FINNEY THOMAS DEMANA WAITS

INSTRUCTOR'S SOLUTIONS MANUAL

Calculus

Graphical, Numerical, Algebraic

FINNEY THOMAS DEMANA WAITS

INSTRUCTOR'S SOLUTIONS MANUAL

Calculus

Graphical, Numerical, Algebraic

DAVID L. WINTER
MARY JEAN WINTER

MICHIGAN STATE UNIVERSITY

Exploration Solutions Contributed by
ALAN LIPP
CHARLES RENO

Addison-Wesley Publishing Company
Reading, Massachusetts • Menlo Park, California • New York
Don Mills, Ontario • Wokingham, England • Amsterdam • Bonn
Sydney • Singapore • Tokyo • Madrid • San Juan • Milan • Paris

ISBN 0–201–56905-1

1 2 3 4 5 6 7 8 9 10–VG–97969594

We would like to acknowledge the important help provided by our solution checkers, Alyson E. Elliott and Harriet E. Winter. We would like to thank Tammy A. Hatfield and Catherine E. Friess for their advice and great technical skill in preparing the manuscript.

TABLE OF CONTENTS

FINNEY THOMAS DEMANA WAITS

INSTRUCTOR'S SOLUTIONS MANUAL

Calculus

Graphical, Numerical, Algebraic

CHAPTER 1

PREREQUISITES FOR CALCULUS

1.1 COORDINATES AND GRAPHS IN THE PLANE

1. In the given viewing rectangle $-5 \leqq x \leqq 5$ and $-10 \leqq y \leqq 10$. The only points satisfying these conditions are $(0,6)$ and $(5,-5)$.

2. $(1.5, 17.5)$, $(10, 100)$

3. One choice: $[-17, 21]$ by $[-12, 76]$

4. One choice: $[-42, 53]$ by $[-89, 31]$

5. For graph reading in the entire viewing rectangle we want (sometimes but not always) to have as many scale marks as possible but still have them far enough apart to be easily distinguishable. For this purpose we suggest 25 or fewer scale marks in each direction. At the same time the distance between them should be a convenient number for measurement, for example, $m10^n$ where m is a positive integer and n is an integer, positive, negative or 0. Here we have $\frac{x\,\mathrm{Max}-x\,\mathrm{Min}}{25} = \frac{50-(-10)}{25} = 2.4$. $x\,\mathrm{Scl} = 5$ would be convenient. $\frac{y\,\mathrm{Max}-y\,\mathrm{Min}}{25} = \frac{50-(-50)}{25} = 4$. Again $y\,\mathrm{Scl} = 5$ would be convenient.

6. $\frac{x\,\mathrm{Max}-x\,\mathrm{Min}}{25} = \frac{1.5-(-1)}{25} = 0.1$. $\frac{y\,\mathrm{Max}-y\,\mathrm{Min}}{25} = \frac{1}{25} = 0.04$. Thus we may take $x\,\mathrm{Scl} = 0.1$, $y\,\mathrm{Scl} = 0.05$.

7. $\frac{x\,\mathrm{Max}-x\,\mathrm{Min}}{25} = 0.04$ so we could take $x\,\mathrm{Scl} = 0.05$, $\frac{y\,\mathrm{Max}-y\,\mathrm{Min}}{25} = \frac{200}{25} = 8$ so $y\,\mathrm{Scl} = 10$ would be convenient.

8. $\frac{x\,\mathrm{Max}-x\,\mathrm{Min}}{25} = 8$ so we take $x\,\mathrm{Scl} = 10$. $\frac{y\,\mathrm{Max}-y\,\mathrm{Min}}{25} = \frac{4}{25} = 0.16$. We may take $y\,\mathrm{Scl} = 0.5$.

9. $y = x + 1$.
x-intercept: $0 = x + 1$, $x = -1$.
y-intercept: $y = 0 + 1 = 1$.

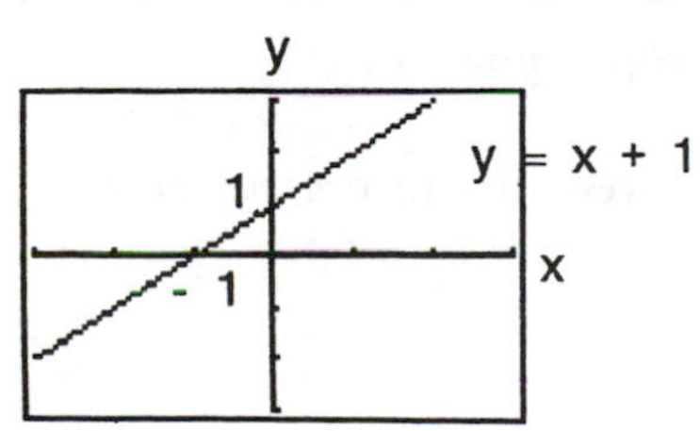

10. Intercepts: $x = 1$, $y = 1$.

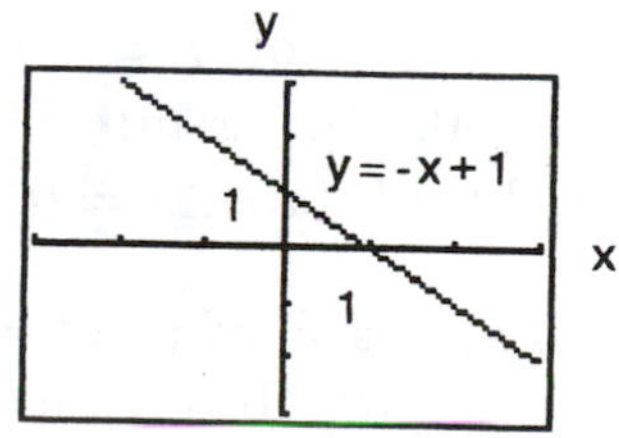

11. $y = -x^2$.
(0, 0) gives the only intercepts.

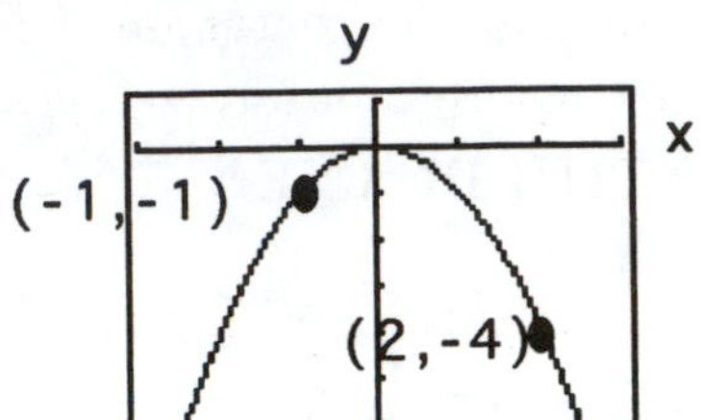

12. $y = 4 - x^2$.
x-intercepts: $x = \pm 2$
y-intercept: $y = 4$.

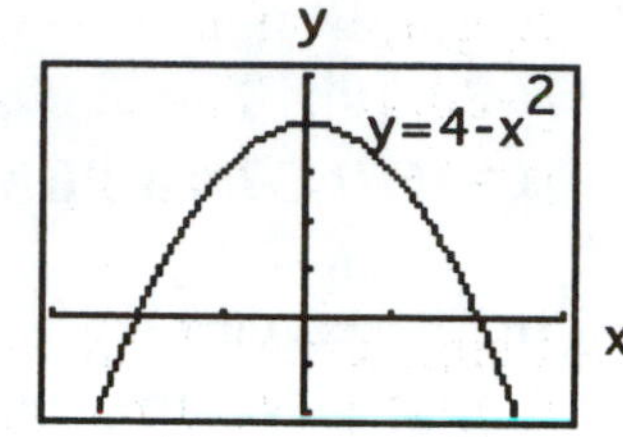

13. $x = -y^2$.
(0, 0) gives the only intercepts.

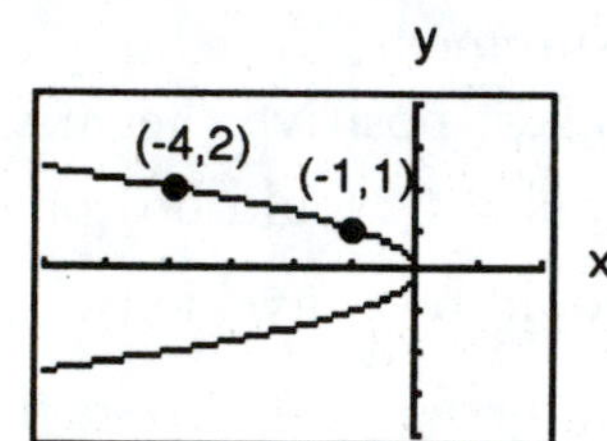

14. $x = 1 - y^2$.
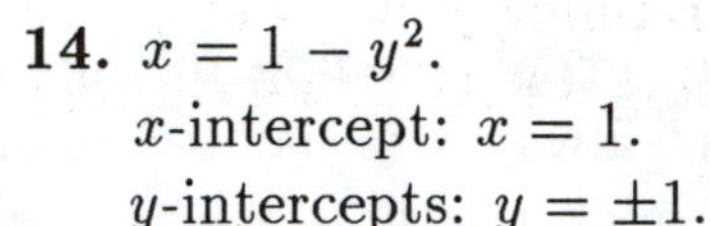
x-intercept: $x = 1$.
y-intercepts: $y = \pm 1$.

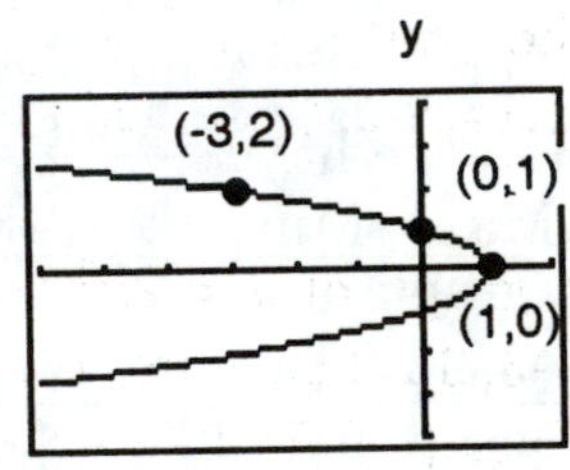

15. (e) **16.** (b) **17.** (e) **18.** (c) **19.** (e) **20.** (d)

21. A complete graph of $y = 3x - 5$ may be viewed in the rectangle $[-2, 4]$ by $[-10, 5]$. x-intercept: $0 = 3x - 5$, $x = 5/3$. y-intercept: $y = 0 - 5 = -5$.

22. A complete graph of $y = 4 - 5x$ may be viewed in the rectangle $[-2, 3]$ by $[-10, 10]$. Intercepts: $x = 4/5$, $y = 4$.

23. A graph of $y = 10 + x - 2x^2$ may be viewed in the rectangle $[-6, 6]$ by $[-30, 15]$. x-intercepts: $0 = 10 + x - 2x^2$ leads to $2x^2 - x - 10 = 0$, $(2x - 5)(x + 2) = 0$, $x = -2,\ 5/2$. y-intercept: $y = 10$.

24. The graph of $y = 2x^2 - 2x - 12$ may be viewed in the rectangle $[-10, 10]$ by $[-30, 70]$. Intercepts: $x = -2, 3$; $y = -12$.

25. A graph of $y = 2x^2 - 8x + 3$ may be viewed in the rectangle $[-7, 10]$ by $[-20, 70]$. x-intercepts: $2x^2 - 8x + 3 = 0$ leads to $x = (4 \pm \sqrt{10})/2$ or $x \approx 0.42, 3.58$. y-intercept: $y = 3$.

26. A graph of $y = -3x^2 - 6x - 1$ may be viewed in the rectangle $[-5, 3]$ by $[-30, 10]$. The x-intercepts, $(-3 \pm \sqrt{6})/3$, are approximately $-1.82, -0.18$. $y = -1$ is the y-intercept.

27. Using TRACE on the graph of $y = x^2 + 4x + 5 = (x+2)^2 + 1$ in the rectangle $[-8, 4]$ by $[-2, 20]$, we see there are no x-intercepts, $y = 5$ is the y-intercept and $(-2, 1)$ is the low point.

28. Graphing $y = -3x^2 + 12x - 8$ in the rectangle $[-1, 5]$ by $[-20, 5]$, we get, using TRACE, x-intercepts: $0.86, 3.14$; y-intercept: $y = -8$ and high point $(2, 4)$.

29. We graph $y = 12x - 3x^3$ in the rectangle $[-4, 4]$ by $[-15, 15]$. $x = \pm 2, 0$ are the x-intercepts, $y = 0$ is the y-intercept. Using TRACE, $(-1.14, -9.24)$ and $(1.14, 9.24)$ are the approximate local low and high points, respectively.

30. A graph of $y = 2x^3 - 2x$ can be obtained in the viewing rectangle $[-2, 2]$ by $[-3, 3]$. The x-intercepts are $-1, 0, 1$ and the y-intercept is 0. Use of TRACE gives $(-0.57, 0.77)$ and $(0.57, -0.77)$ for an approximation of the local high and low point, respectively.

31. A complete graph of $y = -x^3 + 9x - 1$ can be obtained in the viewing rectangle $[-5,,5]$ by $[-35, 35]$. The y-intercept is -1. Use of TRACE yields the following approximations. x-intercepts: $-3.02, 0.08, 2.93$; low point: $(-1.75, -11.39)$; high point: $(1.75, 9.39)$.

32. Graph $y = x^3 - 4x + 3$ in $[-5, 5]$ by $[-5, 10]$. x-intercepts: $-2.30, 1, 1.30$; y-intercept: 3. High point: $(-1.15, 6.08)$, low point: $(1.15, -0.079)$.

33. An idea of a complete graph of $y = x^3 + 2x^2 + x + 5$ can be obtained by using the viewing rectangles $[-3, 2]$ by $[-2, 10]$ and $[-2, 1]$ by $[4, 6]$. The y-intercept is 5 and using TRACE in the first rectangle, we obtain -2.44 as the approximate x-intercept. In the second rectangle, we obtain $(-1, 5)$ and $(-0.33, 4.85)$ as the approximate local high and low points.

34. An idea of a complete graph of $y = 2x^3 - 5.5x^2 + 5x - 5$ can be obtained by using the viewing rectangles $[-2, 4]$ by $[-10, 2]$ and $[0.73, 1.06]$ by $[-3.51, -3.48]$. The y-intercept is -5 and using TRACE in the first rectangle, we obtain 2.14 as the approximate x-intercept. In the second rectangle we obtain $(0.832, -3.495)$ and $(1.00, -3.500)$ as the approximate local high and low points.

35. Every pixel in the first column of pixels has x-coordinate $x\,\text{Min}$. Every pixel in the second column has x-coordinate $x\,\text{Min} + \delta x$. Every pixel in the third column has screen x-coordinate $x\,\text{Min} + 2\delta x$. Continuing inductively, we see that every pixel in the ith column has screen x-coordinate $x\,\text{Min} + (i-1)\delta x$. Similarly every pixel in the jth row, starting from the bottom, has screen y-coordinate $y\,\text{Max} - (j - i)\delta y$.

36. a is at least as close to x_i as it is to $x_i \pm \delta x$. Thus $x_i - \delta x/2 \leqq a \leqq x_i + \delta x/2$ and so $-\delta x/2 \leqq a - x_i \leqq \delta x/2$. That is, $|a - x_i| \leqq \delta x/2$. Similarly $|b - y_j| \leqq \delta y/2$.

37. We assume $N = 127$ and $M = 63$.

$\delta x = \frac{x\,\text{Max} - x\,\text{Min}}{N-1} = \frac{a-(-10)}{126} = 1, \quad a = 116.$

$\delta y = \frac{y\,\text{Max} - y\,\text{Min}}{M-1} = \frac{b-(-10)}{62} = 1, \quad b = 52.$

38. We take $(N, M) = (127, 63)$. $\delta x = \frac{a+10}{126} = 0.1$, $a = 2.6$. $\delta y = \frac{b+10}{62} = 0.1$, $b = -3.8$.

39. We take $N = 127$ and $M = 63$. $\delta x = \frac{a-(-10)}{126} = 0.5$, giving $a = 53$. $\delta y = \frac{b-(-10)}{62} = 2$, giving $b = 114$.

40. We take $(N, M) = (127, 63)$. $\delta x = \frac{a+10}{126} = 2$, $a = 242$. $\delta y = \frac{b+10}{62}$, $b = 610$.

41. Yes, the point $(100, 60)$ is above the curve at $x = 100$.

42. Only b) and c) require a stop.

1.2 SLOPE, AND EQUATIONS FOR LINES

1. $\Delta x = -1 - 1 = -2$, $\Delta y = -1 - 2 = -3$

2. $\Delta x = -1 - (-3) = 2$, $\Delta y = -2 - 2 = -4$

3. $\Delta x = -8 - (-3) = -5$, $\Delta y = 1 - 1 = 0$

4. $\Delta x = 0 - 0 = 0,\ \Delta y = -2 - (4) = -6$

5. $m = \frac{\Delta y}{\Delta x} = \frac{1-(-2)}{2-1} = 3$. The slope of lines perpendicular to AB is $-1/3$.

6. $m = \frac{-2-(-1)}{1-(-2)} = \frac{-1}{3};\ m_\perp = 3$.

7. $m = \frac{3-3}{-1-2} = 0$. The perpendicular lines are vertical and have no slope.

8. A and B determine the vertical line $x = 1$ which has no slope. The lines perpendicular to AB have slope 0.

9. $\sqrt{(x_2 - x_1)^2 + (y_2 - y_1)^2} = \sqrt{(0-1)^2 + (1-0)^2} = \sqrt{2}$

10. $\sqrt{(-1-2)^2 + (0-4)^2} = \sqrt{9+16} = 5$

11. $\sqrt{(-\sqrt{3} - 2\sqrt{3})^2 + (1-4)^2} = \sqrt{(-3\sqrt{3})^2 + (-3)^2} = \sqrt{9 \cdot 3 + 9} = 6$

12. $\sqrt{(1-2)^2 + (-\frac{1}{3} - 1)^2} = \sqrt{1 + \frac{16}{9}} = \frac{5}{3}$

13. $\sqrt{(0-a)^2 + (0-b)^2} = \sqrt{a^2 + b^2}$

14. $\sqrt{(x-0)^2 + (0-y)^2} = \sqrt{x^2 + y^2}$

15. $|-3| = 3$

16. $|2-7| = |-5| = 5$

17. $|-2+7| = |5| = 5$

18. $|1.1 - 5.2| = |-4.1| = 4.1$

19. $|(-2)3| = |-2||3| = 2 \cdot 3 = 6$

20. $\left|\frac{2}{-7}\right| = \frac{|2|}{|-7|} = \frac{2}{7}$

21. a) $x = 2$ b) $y = 3$

22. a) $x = -1$ b) $y \doteq 4/3$

23. a) $x = 0$ b) $y = -\sqrt{2}$

24. a) $x = -\pi$ b) $y = 0$

25. $y - 1 = (1)(x-1)$ or $y = x$

26. $y - (-1) = (-1)(x-1),\ y + 1 = -x + 1$ or $y = -x$

27. $y - 1 = (1)[x - (-1)] = x + 1$ or $y = x + 2$

28. $y - 1 = (-1)[x - (-1)] = -x - 1$ or $y = -x$

29. $y = 2x + b$

30. $y - 0 = (-2)(x - a)$ or $y = -2x + 2a$

31. $m = \frac{3-0}{2-0} = \frac{3}{2},\ y - 0 = \frac{3}{2}(x-0)$ or $y = \frac{3}{2}x$

32. This is the horizontal line $y = 1$. **33.** $x = 1$ **34.** $x = -2$

35. $m = \frac{-2-1}{2-(-2)} = \frac{-3}{4}$. $y - 1 = -\frac{3}{4}[x - (-2)]$ leads to $3x + 4y + 2 = 0$

36. $m = \frac{1-3}{3-1} = -1$. $y - 3 = -(x - 1)$ or $y = -x + 4$

37. $y = mx + b = 3x - 2$. Answer: $y = 3x - 2$

38. $y = -x + 2$ **39.** $y = x + \sqrt{2}$

40. $y = -\frac{1}{2}x - 3$ **41.** $y = -5x + 2.5$

42. $y = \frac{1}{3}x - 1$

43. $3x + 0 = 12$, $x = 4$ is x-intercept. $0 + 4y = 12$, y-intercept is 3.

44. x-intercept $=$ y-intercept $= 2$

45. $4x - 3y = 12$. x-intercept: $4x - 0 = 12$, $x = 3$.
y-intercept: $0 - 3y = 12$, $y = -4$.

46. $2x - 6 = 4$, x-intercept: $x = 2$, y-intercept: $y = -4$.

47. $y = 2x + 4$, x-intercept: $0 = 2x + 4$, $x = -2$. y-intercept: $y = 4$.

48. $x + 2y = -4$. x-intercept: $x = -4$, y-intercept: $y = -2$.

49. For the x-intercept we set $y = 0$: $\frac{x}{3} + 0 = 1$, $x = 3$. For the y-intercept we set $x = 0$: $0 + \frac{y}{4} = 1$, $y = 4$.

50. $\frac{x}{-2} + \frac{y}{3} = 1$. Intercepts: $x = -2$, $y = 3$.

51. $\frac{x}{a} + \frac{y}{b} = 1$. For the x-intercept set $y = 0$: $\frac{x}{a} + 0 = 1$, $x = a$. For the y-intercept set $x = 0$: $0 + \frac{y}{b} = 1$, $y = b$. The point is that if we write a linear equation in the form $\frac{x}{a} + \frac{y}{b} = 1$, we can immediately read off the intercepts, or, if we know the intercepts, we can immediately write down an equation of the line.

52. $\frac{x}{a} + \frac{y}{b} = 2$ leads to $\frac{x}{2a} + \frac{y}{2b} = 1$. Intercepts: $x = 2a$, $y = 2b$.

53. $P(0,0)$, $L : y = -x + 2$. For L, $m = -1$. Hence for a perpendicular line the slope is $-(1/-1) = 1$. The line through $(0,0)$ perpendicular to L is $y - 0 = (1)(x - 0)$ or $y = x$. A complete graph of $y = -x + 2$ and $y = x$ may be seen in the viewing rectangle $[-6, 6]$ by $[-4, 4]$. For the point of intersection of the two lines $y = -x + 2 = x$, $2x = 2$, $x = 1$, $y = x = 1$ and $(1, 1)$ is the point of intersection. The distance from P to L is the distance from P to $(1,1)$: $\sqrt{(1-0)^2 + (1-0)^2} = \sqrt{2}$.

54. $P(0,0)$, $L : x + \sqrt{3}y = 3$. $y = -\frac{1}{\sqrt{3}}x + \sqrt{3}$. Thus L has slope $-\frac{1}{\sqrt{3}}$ and so $\sqrt{3}$ is the slope of a perpendicular, $y = \sqrt{3}x$ is an equation of the perpendicular through P. Graph $y = -\frac{1}{\sqrt{3}}x + \sqrt{3}$ and $y = \sqrt{3}x$ at the same time in the viewing rectangle $[-6, 6]$ by $[-4, 4]$. (Note that in order to maintain the appearance of perpendicularity, we have (x Max $-$ x Min)/(y Max $-$ y Min) $= 3/2$.) For the point of intersection: $y = -\frac{1}{\sqrt{3}}x + \sqrt{3} = \sqrt{3}x$ leads to $-x + 3 = 3x$, $x = 3/2$, $y = \sqrt{3}x = 3\sqrt{3}/2$ and so $(3/2, 3\sqrt{3}/2)$ is the point of intersection. The distance from P to this point is $\sqrt{(\frac{3}{2})^2 + (\frac{3\sqrt{3}}{2})^2} = \sqrt{\frac{36}{4}} = 3$.

55. $P(1,2)$, $L : x + 2y = 3$. $y = -\frac{1}{2}x + \frac{3}{2}$, L has slope $-\frac{1}{2}$. The line through $(1,2)$ perpendicular to L has equation $y - 2 = 2(x - 1)$ or $y = 2x$. We may graph $y = -.5x + 1.5$ and $y = 2x$ in the viewing rectangle $[-6, 6]$ by $[-4, 4]$. For the point of intersection $y = 2x = -\frac{1}{2}x + \frac{3}{2}$, $4x = -x + 3$, $x = \frac{3}{5}$ and $(\frac{3}{5}, \frac{6}{5})$ is the point of intersection. The distance is $\sqrt{(1 - \frac{3}{5})^2 + (2 - \frac{6}{5})^2} = \frac{2\sqrt{5}}{5}$.

56. $P(-2,2)$, $L : 2x + y = 4$. $y = -2x + 4$ so perpendicular through P has equation $y - 2 = \frac{1}{2}(x + 2)$ or $y = \frac{1}{2}x + 3$. The graphs of the two lines may be viewed in $[-7, 8]$ by $[-3, 7]$. For the point of intersection $y = \frac{1}{2}x + 3 = -2x + 4$, $x + 6 = -4x + 8$, $5x = 2$, $x = 2/5$, $y = \frac{1}{2}x + 3 = \frac{1}{5} + 3 = 3.2$. The point of intersection is $(0.4, 3.2)$. The distance is $\sqrt{(-2 - 0.4)^2 + (2 - 3.2)^2} = \sqrt{(\frac{12}{5})^2 + (-\frac{6}{5})^2} = \frac{6\sqrt{5}}{5}$.

57. $P(3,6)$, $L : x + y = 3$. $y = -x + 3$ so L has slope -1. The required perpendicular has equation $y - 6 = (1)(x - 3)$ or $y = x + 3$. Graph $y = -x + 3$ and $y = x + 3$ in the viewing rectangle $[-7, 8]$ by $[-3, 7]$. The point of intersection is the y-intercept $(0, 3)$. The distance is $\sqrt{(0-3)^2 + (3-6)^2} = 3\sqrt{2}$

58. $P(-2,4)$, $L : x = 5$. The perpendicular through P is $y = 4$. The point of intersection is $(5, 4)$ and the distance is 7.

59. $P(2,1)$, $L: y = x + 2$. L has slope 1 so any line parallel to L has slope 1. The line requested has equation $y - 1 = (1)(x - 2)$ or $y = x - 1$. The graphs of $y = x + 2$ and $y = x - 1$ may viewed in the standard viewing rectangle $[-10, 10]$ by $[-10, 10]$.

60. $P(0,0)$, $L: y = 3x - 5$. Desired parallel line has equation $y = 3x$.

61. $P(1,0)$, $L: 2x + y = -2$ or $y = -2x - 2$. Parallel line through $P: y - 0 = -2(x - 1)$ or $y = -2x + 2$.

62. $P(1,1)$, $L: x + y = 1$ or $y = -x + 1$. Requested parallel line: $y = -x + 2$.

63. a) $|x - 3|$ b) $|x - (-2)| = |x + 2|$

64. a) $|y + 1.3|$ b) $|y - 5.5|$

65. The distance between 5 and x is 1.

66. The distance between x and -3 is 5.

67. a) $a = -2$ b) $a \geqq 0$

68. $1 - x \geqq 0$ if and only if $x \leqq 1$. $|1 - x| = x - 1 \Leftrightarrow x \geqq 1$

69. a) From $(0, 69°)$ to $(0.4, 68°)$, $\frac{68°-69°}{0.4-0} = -2.5°/\text{in.}$

b) From $(0.4, 68°)$ to $(4, 10°)$, $\frac{10°-68°}{4-0.4} \approx -16.1°/\text{in.}$

c) From $(4, 10°)$ to $(4.6, 5°)$, $\frac{5°-10°}{4.6-4} = -8\frac{1}{3}°/\text{in.}$

70. Fiber glass is the best insulator (least interior heat reaches the outside), gypsum wall board is the poorest insulator.

71. $p = kd + 1$ and $10.94 = k100 + 1$. Hence $k = 0.0994$ and $p = 0.0994d + 1$. Letting $d = 50$ meters, we find $p = 5.97$ atmospheres.

72. Since the tangent of the angle of arrival is $1/1 = 1$, the same is true of the angle of departure. Thus the line has slope 1 and passes through $(1, 0)$. It has equation $y - 0 = 1(x - 1)$ or $y = x - 1$.

73. $F = \frac{9}{5}C + 32$. Setting $F = C$, we have $C = \frac{9}{5}C + 32$, $-\frac{4}{5}C = 32$, $C = -40°$. $-40°$ Celcius is equivalent to $-40°$ Fahrenheit.

74. We may represent the car on $y = 0.371x$ starting at $(0,0)$. When $y = 14$, $x \approx$ 37.74ft. The car has length $\sqrt{14^2 + (\frac{14}{.371})^2} \approx 40.249$ft.

75. a) $d(t) = 45t + d_0$ where d_0 is the distance of the car from the point P at time $t = 0$.

c) $m = 45$

d) If $t = 0$ corresonds to a specific time, say 1:00 p.m., then negative values of t would correspond to times before 1:00 p.m.

e) The initial distance (from point P) $d_0 = 30$ miles.

76. a) $d(t) = 55t + d_0$ where d_0 is the distance of the car from the point P at time $t = 0$.

c) $m = 55$

d) If $t = 0$ corresponds to a specific time, say 1:00 p.m., then negative values of t would correspond to times before 1:00 p.m.

e) The initial distance (from point P) $d_0 = 40$ miles.

77. Let A, B, C be, respectively, $(-1,1), (2,0), (2,3)$. For one side of the parallelogram we may take the vertical line segment BC of length 3. Thus for the fourth vertex D we take either $(-1,4)$ or $(-1,-2)$. The third possibility for D, $(5,2)$, is found as the intersection of line $BD(y = \frac{2}{3}(x-2))$ and line $CD(y = 3 - \frac{1}{3}(x-2))$.

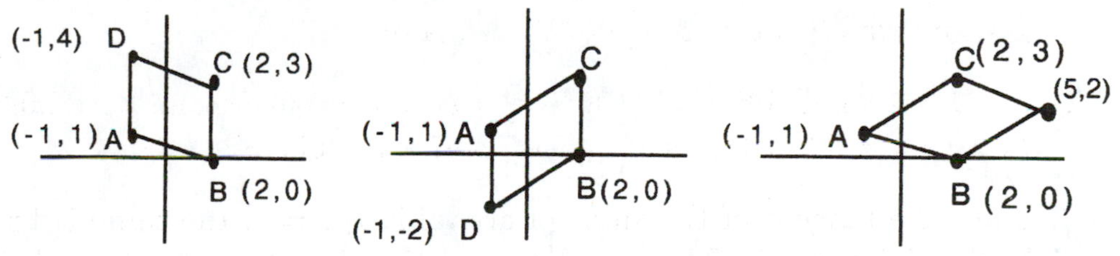

78. $2x + ky = 3$ or $y = -\frac{2}{k}x + \frac{3}{k}$ has slope $-\frac{2}{k}$ while $y = -x + 1$ has slope -1. The lines are perpendicular if $-\frac{2}{k} = -(\frac{1}{-1}) = 1$ or $k = -2$. The lines are parallel if $-\frac{2}{k} = -1$ or $k = 2$.

79. The point of intersection is $(1,1)$. An equation of the line through $(1,1)$ and $(1,2)$ is $x = 1$.

80. We first solve a simpler problem: Show that the distance d from $(0,0)$ to the line $ax+by+c=0$ is $\frac{|c|}{\sqrt{a^2+b^2}}$. We omit the easy verification of this if either a or b is 0 and so can assume both a and b are non-zero. The line has slope $-\frac{a}{b}$ and so the perpendicular line through $(0,0)$ has equation $y=\frac{b}{a}x$. To find the x-coordinate of the point of intersection we set the two y-values equal: $\frac{b}{a}x=-\frac{a}{b}x-\frac{c}{b}$. First we multiply by ab : $b^2x=-a^2x-ac$. Then $x=\frac{-ac}{a^2+b^2}$. $y=\frac{b}{a}x=\frac{-bc}{a^2+b^2}$. $d^2=x^2+y^2=\frac{c^2}{a^2+b^2}$, $d=\frac{|c|}{\sqrt{a^2+b^2}}$.

The graph of $L_2 : A(x+x_1)+B(y+y_1)+C=0$ is the same as that of $L_1 : Ax+By+C=0$ except that it is shifted $-x_1$ units in the x-direction and $-y_1$ units in the vertical direction. If the same rigid transformations are applied to the point (x_1,y_1), it is moved to $(0,0)$. Thus the distance d between (x_1,y_1) and L_1 equals the distance between $(0,0)$ and $L_2 : Ax+By+(Ax_1+Bx_1+C)=0$. By the preceding paragraph $d=\frac{|c|}{\sqrt{a^2+b^2}}=\frac{|Ax_1+By_1+C|}{\sqrt{A^2+B^2}}$.

1.3 RELATIONS, FUNCTIONS, AND THEIR GRAPHS

1. This relation is not a function. The graph does not pass the vertical line test. There are x-values that correspond to two y-values.

2. This is the graph of a function. Each x in the domain corresponds to exactly one y.

3. This is a function. No value of x corresponds to two y-values.

4. Not a function. There are x-values that correspond to more than one y-value.

5. a) $(3,-1)$ b) $(-3,1)$ c) $(-3,-1)$

6. a) $(-2,-2)$ b) $(2,2)$ c) $(2,-2)$

7. a) $(-2,-1)$ b) $(2,1)$ c) $(2,-1)$

8. a) $(-1,1)$ b) $(1,-1)$ c) $(1,1)$

9. a) $(1,\sqrt{2})$ b) $(-1,-\sqrt{2})$ c) $(-1,\sqrt{2})$

10. a) $(-\sqrt{3},\sqrt{3})$ b) $(\sqrt{3},-\sqrt{3})$ c) $(\sqrt{3},\sqrt{3})$

11. a) $(0,-\pi)$ b) $(0,\pi)$ c) $(0,-\pi)$

12. a) $(2,0)$ b) $(-2,0)$ c) $(-2,0)$

13. Domain: $[1, \infty)$, range: $[2, \infty)$

14. Domain: $[-4, \infty)$, range: $[-3, \infty)$

15. Domain: $(-\infty, 0]$, range: $(-\infty, 0]$

16. Domain: $(-\infty, 0]$, range: $[0, \infty)$

17. Domain: $(-\infty, 3]$, range: $[0, \infty)$

18. Domain: $(-\infty, 2]$, range: $(-\infty, 0]$

19. Domain: $(-\infty, 2) \cup (2, \infty)$, range: $(-\infty, 0) \cup (0, \infty)$; any non-zero number is a reciprocal of another number.

20. Domain: $(-\infty, -2) \cup (-2, \infty)$, range: $(-\infty, 0) \cup (0, \infty)$

21. Domain: $(-\infty, \infty)$, range: $[-9, \infty)$. Symmetric about the y-axis.

22. Domain: $(-\infty, \infty)$, range: $(-\infty, 4]$. Symmetric about the y-axis.

23. Domain = range = $(-\infty, \infty)$. No symmetry.

24. Domain = range = $(-\infty, \infty)$. No symmetry.

25. Domain = range = $(-\infty, \infty)$. No symmetry.

26. Domain = range = $(-\infty, \infty)$. No symmetry.

27. Domain = range = $(-\infty, 0) \cup (0, \infty)$. Symmetric about the origin.

28. Domain = $(-\infty, 0) \cup (0, \infty)$, range: $(-\infty, 0)$. Symmetric about the y-axis.

29. Domain: $(-\infty, 0) \cup (0, \infty)$. Since $\frac{1}{x}$ can have any value except 0, $1 + \frac{1}{x}$ can have any value except 1. Range: $(-\infty, 1) \cup (1, \infty)$. No symmetry.

30. Domain: $(-\infty, 0) \cup (0, \infty)$, range: $(1, \infty)$. Symmetric about y-axis.

31. a) No b) No c) $(0, \infty)$

32. a) No b) No c) No d) $(0, 1]$

33. Odd **34.** Even **35.** Neither **36.** Neither **37.** Even

38. Odd **39.** Even **40.** Neither **41.** Odd **42.** Even

43. Symmetric about the y-axis. Graph $y = -x^2$ in the viewing rectangle $[-10, 10]$ by $[-10, 10]$.

44. Symmetric about the x-axis. Graph $y = \sqrt{4 - x}$ and $y = -\sqrt{4 - x}$ simultaneously in the viewing rectangle $[-10, 4]$ by $[-5, 5]$.

45. Symmetric about the y-axis. Graph $y = 1/x^2$ in the viewing rectangle $[-5, 5]$ by $[0, 3]$.

46. Symmetric about the y-axis. Graph $y = 1/(x^2 + 1)$ in the viewing rectangle $[-5, 5]$ by $[0, 1]$.

47. Since $(-x)(-y) = xy$, the equation is unchanged if x is replaced by $-x$ and y by $-y$. Thus the graph is symmetric about the origin. A complete graph of $y = 1/x$ can be obtained in $[-4, 4]$ by $[-4, 4]$.

48. Symmetric about the x-axis. Graph $y = 1/\sqrt{x}$ and $-1/\sqrt{x}$ in the viewing rectangle $[0, 4]$ by $[-4, 4]$.

49. $x^2y^2 = 1$ has graph symmetric about both axes and the origin. Graph $y = 1/|x|$ and $y = -1/|x|$ in the viewing rectangle $[-4, 4]$ by $[-4, 4]$.

50. Symmetric about both axes and the origin. Graph $y = \sqrt{1 - x^2}/2$ and $y = -\sqrt{1 - x^2}/2$ in the viewing rectangle $[-1, 1]$, by $[-1, 1]$.

51. The graph of $y = |x + 3|$ can be obtained by translating the graph of $y = |x|$ three units to the left. Graph $y = \text{abs}(x + 3)$ in the viewing rectangle $[-7, 1]$ by $[0, 4]$.

52. Graph $y = |2 - x| = |x - 2|$ in the viewing rectangle $[-2, 6]$ by $[0, 4]$.

53. Graph $y = \frac{|x|}{x}$ in the viewing rectangle $[-2, 2]$ by $[-2, 2]$. There is not point on the graph when $x = 0$.

54. This is the last graph shifted to the right 1 unit.

55. $y = \frac{x - |x|}{2}$. $y = x$ when $x \leqq 0$ and $y = 0$ when $x \geqq 0$.

56. $y = \frac{x + |x|}{2}$. $y = 0$ when $x \leqq 0$ and $y = x$ when $x \geqq 0$.

57. a) Graph $y = 3 - x + 0\sqrt{1 - x}$ and $y = 2x + 0\sqrt{x - 1}$ in the viewing rectangle $[-10, 10]$ by $[0, 20]$. b) $f(0) = 3$, $f(1) = 2$, $f(2.5) = 5$.

58. a) Graph $y = (1/x) + 0\sqrt{(-x)}$ and $y = x + 0\sqrt{x}$ in the viewing rectangle $[-4, 4]$ by $[-4, 4]$. b) $f(-1) = -1$, $f(0) = 0$, $f(\pi) = \pi$.

59. a) Graph $y = 1 + 0\sqrt{5 - x}$ in the viewing rectangle $[-3, 10]$ by $[0, 2]$. It is understood that the x-axis for $x \geqq 5$ is part of the graph and the point $(5, 1)$ is not. b) $f(0) = 1$, $f(5) = 0$, $f(6) = 0$.

60. a) Graph $y = 1+0\sqrt{-x}$ and $y = \sqrt{x}$ in the viewing rectangle $[-4, 4]$ by $[0, 2]$. $(0, 1)$ is not a point of the graph. b) $f(-1) = 1,\ f(0) = 0,\ f(5) = \sqrt{5}$.

61. a) Graph $y = 4 - x^2 + 0\sqrt{1-x}$, $y = \frac{3}{2}x + \frac{3}{2} + 0\sqrt{x-1} + 0\sqrt{3-x}$ and $y = x + 3 + 0\sqrt{x-3}$ in the viewing rectangle $[-3, 7]$ by $[-5, 10]$. b) $f(0.5) = 3.75,\ f(1) = 3,\ f(3) = 6,\ f(4) = 7$.

62. a) Graph $y = x^2+0\sqrt{-x}$, $y = x^3+0\sqrt{x}+0\sqrt{1-x}$ and $y = 2x-1+0\sqrt{x-1}$ in the viewing rectangle $[-3, 3]$ by $[0, 5]$. b) $f(-1) = 1,\ f(0) = 0,\ f(1) = 1,\ f(2.5) = 4$.

63. a) $1 - |x-1|,\ 0 \leqq x \leqq 2$ b) $f(x) = \begin{cases} 2, & 0 \leqq x < 1 \\ 0, & 1 \leqq x < 2 \\ 2, & 2 \leqq x < 3 \\ 0, & 3 \leqq x \leqq 4 \end{cases}$

64. a) $y = 0,\ 0 \leqq x \leqq \frac{T}{2},\ y = \frac{2}{T}x-1,\ \frac{T}{2} < x \leqq T$ or $y = \frac{1}{T}[|\frac{T}{2}-x|-(\frac{T}{2}-x)],\ 0 \leqq x \leqq T$ b) $y = A,\ 0 \leqq x < \frac{T}{2},\ y = -A,\ \frac{T}{2} \leqq x < T,\ y = A,\ T \leqq x < \frac{3T}{2},\ y = -A,\ \frac{3T}{2} \leqq x \leqq 2T$

65. a) $0 \leq x < 1$ b) $-1 < x \leqq 0$

66. $[x] = \lceil x \rceil$ if and only if x is an integer. If x is between n and $n+1$, $[x] = n$ and $\lceil x \rceil = n+1$.

67. a) Graph of $y = x - [x],\ -3 \leqq x \leqq 3$

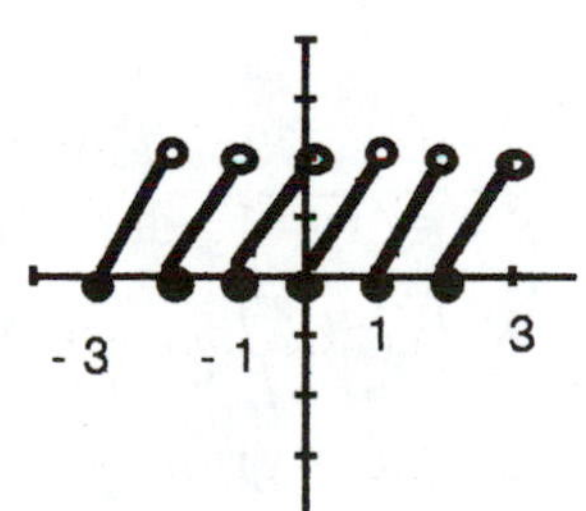

b) Graph of $y = [x] - \lceil x \rceil,\ -3 \leqq x \leqq 3$

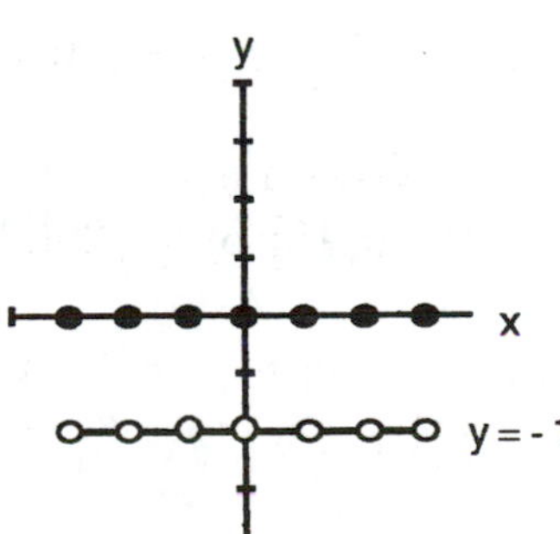

68. When x is negative or zero, $\lceil x \rceil$ is the integer part of the decimal representation of x.

69. Graph $y = abs(x+1) + 2abs(x-3)$ in the viewing rectangle $[-2, 5]$ by $[0, 12]$. For $x \leqq -1$, both $x+1$ and $x-3$ are non-positive and $f(x) = |x+1| + 2|x-3| = -(x+1) - 2(x-3) = -3x+5$. For $-1 < x \leq 3$, $x+1$ is positive and $x-3$ is nonpositive and $f(x) = x+1-2(x-3) = -x+7$. For $x > 3$ both $x+1$ and $x-3$ are positive and $f(x) = x+1+2(x-3) = 3x-5$. Thus

$$f(x) = \begin{cases} -3x+5, & x \leqq -1 \\ -x+7, & -1 < x \leqq 3 \\ 3x-5, & x > 3 \end{cases}$$

70. Graph $y = |x+2| + |x-1|$ in the viewing rectangle $[-4, 3]$ by $[0, 6]$. $y = -2x-1,\ x \leqq -2,\ y = 3,\ -2 < x \leqq 1,\ y = 2x+1,\ x > 1$.

71. Graph $y = |x| + |x-1| + |x-3|$ in the viewing rectangle $[-1, 4]$ by $[0, 7]$. $y = -3x+4,\ x \leqq 0,\ y = -x+4,\ 0 < x \leqq 1,\ y = x+2,\ 1 < x \leqq 3,\ y = 3x-4,\ x > 3$.

72. Graph $y = f(x) = |x+2| + |x| + |x+1|$ in the viewing rectangle $[-3, 1]$ by $[0, 7]$ and in $[-3, 1]$ by $[2, 7]$. $f(x) = -3x-3,\ x \leqq -2,\ f(x) = 1-x,\ -2 < x \leqq -1,\ f(x) = x+3,\ -1 < x \leqq 0,\ f(x) = 3x+3,\ x > 0$

73. $f(x) = x$, domain: $(-\infty, \infty)$. $g(x) = \sqrt{x-1}$, domain: $[1, \infty)$. $f(x)+g(x) = x + \sqrt{x-1}$, domain $(f+g)$: $[1, \infty)$, complete graph in $[0, 5]$ by $[0, 10]$. $f(x) - g(x) = x - \sqrt{x-1}$, domain $(f-g)$: $[1, \infty)$, complete graph in $[0, 5]$ by $[0, 4]$. $f \circ g(x) = f(g(x)) = f(\sqrt{x-1}) = \sqrt{x-1}$, domain $f \circ g : [1, \infty)$, complete graph in $[0, 5)$ by $[0, 2]$. $f(x)/g(x) = x/\sqrt{x-1}$, domain: $(1, \infty)$, complete graph in $[0, 10]$ by $[0, 5]$. $g(x)/f(x) = \sqrt{x-1}/x$, domain: $[1, \infty)$, complete graph in $[0, 10]$ by $[0, 0.5]$, graph starts at $(1, 0)$.

74. $f(x) = \sqrt{x+1}$, domain: $[-1, \infty)$. $g(x) = \sqrt{x-1}$, domain: $[1, \infty)$. $f(x) + g(x) = \sqrt{x+1} + \sqrt{x-1}$, domain: $[1, \infty)$, graph in $[0, 10]$ by $[0, 10]$. $f(x) - g(x) = \sqrt{x+1} - \sqrt{x-1}$, domain $(f-g) : [1, \infty)$, graph in $[0, 10]$ by $[0, 2]$. $f \circ g(x) = f(\sqrt{x-1}) = \sqrt{\sqrt{x-1}+1}$, domain $f \circ g = [1, \infty)$, graph in $[0, 10]$ by $[0, 2]$. $f(x)/g(x) = \sqrt{x+1}/\sqrt{x-1}$, domain $(f/g) : (1, \infty)$, graph in $[0, 10]$ by $[0, 6]$. $g(x)/f(x) = \sqrt{x-1}/\sqrt{x+1}$, domain $(g/f) : [1, \infty)$, graph in $[0, 10]$ by $[0, 2]$.

75. a) $f(g(0)) = f(-3) = 2$ b) $g(f(0)) = g(5) = 22$ c) $f(g(x)) = f(x^2-3) = (x^2-3)+5 = x^2+2$ d) $g(f(x)) = g(x+5) = (x+5)^2 - 3 = x^2 + 10x + 22$ e) $f(f(-5)) = f(0) = 5$ f) $g(g(2)) = g(1) = -2$ g) $f(f(x)) = f(x+5) = (x+5)+5 = x+10$ h) $g(g(x)) = g(x^2-3) = (x^2-3)^2 - 3 = x^4 - 6x^2 + 6$

76. a) $f(g(0)) = f(-1) = 0$ b) $g(f(0)) = g(1) + 0$ c) $f(g(1)) = f(0) = 1$ d) $g(f(1)) = g(2) = 1$ e) $f(g(x)) = f(x-1) = (x-1)+1 = x$ f) $g(f(x)) = g(x+1) = (x+1) - 1 = x$

77. a) $f \circ g(x) = f(g(x)) = f(x-7) = \sqrt{x-7}$
b) $f(g(x)) = f(x+2) = 3(x+2)$
c) $f \circ g(x) = f(g(x)) = \sqrt{g(x) - 5} = \sqrt{x^2 - 5}$ so $g(x) = x^2$
d) $f(g(x)) = f(\frac{x}{x-1}) = \frac{(\frac{x}{x-1})}{(\frac{x}{x-1})-1} = \frac{x}{x-(x-1)} = x$
e) $f(g(x)) = 1 + \frac{1}{g(x)} = x$, $\frac{1}{g(x)} = x - 1$, $g(x) = \frac{1}{x-1}$
f) $f(g(x)) = f(\frac{1}{x}) = x$ so $f(x) = \frac{1}{x}$.

78. a) $f(x+2) - f(2) = \frac{1}{x+2} - \frac{1}{2} = \frac{2-(x+2)}{2(x+2)} = \frac{-x}{2(x+2)}$
b) $F(t+1) - F(1) = 4(t+1) - 3 - (4-3) = 4t$

79. a) $C(10) = 72$ b) $C(30) - C(20)$ is the increase of cost if the production level is raised from 20 to 30 items daily.

80. Domain $\sqrt{x^2}$ = domain $|x| = (-\infty, \infty)$, range $|x| = [0, \infty)$. Domain $(\sqrt{x})^2 = [0, \infty)$ = range $(\sqrt{x})^2$.

81. The two functions are identical.

82. $g(f(x)) = \sqrt{f(x)} = |x|$, $f(x) = x^2$.

83. $g(f(x)) = g(x^2+2x+1) = g((x+1)^2) = |x+1|$. $g(x) = \sqrt{x}$ is one possibility.

84. $f(x) = \sin^2 x$, $g(x) = \sqrt{x}$

85. Graph $y = abs(x+3) + abs(x-2) + abs(x-4)$ in the viewing rectangle $[-4, 5]$ by $[0, 15]$. We see that $d(x)$ is minimized when $x = 2$ so you would put the table next to Machine 2.

86. Graph $y = |x+3| + |x+1| + |x-2| + |x-6|$ in the viewing rectangle $[-4, 7]$ by $[10, 25]$. We see that $d(x)$ has its minimal value of 12 for all x in the interval $-1 \leqq x \leqq 2$. The assembly table should now be put next to Machine 2 or Machine 3 or anywhere between them.

87. Graph $y = 2|x+3| + |x+1| + 3|x-2| + |x-6|$ in the viewing rectangle $[-4, 7]$ by $[15, 40]$. $d(x)$ now has minimum value 17 when $x = 2$. The table should be placed next to Machine 3.

1.4 GEOMETRIC TRANSFORMATIONS: SHIFTS, REFLECTIONS, STRETCHES, AND SHRINKS

1. a) $y = (x+4)^2$ b) $y = (x-7)^2$

2. a) $y = x^2 + 3$ b) $y = x^2 - 5$

3. a) Position 4 b) Position 1 c) Position 2 d) Position 3

4. a) $y = (x+2)^2$ b) $y - 1 = (x-4)^2$ c) $y + 4 = (x+1)^2$ d) $y + 3 = (x-2)^2$

5. Shift the graph of $|x|$ to the left 4 units. Then shift the resulting graph down 3 units.

6. Shift the graph of $y = |x|$ three units to the right and then shift the resulting graph 2 units up.

7. Reflect the graph of $y = \sqrt{x}$ over the y-axis and then stretch the resulting graph vertically by a factor of 3.

8. Reflect the graph of $y = \sqrt{x}$ over the y-axis, shrink vertically by a factor of 0.2, and then reflect the resulting graph over the x-axis.

9. Stretch the graph of $y = \frac{1}{x}$ vertically by a factor of 2 and shift the resulting graph down 3 units.

10. Shrink the graph of $\frac{1}{x}$ by a factor of 0.5. Reflect the resulting graph over the x-axis and then shift the result up 1 unit.

11. Shift the graph of $y = x^3$ right 3 units. Shrink the resulting graph by a factor of 0.5. Reflect the last graph over the x-axis and shift the last graph up 1 unit.

12. Shift the graph of $y = \sqrt{x}$ 2 units to the left (obtaining the graph of $\sqrt{x+2}$). Reflect the resulting graph over the y-axis (obtaining the graph of $\sqrt{-x+2} = \sqrt{2-x}$). Stretch the resulting graph vertically by a factor of 3, and then shift the last result down 5 units.

13. Shift the graph of $y = \frac{1}{x}$ to the right 2 units (obtaining the graph of $\frac{1}{x-2}$). Shift the resulting graph up 3 units.

14. Shift the graph of $\frac{1}{x}$ to the right 3 units. Stretch the resulting graph vertically by a factor of 2. Shift the resulting graph down 5 units.

15. Start with the graph of $y = \sqrt[3]{x}$. Stretch vertically by a factor of 4 and then reflect over the y-axis. We now have the graph of $y = 4\sqrt[3]{-x}$. Now shift to right 2 units obtaining the graph of $y = 4\sqrt[3]{-(x-2)} = 4\sqrt[3]{2-x}$. Finally shift the last graph 5 units down. Viewing rectangle: $[-8, 12]$ by $[-15, 5]$. Domain = range = $(-\infty, \infty)$.

16. Graph $y = 0.5|x+3| - 4$ in the viewing rectangle $[-12, 6]$ by $[-4, 1]$. Domain $= (-\infty, \infty)$, range $= [-4, \infty)$.

17. $y = 5\sqrt[3]{-x} - 1$. Start with the graph of $y = \sqrt[3]{x}$, reflect through the y-axis, stretch vertically by a factor of 5, shift vertically down one unit. Check your result by graphing y in $[-5, 5]$ by $[-10, 10]$. Since $\sqrt[3]{x}$ has domain = range $= (-\infty, \infty)$, the same is true of the present function.

18. Sketch $y = -2\sqrt[4]{1-x} + 3$ in the viewing rectangle $[-9, 2]$ by $[-1, 4]$; recognize that $(1, 3)$ is the point on the graph furthest to the right. Domain $= (-\infty, 1]$, range $= (-\infty, 3]$.

19. Start with the graph of $y = \frac{1}{x^2}$. Shift to left 3 units, reflect over the x-axis and shift up 2 units. Viewing rectangle: $[-10, 5]$ by $[-3, 3]$. Domain $= (-\infty, -3) \cup (-3, \infty)$, range $= (-\infty, 2)$.

20. $y = \frac{1}{(x-2)^2}$. Shift the graph of $y = \frac{1}{x^2}$ to the right 2 units. Domain $= (-\infty, 2) \cup (2, \infty)$, range $= (0, \infty)$.

21. Check your result by graphing $y = -2((x-1)^{(1/3)})^2 + 1$ in the viewing rectangle $[-1, 3]$ by $[-2, 1]$. Domain $= (-\infty, \infty)$, range $= (-\infty, 1]$.

22. Check your result by graphing $y = (x+2)^{3/2} + 2$ in the viewing rectangle $[-2, 4]$ by $[0, 20]$. Domain $= [-2, \infty)$, range $= [2, \infty)$.

23. Check your result by graphing $y = 2[1-x] = 2\ \mathrm{Int}(1-x)$ in the viewing window $[-3, 4]$ by $[-6, 8]$. (Graph this in Dot Mode if possible.) Domain $= (-\infty, \infty)$, range $= \{2n : n = 0, \pm 1, \pm 2, \ldots\}$.

24. Check your result by graphing $y = [x-2] + 0.5$ in the viewing rectangle $[-3, 4]$ by $[-4, 2]$. (Use Dot Mode if possible.) Domain $= (-\infty, \infty)$, range $= \{0.5 + n : n = 0, \pm 1, \pm 2, \ldots\}$.

25. $y = x^2 \to y = 3x^2 \to y = 3x^2 + 4$

26. $y = x^2 \to y = x^2 + 4 \to y = 3(x^2 + 4)$

27. $y = \frac{1}{x} \to y = \frac{1}{x} - 2 \to y = 0.2(\frac{1}{x} - 2)$

28. $y = \frac{1}{x} \to y = \frac{0.2}{x} \to y = \frac{0.2}{x} - 2$

29. $y = |x| \to y = |x+2| \to y = 3|x+2| \to y = 3|x+2| + 5$

30. $y = |x| \to y = |x-3| \to y = 0.3|x-3| \to y = 0.3|x-3| - 1$

31. $y = x^3 \to y = -x^3 \to y = -0.8x^3 \to y = -0.8(x-1)^3 \to y = -0.8(x-1)^3 - 2$

32. $y = x^3 \to y = 2x^3 \to y = -2x^3 \to y = -2(x+5)^3 \to y = -2(x+5)^3 - 6$

33. $y = \sqrt{x} \to y = \sqrt{-x} \to y = 5\sqrt{-x} \to y = 5\sqrt{-(x+y)} \to y = 5\sqrt{-(x+6)} + 5$

34. $y = \sqrt[4]{x} \to y = 0.7\sqrt[4]{x} \to y = 0.7\sqrt[4]{-x} \to y = 0.7\sqrt[4]{-x} \to y = 0.7\sqrt[4]{-(x-8)} \to y = 0.7\sqrt[4]{8-x} - 7$

35. $y = \sqrt{3x} \to y = \sqrt{3(\frac{1}{2}x)} \to y = \sqrt{\frac{3}{2}x} + 1$

36. $y = 4|x| \to y = 4|2x| = 8|x| \to y = 8|x| - 3$

37. Vertical stretch by 4 : $y = x^2 \to y = 4x^2$. Horizontal shrink by 0.5 : $y = x^2 \to y = (2x)^2 = 4x^2$. The resulting curve is the same in both cases. Let $c > 1$ be given. A vertical stretch by c applied to $y = x^n : y = x^n \to y = cx^n$ and a horizontal shrink by $\frac{1}{\sqrt[n]{c}} : y = x^n \to y = (\sqrt[n]{c}x)^n = cx^n$ have the same end result.

38. Vertical stretch by 2 : $y = |x| \to y = 2|x|$. Horizontal shrink by $0.5 = \frac{1}{2} : y = |x| \to y = |2x| = 2|x|$. The resulting curve is the same in both cases. Let $c > 1$ be given. A vertical stretch by c applied to $y = |x| : y = |x| \to y = c|x|$ has the same end result as a horizontal shrink by $\frac{1}{c} : y = |x| \to y = |cx| = c|x|$.

39. In #25 and #26 we obtain, respectively, $y = 3x^2 + 4$ and $y = 3(x^2 + 4) = 3x^2 + 12$. We obtain different geometric results by reversing the order of the transformations. The second graph is 8 units above the first graph.

40. In #27 and #28 we obtain, respectively, $y = 0.2(\frac{1}{x} - 2) = 0.2(\frac{1}{x}) - 0.4$ and $y = 0.2(\frac{1}{x}) - 2$. The second graph is 1.6 units below the first graph.

41. a) Reflect across the y-axis. b) Reflect across the x-axis. c) They are the same: $y = \sqrt[3]{-x} = -\sqrt[3]{x}$.

42. a) Reflect across the y-axis. b) Reflect across the x-axis. c) One is the reflection of the other across the origin.

43. $y = mx + b$

44. $y - y_0 = m(x - x_0)$

45. $(-1,1) \to (2,1) \to (2,3)$. $(0,0) \to (3,0) \to (3,2)$. $(1,1) \to (4,1) \to (4,3)$. The points are $(2,3), (3,2), (4,3)$.

46. $(-1,1) \to (-1,-3) \to (-2,-3)$. $(0,0) \to (0,-4) \to (-1,-4)$. $(1,1) \to (1,-3) \to (0,-3)$. The points are $(-2,-3), (-1,-4), (0,-3)$.

47. $(-1,1) \to (-1,2) \to (-1,-2)$. $(0,0) \to (0,0) \to (0,0)$. $(1,1) \to (1,2) \to (1,-2)$. The points are $(-1,-2), (0,0), (1,-2)$.

48. $(-1,1) \to (-1,0.3) \to (-1,4.3)$. $(0,0) \to (0,0) \to (0,4)$. $(1,1) \to (1,0.3) \to (1,4.3)$. The points are $(-1,4.3), (0,4), (1,4.3)$.

49.

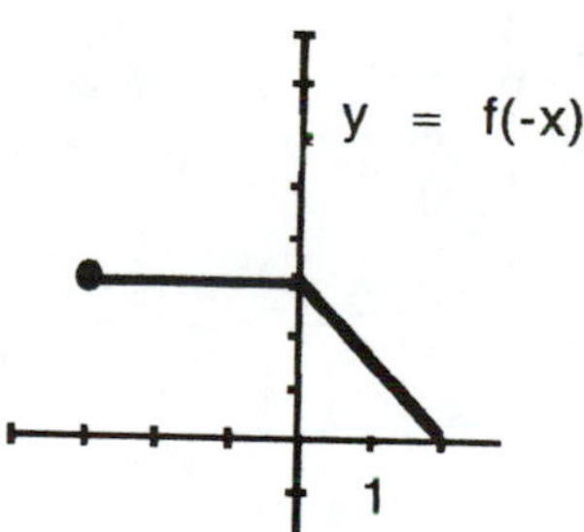

50.

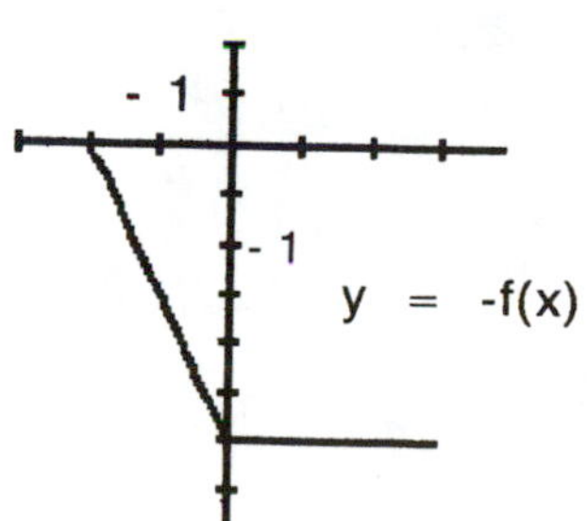

51.

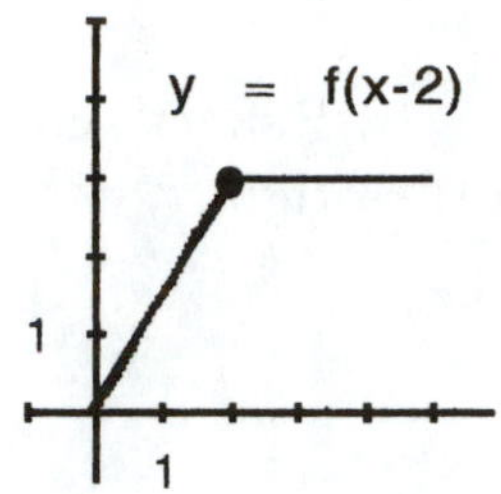

52.

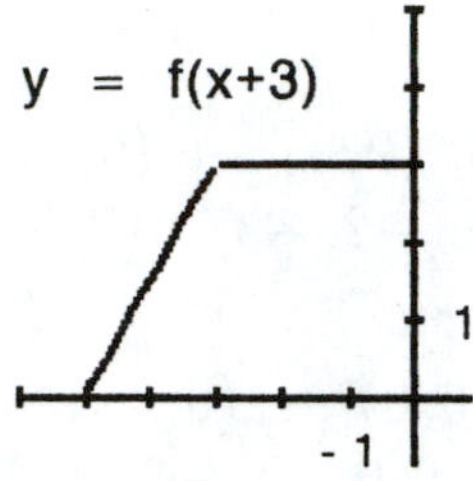

53.

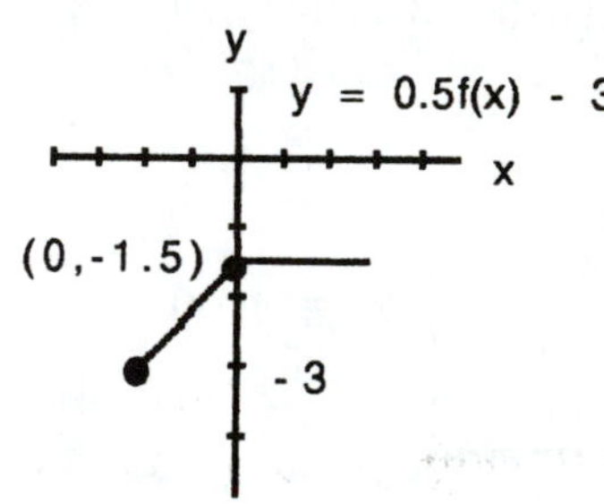

54.

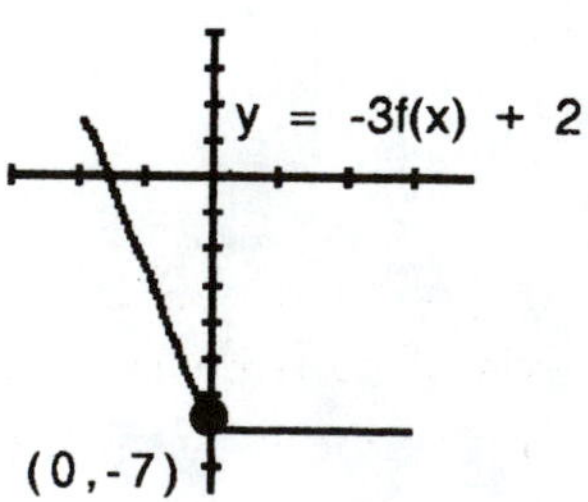

55.

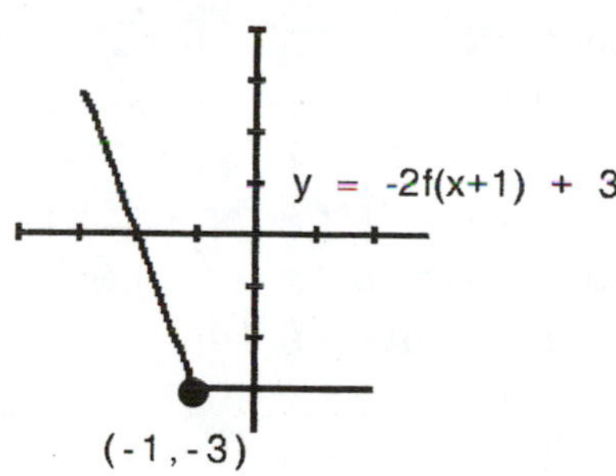

56.

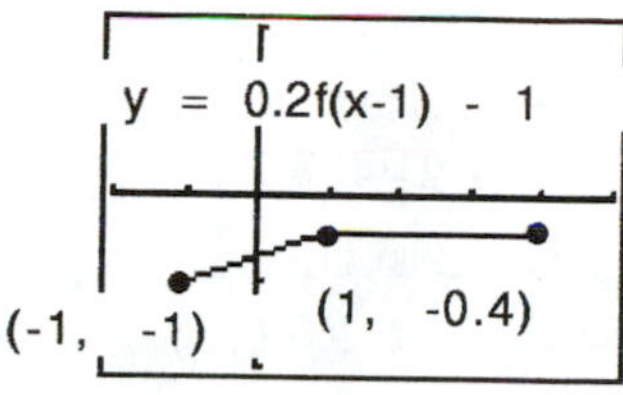

57. Using $y = 2 + \frac{5}{x-2}$, we start with the graph of $y = \frac{1}{x}$. Stretch vertically by a factor of 5, shift right 2 units, shift up 2.

58. By long division $y = \frac{x+1}{x+3} = 1 - \frac{2}{x+3}$. Start with the graph of $y = \frac{1}{x}$. Stretch vertically by a factor of 2, shift left 3, reflect across the x-axis, shift up 1.

59. $y = -2x^2 + 12x - 11 = -2(x^2 - 6x) - 11 = -2(x^2 - 6x + 9) + 2(9) - 11 = -2(x-3)^2 + 7$. Vertex $= (3, 7)$, axis of symmetry: $x = 3$. Check your result by graphing $y = -2x^2 + 12x - 11$ in the viewing rectangle $[-1, 7]$ by $[-16, 7]$.

60. $y = 3x^2 + 12x + 7 = 3(x^2 + 4x + 4) - 12 + 7 = 3(x + 2)^2 - 5$. Vertex: $(-2, -5)$, axis of symmetry: $x = -2$. Check your result by graphing the function in $[-5, 1]$ by $[-5, 10]$.

61. $y = 4x^2 + 20x + 19 = 4[x^2 + 5x + (\frac{5}{2})^2] - 4(\frac{5}{2})^2 + 19 = 4(x + \frac{5}{2})^2 - 6$. Vertex: $(-\frac{5}{2}, -6)$, axis of symmetry: $x = -\frac{5}{2}$. Check your result by graphing the function in $[-5, -0]$ by $[-6, 10]$.

62. $y = -4x^2 + 12x - 3 = -4[x^2 - 3x + (\frac{3}{2})^2] + 4(\frac{3}{2})^2 - 3 = -4(x - \frac{3}{2})^2 + 6$. Vertex: $(\frac{3}{2}, 6)$, axis of symmetry: $x = \frac{3}{2}$. Check your result by graphing the function in $[-1, 4]$ by $[-10, 6]$.

63. $x = y^2 - 6y + 11 = y^2 - 6y + 9 + 2 = (y - 3)^2 + 2$. Start with the graph of $x = y^2$. Shift up 3, shift right 2. $(y - 3)^2 = x - 2$, $y = 3 \pm \sqrt{x - 2}$. Check your sketch by graphing $y = 3 + \sqrt{x - 2}$ and $y = 3 - \sqrt{x - 2}$ in the viewing rectangle $[2, 20]$ by $[-5, 11]$. In parametric mode use $x = t^2 - 6t + 11$, $y = t$, $-5 \leqq t \leqq 11$.

64. $x = y^2 + 4x + 1 = y^2 + 4x + 4 - 3 = (y + 2)^2 - 3$. Shift down 2, shift left 3. $(y + 2)^2 = x + 3$, $y = 2 \pm \sqrt{x + 3}$. Check your sketch by graphing $y = 2 + \sqrt{x + 3}$ and $y = 2 - \sqrt{x + 3}$ in $[-5, 24]$ by $[-4, 8]$. In parametric mode $x = t^2 + 4t + 1$, $y = t$, $-4 \leqq t \leqq 8$.

65. $x = 2y^2 + 4y + 1 = 2(y^2 + 2y) + 1 = 2(y^2 + 2y + 1) - 2 \cdot 1 + 1 = 2(y + 1)^2 - 1$. Shift down 1, stretch horizontally by 2, shift left 1. $\frac{x+1}{2} = (y + 1)^2$, $y = -1 \pm \sqrt{\frac{x+1}{2}}$. Check your sketch by graphing $y = -1 + \sqrt{\frac{x+1}{2}}$ and $y = -1 - \sqrt{\frac{x+1}{2}}$ in $[-1, 10]$ by $[-4, 2]$. In parametric mode $x = 2t^2 + 4t + 1$, $y = t$, $-4 \leqq t \leqq 2$.

66. $x = -3y^2 + 12y - 7 = -3(y^2 - 4y) - 7 = -3(y^2 - 4y + 4) + 3 \cdot 4 - 7 = -3(y - 2)^2 + 5$. Shift up 2, stretch horizontally by 3, reflect across the y-axis, shift right 5. $(y - 2)^2 = \frac{x-5}{-3}$, $y = 2 \pm \sqrt{\frac{5-x}{3}}$. Check your sketch by graphing $y = 2 + \sqrt{\frac{5-x}{3}}$ and $y = 2 - \sqrt{\frac{5-x}{3}}$ in $[-9, 5]$ by $[-1, 5]$. In parametric mode $x = -3t^2 + 12t - 7$, $y = t$, $-1 \leqq t \leqq 5$.

67. $x = -2y^2 + 12y - 13 = -2(y^2 - 6y + 9) + 18 - 13 = -2(y - 3)^2 + 5$. Shift up 3, horizontal stretch by 2, reflection across y-axis, shift right 5. $(y - 3)^2 = \frac{x-5}{-2}$, $y = 3 \pm \sqrt{\frac{5-x}{2}}$. Check your sketch by graphing $y = 3 + \sqrt{\frac{5-x}{2}}$ and $y = 3 - \sqrt{\frac{5-x}{2}}$ in $[-9, 5]$ by $[0, 7]$. In parametric mode $x = -2t^2 + 12t - 13$, $y = t$, $0 \leqq t \leqq 7$.

68. $x = 4y^2 + 16y + 9 = 4(y^2 + 4y + 4) - 16 + 9 = 4(y + 2)^2 - 7$. Shift down 2, horizontal stretch by 4, shift left 7. $(y+2)^2 = \frac{x+7}{4}$, $y = -2 \pm \frac{\sqrt{x+7}}{2}$. Check your sketch by graphing $y = -2 + \frac{\sqrt{x+7}}{2}$ and $y = -2 - \frac{\sqrt{x+7}}{2}$ in $[-7, 30]$ by $[-5, 1]$. In parametric mode $x = 4t^2 + 16t + 9,\ y = t,\ -5 \leqq t \leqq 1$.

69. Two. Only the graphs of $g(x) = (x + 2)^2 + 3$ and $g(x) = (2 - x)^2 + 3$ are possible.

1.5 SOLVING EQUATIONS AND INEQUALITIES GRAPHICALLY

1. $6x^2+5x-6 = 0$. By the quadratic formula or by factoring ($(2x+3)(3x-2) = 0$), we see that $\{-3/2, 2/3\}$ is the solution set. The graph of $y = 6x^2+5x-6$ is a parabola opening upwards with $-3/2$, $2/3$ as x-intercepts.

2. $x^2 - 4x + 1 = 0$. $x = \frac{4\pm\sqrt{16-4}}{2} = \frac{4\pm2\sqrt{3}}{2} = 2 \pm \sqrt{3}$. $\{2 - \sqrt{3}, 2 + \sqrt{3}\}$

3. $4x^3 - 16x^2 + 15x + 2 = 0$. By inspection 2 is a root and so $x - 2$ must be a factor: $(x - 2)(4x^2 - 8x - 1) = 0$ by long division. The solution set is $\{1 - \sqrt{5}/2, 2, 1 + \sqrt{5}/2\}$.

4. $x^3 - x^2 + x = 0$. $x(x^2 - x + 1) = 0$. Since $x^2 - x + 1 = 0$ has no real solution ($b^2 - 4ac = -3 < 0$), $\{0\}$ is the solution set.

5. $2x^2 + 7x - 4 = (2x - 1)(x + 4) = 0$. The solution set is $\{-4, \frac{1}{2}\}$.

6. $\{-0.41, 1.08\}$ or $\{\frac{1\pm\sqrt{5}}{3}\}$

7. $18x^3 - 3x^2 - 14x - 4 = 0$. We determine that $-\frac{1}{2}$ is a root and factor: $2(x+\frac{1}{2})(9x^2-6x-4) = 0$. Using the preceding exercise, we find the solution set to be $\{-0.5, -0.41, 1.08\}$ or $\{-\frac{1}{2}, \frac{1\pm\sqrt{5}}{3}\}$

8. $x^3 - 2x^2 + 2x = x(x^2 - 2x + 2) = 0$. The quadratic has no real roots so the solution set is $\{0\}$.

9. We give a sequence for the smallest solution only. $[-2, -1]$ by $[-1, 1]$, $x\,Scl = 0.1$; $[-2, -1.9]$ by $[-0.1, 0.1]$, $x\,Scl = 0.01$; $[-1.94, -1.93]$ by $[-0.01, 0.01]$, $x\,Scl = 0.001$; $[-1.931, -1.930]$ by $[-0.001, 0.001]$, $x\,Scl = 0.0001$.

10. $x^3 - 4x + 1 = 0$ has 3 solutions. We give a sequence of rectangles for the largest solution only. $[1, 2]$ by $[-1, 1]$, $x\,Scl = 0.1$. $[1.8, 1.9]$ by $[-0.5, 0.5]$, $x\,Scl = 0.01$. $[1.86, 1.87]$ by $[-0.01, 0.01]$, $x\,Scl = 0.001$. $[1.860, 1.861]$ by $[-0.001, 0.001]$, $x\,Scl = 0.0001$. We see that the root lies between 1.8608 and 1.8609.

11. $9x^2 - 6x + 5 = 0$ has no real solution since the discriminant $b^2 - 4ac = (-6)^2 - 4(9)(5) < 0$.

12. $x^2 + x - 12 = 0$. $(x+4)(x-3) = 0$. $\{-4, 3\}$

13. $2x^3 - 8x^2 + 3x + 9 = 0$. Graphing the cubic in $[-10, 10]$ by $[-10, 10]$, $x\,Scl = 1$ suggests $x = 3$ is a solution. This is verified and leads to factoring $(x - 3)(2x^2 - 2x - 3) = 0$. The solution set is $\{3, \frac{1\pm\sqrt{7}}{2}\}$

14. $10 + x + 2x^2 = x^3$. $x^3 - 2x^2 - x - 10 = 0$. A graph of the left-hand side in $[-2, 5]$ by $[-70, 50]$ shows that there is only one real root. Zooming in, we obtain $\{3.25...\}$.

15. Obtain a complete graph of $y = x^3 - 21x^2 + 111x - 71$ in $[0, 15]$ by $[-70, 120]$. To solve the problem we zoom-in to the 3 x-intercepts and obtain $\{0.74, 7.56, 12.70\}$.

16. $y = x^3 + 19x^2 + 90x + 52 = 0$. A graph of y in $[-20, 2]$ by $[-100, 100]$ shows that there are 3 negative roots: approximately $\{-11.67, -6.66, -0.67\}$.

17. We assume that the number x satisfies $2 < x < 6$. The statement a) $0 < x < 4$ about x does not contain precise enough information about x to be able to determine whether the statement is true or false. The same goes for the statement g) which is equivalent to $-2 < x < 6$. Use of rules of inequalities shows that the remaining statements are all equivalent to $2 < x < 6$ whence they are true.

18. All statements are true.

19. $\{-2, 2\}$

20. $|x - 3| = 7$ leads to $x - 3 = -7$ and $x - 3 = 7$ and we get $x = -4$, $x = 10$. The solution set is $\{-4, 10\}$.

21. $|2x + 5| = 4$ leads to $2x + 5 = -4$ and $2x + 5 = 4$. These have solutions $x = -9/2$ and $x = -1/2$. The solution set is $\{-9/2, -1/2\}$.

22. $|1 - x| = 1$. $1 - x = -1$ and $1 - x = 1$. The solution set is $\{0, 2\}$. (The distance between x and 1 is 1.)

23. $|8 - 3x| = 9$. $8 - 3x = -9$ and $8 - 3x = 9$. $x = 17/3$ and $x = -1/3$. $\{-1/3, 17/3\}$.

24. $|\frac{x}{2} - 1| = 1$. $\frac{x}{2} - 1 = -1$ and $\frac{x}{2} - 1 = 1$. $x = 0$ and $x = 4$. $\{0, 4\}$

25. The distance between y and 1 is at most 2: $-1 \leqq y \leqq 3$.

26. $-1 < y + 2 < 1$, $-3 < y < -1$

27. $|3y - 7| < 2$ is equivalent to $-2 < 3y - 7 < 2$, $5 < 3y < 9$, $5/3 < y < 3$.

28. $-10 \leqq \frac{y}{3} \leqq 10$, $-30 \leqq y \leqq 30$

29. The distance between 1 and y is less than $1/10$. $0.9 < y < 1.1$

30. $-1 < \frac{7-3y}{2} < 1$, $-2 < 7 - 3y < 2$, $-9 < -3y < -5$, $3 > y > 5/3$ or $5/3 < y < 3$

31. The midpoint of the interval is $(3 + 9)/2 = 6$ and it is 3 units from the midpoint to an endpoint. Thus $|x - 6| < 3$.

32. The midpoint of the interval is $\frac{-3+9}{2} = 3$ and it is 6 units from the midpoint to either endpoint. $|x - 3| < 6$

33. The midpoint is $(-5 + 3)/2 = -1$ and from there to an endpoint is 4 units. $|x + 1| < 4$.

34. $-7 < x < -1$ has midpoint $\frac{-7+(-1)}{2} = -4$ and radius 3. $|x - (-4)| < 3$ or $|x + 4| < 3$.

35. $|x - 5| < 2$ is equivalent to $-2 < x - 5 < 2$, $3 < x < 7$. Graph $y = abs(x - 5)$ and $y = 2$ and verify that the first curve is below the second in the interval.

36. $|\frac{3x}{2}| < 5$, $\frac{|3x|}{|2|} < 5$, $|3x| < 10$, $|3||x| < 10$, $|x| < \frac{10}{3}$, $-\frac{10}{3} < x < \frac{10}{3}$.

37. $|\frac{4}{x-1}| \leqq 2$ is equivalent to $\frac{|x-1|}{4} \geqq \frac{1}{2}$, $x \neq 1$. $|x - 1| \geq 2$, distance between x and 1 at least 2, leads to $x \leqq -1$ or $x \geqq 3$ but $x \neq 1$. Thus $(-\infty, -1] \cup [3, \infty)$ is the solution set.

38. $|3x + 2| > 3$ has solution set equal to the union of the solution sets of $3x + 2 < -3$ and $3x + 2 > 3$. The first has solution set $(-\infty, -\frac{5}{3})$ and the second has solution set $(\frac{1}{3}, \infty)$. Answer: $(-\infty, -\frac{5}{3}) \cup (\frac{1}{3}, \infty)$.

39. We observe that the graph of $y = |x + 3|$ is below the graph of $y = 5$ in the interval $[-8, 2]$. $|x + 3| = |x - (-3)| \leqq 5$, the distance between x and -3 is at most 5. Hence $-8 \leqq x \leqq 2$.

40. We see that the graph of $y = |\frac{2x}{5}|$ lies below the graph of $y = 1$ in the interval $(-2.5, 2.5)$. $|\frac{2x}{5}| \leqq 1$ leads to $-1 \leqq \frac{2x}{5} \leqq 1$, $-5 \leqq 2x \leqq 5$, $-\frac{5}{2} \leqq x \leqq \frac{5}{2}$. $[-\frac{5}{2}, \frac{5}{2}]$.

41. $|\frac{2}{x+3}| < 1$. $\frac{|x+3|}{2} > 1$, $x \neq -3$, $|x+3| > 2$. We obtain $x+3 < -2$ or $x+3 > 2$, i.e., $x < -5$ or $x > -1$. $(-\infty, -5) \cup (-1, \infty)$ is the solution set.

42. $3x + 2 \leqq -1$ or $3x + 2 \geqq 1$. $(-\infty, -1] \cup [-\frac{1}{3}, \infty)$.

43. $|2 - 3x| < 4$. $-4 < 2 - 3x < 4$, $-6 < -3x < 2$, $2 > x > -2/3$. Answer: $(-2/3, 2)$

44. $-1 \leqq 5 - \frac{x}{2} \leqq 1$, $-6 \leqq -\frac{x}{2} \leqq -4$, $12 \geqq x \geqq 8$. $[8, 12]$.

45. $x^2 + 3x - 10 \leqq 0$, $(x+5)(x-2) \leqq 0$. The graph of the left-hand side is a parabola that opens upward. Thus the solution set will be the interval between the two zeros including the endpoints. Answer: $[-5, 2]$

46. The graph of $y = 4x^2 - 8x + 5$ lies entirely above the x-axis so $4x^2 - 8x + 5 > 0$ is true for all x. We can deduce that this is the solution set by noting that $b^2 - 4ac = 64 - 80 < 0$ (there are no x-intercepts) and the y-intercept is 5 (the parabola must be above the x-axis).

47. Let $y = x^3 - 6x^2 + 5x + 6$. The graph suggests, and we verify, that 2 is a root. $y = (x-2)(x^2 - 4x - 3)$. The other roots are $2 \pm \sqrt{7}$. Answer: $(-\infty, 2 - \sqrt{7}] \cup [2, 2 + \sqrt{7}]$

48. A complete graph of $x^3 - 2x^2 - 5x + 20$ is obtained in $[-10, 10]$ by $[-100, 100]$. There is only one real root. Zooming in, we obtain -2.67. Answer: $(-2.67, \infty)$

49. We graph $y = x^3 - 4x^2 + 3.99x$ in $[-2, 5]$ by $[-2, 2]$. We know this is a complete graph because we a dealing with a cubic function. We determine by Zoom-In those intervals where the graph is above the x-axis. Answer: $(0, 1.9) \cup (2.1, \infty)$.

50. The inequality is algebraically equivalent to $x^3 - 0.2x^2 - 18.14x - 28.8 < 0$. We graph the left-hand side in $[-10, 10]$ by $[-200, 200]$ and Zoom-In to find the intervals in which the graph is below the x-axis: $(-\infty, 4.99)$.

51. $y = 30x - x^2 = x(30 - x)$. The vertex will occur for $x = 15$ midway between the two zeros. Thus $(15, 225)$ is the vertex.

52. Zooming in, we obtain the vertex $(11, 100)$. $y = -(x^2 - 22x) - 21 = -(x^2 - 22x + 121) + 121 - 21 = -(x-11)^2 + 100$. This also implies that $(11, 100)$ is the vertex.

53. b) If the y-range is too large, the graphing utility cannot distinguish between very close values of y.

54. We consider the equivalent equation $y = x^3 - 2x - 2\cos x = 0$. The graph of y in $[-1.5, 1.8]$ by $[-3, 0.1]$ is complete. By zooming in, we confirm there is only one root and it is $1.46\ldots$.

55. We may multiply both sides of the equation by the least common multiple of the denominators of the coefficients. This produces an equivalent equation with integer coefficients. For the given example the l.c.m. is 6 and we obtain the equation $12x^3 + 3x^2 - 4x + 6 = 0$. The possible rational roots of this equation are $\pm\frac{p}{q}$ where $p = 1, 2, 3$ or 6 and $q = 1, 2, 3, 4, 6$ or 12.

56. Non-real roots always occur in conjugate pairs and therefore the real cubic will have 0 or 2 non-real roots. Since there are 3 roots, there will be at least one real root. To find all real roots, first find a complete graph, determine the number of x-intercepts, and zoom in on each of them. The set of x-intercepts is the solution set of the cubic.

57. a) Let y be the length of the adjacent side. $2x + 2y = 100$, $x + y = 50$, $y = 50 - x$, $A = xy$, $A = x(50 - x)$. b) Check you sketch by graphing $y = x(50 - x)$ in $[-10, 60]$ by $[-100, 700]$. c) Domain $= (-\infty, \infty)$, range $= (-\infty, 625]$ d) Both x and y must be positive so $0 < x < 50$. e) Using TRACE for the graph in b) we get the approximate dimensions 13.8ft by 36.2ft. Algebraically we have to solve the system $x + y = 50$, $x(50 - x) = 500$. This gives the dimensions $25 - 5\sqrt{5}$ft by $25 + 5\sqrt{5}$ft. f) In the problem situation the graph is below $y = 500$ for $0 < x < 25 - 5\sqrt{5}$ and for $25 + 5\sqrt{5} < x < 50$.

58. a) $y = 100 - 2x$ is the length of the third side. $A(x) = x(100 - 2x) = 100x - 2x^2$. b) Graph $A(x)$ in $[-20, 70]$ by $[-2000, 1300]$. c) Domain $= (-\infty, \infty)$, range $= (-\infty, 1250]$ d) Problem situation: $0 < x < 50$. Graph $A(x)$ in $[0, 50]$ by $[0, 1250]$. e) We graph $A(x)$ and $y = 500$ and zoom in to find the two points of intersection: $(5.64, 500)$ and $(44.36, 500)$. Since $y = 100 - 2x$, the possible dimensions are 5.64ft (perpendicular to the barn) by 88.72ft and 44.36ft (perpendicular to the barn) by 11.28ft. Algebraically, we set $A(x) = 500$ leading to $x^2 - 50x + 250 = 0$. The quadratic formula yields $x = 25 - 5\sqrt{15} \approx 5.64$ or $25 + 5\sqrt{15} \approx 44.36$ confirming the answer. From d) we see that $A(x) < 500$ has solution set $(0, 5.64) \cup (44.36, 50)$ for $0 < x < 50$.

59. a) $A(x) = (8.5-2x)(11-2x)$ b) Graph $y = A(x)$ in $[0, 10]$ by $[-2, 50]$ c) Domain is $(-\infty, \infty)$. The vertex occurs midway between the x-intercepts at $((8.5/2)+(11/2))/2 = 4.875$. $A(4.875) = -1.5625$. The range is $[-1.5625, \infty)$. d) We must have $8.5 - 2x > 0$ or $0 < x < 8.5/2$, $0 < x < 4.25$. The graph in this interval is the graph of the problem situation. e) $y = (8.5 - 2x)(11 - 2x)$, $y = 60$ has one solution for $0 < x < 4.25$. Zooming in we find its x-coordinate to be $x = 0.95$in.

60. a) $A = (20+2x)(70+2x)-20(70) = 4x^2+180x = 4x(x+45)$ b) Graph $A(x)$ in $[-56, 8]$ by $[-2100, 1000]$ c) $A(x) = 4(x^2+45x) = 4(x^2+45x+\frac{45^2}{4})-45^2 = 4(x+\frac{45}{2})^2 - 2025$. Domain $= (-\infty, \infty)$, range $= [-2025, \infty)$. d) Only $x > 0$ makes sense in the problem situation. The portion of the graph in the first quadrant is the graph of the problem situation. e) From c) $A(x) = 4(x+\frac{45}{2})^2 - 2025 = 500$, $(x+\frac{45}{2})^2 = \frac{2525}{4}$, $x = -\frac{45}{2} + \frac{\sqrt{2525}}{2}$ (x can't be negative) $= 2.62\ldots$ ft.

61. a) $V(x) = 0.1x + 0.25(50 - x)$ b) Graph $y = V(x)$ in $[-20, 100]$ by $[-2, 20]$ c) Domain = range = $(-\infty, \infty)$ d) $0 \leqq x \leqq 50$, x an integer e) $0.1x + 0.25(50 - x) = 9.2$ leads to $10x + 25(50 - x) = 920$, $2x + 5(50 - x) = 184$, $-3x = -66$, $x = 22$. There are 22 dimes and 28 quarters. f) $0.1x + 0.25(50 - x) = 6.25$ does not have an integral solution.

62. $I(x) = 0.065x + 0.08(10000 - x)$ b) Graph $I(x)$ in $[-1000, 11000]$ by $[-100, 1000]$. c) Domain = range = $(-\infty, \infty)$ d) Problem situation: $0 \leqq x \leqq 10000$. Use $[0, 10000]$ for the domain in b). e) The solution of $I(x) = 766.25$ is $x = \$2250$. \$2250 is invested at 6.5% and \$7750 at 8%.

63. a) $V = x(20 - 2x)(25 - 2x)$ b) Graph $y = V$ in $[-2, 17]$ by $[-100, 900]$. c) Domain = range = $(-\infty, \infty)$ d) We must have $x > 0$ and $20 - 2x > 0$ so $0 < x < 10$. Graphing $y = V$ in $[0, 10]$ by $[0, 900]$ gives a graph of the problem situation. e) We zoom in to find the high point to be $(3.68, 820.53)$. For the maximum volume of 820.53in^3 we take $x = 3.68$in.

64. a) $V = x(25 - 2x)(30 - 2x)$ b) Graph V in $[-1, 20]$ by $[-700, 1600]$ c) Domain = range = $(-\infty, \infty)$ d) We must have $25 - 2x > 0$ so $0 < x < 12.5$in. Graph V in $[0, 12.5]$ by $[0, 1600]$. e) $x = 4.52686$in for $V_{\text{max}} = 1512.04\text{in}^3$.

65. a) Let L be the length (horizontal as pictured) of the lid. $2x + 2L = 11$, $2L = 11 - 2x$, $L = 5.5 - x$. b) $V(x) = x(5.5 - x)(8.5 - 2x)$ c) Graph V in $[-1, 7]$ by $[-10, 50]$. The domain and range are both the set of all real numbers. d) The width must be positive: $8.5 - 2x > 0$ which leads to $0 < x < 4.25$. Graph V in $[0, 4.25]$ by $[-10, 50]$. e) In the same viewing window we also graph $y = 25$ and find the x-coordinates of the points of intersection: 0.753in and 2.592in f) $x = 1.585$in. $V_{\text{max}} = V(1.585) = 33.074\text{in}^3$.

66. a) The base has width $2x$. It has length $20 - 2x$. When folded up, the box has height $15 - 2x$. Hence the box has volume $V(x) = 2x(20 - 2x)(15 - 2x)$. b) Graph V in $[-1, 12]$ by $[-750, 864]$. The domain and range are both the set of all real numbers. c) $0 < x < 7.5$. Graph V in $[0, 7.5]$ by $[0, 760]$. d) Graph V and $y = 300$ in the last viewing window. The points of intersection are at $x = 0.574$in and 5.944in. e) Using zoom-in, we obtain $x = 2.829$in, $V_{\text{max}} = V(2.829) = 758.076\text{in}^3$.

1.6 RELATIONS, FUNCTIONS, AND THEIR INVERSES

1. Not one-to-one. Does not pass the Horizontal Line Test.

2. It is one-to-one because it is steadily increasing or it passes the horizontal line test.

3. One-to-one

4. It is not one-to-one because some horizontal lines intersect the graph in two points.

5. The given line is determined by $(-3, 0)$ and $(0, 2)$. The inverse relation has graph a line through $(0, -3)$ and $(2, 0)$.

6. We may assume that this is the graph of $y = x^2 + 1$. The inverse relation is $x = y^2 + 1$. The graph may be seen by graphing $y = \sqrt{x - 1}$ and $y = -\sqrt{x - 1}$ in $[-4, 13]$ by $[-5, 5]$.

7.

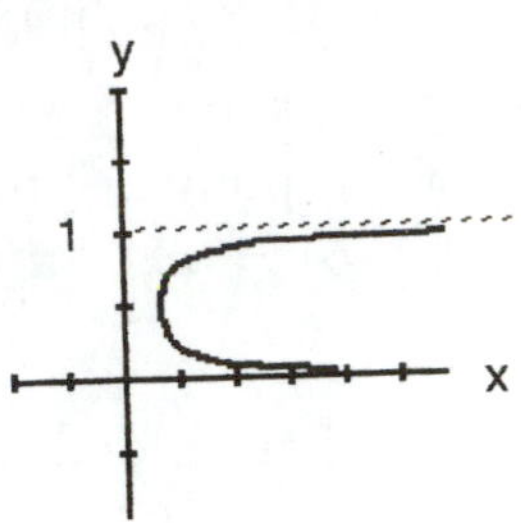

8.

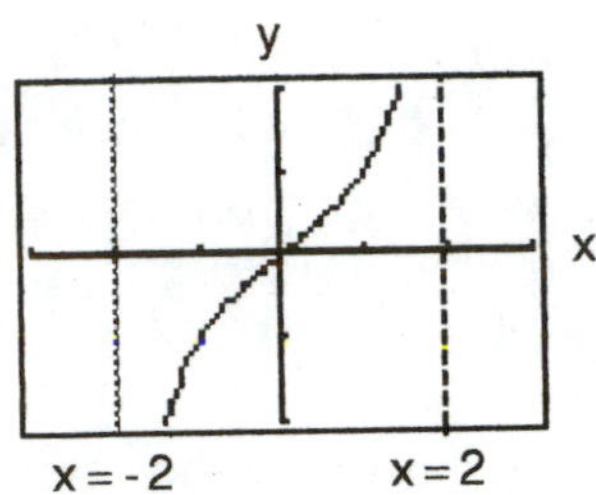

9. $x^2+(y-2)^2=4$. Graph $y=2+\sqrt{4-x^2}$ and $y=2-\sqrt{4-x^2}$ in $[-3.4, 3.4]$ by $[0, 4]$.

10. $(x+2)^2+y^2=9$. Graph $y_1=\sqrt{9-(x+2)^2}$ and $y_2=-y_1$ in $[-7.1, 3.1]$ by $[-3, 3]$.

11. $(x-3)^2+(y+4)^2=25$. Graph $y=-4+\sqrt{25-(x-3)^2}$ and $y=-4-\sqrt{25-(x-3)^2}$ in $[-5.5, 11.5]$ by $[-9, 1]$.

12. $(x-1)^2+(y-1)^2=2$. Graph $y_1=1+\sqrt{2-(x-1)^2}$ and $y_2=1-\sqrt{2-(x-1)^2}$ in $[-1.4, 3.4]$ by $[-0.42, 2.42]$.

13. $x^2+y^2=4$

14. $(x-1)^2+y^2=1$

15. $(x-3)^2+(y-3)^2=9$

16. $(x+1)^2+(y+1)^2=1$

17. $x^2+y^2-6x+8y=-16$. $(x^2-6x)+(y^2+8y)=-16$, $(x^2-6x+9)+(y^2+8y+16)=-16+9+16$, $(x-3)^2+(y+4)^2=9$. Center: $(3, -4)$. Radius: 3. Graph $x=3+3\cos t$, $y=-4+3\sin t$, $0 \leqq t \leqq 2\pi$, in $[-2.1, 8.1]$ by $[-7, -1]$. Domain $=[0, 6]$, range $=[-7, -1]$.

18. $x^2+y^2+2x-4y=11$, $x^2+2x+1+y^2-4x+4=11+1+4$, $(x+1)^2+(y-2)^2=16$. $C=(-1, 2)$, $r=4$. Graph $x=-1+4\cos t$, $y=2+4\sin t$, $0 \leqq t \leqq 2\pi$, in $[-7.8, 5.8]$ by $[-2, 6]$. Domain $=[-5, 3]$, range $=[-2, 6]$.

19. $x^2+y^2+4x+6y+8=0$, $x^2+4x+4+y^2+6y+9=-8+4+9$, $(x+2)^2+(y+3)^2=5$. $C=(-2, -3)$, $r=\sqrt{5}$. Domain $=[-2-\sqrt{5}, -2+\sqrt{5}]$, range $=[-3-\sqrt{5}, -3+\sqrt{5}]$. Graph $x=-2+\sqrt{5}\cos t$, $y=-3+\sqrt{5}\sin t$, $0 \leqq t \leqq 2\pi$, in $[-8.15, 4.15]$ by $[-5.7, 1.5]$.

20. $x^2+y^2-2x-8y+10=0$, $x^2-2x+1+y^2-8y+16=-10+1+16$, $(x-1)^2+(y-4)^2=7$. $C=(1,4)$, $r=\sqrt{7}$. Domain $=[1-\sqrt{7},1+\sqrt{7}]$, range $=[4-\sqrt{7},4+\sqrt{7}]$. Graph $x=1+\sqrt{7}\cos t$, $y=4+\sqrt{7}\sin t$, $0\leqq t\leqq 2\pi$, in $[-5.5,7.5]$ by $[-1,6.7]$.

21. $16x^2-9y^2=144$. $9y^2=16x^2-144=16(x^2-9)$, $y^2=\frac{16}{9}(x^2-9)$. We graph $y=\frac{4}{3}\sqrt{x^2-9}$ and $y=-\frac{4}{3}\sqrt{x^2-9}$ in $[-10,10]$ by $[-10,10]$. The calculator screen shows incorrect gaps in the curve when the slopes of tangent lines become numerically very large.

22. $4x^2+9y^2=36$, $9y^2=4(9-x^2)$, $y=\pm\frac{2}{3}\sqrt{9-x^2}$. Graph $y_1=(2/3)\sqrt{9-x^2}$ and $y_2=-y_1$ in $[-3.4,3.4]$ by $[-2,2]$.

23. a) All points outside the boundary of the unit circle. b) All points in the interior of the circular disk with center $(0,0)$ and radius 2. c) The ring between the two circles.

24. a) Outside and on the unit circle b) Inside and on the circle with center $(0,0)$ and radius 2. c) The annular region between the two circles including the boundaries.

25. $(x+2)^2+(y+1)^2<6$

26. $(x+4)^2+(y-2)^2>16$

27. The inverse relation is given by $x=\frac{3}{y-2}-1$. Its graph is obtained by graphing $x_1=3/(t-2)-1$, $y_1=t$, $-10\leqq t\leqq 10$, tstep $=0.1$, in $[-10,10]$ by $[-10,10]$. We see that the inverse is a function since it passes the vertical line test.

28. The inverse relation is not a function because the function is not one-to-one. For example, $f(0)=f(-5)=0$. The inverse relation is given by $x=y^2+5y$. Graph $x=t^2+5t$, $y=t$, $-7\leqq t\leqq 2$ in $[-6.25,14]$ by $[-8.5,3.5]$.

29. Graph $x=t^3-4t+6$, $y=t$, $-10\leqq t\leqq 10$, tstep $=0.1$, in $[-10,20]$ by $[-15,15]$. This is not a function.

30. The graph steadily rises so the function is one-to-one and its inverse is a function. Graph $x=t^3+t$, $y=t$, $-2\leqq t\leqq 2$ in $[-10,10]$ by $[-5.9,5.9]$.

31. The inverse is given by $x=\ell n\, y^2$ or $e^x=y^2$ which we can graph by graphing $y=e^{x/2}$ and $y=-e^{x/2}$ in $[-2,5]$ by $[-10,10]$. This is not a function.

32. $y = 2^{3-x} = \frac{1}{2^{x-3}}$ steadily decreases so it is one-to-one and the inverse relation is a function. For the inverse relation $x = 2^{3-y}$, graph $x = 2^{3-t}$, $y = t$, $-2 \leqq t \leqq 6$ in $[-2, 32]$ by $[-10, 10]$.

33. $y = f(x) = 2x + 3$. To find f^{-1} we interchange x and $y : x = 2y + 3$, $y = (x-3)/2 = f^{-1}(x)$. $f \circ f^{-1}(x) = f(f^{-1}(x)) = f[(x-3)/2] = 2[(x-3)/2] + 3 = (x-3) + 3 = x$. $f^{-1} \circ f(x) = f^{-1}(2x+3) = [(2x+3) - 3]/2 = x$. Graph $y = 2x + 3$ and $y = (x-3)/2$ in $[-10, 10]$ by $[-10, 10]$.

34. $y = f(x) = 5 - 4x$. $x = 5 - 4y$, $y = \frac{5-x}{4} = f^{-1}(x)$.

35. $y = f(x) = x^3 - 1$. $x = y^3 - 1$, $y^3 = x + 1$, $y = \sqrt[3]{x+1} = f^{-1}(x)$.

36. $y = f(x) = 2 - x^3$. $x = 2 - y^3$, $y = \sqrt[3]{2-x} = f^{-1}(x)$. $f \circ f^{-1}(x) = f(f^{-1}(x)) = f(\sqrt[3]{2-x}) = 2 - (\sqrt[3]{2-x})^3 = 2 - (2-x) = x$. $f^{-1} \circ f(x) = f^{-1}(f(x)) = f^{-1}(2 - x^3) = \sqrt[3]{2 - (2 - x^3)} = \sqrt[3]{x^3} = x$. Graph $y_1 = 2 - x^3$ and $y_2 = (2-x)^{1/3}$ in $[-10, 10]$ by $[-10, 10]$.

37. $y = f(x) = x^2 + 1$, $x \geqq 0$. $x = y^2 + 1$, $y = \pm\sqrt{x-1}$. The domain of f, $x \geqq 0$, must be the range of f^{-1} so we take $f^{-1}(x) = \sqrt{x-1}$.

38. $y = f(x) = x^2$, $x \leqq 0$ (domain $= (-\infty, 0] =$ range f^{-1}). $x = y^2$, $y = -\sqrt{x} = f^{-1}(x)$. Graph $y_1 = x^2 + 0\sqrt{-x}$ and $y_2 = -\sqrt{x}$ in $[-3.5, 5.1]$ by $[-3.9, 7.4]$.

39. $y = f(x) = -(x-2)^2$, $x \leqq 2$. $x = -(y-2)^2$, $y - 2 = \pm\sqrt{-x}$, $y = f^{-1}(x) = 2 - \sqrt{-x}$, $x \leqq 0$. We make sure that the domain $x \leqq 2$ and the range $(-\infty, 0]$ of f become the range $(\infty, 2]$ and the domain $(-\infty, 0]$ of f^{-1}.

40. $y = f(x) = (x+1)^2$, $x \geqq -1$ (domain $f = [-1, \infty) =$ range f^{-1}). $x = (y+1)^2$, $y + 1 = \sqrt{x}$, $y = -1 + \sqrt{x} = f^{-1}(x)$. Graph $y_1 = (x+1)^2 + 0\sqrt{x+1}$ and $y_1 = -1 + \sqrt{x}$ in $[-8, 11]$ by $[-3.8, 7.4]$.

41. $y = f(x) = \frac{1}{x^2}$, $x \geqq 0$. $x = \frac{1}{y^2}$, $y = \pm\frac{1}{\sqrt{x}}$. We have $f^{-1}(x) = \frac{1}{\sqrt{x}}$ because the domain $x \geqq 0$ of f must be the range of f^{-1}.

42. $y = f(x) = \frac{1}{x^3}$. $x = \frac{1}{y^3}$, $y = \frac{1}{\sqrt[3]{x}} = f^{-1}(x)$. Graph $y_1 = x^{-3}$ and $y_2 = x^{(-3^{-1})}$ in $[-8.5, 8.5]$ by $[-5, 5]$.

43. $y = f(x) = \frac{2x+1}{x+3}$, $x \neq -3$. $x = \frac{2y+1}{y+3}$, $(y+3)x = 2y + 1$, $xy + 3x = 2y + 1$, $xy - 2y = 1 - 3x$, $y(x-2) = 1 - 3x$, $y = f^{-1}(x) = \frac{1-3x}{x-2}$, $x \neq 2$.

44. $y = \frac{x+3}{x-2} = f(x)$. $x = \frac{y+3}{y-2}$, $xy - 2x = y + 3$, $(x-1)y = 2x + 3$, $y = \frac{2x+3}{x-1} = f^{-1}(x)$. Graph f and f^{-1} in $[-10, 10]$ by $[-10, 10]$ in dot format.

45. Check your sketch by graphing $y = 2[\ell n(x-4)/\ell n(3)] - 1$ in $[3, 14]$ by $[-10, 4]$ noting that $x = 4$ is a vertical asymptote.

46. Graph $y = -3[\ell n(2-x)/\ell n\, 5] + 1$ in $[-10, 2.7]$ by $[-4, 7]$. $x = 2$ is a vertical asymptote.

47. Check your sketch by graphing $y = -3[\ell n(x+2)/\ell n(0.5)] + 2$ in $[-2, 4]$ by $[-18, 10]$.

48. Graph $y = 2[\ell n(3-x)/\ell n(0.2)] + 1$ in $[-10, 3]$ by $[-2.2, 2.7]$.

49. Graph $y = 5(e^{3x}) + 2$ in $[-2, 1]$ by $[0, 20]$.

50. Graph $y = 3(e^{2-x}) - 1$ in $[-1, 9]$ by $[-2, 25]$.

51. Graph $y = -2(3^x) + 1$ in $[-4, 2]$ by $[-10, 2]$.

52. Graph $y = -5(2^{-x+1}) + 3$ in $[-1, 7]$ by $[-10, 5]$.

53. The graph of $\log x = \log_{10} x$ has the same shape as that in Figure 1.91. We start with the graph of $y = \log x$, reflect it across the x-axis and stretch vertically by a factor of 3. (We now have the graph of $y = -3\log x$.) We then shift left 2 and then up 1. The domain comes from $x + 2 > 0$ so it is $(-2, \infty)$ and the range is $(-\infty, \infty)$.

54. We must have $3 - x > 0$, $3 > x$ so the domain is $(-\infty, 3)$. The range is $(-\infty, \infty)$. Start with the graph of $y = \ell n\, x$. Shift horizontally left 3 units ($y = \ell n(x+3)$). Reflect across the y-axis ($y = \ell n(-x+3) = \ell n(3-x)$). Stretch vertically by a factor of 2. Finally, shift vertically downward 4 units.

55. Start with the graph of $y = 3^x$, shift left 1 ($y = 3^{x+1}$), reflect across the y-axis ($y = 3^{-x+1}$), stretch vertically by a factor of 2 and shift up 1.5. Domain $= (-\infty, \infty)$, range $= (1.5, \infty)$.

56. $5^{x-2} > 0$, $-3(5^{x-2}) < 0$, $-3(5^{x-2}) + 3 < 3$ so the range is $(-\infty, 3)$. The domain is $(-\infty, \infty)$. Shift $y = 5^x$ horizontally 2 units right. Stretch vertically by a factor of 3. Reflect across the x-axis. Shift vertically up 3 units.

57. Shift the graph of $x^2 + y^2 = 9$ three units left and 5 units up. Endpoints of domain are $x = -3 \pm 3$ and the endpoints of the range are $y = 5 \pm 3$. Thus domain $= [-6, 0]$, range $= [2, 8]$.

58. Domain $= [6-5, 6+5] = [1, 11]$. Range $= [-1-5, -1+5] = [-6, 4]$. Shift the graph of $x^2 + y^2 = 5^2$ horizontally 6 units right and vertically one unit down.

59. $f(x) = 2^x$. $f^{-1}(x) = \log_2 x$. Graph $y_1 = 2^x$, $y_2 = \frac{\ell n\, x}{\ell n\, 2}$, $y_3 = x$ in $[-7.5, 14.6]$ by $[-5, 8]$.

60. Graph $y_1 = 0.5^x$, $y_2 = \log_{0.5} x = \frac{\ell n\, x}{\ell n\, 0.5}$, $y_3 = x$ in $[-7.5, 14.6]$ by $[-5, 8]$.

61. Graph $y_1 = \log_3 x = \frac{\ell n\, x}{\ell n\, 3}$, $y_2 = 3^x$, $y_3 = x$ in $[-7.5, 14.6]$ by $[-5, 8]$.

62. Graph $y_1 = \frac{\ell n\, x}{\ell n\, 0.3}$, $y_2 = (0.3)^x$, $y_3 = x$ in $[-2.1, 4.8]$ by $[-1.5, 2.6]$.

63. $e^x + e^{-x} = 3$. Let $u = e^x$. Then $e^{-x} = \frac{1}{e^x} = \frac{1}{u}$. $u + \frac{1}{u} = 3$. $u^2 + 1 = 3u$, $u^2 - 3u + 1 = 0$, $u = \frac{3 \pm \sqrt{9-4}}{2}$, $u = \frac{3 \pm \sqrt{5}}{2} = e^x$. Hence $x = \ell n(\frac{3-\sqrt{5}}{2})$ and $\ell n(\frac{3+\sqrt{5}}{2})$.

64. $2^x + 2^{-x} = 5$. Let $u = 2^x$. $u + u^{-1} = 5$, $u^2 + 1 = 5u$, $u^2 - 5u + 1 = 0$, $u = \frac{5 \pm \sqrt{25-4}}{2} = \frac{5 \pm \sqrt{21}}{2}$. $\ell n\, u = x\, \ell n\, 2 = \ell n((5 \pm \sqrt{21})/2) = \ell n(5 \pm \sqrt{21}) - \ell n\, 2$. $x = -1 + \ell n(5 \pm \sqrt{21})/\ell n\, 2$.

65. $\log_2 x + \log_2(4-x) = 0$. $\log_2[x(4-x)] = 0$, $x(4-x) = 2^0 = 1$, $4x - x^2 = 1$, $x^2 - 4x + 1 = 0$, $x = \frac{4 \pm \sqrt{16-4}}{2} = \frac{4 \pm 2\sqrt{3}}{2} = 2 \pm \sqrt{3}$. In the original equation we must have $x > 0$ and $4 - x > 0$, i.e., $0 < x < 4$. Both these solutions qualify. $\{2 - \sqrt{3}, 2 + \sqrt{3}\}$

66. $\log x + \log(3-x) = 0$ (implying $x > 0$ and $x < 3$ or $0 < x < 3$). $\log(x(3-x)) = 0$, $x(3-x) = 10^0 = 1$, $x^2 - 3x + 1 = 0$, $x = \frac{3 \pm \sqrt{9-4}}{2} = \frac{3 \pm \sqrt{5}}{2}$ (both in $[0, 3]$)

67. Since $R^{-1} = \{(b, a) : (a, b) \text{ is in } R\}$, the hint is the entire proof.

68. We first prove Property 7. $\log_a r^c = \log_a[(a^{\log_a r})^c] = \log_a a^{c \log_a r}$ (property of exponents) $= c \log_a r$ (Property 4). Property 6: $\log_a \frac{r}{s} = \log_a(rs^{-1}) = \log_a r + \log_a s^{-1}$ (Property 5) $= \log_a r + (-1)\log_a s$ (Property 7) $= \log_a r - \log_a s$.

69. $\text{Log}_b a = \frac{\log_a a}{\log_a b} = \frac{1}{\log_a b}$

70. a) $y = a(b^{c-x}) + d$. Domain $= (-\infty, \infty)$. $b^{c-x} > 0$. If $a > 0$, $ab^{c-x} > 0$, $ab^{c-x} + d > d$ and the range is (d, ∞). If $a < 0$, $ab^{c-x} < 0$, $ab^{c-x} + d < d$ and the range is $(-\infty, d)$. b) Domain $= (c, \infty)$, range $= (-\infty, \infty)$.

1.7 A REVIEW OF TRIGONOMETRIC FUNCTIONS

1. $510 \cdot \frac{\pi}{180} = \frac{17\pi}{6} = 8.901$

2. $120° = 120° \frac{\pi}{180°} = \frac{2\pi}{3}$

3. $-42\frac{\pi}{180} = -0.73$

4. $-150\frac{\pi}{180} = -\frac{5\pi}{6}$

5. $6.2(\frac{180}{\pi}) = 355.234°$

6. $-\frac{\pi}{6} = -\frac{\pi}{6}\frac{180°}{\pi} = -30°$

7. $-2(\frac{180}{\pi}) = -114.592°$

8. $\frac{3\pi}{4} = \frac{3\pi}{4}\frac{180°}{\pi} = 135°$

9. $s = r\theta = (2)\frac{5\pi}{8} = \frac{5\pi}{4}$

10. $s = r\theta = 10(\frac{75\pi}{180}) = \frac{25\pi}{6}$

11. $r = \frac{s}{\theta} = \frac{3\pi}{(\pi/6)} = 18$

12. $r = \frac{s}{\theta} = \frac{10}{175\pi/180} = \frac{72}{7\pi}$

13. $\theta = \frac{s}{r} = \frac{7}{14} = \frac{1}{2}$

14. $\theta = \frac{s}{r} = \frac{3\pi/2}{6} = \frac{\pi}{4}$

15. a) $\sin\frac{\pi}{3} = \frac{\sqrt{3}}{2}$, $\cos\frac{\pi}{3} = \frac{1}{2}$, $\tan\frac{\pi}{3} = \sqrt{3}$, $\cot\frac{\pi}{3} = \frac{\sqrt{3}}{3}$, $\sec\frac{\pi}{3} = 2$, $\csc\frac{\pi}{3} = \frac{2\sqrt{3}}{3}$

b) $\sin(-\frac{\pi}{3}) = -\frac{\sqrt{3}}{2}$, $\cos(-\frac{\pi}{3}) = \frac{1}{2}$, $\tan(-\frac{\pi}{3}) = -\sqrt{3}$, $\cot(-\frac{\pi}{3}) = -\frac{\sqrt{3}}{3}$, $\sec(-\frac{\pi}{3}) = 2$, $\csc(-\frac{\pi}{3}) = -\frac{2\sqrt{3}}{3}$

16. a) $\sin(2.5) = 0.5985$, $\cos(2.5) = -0.8011$, $\tan(2.5) = -0.7470$, $\cot(2.5) = -1.3386$, $\sec(2.5) = -1.2482$, $\csc(2.5) = 1.6709$

b) $\sin(-2.5) = -0.5985$, $\cos(-2.5) = -0.8011$, $\tan(-2.5) = 0.7470$, $\cot(-2.5) = 1.3386$, $\sec(-2.5) = -1.2482$, $\csc(-2.5) = -1.6709$

17. a) $\sin(6.5) = 0.2151$, $\cos(6.5) = 0.9766$, $\tan(6.5) = 0.2203$, $\cot(6.5) = 4.5397$, $\sec(6.5) = 1.0240$, $\csc(6.5) = 4.6486$

b) $\sin(-6.5) = -0.2151$, $\cos(-6.5) = 0.9766$, $\tan(-6.5) = -0.2203$, $\cot(-6.5) = -4.5397$, $\sec(-6.5) = 1.0240$, $\csc(-6.5) = -4.6486$

18. a) $\sin(3.7) = -0.5298$, $\cos(3.7) = -0.8481$, $\tan 3.7 = 0.6247$, $\cot(3.7) = 1.6007$, $\sec(3.7) = -1.1791$, $\csc(3.7) = -1.8874$

b) $\sin(-3.7) = 0.5298$, $\cos(-3.7) = -0.8481$, $\tan(-3.7) = -0.6247$, $\cot(-3.7) = -1.6007$, $\sec(-3.7) = -1.1791$, $\csc(-3.7) = 1.8874$

19. a) $\sin\frac{\pi}{2} = 1$, $\cos\frac{\pi}{2} = 0$, $\tan\frac{\pi}{2}$ is undefined, $\cot(\frac{\pi}{2}) = 0$, $\sec\frac{\pi}{2}$ is undefined, $\csc\frac{\pi}{2} = 1$

b) $\sin\frac{3\pi}{2} = -1$, $\cos\frac{3\pi}{2} = 0$, $\tan\frac{3\pi}{2}$ is undefined, $\cot\frac{3\pi}{2} = 0$, $\sec\frac{3\pi}{2}$ is undefined, $\csc\frac{3\pi}{2} = -1$

20. a) $\sin 0 = 0$, $\cos 0 = 1$, $\tan 0 = 0$, $\cot 0$ is undefined, $\sec 0 = 1$, $\csc 0$ is undefined

b) $\sin\pi = 0$, $\cos\pi = -1$, $\tan\pi = 0$, $\cot\pi$ is undefined, $\sec\pi = -1$, $\csc\pi$ is undefined

21. $\sin^{-1}(0.5) = \frac{\pi}{6}$, $30°$

22. $\sin^{-1}(-\frac{\sqrt{2}}{2}) = -\frac{\pi}{4}$, $-45°$

23. $\tan^{-1}(-5) = -1.3734$, $-78.6901°$

24. $\cos^{-1}(0.7) = 0.7954$, $45.5730°$

25. In parametric mode graph $x_1(t) = \cos t$, $y_1(t) = \sin t$, t Min $= 0$, t Max $= 2\pi$, t Step $= 0.1$. Then use TRACE. Then for a given t value, the displayed $x = \cos t$ and $y = \sin t$.

26. In parametric mode and degree mode graph $x_1 = \cos t$, $y_1 = \sin t$, $0 \leqq t \leqq 360$, t Step $= 5$. Use TRACE. For a given t value, the displayed $x = \cos t$ and $y = \sin t$.

27. $[-\pi, 2\pi]$ by $[-1, 1]$, $[-\pi, 2\pi]$ by $[-1, 1]$, $[-1.5\pi, 1.5\pi]$ by $[-2, 2]$, respectively.

28. $[-1.5\pi, 1.5\pi]$ by $[-4, 4]$, $[-\pi, 2\pi]$ by $[-4, 4]$, $[-\pi, 2\pi]$ by $[-4, 4]$, respectively.

29. $[-270°, 450°]$ by $[-3, 3]$, $[-360°, 360°]$ by $[-3, 3]$, $[-180°, 180°]$ by $[-3, 3]$, respectively.

30. $[0, 720]$ by $[-1, 1]$, $[0, 720°]$ by $[-1, 1]$, $[-90, 270]$ by $[-4, 4]$, respectively.

31. Check by graphing the functions in $[-\pi, \pi]$ by $[-3, 3]$.

32. Check by graphing $y_1 = \cos x$, $y_2 = y_1^{-1}$ in $[-1.5\pi, 1.5\pi]$ by $[-4, 4]$.

33. Graph the functions in $[0, 2\pi]$ by $[-1, 1]$.

34. Graph $y_1 = \sin(x/4)$ and $y_2 = \sin x$ in $[0, 8\pi]$ by $[-1, 1]$.

35. $y = 2\cos\frac{x}{3}$. Amplitude $= 2$. Period $= (2\pi)/(1/3) = 6\pi$. To see one period of the function graph it in $[0, 6\pi]$ by $[-2, 2]$. There is a horizontal stretch by a factor of 3.

36. Amplitude $= 2$, period $= \frac{2\pi}{3}$, horizontal shrinking by a factor of $\frac{1}{3}$, vertical stretch by a factor of 2. To see one period of the function graph it in $[0, \frac{2\pi}{3}]$ by $[-2, 2]$.

37. $y = \cot(2x + \frac{\pi}{2}) = \cot[2(x + \frac{\pi}{4})]$. A horizontal shrinking by a factor of $\frac{1}{2}$ is applied to the graph of $y = \cot x$ followed by a horizontal shift left $\frac{\pi}{4}$ units. The period is $\frac{\pi}{2}$. Graphing the function in $[-\frac{\pi}{4}, \frac{3\pi}{4}]$ by $[-2, 2]$ shows two periods of the function.

38. Horizontal shift left $\frac{\pi}{4}$ units, vertical stretch by a factor of 3 followed by a vertical shift 2 units down. Amplitude = 3, period = 2π. Graph $y = 3\cos(x + \pi/4) - 2$ in $[-\frac{\pi}{4}, \frac{7\pi}{4}]$ by $[-5, 1]$ to see one period.

39. Start with the graph of $y = \csc x$, shrink horizontally by a factor of 1/3, shift horizontally left $\pi/3$ units ($y = \csc[3(x + \frac{\pi}{3})]$), stretch vertically by a factor of 3, shift vertically downward 2 units. The period is $2\pi/3$. The domain is the set of all real numbers except the solutions of $\sin(3x + \pi) = 0$, that is except $3x + \pi = n\pi$, $3x = (n-1)\pi$, $x = m\frac{\pi}{3}$, m an integer. The range of $\csc(3x + \pi)$ is $(-\infty, -1) \cup (1, \infty)$. The range of $3\csc(3x + \pi)$ is $(-\infty, -3) \cup (3, \infty)$ and that of $3\csc(3x + \pi) - 2$ is $(-\infty, -5) \cup (1, \infty)$. One period of the graph may be viewed in $[-\frac{\pi}{3}, \frac{\pi}{3}]$ by $[-11, 7]$.

40. Start with the graph of $y = \sin x$. Shift left π units, shrink horizontally by a factor of $\frac{1}{4}$, stretch vertically by a factor of 2 and finally shift vertically up 3 units. Amplitude = 2, period = $\frac{\pi}{2}$, domain = $(-\infty, \infty)$, range = $[1, 5]$. Graph y in $[-\frac{\pi}{4}, \frac{\pi}{4}]$ by $[1, 5]$ to view one period.

41. Start with the graph of $y = \tan x$, shrink horizontally by a factor of 1/3, shift horizontally left $\pi/3$ units, stretch vertically by a factor of 3, reflect across the x-axis, shift vertically upward 2 units. The period is $\pi/3$. The domain is all real numbers except the solutions of $\cos(3x + \pi) = 0$, $3x + \pi = (2m + 1)\pi/2$, $3x = (2m - 1)\pi/2$, $x = (2m - 1)\pi/6$, the odd multiples of $\pi/6$. The range is $(-\infty, \infty)$. One period of the graph may be viewed in $[-\frac{\pi}{6}, \frac{\pi}{6}]$ by $[-11, 15]$.

42. Start with the graph of $y = sinx$. Shift horizontally left $\frac{\pi}{3}$ units, shrink horizontally by a factor of $\frac{1}{2}$, stretch vertically by a factor of 2. Domain = $(-\infty, \infty)$, range = $[-2, 2]$, period = π, amplitude = 2. Graph $y = 2\sin(2x + \frac{\pi}{3})$ in $[-\frac{\pi}{6}, \frac{5\pi}{6}]$ by $[-2, 2]$ to view one period.

43. $\cos x = -0.7$. Two solutions are $\pm\cos^{-1}(-0.7)$, the only solutions in the interval $[-\pi, \pi]$. The solution set is $\{\pm\cos^{-1}(-0.7) + 2n\pi\}$.

44. $\sin x = 0.2$. In $[0, 2\pi)$ the solutions are $\sin^{-1} 0.2$ and $\pi - \sin^{-1} 0.2$. This leads to the solution set $\{x + 2n\pi : x = 0.20 \text{ and } x = 2.94\}$.

45. $\tan x = 4$ has solution set $\{(\tan^{-1} 4) + n\pi\}$

46. $\sin x = -0.2$, $-\sin x = 0.2$, $\sin(-x) = 0.2$. By Exercise 44, $-x = 0.20 + 2n\pi$ and $2.94 + 2n\pi$. So the solutions are $x + 2n\pi$ where $x = -0.20$ and -2.94.

47. To solve $\sin x = 0.2x$, we may first graph $y = 0.2x - \sin x$ in $[-10, 10]$ by $[-2, 2]$. We see that there are 3 solutions (x-intercepts) and we Zoom-In to find them: $\{-2.596, 0, 2.596\}$.

48. To solve $\cos x = 0.2x$, we may first graph $y = 0.2x - \cos x$ in $[-10, 10]$ by $[-2, 2]$. We see that there are 3 solutions and we Zoom-In to find them: $\{-3.837, -1.977, 1.306\}$.

49. We graph $y = \ell n\ x - \sin x$ in $[0, 4]$ by $[-2, 2]$ and Zoom-In to the x-intercept: $x = 2.219$.

50. We graph $y = 0.23x^2 - 0.5 + \cos x$ in $[-5, 5]$ by $[-0.1, 3]$. Zooming-In, we see that there is no x-intercept, hence no solution to the equation.

51. We may use the work in Exercise 47 and solve the equivalent inequality $0.2x - \sin x < 0$. The set of x for which the graph is below the x-axis is $(-\infty, -2.596) \cup (0, 2.596)$.

52. We may use the work in Exercise 46 and solve the equivalent inequality $0.2x - \cos x < 0$. The set of x for which the graph is below the x-axis is $(-\infty, -3.837) \cup (-1.977, 1.306)$.

53. Graph $x_1 = \sqrt{5}\cos t$, $y_1 = \sqrt{5}\sin t$, $0 \leqq t \leqq 2\pi$ in $[-3.8, 3.8]$ by $[-\sqrt{5}, \sqrt{5}]$. We use the "squaring" device on the calculator to produce a viewing window in which the curve appears to be a circle.

54. Graph $x_1 = 2\cos t$, $y_1 = 2\sin t$, $0 \leqq t \leqq 2\pi$. After graphing in $[-2, 2]$ by $[-2, 2]$, we used the "squaring" device on the calculator to produce the viewing rectangle $[-3.4, 3.4]$ by $[-2, 2]$ in which the curve appears to be a circle.

55. Graph $x_1 = 2 + 3\cos t$, $y_1 = -3 + 3\sin t$, $0 \leqq t \leqq 2\pi$, in $[-3.1, 7.1]$ by $[-6, 0]$.

56. Graph $x_1 = -1 + 4\cos t$, $y_1 = -3 + 4\sin t$, $0 \leqq t \leqq 2\pi$ in $[-7.8, 5.8]$ by $[-7, 1]$.

57. $y = 2\sin x + 3\cos x$. $a = 2, b = 3$. $A = \sqrt{a^2 + b^2} = \sqrt{13}$. $\cos\alpha = a/A = 2/\sqrt{13}$, $\sin\alpha = 3/\sqrt{13}$. Thus α is in the first quadrant and we may take $\alpha = \sin^{-1}(3/\sqrt{13})$. $y = A\sin(x + \alpha) = \sqrt{13}\sin(x + \alpha)$, $\alpha = \sin^{-1}(3/\sqrt{13}) = 0.9828$.

58. $y = \sin x + \sqrt{3}\cos x$. $a = 1, b = \sqrt{3}$. $A = \sqrt{1+3} = 2$. $\cos\alpha = a/A = 1/2$, $\sin\alpha = b/A = \sqrt{3}/2$, $\alpha = \pi/3$. $y = A\sin(x+\alpha) = 2\sin(x + \frac{\pi}{3})$.

59. $y = \sin 2x + \cos 2x = \sqrt{2}[\sin 2x\cos(\pi/4) + \cos 2x \sin(\pi/4)] = \sqrt{2}\sin(2x + \frac{\pi}{4})$.

60. $y = 2\sin 3x + 2\cos 3x = 2\sqrt{2}(\sin 3x \cos(\pi/4) + \cos 3x \sin(\pi/4)) = 2\sqrt{2}\sin(3x + \frac{\pi}{4})$.

61. b) and d); c) and e)

62. c), g), h) all have the same graph as $y = -\sin x$. b), e), f) all have the same graph as $y = -\cos x$.

63. $1 + \tan^2\theta = 1 + (\frac{y}{x})^2 = \frac{x^2+y^2}{x^2} = \frac{1}{x^2} = \sec^2\theta$.

64. $1 + \cot^2\theta = 1 + (\frac{x}{y})^2 = \frac{y^2+x^2}{y^2} = \frac{1}{y^2} = \csc^2\theta$.

65. The equation $y = \cos(-x)$ is the same as $y = \cos x$ so the graph of the cosine is symmetric with respect to the y-axis. $-y = \sin(-x) = -\sin x$ is the same as $y = \sin x$ so the graph of the sine is symmetric with respect to the origin. Similarly, the graph of the tangent is symmetric with respect to the origin.

66. The graph of the secant is symmetric with respect to the y-axis. The graphs of the cosecant and cotangent are symmetric with respect to the origin.

67. a) yes, b) $-1 \leqq \cos 2x \leqq 1$, c) $0 \leqq 1 + \cos 2x \leqq 1 + 1$, $0 \leqq \frac{1+\cos 2x}{2} \leqq 1$, d) $0 \leqq \sqrt{\frac{1+\cos 2x}{2}} \leqq \sqrt{1} = 1$. The domain is the set of all reals; the range is the interval $[0,1]$. Alternative approach: $y = \sqrt{(1+\cos 2x)/2} = \sqrt{\cos^2 x} = |\cos x|$.

68. a) Exclude $(2k+1)\frac{\pi}{2}$, b) Exclude $(2k+1)\pi$, c) All real numbers, d) The domain consists of all real numbers except $(2k+1)\pi$. The range is the set of all real numbers.

69. a) 37 b) period $= 2\pi/(2\pi/365) = 365$ c) 101 units to the right d) 25 units upward

70. a) Since the extreme values of the sine function are ± 1, $y_{\max} = 37 + 25 = 62°$, $y_{\min} = -37 + 25 = -12°$.

b) $\frac{62° + (-12°)}{2} = 25°$. This is equal to the vertical shift because the range of the sine function is symmetric about $y = 0$.

71. We obtain $\cos(A - A) = \cos A \cos A + \sin A \sin A$ or $1 = \cos^2 A + \sin^2 A$ for the first equation. For the second equation we obtain $\sin(A - A) = \sin A \cos A - \cos A \sin A$ or $0 = 0$.

72. $\cos(A - \pi/2) = \cos A \cos(\pi/2) + \sin A \sin(\pi/2) = \sin A$. $\sin(A - \pi/2) = \sin A \cos(\pi/2) - \cos A \sin(\pi/2) = -\cos A$.

73. $\cos(A + \frac{\pi}{2}) = \cos A \cos \frac{\pi}{2} - \sin A \sin \frac{\pi}{2} = -\sin A$. If we start the cosine curve, reflect it across the x-axis and shift horizontally $\frac{\pi}{2}$ units to the left, we obtain the sine curve. $\sin(A + \frac{\pi}{2}) = \sin A \cos \frac{\pi}{2} + \cos A \sin \frac{\pi}{2} = \cos A$.

74. $\cos(A + \pi) = \cos A \cos \pi - \sin A \sin \pi = -\cos A$.

$\sin(A + \pi) = \sin A \cos \pi + \cos A \sin \pi = -\sin A$.

$\cos(A - \pi) = \cos A \cos \pi + \sin A \sin \pi = -\cos A$.

$\sin(A - \pi) = \sin A \cos \pi - \cos A \sin \pi = -\sin A$.

75. $\cos 15° = \cos(45° - 30°) = \cos 45° \cos 30° + \sin 45° \sin 30° = \frac{\sqrt{2}}{2}\frac{\sqrt{3}}{2} + \frac{\sqrt{2}}{2}\frac{1}{2} = \frac{\sqrt{2}}{4}(\sqrt{3} + 1)$ or $\frac{\sqrt{6}+\sqrt{2}}{4}$

76. $\sin 75° = \sin(45° + 30°) = \sin 45° \cos 30° + \cos 45° \sin 30° = \frac{\sqrt{2}}{2}\frac{\sqrt{3}}{2} + \frac{\sqrt{2}}{2}\frac{1}{2} = \frac{\sqrt{6}+\sqrt{2}}{4}$

77. $\sin \frac{7\pi}{12} = \sin(\frac{\pi}{4} + \frac{\pi}{3}) = \sin \frac{\pi}{4} \cos \frac{\pi}{3} + \cos \frac{\pi}{4} \sin \frac{\pi}{3} = \frac{\sqrt{2}}{2}\frac{1}{2} + \frac{\sqrt{2}}{2}\frac{\sqrt{3}}{2} = \frac{\sqrt{2}+\sqrt{6}}{4}$

78. $\cos \frac{10\pi}{24} = \cos(\frac{\pi}{4} + \frac{\pi}{6}) = \cos \frac{\pi}{4} \cos \frac{\pi}{6} - \sin \frac{\pi}{4} \sin \frac{\pi}{6} = \frac{\sqrt{2}}{2}\frac{\sqrt{3}}{2} - \frac{\sqrt{2}}{2}\frac{1}{2} = \frac{\sqrt{6}-\sqrt{2}}{4}$

79. $\cos^2 \frac{\pi}{8} = \frac{1+\cos\frac{\pi}{4}}{2} = \frac{1+\frac{\sqrt{2}}{2}}{2} = \frac{2+\sqrt{2}}{4}$

80. $\cos^2 \frac{\pi}{12} = \frac{1+\cos\frac{\pi}{6}}{2} = \frac{1+\frac{\sqrt{3}}{2}}{2} = \frac{2+\sqrt{3}}{4}$

81. $\sin^2 \frac{\pi}{12} = \frac{1-\cos\frac{\pi}{6}}{2} = \frac{1-\frac{\sqrt{3}}{2}}{2} = \frac{2-\sqrt{3}}{4}$

82. $\sin^2 \frac{\pi}{8} = \frac{1}{2}(1 - \sin \frac{\pi}{4}) = \frac{1}{2}(1 - \frac{\sqrt{2}}{2}) = \frac{2-\sqrt{2}}{4}$

83. We discuss only $\cos 2\theta = \cos^2\theta - \sin^2\theta$. We may graph $y_1 = \cos 2x$ and $y_2 = \cos^2 x - \sin^2 x$ in $[0, 2\pi]$ by $[-2, 2]$ and see that we get just one curve. Alternatively we can shift one curve up slightly and see that we get two parallel curves: In the same viewing rectangle graph $y_1 = \cos 2x$ and $y_2 = \cos^2 x - \sin^2 x + 0.5$.

84. Let $A = \sqrt{a^2+b^2}$ and let α be the unique solution in $[0, 2\pi)$ of $\cos\alpha = a/A$, $\sin\alpha = b/A$. $(a/A, b/A)$ is a point on the unit circle so there is such an angle α. Then $A\sin(x+\alpha) = A\sin x\cos\alpha + A\cos x\sin\alpha = a\sin x + b\cos x$.

85. $\tan(A+B) = \frac{\sin(A+B)}{\cos(A+B)} = \frac{\sin A\cos B + \cos A\sin B}{\cos A\cos B - \sin A\sin B} = \frac{\tan A + \tan B}{1-\tan A\tan B}$. The last step was obtained by dividing all terms by $\cos A\cos B$.

86. $\tan(A-B) = \frac{\tan A + \tan(-B)}{1-\tan A\tan(-B)} = \frac{\tan A - \tan B}{1+\tan A\tan B}$

87. a) Let $f(x) = \cot x$. $f(-x) = \frac{\cos(-x)}{\sin(-x)} = \frac{\cos x}{-\sin x} = -f(x)$, proving $f(x)$ is odd.

b) Let $g(x) = \frac{h(x)}{k(x)}$ where $h(x)$ is even and $k(x)$ is odd. $g(-x) = \frac{h(-x)}{k(-x)} = \frac{h(x)}{-k(x)} = -g(x)$ proving $g(x)$ is odd where defined.

c) The graph of $y = \cot(-x) = -\cot x$ can be obtained by reflecting the graph of $y = \cot x$ across the x-axis.

88. a) $\csc\theta = \frac{y}{r}$, $\csc(-\theta) = -\frac{y}{r} = -\csc\theta$.

b) Let $f(x)$ be an odd function and let $F(x) = 1/f(x)$, $F(-x) = 1/f(-x) = 1/[-f(x)] = -1/f(x) = -F(x)$.

c) By reflecting it across the origin.

89. Use the method indicated in the solution of Exercise 87.

90. a) Let $f(x) = \sin^2 x = (\sin x)\sin x$. $f(-x) = (\sin(-x))\sin(-x) = -\sin x(-\sin x) = \sin^2 x$.

b) Let $F(x) = f^2(x)$ where $f(x)$ is odd. $F(-x) = f^2(-x) = [-f(x)]^2 = f^2(x) = F(x)$ so $F(x)$ is even.

c) Let $F(x) = f(x)g(x)$ where f and g are odd. $F(-x) = f(-x)g(-x) = [-f(x)][-g(x)] = f(x)g(x) = F(x)$ so $F(x)$ is even.

91. $f(x) = \sin(60x)$. $60x$ will range between 0 and 2π if x ranges between 0 and $\frac{2\pi}{60}$. Thus $f(x)$ has fundamental period $\frac{\pi}{30}$. Graph $y = \sin(60x)$ in the window $[0, \frac{\pi}{30}]$ by $[-1, 1]$.

92. $f(x) = \cos(60\pi x)$. Fundamental period $= \frac{2\pi}{60\pi} = \frac{1}{30}$. Graph $y = \cos(60\pi x)$ in $[0, \frac{1}{30}]$ by $[-1, 1]$.

93. Since parallel lines have the same slope m and the same angle of inclination α, we may assume that the line passes through the origin. Let (a, b) be a point on the line, $a \neq 0$, $b \geqq 0$. Then $m = \frac{b}{a} = \tan\alpha$.

94. If α is the angle of inclination of a line, then $\tan\alpha = m$, the slope of the line. Thus $\alpha = \tan^{-1} m$ if $m \geqq 0$ and $\alpha = (\tan^{-1} m) + \pi$ if $m < 0$. In the following problems we will convert to degree measure.

$m = \frac{y_2-y_1}{x_2-x_1} = \frac{1-5}{3-1} = -2$. $\alpha = \tan^{-1}(-2) + 180° =$ (in degrees) $116.565°$.

95. $m = \frac{5.5-2}{3+1} = 0.875$. In degrees, $\alpha = \tan^{-1} 0.875 = 41.186°$.

96. $m = 2.5$. In degrees, $\alpha = \tan^{-1}(2.5) = 68.199°$.

97. $2x - 6y = 7$, $6y = 2x - 7$, $y = \frac{1}{3}x - \frac{7}{6}$, $m = \frac{1}{3}$. In degrees, $\alpha = \tan^{-1}\frac{1}{3} = 18.435°$.

PRACTICE EXERCISES, CHAPTER 1

1. a) $(1, -4)$ b) $(-1, 4)$ c) $(-1, -4)$ **2.** a) $(2, 3)$ b) $(-2, -3)$ c) $(-2, 3)$

3. a) $(-4, -2)$ b) $(4, 2)$ c) $(4, -2)$ **4.** a) $(-2, 2)$ b) $(2, -2)$ c) $(2, 2)$

5. a) origin b) y-axis

6. a) $y = x^3$. Symmetric about the origin only. b) $y = x^4$. Symmetric about the y-axis only.

7. a) both axes and the origin b) none of the mentioned symmetries

8. a) $y = x^{1/3}$. Origin only. b) $y = x^{2/3}$. y-axis only.

9. $x = 1$, $y = 3$ **10.** $x = 2$, $y = 0$

11. $x = 0$, $y = -3$ **12.** $x = x_0$, $y = y_0$

13. Using $y - y_0 = m(x - x_0)$, we get $y - 3 = 2(x - 2)$ or $y = 2x - 1$. y-intercept is $y = -1$. x-intercept: $0 = 2x - 1$, $x = \frac{1}{2}$. We may graph the line by drawing the line through $(\frac{1}{2}, 0)$ and $(0, -1)$.

14. $y = 3$. y-intercept 3.

15. $y - 0 = -(x - 1)$. Intercepts: $x = 1$, $y = 1$

16. $y - 1 = (-1)x$, $y = 1 - x$. x-intercept 1, y-intercept 1.

17. $y + 6 = 3(x - 1)$ or $y = 3x - 9$. Intercepts: $x = 3$, $y = -9$.

18. $y - 0 = (1)(x + 2)$, $y = x + 2$. x-intercept -2, y-intercept 2.

19. $y - 2 = -\frac{1}{2}(x+1)$ or $y = -\frac{1}{2}x + \frac{3}{2}$. Intercepts: $x = 3$, $y = \frac{3}{2}$.

20. $y - 1 = \frac{1}{3}(x-3) = \frac{1}{3}x - 1$, $y = \frac{1}{3}x$. x-intercept 0, y-intercept 0.

21. $m = \frac{y_2 - y_1}{x_2 - x_1} = \frac{3+2}{1+2} = \frac{5}{3}$. $y + 2 = \frac{5}{3}(x+2)$, $3y + 6 = 5(x+2) = 5x + 10$, $3y = 5x + 4$.

22. $m = \frac{-2-6}{1+3} = -\frac{8}{4} = -2$. $y + 2 = -2(x-1)$, $y = -2x$

23. $m = \frac{4+1}{4-2} = \frac{5}{2}$. $y + 1 = \frac{5}{2}(x-2) = \frac{5}{2}x - 5$, $y = \frac{5}{2}x - 6$

24. $m = \frac{5-3}{-2-3} = -\frac{2}{5}$. $y - 3 = -\frac{2}{5}(x-3)$ or $5y - 15 = -2x + 6$, $2x + 5y = 21$

25. $y = \frac{1}{2}x + 2$

26. $y = -3x + 3$

27. $y = -2x - 1$

28. $y = 2x$

29. $2x - y = -2$ or $y = 2x + 2$ so $m = 2$. a) $y = 2(x-6)$ or $2x - y = 12$ b) $y = -\frac{1}{2}(x-6)$ or $2y = -x + 6$ or $x + 2y = 6$. c) For the point of intersection of the two lines: $y = 2x + 2 = -\frac{1}{2}x + 3$, $\frac{5}{2}x = 1$, $x = \frac{2}{5}$, $y = \frac{14}{5}$. Distance $= \sqrt{(6 - \frac{2}{5})^2 + (0 - \frac{14}{5})^2} = \frac{14}{5}\sqrt{5}$

30. a) $m = 1$. $y - 1 = (1)(x-3)$, $y = x - 2$ b) $y - 1 = -(x-3) = -x + 3$, $y = -x + 4$. c) Point of intersection: $y = -x + 4 = x + 2$, $2x = 2$, $x = 1$, $y = 3$. Distance $= \sqrt{(3-1)^2 + (1-3)^2} = \sqrt{8} = 2\sqrt{2}$

31. $L : y = -\frac{4}{3}x + 12$, $m = -\frac{4}{3}$. a) $y + 12 = -\frac{4}{3}(x-4)$ or $4x + 3y = -20$ b) $y + 12 = \frac{3}{4}(x-4)$. c) For the point of intersection of L and the perpendicular we obtain $(\frac{228}{25}, -\frac{204}{25})$. The distance is $\frac{32}{5}$.

32. a) $m = -\sqrt{3}$. $y - 1 = -\sqrt{3}(x-0)$, $y = -\sqrt{3}x + 1$. b) $y - 1 = \frac{1}{\sqrt{3}}x$, $y = \frac{1}{\sqrt{3}}x + 1$. c) Point of intersection: $y = \frac{1}{\sqrt{3}}x + 1 = -\sqrt{3}x - 3$, $x + \sqrt{3} = -3x - 3\sqrt{3}$, $4x = -4\sqrt{3}$, $x = -\sqrt{3}$, $y = 0$. Distance $= \sqrt{(-\sqrt{3} - 0)^2 + (0-1)^2} = 2$.

33. Domain = range = $(-\infty, \infty)$

34. Graph $y = |x| - 2$ in $[-6, 6]$ by $[-4, 4]$. Domain = $(-\infty, \infty)$, range = $[-2, \infty)$.

35. The range of $y = 2|x-1|$ is $[0, \infty)$ so the range of $y = 2|x-1| - 1$ is $[-1, \infty)$. The domain is $(-\infty, \infty)$. Check your sketch by graphing $y = 2\,\text{abs}(x-1) - 1$ in $[-2, 4]$ by $[-1, 5]$.

36. Graph one period of $y = \sec x$ in $[-\frac{\pi}{2}, 1.5\pi]$ by $[-4, 4]$. Domain = all reals except $(2k+1)\frac{\pi}{2}$, range = $(-\infty, -1] \cup [1, \infty)$.

37. Domain = $(-\infty, \infty)$, range = $[-1, 1]$.

38. Domain = $(-\infty, \infty)$, range = all integers.

39. Domain = $(-\infty, 0]$, range = $(-\infty, \infty)$. Check your sketch by graphing $y = -\sqrt{-x}$ and $y = \sqrt{-x}$ in $[-9, 0]$ by $[-3, 3]$.

40. Graph $y = -2 + \sqrt{1-x}$ in $[-10, 2]$ by $[-3.5, 3.5]$. Domain = $(-\infty, 1]$, range = $[-2, \infty)$.

41. Domain = range = $(-\infty, \infty)$. Graph $f(x)$ in $[-9, 4]$ by $[-100, 100]$.

42. Graph $y = -1 + \sqrt[3]{1-x}$ in $[-4, 6]$ by $[-4.5, 3.2]$. Domain = range = $(-\infty, \infty)$.

43. Domain = $(1, \infty)$, range = $(-\infty, \infty)$. Graph $y = \ell n(x-1)/\ell n(7) + 1$ in $[1, 3]$ by $[-3, 3]$, recalling that $x = 1$ is a vertical asymptote.

44. Graph $y = 3^{2-x} + 1$ in $[-1, 6]$ by $[0, 28]$. Domain = $(-\infty, \infty)$, range = $(1, \infty)$.

45. $f(x) = \begin{cases} -2x - 1, & x < -3 \\ 5, & -3 \leqq x < 2 \\ 2x + 1, & x \geqq 2 \end{cases}$.

Domain = $(-\infty, \infty)$, range = $[5, \infty)$. Graph $y = \text{abs}(x-2) + \text{abs}(x+3)$ in $[-5, 4]$ by $[4, 9]$.

46. $f(x) = \frac{|x-2|}{x-2} = \begin{cases} -1, & x < 2 \\ 1, & x > 2 \end{cases}$.

Domain = all reals except 2, range = $\{-1, 1\}$.

47. Stretch vertically by a factor of 2, reflect across the x-axis, shift horizontally right one unit, shift vertically 5 units upward.

48. $f(x) = 2\,\ell n(-x+1) + 3$, $g(x) = \ell n\, x$. Shift horizontally left one unit, reflect across the y-axis, vertically stretch by a factor of 2, shift vertically upward 3 units.

49. $f(x) = 3\sin(3x + \pi) = 3\sin[3(x + \frac{\pi}{3})]$. Stretch vertically by a factor of 3, shrink horizontally by a factor of 1/3, shift horizontally $\pi/3$ units left.

50. Shift horizontally left 3 units, reflect across the y-axis, stretch vertically by a factor of 2, reflect across the x-axis, shift vertically 5 units up.

51. $y = x^2$, $y = 2x^2$, $y = -2x^2$, $y = -2(x-2)^2$, $y = -2(x-2)^2 + 3$.

52. $x = y^2 \to 2x = y^2 \to -2x = y^2 \to -2(x+3) = y^2$, $-2(x+3) = (y+2)^2$

53. $y = \frac{1}{x}$, $y = \frac{3}{x}$, $y = \frac{3}{x+2}$, $y = \frac{3}{x+2} + 5$

54. $x^2 + y^2 = 1 \to (x+3)^2 + y^2 = 1 \to (x+3)^2 + (y-5)^2 = 1$

55.

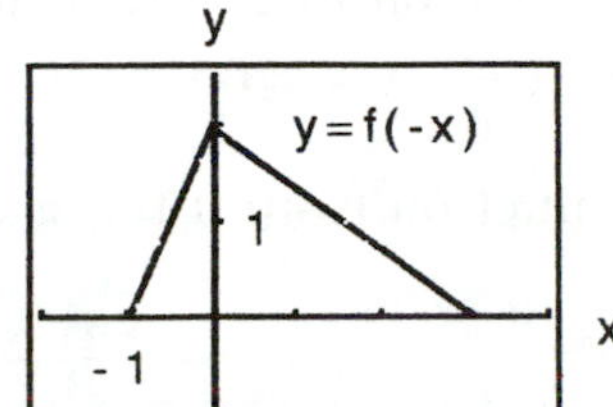

56.

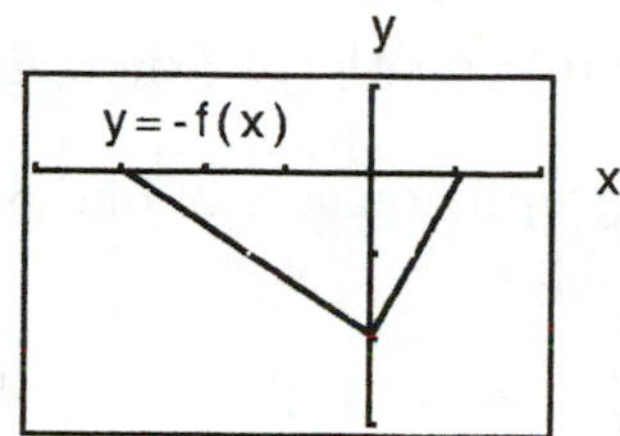

57.

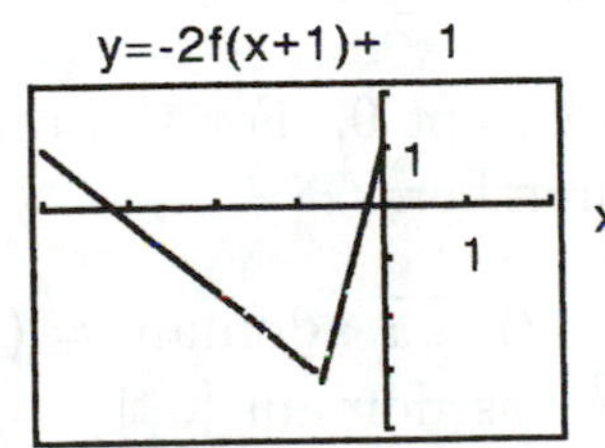

58.

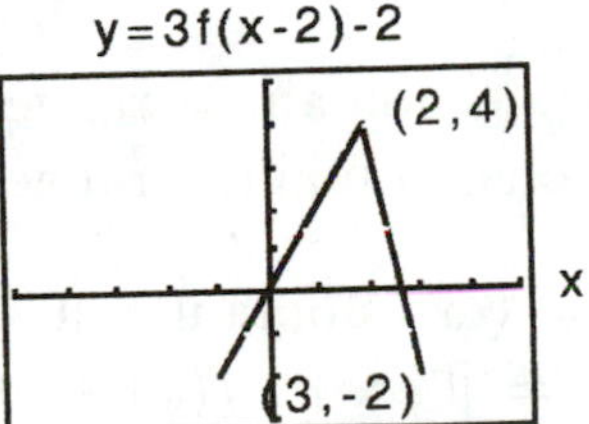

59. $y = -x^2 + 4x - 1 = -(x^2 - 4x) - 1 = -(x^2 - 4x + 4) + 4 - 1 = 3 - (x-2)^2$. $(2, 3)$ is the vertex, $x = 2$ is the axis of symmetry. Check your sketch by graphing y in $[-2, 5]$ by $[-6, 3]$.

60. $x = 2y^2 + 8y + 3 = 2(y^2 + 4y) + 3 = 2(y^2 + 4y + 4) - 8 + 3 = 2(y+2)^2 - 5$. Vertex: $(-5, -2)$, axis of symmetry: $y = -2$. In parametric mode graph $x = 2t^2 + 8t + 3$, $y = t$, $-5 \leqq t \leqq 1$ in $[-5, 10]$ by $[-6.5, 1.9]$.

61. a) $\cos(-x) = \cos x$, even b) even c) even

62. a) odd b) odd c) neither

63. a) even b) odd c) odd

64. a) odd b) odd c) even

65. a) even b) odd c) odd

66. a) odd b) odd c) even

67. Graph y in $[-2,2]$ by $[0,2]$. The function is periodic of period 1.

68. The function has graph the same as the graph of $y=1$ except at $x=n$, an integer, $y=0$. The function is periodic of period 1.

69. Graph $y_1=\sqrt{-x}$ and $y_2=\sqrt{x}$ at the same time in $[-4,4]$ by $[0,2]$.

70. Graph $y_1=-x-2+0\sqrt{x+2}+0\sqrt{-(x+1)}$, $y_2=x+0\sqrt{x+1}+0\sqrt{1-x}$, $y_3=-x+2+0\sqrt{x-1}+0\sqrt{2-x}$ in $[-3,3]$ by $[-1,1]$.

71. The graph consists of one period of the sine function on $[0,2\pi]$ together with all points on the x-axis larger than 2π.

72. The graph consists of one period of the cosine function from 0 to 2π and the x-axis for $2\pi<x<\infty$.

73. $y=\begin{cases} 1-x, & 0\leqq x<1 \\ 2-x, & 1\leqq x\leqq 2 \end{cases}$

74. For the first part $f(x)=\frac{5}{2}x$. In the second part $m=\frac{5-0}{2-4}=-\frac{5}{2}$, $y-0=-\frac{5}{2}(x-4)=-\frac{5}{2}x+10$. $f(x)=\begin{cases} \frac{5}{2}x, & 0\leqq x\leqq 2 \\ -\frac{5}{2}x+10, & 2\leqq x\leqq 4 \end{cases}$

75. For $f(x)$, domain = range = all real numbers except 0. For the remaining functions, domain = range = all positive real numbers.

76. $f(x)=\sqrt{x}$: domain $=[0,\infty)=$ range. $g(x)=\sqrt{1-x}$: domain $=(-\infty,1]$, range $=[0,\infty)$. $f(x)+g(x)=\sqrt{x}+\sqrt{1-x}$ has domain $[0,1]$. $(f+g)^2$ takes on its maximum and minimum values at the same values of x since $f+g\geqq 0$. $[f(x)+g(x)]^2=x+2\sqrt{x}\sqrt{1-x}+1-x=1+2\sqrt{x(1-x)}$. The graph of $x(1-x)$ is a parabola opening downward with maximum 0.25 at $x=0.5$ and minimum 0 at $x=0$ and $x=1$ on $[0,1]$. Hence $f+g$ has maximum $(f+g)(0.5)=\sqrt{2}$ and minimum $(f+g)(0)=1$. Hence $f+g$ has range $[1,\sqrt{2}]$. $(f\cdot g)(x)=\sqrt{x}\sqrt{1-x}$ has domain $[0,1]$, range $[0,0.5]$. $f/g=\sqrt{x}/\sqrt{1-x}$ has domain $[0,1)$ and range $[0,\infty)$. $g/f=\sqrt{1-x}/\sqrt{x}$ has domain $(0,1]$ and range $[0,\infty)$. $f\circ g(x)=f(\sqrt{1-x})=\sqrt{\sqrt{1-x}}$ has domain $(-\infty,1]$ and range $[0,\infty)$. $g\circ f(x)=g(\sqrt{x})=\sqrt{1-\sqrt{x}}$ has domain and range $[0,1]$.

77. $(x-1)^2+(y-1)^2=1$

78. $(x-2)^2+y^2=25$

79. $(x-2)^2+(y+3)^2=\frac{1}{4}$

80. $(x+3)^2+y^2=9$

81. $(3,-5)$, 4

82. Center: $(0,5)$. Radius: $\sqrt{2}$

83. $(-1,7)$, 11

84. $(-4,-1)$, 9

85. a) $x^2+y^2<1$ b) $x^2+y^2\leqq 1$

86. a) $(x-1)^2+(y-1)^2>4$ b) $(x-1)^2+(y-1)^2\geqq 4$

87. $|x-1|=\frac{1}{2}$ leads to $x-1=-\frac{1}{2}$ and $x-1=\frac{1}{2}$. The solution set is $\{\frac{1}{2},\frac{3}{2}\}$.

88. $|2-3x|=1$ leads to $2-3x=-1$ and $2-3x=1$. $3x=3$, $x=1$. $3x=1$, $x=\frac{1}{3}$. Answer: $\{\frac{1}{3},1\}$

89. $\frac{2x}{5}+1=-7$ and $\frac{2x}{5}+1=7$. $\{-20,15\}$

90. $|\frac{5-x}{2}|=7$, $\frac{5-x}{2}=\pm 7$, $5-x=\pm 14$, $x=5\mp 14$, $x=-9,19$.

91. $-\frac{1}{2}\leqq x+2\leqq\frac{1}{2}$, $-\frac{5}{2}\leqq x\leqq -\frac{3}{2}$

92. $|2x-7|\leqq 3$, $-3\leqq 2x-7\leqq 3$, $4\leqq 2x\leqq 10$, $2\leqq x\leqq 5$

93. $-\frac{3}{5}<y-\frac{2}{5}<\frac{3}{5}$, $-\frac{1}{5}<y<1$

94. $|8-\frac{y}{2}|<1$, $-1<8-\frac{y}{2}<1$, $-9<-\frac{y}{2}<-7$, $18>y>14$ or $14<y<18$

95. $\{0.19,2.47,4.34\}$

96. $4x^3-10x^2+9=0$. By zoom-in: $\{-0.82,1.5,1.82\}$. Since 1.5 is a root, $x-\frac{3}{2}$ or $2x-3$ is a factor: $4x^3-10x^2+9=(2x-3)(2x^2-2x-3)$ and we get $\{\frac{3}{2},\frac{1\pm\sqrt{7}}{2}\}$.

97. $2+\log_3(x-2)+\log_3(3-x)=2+\log_3[(x-2)(3-x)]=0$, $\log_3(-x^2+5x-6)=-2$, $-x^2+5x-6=3^{-2}=\frac{1}{9}$, $-9x^2+45x-54=1$, $9x^2-45x+55=0$. From the quadratic formula we get the solution set $\{\frac{15-\sqrt{5}}{6},\frac{15+\sqrt{5}}{6}\}$. Solving with zoom-in, $\{2.127,2.873\}$.

98. $\sin x=-0.7$. $x=\sin^{-1}(-0.7)+2n\pi$ and $x=\pi-\sin^{-1}(-0.7)+2n\pi=(2n+1)\pi+\sin^{-1}(0.7)$. $\{-0.775+2n\pi,0.775+(2n+1)\pi\}$.

99. $|1-2x|<3$, $-3<1-2x<3$, $-4<-2x<2$, $2>x>-1$ or $-1<x<2$.

100. $|\frac{2x-1}{5}|\leqq 1$, $|2x-1|\leqq 5$, $-5\leqq 2x-1\leqq 5$, $-4\leqq 2x\leqq 6$, $-2\leqq x\leqq 3$. We may support this graphically by graphing $y_1=\text{abs}((2x-1)/5)$ and $y_2=1$ and observing that y_2 is above y_1 on, and only on, $-2<x<3$.

101. $|\frac{3}{x-2}| < 1$ is equivalent to $\frac{|x-2|}{3} > 1$, $|x-2| > 3$ which has solution set $(-\infty, -1) \cup (5, \infty)$.

102. $|2-3x| > 1$. $2-3x < -1$ or $2-3x > 1$. $2-3x < -1$ leads to $3x > 3$, $x > 1$. $2-3x > 1$ leads to $3x < 1$, $x < \frac{1}{3}$. Answer: $(-\infty, \frac{1}{3}) \cup (1, \infty)$.

103. The inequality is equivalent to $x^3 - 7x^2 + 12x - 2 < 0$. We seek the set of all x for which the graph of the left-hand side lies below the x-axis. Using the graph and Exercise 95 we get $(-\infty, 0.19) \cup (2.47, 4.34)$.

104. We observe when the graph of $y = 4x^3 - 10x^2 + 9$ is above or touches the x-axis and use the result of Exercise 96. $[-0.82, 1.5] \cup [1.82, \infty)$ or $[\frac{1-\sqrt{7}}{2}, 1.5] \cup [\frac{1+\sqrt{7}}{2}, \infty)$.

105. a) $30(\frac{\pi}{180}) = \frac{\pi}{6}$ b) $22(\frac{\pi}{180}) = 0.122\pi$,
c) $-130(\frac{\pi}{180}) = -0.722\pi$ d) $-150(\frac{\pi}{180}) = -\frac{5\pi}{6}$

106. a) $\frac{3\pi}{2} \to 270°$ b) $-0.9 \to -0.9\frac{(180°)}{\pi} = -51.57°$
c) $2.75 \to 2.75\frac{(180°)}{\pi} = 157.56°$ d) $-\frac{5\pi}{4} = -225°$

107. a) 0.891, 0.454, 1.965, 0.509, 2.205, 1.122
b) −0.891, 0.454, −1.965, −0.509, 2.205, −1.122
c) $\sqrt{3}/2$, $-1/2$, $-\sqrt{3}$, $-\sqrt{3}/3$, -2, $2\sqrt{3}/3$
d) $-\sqrt{3}/2$, $-1/2$, $\sqrt{3}$, $\sqrt{3}/3$, -2, $-2\sqrt{3}/3$

108. a) $\frac{\sqrt{2}}{2}, \frac{\sqrt{2}}{2}, 1, 1, \sqrt{2}, \sqrt{2}$ b) $-\frac{\sqrt{2}}{2}, \frac{\sqrt{2}}{2}, -1, -1, \sqrt{2}, -\sqrt{2}$ c) $0.43, -0.90, -0.47, -2.12, -1.11, 2.33$ d) $-0.43, -0.90, 0.47, 2.12, -1.11, -2.33$

109. Graph the functions in $[0, 2\pi]$ by $[-1, 2]$.

110. Check your result by graphing $y_1 = \cos 2x$, $y_2 = -y_1$, $y_3 = 1 - \cos 2x$, $y_4 = (\sin x)^2$ in $[0, 2\pi]$ by $[-1, 2]$.

111. a) $(\cos \frac{\pi}{6})^2 = (\frac{\sqrt{3}}{2})^2 = \frac{3}{4}$ b) $\cos^2 \frac{\pi}{6} = \frac{1+\cos \frac{\pi}{3}}{2} = \frac{1+\frac{1}{2}}{2} = \frac{3}{4}$

112. a) $\sin^2 \frac{\pi}{4} = (\frac{1}{\sqrt{2}})^2 = \frac{1}{2}$. b) $\sin^2 \frac{\pi}{4} = \frac{1-\cos[2(\frac{\pi}{4})]}{2} = \frac{1-0}{2} = \frac{1}{2}$

113. $f(x) = 2-3x$. $x = 2-3y$, $3y = 2-x$, $f^{-1}(x) = \frac{2-x}{3}$. $f \circ f^{-1}(x) = f(\frac{2-x}{3}) = 2 - 3(\frac{2-x}{3}) = 2-2+x = x$. $f^{-1} \circ f(x) = f^{-1}(2-3x) = \frac{2-(2-3x)}{3} = x$. Graph $y = 2-3x$ and $y = \frac{2-x}{3}$ in $[-10, 10]$ by $[-5, 5]$.

114. $f(x) = (x+2)^2$, $x \geqq -2$. Domain $f = [-2, \infty) = $ range f^{-1}. $y = (x+2)^2$, $x = (y+2)^2$, $y+2 = \pm\sqrt{x}$, $y = -2 \pm \sqrt{x}$. By the above $f^{-1}(x) = -2 + \sqrt{x}$. $f \circ f^{-1}(x) = f(-2+\sqrt{x}) = [(-2+\sqrt{x})+2]^2 = x$. $f^{-1} \circ f(x) = f^{-1}[(x+2)^2] = -2 + \sqrt{(x+2)^2} = -2 + x + 2 = x$. Graph $y_1 = (x+2)^2 + 0\sqrt{x+2}$ and $y_2 = -2 + \sqrt{x}$ in $[-9.5, 13]$ by $[-3.5, 10]$.

115. $y = x^3 - x$. The inverse relation is determined by $x = y^3 - y$. This may be graphed in parametric mode using $x_1 = t^3 - t$, $y_1 = t$, $-2 \leqq t \leqq 2$ in $[-6, 6]$ by $[-2, 2]$. The inverse relation is not a function.

116. $y = \frac{x+2}{x-1}$. $x = \frac{y+2}{y-1}$, $xy - x = y + 2$, $y(x-1) = x + 2$, $y = \frac{x+2}{x-1} = f^{-1}(x)$. Graph $f^{-1}(x) = \frac{x+2}{x-1}$, a function, in $[-10, 10]$ by $[-10, 10]$.

117. $\sin^{-1}(0.7) = 0.775,\ 44.427°$

118. $\tan^{-1}(-2.3) = -1.161$. Changing to degree mode, we read $\tan^{-1}(-2.3) = -66.501°$.

119. Graph $y = |\cos x|$ in $[-\frac{\pi}{2}, \frac{\pi}{2}]$ by $[0, 1]$. The graph is complete because the function has period π.

120. $y = \frac{\cos x + |\cos x|}{2}$ is periodic of period 2π. On $[-\frac{\pi}{2}, \frac{\pi}{2}]$, $y = \cos x$ and on $[\frac{\pi}{2}, \frac{3\pi}{2}]$, $y = \frac{\cos x - \cos x}{2} = 0$. Thus the graph of one period of y, $-\frac{\pi}{2} \leqq x \leqq \frac{3\pi}{2}$, coincides with the graph of $y = \cos x$ on $[-\frac{\pi}{2}, \frac{\pi}{2}]$ and with the x-axis on $[\frac{\pi}{2}, \frac{3\pi}{2}]$.

121. For x in the interval $[-\frac{\pi}{2}, \frac{3\pi}{2}]$, $y = \begin{cases} 0, & -\frac{\pi}{2} \leqq x < \frac{\pi}{2} \\ -\cos x, & \frac{\pi}{2} \leqq x \leqq \frac{3\pi}{2} \end{cases}$. Graphing this part gives a complete graph because the function has period 2π.

122. The graph of one period of y, $-\frac{\pi}{2} \leqq x \leqq \frac{3\pi}{2}$, coincides with the x-axis on $[-\frac{\pi}{2}, \frac{\pi}{2}]$ and with the graph of $\cos x$ on $[\frac{\pi}{2}, \frac{3\pi}{2}]$.

123. a) $A(x) = (\frac{x}{4})^2 + (\frac{100-x}{4})^2$ b) Graph this function in $[-50, 150]$ by $[300, 1000]$ c) The domain is the set of all real numbers. Since $A(0) = A(100)$, $A(50) = 2(\frac{50}{4})^2 = 312.5$ is the minimum of A and the range is $[312.5, \infty)$. d) $0 < x < 100$ e) We must solve $(\frac{x}{4})^2 + (\frac{100-x}{4})^2 = 400$, $x^2 + 10000 - 200x + x^2 = 6400$, $2x^2 - 200x + 3600 = 0$, $x^2 - 100x + 1800 = 0$, $(x-50)^2 = 700$, $x = 50 \pm 10\sqrt{7}$. The two lengths are $50 - 10\sqrt{7}$in and $50 + 10\sqrt{7}$in f) We can only approach the maximum obtained by not cutting and getting $(\frac{100}{4})^2 = 625\text{in}^2$. As noted in c) if both pieces are 50in, we get the minimum area of 312.5in^2.

124. a) $V = x(20 - 2x)(30 - 2x)$

b) Graph V in $[-5, 21]$ by $[-3800, 3200]$.

c) Domain = range = $(-\infty, \infty)$

d) $0 < x < 10$

e) $V_{\text{max}} = V(3.9237) = 1056.31\text{in}^3$

f) $x = 1.70$in by 16.6in by 26.6in or $x = 6.62$in by 6.76in by 16.76in.

125. In the first quadrant $|x| + |y| = 1$ becomes $x + y = 1$ and the graph is the line segment connecting $(0, 1)$ and $(1, 0)$. Since the graph is symmetric with respect to both axes, its graph is the square with vertices $(0, 1), (1, 0), (0, -1), (-1, 0)$.

CHAPTER 2
LIMITS AND CONTINUITY

2.1 LIMITS

1. $\lim_{x\to 2} 2x = 2\cdot 2 = 4$

2. $\lim_{x\to 0} 2x = 2(0) = 0$

3. $\lim_{x\to 1}(3x-1) = (3\cdot 1 - 1) = 2$

4. $\lim_{x\to \frac{1}{3}}(3x-1) = 3(\frac{1}{3}) - 1 = 0$

5. $\lim_{x\to -1} 3x(2x-1) = 3(-1)[2(-1)-1] = -3[-3] = 9$

6. $\lim_{x\to -1} 3x^2(2x-1) = 3(-1)^2[2(-1)-1] = 3(-3) = -9$

7. $\lim_{x\to -2}(x+3)^{171} = (-2+3)^{171} = 1^{171} = 1$

8. $\lim_{x\to -4}(x+3)^{1994} = (-1)^{1994} = 1$

9. $\lim_{x\to 1}(x^3+3x^2-2x-17)$
$= 1^3 + 3(1)^2 - 2(1) - 17$
$= 1+3-2-17$
$= 4-19$
$= -15$

10. $\lim_{x\to -2}(x^3-2x^2+4x+8) = -8-8-8+8 = -16$

11. $\lim_{x\to -1} \frac{x+3}{x^2+3x-1} = \frac{-1+3}{(-1)^2+3(-1)+1} = \frac{2}{1-3+1} = \frac{2}{-1} = -2$

12. $\lim_{y\to 2} \frac{y^2+5y+6}{y+2} = \frac{4+10+6}{4} = \frac{20}{4} = 5$

13. $\lim_{y\to -3} \frac{y^2+4y+3}{y^2-3} = \frac{(-3)^2+4(-3)+3}{(-3)^2-3}$
$= \frac{9-12+3}{9-3}$
$= \frac{0}{6}$
$= 0$

14. $\lim_{x\to -1} \frac{x^3-5x+7}{-x^3+x^2-x+1} = \frac{-1+5+7}{1+1+1+1} = \frac{11}{4}$

15. $\lim_{x\to -2}\sqrt{x-2}$. $\sqrt{x-2}$ is defined only for $x \geqq 2$ so it is not defined for values of x near -2. Thus the limit does not exist.

16. Since division by 0 is not defined, substitution cannot be used. The value of $\frac{1}{x^2}$ increases without bound as $x \to 0$. Therefore the limit does not exist.

17. $\lim_{x\to 0} \frac{|x|}{x}$. Since $\frac{|x|}{x}$ is not defined when $x = 0$, we cannot use substitution. $\lim_{x\to 0^-} \frac{|x|}{x} = \lim_{x\to 0^-} \frac{-x}{x} = \lim_{x\to 0^-}(-1) = -1$. But $\lim_{x\to 0^+} \frac{|x|}{x} = \lim_{x\to 0^+} \frac{x}{x} = 1$. Since the two one-sided limits are not equal, the limit does not exist.

18. $\lim_{x\to 0} \frac{(4+x)^2-16}{x} = \lim_{x\to 0} \frac{8x+x^2}{x} = \lim_{x\to 0}(8+x) = 8$

19. $\lim_{x\to 1} \frac{x-1}{x^2-1} = \lim_{x\to 1} \frac{x-1}{(x-1)(x+1)} = \lim_{x\to 1} \frac{1}{x+1} = \frac{1}{1+1} = \frac{1}{2}$

20. $\lim_{x\to -5} \frac{x^2+3x-10}{x+5} = \lim_{x\to -5} \frac{(x+5)(x-2)}{x+5} = \lim_{x\to -5}(x-2) = -7$

21. $\lim_{t\to 1} \frac{t^2-3t+2}{t^2-1} = \lim_{t\to 1} \frac{(t-1)(t-2)}{(t-1)(t+1)} = \lim_{t\to 1} \frac{t-2}{t+1} = \frac{1-2}{1+1} = -\frac{1}{2}$

22. $\lim_{t\to 2} \frac{t^2-3t+2}{t^2-4} = \lim_{t\to 2} \frac{(t-2)(t-1)}{(t-2)(t+2)} = \lim_{t\to 2} \frac{t-1}{t+2} = \frac{1}{4}$

23. $\lim_{x\to 2} \frac{2x-4}{x^3-2x^2} = \lim_{x\to 2} \frac{2(x-2)}{x^2(x-2)} = \lim_{x\to 2} \frac{2}{x^2} = \frac{2}{2^2} = \frac{1}{2}$

24. $\lim_{x\to 0} \frac{5x^3+8x^2}{3x^4-16x^2} = \lim_{x\to 0} \frac{5x+8}{3x^2-16} = \frac{8}{-16} = -\frac{1}{2}$

25. $\lim_{x\to 0} \frac{\frac{1}{2+x}-\frac{1}{2}}{x} = \lim_{x\to 0} \frac{\frac{2-(2+x)}{2(2+x)}}{x}$
$= \lim_{x\to 0} \frac{\frac{-x}{2(2+x)}}{x}$
$= \lim_{x\to 0} \frac{-1}{2(2+x)} = -\frac{1}{4}$

26. $\lim_{x\to 0} \frac{(2+x)^3-8}{x} = \lim_{x\to 0} \frac{8+12x+6x^2+x^3-8}{x} = \lim_{x\to 0}(12+6x+x^2) = 12$

27. $\lim_{x\to 0} x\sin\frac{1}{x} = 0$ **28.** $\lim_{x\to 0} \frac{\sin x}{x} = 1$ **29.** The limit does not exist.

30. $\lim_{x\to 0} \frac{10^x-1}{x} = 2.30\ldots$ We will later be able to show that this limit is $\ell n\, 10$.

31. $\lim_{x\to 0} \frac{2^x-1}{x} = 0.693\ldots$ We will later be able to show that the limit is $\ell n\, 2$.

32. $\lim_{x\to 0} x\sin(\ell n|x|) = 0$

33. We graph $y = 100(1+0.06/x)^x$ in $[0, 1000]$ by $[106, 106.5]$. As x increases, using TRACE, the limiting value of y, rounded to two decimal places, is \$106.18. After compounding 8 times, the value of the investment has already reached \$106.15. So frequent compounding within one year is not much of an advantage to the investor of \$100.

34. a) $0.00052 \approx 0.001$. To get a one cent difference you would have to invest at least 10 times as much or \$1000. b) We first graph f and g in $[0, 1000]$ by $[108, 108.5]$. The difference is about $0.0009 \approx 0.001$. To pocket a real difference you would have to invest at least \$1000 and hope the bank rounds to the nearest cent.

35. There appear to be no points of the graph of the function very near to $(0, 4)$. The actual graph is a straight line with one point, $(0, 4)$, missing. $y = (x^2 + 4x + 4 - 4)/x = x + 4$ when $x \neq 0$.

36. There appear to be no points of the graph near $(0, 3)$. But if $x \neq 0$, $f(x) = \frac{(x+1)^3 - 1}{x} = x^2 + 3x + 3$. Thus the graph is a parabola with the point $(0, 3)$ missing.

37. $f(x) = \frac{x^3 - 1}{x - 2}$. The graphs indicate $\lim_{x \to 2^+} f(x) = \infty$ and $\lim_{x \to 2^-} f(x) = -\infty$. When $x > 2$, the numerator $\to 7$ while the denominator $\to 0$ but is always positive. When $x < 2$, the denominator is negative and $f(x) \to -\infty$ as $x \to 2^-$.

38. $f(x) = \frac{1 - x^3}{x - 2}$. As $x \to 2$, the numerator $\to -7$. As $x \to 2^+$, $x - 2 > 0$ and $f(x) \to -\infty$. As $x \to 2^-$, $x - 2 < 0$ and $f(x) \to \infty$.

39. Only the following relevant one-sided limits exist: $\lim_{x \to -1^+} f(x) = 1$, $\lim_{x \to 0^-} f(x) = 0$, $\lim_{x \to 0^+} f(x) = 0$, $\lim_{x \to 1^-} f(x) = 1$, $\lim_{x \to 1^+} f(x) = 0$, $\lim_{x \to 2^-} f(x) = 0$. We do therefore have $\lim_{x \to 0} f(x) = 0$. The true statements are a), b), d), e), f).

40. All are true except b) and c).

41. a)

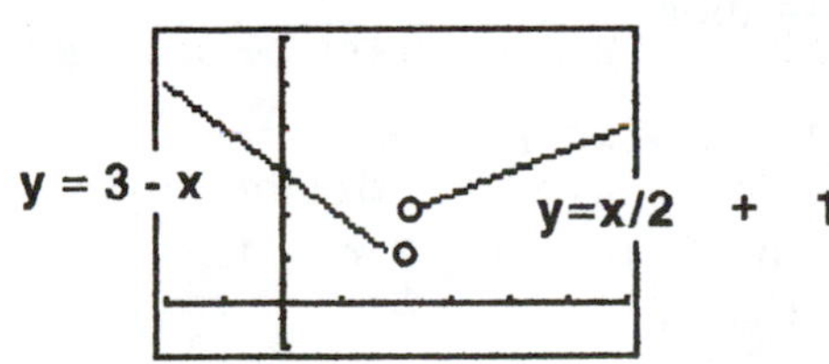

b) $\lim_{x \to 2^+} f(x) = 2$, $\lim_{x \to 2^-} f(x) = 1$

c) Does not exist because right-hand and left-hand limits are not equal.

42. a)

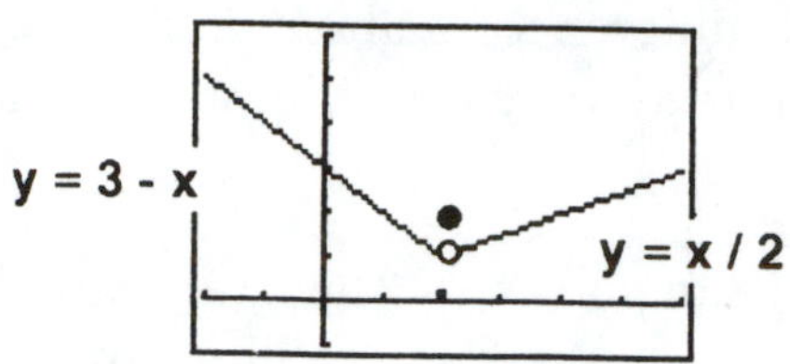

b) $\lim_{x\to 2^+} f(x) = 1$, $\lim_{x\to 2^-} f(x) = 1$.

c) The limit exists and $\lim_{x\to 2} f(x) = 1$

43. a) A complete graph of $f(x)$ can be obtained on a graphing calculator by graphing both $y = (x-1)^{-1} + 0\sqrt{(1-x)}$ and $y = x^3 - 2x + 5 + 0\sqrt{(x-1)}$ in the viewing rectangle $[-3, 5]$ by $[-25, 25]$.

b) $\lim_{x\to 1^+} f(x) = 4$ and $\lim_{x\to 1^-} f(x)$ does not exist.

c) No. For this limit to exist, the two limits in b) must be equal to the same finite number.

44. a) A complete graph of $f(x)$ can be obtained on a graphing calculator by graphing both $y = (2-x)^{-1} + 0\sqrt{(2-x)}$ and $y = 5 - x^2 + 0\sqrt{(x-2)}$ in the viewing rectangle $[-3, 5]$ by $[-10, 10]$.

b) $\lim_{x\to 2^+} f(x) = 1$ and $\lim_{x\to 2^-} f(x)$ does not exist.

c) $\lim_{x\to 2} f(x)$ does not exist because the two limits in b) are not equal to the same finite number.

45. $\text{Lim}_{x\to 2^+} f(x) = \lim_{x\to 2^+}(x^2 + 5x - 3) = 11.$

$\text{Lim}_{x\to 2^-} f(x) = \lim_{x\to 2^-}(a - x^2) = a - 4.$

$\text{Lim}_{x\to 2} f(x)$ will exist if and only if $a - 4 = 11$ or $a = 15$.

46. $\lim_{x\to -1^-} f(x) = \lim_{x\to -1^-}(x^3 - 4x) = -1 + 4 = 3.$

$\lim_{x\to -1^+} f(x) = \lim_{x\to -1^+}(2x + a) = a - 2.$

$\lim_{x\to -1} f(x)$ exists if and only if $a - 2 = 3$ or $a = 5$.

47. a)

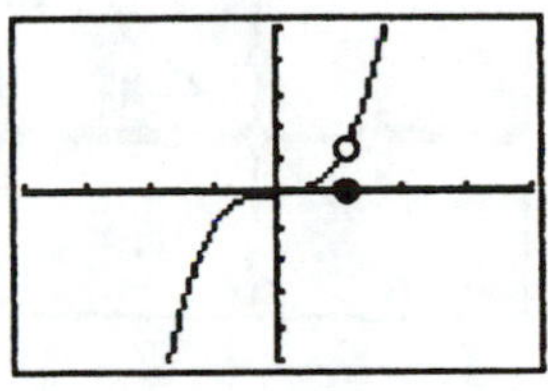

$[-4, 4]$ by $[-6, 6]$

b) $\text{Lim}_{x\to 1^-} f(x) = 1 = \lim_{x\to 1^+} f(x)$

c) $\text{Lim}_{x\to 1} f(x) = 1$.

48. a)

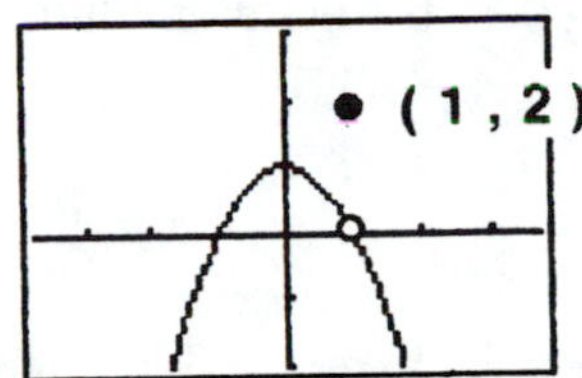

b) $\lim_{x\to 1^+} f(x) = 0$, $\lim_{x\to 1^-} f(x) = 0$

c) Yes, $\lim_{x\to 1} f(x) = 0$

49.

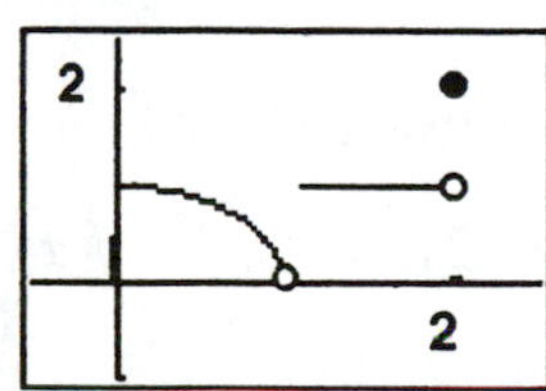

a) All points c except $c = 0, 1, 2$. b) $x = 2$ c) $x = 0$

50.

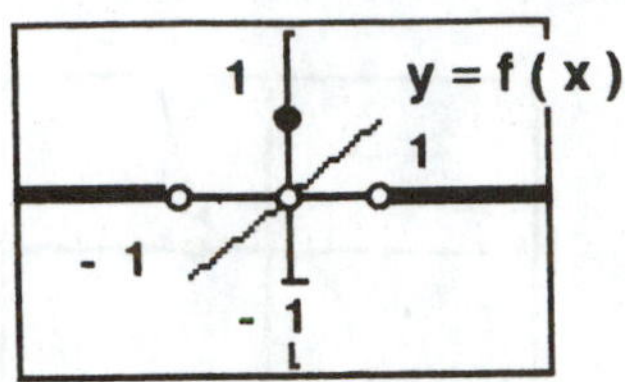

a) $\lim_{x\to c} f(x)$ exists at all points c except $c = \pm 1$. b) None c) None

51. $\lim_{x\to 0^+} [x] = \lim_{x\to 0^+} 0 = 0$

52. $\lim_{x\to 0^-} [x] = -1$

53. $\lim_{x\to 0.5} [x] = \lim_{x\to 0.5} 0 = 0$ because when x is sufficiently close to 0.5, $\lfloor x \rfloor = 0$.

54. $\lim_{x\to 2^-} [x] = 1$

55. $\lim_{x\to 0^+} \frac{x}{|x|} = \lim_{x\to 0^+} \frac{x}{x} = \lim_{x\to 0^+} 1 = 1$ because $|x| = x$ when $x > 0$.

56. $\lim_{x\to 0^-} \frac{x}{|x|} = \lim_{x\to 0^-} \frac{x}{-x} = \lim_{x\to 0^-} (-1) = -1$

57. When $x > a$, $x - a > 0$ and $|x - a| = x - a$. Hence $\lim_{x\to a^+} \frac{|x-a|}{x-a} = \lim_{x\to a^+} \frac{x-a}{x-a} = \lim_{x\to a^+} 1 = 1$.

58. $\lim_{x\to a^-} \frac{|x-a|}{x-a} = \lim_{x\to a^-} \frac{-(x-a)}{x-a} = \lim_{x\to a^-} (-1) = -1$

59. $\lim_{x\to c} f(x) = 5$ and $\lim_{x\to c} g(x) = 2$

a) $\lim_{x\to c} f(x)g(x) = \lim_{x\to c} f(x) \cdot \lim_{x\to c} g(x) = 5 \cdot (2) = 10$

b) $\lim_{x\to c} 2f(x)g(x) = 2 \lim_{x\to c} f(x)g(x) = 2(10) = 20$

60. $\lim_{x\to 4} f(x) = 0$ and $\lim_{x\to 4} g(x) = 3$

a) $\lim_{x\to 4}(g(x) + 3) = 3 + 3 = 6$ b) $\lim_{x\to 4} x f(x) = 4 \cdot 0 = 0$

c) $\lim_{x\to 4} g^2(x) = 3^2 = 9$ d) $\lim_{x\to 4} \frac{g(x)}{f(x)-1} = \frac{3}{0-1} = -3$

61. a) $\lim_{x\to b}(f(x)+g(x)) = \lim_{x\to b} f(x) + \lim_{x\to b} g(x) = 7 + (-3) = 4$

b) $\lim_{x\to b} f(x)\cdot g(x) = \lim_{x\to b} f(x)\cdot \lim_{x\to b} g(x) = 7\cdot(-3) = -21$

c) $\lim_{x\to b} 4g(x) = 4\lim_{x\to b} g(x) = 4(-3) = -12$

d) $\lim_{x\to b} f(x)/g(x) = \lim_{x\to b} f(x)/\lim_{x\to b} g(x) = 7/-3 = -7/3$

62. a) $\lim_{x\to -2}(p(x)+r(x)+s(x)) = 4+0-3 = 1$

b) $\lim_{x\to -2} p(x)\cdot r(x)\cdot s(x) = 4(0)(-3) = 0$

63. $\lim_{x\to 0} x\sin x = \lim_{x\to 0} x \lim_{x\to 0}\sin x = 0\cdot 0 = 0$. The graph of $y = x\sin x$ in the viewing rectangle $[-10, 10]$ by $[-10, 10]$ suggests this limit.

64. A complete graph of $\sin x/x$ can be seen in the viewing rectangle $[-20, 20]$ by $[-1, 1]$. $\lim_{x\to 0}\frac{\sin x}{x} = 1$.

65. A view of the graph of $f(x) = x^2\sin x$ in the square $[-1, 1]$ by $[-1, 1]$ supports $\lim_{x\to 0} f(x) = 0$. $\lim_{x\to 0} x^2\sin x = \lim_{x\to 0} x^2 \lim_{x\to 0}\sin x = 0\cdot 0 = 0$.

66. The graph can be viewed in the rectangle $[-5, 5]$ by $[-3, 3]$ and, for an indication of the behavior as $x\to 0$, in $[-10^{-7}, 10^{-7}]$ by $[-10^8, 10^8]$. As $x\to 0$, values of $|\frac{1}{x}\sin\frac{1}{x}|$ can be made arbitrarily large; the limit does not exist.

67. The magnified graph of the function in the viewing rectangle $[-0.03, 0.03]$ by $[-0.001, 0.001]$ strongly suggests the limit is 0 as $x\to 0$.

68. Although the function is not defined when $x = 0$, when we zoom in on points of the graph near the y-axis, the points of the graph seem to approach $(0, 20.085)$ approximately. Hence $\lim_{x\to 0}(1+x)^{3/x} = L$ where $L \approx 20.085$. $[-0.3, 0.3]$ by $[10, 30]$ is a possible viewing rectangle with which to get started.

69. $\lim_{x\to 0}(1+x)^{4/x} = 54.598$ with error less than 0.01.

70. For the graph of $f(x) = \frac{\ell n(x^2)}{\ell n\, x}$ we apparently get the graph of $y = 2$ for $x > 0$. This is to be expected since $f(x) = \frac{\ell n(x^2)}{\ell n\, x} = \frac{2\,\ell n\, x}{\ell n\, x} = 2$ as long as $x > 0$ and $x \neq 1$. Hence $\lim_{x\to 1} f(x) = \lim_{x\to 1} 2 = 2$.

71. Graph $y = \frac{2^x-1}{x}$ in $[-9, 7]$ by $[-0.3, 3.5]$. Then zoom in to the y-intercept. We see that $\lim_{x\to 0}\frac{2^x-1}{x} = 0.6931$ with error less than 0.0001.

72. Graph $y = \frac{3^x-1}{x}$ in $[-9, 3]$ by $[-0.3, 8]$. Then zoom in to the y-intercept, finding $\lim_{x\to 0}\frac{3^x-1}{x} = 1.0986$ with error less than 0.0002.

74. Graph the function in $[-1,1]$ by $[0.3,0.7]$. As $x \to 0$, the limit seems to be 0.5 until the x reaches the interval $|x| < 0.4$ where the calculator can no longer give meaningful results. If $u = x^{15}$, then $u \to 0$ as $x \to 0$ and $\frac{1-\cos x^{15}}{x^{30}} = \frac{1-\cos u}{u^2}\frac{1+\cos u}{1+\cos u} = \frac{1-\cos^2 u}{u^2}\frac{1}{1+\cos u} = (\frac{\sin u}{u})^2\frac{1}{1+\cos u} \to \frac{1}{2}$ as $x \to 0$.

75. Graph the function in $[1.49\pi, 1.51\pi]$ by $[-0.00001, 0.00001]$ and use TRACE. This suggests the limit is about 9.536×10^{-7}. This is close to the actual limit which can be shown to be $\frac{1}{2^{20}}$. Let $u = x - \frac{3\pi}{2}$, $x = u + \frac{3\pi}{2}$. Then $u \to 0$ as $x \to \frac{3\pi}{2}$ and $\sin x = \sin(u + \frac{3\pi}{2}) = -\cos u$. $\frac{(1+\sin x)^{20}}{(x-\frac{3\pi}{2})^{40}} = \frac{(1-\cos u)^{20}}{u^{40}}\frac{(1+\cos u)^{20}}{(1+\cos u)^{20}} = \frac{(1-\cos^2 u)^{20}}{u^{40}}\frac{1}{(1+\cos u)^{20}} = (\frac{\sin u}{u})^{40}\frac{1}{(1+\cos u)^{20}} \to \frac{1}{2^{20}}$

76. The graph of the function in $[0.9\pi, 1.1\pi]$ by $[-0.00001, 0.00001]$ and TRACE suggest the limit is about 9.5363×10^{-7}. The actual limit is $\frac{1}{2^{20}} \approx 9.5367 \times 10^{-7}$. Let $u = x - \pi$, $x = u + \pi$. Then $u \to 0$ as $x \to \pi$ and $\cos x = \cos(u + \pi) = -\cos u$. $\frac{(1+\cos x)^{20}}{(x-\pi)^{40}} = \frac{(1-\cos u)^{20}}{u^{40}}$. Now look at the solution to Exercise 75 to see that the limit is $\frac{1}{2^{20}}$.

77. The midpoint of OP is $(\frac{a}{2}, \frac{a^2}{2})$ and OP has slope $\frac{a^2}{a} = a$. Thus the perpendicular bisector has equation $y - \frac{a^2}{2} = -\frac{1}{a}(x - \frac{a}{2}) = -\frac{1}{a}x + \frac{1}{2}$ or $y = -\frac{1}{a}x + (\frac{1}{2} + \frac{a^2}{2})$. Hence $b = \frac{1}{2} + \frac{a^2}{2} \to \frac{1}{2}$ as $a \to 0$ or as $P \to 0$. In the viewing window $[-1,1]$ by $[-.5,1]$ graph $y_1 = -A^{-1}x + .5 + \frac{A^2}{2}$, $y_2 = x^2$ and $y_3 = Ax$, successively storing to A the values $A = 0.4, 0.1, 0.01$ and using TRACE to find the y-intercept of y_1. We find the vlaues to be $0.58, 0.505, 0.50005$.

2.2 CONTINUOUS FUNCTIONS

1. a) Yes, $f(-1) = 0$ b) Yes, $\lim_{x\to -1^+} f(x) = 0$ c) Yes d) Yes

2. a) Yes, $f(1) = 1$ b) Yes, $\lim_{x\to 1} f(x) = 2$ c) No d) No

3. Since f is not defined at $x = 2$, it can't be continuous at $x = 2$.

4. All points in $[-1,3]$ except $0, 1, 2$.

5. a) $\lim_{x\to 2} f(x) = 0$ b) Define $g(x) = f(x)$, $x \neq 2$, $g(2) = 0$. Then g is an extension of f which is continuous at $x = 2$.

6. $h(x) = f(x)$, $x \neq 1$, $h(1) = 2$

7. $f(x)$ is continuous at all points except $[-1,2]$ except $x = 0$ and $x = 1$. $\lim_{x\to 0} f(x)$ exists but it is not equal to $f(0)$. $\lim_{x\to 1} f(x)$ does not exist.

8. $f(x)$ is continuous at all points of $[-1, 3]$ except $x = 1$ and $x = 2$. $\lim_{x\to 1} f(x)$ does not exist. $\lim_{x\to 2} f(x) = 1$ but $1 \neq f(2) = 2$.

9. $f(x)$ is continuous at all points except $x = 2$. $\lim_{x\to 2} f(x)$ does not exist.

10. $f(x)$ is continuous at all points except $x = 2$. $\lim_{x\to 2} f(x) = 1$ but $1 \neq f(2) = 2$.

11. $f(x)$ is continuous at all points except $x = 1$. $\lim_{x\to 1^-} f(x)$ does not exist.

12. $f(x)$ is continuous at all points except $x = 2$. $\lim_{x\to 2^-} f(x)$ does not exist.

13. $f(x)$ is continuous at all points except $x = 1$. $\lim_{x\to 1} f(x) = 1 \neq f(1) = 0$.

14. $f(x)$ is continuous at all points except $x = 1$. $\lim_{x\to 1} f(x) = 0 \neq f(1) = 2$.

15. a)

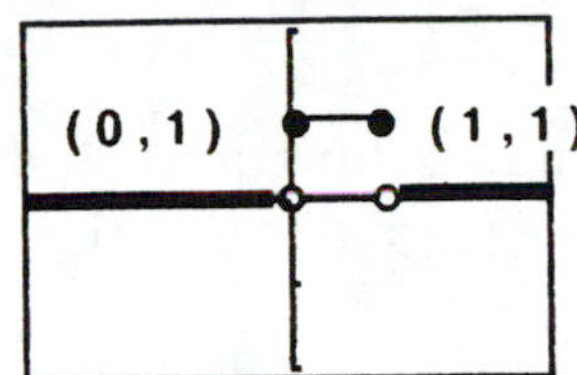

b) f is continuous at all points except $x = 0$ and $x = 1$ where there are jump discontinuities.

16. a)

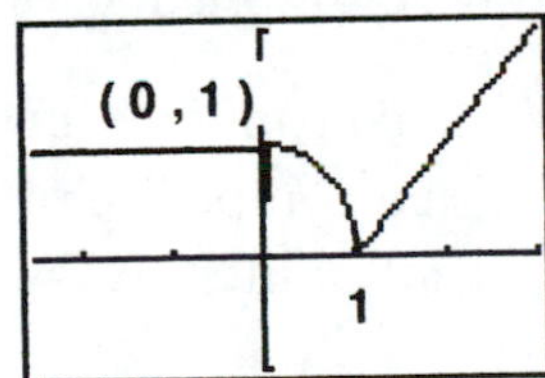

b) f is continuous at all points. In particular, at $x = 0$ and $x = 1$ the right-hand limit and the left-hand limit are equal even though they must be determined by different formulas.

17. $x = 2$ **18.** -2

19. $y = \frac{x+1}{x^2-4x+3} = \frac{x+1}{(x-1)(x-3)}$ is undefined at $x = 1$ and $x = 3$.

20. $y = \frac{x+3}{x^2-3x-10} = \frac{x+3}{(x+2)(x-5)}$ is not continuous only at -2 and 5.

21. $y = \frac{x^3-1}{x^2-1}$ is undefined when $x = \pm 1$. The factor $x - 1$ can be cancelled only when $x \neq 1$; $x = 1$ is a removable discontinuity. y is not continuous only at $x = \pm 1$.

22. No discontinuities

23. $|x - 1|$ is the composite of two continuous functions $|x|$ (Example 6) and $x - 1$. It is therefore continuous by Theorem 6.

24. $y = |2x + 3|$ is the composite of $|x|$ and $2x + 3$ each of which is continuous everywhere. Hence $|2x + 3|$ is continuous everywhere. Answer: No discontinuities.

25. The quotient of two continuous functions is discontinuous only at points where the denominator is 0. Thus $x = 0$ is the only point of discontinuity.

26. An alternative definition of y is

$$y = \begin{cases} -1, & x < 0 \\ 1, & x > 0. \end{cases}$$

$x = 0$ is the only point of discontinuity.

27. $\sqrt{2x + 3}$ will be discontinuous at x if and only if $2x+3 < 0$, that is, $x < -3/2$.

28. $\sqrt[4]{3x - 1}$ is discontinuous at x if and only if $3x - 1 < 0$, that is, $x < 1/3$.

29. The cube root function $x^{1/3}$ is continuous for all x and the same is true of $2x - 1$. Thus $y = \sqrt[3]{2x - 1}$ is the composite of two continuous functions. By Theorem 6 y is continuous at all points. Answer: No discontinuities.

30. $\sqrt[5]{2 - x}$ is the composite of $\sqrt[5]{x}$ and $2 - x$ both of which are continuous for all x. By Theorem 6 there are no discontinuities.

31. $\text{Lim}_{x \to 1} \frac{x^2-1}{x-1} = \lim_{x \to 1}(x + 1) = 2$. Hence $f(1) = \lim_{x \to 1} f(x)$ and f is continuous at $x = 1$.

32. If $x \neq 3$, $g(x) = \frac{x^2-9}{x-3} = x + 3 \to 6$ as $x \to 3$. Thus we must define $g(3) = 6$.

33. $\text{Lim}_{x\to 2} h(x) = \lim_{x\to 2} \frac{(x-2)(x+5)}{x-2} = \lim_{x\to 2}(x+5) = 7$. Thus if $h(2) = 7$, $h(x)$ will be continuous at $x = 2$.

34. $f(x) = \frac{x^3-1}{x^2-1} = \frac{(x-1)(x^2+x+1)}{(x-1)(x+1)} = \frac{x^2+x+1}{x+1} \to \frac{3}{2}$ as $x \to 1$. Therefore we must define $f(1) = 3/2$.

35. $\text{Lim}_{x\to 4} g(x) = \lim_{x\to 4} \frac{(x-4)(x+4)}{(x-4)(x+1)} = \lim_{x\to 4} \frac{x+4}{x+1} = \frac{8}{5}$. If we assign $g(4) = \frac{8}{5}$, the extended function will be continuous at $x = 4$.

36. $g(2) = 1$

37. $\text{Lim}_{x\to 3^-} f(x) = \lim_{x\to 3^-}(x^2-1) = 8$ and $\lim_{x\to 3^+} f(x) = \lim_{x\to 3^+} 2ax = 2a(3) = 6a$. We must have $6a = 8$ or $a = 4/3$. Then we have

$$f(x) = \begin{cases} x^2-1, & x < 3 \\ \frac{8}{3}x, & x \geqq 3. \end{cases}$$

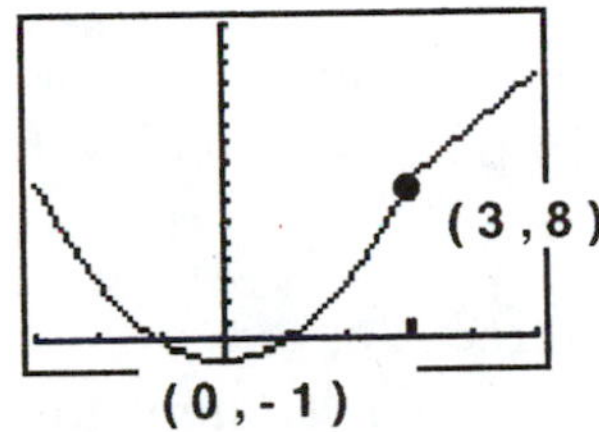

38. $\lim_{x\to \frac{1}{2}^-} g(x) = \lim_{x\to \frac{1}{2}^-} x^3 = \frac{1}{8}$. $\lim_{x\to \frac{1}{2}^+} g(x) = \lim_{x\to \frac{1}{2}^+} bx^2 = \frac{b}{4}$. For continuity we need $\frac{1}{8} = \frac{b}{4}$ or $b = \frac{1}{2}$. Then we have

$$g(x) = \begin{cases} x^3, & x < 1/2 \\ x^2/2, & x \geqq 1/2. \end{cases}$$

For a graph of $g(x)$ on a graphing calculator we graph $y = x^3 + 0\sqrt{(0.5-x)}$ and $y = x^2/2 + 0\sqrt{(x-0.5)}$ in $[-10, 10]$ by $[-10, 10]$. For more detail we can zoom in to the point $(1/2, 1/8)$.

39. $\lim_{x\to 0} \sec x = \lim_{x\to 0} \frac{1}{\cos x} = \frac{\lim_{x\to 0} 1}{\lim_{x\to 0} \cos x} = \frac{1}{\cos 0} = \frac{1}{1} = 1$

40. $\text{Lim}_{x\to 0} \tan x = \lim_{x\to 0} \frac{\sin x}{\cos x} = \frac{\sin 0}{\cos 0} = \frac{0}{1} = 0$

41. $\text{Lim}_{x\to 0}\frac{1+\cos x}{2} = \frac{1+\cos 0}{2} = \frac{1+1}{2} = 1$. We are able to replace x by 0 because $\cos x$ is a continuous function.

42. $\lim_{x\to 0}\sin(\frac{\pi}{2}\cos(\tan x)) = \sin(\frac{\pi}{2}\cos(\tan 0)) = \sin(\frac{\pi}{2}\cos 0) = \sin\frac{\pi}{2} = 1$

43. $\lim_{x\to 0}\cos(1 - \frac{\sin x}{x}) = \cos(1-1) = \cos 0 = 1$

44. $\lim_{x\to 0}\frac{1-\cos x}{x} = \lim_{x\to 0}\frac{1-\cos x}{x}\frac{1+\cos x}{1+\cos x} = \lim_{x\to 0}\frac{1-cos^2 x}{x(1+\cos x)}$
$= \lim_{x\to 0} x(\frac{\sin x}{x})^2\frac{1}{1+\cos x} = 0(1)^2\frac{1}{1+1} = 0.$
$\lim_{x\to 0}\sin(\frac{1-\cos x}{x}) = \sin 0 = 0.$

45. $f(x) = x^3 + 4 = 2,\ x^3 = -2,\ x = -\sqrt[3]{2} \approx -1.2599$

46. $f(x) = 2 - x^3 = 5,\ x^3 = -3,\ x = -\sqrt[3]{3} \approx -1.442$

47. Zooming in to the x-intercept as much as possible we obtain $x = 1.324717957$ with error at most 10^{-9}. In part (b) we obtain the same result.

48. a) Zooming in to the x-intercept of $y = x^3 - 2x + 2$, we obtain $x = -1.7693$ with error at most 10^{-4}. b) The same value is obtained.

49. The maximum value 2 of f is taken on at $x = 2$ and $x = 3$. $\text{Lim}_{x\to 1^-}f(x) = 0$ but the value 0 is not taken on by f for any x; the minimum 0 can be approached arbitrarily closely but cannot be attained. Since f has discontinuities in $[0,4]$, this does not contradict Theorem 7.

50. The function on the domain $(-\infty,\infty)$ takes on neither a maximum value nor a minimum value.

51. The maximum value 1 is not attained but is only approached as x approaches ± 1. The minimum value 0 is attained at $x = 0$. Theorem 7 is not contradicted because the interval $(-1,1)$ is not closed.

52. No. By the theorem, continuity on $0 \leqq x \leqq 1$ is sufficient for the attainment of extreme values. The theorem does not state that continuity is necessary.

53. We are given $f(0) < 0 < f(1)$. By Theorem 8 there exists some c in $[0,1]$ such that $f(c) = 0$. A possible graph is

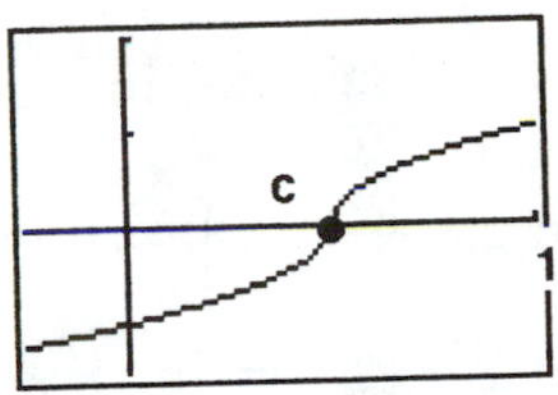

54. Let $f(x) = \cos 3x - x$. Then $f(x)$ is continuous being the difference of two continuous functions. $f(0) = 1$ is positive and $f(\frac{\pi}{2}) = -\pi/2$ is negative. By the Intermediate Value Theorem there exists a c between 0 and $\pi/2$ such that $f(c) = 0$, $\cos 3c - c = 0$ or $\cos 3x = c$. Therefore $\cos 3x = x$ has at least one solution.

55. Let $f(x) = e^{-x} - x$. Then $f(-1) = e + 1 > 0$ and $f(1) = \frac{1}{e} - 1 < 0$. By the Intermediate Value Theorem there is a number c, $-1 < c < 1$, such that $f(c) = 0$, i.e., $e^{-c} = c$. Thus c is a solution of the original equation.

2.3 THE SANDWICH THEOREM AND $(\sin\theta)/\theta$

1. $\lim_{x\to 0} \frac{1}{\cos x} = \frac{1}{\cos 0} = \frac{1}{1} = 1$

2. $\lim_{x\to 0}(2\sin x + 3\cos x) = 2\sin 0 + 3\cos 0 = 3$

3. $\lim_{x\to 0} \frac{1+\sin x}{1+\cos x} = \frac{1+\sin 0}{1+\cos 0} = \frac{1+0}{1+1} = \frac{1}{2}$

4. $\lim_{x\to 0} \frac{x^2+1}{1-\sin x} = \frac{0^2+1}{1-\sin 0} = \frac{1}{1} = 1$

5. $\lim_{x\to 0} \frac{x}{\sin x} = \lim_{x\to 0} \frac{1}{\frac{\sin x}{x}} = \frac{1}{1} = 1$

6. $\lim_{x\to 0} \frac{x}{\tan x} = \lim_{x\to 0} \frac{x\cos x}{\sin x} = \lim_{x\to 0} \frac{\cos x}{(\frac{\sin x}{x})} = \frac{1}{1} = 1$

7. $\lim_{x\to 0} \frac{\sin 2x}{x} = \lim_{x\to 0} 2\frac{\sin 2x}{2x} = 2\lim_{x\to 0} \frac{\sin 2x}{2x} = 2\lim_{\theta\to 0} \frac{\sin\theta}{\theta} = 2\cdot 1 = 2$. Here $\theta = 2x$.

8. $\lim_{x\to 0} \frac{x}{\sin 3x} = \lim_{x\to 0} \frac{1}{3}\frac{3x}{\sin 3x} = \frac{1}{3}\lim_{\theta\to 0} \frac{\theta}{\sin\theta} = \frac{1}{3}\cdot 1 = \frac{1}{3}$ where $\theta = 3x$.

9. $\lim_{x\to 0} = \frac{\tan 2x}{2x} = \lim_{x\to 0} \frac{1}{\cos 2x}\frac{\sin 2x}{2x} = 1\cdot 1 = 1$

10. $\lim_{x\to 0} \frac{\tan 2x}{x} = \lim_{x\to 0} 2\frac{\sin 2x}{2x}\frac{1}{\cos x} = 2\cdot 1\cdot 1 = 2$

11. $\lim_{x\to 0} \frac{\sin x}{2x^2-x} = \lim_{x\to 0} \frac{\sin x}{x}\frac{1}{2x-1} = 1\cdot(-1) = -1$

12. $\lim_{x\to 0} \frac{x+\sin x}{x} = \lim_{x\to 0}(1 + \frac{\sin x}{x}) = 1 + 1 = 2$

13. $\lim_{x\to 0} \frac{\sin^2 x}{x} = \lim_{x\to 0} \sin x \frac{\sin x}{x} = 0\cdot 1 = 0$

14. $\lim_{x\to 0} \frac{\tan^2 x}{2x} = \lim_{x\to 0} \frac{1}{2}\frac{\sin x}{x}\frac{1}{\cos x} \tan x = \frac{1}{2}(1)(1)(0) = 0$

15. $\lim_{x\to 0} \frac{3\sin 4x}{\sin 3x} = 3\lim_{x\to 0} \frac{\sin 4x}{\sin 3x} = 3\lim_{x\to 0} \frac{(4x)\frac{\sin 4x}{4x}}{(3x)\frac{\sin 3x}{3x}} = 3\lim_{x\to 0} \frac{4}{3}\frac{\frac{\sin 4x}{4x}}{\frac{\sin 3x}{3x}} = 3\cdot\frac{4}{3}\cdot \frac{1}{1} = 4.$

16. $\lim_{x\to 0} \frac{\tan 5x}{\tan 2x} = \lim_{x\to 0} \sin 5x \frac{1}{\sin 2x}\frac{\cos 2x}{\cos 5x} = \lim_{x\to 0} \frac{\sin 5x}{5x}\frac{2x}{\sin 2x}\frac{5x}{2x}\frac{\cos 2x}{\cos 5x} = 1\cdot 1\cdot\frac{5}{2}\cdot 1 = \frac{5}{2}$

17. a) Approximately 0.6 b) Very close

c) $\lim_{x\to 0} \frac{\tan 3x}{\sin 5x} = \lim_{x\to 0} \frac{\frac{\sin 3x}{\cos 3x}}{\sin 5x} = \lim_{x\to 0} \frac{\frac{1}{\cos 3x}\frac{\sin 3x}{3x}3x}{\frac{\sin 5x}{5x}5x} = 1\cdot 1\cdot\frac{3}{5} = \frac{3}{5}.$

18. a) 0.6666666666 b) Extremely close

c) $\lim_{x\to 0} \frac{\cot 3x}{\csc 2x} = \lim_{x\to 0} \frac{\cos 3x}{\sin 3x} \sin 2x = \lim_{x\to 0} \cos 3x \frac{3x}{\sin 3x}\frac{\sin 2x}{2x}\frac{2x}{3x} = 1\cdot 1\cdot 1\cdot\frac{2}{3} = \frac{2}{3}.$

19. Since $\lim_{x\to 0} 1 - \frac{x^2}{6} = 1$ and $\lim_{x\to 0} 1 = 1$, it follows from the inequality and the Sandwich Theorem that $\lim_{x\to 0} \frac{\sin x}{x} = 1$.

20. a) We graph $y_1 = 1 - \frac{x^2}{6}$, $y_2 = \frac{x\sin x}{2-2\cos x}$ and $y_3 = 1$ in the viewing window $[-0.12, 0.12]$ by $[0.998, 1.002]$ which supports the inequality.

b) 1 c) Since both the left-hand function and the right-hand function $\to 1$ as $x \to 0$, $\lim_{x\to 0} f(x) = 1$ by the Sandwich Theorem.

21. a) We graph $y_1 = \frac{1}{2} - \frac{x^2}{24}$, $y_2 = \frac{1-\cos x}{x^2}$ and $y_3 = \frac{1}{2}$ (y_1 and y_2 are extremely close near $x = 0$) in the window $[-6, 6]$ by $[-0.1, 0.6]$ supporting the inequality.

b) $\frac{1}{2}$ c) Since both y_1 and $y_3 \to \frac{1}{2}$ as $x \to 0$, $y_2 = f(x) \to \frac{1}{2}$ by the Sandwich Theorem.

22. a) Graph $y_1 = 1$, $y_2 = \frac{\tan x}{x}$ and $y_3 = 1 + x^2$ in $[-0.1, 0.1]$ by $[0.9995, 1.0025]$ to support the inequality. b) 1

c) Since y_1 and y_3 both $\to 1$ as $x \to 0$, $y_2 = f(x) \to 1$ as $x \to 0$.

23. The numerator $\cos x$ approaches 1 as $x \to 0$ while the denominator approaches 0. Thus the fraction can be made arbitrarily large in absolute value if x is sufficiently close to 0. Therefore the fraction cannot approach any finite number and so the limit does not exist.

24. If θ is measured in degrees, its equivalent radian measured is $\frac{\pi}{180}\theta$. Thus $A = \frac{1}{2}r^2\frac{\pi}{180}\theta = \frac{\pi}{360}r^2\theta$.

25. Let $f(\theta) = \frac{\sin\theta}{\theta}$ where the sine function is in degree mode. Then $f(\theta) = \frac{\sin(\frac{\pi}{180}\theta)}{\theta}$ if sine is in radian mode. $\frac{\sin(\frac{\pi}{180}\theta)}{\theta} = \frac{(\frac{\pi}{180})\sin(\frac{\pi}{180}\theta)}{(\frac{\pi}{180}\theta)} \to (\frac{\pi}{180})(1) = \frac{\pi}{180}$ as $\theta \to 0$.

26. $f(x) = \frac{\sin 2x}{x}$.

x	± 0.1	± 0.01	± 0.001
$f(x)$	1.987	1.999	1.9999

Conjecture: $\lim_{x\to 0} f(x) = 2$.

27. $f(x) = \frac{\tan 3x}{x}$.

x	± 0.1	± 0.01	± 0.001
$f(x)$	3.0934	3.0009	3.0000

Conjecture: $\lim_{x\to 0} f(x) = 3$.

28. $f(x) = \frac{\cos -1}{x}$.

x	± 0.1	± 0.01	± 0.001
$f(x)$	∓ 0.04996	∓ 0.00500	∓ 0.00050

Conjecture: $\lim_{x\to 0} f(x) = 0$.

29. $f(x) = \frac{x-\sin x}{x^2}$.

x	-0.01	0.01	-0.001	0.001
$f(x)$	-0.00167	0.00167	-0.00017	0.00017

Conjecture: $\lim_{x\to 0} f(x) = 0$.

30. We first investigate $\lim_{\theta\to 0} \frac{1-\cos\theta}{\sin\theta} \cdot \frac{1-\cos\theta}{\sin\theta} = \frac{(1-\cos\theta)\sin\theta}{\sin^2\theta} = \frac{(1-\cos\theta)\sin\theta}{1-\cos^2\theta} = \frac{\sin\theta}{1+\cos\theta} \to \frac{0}{1+1} = 0$ as $\theta \to 0$. As in the text, using Fig. 2.27, we may arrive at the inequality $\frac{1}{2}\sin\theta < \frac{1}{2}\theta < \frac{1}{2}\tan\theta$. Taking θ as a small positive number, we multiply by 2 and take reciprocals, obtaining $\frac{1}{\sin\theta} > \frac{1}{\theta} > \frac{\cos\theta}{\sin\theta}$. This leads to $\frac{1-\cos\theta}{\sin\theta} > \frac{1-\cos\theta}{\theta} > \cos\theta\frac{(1-\cos\theta)}{\sin\theta}$. Now by the Sandwich Theorem $\lim_{\theta\to 0^+} \frac{1-\cos\theta}{\theta} = 0$. Since the cosine is an even function, we also have $\lim_{\theta\to 0^-} \frac{1-\cos\theta}{\theta} = 0$. Therefore $\lim_{\theta\to 0} \frac{1-\cos\theta}{\theta} = 0$.

2.4 LIMITS INVOLVING INFINITY

1. a) $\lim_{x\to\infty} \frac{2x+3}{5x+7} = \lim_{x\to\infty} \frac{2+\frac{3}{x}}{5+\frac{7}{x}} = \frac{2+0}{5+0} = \frac{2}{5}$

b) $\lim_{x\to-\infty} \frac{2x+3}{5x+3} = \frac{2}{5}$ by the same reasoning.

2. $\lim_{x\to\pm\infty} \frac{2x^3+7}{x^3-x^2+x+7} = \lim_{x\to\pm\infty} \frac{2+\frac{7}{x^3}}{1-\frac{1}{x}+\frac{1}{x^2}+\frac{7}{x^3}} = \frac{2+0}{1-0+0+0} = 2$

3. $\lim_{x\to\infty} \frac{x+1}{x^2+3} = \lim_{x\to\infty} \frac{\frac{1}{x}+\frac{1}{x^2}}{1+\frac{3}{x^2}} = \frac{0+0}{1+0} = \frac{0}{1} = 0$

b) $\lim_{x\to-\infty} \frac{x+1}{x^2+3} = 0$ by the same reasoning.

4. $\lim_{x\to\pm\infty} \frac{3x+7}{x^2-2} = \lim_{x\to\pm\infty} \frac{\frac{3}{x}+\frac{7}{x^2}}{1-\frac{2}{x^2}} = \frac{0+0}{1+0} = 0$

5. a) $\lim_{x\to\infty} \frac{3x^2-6x}{4x-8} = \lim_{x\to\infty} \frac{3x-6}{4-\frac{8}{x}} = \infty$

b) $\lim_{x\to-\infty} \frac{3x^2-6x}{4x-8} = -\infty$ by the same method.

6. $\lim_{x\to\pm\infty} \frac{x^4}{x^3+1} = \lim_{x\to\pm\infty} \frac{x}{1+\frac{1}{x^3}} = \pm\infty$

7. a) $\lim_{x\to\infty} \frac{1}{x^3-4x+1} = \lim_{x\to\infty} \frac{\frac{1}{x^3}}{1-\frac{4}{x^2}+\frac{1}{x^3}} = \frac{0}{1-0+0} = 0$

b) $\lim_{x\to-\infty} \frac{1}{x^3-4x+1} = 0$ by the same method.

8. $\lim_{x\to\pm\infty} \frac{10x^5+x^4+31}{x^6} = \lim_{x\to\pm\infty} \left(\frac{10}{x} + \frac{1}{x^2} + \frac{31}{x^6}\right) = 0$

9. a) $\lim_{x\to\infty} \frac{-2x^3-2x+3}{3x^3+3x^2-5x} = \lim_{x\to\infty} \frac{-2-\frac{2}{x^2}+\frac{3}{x^3}}{3+\frac{3}{x}-\frac{5}{x^2}} = \frac{-2-0+0}{3+0-0} = -\frac{2}{3}$

b) $\lim_{x\to-\infty} \frac{-2x^3-2x+3}{3x^3+3x^2-5x} = -\frac{2}{3}$ by the same method.

10. $\lim_{x\to\pm\infty} \frac{-x^4}{x^4-7x^3+7x^2+9} = \lim_{x\to\pm\infty} \frac{-1}{1-\frac{7}{x}+\frac{7}{x^2}+\frac{9}{x^3}} = \frac{-1}{1-0+0+0} = -1$

11. a) $\lim_{x\to\infty} \left(\frac{-x}{x+1}\right)\left(\frac{x^2}{5+x^2}\right) = \lim_{x\to\infty} \left(\frac{-1}{1+\frac{1}{x}}\right)\left(\frac{1}{\frac{5}{x^2}+1}\right) = (-1)(1) = -1$

b) $\lim_{x\to-\infty} \left(\frac{-x}{x+1}\right)\left(\frac{x^2}{5+x^2}\right) = -1$ by the same proof.

12. $\lim_{x\to\pm\infty} \left(\frac{2}{x}+1\right)\left(\frac{5x^2-1}{x^2}\right) = (1)(5) = 5$

13. $\lim_{x\to2^+} \frac{1}{x-2} = \infty$ because $x-2$ is always positive in this process.

14. $\lim_{x\to2^-} \frac{1}{x-2} = -\infty$

15. $\lim_{x\to 2^+} \frac{x}{x-2} = \infty$ because the numerator approaches 2 and the denominator approaches 0 but is always positive in the process.

16. $\lim_{x\to 2^-} \frac{x}{x-2} = -\infty$

17. $\lim_{x\to -3^+} \frac{1}{x+3} = \infty$ (the denominator is always positive)

18. $\lim_{x\to -3^-} \frac{1}{x+3} = -\infty$

19. $\lim_{x\to -3^+} \frac{x}{x+3} = \lim_{x\to -3^+} x\frac{1}{x+3} = -\infty$ by Theorem 11 or it can be argued that the numerator is negative and approaching -3 while the denominator is positive and approaching 0.

20. $\lim_{x\to -3^-} \frac{x}{x+3} = \lim_{x\to -3^-} x\frac{1}{x+3} = -(-\infty) = \infty$ by Theorem 11 or one could argue that the numerator is approaching -3 while the denominator is approaching 0 through negative values.

21. $\lim_{x\to\pm\infty} f(x) = \lim_{x\to\pm\infty} \frac{x-2}{2x^2+3x-5} = \lim_{x\to\pm\infty} \frac{\frac{1}{x}-\frac{2}{x^2}}{2+\frac{3}{x}-\frac{5}{x^2}} = \frac{0-0}{2+0-0} = 0$. Therefore $y = 0$ is the end behavior asymptote. $2x^2+3x-5 = (2x+5)(x-1)$ so there are vertical asymptotes at $x = -\frac{5}{2}$ and $x = 1$.

22. $\lim_{y\to\pm\infty} T(y) = \lim_{y\to\pm\infty} \frac{2y+3}{4-y^2} = 0$ so $T = 0$ is the end behavior asymptote. Vertical asymptotes at $y = \pm 2$ in the (y, T)-plane.

23. $\lim_{x\to\pm\infty} g(x) = \lim_{x\to\pm\infty} \frac{3x^2-x+5}{x^2-4} = \lim_{x\to\pm\infty} \frac{3-\frac{1}{x}+\frac{5}{x^2}}{1-\frac{4}{x^2}} = 3$ so $y = 3$ is the end behavior asymptote. VA: $x = \pm 2$.

24. $\lim_{x\to\pm\infty} f(x) = \frac{1}{2}$ so $y = \frac{1}{2}$ is the end behavior asymptote. VA: $x = \frac{1\pm\sqrt{3}}{6}$

25. $f(x) = \frac{x^2-2x+3}{x+2} = (x-4) + \frac{11}{x+2}$ so $x-4$ is the end behavior asymptote. VA: $x = -2$.

26. $f(x) = (x-6) + \frac{11}{x+3}$ so $y = x-6$ is the end behavior asymptote. VA: $x = -3$.

27. $g(x) = x^2+2x+2+\frac{5}{x-2}$ so x^2+2x+2 is the end behavior asymptote. VA: $x = 2$.

28. $g(x) = x^2+2+\frac{-x+11}{x^2-4}$ so x^2+2 is the end behavior asymptote. VA: $x = \pm 2$.

29. $\lim_{x\to\pm\infty} \left(\frac{x}{x^2+3} - \frac{1+\frac{2}{x^2}}{\frac{1}{x^2}+\frac{1}{x}-1}\right) = 0-(-1) = 1$ so $y = 1$ is the end behavior asymptote. The quadratic formula yields the roots of $1+x-x^2$ and so the vertical asymptotes: $x = \frac{1\pm\sqrt{5}}{2}$.

30. $g(x) = \frac{x^2-4}{x+1} + \frac{x}{x^2-3x+2} = x - 1 - \frac{3}{x+1} + \frac{x}{x^2-3x+2}$ so $y = x - 1$ is the end behavior asymptote. VA: $x = -1, 1, 2$.

31. To study $y = \frac{1}{x^2-4}$, a convenient viewing rectangle for its graph is $[-3, 3]$ by $[-5, 5]$. From this graph we see a) $\lim_{x\to 2^+} \frac{1}{x^2-4} = \infty$, b) $\lim_{x\to 2^-} \frac{1}{x^2-4} = -\infty$, c) $\lim_{x\to -2^+} \frac{1}{x^2-4} = -\infty$, and d) $\lim_{x\to -2^-} \frac{1}{x^2-4} = \infty$

32. Looking at the graph of $f(x) = \frac{x}{x^2-1}$ in the viewing rectangle $[0, 2]$ by $[-20, 20]$, one sees that a) $\lim_{x\to 1^+} f(x) = \infty$ and b) $\lim_{x\to 1^-} f(x) = -\infty$. Using $[-2, 0]$ by $[-20, 20]$, one sees that c) $\lim_{x\to -1^+} f(x) = \infty$ and d) $\lim_{x\to -1^-} f(x) = -\infty$.

33. The graph of $f(x) = \frac{x^2-1}{2x+4}$ in the viewing window $[-10, 10]$ by $[-10, 10]$ indicates that a) $\lim_{x\to -2^+} f(x) = \infty$, and b) $\lim_{x\to -2^-} f(x) = -\infty$.

34. a) $\lim_{x\to 0^+}(x^2 + \frac{4}{x}) = \infty$, and b) $\lim_{x\to 0^-}(x^2 + \frac{4}{x}) = -\infty$

35. $\lim_{x\to 0^+} \frac{[x]}{x} = \lim_{x\to 0^+} \frac{0}{x} = \lim_{x\to 0^+} 0 = 0$. Here we use the fact that $[x] = 0$ if $0 < x < 1$.

36. $\lim_{x\to 0^-} \frac{[x]}{x} = \lim_{x\to 0^-} \frac{-1}{x} = \infty$

37. $\lim_{x\to \infty} \frac{|x|}{|x|+1} = \lim_{x\to \infty} \frac{x}{x+1} = \lim_{x\to \infty} \frac{1}{1+\frac{1}{x}} = \frac{1}{1+0} = 1$. Here we use the fact that $|x| = x$ for $x \geqq 0$.

38. $\lim_{x\to -\infty} \frac{x}{|x|} = \lim_{x\to -\infty} \frac{x}{-x} = \lim_{x\to -\infty}(-1) = -1$

39. $\lim_{x\to 0^+} \frac{1}{\sin x} = \infty$ because $\sin x$ is a positive number if x is a small positive number $0 < x < \pi$.

40. $\lim_{x\to 0^-} \frac{1}{\sin x} = -\infty$ because $\sin x < 0$ if $-\pi < x < 0$.

41. $\lim_{x\to \frac{\pi}{2}^+} \frac{1}{\cos x} = -\infty$ because $\cos x < 0$ for $\frac{\pi}{2} < x < \frac{3\pi}{2}$.

42. $\lim_{x\to \frac{\pi}{2}^-} \frac{1}{\cos x} = \infty$ because $\cos x > 0$ for $0 < x < \frac{\pi}{2}$.

43. $\lim_{x\to -\infty} f(x) = \lim_{x\to -\infty} \frac{1}{x} = 0$. $\lim_{x\to 0^-} f(x) = \lim_{x\to 0^-} \frac{1}{x} = -\infty$. $\lim_{x\to 0^+} f(x) = \lim_{x\to 0^+}(-1) = -1$. $\lim_{x\to \infty} f(x) = \lim_{x\to \infty}(-1) = -1$.

44. $\lim_{x\to -\infty} f(x) = \lim_{x\to -\infty} \frac{x-2}{x-1} = 1$. $\lim_{x\to 0^-} f(x) = \lim_{x\to 0^-} \frac{x-2}{x-1} = \frac{-2}{-1} = 2$. $\lim_{x\to 0^+} f(x) = \lim_{x\to 0^+} \frac{1}{x^2} = \infty$. $\lim_{x\to \infty} f(x) = \lim_{x\to \infty} \frac{1}{x^2} = 0$.

45. We first prove $\lim_{x\to\infty} \frac{\sin x}{x} = 0$. We know $-1 \leqq \sin x \leqq 1$ for all x. Hence for $x > 0$, $-\frac{1}{x} \leqq \frac{\sin x}{x} \leqq \frac{1}{x}$. But $\lim_{x\to\infty}(-\frac{1}{x}) = 0 = \lim_{x\to\infty} \frac{1}{x}$. By the Sandwich Theorem $\lim_{x\to\infty} \frac{\sin x}{x} = 0$. Therefore $\lim_{x\to\infty}(2 + \frac{\sin x}{x}) = 2 + 0 = 2$.

46. $-1 \leqq \sin x \leqq 1$. Hence for $x < 0$, $-\frac{1}{x} \geqq \frac{\sin x}{x} \geqq \frac{1}{x}$. Using the Sandwich Theorem, we get $\lim_{x\to-\infty} \frac{\sin x}{x} = 0$.

47. $\lim_{x\to\infty}(1 + \cos \frac{1}{x}) = \lim_{\theta\to 0^+}(1 + \cos\theta) = 1 + \cos 0 = 2$.

48. $\lim_{x\to\infty} x \sin \frac{1}{x} = \lim_{\theta\to 0^+} \frac{1}{\theta} \sin\theta = 1$

49. $-1 \leqq \sin 2x \leqq 1$ for all x. Hence for $x > 0$, $-\frac{1}{x} \leqq \frac{\sin 2x}{x} \leqq \frac{1}{x}$. But $\lim_{x\to\infty}(-\frac{1}{x}) = 0 = \lim_{x\to\infty} \frac{1}{x}$. By the Sandwich Theorem, $\lim_{x\to\infty} \frac{\sin 2x}{x} = 0$.

50. $\lim_{x\to\infty} \frac{\cos(1/x)}{1+(1/x)} = \lim_{\theta\to 0^+} \frac{\cos\theta}{1+\theta} = \frac{\cos 0}{1+0} = 1$

51. $\lim_{x\to\pm\infty} \frac{2x^2}{x^2+1} = \lim_{x\to\pm\infty} \frac{2}{1+\frac{1}{x^2}} = \frac{2}{1+0} = 2$ and $\lim_{x\to\pm\infty} \frac{2x^2+5}{x^2} = \lim_{x\to\pm\infty} (2 + \frac{5}{x^2}) = 2 + 0 = 2$. By the Sandwich Theorem, $\lim_{x\to\infty} f(x) = 2$ and $\lim_{x\to-\infty} f(x) = 2$.

52. $\lim_{x\to\pm\infty} 1 = 1 = \lim_{x\to\pm\infty} \frac{x-1}{x}$. By the Sandwich Theorem, $\lim_{x\to\pm\infty} \frac{[x]}{x} = 1$.

53. Each graph satisfies $y \to \infty$ as $x \to \infty$ and $y \to -\infty$ as $x \to -\infty$. As the power of x increases, the vertical steepness of the graph increases for $|x| > 1$.

54. Each graph satisfies $y \to -\infty$ as $x \to \pm\infty$. For $|x| > 1$, the vertical steepness of the graph increases as the power of x increases.

55. $\lim_{x\to\pm\infty} \frac{f(x)}{-\frac{1}{7}} = \lim_{x\to\pm\infty} -7f(x) = -7\lim_{x\to\pm\infty} f(x) = -7(-\frac{1}{7}) = 1$ using the result of Example 8 (the same method can be used for $x \to -\infty$).

56. $\lim_{x\to\pm\infty} \frac{f(x)}{\frac{2}{3}} = \frac{3}{2} \lim_{x\to\pm\infty} f(x) = \frac{3}{2}\frac{2}{3} = 1$ by Example 9.

57. One such function is $f(x) = \begin{cases} x+1, & x \leqq 2 \\ \frac{1}{5-x}, & 2 < x < 5 \\ -1, & x \geqq 5 \end{cases}$. Graph $y_1 = x + 1 + 0\sqrt{2-x}$, $y_2 = \frac{1}{5-x} + 0\sqrt{x-2} + 0\sqrt{5-x}$ and $y_3 = -1 + 0\sqrt{x-5}$ in $[-5, 10]$ by $[-10, 10]$.

58. One such function is $f(x) = \begin{cases} 2 - \frac{15}{x^2+1}, & x \leqq 4 \\ x + \frac{1}{4-x}, & x > 4 \end{cases}$. Graph $y_1 = 2 - \frac{15}{x^2+1} + 0\sqrt{4-x}$, and $y_2 = x + \frac{1}{4-x} + 0\sqrt{x-4}$ in $[-10, 9]$ by $[-20, 20]$.

59. $\lim_{x\to 0} f(x) = \lim_{x\to 0} \frac{1}{x}$ does not exist. $\lim_{x\to 0} g(x) = \lim_{x\to 0} x = 0$. $\lim_{x\to 0} f(x)g(x) = \lim_{x\to 0} \frac{x}{x} = 1$.

60. $\lim_{x\to 0}(-\frac{2}{x^3})$ does not exist, $4x^3 \to 0$ as $x \to 0$, $(-\frac{2}{x^3})4x^3 = -8 \to -8$ as $x \to 0$.

61. $\lim_{x\to 2} \frac{3}{x-2}$ does not exist, $(x-2)^3 \to 0$ as $x \to 2$, $f(x)g(x) = \frac{3}{x-2}(x-2)^3 = 3(x-2)^2 \to 0$ as $x \to 2$.

62. $\frac{5}{(3-x)^4} \to \infty$ as $x \to 3$, $(x-3)^2 \to 0$ as $x \to 3$, $\frac{5}{(3-x)^4}(x-3)^2 = \frac{5}{(x-3)^2} \to \infty$ as $x \to 3$.

63. Let $f(x) = x^3$, $g(x) = \frac{1}{x^2}$. Then $\lim_{x\to 0}(fg) = \lim_{x\to 0} x = 0$.

Let $f(x) = 5x^2$, $g(x) = \frac{1}{x^2}$. Then $\lim_{x\to 0}(fg) = \lim_{x\to 0} 5 = 5$.

Let $f(x) = x^2$, $g(x) = \frac{1}{x^4}$. Then $\lim_{x\to 0}(fg) = \lim_{x\to 0} \frac{1}{x^2} = \infty$.

In each case $\lim_{x\to 0} f(x) = 0$ and $\lim_{x\to 0} g(x) = \infty$.

64. Yes, $\lim_{x\to c}(f \pm g) = \pm\lim_{x\to c} g(x)$ because the values of $f(x) \pm g(x)$ are dominated by $\pm g(x)$ as $x \to c$. The values of $f(x)$ tend toward L but the values of $g(x)$ become numerically arbitrarily large as $x \to c$.

65. $\lim_{x\to\pm\infty} \frac{f(x)}{a_n x^n} = \lim_{x\to\pm\infty} \frac{a_n x^n + a_{n-1}x^{n-1} + a_{n-2}x^{n-2} + \cdots + a_1 x + a_0}{a_n x^n} = \lim_{x\to\pm\infty}(1 + \frac{a_{n-1}}{a_n}\frac{1}{x} + \frac{a_{n-2}}{a_n}\frac{1}{x^2} + \cdots + \frac{a_1}{a_n}\frac{1}{x^{n-1}} + \frac{a_0}{a_n}\frac{1}{x^n}) = 1 + 0 + 0 + \cdots + 0 + 0 = 1$

66. a) $\lim_{x\to\infty} \frac{\ell n(x+1)}{\ell n\ x} = 1$ b) $\lim_{x\to\infty} \frac{\ell n(x+999)}{\ell n\ x} = 1$

c) $\lim_{x\to\infty} \frac{\ell n\ x^2}{\ell n\ x} = 2$ d) $\lim_{x\to\infty} \frac{\ell n\ x}{\log x} = \lim_{x\to\infty} \frac{\ell n\ x}{(\frac{\ell n\ x}{\ell n\ 10})} = \lim_{x\to\infty} \ell n\ 10 = \ell n\ 10$

67. Using graphs, support $(1 + \frac{1}{x})^x \to e$ as $x \to \pm\infty$.

68. Using graphs, we can support $(1 + \frac{5}{x})^x \to e^5$ as $x \to \pm\infty$.

69. Using graphs, we can support $(1 + \frac{0.07}{x})^x \to e^{0.07}$ as $x \to \pm\infty$.

70. Using graphs, we can support $(1 + \frac{1}{x})^{1/x} \to 1$, as $x \to \pm\infty$.

71. $xe^{-x} = \frac{x}{e^x} \to 0$ as $x \to \infty$. $xe^{-x} \to -\infty$ as $x \to -\infty$. This is supported by graphing $y = xe^{-x}$ in the window $[-3, 5]$ by $[-5, 1]$.

72. $y = x^2 e^{-x} = \frac{x^2}{e^x} \to 0$ as $x \to \infty$. $y \to \infty$ as $x \to -\infty$.

73. $y = xe^x \to \infty$ as $x \to \infty$. $y \to 0$ as $x \to -\infty$. This may be supported by graphing y in $[-5, 3]$ by $[-1, 5]$.

74. $y = x^2e^x \to \infty$ as $x \to \infty$. $y \to 0$ as $x \to -\infty$.

75. As θ increases $\sin\theta$ steadily oscillates between the values -1 and $+1$ passing through all intermediate values. No matter how large θ gets, $\sin\theta$ thereafter does not stay arbitrarily close to any fixed value.

76. We assume that the numerator and denominator of the rational function have no common factor of the form $x-a$. Then the number of vertical asymptotes is the number of distinct factors of the denominator of the form $x-a$. For example, $\frac{2x^3-7x}{x^2(x+1)(x-2)^4}$ has the three distinct factors x, $x+1$, $x-2$ in the denominator and has the vertical asymptotes $x=0$, $x=-1$, $x=2$. If the degree of the numerator is larger than the degree of the denominator, the value of the function becomes infinite in absolute value as $x \to \pm\infty$ and there is no horizontal asymptote. If numerator and denominator have the same degree, then there is one horizontal asymptote $y = k \neq 0$ as in Example 12. Finally, if the degree of the denominator exceeds that of the numerator, there is one and only one horizontal asymptote $y=0$ as illustrated in Example 13.

2.5 CONTROLLING FUNCTION OUTPUTS

1. a) $0 < x < 6$. Not equivalent. b) $1 < x-1 < 7$. Adding 1, we get $2 < x < 8$. Equivalent. c) $1 < \frac{x}{2} < 4$. Multiplying by 2, we get $2 < x < 8$. Equivalent. d) $\frac{1}{8} < \frac{1}{x} < \frac{1}{2}$. Taking reciprocals, we get $8 > x > 2$. Equivalent. e) $x > 8$. Not equivalent. f) $|x-5| < 3$. $-3 < x-5 < 3$. Adding 5, we get $2 < x < 8$. Equivalent. g) $4 < x < 10$. Not equivalent. h) $-8 < -x < -2$. Multiplying by -1, we get $8 > x > 2$. Equivalent.

2. All are equivalent except c), d) and h).

3. Change $|x+3| < 1$ to $-1 < x+3 < 1$ to $-4 < x < -2$. Answer: g). Equivalently, we can read $|x+3| < 1$ as $|x-(-3)| < 1$, that is, the distance between x and -3 is less than 1 and thus $-4 < x < -1$.

4. Change $|x-5| < 2$ to $3 < x < 7$. Answer: c)

5. Change $|\frac{x}{2}| < 1$ to $-1 < \frac{x}{2} < 1$ to $-2 < x < 2$. Answer: e)

6. $|1-x| < 2$ or $|x-1| < 2$ yields $-1 < x < 3$. Answer: b)

7. Change $|2x-5| \leqq 1$ to $-1 \leqq 2x-5 \leqq 1$ to $4 \leqq 2x \leqq 6$ to $2 \leqq x \leqq 3$. Answer: h)

8. Change $|2x+4|<1$ to $-1<2x+4<1$ to $-5<2x<-3$ to $-\frac{5}{2}<x<-\frac{3}{2}$. Answer: d)

9. $|\frac{x-1}{5}| \leqq 1$ leads to $|x-1| \leqq 5$, $-5 \leqq x-1 \leqq 5$, $-4 \leqq x \leqq 6$. Answer: i)

10. Change $|\frac{2x+1}{3}|<1$ to $|2x+1|<3$ to $-3<2x+1<3$ to $-4<2x<2$ to $-2<x<1$. Answer: a)

11. $|y-2| \leqq 5$. $-5 \leqq y-2 \leqq 5$, $-3 \leqq y \leqq 7$.

12. $|y+3|<1$. $-1<y+3<1$, $-4<y<-2$.

13. $|2y-5|<1$. $-1<2y-5<1$, $4<2y<6$, $2<y<3$.

14. Answer: $-3<y<-2$

15. Change $|\frac{y}{2}-1| \leqq 1$ to $-1 \leqq \frac{y}{2}-1 \leqq 1$ to $0 \leqq \frac{y}{2} \leqq 2$ to $0 \leqq y \leqq 4$. Answer: $0 \leqq y \leqq 4$

16. Answer: $3<y<5$

17. $|2-y|<\frac{1}{5}$. $-\frac{1}{5}<2-y<\frac{1}{5}$, $-\frac{11}{5}<-y<-\frac{9}{5}$, $\frac{9}{5}<y<\frac{11}{5}$.

18. $|\frac{5-2y}{3}|<1$. $|5-2y|<3$, $-3<5-2y<3$, $-8<-2y<-2$, $1<y<4$.

19. The midpoint of the interval is $\frac{1+8}{2}=\frac{9}{2}$ and it has radius $\frac{9}{2}-1=\frac{7}{2}$. Answer: $|x-\frac{9}{2}|<\frac{7}{2}$

20. The midpoint of the interval is $\frac{-2+7}{2}=\frac{5}{2}$ and it has radius $7-\frac{5}{2}=\frac{9}{2}$. Answer: $|x-\frac{5}{2}|<\frac{9}{2}$

21. The midpoint of the interval is $\frac{-4+1}{2}=-\frac{3}{2}$ and it has radius $1-(-\frac{3}{2})=\frac{5}{2}$. Answer: $|x+\frac{3}{2}|<\frac{5}{2}$

22. The midpoint of the interval is $\frac{-8-1}{2}=-\frac{9}{2}$ and it has radius $-1-(-\frac{9}{2})=\frac{7}{2}$. Answer: $|x+\frac{9}{2}|<\frac{7}{2}$

23. $0.5<x^2<1.5$ yields $\sqrt{0.5}<|x|<\sqrt{1.5}$ or after appropriate rounding $0.71<|x|<1.22$. Thus we obtain $-1.22<x<-0.71$

24. If $0<x<\frac{\pi}{2}$, $0.3<\sin x<0.7$ yields $\sin^{-1}0.3<x<\sin^{-1}0.7$. In the second quadrant we need $\pi-\sin^{-1}0.7<x<\pi-\sin^{-1}0.3$ or, rounding appropriately, $2.37<x<2.83$.

25. For x in the interval $[0,\frac{\pi}{2}]$, $\cos x$ is decreasing and so $0.2<\cos x<0.6$ yields $\cos^{-1}0.6<x<\cos^{-1}0.2$, or rounding appropriately, $0.93<x<1.36$.

26. $1.98 < \tan x < 2.2$ yields $\tan^{-1} 1.98 < x < \tan^{-1} 2.2$ (since $\tan x$ is an increasing function on $[0, \frac{\pi}{2})$) or, rounding appropriately, $1.11 < x < 1.14$.

27. The graph of $y = \cos x$ is symmetric with respect to the y-axis. We need only reflect the corresponding interval in $[0, \frac{\pi}{2}]$ found in Exercise 25 over the y-axis. We obtain $-1.36 < x < -0.93$.

28. By the symmetry of the graph of $y = \tan x$ with respect to the origin, we need only take the reflection of the interval found in Exercise 26: $-1.14 < x < -1.11$.

29. $99.9 < x^2 < 100.1$ yields $\sqrt{99.9} < x < \sqrt{100.1}$, or rounding to thousandths appropriately, $9.995 < x < 10.004$.

30. From Exercise 29 we obtain $-10.004 < x < -9.995$.

31. Change $3.9 < \sqrt{x-7} < 4.1$ to $3.9^2 < x - 7 < 4.1^2$ to $22.21 < x < 23.81$.

32. Change $2 < \sqrt{19-x} < 4$ to $4 < 19 - x < 16$ to $-15 < -x < -3$ to $3 < x < 15$.

33. Change $4 < \frac{120}{x} < 6$ to $\frac{1}{4} > \frac{x}{120} > \frac{1}{6}$ to $30 > x > 20$ or $20 < x < 30$.

34. Change $\frac{1}{2} < \frac{1}{4x} < \frac{3}{2}$ to $2 > 4x > \frac{2}{3}$ to $\frac{1}{2} > x > \frac{1}{6}$ or $\frac{1}{6} < x < \frac{1}{2}$.

35. The graph of $y = \frac{3-2x}{x-1}$ is steadily falling as it passes through the horizontal channel between $y = -3.1$ and $y = -2.9$. Next we solve for x in terms of $y : (x-1)y = 3-2x$, $xy - y = 3 - 2x$, $xy + 2x = y + 3$, $x(y+2) = y+3$, $x = (y+3)/(y+2)$. Thus when $y = -2.9$, $x = (-2.9+3)/(-2.9+2) = -\frac{0.1}{0.9} = -\frac{1}{9}$ and, similarly, when $y = -3.1$, $x = \frac{1}{11}$. Answer: $-\frac{1}{9} < x < \frac{1}{11}$

36. $y = \frac{3x+8}{x+2}$ yields $(x+2)y = 3x + 8$, $xy + 2y = 3x + 8$, $x(y-3) = 8 - 2y$, $x = \frac{8-2y}{y-3}$. When $y = 0.9$, we get $x = -62/21$. When $y = 1.1$, we get $x = -58/19$. Thus $-58/19 < x < -62/21$, or rounding to hundredths appropriately, $-3.05 < x < -2.96$.

37. $10.5 < x^2 - 5 < 11.5$ leads to $15.5 < x^2 < 16.5$, $\sqrt{15.5} < x < \sqrt{16.5}$ since $x_0 > 0$. Thus $3.94 < x < 4.06$ rounding to hundredths appropriately.

38. Since the graph is symmetric with respect to the y-axis, we obtain, using Exercise 37, $-4.06 < x < -3.94$.

39. We graph $y = 4.8$, $y = 5.2$ and $y = x^3 - 9x$ in the same viewing rectangle, $[-4, 5]$ by $[-10, 10]$, for example. We then zoom in on the two points of intersection near $x = -3$. Rounding to hundredths appropriately, we obtain $-2.68 < x < -2.66$.

40. In the same viewing rectangle we graph $y = -5.2$, $y = -4.8$ and $y = x^4 - 10x^2$. Then zoom in on the two points of intersection near $x = 1$. We obtain, rounding to hundredths appropriately, $0.72 < x < 0.74$. In this example it is also possible to solve algebraically. $y+25 = x^4 - 10x^2 + 25$, $y+25 = (x^2-5)^2$, and since x is near 1, $x = \sqrt{5 - \sqrt{y+25}}$. We can now find the endpoints of the desired interval using $y = -4.8$ and $y = -5.2$.

41. $0.4 < e^x < 0.6$ leads to $\ell n\, 0.4 < \ell n\, e^x < \ell n\, 0.6$ since $\ell n\, x$ is an increasing function. Because $\ell n\, e^x = x$, we obtain after rounding appropriately to hundredths $-0.91 < x < -0.52$.

42. $1.9 < \ell n\, x < 2.1$ yields $e^{1.9} < x < e^{2.1}$ or $6.69 < x < 8.16$ rounding to hundredths appropriately.

43. $|f(x) - y_0| < E = 0.5$ is equivalent to $|x + 1 - 4| < 0.5$ or $|x - 3| < 0.5$ which is the desired inequality.

44. $|f(x) - y_0| = |2x - 1 + 5| = 2|x + 2| < 1$ is equivalent to $|x + 2| < \frac{1}{2}$ which is the desired inequality.

45. In the same viewing retangle we graph $y = 2.8$, $y = 3.2$ and $y = 2x^2 + 1$. The latter curve is rising near $x = 1$ and we zoom in on its points of intersection with the horizontal lines near $x = 1$. The curve is in the channel for $0.95 < x < 1.04$ after appropriate rounding. Thus we may take $|x - 1| < 0.04$.

46. $1.98 < \sqrt{2x - 3} < 2.2$ yields $1.98^2 < 2x - 3 < 2.2^2$ and $\frac{3+1.98^2}{2} < x < \frac{3+2.2^2}{2}$. After rounding to hundredths appropriately, we obtain $3.47 < x < 3.92$. Thus we may take $|x - 3.5| < 0.03$.

47. $\lim_{x\to 1} x^2 = 1$, $\lim_{x\to 1} x^2 = 1$, $\lim_{x\to\pi/6} \sin x = 0.5$, $\lim_{x\to 3} \frac{x+1}{x-2} = 4$, respectively.

48. $\lim_{x\to -1} x^2 = 1$, $\lim_{x\to 5\pi/6} \sin x = 0.5$, $\lim_{x\to x_0} \cos x = 0.4$ where $x_0 = \cos^{-1} 0.4$, $\lim_{x\to x_0} \tan x = 2$ where $x_0 = \tan^{-1} 2$, $\lim_{x\to x_0} \cos x = 0.4$ where $x_0 = \cos^{-1} 0.4$, $\lim_{x\to x_0} \tan x = -2$ where $x_0 = \tan^{-1}(-2)$, respectively.

49. $\lim_{x\to 10} x^2 = 100$, $\lim_{x\to -10} x^2 = 100$, $\lim_{x\to 23} \sqrt{x-7} = 4$, $\lim_{x\to 10} \sqrt{19-x} = 3$, $\lim_{x\to 24}(\frac{120}{x}) = 5$, $\lim_{x\to 1/4}(\frac{1}{4x}) = 1$, $\lim_{x\to 0} \frac{3-2x}{x-1} = -3$, $\lim_{x\to -3} \frac{3x+8}{x+2} = 1$, respectively.

50. $\lim_{x\to 4}(x^2-5) = 11$, $\lim_{x\to -4}(x^2-5) = 11$, $\lim_{x\to x_0}(x^3-9x) = 5$ where x_0 is a solution of $x^3 - 9x = 5$ near -3, $\lim_{x\to x_0}(x^4 - 10x^2) = -5$ where x_0 is the solution of $x^4 - 10x^2 = -5$ near 1, $\lim_{x\to x_0} e^x = 0.5$ where $x_0 = \ell n\, 0.5$, $\lim_{x\to e^2} \ell n\, x = 2$, $\lim_{x\to 3}(x+1) = 4$, $\lim_{x\to -2}(2x-1) = -5$, $\lim_{x\to 1}(2x^2 + 1) = 3$, $\lim_{x\to 3.5} \sqrt{2x-3} = 2$, respectively.

51. $8.99 < \pi(x/2)^2 < 9.01$ leads to $\frac{8.99}{\pi} < (\frac{x}{2})^2 < \frac{9.01}{\pi}$, $\frac{4(8.99)}{\pi} < x^2 < \frac{4(9.01)}{\pi}$ and to $\sqrt{\frac{4(8.99)}{\pi}} < x < \sqrt{\frac{4(9.01)}{\pi}}$. Rounding appropriately to thousandths, we obtain $3.384 < x < 3.387$ or, in symmetric form, $|x - x_0| < 0.001$.

52. $4.9 < I < 5.1$ or $4.9 < \frac{V}{R} < 5.1$ leads to $\frac{1}{4.9} > \frac{R}{V} > \frac{1}{5.1}$ and to $\frac{120}{4.9} > R > \frac{120}{5.1}$. Rounding to tenths appropriately, we obtain $23.6 < R < 24.4$ or $|R - 24| < 0.4$.

53. $f(x) = \frac{3x+1}{x-2} = 3 + \frac{7}{x-2} \to 3$ as $x \to \infty$ (the equality can be obtained by long division). The same equality (or a graph) shows that $f(x) > 3$ for $x > 2$. Thus we want to solve $3 + \frac{7}{x-2} < 3.01$, $x > 2$. This leads to $\frac{7}{x-2} < 0.01$, $x - 2 > \frac{7}{0.01}$ (since $x - 2 > 0$), $x > 702$.

54. $f(x) = \frac{2x+5}{5x-7} = \frac{2}{5} + \frac{39}{5(5x-7)}$. Thus $f(x) \to \frac{2}{5}$ as $x \to \infty$ and $f(x) > \frac{2}{5}$ if $x > \frac{7}{5}$. $\frac{2x+5}{5x-7} < 0.41$ leads to $2x + 5 < 0.41(5x - 7)$ (since we may take $x > \frac{7}{5}$), $2x + 5 < 2.05x - 2.87$, $0.05x > 7.87$, $x > 157.4$.

55. By long division $f(x) = \frac{2x^2-x+2}{x^2-4} = 2 + \frac{10-x}{x^2-4}$. This shows $f(x) < 2$ for $x > 10$ and $f(x) = 2 + \frac{(10/x^2)-(1/x^2)}{1-(4/x^2)} \to 2 + 0 = 2$ as $x \to \infty$. $f(x) > 1.99$, $x > 10$ leads to $2x^2 - x + 2 > 1.99(x^2 - 4)$, $0.01x^2 - x + 9.96 > 0$, $x^2 - 100x + 996 > 0$. By the quadratic formula, the larger root of the quadratic is $50 + \sqrt{1504}$. Thus we require $x > 50 + \sqrt{1504} \approx 88.781$.

56. By long division $f(x) = \frac{3x^3-x+1}{2x^3+5} = \frac{3}{2} - \frac{2x+13}{2(2x^3+5)}$. Thus $f(x) \to \frac{3}{2}$ as $x \to \infty$ and $f(x) < \frac{3}{2}$ for $x > 0$. $f(x) > 1.5 - 0.01 = 1.49$ leads to $3x^3 - x + 1 > 1.49(2x^3 + 5)$ (since $2x^3 + 5 > 0$), $0.02x^3 - x - 6.45 > 0$, $2x^3 - 100x - 645 > 0$. By ZOOM-IN, the cubic has the unique root $x = 9.2186...$ Hence we require $x > 9.2186...$

57. In the interval $[0, 2\pi]$, $\sin x = \frac{\sqrt{2}}{2}$ has two solutions $\frac{\pi}{4}$ in the first quadrant and $\pi - \frac{\pi}{4} = \frac{3\pi}{4}$ in the second quadrant, recalling that $\sin(\pi - x) = \sin x$ for all x. We first solve the problem in the first quadrant. We use the result to solve the problem in the second quadrant and then use the periodicity of the sine function to give the complete solution set. Graph $y_1 = \sin x$, $y_2 = \frac{\sqrt{2}}{2} - 0.1$ and $y_3 = \frac{\sqrt{2}}{2} + 0.1$ in $[0.6, 1]$ by $[0.5, 1]$. We then zoom in to the two points of intersection and find that they occur at $x_1 = 0.65241449628$ and $x_2 = 0.93923517764$. Thus in the first quadrant we must have $x_1 < x < x_2$. In the second quadrant we must have $\pi - x_2 < x < \pi - x_1$. The complete solution set is $\{x \mid x_1 + 2n\pi < x < x_2 + 2n\pi \text{ or } (2n+1)\pi - x_2 < x < (2n+1)\pi - x_1\}$.

58. Let h be the width of a stripe in cm. Let ΔV be the volume of liquid in the cylinder at the level of the entire stripe. Then ΔV is the volume of a small right circular cylinder of base radius 5cm and height $h$cm. $\Delta V = \pi 5^2 h$ gives ΔV in cubic centimeters. Since 1 liter = 1000cm^3, we need $\Delta V = \frac{25\pi h}{1000} = \frac{\pi h}{40}$ to have ΔV represented in liters. Thus we must have $\frac{\pi h}{40} < \frac{0.01}{2} = 0.005$ since the middle of the stripe corresponds to exactly one liter. $h < \frac{40(0.005)}{\pi} \approx 0.06$cm.

59. By long division $f(x) = \frac{x-3}{2x+1} = \frac{1}{2} - \frac{7}{2(2x+1)} > \frac{1}{2}$ for $x < -\frac{1}{2}$, and $f(x) \to \frac{1}{2}$ as $x \to -\infty$. $f(x) < 0.51$ leads to $\frac{x-3}{2x+1} < 0.51$, $x - 3 > 0.51(2x+1)$ (since $2x + 1 < 0$), $x - 3 > 1.02x + 0.51$, $-3.51 > .02x$, $x < -175.5$.

60. $f(x) = \frac{3x+4}{2x-1} = \frac{3}{2} + \frac{11}{2(2x-1)} < \frac{3}{2}$ for $x < \frac{1}{2}$ and $f(x) \to \frac{3}{2}$ as $x \to -\infty$. $\frac{3x+4}{2x-1} > 1.49$ leads to $3x + 4 < 1.49(2x - 1)$ (since $2x - 1 < 0$), $0.02x < -5.49$, $x < -274.5$.

61. By long division $f(x) = \frac{x-4}{1-3x} = -\frac{1}{3} - \frac{11}{3(1-3x)} < -\frac{1}{3}$. $\frac{x-4}{1-3x} \to -\frac{1}{3}$ as $x \to -\infty$. $\frac{x-4}{1-3x} > -\frac{1}{3} - \frac{1}{100} = -\frac{103}{300}$ leads to $300(x-4) > -103(1-3x)$ (since $1 - 3x > 0$ for $x < \frac{1}{3}$), $9x < -1200 + 103 = -1097$, $x < -\frac{1097}{9}$.

62. $f(x) = \frac{5-2x}{x+1} = -2 + \frac{7}{x+1} < -2$ when $x < -1$. $f(x) \to -2$ as $x \to -\infty$ and $\frac{5-2x}{x+1} > -2.01$ leads to $5 - 2x < -2.01x - 2.01$ (since $x + 1 < 0$ for $x < -1$), $0.01x < -7.01$, $x < -701$.

2.6 DEFINING LIMITS FORMALLY WITH EPSILONS AND DELTAS

1. $x_0 - a = 5 - 1 = 4$ and $b - x_0 = 7 - 5 = 2$. Since $2 < 4$, we choose $\delta = 2$. Then $|x - x_0| < 2$ or $|x - 5| < 2$ implies $3 < x < 7$ and so $1 < x < 7$.

2. $\delta = 1$

3. $x_0 - a = -3 + \frac{7}{2} = \frac{1}{2}$ and $b - x_0 = -\frac{1}{2} + 3 = \frac{5}{2}$. Since $\frac{1}{2} < \frac{5}{2}$, $\delta = \frac{1}{2}$.

4. $\delta = 1$

5. From the graph we see that if x is between 4.9 and 5.1 on the x-axis, then y on the curve (corresponding to x) is in the range $5.8 < y < 6.2$. Therefore we can take $\delta = 0.1$ because $0 < |x - 5| < \delta = 0.1$ does imply $|f(x) - L| = |y - 6| < 0.2 = \varepsilon$.

6. $\delta = 0.1$

7. From the graph we see that $\sqrt{3} < x < \sqrt{5}$ implies that $|f(x) - L| < \varepsilon$. Rounding the first inequality appropriately to hundredths, we get $1.74 < x < 2.23$. Since 2 is closer to 2.23 than to 1.74, we take $\delta = 2.23 - 2 = 0.23$.

8. $\delta = \min\{-1 + \frac{\sqrt{5}}{2}, -\frac{\sqrt{3}}{2} + 1\} = 0.11$ rounded to hundredths appropriately.

9. From the graph we see that $\frac{9}{16} < x < \frac{25}{16}$ implies $|f(x) - L| < \varepsilon = \frac{1}{4}$. $1 - \frac{9}{16} = \frac{7}{16}$ and $\frac{25}{16} - 1 = \frac{9}{16}$. Thus 1 is closer to $\frac{9}{16}$. We take $\delta = \frac{7}{16}$. Note that $|x - x_0| < \delta$ is the same as $|x - 1| < \frac{7}{16}$ which is equivalent to $\frac{9}{16} < x < \frac{23}{16}$. So $\delta = \frac{7}{16}$ works.

10. $\delta = \min\{3 - 2.61, 3.41 - 3\} = 0.39$

11. $\lim_{x\to 1}(2x + 3) = 5$. $|f(x) - L| = |2x + 3 - 5| = |2x - 2| = |2(x - 1)| = |2||x - 1| = 2|x - 1|$. Thus $|f(x) - L| < 0.01$ is $2|x - 1| < 0.01$ which is equivalent to $|x - 1| < \frac{0.01}{2} = 0.005$. Thus $\delta = 0.005$ will do because $0 < |x - 1| < 0.005 \Rightarrow |f(x) - L| < 0.01 = \varepsilon$.

12. $\lim_{x\to 3}(3 - 2x) = -3$. $|3 - 2x + 3| = |2(3 - x)| = 2|x - 3| < 0.02$ if and only if $|x - 3| < 0.01$. Thus $\delta = 0.01$ will do.

13. $\lim_{x\to 2} \frac{x^2-4}{x-2} = \lim_{x\to 2} \frac{(x-2)(x+2)}{x-2} = \lim_{x\to 2}(x + 2) = 4$. If $x \neq 2$, $\frac{x^2-4}{x-2} = x + 2$ and we may assume this here because x is never equal to 2 as $x \to 2$. Thus $|f(x) - L| = |x + 2 - 4| = |x - 2| < \varepsilon = 0.05$ is equivalent to $0 < |x - 2| < 0.05$ and we may take $\delta = 0.05$.

14. $\lim_{x\to -5} \frac{x^2+6x+5}{x+5} = \lim_{x\to -5} \frac{(x+5)(x+1)}{x+5} = -4$. If $x \neq -5$, $|f(x) - L| = |(x + 1) + 4| = |x + 5|$. So we may take $\delta = \varepsilon = 0.05$.

15. $L = \lim_{x\to 11} \sqrt{x-7} = \sqrt{11-7} = 2$. The inequality $|\sqrt{x-7}-2| < 0.01$ leads successively to $-0.01 < \sqrt{x-7} - 2 < 0.01$, $1.99 < \sqrt{x-7} < 2.01$, $1.99^2 < x - 7 < 2.01^2$, $7 + 1.99^2 < x < 7 + 2.01^2$, $10.9601 < x < 11.0401$. The distance, 0.0399, between $x_0 = 11$ and the left endpoint is less than the distance between x_0 and the right endpoint. Thus we may take $\delta = 0.0399$.

16. $L = \lim_{x\to -3} \sqrt{1-5x} = \sqrt{1+15} = 4$. The inequality $|\sqrt{1-5x} - 4| < 0.5$ leads to $-0.5 < \sqrt{1-5x} - 4 < 0.5$, $3.5 < \sqrt{1-5x} < 4.5$, $3.5^2 < 1 - 5x < 4.5^2$, $-4.5^2 < 5x - 1 < -3.5^2$, $\frac{1-4.5^2}{5} < x < \frac{1-3.5^2}{5}$, $-3.85 < x < 2.25$. The distance, 0.75, between $x_0 = -3$ and the right endpoint is smaller than the distance between x_0 and the left endpoint. Thus we may take $\delta = 0.75$ or any other smaller positive number.

17. $\lim_{x\to 2} \frac{4}{x} = 2$. $|f(x) - L| < \varepsilon$ is equivalent to each of the following: $|\frac{4}{x} - 2| < 0.4 = \frac{2}{5}$, $2 - \frac{2}{5} < \frac{4}{x} < 2 + \frac{2}{5}$, $\frac{8}{5} < \frac{4}{x} < \frac{12}{5}$, $\frac{5}{12} < \frac{x}{4} < \frac{5}{8}$, $\frac{5}{3} < x < \frac{5}{2}$, $2 - \frac{1}{3} < x < 2 + \frac{1}{2}$. The last inequality is satisfied if $|x - 2| < \frac{1}{3}$. We may therefore take $\delta = \frac{1}{3}$.

18. $\lim_{x\to \frac{1}{2}} \frac{4}{x} = 8$. $|f(x) - L| < \varepsilon$ is equivalent to each of the following: $|\frac{4}{x} - 8| < 0.04 = \frac{1}{25}$, $8 - \frac{1}{25} < \frac{4}{x} < 8 + \frac{1}{25}$, $\frac{199}{25} < \frac{4}{x} < \frac{201}{25}$, $\frac{25}{201} < \frac{x}{4} < \frac{25}{199}$, $\frac{100}{201} < x < \frac{100}{199}$, $\frac{1}{2} - \frac{1}{402} < x < \frac{1}{2} + \frac{1}{398}$. Thus we may take $\delta = \frac{1}{402}$ or $\delta = 0.0024$ rounding to ten thousandths appropriately.

19. $|f(x) - 5| = |9 - x - 5| = |4 - x| = |x - 4|$. Thus $|f(x) - 5| < \varepsilon$ is equivalent to $|x - 4| < \varepsilon$. Thus in each case $\delta = \varepsilon$.

20. $|f(x) - 5| = |3x - 12| = 3|x - 4| < \varepsilon$ if and only if $|x - 4| < \varepsilon/3$. Thus $\delta = 0.001$, 0.0001, $\varepsilon/3$, respectively.

21. Let $f(x) = \frac{x+2}{x+1} = \frac{(x+1)+1}{x+1} = 1 + \frac{1}{x+1}$. $\lim_{x\to\infty}(1 + \frac{1}{x+1}) = 1$. Let $\varepsilon > 0$ be given. Then $|f(x) - 1| = \frac{1}{x+1}$ (for $x > -1$) $< \varepsilon$ if and only if $x + 1 > \frac{1}{\varepsilon}$ or $x > \frac{1}{\varepsilon} - 1$. Thus if $N = \frac{1}{\varepsilon} - 1$, $x > N$ implies $|f(x) - 1| < \varepsilon$.

22. Let $f(x) = \frac{x^2}{2x^2-1}$. By long division $f(x) = \frac{1}{2} + \frac{1}{2(2x^2-1)}$. We can see that $\lim_{x\to\infty} f(x) = \frac{1}{2}$. Let $\varepsilon > 0$ be given. $|f(x) - \frac{1}{2}| = \frac{1}{2(2x^2-1)}$ (for $x > \frac{1}{\sqrt{2}}$) $< \varepsilon$ is equivalent to $2x^2 - 1 > \frac{1}{2\varepsilon}$, $2x^2 > 1 + \frac{1}{2\varepsilon}$, $x > \sqrt{\frac{1}{2} + \frac{1}{4\varepsilon}}$. Thus if $N = \sqrt{\frac{1}{2} + \frac{1}{4\varepsilon}}$, $x > N$ implies $|f(x) - \frac{1}{2}| < \varepsilon$.

23. Let $f(x) = \frac{x+2}{x+1} = 1 + \frac{1}{x+1}$. Here since $x \to -1^+$, $x > -1$ or $x + 1 > 0$. Let $N > 1$ be given, $f(x) > N$ is equivalent to $1+\frac{1}{x+1} > N$, $\frac{1}{x+1} > N-1$, $x+1 < \frac{1}{N-1}$. Let $\delta = \frac{1}{N-1}$. Then $-1 < x < -1 + \delta$ implies $f(x) > N$. This proves $\lim_{x\to -1^+} f(x) = \infty$.

24. Let $N > 2$ be given. Here $2x^2 - 1 > 0$ because $x > \frac{\sqrt{2}}{2}$. $f(x) = \frac{x^2}{2x^2-1} > N$ leads to $x^2 > (2x^2 - 1)N$, $x^2(1 - 2N) > -N$, $x^2(2N - 1) < N$, $x^2 < \frac{N}{2N-1}$, $x < \sqrt{\frac{N}{2N-1}}$, $x - \frac{\sqrt{2}}{2} < \sqrt{\frac{N}{2N-1}} - \frac{\sqrt{2}}{2}$. So we choose $\delta = \sqrt{\frac{N}{2N-1}} - \frac{\sqrt{2}}{2}$. Then $\frac{\sqrt{2}}{2} < x < \frac{\sqrt{2}}{2} + \delta$ implies $f(x) > N$. This proves $\lim_{x\to \frac{\sqrt{2}}{2}^+} \frac{x^2}{2x^2-1} = \infty$.

25. $\lim_{x\to 1} \sin x = 0.84 (= \sin 1)$ rounded to hundredths. In the viewing rectangle $[0.99, 1.01]$ by $[0.83, 0.85]$ we graph $y = 0.84$ and $y = \sin x$ (which appears as a straight line). Using the endpoints of $y = \sin x$ in this rectangle, we calculate the slope $m = \frac{0.8468-0.8360}{1.01-0.99} = 0.54$. As in Example 6, $\delta = \varepsilon|m| = \varepsilon/0.54 = 1.85\varepsilon$ rounding δ down to be safe.

26. $\lim_{x\to 1} \tan x = \tan 1 = 1.56$ after rounding. In the viewing rectangle $[0.99, 1.01]$ by $[1.55, 1.57]$ we graph $y = 1.56$ and $y = \tan x$ (which appears as a straight line). Using the endpoints of $y = \tan x$ in the rectangle, we find $m = 3.44$ (rounding up to be safe). $\delta = \frac{\varepsilon}{|m|} = 0.29\varepsilon$ rounding down to be safe.

27. $\lim_{x\to 1} \cos x = \cos 1 = 0.54$ after rounding. In the viewing rectangle $[0.99, 1.01]$ by $[0.53, 0.55]$ we graph $y = 0.54$ and $y = \cos x$. Using two points on the graph of $y = \cos x$, we get for the estimate of m, $m = -0.85$. Thus $\delta = \varepsilon/0.85 = 1.17\varepsilon$.

28. $\lim_{x\to 4} \sec x = \sec 4 = -1.53$ after rounding. We graph $y = -1.53$ and $y = \sec x$ in the viewing rectangle $[3.99, 4.0]$ by $[-1.54, -1.52]$. Our estimate of m is $m = -1.78$. Hence $\delta = \varepsilon/1.78 = 0.56\varepsilon$.

29. $\lim_{x\to 0.5}(x^3 - 4x) = (0.5)^3 - 4(0.5) = -1.88$ after rounding. In the viewing rectangle $[0.49, 0.51]$ by $[-1.89, 1.87]$ we graph $y = -1.88$ and $y = x^3 - 4x$. Using the endpoints to estimate the slope, we get $m = -3.25$. Therefore $\delta = \varepsilon/3.25 = 0.30\varepsilon$ rounding down to be safe.

30. $\lim_{x\to 2.5} 9x - x^3 = 6.875$. In the viewing rectangle $[2.49, 2.51]$ by $[6.874, 6.876]$ we graph $y = 6.875$ and $y = 9x - x^3$. In this viewing rectangle, the latter graph appears as a vertical line and we are unable to calculate a slope. In several smaller viewing rectangles we obtain slope estimates near -10. To be safe we round $|m|$ to be 11. Thus $\delta = \varepsilon/11$ is our estimate.

31. $\lim_{x\to -1} \frac{x}{x^2-4} = \frac{1}{3}$. In the viewing rectangle $[-1.01, -0.99]$ by $[0.32, 0.34]$ we graph $y = 0.33$ and $y = \frac{x}{x^2-4}$. Using the endpoints of the latter graph, we get our estimate of the slope, $m = -0.56$. Thus $\delta = \varepsilon/0.56 = 1.78\varepsilon$ rounding appropriately.

32. $\lim_{x\to -1} \frac{2x}{5-x^2} = -\frac{1}{2}$. In the viewing rectangle $[-1.01, -0.99]$ by $[-0.51, -0.49]$ we graph $y = -0.5$ and $y = \frac{2x}{5-x^2}$. Using the endpoints of the latter graph, we get our estimate of the slope, $m = 0.75$. Hence our estimate is $\delta = \frac{\varepsilon}{|m|} = \frac{4\varepsilon}{3}$.

33. $\sqrt{x-5} < \varepsilon$ is equivalent to $x - 5 < \varepsilon^2$ or $x < 5 + \varepsilon^2$ since $x \geqq 5$. Thus $I = (5, 5+\varepsilon^2)$. Since $5 < x < 5 + \varepsilon^2$ implies $|\sqrt{x-5} - 0| < \varepsilon$, this verifies that $\lim_{x\to 5^+} \sqrt{x-5} = 0$.

34. $\sqrt{4-x} < \varepsilon$ is equivalent to $4 - x < \varepsilon^2$ or $4 - \varepsilon^2 < x$ since $x \leqq 4$. Thus $I = (4-\varepsilon^2, 4)$. This verifies $\lim_{x\to 4^-} \sqrt{4-x} = 0$.

35.

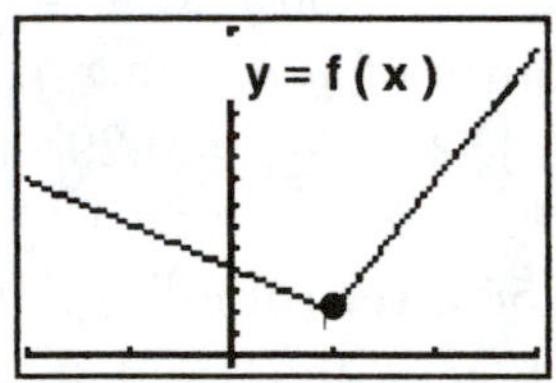

$[-2, 3]$ by $[-1, 16]$

Since $y \geqq 2$, we need only be concerned that $y < 2 + \varepsilon$. For $x < 1$, this is equivalent to $4 - 2x < 2 + \varepsilon$ which leads to $2 < 2x + \varepsilon$ and to $x > 1 - \varepsilon/2$. For $x \geqq 1$, $y < 2+\varepsilon$ is $6x - 4 < 2 + \varepsilon$ which leads to $x < 1 + \varepsilon/6$. The largest δ can be is the smaller of $\varepsilon/2$, $\varepsilon/6$ which is $\varepsilon/6$ and $I = (1 - \varepsilon/6, 1 + \varepsilon/6)$.

36. $f(x) = -1$ for $x < 5$ and $f(x) = 1$ for $x > 5$ and f is not defined for $x = 5$. If $\varepsilon = 4$, $1 - \varepsilon < f(x) < 1 + \varepsilon$ becomes $-3 < f(x) < 4$ which is true for all $x \neq 5$. If $\varepsilon = 2$, $-1 < f(x) < 3$ is true for all $x > 5$. If $\varepsilon = 1$, $0 < f(x) < 2$ is true for all $x > 5$. If $\varepsilon = \frac{1}{2}$, $\frac{1}{2} < f(x) < \frac{3}{2}$ is true for all $x > 5$.

37. $\lim_{x\to 2} f(x) = 5$ means corresponding to any radius $\varepsilon > 0$ about 5, there exists a radius $\delta > 0$ about 2 such that $0 < |x - 2| < \delta$ implies $|f(x) - 5| < \varepsilon$.

38. $\lim_{x\to 0} g(x) = k$ means corresponding to each radius $\varepsilon > 0$ about k, there exists a radius $\delta > 0$ about 0 such that $0 < |x-0| < \delta$ implies $|g(x)-k| < \varepsilon$, x in the domain of g.

39. Since we need $|x^2-4| < \varepsilon < 4$, x cannot be 0 and so $\delta < 2$, $0 < 2-\delta < x$, i.e., $x > 0$. Now $|x^2-4| < \varepsilon$ is equivalent to each of the following: $-\varepsilon < x^2-4 < \varepsilon$, $4-\varepsilon < x^2 < 4+\varepsilon$, $\sqrt{4-\varepsilon} < |x| < \sqrt{4+\varepsilon}$, $\sqrt{4-\varepsilon} < x < \sqrt{4+\varepsilon}$ since $x > 0$. The distance from $x = 2$ to the left endpoint is $2-\sqrt{4-\varepsilon} = (2-\sqrt{4-\varepsilon})\frac{(2+\sqrt{4-\varepsilon})}{2+\sqrt{4-\varepsilon}} = \frac{\varepsilon}{2+\sqrt{4-\varepsilon}}$, and the distance from $x = 2$ to the right endpoint is $\sqrt{4+\varepsilon}-2 = (\sqrt{4+\varepsilon}-2)\frac{(\sqrt{4+\varepsilon}+2)}{\sqrt{4+\varepsilon}+2} = \frac{\varepsilon}{\sqrt{4+\varepsilon}+2}$. Since the second distance has a larger denominator, it is smaller and $\delta = \sqrt{4+\varepsilon}-2$. This verifies $\lim_{x\to 2} x^2 = 4$ or $\lim_{x\to 2}(x^2-4) = 0$. $\delta \to 0$ as $\varepsilon \to 0$. The graph of δ as a function of ε can be viewed by graphing of δ as a function of ε can be viewed by graphing $y = \sqrt{4+x}-2$ in the rectangle $[0,4]$ by $[0,1]$. The endpoints are not included.

40. $|\frac{2}{x-1}-1| < \varepsilon$ is equivalent to the following: $-\varepsilon < \frac{2}{x-1}-1 < \varepsilon$, $1-\varepsilon < \frac{2}{x-1} < 1+\varepsilon$, $\frac{1}{1+\varepsilon} < \frac{x-1}{2} < \frac{1}{1-\varepsilon}$ (since $\varepsilon < 1$), $\frac{2}{1+\varepsilon} < x-1 < \frac{2}{1-\varepsilon}$, $1+\frac{2}{1+\varepsilon} < x < 1+\frac{2}{1-\varepsilon}$. The distance between 3 and the left endpoint is $3-(1+\frac{2}{1+\varepsilon}) = 2(1-\frac{1}{1+\varepsilon}) = \frac{2\varepsilon}{1+\varepsilon}$. The distance between 3 and the right endpoint is $1+\frac{2}{1-\varepsilon}-3 = 2(\frac{1}{1-\varepsilon}-1) = \frac{2\varepsilon}{1-\varepsilon}$. We must use the shorter distance so $\delta = 2\varepsilon/(1+\varepsilon)$. This verifies $\lim_{x\to 3}\frac{2}{x-1} = 1$ or $\lim_{x\to 3}(\frac{2}{x-1}-1) = 0$. $\delta = 2\varepsilon/(1+\varepsilon) \to 0$ as $\varepsilon \to 0$. The graph of δ as a function of ε may be viewed by graphing $y = 2x/(1+x)$ in the rectangle $[0,1]$ by $[0,1]$ and excluding the endpoints since $0 < \varepsilon < 1$. This verifies $\lim_{x\to 3}\frac{2}{x-1} = 1$ or $\lim_{x\to 3}(\frac{2}{x-1}-1) = 0$.

41. One need only reverse the order of the steps in Example 3. Each line implies the preceding line in that list of steps.

42. a) The inequality is supported by graphing $y_1 = \frac{1}{2+x}$, $y_2 = \frac{1}{2}$, $y_3 = \frac{1}{2-x}$ in $[0,2]$ by $[-1,3]$. b) If $\varepsilon > 2$, $\frac{1}{2} < \frac{1}{2-\varepsilon}$ fails because $\frac{1}{2-\varepsilon}$ is negative.

43. Suppose $\varepsilon < 2$. Then $\delta = \frac{\varepsilon}{2(2+\varepsilon)}$. $|x-0.5| < \delta$ implies $-\frac{\varepsilon}{2(2+\varepsilon)} < x-\frac{1}{2} < \frac{\varepsilon}{2(2+\varepsilon)} < \frac{\varepsilon}{2(2-\varepsilon)}$. We may now use the steps in Example 4 in reverse order to obtain $|f(x)-2| < \varepsilon$. Now suppose $\varepsilon \geqq 2$, $\delta = \frac{1}{4}$. $|x-0.5| < \delta = \frac{1}{4}$ implies $\frac{1}{4} < x < \frac{3}{4}$, $\frac{4}{3} < \frac{1}{x} < 4$, $-\frac{2}{3} < \frac{1}{x}-2 < 2$ which implies $|\frac{1}{x}-2| < 2 \leqq \varepsilon$.

44. a) Graph $y_1 = x^3 + 1.001 + 0\sqrt{x}$ and $y_2 = x^3 + 0.009 + 0\sqrt{-x}$ in $[-1,1]$ by $[-0.9,1.6]$. b) Since $\lim_{x\to 0^+} f(x) = 1.001 \neq 0.009 = \lim_{x\to 0^-} f(x)$, $\lim_{x\to 0} f(x)$ does not exist. c) If x_1 is negative and x_2 is positive, then $|f(x_2) - f(x_1)| > 1.001 - 0.009 = 0.992$. For $\varepsilon = 0.4$ the implication fails no matter what $\delta > 0$ is. Because if it did hold, and if x_1 is negative and x_2 is positive, both within δ of $x_0 = 0$, we would have $|f(x_2) - f(x_1)| = |f(x_2) - L + L - f(x_1)| \leqq |f(x_2) - L| + |L - f(x_1)| < \varepsilon + \varepsilon = 2\varepsilon = 0.8$, a contradiction.

45. a)

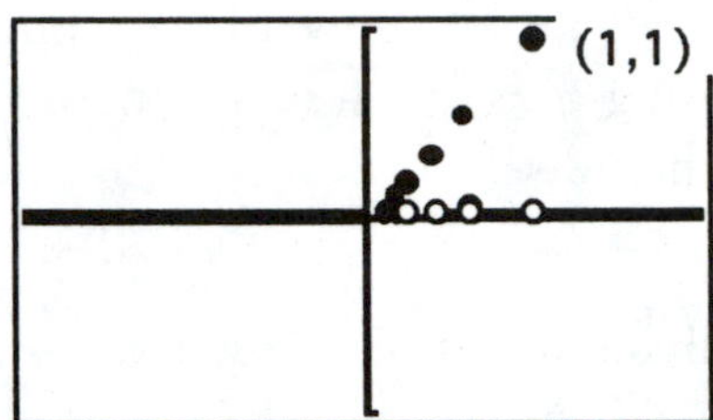

$[-2,2]$ by $[-1,1]$

b) Since $|f(x) - 0| = |f(x)| \leqq |x| = |x - 0|$, we may take $\delta = \varepsilon$ in the definition of limit to prove $\lim_{x\to 0} f(x) = 0$.

46. Let $f(x) = \frac{|x+1|}{x+1}$. Then $f(x) = \begin{cases} -1, & x < -1 \\ 1, & x > -1 \end{cases}$. For any $\delta > 0$ let $x_0 = -1 - \frac{\delta}{2}$. Then $0 < |x_0 + 1| = \frac{\delta}{2} < \delta$, but $|f(x_0) - 1| = |-1 - 1| = 2$. Thus given ε, $0 < \varepsilon \leqq 2$, there is no corresponding δ as in the definition of limit. Hence the limit is not 1 and it therefore does not exist.

PRACTICE EXERCISES, CHAPTER 2

1. Exists **2.** Exists **3.** Exists

4. Does not exist. The right-hand and left-hand limits exist but are not equal.

5. Exists **6.** Exists **7.** Continuous at $x = a$

8. Not continuous at $x = b$. $\lim_{x\to b} f(x) \neq f(b)$

9. Not continuous at $x = c$ since $\lim_{x\to c} f(x)$ does not exist.

10. Continuous at $x = d$

11. $\text{Lim}_{x\to -2} x^2(x+1) = (-2)^2(-2+1) = 4(-1) = -4$

12. $\text{Lim}_{x\to 3}(x+2)(x-5) = 5(-2) = -10$

13. $\text{Lim}_{x\to 3} \frac{x-3}{x^2} = \frac{3-3}{3^2} = \frac{0}{9} = 0$

14. $\text{Lim}_{x\to -1} \frac{x^2+1}{3x^2-2x+5} = \frac{2}{3+2+5} = \frac{1}{5}$

15. $\text{Lim}_{x\to -2} (\frac{x}{x+1})(\frac{3x+5}{x^2+x}) = (\frac{-2}{-1})(\frac{-1}{4-2}) = -1$

16. $\text{Lim}_{x\to 1} (\frac{1}{x+1})(\frac{x+6}{x})(\frac{3-x}{7}) = \frac{1}{2}\frac{7}{1}\frac{2}{7} = 1$

17. $\text{Lim}_{x\to 4} \sqrt{1-2x}$ does not exist.

18. $\text{Lim}_{x\to 5} \sqrt[4]{9-x^2}$ does not exist.

19. $\lim_{x\to 1} \frac{x^2-1}{x-1} = \lim_{x\to 1} x+1 = 2$

20. $\text{Lim}_{x\to -5} \frac{x^2+3x-10}{x+5} = \lim_{x\to -5} \frac{(x+5)(x-2)}{x+5} = \lim_{x\to -5} x-2 = -7$

21. $\text{Lim}_{x\to 2} \frac{x-2}{x^2+x-6} = \lim_{x\to 2} \frac{x-2}{(x-2)(x+3)} = \lim_{x\to 2} \frac{1}{x+3} = \frac{1}{5}$

22. $\text{Lim}_{x\to 1} \frac{x^2-2x+1}{x^3-2x^2+x} = \lim_{x\to 1} \frac{1}{x} = 1$

23. $\text{Lim}_{x\to 0} \frac{(1+x)(2+x)-2}{x} = \lim_{x\to 0} \frac{3x+x^2}{x} = \lim_{x\to 0} 3+x = 3$

24. $\text{Lim}_{x\to 0} \frac{\frac{1}{2+x}-\frac{1}{2}}{x} = \lim_{x\to 0} \frac{2-(2+x)}{2(2+x)}\frac{1}{x} = \lim_{x\to 0} \frac{-1}{2(2+x)} = -\frac{1}{4}$

25. $\text{Lim}_{x\to\infty} \frac{2x+3}{5x+7} = \lim_{x\to\infty} \frac{2+\frac{3}{x}}{5+\frac{7}{x}} = \frac{2+0}{5+0} = \frac{2}{5}$

26. $\text{Lim}_{x\to\infty} \frac{2x^2+3}{5x^2+7} = \lim_{x\to\infty} \frac{2+\frac{3}{x^2}}{5+\frac{7}{x^2}} = \frac{2}{5}$

27. $\text{Lim}_{x\to -\infty} \frac{x^2-4x+8}{3x^3} = \lim_{x\to -\infty} \frac{\frac{1}{x}-\frac{4}{x^2}+\frac{8}{x^3}}{3} = 0$

28. $\text{Lim}_{x\to\infty} \frac{1}{x^2-7x+1} = \lim_{x\to\infty} \frac{\frac{1}{x^2}}{1-\frac{7}{x}+\frac{1}{x^2}} = \frac{0}{1-0+0} = 0$

29. $\text{Lim}_{x\to -\infty} \frac{x^2-7x}{x+1} = \lim_{x\to -\infty} \frac{x(x-7)}{x+1} = \lim_{x\to -\infty} \frac{x(1-\frac{7}{x})}{1+\frac{1}{x}} = -\infty$
because $\lim_{x\to -\infty} \frac{1-\frac{7}{x}}{1+\frac{1}{x}} = 1$.

30. $\text{Lim}_{x\to\infty} \frac{x^4+x^3}{12x^3+128} = \lim_{x\to\infty} \frac{x(x^3+x^2)}{12x^3+128} = \lim_{x\to\infty} x(\frac{1+\frac{1}{x}}{12+\frac{128}{x^3}}) = \infty$

31. $\text{Lim}_{x\to 3^+}\frac{1}{x-3}=\infty$ **32.** $\text{Lim}_{x\to 3^-}\frac{1}{x-3}=-\infty$

33. $\text{Lim}_{x\to 0^+}\frac{1}{x^2}=\infty$ **34.** $\text{Lim}_{x\to 0^-}\frac{1}{|x|}=\infty$

35. $\text{Lim}_{x\to 0}\frac{\sin 2x}{4x}=\lim_{x\to 0}\frac{1}{2}\frac{\sin 2x}{2x}=\frac{1}{2}\cdot 1=\frac{1}{2}$

36. $\text{Lim}_{x\to 0}\frac{x+\sin x}{x}=\lim_{x\to 0}(1+\frac{\sin x}{x})=1+1=2$

37. $\text{Lim}_{x\to 0}\frac{\sin^3 2x}{x^3}=\lim_{x\to 0}(\frac{\sin 2x}{x})^3=\lim_{x\to 0}(2\frac{\sin 2x}{2x})^3=(2\cdot 1)^3=8$

38. $\text{Lim}_{x\to 0}\frac{2\csc 5x}{\csc 3x}=\lim_{x\to 0}\frac{2\sin 3x}{\sin 5x}=\lim_{x\to 0}\frac{2\sin 3x}{3x}\frac{5x}{\sin 5x}\frac{3x}{5x}=2\cdot 1\cdot 1\cdot\frac{3}{5}=\frac{6}{5}$

39. a) the y-intercept is approximately 0.78.

b) The values are all close to 0.78

c) f appears to have a minimal value at $x=0$.

d) $\text{Lim}_{x\to 0}\frac{\sec 2x\csc 9x}{\cot 7x}=\lim_{x\to 0}\frac{1}{\cos 2x}\frac{\sin 7x}{\cos 7x}\frac{1}{\sin 9x}$

$=\lim_{x\to 0}\frac{1}{\cos 2x}\frac{1}{\cos 7x}\frac{\sin 7x}{7x}\frac{1}{\frac{\sin 9x}{9x}}\frac{7x}{9x}=1\cdot 1\cdot 1\cdot 1\cdot\frac{7}{9}=\frac{7}{9}$

40. a) 1.60 b) quite close for x in the range $[-0.1, 0.1]$

c) f appears to have a minimal value at $x=0$.

d) $\text{Lim}_{x\to 0}\frac{\sin 8x\cos 3x}{\sin 5x\cos 8x}=\lim_{x\to 0}\cos 3x\frac{1}{\cos 8x}\frac{\sin 8x}{8x}\frac{5x}{\sin 5x}\frac{8x}{5x}=1\cdot 1\cdot 1\cdot 1\cdot\frac{8}{5}=8/5$

41. a) $\lim_{x\to -2^+}\frac{x+3}{x+2}=\infty$ b) $\lim_{x\to 2^-}\frac{x+3}{x+2}=-\infty$. As $x\to -2$, $x+3$ is positive and $x+2$ is positive in a) but negative in b). The answer may be confirmed by graphing $y=\frac{x+3}{x+2}$.

42. a) $\lim_{x\to 2^+}\frac{x-1}{x^2(x-2)}=\infty$ b) $\lim_{x\to 2^-}\frac{x-1}{x^2(x-2)}=-\infty$

c) $\lim_{x\to 0^+}\frac{x-1}{x^2(x-2)}=\infty$ d) $\lim_{x\to 0^-}\frac{x-1}{x^2(x-2)}=\infty$

43. a)

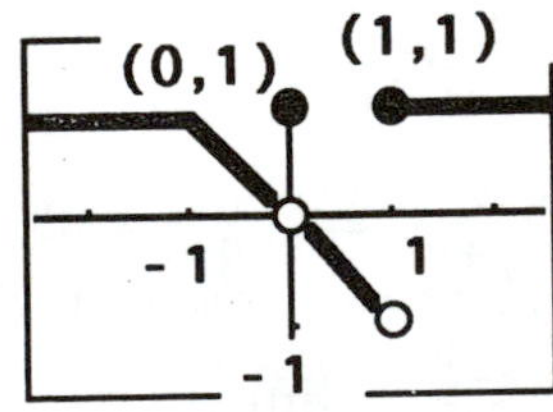

b) $\lim_{x\to -1^+}f(x)=1$, $\lim_{x\to -1^-}f(x)=1$, $\lim_{x\to 0^+}f(x)=0$, $\lim_{x\to 0^-}f(x)=0$, $\lim_{x\to 1^+}f(x)=1$, $\lim_{x\to 1^-}f(x)=-1$

c) $\lim_{x\to -1} f(x) = 1$, $\lim_{x\to 0} f(x) = 0$ but $\lim_{x\to 1} f(x)$ does not exist because the right-hand and left-hand limits of f at 1 are not equal.

d) Only at $x = -1$

44. a)

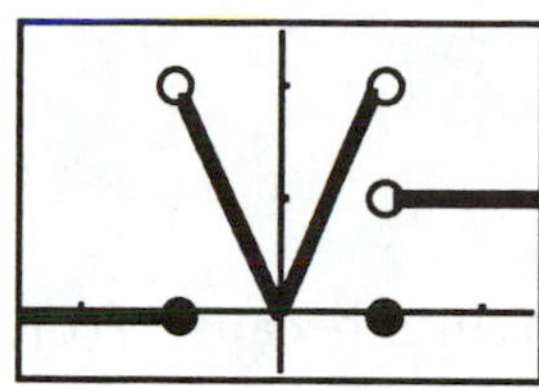

b) $\lim_{x\to -1^-} f(x) = 0$, $\lim_{x\to -1^+} f(x) = 2$, $\lim_{x\to 0^-} f(x) = 0$, $\lim_{x\to 0^+} f(x) = 0$, $\lim_{x\to 1^-} f(x) = 2$, $\lim_{x\to 1^+} f(x) = 1$

c) $\lim_{x\to 0} f(x) = 0$ but $\lim_{x\to -1} f(x)$ and $\lim_{x\to 1} f(x)$ do not exist because the left-hand and right-hand limits at $x = -1$ are not equal and the same is true at $x = 1$.

d) Only at $x = 0$.

45. a) A graph of f may be obtained by graphing the functions $y = abs(x^3 - 4x) + 0\sqrt{1-x}$ and $y = x^2 - 2x - 2 + 0\sqrt{x-1}$ in the viewing rectangle $[-5, 7]$ by $[-4, 10]$.

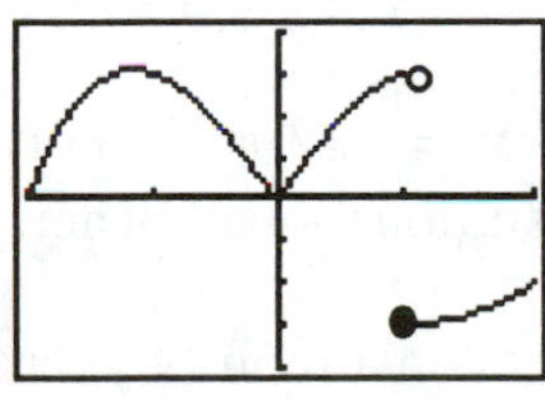

[-2,2] by [-4,4]

b) $\lim_{x\to 1^+} f(x) = \lim_{x\to 1^+}(x^2 - 2x - 2) = -3$.

$\lim_{x\to 1^-} f(x) = \lim_{x\to 1^-} |x^3 - 4x| = 3$

c) f does not have a limit at $x = 1$ because the right-hand and left-hand limits at $x = 1$ are not equal.

d) $x^3 - 4x$ is continuous by 2.2 Example 5 and $|x|$ is continuous by 2.2 Example 8. Thus $|x^3 - 4x|$ is continuous by Theorem 5 and so f is continuous for $x < 1$. For $x > 1$, $f(x) = x^2 - 2x - 2$, a polynomial, is continuous. Thus $f(x)$ is continuous at all points except $x = 1$.

e) f is not continuous at $x = 1$ because the two limits in b) are not equal and so $\lim_{x\to1} f(x)$ does note exist.

46. a) A good idea of the graph is obtained by graphing $y = 1 - \sqrt{3-2x} + 0\sqrt{1.5-x}$ and $y = 1 + \sqrt{2x-3} + 0\sqrt{x-1.5}$ in the rectangle $[-2, 4]$ by $[-3, 5]$.

b) Both limits are equal to 1

c) $\lim_{x\to3/2} f(x) = 1$

d) f is continuous at all points because $f(c) = \lim_{x\to c} f(x)$ for all points c.

e) There are no points of discontinuity because of d)

47. a) A graph of f is obtained by graphing $y = -x + 0\sqrt{1-x}$ and $y = x - 1 + 0\sqrt{x-1}$ in the rectangle $[-2, 4]$ by $[-2, 4]$.

b) $\lim_{x\to1^+} f(x) = \lim_{x\to1^+}(x-1) = 0$. $\lim_{x\to1^-} f(x) = \lim_{x\to1^-} -x = -1$.

c) No value assigned to $f(1)$ makes f continuous at $x = 1$.

48. a) The idea of a complete graph of f can be obtained by graphing $y = 3x^2 + 0\sqrt{1-x}$ and $y = 4 - x^2 + 0\sqrt{x-1}$ in the viewing rectangle $[-2, 4]$ by $[-2, 4]$.

b) $\lim_{x\to1^+} f(x) = \lim_{x\to1^-} f(x) = 3$

c) b) shows that $\lim_{x\to1} f(x) = 3$ so define $f(1) = 3$.

49. f is not defined at $x = \pm2$ so is not continuous at these points. By 2.2 Example 6 f is continuous at all other points.

50. $3x + 2$ and $\sqrt[3]{x}$ are continuous by Examples 5 and 7 of 2.2. Hence the composite, f, is continuous by Theorem 5. There are no discontinuities.

51. $\lim_{x\to\pm\infty} \frac{2x+1}{x^2-2x+1} = \lim_{x\to\pm\infty} \frac{\frac{2}{x}+\frac{1}{x^2}}{1-\frac{2}{x}+\frac{1}{x^2}} = \frac{0+0}{1-0+0} = 0$.

52. $\lim_{x\to\pm\infty} \frac{2x^2+5x-1}{x^2+2x} = \lim_{x\to\pm\infty} \frac{2+\frac{5}{x}-\frac{1}{x^2}}{1+\frac{2}{x}} = \frac{2+0-0}{1+0} = 2$. $y = 2$ is the end behavior asymptote.

53. By long division $h(x) = x^2 - x + \frac{3}{x-3}$. Thus $y = x^2 - x$ is the end behavior asymptote of h.

54. $T(x) = x - \frac{2x^2+1}{x^3-x+1}$. Thus $y = x$ is the end behavior asymptote of T.

55. a) $\lim_{x\to c} 3f(x) = 3\lim_{x\to c} f(x) = 3(-7) = -21$

b) $\lim_{x\to c}(f(x))^2 = \lim_{x\to c} f(x)\lim_{x\to c} f(x) = (-7)(-7) = 49$

c) $\lim_{x\to c} f(x)\cdot g(x) = \lim_{x\to c} f(x)\lim_{x\to c} g(x) = (-7)(0) = 0$

d) $\lim_{x\to c} \frac{f(x)}{g(x)-7} = \frac{\lim_{x\to c} f(x)}{\lim_{x\to c} g(x)-7} = \frac{-7}{0-7} = 1$

e) $\lim_{x\to c} \cos(g(x)) = \cos[\lim_{x\to c} g(x)] = \cos 0 = 1.$

f) $\lim_{x\to c} |f(x)| = |\lim_{x\to c} f(x)| = |-7| = 7.$

56. a) $\lim_{x\to 0} -g(x) = -\sqrt{2}$ b) $\lim_{x\to 0} g(x)\cdot f(x) = \sqrt{2}\cdot\frac{1}{2} = \frac{\sqrt{2}}{2}$

c) $\lim_{x\to 0}[f(x)+g(x)] = \frac{1}{2}+\sqrt{2}$ d) $\lim_{x\to 0}[1/f(x)] = 2$

e) $\lim_{x\to 0}[x+f(x)] = 0+\frac{1}{2} = \frac{1}{2}$ f) $\lim_{x\to 0} \frac{f(x)\cdot\sin x}{x} = \frac{1}{2}\cdot 1 = \frac{1}{2}$

57. Both 0 and $\sqrt{x}$ approach 0 as $x \to 0$. By the Sandwich Theorem $|\sqrt{x}\sin\frac{1}{x}| \to 0$ and hence $\sqrt{x}\sin\frac{1}{x} \to 0$ as $x \to 0$.

58. Both 0 and x^2 approach 0 as $x \to 0$. By the Sandwich Theorem $|x^2\sin\frac{1}{x}| \to 0$ and hence $x^2\sin\frac{1}{x} \to 0$ as $x \to 0$.

59. $\text{Lim}_{x\to\infty}\frac{x+\sin x}{x} = \lim_{x\to\infty}[1+\frac{\sin x}{x}] = 1+0 = 1$

60. $\text{Lim}_{x\to\infty}\frac{x+\sin x}{x+\cos x} = \lim_{x\to\infty}\frac{1+\frac{\sin x}{x}}{1+\frac{\cos x}{x}} = \frac{1+0}{1+0} = 1$

61. $-1 \leqq \sin x \leqq 1$. Hence $-\frac{1}{\sqrt{x}} \leqq \frac{\sin x}{\sqrt{x}} \leqq \frac{1}{\sqrt{x}}$ for $x > 0$. Both $-\frac{1}{\sqrt{x}}$ and $\frac{1}{\sqrt{x}}$ approach 0 as $x \to \infty$. By the Sandwich Theorem $\lim_{x\to\infty}\frac{\sin x}{\sqrt{x}} = 0$.

62. $-\frac{1}{\sqrt{x}} \leqq \frac{\cos x}{\sqrt{x}} \leqq \frac{1}{\sqrt{x}}$. Since $\lim_{x\to\infty} \pm\frac{1}{\sqrt{x}} = 0$, $\lim_{x\to\infty}\frac{\cos x}{\sqrt{x}} = 0$.

63. $\text{Lim}_{x\to 3}\frac{x^2+2x-15}{x-3} = \lim_{x\to 3}\frac{(x-3)(x+5)}{x-3} = \lim_{x\to 3}(x+5) = 8$. In order to have $\lim_{x\to 3} f(x) = f(3) = k$, we should set $k = 8$.

64. $\text{Lim}_{x\to 0^+} x^x = 1$. The limit must be approached from the right-hand side only because x^x is defined only for $x > 0$.

65. a) $\lim_{x\to 0^-} f(x) = 0$ b) $\lim_{x\to 0^+} f(x) = \infty$

c) The limit does not exist. To exist both one-sided limits must exist and be equal.

66. $\text{Lim}_{x\to 0}\frac{\sin x}{2x} = \frac{1}{2}\lim_{x\to 0}\frac{\sin x}{x} = \frac{1}{2}$. Thus the value $\frac{1}{2}$ should be assigned to k.

67. This is not a contradiction because $0 < x < 1$ is not a *closed* interval.

68. Let $y = f(x) = |x|$ on $-1 \leq x < 1$. This function attains its minimum ($f(0) = 0$) and its maximum ($f(-1) = 1$). This is not a contradiction of the theorem which is only a sufficient (not necessary) condition for the attainment of extreme values.

69. True because $0 = f(1) < 2.5 < f(2) = 3$ and so by Theorem 7, $2.5 = f(c)$ for some c in $[1, 2]$.

70. Let $f(x) = x + \cos x$. Then $f(x)$ is continuous. $f(-\pi/2) = -\frac{\pi}{2}$ and $f(0) = 1$. Thus $f(-\pi/2) < 0 < f(0)$. By Theorem 7, $f(c) = 0$ for some c in $[-\pi/2, 0]$.

71. Let $f(x) = x + \log x$. $f(\frac{1}{10}) = \frac{1}{10} - 1 < 0$. $f(1) = 1 > 0$. By the Intermediate Value Theorem there is a number c, $\frac{1}{10} < c < 1$ such that $f(c) = 0$. This means that c is a solution of the given equation.

72. $\text{Lim}_{x\to 1} f(x) = 3$ means given any radius $\varepsilon > 0$ about 3 there exists a radius $\delta > 0$ about 1 such that for all x $0 < |x - 1| < \delta$ implies $|f(x) - 3| < \varepsilon$.

73. $\text{Lim}_{x\to 0} \frac{\sin x}{x} = 1$ means given any radius $\varepsilon > 0$ about 1 there exists a radius $\delta > 0$ about 0 such that, for all x, $0 < |x - 0| < \delta$ implies $|f(x) - 1| < \varepsilon$.

74. Let $f(x) = x^2$. $f(x)$ gets closer to -1 as x approaches 0 but -1 is not equal to $\lim_{x\to 0} f(x)$.

75. This "definition" puts a requirement on $f(x)$ but not necessarily for x near x_0. Thus for any given $\varepsilon > 0$ there is an x such that $|x^2 - 0| < \varepsilon$ but this does not imply $\lim_{x\to x_0} x^2 = 0$ in general.

76. $-1 < \sqrt{x+2} - 4 < 1$ leads successively to $3 < \sqrt{x+2} < 5$, $9 < x + 2 < 25$, $7 < x < 23$, the last interval being the solution of the first part of the problem. The midpoint of this interval is $\frac{7+23}{2} = 15$. The interval can be described by $|x - 15| < 8$.

77. $-\frac{1}{2} < \sqrt{\frac{x+1}{2}} - 1 < \frac{1}{2}$ is equivalent to each of $\frac{1}{2} < \sqrt{\frac{x+1}{2}} < \frac{3}{2}$, $\frac{1}{4} < \frac{x+1}{2} < \frac{9}{4}$, $\frac{1}{2} < x + 1 < \frac{9}{2}$, $-\frac{1}{2} < x < \frac{7}{2}$. The midpoint of the latter interval is $(-\frac{1}{2} + \frac{7}{2})/2 = 3/2$ so that it can be written as $|x - 3/2| < 2$.

78. Let $f(x) = \frac{x-1}{x-3}$. Following the method of 2.5 Example 6, we solve the equations $f(x) = 1.9$ and $f(x) = 2.1$, obtaining 5.22 and 4.82. Thus $1.9 < f(x) < 2.1$ if $4.82 < x < 5.22$. Letting $x_0 = 5$, we have $|x - 5| < 0.18$ implies $|f(x) - 2| < 0.1$.

79. Let $f(x) = \frac{x-1}{x-3}$. Following the method of 2.5 Example 6, we solve the equations $f(x) = -2.1$ and $f(x) = -1.9$ obtaining 2.35 and 2.31. Thus $f(x)$ is within 0.1 unit of -2 if $2.31 < x < 2.35$ or $|x - 7/3| < 0.02$.

80. Let $f(x) = x^3 - 4x$. We graph $f(x)$, $y = 3.9$ and $y = 4.1$ and we use zoom-in to find the x-coordinates of the two points of intersection obtaining 2.38 and 2.39. Thus $f(x)$ is within 0.1 unit of 4 if $2.38 < x < 2.39$. If we take $x_0 = 2.383$ (near the root of $f(x) = 4$), we can say $|x - 2.383| < 0.003$ implies $|f(x) - 4| < 0.1$.

81. Let $f(x) = x^3 - 4x$. We graph $f(x)$, $y = 0.9$ and $y = 1.1$ and use zoom-in to find the x-coordinates of the points of intersection $(-1 < x < 0)$ obtaining -0.28 and -0.228. Thus $f(x)$ is within 0.1 unit of 1 if $-0.280 < x < -0.228$ (rounding appropriately) or if $|x + 0.254| < 0.026$.

82. $|f(x) - 1| = |2x - 4| = 2|x - 2| < \varepsilon$ is equivalent to $|x - 2| < \varepsilon/2$. Thus we must have $0 < \delta \leqq \varepsilon/2$.

83. $|f(x) - 0| + \|x\| = |x| < \varepsilon$ is equivalent to $|x - 0| < \varepsilon$. Thus we must have $0 < \delta \leq \varepsilon$.

84. $\text{Lim}_{x\to\infty} \frac{1-2x}{3x-1} = \lim_{x\to\infty} \frac{\frac{1}{x}-2}{3-\frac{1}{x}} = \frac{-2}{3}$. To confirm this using the definition of this type of limit, we follow Example 6. Let $\varepsilon > 0$ be given. $|\frac{1-2x}{3x-1} + \frac{2}{3}| < \varepsilon$ is equivalent to each of the following. $|\frac{3-6x+6x-2}{3(3x-1)}| < \varepsilon$, $\frac{1}{3(3x-1)} < \varepsilon$ (assuming x is large and positive), $\frac{1}{3\varepsilon} < 3x - 1$, $\frac{1}{3\varepsilon} + 1 < 3x$, $x > \frac{1}{9\varepsilon} + \frac{1}{3} = \frac{3\varepsilon+1}{9\varepsilon}$. So we choose $N = \frac{3\varepsilon+1}{9\varepsilon}$. Then $x > N$ implies $|\frac{1-2x}{3x-1} + \frac{2}{3}| < \varepsilon$. This confirms the limit.

85. As $x \to \frac{1}{3}^+$, $1 - 2x \to \frac{1}{3}$ and $3x - 1 \to 0$ through positive values. Hence $\lim_{x\to\frac{1}{3}^+} \frac{1-2x}{3x-1} = \infty$. With $x > \frac{1}{3}$, all of the following are equivalent. $\frac{1-2x}{3x-1} > N$, $1-2x > 3xN - N$, $1+N > x(3N+2)$, $x < \frac{N+1}{3N+2} = \frac{1}{3} + \frac{1}{3(3N+2)}$, $x - \frac{1}{3} < \frac{1}{3(3N+2)}$. So we choose $\delta = \frac{1}{3(3N+2)}$. Then $0 < x - \frac{1}{3} < \delta$ implies $\frac{1-2x}{3x-1} > N$. Since this is true for any $N > 0$, this confirms the limit.

86. $L = \lim_{x\to3}(5x - 10) = 5 \cdot 3 - 10 = 5$. $|f(x) - 5| = |5x - 15| = 5|x - 3| < \varepsilon = 0.05$ is equivalent to $|x - 3| < 0.01$. Let $\delta = 0.01$. Then for all x $0 < |x - 3| < \delta \Rightarrow |f(x) - 5| < \varepsilon$.

87. $L = \lim_{x\to2}(5x - 10) = 0$. $|f(x) - L| = |5x - 10| = 5|x - 2| < \varepsilon$ is equivalent to $|x - 2| < \varepsilon/5$. Thus $\delta = \varepsilon/5 = 0.01$.

88. $L = \lim_{x\to 9}\sqrt{x-5} = \sqrt{4} = 2$. $|f(x) - L| = |\sqrt{x-5} - 2| < \varepsilon = 1$ is equivalent to $-1 < \sqrt{x-5} - 2 < 1$, $1 < \sqrt{x-5} < 3$, $1 < x - 5 < 9$, $6 < x < 14$, $-3 < x - 9 < 5$. Thus $|x - 9| < 3$ guarantees $|f(x) - L| < \varepsilon$ and so we take $\delta = 3$.

89. $L = \lim_{x\to 2}\sqrt{2x-3} = 1$. $1/2 < \sqrt{2x-3} < 3/2$ yields $\frac{1}{4} < 2x-3 < \frac{9}{4}$, $-\frac{3}{4} < 2x - 4 < \frac{5}{4}$, $-\frac{3}{8} < x - 2 < \frac{5}{8}$. Thus $|x - 2| < \frac{3}{8} \Rightarrow |f(x) - L| < \varepsilon$ so $\delta = 3/8$.

90. $L = \lim_{x\to 5}\frac{x^2-x}{x+2} = \frac{20}{7} = 2.86$. We use the method of 2.6. In the viewing rectangle $[4.98, 5.02]$ by $[2.85, 2.87]$, we find the coordinates of two points on the graph of $f(x)$ and use these to determine the slope m. We find $m = 0.88$ approximately and $\delta = \varepsilon/|m| = \varepsilon/0.88 = 1.13\varepsilon$ rounding down to be safe.

91. $L = \lim_{x\to -5}\frac{x-1}{x^2+3x} = \frac{-6}{10} = -0.6$. In the viewing rectangle $[-5.1, -4.9]$ by $[-0.61, -0.59]$ we graph $f(x)$ and using two points on the graph we find $m = -0.32$. Hence we take $\delta = \varepsilon/|m| = 3.12\varepsilon$ rounding down to be safe.

92. $0.8 < \frac{\sqrt{x}}{2} < 1.2$ is equivalent to $1.6 < \sqrt{x} < 2.4$, $1.6^2 < x < 2.4^2$ or $2.56 < x < 5.76$. $0.9 < \frac{\sqrt{x}}{2} < 1.1$ is equivalent to $1.8 < \sqrt{x} < 2.2$ or $3.24 < x < 4.84$.

93. $9.9995 < 10 + (t - 70) \times 10^{-4} < 10.0005$ is equivalent to $-0.0005 < (t - 70)10^{-4} < 0.0005$, $-5 < t - 70 < 5$ or $65° < t < 75°$.

94. $1000x^2 + x - 10^{-15} = 0$. If $x = 0$, the left-hand side becomes $-10^{-15} \neq 0$, so $x = 0$ is not a solution. By the quadratic formula $x = \frac{-1\pm\sqrt{1+4(10^{-12})}}{2000}$. On a calculator the positive solution reads 10^{-15}. Because of the limits of calculator precision the exact solution cannot be found. We can be sure the error is less than 10^{-14}.

CHAPTER 3

DERIVATIVES

3.1 SLOPES, TANGENT LINES, AND DERIVATIVES

1. a) 0 b) -4 because a drawn-in tangent line at $x = 4$ descends 4 y-units as x increases one unit from 4 to 5.

2. a) 0 b) -3

3. a) 1 b) The spacing between horizontal and vertical dots is 0.5 units. The tangent line at $x = 4$ passes through the points $(4, -\frac{3}{4})$ and $(5, -\frac{3}{2})$. Thus $m = \frac{-1.5-(-.75)}{5-4} = -0.75$.

4. a) Tangent line passes through $(-7, -1)$ and $(1, \frac{3}{2})$ and hence has slope $\frac{1.5+1}{1+7} = \frac{2.5}{8} = \frac{5}{16}$. b) Symmetrically, here, $m = -\frac{5}{16}$.

5. a) April 15 (each vertical line of dots corresponds to the first day of a month). Tangent seems to pass through $(120, 30)$ and $(150, 50)$. Hence $m = \frac{50-30}{150-120} = \frac{2}{3}$ degree per day.

b) Yes. Near January 1 and July 1 c) Positive from mid-January until about July 1, negative from about July 2 until mid-December.

6. a) April 7; 10 min/day b) Yes. Near January 1 and July 1 c) Positive from mid-January until about July 1, negative from about July 2 until mid-December.

7. $f'(x) = \lim_{h\to 0} \frac{f(x+h)-f(x)}{h} = \lim_{h\to 0} \frac{[2(x+h)^2-5]-[2x^2-5]}{h}$
$= \lim_{h\to 0} \frac{2x^2+4xh+2h^2-5-2x^2+5}{h} = \lim_{h\to 0} \frac{4xh+2h^2}{h} = \lim_{h\to 0}(4x+2h) = 4x$

When $x = 3$, $m = f'(3) = 4 \cdot 3 = 12$ and $y = f(3) = 2(3)^2 - 5 = 13$. The tangent at $(3, 13)$ therefore has equation $y - 13 = 12(x - 3)$ or $y = 12x - 23$.

8. $f(x+h)-f(x) = [(x+h)^2 - 6(x+h)] - [x^2 - 6x] = h[2x - 6 + h]$.
$f'(x) = \lim_{h\to 0} \frac{f(x+h)-f(x)}{h} = \lim_{h\to 0}(2x - 6 + h) = 2x - 6$. $m = f'(3) = 0$. $y = f(3) = -9$. The tangent line has equation $y = -9$.

9. $f(x+h)-f(x) = [2(x+h)^2 - 13(x+h)+5] - [2x^2 - 13x + 5] = h(4x - 13 + 2h)$.
$f'(x) = \lim_{h\to 0} \frac{f(x+h)-f(x)}{h} = \lim_{h\to 0}(4x - 13 + 2h) = 4x - 13$. $m = f'(3) = -1$. $f(3) = -16$. Tangent: $y + 16 = (-1)(x-3)$ or $y = -x - 13$.

10. $f(x+h)-f(x) = [-3(x+h)^2 + 4(x+h)] - [-3x^2 + 4x] = h[-6x + 4 - 3h]$.
$f'(x) = \lim_{h\to 0} \frac{f(x+h)-f(x)}{h} = \lim_{h\to 0}(-6x + 4 - 3h) = -6x + 4$. $m = f'(3) = -14$. $f(3) = -15$. Tangent: $y + 15 = -14(x-3)$ or $y = -14x + 27$.

11. $f'(x) = \lim_{h\to 0} \frac{\frac{2}{x+h} - \frac{2}{x}}{h} = \lim_{h\to 0} \frac{1}{h}\frac{2x - 2x - 2h}{(x+h)x} = \lim_{h\to 0} \frac{-2}{(x+h)x} = -\frac{2}{x^2}$.
$f'(3) = -2/3^2 = -2/9$ is the slope and $f(3) = 2/3$. So the tangent at $x = 3$ has equation $y - 2/3 = -(2/9)(x-3)$ or $2x + 9y = 12$.

12. $f'(x) = \lim_{h\to 0} \frac{\frac{1}{x+h+1} - \frac{1}{x+1}}{h} = \lim_{h\to 0} \frac{1}{h}\frac{x+1-x-h-1}{(x+h+1)(x+1)} = \lim_{h\to 0} \frac{-1}{(x+h+1)(x+1)} = -\frac{1}{(x+1)^2}$. $f'(3) = -\frac{1}{(3+1)^2} = -\frac{1}{16}$. $f(3) = 1/4$. An equation of the tangent at $x = 3$ is $y - 1/4 = (-1/16)(x-3)$ or $x + 16y = 7$.

13. $\frac{1}{h}[f(x+h) - f(x)] = \frac{1}{h}[\frac{x+h}{x+h+1} - \frac{x}{x+1}] = \frac{1}{(x+h+1)(x+1)}$.

$f'(x) = \lim_{h\to 0} \frac{1}{(x+h+1)(x+1)} = \frac{1}{(x+1)^2}$. $f'(3) = \frac{1}{16}$. $f(3) = \frac{3}{4}$. Tangent: $y - \frac{3}{4} = \frac{1}{16}(x-3)$ or $16y - x = 9$.

14. $f(x+h) - f(x) = [\frac{1}{2(x+h)+1} - \frac{1}{2x+1}] = \frac{-2h}{(2x+2h+1)(2x+1)}$.

$f'(x) = \lim_{h\to 0} \frac{-2}{(2x+2h+1)(2x+1)} = \frac{-2}{(2x+1)^2}$. $f'(3) = \frac{-2}{49}$. $f(3) = \frac{1}{7}$. Tangent: $y - \frac{1}{7} = \frac{-2}{49}(x-3)$ or $49y + 2x = 13$.

15. $f'(x) = \lim_{h\to 0} \frac{x+h+\frac{9}{x+h} - [x + \frac{9}{x}]}{h} = \lim_{h\to 0} \left\{\frac{h + \frac{9x - 9x - 9h}{(x+h)x}}{h}\right\} = \lim_{h\to 0}[1 + \frac{-9}{(x+h)x}] = 1 - 9/x^2$. $f'(3) = 1 - 9/3^2 = 0$ is the slope and $f(3) = 3 + 9/3 = 6$. An equation of the tangent at $x = 3$ is $y = 6$.

16. $f'(x) = \lim_{h\to 0} \frac{x+h-\frac{1}{x+h} - (x - \frac{1}{x})}{h} = \lim_{h\to 0} \frac{1}{h}(h + \frac{1}{x} - \frac{1}{x+h}) = \lim_{h\to 0} \frac{1}{h}[h + \frac{h}{x(x+h)}]$
$= \lim_{h\to 0}[1 + \frac{1}{x(x+h)}] = 1 + \frac{1}{x^2}$.

At $x = 3$, the slope is $f'(3) = 1 + 1/3^2 = 10/9$, $f(3) = 8/3$ and the tangent has equation $y - 8/3 = (10/9)(x-3)$ or $9y = 10x - 6$.

17. $f(x+h)-f(x) = ([1+\sqrt{x+h}]-[1+\sqrt{x}])$
$= \sqrt{x+h}-\sqrt{x} = (\sqrt{x+h}-\sqrt{x})\frac{\sqrt{x+h}+\sqrt{x}}{\sqrt{x+h}+\sqrt{x}}$
$= \frac{x+h-x}{\sqrt{x+h}+\sqrt{x}} = \frac{h}{\sqrt{x+h}+\sqrt{x}}$. $f'(x) = \lim_{h\to 0}\frac{1}{\sqrt{x+h}+\sqrt{x}} = \frac{1}{2\sqrt{x}}$.
$f'(3) = \frac{1}{2\sqrt{3}}$. $f(3) = 1+\sqrt{3}$. Tangent line: $y-(1+\sqrt{3}) = \frac{1}{2\sqrt{3}}(x-3)$.

18. $f(x+h)-f(x) = [\sqrt{x+h+1}-\sqrt{x+1}]\frac{\sqrt{x+h+1}+\sqrt{x+1}}{\sqrt{x+h+1}+\sqrt{x+1}}$
$= \frac{x+h+1-(x+1)}{\sqrt{x+h+1}+\sqrt{x+1}} = \frac{h}{\sqrt{x+h+1}+\sqrt{x+1}}$.
$f'(x) = \lim_{h\to 0}\frac{1}{\sqrt{x+h+1}+\sqrt{x+1}} = \frac{1}{2\sqrt{x+1}}$. $f'(3) = \frac{1}{4}$. $f(3) = 2$. Tangent line: $y-2 = \frac{1}{4}(x-3)$ or $4y-x=5$.

19. $f'(x) = \lim_{h\to 0}\frac{\sqrt{2(x+h)}-\sqrt{2x}}{h} = \lim_{h\to 0}\frac{\sqrt{2(x+h)}-\sqrt{2x}}{h}\frac{\sqrt{2(x+h)}+\sqrt{2x}}{\sqrt{2(x+h)}+\sqrt{2x}}$
$= \lim_{h\to 0}\frac{2(x+h)-2x}{h(\sqrt{2(x+h)}+\sqrt{2x})} = \lim_{h\to 0}\frac{2}{\sqrt{2(x+h)}+\sqrt{2x}} = \frac{2}{2\sqrt{2x}} = \frac{1}{\sqrt{2x}}$.

$f'(3) = \frac{1}{\sqrt{2\cdot 3}} = \frac{1}{\sqrt{6}}$ is the slope at $x=3$. $f(3) = \sqrt{6}$ is the slope at $x=3$. $f(3) = \sqrt{6}$. The tangent at $x=3$ has equation $y-\sqrt{6} = (1/\sqrt{6})(x-3)$ or $\sqrt{6}y = x+3$.

20. $f'(x) = \lim_{h\to 0}\frac{\sqrt{2(x+h)+3}-\sqrt{2x+3}}{h}\frac{\sqrt{2(x+h)+3}+\sqrt{2x+3}}{\sqrt{2(x+h)+3}+\sqrt{2x+3}} = \lim_{h\to 0}\frac{2x+2h+3-(2x+3)}{h(\sqrt{2(x+h)+3}+\sqrt{2x+3})}$
$= \lim_{h\to 0}\frac{2}{\sqrt{2(x+h)+3}+\sqrt{2x+3}} = \frac{2}{\sqrt{2x+3}+\sqrt{2x+3}} = 1/\sqrt{2x+3}$.

At $x=3$ the slope is $f'(3) = 1/3$, $f(3) = 3$ and the tangent has equation $y-3 = (1/3)(x-3)$ or $3y = x+6$.

21. Using the method of preceding exercises, we find $f'(x) = -2x$. $m = f'(-1) = 2$. Tangent line: $y-3 = 2(x+1)$ or $y = 2x+5$. Graph $y_1 = 4-x^2$ and $y_2 = 2x+5$ in $[-5,5]$ by $[-10,10]$.

22. Using the method of the preceding exercises, we find $f'(x) = 2x-2$. $f'(1) = 0$. Tangent line: $y=1$. Graph $y_1 = (x-1)^2+1$ and $y_2 = 1$ in $[-3,5]$ by $[0,10]$.

23. Proceeding as in Exercise 18 or 19, we find $f'(x) = \frac{1}{2\sqrt{x}}$. $m = f'(1) = \frac{1}{2}$. Tangent line: $y-1 = \frac{1}{2}(x-1)$ or $y = \frac{1}{2}x+\frac{1}{2}$. Graph $y_1 = \sqrt{x}$ and $y_2 = 0.5x+0.5$ in $[-3,5]$ by $[-1,3]$.

24. $f(x+h)-f(x) = \frac{1}{(x+h)^2}-\frac{1}{x^2} = \frac{h(-2x-h)}{(x+h)^2x^2}$. $f'(x) = \lim_{h\to 0}\frac{-2x-h}{(x+h)^2x^2} = \frac{-2x}{x^4} = \frac{-2}{x^3}$. $f'(-1) = 2$. Tangent line: $y-1 = 2(x+1)$ or $y = 2x+3$. Graph $y_1 = \frac{1}{x^2}$ and $y_2 = 2x+3$ in $[-3,3]$ by $[-1,10]$.

25. $f'(\frac{1}{2}) = \lim_{x\to\frac{1}{2}} \frac{f(x)-f(\frac{1}{2})}{x-\frac{1}{2}} = \lim_{x\to\frac{1}{2}} \frac{x^2-x+1-[(\frac{1}{2})^2-\frac{1}{2}+1]}{x-\frac{1}{2}} = \lim_{x\to\frac{1}{2}} \frac{x^2-(\frac{1}{2})^2-(x-\frac{1}{2})}{x-\frac{1}{2}} = \lim_{x\to\frac{1}{2}}(x+\frac{1}{2})-1=0$

26. $f(x)-f(2) = -3x^2+7x+5-[-3(2)^2+7(2)+5] = -3(x^2-2^2)+7(x-2)$. $f' = \lim_{x\to 2}[-3(x+2)+7] = -5$

27. $f(x)-f(-1) = \frac{1}{x+2} - \frac{1}{(-1)+2} = -\frac{(x+1)}{x+2}$. $f'(-1) = \lim_{x\to -1}\frac{f(x)-f(-1)}{x-(-1)} = \lim_{x\to -1}\frac{-1}{x+2} = -1$

28. $f(x)-f(2) = \frac{1}{(x-1)^2} - 1 = \frac{-x^2+2x}{(x-1)^2} = \frac{-x(x-2)}{(x-1)^2}$. $f'(2) = \lim_{x\to 2}\frac{-x}{(x-1)^2} = -2$

29. $f(x)-f(4) = \frac{1}{\sqrt{x}} - \frac{1}{2} = \frac{2-\sqrt{x}}{2\sqrt{x}}\frac{2+\sqrt{x}}{2+\sqrt{x}} = \frac{4-x}{2\sqrt{x}(2+\sqrt{x})}$. $f'(4) = \lim_{x\to 4}\frac{f(x)-f(4)}{x-4} = \lim_{x\to 4}\frac{-1}{2\sqrt{x}(2+\sqrt{x})} = -\frac{1}{16}$

30. $f(x)-f(-2) = \frac{1}{\sqrt{2x+13}} - \frac{1}{3} = \frac{3-\sqrt{2x+13}}{3\sqrt{2x+13}}\frac{3+\sqrt{2x+13}}{3+\sqrt{2x+13}} = \frac{9-(2x+13)}{3\sqrt{2x+13}(3+\sqrt{2x+13})} = \frac{-2x-4}{9\sqrt{2x+13}+3(2x+13)}$. $f'(-2) = \lim_{x\to -2}\frac{f(x)-f(-2)}{x-(-2)} = \lim_{x\to -2}\frac{-2}{9\sqrt{2x+13}+3(2x+13)} = \frac{-2}{27+27} = -\frac{1}{27}$

31. The right-hand derivative at $x=0$ is $\lim_{h\to 0^+}\frac{(0+h)-0}{h} = \lim_{h\to 0^+} 1 = 1$ while the left-hand derivative is $\lim_{h\to 0^-}\frac{(0+h)^2-0^2}{h} = \lim_{h\to 0^-} h = 0$. Since these are unequal, the function is not differentiable at $x=0$.

32. The right-hand and left-hand derivatives at $x=1$ are, respectively, 2 and 0. Since these are unequal, the function is not differentiable at $x=1$.

33. The left-hand derivative is $\lim_{h\to 0^-}\frac{\sqrt{1+h}-1}{h} = \lim_{h\to 0^-}\frac{\sqrt{1+h}-1}{h}\frac{\sqrt{1+h}+1}{\sqrt{1+h}+1} = \lim_{h\to 0^-}\frac{1+h-1}{h(\sqrt{1+h}+1)} = \lim_{h\to 0^-}\frac{1}{\sqrt{1+h}+1} = \frac{1}{2}$. The right-hand derivative at $x=1$ is $\lim_{h\to 0^+}\frac{[2(1+h)-1]-1}{h} = \lim_{h\to 0^+}\frac{2h}{h} = 2$. Since the limits are unequal, f is not differentiable at $x=1$.

34. The left- and right-hand limits at $x=1$ are, respectively, 1 and -1. Hence f is not differentiable at $x=1$.

35. Let $f(x) = y = -x^2$. a) $f'(x) = -2x$ (see Example 3). b) Graph $y_1 = -x^2$ and $y_2 = -2x$ in $[-3,3]$ by $[-10,5]$. c) $y' > 0$ for $x < 0$ and $y' < 0$ for $x > 0$. $f'(0) = 0$. d) y increases on $(-\infty, 0)$ and decreases on $(0,\infty)$. The interval on which $y' > 0 (y' < 0)$ is the interval on which y increases (decreases).

36. a) $f'(x) = \frac{1}{x^2}$ (See Ex. 4). b) Graph $y_1 = -\frac{1}{x}$ and $y_2 = \frac{1}{x^2}$ in $[-3,3]$ by $[-5,5]$. c) $f' > 0$ on $(-\infty,0)$ and $(0,\infty)$. $f'(x)$ is never zero. d) f is increasing on the same intervals.

37. a) $f'(x) = x^2$ (See Ex. 5). b) Graph $y_1 = \frac{x^3}{3}$ and $y_2 = x^2$ in $[-3,3]$ by $[-5,5]$. c) $f' > 0$ on $(-\infty,0)$ and $(0,\infty)$. $f'(0) = 0$. d) f is increasing on the same intervals.

38. a) Proceeding as in Ex. 5, we find $f'(x) = x^3$. b) Graph $y_1 = \frac{x^4}{4}$ and $y_2 = x^3$ in $[-3,3]$ by $[-5,5]$. c) $f' > 0$ on $(0,\infty)$ and $f' < 0$ on $(-\infty,0)$. $f'(0) = 0$. d) f is increasing on the interval where $f' > 0$ and decreasing on the interval where $f' < 0$.

3.2 NUMERICAL DERIVATIVES

1. NDER$(x^2+1,2) = 4$. $f(2) = 5$ and the tangent at $(2,5)$ has equation $y-5 = 4(x-2)$ or $y = 4x-3$. The graphs of $y = x^2+1$ and $y = 4x-3$ can be viewed in the rectangle $[-10,10]$ by $[-10,20]$.

2. NDER$(2x^3-5x-2,1.5) = 8.50$. $f(1.5) = -2.75$ and the tangent at $x = 1.5$ has equation $y+2.75 = 8.5(x-1.5)$. The graph of $f(x)$ and this tangent line can be viewed in the rectangle $[-4,4]$ by $[-50,50]$.

3. NDER$(\sqrt{4-x^2},-1) = 0.58$. $f(-1) = \sqrt{3}$ and the tangent at $(-1,\sqrt{3})$ has equation $y-\sqrt{3} = 0.58(x+1)$. The graph of $y = \sqrt{4-x^2}$ and $y = \sqrt{3}+0.58(x+1)$ can be viewed in the rectangle $[-6,6]$ by $[-4,4]$.

4. NDER$((x-1)^3+1,2.5) = 6.75$. $f(2.5) = 4.375$ and the tangent at $x = 2.5$ has equation $y = 4.375+6.75(x-2.5)$. For a good idea of the graphs one can view them in the two rectangles $[-4,4]$ by $[-10,10]$ and $[-6,6]$ by $[-60,60]$.

5. NDER$(\frac{x^2-4}{x^2+1},2) = 0.80$. Since $f(2) = 0$, the tangent at $x = 2$ has equation $y = 0.8(x-2)$. The graph of $f(x)$ and its tangent may be viewed in the rectangle $[-8,8]$ by $[-8,8]$.

6. NDER$(x\sin x,2) = 0.08$. $f(2) = 2\sin 2 \approx 1.82$ and the tangent at $x = 2$ has equation $y = 1.82+0.08(x-2)$. View the graphs in the rectangle $[-4,4]$ by $[-3,3]$ and then zoom out.

7. a) only. We can draw the graph without lifting our pencil so the function is continuous. But the function is not differentiable at each of the points which are peaks or low points. At these points the left-hand and right-hand derivatives are not equal (there is not a *unique* tangent line).

8. c). The function is continuous and differentiable because its graph has a unique tangent line at each point.

9. c). $x = 0$ is not a point of the domain. At every other point there is a unique tangent line so the function is both continuous and differentiable.

10. d). The function is not continuous (hence not differentiable) at each of the x-intercepts where there is a jump discontinuity.

11 through 18. In these exercises one may evaluate $D(h)$ and $S(h)$ directly using a calculator, or one may first algebraically simplify $D(h)$ and $S(h)$. If the calculator is used, meaningful results may not be obtained when $h = \pm 10^{-15}$ due to the limits of machine accuracy. This answers part c) of these exercises.

11. b) Conjectures: $f'(2) = 10$, $f'(0) = -2$. $S(h)$ is closer in both cases (in fact exact).

12. b) Conjectures: $f'(2) = 28$, $S(h)$ is closer. $f'(0) = 4$, $D(h) = S(h)$.

13. b) $f'(2) = -0.25$, $S(h)$ is closer. $f'(0)$ and $D(h)$ for $a = 0$ are not defined but $S(h) = 0$ for all h.

14. b) $f'(2) = -0.25$, $S(h)$ is closer. When $a = 0$, $f(a)$ and $D(h)$ are not defined, and $f'(0)$ is not defined.

15. a) If $a = 2$, $D(h) = S(h) = 1$ for all the h's considered. If $a = 0$, $D(h) = \begin{cases} -1, & \text{if } h < 0 \\ 1, & \text{if } h > 0 \end{cases}$ and $S(h) = 0$ for all h. b) $f'(2) = 1$; both $S(h)$ and $D(h)$ are exact. $f'(0)$ does not exist.

16. a) If $a = 2$, $D(h) = \begin{cases} -1, & \text{if } h < 0 \\ 1, & \text{if } h > 0 \end{cases}$ and $S(h) = 0$ for all h. If $a = 0$, $D(h) = S(h) = -1$ for the h's considered. b) $f'(2)$ does not exist. $f'(0) = -1$.

17. a) If $a = 2$, $D(h) \to 0$ as $h \to 0$ while $S(h) = 0$ for those h's considered. If $a = 0$, $D(h)$ is undefined for $h < 0$ while $D(h) \to \infty$ as $h \to 0^+$. b) $f'(2) = 0$, $S(h)$ is closer. $f'(0)$ does not exist.

18. a) If $a = 2$, $D(h) \to -\infty$ as $h \to 0^-$ while $D(h)$ is undefined for $h > 0$. If $a = 0$, $D(h) \to 0$ as $h \to 0$ while $S(h) = 0$ for those h's considered. b) $f'(2)$ does not exist. $f'(0) = 0$.

19. Even though the derivative may not exist, the values of $S(h)$ may be defined giving meaningless approximations of the derivative.

20. See the comment for Exercise 19. **21.** See the comment for Exercise 19.

22. $f'(2)$ does not exist. See the comment for Exercise 19.

23. Since $4x - x^2 < 0$ for $x < 0$, $f(x)$ is not defined when $x < 0$ and so the appropriate $S(h)$ is not defined.

24. $f(x)$ is not defined for $x > 2$ and so the appropriate $S(h)$ is not defined.

25. a) We agree that $f'(a) = \text{NDER}(f(x), a)$ (with $h = 0.01$) rounded to two decimal places and obtain $f'(-1) = 0.14$, $f'(0) = -14.94$, $f'(1.5) = -9.69$ and $f'(3.5) = 0.13$. b) $\text{NDER}(f(x), a)$ may give results even though $f'(a)$ does not exist. This was the case for $f'(0)$. The denominator is zero when $x = 0$ and when $x = x_1$ where $x_1 \approx 1.2$. The numerator is undefined when $\tan x$ is undefined. Hence the domain of f consists of all real numbers except $0, x_1$ and $(2n+1)\pi/2$, n an integer.

26. a) Using the understanding of Exercise 23, we obtain $g'(0) = -0.03$, $g'(1.5) = 0.03$ and $g'(5) = -0.06$. b) $g(x)$ is the quotient of differentiable functions and the denominator is never 0. Hence $g(x)$ is differentiable and we are confident about the results of a). Zooming in to various suspicious parts of the graph of $g(x)$ supports the conclusion that there is a unique tangent line at each point.

27. a) $y_1 = -x^2$, $y_2 = \text{NDER}\ y_1$. Graph y_1, y_2 in $[-4, 4]$ by $[-10, 10]$. b) y_1' exists for all x. c) y_1' is positive for negative x, negative for positive x and $y_1'(0)$. e) y_1 is increasing over the interval $(-\infty, 0)$ where y_1' is positive. y_1 is decreasing over the interval $(0, \infty)$ where y_1' is negative.

28. a) Graph $y_1 = -\frac{1}{x}$ and $y_2 = \text{NDER}\ y_1$ in $[-3, 3]$ by $[-8, 9]$. b) y_1' does not exist at $x = 0$ because even y_1 does not exist at $x = 0$. This is suggested by the graph of y_2 since $y_2 \to \infty$ as $x \to 0$. y_1' is positive, tangent lines have positive slope and y_1 is increasing for all x except $x = 0$.

29. a) Graph $y_1 = \sqrt[3]{x-2}$ and $y_2 = \text{NDER}\ y_1$ in $[0, 3]$ by $[-2.2, 3.8]$. b) y_1' does not exist at $x = 2$ because the tangent line is vertical there. This is suggested by the graph because $y_2 \to \infty$ as $x \to 2$. y_1' is positive, tangent lines have positive slope and y_1 is increasing for all x except $x = 2$.

30. This function is the negative of the function in Exercise 27. The graphs may be obtained by reflecting the preceding graphs through the x-axis. Use the preceding solution but change positive to negative and increasing to decreasing. Also here $y_2 \to -\infty$ as $x \to 2$.

31. a) Graph $y_1 = \sqrt{1-x}$ and $y_2 = \text{NDER } y_1$ in $[-2, 1]$ by $[-2.2, 2.5]$. b) y_1' does not exist for $x > 1$ because $(1, \infty)$ is not part of the domain of y_1. y_1' does not exist at $x = 1$ because the graph has a vertical tangent there. This is suggested by the graph because $y_1 \to -\infty$ as $x \to 1^-$. y_1' is negative, slopes of tangent lines are negative and y_1 is decreasing on the interval $(-\infty, 1)$.

32. Graph $y_1 = \sqrt[4]{x-1}$ and $y_2 = \text{NDER } y_1$ in $[0, 2]$ by $[-3, 3]$. b) y_1' does not exist on $(-\infty, 1)$ because this interval is not in the domain of y_1. y_1' does not exist at $x = 1$ because there is a vertical tangent there. This is suggested by the graph because $y_2 \to \infty$ as $x \to 1^+$. y_1' is positive, slopes of tangent lines are positive and y_1 is increasing on $(1, \infty)$.

33. $[0.25895206, 0.25895208]$ by $[0.135, 0.145]$ is one possibility.

34. $[0.13231095, 0.13231302]$ by $[0.184, 0.196]$ is one possibility.

35. a) Let $y_1 = -x^2(x < 0) + (4 - x^2)(x \geqq 0)$. Graph $\text{NDER}(y_1, x)$ in the given window. b) We see the graph of $y = -2x$, $x \neq 0$. c) With $h = 0.01$, $\text{NDER}(f(x), 0) = \frac{f(0+0.01)-f(0-0.01)}{2(0.01)} = \frac{4-(0.01)^2-[-(0.01)^2]}{0.02} = \frac{4}{0.02} = 200$. $f(x)$ has a jump discontinuity at $x = 0$. So $f'(0)$ cannot exist by Theorem 2.

36. a) Let $y_1 = x^2(x < 0) + x(x \geqq 0)$. Graph $y_2 = \text{NDER}(y_1, x)$ in the given window. b) We see the graph of $\begin{cases} 2x, & x < 0 \\ 1, & x \geqq 0 \end{cases}$. c) With $h = 0.01$, $\text{NDER}(f(x), 0) = \frac{f(0+0.01)-f(0-0.01)}{2(0.01)} = \frac{0.01-(-0.01)^2}{0.02} = 0.495$. $f'(0)$ does not exist: the left-hand derivative is 0 and the right-hand derivative at $x = 0$ is 1.

37. a) Let $y_1 = -\frac{x^2}{2} + 0\sqrt{-x}$, $y_2 = \frac{x^2}{2} + 0\sqrt{x}$, $y_3 = \text{NDER}(y_1, x)$, $y_4 = \text{NDER}(y_2, x)$. Graph y_3, y_4 in the given viewing window. b) This is the graph of $y = \begin{cases} -x, & x < 0 \\ x, & x \geqq 0 \end{cases}$. c) $\text{NDER}(f(x), 0) = \frac{f(0+0.01)-f(0-0.01)}{2(0.01)} = \frac{0.0001}{2(0.01)} = 0.005$. $f'(0)$ does not exist because the left-hand derivative and the right-hand derivative at $x = 0$ are not equal.

38. b) $\frac{f(x+h)-f(x-h)}{2h} = \frac{a(x+h)^2+b(x+h)+c-[a(x-h)^2+b(x-h)+c]}{2h} = \frac{4axh+2bh}{2h} = 2ax + b$.

39. Graph $y_1 = (x^3+6x^2+12x)(x<0)+(-x^2)(x>0)$ and $y_2 = \text{NDER}(y_1, x)$ in $[-5,5]$ by $[-10,10]$. Since the two one-sided derivatives at $x=0$ are unequal, $f'(0)$ does not exist.

40. Graph $y_1 = -3\sqrt{-x}(x<0)+(3-0.2x^2)(x>0)$ and $y_2 = \text{NDER}(y_1, x)$ in $[-5,5]$ by $[-10,10]$. Since the left-hand derivative is $\lim_{h\to 0^-} \frac{f(0+h)-f(0)}{h} = \lim_{h\to 0^-} \frac{-3\sqrt{|h|}-3}{h} = \lim_{h\to 0^-} \frac{-3(-h)^{1/2}}{h} - \frac{3}{h} = \lim_{h\to 0^-} 3\frac{1}{(-h)^{1/2}} + \frac{3}{(-h)} = \infty$, $f'(0)$ does not exist.

41. Graph $y_1 = -5(abs\, x)^{1/3}$ and $y_2 = \text{NDER}(y_1, x)$ in $[-5,5]$ by $[-10,10]$. $f'(0)$ does not exist. For example, $\lim_{h\to 0^+} \frac{f(0+h)-f(0)}{h} = \lim_{h\to 0^+} \frac{-5h^{1/3}}{h} = -5\lim_{h\to 0^+} \frac{1}{h^{2/3}} = -\infty$.

42. b) In the given rectangle, if $a \approx 2.7$, the two graphs appear to coincide. c) From part b), $f(x) \approx 2.7^x$. All solutions to this problem are determined in Chapter 7.

43. b) $D_x(\sin x) = \cos x$

44. $f'(0) = \lim_{h\to 0} \frac{\sqrt[5]{0+h}-\sqrt[5]{0}}{h} = \lim_{h\to 0} \frac{1}{h^{4/5}}$ does not exist. But $\text{NDER}(f(x), 0) = 39.81$. Thus the numerical derivative at a point may exist even if the function is not differentiable at the point.

45. By Example 6 of 3.1, $f'(0)$ does not exist. But $\text{NDER}(|x|, 0) = 0$. $\text{NDER}(f(x), a)$ may exist even if $f'(a)$ does not exist.

46. a) Since $f(x) = -x$ if $x<0$ and $f(x) = x$ if $x>0$, f is continuous for $x \neq 0$. But $0 = f(0)$ and $\lim_{x\to 0} f(x) = \lim_{x\to 0} |x| = 0$. So f is also continuous for $x=0$. b) $g(x)$ is continuous because it is a constant, $1/0.02$, times the difference of two continuous functions.

e) $g(x) = \begin{cases} -1, & x < -0.01 \\ 100x, & -0.01 \leqq x < 0.01 \\ 1, & x \geqq 0.01 \end{cases}$

47. $0 < \frac{1}{2} < 1$ but $f(x) = \frac{1}{2}$ for no x, $-1 \leqq x \leqq 1$. Therefore f does not have the Intermediate Value Property on the interval. By the theorem alluded to in the text, f cannot be the derivative of any function on the interval.

48. $f'(0) = \lim_{h\to 0} \frac{f(0+h)-f(0)}{h} = \lim_{h\to 0} \frac{\sin(0+h)-\sin 0}{h} = \lim_{h\to 0} \frac{\sin h}{h} = 1$

49. The range of $[x]$ is the set of all integers so it does not have the Intermediate Value Property. Hence it cannot be the derivative of any function on $(-\infty, \infty)$.

50. It is continuous because $S_f(h) = \frac{1}{2h}[f(x+h) - f(x-h)]$ is a constant times the difference of two continuous functions.

3.3 DIFFERENTIATION RULES

1. $1,\ 0$ **2.** $-1,\ 0$ **3.** $-2x,\ -2$ **4.** $x^2-1,\ 2x$ **5.** $2,\ 0$ **6.** $2x+1,\ 2$

7. $x^2+x+1,\ 2x+1$ **8.** $-1+2x-3x^2,\ 2-6x$

9. $4x^3-21x^2+4x,\ 12x^2-42x+4$ **10.** $15x^2-15x^4,\ 30x-60x^3$

11. $8x-8,\ 8$ **12.** $x^3-x^2+x-1,\ 3x^2-2x+1$

13. $y' = 2x-1,\ y'' = 2,\ y^{(n)} = 0$ for $n \geqq 3$

14. $y' = x^2+x,\ y'' = 2x+1,\ y''' = 2,\ y^{(n)} = 0$ for $n \geqq 4$

15. $y' = 2x^3-3x-1,\ y'' = 6x^2-3,\ y''' = 12x,\ y^{(4)} = 12,\ y^{(n)} = 0$ for $n \geqq 5$

16. $y' = \frac{x^4}{24},\ y'' = \frac{x^3}{6},\ y''' = \frac{x^2}{2},\ y^{(4)} = x,\ y^{(5)} = 1,\ y^{(n)} = 0$ for $n \geqq 6$

17. a) $y' = (x+1)2x + 1\cdot(x^2+1) = 2x^2+2x+x^2+1 = 3x^2+2x+1$
b) $y = x^3+x^2+x+1,\ y' = 3x^2+2x+1$

18. $-3x^2-2x+3$

19. a) $y' = (x-1)(2x+1) + 1\cdot(x^2+x+1) = 2x^2-x-1+x^2+x+1 = 3x^2$
b) $y = x^3-1,\ y' = 3x^2$

20. a) $y' = (x+\frac{1}{x})(1+\frac{1}{x^2}) + (1-\frac{1}{x^2})(x-\frac{1}{x}) = 2x+\frac{2}{x^3}$
b) $y = x^2 - \frac{1}{x^2} = x^2 - x^{-2},\ y' = 2x+2x^{-3} = 2x+\frac{2}{x^3}$

21. a) $y' = (3x-1)(2) + 3(2x+5) = 12x+13$
b) $y = 6x^2+13x-5,\ y' = 12x+13$

22. a) $y' = (5-3x)(-2) - 3(4-2x) = 12x-22$
b) $y = 6x^2-22x+20,\ y' = 12x-22$

23. a) $y' = x^2(3x^2) + 2x(x^3-1) = 5x^4-2x$
b) $y = x^5-x^2,\ y' = 5x^4-2x$

24. a) $y' = x^2(1-\frac{1}{x^2}) + 2x(x+5+\frac{1}{x}) = x^2-1+2x^2+10x+2 = 3x^2+10x+1$
b) $y = x^3+5x^2+x,\ y' = 3x^2+10x+1$

25. $\frac{dy}{dx} = \frac{(x+7)\cdot 1-(x-1)\cdot 1}{(x+7)^2} = \frac{x+7-x+1}{(x+7)^2} = \frac{8}{(x+7)^2}$ **26.** $\frac{-19}{(3x-2)^2}$

27. $y = \frac{x^3+7}{x} = x^2 + 7x^{-1}$. $\frac{dy}{dx} = 2x - 7x^{-2} = 2x - \frac{7}{x^2} = \frac{2x^3-7}{x^2}$

28. $y = \frac{x^2+5x-1}{x^2} = 1 + 5x^{-1} - x^{-2}$. $\frac{dy}{dx} = -5x^{-2} + 2x^{-3} = \frac{-5}{x^2} + \frac{2}{x^3} = \frac{2-5x}{x^3}$

29. $y = \frac{(x-1)(x^2+x+1)}{x^3} = \frac{x^3-1}{x^3} = 1 - x^{-3}$. $\frac{dy}{dx} = 3x^{-4} = \frac{3}{x^4}$

30. $y = \frac{(x^2+x)(x^2-x+1)}{x^4} = \frac{x^4+x}{x^4} = 1 + x^{-3}$. $\frac{dy}{dx} = -3x^{-4} = -3/x^4$.

31. $y = \frac{1-x}{1+x^2}$. $\frac{dy}{dx} = \frac{(1+x^2)(-1)-(1-x)(2x)}{(1+x^2)^2} = \frac{-1-x^2-2x+2x^2}{(1+x^2)^2} = \frac{x^2-2x-1}{(1+x^2)^2}$ **32.** $\frac{-17}{(2x-7)^2}$

33. $y' = \frac{(1-x^3)2x-x^2(-3x^2)}{(1-x^3)^2} = \frac{2x-2x^4+3x^4}{(1-x^3)^2} = \frac{x^4+2x}{(1-x^3)^2}$

34. $y = \frac{x^2-1}{x^2+x-2} = \frac{(x-1)(x+1)}{(x-1)(x+2)} = \frac{x+1}{x+2}$. $y' = \frac{(x+2)\cdot 1-(x+1)\cdot 1}{(x+2)^2} = \frac{1}{(x+2)^2}$.

35. $y = (10)\frac{1}{\sqrt{x}-4}$, $y' = (10)\frac{-1}{(\sqrt{x}-4)^2}\left(\frac{1}{2\sqrt{x}}\right) = \frac{-5}{\sqrt{x}(\sqrt{x}-4)^2}$

36. $y' = \frac{(2\sqrt{x}-7)-x(\frac{1}{\sqrt{x}})}{(2\sqrt{x}-7)^2} = \frac{\sqrt{x}-7}{(2\sqrt{x}-7)^2}$

37. $y' = \frac{(\sqrt{x}+1)(\frac{1}{2\sqrt{x}})-(\sqrt{x}-1)(\frac{1}{2\sqrt{x}})}{(\sqrt{x}+1)^2} = \frac{1}{\sqrt{x}(\sqrt{x}+1)^2}$

38. $y = \frac{1}{x} + 1 - 4x^{-1/2}$, $y' = -\frac{1}{x^2} + 2x^{-3/2}$

39. $y = \frac{1}{(x^2-1)(x^2+x+1)}$. $y' = -\frac{(x^2-1)(2x+1)+2x(x^2+x+1)}{(x^2-1)^2(x^2+x+1)^2} = -\frac{4x^3+3x^2-1}{(x^2-1)^2(x^2+x+1)^2}$

40. $y = \frac{x^2+3x+2}{x^2-3x+2}$, $y' = \frac{(x^2-3x+2)(2x+3)-(x^2+3x+2)(2x-3)}{(x^2-3x+2)^2} = \frac{-6(x^2-2)}{(x^2-3x+2)^2}$.

41. $y = 3x^{-2}$. $y' = -6x^{-3} = -\frac{6}{x^3}$. $y' = -6x^{-3}$ so $y'' = 18x^{-4} = \frac{18}{x^4}$.

42. $y = -\frac{1}{x}$, $y' = \frac{1}{x^2} = x^{-2}$. $y'' = -2x^{-3} = \frac{-2}{x^3}$.

43. $y = \frac{5}{x^4} = 5x^{-4}$. $y' = -20x^{-5} = -\frac{20}{x^5}$. $y' = -20x^{-5}$ so $y'' = 100x^{-6} = \frac{100}{x^6}$.

44. $y = -\frac{3}{x^7} = -3x^{-7}$. $y' = 21x^{-8} = \frac{21}{x^8}$. $y' = 21x^{-8}$ so $y'' = -168x^{-9} = -\frac{168}{x^9}$.

45. $y = x + 1 + x^{-1}$ so $y' = 1 - x^{-2}$, $y'' = 2x^{-3}$ or $y' = 1 - \frac{1}{x^2}$, $y'' = \frac{2}{x^3}$.

46. $y = 12x^{-1} - 4x^{-3} + x^{-4}$ so $y' = -12x^{-2} + 12x^{-4} - 4x^{-5}$ and $y'' = 24x^{-3} - 48x^{-5} + 20x^{-6}$ or $y' = -\frac{12}{x^2} + \frac{12}{x^4} - \frac{4}{x^5}$ and $y'' = \frac{24}{x^3} - \frac{48}{x^5} + \frac{20}{x^6}$.

47. $\text{NDER}(x3^{-0.2x}, 1) = 0.63 = \text{slope}$. $f(1) = e^{-0.2}$. The tangent line at $(1, 3^{-0.2})$ is $y - 3^{-0.2} = 0.63(x-1)$. The result is confirmed by viewing $y = x3^{-0.2x}$ and $y = 3^{-0.2} + 0.63(x-1)$ in the rectangle $[-10, 10]$ by $[-10, 10]$.

48. $\text{NDER}(\frac{\sin x}{x}, \pi) = -0.32$. $f(\pi) = 0$. The tangent line at $(\pi, 0)$ has equation $y - 0 = -0.32(x - \pi)$. The result is confirmed by graphing $y = \sin x/x$ and $y = -0.32(x-\pi)$ in $[2, 4]$ by $[-0.5, 0.5]$.

49. $\text{NDER}(\frac{x+3}{x^3-2x+5}, 0) = 0.44$. $f(0) = 3/5$. The tangent line at $(0, 3/5)$ has equation $y - 3/5 = 0.44x$. The result is confirmed by graphing $y = f(x)$ and $y = (3/5) + 0.44x$ in the rectangle $[-1, 1]$ by $[0.3, 0.9]$.

50. $\text{NDER}(\sqrt[3]{\frac{x-1}{x^2+5}}, 2) = 0.09$. $f(2) = \frac{1}{\sqrt[3]{9}} = \frac{1}{\sqrt[3]{9}}\frac{\sqrt[3]{3}}{\sqrt[3]{3}} = \frac{\sqrt[3]{3}}{3}$. The tangent line at $(2, \sqrt[3]{3}/3)$ has equation $y = (\sqrt[3]{3}/3) + 0.09(x-2)$. The result can be confirmed by graphing $y = f(x)$ and $y = (\sqrt[3]{3}/3) + 0.09(x-2)$ in the rectangle $[-12, 12]$ by $[-1, 1]$.

51. The graph of $y = \text{NDER}(f(x))$ oscillates, appears to cross the x-axis infinitely often and to be symmetric with respect to the origin. The graph of $y = \text{NDER2}(f(x))$ oscillates and appears to cross the x-axis infinitely often and to be symmetric with respect to the y-axis. These graphs can be viewed in $[-10, 10]$ by $[-10, 10]$ and in $[-50, 50]$ by $[-50, 50]$.

52. The graph of $y = \text{NDER}(f(x))$ oscillates, is symmetric with respect to the y-axis and appears to cross the x-axis infinitely often. The graph of $y = \text{NDER2}(f(x))$ oscillates, appears to cross the x-axis infinitely often and to be symmetric with respect to the origin. These graphs can be viewed in the rectangles $[-5, 5]$ by $[-25, 25]$ and $[-50, 50]$ by $[-2500, 2500]$.

53. $y = f'(x)$ or $y = \text{NDER}(f(x))$ can be viewed in the rectangle $[-2, 8]$ by $[-4, 4]$. $y = f''(x)$ or $y = \text{NDER2}(f(x))$ can be viewed in $[-2, 10]$ by $[-4, 10]$.

54. Both $y = \text{NDER}(f(x), x)$ and $y = \text{NDER2}(f(x), x)$ can be viewed in $[-3, 5]$ by $[-5, 5]$.

55. $y = \text{NDER}(f(x))$ can be viewed in the rectangle $[-8, 8]$ by $[-2, 0]$. $y = \text{NDER2}(f(x))$ can be viewed in $[-6, 8]$ by $[-1, 1]$.

56. $y = \text{NDER}(f(x))$ can be viewed in the rectangle $[-10, 10]$ by $[-0.1, 0.2]$. To obtain an idea of the complete graph of $y = \text{NDER2}(f(x))$ it is necessary to view the graph in two rectangles, for example, $[-10, 10]$ by $[-0.1, 0.1]$ and $[-5, 2]$ by $[-1, 1]$.

57. The graph of $y =$ NDER$(f(x))$ oscillates, appears to cross the x-axis infinitely often and to be symmetric with respect to the origin. The three solutions of $f'(x) = 0$ of smallest absolute value are -2.029, 0, 2.029. The graph of $y =$ NDER2$(f(x))$ appears to cross the x-axis infinitely often. The solution set of $f''(x) > 0$ consists of an infinite sequence of intervals. The three closest to the origin are $(-6.578, -3.644)$, $(-1.077, 1.077)$ and $(3.644, 6.578)$ rounding appropriately.

58. The graph of $y = f'(x)$ is symmetric with respect to the y-axis. It crosses the x-axis infinitely many times. The three numerically smallest solutions to $f'(x) = 0$ are -2.29, 0, 2.29. The graph of $y = f''(x)$ is symmetric with respect to the origin, crosses the x-axis infinitely often. The solution set of $f''(x) > 0$ consists of an infinite sequence of intervals, of which the two closest to the origin are $(-3.99, -1.52)$ and $(0, 1.51)$ rounding appropriately.

59. We use zoom-in and the graphs of Exercise 53. $f'(x) = 0$ for $x = -0.313$ and $x = 3.198$. $f''(x) > 0$ for x in the set $(-\infty, -1) \cup (1, \infty)$.

60. $f'(x) = 0$ has solution set $\{-1.287, 3.108\}$. The solution set of $f''(x) > 0$ is the interval $(-2, 2)$.

61. The graph of $f'(x)$ does not cross the x-axis. Hence there is no solution to $f'(x) = 0$. The solution set of $f''(x) > 0$ (that is, those x for which the graph of $y = f''(x)$ is above the x-axis) is $(-3, -0.333) \cup (5, \infty)$.

62. We use the graph of $y =$ NDER$(f(x))$ obtained in Exercise 56 and zoom-in to determine the x-intercepts. This yields that $f'(x) = 0$ when $x = -2.732$ and 0.732. To solve NDER2$(f(x)) > 0$ we start with the graph of $y =$ NDER2$(f(x))$ in $[-10, 10]$ by $[-0.1, 0.1]$ and use zoom-in. The solution set is $(-4.648, -1) \cup (1.804, \infty)$, rounding appropriately to be sure.

63. $f(x) = \frac{2x-5}{3x^2+4}$. $f'(x) = \frac{(3x^2+4)2-(2x-5)6x}{(3x^2+4)^2} = \frac{-6x^2+30x+8}{(3x^2+4)^2} = \frac{-2(3x^2-15x-4)}{(3x^2+4)^2}$. $f''(x) = (-2)\frac{[(3x^2+4)^2(6x-15)-(3x^2-15x-4)2(3x^2+4)6x]}{(3x^2+4)^4} = (-2)\frac{[(3x^2+4)(6x-15)-12x(3x^2-15x-4)]}{(3x^2+4)^3} = \frac{6[6x^3-45x^2-24x+20]}{(3x^2+4)^3}$.

$f''(x)$ changes sign only when $6x^3 - 45x^2 - 24x + 20 = 0$ but we cannot solve this exactly. To solve $f''(x) > 0$ we find the intervals in which the graph of $y =$ NDER2$(f(x))$ lies above the x-axis. The solution set is $(-0.911, 0.460) \cup (7.950, \infty)$.

64. a) Let $f(x)=x$. $\frac{d}{dx}(x)=\lim_{h\to 0}\frac{f(x+h)-f(x)}{h}=\lim_{h\to 0}\frac{(x+h)-x}{h}=\lim_{h\to 0}1=1$.

b) Let $f(x)=-u(x)$. $\frac{d}{dx}(-u)=\lim_{h\to 0}\frac{f(x+h)-f(x)}{h}=\lim_{h\to 0}\frac{-u(x+h)-(-u(x))}{h}=-\lim_{h\to 0}\frac{u(x+h)-u(x)}{h}=-\frac{du}{dx}$.

65. $\frac{d}{dx}(cf(x))=c\frac{df}{dx}+\frac{dc}{dx}f(x)=c\frac{df}{dx}+0(f(x))=c\frac{df}{dx}$

66. $\frac{d}{dx}\left(\frac{1}{f(x)}\right)=\frac{f(x)\frac{d}{dx}(1)-(1)\frac{d}{dx}f(x)}{f^2(x)}=-\frac{f'(x)}{f^2(x)}$

67. a) $\frac{d}{dx}(uv)=uv'+u'v$. The value of this function at $x=0$ is $u(0)v'(0)+u'(0)v(0)=5(2)+(-3)(-1)=13$.

b) $\frac{d}{dx}\left(\frac{u}{v}\right)=\frac{vu'-uv'}{v^2}$. At $x=0$ this becomes $\frac{(-1)(-3)-5(2)}{(-1)^2}=-7$.

c) $\frac{d}{dx}\left(\frac{v}{u}\right)=\frac{uv'-vu'}{u^2}$. At $x=0$ this becomes $\frac{5(2)-(-1)(-3)}{5^2}=\frac{7}{25}$.

d) $\frac{d}{dx}(7v-2u)=7v'-2u'$. At $x=0$ this becomes $7(2)-2(-3)=20$.

68. a) $u(2)v'(2)+u'(2)v(2)=3(2)+(-4)(1)=2$

b) $\frac{v(2)u'(2)-u(2)v'(2)}{v^2(2)}=\frac{(1)(-4)-3(2)}{1^2}=-10$

c) $\frac{u(2)v'(2)-v(2)u'(2)}{u^2(2)}=\frac{3(2)-(1)(-4)}{3^2}=\frac{10}{9}$

d) $3u'(2)-2v'(2)+2(2)$ (from a)) $=-12-4+4=-12$

69. Let $f(x)=x^2+5x$. $f'(x)=2x+5$ and $f'(3)=11$. Answer: c)

70. $2y=3x+12$, $y=\frac{3}{2}x+6$, $y'=\frac{3}{2}$. Answer: c)

71. $y'=3x^2-3$. $y'(2)=9$ is the slope of the tangent. Hence $-1/9$ is the slope of the perpendicular at $x=2$. Thus it has equation $y-3=(-1/9)(x-2)$ or $x+9y=29$.

72. $y'=3x^2+1$. $3x^2+1=4$ yields $x^2=1$, $x=\pm 1$. At $x=1$, $y=2$ and the tangent at $(1,2)$ has equation $y-2=4(x-1)$ or $y=4x-2$. At $x=-1$, $y=-2$ and the tangent at $(-1,-2)$ has equation $y+2=4(x+1)$ or $y=4x+2$. The smallest slope is the smallest value of $y'=3x^2+1$ which 1 and this occurs at $x=0$.

73. $y'=6x^2-6x-12=6(x^2-x-2)=6(x+1)(x-2)=0$ at $x=-1$ and $x=2$. Answer: $(-1,27)$, $(2,0)$

74. $y'=3x^2$. $y'(-2)=3(-2)^2=12$. The line in question has equation $y+8=12(x+2)$ or $y=12x+16$. Thus the y-intercept is 16. To obtain the x-intercept we set $y=0$ and obtain $x=-16/12=-4/3$.

75. $y' = \frac{(x^2+1)4-4x(2x)}{(x^2+1)^2} = \frac{4-4x^2}{(x^2+1)^2} = \frac{4(1-x^2)}{(x^2+1)^2}$. At $x = 0$, $y' = 4$ and the tangent has equation $y = 4x$. At $x = 1$, $y' = 0$ and the tangent has equation $y = 2$.

76. $y' = -\frac{8(2x)}{(4+x^2)^2} = -\frac{16x}{(4+x^2)^2}$. At $x = 2$, $y' = -\frac{32}{8^2} = -\frac{4}{8} = -\frac{1}{2}$ and the tangent has equation $y - 1 = (-1/2)(x - 2)$ or $x + 2y = 4$.

77. $P = nRT(\frac{1}{V-nb}) - an^2V^{-2}$. $\frac{dP}{dV} = nRT\frac{[(V-nb)0-1\cdot1]}{(V-nb)^2} + 2an^2V^{-3} = -\frac{nRT}{(V-nb)^2} + \frac{2an^2}{V^3}$

78. $\frac{ds}{dt} = 9.8t$, $\frac{d^2s}{dt^2} = 9.8$

79. $R = \frac{CM^2}{2} - \frac{M^3}{3}$. $\frac{dR}{dM} = CM - M^2$

80. The derivative of $Kf(x)$ is $Kf'(x)$ so $f'(x)$ and $(Kf(x))'$ have the same zeros. For the case mentioned $x = 1.12$ is the only solution to $f'(x) = 0$. The graphs of $y = K2^x/(2+3^x)$, $K = 1, 2, 3$ can be viewed in $[-10, 15]$ by $[0, 1.5]$.

81. $f'(-x) = \lim_{h\to0} \frac{f(-x+h)-f(-x)}{h} = \lim_{h\to0} \frac{f[-(x-h)]-f(x)}{h} = \lim_{h\to0} \frac{f(x-h)-f(x)}{h} = \lim_{h\to0} -\frac{f(x-h)-f(x)}{-h} = -f'(x)$ so $f'(x)$ is odd.

82. $f'(-x) = \lim_{h\to0} \frac{f(-x+h)-f(-x)}{h} = \lim_{h\to0} \frac{-f(x-h)+f(x)}{h} = \lim_{h\to0} \frac{f(x-h)-f(x)}{-h} = f'(x)$ so $f'(x)$ is even.

83. a) $x = 1.442695\ldots$ b) $x = 1.442698\ldots$ c) very close

84. a) $x = 2.88543\ldots$ b) $x = 2.88541\ldots$ c) very close

85. a) Graph $y = abs(y_1 - f''(x))$ where $f''(x) = 20x^3 - 36x^2 + 6x - 12$ in $[-10, 10]$ by $[-0.1, 0.1]$. Its maximum value (which is the maximum error) is found to be 0.0424.

b) Let $y_2 = \frac{y_1(x+0.01)-y_1(x-0.01)}{0.02}$.

c) Graph $y = abs(y_2 - f^{(3)}(x))$ where $f^{(3)}(x) = 60x^2 - 72x + 6$. Its maximum (and rather constant) value on $[-10, 10]$ is about 0.01 (with dubious accuracy). NOTE: The answers for a) and c) depend on the h value used in the NDER algorithm. We used $\delta = 0.01$ here. On our calculator we had $y_1(x + 0.01) = \text{NDER}(f(x), x), x, x + 0.01)$.

3.4 VELOCITY, SPEED, AND OTHER RATES OF CHANGE

1. a) Let $x_1(t) = t^2 - 3t + 2$, $y_1(t) = 3$, $0 \leqq t \leqq 5$. On a parametric graphing utility with Tstep 0.05 in $[-1, 15]$ by $[-1, 15]$, holding down TRACE, we see the particle moves along the line $y = 3$ first (starting with $t = 0$) to the left and then to the right. b) Using TRACE, we find the positions to be: $(2, 3)$ at $t = 0$, $(0, 3)$ at $t = 1$, $(0, 3)$ at $t = 2$, $(2, 3)$ at $t = 3$. c) Using TRACE, we find the particle changes direction at $(-0.25, 3)$ when $t = 1.5$. The velocity $v = s'(t) = 2t - 3$ at this time is 0 while the acceleration $a = v' = 2$. d) The particle first travels left from $(2, 3)$ to $(-0.25, 3)$ and then right to $(12, 3)$ at $t = 5$. Thus the total distance traveled is $2.25 + 12.25 = 14.5$. e) Along with the graph in a), we graph $x_2(t) = t$, $y_2(t) = t^2 - 3t + 2$. Use simultaneous graphing format. f) Along with the graph in a), we simultaneously graph $x_2(t) = t$, $y_2(t) = 2t - 3 = v$ and $x_3(t) = 5$, $y_3(t) = 2 = a$. The particle is at rest when $v = 0$, that is, when $t = 1.5$sec.

2. a) Let $x_1(t) = 5 + 3t - t^2$, $y_1(t) = 3$, $0 \leqq t \leqq 5$. On a parametric graphing utility with Tstep 0.05 in $[-8, 8]$ by $[-10, 10]$, holding down TRACE, we see the particle moves along the line $y = 3$ first (starting with $t = 0$) to the right and then to the left. b) $(5, 3)$ when $t = 0$, $(7, 3)$ when $t = 1$, $(7, 3)$ when $t = 2$, $(5, 3)$ when $t = 3$. c) The particle changes direction at $(7.25, 3)$ when $t = 1.5$. At this time $v = 0$ and $a = -2$. d) The particle first travels right from $(5, 3)$ to $(7.25, 3)$ and then left to $(-5, 3)$ reached at $t = 5$. Thus the total distance traveled is $2.25 + 12.25 = 14.5$meters. f) The particle is at rest when $v = 3 - 2t = 0$, i.e., when $t = 1.5$sec.

3. For this problem we use $x_1(t) = t^3 - 6t^2 + 7t - 3$, $y_1(t) = 3$, $x_2(t) = t$, $y_2(t) = 3t^2 - 12t + 7 (= v = s')$ and $x_3(t) = t$, $y_3(t) = 6t - 12 (= a(t))$ with $0 \leqq t \leqq 5$, Tstep 0.05 in the viewing rectangle $[-10, 10]$ by $[-15, 25]$. a) The particle first ($t = 0$) moves to the right, then to the left and then to the right again. b) $(-3, 3)$ when $t = 0$, $(-1, 3)$ when $t = 1$, $(-5, 3)$ when $t = 2$, $(-9, 3)$ when $t = 3$sec. c) Using TRACE, we get $(-0.70, 3)$ when $t = 0.7$, $v = 0.07$, $a = -7.8$ and $(-9.30, 3)$ when $t = 3.3$sec $v = 0.07$, $a = 7.8$. d) The particle travels right from $(-3, 3)$ to $(-0.70, 3)$ then left from $(-0.70, 3)$ to $(-9.30, 3)$ then right from $(-9.30, 3)$ to $(7, 3)$. Thus the total distance traveled is $2.3 + 8.6 + 16.3 = 27.2$meters. f) The particle is at rest when $v = 0$. With TRACE our approximation of this is at $t = 0.7$ and $t = 3.3$sec.

4. a) For this problem we use $x_1(t) = 4 - 7t + 6t^2 - t^3$, $y_1(t) = 3$, $x_2(t) = t$, $y_2(t) = -7 + 12t - 3t^2 (= v)$ and $x_3(t) = t$, $y_3(t) = 12 - 6t (= a)$ with $0 \leqq t \leqq 5$, Tstep 0.05 in the viewing rectangle $[-10, 15]$ by $[-25, 15]$. The particle first moves left, then right and then left again. b) $(4,3),(2,3),(6,3),(10,3)$ c) Using TRACE, we get $(1.70,3)$ when $t=0.7$, $v=-0.07$, $a=7.8$ and $(10.30,3)$ when $t=3.3$, $v=-0.07$, $a=-7.8$. d) The particle travels left from $(4,3)$ to $(1.70,3)$, then right from $(1.70,3)$ to $(10.30,3)$ and then left to $(-6,3)$. Thus the total distance traveled is $2.3+8.6+16.3=27.2$meters. f) $v=0$ when (approximately) $t=0.7$ and $t=3.3$.

5. For this problem we use $x_1(t)=t\sin t$, $y_1(t)=3$, $x_2(t)=t$, $y_2(t)=$ NDER$(x_1,t,t)(=v)$ and $x_3(t)=t$, $y_3(t)=$ NDER$(y_2,t,t)(=a(t))$ with $0\leqq t\leqq 15$, Tstep 0.05 in the window $[-15,15]$ by $[-15,15]$. a) The particle first moves right, then left, then right, then left, then right, then left slightly. b) $(0,3)$, $(0.84,3)$, $(1.82,3)$, $(0.42,3)$ c) Using TRACE, we obtain $(1.82,3)$ when $t=2.05$, $v=-0.06$, $a=-2.7$; $(-4.814,3)$ when $t=4.90$, $v=-0.07$, $a=5.19$; $(7.91,3)$ when $t=8$, $v=-0.17$, $a=-8.2$; $(-11.04,3)$ when $t=11.1$, $v=.16$, $a=11.2$; $(14.17,3)$ at $t=14.2$, $v=0.106$, $a=-14.3$ d) The particle travels right from $(0,3)$ to $(1.82,3)$, from $(1.82,3)$ to $(-4.81,3)$, from $(-4.81,3)$ to $(7.91,3)$, from $(7.91,3)$ to $(-11.04,3)$, from $(-11.04,3)$ to $(14.17,3)$ and from $(14.17,3)$ to $(9.75,3)$. Thus the total distance traveled is $1.82+6.63+12.72+18.95+25.21+4.42=69.75$meters. e) We simultaneously graph x_1, y_1 and $x_4=t$, $y_4=t\sin t$. f) The particle is at rest when $v=y_2(t)=0$. With TRACE our approximations of these times are $t=0$, 2.05, 4.9, 8, 11.1 and 14.2.

6. For this problem we use $x_1(t)=5\sin(\frac{2t}{\pi})$, $y_1(t)=3$, $x_2(t)=t$, $y_2(t)=$ NDER$(x_1,t)(=v)$ and $x_3(t)=t$, $y_3(t)=$ NDER$(y_2,t)(=a)$ with $0\leqq t\leqq 2\pi$, Tstep .05 in $[-5,7]$ by $[-4,4]$. a) The particle moves to the right and then to the left. b) $(0,3)$ when $t=0$, $(2.97,3)$ when $t=1$, $(4.78,3)$ when $t=2$ and $(4.72,3)$ when $t=3$. c) Using TRACE, we get $(5.00,3)$ when $t=2.45$, $v=.03$, $a=-2.03$. d) The particle travels from $(0,3)$ to $(5,3)$ and then left to $(-3.71,3)$ a total distance of $5+8.71=13.71$meters. f) By c) the particle is at rest at approximately 2.45sec.

7. On Mars $s=1.86t^2$, $v=2(1.86)t=16.6$ so $t=4.46$sec. On Jupiter $s=11.44t^2$, $v=22.88t=16.6$ so $t=0.73$sec.

8. a) $v=s'=24-1.6t$, $a=v'=-1.6$ b) $v=24-1.6t=0$ when $t=24/1.6=15$sec c) $s(15)=24(15)-0.8(15)^2=180$meters d) $s(t)=24t-0.8t^2=90$ is equivalent to $0.8t^2-24t+90=0$ the smaller solution of which is 4.39sec. e) The rock was aloft twice as long as it took to reach its highest point, hence 30sec.

9. One possibility: Graph $x_1(t) = t$, $y_1(t) = 24t - 0.8t^2$, $0 \leqq t \leqq 30$, in $[0, 30]$ by $[0, 180]$. Then use TRACE and zoom-in if more accuracy is desired.

10. $s = 24t - 4.9t^2$, $v(t) = s'(t) = 24 - 9.8t = 0$ when $t = 24/9.8 = 12/4.9$sec. $s(12/4.9) = 24(12/4.9) - 4.9(12/4.9)^2 = 29.39$meters.

11. $s = 832t - 2.6t^2 = t(832 - 2.6t) = 0$ when $t = 0$ and when $t = 832/2.6 =$ 320sec. Thus it will get back down in 320sec. On earth $s = 832t - 16t^2 = t(832 - 16t) = 0$. When $t = 0$ and when $t = 832/16 = 52$sec so it will get back down in 52sec.

13. $b(t) = 10^6 + 10^4 t - 10^3 t^2$. $b'(t) = 10^4 - (2)10^3 t$. a)$b'(0) = 10^4$ per hour b) $b'(5) = 10^4 - (2)10^3(5) = 0$ c) $b'(10) = 10^4 - (2)10^4 = -10^4$ per hour

14. $Q(t) = 200(30-t)^2 = 200(900-60t+t^2)$. $Q'(t) = 200(-60+2t) = 400(t-30)$. $Q'(10) = 400(-20) = -8000$ so at the end of 10 minutes it is following out at the rate of 8000gal./min. Average rate $= \frac{Q(10)-Q(10)}{10-0} = \frac{200(20)^2-200(30)^2}{10} =$ $-10,000$gal/min.

15. $c(x) = 2000 + 100x - 0.1x^2$. a) The average cost of one washing machine when producing the first 100 washing machines is $c(100)/100 = \$110$. During production of the first 100 machines the average increase in producing one more machine is: average increase $= \frac{c(100)-c(0)}{100-0} = \$90$; the fixed cost $c(0) = \$2000$ is omitted with this method. b) $c'(x) = 100 - 0.2x$, $c'(100) = \$80$ c) $c(101) - c(100) = \$79.90$.

16. $r(x) = 2000(1 - \frac{1}{x+1}) = 2000\frac{x}{x+1}$. a) A graph of $r(x)$, $x \geq 0$ can be viewed in $[0, 50]$ by $[0, 2000]$. Only non-negative integral values of x make sense in this problem situation. b) $r'(x) = 2000\frac{(x+1)\cdot 1 - x\cdot 1}{(x+1)^2} = \frac{2000}{(x+1)^2}$ c) $r'(5) = \frac{2000}{6^2} = \$55.56$ d) $\lim_{x\to\infty} 2000(1 - \frac{1}{x+1}) = 2000(1 - 0) = \$2000$. As more and more desks are made, the revenue gets arbitrarily close to \$2000. Note that the marginal revenue $r'(x) \to 0$ as $x \to \infty$ so that it does not pay to produce many desks.

17. $s = t^3 - 6t^2 + 9t$. $v = s' = 3t^2 - 12t + 9 = 3(t^2 - 4t + 3) = 3(t-1)(t-3) = 0$ when $t = 1$ and 3sec. $a = v' = 6t - 12$. $a(1) = -6$, $a(3) = 6$. Thus $a = -6\text{m/sec}^2$ when $t = 1$sec and $a = 6\text{m/sec}^2$ when $t = 3$sec.

18. $v = 2t^3 - 9t^2 + 12t - 5$, $a = v' = 6t^2 - 18t + 12 = 6(t^2 - 3t + 2) = 6(t-1)(t-2) =$ 0 when $t = 1, 2$. $v(1) = 0$ and $v(2) = -1$. Thus at $t = 1$ the speed is 0m/sec and at $t = 2$sec the speed is 1m/sec.

19. a) We use $-12 \leqq t \leqq 12$, Tstep 0.05 in the viewing rectangle $[-35, 35]$ by $[-3, 10]$. b) For this graph we can use $0 \leqq t \leqq 6.29 \approx 2\pi$, Tstep 0.05 in $[-3, 8]$ by $[-3, 3]$. c) This line segment can be viewed using $0 \leqq t \leqq 6.29$, Tstep 0.05 in $[-6, 10]$ by $[-6, 2]$.

20. Only c) is the graph of a function.

21. a) All have derivative $3x^2$. c) The result of a) suggests that the family consists of all functions of the form x^3+C where C can be any constant. b) From c) $f(x)$ has the form $f(x) = x^3+C$. Since $f(0) = 0$, $f(0) = 0^3+C = 0$ so $C = 0$ and $f(x) = x^3$ is the unique function with the two properties. e) Yes. $g(x) = x^3 + C$ and $g(0) = 0^3 + C = 3$ so $g(x) = x^3 + 3$.

22. a) A graph of $P(x)$ can be obtained in the viewing rectangle $[0, 200]$ by $[0, 11]$. b) Only non-negative integral values of x make sense in the problem situation. c) The graph of $y = P'(x) = \text{NDER}(P(x))$ is the same as the graph of $y = M(x)$ in $[0, 200]$ by $[0, 0.2]$. d) Using only integral values of x and zoom-in and TRACE, we find a maximal marginal profit of 0.1732 at $x = 106$ and 107. Returning to the graph of $P(x)$, using zoom-in and TRACE, we find $P(106) = 4.93$ and $P(107) = 5.10$. We estimate the remaining values of $P'(x)$ using TRACE and the viewing rectangle $[0, 330]$ by $[0, 0.2]$: $P'(50) = 0.01$, $P'(100) = 0.17$, $P'(125) = 0.12$, $P'(150) = 0.03$, $P'(175) = 0.006$, $P'(300) = 1.1 \times 10^{-6}$. e) Since $2^{5-0.1x} = \frac{1}{2^{0.1x-5}} \to 0$ as $x \to \infty$, $\lim_{x\to\infty} P(x) = \frac{10}{1+0} = 10$. The profit can approach 10 (thousand dollars) arbitrarily closely but cannot attain it exactly. But $P(316)$ rounds out to \$10,000.00. f) More information is needed but since $P(240)$ is about \$999 not much is lost by cutting product at $x = 240$.

23. a) 190ft/s b) 2 c) At $8s$ when $v = 0$ d) At $10.8s$ when it was falling at 90ft/s e) From $t = 8s$ to $10.8s$, i.e., $2.8s$ f) Just before burnout, i.e., just before $t = 2s$. The acceleration was constant from $t = 2s$ to $t = 10.8s$ during free fall. Note that the graph of v has constant slope during this time.

24.

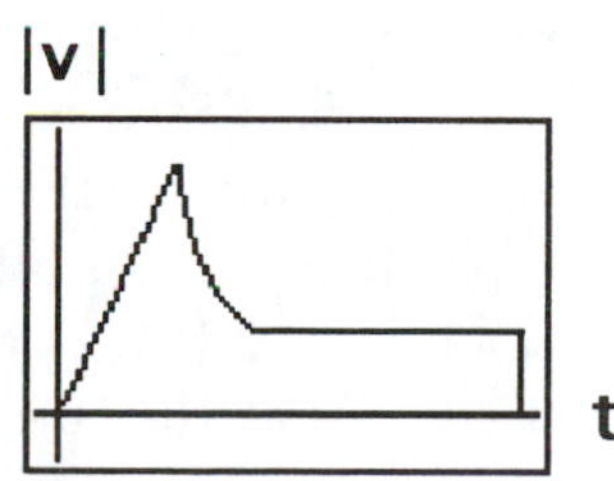

25. a) 0, 0 b) 1700, 1400 c) Rabbits per day and foxes per day

27. (b) **28.** (a) **29.** (d) **30.** (c)

31. a)

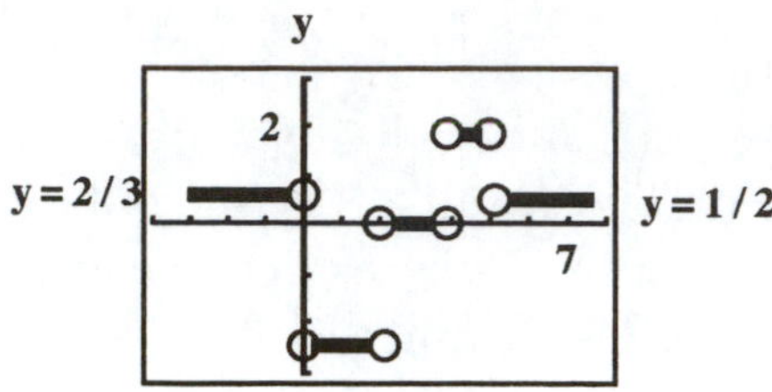

b) $x = 0, 2, 4, 5$

32. a)

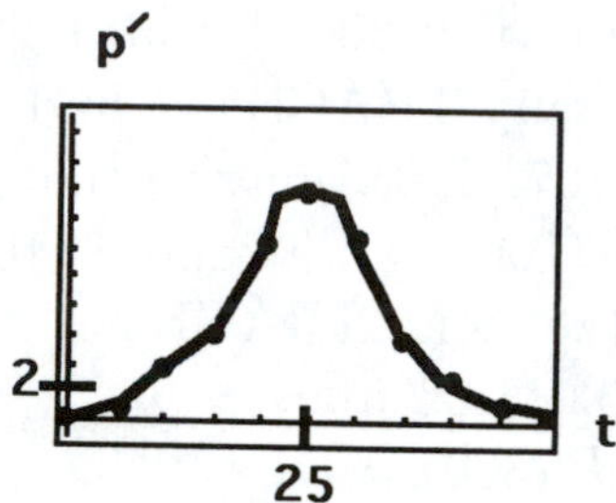

The horizontal axis is measured in days, the vertical axis in fruit flies per day.

b) Growing fastest around day 25, slowest near the beginning day 0 and near the end day 50.

33. We must solve the equation $x_1(t) = 4t^3 - 16t^2 + 15t = 5$ or $4t^3 - 16t^2 + 15t - 5 = 0$. The approximate solution is $t = 2.832$.

34. Our results are approximations using TRACE when we graph $x_1(t) = 2t^3 - 13t^2 + 22t - 5$, $y_1(t) = 2$, $0 \leqq t \leqq 5$, TStep $= 0.05$ in $[-6, 30]$ by $[-1, 3]$. Parts a), c), d), e): We find that the particle starts ($t = 0$) at $(-5, 2)$, moves right until $t = 1.15$ when the particle is at $(6.149, 2)$, then moves left until $t = 3.2$ at which time its location is $(-2.184, 2)$ and then it moves right for $t > 3.2$. The particle is at rest ($v = 0$) at the turning points, i.e., at $t = 1.15$

and $t = 3.2$. We graph v and the speed $|v|$. Here $v = \frac{ds}{dt} = 6t^2 - 26t + 22$.

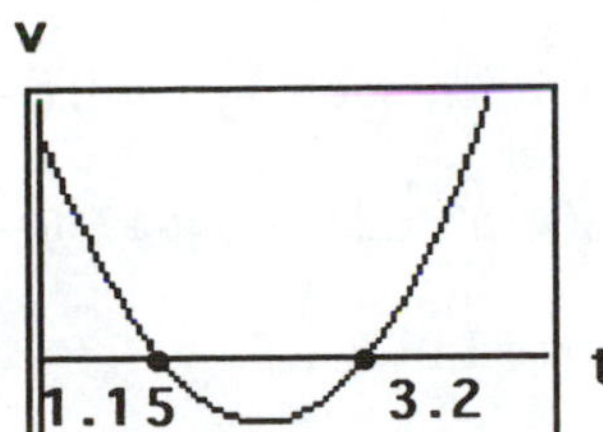

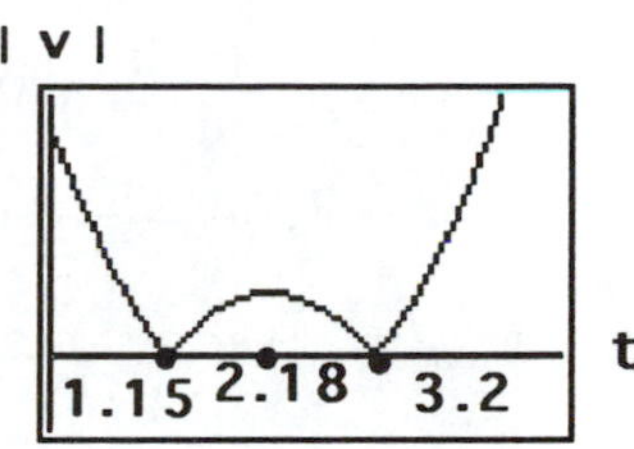

These graphs describe the velocity and speed of the particle. b) $|v|$ is increasing for $1.15 < t < 2.18$ and for $t > 3.2$. $|v|$ is decreasing for $0 < t < 1.15$ and for $2.18 < t < 3.2$. f) $x_1(t) \approx 5$ when $t = 0.75, 1.6, 4.15$.

35.

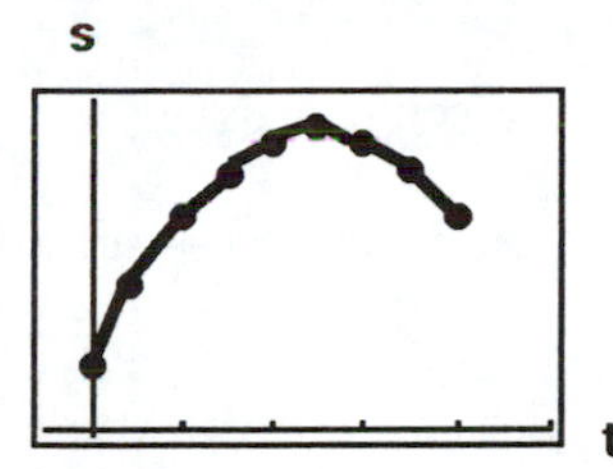

a) $v \approx 18$ft/sec b) $v \approx 0$ft/sec c) $v \approx -12$ft/sec

36.

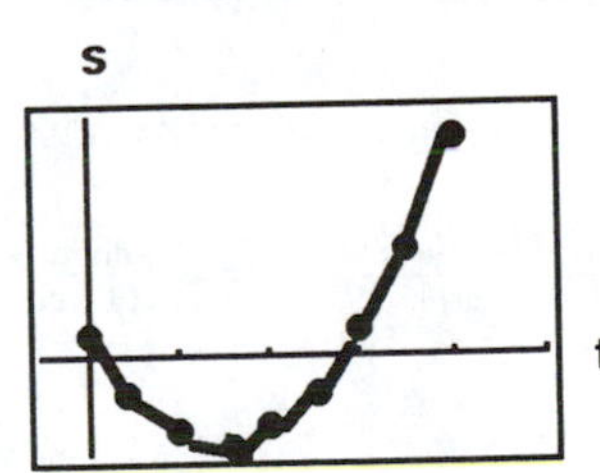

a) $v \approx -6$ftsec b) $v \approx 12$ft/sec c) $v \approx 24$ft/sec

3.5 DERIVATIVES OF TRIGONOMETRIC FUNCTIONS

1. $y = 1 + x - \cos x,\ y' = 1 + \sin x$ **2.** $y = 2\sin x - \tan x,\ y' = 2\cos x - \sec^2 x$

3. $y = \frac{1}{x} + 5\sin x,\ y' = -\frac{1}{x^2} + 5\cos x$ **4.** $y = x^2 - \sec x,\ y' = 2x - \sec x \tan x$

5. $y = \csc x - 5x + 7,\ y' = -\csc x \cot x - 5$ **6.** $y = 2x + \cot x,\ y' = 2 - \csc^2 x$

7. $y = x\sec x,\ y' = x\sec x \tan x + \sec x = \sec x(x\tan x + 1)$

8. $y = x\csc x,\ y' = -x\csc x \cot x + \csc x = \csc x(1 - x\cot x)$

9. $y = x^2 \cot x,\ y' = -x^2\csc^2 x + 2x\cot x = x(2\cot x - x\csc^2 x)$

10. $y = 4 - x^2\sin x,\ y' = -x^2\cos x - 2x\sin x = -x(x\cos x + 2\sin x)$

11. $y = 3x + x\tan x,\ y' = 3 + x\sec^2 x + \tan x$

12. $y = x\sin x + \cos x,\ y' = x\cos x + \sin x - \sin x = x\cos x$

13. $y = \sin x \sec x = \tan x,\ y' = \sec^2 x$

14. $y = \sec x \csc x,\ y' = \sec x(-\csc x \cot x) + \sec x \tan x \csc x = \sec x \csc x(\tan x - \cot x) = \sec^2 x - \csc^2 x$

15. $y = \tan x \cot x = 1,\ y' = 0$

16. $y = \cos x(1 + \sec x) = \cos x + 1,\ y' = -\sin x$

17. $y = \frac{4}{\cos x} = 4\sec x,\ y' = 4\sec x \tan x$

18. $y = 5 + \frac{1}{\tan x} = 5 + \cot x,\ y' = -\csc^2 x$

19. $y = \frac{\cos x}{x},\ y' = \frac{x(-\sin x) - (\cos x)(1)}{x^2} = -\frac{x\sin x + \cos x}{x^2}$

20. $y = \frac{2}{\csc x} - \frac{1}{\sec x} = 2\sin x - \cos x,\ y' = 2\cos x + \sin x$

21. $y = \frac{x}{1+\cos x},\ y' = \frac{(1+\cos x)(1) - x(-\sin x)}{(1+\cos x)^2} = \frac{1+\cos x + x\sin x}{(1+\cos x)^2}$

22. $y = \frac{\sin x + \cos x}{\cos x} = \tan x + 1,\ y' = \sec^2 x$

23. $y = \frac{\cot x}{1+\cot x},\ y' = \frac{(1+\cot x)(-\csc^2 x) - \cot x(-\csc^2 x)}{(1+\cot x)^2} = -\frac{\csc^2 x}{(1+\cot x)^2}$

24. $y = \frac{\cos x}{1+\sin x},\ y' = \frac{(1+\sin x)(-\sin x) - \cos x(\cos x)}{(1+\sin x)^2} = \frac{-\sin x - (\sin^2 x + \cos^2 x)}{(1+\sin x)^2} = -\frac{\sin x + 1}{(1+\sin x)^2} = -\frac{1}{1+\sin x}$

25. $y = \csc x,\ y' = -\csc x \cot x,\ y'' = -[\csc x(-\csc^2 x) + (-\csc x \cot x)\cot x] = \csc^3 x + \csc x \cot^2 x = \csc x(\csc^2 x + \cot^2 x)$

26. a) $y = \sin x,\ y^{(4)} = \sin x$ b) $y = \cos x,\ y^{(4)} = \cos x$

27. $y = \sin x,\ x = 0.\ y' = \cos x.\ f(0) = \cos 0 = 1$ is the slope m of the tangent line. Hence the tangent line has equation $y - 0 = 1(x - 0)$ or $y = x$. For the normal line we need the slope $-1/m = -1$. The normal line has equation $y - 0 = -1(x - 0)$ or $y = -x$. We graph the three functions $y = \sin x,\ y = x$ and $y = -x$ in the viewing rectangle $[-3, 3]$ by $[-2, 2]$. Notice that the horizontal length 6 and the vertical height 4 of the screen are in the ratio 3 to 2 to ensure that perpendicular lines appear to intersect in a right angle.

28. $y = \tan x,\ x = 0.\ y' = \sec^2 x.\ m = \sec^2 0 = 1$. Tangent line: $y - 0 = 1(x - 0)$ or $y = x$. Normal line $y - 0 = (-1/m)(x - 0)$ or $y = -x$. We may graph $y = \tan x,\ y = x$ and $y = -x$ in $[-3, 3]$ by $[-2, 2]$.

29. $y = 2\sin^2 x,\ x = 2.\ y = 2\sin x \sin x$ so by the product rule $y' = 2(\sin x \cos x + \cos x \sin x) = 4\sin x \cos x.\ f'(2) = 4\sin 2\cos 2$. Tangent: $y - 2\sin^2 2 = 4\sin 2\cos 2(x - 2)$. Normal: $y - 2\sin^2 2 = -(4\sin 2\cos 2)^{-1}(x - 2)$. We may view $y = 2(\sin x)^2$, the tangent and the normal in $[0, 4]$ by $[0, 2.7]$.

30. $y = \frac{2+\cot x}{x},\ x = 1.\ y' = \frac{x(-\csc^2 x)-(2+\cot x)(1)}{x^2} = -\frac{x\csc^2 x+\cot x+2}{x^2}.\ m = f'(1) = -(\csc^2 1 + \cot 1 + 2)$. Tangent: $y - (2 + \cot 1) = m(x - 1)$. Normal: $y - (2 + \cot 1) = -(1/m)(x - 1)$. The view of $y = f(x)$, the tangent and the normal in $[-1, 3]$ by $[2, 4.7]$ confirms the result.

31. $D_x \cos x = \lim_{h\to 0} \frac{\cos(x+h)-\cos x}{h} = \lim_{h\to 0} \frac{\cos x \cos h - \sin x \sin h - \cos x}{h} = \lim_{h\to 0}\left[\cos x \frac{(\cos h - 1)}{h} - \sin x \frac{\sin h}{h}\right] = (\cos x)0 - (\sin x)1 = -\sin x$

32. Let $f(x) = \sec x.\ f'(x) = \sec x \tan x$ and $f'(0) = \sec 0 \tan 0 = 1 \cdot 0 = 0$. Let $g(x) = \cos x.\ g'(x) = -\sin x$ so $g'(0) = 0$, that is, the slope of the tangent line is 0.

33. $(\tan x)' = \sec^2 x = \frac{1}{\cos^2 x}$ and $(\cot x)' = -\csc^2 x = -\frac{1}{\sin^2 x}$ cannot be 0 for any value of x.

34. $y = x + \sin x,\ y' = 1 + \cos x = 0$ is equivalent to $\cos x = -1,\ 0 \leqq x \leqq 2\pi$ which has the unique solution $x = \pi$. Thus (π, π) is the only point where there is a horizontal tangent.

35. $y = 2x + \sin x,\ y' = 2 + \cos x = 0$ is equivalent to $\cos x = -2$. Since the latter equation has no solution, the graph has no horizontal tangent.

36. $y = x + \cos x$, $y' = 1 - \sin x = 0$ is equivalent to $\sin x = 1$ which has the unique solution $x = \frac{\pi}{2}$ in the interval. Thus $(\frac{\pi}{2}, \frac{\pi}{2})$ is the only point where there is a horizontal tangent.

37. $y = x + 2\cos x$, $y' = 1 - 2\sin x = 0$ is equivalent to $\sin x = \frac{1}{2}$ which has two solutions $x = \pi/6$, $5\pi/6$ in the interval. Thus there are horizontal tangents at the points $(\pi/6, (\pi/6) + \sqrt{3})$ and $(5\pi/6, (5\pi/6) - \sqrt{3})$.

38. a) $s = 2 - 2\sin t$, $v = \frac{ds}{dt} = -2\cos t$, $a = \frac{d^2s}{dt^2} = 2\sin t$. Speed $= |v| = 2|\cos t|$. $v(\frac{\pi}{4}) = -\sqrt{2}$, speed at $\frac{\pi}{4} = \sqrt{2}$, $a(\frac{\pi}{4}) = \sqrt{2}$. b) $s = \sin t + \cos t$, $v = \cos t - \sin t$, $a = -\sin t - \cos t$. $v(\frac{\pi}{4}) = 0$, $|v(\frac{\pi}{4})| = 0$, $a(\frac{\pi}{4}) = -\sqrt{2}$

39. $y' = -\sqrt{2}\sin x$. $f'(\pi/4) = -\sqrt{2}\sin(\pi/4) = -1 = m$. $y - 1 = (-1)(x - \pi/4)$, $y = -x + 1 + \pi/4$ is an equation of the tangent. $y - 1 = (-1/m)(x - \pi/4) = x - \pi/4$, $y = x + 1 - \pi/4$ is an equation of the normal.

40. $y = \tan x$, $y' = \sec^2 x$. The slope of $y = 2x$ is 2 so we must solve $\sec^2 x = 2$, $-\pi/2 < x < \pi/2$. This leads to $\sec x = \pm\sqrt{2}$, $\cos x = \pm 1/\sqrt{2}$. Thus the solutions for x are $\pm\pi/4$. This yields the answer $(-\pi/4, -1)$, $(\pi/4, 1)$.

41. $y = \cot x - \sqrt{2}\csc x$, $0 < x < \pi$. $y' = -\csc^2 x + \sqrt{2}\csc x \cot x = \csc x(\sqrt{2}\cot x - \csc x)$. Since $\csc x \neq 0$, $y' = 0$ is equivalent to $\sqrt{2}\cot x - \csc x = 0$ which leads to (by multiplication by $\sin x$) $\sqrt{2}\cos x - 1 = 0$, $\cos x = 1/\sqrt{2}$, $x = \pi/4$. $f(\pi/4) = \cot(\pi/4) - \sqrt{2}\csc(\pi/4) = 1 - 2 = -1$. Thus $(\pi/4, -1)$ is the only point where there is a horizontal tangent. The tangent there has equation $y = -1$. This answer is supported by graphing $f(x)$ and $y = -1$ in the viewing window $[0, 3.14]$ by $[-10, 1]$.

42. $y = \tan x + 3\cot x - 3$, $0 < x < \pi/2$. $y' = \sec^2 x - 3\csc^2 x = 1/\cos^2 x - 3/\sin^2 x = (\sin^2 x - 3\cos^2 x)/\cos^2 x \sin^2 x$. Thus $y' = 0$ is equivalent to $\tan^2 x = 3$, $\tan x = \pm\sqrt{3}$. In $0 < x < \pi/2$, this has the unique solution $x = \pi/3$. $f(\pi/3) = \sqrt{3} + 3/\sqrt{3} - 3 = 2\sqrt{3} - 3$. Thus $(\pi/3, 2\sqrt{3} - 3)$ is the only point where there is a horizontal tangent. The tangent there has equation $y = 2\sqrt{3} - 3$. This answer is supported by graphing $f(x)$ and $y = 2\sqrt{3} - 3$ in the viewing window $[0, 1.57]$ by $[0, 10]$.

43. The graph of $\tan x$ and its derivative $\sec^2 x$, $-\pi/2 < x < \pi/2$ may be viewed in the rectangle $[-1.57, 1.57]$ by $[-5, 5]$.

44. The graph of $\cot x$ and its derivative $-\csc^2 x$, $0 < x < \pi$ may be viewed in the rectangle $[0, 3.14]$ by $[-5, 5]$.

45. $\lim_{h\to 0}\frac{1-\cos h}{h^2} = \lim_{h\to 0}\frac{1-\cos h}{h^2}\frac{1+\cos h}{1+\cos h} = \lim_{h\to 0}\frac{1-\cos^2 h}{h^2(1+\cos h)} = \lim_{h\to 0}(\frac{\sin h}{h})^2\frac{1}{1+\cos h} = (1)^2\frac{1}{1+\cos 0} = \frac{1}{2}$

46. $(\sec x)' = (\frac{1}{\cos x})' = \frac{(\cos x)0-1(-\sin x)}{\cos^2 x} = \frac{\sin x}{\cos^2 x} = \frac{1}{\cos x}\frac{\sin x}{\cos x} = \sec x \tan x$

47. $(\cot x)' = (\frac{\cos x}{\sin x})' = \frac{\sin x(-\sin x)-\cos x\cos x}{\sin^2 x} = \frac{-(\sin^2 x+\cos^2 x)}{\sin^2 x} = -\frac{1}{\sin^2 x} = -\csc^2 x$

48. $(\csc x)' = (\frac{1}{\sin x})' = \frac{(\sin x)0-1(\cos x)}{\sin^2 x} = -\frac{\cos x}{\sin^2 x} = -\frac{1}{\sin x}\frac{\cos x}{\sin x} = -\csc x\cot x$

3.6 THE CHAIN RULE

1. $y = \sin(3x+1)$. $y' = 3\cos(3x+1)$

2. $y = \sin(7-5x)$. $y' = -5\cos(7-5x)$

3. $y = \cos(-x/3) = \cos(x/3)$. $y' = -\frac{1}{3}\sin(x/3)$

4. $y = \cos(\sqrt{3}x)$. $y' = -\sqrt{3}\sin(\sqrt{3}x)$

5. $y = \tan(2x - x^3)$. $y' = (2-3x^2)\sec^2(2x-x^3)$

6. $y = \tan[5(2x-1)]$. $y' = 10\sec^2[5(2x-1)]$

7. $y = x\sec(x^2+\sqrt{2})$. $y' = 1 + 2x\sec(x^2+\sqrt{2})\tan(x^2+\sqrt{2})$

8. $y = x\sec(3-8x)$. $y' = -8x\sec(3-8x)\tan(3-8x) + \sec(3-8x) = \sec(3-8x)[1-8x\tan(3-8x)]$

9. $y = -\csc(x^2+7x)$. $y' = (2x+7)\csc(x^2+7x)\cot(x^2+7x)$

10. $y = \sqrt{x} + \csc(1-2x)$. $y' = \frac{1}{2\sqrt{x}} + 2\csc(1-2x)\cot(1-2x)$

11. $y = 5\cot(\frac{2}{x})$. $y' = 5(-\frac{2}{x^2})(-\csc^2(\frac{2}{x})) = \frac{10}{x^2}\csc^2(\frac{2}{x})$

12. $y = \cot(\pi - \frac{1}{x})$. $y' = -\frac{1}{x^2}\csc^2(\pi - \frac{1}{x})$

13. $y = \cos(\sin x)$. $y' = [-\sin(\sin x)]\cos x$

14. $y = \sec(\tan x)$. $y' = [\sec(\tan x)\tan(\tan x)]\sec^2 x$

15. $y = (2x+1)^5$. $y' = 5(2x+1)^4(2) = 10(2x+1)^4$. We support this result by graphing $10(2x+1)^4$ and NDER($(2x+1)^5, x$) in $[-2,1]$ by $[0,810]$ and seeing that the two graphs coincide.

16. $y = (4-3x)^9$. $y' = -27(4-3x)^8$

17. $y = (x^2+1)^{-3}$. $y' = (-3)(x^2+1)^{-4}(2x) = -6x(x^2+1)^{-4}$

18. $y = (x+\sqrt{x})^{-2}$. $y' = -2(x+\sqrt{x})^{-3}(1+\frac{1}{2\sqrt{x}})$

19. $y = (1-\frac{x}{7})^{-7}$. $y' = -7(1-\frac{x}{7})^{-8}(-\frac{1}{7}) = (1-\frac{x}{7})^{-8}$

20. $y = (\frac{x}{2}-1)^{-10}$. $y' = -10(\frac{x}{2}-1)^{-11}(\frac{1}{2}) = -5(\frac{x}{2}-1)^{-11}$

21. $y = (\frac{x^2}{8}+x-\frac{1}{x})^4$. $y' = 4(\frac{x^2}{8}+x-\frac{1}{x})^3(\frac{x}{4}+1+\frac{1}{x^2})$

22. $y = (\frac{x}{5}+\frac{1}{5x})^5$. $y' = 5(\frac{x}{5}+\frac{1}{5x})^4(\frac{1}{5}-\frac{1}{5x^2}) = (1-\frac{1}{x^2})(\frac{x}{5}+\frac{1}{5x})^4$

23. $y = (\csc x + \cot x)^{-1}$. $y' = -(\csc x+\cot x)^{-2}(-\csc x\cot x - \csc^2 x) = \csc x(\csc x+\cot x)^{-2}(\cot x+\csc x) = \csc x(\csc x+\cot x)^{-1}$

24. $y = -(\sec x+\tan x)^{-1}$. $y' = (\sec x+\tan x)^{-2}(\sec x\tan x+\sec^2 x) = \sec x$ $(\sec x+\tan x)^{-1}$

25. $y = \sin^4 x + \cos^{-2} x$. $y' = 4\sin^3 x\cos x - 2\cos^{-3} x(-\sin x) = 2\sin x$ $(2\sin^2 x\cos x + \cos^{-3} x)$

26. $y = \sin^{-5} x - \cos^3 x$. $y' = -5\sin^{-6} x\cos x + 3\cos^2 x\sin x$

27. $y = x^3(2x-5)^4$. $y' = 4x^3(2x-5)^3(2) + 3x^2(2x-5)^4 = x^2(2x-5)^3(8x + 3(2x-5)) = x^2(2x-5)^3(14x-15)$

28. $y = (1-x)(3x^2-5)^5$. $y' = (1-x)5(3x^2-5)^4 6x - (3x^2-5)^5 = (3x^2 - 5)^4(30x-30x^2-3x^2+5) = (3x^2-5)^4(5+30x-33x^2)$

29. $y = (4x+3)^4(x+1)^{-3}$. $y' = (4x+3)^4[-3(x+1)^{-4}] + 4(4x+3)^3 4(x+1)^{-3} = (4x+3)^3(x+1)^{-4}[-3(4x+3)+16(x+1)] = (4x+3)^3(x+1)^{-4}(4x+7)$

30. $y = (2x-5)^{-1}(x^2-5x)^6$. $y' = (2x-5)^{-1}6(x^2-5x)^5(2x-5) - (2x-5)^{-2}(2)(x^2 - 5x)^6 = 6(x^2-5x)^5 - 2(2x-5)^{-2}(x^2-5x)^6 = 2(x^2-5x)^5[3-(2x-5)^{-2}(x^2-5x)]$

31. $y = (\frac{\sin x}{1+\cos x})^2$. $y' = 2(\frac{\sin x}{1+\cos x})\frac{(1+\cos x)\cos x - \sin x(-\sin x)}{(1+\cos x)^2} = \frac{2\sin x}{(1+\cos x)^3}(\cos x + \cos^2 x + \sin^2 x) = \frac{2\sin x}{(1+\cos x)^2}$

32. $y = (\frac{1+\cos x}{\sin x})^{-1} = \frac{\sin x}{1+\cos x}$. $y' = \frac{(1+\cos x)\cos x - \sin x(-\sin x)}{(1+\cos x)^2} = \frac{\cos x + 1}{(1+\cos x)^2} = \frac{1}{1+\cos x}$

33. $y = (\frac{x}{x-1})^{-3} = (\frac{x-1}{x})^3$. $y' = 3(\frac{x-1}{x})^2\frac{x-(x-1)}{x^2} = 3(\frac{x-1}{x})^2\frac{1}{x^2}$

34. $y = (\frac{x}{x-1})^2 - \frac{4}{x-1}$. $y' = 2(\frac{x}{x-1})\frac{x-1-x}{(x-1)^2} + \frac{4}{(x-1)^2} = (\frac{2x}{x-1})(\frac{-1}{(x-1)^2}) + \frac{4}{(x-1)^2} = \frac{2}{(x-1)^2}(2 - \frac{x}{x-1}) = \frac{2}{(x-1)^2}(\frac{2x-2-x}{x-1}) = \frac{2(x-2)}{(x-1)^3}$. $\frac{2(x-2)}{(x-1)^3}$ and NDER(y,x) have the same graph in $[-5,5]$ by $[-10,10]$ supports the result.

35. $y = \sin^3 x \tan 4x$. $y' = \sin^3 x(\sec^2 4x)4 + 3\sin^2 x \cos x \tan 4x = \sin^2 x\ (4\sin x \sec^2 4x + 3\cos x \tan 4x)$

36. $y = \cos^4 x \cot 7x$. $y' = \cos^4 x(-(\csc^2 7x)7) + 4\cos^3 x(-\sin x)\cot 7x = -\cos^3 x\ (7\cos x \csc^2 7x + 4\sin x \cot 7x)$

37. $y = \sqrt{\sin x}$. $y' = \frac{1}{2\sqrt{\sin x}}(\cos x) = \frac{\cos x}{2\sqrt{\sin x}}$. The result is supported by graphing $\frac{\cos x}{2\sqrt{\sin x}}$ and NDER(y,x) in $[-10,10]$ by $[-3,3]$ and seeing that the two graphs coincide.

38. $y = \sqrt{\cos x}$. $y' = \frac{1}{2\sqrt{\cos x}}(-\sin x) = -\frac{\sin x}{2\sqrt{\cos x}}$

39. $y = 4\sqrt{\sec x + \tan x}$. $y' = 4\frac{1}{2\sqrt{\sec x+\tan x}}(\sec x \tan x + \sec^2 x) = \frac{2\sec x(\tan x+\sec x)}{\sqrt{\sec x+\tan x}} = 2\sec x\sqrt{\sec x + \tan x}$. The result is supported by graphing the last function and NDER(y,x) in $[-10,10]$ by $[0,10]$ and seeing that the two graphs coincide.

40. $y = 2\sqrt{\csc x + \cot x}$. $y' = 2\frac{1}{2\sqrt{\csc x+\cot x}}(-\csc x \cot x - \csc^2 x) = -\frac{\csc x(\cot x+\csc x)}{\sqrt{\csc x+\cot x}} = -\csc x\sqrt{\csc x + \cot x}$

41. $y = \frac{3}{\sqrt{2x+1}} = 3(2x+1)^{-1/2}$. $y' = 3(-\frac{1}{2})(2x+1)^{-3/2}(2) = -\frac{3}{(2x+1)^{3/2}}$

42. $y = \frac{x}{\sqrt{1+x^2}}$. $y' = \frac{\sqrt{1+x^2} - x\left(\frac{x}{\sqrt{1+x^2}}\right)}{1+x^2} = \frac{1+x^2-x^2}{(1+x^2)^{3/2}} = \frac{1}{(1+x^2)^{3/2}}$

43. $y = (2x-6)\sqrt{x+5}$. $y' = (2x-6)\frac{1}{2\sqrt{x+5}} + 2\sqrt{x+5} = \frac{2x-6+4(x+5)}{2\sqrt{x+5}} = \frac{6x+14}{2\sqrt{x+5}} = \frac{3x+7}{\sqrt{x+5}}$

44. $y = x\sqrt{x^2-2x}$. $y' = x(\frac{1}{2\sqrt{x^2-2x}})(2x-2) + \sqrt{x^2-2x} = \frac{x(x-1)}{\sqrt{x^2-2x}} + \sqrt{x^2-2x} = \frac{x^2-x+x^2-2x}{\sqrt{x^2-2x}} = \frac{2x^2-3x}{\sqrt{x^2-2x}}$

45. $s = \cos(\frac{\pi}{2} - 3t) = \sin 3t$(complementary angles). $\frac{ds}{dt} = 3\cos 3t$ or $3\sin(\frac{\pi}{2} - 3t)$

46. $s = t\cos(\pi - 4t)$. $\frac{ds}{dt} = 4t\sin(\pi - 4t) + \cos(\pi - 4t)$

47. $s = \frac{4}{3\pi}\sin 3t + \frac{4}{5\pi}\cos 5t$. $\frac{ds}{dt} = \frac{4}{\pi}\cos 3t - \frac{4}{\pi}\sin 5t = \frac{4}{\pi}(\cos 3t - \sin 5t)$

48. $s = \sin(\frac{3\pi}{2}t) + \cos(\frac{7\pi}{4}t)$. $\frac{ds}{dt} = \frac{3\pi}{2}\cos(\frac{3\pi}{2}t) - \frac{7\pi}{4}\sin(\frac{7\pi}{4}t)$

49. $r = \tan(2-\theta)$. $\frac{dr}{d\theta} = [\sec^2(2-\theta)](-1) = -\sec^2(2-\theta)$

50. $r = \sec 2\theta \tan 2\theta$. $\frac{dr}{d\theta} = \sec 2\theta(2\sec^2 2\theta) + 2\sec 2\theta \tan^2 2\theta = 2\sec 2\theta(\sec^2 2\theta + \tan^2 2\theta)$

51. $r = \sqrt{\theta \sin\theta}$. $\frac{dr}{d\theta} = \frac{1}{2\sqrt{\theta\sin\theta}}(\theta\cos\theta + \sin\theta) = \frac{\theta\cos\theta+\sin\theta}{2\sqrt{\theta\sin\theta}}$

52. $r = 2\theta\sqrt{\sec\theta}$. $\frac{dr}{d\theta} = 2\theta\frac{1}{2\sqrt{\sec\theta}}(\sec\theta\tan\theta) + 2\sqrt{\sec\theta} = 2\frac{(\theta\sec\theta\tan\theta+2\sec\theta)}{2\sqrt{\sec\theta}} = \sqrt{\sec\theta}(\theta\tan\theta + 2)$

53. $y = \sin^2(3x-2)$. $y' = 2\sin(3x-2)\cos(3x-2)\cdot 3 = 6\sin(3x-2)\cos(3x-2)$

54. $y = \sec^2 5x$. $y' = 2\sec 5x \sec 5x(\tan 5x)5 = 10(\sec^2 5x)\tan 5x$

55. $y = (1+\cos 2x)^2$. $y' = 2(1+\cos 2x)(-2\sin 2x) = -4(\sin 2x)(1+\cos 2x)$

56. $y = (1-\tan(x/2))^{-2}$. $y' = -2(1-\tan(x/2))^{-3}(-\sec^2(x/2))(1/2) = \sec^2(x/2)(1-\tan(x/2))^{-3}$

57. $y = \sin(\cos(2x-5))$. $y' = \cos(\cos(2x-5))(-\sin(2x-5))2 = -2[\sin(2x-5)]\cos(\cos(2x-5))$

58. $y = (1+\cos^2 7x)^3$. $y' = 3(1+\cos^2 7x)^2(2\cos 7x)(-7\sin 7x) = -42(\sin 7x)(\cos 7x)(1+\cos^2 7x)^2$

59. $y = \cot\sqrt{2x}$. $y' = (-\csc^2\sqrt{2x})(\frac{1}{2\sqrt{2x}})2 = -\frac{\csc^2\sqrt{2x}}{\sqrt{2x}}$

60. $y = \sqrt{\tan 5x}$. $y' = (\frac{1}{2\sqrt{\tan 5x}})(\sec^2 5x)5 = \frac{5\sec^2 5x}{2\sqrt{\tan 5x}}$

61. $y = \tan x$. $y' = \sec^2 x$. $y'' = 2\sec x \sec x \tan x = 2(\sec^2 x)\tan x$

62. $y = \cot x$, $y' = -\csc^2 x$, $y'' = -2\csc x[-\csc x \cot x] = 2(\csc^2 x)\cot x$

63. $y = \cot(3x-1)$. $y' = -3\csc^2(3x-1)$, $y'' = -3(2)\csc(3x-1)[-\csc(3x-1)\cot(3x-1)(3)] = 18\csc^2(3x-1)\cot(3x-1)$

64. $y = 9\tan(x/3)$. $y' = 9(\sec^2(x/3))(1/3) = 3\sec^2(x/3)$. $y'' = 6\sec(x/3)\sec(x/3)(\tan(x/3))(x/3) = 2\sec^2(x/3)\tan(x/3)$

65. We have $f'(u) = 5u^4$ and $g'(x) = \frac{1}{2\sqrt{x}} \cdot (f\circ g)^1(1) = f'(g(1))g'(1) = 5(g(1))^4(\frac{1}{2}) = \frac{5}{2}$

66. $f'(u) = \frac{1}{u^2}$, $g'(x) = \frac{1}{(1-x)^2} \cdot (f\circ g)'(-1) = f'(g(-1))g'(-1) = f'(\frac{1}{2})\frac{1}{4} = 1$

67. $f(u) = \cot \frac{\pi u}{10}$, $f'(u) = -\frac{\pi}{10}\csc^2(\frac{\pi u}{10})$. $g(x) = 5\sqrt{x}$, $g'(x) = \frac{5}{2\sqrt{x}}$. $(f \circ g)'(1) = f'(g(1))g'(1) = f'(5)(\frac{5}{2}) = -\frac{5}{2}\frac{\pi}{10}\csc^2(\frac{5\pi}{10}) = -\frac{\pi}{4}$

68. $f(u) = u + \sec^2 u$, $f'(u) = 1 + 2\sec^2 u \tan u$, $g'(x) = \pi$. $(f \circ g)'(\frac{1}{4}) = f'(g(\frac{1}{4}))g'(\frac{1}{4}) = f'(\frac{\pi}{4})\pi = (1 + 2\sec^2\frac{\pi}{4}\tan\frac{\pi}{4})\pi = 5\pi$

69. $f(u) = \frac{2u}{u^2+1}$. $f'(u) = \frac{2[(u^2+1)\cdot 1 - u(2u)]}{(u^2+1)^2} = \frac{2(1-u^2)}{(u^2+1)^2} \cdot g(x) = 10x^2 + x + 1$, $g'(x) = 20x + 1$. $(f \circ g)'(0) = f'(g(0))g'(0) = f'(1)(1) = 0$

70. $f'(u) = 2(\frac{u-1}{u+1})\frac{(u+1)-(u-1)(1)}{(u+1)^2} = \frac{4(u-1)}{(u+1)^3}$, $g'(x) = -2x^{-3}$. $(f \circ g)'(-1) = f'(g(-1))g'(-1) = f'(0)(2) = -8$

71. a) $\frac{dy}{dx} = \frac{dy}{du}\frac{du}{dx} = -\sin u(6) = -6\sin(6x+2)$

b) $\frac{dy}{dx} = \frac{dy}{du}\frac{du}{dx} = -2\sin 2u(3) = -6\sin[2(3x+1)] = -6\sin(6x+2)$

72. a) $\frac{dy}{dx} = \frac{dy}{du}\frac{du}{dx} = \cos(u+1)(2x) = 2x\cos(x^2+1)$

b) $\frac{dy}{dx} = \frac{dy}{du}\frac{du}{dx} = (\cos u)2x = 2x\cos(x^2+1)$

73. a) $\frac{dy}{dx} = \frac{dy}{du}\frac{du}{dx} = \frac{1}{5}5 = 1$

b) $\frac{dy}{dx} = \frac{dy}{du}\frac{du}{dx} = -\frac{1}{u^2}[-\frac{1}{(x-1)^2}] = \frac{1}{u^2}u^2 = 1$

74. a) $\frac{dy}{dx} = \frac{dy}{du}\frac{du}{dx} = (\cos u)(2\cos 2x)(\sin 2x)] = 2[\cos(\sin 2x)]\cos 2x$.

b) $\frac{dy}{dx} = \frac{dy}{du}\frac{du}{dx} = [\cos(\sin u)\cos u]2 = 2[\cos(\sin 2x)]\cos 2x$

75. $\frac{ds}{dt} = \frac{ds}{d\theta}\frac{d\theta}{dt} = -\sin\theta(5) = -5\sin\theta$. When $\theta = 3\pi/2$, $\frac{ds}{dt} = -5\sin(3\pi/2) = 5$.

76. $\frac{dy}{dt} = (\frac{dy}{dx})_{x=1}(\frac{dx}{dt}) = (2x+7)_{x=1}(\frac{1}{3}) = 3$

77. Slope $= y' = \frac{1}{2}\cos(x/2)$. Since the largest value of the cosine function is 1, the largest value of the slope is 1/2.

78. $y = \sin mx$, $y' = m\cos mx$. At $x = 0$, $y' = m$. The tangent has equation $y - 0 = m(x - 0)$ or $y = mx$

79. $y = 2\tan(\pi x/4)$, $y' = 2\sec^2(\pi x/4)(\pi/4) = (\pi/2)\sec^2(\pi x/4)$. When $x = 1$, $y = 2$, $y' = (\pi/2)2 = \pi$. Tangent: $y - 2 = \pi(x-1)$. Normal: $y - 2 = -\frac{1}{\pi}(x-1)$.

80. $y = \sin 2x$ and $y = -\sin\frac{x}{2}$ intersect at the origin. For the first curve $y' = 2\cos 2x$ and at $x = 0$, $y' = 2$. For the second curve $y' = -\frac{1}{2}\cos\frac{x}{2}$ and at $x = 0$, $y' = -\frac{1}{2}$. Since the slopes of the tangents at the origin, 2 and

$-\frac{1}{2}$, are negative reciprocals of each other, they are perpendicular. Therefore the curves are orthogonal at the origin. Graph the functions in $[-6, 6]$ by $[-3.53, 3.53]$.

81. a) $\left[\frac{d}{dx}\{2f(x)\}\right]_{x=2} = [2f'(x)]_{x=2} = 2f'(2) = 2/3$

b) $f'(3) + g'(3) = 2\pi + 5$

c) $[f(x)g'(x) + f'(x)g(x)]_{x=3} = 3.5 + 2\pi(-4) = 15 - 8\pi$

d) $\left[\frac{g(x)f'(x) - f(x)g'(x)}{g^2(x)}\right]_{x=2} = \frac{2(1/3)-8(-3)}{2^2} = \frac{2/3+24}{4} = 1/6 + 6 = 37/6$

e) $f'(g(2))g'(2) = f'(2)(-3) = -1$

f) $\left[\frac{1}{2\sqrt{f(x)}}f'(x)\right]_{x=2} = \frac{1}{2\sqrt{8}}\left(\frac{1}{3}\right) = \frac{1}{12\sqrt{2}} = \frac{\sqrt{2}}{24}$

g) $\left[\frac{d}{dx}\{g^{-2}(x)\}\right]_{x=3} = [-2g^{-3}(x)g'(x)]_{x=3} = -2(-4)^{-3}5 = \frac{2(5)}{64} = \frac{5}{32}$

h) $\left[\frac{1}{2\sqrt{f^2(x)+g^2(x)}}2f(x)f'(x) + 2g(x)g'(x)\right]_{x=2} = \left[\frac{f(x)f'(x)+g(x)g'(x)}{\sqrt{f^2(x)+g^2(x)}}\right]_{x=2}$ $= \frac{8(1/3)+2(-3)}{\sqrt{8^2+2^2}} = \frac{-5\sqrt{17}}{51}$

82. a) $5f'(1) - g'(1) = -5/3 + 8/3 = 1$

b) $f(0)3g^2(0)g'(0) + f'(0)g^3(0) = (1)3(1)^2(1/3) + 5(1)^3 = 1 + 5 = 6$

c) $\frac{(g(1)+1)f'(1) - f(1)g'(1)}{(g(1)+1)^2} = \frac{(-3)(-1/3)-3(-8/3)}{(-3)^2} = \frac{1+8}{9} = 1$

d) $f'(g(0))g'(0) = f'(1)(1/3) = (-1/3)(1/3) = -1/9$

e) $g'(f(0))f'(0) = g'(1)5 = (-8/3)5 = -40/3$

f) $-2(g(1) + f(1))^{-3}(g'(1) + f'(1)) = -2(3-4)^{-3}(-\frac{8}{3} - \frac{1}{3}) = -6$

g) $f'(0 + g(0))(1 + g'(0)) = f'(1)(4/3) = (-1/3)(4/3) = -4/9$

83. $s = A\cos(2\pi bt)$, $V = s' = -2\pi bA\sin(2\pi bt)$ and $a = v' = -4\pi^2b^2A\cos(2\pi bt)$. Now let $s_1 = A\cos[2\pi(2b)t] = A\cos(4\pi bt)$. Then the new velocity and acceleration are given by $v_1 = -4\pi bA\sin(4\pi bt) = 2(-2\pi bA)\sin(4\pi bt)$ and $a_1 = -16\pi^2b^2A\cos(4\pi bt) = 4(-4\pi^2b^2A)\cos(4\pi bt)$. Thus the amplitude of v is doubled and the amplitude of a is quadrupled.

84. For simplicity we take $A = b = 1$ and $0 \leqq t \leqq 5$. Then $s = \cos(2\pi t)$, $v = -2\pi\sin(2\pi t)$. We simultaneously graph the three curves: $x = -1$, $y = \cos(2\pi t)$; $x = t$, $y = \cos(2\pi t)$; $x = t$, $y = -2\pi\sin(2\pi t)$; $0 \leqq t \leqq 5$, T step 0.05, $-2 \leqq x \leqq 5$, $-7 \leqq y \leqq 7$. Using TRACE, we can study the motion on each curve as t goes from 0 to 5.

85. a) Graph $y = 37\sin\left[\frac{2\pi}{365}(t-101)\right] + 25$ in $[0, 365]$ by $[-12, 62]$.

b) $y' = 37(\frac{2\pi}{365})\cos\left[\frac{2\pi}{365}(t-101)\right]$ is largest when $t - 101 = 0,\ t = 101$ or on April 12 (of a non-leap year)

c) When $t = 101,\ y' = \frac{37(2\pi)}{365} = .6369..^{\circ}F/\text{day}$.

86. a) Graph y in $[0, 365]$ by $[4, 21]$.

b) $y' = 8.5(\frac{2\pi}{365})\cos(\frac{2\pi(t-83)}{365})$ is largest when the cosine factor is 1, i.e., when $\frac{2\pi(t-83)}{365} = 0$ or $t = 83$ which corresponds to March 25 in a non-leap year.

c) $y'(83) = 8.5(\frac{2\pi}{365}) \approx 0.146$ hours per day

3.7 IMPLICIT DIFFERENTIATION AND FRACTIONAL POWERS

1. $y = x^{9/4}$. $y' = (9/4)x^{(9/4)-1} = (9/4)x^{5/4}$

2. $y = x^{-3/5}$. $y' = -(3/5)x^{-8/5}$

3. $y = \sqrt[3]{x} = x^{1/3}$. $y' = (1/3)x^{-2/3} = \frac{1}{3\sqrt[3]{x^2}}$

4. $y = \sqrt[4]{x} = x^{1/4}$. $y' = (1/4)x^{-3/4} = \frac{1}{4\sqrt[4]{x^3}}$

5. $y = (2x+5)^{-1/2}$. $y' = -(1/2)(2x+5)^{-3/2}(2) = -(2x+5)^{-3/2}$

6. $y = (1-6x)^{2/3}$. $y' = (2/3)(1-6x)^{-1/3}(-6) = -4(1-6x)^{-1/3}$

7. $y = x\sqrt{x^2+1} = x(x^2+1)^{1/2}$. $y' = x(1/2)(x^2+1)^{-1/2}(2x) + 1\cdot(x^2+1)^{1/2} = \frac{x^2}{\sqrt{x^2+1}} + \sqrt{x^2+1} = \frac{x^2+x^2+1}{\sqrt{x^2+1}} = \frac{2x^2+1}{\sqrt{x^2+1}}$

8. $y = \frac{x}{\sqrt{x^2+1}}$. $y' = \frac{\sqrt{x^2+1}(1) - x\cdot\frac{x}{\sqrt{x^2+1}}}{x^2+1} = \frac{x^2+1-x^2}{(x^2+1)^{3/2}} = \frac{1}{(x^2+1)^{3/2}}$

9. $x^2y + xy^2 = 6$. $\frac{d}{dx}(x^2y) + \frac{d}{dx}(xy^2) = \frac{d}{dx}(6)$. $(x^2\frac{dy}{dx} + 2xy) + (x2y\frac{dy}{dx} + 1\cdot y^2) = 0$. $x^2y' + 2xyy' + 2xy + y^2 = 0$. $y'(x^2+2xy) = -2xy - y^2$. $y' = -\frac{2xy+y^2}{x^2+2xy}$

10. $x^3 + y^3 = 18xy$. $3x^2 + 3y^2y' = 18xy' + 18y$. $y'(3y^2 - 18x) = 18y - 3x^2$, $y' = \frac{6y-x^2}{y^2-6x}$

11. $2xy + y^2 = x + y$. $2xy' + 2y + 2yy' = 1 + y'$, $y'(2x + 2y - 1) = 1 - 2y$, $y' = (1-2y)/(2x+2y-1)$

12. $x^3 - xy + y^3 = 1$. $3x^2 - xy' - y + 3y^2y' = 0$, $y'(3y^2 - x) = y - 3x^2$, $y' = (y - 3x^2)/(3y^2 - x)$

13. $x^2y^2 = x^2 + y^2$. $x^2 2yy' + 2xy^2 = 2x + 2yy'$, $x^2yy' + xy^2 = x + yy'$, $y'(x^2y - y) = x - xy^2$, $y' = x(1 - y^2)/[y(x^2 - 1)]$

14. $(3x + 7)^2 = 2y^3$. $2(3x + 7)3 = 6y^2y'$, $y^2y' = 3x + 7$, $y' = \frac{3x+7}{y^2}$

15. $y^2 = \frac{x-1}{x+1}$. $2yy' = \frac{(x+1)\cdot 1-(x-1)\cdot 1}{(x+1)^2} = \frac{2}{(x+1)^2}$, $y' = \frac{1}{y(x+1)^2}$

16. $x^2 = \frac{x-y}{x+y}$. $x^3 + x^2y = x - y$, $3x^2 + x^2y' + 2xy = 1 - y'$, $(x^2 + 1)y' = 1 - 3x^2 - 2xy$, $y' = (1 - 3x^2 - 2xy)/(x^2 + 1)$. Immediate differentiation yields $y' = \frac{y}{x} - (x + y)^2$. First solving for y explicitly, we get $y' = (1 - 4x^2 - x^4)/(x^2 + 1)^2$.

17. $y = \sqrt{1 - \sqrt{x}}$. $y' = \frac{1}{2\sqrt{1-\sqrt{x}}}(-\frac{1}{2\sqrt{x}}) = \frac{-1}{4\sqrt{x}\sqrt{1-\sqrt{x}}}$

18. $y = 3(2x^{-1/2} + 1)^{-1/3}$. $y' = 3(-1/3)(2x^{-1/2} + 1)^{-4/3}(-x^{-3/2}) = x^{-3/2}(2x^{-1/2} + 1)^{-4/3}$

19. $y = 3(\csc x)^{3/2}$. $y' = (9/2)(\csc x)^{1/2}(-\csc x \cot x) = -(9/2)\csc^{3/2} x \cot x$

20. $y = [\sin(x + 5)]^{5/4}$. $y' = (5/4)[\sin(x + 5)]^{1/4}\cos(x + 5)$

21. $x = \tan y$. $1 = (\sec^2 y)y'$. $y' = 1/\sec^2 y = \cos^2 y$

22. $x = \sin y$. $1 = (\cos y)y'$. $y' = \sec y$

23. $x + \tan(xy) = 0$. $1 + \sec^2(xy)[xy' + y] = 0$, $xy'\sec^2(xy) + y\sec^2(xy) = -1$, $y' = \frac{-[1+y\sec^2 xy]}{x\sec^2(xy)} = -\frac{[\cos^2(xy)+y]}{x}$

24. $x + \sin y = xy$, $1 + y'\cos y = xy' + y$, $y'(\cos y - x) = y - 1$, $y' = (y - 1)/(\cos y - x)$

25. $y\sin(\frac{1}{y}) = 1 - xy$. $y\cos(\frac{1}{y})(-\frac{1}{y^2})y' + y'\sin(\frac{1}{y}) = -y - xy'$, $y'[-\frac{1}{y}\cos(\frac{1}{y}) + \sin(\frac{1}{y}) + x] = -y$, $y' = y/[\frac{1}{y}\cos(\frac{1}{y}) - \sin(\frac{1}{y}) - x]$

26. $y^2\cos(\frac{1}{y}) = 2x + 2y$. $y^2[-\sin(\frac{1}{y})(-\frac{1}{y^2})y'] + 2yy'\cos(\frac{1}{y}) = 2 + 2y'$, $y'[\sin(\frac{1}{y}) + 2y\cos(\frac{1}{y}) - 2] = 2$, $y' = 2/[\sin(\frac{1}{y}) + 2y\cos(\frac{1}{y}) - 2]$

27. a) $f(x) = \frac{3}{2}x^{2/3} - 3$. $f'(x) = x^{-1/3}$, $f''(x) = -\frac{1}{3}x^{-2/3}$ so a) is not true. b) $f(x) = \frac{9}{10}x^{5/3} - 7$ leads to $f'(x) = \frac{3}{2}x^{2/3}$, $f''(x) = x^{-1/3}$ so b) could be true. c) Since $f''(x) = x^{-1/3}$ leads to $f'''(x) = -\frac{1}{3}x^{-4/3}$, c) is true. d) $f'(x) = \frac{3}{2}x^{2/3} + 6$ leads to $f''(x) = x^{-1/3}$, d) could be true. Answer: b), c), d).

28. a) $g'(t) = 4t^{1/4} - 4 \Rightarrow g''(t) = t^{-3/4}$ so this is possible. b) $g''(t) = t^{-3/4} \Rightarrow g'''(t) = -\frac{3}{4}t^{-7/4}$ so this can't be true. c) $g(t) = t - 7 + \frac{16}{5}t^{5/4} \Rightarrow g'(t) = 1 - 4t^{1/4} \Rightarrow g''(t) = -t^{3/4}$ so this is possible.

29. $x^2 + y^2 = 1$. $2x + 2yy' = 0$, $y' = -x/y$. $y'' = -\frac{y\cdot 1 - xy'}{y^2} = -\frac{y - x(-x/y)}{y^2} = -\frac{y^2+x^2}{y^3} = -\frac{1}{y^3}$

30. $x^{2/3} + y^{2/3} = 1$. $(2/3)x^{-1/3} + (2/3)y^{-1/3}y' = 0$, $y' = -x^{-1/3}/y^{-1/3} = -x^{-1/3}y^{1/3}$. $y'' = -[x^{-1/3}(1/3)y^{-2/3}y' - (1/3)x^{-4/3}y^{1/3}] = -(1/3)\cdot [x^{-1/3}y^{-2/3}(-x^{-1/3}y^{1/3}) - x^{-4/3}y^{1/3}] = (1/3)[x^{-2/3}y^{-1/3} + x^{-4/3}y^{1/3}]$

31. $y^2 = x^2 + 2x$. $2yy' = 2x+2$, $y' = (x+1)/y$. $y'' = \frac{y\cdot 1 - (x+1)y'}{y^2} = \frac{y - (x+1)(x+1)/y}{y^2} = \frac{y^2 - (x+1)^2}{y^3}$

32. $y^2 + 2y = 2x + 1$. $2yy' + 2y' = 2$, $yy' + y' = 1$, $y'(y+1) = 1$, $y' = 1/(y+1)$. $y'' = -\frac{1}{(y+1)^2}y' = -\frac{1}{(y+1)^3}$

33. $y + 2\sqrt{y} = x$. $y' + 2(\frac{1}{2\sqrt{y}})y' = 1$, $y'(1 + \frac{1}{\sqrt{y}}) = 1$, $y'(\frac{\sqrt{y}+1}{\sqrt{y}}) = 1$, $y' = \frac{\sqrt{y}}{\sqrt{y}+1}$. $y'' = \frac{(\sqrt{y}+1)(1/2\sqrt{y})y' - \sqrt{y}(1/(2\sqrt{y}))y'}{(\sqrt{y}+1)^2} = \frac{y'/(2\sqrt{y})}{(\sqrt{y}+1)^2} = \frac{(1/2)}{(\sqrt{y}+1)^3} = \frac{1}{2(\sqrt{y}+1)^3}$

34. $xy + y^2 = 1$. $xy' + y + 2yy' = 0$, $y'(x + 2y) = -y$, $y' = -\frac{y}{x+2y}$. $y'' = -\frac{(x+2y)y' - y(1+2y')}{(x+2y)^2} = -\frac{xy' - y}{(x+2y)^2} = -\frac{x[-y/(x+2y)] - y}{(x+2y)^2} = -\frac{-xy - y(x+2y)}{(x+2y)^3} = \frac{xy + xy + 2y^2}{(x+2y)^3} = \frac{2y(x+y)}{(x+2y)^3}$

35. $x^2 + xy - y^2 = 1$ at $(2,3)$. $2x + xy' + y - 2yy' = 0$, $y'(x - 2y) = -(2x+y)$, $y' = \frac{(2x+y)}{2y-x}$. At $(2,3)$, $y' = \frac{4+3}{6-2} = 7/4$. Tangent: $y - 3 = \frac{7}{4}(x-2)$ or $7x - 4y = 2$. Normal: $y - 3 = -\frac{4}{7}(x-2)$ or $4x + 7y = 29$.

36. $x^2 + y^2 = 25$ at $(3,-4)$. $2x + 2yy' = 0$, $y' = -x/y$. a) Tangent: $y + 4 = [-3/(-4)](x-3) = (3/4)(x-3)$ or $3x - 4y = 25$. b) Normal: $y + 4 = (-4/3)(x-3)$ or $4x + 3y = 0$.

37. $x^2y^2 = 9$ at $(-1,3)$. $2xy^2 + x^2(2yy') = 0$, $x^2yy' = -xy^2$, $y' = -\frac{y}{x}$ (= 3 at $(-1,3)$). a) Tangent: $y - 3 = 3(x+1)$ or $y - 3x = 6$. b) Normal: $y - 3 = (-1/3)(x+1)$ or $x + 3y = 8$.

38. $y^2 - 2x - 4y - 1 = 0$ at $(-2,1)$. $2yy' - 2 - 4y' = 0$. $yy' - 1 - 2y' = 0$, $y'(y-2) = 1$, $y' = 1/(y-2)$ (= -1 at $(-2,1)$). a) Tangent: $y - 1 = -(x+2)$ or $y = -x - 1$. b) Normal: $y - 1 = (1)(x+2)$ or $y = x + 3$

39. $6x^2+3xy+2y^2+17y-6=0$ at $(-1,0)$. $12x+3xy'+3y+4yy'+17y'=0$, $y'(3x+4y+17)=-12x-3y$, $y'=\frac{-3(4x+y)}{3x+4y+17}$ $(=\frac{12}{14}=\frac{6}{7}$ at $(-1,0))$. a) Tangent: $y=\frac{6}{7}(x+1)$ b) Normal: $y=-\frac{7}{6}(x+1)$

40. $x^2-\sqrt{3}xy+2y^2=5$ at $(\sqrt{3},2)$. $2x-\sqrt{3}xy'-\sqrt{3}y+4yy'=0$, $y'[-\sqrt{3}x+4y]=-2x+\sqrt{3}y$, $y'=\frac{2x-\sqrt{3}y}{\sqrt{3}x-4y}$ $(=0$ at $(\sqrt{3},2))$. a) Tangent: $y=2$ b) Normal: $x=\sqrt{3}$

41. $2xy+\pi\sin y=2\pi$ at $(1,\pi/2)$. $2xy'+2y+\pi(\cos y)y'=0$, $y'(2x+\pi\cos y)=-2y$, $y'=\frac{-2y}{2x+\pi\cos y}$ $(=\frac{-\pi}{2+0}=-\frac{\pi}{2}$ at $(1,\frac{\pi}{2}))$. a) Tangent: $y-\frac{\pi}{2}=-\frac{\pi}{2}(x-1)$ or $y=-\frac{\pi}{2}x+\pi$ b) Normal: $y-\frac{\pi}{2}=\frac{2}{\pi}(x-1)$ or $y=\frac{2}{\pi}x+\frac{\pi^2-4}{2\pi}$

42. $x\sin 2y=y\cos 2x$ at $(\frac{\pi}{4},\frac{\pi}{2})$. $x(2\cos 2y)y'+\sin 2y=y(-2\sin 2x)+y'\cos 2x$, $y'(2x\cos 2y-\cos 2x)=-\sin 2y-2y\sin 2x$, $y'=\frac{\sin 2y+2y\sin 2x}{\cos 2x-2x\cos 2y}$ $(=\frac{0+\pi}{0-(\pi/2)(-1)}=2$ at $(\frac{\pi}{4},\frac{\pi}{2}))$. a) Tangent: $y-\frac{\pi}{2}=2(x-\frac{\pi}{4})$ or $y=2x$ b) Normal: $y-\frac{\pi}{2}=-\frac{1}{2}(x-\frac{\pi}{2})$ or $y=-\frac{1}{2}x+\frac{5\pi}{8}$

43. $y=2\sin(\pi x-y)$ at $(1,0)$. $y'=(2\cos(\pi x-y))(\pi-y')$, $y'[1+2\cos(\pi x-y)]=2\pi\cos(\pi x-y)$, $y'=\frac{2\pi\cos(\pi x-y)}{1+2\cos(\pi x-y)}$ $(=\frac{-2\pi}{-1}=2\pi$ at $(1,0))$. a) Tangent: $y=2\pi(x-1)$ b) Normal: $y=-\frac{1}{2\pi}(x-1)$

44. $x^2\cos^2 y-\sin y=0$ at $(0,\pi)$. $x^2(2\cos y)(-\sin y)y'+2x\cos^2 y-(\cos y)y'=0$, $-y'(2x^2\cos y\sin y+\cos y)=-2x\cos^2 y$, $y'=\frac{2x\cos^2 y}{2x^2\cos y\sin y+\cos y}$ $(=0$ at $(0,\pi))$. a) Tangent: $y=\pi$ b) Normal: $x=0$

45. $2xy+\pi\sin y=2\pi$. $2xy'+2y+\pi(\cos y)y'=0$, $y'(2x+\pi\cos y)=-2y$, $y'=-2y/(2x+\pi\cos y)$. At $(1,\pi/2)$, $y'=-\pi/(2+\pi\cdot 0)=-\pi/2$.

46. $x\sin 2y=y\cos 2x$ leads to $x(\cos 2y)2y'+\sin 2y=y(-2\sin 2x)+y'\cos 2x$, $y'(2x\cos 2y-\cos 2x)=-2y\sin 2x-\sin 2y$, $y'=\frac{2y\sin 2x+\sin 2y}{\cos 2x-2x\cos 2y}$. At $(\pi/4,\pi/2)$, $y'=\frac{\pi\cdot 1+0}{0-(\pi/2)(-1)}=\frac{\pi}{\pi/2}=2$. The tangent has equation $y-(\pi/2)=2(x-\pi/4)$ or $y=2x$.

47. a) $y^4=y^2-x^2$ leads to $4y^3y'=2yy'-2x$, $y'(4y^3-2y)=-2x$, $y'=\frac{2x}{2y-4y^3}=\frac{x}{y(1-2y^2)}$. At $(\frac{\sqrt{3}}{4},\frac{\sqrt{3}}{2})$, the slope is $\frac{\sqrt{3}/4}{(\sqrt{3}/2)(1-3/2)}=-1$ and at $(\frac{\sqrt{3}}{4},\frac{1}{2})$ the slope is $\frac{\sqrt{3}/4}{(1/2)(1-1/2)}=\sqrt{3}$ b) Graph $x_1=\sqrt{t^2-t^4}$, $y_1=t$ and $x_2=-x_1$, $y_2=t$, $-1\leqq t\leqq 1$ in $[-0.5,0.5]$ by $[-1,1]$.

48. a) $y^2(2-x)=x^3$ leads to $y^2(-1)+2yy'(2-x)=3x^2$, $2yy'(2-x)=3x^2+y^2$, $y'=\frac{3x^2+y^2}{2y}$. At $(1,1)$ the slope is $\frac{3+1}{2}=2$. The tangent at $(1,1)$

has equation $y - 1 = 2(x - 1)$ or $y = 2x - 1$. The normal at $(1, 1)$ has equation $y - 1 = (-1/2)(x - 1)$ or $x + 2y = 3$. b) Setting $y = tx$, we obtain $t^2x^2(2 - x) = x^3$, $t^2(2 - x) = x$ which leads to $x = 2t^2/(1 + t^2)$, $y = 2t^3/(1 + t^2)$. We graph this curve along with $x = t$, $y = 2t - 1$ and $x = 3 - 2t$, $y = t$, $-2 \leqq t \leqq 2$, Tstep $= 0.05$ in rectangle $[-1, 8]$ by $[-3, 3]$, the ratio of the horizontal length of the rectangle to the vertical length is kept at 3 to 2 to preserve the appearance of perpendicularity. Of course, $y = \pm\sqrt{x^3/(2 - x)}$ may also be used.

49. $x^3y^2 = \cos(\pi y)$. a) Substituting $x = -1$, $y = 1$, we get $(-1)^3(1)^2 = \cos\pi$, a true statement. b) Differentiating both sides with respect to x, we obtain $x^3(2yy') + 3x^2y^2 = -\sin(\pi y)(\pi y')$. This leads to $y'[2x^3y + \pi\sin(\pi y)] = -3x^2y^2$, $y' = \frac{-3x^2y^2}{2x^3y + \pi\sin(\pi y)}$. At $(-1, 1)$, $y' = \frac{-3}{-2+0} = \frac{3}{2}$.

50. a) When $x = 2$, $y^3 - 2y + 1 = 0$. By inspection $y = 1$ is a solution. To find the other solutions use graphing techniques or long division: $y^3 - 2y + 1 = (y - 1)(y^2 + y - 1) = 0$. The quadratic has roots $\frac{1\pm\sqrt{5}}{2}$. Thus the relation cannot be a function because it contains the points $(2, 1)$ and $(2, \frac{1\pm\sqrt{5}}{2})$. b) Differentiating implicity with respect to x, we obtain $3y^2y' - xy' - y = 0$, $y' = \frac{y}{3y^2 - x}$. At $(2, 1)$, $y'(2) = f'(2) = \frac{1}{3-2} = 1$. $y'' = \frac{(3y^2-x)y' - y(6yy'-1)}{(3y^2-x)^2}$ and $f''(2) = y''(2) = \frac{1-(6-1)}{1} = -4$.

51. Differentiating $x^2 + xy + y^2 = 7$, we get $2x + xy' + y + 2yy' = 0$, $y'(x + 2y) = -(2x + y)$, $y' = -(2x + y)/(x + 2y)$. For the x-intercepts we set $y = 0$: $x^2 + x(0) + 0^2 = 7$ and obtain $x = \pm\sqrt{7}$. For these points $y' = -[2(\pm\sqrt{7}) + 0]/[\pm\sqrt{7} + 2(0)] = -2$. This shows that the two tangents are parallel with -2 as their common slope.

52. Differentiating $x^2 + xy + y^2 = 7$, we get as in the preceding exercise $y' = -(2x + y)/(x + 2y)$. Similarly, $\frac{dx}{dy} = -\frac{2y+x}{y+2x}$. a) The tangent is parallel to the x-axis when $y' = 0$, i.e., when $2x + y = 0$ or $y = -2x$. Substituting this into the original equation, we obtain $x^2 - 2x^2 + 4x^2 = 7$, $3x^2 = 7$, $x = \pm\sqrt{7/3}$. Thus the points on the curve where the tangent is horizontal are $(\sqrt{7/3}, -2\sqrt{7/3})$ and $(-\sqrt{7/3}, 2\sqrt{7/3})$. b) The tangent is parallel to the y-axis when $\frac{dx}{dy} = 0$, i.e., when $x = -2y$. This yields $y = \pm\sqrt{7/3}$ and the point $(-2\sqrt{7/3}, \sqrt{7/3})$ and $(2\sqrt{7/3}, -\sqrt{7/3})$.

53. Let C_1 be the curve $2x^2 + 3y^2 = 5$ and let C_2 be the curve $y^2 = x^3$. $4x + 6yy' = 0$ leads to $y' = -\frac{2x}{3y}$ for the slope of C_1. $y^2 = x^3$ leads to $2yy' = 3x^2$, $y' = \frac{3x^2}{2y}$

for the slope of C_2. At $(1,1)$ the slopes are $-\frac{2}{3}$ and $\frac{3}{2}$ (negative reciprocals) so C_1 and C_2 are orthogonal at $(1,1)$. At $(1,-1)$ the slopes are $\frac{2}{3}$ and $-\frac{3}{2}$ so we reach the same conclusion at $(1,-1)$. Graph $y_1 = \sqrt{(5-2x^2)/3}$, $y_2 = -y_1$, $y_3 = x^{1.5}$, $y_4 = -y_3$, $y_5 = -(2/3)x + 5/3$, $y_6 = (3/2)x - 1/2$, $y_7 = (2/3)x - 5/3$, $y_8 = -(3/2)x + 1/2$ in $[-6,6]$ by $[-3.5, 3.5]$.

54. $s = \sqrt{1+4t}$. $v = \frac{ds}{dt} = \frac{1}{2\sqrt{1+4t}}(4) = \frac{2}{\sqrt{1+4t}} = 2(1+4t)^{-1/2}$. $a = \frac{dv}{dt} = 2[(-1/2)(1+4t)^{-3/2}(4)] = \frac{-4}{(1+4t)^{3/2}}$. $v(6) = \frac{2}{\sqrt{25}} = \frac{2}{5}$meters/sec. $a(6) = \frac{-4}{25^{3/2}} = \frac{-4}{125}$m/sec^2.

55. $a = \frac{dv}{dt} = \frac{k}{2\sqrt{s}}\frac{ds}{dt} = \frac{k}{2\sqrt{s}}v = \frac{k}{2\sqrt{s}}(k\sqrt{s}) = \frac{k^2}{2}$

56. To graph $2y = x^2 + \sin y$ we may use $x = \pm\sqrt{2t - \sin t}$, $y = t$, $0 \leqq t \leqq 10$. We see that this is the graph of a function by the vertical line test. We may obtain a parametrization of the derivative by substituting the above values directly into (from Example 2) $y' = 2x/(2-\cos y)$ or by using $\frac{dy}{dx} = \frac{dy}{dt}/(\frac{dx}{dt})$. Either way we obtain $x = \pm\sqrt{2t-\sin t}$, $y = \pm 2\sqrt{2t - \sin t}/(2-\cos t)$. Allow larger values of t to develop an idea of the complete graph. You may later prove the relation is a function by showing that $2y - \sin y$ is an increasing function of y (its derivative, $2 - \cos y$, is positive) and so $2y - \sin y = x^2$ can only relate one y to a given x. There is a solution for any x so the domain is $(-\infty, \infty)$.

57. a) $-1 \leqq -\sin xy \leqq 1$ so $-1 \leqq y^5 \leqq 1$. This implies $-1 \leqq y \leqq 1$. b) $\sin xy = -y^5$ leads to $xy = \sin^{-1}(-y^5) + 2k\pi$ or $xy = \pi - \sin^{-1}(-y^5) + 2k\pi = (2k+1)\pi + \sin^{-1}(y^5)$. c) When we graph $x(t) = t^{-1}\sin^{-1}(-t^5)$, $y(t) = t$, $-1 \leqq t \leqq 1$, we see that $-\frac{\pi}{2} \leqq x < 0$ so the domain is $[-\frac{\pi}{2}, 0)$. The range of the graphed relation is $[-1,0)\cup(0,1]$. d) Graph $x(t) = \frac{\sin^{-1}(-t^5)+2\pi}{t}$, $y(t) = t$, $-1 \leqq t \leqq 1$ in $[-50, 50]$ by $[-2, 2]$. The domain (set of x-values) is $(-\infty, \frac{-5\pi}{2}] \cup [\frac{3\pi}{2}, \infty)$. The range is $[-1,0) \cup (0,1]$.

58. a) When $y = -1/2$, $x = \frac{\sin^{-1}(-(-1/2)^5)}{-1/2} = -2\sin^{-1}(1/32) = -0.06251\ldots$ b) We use $y^5 + \sin xy = 0$ and differentiate implicitly with respect to x. $5y^4y' + (\cos xy)(xy'+y) = 0$, $y'(5y^4 + x\cos xy) = -y\cos xy$, $\frac{dy}{dx} = y' = \frac{-y\cos xy}{5y^4 + x\cos xy}$. At the point $(-0.06251, -0.5)$, this gives $\frac{dy}{dx} = 1.9988\ldots$ c) The tangent line has approximate equation $y = -0.5 + 2(x + 0.06251)$, $y = -0.5 + 2(x+0.06251)$ which can be graphed in $[-2,2]$ by $[-4,4]$.

59. $\frac{dy}{dx} = \frac{y-2x}{2y-x}$ does not exist when $2y - x = 0$ or $x = 2y$. Substituting this into $x^2 - xy + y^2 = 7$, we obtain $4y^2 - 2y^2 + y^2 = 3y^2 = 7$, $y = \pm\sqrt{\frac{7}{3}}$. Since

$x = 2y$, the points we seek are $(-2\sqrt{\frac{7}{3}}, -\sqrt{\frac{7}{3}})$ and $(2\sqrt{\frac{7}{3}}, \sqrt{\frac{7}{3}})$.

60. Graph $y = x^{1/3}$ in $[-2, 2]$ by $[-2, 2]$; domain = range = $(-\infty, \infty)$. Our calculator will not graph $\frac{1}{3}x^{-2/3}$ for negative x. Instead we graph $y' = \frac{1}{3}(absx)^{-2/3}$ in $[-3, 3]$ by $[0, 2]$; domain: $(-\infty, 0) \cup (0, \infty)$, range: $(0, \infty)$.

3.8 LINEAR APPROXIMATIONS AND DIFFERENTIALS

1. $f(x) = x^4$, $a = 1$. $f'(x) = 4x^3$, $f'(1) = 4$. $L(x) = f(a) + f'(a)(x - a) = 1 + 4(x - 1) = 4x - 3$.

2. $f(x) = x^{-1}$, $a = 2$. $L(x) = f(a) + f'(a)(x - a) = \frac{1}{2} - \frac{1}{4}(x - 2) = 1 - \frac{x}{4}$.

3. $f(x) = x^3 - x$, $a = 1$. $f'(x) = 3x^2 - 1$. $L(x) = f(a) + f'(a)(x - a) = 0 + 2(x - 1) = 2(x - 1)$.

4. $f(x) = x^3 - 2x + 3$, $a = 2$. $f(2) = 8 - 4 + 3 = 7$. $f'(x) = 3x^2 - 2$, $f'(2) = 10$. $L(x) = f(a) + f'(a)(x - a) = 7 + 10(x - 2) = 10x - 13$.

5. $f(x) = \sqrt{x}$, $x = 4$. $f(4) = 2$, $f'(x) = \frac{1}{2\sqrt{x}}$, $f'(4) = \frac{1}{4}$. $L(x) = f(a) + f'(a)(x - a) = 2 + \frac{1}{4}(x - 4) = \frac{x}{4} + 1$.

6. $f(x) = \sqrt{x^2 + 9}$ at $x = -4$. $f(-4) = 5$, $f'(x) = \frac{x}{\sqrt{x^2+9}}$, $f'(-4) = \frac{-4}{5}$. $L(x) = f(a) + f'(a)(x - a) = 5 - \frac{4}{5}(x + 4) = -\frac{4x}{5} + \frac{9}{5} = \frac{(9-4x)}{5}$.

7. $a = 0$. $f(x) = x^2 + 2x$, $f'(x) = 2x + 2$, $f(0) = 0$, $f'(0) = 2$. $L(x) = f(a) + f'(a)(x - a) = 0 + 2(x - 0) = 2x$.

8. $a = 1$. $f(x) = \frac{1}{x}$, $f'(x) = -\frac{1}{x^2}$. $f(1) = 1$, $f'(1) = -1$. $L(x) = f(a) + f'(a)(x - a) = 1 - (x - 1) = 2 - x$.

9. $a = -1$. $f(x) = 2x^2 + 4x - 3$, $f'(x) = 4x + 4$, $f(-1) = -5$, $f'(-1) = 0$. $L(x) = f(-1) + f'(-1)(x + 1) = -5$.

10. $a = 8$. $f(x) = 1 + x$, $f'(x) = 1$. $L(x) = f(a) + f'(a)(x - a) = 9 + (x - 8) = 1 + x$.

11. $f(x) = \sqrt[3]{x}$, $a = 8$. $f'(x) = \frac{1}{3}x^{-2/3} = \frac{1}{3(\sqrt[3]{x})^2}$. $f(8) = 2$, $f'(8) = \frac{1}{3(2)^2} = \frac{1}{12}$. $L(x) = f(a) + f'(a)(x - a) = 2 + \frac{1}{12}(x - 8)$.

12. $a = 1$. $f(x) = \frac{x}{x+1}$, $f'(x) = \frac{(x+1)\cdot 1 - x \cdot 1}{(x+1)^2} = \frac{1}{(x+1)^2}$. $f(1) = 1/2$, $f'(1) = 1/4$. $L(x) = (1/2) + (1/4)(x - 1) = (x + 1)/4$.

13. $f(x) = \sin x,\ a = 0.\ f'(x) = \cos x.\ f(0) = 0,\ f'(0) = 1.\ L(x) = 0 + (1)(x - 0) = x.$

14. $f(x) = \cos x,\ a = 0,\ f'(x) = -\sin x.\ f(0) = 1,\ f'(0) = 0.\ L(x) = 1 + 0(x - 0) = 1.$

15. $f(x) = \sin x,\ a = \pi.\ f'(x) = \cos x.\ f(\pi) = 0,\ f'(\pi) = -1.\ L(x) = f(a) + f'(a)(x - a) = 0 - (x - \pi) = \pi - x.$

16. $f(x) = \cos x,\ a = -\pi/2.\ f'(x) = -\sin x.\ f(-\pi/2) = 0,\ f'(-\pi/2) = 1.\ L(x) = f(a) + f'(a)(x - a) = 0 + (1)(x + \pi/2) = x + \pi/2.$

17. $f(x) = \tan x,\ a = \pi/4.\ f'(x) = \sec^2 x,\ f(\pi/4) = 1,\ f'(\pi/4) = 2.\ L(x) = f(a) + f'(a)(x - a) = 1 + 2(x - \pi/4).$

18. $f(x) = \sec x,\ a = \pi/4.\ f'(x) = \sec x \tan x,\ f(\pi/4) = \sqrt{2},\ f'(\pi/4) = \sqrt{2}.\ L(x) = f(a) + f'(a)(x - a) = \sqrt{2} + \sqrt{2}(x - \pi/4) = \sqrt{2}[1 + (x - \pi/4)].$

19. a) $(1 + x)^2 \approx 1 + 2x$ b) $(1 + x)^{-5} \approx 1 - 5x$ c) $2(1 + (-x))^{-1} \approx 2[1 + (-1)(-x)] = 2(1+x)$ d) $(1+x)^6 \approx 1+6x$ and replacing x by $-x$, $(1-x)^6 \approx 1 - 6x$ e) $3(1 + x)^{1/3} \approx 3[1 + (1/3)x] = 3 + x$ f) $(1 + x)^{-1/2} \approx 1 - (1/2)x$

20. a) $(1.002)^{100} = (1+0.002)^{100} \approx 1+100(0.002) = 1.2$. Calculator: $(1.002)^{100} \approx 1.22$

b) $\sqrt[3]{1.009} = (1+0.009)^{1/3} \approx 1+(1/3)(0.009) = 1.003$. Calculator: $\sqrt[3]{1.009} \approx 1.00299$.

21. $f(x) = \sqrt{x+1} + \sin x.\ f(0) = \sqrt{0+1} + \sin 0 = 1+0 = 1.\ f'(x) = (\sqrt{x+1})' + (\sin x)' = \frac{1}{2\sqrt{x+1}} + \cos x.\ f'(0) = \frac{1}{2} + 1 = \frac{3}{2}.\ L(x) = 1 + \frac{3}{2}x$ is the sum of the linearizations of $\sqrt{x+1}$ and $\sin x$.

22. $f(x) = (1 + x)^k,\ f'(x) = k(1 + x)^{k-1}.\ f(0) = 1,\ f'(0) = k.\ L(x) = f(a) + f'(a)(x - a) = 1 + kx.$

23. b) From Example 1, $\sqrt{1+x} \approx 1 + \frac{x}{2}$. Thus when x represents the decimal part of $1 + x$ and the square root of $1 + x$ is taken, the result is nearly the same as halving the decimal part.

c) $\sqrt{2} = \sqrt{1+1} \approx 1 + \frac{1}{2}$ using the linearization. Then $(2^{\frac{1}{2}})^{\frac{1}{2}} \approx (1+\frac{1}{2})^{\frac{1}{2}} \approx 1+\frac{1}{4}$ or $1 + \frac{1}{2^2}$ applying the linearization to $\sqrt{1+\frac{1}{2}}$. Repeating the process, we arrive at $((2^{\frac{1}{2}})^{\frac{1}{2}})^{\frac{1}{2}} \approx 1 + \frac{1}{2^3}$ and in general $2^{1/2^n} \approx 1 + \frac{1}{2^n}$ if the process is carried out n times. These numbers are approaching $1 + 0 = 1$.

d) Starting with $m > 1$, $\sqrt{m} = \sqrt{1+(m-1)} \approx 1 + \frac{m-1}{2}$ and $(m)^{1/2^n} \approx 1 + \frac{m-1}{2^n}$ approaches 1.

24. a) The sequence of numbers increases and approaches 1. b) If $0 < m < 1$, $\sqrt{m} = \sqrt{1+[-(1-m)]} \approx 1 + \frac{[-(1-m)]}{2} = 1 - \frac{(1-m)}{2}$ and, more generally, $m^{1/2^n} \approx 1 - \frac{(1-m)}{2^n}$. So we are subtracting smaller and smaller positive numbers from 1, and the numbers approach 1.

c) The linearization of $(1+x)^{1/10}$ is $1 + \frac{x}{10}$. If $m > 1$, $m^{1/10} = [1+(m-1)]^{1/10} \approx 1 + \frac{m-1}{10}$, and $m^{1/10^n} \approx 1 + \frac{m-1}{10^n}$. Each time we take a tenth root the part of m greater than 1 is divided by 10. Also if $0 < m < 1$, $m^{1/10^n} = [1+(-(1-m))]^{1/10^n} \approx 1 - \frac{(1-m)}{10^n}$.

25. $f(x) = x^2 + 2x$, $x_0 = 0$, $dx = 0.1$. a) $\Delta f = f(x_0 + dx) - f(x_0) = f(0.1) - f(0) = 0.01{+}0.2{-}0 = 0.21$. b) $f'(x) = 2x{+}2$. $df = f'(x_0)dx = 2(0.1) = 0.2$. c) $|\Delta f - df| = |0.21 - 0.2| = 0.01$.

26. $f(x) = 2x^2 + 4x - 3$, $x_0 = -1$, $dx = 0.1$. a) $\Delta f = f(x_0 + dx) - f(x_0) = 2(-0.9)^2 + 4(-0.9) - 3 - [2 - 4 - 3] = 0.02$. b) $f'(x) = 4x + 4$. $df = f'(x_0)dx = 0(0.1) = 0$ c) $|\Delta f - df| = 0.02$.

27. $f(x) = x^3 - x$, $x_0 = 1$, $dx = 0.1$. a) $\Delta f = f(x_0 + dx) - f(x_0) = (1.1)^3 - 1.1 - 0 = 0.231$. b) $f'(x) = 3x^2 - 1$. $df = f'(x_0)dx = 2(0.1) = 0.2$. c) $|\Delta f - df| = |0.231 - 0.2| = 0.031$.

28. $f(x) = x^4$, $x_0 = 1$, $dx = 0.1$. a) $\Delta f = f(x_0 + dx) - f(x_0) = 1.1^4 - 1 = 0.4641$. b) $f'(x) = 4x^3$, $df = f'(x_0)dx = 4(0.1) = 0.4$ c) $|\Delta f - df| = 0.0641$.

29. $f(x) = x^{-1}$, $x_0 = 0.5$, $dx = 0.1$. a) $\Delta f = f(x_0 + dx) - f(x_0) = (0.6)^{-1} - (0.5)^{-1} = (5/3) - 2 = -1/3$. b) $f'(x) = -1/x^2$, $df = f'(x_0)dx = (-1/0.25)(0.1) = -4(0.1) = -0.4 = -2/5$ c) $|\Delta f - df| = |-1/3 - (-2/5)| = 1/15$.

30. $f(x) = x^3 - 2x + 3$, $x_0 = 2$, $dx = 0.1$. a) $\Delta f = f(x_0 + dx) - f(x_0) = (2.1)^3 - 2(2.1) + 3 - [2^3 - 2(2) + 3] = 1.061$. b) $f'(x) = 3x^2 - 2$, $df = f'(x_0)dx = 10(0.1) = 1$. c) $|\Delta f - df| = 0.061$.

31. $dV = \frac{dV}{dr}dr = 4\pi r^2 dr$ and, at r_0, $dV = 4\pi r_0^2 dr$.

32. $S'(r) = 8\pi r$. $dS = 8\pi r_0 dr$.

33. $V(x) = x^3$. $V'(x) = 3x^2$. $dV = 3x_0^2 dx$.

34. $S(x) = 6x^2$. $S'(x) = 12x$. $dS = 12x_0dx$.

35. $V(r) = \pi r^2h$. $V'(r) = 2\pi rh$. $dV = 2\pi r_0hdr$.

36. $S(h) = 2\pi rh$. $S'(h) = 2\pi r$. $dS = 2\pi rdh$.

37. a) $A(r) = \pi r^2$, $A'(r) = 2\pi r$. $dA = A'(2.00)dr = \pi 4(0.02) = 0.08\pi \approx 0.2513m^2$ b) $\frac{0.2513}{4\pi}(100) \approx 2.000\%$.

38. $C = 2\pi r = \pi d$. $d(C) = C/\pi$. $d'(C) = 1/\pi$. When $d = 10$, $C = 10\pi$. $d(d)$ (the differential of d) $= d'(10\pi)dC = (1/\pi)2 = 2/\pi \approx 0.6366$. Thus the diameter grew about 0.6366 in. The cross section area is $A = \pi r^2 = \pi(C/2\pi)^2 = C^2/4\pi$. $A'(C) = C/2\pi$. $dA = A'(C_0dC = A'(10\pi)2 = (10\pi/2\pi)2 = 10$. Thus the tree's cross section grew about 10 in^2.

39. $V = x^3$, $V'(x) = 3x^2$, $dV = V'(x_0)dx = V'(10)(\frac{10}{100}) = 3(10)^2(\frac{1}{10}) = 30$ is the estimated error. The estimated percentage error is $\frac{30}{100}(100) = 3\%$.

40. $A(x) = x^2$. The estimated percentage error in calculating the area is $(\frac{dA}{A})100 = \frac{(2x_0dx)100}{x_0^2} = \frac{200dx}{x_0}$. Thus $\frac{200dx}{x_0} = 2$ implies $dx = \frac{x_0}{100}$ so the side should be measured with an error equal to at most 1% of its length.

41. $V = \frac{4}{3}\pi r^3 = \frac{4}{3}\pi(\frac{d}{2})^3 = \frac{\pi}{6}d^3$. $V'(d) = \frac{\pi d^2}{2}$. The estimated error in the volume calculation is $dV = V'(d_0)d(d) = V'(d_0)$ (differential of d) $= V'(100)(1) = 5000\pi$. The estimated percentage error is $\frac{dV}{V}(100) = \frac{5000\pi(100)}{(\pi/6)100^3} = \frac{(100^2/2)\pi 100}{(\pi/6)100^3} = \frac{6}{2} = 3\%$.

42. $V = \frac{4}{3}\pi r^3 = \frac{4}{3}\pi(\frac{d}{2})^3 = \frac{\pi d^3}{6}$. The estimated error in calculating V is $dV = V'(d_0)u$ where u is the differential of d. $V'(d) = \frac{\pi d^2}{2}$ so $dV = \frac{\pi d_0^2}{2}u$. The estimated percentage error being 3% means $\frac{(\pi d_0^2u/2)100}{(\pi d_0^3/6)} = \frac{300u}{d_o} = 3$ or $u = \frac{d_0}{100}$, i.e., the error in the measurement of the diameter should be at most 1% of its length.

43. We have $V = \pi h^3$, $dV = 3\pi h^2dh$. We require $dV/V \leqq 0.01$, $3\pi h^2dh/\pi h^3 = 3dh/h \leqq 0.01$ or $dh/h \leqq 0.01/3$. Thus h should be measured with an error of no more than 1/3 of 1%.

44. a) $V = \pi r^2h = \pi(d/2)^210 = (5/2)\pi d^2$. $dV = 5\pi du$ where u is the differential of d. It is required that $dV/V \leqq 0.01$. Hence $5\pi du/((5/2)\pi d^2) \leqq 0.01$, $2u/d \leqq 0.01$ or $u/d \leqq 0.01/2$. Thus the diameter must be measured to within 1/2 of 1% of its true value.

b) Let w be the tank's exterior diameter. Let S = surface area of side of tank $= 2\pi rh = 10\pi w$. $dS = 10\pi dw$. It is required that $dS/S \leqq 0.05$, $10\pi dw/10\pi w \leqq 0.05$, $dw/w \leqq 0.05$. Thus w should be measured with an error of no more than 5%.

45. Let weight of coin $= W = kV$ when k is the density and V is the volume. $V = \pi r^2 h$. Thus $W = kh\pi r^2$, $dW = 2kh\pi r dr$. It is required that $dW \leqq (1/1000)W$, $2kh\pi r dr \leqq (1/1000)kh\pi r^2$ or $dr \leqq (1/2000)r$. The variation of the radius should not exceed 1/2000 of its ideal value, that is, 0.05% of the ideal value.

46. As shown in Example 10, the relative change in V is four times the relative change in r. Thus for a 50% increase in V a $50/4 = 12.5\%$ increase in r is required.

47. $s = 16t^2$, $ds = 32t dt$. $ds/s = 32t dt/16t^2 = 2dt/t$. Thus the relative error in s is twice the relative error in measuring t.

48. $W = a + b/g$, $dW = (-b/g^2)dg$. $dW_{\text{moon}}/dW_{\text{earth}} = (-b/5.2^2)dg/(-b/32^2)dg = 32^2/5.2^2 \approx 37.9$.

49. a) $g(x) = \sqrt{x} + \sqrt{1+x} - 4$. $g(3) = \sqrt{3} + \sqrt{4} - 4 = \sqrt{3} - 2 < 0$. $g(4) = \sqrt{4} + \sqrt{5} - 4 = \sqrt{5} - 2 > 0$.

b) We use the linearization of $g(x)$: $L(x) = g(3) + g'(3)(x-3)$. $g'(x) = 1/(2\sqrt{x}) + 1/(2\sqrt{x+1})$. $g'(3) = 1/(2\sqrt{3}) + 1/2\sqrt{4} = 1/(2\sqrt{3}) + 1/4$. $L(x) = (\sqrt{3} - 2) + (1/(2\sqrt{3}) + 1/4)(x-3) = 0$ leads to $[\frac{1}{2\sqrt{3}} + \frac{1}{4}](x-3) = 2 - \sqrt{3}$, $\frac{2+\sqrt{3}}{4\sqrt{3}}(x-3) = 2 - \sqrt{3}$, $x - 3 = \frac{4\sqrt{3}}{2+\sqrt{3}}(2-\sqrt{3})\frac{2-\sqrt{3}}{2-\sqrt{3}} = \frac{4\sqrt{3}(4-4\sqrt{3}+3)}{4-3} = 28\sqrt{3} - 48$, $x = 28\sqrt{3} - 45 \approx 3.497$.

c) We find $g(28\sqrt{3} - 45) \approx -0.009$. d) $x = 3.516$ with error at most 0.01.

e) $\sqrt{x} + \sqrt{1+x} - 4 = 0$, leads to $\sqrt{1+x} = 4 - \sqrt{x}$, $1 + x = 16 - 8\sqrt{x} + x$, $\sqrt{x} = \frac{15}{8}$, $x = \frac{225}{64}$.

50. b) The linearization $L_1(x)$ of $2\cos x$ at $x = \pi/4$ is $L_1(x) = 2\cos(\pi/4) - 2\sin(\pi/4)(x - \pi/4) = \sqrt{2}(1 - x + \pi/4)$. The linearization $L_2(x)$ of $\sqrt{1+x}$ at $x = 0.69$ is $L_2(x) = \sqrt{1.69} + (1/2\sqrt{1.69})(x - 0.69)$.

c) We change $2\cos x = \sqrt{1+x}$ to $L_1(x) = L_2(x)$: $\sqrt{2}(1 + \pi/4 - x) = \sqrt{1.69} + (1/2\sqrt{1.69})(x - 0.69)$. Solving this linear equation for x, we obtain $x = \frac{\sqrt{2}\sqrt{1.69}(\pi+4) - 5.38}{2(2\sqrt{2}\sqrt{1.69}+1)} \approx 0.8285$. With $f(x) = 2\cos x - \sqrt{1+x}$, we find $f(0.8285) \approx -2.6 \times 10^{-4}$.

d) $x = 0.82836$ e) We have no method to find an exact solution.

51. $y = x^3 - 3x$. $dy = (3x^2 - 3)dx = 3(x^2 - 1)dx$.

52. $y = x\sqrt{1-x^2}$. $dy = \left[x\frac{1\cdot(-2x)}{2\sqrt{1-x^2}} + \sqrt{1-x^2}\right]dx = \left[\frac{-x^2}{\sqrt{1-x^2}} + \sqrt{1-x^2}\right]dx = \left[\frac{-x^2+1-x^2}{\sqrt{1-x^2}}\right]dx = \frac{(1-2x^2)dx}{\sqrt{1-x^2}}$.

53. $y = \frac{2x}{1+x^2}$. $dy = 2\frac{[(1+x^2)\cdot 1 - x(2x)]dx}{(1+x^2)^2} = \frac{2(1-x^2)dx}{(1+x^2)^2}$.

54. $y = (3x^2-1)^{3/2}$. $dy = \frac{3}{2}(3x^2-1)^{1/2}(6x)dx = 9x(3x^2-1)^{1/2}dx$.

55. $y + xy - x = 0$ leads to $y(1+x) - x = 0$, $y = \frac{x}{1+x}$, $dy = \frac{[(1+x)\cdot 1 - x(1)]dx}{(1+x)^2} = \frac{dx}{(1+x)^2}$.

56. $xy^2 + x^2y - 4 = 0$. $2xyy' + y^2 + x^2y' + 2xy = 0$, $y'(2xy + x^2) = -y^2 - 2xy$, $dy = -\frac{(y^2+2xy)dx}{2xy+x^2}$.

57. $y = \sin(5x)$. $dy = [\cos(5x)]5dx = 5\cos(5x)dx$.

58. $y = \cos(x^2) = -\sin(x^2)(2x)dx = -2x\sin(x^2)dx$.

59. $y = 4\tan(x/2)$. $dy = 4\sec^2(x/2)(1/2)dx = 2\sec^2(x/2)dx$.

60. $y = \sec(x^2-1)$. $dy = \sec(x^2-1)\tan(x^2-1)(2x)dx = 2x\sec(x^2-1)\tan(x^2-1)dx$.

61. $y = 3\csc(1-(x/3))$. $dy = -3\csc(1-(x/3))\cot(1-(x/3))(-1/3)$. $dy = -3\csc(1-(x/3))\cot(1-(x/3))(-1/3)dx = \csc(1-(x/3))\cot(1-(x/3))dx$.

62. $y = 2\cot\sqrt{x}$. $dy = -2(\csc^2\sqrt{x})\frac{dx}{2\sqrt{x}} = -\frac{(\csc^2\sqrt{x})dx}{\sqrt{x}}$.

63. a) We graphed $y = \sqrt{1+x}$ in $[-0.17, 0.17]$ by $[0.91, 1.08]$. b) We use TRACE to estimate the coordinates of the extreme points of the graph in a): $m = \frac{y_2-y_1}{x_2-x_1} = \frac{1.08-0.91}{0.17-(-0.17)} = 0.5$. $f'(x) = \frac{1}{2\sqrt{1+x}}$. $f'(0) = 0.5$. c) Also graph $y = 1 + \frac{x}{2}$. The graphs nearly coincide (especially near $x = 0$) in the above window.

64. Use the methods of Exercise 63. Here $f'(x) = \frac{(x^2-4)-x(2x)}{(x^2-4)^2} = \frac{-(x^2+4)}{(x^2-4)^2}$. $f'(3) = \frac{-13}{25}$. $L(x) = f(3) + f'(3)(x-3) = \frac{3}{5} - \frac{13}{25}(x-3)$.

65. $\lim_{x\to 0}\frac{\sqrt{1+x}}{1+(x/2)} = \frac{\sqrt{1+0}}{1+0} = 1$.

66. $\lim_{x\to 0}\frac{\tan x}{x} = \lim_{x\to 0}\frac{1}{x}\frac{\sin x}{\cos x} = \lim_{x\to 0}\frac{1}{\cos x}\frac{\sin x}{x} = (1)(1) = 1$.

67. $E(x) = f(x) - g(x) = f(x) - m(x-a) - c$. $E(a) = 0$ implies $f(a) - 0 - c = 0$ and so $c = f(a)$. $E(x) = f(x) - f(a) - m(x-a)$. $0 = \lim_{x\to a} \frac{E(x)}{x-a} = \lim_{x\to a} \frac{f(x)-f(a)}{x-a} - m = f'(a) - m$. Thus $m = f'(a)$ and $g(x) = L(x)$.

PRACTICE EXERCISES, CHAPTER 3

1. $y = x^5 - \frac{1}{8}x^2 + \frac{1}{4}x$. $y' = 5x^4 - \frac{1}{4}x + \frac{1}{4}$.

2. $y = 3 - 7x^3 + 3x^7$. $y' = -21x^2 + 21x^6 = 21x^2(x^4 - 1)$.

3. $y = (x+1)^2(x^2+2x)$. $y' = (x+1)^2(2x+2)+2(x+1)(x^2+2x) = 2(x+1)[(x+1)^2 + x^2 + 2x] = 2(x+1)(2x^2+4x+1)$.

4. $y = (2x-5)(4-x)^{-1} = \frac{2x-5}{4-x}$. $y' = \frac{(4-x)2-(2x-5)(-1)}{(4-x)^2} = \frac{8-2x+2x-5}{(4-x)^2} = \frac{3}{(4-x)^2}$.

5. $y = 2\sin x \cos x = \sin 2x$. $y' = 2\cos 2x$.

6. $y = \sin x - x\cos x$. $y' = \cos x - (-x\sin x + \cos x) = x\sin x$.

7. $y = \frac{x}{x+1}$. $y' = \frac{(x+1)\cdot 1 - x\cdot 1}{(x+1)^2} = \frac{1}{(x+1)^2}$.

8. $y = \frac{2x+1}{2x-1}$. $y' = \frac{(2x-1)(2)-(2x+1)2}{(2x-1)^2} = \frac{4x-2-4x-2}{(2x-1^2} = \frac{-4}{(2x-1)^2}$.

9. $y = (x^3+1)^{-4/3}$. $y' = (-4/3)(x^3+1)^{-7/3}(3x^2) = -4x^2(x^3+1)^{-7/3}$.

10. $y = (x^2-8x)^{-1/2}$. $y' = (-1/2)(x^2-8x)^{-3/2}(2x-8) = -(x-4)(x^2-8x)^{-3/2}$.

11. $y = \cos(1-2x)$. $y' = [-\sin(1-2x)](-2) = 2\sin(1-2x)$.

12. $y = \cot\frac{2}{x}$. $y' = (-\csc^2(\frac{2}{x}))(-\frac{2}{x^2}) = \frac{2}{x^2}\csc^2(\frac{2}{x})$.

13. $y = (x^2+x+1)^3$. $y' = 3(x^2+x+1)^2(2x+1)$.

14. $y = (-1-\frac{x}{2}-\frac{x^2}{4})^2 = (1+\frac{x}{2}+\frac{x^2}{4})^2$. $y' = 2(1+\frac{x}{2}+\frac{x^2}{4})(\frac{1}{2}+\frac{x}{2}) = (1+x)(1+\frac{x}{2}+\frac{x^2}{4})$.

15. $y = \sqrt{2u+u^2}$, $u = 2x+3$. $\frac{dy}{dx} = \frac{dy}{du}\frac{du}{dx} = \left(\frac{2+2u}{2\sqrt{2u+u^2}}\right)(2) = \frac{2(1+u)}{\sqrt{2u+u^2}} = \frac{2(1+2x+3)}{\sqrt{2(2x+3)+(2x+3)^2}} = \frac{4(x+2)}{\sqrt{4x^2+16x+15}}$.

16. $y = \frac{-u}{1+u}$, $u = \frac{1}{x}$. $\frac{dy}{dx} = \frac{dy}{du}\frac{du}{dx} = \frac{-1}{(1+u)^2}(-\frac{1}{x^2})$ (by Exercise 7) $= \frac{1}{(1+u)^2}u^2 = (\frac{u}{1+u})^2 = \frac{1}{(x+1)^2}$.

17. $xy + y^2 = 1$. $xy' + y + 2yy' = 0$, $y'(x+2y) = -y$, $y' = \frac{-y}{x+2y}$.

18. $xy + 2x + 3y = 1$. $y(x+3) = 1 - 2x$, $y = \frac{1-2x}{x+3}$. $y' = \frac{(x+3)(-2)-(1-2x)}{(x+3)^2} = \frac{-2x-6-1+2x}{(x+3)^2} = \frac{-7}{(x+3)^2}$.

19. $x^2+xy+y^2-5x = 2$. $2x+xy'+y+2yy'-5 = 0$, $y'(x+2y) = 5-2x-y$, $y' = \frac{5-2x-y}{x+2y}$.

20. $x^3+4xy-3y^2 = 2$. $3x^2+4xy'+4y-6yy' = 0$, $y'(4x-6y) = -3x^2-4y$, $y' = \frac{3x^2+4y}{2(3y-2x)}$.

21. $5x^{4/5} + 10y^{6/5} = 15$. $4x^{-1/5} + 12y^{1/5}y' = 0$, $y' = \frac{-4x^{-1/5}}{12y^{1/5}} = \frac{-1}{3(xy)^{1/5}}$.

22. $\sqrt{xy} = 1$. $\frac{1}{2\sqrt{xy}}(xy'+y) = 0$, $xy'+y = 0$, $y' = -\frac{y}{x}$.

23. $y^2 = \frac{x}{x+1}$. $2yy' = \frac{1}{(x+1)^2}$ by Exercise 7. $y' = \frac{1}{2y(x+1)^2}$.

24. $y^2 = \sqrt{\frac{1+x}{1-x}}$. $2yy' = \frac{1}{2}\sqrt{\frac{1-x}{1+x}}\frac{(1-x)-(1+x)(-1)}{(1-x)^2} = \frac{1}{2}\sqrt{\frac{1-x}{1+x}}\frac{2}{(1-x)^2}$, $y' = \frac{1}{2y(1-x)^2}\sqrt{\frac{1-x}{1+x}} = \frac{1}{2y^3(1-x)^2}$.

25. $y^2 = \frac{(5x^2+2x)^{3/2}}{3}$. $2yy' = \frac{1}{2}(5x^2+2x)^{1/2}(10x+2) = (5x+1)(5x^2+2x)^{1/2}$. $y' = \frac{(5x+1)(5x^2+2x)^{1/2}}{2y}$.

26. $y = \frac{3}{(5x^2+2x)^{3/2}} = 3(5x^2+2x)^{-3/2}$. $y' = (-9/2)(5x^2+2x)^{-5/2}(10x+2) = \frac{-9(5x+1)}{(5x^2+2x)^{5/2}}$.

27. $y = \sqrt{x}+1+\frac{1}{\sqrt{x}} = \sqrt{x}+1+x^{-1/2}$. $y' = \frac{1}{2\sqrt{x}} - \frac{1}{2}x^{-3/2} = \frac{1}{2\sqrt{x}} - \frac{1}{2x\sqrt{x}} = \frac{x-1}{2x\sqrt{x}}$.

28. $y = x\sqrt{2x+1}$. $y' = x\frac{1}{2\sqrt{2x+1}}(2) + \sqrt{2x+1} = \frac{x}{\sqrt{2x+1}} + \frac{2x+1}{\sqrt{2x+1}} = \frac{3x+1}{\sqrt{2x+1}}$.

29. $y = \sec(1+3x)$. $y' = \sec(1+3x)\tan(1+3x)\cdot 3 = 3\sec(1+3x)\tan(1+3x)$.

30. $y = \sec^2(1+3x)$. $y' = 2\sec(1+3x)\sec(1+3x)\tan(1+3x)(3) = 6\sec^2(1+3x)\tan(1+3x)$.

31. $y = \cot x^2$. $y' = (-\csc^2 x^2)2x = -2x(\csc x^2)^2$.

32. $y = x^2\cos 5x$. $y' = x^2(-\sin 5x)5 + 2x\cos 5x = x(2\cos 5x - 5x\sin 5x)$.

33. $y = \sqrt{\frac{1-x}{1+x^2}} = \sqrt{u}$ where $u = \frac{1-x}{1+x^2}$. $\frac{dy}{dx} = \frac{dy}{du}\frac{du}{dx} = \frac{1}{2\sqrt{u}}\frac{(1+x^2)(-1)-(1-x)2x}{(1+x^2)^2} = \frac{1}{2}\sqrt{\frac{1}{u}}\frac{-1-x^2-2x+2x^2}{(1+x^2)^2} = \frac{1}{2}\sqrt{\frac{1+x^2}{1-x}}\frac{x^2-2x-1}{(1+x^2)^2} = \frac{x^2-2x-1}{(1-x)^{1/2}(1+x^2)^{3/2}}$.

34. $y^2 = \frac{x^2-1}{x^2+1}$. $2yy' = \frac{(x^2+1)2x-(x^2-1)2x}{(x^2+1)^2} = \frac{2x^3+2x-2x^3+2x}{(x^2+1)^2} = \frac{4x}{(x^2+1)^2}$. Therefore $y' = \frac{1}{2y}\frac{4x}{(x^2+1)^2} = \frac{2x}{y(x^2+1)^2}$.

35. a)

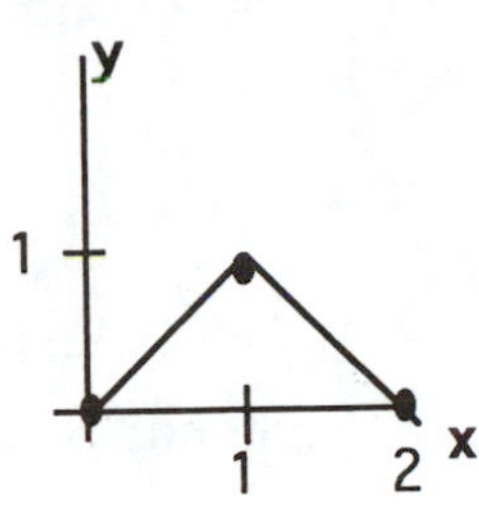

b) Yes. $f(x)$ is continuous at $x = 1$ because $\lim_{x\to1} f(x) = f(1)$.

c) f is not differentiable at $x = 1$ because its left-hand derivative (1) is not equal to its right-hand derivative (-1) at $x = 1$.

36. a) $\lim_{h\to0^-} \frac{f(0+h)-f(0)}{h} = \lim_{h\to0^-} \frac{\sin 2h-0}{h} = 2\lim_{h\to0^-} \frac{\sin 2h}{2h} = 2$. $\lim_{h\to0^+} \frac{f(0+h)-f(0)}{h} = \lim_{h\to0^+} \frac{mh-0}{h} = m$.

b) f is differentiable at $x = 0$ if and only if $m = 2$.

37. $y = 2x^3 - 3x^2 - 12x + 20$. $y' = 6x^2 - 6x - 12 = 6(x^2 - x - 2) = 6(x+1)(x-2)$. The tangent is horizontal if and only if $y' = 0$. Thus $x = -1, 2$. The points on the curve are $(-1, 27)$ and $(2, 0)$.

38. Let $y = f(x) = x^2 + 2x - 3$. $f'(x) = 2x + 2$. $f'(1) = 4$. The normal to the curve at $(1, 0)$ has equation $y - 0 = (-1/4)(x-1)$. Solving simultaneously, we have $x^2 + 2x - 3 = (-1/4)(x-1)$, $4x^2 + 8x - 12 = -x + 1$, $4x^2 + 9x - 13 = 0$ or $(x-1)(4x+13) = 0$, $x = 1, -13/4$. $f(-13/4) = 17/16$. Thus the normal at $(1, 0)$ also intersects the curve at $(-13/4, 17/16)$. The result may be confirmed by graphing $y = x^2 + 2x - 3$ and $y = (-1/4)(x-1)$ and using TRACE.

39. $s(t) = 10\cos(t + \pi/4)$. b) $s(0) = 10\cos(0 + \pi/4) = 5\sqrt{2}$. c) The smallest and largest values of the cosine function are -1 and 1 so the largest and smallest values of s are -10 and 10. d) $v = \frac{ds}{dt} = -10\sin(t + \pi/4)$ and $a = \frac{dv}{dt} = -10\cos(t + \pi/4)$. $s(3\pi/4) = -10$ and $s(7\pi/4) = 10$. $v(3\pi/4) = 0$, $a(3\pi/4) = 10$ and $v(7\pi/4) = 0$, $a(7\pi/4) = -10$. e) $s(t)$ will first be 0 when $t + \pi/4 = \pi/2$ or $t = \pi/4$. $v(\pi/4) = -10$, speed $= |v(\pi/4)| = 10$, $a(\pi/4) = 0$.

40. b) $v = \frac{ds}{dt} = 64 - 32t = 0$ when $t = 2$. So maximum height is reached when $t = 2\,\text{sec}$. $v(0) = 64$ ft/sec. c) $s(t) = 64t - 2.6t^2$, $v(t) = 64 - 5.2t$. $v = 0$ when $t = 64/5.2 \approx 12.308\text{sec}$ $\approx$ time maximum height is reached. The maximum height is $s(64/5.2) \approx 393.846$ ft.

41. a) $s = 490t^2$. $490t^2 = 160$, $t^2 = 16/49$, $t = (4/7)$ sec. Average velocity $= \frac{s(t_2)-s(t_1)}{t_2-t_1} = \frac{s(4/7)-s(0)}{(4/7)-0} = \frac{490(4/7)^2}{(4/7)} = 280$ cm/sec.

b) $v = \frac{ds}{dt} = 980t$, $a = 980$. $v(4/7) = 560$ cm/sec, $a(4/7) = 980$ cm/sec^2.

42.

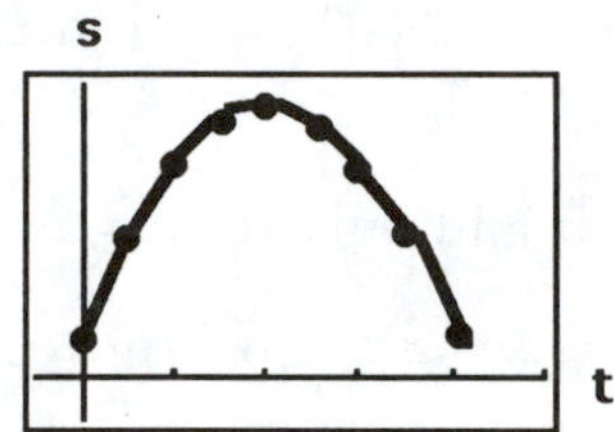

For our estimates we use Equation (7) of 3.1 with $h = 0.5$: $f'(a) \approx \frac{f(a+0.5)-f(a-0.5)}{2(0.5)} = f(a+0.5) - f(a-0.5)$.

a) $v(1.0) = f(1.5) - f(0.5) = 32$ ft/sec. b) $v(2.5) = f(3.0) - f(2.0) = -16$ ft/sec. c) $v(2.0) = v(2.5) - v(1.5) = 0$ ft/sec.

43. The steadily increasing distance is shown in (iii). Its rate of increase (velocity) which is always positive is shown in (i). (ii) is the graph of the rate of change of (i): it is positive (above t-axis) between $t = 0$ and $t = 1$ where (i) is increasing, and negative between $t = 1$ and $t = 2$ where (i) is decreasing. Thus (i) shows velocity and (ii) shows acceleration. a) (iii) b) (i) c) (ii)

44. a)

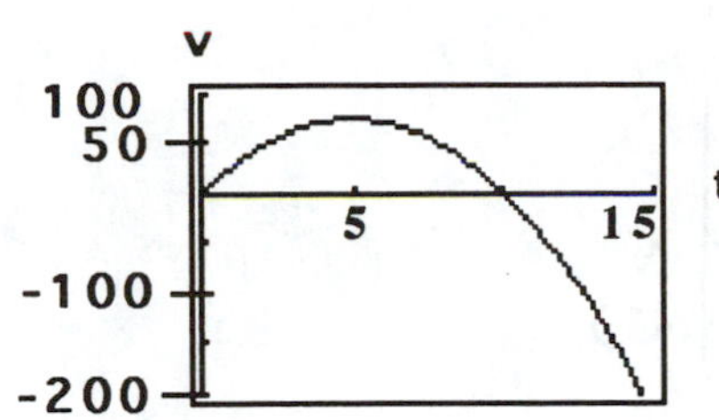

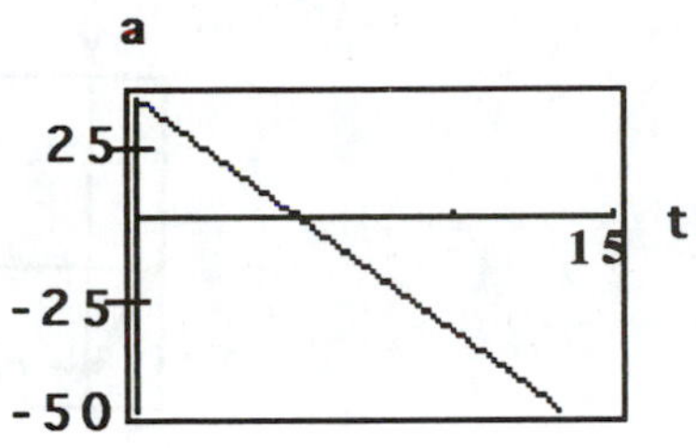

b) $|v|$ is least at $t = 0$ and $t = 10$. $|v|$ is highest at $t = 15$.

c) Compare with the graphs of $s'(t) = 30t - 3t^2$ and $s''(t) = 30 - 6t$ in the viewing window $[0, 15]$ by $[-200, 100]$.

45. The graph of f looks like this:

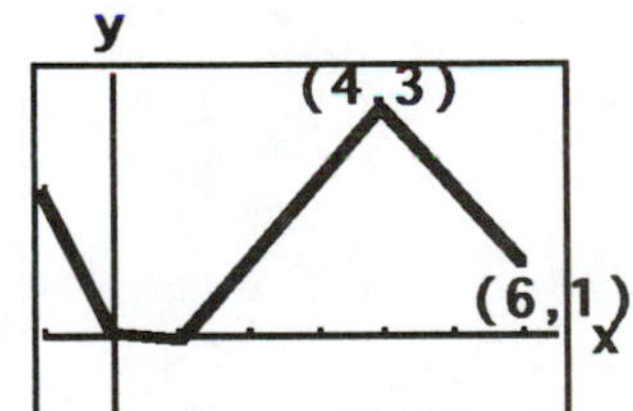

Note that for $0 \le x \le 1$, $f' = 0$

and thus $f =$ constant.

46. The graph of the solution to #45 is lowered by 2 units.

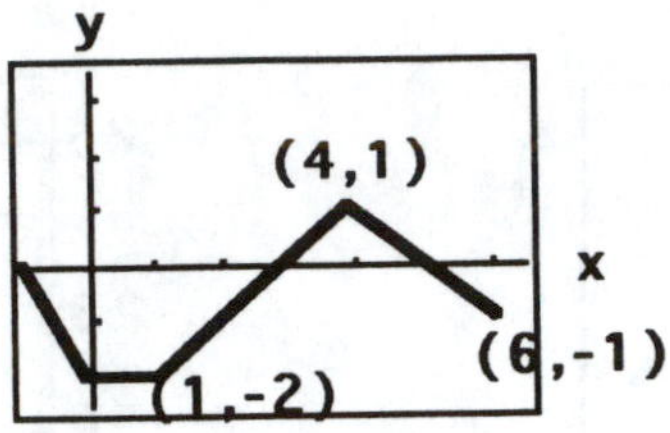

47.

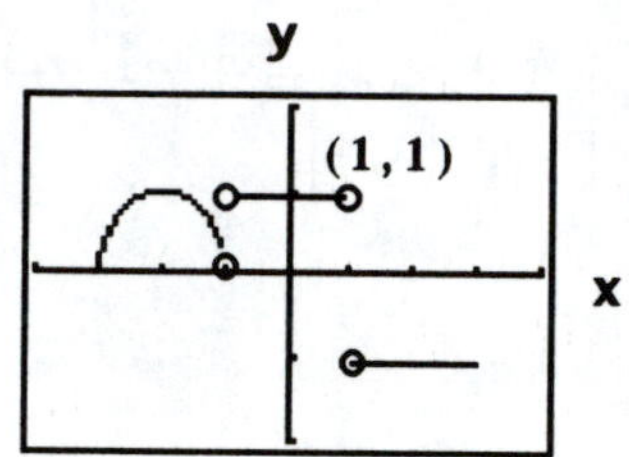

48.

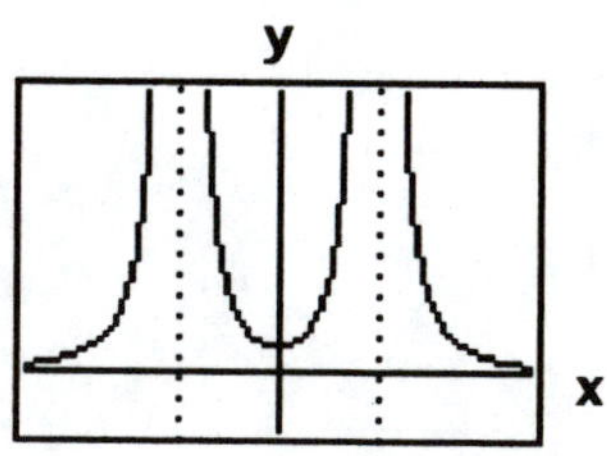

49. $V = \pi[10 - (x/3)]x^2$. $dV/dx = \pi[(10 - (x/3))2x - (1/3)x^2] = \pi(20x - x^2)$.

50. $p = [3 - (x/40)]^2$. $r(x) = xp = x[3 - (x/40)]^2$, $0 \leqq x \leqq 60$. $dr/dx = x(2)[3 - (x/40)](-1/40) + [3 - (x/40)]^2 = [3 - (x/40)][-x/20 + 3 - x/40] = [3 - (x/40)][3 - 3(x/40] = 0$ when $x = 40$ and when $x = 120$ (not in the domain). $p(40) = [3 - (40/40)]^2 = 4$ dollars.

51. $y = 4 + \cot x - 2\csc x$, $y' = -\csc^2 x + 2\csc x \cot x = \csc x(2\cot x - \csc x)$. a) Since $\csc x \neq 0$, $y' = 0$ when $2\cot x = \csc x$ which leads to $2\cos x = 1$, $\cos x = \frac{1}{2}$, $x = \pm\frac{\pi}{3} + 2k\pi$. Evaluating y at these points, we get the points $(\pm\frac{\pi}{3} + 2k\pi,\ 4 \mp \sqrt{3})$. b) $f'(\frac{\pi}{2}) = -1$ and so the tangent at p has equation $y - 2 = -(x - \frac{\pi}{2})$.

52. Let $y = f(x) = 1 + \sqrt{2}\csc x + \cot x$. $y' = f'(x) = -\sqrt{2}\csc x\cot x - \csc^2 x = -\csc x(\sqrt{2}\cot x + \csc x)$. a) $y' = 0$ when $\sqrt{2}\cot x = -\csc x$ or $\cos x = -\sqrt{2}/2$ or when $x = \pm\frac{3\pi}{4} + 2k\pi$. Evaluating y at these points, we get the points $(\frac{3\pi}{4} + 2k\pi, 2)$ and $(-\frac{3\pi}{4} + 2k\pi, 0)$. b) $f'(\frac{\pi}{4}) = -\sqrt{2}(\sqrt{2} + \sqrt{2}) = -4$. Hence the tangent at P has equation $y - 4 = -4(x - \frac{\pi}{4})$ or $y + 4x = \pi + 4$.

53. $y = f(x) = \sin(x - \sin x)$, $y' = f'(x) = \cos(x - \sin x)[1 - \cos x]$. $y' = 0$ when $\cos(x - \sin x) = 0$ or $1 - \cos x = 0$. From the first equation we get $x - \sin x = (2n + 1)\pi/2$. At a solution x of this equation we see that $f(x) = \sin(x - \sin x) = \sin[(2n+1)\pi/2] = \pm 1$ so $(x, f(x))$ is not on the x-axis. The second equation yields $\cos x = 1$, $x = 2n\pi$. $f(2n\pi) = \sin(2n\pi - 0) = 0$. Thus $(2n\pi, 0)$, $n = 0, \pm 1, \pm 2, \ldots$ are points on the curve where the tangent is horizontal.

54. a) $\tan\theta = x/1$, $x = \tan\theta$. b) $dx/dt = (\sec^2\theta)d\theta/dt = (-3/5)\sec^2\theta$ km/sec c) At A, $\theta = 0$ and $dx/dt = (-3/5)$ km/sec $= (-3000/5)$ m/sec $= -600$ m/sec. Thus the speed of the light along the shore at point A is 600 m/sec. d) $\frac{0.6 \text{ radian}}{\text{sec}} = (0.6 \text{ radian})\left(\frac{1 \text{ rev}}{2\pi \text{ radian}}\right)\frac{1}{\text{sec}}\frac{60 \text{ sec}}{\text{min}} = \frac{18}{\pi}$rev/min ≈ 5.73 rev/min.

55. a) $(5f(x) - g(x))'_{x=1} = (5f'(x) - g'(x))_{x=1} = 5f'(1) - g'(1) = 1$.

b) $(f(x)g^3(x))'_{x=0} = f(0)3g^2(0)g'(0) + f'(0)g^3(0) = 6$.

c) $\left(\frac{f(x)}{g(x)+1}\right)'_{x=1} = \frac{(g(1)+1)f'(1) - f(1)g'(1)}{(g(1)+1)^2} = 1$.

d) $(f(g(x)))'_{x=0} = f'(g(0))g'(0) = f'(1)(1/3) = -1/9$.

e) $(g(f(x)))'_{x=0} = g'(f(0))f'(0) = g'(1)5 = -40/3$.

f) $((x + f(x))^{3/2})'_{x=1} = (3/2)(1 + f(1))^{1/2}(1 + f'(1)) = 2$.

g) $(f(x + g(x)))'_{x=0} = f'(0 + g(0))(1 + g'(0)) = -4/9$.

56. No. The Chain Rule gives a sufficient condition for the composites to be differentiable. It does not state that it is necessary for f and g to be differentiable.

57. Differentiating both sides of the identity $\sin(x+a) = \sin x \cos a + \cos x \sin a$ with respect to x, we obtain the identity $\cos(x+a) = \cos x \cos a - \sin x \sin a$. We cannot do the same with $x^2 - 2x - 8 = 0$ because this is not an identity between two functions.

58. $y = 3\sin 2x$, $x = t^2 + \pi$. $\frac{dy}{dt} = \frac{dy}{dx}\frac{dx}{dt} = (6\cos 2x)2t$. At $t = 0, x = \pi$ and $\frac{dy}{dt} = (6\cos 2\pi)(2\cdot 0) = 0$.

59. $s = t^2 + 5t$, $t = (u^2 + 2u)^{1/3}$. $\frac{ds}{du} = \frac{ds}{dt}\frac{dt}{du} = (2t+5)\frac{1}{3}(u^2+2u)^{-2/3}(2u+2)$. If $u = 2$, $t = 2$ and $\frac{ds}{du} = (9)\frac{1}{3}(8)^{-2/3}(6) = \frac{9}{2}$.

60. $w = \sin(\sqrt{r} - 2)$, $r = 8\sin(s + \pi/6)$. $\frac{dw}{ds} = \frac{dw}{dr}\frac{dr}{ds} = \frac{1}{2\sqrt{r}}\cos(\sqrt{r}-2)8\cos(s + \pi/6)$. When $s = 0$, $r = 4$ and $\frac{dw}{ds} = \frac{1}{4}(\cos 0)8\cos(\pi/6) = \sqrt{3}$.

61. $y = \sqrt{x}$, $y' = 1/(2\sqrt{x})$. At $x = 4$, $y' = 1/4$ and the tangent has equation $y - 2 = (1/4)(x-4)$ which intersects the coordinates at $(0,1)$ and $(-4,0)$.

62. Let $y = f(x) = \sqrt{x}$, $y' = f'(x) = 1/(2\sqrt{x})$. $y' = 1$ leads to $x = 1/4$, $f(1/4) = 1/2$. Thus $y = 1/2$ is the desired line.

63. a) $x^2 + 2y^2 = 9$, $2x + 4yy' = 0$, $y' = -x/(2y)$. At $(1,2)$, $m = y' = -1/4$. Tangent: $y - 2 = (-1/4)(x-1)$ or $x + 4y = 9$. Normal: $y - 2 = 4(x-1)$ or $4x - y = 2$. b) $x^3 + y^2 = 2$, $3x^2 + 2yy' = 0$, $y' = -3x^2/(2y)$. At $(1,1)$, $m = y' = -3/2$. Tangent: $y - 1 = (-3/2)(x-1)$ or $3x + 2y = 5$. Normal: $y - 1 = (2/3)(x-1)$ or $3y - 2x = 1$. c) $xy + 2x - 5y = 2$, $xy' + y + 2 - 5y = 0$, $y'(x-5) = -(y+2)$, $y' = (y+2)/(5-x)$. At $(3,2)$, $m = y' = 2$. Tangent: $y - 2 = 2(x-3)$ or $y = 2x - 4$. Normal: $y - 2 = (-1/2)(x-3)$ or $x + 2y = 7$.

64. I $f(x) = \frac{9}{28}x^{7/3}$, $f'(x) = \frac{3}{4}x^{4/3}$, $f''(x) = x^{1/3}$ so I is possible. II $f'(x) = \frac{9}{28}x^{7/3} - 2$, $f''(x) = \frac{3}{4}x^{4/3}$ so II is not possible. III $f'(x) = \frac{3}{4}x^{4/3} + 6$, $f''(x) = x^{1/3}$ so III is possible. IV $f(x) = \frac{3}{4}x^{4/3} - 4$, $f'(x) = x^{1/3}$, $f''(x) = \frac{1}{3}x^{-2/3}$ so IV is not possible. Answer: d).

65. From the figure we see that if we find the tangent to the circle at $(12,-9)$ and then find its x-value when $y = -15 - 8 = -23$ we will then have half the desired width. $x^2 + y^2 = 225$, $2x + 2yy' = 0$, $y' = -x/y$. At $(12,-9)$, $y' = 4/3$ and the tangent has equation $y + 9 = (4/3)(x - 12)$. When $y = -23$, $x = 3/2$ and so the width is $2(3/2) = 3$ ft.

66. $y = \frac{1}{2r\ell}\sqrt{\frac{T}{\pi d}}$. $y'(r) = \frac{1}{2\ell}\sqrt{\frac{T}{\pi d}}(-\frac{1}{r^2}) = -\frac{1}{2r^2\ell}\sqrt{\frac{T}{\pi d}}$. $y'(\ell) = -\frac{1}{2r\ell^2}\sqrt{\frac{T}{\pi d}}$. $y'(d) = \frac{1}{2r\ell}\frac{1}{2}\sqrt{\frac{\pi d}{T}}(-\frac{T}{\pi d^2}) = -\frac{1}{4r\ell}\sqrt{\frac{T}{\pi d^3}}$. $y'(T) = \frac{1}{2r\ell}\frac{1}{2}\sqrt{\frac{\pi d}{T}}\frac{1}{\pi d} = \frac{1}{4r\ell}\sqrt{\frac{1}{\pi dT}} = \frac{1}{4r\ell\sqrt{\pi dT}}$.

67. a) $x^3 + y^3 = 1$. $3x^2 + 3y^2y' = 0$, $y' = -x^2/y^2$. $y'' = -\frac{y^2(2x)-x^2(2yy')}{y^4} = -\frac{2xy^2-2x^2y(-x^2/y^2)}{y^4} = -\frac{2xy^3+2x^4}{y^5} = \frac{-2x(x^3+y^3)}{y^5} = -\frac{2x}{y^5}$.

b) $y^2 = 1 - \frac{2}{x}$. $2yy' = \frac{2}{x^2}$, $y' = \frac{1}{x^2y}$. $y'' = \frac{0-(x^2y'+2xy)}{x^4y^2} = \frac{-((1/y)+2xy)}{x^4y^2} = \frac{-(1+2xy^2)}{x^4y^3} = -\frac{(1+2x(1-2/x))}{x^4y^3} = \frac{3-2x}{x^4y^3}$.

68. a) $x^2 - y^2 = 1$. $2x - 2yy' = 0$, $y' = x/y$. b) $y'' = \frac{y-xy'}{y^2} = \frac{y-x^2/y}{y^2} = \frac{y^2-x^2}{y^3} = \frac{-1}{y^3}$.

69. a) $y = \sqrt{2x+7}$. $y' = \frac{1}{2\sqrt{2x+7}}(2) = \frac{1}{\sqrt{2x+7}}$. $y'' = \frac{0-1/\sqrt{2x+7}}{2x+7} = -\frac{1}{(2x+7)^{3/2}}$.
b) $x^2 + y^2 = 1$. $2x + 2yy' = 0$, $y' = -x/y$. $y'' = -\frac{y-xy'}{y^2} = -\frac{y+x^2/y}{y^2} = -\frac{x^2+y^2}{y^3} = \frac{-1}{y^3}$.

70. $y^3 + y = 8x - 6$. $3y^2y' + y' = 8$, $y' = \frac{8}{3y^2+1}$. $y'' = \frac{0-8(6yy')}{(3y^2+1)^2} = \frac{-48y(8)}{(3y^2+1)^3}$. At $(1,1)$, $y'' = -6$.

71. a) Let $f(x) = \tan x$. $f'(x) = \sec^2 x$. $L(x) = f(a) + f'(a)(x-a) = \tan(-\pi/4) + \sec^2(-\pi/4)(x+\pi/4) = -1 + 2(x+\pi/4) = 2x + (\pi/2) - 1$. Graph $y = \tan x$ and $y = 2x + (\pi/2) - 1$ in the viewing rectangle $[-5,5]$ by $[-8,8]$.

b) Let $f(x) = \sec x$. $f'(x) = \sec x \tan x$. $L(x) = f(a) + f'(a)(x-a) = \sec(-\pi/4) + \sec(-\pi/4)\tan(-\pi/4)(x+\pi/4) = \sqrt{2} - \sqrt{2}(x+\pi/4)$. Graph $y = \sec x$ and $y = \sqrt{2} - \sqrt{2}(x+\pi/4)$ in the viewing rectangle $[-8,8]$ by $[-10,10]$.

72. Let $f(x) = \frac{1}{1+\tan x}$. $f'(x) = \frac{0-\sec^2 x}{(1+\tan x)^2} = \frac{-\sec^2 x}{(1+\tan x)^2}$. $L(x) = f(a)+f'(a)(x-a) = \frac{1}{1+0} - \frac{1}{(1+0)^2}(x-0) = 1 - x$.

73. $f(x) = \sqrt{1+x} + \sin x - 0.5$. a) $f(-\pi/4) = \sqrt{1-\pi/4} - \frac{\sqrt{2}}{2} - \frac{1}{2} < 0$. $f(0) = \sqrt{1} + 0 - 0.5 = 0.5 > 0$. b) (The linearization of a sum is the sum of the linearizations.) $f'(x) = \frac{1}{2\sqrt{1+x}} + \cos x$. $L(x) = f(0)+f'(0)(x-0) = 0.5+1.5x$. We solve $0.5 + 1.5x = 0$ and obtain $x = -1/3$. c) $f(-1/3) \approx -0.01$. d) By zoom-in, $x = -0.326$. We cannot find the exact solution with the tools at hand.

74. $f(x) = \frac{2}{1-x} + \sqrt{1+x} - 3.1$. b) $f'(x) = \frac{2}{(1-x)^2} + \frac{1}{2\sqrt{1+x}}$. $L(x) = f(0) + f'(0)(x - 0) = -0.1 + 2.5x$. $L(x) = 0$ leads to $x = 0.04$. c) $f(0.04) \approx 0.003$. d) With zoom-in: $\{0.0388196\ldots, 10.031939\ldots\}$. We cannot find the exact solutions with the tools at hand.

75. $V = \frac{1}{3}\pi r^2 h$. $dV = \frac{2}{3}\pi r h dr$. An estimate of the change in question is $dV = \frac{2}{3}\pi r_0 h dr$.

76. $S = \pi r\sqrt{r^2+h^2}$, $dS = \pi r \frac{1}{2\sqrt{r^2+h^2}} 2h dh = \frac{\pi r h dh}{\sqrt{r^2+h^2}}$. An estimate of the change in question is $\frac{\pi r h_0 dh}{\sqrt{r^2+h_0^2}}$.

77. a) Surface area $= S = 6x^2$. An estimate of the error in measuring S is $dS = 12x dx$. We require $\frac{dS}{S} \leqq 0.02$, $\frac{12x dx}{6x^2} = \frac{2}{x} dx \leqq 0.02$ or $\frac{dx}{x} \leqq 0.01$. Thus the edge should be measured with an error of no more than 1%.

b) $V = x^3$. $dV = 3x^2 dx$. $\frac{dV}{V} = \frac{3x^2 dx}{x^3} = 3\frac{dx}{x} \leqq 0.03$. Therefore the volume can be measured with an estimated error of no more than 3%.

78. $C = 2\pi r$, $dC = 2\pi dr$, $2\pi dr = 0.4$. a) $dr = 0.2/\pi \approx 0.06$. But $10 = 2\pi r$, $r = 5/\pi$. So $\frac{dr}{r} = \frac{0.2/\pi}{5/\pi} = 0.04$. The calculated value of r has a possible error of 4%.

b) $S = 4\pi r^2$, $dS = 8\pi r dr$, $\frac{dS}{S} = \frac{8\pi r dr}{4\pi r^2} = 2\frac{dr}{r} = 0.08$. The calculated value of S has a possible error of 8%.

c) $V = \frac{4}{3}\pi r^3$, $dV = 4\pi r^2 dr$, $\frac{dV}{V} = \frac{4\pi r^2 dr}{(4/3)\pi r^3} = 3\frac{dr}{r} = 0.12$. The calculated value of V has a possible error of 12%.

79. $1° = 1° \frac{\pi}{180°}$ radian $= \frac{\pi}{180}$ radian is the possible error in measuring θ. $h = 100\tan\theta$, $dh = 100\sec^2\theta d\theta$. An estimate of the possible error in calculating h is $dh = 100\sec^2(\pi/6)(\pi/180) = (5/9)\pi(4/3) = (20/27)\pi \approx 2.33$ ft.

80. We have $da = (1/12)$ ft. $\tan\theta = h/(20+a)$, $h = (20+a)\tan\theta = (20+a)6/a = 6(20/a+1)$. When $a = 15$, $h = 6(4/3+1) = 14$ ft. $dh = 6(-20/a^2)da = -120da/a^2$. With the given values, the possible error $|dh|$ is $|-2/45| \approx 0.04$ ft.

81.

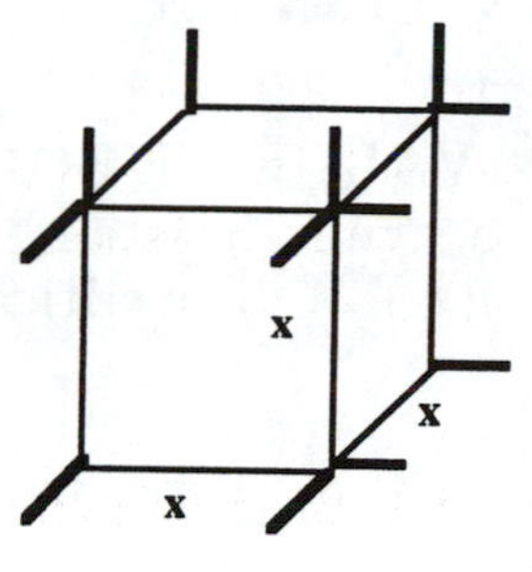

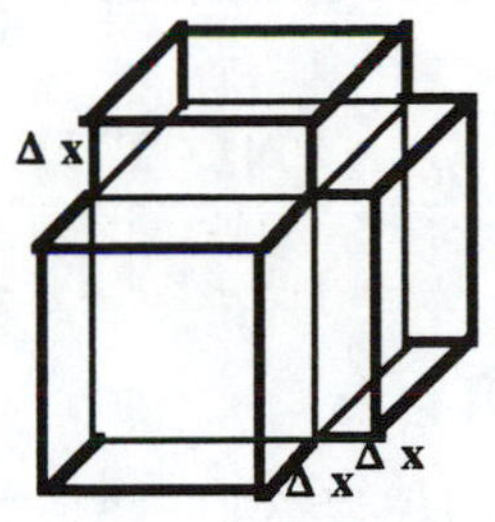

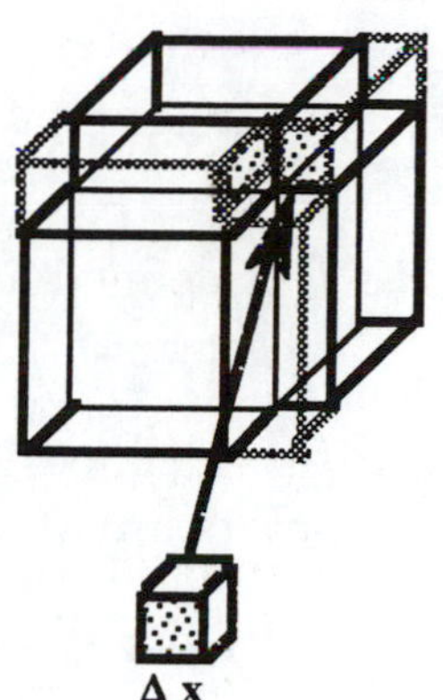

CHAPTER 4

APPLICATIONS OF DERIVATIVES

4.1 MAXIMA, MINIMA, AND THE MEAN VALUE THEOREM

1. f' is defined for all x; $f' = 3x^2 + 2x - 8 = (3x-4)(x+2)$. The critical points are at $x = 4/3$ and $x = -2$.

2. f' is defined for all x; $f' = 3x^2 - 4x - 15 = (x-3)(3x+5)$. The critical points are at $x = -5/3$, $x = 3$.

3. $F(x) = x^{1/3} \Rightarrow F'(x) = \frac{1}{3}x^{-2/3}$; F' does not exist at $x = 0$.

4. $F'(x) = \cos x$ which exists for all x, $-\frac{\pi}{2} \le x \le \frac{5\pi}{6}$; $F' = 0$ at $x = -\frac{\pi}{2}, \frac{\pi}{2}$.

5. g' is not defined at $x = -1$ and $x = 3$. $g' = 0$ at $x = 1$. The critical points are $-1, 1, 3$.

6. g' is not defined at $x = -5$ and $x = 1$; $g'(2) = 0$. The critical points are $-5, -2, 1$.

7. h' is not defined at $x = 0$; where defined, $h' \neq 0$. The only critical point of h is at $x = 0$.

8. On the open intervals $h'(x) = \begin{cases} -x & x < 0 \\ x & 0 < x < 2 \\ 2 & 2 < x \end{cases}$; taking limits gives $h'(x) = \begin{cases} -x & x < 0 \\ x & 0 \le x \le 2 \\ 2 & 2 < x \end{cases}$. $h''(x) = \begin{cases} -1 & x < 0 \\ 1 & 0 < x < 2 \\ 0 & 2 < x \end{cases}$. $h' = 0$ at $x = 0$ which is an inflection point.

9. a) Let F denote the zeros of y, D a zero of its derivative. The screen ranges and orders of zeros are

i $[-6, 6]$ by $[-6, 25]$

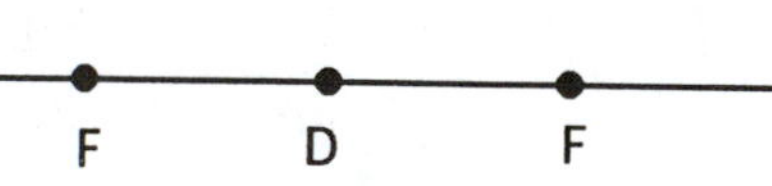

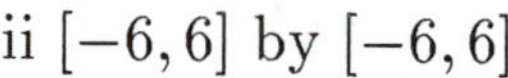

ii $[-6, 6]$ by $[-6, 6]$

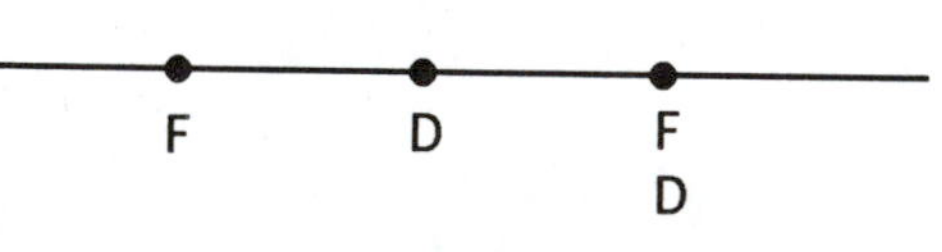

iii $[-2, 40]$ by $[-500, 500]$

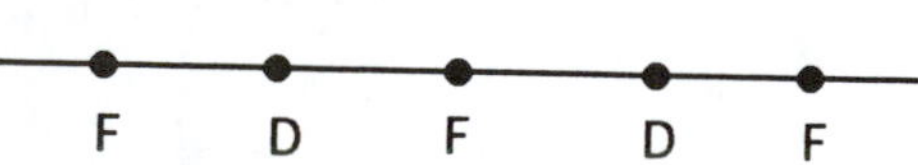

iv $[-6, 6]$ by $[-6, 6]$

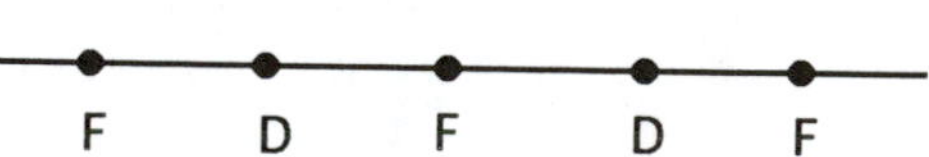

Between consecutive F's there is a D

b) Let $f(x) = x^n + a_{n-1}x^{n-1} \cdots + a_0$; $f'(x) = nx^{n-1} + \cdots + a_1$. Let x_1, x_2 be consecutive zeros of f. By Rolle's Theorem, there exists c, $x_1 < c < x_2$ such that $f'(c) = 0$.

10. $f(1) = 0$, but $\lim_{x \to 1} f(x) = 1$. f is not continuous on $[0, 1]$. Hence, Rolle's Theorem does not apply.

11. (b) Let $f(x) = \sin x$. Between every two zeros of $f(x)$ there is a zero of $f'(x) = \cos x$.

12. Use the window $[-5, 5]$ by $[-13, 3]$. The positive zero of $f'(x)$ lies between two positive zeros of $f(x)$; its existence is guaranteed by Rolle's Theorem since f is continuous and differentiable if $x > -1$. The existence of the negative zero of f' cannot be proved by use of Rolle's Theorem; it does not lie *between* zeros of f.

13. a) Use the window $[-5, 5]$ by $[-1, 15]$

b) $f'(0)$ does not exist; the graph does not have a horizontal tangent at $(0, 0)$

c) $f'(3)$ does not exist; the graph does not have a horizontal tangent at $(3, 0)$

d) f has a local minimum at each of $(-3, 0), (0, 0)$ and $(3, 0)$. f has a local maximum at $(\pm 1.73, 10.39)$

e) f' is defined at ± 1.73; the tangent line at each local maximum is horizontal f' is not defined at $\pm 3, 0$

f) f is increasing on $[-3, -1.73] \cup [0, 1.73] \cup [3, \infty)$. f is decreasing on $(-\infty, -3] \cup [-1.73, 0] \cup [1.73, 3]$

14. Graph $g(x) = ((x-2)^2)$^(1/3) on $[-20, 20]$ by $[-20, 20]$; $g'(2)$ does not exist (cusp); g has a local minimum at 2; no conflict as $g'(2)$ does not exist; e) increasing $[2, \infty)$, decreasing $(-\infty, 2]$

15. $f(0) = 0$, $f(6) = -6$. $f'(x) = 5 - 2x$; set $f'(x) = 0$ to find the critical point $x = 5/2$. $f(5/2) = 6.25$. f has a local maximum at $(5/2, 6.25)$. f is increasing on $[0, 5/2]$; f is decreasing on $[5/2, 6]$. The absolute minimum of f on $[0, 6]$ is at $(6, -6)$. The absolute maximum of f on $[0, 6]$ is at $(5/2, 6.25)$. f has a local minimum at $(0, 0)$.

16. $f(-4) = 8$; $f(4) = 0$; $f' = 2x - 1 \Rightarrow f$ decreasing on $[-4, 1/2]$, increasing on $[1/2, 4]$: $f(1/2) = -12.25$, a local minimum. On $[-4, 4]$, $-12.25 \le f(x) \le 8$. $f(4)$ is also a local maximum.

17. $f(4) = 0$, $f(8) = 2$; $f'(x) = \frac{1}{2\sqrt{x-4}}$, not defined at $x = 4$. Since $f' \neq 0$ between 4 and 8 there are no interior local extrema. The absolute minimum occurs at $(4,0)$; the absolute maximum occurs at $(8,2)$; f is increasing on $[4,8]$.

18. $f(-2) = 4$, $f(5) = 1.35\ldots$; $f' = -\frac{1}{2\sqrt{x+2}}$; f' is never 0, f' does not exist at $x = -2$, $f < 0 \Rightarrow f$ decreasing on $[-2,5]$; $1.35 \leq f \leq 4$.

19. $y(-3) = 0$; $y(3) = 0$. $y'(x) = 4x^3 - 20x = 4x(x^2 - 5)$. Critical points are $x = 0, \pm\sqrt{5}$: $y(0) = 9$, $y(\sqrt{5}) = y(-\sqrt{5}) = -16$. There are local minima at $(\pm\sqrt{5}, -16)$; local maximum at $(0,9)$. Absolute minima at $(\pm\sqrt{5}, -16)$; absolute maximum at $(0,9)$. $(-3,0)$ and $(3,0)$ are local maxima. f is increasing on $[-\sqrt{5}, 0] \cup [\sqrt{5}, 3]$. f is decreasing on $[-3, -\sqrt{5}] \cup [0, \sqrt{5}]$.

20. $g(-3) = -40$, $g(3) = -40$; $g' = -4x^3 + 10x = -2x(2x^2 - 5)$; $g' > 0$ (g increasing) on $[-3, -\sqrt{\frac{5}{2}}]$ and $[0, \sqrt{\frac{5}{2}}]$; $g' < 0$ (g decreasing) on $[-\sqrt{\frac{5}{2}}, 0]$ and $[\sqrt{\frac{5}{2}}, 3]$; $g(-\sqrt{\frac{5}{2}}) = 2.25$, $g(0) = -4$, $g(\sqrt{\frac{5}{2}}) = 2.25$; local maxima at $-\sqrt{\frac{5}{2}}$, local minimum at 0; $-40 \leq g \leq 2.25$.

21. Graph h and NDER on $[-5,5]$ by $[-5,5]$. h has absolute maximum at $x = 5$, absolute minimum at $y = -5$; local minimum at $(0.559, -2.639)$, local maximum at $(-1.126, -0.362)$. Increasing on $[-5, -1.126] \cup [0.559, 5]$; decreasing on $[-1.126, 0.559]$.

22. Graph h and its NDER on $[0, 2\pi]$ by $[-6,6]$. Local minima at $(0,2)$, $(1.571, -1)$, $(4.712, -3)$; local maxima at $(0.125, 2.063)$, $(3.016, 2.063)$, $(2\pi, 2)$. On $[0, 2\pi]$ the absolute minimum is -3, the absolute maximum is 2.063. Increasing on $[0, 0.125] \cup [1.571, 3.016] \cup [4.712, 2\pi]$; decreasing on $[0.125, 1.571] \cup [3.016, 4.712]$.

23. $f'(x) = 2x + 2$; $\frac{f(1)-f(0)}{1-0} = \frac{2-(-1)}{1} = 3$. By the M.V.T., $2c + 2 = 3$. $c = \frac{1}{2}$.

24. $f(x) = x^{2/3} \Rightarrow \frac{f(1)-f(0)}{1-0} = f'(c) = 1$; $f'(x) = \frac{2}{3}x^{-1/3} \Rightarrow \frac{2}{3}c^{-1/3} = 1$ on $c = (\frac{2}{3})^3 = \frac{8}{27}$.

25. $f'(x) = 1 - \frac{1}{x^2}$; $\frac{f(2)-f(1/2)}{2-1/2} = \frac{2.5-2.5}{1.5} = 0$. By the M.V.T., $1 - \frac{1}{c^2} = 0$. $c = \pm 1$. The value of c in $(\frac{1}{2}, 2)$ is $c = 1$.

26. $f(x) = \sqrt{x-1} \Rightarrow \frac{f(3)-f(1)}{3-1} = \frac{\sqrt{2}}{2} = f'(c)$; $f'(x) = \frac{1}{2\sqrt{x-1}} \Rightarrow \frac{1}{2\sqrt{c-1}} = \frac{\sqrt{2}}{2} \Rightarrow c = \frac{3}{2}$.

27. Let $s(t)$ = distance traveled in t hours. $s'(t) = v(t)$, the velocity. By the M.V.T., there exists c such that $v(c) = \frac{s(2)-s(0)}{2-0} = \frac{159-0}{2} = 79.5$. At time $t = c$, the trucker was going 79.5 mph.

28. If $T(t)$ = temperature at time t in seconds, $\frac{T(20)-T(0)}{20-0} = \frac{202}{20} = 10.1°\text{F/sec}$. By the M.V.T, there exists c such that $T'(c) = 10.1$.

29. Let $s(t)$ = sea miles covered in t hours. By the M.V.T. there exists c, $0 < c < 24$ such that $s'(c) = v(c) = \frac{s(24)-s(0)}{24-0} = \frac{184-0}{24-0} = 7.66$ sea miles/hr $= 7.66$ knot.

30. Let $d(t)$ = distance from start at time t. $d(0) = 0$, $d(2.2) = 26.2$; if $v(t) = d'(t), v(0) = v(2.2) = 0$. The maximum velocity must exceed 11mph (because $11*2.2 < 26.2$), assume $v(t_M) = v_{\max}$. By the Intermediate Value Theorem, $v = 11$ at least once on $(0, t_M)$ and at least once on $(t_M, 2, 2)$.

31. $[-2, 9]$ by $[-7, 3]$

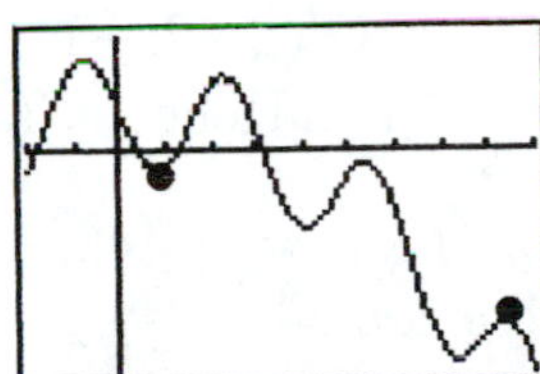

32. Let $g(x) = f(x) - K$. Then $g(a) = g(b) = K - K = 0$. g satisfies the hypotheses of Rolle's Theorem. Hence there is a c, $a < c < b$ such that $g'(c) = 0$. However $g'(x) = f'(x) - 0$; hence $f'(c) = 0$.

33. $f(-4) = f(4) = 4\sin 4$; $f'(x) = \sin x + x\cos x$. $f'(0) = 0$, $f'(\pm 2.029) = 0$.

34. $f(5) = f(-5) = 25\cos 5$; $f'(x) = x(2\cos x - x\sin x)$; clearly $f'(0) = 0$. Graphing shows $f' = 0$ at $x = -3.644, -1.077, 1.077, 3.644$.

35. $f = x^2 - 2x$; $f(-2) = f(4) = 8$; $f' = 2x - 2 = 2(x-1)$; $f'(1) = 0$

36. $f = 3x - x^2$; $f(0.5) = f(2.5) = 1.25$; $f' = 3 - 2x$, $f'(1.5) = 0$

37. $y' = \frac{-1}{x^2} < 0$ wherever it is defined. By Corollary 1 to the M.V.T., y is a decreasing function of x on any interval on which it is defined. NOTE:

$y(-1) = -1$, $y(1) = 1$, but y is *not* defined on $[-1, 1]$, so this is not a contradiction.

38. $y' = -2x^{-3} = \frac{-2}{x^3} > 0$ if $x < 0$. Hence y^∞ increasing when $x < 0$. Similarly $y' < 0$ for $x > 0 \Rightarrow y$ is decreasing.

39. If $f(0) = f(1)$, there would exist c such that $f'(c) = \frac{f(1)-f(0)}{1-0} = 0$, which is precluded.

40. Let $f(x) = \sin x$. By the M.V.T., there exists c, between a and b, such that $\frac{\sin b - \sin a}{b-a} = \cos c$. Using absolute values and $|\cos t| \le 1$, $\frac{|\sin b - \sin a|}{|b-a|} \le 1$.

41. $\frac{f(b)-f(a)}{b-a} < 0$. By the M.V.T., there exists c such that $f'(c) = (f(b) - f(a))/(b-a)$.

42. f has at least one zero by the Intermediate Value Theorem. If f were to have two zeros, by Rolle's Theorem f' would have a zero between them. But $f' \neq 0$ on (a, b). Hence f has exactly one zero.

43. $f(-2) = 11$; $f(-1) = -1$. On $(-2, -1)$, $f'(x) = 4x^3 + 3$. The only zero of $f'(x)$ is at $-(.75)^{1/3} \approx -0.91$. Hence, $f'(x) \neq 0$ on $(-2, 1)$.

44. $f(0) = 1$, $f(1) = -3$, $f'(x) = -3x^2 - 3 = -3(x^2 + 1)$ which is never 0. Hence f has at most one zero.

45. $f(1) = -1$, $f(3) = \frac{7}{3}$. $f'(x) = 1 + \frac{2}{x^2}$, which is never zero.

46. $f(-\pi) = -2\pi + 1$; $f(\pi) = 2\pi - 1 = -(-2\pi + 1)$. $f'(x) = 2 + \sin x > 0$ everywhere.

47. Let $x = b$, where b is an arbitrary real number, $b \neq 0$. $\frac{f(b)-f(0)}{b-0} = \frac{f(b)-3}{b-0} = f'(c) = 0$. Thus $f(b) - 3 = 0$, or $f(b) = 3$. Since b was an arbitrary real number, $f(x) = 3$ for all x.

48. $\frac{f(x)-f(0)}{x-0} = 2 \Rightarrow f(x) = 2x + f(0) = 2x + 5$.

49. a) $f' > 0$ where f increases: $(2, 5)$; b) $f' < 0$ where f decreases: $(-2, 2)$; $f'(2) = 0$.

50. a) $f' > 0$ on $(-1.25, 1.25)$; b) $f' < 0$ on $(-2, -1.25)$ and $(1.25, 2.5)$ c) $f'(-1.25) = 0$, $f'(1.25) = 0$.

51. Let $f(x) = \sqrt{(x-1)(4-x)}$; at $x = 1$ and $x = 4$ the tangent lines are vertical.

52. a)

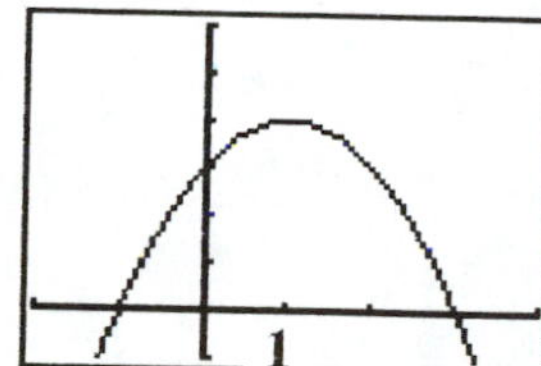

b)

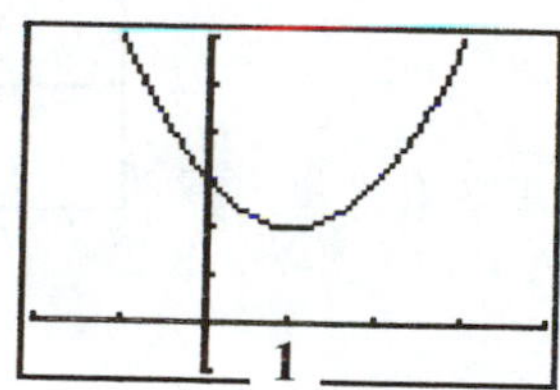

c)

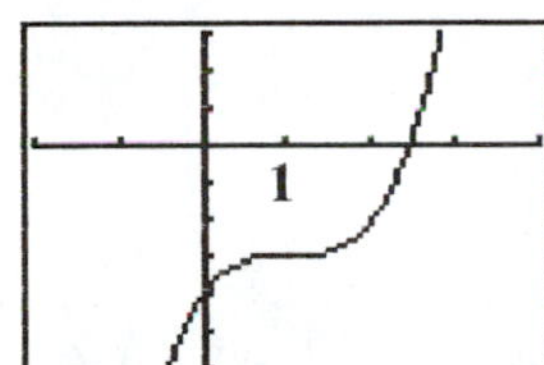

d)

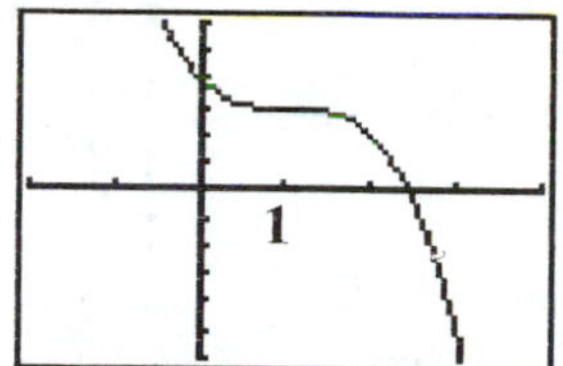

53.

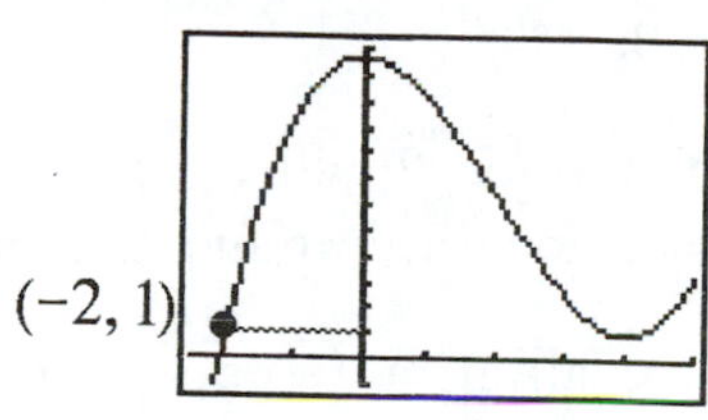

54.

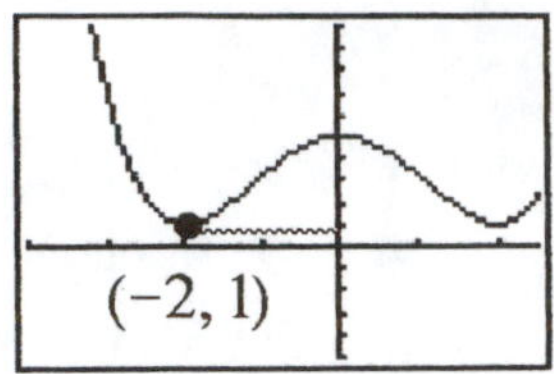

55. Graphing $y = xe^{-x}$ on $[0,3]$ by $[-.5,.5]$. a) If $a = 1.000$, the rectangle has width 0. b) For each value of $\underline{a}$, use trace to find $\underline{b}$ such that $f(b) = f(a)$.

a	$f(a)$	b	area
0.5	0.303	1.762	0.382
0.8	0.359	1.238	0.157
1.0	0.368	1.0	0
1.2	0.361	0.817	0.138
1.5	0.335	0.627	0.292

c) When $a = .25$, area $\approx .455$, when $a = .30$, area $= .459$, when $a = .35$, area $\approx .450$. The maximum appears to occur near $x = 0.3$, area ≈ 0.459.

4.2 ANALYZING HIDDEN BEHAVIOR

1. $f' = 0$ at $x = \pm 1$; $f' > 0$ on $(-\infty, -1)$ and $(1, \infty)$; $f' < 0$ on $(-1, 1)$.

2. $f' > 0$ on $(-1.5, 0)$ and $(1.5, \infty)$; $f' < 0$ on $(-\infty, -1.5)$ and $(0, 1.5)$; $f'(-1.5) = f'(0) = f'(1.5) = 0$.

3. $f' > 0$, so graph of f is rising on $(-\infty, 2] \cup [0, 2]$; f is falling on $[-2, 0] \cup [2, \infty)$ local maxima at $x = -2, x = 2$; local minimum at $x = 0$.

4. f is rising on $(-2, 0)$ and $(0, 2)$, falling on $(-\infty, 2)$ and $(2, \infty)$; local minimum at $x = -2$, local maximum at $x = 2$.

5. $y' = 2x - 1$; $y'' = 2$; $y' = 0$ at $x = 1/2$; $y'' \neq 0$. The graph is falling for $x < 1/2$, rising for $x \geq 1/2$. The graph is always concave up. There is a local minimum at $(1/2, -5/4)$.

6. $y = 4x^2 + 8x + 1 \Rightarrow y' = 8x + 8 = 8(x+1)$, $y'' = 8$; $y' > 0$ (graph of y rising) for $x \geq -1$, $y' < 0$ (graph of y falling) for $x < -1$, local minimum at $x = -1$, $y = -3$; $y'' > 0 \Rightarrow y$ is concave up.

7. $y' = 3x^2 - 12x + 9 = 3(x-3)(x-1)$; $y'' = 6x - 12$

x	y	y'	y''	concavity	ext/inf	
$(-\infty, 1)$		pos	neg	down		rising
$x = 1$	5	0	neg	down	max	
$(1, 2)$		neg	neg	down		falling
$x = 2$	3	neg	0	–	inf	falling
$(2, 3)$		neg	pos	up		falling
$x = 3$	1	0	pos	up	min	
$(3, \infty)$		pos	pos	up		rising

8. $y = -2x^3 + 6x^2 - 3$, $y' = -6x^2 + 12x = -6x(x-2)$, $y'' = -12x + 12 = -12(x-1)$; $y' > 0$ (graph rising) on $(0, 2)$; $y' < 0$ (graph falling) on $(-\infty, 0)$ and $(2, \infty)$; local minimum at $(0, -3)$, local maximum at $(2, 5)$; inflection point at $(1, 1)$. Concave up for $x < 1$, concave down for $x > 1$.

9. $y' = 8x^3 - 8x = 8x(x^2 - 1) = 8x(x-1)(x+1)$

$y'' = 24x^2 - 8 = 8(3x^2 - 1) = 24(x^2 - 1/3) = 24(x - \frac{1}{\sqrt{3}})(x + \frac{1}{\sqrt{3}})$

x	y	y'	y''	concavity	ext/inf	
$(-\infty, -1)$		–	+	up		falling
$x = -1$	-1	0	+	up	min	—
$(-1, -1/\sqrt{3})$		+	+	up		rising
$x = -1/\sqrt{3}$	-1/9	+	0		inf	rising
$(-1/\sqrt{3}, 0)$		+	–	down		rising
$x = 0$	1	0	–	down	max	—
$(0, 1/\sqrt{3})$		–	–	down		falling
$x = 1/\sqrt{3}$	-1/9	–	0		inf	falling
$(1/\sqrt{3}, 1)$		–	+	up		falling
$x = 1$	-1	0	+	up	min	—
$(1, \infty)$		+	+	up		rising

10. $y = x^4 - 2x^2$, $y' = 4x^3 - 4x = 4x(x^2 - 1) = 4x(x-1)(x+1)$; $y'' = 12x^2 - 4 = 4(3x^2 - 1) = 12(x - \sqrt{1/3})(x + \sqrt{1/3})$; falling on $(-\infty, -1]$ and $[0, 1]$; rising on

$[-1,0]$ and $[1,\infty)$; local minima at $(\pm 1,-1)$; local maximum at $(0,0)$; inflection points at $(\pm\sqrt{1/3},-0.5\bar{5})$. Concave down on $(-1/\sqrt{3},1/\sqrt{3})$; concave up elsewhere.

11. The viewing rectangle $[-3,3]$ by $[-5,15]$ shows the global behavior. Zoom in to determine the behavior of the function at the "bend". ($[0,2]$ by $[10.5,11.3]$ indicates the falling behavior.) $y' = 6x^2 - 10x + 4 = 2(3x^2 - 5x + 2) = 2(x-1)(3x-2)$. $y'' = 12x - 10 = 2(6x-5)$. The graph is concave down on $(-\infty,5/6)$, concave up on $(5/6,\infty)$. $(5/6,11.019)$ is an inflection point. By the second derivative test, there is a local maximum at $(1,11)$ and a local minimum at $(1.5,11.5)$. Falling on $[2/3,1]$, rising elsewhere.

12. $y = 4x^3+21x^2+36x-20$, $y' = 12x^2+42x+36 = 6(2x^2+7x+6) = 12(x+2)(x+3/2)$. $y'' = 24x+42 = 6(4x+7)$; rising on $(-\infty,-2]$ and $[-3/2,\infty)$; falling on $[-2,-3/2]$; local maximum at $(-2,-40)$, local minimum at $(-3/2,-40.25)$ inflection point at $(-1.75),-40.125)$. Concave down for $x < 1.75$, concave up for $x > 1.75$.

13. The viewing rectangle $[-3,3]$ by $[-12,4]$ indicates the behavior. $y' = 12x^3 - 2x = 2x(6x^2-1) = 12x(x-1/\sqrt{6})(x+1/\sqrt{6})$. $y'' = 36x^2 - 2 = 36(x - 1/\sqrt{18})(x+1/\sqrt{18})$. There are inflection points at $(\pm 1/\sqrt{18},-10.046))$; local minima at $(\pm 1/\sqrt{6},-121/12)$ and a local maximum at $(0,-10)$. Concave up for $x < -1/\sqrt{18}$ and $x > 1/\sqrt{18}$, concave down on $(-1/\sqrt{18},1/\sqrt{18})$. Falling on $(-\infty,-1/\sqrt{6}]$ and $[0,1/\sqrt{6}]$; rising elsewhere.

14. $y = 20 + 2x^2 - 9x^4$, $y' = 4x - 36x^3 = 4x(1-9x^2) = 4x(1-3x)(1+3x)$, $y'' = 4 - 108x^2 = 4(1-27x^2)$; increasing on $(-\infty,-1/3]$ and $[0,1/3]$; decreasing on $[-1/3,0]$ and $[1/3,\infty)$; local minimum at $(0,20)$; local maxima at $(\pm 1/3,20.111)$, inflection points at $(\pm 0.192,20.06)$. Concave up on $(-0.192,0.192)$; concave down otherwise.

15. The viewing rectangle $[0,2\pi]$ by $[0,7]$ indicates the behavior of the function. $y' = 1+\cos x$; $y' = 0$ at (π,π). For $x \neq \pi$, $y'(x) > 0$ and the graph is rising. $y'' = -\sin x$; $y''(\pi) = 0$, $y''(\pi^-) < 0$, $y''(\pi^+) > 0$, so (π,π) is an inflection point $y(0) = 0$, $y(2\pi) = 2\pi$. Always rising; minimum at $(0,0)$, maximum at $(2\pi,2\pi)$. Concave down for $0 < x < \pi$, concave up for $\pi < x < 2\pi$.

16. $y = x - \sin x$, $y' = 1 - \cos x$, $y'' = \sin x$; on $(0,2\pi)$, $y' > 0$ so y is always rising; local minimum at $(0,0)$, local maximum at $(2\pi,2\pi)$; inflection point at (π,π). Concave up for $x < \pi$, down for $x > \pi$.

17. The viewing rectangle $[-4,4]$ by $[-25,25]$ shows the complete graph. To find the local extrema, graph $y' = 4x^3 - 16x + 4 = 4(x^3 - 4x + 1)$. $y'' = 4(3x^2 - 4) = 12(x - \frac{2}{\sqrt{3}})(x + \frac{2}{\sqrt{3}})$. The zeros of y' are, approximately $x = -2.115, 0.254, 1.861$. Using technology to evaluate y: local minima at $(-2.115, -22.236)$ and $(1.861, -6.268)$; local maximum at $(0.254, 2.504)$; inflection points at $(-2/\sqrt{3}, -11.508)$ and $(2/\sqrt{3}, -2.270)$. Concave down on $(-2/\sqrt{3}, 2/\sqrt{3})$, concave up elsewhere. Falling on $(-\infty, -2.115] \cup [0.254, 1.861]$, rising elsewhere.

18. Graph on $[-4,4]$ by $[-10,20]$; use FMIN, FMAX, and INFLC to analyze the graph. Concave up between inflection points at $(0,1)$ and $(2,9)$; local maxima at $(-0.532, 2.446)$ and $(2.880, 16.234)$; local minimum at $(0.653, -0.680)$; rising on $(-\infty, -0.532] \cup [0.653, 2.880]$; falling elsewhere.

19. $y' = -4x^3 + 4x - 3$; $y'' = -12x^2 + 4 = -4(3x^2 - 1)$. The only zero of y' is, approximately -1.263. Using technology, we find: rising on $(-\infty, -1.263]$, local maximum at $(-1.263, 2.435)$; falling thereafter; inflection points at $(-1/\sqrt{3}, 0.288)$ and $(1/\sqrt{3}, -3.176)$; concave up on $(-1/\sqrt{3}, 1/\sqrt{3})$; concave down elsewhere. The viewing rectangle $[-3,3]$ by $[-6,6]$ shows a complete graph.

20. $y = 2x^4 - x^2 - 3x + 5$, $y' = 8x^3 - 2x - 3$ which has only one zero; the graph has no hidden behavior. Graph on $[-1.5, 1.5]$ by $[0, 10]$; local minimum at $(0.836, 2.770)$, inflection points at $(-0.289, 5.797)$ and $(0.289, 4.065)$. Concave down $(-0.289, 0.289)$; falling for $x \leq 0.836$.

21. Graph on $[-4,4]$ by $[-6,6]$. Graph is always rising; inflection point at $(0,3)$; concave up for $x < 0$, concave down for $x > 0$.

22. Graph on $[-4,4]$ by $[-2,8]$. Concave down for $x < 0$; inflection point at $(0,5)$; concave up for $x > 0$. Always falling.

23. Defined for $x \geq 0$; always rising and concave down; local minimum at $(0,-1)$.

24. Defined for $x \geq 0$; always rising and concave down; local minimum at $(0,0)$.

25. The viewing rectangle $[-3,3]$ by $[0,3]$ shows a complete graph. The function is always increasing, concave up, for $x < 2$, inflection point at $(2, 2.5)$.

26. Graph on $[-5,10]$ by $[-10,0]$; graph is always falling, inflection point at $(2,-3.5)$. Concave down for $x < 2$.

27. Graph on $[-4,4]$ by $[-6,6]$. $f(0)=1$ is a local minimum. Always rising, concave down for $x<0$, concave up for $x>0$. $x=0$ is not an inflection point since there is no tangent line at $x=0$.

28. Local maximum at $(1,2)$; no concavity for $x<1$, concave down for $x>1$. Rising for $x<1$, falling for $x>1$.

29. Graph $y=(x\tilde{}(1/3))^2(3-x)$ on $[-3,4]$ by $[-6,15]$. Local minimum and inflection point at $(0,0)$; local maximum at $(1.200, 2.033)$; $y''(-0.6)=0$, $y''(-0.7)>0 \Rightarrow$ concave up for $x<-0.6$, concave down on $(-0.6,0)$ and $(0,\infty)$; concave down for $x>0$. Rising on $[0,1.2]$.

30. Defined for $x\geq 0$; graph on $[0,10]$ by $[-6,15]$. Local minimum at $(0,0)$; local maximum at $(2.143, 5.060)$; always concave down. Rising on $[0,2.143]$.

31. Graphing this does not show the concavity: $y'=\frac{4}{3}x^{1/3}(1-\frac{1}{x})$ and $y''=\frac{4}{9}x^{-2/3}[1+\frac{2}{x}]$ tell us that $(0,0)$ and $(-2,7.560)$ are inflection points; local minimum at $(1,-3)$; concave up for $x>0$ and $x<-2$, concave down for $-2<x<0$. Falling on $(-\infty,1]$, rising on $[1,\infty)$.

32. Defined for $x\geq 0$; local minimum at $(0,0)$; inflection point at $(1.800, 5.560)$; concave down for $x<1.800$, concave up for $x>1.800$. Confirm by calculating $y''=\frac{1}{16}x^{-3/4}[5-9/x]$. Always rising.

33. $s(t)=t^2-4t+3=(t-3)(t-1)$, $\nu(t)=2t-4=2(t-2)$, $a(t)=2$.

time	$a(t)$	$\nu(t)$	speed $=\lvert\nu\rvert$	direction
$t<2$	pos	neg, inc	slowing	to the left
$t=2$	pos	0		stopped
$2<t$	pos	pos, inc	gaining	to the right

34. $s=6-2t-t^2$, $\nu=-2-2t=-2(t+1)$, $a=-2$.

time	$a(t)$	$\nu(t)$	speed $=\lvert\nu\rvert$	direction
$t<-3.646$	neg	pos, dec	dec	to the left
$-3.646<t<-1$	neg	pos, dec	dec	to the right
$t=-1$	neg	0, dec	0	stopped
$-1<t$	neg	neg, dec	inc	to the left

35. $s = t^3 - 3t + 3,\ \nu = 3t^2 - 3 = 3(t-1)(t+1),\ a = 6t.$

time	$a(t)$	$\nu(t)$	speed	direction
$t < -1$	neg	pos, dec	dec	to the right
$t = -1$	neg	0	0	stopped
$-1 < t < 0$	neg	neg, dec	inc	to the left
$0 < t < 1$	pos	neg, inc	dec	to the left
$1 < t$	pos	pos	inc	to the right

36. $s = 3t^2 - 2t^3 = t^2(3-2t),\ \nu = 6t - 6t^2 = 6t(1-t),\ a = 6 - 12t.$

time	$a(t)$	$\nu(t)$	speed	direction
$t < 0$	pos	neg, inc	dec	to the left
$0 < t < 1/2$	pos	pos, inc	inc	to the right
$1/2 < t < 1$	neg	pos, dec	dec	to the right
$1 < t$	neg	neg, dec	inc	to the left

37. $y' = 0$ at $1, 2$;

$$\begin{aligned} y'' &= 2(x-1)(x-2) + (x-1)^2 \\ &= (x-1)[2x - 4 + x - 1] \\ &= (x-1)[3x-5] \end{aligned}$$

x	y'
1^-	neg
1	0
1^+	neg
2^-	neg
2	0
2^+	pos

x	y''
1^-	pos
1	0
1^+	neg
$5/3^-$	neg
$5/3$	0
$5/3^+$	pos

y has a local minimum at $x = 2$; y has inflection points at $x = 1,\ 5/3$.

38. Graph y' on $[-5, 8]$ by $[-7, 5]$; y' changes sign at 2 and at 4; local maximum at 2, local minimum at 4. Inflection points correspond to local extremes of y': $x = 1,\ x = 1.634,\ x = 3.366$.

39. $y' = 1 - \frac{1}{x^2}$; $y'' = \frac{2}{x^3}$; $y'(1) = 0,\ y''(1) > 0$; $y'(-1) = 0$; $y''(-1) < 0$. y has a local maximum at $(-1, -2)$ and a local minimum at $(1, 2)$.

40. $y' = \frac{1}{2} - \frac{2}{(2x-1)^2},\ y'' = \frac{8}{(2x-1)^3}$; y'' changes sign only at $x = 1/2$, where y is not defined $\Rightarrow$ no inflection points. $y' = 0 \Rightarrow (2x-1)^2 = 4$ or $2x - 1 = \pm 2 \Rightarrow x =$

$3/2$ or $x = -1/2$, $y''(3/2) > 0 \Rightarrow$ local minimum at $(3/2, 1.25)$; $y''(-1/2) < 0 \Rightarrow$ local maximum at $(-1/2, -.75)$.

41. No, f might have an inflection point at c. Example: $f(x) = x^3 + 1$ has $f' = 0$ at 0, but $x = 0$ is not a local extremum.

42. No, f could have a local max or min at c, e.g., $f(x) = (x-1)^4$.

43. $y'' = 2a$ which is never 0. The statement is true.

44. $y'' = 6ax + 2b$ equals 0 and changes sign at $x = -b/3a$. True.

45. a) The velocity is zero when the tangent to $y = s(t)$ is horizontal: at $t = 2$, $t = 6$, $t = 9.5$.

b) The acceleration is zero when the graph has an inflection point: at $t = 4$, $t = 8$, $t = 11.5$.

46. a) A horizontal tangent implies $D = 0$ at $t = 0, 4, 12$, and 16, b) $s'' = 0$ (acceleration $= 0$) at the inflection points: $t = 1.5, 5.5, 7, 10.5, 13.5$.

47.

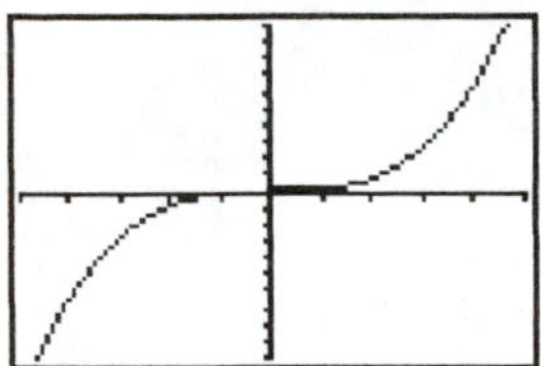

48.

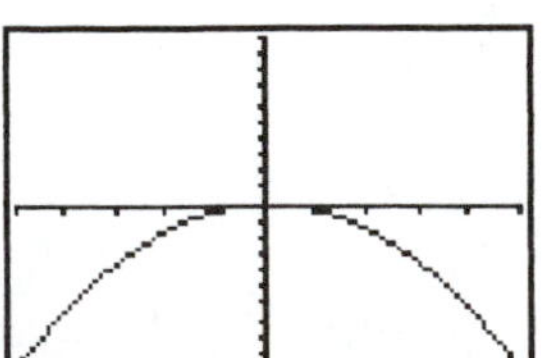

49.

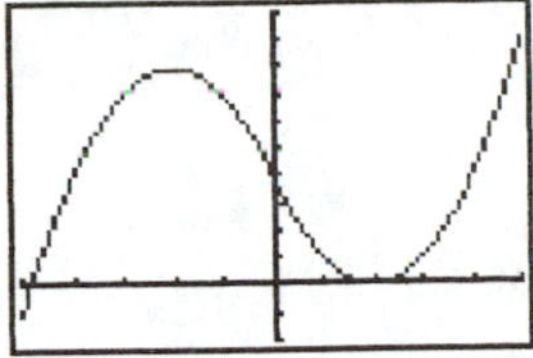

50.

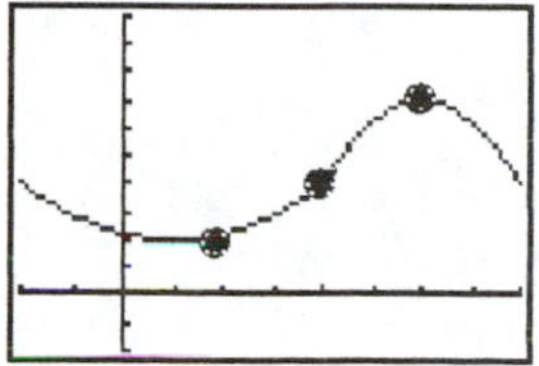

51. Look at both graphs in the viewing rectangle $[-1,1]$ by $[-1,1]$.

52. Graph $\sin x$ and $L(x) = -x + \pi$ on $[\pi - 2, \pi + 2]$ by $[-1.5, 1.5]$.

53. Look at both graphs in $[-0.5, 0.5]$ by $[-5, 5]$.

54. Graph $f(x)$ and $L(x)$ on $[\sqrt{3} - 2, \sqrt{3} + 2]$ by $[\sqrt{3} - 2, \sqrt{3} + 1]$.

55. a) $f = x^3 - 9x \Rightarrow f' = 3x^2 - 9 \Rightarrow f'' = 6x$ which changes sign at $x = 0$.

b) $y - f(a) = (x-a)f'(a) \Rightarrow y = a^3 - 9a + (x-a)(3a^2-9) = (-2a^3) + x(3a^2-9)$.

c) Graph on $[-4, 4]$ by $[-12, 12]$; y_1 lies above y_2; $f''(1) > 0 \Rightarrow$ the graph is concave up at $x = 1$.

56. The graph looks linear as one ZOOMs. The inflection point is at $(-0.317, -1.299)$.

57. Graph in $[0, 300]$ by $[-20, 150]$ using $0 \le t \le 10$.

58. The ball is in the air until $y = 0$, i.e., $t = \frac{V_0 \sin\alpha}{16}$; the range is $x = \frac{V_0^2 \sin\alpha}{16} \cos\alpha$; maximum height occurs when $y'(t) = 0$, i.e., $t = \frac{V_0 \sin\alpha}{32} \cdot y(\frac{V_0 \sin\alpha}{32}) = \frac{V_0^2 \sin^2\alpha}{32} - \frac{16V_0^2 \sin^2\alpha}{32^2} = \frac{V_0^2 \sin^2\alpha}{32}[\frac{1}{2}]$.

59. Setting range = maximum height $\Rightarrow \frac{V_0^2 \sin\alpha\cos\alpha}{16} = \frac{V_0^2 \sin^2\alpha}{64} \Rightarrow \cos\alpha = \frac{1}{4}\sin\alpha \Rightarrow$ $\tan\alpha = 4 \Rightarrow \alpha = 75.964°$.

4.3 POLYNOMIAL FUNCTIONS, NEWTON'S METHOD, AND OPTIMIZATION

1. The procedure: $0 \to$ Ans, followed by repeating the iteration: Ans $-$ (Ans^2 + Ans $-$ 1) $\div$ (2 Ans + 1) leads to the sequence:

1
0.6666666667
0.619047619
0.6180344478
0.6180339888
0.6180339888

and we conclude 0.6180339888 is a solution. Starting with -5, gives the solution -1.618033989.

2. $x_{n+1} = x_n - (x_n^3 + x_n - 1)/(3x_n^2 + x_n)$. Unique solution is $+0.682327\ldots$

3. A graph in $[-10, 10]$ by $[-5, 5]$ indicates there are two solutions. The iteration for Newton's method is ANS $-$ (ANS^4 + ANS $-$ 3)/(4 ANS^3 + 1). A starting value of $x = 1$ leads to $x = 1.16403514$. A starting value of $x = -1$ leads to $x = -1.452626879$.

If $y1 = x^4 + x + 3$, $y2 = \text{NDER}(y1, x)$, the sequence: $1 \to x$ (enter), followed by $x - y1/y2 \to x$ (enter) (enter) (enter) ... will display the first set of iterates.

4. $x_{n+1} = x_n - (2x_n - x_n^2 + 1)/(2 - 2x_n)$; $x_0 = -2 \to x = -.4142135\ldots$, $x_0 = 3 \to x = 2.4142135\ldots$.

5. Graphing the function in $[-10, 10]$ by $[-10, 10]$ indicates there are two solutions. The iteration for Newton's method is $x - (x^4 - 2x^3 - x^2 - 2x + 2)/(4x^3 - 6x^2 - 2x - 2) \to x$. Starting with $x = 1$ leads to 0.6301153962. Starting with $x = 5$ leads to 2.5732719864.

6. $x_{n+1} = x_n - (2x_n^4 - 4x_n^2 + 1)/(8x_n^3 - 8x_n)$; $x_0 = 1 \to x = 1.306565296$, $x_0 = 0.5 \to x = 0.5411961001$, $x_0 = -0.5 \to x = -0.5411961001$, $x_0 = -1 \to x = -1.30656296$.

7. A graph in $[-10, 10]$ by $[-10, 10]$ indicates there are two solutions. The iteration is: $x - (9 - \frac{3}{2}x^2 + x^3 - \frac{1}{4}x^4)/(-3x + 3x^2 - x^3) \to x$. If $x_0 = 1$, Newton's method converges to 3.216451347. If $x_0 = -1$, Newton's method converges to -1.564587289.

8. $x_{n+1} = x_n - (x_n^5 - 5x_n^3 + 4x_n + 5)/(5x_n^4 - 15x_n^2 + 4)$; $x_0 = -4 \to x = -2.153634878$.

9. The iteration is $x_{n+1} = x_n - (x_n^4 - 2)/4x^3$ $x_0 = 1 \Rightarrow x_1 = 1.25$, $x_2 = 1.1935$, $x \to 1.89207115$.

10. $x_1 = -1.2499..$, $x_2 = -1.193500..$, $x \to -1.189207115$.

11. If $f(x_0) = 0$, then $x_1 = x_0 - f(x_0)/f'(x_0) = x_0 - 0/f'(x_0)$. If $f'(x_0) \neq 0$ then x_1 and all subsequent approximations are equal to x_0. If $f'(x_0) = 0$, a calculator may show a "math error".

12. If $x_0 = 0$, the tangent line is horizontal, i.e. there is no x_1. Attempts to use a program leads to an error message of "division by 0". On the other hand, $x_0 = 3 \to x = -4.7123889 = -3\pi/2$. You must have x_0 very close to $\pi/2$.

13.

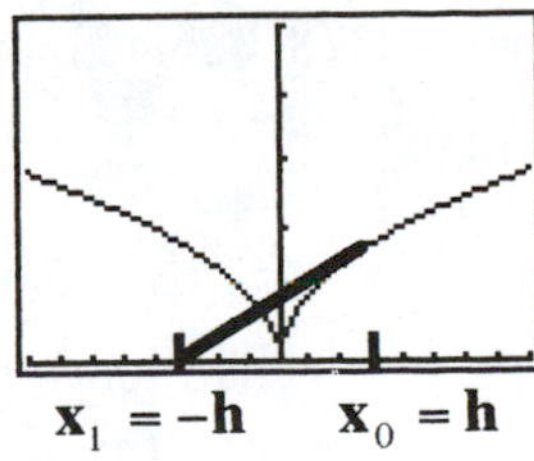

If $x_0 = h > 0$, $f(x) = \sqrt{x}$ and $x_1 = x_0 - \frac{\sqrt{x_0}}{\frac{1}{2\sqrt{x_0}}} = x_0 - 2x_0 = -x_0$. If $x_0 = -h < 0$, $f(x) = \sqrt{-x}$ and $x_1 = x_0 - \frac{\sqrt{-x_0}}{\frac{-1}{2\sqrt{-x_0}}} = x_0 + 2\sqrt{x_0^2} = x_0 + 2|x_0| = -h + 2h = h$.

14.

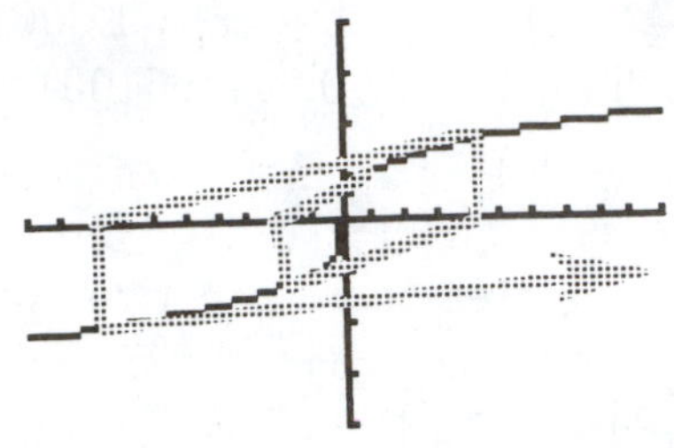

$x_0 = 1 \to x_1 \approx -2,\ x_2 \approx 4,$
$x_3 \approx -8,\ x_4 \approx 16,$
$x_{n+1} = x_n - x_n^{1/3}/(\frac{1}{3}x_n^{-2/3})$
$= x_n - 3x_n = -2x_n$
$|x_n| \to \infty$ as $x_n \to \infty$

15. $y' = 3x^2 - 6x + 5$ which is always > 0. The graph is always rising. $y'' = 6x - 6 = 6(x-1)$; the graph is concave down for $x < 1$, concave up for $x > 1$, has an inflection point at $(1, -1)$. The viewing rectangle $[-5, 5]$ by $[-10, 10]$ shows a complete graph. There is one real root.

16. $y' = x^2 - 4x + 4$, $y'' = 2x - 4$; $y' = 0$ at $x = 2$, $y'' = 0$ at $x = 2$, $y'(2^-) > 0$, $y'(2^+) > \Rightarrow$ no local extremum, $y''(2^-) < 0$, $y''(2^+) > 0 \Rightarrow$ inflection point. Graph is always rising, concave down until inflection point at $(2, 10.667)$. One real root.

17. $y' = 3x^2 - 4x - 3$; $y' = 0$ at $(2 \pm \sqrt{13})/3$, $y'' = 6x - 4$; $y'' < 0$ for $x < 2/3$, $y'' > 0$ for $x > 2/3$; local maximum at $(\frac{2-\sqrt{13}}{3}, 8.879)$; local minimum at $(\frac{2+\sqrt{13}}{3}, 1.9354)$; inflection point at $(\frac{2}{3}, 5.407)$. A viewing rectangle of $[-5, 5]$ by $[-10, 10]$ shows a complete graph. There is one real root. Rising on $(-\infty, \frac{2-\sqrt{13}}{3}] \cup [\frac{2+\sqrt{13}}{3}, \infty)$; falling on $[\frac{2-\sqrt{13}}{3}, \frac{2+\sqrt{13}}{3}]$; concave down for $x < \frac{2}{3}$, concave up for $x > \frac{2}{3}$.

18. $y' = 3x^2 + 20x - 23 = (x-1)(3x+23)$, $y'' = 6x + 20$. Rising on $(-\infty, -23/3]$ and $[1, \infty)$, falling on $[23/3, 1]$, concave down on $(-\infty, -10/3)$, concave up on $(-10/3, \infty)$. Local maximum at $(-23/3, 325.48)$, local minimum at $(1, 0)$; inflection point at $(-10/3, 162.7)$. Two distinct roots.

19. $y' = 1 - 12x^2$; $y' = 0$ at $\pm\frac{1}{\sqrt{12}}$, $y'' = -24x$. Local minimum at $(-1/\sqrt{12}, 11.808)$: local maximum at $(1/\sqrt{12}, 12.192)$ inflection point at $(0, 12)$. The viewing rectangle $[-5, 5]$ by $[-10, 10]$ shows a complete graph. There is one real root. Rising on $[-1\sqrt{12}, 1/\sqrt{12}]$, falling elsewhere. Concave up for $x < 0$, concave down for $x > 0$.

20. $y'' = 6x^2 - 2x - 14 = 2(3x^2 - x - 7)$, $y'' = 12x - 2$. Rising to a local maximum at $(-1.37, 0.160)$, falling to a local minimum at $(1.70, -28.864)$, inflection point at $(1/6, -14.352)$; three real roots. Concave down for $x < 1/6$, concave up for $x > 0$.

21. The viewing rectangle $[-10, 10]$ by $[-60, 60]$ appears to show a complete graph. Analytically, $y' = -3 - x^2 < 0$. There are no local extremes, the graph is always falling. $y'' = -2x \Rightarrow (0, 20)$ is an inflection point. The graph was complete. Concave up for $x < 0$, concave down for $x > 0$. One real root.

22. Graph on $[-3, 3]$ by $[-60, 10]$; $y' = -27x^2 + 4$, $y'' = -54x$, falling to local minimum at $(-2/\sqrt{27}, -16.03)$, then rising to local maximum at $(2/\sqrt{27}, -13.97)$, then falling; inflection point at $(0, -15)$. Concave up for $x < 0$, concave down for $x > 0$.

23. The viewing rectangle $[-5, 5]$ by $[0, 20]$ appears to show a complete graph. To be sure that a slight "wiggle" hasn't been missed, look at $y' = 4x^3 + 2x + 1$ and $y'' = 12x^2 + 2 > 0$. The graph is always concave up. It is falling to a local minimum at $(-0.383, 7.785)$, then rising. The function is never zero.

24. $y' = 4x^3 - 24x^2 + 34x - 10$, $y'' = 12x^2 - 48x + 34$. Graph on $[-3, 6]$ by $[-30, 10]$ to see a complete graph; zoom in to $[.85, 2.4]$ by $[-2.5, 2.5]$ to ascertain that there are 4 real zeros. Local minima at $(4.056, -25.057)$ and $(0.399, -2.766)$; local maximum at $(1.545, 0.324)$; inflection points at $(0.920, -1.325)$ and $(3.080, -14.286)$. Falling on $(-\infty, 0.399] \cup [1.545, 4.056]$, rising elsewhere; concave down on $(0.920, 3.080)$, concave up on $(-\infty, 0.920] \cup [3.080, \infty]$.

25. The rectangle $[-1, 6]$ by $[-30, 10]$ gives a complete graph. Rising to a local maximum at $(0.259, 0.001)$, falling to a local minimum at $(2.574, -24.825)$, then rising; concave down before the inflection point at $(1.417, -12.412)$, then concave up. There is only one zero.

26. $[-10, 5]$ by $[-40, 15]$ shows a complete graph; there are 3 real zeros; falling to a local minimum at $(-3.082, -27,041)$, rising to a local maximum at $(1.082, 9.041)$; inflection point at $(-1, -9)$; concave up for $x < 1$, concave down for $x > 1$.

27. The rectangle $[-5, 5]$ by $[-17, 17]$ suggests a complete graph, but the behavior of the function between $x = -1$ and $x = 1$ is not clear. A graph of the derivative, $y' = 8x^2 - 4x^3 = 4x^2(2 - x)$, shows that $y' \geq 0$ on $[-1, 1]$, i.e., there is no hidden local maximum. $(2.000, 15.333)$ is a local maximum. Rising on $(-\infty, 2.000]$, falling on $[2.000, \infty)$; Concave up between inflection points at $(0, 10)$ and $(1.333, 13.160)$, concave down elsewhere. Two real roots.

28. Graph y and y' on $[-5, 5]$ by $[-12, 4]$, $y' > 0$ for $x > -1.5$ shows there is no hidden behavior; two real roots; falling to a minimum at $(-1.793, -11.397)$, then rising. y' always increasing $\Rightarrow y$ is always concave up.

29. Use $[-3, 3]$ by $[0, 34]$. Falling to a local minimum at $(-1.107, 17.944)$; rising to a local maximum at $(0.270, 20.130)$; falling to a local minimum at $(0.837, 19.927)$; rising thereafter. Concave down between inflection points $(-0.577, 18.867)$ and $(0.577, 20.022)$; concave up elsewhere. No real roots.

30. $[-5, 5]$ by $[-10, 30]$ shows a complete graph; 2 real roots. Rising to a maximum at $(-0.667, 22.222)$, falling to a local minimum at $(0.667, 18.667)$, rising to a local maximum at $(1.000, 18.750)$, falling thereafter. Concave up between inflection points at $(-0.176, 20.747)$ and $(0.843, 18.710)$.

31. Use $[-8, 4]$ by $[-20, 35]$. The graph is rising to a local maximum at $(-4.023, 32.012)$; falling to a local minimum at $(-1.514, -0.189)$; rising to a local maximum at $(0.287, 13.095)$; falling thereafter. Concave up between inflection points $(-3, 18)$ and $(-0.5, 7.063)$; concave down elsewhere. Four real roots.

32. $[-2, 7]$ by $[-10, 30]$ shows a complete graph. Falling to a local minimum at $(-0.236, -6.000)$, then rising to a local maximum at $(2.000, 19.000)$, falling to a local minimum at $(4.236, -6.000)$ then rising. Concave down between inflection points at $(0.709, 5.111)$ and $(3.291, 5.111)$, concave up elsewhere. Four real roots.

33. Use $[-4, 4]$ by $[-10, 55]$. Falling to local minimum at $(-2.601, -7.580)$; rising to local maximum at $(-1.097, 21.196)$; falling to local minimum at $(0.534, -0.495)$; rising to $(2.364, 53.006)$, falling thereafter. Concavity: up until $(-2.016, 4.530)$, down until $(-0.266, 10.206)$, up until $(1.681, 31.029)$, down thereafter. Five real roots.

34. $[-3, 6]$ by $[-40, 20]$ shows a complete graph. Rising to a local maximum at $(0, 1)$, falling to a local minimum at $(2.359, -31.191)$, rising thereafter; inflection point at $(1.678, -19.587)$; concave down for $x < 1.678$, concave up for $x > 1.678$. Three real roots.

35. Use $[-2, 2]$ by $[-50, 50]$; graph y and y'. Since $y' > 0$, the graph is always rising. Inflection point at $(-0.288, -2.922)$. For $x < -0.288$, the graph is concave down. One real root.

36. Graphing y' on $[-6, 6]$ on $[-20, 20]$ shows interesting behavior on $4 < x < 5$. $[-3, 7]$ by $[-20, 80]$ gives a complete graph; $[4, 5]$ by $[43, 44]$ shows the hidden interval on which y decreases. Rising to a local maximum at $(-0.929, 12.856)$, falling to a local minimum at $(0.835, -6.548)$, rising to a local maximum at $(4.340, 43.742)$, falling to a local minimum at $(4.755, 43.500)$, thereafter rising. Concave down on $(-\infty, -0.206)$ and $(2.399, 4.557)$, concave up elsewhere. Inflection points at $(-0.206, 4.424)$, $(2.399, 17.322)$, and $(4.557, 43.617)$. Three real roots.

37. Let x be one number; $20 - x$ is the other. $S(x) = x^2 + (20 - x)^2$, $0 \leq x \leq 20$. $S'(x) = 2x + 2(20 - x)(-1) = 4x - 40$. $S''(x) = 4$. $S'(10 = 0$. The maximum value of $S(x)$ is the maximum of $S(10) = 20$, $S(0) = 400$ and $S(20) = 400$. The numbers are 0 and 20.

38. $A = \ell w$, where $2(\ell + w) = 8$; $A = w(4 - w)$; $A'(w) = 4 - 2w$, $A'' = 4 > 0$, $A'(2) = 0$, $A'(2) > 0 \Rightarrow$ maximum area occurs when $w = \ell = 2$ ft.

39.

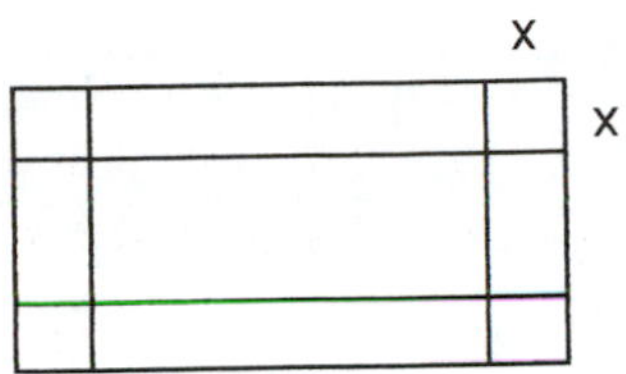

$V(x) = x(8 - 2x)(15 - 2x)$, $0 \leq x \leq 4$. $V'(x) = (8 - 2x)(15 - 2x) - 2x(15 - 2x) - 2x(8 - 2x) = 4x^2 - 46x + 120 + 4x^2 - 30x + 4x^2 - 16x = 12x^2 - 92x + 120$. Solving $V'(x) = 0$ gives $x = \frac{92 \pm 52}{24} = 6$ or $5/3$. 6 is not in the domain of V. Since $V(0) = V(4) = 0$, $V(5/3)$ is the maximum volume. The box has dimensions $5/3 \times 14/3 \times 35/3$ inches.

40.

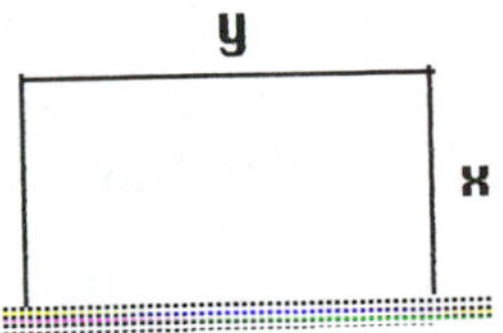

$2x + y = 800$; $A = xy = x(800 - 2x)$; $A' = 800 - 4x$; $A'' = -4 < 0$. $A'(200) = 0$, $A''(200) < 0 \Rightarrow A\,\text{max} = A(200) = 80000\text{m}^2$.

41. a) Since OAB is an isosceles right triangle, B has coordinates $(0,1)$ and the equation of the line AB is $y = -(x-1) = 1-x$

b) $A(x) = 2x(1-x), \quad 0 \le x \le 1$

c) $A'(x) = 2(1-x) + 2x(-1) = -4x + 2$. $A'(1/2) = 0$. Since $A(0) = 0$, and $A(1) = 0$, the largest area is $A(1/2) = 1/2$.

42. $A = 2x(12 - x^2)$; $A' = 24 - 6x^2$; $A'' = -12x$. $A'(2) = 0$, $A''(2) < 0 \Rightarrow \max A = A(2) = 32$.

43. $v = -32t + 100$

a) $v(0) = 100$ft/sec.

b) $v = 0$ when $t = 100/32$ seconds; $s(100/32) = 356.25$ feet.

c) $s = 0$ when $t = (-100 - \sqrt{(100^2 + 64 \cdot 200)})/(-32) = 7.844$. $v(7.844) \approx -150.997$ feet/second.

44. $s'(t) = -32t + 96 = -32(t-3)$; $s'(0) = 96$ ft/sec; $s'(t) = 0$ at $t = 3$, $s(3) = 256$ ft; $s(t) = -16(t^2 - 6t - 7) = -16(t-7)(t+1)$, $s = 0$ at $t = 7$, $s'(7) = -128$ ft/sec.

45. Let one side of the square be x. The length of the box is $108 - 4x$ and its volume is $v(x) = x^2(108 - 4x)$, $0 \le x \le 27$. $v'(x) = 2x(108 - 4x) - 4x^2 = 4x(54 - 2x - x) = 12x(18 - x)$. Since $v(0) = 0$ and $v'(18) < 0$, v has a maximum at $x = 18$. The dimensions of the box are $18 \times 18 \times 36$ inches.

46. a) $D^2 = (x-2)^2 + (y+1/2)^2 = (x-2)^2 + (x^2+1/2)^2$; $2DD' = 2(x-2) + 2(x^2 + 1/2)(2x)$; $D' = 0 \Rightarrow (x-2) + (x^2+1/2)(2x) = 0$ or $x - 2 + 2x^3 + x = 0$, or $x + x^3 - 1 = 0$ which is equivalent to $x = 1/(x^2+1)$. b) For $f(x) = x^3 + x - 1$, Newton's method is $x_{n+1} = x_n - (x_n^3 + x_n - 1)/(3x_n^2 + 1)$; $x_0 = 1 \Rightarrow x_4 = 0.68233$.

47. a) A circle with circumference x has area $\frac{x^2}{2\pi^2}$; in this problem, $2x + 2y = 36$, or $y = 18 - x$. The volume of the cylinder is $v = \frac{x^2 y}{2\pi^2} = x^2(18-x)/(2\pi^2)$, $0 \le x \le 18$. $v'(x) = \frac{1}{2\pi^2}[2x(18-x) - x^2] = \frac{x}{2\pi^2}[36 - 2x - x]$. Since $v(0) = v(18) = 0$, the maximum value of v occurs when $x = 12$ and $y = 6$. (The maximum value is 4.05cm^3).

b) From the diagram, $v = \pi x^2 y = \pi x^2(18 - x)$. As above, the maximum occurs when $x = 12$, $y = 6$; the maximum value is 251.33cm^3.

48.

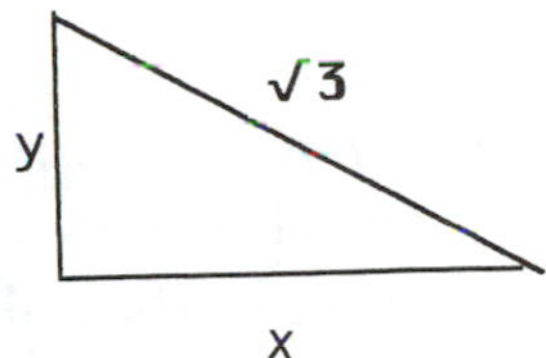

Revolve the line about $\log x$; $v = \frac{1}{3}\pi r^2 h = \frac{1}{3}\pi y^2 x = \frac{1}{3}\pi(3 - x^2)x$; $v' = \frac{\pi}{3}(3 - 3x^2) = \pi(1 - x^2)$; $v'' = -2\pi x$; $v'(1) = 0$, $v''(1) < 0$. $y = r = \sqrt{3 - 1} = \sqrt{2}$, $v(1) = 2\pi/3$ is the maximum area.

49. Exactly 1/4 of the volume of the buoy is submerged, i.e., $\frac{1}{4}(\frac{4}{3}\pi) = \frac{\pi}{3}$ is submerged. The volume of a spherical segment of height x and radius r is $v = \frac{\pi x}{6}(3r^2 + x^2)$. From the diagram, $r^2 + (1 - x)^2 = 1^2$. To find x, solve the equation $v = \frac{\pi}{3}$. $\frac{\pi x}{6}(3 - 3(1 - x)^2 + x^2) = \frac{\pi}{3}$. $x(3 + x^2 - 3 + 6x - 3x^2) = 2$ or $x^3 - 3x^2 + 1 = 0$. Using technology, the root between 0 and 1 is 0.653.

50. The volume submerged is $\frac{1}{3}(\frac{4}{3}\pi) = \frac{4\pi}{9}$. As in problem 49, the volume of the spherical segment is $\frac{\pi x}{6}(3 - 3(1 - x^2) + x^2)$. Equating the two leads to the equation $3x^3 - 9x^2 + 4 = 0$ which has solution $x = 0.774$.

51. $f'(x) = 3x^2 + 2ax + b$

a) Since $f'(-1) = f'(3) = 0$, $f'(x) = c(x + 1)(x - 3) = cx^2 - 2cx - 3c$. Equating coefficients gives $c = 3$, $a = -3$, $b = -9$. Hence $f'(x) = 3x^2 - 6x - 9$; $f''(x) = 6x - 6$; $f''(-1) < 0$, hence f has a maximum at $x = -1$; $f''(3) > 0$, hence f has a minimum at 3.

b) $f'' = 6x + 2a$. Since $f''(1) = 0$, $a = -3$. $f'(x) = 3x^2 - 6x + b$. If $f'(4) = 0$, $b = -24$. $f''(4) > 0 \Rightarrow f(4)$ is a minimum.

52. $x'(t) = v(t) = (t-4)^4 + 4(t-1)(t-4)^3 = (t-4)^3[t-4+4t-4] = (t-4)^3[5t-8]$; a) $v = 0$ at $t = 4$, $t = 8/5$; b) $v < 0$ for $8/5 < t < 4$, hence the particle moves left, c) the maximum negative value of v occurs when $v' = 0$: $v' = 3(t-4)^2[5t-8] + 5(t-4)^3$; $v'(11/5) = 0$, $v(11/5) = -17.496$; $\max|v| = 17.496$.

53. $h = (y + 3)$, radius of cone $= x = \sqrt{9 - y^2}$, $v = \frac{1}{3}\pi(9 - y^2)(y + 3)$; $v' = \frac{\pi}{3}[(-2y)(y + 3) + (9 - y^2)]$; $v' = 0 \Rightarrow -2y^2 - 6y + 9 - y^2 = 0$ or $3y^2 + 6y - 9 = 0 \Rightarrow y^2 + 2y - 3 = 0 \Rightarrow y = 1$; $x = \sqrt{8} = 2\sqrt{2}$, $v = \frac{1}{3}\pi(8)4 = 32\pi/3 =$ maximum volume.

54. From the diagram, $s(w) = kwd^2 = kw(144 - w^2)$, $0 \leq w \leq 12$. $s'(w) = k[144 - w^2 - 2w^2]$. $s'(w) = 0$ at $w = \sqrt{48} = 4\sqrt{3}$ inches. The strongest beam is $4\sqrt{3}$ wide $\times 4\sqrt{6}$ deep.

55. Let F be the stiffness; $F(d) = kwd^2 = k(\sqrt{144 - d^2})d^3$. $[F(d)]^2 = k^2d^6(144 - d^2)$. Differentiating gives $2F(d)F'(d) = k^2[6d^5(144 - d^2) - 2d^7]$. Setting $F'(d) = 0$ leads to $3(144 - d^2) = d^2$. The dimensions of the stillest beam are $d = 6 \times w = 6\sqrt{3}$ inches.

56. If $v(t) = kax(t) - k[x(t)]^2$, then $v'(t) = ka\frac{dx}{dt} - k \cdot 2x(t)\frac{dx}{dt}$. Set $v'(t) = 0$ to get $a = 2x(t)$. $v'(t) = 0$ when $x(t) = a/2$; i.e., the rate is a maximum when exactly half the original tin has crumbled.

57. To maximize $R(p)$, calculate $R'(p) : R'(p) = r(\frac{K-p}{K}) - \frac{rp}{K} = r(K - 2p)/K$; $R'(p) = 0$ when $p = K/2$, one half the carrying capacity. $R''(p) = \frac{-2r}{K} < 0$ so rate is maximized at $p = K/2$.

58. $v' = c[-r^2 + 2r(r_0 - r)] = cr[2r_0 - 3r]$. $\frac{r_0}{2} \leq r < r_0$. $v' = 0$ when $r = \frac{2}{3}r_0$ or when $r = 0$. Since $r = 0$ is not allowed, v has its maximum at $\frac{r_0}{2}$, $\frac{2}{3}r_0$, or r_0 $v(\frac{r_0}{2}) = c\frac{r_0^3}{8}$, $v(\frac{2}{3}r_0) = c\frac{4}{27}r_0$, $v(r_0) = 0$. Hence, the maximum value occurs when $r = \frac{2}{3}r_0$.

59. $R = M^2(\frac{c}{2} - \frac{M}{3})$; $\frac{dR}{dM} = cM - M^2$, $\frac{d^2R}{dM^2} = c - 2M$; $\frac{d^3R}{dM^3} = -2$; $\frac{dR}{dM}$ is maximized if $\frac{d^2R}{dM^2} = 0$ and $\frac{d^3R}{dM^3} < 0$; this occurs at $M = c/2$.

60. deg $p'(x) = 6 \Rightarrow$ degree of $p = 7$. $p'(-1^-) > 0$, $p'(-1)^+ < 0 \Rightarrow p$ has a local maximum at $x = -1$. $p'(5^-) < 0$, $p'(5^+) > 0 \Rightarrow p$ has a local minimum at $x = 5$. p' has a local maximum at $x = -1$, so p has an inflection point at $x = 1$. There are inflection points at $x = -0.48$ and 2.81. The graph of p is concave up on $(-0.48, 1)$ and $(2.81, \infty)$. Schematically, the graph behaves like this:

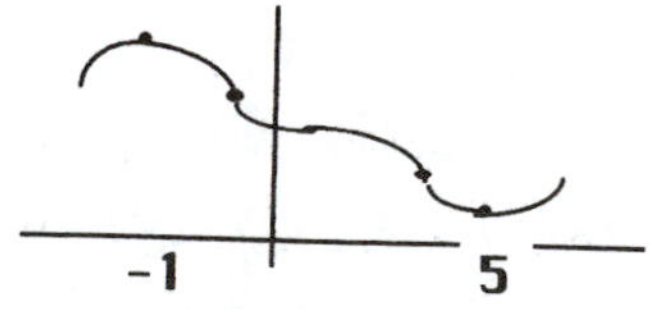

61. The iteration is $x - \frac{(x-1)^{40}}{40(x-1)^{39}}$. The stopping value depends on the computer/calculator used, but is usually about 1.003.

63. a) $f(x) = 0$ can be written as $x^3 = 3x + 1$ and as $x^3 - 3x = 1$. $f(x) = g'(x)$, hence $f = 0$ when $g' = 0$.

b) Use $[-2, 2]$ by $[-1, 1]$.

c) Using technology, the roots are: $1.879, -0.347, -1.532$.

64. Set $f(x) = 2x - \tan x$; use a solver, or zoom, to find $x = 1.16556$. (If you use Newton's Method, start with $x_0 > 1.15$).

65. Use the iteration $x - (\tan x) \times (\cos x)^2 \rightarrow x$. Start with $x_0 = 3$. After two iterations $x = 3.14159$.

66. Newton's Method gives $x_{n+1} = x_n - (x_n - 1 - 0.5\sin x)/(1 - 0.5\cos x)$. $x_0 = 1.6 \Rightarrow x_3 = 1.49870$.

67. b) $v(x) = 2x(24 - 2x)(18 - 2x)$ c) $[-10, 10]$ by $[-1500, 1500]$ shows a complete graph; d) $0 < x < 9$, $[0, 9]$ by $[0, 1500]$ is a graph of the problem situation; e) $x = 3.394$ in; $v = 1309.955$ in^3; analytically $v' = 2(24 - 2x)(18-2x) - 4x(18-2x) - 4x(24-2x) = 24x^2 - 336x + 864 = 24(x^2 - 14x + 36)$. $v' = 0$ at $\frac{14-\sqrt{52}}{2} = 3.3944$; f) $v(x) = 1120$ becomes $x(12 - x)(9 - x) = 140$; since a factor of 10 is needed, try $x = 2$. Looking at the graph suggests $x = 5$ also works.

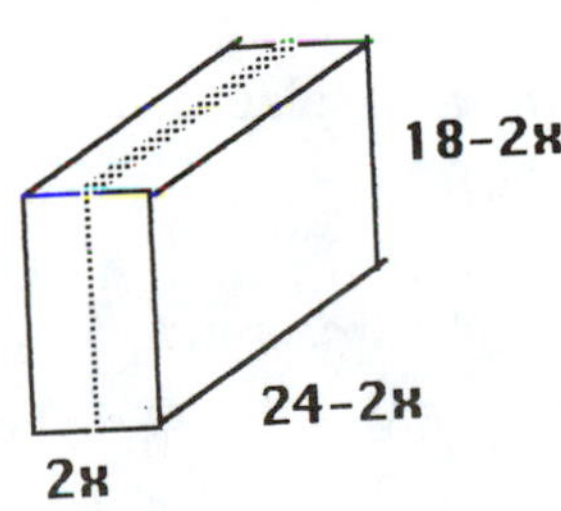

68. If the squares are cut from the 10 inch end, the volume is $v = x(10 - 2x)(\frac{15}{2} - x)$, $0 < x < 5$. Graph on $[0, 5]$ by $[0, 100]$; $v_{\max} = 66.019$ in^3 at $x = 1.962$ in. Analytically, $v' = (10 - 2x)(\frac{15}{2} - x) - 2x(\frac{15}{2} - x) - x(10 - 2x) = 6x^2 - 50x + 75$, $v'(1.96187) = 0$.

69. $v = x(15 - 2x)(5 - x)$. Graph on $[0, 5]$ by $[0, 100]$; $v_{\max} = 66.019$ in^3 at $x = 1.962$ in.

70. a) If $f'(x) = (x-c)^{2k}g(x)$ with $g(c) \neq 0$, f' does not change sign as x increases from c^- to c^+; hence f does not have a local extremum at $x = c$. The number of zeros of f' is $2k + z_g$ where z_g is the number of zeros of $g(x)$. Only the z_g zeros of $g(x)$ are possible extrema of f; $2k$ have been eliminated.

b) If $f'(x) = (x-c)^{2k+1}g(x)$, f' will change sign at $x = c$. Hence f has an extremum there. The number of zeros of f' is $2k+1+z_g$; only the remaining z_g zeros of g are also possible extrema of f.

71. f and g will have extrema at the same values of x-although their nature may be reversed.

72. Assume the extrema, y_1, y_2, y_3, occur as: local maximum, local minimum, local maximum (if not, the following argument applies to $-f$). Either $y_2 < y_3 \leq y_1$ or $y_2 < y_1 \leq y_3$; let $c = \frac{y_2 + \min(y_1, y_3)}{2}$

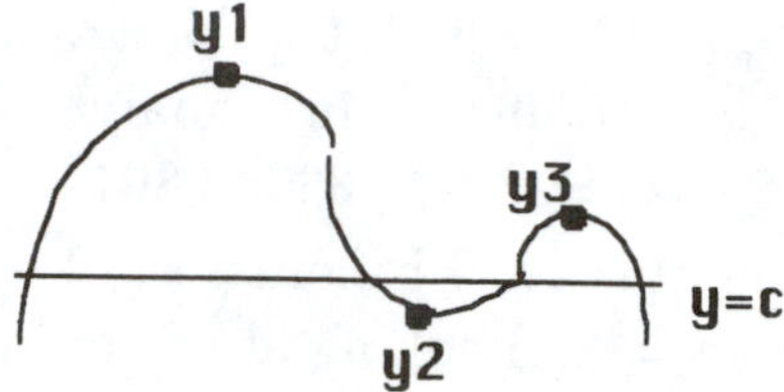

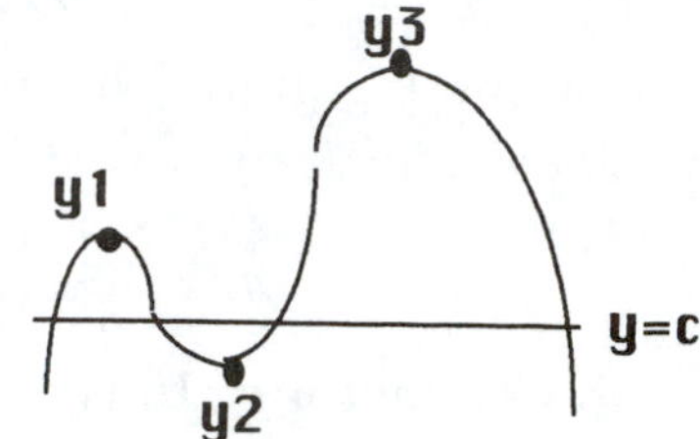

Then $y = c$ intersects the graph at 4 places (x_1, x_2, x_3, x_4), i.e. $f(x) - c = a(x-x_1)(x-x_2)(x-x_3)(x-x_4)$.

73. $y' = 3ax^2 + 2bx + c$; at $x = 0$, $y' = c$. Since the path is tangential to the runway at the point of contact, $c = 0$.

74. Physically, $y'(-L) = 0$, i.e., the airplane has been flying at a constant altitude when it begins the landing path. $y'(-L) = 3aL^2 - 2bL = 0$ ($c = 0$ by #73).

75. $y(0) = 0 \Rightarrow d = 0$. #73 $\Rightarrow c = 0$, #74 $\Rightarrow 3aL - 2b = 0$ since $L \neq 0$, $y(-L) = H \Rightarrow -aL^3 + bL^2 = H$. Solving for a and $b \Rightarrow a = 2H/L^3$, $b = 3H/L^2$. Hence $y = 2H\frac{x^3}{L^3} + 3H\frac{x^2}{L^2}$.

76. Using miles, graph $y = \frac{25000}{5280}[2(\frac{x}{90})^3 + 3(\frac{x}{90})^2]$ in the window $[-90, 0]$ by $[-1, 25000/5280]$.

4.4 RATIONAL FUNCTIONS AND ECONOMIC APPLICATIONS

1. Rewrite as $y = x - \frac{1}{x}$; then $y' = 1 + \frac{1}{x^2}$, $y'' = -\frac{2}{x^3}$. $y' > 0$ for $x \neq 0$, y'' changes sign at $x = 0$. The graph is always rising, vertical asymptote at $x = 0$, concave up for $x < 0$, concave down $x > 0$. The rectangle $[-5, 5]$ by $[-5, 5]$ shows a complete graph.

2. $y = \frac{1}{2}\frac{x^2+4}{x} = \frac{1}{2}[x + \frac{4}{x}]$; $y' = \frac{1}{2}[1 - \frac{4}{x^2}]$, $y'' = \frac{4}{x^3}$; $y' = 0$ at $x = \pm 2$, $y'' \neq 0$ for all x. $[-6, 6]$ by $[-4, 4]$ shows both the behavior at $x = 0$ and end behavior like $y = x/2$. Local maximum at $(-2, -2)$; local minimum at $(2, 2)$. Rising on $(-\infty, -2] \cup [2, \infty)$, falling on $[-2, 0) \cup (0, 2]$; concave down for $x < 0$, concave up for $x > 0$.

3. Rewrite as $y = x^2 + \frac{1}{x^2}$; $y' = 2x - \frac{2}{x^3}$, $y'' = 2 + \frac{6}{x^4}$. Concave up if $x \neq 0$. Local minima at $(\pm 1, 2)$. $[-5, 5]$ by $[0, 5]$ shows a complete graph. Always concave up. Rising on $[-1, 0) \cup [1, \infty)$, falling on $(-\infty, -1] \cup (0, 1]$.

4. Rewrite as $y = x + \frac{1}{x^2}$; $y' = 1 - \frac{2}{x^3}$, $y'' = \frac{6}{x^4}$; local minimum at $x = 2^{1/3}$, no inflection points. Graph on $[-6, 6]$ by $[-4, 4]$.

5. $y = \frac{x}{x^2-4} \Rightarrow y' = \frac{x^2-4-2x^2}{(x^2-4)^2} = -\frac{(4+x^2)}{(x^2-4)^2} < 0$ when defined so the graph is always falling. $y'' = \left[\frac{-2x(x^2-4)^2+(4+x^2)2(x^2-4)\cdot 2x}{(x^2-4)^4}\right]$; clearly $y''(0) = 0$. $[-5, 5] \times [-5, 5]$ appears to be a complete graph. Concavity is: down $(-\infty, -2)$, up $(-2, 0)$, down $(0, 2)$, up $(2, \infty)$. Always falling. Concavity: down $(-\infty, -2)$, up $(-2, 0)$, down $(0, 2)$, up $(2, \infty)$.

6. Graph on $[-4, 4]$ by $[-4, 4]$; the graph suggests that y' is never 0; analysis gives $y' = \frac{(-2x^2+5x-4)}{x^3(x-2)^2}$; the discriminant of the numerator is $25 - 4(-4)(-2) < 0$; hence y' is never 0. No local extrema; the graph of y' has a maximum at $x = 1.223$, hence y has an inflection point at $(1.223, -0.192)$. Rising on $(-\infty, 0)$; falling on $(0, 2) \cup (2, \infty)$; concave up for $x < 0$, $x > 2$ and on $(0, 1.223)$; concave down on $(1.223, 2)$.

7. $[-5, 5]$ by $[-5, 5]$ shows a complete graph of $y = \frac{1}{x^2-1}$; $y' = \frac{-2x}{(x^2-1)^2}$. Graph is rising for $x < 0$, falling for $x > 0$. Local maximum at $(0, -1)$. A graph of y' shows that y' has no local extremes; hence y has no inflection points. Concave up for $|x| > 1$, concave down on $(-1, 1)$. Rising on $(-\infty; -1) \cup (-1, 0]$. Falling on $[0, 1) \cup (1, \infty)$.

8. Not defined at $x = \pm 1$. Graph y, y', y'' on $[-3,3]$ by $[-10,10]$; at $x = 0$, y' changes from positive to negative $\Rightarrow$ y has a local maximum at 0. For $x < -1$, y'' is never zero $\Rightarrow$ no inflection points. Local maximum at $(0,0)$; no inflection points; graph on $[-3,3]$ by $[-10,10]$. Rising on $(-\infty,-1)\cup(-1,0]$; falling on $[0,1)\cup(1,\infty)$; concave up for $|x| > 1$, concave down on $(-1,1)$.

9. Rewrite as $y = \frac{1}{x^2-1} - 1$. The graph will be the graph in #7, decreased by 1. Local maximum at $(0,-2)$.

10. Not defined at $x = \pm\sqrt{2}$ which are vertical asymptotes. Graph y, y', y'' on $[-3,3]$ by $[-10,10]$; y' changes from negative to positive $\Rightarrow$ y has a local minimum at $x = 0$; y'' is never zero $\Rightarrow$ no inflection points. Concave up on $(-\sqrt{2},\sqrt{2})$, concave down for $|x| > \sqrt{2}$, rising on $[0,\sqrt{2})\cup(\sqrt{2},\infty)$, falling on $(-\infty,-\sqrt{2})\cup(-\sqrt{2},0]$.

11. Rewrite as $y = x+1-\frac{3}{x-1}$; $y' = 1+\frac{3}{(x-1)^2}$, $y'' = \frac{-6}{(x-1)^3}$, $[-10,10]\times[-10,10]$ shows a complete graph with end behavior $y = x+1$. No local extrema, no inflection points. Rising on $(-\infty,1)\cup(1,\infty)$; concave up for $x < 1$, concave down for $x > 1$.

12. Not defined at $x = -1$. Graph y, y', y'' on $[-3,3]$ by $[-10,10]$. $y' < 0$ for all x in the domain $\Rightarrow$ y is decreasing on $(-\infty,-1)\cup(-1,\infty)$. y'' changes sign only at $x = -1$ $\Rightarrow$ no inflection points. Concavity changes from down to up at $x = -1$.

13. $[-5,4]$ by $[-10,25]$ gives a complete graph of both y and y'; local minimum at $(0.575, 0.144)$. Inflection point at $(-3,11)$. Concave up on $(-\infty,-3)\cup(-2,\infty)$, concave down on $(-3,-2)$; falling on $(-\infty,-2)\cup(-2,0.575]$, rising on $[0.575,\infty)$.

14. Not defined at $x = 0$. Graph y, y', y'' on $[-3,3]$ by $[-10,10]$. Local minimum at $(2,3)$. No inflection points. Concave up on $(-\infty,0)\cup(0,\infty)$; rising on $(-\infty,0)\cup[2,\infty)$, falling on $(0,2]$.

15. Graph y and y' in $[-4,4]$ by $[-0.5,1.5]$. Find the zeros and local extrema of y' to locate the local extrema and inflection points of y. $(-2.414,-0.207)$ is a local minimum; $(-0.268, 0.683)$ is an inflection point; $(0.414, 1.207)$ is a local maximum; $(1,1)$ is an inflection point. Rising on $[-2.414, 0.414]$, falling on $(-\infty,-2.414]$ and $[0.414,\infty)$; concave down on $(-0.268,1)$, concave up elsewhere.

16. Not defined at $x = 0, \pm 2$. Graph y, y', y'' on $[-5, 5]$ by $[-10, 10]$. Local minimum at $(-0.729, 0.606)$; local maximum at $(0.729, -0.606)$. No inflection points. Falling on $(-\infty, -2) \cup (-2, -0.729] \cup [0.729, 2) \cup (2, \infty)$; rising on $[-0.729, 0) \cup (0, 0.729]$; concave down on $[-\infty, -2) \cup (0, 2)$, concave up on $(-2, 0) \cup (2, \infty)$.

17. Graph y and y' in $[-4, 4]$ by $[-10, 10]$. $(-0.475, -3.331)$ local maximum, $(0.490, 0.800)$ local minimum. Rising on $(-\infty, -2) \cup (-2, -0.475] \cup [0.490, \infty)$; falling on $[-0.475, 0) \cup (0, 0.490]$. Concave up on $(-\infty, -2) \cup (0, \infty)$; concave down on $(-2, 0)$.

18. Not defined at $x = 0, \pm\sqrt{3}$. To see the end behavior of $y = 3x$, graph on $[-5, 5]$ by $[-10, 20]$. Local minimum at $(0.324, 4.499)$; local maximum at $(-0.327, 0.400)$; falling only on $[-0.327, 0) \cup (0, 0.323]$, rising on $(-\infty, -0.327] \cup [0.323, 3) \cup (3, \infty)$; concave up on $(0, 3)$, concave down on $(-\infty, 0)$ and $(3, \infty)$.

19. Two views are needed for a complete graph: $[-2, 6]$ by $[-10, 30]$ shows the end behavior, while $[-2, 2]$ by $[-2, 2]$ shows the local behavior of the left-hand branch. Local minima at $(0.243, -1.589)$ and $(2.543, 18.459)$. Local maximum at $(1.214, -0.869)$; inflection point at $(0.855, -1.158)$. Rising on $[0.243, 1.214] \cup [2.543, \infty)$; falling on $(-\infty, 0.243] \cup [1.214, 2) \cup (2, 2.543]$. Concave up on $(-\infty, 0.855) \cup (2, \infty)$, concave down on $(0.855, 2)$.

20. Not defined at $x = -3$. Graph on $[-5, 5]$ by $[-30, 10]$. Local maxima at $(-0.121, 1.680)$ and $(-3.532, -15.234)$; local minimum at $(-2.347, -1.446)$; inflection point at $(-1.740, -0.440)$. Concave down on $(-\infty, -3) \cup (-1.740, \infty)$, concave up on $(-3, -1.740)$; rising on $(-\infty, -3.532] \cup [-2.347, -0.121]$, falling on $[-3.532, -3) \cup (-3, -2.347] \cup [-0.121, \infty)$.

21. Graph both y and y' in $[-5, 5]$ by $[-15, 15]$. The graph of y' shows that y has two inflection points, at $(0, -0.5)$ and $(-3.005, 10.792)$, and a local minimum at $(1.666, 2.884)$. Falling on $(-\infty, -2) \cup (-2, 1) \cup (1, 1.666]$; rising on $[1.666, \infty)$; concave down on $(-3.005, -2)$ and $(0, 1)$, concave up on $(-\infty, -3.005)$, $(-2, 0)$ and $(1, \infty)$.

22. Not defined at $x = 0, -1$. Graph on $[-4, 4]$ by $[-20, 30]$; local minimum at $(-0.457, 13.892)$; inflection points at $(1.070, 2.486)$ and $(-2.206, 5.572)$. Falling on $(-\infty, -1) \cup (-1, -0.457]$, rising on $[-0.457, 0)$ and $(0, \infty)$; concave up on $(-\infty, -2.206) \cup (-1, 0) \cup (1.070, \infty)$, concave down on $(-2.206, -1) \cup (0, 1.070)$

23. Three views are necessary to show the complete graph: $[-5, 5]$ by $[-100, 100]$ shows the end behavior, $[-3.2, -2.8]$ by $[-100, 100]$ and $[2.8, 3.2]$ by $[-100, 100]$ show a local maximum at $(2.919, 45.572)$ and a local minimum at $(3.077, 62.540)$. Inflection points at $(-3.257, -67.885)$, $(0.004, -0.222)$ and $(2.727, 39.274)$. Rising on $(-\infty, -3) \cup (-3, 2.919] \cup [3.077, \infty)$, falling on $[2.919, 3) \cup (3, 3.077)$; concave up for $-3.257 < x < -3$, $0.004 < x < 2.727$, and $x > 3$; concave down on $(-\infty, -3.257) \cup (-3, 0.004) \cup (2.919, 3)$.

24. Not defined at $x = -5$, $x = 2$. $[-6, 3]$ by $[-150, 20]$ gives a general view but does not show behavior near $x = -5$. Local maximum at $(-0.535, 0.688)$; local minimum at $(0.460, 0.168)$; inflection points at $(-5.209, -135.337)$, $(-0.036, 0.426)$ and $(2.486, 11.099)$. Rising on $(-\infty, -5) \cup (-5, 2) \cup (2, \infty)$; concave downon $(-\infty, -5.209)$, $(-5, -0.036)$, and $(2, 2.486)$; concave up on $(-5.209, -5)$, $(-0.036, 2)$ and $(2.486, \infty)$.

25. Graph y and y' on $[-4, 4]$ by $[-1, 2]$. Local maximum at $(0, 2)$; inflection points at $(\pm 1.155, 1.5)$. Rising on $(-\infty, 0]$, falling on $[0, \infty)$; concave down for $|x| < 1.155$, concave up for $|x| > 1.55$.

26. Graph on $[-15, 15]$ by $[-1.4, 1.3]$; local minimum at $(-2, -1)$, local maximum at $(2, 1)$; inflection points at $(0, 0)$, $(\pm 3.464, \pm 0.866)$. Rising on $[-2, 2]$, falling elsewhere, concave down on $(-\infty, -3.464) \cup (0, 3.464)$; concave up elsewhere.

27. $P = 2x + 2y$, where $xy = 16$. $P(x) = 2x + \frac{32}{x}$, $x > 0$; $P'(x) = 2 - \frac{32}{x^2}$: $P''(x) = \frac{64}{x^3}$. Set $P'(x) = 0$ to obtain $x = 4$. Since $P''(4) > 0$, $P(4)$ is a minimum. $P(4) = 16$.

28. Let x = side of base. Then $V = x^2 h = 50$; $SA = S(x) = x^2 + 4xh = x^2 + 4x(50/x^2) = x^2 + 200/x$, $S' = 2x - 200/x^2$; critical points are $x = 0$ and $x = \sqrt[3]{100}$. Since $S'' = 2 + 400/x^3 > 0$, S has a minimum at $x = 4.642$, $h = 2.321$.

29.

x

y y

x

$F = 3x + 2y$ where $xy = 216$.

$F(x) = 3x + \frac{432}{x}$. $F'(x) = 3 - \frac{432}{x^2}$.

Set $F'(x) = 0$ to obtain $x^2 = 432/3$, i.e., $x = 12$. Since $F''(12) > 0$, $F(12)$ is a minimum. $y = 18$. $F(12) = 72$ m (the "equal parts" condition does not affect the answer).

30. See #28. $SA = S(x) = x^2 + 4x(\frac{500}{x^2})$. $S' = 2x - \frac{(2000)}{x^2}$; critical points are $x = 0$, $x = 10$. Minimum SA occurs at $x = 10$, $h = 5$.

31. From the problem, $x^2y = 1,125$. Thus $C = 5(x^2 + 4x \cdot \frac{1,125}{x^2}) + 10x \cdot \frac{1,125}{x^2} = 5(x^2 + \frac{5,400}{x}) + \frac{11,250}{x} = 5x^2 + \frac{33,750}{x}$. $C'(x) = 10x - \frac{33,750}{x^2}$. Set $C'(x) = 0$: $x^3 = 3,375 \Rightarrow x = 15$ ft, $x = 15 \Rightarrow y = 5$ ft.

32.

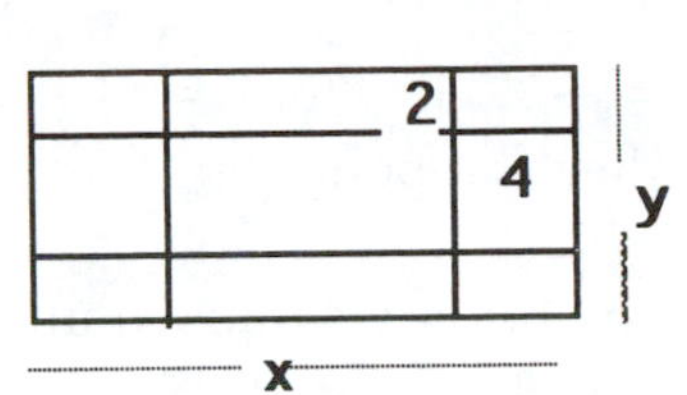

$(x-8)(y-4) = 50$

$A = xy = x[\frac{50}{x-8} + 4]$

$A' = [\frac{50}{x-8} + 4] - \frac{50x}{(x-8)^2} = \frac{50(x-8)+4(x-8)^2-50x}{(x-8)^2}$

$= -\frac{400+4(x-8)^2}{(x-8)^2}$; critical points are at 8, and $x - 8 = \pm 10$. Graphing $A(x)$ on $[0, 30]$ by $[0, 300]$ shows that A is minimized at $x = 18$ in, $y = 9$ in.

33. For this can, $A = 2\pi rh + \pi r^2$ where $\pi r^2 h = 1000$, $A = 2\pi r(\frac{1000}{\pi r^2}) + \pi r^2 = \frac{2000}{r} + \pi r^2$, $\frac{dA}{dr} = \frac{-2000}{r^2} + 2\pi r$. Set $\frac{dA}{dr} = 0$, giving $2000 = 2\pi r^3$ or $r = \frac{10}{\sqrt[3]{\pi}}$, $h = \frac{1000}{\pi r^2} = \frac{1000}{100\cdot\pi}\pi^{2/3} = \frac{10}{\sqrt[3]{\pi}} = r$.

34. $A = 8r^2 + 2\pi rh = 8r^2 + 2\pi r[\frac{1000}{\pi r^2}] = 8r^2 + \frac{2000}{r}$, $0 \le r \le 1$. $A' = 16r - 2000/r^2$; critical points are 0, $\sqrt[3]{\frac{2000}{16}} = 5$, which is not in the domain (the side of a square is 2 units). Hence the minimum occurs at $r = 1$ cm. $h = \frac{1000}{\pi r^2} = \frac{1000}{\pi}$ cm; $\frac{h}{r} = \frac{1000}{\pi}$.

35. Draw $\overline{RT}$. $\overline{PB} = 8.5 - x$, $\overline{QB}$ $= \sqrt{x^2 - (8.5-x)^2} = s$, $\overline{TQ} = \sqrt{L^2 - x^2} - s$.

Use triangle RQT to obtain: $L^2 - x^2 = (8.5)^2 + (\sqrt{L^2 - x^2} - s)^2 \Rightarrow L^2 = x^2 + \frac{17x^2}{4x-17} = \frac{4x^3}{4x-17}$.

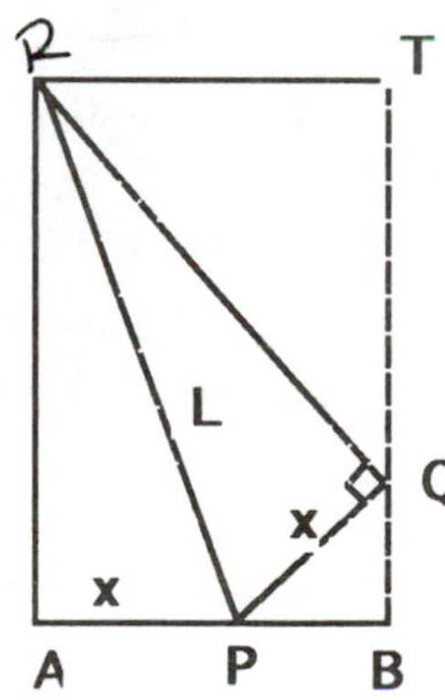

c. Differentiating gives $2LL' = \frac{12x^2(4x-17) - 4x^3(4)}{(4x-17)^2} = \frac{4x^2[12x-51-4x]}{(4x-17)^2}$, $L' = \frac{1}{2L}\frac{4x^2}{(4x-17)^2}[8x - 51]$.

$L' = 0$ at $x = 51/8$, $L'(51/8^-) < 0$, $L'(51/8^+) > 0$,

so there is a local minimum of L at $51/8$.

d. $L(51/8) = 11.04$ inches.

36. Let $d(x) = g(x) - f(x)$ be the distance between the graphs; $d(a) = d(b) = 0$. At a local maximum $c, d'(c) = 0$. $d'(c) = g'(c) = f'(c) = 0 \Rightarrow g'(c) = f'(c)$, i.e., the tangent lines are parallel.

37. $f'(x) = 2x - \frac{a}{x^2}$; $f''(x) = 2 + \frac{2a}{x^3}$.

a. For f to have a local minimum at $x = 2$, $f'(2) = 0$ and $f''(2) > 0$. Set $f'(2) = 0 \Rightarrow 0 = 4 - \frac{a}{4} \Rightarrow a = 16$. Check that $f''(16) > 0$.

b. Set $f''(1) = 0 \Rightarrow 0 = 2 + 2a \Rightarrow a = -1$. If $f''(x) = 2 - 2/x^3$ then $f''(1^-) < 0$, $f''(1^+) > 0$ and $x = 1$ is an inflection point.

38. $f' = 2x - a/x^2$, $f'' - 2 + 2a/x^3 > 0$. The function is concave up for $x > 0$; hence for $x > 0$ there is no local maximum. For $x < 0$, $f' < 0$; i.e., f is always decreasing.

39. Method 1 – using algebra:

$$\begin{aligned} 6x^2 + 8x + 19 &= 6(x^2 + \frac{8}{6}x + \frac{4}{9}) + 19 - \frac{4}{9} \cdot 6 \\ &= 6(x + \frac{2}{3})^2 + \frac{147}{9} > 0. \end{aligned}$$

Method 2 – using calculus: To show $y > 0$ for all x, find the minimum value of y and show that $y \geq \min y > 0$. $y' = 12x + 8$, $y'' = 12$, $y' = 0$ at $x = -2/3$; since $y'' > 0$, this is a minimum $y(-2/3) = \frac{147}{9} > 0$.

40. $f' = 2x - 1 \Rightarrow f'(\frac{1}{2}) = 0$; $f'' = 2$. f has a minimum at $1/2$, i.e., $f(x) \geq f(1/2) > 0$.

41. The profit is $P = Nx - Nc = (x - c)[\frac{a}{(x-c)} + b(100 - x)] = a + b(x - c)(100 - x)$. $P'(x) = b[100 - x - (x - c)] = b[100 + c - 2x]$; $P''(x) = -2b < 0$. P is a minimum when $x = 50 + c/2$.

42. Assume x people go on the tour, $50 \leq x \leq 80$. Then $P(x) = R(x) - C(x) = [200 - 2(x - 50)]x - 6000 - 32x = 200x - 2x^2 + 100x - 6000 - 32x$. $P'(x) = 200 - 4x + 100 - 32$, $P'(67) = 0$. Checking $P(50)$, $P(67)$, $P(80)$ shows $P(67)$ is the maximum profit.

43. $A'(q) = \frac{-km}{q^2} + \frac{h}{2} = \frac{-2km + hq^2}{2q^2}$; $A''(q) = \frac{2km}{q^3} > 0$, $A'(q) = 0$ when $q = \sqrt{(2km/h)}$. $A'' > 0 \Rightarrow$ this is a local minimum.

44. $A(q) = \frac{m(k+bq)}{q} + cm + \frac{hq}{2} = b + \frac{km}{q} + cm + \frac{hq}{2}$. This is the previous cost function increased by a constant. It will have a local minimum for the same value of q, $q = (2km/h)^{1/2}$.

45. The profit is $p(x) = r(x) - c(x) = 6x - x^3 + 6x^2 - 15x$. $p'(x) = 6 - 3x^2 + 12x - 15 = -3(x^2 - 4x + 3)$. $p''(x) = -6x + 12$. The critical points of p are $x = 1$ and $x = 3$. At $x = 1$, since $p''(1) > 0$, p has a local minimum. $p''(3) < 0$; the maximum value of p is $p(3) = 0$. p is never positive, i.e., there is never a profit.

46. The solution to $c(x)/x = c'(x)$, or $c(x) = xc'(x)$ will minimize the average cost: $(x^3 - 20x^2 + 20,000x + 1000) = x(3x^2 - 40x + 20,000) \Rightarrow x = 12.972$. Assuming the solution must be an integer, compare $c(x)/x$ at $x = 12$ and $x = 13$. $c(13)/13 = 1985.923$; $c(12)/12 = 1987.333$. Hence, 13 items is the optimal number.

47. If $f(x) = \frac{p(x)}{q(x)} = \frac{p}{q}$, then $f' = \frac{p'q - q'p}{q^2}$. The zeros of $q^2(x)$ are the zeros of $q(x)$. If $q(x) \neq 0$, x is in the domain of f; if $q(\bar{x}) = 0$, then $x = \bar{x}$ is a vertical asymptote of f. Moreover, if $q(x) \neq 0$, x is in the domain of f'. The tricky part is to show that $q(\bar{x}) = 0 \Rightarrow \bar{x}$ not in the domain of f' (i.e., that the numerator does not cancel the denominator). This requires writing $q(x) = (x - \bar{x})^m r(x)$, where $r(\bar{x}) \neq 0$. Then

$$f'(x) = \frac{p'(x-\bar{x})^m r - p(m(x-\bar{x})^{m-1}r + (x-\bar{x})^m r')}{(x-\bar{x})^{2m} r^2}$$

Divide by $(x - \bar{x})^{m-1}$,

$$f'(x) = \frac{(p'r - pr')(x-\bar{x}) - pmr}{(x-\bar{x})^{m+1} r^2}$$

at $x = \bar{x}$, the denominator is zero but the numerator is $p(\bar{x})mr(\bar{x}) \neq 0$. Hence a rational function and its derivative have the same domain and same vertical asymptotes. Since f'' is the derivative of f', the assertion extends to f''.

48. $f(x) = q(x) + r(x)$ where $r(x) \to 0$ as $x \to \pm\infty$. Since f is rational, r is differentiable. As $r \to 0$, a constant, $r' \to 0$ also.

49. Let the specified volume be V; let the cylinder have radius r and height h. Then $V = \pi r^2 h$; the material M is given by $M = 2\pi rh + 2\pi r^2 = \frac{2\pi rV}{\pi r^2} + 2\pi r^2 = 2[\frac{V}{r} + \pi r^2]$. $M' = 0$ when $\frac{V}{r^2} = 2\pi r$, i.e., $2r = \frac{V}{\pi r^2} = h$. The height will equal the diameter (twice the radius).

50. Let the can have radius r and height h; $V = \pi r^2 h = 1 \Rightarrow h = \frac{1}{\pi r^2}$. Cost $=$ cost of materials $+$ cost of soldering $= 0.80[2\pi r^2 + 2\pi r(\frac{1}{\pi r^2})] + 0.20[\frac{1}{\pi r^2} + 2(2\pi r)]$. Graph on $[0, 3]$ by $[0, 10]$ to find the minimum cost $\approx \$5.97$, which occurs when $r = 0.498$. Analytically, $C'(r) = .8(4\pi r - 2/r^2) + 0.2(-2/\pi r^3) + 4\pi)$. Solving the equation $C'(r) = 0$ graphically shows C has a minimum at $r = 0.497$ ft. ($h = 1.288$ ft)

4.5 RADICAL AND TRANSCENDENTAL FUNCTIONS

Exercises 1-47. For all problems, local extrema and inflection points are given. If intervals for rising, falling and/or concavity are not specified and there are no discontinuities, the graph rises as y goes from a local minimum to a local maximum, etc. Concavity changes at an inflection point. Normally, if y is differentiable at a local maximum (minimum) $x = c$, the graph is concave down (up) on a neighborhood of c.

1. $y = x^{1/3} \Rightarrow y' = \frac{1}{3}x^{-2/3}$ and $y'' = \frac{-2}{9}x^{-5/3}$. The derivative is not defined at $x = 0$; elsewhere $y' > 0$ and y is increasing. For $x < 0$, $y'' > 0$ and y is concave up, for $x > 0$, $y'' < 0$ and y is concave down. $[-5, 5]$ by $[-5, 5]$ confirms this analysis. Inflection point at $(0, 0)$.

2. $y = x^{4/3}$, $y' = \frac{4}{3}x^{1/3}$, $y'' = \frac{4}{9}x^{-2/3}$; $y' > 0$ for $x > 0$, $y' < 0$ for $x < 0$, $y'(0) = 0$. $y'' > 0 \Rightarrow y$ is always concave up. y has a local minimum at $(0, 0)$; y is decreasing for $x < 0$, increasing for $x > 0$.

3. $y = x^{3/2}$, $y' = \frac{3}{2}x^{1/2}$, $y'' = \frac{3}{4}x^{-1/2}$; y defined for $x \geq 0$, $y' > 0 \Rightarrow y$ is always increasing, $y'' > 0 \Rightarrow y$ is concave up; local minimum $(0, 0)$.

4. $y = (1 - x)^{0.2} \Rightarrow y' = -0.2(1 - x)^{-0.8}$, $y'' = 0.16(1 - x)^{-1.8}$. Inflection point at $(1, 0)$. Graph in $[-5, 5]$ by $[-5, 5]$.

5. $y = (2x+3)^{1/3} \Rightarrow y' = \frac{1}{2}(2x+3)^{-1/2}2$ and $y'' = -(2x+3)^{-3/2}$. For $x < -\frac{3}{2}$, y is not defined. For $x > -\frac{3}{2}$, y is rising and concave down. $[-5, 5]$ by $[-5, 5]$ shows the complete graph; $(-1.5, 0)$ is a local minimum.

6. Graph on $[0, 5]$ by $[0, 10]$. Not defined for $x < 2$; $y' = .25(x - 2)^{-.75} > 0$. Always increasing and concave down; $(2, 5)$ is a local minimum.

7. $[-1, 3]$ by $[0, 6]$ gives a good view of slightly more than two complete periods; $[-0.5, 2]$ by $[0, 6]$ shows one period. The "bottoms" correspond to $3x + 5 =$

$3\pi/2$ and $3x + 5 = 7\pi/2$, i.e., $-0.144 \leq x \leq 1.999$ describes a complete period. The graph has a local maximum at $(0.951, 5)$. To find the inflection points analytically, look for the extrema of $y' = 6\cos(3x + 5)$. These occur at $3x + 5 = 2\pi$ and 3π. The inflection points are $(0.428, 3)$, $(1.147, 3)$.

8. Graph y, y' on $[0, \pi]$ by $[-6, 6]$. $y' = 0$ at $\pi/2$ falling on $[0, \pi/2]$, rising on $[\pi/2, \pi]$. $y'' = 12\cos(2x + \pi)$; $y'' = 0$ at $x = \pi/4$, $3\pi/4$; local maximum at $(0, 4)$, local minimum at $(\pi/2, -2)$.

9. $y = \sin 3x + \cos 3x \Rightarrow y' = 3(\cos 3x - \sin 3x)$ three successive zeros of y' will determine a complete period. $\cos 3x = \sin 3x$ at $3x = \pi/4$, $5\pi/4$, $9\pi/4$; the function has a local maximum of $\sqrt{2}/2 + \sqrt{2}/2 = \sqrt{2}$ at $x = \pi/12$ and $x = 9\pi/12$. It has a local minimum of $-\sqrt{2}$ at $x = 5\pi/12$. Inflection points are found by locating the extremes of y'; they are $(3\pi/12, 0)$ and $(7\pi/12, 0)$.

10. $y' = 10\cos 2x - 6\sin 2x$; $y' = 0$ at $\tan 2x = 10/6$, or $x = 2.086, 0.515$. Rising on $[0, 0.515]$, falling on $[0.515, 2.086]$, rising $[2.086, \pi]$. Inflection point at $[1.3, 0]$; local maximum at $(0.515, 5.831)$, local minimum at $(2.086, -5.831)$.

11. Graph $y = (x\text{\^{}}(1/3))\text{\^{}}5$ on $[-5, 5]$ by $[-5, 5]$. The graph appears to be always increasing, concave down for $x < 0$, inflection point at $x = 0$, concave up for $x > 0$. Analytically, $y' = \frac{5}{3}x^{2/3}$, $y'' = \frac{10}{9}x^{-1/3}$; $y' \geq 0$ for all x, y'' changes sign at $x = 0$.

12. Graph $y = x\text{\^{}}^{1.25}$ on $[0, 5]$ by $[0, 5]$; the graph appears to be increasing from a minimum at $(0, 0)$ and concave up. $y' = \frac{5}{4}x^{1/4}$, $y'' = \frac{5}{16}x^{-3/4}$, which are both positive for $x > 0$.

13. $y = \sqrt{2x - 3}$; graph on $[0, 5]$ by $[0, 5]$. The graph appears to be concave down and increasing from a minimum at $(\frac{3}{2}, 0)$; $y' = \frac{2}{2\sqrt{2x-3}}$; $y'' = (-\frac{1}{2})\frac{(2)1}{(2x-3)^{3/2}}$ which is < 0 for $x > \frac{3}{2}$.

14. $y' = -\frac{(3-x)^{-2/3}}{3} < 0$ for $x \neq 3$. The function is always decreasing. There is a vertical tangent at $x = 3$; the concavity changes at $x = 3$; $(3, 5)$ is an inflection point.

15. Graphing y' on $[3, 5]$ by $[-10, 10]$ shows the interval $3 \leq x \leq 5$ includes all possible extrema and inflection points (the domain is $[4, \infty)$) $[0, 50]$ by $[-3, 4]$ shows the function, which increases from a minimum at $(4, -2)$, concave down.

16. Graph on $[-5, 8]$ by $[-4, 4]$. Using NDER graph y' also. Concavity changes at $(0, 0)$, $(3, 0)$; $y' = 0$ at $x = 1.5$; local minimum at $(1.500, -1.310)$.

17. $y = \frac{3}{\sin(2x+\pi)} - 5$. A graph in $[-\pi, \pi]$ by $[-10, 6]$ shows that $(0, \pi)$ is a complete period. y has a minimum of -2 when $2x + \pi = 5\pi/2$, i.e. at $(3\pi/4, -2)$ and a maximum of -8 when $2x + \pi = 3\pi/2$, i.e. at $x = \pi/4$. For this function, a knowledge of trigonometry is more helpful than calculus. Concave down on $(0, \pi/2)$, concave up on $(\pi/2, \pi)$.

18. Graph y and y' in $[-\pi/2, \pi/2]$ by $[-20, 20]$; $-\pi/6 \le x \le \pi/2$ is a complete period. $y' = 0$ at $x = 0$, $x = \pi/3$; local minimum at $(0, 9)$; local maximum at $(\pi/3, 5)$. Concave up on $(-\pi/6, \pi/6)$, concave down on $(\pi/6, \pi/2)$.

19. Graph y and y' (use NDER) on $[-1, 1]$ by $[-3, 3]$. There will be an inflection point at $x = 0$ and where y' has a local minimum, i.e. at $(0, 0)$ and also at $(0.268, 0.716)$. y has a local minimum at $(-0.341, -0.583)$.

20. Graph $y = (x^2)$^(1/3)$-x$ on $[-1.5, 1.5]$ by $[-0.5, 1.5]$; graph y' on $[-1.5, 1.5]$ by $[-4, 2]$. Local maximum at $(0.296, 0.148)$; local minimum at $(0, 0)$. $y'' = -\frac{2}{9}x^{-4/3} < 0$ for $x \ne 0 \Rightarrow$ graph is concave down for $x < 0$ and $x > 0$.

21. Graph y = $((x)$^(1/5))^3 in $[-5, 5]$ by $[-1, 4]$; always increasing; $y'' = -\frac{6}{25}(x+2)^{-7/5} \Rightarrow y'' > 0$ for $x < -2 \Rightarrow$ concave up; concave down for $x > -2$; $(-2, 0)$ is an inflection point.

22. Graph $y = ((3-x)^2)$^(1/3) and y' on $[-2, 4]$ by $[-3, 3]$; local minimum at $(3, 0)$; concave down on $(-\infty, 3)$ and $(3, \infty)$.

23. Graph on $[-\pi, \pi]$ by $[-2, 2]$. Concave up on $(-\pi/4, \pi/4)$ with minimum at $(0, 1)$; concave down on $(\pi/4, 3\pi/4)$ with maximum at $(\pi/2, -1)$.

24. Graph y, y' on $[\pi/6, \pi/6 + \pi/3]$ by $[-3, 10]$; inflection point at $(\pi/3, 0)$; concave down on $(\pi/6, \pi/3)$.

25. $[-5, 5]$ by $[0, 100]$ shows a complete graph. The graph is always rising, always concave up.

26. Graph y on $[-6, 6]$ by $[0, 10]$; y' on $[-6, 6]$ by $[-1, 3]$. Always rising and concave up.

27. Graph $y = 3(\ell n(x+1))/\ell n\, 2$ in $[-5, 5]$ by $[-10, 10]$ the graph is always rising, always concave down.

28. Graph $y = 2\,\ell n(2-x)/\ell n\, 3$, and y' on $[-6, 6]$ by $[-3, 3]$. Falling and concave down for $x < 2$.

29. Use $[-5,5]$ by $[0,4]$. Concave down $x \leq 0$ and $x \geq 2$. Not defined on $(0,2)$; local minima at $(0,0)$ and $(2,0)$.

30. Graph $y = (x^3 - 4x)$^(1/3) and y' on $[-4,4]$ by $[-3,3]$; local maximum at $(-1.155, 1.455)$; local minimum at $(1.155, -1.455)$; inflection points at $(-2,0), (0,0), (2,0)$.

31. $[-\pi/2, 3\pi/2]$ by $[-8,8]$ shows one period. Inflection points at $(0,0)$ and $(\pi, 0)$. Rising on $(-\pi/2, \pi/2)$, falling on $(\pi/2, 3\pi/2)$¡ concave up on $(0, \pi/2) \cup (\pi/2, \pi)$.

32. Graph $y = \cos x/(\sin x)^2$ and y' on $[-\pi, \pi]$ by $[-5, 10]$; inflection points at $(-\pi/2, 0), (\pi/2, 0)$. This is the same graph as #31, shifted $\pi/2$ units to the left.

33. Graph y and y' in $[-\pi/2, 3\pi/2]$ by $[-3,3]$, using a horizontal scale of $\pi/16$.

Local minima at	**Local maxima at**	**Inflection points at**
(−0.968, −1.906)	(0.216, 1.216)	(−0.413, −0.408)
(1.228, −0.223)	(1.914, 0.223)	(0.673, 0.542)
(2.925, −1.216)	(4.109, 1.906)	(1.571, 0)
		(2.469, −0.542)
		(3.554, 0.408)
		(4.712, 0)

34. $[0, 24\pi]$ by $[-2, 2]$ shows two complete periods;

Local maxima at	**Local minima at**	**Inflection points at**
(3.621, 1.906)	(10.724, −1.216)	(0, 0)
(16.792, 0.223)	(20.907, −0.223)	(6.949, 0.408)
(26.975, 1.216)	(34.078, −1.906)	(18.850, 0)
		(13.463, −0.542)
		(24.237, 0.542)
		(30.751, −0.408)

35. Looking at the graph in $[-2\pi, 2\pi]$ by $[-10, 10]$ indicates that $[-\pi, \pi]$ by $[-6, 6]$ will show a complete period. Graph y to find that y has: a local minimum at $(-1.298, -4.132)$, an inflection point at $(-0.858, -3.890)$, a local maximum at $(-0.578, -3.718)$, a local minimum at $(0.578, 3.718)$, an inflection point at $(0.858, 3.890)$, and a maximum at $(1.298, -4.132)$.

36. Graph y, y' on $[0, 2\pi]$ by $[-4, 4]$;

Local maxima at	**Local minima at**	**Inflection points at**
(1.977, 3.186)	(2.407, 3.118)	(2.220, 3.143)
(3.876, −3.118)	(4.306, −3.186)	(3.143, −3.149)

37. Graph y and y' on $[0, 5]$ by $[-2, 5]$. Using trace, y has a minimum at $(0.368, -0.368)$. Analytically, $y' = \ell n\, x + 1$; $y' = 0$ at $x = 1/e$. $y(1/e) = -1/e$.

38. Graph y on $[0, 3]$ by $[-1, 4]$; graph y, y' on $[0, 1]$ by $[-0.6, 0]$; local minimum at $(0.607, -0.184)$; inflection point at $(0.223, -0.075)$.

39. Graph y and y' on $[-3, 3]$ by $[-1, 2]$. Using trace, y has a minimum at $(\pm .60, -0.18)$. Analytically, for $x > 0$. $y' = 2x\, \ell n\, x - x = x(\ell n\, x^2 - 1)$; $y' = 0$ at $x = \sqrt{(1/e)}$. y has a local minimum at $(\pm\sqrt{(1/e)}, -0.18)$. y is not defined at 0, but, if we define $y(0) = 0$, then y has a local maximum at $(0, 0)$.

40. Graph y, y' on $[0, 20]$ by $[-0.15, 0.4]$; local maximum at $(e, 0.368)$; inflection point at $(4.482, 0.335)$.

41. Use $[-3, 3]$ by $[-1, 6]$. The graph of y' shows that there is no hidden behavior near $x = 0$. $(0, 1/2)$ is a minimum. Analytically, $y' = 2^{x^2-1}(\ell n\, 2)(2x)$.

42. Graph y, y' on $[-2, 2]$ by $[-1, 1]$; local maximum at $(0, 1)$; inflection points at $(-0.675, 0.607)$, $(0.675, 0.607)$.

43. Graph y and y' on $[-3, 3]$ by $[-1, 1]$. Local extrema at $(\mp 0.850, \mp 0.515)$; inflection points at $(\mp 1.471, \mp 0.328)$ and $(0, 0)$.

44. Graph y on $[-4, 1]$ by $[-0.5, 2]$; graph y, y' on $[-4, 1]$ by $[-0.2, 0.2]$; local minimum at $(-0.910, -0.335)$; inflection point at $(-1.820, -0.246)$.

45. Graphing the function on $[-4\pi, 4\pi]$ by $[-20, 20]$ shows the symmetry with respect to $x = 0$. Graph y and y' on $[0, 4\pi]$ by $[-15, 12]$ and tracing shows local maxima at $(\pm 4.493, -4.603)$ and $(\pm 10.904, -10.950)$, local minima at $(\pm 7.725, 7.790)$. The function is not defined at $x = 0$, but if we define $y(0) = 1$, y is continuous at 0 and has a minimum at $(0, 1)$. Concavity changes at $x = k\pi$, $k = \pm 1, \pm 2, \ldots$.

46. Graph y on $[-4\pi, 4\pi]$; graph y, y' on $[-4\pi, 4\pi]$ by $[-5, 5]$;

Local maxima at	Local minima at	Inflection points at
(2.289, 3.945)	(−2.289, −3.945)	x = 0
(−5.087, 24.083)	(+5.087, −24.083)	x = ±1.520
(8.096, 63.635)	(−8.096, −63.635)	x = ±3.994
(−11.173, 122.876)	(11.173, −122.876)	etc.

47. Graph y and y' on $[-2, 6]$ by $[-1, 1]$. There are inflection points at $(-2.626, -0.500)$, $(1.090, 0.089)$ and $(5.333, 0.182)$. If $y(0)$ is defined to be 0, y has a minimum at $(0, 0)$. Concave down for $-2.626 < x < -1$ and $1.090 < x < 5.333$, concave up on $(-\infty, -2.626) \cup (-1, 1.090) \cup (5.333, \infty)$ (assuming $y(0) = 0$).

48. Graph y, y' on $[-2, 4]$ by $[-4, 4]$. (Assume $y(0)$ is defined to be 0.) Local maximum at $(-1.743, 0.180)$; inflection point at $(-3.447, 0.104)$; concave up $x < -3.447$ and $x > 2$, concave down $-3.447 < x < 2$.

49. Graph $y = (x\text{^}3)\text{^}(1/5)$ on $[-5, 5]$ by $[-3, 3]$ in order to see the entire function. $\min y = -2.627$, $\max y = 2.627$.

50. Graph $y = (x\text{^}(1/3))\text{^}4$ on $[-3, 4]$ by $[--2, 7]$. Absolute minimum is 0, absolute maximum is $y(4) = 6.350$.

51. Graph y on $[-6, 6]$ by $[-200, 20]$. To find the maximum and minimum, graph $y' = e^x \sin x + e^x \cos x = e^x(\sin x + \cos x)$ on $[0, 6]$ by $[-10, 10]$. $y' = 0$ at $3\pi/4$ (2.356), where $y = 7.460$; and at $7\pi/4$ (5.498) where $y = -172.64$.

52. Maximum occurs at -5, $y(-5) = 42.099$; minimum is $y(-3.927) = -35.889$.

53-57

problem	**function**	**window for complete graph**	**inflection point at**	**local maximum**	**local minimum**
53.	$\sinh x$	[-3,3] by [-3,3]	(0,0)		
54.	$\tanh x$	[-3,3] by [-3,3]	(0,0)		
55.	$\coth x$	[-3,3] by [-3,3]	none		concavity changes at 0
56.	$\operatorname{sech} x$	[-3,3] by [-3,3]		(0,1)	
57.	$\operatorname{csch} x$	[-3,3] by [-3,3]	none		

58. Let x and $20 - x$ be the numbers, $x \geq 0$. Let $S = 20 - x + \sqrt{x}$. Then $S'(x) = -1 + \frac{1}{2\sqrt{x}}$; $S' = 0$ at $x = 1/4$. The numbers are 79/4 and 1/4.

59. If x and y are the legs, then $x^2 + y^2 = 25$ and $A = \frac{1}{2}xy = \frac{1}{2}x\sqrt{25 - x^2}$. Graphing A on $[0, 5]$ by $[0, 7]$ shows maximum area is 6.25 cm^2.

60. $A'(\theta) = (1/2)ab\cos\theta$. $0 < \theta < \pi$. $A' = 0$ at $\theta = \pi/2$. Since $A''(\theta) = -(1/2)ab\sin\theta < 0$, $\theta = \pi/2$ provides a local maximum. As $\theta \to 0$, and also as $\theta \to \pi$, $A(\theta) \to 0$; hence $A(\pi/2)$ is an absolute maximum.

61. $x^2 + y^2 = 5 \Rightarrow s = 2x + y = 2x + \sqrt{5 - x^2}$. Graphing s on $[0, 3]$ by $[0, 6]$ shows $\max s = 5$ (when $x = 2$).

62. The area is $A = \frac{1}{2}ab$, where $a^2 + b^2 = 20^2$. Hence $A = \frac{1}{2}a\sqrt{20^2 - a^2}$. $A'(a) = \frac{1}{2}\left[\sqrt{20^2 - a^2} + a\frac{1}{2\sqrt{20^2 - a^2}}(-2a)\right]$. $A' = 0$ when $(20^2 - a^2) = a^2$, or, $a^2 = 20^2/2$; if $a^2 = 20^2/2$, then $b^2 = 20^2/2$ and A has a maximum when $a = b$.

63. From the diagram, the height of the trough is $h = 1 \cdot \cos\theta$; the triangles form a rectangle of width $\sin\theta$. The area of a cross section is $A(\theta) = \cos\theta\sin\theta + (\cos\theta) \cdot 1$, $0 \le \theta < \pi/2$; volume is maximized when $A(\theta)$ is a maximum. $A'(\theta) = -\sin^2\theta + \cos^2\theta - \sin\theta = 1 - 2\sin^2\theta - \sin\theta = -(2\sin\theta - 1)(\sin\theta + 1)$: $A'(\pi/6) = 0$, $A' > 0$ for $0 < \theta < \pi/6$, $A' < 0$ for $\pi/6 < \theta < \pi/2$. The volume is maximized at $\pi/6$.

64. Following the hint, let $f(x) = (\text{distance})^2 = (x - 3/2)^2 + (\sqrt{x} - 0)^2 = (x - 3/2)^2 + x$. $f'(x) = 2(x - 3/2) + 1$; $f' = 0$ when $x = 1$. Since $f(1) = 1.25$, the minimum distance in $\sqrt{1.25} = 1.12$.

65. Sketch the semicircles $y = \sqrt{16 - x^2}$ and $y = \sqrt{4 - x^2}$; they are 2 units apart. The point $(1, \sqrt{3})$ lies on the second semicircle.

66. $f(x) = 3 + 4\cos x + \cos 2x \Rightarrow f'(x) = -4\sin x - 2\sin 2x$ and $f''(x) = -4\cos x - 4\cos 2x$. Set $f'(x) = 0$; $-4\sin x - 2(2\sin x\cos x) = 0$; $\sin x = 0$ or $1 + \cos x = 0$, i.e. $x = 0$ or $x = \pi$. $f(0) = 8$; $f(\pi) = 3 - 4 + (1) = 0$. $f(x) \ge \min f(x) = 0$. Therefore $f(x)$ is never negative.

67. $f' = (3^x\ell n\, 3 - 3^{-x}\ell n\, 3)/2 = \frac{\ell n\, 3}{2}[3^x - 3^{-x}]$; a graph of f' shows f has a minimum at $x = 0$ where f' changes from negative to positive). Hence $f(x) \ge f(0) = 0$. f is never negative.

68. Let B' be the reflection of B in the mirror AOB is shortest when AOB is a straight line, i.e., when $\pi/2 - \theta_1 = \alpha$. But $\theta_2 + \alpha = \pi/2$; hence $\theta_1 = \theta_2$.

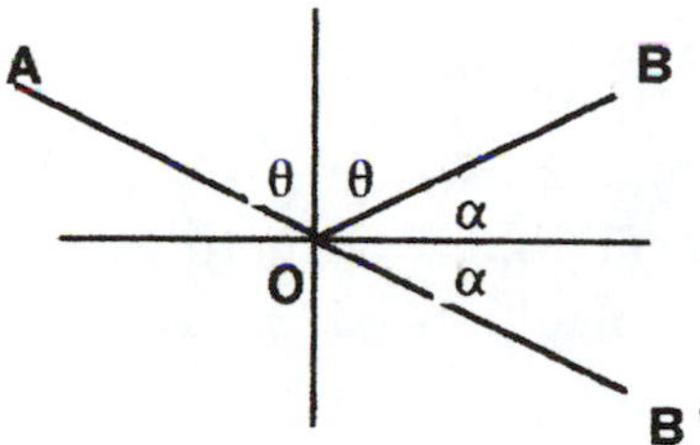

69. Let P represent the closest point on the shore, C the city, and x the distance from P where the boat lands. The necessary time is $T(x) = \frac{\sqrt{4+x^2}}{2} + \frac{6-x}{5}$, $0 \leq x \leq 6$. Graph $T(x)$ on $[0,6]$ by $[0,4]$; finding its minimum value shows $x = 0.873$ miles.

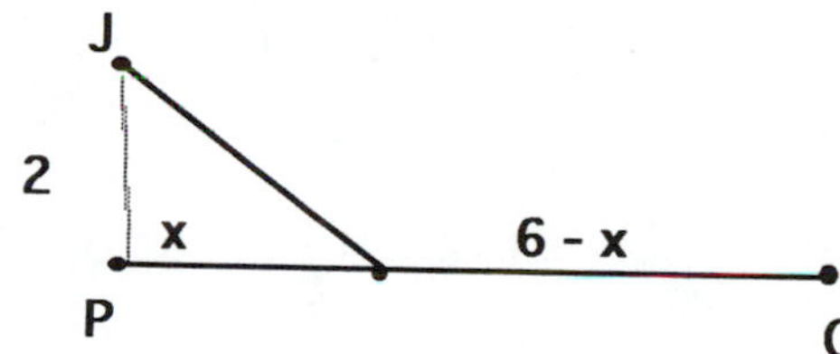

70.

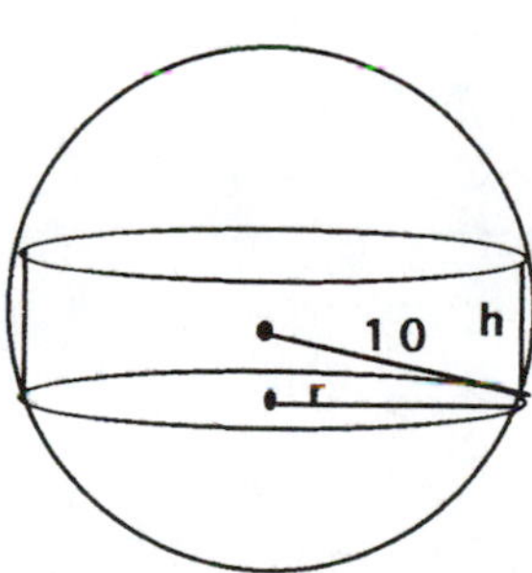

$V = \pi r^2 h$ where $(\frac{h}{2})^2 + r^2 = 100 \Rightarrow r^2 = 100 - \frac{h^2}{4}$. $V = \pi(100h - \frac{h^3}{4})$; $V' = \pi(100 - \frac{3h^2}{4})$. $V' = 0$ when $h = \frac{20}{\sqrt{3}}$ cm and $r = \frac{1}{2}\sqrt{400 - h^2} = \frac{1}{2} \cdot 20\sqrt{1 - 1/3} = \frac{10\sqrt{2}}{\sqrt{3}}$ cm. Since $V'' = \frac{-6}{4}\pi h < 0$, this is a maximum.

71. Height of rectangle is $2\sin\theta$, width is $2(2\cos\theta)$. Hence $A(\theta) = 8\sin\theta\cos\theta = 4\sin 2\theta$; $A'(\theta) = -8\cos 2\theta$, $A' = 0$ when $2\theta = \pi/2$ or $\theta = \pi/4$; $A(\pi/4) = 8(\frac{1}{\sqrt{2}})(\frac{1}{\sqrt{2}}) = 4$.

72. a) $V = (\pi/3)r^2h = (\pi/3)(\frac{2a\pi - x}{2\pi})^2\sqrt{a^2 - r^2} = (\pi/3)(\frac{2a\pi - x}{2\pi})^2[a^2 - (\frac{2a\pi - x}{2\pi})^2]^{1/2}$.

b) Set $y1 = (2A * \pi - x)/(2\pi)$, $y2 = \sqrt{(A^2 - y1^2)}$, $y3 = (\pi/3) * y1^2 * y2$; for the given values of A, graph only $y3$. Evaluate $y1$ (radius) and $y2$ (height) at the maximum of $y3$. Also calculuate $y1/y2$ to 3 place accuracy

a	r	h	r/h
4	3.666 ...	2.309 ...	1.414
5	4.082 ...	2.887 ...	1.414
6	4.899	3.464	1.414 ...
8	6.532	4.619	1.414

c) evidently $r = h\sqrt{2}$

d) $V = (\pi/3)r^2h \Rightarrow V' = 0$ when $2rr'h + r^2h' = 0$. Since $h^2 + r^2 = a^2$, $2rr' + 2hh' = 0$; substituting for $2rr'$ gives: $-2h^2h' + r^2h' = 0$ or $r^2 = 2h^2$.

73. $A(x) = \text{base} \times \text{height} = 2x(8\cos 0.3x)$. Graphing on $[0, \frac{\pi}{2(.3)}]$ by $[0, 30]$ shows a maximum area of 29.925.

74 a)

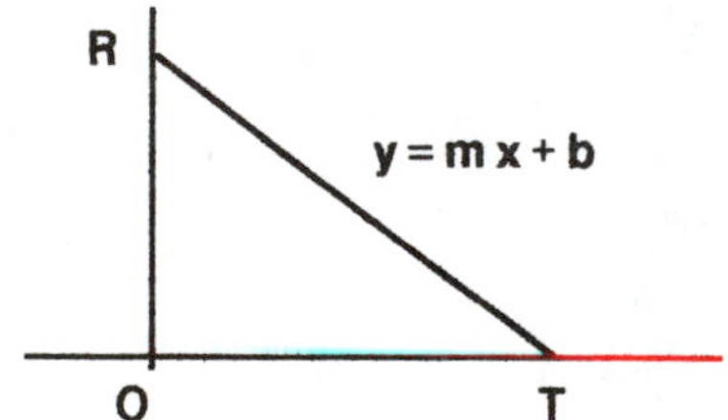

For the line $y = mx + b$, the area if triangle ORT is $\frac{1}{2}(b)(\frac{-b}{m})$. The area of triangle RST, where $m = f'(x)$ is thus $A(x) = \frac{-b^2}{f'(x)}$. To find b one must determine the y-intercept of the line through $(\bar{x}, f(\bar{x}))$ with slope $f'(\bar{x})$; this line has equation $y - f(\bar{x}) = f'(\bar{x})(x - \bar{x})$. Setting $x = 0$ gives the y intercept of $f(\bar{x}) - \bar{x}f'(\bar{x})$. Hence the area of triangle RST is $A(x) = \frac{-[f(x) - xf'(x)]^2}{f'(x)} = -f'(x)[\frac{f}{f'} - x]^2$.

b) Set $y1 = 5 + 5(1 - x^2/10^2)$^.5; $y2 =$ NDER $1(y1, x)$ and $y3 = -y2(x - y1/y2)^2$; graph $y3$ on $[0, 15]$ by $[0, 1000]$. The asymptotes correspond to triangles with infinitely large bases or heights. Meaningful value of x are from 0 to the right-hand vertex of the ellipse.

c) Use trace to find the minimum area. Evaluating $y1 - xy2$ at the maximum gives $h = 15 = 3 \times 5$.

d) (This is more easily done using a computer algebra system.) $A(x) = \frac{-h^2(x)}{f'(x)} = \frac{-[f(x) - xf'(x)]^2}{f'(x)}$ where $f(x) = B + \frac{B}{C}\sqrt{C^2 - x^2}$ and $f'(x) = -\frac{B}{C}\frac{x}{\sqrt{C^2-x^2}}$. $A'(x) = \frac{-2hh'}{f'} + \frac{h^2 f''}{(f')^2} = \frac{h}{f'}[-2\{f' - f' - xf''\} + \frac{hf''}{f'}] = \frac{hf''}{f'}[2x + \frac{f}{f'} - x]$. Hence $A' = 0$ when $-x = \frac{f}{f'}$, i.e. $-xf' = f$. Substituting gives $\frac{B}{C}\frac{x^2}{\sqrt{C^2-x^2}} = B + \frac{B}{C}\sqrt{C^2 - x^2}$, or $x^2 = C\sqrt{C^2 - x^2} + C^2 - x^2$; $2x^2 - C^2 = C\sqrt{C^2 - x^2} \Rightarrow 4x^4 - 4C^2x^2 + C^4 = C^4 - C^2x^2$ or $4x^4 - 3C^2x^2 = 0 \Rightarrow x^2 = \frac{3}{4}C^2 \Rightarrow h_{opt} = B + \frac{B}{C}\cdot\frac{C}{2} + \frac{B}{C}\frac{\frac{3}{4}C^2}{\frac{1}{2}C} = 3B$.

4.6 RELATED RATES OF CHANGE

1. $A = \pi r^2 \Rightarrow dA/dt = 2\pi r\ dr/dt$

2. $dS/dt = 8\pi r\ dr/dt$

3. $dV/dt = (2/3)\pi rh\ dr/dt$

4. $dV/dt = (1/3)\pi \frac{r^2 dh}{dt}$

5. a) $\frac{dP}{dt} = I^2 \frac{dR}{dt} + 2IR\frac{dI}{dt}$

b) $\frac{dR}{dt} = 0 \Rightarrow \frac{dP}{dt} = 2IR\frac{dI}{dt}$

6. a) $\frac{ds}{dt} = \frac{2x\frac{dx}{dt} + 2y\frac{dy}{dt}}{2\sqrt{x^2+y^2}} = \frac{1}{s}(x\frac{dx}{dt} + y\frac{dy}{dt})$

b) $\frac{ds}{dt} = 0 \Rightarrow \frac{dx}{dt} = \frac{-y}{x}\frac{dy}{dt}$.

7. $\frac{ds}{dt} = \frac{1}{2\sqrt{x^2+y^2+z^2}}(2x\frac{dx}{dt} + 2y\frac{dy}{dt} + 2z\frac{dz}{dt}) = \frac{x}{s}\frac{dx}{dt} + \frac{y}{s}\frac{dy}{dt} + \frac{z}{s}\frac{dz}{dt}$

8. $\frac{dA}{dt} = \frac{1}{2}\left[\frac{da}{dt}\cdot b\sin\theta + a\frac{db}{dt}\sin\theta + (ab\cos\theta)\frac{d\theta}{dt}\right]$

9. $A = \pi r^2$; $dA/dt = 2\pi r\ dr/dt$. When $r = 50$ cm and $dr/dt = 0.01$ cm/min, $dA/dt = 2\pi 50(0.01) = \pi$ cm^2/min.

10. a) $\frac{dV}{dt} = 1$ volt/sec, b) $\frac{dI}{dt} = -(1/3)$ amp/sec, c) $\frac{dV}{dt} = I\frac{dR}{dt} + R\frac{dI}{dt} \Rightarrow 1 = I\frac{dR}{dt} + R(-1/3)$, d) $1 = 2\frac{dR}{dt} + (\frac{12}{2})(-\frac{1}{3}) \Rightarrow \frac{dR}{dt} = \frac{3}{2}$ ohms/sec; R is increasing.

11. $A = \ell w \Rightarrow dA/dt = \ell dw/dt + w d\ell/dt = \ell(+2) + w(-2)$ cm^2/sec. $P = 2(\ell + w) \Rightarrow dP/dt = 2(d\ell/dt + dw/dt) = 2(-2+2) = 0$ cm/sec. A diagonal has length $D = \sqrt{\ell^2 + w^2}$; hence

$$\begin{aligned} dD/dt &= \frac{1}{2\sqrt{\ell^2+w^2}}(2\ell\ d\ell/dt + 2w\ dw/dt) \\ &= \frac{\ell\ d\ell/dt + w\ dw/dt}{\sqrt{\ell^2+w^2}} = \frac{\ell(-2) + w(2)}{\sqrt{\ell^2+w^2}} \text{ cm/sec} \end{aligned}$$

Evaulating these rates at $\ell = 12$ and $w = 15$ gives $dA/dt = 12(+2)+5(-2) = +14$ cm^2/sec, increasing; $dP/dt = 0$ cm/sec, constant; $dD/dt = \frac{12(-2)+5(2)}{\sqrt{12^2+5^2}} = \frac{-14}{\sqrt{169}} = -14/13$ cm/sec., decreasing.

12. a) $V = xyz \Rightarrow \frac{dV}{dt} = \frac{dx}{dt}yz + xz\frac{dy}{dt} + xy\frac{dz}{dt}$; at (4,3,2) $\frac{dV}{dt} = 6\cdot 1 + 8(-2) + 12\cdot 1 = 2$ m^3/sec.

b) $SA = 2(xy + xz + yz) \Rightarrow \frac{dSA}{dt} = 2(x\frac{dy}{dt} + y\frac{dx}{dt} + x\frac{dz}{dt} + z\frac{dx}{dt} + y\frac{dz}{dt} + z\frac{dy}{dt})$; at (4,3,2) $\frac{dSA}{dt} = 2(4(-2) + 3\cdot 1 + 4\cdot 1 + 2\cdot 1 + 3\cdot 1 + 2(-2)) = 0$.

c) $\frac{ds}{dt} = \frac{1}{2s}(2x\frac{dx}{dt} + 2y\frac{dy}{dt} + 2z\frac{dz}{dt})$; at (4,3,2) $\frac{ds}{dt} = \frac{4\cdot1+3(-2)+2\cdot1}{\sqrt{29}} = 0$.

13. $D^2 = x^2 + y^2$ where plane A is x miles from the intersection point of the courses. $2D\frac{dD}{dt} = 2x\frac{dx}{dt} + 2y\frac{dy}{dt} = 2[x(-520) + y(-520)]$; when $x = 5$, $y = 12$, $D = 13$, so $13\frac{dD}{dt} = -520[5+12]$ or $\frac{dD}{dt} = -680$ miles/hr. At any time $\frac{dD}{dt} = \frac{-520(x+y)}{\sqrt{x^2+y^2}}$ mph.

14. $x^2 + y^2 = 13^2$, so that $2x\frac{dx}{dt} + 2y\frac{dy}{dt} = 0$, $\frac{dy}{dt} = -\frac{x}{y}\frac{dx}{dt} = -\frac{x}{y}(5)$ ft/sec. When $x = 12$, $y = \sqrt{13^2 - 12^2} = 5$, and $\frac{dy}{dt} = -\frac{12}{5}\cdot 5 = -12$ ft/sec. $A = \frac{1}{2}xy$, so that $dA/dt = \frac{1}{2}[x\frac{dy}{dt} + y\frac{dx}{dt}] = \frac{1}{2}[12(-12) + 5\cdot 5] = -119/2$ ft^2/sec.

15. $V = \frac{4}{3}\pi r^3 \Rightarrow \frac{dV}{dt} = 4\pi r^2\frac{dr}{dt} \Rightarrow \frac{dr}{dt} = \frac{dV/dt}{4\pi r^2}$; $\frac{dV}{dt} = -0.08$ mL $= -0.08$ cm^3; when $r = 10$ mm $=$ 1cm, $\frac{dr}{dt} = \frac{-0.08}{4\pi\cdot 1^2} = -.006366$ cm/min $= -0.06366$ mm/min.

16. $V = \pi r^2 6$ so that $\frac{dV}{dt} = 12\pi r\frac{dr}{dt} = 12\pi r(\frac{.001}{3}) = .004\pi r$ in^3/min. When the diameter is 3.80, $r = 1.90$ and $\frac{dV}{dt} = .004\pi(1.90) = 0.0239$ in^3/min.

17. Since $h = \frac{3}{8}(2r)$, $V = \frac{\pi}{3}r^2h = \frac{\pi}{3}r^2(\frac{3}{8}2r) = \frac{\pi}{4}r^3$; $\frac{dV}{dt} = \frac{3\pi}{4}r^2\frac{dr}{dt}$. When $h = 4$ m, $r = \frac{4}{3}\cdot 4 = \frac{16}{3}$ m; a) $\frac{dh}{dt} = \frac{3}{4}\frac{dr}{dt} = \frac{\frac{3}{4}\frac{dV}{dt}}{\frac{3}{4}\pi r^2} = \frac{10}{\pi(\frac{16}{3})^2}$ m/min $= \frac{90\cdot 100}{256\pi}$ cm/min $= 11.191$ cm/min.

b) $\frac{dr}{dt} = \frac{4}{3}\frac{dh}{dt} = \frac{4}{3}\cdot\frac{1125}{32\pi}$ cm/min $= 14{,}921$ cm/min.

18. $V = \frac{\pi}{3}r^2h$; $\frac{r}{h} = \frac{45}{6} \Rightarrow h = \frac{6r}{45}$ and $V = \frac{\pi\cdot 6}{3\cdot 45}r^3$; $\frac{dV}{dt} = \frac{6\pi}{45}r^2\frac{dr}{dt}$. When $h = 5$, $r = 5\cdot 45/6$, $\frac{dr}{dt} = \frac{-50}{\frac{6\pi}{45}(\frac{5\cdot 45}{6})^2} = \frac{-4}{15\pi}$. a) $\frac{dh}{dt} = \frac{6}{45}\frac{dr}{dt} = \frac{-6}{45}\cdot\frac{4}{15\pi} = \frac{-8}{225\pi}$ m/min ≈ -1.132 cm/min. b) $\frac{-4\cdot 100}{15\pi} = \frac{-400}{15\pi}$ cm/min.

19. $V = (\frac{\pi}{3})y^2(3\cdot 13 - y) \Rightarrow \frac{dV}{dt} = \frac{\pi}{3}[78y - 3y^2]\frac{dy}{dt}$.

a) $\frac{dV}{dt} = -6$, $y = 8 \Rightarrow \frac{dy}{dt} = -1.362$ cm/min

b) From the diagram, for depth d,

$x^2 = 13^2 - (13-d)^2 = 26d - d^2$.

Hence, for depth $y, r = \sqrt{26y - y^2}$.

c) $\frac{dr}{dt} = \frac{1}{2\sqrt{26y-y^2}}(26-2y)(\frac{dy}{dt}) = -0.553$ cm/min.

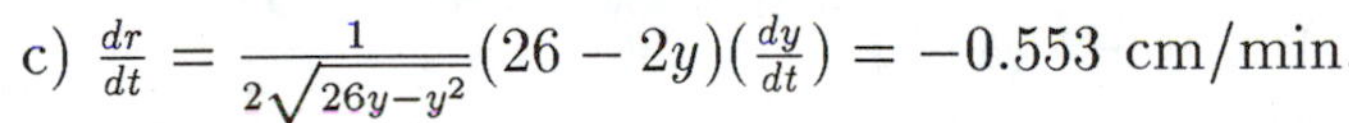

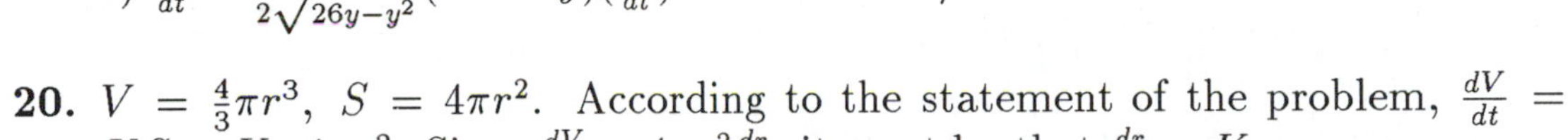

20. $V = \frac{4}{3}\pi r^3$, $S = 4\pi r^2$. According to the statement of the problem, $\frac{dV}{dt} = KS = K\cdot 4\pi r^2$. Since $\frac{dV}{dt} = 4\pi r^2\frac{dr}{dt}$, it must be that $\frac{dr}{dt} = K$.

21. $V = \frac{4}{3}\pi r^3$, $S = 4\pi r^2 \Rightarrow \frac{dV}{dt} = 4\pi r^2\frac{dr}{dt}$, $\frac{dS}{dt} = 8\pi r\frac{dr}{dt}$. $V = 100\pi$; when $r = 5$, $\frac{dr}{dt} = \frac{100\pi}{4\pi\cdot 25} = 1$ ft/min, $\frac{dS}{dt} = 8\pi 5$ ft $\cdot 1$ ft/min $= 40\pi$ ft^2/min.

22. Let x be the distance from the bow to the foot of the dock, and r be the length of the rope still out. Then $x^2 + 6^2 = r^2$ and $x\frac{dx}{dt} + 0 = r\frac{dr}{dt}$. When $r = 10$, $x = \sqrt{100-36} = 8$, and $8\frac{dx}{dt} = 10(-2)$, i.e., $\frac{dx}{dt} = -5/2$ ft/sec. The rate at which x decreases is the rate at which the boat approaches the dock, 2.5 ft/sec.

23. $[s(t)]^2 = x^2(t) + y^2(t)$. $2s(t)\frac{ds}{dt} = 2x(t)\frac{dx}{dt} + 2y(t)\frac{dy}{dt} = 2x\cdot 17 + 26\cdot 1$. When $t = 3$, $x = 17\cdot 3 = 51$, $y = 65 + 3 = 68$, $s = \sqrt{51^2 + 68^2} = 85$. Hence $\frac{ds}{dt} = \frac{1}{85}[51\cdot 17 + 68\cdot 1] = 11$ ft/sec.

24. The volume of coffee in the pot is $V_{\text{pot}} = \pi\cdot 3^2\cdot h$; $\frac{dV_{\text{pot}}}{dt} = 9\pi\frac{dh}{dt}$. It is given $dV_{\text{pot}}/dt = 10$ in^3/min, hence $10 = 9\pi\frac{dh}{dt}$, i.e., $\frac{dh}{dt} = \frac{10}{9\pi}$ in/min. Assume that the level of coffee in the filter is ℓ, and the diameter across the top of the coffee is d. Then $\ell/d = 6/6 = 1$ and $V_{\text{filter}} = \frac{1}{3}\pi(\frac{d}{2})^2\ell = \frac{1}{3}\pi(\frac{\ell}{2})^2\ell = \frac{1}{12}\pi\ell^3$. $dV_f/dt = \frac{1}{4}\pi\ell^2\cdot\frac{d\ell}{dt} = -10$. When $\ell = 5$, $\frac{d\ell}{dt} = \frac{-10\cdot 4}{\pi\cdot 25} = -8/(5\pi)$ in/min.

25. $\frac{dr}{dt} = -0.2$, $\frac{dV}{dt} = 4Kr^3\frac{dr}{dt}$; $\frac{\frac{dV}{dt}}{V} = \frac{4Kr^3\frac{dr}{dt}}{Kr^4} = 4\frac{\frac{dr}{dt}}{r} = -0.8$. The volume decreases 80% per minute.

26. $y = Q/D$, or $Dy = Q$. Differentiating with respect to t gives: $Ddy/dt + ydD/dt = 0$, since Q does not change. When $Q = 233$, $D = 41$, $y = Q/D = 233/41$, and $dD/dt = -2$. $41\frac{dy}{dt} + \frac{233}{41}(-2) = 0$ or $\frac{dy}{dt} = \frac{2\cdot 233}{41\cdot 41} = 0.2772$ units/min.

27. Let $P(x, y)$ be a point on the curve. Then $y/x = \tan\theta$, where θ is the angle of inclination of the line OP. Since $y = x^2$, we have $x = \tan\theta$; $dx/dt = \sec^2\theta\ d\theta/dt$ or $d\theta/dt = (\cos^2\theta)dx/dt = 10\cos^2\theta$. When $x = 3$, $y = 9$, $\cos\theta = 3/\sqrt{9+81} = 1/\sqrt{10}$ and $d\theta/dt = 10/10 = 1$ rad/sec. As $x \to \infty$, $\frac{d\theta}{dt} = 10\frac{x^2}{x^4+x^2}$. $\text{Lim}_{x\to\infty} d\theta/dt = 0$.

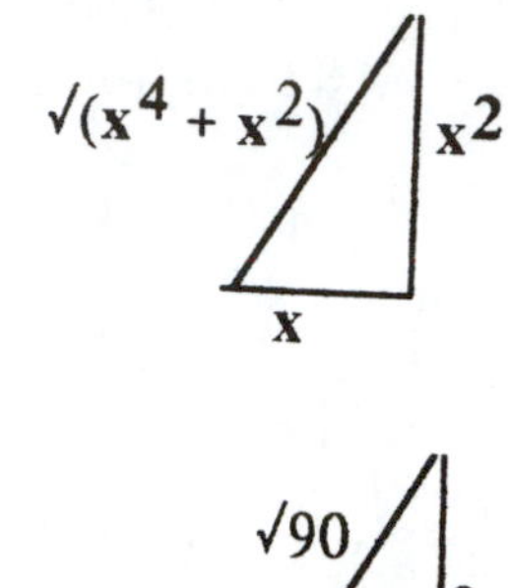

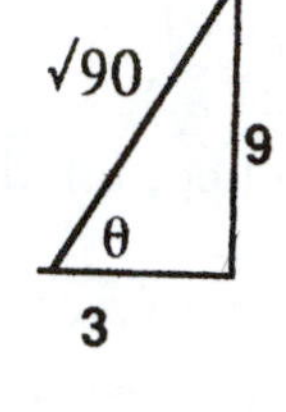

28. $y = \sqrt{-x} \Rightarrow \frac{dy}{dt} = \frac{-1}{2\sqrt{-x}}\frac{dx}{dt}$; $x = -4 \Rightarrow y = 2$; from the diagram, $\tan\theta = y/x \Rightarrow$ $\sec^2\theta\frac{d\theta}{dt} = \frac{x\frac{dy}{dt} - y\frac{dx}{dt}}{x^2}$; at $(-4, 2)$ $\frac{dx}{dt} = -8$, $\frac{dy}{dt} = \frac{8}{4} = 2$, $\sec^2\theta = (\frac{\sqrt{20}}{-4})^2$ $\Rightarrow \frac{d\theta}{dt} = (\frac{16}{20})\frac{-8-(-16)}{16} = \frac{2}{5}$ rad/sec.

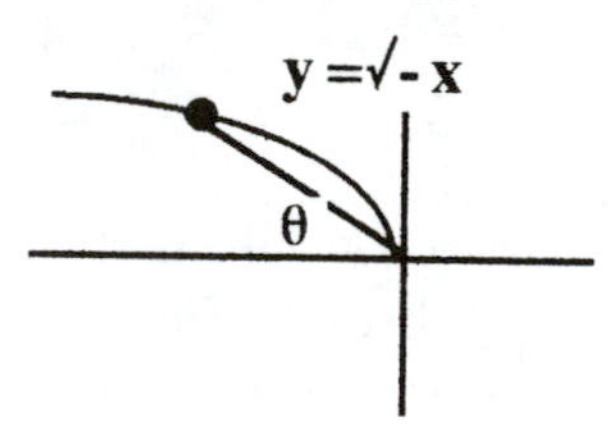

29. a) $\frac{dr}{dt} = 9\frac{dx}{dt}$, $\frac{dc}{dt} = 3x^2\frac{dx}{dt} - 12x\frac{dx}{dt} + 15\frac{dx}{dt} = (3x^2 - 12x + 15)\frac{dx}{dt}$; when $x = 2$, $\frac{dx}{dt} = 0.1$, $\frac{dr}{dt} = 0.9$, $\frac{dc}{dt} = (3)(0.1) = 0.3$; $\frac{dp}{dt} = 0.9 - 0.3 = 0.6$.
b) $\frac{dr}{dt} = 70\frac{dx}{dt}$, $\frac{dc}{dt} = (3x^2 - 12x - \frac{45}{x^2})\frac{dx}{dt}$; when $\frac{dx}{dt} = 0.05$ and $x = 1.5$, $\frac{dr}{dt} = 3.5$, $\frac{dc}{dt} = -1.5625$; $\frac{dp}{dt} = 3.5 - (-1.5625) = 5.0625$.

30. Let s be the length of the shadow and let x be the man's distance from the light, $\frac{16}{x+s} = \frac{6}{s}$ or $16s = 6x + 6s$, i.e., $5s = 3x$, $5ds/dt = 3dx/dt$ or $ds/dt = \frac{3}{5}(-5) = -3$ ft/sec. The length of the shadow is decreasing at a rate of 3 ft/sec. The tip of his shadow is given by $x + s$. Since $\frac{d}{dt}(x+s) = \frac{dx}{dt} + \frac{ds}{dt} = -5 - 3 = -8$, the tip of the shadow is moving at a rate of 8 ft/sec. The distance of 10 feet does not enter into the problem.

31. Let ℓ = distance from pole to the ball's shadow, h the height of the ball.
$h = 50 - 16t^2$, $\frac{50-h}{30} = \frac{50}{\ell}$, or $\ell(50-h) = 1500$
$\frac{d\ell}{dt}(50-h) + \ell(-\frac{dh}{dt}) = 0$; $\frac{dh}{dt} = -32t$,
$t = \frac{1}{2} \Rightarrow h = 46 \Rightarrow \ell = 375$,
$\frac{d\ell}{dt} = \frac{(-32)(1/2)(375)}{4} = -1500$ ft/sec.

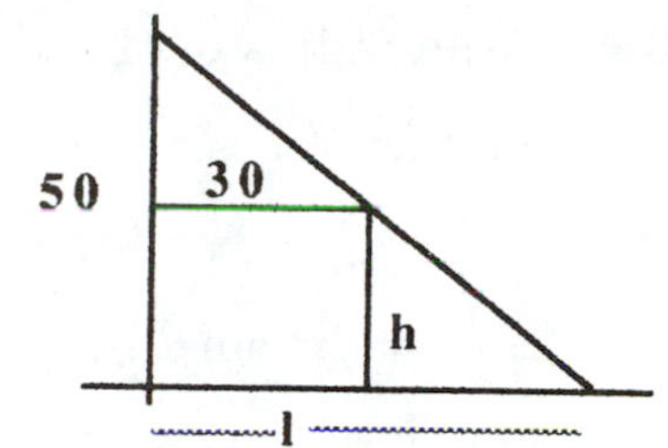

32. Let s be the length of the string (or distance of the kite from the girl.
$x^2 + 300^2 = s^2$ and $dx/dt = 25$ ft/sec.
When $s = 500$, $x = 400$. Hence
$xdx/dt = sds/dt$ or
$\frac{400(25)}{500} = ds/dt = 20$ ft/sec.

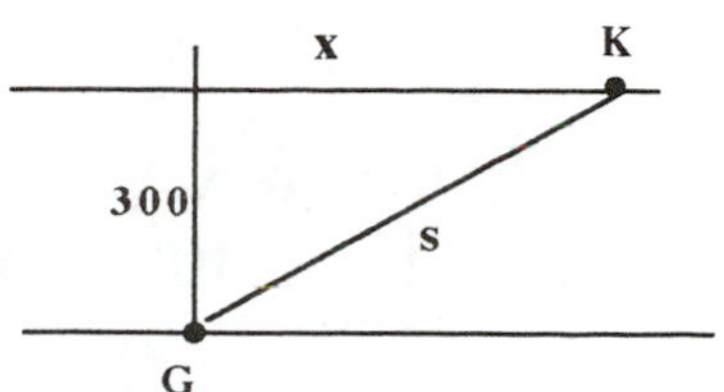

33. When the ice is 2 in thick, the radius of the ice ball is $(4+2) = 6$ in. $V = \frac{4}{3}\pi r^3$, $S = 4\pi r^2$. $\frac{dV}{dt} = 4\pi r^2 \frac{dr}{dt}$, $\frac{dS}{dt} = 8\pi r \frac{dr}{dt}$. When $\frac{dV}{dt} = -10$ and $r = 6$, $\frac{dr}{dt} = \frac{-10}{4\pi \cdot 36} = \frac{-5}{72\pi}$ in/min. $\frac{dS}{dt} = 8\pi \cdot 6(\frac{-5}{72\pi}) = \frac{-10}{3}$ in^2/min.

34. From the diagram, $3^2 + x^2 = s^2$. If the car is going ν mph, then $\frac{dx}{dt} = -\nu - 120$. When $s = 5$, $x = 4$ and $ds/dt = -160$. Differentiating with respect to t gives: $x\frac{dx}{dt} = s\frac{ds}{dt}$; $4(-\nu - 120) = 5(-160)$, or $\nu = 80$ mph.

35. $0.27° = 0.27\pi/180$ radians. Let s be the length of the shadow; $\frac{80}{s} = \tan\theta \Rightarrow \frac{-80}{s^2}\frac{ds}{dt} = \sec^2\theta \frac{d\theta}{dt}$; $\frac{ds}{dt} = \frac{-s^2 \sec^2\theta}{80}\frac{d\theta}{dt}$. When $s = 60$, $\frac{d\theta}{dt} = 0.27\pi/180$, $\sec^2\theta = \tan^2\theta + 1 = (\frac{80}{60})^2 + 1$. Hence $\frac{ds}{dt} = \frac{-60^2(\frac{80^2}{60^2}+1)}{80} \cdot \frac{0.27\pi}{180} = 0.589$ ft/min ≈ 7.1 in/min.

36. From the diagram, $\tan\theta = \frac{A}{B} \Rightarrow \sec^2\theta \frac{d\theta}{dt} = \frac{B\frac{dA}{dt} - A\frac{dB}{dt}}{B^2}$. When $A = 10$, $B = 20$, $\sec\theta = \frac{10\sqrt{5}}{20} \Rightarrow \frac{d\theta}{dt} = \frac{4}{5}\frac{(-40)-10}{20^2} = -\frac{1}{10}$ rad/sec $= -\frac{1}{10}\frac{\frac{180}{\pi}\text{ deg}}{\frac{1}{60}\text{ min}} \approx -344°$/min.

37. By the law of cosines: $AB^2 = OA^2 + OB^2 - 2 \cdot OA \cdot OB \cdot \cos 120$ or $AB^2 = OA^2 + OB^2 + OA \cdot OB$. Hence

$2 \cdot AB \cdot dAB/dt = 2OA \cdot dOA/dt + 2OB \cdot dOB/dt + OA \cdot dOB/dt + OB \cdot dOA/dt$.

When $OA = 5$ and $OB = 3$, $AB = \sqrt{(25 + 9 + 15)} = \sqrt{(49)} = 7$ and

$$2 \cdot 7 \cdot dAB/dt = 2 \cdot 5 \cdot 14 + 2 \cdot 3 \cdot 21 + 5 \cdot 21 + 3 \cdot 14$$

or $dAB/dt = 29.5$ knots.

38. If the radius of the clock face is r, by the law of cosines, $x^2 = 2r^2 - 2r^2\cos\theta$, where $\frac{d\theta}{dt} - 2\pi$ rad/min. $2x\frac{dx}{dt} = 2r^2\sin\theta\frac{d\theta}{dt}$. At 4 pm, $\theta = 120°$, $x = r\sqrt{3}$, $\frac{dx}{dt} = \frac{r^2}{r\sqrt{3}}\frac{\sqrt{3}}{2}\cdot 2\pi = \pi r$ units/min.

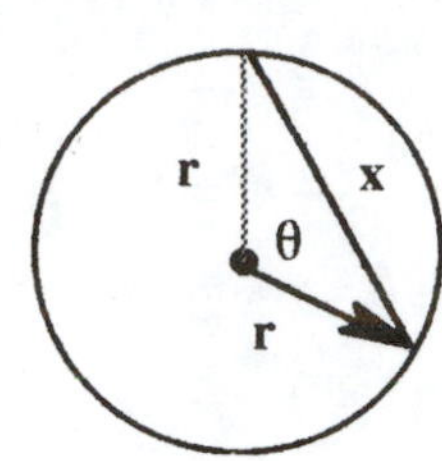

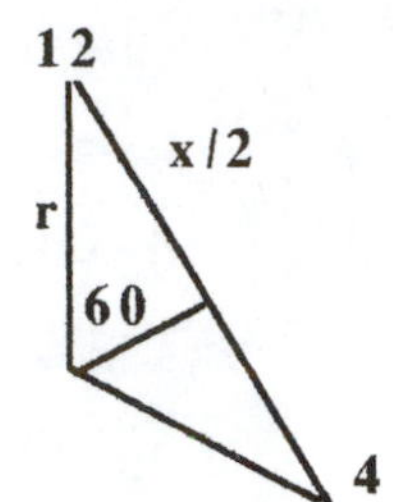

4.7 ANTIDERIVATIVES, INITIAL VALUE PROBLEMS, AND MATHEMATICAL MODELING

1. a) $x^2 + C$, b) $\frac{x^3}{3} + C$ c) $\frac{x^3}{3} - x^2 + x + C$

2. a) $3x^2 + C$, b) $\frac{x^6}{6} + C$, c) $\frac{x^6}{6} - 3x^2 + 3x + C$

3. a) $y' = -3x^{-4} \Rightarrow y = -3x^{-3}/(-3) + C = x^{-3} + C$; b) $y' = x^{-4} \Rightarrow y = x^{-3}/(-3) + C$ c) $y' = x^{-4} + 2x + 3 \Rightarrow y = -x^{-3}/3 + x^2 + 3x + C$

4. a) $-x^{-2} + C$, b) $\frac{-x^{-2}}{4} + \frac{x^3}{3} + C$, c) $\frac{x^{-2}}{2} + \frac{x^2}{2} - x + C$

5. a) $y' = x^{-2} \Rightarrow y = -x^{-1} + C$, b) $y = \frac{-5}{x} + C$ c) $y = 2x + \frac{5}{x} + C$

6. a) $x^{-2} + C$, b) $\frac{-x^{-2}}{4} + C$, c) $\frac{x^4}{4} + \frac{x^{-2}}{2} + C$

7. a) $y' = \frac{3}{2}x^{1/2} \Rightarrow y = \frac{(\frac{3}{2})x^{1/2+1}}{1/2+1} + C = \frac{(\frac{3}{2})}{(\frac{3}{2})}x^{3/2} + C = x^{3/2} + C$ b) $y' = 4x^{1/2} \Rightarrow y = \frac{4}{(\frac{3}{2})}x^{3/2} + C = \frac{8}{3}x^{3/2} + C$ c) $y' = x^2 - 4x^{1/2} \Rightarrow y = x^3/3 - (8/3)x^{3/2} + C$

8. a) $y' = \frac{4}{3}x^{1/3} \Rightarrow y = \frac{\frac{4}{3}x^{4/3}}{\frac{4}{3}} + C = x^{4/3} + C$, b) $y' = \frac{1}{3}x^{-1/3} \Rightarrow y = \frac{\frac{1}{3}x^{2/3}}{\frac{2}{3}} + C = \frac{1}{2}x^{2/3} + C$ c) $y' = x^{1/3} + x^{-1/3} \Rightarrow y = \frac{3}{4}x^{4/3} + \frac{3}{2}x^{2/3} + C$

9. a) $y' = \frac{2}{3}x^{-1/3} \Rightarrow y = \frac{\frac{2}{3}}{1-\frac{1}{3}}x^{-1/3+1} + C = x^{2/3} + C$ b) $y' = \frac{1}{3}x^{-2/3} \Rightarrow y = \frac{\frac{1}{3}}{\frac{1}{3}}x^{1/3} + C = x^{1/3} + C$ c) $y' = -\frac{1}{3}x^{-4/3} \Rightarrow y = \frac{-\frac{1}{3}}{-\frac{4}{3}+1}x^{-1/3} + C = x^{-1/3} + C$

10. a) $y = \frac{1}{2}\frac{x^{-1/2+1}}{1/2} + C = x^{1/2} + C$, b) $y = -\frac{1}{2}\frac{x^{-1/2}}{-\frac{1}{2}} + C = x^{-1/2} + C$ c) $y = -\frac{3}{2}\frac{x^{-5/2+1}}{-\frac{3}{2}} + C = x^{-3/2} + C$

11. a) $y' = \sin(3x) \Rightarrow y = \frac{\cos(3x)}{3} + C$ b) $y' = 3\sin x \Rightarrow y = -3\cos x + C$ c) $y' = 3\sin x - \sin(3x) \Rightarrow y' = -3\cos x + \frac{\cos(3x)}{3} + C$

12. a) $y = \pi\frac{\sin(\pi x)}{\pi} + C = \sin(\pi x) + C$, b) $y = \sin(\frac{\pi x}{2}) + C$, c) $y = \frac{2}{\pi}\sin(\frac{\pi x}{2}) + C$

13. a) $y' = \sec^2 x \Rightarrow y = \tan x + C$ b) $y' = 5\sec^2(5x) \Rightarrow y = 5\frac{\tan(5x)}{5} + C = \tan(5x) + C$ c) $y' = \sec^2(5x) \Rightarrow y = \frac{\tan(5x)}{5} + C$

14. a) $y = -\cot x + C$, b) $y = -\cot(7x) + C$, c) $y = -\frac{1}{7}\cot(7x) + C$

15. a) $y' = \sec x \tan x \Rightarrow y = \sec x + C$ b) $y' = 2\sec 2x \tan 2x \Rightarrow y = \sec 2x + C$ c) $y' = 4\sec 2x \tan 2x \Rightarrow y = \frac{4\sec 2x}{2} + C = 2\sec 2x + C$

16. a) $y = -\csc x + C$, b) $y = -2\csc(4x) + C$, c) $y = \frac{-1}{4}\csc(4x) + C$

17. $y' = (\sin x - \cos x)^2 = \sin^2 x - 2\sin x\cos x + \cos^2 x = 1 - 2\sin x\cos x = 1 - \sin 2x \Rightarrow y = x + \frac{\cos 2x}{2} + C$

18. Using the hint, $y' = 1 + 4\cos x + 4\cos^2 x = 1 + 4\cos x + 2(1 + \cos(2x)) = 3 + 4\cos x + 2\cos(2x)$. Hence $y = 3x + 4\sin x + \sin(2x) + C$.

19. a) $1 - \sqrt{x} + C$ or, since $1 + C$ is a constant, the general antiderivative can be written $-\sqrt{x} + C$ b) $x + C$ c) $\sqrt{x} + C$ d) $-x + C$ e) $x - \sqrt{x} + c$ f) $-3\sqrt{x} - 2x + C$ g) $\frac{x^2}{2} - \sqrt{x} + C$ h) $x - 4x + C = -3x + C$

20. a) $e^x + C$ b) $x\sin x + C$ c) $-e^x + C$ d) $-x\sin x + C$ e) $e^x + x\sin x + C$ f) $3e^x - 2x\sin x + C$ g) $\frac{x^2}{2} + e^x + C$ h) $x\sin x - 4x + C$

21. $y' = 2x \Rightarrow y = x^2 + C$; $y(1) = 4 \Rightarrow y = x^2 + 3$; graph b.

22. $y' = -x \Rightarrow y'(-1) = +1 > 0$. $\underline{b}$ is the only graph which is increasing at $(-1, 1)$.

23. $y' = 2x - 7 \Rightarrow y(-x) = x^2 - 7x + C$. Evaluating at $x = 2$ gives $0 = 2^2 - 7 \cdot 2 + C$, $C = 10$, $y = x^2 - 7x + 10$.

24. $y = 10x - \frac{x^2}{2} + C'$; $-1 = 10 \cdot 0 - 0^2/2 + C \Rightarrow C = -1$, $y = 10x - \frac{x^2}{2} - 1$

25. $y' = x^2 + 1 \Rightarrow y = x^3/3 + x + C$; $y(0) = 1 \Rightarrow 1 = 0 + 0 + C$; $y = x^3/3 + x + 1$

26. $y = \frac{x^3}{3} + \frac{2}{3}x^{3/2} + C$; $1 = \frac{1}{3} + \frac{2}{3} + C \Rightarrow C = 0$; $y = \frac{x^3}{3} + \frac{2}{3}x^{3/2}$

27. $y' = -5x^{-2} \Rightarrow y = \frac{-5}{-1}x^{-1} + C = 5x^{-1} + C$; $3 = 5(5^{-1}) + C \Rightarrow C = 2$. $y = 5/x + 2$

28. $y = -x^{-1} + \frac{x^2}{2} + C$; $1 = -\frac{1}{2} + \frac{2^2}{2} + C \Rightarrow C = -\frac{1}{2}$; $y = \frac{x^2}{2} - \frac{1}{x} - \frac{1}{2}$

29. $y' = 3x^2 + 2x + 1 \Rightarrow y = x^3 + x^2 + x + C$; $0 = 1 + 1 + 1 + C \Rightarrow C = -3$, $y = x^3 + x^2 + x - 3$

30. $y = 3x^3 - 2x^2 + 5x + C$; $0 = -3 - 2 - 5 + C \Rightarrow C = 10$, $y = 3x^3 - 2x^2 + 5x + 10$

31. $y' = 1 + \cos x \Rightarrow y = x + \sin x + C$; $4 = 0 + \sin 0 + C \Rightarrow C = 4$. $y = x + \sin x + 4$

32. $y = \sin x - \cos x + C$; $1 = \sin \pi - \cos \pi + C = 1 + C \Rightarrow C = 0$; $y = \sin x - \cos x$

33. $y'' = 2 - 6x \Rightarrow y' = 2x - 3x^2 + C$; since $y' = 4$ when $x = 0$, $4 = 0 - 0 + C$. Hence $y' = 2x - 3x^2 + 4$ giving $y = x^2 - x^3 + 4x + C$; since $y = 1$ when $V = 0$, $1 = C$. Hence $y = x^2 - x^3 + 4x + 1$.

34. $\frac{d^2y}{dx^2} = 6x + C$; $-8 = 6 \cdot 0 + C \Rightarrow C = -8$; $\frac{d^2y}{dx^2} = 6x - 8 \Rightarrow \frac{dy}{dx} = 3x^2 - 8x + K$; $0 = 0 + K \Rightarrow K = 0$; $\frac{dy}{dx} = 3x^2 - 8x \Rightarrow y = x^3 - 4x^2 + C$; $5 = 0 - 0 + C \Rightarrow C = 5$; $y = x^3 - 4x^2 + 5$.

35. $v = s' = 9.8t \Rightarrow s = \frac{9.8}{2}t^2 + C = 4.9t^2 + C$. $s(0) = 10 \Rightarrow 10 = C$. Hence $s = 4.9t^2 + 10$.

36. $s = -\cos t + C$; $0 = -\cos 0 + C \Rightarrow C = 1$; $s = 1 - \cos t$

37. $a = v' = 32 \Rightarrow v = 32t + C$; $v(0) = 20 \Rightarrow C = 20$. Hence $v = 32t + 20$. $s' = v = 32t + 20 \Rightarrow s = 16t^2 + 20t + C$; $s(0) = 0 \Rightarrow C = 0$. Hence $s = 16t^2 + 20t$.

38. $v = -\cos t + C$; $-1 = -\cos 0 + C \Rightarrow C = 0$; $v = -\cos t \Rightarrow s = -\sin t + K$; $1 = -\sin 0 + K \Rightarrow K = 1$; $s = 1 - \sin t$

39. The slope, y', satisfies the equation $y' = 3x^{1/2}$. Hence, $y = \frac{3}{3/2}x^{3/2} + C = 2x^{3/2} + C$. Since $y = 4$ when $x = 9$, $4 = 2 \cdot 27 + C$, or $C = -50$. The curve is given by $y = 2x^{3/2} - 50$.

40. The problem requires that $y = 1$ and $\frac{dy}{dx} = 0$ at $x = 0$. Hence $\frac{dy}{dx} = 3x^2 + C$; $0 = 0 + C \Rightarrow C = 0$; $\frac{dy}{dx} = 3x^2 \Rightarrow y = x^3 + C$; $1 = 0 + C \Rightarrow y = x^3 + 1$. There are no undetermined constants; the answer is unique.

41. $r' = 3x^2 - 6x + 12 \Rightarrow r = x^3 - 3x^2 + 12x + C$. $r(0) = 0 \Rightarrow C = 0 \Rightarrow r = x^3 - 3x^2 + 12x$

42. $c'(x) = 3x^2 - 12x + 15 \Rightarrow c = x^3 - 6x^2 + 15x + K$; $c(0) = 400 \Rightarrow K = 400$. $c(x) = x^3 - 6x^2 + 15x + 400$

43. $a = -1.6 \Rightarrow v = -1.6t + C$. If the rock is dropped (not thrown), $v(0) = 0$. Hence $v = -1.6t$ m/sec. At $t = 30$ sec, $v = -48$ m/sec. The velocity is 48 m/sec in a downwards direction.

44. Assume the acceleration remains constant. $v'(t) = 20 \Rightarrow v(t) = 20t + C$; $v(0) = 0 \Rightarrow v(t) = 20t$; $v(60) = 1200$ m/sec.

45. The problem assumes that the diver does not jump or take a running start, i.e. $v(0) = 0$. $a = v' = -9.8 \Rightarrow v = -9.8t + C$. By assumption, $C = 0$, $s' = v = -9.8t \Rightarrow s = -4.9t^2 + C$. The height of the board is 10 m, hence $s(0) = 10$, $s = -4.9t^2 + 10$. You enter the water when $s = 0$. Solving $0 = -4.9t^2 + 10$ gives $t = \sqrt{10/4.9}$. $v(\sqrt{10/4.9}) = -9.8\sqrt{10/4.9} = -14$ m/sec.

46. $a = v'(t) = -3.72 \Rightarrow v(t) = -3.72t + C$. $v(0) = 93$. Hence $v(t) = s'(t) = 93 - 3.72t$; $s(t) = 93t - \frac{3.72}{2}t^2 + 0$. $v = 0$ at $t = 93/3.72$ sec, $s(\frac{93}{3.72}) = 1162.5$ m

47. The general antiderivative of $y^{-1/2}\frac{dy}{dt}$ is $\frac{y^{1/2}}{\frac{1}{2}} + C_1 = 2y^{1/2} + C_1$. The general antiderivative of $-k$ is $-kt + C_2$. Hence $2y^{1/2} + C_1 = -kt + C_2$. $2y^{1/2} = -kt + (C_2 - C_1) = -kt + C$, $y^{1/2} = -\frac{kt}{2} + \frac{C}{2} = \frac{C-kt}{2}$, $y = \frac{(C-kt)^2}{4}$.

48. $y = \frac{(C-kt)^2}{4} = \frac{(C-0.1t)^2}{4}$; a) $9 = \frac{C^2}{4} \Rightarrow C = 6$. $y = (6 - 0.1t)^2/4$; b) setting $y = 0$ gives $t = 60$ min.

49. The least common multiple of 8 and 12 is 24; every 24 seconds Renée and Sherrie will return to points R and S. Graph (x_3, y_3) using $x\,$Min $= t\,$Min $= 0$, $x\,$Max $= t\,$Max $= 48$, $y\,$Min $= 0$, $y\,$Max $= 60$. The screen shows two periods of $D(t)$. c) Using zoom, the maximum distance is 45.821 at $t = 15.925$, the minimum is 4.109 at $t = 21.5$ d) Try to find the critical points of $D(t)$: $D^2(t) = (x_1 - x_2)^2 + (y_1 - y_2)^2$; $2DD'(t) = 2(x_1 - x_2)(x_1' - x_2') + 2(y_1 - y_2)(y_1' - y_2')$: Set $D'(t) = 0$ and substitute for the derivatives of x_1, x_2, y_1 and y_2. the algebra is almost impossible.

50. The period remains the same (24). For Renée's position (x_1, y_1) to be $(0,0)$ at $t = 0$ the argument of the sin and cosine terms must be $3\pi/2$ when $t = 0$. Hence $x_1(t) = 20\cos(\frac{\pi}{6}(t+9))$, $y_1(t) = 20 + 20\sin(\frac{\pi}{6}(t+9))$. Similarly, for Sherrie, $x_2(t) = 15 + 15\cos(\frac{\pi(t+6)}{4})$, $y_2(t) = 15 + 15\sin(\frac{\pi(t+6)}{4})$. $D\,$max $=$ 48.537 at 9.375: $D\,$min $= 1.817$ at $t = 3.86$.

51. The diagram shows that $y(1) = 1$ and $y'(1) = 2$. The only initial value problem listed satisfying both these conditions is (d).

52. $y(0) = \frac{2}{3}(3)^{3/2} + 2 \neq 2$ so y does not pass through $(0, 2)$. However, y does satisfy the differential equation.

53. By the sign of y', y decreases for $x < 0$ and increases for $y > 0$; $y'' = 2 > 0 \Rightarrow y$ is always concave up. Actual solutions are of the form $y = x^2 + C$.

54. The solution curves are $y = -x^2 + 2x + C$.

55. $y' = 1 - 3x^2 \Rightarrow y$ decreasing for $x < -\frac{1}{\sqrt{3}}$ increasing on $(-\frac{1}{\sqrt{3}}, \frac{1}{\sqrt{3}})$, decreasing thereafter. $y'' = -6x \Rightarrow y$ concave up for $x < 0$, concave down for $x > 0$. The solution curves are $y = x - x^3 + C$.

56. The solution curves are $y = \frac{x^3}{3} + C$.

57. $y' > 0 \Rightarrow y$ is always increasing. $y'' = \frac{4x^3}{2\sqrt{1+x^4}}$ shows y has an inflection point at $x = 0$. To see the shape of the graph of an antiderivative, put your calculator in Dif Eq Mode, select GRAPH and set the ranges $t : [-2, 2]$, $x[-2, 2]$, $y[-4, 4]$. The calculator graphs an antiderivative of $Q'(t)$. Enter, for this problem, $Q'(t) = \sqrt{}(1 + t\text{^}4)$, then select INITC to specify the value of y when $x = x\min = t\min$. For this problem $y(-2) = -4$ will show the graph of one antiderivative of $\sqrt{}(1 + t\text{^}4)$.

58. Inflection point at $x = 0$. To graph an antiderivative of y', set the ranges: $t[-0.9, 0.9]$, $x[-2, 2]$, $y[-4, 4]$. Try INITC of $QI1 = -1$.

59. $y' \to 0$ as $x \to \pm\infty \Rightarrow y$ approaches a constant. $y'(0) = 0$ and changes from negative to positive there $\Rightarrow$ minimum at $(0, 0)$. To see an antiderivative, use Dif Eq. setting the ranges: $t[-8, 8]$, $x[-8, 8]$, $y[-6, 6]$. Try $y(-8) = 3$ to see the shape of the curve.

60. Always decreasing, inflection point $x = 0$. Set all ranges $-t, x, y-$ to $[-8, 8]$. Try $y(-8) = 4$. As $|x| \to \infty$ y approaches line of slope -1.

61.

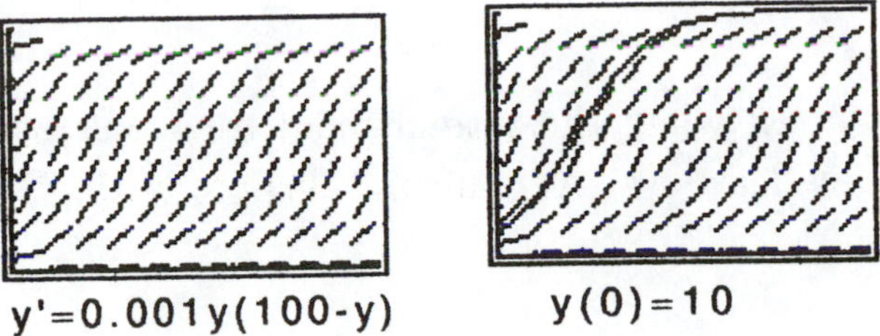

PRACTICE EXERCISES, CHAPTER 4

1. $y' = \frac{(x+1)-(x)}{(x+1)^2} = \frac{1}{(x+1)^2} > 0$. Hence y is always increasing.

2. $y = \sin^2 t - 3t \Rightarrow y' = 2\sin t\cos t - 3 < 0$ for all t.

3. $y = x^3 + 2x \Rightarrow y' = 3x^2 + 2$ which is defined for all x but is never 0. y' is always > 0 and y is always increasing.

4. $f' = 3x^2 + 2 + \sec^2 x > 0$; the function is always increasing.

5. By the M.V.T. the increase in f can be written $f(6) - f(0) = (6-0)f'(c) \leq 6 \cdot 2 = 12$, the maximum increase.

6. Let $y = x^4 + 2x^2 - 2$. $y(0) = -2$, $y(1) = 1$. Hence y has at least one zero on $[0,1]$. $y' = 4x^3 + 4x = 4x(x^2+1) > 0$ for $0 < x < 1$. Hence y is strictly increasing and has at most one zero.

7. Graph y' in $[-3,3]$ by $[-5,30]$. $y'(-1^{-1}) < 0$, $y'(-1) = 0$ $y'(-1^{+}) > 0$ means y changes from a decreasing to an increasing function, i.e. y has a local minimum at $x = -1$. The inflection points of y are at the local extremes of y', i.e. at $x = 0$ and $x = 2$.

8. Graphing y' on $[-4,4]$ by $[-30,10]$ shows $y' < 0$ for $x < -1$, $y' > 0$ on $(-1,1)$, $y' < 0$ on $(0,2)$, $y' > 0$ for $x > 2$. Hence y has local minima at $x = -1$ and $x = 2$; y has a local maximum at $x = 0$. The extremes of y' occur at points of inflection of y: $x = -0.549$ and $x = 1.215$.

9. a) For $y' < 0$ and $y'' < 0$, y must be decreasing and concave down. These conditions are met at T. b) $y' < 0$ and $y'' > 0$ means y is decreasing and concave up: P.

10. The graph is steepest when the rate changes from increasing to decreasing approximately day 23.

Exercises 11 - 30: See comment at beginning of solutions to Exercises, Section 4.5.

11. $y' = -3x^2 - 6x - 4$; $y'' = -6x - 6 = -6(x+1)$. y has an inflection point at $x = -1$, $y = 0$. Set $y' = 0$: $x = (6 \pm \sqrt{36-48})/(-6)$; there are no local extrema. $[-6, 6]$ by $[-4, 4]$ shows a complete graph.

12. $y' = 3x^2 - 18x - 21 = 3(x^2 - 6x - 7) = 3(x-7)(x+1)$; $y'' = 3(2x-6)$, $y'(7) = 0$, $y''(7) > 0 \Rightarrow (7, -256)$ is a minimum, $y'(-1) = 0$, $y''(-1) < 0 \Rightarrow (-1, 0)$ is a maximum; $y''(3) = 0$, y'' changes sign at $3 \Rightarrow (3, -128)$ is an inflection point. $[-4, 10]$ by $[-300, 5]$ shows a complete graph.

13. $y = (x-2)^{1/3} \Rightarrow y' = \frac{1}{3}(x-2)^{-2/3}$, $y'' = \frac{-2}{9}(x-2)^{-5/3}$, y'' changes sign at $x = 2$. y' is never 0; y' is not defined at 2, but $(2, 0)$ is an inflection point. $[-6, 6]$ by $[-4, 4]$ shows the graph.

14. y is defined for $x \leq 1$, $y(1) = 0$; $y' = -(1-x)^{-3/4}$ which is never 0; $y'' = -\frac{3}{4}(1-x)^{-7/4} < 0$ for $x < 1$. The graph is always concave down. $[-4, 4]$ by $[-4, 4]$ shows a complete graph.

15. $[-6, 6]$ by $[-4, 4]$ appears to show a complete graph. $y' = 1 - 2x - 4x^3$; $y'' = -2 - 12x^2 = -2(1 + 6x^2)$. There are no inflection points. The unique zero of y', between 0 and 1, corresponds to the plotted maximum of $(0.385, 1.215)$.

16. $[-4, 4]$ by $[-4, 30]$ appears to show a complete graph. There are no local extrema; there appears to be an inflection point near $x = 0$. Using calculus, $y' = 2x^2 + 5$, $y'' = 4x$. $(0, 20)$ is the inflection point.

17. Use a grapher to graph y and y' in $[-6, 6]$ by $[-50, 50]$. To determine whether y is ever increasing, look at $y' = -8x^2 + 8x - 2$. If y' is positive, the discriminant of the quadratic will be positive; however $64 - 4 \cdot 2 \cdot 8 = 0$. We have not missed any hidden behavior of y. Inflection point at $(\frac{1}{2}, -12.333)$.

18. Use $[-4,4]$ by $[-4,25]$ then zoom; graph y, NDER(y), and NDER2(y). Local maxima at $(0.163, 20.073)$ and $(2.107, 22.056)$, local minimum at $(0.730, 19.870)$, inflection points at $(0.423, 19.978)$ and $(1.577, 21.133)$.

19. Use $[-6,6]$ by $[-8,3]$; graph both y and y'. Zooming in on y' near its local maximum shows that y' does turn positive, and hence, that y is increasing between $x = -0.12$ and $x = 0$. To see the increasing behavior of y, use $[-0.5, 0.5]$ by $[0.99, 1.01]$. Local minima at $(-0.118, 0.9976)$ and $(2.118, -6.456)$, local maximum at $(0,1)$, inflection points at $(-0.598, 0.9988)$ and $(1.393, -3.414)$.

20. Graph y, NDER(y), NDER2(y) on $[-2,2]$ by $[-10,10]$; zoom in to look at y' near $x = 0$ to investigate hidden behavior. $y' > 0$ except at $x = 0$ means that y has only an inflection point at $(0,4)$. In fact, $y' = 20x^2(x^2 + x + 1) \geq 0$.

21. Graphing y and y' in $[-5,5]$ by $[-15,25]$ shows the end behavior of y. On the interval $-1.5 \leq x \leq 0$, the graph of y' is steadily increasing. This means there is no concealed behavior in the graph of y. Local minimum at $(-0.578, 0.972)$, inflection point at $(1.079, 13.601)$, local maximum at $(1.692, 20.517)$.

22. $[-6,4]$ by $[-40,10]$ gives the overall shape, but the graph of y' in that window shows there are 4 local extrema. Use $[-5,-3]$ by $[-40,-30]$ and $[0,1]$ by $[-0.5, 2.5]$ to see them. Local maximum at $(-3.791, -34, 193)$, inflection point at $(-3.562, -34.348)$, local minimum at $(-3.303, -34.518)$, inflection point at $(-1.5, -16.55)$, local maximum at $(0.303, 1.418)$, inflection point at $(0.562, 1.248)$, local minimum at $(0.791, 1.093)$.

23. Graph y and y' in $[-5,5]$ by $[-15,20]$. Tracing along y' shows that y has an inflection point at $(3.710, -3.420)$. Local maximum at $(0.215, -2.417)$.

24. $[-4,4]$ by $[-10,10]$ shows a complete graph. Local maximum at $(-1.805, -6.483)$, local minimum at $(-0.209, 3.525)$.

25. If $y = \log_3|x|$, then $3^y = |x|$ and $y = \frac{\ell n|x|}{\ell n\, 3}$. The graph will be symmetric about the y-axis. Graphing y and y' in $[-1,4]$ by $[-5,5]$ shows the complete behavior of y. The graph of y' gets closer and closer to zero as $x \to \infty \Rightarrow$ the graph of y grows more and more slowly as $x \to \infty$.

26. $[-4,4]$ by $[-5,20]$ gives the general shape. Investigating the graphs of y, y', y'' on $[-0.5, 1.5]$ by $[-0.6, 0.6]$ shows there is no hidden behavior. Local minimum at $(1,0)$, $y'' = e^{x-1} > 0 \Rightarrow$ always concave up.

27. Use $[1.9, 4]$ by $[-5, 5]$. Defined for $x > 2$, always rising, concave down on $(2, 4)$. Inflection point at $(4, 2.773)$.

28. $[0, 4\pi]$ by $[-2, 2]$ shows two complete periods. On $[0, 2\pi]$:

Local maxima at	**Local minima at**	**Inflection points at**
$(0.176, 1.266)$	$(0.994, -0.513)$	$(0.542, 0.437)$
$(1.571, 0) = (\pi/2, 0)$	$(2.148, -0.513)$	$(1.266, -0.267)$
$(2.965, 1.266)$	$(3.834, -1.806)$	$(1.876, -0.267)$
$(4.712, 2.000)$	$(5.591, -1.806)$	$(2.600, 0.437)$
		$(3.425, -0.329)$
		$(4.281, 0.120)$
		$(5.144, 0.120)$
		$(6.000, -0.329)$

29. $y = [x(1-x)]^{1/4}$. The domain of y is $0 \le x \le 1$; $y \ge 0$. Use $[-1, 2]$ by $[-1, 2]$. Local maximum at $(0.500, 0.707)$; local minima at $(0, 0)$ and $(1, 0)$.

30. Use $[-4, 3]$ by $[-6, 10]$. Inflection point at $(-2, 0)$.

31. The question asks for the solution to $-x^3 + 3x + 4 = 0$. Let $f(x) = -x^3 + 3x + 4$. $f(2) = 2$, $f(3) = -14 < 0$. For Newton's method, the iteration is $x_{n+1} = x_n - \frac{-x^3+3x+4}{-3x^2+3}$. Starting with $x_0 = 1$, leads to division by 0, a "math error". Starting with $x_0 = 2$ leads quickly to $x = 2.1958$.

32. The iteration is: $x_{n+1} = x_n - (\sec x_n - 4)/\sec x_n \tan x_n = x_n - \frac{\frac{1-4\cos x_n}{\cos x_n}}{\tan x_n} \cdot \cos x_n = x_n - (1 - 4\cos x_n)/\tan x_n$. x_0 cannot be 0 or $\pi/2$. Starting with $x_0 = 0.5$ leads to 4.965 which is not in the interval. $x_0 = 1.5$ leads to 1.318.

33. Let's blindly start with $x_0 = 0$ and use Newton's method: $x_{n+1} = x_n - (2\cos x_n - \sqrt{1 + x_n})/(-2\sin x_n - \frac{1}{2\sqrt{1+x_n}})$. The iterates are:

$$\begin{aligned} x_1 &= 2 \\ x_2 &= 0.783\ldots \\ x_3 &= 0.829\ldots \\ x_4 &= 0.828\ldots \\ x_5 &= 0.828\ldots \end{aligned}$$

34. Set $y1 = 8x^4 - 14x^3 - 9x^2 + 11x - 1$; $y_2 = \text{NDER}(y)$. Use Newton's method: $x - y1/y2 \to x$ for various starting values: $x_0 = 0 \to 0.100$, $x_0 = 2 \to 1.984$, $x_0 = -1 \to -0.977$, $x_0 = 0.5 \to 0.643$. Confirm by comparing the graphs of $y1$ and $y3 = 8(x-0.100)(x-1.984)(x+0.977)(x-0.643)$ on $[-2,2]$ by $[-6,6]$. The graphs are very close.

35. b) $x - f(x)/f'(x) = x - \frac{\frac{1}{x}-3}{-\frac{1}{x^2}} = x + x^2(\frac{1}{x} - 3) = x + x - 3x^2 = x(2-3x)$

36. Let $f(x) = x^3 + x - 1$; $f(0) = -1$, $f(1) = 1$. There is at least one solution between 0 and 1. Since $f'(x) = 3x^2 + x > 0$, $f(x) = 0$ has at most one solution. Therefore there is exactly one solution. The iteration for Newton's method is

$$x_{n+1} = x_n - f(x_n)/f'(x_n) = x_n - \frac{(x_n^3 + x_n - 1)}{3x_n^2 + 1}$$

Starting with $x_0 = 0$, the iterates are

$$\begin{aligned} x_1 &= 1 \\ x_2 &= 0.75 \\ x_3 &= 0.68604 \\ x_4 &= 0.68234 \\ x_5 &= 0.68233 \end{aligned}$$

The solution, to three decimal places is 0.682.

37. Graph $f(x)$ on $[-6,1]$ by $[-80,30]$, $f(-6) = -74$ is the minimum value. The maximum value of 16.25 is taken on at -4.550.

38. Graph on $[0,11]$ by $[-5,10]$. $f(2.840) = 0.730$ is the minimum value; $f(6.481) = 3.526$ is the maximum.

39. $A = \frac{1}{2}rs = \frac{1}{2}r(100-2r) = 50r - r^2$, $0 \le r \le 50$. $A' = 50 - 2r \Rightarrow A' = 0$ when $r = 25$. $A'' = -2 < 0$, so this gives a maximum for A ($A(0) = A(50) = 0$). When $r = 25$, $s = 50$.

40. For $x > 0$, $A(x) = \frac{1}{2}bh = \frac{1}{2}(2x)(27 - x^2)$; $A'(x) = 27 - x^2 + x(-2x) = 27 - 3x^2 = 3(9 - x^2) = 3(3-x)(3+x)$. $A'(3) = 0$, $A'(3^-) > 0$. $A'(3^+) < 0$, hence $A(3) = 54$ is a local maximum. Since $A = 0$ at the endpoints of the domain $0 \le x \le \sqrt{27}$, the maximum value of the area is 54 square units.

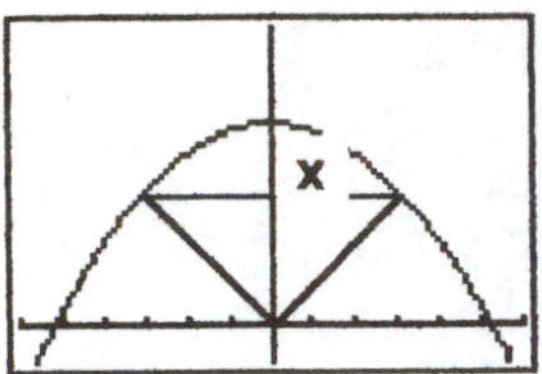

41. Let the sides of the base be x, let the height be h. Then $V = s^2h$, $108 =$ (area of base) $+4$(area of one side) $= s^2+4sh$. Solving for h gives $h = \frac{108-s^2}{4s}$. $V = \frac{s^2(108-s^2)}{4s} = \frac{1}{4}s(108-s^2) = \frac{1}{4}[108s-s^3]$. $V' = \frac{1}{4}[108-3s^2]$; $V' = 0$ when $s^2 = \frac{108}{3} = 36$. The maximum volume occurs when $s = 6$ ft, $h = \frac{108-36}{24} = 3$ ft.

42. Let the base of the vat have side x, let its height be h. Then $V = x^2h = 32 \Rightarrow h = 32/x^2$. The weight is proportional to the surface area $S = x^2 + 4xh = x^2 + \frac{4\cdot 32}{x}$. $S' = 2x - \frac{128}{x^2} = \frac{2}{x^2}(x^3-64)$. $S'(4) = 0$, $S'(4^-) < 0$, $S'(4^+) > 0 \Rightarrow S(4) = 48$ is the minimum value for S. The vat should be of size $4 \times 4 \times 2$ ft high.

43. From the diagram $(\frac{h}{2})^2 + r^2 = (\sqrt{3})^2$, or $r^2 = 3 - h^2/4$. $V = \pi r^2 h = \pi(3-\frac{h^2}{4})h$, $0 \le h \le 2\sqrt{3}$, $V' = \pi[-\frac{-2h}{4}\cdot h + (3-\frac{h^2}{4})] = \frac{\pi}{4}[12-3h^2]$. $V' = 0$ for $h = 2$; for $h > 0$, $V'' < 0$ so $h = 2$ gives the maximum volume. $r^2 = 3 - 1 = 2 \Rightarrow r = \sqrt{2}$.

44. If the interior cone has radius x and height y, then $\frac{12}{6} = \frac{12-y}{x}$ or $y = 12-2x$. Hence, the volume of the interior cone is $V = \frac{1}{3}\pi x^2(12-2x)$. $V' = \frac{\pi}{3}[2x(12-2x)-2x^2] = \frac{2\pi x}{3}[12-2x-x] = \frac{2\pi x}{3}[12-3x]$: Maximum occurs when radius $= x = 4$, height $= 12-8 = 4$.

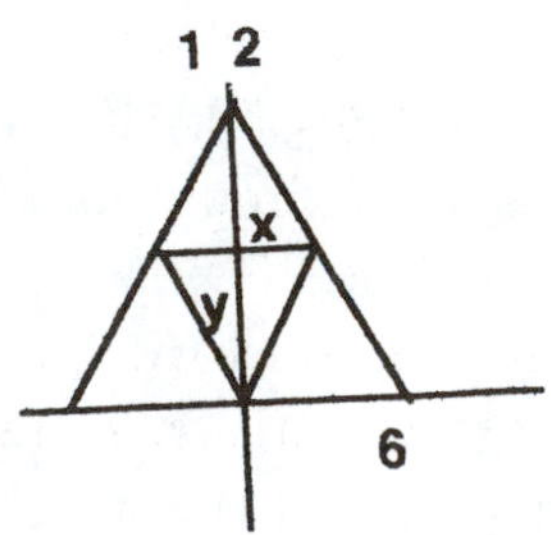

45. The cost, $C = x \cdot 40000 + (20 - y)30000$, where $y^2 + 12^2 = x^2$. $C = 10000[4\sqrt{y^2+12^2} + 3(20-y)]$. $C' = 10000[\frac{4 \cdot y}{\sqrt{y^2+12^2}} - 3]$; setting $C' = 0$ gives $4y = 3\sqrt{y^2+12^2} \Rightarrow 16y^2 = 9(y^2+12^2)$, or $y = 36/\sqrt{7}$; $x^2 = \frac{9 \cdot 12^2}{7} + 12^2 = \frac{16 \cdot 12^2}{7} \Rightarrow x = 48/\sqrt{7}$. Alternatively, graphing $C/10000$ on $[0, 40]$ by $[-5, 150]$ gives $y = 13.607$, $x = \sqrt{y^2+12^2} = 18.143$.

46. The track length is $400 = 2x + 2\pi r$, the area of the rectangle is $A = x(2r)$. Substituting $x = 200 - \pi r$ into A and maximizing. $A = 400r - 2\pi r^2$ gives $r = \frac{100}{\pi}$, $x = 100$.

47. The total profit is $P = K(2x + y)$, where K is a constant; $P = K(2x + \frac{40-10x}{5-x}) = K[\frac{40-2x^2}{5-x}]$; $P' = 0$ when $-4x(5-x) - (40-2x^2)(-1) = 0$, i.e. at $x = 2.764$. Since $P(2.764) = 11.048K$, $P(0) = 8K$, $P(4) = 8K$, for maximum profit you should make 276 Grade A tires, 553 Grade B tires.

48. The distance is $d(t) = \sin t - \sin(t + \pi/3)$; $d'(t) = \cos t - \cos(t + \pi/3)$. a) Using technology, the maximum value of d is 1 at $x = 2.618 = 5\pi/6$, and $11\pi/6$. b) They collide at $t = \pi/3, 4\pi/3, 7\pi/3, \ldots$.

49. If the squares have side x, the volume of the box is $V = x(16-2x)(10-2x)$. $V' = (16-2x)(10-2x) - 2x(16-2x) - 2x(10-2x) = 12x^2 - 104x + 160 = (x-2)(12x-80)$. $x = 80/12 > 6$ is impossible for this problem. $V(2) = 144$ in^3 is the maximum volume. The dimensions are $12 \times 6 \times 2$ inches.

50. a) Following the diagram in Fig. 1.95, $V(x) = 2x(22-2x)(17-2x)$; since the factors must be positive $0 < x < 8.5$. The window $[0, 9] \times [0, 2000]$ shows the graph of V; $V\max = 1058.37$ at $x = 3.171$. Analytically, $V' = 2(22-2x)(17-2x) - 4x(22-2x) - 4x(17-2x) = 24x^2 - 312x + 748$. Using the quadratic equation, $V' = 0$ when $x = 3.171$. e) Solve the equation $2x(22-2x)(17-2x) = 400$; $x = 0.610$.

b) $V_1(x) = x(12-2x)(1-x)$ or $V_2 = x(20-2x)(6-x)$ depending on whether the squares are cut from a short edge or from a long edge. In each case, $0 < x < 6$. $V_1(2.427) = 131.341$ is a maximum; $V_2(2.427) = 131.341$ is a maximum. To confirm, graph V_i' and locate its root.

51. $s'(t) = 3t^2 + 2t - 6$. $s' = 0$ at $t = \frac{-2 \pm \sqrt{4+72}}{6} = \frac{-2 \pm \sqrt{76}}{6}$. $s'(0) < 0$, so for $0 < t < (-2+\sqrt{76})/6$ the particle moves to the left. It then moves to the right. At $t = (-2+\sqrt{76})/6$, the particle is 0.94 units to the right of the origin.

52. $s'(t) = 4 - 6t - 3t^2$; the particle moves to the right $0 \le t \le 0.524$, after which time it moves to the left. It passes through the origin at $t = 1.398$.

53. $A = \pi r^2$ and $dr/dt = -2/\pi$. $dA/dt = 2\pi r\, dr/dt = -4r$. $\frac{dA}{dt}|_{r=10} = -40\ \text{m}^2/\text{sec}$.

54. The distance from the origin, $s(t)$, satisfies the equation $s^2(t) = x^2(t) + y^2(t)$; hence $2s(t)s'(t) = 2x(t)x'(t) + 2y(t)y'(t)$. At $(5, 12)$ $s(t) = \sqrt{25 + 144} = 13$. Hence at $(5, 12)$, $13s'(t) = 5(-1) + 12(-5)$, or $s'(t) = -65/13 = 5$ m/sec.

55. Let s be the side of the cube. $V = s^3$ and $\frac{dV}{dt} = 1200\ \text{cm}^3/\text{min}$. $\frac{dV}{dt} = 3s^2\frac{ds}{dt}$. Hence $\frac{ds}{dt} = \frac{1}{3s^2}\frac{dV}{dt} = \frac{1200}{3s^2}|_{s=20} = \frac{1200}{3 \cdot 20 \cdot 20} = 1$ cm/min.

56. $s^2(t) = x^2(t) + y^2(t) = x^2(t) + x^3(t)$; $2s \cdot s' = 2xx' + 3x^2x'$. When $x = 3$, $s = \sqrt{9 + 27} = 6$; at $x = 3$, $s' = 11$, $2 \cdot 6 \cdot 11 = (6 + 27)x'$, or $x' = 132/33 = 4$ units/sec.

57. a) From the figure, $h/r = 10/4$ or $r = 2h/5$ b) $V = \frac{1}{3}\pi r^2 h = \frac{1}{3}\pi\frac{4h^3}{25}$; $dV/dt = \frac{4}{25}\pi h^2 dh/dt = -5$ when $h = 6$, $dh/dt = \frac{-25 \cdot 5}{4\pi \cdot 36} = \frac{-125}{144\pi}$ ft/min.

58. $s^2 = x^2 + y^2$; $ss' = x \cdot x' + y \cdot y' = x(-36) + y(-50)$, or $s' = \frac{-36x - 50y}{s}$. When $x = 5$, $y = 12$, $s = 13$ and $s' = -60$ mph.

59. Let $x =$ distance from the car to the point on the track which is right in front of you. $\tan\theta = x/132 \Rightarrow \sec^2\theta\frac{d\theta}{dt} = \frac{1}{132}\frac{dx}{dt} = -\frac{264}{132}$ or $\frac{d\theta}{dt} = -2\cos^2\theta$. When the car is in front of you, $\theta = 0$, $\cos\theta = 1$ and $\frac{d\theta}{dt} = -2$ rad/sec.

60. $\frac{ds}{dt} = r\frac{d\theta}{dt}$; when $r = 1.2$, $\frac{d\theta}{dt} = \frac{6\text{ft/sec}}{1.2\text{ft}} = 5$ rad/sec.

61. Let $H(x) = 3x$; then $H'(x) = 3$. If $g'(x) = 3$, for some function $g(x)$, then $g(x) - H(x) = K$, a constant. Thus, $g(x) = H(x) + K = 3x + K$ and $g(x)$ is one of the functions $F(x) = 3x + C$. All functions with derivative 3 are of the form $F(x) = 3x + C$.

62. The corollary says the functions must differ by a constant: $\frac{x}{x+1} - (-\frac{1}{x+1}) = 1$, a constant.

63. a) C b) $x + C$, c) $\frac{x^2}{2} + C$ d) $\frac{x^3}{3} + C$ e) $x^{11}/11 + C$ f) $-x^{-1} + C$ g) $-x^{-4}/4 + C$ h) $\frac{2}{7}x^{7/2} + C$ i) $\frac{3}{7}x^{7/3} + C$ j) $\frac{4}{7}x^{7/4} + C$ k) $\frac{2}{3}x^{3/2} + C$ l) $2x^{1/2} + C$ m) $\frac{7}{4}x^{4/7} + C$ n) $\frac{-3}{4}x^{-4/3} + C$

64. a) $-\cos x + C$ b) $\sin x + C$ c) $\sec x + C$ d) $\cot x + C$ e) $\tan x + C$ f) $\csc x + C$

65. $y' = 3x^2 + 5x - 7 \Rightarrow y = x^3 + (5/2)x^2 - 7x + C$

66. $y' = x^{-2} + x + 1 \Rightarrow y = -\frac{1}{x} + \frac{x^2}{2} + x + C$

67. $y' = x^{1/2} + x^{-1/2} \Rightarrow y = \frac{x^{3/2}}{3/2} + \frac{x^{1/2}}{1/2} + C = (2/3)x^{3/2} + 2x^{1/2} + C$

68. $y' = x^{1/3} + x^{1/4} \Rightarrow y = \frac{x^{4/3}}{4/3} + \frac{x^{5/4}}{5/4} + C = \frac{3}{4}x^{4/3} + \frac{4}{5}x^{5/4} + C$

69. $y' = 3\cos 5x \Rightarrow y = 3(\frac{1}{5})\sin 5x + C$

70. $-16\cos(x/2) + C$

71. $y' = 3\sec^2 3x \Rightarrow y' = 3(\frac{1}{3})\tan 3x + C = \tan 3x + C$

72. $-2\cot 2x + C$

73. $y' = \frac{1}{2} - \cos x \Rightarrow y = (1/2)x - \sin x + C$

74. $\frac{1}{2}x^6 + 2\sin 8x + C$

75. $y' = \sec\frac{x}{3}\tan\frac{x}{3} + 5 \Rightarrow y = 3\sec(x/3) + 5x + C$

76. $x + \csc x + C$

77. $y' = \tan^2 x = \sec^2 x - 1 \Rightarrow y = \tan x - x + C$

78. $y' = \cot^2 x = \csc^2 x - 1 \Rightarrow y = -\cot x - x + C$

79. $y' = 2\sin^2 x = 1 - \cos 2x \Rightarrow y = x - \frac{1}{2}\sin 2x + C$

80. $y' = \cos 2x \Rightarrow y = \frac{1}{2}\sin 2x + C$

81. $y' = 1 + x + \frac{x^2}{2} \Rightarrow y = x + \frac{x^2}{2} + \frac{x^3}{6} + C$. Since $y(0) = 1$, $1 = C$. $y = x + \frac{x^2}{2} + \frac{x^3}{6} + 1$. Or in the powers of x, $y = 1 + x + \frac{x^2}{2} + \frac{x^3}{6}$.

82. $y = x^4 - \frac{21}{3}x^3 + 7x^2 - 7x + C$; $1 = 1 - 7 + 7 - 7 + C \Rightarrow C = 7$. $y = x^4 - 7x^3 + 7x^2 - 7x + 7$

83. $y' = \frac{x^2+1}{x^2} = 1 + x^{-2} \Rightarrow y = x - x^{-1} + C$. Since $y(1) = -1$, $-1 = 1 - 1/1 + C$, or $C = -1$; $y = x - 1/x - 1$

84. $y' = x^2 + 2 + x^{-2} \Rightarrow y = \frac{1}{3}x^3 + 2x - \frac{1}{x} + C$; $1 = \frac{1}{3} + 2 - 1 + C \Rightarrow C = -\frac{1}{3}$; $y = \frac{1}{3}x^3 + 2x - \frac{1}{x} - \frac{1}{3}$

85. $y'' = -\sin x \Rightarrow y' = \cos x + C$. Since $dy/dx = 1$ when $x = 0$, $C = 0$. Hence $y' = \cos x$, which gives $y = \sin x + C$. Since $y(0) = 0$, this $C = 0$ also and $y = \sin x$ is the solution.

86. $y'' = \cos x \Rightarrow y' = \sin x + C_1$; $y'(0) = 0 \Rightarrow C_1 = 0$. $y' = \sin x \Rightarrow y = -\cos x + C_2$; $y(0) = -1 \Rightarrow C_2 = 0$. $y = -\cos x$.

87. a) $y'' = 0$ for all $x \Rightarrow y' = C$ b) $\frac{dy}{dx} = 1$ when $x = 0 \Rightarrow C = 1$, i.e., $y' = 1$ c) $y' = 1 \Rightarrow y = x + C$. Since $y(0) = 0$, $C = 0$. $y = x$ satisfies all the conditions.

88. $y' = 3x^2 + 2 \Rightarrow y = x^3 + 2x + C$; $y(1) = -1 \Rightarrow -1 = 1 + 2 +$; $C = -4$; $y = x^3 + 2x - 4$.

89. Let $s(t)$ be the height of the shovelful of dirt at time t and let the bottom of the hole correspond to $s(0) = 0$. The problem states that $s'(0) = v(0) = 32$. By the law of gravity, $s''(t) = -32$ for all t. Hence $s'(t) = -32t + C$; $32 = -32 \cdot 0 + C$, so $s'(t) = -32t + 32$. $s(t) = -16t^2 + 32t + C$. $s(0) = 0 \Rightarrow C = 0$, so the height of the dirt is $s(t) = -16t^2 + 32t$. Set $s'(t) = 0$ to find the value of t at which s has a maximum: $t = 1$. $s(1) = -16 + 32 = 16$. Duck!

90. $a(t) = 2 + 6t \Rightarrow v(t) = 2t + 3t^2 + C_1$; $v(0) = 4 \Rightarrow C_1 = 4$; $v(t) = 2t + 3t^2 + 4 \Rightarrow s(t) = t^2 + t^3 + 4t + C_2$. We are not given $s(0)$; however, $s(1) - s(0) = [1^2 + 1^3 + 4 + C_2] - [0 + C_2] = 6$.

91. For $-\sqrt{13} < x < \sqrt{13}$, $y' > 0$ and the function is increasing. There is a local minimum at $-\sqrt{13} \approx -3.6$ and a local maximum at $\sqrt{13} \approx 3.6$. $y'' = \frac{-x}{\sqrt{x^2+3}}$ is 0 at $x = 0$; there is an inflection point at 0. Graph the slope field on $[-6, 0]$ by $[-4, 4]$ and $[0, 6]$ by $[-4, 4]$; the solution behaves like

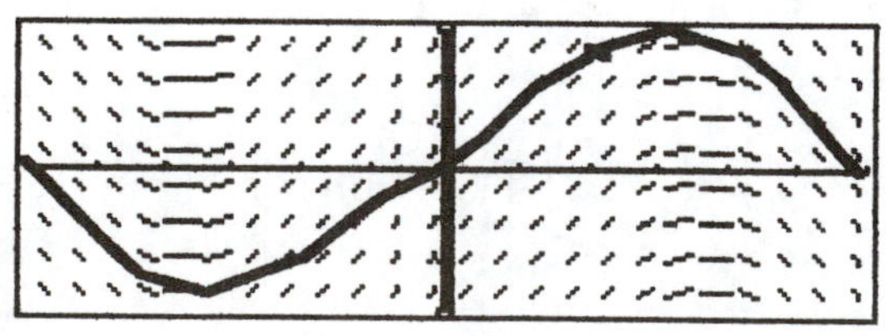

92. Since $y' \geq 0$, y is always increasing. $y'' = \frac{x}{\sqrt{x^2+1}}$ changes sign at $x = 0$.

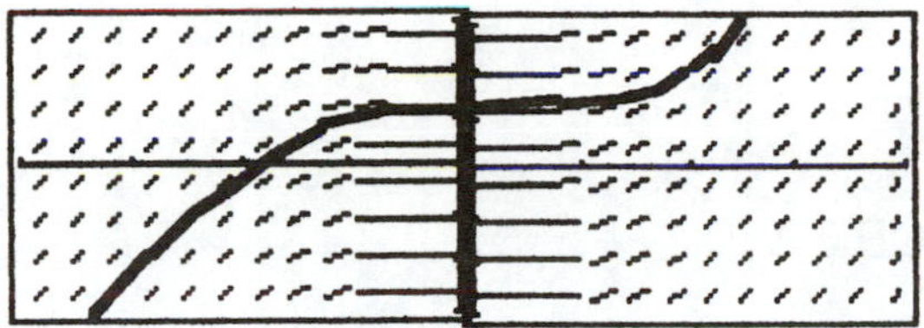

CHAPTER 5
INTEGRATION

5.1 CALCULUS AND AREA

1. a)

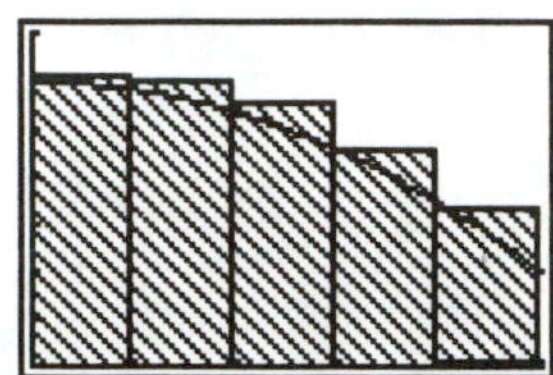  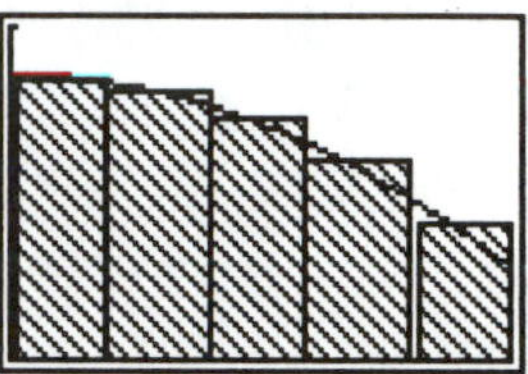

b) $\text{LRAM}_5(6 - x^2) = (6 - 0^2)\Delta x + (6 - (0.4)^2)\Delta x + (6 - (0.8)^2)\Delta x + (6 - (1.2)^2)\Delta x + (6 - (1.6)^2)\Delta x = 10.08$(where $\Delta x = 0.4$).

$\text{RRAM}_5(6-x^2) = (6-0.4^2)\Delta x+(6-0.8)^2)\Delta x+(6-1.2^2)\Delta x+(6-1.6^2)\Delta x+ (6 - 2^2)\Delta x = 8.48.$

$\text{MRAM}_5(6-x^2) = (6-0.2^2)\Delta x+(6-0.6^2)\Delta x+(6-1^2)\Delta x+(6-1.4^2)\Delta x+ (6 - 1.8^2)\Delta x = 9.36$

2. a)

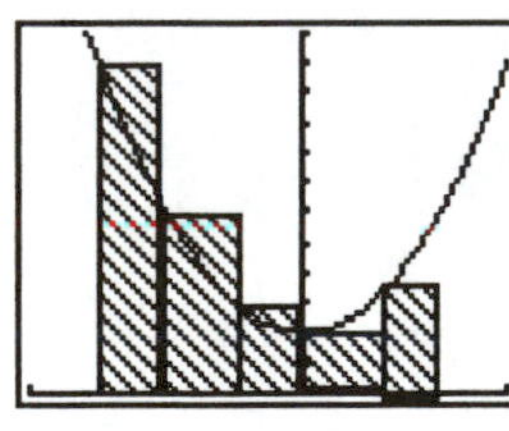 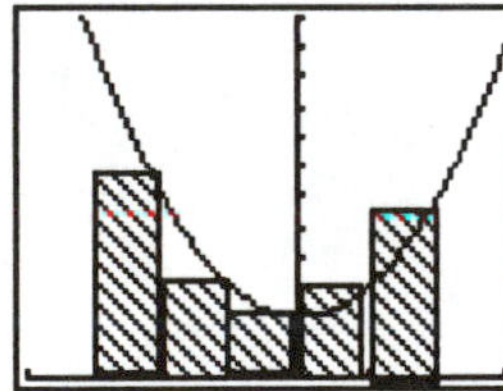 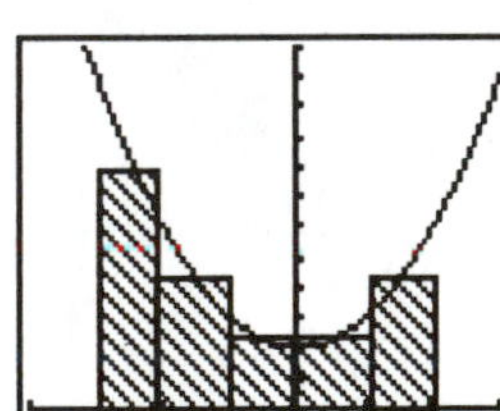

b) $\text{LRAM}_5(x^2+2) = ((-3)^2+2)\cdot 1+((-2)^2+2)\cdot 1+((-1)^2+2)\cdot 1+((0)^2+ 2)\cdot 1 + ((1)^2 + 2)\cdot 1 = 25.$

$\text{RRAM}_5(x^2 + 2) = ((-2)^2 + 2)\cdot 1 + ((-1)^2 + 2)\cdot 1 + ((0)^2 + 2)\cdot 1 + ((1)^2 + 2)\cdot 1 + ((2)^2 + 2)\cdot 1 = 20.$

$\text{MRAM}_5(x^2 + 2) = ((-2.5)^2 + 2)\cdot 1 + ((-1.5)^2 + 2)\cdot 1 + ((-0.5)^2 + 2)\cdot 1 + ((0.5)^2 + 2)\cdot 1 + ((1.5)^2 + 2)\cdot 1 = 21.25.$

3. a)

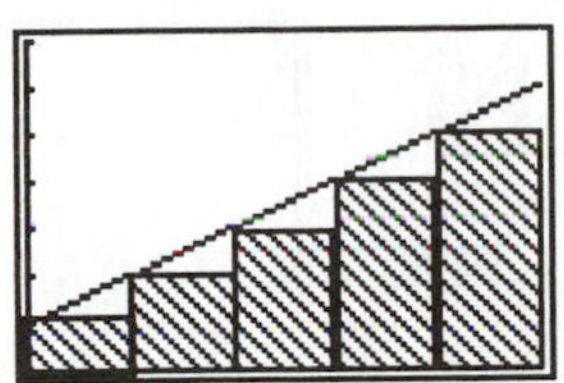
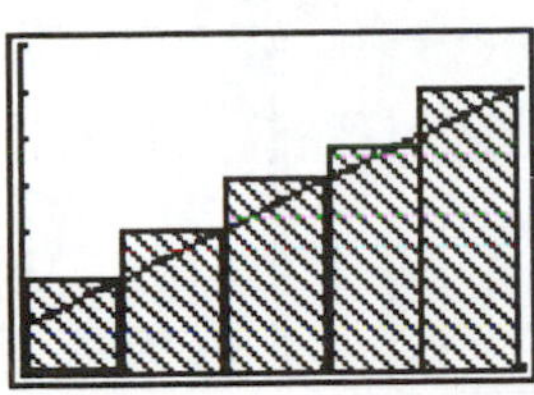
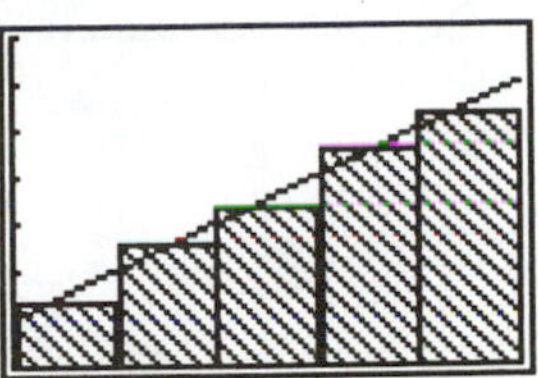

b) $\text{LRAM}_5(x+1) = [(0+1)+(1+1)+(2+1)+(3+1)+(4+1)]\Delta x = 15$(where $\Delta x = 1$).

$\text{RRAM}_5(x+1) = [(1+1)+(2+1)+(3+1)+(4+1)+(5+1)]\Delta x = 20.$

$\text{MRAM}_5(x+1) = [(0.5+1)+(1.5+1)+(2.5+1)+(3.5+1)+(4.5+1)]\Delta x = 17.5.$

4. a)

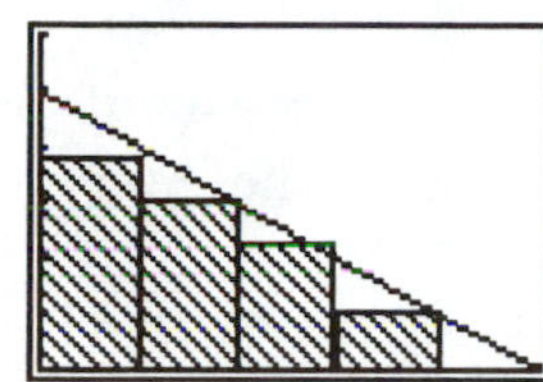
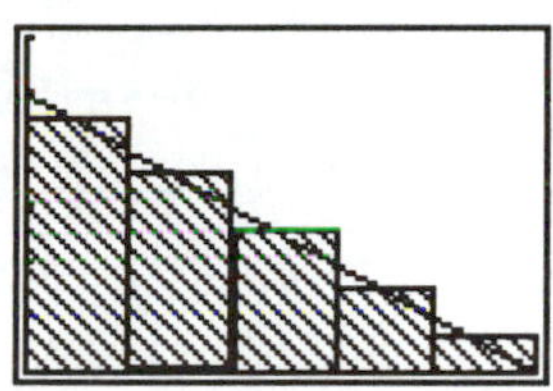

b) $\text{LRAM}_5(5-x) = 5\cdot 1 + 4\cdot 1 + 3\cdot 1 + 2\cdot 1 + 1\cdot 1 = 15.$

$\text{RRAM}_5(5-x) = 4+3+2+1+0 = 10.$

$\text{MRAM}_5(5-x) = 4.5+3.5+2.5+1.5+0.5 = 12.5.$

5. a)

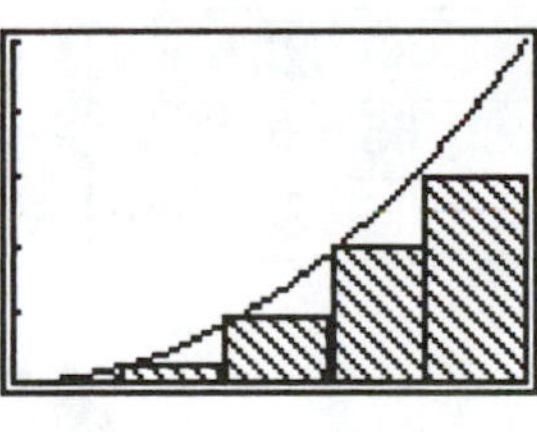
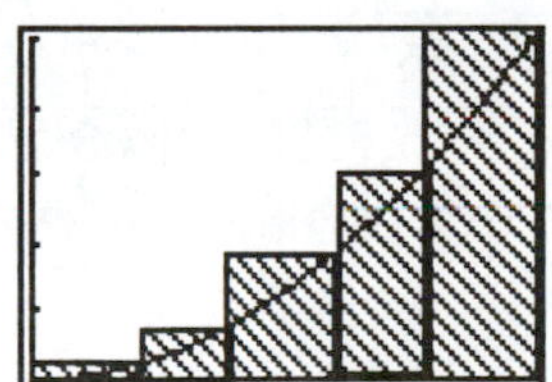
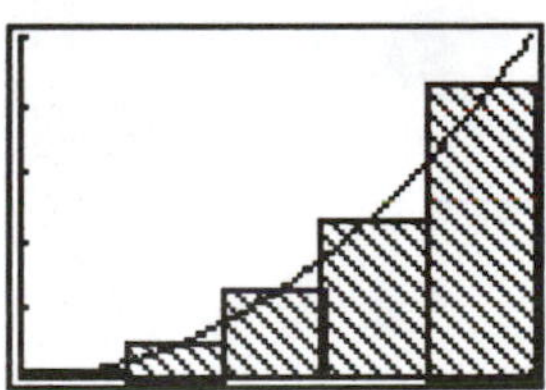

b) $\text{LRAM}_5(2x^2) = [2(0)^2 + 2(1)^2 + 2(2^2) + 2(3^2) + 2(4^2)]\Delta x = 60$ (where $\Delta x = 1$).

$\text{RRAM}_5(2x^2) = [2(1^2) + 2(2^2) + 2(3^2) + 2(4^2) + 2(5^2)]\Delta x = 110.$

$\text{MRAM}_5(2x^2) = [2(0.5^2) + 2(1.5^2) + 2(2.5^2) + 2(3.5^2) + 2(4.5^2)]\Delta x = 82.5.$

6. a)

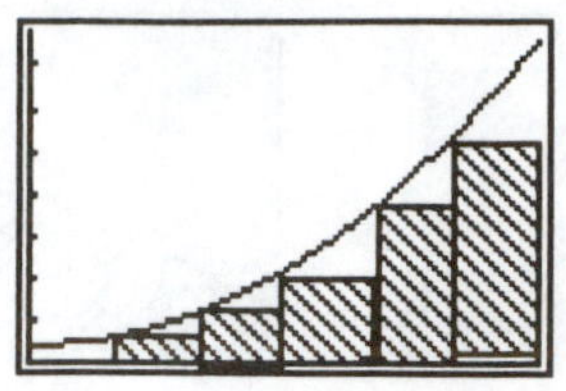 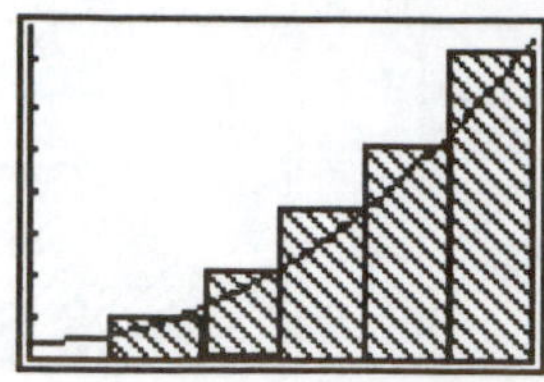 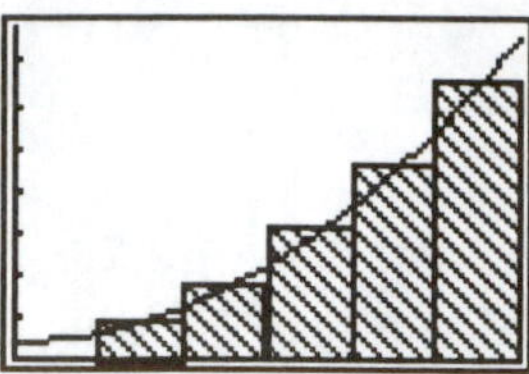

b) $\text{LRAM}_5(x^2+2) = (1^2+2)(1) + (2^2+2)(1) + (3^2+2)(1) + (4^2+2)(1) + (5^2+2)(1) = 65$.

$\text{RRAM}_5(x^2+2) = (2^2+2)(1) + (3^2+2)(1) + (4^2+2)(1) + (5^2+2)(1) + (6^2+2)(1) = 100$.

$\text{MRAM}_5(x^2+2) = (1.5^2+2)(1) + (2.5^2+2)(1) + (3.5^2+2)(1) + (4.5^2+2)(1) + (5.5^2+2)(1) = 81.25$.

7. We may verify $f(x) = x^2 - x + 3 \geqq 0$ on $[0, 3]$ by seeing that its graph does not fall below the x-axis on the interval.

n	$\text{LRAM}_n f$	$\text{RRAM}_n f$	$\text{MRAM}_n f$
10	12.645	14.445	13.4775
100	13.41045	13.59045	13.499775
1000	13.4910045	13.5090045	13.49999775

8. $f(x) = 2x^2 - 5x + 6$. $b^2 - 4ac = 25 - 4(2)6 < 0$. Hence $f(x)$ had no real zero and $f(x) > 0$ for all x.

n	$\text{LRAM}_n f$	$\text{RRAM}_n f$	$\text{MRAM}_n f$
10	35	37.5	35.625
100	35.7125	35.9625	35.83125
1000	35.820875	35.845875	35.8333125

9.

n	$\text{LRAM}_n f$	$\text{RRAM}_n f$	$\text{MRAM}_n f$
10	268.125	393.125	325.9375
100	321.28125	333.78125	327.48438..
1000	326.87531..	328.12531..	327.49984..

10. $f(x) = x^3 + x^2 + 2x + 3$ and $f(-1) = 1$. $f'(x) = 3x^2 + 2x + 2 > 0$ for all $x(b^2 - 4ac < 0)$ so f is an increasing function. Therefore $f(x) \geqq 1 > 0$ on $[-1, 3]$.

n	$\text{LRAM}_n f$	$\text{RRAM}_n f$	$\text{MRAM}_n f$
10	40.96	58.56	49.12
100	48.4576	50.2176	49.3312
1000	49.245376	49.421376	49.333312

11.

n	$\text{LRAM}_n f$	$\text{RRAM}_n f$	$\text{MRAM}_n f$
10	1.98352..	1.98352..	2.00825..
100	1.99984..	1.99984	2.00008..
1000	1.99999..	1.99999..	2.00000..

12.

n	$\text{LRAM}_n f$	$\text{RRAM}_n f$	$\text{MRAM}_n f$
10	1.07648..	0.91940..	1.00102..
100	1.00783..	0.99187..	0.99988..
1000	1.00078..	0.99921..	0.99999..

13.

n	$\text{LRAM}_n f$	$\text{RRAM}_n f$	$\text{MRAM}_n f$
10	1.77264..	1.77264..	1.77227..
100	1.77245..	1.77245..	1.77245..
1000	1.77245..	1.77245..	1.77245..

14. $f(x) = 2 + \frac{\sin x}{x}$. We assume $f(0)$ is defined as $2 + 1 = 3$. Graph f in $[-3, 4]$ by $[0, 3]$ to see that $f(x) > 0$ on $[-3, 4]$.

n	$\text{LRAM}_n f$	$\text{RRAM}_n f$	$\text{MRAM}_n f$
10	17.67064..	17.50527..	17.61632
100	17.61493..	17.59839..	17.60694..
1000	17.60768..	17.60602..	17.60685..

15. 17.5, 83, 13.5, 327.5, 2, respectively.

16. 12.5, 82, 35.83, 49.33, 1, respectively.

19. $\sum_{k=1}^{4} \frac{1}{k} = \frac{1}{1} + \frac{1}{2} + \frac{1}{3} + \frac{1}{4} = \frac{25}{12}$

20. $\sum_{k=1}^{4} \frac{12}{k} = \frac{12}{1} + \frac{12}{2} + \frac{12}{3} + \frac{12}{4} = 12 + 6 + 4 + 3 = 25$

21. $\sum_{k=1}^{3}(k+2) = (1+2) + (2+2) + (3+2) = 12$

22. $\sum_{k=1}^{5}(2k-1) = 1 + 3 + 5 + 7 + 9 = 25$

23. $\sum_{k=0}^{4} \frac{k}{4} = \frac{0}{4} + \frac{1}{4} + \frac{2}{4} + \frac{3}{4} + \frac{4}{4} = \frac{5}{2}$

24. $\sum_{k=-2}^{2} 3k = 3(-2) + 3(-1) + 3(0) + 3(1) + 3(2) = 0$

25. $\sum_{k=1}^{4} \cos k\pi = \cos(1 \cdot \pi) + \cos 2\pi + \cos 3\pi + \cos 4\pi = 0$

26. $\sum_{k=1}^{3} \sin \frac{\pi}{k} = \sin \frac{\pi}{1} + \sin \frac{\pi}{2} + \sin \frac{\pi}{3} = 0 + 1 + \frac{\sqrt{3}}{2} = \frac{2+\sqrt{3}}{2}$

27. $\sum_{k=1}^{4}(-1)^k = (-1)^1 + (-1)^2 + (-1)^3 + (-1)^4 = 0$

28. $\sum_{k=1}^{4}(-1)^{k+1} = (-1)^2 + (-1)^3 + (-1)^4 + (-1)^5 = 0$ **29.** All

30. a) $\sum_{k=-1}^{1} \frac{(-1)^k}{k+2} = \frac{-1}{1} + \frac{1}{2} - \frac{1}{3}$ b) $\sum_{k=0}^{2} \frac{(-1)^k}{k+1} = \frac{1}{1} - \frac{1}{2} + \frac{1}{3}$ c) $\sum_{k=1}^{3} \frac{(-1)^k}{k} = -\frac{1}{1} + \frac{1}{2} - \frac{1}{3}$ d) $\sum_{k=2}^{4} \frac{(-1)^{k-1}}{k-1} = -\frac{1}{1} + \frac{1}{2} - \frac{1}{3}$ Answer: b) is not equivalent to the others.

31. $1 + 2 + 3 + 4 + 5 + 6 = \sum_{k=1}^{6} k$

32. $1 + 4 + 9 + 16 = \sum_{n=1}^{4} n^2$

33. $\sum_{k=1}^{4} \frac{1}{2^k}$

34. $1 + \frac{1}{2} + \frac{1}{3} + \frac{1}{4} + \frac{1}{5} = \sum_{k=1}^{5} \frac{1}{k}$

35. $\sum_{k=1}^{5}(-1)^{k+1} \frac{k}{5}$

36. $-\frac{1}{5} + \frac{2}{5} - \frac{3}{5} + \frac{4}{5} - \frac{5}{5} = \sum_{k=1}^{5}(-1)^k \frac{k}{5}$

37. $\sum_{k=1}^{10} k = \frac{10(11)}{2} = 55$

38. $\sum_{k=1}^{7} 2k = 2\sum_{k=1}^{7} k = 2[\frac{7(7+1)}{2}] = 56$

39. $\sum_{k=1}^{6} -k^2 = -\sum_{k=1}^{6} k^2 = -\frac{6(6+1)(2 \cdot 6+1)}{6} = -91$

40. $\sum_{k=1}^{6}(k^2+5) = \sum_{k=1}^{6} k^2 + \sum_{k=1}^{6} 5 = \frac{6(6+1)(2 \cdot 6+1)}{6} + 5 \cdot 6 = 91 + 30 = 121$

41. $\sum_{k=1}^{5} k(k-5) = \sum_{k=1}^{5}(k^2-5k) = \sum_{k=1}^{5} k^2 - 5\sum_{k=1}^{5} k = \frac{5(6)11}{6} - \frac{5(5)(6)}{2} = -20$

42. $\sum_{k=1}^{7}(2k-8) = 2\sum_{k=1}^{7} k - \sum_{k=1}^{7} 8 = 2[\frac{7(7+1)}{2}] - 8\cdot 7 = 0$

43. $\sum_{k=1}^{100} k^3 - \sum_{k=1}^{99} k^3 = 100^3 + \sum_{k=1}^{99} k^3 - \sum_{k=1}^{99} k^3 = 1,000,000$

44. $(\sum_{k=1}^{7} k)^2 - \sum_{k=1}^{7} k^3 = [\frac{7(7+1)}{2}]^2 - [\frac{7(7+1)}{2}]^2 = 0$

45. a) $\sum_{k=1}^{n} 3a_k = 3\sum_{k=1}^{n} a_k = 3(-5) = -15$

b) $\sum_{k=1}^{n} \frac{b_k}{6} = \frac{1}{6}\sum_{k=1}^{n} b_k = 1$

c) $\sum_{k=1}^{n}(a_k+b_k) = \sum_{k=1}^{n} a_k + \sum_{k=1}^{n} b_k = -5+6 = 1$

d) $\sum_{k=1}^{n}(a_k-b_k) = \sum_{k=1}^{n} a_k - \sum_{k=1}^{n} b_k = -11$

e) $\sum_{k=1}^{n}(b_k-2a_k) = \sum_{k=1}^{n} b_k - 2\sum_{k=1}^{n} a_k = 16$

46. a) $\sum_{k=1}^{n} 8a_k = 8\sum_{k=1}^{n} a_k = 0$

b) $\sum_{k=1}^{n} 250b_k = 250\sum_{k=1}^{n} b_k = 250$

c) $\sum_{k=1}^{n}(a_k+1) = \sum_{k=1}^{n} a_k + \sum_{k=1}^{n} 1 = 0 + n = n$

d) $\sum_{k=1}^{n}(b_k-1) = \sum_{k=1}^{n} b_k - \sum_{k=1}^{n} 1 = 1 - n$

47. $\frac{6\cdot 1}{1+1} + \frac{6\cdot 2}{2+1} + \frac{6\cdot 3}{3+1} + \frac{6\cdot 4}{4+1} + \frac{6\cdot 5}{5+1}$. $\sum_{k=1}^{100} \frac{6k}{k+1} = 574.816..$

48. $\frac{1-1}{1} + \frac{2-1}{2} + \frac{3-1}{3} + \frac{4-1}{4} + \frac{5-1}{5}$. $\sum_{k=1}^{100} \frac{k-1}{k} = 94.8126..$

49. $1(1-1)(1-2) + 2(2-1)(2-2) + 3(3-1)(3-2) + 4(4-1)(4-2) + 5(5-1)(5-2)$. $\sum_{k=1}^{500} k(k-1)(k-2) = 15,562,437,750$

50. $(1-0)(2-0) + (1-1)(2-1) + (1-3)(2-3) + (1-4)(2-4) + (1-5)(2-5)$. $\sum_{k=0}^{500}(1-k)(2-k) = 41,417,000$

51. $\sum_{k=1}^{12} k = 12(13)/2 = 78$

52. $\sum_{k=1}^{n}(2k-1) = 2\sum_{k=1}^{n} k - \sum_{k=1}^{n} 1 = n(n+1) - n = n^2$

53. $\sum_{k=1}^{n} k + \sum_{k=1}^{n-1} k = \frac{n(n+1)}{2} + \frac{(n-1)n}{2} = \frac{n}{2}(n+1+n-1) = n^2$

54. In $\text{LRAM}_n x^2$ each of the rectangles used is an inscribed rectangle and underestimates the area in its interval under the curve $y = x^2$. In $\text{RRAM}_n x^2$ each of the rectangles used is a superscribed rectangle and overestimates the area in its interval under the curve $y = x^2$.

55. Because $0 < x_{k-1} < x_k$ in the partition of $[a, b]$, it follows that $\frac{1}{x_{k-1}} > \frac{1}{x_k}$. Therefore $\text{LRAM}_n(1/x) = (\frac{1}{x_0} + \frac{1}{x_1} + \frac{1}{x_2} + \cdots + \frac{1}{x_{n-1}})\Delta x > (\frac{1}{x_1} + \frac{1}{x_2} + \cdots + \frac{1}{x_n})\Delta x = \text{RRAM}_n(1/x)$.

56. Let A_k be the area of the region under $y = f(x)$, over the x-axis from $x = x_{k-1}$ to $x = x_k$. Since $f(x)$ is increasing, $f(x_{k-1})\Delta x < A_k < f(x_k)\Delta x$ and therefore $\sum_{k=1}^n f(x_{k-1})\Delta x < \sum_{k=1}^n A_k < \sum_{k=1}^n f(x_k)\Delta x$ which is to say $\text{LRAM}_n f < A_a^b f < \text{RRAM}_n f$.

57. Let A_k be the area of the region under $y = f(x)$ over the x-axis from $x = x_{k-1}$ to $x = x_k$. Since $f(x)$ is decreasing and nonnegative, $f(x_{k-1})\Delta x > A_k > f(x_k)\Delta x$ and therefore $\sum_{k=1}^n f(x_{k-1})\Delta x > \sum_{k=1}^n A_k > \sum_{k=1}^n f(x_k)\Delta x$ which is to say $\text{LRAM}_n f > A_a^b f > \text{RRAM}_n f$ or $\text{RRAM}_n f < A_a^b f < \text{LRAM}_n f$.

58. a) $\text{RRAM}_5 x^2 = 55$. $\text{LRAM}_5 x^2 + 5^2 \cdot 1 - 0^2 \cdot 1 = 30 + 25 - 0 = 55$.

b) $\text{RRAM}_n f = \left[\sum_{k=1}^{n-1} f(x_k)\Delta x\right] + f(x_n)\Delta x + (f(x_0)\Delta x - f(x_0)\Delta x)$ $= [f(x_0)\Delta x + \sum_{k=1}^{n-1} f(x_k)\Delta x] + f(x_n)\Delta x - f(x_0)\Delta x = \text{LRAM}_n f + f(x_n)\Delta x - f(x_0)\Delta x$.

59. The results of Example 2 disprove the statement.

60. $\sin x_0 = \sin 0 = 0$ and $\sin x_n = \sin \pi = 0$. The explanation now follows from Exercise 58 b).

61. Since $x = \frac{a+b}{2}$ is a line of symmetry, $f(a) = f(b)$ or $f(x_0) = f(x_n)$. The result now follows from #58 b).

62. $\text{RRAM}_n(x + 1) = \Delta x(1 \cdot \frac{5}{n} + 1) + \Delta x(2 \cdot \frac{5}{n} + 1) + \cdots + \Delta x(n \cdot \frac{5}{n} + 1) = \Delta x \sum_{k=1}^n (k(\frac{5}{n}) + 1) = \frac{5}{n}\left[\frac{5}{n}\sum_{k=1}^n k + \sum_{k=1}^n 1\right] = \frac{5}{n}\left[\frac{5}{n}\frac{n(n+1)}{2} + n\right] = 5\left[\frac{5}{2}\frac{n+1}{n} + 1\right]$. $\text{Lim}_{n\to\infty}\text{RRAM}_n(x + 1) = 5\left[\frac{5}{2} \cdot 1 + 1\right] = \frac{35}{2} = A_0^5(x + 1)$.

63. $\Delta x = \frac{5}{n}$, $x_k = \frac{5k}{n}$ $\cdot$ $\text{RRAM}_n(2x^2) = 2\sum_{k=1}^n (\frac{5k}{n}^2 \frac{5}{n} = (\frac{250}{n^3})\frac{n(n+1)(2n+1)}{6} = \frac{125}{3}(\frac{n+1}{n})(\frac{2n+1}{n}) \to \frac{125}{3}(1)(2) = \frac{250}{3}$.

64. Here $\Delta x = \frac{6-1}{n} = \frac{5}{n}$, $x_k = 1 + \frac{5k}{n}$. $\text{RRAM}_n(x^2 + 2) = \Delta x \sum_{k=1}^n [(1 + \frac{5k}{n})^2 + 2] = \frac{5}{n}\sum_{k=1}^n (\frac{10k}{n} + \frac{25k^2}{n^2} + 3) = \frac{5}{n}[\frac{10}{n}(\sum_{k=1}^n k) + \frac{25}{n^2}(\sum_{k=1}^n k^2) + 3n] = \frac{50}{n^2}\frac{n(n+1)}{2} + \frac{125}{n^3}\frac{n(n+1)(2n+1)}{6} + 15$. Thus $\text{RRAM}_n = 25(1 + \frac{1}{n}) + \frac{125}{6}(1 + \frac{1}{n})(2 + \frac{1}{n}) + 15$. $A_1^6(x^2 + 2) = \lim_{n\to\infty} \text{RRAM}_n(x^2 + 2) = 25 + \frac{125}{6}(2) + 15 = \frac{245}{3}$.

65. $\Delta x = \frac{3}{n}$, $x_k = \frac{3k}{n} \cdot \text{RRAM}_n(x^2 - x + 3) = \sum_{k=1}^{n}\left[\left(\frac{3k}{n}\right)^2 \frac{3}{n} - \frac{3k}{n}\frac{3}{n} + 3\left(\frac{3}{n}\right)\right] = \frac{27}{n^3}\sum_{k=1}^{n} k^2 - \frac{9}{n^2}\sum_{k=1}^{n} k + \frac{9}{n}(n) = \frac{27}{n^3}\frac{n(n+1)(2n+1)}{6} - \frac{9}{n^2}\frac{n(n+1)}{2} + 9 = \frac{9}{2}\left(\frac{n+1}{n}\right)\left(\frac{2n+1}{n}\right) - \frac{9}{2}\left(\frac{n+1}{n}\right) + 9 \to \frac{9}{2}(1)(2) - \frac{9}{2}(1) + 9 = \frac{27}{2}$.

66. $\Delta x = \frac{5}{n}$, $x_k = \frac{5k}{n} \cdot \text{RRAM}_n(2x^3 + 3) = \Delta x(2(\frac{5}{n})^3 + 3) + \Delta x(2(\frac{5\cdot 2}{n})^3 + 3) + \cdots + \Delta x(2(\frac{5n}{n})^3 + 3) = \frac{5}{n}\sum_{k=1}^{n}\left[2\frac{5^3k^3}{n^3} + 3\right] = \frac{5}{n}\left[\frac{2(5^3)}{n^3}\sum_{k=1}^{n} k^3 + \sum_{k=1}^{n} 3\right] = \frac{5}{n}\left[\frac{2(5^3)}{n^3}\left(\frac{n(n+1)}{2}\right)^2 + 3n\right] = 5\left[\frac{2(5^3)}{4}\left(\frac{n}{n}\cdot\frac{n+1}{n}\right)^2 + 3\right] = 5\left[\frac{125}{2}\left(\frac{n+1}{n}\right)^2 + 3\right]$. $A_0^5(2x^3 + 3) = \lim_{n\to\infty} \text{RRAM}_n(2x^3 + 3) = 5\left[\frac{125}{2}(1)^2 + 3\right] = \frac{655}{2} = 327.5$.

67. $\Delta x = \frac{4}{n}$, $x_k = -1 + \frac{4k}{n}$. $\text{RRAM}_n f = \sum_{k=1}^{n}[(-1 + \frac{4k}{n})^3 + (-1 + \frac{4k}{n})^2 2(-1 + \frac{4k}{n}) + 3]\frac{4}{n} = \frac{4}{n}\sum_{k=1}^{n}[1 + \frac{12k}{n} - \frac{32k^2}{n^2} + \frac{64k^3}{n^3}] = \frac{4}{n}[n + \frac{12}{n}\frac{n(n+1)}{2} - \frac{32}{n^2}\frac{n(n+1)(2n+1)}{6} + \frac{64}{n^3}\frac{n^2(n+1)^2}{4}] = 4 + 24(\frac{n+1}{n}) - \frac{64}{3}(\frac{n+1}{n})(\frac{2n+1}{n}) + 64(\frac{n+1}{n})^2 \to 4 + 24 - \frac{128}{3} + 64 = \frac{148}{3}$.

68. $\Delta x = (5 - 0)/n$, $x_k = (k - 1)\Delta x$ for the left-hand endpoint. $\text{LRAM}_n x^2 = \sum_{k=1}^{n} x_k^2\Delta x = \sum_{k=1}^{n}(k-1)^2(\Delta x)^3 = \frac{125}{n^3}\sum_{k=1}^{n}(k-1)^2 = \frac{125}{n^3}\sum_{k=0}^{n-1} k^2 = \frac{125}{n^3}\frac{(n-1)n[2(n-1)+1]}{6}$. $\lim_{n\to\infty} \text{LRAM}_n x^2 = \lim_{n\to\infty} \frac{125}{6}\frac{n-1}{n}\frac{n}{n}\frac{2(n-1)+1}{n} = \frac{125}{6}(1)(1)(2) = \frac{125}{3}$.

69. a) $\sum_{k=1}^{n}(2k-1)^2 = 1^2 + 2^2 + \cdots + (2n)^2 - [2^2 + 4^2 + 6^2 + \cdots + (2n)^2] = \sum_{k=1}^{2n} k^2 - \sum_{k=1}^{n}(2k)^2 = \sum_{k=1}^{2n} k^2 - 4\sum_{k=1}^{n} k^2 = \frac{2n(2n+1)(4n+1)}{6} - \frac{4n(n+1)(2n+1)}{6} = \frac{2n(2n+1)}{6}[4n + 1 - 2(n+1)] = \frac{(2n-1)n(2n+1)}{3}$.

b) $\sum_{k=1}^{n}(2k-1)^3 = \sum_{k=1}^{2n} k^3 - \sum_{k=1}^{n}(2k)^3 = \sum_{k=1}^{2n} k^3 - 2^3\sum_{k=1}^{n} k^3 = [\frac{2n(2n+1)}{2}]^2 - 8[\frac{n(n+1)}{2}]^2 = n^2(2n+1)^2 - 2n^2(n+1)^2 = n^2(2n^2 - 1)$.

70. $\Delta x = 5/n$, $x_k = [(k-1)\Delta x + k\Delta x]/2 = (2k-1)\Delta x/2$. $\text{MRAM}_n x^2 = \sum_{k=1}^{n}\left[\frac{(2k-1)\Delta x}{2}\right]^2\Delta x = \frac{125}{4n^3}\sum_{k=1}^{n}(2k-1)^2 = \frac{125}{4n^3}\frac{(2n-1)n(2n+1)}{3}$. $\lim_{n\to\infty} \text{MRAM}_n x^2 = \lim_{n\to\infty} \frac{125}{3}\frac{(2n-1)n(2n+1)}{4(n)(n)(n)} = \frac{125}{3}\frac{1}{4}(2)(1)(2) = \frac{125}{3}$.

71. $\text{LRAM}_n x^3 = \sum_{k=0}^{n-1} x_k^3\,\Delta x = \sum_{k=0}^{n-1}(\frac{5k}{n})^3\frac{5}{n} = \frac{5^4}{n^4}\sum_{k=0}^{n-1} k^3 = \frac{5^4}{n^4}\sum_{k=1}^{n-1} k^3 = \frac{5^4}{n^4}[\frac{(n-1)n}{2}]^2 = \frac{625}{4}(\frac{n-1}{n})^2 \to \frac{625}{4} = 156.25$.

72. $\Delta x = \frac{5}{n}$. $x_k = [(k-1)\Delta x + k\Delta x]/2 = (2k-1)\Delta x/2$. $\text{MRAM}_n x^3 = \sum_{k=1}^{n}[\frac{(2k-1)\Delta x}{2}]^3\Delta x = \frac{625}{8n^4}\sum_{k=1}^{n}(2k-1)^3 = \frac{625}{8n^4}n^2(2n^2-1) = \frac{625}{8}(2 - \frac{1}{n^2}) \to \frac{625}{4} = 156.25$ as $n \to \infty$.

73. In the solution to Exercise 71, we replace the 5 by x. $\text{LRAM}_n t^3 = \frac{x^4}{4}(\frac{n-1}{n})^2$. $A_0^x(t^3) = \lim_{n\to\infty} \frac{x^4}{4}(\frac{n-1}{n})^2 = \frac{x^4}{4}$.

74. If we count the black dots, we get $1+3+5+7=\sum_{k=1}^{4}(2k-1)$. If we count the dots in the square we get 4^2 so $\sum_{k=1}^{4}(2k-1)=4^2$.

75. The explanation is below the picture.

76. Number of dots in square $(n^2)+$ number of dots in triangle $(\sum_{k=1}^{n}k)$ = number of dots in trapezoid $(\sum_{k=n+1}^{2n}k)$.

5.2 DEFINITE INTEGRALS

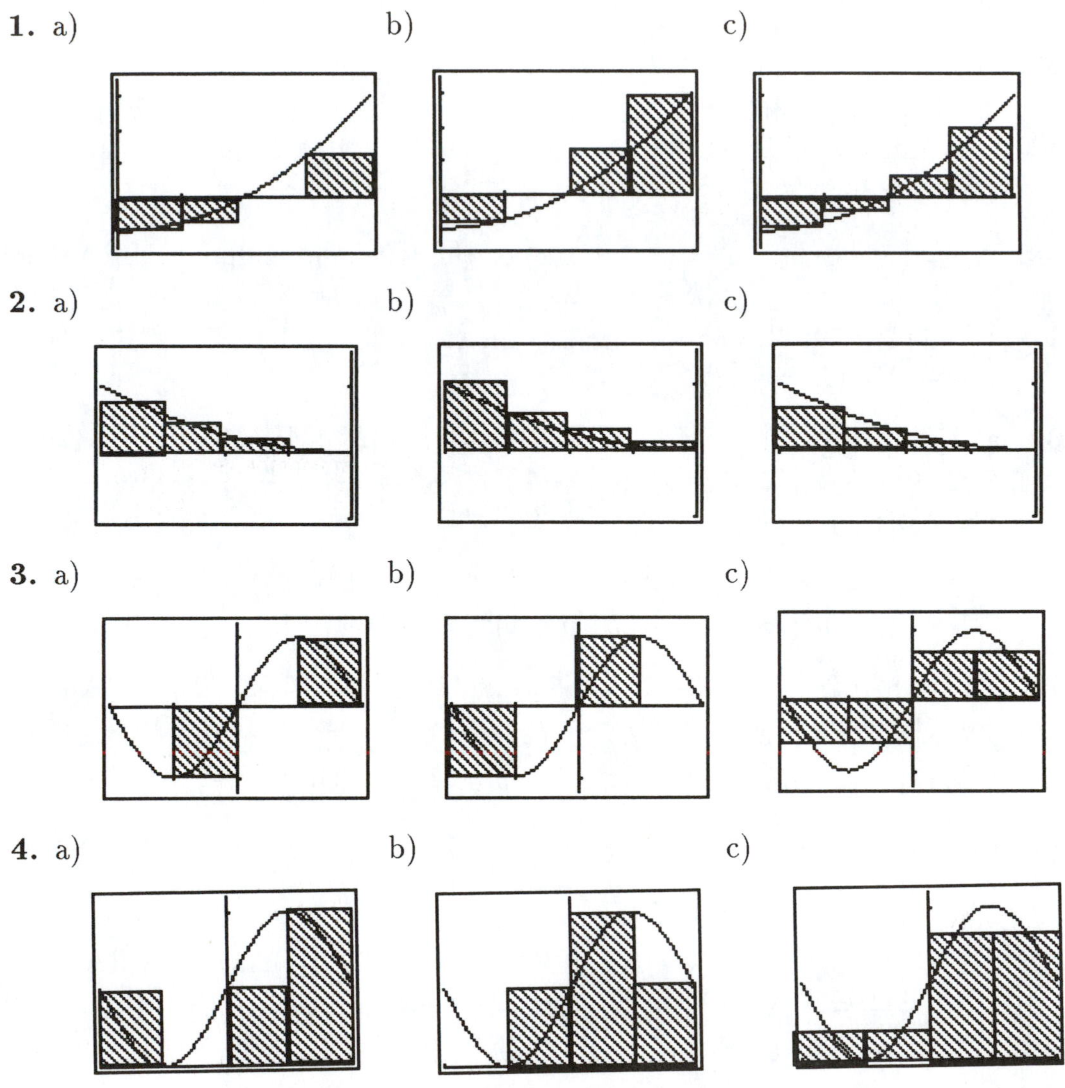

5. $\Delta x_1=1.2$, $\Delta x_2=1.5-1.2=0.3$, $\Delta x_3=2.3-1.5=0.8$, $\Delta x_4=0.3$, $\Delta x_5=0.4$. The largest of these is 0.8 so $\|P\|=0.8$.

6. $\Delta x_1 = -1.6 - (-2) = 0.4$, $\Delta x_2 = -0.5 - (-1.6) = 1.1$, $\Delta x_3 = 0.5$, $\Delta x_4 = 0.8$, $\Delta x_5 = 0.2$. The largest of these is $\|P\| = 1.1$.

7. $\int_0^2 x^2 dx$ **8.** $\int_{-1}^0 2x^3 dx$ **9.** $\int_{-7}^5 (x^2 - 3x)dx$ **10.** $\int_1^4 \frac{1}{x} dx$

11. $\int_2^3 \frac{1}{1-x} dx$ **12.** $\int_0^1 \sqrt{4 - x^2} dx$ **13.** $\int_0^4 \cos x \, dx$ **14.** $\int_{-\pi}^{\pi} \sin^3 x \, dx$

15. Graph $y = x^2 - 4$ in $[0, 2]$ by $[-5, 1]$. We see that $y = f(x) \leqq 0$ on the interval. $A = -\int_0^2 (x^2 - 4)dx = 5.333$.

16. $f(x) \geqq 0$ on $[a, b]$. $A = \int_0^3 (9 - x^2)dx = 18$.

17. $f(x) = \sqrt{25 - x^2} \geqq 0$. $A = \int_0^5 \sqrt{25 - x^2} dx = 19.634$.

18. $f(x) = \sqrt{36 - 4x^2} = 2\sqrt{9 - x^2} \geqq 0$. $A = \int_{-3}^3 2\sqrt{9 - x^2} dx = 28.274$.

19. $f(x) = \tan x \geqq 0$ for $0 \leqq x \leqq \frac{\pi}{4}$. $A = \int_0^{\pi/4} \tan x \, dx = 0.346$.

20. $f(x) = \cos x \leqq 0$ for $\frac{\pi}{2} \leqq x \leqq \frac{3\pi}{2}$. $A = -\int_{\pi/2}^{3\pi/2} \cos x \, dx = 2$.

21. $f(x) = x^2 e^{-x^2}$, $[0, 3]$.

	LRAM_nf	RRAM_nf	MRAM_nf
n = 100	0.44290145	0.44293477	0.44291878
n = 1000	0.44291688	0.44292022	0.44291856
NINT(f)		0.442918559	

22. $f(x) = \sin(x^2)$, $[0, 2\pi]$

	LRAM_nf	RRAM_nf	MRAM_nf
n = 100	0.61053958	0.67201053	0.64257251
n = 1000	0.63905608	0.64520317	0.64214246
NINT(f)		0.64213818	

23. $f(x) = \frac{\sin x}{x}$, $[1, 10]$.

	LRAM_nf	RRAM_nf	MRAM_nf
n = 100	0.75272914	0.67210056	0.71218935
n = 1000	0.71629745	0.70823459	0.71226377
NINT(f)		0.71226452	

24. $f(x) = \frac{1-\cos x}{x^2}$, $[1, 2\pi]$.

	LRAM_nf	RRAM_nf	MRAM_nf
n = 100	0.94392766	0.91964098	0.93175713
n = 1000	0.93298071	0.93055204	0.93176610
NINT(f)		0.93176619	

25. $f(x) = x \sin x$, $[-1, 2]$

	LRAM_nf	RRAM_nf	MRAM_nf
n = 100	2.0282123	2.0575260	2.0427050
n = 1000	2.0412951	2.0442265	2.0427592
NINT(f)		2.0427597	

26. $f(x) = e^{-x^2}$, $[0, 5]$

	LRAM_nf	RRAM_nf	MRAM_nf
n = 100	0.91122692	0.86122692	0.86122692
n = 1000	0.88872692	0.88372692	0.88622692
NINT(f)		0.88622692	

27. $\text{NINT}((2 - x - 5x^2, x, -1, 3) = -42.666$.

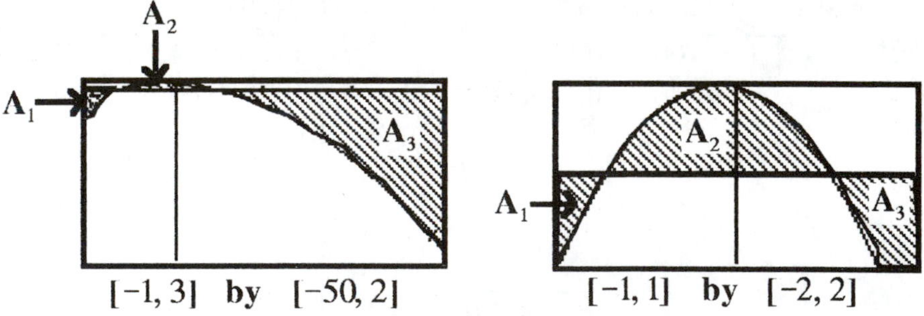

The definite integral has the value $A_2 - A_1 - A_3$. Since A_1 and A_3 are below the x-axis, they each contribute a negative value to the integral.

28. $\text{NINT}(x^2e^{-x^2}, x, 0, 3) = 0.443$. Graph $y = x^2e^{-x^2}$ in $[0, 3]$ by $[0, 0.5]$. The integral gives the area of the region between the curve and the x-axis for $0 \leqq x \leqq 3$.

29. $\text{NINT}(\sin(x^2), x, 0, 2\pi) = 0.642$. Graph $y = \sin(x^2)$ in $[0, 2\pi]$ by $[-1, 1]$. The integral is the sum of the signed areas (positive if above the x-axis, negative if below the x-axis) between the x-axis and the curve.

30. $\text{NINT}(\frac{\sin x}{x}, x, 0, 10) = 1.658$. Graph $y = \frac{\sin x}{x}$ in $[0, 10]$ by $[-1, 1]$. The value of the integral is the sum of the signed areas between the x-axis and the curve.

31. $\int_{-1}^{1} \sqrt{1-x^2}dx = \frac{\pi}{2}$ (area of upper half of unit circle). $\text{NINT}(\sqrt{1-x^2}, x, -1, 1) = 1.571$.

32. This is the area of the quarter of the circle lying in the first quadrant. $\pi r^2/4 = \pi$. $\text{NINT}(\sqrt{4-x^2}, x, 0, 2) = 3.142$.

33. $\int_{-1}^{1}(1-|x|)dx =$ area of rectangle with vertices $(-1, 0)$, $(-1, 1)$, $(1, 1)$, $(1, 0)$ minus the area of the triangle with vertices $(0, 0)$, $(-1, 1)$, $(1, 1)$. $A = 2(1) - \frac{1}{2}(2)1 = 1$. $\text{NINT}(1 - |x|, x, -1, 1) = 1$.

34. This region may be regarded as the rectangle with vertices $(-1, 0)$, $(1, 0)$, $(-1, 1)$, $(1, 1)$ surmounted by a semicircle of radius 1. Hence its area is $2 \cdot 1 + \pi(1^2)/2 = 2 + \pi/2$. $\text{NINT}(1 + \sqrt{1-x^2}, x, -1, 1) = 3.571$.

35. For $[0, 5]$: $\Delta x = 5/n$. $x_k = k\Delta x - \frac{\Delta x}{2} = (2k-1)\frac{\Delta x}{2}$. $\text{MRAM}_n\text{f} = \frac{1}{8}\sum_{k=1}^{n}(2k-1)^3(\Delta x)^4 = \frac{5^4}{8n^4}n^2(2n^2-1) = \frac{625}{8}(2 - \frac{1}{n^2})$. $\text{Lim}_{n\to\infty}\text{MRAM}_n\text{f} = \frac{625}{4}$. For $[0, a]$: $\Delta x = a/n$, $x_k = k\Delta x - \frac{\Delta x}{2} = (2k-1)\frac{\Delta x}{2}$. $\text{MRAM}_n\text{f} = \frac{1}{8}\sum_{k=1}^{n}(2k-1)^3\Delta x^4 = \frac{a^4}{8n^4}n^2(2n^2-1) = \frac{a^4}{8}(2 - \frac{1}{n^2})$. $\text{Lim}_{n\to\infty}\text{MRAM}_n\text{f} = \frac{a^4}{4}$.

36. $\Delta x = \frac{2}{n}$. We use $c_k = x_k = \frac{2k}{n}$. $\text{RRAM}_n\text{f} = \sum_{k=1}^{n} f(c_k)\Delta x = \sum_{k=1}^{n}[(\frac{2k}{n})^3 + 1]\frac{2}{n} = \frac{8}{n^3}[\sum_{k=1}^{n} k^3]\frac{2}{n} + [\sum_{k=1}^{n} 1]\frac{2}{n} = \frac{16}{n^4}\frac{n^2(n+1)^2}{4} + 2 = 4(\frac{n+1}{n})^2 + 2 \to 4 + 2 = 6$.

37. $|x| = x$ if $x \geqq 0$ and $|x| = -x$ if $x < 0$. Thus $\frac{|x|}{x} = 1$ if $x > 0$ and $\frac{|x|}{x} = -1$ if $x < 0$. $x = 0$ is the only discontinuity. Graph $y_1 = 1 + 0\sqrt{x}$ and $y_2 = -1 + 0\sqrt{-x}$ in $[-2, 3]$ by $[-2, 2]$. The value of the integral is the net area between the graph and the x-axis from $x = -2$ to $x = 3$ counting any area below the x-axis as a negative contribution. Thus integral $= -2+3 = 1$.

38. In dot format graph $y = 2\ \text{int}(abs(x-3))$ in $[-6,5]$ by $[-2,19]$. The discontinuities occur at $x = n$ for every integer n. The value of the integral will be the sum of the areas of the rectangular strips defined by the graph $= 16+14+12+\cdots+2+2 = 2+2\sum_{k=1}^{8} k = 2+8(8+1) = 74$.

39. Graph $y = \frac{x^2-1}{x+1}$ in $[-3,4]$ by $[-5,4]$. Since $y = \frac{(x-1)(x+1)}{x+1} = x-1$ when $x \neq -1$, the graph is a straight line with a hole at $x = -1$, the only discontinuity. On $[-3,4]$ the area between the curve and the x-axis consists of two triangles, one below and one above the x-axis separated at the point $(1,0)$. Thus the net area $= I = -[\frac{1}{2}(4)(4)] + \frac{1}{2}(3)(3) = -\frac{7}{2}$.

40. Graph $y = \frac{9-x^2}{x-3}$ in $[-5,6]$ by $[-10,3]$. We obtain the graph of $y = -(x+3)$ if we do not evaluate at $x = 3$, the only discontinuity. The net area $= I = \frac{1}{2}(2)(2) - \frac{1}{2}(9)9 = -\frac{77}{2} = -38.5$.

41. a) Since $\lim_{x\to0} \frac{\sin x}{x} = 1$, $g(x) = \frac{\sin x}{x}$, $x \neq 0$, $g(0) = 1$ is the continuous extension of y. b) 1.848 c) 3.697

42. a) Since $\lim_{x\to0} \frac{1-\cos x}{x^2} = \lim_{x\to0} \frac{2\sin^2\frac{x}{2}}{x^2} = \lim_{x\to0} \frac{2\sin^2\frac{x}{2}}{4(\frac{x}{2})^2} = \frac{1}{2}$, the continuous extension of y is $g(x) = \frac{1-\cos x}{x^2}$, $x \neq 0$, $g(0) = \frac{1}{2}$. b) 1.185 c) $2.658 \times 10^{-8} \approx 0$.

43. Let $f(x) = \frac{\sin x}{x}$. $f(-x) = \frac{\sin(-x)}{-x} = \frac{-\sin(x)}{-x} = \frac{\sin x}{x} = f(x)$. Thus $f(-x) = f(x)$, $f(x)$ is an even function, its graph is symmetric about the y-axis. Therefore the net area between the curve and the x-axis from $x = -3$ to $x = 0$ equals that from $x = 0$ to $x = 3$, and is $\frac{1}{2}$ that from $x = -3$ to $x = 3$.

44. Let $f(x) = \frac{|x|}{x}\left(\frac{1-\cos x}{x^2}\right)$. $f(-x) = \frac{|-x|}{-x}\left(\frac{1-\cos(-x)}{(-x)^2}\right) = \frac{|x|}{-x}\left(\frac{1-\cos x}{x^2}\right) = -f(x)$. Thus $f(-x) = -f(x)$, $f(x)$ is an odd function, its graph is symmetric about the origin. For any positive number a, any positive area on $[-a,0]$ defined by the graph of f is matched by a negative area on $[0,a]$. Any negative area on $[-a,0]$ is matched by a positive area on $[0,a]$. Since $I = \int_{-a}^{a} f(x)dx$ is the net area between the graph and the x-axis on $[-a,a]$, $I = 0$.

45. They are equal for each of the three functions. Conjecture: If $a \leqq c \leqq b$, then $\int_a^b f(x)dx = \int_a^c f(x)dx + \int_c^b f(x)dx$ when these integrals exist.

46. Let $f(x) = x^2e^{-x}$. $f(x) \geqq 0$ for all x. The area $\int_0^{10} f(x)dx$ includes and is greater than the area $\int_0^3 f(x)dx$.

47. From the graph of $f(x)$ we see that $\int_{-1}^{0} f(x)dx$ and $\int_2^3 f(x)dx$ are negative numbers because they are related to areas below the x-axis while $\int_0^2 f(x)dx$ is positive. We therefore take $a = c = -1$, $b = 1$.

48. The x-intercepts of the graph of $f(x)$ on $[a, b]$ determine regions between the x-axis and the graph of $f(x)$. Let $A_1, A_2, \ldots, A_s$ be the area of those regions so determined which are above the x-axis and let $B_1, B_2, \ldots, B_t$ be the areas of those below the x-axis. Then $\int_a^b f(x)dx = A_1 + A_2 + \cdots + A_s - B_1 - B_2 - \cdots - B_t =$ net area.

49. In work with $\text{RRAM}_n(x^2)$ on $[0, 1]$, we have $\Delta x = \frac{1-0}{n} = \frac{1}{n}$ and $x_k = \frac{k}{n}$. Thus $\text{RRAM}_n(x^2) = (\frac{1}{n})^2\frac{1}{n} + (\frac{2}{n})^2\frac{1}{n} + \cdots + (\frac{n}{n})^2\frac{1}{n} = S_n$. So $\lim_{n\to\infty} S_n = \lim_{n\to\infty}$ $\text{RRAM}_n(x^2) = \int_0^1 x^2 dx = \text{NINT}(x^2, 0, 1) = 0.333$.

50. $S_n = \sum_{k=1}^n \frac{8k^2}{n^3} = \sum_{k=1}^n 8(\frac{k}{n})^2\frac{1}{n} = \text{RRAM}_n(8x^2)$ on $[0, 1]$. $\text{Lim}_{n\to\infty} S_n = \int_0^1 8x^2 dx = \text{NINT}(8x^2, 0, 1) = 2.667$.

51. $S_n = \sum_{k=1}^n \frac{n+k}{n^2} = \sum_{k=1}^n (\frac{n+k}{n})\frac{1}{n} = \sum_{k=1}^n (1 + \frac{k}{n})\frac{1}{n} = \text{RRAM}_n(1 + x)$ on $[0, 1]$. $\lim_{n\to\infty} S_n = \int_0^1 (1 + x)dx = \text{NINT}(1 + x, 0, 1) = 1.5$.

52. $S_n = \sum_{k=1}^n \frac{1}{n+k} = \sum_{k=1}^n \frac{n}{n+k}\frac{1}{n} = \sum_{k=1}^n \frac{1}{1+(k/n)}\frac{1}{n} = \text{RRAM}_n(\frac{1}{1+x})$ on $[0, 1]$. $\text{Lim}_{n\to\infty} S_n = \int_0^1 \frac{1}{1+x} dx = \text{NINT}(\frac{1}{1+x}, 0, 1) = 0.693$.

53. $S_n = \sum_{k=1}^n \frac{(2n+k)^2}{n^3} = \sum_{k=1}^n (\frac{2n+k}{n})^2 \frac{1}{n} = \sum_{k=1}^n (2 + \frac{k}{n})^2 \frac{1}{n} = \text{RRAM}_n((2 + x)^2)$ on $[0, 1]$. $\text{Lim}_{n\to\infty} S_n = \int_0^1 (2 + x)^2 dx = \text{NINT}((2 + x)^2, 0, 1) = 6.333$.

5.3 ANTIDERIVATVES AND DEFINITE INTEGRALS

1. $F(x) = \frac{x^3}{3}$. $F(2) - F(0) = \frac{8}{3}$.

2. $F(x) = \frac{x^3}{3}$. $F(1) - F(-2) = 3$.

3. $F(x) = \frac{2}{3}x^{3/2}$. $F(4) - F(0) = \frac{16}{3}$.

4. $F(x) = -\frac{1}{x}$. $F(2) - F(\frac{1}{2}) = -\frac{1}{2} - (-2) = \frac{3}{2}$.

5. $F(x) = 2x - \frac{2}{3}x^{3/2}$. $F(4) - F(0) = 8 - \frac{2}{3}(4)^{3/2} = 8 - \frac{16}{3} = \frac{8}{3}$.

6. $F(x) = x - \frac{x^3}{3}$. $F(1) - F(-1) = \frac{2}{3} - (-\frac{2}{3}) = \frac{4}{3}$.

7. $F(x) = \frac{x^4}{4} - x^3 + 4x$. $F(2) - F(-1) = 4 - 8 + 8 - [\frac{1}{4} + 1 - 4] = \frac{27}{4}$.

8. $F(x) = \frac{2}{5}x^5 - \frac{4}{3}x^3 + 2x$. $F(1) - F(-1) = \frac{32}{15}$.

9. $F(x) = \sin x$. $F(\frac{\pi}{2}) - F(0) = \sin\frac{\pi}{2} - \sin 0 = 1$.

10. $F(x) = \sin 2x$. $F(\frac{\pi}{4}) - F(0) = 1$.

11. $F(x) = -\frac{1}{2}\cos 2x$. $F(\frac{\pi}{2}) - F(0) = -\frac{1}{2}[\cos\pi - \cos 0] = 1$.

12. $F(x) = -\frac{1}{2}\cos x$. $F(\pi) - F(0) = 1$.

13. $F(x) = \frac{\sin\pi x}{\pi}$. $F(\frac{1}{2}) - F(-\frac{1}{2}) = \frac{1}{\pi}[\sin\frac{\pi}{2} - \sin(-\frac{\pi}{2})] = \frac{2}{\pi}$.

14. $F(x) = x - \cos x$. $F(\pi) - F(0) = \pi - (-1) - [0 - 1] = 2 + \pi$.

15. $F(x) = \tan x$. $F(\frac{\pi}{3}) - F(-\frac{\pi}{4}) = \tan\frac{\pi}{3} - \tan(-\frac{\pi}{4}) = \sqrt{3} + 1$.

16. $F(x) = \sec x$. $F(\frac{\pi}{3}) - F(0) = 2 - 1 = 1$.

17. $F(x) = \frac{x^{n+1}}{n+1}$ is antiderivative of x^n. $A = F(b) - F(0) = \frac{b^{n+1}}{n+1}$.

18. $F(x) = \frac{h}{2b}x^2$. $F(b) - F(0) = \frac{bh}{2}$.

19. a) 0 b) $\int_5^1 g(x)dx = -\int_1^5 g(x)dx = -8$ c) $\int_1^2 3f(x)dx = 3\int_1^2 f(x)dx = -12$ d) $\int_2^5 f(x)dx = \int_2^1 f(x)dx + \int_1^5 f(x)dx = -\int_1^2 f(x)dx + 6 = 10$ e) $\int_1^5[f(x) - g(x)]dx = \int_1^5 f(x)dx - \int_1^5 g(x)dx = 6 - 8 = -2$ f) $\int_1^5[4f(x) - g(x)]dx = \int_1^5 4f(x)dx - \int_1^5 g(x)dx = 4\int_1^5 f(x)dx - \int_1^5 g(x)dx = 4(6) - 8 = 16$

20. a) $\int_1^9 -2f(x)dx = -2\int_1^9 f(x)dx = 2$ b) $\int_7^9[f(x) + h(x)]dx = \int_7^9 f(x)dx + \int_7^9 h(x)dx = 5+4 = 9$ c) $\int_7^9[2f(x) - 3h(x)]dx = 2\int_7^9 f(x)dx - 3\int_7^9 h(x)dx = 10-12 = -2$ d) $\int_9^1 f(x)dx = -\int_1^9 f(x)dx = 1$ e) $\int_1^7 f(x)dx = \int_1^9 f(x)dx - \int_7^9 f(x)dx = -1 - 5 = -6$ f) $\int_9^7[h(x) - f(x)]dx = \int_9^7 h(x)dx - \int_9^7 f(x)dx = -\int_7^9 h(x)dx + \int_7^9 f(x)dx = -4 + 5 = 1$

21. a) $\int_1^2 f(u)du = 5$ b) $\int_2^1 f(t)dt = -\int_1^2 f(t)dt = -5$ c) $\int_2^3 f(y)dy = \int_1^3 f(y)dy - \int_1^2 f(y)dy = 2 - 5 = -3$

22. a) $\int_0^3 f(t)dt = 3$ b) $\int_4^0 f(w)dw = -\int_0^4 f(w)dw = -7$ c) $\int_3^4 f(y)dy = \int_0^4 f(y)dy - \int_0^3 f(y)dy = 7 - 3 = 4$

23. An antiderivative of x^3 is $\frac{x^3}{3}$. Hence $\int_{-3}^5 x^2dx = \frac{5^3}{3} - \frac{(-3)^3}{3} = \frac{125+27}{3} = \frac{152}{3}$. $\int_{-3}^5(2 - x^2)dx = \int_{-3}^5 2\,dx - \int_{-3}^5 x^2dx = 2[5 - (-3)] - \frac{152}{3} = 16 - \frac{152}{3} = -\frac{104}{3}$. The average value is $\frac{1}{5-(-3)}\int_{-3}^5(2 - x^2)dx = \frac{1}{8}(-\frac{104}{3}) = -\frac{13}{3}$.

24. $\frac{x^4}{4} - x$ is an antiderivative of $f(x)$. $\frac{1}{3-(-1)}\int_{-1}^3(x^3 - 1)dx = \frac{1}{4}[(\frac{3^4}{4} - 3) - (\frac{(-1)^4}{4} - (-1))] = \frac{1}{4}[\frac{81}{4} - 3 - \frac{1}{4} - 1] = \frac{1}{4}[\frac{80}{4} - 4] = 4$.

25. $\frac{2}{3}x^{3/2}$ is an antiderivative of $\sqrt{x}$. $\frac{1}{3-0}\int_0^3 \sqrt{x}\,dx = \frac{1}{3}[\frac{2}{3}3^{3/2} - \frac{2}{3}0^{3/2}] = \frac{2}{\sqrt{3}}$.

26. We use $\frac{1}{\pi-(-\pi)}\int_{-\pi}^{\pi}\cos^2 x\,dx = \frac{1}{2\pi}$ NINT$(\cos^2 x, x, -\pi, \pi) = \frac{1}{2}$.

27. We use $\frac{1}{5-0}$ NINT$(x\sin x, x, 0, 5) = -0.475...$

28. $\frac{1}{4-1}\int_1^4 2e^{-x^2}dx = \frac{2}{3}$ NINT$(e^{-x^2}, x, 1, 4) = 0.0929...$

29. Min$(\frac{1}{1+x^2}) = \frac{1}{1+1^2} = \frac{1}{2}$ and max $\frac{1}{1+x^2} = \frac{1}{1+0^2} = 1$ for $0 \leqq x \leqq 1$. By Rule 7, $\frac{1}{2}(1-0) \leqq \int_0^1 \frac{1}{1+x^2}dx \leqq 1(1-0)$ or $\frac{1}{2} \leqq \int_0^1 \frac{1}{1+x^2}dx \leqq 1$.

30. $\frac{1}{1+(\frac{1}{2})^2}(\frac{1}{2}) \leqq \int_0^{1/2}\frac{1}{1+x^2}dx \leqq \frac{1}{1+0^2}(\frac{1}{2})$ or $\frac{2}{5} \leqq \int_0^{1/2}\frac{1}{1+x^2}dx \leqq \frac{1}{2}$. $\frac{1}{1+1^2}(\frac{1}{2}) \leqq \int_{1/2}^1 \frac{1}{1+x^2}dx \leqq \frac{1}{1+(\frac{1}{2})^2}(\frac{1}{2})$ or $\frac{1}{4} \leqq \int_{1/2}^1 \frac{1}{1+x^2}dx \leqq \frac{2}{5}$. Adding these, we get $\frac{2}{5}+\frac{1}{4} \leqq \int_0^{1/2}\frac{1}{1+x^2}dx + \int_{1/2}^1\frac{1}{1+x^2}dx \leqq \frac{1}{2}+\frac{2}{5}$ or $\frac{13}{20} \leqq \int_0^1 \frac{1}{1+x^2}dx \leqq \frac{9}{10}$.

31. Since $f(x) = x^3+1$ is continuous on $[0,1]$, the Mean Value Theorem tells us that there is at least one such point c. $(c^3+1)(1-0) = \frac{5}{4}$, $c^3 = \frac{1}{4}$, $c = \frac{1}{\sqrt[3]{4}}$. Graph $y_1 = x^3+1$ and $y_2 = f(c) = \frac{5}{4}$ in $[0,1]$ by $[-1,2]$.

32. $f(c)(b-a) = 18$, $(9-c^2)(3) = 18$, $9-c^2 = 6$, $c = \pm\sqrt{3}$. $c = \sqrt{3}$ is in the interval. Graph $y_1 = 9-x^2$, $y_2 = 9-c^2 = 6$ in $[0,3]$ by $[-1,9]$.

33. $f(c)(b-a) = \frac{\pi}{4}$, $\frac{1}{c^2+1}(1-0) = \frac{\pi}{4}$, $c^2+1 = \frac{4}{\pi}$, $c = \sqrt{\frac{4}{\pi}-1} \approx 0.52$. Graph $y_1 = \frac{1}{x^2+1}$, $y_2 = f(c) = \frac{\pi}{4}$ in $[0,1]$ by $[-0.5,1]$.

34. $f(c)(b-a) = \frac{\pi}{3}$, $\frac{1}{\sqrt{1-c^2}}(1) = \frac{\pi}{3}$, $1-c^2 = \frac{9}{\pi^2}$, $c^2 = 1-\frac{9}{\pi^2}$, $c = \pm\sqrt{1-9/\pi^2} \approx \pm 0.30$. Graph $y_1 = 1/\sqrt{1-x^2}$, $y_2 = f(c) = \frac{\pi}{3}$ on $[-\frac{1}{2},\frac{1}{2}]$ by $[-\frac{1}{2},\frac{3}{2}]$.

35. We are given that $\int_{-1}^2 f(t)dt = (2-(-1))5 = 15$ and $\int_2^7 f(t)dt = (7-2)3 = 15$. $\int_{-1}^7 f(t)dt = \int_{-1}^2 f(t)dt + \int_2^7 f(t)dt = 15+15 = 30$. Therefore, av. value of f on $[-1,7] = \frac{1}{7-(-1)}\int_{-1}^7 f(t)dt = \frac{30}{8} = \frac{15}{4}$.

36. $\frac{1}{10-(-1)}\int_{-1}^{10} g(s)ds = \frac{1}{11}[\int_{-5}^{10} g(s)ds - \int_{-5}^{-1} g(s)ds] = \frac{1}{11}[15(\frac{1}{15}\int_{-5}^{10} g(s)ds) - 4(\frac{1}{4}\int_{-5}^{-1} g(s)ds)] = \frac{1}{11}[15(50) - 4(10)] = \frac{710}{11}$

37. This is an immediate consequence of the Mean Value Theorem for Definite Integrals.

38. $0 \leqq \sin^2 x \leqq 1$. By Rule 7, $0 \cdot 1 \leqq \int_0^1 \sin^2 x dx \leqq 1 \cdot 1$ or $0 \leqq \int_0^1 \sin^2 x\,dx \leqq 1$ so that the value of $\int_0^1 \sin^2 x\,dx$ cannot possibly be 2.

5.4 THE FUNDAMENTAL THEOREM OF CALCULUS

1. $\int_0^3 (4-x^2)dx = \left[4x - \frac{x^3}{3}\right]_0^3 = 4\cdot 3 - \frac{3^3}{3} - (4\cdot 0 - \frac{0^3}{3}) = 3.$

2. $\int_0^1 (x^2 - 2x + 3)dx = \left[\frac{x^3}{3} - x^2 + 3x\right]_0^1 = \frac{1}{3} - 1 + 3 = \frac{7}{3}.$

3. $\int_0^1 (x^2 + \sqrt{x})dx = \frac{x^3}{3} + \frac{2}{3}x^{3/2}\,|_0^1 = \frac{1}{3} + \frac{2}{3} = 1.$

4. $\int_0^5 x^{3/2}dx = \frac{2}{5}\left[x^{5/2}\right]_0^5 = \frac{2}{5}5^2\sqrt{5} = 10\sqrt{5}.$

5. $\int_1^{32} x^{-6/5}dx = -5x^{-1/5}\,|_1^{32} = -5\left(\frac{1}{32^{1/5}} - 1\right) = -5\left(\frac{1}{2} - 1\right) = \frac{5}{2}.$

6. $\int_{-2}^{-1} \frac{2}{x^2}dx = -2\left[\frac{1}{x}\right]_{-2}^{-1} = -2\left[-1 - (-\frac{1}{2})\right] = 1.$

7. $\int_0^\pi \sin x dx = -\cos x\,|_0^\pi = -(\cos\pi - \cos 0) = -(-1-1) = 2.$

8. $\int_0^\pi (1 + \cos x)dx = [x + \sin x]_0^\pi = \pi.$

9. $\int_0^{\pi/3} 2\sec^2 x dx = 2\tan x\,\Big|_0^{\pi/3} = 2(\tan\pi/3 - \tan 0) = 2\left(\sqrt{3} - 0\right) = 2\sqrt{3}.$

10. $\int_{\pi/6}^{5\pi/6} \csc^2 x dx = -[\cot x]_{\pi/6}^{5\pi/6} = -\left[\cot\frac{5\pi}{6} - \cot\frac{\pi}{6}\right] = -\left[-\sqrt{3} - \sqrt{3}\right] = 2\sqrt{3}.$

11. $\int_{\pi/4}^{3\pi/4} \csc x \cot x dx = -\csc x\,\Big|_{\pi/4}^{3\pi/4} = -\left(\csc\frac{3\pi}{4} - \csc\frac{\pi}{4}\right) = -\left(\sqrt{2} - \sqrt{2}\right) = 0.$

12. $\int_0^{\pi/3} 4\sec x \tan x dx = [4\sec x]_0^{\pi/3} = 4[2-1] = 4.$

13. $\int_{-1}^1 (r+1)^2 dr = \frac{1}{3}(r+1)^3]_{-1}^1 = \frac{1}{3}[(1+1)^3 - (-1+1)^3] = \frac{8}{3}.$ Alternatively, $\int_{-1}^1 (r^2 + 2r + 1) = \frac{r^3}{3} + r^2 + r]_{-1}^1 = (\frac{1}{3} + 1 + 1) - (-\frac{1}{3} + 1 - 1) = \frac{1}{3} + 2 + \frac{1}{3} = \frac{8}{3}.$

14. $\int_9^4 \frac{1-\sqrt{u}}{\sqrt{u}}du = \int_9^4 (u^{-1/2} - 1)du = [2u^{1/2} - u]_9^4 = [2(2) - 4] - [2(3) - 9] = 3.$

15. $A = \int_0^2 (2-x)dx - \int_2^3 (2-x)dx = \left[2x - \frac{x^2}{2}\right]_0^2 - \left[2x - \frac{x^2}{2}\right]_2^3 = (4-2) - \left[6 - \frac{9}{2} - (4-2)\right] = 2 - \left[-\frac{1}{2}\right] = \frac{5}{2}.$

16. $y = 3x^2 - 3,\ -2 \leqq x \leqq 2.$ $y = 3(x^2 - 1) < 0$ when $|x| < 1$. Since y is an even function, we may take twice the area to the right of the y-axis. $A = 2\left\{-3\int_0^1 (x^2-1)dx + 3\int_1^2 (x^2-1)dx\right\} = 2\left\{-3\left[\frac{x^3}{3} - x\right]_0^1 + 3\left[\frac{x^3}{3} - x\right]_1^2\right\}$ $= 2\left\{-3\left[\frac{1}{3} - 1\right] + 3\left[\frac{8}{3} - 2\right] - 3\left[\frac{1}{3} - 1\right]\right\} = 12.$

17. $y = x^3 - 3x^2 + 2x = x(x^2 - 3x + 2) = x(x-1)(x-2)$, $0 \leqq x \leqq 2$. Graphically or algebraically, we see that $y \geqq 0$ for $0 \leqq x \leqq 1$ and $y \leqq 0$ for $1 \leqq x \leqq 2$. Thus $A = \int_0^1 f(x)dx - \int_1^2 f(x)dx = \left[\frac{x^4}{4} - x^3 + x^2\right]_0^1 - \left[\frac{x^4}{4} - x^3 + x^2\right]_1^2 = \left[\frac{1}{4} - 1 + 1\right] - \left[\left(\frac{16}{4} - 8 + 4\right) - \left(\frac{1}{4} - 1 + 1\right)\right] = \frac{1}{4} - \left[\frac{15}{4} - 4\right] = 4 - \frac{14}{4} = \frac{1}{2}$.

18. $y = x^3 - 4x$, $-2 \leqq x \leqq 2$. Since this is an odd function, its graph is symmetric about the origin. $y = x(x^2 - 4) \leqq 0$ for $0 \leqq x \leqq 2$. Thus $A = -2\int_0^2 (x^3 - 4x)dx = -2\left[\frac{x^4}{4} - 2x^2\right]_0^2 = 8$.

19. Let $f(x) = \frac{x^2-1}{x+1}$. If $x \neq -1$, $f(x) = \frac{(x-1)(x+1)}{x+1} = x - 1$. Thus $x - 1$ is a continuous extension of $f(x)$ and the discontinuity at $x = -1$ is removable. In this case $\int_{-2}^3 f(x)dx = \int_{-2}^3 (x-1)dx = \left[\frac{x^2}{2} - x\right]_{-2}^3 = \frac{3}{2} - \frac{8}{2} = -\frac{5}{2}$.

20. Let $f(x) = \frac{9-x^2}{3x-9} = \frac{(3-x)(3+x)}{3(x-3)} = -\frac{(x-3)(x+3)}{3(x-3)} = -\frac{x+3}{3}$. $f(x)$ has a removable discontinuity at $x = 3$ so Theorem 4 cannot be used directly. But we may use the continuous extension $-\frac{x+3}{3}$. $\int_0^5 f(x)dx = -\frac{1}{3}\int_0^5 (x+3)dx = -\frac{1}{3}\left[\frac{x^2}{2} + 3x\right]_0^5 = -\frac{1}{3}\frac{55}{2} = -\frac{55}{6}$.

21. Tan x has a discontinuity at $x = \frac{\pi}{2}$ so Theorem 4 does not apply. Tan x is unbounded on $[0, \frac{\pi}{2})$. The integral does not exist.

22. In $[0, 2]$, $x \neq -1$, so that $f(x) = \frac{x+1}{x^2-1} = \frac{1}{x-1}$ on the interval. However, $f(x)$ is unbounded and the integral does not exist.

23. $f(x) = \frac{\sin x}{x}$ has a removable discontinuity at $x = 0$ since $\frac{\sin x}{x} \to 1$ as $x \to 0$. NINT $\left(\frac{\sin x}{x}, x, -1, 2\right) = 2.551..$

24. $f(x) = \frac{1-\cos x}{x^2}$ has a removable discontinuity at $x = 0$: $\frac{1-\cos x}{x^2} = \frac{2\sin^2(\frac{x}{2})}{4(\frac{x}{2})^2} \to \frac{1}{2}$ as $x \to 0$. We cannot find an explicit antiderivative. However, NINT$(f(x), -2, 3) = 2.082..$

25. $\int_0^1 x^2dx + \int_1^2 (2-x)dx = \frac{x^3}{3}\Big]_0^1 + \left[2x - \frac{x^2}{2}\right]_1^2 = \frac{1}{3} + (4-2) - (2 - \frac{1}{2}) = \frac{1}{3} + \frac{1}{2} = \frac{5}{6}$.

26. $A = \int_0^1 x^{1/2}dx + \int_1^2 x^2dx = \left[\frac{2}{3}x^{3/2}\right]_0^1 + \frac{1}{3}\left[x^3\right]_1^2 = \frac{2}{3} + \frac{1}{3}[8-1] = 3$.

27. Area of rectangle − area under curve $= 2\pi - \int_0^\pi (1 + \cos x)dx = 2\pi - [x + \sin x]_0^\pi = 2\pi - [\pi + 0 - (0+0)] = \pi$.

28. $A = \int_{\pi/6}^{5\pi/6} \sin x dx$ − (area of rectangle) $= [-\cos x]_{\pi/6}^{5\pi/6} - \left(\frac{5\pi}{6} - \frac{\pi}{6}\right)\left(\sin\frac{\pi}{6}\right) = -\left[-\frac{\sqrt{3}}{2} - \frac{\sqrt{3}}{2}\right] - \frac{2\pi}{3}\left(\frac{1}{2}\right) = \sqrt{3} - \frac{\pi}{3}$.

29. $F(x) = \int_0^x (t-2)dt = [\frac{t^2}{2} - 2t]_0^x = \frac{x^2}{2} - 2x$. The two graphs are the same in the standard viewing rectangle $[-10, 10]$ by $[-10, 10]$. $F(0.5) = -0.875$ and we get the same value after zooming in. $F(1) = -1.5$ compared to -1.516 as one approximation. $F(1.5) = -1.875$ compared to -1.879 as one approximation. $F(2) = -2$ compared to -2. $F(5) = 2.5$ compared to 2.53 as one approximation.

30. $F(x) = \int_0^x (t^3+1)dt = [\frac{t^4}{4}+t]_0^x = \frac{x^4}{4}+x$. The two graphs are indistinguishable in $[-2, 2]$ by $[-0.75, 6]$. The values of $F(x)$ and $\text{NINT}(t^3+1, t, 0, x)$ agree when accurately calculated.

31. $F(x) = \int_0^x (t^2 - 3t + 6)dt = [\frac{t^3}{3} - \frac{3}{2}t^2 + 6t]_0^x = \frac{x^3}{3} - \frac{3}{2}x^2 + 6x$. The two graphs are indistinguishable in the viewing rectangle $[-15, 15]$ by $[-1,000, 1,000]$. The values of $F(x)$ and $\text{NINT}(f(t), t, 0, x)$ agree when accurately calculated.

32. $F(x) = \int_0^x 3\sin t\; dt = -3[\cos t]_0^x = -3[\cos x - 1] = 3[1 - \cos x]$. The two graphs are indistinguishable in $[-2\pi, 2\pi]$ by $[0, 6]$. The values of $F(x)$ and $\text{NINT}(f(t), t, 0, x)$ agree when accurately calculated.

33. Graph $y = \text{NINT}(t^2 \sin t, t, 0, x)$ in the viewing window $[-3, 3]$ by $[0, 9]$.

34. Graph $y = \text{NINT}(\sqrt{1+t^2}, t, 0, x)$ in the viewing window $[0, 5]$ by $[0, 20]$.

35. Graph $y = \text{NINT}(5e^{-0.3t^2}, t, 0, x)$ in the viewing window $[0, 5]$ by $[0, 10]$.

36. Graph $y = \text{NINT}(t\sin(t^3), t, 0, x)$ in $[0, \pi]$ by $[-1, 1]$.

37. We graph $y = \text{NDER}(\text{NINT}(4 - t^2, t, 0, x), x, x)$ and $y = 4 - x^2$ in $[-5, 5]$ by $[-21, 4]$ and see that the graphs are identical. This visually supports $D_x(\int_0^x (4 - t^2)dt) = 4 - x^2$.

38. Graph $y = \text{NDER}(\text{NINT}(t\sin t, t, 0, x), x, x)$ and $y = x\sin x$ in $[0, 2\pi]$ by $[-2\pi, 2\pi]$ and see that the graphs are identical. This visually supports $D_x \int_0^x t\sin t\; dt = x\sin x$.

39. We know the same K works for all x because the two integrals differ by a constant. Let $x = b = 2$. Then the equation becomes $\int_{-1}^2 f(t)dt + K = \int_2^2 f(t)dt$. Evaluating, we get $\frac{3}{2} + K = 0$, $K = -\frac{3}{2}$.

40. The same K must work for all x. Let $x = b = 2$. Then the equation becomes $\int_0^2 \sin^2 t\; dt + K = \int_2^2 \sin^2 t\; dt = 0$. $K = -\text{NINT}(\sin^2 t, t, 0, 2) = -1.189...$

41. We solve $\int_0^x e^{-t^2} dt = 0.6$. Graph $y = \text{NINT}(e\text{^}(-t^2), t, 0, x)$ and $y = 0.6$ in $[0, 20]$ by $[0, 2]$. Zooming in to the point of intersection, we find $x = 0.70$. Remark: On the TI-85 one may enter the functions and range and then use the ISECT function.

42. Graph $y = \text{NINT}(\frac{\sin t}{t}, t, 0, x)$ and $y = 1.8$ in $[-10, 10]$ by $[-3, 3]$ and find the points of intersection, $x = 2.60, 3.76$.

43. $\sqrt{1 + x^2}$

44. $\frac{1}{x}$, $x > 0$

45. $y = \int_0^{\sqrt{x}} \sin(t^2) dt$. Let $u = \sqrt{x}$, $y = \int_0^u \sin(t^2) dt$. $\frac{dy}{dx} = \frac{dy}{du}\frac{du}{dx} = \sin(u^2)\frac{1}{2\sqrt{x}} = \frac{\sin x}{2\sqrt{x}}$.

46. $y = \int_0^u \cos t \, dt$ where $u = 2x$. $\frac{dy}{dx} = \frac{dy}{du}\frac{du}{dx} = (\cos u)2 = 2\cos(2x)$.

47. d)

48. c)

49. b)

50. a)

51. $x = a$

52. Differentiating both sides of the equation, we get $f(x) = 2x - 2$.

53. Differentiating both sides of the given equation, we obtain $f(x) = \cos \pi x - \pi x \sin \pi x$. $f(4) = \cos 4\pi - 4\pi \sin 4\pi = 1$.

54. We have $f(0) = 2$, $f'(x) = \frac{10}{1+x}$, $f'(0) = 10$. Hence $L(x) = f(0) + f'(0)(x - 0) = 2 + 10x$.

55. If we start with $x = 0$ on the graph of $y = \sin kx$, the first arch will be completed when $kx = \pi$ or $x = \pi/k$. $A = \int_0^{\pi/k} \sin(xk) dx = -\frac{\cos(kx)}{k}\Big]_0^{\pi/k} = -\frac{1}{k}[\cos(k(\pi/k)) - \cos 0] = -\frac{1}{k}[-1 - 1] = \frac{2}{k}$.

56. a) $y = 6 - x - x^2 = (2 - x)(3 + x)$. $A = \int_{-3}^{2}(6 - x - x^2)dx = [6x - \frac{x^2}{2} - \frac{x^3}{3}]_{-3}^{2} = (12 - 2 - \frac{8}{3}) - (-18 - \frac{9}{2} + 9) = \frac{125}{6}$. b) y_{max} occurs where $y' = -1 - 2x = 0$, i.e., at $x = -\frac{1}{2}$. So height $= f(-\frac{1}{2}) = \frac{25}{4}$. c) $\frac{2}{3}$ (base)(height) $= \frac{2}{3}(5)\frac{25}{4} = \frac{125}{6} = A$.

57. a) $c(100) - c(1) = \int_1^{100} \frac{dx}{2\sqrt{x}} = \sqrt{x}]_1^{100} = 9$ dollars. b) $c(400) - c(100) = \int_{100}^{400} \frac{dx}{2\sqrt{x}} = \sqrt{x}]_{100}^{400} = 20 - 10 = 10$ dollars.

58. $\int_0^3[2 - \frac{2}{(x+1)^2}]dx = [2x + \frac{2}{x+1}]_0^3 = (6 + \frac{1}{2}) - 2 = 4.5$ thousand dollars or $\text{NINT}(2 - \frac{2}{(x+1)^2}, x, 0, 3) = 4.5$ thousand dollars.

59. $I_{av} = \frac{1}{30-0}\int_0^{30}(450 - \frac{x^2}{2})dx = \frac{1}{30}[450x - \frac{x^3}{6}]_0^{30} = \frac{1}{30}[450(30) - \frac{30^3}{6}] = 300$. Average daily holding cost $= 0.02(300) = 6$ dollars per day.

60. During the 60 days after a shipment is received the average value of $I(x)$ is $\frac{1}{60}\int_0^{60}[600 - 20\sqrt{15}x^{1/2}]dx = \frac{1}{60}[600x - 20\sqrt{15}(\frac{2}{3})x^{3/2}]_0^{60} = \frac{1}{60}[600(60) - \frac{40\sqrt{15}}{3}(\sqrt{4}\sqrt{15})^3] = 200$ cases. Average daily holding cost = (average number of cases per day)(holding cost per case) $= 200(\frac{1}{2}) = 100$ cents $= \$1.00$.

61. a) Compare your drawing with the result of graphing $y = (\cos x)/x$ in $[-15, 15]$ by $[-1, 1]$. The $x-$ and $y-$axes are asymptotes. b) Graph $y =$ NINT$((\cos t)/t, t, 1, x)$ in $[0, 15]$ by $[-1, 1]$. c) Because $f(0)$ is undefined. d) For $x > 0$, $g(x)$ and $h(x)$ have the same derivative $f(x)$ and so they differ by an additive constant. This is confirmed if one graph can be obtained from the other by a vertical shift. Along with the function in b), graph $y =$ NINT$((\cos t)/t, t, 0.5, x)$ in $[0.01, 3]$ by $[-3, 3]$ to see that this is the case. Alternatively, $\int_{0.5}^{x} f(t)dt = \int_{0.5}^{1} f(t)dt + \int_1^x f(t)dt \approx 0.5 + \int_1^x f(t)dt$.

62. Let $I = \int \frac{\sin 2x}{x}dx$. Let $u = 2x$, $du = 2dx$, $dx = \frac{1}{2}du$. $I = \frac{1}{2}\int \frac{\sin u}{(u/2)}du = \int \frac{\sin u}{du}du$. Since $\int \frac{\sin u}{u}du$ has no explicit finite formula in terms of elementary functions, the same must be true of I. $f(x) = \int_0^x \frac{\sin 2t}{t}dt$. Let $v = 2t$, $dv = 2dt$, $t = \frac{v}{2}$, $dt = \frac{dv}{2}$. $f(x) = \int_0^{2x} \frac{\sin v}{(v/2)}\frac{dv}{2} = \int_0^{2x} \frac{\sin v}{v}dv$. Let $g(x) = \int_0^x \frac{\sin t}{t}dt$. We see that $f(\frac{1}{2}x) = g(x)$. Thus the graph of $g(x)$ can be obtained from the graph of $f(x)$ by stretching the graph of $f(x)$ horizontally by a factor of 2. This is supported by graphing $f(x) \approx$ NINT$(\frac{\sin 2t}{2}, t, 0.1, x)$ and $g(x) \approx$ NINT$(\frac{\sin t}{t}, t, 0.1, x)$ in $[0.1, 2]$ by $[0, 2]$.

63. $F(x) = \int_1^{x^2} \sqrt{1-t^2}dt$. a) The integrand has domain $[-1, 1]$. x^2 is in $[-1, 1]$ if x is. Hence $F(x)$ has domain $[-1, 1]$. b) $F'(x) = \sqrt{1-(x^2)^2}(2x) = 2x\sqrt{1-x^4}$. $F'(x) = 0$ when $x = 0, \pm 1$. $F''(x) = 2x\frac{(-4x^3)}{2\sqrt{1-x^4}} + 2\sqrt{1-x^4} = \frac{-4x^4+2(1-x^4)}{\sqrt{1-x^4}} = \frac{2(1-3x^4)}{\sqrt{1-x^4}}$. We see that $F'' < 0$ for $-1 < x < -\frac{1}{\sqrt[4]{3}}$ and for $\frac{1}{\sqrt[4]{3}} < x < 1$ and $F''(x) > 0$ for $-\frac{1}{\sqrt[4]{3}} < x < \frac{1}{\sqrt[4]{3}}$. Thus F' decreases on $(-1, -\frac{1}{\sqrt[4]{3}})$ and $(\frac{1}{\sqrt[4]{3}}, 1)$ and increases on $(-\frac{1}{\sqrt[4]{3}}, \frac{1}{\sqrt[4]{3}})$. F' has a local maximum at $x = 1$, local minimum at $x = -\frac{1}{\sqrt[4]{3}}$, local maximum at $x = \frac{1}{\sqrt[4]{3}}$ and local minimum at $x = 1$. d) The zeros of $F'(x)$ tell us where the graph has a horizontal tangent and possible local extremes. The graph of F is concave up where F' is increasing and concave down where F' is decreasing. The local extrema of F' at $x = \pm\frac{1}{\sqrt[4]{3}}$ correspond to inflection points of F. Graph NINT$(\sqrt{1-t^2}, t, 1, x^2)$ in $[-1, 1]$ by $[-2, 1]$.

64. $F(x) = \int_0^{x^2} \sqrt{1-t^2}dt = \int_0^1 \sqrt{1-t^2}dt + \int_1^{x^2} \sqrt{1-t^2}dt = \frac{\pi}{4} + \int_1^{x^2} \sqrt{1-t^2}dt$. Thus the graph here can be obtained from the graph of Exercise 63 by shifting it up $\frac{\pi}{4}$. The data on the derivatives is the same.

65. $F(x) = \int_1^{2x} \frac{dt}{\sqrt{1-t^2}}$. a) Since we must have $|t| < 1$, we get $|2x| < 1$, and so F has domain $(-\frac{1}{2}, \frac{1}{2})$. Later work will show that F has a continuous extension to a function with domain $[-\frac{1}{2}, \frac{1}{2}]$. b) By Theorem 4 (extended form), $F'(x) = \frac{1}{\sqrt{1-(2x)^2}}(2) = \frac{2}{\sqrt{1-4x^2}}$. c) $F'(x) > 0$ for all x in the domain (F is increasing). By inspection, F' is decreasing on $(-\frac{1}{2}, 0)$ and increasing on $(0, \frac{1}{2})$. Hence $x = 0$ is a local minimum of F', the only local extreme in $(-\frac{1}{2}, \frac{1}{2})$. d) c) tells us that the graph of F is concave down on $(-\frac{1}{2}, 0)$, concave up on $(0, \frac{1}{2})$ and that F has an inflection point at $x = 0$. In parametric mode (to save time) graph $x = t$, $y = \text{NINT}(\frac{1}{\sqrt{1-s^2}}, s, 1, 2t)$, $-0.499 \leqq t \leqq 0.499$, t step $= 0.05$ in $[-0.499, 0.499]$ by $[-\pi, 0]$.

66. $F(x) = \int_1^{3x} \frac{dt}{\sqrt{1-t^2}}$. Since $G(x) = \int_1^{2x} \frac{dt}{\sqrt{1-t^2}}$ is the function of the preceding exercise and $F(x) = G(\frac{3}{2}x)$, the graph of F can be obtained from that of the preceding exercise by horizontally shrinking by a factor $\frac{2}{3}$. Note that in Chapter 7 we will be able to see that $\sin^{-1} x = \int_0^x \frac{dt}{\sqrt{1-t^2}}$. From this it follows that the function of Exercise 65 is $-\frac{\pi}{2} + \sin^{-1} 2x$ and the present function is $-\frac{\pi}{2} + \sin^{-1} 3x$.

67. $F(x) = \int_{x^2}^{x^3} \cos(2t)dt = \frac{1}{2}\sin(2t)]_{x^2}^{x^3} = \frac{1}{2}[\sin(2x^3) - \sin(2x^2)]$. $F'(x) = \frac{1}{2}[6x^2\cos(2x^3) - 4x\cos(2x^2)] = 3x^2\cos(2x^3) - 2x\cos(2x^2)$. This is supported by graphing, in parametric mode, $x_1 = t$, $y_1 = \text{NDER}(\text{NINT}(\cos(2s), s, t^2, t^3), t, t)$ and $x_2 = t$, $y_2 = 3t^2\cos(2t^3) - 2t\cos(2t^2)$, $-1.5 \leqq t \leqq 1.5$, t Step $= 0.05$, in $[-1.5, 1.5]$ by $[-12, 12]$.

68. $F(x) = \int_{\sin x}^{\cos x} t^2 dt = \frac{1}{3}t^3]_{\sin x}^{\cos x} = \frac{1}{3}\cos^3 x - \frac{1}{3}\sin^3 x$. $F'(x) = -\cos^2 x \sin x - \sin^2 x \cos x = -\sin x \cos x(\sin x + \cos x)$. This is supported by graphing, in parametric mode, $x_1 = t$, $y_1 = \text{NDER}(\text{NINT}(s^2, s, \sin t, \cos t), t, t)$ and $x_2 = t$, $y_2 = -\sin t \cos t(\sin t + \cos t)$, $0 \leqq t \leqq 2\pi$, t Step $= 0.05$ in $[0, 2\pi]$ by $[-1, 1]$.

5.5 INDEFINITE INTEGRALS

1. $\int x^3 dx = \frac{x^4}{4} + C$. $(\frac{x^4}{4} + C)' = x^3$

2. $\int 7\,dx = 7x + C$. $(7x + C)' = 7$

3. $\int (x+1)dx = \frac{x^2}{2} + x + C$. $(\frac{x^2}{2} + x + C)' = x + 1$

4. $\int(6-6x)dx = 6x - 3x^2 + C$

5. $\int 3\sqrt{x}dx = 3\int\sqrt{x}dx = 3(\frac{2}{3})x^{3/2} + C = 2x^{3/2} + C$

6. $\int \frac{4}{x^2}dx = -\frac{4}{x} + C$

7. $\int x^{-1/3}dx = \frac{3}{2}x^{2/3} + C$

8. $\int(1-4x^{-3})dx = x - 4\frac{x^{-2}}{(-2)} + C = x + 2x^{-2} + C$

9. $\int(5x^2+2x)dx = \frac{5}{3}x^3 + x^2 + C$

10. $\int(\frac{x^2}{2}+\frac{x^3}{3})dx = \frac{1}{2}\frac{x^3}{3} + \frac{1}{3}\frac{x^4}{4} + C = \frac{x^3}{6} + \frac{x^4}{12} + C$

11. $\int(2x^3-5x+7)dx = \frac{1}{2}x^4 - \frac{5}{2}x^2 + 7x + C$. We graph the integrand and NDER$(\frac{1}{2}x^4 - \frac{5}{2}x^2 + 7x, x)$ in $[-3,3]$ by $[-50,50]$ and see that two graphs are identical.

12. $\int(1-x^2-3x^5)dx = x - \frac{x^3}{3} - \frac{x^6}{2} + C$

13. $\int 2\cos x dx = 2\sin x + C$

14. $\int 5\sin\theta d\theta = -5\cos\theta + C$

15. $\int \sin\frac{x}{3}dx = -3\cos\frac{x}{3} + C$

16. $\int 3\cos 5x dx = \frac{3}{5}\sin 5x + C$

17. $\int 3\csc^2 x\, dx = -3\cot x + C$

18. $\int \frac{\sec^2 x}{3}dx = \frac{\tan x}{3} + C$

19. $\int \frac{\csc x \cot x}{2}dx = -\frac{1}{2}\csc x + C$

20. $\int \frac{2}{5}\sec x \tan x\, dx = \frac{2}{5}\sec x + C$

21. $\int(4\sec x\tan x - 2\sec^2 x)dx = 4\sec x - 2\tan x + C$. With $C=0$, for example, we may graph the result and also $y = \text{NINT}(4\sec t\tan t - 2\sec^2 t, t, 0, x)$ in the relatively small window $[-1.5, 1.5]$ by $[-1, 20]$. Use $tol = 1$ to save time. We see that one graph can be obtained from the other by a vertical shift, supporting that the functions differ by a constant.

22. $\int \frac{1}{2}(\csc^2 x - \csc x\cot x)dx = \frac{1}{2}(-\cot x + \csc x) + C$

23. $\int(\sin 2x - \csc^2 x)dx = -\frac{1}{2}\cos 2x + \cot x + C$

24. $\int(2\cos 2x - 3\sin 3x)dx = \sin 2x + \cos 3x + C$

25. $\int 4\sin^2 y\, dy = 2\int(1-\cos 2y)dy = 2(y - \frac{1}{2}\sin 2y) + C = 2y - \sin 2y + C$

26. $\int \frac{\cos^2 x}{7}dx = \frac{1}{14}\int(1+\cos 2x)dx = \frac{1}{14}(x + \frac{\sin 2x}{2}) + C$

27. $\int \sin x\cos x dx = \frac{1}{2}\int 2\sin x\cos x dx = \frac{1}{2}\int\sin 2x dx = -\frac{1}{4}\cos 2x + C$

28. $\int(1-\cos^2 t)dt = \int\sin^2 t\, dt = \frac{t}{2} - \frac{\sin 2t}{4} + C$ by Example 8

29. $\int(1+\tan^2\theta)d\theta = \int \sec^2\theta\, d\theta = \tan\theta + C$

30. $\int \frac{1+\cot^2 x}{2}dx = \frac{1}{2}\int \csc^2 x\, dx = -\frac{\cot x}{2} + C$

31. $\lim_{x\to 0}\frac{1-\cos x}{x^2} = \lim_{x\to 0}\frac{1-\cos x}{x^2}(\frac{1+\cos x}{1+\cos x}) = \lim_{x\to 0}\frac{1-\cos^2 x}{x^2(1+\cos x)} = \lim_{x\to 0}(\frac{\sin x}{x})^2\frac{1}{1+\cos x} = (1)^2\frac{1}{1+1} = \frac{1}{2}$. Thus the continuous extension g of f may be written as $g(x) = \frac{1-\cos x}{x^2}$, $-10 \leqq x \leqq 10$, $x \neq 0$, $g(0) = \frac{1}{2}$. Graph $y = \int_0^x f(t)dt =$ NINT$(f(t), t, 0, x)$ in $[-10, 10]$ by $[-2, 2]$. For quicker results, use parametric mode, $x = t$, $y =$ NINT$((1-\cos s)/s^2, s, 0, t)$, t step= 0.3, $-10 \leqq t \leqq 10$, in the above window.

32. We see that $f(0.10 = 1.270\ldots$, $f(0.01) = 1.047\ldots$, $f(0.001) = 1.006\ldots$. This suggests $\lim_{x\to 0^+} f(x) = 1$. (In a later chapter when we study l'Hospital's rule, we will be able to prove this.) Thus $g(x) = f(x)$, $x \neq 0$, $g(0) = 1$ is the continuous extension of $f(x)$ on the interval. Graph NINT$((1+t^{-1})^t, t, 0, x)$ in $[0, 5]$ by $[0, 12]$.

33. $[\frac{(7x-2)^4}{28} + C]' = \frac{4(7x-2)^3(7)}{28} = (7x-2)^3$

34. $(\frac{\tan 5x}{5} + C)' = \sec^2 5x$

35. $[-\frac{1}{x+1} + C]' = -[(x+1)^{-1}]' = -[(-1)(x+1)^{-2}] = \frac{1}{(x+1)^2}$

36. $(\frac{x}{x+1} + C)' = \frac{(x+1)\cdot 1 - x\cdot 1}{(x+1)^2} = \frac{1}{(x+1)^2}$

37. a) $(\frac{x^2}{2}\sin x)' = \frac{x^2}{2}\cos x + x\sin x$. Wrong b) $(-x\cos x)' = x\sin x - \cos x$. Wrong c) $(-x\cos x + \sin x)' = x\sin x - \cos x + \cos x = \sin x$. Right

38. a) $[\frac{(2x+1)^3}{3} + C]' = \frac{1}{3}(3)(2x+1)^2(2) = 2(2x+1)^2$, wrong b) $[(2x+1)^3 + C]' = 6(2x+1)^2$ wrong c) right by the work in in b)

39. $\frac{dy}{dx} = 3\sqrt{x}$, $y = \int 3\sqrt{x}dx = 3(\frac{2}{3})x^{3/2} + C = 2x^{3/2} + C$. When $x = 9$, $y = 2(9)^{3/2} + C = 54 + C = 4$, $C = -50$. $y = 2x^{3/2} - 50$

40. $y = \int_4^x \frac{1}{4}x^{-1/2}dx = [\frac{1}{2}(2x^{1/2})]_4^x = \sqrt{x} - 2$

41. $y = \int_0^x 2^t dt + C$. When $x = 0$, $y = \int_0^0 2^t dt + C = C = 2$. $y = \int_0^x 2^t dt + 2$. Graph $y = 2 +$ NINT$(2^t, t, 0, x)$ in $[-7.5, 9.5]$ by $[0, 10]$.

42. Graph $y = 0.5 + \int_0^x \frac{(1-\cos t)}{t}dt = 0.5 +$ NINT$((1-\cos t)/t, t, 0, x)$ in $[-20, 20]$ by $[-1, 5]$. Analytic means are unavailable.

43. $\frac{d^2y}{dx^2} = 0$ leads to $\frac{dy}{dx} = C$ and so $\frac{dy}{dx} = 2$. $y = \int 2\, dx = 2x + C_1$. When $x = 0$, $y = 0$ so $C_1 = 0$ and $y = 2x$.

44. $y' = 1 + \int_1^x 2t^{-3}dt = 1 + \frac{2}{-2}[t^{-2}]_1^x = 1 - [x^{-2} - 1] = 2 - x^{-2}$. $y = 1 + \int_1^x (2 - t^{-2})dt = 1 + [2t + t^{-1}]_1^x = 1 + 2x + \frac{1}{x} - (2+1) = 2x + \frac{1}{x} - 2$

45. $\frac{dy}{dx} = \int \frac{3x}{8} dx = \frac{3}{16}x^2 + C_1$. When $x = 4$, $\frac{dy}{dx} = \frac{3}{16}(4^2) + C_1 = 3$ so $C_1 = 0$ and $\frac{dy}{dx} = \frac{3x^2}{16}$. $y = \int \frac{3}{16}x^2 dx = \frac{1}{16}x^3 + C_2$. When $x = 4$, $4 = y = \frac{1}{16}(4^3) + C_2 = 4 + C_2$ so $C_2 = 0$ and $y = \frac{x^3}{16}$.

46. $y'' = -8 + \int_0^x 6\,dt = -8 + [6t]_0^x = -8 + 6x$. $y' = \int_0^x (-8+6t)dt = [-8t + 3t^2]_0^x = -8x + 3x^2$. $y = 5 + \int_0^x (-8t + 3t^2)dt = 5 - 4x^2 + x^3$

47. Step 1. $v = \frac{ds}{dt} = \int -k\,dt = -kt + C_1 = -kt + 88$ since $v = 88$ when $t = 0$. $s = \int(-kt + 88)dt = -k\frac{t^2}{2} + 88t + C_2 = -k\frac{t^2}{2} + 88t$ since $s = 0$ when $t = 0$. Step 2. $\frac{ds}{dt} = 0$ when $t = 88/k$. Step 3. $s = -\frac{k}{2}(\frac{88}{k})^2 + 88(\frac{88}{k}) = 242$ leads to $\frac{88^2}{2k} = 242$, $k = \frac{88^2}{484} = 16\text{ft/sec}^2$.

48. a) $v = 4 + \int_1^t [15u^{1/2} - 3u^{-1/2}]du = 4 + [15(\frac{2}{3}u^{3/2} - 3(2)u^{1/2}]_1^t = 4 + 10t^{3/2} - 6t^{1/2} - [10 - 6] = 10t^{3/2} - 6t^{1/2}$. b) $s = \int_1^t (10u^{3/2} - 6u^{1/2})du = [10(\frac{2}{5})u^{5/2} - 6(\frac{2}{3})u^{3/2}]_1^t = 4t^{5/2} - 4t^{3/2} - [4 - 4] = 4(t^{5/2} - t^{3/2})$

49. $\frac{ds}{dt} = \int 5.2dt = 5.2t + C_1 = 5.2t$ since $\frac{ds}{dt} = 0$ when $t = 0$. $s = \int 5.2t\,dt = \frac{5.2}{2}t^2 + C_2 = 2.6t^2$ since $s = 0$ when $t = 0$. $2.6t^2 = 4$ leads to $\sqrt{2.6}t = 2$, $t = 2/\sqrt{2.6} = 1.24$sec approximately.

50. $v = v_0 + \int_0^t g\,du = v_0 + [gu]_0^t = v_0 + gt$. $s = s_0 + \int_0^t (v_0 + gu)du = s_0 + [v_0u + g\frac{u^2}{2}]_0^t = s_0 + v_0t + \frac{1}{2}gt^2$

51. $\int \cos^2 x\,dx = \frac{1}{2}\int(1 + \cos 2x)dx = \frac{1}{2}(x + \frac{\sin 2x}{2}) + C = \frac{x}{2} + \frac{\sin 2x}{4} + C$

52. Since $y(0) = 0$, there is always a discontinuity at the lower endpoint $t = 0$. Theorem 4 does not apply because the integrand is not continuous on $[0, 1]$. It can be applied if we can replace f by a continuous extension.

54. A) $f(t) = \sqrt{\frac{(x'(t))^2+(y'(t))^2}{-2y(t)}} = \sqrt{\frac{\pi^2+4}{4t}}$. $\lim_{t\to 0^+} f(t) = \infty$. B) $f(t) = \sqrt{\frac{\pi^2+16(t-1)^2}{-4(t-1)^2+4}} \to \infty$ as $t \to 0^+$. C) $f(t) = \sqrt{\frac{[\pi-\pi\cos(\pi t)]^2+\pi^2\sin^2(\pi t)}{-2[\cos(\pi t)-1]}} = \sqrt{\frac{2\pi^2[1-\cos(\pi t)]}{2[1-\cos(\pi t)]}} = \pi \to \pi$ as $t \to 0^+$. D) $f(t) = \sqrt{\frac{\pi^2+(-1/\sqrt{t})^2}{4\sqrt{t}}} = \sqrt{\frac{t\pi^2+1}{4t^{3/2}}} \to \infty$ as $t \to 0^+$.

55. A) From the preceding exercise $f(t) = \frac{\sqrt{\pi^2+4}}{2}t^{-1/2}$. $\int_x^1 f(t)dt = \frac{\sqrt{\pi^2+4}}{2}[2t^{1/2}]_x^1 = \sqrt{\pi^2+4}(1 - \sqrt{x})$. $T = \lim_{x\to 0^+} \frac{1}{\sqrt{g}}\sqrt{\pi^2+4}(1 - \sqrt{x}) = \frac{\sqrt{\pi^2+4}}{\sqrt{g}}$. C) $T = \lim_{x\to 0^+} \frac{1}{\sqrt{g}}\int_x^1 \pi\,dt = \lim_{x\to 0^+} \frac{\pi(1-x)}{\sqrt{g}} = \frac{\pi}{\sqrt{g}}$

56. We use the expressions for $f(t)$ obtained in Exercise 54.

		$h = 0.1$	0.01	0.001
B	$T =$	$2.15/\sqrt{g}$	$2.92/\sqrt{g}$	$3.16/\sqrt{g}$
D	$T =$	$1.95/\sqrt{g}$	$2.53/\sqrt{g}$	$2.81/\sqrt{g}$

Since the initial velocity is 0, the closer to time $t = 0$ that we start, the larger T will be especially since the new initial velocity will be closer to 0 giving a slower start.

57. In Exercise 55 we found that T for track C is $\pi/\sqrt{g}$. The times T found in Exercise 56 are all less than this.

5.6 INTEGRATION BY SUBSTITUTION - RUNNING THE CHAIN RULE BACKWARD

1. $u = 3x$, $du = 3\,dx$, $dx = \frac{1}{3}du$. $\int \sin 3x\,dx = \int \sin u \frac{1}{3}\,du = \frac{1}{3}\int \sin u\,du = \frac{1}{3}(-\cos u) + C = -\frac{1}{3}\cos 3x + C$

2. $u = 2x^2$, $du = 4x\,dx$, $x\,dx = \frac{1}{4}\,du$. $\int x\sin(2x^2)dx = \frac{1}{4}\int \sin u\,du = -\frac{1}{4}\cos u + C = -\frac{1}{4}\cos(2x^2) + C$

3. $u = 2x$, $du = 2\,dx$, $dx = du/2$. $\int \sec 2x \tan 2x\,dx = \frac{1}{2}\int \sec u \tan u\,du = \frac{1}{2}\sec u + C = \frac{1}{2}\sec 2x + C$

4. $u = 1 - \cos\frac{t}{2}$, $du = \frac{1}{2}(\sin\frac{t}{2})dt$. $\int(1 - \cos\frac{t}{2})^2 \sin\frac{t}{2}dt = 2\int u^2du = \frac{2}{3}u^3 + C = \frac{2}{3}(1 - \cos\frac{t}{2})^3 + C$

5. $u = 7x - 2$, $du = 7dx$, $dx = du/7$. $\int 28(7x-2)^3dx = \frac{28}{7}\int u^3du = 4\frac{u^4}{4} + C = (7x-2)^4 + C$

6. $u = x^4 - 1$, $du = 4x^3dx$, $x^3dx = \frac{1}{4}du$. $\int 4x^3(x^4-1)^2dx = \int u^2du = \frac{u^3}{3} + C = \frac{(x^4-1)^3}{3} + C$

7. $u = 1 - r^3$, $du = -3r^2dr$, $r^2dr = -\frac{1}{3}du$. $\int \frac{9r^2dr}{\sqrt{1-r^3}} = (-\frac{1}{3})9\int \frac{du}{u^{1/2}} = -3\int u^{-1/2}du = -3\frac{u^{1/2}}{1/2} + C = -6\sqrt{1-r^3} + C$

8. $u = y^4 + 4y^2 + 1$, $du = (4y^3 + 8y^2)dy$, $(y^3 + 2y^2)dy = \frac{1}{4}du$. $\int 12(y^4 + 4y^2 + 1)^2(y^3 + 2y)dy = \frac{12}{4}\int u^2du = u^3 + C = (y^4 + 4y^2 + 1)^3 + C$

9. a) $u = \cot 2\theta$, $du = -2\csc^2 2\theta\, d\theta$. $\int \csc^2 2\theta \cot 2\theta\, d\theta = -\frac{1}{2}\int u\, du = -\frac{1}{2}\frac{u^2}{2} + C = -\frac{\cot^2 2\theta}{4} + C$ b) $u = \csc 2\theta$, $du = -2\csc 2\theta \cot 2\theta\, d\theta$. $\int \csc^2 2\theta \cot 2\theta\, d\theta = -\frac{1}{2}\int u\, du = -\frac{1}{2}\frac{u^2}{2} + C = -\frac{\csc^2 2\theta}{4} + C$. The two answers can be seen to be equivalent by using the identity $\csc^2 x = 1 + \cot^2 x$.

10. a) $u = 5x$, $du = 5\,dx$, $dx = \frac{1}{5}du$. $\int \frac{dx}{\sqrt{5x}} = \frac{1}{5}\int u^{-1/2}du = \frac{2}{5}u^{1/2} + C = \frac{2}{5}\sqrt{5x} + C$. b) $u = \sqrt{5x}$, $du = \frac{5\,dx}{2\sqrt{5x}}$, $\frac{dx}{\sqrt{5x}} = \frac{2}{5}du$. $\int \frac{dx}{\sqrt{5x}} = \frac{2}{5}\int du = \frac{2}{5}u + C = \frac{2}{5}\sqrt{5x} + C$

11. $u = 2x + 1$, $du = 2\,dx$, $dx = \frac{1}{2}du$. Note that $u = 1$ when $x = 0$ and $u = 2$ when $x = \frac{1}{2}$ so the u-limits are 1 to 2. $\int_0^{1/2} \frac{dx}{(2x+1)^3} = \frac{1}{2}\int_1^2 u^{-3}du = \frac{1}{2}\frac{u^{-2}}{-2}\Big]_1^2 = -\frac{1}{4}[\frac{1}{4} - 1] = \frac{1}{4}\frac{3}{4} = \frac{3}{16}$

12. $u = 5x + 4$, $du = 5\,dx$, $dx = \frac{1}{5}du$. $\int_0^1 \sqrt{5x+4}\,dx = \frac{1}{5}\int_4^9 u^{1/2}du = \frac{1}{5}(\frac{2}{3})[u^{3/2}]_4^9 = \frac{2}{15}(27 - 8) = \frac{38}{15}$

13. $u = \cos 2x$, $du = -2\sin 2x\, dx$, $\sin 2x\, dx = -\frac{1}{2}du$. $\int_0^{\pi/6} \frac{\sin 2x}{\cos^2 2x}dx = -\frac{1}{2}\int_1^{1/2} \frac{du}{u^2} = -\frac{1}{2}\int_1^{1/2} u^{-2}du = -\frac{1}{2}\frac{u^{-1}}{-1}\Big]_1^{1/2} = \frac{1}{2}[2 - 1] = \frac{1}{2}$

14. $u = \sin\theta$, $du = \cos\theta\, d\theta$. $\int_{\pi/6}^{\pi/2} \sin^2\theta \cos\theta\, d\theta = \int_{1/2}^1 u^2 du = [\frac{u^3}{3}]_{1/2}^1 = \frac{1}{3}(1 - \frac{1}{8}) = \frac{7}{24}$

15. $u = 1 - x^2$, $du = -2x\, dx$, $x\, dx = -\frac{1}{2}du$. $\int_{-1}^1 x\sqrt{1-x^2}dx = -\frac{1}{2}\int_0^0 \sqrt{u}\, du = 0$

16. $u = 1 + \sqrt{y}$, $du = \frac{dy}{2\sqrt{y}}$. $\int_1^4 \frac{dy}{2\sqrt{y}(1+\sqrt{y})^2} = \int_2^3 u^{-2}du = [-\frac{1}{u}]_2^3 = -\frac{1}{3} + \frac{1}{2} = \frac{1}{6}$

17. $u = 2 + \sin x$, $du = \cos x\, dx$. $\int_{-\pi/2}^{\pi/2} \frac{\cos x}{(2+\sin x)^2}dx = \int_1^3 u^{-2}du = -u^{-1}|_1^3 = -(\frac{1}{3} - 1) = \frac{2}{3}$

18. $u = \sqrt{x}$, $du = \frac{dx}{2\sqrt{x}}$. $\int_{\pi^2/4}^{\pi^2} \frac{\sin\sqrt{x}}{\sqrt{x}}dx = 2\int_{\pi/2}^{\pi} \sin u\, du = -2[\cos u]_{\pi/2}^{\pi} = -2(-1 - 0) = 2$

19. Let $u = 1 - x$, $du = -dx$. $\int \frac{dx}{(1-x)^2} = -\int u^{-2}du = u^{-1} + C = \frac{1}{1-x} + C$

20. Let $I = \int \frac{4y}{\sqrt{2y^2+1}}dy$. Let $u = 2y^2 + 1$, $du = 4y\, dy$. $I = \int u^{-1/2}du = 2u^{1/2} + C = 2\sqrt{2y^2+1} + C$

21. Let $u = x + 2$, $du = dx$. $\int \sec^2(x+2)dx = \int \sec^2 u\, du = \tan u + C = \tan(x+2) + C$

22. Let $I = \int \sec^2(\frac{x}{4})dx$. Let $u = \frac{x}{4}$, $du = \frac{1}{4}dx$, $dx = 4\,du$. $I = 4\int \sec^2 u\,du = 4\tan u + C = 4\tan(\frac{x}{4}) + C$.

23. $u = r^2 - 1$, $du = 2r\,dr$, $r\,dr = \frac{1}{2}du$. $\int 8r(r^2-1)^{1/3}dr = \frac{8}{2}\int u^{1/3}du = 4(\frac{3}{4})u^{4/3} + C = 3(r^2-1)^{4/3} + C$

24. Let $I = \int x^4(7-x^5)^3dx$. Let $u = 7 - x^5$, $du = -5x^4dx$, $x^4dx = -\frac{1}{5}du$. $I = -\frac{1}{5}\int u^3du = -\frac{1}{20}u^4 + C = -\frac{1}{20}(7-x^5)^4 + C$

25. $u = \theta + \frac{\pi}{2}$, $du = d\theta$. $\int \sec(\theta+\frac{\pi}{2})\tan(\theta+\frac{\pi}{2})d\theta = \int \sec u \tan u\,du = \sec u + C = \sec(\theta + \frac{\pi}{2}) + C$

26. Let $I = \int \sqrt{\tan x}\sec^2 x\,dx$. Let $u = \tan x$, $du = \sec^2 x\,dx$. $I = \int u^{1/2}du = \frac{2}{3}u^{3/2} + C = \frac{2}{3}(\tan x)^{3/2} + C$

27. $u = 1 + x^4$, $du = 4x^3dx$, $x^3dx = \frac{1}{4}du$. $\int \frac{6x^3}{\sqrt[4]{1+x^4}}dx = \frac{6}{4}\int u^{-1/4}du = \frac{3}{2}(\frac{4}{3})u^{3/4} + C = 2(1+x^4)^{3/4} + C$

28. Let $I = \int(s^3+2s^2-5s+6)^2(3s^2+4s-5)ds$. Let $u = s^3+2s^2-5s+6$, $du = (3s^2+4s-5)ds$. $I = \int u^2du = \frac{u^3}{3} + C = \frac{(s^3+2s^2-5s+6)^3}{3} + C$

29. $u = y+1$, $du = dy$. a) $\int_0^3 \sqrt{y+1}dy = \int_1^4 u^{1/2}du = \frac{2}{3}u^{3/2}]_1^4 = \frac{2}{3}[4^{3/2} - 1^{3/2}] = \frac{2}{3}[8-1] = \frac{14}{3}$ b) $\int_{-1}^0 \sqrt{y+1}dy = \int_0^1 u^{1/2}du = \frac{2}{3}u^{3/2}]_0^1 = \frac{2}{3}(1-0) = \frac{2}{3}$

30. a) Let $I = \int_0^1 r\sqrt{1-r^2}dr$. Let $u = 1 - r^2$, $du = -2r\,dr$, $r\,dr = -\frac{1}{2}du$. $I = -\frac{1}{2}\int_1^0 u^{1/2}du = -\frac{1}{2}(\frac{2}{3})[u^{3/2}]_1^0 = -\frac{1}{3}(0-1) = \frac{1}{3}$. b) Let $I = \int_{-1}^1 r\sqrt{1-r^2}dr$. Using the same substitution, we obtain $I = -\frac{1}{2}\int_0^0 u^{1/2}du = 0$.

31. $u = \tan x$, $du = \sec^2 x\,dx$. a) $\int_0^{\pi/4} \tan x\sec^2 x\,dx = \int_0^1 u\,du = \frac{u^2}{2}]_0^1 = \frac{1}{2}$. b) $\int_{-\pi/4}^0 \tan x\sec^2 x\,dx = \int_{-1}^0 u\,du = \frac{1}{2}u^2]_{-1}^0 = \frac{1}{2}[0^2 - (-1)^2] = -\frac{1}{2}$.

32. Let $u = 1+x^4$, $du = 4x^3dx$, $x^3dx = \frac{1}{4}du$. a) $\int_0^1 x^3(1+x^4)^3dx = \frac{1}{4}\int_1^2 u^3du = \frac{1}{16}[u^4]_1^2 = \frac{15}{16}$ b) $\int_{-1}^1 x^3(1+x^4)dx = \frac{1}{16}[u^4]_2^2 = 0$

33. $u = x^4 + 9$, $du = 4x^3dx$, $x^3dx = \frac{1}{4}du$. a) $\int_0^1 \frac{x^3}{\sqrt{x^4+9}}dx = \frac{1}{4}\int_9^{10} u^{-1/2}du = \frac{1}{4}(2)u^{1/2}]_9^{10} = \frac{1}{2}(10^{1/2} - 9^{1/2}) = \frac{1}{2}(\sqrt{10}-3)$ b) $\int_{-1}^0 \frac{x^3}{\sqrt{x^4+9}}dx = \frac{1}{4}\int_{10}^9 u^{-1/2}du = -\frac{1}{4}\int_9^{10} u^{-1/2}du =$ (from a)) $-\frac{1}{2}(\sqrt{10}-3) = \frac{1}{2}(3-\sqrt{10})$

34. Let $u = 1+x^2$, $du = 2x\,dx$, $x\,dx = \frac{1}{2}du$. a) $\int_{-1}^1 \frac{x}{(1+x^2)^2}dx = \frac{1}{2}\int_2^2 u^{-2}du = 0$ b) $\int_0^1 \frac{x}{(1+x^2)^2}dx = \frac{1}{2}\int_1^2 u^{-2}du = [-\frac{1}{2u}]_1^2 = -\frac{1}{4} + \frac{1}{2} = \frac{1}{4}$

35. $u = x^2+1$, $du = 2x\,dx$, $x\,dx = \frac{1}{2}du$. a) $\int_0^{\sqrt{7}} x(x^2+1)^{1/3}dx = \frac{1}{2}\int_1^8 u^{1/3}du = \frac{1}{2}\frac{3}{4}u^{4/3}]_1^8 = \frac{3}{8}(16-1) = \frac{45}{8}$. b) $\int_{-\sqrt{7}}^0 x(x^2+1)^{1/3}dx = \frac{1}{2}\int_8^1 u^{1/3}du = -\frac{1}{2}\int_1^8 u^{1/3}du = -\frac{45}{8}$ from a)

36. Let $u = \cos x$, $du = -\sin x\,dx$. a) $\int_0^{\pi} 3\cos^2 x \sin x\,dx = -3\int_1^{-1} u^2du = [-u^3]_1^{-1} = 2$ b) $\int_{2\pi}^{3\pi} 3\cos^2 x\sin x\,dx = [-u^3]_1^{-1} = 2$

37. $u = 1-\cos 3x$, $du = 3\sin 3x\,dx$. a) $\int_0^{\pi/6}(1-\cos 3x)\sin 3x\,dx = \frac{1}{3}\int_0^1 u\,du = \frac{1}{3}\frac{u^2}{2}]_0^1 = \frac{1}{6}$ b) $\int_{\pi/6}^{\pi/3}(1-\cos 3x)\sin 3x\,dx = \frac{1}{3}\int_1^2 u\,du = \frac{1}{3}\frac{u^2}{2}]_1^2 = \frac{1}{6}(4-1) = \frac{1}{2}$

38. $u = x^2+1$, $du = 2x\,dx$. a) $\int_0^{\sqrt{3}} \frac{4x}{\sqrt{x^2+1}}dx = 2\int_1^4 u^{-1/2}du = 4[u^{1/2}]_1^4 = 4$ b) $\int_{-\sqrt{3}}^{\sqrt{3}} \frac{4x}{\sqrt{x^2+1}}dx = 4[u^{1/2}]_3^3 = 0$

39. $u = 2+\sin x$, $du = \cos x\,dx$. a) $\int_0^{2\pi} \frac{\cos x}{\sqrt{2+\sin x}}dx = \int_2^2 u^{-1/2}du = 0$. b) $\int_{-\pi}^{\pi} \frac{\cos x}{\sqrt{2+\sin x}}dx = \int_2^2 u^{-1/2}du = 0$

40. $u = 3+\cos x$, $du = -\sin x\,dx$. a) $\int_{-\pi/2}^0 \frac{\sin x}{(3+\cos x)^2}dx = -\int_3^4 u^{-2}du = [\frac{1}{u}]_3^4 = \frac{1}{4}-\frac{1}{3} = -\frac{1}{12}$ b) $\int_0^{\pi/2} \frac{\sin x\,dx}{(3+\cos x)^2} = [\frac{1}{u}]_4^3 = \frac{1}{12}$

41. $u = t^5+2t$, $du = (5t^4+2)dt$. $\int_0^1 \sqrt{t^5+2t}(5t^4+2)dt = \int_0^3 u^{1/2}du = \frac{2}{3}u^{3/2}]_0^3 = \frac{2}{3}3^{3/2} = 2\sqrt{3}$

42. Let $I = \int_0^3 t\sqrt{1+t}\,dt$, $u = 1+t$, $t = u-1$, $du = dt$. Then $I = \int_1^4(u-1)u^{1/2}du = \int_1^4(u^{3/2}-u^{1/2})du = \frac{2}{5}u^{5/2}-\frac{2}{3}u^{3/2}]_1^4 = (\frac{64}{5}-\frac{16}{3})-(\frac{2}{5}-\frac{2}{3}) = \frac{116}{15}$

43. $u=\cos 2x$, $du=-2\sin 2x\,dx$. $\int_0^{\pi/2}\cos^3 2x\sin 2x\,dx = -\frac{1}{2}\int_1^{-1}u^3du = -\frac{1}{2}\frac{u^4}{4}]_1^{-1} = -\frac{1}{8}((-1)^4-(1)^4) = 0$

44. $I = \int_{-\pi/4}^{\pi/4}\tan^2 x\sec^2 x\,dx$. $u = \tan x$, $du = \sec^2 x\,dx$. $I = \int_{-1}^1 u^2du = [\frac{u^3}{3}]_{-1}^1 = \frac{2}{3}$

45. $u = 5-4\cos t$, $du = 4\sin 5\,dt$. $\int_0^{\pi}\frac{8\sin t}{\sqrt{5-4\cos t}}dt = \frac{8}{4}\int_1^9 u^{-1/2}du = 2(2u^{1/2})]_1^9 = 4(3-1) = 8$

46. $I = \int_0^{\pi/4}(1-\sin 2t)^{3/2}\cos 2t\,dt$. $u = 1-\sin 2t$, $du = -2\cos 2t\,dt$. $I = -\frac{1}{2}\int_1^0 u^{3/2}du = -\frac{1}{2}(\frac{2}{5})[u^{5/2}]_1^0 = \frac{1}{5}$

47. $u = 5x^3+4$, $du = 15x^2dx$. $\int_0^1 15x^2\sqrt{5x^3+4}dx = \int_4^9 u^{1/2}du = \frac{2}{3}u^{3/2}]_4^9 = \frac{2}{3}(9^{3/2}-4^{3/2}) = \frac{2}{3}(27-8) = \frac{38}{3}$

48. $I = \int_0^1 (y^3 + 6y^2 - 12y + 5)(y^2 + 4y - 4)dy$. $u = y^3 + 6y^2 - 12y + 5$, $du = (3y^2 + 12y - 12)dy$, $\frac{1}{3}du = (y^2 + 4y - 4)dy$. $I = \frac{1}{3}\int_5^0 u\,du = \frac{1}{6}[u^2]_5^0 = -\frac{25}{6}$

49. $u = 4 - x^2$, $du = -2x\,dx$, $x\,dx = -\frac{1}{2}du$. $A = -\int_{-2}^0 x\sqrt{4-x^2}dx + \int_0^2 x\sqrt{4-x^2}dx = -(-\frac{1}{2})\int_0^4 u^{1/2}du - \frac{1}{2}\int_4^0 u^{1/2}du = \frac{1}{2}\int_0^4 u^{1/2}du + \frac{1}{2}\int_0^4 u^{1/2}du = \int_0^4 u^{1/2}du = \frac{2}{3}u^{3/2}]_0^4 = \frac{16}{3}$

50. $A = \int_0^\pi (1 - \cos x)\sin x\,dx$. Let $u = 1 - \cos x$, $du = \sin x\,dx$. $A = \int_0^2 u\,du = [\frac{u^2}{2}]_0^2 = 2$

51. Let $u = 3t^2 - 1$, $du = 6t\,dt$. $s = \int 24t(3t^2 - 1)^3 dt = \frac{24}{6}\int u^3 du = 4\frac{u^4}{4} + C = (3t^2 - 1)^4 + C$. If $t = 0$, $s = (0 - 1)^4 + C = 1 + C = 0$ so $C = -1$. Thus $s = (3t^2 - 1)^4 - 1$.

52. $y = \int_0^x 4t(t^2 + 8)^{-1/3}dt$. Let $u = t^2 + 8$, $du = 2t\,dt$. $y = 2\int_8^{x^2+8} u^{-1/3}du = 2(\frac{3}{2})[u^{2/3}]_8^{x^2+8} = 3[(x^2 + 8)^{2/3} - 4]$

53. Let $u = t + \pi$, $du = dt$. $s = \int 6\sin(t + \pi)dt = 6\int \sin u\,du = -6\cos u + C = -6\cos(t + \pi) + C$. When $t = 0$, $s = -6\cos\pi + C = 6 + C = 0$ so $C = -6$. Thus $s = -6\cos(t + \pi) - 6$.

54. $\frac{d^2s}{dt^2} = -4\sin(2t - \frac{\pi}{2}) = 4\sin(\frac{\pi}{2} - 2t) = 4\cos 2t$. $\frac{ds}{dt} = 100 + 4\int_0^t \cos 2v\,dv = 100 + 4[\frac{\sin 2v}{2}]_0^t = 100 + 2\sin 2t$. $s = 100t - \cos 2t + C$. $0 = s(0) = -1 + C$, $C = 1$. $s = 100t - \cos 2t + 1$ or $s = 100t + \sin(2t - \frac{\pi}{2}) + 1$

55. Let $I = \int_0^{\pi/4} \frac{18\tan^2 x\sec^2 x}{(2+\tan^3 x)^2}dx$. a) $u = \tan x$, $du = \sec^2 x\,dx$. $I = 18\int_0^1 \frac{u^2\,du}{(2+u^3)^2}$. $v = u^3$, $dv = 3u^2du$, $u^2du = \frac{1}{3}dv$. $I = \frac{18}{3}\int_0^1 \frac{dv}{(2+v)^2} = 6\int_0^1 \frac{dv}{(2+v)^2}$. $w = 2 + v$, $dw = dv$. $I = 6\int_2^3 w^{-2}dw = -6\frac{1}{w}]_2^3 = -6(\frac{1}{3} - \frac{1}{2}) = 1$ b) $u = \tan^3 x$, $du = 3\tan^2 x\sec^2 x\,dx$. $I = \frac{18}{3}\int_0^1 \frac{du}{(2+u)^2} = 6\int_2^3 v^{-2}dv = 1$ (from end of a)). c) $u = 2 + \tan^3 x$, $du = 3\tan^2 x\sec^2 x\,dx$. $I = \frac{18}{3}\int_2^3 u^{-2}du = 1$ as before.

56. a) $u = x - 1$, $du = dx$. $I = \int \sqrt{1 + \sin^2 u}\sin u\cos u\,du$. $v = \sin u$, $dv = \cos u\,du$. $I = \int \sqrt{1 + v^2}v\,dv$. $w = 1 + v^2$, $dw = 2v\,dv$. $I = \frac{1}{2}\int w^{1/2}dw = \frac{1}{2}\frac{2}{3}w^{3/2} + C = \frac{1}{3}(1 + v^2)^{3/2} + C = \frac{1}{3}(1 + \sin^2 u)^{3/2} + C = \frac{1}{3}[1 + \sin^2(x - 1)]^{3/2} + C$
b) $u = \sin(x - 1)$, $du = \cos(x - 1)dx$. $I = \int \sqrt{1 + u^2}u\,du$. $v = 1 + u^2$, $dv = 2u\,du$. $I = \frac{1}{2}\int v^{1/2}dv = \frac{1}{3}v^{3/2} + C = \frac{1}{3}(1 + u^2)^{3/2} + C = \frac{1}{3}[1 + \sin^2(x - 1)]^{3/2} + C$
c) $u = 1 + \sin^2(x - 1)$, $du = 2\sin(x - 1)\cos(x - 1)dx$. $I = \frac{1}{2}\int u^{1/2}du + C = \frac{1}{2}(\frac{2}{3})u^{3/2} + C = \frac{1}{3}[1 + \sin^2(x - 1)]^{3/2} + C$

57. The answers can be seen to be equivalent: $\sin^2 x + C_1 = (1 - \cos^2 x) + C_1 = -\cos^2 x + (1 + C_1) = -\cos^2 x + C_2 = -(\frac{1+\cos 2x}{2}) + C_2 = -\frac{\cos 2x}{2} + (-\frac{1}{2} + C_2) = -\frac{\cos 2x}{2} + C_3$. The graph of any one of the antiderivatives can be obtained from the graph of any other antiderivative by a vertical shift verifying that they differ by an additive constant.

5.7 NUMERICAL INTEGRATION: TRAPEZOIDAL RULE & SIMPSON'S METHOD

1. Let $f(x) = x$. $h = \frac{2-0}{4} = \frac{1}{2}$. a) $T = \frac{1}{2}(\frac{1}{2})[f(0)+2f(\frac{1}{2})+2f(1)+2f(\frac{3}{2})+f(2)] = \frac{1}{4}[0+1+2+3+2] = 2$ b) $S = \frac{1}{3}(\frac{1}{2})[f(0)+4f(\frac{1}{2})+2f(1)+4f(1.5)+f(2)] = 2$ c) $\int_0^2 x\ dx = \frac{x^2}{2}]_0^2 = 2$

2. a) $T = \frac{1}{2}(\frac{1}{2})[0 + 2(\frac{1}{2})^2 + 2(1)^2 + 2(\frac{3}{2})^2 + 2^2] = \frac{11}{4}$ b) $S = \frac{1}{3}(\frac{1}{2})[0 + 4(\frac{1}{2})^2 + 2(1)^2 + 4(\frac{3}{2})^2 + 2^2] = \frac{8}{3}$ c) $\int_0^2 x^2\ dx = [\frac{x^3}{3}]_0^2 = \frac{8}{3}$.

3. Let $f(x) = x^3$, $h = \frac{2-0}{4} = \frac{1}{2}$. a) $T = \frac{1}{4}[f(0) + 2f(\frac{1}{2}) + 2f(1) + 2f(\frac{3}{2}) + f(2)] = 4.25$ b) $S = \frac{1}{3}(\frac{1}{2})[f(0) + 4f(\frac{1}{2}) + 2f(1) + 4f(\frac{3}{2}) + f(2)] = 4$ c) $\int_0^2 x^3\ dx = \frac{x^4}{4}]_0^2 = \frac{16}{4} - \frac{0}{4} = 4$.

4. a) $T = \frac{1}{2}\frac{1}{4}[1 + 2(\frac{4}{5})^2 + 2(\frac{2}{3})^2 + 2(\frac{4}{7})^2 + \frac{1}{2^2}] = 0.50899$ b) $S = \frac{1}{3}\frac{1}{4}[1 + 4(\frac{4}{5})^2 + 2(\frac{2}{3})^2 + 4(\frac{4}{7})^2 + \frac{1}{2^2}] = 0.50042$ c) $\int_1^2 \frac{1}{x^2}dx = [-\frac{1}{x}]_1^2 - [\frac{1}{2} - 1] = 0.5$.

5. a) $T = \frac{1}{2}(1)[\sqrt{0} + 2\sqrt{1} + 2\sqrt{2} + 2\sqrt{3} + \sqrt{4}] = \frac{1}{2}[2 + 2\sqrt{2} + 2\sqrt{3} + 2] = 2 + \sqrt{2} + \sqrt{3} = 5.146...$ b) $S = \frac{1}{3}(1)[\sqrt{0} + 4\sqrt{1} + 2\sqrt{2} + 4\sqrt{3} + \sqrt{4}] = \frac{1}{3}(6 + 2\sqrt{2} + 4\sqrt{3}) = 5.252...$ c) $\int_0^4 \sqrt{x}\ dx = \frac{2}{3}x^{3/2}\Big|_0^4 = \frac{16}{3} = 5.333...$

6. a) $T = \frac{1}{2}\frac{\pi}{4}[\sin 0 + 2\sin\frac{\pi}{4} + 2\sin\frac{\pi}{2} + 2\sin\frac{3\pi}{4} + \sin\pi] = \frac{\pi}{8}[0 + \sqrt{2} + 2 + \sqrt{2} + 0] = \frac{\pi}{8}(2 + 2\sqrt{2}) = \frac{\pi}{4}(1 + \sqrt{2}) = 1.89612$ b) $S = \frac{1}{3}\frac{\pi}{4}[\sin 0 + 4\sin\frac{\pi}{4} + 2\sin\frac{\pi}{2} + 4\sin\frac{3\pi}{4} + \sin\pi] = \frac{\pi}{12}[0 + 2\sqrt{2} + 2 + 2\sqrt{2} + 0] = \frac{\pi}{12}[2 + 4\sqrt{2}] = \frac{\pi}{6}[1 + 2\sqrt{2}] = 2.00456$ c) $\int_0^\pi \sin x\ dx = [-\cos x]_0^\pi = 2$.

7. $\int_{-1}^3 e^{-x^2}\ dx$

n	TRAP	SIMP	LRAM	RRAM	MRAM
10	1.62316	1.63322	1.69671	1.54961	1.63799
100	1.63293	1.63303150	1.64029	1.62558	1.63308
1000	1.6330305	1.63303148	1.63377	1.63229	1.63303

NINT yields 1.63303148105.

8. $\int_2^5 \sqrt{x^2-2}\ dx$

n	TRAP	SIMP	LRAM	RRAM	MRAM
10	9.519	9.52128	9.01	10.023	9.5227
100	9.52133	9.52135547	9.47	9.57	9.521369
1000	9.5213552	9.52135548278	9.516	9.526	9.5213556

NINT yields 9.52135548276.

9. $\int_{-5}^5 x \sin x\ dx$

n	TRAP	SIMP	LRAM	RRAM	MRAM
10	-4.682	-4.73	-4.68	-4.68	-4.79
100	-4.7537	-4.754469	-4.7537	-4.7537	-4.7549
1000	-4.75446	-4.7544704038	-4.75446	-4.75446	-4.754474

NINT yields -4.75447040396.

10. $\int_{-2}^2 \sin x^2\ dx$

n	TRAP	SIMP	LRAM	RRAM	MRAM
10	1.535	1.63	1.535	1.535	1.649
100	1.6088	1.6095547	1.60886	1.60886	1.6099
1000	1.609546	1.60955297886	1.609546	1.609546	1.609556

NINT yields 1.60955297869.

11. $\int_{3\pi/4}^{4.5} \frac{\tan x}{x} dx$

n	TRAP	SIMP	LRAM	RRAM	MRAM
10	0.257	0.246	0.101	0.413	0.238
100	0.244	0.243771	0.228	0.259	0.244
1000	0.2437718	0.2437703542	0.242	0.245	0.2437696

NINT yields 0.243770354155.

12. $\int_1^{2\pi} \frac{2 \sin x}{x} dx$

n	TRAP	SIMP	LRAM	RRAM	MRAM
10	0.966	0.94386	1.41	0.52	0.933
100	0.94435	0.94413698	0.989	0.8999	0.94403
1000	0.944139	0.944137011524	0.9486	0.9397	0.9441359

NINT yields 1.944137011531.

13. $f(x) = x^{-1}$, $f'(x) = -x^{-2}$, $f''(x) = \frac{2}{x^3} \leqq \frac{2}{1^3}$ on $[1,2]$ so we may take $M = 2$. $|E_T| \leqq \frac{b-a}{12}h^2M = \frac{1}{12}(\frac{1}{10})^2 2 = \frac{1}{600} = 0.0016666\ldots$.

14. $f(x) = \frac{1}{x}$, $f'(x) = -x^{-2}$, $f''(x) = 2x^{-3}$, $f'''(x) = -6x^{-4}$, $f^{(4)}(x) = 24x^{-5} = \frac{24}{x^5} \leqq 24$ on $[1,2]$. $|E_S| \leqq \frac{b-a}{180}h^4M = \frac{1}{180}(0.1)^4 24 = 0.00001333\ldots$.

15. With $f(x) = x$, we have $f''(x) = f^{(4)}(x) = 0$ and so we may take $M = 0$ in both Eq. (6) and Eq. (7). Hence $E_T = E_S = 0$ for all possible n. a) $n = 1$ b) $n = 2$ (n is always even in S)

16. $f(x) = x^2$, $f'(x) = 2x$, $f''(x) = 2$, $f''(x) = f^{(4)}(x) = 0$. $|E_T| \leqq \frac{b-a}{12}h^2M = \frac{2}{12}(\frac{2}{n})^2 2 = \frac{4}{3n^2} < \frac{1}{10^4}$ leads to $n > (\frac{4(10^4)}{3})^{1/2} = 115.47$. We may take $n = 116$ for a). In b) $M = 0$, Simpson's Rule is exact and we may take $n = 2$.

17. Let $f(x) = x^3$, $f'(x) = 3x^2$, $f''(x) = 6x$. For $0 \leqq x \leqq 2$, $|f''(x)| \leqq 6 \cdot 2 = 12$ so we may take $M = 12$ in Eq. (6). $h = (b/a)/n = (2-0)/n = 2/n$. By Eq. (4), $|E_T| \leqq \frac{2}{12}(\frac{2}{n})^2 12 = \frac{8}{n^2}$. In a) we wish to have $\frac{8}{n^2} < \frac{1}{10^4}$ or $\frac{n^2}{8} > 10^4$, $n^2 > (8)10^4$, $n > 2\sqrt{2}10^2 = 200\sqrt{2} = 282.84\ldots$. Thus, we may take $n = 283$ in a). Since $f^{(4)}(x) = 0$, we may take $M = 0$ in Eq. (7) and S is the exact value of the integral for all even $n > 0$. b) $n = 2$

18. $f(x) = x^{-2}$, $f'(x) = -2x^{-3}$, $f''(x) = 6x^{-4}$, $f'''(x) = -24x^{-5}$, $f^{(4)}(x) = 120x^{-6}$. a) $|f''(x)| = \frac{6}{x^4} \leqq 6 = M$ on $[1,2]$. $|E_T| \leqq \frac{1}{12}(\frac{1}{n})^2 6 = \frac{1}{2n^2} < \frac{1}{10^4}$ leads to $n^2 > \frac{10^4}{2}$, $n > \frac{10^2}{\sqrt{2}} = 70.71$. We may take $n = 71$. b) Here $M = 120$. $|E_S| \leqq \frac{1}{180}(\frac{1}{n})^4 120 = \frac{2}{3n^4} < \frac{1}{10^4}$ leads to $n^4 > \frac{2(10^4)}{3}$, $n > (2/3)^{1/4}10 = 9.04$ so we may take $n = 10$.

19. $f(x) = x^{1/2}$, $f'(x) = \frac{1}{2}x^{-1/2}$, $f''(x) = -\frac{1}{4}x^{-3/2}$, $f'''(x) = \frac{3}{8}x^{-5/2}$, $f^{(4)}(x) = -\frac{15}{16}x^{-7/2}$. $|f''(x)| = |\frac{1}{4x^{3/2}}| \leqq \frac{1}{4}$ for $1 \leqq x \leqq 4$ and so we may take $M = \frac{1}{4}$ in Eq. (6). For a) we wish to have $|E_T| \leqq \frac{b-a}{12}h^2M = \frac{3}{12}(\frac{3}{n})^2\frac{1}{4} = \frac{9}{16n^2} < \frac{1}{10^4}$ or $\frac{16n^2}{9} > 10^4$, $n^2 > \frac{9}{16}10^4$, $n > \frac{3}{4}(10^2) = 75$. Thus in a) we take $n = 76$.

$|f^{(4)}(x)| = |\frac{15}{16x^{7/2}}| \leqq \frac{15}{16}$ on $[1,4]$ so we may take $M = \frac{15}{16}$ in Eq. (7). $|E_S| \leqq \frac{b-a}{180}h^4M = \frac{3}{180}(\frac{3}{n})^4\frac{15}{16} = \frac{81}{64n^4} < \frac{1}{10^4}$ leads to $\frac{64n^4}{81} > 10^4$, $n > (\frac{81}{64}(10^4))^{1/4} = 10.6\ldots$. Since n is even in Simpson's Rule, we take $n = 12$ in b).

20. In both cases $M = 1$. a) $|E_T| \leqq \frac{\pi}{12}(\frac{\pi}{n})^2 = \frac{\pi^3}{12n^2} < \frac{1}{10^4}$ leads to $n^2 > \frac{10^4\pi^3}{12}$, $n > \sqrt{\frac{10^4\pi^3}{12}} = 160.74$. We may take $n = 161$. b) $|E_S| \leqq \frac{\pi}{180}\frac{\pi^4}{n^4} < \frac{1}{10^4}$ leads to $n^4 > \frac{\pi(10\pi)^4}{180}$, $n > (\frac{\pi}{180})^{1/4}10\pi = 11.42$. We may take $n = 12$.

21. $3.1379\ldots$, $3.14029\ldots$

22. 1.08942941322..., 1.08942941322...

23. 1.3669..., 1.3688...

24. 0.828116396058..., 0.828116333053...

25. a) 0.057... and 0.0472... b) Let $y_1 = \sin x/x$, $y_2 = \text{der2}(y_1, 2, 2)$, $y_3 = (\frac{1.5\pi}{12})(\frac{1.5\pi}{10})^2 abs y_2$, $y_4 = \text{NDER}(\text{NDER}(y_2, x, x), x, x)$, $y_5 = \frac{1.5\pi}{180}(\frac{1.5\pi}{10})^4 abs y_4$. We graph y_3 in $[\frac{\pi}{2}, 2\pi]$ by $[0, 0.03]$ and y_5 in $[\frac{\pi}{2}, 2\pi]$ by $[0, 3 \times 10^{-4}]$. Max $y_3 = 2.168\ldots \times 10^{-2}$, $\max y_5 = 2.26\ldots \times 10^{-4}$ c) $\max E_T(x) = 5.42\ldots \times 10^{-3}$, $\max E_S(x) = 1.41\ldots \times 10^{-5}$ d) $\max E_T(x) = 8.67 \times 10^{-4}$, $\max E_S(x) = 3.62\ldots \times 10^{-7}$ e) We cannot find the exact value, but we can approximate the integral as closely as we like by increasing n. With $n = 50$, Simpson's Rule gives the value 0.0473894... . By d) the error is at most 3.62×10^{-7}.

26. Refer to the method of Exercise 25. a) 0.64..., 0.73... b) $\max E_T = 0.77095\ldots$, $\max E_S = 0.0948\ldots$ c) $\max E_T = 0.1927\ldots$, $\max E_S = 0.0059\ldots$ d) $\max E_T = 0.0308\ldots$, $\max E_S = 1.5\ldots \times 10^{-4}$ e) Let $I = \int_1^{\pi} x \sin x^2 dx$, $u = x^2$, $du = 2x dx$, $x dx = \frac{1}{2} du$. $I = \frac{1}{2}\int_1^{\pi^2} \sin u\, du = [-\frac{1}{2}\cos u]_1^{\pi^2} = -\frac{1}{2}[\cos \pi^2 - \cos 1]$ (exact value) $\approx 0.721493833901\ldots$.

27. Refer to the method of Exercise 25. a) 3.6664... and 3.65348218... b) $\max E_T = 0.2466\ldots$, $\max E_S = 2.55\ldots \times 10^{-4}$ c) $\max E_T = 6.16\ldots \times 10^{-2}$, $\max E_S = 1.59\ldots \times 10^{-5}$ d) $\max E_T = 9.86\ldots \times 10^{-3}$, $\max E_S = 4.08\ldots \times 10^{-7}$ e) Simpson's Rule with $n = 50$ yields 3.6534844... with error at most $4.08\ldots \times 10^{-7}$.

28. $f(x) = \sin(x^2)$. After some simplification, we obtain $f''(x) = -4x^2 \sin(x^2) + 2\cos(x^2)$, $y_1 = f^{(4)}(x) = 16x^4 \sin(x^2) - 48x^2 \cos(x^2) = 12\sin(x^2)$. Let $y_2 = \text{NDER2}(f''(x), x)$. Graph y_1 and y_2 in $[-1, 1]$ by $[-28.5, 0]$; the graphs are indistinguishable. Evaluating directly, we obtain

x	-0.2	0.2	0.4	0.6
y_1	-2.3973	-2.3973	-9.4285	-19.6691
y_2	-2.4012	-2.4012	-9.4314	-19.6679
$\lvert y_1 - y_2\rvert$	0.0039	0.0039	0.0029	0.0012

Our calculator finds $\max |y_1 - y_2| = 0.004$ in $-1 \leqq x \leqq 1$. If E_S turns out to be very small, say 10^{-7}, it makes little difference whether y_1 or y_2 is used to compute E_S.

29. $f(x) = x^2 + \sin x$. On $[1, 2]$, $\text{LRAM}_{50} f = 3.2591$, $\text{RRAM}_{50} f = 3.3205$ and $T_{50} f = 3.2898$. We find $2T_{50} f = \text{LRAM}_{50} f + \text{RRAM}_{50} f$ (see Exercise 31).

30. $f(x) = x^3 - \cos x$ on $[1, 2]$. $T_{50}f = 3.6825$, $\text{MRAM}_{25}f = 3.6816$ and $S_{50}f = 3.6822$. We find $S_{50}f = (\text{MRAM}_{25}f + 2T_{50}f)/3$ (see Exercise 32).

31. $\frac{\text{LRAM}_{10}f + \text{RRAM}_{10}f}{2} = [(y_0 + y_1 + \cdots + y_9)h + (y_1 + y_2 + \cdots + y_{10})h]/2 = \frac{h}{2}(y_0 + 2y_1 + 2y_2 + \cdots + 2y_9 + y_{10}) = T_{10}f$.

32. Let $y_0, y_1, \ldots, y_{10}$ be the f-values required for $T_{10}f$ and $S_{10}f$. Then y_1, y_3, y_5, y_7, y_9 are the f-values required for $\text{MRAM}_5 f$. Let $h = \frac{b-a}{2n}$. $\frac{[\text{MRAM}_5 f + 2T_{10}f]}{3} = \frac{(y_1 + y_3 + \cdots + y_9)(2h) + (y_0 + 2y_1 + 2y_2 + \cdots + 2y_9 + y_{10})h}{3} = \frac{h}{3}(y_0 + 4y_1 + 2y_2 + 4y_3 + 2y_4 + \cdots + 4y_9 + y_{10}) = S_{10}f$.

33. We estimate the surface area of the pond using Simpson's Rule. $S = \frac{200}{3}[0+4(520)+2(800)+400(1000)+2(1140)+4(1160)+2(1110)+4(860)+0] = 1350666.667$ ft^2. The volume of the point is $20S$ ft^3. The number of fish initially is $\frac{20S}{1000}$ and the maximum number to be caught is $(\frac{3}{4})\frac{20S}{1000}$. Thus the maximum number of licenses to be sold is $(\frac{1}{20})(\frac{3}{4})\frac{20S}{1000} = 1013$.

34. Let V be the volume of the tank. Then $42V = 5000$, $V = \frac{5000}{42} = \frac{2500}{21}$ ft^3. since the cross-section area A is constant, $V = AL$ where L is the length of the tank. By Simpson's rule $A \approx \frac{1}{3}[1.5 + 4(1.6) + 2(1.8) + 4(1.9) + 2(2.0) + 4(2.1) + 2.1] = 11.2$. Hence $L = V/A \approx \frac{2500}{21(11.2)} = 10.63$ ft.

35. $S = \frac{1}{3}(\frac{24}{6})[0 + 4(18.75) + 2(24) + 4(26) + 2(24) + 4(18.75) + 0] = 466.66\ldots$ in^2.

36. a) $A \approx T = \frac{1}{2}[0 + 2(\text{sum of values given}) + 0] = 541.5$. b) Cardiac Output $= \frac{5.6(60)}{541.5} = 0.62\ldots L/\text{min}$.

37. We use the odd-numbered hours. $n = 12$ and $h = 2$. $T = \frac{h}{2}[1.88 + 2(2.02) + 2(2.25) + 2(3.60) + 2(3.05) + 2(2.38) + 2(2.02) + 2(1.72) + 2(1.97) + 2(2.68) + 2(2.65) + 2(2.21) + 1.88] = 56.86$ kwh per customer.

38. $T = \frac{1}{2}[1.69 + 2(\text{sum of intermediate terms}) + 2.22] = 51.605$ kwh.

39. In parametric mode graph $x = \pi t - \sin(\pi t)$, $y = 1 + \cos(\pi t)$, $0 \leqq t \leqq 1$, t step $= 0.05$ in $[0, \pi]$ by $[0, 2]$. The fact that the graph appears to pass the vertical line test, supports that the relation is a function. $\frac{dx}{dt} = \pi - \pi\cos(\pi t) = \pi(1 - \cos \pi t) > 0$ for $0 < t \leqq 1$. This shows that x is an increasing function of t on the interval. Hence no value of x is repeated and so the relation is a function. Since x is an increasing function of t, we see that as t goes from 0 to 1, x goes from 0 to π, so the domain is $[0, \pi]$. Similarly, we see that y is a decreasing function of t and that the range is $[0, 2]$.

40. $T = \sum_{i=1}^{4} \frac{1}{2}(y_{i-1}+y_i)(x_i-x_{i-1}) = \frac{1}{2}[(2+1.707)(0.0783-0)+(1.707+1)(0.571-0.0783) + (1+0.293)(1.649-0.571) + (0.293+0)(3.142-1.649)] = 1.728$. By the method shown in Exercise 41 we get $T = 1.727379\ldots$.

41. Let $t_k = \frac{k}{n} = \frac{k}{10} = (0.1)k$, $k = 0, 1, 2, \ldots, 10$. Each t_k determines a point (x_k, y_k) on the curve. $n = 10$ trapezoids are determined, the kth one having area $A_k = \frac{1}{2}(y_{k-1}+y_k)(x_k - x_{k-1}) = \frac{1}{2}[1 + \cos((k-1)(0.1)\pi) + 1 + \cos(k(0.1)\pi)] \times [\pi k(0.1) - \sin(k(0.1)\pi) - ((k-1)(0.1)\pi - \sin((k-1)(0.1)\pi)] = \frac{1}{2}[2 + \cos((k-1)(0.1)\pi) + \cos((0.1)k\pi)] \times [(0.1)\pi - \sin((0.1)k\pi) + \sin((k-1)(0.1)\pi)]$. An approximation for the area is $T = \sum_{k=1}^{10} A_k = 1.5965087\ldots$. Your calculator may have a "sum seq" feature that can be used.

42. $\frac{dx}{dt} = \pi - \pi\cos(\pi t) = \pi - \pi(y-1) = 2\pi - \pi y$. $\frac{dy}{dt} = -\pi\sin(\pi t) = -\pi\sqrt{1-\cos^2(\pi t)} = -\pi\sqrt{1-(y-1)^2}$. After simplification, $\frac{dx}{dy} = \frac{(dx/dt)}{(dy/dt)} = \frac{y-1}{\sqrt{1-(y-1)^2}} - \frac{1}{\sqrt{1-(y-1)^2}}$. Integrating, we obtain $x = -\sqrt{1-(y-1)^2} - \sin^{-1}(y-1) + C$. Using values at $t = 0$, we find $C = \frac{\pi}{2}$, $x = \frac{\pi}{2} - \sqrt{2y-y^2} - \sin^{-1}(y-1) = g(y)$. $y = g(x) = \frac{\pi}{2} - \sqrt{2x-x^2} - \sin^{-1}(x-1)$. Graph $g(x)$ in $[0, 2]$ by $[0, \pi]$. f and g are inverses of each other because, in general, if a relation is the solution set of an equation, the inverse relation is the solution set of the equation obtained from the original equation by interchanging x and y.

43. $g(x) = \frac{\pi}{2} - \sqrt{2x-x^2} - \sin^{-1}(x-1)$. $\text{NINT}(g(x), x, 0, 2) = 1.57079603391$. The region under the graph of g from 0 to 2 is the reflection of the region under the graph of f from 0 to π across the line $y = x$.

44. With $n = 100$ used in the method of Exercise 41 we obtain 1.57105469968 which is within 0.001 of the NINT value.

45. $\text{LRAM}_n f + \text{RRAM}_n f = \sum_{k=0}^{n-1} y_k h + \sum_{k=1}^{n} y_k h = h[y_0 + 2\sum_{k=1}^{n-1} y_k + y_n] = 2T_n f$ which is equivalent to the first formula. Let $y_0, y_1, \ldots, y_{2n}$ be the f-values corresponding to the regular partition of the interval into $2n$ subintervals. Then $y_1, y_3, y_5, \ldots, y_{2n-1}$ are the f-values needed for MRAM_n for a partition of the interval into n subintervals. With $h = \frac{b-a}{2n}$, $(\text{MRAM}_n f + 2T_{2n} f)/3 = \frac{1}{3}[(y_1 + y_3 + \cdots + y_{2n-1})(2h) + (y_0 + 2y_1 + 2y_2 + \cdots + 2y_{2n-1} + y_{2n})h] = \frac{h}{3}[y_0 + 4y_1 + 2y_2 + 4y_3 + \cdots + 2y_{2n-2} + 4y_{2n-1} + y_{2n}] = S_{2n} f$.

PRACTICE EXERCISES, CHAPTER 5

1. a)

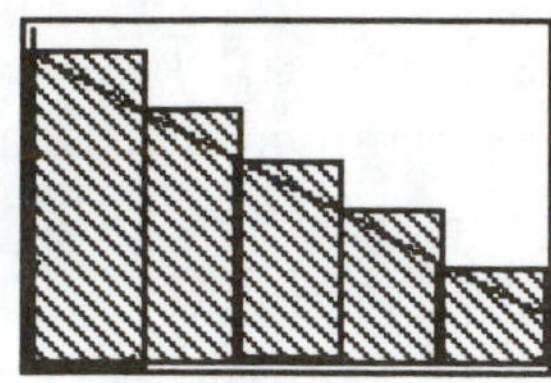
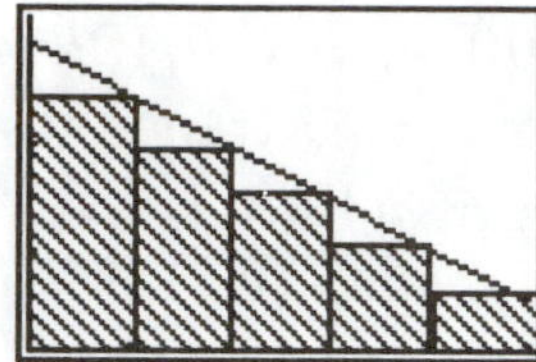
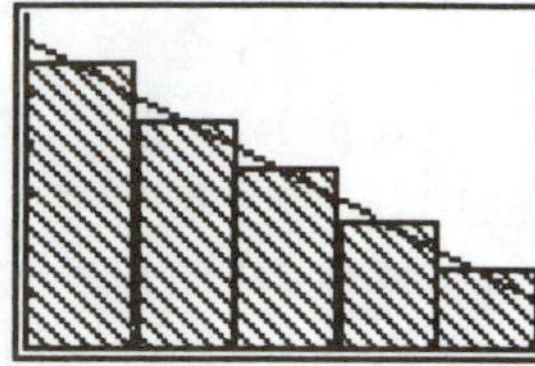

b) $\text{LRAM}_5 f = f(0)\Delta x + f(1)\Delta x + f(2)\Delta x + f(3)\Delta x + f(4)\Delta x = 6\cdot1+5\cdot1+4\cdot1+3\cdot1+2\cdot1 = 20$. $\text{RRAM}_5 f = [f(1)+f(2)+f(3)+f(4)+f(5)]\Delta x = 15$. $\text{MRAM}_5 f = [f(\frac{1}{2}) + f(\frac{3}{2}) + f(\frac{5}{2}) + f(\frac{7}{2}) + f(\frac{9}{2})]\Delta x = 17.5$

2. a)

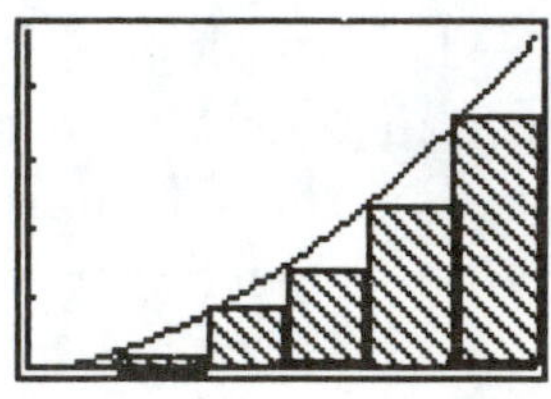
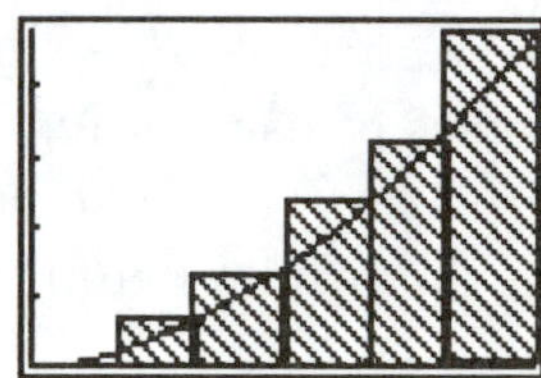
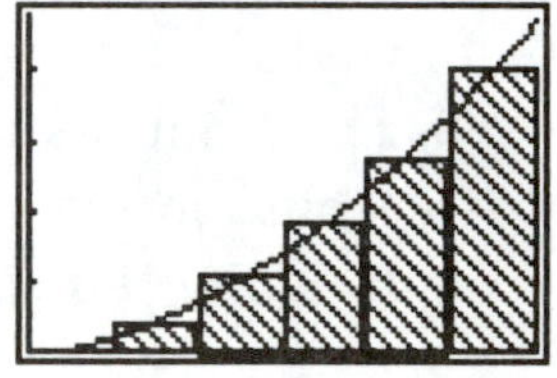

b) $\text{LRAM}_5 = [2+7+14+23+34]\Delta x = 80$. $\text{RRAM}_5 = [7+14+23+34+47]\Delta x = 125$. $\text{MRAM}_5 = [4.25+10.25+18.25+28.25+40.25]\Delta x = 101.25$

3.

	n=10	n=100	n=1000
LRAM_n	22.695	23.86545	23.9865045
RRAM_n	25.395	24.13545	24.0135045
MRAM_5	23.9775	23.999775	23.99999775

4. $f(x) = x^3 - 2x + 2$, $f'(x) = 3x^2 - 2$. We find that, on $[0,2]$, $\min f = f(\sqrt{2/3}) = 0.91 > 0$ so f is certainly non-negative on $[0,2]$.

	n=10	n=100	n=1000
LRAM_n	3.64	3.9604	3.996004
RRAM_n	4.44	4.0404	4.004004
MRAM_n	3.98	3.9998	3.999998

5.

	n=10	n=100	n=1000
LRAM_n	3.9670	3.99967	3.9999967
RRAM_n	3.9670	3.99967	3.9999967
MRAM_n	4.0165	4.00016	4.0000016

6.

	n=10	n=100	n=1000
LRAM_n	.886319	.8862269	.886226925
RRAM_n	.886319	.8862269	.886226925
MRAM_n	.886135	.8862269	.886226925

7. a) $\sum_{k=1}^{10}(k+2) = \sum_{k=1}^{10} k + \sum_{k=1}^{10} 2 = \frac{10(11)}{2} + (10)2 = 75$ b) $\sum_{k=1}^{10}(2k-12) = 2\sum_{k=1}^{10} k - \sum_{k=1}^{10} 12 = 2\frac{(10)(11)}{2} - 12(10) = -10$

8. a) $\sum_{k=1}^{6}(k^2 - \frac{1}{6}) = \sum_{k=1}^{6} k^2 - \frac{1}{6}\sum_{k=1}^{6} 1 = \frac{6(7)(2\cdot6+1)}{6} - 1 = 90.$ b) $\sum_{k=1}^{6} k(k+1) = \sum_{k=1}^{6} k^2 + \sum_{k=1}^{6} k = \frac{6(7)13}{6} + \frac{6(7)}{2} = 91 + 21 = 112$

9. a) $\sum_{k=1}^{5}(k^3 - 45) = \sum_{k=1}^{5} k^3 - \sum_{k=1}^{5} 45 = [\frac{5(5+1)}{2}]^2 - 5(45) = 0$ b) $\sum_{k=1}^{6}(\frac{k^3}{7} - \frac{k}{7}) = \frac{1}{7}) = \frac{1}{7}[\sum_{k=1}^{6} k^3 - \sum_{k=1}^{6} k] = \frac{1}{7}[(\frac{6\cdot7}{2})^2 - \frac{6\cdot7}{2}] = \frac{21}{7}[21-1] = 60$

10. a) $\sum_{k=1}^{3} 2^{k-1} = 1+2+2^2 = 7$ b) $\sum_{k=0}^{4}(-1)^k \cos k\pi = \cos 0 - \cos\pi + \cos 2\pi - \cos 2\pi + \cos 4\pi = 5$ c) $\sum_{k=-1}^{2} k(k+1) = -1(0) + 0(1) + 1(2) + 2(3) = 8$ d) $\sum_{k=1}^{4} \frac{(-1)^{k+1}}{k(k+1)} = \frac{1}{1(2)} - \frac{1}{2(3)} + \frac{1}{3(4)} - \frac{1}{4(5)} = \frac{11}{30}$

11. a) $\sum_{k=0}^{3} 2^k$ b) $\sum_{k=0}^{4} \frac{1}{3^k}$ c) $\sum_{k=1}^{5}(-1)^{k+1}k$ d) $\sum_{k=1}^{3} \frac{5}{2^k}$

12. $\int_1^2 \frac{1}{x}dx$ **13.** $\int_0^1 e^x dx$

14. $x_k = 0 + k\Delta x = \frac{4k}{n}$. $\text{RRAM}_n f = \sum_{k=1}^{n} f(x_k)\Delta x = \Delta x \sum_{k=1}^{n}((\frac{4k}{n})^2 + 2(\frac{4k}{n}) + 3) = \frac{4}{n}[\frac{16}{n^2}\sum_{k=1}^{n} k^2 + \frac{8}{n}\sum_{k=1}^{n} k + 3n] = \frac{4}{n}[\frac{16}{n^2}\frac{n(n+1)(2n+1)}{6} + \frac{8}{n}\frac{n(n+1)}{2} + 3n] = \frac{32}{3}\frac{n}{n}(\frac{n+1}{n})(\frac{2n+1}{n}) + 16\frac{n}{n}\frac{n+1}{n} + 12 = \frac{32}{3}(1+\frac{1}{n})(2+\frac{1}{n}) + 16(1+\frac{1}{n}) + 12$. $\lim_{n\to\infty} \text{RRAM}_n f = \frac{32}{3}(1)(2) + 16(1) + 12 = \frac{64}{3} + 28 = \frac{148}{3}$

15. $x_k = 0 + k\Delta x = k(b-a)/n = 5k/n$. $\text{RRAM}_n f = \sum_{k=1}^{n} f(x_k)\Delta x = \Delta x \sum_{k=1}^{n}(2x_k^3 + 3x_k) = \frac{5}{n}[2\sum_{k=1}^{n} \frac{5^3k^3}{n^3} + 3\sum_{k=1}^{n} \frac{5k}{n}] = \frac{10}{n}\frac{5^3}{n^3}[\frac{n(n+1)}{2}]^2 + 3(\frac{5}{n})^2\frac{n(n+1)}{2} = \frac{1250}{4}\frac{n^2(n+1)^2}{n^4} + \frac{75}{2}\frac{n(n+1)}{n^2} = \frac{625}{2}(\frac{n+1}{n})^2 + \frac{75}{2}\frac{n+1}{n} = \frac{625}{2}(1+\frac{1}{n})^2 + \frac{75}{2}(1+\frac{1}{n})$. $\lim_{n\to\infty} \text{RRAM}_n f = \frac{625}{2} + \frac{75}{2} = 350$

16. a) True b) $\int_{-2}^{5}(f(x)+g(x))dx = \int_{-2}^{5} f(x)dx + \int_{-2}^{5} g(x)dx = \int_{-2}^{2} f(x)dx + \int_{2}^{5} f(x)dx + 2 = 4+3+2 = 9$. b) is True c) Since $7 = \int_{-2}^{5} f(x)dx > \int_{-2}^{5} g(x)dx = 2$, c) is false because $g(x) \geqq f(x)$ on $[a,b]$ implies $\int_a^b g(x)dx \geqq \int_a^b f(x)dx$.

17. a) $\int_0^1 f(t)dt = \pi$ b) $\int_1^0 f(y)dy = -\int_0^1 f(y)dy = -\pi$ c) $\int_0^1 -3f(z)dz = -3\int_0^1 f(z)dz = -3\pi$

18. $\int_{\pi/4}^{3\pi/4}(2-\csc^2 x)dx = [2x+\cot x]_{\pi/4}^{3\pi/4} = (\frac{3\pi}{2}+\cot\frac{3\pi}{4}) - (\frac{\pi}{2}+\cot\frac{\pi}{4}) = \pi - 2$

19. Area=area of rectangle − area under curve$=\pi - \int_0^\pi \sin x\,dx = \pi - [-\cos x]_0^\pi = \pi + [-1-1] = \pi - 2 \approx 1.14159$. $\pi - \text{NINT}(\sin x, x, 0, \pi) \approx 1.14159$

20. $\int_0^1(e-e^x)dx = [ex-e^x]_0^1 = (e-e^1)-(0-e^0) = 1$

21. Total area $= \int_0^6 |4-x|dx = \int_0^4(4-x)dx + \int_4^6(x-4)dx = [4x-\frac{x^2}{2}]_0^4 + [\frac{x^2}{2}-4x]_4^6 = 8+(18-24)-(8-16) = 10$

22. $\cos x \leqq 0$ in $[-\pi,-\frac{\pi}{2}]$, $\cos x \geqq 0$ in $[-\frac{\pi}{2},\frac{\pi}{2}]$, $\cos x \leqq 0$ in $[\frac{\pi}{2},\pi]$. Hence, total area $= -\int_{-\pi}^{-\pi/2}\cos x\,dx + \int_{-\pi/2}^{\pi/2}\cos x\,dx - \int_{\pi/2}^{\pi}\cos x\,dx = -\sin x]_{-\pi}^{-\pi/2} + \sin x]_{-\pi/2}^{\pi/2} - \sin x]_{\pi/2}^{\pi} = 1+2+1 = 4$

23. $\int_{-1}^{1}(3x^2-4x+7)dx = [x^3-2x^2+7x]_{-1}^{1} = (1-2+7)-(-1-2-7) = 6+10 = 16$

24. $\int_0^1(8s^3-12s^2+5)ds = 2s^4-4s^3+5s|_0^1 = 2-4+5 = 3$

25. $\int_1^2 \frac{4}{x^2}dx = [-\frac{4}{x}]_1^2 = -[\frac{4}{2}-\frac{4}{1}] = 2$

26. $\int_1^{27} x^{-4/3}dx = -3x^{-1/3}]_1^{27} = -3(\frac{1}{3}-1) = 2$.

27. $\int_1^4 \frac{dt}{t\sqrt{t}} = \int_1^4 t^{-3/2}dt = [-2t^{-1/2}]_1^4 = -2[4^{-1/2}-1] = 1$

28. $\int_0^2 3\sqrt{4x+1}dx$. Let $u = 4x+1$, $du = 4dx$, $dx = \frac{1}{4}du$. $\int_0^2 3\sqrt{4x+1}dx = \frac{3}{4}\int_1^9 u^{1/2}du = \frac{3}{4}(\frac{2}{3})u^{3/2}]_1^9 = \frac{1}{2}(27-1) = 13$. $\text{NINT}(3\surd(4x+1), x, 0, 2) = 12.999\ldots$

29. $\int_0^\pi \sin 5\theta\,d\theta = [-\frac{\cos 5\theta}{5}]_0^\pi = -\frac{1}{5}[\cos 5\pi - \cos 0] = \frac{2}{5}$

30. $\int_0^\pi \cos 5t\,dt = \frac{\sin 5t}{5}]_0^\pi = 0-0 = 0$

31. $\int_0^{\pi/3}\sec^2\theta\,d\theta = [\tan\theta]_0^{\pi/3} = \sqrt{3}$

32. $\int_{\pi/4}^{3\pi/4}\csc^2 x\,dx = -\cot x]_{\pi/4}^{3\pi/4} = -[\cot(\frac{3\pi}{4}) - \cot(\frac{\pi}{4})] = -(-1-1) = 2$

33. $I = \int_{\pi}^{3\pi} \cot^2 \frac{x}{6} dx = \int_{\pi}^{3\pi}(\csc^2 \frac{x}{6} - 1)dx$. Let $u = \frac{x}{6}$, $du = \frac{1}{6}dx$. $I = 6\int_{\pi/6}^{\pi/2}(\csc^2 u - 1)du = -6[\cot u + u]_{\pi/6}^{\pi/2} = -6[\cot\frac{\pi}{2} + \frac{\pi}{2} - \cot\frac{\pi}{6} - \frac{\pi}{6}] = -6[0 + \frac{\pi}{2} - \sqrt{3} - \frac{\pi}{6}] = 6(\sqrt{3} - \frac{\pi}{3}) = 6\sqrt{3} - 2\pi$

34. $\int_0^{\pi} \tan^2 \frac{\theta}{3} d\theta = \int_0^{\pi}(\sec^2 \frac{\theta}{3} - 1)d\theta = 3\tan\frac{\theta}{3} - \theta]_0^{\pi} = 3\tan\frac{\pi}{3} - \pi = 3\sqrt{3} - \pi = 2.054559\ldots$. NINT$((\tan(x/3))^2, x, 0, \pi) = 2.054559\ldots$

35. Let $\int_0^1 \frac{36dx}{(2x+1)^3} = I$. Let $u = 2x + 1$, $du = 2\,dx$. $I = \frac{36}{2}\int_1^3 u^{-3}du = \frac{18}{-2}[u^{-2}]_1^3 = -9[\frac{1}{9} - 1] = -1 + 9 = 8$

36. $\int_1^2 (x + x^{-2})dx = \frac{x^2}{2} - x^{-1}]_1^2 = 2 - \frac{1}{2} - (\frac{1}{2} - 1) = 2$

37. $\int_{-\pi/3}^{0} \sec x \tan x\, dx = [\sec x]_{-\pi/3}^{0} = 1 - 2 = -1$

38. $\int_{\pi/4}^{3\pi/4} \csc x \cos x\, dx = -\csc x]_{\pi/4}^{3\pi/4} = -(\csc\frac{3\pi}{4} - \csc\frac{\pi}{4}) = -(\sqrt{2} - \sqrt{2}) = 0$. NINT$((\sin x \tan x)^{-1}, x, \frac{\pi}{4}, \frac{3\pi}{4}) = -1.9 \times 10^{-13}$.

39. $I = \int_0^{\pi/2} 5(\sin x)^{3/2} \cos x\, dx$. Let $u = \sin x$, $du = \cos x\, dx$. $I = 5\int_0^1 u^{3/2}du = 5(\frac{2}{5})[u^{5/2}]_0^1 = 2$

40. Let $u = 1 - x^2$, $du = -2x\, dx$. $\int_{-1}^{1} 2x\sin(1 - x^2)dx = -\int_0^0 \sin u\, du = 0$

41. $\int_4^8 \frac{1}{5}dt = \text{NINT}(\frac{1}{t}, 4, 8) = 0.6931\ldots$

42. $\int_0^2 \frac{2}{x+1}dx = \text{NINT}(\frac{2}{x+1}, 0, 2) = 2.1972\ldots$

43. $\int_0^2 \frac{x\,dx}{x^2+5} = \text{NINT}(\frac{x}{x^2+5}, 0, 2) = 0.2938\ldots$

44. Let $u = 3 - \sin x$, $du = -\cos x\, dx$. $\int_0^{\pi} \frac{\cos x}{3-\sin x}dx = -\int_3^3 \frac{du}{u} = 0$

45. $\int_2^3 (t - \frac{2}{t})(t + \frac{2}{t})dt = \int_2^3 (t^2 - \frac{4}{t^2})dt = \int_2^3 (t^2 - 4t^{-2})dt = \frac{t^3}{3} + 4t^{-1}]_2^3 = 9 + \frac{4}{3} - (\frac{8}{3} + 2) = 7 - \frac{4}{3} = \frac{17}{3}$. NINT$((x - 2/x)(x + 2/x), x, 2, 3) = 5.6666\ldots$

46. $\int_{-1}^{0} (1-3w)^2 dw = \int_{-1}^{0}(1-6w+9w^2)dw = w-3w^2+3w^3]_{-1}^{0} = 0-(-1-3-3) = 7$

47. $\int_{-4}^{0} |x|dx = -\int_4^0 x\, dx = -[\frac{x^2}{2}]_{-4}^{0} = -[0 - \frac{16}{2}] = 8$

48. $\int_{1/2}^{4} \frac{x^2+3x}{x}dx = \int_{1/2}^{4}(x + 3)dx = \frac{x^2}{2} + 3x]_{1/2}^{4} = 8 + 12 - (\frac{1}{8} + \frac{3}{2}) = \frac{147}{8}$

49. $I = \int_{-\pi/2}^{\pi/2} 15\sin^4 3x \cos 3x\, dx$, $u = \sin 3x$, $du = 3\cos 3x\, dx$. $I = \frac{15}{3}\int_1^{-1} u^4 du = [u^5]_1^{-1} = -2$

50. $I=\int_0^{\pi/2}\frac{3\sin x\cos x}{\sqrt{1+3\sin^2 x}}dx$, $u=1+3\sin^2 x$, $du=6\sin x\cos x\,dx$. $I=\frac{1}{2}\int_1^4 u^{-1/2}du=[u^{1/2}]_1^4=1$

51. We graph $y = \text{NINT}((\ell n(5t))/t,t,1,x)$ and $y = \frac{1}{2}(\ell n\,5x)^2$ in $[0.01,7]$ by $[-1.3,6.4]$. This suggests that the two functions differ by a constant because their graphs appear to be obtainable, one from the other, by a vertical shift. We may also graph $\frac{\ell n\,5x}{x}$ and $\text{NDER}(\frac{1}{2}(\ell n\,5x)^2,x)$ in the above rectangle and see that the graphs coincide.

52. We graph $y = \text{NINT}(t^2\ell n\,t,t,1,x)$ and $y = x^3\ell n\,x/3 - x^3/9 + 10$ in $[2,7]$ by $[-10,200]$. This suggests that the two functions differ by a constant because their graphs appear to be related by a vertical shift. We may also graph $x^2\ell n\,x$ and $\text{NDER}(\frac{x^3}{3}\ell n\,x - \frac{x^3}{9},x)$ and see that the graphs appear to coincide.

53. Graph $e^x\sin x$ and $\text{NDER}(\frac{e^x}{2}(\sin x - \cos x),x)$ in the viewing window $[-3,3]$ by $[-1,8]$ and see that the two graphs coincide.

54. Graph $\text{NINT}(te^t,t,0,x)$ and $xe^x - e^x + 5$ in $[-10,3]$ by $[-1,10]$ and observe that one graph may be obtained from the other by a vertical shift. This implies that the two functions differ by a constant.

55. Graph $\text{NINT}((\sin(3t))/t,t,1,x)$ in $[0.01,8]$ by $[-2,0.004]$. It was mentioned earlier that $\frac{\sin x}{x}$ has no explicit antiderivative in terms of elementary functions. Let $I = \int\frac{\sin 3x}{x}dx$. Let $u = 3x$, $du = 3dx$, $dx = \frac{1}{3}du$. Then $I = \frac{1}{3}\int\frac{\sin u}{u/3}du = \int\frac{\sin u}{u}du$. Thus there is no explicit antiderivative of $\frac{\sin 3x}{x}$ otherwise $\sin u/u$ and so $\sin x/x$ would have an explicit antiderivative.

56. Graph $\text{NINT}(0.25e^{-(0.5t^2)},t,0,x)$ in $[-10,10]$ by $[-1,1]$. It was mentioned earlier that e^{-x^2} has no explicit antiderivative in terms of elementary functions. Let $I = \int\frac{1}{4}e^{-x^2/2}dx$. Let $u = x/\sqrt{2}$ and $du = dx/\sqrt{2}$, $dx = \sqrt{2}du$. Then $I = \frac{\sqrt{2}}{4}\int e^{-u^2}du$. $\int e^{-u^2}du = \frac{4}{\sqrt{2}}I$. Since $\int e^{-u^2}du$ has no elementary formula, I cannot have an elementary formula.

57. Solve $I = \int_0^x(t^3 - 2t + 3)dt = 4$. $I = [\frac{t^4}{4} - t^2 + 3t]_0^x = \frac{x^4}{4} - x^2 + 3x = 4$, $x^4 - 4x^2 + 12x - 16 = 0$. We graph the left-hand side of the last equation and zoom in to its x-intercepts obtaining $x = -3.091$, 1.631. Alternatively, we may graph $y_1 = \text{NINT}(t^3 - 2t + 3,t,0,x)$ and $y_2 = 4$ and find the points of intersection.

58. $\int_1^x \frac{1}{2}e^{-t^2}\,dt = 0.05$. Multiplying by 2, we have $\int_1^x e^{-t^2}\,dt = 0.1$. We graph $y_1 = \text{NINT}(e^{-t^2}, t, 1, x)$ and $y_2 = 0.1$ in $[0, 5]$ by $[-1, 0.2]$ and then zoom in to the point of intersection: $x = 1.421$

59. We graph $y = \text{NINT}(\cos t/t, t, 1, x)$ in $[0.01, 20]$ by $[-1, 0.5]$. We see that $y < 1$ and so there is no solution.

60. Since t cannot be 0 in the integrand, the domain may be taken to be $(0, \infty)$. We may graph $y = \text{NINT}(\cos t/t, t, 1, x)$ in $[0.01, 20]$ by $[-1, 0.5]$.

61. a) and c)

62. Both integrations are correct. $\frac{\sec^2 x}{2} = \frac{1+\tan^2 x}{2} = \frac{\tan^2 x}{2} + \frac{1}{2}$ shows that the two results differ by a constant.

63. By the Fundamental Theorem of Calculus the function $F(x)$ defined by $F(x) = \int_0^x f(t)dt$, $0 \leqq x \leqq 1$, is differentiable and $F'(x) = f(x)$. On the other hand we are given $F(x) = \sin x$ and so $F'(x) = \cos x$. Therefore $f(x) = \cos x$, $0 \leqq x \leqq 1$.

64. $g'(x) = f(x)$ by the Fundamental Theorem so a) is true. b) is true because differentiability implies continuity. $g'(1) = f(1) = 0$ so c) is true. $g''(x) = f'(x) > 0$ for all x so $g''(1) > 0$ which proves e) is true and d) and f) are false. Since $g'(1) = 0$ and $g''(1) > 0$ (g' is increasing at $x = 1$), g) is true.

65. $\int_0^1 \sqrt{1 + x^4}dx = F(1) - F(0)$

66. $y' = 2x + \frac{1}{x}$, $y'' = 2 - \frac{1}{x^2}$ as required. When $x = 1$, $y = 1^2 + \int_1^1 \frac{1}{t}dt + 1 = 1 + 0 + 1 = 2$ and $y' = 2(1) + \frac{1}{1} = 3$.

67. $\frac{d^2 s}{dt^2} = \pi^2 \cos \pi t$. $\frac{ds}{dt} = \pi \sin \pi t + v_0 = \pi \sin \pi t + 8$. $s = -\cos \pi t + 8t + C$, $0 = s(0) = -1 + C$, $C = 1$. $s = -\cos \pi t + 8t + 1$. $s(1) = -\cos \pi + 8 + 1 = 10\text{m}$.

68. a) $y^{(4)} = \cos x$, $y''' = \sin x + C_1$. When $x = 0$, $y''' = 0$ so $C_1 = 0$, $y''' = \sin x$. $y'' = -\cos x + C_2$. When $x = 0$, $y'' = -\cos 0 + C_2 = 1$, $C_2 = 2$ and so $y'' = -\cos x + 2$. $y' = -\sin x + 2x + C_3$. When $x = 0$, $y' = -0 + 0 + C_3 = 2$ so $y' = -\sin x + 2x + 2$. $y = \cos x + x^2 + 2x + C_4$. When $x = 0$, $y = \cos 0 + 0 + 0 + C_4 = 3$ so $C_4 = 2$ and $y = \cos x + x^2 + 2x + 2$. b) We see that $y = \int_0^x \frac{4t}{(1+t^2)^2}dt$ is the solution (note that $y = 0$ if $x = 0$). Let $u = 1 + t^2$, $du = 2t\,dt$, $2du = 4t\,dt$. Then $y = 2\int_1^{1+x^2} u^{-2}du$ ($u = 1$ when $t = 0$ and $u = 1 + x^2$ when $t = x$). $y = -2u^{-1}]_1^{1+x^2} = -2(\frac{1}{1+x^2} - 1) = -2[\frac{1-(1+x^2)}{1+x^2}] = \frac{2x^2}{1+x^2}$.

69. $a = \frac{d^2s}{dt^2} = -k$. $v = \frac{ds}{dt} = -kt + v_0 = -kt + 44$. $s = -\frac{k}{2}t^2 + 44t + s_0 = -\frac{k}{2}t^2 + 44t$. $v(t^*) = -kt^* + 44 = 0$, $t^* = \frac{44}{k}$. $s(t^*) = -\frac{k}{2}(\frac{44}{k})^2 + 44(\frac{44}{k}) = -\frac{1}{2}\frac{44^2}{k} + \frac{44^2}{k} = \frac{1}{2}\frac{44^2}{k} = 45$, $k = \frac{44^2}{90} = 21.511\text{ft/sec}^2$

70. $\frac{dy}{dx} = 2x$, $y = x^2 + C$. When $x = 1$, $4 = y = 1^2 + C$, $C = 3$, $y = x^2 + 3$. Thus b) is the correct graph.

71. a) Because of periodicity we may use any interval of length 1 and we choose $[0,1]$. $(V^2)_{av} = (\int_0^1 V^2 dt)/1 = (V_{\max})^2 \int_0^1 \sin^2(120\pi t)dt = \frac{(V_{\max})^2}{2}\int_0^1(1 - \cos(240\pi t))dt = \frac{(V_{\max})^2}{2}[t - \frac{\sin(240\pi t)}{240\pi}] = \frac{V_{\max}^2}{2}$. $V_{rms} = \sqrt{(V^2)_{av}} = \sqrt{\frac{(V_{\max})^2}{2}} = \frac{V_{\max}}{\sqrt{2}}$ verifying (1). b) $V_{\max} = \sqrt{2}V_{rms} = \sqrt{2}(240) = 339.411$ volts.

72. We use av. value of $f(x) = \frac{1}{b-a}\int_a^b f(t)dt = \frac{1}{365-0}\int_0^{365}\{37\sin[\frac{2\pi}{365}(t-101)] + 25\}dt = \frac{1}{365}\text{NINT}(37\sin(\frac{2\pi}{365}(t-101)) + 25, t, 0, 365) = 25°$. We remark that this can be calculated without a calculator if one uses the periodicity of the cosine to cancel out terms.

73. Average value of f' on $[a,b] = \frac{1}{b-a}\int_a^b f'(x)dx = \frac{1}{b-a}[f(b) - f(a)] = \frac{f(b)-f(a)}{b-a}$.

74. a) Let $A = \frac{1}{b-a}\int_a^b f(x)dx = \frac{1}{3-0}\int_0^3 \sqrt{3x}dx$. Let $u = 3x$, $du = 3dx$, $\frac{1}{3}du = dx$. $A = \frac{1}{3}\int_0^9 u^{1/2}\frac{1}{3}du = \frac{1}{9}\int_0^9 u\, du = \frac{1}{9}\frac{2}{3}u^{3/2}]_0^9 = \frac{2}{27}9^{3/2} = 2$ b) $\frac{1}{a-0}\int_0^a \sqrt{ax}dx = \frac{1}{a}\frac{2}{3a}(ax)^{3/2}]_0^a = \frac{2}{3a^2}(a^2)^{3/2} = \frac{2a}{3}$

75. Let $f(x) = \frac{1}{x}$. Then we may find that $f^{(4)}(x) = \frac{24}{x^5} \leqq 24 = M$ on $[1,3]$. $|E_S| \leqq \frac{b-a}{180}h^4 M = \frac{3-1}{180}h^4(24) = \frac{4}{15}h^4 \leqq 10^{-4}$. $h^4 \leqq \frac{15}{4}\frac{1}{10^4}$, $h \leqq \frac{1}{10}\sqrt[4]{3.75} = 0.13915788\ldots$ But $h = \frac{b-a}{n} = \frac{2}{n} \leqq 0.13915788\ldots$ which implies $n \geq \frac{2}{0.13915788} = 14.372$. Since n must be even, we have $n \geqq 16$, $h = \frac{b-a}{n} \leqq \frac{2}{16} = \frac{1}{8}$.

76. $h = \frac{b-a}{n} = \frac{1-0}{n} = \frac{1}{n}$. $|E_T| \leqq \frac{b-a}{12}h^2 M \leqq \frac{1}{12}\frac{1}{n^2}8 = \frac{2}{3n^2} \leqq \frac{1}{1000}$. This is equivalent to $\frac{3n^2}{2} \geqq 1000$, $n \geqq \sqrt{\frac{2000}{3}} \approx 25.8$. Thus 26 subdivisions would be needed using the Trapezoidal Rule.

77. $T = ((\pi/6)/2)(2\sin^2 0 + 4\sin^2\frac{\pi}{6} + 4\sin^2\frac{\pi}{3} + 4\sin^2\frac{\pi}{2} + 4\sin^2(\frac{2\pi}{3}) + 4\sin^2(\frac{5\pi}{6}) + 2\sin^2\pi) = 3.14159265359\ldots$ $|E_T| = |\pi - T| \leq 10^{-11}$.

$S = \frac{\pi}{18}(2\sin^2 0 + 8\sin^2\frac{\pi}{6} + 4\sin^2\frac{\pi}{3} + 8\sin^2\frac{\pi}{2} + 4\sin^2(\frac{2\pi}{3}) + 8\sin^2(\frac{5\pi}{6}) + 2\sin^2\pi) = 3.14159265359\ldots$ $|E_S| < 10^{-11}$. T and S both agree with π up to the limits of calculator accuracy.

78. $h = \frac{b-a}{n} = \frac{2-1}{n} = \frac{1}{n}$. $|E_S| \leqq \frac{b-a}{180}h^4M \leqq \frac{1}{180}\frac{1}{n^4}3 = \frac{1}{60n^4} < \frac{1}{10^5}$. This is equivalent to $60n^4 > 10^5$, $n > (\frac{10^5}{60})^{1/4} \approx 6.4$. Since n must be even, we take $n = 8$.

79. By Simpson's Rule the approximate area of the lot is $\frac{15}{3}[0 + 4(36) + 2(54) + 4(51) + 2(49.5) + 4(54) + 2(64.4) + 4(67.5) + 42] = 6059\text{ft}^2$. The cost for the job is $0.10(6059) + (2.00)6059 = \$12,723.90$.

CHAPTER 6

APPLICATIONS OF DEFINITE INTEGRALS

6.1 AREAS BETWEEN CURVES

1. $A = \int_{-2}^{3}[(3x+7)-(x^2+2x+1)]dx = [3\frac{x^2}{2}+7x-\frac{x^3}{3}-x^2-x]_{-2}^{3} = \frac{125}{6}$.

2. $A = \int_0^{3.5}[(-x^2+4x+2)-(x^2-3x+2)]dx = \int_0^{3.5}(-2x^2+7x)dx = [-\frac{2x^3}{3}+\frac{7x^2}{2}]_0^{3.5} = \frac{343}{24}$.

3. $A = \int_{-3}^{2}[(2-y)-(y^2-4)]dy = [6y-\frac{y^2}{2}-\frac{y^3}{3}]_{-3}^{2} = \frac{125}{6}$.

4. $A = \int_0^1[2\sqrt{y}-y]dx = [\frac{2y^{\frac{3}{2}}}{\frac{3}{2}}-\frac{y^2}{2}]_0^1 = \frac{4}{3}-\frac{1}{2} = \frac{5}{6}$.

5. $\int_0^\pi(1-\cos^2 x)dx = \int_0^\pi \sin^2 x\ dx = \frac{1}{2}\int_0^\pi(1-\cos 2x)dx = \frac{1}{2}[x-\frac{1}{2}\sin 2x]_0^\pi = \frac{1}{2}[\pi-0]-\frac{1}{2}[0] = \frac{\pi}{2}$.

6. $\int_{\pi/4}^{5\pi/4}(\sin x-\cos x)dx = [-\cos x-\sin x]_{\pi/4}^{5\pi/4} = [-(-\frac{\sqrt{2}}{2})-(-\frac{\sqrt{2}}{2})]-[-\frac{\sqrt{2}}{2}-\frac{\sqrt{2}}{2}] = \frac{4\sqrt{2}}{2} = 2\sqrt{2}$.

7. The curves intersect at $(-2,2)$ and $(2,2)$, $\int_{-2}^{2}(2-(x^2-2))dx = \int_{-2}^{2}(4-x^2)dx = 2\int_0^2(r-x^2)dx$ by symmetry $= 2[4x-\frac{x^3}{3}]_0^2 = 2[8-\frac{8}{3}] = \frac{32}{3}$.

8. $\int_0^2(2x-x^2)dx = [x^2-\frac{x^3}{3}]_0^2 = 4-\frac{8}{3} = \frac{4}{3}$.

9.

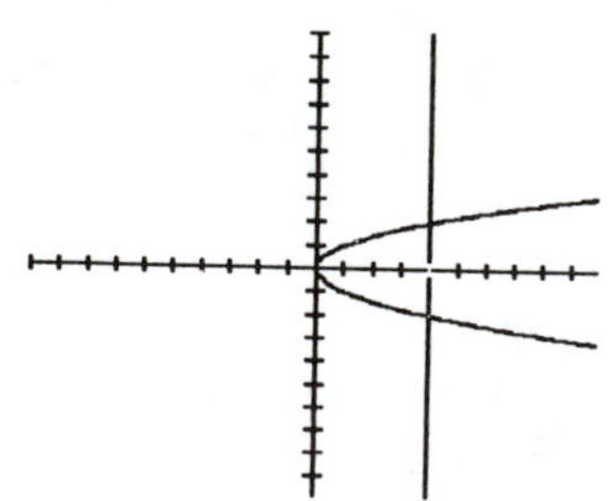

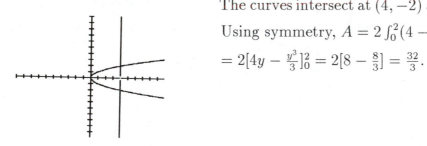

The curves intersect at $(4,-2)$ and $(4,2)$. Using symmetry, $A = 2\int_0^2(4-y^2)dy$ $= 2[4y-\frac{y^3}{3}]_0^2 = 2[8-\frac{8}{3}] = \frac{32}{3}$.

10. The curves intersect at $(-1,-3)$ and $(3,-3)$. $\int_{-1}^{3}(2x-x^2-(-3))dx = \frac{32}{3}$.

11. $\int_0^1(x-x^2)dx = \frac{1}{6}$.

12.

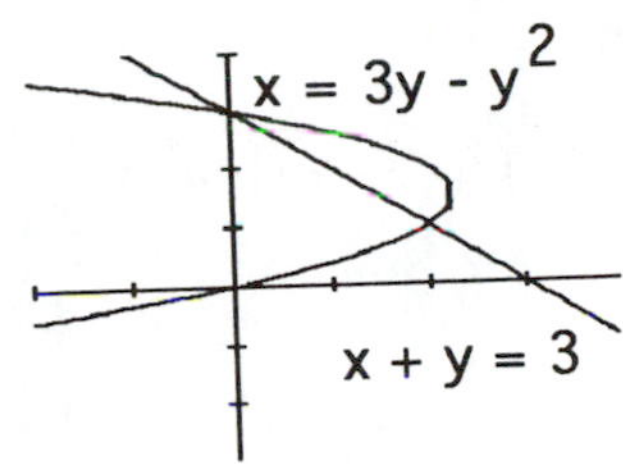

The curves intersect at $(2,1)$ and $(0,3)$. $\int_1^3 (3y - y^2 - (3 - y)dy = \frac{4}{3}$.

13.

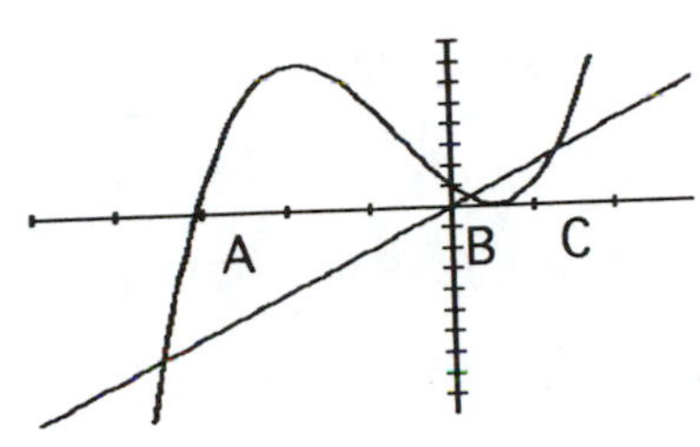

Two regions are enclosed by the curves. Use technology to find the x-coordinates: $A: -3.50700,\ B: 0.22187,\ C: 1.28514$
$\int_A^B (x^2 + 2x^2 - 3x + 1 - 2x)dx$
$+ \int_B^C (2x - (x^3 + 2x^2 - 3x + 1))dx$
$= 25.299838 + 0.853568 = 26.15341.$

14. $\int_0^{\pi/4} (3 - 2x - 2\cos 2x)dx + \int_{\pi/4}^{1.5} 3 - 2x\, dx = 3\pi/4 - \pi^2/16 - 1 + [3x - x^2]_{\pi/4}^{1.5} = 1.250.$

15. $\int_0^3 (x - (x^2 - 2x))dx = 9/2.$

16. $2\int_0^3 (10 - y^2 - 1)dy = 36.$

17. Use technology to find the right-hand value of $B = 0.41936\ldots$. Use technology to compute NINT$(e\text{^}(-x^2) - 2x, x, 0, B) = 0.22016$.

18. Use technology to find the left-hand limit of $x = -0.567143$. Use technology to find the answer of 8.661.

19. $\int_1^2 (2 - (x - 2)^2 - x)dx = 1/6$. (With the change of variable $t = x - 2$, the integral becomes $\int_{-1}^0 (-t - t^2)dt$.)

20. $\int_{-1}^1 (7 - 2x^2 - (x^2 + 4))dx = 2\int_0^1 (3 - 3x^2)dx = 6\int_0^1 (1 - x^2)dx = 4.$

21. Graph using $[-2, 5]$ by $[-10, 10]$. Use technology to find the x-coordinates of the points of intersection $A = -1.39138,\ B = 0.22713,\ C = 3.16425$. $\int_A^B (x^3 - 2x^2 - 3x + 1 - x)dx + \int_B^C (x - (x^3 - 2x^2 - 3x + 1))dx = 2.64735 + 13.03641 = 15.68376.$

22. The curves intersect when $x = 0$ and $x = \pm 1.89549 = \pm A$. Using technology, $2\text{NINT}(2 - x^2 - 2\cos 2x, x, 0, A) = 4.25108$.

23. This area can be found by both methods. $A = \int_0^1 (y^2 - y^3)dy = [\frac{y^3}{3} - \frac{y^4}{4}]_0^1 = \frac{1}{12}$ and $A = \int_0^1 (x^{1/3} - x^{1/2})dx = [\frac{x^{4/3}}{4/3} - \frac{x^{3/2}}{3/2}]_0^1 = \frac{3}{4} - \frac{2}{3} = \frac{1}{12}$.

24. $2\int_0^1 (4 - 4x^2 - (x^4 - 1))dx = \frac{104}{15}$.

25. $\int_0^1 12(y^2 - y^3)dy = 1$.

26. $2\int_0^2 (2x^2 - (x^4 - 2x^2))dx = \frac{128}{15}$.

27. The curves meet at $x = \pi/4$. $\int_0^{\pi/4} (\cos x - \sin x)dx = [\sin x + \cos x]_0^{\pi/4} = [\frac{2\sqrt{2}}{2}] - [1] = \sqrt{2} - 1$.

28.

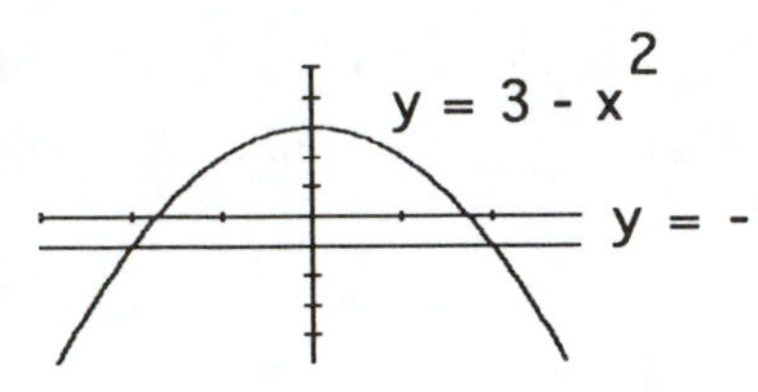

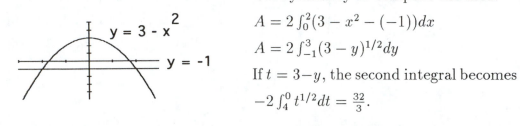

Use symmetry to compute the area.

$A = 2\int_0^2 (3 - x^2 - (-1))dx$

$A = 2\int_{-1}^3 (3 - y)^{1/2}dy$

If $t = 3 - y$, the second integral becomes $-2\int_4^0 t^{1/2}dt = \frac{32}{3}$.

29.

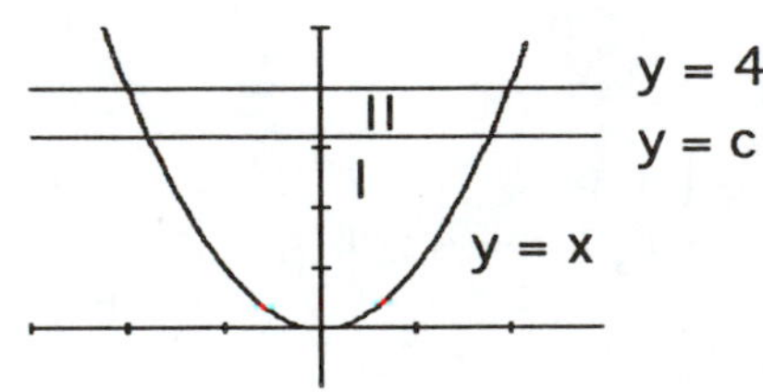

By symmetry, regions I and II will have equal areas

a) $\int_0^c y^{1/2}dy = \int_c^4 y^{1/2}dy$

b) $\int_0^{\sqrt{c}} (c - x^2)dx = (4 - c)\sqrt{c} + \int_{\sqrt{c}}^2 (4 - x^2)dx$

Integrating equation a) leads to $c^{3/2} = 4^{3/2} - c^{3/2}$ or $c = 2^{4/3}$.

30. The ratio is $\frac{2 \cdot \frac{1}{2} \cdot a \cdot a^2}{2\int_0^a (a^2 - x^2)dx} = \frac{a^3}{2(a^3 - \frac{a^3}{3})} = \frac{3}{4}$.

31. a) $A(t) = \int_0^{t/\sqrt{2}} (2x^2 - x^2)dx + \int_{t/\sqrt{2}}^t (t^2 - x^2)dx = \frac{x^3}{3}]_0^{t/\sqrt{2}} + [t^2x - \frac{x^3}{3}]_{t/\sqrt{2}}^t = \frac{2}{3}t^3(1 - \frac{1}{\sqrt{2}})$. b) $A(t) = \int_0^{t^2} (\sqrt{y} - \sqrt{\frac{y}{2}})dy = (1 - \frac{1}{\sqrt{2}})\int_0^{t^2} y^{1/2}dy = (1 - \frac{1}{\sqrt{2}})\frac{2}{3} \cdot t^3$. c) $R(t) = t \cdot t^2 \Rightarrow A(t)/R(t) = (1 - \frac{1}{\sqrt{2}})\frac{2}{3} = 0.195\ldots$.

32. a) $A(t) = \int_0^t (2x^2 - x^2)dx = \frac{t^3}{3}$ b) $A(t) = \int_0^{t^2} (\sqrt{y} - \sqrt{\frac{y}{2}})dy + \int_{t^2}^{2t^2} (t - \sqrt{\frac{y}{2}})dy = (1 - \frac{1}{\sqrt{2}})\frac{2}{3}t^3 + [ty - \frac{1}{\sqrt{2}}\frac{2}{3}y^{3/2}]_{t^2}^{2t^2} = (1 - \frac{1}{\sqrt{2}})\frac{2}{3}t^3 + [2t^3 - \frac{2}{3\sqrt{2}} \cdot 2\sqrt{2}t^3] - [t^3 - \frac{2}{3\sqrt{2}}t^3] = \frac{t^3}{3}$ c) $R(t) = 2t^3 \Rightarrow A(t)/R(t) = \frac{1}{6}$

33. We are given: $\int_a^b f(x)dx = 4$, therefore $\int_a^b (2f(x) - f(x))dx = \int_a^b f(x)dx = 4$

34. A region is bounded if it is entirely contained in a rectangle $|x| < M$, $|y| < N$. The graph of $y = e^{-x^2}$ approaches the x-axis asymptotically, but for all <u>finite</u> values of x, lies above it. Hence no M can be found.

6.2 VOLUMES OF SOLIDS OF REVOLUTION - DISKS AND WASHERS

1. $V = \pi \int_0^2 (2 - x)^2 dx = \pi \int_0^2 (x - 2)^2 dx = \pi[\frac{(x-2)^3}{3}]_0^2 = \pi[0 - \frac{(-2)^3}{3}] = \frac{8\pi}{3}$

2. $V = \pi \int_0^2 (x^2)^2 dx = \pi \frac{x^5}{5}]_0^2 = 32\pi/5$

3. $V = 2\pi \int_0^3 (\sqrt{9 - x^2})^2 dx = 2\pi \int_0^3 (9 - x^2)dx = 2\pi[9x - \frac{x^3}{3}]_0^3 = 36\pi$

4. $V = \pi \int_0^1 (x - x^2)^2 dx = \pi \int_0^1 (x^2 - 2x^3 + x^4)dx = \pi/30$

5. $V = \pi \int_0^2 (x^3)^2 dx = \pi[\frac{x^7}{7}]_0^2 = 128\pi/7$

6. $V = \pi \int_0^{\ln 2} (e^x)^2 dx = \frac{\pi}{2}[e^{2x}]_0^{\ln 2} = \frac{\pi}{2}[4 - 1] = 3\pi/2$

7. $V = \pi \int_0^{\pi/2} (\sqrt{\cos x})^2 dx = \pi[\sin x]_0^{\pi/2} = \pi$

8. $V = \pi \int_{-\pi/4}^{\pi/4} (\sec x)^2 dx = \pi[\tan x]_{-\pi/4}^{\pi/4} = 2\pi$

9. $V = \pi \int_0^2 (2y)^2 dy = 4\pi[\frac{y^3}{3}]_0^2 = 32\pi/3$

10. $V = \pi \int_0^4 (\sqrt{4 - y})^2 dy = \pi \int_0^4 (4 - y)dy = 8\pi$

11.

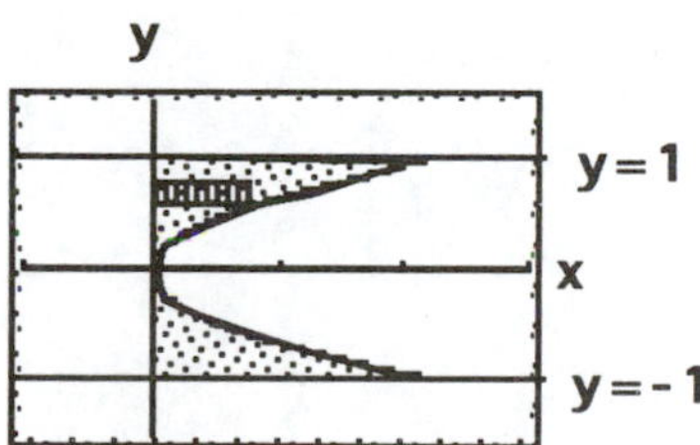

Using symmetry,
$V = 2\pi \int_0^1 (\sqrt{5}y^2)^2 dy$
$= 10\pi[\frac{y^5}{5}]_0^1 = 2\pi$

12. $V = 2\pi \int_0^1 (1-y^2)^2 dy = 16\pi/15$

13. $V = \pi \int_0^2 (y^{3/2})^2 dy = \frac{\pi \cdot 2^4}{4} = 4\pi$

14. $V = \pi \int_0^{\pi/2} (\sqrt{2 \sin 2y})^2 dy = \pi \int_0^{\pi/2} 2 \sin 2y dy = \pi[-\cos 2y]_0^{\pi/2} = 2\pi$

15. $V = \pi \int_0^3 [\frac{2}{\sqrt{y+1}}]^2 dy = 4\pi \int_0^3 \frac{1}{y+1} dy = 4\pi[\ln(y+1)]_0^3 = 4\pi \ \ln 4$

16. $V = \pi \int_0^1 [\frac{2}{y+1}]^2 dy = 4\pi[\frac{-1}{y+1}]_0^1 = 2\pi$

17. $V = \pi \int_0^1 (1^2 - x^2) dx = 2\pi/3$

18. $V = \pi \int_0^1 ((2x)^2 - x^2) dx = \pi \int_0^1 3x^2 dx = \pi[x^3]_0^1 = \pi$

19. $V = \pi \int_0^2 (4^2 - (x^2)^2) dx = 128\pi/5$

20. $V = 2\pi \int_0^1 (4^2 - (x^2+3)^2) dx = 48\pi/5$

21. $V = \pi \int_{-1}^2 ((x+3)^2 - (x^2+1)^2) dx = \pi \int_{-1}^2 (x^2 + 6x + 9 - x^4 - 2x^2 - 1) dx =$
$\pi \int_{-1}^2 (8 + 6x - x^2 - x^4) dx = \pi[8x + 3x^2 - x^3/3 - x^5/5]_{-1}^2 = \frac{117\pi}{5}$

22. $V = \pi \int_{-1}^2 ((4-x^2)^2 - (2-x)^2) dx = 108\pi/5$

23. $V = 2\pi \int_0^{\pi/4} ((\sqrt{2})^2 - \sec^2 x) dx = 2\pi[2x - \tan x]_0^{\pi/4} = 2\pi(\frac{\pi}{2} - 1) = \pi^2 - 2\pi$

24. $V = \pi \int_1^4 (2^2 - (2/\sqrt{x})^2) dx = 4\pi \int_1^4 (1 - \frac{1}{x}) dx = 4\pi[x - \ln x]_1^4 = 4\pi(3 - \ln 4)$

25. $V = \pi \int_0^1 ((y+1)^2 - 1^2) dy = 4\pi/3$

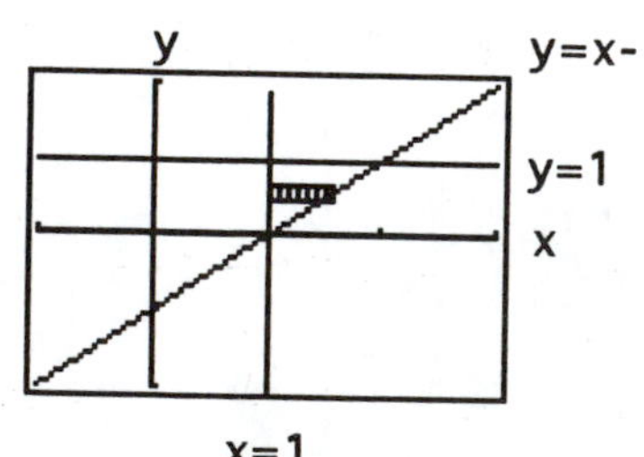

26. 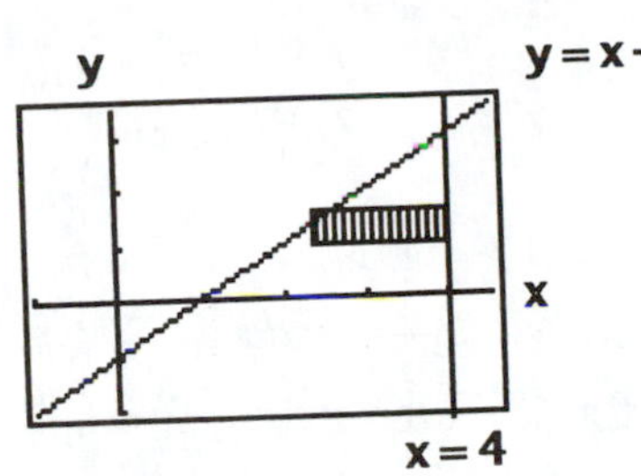

$V = \pi \int_0^3 [4^2 - (y+1)^2]dy = 27\pi$

27. $V = \pi \int_0^4 (2^2 - (\sqrt{y})^2)dy = \pi[4y - \frac{y^2}{2}]_0^4 = 8\pi$

28. $V = \pi \int_0^1 (y^2 - (y^2)^2)dy = 2\pi/15$

29. $V = 2\pi \int_0^5 (\sqrt{25 - y^2})^2 dy = 2 \cdot 250\pi/3 = 500\pi/3$

30. $V = 2\pi \int_0^3 ((\sqrt{25 - y^2})^2 - 4^2)dy = 36\pi$

31. $V = 2\pi \int_0^{\pi/2} (1^2 - (\sqrt{\cos x})^2)dx = 2x[x - \sin x]_0^{\pi/2} = 2\pi(\frac{\pi}{2} - 1) = \pi^2 - 2\pi$

32. $V = \pi \int_0^{2 \ln 2} (2^2 - (e^{x/2})^2)dx = \pi \int_0^{2 \ln 2} (4 - e^x)dx = \pi[4x - e^x]_0^{2 \ln 2} = \pi(8 \ln 2 - 4 + 1) = \pi(8 \ln 2 - 3)$

33. $V = \pi \int_0^1 (1^2 - (\sqrt{1 - y^2})^2)dy = \pi/3$

34. a) $V = \pi \int_1^2 ((4/x^2)^2 - 1^2)dx = 11\pi/3$ b) $V = \pi \int_1^4 ((\sqrt{\frac{4}{y}})^2 - 1^2)dy = \pi[4 \ln y - y]_1^4 = \pi(4 \ln 4 - 3)$

35. a) $V = \pi \int_0^4 (2^2 - (\sqrt{x})^2)dx = 8\pi$ b) $V = \pi \int_0^2 (y^2)^2 dy = 32\pi/5$ c) $V = \pi \int_0^4 (2 - \sqrt{x})^2 dx = 8\pi/3$ d) $V = \pi \int_0^2 (4^2 - (4 - y^2)^2)dy = 224\pi/15$

36. a) $V = \pi \int_0^2 (1 - \frac{y}{2})^2 dy = 2\pi/3$ b) $V = \pi \int_0^2 ((2 - \frac{y}{2})^2 - 1^2)dy = \pi[3y - y^2 + \frac{y^3}{12}]_0^2 = 8\pi/3$

37. a) $V = 2\pi \int_0^1 (1 - x^2)^2 dx = 2\pi \int_0^1 (1 - 2x^2 + x^4)dx = 2\pi[x - \frac{2x^3}{3} + \frac{x^5}{5}]_0^1 = 16\pi/15$ b) $V = 2\pi \int_0^1 ((2 - x^2)^2 - 1^2)dx = 2\pi \int_0^1 (3 - 4x^2 + x^4)dx = 2\pi[3x - \frac{4x^3}{3} + \frac{x^5}{5}]_0^1 = 56\pi/15$ c) $V = 2\pi \int_0^1 (2^2 - (x^2 + 1)^2)dx = 2\pi \int_0^1 (3 - x^4 - 2x^2)dx = 2\pi[3x - \frac{x^5}{5} - \frac{2x^3}{3}]_0^1 = 64\pi/15$

38. $V = \pi \int_0^h (r)^2 dx = \pi r^2 h$

39. The line has equation $y = -\frac{h}{r}x + h$, or $x = -\frac{r}{h}y + r$. $V = \pi \int_0^h = (-\frac{r}{h}y + r)^2 dy = \pi \int_0^h (\frac{r^2}{h^2}y^2 - \frac{2r^2}{h}y + r^2)dy = \pi[\frac{r^2}{h^2}\frac{y^3}{3} - \frac{r^2}{h}y^2 + r^2 y]_0^h = \pi r^2(\frac{h}{3} - h + h) = \frac{\pi r^2 h}{3}$

40. $V(c) = 2\pi \int_0^{\sin^{-1} c}(c - \sin x)^2 dx + 2\pi \int_{\sin^{-1} c}^{\pi/2}(\sin x - c)^2 dx = 2\pi \int_0^{\pi/2}(\sin x - c)^2 dx = 2\pi \int_0^{\pi/2}(\sin^2 x - 2c \sin x + c^2)dx = 2\pi \int_0^{\pi/2} \sin^2 x\, dx - 4\pi c \int_0^{\pi/2} \sin x\, dx + 2\pi c^2 \int_0^{\pi/2} dx$; $\frac{d}{dc}V(c) = -4\pi \int_0^{\pi/2} \sin x\, dx + 4\pi c \int_0^{\pi/2} dx = -4\pi + 4\pi \cdot c \cdot \pi/2$ Setting $V'(c) = 0$ gives $c = 2/\pi$. Since $V''(c) = 4\pi \int_0^{\pi/2} dx > 0$, this will be a minimum value of V. $V(2/\pi) = 2\pi \int_0^{\pi/2} \sin^2 x\, dx - 8[-\cos x]_0^{\pi/2} + \frac{8}{\pi} \int_0^{\pi/2} dx = 2\pi \int_0^{\pi/2} \sin^2 x\, dx - 8 + 4 = 2\pi \cdot \pi/4 - 4 \approx 0.935$

41. a) $x = -2$, $x = 0.59375\ldots$ Store the value of x in A.

b) $V = \pi \int_{-2}^{0.59375\ldots}[5\cos(0.5x + 1)]^2 - [x^2 + 1]^2 dx = \pi$ NINT$((5\cos(0.5x + 1))^2 - (x^2 + 1)^2, x, -2, A) = 76.815307\ldots$

42. Following the approach of problem 40. $V(c) = \pi \int_0^{\pi}(\sin(x\sqrt{x^2 + 3}) - c)^2 dx$; $V'(c) = 0$ when $2\pi \int_0^{\pi} \sin(x\sqrt{x^2 + 3})dx = 2c\pi \int_0^{\pi} dx$;

$c = \frac{1}{\pi} \int_0^{\pi} \sin(x\sqrt{x^2 + 3})dx = 0.1761$. $V(c) \approx 4.6562$.

43. Revolve the area bounded by $y = -7$, $x = 0$, $x^2 + y^2 = 16^2$ about the y-axis $V = \pi \int_{-16}^{-7}(16^2 - y^2)dy = 1053\pi \approx 3.3L$

44. $V = \pi \int_0^6 [\frac{x}{12}\sqrt{36 - x^2}]^2 dx = \frac{\pi}{144} \int_0^6 x^2(36 - x^2)dx = 36\pi/5$, about 192.27gm

6.3 CYLINDRICAL SHELLS – AN ALTERNATIVE TO WASHERS

1. $V = 2\pi \int_0^2$ (radius)(height) $dx = 2\pi \int_0^2 x(x - (-x/2))dx = 8\pi$.

2. $V = 2\pi \int_0^4 x\sqrt{x}\, dx = 2\pi \int_0^4 x^{3/2} dx = 128\pi/5$.

3. $V = 2\pi \int_0^1$ (radius)(height) $dx = 2\pi \int_0^1 x(x^2 + 1)dx = 3\pi/2$.

4. $V = 2\pi \int_0^1 x(\sqrt{x} - (2x - 1))dx = 7\pi/15$.

5.

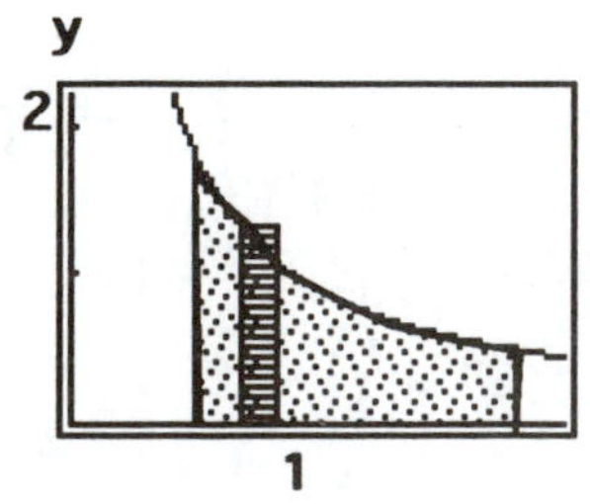

$V = 2\pi \int_{1/2}^2 x(\frac{1}{x})dx = 2\pi \cdot \frac{3}{2} = 3\pi$.

6. $V = 2\pi \int_{1/2}^{2} x(\frac{1}{x^2})dx = 2\pi[\ln x]_{1/2}^{2} = 2\pi[\ln 2 - \ln \frac{1}{2}] = 4\pi \ln 2.$

7.

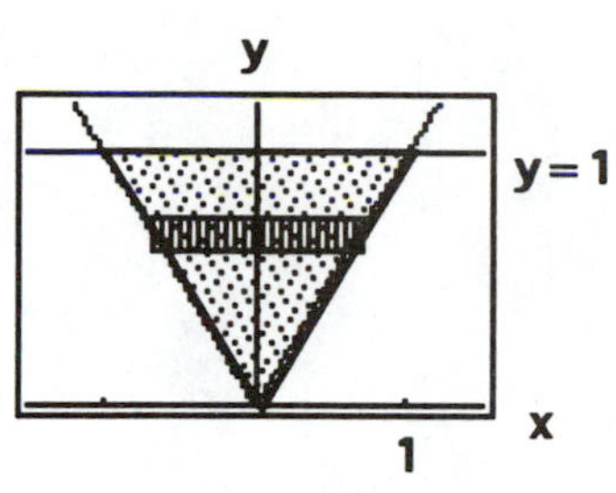

Using symmetry,

$V = 2(2\pi) \int_0^1 y(y)dy = 4\pi/3.$

8.

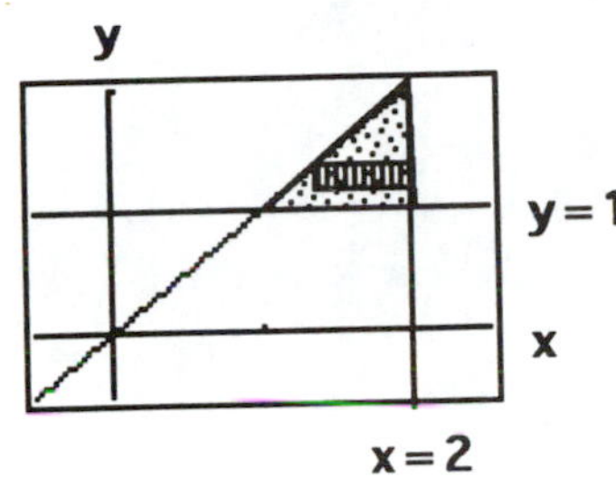

$V = 2\pi \int_0^2 y(2-y)dy = 4\pi/3.$

9.

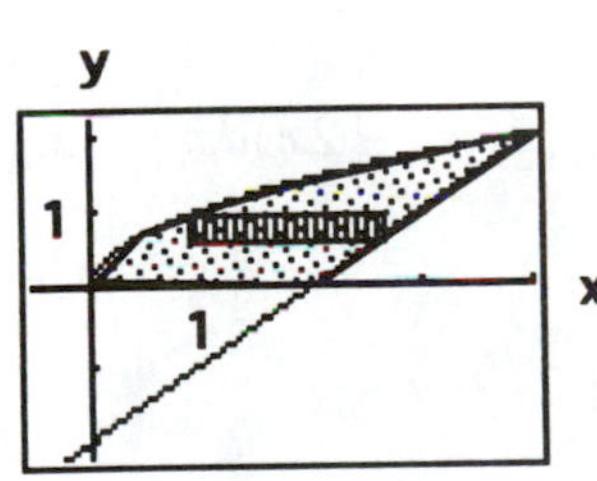

$V = 2\pi \int_0^2 y(y+2-y^2)dy = 16\pi/3.$

10.

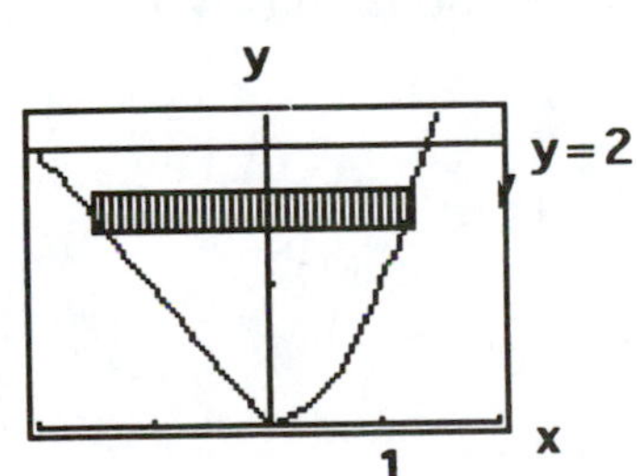

$V = 2\pi \int_0^2 y(\sqrt{y}+y)dy =$

$2\pi[y^{5/2} \cdot \frac{2}{5} + \frac{y^3}{3}]_0^2 =$

$2\pi(8\sqrt{2}/5 + 8/3) = 16\pi(3\sqrt{2}+5)/15.$

11. $V = 2\pi \int_0^2 y(2y - y^2)dy = 8\pi/3.$

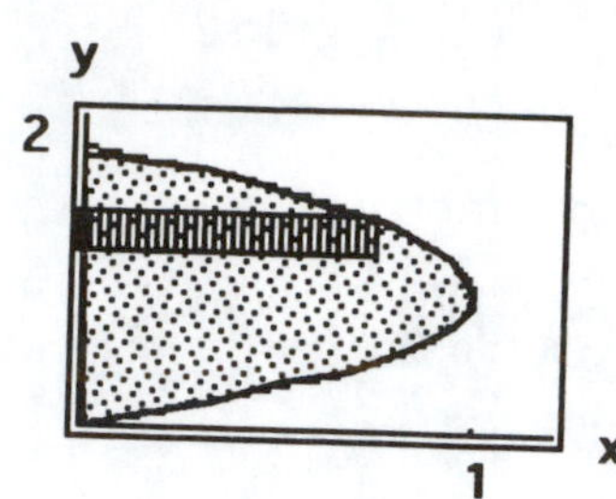

12. $V = 2\pi \int_0^1 y(2y - y^2 - y)dy = \pi/6.$

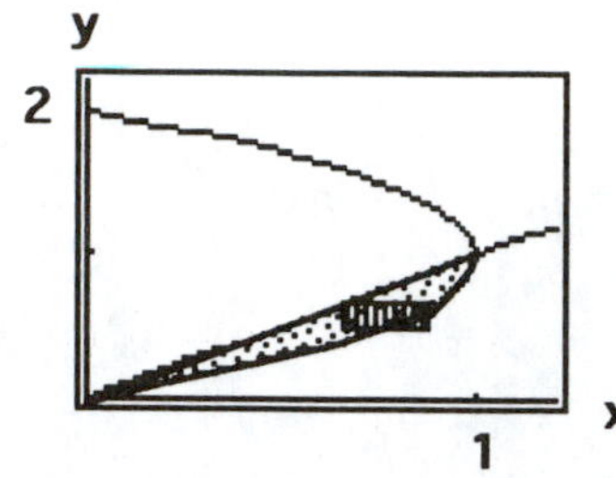

13. Use shells. $V = 2\pi \int_0^{\sqrt{3}} x\sqrt{x^2+1}dx$. Let $u = x^2 + 1$, then $du = 2xdx$. $V = \pi \int_1^4 u^{1/2}du = \frac{\pi u^{3/2}}{3/2}\Big|_1^4 = \frac{2\pi}{3}[8 - 1] = 14\pi/3$.

14. Use shells. $V = \pi \int_{\ln 2}^{2\ln 2} x[\frac{e^x}{x} - 2]dx = 2\pi \int_{\ln 2}^{\ln 4}(e^x - 2x)dx = 2\pi(2 + (\ln 2)^2 - (\ln 4)^2) \approx 3.510$.

15. Use shells. $V = 2\pi \int_0^1 y(12(y^2 - y^3))dy = 6\pi/5$.

16. Use shells. $V = 2\pi \int_0^2 y(\frac{y^2}{2} - (\frac{y^4}{4} - \frac{y^2}{2}))dy = 8\pi/3$.

17.

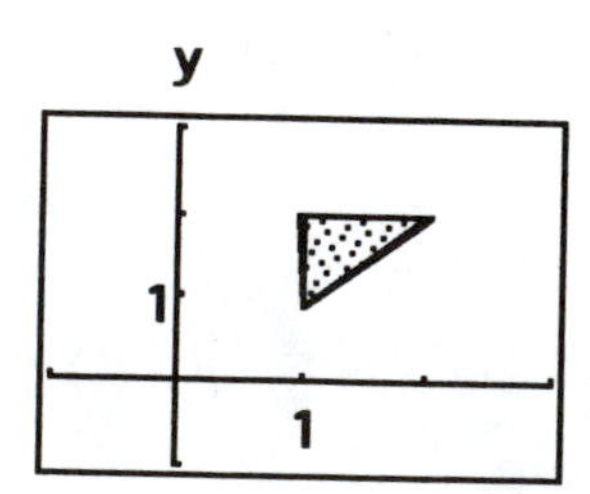

The region is bounded by $y = x$, $x = 1$ and $y = 2$

a) Use shells.

$V = 2\pi \int_1^2 y(y-1)dy = 2\pi[\frac{y^3}{3} - \frac{y^2}{2}]_1^2 = \frac{5\pi}{3}$.

b) Use shells.

$V = 2\pi \int_1^2 x(2-x)dx = 4\pi/3$.

18. 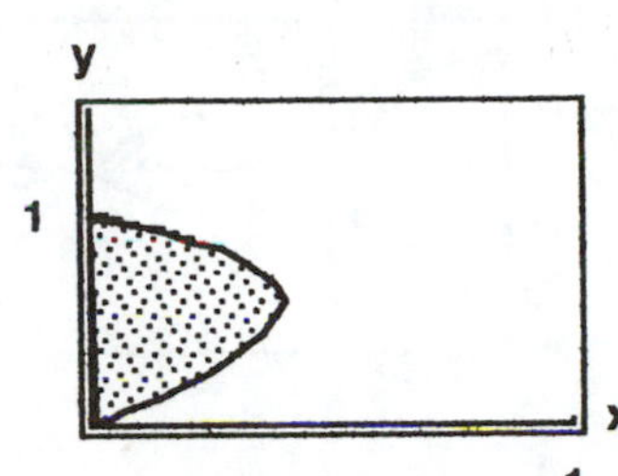

a) Use shells.

$V = 2\pi \int_0^1 y(y - y^3)dy = 4\pi/15.$

b) Use washers.

$V = \pi \int_0^1 (y - y^3)^2 dy = 8\pi/105.$

19. 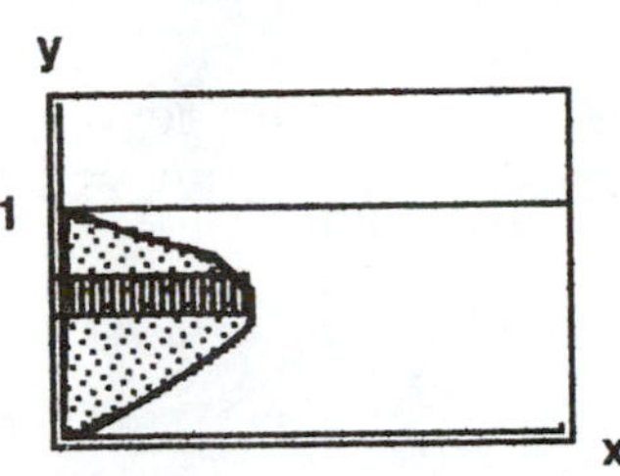

a) Use shells.

$V = 2\pi \int_0^1 y(1 - (y - y^3))dy = 11\pi/15.$

b) Use washers.

$V = \pi \int_0^1 (1^2 - (y - y^3)^2)dy = 97\pi/105.$

c) Use washers.

$V = \pi \int_0^1 (1 - (y - y^3))^2 dy = 121\pi/210.$

d) Use shells. $V = 2\pi \int_0^1 (1 - y)(1 - (y - y^3))dy = 23\pi/30.$

20. a) Use washers. $V = \pi \int_0^4 ((\frac{x+4}{2})^2 - x^2)dx = 16\pi.$

b) Use shells. $V = 2\pi \int_0^4 x(\frac{x+4}{2} - x)dx = 32\pi/3.$

c) Use shells. $V = 2\pi \int_0^4 (4 - x)(\frac{x+4}{2} - x)dx = 64\pi/3.$

d) Use washers. $V = \pi \int_0^4 ((8 - x)^2 - (8 - \frac{x+4}{2})^2)dx = 48\pi.$

21. The curves intersect at (2,8). a) Using shells, $V = 2\pi \int_0^8 y(y^{1/3} - \frac{y}{4})dy$. Using washers, $V = \pi \int_0^2 ((4x)^2 - (x^3)^2)dx = 512\pi/21.$

b) Using washers, $V = \pi \int_0^2 ((8 - x^3)^2 - (8 - 4x)^2)dx = 832\pi/21.$

22. a) Using shells, $V = 2\pi \int_0^2 y(\sqrt{8y} - y^2)dy$. Using washers, $V = \pi \int_0^4 ((\sqrt{x})^2 - (x^2/8)^2)dx = 24\pi/5.$

b) Using shells, $V = 2\pi \int_0^4 x(\sqrt{x} - x^2/8)dx = 48\pi/5.$

23. a) Use shells. $V = 2\pi \int_0^1 x(2x - x^2 - x)dx = \pi/6.$

b) Use shells. $V = 2\pi \int_0^1 (1 - x)(2x - x^2 - x)dx = \pi/6.$

24. a) Use shells. $V = 2\pi \int_0^2 y(y^2)dy = 8\pi.$

b) Use shells. $V = 2\pi \int_0^4 x(2 - \sqrt{x})dx = 32\pi/5.$

c) Use shells. $V = 2\pi \int_0^4 (4 - x)(2 - \sqrt{x})dx = 224\pi/15.$

d) Use shells. $V = 2\pi \int_0^2 (2 - y)y\ dy = 8\pi/3.$

25. Following the approach of Exploration 2,

a) The solid is generated by revolving the regions OBD, ODCE, and ECA about the x-axis

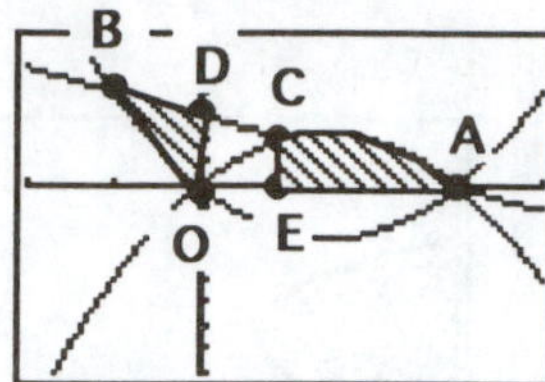

$V = \pi \int_{-1}^{0} [(-x+3)^2 - (x^2-3x)^2]dx + \pi \int_0^1 (-x+3)^2 dx + \pi \int_1^3 (3x-x^2)^2 dx = \pi[7.6333 + 6.3333 + 6.4] = 20.367\pi.$

b) 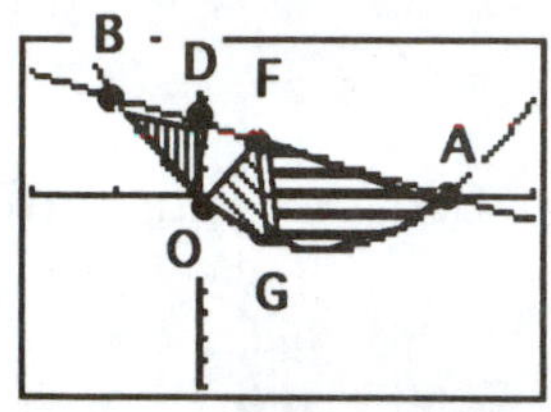

The solid is generated by revolving the regions OBD, OFG, and FGA about the y-axis

$V = 2\pi \int_{-1}^{0} -x(-x+3-(x^2-3x))dx + 2\pi \int_0^{0.6458} x(x^2+3x-(x^2-3x))dx + 2\pi \int_{0.6458}^{3} x(-x+3-(x^2-3x))dx = 2\pi[0.58333 + 0.53867 + 10.4883] = 2\pi(11.610)$

26. a) $V = \pi \int_{-1}^{1} (-x^2+x+2)^2 dx$
$+\pi \int_1^2 (-x-1)^2 dx$
$+\pi \int_2^3 ((-x-1)^2 - (-x^2+x+2)^2)dx$
$= \pi[6.4 + 6.3333 + 7.633] = 20.367\pi.$

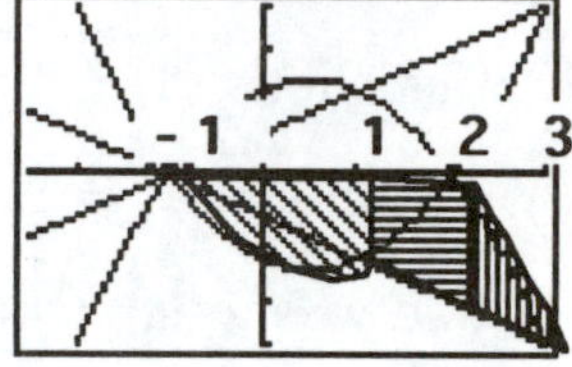

b) The solid is generated by the region to the right of the y-axis. Using shells
$V = 2\pi \int_0^3 x(-x^2+x+2-(-x-1))dx$
$= 2\pi \int_0^3 x(-x^2+2x+3)dx = 2\pi(11.25).$

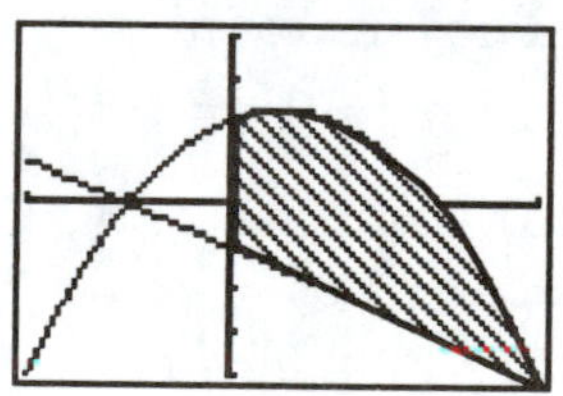
[-2, 3] by [-4, 4]

27. $V = 2 \cdot 2\pi \int_1^3 x\sqrt{9-x^2}dx$
$= \frac{4\pi \cdot 16\sqrt{2}}{3} = 94.782.$

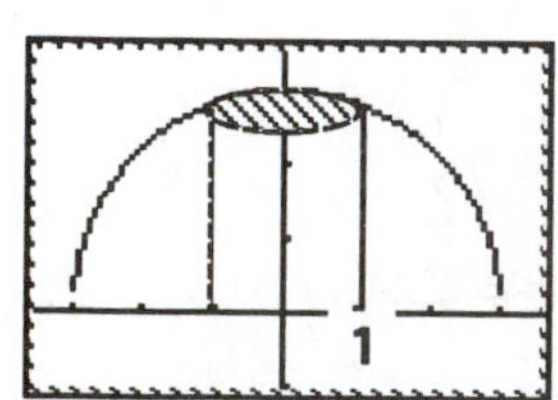

28. $V = 2 \cdot 2\pi \int_{1/2}^{2} x \cdot 3\sqrt{1-(x^2/4)}dx$
$= 45.627$ using NINT.

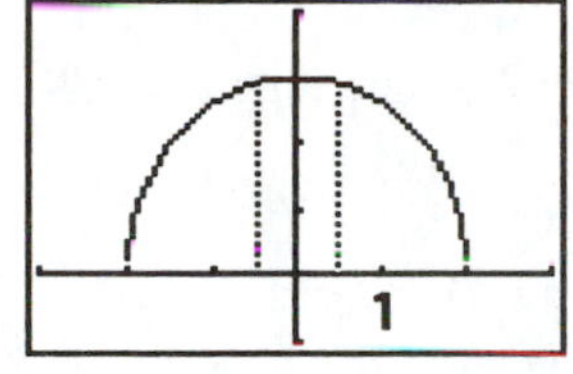

29. Use shells. Graph on $[0,3]$ by $[-1,2]$.

a) $V = 2\pi \int_0^b$ radius $\cdot$ height $dx = 2\pi \int_0^b (2-x)(1-(1-\frac{\sin 4x}{4x}))dx = 2\pi \int_0^b (2-x)\frac{(\sin(4x))}{4x}dx$.

b) The curves intersect when $1 = \frac{\sin 4x}{4x} = 1$, i.e., at $x = \pi/4$.

c) $V = 2\pi \int_0^{\pi/4} \frac{(2-x)\sin(4x)}{4x}dx = 5.033$, using NINT. Because the integrand is not defined at $x = 0$, it may be necessary to use a lower limit of 0.000001.

30. Not always true: using washers, $V = \pi \int_a^b (r_{\text{outer}}^2 - r_{\text{inner}}^2)dx$; If $f > g > 0$ on (a,b), $V = \pi \int_a^b (f^2 - g^2)dx \neq \pi \int_a^b (f-g)^2 dx$.

b) Always true: using shells, $V =$
$2\pi \int_b^c$radius $\cdot$ height $dx = 2\pi \int_b^c (x-a)(g(x)-f(x))dx$.

c) Not always true; consider this possibility:

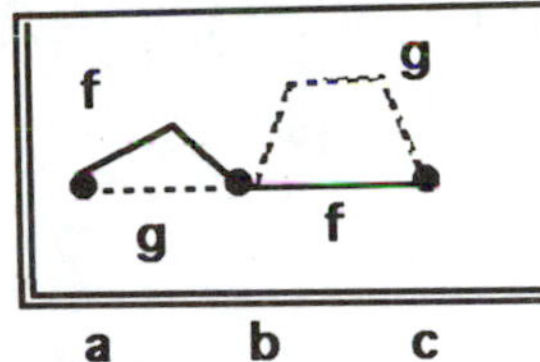

6.4 LENGTHS OF CURVES IN THE PLANE

1. $\frac{dy}{dx} = \frac{1}{3} \cdot \frac{3}{2}(x^2+2)^{1/2}2x = x(x^2+2)^{1/2}$; $L = \int_0^3 \sqrt{1+x^2(x^2+2)}\ dx = \int_0^3 \sqrt{x^4+2x^2+1}\ dx = \int_0^3 (x^2+1)dx = [\frac{x^3}{3}+x]_0^3 = 12$.

2. $\frac{dy}{dx} = \frac{3}{2}x^{1/2}$; $L = \int_0^4 \sqrt{1+\frac{9}{4}x}\ dx = \frac{8}{27}(10^{3/2}-1)$.

3. $y = (\frac{9x^2}{4})^{1/3}$ is not differentiable at $x = 0$. Use $x = \frac{2}{3}y^{3/2}$. Then $\frac{dx}{dy} = y^{1/2}$ and $L = \int_0^3 \sqrt{1+y}\ dy = \frac{2}{3}[(1+y)^{3/2}]_0^3 = \frac{2}{3}[8-1] = \frac{14}{3}$.

4. The endpoints are $(0,0)$ and $(4, 4^{2/3})$. Rewrite to find x as a function of y: $x = y^{3/2}$; $x' = \frac{3}{2}y^{1/2}$. $L = \int_0^{4^{2/3}} \sqrt{1+\frac{9}{4}\cdot y}\ dy = \frac{4}{9}\cdot\frac{2}{3}[(1+\frac{9}{4}y)^{3/2}]_0^{4^{2/3}} = \frac{8}{27}[(1+9\cdot 4^{-1/3})^{3/2}-1] = 4.807$.

5. $1+(\frac{dy}{dx})^2 = 1+(x^2-\frac{1}{4x^2})^2 = (x^2+\frac{1}{4x^2})^2$; $L = \int_1^3 (x^2+\frac{1}{4x^2})dx = [x^3/3 - 1/(4x)]_1^3 = \frac{53}{6}$.

6. $1+(\frac{dy}{dx})^2 = 1+\frac{1}{4}(x^{1/2}-x^{-1/2})^2 = \frac{1}{4}(x^{1/2}+x^{-1/2})^2$; $L = \frac{1}{2}\int_1^9 (x^{1/2}+x^{-1/2})dx = \frac{32}{3}$.

7. $1+(\frac{dx}{dy})^2 = 1+(y^3-\frac{1}{4y^3})^2 = (y^3+\frac{1}{4y^3})^2$; $L = \int_1^2(y^3+\frac{1}{4}y^{-3})dy = [\frac{y^4}{4}-\frac{1}{8}y^{-2}]_1^2 = \frac{123}{32}$.

8. $1 + (\frac{dx}{dy})^2 = 1 + \frac{1}{4}(y^2 - y^{-2})^2 = \frac{1}{4}(y^2 + y^{-2})^2$; $L = \frac{1}{2}\int_2^3(y^2 + y^{-2})dy = \frac{1}{2}[y^3/3 - 1/y]_2^3 = \frac{13}{4}$.

9. $1 + (\frac{dy}{dx})^2 = 1 + [\frac{e^x-e^{-x}}{2}]^2 = \frac{1}{4}(e^x + e^{-x})^2$; $L = \frac{1}{2}\int_{-\ln 2}^{\ln 2}(e^x + e^{-x})dx = \frac{1}{2}[e^x - e^{-x}]_{-\ln 2}^{\ln 2} = \frac{1}{2}[(2 - \frac{1}{2}) - (\frac{1}{2} - 2)] = \frac{3}{2}$.

10. $1 + (\frac{dx}{dy})^2 = 1 + (2y - \frac{1}{8y})^2 = (2y + 1/(8y))^2$; $L = \int_1^3(2y + y^{-1}/8)dy = [y^2 + \frac{1}{8}\ln y]_1^3 = 8 + \frac{1}{8}\ln 3$.

11. $1+(\frac{dy}{dx})^2 = 1+\sec^2 x\tan^2 x$; $L = \int_{-\pi/3}^{\pi/3}\sqrt{1 + \sec^2 x\tan^2 x}\, dx = 3.1385$, using NINT.

12. $1 + (\frac{dy}{dx})^2 = 1 + \cos^2 x$; $L = \int_0^{2\pi}\sqrt{1 + \cos^2 x}\, dx = 7.640$, using NINT.

13. $y' = (e^2 - e^{-x})/2$; $1+(y')^2 = \frac{4+(e^{2x}-2+e^{-2x})}{4} = \frac{(e^x+e^{-x})^2}{4}$; $L = \int_{-3}^{3}\frac{e^x+e^{-x}}{2}dx = \frac{1}{2}[e^x - e^{-x}]_{-3}^{3} = e^3 - e^{-3} = 20.036$.

14. $y' = \frac{(2^x-2^{-x})}{2}\ln 2$; $1 + (y')^2 = 1 + (2^x - 2^{-x})^2(\frac{\ln 2}{2})^2$; $L = \int_{-3}^{3}\frac{1}{2}\sqrt{4 + (2^x - 2^{-x})^2(\ln 2)^2}\, dx = 9.138$.

15. $y = (1 - x^{2/3})^{3/2} \Rightarrow \frac{dy}{dx} = \frac{3}{2}(1 - x^{2/3})^{1/2}(-\frac{2}{3}x^{-1/3}) = -x^{-1/3}(1 - x^{2/3})^{1/2}$; $1 + (\frac{dy}{dx})^2 = 1 + x^{-2/3}(1 - x^{2/3}) = x^{-2/3}$. This function is not continuous on $[0,1]$. It is, however, continuous on $[\sqrt{2}/4, 1]$, which is where the curve and the line $y = x$ intersect. $L = 8\int_{\sqrt{2}/4}^{1} x^{-1/3}dx = \frac{8x^{2/3}}{2/3}\Big|_{\sqrt{2}/4}^{1} = \frac{3\cdot 8}{2}[1 - (2^{-3/2})^{2/3}] = 4\cdot 3(\frac{1}{2}) = 6$.

16. $\frac{dy}{dx} = (\cos 2x)^{1/2}$; $L = \int_0^{\pi/4}\sqrt{1 + \cos 2x}\, dx = \sqrt{2}\int_0^{\pi/4}\cos x dx = \sqrt{2}[\sin x]_0^{\pi/4} = \sqrt{2}\,\frac{\sqrt{2}}{2} = 1$.

17. $1 + (\frac{dy}{dx})^2 = 1 + 1/(4x)$, $\frac{dy}{dx} = \frac{1}{2\sqrt{x}}$, $y = \int \frac{1}{x}x^{-1/2}\, dx = x^{1/2} + C$. Since $y(0) = 0$, $C = 0$ and the curve is $y = x^{1/2}$, $0 \le x \le 4$.

18. $\frac{dy}{dx} = e^x$; $y = e^x + C$. Since $y(0) = 1$, $C = 0$, $y = e^x$.

19. $\frac{dy}{dx} = 1/x$. $y = \ln x + C$. Since $y(1) = 3$, $C = 3$ and the curve is $y = \ln x + 3$.

20. The left-hand integral is the area of the unit circle, $\pi 1^2 = \pi$. If $y = \sqrt{1-x^2}$, then $1 + (\frac{dy}{dx})^2 = 1 + (\frac{-x}{\sqrt{1-x^2}})^2 = 1 + \frac{x^2}{1-x^2} = \frac{1}{1-x^2}$. Hence the value of $\int_{-1}^{1} \frac{1}{\sqrt{1-x^2}} dx$ is half the circumference of the unit circle, $\frac{1}{2}(2\pi) = \pi$.

21. $\frac{dy}{dx} = \frac{-2x}{1-x^2}$. $L = \int_0^{1/2} \sqrt{1 + \frac{4x^2}{(1-x^2)^2}} dx = \int_0^{1/2} \frac{\sqrt{1+2x^2+x^4}}{(1-x^2)} dx = \int_0^{1/2} \frac{1+x^2}{1-x^2} dx = 0.598612\ldots$.

22. $\frac{dy}{dx} = -\frac{\sin x}{\cos x}$; $L = \int_0^{\pi/3} \sqrt{1 + \tan^2 x}\ dx = \int_0^{\pi/3} \sec x\, dx = \text{NINT}(1/\cos x, x, 0, \pi/3) = 1.316957\ldots$.

23. The width of the material equals the length of the curve which is found by evaluating $\int_0^{20} [1 + \frac{9\pi^2}{400} \cos^2 \frac{3\pi}{20} x]^{1/2} dx = 21.068$ inches, using NINT.

24. Area $= 300 \int_{-25}^{25} [1 + \frac{\pi^2}{4} \sin^2 \frac{\pi x}{50}]^{1/2} dx$. Cost $=$ Area$*1.75 = \$38,422$.

25. The path is a hyperbola; the asteroid will be close to earth at the vertex $(\sqrt{5}, 0)$. The distance travelled is the arc length $L = \int_{\sqrt{5}}^{10} [1 + (\frac{dy}{dx})^2]^{1/2} dx$. Either solve explicitly for y and differentiate, or use the following method: Since $y^2 = 0.2x^2 - 1$, $2yy' = 0.4x$ and $y' = \frac{0.4x}{2y} = \frac{0.2x}{y}$; $(y')^2 = \frac{0.04x^2}{y^2} = \frac{0.04x^2}{0.2x^2-1}$. $L = \text{NINT}(\sqrt{}(1 + 0.04x^2/(0.2x^2 - 1)), x, \sqrt{5}, 10) = 9.0333173\ldots$.

26. Even NINT will not compute $\int_{-\sqrt{500}}^{\sqrt{500}} (1 + (\frac{dy}{dx})^2)^{1/2} dx$ because $\frac{dy}{dx}$ is infinite at the endpoints. To approximate L_1, calculate $2 \int_0^{\sqrt{500}-0.000001} (1+(\frac{dy}{dx})^2)^{1/2} dx = 52.69679$. If the starting point of track 2 is at t, then the equation $\int_x^{\sqrt{750}-0.000001} (1 + (dy_2/dt)^2)^{1/2} dt = L_1$ must be solved for x. The problem suggests $x < 0$; experimental values of x place it between -20.0 and -19.9.

27. The pool is shaped like a teardrop. Use $[0, 20]$ by $[-30, 30]$. The limits are $x = 0$ and $x = 20$, $y = 2x(2-0.1x)^{1/2}$; $\frac{dy}{dx} = 2(2-0.1x)^{1/2} - \frac{2x(0.1)}{2(2-0.1x)^{1/2}}$; $\frac{dy}{dx} = \frac{2(2-0.1x)-x(0.1)}{(2-0.1x)^{1/2}} = \frac{4-0.3x}{(2-0.1x)^{1/2}}$; $ds^2 = 1+(\frac{dy}{dx})^2 = 1+\frac{(4-0.3x)^2}{(2-0.1x)}$; $L = 2 \int_0^{19.999999} ds = 100.89\ldots$.

28. See #27. Here, $y' = \frac{0.1\sqrt{K}}{2\sqrt{2-0.1x}}[4 - 0.3x] \Rightarrow ds = [1 + \frac{K(4-0.3x)^2(0.01)}{4(2-0.1x)}]$ and we must solve the equation $80 = 2 \int_0^{19.9999} [1 + \frac{K(4-0.3x)^2(0.01)}{4(2-0.1x)}]^{1/2} dx$. If $K = 212$, the value of the integral is 79.838.

6.5 AREAS OF SURFACES OF REVOLUTION

1. $S = 2\pi \int_0^4 y ds = 2\pi \int_0^4 \frac{x}{2}\sqrt{1+(\frac{1}{2})^2}dx = \frac{\pi\sqrt{5}}{2}[\frac{x^2}{2}]_0^4 = 4\pi\sqrt{5}$

[(base circumference)/2] $\times$ slant height $= \frac{\pi \cdot 4}{2}\sqrt{4^2+2^2} = 4\pi\sqrt{5}$.

2. $S = 2\pi \int x ds = 2\pi \int_{x=0}^{x=4} x\sqrt{1+(\frac{1}{2})^2}dx = \pi\sqrt{5}\int_0^4 x dx = 8\sqrt{5}\pi$.

3. $S = 2\pi \int y ds = 2\pi \int_1^3 (\frac{x}{2}+\frac{1}{2})\sqrt{1+(\frac{1}{2})^2}dx = \frac{\pi\sqrt{5}}{2}\int_1^3 (x+1)dx = \frac{\pi\sqrt{5}}{2}[\frac{x^2}{2}+x]_1^3 = 3\pi\sqrt{5}$.

4. $S = 2\pi \int_1^3 x(\sqrt{1+(\frac{1}{4})^2}dx = \frac{\sqrt{5}\pi}{2}[x^2]_1^3 = 4\pi\sqrt{5}$.

5. $S = 2\pi \int_0^2 \frac{x^3}{9}\sqrt{1+(\frac{x^2}{3})^2}dx = (2\pi/27)\int_0^2 x^3\sqrt{9+x^4}dx$. Let $u = x^4+9$, then $du = 4x^3dx$ and $S = \frac{(2\pi/27)}{4}\int_{x=0}^{x=2} u^{1/2}du = \frac{\pi}{54}\int_9^{25} u^{1/2}du = \frac{\pi}{54}\cdot\frac{2}{3}[u^{3/2}]_9^{25} = \frac{\pi}{27\cdot 3}[125-27] = 98\pi/81$.

6. $S = 2\pi \int_0^1 \frac{y^3}{3}\sqrt{1+(y^2)^2}dy = (2\pi/3)\int_0^1 y^3\sqrt{1+y^4}dy$. Let $u = y^4+1$ then $du = 4y^3dy$. $S = (2\pi/3)(\frac{1}{4})\int_{u=1}^{u=2} u^{1/2}du = \frac{2}{3}(\frac{2\pi}{3})(\frac{1}{4})[u^{3/2}]_1^2 = (\sqrt{8}-1)\pi/9$.

7. $S = 2\pi \int_{3/4}^{15/4} \sqrt{x}\sqrt{1+(\frac{1}{2\sqrt{x}})^2}dx = \pi\int_{3/4}^{15/4}\sqrt{4x+1}dx = 28\pi/3$.

8. $\frac{dy}{dx} = \frac{1}{2\sqrt{2x-x^2}}(2-2x) = \frac{1-x}{\sqrt{2x-x^2}}$. $S = 2\pi\int_0^2 \sqrt{2x-x^2}\ \sqrt{1+\frac{(1-x)^2}{2x-x^2}}dx =$

$2\pi\int_0^2\sqrt{2x-x^2+(1-x)^2}dx = 2\pi\int_0^2 1dx = 4\pi$.

9. $S = 2\pi\int_1^5 \sqrt{x+1}\ \sqrt{1+(\frac{1}{2\sqrt{x+1}})^2}dx = \pi\int_1^5\sqrt{4x+5}dx = \frac{\pi}{4}\cdot\frac{2}{3}[(4x+5)^{3/2}]_1^5 = (\pi/6)[125-27] = 49\pi/3$.

10. $S = 2\pi\int$ radius $\cdot$ arc length $= 2\pi\int_1^4 x \cdot$ arc length dx. Since $y = 2\sqrt{4-x}$, $y' = \frac{-1}{\sqrt{4-x}}$, $\sqrt{1+(y')^2} = \sqrt{1+1/(4-x)}$. $S = 2\pi\int_1^4 x\sqrt{\frac{5-x}{4-x}}\,dx = 2\pi\ \text{NINT}(x\sqrt{((5-x)/(4-x))}, x, 1, 4) =$ "div by zero".

If the 4 is replaced by 3.999 the integral is 13.311...

If the 4 is replaced by 3.9999999 the integral is 13.3888...

If the 4 is replaced by 3.9999999999 the integral is 13.3912...

A close approximation to the surface area is 84.1395...

11. $S = 2\pi\int_0^1 x\sqrt{1+x^2}dx = \pi\cdot\frac{2}{3}[(1+x^2)^{3/2}]_0^1 = 2\pi(2\sqrt{2}-1)/3$.

12. $S = 2\pi \int_0^{\ln 2} \frac{e^y+e^{-y}}{2}\sqrt{1+(\frac{e^y-e^{-y}}{2})^2}dy = \frac{\pi}{2}\int_0^{\ln 2}(e^y+e^{-y})^2dy = \frac{\pi}{2}\int_0^{\ln 2}(e^{2y}+2+e^{-2y})dy = \frac{\pi}{2}[\frac{1}{2}e^{2y}+2y-\frac{1}{2}e^{-2y}]_0^{\ln 2} = \frac{\pi}{2}[\frac{1}{2}\cdot 4+2\ln 2-\frac{1}{2}\cdot\frac{1}{4}] - \frac{\pi}{2}[\frac{1}{2}+0-\frac{1}{2}] = \pi(\frac{15}{16}+\ln 2)$.

13. $S = 2\pi\int_1^2 y\sqrt{1+(y^3-\frac{1}{4y^3})^2}dy = 2\pi\int_1^2 y\sqrt{(y^3+\frac{1}{4y^3})^2}dy = 2\pi\int_1^2(y^4+\frac{1}{4y^2})dy = 253\pi/20$.

14. $S = 2\pi\int_{x=0}^{x=3} x ds = 2\pi\int_0^3 x\sqrt{1+[(\frac{x^2+2}{2})^{1/2}(2x)]^2}\,dx = 2\pi\int_0^3 x\sqrt{x^4+2x^2+1}\,dx = 2\pi\int_0^3(x^3+x^2)dx = \frac{117}{2}\pi$.

15. $S = \int_0^{1/2} 2\pi x^3(1+9x^4)^{1/2}dx$; let $\omega = (1+9x^4)$, $d\omega = 36x^3dx$. $S = \frac{2\pi}{36}\int_{x=0}^{x=1/2}\omega^{1/2}d\omega = \frac{2\pi}{36}\frac{(1+9x^4)^{3/2}}{3/2}\Big|_0^{1/2} = \frac{\pi}{27}[(\frac{5}{4})^3-1]$.

16. Using symmetry, $S = 4\pi\int_0^1\sqrt{1-x^2}\sqrt{1+(x/\sqrt{1-x^2})^2}dx = 4\pi\int_0^1 1dx = 4\pi$.

17. $S = 2\pi\int y ds = 2\pi\int_{-\pi/2}^{\pi/2}\cos x(1+\sin^2 x)^{1/2}dx = 4.591\pi$.

18. $S = 4\pi\int_0^1(1-x^{2/3})^{3/2}\sqrt{1+[\frac{3}{2}(1-x^{2/3})^{1/2}(-\frac{2}{3}x^{-1/3})]^2}\,dx = 4\pi\int_0^1(1-x^{2/3})^{3/2}\cdot\sqrt{1+(1-x^{2/3})x^{-2/3}}\,dx = 4\pi\int_0^1(1-x^{2/3})^{3/2}x^{-1/3}\,dx = -4\cdot\frac{3\pi}{2}\int_1^0 t^{3/2}dt = 12\pi/5$. Let $t = 1-x^{2/3}$.

19. For one wok, the area of the outside is $S = 2\pi\int_{y=-16}^{y=-7} x ds$. Since $dx = \frac{-y}{x}dy$, $ds = \frac{16}{x}dy$. $S = 2\pi\int_{-16}^{-7} x\cdot\frac{16}{x}dy = 32\pi\cdot 9 = 904.779$ cm^2; the amount of paint is $904.779(.1)$ cm^3 = 0.0904779 liters per side per wok. For 5000 woks, 452.390 liters of each color will be needed.

20. $S_{AB} = 2\pi\int_a^{a+h}\sqrt{r^2-x^2}\sqrt{1+(\frac{-x}{\sqrt{r^2-x^2}})^2}\,dx = 2\pi\int_a^{a+h}\sqrt{r^2}\,dx = 2\pi r[a+h-a] = 2\pi rh$ which is independent of a;

21. $S = 2\pi\int_{x=0}^{x=40} y ds = 2\pi\int_0^{40} 10x^{13/30}\sqrt{1+(\frac{13}{3}x^{-17/30})^2}\,dx = 11,899.571$. 11,900 tiles are needed.

22. Since $y' = \frac{1}{\sqrt{x}}$, solve $50 = 2\pi\int_0^c 2\sqrt{x}\sqrt{1+\frac{1}{x}}\,dx = 4\pi\int_0^c\sqrt{x+1}\,dx = \frac{4\pi(x+1)^{3/2}}{3/2}\Big|_0^c$. Algebraic manipulation gives $\frac{150}{8\pi} = (c+1)^{3/2}-1$. $c = -1+[\frac{150}{8\pi}+1]^{2/3} = 2.648\ldots$

6.6 WORK

1. $W = \int_0^{20} 40(\frac{20-x}{20})dx = 2\int_0^{20}(20-x)dx = 400$ ft · lb.

2. The force varies steadily from 16 to 8 $\Rightarrow$ $F(x) = 16 - \frac{2}{5}x$. $W = \int_0^{20}(16 - \frac{2}{5}x)dx = 240$ ft · lb.

3. $W = \int_0^{50} 0.74(50-x)dx = 0.74[50x - \frac{x^2}{2}]_0^{50} = 925\ N \cdot m.$

4. $W = \int_0^{170}(2 - \frac{2}{170}x)dx = 170$ ft · oz $= \frac{170}{16}$ ft · lb $= 10.625$ ft · lb.

5. $W = \int_0^{180} 4(180-x)dx = 64,800$ ft · lb.

6. $W = \int_0^{18}(144 - \frac{72}{18}x)dx = 1944$ ft · lb.

7. $6 = K(0.4)$, so $K = 15N/m$. $W = \int_0^{0.4} 15\ x dx = 1.2\ N \cdot m.$

8. $F = 90x \Rightarrow W = \int_0^5 90\ x dx = 1125\ N \cdot m.$

9. $10,000 = K(1)$. a) $W = \int_0^{1/2} 10,000\ x dx = 1250$ in · lb $= 104\frac{1}{6}$ ft · lb.

b) $W = \int_{1/2}^1 10,000\ x dx = 3750$ in · lb $= 312.5$ ft · lb.

10. $F = kx \Rightarrow 150 = (-1/16)k \Rightarrow F = -2400x$; if $x = -1/8$, $F = 300$ lb. $W = \int_0^{-1/8}(-2400x)dx = 18.75(\frac{1}{12}$ ft · lb$) = 1.5625$ ft · lb.

11. $W = \int_0^{10} 25\pi w y dy = 62.5 \times 25 \times \pi \cdot 10^2/2 = 245,436.926$ ft · lb.

12. $W = 25\pi 62.5 \int_5^{10}(y+4)dy = 282,252.465$ ft · lb.

13. $W = 51.2\int_0^{30} \pi(10^2)(30-y)dy = 7,238,229.473$ ft · lb.

14. a) $W = \pi\int_0^8 \frac{64.5}{4}(10-y)y^2\ dy = 34,582.652$ ft · lb.

b) $W = \pi\int_0^8 \frac{57}{4}(13-y)y^2\ dy = 17,024\pi = 53,482.473$ ft · lb.

15. a) $W = \int_0^{20} 62.5(20-y)120dy = 120(62.5)[20y - \frac{y^2}{2}]_0^{20} = 1,500,000$ ft · lb.

b) $1,500,000/250 = 6000$ sec $= 100$ minutes.

c) $W = \int_{10}^{20} 62.5(20-y)120dy = 375,000$ ft · lb are required to lower the level 10 feet. The pump will require $375,000/250 = 1500$ sec $= 25$ minutes.

16. a) $W = 62.5\int_{10}^{20} y(20)(12)dy = 2,250,000$ ft · lb.

b) 2 hrs 46.7 minutes

c) $W = 62.5\int_{10}^{15} 240y dy = 937,500$ ft · lb; 1 hr 9.4 minutes.

17. $W = \omega \int_0^{16}(16-y)\pi(\sqrt{y})^2\,dy = \omega\pi \int_0^{16}(16y-y^2)dy = \frac{2048}{3}\pi\omega = 21,446,605.85$ $N \cdot m$.

18. $W = 62.5\pi \int_0^5 (25-y^2)(y+4)dy = 96,129.463$ ft $\cdot$ lb.

19. The water weighs $\pi \cdot 2^2 \cdot 6 \cdot 62.5 = 4712.389$ pounds. If it is pumped to the top, $W = 4712.389 \cdot (15+6) = 98,960.17$ ft $\cdot$ lb. If it is pumped to the level of the valve and then pumped in, $W = 4712.389 \cdot 15 + 62.5(2^2)\pi \int_0^6 y\,dy = 84,823$ ft$\cdot$lb through the valve is faster.

20. $W = \frac{4}{9}\int_0^7 \pi(\frac{y+17.5}{14})^2(8-y)dy = 91.324$ in $\cdot$ oz.

21. $W = 56\int_0^{10} \pi(\sqrt{100-y^2})^2(12-y)dy = 56\pi \int_0^{10}(1200-12y^2-100y+y^3)dy = 56\pi(5500) = 967,610.537$ ft $\cdot$ lb. Cost is \$4838.05.

22. W = Work to raise tank water to base + Work to fill tank + Work to fill pipe $= 62.5(360)\pi 10^2 \cdot 25 + 62.5\pi\int_0^{25}(25-y)(100)dy + 62.5\pi\int_0^{360}(\frac{1}{6})^2(360-y)dy = 176,714,586.764 + 6,135,923.152 + 353,429.174 = 183,203,939$ ft $\cdot$ lb; 30 hrs 54 minutes.

23. $\int_a^b \frac{mMG}{r^2}dr = -mMG[\frac{1}{r}]_1^b = -1000 \times 5.975 \times 10^{24} \times 6.6720 \times 10^{-11}[\frac{1}{3578} \times 10^{-4} - \frac{1}{637} \times 10^{-4}] = -1000 \times 5.975 \times 6.6720[\frac{1}{3578} - \frac{1}{637}] \times 10^9 = 0.051441 \times 10^3 \times 10^9 = 5.1441 \times 10^{-2+3+9} = 5.1441 \times 10^{10}$ $N \cdot m$.

24. a) When the electron is at x, $-1 \le x \le 0$, the force is $23 \times 10^{-29}(1-x)^{-2}$; $W = \int_{-1}^0 23 \cdot 10^{-29}(1-x)^{-2}dx = 11.5 \times 10^{-29}$ $N \cdot m = 1.15 \times 10^{-3} N \cdot m$.

b) The forces are added, $W = 23 \cdot 10^{-29}\int_5^3[\frac{1}{(x+1)^2} + \frac{1}{(x-1)^2}]dx = -7.67 \times 10^{-29}$ $N \cdot m$, the negative sign indicates the direction.

6.7 FLUID PRESSURES AND FLUID FORCES

1. $F = \int_0^3 (62.5)2(7-y)y\,dy = 125\int_0^3(7y-y^2)dy = 2812.5$ lb.

2. Full tank: $F = 100\int_0^{90}$ depth $\cdot$ circumference $dy = 100\int_0^{90}(90-y)(90\pi)dy = 114,511,052$ lb. Half tank: $F = 100\int_0^{45}(45-y)90\pi dy = 28,627,763$ lb.

3. $F = 62.5\int_0^3 2x(3-y)dy = 125\int_0^3 \frac{2y}{3}(3-y)dy = 375$ lb. The length does not affect the force.

4. $F = 62.5\int_0^2 2x(2-y)dy = 125\int_0^2 \frac{2y}{3}(2-y)dy = 111.1$ lb.

5. The side of the plate lies along the line through $B(2,-1)$ and $(0,-5)$: $y = 2x-5$. $F = 62.5\int_{-5}^{-1} 2(\frac{y+5}{2})(-y)dy = 62.5\int_{-5}^{-1}(5+y)(-y)dy = 62.5\cdot\frac{56}{3} = 1166.67$ lb.

6. When revolved, the vertex is at $(0,3)$. The plate lies along the line through $(0,3)$ and $B(2,-1)$: $y = 3-2x$. $F = 62.5\int_{-1}^{0} 2[\frac{3-y}{2}](-y)dy = 114.583$ lb.

7. $F = 62.5\int_{-1}^{0} 2\sqrt{1-y^2}(-y)dy = +62.5\int_0^1 u^{1/2}du = 62.5[\frac{2}{3}] = 41.67$ lb. Let $u = 1-y^2$.

8. Sides: $F = 62.5\int_0^{2-1/6}(2-\frac{1}{6}-y)4dy = 420.139$ lb $=$ force on one side; force the two sides $= 840.278$ lb.

Ends: $F = 62.5\int_0^{11/6}(\frac{11}{6}-y)2dy = 210.069$ lb; force on the ends is 420.139 lb.

9. $F = \omega\int_{-33.5}^{-0.5} 63(-y)dy = -\omega\cdot 63[\frac{y^2}{2}]_{-33.5}^{-0.5} = \omega\cdot 63[561]$, $\omega = 64$ lb/ft^3 $=$ 0.03073 lb/in^3, $F = 1309$ lb/in^3.

10. $F = 64.5\int_0^{7.75/12}(\frac{7.75}{12}-y)(\frac{3.75}{12})dy = 4.204$ lb.

11. $F = 64.5\int_{-3}^{0} 2\sqrt{9-y^2}(-y)dy = 64.5\int_0^9 u^{1/2}du = 64.5(\frac{2}{3})9^{3/2} = 1161$ lb. Let $u = 9-y^2$.

12. a) $F = 50\int_0^{\text{gate height}}(2-y)2x\,dy = 50\int_0^1(2-y)2\sqrt{y}\,dy = 93.333$ lb.

b) Solve $160 = 50\int_0^1(h-y)2\sqrt{y}\,dy \Rightarrow h = \frac{160+50\int_0^1 2y^{3/2}dy}{50\int_0^1 2\sqrt{y}\,dy} = 3$ ft.

13. Assume the depth of the water is h. Then $V = 30\cdot h(\frac{2h}{5}) = 12h^2$. $F = 62.5\int_0^h 2(2y/5)(h-y)dy = 50\int_0^h(hy-y^2)dy = 50[\frac{hy^2}{2}-\frac{y^3}{3}]_0^h = 50\cdot\frac{h^3}{6}$. $F = 6667 \Rightarrow h = 9.23$, $V = 1034.16$ ft^3.

14. a) After 9 hrs, depth $= 9000/1500 = 6$ feet. $F = 62.5\int_0^1(6-y)2y\,dy = 333.33$ lb.

b) Solve $520 = 62.5\int_0^1(h-y)2y\,dy \Rightarrow h = (520+125\int_0^1 y^2dy)/(125\int_0^1 y\,dy) = 8.987$ ft.

15. Let the plate have width w, height h. Assume the top of the plate is d units below the surface of the fluid which is on the x-axis. The force on one side of the plate is $F = \omega\int_{-h-d}^{-3} w(-y)dy = \omega w[\frac{h^2+2dh}{2}]$. The average of the pressure at the top and bottom is $\frac{\omega d+\omega(d+h)}{2}$. Multiplied by the area, this is $\frac{\omega}{2}(h+2d)hw)$.

6.8 CENTERS OF MASS

1. $80 \cdot 5 = 100 \cdot x$. $x = 4$ ft.

2. The center of mass of each rod is at its center. Since they are equal, the center of mass will be halfway between them at $(L/2, L/2)$.

3. $M_0 = \int_0^2 \delta(x)x dx = 4\int_0^2 x dx = 4[\frac{x^2}{2}]_0^2 = 8$. $M = \int_0^2 \delta dx = 4\int_0^2 dx = 8$. $\bar{x} = \frac{8}{8} = 1$.

4. $M_0 = \int_0^3 x(1 + x/3)dx = 7.5$; $M = \int_0^3(1 + x/3)dx = 4.5$; $\bar{x} = \frac{7.5}{4.5} = \frac{5}{3}$.

5. $M_0 = \int_0^4 x\delta(x)dx = \int_0^4 x(1 + \frac{x}{4})^2 dx = \int_0^4 x(1 + \frac{x}{2} + \frac{x^2}{16})dx = [\frac{x^2}{2} + \frac{x^3}{6} + \frac{x^4}{64}]_0^4 = \frac{68}{3}$. $M = \int_0^4 (1+\frac{x}{4})^2 dx = \frac{4}{3}[(1+\frac{x}{4})^3]_0^4 = \frac{4}{3}(8-1) = \frac{28}{3}$. $\bar{x} = M_0/M = 68/28 = 17/7$.

6. $M_0 = \int_0^3 2x dx + \int_3^6 x dx = 22.5$; $M = \int_0^3 2dx + \int_3^6 1dx = 9$; $\bar{x} = 2.5$.

7.

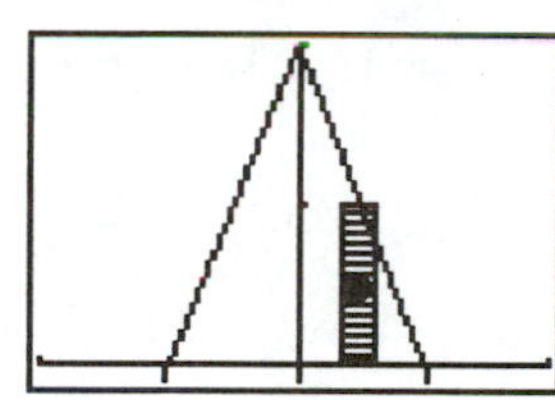

$\bar{x} = 0$. Using vertical strips and symmetry,

$M_x = 2\delta\int_{x=0}^{x=1} \tilde{y} dA =$

$2\delta\int_0^1 \frac{(-2x+2)}{2}(-2x + 2)dx =$

$4\delta\int_0^1 (1 - x)^2 dx = 4\delta[\frac{(1-x)^3}{-3}]_0^1 = 4\delta/3$.

$M = 2\delta\int_0^1 dA = 2\delta\int_0^1(-2x + 2)dx =$

$4\delta\int_0^1(1 - x)dx = 4\delta[\frac{(1-x)^2}{-2}]_0^1 = 2\delta$. $\bar{y} = 2/3$.

8.

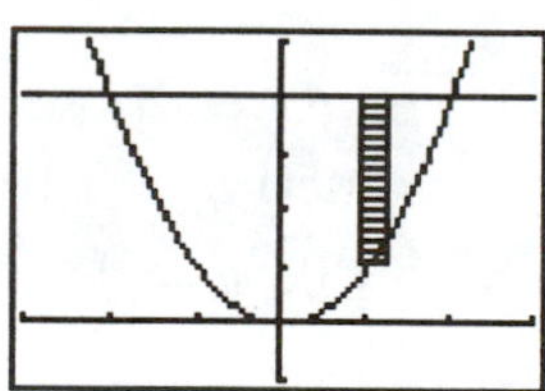

$\bar{x} = 0$; $M_x = \delta\int_{-2}^2 (\frac{4+x^2}{2})(4 - x^2)dx = 25.6\delta$;

$M = \delta\int_{-2}^2 (4 - x^2)dx = 10^{2/3}$

$\bar{y} = \frac{25.6\delta}{10^{2/3}\delta} = 2.4$.

9.

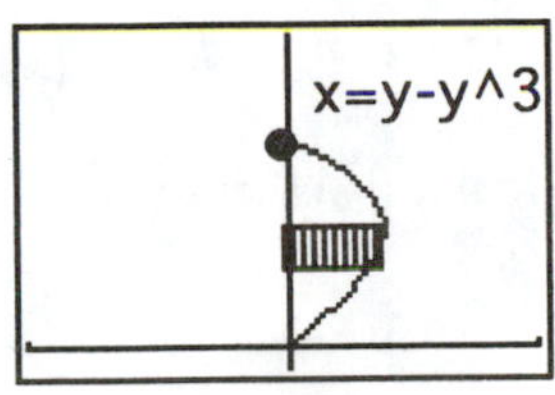

$M_x = \delta\int_0^1 y(y - y^3)dy$

$= \delta[\frac{y^3}{3} - \frac{y^5}{5}]_0^1 = 2\delta/15$.

$M_y = \delta\int_0^1 (\frac{y-y^3}{2})(y - y^3)dy = 4\delta/105$.

$M = \delta\int_0^1 (y - y^3)dy = \delta[\frac{y^2}{2} - \frac{y^4}{4}]_0^1 = \delta/4$.

$\bar{x} = M_y/M = 16/105$; $\bar{y} = M_x/M = 8/15$.

10. $M_y = \delta\int_0^2 x(x - x^2 + x)dx = \frac{4\delta}{3}$; $M_x = \delta\int_0^2 \frac{x-x^2-x}{2}(2x - x^2)dx = \frac{-4\delta}{5}$; $M = \delta\int_0^2(2x - x^2)dx = \frac{4\delta}{3} \Rightarrow \bar{x} = 1,\ \bar{y} = -\frac{3}{5}$.

11.

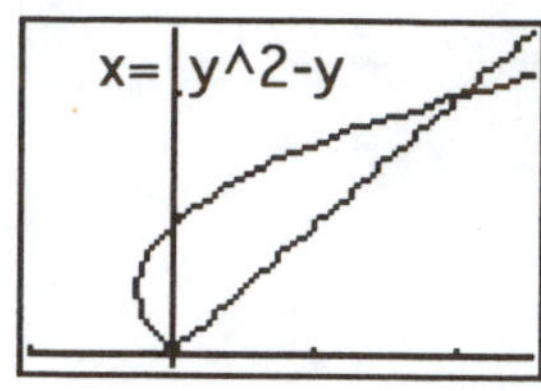

$M_x = \delta\int_0^2 y dA = \delta\int_0^2 y[y - (y^2 - y)]dy$
$= \delta\int_0^2(2y^2 - y^3)dy = \delta[\frac{2y^3}{3} - \frac{y^4}{4}]_0^2 = 4\delta/3.$
$M_y = \delta\int_0^2 \tilde{x} dA = \delta\int_0^2 \frac{[y+(y^2-y)]}{2}[y-(y^2-y)]dy$
$= \frac{\delta}{2}\int_0^2[y^2 - (y^2 - y)^2]dy = \frac{\delta}{2}\int_0^2(2y^3 - y^4)dy$
$= 4\delta/5.$
$M = \delta\int_0^2[y-(y^2-y)]dy = \delta[y^2-\frac{y^3}{3}]_0^2 = 4\delta/3,$
$\bar{x} = M_y/M = 3/5,\ \bar{y} = M_x/M = 1.$

12. $M_y = \delta\int_{-5}^5 x(25 - x^2)dx = 0$; $M_x = \delta\int_{-5}^5 \frac{25-x^2}{2}(25 - x^2)dx = \frac{5000}{3}\delta$; $M = \delta\int_{-5}^5(25 - x^2)dx = \frac{500}{3}\delta \Rightarrow \bar{x} = 0,\ \bar{y} = 10.$

13. By symmetry $\bar{x} = 0$. $M_x = \delta\int_{-\pi/2}^{\pi/2} \frac{\cos x}{2}\cos x dx = \frac{\delta}{2}\frac{1}{2}[x + \frac{1}{2}\sin 2x]_{-\pi/2}^{\pi/2} = \frac{\delta}{4}(\pi)$. $M = \delta\int_{-\pi/2}^{\pi/2}\cos x dx = \delta[\sin x]_{-\pi/2}^{\pi/2} = 2\delta,\ \bar{y} = M_x/M = \pi/8.$

14. $M_y = 0$; $M_x = \delta\int_{-\pi/4}^{\pi/4} \frac{\sec x}{2}\cdot\sec x dx = 1$; $M = \delta\int_{-\pi/4}^{\pi/4}\sec x dx = 1.763 \Rightarrow \bar{x} = 0,\ \bar{y} = \frac{1}{1.763}\delta \approx 0.567.$

15. $M = \delta\int_0^2[(2x - x^2) - (2x^2 - 4x)]dx = \delta\int_0^2(6x - 3x^2)dx = 4\delta$; $M_x = \delta\int_0^2 \frac{(2x-x^2)+(2x^2-4x)}{2}(6x - 3x^2)dx = \frac{\delta}{2}\int_0^2(x^2 - 2x)(6x - 3x^2)dx = \frac{\delta}{2}\int_0^2(6x^3 - 3x^4-12x^2+6x^3)dx = \frac{-\delta}{2}\cdot\frac{16}{5} = \frac{-8\delta}{5}$. $M_y = \delta\int_0^2 x(6x-3x^2)dx = \delta[2x^3-3\frac{x^4}{4}]_0^2 = 4\delta$. $\bar{x} = M_y/M = 1$; $\bar{y} = M_x/M = -2/5.$

16. a) $M_y = \delta\int_0^3 x\sqrt{9 - x^2}\,dx = 9\delta$; $M = \delta\int_0^3\sqrt{9 - x^2}\,dx =$ (area of quarter circle) $\cdot\,\delta = \frac{9\pi}{4}\delta \Rightarrow \bar{x} = \frac{4}{\pi}$; $\bar{y} = \bar{x} = \frac{4}{\pi}$ by symmetry. b) $M_y = 0$; $M_x = \delta\int_{-3}^3 \frac{\sqrt{9-x^2}}{2}\sqrt{9 - x^2}\,dx = 18\delta$; $M = \frac{9\pi}{2}\delta \Rightarrow \bar{x} = 0,\ \bar{y} = \frac{4}{\pi}$.

17. Area $= \frac{1}{4}$ (area of square of side 6 - area of circle) $= \frac{1}{4}(36 - \pi\cdot 3^2) = \frac{9}{4}(4 - \pi)$; $M = \frac{9}{4}(4 - \pi)\delta$; $M_x = \delta\int_0^3 \frac{3+\sqrt{9-x^2}}{2}\cdot(3 - \sqrt{9 - x^2})\,dx = \frac{\delta}{2}\int_0^3(9 - (9 - x^2))dx = \frac{\delta}{2}[\frac{x^3}{3}]_0^3 = \frac{9\delta}{2}$; $\bar{y} = M_x/M = \frac{2}{4-\pi}$. The center of mass of a symmetric region lies on its axis of symmetry. Hence $\bar{x} = \bar{y} = (\frac{2}{4-\pi})$.

18. $M_x = 0$ by symmetry; $M_y = \delta\int_1^2 x(\frac{2}{x^2})dx = (\ln 4)\delta$; $M = \delta\int_1^2 \frac{2}{x^2}dx = \delta \Rightarrow \bar{x} = \ln 4,\ \bar{y} = 0.$

19. $\bar{y} = \frac{1}{3}(3 - 0)$. $\bar{x} = 0$.

20. The line $y = (x-1)(-\frac{1}{2})$, through $(0, \frac{1}{2})$ and $(1,0)$, intersects the line $y = (x - \frac{1}{2})(-2)$, through $(0,1)$ and $(\frac{1}{2}, 0)$, at $(\frac{1}{3}, \frac{1}{3})$.

21. By symmetry, the centroid lies on the line $y = x$. One median coincides with the line joining $(0, a)$ and $(\frac{a}{2}, 0) : y = -2x + a$. The lines intersect at $(\frac{a}{3}, \frac{a}{3})$.

22. The lines $y = \frac{-b/2}{a}x + \frac{b}{2}$ and $y = \frac{-b}{a/2}x + b$ intersect at $(\frac{a}{3}, \frac{b}{3})$.

23. $M = \int_1^2 \delta dA = \int_1^2 x^3 \cdot \frac{2}{x^2} dx = [x^2]_1^2 = 3$. $M_x = \int_1^2 \frac{1}{2}(\frac{2}{x^2})x^3 \cdot \frac{2}{x^2} dx = 2\int_1^2 \frac{dx}{x} = 2\ln 2 = \ln 4$. $M_y = \int_1^2 x \cdot x^3 \cdot \frac{2}{x^2} dx = 2\int_1^2 x^2 dx = 2[\frac{x^3}{3}]_1^2 = \frac{14}{3}$. $\bar{x} = 14/9$, $\bar{y} = (\ln 4)/3$.

24. $M_x = \int_0^1 12x(\frac{x+x^2}{2})(x - x^2)dx = \frac{1}{2}$; $M_y = \int_0^1 12x \cdot x \cdot (x - x^2)dx = \frac{3}{5}$; $M = \int_0^1 12x(x - x^2)dx = 1 \Rightarrow \bar{x} = \frac{3}{5}$, $\bar{y} = \frac{1}{2}$.

25. By symmetry, the centroid is at $(2, 2)$. The area of the square is $(\sqrt{8})^2$. $V = 2\pi \cdot 2 \cdot 8 = 32\pi$. The perimeter of the square is $4\sqrt{8}$. $S = 2\pi \cdot 2 \cdot 4\sqrt{8} = 32\sqrt{2}\,\pi$.

26. By problem 22, the centroid is at $(1, 2)$; the distance travelled is $2\pi \cdot 4 \Rightarrow A = 2\pi \cdot 4 \cdot \text{area } = 8\pi\frac{1}{2} \cdot 3 \cdot 6 = 72\pi$.

27. The centroid is at $(2, 0)$. The area of the circle is π. Hence $V = 2\pi \cdot 2 \cdot \pi = 4\pi^2$.

28. Generate the cone by revolving about the y-axis the triangular region bounded by the coordinate axes and the line through $A(0, h)$ and $B(r, 0)$. The centroid is at $(\frac{r}{3}, \frac{h}{3})$; the area of $OAB = \frac{1}{2}rh$; length of $AB = \sqrt{r^2 + h^2}$. The arc's center of mass is at $(\frac{r}{2}, \frac{h}{2})$. By the theorems of Pappus, $V = 2\pi\frac{r}{3}(\frac{1}{2}rh)$ and $S = \sqrt{r^2 + h^2} \cdot 2\pi \cdot \frac{r}{2} \Rightarrow V = \frac{\pi}{3}r^2 h$, $S = \pi r\sqrt{r^2 + h^2}$.

29. By symmetry $\bar{x} = 0$. The length of the semicircle is πa. By Theorem 2, $4\pi a^2 = 2\pi\bar{y}(\pi a)$. $\bar{y} = 2a/\pi$.

30. $S =$ (arc length) distance travelled $= \pi a \cdot 2\pi(a - \frac{2a}{\pi}) = 2\pi a^2(\pi - 2)$.

31. The area of the semicircle is $\pi a^2/2$. Hence $2\pi\bar{y} \cdot \pi a^2/2 = (4/3)\pi a^3$ implies $\bar{y} = 4a/(3\pi)$. By symmetry, $\bar{x} = 0$.

32. $V =$ (distance travelled)(area) $= 2\pi(\frac{4a}{3\pi} + a)\frac{\pi a^2}{2} = \pi a^3\frac{(4+3\pi)}{3}$.

33.

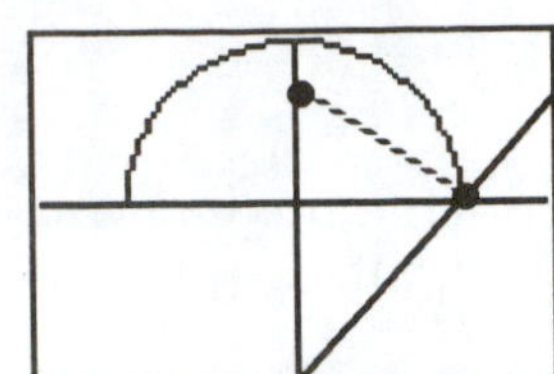

By Problem 31, the centroid is located at $(0, 4a/3\pi)$. To find the distance the centroid travels, first find the line through $(0, 4a/3\pi)$ with slope $-1 : y = -x + 4a/3\pi$. This line meets $y = x - a$ at $(\frac{a(4+3\pi)}{6\pi}, \frac{a(4-3\pi)}{6\pi})$. The distance from the centroid to the axis of rotation: $\sqrt{[\frac{a(4+3\pi)}{6\pi}]^2 + [\frac{a(4-3\pi)}{6\pi} - \frac{8a}{6\pi}]^2} = \sqrt{2[\frac{a(4+3\pi)}{6\pi}]^2} = \frac{a(4+3\pi)}{6\pi}\sqrt{2}$; this is the radius of the circle swept out by the centroid. By Pappus, $V = \text{area} \cdot \text{distance} = \frac{\pi a^2}{2} \cdot 2\pi \frac{a(4+3\pi)}{6\pi}\sqrt{2} = \frac{\pi a^3}{6}(4 + 3\pi)\sqrt{2}$.

34. $S = \pi a \cdot 2\pi \cdot$ (perpendicular distance from $(0, \frac{2a}{\pi})$ to $y = x - a$); the perpendicular line through **cm** is $y - 0 = -(x - \frac{2a}{\pi})$ which intersects $y = x - a$ at $(\frac{a}{2} + \frac{a}{\pi}, -\frac{a}{2} + \frac{a}{\pi}) \Rightarrow S = 2\pi^2 a\sqrt{(\frac{a}{2} + \frac{a}{\pi})^2 + (-\frac{a}{2} - \frac{a}{\pi})^2} = 2\pi^2 a(\frac{a}{2} + \frac{a}{\pi})\sqrt{2} = \sqrt{2}\pi a^2(\pi + 2)$.

6.9 THE BASIC IDEA. OTHER MODELING APPLICATIONS

1. Cross-sections are squares with diagonal of length $2\sqrt{x}$. The area of the square is $\frac{(2\sqrt{x})^2}{2} = 2x$ Hence $V = \int_0^4 2x dx = [x^2]_0^4 = 16$.

2. A square with diagonal $= 2\sqrt{1 - x^2}$ has area $4(1 - x^2)/2$. By symmetry, $V = 2\int_0^1 2(1 - x^2)dx = 8/3$.

3. The side of each square is $2\sqrt{1 - x^2}$. $V = \int_{-1}^1 (2\sqrt{1 - x^2})^2 dx = 4\int_{-1}^1 (1 - x^2)dx = 4[x - \frac{x^3}{3}]_{-1}^1 = \frac{16}{3}$.

4. A disk with diameter $2 - 2x^2$ has area $\pi(1 - x^2)^2 \Rightarrow V = 2\pi\int_0^1 (1 - x^2)dx = 16\pi/15$.

5. The radius of each disk is $\frac{1}{\sqrt{x}}$. $V = \int_1^2 \pi(\frac{1}{x})dx = [\pi \ln x]_1^2 = \pi \ln 2$.

6. radius $= \frac{\sqrt{5}x^2}{2} \Rightarrow$ area of disk$= \frac{5\pi x^4}{4} \Rightarrow V = \int_0^2 \frac{5\pi x^4 dx}{4} = 8\pi$

7. The base and height of each triangle are $2\sqrt{1 - x^2}$. Hence $A(x) = \frac{1}{2}2\sqrt{1 - x^2} \cdot 2\sqrt{1 - x^2} = 2 - 2x^2$. $V = \int_{-1}^1 (2 - 2x^2)dx = 2\int_0^1 (2 - 2x^2)dx = 2[2x - \frac{2x^3}{3}]_0^1 = \frac{8}{3}$.

8. Area of cross-section is $\frac{\sqrt{3}}{4}(2\sqrt{\sin x})^2 \Rightarrow V = \sqrt{3}\int_0^\pi \sin x\, dx = 2\sqrt{3}$.

9.

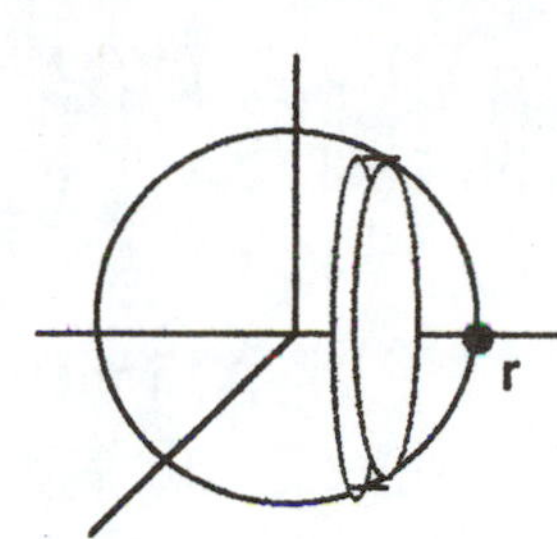

$V = 2\int_0^r$ area of slice dx
$= 2\int_0^r \pi(\sqrt{r^2 - x^2})^2 dx$
$= 2\pi \int_0^r (r^2 - x^2)dx = 2\pi(r^3 - \frac{r^3}{3}) = \frac{4\pi r^3}{3}$

10.

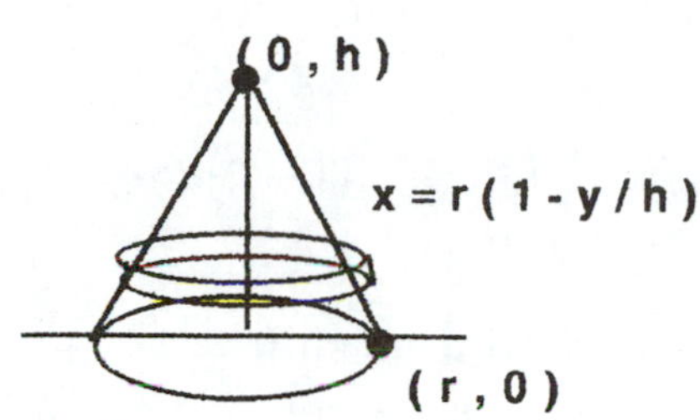

$V = \int_0^h \pi x^2 dy = \pi r^2 \int_0^h (1 - \frac{y}{h})^2 dy$. Let
$u = (-\frac{y}{h})$;
$V = -\frac{h\pi r^2}{3}(1 - \frac{y}{h})^3]_0^h = \frac{\pi r^2 h}{3}$

11. Twisted or not, the area of each square is s^2. $V = \int_0^h s^2 dx = s^2 h$. V is independent of the number of revolutions of the square.

12. Let $A_1(x)$ be the area of one cross-section, $A_2(x)$ the other. By assumption $A_1(x) = A_2(x) \Rightarrow V(\text{first solid}) = \int_a^b A_1(x)dx = \int_a^b A_2(x)dx = V(\text{second solid})$.

13. b) Total distance $= \int_0^{2\pi} |5\cos t|dt = 5\int_0^{\pi/2}\cos t dt - 5\int_{\pi/2}^{3\pi/2}\cos t\, dt + 5\int_{3\pi/2}^{2\pi}\cos t = 5[\sin t]_0^{\pi/2} - 5[\sin t]_{\pi/2}^{3\pi/2} + 5[\sin t]_{3\pi/2}^{2\pi} = 5[1-0] - 5[-1-(+1)] + 5(0-(-1)] = 20$ meters. c) $\int_0^{2\pi} 5\cos t\, dt = 5[\sin t]_0^{2\pi} = 0$ m

14. b) Total distance $= \int_0^2 |\sin \pi t|dt = \int_0^1 \sin \pi t\, dt - \int_1^2 \sin \pi t\, dt = \frac{4}{\pi}$ m.

c) $\int_0^2 \sin \pi t\, dt = 0$ m.

15. b) Total distance $= \int_0^{\pi/2} |6\sin 3t|dt = 6\int_0^{\pi/3}\sin 3t\, dt - 6\int_{\pi/3}^{\pi/2}\sin 3t\, dt = -\frac{6}{3}[\cos 3t]_0^{\pi/3} + \frac{6}{3}[\cos 3t]_0^{\pi/2} = -2[-1-(1)] + 2[0-(-1)] = 6$ meters

c) $\int_0^{\pi/2}(6\sin 3t)dt = -\frac{6}{3}[\cos 3t]_0^{\pi/2} = -2[0-1] = 2$ meters.

16. b) Total distance $= \int_0^\pi |4\cos 2t|dt = \int_0^{\pi/4} 4\cos 2t\,dt - \int_{\pi/4}^{3\pi/4} 4\cos 2t\,dt + \int_{3\pi/4}^{\pi} 4\cos 2t\,dt = 4\int_0^{\pi/4} 4\cos 2t\,dt$ (from the graph) $= 4\cdot 2 = 8$ m c) From the graph, the shift is 0 m.

17. b) $\int_0^{10} |49 - 9.8t|dt = 2\int_0^5 (49 - 9.8t)dt = 2[49t - \frac{9.8t^2}{2}]_0^5 = 245$ meters.

c) By symmetry, 0 meters.

18. b) $\int_0^{10} |8 - 1.6t|dt = \int_0^5 (8 - 1.6t)dt + \int_5^{10}(1.6t - 8)dt = 40$ m.

c) $\int_0^{10}(8 - 1.6t)dt = 0$ m

19. b) $\int_0^2 |6(t-1)(t-2)|dt = \int_0^1 (6t^2 - 18t + 12)dt - \int_1^2 (6t^2 - 18t + 12)dt = [2t^3 - 9t^2 + 12t]_0^1 - [2t^3 - 9t^2 + 12t]_1^2 = [5] - [16 - 36 + 24 - (2 - 9 + 12)] = [5] - [4 - (5)] = 5 - (-1) = 6$ meters. c) $\int_0^2 (6t^2 - 18t + 12)dt = 5 - 1 = 4$ meters.

20. b) $\int_0^3 |6(t-1)(t-2)|dt = 6\int_0^1 (t-1)(t-2)dt - 6\int_1^2 (t-1)(t-2)dt + 6\int_2^3 (t-1)(t-2)dt = 11$ m. c) $\int_0^3 6(t-1)(t-2)dt = 9$ m.

21. a) distance travelled $= 1\cdot 1 + 0\cdot 1 + 1\cdot 1 + 0\cdot 1 = 2$; position shift $= 1\cdot 1 + 0\cdot 1 + 1\cdot 1 + 0\cdot 1 = 2$. The velocity is always non-negative; hence position shift = distance travelled. b) distance travelled $= 1\cdot 1 + 1\cdot 1 + 1\cdot 1 + 1\cdot 1 = 4$; position shift $= 1\cdot 1 + (-1)\cdot 1 + 1\cdot + (-1)\cdot 1 = 0$ c) distance travelled $= 4\cdot\frac{1}{2}(1)(2) = 4$; position shift $= 4$ d) distance travelled $= \frac{1}{2}(2)(1) + 0 + \frac{1}{2}(2)(1) = 2$; position shift $= 2$

22. Distance $= \int_0^{10} |v(t)|dt \approx \frac{1}{3}[0 + 4\cdot 12 + 2\cdot 22 + 4\cdot 10 \oplus 2\cdot 5 \oplus 4\cdot 13 \oplus 2\cdot 11 \oplus 4\cdot 6 + 2\cdot 2 = 268/3$; for the shift, change the $\oplus$ to $\ominus$; shift $= \frac{52}{3}$ inches.

23. $\int_0^{\sqrt{3}} \frac{2\pi x}{\sqrt{3}} dx = \frac{\pi\cdot x^2}{\sqrt{3}}\Big|_0^{\sqrt{3}} = \frac{3\pi}{\sqrt{3}} \approx 1.73\pi$ instead of 2π

24. $\int_0^h 2\pi r\,dx = 2\pi r[h - 0] = 2\pi rh.$

PRACTICE EXERCISES, CHAPTER 6

1. $A = \int_1^2 (x - \frac{1}{x^2})dx = [\frac{x^2}{2} + \frac{1}{x}]_1^2 = 1$

2. $A = \int_1^2 (x - x^{-1/2})dx = [\frac{x^2}{2} - 2x^{1/2}]_1^2 = 2 - 2\sqrt{2} - 0.5 + 2 = \frac{7}{2} - 2\sqrt{2}$

3. $A = \int_{-2}^1 (3 - x^2 - (x+1))dx = \int_{-2}^1 (2 - x - x^2)dx = [2x - \frac{x^2}{2} - \frac{x^3}{3}]_{-2}^1 = \frac{9}{2}$

4. $A = \int_0^1 (1 - \sqrt{x})^2 dx = \int_0^1 (1 + x - 2\sqrt{x})dx = \frac{1}{6}$

5. $A = \int_0^3 2y^2 dy = [\frac{2y^3}{3}]_0^3 = 18$

6. $A = \int_{-4}^5 (\frac{y+16}{4} - \frac{y^2-4}{4})dy = 243/8$

7. $A = \int_{-1}^2 (\frac{y+2}{4} - \frac{y^2}{4})dy = \frac{1}{4}\int_{-1}^2 (y+2-y^2)dy = \frac{1}{4}[\frac{y^2}{2} + 2y - \frac{y^3}{3}]_{-1}^2 = \frac{9}{8}$

8. $A = \int_0^4 \sqrt{4-x}dx = -\int_4^0 \sqrt{t}dt = 16/3$

9. $A = \int_0^{\pi/4}(x - \sin x)dx = [\frac{x^2}{2} + \cos x]_0^{\pi/4} = \frac{\pi^2}{32} + \frac{\sqrt{2}}{2} - 1 = 0.0155$

10. $A = 2\int_0^{\pi/2} 1 - \sin x\, dx = \pi - 2$

11. $A = \int_0^\pi (2\sin x - \sin 2x)dx = [-2\cos x + \frac{1}{2}\cos 2x]_0^\pi = 4$

12. $A = 2\int_0^{\pi/3}(8\cos x - \sec^2 x)dx = 2[8\sin x - \tan x]_0^{\pi/3} = 6\sqrt{3}$

13. $A = \int_1^2 (\sqrt{y} - (2-y))dy = [\frac{2y^{3/2}}{3} - 2y + \frac{y^2}{2}]_1^2 = (8\sqrt{2} - 7)/6$

14. $A = \int_0^4 (\sqrt{x} - 1)dx + \int_4^5 (6 - x - 1)dx = \int_1^2 (6 - y - y^2)dy = 13/6$

15. $A = \int_{-\ln 2}^{\ln 2} 2e^{-x}dx = -2[e^{-x}]_{-\ln 2}^{\ln 2} = -2[e^{-\ln 2} - e^{\ln 2}] = -2[\frac{1}{2} - 2] = 3$

16. $A = 2\int_1^4 \frac{2}{y}dy = 4\ \ln 4$

17. a) $V = \pi\int_{-1}^1 (3x^4)^2 dx = 2\pi\cdot 9\int_0^1 x^8 dx = 2\pi\cdot 9[\frac{x^9}{9}]_0^1 = 2\pi$ b) The solid is generated by revolving the region between $x = 0$ and $x = 1$. $V = 2\pi\int_0^1 x\cdot 3x^4 dx = 6\pi[\frac{x^6}{6}]_0^1 = \pi$

18. a) $V = \pi\int_0^3 (x^2)^2 dx = \frac{243\pi}{5}$ b) $V = \pi\int_0^9 (9 - (\sqrt{y})^2)dy = \frac{81\pi}{2}$

19. a) $V = \pi\int_1^5 y^2 dx = \pi\int_1^5 (x-1)dx = \pi[\frac{x^2}{2} - x]_1^5 = 8\pi$ b) $V = 4\pi\int_1^5 x\sqrt{x-1}dx = 4\pi\int_0^4 (u+1)u^{1/2}du = 4\pi[\frac{2u^{5/2}}{5} + \frac{2u^{3/2}}{3}]_0^4 = 4\pi[\frac{64}{5} + \frac{16}{3}] = \frac{1088\pi}{15}$, where $u = x - 1$. c) $V = 4\pi\int_1^5 (5-x)\sqrt{x-1}dx = 4\pi\int_0^4 (4-u)u^{1/2}du = 4\pi[\frac{4u^{3/2}\cdot 2}{3} - \frac{2u^{5/2}}{5}]_0^4 = 512\pi/15$, where $u = x - 1$.

20. a) $V = \pi\int_0^4 (4x - x^2)dx = 32\pi/3$ b) $V = \pi\int_0^4 (y^2 - (\frac{y^2}{4})^2)dy = 128\pi/15$ c) $V = \pi\int_0^4 ((4 - \frac{y^2}{4})^2 - (4-y)^2)dy = 64\pi/5$ d) $V = \pi\int_0^4 ((4-x)^2 - (4 - \sqrt{4x})^2)dx = 32\pi/3$

21. $V = \pi\int_0^{\pi/3} (\tan x)^2 dx = \pi\int_0^{\pi/3}(\sec^2 x - 1)dx = \pi[\sqrt{3} - \frac{\pi}{3}]$

22. $V = \pi\int_0^\pi (2 - \sin x)^2 dx = \pi\int_0^\pi (4 - 4\sin x + \sin^2 x)dx = \pi[4x + 4\cos x + \frac{1}{2}(x - \frac{1}{2}\cos 2x)]_0^\pi = 9\pi/2 - 8$

23. $V = \pi \int_1^{16}(\frac{1}{\sqrt{x}})^2 dx = \pi[\ln x]_1^{16} = \pi \ln 16$

24. $V = \pi \int_0^{\ln 3}((e^{x/2})^2 - 1^2)dx = \pi \int_0^{\ln 3}(e^x - 1)dx = \pi(2 - \ln 3)$

25. Revolve the strip bounded by $x = 0, x = \sqrt{3}, x^2 + y^2 = 4$ about the y-axis. $V = 4\pi \int_0^{\sqrt{3}} x\sqrt{4 - x^2}dx = -\frac{4\pi}{3}[(4 - x^2)^{3/2}]_0^{\sqrt{3}} = \frac{28\pi}{3}$

26. $V = 2\pi \int_0^{11/2} 12(1 - \frac{4x^2}{|2|})dx = 88\pi = 276$ cubic inches

27. $L = \int_0^3 \sqrt{1 + [\frac{1}{2}x^{-1/2} - \frac{1}{2}x^{1/2}]^2}dx = \int_0^3[\frac{1}{2}x^{-1/2} + \frac{1}{2}x^{1/2}]dx = \frac{1}{2}[2x^{1/2} + \frac{2x^{3/2}}{3}]_0^3 = 2\sqrt{3}$

28. $dx = \frac{2}{3}y^{-1/3}dy \Rightarrow L = \int_1^8 \sqrt{1 + \frac{4}{9}y^{-2/3}}dy = \frac{1}{3}\int_1^8(\sqrt{9y^{2/3} + 4})y^{-1/3}dy = \frac{1}{18}\int_{13}^{40} u^{1/2}du = 7.634$ where $u = 9y^{2/3} + 4$.

29. $y' = \frac{x^3}{2} - \frac{1}{2x^3}$; $1 + (y')^2 = 1 + \frac{x^6}{4} - \frac{1}{2} + \frac{1}{4x^6} = (\frac{x^3}{2} + \frac{1}{2x^3})^2$; $s = \frac{1}{2}\int_1^3(x^3 + x^{-3})dx = \frac{1}{2}[\frac{x^4}{4} - \frac{1}{2x^2}]_1^3 = \frac{92}{9}$

30. $y' = \frac{3x^2}{2} - \frac{1}{6x^2}$; $1 + (y')^2 = 1 + \frac{9x^4}{4} - \frac{1}{2} + \frac{1}{36x^4} = (\frac{3x^2}{2} + \frac{1}{6x^2})^2$; $s = \int_1^2(\frac{3x^2}{2} + \frac{1}{6x^2})dx = [\frac{1}{2}x^3 - \frac{1}{6x}]_1^2 = \frac{43}{12}$

31. $1 + (y')^2 = 1 + [\frac{2}{2\sqrt{2x+1}}]^2 = \frac{2x+2}{2x+1}$; $S = 2\pi \int_0^{12} \sqrt{2x+1}\frac{\sqrt{2x+2}}{\sqrt{2x+1}}dx = 2\pi \int_0^{12}(2x + 2)^{1/2}dx = \pi[\frac{2(2x+2)^{3/2}}{3}]_0^{12} = \frac{2\pi}{3}[26^{3/2} - 2^{3/2}] \approx 86.5\pi$

32. $ds = [1 + (\frac{x^2}{3})^2]^{1/2}dx \Rightarrow S = 4\pi \int_0^1 \frac{x^3}{9}\sqrt{\frac{9+x^4}{9}}dx = \frac{\pi}{27}\int_9^{10} \sqrt{u}\, du \approx 0.114\pi$ where $u = 9 + x^4$

33. $1 + (y')^2 = 1 + [\frac{x^{1/2}}{2} - \frac{1}{2}x^{-1/2}]^2 = [\frac{x^{1/2}}{2} + \frac{x^{-1/2}}{2}]^2$. The graph lies below the x axis. Hence $S = 2\pi \int_0^3 -(\frac{x^{3/2}}{3} - x^{1/2})\frac{1}{2}(x^{1/2} + x^{-1/2})dx = -\pi \int_0^3(\frac{1}{3}x^2 + \frac{1}{3}x - x - 1)dx = -\pi[\frac{x^3}{9} - \frac{x^2}{3} - x]_0^3 = -\pi[-3] = 3\pi$

34. From #30, $ds = \frac{4e^x + e^{-x}}{4}dx \Rightarrow S = 2\pi \int_{-\ln 4}^{\ln 2} y[\frac{4e^x + e^{-x}}{4}]dx = \frac{\pi}{8}\int_{-\ln 4}^{\ln 2}[4e^x + e^{-x}]^2 dx = 21.995$

35. $1 + (y')^2 = 1 + x^2(x^2 + 2) = (x^2 + 1)^2$. $S = 2\pi \int_0^1 x(x^2 + 1)dx = 2\pi[\frac{x^4}{4} + \frac{x^2}{2}]_0^1 = 3\pi/2$

36. $dy = 2x dx \Rightarrow (dx)^2 = \frac{(dy)^2}{4x^2} = \frac{(dy)^2}{4y} \Rightarrow S = 2\pi \int_0^2 \sqrt{y}\sqrt{1 + \frac{1}{4y}}dy = \pi \int_0^2 \sqrt{4y + 1}dy = \frac{13\pi}{3}$

37. The work needed to haul up the equipment is $10(9.8)(40) = 3920N \cdot m$. For the rope, $W = \int_0^{40} 0.8(40-x)dx = [32x - 0.4x^2]_0^{40} = 640N \cdot m$. Total work $= 3920 + 640 = 4560N \cdot m$.

38. $W = \int$ Force $\cdot$ distance $= \int_0^{50} 8[800 - \frac{400}{50}t]95dt = 22,800,000$ ft $\cdot$ lb.

39. $K = 20$ lb/ft. $W = \int_0^1 20x\,dx = [10x^2]_0^1 = 10$ ft $\cdot$ lb. To stretch an additional foot requires $W = \int_1^2 20xdx = [10x^2]_1^2 = 30$ ft $\cdot$ lb.

40. $2N = K \cdot 2$ cm $\Rightarrow F = 1 \cdot (\text{cm}/N)x$; a $4 - N$ force will stretch the band 4 cm. $W = \int_0^4 x\,dx = 8$ N $\cdot$ cm.

41. $W = \omega \int_0^8 (14-y)\pi r^2 dy = \omega \int_0^8 (14-y)\pi(\frac{10y}{8})^2 dy = \omega\pi\, 6400/3$ ft $\cdot$ lb, where $\omega = 62.5$

42. $W = \omega \int_0^5 (8-y)\pi(\frac{5}{4}y)^2 dy = \frac{25\pi\omega}{16}\int_0^5 (8y^2 - y^3)dy = \frac{25\pi\omega}{16}[\frac{8y^3}{3} - \frac{y^4}{4}]_0^5 = 53125\omega\pi/192$ ft $\cdot$ lb where $\omega = 62.5$

43. $F = 2\omega \int_0^2 (2-y)2ydy = 4 \cdot \omega \int_0^2 (2y - y^2)dy = 4\omega[y^2 - \frac{y^3}{3}]_0^2 = \frac{16\omega}{3} = \frac{16(62.5)}{3} =$ 333.3 lb

44. $F = \omega \int_0^{10/12} (\frac{5}{6} - y)2(y+2)dy = 2\omega \int_0^{5/6} (\frac{5}{6}y + \frac{10}{6} - y^2 - 2y)dy = 2\omega[\frac{5y^2}{12} + \frac{10y}{6} - \frac{y^3}{3} - \frac{2y^2}{2}]_0^{5/6} = \frac{1025}{648}\omega = 118.63$ lb

45. $F = 2\int_0^4 (9-y)62.5\frac{\sqrt{y}}{2}dy = 62.5\int_0^4 (9y^{1/2} - y^{3/2})dy = 62.5[\frac{2}{3} \cdot 9y^{3/2} - \frac{2}{5}y^{5/2}]_0^4 = 62.5 \times \frac{176}{5} = 2200$ lb

46. Olive oil has weight density 57 lb/ft^3 $= \frac{57}{12^3}$ lb/in^3. The force on the base is $\frac{57}{12^3}(10 \cdot 5.75 \cdot 3.5) = 6.638$ lb. Against an end panel $F = \int_0^{10} \frac{57}{12^3}(10 - y)(3.75dy) = 5.773$ lb; against a side panel, $F = \frac{57}{12^3}\int_0^{10}(10-y)(5.75dy) = 9.484$ lb.

47. $M_x = \int_{-1}^1 \tilde{y}dm = \delta\int_{-1}^1 [\frac{3-x^2+2x^2}{2}](3 - x^2 - 2x^2)dx = \frac{\delta}{2}\int_{-1}^1 (3+x^2)(3-3x^2)dx = \frac{3\delta}{2}\int_{-1}^1 [3 - 2x^2 - x^4]dx = \frac{3\delta}{2} \cdot 2[3x - \frac{2x^3}{3} - \frac{x^5}{5}]_0^1 = \frac{32}{5}\delta$. $M = \int_{-1}^1 \delta dA = 2\delta\int_0^1 (3 - 3x^2)dx = 6\delta[x - \frac{x^3}{3}]_0^1 = 48$. $\bar{y} = M_x/M = 8/5$. By symmetry, $\bar{x} = 0$.

48. $\bar{x} = 0$ by symmetry: $M = 2\int_0^2 x^2 dx = \frac{16}{3}$, $M_x = 2\int_0^2 (\frac{x^2}{2})x^2 dx = \frac{32}{5} \Rightarrow \bar{y} = \frac{32/5}{16/3} = 6/5$.

49. $M_x = \int \tilde{y}dm = \delta\int_0^4 y(2\sqrt{y})dy = 2\delta\frac{2}{5}[y^{5/2}]_0^4 = \frac{128}{5}\delta$; $M_y = \int \tilde{x}dm = \delta\int_0^4 x(4 - \frac{x^2}{4})dx = \delta[2x^2 - \frac{x^4}{16}]_0^4 = 16\delta$; $M = \delta\int_0^4 (4 - \frac{x^2}{4})dx = \delta[4x - \frac{x^3}{12}]_0^4 = \frac{32}{3}\delta$; $\bar{x} = M_y/M = 3/2$; $\bar{y} = M_x/M = 12/5$

50. $M = 3\int_0^4(\sqrt{x} - \frac{x}{2})dx = 4$; $M_x = 3\int_0^4 \frac{1}{2}(\sqrt{x} + \frac{x}{2})(\sqrt{x} - \frac{x}{2})dx = 4$; $M_y = 3\int_0^4 x(\sqrt{x} - \frac{x}{2})dx = 32/5$; $\bar{x} = \frac{M_y}{M} = \frac{8}{5}$; $\bar{y} = \frac{4}{4} = 1$

51. $M_x = \int \tilde{y}dm = \delta \int_0^1 y \cdot 2(y+2)dy = 2\delta[\frac{y^3}{3} + y^2]_0^1 = 8\delta/3$; $M = 2\delta\int_0^1(y+2)dy = 2\delta[\frac{y^2}{2} + 2y]_0^1 = 5\delta$; $\bar{y} = M_x/M = 8/15$. By symmetry, $\bar{x} = 0$.

52. By symmetry, $\bar{x} = 0$; $M = 2\int_0^4(2 - x/2)dx = 8$, $M_x = 2\int_0^4 \frac{1}{2}(2 + \frac{x}{2})(2 - \frac{x}{2})dx = 32/3$; $\bar{y} = \frac{32/3}{8} = \frac{4}{3}$.

53. $M_x = \int \tilde{y}dm = \delta\int_1^{16} \frac{1}{2\sqrt{x}} \cdot \frac{1}{\sqrt{x}}dx = \frac{1}{2}\delta[\ln x]_1^{16} = \delta \ln 4 = 2\delta \ln 2$; $M_y = \int \tilde{x}dm = \delta\int_1^{16} x\frac{1}{\sqrt{x}}dx = \delta \cdot \frac{2}{3}[x^{3/2}]_1^{16} = 42\delta$; $M = \int dm = \delta\int_1^{16}\frac{1}{\sqrt{x}}dx = 2\delta[x^{1/2}]_1^{16} = 6\delta$. $\bar{x} = M_y/M = 7$; $\bar{y} = M_x/M = \frac{1}{3}\ln 2$

54. $M_x = \int_1^{16} \frac{1}{2\sqrt{x}} \frac{1}{\sqrt{x}} \frac{4}{\sqrt{x}}dx = 2\int_1^{16} x^{-3/2}dx = 2(-2)[x^{-1/2}]_1^{16} = 3$; $M_y = \int_1^{16} x\frac{1}{\sqrt{x}}\frac{4}{\sqrt{x}}dx = 4\int_1^{16} dx = 4 \cdot 15 = 60$; $M = \int_1^{16}\frac{1}{\sqrt{x}}\frac{4}{\sqrt{x}}dx = 4[\ln x]_1^{16} = 4\ln 16$. $\bar{x} = M_y/M = 15/\ln 16$; $\bar{y} = M_x/M = 3/(4\ln 16)$

55. The radius of each cross-section is $\frac{1}{2}[x^2 - \sqrt{x}]$. $V = \int_0^1 \pi[\frac{1}{2}(x^2 - \sqrt{x})]^2dx = \frac{\pi}{4}\int_0^1(x^4 - 2x^{5/2} + x)dx = \frac{\pi}{4}[\frac{x^5}{5} - 2 \cdot \frac{2}{7}x^{7/2} + \frac{x^2}{x}]_0^1 = \frac{9\pi}{280}$.

56. The area of a cross-section is $\frac{1}{2} \cdot \frac{\sqrt{3}}{2}(2\sqrt{x} - x)^2 \Rightarrow V = \frac{\sqrt{3}}{4}\int_0^4(2\sqrt{x} - x)^2dx = \frac{\sqrt{3}}{4}\int_0^4(4x - 4x^{3/2} + x^2)dx = \frac{\sqrt{3}}{4}[2x^2 - \frac{8}{5}x^{5/2} + \frac{x^3}{3}]_0^4 = \frac{8\sqrt{3}}{15}$.

57. The radius of each cross-section is $\frac{1}{2}[2\sin x - 2\cos x]$. $V = \pi\int_{\pi/4}^{5\pi/4}[\sin x - \cos x]^2dx = \pi\int_{\pi/4}^{5\pi/4}[\sin^2 x - 2\sin x\cos x + \cos^2 x]dx = \pi\int_{\pi/4}^{5\pi/4}(1 - 2\sin x\cos x)dx = \pi[x + \frac{\cos^2 x}{2}]_{\pi/4}^{5\pi/4} = \pi^2$.

58. $V = \int_0^6(\sqrt{6} - \sqrt{x})^4dx = 14.4$, using technology. Analytically, $V = \int_0^6(36 - 4 \cdot 6\sqrt{6}\sqrt{x} + 6 \cdot 6 \cdot x - 4 \cdot \sqrt{6}x^{3/2} + x^2)dx = \frac{72}{5}$.

59. Each cross-section is an isosceles triangle with acute angles 45°. $A(y) = \frac{1}{2}$ base $\cdot$ height $= \frac{1}{2}(\text{base})^2 = \frac{1}{2}x^2 = \frac{1}{2}(9 - y^2)$; $V = \int_{-3}^3 \frac{1}{2}(9 - y^2)dy = 2 \cdot \frac{1}{2}\int_0^3(9 - y^2)dy = [9y - \frac{y^3}{3}]_0^3 = 18$.

60. The cross-sections at each value of x have area $\pi(\frac{x - \frac{x}{2}}{2})^2 = \pi x^2/16$; the altitude is 12. The cone can be generated by revolving $y = 3x/12 = x/4$, $0 \le x \le 12$, about the x-axis. The cross-section at x has area $\pi(\frac{x}{4})^2 = \pi x^2/16$. By Cavalieri's Theorem, the solids have the same volume.

61. b) $\int_0^6 |v(t)|dt = \int_0^2 (1 - t/2)dt + \int_2^6 (t/2 - 1)dt = [t - t^2/4]_0^2 + [t^2/4 - t]_2^6 = 5$ ft c) $\int_0^6 v(t)dt = \int_0^6 (t/2 - 1)dt = [t^2/4 - t]_0^6 = 3$ ft.

62. b) $\int_0^6 |v(t)|dt = \int_0^2 v(t)dt - \int_2^6 v(t)dt = 64/3$; analytically the integral is simplified by letting $\omega = t - 2 \Rightarrow \int_0^6 |v(t)|dt = \int_{-2}^4 |\omega(\omega - 4)|d\omega = \int_{-2}^0 (\omega^2 - 4\omega)d\omega - \int_0^4 (\omega^2 - 4\omega)d\omega = \frac{8}{3} + 8 - \frac{64}{3} + 32 = 64/3$ ft. c) $\int_0^6 (t-2)(t-6)dt = \int_{-2}^4 (\omega^2 - 4\omega)d\omega = 0$ ft.

63. b) $\int_0^{3\pi/2} |v(t)|dt = \int_0^{\pi/2} 5\cos t\, dt - \int_{\pi/2}^{3\pi/2} 5\cos t\, dt = [5\sin t]_0^{\pi/2} - [5\sin t]_{\pi/2}^{3\pi/2} =$ 15ft. c) $\int_0^{3\pi/2} 5\cos t\, dt = [5\sin t]_0^{3\pi/2} = -5$ ft.

64. b) $\int_0^{3/2} |v(t)|dt = \int_0^1 \pi\sin(\pi t)dt - \int_1^{3/2} \pi\sin(\pi t)dt = 1.5\cdot$ (area under one 'bump' of sine curve) $= 1.5(2) = 3$ ft. c) $2 - 1 = 1$ ft.

CHAPTER 7

THE CALCULUS OF TRANSCENDENTAL FUNCTIONS

7.1 THE NATURAL LOGARITHM FUNCTION

1. $\ln 4/9 = \ln 4 - \ln 9 = \ln 2^2 - \ln 3^2 = 2\ln 2 - 2\ln 3 = 2(\ln 2 - \ln 3)$.

2. $\ln 12 = \ln(2^2 3) = \ln 2^2 + \ln 3 = 2\ln 2 + \ln 3$.

3. $\ln(1/2) = \ln 1 - \ln 2 = -\ln 2$.

4. $\ln(1/3) = \ln 1 - \ln 3 = -\ln 3$.

5. $\ln 4.5 = \ln(9/2) = \ln 3^2 - \ln 2 = 2\ln 3 - \ln 2$.

6. $\ln \sqrt[3]{9} = \frac{1}{3}\ln 3^2 = \frac{2}{3}\ln 3$.

7. $\ln 3\sqrt{2} = \ln 3 + \ln 2^{1/2} = \ln 3 + \frac{1}{2}\ln 2$.

8. $\ln \sqrt{13.5} = \ln(\frac{27}{2})^{1/2} = \frac{1}{2}[\ln 27 - \ln 2] = \frac{1}{2}[\ln 3^3 - \ln 2] = \frac{1}{2}[3\ln 3 - \ln 2]$.

9. $y = \ln(x^2) = 2\ln x$. $\frac{dy}{dx} = \frac{2}{x}$. The result is supported by graphing $2/x$ and NDER$(\ln(x^2), x, x)$ in $[-5, 5]$ by $[-5, 5]$.

10. $y = (\ln x)^2$. $\frac{dy}{dx} = 2(\ln x)\frac{1}{x} - \frac{2\ln x}{x}$.

11. $y = \ln(1/x) = \ln x^{-1} = -\ln x$. $\frac{dy}{dx} = -\frac{1}{x}$.

12. $y = \ln(10/x) = \ln 10 - \ln x$. $\frac{dy}{dx} = -\frac{1}{x}$.

13. $y = \ln(x+2)$. $\frac{dy}{dx} = \frac{1}{x+2}(x+2)' = \frac{1}{x+2}$.

14. $y = \ln(2x+2)$. $\frac{dy}{dx} = \frac{1}{2x+2}(2) = \frac{1}{x+1}$.

15. $y = \ln(2 - \cos x)$. $\frac{dy}{dx} = \frac{\sin x}{2 - \cos x}$.

16. $y = \ln(x^2+1)$. $\frac{dy}{dx} = \frac{1}{x^2+1}(2x) = \frac{2x}{x^2+1}$.

17. $y = \ln(\ln x)$. $\frac{dy}{dx} = \frac{\frac{1}{x}}{\ln x} = \frac{1}{x\ln x}$.

18. $y = x\ln x - x$. $\frac{dy}{dx} = x\frac{1}{x} + \ln x - 1 = \ln x$.

19. $y = \sqrt{x(x+1)} = [x(x+1)]^{1/2}$. $\ln y = \ln[x(x+1)]^{1/2} = \frac{1}{2}\ln[x(x+1)] = \frac{1}{2}[\ln x + \ln(x+1)]$. $\frac{y'}{y} = \frac{1}{2}[\frac{1}{x} + \frac{1}{x+1}]$, $y' = \frac{y}{2}[\frac{x+1+x}{x(x+1)}] = \sqrt{x(x+1)}\ \frac{2x+1}{2x(x+1)} = \frac{2x+1}{2\sqrt{x(x+1)}}$. See the remark in the solution of Exercise 21.

20. $y = (\frac{x}{x+1})^{1/2}$. $\ln y = \frac{1}{2}\ln\frac{x}{x+1} = \frac{1}{2}(\ln x - \ln(x+1))$. $\frac{y'}{y} = \frac{1}{2}(\frac{1}{x} - \frac{1}{x+1})$. $y' = \frac{1}{2}\sqrt{\frac{x}{x+1}}\ (\frac{1}{x(x+1)}) = \frac{1}{2\sqrt{x}\ (x+1)^{3/2}}$.

21. $y = (x+3)^{1/2}\sin x$. $\ln y = \ln(x+3)^{1/2} + \ln\sin x$, $\ln y = \frac{1}{2}\ln(x+3) + \ln\sin x$. $\frac{y'}{y} = \frac{1}{2}\frac{1}{x+3} + \frac{\cos x}{\sin x}$, $y' = (x+3)^{1/2}\sin x[\frac{1}{2(x+3)} + \frac{\cos x}{\sin x}] = \frac{\sin x}{2\sqrt{x+3}} + \sqrt{x+3}\ \cos x = \frac{\sin x + 2(x+3)\cos x}{2\sqrt{x+3}}$. Remark: To be sure we are "taking log's" of positive numbers only, we could proceed as follows. Since $y > 0$, $y = |y| = |x+3|^{1/2}|\sin x|$ in the domain of y. $\ln|y| = \frac{1}{2}\ln|x+3| + \ln|\sin x|$. Differentiating the last equation yields $\frac{y'}{y} = \frac{1}{2(x+3)} + \frac{\cos x}{\sin x}$ and we may proceed as above. Other exercises in this section may be treated in a similar fashion.

22. $y = \frac{\tan x}{\sqrt{2x+1}}$. $\ln y = \ln\tan x - \frac{1}{2}\ln(2x+1)$. $\frac{y'}{y} = \frac{\sec^2 x}{\tan x} - \frac{1}{2}\frac{2}{2x+1}$. $y' = \frac{\tan x}{\sqrt{2x+1}}\ [\frac{\sec^2 x}{\tan x} - \frac{1}{2x+1}]$.

23. $y = x(x+1)(x+2)$, $x > 0$. $\ln y = \ln x + \ln(x+1) + \ln(x+2)$, $\frac{y'}{y} = \frac{1}{x} + \frac{1}{x+1} + \frac{1}{x+2}$, $y' = x(x+1)(x+2)(\frac{1}{x} + \frac{1}{x+1} + \frac{1}{x+2}) = (x+1)(x+2) + x(x+2) + x(x+1) = 3x^2 + 6x + 2$.

24. $y = \frac{1}{x(x+1)(x+2)}$. $\ln y = -\ln x - \ln(x+1) - \ln(x+2)$. $\frac{y'}{y} = -\frac{1}{x} - \frac{1}{x+1} - \frac{1}{x+2}$. $y' = \frac{-1}{x(x+1)(x+2)}\ [\frac{1}{x} + \frac{1}{x+1} + \frac{1}{x+2}]$.

25. $y = \frac{x+5}{x\cos x}$, $x > 0$. $\ln y = \ln(x+5) - \ln x - \ln\cos x$. $\frac{y'}{y} = \frac{1}{x+5} - \frac{1}{x} + \frac{\sin x}{\cos x}$, $y' = \frac{x+5}{x\cos x}\ [\tan x - \frac{5}{x(x+5)}]$.

26. $y = \frac{x\sin x}{\sqrt{\sec x}}$. $\ln y = \ln x + \ln\sin x - \frac{1}{2}\ln\sec x$. $\frac{y'}{y} = \frac{1}{x} + \frac{\cos x}{\sin x} - \frac{1}{2}\frac{\sec x\tan x}{\sec x}$, $y' = \frac{x\sin x}{\sqrt{\sec x}}\ [\frac{1}{x} + \cot x - \frac{1}{2}\tan x]$.

27. $y = \frac{x(x^2+1)^{1/2}}{(x+1)^{2/3}}$. $\ln y = \ln x + \frac{1}{2}\ln(x^2+1) - \frac{2}{3}\ln(x+1)$. $\frac{y'}{y} = \frac{1}{x} + \frac{1}{2}\frac{2x}{x^2+1} - \frac{2}{3}\frac{1}{x+1}$, $y' = \frac{x\sqrt{x^2+1}}{(x+1)^{2/3}}\ [\frac{1}{x} + \frac{x}{x^2+1} - \frac{2}{3(x+1)}]$.

28. $y = \sqrt{\frac{(x+1)^{10}}{(2x+1)^5}}$. $\ln y = \frac{1}{2}[10\ln(x+1) - 5\ln(2x+1)]$, $\frac{y'}{y} = \frac{5}{x+1} - \frac{5}{2}\frac{2}{2x+1}$, $y' = 5\sqrt{\frac{(x+1)^{10}}{(2x+1)^5}}\ [\frac{1}{x+1} - \frac{1}{2x+1}]$.

29. $y = (\frac{x(x-2)}{x^2+1})^{1/3}$. $\ln y = \frac{1}{3}[\ln x + \ln(x-2) - \ln(x^2+1)]$, $\frac{y'}{y} = \frac{1}{3}[\frac{1}{x} + \frac{1}{x-2} - \frac{2x}{x^2+1}]$. $y' = \frac{2}{3}\sqrt[3]{\frac{x(x-2)}{x^2+1}}\ \frac{(x^2+x-1)}{x(x-2)(x^2+1)}$.

30. $y = \sqrt[3]{\frac{x(x+1)(x-2)}{(x^2+1)(2x+3)}}$. $\ln y = \frac{1}{3}[\ln x + \ln(x+1) + \ln(x-2) - \ln(x^2+1) - \ln(2x+3)]$. $\frac{y'}{y} = \frac{1}{3}[\frac{1}{x} + \frac{1}{x+1} + \frac{1}{x-2} - \frac{2x}{x^2+1} - \frac{2}{2x+3}]$, $y' = \frac{1}{3}\sqrt[3]{\frac{x(x+1)(x-2)}{(x^2+1)(2x+3)}}\ [\frac{1}{x} + \frac{1}{x+1} + \frac{1}{x-2} - \frac{2x}{x^2+1} - \frac{2}{2x+3}]$.

31. $\int_{-3}^{-2} \frac{dx}{x} = \ln|x|\ ]_{-3}^{-2} = \ln 2 - \ln 3$.

32. $\int_{-9}^{-4} \frac{dx}{2x} = \frac{1}{2}\int_{-9}^{-4} \frac{dx}{x} = \frac{1}{2}\ln|x|\ ]_{-9}^{-4} = \frac{1}{2}[\ln 4 - \ln 9] = \ln(\frac{4}{9})^{1/2} = \ln\frac{2}{3}$.

33. $I = \int_{-1}^{0} \frac{3dx}{3x-2}$. Let $u = 3x-2$, $du = 3dx$ and note that $u = -5$ when $x = -1$ and $u = -2$ when $x = 0$. $I = \int_{-5}^{-2} \frac{du}{u} = \ln|u|\ ]_{-5}^{-2} = \ln 2 - \ln 5$.

34. $\int_{-1}^{0} \frac{dx}{2x+3} = \frac{1}{2}\ln|2x+3|\ ]_{-1}^{0} = \frac{1}{2}[\ln 3 - \ln 1] = \frac{1}{2}\ln 3$.

35. $\int_3^4 \frac{dx}{x-5} = \ln|x-5|\ ]_3^4$ (by inspection) $= \ln 1 - \ln 2 = -\ln 2$.

36. $\int_2^5 \frac{3x}{1-x} = -\ln|1-x|\ ]_2^5 = -[\ln 4 - \ln 1] = -\ln 4$.

37. Let $I = \int_0^3 \frac{2xdx}{x^2-25}$, $u = x^2 - 25$, $du = 2xdx$. Then $I = \int_{-25}^{-16} \frac{du}{u} = \ln|u|\ ]_{-25}^{-16} = \ln 16 - \ln 25 = \ln(16/25) = \ln(4/5)^2 = 2\ln(4/5) = 2\ln(0.8)$.

38. $\int_0^1 \frac{8xdx}{4x^2-5} = \ln|4x^2-5|\ ]_0^1 = \ln 1 - \ln 5 = -\ln 5$.

39. $\int_0^3 \frac{1}{x+1}dx$, $u = x+1$, $du = dx$. $\int_1^4 \frac{du}{u} = \ln u]_1^4 = \ln 4 - \ln 1 = \ln 4$.

40. $I = \int_0^4 \frac{2xdx}{x^2+9}$, $u = x^2+9$. $du = 2xdx$. $I = \int_9^{25} \frac{du}{u} = [\ln u]_9^{25} = \ln 25 - \ln 9 = \ln\frac{25}{9}$.

41. Let $I = \int_0^{\pi} \frac{\sin x}{2-\cos x}dx$. Let $u = 2 - \cos x$. Then $du = \sin x dx$, $u = 1$ when $x = 0$ and $u = 3$ when $x = \pi$, $I = \int_1^3 \frac{du}{u} = \ln|u|\ ]_1^3 = \ln 3 - \ln 1 = \ln 3$.

42. $\int_0^{\pi/3} \frac{4\sin x}{1-4\cos x}dx = \ln|1-4\cos x|\ ]_0^{\pi/3} = \ln|1-4(\frac{1}{2})| - \ln|1-4(1)| = \ln 1 - \ln 3 = -\ln 3$.

43. Let $I = \int_1^2 \frac{2\ln x}{x}dx$, $u = \ln x$, $du = \frac{1}{x}$. Then $I = 2\int_0^{\ln 2} udu = u^2]_0^{\ln 2} = (\ln 2)^2$.

44. Let $I = \int_3^4 \frac{5\ln x}{x}dx$, $u = \ln x$, $du = \frac{dx}{x}$. Then $I = 5\int_{\ln 3}^{\ln 4} udu = \frac{5u^2}{2}]_{\ln 3}^{\ln 4} = \frac{5}{2}[(\ln 4)^2 - (\ln 3)^2]$.

45. $I = \int_0^{\pi/2} \tan \frac{x}{2} dx = \int_0^{\pi/2} \frac{\sin \frac{x}{2}}{\cos \frac{x}{2}} dx$. Let $u = \cos \frac{x}{2}$. Then $du = -\frac{1}{2}(\sin \frac{x}{2})dx$, $-2du = (\sin \frac{x}{2})dx$. $I = -2 \int_1^{\sqrt{2}/2} \frac{du}{u} = -2 \ln |u| \,]_1^{\sqrt{2}/2} = -2[\ln(\sqrt{2})^{-1} - \ln 1] = 2 \ln 2^{1/2} = \ln 2$.

46. $\int_{-\pi/2}^{-\pi/4} \cot x dx = \ln |\sin x| \,]_{-\pi/2}^{-\pi/4} = \ln \frac{\sqrt{2}}{2} - \ln 1 = -\frac{1}{2} \ln 2$.

47. Let $I = \int_{\pi/2}^{\pi} 2 \cot \frac{x}{3} dx = 2 \int_{\pi/2}^{\pi} \frac{\cos \frac{x}{3}}{\sin \frac{x}{3}} dx$. Let $u = \sin \frac{x}{3}$, $du = \frac{1}{3}(\cos \frac{x}{3})dx$. Then $I = 6 \int_{1/2}^{\sqrt{3}/2} \frac{du}{u} = 6 \ln |u| \,]_{1/2}^{\sqrt{3}/2} = 6[(\ln \sqrt{3} - \ln 2) - (\ln 1 - \ln 2)] = 6(\frac{1}{2}) \ln 3 = 3 \ln 3$.

48. $\int_0^{\pi/12} 6 \tan 3x \; dx = -2 \ln |\cos 3x| \,]_0^{\pi/12} = -2[\ln \frac{\sqrt{2}}{2} - \ln 1] = \ln(\frac{1}{\sqrt{2}})^{-2} = \ln 2$.

49. $\int \frac{2dx}{x} = 2 \ln |x| + C = \ln x^2 + C$.

50. $\int \frac{dx}{3x+1} = \frac{1}{3} \ln |3x + 1| + C$.

51. Let $I = \int \frac{xdx}{x^2+4}$, $u = x^2 + 4$, $du = 2xdx$, $\frac{du}{2} = xdx$. Then $I = \frac{1}{2} \int \frac{du}{u} = \frac{1}{2} \ln |u| + C = \frac{1}{2} \ln(x^2 + 4) + C$.

52. Let $I = \int \frac{\sin x \; dx}{1+2\cos x}$, $u = 1 + 2\cos x$, $du = -2 \sin x \; dx, -\frac{1}{2}du = \sin x \; dx$. Then $I = \frac{1}{2} \int \frac{du}{u} = -\frac{1}{2} \ln |u| + C = -\frac{1}{2} \ln |1 + 2\cos x| + C$.

53. Let $I = \int \tan \frac{x}{3} dx = \int \frac{\sin(x/3)dx}{\cos(x/3)}$, $u = \cos \frac{x}{3}$, $du = -\frac{1}{3} \sin \frac{x}{3} dx$, $-3du = \sin \frac{x}{3} dx$. Then $I = -3 \int \frac{du}{u} = -3 \ln |u| + C = -3 \ln |\cos \frac{x}{3}| + C$.

54. Let $I = \int \cot 2x dx = \int \frac{\cos 2x}{\sin 2x} dx$, $u = \sin 2x$, $du = 2\cos 2x dx$, $\frac{1}{2}du = \cos 2x dx$. Then $I = \frac{1}{2} \int \frac{du}{u} = \frac{1}{2} \ln |u| + C = \frac{1}{2} \ln |\sin 2x| + C$.

55. $\lim_{x \to \infty} \ln \frac{1}{x} = \lim_{u \to 0+} \ln u = -\infty$.

56. $\lim_{x \to 0+} \ln \frac{1}{x} = \lim_{u \to \infty} \ln u = \infty$ using $u = \frac{1}{x}$.

57. $\lim_{x \to 0} \ln |x| = -\infty$.

58. $\lim_{x \to 0} \ln |\frac{1}{x}| = \lim_{u \to \infty} \ln u = \infty$. Also $\lim_{x \to 0} \ln \frac{1}{|x|} = \lim_{x \to 0}[-\ln |x| \,] = \infty$.

59. $\int_0^{\pi/3} \tan x \; dx = -\ln |\cos x| \Big|_0^{\pi/3} = -[\ln |\cos(\pi/3)| - \ln |\cos 0| \,] = -[\ln(1/2) - \ln 1] = -\ln 2^{-1} = \ln 2$.

60. $\pi \int_{\pi/6}^{\pi/2} (\sqrt{\cot x})^2 dx = \pi \ln |\sin x| \Big|_{\pi/6}^{\pi/2} = \pi[\ln 1 - \ln \frac{1}{2}] = \pi[0 - \ln 2^{-1}] = \pi \ln 2$.

61. Suppose $f(x)$ is an increasing function on its domain. Let $x_1 < x_2$ be distinct points of the domain. Then $f(x_1) < f(x_2)$ and so $f(x_1) \neq f(x_2)$. Therefore $f(x)$ is one-to-one. The proof for decreasing functions is similar. Since $\frac{d}{dx}(\ln x) = \frac{1}{x} > 0$ for $x > 0$, $\ln x$ is an increasing function. Therefore $\ln x$ is one-to-one.

62. $y = \frac{(x^2+1)(x+3)^{1/2}}{x-1}$. a) We require $x + 3 \geqq 0$ and $x \neq 1$. Thus the domain is $[-3,1) \cup (1,\infty)$. b) Graph y in $[-3,4]$ by $[-3,15]$. c) $(-3,1) \cup (1,\infty)$.

63. a) Graph $y_1 = \sqrt{x+3}\ \sin x$ in $[-4,20]$ by $[-5,5]$. b) In the same viewing rectangle graph $y_2 = \text{NDER}(y_1, x)$ and $y_3 = \frac{dy}{dx}$ (found in Exercise 21). The fact that y_2 and y_3 coincide supports that y_3 is valid where $y < 0$.

64. a) Graph $y_1 = x(x+1)(x+2)$ in $[-5,3]$ by $[-5,5]$. Proceed now as in Exercise 63.

65. Graph $y_1 = \frac{x+5}{x \cos x}$ in $[-20,20]$ by $[-20,20]$. Proceed now as in Exercise 63.

66. Graph $y_1 = \frac{x\sqrt{x^2+1}}{(x+1)^{2/3}} = x\sqrt{x^2+1}/((x+1)^2)^{1/3}$ in $[-7.6,5.7]$ by $[-16,9.7]$. Proceed now as in Exercise 63.

67. Graph $y_1 = (x(x-2)/(x^2+1))^{1/3}$ in $[-20,20]$ by $[-1.1,1.2]$. Proceed now as in Exercise 63.

68. $y = fg$. $|y| = |fg| = |f||g|$. $\ln|y| = \ln|f| + \ln|g|$. $\frac{y'}{y} = \frac{f'}{f} + \frac{g'}{g}$, $y' = fg(\frac{f'}{f} + \frac{g'}{g}) = f'g + fg'$ or $\frac{dy}{dx} = \frac{df}{dx}g + f\frac{dg}{dx}$.

69. $y = \frac{f}{g}$. $|y| = \frac{|f|}{|g|}$. $\ln|y| = \ln|f| - \ln|g|$. $\frac{y'}{y} = \frac{f'}{f} - \frac{g'}{g}$, $y' = \frac{f}{g}(\frac{f'}{f} - \frac{g'}{g}) = \frac{f'}{g} - \frac{fg'}{g^2} = \frac{gf'-fg'}{g^2}$ or $\frac{dy}{dx} = \frac{\frac{df}{dx}g - f\frac{dg}{dx}}{g^2}$.

70. a) Graph $y = f(x) = \text{NINT}(\tan t, t, 0, x)$ in $[-\frac{\pi}{2}, \frac{\pi}{2}]$ by $[-3,3]$. (Use $tol = 1$.) Conjecture: $\lim_{x\to(-\pi/2)^+} f(x) = \lim_{x\to(\pi/2)^-} f(x) = \infty$. b) Near $x = 2$ there is rather wild oscillating behavior due to the discontinuity of the integrand at $x = \frac{\pi}{2}$. c) Except for the vertical asymptote at $x = \frac{\pi}{2}$ the graph of the antiderivative $-\ln|\cos x|$ is smooth. The graphs are the same on $(-\frac{\pi}{2}, \frac{\pi}{2})$.

71. a) Graph $y = f(x) = \text{NINT}(\cot t, t, \pi/2, x)$ in $[0,\pi]$ by $[-5,5]$. (Use $tol = 1$.) Conjecture: $\lim_{x\to 0^+} f(x) = \lim_{x\to\pi^-} f(x) = -\infty$. b) Near $x = 4$ there is rather wild oscillating behavior due to the discontinuity of the integrand at $x = \pi$. c) Except for the vertical asymptote at $x = \pi$ the graph of the antiderivative $\ln|\sin x|$ is smooth. The two graphs are identical on $(0,\pi)$.

72. a) Graph y_1 and y_2 in $[-5,5]$ by $[-3,5]$ for several values of a. The graph of y_1 can be obtained from the graph of y_2 by a vertical shift. b) There is no value of a for which $y_1 = y_2$. This can be confirmed analytically. $y_1 = \ln(t^2+3)\,]_a^x = \ln(x^2+3) - \ln(a^2+3)$. For no value of a does this become $y_2 = \ln|t|\,]_1^{x^2+3} = \ln(x^2+3) - \ln 1 = \ln(x^2+3)$. Thus $\int_a^x f(t)dt$ may not give the entire family of antiderivatives of $f(x)$. c) y_1 is an antiderivative of $\frac{2x}{x^2+3}$. By part a), $y_1 = y_2 + C$. But $y_2 = \int_1^{x^2+3} \frac{dt}{t} = \ln|t|\,]_1^{x^2+3} = \ln(x^2+3) - \ln 1 = \ln(x^2+3)$. Therefore $y_1 = \ln(x^2+3) + C$ is the antiderivative of $\frac{2x}{x^2+3}$.

7.2 THE EXPONENTIAL FUNCTION

1. $e^{\ln 7} = 7$

2. $e^{-\ln 7} = e^{\ln 7^{-1}} = 7^{-1} = \frac{1}{7}$

3. $\ln e^2 = 2$

4. $e^{3\ln 2} = e^{\ln 2^3} = 8$

5. $e^{2+\ln 3} = e^2 e^{\ln 3} = 3e^2$

6. $e^{-2\ln 3} = e^{\ln 3^{-2}} = \frac{1}{9}$

7. $e^{2k} = 4$. $2k = \ln e^{2k} = \ln 4$. $k = \frac{1}{2}\ln 4 = \ln 2$

8. $e^{5k} = \frac{1}{4}$. $\ln e^{5k} = \ln\frac{1}{4}$, $5k = \ln 1 - \ln 4 = -\ln 4$, $k = \frac{-\ln 4}{5}$

9. $100e^{10k} = 200$. $e^{10k} = 2$, $10k = \ln 2$, $k = \frac{\ln 2}{10}$

10. $100e^k = 1$, $e^k = 0.01$, $k = \ln 0.01$

11. $2^{k+1} = 3^k$. $(k+1)\ln 2 = k\ln 3$, $k[\ln 3 - \ln 2] = \ln 2$, $k = \frac{\ln 2}{\ln\frac{3}{2}}$

12. $4^{(k-1)/2} = 3^k$ leads to $\frac{(k-1)}{2}\ln 4 = k\ln 3$, $-\frac{1}{2}\ln 2^2 = k(\ln 3 - \frac{1}{2}\ln 2^2)$, $-\ln 2 = k(\ln 3 - \ln 2)$, $k = \frac{\ln 2}{\ln 2 - \ln 3}$.

13. $e^t = 1$. $t = \ln e^t = \ln 1 = 0$.

14. $e^{kt} = \frac{1}{2}$, $kt = \ln\frac{1}{2} = -\ln 2$, $t = \frac{-\ln 2}{k}$

15. $e^{-0.3t} = 27$. $-0.3t = \ln 27$, $t = -\frac{\ln 27}{0.3} = -\frac{3\ln 3}{0.3}\frac{10}{10} = -10\ln 3$

16. $e^{-0.01t} = 1000$, $-0.01t = \ln 1000$, $t = -\frac{\ln 1000}{0.01} = -100\ln 1000 = -100\ln 10^3 = -300\ln 10$

17. $2^{e^t} = 2 - t$. $y = 2^{e^t}$ is an increasing function while $y = 2 - t$ is a linear decreasing function. There is therefore at most one point of intersection. By inspection $(0,2)$ is one such point. Therefore $t = 0$ is the only solution.

18. $e^{-2^t} = t + 2$, $e^{-2^t} - t - 2 = 0$. We graph $y = e^{-2^x} - x - 2 = $ e^(−2^x) $- x - 2$ and zoom in to the x-intercept $x = -1.328\ldots$.

19. $\ln y = 2t = 4.\ y = e^{2t+4}$

20. $\ln y = -t+5,\ y = e^{-t+5}$

21. $\ln(y-40) = 5t.\ y-40 = e^{5t},\ y = 40+e^{5t}$

22. $\ln(1-2y) = t,\ 1-2y = e^t,\ y = \frac{1-e^t}{2}$

23. $5+\ln y = 2^{x^2+1}.\ \ln y = 2^{x^2+1}-5,\ y = e^{(2^{x^2+1}-5)}$

24. $\ln(2^y-1) = x^2-3,\ 2^y-1 = e^{x^2-3},\ 2^y = e^{x^2-3}+1,\ y\ln 2 = \ln(e^{x^2-3}+1),\ y = \frac{\ln(e^{x^2-3}+1)}{\ln 2}$

25. $y = 2e^x.\ dy/dx = 2e^x$

26. $y = e^{2x}.\ \frac{dy}{dx} = (e^{2x})2 = 2e^{2x}$

27. $y = e^{-x}.\ dy/dx = e^{-x}(-1) = -e^{-x}$

28. $y = e^{-5x}.\ \frac{dy}{dx} = -5e^{-5x}$

29. $y = e^{2x/3}.\ \frac{dy}{dx} = \frac{2}{3}e^{2x/3}$

30. $y = e^{-x/4}.\ \frac{dy}{dx} = -\frac{1}{4}e^{-x/4}$

31. $y = xe^2 - e^x.\ \frac{dy}{dx} = e^2 - e^x$

32. $y = x^2e^x - xe^x.\ \frac{dy}{dx} = x^2e^x + 2xe^x - xe^x - e^x = x^2e^x + xe^x - e^x$

33. $y = e^{\sqrt{x}}.\ \frac{dy}{dx} = e^{\sqrt{x}}\frac{1}{2\sqrt{x}} = \frac{e^{\sqrt{x}}}{2\sqrt{x}}$

34. $y = e^{x^2}.\ \frac{dy}{dx} = 2xe^{x^2}$

35. $\int_1^{e^2}\frac{dx}{x} = \ln x]_1^{e^2} = \ln e^2 - \ln 1 = 2-0 = 2$

36. $\int_1^e \frac{2}{x}dx = 2\int_1^e\frac{dx}{x} = 2[\ln x]_1^e = 2$

37. $\int_{\ln 2}^{\ln 3} e^x dx = e^x]_{\ln 2}^{\ln 3} = e^{\ln 3} - e^{\ln 2} = 3-2 = 1$

38. $I = \int_{-1}^1 e^{(x+1)}dx$. Let $u = x+1,\ du = dx.\ I = \int_0^2 e^u du = [e^u]_0^2 = e^2-1.$

39. $\int_{\ln 3}^{\ln 5} e^{2x}dx,\ u = 2x,\ du = 2\,dx.\ \frac{1}{2}\int_{2\ln 3}^{2\ln 5} e^u du = \frac{1}{2}e^u]_{\ln 9}^{\ln 25} = \frac{1}{2}(e^{\ln 25} - e^{\ln 9}) = \frac{1}{2}(25-9) = 8.$

40. $I = \int_0^{\ln 2} e^{-x}dx,\ u = -x.\ du = -dx.\ I = -\int_0^{-\ln 2} e^u du = -[e^u]_0^{\ln 2^{-1}} = -[2^{-1}-1] = \frac{1}{2}$

41. $\int_0^1(1+e^x)e^x dx,\ u = 1+e^x,\ du = e^x dx.\ \int_2^{1+e} u\,du = \frac{u^2}{2}]_2^{1+e} = \frac{1}{2}[(1+e)^2-2^2] = \frac{1}{2}(e^2+2e-3)$

42. Let $I = \int_{-1}^1 \frac{e^x}{1+e^x}dx,\ u = 1+e^x,\ du = e^x dx.$ Then $I = \int_{1+e^{-1}}^{1+e}\frac{du}{u} = \ln|u|]_{1+e^{-1}}^{1+e} = \ln(1+e) - \ln(1+e^{-1}) = \ln(1+e) - \ln(\frac{e+1}{e}) = \ln(1+e) - \ln(e+1) + \ln e = 1$

43. $\int_2^4 \frac{dx}{x+2},\ u = x+2,\ du = dx.\ \int_4^6\frac{du}{u} = \ln u]_4^6 = \ln 6 - \ln 4 = \ln(6/4) = \ln(3/2)$

44. $I = \int_{-1}^{0} \frac{8dx}{2x+3}$. Let $u = 2x+3$, $du = 2dx$. $I = 4\int_1^3 \frac{du}{u} = 4[\ln u]_1^3 = 4\ln 3$

45. $\int_{-1}^{1} 2xe^{-x^2}dx$, $u = -x^2$, $du = -2xdx$. $-\int_{-1}^{-1} e^u du = 0$

46. $I = \int_0^1 \frac{xdx}{4x^2+1}$. Let $u = 4x^2+1$, $du = 8xdx$, $xdx = \frac{1}{8}du$. $I = \frac{1}{8}\int_1^5 \frac{du}{u} = \frac{1}{8}[\ln u]_1^5 = \frac{\ln 5}{8}$

47. $\int_1^4 \frac{e^{\sqrt{x}}dx}{2\sqrt{x}}$, $u = \sqrt{x}$, $du = \frac{dx}{2\sqrt{x}}$. $\int_1^2 e^u du = e^u]_1^2 = e^2 - e$

48. Let $I = \int_e^{e^2} \frac{1}{x\ln x}dx$. Let $u = \ln x$, $du = \frac{1}{x}dx$. $I = \int_1^2 \frac{du}{u} = \ln|u|]_1^2 = \ln 2$

49. Let $I = \int 2e^x\cos(e^x)dx$, $u = e^x$, $du = e^x dx$. Then $I = 2\int \cos u du = 2\sin u + C = 2\sin(e^x) + C$.

50. $\int 3e^x \sin(e^x)dx = -3\cos(e^x) + C$

51. Let $I = \int \frac{e^x dx}{1+e^x}$, $u = 1+e^x$, $du = e^x dx$. Then $I = \int \frac{du}{u} = \ln|u| + C = \ln(1+e^x) + C$.

52. $\int \frac{dx}{1-e^{-x}} = \int \frac{e^x}{e^x}\frac{dx}{1-e^{-x}} = \int \frac{e^x dx}{e^x - 1} = \ln|e^x - 1) + C$

53. Let $I = \int \frac{\tan(\sqrt{x})dx}{\sqrt{x}}$, $u = \sqrt{x}$, $du = \frac{dx}{2\sqrt{x}}$, $\frac{dx}{\sqrt{x}} = 2du$. Then $I = 2\int \tan u\, du = -2\ln|\cos u| + C = -2\ln|\cos(\sqrt{x})| + C$

54. Let $I = \int \frac{\cot(\sqrt{2x})dx}{\sqrt{x}}$, $u = \sqrt{2x}$, $du = \frac{2dx}{2\sqrt{2x}} = \frac{dx}{\sqrt{2}\sqrt{x}}$, $\sqrt{2}du = \frac{dx}{\sqrt{x}}$. Then $I = \sqrt{2}\int \cot u\, du = \sqrt{2}\ln|\sin u| + C = \sqrt{2}\ln|\sin(\sqrt{2x})| + C$.

55. a) Let $y = f(x) = 2x+3$. The inverse relation is determined by $x = 2y + 3$, $2y = x - 3$, $y = \frac{x-3}{2}$, $f^{-1}(x) = \frac{x-3}{2}$. b) We may graph $y = 2x+3$ and $y = \frac{x-3}{2}$ together in $[-10, 10]$ by $[-10, 10]$. c) $\frac{df}{dx}|_{x=a} = 2$, $\frac{df^{-1}}{dx}|_{x=f(a)} = \frac{1}{2}$ which verifies the Equation.

56. $f(x) = 5 - 4x$, $a = \frac{1}{2}$. a) $x = 5 - 4y$, $4y = 5 - x$, $y = \frac{5-x}{4}$, $f^{-1}(x) = \frac{5-x}{4}$. b) Graph f and f^{-1} in $[-7.4, 10.1]$ by $[-2.9, 7.4]$. c) $\frac{df}{dx}|_{x=\frac{1}{2}} = -4$. $f(a) = f(\frac{1}{2}) = 3$. $\frac{df^{-1}}{dx}|_{x=3} = -\frac{1}{4}$, verifying the Equation.

57. $y = f(x) = \frac{1-2x}{x+2}$. $x = \frac{1-2y}{y+2}$, $xy + 2x = 1 - 2y$, $xy + 2y = 1 - 2x$, $y(x+2) = 1 - 2x$, $y = \frac{1-2x}{x+2}$, $f^{-1}(x) = \frac{1-2}{x+2}$. Thus $f = f^{-1}$, f is self-inverse. b) We see that the graph of f is symmetric about the line $y = x$ confirming that it is self-inverse. c) $f' = (f^{-1})' = \frac{(x+2)(-2)-(1-2x)}{(x+2)^2} = \frac{-5}{(x+2)^2}$. $f'(a) = \frac{5}{(a+2)^2}$. $f(a) = \frac{1-2a}{a+2}$, $\frac{df^1}{dx}|_{f(a)} = \frac{-5}{(\frac{1-2a}{a+2}+2)^2} = \frac{-5(a+2)^2}{(1-2a+2a+4)^2} = -\frac{(a+2)^2}{5}$, verifying the Equation.

58. $y = f(x) = \frac{x-5}{x-3}$, $a = 2$. a) $x = \frac{y-5}{y-3}$, $xy - 3x = y - 5$, $y(x-1) = 3x - 5$, $y = f^{-1}(x) = \frac{3x-5}{x-1}$. b) Graph f and f^{-1} in $[-6.5, 9.4]$ by $[-6.1, 7.7]$. c) $f'(x) = \frac{(x-3)-(x-5)}{(x-3)^2} = \frac{2}{(x-3)^2}$. $f'(2) = 2$. $(f^{-1}(x))' = \frac{(x-1)3-(3x-5)}{(x-1)^2} = \frac{2}{(x-1)^2}$. $f(2) = \frac{-3}{-1} = 3$. $(f^{-1}(x))'|_{x=3} = \frac{2}{2^2} = \frac{1}{2}$, verifying Eq. (19).

59. $f'(x) = 2x - 4$. $\frac{df^{-1}}{dx}|_{x=f(4)=-3} = \frac{1}{f'(4)} = \frac{1}{4}$

60. $f'(x) = 2x - 4$. $\frac{df^{-1}}{dx}|_{x=2=f(5)} = \frac{1}{f'(5)} = \frac{1}{6}$

61. $f(x) = \frac{2x+3}{x+3} = 2 - \frac{3}{x+3}$ whence $f'(x) = \frac{3}{(x+3)^2}$. $f^{-1}(x) = \frac{-3x+3}{x-2} = -3 - \frac{3}{x-2}$ whence $\frac{df^{-1}(x)}{dx} = \frac{3}{(x-2)^2}$. $f'(-5) = \frac{3}{(-2)^2} = \frac{3}{4}$. $(f^{-1})'(3.5) = \frac{3}{(3/2)^2} = \frac{4}{3}$.

62. $\lim_{x\to\infty} e^{-x} = \lim_{x\to\infty} \frac{1}{e^x} = 0$

63. $\lim_{x\to-\infty} e^{-x} = \infty$

64. $\lim_{x\to-\infty} \ln(2 + e^x) = \lim_{x\to-\infty} \ln(2 + \frac{1}{e^{-x}}) = \ln(2+0) = \ln 2$

65. $\lim_{x\to\infty} \int_x^{2x} \frac{1}{t} dt = \lim_{x\to\infty} \ln|t|]_x^{2x} = \lim_{x\to\infty}[\ln|2x| - \ln|x|] = \lim_{x\to\infty} \frac{\ln 2|x|}{|x|} = \ln 2$

66. $\int_0^x \frac{1}{\sqrt{2\pi}} e^{(-t^2/2)} dt = 0.3$. Graph $y = \text{NINT}(\frac{1}{\sqrt{2\pi}} e^{(-t^2/2)}, t, 0, x)$ and $y = 0.3$ in $[-3, 3]$ by $[-2, 2]$ and zoom in to the point of intersection: $x = 0.84162123357$.

67. We graph $y = \text{NINT}(2^t \ln t, t, 1, x) - 0.1$ in $[0.001, 3]$ by $[-1, 10]$ and zoom in to the x-intercepts getting $\{0.679\ldots, 1.3086\ldots\}$.

68. $f'(x) = e^x$, $f(0) = f'(0) = e^0 = 1$. We are finding $L(x) = f(0) + f'(0)(x - 0) = 1 + x$.

69. $f(x) = x + e^{4x}$, $f'(x) = 1 + 4e^{4x}$. $L(x) = f(0) + f'(0)(x-0) = 1 + 5x$

70. $f(x) = x^2 \ln(1/x) = -x^2 \ln x$. $f'(x) = -x^2(\frac{1}{x}) - 2x \ln x = -x(1 + 2\ln x)$. The factor $-x$ is always negative since $x > 0$ here. The factor $1 + 2\ln x$ is first negative, then positive, being 0 and changing sign at $x = e^{-1/2}$. Thus $f'(x)$ is first positive and then negative, being 0 and changing sign at $x = e^{-1/2}$. Therefore $f(x)$ increases on $(0, e^{-1/2})$ and decreases on $(e^{-1/2}, \infty)$ and so has maximum value $f(e^{-1/2}) = e^{-1} \ln e^{1/2} = \frac{1}{2e}$.

71. Since $f(x)$ is periodic of period 2π, we need only investigate the extreme values of f on $[0, 2\pi]$. Since e^x is an increasing function, we can use the extreme values of $\sin x$. Thus the maximum and minimum values of f are, respectively, e and e^{-1}.

72. $y = e^{\sin x} + C$. When $x = 0$, $0 = y = e^0 + C$, $C = -1$, $y = e^{\sin x} - 1$.

73. $\frac{dy}{dx} = 1 + \frac{1}{x}$, $y(1) = 3$. $y(x) = 3 + \int_1^x (1 + \frac{1}{t})dt = 3 + [t + \ln|t|]_1^x = 3 + [x + \ln|x|] - 1 = 2 + x + \ln|x|$

74. $v = \int \frac{4dt}{(4-t)^2}$. Let $u = 4-t$, $du = -dt$. $v = -4\int u^{-2}du = 4u^{-1} + C = \frac{4}{4-t} + C$. When $t = 0$, $2 = v = \frac{4}{4-0} + C = 1 + C$, $C = 1$ and $v = \frac{4}{4-t} + 1$. The desired distance is $\int_1^2 (\frac{4}{4-t} + 1)dt = [-4\ln|4-t| + t]_1^2 = -4\ln 2 + 2 - (-4\ln 3 + 1) = 4\ln(3/2) + 1$ meters.

75. The first integral is the area under the curve $y = \ln x$ from 1 to a. The second integral is the area in the rectangle to the left of the curve. The area, $a \ln a$, of the rectangle is the sum of these two areas.

76. $\frac{e^{x_1}}{e^{x_2}} = \frac{e^{x_1 - x_2 + x_2}}{e^{x_2}} = \frac{e^{x_1 - x_2} e^{x_2}}{e^{x_2}} = e^{x_1 - x_2}$

77. $\frac{1}{e^x} = \frac{e^0}{e^x} = e^{0-x} = e^{-x}$

78. $y = y_0 e^{kt} = e^{kt}$. When $t = 0.5$ hr, $y = 2 = e^{0.5k}$, and $\ln 2 = 0.5k$, $k = 2\ln 2 = \ln 4$. When $t = 24$, $y = e^{24\ln 4} = 4^{24} \approx 2.81 \times 10^{14}$

79. Using $y = y_0 e^{kt}$, when $t = 3$, we get $10000 = y_0 e^{3k}$ and when $t = 5$, $40000 = y_0 e^{5k}$. Dividing the second relation by the first, we have $4 = e^{2k}$, $\ln 4 = 2k$, $k = \frac{1}{2}\ln 4 = \frac{1}{2}\ln 2^2 = \ln 2$. $10000 = y_0 e^{3k}$ leads to $y_0 = \frac{10000}{e^{\ln 2^3}} = \frac{10000}{8} = 1250$ for the initial population.

80. $y = 10000e^{kt}$, $7500 = 10000e^{k(1)}$, $e^k = 0.75$, $k = \ln(0.75)$. a) $1000 = 10000e^{kt}$, $e^{kt} = 0.1$, $kt = \ln 0.1$, $t = \frac{\ln 0.1}{k} = \frac{\ln 0.1}{\ln(.75)} \approx 8.00$ years b) $1 = 10000e^{kt}$, $e^{kt} = 0.0001$, $t = \frac{\ln 0.0001}{\ln 0.75} \approx 32.02$ years

81. a) $A(t) = A_0 e^{rt}$ and here $r = 1$ so the amount in the account after t years is $A(t) = A_0 e^t$. b) $A(t) = 3A_0 = A_0 e^t$, $3 = e^t$, $t = \ln 3 = 1.0986$ years (rounded). c) $A(1) = A_0 e^1 = eA_0$, e times the original amount.

82. $A(100) = 90A_0 = A_0 e^{r(100)}$, $e^{100r} = 90$, $100r = \ln 90$, $r = \frac{\ln 90}{100} \approx 0.0450$ or 4.50%.

83. $A(t) = A_0 e^{rt}$ and we wish to find r so that $131A_0 = A(100) = A_0 e^{100r}$, $131 = e^{100r}$, $\ln|3| = 100r$, $r = \frac{\ln 131}{100} = 0.04875\ldots$. The interest rate should be $4.875\ldots\%$ per year.

84. $p(t) = p_0 e^{kt}$, $p(10) = 246,605,103e^{0.00609(10)} \approx 262,090,086$

85. a) $y = \frac{70}{i} = \frac{70}{5} = 14$ years. $y = \frac{70}{7} = 10$ years b) $y = \frac{70}{i}$. $5 = \frac{70}{i}$ leads to $i = 14\%$. $20 = \frac{70}{i}$ leads to $i = 3.5\%$.

86. $y = \frac{70}{i} = \frac{70}{3.1} = 22.58$. About December, 2015.

87. $p(t) = 144.4e^{0.031t}$. $2(144.4) = 144.4e^{0.031t}$, $2 = e^{0.031t}$, $\ln 2 = 0.031t$, $t \approx 22.36$ years. $22\frac{1}{3}$ years from May, 1993 would be August, 2015.

88. 1995 CPI $= 144.4e^{0.031(2)} = 153.6$. 1997 CPI $= 144.4e^{0.031(4)} = 163.5$. 1999 CPI $= 144.4e^{0.031(6)} = 173.9$. The respective purchasing powers of the dollar are $100/153.6 = 0.65$, $100/163.5 = 0.61$, $100/173.9 = 0.58$.

89. $1.04 = \frac{p(t+1)}{p(t)} = \frac{p_0 e^{k(t+1)}}{p_0 e^{kt}} = e^{kt+k-kt} = e^k$. $k = \ln 1.04 \approx 0.039$. The inflation rate was about 3.9%.

90. $p_0 e^{8t} = 2p_0$, $e^{8t} = 2$, $8t = \ln 2$, $t = \frac{\ln 2}{8}$ years $= 365\frac{\ln 2}{8}$ days $= 32$ days

91. The required number of years t satisfies $\frac{100}{p(t)} = \frac{1}{2}\frac{100}{p_0}$ or $p(t) = 2p_0$, $p_0 e^{(r/100)t} = 2p_0$. This leads to $e^{(r/100)t} = \frac{r}{100}t = \ln 2$, $t = \frac{100 \ln 2}{r}$ years.

92. $\frac{1}{5}p_0 = p_0 e^{-0.10t}$, $0.2 = e^{-0.1t}$, $\ln 0.2 = -t/10$, $t = -10 \ln 0.2 = 16.09$; in 16.09 years.

93. a) $p(x) = p_0 e^{-0.01x}$. $p(100) = 20.09 = p_0 e^{-1}$ yielding $p_0 = 20.09e$ and $p(x) = 20.09e^{1-0.01x}$. b) $p(10) = 20.09e^{1-0.1} = 20.09e^{0.9} = \$49.41$ (rounded). $p(90) = 20.09e^{1-0.9} = 20.09e^{0.1} = \$22.20$ (rounded). c) $r(x) = xp(x) = 20.09xe^{1-0.01x}$. $r'(x) = 20.09[e^{1-0.01x} - 0.01xe^{1-0.01x}] = 20.09e^{1-0.01x}(1 - 0.01x)$. We see from this that r increases ($r' > 0$) until x reaches 100 and from then on decreases. Thus $r(100)$ is the maximum value of r. d) Graph $y = r(x) = 20.09xe^{1-0.01x}$ in $[0, 200]$ by $[0, 2100]$.

94. a) $1.5p_0 = p_0 e^{0.03t}$, $0.03t = \ln 1.5$, $t = \ln 1.5/0.03 = 13.52$ years at 3%. $1.5p_0 = p_0 e^{0.05t}$, $t = \ln 1.5/0.05 = 8.11$ years at 5%. $t = \ln 1.5/0.064 = 6.34$ years at 6.4%. b) $70/3 = 23.3$ years at 3%. $70/5 = 14$ years at 5%. $70/6.4 = 10.9$ years at 6.4%.

95. a) $0.04 \int_0^9 \frac{d\tau}{1+\tau} = 0.04 \ln(1+\tau)]_0^9 = 0.04[\ln 10 - \ln 1] = 0.04 \ln 10$ b) $p(9) = 100e^{0.04 \ln 10} = 109.65$ (rounded) c) The exponent $\int_0^t k(\tau)d\tau = (\frac{1}{t-0}\int_0^t k(\tau)d\tau)t = kt$ where k is the average value of $k(\tau)$ from $\tau = 0$ to $\tau = t$. Hence we will always get the same result in b) and c).

96. a) $p(t) = 100e^{\int_0^t (1+1.3\tau)d\tau} = 100e^{t+1.3t^2/2}$ b) $p(1) = 100e^{1+1.3/2} = 520.70$. $p(2) = 100e^{2+1.3(2)} = 100e^{4.6} = 9948.43$. The respective percentage increases in the associated consumer price index are $520.70 - 100 = 420.70\%$ and $9948.43 - 100 = 9848.43\%$.

97. a) $P(0) = \frac{1000}{1+e^{4.8}} \approx 8$ b) $\lim_{t\to\infty} e^{4.8-0.7t} = e^{4.8}\lim_{t\to\infty}\frac{1}{e^{0.7t}} = 0$. Hence $\lim_{t\to\infty} P(t) = 1000$ is the limiting maximum of the population. c) By b) $P(t) = 1200$ is impossible. $P(t) = 700$ leads to $1000 = 700(1+e^{4.8-0.7t})$, $10 = 7 + 7e^{4.8}e^{-0.7t}$, $e^{-0.7t} = \frac{3}{7e^{4.8}}$, $e^{0.7t} = \frac{7e^{4.8}}{3}$, $t = \frac{1}{0.7}\ln\frac{7e^{4.8}}{3} \approx 8.03$ months d) Determining this analytically would be quite messy. Instead we graph NDER$(P(x), x)$ in $[0, 15]$ by $[0, 200]$ and zoom in to the maximum. We find $\max P' = 173.919$rabbits per month which occurs when $t = 7.082$ months.

98. a) $P(0) = \frac{200}{1+e^{5.3}} \approx 1$ b) $P(t) \to 200$ as $t \to \infty$ c) $P(t) = 150$ leads to $200 = 150(1 + e^{5.3-t})$, $\frac{4}{3} = 1 + e^{5.3}e^{-t}$, $\frac{1}{3e^{5.3}} = e^{-t}$, $e^t = 3e^{5.3}$, $t = (\ln 3) + 5.3 \approx 6.4$ days. d) Graph NDER$(\frac{200}{1+e^{5.3-x}}, x)$ in $[0, 12]$ by $[0, 75]$ and zoom in to the maximum. Max $P' = 49.971$ students/day which occurs after 5.252 days.

99. a) We know $D_xF^{-1}(1) = \frac{1}{F'(a)} = \frac{1}{f(a)}$ where a is the number with $F(a) = 1$. So we must find an estimate for a. That is, we must estimate the x-coordinate of the point of intersection of the graphs of $y_1 = F(x) = \text{NINT}(e^{-0.7t^2}, t, 0, x)$ and $y_2 = 1$. Using ZOOM-IN after graphing in $[0, 2]$ by $[0, 2]$, we find $a = 1.616$. Hence $D_xF^{-1}(1) = \frac{1}{f(a)} \approx e^{0.7(1.616)^2} = 6.222$. b) In parametric mode, graph $x_1 = t$, $y_1 = F(t) = \text{NINT}(e^{-0.7s^2}, s, 0, t)$ and the inverse, $x_2 = y_1(t), y_2 = t$, in $[0, 2]$ by $[0, 2]$. Using TRACE to find the coordinates of two points on $y = F^{-1}(x)$, one to the left and one to the right of $x = 1$, we estimate the slope, $D_xF^{-1}(1)$, and obtain 6.348. This estimate supports the estimate found in a).

7.3 OTHER EXPONENTIAL AND LOGARITHMIC FUNCTIONS

1. $y = x^{\pi}$. $\frac{dy}{dx} = \pi x^{\pi-1}$

2. $y = x^{1+\sqrt{2}}$. $\frac{dy}{dx} = (1+\sqrt{2})x^{\sqrt{2}}$

3. $y = x^{-\sqrt{2}}$. $\frac{dy}{dx} = -\sqrt{2}\,x^{-\sqrt{2}-1}$

4. $y = x^{1-e}$. $\frac{dy}{dx} = (1-e)x^{-e}$

5. $y = 8^x$. $\frac{dy}{dx} = 8^x \ln 8$

6. $y = 9^{-x}$. $\frac{dy}{dx} = -9^{-x}\ln 9$

7. $y = 3^{\csc x}$. $\frac{dy}{dx} = (\ln 3)3^{\csc x}(-\csc x \cot x) = -\csc x \cot x(3^{\csc x})\ln 3$

8. $y = 3^{\cot x}$. $\frac{dy}{dx} = -(\csc^2 x)3^{\cot x}\ln 3$

9. $y = x^{\ln x}$, $x > 0$. $\ln y = (\ln x)\ln x = (\ln x)^2$. $\frac{y'}{y} = 2(\ln x)\frac{1}{x}$, $y' = y(2\frac{\ln x}{x}) = 2(\frac{\ln x}{x})x^{\ln x}$.

10. $y = x^{(1/\ln x)}$. $\ln y = \frac{1}{\ln x}(\ln x) = 1$ so $y = e$, $x > 0$, $x \neq 1$, and $\frac{dy}{dx} = 0$.

11. $y = (x+1)^x$. $\ln y = x\ln(x+1)$, $\frac{y'}{y} = \frac{x}{x+1} + \ln(x+1)$, $y' = y[\frac{x}{x+1} + \ln(x+1)] = (x+1)^x[\frac{x}{x+1} + \ln(x+1)]$.

12. $y = (x+2)^{x+2}$. $\ln y = (x+2)\ln(x+2)$. $\frac{y'}{y} = (x+2)\frac{1}{(x+2)} + \ln(x+2)$, $y' = (x+2)^{x+2}[1 + \ln(x+2)]$.

13. $y = x^{\sin x}$. $\ln y = \sin x(\ln x)$, $\frac{y'}{y} = (\sin x)\frac{1}{x} + (\cos x)\ln x$, $y' = x^{\sin x}[\frac{\sin x}{x} + (\cos x)\ln x]$.

14. $y = (\sin x)^{\tan x}$. $\ln y = (\tan x)\ln\sin x$, $\frac{y'}{y} = (\tan x)\frac{\cos x}{\sin x} + (\sec^2 x)\ln\sin x = 1 + (\sec^2 x)\ln\sin x$, $y' = (\sin x)^{\tan x}[1 + (\sec^2 x)\ln\sin x]$.

15. $y = \log_4 x^2 = 2\log_4 x = 2\frac{\ln x}{\ln 4}$. $\frac{dy}{dx} = \frac{2}{(\ln 4)x} = \frac{2}{2(\ln 2)x} = \frac{1}{x\ln 2}$.

16. $y = \log_5\sqrt{x} = \frac{\ln\sqrt{x}}{\ln 5}$. $\frac{dy}{dx} = \frac{1}{\ln 5}\frac{1}{\sqrt{x}}\frac{1}{2\sqrt{x}} = \frac{1}{2x\ln 5}$, $x > 0$.

17. $y = \log_2(3x+1) = \frac{\ln(3x+1)}{\ln 3}$. $\frac{dy}{dx} = \frac{3}{(\ln 3)(3x+1)}$.

18. $y = \log_{10}\sqrt{x+1} = \frac{\ln(x+1)^{1/2}}{\ln 10} = \frac{\ln(x+1)}{2\ln 10}$. $\frac{dy}{dx} = \frac{1}{2(\ln 10)(x+1)}$.

19. $y = \log_2(1/x) = -\log_2 x = -\frac{\ln x}{\ln 2}$. $\frac{dy}{dx} = -\frac{1}{x\ln 2}$, $x > 0$.

20. $y = \frac{1}{\log_2 x} = \frac{1}{\frac{\ln x}{\ln 2}} = (\ln 2)\frac{1}{\ln x}$. $\frac{dy}{dx} = (\ln 2)\frac{-1}{(\ln x)^2}\frac{1}{x} = -\frac{\ln 2}{x(\ln x)^2}$.

21. $y = (\ln 2)\log_2 x = \ln 2\frac{\ln x}{\ln 2} = \ln x$. $\frac{dy}{dx} = \frac{1}{x}$.

22. $y = \log_3(1 + x\ln 3) = \frac{1}{\ln 3}\ln(1 + x\ln 3)$. $\frac{y'}{y} = \frac{1}{\ln 3}\frac{\ln 3}{1+x\ln 3} = \frac{1}{1+x\ln 3}$. $y' = \frac{\log_3(1+x\ln 3)}{1+x\ln 3}$.

23. $y = \log_{10} e^x = \frac{\ln e^x}{\ln 10} = \frac{x}{\ln 10}$. $\frac{dy}{dx} = \frac{1}{\ln 10}$.

24. $y = \ln 10^x = x\ln 10$. $\frac{dy}{dx} = \ln 10$.

25. $\int_0^1 3x^{\sqrt{3}}dx = 3\frac{x^{\sqrt{3}+1}}{\sqrt{3}+1}\Big]_0^1 = \frac{3}{\sqrt{3}+1} = \frac{3}{\sqrt{3}+1}\frac{\sqrt{3}-1}{\sqrt{3}-1} = \frac{3(\sqrt{3}-1)}{3-1} = \frac{3}{2}(\sqrt{3}-1)$.

26. $\int_0^1 x^{\sqrt{2}}dx = \frac{x^{\sqrt{2}+1}}{\sqrt{2}+1}\Big]_0^1 = \frac{1}{\sqrt{2}+1}$.

27. $\int_0^1 5^x dx = \frac{5^x}{\ln 5}\Big]_0^1 = \frac{5-5^0}{\ln 5} = \frac{4}{\ln 5}$.

28. $\int_1^e x^{\ln 2-1}dx = \frac{x^{\ln 2}}{\ln 2}\Big]_1^e = \frac{e^{\ln 2}-1}{\ln 2} = \frac{1}{\ln 2}$.

29. Let $I = \int_0^1 2^{-x}dx$. Let $u = -x,\ du = -dx$. Then $I = -\int_0^{-1} 2^u du = \frac{-2^u}{\ln 2}\Big]_0^{-1} = \frac{-(2^{-1}-2^0)}{\ln 2} = \frac{1}{2\ln 2} = \frac{1}{\ln 4}$.

30. $\int_{-1}^1 2^{(x+1)}dx = \frac{2^{(x+1)}}{\ln 2}\Big]_{-1}^1 = \frac{2^2-2^0}{\ln 2} = \frac{3}{\ln 2}$.

31. Let $I = \int_{-1}^0 4^{-x}\ln 2\ dx$. Let $u = -x,\ du = -dx$. Then $I = -\ln 2\int_1^0 4^u du = -\ln 2\frac{4^u}{\ln 4}\Big|_1^0 = -\frac{\ln 2}{2\ln 2}(1-4) = \frac{3}{2}$.

32. $\int_{-2}^0 5^{-x}dx = \frac{-5^{-x}}{\ln 5}\Big]_{-2}^0 = \frac{-[5^0-5^2]}{\ln 5} = \frac{24}{\ln 5}$.

33. Let $I = \int_1^{\sqrt{2}} x2^{x^2}dx$. Let $u = x^2,\ du = 2xdx,\ \frac{1}{2}du = xdx$. Then $I = \frac{1}{2}\int_1^2 2^u du = \frac{1}{2}\frac{2^u}{\ln 2}\Big]_1^2 = \frac{2^2-2^1}{2\ln 2} = \frac{1}{\ln 2}$.

34. $\int_0^{\pi/2} 2^{\cos x}\sin x\ dx = \frac{-2^{\cos x}}{\ln 2}\Big]_0^{\pi/2} = \frac{-[2^{\cos \pi/2}-2^{\cos 0}]}{\ln 2} = \frac{1}{\ln 2}$.

35. Let $I = \int_1^{10}\frac{\log_{10}x}{x}dx = \frac{1}{\ln 10}\int_1^{10}\frac{\ln x}{x}dx$. Let $u = \ln x$. Then $du = \frac{dx}{x}$, $I = \frac{1}{\ln 10}\int_0^{\ln 10} udu = \frac{1}{\ln 10}\frac{u^2}{2}\Big]_0^{\ln 10} = \frac{1}{2}\frac{(\ln 10)^2}{\ln 10} = \frac{\ln 10}{2}$.

36. $\int_1^4 \frac{\log_2 x}{x}dx = \frac{1}{\ln 2}\int_1^4\frac{\ln x}{x}dx = \frac{1}{\ln 2}\frac{1}{2}(\ln x)^2]_1^4 = \frac{1}{2\ln 2}[(\ln 4)^2-(\ln 1)^2] = \ln 4$.

37. Let $I = \int_0^2\frac{\log_2(x+2)}{x+2}dx = \frac{1}{\ln 2}\int_0^2\frac{\ln(x+2)}{(x+2)}dx$. Let $u = \ln(x+2)$. Then $du = \frac{dx}{x+2}$ and $I = \frac{1}{\ln 2}\int_{\ln 2}^{\ln 4} udu = \frac{1}{\ln 2}\frac{u^2}{2}\Big|_{\ln 2}^{\ln 4} = \frac{(\ln 4)^2-(\ln 2)^2}{2\ln 2} = \frac{(2\ln 2)^2-(\ln 2)^2}{2\ln 2} = \frac{3(\ln 2)^2}{2\ln 2} = \frac{3}{2}\ln 2$.

38. $\int_{1/10}^{10}\frac{\log_{10}(10x)}{x}dx = \frac{1}{\ln 10}\int_{1/10}^{10}\frac{\ln(10x)}{x}dx = \frac{1}{\ln 10}\frac{1}{2}(\ln(10x))^2]_{1/10}^{10} = \frac{1}{2\ln 10}[(\ln 100)^2-(\ln 1)^2] = \frac{1}{2\ln 10}(2\ln 10)^2 = 2\ln 10$.

39. $\int_0^9\frac{2\log_{10}(x+1)}{x+1}dx = \frac{2}{\ln 10}\int_0^9\frac{\ln(x+1)}{(x+1)}dx = \frac{2}{\ln 10}\frac{[\ln(x+1)]^2}{2}\Big]_0^9 = \frac{1}{\ln 10}[(\ln 10)^2-(\ln 1)^2] = \ln 10$.

40. $\int_2^3\frac{2\log_2(x-1)}{x-1}dx = \frac{1}{\ln 2}\int_2^3\frac{2\ln(x-1)}{x-1}dx = \frac{1}{\ln 2}(\ln(x-1))^2]_2^3 = \frac{1}{\ln 2}[(\ln 2)^2-(\ln 1)^2] = \ln 2$.

41. Let $I = \int 2^{\sin x}\cos x dx$, $u = \sin x$, $du = \cos x dx$. Then $I = \int 2^u du = \frac{2^u}{\ln 2} + C = \frac{2^{\sin x}}{\ln 2} + C$.

42. Let $I = \int \frac{3^{\sqrt{x}}dx}{\sqrt{x}}$, $u = \sqrt{x}$, $du = \frac{dx}{2\sqrt{x}}$, $\frac{dx}{\sqrt{x}} = 2du$. Then $I = 2\int 3^u du = 2\frac{3^u}{\ln 3} + C = \frac{2(3^{\sqrt{x}})}{\ln 3} + C$.

43. Let $I = \int \frac{\log_3(x-2)dx}{x-2} = \frac{1}{\ln 3}\int \frac{\ln(x-2)dx}{x-2}$, $u = \ln(x-2)$, $du = \frac{dx}{x-2}$. Then $I = \frac{1}{\ln 3}\int u du = \frac{u^2}{2\ln 3} + C = \frac{(\ln(x-2))^2}{2\ln 3} + C$.

44. Let $I = \int \frac{\log_5(2x-1)dx}{1-2x} = \frac{1}{\ln 5}\int \frac{\ln(2x-1)dx}{1-2x}$, $u = \ln(2x-1)$, $du = \frac{2dx}{2x-1}$, $-\frac{1}{2}du = \frac{dx}{1-2x}$. Then $I = -\frac{1}{2\ln 5}\int u du = -\frac{u^2}{4\ln 5} + C = -\frac{(\ln(2x-1))^2}{4\ln 5} + C$.

45. $y = x^{-\sqrt{3}} = \frac{1}{x^{\sqrt{3}}}$. As $x \to 0^+$, $y \to \infty$ and as $x \to \infty$, $y \to 0$. $y' = -\sqrt{3}\, x^{-\sqrt{3}-1} = \frac{-\sqrt{3}}{x^{\sqrt{3}+1}} < 0$ since the domain of the function is $x > 0$. Hence the graph is falling on $(0, \infty)$. $y'' = \sqrt{3}(\sqrt{3}+1)x^{-\sqrt{3}-2} = \frac{\sqrt{3}(\sqrt{3}+1)}{x^{\sqrt{3}+2}} > 0$ on $(0, \infty)$. The curve is concave up for $x > 0$. There are no local extrema or inflection points. Check your graph by graphing y in $[0, 10]$ by $[0, 3]$.

46. $y = x^{\sqrt{7}}$. $y' = \sqrt{7}\, x^{\sqrt{7}-1}$, $y'' = (\sqrt{7}-1)\sqrt{7}\, x^{\sqrt{7}-2}$. The domain here is $x > 0$. In this domain y, y', y'' are all positive. Hence y steadily increases and the graph is concave up. Graph y in $[0, 10]$ by $[0, 20]$.

47. Graph $y_1 = x^{\sqrt{x}}$ in $[-1, 3]$ by $[-2, 8]$. The domain is $(0, \infty)$. $\ln y = \sqrt{x}\, \ln x$, $\frac{y'}{y} = \frac{\sqrt{x}}{x} + \frac{\ln x}{2\sqrt{x}}$, $y' = y(\frac{1}{\sqrt{x}} + \frac{\ln x}{2\sqrt{x}}) = x^{\sqrt{x}}\frac{(2+\ln x)}{2\sqrt{x}}$. $y' = 0$ when $\ln x = -2$ or when $x = e^{-2}$. $y' < 0$ on $(0, e^{-2})$ and $y' > 0$ on (e^{-2}, ∞). Therefore y is decreasing on $(0, e^{-2}]$, has a relative minimum at $x = e^{-2}$, and is increasing on $[e^{-2}, \infty)$. It is more practical to deal with y'' graphically. Also graph separately $y' = y_2 =$ NDER(y_1, x) and $y'' = y_3 =$ NDER(y_2, x) in the above window. We see that $y'' > 0$ for all $x > 0$ (y' is increasing steadily) and so the graph of y is concave up for all such x.

48. $y = x^{\ln x}$. $\ln y = (\ln x)^2$, $\frac{y'}{y} = \frac{2\ln x}{x}$, $y' = 2(x^{\ln x})\frac{\ln x}{x}$. The domain is $(0, \infty)$. As x increases through 1, $\ln x$ (and so y') passes through negative, 0 and then positive values. Hence the graph falls on $(0, 1]$ to the point $(1, 1)$, a local minimum, and then rises on $[1, \infty)$. The graph of NDER(y', x) in $[0, 10]$ by $[-5, 5]$ shows that $y'' > 0$ and so the graph is always concave up. Graph y in $[0, 10]$ by $[0, 20]$.

49. $y = 2^{\sec x}$ has period 2π and the graph is symmetric with respect to the vertical lines $x = n\pi$. In order to include the relative extrema in the interior we work in the interval $[-\frac{\pi}{2}, \frac{3\pi}{2})$. Graph y in $[-\frac{\pi}{2}, \frac{3\pi}{2}]$ by $[0, 4]$. $y' = 2^{\sec x}(\ln 2)\sec x \tan x$ and we use $y'' =$ NDER(y', x, x). Rel. min. at $(0, 2)$, rel. max. at $(\pi, \frac{1}{2})$. A root of $y'' = 0$ is $v = 1.90392136$. Inflection points at $(v, 0.12)$ and $(2\pi - v, 0.12)$. y is rising on $[0, \frac{\pi}{2})$ and $(\frac{\pi}{2}, \pi]$, falling on $(-\frac{\pi}{2}, 0]$ and $[\pi, \frac{3\pi}{2})$. It is concave up on $(-\frac{\pi}{2}, \frac{\pi}{2})$, $(\frac{\pi}{2}, v)$ and $(2\pi - v, \frac{3\pi}{2})$. It is concave down on $(v, 2\pi - v)$.

50. $y = 2^{\tan x}$. $y' = 2^{\tan x}(\ln 2)\sec^2 x > 0$ so y is increasing within its period $-\frac{\pi}{2} < x < \frac{\pi}{2}$, with $x = \frac{\pi}{2}$ being a vertical asymptote. After factoring, $y'' = 2^{\tan x}(\ln 2)\sec^2 x[2\tan x + (\ln 2)\sec^2 x]$. Working with $2\tan x + (\ln 2)\sec^2 x$ graphically, we solve $y'' > 0$ and $y'' < 0$. We find that the graph of y is concave up on $(-\frac{\pi}{2}, -1.188]$ and $[-0.383, \frac{\pi}{2})$, and concave down on $[-1.188, -0.383]$. Graph one period of y in $[-\frac{\pi}{2}, \frac{\pi}{2}]$ by $[0, 10]$.

51. $y = \log_7 \sin x = \ln \sin x / \ln 7$ is periodic of period 2π. We consider only the interval $[0, 2\pi)$. In here we note that the domain is $(0, \pi)$. Graph y in $[0, 2\pi]$ by $[-3, 0]$. $y' = \frac{1}{\ln 7}\frac{\cos x}{\sin x} = \frac{1}{\ln 7}\cot x$. The graph is rising on $(0, \frac{\pi}{2}]$ and falling on $[\frac{\pi}{2}, \pi)$. Rel. max at $(\frac{\pi}{2}, 0)$. $y'' = -\frac{1}{\ln 7}\csc^2 x$. Thus the curve is concave down on $(0, \frac{\pi}{2})$ and $(\frac{\pi}{2}, \pi)$ and there is no point of inflection.

52. $y = \log_5(x+1)^2 = 2\log_5(x+1) = \frac{2}{\ln 5}\ln(x+1) \approx 1.243\ln(x+1)$. The graph of y can be obtained from the graph of $\ln x$ by shifting it horizontally one unit to the left and stretching it vertically by a factor of 1.243. It is rising and concave down on $(-1, \infty)$ with $x = -1$ being a vertical asymptote. Graph $y = (2/\ln 5)\ln(x+1)$ in $[-1, 5]$ by $[-2, 4]$.

53. We have no analytic method for this integral. $\text{NINT}(2^{x^2}, x, 1, 2) = 6.052$.

54. $\text{NINT}(2^{\cos 2}, x, 0, \frac{\pi}{2}) = 2.496$

55. $\text{NINT}(x^{\ln x}, x, 1, 3) = 3.591$

56. $\text{NINT}(x^x, x, 0, 3) = 14.5085$

57. The equation may be written $7 + 5 = x$ $(a^{\log_a v} = v$ for all $v > 0)$, $x = 12$.

58. The equation may be written as $3 - 5 = x^2 - 3x$, $x^2 - 3x + 2 = 0$, $(x-1)(x-2) = 0$. The solution set is $\{1, 2\}$.

59. The graphs of the two functions in $[-2, 5]$ by $[0, 20]$ show that there are 3 solutions (points of intersection). They are $(-0.77, 0.58), (2, 4)$ and $(4, 16)$.

60. By inspection $(10, 10^{10})$ is one point of intersection. To see the other two we may graph the two functions in the window $[-2, 3]$ by $[0, 50]$. Using zoom-in, we find the solutions: $\{(-0.827, 0.149), (1.371, 23.512), (10, 10^{10})\}$.

61. a) $\lim_{x\to\infty}\log_2 x = \lim_{x\to\infty}\frac{\ln x}{\ln 2} = \frac{1}{\ln 2}\lim_{x\to\infty}\ln x = \infty$
b) $\lim_{x\to\infty}\log_2(1/x) = \lim_{x\to\infty}\log_2 x^{-1} = -\lim_{x\to\infty}\log_2 x = -\infty$ by part a).

62. a) $\lim_{x\to 0^+}\log_{10} x = \lim_{x\to 0^+}\frac{\ln x}{\ln 10} = -\infty$
b) $\lim_{x\to 0^+}\log_{10}(1/x) = -\lim_{x\to 0^+}\log_{10} x = \infty$

63. a) $\lim_{x\to\infty} 3^x = \infty$ b) $\lim_{x\to\infty} 3^{-x} = \lim_{x\to\infty}\frac{1}{3^x} = 0$

64. a) $\lim_{x\to-\infty} 3^x = \lim_{x\to-\infty}\frac{1}{3^{-x}} = 0$ b) $\lim_{x\to-\infty} 3^{-x} = \infty$

65. a) $\frac{\ln x}{\log x} = \frac{\ln x}{\ln x/\ln 10} = \ln 10.$ $\lim_{x\to\infty}\ln 10 = \ln 10$ b) $\frac{\log_2 x}{\log_3 x} = \frac{\ln x/\ln 2}{\ln x/\ln 3} = \frac{\ln 3}{\ln 2}.$ $\lim_{x\to\infty}\frac{\ln 3}{\ln 2} = \frac{\ln 3}{\ln 2}$

66. a) $\frac{\log_9 x}{\log_3 x} = \frac{(\ln x/\ln 9)}{(\ln x/\ln 3)} = \frac{\ln 3}{\ln 3^2} = \frac{1}{2}$ b) $\frac{\log_{\sqrt{10}} x}{\log_{\sqrt{2}} x} = \frac{(\ln x/\ln\sqrt{10})}{(\ln x/\ln\sqrt{2})} = \frac{0.5\ln 2}{0.5\ln 10} = \frac{\ln 2}{\ln 10}$

67. Check your result by graphing $y = x^{\sin x}$ in $[0, 40]$ by $[0, 40]$.

68. Graph y in $[0, 6\pi]$ by $[0, 20]$.

69. In each case we have $x^\beta < x^{\sqrt{3}} < x^\alpha$ for $0 < x < 1$ and $x^\alpha < x^{\sqrt{3}} < x^\beta$ for $x > 1$ where $0 < \alpha < \sqrt{3} < \beta$. The closer α and β are, the more we must zoom in to distinguish the curves. For $x > 0$, $(1, 1)$ is the only point of intersection.

70. a) $y = 2^{\ln x}$. $\frac{dy}{dx} = 2^{\ln x}(\ln 2)(\frac{1}{x}) = (\ln 2)\frac{2^{\ln x}}{x}$ b) $y = \ln 2^x = x\ln 2$. $\frac{dy}{dx} = \ln 2$
c) $y = \ln x^2 = 2\ln x$. $\frac{dy}{dx} = \frac{2}{x}$ d) $y = (\ln x)^2$. $\frac{dy}{dx} = \frac{2\ln x}{x}$

71. a) Let $x_1 < x_2$ be in the domain of $f\circ g$. Since g is increasing, $g(x_1) < g(x_2)$. Since f is increasing, $f(g(x_1)) < f(g(x_2))$. But this is the same as $f\circ g(x_1) < f\circ g(x_2)$. Therefore $f\circ g$ is an increasing function. b) Let $f(x) = e^x$ and $g(x) = \sqrt{3}\ln x$. Then $f(x)$ is increasing for all x and $g(x)$ is increasing for $x > 0$. By part a), $f\circ g(x) = f(g(x)) = e^{\sqrt{3}\ln x}$ is an increasing function.

72. $\frac{d(a^x)}{dx} = a^x\ln a = a^x$ if and only if $\ln a = 1 = \ln e$ if and only if $a = e$.

73. Let $u = [H_30^+]$. $-\log_{10} u = 7.37$ leads to $\log_{10} u = -7.37$, $u = 10^{-7.37} = 4.27\times 10^{-8}$. For the other bound $u = 10^{-7.44} = 3.63\times 10^{-8}$.

74. $pH = -\log_{10}[H_30^+] = -\log_{10}(4.8\times 10^{-8}) = -[\log_{10} 4.8 + \log_{10} 10^{-8}] = -[0.68 - 8] = 7.32.$

75. k must satisfy $10\log_{10}(kI \times 10^{12}) = 10\log_{10}(I \times 10^{12}) + 10$, $\log_{10} k + \log_{10} I + 12 = \log_{10} I + 12 + 1$, $\log_{10} k = 1$, $k = 10$.

76. $\log_2 x = \frac{\log_{10} x}{\log_{10} 2} = (\log_{10} x)/(\ln 2/\ln 10) = \frac{\ln 10}{\ln 2}\log_{10} x$.

77. $\log_b = \frac{\log_a x}{\log_a b} = (\log_a x)/(\ln b/\ln a) = \frac{\ln a}{\ln b}\log_a x$.

7.4 THE LAW OF EXPONENTIAL CHANGE REVISITED

1. a) $0.99y_0 = y_0 e^{1000k}$, $0.99 = e^{1000k}$, $\ln 0.99 = 1000k$, $k = \frac{\ln 0.99}{1000}$.
b) $0.9y_0 = y_0 e^{kt}$, $\ln 0.9 = kt$, $t = \frac{\ln 0.9}{k} = 1000\frac{\ln 0.9}{\ln 0.99} \approx 10,483$ years. To support graphically, let $y_0 = 1$ and investigate the point of intersection of $y = e^{kt} = e \wedge ((\ln 0.99/1000)x)$ and $y = 0.9$. c) $y = y_0 e^{20,000k} = y_0 e^{20 \ln 0.99} = y_0(0.99)^{20} \approx 0.82y_0$, about 82%.

2. a) $p = p_0 e^{kh} = 1013e^{kh}$. $p(20) = 90 = 1013e^{20k}$, $e^{20k} = \frac{90}{1013}$, $k = \frac{1}{20}\ln\frac{90}{1013} \approx -0.121043091696$

b) $p(50) = 1013e^{50k} \approx 2.383$ millibars c) $p(h) = 1013e^{kh} = 900$, $e^{kh} = \frac{900}{1013}$, $h = \frac{1}{k}\ln\frac{900}{1013} \approx 0.977$ km.

3. From equations (1) and (2), $y = 100e^{-0.6t}$. When $t = 1$ hour, $y = 100e^{-0.6} \approx 54.88$ grams.

4. Let $A(t)$ be the amount of sugar at time t. Then $A(t) = A_0 e^{kt} = 1000e^{kt}$. $A(10) = 800 = 1000e^{10k}$. $0.8 = e^{10k}$, $k = 0.1\ln(0.8)$. $A(24) = 1000e^{24k} = 585.350$ kg.

5. Let t be the required number of days. Then $0.9y_0 = y_0 e^{-0.18t}$, $\ln 0.9 = -0.18t$, $t = \frac{\ln 0.9}{0.18} = 0.59$ day.

6. $A = A_0 e^{-kt}$ where A_0 is the amount on the day the sample arrives and $k = \frac{\ln 2}{140}$. $0.05A_0 = A_0 e^{-kt}$, $\ln 0.05 = -kt$, $t = -\frac{\ln 0.05}{k} = -140\frac{\ln 0.05}{\ln 2} = 605.070$, about 605 days.

7. a) We use $m = 66 + 7 = 73$. Then distance coasted $= \frac{v_0 m}{k} = \frac{9(73)}{3.9} = 168.46\ldots m$. b) $v = v_0 e^{-(k/m)t} = 9e^{-3.9t/73} = 1$ m/sec. $-3.9t/73 = \ln(1/9) = -\ln 9$, $t = 73(\ln 9)/3.9 = 41.1$ sec.

8. a) Distance coasted $= \frac{v_0 m}{k} = \frac{9(51\times 10^6)}{59\times 10^3} = 7780m$. b) We solve $1 = \frac{ds}{dt} = v_0 e^{-(k/m)t}$ for t: $v_0^{-1} = e^{(-k/m)t}$, $\ln v_0 = \frac{kt}{m}$, $t = \frac{m \ln v_0}{k} = \frac{(51\times 10^6)\ln 9}{59\times 10^3} = 1899$ sec or about 31.65 minutes.

9. By (8), $T = T_s + (T_0 - T_s)e^{-kt} = 20 + (90 - 20)e^{-kt}$, $T = 20 + 70e^{kt}$. When $t = 10$, $60 = 20 + 70e^{10k}$, $\frac{4}{7} = e^{10k}$, $10k = \ln(4/7)$, and $k = \ln(4/7)/10$ in $T = 20 + 70e^{kt}$. a) $35 = 20 + 70e^{kt}$, $15/70 = e^{kt}$, $kt = \ln(3/14)$, $t = \frac{10\ln(3/14)}{\ln(4/7)} \approx 27.53$ min. so $27.53 - 10 = 17.53$ min. longer. b) Here $T = T_s + (T_0 - T_s)e^{-kt} = -15 + (90+15)e^{kt} = -15 + 105e^{kt} = 35$. $e^{kt} = 50/105 = 10/21$, $kt = \ln(10/21)$, $t = (\frac{1}{k})\ln(10/21) = \frac{10\ln(10/21)}{\ln(4/7)} \approx 13.26$ min.

10. $T - T_s = (T_0 - T_s)e^{-kt}$, $T - 30 = (T_0 - 30)e^{-kt}$. When $t = 10:\ 0 - 30 = (T_0 - 30)e^{-10k}$. When $t = 20:\ 15 - 30 = (T_0 - 30)e^{-20k} = (T_0 - 30)(e^{-10k})^2$ or $-15 = (T_0 - 30)(e^{-10k})^2 = (T_0 - 30)(\frac{-30}{T_0 - 30})^2 = \frac{900}{T_0 - 30}$, $-15(T_0 - 30) = 900$, $T_0 = -30°$ F.

11. T_s = temperature of refrigerator. $T = T_s + (T_0 - T_s)e^{-kt} = T_s + (46 - T_s)e^{-kt}$. $39 = T_s + (46 - T_s)e^{-10k}$ and $33 = T_s + (46 - T_s)e^{-20k} = T_s + (46 - T_s)(e^{-10k})^2$. From the first equation $(e^{-10k})^2 = (\frac{39 - T_s}{46 - T_s})^2 = \frac{33 - T_s}{46 - T_s}$ from the second equation. Multiplying by $(46 - T_s)^2$, we find $T_s = -3°C$.

12. $T - T_s = (T_0 - T_s)e^{-kt}$. At $t = 0$ (right now), $60 = (T_0 - T_s)$. At $t = -20$, $70 = (T_0 - T_s)e^{20k}$. When $t = 15$, $T - T_s = (T_0 - T_s)e^{-15k} = (T_0 - T_s)(e^{20k})^{-\frac{15}{20}} = 60(\frac{70}{T_0 - T_s})^{-3/4} = 60(\frac{70}{60})^{-3/4} = 53.449°$ C, i.e., 15 minutes from now it will be 53.449° C above room temperature. When $t = 120$, $T - T_s = 60(\frac{70}{60})^{-120/20} = 23.794°$ C, i.e., two hours from now it will be 23.794° C above room temperature. $10 = 60(\frac{70}{60})^{-t/20}$, $\frac{1}{6} = (\frac{7}{6})^{-t/20}$, $-\ln 6 = -\frac{t}{20}\ln\frac{7}{6}$, $t = 20\frac{\ln 6}{\ln(7/6)} = 232.469$ minutes $= 3.874$ hours is the time when it will be 10° C above room temperature.

13. $\frac{dV}{dt} = -\frac{1}{40}V$, $\frac{1}{V}\frac{dV}{dt} = -\frac{1}{40}$ so $\ln|V| = -\frac{1}{40}t + C$, $V = e^{-t/40}e^C$, taking V to be positive and when $t = 0$, $V = e^C = V_0$. Hence $V = V_0e^{-t/40}$. If t is the desired time, then $0.1V_0 = V_0e^{-t/40}$, $\ln 0.1 = -\frac{t}{40}$, $t = -40\ln 0.1 \approx 92.10$ sec.

14. $y = y_0e^{-kt}$. 95% of y_0 will have disintegrated when we have left $y = 0.05y_0 = y_0e^{-kt}$, $\ln 0.05 = -kt$, $t = (-\ln 0.05)\frac{1}{k} = 2.996(\frac{1}{k}) \approx 3(\frac{1}{k})$.

15. From Example 3, $k = \ln 2/5700$. If t is the age $0.445y_0 = y_0e^{-kt}$, $\ln 0.445 = -kt$, $t = -\frac{1}{k}\ln 0.445 = -\frac{5700\ln 0.445}{\ln 2} \approx 6658.30$ years.

16. a) $0.17y_0 = y_0e^{kt}$, $\ln 0.17 = kt$, $t = 5700\frac{\ln 0.17}{\ln 2} = -14,571$ years. The estimate of the year the animal died is $14,571 - 2000 = 12,571$ B.C. b) $t = 5700\frac{\ln 0.18}{\ln 2} = -14,101$. In this case the estimate is 12,101 B.C. c) $t = 5700\frac{\ln 0.16}{\ln 2} = -15,070$. Now the estimate is 13,070 B.C. d) Let r

be the percentage of the original amount of C-14 remaining in the bone. The value of t corresponding to the time the animal died satisfies $\frac{r}{100}y_0 = y_0e^{kt}$, $kt = \ln(\frac{r}{100})$. The age of the bone is $|t| = |\ln(\frac{r}{100})/k|$. To graph the age of the bone as a function of the percentage of C-14 remaining, graph $y = abs((5700/\ln 2)\ln(\frac{x}{100}))$ in $[16, 18]$ by $[14101, 15070]$.

17. $y = y_0e^{-kt}$ where $k = \ln 2/5700$ as in Example 2. If t is the age of the painting, then $0.995y_0 = y_0e^{-kt}$, $\ln 0.995 = -kt$, $t = -5700\ln 0.995/\ln 2 \approx 41.22$ years.

18. Assume the room temperature is $T_s = 70°\text{F}$ and that the liquid originally has temperature $T_0 = 180°\text{F}$. Then $T = T_s + (T_0 - T_s)e^{-kt}$ becomes $T = 70 + 110e^{-kt}$. k may be found using your data.

19. a) f) $V(t) = 7e^{-2.5t/50} = 1$, $-2.5t/50 = \ln(1/7) = -\ln 7$, $t = 50(\ln 7)/2.5 = 38.918$ sec. g) $s(3) = \frac{350}{2.5}(1-e^{-(2.5/50)3}) = 19.50088\ldots m$. h) $s(t) = 100$ leads to $\frac{350}{2.5}(1 - e^{-(2.5/50)t}) = 100$, $1 - e^{-(2.5/50)t} = \frac{250}{350} = \frac{5}{7}$, $-(2.5/50)t = \ln\frac{2}{7}$, $t = -\frac{50}{2.5}\ln\frac{2}{7} = 25.055\ldots$sec. b) $s(t) = \frac{350}{2.5}(1 - e^{-(2.5/50)t}) \leq \frac{350}{2.5} = 140$. So Jenny will never coast 141 m from the finish line.

20. The differential equation may be rewritten as $\frac{1}{y}\frac{dy}{dt} = k$ which is the same as $\frac{d}{dt}\ln|y| = k$. This implies that $\ln|y| = kt + C$, $|y| = e^{kt+C} = e^Ce^{kt}$. Setting $t = 0$, we obtain $|y_0| = e^C$, $|y| = |y_0|e^{kt}$. If $y_0 = 0$, then we obtain $y = 0$ for all t. If $y_0 \neq 0$, then $|y| \neq 0$ for all t and so $y \neq 0$ for all t. Since y is continuous (it has derivative ky), by the Intermediate Value Theorem, y is either always positive or always negative. Hence it has the same sign as y_0. It follows that $y = y_0e^{kt}$. One may verify directly that this is a solution of the initial value problem. It is not too difficult to show that it is the only solution.

21. We first find the amount of time, t_1, required for the population to grow from 5,000 to 10,000. $y = y_0e^{kt}$, $10000 = 5000e^{t_1/4}, 2 = e^{t_1/4}$, $\ln 2 = t_1/4$, $t_1 = 4\ln 2$. Next we find the amount of time, t_2, required for the population to grow from 10,000 to 25,000. $y = 10000e^{t/12}$, $25000 = 10000e^{t_2/12}$, $2.5 = e^{t_2/12}$, $\ln 2.5 = t_2/12$, $t_2 = 12\ln 2.5$. Answer: $t_1 + t_2 = 13.768$ years from now.

7.5 INDETERMINATE FORMS AND L'HÔPITAL'S RULE

1. $\lim_{x\to 2}\frac{x-2}{x^2-4} = \lim_{x\to 2}\frac{1}{2x} = \frac{1}{4}$

2. $\lim_{x\to 2}\frac{x^2-4}{x-2}=\lim_{x\to 2}\frac{2x}{1}=4$

3. $\lim_{x\to 1}\frac{x^3-1}{4x^3-x-3}$(form $\frac{0}{0}$) $=\lim_{x\to 1}\frac{3x^2}{12x^2-1}=\frac{3}{11}$

4. $\lim_{x\to 0}\frac{1-\cos x}{x^2}=\lim_{x\to 0}\frac{\sin x}{2x}=\lim_{x\to 0}\frac{\cos x}{2}=\frac{1}{2}$

5. $\lim_{t\to 0}\frac{\sin t^2}{t}=\lim_{t\to 0}\frac{2t\cos t}{1}=0$

6. $\lim_{x\to 0}\frac{\sin 5x}{x}=\lim_{x\to 0}\frac{5\cos x}{1}=5$

7. $\lim_{x\to\infty}\frac{3x^2-1}{2x^2-x+1}(\frac{\infty}{\infty})=\lim_{x\to\infty}\frac{6x}{4x-1}(\frac{\infty}{\infty})=\lim_{x\to\infty}\frac{6}{4}=\frac{3}{2}$

8. $\lim_{t\to\infty}\frac{6t+5}{3t-8}=\lim_{t\to\infty}\frac{6}{3}=2$

9. Graph $y=\frac{2x-\pi}{\cos x}$ in $[\frac{\pi}{2}-0.1,\frac{\pi}{2}+0.1]$ by $[-3,3]$. Use of TRACE suggests $y\to -2$ as $x\to\frac{\pi}{2}$. $\lim_{x\to\pi/2}\frac{2x-\pi}{\cos x}(\frac{0}{0})=\lim_{x\to\frac{\pi}{2}}\frac{2}{-\sin x}=-2$.

10. $\lim_{x\to 0}\frac{(1/2)^x-1}{x}=\lim_{x\to 0}\frac{(1/2)^x\ln(1/2)}{1}=\ln(1/2)$

11. $\lim_{x\to\infty}\frac{5x^2-3x}{7x^2+1}=\lim_{x\to\infty}\frac{10x-3}{14x}=\lim_{x\to\infty}\frac{10}{14}=\frac{5}{7}$

12. $\lim_{t\to 0}\frac{\cos t-1}{t^2}=\lim_{t\to 0}\frac{-\sin t}{2t}=\lim_{t\to 0}\frac{-\cos t}{2}=-\frac{1}{2}$

13. $\lim_{x\to\pi/2}\frac{1-\sin x}{1+\cos 2x}=\lim_{x\to\pi/2}\frac{-\cos x}{-2\sin 2x}=\lim_{x\to\pi/2}\frac{-\sin x}{4\cos 2x}=\frac{-1}{4(-1)}=\frac{1}{4}$

14. $\lim_{x\to\pi/2}(\frac{\pi}{2}-x)\tan x=\lim_{x\to\pi/2}\frac{(\pi/2)-x}{\cot x}(\frac{0}{0})=\lim_{x\to\pi/2}\frac{-1}{-\csc^2 x}=1$

15. $\lim_{x\to 0^+}\frac{2x}{x+7\sqrt{x}}=\lim_{x\to 0^+}\frac{2}{1+7/(2\sqrt{x})}=0$ since $\lim_{x\to 0^+}\frac{7}{2\sqrt{x}}=\infty$.

16. $\lim_{x\to\infty}\frac{x-2x^2}{3x^2+5x}=\lim_{x\to\infty}\frac{1-4x}{6x+5}=\lim_{x\to\infty}\frac{-4}{6}=-\frac{2}{3}$

17. $\lim_{t\to 0}\frac{10(\sin t-t)}{t^3}=\lim_{t\to 0}\frac{10(\cos t-1)}{3t^2}=\lim_{t\to 0}\frac{-10\sin t}{6t}=-\frac{5}{3}\lim_{t\to 0}\frac{\cos t}{1}=-\frac{5}{3}$

18. $\lim_{x\to 0}\frac{x(1-\cos x)}{x-\sin x}=\lim_{x\to 0}\frac{x\sin x+1-\cos x}{1-\cos x}=\lim_{x\to 0}\frac{x\cos x+\sin x+\sin x}{\sin x}=$ $\lim_{x\to 0}\frac{-x\sin x+\cos x+2\cos x}{\cos x}=\frac{0+1+2}{1}=3$

19. $\lim_{x\to 0}(\frac{1}{\sin x}-\frac{1}{x})=\lim_{x\to 0}\frac{x-\sin x}{x\sin x}(\frac{0}{0})=\lim_{x\to 0}\frac{1-\cos x}{x\cos x+\sin x}=$ $\lim_{x\to 0}\frac{\sin x}{-x\sin x+\cos x+\cos x}=\frac{0}{2}=0$

20. $\lim_{x\to 0^+}(\frac{1}{x}-\frac{1}{\sqrt{x}})=\lim_{x\to 0^+}\frac{\sqrt{x}-x}{x^{3/2}}=\lim_{x\to 0^+}\frac{(1/2)x^{-1/2}-1}{(3/2)x^{1/2}}=\infty$ (form $\frac{\infty}{0}$ as $x\to 0^+$)

21. Let $y = x^{(1/\ln x)}$. Then $\lim_{x\to 0^+} \ln y = \lim_{x\to 0^+} \frac{1}{\ln x} \ln x = \lim_{x\to 0^+} 1 = 1$. Therefore, $\lim_{x\to 0^+} y = e^1 = e$.

22. Let $y = x^{1/x}$. $\ln y = \frac{1}{x}\ln x \to -\infty$ as $x \to 0^+$ because it has the form $(\infty)(-\infty)$ as $x \to 0^+$. Therefore $y \to 0$ as $x \to 0^+$.

23. Let $y = (e^x + x)^{1/x}$. Then $\lim_{x\to 0} \ln y = \lim_{x\to 0} \frac{1}{x}\ln(e^x + x)(\frac{0}{0}) = \lim_{x\to 0} \frac{\frac{e^x+1}{e^x+x}}{1} = \frac{e^0+1}{e^0+0} = 2$. Therefore, $\lim_{x\to 0} y = e^2$.

24. Let $y = x^{1/(x-1)}$. Then $\lim_{x\to 1} \ln y = \lim_{x\to 1} \frac{\ln x}{x-1} = \lim_{x\to 1} \frac{1}{x} = 1$. Hence $\lim_{x\to 1} y = e^1 = e$.

25. Let $y = (\frac{1}{x^2})^x$. Then $\lim_{x\to 0} \ln y = \lim_{x\to 0} x\ln(1/x^2) = -\lim_{x\to 0} x\ln(x^2) = -\lim_{x\to 0} \frac{\ln x^2}{\frac{1}{x}} = -\lim_{x\to 0} \frac{(2x/x^2)}{(-1/x^2)} = \lim_{x\to 0} 2x = 0$. Therefore $\lim_{x\to 0} y = e^0 = 1$.

26. Let $y = (\ln|\frac{1}{x}|)^x$. Then $\lim_{x\to 0} \ln y = \lim_{x\to 0} x\ln(\ln|\frac{1}{x}|) = \lim_{x\to 0} \frac{\ln(\ln|\frac{1}{x}|)}{1/x} = \lim_{x\to 0} \frac{(1/\ln|\frac{1}{x}|)x(-1/x^2)}{(-1/x^2)}$ (Recall $\frac{d}{dx}\ln|u| = \frac{1}{u}\frac{du}{dx}$) $= \lim_{x\to 0} x\frac{1}{\ln|\frac{1}{x}|} = 0 \cdot 0 = 0$. Hence $\lim_{x\to 0} y = e^0 = 1$. This is supported by graphing y in $[-1, 1]$ by $[-1, 4]$.

27. $\lim_{x\to 0^+} x^{\sqrt{2}} = \lim_{x\to 0^+} e^{\sqrt{2}\ln x} = 0$ because $\lim_{x\to 0^+} \ln x = -\infty$. If we define the function to be 0 at 0, then the function is continuous on $[0, \infty)$.

28. $y = x^{-\sqrt{3}} = \frac{1}{x^{\sqrt{3}}} \to \infty$ as $x \to 0^+$. We can't make the function continuous at $x = 0$.

29. $y = x^{\ln x} = e^{\ln x^{\ln x}} = e^{(\ln x)^2} \to \infty$ as $x \to 0^+$. y cannot be defined at 0 in a way that it would be continuous at $x = 0$.

30. $y = x^{x+1}$. $\lim_{x\to 0^+} \ln y = \lim_{x\to 0^+} (x+1)\ln x = -\infty$. Therefore $\lim_{x\to 0^+} y = \lim_{x\to 0^+} e^{\ln y} = 0$. Yes, define $y(0) = 0$.

31. $\lim_{x\to 0^+} \ln f(x) = \lim_{x\to 0^+} x\ln x = \lim_{x\to 0^+} \frac{\ln x}{\frac{1}{x}}(-\frac{\infty}{\infty}) = \lim_{x\to 0^+} \frac{\frac{1}{x}}{-\frac{1}{x^2}} = -\lim_{x\to 0^+} x = 0$. Therefore, $\lim_{x\to 0^+} f(x) = e^0 = 1$. Thus the function $F(x)$ defined by $F(x) = f(x)$, $x > 0$ and $F(0) = 1$ is continuous on $[0, \infty)$.
b) In part a) we saw that $f(x) = x^x \to 1$ as $x \to 0^+$. Thus $f'(x) = (1 + \ln x)x^x$ has the form $(-\infty \cdot 1)$ as $x \to 0^+$ and so $\lim_{x\to 0^+} f'(x) = -\infty$.

32. $f(x) = x^x$. $f'(0)$ does not exist because $f(0)$ is not even defined. We may use L'Hôpital's rule to find $\lim_{x\to 0^+} f(x) = 1$ and try to work with the continuous extension $g(x)$ of $f(x)$ where

$$\begin{cases} g(x) = f(x), & x > 0 \\ g(0) = 1 \end{cases}$$

to see if $g(x)$ has a right-hand derivative at $x = 0$. $\lim_{h\to 0^+} \frac{g(0+h)-g(0)}{h} = \lim_{h\to 0^+} \frac{h^h-1}{h} = \lim_{h\to 0^+} h^h(1 + \ln h) = -\infty$ by L'Hôpital's rule. Thus even $g(x)$ does not have a right-hand derivative at $x = 0$.

33. b) is correct. a) is incorrect because L'Hôpital's rule does not apply to the limit form $\frac{0}{6}$; it is not an indeterminate form.

34. $\lim_{x\to-\infty} x \sin\frac{1}{x} = \lim_{x\to-\infty} \frac{\sin\frac{1}{x}}{\frac{1}{x}} = \lim_{x\to-\infty} \frac{(\cos\frac{1}{x})(-\frac{1}{x^2})}{(-\frac{1}{x^2})} = \lim_{x\to-\infty} \cos\frac{1}{x} = 1$

35. a) The domain of $f(x) = (1+\frac{1}{x})^x$ is the solution set of $1 + \frac{1}{x} > 0$ or $\frac{x+1}{x} > 0$. Numerator and denominator have the same sign in $(-\infty, -1) \cup (0, \infty)$. b) $\lim_{x\to-1^-} f(x) = \lim_{x\to-1^-} e^{x\ln(1+\frac{1}{x})} = \infty$ because the exponent has limit form $(-1)(-\infty)$. c) $\lim_{x\to 0^+} \ln f(x) = \lim_{x\to 0^+} x\ln(1+\frac{1}{x}) = \lim_{x\to 0^+} \frac{\ln(1+\frac{1}{x})}{\frac{1}{x}} (\frac{\infty}{\infty}) = \lim_{x\to 0^+} \frac{\frac{1}{1+\frac{1}{x}}(-\frac{1}{x^2})}{-\frac{1}{x^2}} = \lim_{x\to 0^+} \frac{1}{1+\frac{1}{x}} = 0$. Hence $\lim_{x\to 0^+} f(x) = e^0 = 1$. d) $\lim_{x\to\pm\infty} f(x) = e$ was proved in Example 6. $f'(x) = (1+\frac{1}{x})^x[\ln(1+\frac{1}{x}) - \frac{1}{x+1}]$. The second factor $\to 0$ as $x \to \pm\infty$ and it is rising for $x < 0$ and falling for $x > 0$. Thus it, and so $f'(x)$, is positive for all x in the domain. Hence the graph of f is always rising. The graph of NDER$(f'(x), x)$ in $[-4, 4]$ by $[-3, 3]$ shows that $f''(x) > 0$ for $x < 0$ and $f''(x) < 0$ for $x > 0$, confirming the concavity shown in the figure.

36. $\frac{\sec x}{\tan x} = \frac{(1/\cos x)}{(\sin x/\cos x)} = \frac{1}{\sin x} \to 1$ as $x \to (\pi/2)^-$. Similarly $\frac{\tan x}{\sec x} \to 1$ as $x \to (\pi/2)^+$. Use of L'Hôpital's rule before simplifying does not lead to a simpler limit.

37. $f(x) = (1+\frac{2}{x})^x$. By the method of Example 6, $\lim_{x\to\pm\infty} f(x) = e^2$. As in Exercise 35, the domain is $(-\infty, -2) \cup (0, \infty)$ and $\lim_{x\to 0^+} f(x) = 1$. Graph f in $[-10, 10]$ by $[0, 20]$. The graph resembles the one in Fig. 7.30 and may be confirmed as in Exercise 35.

38. $f(x) = (1+\frac{3}{x})^x$. By the method of Example 6, $\lim_{x\to\pm\infty} f(x) = e^3$. As in Exercise 35, the domain is $(-\infty, -3) \cup (0, \infty)$ and $\lim_{x\to 0^+} f(x) = 1$. Graph f in $[-20, 20]$ by $[0, 100]$.

39. $f(x) = x^{(1/\ln x)}$. $\ln f(x) = \frac{1}{\ln x}\ln x = 1$ for $x > 0$, $x \neq 1$, and so $f(x) = e$ for $x > 0$, $x \neq 1$.

40. $f(x) = (1+x)^{1/x}$. f has domain $(-1, \infty)$ and $\lim_{x\to -1^+} f = \infty$. Using the method of Example 6, we can see that $\lim_{x\to 0} f = e$. Similarly, $f \to 1$ as $x \to \infty$. Graph f in $[-2, 3]$ by $[-10, 50]$.

41. $\lim_{x\to 0} \frac{3^{\sin x}-1}{x}(\frac{0}{0}) = \lim_{x\to 0} \frac{3^{\sin x}(\cos x)\ln 3}{1} = \ln 3$. Graph $f(x) = \frac{3^{\sin x}-1}{x}$ in $[-2\pi, 2\pi]$ by $[-1, 2]$. f is continuous on the interval except at $x = 0$ where it has a removable discontinuity.

42. $\lim_{x\to 0} \frac{2^{\cos x}-2}{x} = \lim_{x\to 0} 2^{\cos x}(-\sin x)\ln 2 = 0$. Graph $f(x) = \frac{2^{\cos x}-2}{x}$ in $[-4\pi, 4\pi]$ by $[-1, 1]$. f is continuous on the interval except at $x = 0$ where it has a removable discontinuity.

43. We have $A(t) = \int_0^t e^{-x}dx = -e^{-x}|_0^t = 1 - e^{-t}$, $V(t) = \pi\int_0^t e^{-2x}dx = \frac{\pi}{2}(1 - e^{-2t}) = \frac{\pi}{2}(1-e^{-t})(1+e^{-t})$, $V(t)/A(t) = \frac{\pi}{2}(1+e^{-t})$. a) $\lim_{t\to\infty} A(t) = 1$ b) $\lim_{t\to\infty} V(t)/A(t) = \frac{\pi}{2}$ c) $\lim_{t\to 0^+} V(t)/A(t) = \frac{\pi}{2}(1+1) = \pi$

44. No. L'Hôpital's rule applies only to indeterminate forms such as $\frac{0}{0}$ or $\frac{\infty}{\infty}$. $\frac{f(x)}{g(x)}$ is not indeterminate as $x \to 0$.

45. a) Let $y = (1+\frac{r}{k})^{kt}$. $\lim_{k\to\infty} \ln y = \lim_{k\to\infty} \frac{\ln(1+\frac{r}{k})}{\frac{1}{kt}}(\frac{0}{0}) = \lim_{k\to\infty} \frac{\frac{1}{1+\frac{r}{k}}(-\frac{r}{k^2})}{\frac{1}{t}(-\frac{1}{k^2})} = \lim_{k\to\infty} \frac{rt}{1+\frac{r}{k}} = rt$ whence $\lim_{k\to\infty} y = e^{rt}$. b) $100e^{0.06} = 106.184$ c) $e^{0.06} \approx 1.06$ yields $1{,}000{,}000e^{0.06} = 1{,}060{,}000$ d) $1{,}000{,}000e^{0.06} = 1061836.55$ e) In computing the product $1000000e^{0.06}$, we find a significant difference if we round off before or after the computation. Investors/bankers should pay careful attention as to when the rounding off takes place.

7.6 THE RATES AT WHICH FUNCTIONS GROW

1. a) $\lim_{x\to\infty} \frac{x+3}{e^x} = \lim_{x\to\infty} \frac{1}{e^x}$ (by L'Hôpital's rule) $= 0$. $x + 3$ grows slower than e^x as $x \to \infty$. b) $\lim_{x\to\infty} \frac{x^3-3x+1}{e^x} = \lim_{x\to\infty} \frac{3x^2-3}{e^x} = \lim_{x\to\infty} \frac{6x}{e^x} = \lim_{x\to\infty} \frac{6}{e^x} = 0$. Slower. c) $\lim_{x\to\infty} \frac{\sqrt{x}}{e^x} = \lim_{x\to\infty} \frac{1}{2\sqrt{x}e^x} = 0$. Slower. d) $\lim_{x\to\infty} \frac{e^x}{4^x} = \lim_{x\to\infty}(\frac{e}{4})^x = 0$ since $0 < \frac{e}{4} < 1$. Faster than e^x as $x \to \infty$. e) $\lim_{x\to\infty} \frac{2.5^x}{e^x} = \lim_{x\to\infty}(\frac{2.5}{3})^x = 0$ since $0 < \frac{2.5}{e} < 1$. Slower. f) $\lim_{x\to\infty} \frac{\ln x}{e^x} = \lim_{x\to\infty} \frac{1}{xe^x} = 0$. Slower. g) $\lim_{x\to\infty} \frac{\log_{10} x}{e^x} = \lim_{x\to\infty} \frac{\ln x}{(\ln 10)e^x} = \lim_{x\to\infty} \frac{1}{(\ln 10)xe^x} = 0$. Slower. h) $\lim_{x\to\infty} \frac{e^{-x}}{e^x} =$

$\lim_{x\to\infty} \frac{1}{e^{2x}} = 0$. Slower. i) $\lim_{x\to\infty} \frac{e^{x+1}}{e^x} = \lim_{x\to\infty} e = e$. Same rate. j) $\lim_{x\to\infty} \frac{(1/2)e^x}{e^x} = \lim_{x\to\infty}(1/2) = 1/2$. Same rate.

2. a) $\frac{10x^4+30x+1}{e^x} \to 0$ as $x \to \infty$ applying L'Hôpital's rule several times. Slower. b) $\lim_{x\to\infty} \frac{x\ln x - x}{e^x} = \lim_{x\to\infty} \frac{x(1/x)+\ln x - 1}{e^x} = \lim_{x\to\infty} \frac{\ln x}{e^x} = \lim_{x\to\infty} \frac{1}{xe^x} = 0$. Slower. c) For $x > 0$, $\sqrt{1+x^4} < 1 + x^4$ and $1 + x^4$ grows slower than e^x. Slower. d) $(x^{1000})' = 1000x^{999}$, $(x^{1000})'' = 1000(999)x^{998}$. Continuing, we deduce that $(x^{1000})^{(1000)} = 1000!$. Thus applying L'Hôpital's rule 1000 times, we find $\lim_{x\to\infty} \frac{x^{1000}}{e^x} = \lim_{x\to\infty} \frac{1000!}{e^x} = (1000!) \lim_{x\to\infty} \frac{1}{e^x} = (1000!)(0) = 0$. Slower. e) $[(e^x + e^{-x})/2]/e^x = \frac{1}{2} + \frac{1}{2e^{2x}} \to \frac{1}{2}$ as $x \to \infty$. Same rate. f) $\frac{xe^x}{e^x} = x \to \infty$ as $x \to \infty$. Faster. g) $\frac{e^{-1}}{e^x} \leq \frac{e^{\cos x}}{e^x} \leq \frac{e^1}{e^x}$ and $\frac{e^{-1}}{e^x} \to 0$, $\frac{e}{e^x} \to 0$ as $x \to \infty$. By the Squeeze Theorem, $\lim_{x\to\infty} \frac{e^{\cos x}}{e^x} = 0$. Slower. h) $\frac{e^{x-1}}{e^x} = e^{-1} \to e^{-1}$ as $x \to \infty$. Same rate.

3. a) $\frac{x^2+4x}{x^2} = 1 + \frac{4}{x} \to 1$ as $x \to \infty$. $x^2 + 4x$ grows at the same rate as x^2 as $x \to \infty$. b) $\frac{x^2}{x^3+3} = \frac{1}{x+(3/x^2)} \to 0$ as $x \to \infty$. Faster. c) $\frac{x^2}{x^5} = \frac{1}{x^3} \to 0$ as $x \to \infty$. Faster. d) $\frac{15x+3}{x^2} = \frac{15}{x} + \frac{3}{x^2} \to 0$. Slower. e) $\frac{\sqrt{x^4+5x}}{x^2} = \frac{\sqrt{x^4+5x}}{\sqrt{x^4}} = \sqrt{1+\frac{5}{x^3}} \to 1$ as $x \to \infty$. Same rate. f) $\frac{(x+1)^2}{x^2} = (\frac{x+1}{x})^2 = (1+\frac{1}{x})^2 \to 1$ as $x \to \infty$. Same rate. g) $\lim_{x\to\infty} \frac{\ln x}{x^2} = \lim_{x\to\infty} \frac{(1/x)}{2x} = \lim_{x\to\infty} \frac{1}{2x^2} = 0$. Slower. h) $\lim_{x\to\infty} \frac{\ln(x^2)}{x^2} = \lim_{x\to\infty} = \frac{2\ln x}{x^2} = 0$ (as in g)). Slower. i) $\frac{\ln(10^x)}{x^2} = \frac{x\ln 10}{x^2} = \frac{x\ln 10}{x^2} = \frac{\ln 10}{x} \to 0$. Slower. j) $\lim_{x\to\infty} \frac{2^x}{x^2} = \lim_{x\to\infty} \frac{2^x \ln 2}{2x} = \lim_{x\to\infty} \frac{2^x(\ln 2)^2}{2} = \infty$. Faster.

4. a) $\frac{\log_3 x}{\ln x} = \frac{\ln x}{(\ln 3)\ln x} = \frac{1}{\ln 3} \to \frac{1}{\ln 3}$ as $x \to \infty$. Same rate. b) $\frac{\log_2 x^2}{\ln x} = \frac{2\log_2 x}{\ln x} = \frac{2\ln x}{(\ln 2)\ln x} = \frac{2}{\ln 2} \to \frac{2}{\ln 2}$. Same rate. c) $\frac{\log_{10}\sqrt{x}}{\ln x} = \frac{\log_{10} x}{2\ln x} = \frac{\ln x}{2(\ln 10)\ln x} = \frac{1}{2\ln 10} \to \frac{1}{2\ln 10}$. Same rate. d) $\frac{(1/x)}{\ln x} = \frac{1}{x\ln x} \to 0$ as $x \to \infty$. Slower. e) $\frac{(1/\sqrt{x})}{\ln x} = \frac{1}{\sqrt{x}\ln x} \to 0$ as $x \to \infty$. Slower. f) $\frac{e^{-x}}{\ln x} = \frac{1}{e^x \ln x} \to 0$ as $x \to \infty$. Slower. g) $\lim_{x\to\infty} \frac{x}{\ln x} = \lim_{x\to\infty} \frac{1}{(1/x)} = \lim_{x\to\infty} x = \infty$. Faster. h) $\frac{5\ln x}{\ln x} = 5 \to 5$. Same rate. i) $\frac{2}{\ln x} \to 0$ as $x \to \infty$. Slower. j) $\frac{-1}{\ln x} \leq \frac{\sin x}{\ln x} \leq \frac{1}{\ln x}$ and the Squeeze Theorem show $\frac{\sin x}{\ln x} \to 0$ as $x \to \infty$. Slower.

5. $\frac{e^x}{e^{x/2}} = e^{x/2} \to \infty$ as $x \to \infty$. $\frac{(\ln x)^x}{e^x} = (\frac{\ln x}{e})^x \to \infty$ as $x \to \infty$. $\frac{x^x}{(\ln x)^x} = (\frac{x}{\ln x})^x \to \infty$ as $x \to \infty$ because $\frac{x}{\ln x} \to \infty$ as $x \to \infty$. The order from slowest to fastest is $e^{x/2}, e^x, (\ln x)^x, x^x$.

6. Since $0 < \ln 2 < 1$, $(\ln 2)^x \to 0$ as $x \to \infty$. Hence $\frac{(\ln 2)^x}{x^2} \to 0$ as $x \to \infty$. $\lim_{x\to\infty} \frac{x^2}{2^x} = \lim_{x\to\infty} \frac{2x}{2^x \ln 2} = \lim_{x\to\infty} \frac{2}{2^x (\ln 2)^2} = 0$. $\frac{2^x}{e^x} = \left(\frac{2}{e}\right)^x \to 0$ as $x \to \infty$ since $0 < \frac{2}{e} < 1$. Answer: $(\ln 2)^x, x^2, 2^x, e^x$.

7. $\lim_{x\to\infty} \frac{\sqrt{10x+1}}{\sqrt{x}} = \lim_{x\to\infty} \sqrt{\frac{10x+1}{x}} = \lim_{x\to\infty} \sqrt{10+(1/x)} = \sqrt{10}$. $\lim_{x\to\infty} \frac{\sqrt{x+1}}{\sqrt{x}} = \lim_{x\to\infty} \sqrt{\frac{x+1}{x}} = \lim_{x\to\infty} \sqrt{1+(1/x)} = \sqrt{1} = 1$.

8. $\frac{\sqrt{x^4+x}}{x^2} = \sqrt{\frac{x^4+x}{x^4}} = \sqrt{1+\frac{1}{x^3}} \to \sqrt{1+0} = 1$ as $x \to \infty$. So $\sqrt{x^4+x}$ and x^2 grow at the same rate. Similarly, $\sqrt{x^4-x^3}$ and x^2 grow at the same rate.

9. $\frac{\sqrt{x^4+x}}{x^2} = \sqrt{\frac{x^4+x}{x^4}} = \sqrt{1+\frac{1}{x^2}} \to \sqrt{1+0} = 1$ as $x \to \infty$. So $\sqrt{x^4+x}$ and x^2 grow at the same rate. $\frac{\sqrt[3]{x^6+x}}{x^2} = \sqrt[3]{\frac{x^6+x}{x^6}} = \sqrt[3]{1+\frac{1}{x^5}} \to 1$ as $x \to \infty$. So $\sqrt[3]{x^6+x}$ and x^2 grow at the same rate.

10. $\frac{\sqrt[4]{x^6+x}}{x^{3/2}} = \sqrt[4]{\frac{x^6+x}{x^6}} = \sqrt[4]{1+\frac{1}{x^5}} \to 1$ as $x \to \infty$. Hence $\sqrt[4]{x^6+x}$ and $x^{3/2}$ grow at the same rate as $x \to \infty$. $\frac{\sqrt{x^3-4x}}{x^{3/2}} = \sqrt{\frac{x^3-4x}{x^3}} = \sqrt{1-\frac{4}{x^2}} \to 1$ as $x \to \infty$. Hence $\sqrt{x^3-4x}$ and $x^{3/2}$ grow at the same rate as $x \to \infty$.

11. a) $\frac{x}{x} \to 1 \neq 0$. False b) $\frac{x}{x+5} = \frac{1}{1+(5/x)} \to 1 \neq 0$ as $x \to \infty$. False c) True by the preceding work. d) $\frac{x}{2x} = \frac{1}{2}$. True e) $\frac{e^x}{e^{2x}} = \frac{1}{e^x} \to 0$ as $x \to \infty$. True f) $\frac{x+\ln x}{x} = 1 + \frac{\ln x}{x} \to 1+0$ as $x \to \infty$ by L'Hôpital's rule. True g) $\frac{\ln x}{\ln 2x} = \frac{\ln x}{\ln 2 + \ln x} = \frac{1}{(\ln 2/\ln x)+1} \to \frac{1}{0+1} = 1$ as $x \to \infty$. False h) $\frac{\sqrt{x^2+5}}{x} = \sqrt{\frac{x^2+5}{x^2}} = \sqrt{1+5/x^2} \to 1$ as $x \to \infty$. True

12. a) $\left(\frac{1}{x+3}\right)/\left(\frac{1}{x}\right) = \frac{x}{x+3} = 1 - \frac{3}{x+3} > 1$ for $x > 0$. True. b) $\left(\frac{1}{x}+\frac{1}{x^2}\right)/\left(\frac{1}{x}\right) = 1 + \frac{1}{x} \leq 2$ for $x \geq 1$. True. c) $\left(\frac{1}{x}-\frac{1}{x^2}\right)/\left(\frac{1}{x}\right) = 1 - \frac{1}{x} \to 1$ as $x \to \infty$. False. d) $\frac{2+\cos x}{2} = 1 + \frac{\cos x}{2} \leq 1 + \frac{1}{2} = \frac{3}{2}$ for all x. True. e) $\frac{e^x+x}{e^x} = 1 + \frac{x}{e^x} \to 1$ as $x \to \infty$. So $1 + \frac{x}{e^x} \leq 2$ for x sufficiently large. True. f) $\frac{x \ln x}{x^2} = \frac{\ln x}{x} \to 0$ as $x \to \infty$. True. g) $\lim_{x\to\infty} \frac{\ln(\ln x)}{\ln x} = \lim_{x\to\infty} \frac{\left(\frac{1}{\ln x}\right)\frac{1}{x}}{\frac{1}{x}} = \lim_{x\to\infty} \frac{1}{\ln x} = 0$. Since $\ln(\ln x) = o(\ln x)$, $\ln(\ln x) = O(\ln x)$. True. h) $\lim_{x\to\infty} \frac{\ln x}{\ln(x^2+1)} = \lim_{x\to\infty} \frac{\frac{1}{x}}{\left(\frac{2x}{x^2+1}\right)} = \lim_{x\to\infty} \frac{x^2+1}{2x^2} = \frac{1}{2}$. False.

13. By induction, the nth derivative of x^n is $n!$ for any positive integer n. By applying l'Hôpital's rule n times $\lim_{x\to\infty} \frac{x^n}{e^x} = \lim_{x\to\infty} \frac{n!}{e^x} = 0$.

14. We use induction on the degree n of the polynomial. If $n = 0$, the polynomial is a constant c and $\frac{c}{e^x} \to 0$ as $x \to \infty$. The statement is true when $n = 0$. Assume the statement is true for $n = k$. Let $p(x)$ be a polynomial of degree $k + 1$. Then $p'(x)$ is a polynomial of degree k and by l'Hôpital's rule, $\lim_{x\to\infty} \frac{p(x)}{e^x} = \lim_{x\to\infty} \frac{p'(x)}{e^x} = 0$. Therefore the statement is true for $n = k + 1$. By induction the statement is true for all n.

15. a) $\lim_{x\to\infty} \frac{\ln x}{x^{1/n}} = \lim_{x\to\infty} \frac{\frac{1}{x}}{\frac{1}{n}x^{(1/n)-1}} = \lim_{x\to\infty} \frac{n}{x^{1/n}} = 0$ b) $\ln x = x^{10^{-6}}$ implies $\ln(\ln x) = 10^{-6} \ln x$ and $\ln(\ln(\ln x)) = \ln(\ln x) - 6 \ln 10$. Let $u = \ln(\ln x)$. Then the last equation becomes $\ln u = u - 6 \ln 10$. Graph $y_1 = \ln x$ and $y_2 = x - 6 \ln 10$ in $[0, 20]$ by $[-2, 7]$ and zoom in to the point of intersection. Its x-coordinate is $u = 16.6265089014 = \ln(\ln x)$. Hence $\ln x = e^u$ and $x = e^{e^u}$.

16. Let $p(x)$ be a polynomial of degree $n \geqq 1$. By l'Hôpital's rule, $\lim_{x\to\infty} \frac{\ln x}{p(x)} = \lim_{x\to\infty} \frac{1}{xp'(x)} = 0$. Therefore $\ln x$ grows slower than $p(x)$ as $x \to \infty$.

17. $\lim_{x\to\infty} \frac{(2\sqrt{x}-1)^2}{4x} = \lim_{x\to\infty} \frac{4x-4\sqrt{x}+1}{4x} = \lim_{x\to\infty}(1 - \frac{1}{\sqrt{x}} + \frac{1}{4x}) = 1$

18. $\lim_{x\to\pm\infty} \frac{\sqrt{x^2+5}}{|x|} = \lim_{x\to\pm\infty} \sqrt{\frac{x^2+5}{x^2}} = \lim_{x\to\pm\infty} \sqrt{1 + \frac{5}{x^2}} = 1.$

19. $\lim_{x\to\infty} \frac{e^x+x^2}{e^x} = \lim_{x\to\infty}(1 + \frac{x^2}{e^x}) = 1$ as in Exercise 1. $\lim_{x\to-\infty} \frac{e^x+x^2}{x^2} = \lim_{x\to-\infty}(\frac{1}{x^2e^{-x}} + 1) = 1.$

20. $\lim_{x\to\pm\infty} \frac{f+g}{f} = \lim_{x\to\pm\infty}(1 + \frac{g}{f}) = 1$

21. Since $0 \leqq |\frac{\sin x}{x}| \leqq \frac{1}{|x|} \to 0$ as $x \to \pm\infty$, $\lim_{x\to\pm\infty} \frac{\sin x}{x} = 0$. Hence $\lim_{x\to\pm\infty} \frac{f(x)}{g(x)} = \lim_{x\to\pm\infty}(1 + \frac{\sin x}{x}) = 1$ and g is an end behavior model for f.

22. $\frac{x^2-\cos x}{x^2} = 1 - \frac{\cos x}{x^2} \to 1 - 0 = 1$ as $x \to \pm\infty$. Therefore g is an end behavior model for f.

23. $\lim_{x\to\pm\infty} \frac{2x^3-3x^2+x-1}{2x^3} = \lim_{x\to\pm\infty}(1 - \frac{3}{2x} + \frac{1}{2x^2} - \frac{1}{2x^3}) = 1$. g is an end behavior model for f.

24. $\frac{f(x)}{g(x)} = \frac{2}{3}\frac{3x^4-x^3+x-1}{2x^4+x^3-1} \to \frac{2}{3}\frac{3}{2} = 1$ as $x \to \pm\infty$. Therefore g is an end behavior model for f.

25. $\lim_{x\to\infty} \frac{f(x)}{g(x)} = \lim_{x\to\infty} \frac{2^x+x}{x} = \lim_{x\to\infty} \frac{2^x \ln 2+1}{1} = \infty$ (l'Hôpital's rule). $\lim_{x\to-\infty} \frac{2^x+x}{x} = \lim_{x\to-\infty}(\frac{1}{x2^{-x}} + 1) = 1$. g is only a left end behavior model for f.

26. $\frac{f(x)}{g(x)} = 1 + \frac{x}{2^x} \to 1$ if $x \to \infty$, $\to \infty$ if $x \to -\infty$. Therefore g is a right end behavior model for f.

27. $\lim_{x\to\pm\infty} \frac{\sqrt{x^4+2x-1}}{x^2} = \lim_{x\to\pm\infty} \sqrt{\frac{x^4+2x-1}{x^4}} = \lim_{x\to\pm\infty} \sqrt{1+\frac{2}{x^3}-\frac{1}{x^4}} = \sqrt{1+0-0} = 1$. g is an end behavior model for f.

28. $\frac{f(x)}{g(x)} = \frac{\sqrt{x^3+1}}{x^{3/2}} = \sqrt{\frac{x^3+1}{x^3}} = \sqrt{1+\frac{1}{x^3}} \to 1$ as $x \to \infty$. Therefore g is a right end behavior model for f.

29. $\lim_{x\to\pm\infty} \frac{\sqrt[3]{x^2-2x-1}}{\sqrt[3]{x^2}} = \lim_{x\to\pm\infty} \sqrt[3]{1-\frac{2}{x}-\frac{1}{x^2}} = 1$. g is an end behavior model for f.

30. $\frac{\sqrt[4]{x+2}}{x^{1/4}} = \sqrt[4]{\frac{x+2}{x}} = \sqrt{1+\frac{2}{x}} \to 1$ as $x \to \infty$. Therefore g is a right end behavior model for f.

31. We show only that f_1/g_1 is a right end behavior model for f/g. The other proofs are similar. $\lim_{x\to\infty} \frac{(f/g)}{(f_1/g_1)} = \lim_{x\to\infty} \frac{f}{f_1}\frac{g_1}{g} = 1 \cdot 1 = 1$.

32. We show only that $f_1 g_1$ is a left end behavior model for fg. The other proofs are similar. $\lim_{x\to-\infty} \frac{f_1 g_1}{fg} = \lim_{x\to-\infty} \frac{f_1}{f}\frac{g_1}{g} = 1 \cdot 1 = 1$.

33. Repeat Exercise 31 replacing $x \to \infty$ by $x \to \pm\infty$.

34. This is an immediate consequence of the definitions.

35. By assumption $0 = \lim_{x\to\infty} \frac{x}{a_n x^n + a_{n-1}x^{n-1} + \cdots + a_0}$. Dividing numerator and denominator by x^n, we have $\lim_{x\to\infty} \frac{x^{1-n}}{a_n + \frac{a_{n-1}}{x} + \cdots + \frac{a_0}{x^n}} = 0$. Since the limit of the denominator is a_n, we must have $\lim_{x\to\infty} x^{1-n} = 0$. This is true if and only if $1-n < 0$ or $n > 1$.

36. Since $p(x)$ and $q(x)$ grow at the same rate, they have the same degree. If their respective leading coefficients are p_n and q_n, then $\lim_{|x|\to\infty} \frac{p(x)}{q(x)} = \frac{p_n}{q_n}$.

37. The first is the most efficient because the number of steps for each of the others grows faster than n as $n \to \infty$.

38. $O(10^6)$ steps in a sequential search (as much as one million steps). In a binary search it may take as many as $\log_2 10^6 = 6\log_2 10 = 6\ln 10/\ln 2 \approx 20$ steps.

39. We are given that $\lim_{x\to\infty} \frac{f}{g} = L$ where L is a non-zero finite number. Hence $\frac{f(x)}{g(x)} \leqq L+1$ for x sufficiently large. Therefore $f = O(g)$. $\lim_{x\to\infty} \frac{g}{f} =$

$\lim_{x\to\infty}(\frac{f}{g})^{-1} = L^{-1}$. Hence $\frac{g}{f} \leq \frac{1}{L} + 1$ for x sufficiently large. Therefore $g = O(f)$.

40. $\frac{|E_S|}{h^4} \leq \frac{b-a}{180} h^4 M/h^4 = \frac{b-a}{180} M$ as $h \to 0$. $\frac{|E_T|}{h^2} \leq \frac{b-a}{12} h^2 M/h^2 = \frac{b-a}{12} M$ as $h \to 0$.

7.7 THE INVERSE TRIGONOMETRIC FUNCTIONS

1. a) $\frac{\pi}{4}$ b) $\frac{\pi}{3}$ c) $\frac{\pi}{6}$

2. a) $-\frac{\pi}{4}$ b) $-\frac{\pi}{3}$ c) $-\frac{\pi}{6}$

3. a) $-\frac{\pi}{6}$ b) $-\frac{\pi}{4}$ c) $-\frac{\pi}{3}$

4. a) $\frac{\pi}{6}$ b) $\frac{\pi}{4}$ c) $\frac{\pi}{3}$

5. a) $\frac{\pi}{3}$ b) $\frac{\pi}{4}$ c) $\frac{\pi}{6}$

6. a) $\frac{2\pi}{3}$ b) $\frac{3\pi}{4}$ c) $\frac{5\pi}{6}$

7. a) $\frac{3\pi}{4}$ b) $\frac{5\pi}{6}$ c) $\frac{2\pi}{3}$

8. a) $\frac{\pi}{4}$ b) $\frac{\pi}{6}$ c) $\frac{\pi}{3}$

9. a) $\frac{\pi}{4}$ b) $\frac{\pi}{3}$ c) $\frac{\pi}{6}$

10. a) $-\frac{\pi}{4}$ b) $-\frac{\pi}{3}$ c) $-\frac{\pi}{6}$

11. a) $\frac{3\pi}{4}$ b) $\frac{5\pi}{6}$ c) $\frac{2\pi}{3}$

12. a) $\frac{\pi}{4}$ b) $\frac{\pi}{6}$ c) $\frac{\pi}{3}$

13. $\alpha = \sin^{-1}(1/2) = \pi/6$. $\cos\alpha = \frac{\sqrt{3}}{2}$, $\tan\alpha = \frac{1}{\sqrt{3}}$, $\sec\alpha = \frac{2}{\sqrt{3}}$, $\csc\alpha = 2$

14. $\alpha = \cos^{-1}(-1/2) = \frac{2\pi}{3}$. $\sin\alpha = \frac{\sqrt{3}}{2}$, $\tan\alpha = -\sqrt{3}$, $\sec\alpha = -2$, $\csc\alpha = \frac{2}{\sqrt{3}}$

15. $\alpha = \tan^{-1}(4/3)$ implies $\tan\alpha = 4/3$ and α is in the first quadrant. Cot $\alpha = 3/4$. $\mathrm{Sec}^2\alpha = 1 + \tan^2\alpha = 1 + \frac{16}{9} = \frac{25}{9}$ and so $\sec\alpha = \frac{5}{3}$, $\cos\alpha = \frac{3}{5}$. Sin $\alpha = (\cos\alpha)\tan\alpha = \frac{3}{5}\frac{4}{3} = \frac{4}{5}$. Csc $\alpha = \frac{5}{4}$.

16. $\alpha = \sec^{-1}(-\sqrt{5})$ implies $\sec\alpha = -\sqrt{5}$ and $\frac{\pi}{2} < \alpha < \pi$. $\cos\alpha = -\frac{1}{\sqrt{5}}$. $\sin^2\alpha = 1 - \cos^2\alpha = 1 - \frac{1}{5} = \frac{4}{5}$ and so $\sin\alpha = \frac{2}{\sqrt{5}}$, $\csc\alpha = \frac{\sqrt{5}}{2}$. Tan $\alpha = (\sin\alpha)/(\cos\alpha) = -2$, cot $\alpha = -\frac{1}{2}$.

17. $\sin(\cos^{-1}\frac{\sqrt{2}}{2}) = \sin\frac{\pi}{4} = \frac{\sqrt{2}}{2}$

18. $\sec(\cos^{-1}\frac{1}{2}) = \sec\frac{\pi}{3} = 2$

19. $\tan(\sin^{-1}(-\frac{1}{2})) = \tan(-\frac{\pi}{6}) = -\frac{1}{\sqrt{3}}$

20. $\cot[\sin^{-1}(-\frac{\sqrt{3}}{2})] = \cot(-\frac{\pi}{3}) = -\cot\frac{\pi}{3} = -\frac{\sqrt{3}}{3}$

21. $\csc(\sec^{-1}2) + \cos(\tan^{-1}(-\sqrt{3})) = \csc\frac{\pi}{3} + \cos(-\frac{\pi}{3}) = \frac{2}{\sqrt{3}} + \frac{1}{2}$

22. $\tan(\sec^{-1}1) + \sin(\csc^{-1}(-2)) = \tan 0 + \sin(\sin^{-1}(-\frac{1}{2})) = 0 - \frac{1}{2} = -\frac{1}{2}$

23. $\sin[\sin^{-1}(-\frac{1}{2}) + \cos^{-1}(-\frac{1}{2})] = \sin(-\frac{\pi}{6} + \frac{2\pi}{3}) = \sin\frac{\pi}{2} = 1$

24. $\cot[\sin^{-1}(-\frac{1}{2}) - \sec^{-1} 2] = \cot(-\frac{\pi}{6} - \frac{\pi}{3}) = \cot(-\frac{\pi}{2}) = 0$

25. $\sec(\tan^{-1} 1 + \csc^{-1} 1) = \sec(\frac{\pi}{4} + \frac{\pi}{2}) = \sec(\frac{3\pi}{4}) = -\sqrt{2}$

26. $\sec(\cot^{-1}\sqrt{3} + \csc^{-1}(1)) = \sec(\frac{\pi}{6} + \frac{\pi}{2}) = \sec\frac{2\pi}{3} = -2$

27. $\sec^{-1}(\sec(-\frac{\pi}{6})) = \sec^{-1}(\frac{2}{\sqrt{3}}) = \cos^{-1}(\frac{\sqrt{3}}{2}) = \frac{\pi}{6}$

28. $\cot^{-1}(\cot(-\frac{\pi}{4})) = \cot^{-1}(-1) = \frac{\pi}{2} - \tan^{-1}(-1) = \frac{\pi}{2} + \frac{\pi}{4} = \frac{3\pi}{4}$

29. $\sec[\tan^{-1}\frac{x}{2}]$. Let $\theta = \tan^{-1}\frac{x}{2}$. Then $-\frac{\pi}{2} < \theta < \frac{\pi}{2}$ and $\tan\theta = \frac{x}{2}$. Since $-\frac{\pi}{2} < \theta < \frac{\pi}{2}$, $\sec\theta > 0$. $\sec^2\theta = 1 + \tan^2\theta = 1 + \frac{x^2}{4} = \frac{x^2+4}{4}$. $\sec\theta = \frac{\sqrt{x^2+4}}{2}$.

30. Let $\theta = \tan^{-1} 2x$. Then $-\frac{\pi}{2} < \theta < \frac{\theta}{2}$ (so $\sec\theta > 0$) and $\tan\theta = 2x$. $\sec^2\theta = 1 + \tan^2\theta = 1 + 4x^2$. $\sec\theta = \sqrt{4x^2+1}$.

31. Let $\theta = \sec^{-1} 3y = \cos^{-1}\frac{1}{3y}$. Then $0 \leqq \theta < \frac{\pi}{2}$ if $3y \geqq 1$, $\frac{\pi}{2} < \theta \leqq \pi$ if $3y \leqq -1$ and $\cos\theta = \frac{1}{3y}$, $\sec\theta = 3y$. $\tan^2\theta = \sec^2\theta - 1 = 9y^2 - 1$. $\tan\theta = \sqrt{9y^2-1}$ if $y \geqq \frac{1}{3}$, $\tan\theta = -\sqrt{9y^2-1}$ if $y \leqq -\frac{1}{3}$.

32. $\tan[\sec^{-1}\frac{y}{5}] = \begin{cases} -\sqrt{y^2-25}/5 & \text{if } y \leqq -5 \\ \sqrt{y^2-25}/5 & \text{if } y \geqq 5 \end{cases}$ (See the preceding solution.)

33. Let $\theta = \sin^{-1} x$. Then $-\frac{\pi}{2} \leqq \theta \leqq \frac{\pi}{2}$ (so $\cos\theta \geqq 0$) and $\sin\theta = x$. $\cos^2\theta = 1 - \sin^2\theta = 1 - x^2$. $\cos\theta = \sqrt{1-x^2}$.

34. Let $\theta = \cos^{-1} x$. Then $0 \leqq \theta \leqq \frac{\pi}{2}$ if $x \geqq 0$ and $\frac{\pi}{2} < x \leqq \pi$ if $x < 0$. Thus $\tan\theta > 0$ if $x > 0$ and $\tan\theta < 0$ if $x < 0$. $\sin^2\theta = 1 - \cos^2 = 1 - x^2$, $\sin\theta = \sqrt{1-x^2}$ for $0 \leqq \theta \leqq \pi$. $\tan\theta = \frac{\sin\theta}{\cos\theta}$. $\tan\theta$ will have the correct sign if we write $\tan\theta = \frac{\sqrt{1-x^2}}{x}$.

35. We claim $\sin(\tan^{-1} u) = \frac{u}{\sqrt{1+u^2}}$ for any u. Let $\theta = \tan^{-1} u$. Then $-\frac{\pi}{2} < \theta < \frac{\pi}{2}$ (so $\cos\theta > 0$) and $\tan\theta = u$. $\sec^2\theta = \tan^2\theta + 1 = u^2 + 1$, $\sec\theta = \sqrt{u^2+1}$. $\sin\theta = \frac{\sin\theta}{\cos\theta}\cos\theta = u\frac{1}{\sqrt{u^2+1}} = \frac{u}{\sqrt{u^2+1}}$. Hence $\sin[\tan^{-1}\sqrt{x^2-2x}] = \frac{u}{\sqrt{1+u^2}} = \frac{\sqrt{x^2-2x}}{\sqrt{1+(\sqrt{x^2-2x})^2}} = \frac{\sqrt{x^2-2x}}{\sqrt{x^2-2x+1}} = \frac{\sqrt{x^2-2x}}{|x-1|}$.

36. By the preceding solution, $\sin(\tan^{-1}\frac{x}{\sqrt{x^2+1}}) = \frac{u}{\sqrt{1+u^2}} = \frac{x/\sqrt{x^2+1}}{\sqrt{1+x^2/(x^2+1)}} = \frac{x/\sqrt{x^2+1}}{\sqrt{2x^2+1}/\sqrt{x^2+1}} = \frac{x}{\sqrt{2x^2+1}}$

37. We claim $\cos(\sin^{-1} u) = \sqrt{1-u^2}$ for any u with $|u| \leqq 1$. Let $\theta = \sin^{-1} u$. Then $-\frac{\pi}{2} \leqq \theta \leqq \frac{\pi}{2}$ (so $\cos\theta \geqq 0$) and $\sin\theta = u$. $\cos^2\theta = 1 - \sin^2\theta = 1 - u^2$, $\cos\theta = \sqrt{1-u^2}$. Hence $\cos(\sin^{-1}\frac{2y}{3}) = \sqrt{1 - \frac{4y^2}{9}} = \sqrt{9-4y^2}/3$.

38. By the preceding proof, $\cos(\sin^{-1}\frac{y}{5}) = \sqrt{1-(\frac{y}{5})^2} = \sqrt{25-y^2}/5$.

39. We claim $\sin(\sec^{-1} u) = \frac{\sqrt{u^2-1}}{|u|}$ for all u, $|u| \geqq 1$. Let $\theta = \sec^{-1} u = \cos^{-1}(1/u)$. Then $0 \leqq \theta \leqq \pi$ (so $\sin\theta \geqq 0$) and $\cos\theta = \frac{1}{u}$. $\sin^2\theta = 1 - \cos^2\theta = 1 - \frac{1}{u^2} = \frac{u^2-1}{u^2}$, so $\sin\theta = \frac{\sqrt{u^2-1}}{|u|}$. It follows that $\sin(\sec^{-1}\frac{x}{4}) = \frac{\sqrt{\frac{x^2}{4^2}-1}}{(|x|/4)} = \frac{\sqrt{x^2-16}}{|x|}$.

40. By the preceding solution, $\sin(\sec^{-1}\frac{\sqrt{x^2+1}}{x}) = \frac{\sqrt{\frac{x^2+4}{x^2}-1}}{(\sqrt{x^2+4}/|x|)} = \frac{2/|x|}{\sqrt{x^2+4}/|x|} = \frac{2}{\sqrt{x^2+4}}$.

41. $\lim_{x\to 1^-} \sin^{-1} x = \frac{\pi}{2}$

42. $\cos^{-1}(-1) = \pi$ and $\cos^{-1} x$ is continuous. Hence $\lim_{x\to -1^+} \cos^{-1} x = \pi$.

43. $\lim_{x\to\infty} \tan^{-1} x = \frac{\pi}{2}$ **44.** $\lim_{x\to-\infty} \tan^{-1} x = -\frac{\pi}{2}$ **45.** $\lim_{x\to\infty} \sec^{-1} x = \frac{\pi}{2}$

46. $\lim_{x\to-\infty} \sec^{-1} x = \lim_{x\to-\infty} \cos^{-1}(\frac{1}{x}) = \lim_{\theta\to 0^-} \cos^{-1}\theta = \frac{\pi}{2}$

47. $\lim_{x\to\infty} \csc^{-1} x = 0$

48. $\lim_{x\to-\infty} \csc^{-1} x = \lim_{x\to-\infty} \sin^{-1}(\frac{1}{x}) = \lim_{\theta\to 0^-} \sin^{-1}\theta = 0$

49. $\alpha =$ large angle $-$ small angle $= \cot^{-1}\frac{x}{15} - \cot^{-1}\frac{x}{3}$

50. $V = \pi\int_0^{\pi/3}(2^2 - \sec^2 y)dy = \pi[4y - \tan y]_0^{\pi/3} = \pi[\frac{4\pi}{3} - \sqrt{3}]$

51. Let θ be the indicated angle. Then $r = 3\sin\theta$, $h = 3\cos\theta$. $V = \frac{1}{3}\pi r^2 h = \frac{1}{3}\pi 9(\sin^2\theta)3\cos\theta = 9\pi\sin^2\theta\cos\theta$. $\frac{dV}{d\theta} = 9\pi(-\sin^3\theta + 2\sin\theta\cos^2\theta) = 9\pi\sin\theta(2\cos^2\theta - \sin^2\theta) = 9\pi\sin\theta(3\cos^2\theta - 1)$. $\frac{dV}{d\theta} = 0$ leads, under the given circumstances, to $\cos\theta = \frac{1}{\sqrt{3}}$, $\theta = \cos^{-1}\frac{1}{\sqrt{3}} = 0.955$ radians or $54.736°$.

52. $65° + (90° - \beta) + (90° - \alpha) = 180°$ (the left-most straight line). Hence $\alpha + \beta = 65°$, $\alpha = 65° - \beta = 65° - \tan^{-1}\frac{21}{50}$ (in degree mode) $= 42.218°$.

53. $\cot^{-1} 2 = \frac{\pi}{2} - \tan^{-1} 2 = 0.464$, $\sec^{-1}(1.5) = \cos^{-1}(1/1.5) = 0.841$, $\csc^{-1}(1.5) = \sin^{-1}(1/15) = 0.730$

54. In the diagram three angles are indicated with arrows. The lowest is $\tan^{-1} 1$, the next highest is $\tan^{-1} 2$ and the third is $\tan^{-1} 3$. Their sum is clearly π.

55. Let $\theta = \pi - \cos^{-1} x$. $0 \leqq \cos^{-1} x \leqq \pi$ leads to $0 \geqq -\cos^{-1} x \geqq -\pi$ and $\pi \geqq \pi - \cos^{-1} x \geqq 0$ or 1) $0 \leqq \theta \leqq \pi$. $\cos(\pi - \cos^{-1} x) = (\cos \pi)\cos(\cos^{-1} x) + \sin \pi \sin(\cos^{-1} x) = (-1)x + 0$ and so 2) $\cos\theta = -x$. 1) and 2) together prove $\theta = \cos^{-1}(-x)$ as was to be shown.

56. Let $\theta = \frac{\pi}{2} - \cos^{-1} x$. $0 \leqq \cos^{-1} x \leqq \pi$ leads to $0 \geqq -\cos^{-1} x \geqq -\pi$ and then $\frac{\pi}{2} \geqq \frac{\pi}{2} - \cos^{-1} x \geqq -\frac{\pi}{2}$ or 1) $-\frac{\pi}{2} \leqq \theta \leqq \frac{\pi}{2}$. $\text{Sin}(\frac{\pi}{2} - \cos^{-1} x) = (\sin\frac{\pi}{2})\cos(\cos^{-1} x) - (\cos\frac{\pi}{2})\sin(\cos^{-1} x) = x$ and so 2) $\sin\theta = x$. 1) and 2) imply $\theta = \sin^{-1} x$ which is equivalent to $\sin^{-1} x + \cos^{-1} x = \frac{\pi}{2}$.

57. Let $\theta = -\sin^{-1} x$. Then $-\frac{\pi}{2} \leqq \theta \leqq \frac{\pi}{2}$ and $\sin\theta = \sin(-\sin^{-1} x) = -\sin(\sin^{-1} x) = -x$. These two facts imply $\theta = \sin^{-1}(-x)$.

58. $-\frac{\pi}{2} < -\tan^{-1} x < \frac{\pi}{2}$ and $\tan(-\tan^{-1} x) = -\tan(\tan^{-1} x) = -x$. These two facts imply $-\tan^{-1} x = \tan^{-1}(-x)$.

59. $\lim_{x\to\infty} \frac{\tan^{-1} x}{(\pi/2)} = \frac{(\pi/2)}{(\pi/2)} = 1$. $\lim_{x\to-\infty} \frac{\tan^{-1} x}{(-\pi/2)} = \frac{(-\pi/2)}{(-\pi/2)} = 1$.

60. $y = \sin(\sin^{-1} x) = x$, $-1 \leqq x \leqq 1$. Domain = range = $[-1, 1]$.

61. Graph $y = \sin^{-1}(1/2x) = \sin^{-1}((2x)^{-1})$ in $[-3.5, 3.5]$ by $[-\frac{\pi}{2}, \frac{\pi}{2}]$

62. Graph $y = 3\tan^{-1} x$ in $[-10, 10]$ by $[-1.5\pi, 1.5\pi]$.

63. Graph $y = 2\cos^{-1}(1/3x) = 2\cos^{-1}((3x)^{-1})$ in $[-3, 3]$ by $[0, 2\pi]$.

64. Graph $y = \cot^{-1}(x+2) = \frac{\pi}{2} - \tan^{-1}(x+2)$ in $[-10, 10]$ by $[0, \pi]$.

65. Graph $y = 3 + \cos^{-1}(x-2)$ in $[1, 3]$ by $[3, 3+\pi]$.

66. Graph $y = -2 + \sin^{-1}(x-3)$ in $[-1, 5]$ by $[-4, 1]$.

7.8 DERIVATIVES OF INVERSE TRIGONOMETRIC FUNCTIONS; RELATED INTEGRALS

1. $y = \cos^{-1} x^2$. $y' = -\frac{1}{\sqrt{1-(x^2)^2}}(2x) = -\frac{2x}{\sqrt{1-x^4}}$. To confirm graphically we graph this last result and NDER(y, x) in $[-1, 1]$ by $[-5, 5]$ and see that the two graphs match.

2. $y = \cos^{-1}(\frac{1}{x})$. $y' = -\frac{1}{\sqrt{1-(\frac{1}{x})^2}}(-\frac{1}{x^2}) = \frac{\sqrt{x^2}}{x^2\sqrt{x^2-1}} = \frac{|x|}{|x|^2\sqrt{x^2-1}} = \frac{1}{|x|\sqrt{x^2-1}}$

3. $y = 5\tan^{-1} 3x$. $\frac{dy}{dx} = 5\frac{1}{1+(3x)^2}(3) = \frac{15}{1+9x^2}$

4. $y = \cot^{-1}\sqrt{x} = \frac{\pi}{2} - \tan^{-1}\sqrt{x}$. $y' = -\frac{1}{(1+x)}\frac{1}{2\sqrt{x}} = -\frac{1}{2\sqrt{x}(1+x)}$

5. $y = \sin^{-1}(x/2)$. $y' = \frac{1}{\sqrt{1-(x/2)^2}}(\frac{1}{2}) = \frac{1}{2\sqrt{\frac{4-x^2}{4}}} = \frac{1}{\sqrt{4-x^2}}$

6. $y = \sin^{-1}(1-x)$. $y' = -\frac{1}{\sqrt{1-(1-x)^2}} = -\frac{1}{\sqrt{2x-x^2}}$

7. $y = \sec^{-1}(5x)$. $y' = \frac{5}{|5x|\sqrt{25x^2-1}} = \frac{1}{|x|\sqrt{25x^2-1}}$

8. $y = \frac{1}{3}\tan^{-1}(\frac{x}{3})$. $y' = \frac{1}{3}\frac{1}{1+\frac{x^2}{9}}(\frac{1}{3}) = \frac{1}{9+x^2}$

9. $y = \csc^{-1}(x^2+1)$. $y' = \frac{-2x}{|x^2+1|\sqrt{(x^2+1)^2-1}} = \frac{-2x}{(x^2+1)\sqrt{x^4+2x^2}}$

10. $y = \cos^{-1}(2x)$. $y' = -\frac{2}{\sqrt{1-4x^2}}$.

11. $y = \csc^{-1}\sqrt{x} + \sec^{-1}\sqrt{x} = \sin^{-1}\frac{1}{\sqrt{x}} + \cos^{-1}\frac{1}{\sqrt{x}} = \frac{\pi}{2}$, a constant function. $y' = 0$.

12. $y = \csc^{-1}(\frac{1}{x})$, $x > 0$. $y' = -\frac{(-\frac{1}{x^2})}{|\frac{1}{x}|\sqrt{\frac{1}{x^2}-1}} = \frac{1}{\sqrt{1-x^2}}$

13. $y = \cot^{-1}\sqrt{x-1}$. $y' = -\frac{\frac{1}{2}(x-1)^{-1/2}}{1+(\sqrt{x-1})^2} = \frac{-1}{2x\sqrt{x-1}}$

14. $y = x\sqrt{1-x^2} - \cos^{-1}x$. $y' = \frac{-x^2}{\sqrt{1-x^2}} + \sqrt{1-x^2} + \frac{1}{\sqrt{1-x^2}} = \frac{-x^2+(1-x^2)+1}{\sqrt{1-x^2}} = \frac{2(1-x^2)}{\sqrt{1-x^2}} = 2\sqrt{1-x^2}$

15. $y = \sqrt{x^2-1} - \sec^{-1}x$. $y' = \frac{1}{2}(x^2-1)^{-1/2}(2x) - \frac{1}{|x|\sqrt{x^2-1}} = \frac{x}{\sqrt{x^2-1}} - \frac{1}{|x|\sqrt{x^2-1}} = \frac{x|x|-1}{|x|\sqrt{x^2-1}}$

16. $y = \cot^{-1}(\frac{1}{x}) - \tan^{-1}x = \frac{\pi}{2} - \tan^{-1}\frac{1}{x} - \tan^{-1}x$. $y' = \frac{\frac{1}{x^2}}{1+\frac{1}{x^2}} - \frac{1}{1+x^2} = 0$, $x \neq 0$. (Remark: $y = \pi$ for $x < 0$ and $y = 0$ for $x > 0$.)

17. $y = 2x\tan^{-1}x - \ln(x^2+1)$. $y' = 2(x\frac{1}{1+x^2} + \tan^{-1}x) - \frac{2x}{x^2+1} = 2\tan^{-1}x$

18. $y = \ln(x^2+1) - 2x + 2\tan^{-1}x$. $y' = \frac{2x}{x^2+1} - 2 + \frac{2}{x^2+1} = \frac{2x(1-x)}{x^2+1}$

19. $\int_0^{1/2} \frac{dx}{\sqrt{1-x^2}} = \sin^{-1}x]_0^{1/2} = \sin^{-1}(1/2) - \sin^{-1}0 = \frac{\pi}{6}$

20. $\int_{-1}^{1} \frac{dx}{1+x^2} = \tan^{-1}x]_{-1}^{1} = \frac{\pi}{2}$

21. $\int_{\sqrt{2}}^{2} \frac{dx}{x\sqrt{x^2-1}} = \sec^{-1}x]_{\sqrt{2}}^{2} = \cos^{-1}(1/x)]_{\sqrt{2}}^{2} = \cos^{-1}(1/2) - \cos^{-1}(1/\sqrt{2}) = \frac{\pi}{3} - \frac{\pi}{4} = \frac{\pi}{12}$

22. $\int_{-2}^{-\sqrt{2}} \frac{dx}{x\sqrt{x^2-1}} = \sec^{-1}(|x|)]_{-2}^{-\sqrt{2}} = \frac{\pi}{4} - \frac{\pi}{3} = -\frac{\pi}{12}$

23. $\int_{-1}^{0} \frac{4dx}{1+x^2} = 4\tan^{-1}x]_{-1}^{0} = 4[\tan^{-1}0 - \tan^{-1}(-1)] = 4[0 - (-\frac{\pi}{4})] = \pi$

24. $\int_{\sqrt{3}/3}^{\sqrt{3}} \frac{6dx}{1+x^2} = 6\tan^{-1}x]_{\sqrt{3}/3}^{\sqrt{3}} = 6[\frac{\pi}{3} - \frac{\pi}{6}] = \pi$

25. Let $I = \int_0^{\sqrt{2}/2} \frac{xdx}{\sqrt{1-x^4}}$. Let $u = x^2$. Then $u^2 = x^4$, $du = 2xdx$, $I = \frac{1}{2}\int_0^{1/2} \frac{du}{\sqrt{1-u^2}} = \frac{1}{2}\sin^{-1}u]_0^{1/2} = \frac{1}{2}[\frac{\pi}{6} - 0] = \frac{\pi}{12}$.

26. Let $I = \int_0^{1/4} \frac{dx}{\sqrt{1-4x^2}}$, $u = 2x$, $du = 2dx$. Then $I = \frac{1}{2}\int_0^{1/2} \frac{du}{\sqrt{1-u^2}} = \frac{1}{2}\sin^{-1}u]_0^{1/2} = \frac{\pi}{12}$.

27. Let $I = \int_{1/\sqrt{3}}^{1} \frac{dx}{x\sqrt{4x^2-1}}$. Let $u = 2x$, $du = 2dx$. $I = \frac{1}{2}\int_{2/\sqrt{3}}^{2} \frac{2du}{u\sqrt{u^2-1}} = \sec^{-1}|u|]_{2/\sqrt{3}}^{2} = \cos^{-1}(1/2) - \cos^{-1}(\sqrt{3}/2) = \frac{\pi}{3} - \frac{\pi}{6} = \frac{\pi}{6}$.

28. Let $I = \int_0^1 \frac{xdx}{1+x^4}$, $u = x^2$, $du = 2xdx$. Then $I = \frac{1}{2}\int_0^1 \frac{du}{1+u^2} = \frac{1}{2}\tan^{-1}u]_0^1 = \frac{\pi}{8}$.

29. Let $I = \int_0^1 \frac{4xdx}{\sqrt{4-x^4}} = \frac{4}{2}\int_0^1 \frac{xdx}{\sqrt{1-(x^4/4)}}$. Let $u = \frac{x^2}{2}$, $du = xdx$. $I = 2\int_0^{1/2} \frac{du}{\sqrt{1-u^2}} = 2\sin^{-1}u]_0^{1/2} = \frac{\pi}{3}$.

30. Let $I = \int_0^1 \frac{dx}{\sqrt{4-x^2}}$, $u = \frac{x}{2}$, $du = \frac{1}{2}dx$. Then $I = 2\int_0^{1/2} \frac{du}{\sqrt{4-4u^2}} = \int_0^{1/2} \frac{du}{\sqrt{1-u^2}} = \sin^{-1}u]_0^{1/2} = \frac{\pi}{6}$.

31. Let $I = \int \frac{dx}{\sqrt{9-x^2}}$, $x = 3u$, $dx = 3du$. Then $I = \int \frac{3du}{\sqrt{9(1-u^2)}} = \int \frac{du}{\sqrt{1-u^2}} = \sin^{-1}u + C = \sin^{-1}\frac{x}{3} + C$.

32. Let $I = \int \frac{dx}{\sqrt{1-4x^2}}$, $u = 2x$, $du = 2dx$. Then $I = \frac{1}{2}\int \frac{du}{\sqrt{1-u^2}} = \frac{1}{2}\sin^{-1}u + C = \frac{1}{2}\sin^{-1}(2x) + C$.

33. Let $I = \int \frac{dx}{17+x^2}$, $x = \sqrt{17}u$, $dx = \sqrt{17}du$. Then $I = \int \frac{\sqrt{17}du}{17+17u^2} = \frac{1}{\sqrt{17}} \int \frac{du}{1+u^2} = \frac{1}{\sqrt{17}} \tan^{-1} u + C = \frac{1}{\sqrt{17}} \tan^{-1} \frac{x}{\sqrt{17}} + C$.

34. Let $I = \int \frac{dx}{9+3x^2} = \frac{1}{3} \int \frac{dx}{3+x^2}$, $x = \sqrt{3}u$, $dx = \sqrt{3}du$. Then $I = \frac{1}{3} \int \frac{\sqrt{3}du}{3+3u^2} = \frac{1}{3\sqrt{3}} \int \frac{du}{1+u^2} = \frac{1}{3\sqrt{3}} \tan^{-1} u + C = \frac{1}{3\sqrt{3}} \tan^{-1} \frac{x}{\sqrt{3}} + C$.

35. Let $I = \int \frac{dx}{x\sqrt{25x^2-2}}$, (we need $2u^2 = 25x^2$) $u = \frac{5}{\sqrt{2}}x$, $u^2 = \frac{25x^2}{2}$, $x = \frac{\sqrt{2}}{5}u$, $dx = \frac{\sqrt{2}}{5}du$. Then $I = \int \frac{(\sqrt{2}/5)du}{(\sqrt{2}/5)u\sqrt{2u^2-2}} = \frac{1}{\sqrt{2}} \int \frac{du}{u\sqrt{u^2-1}} = \frac{1}{\sqrt{2}} \sec^{-1} |u| + C = \frac{1}{\sqrt{2}} \sec^{-1}(\frac{5}{\sqrt{2}}|x|) + C$.

36. Let $I = \int \frac{dx}{x\sqrt{5x^2-4}}$, (we need $4u^2 = 5x^2$), $u = \frac{\sqrt{5}}{2}x$, $du = \frac{\sqrt{5}}{2}dx$. Then $I = \int \frac{(2/\sqrt{5})du}{(2/\sqrt{5})u\sqrt{4u^2-4}} = \frac{1}{2} \int \frac{du}{u\sqrt{u^2-1}} = \frac{1}{2} \sec^{-1} |u| + C = \frac{1}{2} \sec^{-1}(\frac{\sqrt{5}}{2}|x|) + C$.

37. Let $I = \int \frac{ydy}{\sqrt{1-y^4}}$, $u = y^2$, $du = 2ydy$, $\frac{1}{2}du = ydy$. Then $I = \frac{1}{2} \int \frac{du}{\sqrt{1-u^2}} = \frac{1}{2} \sin^{-1} u + C = \frac{1}{2} \sin^{-1}(y^2) + C$.

38. Let $I = \int \frac{\sec^2 ydy}{\sqrt{1-\tan^2 y}}$, $u = \tan y$, $du = \sec^2 ydy$. Then $I = \int \frac{du}{\sqrt{1-u^2}} = \sin^{-1} u + C = \sin^{-1}(\tan y) + C$.

39. Let $I = \int_{\sqrt[4]{2}}^{\sqrt{2}} \frac{2xdx}{x^2\sqrt{x^4-1}}$, $u = x^2$, $du = 2xdx$. $I = \int_{\sqrt{2}}^{2} \frac{du}{u\sqrt{u^2-1}} = \sec^{-1} |u|]_{\sqrt{2}}^{2} = \cos^{-1} \frac{1}{2} - \cos^{-1} \frac{1}{\sqrt{2}} = \frac{\pi}{3} - \frac{\pi}{4} = \frac{\pi}{12}$.

40. $I = \int_0^2 \frac{dx}{1+(x-1)^2}$, $u = x - 1$, $du = dx$. $I = \int_{-1}^{1} \frac{du}{1+u^2} = \tan^{-1} u]_{-1}^{1} = \frac{\pi}{2}$

41. Let $I = \int_1^{\sqrt{3}} \frac{2dx}{(1+x^2)\tan^{-1} x}$, $u = \tan^{-1} x$, $du = \frac{dx}{1+x^2}$. $I = 2\int_{\pi/4}^{\pi/3} \frac{du}{u} = 2\ln |u|]_{\pi/4}^{\pi/3} = 2[\ln(\pi/3) - \ln(\pi/4)] = 2\ln(4/3)$.

42. $I = \int_0^{\ln\sqrt{3}} \frac{e^x dx}{1+e^{2x}}$, $u = e^x$, $du = e^x dx$. $I = \int_1^{\sqrt{3}} \frac{du}{1+u^2} = \tan^{-1} u]_1^{\sqrt{3}} = \frac{\pi}{3} - \frac{\pi}{4} = \frac{\pi}{12}$

43. Let $I = \int_2^4 \frac{dx}{2x\sqrt{x-1}}$, $u^2 = x$, $u = \sqrt{x}$, $du = \frac{dx}{2\sqrt{x}}$, $\frac{du}{u} = \frac{dx}{2x}$. $I = \int_{\sqrt{2}}^{2} \frac{du}{u\sqrt{u^2-1}} = \sec^{-1} u]_{\sqrt{2}}^{2} = \cos^{-1}(\frac{1}{2}) - \cos^{-1}(\frac{1}{\sqrt{2}}) = \frac{\pi}{3} - \frac{\pi}{4} = \frac{\pi}{12}$.

44. $I = \int_{-\pi/2}^{\pi/2} \frac{2\cos xdx}{1+\sin^2 x}$, $u = \sin x$, $du = \cos xdx$. $I = 2\int_{-1}^{1} \frac{du}{1+u^2} = 2\tan^{-1} u]_{-1}^{1} = 2\frac{\pi}{2} = \pi$

45. $\lim_{x\to 0} \frac{\sin^{-1} x}{x} = \lim_{x\to 0} \frac{\frac{1}{\sqrt{1-x^2}}}{1} = 1$

46. $\lim_{x\to 0} \frac{\sin^{-1} x}{x^3} = \lim_{x\to 0} \frac{1}{3x^2\sqrt{1-x^2}} = \infty$

47. $\lim_{x\to 0} \frac{\tan^{-1} x}{x} = \lim_{x\to 0} \frac{\frac{1}{1+x^2}}{1} = 1$

48. $\lim_{x\to 0} \frac{\tan^{-1} x}{x^3} = \lim_{x\to 0} \frac{1}{3x^2(1+x^2)} = \infty$

49. $V = \pi \int_{-\frac{\sqrt{3}}{3}}^{\sqrt{3}} \frac{dx}{1+x^2} = \pi \tan^{-1} x\Big]_{-\sqrt{3}/3}^{\sqrt{3}} = \pi[\frac{\pi}{3} - (-\frac{\pi}{6})] = \frac{\pi^2}{2}$

50. $y = \sqrt{1-x^2}$, $y' = \frac{-x}{\sqrt{1-x^2}}$, $1 + (y')^2 = \frac{1}{1-x^2}$. $s = \int_{-1/2}^{1/2} \frac{dx}{\sqrt{1-x^2}} = \sin^{-1} x\Big]_{-1/2}^{1/2} = \frac{\pi}{3}$

51. We wish to maximize $\alpha = \cot^{-1} \frac{x}{15} - \cot^{-1} \frac{x}{3}$. $\frac{d\alpha}{dx} = -\frac{1}{15}\frac{1}{1+(x/15)^2} + \frac{1}{3}\frac{1}{1+(x/3)^2} = -\frac{15}{225+x^2} + \frac{3}{9+x^2} = 12[\frac{45-x^2}{(9+x^2)(225+x^2)}]$. From this we see that α rises from $x = 0$ to $x = \sqrt{45} = 3\sqrt{5}$ and then falls from larger x. You should sit $3\sqrt{5}$ft ≈ 6.71ft from the wall. You may confirm graphically by zooming in to the maximum of the graph of $\alpha = \tan^{-1} \frac{15}{x} - \tan^{-1} \frac{3}{x}$, $x > 0$.

52. By symmetry, $\bar{y} = 0$. $A = 2\int_0^1 \frac{dx}{1+x^2} = 2\tan^{-1} x\Big]_0^1 = \frac{\pi}{2}$. $M_y = \int_0^1 \frac{2x\,dx}{1+x^2} = \ln(1+x^2)\Big]_0^1 = \ln 2$. $\bar{x} = (\ln 2)/(\pi/2) = (\ln 4)/\pi$.

53. We start with the straight line angle $\pi = \cot^{-1} \frac{x}{1} + \theta + \cot^{-1} \frac{2-x}{1}$, $\theta = \pi - \cot^{-1} x - \cot^{-1}(2-x)$. $\frac{d\theta}{dx} = \frac{1}{1+x^2} - \frac{1}{1+(2-x)^2} = \frac{4(1-x)}{(1+x^2)[1+(2-x)^2]}$. Thus θ rises for $0 < x < 1$ and falls for $x > 1$. Hence θ is a maximum for $x = 1$. When $x = 1$, $\theta = \pi - \cot^{-1} 1 - \cot^{-1} 1 = \pi - \frac{\pi}{4} - \frac{\pi}{4} = \frac{\pi}{2}$.

54. In each case we use $L(x) = f(0) + f'(0)x$. For $\sin^{-1} x$, $L(x) = 0 + x = x$. For $\cos^{-1} x$, $L(x) = \frac{\pi}{2} - x$. For $\tan^{-1} x$, $L(x) = 0 + x = x$. For $\cot^{-1} x = \frac{\pi}{2} - \tan^{-1} x$, $L(x) = \frac{\pi}{2} - x$.

55. $y = \sec^{-1}|x| + C$. When $x = 2$, $\pi = \sec^{-1} 2 + C$, $\pi = \frac{\pi}{3} + C$, $C = \frac{2\pi}{3}$. $y = \sec^{-1}|x| + \frac{2\pi}{3}$ or $y = \sec^{-1} x + \frac{2\pi}{3}$ since $x > 0$.

56. $y = 1 + \int_0^x \frac{dt}{\sqrt{1-t^2}} = 1 + \sin^{-1} x$

57. $y = \cos^{-1} x + C$. When $x = -\sqrt{2}/2$, $\pi/2 = \cos^{-1}(-\sqrt{2}/2) + C = 3\pi/4 + C$, $C = -\pi/4$. $y = \cos^{-1} x - \frac{\pi}{4}$

58. $y = \frac{\pi}{2} + \int_0^x \frac{-dt}{1+t^2} = \frac{\pi}{2} - \tan^{-1} x$

60. $4\text{NINT}(\frac{1}{1+x^2}, x, 0, 1) = 3.14159265358$ (agrees with π up to the last decimal place).

61. Both answers can be correct because they differ by a constant: $\sin^{-1} x = \frac{\pi}{2} - \cos^{-1} x$.

62. a) $f'(x) = \frac{1}{\sqrt{1-(\frac{x-1}{x+1})^2}} \frac{(x+1)-(x-1)}{(x+1)^2} = \sqrt{\frac{(x+1)^2}{4x}} \frac{2}{(x+1)^2} = \frac{1}{\sqrt{x}(x+1)}$ for $x \geqq 0$. $g'(x) = 2\frac{1}{1+x}\frac{1}{2\sqrt{x}} = \frac{1}{\sqrt{x}(x+1)}$. Therefore $f(x) = g(x) + C$. b) $f(0) = g(0) + C$, $-\frac{\pi}{2} = 0 + C$, $C = -\frac{\pi}{2}$.

63. $\frac{d(\cos^{-1} u)}{dx} = \frac{d}{dx}(\frac{\pi}{2} - \sin^{-1} u) = -\frac{d(\sin^{-1} u)}{dx} = -\frac{du/dx}{\sqrt{1-u^2}}$

64. Let $y = \tan^{-1} x$. Then $-\frac{\pi}{2} < y < \frac{\pi}{2}$ and $x = \tan y$. Differentiating implicitly with respect to x, we obtain $1 = \sec^2 y \frac{dy}{dx}$, $\frac{dy}{dx} = \frac{1}{\sec^2 y} = \frac{1}{1+\tan^2 y} = \frac{1}{1+x^2}$. The formula now follows from the chain rule.

65. $\frac{d(\cot^{-1} u)}{dx} = \frac{d}{dx}(\frac{\pi}{2} - \tan^{-1} u) = -\frac{d}{dx}(\tan^{-1} u) = -\frac{du/dx}{1+u^2}$ by the preceding exercise.

66. $\frac{d}{dx} \csc^{-1} u = \frac{d}{dx} \sin^{-1}(\frac{1}{u}) = \frac{1}{\sqrt{1-\frac{1}{u^2}}}(-\frac{1}{u^2})\frac{du}{dx} = -\frac{|u|}{|u|^2\sqrt{u^2-1}}\frac{du}{dx} = -\frac{1}{|u|\sqrt{u^2-1}}\frac{du}{dx}$, $|u| > 1$.

67. Graph $y_1 = \text{NDER}(\sin^{-1} x, x)$ and $y_2 = \frac{1}{\sqrt{1-x^2}}$ in $[-1, 1]$ by $[0, 5]$ and see that the graphs match. Graph $\text{NDER}(\cos^{-1} x, x)$ and $\frac{-1}{\sqrt{1-x^2}}$ in $[-1, 1]$ by $[-5, 0]$ and see that the graphs coincide.

68. Graph $y_1 = \text{NDER}(\tan^{-1} x, x)$ and $y_2 = \frac{1}{1+x^2}$ in $[-5, 5]$ by $[-0.5, 1.5]$ and see that the graphs coincide. Graph $y_1 = \text{NDER}(\frac{\pi}{2} - \tan^{-1} x, x)$ and $y_2 = -\frac{1}{1+x^2}$ in $[-5, 5]$ by $[-1.5, 0.5]$ and see that the graphs coincide.

69. Graph $y_1 = \text{NDER}(\cos^{-1}(1/x), x)$ and $\frac{1}{|x|\sqrt{x^2-1}}$ in $[-3, 3]$ by $[0, 3]$ and see that the graphs coincide. Similarly for $\text{NDER}(\sin^{-1}(1/x), x)$ and $\frac{-1}{|x|\sqrt{x^2-1}}$ in $[-3, 3]$ by $[-3, 0]$.

70. Graph $y = \tan^{-1} \sqrt{x+1}$ in $[-1, 10]$ by $[0, \frac{\pi}{2}]$.

71. Graph $y = \sec^{-1}(3x) = \cos^{-1}(\frac{1}{3x})$ in $[-5, 5]$ by $[0, \pi]$. Other windows can show that $y = 0$ when $x = \frac{1}{3}$ and $y = \pi$ when $x = -\frac{1}{3}$.

72. $y = \csc^{-1} \sqrt{x} + \sec^{-1} \sqrt{x} = \frac{\pi}{2}$, $x \geqq 1$. Graph $y = \sin^{-1}(1/\sqrt{x}) + \cos^{-1}(1/\sqrt{x})$ in $[-1, 10]$ by $[-1, 8]$.

73. Graph $y = \cot^{-1} \sqrt{x^2-1} = \frac{\pi}{2} - \tan^{-1} \sqrt{x^2-1}$ in $[-10, 10]$ by $[-1, 2]$.

7.9 HYPERBOLIC FUNCTIONS

1. $\sinh x = -\frac{3}{4}$. $x = \sinh^{-1}(-0.75) = -0.693$. Alternatively one may proceed as follows. Let $u = e^x$. Then $\frac{1}{u} = e^{-x}$. $\sinh x = -\frac{3}{4}$ becomes $\frac{e^x - e^{-x}}{2} = -\frac{3}{4}$, $u - \frac{1}{u} = -\frac{3}{2}$, $u^2 - 1 = -\frac{3}{2}u$, $2u^2 - 2 = -3u$, $2u^2 + 3u - 2 = 0$, $(2u-1)(u+2) = 0$, $u = \frac{1}{2}$, $u = -2$ so $e^x = \frac{1}{2}$, $e^x = -2$ (no solution), $x = \ln(\frac{1}{2}) = -\ln 2$. Answer: $x = -\ln 2$. This method may be used in the next 5 exercises.

2. $\operatorname{csch} x = \frac{1}{\sinh x} = \frac{4}{3}$, $\sinh x = \frac{3}{4}$, $x = 0.693$ by Exercise 1.

3. $\cosh x = 2$. $x = \pm \cosh^{-1} 2 = \pm 1.317$.

4. $\tanh x = 0.5$. $x = \tanh^{-1} 0.5 = 0.549$.

5. $\operatorname{sech} x = 0.7$. $x = \pm \operatorname{sech}^{-1} 0.7 = \pm \cosh^{-1}(0.7^{-1}) = \pm 0.896$.

6. $\coth x = -2$. $x = \coth^{-1}(-2) = \tanh^{-1}(-\frac{1}{2}) = -0.549$.

7. $2\cosh(\ln x) = 2\frac{e^{\ln x} + e^{-\ln x}}{2} = x + \frac{1}{x}$, $x > 0$. To confirm graph both $2\cosh(\ln x)$ and $x + \frac{1}{x}$ in $[0, 4]$ by $[0, 5]$ and see that the two graphs coincide.

8. $\sinh(2\ln x) = 2\sinh(\ln x)\cosh(\ln x) = 2\frac{e^{\ln x} - e^{-\ln x}}{2}\frac{e^{\ln x} + e^{-\ln x}}{2} = \frac{1}{2}(x - \frac{1}{x})(x + \frac{1}{x}) = \frac{1}{2}(x^2 - \frac{1}{x^2})$, $x > 0$. Confirm by graphing both functions in $[0, 5]$ by $[-5, 12]$.

9. $\cosh 5x + \sinh 5x = \frac{e^{5x} + e^{-5x} + e^{5x} - e^{-5x}}{2} = e^{5x}$.

10. $(\sinh x + \cosh x)^4 = (e^x)^4 = e^{4x}$.

11. $\cosh 3x - \sinh 3x = \frac{1}{2}(e^{3x} + e^{-3x} - e^{3x} + e^{-3x}) = e^{-3x}$.

12. $\ln(\cosh x + \sinh x) + \ln(\cosh x - \sinh x) = \ln(\cosh^2 x - \sinh^2 x) = \ln 1 = 0$.

13. a) $\sinh 2x = \sinh(x + x) = \sinh x \cosh x + \cosh x \sinh x = 2\sinh x \cosh x$

b) $\cosh 2x = \cosh(x + x) = \cosh x \cosh x + \sinh x \sinh x = \cosh^2 x + \sinh^2 x$.

14. $\cosh^2 x - \sinh^2 x = (\frac{e^x + e^{-x}}{2})^2 - (\frac{e^x - e^{-x}}{2})^2 = \frac{e^{2x} + 2 + e^{-2x}}{4} - \frac{e^{2x} - 2 + e^{-2x}}{4} = \frac{4}{4} = 1$.

15. Graph $y = \sinh 3x$ in $[-3, 3]$ by $[-3, 3]$. This graph may be obtained from the graph of $y = \sinh x$ by horizontally shrinking it by a factor of $\frac{1}{3}$.

16. $y = \frac{1}{2}\sinh(2x + 1) = \frac{1}{2}\sinh[2(x + \frac{1}{2})]$. Graph in the viewing window $[-3.5, 3]$ by $[-10, 10]$. The graph can be obtained from the graph of $y = \sinh x$ using the following sequence of transformations: $\sinh x \rightarrow \sinh 2x \rightarrow \sinh[2(x + \frac{1}{2})] \rightarrow \frac{1}{2}\sinh[2(x + \frac{1}{2})]$.

17. Graph $y = 2\tanh\frac{x}{2}$ in $[-4,4]$ by $[-2,2]$. The graph can be obtained from the graph of $y = \tanh x$ by stretching vertically and horizontally by a factor of 2.

18. Graph $y = x - \tanh x$ in $[-2.5, 2.5]$ by $[-2.5, 2.5]$. y is an odd function so the graph is symmetric with respect to the origin. $y' = 1 - \text{sech}^2 x \geqq 0$ and $y' = 0$ only for $x = 0$.

19. Graph $y = \ln(\text{sech}\, x)$ in $[-10,10]$ by $[-10,0]$. $y' = -\frac{\text{sech}\, x \tanh x}{\text{sech}\, x} = -\tanh x$ which is positive for $x < 0$ and negative for $x > 0$. $y'' = -\text{sech}^2\, x$ which shows that the graph of y is concave down for all x.

20. Graph $y = \ln(\text{csch}\, x)$ in $[0,3]$ by $[-10,10]$. Since $\lim_{x\to 0^+} \text{csch}\, x = \infty$ and $\lim_{x\to\infty} \text{csch}\, x = 0$, $\lim_{x\to 0^+} y = \infty$ and $\lim_{x\to\infty} y = -\infty$.

21. Graph $y = \sinh^{-1}(2x)$ in $[-5,5]$ by $[-5,5]$. The graph may be obtained from the graph of $y = \sinh^{-1} x$ by shrinking horizontally by a factor of $\frac{1}{2}$.

22. Graph $y = 2\cosh^{-1}\sqrt{x}$ in $[-1,5]$ by $[-1,5]$.

23. Graph $y = (1-x)\tanh^{-1} x$ in $[-1,1]$ by $[-4,0.5]$. $\lim_{x\to -1^+}(1-x)\tanh^{-1} x = -\infty$ (has the form $2(-\infty)$). $\lim_{x\to 1^-}(1-x)\tanh^{-1} x = \lim_{x\to 1^-} \frac{\tanh^{-1} x}{\frac{1}{1-x}} (\frac{\infty}{\infty}) = \lim_{x\to 1^-} \frac{\frac{1}{1-x^2}}{\frac{1}{(1-x)^2}} = \lim_{x\to 1^-} \frac{1-x}{1+x} = 0$. $y' = \frac{1}{1+x} - \tanh^{-1} x$ and, using technology, we find that the graph of y rises on $(-1, 0.564]$ to a maximum at $x = 0.564$ and then falls on $[0.564, 1)$. $y'' = \frac{-2+x}{1-x^2}$ which is negative in the domain $(-1,1)$ and so the graph of y is concave down on $(-1,1)$.

24. Graph $y = (1-x^2)\coth^{-1} x = (1-x^2)\tanh^{-1}(\frac{1}{x})$ in $[-10,10]$ by $[-10,10]$.

25. Graph $y = x\,\text{sech}^{-1}\, x = x\cosh^{-1}(x^{-1})$ in $[0,1]$ by $[0,1]$. Since $(0,1]$ is the domain of $\text{sech}^{-1}\, x$, it is the domain of y. $y' = -\frac{1}{\sqrt{1-x^2}} + \text{sech}^{-1}\, x$. From this, using technology, we find y is rising for $0 < x \leqq 0.552$ and falling for $0.552 \leqq x \leqq 1$. $y'' = -\frac{1}{(1-x^2)^{3/2}} - \frac{1}{x\sqrt{1-x^2}} < 0$ for $0 < x < 1$ so the curve is concave down for $0 < x < 1$.

26. Graph $y = x^2\text{csch}^{-1}\, x^2 = x^2\sinh^{-1}(\frac{1}{x^2})$ in $[-10,10]$ by $[-1,2]$.

27. $y = \ln(\text{csch}\, x + \coth x)$. $y' = \frac{-\text{csch}\, x \coth x - \text{csch}^2 x}{\text{csch}\, x + \coth x} = \frac{-\text{csch}\, x(\coth x + \text{csch}\, x)}{\text{csch}\, x + \coth x} = -\text{csch}\, x$.

28. $y = x\cosh x - \sinh x$. $y' = x\sinh x + \cosh x - \cosh x = x\sinh x$.

29. $y = \frac{1}{2}\ln|\tanh|$. $y' = \frac{1}{2}\frac{\operatorname{sech}^2 x}{\tanh x} = \frac{1}{2\cosh^2 x \frac{\sinh x}{\cosh x}} = \frac{1}{2\cosh x \sinh x} = \frac{1}{\sinh(2x)}$ (by Exercise 13 a)) $= \operatorname{csch}(2x)$.

30. $y = \tan^{-1}(\sinh x)$. $y' = \frac{1}{1+\sinh^2 x}\cosh x = \frac{\cosh x}{\cosh^2 x} = \operatorname{sech} x$.

31. a) $y = \cosh^2 x$. $y' = 2\cosh x \sinh x = \sinh(2x)$. b) $y = \sinh^2 x$. $y' = 2\sinh x \cosh x = \sinh(2x)$. c) $y = \frac{1}{2}\cosh 2x$. $y' = \frac{1}{2}(\sinh 2x)2 = \sinh(2x)$.

32. $y = (x^2+1)\operatorname{sech}(\ln x) = \frac{x^2+1}{\cosh(\ln x)} = \frac{2(x^2+1)}{x+\frac{1}{x}} = 2x$. $y' = 2$ for all $x > 0$.

33. $y = \sinh^{-1}(\tan x)$. $y' = \frac{\sec^2 x}{\sqrt{1+\tan^2 x}} = \frac{|\sec x|^2}{|\sec x|} = |\sec x|$.

34. $y = \cosh^{-1}(\sec x)$, $0 < x < \frac{\pi}{2}$. $y' = \frac{1}{\sqrt{\sec^2 x - 1}}(\sec x \tan x) = \sec x$, $0 < x < \frac{\pi}{2}$.

35. $y = \tanh^{-1}(\sin x)$, $-\frac{\pi}{2} < x < \frac{\pi}{2}$. $y' = \frac{\cos x}{1-\sin^2 x} = \sec x$.

36. $y = \coth^{-1}(\sec x)$, $-\frac{\pi}{2} < x < \frac{\pi}{2}$. $y' = \frac{1}{1-\sec^2 x}(\sec x \tan x) = -\frac{\sec x}{\tan x} = -\csc x$, $-\frac{\pi}{2} < x < \frac{\pi}{2}$, $x \neq 0$.

37. $y = \operatorname{sech}^{-1}(\sin x)$, $0 < x < \frac{\pi}{2}$. $y' = \frac{-\cos x}{\sin x\sqrt{1-\sin^2 x}} = -\csc x$, $0 < x < \frac{\pi}{2}$.

38. $y = \operatorname{csch}^{-1}(\tan x)$, $0 < x < \frac{\pi}{2}$. $y' = -\frac{\sec^2 x}{|\tan x|\sqrt{1+\tan^2 x}} = -\frac{\sec^2 x}{\tan x \sec x} = -\frac{\sec x}{\tan x} = -\csc x$.

39. $\int_{-1}^{1} \cosh 5x \, dx = \frac{\sinh 5x}{5}\Big]_{-1}^{1} = \frac{1}{5}(\sinh(5) - \sinh(-5)) = \frac{2\sinh 5}{5}$.

40. $\int_{-1}^{0} \cosh(2x+1)dx = \frac{1}{2}\sinh(2x+1)\Big]_{-1}^{0} = \frac{1}{2}[\sinh 1 - \sinh(-1)] = \sinh 1$.

41. $\int_{-3}^{3} \sinh x \, dx = 0$ since $\sinh x$ is an odd function.

42. $\int_{-\pi}^{\pi} \tanh 2x \, dx = 0$ because $\tanh 2x$ is an odd function.

43. $\int_0^{1/2} 4e^x \cosh x \, dx = 2\int_0^{1/2}(e^{2x}+1)dx = 2\left[\frac{e^{2x}}{2} + x\right]_0^{1/2} = 2\left[\frac{e}{2} + \frac{1}{2} - \frac{1}{2}\right] = e$.

44. $\int_0^{1/2} 4e^{-x}\sinh x \, dx = \int_0^{1/2} 2(1-e^{-2x})dx = 2x + e^{-2x}]_0^{1/2} = 1 + e^{-1} - 1 = \frac{1}{e}$.

45. Let $I = \int_1^2 \frac{\cosh(\ln x)}{x}dx$. Let $u = \ln x$, $du = \frac{1}{x}dx$. $I = \int_0^{\ln 2}\cosh u \, du = \sinh(\ln 2) - \sinh 0 = \frac{e^{\ln 2} - e^{\ln 2^{-1}}}{2} = \frac{2-\frac{1}{2}}{2} = \frac{3}{4}$.

46. $\int_0^{\ln 2} \frac{\sinh x \, dx}{\cosh x} = \ln\cosh x]_0^{\ln 2} = \ln(\cosh\ln 2) - \ln(\cosh 0) = \ln\frac{2+\frac{1}{2}}{2} - \ln 1 = \ln\frac{5}{4}$.

47. $\int_0^{\ln 3} \operatorname{sech}^2 x\, dx = \tanh x]_0^{\ln 3} = \frac{4}{5}$.

48. $\int_0^{\ln 2} \tanh^2 x\, dx = \int_0^{\ln 2}(1 - \operatorname{sech}^2 x)dx = [x - \tanh x]_0^{\ln 2} = \ln 2 - \tanh(\ln 2) = \ln 2 - \frac{2-\frac{1}{2}}{2+\frac{1}{2}} = (\ln 2) - \frac{3}{5}$.

49. Let $I = \int_0^4 \frac{\cosh\sqrt{x}}{\sqrt{x}} dx$. Let $u = \sqrt{x}$, $du = \frac{1}{2\sqrt{x}} dx$. $I = 2\int_0^2 \cosh u\, du = 2\sinh u]_0^2 = 2\sinh 2$.

50. $\int_{\ln 2}^{\ln 3} \operatorname{csch}^2 x\, dx = [-\coth x]_{\ln 2}^{\ln 3} = \coth(\ln 2) - \coth(\ln 3) = \frac{2+\frac{1}{2}}{2-\frac{1}{2}} - \frac{3+\frac{1}{3}}{3-\frac{1}{3}} = \frac{5}{3} - \frac{10}{8} = \frac{5}{12}$.

51. $\int \sinh 2x\, dx = \frac{\cosh 2x}{2} + C$.

52. Let $I = \int 4\cosh(3x - \ln 2)dx$, $u = 3x - \ln 2$, $du = 3dx$. Then $I = \frac{4}{3}\int \cosh u\, du = \frac{4}{3}\sinh u + C = \frac{4}{3}\sinh(3x - \ln 2) + C$.

53. $\int 2e^{2t}\cosh t\, dt = 2\int e^{2t}\frac{e^t+e^{-t}}{2} dt = \int(e^{3t} + e^t)dt = \frac{e^{3t}}{3} + e^t + C$.

54. $\int 8e^{-t}\sinh t\, dt = \int 8e^{-t}\frac{e^t-e^{-t}}{2} dt = 4\int(1 - e^{-2t})dt = 4(t + \frac{e^{-2t}}{2}) + C = 2(2t + e^{-2t}) + C$.

55. Let $I = \int \tanh\frac{x}{7} dx$, $u = \frac{x}{7}$, $du = \frac{dx}{7}$. Then $I = 7\int \tanh u\, du = 7\int \frac{\sinh u}{\cosh u} du$. Let $v = \cosh u$, $dv = \sinh u\, du$. $I = 7\int \frac{dv}{v} = 7\ln|v| + C = 7\ln|\cosh u| + C = 7\ln(\cosh\frac{x}{7}) + C$.

56. Let $I = \int \coth\sqrt{2}\, x\, dx$, $u = \sqrt{2}\, x$, $du = \sqrt{2}\, dx$. $I = \frac{1}{\sqrt{2}}\int \coth u\, du = \frac{1}{\sqrt{2}}\int \frac{\cosh u}{\sinh u} du = \frac{1}{\sqrt{2}}\int \frac{dv}{v}$ $(v = \sinh u)$ $= \frac{1}{\sqrt{2}}\ln|v| = \frac{1}{\sqrt{2}}\ln|\sinh\sqrt{2}\, x| + C$.

57. Since $\frac{d}{dt}(-2\operatorname{sech}\sqrt{t}) = -2\left[-\operatorname{sech}\sqrt{t}\tanh\sqrt{t}\,\frac{1}{2\sqrt{t}}\right] = \frac{\operatorname{sech}\sqrt{t}\tanh\sqrt{t}}{\sqrt{t}}$, $\int \frac{\operatorname{sech}\sqrt{t}\tanh\sqrt{t}}{\sqrt{t}} dt = -2\operatorname{sech}\sqrt{t} + C$.

58. Since $\frac{d}{dt}[-\operatorname{csch}(\ln t)] = \frac{\operatorname{csch}(\ln t)\coth(\ln t)}{t}$, $\int \frac{\operatorname{csch}(\ln t)\coth(\ln t)}{t} dt = -\operatorname{csch}(\ln t) + C$.

59. One may also verify an identity by showing that both sides have the same derivative (hence they differ by a constant) and then showing that both sides have the same value at a convenient value of x (so the constant is 0).

61. a) $\int_0^1 \frac{dx}{\sqrt{1+x^2}} = \sinh^{-1} x\Big]_0^1 = \sinh^{-1}(1)$ b) $\sinh^{-1}(1) = \ln(1 + \sqrt{2})$.

62. a) $\int_{3/5}^{4/5} \frac{dx}{x\sqrt{1-x^2}} = -\text{sech}^{-1} x\Big]_{3/5}^{4/5} = \text{sech}^{-1}(3/5) - \text{sech}^{-1}(4/5)$ b) $\text{sech}^{-1}(3/5) - \text{sech}^{-1}(4/5) = \ln\left(\frac{1+\sqrt{1-(3/5)^2}}{(3/5)}\right) - \ln\left(\frac{1+\sqrt{1-(4/5)^2}}{(4/5)}\right) = \ln 3 - \ln 2 = \ln(3/2)$.

63. a) $\int_{5/4}^{5/3} \frac{dx}{\sqrt{x^2-a}} = \cosh^{-1} x\Big]_{5/4}^{5/3} = \cosh^{-1}\left(\frac{5}{3}\right) - \cosh^{-1}\left(\frac{5}{4}\right)$ b) $\ln\left(\frac{5}{3} + \sqrt{\frac{25}{9} - 1}\right) - \ln\left(\frac{5}{4} + \sqrt{\frac{25}{16} - 1}\right) = \ln \frac{\frac{5}{3}+\frac{4}{3}}{\frac{5}{4}+\frac{3}{4}} = \ln \frac{3}{2}$.

64. a) $\int_0^{1/2} \frac{dx}{1-x^2} = \tanh^{-1} x\Big]_0^{1/2}$ (since $|x| < 1$) $= \tanh^{-1}\left(\frac{1}{2}\right)$ b) $\tanh^{-1}\left(\frac{1}{2}\right) = \frac{1}{2} \ln \frac{1+\frac{1}{2}}{1-\frac{1}{2}} = \frac{1}{2} \ln 3$.

65. a) $\int_{5/4}^{2} \frac{dx}{1-x^2} = \coth^{-1} x\Big]_{5/4}^{2} = \coth^{-1} 2 - \coth^{-1}\left(\frac{5}{4}\right)$ b) $\frac{1}{2} \ln \frac{2+1}{2-1} - \frac{1}{2} \ln \frac{\frac{5}{4}+1}{\frac{5}{4}-1} = \frac{1}{2} \ln 3 - \frac{1}{2} \ln 9 = \frac{1}{2} \ln \frac{1}{3} = -\frac{\ln 3}{2}$.

66. a) Let $I = \int_0^{2\sqrt{3}} \frac{dx}{\sqrt{4+x^2}}$, $u = \frac{x}{2}$, $du = \frac{1}{2} dx$. Then $I = 2\int_0^{\sqrt{3}} \frac{du}{\sqrt{4+4u^2}} = \int_0^{\sqrt{3}} \frac{du}{\sqrt{1+u^2}} = \sinh^{-1} u\Big]_0^{\sqrt{3}} = \sinh^{-1} \sqrt{3}$ b) $\sinh^{-1} \sqrt{3} = \ln(\sqrt{3} + 2)$.

67. a) Let $I = \int_1^2 \frac{dx}{x\sqrt{4+x^2}}$. Let $x = 2u$, $dx = 2du$. $I = \int_{1/2}^{1} \frac{2du}{2u\sqrt{4(1+u^2)}} = \frac{1}{2}\int_{1/2}^{1} \frac{du}{u\sqrt{1+u^2}} = -\frac{1}{2} \sinh^{-1}\left(\frac{1}{|u|}\right)\Big]_{1/2}^{1} = \frac{1}{2}\left[\sinh^{-1} 2 - \sinh^{-1}(1)\right]$.
b) $\frac{1}{2}\left[\sinh^{-1} 2 - \sinh^{-1}(1)\right] = \frac{1}{2}\left[\ln(2+\sqrt{5}) - \ln(1+\sqrt{2})\right] = \frac{1}{2} \ln \frac{2+\sqrt{5}}{1+\sqrt{2}}$.

68. Let $I = \int_0^{\pi} \frac{\cos x \; dx}{\sqrt{1+\sin^2 x}}$, $u = \sin x$, $du = \cos x \; dx$. Then $I = \int_0^0 \frac{du}{\sqrt{1+u^2}} = 0$.

69. $\int_1^3 \frac{\sinh x}{x} \, dx = \text{NINT}(\frac{\sinh x}{x}, x, 1, 3) \approx 3.916$. We used Tol $= 1$.

70. $\int_{-1}^{3} \cosh(x^2) dx = \text{NINT}(\cosh(x^2), x, -1, 3) = 723.8204$. There is no explicit antiderivative.

71. $\int_1^4 \frac{\cosh x - 1}{x} \, dx = \text{NINT}(\frac{\cosh x - 1}{x}, x, 1, 4) \approx 7.589$.

72. $\int_{-2}^{2} \frac{\sinh^{-1} x}{x} \, dx = \text{NINT}(\frac{\sinh^{-1} x}{x}, x, -2, 2) = 3.5069\ldots$. To avoid division by 0 in the integration process, we replaced the limit 2 by 2.000001.

73. Graph $y = \text{NINT}(\frac{\sinh t}{t}, t, 1, x)$ in $[-4, 4]$ by $[-10, 10]$.

74. Graph $y = \int_0^x \cosh(t^2) dt = \text{NINT}(\cosh(t^2), t, 0, x)$ in $[-3, 3]$ by $[-10, 10]$. We used Tol $= 1$.

75. Graph NINT$((\cosh t - 1)/t, t, 1, x)$ in $[-5, 5]$ by $[-1, 10]$.

76. Graph $y = \int_1^x \frac{\sinh^{-1} t}{t}\, dt =$ NINT$((\sinh^{-1} t)/t, t, 1, x)$ in $[-2, 2]$ by $[-3, 3]$.

77. $V = \pi \int_0^2 (\cosh^2 x - \sinh^2 x) dx = \pi \int_0^2 1 dx = 2\pi$.

78. $V = 2\pi \int_0^{\ln\sqrt{3}} \operatorname{sech}^2 x\, dx = 2\pi \tanh x \Big]_0^{\ln\sqrt{3}} = 2\pi \frac{\sqrt{3} - \frac{1}{\sqrt{3}}}{\sqrt{3} + \frac{1}{\sqrt{3}}} = 2\pi \frac{3-1}{3+1} = \pi$.

79. By symmetry $\bar{x} = 0$. $A = 2\int_0^{\ln\sqrt{3}} \operatorname{sech} x\, dx =$ 2NINT$((\cosh t)^{-1}, t, 0, \ln\sqrt{3}) \approx$ 1.0472. $M_x = \frac{1}{2}\int_{-\ln\sqrt{3}}^{\ln\sqrt{3}} \operatorname{sech}^2 x\, dx = \int_0^{\ln\sqrt{3}} \operatorname{sech}^2 x\, dx = \tanh x]_0^{\ln\sqrt{3}} = \tanh(\ln\sqrt{3}) = \frac{\sqrt{3} - \frac{1}{\sqrt{3}}}{\sqrt{3} + \frac{1}{\sqrt{3}}} = \frac{3-1}{3+1} = \frac{1}{2}$. $\bar{y} = \frac{M_x}{A} \approx 0.477$.

80. $V = \pi \int_0^{\ln\sqrt{199}} (1 - \tanh x)^2 dx = \pi \int_0^{\ln\sqrt{199}} (1 - 2\tanh x + \tanh^2 x) dx = \pi \int_0^{\ln\sqrt{199}} (1 - 2\tanh x + 1 - \operatorname{sech}^2 x) dx = \pi\,[2x - 2\ln\cosh x - \tanh x]_0^{\ln\sqrt{199}} = \pi\left[(2\ln\frac{199}{100}) - \frac{99}{100}\right]$.

81. Let $f_1(x) = \frac{f(x)+f(-x)}{2}$ and $f_2(x) = \frac{f(x)-f(-x)}{2}$. $f_1(-x) = \frac{f(-x)+f(x)}{2} = f_1(x)$ so $f_1(x)$ is even. $f_2(-x) = \frac{f(-x)-f(x)}{2} = -\frac{f(x)-f(-x)}{2} = -f_2(x)$ so $f_2(x)$ is odd.

82. a) If $f(x)$ is even, $f(-x) = f(x)$ and the equation becomes $f(x) = f(x)$. b) If $f(x)$ is odd, $f(-x) = -f(x)$ and the equation becomes $f(x) = f(x)$.

83. $v = \sqrt{\frac{mg}{k}} \tanh(\sqrt{\frac{gk}{m}} t)$. a) $\frac{dv}{dt} = \sqrt{\frac{mg}{k}} \sqrt{\frac{gk}{m}} \operatorname{sech}^2(\sqrt{\frac{gk}{m}} t) = g \operatorname{sech}^2(\sqrt{\frac{gk}{m}} t)$. $m\frac{dv}{dt} = mg \operatorname{sech}^2(\sqrt{\frac{gk}{m}} t)$. $mg - kv^2 = mg - mg\tanh^2(\sqrt{\frac{gk}{m}} t) = mg \operatorname{sech}^2(\sqrt{\frac{gk}{m}} t)$. Hence $m\frac{dv}{dt} = mg - kv^2$ and v satisfies the differential equation. When $t = 0$, $v = \sqrt{\frac{mg}{k}} \tanh(0) = 0$. b) Since $\lim_{x\to\infty} \tanh x = 1$, $\lim_{t\to\infty} v = \sqrt{\frac{mg}{k}}$. c) $\sqrt{\frac{160}{0.005}}$ ft/sec. $= 178.885$ ft/sec.

84. a) $v = \frac{ds}{dt} = -ak\sin kt + bk\cos kt$, $\frac{d^2 s}{dt^2} = -k^2(a\cos kt + b\sin kt) = -k^2 s$. $\frac{d^2 s}{dt^2}$ and s have opposite signs so the acceleration is directed toward the origin. b) $\frac{ds}{dt} = ak\sinh kt + bk\cosh bt$, $\frac{d^2 s}{dt^2} = k^2(a\cosh kt + b\sinh kt) = k^2 s$. Now $\frac{d^2 s}{dt^2}$ and s have the same sign so the acceleration is directed away from the origin.

85. $y = 10\cosh(\frac{x}{10})$. $y' = \sinh\frac{x}{10}$. $L = 2\int_0^{10\ln 10} \sqrt{1 + \sinh^2 x}\, dx = 2\int_0^{10\ln 10} \cosh x\, dx = 2\sinh x]_0^{10\ln 10} = 2\sinh(10\ln 10) = 10^{10} - \frac{1}{10^{10}}$. $L =$

$2\int_0^{10\ln 10}\sqrt{1+\sinh^2(\frac{x}{10})}\ dx = 2\int_0^{10\ln 10}\cosh(\frac{x}{10})dx = 20\sinh(\frac{x}{10})\Big]_0^{10\ln 10} =$ $20\sinh(\ln 10) = 20\frac{10-\frac{1}{10}}{2} = 99.$

86. $A = 10\int_{-10\ln 10}^{10\ln 10}\cosh(x/10)\ dx = 20\int_0^{10\ln 10}\cosh(x/10)\ dx =$ $20\left[10\sinh(x/10)\right]_0^{10\ln 10} = 200\sinh(\ln 10) = 200\frac{10-\frac{1}{10}}{2} = 990.$

87. $2\pi\int_0^{\ln 8}2\cosh(x/2)\sqrt{1+\sinh^2(\frac{x}{2})}\ dx = 4\pi\int_0^{\ln 8}\cosh^2(\frac{x}{2})dx = 2\pi\int_0^{\ln 8}(\cosh x +$ $1)dx = 2\pi\left[\sinh x + x\right]_0^{\ln 8} = 2\pi\left(\frac{8-\frac{1}{8}}{2}+\ln 8\right) = \pi\left(\frac{63}{8}+2\ln 8\right).$

88. $y = a\cosh(x/a)$. $y' = \sinh(x/a)$, $y'' = (1/a)\cosh(x/a)$. $(1/a)\sqrt{1+(y')^2} =$ $(1/a)\sqrt{1+\sinh^2(x/a)} = (1/a)\cosh(x/a)$. Thus $y'' = (1/a)\sqrt{1+(y')^2}$. It is easy to see that the initial conditions are satisfied by y.

89. a) $A(u)$ = area of triangle $-\int_1^{\cosh u}\sqrt{x^2-1}\ dx = \frac{1}{2}\cosh u\sinh u -$ $\int_1^{\cosh u}\sqrt{x^2-1}\ dx$. b) $A'(u) = \frac{1}{2}(\cosh^2 u+\sinh^2 u) - \sqrt{\cosh^2 u - 1}\ \sinh u =$ $\frac{1}{2}(\cosh^2 u - \sinh^2 u) = \frac{1}{2}$. c) $A(u) = \frac{1}{2}u + C$. Since $A(0) = 0$, $C = 0$ and $A(u) = \frac{1}{2}u$, $u = 2A(u)$.

PRACTICE EXERCISES, CHAPTER 7

1. $y = \ln\sqrt{x} = \frac{1}{2}\ln x$. $\frac{dy}{dx} = \frac{1}{2x}$. Support by graphing NDER$(\ln\sqrt{x}, x)$ in $[0,2]$ by $[0,10]$.

2. $y = \ln(\frac{e^x}{2}) = \ln e^x - \ln 2 = x - \ln 2$. $\frac{dy}{dx} = 1$. Confirm by graphing NDER$(\ln(e^x/2), x)$.

3. $y = \ln(3x^2+6)$. $\frac{dy}{dx} = \frac{6x}{3x^2+6} = \frac{2x}{x^2+2}$.

4. $y = \ln(1+e^x)$. $\frac{dy}{dx} = (\frac{1}{1+e^x})e^x = \frac{e^x}{1+e^x}$. We may confirm this graphically by graphing $\frac{e^x}{1+e^x}+2$ and NDER$(\ln(1+e^x), x)$ and seeing that the first curve may be obtained from the second by vertically shifting the latter curve 2 units upward.

5. $y = \frac{1}{e^x} = e^{-x}$. $y' = e^{-x} = -\frac{1}{e^x}$.

6. $y = xe^{-x}$. $\frac{dy}{dx} = -xe^{-x}+e^{-x} = e^{-x}(1-x)$.

7. $y = e^{(1+\ln x)} = e^1 e^{\ln x} = ex$. $\frac{dy}{dx} = e$.

8. $\frac{dy}{dx} = \ln x$. This can be confirmed by graphing $\ln x + 2$ and NDER(NINT($\ln t, t, 1, x$), x) in $[1, 10]$ by $[-5, 5]$ and observing that one graph can be obtained from the other by a vertical shift.

9. $(\ln(\cos x))' = \frac{-\sin x}{\cos x} = -\tan x$.

10. $y = \ln(\sin x)$. $y' = \frac{\cos x}{sin x} = \cot x$.

11. $(\ln(\cos^{-1} x))' = \frac{-\frac{1}{\sqrt{1-x^2}}}{\cos^{-1} x} = -\frac{1}{\sqrt{1-x^2}\cos^{-1} x}$.

12. $y = \ln(\sin^{-1} x)$. $y' = \frac{1}{\sqrt{1-x^2}\sin^{-1} x}$.

13. $(\log_2(x^2))' = (\frac{\ln(x^2)}{\ln 2})' = \frac{1}{\ln 2}\frac{2x}{x^2} = \frac{2}{(\ln 2)x}$.

14. $y = \log_5(x-7) = \frac{\ln(x-7)}{\ln 5}$. $y' = \frac{1}{(\ln 5)(x-7)}$.

15. $(8^{-x})' = 8^{-x}(\ln 8)(-1) = -(\ln 8)8^{-x}$.

16. $y = 9^x$. $y' = 9^x \ln 9$.

17. $(\sin^{-1}(\sqrt{1-x})' = \frac{1}{\sqrt{1-(1-x)}}\frac{-1}{2\sqrt{1-x}} = \frac{-1}{2\sqrt{x(1-x)}}$.

18. $(\tan^{-1}(\tan 2x))' = \frac{2\sec^2 2x}{1+\tan^2 2x} = 2$. $(\tan^{-1}(\tan 2x) = 2x + k\pi$ for some integer k.)

19. $y = \cos^{-1}(1/x) - \csc^{-1} x = \cos^{-1}(1/x) - \sin^{-1}(1/x)$, $x > 0$. $y' = \frac{(1/x^2)}{\sqrt{1-(1/x)^2}} + \frac{(1/x^2)}{\sqrt{1-(1/x)^2}} = \frac{2}{|x|\sqrt{x^2-1}} = \frac{2}{x\sqrt{x^2-1}}$, $x > 0$.

20. $y = (1+x^2)\cot^{-1}(2x)$. $y' = (1+x^2)\frac{-2}{1+4x^2} + 2x\cot^{-1}(2x) = 2\left[x\cot^{-1}(2x) - \frac{1+x^2}{1+4x^2}\right]$.

21. $(2\sqrt{x-1}\sec^{-1}\sqrt{x})' = 2\sqrt{x-1}\,\frac{\frac{1}{2\sqrt{x}}}{\sqrt{x}\,\sqrt{x-1}} + \frac{1}{\sqrt{x-1}}\sec^{-1}\sqrt{x} = \frac{1}{x} + \frac{\sec^{-1}\sqrt{x}}{\sqrt{x-1}}$.

22. $y = \csc^{-1}(\sec x)$, $0 < x < \frac{\pi}{2}$. $y' = \frac{-\sec x \tan x}{|\sec x|\sqrt{sec^2 x - 1}} = -1$, $0 < x < \frac{\pi}{2}$. Alternatively, $y = \frac{\pi}{2} - \sec^{-1}(\sec x) = \frac{\pi}{2} - x$, $y' = -1$, $0 < x < \frac{\pi}{2}$.

23. a) $\ln e^{2x} = 2x$ ($\ln e^u = u$ for any u) b) $\ln 2e = \ln 2 + \ln e = 1 + \ln 2$, c) $\ln\frac{1}{e} = \ln e^{-1} = -1$

24. a) $e^{2\ln 2} = e^{\ln 2^2} = 4$ b) $e^{-\ln 4} = e^{\ln 4^{-1}} = \frac{1}{4}$ c) $e^{\ln(\ln x)} = \ln x$

25. $\ln(y^2+y)-\ln y = x,\ \ln(\frac{y^2+y}{y}) = x,\ \ln(y+1) = x,\ y+1 = e^x,\ y = e^x - 1$

26. $\ln(y-4) = -4t.\ \ y-4 = e^{-4t},\ y = 4 + e^{-4t}.$

27. $e^{2y} = 4x^2,\ x > 0.\ \ 2y = \ln 4x^2,\ y = \frac{1}{2}\ln 4x^2,\ y = \ln(4x^2)^{1/2} = \ln 2x.$

28. $e^{-0.1y} = \frac{1}{2}.\ \ -0.1y = \ln\frac{1}{2} = \ln 2^{-1} = -\ln 2,\ y = 10\ln 2.$

29. $y = \frac{2(x^2+1)}{\sqrt{\cos 2x}}.\ \ln y = \ln 2 + \ln(x^2+1) - \frac{1}{2}\ln\cos 2x.\ \frac{y'}{y} = \frac{2x}{x^2+1} - \frac{1}{2}\frac{(-2\sin 2x)}{\cos 2x},\ y' = \frac{2(x^2+1)}{\sqrt{\cos 2x}}(\frac{2x}{x^2+1} + \tan 2x).$

30. $y = \sqrt[10]{\frac{3x+4}{2x-4}}.\quad \ln y = \frac{1}{10}\ln\frac{3x+4}{2x-4} = \frac{1}{10}[\ln(3x+4) - \ln(2x-4)].\quad \frac{y'}{y} = \frac{1}{10}\left[\frac{3}{3x+4} - \frac{2}{2x-4}\right],\ y' = \frac{1}{10}\left[\frac{3}{3x+4} - \frac{2}{2x-4}\right]\sqrt[10]{\frac{3x+4}{2x-4}}$

31. $y = \left(\frac{(x+5)(x-1)}{(x-2)(x+3)}\right)^5.\ \ln y = 5[\ln(x+5) + \ln(x-1) - \ln(x-2) - \ln(x+3)],\ \frac{y'}{y} = 5\left[\frac{1}{x+5} + \frac{1}{x-1} - \frac{1}{x-2} - \frac{1}{x+3}\right],\ y' = 5\left(\frac{(x+5)(x-1)}{(x-2)(x+3)}\right)^5\left[\frac{1}{x+5} + \frac{1}{x-1} - \frac{1}{x-2} - \frac{1}{x+3}\right].$

32. $y = x^{\ln x},\ x > 1.\ \ \ln y = (\ln x)^2,\ \ \frac{y'}{y} = 2(\ln x)\frac{1}{x},\ y' = 2\left(\frac{\ln x}{x}\right)x^{\ln x},\ x > 1.$

33. $y = (1+x^2)e^{\tan^{-1}x}.\ \ \ln y = \ln(1+x^2) + \tan^{-1}x,\ \frac{y'}{y} = \frac{2x}{1+x^2} + \frac{1}{1+x^2}.\ \ y' = e^{\tan^{-1}x}(2x+1).$

34. $y = \frac{2x2^x}{\sqrt{x^2+1}}.\ \ \ln y = \ln 2 + \ln x + x\ln 2 - \frac{1}{2}\ln(x^2+1).\ \ \frac{y'}{y} = \frac{1}{x} + \ln 2 - \frac{x}{x^2+1}.\ y' = \frac{2x2^x}{\sqrt{x^2+1}}\left[\frac{1}{x} + \ln 2 - \frac{x}{x^2+1}\right].$

35. $(x - \coth x)' = 1 + \operatorname{csch}^2 x = \coth^2 x$

36. $y = x\sinh x - \cosh x.\ \ y' = x\cosh x + \sinh x - \sinh x = x\cosh x.$

37. $[\ln(\operatorname{csch} x) + x\coth x]' = \frac{-\operatorname{csch} x\coth x}{\operatorname{csch} x} - x\operatorname{csch}^2 x + \coth x = -x\operatorname{csch}^2 x.$

38. $[\ln(\operatorname{sech} x) + x\tanh x]' = -\frac{\operatorname{sech} x\tanh x}{\operatorname{sech} x} + x\operatorname{sech}^2 x + \tanh x = x\operatorname{sech}^2 x.$

39. $(\sin^{-1}(\tanh x))' = \frac{\operatorname{sech}^2 x}{\sqrt{1-\tanh^2 x}} = \operatorname{sech} x.$

40. $[\tan^{-1}(\sinh x)]' = \frac{\cosh x}{1+\sinh^2 x} = \operatorname{sech} x.$

41. $(\sqrt{1+x^2}\sinh^{-1}x)' = \sqrt{1+x^2}\,\frac{1}{\sqrt{1+x^2}} + \frac{x}{\sqrt{1+x^2}}\sinh^{-1}x = 1 + \frac{x\sinh^{-1}x}{\sqrt{1+x^2}}.$

42. $\left[\sqrt{x^2-1}\cosh^{-1}x\right]' = \sqrt{x^2-1}\,\frac{1}{\sqrt{x^2-1}} + \frac{x}{\sqrt{x^2-1}}\cosh^{-1}x = 1 + \frac{x\cosh^{-1}x}{\sqrt{x^2-1}}.$

43. $(1 - \tanh^{-1}(1/x))' = -\frac{(-1/x^2)}{1-(1/x)^2} = \frac{1}{x^2-1}$.

44. $[\coth^{-1}(\csc x)]' = \frac{-\csc x \cot x}{1-\csc^2 x} = \frac{\csc x \cot x}{\cot^2 x} = \sec x$.

45. $(\operatorname{sech}^{-1}(\cos 2x))' = \frac{-(-2\sin 2x)}{(\cos^2 x)\sqrt{1-\cos^2 2x}} = 2\sec 2x$.

46. $[\operatorname{csch}^{-1}(\cot x)]' = \frac{\csc^2 x}{|\cot x|\sqrt{1+\cot^2 x}} = \frac{\csc x}{\cot x} = \sec x,\ 0 < x < \frac{\pi}{2}$.

47. Graph $y = e^{\tan^{-1} x}$ in $[-30, 30]$ by $[0, 5]$. Note that $\lim_{x\to\pm\infty} y = e^{\pm\pi/2}$, so that there are two horizontal asymptotes. $y' = \frac{e^{\tan^{-1} x}}{1+x^2} > 0$ so y is always increasing. $y'' = (1-2x)\frac{e^{\tan^{-1} x}}{(1+x^2)^2}$ so that the curve is concave up on $(-\infty, \frac{1}{2})$, concave down on $(\frac{1}{2}, \infty)$ and has an inflection point at $x = \frac{1}{2}$.

48. Graph $y = e^{\cot^{-1} x} = e^{\pi/2-\tan^{-1} x}$ in $[-50, 30]$ by $[-5, 24]$. Note that $y \to e^0 = 1$ as $x \to \infty$ and $y \to e^{\pi}$ as $x \to -\infty$.

49. Graph $y = x\tan^{-1} x - \frac{1}{2}\ln x$ in $[0, 4]$ by $[0, 5]$. $y' = \frac{x}{1+x^2} + \tan^{-1} x - \frac{1}{2x}$. Technology yields that y decreases on $(0, 0.544]$ and increases on $[0.544, \infty)$ with a minimum at $(0.544, 0.578)$. $y'' = \frac{2}{(1+x^2)^2} + \frac{1}{2x^2}$ so the curve is concave up.

50. $y = x\cos^{-1} x - \sqrt{1-x^2}$, $y' = -\frac{x}{\sqrt{1-x^2}} + \cos^{-1} x + \frac{x}{\sqrt{1-x^2}} = \cos^{-1} x \geqq 0$ for $-1 \leqq x \leqq 1$ and $y' = 0$ only for $x = 1$. Thus y is increasing on $[-1, 1]$ and $y = 0$ is the tangent line at the point $(1, 0)$. Graph y in the viewing window $[-2, 2]$ by $[-5, 1]$.

51. $\int_{-1}^{1} \frac{dx}{3x-4} = \frac{1}{3}\ln|3x-4|\Big]_{-1}^{1} = \frac{1}{3}(\ln 1 - \ln 7) = -\frac{\ln 7}{3}$.

52. $I = \int_1^e \frac{\sqrt{\ln x}}{x}\,dx$. Let $u = \ln x$, $du = \frac{dx}{x}$. Then $I = \int_0^1 u^{1/2} du = \frac{2}{3}u^{3/2}\Big]_0^1 = \frac{2}{3}$.

53. $\int_{\ln 3}^{\ln 4} e^x dx = [e^x]_{\ln 3}^{\ln 4} = e^{\ln 4} - e^{\ln 3} = 4 - 3 = 1$.

54. Let $u = 2x$, $du = 2dx$. $\int_0^{\ln 3} e^{2x} dx = \frac{1}{2}\int_0^{2\ln 3} e^u du = \frac{1}{2}e^u\Big]_0^{\ln 3^2} = \frac{1}{2}\left[e^{\ln 9} - e^0\right] = \frac{1}{2}[9-1] = 4$.

55. $\int_0^{\pi/4} e^{\tan x}\sec^2 x\,dx = [e^{\tan x}]_0^{\pi/4} = e^1 - e^0 = e - 1$.

56. Let $u = \sec x$, $du = \sec x\tan x\,dx$. $\int_0^{\pi/3} e^{\sec x}\sec x\tan x\,dx = \int_1^2 e^u du = e^u\Big]_1^2 = e^2 - e$.

57. $\int_0^{\pi} \tan \frac{x}{3} dx = -3 \ln|\cos \frac{x}{3}|\Big]_0^{\pi} = -3\left[\ln \frac{1}{2} - \ln 1\right] = 3\ln 2 = \ln 8.$

58. $\int_{1/6}^{1/4} 2\cot \pi x \, dx = \frac{2}{\pi}\ln(\sin \pi x)\Big]_{1/6}^{1/4} = \frac{2}{\pi}\left[\ln \frac{1}{\sqrt{2}} - \ln \frac{1}{2}\right] = \frac{2}{\pi}\ln\sqrt{2} = \frac{\ln 2}{\pi}.$

59. $I = \int_0^4 \frac{2x \, dx}{x^2-25}$. Let $u = x^2 - 25$, $du = 2x \, dx$. $I = \int_{-25}^{-9} \frac{du}{u} = \ln|u|\big]_{-25}^{-9} = \ln 9 - \ln 25 = \ln(9/25)$.

60. $\int_{-\pi/3}^{\pi/3} \frac{\sec x + \tan x}{\sec x} dx = \int_{-\pi/3}^{\pi/3}(1 + \sin x)dx = \frac{2\pi}{3}$. ($\int_{-\pi/3}^{\pi/3} \sin x \, dx = 0$ since the sine is an odd function.)

61. Let $I = \int_0^{\pi/4} \frac{\sec x \tan x + \sec^2 x}{\sec x + \tan x} dx$. Let $u = \sec x + \tan u$. Then $du = (\sec x \tan x + \sec^2 x)dx$, $I = \int_1^{1+\sqrt{2}} \frac{du}{u} = \ln|u|\big]_1^{1+\sqrt{2}} = \ln(1+\sqrt{2})$.

62. $\int_{-\pi/2}^{\pi/2} \frac{\cos x dx}{2 - \sin x} = -\ln(2 - \sin x)\big]_{-\pi/2}^{\pi/2} = -[\ln 1 - \ln 3] = \ln 3.$

63. Let $I = \int_1^8 \frac{\log_4 x}{x} dx = \frac{1}{\ln 4}\int_1^8 \frac{\ln x}{x} dx$. Let $u = \ln x$. Then $du = \frac{dx}{x}$, $I = \frac{1}{\ln 4}\int_0^{\ln 8} u \, du = \frac{1}{\ln 4}\frac{u^2}{2}\Big]_0^{\ln 8} = \frac{(\ln 8)^2}{2\ln 4} = \frac{(\frac{3}{2}\ln 4)^2}{2\ln 4} = \frac{9}{8}\ln 4 = \frac{9}{4}\ln 2.$

64. Let $I = \int_1^e \frac{8 \ln 3 \log_3 x \, dx}{x} = 8\int_1^e \frac{\ln x \, dx}{x}$, $u = \ln x$, $du = \frac{dx}{x}$. $I = 8\int_0^1 u \, du = 4u^2\big]_0^1 = 4.$

65. Let $I = \int_0^1 x \, 3^{x^2} dx$. Let $u = 3^{x^2}$, $du = 2(\ln 3)x \, 3^{x^2} dx$. $I = \frac{1}{2\ln 3}\int_1^3 du = \frac{1}{2\ln 3}(2) = \frac{1}{\ln 3}.$

66. $\int_0^{\pi/4} 2^{\tan x} \sec^2 x \, dx = \frac{2^{\tan x}}{\ln 2}\Big]_0^{\pi/4} = \frac{1}{\ln 2}.$

67. $\int_{-1/2}^{1/2} \frac{3dx}{\sqrt{1-x^2}} = 6\int_0^{1/2} \frac{dx}{\sqrt{1-x^2}}$ (integral of even function) $= 6\sin^{-1} x\Big]_0^{1/2} = 6\frac{\pi}{6} = \pi.$

68. $\int_1^{1+\sqrt{2}/2} \frac{dx}{\sqrt{1-(x-1)^2}} = \sin^{-1}(x-1)\Big]_1^{1+\sqrt{2}/2} = \frac{\pi}{4}.$

69. $\int_{-1}^1 \frac{dx}{1+x^2} = \tan^{-1} x\big]_{-1}^1 = \frac{\pi}{4} - (-\frac{\pi}{4}) = \frac{\pi}{2}.$

70. Let $I = \int_1^3 \frac{2dx}{\sqrt{x}(1+x)}$, $u = \sqrt{x}$, $du = \frac{dx}{2\sqrt{x}}$. $I = 2\int_1^{\sqrt{3}} \frac{2du}{(1+u^2)} = 4\tan^{-1} u\big]_1^{\sqrt{3}} = 4\left[\frac{\pi}{3} - \frac{\pi}{4}\right] = \frac{\pi}{3}.$

71. Let $I = \int_{1/2}^{3/4} \frac{dx}{\sqrt{x}\sqrt{1-x}}$. Let $u = \sqrt{x}$, $du = \frac{dx}{2\sqrt{x}}$. $I = 2\int_{1/\sqrt{2}}^{\sqrt{3}/2} \frac{du}{\sqrt{1-u^2}} = 2\sin^{-1} u\Big]_{1/\sqrt{2}}^{\sqrt{3}/2} = 2\left(\frac{\pi}{3} - \frac{\pi}{4}\right) = \frac{\pi}{6}.$

72. Let $I = \int_{\sqrt{2}/3}^{2/3} \frac{dx}{x\sqrt{9x^2-1}}$, $u = 3x$, $du = 3dx$. Then $I = \frac{1}{3}\int_{\sqrt{2}}^{2} \frac{3du}{u\sqrt{u^2-1}} = \sec^{-1} u\big]_{\sqrt{2}}^{2} = \frac{\pi}{3} - \frac{\pi}{4} = \frac{\pi}{12}$.

73. $\int_0^{\ln 2} 4e^x \cosh x\, dx = 4\int_0^{\ln 2} e^x \frac{(e^x+e^{-x})}{2} dx = 2\int_0^{\ln 2}(e^{2x}+1)dx = 2\left(\frac{e^{2x}}{2}+x\right)_0^{\ln 2} = 2\left[\frac{4}{2} + \ln 2 - \frac{1}{2}\right] = 3 + \ln 4$.

74. $\int_0^{\ln 2} \frac{\sinh x\, dx}{1+\cosh x} = \ln(1+\cosh x)]_0^{\ln 2} = \ln\left(1 + \frac{2+\frac{1}{2}}{2}\right) - \ln 2 = \ln\frac{9}{4} - \ln 2 = \ln\frac{9}{8}$.

75. $\int_{-\ln 3}^{\ln 3} 3\sqrt{\cosh 2x + 1}\ \ dx = 6\int_0^{\ln 3}\sqrt{2\cosh^2 x} = 6\sqrt{2}\int_0^{\ln 3}\cosh x\ \ dx = 6\sqrt{2}\sinh x\Big]_0^{\ln 3} = 6\sqrt{2}\ \frac{3-\frac{1}{3}}{2} = 8\sqrt{2}$.

76. $\int_1^2 \frac{5\ \operatorname{sech}^2(\ln x)dx}{x} = 5\tanh(\ln x)]_1^2 = 5\left[\tanh(\ln 2) - 0\right] = 5\frac{2-\frac{1}{2}}{2+\frac{1}{2}} = 3$.

77. $\int_2^4 10\ \operatorname{csch}^2 x \coth x\ dx = -10\ \frac{\operatorname{csch}^2\ x}{2}\Big]_2^4 = 5(\operatorname{csch}^2\ 2 - \operatorname{csch}^2\ 4)$.

78. $\int_0^{\ln\sqrt{2}} 4\ \operatorname{sech}^4 \tanh x\ dx = -\operatorname{sech}^4\ x\Big]_0^{\ln\sqrt{2}} = 1 - \left(\frac{2}{\sqrt{2}+\frac{1}{\sqrt{2}}}\right)^4 = \frac{17}{81}$.

79. Let $I = \int \sec^2(x)e^{\tan x}dx$, $u = \tan x$, $du = \sec^2 x\ dx$. Then $I = \int e^u du = e^u + C = e^{\tan x} + C$.

80. $\int \csc^2(x)e^{\cot x}\ dx = -e^{\cot x} + C$.

81. Let $I = \int \frac{\tan(\ln v)}{v}\ dv$, $u = \ln v$, $du = \frac{dv}{v}$. Then $I = \int \tan u\ du = -\ln|\cos u| + C = -\ln|\cos(\ln v)| + C$.

82. Let $I = \int \frac{dv}{v\ln v}$, $u = \ln v$, $du = \frac{dv}{v}$. Then $I = \int \frac{du}{u} = \ln|u| + C = \ln(|\ln v|) + C$.

83. Let $I = \int (x)3^{x^2} dx$, $u = x^2$, $du = 2x\ dx$, $\frac{1}{2}du = x\ dx$. Then $I = \frac{1}{2}\int 3^u\ du = \frac{1}{2}3^u/\ln 3 + C = \frac{3^{x^2}}{2\ln 3} + C$.

84. $\int 2^{\tan x}\sec^2 x\ dx = \frac{2^{\tan x}}{\ln 2} + C$.

85. Let $I = \int \frac{dy}{y\sqrt{4y^2-1}}$, $u = 2y$, $y = \frac{u}{2}$, $dy = \frac{1}{2}du$. Then $I = \frac{1}{2}\int \frac{du}{(u/2)\sqrt{u^2-1}} = \int \frac{du}{u\sqrt{u^2-1}} = \sec^{-1}|u| + C = \sec^{-1}|2y| + C$.

86. Let $I = \int \frac{24dy}{y\sqrt{y^2-16}}$, $y = 4u$, $u = \frac{1}{4}y$, $du = \frac{1}{4}dy$. Then $I = \int \frac{24(4du)}{4u\sqrt{16u^2-16}} = 6\int \frac{du}{u\sqrt{u^2-1}} = 6\sec^{-1}|u| + C = 6\sec^{-1}(|y|/4) + C$.

87. a) Let $I = \int_0^{\pi/2} \frac{\sin x\ dx}{\sqrt{1+\cos^2 x}}$. Let $u = \cos x$, $du = -\sin x\ dx$. $I = -\int_1^0 \frac{du}{\sqrt{1+u^2}} = \sinh^{-1} 1$. b) $\ln(1+\sqrt{2})$

88. a) Let $I = \int_{\sqrt{2}}^{\sqrt{17}} \frac{2x\ dx}{\sqrt{x^4-1}}$, $u = x^2$, $du = 2x\ dx$. $I = \int_2^{17} \frac{du}{\sqrt{u^2-1}} = \cosh^{-1} u\Big]_2^{17} = \cosh^{-1} 17 - \cosh^{-1} 2$. b) $\cosh^{-1} 17 - \cosh^{-1} 2 = \ln(17+\sqrt{288}) - \ln(2+\sqrt{3}) = \ln \frac{17+12\sqrt{2}}{2+\sqrt{3}}$.

89. a) $\int_{1/5}^{1/2} \frac{4\tanh^{-1} x}{1-x^2} dx = 2(\tanh^{-1} x)^2\Big]_{1/5}^{1/2} = 2\left[(\tanh^{-1}(\frac{1}{2}))^2 - (\tanh^{-1}(\frac{1}{5}))^2\right]$ b) $2\left[(\frac{1}{2}\ln 3)^2 - (\frac{1}{2}\ln\frac{3}{2})^2\right] = \frac{1}{2}\left[(\ln 3)^2 - (\ln\frac{3}{2})^2\right] = \frac{1}{2}\left[\ln 3 - \ln\frac{3}{2}\right]\left[\ln 3 + \ln\frac{3}{2}\right] = \frac{1}{2}(\ln 2)\ln\frac{9}{2}$.

90. a) Let $I = \int_{\pi/6}^{\pi/4} \frac{\cos x\ dx}{\sin x\sqrt{1+\sin^2 x}}$, $u = \sin x$, $du = \cos x\ dx$. Then $I = \int_{1/2}^{1/\sqrt{2}} \frac{du}{u\sqrt{1+u^2}} = -\operatorname{csch}^{-1} u\Big]_{1/2}^{1/\sqrt{2}} = \operatorname{csch}^{-1}\frac{1}{2} - \operatorname{csch}^{-1}\frac{1}{\sqrt{2}}$. b) $\operatorname{csch}^{-1}\frac{1}{2} - \operatorname{csch}^{-1}\frac{1}{\sqrt{2}} = \ln(2+2\sqrt{1+1/4}) - \ln(\sqrt{2}+\sqrt{3}) = \ln\frac{2+\sqrt{5}}{\sqrt{2}+\sqrt{3}}$.

91. a) $\int_{3/5}^{4/5} \frac{2\operatorname{sech}^{-1} x}{x\sqrt{1-x^2}} dx = -(\operatorname{sech}^{-1} x)^2\Big]_{3/5}^{4/5} = (\operatorname{sech}^{-1}(\frac{3}{5}))^2 - (\operatorname{sech}^{-1}(\frac{4}{5}))^2$ b) $(\ln 3)^2 - (\ln 2)^2 = (\ln 3 - \ln 2)(\ln 3 + \ln 2) = (\ln\frac{3}{2})\ln 6$.

92. a) $\int_{\sqrt{8}}^{\sqrt{3}} \frac{e^{\coth^{-1} x}}{1-x^2} dx = e^{\coth^{-1} x}\Big]_{\sqrt{8}}^{\sqrt{3}} = e^{\coth^{-1}\sqrt{3}} - e^{\coth^{-1}\sqrt{8}}$ b) $e^{\coth^{-1}\sqrt{3}} - e^{\coth^{-1}\sqrt{8}} = e^{\ln(\frac{\sqrt{3}+1}{\sqrt{3}-1})^{1/2}} - e^{\ln(\frac{\sqrt{8}+1}{\sqrt{8}-1})^{1/2}} = \sqrt{\frac{\sqrt{3}+1}{\sqrt{3}-1}} - \sqrt{\frac{\sqrt{8}+1}{\sqrt{8}-1}}$.

93. Let $I = \int_1^6 \log_5 x\ dx = \frac{1}{\ln 5}\int_1^6 \ln x\ dx$. Let $u = \ln x$, $dv = dx$, $du = \frac{1}{x}dx$, $v = x$. $(\ln 5)I = x\ln x\Big]_1^6 - \int_1^6 x\frac{1}{x}dx = 6\ln 6 - 5$, $I = \frac{6\ln 6 - 5}{\ln 5}$.

94. $\int_0^1 3^{x^2} dx = \text{NINT}(3^{x^2}, x, 0, 1) = 1.5266$.

95. $\int_1^5 \frac{e^{\sinh^{-1} x}}{1+x^2} dx \approx \text{NINT}\left(\frac{e^{\sinh^{-1} x}}{1+x^2}, x, 1, 5\right) \approx 2.714$.

96. $\int_2^5 \frac{e^{\cosh^{-1} x}}{1+\sqrt{x}} dx = \text{NINT}\left(e^{\cosh^{-1} x}/(1+\sqrt{x}), x, 2, 5\right) = 7.08085$.

97. Graph $\int_1^x \frac{\tanh t}{t} dt = \text{NINT}\left(\frac{\tanh t}{t}, t, 1, x\right)$ in $[-10, 10]$ by $[-5, 3]$.

98. Graph $y = \int_1^x t\coth t\ dt = \text{NINT}(t(\tanh t)^{-1}, t, 1, x)$ in the "squared" window $[-17, 17]$ by $[-10, 10]$.

99. $f(x) = e^x + x$, $f'(x) = e^x + 1$. $df^{-1}/dx\big]_{x=f(\ln 2)} = \frac{1}{f'(\ln 2)} = \frac{1}{3}$.

100. $A = 2\int_1^e \frac{\ln x\, dx}{x} = (\ln x)^2]_1^e = 1.$

101. $\ln 5x - \ln 3x = \ln \frac{5x}{3x} = \ln \frac{5}{3}.$

102. $A_1 = \int_1^2 \frac{dx}{x} = \ln 2 - \ln 1 = \ln 2.$ $A_2 = \int_{10}^{20} \frac{dx}{x} = \ln 20 - \ln 10 = \ln \frac{20}{10} = \ln 2.$

103. For any positive $a \neq 1$, $\int_1^e \frac{2\log_a x}{x} = \frac{2}{\ln a}\int_1^e \frac{\ln x}{x}dx = \frac{1}{\ln a}(\ln x)^2\Big]_1^e = \frac{1}{\ln a}$. Thus the two areas are $\frac{1}{\ln 2}$ and $\frac{1}{\ln 4}$ and their ratio is 2.

104. a) $\log_{10} 5 = \frac{\ln 5}{\ln 10} = 0.698970004336$ b) $\log_2 3 = \frac{\ln 3}{\ln 2} = 1.58496250072$
c) $\log_7 2 = \frac{\ln 2}{\ln 7} = 0.356207187108$

105. $y = y_0e^{-kt}$ and $k = \ln 2/5700$ from Example 3 of Section 7.3. Let t be the desired age. Then $0.1y_0 = y_0e^{-kt}$, $-kt = \ln(0.1)$, $t = -5700\frac{\ln(0.1)}{\ln 2} \approx 18935$ years.

106. a) $y = y_0e^{-kt}$, $k = \frac{\ln 2}{2.645} = 0.262$ b) $1/k = 3.816$ years c) By Exercise 14 of 7.4, $t = 3/k = 11.448$ years.

107. We wish to determine k in the model $y = y_0e^{kt}$. Let $t = 0$ correspond to 1924. Then $y_0 = 250$ and $t = 64$ corresponds to 1988. When $t = 64$, $7500 = 250e^{64k}$, $e^{64k} = 30$, $64k = \ln 30$, $k = (\ln 30)/64 \approx 0.053$. Thus the rate of appreciation is about 5.3%.

108. $L(x) = L_0e^{-kt}$. $L(18) = 0.5L_0 = L_0e^{-18k}$. $0.5 = e^{-18k}$, $\ln 0.5 = -18k$, $k = -\frac{\ln 0.5}{18}$. Let x be the desired depth. $L(x) = 0.1L_0 = L_0e^{-kx}$. $0.1 = 3^{-kx}$, $\ln 0.1 = -kx$, $x = \frac{\ln 0.1}{-k} = 18\frac{\ln 0.1}{\ln 0.5} = 59.795$ ft.

109. $p = p_0e^{0.04t}$. Let t be the required time. Then $\frac{4}{3}p_0 = p_0e^{0.04t}$, $\ln(4/3) = 0.04t$, $t = \ln(4/3)/0.04 \approx 7.19$ years.

110. a) $p(t) = p_0e^{kt} = 175.8e^{kt}$ where $t = 0$ corresponds to 1980. $p(6) = 211.9 = 175.8e^{6k}$, $e^{6k} = \frac{211.9}{175.8}$, $6k = \ln\frac{211.9}{175.8}$, $k = \frac{1}{6}\ln\frac{211.9}{175.8} = 0.0311$, 3.11%
b) $p(12) = 175.8e^{12k} = 255.414$ using the exact value of k.

111. a) $p = p_0e^{kt} = 295.5e^{kt}$ using $t = 0$ for 1980. When $t = 6$, $p = 480.1 = 295.5e^{6k}$. This leads to $k = 0.08\ldots$ or an inflation rate of about 8%.
b) $p(12) = 295.5e^{(0.08)12} \approx 771.8$.

112. a) $P(0) = \frac{150}{1+e^{4.3}} \approx 2$ b) $P(t) \to 150$ as $t \to \infty$ c) $P(t) = 125$ leads to $150 = 125(1+e^{4.3-t})$, $\frac{6}{5} = 1+e^{4.3}e^{-t}$, $\frac{1}{5e^{4.3}} = e^{-t}$, $e^t = 5e^{4.3}$, $t = 4.3+\ln 5 \approx 5.9$ days d) Graph NDER$(P(x), x)$ in $[0, 10]$ by $[0, 50]$ and zoom in to the maximum. Max $P' = 37.5$ students/day which occurs after 4.3 days.

113. $A_0 e^{-kt} = \frac{1}{2} A_0$ leads to $e^{-kt} = \frac{1}{2}$, $-kt = \ln 1 - \ln 2 = -\ln 2$, $t = (\ln 2)/k$.

114. a) $t \approx \frac{70}{5} = 14$ years b) $t \approx \frac{70}{7} = 10$ years

115. $T - T_s = (T_0 - T_s)e^{-kt}$. After 15 minutes, $180 - 40 = (220 - 40)e^{-15k}$, $e^{-15k} = \frac{7}{9}$, $k = -\frac{\ln \frac{7}{9}}{15}$. Let t be the total number of minutes to cool to 70°. $70 - 40 = (220 - 40)e^{-kt}$, $t = \frac{\ln 6}{k} \approx 106.943$ minutes. Thus the answer is $106.94 - 15 = 91.943$ minutes.

116. We will assume $c > y$. a) $\frac{dy}{dt} = k\frac{A}{V}(c - y)$. $\frac{dy}{(c-y)} = \frac{kA}{V}dt$, $\int \frac{dy}{c-y} = \frac{kA}{V}\int dt + C_0$, $-\ln(c - y) = \frac{kAT}{V} + C_0$, $c - y = C_1 e^{-(kA/V)t}$. Setting $t = 0$, we obtain $C_1 = c - y_0$, $y = c - (c - y_0)e^{-(kA/V)t}$. b) As $t \to \infty$, $y \to c - (c - y_0)0 = c$.

117. $\lim_{t\to 0} \frac{t - \ln(1+2t)}{t^2} = \lim_{t \to 0} \frac{1 - \frac{2}{1+2t}}{2t}$. From this we see that the limit is ∞ as $t \to 0^-$ and $-\infty$ as $t \to 0^+$. This can be confirmed by graphing the original function in $[-0.5, 1]$ by $[-50, 50]$.

118. $\lim_{x\to 0} \frac{\sin x}{e^x - x - 1} = \lim_{x \to 0} \frac{\cos x}{e^x - 1}$ does not exist. $\lim_{x\to 0^-} \frac{\cos x}{e^x - 1} = -\infty$ while $\lim_{x \to 0^+} \frac{\cos x}{e^x - 1} = \infty$. L'Hospital's rule can be applied to one-sided limits and to the case where the limit is ∞ or $-\infty$.

119. $\lim_{x\to 0} \frac{x \sin x}{1 - \cos x} = \lim_{x\to 0} \frac{x \cos x + \sin x}{\sin x}$ (still $\frac{0}{0}$) $= \lim_{x \to 0} \frac{-x \sin x + 2\cos x}{\cos x} = 2$.

120. $\lim_{x\to 1} \frac{\log_4 x}{\log_2 x} = \lim_{x \to 1} \frac{\ln 2}{\ln 4} \frac{\ln x}{\ln x} = \frac{\ln 2}{\ln 4} = \frac{\ln 2}{2 \ln 2} = \frac{1}{2}$.

121. $\lim_{x \to 0} \frac{2^{\sin x} - 1}{e^x - 1} = \lim_{x \to 0} \frac{2^{\sin x}(\cos x) \ln 2}{e^x} = \ln 2$. Defining $f(0) = \ln 2$ removes the discontinuity.

122. $\lim_{x \to 0^+} x \ln x = \lim_{x \to 0^+} \frac{\ln x}{\frac{1}{x}} = \lim_{x\to 0^+} \frac{\frac{1}{x}}{-\frac{1}{x^2}} = \lim_{x \to 0^+}(-x) = 0$. Thus if $g(0) = 0$, $g(x) = f(x)$ for $0 < x \leqq 2$, then g is an extension of f continuous on $[0, 2]$. Graph $y = x \ln x$ in $[0, 2]$ by $[-2, 3]$.

123. Let $y = x^{\frac{1}{x}}$. $\lim_{x\to\infty} \ln y = \lim_{x \to \infty} \frac{\ln x}{x} = \lim_{x\to\infty} \frac{1}{x} = 0$. Therefore $\lim_{x \to \infty} y = e^0 = 1$.

124. Let $y = x^{1/x^2}$, $\ln y = \frac{1}{x^2} \ln x$. $\lim_{x\to\infty} \ln y = \lim_{x\to\infty} \frac{\ln x}{x^2} = \lim_{x \to \infty} \frac{(1/x)}{2x} = \lim_{x\to\infty} \frac{1}{2x^2} = 0$. Therefore $\lim_{x\to\infty} y = e^0 = 1$.

125. $\lim_{x\to\infty}(1 + \frac{3}{x})^x = e^3$ is found by the method of Example 6 of 7.5. See also Exercise 45 of that section.

126. Let $y = (1+\frac{3}{x})^x$. $\lim_{x\to 0}\ln y = \lim_{x\to 0} x\ln(1+\frac{3}{x}) = \lim_{x\to 0}\frac{\ln(1+\frac{3}{x})}{\frac{1}{x}} = \lim_{x\to 0}\frac{[1/(1+3/x)](-3/x^2)}{(-1/x^2)} = \lim_{x\to 0}\frac{3}{(1+3/x)} = 3$. Therefore $\lim_{x\to 0} y = e^3$.

127. In all three cases the functions grow at the same rate as $x\to\infty$ because $\lim_{x\to\infty}\frac{f(x)}{g(x)} = L$, L finite and not 0.

128. a) $\lim_{x\to\infty}\frac{\ln 2x}{\ln x^2} = \lim_{x\to\infty}\frac{\ln 2+\ln x}{2\ln x} = \lim_{x\to\infty}\left[\frac{\ln 2}{2}\frac{1}{\ln x}+\frac{1}{2}\right] = \frac{1}{2}$. $\lim_{x\to\infty}\frac{\ln x^2}{\ln(x+2)} = \lim_{x\to\infty}\frac{2\ln x}{\ln(x+2)} = \lim_{x\to\infty}\frac{2(1/x)}{1/(x+2)} = \lim_{x\to\infty}\frac{2(x+2)}{x} = \lim_{x\to\infty}2(1+\frac{2}{x}) = 2$. So all three grow at the same rate. b) $\frac{x^{\ln x}}{x^{\log_2 x}} = \frac{1}{x^{\ln x/\ln 2-\ln x}} = \frac{1}{x^{(\ln x)(1/\ln 2-1)}}\to 0$ as $x\to\infty$. So $x^{\log_2 x}$ grows faster than $x^{\ln x}$ as $x\to\infty$. c) $(1/2)^x/(1/3)^x = (3/2)^x\to\infty$ as $x\to\infty$. So $(1/2)^x$ grows faster than $(1/3)^x$ as $x\to\infty$.

129. a) $\frac{\frac{1}{x^2}+\frac{1}{x^4}}{\frac{1}{x^2}} = 1+\frac{1}{x^2}\leqq 2$ for $x\geqq 1$. True. b) $\frac{\frac{1}{x^2}+\frac{1}{x^4}}{\frac{1}{x^4}} = x^2+1\to\infty$ as $x\to\infty$. False. c) $\frac{\sqrt{x^2+1}}{x} = \sqrt{\frac{x^2+1}{x^2}}$ (for $x>0$) $= \sqrt{1+\frac{1}{x^2}}\leqq\sqrt{2}$ for $x\geqq 1$. True.

130. a) $\lim_{x\to\infty}\frac{\ln x}{x} = \lim_{x\to\infty}\frac{(1/x)}{1} = 0$ so a) is true. b) $\lim_{x\to\infty}\frac{\ln\ln x}{\ln x} = \lim_{x\to\infty}\frac{(1/\ln x)(1/x)}{(1/x)} = 0$ so b) is true. c) $\lim_{x\to\infty}\frac{x}{x+\ln x} = \lim_{x\to\infty}\frac{1}{1+(\ln x/x)} = \frac{1}{1+0} = 1$ so c) is false.

131. $g(x)$ is not defined for $x<0$ and $f(x)$ is not defined for $x<-1$. $\lim_{x\to\infty}\frac{f(x)}{g(x)} = \lim_{x\to\infty}\frac{\sqrt[4]{x^3-x}}{\sqrt[4]{x^3}} = \lim_{x\to\infty}\sqrt[4]{1-\frac{1}{x^2}} = 1$. g is a right end behavior model for f.

132. $\frac{f(x)}{g(x)} = \frac{\sqrt[3]{x^2-2x+1}}{\sqrt[3]{x^2}} = \sqrt[3]{1-\frac{2}{x}+\frac{1}{x^2}}\to 1$ as $x\to\pm\infty$. g is an end behavior model for f.

133. $\frac{f(x)}{g(x)} = \frac{2x^3-x+1}{(4x^2-x+1)(0.5x)} = \frac{2x^3-x+1}{2x^3-0.5x+0.5x} = \frac{2-\frac{1}{x^2}+\frac{1}{x^3}}{2-\frac{0.5}{x}+\frac{0.5}{x^2}}\to\frac{2}{2} = 1$ as $x\to\pm\infty$. g is an end behavior model for f.

134. $\frac{f(x)}{g(x)} = \frac{x^5-1}{(x^2+x-1)(x^3)} = \frac{x^5-1}{x^5+x^4-x^3} = \frac{1-\frac{1}{x^5}}{1+\frac{1}{x}-\frac{1}{x^2}}\to 1$ as $x\to\pm\infty$. g is an end behavior model for f.

135. $\frac{f(x)}{g(x)} = \frac{2^x+2^{-x}}{2^x} = 1+\frac{1}{2^{2x}}\to 1$ as $x\to\infty$. ($\to\infty$ as $x\to-\infty$) g is a right end behavior model for f.

136. $\frac{f(x)}{g(x)} = \frac{2^x+2^{-x}}{2^{-x}} = 2^{2x}+1\to 1$ as $x\to-\infty$. g is a left end behavior model for f.

137. Let $f(x) = \tan^{-1} x + \tan^{-1} \frac{1}{x}$. $f'(x) = \frac{1}{1+x^2} + \frac{(-\frac{1}{x^2})}{1+(\frac{1}{x})^2} = \frac{1}{1+x^2} - \frac{1}{x^2+1} = 0$ so $f(x)$ is constant. $f(1) = \frac{\pi}{2}$ but $f(-1) = -\frac{\pi}{2}$. Hence $f(x) = \frac{\pi}{2}$ for $x > 0$ and $f(x) = -\frac{\pi}{2}$ for $x < 0$.

138. Draw the horizontal line segment from $(0, \frac{\pi}{2})$ to $(1, \frac{\pi}{2})$ to complete a second congruent rectangle. The area, A_1, in this rectangle above $y = \sin^{-1} x$ is equal to the area, A_2, under $y = \sin x$ from $x = 0$ to $\frac{\pi}{2}$. $A_1 = \frac{\pi}{2} - \int_0^1 \sin^{-1} x \, dx = A_2 = \int_0^{\pi/2} \sin x \, dx$.

139. $\theta = \pi - \cot^{-1} \frac{x}{60} - \cot^{-1} \frac{50-x}{30}$, $\frac{d\theta}{dx} = \frac{1}{60} \frac{1}{1+(\frac{x}{60})^2} - \frac{1}{30} \frac{1}{1+(\frac{50-x}{30})^2} = \frac{30(x^2-200x+3200)}{(3600+x^2)[900+(50-x)^2]}$. $\frac{d\theta}{dx} = 0$ when $x = 100 \pm 20\sqrt{17}$. For the problem situation the solution is $x = 100 - 20\sqrt{17} \approx 17.538$ m.

140. a) $\frac{dL}{d\theta} = k\left[\frac{b \csc^2 \theta}{R^4} - \frac{b \csc\theta \cot\theta}{r^4}\right] = kb \csc\theta \left[\frac{\csc\theta}{R^4} - \frac{\cot\theta}{r^4}\right] = 0 \Rightarrow \frac{\cot\theta}{\csc\theta} = \frac{r^4}{R^4}$, $\cos\theta = \frac{r^4}{R^4}$, $\theta = \cos^{-1} \frac{r^4}{R^4}$. b) $\theta = \cos^{-1}((\frac{5}{6})^4) = 61°$ to the nearest degree.

141. $y = -\int_1^x \frac{dt}{t\sqrt{1-t^2}} + \int_1^x \frac{t\,dt}{\sqrt{1-t^2}}$. (Note $y = 0$ if $x = 1$.) $y = \operatorname{sech}^{-1}|t| - \sqrt{1-t^2}\Big|_1^x = \operatorname{sech}^{-1}|x| - \sqrt{1-x^2} - \operatorname{sech}^{-1} 1 = \operatorname{sech}^{-1}|x| - \sqrt{1-x^2} = \cosh^{-1}(\frac{1}{|x|}) - \sqrt{1-x^2}$. You may graph this to confirm the given figure.

CHAPTER 8

TECHNIQUES OF INTEGRATION

8.1 FORMULAS FOR ELEMENTARY INTEGRALS

1. Let $u = 8x^2 + 1$; $du = 16x^2 dx$. The integral becomes $\int \frac{du}{\sqrt{u}} = \int u^{-1/2} du = \frac{u^{1/2}}{1/2} + C = 2(8x^2+1)^{1/2} + C$.

2. Let $u = 1 + 3\sin x$; $du = 3\cos x\,dx$. The integral becomes $\int u^{-1/2} du = 2u^{1/2} + C = 2\sqrt{1+3\sin x} + C$.

3. Let $u = 8x^2 + 2 : 0 \le x \le 1 \Rightarrow 2 \le u \le 10$, $du = 16x dx$. The integral becomes $\int_2^{10} \frac{du}{u} \approx 1.60944$.

4. Let $u = \sqrt{x}$; $du = \frac{1}{2\sqrt{x}} dx$; $\int_4^9 \frac{dx}{x-\sqrt{x}} = \int_4^9 \frac{1}{\sqrt{x}} \frac{dx}{\sqrt{x}-1} = 2\int_2^3 \frac{du}{u-1} = 2\ln 2 = \ln 4$.

5. Let $u = x^2$, $du = 2x dx \Rightarrow 4x dx = 2du$. The integral becomes $2\int \tan u du = -2\ln|\cos u| + C = \ln(\cos^2 x^2) + C$.

6. $\int \cot(3-7x)dx = -\frac{1}{7}\int \cot u du = -\frac{1}{7}\ln|\sin(3-7x)| + C$.

7. Let $u = x/3$; $du = \frac{1}{3}dx$ or $dx = 3du$, $-\pi \le x \le \pi \Rightarrow -\pi/3 \le u \le \pi/3$. The integral becomes $\int_{-\pi/3}^{\pi/3} \sec u(3du) = 3\int_{-\pi/3}^{\pi/3} \sec u du \approx 7.902$.

8. $\int x\sec(x^2-5)dx = \frac{1}{2}\int \sec u du = \frac{1}{2}\ln|\sec u + \tan u| + C = \frac{1}{2}\ln|\sec(x^2-5) + \tan(x^2-5)| + C$, where $u = x^2 - 5$.

9. The integral becomes $\int_{\pi/2}^{3\pi/4} \csc u du \approx 0.881$, where $u = x - \pi$.

10. $\int \frac{1}{x^2}\csc(\frac{1}{x})dx = -\int \csc u du = \ln|\csc u + \cot u| + C = \ln|\csc(\frac{1}{x}) + \cot(\frac{1}{x})| + C$, where $u = \frac{1}{x}$.

11. Let $u = e^x + 1$. The integral becomes $\int \csc u du = -\ln|\csc(e^x+1) + \cot(e^x+1)| + C$.

12. Let $u = 3 + \ln x$. The integral becomes $\int \cot u du = \ln|\sin(3+\ln x)| + C$.

13. The integral becomes: $\int_0^{\ln 2} e^u du = 2 - 1 = 1$, where $u = x^2$.

14. $\int_{\pi/2}^{\pi} \sin x e^{\cos x} dx = -\int_0^{-1} e^u du = -[e^{-1} - e^0] = 1 - \frac{1}{e}$.

15. The integral becomes: $\int_0^1 3^u du \approx 1.82048$.

16. $\int_1^2 \frac{2^{\ln x}}{x} dx = \int_0^{\ln 2} 2^u du = \frac{1}{\ln 2} 2^u]_0^{\ln 2} = \frac{1}{\ln 2}[2^{\ln 2} - 1]$; where $u = \ln x$.

17. $u = \sqrt{y} \Rightarrow du = \frac{1}{2\sqrt{y}} dy$ so $\frac{dy}{\sqrt{y}} = 2du$; $1 \le y \le 3 \Rightarrow 1 \le u \le \sqrt{3}$. The integral becomes: $\int_1^{\sqrt{3}} \frac{2 \cdot 6du}{1+u^2} = 12 \int_1^{\sqrt{3}} \frac{du}{1+u^2} \approx 3.14159$ (which is obviously π).

18. $\int_{-1}^0 \frac{4dx}{1+(2x+1)^2} = \int_{-1}^1 \frac{2du}{1+u^2} = 2\tan^{-1} u]_{-1}^1 = 2(\pi/4 + \pi/4) = \pi$, where $u = 2x+1$.

19. $u = 3x \Rightarrow du = 3dx \Rightarrow dx = \frac{1}{3}du$. The integral becomes $\int_0^{1/2} \frac{\frac{1}{3}du}{\sqrt{1-u^2}} = \frac{1}{3}\int_0^{1/2} \frac{du}{\sqrt{1-u^2}} = \frac{1}{3}\sin^{-1} u|_0^{1/2} \approx 0.1743 = \frac{\pi}{18}$.

20. $\int_0^1 \frac{dx}{\sqrt{4-x^2}} = \frac{1}{2}\int_0^1 \frac{dx}{\sqrt{1-(x/2)^2}} = \int_0^{1/2} \frac{du}{\sqrt{1-u^2}} = \sin^{-1} u]_0^{1/2} = \frac{\pi}{6}$.

21. Let $u = 5x$; $2/(5\sqrt{3}) \le x \le \frac{2}{5} \Rightarrow 2/\sqrt{3} \le u \le 2$; $x = u/5 \Rightarrow dx = \frac{1}{5}du$. The integral becomes $\int_{2/\sqrt{3}}^2 \frac{6 \cdot \frac{1}{5}du}{\frac{1}{5}u\sqrt{u^2-1}} = 6\int_{2/\sqrt{3}}^2 \frac{du}{u\sqrt{u^2-1}} \approx 3.14159$, (π).

22. $\int_6^{-3\sqrt{2}} \frac{dx}{x\sqrt{x^2-9}} = \int_{-6}^{-3\sqrt{2}} \frac{1}{3x}\frac{dx}{\sqrt{(x/3)^2-1}} = \int_{-2}^{-\sqrt{2}} \frac{du}{3u\sqrt{u^2-1}} = \frac{1}{3}\sec^{-1}|u|]_{-2}^{-\sqrt{2}} = \frac{1}{3}[\sec^{-1}\sqrt{2} - \sec^{-1} 2] = \frac{1}{3}[\frac{\pi}{4} - \frac{\pi}{3}] = \frac{-\pi}{36}$, where $3u = x$.

23. $(-x^2 + 4x - 3) = -(x^2 - 4x) - 3 = -(x^2 - 4x + 4) - 3 + 4 = 1 - (x-2)^2$. The integral becomes $\int \frac{dx}{\sqrt{1-(x-2)^2}}$. Let $u = x - 2$. Then $du = dx$; the integral is $\int \frac{du}{\sqrt{1-u^2}} = \sin^{-1} u + C = \sin^{-1}(x-2) + C$.

24. $\int \frac{dx}{\sqrt{2x-x^2}} = \int \frac{dx}{\sqrt{1-1+2x-x^2}} = \int \frac{dx}{\sqrt{1-(x-1)^2}} = \sin^{-1}(x-1) + C$.

25. $x^2 - 2x + 2 = x^2 - 2x + 1 + 1 = (x-1)^2 + 1$. Let $u = x - 1$. Then $\int_1^2 \frac{8dx}{(x-1)^2+1} = \int_0^1 \frac{8du}{u^2+1} \approx 6.28319$ (2π).

26. $\int_2^4 \frac{2dx}{x^2-6x+10} = \int_2^4 \frac{2dx}{x^2-6x+9+1} = \int_2^4 \frac{2dx}{(x-3)^2+1} = 2\tan^{-1}(x-3)]_2^4 = 2[\tan^{-1}(1) - \tan^{-1}(-1)] = 2 \cdot [\pi/4 + \pi/4] = \pi$.

27. $x^2 + 2x = x^2 + 2x + 1 - 1 = (x+1)^2 - 1$. Let $u = x + 1$. The integral becomes $\int \frac{du}{u\sqrt{u^2-1}} = \sec^{-1}|u| + C = \sec^{-1}|x+1| + C$.

28. $\int \frac{dx}{(x-2)\sqrt{x^2-4x+3}} = \int \frac{dx}{(x-2)\sqrt{(x-2)^2-1}} = \sec^{-1}|x-2| + C$.

29. $\int(\csc x - \cot x)^2 dx = \int(\csc^2 x + \cot^2 x - 2\csc x \cot x)dx = \int(2\csc^2 x - 1)dx - 2\int \csc x \cot x\, dx = -2\cot x - x + 2\csc x + C$. $[-2\cot x - x + 2\csc x]_{\pi/4}^{3\pi/4} = [2 - 3\pi/4 + \sqrt{2}] - [-2 - \pi/4 + \sqrt{2}] = [4 - \pi/2]$.

30. $\int_0^{\pi/4}(\sec x + 4\cos x)^2 dx = \int_0^{\pi/4} \sec^2 x + 8 + 16\cos^2 x\, dx = [\tan x + 8x + 16[\frac{x}{2} - \frac{\cos 2x}{4}]]_0^{\pi/4} = 1 + \frac{8\pi}{4} + \frac{8\pi}{4} - [-4] = 5 + 4\pi$.

31. $\int(\csc x - \sec x)(\sin x + \cos x)dx = \int(\frac{1}{\sin x} - \frac{1}{\cos x})(\sin x + \cos x)dx = \int(1 + \cot x - \tan x - 1)dx = \ln|\sin x| - \ln|\sec x| + C = \ln|\sin x| + \ln|\cos x| + C$. $[\ln|(\sin x)(\cos x)|]_{\pi/6}^{\pi/3} = [\ln(\frac{1}{2}\frac{\sqrt{3}}{2})] - [\ln(\frac{\sqrt{3}}{2}\cdot\frac{1}{2})] = 0$.

31. $\int_0^{\pi/2}(\sin 3x \cos 2x - \cos 3x \sin 2x)dx = \int_0^{\pi/2} \sin(3x - 2x)dx = -\cos x]_0^{\pi/2} = 1$.

32. $\int_0^{\pi/2}(\sin 3x \cos 2x - \cos 3x \sin 2x)dx = \int_0^{\pi/2} \sin(3x - 2x)dx = -\cos x]_0^{\pi/2} = 1$.

33. $\int_0^{\sqrt{3}/2} \frac{1}{\sqrt{1-x^2}}dx - \int_0^{\sqrt{3}/2} \frac{x}{\sqrt{1-x^2}}dx = \sin^{-1} x|_0^{\sqrt{3}/2} + \frac{1}{2}\int_0^{\sqrt{3}/2}(-2x)(1-x^2)^{-1/2}dx = \pi/3 + \frac{1}{2}[\frac{(1-x^2)^{1/2}}{1/2}]_0^{\sqrt{3}/2} = \frac{\pi}{3} + [\frac{1}{2} - 1] = \frac{\pi}{3} - \frac{1}{2}$.

34. $\int_2^5 \frac{x+2\sqrt{x-1}}{2x\sqrt{x-1}}dx = \int_2^5 \frac{1}{2\sqrt{x-1}}dx + \int_2^5 \frac{dx}{x} = [(x-1)^{1/2} + \ln|x|]_2^5 = 2 + \ln 5 - 1 - \ln 2 = 1 + \ln 2.5$.

35. $\int_0^{\pi/4} \frac{1}{\cos^2 x}dx + \int_0^{\pi/4} \frac{\sin x}{\cos^2 x}dx = \int_0^{\pi/4} \sec^2 x dx + \int_{x=0}^{x=\pi/4} \frac{-du}{u^2} = [\tan x]_0^{\pi/4} + [\frac{1}{u}]_{x=0}^{x=\pi/4} = 1 + [\frac{1}{\cos x}]_0^{\pi/4} = 1 + [\sqrt{2} - 1] = \sqrt{2}$; where $u = \cos x$.

36. $\int_0^{1/2} \frac{2-8x}{1+4x^2}dx = \int_0^{1/2} \frac{2dx}{1+(2x)^2} - \int_0^{1/2} \frac{8xdx}{1+4x^2} = [\tan^{-1} 2x - \ln(1+4x^2)]_0^{1/2} = \pi/4 - \ln 2$.

37. $y = \ln(\cos x) \Rightarrow y' = \frac{1}{\cos x}(-\sin x)$; $L = \int_0^{\pi/3} \sqrt{1+(y')^2}dx = \int_0^{\pi/3} \sqrt{1+\tan^2 x}dx = \int_0^{\pi/3} \sqrt{\sec^2 x}dx = \int_0^{\pi/3} \sec x\, dx \approx 1.31696$.

38. $y' = \frac{1}{\sec x}(\sec x \tan x) = \tan x \Rightarrow ds = \sqrt{1+\tan^2 x}dx = \sec x dx \Rightarrow L = \int_0^{\pi/4} \sec x\, dx = \ln|\sec x + \tan x||_0^{\pi/4} = \ln|\sqrt{2}+1| - \ln 1 = \ln(\sqrt{2}+1)$.

39.

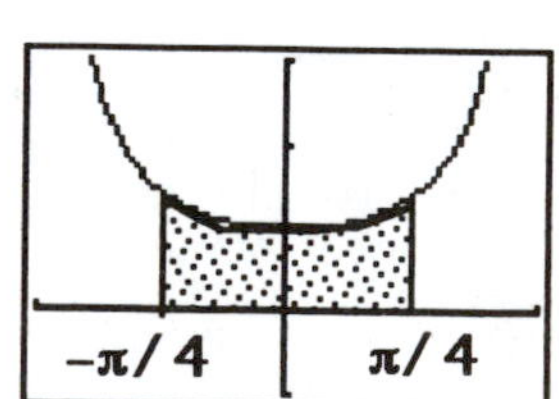

By symmetry $\bar{x} = 0$.
$M_x = \int_{-\pi/4}^{\pi/4} \frac{1}{2}(\sec x)(\sec x)dx = \int_0^{\pi/4} \sec^2 x\, dx = [\tan x]_0^{\pi/4} = 1$; $M = 2\int_0^{\pi/4} \sec x\, dx = 2[\ln|\sec x + \tan x|]_0^{\pi/4} = 2\ln(\sqrt{2}+1) - 0$; $\bar{y} = \frac{1}{2\ln(1+\sqrt{2})} = \frac{1}{\ln(1+\sqrt{2})^2} = \frac{1}{\ln(3+2\sqrt{2})}$.

40. $A = 2\int_0^{\pi/4}(2\cos x - \sec x)dx = [2\sin x - \ln|\sec x + \tan x|]_0^{\pi/4} = 2[2\cdot\frac{\sqrt{2}}{2} - \ln|\sqrt{2}+1|] = 2(\sqrt{2} - \ln(1+\sqrt{2}))$.

41. $D = 25\int_{30^\circ}^{45^\circ}\sec x\,dx = 25\ln|\sec x + \tan x|]_{30^\circ}^{45^\circ} = 8.30169$ cm.

42. $D = 25\int_{45^\circ}^{60^\circ}\sec x\,dx = 25\ln|\sec x + \tan x|]_{45^\circ}^{60^\circ} = 10.8896$ cm.

43. Let $u = \sin x$; $du = \cos x\,dx$; $0 \le x \le \pi/2 \Rightarrow 0 \le u \le 1$. The integral becomes $3\int_0^1 u^{1/2}du = \frac{3u^{3/2}}{\frac{3}{2}}]_0^1 = 2$.

44. Let $u = \cot x$; $\int_{\pi/6}^{\pi/2}\cot^3 x\csc^2 x\,dx = -\int_{\sqrt{3}}^0 u^3du = \frac{9}{4}$.

45. Let $u = 2+\cos x$. $-\pi \le x \le 0 \Rightarrow 1 \le u \le 3$; $du = -\sin x\,dx$. The integral becomes $-\int_1^3\frac{du}{u} = -\ln|u|]_1^3 = -[\ln 3 - \ln 1] = \ln(\frac{1}{3}) \approx -1.0986$.

46. Let $u = 4x^2+1$: $\int_0^2\frac{x\,dx}{4x^2+1} = \frac{1}{8}\int_1^{17}\frac{du}{u} = \frac{1}{8}\ln 17$.

47. Let $u = \pi x$; then $du = \pi\,dx$ or $x = (\frac{1}{\pi}du)$. The integral becomes $\frac{1}{\pi}\int_0^{\pi/4}\sec u\,du = \frac{1}{\pi}\ln|\sec u + \tan u|]_0^{\pi/4} = \frac{1}{\pi}[\ln(\sqrt{2}+1) - \ln(1+0)] = \frac{1}{\pi}\ln(\sqrt{2}+1) = 0.28055$.

48. Let $u = \pi x$: $\int_{1/4}^{3/4}\csc\pi x\,dx = \frac{1}{\pi}\int_{\pi/4}^{3\pi/4}\csc u\,du = \frac{-1}{\pi}\ln|\csc u + \cot u|]_{\pi/4}^{3\pi/4} = \frac{-1}{\pi}[\ln|\sqrt{2}-1| - \ln|\sqrt{2}+1|] = \frac{1}{\pi}\ln(3+2\sqrt{2})$.

49. Let $u = \tan x$; then $0 \le u \le \sqrt{3}$ and $du = \sec^2 x\,dx$. The integral becomes $\int_0^{\sqrt{3}}e^u du = e^u]_0^{\sqrt{3}} = e^{\sqrt{3}} - 1 = 4.65223$.

50. Let $u = \sqrt{x}$: $\int_{\ln^2 2}^{\ln^2 3}\frac{e^{\sqrt{x}}dx}{\sqrt{x}} = 2\int_{\ln 2}^{\ln 3}e^u du = 2$.

51. Let $u = \sqrt{x}$; then $du = \frac{1}{2\sqrt{x}}dx$. The integral becomes $\int_1^2 2^u du = \frac{1}{\ln 2}[2^u]_1^2 = \frac{1}{\ln 2}[4-2] = \frac{2}{\ln 2} = 2.8854$.

52. Let $u = 2x$: $\int_0^1 10^{2x}dx = \frac{1}{2}\int_0^2 10^u du = \frac{1}{2}\cdot\frac{1}{\ln 10}[100-1] = \frac{99}{\ln 100}$.

53. Let $u = 3x$; $du = 3dx$ and $0 \le u \le \sqrt{3}$. The integral becomes $3\int_0^{\sqrt{3}}\frac{du}{1+u^2} = 3[\tan^{-1}u]_0^{\sqrt{3}} = 3\cdot[\pi/3 - 0] = \pi$.

54. Let $u = e^x$: $\int_0^{\ln\sqrt{3}}\frac{e^x dx}{1+e^{2x}} = \int_1^{\sqrt{3}}\frac{du}{1+u^2} = \tan^{-1}u]_1^{\sqrt{3}} = [\pi/3 - \pi/4] = \frac{\pi}{12}$.

55. Let $u = 2x$. The integral becomes $\int_0^{1/2}\frac{du}{\sqrt{1-u^2}} = [\sin^{-1}u]_0^{1/2} = \frac{\pi}{6} - 0 = \frac{\pi}{6}$.

56. Let $u = x^2$: $\int_0^{1/\sqrt{2}} \frac{2x\,dx}{\sqrt{1-x^4}} = \int_0^{1/2} \frac{du}{\sqrt{1-u^2}} = \sin^{-1} u]_0^{1/2} = \pi/6$.

57. Let $u = 2x$; then $du = 2dx$. $\int_{1/\sqrt{2}}^{1} \frac{dx}{x\sqrt{4x^2-1}} = \int_{1/\sqrt{2}}^{1} \frac{2dx}{2x\sqrt{4x^2-1}} = \int_{2/\sqrt{2}}^{2} \frac{du}{u\sqrt{u^2-1}} = [\sec^{-1}|u|]_{\sqrt{2}}^{2} = \sec^{-1} 2 - \sec^{-1}\sqrt{2} = \frac{\pi}{3} - \frac{\pi}{4} = \frac{\pi}{12}$.

58. Let $u = e^{-x}$: $\int_{\ln(\frac{2}{\sqrt{3}})}^{\ln 2} \frac{3^{-x}dx}{\sqrt{e^{2x}-1}} = -\int_{\sqrt{3}/2}^{1/2} \frac{du}{\sqrt{\frac{1}{u^2}-1}} = -\int_{\sqrt{3}/2}^{1/2} \frac{u\,du}{\sqrt{1-u^2}} = \frac{1}{2}\int_{u=\frac{\sqrt{3}}{2}}^{1/2} w^{-1/2}dw = \frac{1}{2}[2w^{1/2}]_{u=\sqrt{3}/2}^{1/2} = \sqrt{1-u^2}]_{\sqrt{3}/2}^{1/2} = \frac{\sqrt{3}}{2} - \frac{1}{2}$, where $w = \sqrt{1-u^2}$.

59. Graph $y1 = \int_0^x \frac{dt}{t^2+2t+2} = \text{NINT}(1/(t^2+2t+2), t, 0, x)$ and $y2 = \tan^{-1}(x+1)$ on $[-2,2]$ by $[-2,2]$. Using trace or trigonometry we see that $y2(0) = \pi/4 \approx .78$. Thus $C = -\pi/4$, and $\tan^{-1}(x+1) - \pi/4 = \int_0^x \frac{dt}{t^2+2t+1}$

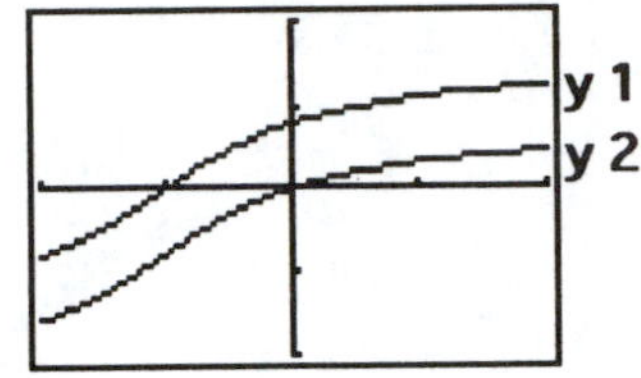

60. Graph $y1 = \int_0^x \frac{dt}{1+4t^2}$ and $y2 = \frac{1}{2}\tan^{-1}(2x)$ on $[-2,2]$ by $[-2,2]$. The graphs are identical. Hence $C = 0$.

8.2 INTEGRATION BY PARTS

1. Let $u = x$, $dv = \sin x\,dx$. Then $du = dx$, $v = -\cos x$; $\int x \sin x\,dx = x(-\cos x) - \int(-\cos x)dx = -x\cos x + \int \cos x\,dx = -x\cos x + \sin x + C$. To support graphically, compare the graphs of $y = -x\cos x + \sin x + 2$ and $y = \int_0^x t\sin t\,dt = \text{NINT}(t\sin t, t, 0, x)$ on $[-5,5]$ by $[-5,5]$.

2. Let $u = x$, $dv = \cos 2x dx$; $du = dx$, $v = \frac{1}{2}\sin 2x \Rightarrow \int x\cos 2x dx = \frac{x}{2}\sin 2x - \frac{1}{2}\int \sin 2x dx = \frac{x}{2}\sin 2x + \frac{1}{4}\cos 2x + C$.

3. Let $u = x^2$, $dv = \sin$; $du = 2x dx$, $v = -\cos x$; $\int x^2 \sin x\,dx = -x^2\cos x + 2\int x\cos x\,dx = -x^2\cos x + 2[x\sin x - \int \sin dx] = -x^2\cos x + 2x\sin x + 2\cos x + C$, where $u = x$, $dv = \cos x\,dx$ for the second integration by parts. Support this answer with an NINT computation, or by comparing the graph of $y = x^2\sin x$ to the graph of $y = \text{NDER } 1(-x^2\cos x + 2x\sin x + 2\cos x)$.

Since the graphs are the same, the answer is correct; this method is often quicker.

4. Using tabular integration, $\int x^2 \cos x\, dx = x^2 \sin x - (2x)(-cos x) + 2(-\sin x) + C$

x^2	$+$ ↘	$\cos x$
$2x$	$-$ ↘	$\sin x$
2	$+$ ↘	$-\cos x$
		$-\sin x$

5. Let $u = \ln x$, $dv = x\, dx$; $du = \frac{dx}{x}$, $v = \frac{x^2}{2} \Rightarrow \int x \ln x\, dx = \frac{x^2}{2} \ln x - \int \frac{x^2}{2} \frac{dx}{x} = \frac{x^2}{2} \ln x - \frac{1}{4}x^2 + C$.

6. Let $u = \ln x$, $dv = x^3 dx$; $du = \frac{dx}{x}$, $v = \frac{x^4}{4}$. $\int x^3 \ln x\, dx = \frac{x^4}{4} \ln x - \int \frac{x^3}{4} dx = \frac{x^4}{4}(\ln x - \frac{1}{4}) + C$.

7. Let $u = \tan^{-1} x$, $dv = x$; $du = \frac{1}{1+x^2}$, $v = x$; $\int \tan^{-1} x\, dx = x \tan^{-1} x - \int \frac{x}{1+x^2} dx = x \tan^{-1} x - \frac{1}{2} \int \frac{2x}{1+x^2} dx = x \tan^{-1} x - \frac{1}{2} \ln(1 + x^2) + C$.

8. Let $u = \sin^{-1} x$, $dv = dx$; $du = \frac{dx}{\sqrt{1-x^2}}$, $v = x$. $\int \sin^{-1} x\, dx = x \sin^{-1} x - \int x(1-x^2)^{-1/2} dx = x \sin^{-1} x + \frac{1}{2} \int t^{-1/2} dt = x \sin^{-1} x + (1-x^2)^{1/2} + C$, where $t = 1 - x^2$.

9. Let $u = x$, $dv = \sec^2 x\, dx$; $du = dx$, $v = \tan x$; $\int x \sec^2 x\, dx = x \tan x - \int \tan x\, dx = x \tan x + \ln|\cos x| + C$.

10. Let $u = 2x$, $dv = 2\sec^2 2x dx$; $du = 2dx$, $v = \tan 2x$; $\int 4x \sec^2 3x dx = 2x \tan 2x - \int 2 \tan 2x dx = 2x \tan 2x - \ln|\sec 2x| + C$.

11. Evaluating $\int x^3 e^x dx$ requires using integration by parts three times. This is most easily done by tabular integration (see Example 7)

f and its derivatives		g and its integrals
x^3	(+) ↘	e^x
$3x^2$	(−) ↘	e^x
$6x$	(+) ↘	e^x
6	− ↘	e^x
0		e^x

$\int x^3 e^x dx = x^3e^x - 3x^2e^x + 6xe^x - 6e^x + C.$

12. By tabular integration: $\int x^4 e^{-x} dx = -e^{-x}(x^4 + 4x^3 + 12x^2 + 24x + 24) + C$

x^4		e^{-x}
	(+) ↘	
$4x^3$		$-e^{-x}$
	(−) ↘	
$12x^2$		e^{-x}
	(+) ↘	
$24x$		$-e^{-x}$
	− ↘	
24		$-e^{-x}$
	+ ↘	
		$-e^{-x}$

13. Again, use tabular integration:

f and its derivatives		g and its integrals
$x^2 - 5$		e^x
	(+) ↘	
$2x - 5$		e^x
	(−) ↘	
2		e^x
	(+) ↘	
0		e^x

$\int (x^2 - 5x)e^x dx = (x^2 - 5x)e^x - (2x - 5)e^x + 2e^x + C = (x^2 - 7x + 7)e^x + C.$

14. $\int x^2e^x + \int xe^x + \int e^x dx = (x^2-2x+2)e^x + (x-1)e^x + e^x + C = (x^2-x+2)e^x + C,$ by examples 3, 7.

15. Use tabular integration:

f		g
x^5	(+) ↘	e^x
$5x^4$	·	e^x
$20x^3$	·	e^x
$60x^2$	·	e^x
$120x$	·	e^x
120	(−) ↘	e^x
		e^x

$\int x^5e^xdx = x^5e^x - 5x^4e^x + 20x^3e^x - 60x^2e^x + 120xe^x - 120e^x + C.$

16. $\int x^2e^{4x}dx = \frac{x^2}{4}e^{4x} - \frac{2x}{16}e^{4x} + \frac{2}{64}e^{4x} + C = (x^2 - \frac{x}{2} + \frac{1}{8})\frac{e^{4x}}{4} + C$ by tabular integration.

17. Use tabular integration $\int_0^{\pi/2} x^2\sin 2xdx = [x^2(\frac{-\cos 2x}{2}) - 2x(\frac{-\sin 2x}{4}) + 2(\frac{\cos 2x}{8})]_0^{\pi/2} = [\frac{\pi^2}{4}(\frac{-(-1)}{2}) - 2\frac{\pi}{2}(0) + \frac{2(-1)}{8}] - [0 - 0 + 2(\frac{1}{8})] = \frac{\pi^2-4}{8} = 0.7337$

f and its derivatives		g and its integrals
x^2	(+) ↘	$\sin 2x$
$2x$	(−) ↘	$-\frac{1}{2}\cos 2x$
2	(+) ↘	$-\frac{1}{4}\sin 2x$
0		$\frac{1}{8}\cos 2x$

Use NINT to confirm.

18. $\int_0^{\pi/2} x^3\cos 2xdx = [x^3\frac{\sin 2x}{2} - 3x^2(\frac{-\cos 2x}{4}) + 6x(\frac{-\sin 2x}{8}) - 6(\frac{\cos 2x}{16})]_0^{\pi/2} = (0 - \frac{3\pi^2}{4}\cdot\frac{1}{4} + 0 + \frac{6(-1)}{16}) - (\frac{6}{16}) = \frac{-1}{16}(3\pi^2 - 12) = \frac{-3}{16}(\pi^2 - 4)$

19. $\int_1^2 x\sec^{-1}x\,dv = [\frac{x^2}{2}\sec^{-1}x]_1^2 - \frac{1}{2}\int_1^2 \frac{x}{\sqrt{x^2-1}}dx = [\frac{4}{2}\sec^{-1}2 - \frac{1}{2}\sec^{-1}1] - \frac{1}{4}\int_1^2(x^2-1)^{-1/2}2xdx$; (let $u = x^2 - 1$; then $du = 2xdx$, $x = 1 \Rightarrow u = 0$, $x = 2 \Rightarrow u = 3$) $= [2\cdot\frac{\pi}{3} - \frac{1}{2}\cdot 0] - \frac{1}{4}\int_0^3 u^{-1/2}du = \frac{2\pi}{3} - \frac{\frac{1}{4}[u^{1/2}]_0^3}{1/2} = \frac{2\pi}{3} - \frac{1}{2}\sqrt{3} = 1.228$, where $u = \sec^{-1}x$, $dv = xdx$, $du = \frac{dx}{|x|\sqrt{x^2-1}} = \frac{dx}{x\sqrt{x^2-1}}$ since $1 \le x \le 2$. To confirm, evaluate $\text{NINT}(t\cos^{-1}(1/t), t, 1, 2)$.

20. Let $u = \sec^{-1}\sqrt{x}$, $dv = dx$; $du = \frac{1}{\sqrt{x}\sqrt{(\sqrt{x})^2-1}}(\frac{1}{2\sqrt{x}})dx$, $v = x$; $\int_1^4 \sec^{-1}\sqrt{x}dx = x\sec^{-1}\sqrt{x}]_1^4 - \frac{1}{2}\int_1^4 (x-1)^{-1/2}dx = [x\sec^{-1}\sqrt{x} - (x-1)^{1/2}]_1^4 = 4\sec^{-1}2 - \sqrt{3} = \frac{4\pi}{3} - \sqrt{3}$.

21. Use integration by parts twice: $\int_{-1}^3 e^x \sin x\, dx = e^x \sin x|_{-1}^3 - \int_{-1}^3 e^x \cos x\, dx = e^x \sin x|_{-1}^3 - \{e^x \cos x|_{-1}^3 + \int_{-1}^3 e^x \sin x\, dx\}$. Thus $2\int_{-1}^3 e^x \sin x\, dx = (e^x \sin x - e^x \cos x)|_{-1}^3 = 23.227$ or, $\int_{-1}^3 e^x \sin x\, dx = 11.614$.

22. Let $I = \int e^{-x} \cos x\, dx$, $u = \cos x$, $dv = e^{-x}dx$. Then $I = -e^{-x}\cos x + \int e^{-x}(-\sin x)dx = -e^{-x}\cos x - \int e^{-x}\sin dx$; now let $u = \sin x$, $dv = e^{-x}dx$; $I = -e^{-x}\cos x - \{-e^{-x}\sin x + \int e^{-x}\cos x\, dx\}$ or $2I = e^{-x}(\sin x - \cos x) + C \Rightarrow I = \frac{e^{-x}}{2}(\sin x - \cos x) + C$. Hence $\int_{-3}^2 e^{-x}\cos x\, dx = \frac{e^{-x}}{2}(\sin x - \cos x)|_{-3}^2 = -8.435$.

23. Let $I = \int e^{2x}\cos 3x dx$ and use integration by parts twice. First let $u = \cos 3x$, $dv = e^{2x}dv$, $du = -3\sin 3x$, $v = \frac{1}{2}e^{2x}$; $I = (e^{2x}\cos 3x)/2 + \frac{3}{2}\int e^{2x}\sin 3x dx = (e^{2x}\cos 3x)/2 + \frac{3}{2}[(e^{2x}\sin 3x)/2 - \frac{3}{2}I]$, where $u = \sin 3x$, $dv = e^{2x}dx$, $du = 3\cos 3x dx$, $v = \frac{1}{2}e^{2x}$ in the second integration by parts. $(1 + \frac{9}{4})I = (\cos 3x + \frac{3}{2}\sin 3x)e^{2x}/2$. $I = \frac{4}{13}(\cos 3x + \frac{3}{2}\sin 3x)\frac{e^{2x}}{2} + C$. Hence, $\int_{-2}^3 e^{2x}\cos 3x dx = \frac{2e^{2x}}{13}(\cos 3x + \frac{3}{2}\sin 3x)|_{-2}^3 = 18.186$.

24. Let $2x = t$, then $I = \frac{1}{2}\int e^{-t}\sin t\, dt = \frac{1}{2}J$. Let $u = \sin t$, $dv = e^{-t}dt$, $J = -e^{-t}\sin t + \int e^{-t}\cos t\, dt = -e^{-t}\sin t + [-e^{-t}\cos t - \int e^{-t}\sin t\, dt]$ or $J = \frac{-e^{-t}(\sin t + \cos t)}{2} + C$. Hence $I = \frac{-e^{-2x}(\sin 2x + \cos 2x)}{4} + C$, $\int_{-3}^2 e^{-2x}\sin 2x dx = \frac{-e^{-2x}(\sin 2x+\cos 2x)}{4}|_{-3}^2 = 125.028$.

25.

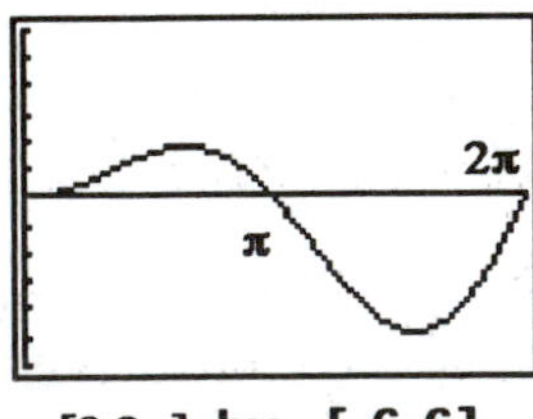

[0,2π] by [-6,6]

From problem 1,

$$\int x \sin x\, dx = \sin x - x\cos x + C$$

a) $\int_0^\pi x \sin x\, dx = [\sin x - x\cos x]_0^\pi = \pi$

b) $\int_\pi^{2\pi} x\sin x\, dx = [\sin x - x\cos x]_\pi^{2\pi} = -2\pi - \pi = -3\pi$

Since area, in itself, is taken to be positive, the answer to b) is 3π.

26. $V = 2\pi \int_0^{\pi/2} x \cos x \, dx = 2\pi[x \sin x + \cos x]_0^{\pi/2}$ by example 1, $V = 2\pi[\frac{\pi}{2} - 1]$.

27.

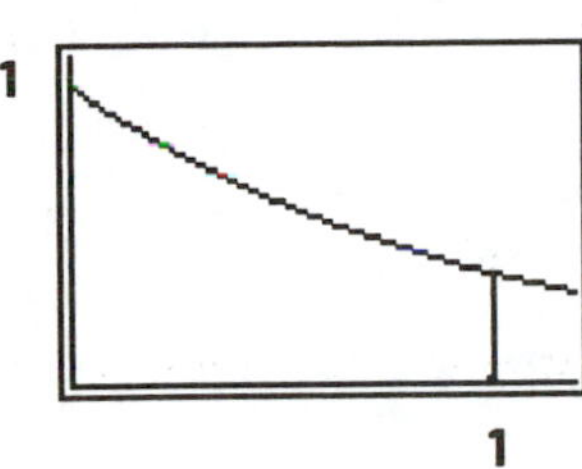

Using cylindrical shells, and tabular integration by parts, $V = 2\pi \int$ radius $\cdot$ height dx $= 2\pi \int_0^1 xe^{-x} dx = 2\pi[x(-e^{-x}) - 1 \cdot (e^{-x})]_0^1$

f		g
x	(+) ↘	e^{-x}
1	(−) ↘	$-e^{-x}$
0		e^{-x}

$= 2\pi[(-e^{-1} - e^{-1}) - (-1)] = 2\pi[1 - 2/e]$

28. a) $V = \pi \int_0^{\pi} x^2 \sin^2 x \, dx = \frac{\pi}{2} \int_0^{\pi} x^2(1 - \cos 2x)dx = \frac{\pi}{2}\frac{x^3}{3}\Big]_0^{\pi} - \frac{\pi}{2}\int_0^{\pi} x^2 \cos 2x dx = \frac{\pi}{2}[\frac{x^3}{3} - \frac{x^2}{2} \sin 2x - \frac{x \cos 2x}{2} + \frac{\sin 2x}{4}]_0^{\pi} = \frac{\pi}{2}[\frac{\pi^3}{3} - \frac{\pi}{2}]$ b) $V = 2\pi \int_0^{\pi} (\pi - x)x \sin x \, dx = 2\pi^2 \int_0^{\pi} x \sin x \, dx - 2\pi \int_0^{\pi} x^2 \sin x \, dx = 2\pi^2(-x \cos x + \sin x)]_0^{\pi} - 2\pi(-x^2 \cos x + 2x \sin x + 2 \cos x)]_0^{\pi} = 2\pi^2(\pi) - 2\pi(\pi^2 - 2 - 2) = 8\pi$ (by problems 1 and 3).

29. $M_y = \int_0^{\pi} x(\sin x)(1+x)dx = \int_0^{\pi}(x^2+x) \sin x \, dx$. Use tabular integration to get $= [(x^2+x)(-\cos x) - (2x+1)(-\sin x) + 2 \cos x]_0^{\pi} = [\pi^2+\pi-2] - [2] = \pi^2+\pi-4$.

30. Let $u = \tan^{-1} x$, $dv = x \, dx$; $du = \frac{dx}{x^2+1}$, $v = \frac{x^2}{2} + \frac{1}{2}$; $\int x \tan^{-1} x \, dx = \frac{1}{2}(x^2 + 1) \tan^{-1} x - \frac{1}{2} \int (x^2 + 1)\frac{dx}{x^2+1} = \frac{1}{2}(x^2 + 1) \tan^{-1} x - \frac{1}{2}x + C$.

31. a) Let $u = (\ln x)^n$ and $dv = dx$. $du = n(\ln x)^{n-1}(\frac{1}{x})dx$; $v = x$. Then $\int(\ln x)^n dx = x(\ln x)^n - n\int(\ln x)^{n-1}(\frac{1}{x})x \, dx = x(\ln x)^n - n\int(\ln x)^{n-1}dx$.

b) $\int_1^3(\ln x)^4 x = [x(\ln x)^4 - 4[x(\ln x)^3 - 3[x(\ln x)^2 - 2[x(\ln x)^1 - x(\ln x)^0]]]]^3 = [x(\ln x)^4 - 4x(\ln x)^3 + 12x(\ln x)^2 - 24x(\ln x) + 24x]_1^3 = [3(\ln 3)^4 - 12(\ln 3)^3 + 36(\ln 3)^2 - 72 \ln 3 + 72 - 24]$. To evaluate: store $\ln 3$ in x, then evaluate $3x^4 - 12x^3 + 36x^2 - 72x + 72 - 24 \approx 0.8086\ldots$.

c) $\text{NINT}((\ln x)^4 - 4, x, 1, 3) = 0.8086\ldots$.

32. a) $\int \tan^n x \, dx = \int \tan^{n-2} x \tan^2 x \, dx = \int \tan^{n-2} x(\sec^2 x - 1)dx = \frac{\tan^{n-1}(x)}{n-1} - \int \tan^{n-2} x \, dx$ b) $\int_{\pi/6}^{\pi/3} \tan^4 x \, dx = \frac{\tan^3 x}{3}\Big]_{\pi/6}^{\pi/3} - \int_{\pi/6}^{\pi/3}(\sec^2 x - 1)dx = [\frac{\tan^3 x}{3} + x - \tan x]_{\pi/6}^{\pi/3} = (\sqrt{3} + \frac{\pi}{3} - \sqrt{3}) - (\frac{\sqrt{3}}{27} + \frac{\pi}{6} - \frac{\sqrt{3}}{3}) = \frac{\pi}{6} + \sqrt{3}(\frac{8}{27})$

33. a) $\int_{\ln 2}^{\ln 3} e^x dx = 3 - 2 = 1$; b) $\int_2^3 \ln y \, dy = [y \ln y - y]_2^3 = \ln \frac{27}{4} - 1$

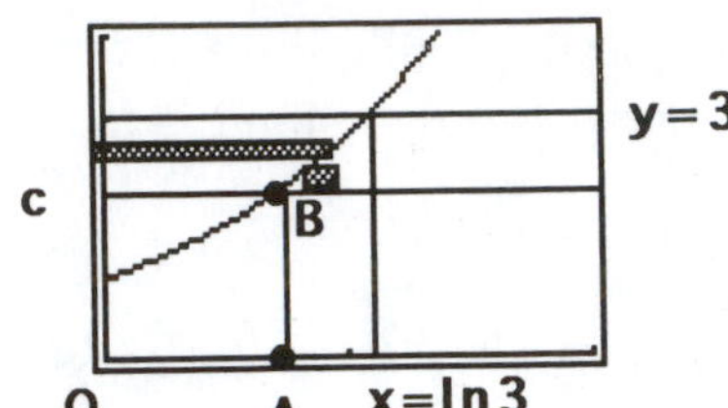

c) $(a)+(b)+\text{area}(OABC) = \text{area of rectangle} = 3\ln 3$, or $(a)+(b) = 3\ln 3 - 2\ln 2$

d) Let $y = e^x = u$, $x = v$; $(a) = \int u \, dv$, $(b) = \int x \, dy = \int v \, du$

34.

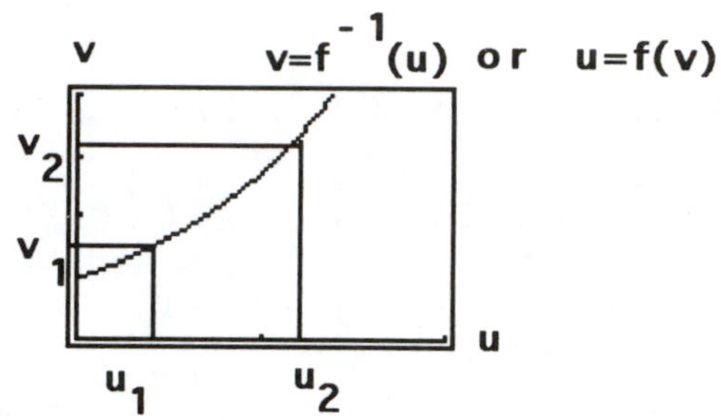

8.3 INTEGRALS INVOLVING TRIGONOMETRIC FUNCTIONS

1. $\int \sin^5 x \, dx = \int \sin x (\sin^4 x) dx = \int \sin x (\sin^2 x)^2 dx = \int \sin x (1 - \cos^2 x)^2 dx = \int \sin x (1 - 2\cos^2 x + \cos^4 x) dx = -\cos x + 2\frac{\cos^3 x}{3} - \frac{\cos^5 x}{5} + C$

2. $\int \sin^5 \frac{x}{2} dx = 2\int \sin^{-5} u \, du$ (where $u = \frac{x}{2}$) $= 2[-\cos \frac{x}{2} + \frac{2\cos^3(\frac{x}{2})}{3} - \frac{\cos^5(\frac{x}{2})}{5}] + C$ by exercise 1.

3. $\int \cos^3 x \, dx = \int \cos x (1 - \sin^2 x) dx = -\frac{\sin^3 x}{3} + \sin x + C$

4. Let $u = 3x$. $\int 3\cos^5 3x dx = \int \cos^5 u \, du = \int (1 - \sin^2 u)^2 \cos u \, du = \int (1 - 2\sin^2 u + \sin^4 u) \cos u \, du = \sin u - \frac{2\sin^3 u}{3} + \frac{\sin^5 u}{5} + C = \sin(3x) - \frac{2\sin^3(3x)}{3} + \frac{\sin^5(3x)}{5} + C$.

5. $\int \sin^7 y \, dy = \int \sin y [\sin^2 y]^3 dy = \int [1 - \cos^2 y]^3 \sin y \, dy = \int [1 - 3\cos^2 y + 3\cos^4 y - \cos^6 y] \sin y \, dy = -\cos y + \frac{3\cos^3 y}{3} - \frac{3\cos^5 y}{5} + \frac{\cos^7 y}{7} + C$

6. $\int 7\cos^7 t \, dt = 7\int (1 - \sin^2 t)^3 \cos t \, dt = 7[\sin t - \sin^3 t + \frac{3\sin^5 t}{5} - \frac{\sin^7 t}{7}] + C$

7. $\int 8\sin^4 x\,dx = 3x - 2\sin 2x + \frac{1}{4}\sin 4x + C$ since $\sin^4 x = (\sin^2 x)^2 = [(1 - \cos 2x)/2]^2 = \frac{1}{4}[1 - 2\cos 2x + \cos^2 2x] = \frac{1}{4}[1 - 2\cos 2x + \frac{1}{2}(1 + \cos 4x)] = \frac{1}{4}[\frac{3}{2} - 2\cos 2x + \frac{1}{2}\cos 4x]$

8. $\int 8(\cos^2 2\pi x)^2 dx = 8\int[\frac{1+\cos 4\pi x}{2}]^2 dx = 2\int(1 + 2\cos 4\pi x + \cos^2 4\pi x)dx = \int(2 + 4\cos 4\pi x + 1 + \cos 8\pi x)dx = 3x + \frac{1}{\pi}\sin 4\pi x + \frac{1}{8\pi}\sin 8\pi x + C$

9. $\int 16\sin^2 x\cos^2 x\,dx = 4\int(2\sin x\cos x)^2 dx = 4\int(\sin 2x)^2 dx = 2\int(1 - \cos 4x)dx = 2[x - \frac{\sin 4x}{4}] + C$

10. $\int 8\sin^4 y\cos^2 y\,dy = 2\int\sin^2 y\sin^2 2y dy = \int(1 - \cos 2y)\sin^2 2y dy = \int\sin^2 2y dy - \int\sin^2 2y\cos 2y dy = \frac{1}{2}(y - \frac{\sin 4y}{4}) - \frac{1}{2}\frac{\sin^3 2y}{3} + C$

11. $\int 35\sin^4 x\cos^3 x\,dx = \int 35\sin^4 x(1 - \sin^2 x)\cos x\,dx = 35[\frac{\sin^5 x}{5} - \frac{\sin^7 x}{7}] + C$

12. $\int \sin 2x\cos^2 2x dx = \frac{-1}{2}\int u^2 du$, (where $u = \cos 2x$), $= -\frac{1}{2}\frac{\cos^3 2x}{3} + C$

13. $\int 8\cos^3 2\theta\sin 2\theta d\theta = -4\int\cos^3 2\theta(-\sin 2\theta)2d\theta = -4\frac{\cos^4 2\theta}{4} + C$

14. $\int\sin^2 2\theta(1 - \sin^2 2\theta)\cos 2\theta d\theta = \frac{1}{2}\int\sin^2 2\theta\cos 2\theta(2d\theta) = -\frac{1}{2}\int\sin^4 2\theta\cos 2\theta(2d\theta) = \frac{1}{2}[\frac{\sin^3 2\theta}{3} - \frac{\sin^5 2\theta}{5}] + C$

15. $\frac{1-\cos x}{2} = \sin^2\frac{x}{2}$. Since $0 \le x \le 2\pi \Rightarrow 0 \le \frac{x}{2} \le \pi \Rightarrow \sin\frac{x}{2} \ge 0$, we may take a square root. The integral becomes $\int_0^{2\pi}\sin\frac{x}{2}dx = [2(-\cos(\frac{x}{2})]_0^{2\pi} = 2 - (-2) = 4$.

16. $\int_0^\pi\sqrt{\frac{(1-\cos 2x)2}{2}}dx = \int_0^\pi\sqrt{2(\sin^2 x)}dx = \sqrt{2}\int_0^\pi|\sin x|dx = \sqrt{2}\int_0^\pi\sin x\,dx = 2\sqrt{2}$

17. $1 - \sin^2 t = \cos^2 t$. Since the integrand is non-negative, $\int_0^\pi\sqrt{1 - \sin^2 t}\,dt = \int_0^\pi|\cos t|dt = \int_0^{\pi/2}\cos t\,dt + \int_{\pi/2}^\pi(-\cos t)dt = 2$ (the area under one loop of the cosine curve is 2).

18. $\int_0^\pi\sqrt{1 - \cos^2\theta}\,d\theta = \int_0^\pi|\sin x|dx = \int_0^\pi\sin x\,dx = 2$

19. $\int_{-\pi/4}^{\pi/4}\sqrt{1 + \tan^2 x}\,dx = 2\int_0^{\pi/4}\sqrt{1 + \tan^2 x}\,dx = 2\int_0^{\pi/4}\sec x\,dx = [2\ln|\sec x + \tan x|]_0^{\pi/4} = 2\ln|\sqrt{2} + 1| = \ln(\sqrt{2} + 1)^2 = \ln(3 + 2\sqrt{2})$

20. $\int_{-\pi/4}^{\pi/4}\sqrt{\sec^2 x - 1}\,dx = \int_{-\pi/4}^{\pi/4}|\tan x|dx = 2\int_0^{\pi/4}\tan x\,dx = 2\ln|\sec x|]_0^{\pi/4} = 2(\ln\sqrt{2} - \ln 1) = \ln 2$

21. On the interval $0 \le \theta \le \pi/2$ the integral becomes $\sqrt{2}\int_0^{\pi/2} \theta \sin\theta\, d\theta = \sqrt{2}[\theta\cos\theta - (-\sin\theta)]_0^{\pi/2} = \sqrt{2}$

$$\begin{array}{ccc} \theta & & \sin\theta \\ & (+)\searrow & \\ 1 & & -\cos\theta \\ & (-)\searrow & \\ 0 & & -\sin\theta \end{array}$$

22. $\int_{-\pi}^{\pi}(1-\cos^2 t)^{3/2}dt = \int_{-\pi}^{\pi}|\sin t|^3 dt = 2\int_0^{\pi}\sin^3 t\, dt = 2\int_0^{\pi}(1-\cos^2 t)\sin t\, dt = 2(-\cos t + \frac{\cos^3 t}{3})]_0^{\pi} = \frac{8}{3}$

23. First, find the indefinite integral $I = \int \sec^3 x\, dx = \int \sec^2 x \sec x\, dx = \sec x\tan x - \int \sec x\tan^2 x\, dx$; $u = \sec x$, $dv = \sec^2 x$, $du = \sec x\tan x$, $v = \tan x$; $I = \sec x\tan x - \int\sec x(\sec^2 x - 1)dx = \sec x\tan x - I + \int \sec x\, dx$; $2I = \sec x \tan + \ln|\sec x + \tan x|$: $I = [\sec x\tan x + \ln|\sec x + \tan x|]/2 + C$. The problem becomes $2I]_{-\pi/3}^{0} = [0 - 2(-\sqrt{3}) + \ln(2-\sqrt{3})] = 2\sqrt{3} - \ln(2-\sqrt{3})$.

24. Let $u = e^x$; $\int_0^{\ln\pi/4} e^x(\sec^3 e^x)dx = \int \sec^3 u\, du = \frac{1}{2}[\sec u\tan u + \ln|\sec u + \tan u|]_1^{\pi/4} = -0.907$.

25. $\int_0^{\pi/4}\sec^2\theta\sec^2\theta\, d\theta = \int_0^{\pi/4}(\tan^2\theta + 1)\sec^2\theta\, d\theta = \int_0^{\pi/4}\tan^2\theta\sec^2\theta\, d\theta + \int_0^{\pi/4}\sec^2\theta\, d\theta = [\frac{\tan^3\theta}{3} + \tan\theta]_0^{\pi/4} = \frac{4}{3}$

26. Let $u = 3x$; $\int_0^{\pi/12} 3\sec^4 3x\, dx = \int_0^{\pi/4}\sec^4 u\, du = \frac{4}{3}$ by exercise 25.

27. $\int_{\pi/4}^{\pi/2}\csc^2\theta\csc^2\theta\, d\theta = \int_{\pi/4}^{\pi/2}(1+\cot^2\theta)\csc^2\theta\, d\theta = [-\cot\theta - \frac{\cot^3\theta}{3}]_{\pi/4}^{\pi/2} = 4/3$

28. Let $u = \theta/2$; $\int_{\pi/2}^{\pi} 3\csc^4\frac{\theta}{2}d\theta = 6\int_{\pi/4}^{\pi/2}\csc^4 u\, du = 6\cdot\frac{4}{3} = 8$ by exercise 27.

29. $4\int_0^{\pi/4}\tan^3 x\, dx = 4\int_0^{\pi/4}(\sec^2 x - 1)\tan x\, dx = 4\int_0^{\pi/4}(\tan x\sec^2 x - \tan x)dx = 4[\frac{\tan^2 x}{2} + \ln|\cos x|]_0^{\pi/4} = 4[\frac{1}{2} + \ln\frac{1}{\sqrt{2}} - 0] = 2 + \ln(\frac{1}{\sqrt{2}})^4 = 2 + \ln\frac{1}{4} = 2 - \ln 4$

30. $\int_{-\pi/4}^{\pi/4} 6\tan^4 x\, dx = 12\int_0^{\pi/4}\tan^2 x(\sec^2 x - 1)dx = 12\int_0^{\pi/4}(\tan^2 x\sec^2 x - \sec^2 x + 1)dx = 12[\frac{\tan^3 x}{3} - \tan x + x]_0^{\pi/4} = 12[\frac{1}{3} - 1 + \pi/4] = 3\pi - 8$

31. $\int_{\pi/6}^{\pi/3}(\csc^2 x - 1)\cot x\, dx = [-\frac{\cot^2 x}{2} - \ln|\sin x|]_{\pi/6}^{\pi/3} = -\{[\frac{1/3}{2} + \ln\frac{\sqrt{3}}{2}] - [\frac{3}{2} + \ln(\frac{1}{2})]\} = -\{-\frac{4}{3} + \ln\frac{\sqrt{3}/2}{1/2}\} = \frac{4}{3} - \ln\sqrt{3}$

32. $8\int_{\pi/4}^{\pi/2}\cot^4 t\,dt = 8\int_{\pi/4}^{\pi/2}\cot^2 t(\csc^2 t-1)dt = 8\int_{\pi/4}^{\pi/2}(\cot^2 t\csc^2 t-\csc^2 t+1)dt = 8[-\frac{\cot^3 t}{3}+\cot t+t]_{\pi/4}^{\pi/2} = 8[+\frac{\pi}{2}-(-\frac{1}{3}+1+\frac{\pi}{4})] = 2\pi - 16/3$

33. $\sin 3x\cos 2x = \frac{1}{2}[\sin x+\sin 5x]$. Hence $\int_{-\pi}^{0}\sin 3x\cos 2x dx = \frac{1}{2}[-\cos x - \frac{1}{5}\cos 5x]_{-\pi}^{0} = \frac{1}{2}[-\frac{6}{5}-\frac{6}{5}] = -\frac{6}{5}$

34. $\int_0^{\pi/2}\sin 2x\cos 3x dx = \frac{1}{2}\int_0^{\pi/2}[\sin(2-3)x+\sin(2+3)x]dx = \frac{1}{2}[+\cos(-x) - \frac{1}{5}\cos 5x]_0^{\pi/2} = \frac{1}{2}[0-(1-\frac{1}{5})] = -\frac{2}{5}$

35. $\int_{-\pi}^{\pi}\sin 3x\sin 3x dx = 2\int_0^{\pi}\sin 3x\sin 3x dx = \frac{2}{2}\int_0^{\pi}(\cos\cdot 0x - \cos 6x)dx = [x - \frac{\sin 6x}{x}]_0^{\pi} = \pi$

36. $\int_0^{\pi/2}\sin x\cos x\,dx = \frac{1}{2}\int_0^{\pi/2}\sin 2x dx = \frac{1}{4}(-\cos 2x)]_0^{\pi/2} = \frac{1}{2}$

37. $\int_0^{\pi}\cos 3x\cos 4x dx = \frac{1}{2}\int_0^{\pi}(\cos x+\cos 7x)dx = \frac{1}{2}[\sin x + \frac{\sin 7x}{7}]_0^{\pi} = 0$

38. $\int_{-\pi/2}^{\pi/2}\cos x\cos 7x dx = \frac{2}{2}\int_0^{\pi/2}(\cos 6x+\cos 8x)dx = [\frac{\sin 6x}{6}+\frac{\sin 8x}{8}]_0^{\pi/2} = 0$

39. (a), (b), (c), (e), (g) and (i) are odd functions. The integrals of these functions over an interval centered at 0 are all zero. The others are even, non-negative functions; their integrals are not zero.

40. The integrands for a, b, c, d, e, h, i are odd functions, hence the integrals are zero; the integrand for g is even and positive, hence $2\int_0^{\pi/2}\sin^2 x\cos x\,dx > 0$; f becomes $2\int_0^{\pi}\cos^5 x\,dx = 2\cdot 0$.

41. d) $\int_{-\pi/2}^{\pi/2} x\sin x\,dx = 2\int_0^{\pi/2} x\sin x\,dx = 2[-x\cos x - (-\sin x)]_0^{\pi/2} = 2[1] = 2$

x	+ ↘	$\sin x$
1	− ↘	$-\cos x$
0		$-\sin x$

f) $\int_{-\pi/2}^{\pi/2}\cos^3 x\,dx = 2\int_0^{\pi/2}(1-\sin^2 x)\cos x\,dx = 2[\sin x - \frac{\sin^3 x}{3}]_0^{\pi/2} = \frac{4}{3}$

h) $2\int_0^{\pi/2}\sin x\sin 2x dx = 2\int_0^{\pi/2}\cos(1x) - \cos 3x dx = 2[\sin x - \frac{\sin 3x}{3}]_0^{\pi/2} = \frac{4}{3}$

42. g) $2\int_0^{\pi/2}\sin^2 x\cos x\,dx = \frac{2}{3}\sin^3 x]_0^{\pi/2} = \frac{2}{3}$ f) $2\int_0^{\pi}\cos^5 x\,dx = 2\int_0^{\pi}\cos x(1-\sin^2 x)^2 = 2\int_0^{\pi}(\cos x - 2\sin^2 x\cos x + \sin^4 x\cos x)dx = 2[\sin x - \frac{2\sin^3 x}{3} + \frac{\sin^5 x}{5}]_0^{\pi} = 0$

43. $I = \int \csc x\,dx = \int \frac{\csc x(\csc x+\cot x)}{\csc x+\cot x}dx$. If $u = \csc x + \cot x$, then $du = (-\csc x \cot x - \csc^2 x)dx$. Hence $I = -\int \frac{du}{u} = -\ln|u| + C = -\ln|\csc x + \cot x| + C$.

44. Let $I = \int \csc^3 x\,dx$, let $u = \csc x$, $dv = \csc^2 x\,dx$; $du = -\csc x \cot x\,dx$, $v = -\cot x \Rightarrow I = -\csc x \cot x - \int \cot^2 x \csc x\,dx = -\csc x \cot x - \int(\csc^2 x - 1)\csc x\,dx = -\csc x \cot x + \int \csc x\,dx - I$; $2I = -\csc x \cot x - \ln|\csc x + \cot x| + C$; $I = -\frac{1}{2}[\csc x \cot x + \ln|\csc x + \cot x|] + C$.

45. a) f even

areas are

doubled

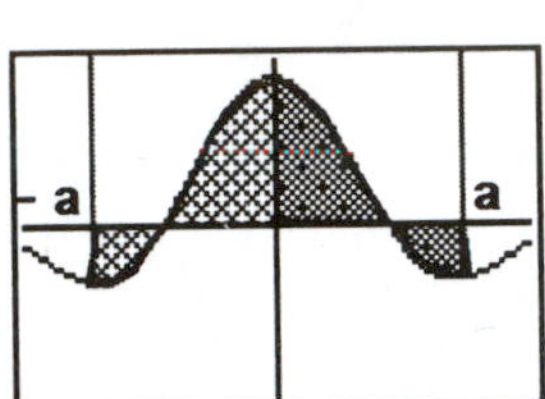

b) f odd

areas are

cancelled

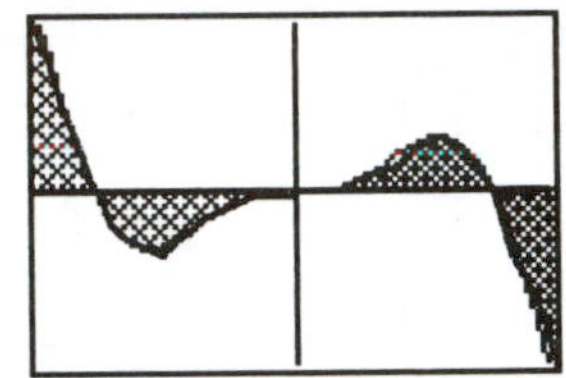

46. Let $t = -x$, $dt = -dx$; $\int_{-a}^{0} f(x)dx = -\int_{a}^{0} f(-t)dt = +\int_{a}^{0} f(t)dt = -\int_{0}^{a} f(t)dt = -\int_{0}^{a} f(x)dx$.

47. $f = x \cdot \sin x = \text{odd} \cdot \text{odd} = \text{even}$; $\int_{-\pi/2}^{\pi/2} x \sin x\,dx = 2\int_0^{\pi/2} x \sin x\,dx = 2(-x\cos x + \sin x)|_0^{\pi/2} = 2 \cdot 1 = 2$

48. $x^2 \sin x = \text{even} \cdot \text{odd} = \text{odd}$. $\int_{-\pi/2}^{\pi/2} x^2 \sin x\,dx = 0$; $\int_0^{\pi/2} x^2 \sin x\,dx = [-x^2\cos x + 2x\sin x + 2\cos x]|_0^{\pi/2}$ by (8.2 #3) $= \pi - 2$

49. $\cos x \sin 3x = \text{even} \cdot \text{odd} = \text{odd}$. $\int_{-\pi/2}^{\pi/2} f(x)dx = 0$, $\int_0^{\pi/2} \cos x \sin 3x dx = 0.5$

50. $\frac{\sin x}{\cos^2 x} = \frac{\text{odd}}{\text{even}} = \text{odd}$; $\int_{-\pi/4}^{\pi/4} f(x)dx = 0$; $\int_0^{\pi/4} \frac{\sin x}{\cos^2 x}dx = \frac{1}{\cos x}|_0^{\pi/4} = \sqrt{2} - 1$

51. Let $t = \sin x$; $-t = \sin(-x)$, $-\frac{\pi}{2} \le x \le \pi/2 \Rightarrow -1 \le t \le 1$; $f(-x) = \cos(\sin(-x)) = \cos(-t) = \cos t \Rightarrow f$ is even, $\int_{-\pi/2}^{\pi/2} \cos(\sin x)dx = 2\int_0^{\pi/2} \cos(\sin x)dx = 2 \cdot 1.202 = 2.404$.

52. $f(x)$ is even; $\int_{-\pi/4}^{\pi/4} \sqrt{1 + \tan^2 x}\,dx = 2 \cdot \int_0^{\pi/4} \sqrt{1 + \tan^2 x}\,dx = 2 \cdot 0.88137 = 1.301$

8.4 TRIGONOMETRIC SUBSTITUTIONS

1. $\int_{-2}^{2} \frac{dx}{4+x^2} = 2\int_0^2 \frac{dx}{2^2+x^2} = 2[\frac{1}{2}\tan^{-1}\frac{x}{2}]_0^2 = \frac{\pi}{4}$ by formula (8).

2. $\int_0^2 \frac{x}{8+2x^2} = \frac{1}{2}\int_0^2 \frac{dx}{4+x^2} = \frac{1}{4}\cdot 2\int_0^2 \frac{dx}{4+x^2} = \frac{1}{4}\cdot\frac{\pi}{4} = \frac{\pi}{16}$, by exercise 1.

3. $\int_0^{3/2} \frac{dx}{\sqrt{9-x^2}} = \int_0^{3/2} \frac{dx}{\sqrt{3^2-x^2}} = [\sin^{-1}\frac{x}{3}]_0^{3/2} = \frac{\pi}{6}$

4. $\int_0^{\frac{1}{2\sqrt{2}}} \frac{2dx}{\sqrt{1-4x^2}} = [[\sin^{-1} 2x]_0^{\frac{1}{2\sqrt{2}}} = \frac{\pi}{4}$

5. $\int \frac{dx}{\sqrt{x^2-4}} = \int \frac{dx}{\sqrt{x^2-2^2}}$. Set $x = 2\sec\theta$. Then $dx = 2\sec\theta\tan\theta\, d\theta$, $x^2-4 = 4\tan^2\theta$. The integral becomes $\int \frac{2\sec\theta\tan\theta\, d\theta}{|2\tan\theta|} = \ln|\sec\theta+\tan\theta| + C' = \ln|\frac{x}{2} + \sqrt{\frac{x^2-4}{4}}| + C' = \ln|x+\sqrt{x^2-4}| + C$.

6. Let $3x = \sec\theta$; $\int \frac{3dx}{\sqrt{9x^2-1}} = \int \frac{d(3x)}{\sqrt{(3x)^2-1}} = \int \frac{\sec\theta\tan\theta\, d\theta}{\sqrt{\sec^2\theta-1}} = \int \sec\theta\, d\theta = \ln|\sec\theta+\tan\theta| + C = \ln|3x+\sqrt{9x^2-1}| + C$

7. Let $x = 5\sin\theta$. $\int\sqrt{25-x^2}dx = \int 5|\cos\theta|\cdot 5\cos\theta\, d\theta = 25\int\cos^2\theta\, d\theta = 25[\frac{\theta}{2} + \frac{\sin 2\theta}{4}] + C = \frac{25}{2}[\sin^{-1}(\frac{x}{5}) + \frac{2}{4}(\frac{x}{5})\sqrt{1-(\frac{x}{5})^2}] + C = [\frac{25}{2}\sin^{-1}(\frac{x}{5}) + \frac{x}{2}\sqrt{25-x^2}] + C$.

8. Let $3x = \sin\theta$. $\int\sqrt{1-9x^2}dx = \int|\cos\theta|(\frac{1}{3})\cos\theta\, d\theta = \frac{1}{3}[\frac{\theta}{2} + \frac{\sin 2\theta}{4}] + C = [\frac{1}{6}\sin^{-1}(3x) + \frac{1}{3}\cdot\frac{1}{4}\cdot 2(3x)\sqrt{1-9x^2}] + C = [\frac{1}{6}\sin^{-1}(3x) + \frac{x}{2}\sqrt{1-9x^2}] + C$.

9. Let $x = \sin\theta$. $\int \frac{4x^2}{(1-x^2)^{3/2}}dx = \int \frac{4\sin^2\theta\cos\theta\, d\theta}{|\cos\theta|^3} = 4\int\tan^2\theta\, d\theta = 4[\tan\theta - \theta] + C = 4[\frac{x}{\sqrt{1-x^2}} - \sin^{-1}x] + C$.

10. Let $x = 2\sin\theta$. $\int \frac{dx}{(4-x^2)^{3/2}} = \int \frac{2\cos\theta\, d\theta}{8|\cos^3\theta|} = \frac{1}{4}\int\sec^2\theta\, d\theta = \frac{1}{4}\tan\theta + C = \frac{x}{4\sqrt{4-x^2}} + C$.

11. $\int_0^{\sqrt{3}/2} \frac{2dy}{1+4y^2} = \int_0^{\sqrt{3}/2} \frac{2dy}{1+(2y)^2} = [\frac{1}{1}\tan^{-1}\frac{2y}{1}]_0^{\sqrt{3}/2} = \pi/3$ by (8).

12. Let $3y = \tan\theta$; $\int_0^{1/3} \frac{3dy}{\sqrt{1+(3y)^2}} = \int_0^{\pi/4} \frac{\sec^2\theta\, d\theta}{\sqrt{1+\tan^2\theta}} = \int_0^{\pi/4}\sec\theta\, d\theta = [\ln|\sec\theta+\tan\theta|]_0^{\pi/4} = \ln(\sqrt{2}+1) - \ln 1 = \ln(\sqrt{2}+1)$.

13. Let $u = -x^2+2x$; $du = (-2x+2)dx = -2(x-1)dx$; $\int \frac{(x-1)dx}{\sqrt{2x-x^2}} = \frac{1}{-2}\int\frac{du}{\sqrt{u}} = -\frac{1}{2}\int u^{-1/2}du = \frac{-\frac{1}{2}u^{1/2}}{\frac{1}{2}} + C = -\sqrt{2x-x^2} + C$.

14. Let $u = 5 + 4x - x^2$, $du = (4 - 2x)dx = -2(x-2)dx$; $\int \frac{x-2}{\sqrt{5+4x-x^2}}dx = -\frac{1}{2}\int u^{-1/2}du = -u^{1/2} + C = -\sqrt{5+4x-x^2} + C$.

15. $\int \frac{dx}{\sqrt{x^2-2x}} = \int \frac{dx}{\sqrt{x^2-2x+1-1}} = \int \frac{dx}{\sqrt{(x-1)^2-1}}$.

Let $(x-1) = \sec\theta$. The integral becomes

$= \int \frac{\sec\theta\tan\theta\, d\theta}{|\tan\theta|} = \ln|\sec\theta + \tan\theta| + C$

$= \ln|x - 1 + \sqrt{x^2-2x}| + C$.

16. $\int \frac{dx}{\sqrt{x^2+2x}} = \int \frac{dx}{\sqrt{(x+1)^2-1}} = \int \frac{\sec\theta\tan\theta\, d\theta}{\tan\theta} = \int \sec\theta\, d\theta = \ln|\sec\theta + \tan\theta| + C = \ln|x+1+\sqrt{x^2+2x}| + C$ where $x+1 = \sec\theta$.

17. $\int_0^{3\sqrt{2}/4} \frac{dx}{\sqrt{9-4x^2}} = \frac{1}{2}\int_0^{3\sqrt{2}/4} \frac{2dx}{\sqrt{3^2-(2x)^2}} = [\frac{1}{2}\sin^{-1}\frac{2x}{3}]_0^{3\sqrt{2}/4} = \frac{1}{2}\sin^{-1}\frac{\sqrt{2}}{2} = \frac{1}{2}\cdot\frac{\pi}{4} = \pi/8$, by (9).

18. Let $x = 5\sin u$; $\int_0^5 \sqrt{25-x^2}dx = \int_0^{\pi/2}\sqrt{25-25\sin^2 u}\,5\cos u\, du = \frac{25}{2}[u + \frac{\sin 2u}{2}]_0^{\pi/2} = \frac{25\pi}{4}$.

19. Let $2z = \sec\theta$. Then $z = \frac{1}{2}\sec\theta \Rightarrow dz = \frac{1}{2}\sec\theta\tan\theta\, d\theta$ and $4z^2 - 1 = \sec^2\theta - 1 = \tan^2\theta$. Since $1/\sqrt{3} \le z \le 1$, $2/\sqrt{3} \le 2z \le 2$ and $\pi/6 \le \theta \le \pi/3$. On this range, $\tan\theta > 0$. The integral becomes $\int_{\pi/6}^{\pi/3} \frac{\sec\theta\tan\theta\, d\theta}{\frac{1}{2}\sec\theta\tan\theta} = 2(\pi/3 - \pi/6) = \pi/3$.

20. Let $x = 4\sec\theta$; $\int \frac{24dx}{x\sqrt{x^2-16}} = \int \frac{24\cdot 4\cdot\sec\theta\tan\theta\, d\theta}{4\cdot\sec\theta\cdot 4\tan\theta} = 6\theta + C$; $\int_{8/\sqrt{3}}^{8} \frac{24dx}{x\sqrt{x^2-16}} = 6\sec^{-1}\frac{x}{4}]_{8/\sqrt{3}}^{8} = 6(\sec^{-1}2 - \sec^{-1}\frac{2}{\sqrt{3}}) = 6(\pi/3 - \pi/6) = \pi$.

21. $\int_0^2 \frac{dx}{\sqrt{2^2+x^2}} = \int_0^{\pi/4} \frac{2\sec^2\theta\, d\theta}{2\sec\theta} = \int_0^{\pi/4}\sec\theta\, d\theta = [\ln(\sec\theta + \tan\theta)]_0^{\pi/4} = \ln(\sqrt{2}+1)$, where $x = 2\tan\theta$.

22. Let $x = \tan\theta$; $\int_0^1 \frac{x^3dx}{\sqrt{x^2+1}} = \int_0^{\pi/4} \frac{\tan^3\theta\sec^2\theta\, d\theta}{\sec\theta} = \int_0^{\pi/4}\tan^2\theta(\sec\theta\tan\theta\, d\theta) = \int_0^{\pi/4}(\sec^2\theta - 1)d(\sec\theta) = [\frac{\sec^3\theta}{3} - \sec\theta]_0^{\pi/4} = [\frac{(\sqrt{2})^3}{3} - \sqrt{2}] - [\frac{1}{3} - 1] = \frac{2-\sqrt{2}}{3}$.

23. Let $(x-1) = u$. Then $1 \le x \le 2 \Rightarrow 0 \le u \le 1$. The integral becomes, by (9), $\int_0^1 \frac{6du}{\sqrt{2^2-u^2}} = 6[\sin^{-1}\frac{u}{2}]_0^1 = 6\cdot\frac{\pi}{6} = \pi$.

24. Let $x = \sin\theta$; $\int_{1/2}^{1} \frac{\sqrt{1-x^2}}{x^2}dx = \int_{\pi/6}^{\pi/2} \frac{\cos\theta\cos\theta\, d\theta}{\sin^2\theta} = \int_{\pi/6}^{\pi/2}\cot^2\theta\, d\theta = \int_{\pi/6}^{\pi/2}(\csc^2\theta - 1)d\theta = [-\cot\theta - \theta]_{\pi/6}^{\pi/2} = -0 - \pi/2 + \sqrt{3} + \pi/6 = 0.685\ldots$.

25. $\int_1^3 \frac{dy}{y^2-2y+5} = \int_1^3 \frac{dy}{y^2-2y+1+4} = \int_1^3 \frac{dy}{(y-1)^2+2^2}$; let $u = y-1$; then $0 \le u \le 2$, $du = dy$, and the integral $= \int_0^2 \frac{du}{u^2+2^2} = [\frac{1}{2}\tan^{-1}\frac{u}{2}]_0^2 = \frac{\pi}{8}$.

26. $\int_1^4 \frac{dy}{y^2-2y+10} = \int_1^4 \frac{dy}{(y-1)^2+3^2} = \int_0^3 \frac{du}{u^2+3^2} = \frac{1}{3}\tan^{-1}\frac{3}{3} = \frac{\pi}{12}$ where $u = y - 1$.

27. Let $u = x^2 + 4x + 13$; then $du = (2x + 4)dx$. $\int_{-2}^{2} \frac{(x+2)dx}{\sqrt{x^2+4x+13}} = \frac{1}{2}\int_9^{25} u^{-1/2}du = [\frac{\frac{1}{2}u^{1/2}}{\frac{1}{2}}]_9^{25} = 2$. Another method of solution is to write the integrand as $\frac{(x+2)}{\sqrt{(x+2)^2+9}}$ and make the substitution $x + 2 = 3\tan\theta$. This gives: $\int_{x=-2}^{x=2} \frac{3\tan\theta 3\sec^2\theta\, d\theta}{3\sec\theta} = [3\sec\theta]_{x=-2}^{x=2} = [\sqrt{(x+2)^2+9}]_{x=-2}^{x=2} = 2$.

28. $\int_0^1 \frac{(1-x)dx}{\sqrt{8+2x-x^2}} = \int_0^1 \frac{(1-x)dx}{\sqrt{3^2-(x-1)^2}} = -\int_{x=0}^{x=1} \frac{3\sin\theta \cdot 3\cos\theta\, d\theta}{3\cos\theta} = 3[\cos\theta]_{x=0}^{x=1} = \sqrt{9}-\sqrt{8} = 3 - 2\sqrt{2}$ where $x - 1 = 3\sin\theta$.

29. $A = \int_0^1 \frac{2dx}{x^2-4x+5} = 2\int_0^1 \frac{dx}{(x-2)^2+1} = [2\tan^{-1}\frac{x-2}{1}]_0^1 = 2[\tan^{-1}(-1) - \tan^{-1}(-2)] = 0.64350\ldots$.

30. $V = \pi\int_{-2}^{11} \frac{20^2}{x^2-2x+17}dx = [400\pi\int_{-2}^{11} \frac{dx}{(x-1)^2+4^2} = 400\pi(\frac{1}{4})\tan^{-1}\frac{x-1}{4}]_{-2}^{11} = 100\pi[\tan^{-1}\frac{5}{2} - \tan^{-1}(-\frac{3}{4})] = 576.102$.

31. Let $x = \sin\theta$. Then $dx = \cos\theta\, d\theta$. The integral becomes $\int \frac{4\sin^2\theta\cos\theta\, d\theta}{(\cos^2\theta)^{3/2}} = 4\int \frac{\sin^2\theta}{\cos^2\theta}d\theta$

$= 4\int \frac{1-\cos^2\theta}{\cos^2\theta}d\theta = 4\int(\sec^2\theta - 1)d\theta$

$= 4[\tan\theta - \theta] + C = 4[\frac{x}{\sqrt{1-x^2}} - \sin^{-1}x] + C$.

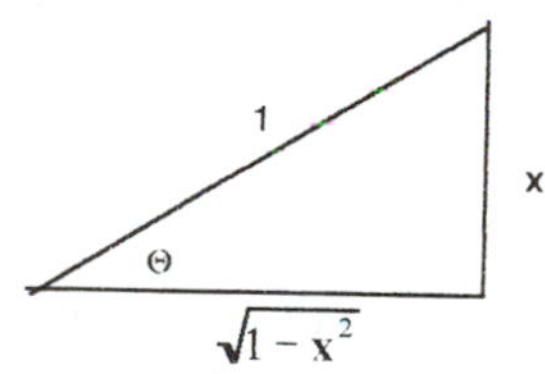

32. Let $x = 2\sin\theta$; $\int_0^1 \frac{4dx}{(4-x^2)^{3/2}} = \int_0^{\pi/6} \frac{4\cdot 2\cos\theta\, d\theta}{2^3\cos^3\theta} = \int_0^{\pi/6}\sec^2\theta\, d\theta = \tan\theta]_0^{\pi/6} = \frac{1}{\sqrt{3}}$.

33. Let $x = 3\sin\theta$; $A = \frac{1}{3}\int_0^3 \sqrt{9-x^2}dx = \frac{1}{3}\int_0^{\pi/2} 3\cos\theta\, 3\cos\theta\, d\theta = 3\int_0^{\pi/2}\cos^2\theta\, d\theta = \frac{3}{2}[\theta + \frac{\sin 2\theta}{2}]_0^{\pi/2} = \frac{3\pi}{4}$.

34. $A = \frac{1}{4-2}\int_2^4 \frac{4dx}{x^2-4x+8} = 2\int_2^4 \frac{dx}{(x-2)^2+4} = 2\cdot\frac{1}{2}[\tan^{-1}\frac{x-2}{2}]_2^4 = (\pi/4 - 0) = \pi/4$.

35. The substitution $x = 3\sin\theta$ would mean that $\sin\theta = 4/3$ at the left hand limit, which is impossible. Rewrite the integral as $-\int_4^5 \frac{dx}{x^2-9}$ and let $x = 3\sec\theta$. This leads to $-\int_{x=4}^{x=5} \frac{3\sec\theta\tan\theta\, d\theta}{9\tan^2\theta} = -\frac{1}{3}\int_{x=4}^{x=5}\csc\theta\, d\theta$ which can be done analytically. To confirm, use NINT: $(-1/3)\text{NINT}(1/\sin t, t, \cos^{-1}(3/4), \cos^{-1}(3/5)) = -0.093269\ldots$; $\text{NINT}(1/(9-x^2), x, 4, 5) = -0.093269\ldots$.

36. Let $f(x) = \frac{1}{\sqrt{2x-x^2}} = \frac{1}{\sqrt{1-(x^2-2x+1)}} = \frac{1}{\sqrt{1-(x-1)^2}}$. The domain of f is determined by the inequality $(x-1)^2 < 1$ or $2x - x^2 > 0$. The solution to this inequality is $0 < x < 2$. The domain of $\sin^{-1}t$ is $-1 \le t \le 1$. The domain of $\sin^{-1}(x-1)$ is $0 \le x \le 2$.

8.5 RATIONAL FUNCTIONS AND PARTIAL FRACTIONS

1. $\frac{5x-13}{(x-3)(x-2)} = \frac{A}{x-3} + \frac{B}{x-2}$ or $5x - 13 = A(x-2) + B(x-3)$ for all x. Set $x = 3$ to obtain $15 - 13 = A \cdot 1$ or $A = 2$; set $x = 2$ to obtain $10 - 13 = B(-1)$ or $B = 3$: $\frac{2}{x-3} + \frac{3}{x-2}$.

2. $\frac{5x-7}{x^2-3x+2} = \frac{A}{x-1} + \frac{B}{x-2} = \frac{2}{x-1} + \frac{3}{x-2}$.

3. $\frac{x+4}{(x+1)^2} = \frac{A}{x+1} + \frac{B}{(x+1)^2}$, or $x+4 = A(x+1)+B$. Rewrite, equating coefficients of corresponding terms; $x+4 = Ax + (A+B)$, and solve to find $A = 1, A+B = 4 \Rightarrow B = 3$. $\frac{1}{x+1} + \frac{3}{(x+1)^2}$.

4. $\frac{2x+2}{x^2-2x+1} = \frac{A}{x-1} + \frac{B}{(x-1)^2} = \frac{2}{x-1} + \frac{4}{(x-1)^2}$.

5. $\frac{x+1}{x^2(x-1)} = \frac{A}{x} + \frac{B}{x^2} + \frac{C}{x-1} = \frac{-2}{x} - \frac{1}{x^2} + \frac{2}{x-1}$.

6. $\frac{x}{x^3-x^2-6x} = \frac{1}{x^2-x-6} = \frac{1}{(x-3)(x+2)} = \frac{1/5}{(x-3)} + \frac{-1/5}{(x+2)}$.

7. Use long division, $x^2 - 5x + 6 \,\overline{\big)\, x^2 + \quad 8}$ $\;\overline{1\,R\,5x+2}$ to find $\frac{x^2+8}{x^2-5x+6} = 1 + \frac{5x+2}{(x-3)(x-2)} = 1 + \frac{A}{x-3} + \frac{B}{x-2} \Rightarrow 5x + 2 = A(x-2) + B(x-3)$. Set $x = 3$, then set $x = 2$ to get $17 = A$, $12 = -B$: $1 + \frac{17}{x-3} - \frac{12}{x-2}$.

8. Use long division to find $\frac{x^3+1}{x^2+4} = x + \frac{-4x+1}{x^2+4}$; x^2+4 is an irreducible quadratic.

9. $\int_0^{1/2} \frac{dx}{1-x^2} = \frac{1}{2}\int_0^{1/2}(\frac{1}{1-x} + \frac{1}{1+x})dx = \frac{1}{2}[-\ln(1-x)+\ln(1+x)]_0^{1/2} = \frac{1}{2}[-\ln(1/2)+\ln(3/2)] = \frac{1}{2}\ln 3$.

10. $\int_1^2 \frac{dx}{x^2+2x} = \frac{1}{2}\int_1^2[\frac{1}{x} - \frac{1}{x+2}]dx = \frac{1}{2}[\ln|x| - \ln|x+2|]_1^2 = \frac{1}{2}\ln(\frac{3}{2})$.

11. $\frac{y}{y^2-2y-3} = \frac{A}{(y-3)} + \frac{B}{(y+1)} = \frac{3/4}{y-3} + \frac{1/4}{y+1}$; $\int_4^8 \frac{ydy}{y^2-2y-3} = [\frac{3}{4}\ln|y-3| + \frac{1}{4}\ln y+1|]_4^8 = \frac{3}{4}\ln 5 + \frac{1}{4}\ln 9 - 0 - \frac{1}{4}\ln 5 = \frac{1}{2}\ln 5 + \frac{1}{2}\ln 3 = \frac{1}{2}\ln 15$.

12. $f(y) = \frac{y+4}{y(y+1)} = \frac{A}{y} + \frac{B}{y+1} = \frac{4}{y} + \frac{-3}{y+1} \Rightarrow \int_1^2 f(y)dy = [4\ln|y| - 3\ln|y+1|]_1^2 = 4\ln 2 - 3\ln 3 + 3\ln 2 = 7\ln 2 - 3\ln 3$.

13. Let $f(t) = \frac{1}{t^3+t^2-2t} = \frac{1}{t(t^2+t-2)} = \frac{A}{t} + \frac{B}{t+2} + \frac{C}{t-1} = \frac{-1/2}{t} + \frac{1/6}{t+2} + \frac{1/3}{t-1} \Rightarrow \int f(t)dt = -\frac{1}{2}\ln|t| + \frac{1}{6}\ln|t+2| + \frac{1}{3}\ln|t-1| + C = \frac{1}{6}\ln|\frac{(t+2)(t-1)^2}{t^3}| + C$.

14. Let $f(t) = \frac{t+3}{2t^3-8t} = \frac{t+3}{2t(t^2-4)} = \frac{A}{2t} + \frac{B}{t-2} + \frac{C}{t+2}$; $t + 3 = A \cdot (t^2 - 4) + B \cdot 2t \cdot (t + 2) + C(2t)(t - 2) \Rightarrow A = \frac{-3}{4}$, $B = \frac{5}{16}$, $C = \frac{1}{16} \Rightarrow \int f(t)dt = \frac{-3}{8} \ln|2t| + \frac{5}{16} \ln|t - 2| + \frac{1}{16} \ln|t + 2| + C$.

15. $\frac{x^3}{x^2+1} = \frac{x^3+x-x}{x^2+1} = x - \frac{x}{x^2+1}$; $\int_0^{2\sqrt{2}} \frac{x^3\,dx}{x^2+1} = \int_0^{2\sqrt{2}} x\,dx - \int_0^{2\sqrt{2}} \frac{x\,dx}{x^2+1} = [\frac{x^2}{2}]_0^{2\sqrt{2}} - \frac{1}{2}\int_0^{2\sqrt{2}} \frac{2x\,dx}{x^2+1} = 4 - [\frac{1}{2}\ln(x^2+1)]_0^{2\sqrt{2}} = 4 - \ln 3$.

16. $\int_0^1 \frac{x^4+2x}{x^2+1}dx = \int_0^1 (x^2 - 1 + \frac{2x+1}{x^2+1})dx = [\frac{x^3}{3} - x + \ln(x^2+1) + \tan^{-1} x]_0^1 = \ln 2 + \pi/4 - 2/3$.

17. $\frac{5x^2}{x^2+1} = \frac{5x^2+5-5}{x^2+1} = 5 - \frac{5}{x^2+1}$; $\int_0^{\sqrt{3}} \frac{5x^2}{x^2+1}dx = 5\int_0^{\sqrt{3}}[1 - \frac{1}{x^2+1}]dx = 5[x - \tan^{-1} x]_0^{\sqrt{3}} = 5[\sqrt{3} - \frac{\pi}{3}]$.

18. $\int_1^5 \frac{y^3+4y^2}{y^3+y}dy = \int_1^5 (1 + \frac{4y^2-y}{y^3+y})dy = \int_1^5 (1 + \frac{4y}{y^2+1} - \frac{1}{y^2+1})dy = [y + 2\ln(y^2+1) - \tan^{-1} y]_1^5 = 8.5419$.

19. Let $f(x) = \frac{1}{(x^2-1)^2} = \frac{A}{x-1} + \frac{B}{(x-1)^2} + \frac{C}{x+1} + \frac{D}{(x+1)^2} \Rightarrow 1 = A(x-1)(x+1)^2 + B(x+1)^2 + C(x+1)(x-1)^2 + D(x-1)^2$; setting $x = 1, x = -1$ gives $B = 1/4, D = 1/4$. Differentiating gives $0 = A(x+1)^2 + 2A(x-1)(x+1) + 2B(x-1) + C(x-1)^2 + 2C(x-1)(x+1) + 2D(x-1)$; setting $x = 1, x = -1$ gives $A = -1/4, C = 1/4$. Hence $\int f(x)dx = \frac{-1}{4}\ln|x-1| - \frac{1}{4}\frac{1}{(x-1)} + \frac{1}{4}\ln|x+1| - \frac{1}{4}\frac{1}{(x+1)} + C$.

20. Let $f(x) = \frac{x^2}{(x-1)(x^2+2x+1)} = \frac{x^2}{(x-1)(x+1)^2} = \frac{A}{x-1} + \frac{B}{(x+1)} + \frac{C}{(x+1)^2}$; $x^2 = A(x+1)^2 + B(x-1)(x+1) + C(x-1) = x^2(A+B) + \cdots + A - B - C$. $x = 1 \Rightarrow A = 1/4$; equating the coefficients of x^2 gives $1 = A + B$, or $B = 3/4$. $A - B - C = 0 \Rightarrow C = -1/2$. Hence $\int f(x)dx = \frac{1}{4}\ln|x-1| + \frac{3}{4}\ln|x+1| + \frac{1}{2(x+1)} + C$.

21. $\int \frac{x+4}{x^2+5x-6}dx = \frac{1}{7}\int(\frac{2}{x+6} + \frac{5}{x-1})dx = \frac{1}{7}[2\ln|x+6| + 5\ln|x-1|] + C = \frac{1}{7}\ln|(x+6)^2(x-1)^5| + C$.

22. $\int \frac{2x+1}{x^2-7x+12}dx = \int[\frac{-7}{x-3} + \frac{9}{x-4}]dx = \ln|\frac{(x-4)^9}{(x-3)^7}| + C$.

23. $\int \frac{2}{x^2-2x+2}dx = 2\int \frac{dx}{(x-1)^2+1} = 2\tan^{-1}(x-1) + C$.

24. $\int \frac{3}{x^2-4x+5}dx = 3\int \frac{dx}{(x-2)^2+1} = 3\tan^{-1}(x-2) + C$.

25. $\frac{x^2-2x-2}{x^3-1} = \frac{A}{x-1} + \frac{Bx+C}{x^2+x+1} \Rightarrow x^2 - 2x - 2 = A(x^2+x+1) + (Bx+C)(x-1)$ evaluating at $x = 1 \Rightarrow A = -1$, evaluating at $x = 0 \Rightarrow C = 1$, evaluating at $x = 2 \Rightarrow B = 2$. $\int \frac{x^2-2x-2}{x^3-1}dx = -\int \frac{dx}{x-1} + \int \frac{2x+1}{x^2+x+1}dx = -\ln|x-1| + \ln|x^2+x+1| + C = \ln|\frac{x^2+x+1}{x-1}| + C$.

26. $\frac{x^2-4x+4}{x^3+1} = \frac{A}{x+1} + \frac{Bx+C}{x^2-x+1} = \frac{3}{x+1} + \frac{-2x+1}{x^2-x+1}$; $\int \frac{x^2-4x+4}{x^3+1}dx = 3\ln|x+1| - \ln|x^2-x+1| + C$.

27. $\frac{2x^4+x^3+16x^2+4x+32}{(x^2+4)^2} = 2 + \frac{x^3+4x}{(x^2+4)^2} = 2 + \frac{x}{x^2+4}$. Hence, the integral is $2x + \frac{1}{2}\ln(x^2+4) + C$.

28. $\frac{x^4-3x^3+2x^2-3x+1}{(x^2+1)^2} = 1 + \frac{-3x^3-3x}{(x^2+1)^2} = 1 - \frac{3x}{x^2+1}$; the integral is $x - \frac{3}{2}\ln(x^2+1)$

29. $\int_0^1 \frac{x}{x+1}dx = \int_0^1 (1 - \frac{1}{x+1})dx = [x - \ln|x+1|]|_0^1 = 1 - \ln 2 = 0.307$.

30. $\int_0^1 \frac{x^2+1-1}{x^2+1}dx = \int_0^1 (1 - \frac{1}{x^2+1})dx = [x - \tan^{-1} x]_0^1 = 1 - \pi/4$.

31. $\int_{\sqrt{2}}^3 \frac{2x^3}{x^2-1}dx = \int_{\sqrt{2}}^3 (2x + \frac{2x}{x^2-1})dx = [x^2 + \ln|x^2-1|]|_{\sqrt{2}}^3 = 9 + \ln 8 - (2) = 7 + \ln 8$.

32. $\int_{-1}^3 \frac{4x^2-7}{2x+3}dx = \int_{-1}^3 (2x - 3 + \frac{2}{2x+3})dx = [x^2 - 3x + \ln|2x+3|]|_{-1}^3 = -4 + \ln 9$.

33. $V = \pi\int_{1/2}^{5/2} \frac{9}{3x-x^2}dx = 9\pi\int_{1/2}^{5/2} \frac{dx}{(3-x)x} = \frac{9\pi}{3}\int_{1/2}^{5/2} (\frac{1}{3-x} + \frac{1}{x})dx$; $\frac{1}{(3-x)(x)} = \frac{A}{3-x} + \frac{B}{x}$; $B = 1/3$, $A = 1/3$; $V = 3\pi[-\ln|3-x| + \ln|x|]|_{1/2}^{5/2} = 3\pi[[-\ln\frac{1}{2} + \ln\frac{5}{2}] - [-\ln\frac{5}{2} + \ln\frac{1}{2}]] = 3\pi[2\ln 5/2 - 2\ln 1/2] = 3\pi\ln 25$.

34. $dy = \frac{-2x}{1-x^2}dx \Rightarrow L = \int_0^{1/2}\sqrt{1 + \frac{4x^2}{(1-x^2)^2}}dx = \int_0^{1/2} \frac{1+x^2}{1-x^2}dx = \int_0^{1/2}(-1 + \frac{2}{1-x^2})dx = \int_0^{1/2}(-1 + \frac{1}{1-x} + \frac{1}{1+x})dx = [-x - \ln|1-x| + \ln|1+x|]|_0^{1/2} = \ln 3 - 1/2$.

35. (a) Rewrite the equation as (*) $\frac{250dx}{x(1000-x)} = dt$. To find the antiderivative of the left-hand-side, use the method of partial fractions to write $\frac{250}{x(1000-x)} = \frac{A}{x} + \frac{B}{1000-x} = \frac{1/4}{x} + \frac{1/4}{1000-x}$. Hence $\int \frac{250}{x(1000-x)}dx = \frac{1}{4}[\ln x - \ln(1000-x)] + C' = \frac{1}{4}\ln[\frac{Cx}{1000-x}]$. Since the integral of $dt = t$; $\frac{1}{4}\ln\frac{Cx}{1000-x} = t$ or $\frac{Cx}{1000-x} = e^{4t}$; when $t = 0$, $x = 2$; hence $\frac{C\cdot 2}{998} = 1$ or $C = 499$. Solve for x: $499x = (1000-x)e^{4t} \Rightarrow x = \frac{1000e^{4t}}{499+e^{4t}}$.

(b) Set $x = 500$ and solve for t: $t = \frac{1}{4}\ln(499) = 1.55$ days.

36. (a) (i) Assume $a = b$; $\frac{dx}{(a-x)^2} = kdt \Rightarrow \int \frac{dx}{(x-a)^2} = \int kdt \Rightarrow \frac{(x-a)^{-1}}{-1} = kt + C'$, or $x - a = \frac{-1}{kt+C}$. When $t = 0$, $x = 0 \Rightarrow -a = -\frac{1}{C}$ on $C = 1/a$. $x = a - \frac{a}{akt+1} = \frac{a^2kt}{akt+1}$.

(ii) If $a \neq b$, $\int \frac{dx}{(a-x)(b-x)} = \int(\frac{A}{a-x} + \frac{B}{b-x})dx = \frac{1}{b-a}\int(\frac{1}{a-x} - \frac{1}{b-x})dx$. As in (i), integrating gives $\frac{1}{(b-a)} \ln|\frac{b-x}{a-x}| = kt + C'$ or, $\ln|\frac{b-x}{a-x}| = (b-a)kt + C'' \Rightarrow \frac{b-x}{a-x} = Ce^{(b-a)kt} = \frac{b}{a}e^{(b-a)kt}$ when C is evaluated. Solving for x gives $x = \frac{ab[e^{(b-a)kt}-1]}{be^{(b-a)kt}-a}$.
(b) $[0,2]$ by $[0,15]$ shows a complete graph.

37. $\int_0^{\pi/2} \frac{dx}{1+\sin x} = \int_0^1 \frac{2dz}{(1+z^2)(1+\frac{2z}{1+z^2})} = 2\int_0^1 \frac{dz}{1+z^2+2z} = 2\int_0^1 \frac{dz}{(z+1)^2} = 2[\frac{(z+1)^{-1}}{-1}]_0^1 = 2(-\frac{1}{2}+1) = 1$

38. $\int_{\pi/3}^{\pi/2} \frac{\pi}{2}\frac{dx}{1-\cos x} = \frac{\pi}{2}\int_{1/\sqrt{3}}^1 \frac{2dz}{(1+z^2)[2-\frac{2}{(-1+z^2)}]} = \frac{\pi}{2}\int_{1/\sqrt{3}}^1 \frac{dz}{z^2} = [-\frac{\pi}{2}z^{-1}]_{1/\sqrt{3}}^1 = \frac{\pi}{2}(\sqrt{3}-1)$

39. $\int \frac{dx}{1-\sin x} = 2\int \frac{dz}{(z-1)^2} = \frac{2}{-1}\frac{1}{(z-1)} + C = \frac{2}{1-\tan(x/2)} + C$

40. $\int \frac{dx}{2+\cos x} = \int \frac{2dz}{(1+z^2)[1+\frac{2}{1+z^2}]} = 2\int \frac{dz}{z^2+3} = \frac{2}{\sqrt{3}}\tan^{-1}\frac{z}{\sqrt{3}} + C = \frac{2}{\sqrt{3}}\tan^{-1}[\frac{\tan(\frac{x}{2})}{\sqrt{3}}] + C$

41. $\int \frac{\cos x\, dx}{1-\cos x} = \int \frac{1-z^2}{1+z^2}\frac{1}{[1-\frac{1-z^2}{1+z^2}]}\frac{2dz}{1+z^2} = 2\int \frac{(1-z^2)dx}{(1+z^2-1+z^2)(1+z^2)} = 2\int \frac{(1-z^2)}{2z^2(1+z^2)}dz = 2\int \frac{1+z^2-2z^2}{2z^2(1+z^2)}dz = \int \frac{dz}{z^2} - 2\int \frac{dz}{1+z^2} = -\frac{1}{z} - 2\tan^{-1} z + C = -\cot(x/2) - 2(x/2) + C = -\cot(x/2) - x + C$

42. $\int \frac{dx}{1+\sin x+\cos x} = \int \frac{2dz}{(1+z^2)(\frac{2}{1+z^2}+\frac{2z}{1+z^2})} = \int \frac{dz}{1+z} = \ln|1+z| + C = \ln|1+\tan(\frac{x}{2})| + C$

43. $\int \frac{dx}{\sin x-\cos x} = \int \frac{2dz}{(1+z^2)}\frac{1}{\frac{2z-(1-z^2)}{(1+z^2)}} = 2\int \frac{dz}{z^2+2z-1} = 2\int \frac{dz}{z^2+2z+1-2} = 2\int \frac{dz}{(z+1)^2-2} = (-2)\frac{1}{2\sqrt{2}}\ln|\frac{z+1+\sqrt{2}}{z+1-\sqrt{2}}| + C = \frac{1}{\sqrt{2}}\ln|\frac{z+1-\sqrt{2}}{z+1+\sqrt{2}}| + C = \frac{1}{\sqrt{2}}\ln|\frac{\tan(x/2)+1-\sqrt{2}}{\tan(x/2)+1+\sqrt{2}}| + C$

44. $\int_{\pi/2}^{2\pi/3} \frac{dx}{\sin x+\tan x} = \int_{\pi/2}^{2\pi/3} \frac{dx}{\sin x(1+\frac{1}{\cos x})} = \int_1^{\sqrt{3}} \frac{2dz}{(1+z^2)\frac{2z}{(1+z^2)}(1+\frac{1+z^2}{1-z^2})} = \int_1^{\sqrt{3}} \frac{1-z^3}{2z}dz = [\frac{1}{2}\ln|z| - \frac{z^2}{4}]_1^{\sqrt{3}} = \frac{1}{2}\ln\sqrt{3} - \frac{3}{4} + \frac{1}{4} = \frac{1}{2}(\ln\sqrt{3} - 1)$

8.6 IMPROPER INTEGRALS

1. $\int_0^\infty \frac{dx}{x^2+4} = \lim_{b\to\infty} \frac{1}{2}[\tan^{-1}\frac{x}{2}]_0^b = \frac{1}{2}\lim_{b\to\infty}\tan^{-1}\frac{b}{2} - \frac{1}{2}\tan^{-1}0 = \frac{1}{2}\cdot\frac{\pi}{2} = \frac{\pi}{4}$.

2. $\int_0^1 \frac{dx}{\sqrt{x}} = \lim_{b\to 0^+} 2\sqrt{x}]_b^1 = 2 - \lim_{b\to 0^+} 2\sqrt{b} = 2$.

3. $\int_{-1}^1 \frac{dx}{x^{2/3}} = \int_{-1}^0 \frac{dx}{x^{2/3}} + \int_0^1 \frac{dx}{x^{2/3}} = \lim_{b\to 0^-}\int_{-1}^b \frac{dx}{x^{2/3}} + \lim_{c\to 0^+}\int_c^1 \frac{dx}{x^{2/3}} = \lim_{b\to 0^-}[3x^{1/3}]_{-1}^b + \lim_{c\to 0^+}[3x^{1/3}]_c^1 = 0 - (-3) + 3 - 0 = 6$.

4. $\int_1^\infty \frac{dx}{x^{1.001}} = \lim_{b\to\infty} \int_1^b \frac{dx}{x^{1.001}} = \lim_{b\to\infty}[x^{-0.001}/(-0.001)]_1^b = 0 + \frac{1}{0.001} = 1000.$

5. $\int_0^4 \frac{dx}{\sqrt{4-x}} = \lim_{b\to 4-} \int_0^b (4-x)^{-1/2}dx = \lim_{b\to 4-} -[\frac{2(4-x)^{1/2}}{}]_0^b = 4.$

6. $\int_0^1 \frac{dx}{\sqrt{1-x^2}} - \lim_{b\to 1-} \int_0^b \frac{dx}{\sqrt{1-x^2}} = \lim_{b\to 1-}[\sin^{-1} x]_0^b = \frac{\pi}{2}.$

7. $\int_0^1 \frac{dx}{x^{0.999}} = \lim_{b\to 0+} \int_b^1 x^{-0.999}dx = \lim_{b\to 0+} \frac{x^{0.001}}{0.001}]_b^1 = 1000.$

8. $\int_{-\infty}^2 \frac{dx}{4-x} = \lim_{c\to-\infty} \int_c^2 \frac{dx}{4-x} = \lim_{c\to-\infty}[-\ln|4-x|]_c^2 =$
$\lim_{c\to-\infty}[-\ln 2 + \ln|4-c|] = \infty$, diverges.

9. $\int_2^\infty \frac{2}{x^2-x}dx = \lim_{b\to\infty} \int_2^b \frac{2}{x^2-x}dx = \lim_{b\to\infty} \int_2^b[\frac{-2}{x}+\frac{2}{x-1}]dx = \lim_{b\to\infty}[-2\ln|x|+$
$2\ln|x-1|]_2^b = \lim_{b\to\infty}[2\ln|\frac{x-1}{x}|]_2^b = 0 - 2\ln(\frac{1}{2}) = \ln 4.$

10. $I = \int_0^\infty \frac{dx}{(1+x)\sqrt{x}} = \int_0^1 \frac{dx}{(1+x)\sqrt{x}} + \int_1^\infty \frac{dx}{(1+x)\sqrt{x}} = \lim_{c\to 0+} \int_c^1 \frac{dx}{(1+x)\sqrt{x}} +$
$\lim_{b\to\infty} \int_1^b \frac{dx}{(1+x)\sqrt{x}}$. Let $u = \sqrt{x}$, $du = \frac{dx}{2\sqrt{x}} \Rightarrow \int \frac{dx}{(1+x)\sqrt{x}} = 2\int \frac{du}{1+u^2} =$
$2\tan^{-1} u$. Hence $I = \lim_{c\to 0+}[2\tan^{-1}\sqrt{x}]_c^1 + \lim_{b\to\infty}[2\tan^{-1}\sqrt{x}]_1^b = 2\cdot\frac{\pi}{4} -$
$0 + 2\cdot\frac{\pi}{2} - 2\cdot\frac{\pi}{4} = \pi.$

11. $\int_1^\infty \frac{dx}{x^{1/3}} = \lim_{b\to\infty} \frac{x^{2/3}}{2/3}]_1^b$; diverges.

12. $\int_1^\infty \frac{dx}{x^3} = \lim_{b\to\infty}[\frac{x^{-2}}{-2}]_1^b$; converges (see the paragraph after Explanation 3).

13. $\int_1^\infty \frac{dx}{x^3+1}$ converges by the Domination Test; $\frac{1}{x^3+1} < \frac{1}{x^3}$; see #12.

14. $\int_0^\infty \frac{dx}{x^3} = \int_0^1 \frac{dx}{x^3} + \int_1^\infty \frac{dx}{x^3}$; the first integral $= \lim_{c\to 0+}[\frac{x^{-2}}{2}]_c^1$ which diverges; diverges.

15. $\int_0^\infty \frac{dx}{x^{3/2}+1} = \int_0^1 \frac{dx}{x^{3/2}+1} + \int_1^\infty \frac{dx}{x^{3/2}+1}$; the first integral is finite, the second is dominated by $\int_1^\infty \frac{dx}{x^{3/2}}$ which converges; converges.

16. $\int_0^\infty \frac{dx}{1+e^x} = \int_1^\infty \frac{du}{u+u^2}$, where $u = e^x$; this integral is dominated by $\int_1^\infty \frac{du}{u^2}$ which converges; converges.

17. $\int_0^{\pi/2} \tan x\, dx = \lim_{b\to\frac{\pi}{2}-} \int_0^b \frac{\sin x}{\cos x}dx = \lim_{b\to\frac{\pi}{2}-}[-\ln|\cos x|]_0^b$; diverges.

18. $\int_{-1}^1 \frac{dx}{x^2} = \int_{-1}^0 \frac{dx}{x^2} + \int_0^1 \frac{dx}{x^2}$; both integrals diverge; diverges.

19. $\int_{-1}^1 \frac{dx}{x^{2/5}} = \int_{-1}^0 \frac{dx}{x^{2/5}} + \int_0^1 \frac{dx}{x^{2/5}}$; both integrals converge; converges.

20. $\int_0^\infty \frac{dx}{\sqrt{x}} = \int_0^1 \frac{dx}{\sqrt{x}} + \int_1^\infty \frac{dx}{\sqrt{x}}$; the second integral diverges; diverges.

21. $\int_2^\infty \frac{dx}{\sqrt{x-1}} = \lim_{b\to\infty}[2(x-1)^{1/2}]_2^b$; diverges.

22. $\int_1^\infty \frac{5}{x}dx = \lim_{b\to\infty}[5\ln|x|]_1^b$; diverges.

23. $\int_0^2 \frac{dx}{1-x^2} = \int_0^1 \frac{dx}{1-x^2} + \int_1^2 \frac{dx}{1-x^2} = \lim_{b\to 1^-} \frac{1}{2}\int_0^b[\frac{1}{1-x} + \frac{1}{1+x}]dx + \lim_{c\to 1^+} \frac{1}{2}\int_c^2[\frac{1}{1-x} + \frac{1}{1+x}]dx = \lim_{b\to 1^-}[\frac{1}{2}\ln|\frac{1+x}{1-x}|]_0^b + \lim_{c\to 1^+}[\frac{1}{2}\ln|\frac{1+x}{1-x}|]_c^2$; both integrals diverge; diverges.

24. $\int_2^\infty \frac{dx}{(x+1)^2}$ is dominated by $\int_2^\infty \frac{dx}{x^2}$ which converges.

25. $\int_0^\infty \frac{dx}{\sqrt{x^6+1}} = \int_0^1 \frac{dx}{\sqrt{x^6+1}} + \int_1^\infty \frac{dx}{\sqrt{x^6+1}}$; the first integral is finite, the second is dominated by $\int_1^\infty \frac{dx}{x^3}$ which converges; converges.

26. $\int_{-1}^1 \frac{dx}{x^{1/3}} = \int_{-1}^0 \frac{dx}{x^{1/3}} + \int_0^1 \frac{dx}{x^{1/3}}$; both integrals converge.

27. $\int_0^\infty x^2 e^{-x}dx = \lim_{b\to\infty}[e^{-x}(-x^2-2x-2)]_0^b = \lim_{b\to\infty}[\frac{-b^2-2b-2}{e^b}] + 2 = 2$; converges.

28. $\int_1^\infty \frac{\sqrt{x+1}}{x^2}dx$; by the limit Comparison Test, $\lim_{x\to\infty} \frac{\frac{\sqrt{x+1}}{x^2}}{\frac{1}{x^{3/2}}} = 1 \Rightarrow$ the integral converges.

29. $\frac{1}{x} \le \frac{2+\cos x}{x}$; $\int_\pi^\infty \frac{dx}{x}$ diverges $\Rightarrow \int_\pi^\infty \frac{2+\cos x}{x}dx$ diverges by the Domination Test.

30. $\int_1^\infty \frac{\ln x}{x}dx = \int_0^\infty u\,du$, where $u = \ln x$; diverges.

31. Diverges by the Limit Comparison Test since $\int_6^\infty \frac{1}{\sqrt{x}}dx$ diverges and $\lim_{x\to\infty} \frac{\frac{1}{\sqrt{x}}}{\frac{1}{\sqrt{x+5}}} = 1$.

32. Diverges by the Limit Comparison Test, (cf #31), since $\lim_{x\to\infty} \frac{\frac{1}{\sqrt{x}}}{\frac{1}{\sqrt{2x+10}}} = \sqrt{2}$.

33. Converges by the Limit Comparison test since $\int_2^\infty \frac{dx}{x^2}$ converges and $\lim_{x\to\infty} \frac{\frac{1}{x^2}}{\frac{2}{x^2-1}} = \frac{1}{2}$.

34. $\int_1^\infty \frac{1}{e^{\ln x}}dx = \int_1^\infty \frac{dx}{x}$ which diverges.

35. $2 \le x \Rightarrow \ln x \le x \Rightarrow \frac{1}{x} \le \frac{1}{\ln x}$; diverges by the Domination Test since $\int_2^\infty \frac{dx}{x}$ diverges.

36. Converges by the the Limit Comparison Test since $\int_1^\infty \frac{1}{\sqrt{e^x}}dx = \int_1^\infty e^{-x/2}dx = \lim_{b\to\infty}[-2e^{-x/2}]_1^b$ converges and $\lim_{x\to\infty} \frac{\frac{1}{\sqrt{e^x}}}{\frac{1}{\sqrt{e^x-x}}} = \lim_{x\to\infty} \sqrt{\frac{e^x-x}{e^x}} = 1$.

37. Converges by the Limit Comparison Test since $\int_1^\infty \frac{dx}{e^x}$ converges and $\lim_{x\to\infty} \frac{\frac{1}{e^x}}{\frac{1}{e^x-2^x}} = \lim_{x\to\infty}(1-(\frac{2}{e})^x) = 1$.

38. Converges by the Limit Comparison Test since $\int_2^\infty \frac{dx}{x^3}$ converges and $\lim_{x\to\infty} \frac{\frac{1}{x^3}}{\frac{1}{x^3-5}} = 1$.

39. $\int_0^\infty \frac{dx}{\sqrt{x+x^4}} = \int_0^1 \frac{dx}{\sqrt{x+x^4}} + \int_1^\infty \frac{dx}{\sqrt{x+x^4}}$; $\frac{1}{\sqrt{x+x^4}} < \frac{1}{\sqrt{x^4}} = \frac{1}{x^2}$ and $\int_1^\infty \frac{dx}{x^2}$ converges $\Rightarrow$ the second integral converges. $\frac{1}{\sqrt{x+x^4}} < \frac{1}{\sqrt{x}}$ and $\int_0^1 \frac{1}{\sqrt{x}}$ converges $\Rightarrow$ the first integral converges. Converges

40. Since $e^{-x}\cos x$ can be negative, the usual tests cannot be directly applied. $\int_0^\infty e^{-x}\cos x\,dx = \lim_{b\to\infty}\int_0^b e^{-x}\cos x\,dx$; $|\int_0^b e^{-x}\cos x\,dx| < \int_0^b |e^{-x}\cos x|dx \le \int_0^b e^{-x}dx$ which converges.

41. $\int_0^\infty e^{-2x}dx = \lim_{b\to\infty}[-\frac{1}{2}e^{-2x}]_0^b$; converges.

42. $\int_{-5}^\infty e^{-3x}dx = \lim_{b\to\infty}[-\frac{1}{3}e^{-3x}]_{-5}^b$; converges.

43. $\int_{-3}^\infty x^2e^{-2x}dx = \lim_{b\to\infty}[e^{-2x}(-\frac{1}{2}x^2 - \frac{1}{2}x - \frac{1}{4})]_{-3}^b$ converges (use L'Hôpital's Rule).

44. $\int_{-2}^\infty xe^{-x}dx = \lim_{b\to\infty}[e^{-x}(-x-1)]_{-2}^b$; converges.

45. a)

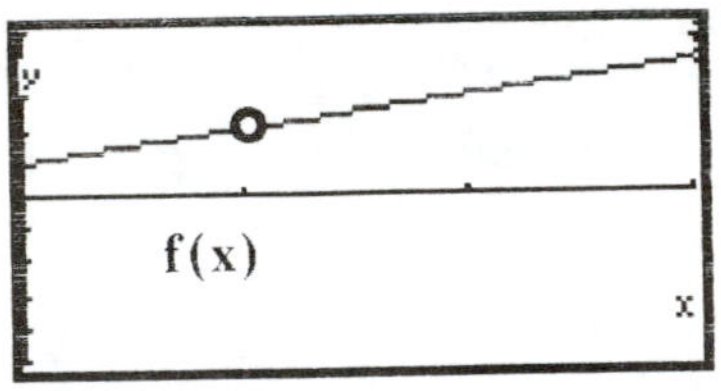

f(x)

b) $\int_0^3 \frac{x^2-1}{x-1}dx = \int_0^1 \frac{x^2-1}{x-1}dx + \int_1^3 \frac{x^2-1}{x-1}dx$
$= \lim_{b\to1^-}[\frac{x^2}{2}+x]_0^b + \lim_{b\to1^+}[\frac{x^2}{2}+x]_b^3$
$= 7.5$.

46. a)

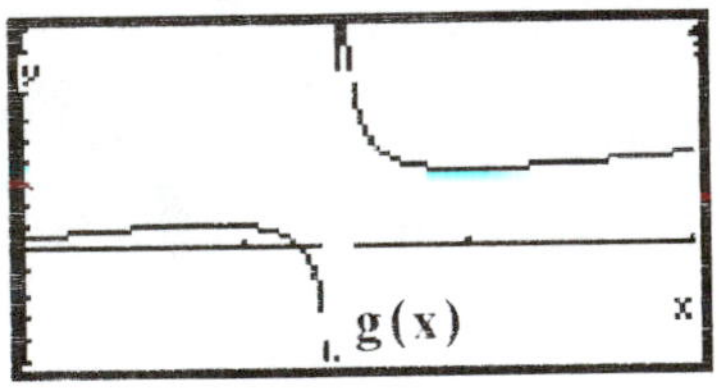

g(x)

b) $\int_0^3 g(x)dx = \int_0^{\sqrt{2}} g(x)dx + \int_{\sqrt{2}}^3 g(x)dx$;
$g(x) = x + 1 + \frac{1}{x^2-2} =$
$x + 1 + \frac{1}{2\sqrt{2}}[\frac{1}{x-\sqrt{2}} - \frac{1}{x+\sqrt{2}}]$.

Hence $\int_0^3 g(x)dx = \lim_{b\to\sqrt{2}^-}[\frac{x^2}{2} + x + \frac{1}{2\sqrt{2}}\ln|\frac{x-\sqrt{2}}{x+\sqrt{2}}|]_0^b + \lim_{b\to\sqrt{2}^+}[\frac{x^2}{2} + x + \frac{1}{2\sqrt{2}}\ln|\frac{x-\sqrt{2}}{x+\sqrt{2}}|]_b^3$; both integrals diverge.

47. $x > 3 \Rightarrow e^{-x\cdot x} < e^{-3\cdot x} \Rightarrow \int_3^\infty e^{-x^2}dx < \int_3^\infty e^{-3x}dx = \lim_{b\to\infty}[-\frac{1}{3}e^{-3x}]_3^b = \frac{1}{3}e^{-9} = 0.000042$. Using NINT, $\int_0^3 e^{-x^2}dx = 0.886217\ldots$; thus $\int_0^\infty e^{-x^2}dx = 0.886217$ with an error of at most 0.000042.

48. $\int_0^\infty \pi(\frac{1}{x})^2dx = \lim_{b\to\infty}[\pi\frac{x^{-1}}{-1}]_0^b = \pi$.

49. $\int_1^\infty x^{-p}dx = \lim_{b\to\infty}[\frac{x^{-p+1}}{1-p}]_1^b = -\frac{1}{1-p} = \frac{1}{p-1}$ if $p > 1$; if $p < 1$, $-p+1 > 1$ and the integral is infinite.

50. $\int_0^1 \frac{dx}{x^p} = \lim_{b\to 0^-}[\frac{x^{-p+1}}{-p+1}]_b^1 = \frac{1}{1-p}$ when $p < 1$; if $p > 1$ the integral is infinite.

51. Let $u = \ln x$; $du = \frac{dx}{x} \Rightarrow \int_1^2 \frac{dx}{x(\ln x)^p} = \int_0^{\ln 2} \frac{du}{u^p}$; the integral converges when $p < 1$ and diverges if $p \geq 1$. (cf #50).

52. Let $u = \ln x$; $du = \frac{dx}{x} \Rightarrow \int_2^\infty \frac{dx}{x(\ln x)^p} = \int_{\ln 2}^\infty \frac{du}{u^p}$ which converges for $p > 1$, diverges for $p \leq 1$ (cf #49).

53. $A = \lim_{b\to\infty} \int_0^b e^{-x}dx = \lim_{b\to\infty}[-\frac{1}{e^b} + \frac{1}{e^0}] = 1$.

54. $M = 1$ from #53. $M_x = \int_0^\infty \frac{e^{-x}}{2} \cdot e^{-x}dx = \frac{1}{2}\int_0^\infty e^{-2x}dx = -\frac{1}{4}\lim_{b\to\infty}[e^{-2x}]_0^b = \frac{1}{4}$; $M_y = \int_0^\infty xe^{-x}dx = \lim_{b\to\infty}[e^{-x}(-x-1)]_0^b = 1$; $\bar{x} = \frac{M_y}{M} = 1$, $\bar{y} = \frac{M_x}{M} = \frac{1}{4}$.

55. (use shells) $V = 2\pi \int xe^{-x}dx = 2\pi \cdot 1$ (see #54).

56. (use disks) $V = \pi \int_0^\infty (e^{-x})^2 dx = -\frac{\pi}{2}\lim_{b\to\infty}[e^{-2x}]_0^b = \frac{\pi}{2}$.

57. $A = \int_0^{\pi/2}(\sec x - \tan x)dx$

$= \lim_{b\to(\frac{\pi}{2})^-} \int_0^b (\sec x - \tan x)dx$

$= \lim_{b\to(\frac{\pi}{2})^-} [\ln|\sec x \tan x| + \ln|\cos x|]_0^b$

$= \lim_{b\to(\frac{\pi}{2})^-} [\ln|1 + \sin x|]_0^b = \ln 2$.

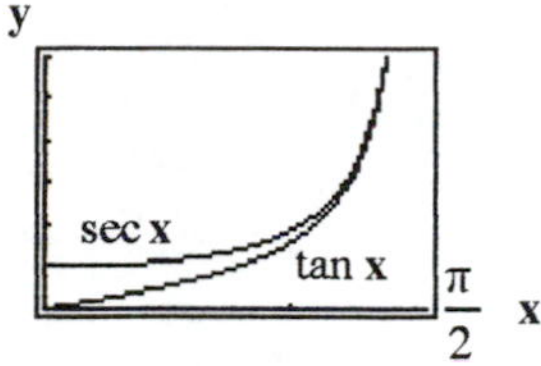

58. By symmetry $\int_{-\infty}^\infty \frac{1}{1+x^2}dx = 2\int_0^\infty \frac{1}{1+x^2}dx = \lim_{b\to\infty} 2\int_0^b \frac{1}{1+x^2}dx = \lim_{b\to\infty} 2[\tan^{-1} x]_0^b = 2 \cdot \frac{\pi}{2} = \pi$ which is the area of the unit disk.

59. $\int_2^\infty \frac{dx}{\sqrt{x^2+1}}$ diverges by the Limit Comparison Theorem since $\lim_{x\to\infty} \frac{\frac{1}{x}}{\frac{1}{\sqrt{x^2+1}}} = 1$ and $\int_2^\infty \frac{dx}{x}$ diverges.

60. By symmetry $\int_{-\infty}^\infty \frac{dx}{\sqrt{x^6+1}} = 2\int_0^\infty \frac{dx}{\sqrt{x^6+1}}$ which converges (see #25).

61. Let $u = e^x$; $\int_{-\infty}^\infty \frac{dx}{e^x+e^{-x}} = \int_0^\infty \frac{du}{u^2+1} = \lim_{b\to\infty} \tan^{-1} u|_0^b = \frac{\pi}{2}$; converges.

62. $\int_{-\infty}^\infty \frac{e^{-x}}{x^2+1}dx = \int_{-\infty}^{-1} \frac{e^{-x}}{x^2+1}dx + \int_{-1}^\infty \frac{e^{-x}}{x^2+1}dx$; we show the first integral diverges: $\int_{-\infty}^{-1} \frac{e^{-x}dx}{x^2+1} = -\int_\infty^1 \frac{e^z dz}{z^2+1} = \int_1^\infty \frac{e^z dz}{z^2+1}$; however $\frac{e^z}{z^2+1} > \frac{1}{z}$ for $z > 1$ (verify with a graph). Hence the original integral diverges.

63. By the Limit Comparison Test, $\int_0^\infty \frac{2x}{x^2+1}dx$ and $\int_0^\infty \frac{2}{x}dx$ both diverge. However $\int_{-b}^{b} \frac{2x}{x^2+1}dx = 0$ for all finite b since the integrand is odd.

64. The triangle formed by the x-axis, $y = 1 + \frac{2}{3}x$, and $y = 1 - \frac{2}{3}x$ has area $\frac{1}{2} \cdot 3 \cdot 1 = 1.5$

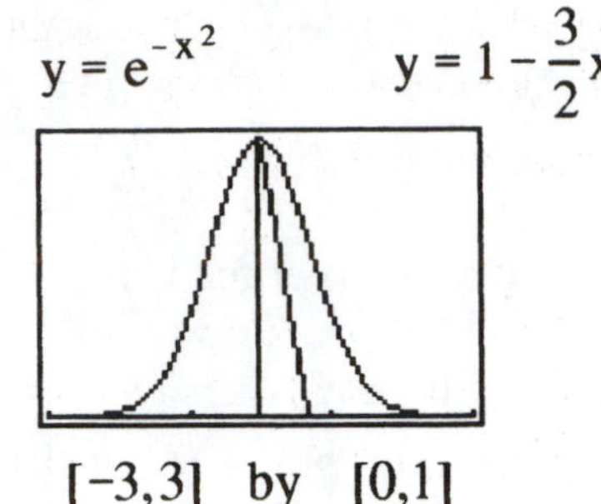

[-3,3] by [0,1]

8.7 DIFFERENTIAL EQUATIONS

1. a) $y = x^2$, $y' = 2x$, $y'' = 2 \Rightarrow xy'' - y' = x \cdot 2 - 2x = 0$; b) $y = 1$, $y' = 0$, $y'' = 0 \Rightarrow xy'' - y' = x \cdot 0 - 0 = 0$; c) $y = c_1x^2 + c_2$, $y' = 2c_1x$, $y'' = 2c_1 \Rightarrow xy'' - y' = x \cdot 2c_1 - 2c_1x = 0$.

2. a) $y' + \frac{1}{x}y = \frac{1}{2} + \frac{1}{x} \cdot \frac{x}{2} = 1$; $y' + \frac{1}{x}y = (-\frac{1}{x^2} + \frac{1}{2}) + \frac{1}{x}(\frac{1}{x} + \frac{x}{2}) = 1$; c) $y' + \frac{1}{x}y = (-\frac{C}{x^2} + \frac{1}{2}) + \frac{1}{x}(\frac{C}{x} + \frac{x}{2}) = 1$.

3. a) $y = e^{-x}$, $y' = e^{-x}$, $y'' = e^{-x} \Rightarrow 2y' + 3y = 2(-e^{-x}) + 3e^{-x} = e^{-x}$; b) $y = e^{-x} + e^{-(3/2)x}$, $y' = -e^{-x} - \frac{3}{2}e^{-(3/2)x}$, $y'' = e^{-x} + \frac{9}{4}e^{-(3/2)x} \Rightarrow 2y' + 3y = 2(-e^{-x} - \frac{3}{2}e^{-(3/2)x}) + 3(e^{-x} + e^{-(3/2)x}) = e^{-x}$; c) $y = e^{-x} + Ce^{-(3/2)x}$, $y' = -e^{-x} - \frac{3}{2}Ce^{-(3/2)x}$, $y'' = e^{-x} + \frac{9}{4}Ce^{-(3/2)x} \Rightarrow 2y' + 3y = 0$.

4. a) $yy'' = 1 \cdot 0 = 0$; $2(y')^2 - 2y' = 0 - 0 = 0$; b) $yy'' = \tan x(2\sec^2 x \tan x) = 2\sec^2 x \tan^2 x$; $2(y')^2 - 2y' = 2\sec^4 x - 2\sec^2 x = 2\sec^2 x(\sec^2 x - 1) = 2\sec^2 x \tan^2 x$.

5. $y = 160x - 16x^2 \Rightarrow y' = 160 - 32x$, $y'' = -32$; $y'' = -32$, thus y is a solution of the differential equation. $y(5) = 160 \cdot 5 - 16 \cdot 25 = 400$; $y'(5) = 160 - 32 \cdot 5 = 0$.

6. $y = 2(1 - 3^{-3t/2})$, $y' = 3e^{-3t/2}$; $y(0) = 2(1 - 1) = 0$; $2\frac{dy}{dt} + 3y = 6e^{-3t/2} + 3(2 - 2e^{-3t/2}) = 6$.

7. $y = 3\cos 2t - \sin 2t \Rightarrow y' = -6\sin 2t - 2\cos 2t$, $y'' = -12\cos 2t + 4\sin 2t$; $y'' + 4y = -12\cos 2t + 4\sin 2t + 12\cos 2t - 4\sin 2t = 0$; $y(0) = 3 \cdot 1 - 0 = 3$, $y'(0) = -6 \cdot 0 - 2 \cdot 1 = -2$.

8. $y = 2 - \ln\cos(x-1)$, $y' = \frac{\sin(x-1)}{\cos(x-1)} = \tan(x-1)$, $y'' = \sec^2(x-1)$; $y(1) = 2 - \ln 1 = 2$, $y'(1) = \tan 0 = 0$; $y'' - (y')^2 = \sec^2(x-1) - \tan^2(x-1) = 1$.

9. a) $\frac{1}{1000+0.10x}\frac{dx}{dt}\cdot dt = 1\cdot dt \Rightarrow \int \frac{1}{1000+0.10x}dx = \int 1\cdot dt \Rightarrow \frac{1}{0.10}\ln(1000+0.10x) = t + C' \Rightarrow \ln(1000+0.10x) = 0.10(t+C') = 0.10t + C'' \Rightarrow 1000 + 0.10x = Ce^{0.10t}$. Using $x(0) = 1000$ gives $1000 + 100 = C$, or $x = \frac{1100e^{0.10t}-1000}{0.10} = 11000e^{0.10t} - 10000$ b) $100000 = 11000e^{0.10t} - 10000 \Rightarrow 10 = e^{0.10t} \Rightarrow t = \frac{\ln 10}{0.10} = 23.026$ yrs.

10. $\frac{dT}{dt} = k(T - T_s) = k(T-20)$ in this problem. $\frac{dT}{T-20} = kdt \Rightarrow \ln(T-20) = kt + C' \Rightarrow T - 20 = Ce^{kt}$; $T(0) = 100° \Rightarrow C = 80 \Rightarrow T = 20 + 80e^{kt}$; $T(20) = 40 \Rightarrow 40 = 20 + 80e^{k\cdot 20} \Rightarrow \frac{1}{4} = e^{k\cdot 20} \Rightarrow k\cdot 20 = \ln(\frac{1}{4}) \Rightarrow k = \frac{-\ln 4}{20}$. To find t, such that $T(t) = 60$, set $T = 20 + 80e^{-\frac{\ln 4}{20}\cdot t}$ equal to 60: $60 = 20 + 80e^{-\frac{\ln 4}{20}t} \Rightarrow \frac{-\ln 4}{20}t = \ln(\frac{1}{2})$; $t = \frac{20\ln(1/2)}{-\ln 4} = \frac{20\ln 2}{\ln 4} = 10$ minutes.

11. $\frac{dy}{y-95} = 0.017dx \Rightarrow \ln|y-95| = 0.017x + C' \Rightarrow |y-95| = Ce^{0.017x}$; $y(0) = 30 \Rightarrow |30-95| = 65 = Ce^0$, so that $|y-95| = 65e^{0.017x}$. Since $y(0) < 95$, for x near 0, $y < 95$ and $|y-95| = 95 - y \Rightarrow y = 95 - 65e^{0.017x}$. The graph of y agrees with that produced by IMPEULG, with $y1 = 0.017(y-95)$, $y(0) = 30$.

12. $\frac{dx}{x-20} = 0.013dt \Rightarrow \ln|x-20| = 0.013t + C' \Rightarrow |x-20| = Ce^{0.013t}$. $|6-20| = 14 = Ce^0 \Rightarrow 20 - x = 14e^{0.013t}$ or $x = 20 - 14e^{0.013t}$.

13. Using partial fractions, $\frac{dy}{y(350-y)} = 0.00125dt$ becomes: $\frac{1}{350}[\frac{1}{y} + \frac{1}{350-y}]dy = 0.00125dt \Rightarrow \frac{1}{350}[\ln|y| - \ln|350-y|] = 0.00125t + C' \Rightarrow \ln|\frac{y}{350-y}| = 0.4375t + C'' \Rightarrow \frac{y}{350-y} = Ce^{0.4375t}$; $\frac{30}{350-30} = C = \frac{3}{32}$. Finally $y = (350-y)\frac{3}{32}e^{0.4375t} \Rightarrow y = \frac{\frac{350\cdot 3}{32}e^{0.4375t}}{1+\frac{3}{32}e^{0.4375t}} = \frac{1050e^{0.4375t}}{32+3e^{0.4375t}}$.

14. $\frac{dx}{x(500-x)} = 0.005dt$. By the method of partial fractions, $\frac{1}{500}[\frac{1}{x} + \frac{1}{500-x}]dx = 0.005dt \Rightarrow \ln|\frac{x}{500-x}| = (500)(0.005t) + C' \Rightarrow |\frac{x}{500-x}| = Ce^{2.5t}$; $x(0) = 20 \Rightarrow |\frac{x}{500-x}| = \frac{x}{500-x}$ and $\frac{20}{500-20} = C = \frac{1}{24}$; $x = (500-x)\frac{1}{24}e^{2.5t} \Rightarrow x = \frac{\frac{500}{24}e^{2.5t}}{1+\frac{1}{24}e^{2.5t}} = \frac{500e^{2.5t}}{24+e^{2.5t}}$.

15. $y' = e^{-2y} \Rightarrow e^{2y}dy = dx \Rightarrow \frac{1}{2}e^{2y} = x + C$; $y(0) = 0 \Rightarrow \frac{1}{2}\cdot 1 = C \Rightarrow e^{2y} = 2(x + \frac{1}{2})$ or $y = \frac{1}{2}\ln(2x+1)$.

16. $y' = e^{2x} - x \Rightarrow y = \frac{1}{2}e^{2x} - \frac{x^2}{2} + C$, $y(0) = 1 \Rightarrow 1 = \frac{1}{2} + C \Rightarrow C = \frac{1}{2}$; $y = \frac{1}{2}(e^{2x} - x^2 + 1)$.

17. $y' = \frac{1}{x^2+1} \Rightarrow y = \arctan x + C$, $y(0) = 1 \Rightarrow 1 = 0 + C$; $y = \arctan x + 1$.

18. $y' = \sqrt{1-y^2} \Rightarrow \frac{dy}{\sqrt{1-y^2}} = dx \Rightarrow \arcsin y = x + C \Rightarrow y = \sin(x+C)$; $y(0) = 1 \Rightarrow 1 = \sin C \Rightarrow C = \frac{\pi}{2}$; $y = \sin(x + \frac{\pi}{2})$.

19. $y' = xe^x \Rightarrow y = xe^x - e^x + C$; $y(0) = 0 \Rightarrow C = 1$; $y = xe^x - e^x + 1$.

20. $y \ln y \, dy = dx \Rightarrow \frac{y^2}{2} \ln y - \frac{y^2}{4} = x + C$ (integrate by parts, let $u = \ln y$, $dv = y$); $y(0) = 1 \Rightarrow 1 \cdot 0 - \frac{1}{4} = C$; the solution satisfies $\frac{y^2}{2} \ln y - \frac{y^2}{4} = x - \frac{1}{4}$.

21. By example 2, $y = C_1 \cos x + C_2 \sin x$; $y' = -C_1 \sin x + C_2 \cos x$; $y(0) = 0$ and $y'(0) = 1$ give $0 = C_1 + 0$, $1 = 0 + C_2$; hence $y = \sin x$.

22. By #21, $y = C_1 \cos x + C_2 \sin x$; $y' = -C_1 \sin x + C_2 \cos x$; $y(0) = 1$, $y'(0) = 0$ give: $1 = C_1 + 0$, $0 = 0 + C_2$; hence $y = \cos x$.

23. $y = C_1 e^x \cos x + C_2 e^x \sin x$; $y' = C_1(e^x \cos x - e^x \sin x) + C_2(e^x \sin x + e^x \cos x)$; $y'' = C_1(-2e^x \sin x) + C_2(2e^x \cos x)$; $y'' - 2y' + 2y = C_1(-2e^x \sin x) + C_2(2e^x \cos x) - 2C_1 e^x \cos x + 2C_1 e^x \sin x - 2C_2 e^x \sin x - 2C_2 e^x \cos x + 2C_1 e^x \cos x + 2C_2 e^x \sin x = e^x \sin x(-2C_1 + 2C_1 - 2C_2 + 2C_2) + e^x \cos x(2C_2 - 2C_1 - 2C_2 + 2C_1) = 0$.

24. a) From 23, $y(0) = 0 \Rightarrow 0 = C_1 \cdot 1 + C_2 \cdot 0$; $y'(0) = 1 \Rightarrow 1 = C_1(1-0) + C_2(0+1); \Rightarrow C_1 = 0$, $C_2 = 1$; $y = e^x \sin x$ b) From 23, $y(0) = 1 \Rightarrow 1 = C_1 \cdot 1 + C_2 \cdot 0$, $y'(0) = 0 \Rightarrow 0 = C_1 + C_2; \Rightarrow C_1 = 1$, $C_2 = -1$; $y = e^x \cos x - e^x \sin x$.

25. $y(0) = 1$; $y(0.2) \approx y_1 = 1 + 0.2 \cdot 1 = 1.2$; $y(0.4) \approx y_2 = 1.2 + 0.2(1.2) = 1.44$; $y(0.6) \approx y_3 = 1.44 + 0.2(1.44) = 1.728$; $y(0.8) \approx y_4 = 1.728$; $y(1) \approx y_5 = 2.0736$. If $\frac{dy}{dt} = y$, $y(0) = 1$, then $\frac{dy}{y} = dt \Rightarrow \ln|y| = t + C' \Rightarrow y = Ce^t$; $y(0) = 1 \Rightarrow C = 1 \Rightarrow y(t) = e^t$; $y(1) = e^1 = 2.71828$.

26. $y_1 = y_0 + \frac{1}{n} y_0 = y_0(1 + \frac{1}{n})$; $y_2 = y_1 + \frac{1}{n} y_1 = y_1(1 + \frac{1}{n}) = y_0(1 + \frac{1}{n})^2; \ldots; y(1) \approx y_n = y_0(1 + \frac{1}{n})^n$. $\lim_{n \to \infty}(1 + \frac{1}{n})^n = e^1$.

27. Using IMPEULT, gives $y(1) \approx y_5 = 2.7027\ldots$.

28. To compute $y_5 \approx y(1)$ by hand, complete by hand the table below. Here $f(x, y) = y$

	$n = 0$	$n = 1$	$n = 2$	$\ldots$	$n = 5$
x_n	0	0.2	0.4		1.0
y_n	1	1.2214	1.4918		2.71825...
K_1	$0.2(1) = 0.2$	0.24428			
K_2	$0.2(1 + \frac{0.2}{2}) = 0.22$	0.268708			
K_3	$0.2(1 + \frac{0.22}{2}) = 0.222$	0.2711508			
K_4	$0.2(1 + 0.222) = 0.2444$	0.29851			
	$(K_1 + 2K_2 + 2K_3 + K_4)/6 = 0.2214$	0.27041			

Or, use the RUNKUT program to find $y_5 = 2.7182511366\ldots$ which is quite close to e.

29. First solve $y' = y^2$, $y(0) = 1$: $y^{-2}dy = dx \Rightarrow -y^{-1} = x + C$; when $x = 0 - 1^{-1} = 0 + C$, so that $-\frac{1}{y} = x - 1$ or $y = \frac{1}{1-x}$ which becomes infinite at $x = 1$. To compare the functions, set $y1 = x^2 + y^2$, $y2 = 1/(1-x)$; set the range to $[0, 1]$ by $[-10, 100]$. Run the program RUNKUTG — take $h = 0.02$ to obtain the graph of $y1$. While the dotted graph is showing, use DrawF to graph $y2$.

30. a) $\frac{dy}{1+y^2} = dx \Rightarrow \tan^{-1} y = x + c$; $y(0) = 0 \Rightarrow c = 0$; $\tan^{-1} y = x$ on $y = \tan x$. As $x \to \pi/2$, y becomes infinite.

b) Let $y = -u'/u$. Then $y' = u''/u + (u'/u)^2$ and the equation becomes $u'' = -u$. By Example 2, the function $u = C_1 \cos x + C_2 \sin x$ is a solution of the new equation. Since $y(0) = 0$, we must have $u'(0) = 0$. But $u'(0) = C_2 \cos(0) = C_2$. Hence $C_2 = 0$ and $u = C_1 \cos x$. Then $y = -u'/u = -\frac{-C_1 \sin x}{C_1 \cos x} = \tan x$.

The solutions to both $y' = x^2 + y^2$ and $y' = 1 + y^2$ become infinite.

31. The computed values of $y(2)$ are:

Euler, 1.271428571;

Improved Euler, 1.28571428571;

Runge-Kutta, 1.28571428571

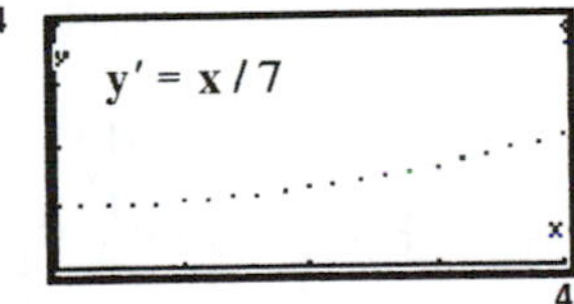

32. The computed values of $y(4)$ are:

Euler, 0.80491664235;

Improved Euler, 0.83868770875;

Runge-Kutta, 0.838119526483

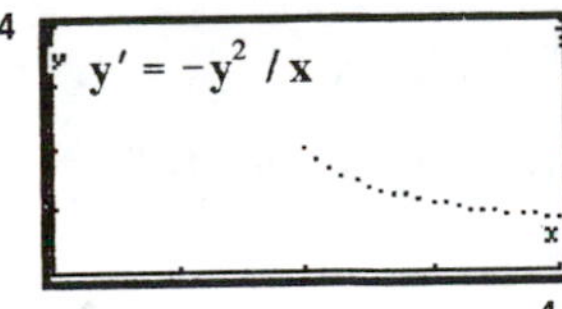

33. The computed values of $y(2)$ are:

Euler, 2.07334340632;

Improved Euler, 2.21157126441;

Runge-Kutta, 2.20283462521

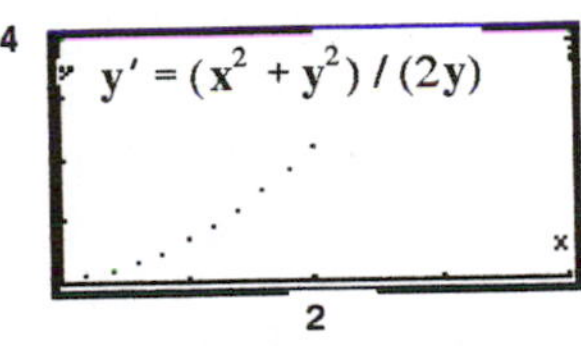

34. The computed values of $y(2)$ are:

Euler, 2.13448296411;

Improved Euler, 2.20482251496;

Runge-Kutta, 2.20273270952

35. a) $y(1) \approx 0.310268270416$ if $h = 0.05$ (20 steps); b) $y(1) \approx 0.310268299767$ if $h = 0.025$ (40 steps).

36. $\int_0^1 x^2e^{-x}dx = \begin{cases} 0.160602806363 & \text{20 steps} \\ 0.160602794907 & \text{40 steps} \end{cases}$;
$\int_0^1 x^2e^{-x}dx = -x^2e^{-x}|_0^1 + 2\int_0^1 xe^{-x}dx = -\frac{1}{e} + 2[-xe^{-x}|_0^1 + \int_0^1 e^{-x}dx] = 2 - 5/e = 0.16060279414$.

37. $y = \int_a^x -2t\sec^2(a^2 - t^2)dt$, $y(a) = 0 \Rightarrow y = \tan(a^2 - x^2)$; $y' = -2x\sec^2(a^2 - x^2) = -2x(\tan^2(a^2 - x^2) + 1) = -2x(y^2 + 1)$.

38. Graph the computed solution on $[1, 3]$ by $[-5, 5]$. An error message may occur, this will not affect the graph.

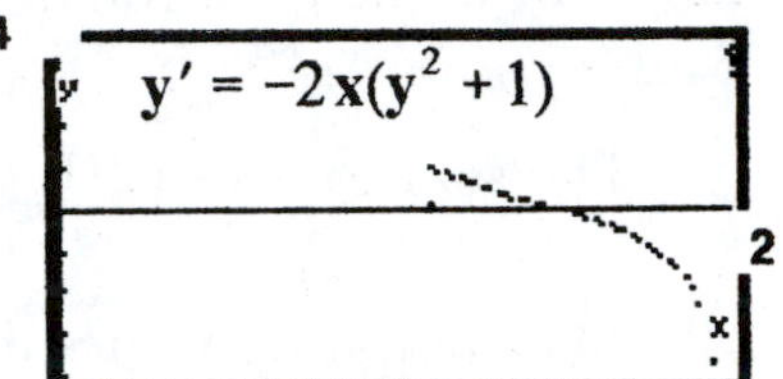

39. $y = \tan(a^2 - x^2)$, $y(1) = 1 \Rightarrow 1 = \tan(a^2 - 1) \Rightarrow a^2 - 1 = \arctan 1 = \frac{\pi}{4} \Rightarrow a = \sqrt{\frac{\pi}{4} + 1} \approx 1.336$.

40. $y = \lim_{b\to\infty} \int_x^b 2t(1 + t^2)^{-2}dt = \lim_{b\to\infty}[\frac{(1+t^2)^{-2}}{-1}]_x^b = \frac{1}{1+x^2}$; $y' = \frac{-2x}{(1+x^2)^2} = -2xy^2$.

41. Using step size 0.5, $N = 6$, the corresponding entries for $x = 3$ are:

	x	y(R-K)	y(true)	Difference
$y' - x - y$, $y(0) = 1$	3	2.09981094687	$\cdots$	2.37×10^{-4}
$y' = x - y$, $y(0) = -2$	3	1.95009452656	$\cdots$	-1.18×10^{-4}

The errors are much greater because the step size is five times as large.

42. Let $y(t)$ = number of guppies at time t. a) $\frac{dy}{dt} = 0.0015y(150-y)$; $y(0)=6$; $\frac{dy}{y(150-y)} = 0.0015dt$. Using partial fractions, $\frac{1}{150}[\frac{1}{y} + \frac{1}{150-y}]dy = 0.0015dt \Rightarrow \ln|\frac{y}{150-y}| = (150 \cdot 0.0015)t + C'$; $\frac{y}{150-y} = Ce^{0.225t}$; $y(0)=6 \Rightarrow \frac{6}{150-6} = C = \frac{1}{24}$; $y = \frac{(150-y)}{24}e^{0.225t} \Rightarrow y = \frac{150e^{0.225t}}{24+e^{0.225t}}$. e) Using a solver, $t = 17.2$ when $y = 100$, $t = 21.3$ when $y = 125$. f) At $t = 40$, $y > 149.5$.

43. a) $\frac{dy}{y(250-y)} = 0.0004dt$; $\frac{1}{250}[\frac{1}{y} - \frac{1}{250-y}] = 0.0004t + C'$; $\ln\frac{y}{250-y} = 0.1t + C'$; $\frac{y}{250-y} = Ce^{0.1t}$; $y(0) = 28 \Rightarrow C = \frac{28}{222} = \frac{14}{111}$; $y = (250-y)\frac{14}{111}e^{0.1t}$; $111y = (250-y)14e^{0.1t}$; $y(111+14e^{0.1t}) = 250 \cdot 14e^{0.15}$; $y = \frac{3500e^{0.1t}}{111+14e^{0.1t}}$

e) $y = 100$ at $t = 16.65$ years, $y = 200$ at $t = 34.57$ years

f) at 92 years, $y > 249$.

44. $A = \int_0^\infty \frac{a^3}{a^2+x^2}dx = \lim_{b\to\infty}\int_0^b \frac{a^3}{a^2+x^2}dx = \frac{a^3}{a}\lim_{b\to\infty}\tan^{-1}\frac{x}{a}\Big|_0^b = a^2\frac{\pi}{2}$.

45. $\frac{\pi}{4}a^2 = 1 \Rightarrow a = \pm\frac{2}{\sqrt{\pi}}$.

46. All paths go through $(0,0)$ at $t=0$ and $(\pi,-2)$ at $t=1$.
a) $x_1(t) = \pi t$, $y_1(t) = -2t$; $\frac{dy}{dx} = \frac{dy/dt}{dx/dt} = \frac{-2}{\pi} < 0 \Rightarrow y_1$ decreasing.
$x_2(t) = \pi t$, $y_2(t) = -2\sqrt{t}$; $\frac{dy}{dx} = \frac{-\frac{1}{\sqrt{t}}}{\pi} < 0 \Rightarrow y$ decreasing.
$x_3(t) = \pi t - \sin\pi t$, $y_3(t) = -1 + \cos\pi t$; $\frac{dy}{dx} = \frac{-\pi\sin\pi t}{\pi - \pi\cos\pi t} = \frac{-\text{pos}}{\text{pos}} < 0$ for $0 < t < 1$.
$x_4(t) = \pi t$, $y_4(t) = 2(t-1)^2 - 2$; $\frac{dy}{dx} = \frac{4(t-1)}{\pi} < 0$, $0 < t < 1$.
$x_5(t) = \pi t$, $y_5(t) = -2\sqrt{2t-t^2}$; $\frac{dy}{dx} = \frac{-\frac{2-2t}{\sqrt{2t-t^2}}}{\pi} < 0$, for $0 < t < 1$.
c) path:
1) $T = \frac{1}{\sqrt{g}}\int_0^1\sqrt{\frac{\pi^2+4}{-2(-2t)}}\,dt$, improper
2) $T = \frac{1}{\sqrt{g}}\int_0^1\sqrt{\frac{\pi^2+\frac{1}{t}}{4\sqrt{t}}}\,dt$, improper
3) $T = \frac{1}{\sqrt{g}}\int_0^1\sqrt{\frac{(\pi-\pi\cos\pi t)^2+(\pi\sin\pi t)^2}{2(1-\cos\pi t)}}\,dt$, improper
4) $T = \frac{1}{\sqrt{g}}\int_0^1\sqrt{\frac{\pi^2+16(t-1)^2}{-4((t-1)^2-1)}}\,dt$, improper

5) $T = \frac{1}{\sqrt{g}}\int_0^1\sqrt{\frac{\pi^2+\frac{(2-2t)^2}{2t-t^2}}{4\sqrt{2t-t^2}}}\,dt$, improper
d) $T = \frac{1}{\sqrt{g}}\int_0^1\sqrt{\frac{\pi^2+4}{-2(-2t)}}dt = \frac{\sqrt{\pi^2+4}}{2\sqrt{g}}\int_0^1 t^{-1/2}dt = \frac{\sqrt{\pi^2+4}}{\sqrt{g}}$

e) Using NINT, and a lower limit of 0.0000001,
1) $\frac{1}{\sqrt{g}}3.7230\ldots \approx \frac{\sqrt{\pi^2+4}}{\sqrt{g}}$, 2) $\frac{1}{\sqrt{g}}3.1368\ldots$, 3) $\frac{1}{\sqrt{g}}3.1415923\ldots$,
4) $\frac{1}{\sqrt{g}}3.27519\ldots$, 5) $\frac{1}{\sqrt{g}}3.1110\ldots$.
f) #46d, curve 3, was very close to $\pi/\sqrt{g}$, using NINT. Analytically, $\int_0^1 [\frac{(\pi-\pi\cos\pi t)^2+(\pi\sin\pi t)^2}{2(1-\cos\pi t)}]^{1/2}dt = \pi\int_0^1[\frac{1-2\cos\pi t+\cos^2\pi t+\sin^2\pi t}{2(1-\cos\pi t)}]^{1/2}dt = \pi\int_0^1[\frac{2-2\cos\pi t}{2(1-\cos\pi t)}]^{1/2}dt = \pi\int_0^1 dt = \pi$. By ignoring $\int_0^{0.0000001}\sqrt{\frac{(x'(t))^2+(y'(t))^2}{-2y(t)}}dt$, an interval on which the integrand becomes infinitely large, we are omitting a potentially sizable component.

47. Your company invents a new game. Let $P(t)$ be the number sold at time t. At first the sales are exponential, but eventually the market approaches saturation. If $\frac{dP}{dt} = aP(b-P)$, the inflection point will be when $\frac{dP}{dt}$, the rate at which the game is selling, is the greatest.

8.8 COMPUTER ALGEBRA SYSTEMS

1. (Using *Mathematica*)

```
In[2]:= Integrate[Exp[-x^2],{x,0,Infinity}]

        Sqrt[Pi]
Out[2]= --------
           2
```

2.

```
In[3]:= Integrate[x ArcCos[x],x]

                        2        2
         -(x Sqrt[1 - x ])      x  ArcCos[x]   ArcSin[x]
Out[3]= -----------------  +  ------------  + ---------
                4                   2             4
```

3.

```
In[5]:= Integrate[1/(x Sqrt[x-3]),{x,6,9}]

            -Pi          2 ArcTan[Sqrt[2]]
Out[5]= ---------  +  -----------------
         2 Sqrt[3]         Sqrt[3]
```

4.

```
In[6]:= Integrate[x ArcTan[2x],{x,0,1/2}]

         -2 + Pi
Out[6]= -------
           16
```

5.

```
In[7]:= Integrate[1/(9-x^2)^2,x]

            -x            Log[-3 + x]    Log[3 + x]
Out[7]= ------------  -  -----------  +  ----------
                   2         108            108
        18 (-9 + x )
```

6.

```
In[8]:= Integrate[(Sqrt[4 x+9])/x^2,{x,4,10}]

                          5                7
                4 ArcTanh[-]     4 ArcTanh[-]
         11               3                3
Out[8]= --  +  -----------   -   ------------
         20         3                  3
```

7.

```
In[9]:= Integrate[1/(x^2 Sqrt[7+x^2]),{x,3,11}]

         4    8 Sqrt[2]
Out[9]= --  -  ---------
         21       77
```

8.

```
In[10]:= Integrate[1/(x^2 Sqrt[7-x^2]),x]

                      2
           -Sqrt[7 - x ]
Out[10] = -------------
               7 x
```

9.

```
In[11]:= Integrate[(Sqrt[x^2-2])/x,{x,-2,-Sqrt[2]}]

            Pi         4 + Pi
Out[11]= -------  -  ---------
         Sqrt[2]     2 Sqrt[2]
```

10.

```
In[12]:= Integrate[1/(5+4Sin[2x]),{x,-Pi/12,Pi/4}]

         Pi
Out[12]= --
         18
```

11.

```
In[13]:= Integrate[1/(4+5Sin[2x]),x]

         -Log[2 Cos[x] + Sin[x]]     Log[Cos[x] + 2 Sin[x]]
Out[13]= ----------------------- +  ----------------------
                    6                         6
```

12.

```
In[14]:= Integrate[x/Sqrt[x-2],{x,3,6}]

         26
Out[14]= --
         3
```

13.

```
In[15]:= Integrate[x Sqrt[2x-3],x]

                                        2
                         3    x    2 x
Out[15]= Sqrt[-3 + 2 x] (-(-) - - + ----)
                         5    5     5
```

14.

```
In[16]:= Integrate[(Sqrt[3x-4])/x,x]

                                           Sqrt[-4 + 3 x]
Out[16]= 2 Sqrt[-4 + 3 x] - 4 ArcTan[--------------]
                                                 2
```

15.

```
In[17]:= Integrate[x^10 Exp[-x],{x,0,Infinity}]

Out[17]= 3628800
```

16.

```
In[1]:= Integrate[x^2 ArcTan[x],{x,0,1}]

           -2 + Pi + 2 Log[2]
Out[1]=    ------------------------
                  12
```

17.

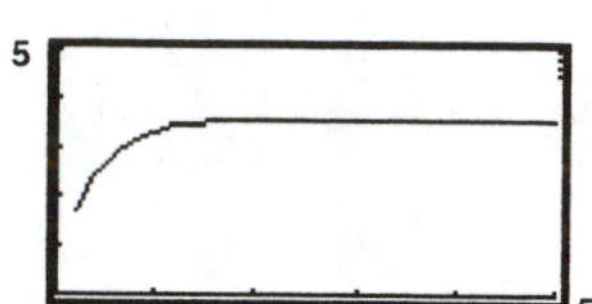

18.

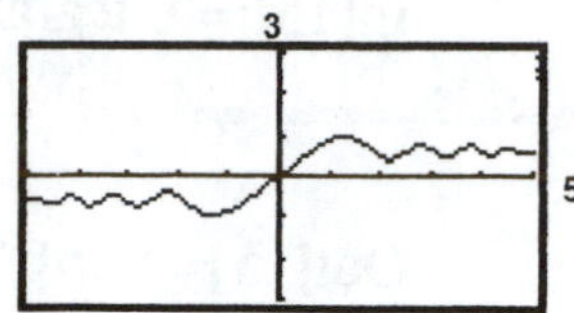

19.

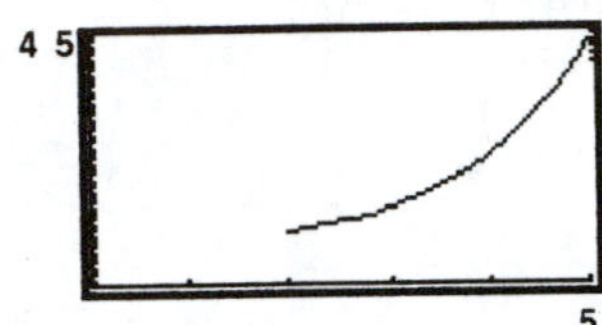

20.

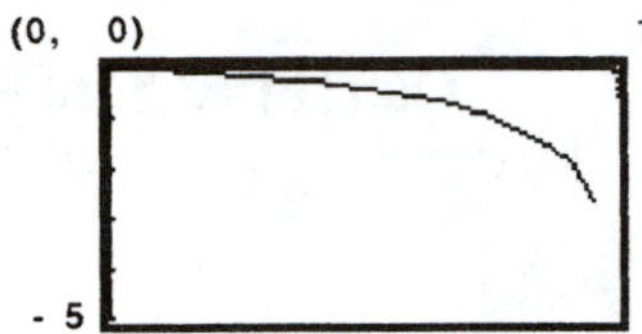

21.

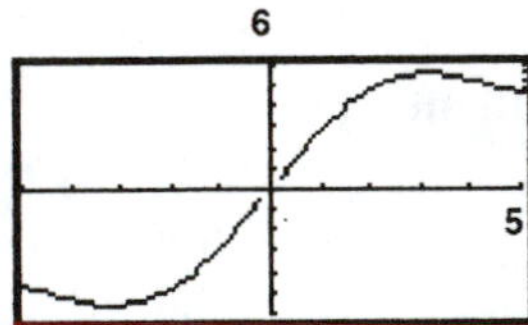

22.

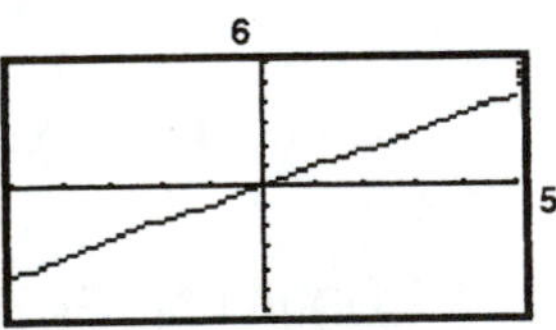

23.

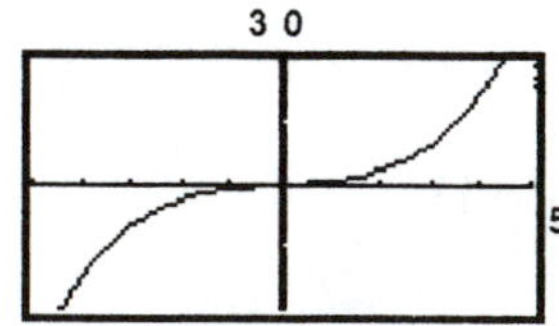

24.

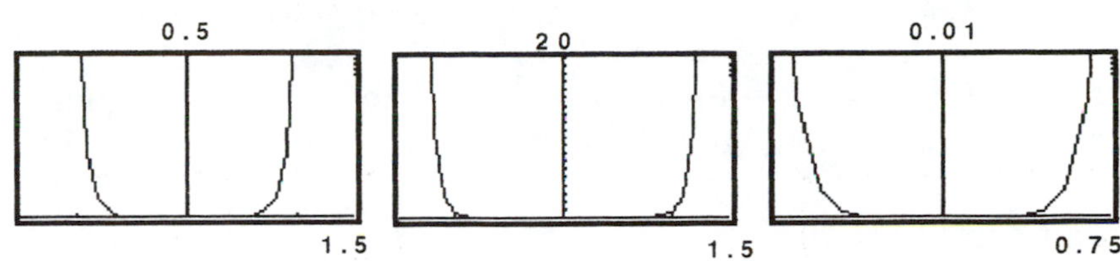

25. Since $y' = \sin^{-1}\sqrt{x}$, $y(0) = 0$, first use IMPEULT on $0 \le x \le 3$ to locate the solution. Set $y1 = \sin^{-1}\sqrt{x}$, take $h = 0.1$. The solution appears to be between 0.7 and 0.8. Graph $y1 = 0.5$, $y2 = \text{NINT}(\sin^{-1}\sqrt{t}, t, 0, x)$ on $[0.7, 0.8]$ by $[0.4, 0.6]$. When the graphs are completed, use trace to find $x = 0.7705$.

26. Use IMPEUG on the initial value problem $y' = \sqrt{1 + \sin^2 x}$, $y(\pi/4) = 0$ to locate x between 1.0 and 1.18. Then graph $y1 = 0.5$ and $y2 = \text{NINT}(\sqrt{}(1 + (\sin T)^2), T, \pi/4, x)$. Use trace to find $x = 1.1714$.

27. Use IMPEUT on $y' = \sqrt{}(1+x^4)$ with stepsize $= 0.01$ to locate x between 0.49 and 0.50. Graph $y1 = 0.5$, $y2 = \text{NINT}(\sqrt{}(1 + T\text{^}4), T, 0, x)$ on $[0.49, 0.50]$ by $[0.4, 0.6]$ to find $x = 0.498730$.

28. The graphs cross at $x = 0.971$, approximately.

29. Graph $f' = y = xe^{-x}$ on $[0, 5]$ by $[0, 0.5]$. Since $f' > 0$, f is always increasing, f' increasing on $[0, 1] \Rightarrow f$ is concave up; there is an inflection point at $x = 1$; $f' \to 0$ as $x \to \infty \Rightarrow f$ approaches a constant value.

30. From the graph of y, the graph of the integral is increasing and concave down until $x = 0.840$, it has an inflection point at $x = 2.166$, it continues to decrease until $x = 3.933$. The critical points are, approximately, $(0.840, 1.234)$, $(2.166, 0.701)$, and $(3.933, 0.034)$.

31. Since $\delta = 1$, $M = \int_0^3 \frac{1}{\sqrt{x+1}}dx$; $M_y = \int_0^3 \frac{x}{\sqrt{x+1}}dx$, $M_y = \int_0^3 \frac{2}{x+1}dx$. Using NINT, $\bar{x} = M_y/M \approx 1.33333333143$, $\bar{y} = 1.38629436004$.

32. $M_y = \int \tilde{x}dm = \int_0^3 \frac{18}{2x+3}dA = \int_0^3 \frac{18}{2x+3} \cdot \frac{36}{2x+3} \cdot dx = 18 \cdot 36[-(2x+3)^{-1}]_0^3 = 72.$

33. Let $y = \sqrt{\frac{2a-x}{x}} = (2a-x)^{1/2}x^{-1/2}$; $y' = \frac{-1}{2}(2a-x)^{-1/2}x^{-1/2} + (2a-x)^{1/2}(-\frac{1}{2})x^{-3/2} = \frac{-1}{2}x^{-1/2}[\frac{1}{\sqrt{2a-x}} + \frac{\sqrt{2a-x}}{x}] = \frac{-1}{2\sqrt{x}}[\frac{x+(2a-x)}{x\sqrt{2a-x}}] = \frac{-a}{x\sqrt{2ax-x^2}}$. Hence $\frac{d}{dx}[-\frac{1}{a}\sqrt{\frac{2a-x}{x}}] = \frac{1}{x\sqrt{2ax-x^2}}$.

34. Let $y = \frac{1}{a}\tan\frac{ax}{2}$. Then $y' = \frac{a}{2a}\sec^2\frac{ax}{2} = \frac{1}{2\cos^2(ax/2)} = \frac{1}{1+\cos ax}$.

35. If $u = ax + b$, $x = (u-b)/a$, $dx = \frac{1}{a}du$. The integral becomes $\frac{1}{a^2}\int\frac{(u-b)du}{u^2} = \frac{1}{a^2}\int[\frac{1}{u} - \frac{b}{u^2}]du = \frac{1}{a^2}[\ln|u| + \frac{b}{u}] + C = \frac{1}{a^2}[\ln|ax+b| + \frac{b}{ax+b}] + C$.

36. The integral becomes

$$\int\frac{a\sec u\tan u\,du}{a^2\sec^2 u\sqrt{a^2\tan^2 u}} = \frac{1}{a^2}\int\frac{du}{\sec u}du$$
$$= \frac{1}{a^2}\int\cos u\,du = \frac{1}{a^2}\sin u + C$$
$$= \frac{1}{a^2}\frac{\sqrt{x^2-a^2}}{x} + C.$$

PRACTICE EXERCISES, CHAPTER 8

1. Let $u = 1+\sin x$, $du = \cos x\,dx$, $0 \le x \le \frac{\pi}{2}$, $1 \le u \le 2$; $\int_0^{\pi/2}\frac{\cos x\,dx}{(1+\sin x)^{1/2}} = \int_1^2 u^{-1/2}du = [\frac{u^{1/2}}{\frac{1}{2}}]_1^2 = 2[\sqrt{2}-1]$.

2. Let $u = \sqrt{x}$; $\int_{\pi^2/16}^{\pi^2/9}\frac{\tan\sqrt{x}}{2\sqrt{x}}dx = \int_{\pi/4}^{\pi/3}\tan u\,du = [\ln|\sec u|]_{\pi/4}^{\pi/3} = [\ln 2 - \ln\sqrt{2}] = \frac{1}{2}\ln 2$.

3. $\int_{-1}^{1}\frac{2y}{y^4+1}dy = 0$. The integrand is odd.

4. Let $u = \sec x$; $\int\sec x\tan x e^{\sec x}dx = \int e^u du = e^u + C = e^{\sec x} + C$.

5. Let $u = \sin^{-1}x$, $du = (1/\sqrt{1-x^2})dx$; $0 \le x \le \sqrt{2}/2 \Rightarrow 0 \le u \le \frac{\pi}{4}$; $\int_0^{\sqrt{2}/2}\frac{\sin^{-1}x}{\sqrt{1-x^2}}dx = \int_0^{\pi/4}u\,du = [\frac{u^2}{2}]_0^{\pi/4} = \frac{\pi^2}{32}$.

6. Let $u = 2+\ln x$; $\int_1^e\frac{dx}{x(2+\ln x)} = \int_2^3\frac{du}{u} = \ln 3 - \ln 2 = \ln 1.5$.

7. $\int_{\pi/4}^{\pi/3}\frac{dx}{2\sin x\cos x} = \int_{\pi/4}^{\pi/3}\frac{1}{\sin 2x}dx = \int_{\pi/4}^{\pi/3}\csc 2x\,dx = \frac{1}{2}[\ln|\csc 2x + \cot 2x|]_{\pi/4}^{\pi/3} = \frac{1}{2}[\ln|\csc\frac{2\pi}{3} + \cot\frac{2\pi}{3}| - \ln|\csc\frac{\pi}{2} + \cot\frac{\pi}{2}|] = \frac{1}{2}[\ln|\frac{2\sqrt{3}}{3} - \frac{\sqrt{3}}{3}| - \ln|1+0|] = \frac{1}{4}\ln 3$.

8. $\int_0^{\pi/6}\frac{2dt}{\cos^2 t - \sin^2 t} = \int_0^{\pi/6}2\sec 2t\,dt = [\ln|\sec 2t + \tan 2t|]_0^{\pi/6} = \ln(2+\sqrt{3}) - \ln 1 = \ln(2+\sqrt{3})$.

9. $\int\frac{x+4}{x^2+1}dx = \frac{1}{2}\int\frac{2x}{x^2+1}dx + 4\int\frac{dx}{x^2+1} = \frac{1}{2}\ln(x^2+1) + 4\tan^{-1}x + C$.

10. $\int\frac{x+2}{\sqrt{1-x^2}}dx = \int\frac{x}{\sqrt{1-x^2}}dx + \int\frac{2}{\sqrt{1-x^2}}dx = -(1-x^2)^{1/2} + 2\sin^{-1}x + C$.

11. Let $u = \ln x$; $dv = x^2 dx$; $du = \frac{dx}{x}$; $v = \frac{x^3}{3}$. $\int x^2 \ln x\, dx = \frac{x^3}{3} \ln x - \int \frac{x^2}{3} dx = \frac{x^3}{3} \ln x - \frac{x^3}{9} + C$.

12. Let $u = x + 1$; $\int_0^1 \ln(x+1) dx = \int_1^2 \ln u\, du = [u \ln u - u]_1^2 = \ln(4) - 1$.

13. $\int x^5 \sin x\, dx = -x^5 \cos x + 5x^4 \sin x + 20x^3 \cos x - 60x^2 \sin x + 120x(-\cos x) + 120 \sin x + C$, using tabular integration.

14. Let $u = \tan^{-1} x$, $dv = \frac{dx}{x^2}$; $du = \frac{dx}{1+x^2}$, $v = \frac{-1}{x} \Rightarrow \int \frac{2\tan^{-1} x}{x^2} dx = 2[\frac{-\tan^{-1} x}{x} + \int \frac{dx}{x(1+x^2)}] = 2[\frac{-\tan^{-1} x}{x} + \int [\frac{1}{x} - \frac{x}{1+x^2}] dx] = \frac{-2\tan^{-1} x}{x} + 2\ln|x| - \ln(1+x^2) + C$.

15. Let $u = \cos 2x$; $dv = e^x dv$; $du = -2\sin 2x$; $v = e^x \Rightarrow I = \int e^x \cos 2x\, dx = e^x \cos 2x + 2\int e^x \sin 2x dx = e^x \cos 2x + 2[e^x \sin 2x - 2I]$, after another integration by parts, $I = \frac{1}{5}[e^x \cos 2x + 2e^x \sin 2x] + C$.

16. Let $u = \sin x$, $dv = e^{-x}$; $I = \int e^{-x} \sin x\, dx = -e^{-x} \sin x + \int e^{-x} \cos x\, dx$; let $u = \cos x$, $dv = e^{-x} \Rightarrow I = -e^{-x} \sin x + [-e^{-x} \cos x - \int e^{-x} \sin dx] \Rightarrow I = \frac{-e^{-x}}{2}[\sin x + \cos x] + C$.

17. $\int \sin^3 y\, dy = \int (1 - \cos^2 y) \sin y\, dy = -\cos y + \frac{\cos^3 y}{3} + C$.

18. $\int \sin^3 y \cos^2 y\, dy = \int \sin y (1 - \cos^2 y) \cos^2 y\, dy = \int \sin y (\cos^2 y - \cos^4 y) dy = \frac{-\cos^3 y}{3} + \frac{\cos^5 y}{5} + C$.

19. $\int \sin^4 x (1 - \sin^2 x) dx = \int \sin^4 x\, dx - \int \sin^6 x\, dx$; use formula (60) $= \int \sin^4 x\, dx - [-\frac{\sin^5 x \cos x}{6} + \frac{5}{6} \int \sin^4 x\, dx] = \frac{\sin^5 x \cos x}{6} + \frac{1}{6} \int \sin^4 x\, dx = \frac{\sin^5 x \cos x}{6} + \frac{1}{6}[-\frac{\sin^3 x \cos x}{4} + \frac{3}{4} \int \sin^2 x\, dx] = \frac{\sin^5 x \cos x}{6} - \frac{1}{24} \sin^3 x \cos x + \frac{1}{8} \cdot \frac{1}{2}[x - \frac{\sin 2x}{2}] + C$.

20. $\int \sin^3 x \cos^3 x\, dx = \int \sin^3 x (1 - \sin^2 x) \cos x\, dx = \frac{\sin^4 x}{4} - \frac{\sin^6 x}{6} + C$.

21. $\int_0^\pi \sqrt{\frac{1+\cos 2x}{2}} = \int_0^\pi \sqrt{\cos^2 x} dx = \int_0^\pi |\cos x| dx = 2$.

22. $\int_{\pi/4}^{3\pi/4} \sqrt{\cot^2 t + 1} dt = \int_{\pi/4}^{3\pi/4} |\csc t| dt = \int_{\pi/4}^{3\pi/4} \csc t\, dt = -\ln|\csc t + \cot t|]_{\pi/4}^{3\pi/4} = -\ln|\sqrt{2} - 1| + \ln|\sqrt{2} + 1| = \ln(3 + 2\sqrt{2})$.

23. $\int_0^{\pi/3} \tan^3 t\, dt = \int_0^{\pi/3} (\sec^2 t - 1) \tan t\, dt = \int_0^{\pi/3} \sec t (\sec t \tan t) dt - \int_0^{\pi/3} \tan t\, dt = [\frac{\sec^2 t}{2} + \ln|\cos t|]_0^{\pi/3} = 2 + \ln\frac{1}{2} - [\frac{1}{2} + 0] = \frac{3}{2} - \ln 2$.

24. $\int_{-\pi/4}^{\pi/4} 6\sec^4 t\, dt = 12 \int_0^{\pi/4} (\tan^2 t + 1) \sec^2 t\, dt = 12[\frac{\tan^3 t}{3} + \tan t]_0^{\pi/4} = 12 \cdot \frac{4}{3} = 16$.

25. Let $z = 4\tan\theta$; $0 \le z \le 3 \Rightarrow 0 \le \theta \le \arctan(3/4)$; $\int_0^3 \frac{dz}{(16+z^2)^{3/2}} = \frac{1}{16} \int_0^{\arctan(3/4)} \frac{\sec^2\theta\, d\theta}{\sec^3\theta} = \frac{1}{16} \int_0^{\arctan(3/4)} \cos\theta\, d\theta = [\frac{\sin\theta}{16}]_0^{\arctan(3/4)} = \frac{1}{16} \cdot \frac{3}{5} = \frac{3}{80}$.

26. Let $u = \sqrt{z^2+1}$; $\int_0^{\sqrt{3}} \frac{z^3+z}{\sqrt{1+z^2}}dz = \int_0^{\sqrt{3}}(z^2+1)\frac{2z}{2\sqrt{z^2+1}}dz = \int_1^2 u^2du = [\frac{u^3}{3}]_0^2 = 7/3$.

27. Let $x = \sin\theta$; $\int \frac{dx}{x^2\sqrt{1-x^2}} = \int \frac{\cos\theta\, d\theta}{\sin^2\theta\cos\theta}$
$= \int \csc^2\theta\, d\theta = -\cot\theta + C = -\frac{\sqrt{1-x^2}}{x} + C$

28. Let $x = \sin u$; $\int \frac{x^2dx}{\sqrt{1-x^2}} = \int \frac{\sin^2 u\cos u}{\cos u}du = \int \sin^2 u\, du = \frac{u}{2} - \frac{\sin u\cos u}{2} + C = \frac{\sin^{-1}x}{2} - \frac{x\sqrt{1-x^2}}{2} + C$.

29. Let $x = \sec\theta$; $\int_{5/4}^{5/3}\frac{12dx}{(x^2-1)^{3/2}} = \int_{x=5/4}^{x=5/3}\frac{12\sec\theta\tan\theta\, d\theta}{(\tan^2\theta)^{3/2}} = 12\int_{x=5/4}^{x=5/3}\frac{\sec\theta}{\tan^2\theta}d\theta = 12\int_{x=5/4}^{x=5/3}\frac{\cos\theta}{\sin^2\theta}d\theta = \frac{12(\sin\theta)^{-1}}{-1}]_{x=5/4}^{x=5/3} = -12[\frac{1}{4/5} - \frac{1}{3/5}] = 5$.

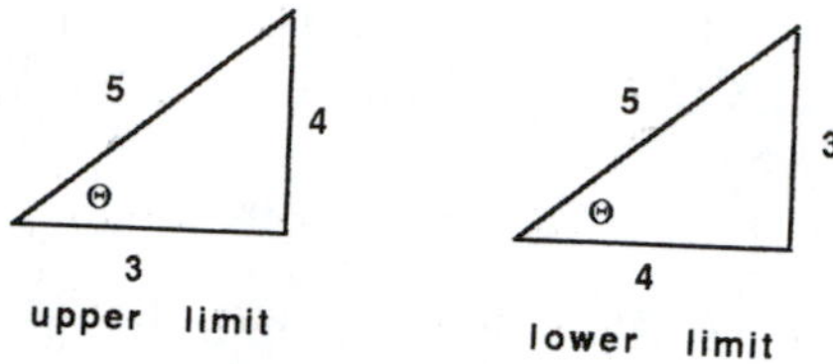

30. Let $x = 3\sec u$; $\int_5^6 \frac{dx}{\sqrt{x^2-9}} = \int_{x=5}^{x=6}\frac{3\sec u\tan u\, du}{3\tan u} = [\ln|\sec u + \tan u|]_{x=5}^{x=6} = [\ln|\frac{x}{3} + \frac{\sqrt{x^2-9}}{3}|]_5^6 = \ln(\frac{2+\sqrt{3}}{3})$.

31. $\int_{1/3}^1 \frac{3dx}{9x^2-6x+1+4} = \int_{1/3}^1\frac{3dx}{(3x-1)^2+2^2} = \int_0^2\frac{dt}{t^2+2^2} = \frac{1}{2}[\tan^{-1}\frac{t}{2}]_0^2$; let $t = 3x-1$; $= \frac{1}{2}[\frac{\pi}{4} - 0] = \frac{\pi}{8}$.

32. $I = \int_{-1}^{-1/2}\frac{dx}{\sqrt{-2x-x^2}} = \int_{-1}^{-1/2}\frac{dx}{\sqrt{1-(x+1)^2}}$; let $x+1 = \sin u$; $I = \int_0^{+\pi/6}\frac{\cos u}{|\cos u|}du = \frac{\pi}{6}$.

33. $\int_0^1 \frac{dx}{(x+1)\sqrt{x^2+2x}} = \int_0^1\frac{dx}{(x+1)\sqrt{(x+1)^2-1}}$; let $t = x+1$; $= \int_1^2\frac{dt}{t\sqrt{t^2-1}} = [\sec^{-1}|t|]_1^2 = \pi/3$.

34. $\int_{-2}^{-1}\frac{2dx}{x^2+4x+5} = \int_{-2}^{-1}\frac{2dx}{(x+2)^2+1} = 2[\tan^{-1}(x+2)]_{-2}^{-1} = \frac{\pi}{2}$.

35. Reduce the integrand: $\frac{x^3+x^2}{x^2+x-2} = x + \frac{2x}{x^2+x-2} = x + \frac{2x}{(x+2)(x-1)} = x + \frac{A}{x+2} + \frac{B}{x-1}$. Since $2x = A(x-1) + B(x+2)$, $-3A = -4$, $3B = 2$. The problem becomes $\int_2^6(x + \frac{4/3}{x+2} + \frac{2/3}{x-1})dx = [\frac{x^2}{2} + \frac{4}{3}\ln|x+2| + \frac{2}{3}\ln|x-1|]_2^6 = [18 + \frac{4}{3}\ln 8 + \frac{2}{3}\ln 5] - [2 + \frac{4}{3}\ln 4 + \frac{2}{3}\ln 1] = 16 + \frac{4}{3}\ln\frac{8}{4} + \frac{2}{3}\ln 5 = 16 + \frac{1}{3}\ln(16\cdot 25) = 16 + (\ln 400)/3$.

36. $\int_2^3 \frac{x^3+1}{x^3-x}dx = \int_2^3(1+\frac{x+1}{x(x^2-1)})dx = \int_2^3(1-\frac{1}{x}+\frac{1}{x-1})dx = [x-\ln|x|+\ln|x-1|]_2^3 = 1+\ln(\frac{4}{3})$.

37. $\int \frac{xdx}{(x-1)^2} = \int \frac{x-1+1}{(x-1)^2}dx = \int(\frac{1}{x-1} + \frac{1}{(x-1)^2})dx = \ln|x-1| - \frac{1}{x-1} + C$.

38. $\int \frac{8dx}{x^3(x+1)} = \int(\frac{1}{x} - \frac{2}{x^2} + \frac{4}{x^3} - \frac{1}{x+2})dx = \ln|x| + \frac{2}{x} - \frac{2}{x^2} - \ln|x+2| + C$.

39. $\frac{4}{x^3+4x} = \frac{4}{x(x^2+4)} = 4[\frac{A}{x} + \frac{Bx+C}{x^2+4}]$. Since $1 = A(x^2+4)+(Bx+C)x$, $x=0 \Rightarrow A = \frac{1}{4}$; $1 = x^2(A+B)+x\cdot C+4A \Rightarrow B=-A=-\frac{1}{4}$, $C=0$. The problem becomes: $4\int[\frac{1/4}{x} + \frac{-1/4x}{x^2+4}]dx = \int[\frac{1}{x} - \frac{x}{x^2+4}]dx = \ln|x| - \frac{1}{2}\ln|x^2+4| + C = \ln\frac{|x|}{\sqrt{x^2+4}} + C$.

40. $\int_0^1 \frac{x}{x^4-16}dx = \frac{1}{2}\int_0^1 \frac{2x}{(x^2)^2-4^2} = \frac{1}{2}\int_0^1 \frac{du}{u^2-4^2} = \frac{1}{2}\int_0^1 \frac{1/8}{u-4} - \frac{1/8}{u+4}du = \frac{1}{16}[\ln|\frac{u-4}{u+4}|]_0^1 = [\frac{1}{16}\ln(\frac{3}{5})$.

41. First, rewrite the integrand $\frac{2}{x(x-1)} = \frac{-1}{x} + \frac{1}{x-2}$. Then write a definite integral with finite upper limit: $\lim_{b\to\infty}\int_3^b[\frac{-1}{x} + \frac{1}{x-2}]dx = \lim_{b\to\infty}[-\ln|x| + \ln|x-2|]_3^b = \lim_{b\to\infty}[\ln|\frac{x-2}{x}|]_3^b = \lim_{b\to\infty}[\ln|1-\frac{2}{x}|]_3^b = \lim_{b\to\infty}[\ln|1-\frac{2}{b}| - \ln|1-\frac{2}{3}|] = -\ln\frac{1}{3} = \ln 3$.

42. Let $u = \ln x$, $dv = dx/x^2$; $\int \frac{\ln x}{x^2}dx = \frac{-\ln x}{x} + \int\frac{dx}{x^2} = \frac{-\ln x}{x} - \frac{1}{x} + C$. Hence $\lim_{b\to\infty}\int_1^b \frac{\ln x}{x^2}dx = \lim_{b\to\infty}[\frac{-\ln b}{b} - \frac{1}{b} + 1] = 1$ since $\lim_{b\to\infty}\frac{\ln b}{b} = 0$.

43. $M_x = \int_1^e[\frac{1+\ln x}{2}]2[1-\ln x]dx = \int_1^e[1-(\ln x)^2]dx = \int_1^e dx - \int_1^e(\ln x)^2dx = [(e-1) - [x(\ln x)^2 - 2\int_1^e \ln x\, dx]_1^e$; let $u = (\ln x)^2$, $du = \frac{2\ln x}{x}dx$; $dv = dx$, $v = x$; $= (e-1) - [x(\ln x)^2 - 2[x\ln x - x]]_1^e = (e-1) - [e\cdot 1 - 2(e-e) - (2)] = 1$.

44. $A = \int_1^2 \ln x\, dx = [x\ln x - x]_1^2 = \ln(4) - 1$.

45. Use shells. $V = 2\pi\int_0^1 x\cdot 3x\sqrt{1-x}\,dx = 6\pi\int_0^1 x^2(1-x)^{1/2}dx$; let $u = 1-x$; $= 6\pi\int_1^0(1-u)^2u^{1/2}(-du) = 6\pi\int_0^1(u^{1/2}-2u^{3/2}+u^{5/2})du = 6\pi[\frac{2}{3}-2\cdot\frac{2}{5}+\frac{2}{7}] = 32\pi/35$.

46. $L = \int_0^{\sqrt{3}/2}\sqrt{1+(2x)^2}dx$; let $2x = \tan u$; $L = \frac{1}{2}\int_0^{\pi/3}\sqrt{\tan^2 u+1}\sec^2 u\, du = \frac{1}{2}\int_0^{\pi/3}\sec^3 u\, du = \frac{1}{2}[\frac{1}{2}\sec u\tan u + \frac{1}{2}\ln|\sec u + \tan u|]_0^{\pi/3}$ by Example 6, 8.3. $L = [\frac{1}{4}\sqrt{3}\cdot 2 + \frac{1}{4}\ln|2+\sqrt{3}|] = (2\sqrt{3} + \ln(2+\sqrt{3}))/4$.

47. $y' = \frac{(3x^2+5)}{2\sqrt{x^3+5x}}$; $s = \int_1^8\sqrt{1+\frac{(3x^2+5)^2}{4(x^3+5x)}}dx$. Use NINT to find $S = 22.25369$. If y is graphed in $[0,8]$ by $[-10,30]$, the graph is very close to being the hypoteneuse of a right triangle with sides $8-2=6$ and 21.7, which is about 22.

48. $M = 2\int_0^1 \frac{1}{\sqrt{1-x^2}}dx = 2\lim_{b\to 1^-}\int_0^b \frac{1}{\sqrt{1-x^2}}dx = \lim_{b\to 1^-} 2[\sin^{-1} x]_0^b = \pi$; $M_y = \int_0^1 \frac{2x}{\sqrt{1-x^2}}dx = \lim_{b\to 1^-}\int_0^b \frac{2x}{\sqrt{1-x^2}}dx = \lim_{b\to 1^-}[-2\sqrt{1-x^2}]_0^b = 2$; $\bar{x} = \frac{2}{\pi}$, $\bar{y} = 0$, by symmetry.

49. $V = \pi\int_0^1(-\ln x)^2dx = \pi\lim_{b\to 0^+}\int_b^1(\ln x)^2dx$ (see problem 43) $= \pi\lim_{b\to 0^+}[x(\ln x)^2 - 2(x\ln x - x)]_b^1 = \pi\lim_{b\to 0^+}[2 - b(\ln b)^2 + 2b\ln b - 2b] = \pi[2-0+0-0] = 2\pi$.

50. Graph on $[0,2]$ by $[-2,2]$; since $r^2 \geq 0$, $V = \pi\int_0^2 x^2(\ln x)^2dx = \lim_{b\to 0^+}\pi\int_b^2 x^2(\ln x)^2dx$. Using integration by parts twice: $\int x^2(\ln x)^2dx = \frac{x^3}{3}(\ln x)^2 - \frac{2}{3}\int x^2\ln x\,dx = \frac{x^3}{3}(\ln x)^2 - \frac{2}{3}[\frac{x^3}{3}\ln x - \int\frac{x^2}{3}]$. Hence $V = \lim_{b\to 0^+}\pi[\frac{x^3}{3}(\ln x)^2 - \frac{2}{9}x^3\ln x + \frac{2}{3}\cdot\frac{x^3}{9}]_b^2 = \pi[\frac{8}{3}(\ln 2)^2 - \frac{16}{9}\ln 2 + \frac{16}{27}]$.

51. $\int_{-\infty}^0 \frac{4x}{x^2+1}dx + \int_0^\infty \frac{4x}{x^2+1}dx$ is the sum of divergent integrals. Therefore, $\int_{-\infty}^\infty \frac{4x}{x^2+1}dx$ diverges.

52. $\int_{-\infty}^\infty \frac{8dx}{2x^2+3} = 2\int_0^\infty \frac{8dx}{2x^2+3}$ which becomes $\lim_{b\to\infty}[C\tan^{-1}Kx]_0^b$ for some constants C, K. The integral converges.

53. Rewrite the integral as $\lim_{c\to -\infty}\int_c^0 \frac{e^{-x}dx}{e^{-x}+e^x} + \lim_{b\to\infty}\int_0^b \frac{e^{-x}dx}{e^{-x}+e^x}$. We will show that the first is divergent: let $u = -e^{-x}$, then $\frac{1}{u} = -e^x$, $du = e^{-x}dx$, and, since $\frac{e^{-x}dx}{e^{-x}+e^x} = \frac{-e^{-x}dx}{-e^{-x}-e^x}$ the integral becomes $\lim_{c\to-\infty}\int_{-e^{-c}}^{-1}\frac{-du}{u+1/u} = \lim_{c\to-\infty}\int_{-1}^{-e^{-c}}\frac{udu}{u^2+1} = \lim_{c\to-\infty}[\frac{1}{2}\ln(u^2+1)]_{-1}^{-e^{-c}}$ as $c\to-\infty$, the logarithm becomes infinite.

54. For $0 < x$, $0 < \frac{x^3}{1+e^x} < \frac{x^3}{e^x}$; by the Domination test, if $\int_0^\infty \frac{x^3}{e^x}dx$ converges, so will $\int_0^\infty \frac{x^3}{1+e^x}dx$. By tabular integration by parts: $\lim_{b\to\infty}\int_0^b x^3e^{-x}dx = \lim_{b\to\infty} -e^{-x}[x^3+3x^2+6x+6]_0^b$ which is finite.

55. Rewrite as $\frac{dy}{dx} = \frac{e^{y-2}}{e^{x+2y}} = e^{-(2+x+y)}$, $y(0) = -2$. Since $y(2)$ is asked for, set the range to $[0,2]$ by $[-5,5]$. Run IMPEULG with $h = 0.1$. To estimate $y(2)$, use IMPEULT. $y(2) \approx -1.377$. Problems 56-62 are done in a similar fashion. you may need to adjust the Y-range values.

56. $\frac{dy}{dx} = \frac{-y\ln y}{(1+x^2)}$; with $h = 0.1$, $y(2) = 1.393$.

57. Taking $h = 0.1$, $y(6) = -7.349$.

58. $y(5) = 0.144$

59. $\frac{dy}{dx} = \frac{-(x^2+y)}{e^y+x}$; taking $h = 0.1$ gives $y(3) = -2.691$.

60. $\frac{dy}{dx} = \frac{-y(e^x+\ln y)}{x+y}$; taking $h = (1-\ln 2)/10$ and 10 steps gives $y(1) \approx 0.675$.

61. $\frac{dy}{dx} = \frac{x-2y}{x+1}$; $y(3) = 0.907$.

62. $\frac{dy}{dx} = \frac{x^2+1-2y}{x}$; taking $h = 0.1$ gives $y(6) = 9.508$.

63. $\frac{dy}{y-22} = -0.15dx \Rightarrow \ln|y-22| = -0.15x + C' \Rightarrow |y-22| = Ce^{-0.15x}$. $y(0) = 50 \Rightarrow |y-22| = y-22 \Rightarrow y-22 = Ce^{-0.15x}$. Applying the initial condition gives $y = 28e^{-0.15x} + 22$.

64. $\frac{dx}{x-55} = 0.03dt \Rightarrow \ln|x-55| = 0.03t + C' \Rightarrow |x-55| = Ce^{0.03t}$; $x(0) = 50 \Rightarrow |x-55| = 55-x$. Hence $55-x = Ce^{0.03t}$; $5 = C$ and $x = 55 - 5e^{0.03t}$.

65. $\frac{dP}{(500-P)P} = 0.002dt \Rightarrow \frac{1}{500}[\frac{1}{500-P} + \frac{1}{P}]dP = 0.002dt$; $[\frac{dP}{500-P} + \frac{dP}{P}] = 1 \cdot dt \Rightarrow -\ln|500-P| + \ln|P| = t + C' \Rightarrow |\frac{P}{500-P}| = Ce^t$. $P(0) = 20 \Rightarrow \frac{P}{500-P} = Ce^t$ where $C = \frac{20}{480} = \frac{1}{24}$. Solving for P gives $P = \frac{500e^t}{24+e^t}$.

66. $\frac{dP}{P(200-P)} = 0.0055dt \Rightarrow [\frac{1}{200-P} + \frac{1}{P}] = 1.1dt \Rightarrow |\frac{P}{200-P}| = Ce^{1.1t}$. $P(0) = 50 \Rightarrow \frac{50}{150} = \frac{1}{3} = C$. Solving $\frac{P}{200-P} = \frac{1}{3}e^{1.1t}$ gives $P = \frac{200e^{1.1t}}{3+e^{1.1t}} = \frac{200}{1+3e^{-1.1t}}$.

67. $\frac{dx}{x(a-x)} = kdt \Rightarrow \frac{1}{a}[\frac{1}{x} + \frac{1}{a-x}]dx = kdt$ or $\frac{dx}{x} + \frac{dx}{a-x} = kadt$. Assume $x(t_0) = x_0$. Then $\ln|\frac{x}{a-x}| = kat + C'$, or, $|\frac{x}{a-x}| = Ce^{akt}$. Assuming $x_0 < a$, this leads to $x = (a-x)[\frac{x_0}{a-x_0}]e^{akt}$. Multiplying by $(a-x_0)$ gives $x[(a-x_0) + x_0e^{akt}] = ax_0e^{akt}$. Before solving for x, multiply by e^{-akt} to get $x = \frac{ax_0}{x_0+(a-x_0)e^{-akt}}$.

CHAPTER 9

INFINITE SERIES

9.1 LIMITS OF SEQUENCES OF NUMBERS

1. $a_1 = 0,\ a_2 = -0.25,\ a_3 = -0.22222\ldots = -2/9,\ a_4 = -0.1875 = -3/16$

2. $a_1 = 1,\ a_2 = \frac{1}{2},\ a_3 = \frac{1}{6},\ a_4 = \frac{1}{24}$

3. $a_1 = (-1)^2/(2\cdot 1 - 1) = 1/1 = 1,\ a_2 = -1/3,\ a_3 = 1/5,\ a_4 = -1/7$

4. $a_1 = 2 + (-1) = 1,\ a_2 = 2 + (-1)^2 = 3,\ a_3 = 1,\ a_4 = 3$

5. $x_1 = 1,\ x_2 = 1 + \frac{1}{2} = \frac{3}{2},\ x_3 = \frac{3}{2} + \frac{1}{4} = \frac{7}{4},\ x_4 = \frac{7}{4} + \frac{1}{8} = \frac{15}{8},\ x_5 = \frac{31}{16},\ x_6 = \frac{63}{32},\ x_7 = \frac{127}{64},\ x_8 = \frac{255}{128},\ x_9 = \frac{511}{256},\ x_{10} = \frac{1023}{512}$

6. $x_1 = 1,\ x_2 = \frac{1}{2},\ x_3 = \frac{\left(\frac{1}{2}\right)}{3} = \frac{1}{6},\ x_4 = \frac{\left(\frac{1}{6}\right)}{4} = \frac{1}{24}, \ldots, x_{10} = \frac{1}{10!}$

7. $x_1 = 2,\ x_2 = \frac{2}{2} = 1, x_3 = \frac{1}{2},\ x_4 = \frac{1}{2^2}, \ldots, x_{10} = \frac{1}{2^8}$

8. $x_1 = -2,\ x_2 = 1\cdot(-2)/2 = -1,\ x_3 = 2(-1)/3 = -\frac{2}{3},\ x_4 = \frac{3\left(-\frac{2}{3}\right)}{4} = -\frac{2}{4} = -\frac{1}{2},\ x_5 = \frac{-4\left(-\frac{2}{4}\right)}{5} = -\frac{2}{5}$; the pattern is $x_n = -\frac{2}{n}$; $x_{10} = -\frac{2}{10}$

9. $x_1 = 1,\ x_2 = 1,\ x_3 = 2,\ x_4 = 3,\ x_5 = 5,\ x_6 = 8,\ x_7 = 13,\ x_8 = 21,\ x_9 = 34,\ x_{10} = 55$

10. $x_1 = 1,\ x_2 = 1,\ x_3 = 1 + 1 = 2,\ x_4 = 1 + 1 + 2 = 4,\ x_5 = 1 + 1 + 2 + 4 = 8$; $x_{10} = 2^{10-2} = 2^8$

11. $\lim_{n\to\infty} (0.1)^n = 0,\ \lim_{n\to\infty} 2 = 2 \Rightarrow \lim_{n\to\infty} a_n = 2$; the sequence converges to 2.

12. $\{a_n\} = \{0, 2, 0, 2, \ldots\}$; diverges

13. $\lim_{n\to\infty} a_n = \lim_{n\to\infty} 5 = 5$; converges

14. $\lim_{n\to\infty} a_n = \lim_{n\to\infty} 10 = 10$; converges

15. $\lim_{n\to\infty} a_n = \infty$; the sequence diverges

16. $\lim_{n\to\infty} a_n = \infty$; diverges

17. $a_n = \frac{1+2n-4n}{1+2n} = 1 - 4\left(\frac{n}{1+2n}\right) = 1 - 4\left(\frac{1}{\frac{1}{n}+2}\right)$. Since $\lim_{n\to\infty} 1/n = 0$, $\lim_{n\to\infty} a_n = 1 - 4/2 = -1$. The sequence converges.

18. $\lim_{n\to\infty} a_n = \lim_{n\to\infty} \frac{2+\frac{1}{n}}{\frac{1}{n}-3} = -\frac{2}{3}$; converges

19. $a_n = n\left[\frac{n-2+\frac{1}{n}}{n-1}\right]$; the bracketed factor converges to 1. Hence $\lim_{n\to\infty} a_n = \infty$. The sequence diverges.

20. $\lim_{n\to\infty} a_n = \lim_{n\to\infty} \frac{\frac{1}{n}+\frac{3}{n^2}}{1+\frac{5}{n}+\frac{6}{n^2}} = 0$; converges

21. $a_n = \frac{-5+\frac{1}{n^4}}{1+\frac{8}{n}}$; $\lim_{n\to\infty} a_n = -5$; the sequence converges

22. $\lim_{n\to\infty} a_n = \lim_{n\to\infty} \frac{n^3}{n^2}\frac{(\frac{1}{n^3}-1)}{\frac{70}{n^2}-4} = +\infty$; diverges

23. $a_n = 1 + \frac{(-1)^n}{n}$; $\lim_{n\to\infty} a_n = 1$; converges

24. $\{a_n\} = \{0, \frac{1}{2}, -\frac{2}{3}, \frac{3}{4}, -\frac{4}{5}, \frac{5}{6}, \ldots\}$; half the terms approach 1, the others approach -1; diverges

25. $\lim_{n\to\infty}\left(\frac{n+1}{2n}\right) = \lim_{n\to\infty}\left(\frac{1+\frac{1}{n}}{2}\right) = \frac{1}{2}$; $\lim_{n\to\infty}\left(1-\frac{1}{n}\right) = 1$. The limits of both factors exist. Hence $\lim_{n\to\infty} a_n = \frac{1}{2}\cdot 1 = \frac{1}{2}$.

26. $\lim_{n\to\infty} a_n = \lim_{n\to\infty}\left(2-\frac{1}{2^n}\right)\lim_{n\to\infty}\left(3+\frac{1}{2^n}\right) = 2\cdot 3 = 6$; converges

27. $|a_n| = \frac{1}{2n-1}$; hence $\lim_{n\to\infty}|a_n| = 0$. This implies $\lim_{n\to\infty} a_n = 0$.

28. $\lim_{n\to\infty} a_n = 0$

29. $\lim_{n\to\infty}\frac{\sin n}{n} = 0$.

30. $\lim_{n\to\infty} a_n = \lim_{n\to\infty}\frac{\sin^2 n}{2^n}$; $0 < \frac{\sin^2 n}{2^n} \le \frac{1}{2^n}$; by the Sandwich theorem $\lim_{n\to\infty} a_n = 0$.

31. $\lim_{n\to\infty}\frac{2n}{n+1} = 2$. Since $\sqrt{\frac{2x}{x+1}}$ is a continuous function of x, $\lim_{n\to\infty}\sqrt{\frac{2n}{n+1}} = \sqrt{\lim_{n\to\infty}\frac{2n}{n+1}} = \sqrt{2}$.

32. By Theorem 3, $\lim_{n\to\infty}\sin\left(\frac{\pi}{2}+\frac{1}{n}\right) = \sin\left(\lim_{n\to\infty}\left(\frac{\pi}{2}+\frac{1}{n}\right)\right) = \sin\frac{\pi}{2} = 1$.

33. Since $\tan^{-1}x$ is a continuous function, $\lim_{n\to\infty}\tan^{-1}n = \tan^{-1}(\lim_{n\to\infty} n) = \pi/2$.

34. $\lim_{n\to\infty} a_n = \lim_{n\to\infty} \ln(\frac{n}{n+1}) = \ln\left(\lim_{n\to\infty}\left(\frac{n}{n+1}\right)\right) = \ln 1 = 0.$

35. Graphing $y = x/2^x$ on $[0,100]$ by $[-2,2]$ suggests the sequence converges to 0. Using L'Hôpital's rule gives $\lim_{n\to\infty} \frac{n}{2^n} = \lim_{n\to\infty} \frac{1}{2^n \ln 2} = 0.$

36. $\lim_{n\to\infty} \frac{3^n}{n^3} = +\infty$; apply l'Hôpital's rule to $\lim_{n\to\infty} \frac{3^x}{x^3}$ three times.

37. Graphing $y = \frac{\ln(x+1)}{x}$ suggests that $\{a_n\}$ converges to 0. Analytically, $\lim_{n\to\infty} a_n = \lim_{n\to\infty} \frac{\frac{1}{x+1}}{1} = 0.$

38. $\lim_{n\to\infty} a_n = \lim_{n\to\infty} \frac{\ln n}{\ln 2 + \ln n} = \lim_{n\to\infty} \frac{1}{\frac{\ln 2}{\ln n}+1} = 1.$

39. Graphing $y = 8^{1/x}$ on $[0,100]$ by $[-2,2]$ suggests $\{a_n\}$ converges to 1. Analytically, $\lim_{n\to\infty} 8^{1/n} = 8^{\lim(1/n)} = 8^0 = 1.$

40. $\lim_{n\to\infty}(0.03)^{1/n} = 1$ by Table 9.1.

41. $\lim_{n\to\infty}\left(1+\frac{7}{n}\right)^n = e^7$, using Table 9.1, #5.

42. $\lim_{n\to\infty}\left(1-\frac{1}{n}\right)^n = \lim_{n\to\infty}\left(1+\frac{-1}{n}\right)^n = e^{-1}$

43. $\lim_{n\to\infty}(0.9)^n = 0$; the sequence diverges

44. $\lim_{n\to\infty} \frac{(n+1)!}{n!} = \lim_{n\to\infty}(n+1) = +\infty$; diverges

45. Graphing $y = (10x)^{(1/x)}$ suggests that $\{a_n\}$ converges to 1. Analytically, let $w = (10x)^{1/x}$; $\ln w = \frac{1}{x}\ln(10x)$; $\lim_{x\to\infty} \ln w = \lim_{x\to\infty} \frac{\frac{10}{10x}}{1} = 0$. $\ln w \to 0 \Rightarrow w \to 1$.

46. $\lim_{n\to\infty} \sqrt[n]{n^2} = \lim_{n\to\infty} (\sqrt[n]{n})^2 = (\lim_{n\to\infty} \sqrt[n]{n})^2 = 1^2 = 1$

47. Let $w = (\frac{3}{x})^{1/x}$; $\ln w = \frac{\ln(3/x)}{x} = \frac{\ln 3 - \ln x}{x}$; $\lim_{x\to\infty} \ln w = \lim_{x\to\infty} \frac{-\frac{1}{x}}{1} = 0$. Thus, $\{a_n\}$ converges to 1.

48. Let $k = n+4$; $\lim_{n\to\infty} a_n = \lim_{k\to\infty} k^{1/k} = 1$, by Table 9.1.

49. Recall that $\lim_{n\to\infty} n^{1/n} = 1$. Since the numerator grows large without bound, $\lim_{n\to\infty} a_n = \lim_{n\to\infty} \frac{\ln n}{1} = \infty$; $\{a_n\}$ diverges.

50. $\lim_{n\to\infty} a_n = \lim_{n\to\infty} 4\sqrt[n]{n} = 4$, by Table 9.1.

51. $\lim_{n\to\infty} x^n = 0$ if $|x| < 1$. $\{(\frac{1}{3})^n\}$ converges to 0.

52. $\lim_{n\to\infty} a_n = \lim_{n\to\infty} \frac{1}{(\sqrt{2})^n} = 0$

53. By the squeeze theorem, $0 < \frac{1}{n!} < \frac{1}{n} \Rightarrow \{\frac{1}{n!}\}$ converges to 0.

54. By Table 9.1, #6, $\lim_{n\to\infty} \frac{(-4)^n}{n!} = 0$

55. Graphing the function $y = (\frac{1}{x})^{1/\ln x}$ and using trace indicates that the sequence converges to 0.3678794. Analytically, let $w = (\frac{1}{x})^{1/\ln x}$; $\ln w = \frac{1}{\ln x}\ln(\frac{1}{x}) = \frac{-\ln x}{\ln x} = -1$. $\lim_{x\to\infty} w = e^{-1} = 0.3678\ldots$.

56. $\lim_{n\to\infty} a_n = \lim_{n\to\infty} \frac{1}{\frac{6^n}{n!}} = +\infty$; the sequence diverges

57. $a_{10} = 3.6\ E - 54$; $a_{100} = 9.3\ E - 443$ certainly suggests that $\lim_{n\to\infty} a_n = 0$.

58. $\lim_{n\to\infty} a_n = \lim_{n\to\infty} \frac{(3^2)^n}{n!} = 0$, by Table 9.1.

59. Graph $y_1 = \text{abs}(0.5\hat{\ }(1/x) - 1)$ and $y_2 = 0.001$ on $0 \le x \le 1000$; find their intersection (at $x = 692.8$). For $n \ge 693$, the inequality will hold.

60. Find the intersection of $y = \text{abs}(x\hat{\ }(1/x) - 1)$ and $y = 0.001$, on $0 \le x \le 10000$; the curves intersect at $x = 9123.126$. The inequality holds for $n > N = 9123$.

61. Graph $y = 0.9\hat{\ }x$ and $y = 0.001$ on $[0, 100]$; the curves intersect at $x = 65.563$. $n \ge N = 66 \Rightarrow 0.9^n < 10^{-3}$.

62. If $n \ge 15$, $\frac{2^n}{n!} < 10^{-7}$.

63. The sequence can be written $2^{\frac{1}{2}}, 2^{\frac{1}{4}}, 2^{\frac{1}{8}}, \ldots$. Graph $y = 2\hat{\ }(1/2\hat{\ }x)$ and $y = 1$ on $[0, 6]$ by $[0, 2]$.

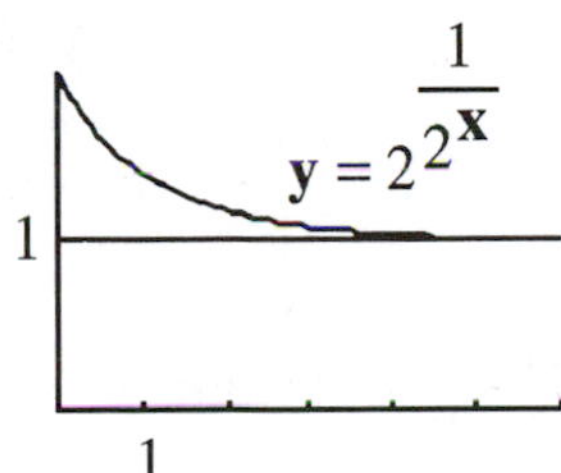

64. Graph $\frac{\ln x}{x}$ in $[0, 100]$ by $[0, 1]$.

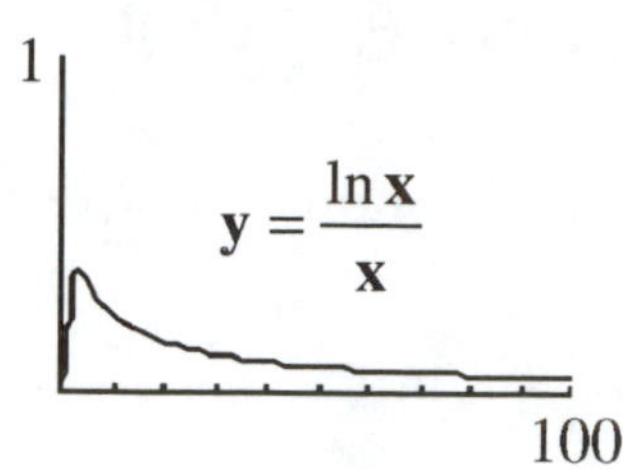

65. Equation 5. Let $a_n(x) = (1+\frac{x}{n})^n$; $a_{10}(0.5) = 1.629$, $a_{100}(0.5) = 1.647$, $e^{0.5} = 1.649$; $a_{10}(2) = 6.192$, $a_{100}(x) = 7.245$, $a_{1000}(2) = 7.374$, $e^2 = 7.389$.

66. Equation 4. Let $a_n(x) = x^n$. $a_{10}(0.99) = 0.904$, $a_{100}(0.99) = 0.366$, $a_{1000}(0.99) = 0.00004$.

67. $f(x) = \sin x - x^2$, $x_{n+1} = x_n - \frac{f(x_n)}{f'(x_n)} \Rightarrow a_{n+1} = a_n - \frac{\sin a_n - a_n^2}{\cos a_n - 2a_n}$,
$a_1 = 1 \to 0.891, 0.877, 0.877, \ldots$; $a_1 = 2 \to 1.300, 0.989, 0.889, \ldots$;
$a_1 = -1 \to -0.275, -0.045, -0.002, \ldots$.

68. $a_{n+1} = a_n - \frac{\cos a_n - a_n^2}{-\sin a_n - 2a_n}$, $a_1 = 1 \to x = 0.824$, $a_1 = -1 \to x = -0.824$, $a_1 = -2 \to x = -0.824$

69. The sequence stabilizes at 1.57079632679 at x_4.

71. Graph $y = 0.5x$ and $y = \sin x$ on $[-5, 5]$ by $[-2, 2]$. There are two solutions, $0.5x = \sin x \Rightarrow x = 2\sin x$ or $x_{n+1} = 2\sin x_n$. The iteration will converge to the solutions. $x_0 = 1 \Rightarrow \{x_n\}$ converges to $1.895494\ldots$. $x_0 = -1 \Rightarrow \{x_n\}$ converges to $-1.895494\ldots$.

72. Graphing $y = 2x$ and $y = \tan x$ shows there are two non-zero solutions. Let $x_0 = 1$; the iteration $x_{n+1} = \frac{1}{2}\tan x$ converges to 0. Rewrite $\tan x = 2x$ as $x = \tan^{-1}(2x)$. The iteration $x_{n+1} = \tan^{-1}(2x_n)$ leads to $x = 1.165561\ldots$.

9.2 INFINITE SERIES

1. $s_n = 2(1 + \frac{1}{3} + \cdots + \frac{1}{3^{n-1}}) = \frac{2(1-(\frac{1}{3})^n)}{1-\frac{1}{3}} \to 3$, $s_{20} = 2.99999999914$

2. $s_n = \frac{9}{100}\left[1 + \frac{1}{100} + \cdots + \frac{1}{100^{n-1}}\right] = \frac{9}{100}\left[\frac{1-\frac{1}{100^n}}{1-\frac{1}{100}}\right]$; $\lim_{n\to\infty} s_n = \frac{\frac{9}{100}}{\frac{99}{100}} = \frac{1}{11}$

3. $s_n = \frac{1-(1/2)^n}{1-(-1/2)} \to 2/3$. $s_{20} = 0.6666660308$

4. $r = -2, a_1 = 1 \Rightarrow s_n = \frac{1\cdot(1-(-2)^n)}{1+2}$; the series diverges

5. Recall that $\frac{1}{n(n+1)} = \frac{1}{n} - \frac{1}{n+1}$. $s_n = \left[\frac{1}{2} - \frac{1}{3}\right] + \left[\frac{1}{3} - \frac{1}{4}\right] + \cdots + \left[\frac{1}{n+1} - \frac{1}{n+2}\right] = \frac{1}{2} - \frac{1}{n+2} \to \frac{1}{2}$

6. $\frac{5}{n(n+1)} = 5\left[\frac{1}{n} - \frac{1}{n+1}\right] \Rightarrow s_n = 5\left[\frac{1}{1} - \frac{1}{2}\right] + \left[\frac{1}{2} - \frac{1}{3}\right] + \cdots + 5\left[\frac{1}{n} - \frac{1}{n+1}\right] = 5\left[1 - \frac{1}{n+1}\right]$; $\lim_{n\to\infty} s_n = 5$

7. $s_1 = \frac{1}{4^0}$, $s_2 = \frac{1}{4^0} + \frac{1}{4^1}$, etc. $s_5 = 1.3320\ldots$, $s_n = 1 + \frac{1}{4} + \frac{1}{4^2} + \cdots + \frac{1}{4^{n-1}} = \frac{1-(1/4)^n}{1-1/4} = 4/3$

8. $\frac{1}{16} + \frac{1}{64} + \frac{1}{256} + \cdots = \frac{1}{16}\left[\frac{1}{1-1/4}\right] = \frac{1}{12}$

9. $s_1 = \frac{7}{4}, \ldots, s_5 = 2.33105\ldots$, $s_n = 7(\frac{1}{4} + \frac{1}{16} + \cdots + \frac{1}{4^n}) = 7\cdot\frac{1}{4}\left[\frac{1-(1/4)^{n+1}}{1-1/4}\right] = 7/3$

10. $5 - \frac{5}{4} + \frac{5}{16} - \frac{5}{64} + \cdots = 5\left[\frac{1}{1-(-1/4)}\right] = \frac{5}{5/4} = 4$

11. $s_1 = 5 + 1$; $s_5 = 11.181327\ldots$. Break up s_n into two parts: $s_n = 5(1 + \frac{1}{2} + \cdots + \frac{1}{2^{n-1}}) + (1 + \frac{1}{3} + \cdots + \frac{1}{3^{n-1}}) = 5(1-(\frac{1}{2})^n)/(1-\frac{1}{2}) + (1-(\frac{1}{3})^n)/(1-\frac{1}{3}) \to 10 + 3/2 = 11.5$

12. $\sum_{n=0}^{\infty} \frac{5}{2^n} = 5\left[\frac{1}{1-1/2}\right] = 10$; $\sum_{n=0}^{\infty} \frac{1}{3^n} = 1\left[\frac{1}{1-1/3}\right] = \frac{3}{2} \Rightarrow \sum_{n=0}^{\infty}\left(\frac{5}{2^n} - \frac{1}{3^n}\right) = 10 - \frac{3}{2} = \frac{17}{2}$, by Theorem 5.

13. $s_n = (1 + \frac{1}{2} + \cdots + \frac{1}{2^{n-1}}) + (1 - \frac{1}{5} + \frac{1}{25} + \cdots + \frac{(-1)^{n-1}}{5^{n-1}}) = \frac{1-(1/2)^n}{1-1/2} + \frac{1-(-1)^n/5^n}{1-(-1/5)} = 2 + \frac{5}{6} = \frac{17}{6}$

14. $\sum_{n=0}^{\infty}(\frac{2^{n+1}}{5^n}) = 2\sum_{n=0}^{\infty}(\frac{2}{5})^n = 2\left[\frac{1}{1-2/5}\right] = \frac{10}{3}$

15. $\frac{4}{(4n-3)(4n+1)} = \frac{1}{4n-3} - \frac{1}{4n+1}$. Hence $s_n = (\frac{1}{1} - \frac{1}{5}) + (\frac{1}{5} - \frac{1}{9}) + (\frac{1}{9} - \frac{1}{13}) + \cdots \to 1$.

16. If each term is multiplied by 4, the series in Problem 15 results. Hence, $\lim_{n\to\infty} s_n = \frac{1}{4}(1) = \frac{1}{4}$.

17. See #15. $s_n = (\frac{1}{9} - \frac{1}{13}) + (\frac{1}{13} - \frac{1}{17}) + \cdots \to \frac{1}{9}$. $s_{50} = 0.10613\ldots$; $s_{100} = 0.10861\ldots$.

18. $\frac{2n+1}{n^2(n+1)^2} = \frac{1}{n^2} - \frac{1}{(n+1)^2}$; the series collapses so that $s_n = \sum_{k=1}^{n} a_k = 1 - \frac{1}{(n+1)^2}$; the series converges to 1.

19. This is a geometric series with $r = 1/\sqrt{2}$, $0 < r < 1$. $\lim_{n\to\infty} s_n = \frac{1}{1-1/\sqrt{2}} = \frac{\sqrt{2}}{\sqrt{2}-1} = \frac{\sqrt{2}(\sqrt{2}+1)}{1} = 2 + \sqrt{2}$. ($s_{20} = 3.410879\ldots$)

20. $\lim_{n\to\infty} a_n = \lim_{n\to\infty} \ln \frac{1}{n} = -\infty$; diverges

21. $s_n = -3\sum_{k=1}^{n}(-\frac{1}{2})^k \to -3(-\frac{1}{2})(1)/(1+\frac{1}{2}) = 1$. This is a multiple of a convergent geometric series.

22. $\sum_{n=1}^{\infty}(\sqrt{2})^n = \sum_{n=1}^{\infty}(2^{1/2})^n$, a geometric series with $|r| > 1$; diverges

23. The series is $1-1+1-1+1\ldots$; the nth term fails to approach 0; the series diverges

24. $\sum_{n=0}^{\infty} \frac{\cos n\pi}{5^n} = 1 - \frac{1}{5} + \frac{1}{5^2} - \frac{1}{5^3} + \cdots$, a geometric series with $r = -\frac{1}{5}$; converges to $1 \cdot \frac{1}{1+\frac{1}{5}} = \frac{5}{6}$

25. Convergent geometric series: $\sum_{n=0}^{\infty}(\frac{1}{e^2})^n \to \frac{1}{1-1/e^2} = \frac{e^2}{e^2-1}$, $s_{30} = 1.15651764275$

26. $a_n = (n^2+1)/n = n + 1/n$; $\lim_{n\to\infty} a_n \neq 0$, diverges

27. Diverges; $\lim_{n\to\infty} a_n \neq 0$.

28. $\sum_{n=1}^{\infty} \frac{2}{10^n} = 2\sum_{n=1}^{\infty}(\frac{1}{10})^n = \frac{2}{10}(\frac{1}{1-1/10}) = \frac{2}{9}$

29. This is the sum of two convergent, geometric series; $\sum_{n=0}^{\infty}(\frac{2}{3})^n - \sum_{n=0}^{\infty}(\frac{1}{3})^n \to \frac{1}{1-2/3} - \frac{1}{1-1/3} = 3 - \frac{3}{2} = \frac{3}{2}$ ($s_{10} = 1.4480008\ldots$)

30. $a_n = (1 - \frac{1}{n})^n \Rightarrow \lim_{n\to\infty} a_n = \frac{1}{e} \neq 0$; diverges

31. By #6, Table 9.1, $\lim_{n\to\infty} \frac{x^n}{n!} = 0$. Hence $\lim_{n\to\infty} a_n = \infty$. The series diverges.

32. $\sum_{n=0}^{\infty} \frac{1}{x^n} = \sum_{n=0}^{\infty}(\frac{1}{x})^n$ a geometric series with $r = \frac{1}{x}$; $|r| = |\frac{1}{x}| = \frac{1}{|x|} < 1 \Rightarrow$ converges to $\frac{1}{1-1/x} = \frac{x}{x-1}$

33. $\sum_{n=0}^{\infty}(-1)^n x^n = 1 - x + x^2 - \cdots = 1(1 - x + x^2 - \cdots)$. $a = 1$, $r = -x$

34. $(1 - x^2 + x^4 - \cdots) = 1(1 - x^2 + x^4 - \cdots)$. $a = 1$, $r = -x^2$

35. a) The sum of the series is $\frac{1}{6} = 0.1666666666\ldots$. If the series is written as $a_0 + a_1 + \cdots$, then $a_n = \frac{1}{3^{n+2}}$, $s_{17} = 0.166666666235$, i.e., 18 terms needed.

b) The sum of the series is $\frac{8}{3}$. If $a_n = 4(-\frac{1}{2})^n$, then $s_n = a_0 + a_1 + \cdots + a_n$; $s_{30} = 2.66666666791$, i.e., 31 terms needed.

36. Analytically, $|s_k - 1| = \frac{1}{k+1}$. If $\frac{1}{k+1} < 0.01$, $k > 100 - 1 = 99$. The smallest value of k is 100. Using PARTSUMT to compute s_{99} and s_{100} confirms this.

37. The ball travels $4+2(0.75)4+2(0.75)^2 4+\cdots$ m $= 8[1+0.75+0.75^2+\cdots]-4 = 8\left[\frac{1}{1-0.75}\right] - 4 = 28$ m.

38. The dropped ball takes $\sqrt{\frac{4}{4.9}}$ seconds to hit the floor. On each succeeding bounce, the time to the top of the bounce is the same as the time to the next hit $= \sqrt{\frac{\text{height}}{4.9}}$. Hence time $= \sqrt{\frac{4}{4.9}} + 2\left[\sqrt{\frac{4(0.75)}{4.9}} + \sqrt{\frac{4(0.75)^2}{4.9}} + \cdots\right] = 2\sqrt{\frac{4}{4.9}}\left[1 + (0.75)^{1/2} + (0.75)^{2/2} + (0.75)^{3/2} + \cdots\right] - \sqrt{\frac{4}{4.9}} = 2\sqrt{\frac{4}{4.9}}\left[\frac{1}{1-\sqrt{3/4}}\right] - \sqrt{\frac{4}{4.9}} = 12.5842$ sec.

39. Let $x = 0.234234\ldots$, then $1000x = 234.234234\ldots$ and $999x = 234 \Rightarrow x = \frac{234}{999} = \frac{26}{111}$

40. $x = 1.24123123123\ldots = 1.24 + 123(0.001001001001\ldots) = \frac{124}{100} + \frac{123}{100}\left[\frac{1}{1000} + \frac{1}{(1000)^2} + \frac{1}{(1000)^3} + \cdots\right] = \frac{124}{100} + \frac{123}{100}\frac{\frac{1}{1000}}{(1-\frac{1}{1000})} = \frac{124}{100} + \frac{123}{999(100)} = \frac{123{,}999}{99{,}900}$

41. The denominator of the first term must be $2\cdot 3$. If $n = -2$, the $n+4 = 2$, $n+5 = 3$. Hence a) $\sum_{n=-2}^{\infty} \frac{1}{(n+4)(n+5)}$, b) $\sum_{n=0}^{\infty} \frac{1}{(n+2)(n+3)}$, c) $\sum_{n=5}^{\infty} \frac{1}{(n-3)(n-2)}$

42. a) If $n = 1$, then $n - 2 = -1$; let $k = n - 2$. $n = k + 2$, $n + 1 = k + 3$, $\sum_{n=1}^{\infty} \frac{5}{(n+2)(n+1)} = \sum_{k=-1}^{\infty} \frac{5}{(k+2)(k+3)} = \sum_{n=-1}^{\infty} \frac{5}{(n+2)(n+3)}$,

b) If $n = 1$, $n+2 = 3$; let $k = n+2$. As in (a) $\sum_{n=1}^{\infty} \frac{5}{n(n+1)} = \sum_{n=3}^{\infty} \frac{5}{(n-2)(n-1)}$,

c) $n + 19 = 20 \Rightarrow \sum_{n=1}^{\infty} \frac{5}{n(n+1)} = \sum_{n=20}^{\infty} \frac{5}{(n-19)(n-18)}$

43. Corollary 1. Let $\sum a_n$ be divergent, let $b_n = ca_n$. Then $s_k = \sum_{n=1}^{k} b_n = \sum_{n=1}^{k} ca_n = c\sum_{n=1}^{k} a_n$, which diverges.

Corollary 2. Let $s_k = \sum_{n=1}^{k}(a_n \pm b_n) = \sum_{n=1}^{k} a_n \pm \sum_{n=1}^{k} b_n$. The first series converges, the second diverges. Hence $\{s_k\}$ diverges.

44. $\sum_{n=1}^{m}(a_n + b_n) = \sum_{n=1}^{m} a_n + \sum_{n=1}^{m} b_n \to A + B$; $\sum_{n=1}^{m}(a_n - b_n) = \sum_{n=1}^{m} a_n - \sum_{n=1}^{m} b_n \to A - B$; $\sum_{n=1}^{m} ka_n = k\sum_{n=1}^{m} a_n \to kA$.

45. The partial sums of the series can be written $s_k = 1 + \frac{1}{2} + \frac{1}{4} + \frac{1}{8} + \cdots + \frac{1}{2^k} = \sum_{n=0}^{k} \frac{1}{2^n}$. $s_{10} = 1.9990\ldots$, $s_{15} = 1.99999694\ldots$, $s_{20} = 2.000000$; the sum appears to be 2.

46. The partial sums can be written $s_k = 2500 + \frac{2500}{1.05} + \frac{2500}{1.05^2} + \cdots + \frac{2500}{105^k} = \sum_{n=0}^{k} \frac{2500}{(1.05)^n}$; $s_{10} = 21,804.337$, $s_{100} = 52,119.776$, $s_{1000} = 52,500$.

47. $s_k = 0 + 1 + 3 + 5 + \cdots (2k-1)$, which diverges.

48. $s = 1 + \frac{1}{1} + \frac{1}{2} + \frac{\left(\frac{1}{2}\right)}{3} + \frac{\left(\frac{1}{6}\right)}{4} + \frac{\left(\frac{1}{24}\right)}{5} + \cdots = 1 + \frac{1}{1} + \frac{1}{2\cdot 1} + \frac{1}{3\cdot 2\cdot 1} + \frac{1}{4\cdot 3\cdot 2\cdot 1} + \frac{1}{5!} + \cdots$, $s_5 = 2.71667$, $s_{10} = 2.718281\cdots$; the series appears to converge to e

49. Present value $= \sum_{n=1}^{\infty} a_n$, where $a_n = (1.07)^{-1}a_{n-1}$, $a_0 = 5000$. $\sum_{n=1}^{100} a_n = 71,346.25$; $\sum_{n=1}^{1000} a_n = 71,428.571$; $\sum_{n=1}^{1500} a_n = 71,428.571$.

50. Present value $= \lim_{N\to\infty} \sum_{n=1}^{N} a_n = (1.05)^{-1}a_{n-1}$, $a_0 = 1000$. Taking N to be 1500 gives a present value of \$20,000.

51. a) The money needed is $a_1 + a_2 + \cdots + a_{50}$; $a_0 = 500$, $a_{n+1} = 1.07^{-1}a_n$, $N = 50$. \$6,900.38 is needed, b) \$7,134.63

52. a) $a_0 = 2000$, $N = 20$, $a_{n+1} = 1.07^{-1}a_n \Rightarrow$ \$21,188.03 is needed.

b) $a_0 = 2000$, $N = 100$, $a_{n+1} = 1.07^{-1}a_n \Rightarrow$ \$28,538.51 is needed.

53. If a square has side s (area s^2), the next square has side $\sqrt{2(\frac{s}{2})^2} = \frac{s}{\sqrt{2}}$; (area $\frac{s^2}{2}$). The next smaller square has side $\frac{s/\sqrt{2}}{\sqrt{2}} = \frac{s}{2}$; (area $\frac{s^2}{4}$). $s = 2 \Rightarrow$ the first square has area 4. The sum of the areas is $4(1 + \frac{1}{2} + \frac{1}{4} + \cdots) = 4\left[\frac{1}{1-1/2}\right] = 8 \text{ m}^2$.

s / 2

s / 2

s / √2

54. Letting each term correspond to a row of semicircles;
$A = \frac{1}{2}\left[2\pi(\frac{1}{2})^2 + 4\pi(\frac{1}{4})^2 + 8\pi(\frac{1}{8})^2 + \cdots\right] = \frac{\pi}{2}\left[\frac{1}{2} + \frac{1}{4} + \frac{1}{8} + \cdots\right] = \frac{\pi}{4}\left[1 + \frac{1}{2} + \frac{1}{4} + \cdots\right] = \frac{\pi}{2}$.

55. Let $a_n = n$, $b_n = -n$. $\sum a_n$, $\sum b_n$ diverge, but $\sum(a_n + b_n) = \sum 0$ converges.

56. Let $a_n = \frac{1}{2^n}$, $b_n = \frac{1}{3^n}$. Then $\frac{a_n}{b_n} = (\frac{3}{2})^n$ and $\sum(\frac{a_n}{b_n})$ diverges.

57. Let $a_n = \frac{1}{3^n}, \sum_{n=0}^{\infty} a_n = \frac{3}{2}$; let $b_n = \frac{1}{2^n}$, $\sum_{n=0}^{\infty} b_n = 2$, $\frac{a_n}{b_n} = (\frac{2}{3})^n$, $\sum(\frac{2}{3})^n = \frac{1}{1-2/3} = 3$, but $A/B = \frac{3}{4}$.

58. $\sum a_n$ converges $\Rightarrow \lim_{n\to\infty} a_n = 0 = \lim_{n\to\infty}(\frac{1}{a_n}) \neq 0$

9.3 SERIES WITHOUT NEGATIVE TERMS: COMPARISON AND INTEGRAL TESTS

1. Converges; geometric series with $r = 1/10$; sum is $10/9$

2. $\sum_{n=1}^{\infty} -\frac{1}{8^n} = -\sum(\frac{1}{8})^n = -\frac{1/8}{1-1/8} = -\frac{1}{7}$, a geometric series

3. $\lim_{n\to\infty} a_n = \lim_{n\to\infty}(1 - \frac{2}{n+2}) \neq 0$; diverges.

4. Compare with $\sum \frac{1}{n}$; $\lim_{n\to\infty} a_n/b_n = \lim_{n\to\infty} \frac{5/n}{1/n}$; both diverge.

5. $\frac{\sin^2 n}{2^n} > 0$; $\frac{\sin^2 n}{2^n} \leq \frac{1}{2^n}$. Converges by the comparison test; sum $\approx s_{100} = s_{666} = 0.637$

6. $0 \leq \frac{1+\cos n}{n^2} \leq \frac{2}{n^2}$. Converges by comparison test; $s_{100} \approx s_{666} = 1.968$

7. For $n \geq 3$, $\ln n > 1$; $\frac{\ln n}{n} > \frac{1}{n}$. Diverges by comparison test.

8. Diverges; by l'Hôpital's rule, $\lim_{n\to\infty} a_n = \lim_{n\to\infty} \frac{\frac{1}{2\sqrt{n}}}{\frac{1}{n}} = \lim_{n\to\infty} \frac{\sqrt{n}}{2} \neq 0$.

9. $\sum_1^{\infty} \frac{1}{n^{3/2}}$ converges by the integral test, since $\int_1^{\infty} \frac{dx}{x^{3/2}}$ converges.

10. $a_n = (\frac{2}{3})^n$; geometric series with $r = 2/3$. Converges to $\frac{2/3}{1-2/3} = 2$.

11. Diverges by the Limit Comparison Test with $\sum \frac{1}{n}$; $\lim_{n\to\infty} \frac{\frac{2}{n+1}}{\frac{1}{n}} = 2 \neq 0$.

12. For $n \geq 1$, $n \geq \ln n$; $1 + n \geq 1 + \ln n$; $\frac{1}{1+n} \leq \frac{1}{1+\ln n} = a_n$; $\lim_{n\to\infty} \frac{\frac{1}{n+1}}{\frac{1}{n}} = 1$; $\sum \frac{1}{n+1}$ and thus $\sum a_n$ diverge.

13. Diverges by the Limit Comparison Test with $\sum \frac{1}{n}$; $\lim_{n\to\infty} \frac{\frac{1}{n}}{\frac{1}{2n-1}} = 2 \neq 0$.

14. $\lim_{n\to\infty} a_n = \lim_{n\to\infty} \frac{2^n}{n+1} + \lim_{n\to\infty} \frac{2^n \ln 2}{1} = \infty$. Diverges.

15. Diverges; $\lim_{n\to\infty} a_n = e \neq 0$

16. Converges by the Comparison Test since $0 \leq (\frac{n}{3n+1})^n < (\frac{n}{3n})^n = (\frac{1}{3})^n$.

17. Diverges by limit comparison with $\sum \frac{1}{n}$: $\lim_{n\to\infty} \frac{\frac{n}{n^2+1}}{\frac{1}{n}} = 1$.

18. Converges; $\lim_{n\to\infty} \frac{\frac{\sqrt{n}}{n^2+1}}{\frac{1}{n^{3/2}}} = 1$; $\sum \frac{1}{n^{3/2}}$ converges by the integral test since $\int_1^\infty \frac{dx}{x^{3/2}}$ converges.

19. Compare with $\sum \frac{1}{n^{3/2}}$ which is a convergent p-series. $\lim_{n\to\infty} \frac{\frac{1}{\sqrt{n^3+2}}}{\frac{1}{\sqrt{n^3}}} = \lim_{n\to\infty} \sqrt{\frac{n^3}{n^3+2}} = 1$. Converges

20. Diverges by the Comparison Test; $n > 1 \Rightarrow \sqrt[n]{n} > 1 \Rightarrow \frac{1}{n\sqrt[n]{n}} > \frac{1}{n}$; $\sum \frac{1}{n}$ diverges.

21. $\sum_{n=1}^{\infty} \frac{1+n}{n\cdot 2^n} = \sum_{n=1}^{\infty} \frac{1}{n\cdot 2^n} + \sum_{n=1}^{\infty} \frac{1}{2^n}$; the second series is a convergent geometric series; the first converges by the comparison test: $\frac{1}{n2^n} < \frac{1}{2^n}$.

22. $\sum(\frac{1}{\ln 2})^n$ is a geometric series wtih $r = (\frac{1}{\ln 2}) > 1$; diverges.

23. Converges by the Comparison Test; $\frac{1}{3^{n+1}+1} < \frac{1}{3^{n-1}}$; $\sum_{n=1}^{\infty} \frac{1}{3^{n-1}} = 3\sum_1^\infty \frac{1}{3^n}$ which is a geometric series with $r = (\frac{1}{3})$.

24. Converges by the Limit Comparison Test with $\sum \frac{1}{n^2}$.

25. $n = 14$.

26. $5 + \frac{2}{3} + 1 + \frac{1}{7} - 1 + 1 - 1 + 1 + \frac{1}{2} + \frac{1}{3!} + \frac{1}{4!} + \cdots = (5 + \frac{2}{3} + 1 + \frac{1}{7} - 2) + \sum_{k=0}^{\infty} \frac{1}{k!} = (4 + \frac{2}{3} + \frac{1}{7}) + e = \frac{101}{21} + e$.

27. $n \approx 365 * 24 * 60^2 * 13 \times 10^9 \approx 4.09968 \times 10^{17}$; $\ln(n+1) = 40.5548\cdots \leq s_n \leq 41.5548$.

28. $\sum_1^\infty \frac{1}{nx} = (\frac{1}{x})\sum_1^\infty \frac{1}{n}$; if $x \neq 0$, $(\frac{1}{x})$ is finite and the series diverges. For $x = 0$, the terms $a_n = \frac{1}{nx}$ are not defined.

29. $0 \leq \frac{a_n}{n} \leq a_n$; by the comparison test, $\sum(a_n/n)$ converges.

30. Since $\sum a_n$ converges, $\lim_{n\to\infty} a_n = 0$. There exists N such that $n \geq N \Rightarrow 0 \leq a_n < 1$. If $n \geq N$, $0 \leq a_n b_n < 1 \cdot b_n = b_n$. By the Comparison Theorem, since $\sum b_n$ converges, $\sum a_n b_n$ must converge.

31. If $\{s_n\}$ is nonincreasing, with $M \leq s_n$ for every n, then $\{-s_n\}$ is nondecreasing, $-s_n \leq -M \Rightarrow \{-s_n\}$ is bounded above. A bounded, nondecreasing sequence converges. If $\{-s_n\} \to s$ then $\{s_n\} \to -s$. Similarly, if $\{s_n\}$ is nonincreasing and <u>not</u> bounded below, the sequence $\{-s_n\}$ diverges.

9.4 SERIES WITH NONNEGATIVE TERMS: RATIO AND ROOT TESTS

1. Converges, by the ratio test: $\lim_{n\to\infty}\left|\frac{(n+1)^2}{2^{n+1}}\cdot\frac{2^n}{n^2}\right| = \frac{1}{2} < 1$. The partial sums stabilize at 6.

2. Diverges, by the ratio test: $\lim_{n\to\infty}\frac{(n+1)!}{10^{n+1}}\frac{10^n}{n!} = \lim_{n\to\infty}\frac{n+1}{10} = \infty$.

3. Converges, by the ratio test: $\lim_{n\to\infty}\frac{(n+1)^{10}}{10^{n+1}}\cdot\frac{10^n}{n^{10}} = \frac{1}{10} < 1$. $s_{20} = 376.1794\ldots$; s_{40} is essentially the same.

4. Converges, by the ratio test: $\lim_{n\to\infty}\frac{(n+1)^2e^{-(n+1)}}{n^2e^n} = \frac{1}{e} < 1$. The partial sums stabilize at 1.99229.

5. Diverges, by the ratio test: $\lim_{n\to\infty}\frac{(n+1)!e^{-(n+1)}}{n!e^{-n}} = \lim_{n\to\infty}\frac{n+1}{e} = \infty$.

6. Converges, this is a power series. $s = -\frac{2}{3}\frac{1}{1+2/3} = -0.4$, $s_{20} = -0.39987\ldots$

7. $a_n = (1-\frac{2}{n})^n$; $a_n \to e^{-2} \neq 0$. The series diverges.

8. Since $0 < a_n \leq \frac{3}{(1.25)^n}$, and $3\sum(\frac{1}{1.25})^n$ converges, the series converges by the comparison test. Because of the alternating sign, $\{s_n\}$ is slow to stabilize. $s_{100} = 7.55555555402$, $s_{500} = 7.55555555556$

9. Diverges: $\lim_{n\to\infty} a_n = \lim_{n\to\infty}(1-\frac{3}{n})^n = e^{-3} \neq 0$.

10. Diverges: $\lim_{n\to\infty}(1-\frac{1}{n^2})^n = \lim_{n\to\infty}\left[(1-\frac{1}{n^2})^{n^2}\right]^{1/n} = e^0 = 1$.

11. Diverges: $\sum\frac{1}{n} - \sum\frac{1}{n^2}$ is the difference between a divergent series and a convergent series.

12. Converges, by the comparison test: $0 < \frac{\ln n}{n^3} \leq \frac{n}{n^3} = \frac{1}{n^2}$ and $\sum 1/n^2$ converges. Estimated sum is between 0.1981 and 0.2. ($s_{200} = 0.19805\ldots$, $s_{500} = 0.19811\ldots$, $s_{2000} = 0.198125\ldots$)

13. For $n \geq 3$, $\ln n > 1$. Series diverges by the comparison test: $n \geq 3 \Rightarrow a_n > \frac{1}{n}$.

14. Converges, by the ratio test: $\lim_{n\to\infty}\frac{(n+1)\ln(n+1)}{2^{n+1}}\cdot\frac{2^n}{n\ln n} = \frac{1}{2} < 1$. Estimated sum is $1.786\ldots$.

15. Converges, by the ratio test: $\lim_{n\to\infty}\frac{(n+2)(n+3)}{(n+1)!}\frac{n!}{(n+1)(n+2)} = \lim_{n\to\infty}\frac{n+2}{(n+1)^2} = 0$. Estimated value is $17.0280\ldots$.

16. Converges, by the ratio test: $\lim_{n\to\infty} \frac{e^{-(n+1)}(n+1)^3}{e^{-n}n^3} = \frac{1}{e} < 1$. Estimated sum is $6.0065\ldots$.

17. Converges, by the ratio test: $\lim_{n\to\infty} \frac{(n+4)!}{3!(n+1)!3^{n+1}} \cdot \frac{3!n!3^n}{(n+3)!} = \lim_{n\to\infty} \frac{n+4}{3(n+1)} = \frac{1}{3} < 1$. Estimated sum is 4.0625.

18. Converges, by the ratio test: $\frac{a_{n+1}}{a_n} = \frac{1}{(2(n+1)+1)!} \frac{(2n+1)!}{1} = \frac{1}{(2n+3)(2n+2)}$. Estimated value is $0.17520\ldots$.

19. The ratio test can be applied to $\sum_{n=1}^{\infty} \frac{n^2}{2^n}$, which is a series with nonnegative terms: $\lim_{n\to\infty} \frac{(n+1)^2}{2^{n+1}} \cdot \frac{2^n}{n^2} = \frac{1}{2} < 1$. The series converges; the estimated value is -6.

20. Converges, by the ratio test: $\lim_{n\to\infty} \frac{(n+1)!}{(n+1)^{n+1}} \cdot \frac{n^n}{n!} = \lim_{n\to\infty}(n+1)(\frac{n}{n+1})^n \frac{1}{n+1} = \lim_{n\to\infty} \frac{1}{(\frac{n+1}{n})^n} = \frac{1}{e} < 1$. Estimated sum is $1.87985386\ldots$.

21. When there are n's in the exponents, and no factorials, the root test is likely to be successful. $\lim_{n\to\infty} \sqrt[n]{a_n} = \lim_{n\to\infty} \frac{\sqrt[n]{n}}{\ln n} = 0$. The series converges; estimated sum is $8.25271035\ldots$.

22. Diverges, by the comparison test. For $n \geq 2$, $\ln n < \sqrt{n}$; hence $\frac{1}{(\ln n)^2} > \frac{1}{n}$.

23. Converges, by the comparison test: $a_n = \frac{1}{(n+1)(n+2)} < \frac{1}{n^2}$; $\sum \frac{1}{n^2}$ converges. Since $s_{500} = 0.498007$, and $s_{2000} = 0.499500$, a reasonable estimate is that $s = 0.5$.

24. Converges, by the ratio test: $\lim_{n\to\infty} \frac{(n+1)!}{(2n+3)!} \frac{(2n+1)!}{n!} = 0$. Estimated sum is $0.18459302\ldots$.

25. Diverges, by the ratio test: $\lim_{n\to\infty} \frac{3^{n+1}}{(n+1)^3 2^{n+1}} \cdot \frac{n^3 2^n}{3^n} = \frac{3}{2} > 1$.

26. By #20, $\sum \frac{n!}{n^n}$ converges. Hence $\frac{n!}{n^n} \to 0$ and for n sufficiently large, $\frac{n!}{n^n} < 1$ and $(\frac{n!}{n^n})^n < \frac{n!}{n^n}$. Once we observe that $\frac{1}{n^{n^2}} < \frac{1}{n^n}$, then $a_n \leq (\frac{n!}{n^n})^n < \frac{n!}{n^n}$ and the series converges, by the comparison test. The estimated sum is 1.26105.

27. Converges, by the ratio test: $\lim_{n\to\infty} \frac{a_{n+1}}{a_n} = \lim_{n\to\infty} \frac{1+\sin n}{n} = 0$. The series stabilizes at 2.680118.

28. Diverges, by the ratio test: $\lim_{n\to\infty} \frac{a_{n+1}}{a_n} = \lim_{n\to\infty} \frac{3n-1}{2n+5} = \frac{3}{2}$.

29. The ratio test is inconclusive: $\lim_{n\to\infty} \frac{a_{n+1}}{a_n} = 1$. Generating the first few terms gives $a_1 = 3$, $a_2 = \frac{1}{2} \cdot 3$, $a_3 = \frac{2}{3} \cdot \frac{3}{2} = \frac{3}{3}$, $a_4 = \frac{3}{4} \cdot \frac{3}{3} = \frac{3}{4}, \ldots$, $a_n = \frac{3}{n}$, the series diverges.

30. Converges, by the ratio test: $\lim_{n\to\infty} \frac{a_{n+1}}{a_n} = 0$. Estimated sum is 14.778112.

31. Converges, by the ratio test: $\lim_{n\to\infty} \frac{1+\ln n}{n} = 0$. Estimated sum is -2.119527.

32. For $n > e^{10}$, i.e., $n > 22026$, $\ln n > 10 \Rightarrow a_{n+1} > 1 \cdot a_n$. In other words, for $n > e^{10}$ all a_n exceed a_{22027}; $\lim_{n\to\infty} a_n \neq 0$. The series diverges.

33. Converges, by the ratio test: $\lim_{n\to\infty} \frac{2^{n+1}(n+1)!(n+1)!}{(2n+2)!} \frac{(2n)!}{2^n n! n!} = \lim_{n\to\infty} \frac{2(n+1)^2}{(2n+1)(2n+2)} = \frac{1}{2}$. The sum stabilizes at $2.5707963\ldots$.

34. Diverges, by the ratio test: $\lim_{n\to\infty} \frac{(3n+3)!}{(n+1)!(n+2)!(n+3)!} \frac{n!(n+1)!(n+2)!}{(3n)!} = \lim_{n\to\infty} \frac{(3n+1)(3n+2)(3n+3)}{(n+3)(n+2)(n+1)} = 27 > 1$.

35. Following the hint $a_1 = 1$, $a_2 = \frac{2\cdot 1}{4\cdot 3}$, $a_3 = \frac{3\cdot 2\cdot 2\cdot 1}{5\cdot 4\cdot 4\cdot 3}$, $a_4 = \frac{3\cdot 2\cdot 2\cdot 1}{6\cdot 5\cdot 5\cdot 4}$, $a_5 = \frac{3\cdot 2\cdot 2\cdot 1}{7\cdot 6\cdot 6\cdot 5}$. In general, $a_n = \frac{3\cdot 2\cdot 2\cdot 1}{(n+2)(n+1)^2 n} = \frac{12}{(n+2)(n+1)^2 n} < \frac{12}{n^4}$. The series converges, by comparison test. The estimated sum is $1.26079119\ldots$.

36. Root test: $\lim_{n\to\infty} \left[\frac{1}{n^p}\right]^{1/n} = \lim_{n\to\infty} \frac{1}{n^{p/n}} = \lim_{n\to\infty} n^{-p/n} = \lim_{n\to\infty} e^{-(p/n)\ln n} = e^{\lim(\frac{-p\ln n}{n})} = e^0 = 1$. Ratio test: $\lim_{n\to\infty} \frac{1}{(n+1)^p} n^p = \lim_{n\to\infty} (\frac{n}{n+1})^p = \left[\lim_{n\to\infty} (\frac{n}{n+1})\right]^p = 1$. Both tests are inconclusive.

9.5 ALTERNATING SERIES AND ABSOLUTE CONVERGENCE

1. Converges by absolute convergence theorem since $\sum \frac{1}{n^2}$ converges. The absolute value of the error, $|s_n - L| < \frac{1}{(n+1)^2}$. Solving $\frac{1}{(n+1)^2} < 0.001$ gives $\sqrt{\frac{1}{0.001}} < n+1$ or $n \geq 31$. $s_{32} = 0.82199\ldots < L < 0.82297\ldots = s_{31}$.

2. Converges by the Alternating Series Theorem since (i), $f(x) = 1/\ln x \Rightarrow f'(x) = -\frac{1}{x(\ln x)^2} < 0 \Rightarrow f(x)$ is decreasing, and (ii), $1/\ln n \to 0$. To find n such that $\frac{1}{\ln(n+1)} < 0.001$, rewrite the inequality as $\frac{1}{0.001} < \ln(n+1)$. $e^{\frac{1}{0.001}} = n+1 \Rightarrow n \approx 1.97 \times 10^{434}$.

3. Diverges; $\lim_{n\to\infty} a_n \neq 0$.

4. Diverges; $\lim_{n\to\infty} |a_n| = \lim_{n\to\infty} \frac{10^n}{n^{10}} \neq 0$.

5. Converges by Alternating Series Theorem. Let $f(x) = \frac{\sqrt{x}+1}{x+1}$; $\lim_{n\to\infty} f(x) = \lim_{n\to\infty} a_n = 0$; to show f is decreasing, compute $f'(x) = \frac{(1-x-2\sqrt{x})}{2\sqrt{x}(x+1)^2} < 0$. To find sums within 0.001, try this approach: for large values of n, $\frac{\sqrt{n+1}}{n+1} \approx \frac{\sqrt{n}}{n} = \frac{1}{\sqrt{n}}$. Solving $\frac{1}{\sqrt{n+1}} < 0.001$ gives $n \approx 1{,}000{,}000$.

6. Converges by absolute convergence theorem. $s_{102} = 0.76466\ldots$, $s_{103} = 0.7656\ldots$; s lies between these bounds. $\frac{s_{102}+s_{103}}{2} = 0.7651$ is a good approximation.

7. Converges: $a_n = f(n)$, where $f(x) = (\ln x)/x$. $f'(x) = \frac{1-\ln x}{x^2} < 0$ for $x > e$. Hence $a_n > a_{n+1}$. $\text{Lim}_{x\to\infty} \frac{\ln x}{x} = \lim_{x\to\infty} \frac{\frac{1}{x}}{1} = 0$. The series converges very slowly; look at the graph of $(\ln x)/x$ on $[0, 10{,}000]$ by $[0, 0.01]$. The series needs at least $10{,}000$ terms to be within 0.001 of its sum.

8. $\ln(n)/\ln(n^2) = \frac{1}{2}$; $a_n \not\to 0$; diverges

9. $|a_n| = \frac{3\sqrt{n+1}}{\sqrt{n}+1} = \frac{3\sqrt{n}\sqrt{1+1/n}}{\sqrt{n}(1+1/\sqrt{n})} \to 3$. Diverges.

10. ~~Converges~~ diverges: $f(x) = \ln(1 + 1/x) \Rightarrow f'(x) = \frac{-\frac{1}{x}}{(1+\frac{1}{x})^2} < 0$; $\lim_{x\to\infty} f(x) = 0$. $|a_{1000}| < 0.001$, $a_{1000} > 0 \Rightarrow s_{999} < L < s_{1000}$.

11. Converges absolutely since $\sum |a_n|$ is a convergent p–series.

12. Converges conditionally since $\frac{\frac{1}{\sqrt{n+1}}}{\frac{1}{\sqrt{n}}} = \sqrt{\frac{n}{n+1}} < 1$ and $\lim_{n\to\infty} \frac{1}{\sqrt{n}} = 0$, but $\sum \frac{1}{\sqrt{n}}$ is a divergent p–series.

13. Converges absolutely; let $c_n = \frac{1}{n^2}$; then $\lim_{n\to\infty} \frac{|a_n|}{c_n} = \lim_{n\to\infty} \frac{n^3}{n^3+1} = 1$.

14. Diverges; by the Ratio Test: $\lim_{n\to\infty} \frac{a_{n+1}}{a_n} = \lim_{n\to\infty} \frac{(n+1)!}{2^{n+1}} \cdot \frac{2^n}{n!} = \infty$.

15. Converges conditionally: $|a_n| = \frac{1}{n+3} > |a_{n+1}| = \frac{1}{n+4}$; $a_n \to 0$. However $\sum |a_n|$ diverges by the Limit Comparison Test when compared to $\sum \frac{1}{n}$.

16. Diverges: $\lim_{n\to\infty} |a_n| = 1 \neq 0$.

17. Converges absolutely: $0 \le |(-1)^n \frac{\sin n}{n^2}| \le \frac{1}{n^2}$; $\sum \frac{1}{n^2}$ is a convergent p–series.

18. Converges conditionally since $\sum (-1)^n \frac{1}{\ln n^3} = \frac{1}{3} \sum (-1)^n \frac{1}{\ln n}$ which converges – See Exercise #2. Does not converge absolutely since $\frac{1}{\ln n^3} > \frac{1}{n}$ for $n \ge 5$.

19. Converges conditionally: $|a_n| = \frac{1+n}{n^2} > \frac{2+n}{(n+1)^2} = |a_{n+1}|$. This can be established directly by comparing $(n+1)^3 = n^3 + 3n^2 + 3n + 1 > n^3 + 2n^2$, or by setting $f(x) = \frac{1+x}{x^2}$ and showing that $f'(x) = \frac{x^2 - 2x(1+x)}{x^4} < 0$. $|a_n| \to$

0, $\sum |a_n| = \sum(\frac{1}{n^2} + \frac{1}{n}) = \sum \frac{1}{n^2} + \sum \frac{1}{n}$, the sum of a convergent and a divergent series, is divergent.

20. Converges absolutely: $|a_n| = \left|\frac{2^{n+1}}{n+5^n}\right| < \frac{2\cdot 2^n}{5^n} = 2(\frac{2}{5})^n$; $\sum 2(\frac{2}{5})^n$ converges.

21. Converges absolutely by the Ratio Test: $\left|\frac{a_{n+1}}{a_n}\right| = \frac{(n+1)^2}{n^2}\frac{2}{3} \to \frac{2}{3}$.

22. Diverges: $\lim_{n\to\infty} |a_n| = \lim_{n\to\infty} 10^{\frac{1}{n}} = 1 \neq 0$.

23. $\int_0^\infty \frac{\tan^{-1} x}{x^2+1} dx = \int_0^{\pi/2} u du$ where $u = \tan^{-1} x$. The integral converges; $\sum a_n$ is absolutely convergent.

24. Converges conditionally since $\frac{1}{n \ln n} > \frac{1}{(n+1)\ln(n+1)}$ and $\lim_{n\to\infty} a_n = 0$. $\sum \frac{1}{n \ln n}$ diverges by the Integral Test: $\int_2^\infty \frac{dx}{x \ln x} = \lim_{b\to\infty} [\ln(\ln x)]_2^\infty = \infty$.

25. Diverges: $a_n = \frac{1}{2n}$, $\frac{1}{2}\sum \frac{1}{n}$ diverges.

26. Converges absolutely: $|a_n| = \frac{1}{n10^n} < \frac{1}{10^n}$ and $\sum(\frac{1}{10})^n$ converges.

27. Diverges: $\lim_{n\to\infty} |a_n| = 1 \neq 0$.

28. Converges conditionally: $\frac{1}{1+\sqrt{n}} > \frac{1}{1+\sqrt{n+1}}$, $\frac{1}{1+\sqrt{n}} \to 0$. $\sum |a_n|$ diverges since $\frac{1}{1+\sqrt{n}} > \frac{1}{\sqrt{n}}$ and $\sum \frac{1}{\sqrt{n}}$ diverges.

29. $a_n = \frac{-1}{(n+1)^2}$; $0 < -a_n < \frac{1}{n^2}$. The series $\sum -a_n$ converges, hence $\sum a_n$ converges.

30. Converges absolutely by the Ratio Test: $\frac{|a_{n+1}|}{|a_n|} = \frac{100}{n+1} \to 0$.

31. Converges absolutely since $|5^{-n}| = (\frac{1}{5})^n$; $\sum(\frac{1}{5})^n$ is a geometric series with $|r| < 1$.

32. $\sum_{n=2}^\infty (-1)^n \left(\frac{\ln n}{\ln n^2}\right)^n = \sum_{n=2}^\infty (-1)^n \left(\frac{1}{2}\right)^n$ which converges absolutely (geometric series with $|r| < 1$).

33. $0 \leq |a_n| < \frac{1}{n^{3/2}}$ since $|\cos x| \leq 1$. The series is absolutely convergent.

34. Converges conditionally: $\sum_{n=1}^\infty \frac{\cos n\pi}{n} = \sum_{n=1}^\infty (-1)^n \frac{1}{n}$, the alternating harmonic series.

35. The series converges conditionally because $|a_n| = \frac{1}{\sqrt{n}+\sqrt{n+1}} > \frac{1}{\sqrt{n+1}+\sqrt{n+2}} = |a_{n+1}|$ and $a_n \to 0$. To show that the series does not converge absolutely, write $|a_n| = \frac{1}{\sqrt{n}(1+\sqrt{1+1/n})}$ and compare it with the divergent series $\sum \frac{1}{\sqrt{n}} = \sum d_n$. By the Limit Comparison Test $\frac{|a_n|}{d_n} = \frac{1}{1+\sqrt{1+1/n}} \to \frac{1}{2}$.

36. Converges absolutely by the Ratio Test: $\lim_{n\to\infty} \frac{[(n+1)!]^2}{(2(n+1))!}\frac{(2n)!}{(n!)^2} = \lim_{n\to\infty} \frac{(n+1)^2}{(2n+2)^2} = \frac{1}{4}$.

37. By Theorem 8, $|s_4 - s| < \frac{1}{5} < 0.2$

38. $|\text{error}| \le \left|(-1)^6 \frac{1}{10^5}\right| = 0.00001$

39. $|\text{error}| < |a_5| = \frac{(0.01)^5}{5} = 2E - 11 = 2 \times 10^{-11}$

40. $|\text{error}| \le |\text{fifth term}| = |t^4| = t^4$

41. The first omitted term must satisfy $\frac{1}{(2n)!} < 5 \cdot 10^{-6}$, i.e. $10^6 < 5 \cdot (2n)!$. This is true for $n = 5$. Thus s_4 will suffice. Using the program PARTSUM, $s_4 = 0.5403023\ldots$.

42. $|\text{error}| < |\text{first omitted term}| = \frac{1}{n!} < 5 \times 10^{-6} \Rightarrow \frac{10^6}{5} < n! \Rightarrow n = 9$, thus $1 - 1 + \frac{1}{2} - \frac{1}{3!} + \frac{1}{4!} - \frac{1}{5!} + \frac{1}{6!} - \frac{1}{7!} + \frac{1}{8!} = 0.36788194$.

43. a) The absolute values of the terms are not not strictly decreasing, b) If the terms are rearranged the series becomes $\left[\frac{1}{3} + \frac{1}{9} + \frac{1}{27} + \cdots\right] - \left[\frac{1}{2} + \frac{1}{4} + \frac{1}{8} + \cdots\right]$. It is, in fact, permitted to rearrange the terms of an absolutely convergent series without affecting its sum. Hence the given series converges to $\frac{1}{2} - 1 = -\frac{1}{2}$.

44. $s_{20} = \sum_{n=1}^{20} \frac{(-1)^{n+1}}{n} = \text{sum}(\text{seq}((-1)\hat{\ }(x+1)/x, x, 1, 20, 1)) = 0.668771\ldots$; hence $s_{20} + \frac{1}{2} \cdot \frac{1}{21} = 0.69258$

45. Assume the series is written $a_1 - a_2 + a_3 - a_4 \ldots$ where all the a_i are positive. If $a_n > 0$ the series can be written $(a_1 - a_2 + \cdots + a_n) - a_{n+1} + a_{n+2} - \cdots$ and the remainder is $-(a_{n+1} - a_{n+2}) - (a_{n+3} - a_{n+4}) - \ldots$. All the terms enclosed in parentheses are positive. Hence the remainder is negative, as is the $n + 1^{st}$ term. If $a_n < 0$ we have $(a_1 - a_2 + \cdots - a_n) + [(a_{n+1} - a_{n+2}) + (a_{n+3} - a_{n+4}) + \cdots]$ and both the remainder and the $(n+1)^{st}$ term are positive.

46. For the first series, $s_2 = 1 - \frac{1}{2}$, $s_4 = 1 - \frac{1}{3}$, $s_6 = 1 - \frac{1}{4}, \Rightarrow s_{2n} = 1 - \frac{1}{n+1}$; $s_{2n+1} = 1$; for the second series $\sum_{k=1}^{n} \frac{1}{k(k+1)} = \sum_{k=1}^{n} \left[\frac{1}{k} - \frac{1}{k+1}\right] = s_{2n}$ of the first series. Both series converge to 1.

47.

	x	Approximate Value of n	Underestimate	Overestimate
a)	0.2	$\frac{0.2^n}{n} < 0.001 \Rightarrow n \geq 4$	$s_4 = 0.18226$	$s_5 = 0.18233$
b)	0.5	$\frac{0.5^n}{n} < 0.001 \Rightarrow n \geq 8$	$s_8 = 0.4053$	$s_9 = 0.4055$
c)	0.8	$\frac{0.8^n}{n} < 0.001 \Rightarrow n \geq 19$	$s_{20} = 0.5875$	$s_{19} = 0.58811$
d)	0.9	$\frac{0.9^n}{n} < 0.001 \Rightarrow n \geq 34$	$s_{34} = 0.64147$	$s_{35} = 0.64218$

48. a) The series is not an alternating series: $s_1, s_2, s_3 > 0$, $s_4, s_5 < 0$, etc.

b) $s_{100} = 1.013856\ldots$, $s_{500} = 1.0139614\ldots$, $s_{1000} = 1.0139590\ldots$, $s_{2000} = 1.01395933\ldots$, c) It appears probable that the sum begins: $1.013959\ldots$. It's possible the sum is 1.01396. d) It's probably within 0.00001 of the sum.

9.6 POWER SERIES

1. $\lim_{n\to\infty} \left|\frac{(x+1)^{n+1}}{(x+1)^n}\right| < 1 \Rightarrow |x+1| < 1 \Rightarrow -2 < x < 0$; when $x = -2$ we have $\sum(-1)^n(-1)^n = \sum 1$, divergent. At $x = 0$ we have $\sum(-1)^n$, divergent.

2. $\lim_{n\to\infty} \left|\frac{(n+1)x^{n+1}}{(n+3)} \cdot \frac{(n+2)}{nx^n}\right| = \lim_{n\to\infty} \left|\frac{(n+1)(n+2)}{(n+3)n}\right| |x| = 1 \cdot |x| < 1$ for $-1 < x < 1$. For $x = \pm 1$, $\lim_{n\to\infty} |a_n| = 1 \Rightarrow$ the series diverges. The series converges for $-1 < x < 1$. It converges absolutely on all intervals $|x| < c$, $0 < c < 1$, or, more informally for $-1 < x < 1$.

3. $\lim_{n\to\infty} \left|\frac{x^{n+1}}{(n+1)\sqrt{n+1}} \frac{n\sqrt{n}}{x^n}\right| < 1 \Rightarrow |x| < 1 \Rightarrow -1 < x < 1$; at $x = -1$, we have $\sum \frac{(-1)^n}{n^{3/2}}$ which is convergent by the alternating series test; at $x = 1$, we have $\sum \frac{1}{n^{3/2}}$, a convergent p-series. Hence the series is absolutely convergent on $[-1, 1]$.

4. $\lim_{n\to\infty} \left|\frac{(x-1)^{n+1}}{\sqrt{n+1}} \cdot \frac{\sqrt{n}}{(x-1)^n}\right| = |x-1| < 1$ when $0 < x < 2$; when $x = 0$ the series converges by the Alternating Series Theorem; when $x = 2$, the series diverges by the integral test: $\int_1^\infty \frac{dx}{\sqrt{x}} = \infty$. a) $0 \leq x < 2$, b) $0 < x < 2$

5. $\lim_{n\to\infty} \left|\frac{x^{2n+3}}{(n+1)!} \frac{n!}{x^{2n+1}}\right| < 1 \Rightarrow 0 < 1$, true for all x.

6. $\lim_{n\to\infty} \left|\frac{(x-3)^{2n+3}}{(n+1)!} \frac{n!}{(x-3)^{2n+1}}\right| = \lim_{n\to\infty} \frac{1}{n+1}|x-3|^2 < 1$ for all x. a) all x, b) all x

7. $\lim_{n\to\infty}\left|\frac{x^{n+1}}{\sqrt{(n+1)^2+3}}\cdot\frac{\sqrt{n^2+3}}{x^n}\right| < 1 \Rightarrow |x|\lim_{n\to\infty}\sqrt{\frac{n^2+3}{(n+1)^2+3}} < 1 \Rightarrow |x|\cdot 1 < 1 \Rightarrow$ $-1 < x < 1$; at $x = -1$ the (alternating) series converges; at $x = 1$ we have $\sum\frac{1}{\sqrt{n^2+3}}$ which diverges.

8. $\lim_{n\to\infty}\left|\frac{x^{n+1}}{\sqrt{n^2+2n+4}}\frac{\sqrt{n^2+3}}{x^n}\right| = |x| < 1$ when $-1 < x < 1$; when $x = -1$, $\sum\frac{(-1)^{2n}}{\sqrt{n^2+3}} = \sum\frac{1}{\sqrt{n^2+3}}$ which diverges by the Integral Test; when $x = 1$, the series is $\sum\frac{(-1)^n}{\sqrt{n^2+3}}$ which converges.

9. $\lim_{n\to\infty}\left|\frac{(n+1)x^{n+1}}{(n+1)^2+1}\cdot\frac{n^2+1}{(n+1)x^n}\right| < 1 \Rightarrow |x|\lim_{n\to\infty}\frac{n^2+1}{(n+1)^2+1} < 1 \Rightarrow |x| < 1 \Rightarrow$ $-1 < x < 1$; when $x = -1$, the (alternating) series converges; when $x = 1$ we have $\sum\frac{n}{n^2+1}$ which diverges.

10. $\lim_{n\to\infty}\left|\frac{(n+1)(x-3)^{n+1}}{n(x-3)^n}\right| = |x-3| < 1$ when $2 < x < 4$; when $x = 4$ and when $x = 2$, $|a_n| = n \Rightarrow$ the series diverge. a) $2 < x < 4$, b) $2 < x < 4$

11. $\lim_{n\to\infty}\left|\frac{\sqrt{n+1}x^{n+1}}{3^{n+1}}\cdot\frac{3^n}{\sqrt{n}x^n}\right| < 1 \Rightarrow |x|\cdot\frac{1}{3} < 1 \Rightarrow -3 < x < 3$; at $x = -3$ we have $\sum(-1)^n\sqrt{n}$ which diverges ($\lim a_n \neq 0$); at $x = 3$ we have $\sum\sqrt{n}$ which diverges.

12. $\lim_{n\to\infty}\left|\frac{(n+1)^{\frac{1}{n+1}}(x-1)^{n+1}}{n^{1/n}(x-1)^n}\right| = |x-1|\lim_{n\to\infty}\frac{\sqrt[n+1]{n+1}}{\sqrt[n]{n}} = |x-1| < 1$ when $0 < x < 2$; when $x = 2$ and when $x = 0$, $|a_n| \to 1$ so both series diverge. a) $0 < x < 2$, b) $0 < x < 2$

13. $\lim_{n\to\infty}\left|\frac{(1+\frac{1}{n+1})^{n+1}x^{n+1}}{(1+\frac{1}{n})^n x^n}\right| < 1 \Rightarrow |x|\cdot\frac{e}{e} < 1 \Rightarrow -1 < x < 1$, at $x = -1$ and $x = 1$, the series diverge ($\lim|a_n| = e$)

14. $\lim_{n\to\infty}\left|\frac{\ln(n+1)x^{n+1}}{\ln(n)x^n}\right| = |x| < 1$ when $-1 < x < 1$; when $x = \pm 1$, $|a_n| = \ln n$, $\lim|a_n| \neq 0$ so the series diverge. a) $-1 < x < 1$, b) $-1 < x < 1$

15. $\lim_{n\to\infty}\left|\frac{x^{n+1}}{x^n}\right| < 1 \Rightarrow |x| < 1 \Rightarrow -1 < x < 1$; diverges for $x = \pm 1$. For $-1 \leq x < 1$, the series converges to $\frac{1}{1-x}$. Use a calculator to compare P_{20} and $\frac{1}{1-x}$: set $y_1 = \text{sum seq}(x\hat{}N, N, 0, 20, 1)$. $y_2 = 1/(1-x)$; $y_3 = \text{abs}(y_1 - y_2)$ and graph y_3 on $[-1, 1]$ by $[0, 0.01]$. Investigation shows $-0.8 \leq x \leq 0.7 \Rightarrow |\text{sum} - P_{20}| < 0.01$.

16. We must have $|x+5| < 1 \Rightarrow -6 < x < -4$, at $x = -6$; and $x = -4$ the series diverges. When it converges, the series converges to $\frac{1}{1-(x+5)} = \frac{-1}{4+x}$. Using the results of #15, P_{20} will approximate the sum with error at most 0.01 for $-0.8 < x+5 < 0.7$, i.e., $-5.8 < x < -4.3$.

17. Absolute convergence when $\left|\frac{x-2}{10}\right| < 1 \Rightarrow -8 < x < 12$. The series converges to $\frac{1}{1-(\frac{x-2}{10})} = \frac{10}{12-x}$. Error is less than 0.0.1 when $-0.8 < \frac{x-2}{10} < 0.7 \Rightarrow -6 < x < 9$.

18. $|2x| < 1 \Rightarrow -\frac{1}{2} < x < \frac{1}{2}$. The series converges to $\frac{1}{1-2x}$. From #15, $-0.8 < 2x < 0.7$ or $-0.4 < x < 0.35$.

19. a) Use sumseq, and the identity $(-1)^{2n-1} = -1$ to calculate $P_{20} = \text{sumseq}(-1/N, N, 1, 20, 1) = -3.597\ldots$, $P_{30} = -3.994\ldots$, $P_{50} = -4.4992$, $P_{100} = -5.187\ldots$. These are the partial sums of the harmonic series, which diverges. b) $P_{10}(-0.9) = -2.1187$; $\ln(1-0.9) = -2.302$. The maximum error is 0.19. c)

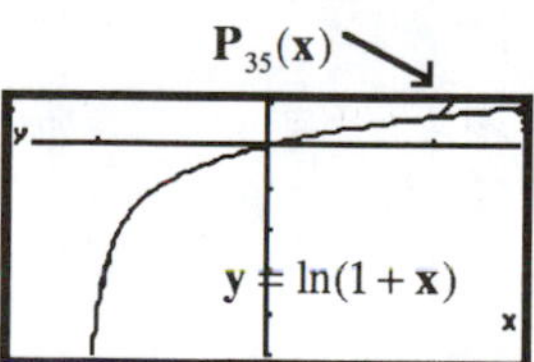

20. a) Using PARTSUM: for $x = 1$, $P_{100} = 0.782$, $P_{1000} = 0.7851$ for $x = -1$, $P_{100} = -0.783$, $P_{1000} = 4.435$, $P_{10000} = -0.7851$. Since $\tan^{-1}(\pm 1) = \pm\pi/4 = \pm 0.7851$, there is good reason to expect convergence for $x = \pm 1$. b) The error is less than the first omitted term; the error is greatest when $|x| = 1$. Solving $\frac{1}{2(n+1)-1} < \frac{1}{100}$ gives $n = 50$. c) For a given value of n, the error in Example 5 was $\frac{1}{n}$; here it is $\frac{1}{2n-1} \approx \frac{1}{2}(\frac{1}{n})$.

	a) Convergent	b) Absolutely Convergent	c) Alternating on	d) Error $<$ 0.01
21.	$[1,3)$	$(1,3)$	$(1,2)$	P_{30} on $[1.1,2]$ Compare with $\ln(3-1.1)$
22.	$(-3,-1]$	$(-3,-1)$	$(-2,-1)$	P_{30} on $[-2.9,-2]$ Compare with $\ln(3-2.9)$
23.	$(2,4)$	$(2,4)$	$(2,3)$	P_{110} on $[2.1,3]$ Compare with $(2.1-3)/(4-2.1)^2$; series is $(x-3)D_x\sum(x-3)^n$
24.	$[1,3)$	$(1,3)$	$(1,2)$	P_{110} on $[1.1,2]$ Series is $(x-2)D_x\sum(x-2)^n$

25. a) Using sum seq $(((-1)\text{^}N(\pi/4)\text{^}(2n+1))/(2N+1)!, N, 0, 5, 1)$ etc. we get

x	$P_5(x)$	$P_{11}(x)$	$P_{17}(x)$
$\pi/4$	$0.707106\ldots$	$0.707106\ldots$	$0.707106\ldots$
$-\pi/4$	$-0.707106\ldots$		
$\pi/2$	-0.99999994	1	
$-\pi/2$	-0.99999994	-1	
2π	-3.19507	$-5.494\,E-06$	$-4E-13$
-2π	3.19507		

b) For positive and negative values of x, the sums are the same. Using the Alternating series estimate and graphing we find:

$|\text{error } P_{11}(x)| < \left|\frac{x^{2\cdot12+1}}{(2\cdot12+1)!}\right| = \frac{x^{25}}{25!} < 0.01 \Rightarrow |x| < 8.4;$

$|\text{error } P_{21}(x)| < 0.01 \Rightarrow \frac{x^{2\cdot22+1}}{45!} < 0.01 \Rightarrow |x| < 15.9;$

$|\text{error } P_{31}(x)| < 0.01 \Rightarrow |x| < 23.3.$

26. a) $\cos x = D_x \sin x = D_x \sum_0^\infty \frac{(-1)^n x^{2n+1}}{(2n+1)!} = \sum_0^\infty \frac{(-1)^n(2n+1)x^{2n}}{(2n+1)!} = \sum_0^\infty \frac{(-1)^n x^{2n}}{(2n)!}$, b) all x, since the series for $\sin x$ converges for all x

27. $\left|\frac{x^{n+1}}{(n+1)!} \cdot \frac{n!}{x^n}\right| = \frac{|x|}{n+1} \to 0$; the series converges for all x. Experimenting with $-10 \to x : e\hat{\ }x - \text{sum seq}(x\hat{\ }N/N!, N, 0.20.1) \Rightarrow -7 < x < 7$.

28. The sum is $-e$. This is the same series as #27.

29. a) The series is $\sum \frac{(3x)^n}{n!}$ which sums to e^{3x}, it converges for all x;

b) $-7 < 3x < 7 \Rightarrow -7/3 < x < 7/3$

30. This series is $D_x \sum_0^\infty x^n = D_x \frac{1}{1-x}$ which converges for $-1 < x < 1$. For $x < 0$, the error, $|\text{sum} - P_{20}(x)| < 21|x|^{21} < 0.01 \Rightarrow |x| < 0.69$.

31. a) $s_n = x \cdot \frac{1-x^{n+1}}{1-x} \to \frac{x}{1-x}$, b) $1 + \sum_1^\infty x^n = \frac{1}{1-x}$; sum $= \frac{1}{1-x} - 1 = \frac{x}{1-x}$,

c) $|x| < 1$, d) for $x < 0$ the series alternates; $|P_{20}(x) - \frac{1}{1-x}| < |x^{21}|$;

solving $|x|^{21} < 0.01 \Rightarrow |x| < 0.80$. Evaluating $-0.80 \to x : x/(1-x)-$

$\text{sumseq}(x\hat{\ }N, N, 1, 20, 1)$ for positive values of x shows the error < 0.01 when $-0.80 < x < 0.74$.

32. a) All x, by the Ratio Test; b) $\frac{(-1)^0 x^1}{1} + \sum_1^\infty \frac{(-1)^n x^{2n+1}}{(2n+1)!} = \sin x$, the sum is $\sin x - x$; c) by #25 b), $-15.9 < x < 15.9$.

33. a) $-\ln(1-x) = -\sum_{n=1}^\infty \frac{(-1)^{n-1}(-x)^n}{n} = -\sum_{n=1}^\infty -1 \frac{(-1)^n(-x)^n}{n} = \sum_{n=1}^\infty \frac{x^n}{n}$; b) $-\ln(3-x) = -\ln(1-(x-2)) = \sum \frac{(x-2)^n}{n}$ by a); c) $-\ln(3+x) = -\ln(1+(2+x)) = -\sum_{n=1}^\infty \frac{(-1)^{n-1}(2+x)^n}{n}$ by a) $= \sum_{n=1}^\infty \frac{(-1)^n(x+2)^n}{n}$

34. a) $\frac{d}{dx}\ln(1+x) = \sum_{n=1}^\infty \frac{n(-1)^{n-1}x^{n-1}}{n} \Rightarrow \frac{d^2}{dx^2}\ln(1+x) = \sum_{n=1}^\infty (n-1)(-1)^{n-1}x^{n-2} = \sum_{n=2}^\infty (n-1)(-1)^{n-1}x^{n-2}$. Hence $x\frac{d^2}{dx^2}\ln(1+x) = \sum_{n=2}^\infty (n-1)(-1)^{n-1}x^{n-1} = \sum_{k=1}^\infty k(-1)^k x^k$ where $k = n-1$; b) replacing x by $-x \Rightarrow \sum_{k=1}^\infty k(x)^k = -\frac{(-x)}{(1-x)^2}$; c) replacing x by $x-3$ in b) $\Rightarrow \frac{x-3}{(1-x+3)^2} = \sum_{n=1}^\infty n(x-3)^n$; d) for a) and b): $-1 < x < 1$; for c): $-2 < x < 4$

35. On a grapher, set $y1 = 1 + x + x^2/2$, $y2 = y1 + x^3/6$, $y3 = y2 + x^4/24$, $y4 = y1 - y2$. Graph the polynomials on $[-2, 2]$ by $[-0.01, 0.01]$. Use Trace to find the results: a) $(-0.38, 0.38)$ b) $(-0.698, 0.698)$ For c), observe that $|P_9(x) - P_{10}(x)| = x^{10}/10!$; graphing that on $[-3, 3]$ by $[-0.01, 0.01]$ gives $(-2.85, 2.85)$ d) the series converges for all x.

36. Look at $|P_n(x) - P_{n+1}(x)|$; for $x = 0$ this difference is always 0. For $|x|$ "near zero", the difference should be small, less than 0.01 for example. However, if as n increases, the definition of "near zero" gets smaller and smaller, one can argue that, eventually only $x = 0$ will be left.

37. $\lim_{n\to\infty} \frac{(n+1)^{n+1}x^{n+1}}{n^n x^n} = |x| \lim_{n\to\infty} \left(\frac{n+1}{n}\right)^n (n+1) = |x| \cdot e \lim_{n\to\infty}(n+1)$; this series converges if and only if $x = 0$.

38. $\lim_{n\to\infty} \left|\frac{(n+1)!(x-4)^{n+1}}{n!(x-4)^n}\right| = |x-4| \lim_{n\to\infty}(n+1)$; the series only converges when $x = 4$.

39. Let $f(x) = 1 - \frac{1}{2}(x-3) + \frac{1}{4}(x-3)^2 + \cdots = \sum_{n=0}^{\infty} \left(\frac{-(x-3)}{2}\right)^n$; this is a geometric series with $|r| = \frac{|x-3|}{2}$ which converges, for $1 < x < 5$, to $\frac{1}{1+\frac{(x-3)}{2}} = \frac{2}{x-1}$; when $x = 1$ the series is $\sum 1^n$ which diverges; at $x = 5$, the series is $\sum (-1)^n$ which diverges. $f' = \sum_{n=1}^{\infty} \frac{n}{2}\left[\frac{-(x-3)}{2}\right]^{n-1} = \sum_{k=0}^{\infty} \frac{(k+1)}{2}\left[\frac{-(x-3)}{2}\right]^k$; by calculus the sum of this series is $\frac{-2}{(x-1)^2}$ which also converges for $1 < x < 5$.

40. $\int f(x)dx = x - \frac{(x-3)^2}{4} + \frac{(x-3)^3}{12} + \cdots + \frac{(-1)^n(x-3)^{n+1}}{2^n(n+1)} + \cdots = \int \frac{2}{x-1}dx = 2\ln|x-1| + C$; find C by evaluating at $x = 3: 3 = 2\ln 2 + C \Rightarrow C = 3 - \ln 4$. The series $\sum_{n=0}^{\infty} \frac{(-1)^n(x-3)^{n+1}}{2^n(n+1)}$ will converge absolutely for $1 < x < 5$, checking the endpoints shows at $x = 1$, $\sum \frac{(-2)^{2n+1}}{2^n(n+1)}$ diverges but at $x = 5$, $\sum \frac{(-1)^n \cdot 2}{n+1}$ converges.

41. a) Graph $P_7(x) - \tan x$: error is < 0.01 on $-0.873 < x < 0.873$; b) $\ln|\sec x| = \int \tan x\, dx = \frac{x^2}{2} + \frac{x^4}{12} + \frac{x^6}{45} + \frac{17x^8}{8 \cdot 315} + C$; evaluating at $x = 0$ gives $C = 0$; the series converges absolutely for the same values as the series for $\tan x$; $-\frac{\pi}{2} < x < \frac{\pi}{2}$; c) differentiating the series for $\tan x$ gives $\sec^2 x = 1 + x^2 + \frac{2x^4}{3} + \cdots$; d) $(1 + \frac{x^2}{2} + \frac{5}{24}x^4 + \cdots)(1 + \frac{x^2}{2} + \frac{5}{24}x^4 + \cdots) = 1 + \frac{2x^2}{2} + x^4(\frac{2 \cdot 5}{24} + \frac{1}{4}) + \cdots = 1 + x^2 + \frac{2x^4}{3} + \cdots$.

42. a) Graphing $P_6(x) - \sec x$ on $[-\pi/2, \pi/2]$ by $[-0.01, 0.01]$ gives $|x| < 0.823$. b) By Theorem 12, we can integrate term by term: $\ln|\sec x + \tan x| + C = x + \frac{x^3}{6} + \frac{x^5}{24} + \frac{61}{7 \cdot 720}x^7 + \cdots$; evaluating at $x = 0 \Rightarrow C = 0$. The series will converge if $|x| < \pi/2$. c) By Theorem 12, differentiate term by term: $\sec x \tan x = x + \frac{20x^3}{24} + \frac{61 \cdot 6}{720}x^5 + \cdots = x + \frac{5}{6}x^3 + \ldots$; converges for $|x| < \pi/2$. d) $\sec x \tan x = (1 + \frac{x^2}{2} + \frac{5}{24}x^4 + \cdots)(x + \frac{x^3}{3} + \frac{2x^5}{15} + \cdots) = x + x^3(\frac{1}{3} + \frac{1}{2}) + x^5(\frac{5}{24} + \frac{1}{6} + \frac{2}{15}) + \cdots = x + \frac{5}{6}x^3 + \frac{61}{120}x^5 + \cdots$.

43. a) The graphs are nearly vertical near $x = \pm\pi/2$; b) Graph s_{30} and the lines $y = \pm 0.99$ on $[-\pi, \pi]$ by $[-1.01, 1.01]$; use Trace to estimate $|s_{30} - f(x)| < 0.01$ for $0 \le |x| < 1.172$ or $1.97 < |x| < \pi$. (This is called a "square wave".)

44. a) The graphs are nearly vertical at $x = 0$, b) (see #43.) Graphing shows $|s_{30} - f(x)| < 0.01$ when $1.4707 < |x| < 2.319$.

45. $|a_n| = \frac{|\sin(n!x)|}{n^2} \leq \frac{1}{n^2}$; $\sum \frac{1}{n^2}$ converges so the series converges absolutely for all x.

46. $12 \cdot 6 \approx 4\pi$, $s(x)$ appears to be a sawtooth curve;

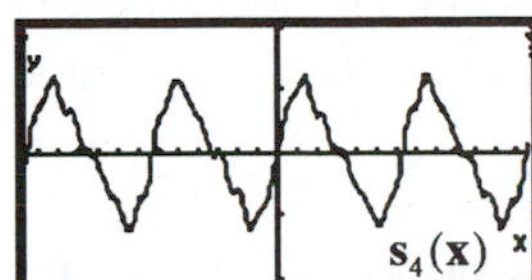

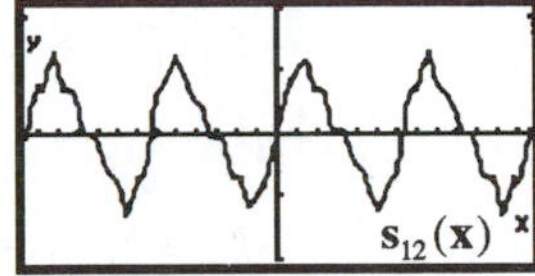

47. The values appear to differ in the hundredths or thousandths place.

48. $|s(x) - s_{13}(x)| = \left|\sum_{n=14}^{\infty} \frac{\sin(n!x)}{n^2}\right| \leq \sum_{n=14}^{\infty} \left|\frac{\sin(n!x)}{n^2}\right| \leq \sum_{n=14}^{\infty} \frac{1}{n^2} = \sum_{n=1}^{\infty} \frac{1}{n^2} - \sum_{n=1}^{13} \frac{1}{n^2} = \frac{\pi^2}{6} - 1.5709 = 0.0740 < 0.075$.

49. Zooming in near a peak reveals a host of subpeaks and valleys.

50. The graph of the derivative of $s_{13}(x)$ is a set of vertical lines.

51. The graph of the exact derivative of $s_{13}(x)$ is a set of vertical lines. This suggests that $s'_{13}(x)$ alternates between very large positive and negative numbers, and that $s(x)$ will not have a derivative.

52. The series for $\tan^{-1} x$ was given in Exercise 20. At $x = 1$, $|\tan^{-1} 1 - s_n| = |\frac{\pi}{4} - s_n| \leq \frac{1}{2(n+1)-1} = \frac{1}{2n+1} < 10^{-3}$ when $n \geq 500$.

53. If $|x| = 1 + C$, by the Ratio Test, $\lim_{n\to\infty} \left|\frac{a_{n+1}}{a_n}\right| = 1 + C$, hence the series diverges. Plotting the points $(k, \sum_{n=1}^{k} a_n)$ will show the divergence.

54. $\frac{(\frac{1}{18})^n}{2n-1} < 10^{-6}$ for $n = 5$; $48 * s_5 = 48(0.0554985) = 2.6639282\ldots$;

$\frac{(\frac{1}{57})^n}{2n-1} < 10^{-6}$ for $n = 5$; $32 * s_5 = 32(0.01754206) = 0.5613459\ldots$;

$\frac{(\frac{1}{239})^n}{2n-1} < 10^{-6}$ for $n = 3$; $-20{*}s_3 = -20(0.004184) = -0.0836815\ldots$. Adding these gives 3.14159265359, i.e., π to 12 places. The increased accuracy is because, for the specified values of n, the errors are much smaller than 10^{-6}.

9.7 TAYLOR SERIES AND MACLAURIN SERIES

1. $f(x) = \ln x$, $f'(x) = \frac{1}{x}$, $f''(x) = -\frac{1}{x^2}$, $f'''(x) = \frac{2}{x^3}$; $f(1) = 0$, $f'(1) = 1$, $f''(1) = -1$, $f'''(1) = 2$; $P_1(x) = (x-1)$; $P_2(x) = (x-1) - (x-1)^2/2$; $P_3(x) = (x-1) - (x-1)^2/2 + 2(x-1)^3/6$. Graph $P_3 - f$ on $[0,3]$ by $[-0.01, 0.01]$; $|P_3 - f| < 0.01$ when $0.60 < x < 1.47$.

2. $f(x) = \ln(1+x)$, $f'(x) = \frac{1}{1+x}$, $f''(x) = \frac{-1}{(1+x)^2}$, $f'''(x) = \frac{+2}{(1+x)^3}$; $f(0) = 0$, $f'(0) = 1$, $f''(0) = -1$, $f'''(0) = 2 \Rightarrow P_0(x) = 0$, $P_1(x) = 1(x)$, $P_2(x) = (x) + \frac{(-1)}{2!}x^2$; $P_3(x) = (x) - \frac{1}{2}(x)^2 + \frac{2}{6}(x)^3$. Graphing $P_3 - \ln(1+x)$ shows $|\text{error}| < 0.01$ for $|x| < 0.481$.

3. $f(x) = \frac{1}{x}$, $f'(x) = -\frac{1}{x^2}$, $f''(x) = \frac{2}{x^3}$, $f'''(x) = -\frac{6}{x^4}$; $f(2) = \frac{1}{2}$, $f'(2) = -\frac{1}{4}$, $f''(2) = \frac{1}{4}$, $f'''(2) = \frac{-3}{8}$; $P_1 = \frac{1}{2} - \frac{1}{4}(x-2)$; $P_2 = \frac{1}{2} - \frac{1}{4}(x-2) + \frac{1}{4}(x-2)^2/2$; $P_3 = \frac{1}{2} - \frac{1}{4}(x-2) + \frac{1}{8}(x-2)^2 - \frac{3}{8}(x-2)^3/6 = \frac{1}{2} - \frac{1}{4}(x-2) + \frac{1}{8}(x-2)^2 - \frac{1}{16}(x-2)^3$. $|P_3 - f| < 0.01$ when $1.34 < x < 2.78$.

4. $f(x) = \frac{1}{x+2}$, $f'(x) = \frac{-1}{(x+2)^2}$, $f''(x) = \frac{2}{(x+2)^3}$, $f'''(x) = \frac{-6}{(x+2)^4}$; $f(0) = \frac{1}{2}$, $f'(0) = -\frac{1}{4}$, $f''(0) = \frac{1}{4}$, $f'''(0) = \frac{-6}{16}$. $P_3(x) = \frac{1}{2} - \frac{x}{4} + \frac{1}{2!}\frac{x^2}{4} - \frac{6}{16}\frac{x^3}{3!} = \frac{1}{2} - \frac{x}{4} + \frac{x^2}{8} - \frac{1}{16}x^3$; graph $P_3(x) - f(x)$ to find $|x - 0| < 0.820 \Rightarrow |\text{error}| < 0.01$.

5. $f(x) = \sin x$, $f'(x) = \cos x$, $f''(x) = -\sin x$, $f'''(x) = -\cos x$; $f(\frac{\pi}{4}) = \frac{1}{\sqrt{2}}$, $f'(\frac{\pi}{4}) = \frac{1}{\sqrt{2}}$, $f''(\frac{\pi}{4}) = -\frac{1}{\sqrt{2}}$, $f'''(\frac{\pi}{4}) = -\frac{1}{\sqrt{2}}$. $P_1 = \frac{1}{\sqrt{2}} + \frac{1}{\sqrt{2}}(x - \frac{\pi}{4})$;
$P_2 = \frac{1}{\sqrt{2}}\left[1 + (x - \frac{\pi}{4})\right] - \frac{1}{\sqrt{2}}(x - \frac{\pi}{4})^2/2$;
$P_3 = \frac{1}{\sqrt{2}}\left[1 + (x - \frac{\pi}{4}) - (x - \frac{\pi}{4})^2/2 - (x - \frac{\pi}{4})^3/6\right]$. $|P_3 - f| < 0.01$ when $-0.008 < x < 1.515$.

6. $f(x) = \cos x$, $f'(x) = -\sin x$, $f''(x) = -\cos x$, $f'''(x) = \sin x$; $f(\frac{\pi}{4}) = \frac{\sqrt{2}}{2}$, $f'(\frac{\pi}{4}) = -\frac{\sqrt{2}}{2}$, $f''(\frac{\pi}{4}) = -\frac{\sqrt{2}}{2}$, $f'''(\frac{\pi}{4}) = \frac{\sqrt{2}}{2}$; $P_3(x) = \frac{\sqrt{2}}{2} - \frac{\sqrt{2}}{2}(x - \frac{\pi}{4}) - \frac{\sqrt{2}}{2}\frac{1}{2!}(x - \frac{\pi}{4})^2 + \frac{\sqrt{2}}{2}\frac{1}{3!}(x - \frac{\pi}{4})^3$; $0.045 < x < 1.587 \Rightarrow |\text{error}| < 0.01$.

7. $f(x) = x^{\frac{1}{2}}$, $f'(x) = \frac{1}{2}x^{-\frac{1}{2}}$, $f''(x) = -\frac{1}{4}x^{-\frac{3}{2}}$, $f'''(x) = \frac{3}{8}x^{-\frac{5}{2}}$; $f(4) = 2$, $f'(4) = \frac{1}{4}$, $f''(4) = \frac{-1}{32}$, $f'''(4) = \frac{3}{256} \Rightarrow P_3(x) = 2 + \frac{1}{4}(x-4) - \frac{1}{32}(x-4)^2/2 + \frac{3}{256}(x-4)^3/6$; $|P_3 - f| < 0.01$ when $1.91 < x < 6.60$.

8. $f(x) = (x+4)^{\frac{1}{2}}$, $f'(x) = \frac{1}{2}(x+4)^{-\frac{1}{2}}$, $f''(x) = -\frac{1}{4}(x+4)^{-\frac{3}{2}}$, $f'''(x) = \frac{3}{8}(x+4)^{-\frac{5}{2}}$; $f(0) = 2$, $f'(0) = \frac{1}{4}$, $f''(0) = \frac{-1}{32}$, $f'''(0) = \frac{3}{256}$; $P_3(x) = 2 + \frac{1}{4}x - \frac{1}{32}\frac{1}{2!}x^2 + \frac{3}{256}\frac{1}{3!}x^3 = 2 + \frac{x}{4} - \frac{x^2}{64} + \frac{x^3}{512}$; $|\text{error}| < 0.01$ when $|x| < 2.624$.

9. $e^x = \sum_{n=0}^{\infty} x^n/n! \Rightarrow e^{-x} = \sum_{n=0}^{\infty}(-x)^n/n!$. $|R_{10}(x)| \le \frac{|e^x||x^{11}|}{11!} \le 0.01$ for $-5 \le x \le 2.5$.

10. $e^x = \sum_{k=0}^{\infty} \frac{x^k}{k!} \Rightarrow e^{x/2} = \sum_{k=0}^{\infty} \frac{x^k}{2^k k!}$; graphing shows $|P_{10}(x) - e^{x/2}| < 0.01$ for $|x| < 6.605$, $|R_{10}(x)| \le \frac{x^{11}}{11!}|f^{(n+1)}(c)| = \frac{x^{11}}{2''\cdot 11!}e^{c/2} < \frac{x^{11}e^{x/2}}{2^{11}\cdot 11!} < 0.01$ for $|x| < 5.1$.

11. $\sin 3x = \sum_{n=0}^{\infty} \frac{(-1)^n(3x)^{2n+1}}{(2n+1)!}$; $P_{10} = \sum_{n=0}^{4} \frac{(-1)^n(3x)^{2n+1}}{(2n+1)!}$; graphically, $|\sin 3x - P_{10}(x)| < 0.01$ when $|x| < 1.083$; analytically, $|R_{10}| < 0.01$, when $\frac{3^{11}|x|^{11}}{11!} \cdot 1 < 0.01$, i.e., $|x| < 0.95$.

12. $5\cos \pi x = 5\sum_{k=0}^{\infty} \frac{(-1)^k(\pi x)^{2k}}{(2k)!}$; $P_{10} = 5\sum_{k=0}^{5} \frac{(-1)^k(\pi x)^{2k}}{(2k)!}$; graphically, $|P_{10}(x) = 5\cos \pi x| < 0.01$ when $|x| < 1.007$; $|R_{10}(x)| < 0.01$ when $\frac{5|x|^{12}}{12!}\pi^{12} \cdot 1 < 0.01$, i.e., $|x| < 1.003$.

13. $\cos(-x) = \sum_{n=0}^{\infty} (-1)^n(-x)^{2n}/(2n)! = \sum_{n=0}^{\infty} (-1)^n x^{2n}/(2n)!$; $|R_{10}| \le 1 \cdot |x|^{11}/(11)! \le 0.01$ for $|x| < 3.22$; graphically, $|P_{10}(x) - \cos(-x)| < 0.01$ for $|x| < 3.624$.

14. $x \sin x = \sum_{k=0}^{\infty} \frac{(-1)^k x^{2k+2}}{(2k+1)!}$; $P_{10}(x) = \sum_{k=0}^{4} \frac{(-1)^k x^{2k+2}}{(2k+1)!}$; $|P_{10}(x) - x\sin x| < 0.01$ when $|x| < 2.942$; $|R_{10}| < 0.01$ when $\frac{|x|^{12}}{11!} \cdot 1 < 0.01$, i.e., $|x| < 2.929$.

15. $\cosh x = \frac{1}{2}\left[1 + x + \frac{x^2}{2!} + \frac{x^3}{3!} + \cdots + 1 - x + \frac{x^2}{2!} - \frac{x^3}{3!} + \cdots\right] = \left[1 + \frac{x^2}{2!} + \frac{x^4}{4!} + \cdots\right] = \sum_{n=0}^{\infty} \frac{x^{2n}}{(2n)!}$; $P_{10}(x) = \sum_{n=0}^{5} \frac{x^{2n}}{(2n)!}$; graphically, $|P_{10}(x) - \cosh x| < 0.01$ when $|x| < 3.581$; $|R_{10}(x)| < 0.01$ when $\left|\frac{x^{12}}{12!}\cosh c\right| < 0.01$; if we assume $|x| < 4$, then $|\cosh c| < \cosh 4 < 27.4$; $\left[\frac{(0.01)(12!)}{27.4}\right]^{1/12} = 2.734$.

16. $\sinh x = \frac{1}{2}\left[\sum_{k=0}^{\infty} \frac{(x)^k}{k!} - \sum_{k=0}^{\infty} \frac{(-x)^k}{k!}\right] = \sum_{j=0}^{\infty} \frac{x^{2j+1}}{(2j+1)!}$; $P_{10}(x) = \sum_{j=0}^{4} \frac{x^{2j+1}}{(2j+1)}$; graphically, $|P_{10}(x) - \sinh x| < 0.01$ when $|x| < 3.210$; $|R_{10}| = \left|\frac{x^{11}}{11!}\sinh c\right| \le \frac{|x^{11}|}{11!} \cdot 17$ if we assume $|x| < 3.5$; thus $|R_{10}| < 0.01$ when $|x| < 2.496$.

17. $\cos x - \left[1 - \frac{x^2}{2}\right] = \left[1 - \frac{x^2}{2!} + \frac{x^4}{4!} - \frac{x^6}{6!} + \cdots\right] - \left[1 - \frac{x^2}{2}\right] = \sum_{n=2}^{\infty} (-1)^n x^{2n}/(2n)!$; $P_{10} = \sum_{n=2}^{5} \frac{(-1)^n x^{2n}}{(2n)!}$; graphically, $|P_{10} - \text{function}| < 0.01$ when $|x| < 3.624$; $|R_{10}| = \frac{|x^{12}|}{12!}|\sin c| \le \frac{|x^{12}|}{12!} < 0.01$ when $|x| < 3.603$.

18. $\cos^2 x = \frac{1}{2}(1 + \cos 2x) = \frac{1}{2}\left(1 + \sum_{k=0}^{\infty} \frac{(-1)^k(2x)^{2k}}{(2k)!}\right) = \frac{1}{2} + \frac{1}{2}\sum_{k=0}^{\infty} \frac{(-1)^k(2x)^{2k}}{(2k)!}$; $P_{10}(x) = \frac{1}{2} + \frac{1}{2}\sum_{k=0}^{5} \frac{(-1)^k(2x)^{2k}}{(2k)!}$; graphically, $|P_{10}(x) - \cos^2 x| < 0.01$ when $|x| < 1.921$; $|R_{10}(x)| \le \frac{1}{2}\frac{(2x)^{12}}{12!} \cdot 1 < 0.01$ when $|x| < 1.908$.

19. $f(x) = \frac{1}{1+x}$; $f'(x) = \frac{-1}{(1+x)^2}$; $f''(x) = \frac{2}{(1+x)^3}$; $f'''(x) = \frac{-6}{(1+x)^4} \Rightarrow f(0) = 1$, $f'(0) = -1$, $f''(0) = 2$, $f'''(c) = \frac{-6}{(1+c)^4}$; $\frac{1}{1+x} = P_2(x) + R_2(x) = 1 - x + \frac{2x^2}{2} - \frac{6x^3}{(1+c)^4}\frac{1}{3!} = [1 - x + x^2] - x^3/(1+c)^4$, where c is between 0 and x.

20. $f(x) = (1+x)^{1/2}$, $f'(x) = \frac{1}{2}(1+x)^{-1/2}$, $f''(x) = -\frac{1}{4}(1+x)^{-3/2}$, $f'''(x) = \frac{3}{8}(1+x)^{-5/2}$; $f(0) = 1$, $f'(0) = \frac{1}{2}$, $f''(0) = -\frac{1}{4}$, $f'''(c) = \left(\frac{3}{8}\right)\frac{1}{(1+c)^{5/2}}$; $\sqrt{1+x} = 1 + \frac{1}{2}\cdot x + \frac{1}{2!}\left(-\frac{1}{4}\right)x^2 + x^3\frac{1}{3!}\frac{3}{8}\cdot\frac{1}{(1+c)^{5/2}}$, for c between 0 and x.

21. $f(x) = \ln(1+x)$; $f'(x) = \frac{1}{1+x}$; $f''(x) = \frac{-1}{(1+x)^2}$; $f'''(x) = \frac{2}{(1+x)^3}$; $\Rightarrow f(0) = 0$, $f'(0) = 1$, $f''(0) = -1$, $f'''(c) = 2/(1+c)^3$. $\ln(1+x) = P_2(x) + R_2(x) = [+x - x^2/2] + [2x^3/(1+c)^3]/3!$

22. $f(x) = (1+x)^k$, $f'(x) = k(1+x)^{k-1}$, $f''(x) = k(k-1)(1+x)^{k-2}$, $f'''(x) = k(k-1)(k-2)(1+x)^{k-3}$; $f(0) = 1$, $f'(0) = k$, $f''(0) = k(k-1)$, $f'''(c) = k(k-1)(k-2)(1+c)^{k-3}$. $(1+x)^k = 1 + x\cdot k + \frac{x^2}{2!}k(k-1) + \frac{x^3}{3!}k(k-1)(k-2)(1+c)^{k-3}$, c between 0 and x.

23. From Example 5, $\sin x = x - \frac{x^3\cos c}{3!}$.

24. $f(x) = \cos x \Rightarrow f(0) = 1$, $f'(0) = 0$, $f''(0) = -1$, $f'''(c) = \sin c$; $\cos x = 1 + \frac{x^2}{2!}(-1) + \frac{x^3}{3!}\sin c$; however $f'''(0) = 0$, so $P_3(x)$ is also quadratic: $\cos x = 1 - \frac{x^2}{2} + \frac{x^4}{4!}\cos c$, c between 0 and 1.

25. All derivatives of f, evaluated at a, are $f^{(k)}(a) = e^a$.
$e^x = \sum_{n=0}^{\infty}(x-a)^n e^a/n! = e^a[\sum_{n=0}^{\infty}(x-a)^n/n!]$.

26. $f(x) = f'(x) = f''(x)\cdots = e^x$; $f(1) = f'(1) = \cdots = e^1$;
$e^x = \sum_{k=0}^{\infty}\frac{(x-1)^k}{k!}f^{(k)}(1) = e\sum_{k=0}^{\infty}(x-1)^k/k!$ as in #25.

27. $x - x^3/6$ is actually P_4. Using the Taylor remainder, $|R_4| = \frac{|x|^5}{5!}|\cos c| \leq \frac{|x|^5}{120} \leq 5\times 10^{-4}$ for $|x| \leq (5\cdot 120\times 10^{-4})^{1/5} = 0.56\ldots$.

28. error $= \cos x - (1 - \frac{x^2}{2}) = \frac{x^4}{4!}\cos c$, $0.5 < |c| < 0$. error > 0, so $1 - \frac{x^2}{2}$ is too small, $|\text{error}| \leq \frac{(0.5)^4}{4!}1 = 0.0026$

29. $|\sin x - x| = |R_3(x)| \leq \frac{|x|^3}{6}\cdot 1 \leq 10^{-9}/6 < 1.67\times 10^{-10}$. For $x > 0$, $x > \sin x$.

30. $|\text{error}| = \left|\frac{x^2}{2!}\frac{1}{4(c+1)^{3/2}}\right| \leq \frac{(0.01)^2}{8}\frac{1}{4(c+1)^{3/2}} \leq \frac{(0.01)^2}{32} = 3.125\times 10^{-6}$.

31. $|R_2| = \frac{|x|^3}{3!}e^c \leq \frac{(0.1)^3 e^{0.1}}{6} = 1.84\times 10^{-4}$.

32. By the Alternating Series Theorem, |error| < |first omitted term| $= \left|\frac{x^3}{6}\right| <$ 0.00017. By the Remainder Estimation Theorem, |error| $= \left|\frac{x^3}{6}e^c\right| < \left|\frac{x^3}{6}\right||1|$, where $-0.1 < c < 0$. The estimates are the same.

33. $|R_4| = \frac{|x|^5|\cos c|}{5!} \le \frac{(0.5)^5(1)}{5!} = 2.6 \times 10^{-4}$.

34. |error| $= \frac{h^2}{2}e^c < \frac{h^2}{2}1.01 < h(0.00502) \le 0.6h$.

35. This is the series for $\sin(0.1) = 0.0998334\ldots$.

36. $s = 1 - \frac{(\frac{\pi}{4})^2}{2!} + \frac{(\frac{\pi}{4})^4}{4!} + \cdots = \cos\frac{\pi}{4} = \frac{1}{\sqrt{2}}$.
sum(seq(((−1)^x)($\frac{\pi}{4}$)^(2x)/(2x)!, x, 0, 3, 1) $- \frac{1}{\sqrt{2}} = -3.6\ E - 6$.

37. $\sin x = x - \frac{x^3}{3!} + \frac{x^5}{5!} - \frac{x^7}{7!} + \cdots$; differentiating each term gives $1 - \frac{x^2}{2!} + \frac{x^4}{4!} - \frac{x^6}{6!} + \cdots = \cos x$. Formally, $D_x e^x = D_x \sum_{n=0}^{\infty} x^n/n! = \sum_{n=0}^{\infty} D_x x^n/n! = \sum_{n=1}^{\infty} x^{n-1}/(n-1)! = \sum_{k=0}^{\infty} x^k/k!$.

38. $\int \sin x\,dx = \sum_{k=0}^{\infty} \frac{(-1)^k}{(2k+1)!}\int x^{2k+1}dx = \sum_{k=0}^{\infty} \frac{(-1)^k x^{2k+2}}{(2k+2)(2k+1)!} + c = \sum_{j=1}^{\infty} \frac{x^{2j}}{(2j)!} + c$ (where $j = k+1$) $= \cos x - 1 + c$; i.e., $\int$ (series for $\sin x$)$dx -$(series for $\cos x$) = constant. The other parts are similar.

39. $e^x = 1 + x + \frac{x^2}{2} + \frac{x^3}{6} + \frac{x^4}{24} + \frac{x^5}{5!} + \cdots$, $\sin x = x - \frac{x^3}{3!} + \frac{x^5}{5!} + \cdots$; $e^x \sin x = x + x^2 + x^3(\frac{1}{2} - \frac{1}{3!}) + x^4(\frac{1}{6} - \frac{1}{3!}) + x^5(\frac{1}{24} - \frac{1}{2\cdot 3!} + \frac{1}{5!}) + x^6(\frac{1}{5!} - \frac{1}{6\cdot 3!} + \frac{1}{5!}) + \cdots = x + x^2 + \frac{1}{3}x^3 + 0\cdot x^4 + x^5(-\frac{1}{30}) + x^6(-\frac{1}{90}) + \cdots$

40. $e^x \cos x = (1+x+\frac{x^2}{2}+\frac{x^3}{3!}+\frac{x^4}{4!}+\frac{x^5}{5!}+\cdots)(1-\frac{x^2}{2}+\frac{x^4}{4!}+\cdots) = 1+x+x^2(\frac{1}{2!}-\frac{1}{2!})+x^3(\frac{1}{3!}-\frac{1}{2})+x^4(\frac{1}{4!}-\frac{1}{2!2!}+\frac{1}{4!})+x^5(\frac{1}{4!}-\frac{1}{2\cdot 3!}+\frac{1}{5!})+\cdots = 1+x-\frac{x^3}{3}-\frac{x^4}{6}+\frac{1}{30}x^5+\cdots$.

41. $P_3 = x - \frac{x^3}{6} = \sin x - R_3 = \sin x - \frac{x^4}{4!}\sin c < \sin x$ if $0 < |x| < 1$; similarly $\sin x < x$ for $0 < |x| < 1$. Hence $x - \frac{x^3}{6} < \sin x < x \Rightarrow 1 - \frac{x^2}{6} < \frac{\sin x}{x} < 1$ if $x > 0$. If $x < 0$, $x = -t$, say, then $\frac{\sin x}{x} = \frac{-\sin t}{-t} = \frac{\sin t}{t}$ and $1 - \frac{t^2}{6} < \frac{\sin t}{t} < 1$, or $1 - \frac{x^2}{6} < \frac{\sin x}{x} < 1$.

42. $\cos x = 1 - \frac{x^2}{2} + \frac{x^4}{24} - \cdots$; by the Alternating Series Theorem, $1 - \frac{x^2}{2} < \cos x < 1 - \frac{x^2}{2} + \frac{x^4}{24}$ or $-\frac{1}{2} < \frac{\cos x - 1}{x^2} < \frac{x^2}{24} - \frac{1}{2}$, i.e., $\frac{1}{2} - \frac{x^2}{24} < \frac{1-\cos x}{x^2} < \frac{1}{2}$.

43. a) $e^{i\pi} = \cos\pi + i\sin\pi = -1$, b) $e^{i\pi/4} = \cos\frac{\pi}{4} + i\sin\frac{\pi}{4} = \frac{1}{\sqrt{2}}(1+i)$; c) $e^{-i\pi/2} = \cos(-\frac{\pi}{2}) + i\sin(-\frac{\pi}{2}) = -i$

44. $\frac{e^{i\theta}+e^{-i\theta}}{2} = \frac{\cos\theta + i\sin\theta + \cos(-\theta) + i\sin(-\theta)}{2} = \frac{\cos\theta + \cos\theta + i\sin\theta - i\sin\theta}{2} = \cos\theta$; $\frac{e^{i\theta}-e^{-i\theta}}{2i} = \frac{1}{2i}[\cos\theta + i\sin\theta - (\cos(-\theta) + i\sin(-\theta)] = \frac{1}{2i}(2i\sin\theta) = \sin\theta$.

45. $e^{i\theta} = 1 + i\theta + \frac{(i)^2\theta^2}{2} + \frac{(i)^3\theta^3}{3!} + \frac{(i)^4\theta^4}{4!} + \cdots$, $e^{-i\theta} = 1 - i\theta + \frac{(i)^2\theta^2}{2} + \frac{(-i)^3\theta^3}{3!} + \frac{(i)^4\theta^4}{4!} + \cdots$, $\frac{1}{2}[e^{i\theta} + e^{-i\theta}] = \frac{1}{2}[2 + \frac{2(i)^2\theta^2}{2!} + \frac{2(i)^4\theta^4}{4!} + \cdots] = \frac{2}{2}[1 - \frac{\theta^2}{2!} + \frac{\theta^4}{4!} + \cdots] = \cos\theta$; $\frac{1}{2i}[e^{i\theta} - e^{-i\theta}] = \frac{1}{2i}[2i\theta - \frac{i\theta^3}{3!} + \cdots] = \sin\theta$.

46. $\frac{d}{dx}e^{(a+ib)x} = \frac{d}{dx}e^{ax}\cos bx + i\frac{d}{dx}e^{ax}\sin bx = ae^{ax}\cos bx - be^{ax}\sin bx + iae^{ax}\sin bx + ibe^{ax}\cos bx = a(e^{ax}(\cos bx + i\sin bx)) + ib(e^{ax}(\cos bx + i\sin bx)) = (a+ib)e^{ax}(\cos bx + i\sin bx) = (a+ib)e^{(a+ib)x}$.

47. $\int e^{(a+ib)x}dx = \int e^{ax}\cos bx\ dx + i\int e^{ax}\sin bx\ dx = \frac{1}{a^2+b^2}[a - ib][e^{ax}][\cos bx + i\sin bx] + c = \frac{e^{ax}}{a^2+b^2}[a\cos bx + b\sin bx] + \frac{ie^{ax}}{a^2+b^2}[a\sin bx - b\cos bx] + c_1 + ic_2$. Equating real and imaginary parts gives $\int e^{ax}\cos bx\ dx = \frac{e^{ax}}{a^2+b^2}[a\cos bx + b\sin bx] + c_1$ and $\int e^{ax}\sin bx\ dx = \frac{e^{ax}}{a^2+b^2}[a\sin bx - b\cos bx] + c_2$.

9.8 FURTHER CALCULATIONS WITH TAYLOR SERIES

1. $\cos \approx \cos 1 + (x-1)(-\sin 1) + \frac{(x-1)^2}{2}(-\cos 1) + \frac{(x-1)^3}{6}\sin 1$. $|\text{error}| \le \frac{|x-1|^4}{4!}\cdot 1$.

2. $f(x) = \sin x$; take $a = 2\pi$: $\sin x = (x-2\pi) - \frac{(x-2\pi)^3}{3!} + \frac{(x-2\pi)^5}{5!} - \frac{(x-2\pi)^7}{7!} + \frac{(x-2\pi)^9}{9!}\cos c$; for $x = 6.3$, $|\text{error}| = \frac{(6.3-2\pi)^9}{9!}|\cos c| \le 3\times 10^{-22}$.

3. $e^x \approx e^{0.4} + (x-0.4)e^{0.4} + \frac{(x-0.4)^2}{2}e^{0.4} + \frac{(x-0.4)^3}{6}e^{0.4}$; $|\text{error}| \le \frac{|x-0.4|^4}{4!}e^{0.4}$.

4. $\ln x = \ln 1 + (x-1)\frac{1}{1} + \frac{(x-1)^2}{2!}\left(\frac{-1}{1}\right) + \frac{(x-1)^3}{3!}\frac{2}{1} + \frac{(x-1)^4}{4!}\left(\frac{-6}{1}\right) + \frac{(x-1)^5}{5!}\frac{(24)}{c^5} = (x-1) - \frac{(x-1)^2}{2} + \frac{(x-1)^3}{3} - \frac{(x-1)^4}{4} + \frac{(x-1)^5}{5}\cdot\frac{1}{c^5}$; for $x = 1.3$, $|\text{error}| < \frac{(0.3)^5}{5!}\frac{24}{c^5} < \frac{(0.3)^5}{5!}\frac{24}{1} < 5\times 10^{-4}$, since $1 < c < 1.3$.

5. $\cos x \approx \cos 69 + (x-69)(-\sin 69) + \frac{(x-69)^2}{2}(-\cos 69) + \frac{(x-69)^3}{6}\sin 69$; $|\text{error}| \le \frac{|x-69|^4}{4!}\cdot 1$.

6. Expand about $x = 1$; $\tan^{-1}x = \frac{\pi}{4} + \frac{1}{2}(x-1) + \frac{1}{2!}(x-1)^2(-\frac{1}{2}) + \frac{(x-1)^3}{3!}\cdot\frac{1}{2} + R = \frac{\pi}{4} + \frac{(x-1)}{2} - \frac{(x-1)^2}{4} + \frac{(x-1)^3}{12} + R$; $\tan^{-1}2 - (\frac{\pi}{4} + \frac{1}{2} - \frac{1}{4} + \frac{1}{12}) \approx 0.988$.

7. $f = (1+x)^3$; $f' = 3(1+x)^2$, $f'' = 6(1+x)$; $f''' = 6$, $f^{\text{iv}} \equiv 0$; $f(0) = 1$, $f'(0) = 3$, $f''(0) = 6$, $f'''(0) = 6 \Rightarrow (1+x)^3 = 1 + x\cdot 3 + \frac{x^2}{2}6 + \frac{x^3}{6}\cdot 6 + \frac{x^4}{4!}\cdot 0$.

8. If m is an integer, for $k \ge m+1$ $\binom{m}{k} = \frac{m(m-1)\cdots(m+1-k)}{k!} = 0$; hence $1 + \sum_{k=1}^{\infty}\binom{m}{k}x^k = \sum_{k=0}^{m}\binom{m}{k}x^k$.

9. – 14. Enter the functions: $y1 = 1 + Mx + (M(M-1)/2)x^2 + (M(M-1)(M-2)/6)x^3 + (M(M-1)(M-2)(M-3)/24)x^4 + (M(M-1)(M-2)(M-3)(M-4)/120)x^5$ and $y2 = (1+x)$^M. Store the desired value of M, then graph $y1 - y2$ on $[a, b]$ by $[-0.01, 0.01]$; for part b) graph $y1$ and $y2$

	x-Range	Viewing Window
9.	$(-0.88, 1.14)$	$(-2, 6)$ by $(-2, 15)$
10.	$(-1, 1.27)$	$(-1, 10)$ by $(0, 100)$
11.	$(-0.35, 0.4)$	$(-1, 6)$ by $(0, 3)$
12.	$(-0.35, 0.41)$	$(-1, 6)$ by $(0, 2)$
13.	$(-0.27, 0.26)$	$(-1, 4)$ by $(0, 1.5)$
14.	$(-0.33, 0.33)$	$(-1, 3)$ by $(0, 1.5)$

15. $\frac{\sin t}{t} - P_{21} = R$, where $|R| = \frac{t^{22}}{(23)!}|\cos c| \le \frac{t^{22}}{23!}$; $\left|\int_0^x \frac{\sin t}{t}dt - \int_0^x P_{21}(t)dt\right| \le \int_0^x \left|\frac{t^{22}}{23!}\right|dt = \frac{x^{23}}{(23)(23)!}$; $\frac{5^{23}}{(23)(23)!} < 2.005 \times 10^{-8}$, $\frac{8^{23}}{(23)(23)!} < 9.93 \times 10^{-4}$, $\frac{10^{23}}{(23)(23)!} < 0.169$.

16. By the Alternating Series Theorem, $|\text{error}| < \frac{|x|^{23}}{23}$; $\frac{|x|^{23}}{23} < 0.001$ when $|x| < 0.848$.

17. $\frac{1-\cos t}{t^2} = \frac{\frac{t^2}{2!} - \frac{t^4}{4!} + \frac{t^6}{6!} + \cdots}{t^2} = -\sum_{k=1}^{\infty} \frac{(-1)^k t^{2k-2}}{(2k)!} = \sum_{k=1}^{\infty} \frac{(-1)^{k+1} t^{2k-2}}{(2k)!}$;

$\int_0^x \frac{1-\cos t}{t^2}dt = \sum_{k=1}^{\infty} \frac{(-1)^{k+1} x^{2k-1}}{(2k-1)(2k)!}$.

18. a) $\int_0^x \frac{1-e^{-t}}{t}dt = \int_0^x \frac{1}{t}(t - \frac{t^2}{2!} + \frac{t^3}{3!} - \frac{t^4}{4!} + \cdots)dt = x - \frac{x^2}{2 \cdot 2!} + \frac{x^3}{3 \cdot 3!} - \frac{x^4}{4 \cdot 4!} + \cdots$, b) Graph both in $[-4, 8]$ by $[-20, 10]$, c) By the Alternating Series Theorem, for $x > 0$, $|\text{error}| \le \frac{|x|^{n+1}}{(n+1)(n+1)!}$. If $|x| \le 1$, $n + 1 \ge 9 \Rightarrow \frac{|x|^{n+1}}{(n+1)(n+1)!} < 10^{-6}$.

19. $\int_0^{0.1} P_{10}(x)dx = \int_0^{0.1} \sum_{k=1}^{6} \frac{(-1)^{k+1} x^{2k-1}}{(2k-1)! x}dx = \sum_1^6 \frac{(-1)^{k+1}}{(2k-1)!} \int_0^{0.1} x^{2k-2}dx = \sum_1^6 \frac{(-1)^{k+1}}{(2k-1)(2k-1)!}(0.1)^{2k-1} = 0.0999444612$; using NINT gives 0.0999444601.

20. $\int_0^{0.1} e^{-x^2}dx = \int_0^{0.1}[1 - x^2 + \frac{x^4}{2!} - \frac{x^6}{3!} + \frac{x^8}{4!} - \frac{x^{10}}{5!} + \cdots]dx =$

$\left[x - \frac{x^3}{3} + \frac{x^5}{5 \cdot 2!} - \frac{x^7}{7 \cdot 3!} + \cdots\right]_0^{0.1} = \sum_{k=0}^{\infty} \frac{(-1)^k x^{(2k+1)}}{(2k+1)k!}\Big]_0^{0.1}$; $\int_0^{0.1} P_{10}(x)dx =$

$\sum_{k=0}^{5} \frac{(-1)^k (0.1)^{2k+1}}{(2k+1)k!} = 0.099667664$.

21. $\int_0^{0.1} \frac{1-\cos x}{x^2} dx = -\int_0^{0.1} \sum_{n=1}^{\infty} \frac{(-1)^n x^{2n-2}}{(2n)!} dx$; $\int_0^{0.1} P_{10}(x) dx =$
$\sum_{n=1}^{5} \int_0^{0.1} \frac{(-1)^{n+1} x^{2n-2}}{(2n)!} dx = \sum_{n=1}^{5} \frac{(-1)^{n+1} x^{2n-1}}{(2n-1)(2n)!} \Big]_0^{0.1} = 0.49986114\ldots$.

22. $\int_0^{0.1} (1+x^4)^{1/2} dx = \int_0^{0.1} \sum_{k=0}^{\infty} \binom{\frac{1}{2}}{k} x^{4k} dx$, where $\binom{\frac{1}{2}}{k} = \frac{1}{2}(\frac{1}{2}-1)(\frac{1}{2}-2)\cdots(\frac{1}{2}-k+1)/k!$; $\int_0^{0.1} \sum_{k=0}^{2} \binom{\frac{1}{2}}{k} x^{4k} dx = \sum_{k=0}^{2} \binom{\frac{1}{2}}{k} \frac{x^{4k+1}}{4k+1} \Big|_0^{0.1} = \frac{0.1}{1} + \frac{1}{2}\frac{(0.1)^5}{5} + \frac{\frac{1}{2}(\frac{1}{2}-1)}{2}\frac{(0.1)^9}{9} = 0.100000999986$; NINT gives 0.100001.

23. $\ln(1+x) = 1 + x + \frac{x^2}{2} + \frac{x^3}{3} + \frac{x^4}{4} + \cdots$, $\ln(1-x) = 1 - x + \frac{x^2}{2} - \frac{x^3}{3} + \frac{x^2}{4} + \cdots$, $\ln(1+x) - \ln(1-x) = \ln\frac{1+x}{1-x} = 2[x + \frac{x^3}{3} + \frac{x^5}{5} + \cdots]$.

24. From the table, $x = 0.1$ so $|\text{error after } n \text{ terms}| \le \frac{|0.1|^{n+1}}{n+1}$; solving $(0.1)^{n+1} < (n+1)10^{-08}$ gives $n \ge 7$.

PRACTICE EXERCISES, CHAPTER 9

1. $\lim_{n\to\infty} a_n = \lim_{n\to\infty} 1 + \lim_{n\to\infty} \frac{(-1)^n}{n} = 1$; converges to 1.

2. $\lim_{n\to\infty} a_n = \lim_{n\to\infty} \left(\frac{1}{2^n} - \frac{2^n}{2^n}\right) = -1$; converges to -1.

3. This sequence is $1, 0, -1, 0, 1, 0, -1, 0, \ldots$; diverges.

4. $\lim_{n\to\infty} a_n = \lim_{n\to\infty} \left[(\frac{4}{n})^{n/4}\right]^2 = 1^2 = 1$; converges to 1.

5. $\lim_{n\to\infty} a_n = 2\lim_{n\to\infty} \frac{\ln n}{n} = 2\lim_{n\to\infty} \frac{\frac{1}{n}}{1} = 0$; converges to 0.

6. $\lim_{n\to\infty} a_n = \lim_{n\to\infty} \left(1 + \frac{5}{n}\right)^n = e^5$; converges to e^5.

7. $\lim_{n\to\infty} a_n = \lim_{n\to\infty} \frac{3}{\sqrt[n]{n}} = \frac{3}{1} = 3$; converges to 3.

8. $\lim_{n\to\infty} \frac{1}{3^{2n-1}} = 0$; converges to 0.

9. $\lim_{n\to\infty} a_n = \lim_{n\to\infty} \frac{(-4)^n}{n!} = 0$, $\left(\frac{x^n}{n!} \to 0 \text{ for all } x\right)$; converges to 0.

10. $\lim_{n\to\infty} \frac{\ln(2n+1)}{n} = \lim_{n\to\infty} \frac{\frac{2}{2n+1}}{1} = 0$, by l'Hôpital's rule; converges to 0.

11. $\lim_{n\to\infty} a_n = \lim_{n\to\infty} (n+1) = +\infty$; diverges.

12. $\lim_{n\to\infty} a_n = \lim_{n\to\infty} \frac{1}{2}\frac{2n^2-2n+n-n}{2n^2+n} = \frac{1}{2}\lim_{n\to\infty}\left(1 - \frac{3n}{2n^2+n}\right) = \frac{1}{2}$; converges to $\frac{1}{2}$.

13. $\sum^k \ln\left(\frac{n}{n+1}\right) = \ln\left(\prod^k \frac{n}{n+1}\right) = \ln\left(\frac{1}{2}\cdot\frac{2}{3}\cdot\frac{3}{4}\cdot \cdots \frac{k}{k+1}\right) = \ln 1 - \ln k$; $\lim_{k\to\infty} \ln 1 - \ln k = -\infty$; diverges.

14. $\sum_{n=2}^{\infty} \frac{-2}{n(n+1)} = -2\sum_{n=2}^{\infty}\left(\frac{1}{n} - \frac{1}{n+1}\right) = -2\cdot\frac{1}{2} = -1$; $\sum_{n=2}^{100} a_n = -0.980$.

15. $1 + \left(\frac{1}{e}\right) + \left(\frac{1}{e}\right)^2 + \cdots$ converges to $\frac{1}{1-1/e} = \frac{e}{e-1}$.

16. $\sum_{n=1}^{\infty}(-1)^n\frac{3}{4^n} = 3\sum_{n=1}^{\infty}\left(-\frac{1}{4}\right)^n = 3\cdot\left(-\frac{1}{4}\right)\frac{1}{1-(-1/4)} = -\frac{3}{5}$; $S_{100} = -0.6$.

17. $a_n = 1.05^{-1}a_{n-1} = 1.05^{-2}a_{n-2} = \cdots = (1.05)^{-n}a_0$; $\sum_{n=0}^{\infty} a_n = a_0\sum_{n=0}^{\infty}\left(\frac{1}{1.05}\right)^n = 125\left(\frac{1}{1-1/1.05}\right) = 2625$.

18. $a_n = a_{n-1}/(1+n) \Rightarrow a_1 = \frac{a_0}{2}$, $a_2 = \frac{a_1}{3} = \frac{a_0}{3!}$, $a_3 = \frac{a_2}{4} = \frac{a_0}{4!}$, etc. $\sum_{n=0}^{\infty} a_n = \sum_{n=0}^{\infty}\frac{a_0}{(n+1)!} = a_0(1 + \frac{1}{2} + \frac{1}{3!} + \frac{1}{4!} + \cdots) = 1\cdot(e-1)$.

19. $\sum_{n=1}^{\infty}\frac{1}{n^p}$ diverges if $p \le 1$; here $p = \frac{1}{2}$; diverges.

20. $\sum_{n=1}^{\infty} -\frac{5}{n} = -5\sum_{n=1}^{\infty}\frac{1}{n}$ which diverges.

21. $\lim_{n\to\infty} a_n = 0$; by the Alternating Series Test, the series converges. By #17, it does not converge absolutely; conditionally convergent. $|\text{error of } S_{9999}| < |a_{10000}| = \frac{1}{100} = 0.01$; on the other hand $S_{1001} = -0.621 < S < S_{1000} = -0.589$. Taking the average, $S = 0.605$ with an error less than 0.02.

22. $\sum_{n=1}^{\infty}\frac{1}{2n^3} = \frac{1}{2}\sum_{n=1}^{\infty}\frac{1}{n^3}$ which is convergent. $S_{1000} = 0.601028\ldots$, $|s - S_{1000}| < \left|\frac{1}{2}\int_{1000}^{\infty}\frac{dx}{x^3}\right| = 2.5\times 10^{-7}$.

23. The series converges by the Alternating Series Test ($\ln x$ is increasing $\Rightarrow$ $1/\ln x$ is decreasing). $\ln(n+1) < n+1 \Rightarrow \frac{1}{\ln(n+1)} > \frac{1}{n+1}$; since $\sum\frac{1}{n+1}$ diverges so does $\sum 1/\ln(n+1)$; conditionally convergent. $S_{1000} = -0.85\ldots$, $S_{1001} = -0.997$; since $|a_{10001}| = 0.087$; the series converges very slowly.

24. $\sum_{n=2}^{\infty}\frac{1}{n(\ln n)^2}$ converges by the Integral Test: $\int_2^{\infty}\frac{dx}{x(\ln x)^2} = \lim_{b\to\infty}\left.\frac{-1}{\ln x}\right]_2^b = \ln 2$. $S_{1000} = 1.965$; the error is less than $\int_{1000}^{\infty}\frac{dx}{x(\ln x)^2} = 0.145$.

25. By the Limit Comparison Test, $\lim_{n\to\infty} = \left(\frac{1}{n\sqrt{n^2+11}}\right)/(1/n^2) = \lim_{n\to\infty}\frac{n}{\sqrt{n^2+1}} = 1$; since $\sum\frac{1}{n^2}$ converges, the series converges absolutely. $S_{100} = -0.55155$; $|\text{error}| \le \frac{1}{11\sqrt{122}} = 0.008$.

26. $|a_n| = \frac{3n^2}{n^3+1}$; $\frac{|a_n|}{\frac{1}{n}} = 3$; the series does not converge absolutely by the Limit Comparison Test; it converges conditionally by the Alternating Series Test. The converge is very slow - (see #18).

27. $\sum_{n=1}^{\infty} \frac{n+1}{n!} = \sum_{n=1}^{\infty} \frac{1}{(n-1)!} + \sum_{n=1}^{\infty} \frac{1}{n!} = \sum_{k=0}^{\infty} \frac{1}{k!} + \sum_{n=1}^{\infty} \frac{1}{n!} = e^1 + (e^1 - 1) = 2e^1 - 1 = 2e - 1 = 4.4365\ldots$.

28. Fails to converge absolutely or conditionally as $\lim_{n\to\infty} |a_n| = \frac{1}{2} \neq 0$.

29. Since $e^{-3} = \sum_{n=0}^{\infty} \frac{(-3)^n}{n!} = 1 + \sum_{n=1}^{\infty} \frac{(-3)^n}{n!}$, the series is convergent; sum is $e^{-3} - 1$.

30. $\sum_1^{\infty} a_n = \sum \left(\frac{6}{n}\right)^n$; by the Root Test $\lim_{n\to\infty} a_n^{1/n} = 0$; the series converges. $S_{60} = 32.024\ldots$, the error $\sum_{n=61}^{\infty} \left(\frac{6}{n}\right)^n \leq \sum_{n=61}^{\infty} (0.1)^n = (0.1)^{61}\frac{1}{1-0.1}$ is very small.

31. $\lim_{n\to\infty} \left|\frac{a_{n+1}}{a_n}\right| = \lim_{n\to\infty} \left|\frac{(x+2)^{n+1}}{3^{n+1}(n+1)} \cdot \frac{3^n n}{(x+2)^n}\right| = \frac{|x+2|}{3} \lim_{n\to\infty} \frac{n}{n+1} = \frac{|x+2|}{3} < 1$ for $-3 < x+2 < 3$ or $-5 < x < 1$. When $x = -5$ we have $\sum \frac{(-1)^n 3^n}{3^n n} = \sum (-1)^n/n$ which converges; when $x = 1$ we have $\sum \frac{3^n}{3^n n} = \sum \frac{1}{n}$ which diverges. S_{20} will have its greatest error at -3; solving $\frac{(x+2)^{21}}{3^{21}\cdot 21} < 0.01$ gives $|x+2| \leq 2.78$; if $x + 2 = -2 + 0.78$, $x = -4.78$.

32. $\lim_{n\to\infty} \left|\frac{x^{n+1}}{\sqrt{n+1}} \cdot \frac{\sqrt{n}}{x^n}\right| = |x|$; the series converges absolutely for, at least, $-1 < x < 1$. If $x = 1$, $\sum \frac{1}{n^{1/2}}$ diverges by the Integral Test; if $x = -1$, $\sum \frac{(-1)^n}{n^{1/2}}$ converges by the Alternating Series Test. a) $-1 \leq x < 1$, b) $-1 < x < 1$.

33. $\lim_{n\to\infty} \left|\frac{x^{n+1}}{(n+1)^{n+1}} \cdot \frac{n^n}{x^n}\right| = |x| \lim_{n\to\infty} \left|\frac{n^n}{(n+1)^{n+1}}\right| = |x| \lim_{n\to\infty} \left|\left(\frac{n}{n+1}\right)^n \cdot \frac{1}{n+1}\right| = |x|(\frac{1}{e}) \cdot 0 = 0 \Rightarrow$ the series converges absolutely for all x.

34. $\lim_{n\to\infty} \left|\frac{n+2}{2n+3} \frac{(x-1)^{n+1}}{2^{n+1}} \cdot \frac{2^n(2n+1)}{(n+1)(x-1)^n}\right| = \frac{|(x-1)|}{2} < 1$ when $-1 < x < 3$. When $x = -1$, $a_n = \frac{n+1}{2n+1}\frac{(-2)^n}{2^n}$; $\lim_{n\to\infty} a_n \neq 0$, at $x = -1$ the series diverges. At $x = 3$, $a_n = \frac{n+1}{2n+1}\frac{2^n}{2^n} \to \frac{1}{2}$; the series diverges. Hence the series converges, and converges absolutely for $-1 < x < 3$.

35. $\lim_{n\to\infty} \left|\frac{a_{n+1}}{a_n}\right| = \lim_{n\to\infty} \left|\frac{(x-1)^{n+1}}{(n+1)^2} \cdot \frac{n^2}{(x-1)^n}\right| = |x-1| \lim_{n\to\infty} \frac{n^2}{(n+1)^2} = |x-1| \cdot 1 \Rightarrow$ the series converges for $|x-1| < 1$, or $0 < x < 2$; at $x = 0$ we have $\sum \frac{(-1)^{n-1}}{n^2}(-1)^n = \sum \frac{(-1)}{n^2}$ which converges; at $x = 2$, $\sum \frac{(-1)^{n-1}}{n^2}$ converges by the alternating series test.

36. $\lim_{n\to\infty}\left|\frac{(x-1)^{2n}}{(2n+1)!}\cdot\frac{(2n-1)!}{(x-1)^{2n-2}}\right| = \lim_{n\to\infty}\left|\frac{(x-1)^2}{(2n+1)(2n)}\right| = 0$ for all x; a) for all x, b) for all x.

37. This is the series for $\frac{1}{1+x}$, where $x = \frac{1}{4}$; sum is $\frac{1}{1+1/4} = 0.8$.

38. The series is $\sum_{n=1}^{\infty} \frac{(-1)^{n-1}(\frac{2}{3})^n}{n} = \ln(1+\frac{2}{3})$; $f(x) = \ln(1+x)$.

39. This is the series for $\sin x$, at $x = \pi$; $\sin\pi = 0$.

40. This is the series for $\cos x$ at $x = \left(\frac{\pi}{3}\right)$; sum is $\cos\frac{\pi}{3} = \frac{1}{2}$.

41. This is the series for e^x at $x = \ln 2$, sum is $e^{\ln 2} = 2$.

42. $\frac{1}{\sqrt{3}} - \frac{1}{9\sqrt{3}} + \frac{1}{45\sqrt{3}} - \cdots = \frac{(\frac{1}{\sqrt{3}})}{1} - \frac{(\frac{1}{\sqrt{3}})^3}{3} + \frac{(\frac{1}{\sqrt{3}})^5}{5} - \cdots = \tan^{-1}(\frac{1}{\sqrt{3}}) = \frac{\pi}{6}$.

43. $f = (3+x^2)^{1/2}$; $f' = \frac{1}{2}(3+x^2)^{-1/2}(2x)$; $f'' = (3+x^2)^{-1/2} + x(-\frac{1}{2})(3+x^2)^{-3/2}\ (2x) = (3+x^2)^{-1/2} - x^2(3+x^2)^{-3/2}$; $f''' = -\frac{1}{2}(3+x^2)^{-3/2}2x - 2x(3+x^2)^{-3/2} + \frac{3x^2}{2}(3+x^2)^{-5/2}2x = -3x(3+x^2)^{-3/2} + 3x^3(3+x^2)^{-5/2}$; $f(-1) = 2$, $f'(-1) = -\frac{1}{2}$, $f''(-1) = \frac{3}{8}$, $f'''(-1) = \frac{3}{8} - \frac{3}{32} = \frac{9}{32}$. $f(x) = 2 + (-\frac{1}{2})(x+1) + (\frac{3}{8})\frac{1}{2}(x+1)^2 + (\frac{9}{32})\frac{1}{6}(x+1)^3 + \cdots$. Graphing abs ($\sqrt{3+x^2}$ - first four non-zero terms) shows the error is less than 0.01 for $-2.143 < x < 0.238$.

44. $f = (1-x)^{-1}$; $f' = (1-x)^{-2}$, $f'' = 2(1-x)^{-3}$, $f''' = 6(1-x)^{-4}$; $f(2) = -1$, $f'(2) = 1$, $f''(2) = -2$, $f'''(2) = 6$; $P_4 = -1 + (x-2)(1) + (-2)\frac{(x-2)^2}{2!} + 6\frac{(x-2)^3}{3!} = -1 + (x-2) - (x-2)^2 + (x-2)^3$. Graph abs $(f - P_4)$ on $(1.01, 3)$ by $(0, 0.2)$; $1.710 < x < 2.340$.

45. $(\sin x)2(\cos x) = 2(x - \frac{x^3}{3!} + \frac{x^5}{5!} - \frac{x^7}{7!})(1 - \frac{x^2}{2!} + \frac{x^4}{4!} - \frac{x^6}{6!} + \cdots) = 2\left[x + x^3(-\frac{1}{3!} - \frac{1}{2!}) + x^5(\frac{1}{5!} + \frac{1}{2!3!} + \frac{1}{4!}) + x^7(-\frac{1}{7!} - \frac{1}{2!5!} - \frac{1}{3!4!} - \frac{1}{6!})\right] = 2x - 2\cdot\frac{4}{6}x^3 + 2\left[\frac{1}{120} + \frac{1}{12} + \frac{1}{24}\right]x^5 - \frac{2}{7!}\left[1 + \frac{7\cdot 6}{2} + \frac{7\cdot 6\cdot 5}{6} + 7\right]x^7 + \cdots = 2x - \frac{2^3x^3}{3!} + \frac{2^5x^5}{5!} - \frac{2^7x^7}{7!} + \cdots$.

46. $\sin^2 x = \frac{1}{2}\left[1 - \sum_{n=0}^{\infty}\frac{(-1)^n(2x)^{2n}}{2n!}\right] = \frac{1}{2}\left(-\sum_{n=1}^{\infty}\frac{(-1)^n(2x)^{2n}}{(2n)!}\right) = \frac{1}{2}\sum_{n=1}^{\infty}\frac{(-1)^{n+1}(2x)^{2n}}{(2n)!} = \frac{(2x)^2}{2\cdot 2!} - \frac{(2x)^4}{2\cdot 4!} + \frac{(2x)^6}{2\cdot 6!} - \frac{(2x)^8}{2\cdot 8!} + \cdots = \frac{2x^2}{2!} - \frac{2^3x^4}{4!} + \frac{2^5x^6}{6!} - \frac{2^7x^8}{8!} + \cdots$.

47. $\int_0^{1/2} e^{-x^3}dx = \int_0^{1/2}(1 - x^3 + \frac{x^6}{2!} - \frac{x^9}{3!} + \frac{x^{12}}{4!})dx + \int_0^{1/2} R_4\ dx$ where $|R_4| \le \int_0^{1/2}\frac{x^{14}}{5!}\ dx = \left.\frac{x^{15}}{15\cdot 5!}\right|_0^{1/2} = 1.69\ E{-}08$ which is a little too large for the problem. For error $< 10^{-8}$ take $\int_0^{1/2}(1 - x^3 + \frac{x^6}{2!} - \frac{x^9}{3!} + \frac{x^{12}}{4!} - \frac{x^{15}}{5!})dx = 0.48491714$ (NINT).

48. $\int_0^{1/2} \frac{\tan^{-1} x}{x} dx = \int_0^{1/2}(1 - \frac{x^2}{3} + \frac{x^4}{5} - \frac{x^6}{7} + \frac{x^8}{9} = \cdots + \frac{x^{16}}{17})dx + \int_0^{1/2} R\, dx$, where $\left|\int_0^{1/2} R\, dx\right| \leq \int_0^{1/2} |R|\, dx \leq \int_0^{1/2} \frac{x^{18}}{19} dx = 0.5\ E - 9$. The first integral has value 0.487222362.

49. a) The MacLaurin series is $\sum_{n=0}^{\infty} \frac{x^n}{n!}\ 0 = 0$ for all x. It converges for all x; it converges to $f(x)$ only at $x = 0$. b) $e^{-1/x^2} = 0 + R_n(x)$; hence $R_n = e^{-1/x^2}$.

50. The iterations $x = \frac{\sin x}{x^2}$ and $x = \sin^{-1}(x^3)$ diverge; however, taking $x = \sqrt{\frac{\sin x}{x}}$ leads to convergence at $x = 0.9286$.

CHAPTER 10

PLANE CURVES, PARAMETRIZATIONS, AND POLAR COORDINATES

10.1 CONIC SECTIONS AND QUADRATIC EQUATIONS

1. $y = \frac{x^2}{4p} = \frac{x^2}{16}$

2. $y = \frac{x^2}{4p} = x^2$. $y = x^2$

3. $y = -\frac{x^2}{4p} = -\frac{x^2}{12}$

4. $y = -\frac{x^2}{4p} = -\frac{x^2}{2}$. $y = -\frac{x^2}{2}$

5. $x = -\frac{y^2}{4p} = -\frac{y^2}{12}$. $x = -\frac{y^2}{12}$

6. $x = \frac{y^2}{4p} = \frac{y^2}{8}$. $x = \frac{y^2}{8}$

7. $y = 4x^2 = \frac{x^2}{4(\frac{1}{16})}$, $p = \frac{1}{16}$. Focus: $(0, \frac{1}{16})$, directrix: $y = -\frac{1}{16}$.

8. $y = \frac{x^2}{3} = \frac{x^2}{4p}$. $4p = 3$, $p = \frac{3}{4}$. Focus: $(0, \frac{3}{4})$. Directrix: $y = -\frac{3}{4}$.

9. $y = -3x^2 = \frac{x^2}{4(-\frac{1}{12})}$. Focus: $(0, -\frac{1}{12})$. Directrix: $y = \frac{1}{12}$.

10. $y = -\frac{x^2}{4} = \frac{x^2}{4p}$, $p = -1$. Focus: $(0, -1)$. Directrix: $y = 1$.

11. Graph $y = \frac{x^2}{2}$ in $[-10.6, 10.6]$ by $[0, 12.5]$ to check you result. (We have used the "screen-squaring" feature of our calculator to help determine the viewing rectangle.)

12. Graph $y = -\frac{x^2}{6}$ in $[-4.25, 4.25]$ by $[-4, 1]$.

13. Graph $y = \sqrt{8x}$ and $y = -\sqrt{8x}$ together in $[-9.4, 14.4]$ by $[-7, 7]$ to check your result.

14. $x = -y^2/4$, $y^2 = -4x$, $y = \pm 2\sqrt{-x}$. Graph $y_1 = 2\sqrt{-x}$ and $y_2 = -y_1$ in $[-11.7, 8.7]$ by $[-6, 6]$.

15. $\frac{x^2}{4} + \frac{y^2}{9} = 1$. $\frac{y^2}{9} = 1 - \frac{x^2}{4} = \frac{4-x^2}{4}$, $y^2 = \frac{9}{4}(4 - x^2)$, $y = \pm\frac{3}{2}\sqrt{4 - x^2}$. Graph $y = \frac{3}{2}\sqrt{4 - x^2}$ and $y = -\frac{3}{2}\sqrt{x - x^2}$ in $[-5.1, 5.1]$ by $[-3, 3]$.

16. Check your result by graphing $y_1 = \sqrt{1 - x^2/2}$ and $y_2 = -y_1$ in $[-2, 2]$ by $[-1.2, 1.2]$.

17. $\frac{y^2}{4} - x^2 = 1$. Graph $y_1 = 2\sqrt{1 + x^2}$ and $y_2 = -y_1$ in $[-13.6, 13.6]$ by $[-8, 8]$.

18. $\frac{x^2}{4} - \frac{y^2}{9} = 1$, $\frac{y^2}{9} = \frac{x^2}{4} - 1 = \frac{x^2-4}{4}$, $y^2 = \frac{9}{4}(x^2 - 4)$. Check your result by graphing $y_1 = \frac{3}{2}\sqrt{x^2 - 4}$ and $y_2 = -y_1$ in $[-17, 17]$ by $[-10, 10]$.

19. $64x^2 - 36y^2 = 2304$, $\frac{64x^2}{2304} - \frac{36y^2}{2304} = 1$, $\frac{x^2}{6^2} - \frac{y^2}{8^2} = 1$. $c = \sqrt{6^2 + 8^2} = 10$, $e = \frac{c}{a} = \frac{10}{a} = \frac{5}{3}$. Graph $y_1 = (4/3)\sqrt{x^2 - 36}$, $y_2 = -y_1$, $y_3 = (4/3)x$ and $y_4 = -y_3$ in $[-34, 34]$ by $[-20, 20]$. Foci: $(\pm 10, 0)$.

20. $16x^2 + 25y^2 = 400$, $\frac{x^2}{25} + \frac{y^2}{16} = 1$. $c = \sqrt{25 - 16} = 3$; $e = \frac{3}{5}$. Foci: $(\pm 3, 0)$. Check your sketch by graphing $y_1 = \frac{4}{5}\sqrt{25 - x^2}$ and $y_2 = -y_1$ in $[-6.8, 6.8]$ by $[-4, 4]$. Intercepts: $(\pm 5, 0)$, $(0, \pm 4)$.

21. $8y^2 - 2x^2 = 16$, $\frac{y^2}{2} - \frac{x^2}{8} = 1$. $c = \sqrt{2 + 8} = \sqrt{10}$. $e = \frac{\sqrt{10}}{\sqrt{2}} = \sqrt{5}$. Foci: $(0, \pm\sqrt{10})$, asymptotes: $y = \pm\frac{\sqrt{2}x}{2\sqrt{2}} = \pm\frac{x}{2}$. Graph $y_1 = \frac{\sqrt{x^2+8}}{2}$, $y_2 = -y_1$, $y_3 = \frac{x}{2}$ and $y_4 = -\frac{x}{2}$ in $[-8.5, 8.5]$ by $[-5, 5]$.

22. $7x^2 + 16y^2 = 112$, $\frac{x^2}{16} + \frac{y^2}{7} = 1$. $c = \sqrt{16 - 7} = 3$. Foci: $(\pm 3, 0)$, $e = \frac{3}{4}$. Check your sketch by graphing $y_1 = \frac{\sqrt{7}}{4}\sqrt{16 - x^2}$ and $y_2 = -y_1$ in $[-4.5, 4.5]$ by $[-2.65, 2.65]$. Intercepts: $(\pm 4, 0)$, $(0, \pm\sqrt{7})$.

23. $169x^2 + 25y^2 = 4225$, $\frac{x^2}{25} + \frac{y^2}{169} = 1$. $c = \sqrt{169 - 25} = 12$, $e = \frac{c}{a} = \frac{12}{13}$, foci: $(0, \pm 12)$. Graph $y_1 = \frac{13}{5}\sqrt{25 - x^2}$ and $y_2 = -y_1$ in $[-22, 22]$ by $[-13, 13]$.

24. $y^2 - 3x^2 = 3$, $\frac{y^2}{3} - x^2 = 1$. $c = \sqrt{3 + 1} = 2$, $e = \frac{2}{\sqrt{3}}$. Foci: $(0, \pm 2)$, asymptotes: $y = \pm\sqrt{3}x$. Graph $y_1 = \sqrt{3x^2 + 3}$, $y_2 = -y_1$, $y_3 = \sqrt{3}x$, $y_4 = -y_3$ in $[-8.5, 8.5]$ by $[-5, 5]$. Vertices: $(0, \pm\sqrt{3})$.

25. $8x^2 - 2y^2 = 16$, $\frac{x^2}{2} - \frac{y^2}{8} = 1$. $c = \sqrt{2 + 8} = \sqrt{10}$, $e = \frac{c}{a} = \frac{\sqrt{10}}{\sqrt{2}} = \sqrt{5}$. Foci: $(\pm\sqrt{10}, 0)$, asymptotes: $y = \pm\frac{\sqrt{8}}{\sqrt{2}}x = \pm 2x$. Graph $y = 2\sqrt{x^2 - 2}$, $y = -2\sqrt{x^2 - 2}$, $y = 2x$ and $y = -2x$ in $[-17, 17]$ by $[-10, 10]$.

26. $6x^2 + 9y^2 = 54$. $\frac{x^2}{9} + \frac{y^2}{6} = 1$. $c = \sqrt{9 - 6} = \sqrt{3}$, $e = \frac{\sqrt{3}}{3}$. Foci: $(\pm\sqrt{3}, 0)$. Graph $y_1 = \sqrt{54 - 6x^2}/3$ and $y_2 = -y_1$ in $[-4.2, 4.2]$ by $[-2.5, 2.5]$. Intercepts: $(\pm 3, 0)$, $(0, \pm\sqrt{6})$.

27. $9x^2 + 10y^2 = 90$, $\frac{x^2}{10} + \frac{y^2}{9} = 1$. $c = \sqrt{10 - 9} = 1$, $e = \frac{c}{a} = \frac{1}{\sqrt{10}}$, foci: $(\pm 1, 0)$. Graph $y_1 = \frac{3}{\sqrt{10}}\sqrt{10 - x^2}$ and $y_2 = -y_1$ in $[-5.1, 5.1]$ by $[-3, 3]$.

28. $y^2 - x^2 = 8$. $\frac{y^2}{8} - \frac{x^2}{8} = 1$. $c = \sqrt{8 + 8} = 4$, $e = \frac{4}{\sqrt{8}} = \sqrt{2}$. Foci: $(0, \pm 4)$, asymptotes: $y = \pm x$. Graph $y_1 = \sqrt{x^2 + 8}$, $y_2 = -y_1$, $y_3 = x$, $y_4 = -x$ in $[-13.6, 13.6]$ by $[-8, 8]$. Vertices: $(0, \pm\sqrt{8})$.

29. $x^2 - y^2 = 1$, $c = \sqrt{1+1} = \sqrt{2}$, $e = \frac{c}{a} = \sqrt{2}$, foci: $(\pm\sqrt{2}, 0)$. Graph $y = \sqrt{x^2-1}$, $y = -\sqrt{x^2-1}$, $y = x$ and $y = -x$ in $[-8.2, 8.2]$ by $[-4.9, 4.9]$.

30. $2x^2 + y^2 = 4$, $\frac{x^2}{2} + \frac{y^2}{4} = 1$. $c = \sqrt{4-2} = \sqrt{2}$, $e = \frac{\sqrt{2}}{2}$. Foci: $(0, \pm\sqrt{2})$. Graph $y_1 = \sqrt{4-2x^2}$ and $y_2 = -y_1$ in $[-3.4, 3.4]$ by $[-2, 2]$. Intercepts: $(0, \pm 2)$, $(\pm\sqrt{2}, 0)$.

31. $y^2 - x^2 = 4$, $\frac{y^2}{4} - \frac{x^2}{4} = 1$, $c = \sqrt{4+4} = 2\sqrt{2}$, $e = \frac{2\sqrt{2}}{2} = \sqrt{2}$, foci: $(0, \pm 2\sqrt{2})$. Graph $y = \sqrt{x^2+4}$, $y = -\sqrt{x^2+4}$, $y = x$ and $y = -x$ in $[-8.2, 8.2]$ by $[-4.9, 4.9]$.

32. $2x^2 + y^2 = 2$, $x^2 + \frac{y^2}{2} = 1$. $c = \sqrt{2-1} = 1$. $e = \frac{1}{\sqrt{2}}$. Foci: $(0, \pm 1)$. Graph $y_1 = \sqrt{2-2x^2}$ and $y_2 = -y_1$ in $[-2.4, 2.4]$ by $[-\sqrt{2}, \sqrt{2}]$. Intercepts: $(0, \pm\sqrt{2})$, $(\pm 1, 0)$.

33. $3x^2 + 2y^2 = 6$, $\frac{x^2}{2} + \frac{y^2}{3} = 1$, $c = \sqrt{3-2} = 1$, $e = \frac{1}{\sqrt{3}}$. Graph $y = \sqrt{1.5}\sqrt{2-x^2}$ and $y = -\sqrt{1.5}\sqrt{2-x^2}$ in $[-2.9, 2.9]$ by $[-\sqrt{3}, \sqrt{3}]$. Foci: $(0, \pm 1)$.

34. $9x^2 - 16y^2 = 144$, $\frac{x^2}{16} - \frac{y^2}{9} = 1$. $c = \sqrt{16+9} = 5$. Foci: $(\pm 5, 0)$, $e = \frac{5}{4}$, asymptotes: $y = \pm\frac{3}{4}x$. Graph $y_1 = 0.75\sqrt{x^2-16}$, $y_2 = -y_1$, $y_3 = 0.75x$, $y_4 = -y_3$ in $[-17, 17]$ by $[-10, 10]$. Vertices: $(\pm 4, 0)$.

35.

	$y = -x^2/4$	$x = -y^2/4$
Focal axis:	The y-axis	The x-axis
Focus:	$(0, -1)$	$(-1, 0)$
Vertex:	$(0, 0)$	$(0, 0)$
Directrix:	$y = 1$	$x = 1$

36.

	$8y^2 - 2x^2 = 16$ $\frac{y^2}{2} - \frac{x^2}{8} = 1$	$8x^2 - 2y^2 = 16$ $\frac{x^2}{2} - \frac{y^2}{8} = 1$
Focal axis:	The y-axis	The x-axis
Center-to-focus distance: $c = \sqrt{a^2+b^2}$	$c = \sqrt{2+8} = \sqrt{10}$	$c = \sqrt{2+8} = \sqrt{10}$
Foci:	$(0, \pm c) = (0, \pm\sqrt{10})$	$(\pm c, 0) = (\pm\sqrt{10}, 0)$
Vertices:	$(0, \pm a) = (0, \pm\sqrt{2})$	$(\pm a, 0) = (\pm\sqrt{2}, 0)$
Center:	$(0, 0)$	$(0, 0)$
Asymptotes:	$y = \pm\frac{a}{b}x = \pm\sqrt{\frac{2}{8}}x = \pm\frac{x}{2}$	$y = \pm\frac{b}{a}x = \pm 2x$

37. Volume of parabolic solid $= 2\pi \int_0^{b/2} x(h - \frac{4h}{b^2}x^2)dx = 2\pi h \int_0^{b/2}(x - \frac{4}{b^2}x^3)dx = 2\pi h[\frac{x^2}{2} - \frac{x^4}{3b^2}]_0^{b/2} = \frac{\pi h b^2}{8}$. $\frac{3}{2}$ (volume of cone) $= \frac{3}{2}(\frac{1}{3}\pi r^2 h) = \frac{1}{2}\pi(\frac{b}{2})^2 h = \frac{\pi h b^2}{8}$.

38. $\frac{dy}{dx} = \frac{w}{H}x$. $y = \frac{wx^2}{2H} + C$. But $C = 0$ by the initial condition. Hence $y = \frac{wx^2}{2H}$ and the curve is a parabola.

39. We require $e = \frac{c}{a} = \frac{4}{5}$. Let us try to find a solution with $c = 4$, $a = 5$. $4 = c = \sqrt{a^2 - b^2} = \sqrt{25 - b^2}$ yielding $b = 3$. Thus the graph of $\frac{x^2}{25} + \frac{y^2}{9} = 1$ is an example of an ellipse with $e = \frac{4}{5}$.

40. From Table 10.2, $e = \frac{1}{4} = \frac{c}{a}$. $a = 4c$, $c^2 = a^2 - b^2$, $b^2 = a^2 - c^2$. If we set $c = 1$, we get $a = 4$, $b^2 = 15$, $b = \sqrt{15}$. The orbit has the shape of $\frac{x^2}{a} + \frac{y^2}{b^2} = \frac{x^2}{16} + \frac{y^2}{15} = 1$. Graph $y_1 = \frac{\sqrt{15}}{4}\sqrt{16 - x^2}$ and $y_2 = -y_1$ in $[-6.58, 6.58]$ by $[-\sqrt{15}, \sqrt{15}]$.

41. Because of the symmetry with respect to both axes and the origin, the coordinate axes break the rectangle into four smaller rectangles, one lying in each of the four quadrants. The rectangle in the first quadrant has vertices $(0,0), (0,1), (2,0)$ and $(x, y) = (x, \frac{\sqrt{4-x^2}}{2})$ because it lies on the upper half of the ellipse. The rectangle in the first quadrant has area $A(x) = xy = \frac{1}{2}x\sqrt{4 - x^2}$. $A'(x) = \frac{1}{2}[-\frac{x^2}{\sqrt{4-x^2}} + \sqrt{4 - x^2}] = \frac{2-x^2}{\sqrt{4-x^2}}$. $A(x)$ is maximized when $x = \sqrt{2}$ and $A(\sqrt{2}) = 1$. The dimensions of the original rectangle of maximal area are $2x = 2\sqrt{2}$ by $2y = \sqrt{4 - x^2} = \sqrt{2}$ and it has area $4A(\sqrt{2}) = 4$.

42. By symmetry $\bar{x} = 0$. For the top half of the ellipse $y = \frac{4}{3}\sqrt{9 - x^2}$ and this top half has area $\frac{\pi ab}{2} = 6\pi$. $\bar{y} = \frac{1}{6\pi}\int_{-3}^{3}\frac{1}{2}y^2 dx = \frac{1}{6\pi}(\frac{1}{2})\frac{16}{9}\int_{-3}^{3}(9 - x^2)dx = \frac{1}{6\pi}(\frac{16}{9})\int_0^3(9 - x^2)dx = \frac{8}{27\pi}[9x - \frac{x^3}{3}]_0^3 = \frac{16}{3\pi}$. $(\bar{x}, \bar{y}) = (0, \frac{16}{3\pi})$.

43. Volume $= 2$(Volume of top half) $= 2\pi\int_0^3(1 + y^2)dy = 2\pi[y + \frac{y^3}{3}]_0^3 = 24\pi$.

44. $A = 2\pi\int y\, ds = 2\pi\int_0^{\sqrt{2}} y\sqrt{1 + (y')^2}dx$. $y = \sqrt{x^2 + 1}$, $y' = \frac{x}{\sqrt{x^2+1}}$, $1 + (y')^2 = \frac{x^2+1}{x^2+1} + \frac{x^2}{x^2+1} = \frac{2x^2+1}{x^2+1}$. $y\sqrt{1 + (y')^2} = \sqrt{2x^2 + 1}$. $A = 2\pi\int_0^{\sqrt{2}}\sqrt{2x^2 + 1}dx$. Let $u = \sqrt{2}x$, $du = \sqrt{2}dx$. Then $A = \frac{2\pi}{\sqrt{2}}\int_0^2\sqrt{u^2 + 1}du = \sqrt{2}\pi[\frac{u}{2}\sqrt{u^2 + 1} + \frac{1}{2}\sinh^{-1}u]_0^2 = \sqrt{2}\pi[\sqrt{5} + \frac{1}{2}\sinh^{-1}2]$.

45. a) $y^2 = 4px$. $2yy' = 4p$, $y' = \frac{2p}{y}$. Tan β is the slope of L. The slope of L is the slope of the parabola at (x_0, y_0), so $\tan\beta = \frac{2p}{y_0}$. b) It is clear that $\tan\phi = \frac{opp}{adj} = \frac{y_0}{x_0 - p}$. c) We see that $\beta + \alpha + (\frac{\pi}{2} - \phi) = \frac{\pi}{2}$ and so $\alpha = \phi - \beta$. Tan $\alpha = \tan(\phi - \beta) = \frac{\tan\phi - \tan\beta}{1 + \tan\phi\tan\beta} = \frac{\frac{y_0}{x_0 - p} - \frac{2p}{y_0}}{1 + \frac{y_0}{x_0 - p}\frac{2p}{y_0}} = \frac{2p}{y_0}$ using $y_0^2 = 4px_0$.

46. Since r_A and r_B have the same derivative, they differ by a constant: $r_A - r_B = C$. Thus the difference of the distance r_A of P to A and the distance r_B of P to B is a constant. Therefore the path of P lies on a hyperbola.

10.2 THE GRAPHS OF QUADRATIC EQUATIONS IN x AND y

1. $x^2 - y^2 - 1 = 0$. $B^2 - 4AC = 0^2 - 4(1)(-1) = 4 > 0$: the equation represents a hyperbola. Graph $y_1 = \sqrt{x^2 - 1}$ and $y_2 = -y_1$ in $[-17, 17]$ by $[-10, 10]$.

2. $25x^2 + 9y^2 - 225 = 0$. $B^2 - 4AC = -4(25)9 < 0$, ellipse. Graph $y_1 = \frac{5}{3}\sqrt{9 - x^2}$ and $y_2 = -y_1$ in $[-8.5, 8.5]$ by $[-5, 5]$.

3. $y^2 - 4x - 4 = 0$. $B^2 - 4AC = 0^2 - 4(0)(1) = 0$. Parabola. Graph $y_1 = 2\sqrt{x + 1}$ and $y_2 = -y_1$ in $[-5.8, 7.8]$ by $[-4, 4]$. Shift the graph of $x = \frac{y^2}{4}$ horizontally left one unit.

4. $x^2 + y^2 - 10 = 0$. $B^2 - 4AC = -4(1)(1) < 0$, ellipse (circle). Graph $y_1 = \sqrt{10 - x^2}$ and $y_2 = -y_1$ in $[-5.37, 5.37]$ by $[-\sqrt{10}, \sqrt{10}]$.

5. $x^2 + 4y^2 - 4x - 8y + 4 = 0$. $B^2 - 4AC = 0^2 - 4(1)(4) = -16 < 0$, ellipse. $4y^2 - 8y + (x^2 - 4x + 4) = 0$. Let $y_1 = \sqrt{64 - 16(x^2 - 4x + 4)}$. We graph $y_2 = (8 + y_1)/8$ and $y_3 = (8 - y_1)/8$ in $[0, 4]$ by $[-3.8, 2.3]$.

6. $2x^2 - y^2 + 4xy - 2x + 3y = 6$. $B^2 - 4AC = 4^2 - 4(2)(-1) > 0$, hyperbola. $\theta = \frac{1}{2}\tan^{-1}\frac{B}{A-C} = \frac{1}{2}\tan^{-1}\frac{4}{3} = 0.464$. $y^2 - (4x + 3)y + (-2x^2 + 2x + 6) = 0$. Let $y_1 = \sqrt{(4x + 3)^2 - 4(-2x^2 + 2x + 6)}$. Graph $y_2 = \frac{1}{2}(4x + 3 + y_1)$ and $y_3 = \frac{1}{2}(4x + 3 - y_1)$ in $[-17, 17]$ by $[-10, 10]$.

7. $x^2 + 4xy + 4y^2 - 3x = 6$. $B^2 - 4AC = 16 - 4(1)4 = 0$, parabola. $4y^2 + 4xy + (x^2 - 3x - 6) = 0$. Let $y_1 = \sqrt{(4x)^2 - 16(x^2 - 3x - 6)}$. Graph $y_2 = (-4x + y_1)/8$ and $y_3 = (-4x - y_1)/8$ in $[-8.4, 13.6]$ by $[-9.1, 3.9]$. $\theta = \frac{1}{2}\tan^{-1}(\frac{B}{A-C}) = \frac{1}{2}\tan^{-1}(\frac{4}{1-4}) = \frac{1}{2}\tan^{-1}(-\frac{4}{3}) = -0.464$.

8. $x^2 + y^2 + 3x - 2y = 10$. $B^2 - 4AC = -4(1)(1) < 0$, ellipse (circle). $y^2 - 2y + (x^2 + 3x - 10) = 0$. Let $y_1 = \sqrt{4 - 4(x^2 + 3x - 10)}$. Graph $y_2 = \frac{1}{2}(2 + y_1)$ and $y_3 = \frac{1}{2}(2 - y_1)$ in $[-8.6, 5.6]$ by $[-3.2, 5.2]$.

9. $xy + y^2 - 3x = 5$. $B^2 - 4AC = 1 - 0 = 1 > 0$, hyperbola. $y^2 + xy + (-3x - 5) = 0$. Let $y_1 = \sqrt{x^2 + 12x + 20}$. Graph $y_2 = (-x + y_1)/2$ and $y_3 = (-x - y_1)/2$ in $[-49, 43]$ by $[-27.4, 26.8]$, $\theta = \frac{1}{2}\tan^{-1}(\frac{B}{A-C}) = \frac{1}{2}\tan^{-1}(-1) = -\frac{\pi}{8}$.

10. $3x^2 + 6xy + 3y^2 - 4x + 5y = 12$. $B^2 - 4AC = 6^2 - 4(3)(3) = 0$, parabola. $\theta = \frac{1}{2}\cot^{-1}\frac{3-3}{6} = \frac{1}{2}\frac{\pi}{2} = \frac{\pi}{4}$. $3y^2 + (6x + 5)y + (3x^2 - 4x - 12) = 0$. Let $y_1 = \sqrt{(6x + 5)^2 - 4(3)(3x^2 - 4x - 12)}$. Graph $y_2 = \frac{1}{6}[-(6x + 5) + y_1]$ and $y_3 = \frac{1}{6}[-(6x + 5) - y_1]$ in $[-7.4, 15.1]$ by $[-10, 3.2]$.

11. $x^2 - y^2 = 1$. $B^2 - 4AC = 0 - 4(1)(-1) = 4 > 0$, hyperbola. Graph $y = \sqrt{x^2 - 1}$ and $y = -\sqrt{x^2 - 1}$ in $[-8.5, 8.5]$ by $[-5, 5]$.

12. $2x^2 + 3y^2 - 4x = 7$. $B^2 - 4AC = -4(2)(3) < 0$, ellipse. Graph $y_1 = \sqrt{(7 + 4x - 2x^2)/3}$ and $y_2 = -y_1$ in $[-2.8, 4.9]$ by $[-2.3, 2.3]$.

13. $xy = 1$. $B^2 - 4AC = 1^2 - 4(0)(0) = 1 > 0$, hyperbola. Graph $y = \frac{1}{x}$ in $[-5.1, 5.1]$ by $[-3, 3]$. Since $A = C$, $\theta = \frac{\pi}{4}$.

14. $xy = -3$. $B^2 - 4AC = 1 > 0$, hyperbola. Graph $y = -\frac{3}{x}$ in $[-10, 10]$ by $[-5.9, 5.9]$. $\theta = \frac{\pi}{4}$.

15. $2x^2 + xy + x - y + 1 = 0$. $B^2 - 4AC = 1^2 - 4(2)(0) = 1 > 0$, hyperbola. Solving for y, we obtain $y = -(2x^2 + x + 1)/(x - 1)$. Graph y in dot format in $[-8, 8.7]$ by $[-19, 9.1]$. $\theta = \frac{1}{2}\tan^{-1}\frac{B}{A-C} = \frac{1}{2}\tan^{-1}\frac{1}{2} = 0.232$.

16. $x^2 + 2xy - 2x + 3y - 3 = 0$. $B^2 - 4AC = 4 - 4(1)(0) = 4 > 0$, hyperbola. Graph $y = -(x^2 - 2x - 3)/(2x + 3)$ in dot format in $[-8, 8.7]$ by $[-19, 20]$. $\theta = \frac{1}{2}\tan^{-1}\frac{B}{A-C} = \frac{1}{2}\tan^{-1}\frac{2}{1} = 0.554$.

17. $x^2 - 3xy + 3y^2 + 6y = 7$. $B^2 - 4AC = 9 - 4(1)3 = -3 < 0$, ellipse. $3y^2 + (6 - 3x)y + (x^2 - 7) = 0$. Let $y_1 = \sqrt{(6 - 3x)^2 - 4(3)(x^2 - 7)}$. Graph $y_2 = \frac{-(6-3x)+y_1}{6}$ and $y_3 = \frac{-(6-3x)-y_1}{6}$ in $[-20.3, 8.2]$ by $[-12.9, 3.9]$. $\theta = \frac{1}{2}\tan^{-1}\frac{-3}{1-3} = \frac{1}{2}\tan^{-1}(1.5) = 0.491$.

18. $25x^2 - 4y^2 - 350x = 0$. $B^2 - 4AC = -4(25)(-4) > 0$, hyperbola. Graph $y_1 = \frac{5}{2}\sqrt{x^2 - 14}$ and $y_2 = -y_1$ in $[-17, 17]$ by $[-10, 10]$.

19. $6x^2+3xy+2y^2+17y+2=0$. $B^2-4AC=9-48<0$, ellipse. $2y^2+(3x+17)y+(6x^2+2)=0$. Let $y_1=\sqrt{(3x+17)^2-4(2)(6x^2+2)}$. Graph $y_2=\frac{-(3x+17)+y_1}{4}$ and $y_3=\frac{-(3x+17)-y_1}{4}$ in $[-11.9,14.4]$ by $[-12.6,2.9]$. $\theta=\frac{1}{2}\tan^{-1}\frac{3}{6-2}=0.322$.

20. $3x^2+12xy+12y^2+435x-9y+72=0$. $B^2-4AC=12^2-4(3)(12)=0$, parabola. $\theta=\frac{1}{2}\tan^{-1}\frac{12}{3-12}=\frac{1}{2}\tan^{-1}(-\frac{4}{3})=-0.464$. $12y^2+(12x-9)y+(3x^2+435x+72)=0$. Let $y_1=\sqrt{(12x-9)^2-4(12)(3x^2+435x+72)}$. Graph $y_2=\frac{1}{24}[-(12x-9)+y_1]$ and $y_3=\frac{1}{24}[-(12x-9)-y_1]$ in $[-2.4,1.3]$ by $[-6.1,5.7]$.

21. $xy=2$. $B^2-4AC=1^2-0>0$, hyperbola. $y=\frac{2}{x}$ may be graphed in $[-8.5,8.5]$ by $[-5,5]$. $\theta=\frac{1}{2}\cot^{-1}(\frac{A-C}{B})=\frac{1}{2}\cot^{-1}0=\frac{\pi}{4}$. $x'=\frac{\sqrt{2}}{2}(x-y)$, $y'=\frac{\sqrt{2}}{2}(x+y)$. (x',y') satisfies $xy=2$ so we obtain $x'y'=2$, $\frac{\sqrt{2}}{2}(x-y)\frac{\sqrt{2}}{2}(x+y)=2$, $x^2-y^2=4$.

22. $x^2+xy+y^2=1$. $B^2-4AC=1-4<0$, ellipse. With $A=C$, we again obtain $\theta=\frac{\pi}{4}$. $x'=\frac{\sqrt{2}}{2}(x-y)$, $y'=\frac{\sqrt{2}}{2}(x+y)$. $x'^2+x'y'+y'^2=1$, $\frac{1}{2}(x^2-2xy+y^2)+\frac{1}{2}(x^2-y^2)+\frac{1}{2}(x^2+2xy+y^2)=1$, $\frac{3}{2}x^2+\frac{1}{2}y^2=1$ is the transformed equation. $y^2+xy+(x^2-1)=0$. Let $y_1=\sqrt{x^2-4(x^2-1)}$. Graph $y_2=\frac{1}{2}(-x+y_1)$ and $y_3=\frac{1}{2}(-x-y_1)$ in $[-2.4,2.4]$ by $[-1.4,1.4]$.

23. $x^2-\sqrt{3}xy+2y^2=1$. $B^2-4AC=3-4(1)2=-5<0$, ellipse. $2y^2-\sqrt{3}xy+(x^2-1)=0$. Let $y_1=\sqrt{3x^2-4(2)(x^2-1)}$. Graph $y_2=(\sqrt{3}x+y_1)/4$ and $y_3=(\sqrt{3}x-y_1)/4$ in $[-2.6,2.1]$ by $[-1.5,1.3]$. $\theta=\frac{1}{2}\tan^{-1}\frac{B}{A-c}=\frac{1}{2}\tan^{-1}(\sqrt{3})=\frac{\pi}{6}$. $x'=x\cos\frac{\pi}{6}-y\sin\frac{\pi}{6}=\frac{\sqrt{3}x-6}{2}$, $y'=x\sin\frac{\pi}{6}+y\cos\frac{\pi}{6}=\frac{x+\sqrt{3}y}{2}$. Substituting this into the original equation, $x'^2-\sqrt{3}x'y'+2y'^2=1$, and simplifying, we obtain $x^2+5y^2=2$.

24. $x^2-2xy+y^2=2$, $(x-y)^2=2$, $(x-y)^2-(\sqrt{2})^2=0$, $(x-y-\sqrt{2})(x-y+\sqrt{2})=0\Rightarrow$ the graph consists of the two straight lines $y=x\pm\sqrt{2}$. $B^2-4AC=0$ so this is a degenerate parabola. Since $A=C$, we get $\theta=\frac{\pi}{4}$. $x'=\frac{\sqrt{2}}{2}(x-y)$, $y'=\frac{\sqrt{2}}{2}(x+y)$. $y'-x'=\sqrt{2}y=\pm\sqrt{2}$. Hence the equivalent graph consists of the graphs of $y=1$ and $y=-1$.

25. $x^2-3xy+y^2=5$. $B^2-4AC=9-4=5>0$, hyperbola. Let $y_1=\sqrt{9x^2-4(x^2-5)}$. Graph $y_2=(3x+y_1)/2$ and $y_3=(3x-y_1)/2$ in $[-17,17]$ by $[-10,10]$. $\theta=\frac{\pi}{4}$ since $A=C$. Substituting $x'=\frac{\sqrt{2}}{2}(x-y)$, $y'=\frac{\sqrt{2}}{2}(x+y)$, we obtain $5y^2-x^2=10$.

26. $xy - y - x + 1 = 0$. $B^2 - 4AC = 1 > 0$, hyperbola. But the equation can be factored: $(x-1)(y-1) = 0$ so its graph is the degenerate hyperbola consisting of the two lines $x = 1$, $y = 1$. Since $A = C$, $\theta = \frac{\pi}{4}$, $x' = \frac{x-y}{\sqrt{2}}$, $y' = \frac{x+y}{\sqrt{2}}$ and the transformed equation is $\frac{x^2-y^2}{2} - \frac{x+y}{\sqrt{2}} - \frac{x-y}{\sqrt{2}} + 1 = 0$, $y^2 = (x - \sqrt{2})^2$ or $y = \pm(x - \sqrt{2})$.

27. $3x^2 + 2xy + 3y^2 = 19$. $B^2 - 4AC = 4 - 4(3)3 < 0$, ellipse. Let $y_1 = \sqrt{4x^2 - 12(3x^2 - 19)}$. Graph $y_2 = (-2x + y_1)/6$ and $y_3 = (-2x - y_1)/6$ in $[-5.1, 5.1]$ by $[-3, 3]$. Since $A = C = 3$, $\theta = \frac{\pi}{4}$, $x' = \frac{\sqrt{2}}{2}(x-y)$, $y' = \frac{\sqrt{2}}{2}(x+y)$. After substituting and simplifying, we obtain $4x^2 + 2y^2 = 19$.

28. $3x^2 + 4\sqrt{3}xy - y^2 = 7$. $B^2 - 4ABC = 48 - 4(3)(-1) > 0$, hyperbola. $\theta = \frac{1}{2}\tan^{-1}\frac{4\sqrt{3}}{3+1} = \frac{1}{2}\tan^{-1}\sqrt{3} = \frac{\pi}{6}$. $x' = x\cos\frac{\pi}{6} - y\sin\frac{\pi}{6} = \frac{\sqrt{3}x-y}{2}$, $y' = x\sin\frac{\pi}{6} + y\cos\frac{\pi}{6} = \frac{x+\sqrt{3}y}{2}$. $3x'^2 + 4\sqrt{3}x'y' - y'^2 = 7$ leads to $5x^2 - 3y^2 = 7$. We return to the original equation, $y^2 - 4\sqrt{3}xy + (7 - 3x^2) = 0$. Let $y_1 = \sqrt{48x^2 - 4(7 - 3x^2)}$. Graph $y_2 = \frac{1}{2}(4\sqrt{3}x + y_1)$ and $y_3 = \frac{1}{2}(4\sqrt{3}x - y_1)$ in $[-5, 5]$ by $[-30, 30]$.

29. The equations give the coordinates of a point after it is rotated through an angle θ. To get the original coordinates back, we can use the same equations to rotate (x', y') through an angle $-\theta$: $x = x'\cos(-\theta) - y'\sin(-\theta) = x'\cos\theta + y'\sin\theta$, $y = x'\sin(-\theta) + y'\cos(-\theta) = -x'\sin\theta + y'\cos\theta$.

Exercises 30-34. Some results of this section may be combined to arrive at the following statement. If x is replaced by $x\cos\theta + y\sin\theta$ and y is replaced by $-x\sin\theta + y\cos\theta$ throughout an equation, the graph of the new equation is the graph of the original equation rotated through an angle θ.

30. We change the equation setting $C = A$ and replacing x by $x\cos(\pi/4) + y\sin(\pi/4) = (x + y)/\sqrt{2}$ and y by $-x\sin(\pi/4) + y\cos(\pi/4) = (y - x)/\sqrt{2}$: $A(x+y)^2/2 + B(y^2-x^2)/2 + A(y-x)^2/2 + D(x+y)/\sqrt{2} + E(y-x)/\sqrt{2} + F = 0$. It is easily seen that the xy-term drops out.

31. $2x^2 - y^2 - 8 = 0$, $\theta = \frac{\pi}{4}$. x is replaced by $x\cos(\pi/4) + y\sin(\pi/4) = (x+y)/\sqrt{2}$ and y is replaced $-x\sin(\pi/4) + y\cos(\pi/4) = (y - x)/\sqrt{2}$: $2(x + y)^2/2 - (y - x)^2/2 - 8 = 0$, $x^2 + 2xy + y^2 + (-y^2 + 2xy - x^2)/2 - 8 = 0$, $\frac{x^2}{2} + 3xy + \frac{y^2}{2} - 8 = 0$ or $x^2 + 6xy + y^2 - 16 = 0$.

32. $x^2+2y^2-10=0$, $\theta=\pi/6$. x is replaced by $x\cos(\pi/6)+y\sin(\pi/6)=(\sqrt{3}x+y)/2$ and y is replaced by $-x\sin(\pi/6)+y\cos(\pi/6)=(\sqrt{3}y-x)/2$: $(\sqrt{3}x+y)^2/4+2(\sqrt{3}y-x)^2/4-10=0$. This reduces to $5x^2-2\sqrt{3}xy+7y^2-40=0$.

33. $4x^2-y^2-10=0$, $\theta=\pi/3$. x is replaced by $x\cos(\pi/3)+y\sin(\pi/3)=(x+\sqrt{3}y)/2$ and y is replaced by $-x\sin(\pi/3)+y\cos(\pi/3)=(y-\sqrt{3}x)/2$: $4(x+\sqrt{3}y)^2/4-(y-\sqrt{3}x)^2/4-10=0$. This reduces to $x^2+10\sqrt{3}xy+11y^2-40=0$.

34. $2x^2+8y^2-7=0$, $\theta=\pi/4$, x is replaced by $x\cos(\pi/4)+y\sin(\pi/4)=(x+y)/\sqrt{2}$ and y is replaced by $-x\sin(\pi/4)+y\cos(\pi/4)=(y-x)/\sqrt{2}$: $2(x+y)^2/2+8(y-x)^2/2-7=0$. This reduces to $5x^2-6xy+5y^2-7=0$.

35. a) and b) are combined. The distance between $(-3,-3)$ and $(3,3)$ is $6\sqrt{2}$. Since the sum of the lengths of two sides of a triangle exceeds the length of the 3rd side, for any point (x,y), $\sqrt{(x+3)^2+(y+3)^2}+\sqrt{(x-3)^2+(y-3)^2}\geqq 6\sqrt{2}$. Thus $\sqrt{(x+3)^2+(y+3)^2}+\sqrt{(x-3)^2+(y-3)^2}=6$ is impossible. (Call this fact (A)). The equation $\sqrt{(x+3)^2+(y+3)^2}-\sqrt{(x-3)^2+(y-3)^2}=\pm 6$ is equivalent to $\sqrt{(x+3)^2+(y+3)^2}=\sqrt{(x-3)^2+(y-3)^2}\pm 6$. The last equation is equivalent to $(x+3)^2+(y+3)^2=(\sqrt{(x-3)^2+(y-3)^2}\pm 6)^2$ using fact (A). The last equation is equivalent to $(x-3)+y=\pm\sqrt{(x-3)^2+(y-3)^2}$. This is equivalent to $[(x-3)+y]^2=(x-3)^2+(y-3)^2$ which in turn is equivalent to $2xy=9$.

36. If $\frac{B'}{A'-C'}\to\infty$, $\tan^{-1}\frac{B'}{A'-C'}\to\frac{\pi}{2}$ and $\frac{1}{2}\tan^{-1}\frac{B'}{A'-C'}\to\frac{\pi}{4}$.

10.3 PARAMETRIC EQUATIONS FOR PLANE CURVES

1. $x=\cos t$, $y=\sin t$, $0\leqq t\leqq\pi$. Graph $x_1=\cos t$, $y_1=\sin t$, $0\leqq t\leqq\pi$ in $[-1.7,1.7]$ by $[-1,1]$. $x^2+y^2=\cos^2 t+\sin^2 t=1$. The upper half of the unit circle is traced out in the counterclockwise direction.

2. Graph $x_1=\cos 2t$, $y=\sin 2t$, $0\leqq t\leqq\pi$ in $[-1.7,1.7]$ by $[-1,1]$. The unit circle $x^2+y^2=1$ is traced out in the counterclockwise direction.

3. Graph $x_1=\sin 2\pi t$, $y_1=\cos 2\pi t$, $0\leqq t\leqq 1$ in $[-1.7,1.7]$ by $[-1,1]$. The unit circle is traced out once in the clockwise direction starting at $(0,1)$.

4. Graph $x_1 = \cos(\pi - t)$, $y_1 = \sin(\pi - t)$, $0 \leqq t \leqq \pi$ in $[-1.7, 1.7]$ by $[-1, 1]$. The top half of the unit circle $x^2 + y^2 = 1$ is traced out in the clockwise direction.

5. $x = 4\cos t$, $y = 2\sin t$, $0 \leqq t \leqq 2\pi$. Graph this with t-step $= 0.1$ in $[-4, 4]$ by $[-2.4, 2.4]$. $(\frac{x}{4})^2 + (\frac{y}{2})^2 = \cos^2 t + \sin^2 t = 1$. The ellipse $\frac{x^2}{16} + \frac{y^2}{4} = 1$ is traced out once in the counterclockwise direction.

6. Graph $x = 4\sin t$, $y = 2\cos t$, $0 \leqq t \leqq \pi$ in $[-4, 4]$ by $[-2.35, 2.35]$. $(\frac{x}{4})^2 + (\frac{y}{2})^2 = \sin^2 t + \cos^2 t = 1$. The right-hand side of the ellipse $\frac{x^2}{16} + \frac{y^2}{4} = 1$ is traced out in the clockwise direction.

7. $x = 4\cos t$, $y = 5\sin t$, $0 \leqq t \leqq \pi$. Graph in $[-8.5, 8.5]$ by $[-5, 5]$. The upper half of the ellipse $\frac{x^2}{16} + \frac{y^2}{25} = 1$ is traced out in the counterclockwise direction.

8. Graph $x_1 = 4\sin t$, $y_1 = 5\cos t$, $0 \leqq t \leqq 2\pi$ in $[-8.5, 8.5]$ by $[-5, 5]$. The entire ellipse $\frac{x^2}{16} + \frac{y^2}{25} = 1$ is traced out in the clockwise direction starting at $(0, 5)$.

9. $x = 3t$, $y = 9t^2$, $-\infty < t < \infty$. Graph for $-1 \leqq t \leqq 1$ in $[-7.6, 7.6]$ by $[0, 9]$. $y = 9t^2 = x^2$. The parabola $y = x^2$ is traced out from left to right.

10. Graph $x_1 = -\sqrt{t}$, $y_1 = t$, $t \geqq 0$ (use $0 \leqq t \leqq 9$) in $[-9.2, 6.2]$ by $[0, 9]$. The curve $x = -\sqrt{y}$ (the left side of $y = x^2$) is traced out from right to left.

11. $x = t$, $y = \sqrt{t}$, $t \geqq 0$. Graph for $0 \leqq t \leqq 10$ in $[0, 10]$ by $[-1.4, 4.5]$. $x = t = y^2$. The upper half of the parabola $y^2 = x$ is traced out from left to right.

12. Graph $x_1 = \sec^2 t - 1$, $y_1 = \tan t$, $-\frac{\pi}{2} < t < \frac{\pi}{2}$ in $[-2.6, 5.1]$ by $[-2.3, 2.3]$. $x = \sec^2 t - 1 = \tan^2 y = y^2$. The parabola $x = y^2$ is traced out, the bottom half from right to left, then the top from left to right.

13. $x = -\sec t$, $y = \tan t$, $-\pi/2 < t < \pi/2$. Graph in $[-17, 17]$ by $[-10, 10]$. $x^2 - 1 = \sec^2 t - 1 = \tan^2 t = y^2$. The left branch of the hyperbola $x^2 - y^2 = 1$ is traced out from bottom to top.

14. Graph $x = \csc t$, $y = \cot t$, $0 < t < \pi$ in $[-1.9, 4.9]$ by $[-1, 3]$. $x^2 - y^2 = \csc^2 t - \cot^2 t = 1$. The right branch of the hyperbola $x^2 - y^2 = 1$ is traced out from top to bottom.

15. $x = 2t - 5$, $y = 4t - 7$, $-\infty < t < \infty$. Graph for $-5 \leqq t \leqq 5$ in $[-16.6, 17.4]$ by $[-7, 13]$. $t = \frac{x+5}{2} = \frac{y+7}{4}$. The graph of the line $y = 2x + 3$ is traced out from left to right.

16. Graph $x = 1 - t$, $y = 1 + t$, $-\infty < t < \infty$ (use $-3 \leqq t \leqq 3$) in $[-4, 6]$ by $[-2, 4]$. $x + y = (1 - t) + (1 + t) = 2$. The line $x + y = 2$ is traced out from right to left.

17. $x = t$, $y = 1 - t$, $0 \leqq t \leqq 1$. Graph in $[-2, 3]$ by $[-1, 2]$. $y = 1 - x$. The line segment from $(0, 1)$ to $(1, 0)$ is traced out from left to right.

18. Graph $x = 3t$, $y = 2 - 2t$, $0 \leqq t \leqq 1$ in $[-1.9, 4.9]$ by $[-1, 3]$. $\frac{x}{3} = t = \frac{2-y}{2}$ leads to $y = -\frac{2}{3}x + 2$. The segment of the line $y = -\frac{2}{3}x + 2$, $(0, 2)$ to $(3, 0)$, is traced out from left to right.

19. $x = t$, $y = \sqrt{1 - t^2}$, $-1 \leqq t \leqq 1$. Graph in $[-1, 1]$ by $[-0.1, 1.1]$. The upper half of the unit circle, $y = \sqrt{1 - x^2}$, is traced out from left to right.

20. Graph $x = t$, $y = \sqrt{4 - t^2}$, $0 \leqq t \leqq 2$ in $[-0.7, 2.7]$ by $[0, 2]$. $y^2 = 4 - t^2 = 4 - x^2$ or $x^2 + y^2 = 4$. The quarter-circle of radius 2 with center $(0, 0)$ is traced out from $(0, 2)$ to $(2, 0)$ in the clockwise direction.

21. $x = t^2$, $y = \sqrt{t^4 + 1}$, $t \geqq 0$. Graph for $0 \leqq t \leqq 4$ in $[-6.9, 21.9]$ by $[0, 17]$. $y = \sqrt{x^2 + 1}$. The top half of the hyperbola $y^2 - x^2 = 1$ for $x \geqq 0$ is traced out from left to right.

22. Graph $x = \sqrt{t + 1}$, $y = \sqrt{t}$, $t \geq 0$ (use $0 \leqq t \leqq 20$) in $[-1.83, 5.83]$ by $[0, 4.5]$. $x^2 = t + 1 = y^2 + 1$ or $x^2 - y^2 = 1$. The upper half of the right branch of the hyperbola $x^2 - y^2 = 1$ is traced out from left to right starting at $(1, 0)$.

23. $x = \cosh t$, $y = \sinh t$, $-\infty < t < \infty$. Graph for $-3 \leqq t \leqq 3$ in $[-17, 17]$ by $[-10, 10]$. $x^2 - y^2 = \cosh^2 t - \sinh^2 t = 1$. The right branch of the hyperbola $x^2 - y^2 = 1$ is traced out from bottom to top.

24. Graph $x = 2\sinh t$, $y = 2\cosh t$, $-\infty < t < \infty$ (use $-2.3 \leqq t \leqq 2.3$) in $[-10, 10]$ by $[-10, 10]$. $1 = \cosh^2 t - \sinh^2 t = (\frac{y}{2})^2 - (\frac{x}{2})^2$ or $\frac{y^2}{4} - \frac{x^2}{4} = 1$. The upper branch of the hyperbola $\frac{y^2}{4} - \frac{x^2}{4} = 1$ is traced out from left to right.

25. a) $x = a\cos t$, $y = -a\sin t$, $0 \leqq t \leqq 2\pi$ b) $x = a\cos t$, $y = a\sin t$, $0 \leqq t \leqq 2\pi$ c) $x = a\cos(2t)$, $y = -a\sin(2t)$, $0 \leqq t \leqq 2\pi$ d) $x = a\cos t$, $y = a\sin t$, $0 \leqq t \leqq 4\pi$

26. $a = \pi$. The curve is also closed for $a = 2\pi$ and this gives the complete curve.

27. $x = 5\sin 2t$, $y = 5\sin 3t$. Take $a = 2\pi$. We obtain more and more of the graph until t reaches 2π. Note that $\sin[2(t + 2\pi)] = \sin 2t$ and $\sin[3(t + 2\pi)] = \sin 3t$. We also get a closed curve for $a = \pi$.

28. $x = (5\sin 3t)\cos t$, $y = (5\sin 3t)\sin t$. Take $a = \pi$ for a complete graph. If in doubt, use the TRACE function. The curve is closed for $a = \pi/3$.

29. $x = (5\sin 2t)\cos t$, $y = (5\sin 2t)\sin t$, $0 \leqq t \leqq a = 2\pi$ for a complete graph. If $a = \pi/2$, a closed curve is obtained.

30. Use $a = 4\pi$ for a complete graph. Note that $\sin[2.5(t + 4\pi)] = \sin(2.5t + 10\pi) = \sin 2.5t$. A closed curve is obtained for $a = \pi/2.5$.

31. Use $a = 4\pi$ for a complete graph. Note that $\sin[1.5(t+4\pi)] = \sin(1.5t+6\pi) = \sin 1.5t$. A closed curve is obtained if $a = \pi/1.5$.

32. No such a exists. Note that $x^2 + y^2 = t^2(\cos^2 t + \sin^2 t) = t^2$. Thus the distance from (x, y) to $(0, 0)$ is $|t|$. So for $t > 0$, (x, y) can't get back to $(0, 0)$.

33. No such a exists. Note that $x^2 + y^2 = t^2(\sin^2 t + \cos^2 t) = t^2$. Thus the distance from (x, y) to $(0, 0)$ is $|t|$. So for $t > 0$, (x, y) can't get back to $(0, 0)$.

34. The following use $0 \leqq t \leqq 2\pi$. a) $x = a\cos t$, $y = -b\sin t$ b) $x = a\cos t$, $y = b\sin t$ c) $x = a\cos 2t$, $y = -b\sin 2t$ d) $x = a\cos 2t$, $y = b\sin 2t$.

35.

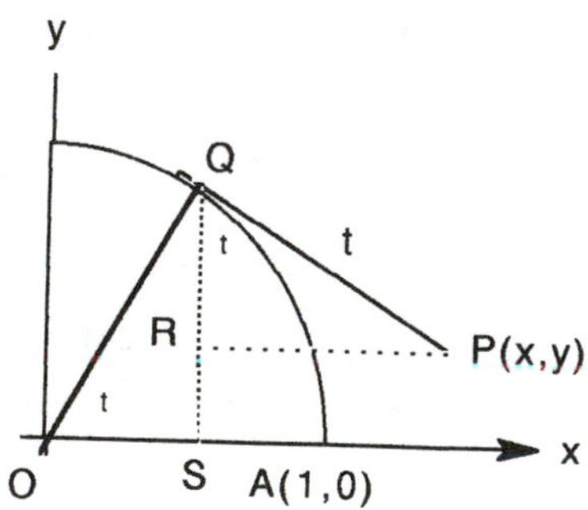

$\angle OQP$ is known to be a right angle. Segment QP = length arc $AQ = r\theta = 1 \cdot t = t$. $x = OS + RP = \cos t + t\sin t$, $y = SQ - RQ = \sin t - t\cos t$.

36. Note that angle $OAQ = t$. $\cot t = AQ/2$, $x = AQ = 2\cot t$. $y = 2 - AB\sin t = 2 - \frac{(AQ)^2}{OA}\sin t = 2 - \frac{(2\cot t)^2}{2\csc t}\sin t = 2(1 - \cos^2 t) = 2\sin^2 t$. We may take $x = 2\cot 2$, $y = 2\sin^2 t$, $0 < t < \pi$.

37. We wish to minimize the square of the distance $D(t) = (x-2)^2 + (y-\frac{1}{2})^2 = (t-2)^2 + (t^2-\frac{1}{2})^2$. $D'(t) = 2(t-2) + 2(t^2-\frac{1}{2})2t = 4(t^3-1)$. $D'(t) < 0$ for $t < 1$ and $D'(t) > 0$ for $t > 1$. $D(t)$ therefore has its minimal value when $t = 1$ and so $(1,1)$ is the closest point.

38. Let $f(t) = (x-\frac{3}{4})^2+y^2 = (2\cos t-\frac{3}{4})^2+\sin^2 t$. $f'(t) = 2(2\cos t-\frac{3}{4})(-2\sin t)+2\sin t\cos t = 2\sin t[-4\cos t+\frac{3}{2}+\cos t] = 6\sin t[\frac{1}{2}-\cos t]$. On $[0,2\pi]$, the critical points (which include the endpoints) are $0, \frac{\pi}{3}, \pi, \frac{5\pi}{3}, 2\pi$. Testing f at each of these values, we find that the smallest value of f is attained when $t = \frac{\pi}{3}$ and $\frac{5\pi}{3}$. This corresponds to the points $(1, \frac{\sqrt{3}}{2})$ and $(1, -\frac{\sqrt{3}}{2})$ on the ellipse.

39. $x = x_0 + (x_1 - x_0)t$, $y = y_0 + (y_1 - y_0)t$. a) By taking $t = 0$ and $t = 1$, we see that the curve passes through the two points. $t = \frac{x-x_0}{x_1-x_0} = \frac{y-y_0}{y_1-y_0}$, a linear equation in x and y and so the curve is a line. b) Take $(x_0, y_0) = (0,0)$: $x = x_1 t$, $y = y_1 t$. c) Take $(x_0, y_0) = (-1, 0)$ and $(x_1, y_1) = (0,1)$: $x = -1 + t$, $y = t$.

40. Use the viewing window $[-4, 4]$ by $[-2.36, 2.36]$ for each of these graphs.

41. a) Use $[-18, 32]$ by $[-15, 15]$. b) Use $[-0.13, 2]$ by $[-0.55, 0.55]$. c) Use $[-50, 1]$ by $[-55, 15]$.

42. $[-3, 9]$ by $[-2.6, 4.5]$ is a possible viewing window.

43. The three graphs can be compared in the viewing window $[0, 4\pi]$ by $[-2.7, 4.7]$.

44. The two graphs in a) and b) may be compared in the viewing window $[-1.7, 1.7]$ by $[-1, 1]$. c) Graph $x_1 = \cos t$, $y_1 = \sin t$, $x_2 = 0.75 + 0.25\cos t$, $y_2 = 0.25\sin t$, $0 \leqq t \leqq 2\pi$ in $[-1.7, 1.7]$ by $[-1, 1]$. If the small circle rolls through 2π, it travels $2\pi r = \pi/2$ which is exactly $1/4$ of the circumference of the unit circle. Graphing all three curves together is also helpful.

45. Graph in $[-5.1, 5.9]$ by $[-3.2, 3.2]$. The new equations amount to $x = -2\cos t + \cos(2t)$, $y = -2\sin t + \sin(2t)$. Graph in the same window. The original curve had three cusps. The new curve appears to be a cardioid.

46. Compare the graphs in the viewing window $[-6.8, 6.8]$ by $[-4, 4]$.

47. Graph a), b), c) in $[0, 128]$ by $[-21, 54]$. In d) the curve is part of the y-axis traced from $(0,0)$ to $(0,64)$ and back down to $(0,0)$.

48. The curves a), b), c) may be graphed in $[-18.7, 18.7]$ by $[-11, 11]$. Graph d) in $[-11.5, 11.5]$ by $[-6.75, 6.75]$. Graph e) in $[-1.5, 1.5]$ by $[-1.5, 1.5]$. In e) $0 \leqq t \leqq \pi$ is sufficient.

49. Suppose a curve is given parametrically as $x = x(t)$, $y = y(t)$. Let the curve be rotated counterclockwise through an angle θ. According to Section 10.2 the new curve is given by $x_1 = x(t)\cos\theta - y(t)\sin\theta$, $y_1 = x(t)\sin\theta + y(t)\cos\theta$. In this problem, $y = x^2 + 1$, $\theta = 30°$. Thus, $x = t$, $y = t^2 + 1$. The rotated curve is given by $x_1 = (\sqrt{3}t - (t^2+1))/2$, $y_1 = (t + \sqrt{3}(t^2+1))/2$. Both curves may be graphed for $-4 \leqq t \leqq 4$ in $[-16.1, 16.1]$ by $[-2, 17]$. If we rotate a curve through an angle θ, we can obtain an equation of the new curve by replacing x by $x\cos\theta + y\sin\theta$ and replacing y by $-x\sin\theta + y\cos\theta$. Here, with $\theta = 30°$, x is replaced by $(\sqrt{3}x + y)/2$ and y is replaced by $(-x + \sqrt{3}y)/2$. After simplifying, we obtain $3x^2 + 2\sqrt{3}xy + y^2 + 2x - 2\sqrt{3}y + 4 = 0$.

50. $y = x^2 + 1$, $\theta = 90°$. The new curve is clearly $x = -y^2 - 1$. Graph $x_1 = t$, $y_1 = t^2 + 1$, $x_2 = -t^2 - 1$, $y_2 = t$, $-3 \leqq t \leqq 3$ in $[-21, 16]$ by $[-11, 11]$.

51. $y^2 = x$, $\theta = \frac{\pi}{2}$. The rotated curve is clearly $y = x^2$. Graph $x_1 = t^2$, $y_1 = t$, $x_2 = t$, $y_2 = t^2$, $-3 \leqq t \leqq 3$ in $[-15.3, 15.3]$ by $[-9, 9]$.

52. $y^2 = x$, $\theta = \frac{\pi}{4}$. Graph $x_1 = t^2$, $y_1 = t$, $x_2 = t^2\cos\frac{\pi}{4} - t\sin\frac{\pi}{4} = (t^2 - t)/\sqrt{2}$, $y_2 = t^2\sin\frac{\pi}{4} + t\cos\frac{\pi}{4} = (t^2 + t)/\sqrt{2}$, $= 3 \leqq t \leqq 3$ in $[-15.3, 15.3]$ by $[-9, 9]$. Using the formulas in the solution of Exercise 49, we find an equation of the rotated curve to be $x^2 - 2xy + y^2 = \sqrt{2}x + \sqrt{2}y$.

53. $\frac{x^2}{4} + \frac{y^2}{9} = 1$, $\theta = \frac{\pi}{3}$. Graph $x_1 = 2\cos t$, $y_1 = 3\sin t$, $x_2 = 2\cos t\cos\frac{\pi}{3} - 3\sin t\sin\frac{\pi}{3} = \cos t - (3\sqrt{3}/2)\sin t$, $y_2 = 2\cos t\sin\frac{\pi}{3} + 3\sin t\cos\frac{\pi}{3} = \sqrt{3}\cos t + (3/2)\sin t$, $0 \leqq t \leqq 2\pi$ in $[-5.1, 5.1]$ by $[-3, 3]$. Using the formulas in the solution of Exercise 49, we find an equation of the rotated curve to be $21x^2 + 10\sqrt{3}xy + 31y^2 = 144$.

54. $\frac{x^2}{4} + \frac{y^2}{9} = 1$, $\theta = \frac{\pi}{4}$. Graph $x_1 = 2\cos t$, $y_1 = 3\sin t$, $x_2 = 2\cos t\cos\frac{\pi}{4} - 3\sin t\sin\frac{\pi}{4} = (2\cos t - 3\sin t)/\sqrt{2}$, $y_2 = 2\cos t\sin\frac{\pi}{4} + 3\sin t\cos\frac{\pi}{4} = (2\cos t + 3\sin t)/\sqrt{2}$, $0 \leqq t \leqq 2\pi$ in $[-5.1, 5.1]$ by $[-3, 3]$. Using the formulas in the solution of Exercise 49, we find an equation of the rotated curve to be $13x^2 + 10xy + 13y^2 = 72$.

55. $\frac{x^2}{16} - \frac{y^2}{25} = 1$, $\theta = 60°$. Graph $x_1 = 4\sec t$, $y_1 = 5\tan t$, $x_2 = 4\sec t\cos 60° - 5\tan t\sin 60° = (4\sec t - 5\sqrt{3}\tan t)/2$, $y_2 = 4\sec t\sin 60° + 5\tan t\cos 60° = (4\sqrt{3}\sec t + 5\tan t)/2$, $0 \leqq t \leqq 2\pi$ in $[-34, 34]$ by $[-20, 20]$ (dot format is suggested). Using the formulas in the solution of Exercise 49, we find an equation of the rotated curve to be $-23x^2 + 82\sqrt{3}xy + 59y^2 = 1600$.

56. $\frac{x^2}{16} - \frac{y^2}{25} = 1$, $\theta = 30°$. Graph $x_1 = 4\sec t$, $y_1 = 5\tan t$, $x_2 = 4\sec t\cos 30° - 5\tan t\sin 30° - (4\sqrt{3}\sec t - 5\tan t)/2$, $y_2 = 4\sec t\sin 30° + 5\tan t\cos 30° = (4\sec t + 5\sqrt{3}\tan t)/2$, $0 \leqq t \leqq 2\pi$ in $[-34, 34]$ by $[-20, 20]$ (dot format is suggested). Using the formulas in the solution of Exercise 49, we find an equation of the rotated curve to be $59x^2 + 82\sqrt{3}xy - 23y^2 = 1600$.

10.4 THE CALCULUS OF PARAMETRIC EQUATIONS

1. $x = 2\cos t$, $y = 2\sin t$, $t = \pi/4$. $\frac{dy}{dx} = \frac{dy/dt}{dx/dt} = \frac{2\cos t}{-2\sin t} = -\cot t$. $m = \frac{dy}{dx}\big|_{t=\pi/4} = -\cot(\pi/4) = -1$. $y - y_0 = m(x - x_0)$ becomes $y - \sqrt{2} = (-1)(x - \sqrt{2})$ or $y = -x + 2\sqrt{2}$. $\frac{d^2y}{dx^2} = \frac{dy'/dt}{dx/dt} = \frac{\csc^2 t}{-2\sin t} = -\frac{1}{2\sin^3 t}$. $\frac{d^2y}{dx^2}\big|_{t=\pi/4} = -\sqrt{2}$. Graph $x_1 = 2\cos t$, $y_1 = 2\sin t$, $x_2 = t$, $y_2 = -t + 2\sqrt{2}$, $-3.5 \leqq t \leqq 9$, Tstep $= 0.05$ in $[-10.5, 10, 5]$ by $[-6.1, 6.1]$.

2. $x = \sin 2\pi t$, $y = \cos 2\pi t$, $t = -\frac{1}{6}$. $\frac{dy}{dx} = \frac{-2\pi\sin 2\pi t}{2\pi\cos 2\pi t} = -\tan 2\pi t$. $m = -\tan(\frac{-\pi}{3}) = \sqrt{3}$. Tangent line: $y - \frac{1}{2} = \sqrt{3}(x + \frac{\sqrt{3}}{2})$ or $y = \sqrt{3}x + 2$. $\frac{d^2y}{dx^2} = \frac{-2\pi\sec^2 2\pi t}{2\pi\cos 2\pi t} = -\sec^3 2\pi t (= -8$ at $t = -\frac{1}{6})$. Graph $x_1 = \sin 2\pi t$, $y_1 = \cos 2\pi t$, $x_2 = t$, $y_2 = \sqrt{3}t + 2$, $-2.2 \leqq t \leqq -0.4$ in $[-2.7, 2.6]$ by $[-1.74, 1.39]$.

3. $\frac{dy}{dx} = \frac{-2\sin t}{4\cos t} = -\frac{1}{2}\tan t$, $m = \frac{dy}{dx}\big|_{x=\pi/4} = -\frac{1}{2}$. Tangent line: $y = -\sqrt{2} = -\frac{1}{2}(x - 2\sqrt{x})$ or $y = -\frac{1}{2}x + 2\sqrt{2}$. $\frac{d^2y}{dx^2} = \frac{dy'/dt}{dx/dt} = \frac{-\frac{1}{2}\sec^2 t}{4\cos t} = -\frac{1}{8}\sec^3 t$. $\frac{d^2y}{dx^2}\big|_{t=\pi/4} = -\frac{\sqrt{2}}{4}$. Graph $x_1 = 4\sin t$, $y_1 = 2\cos t$, $x_2 = t$, $y_2 = -\frac{1}{2}x + 2\sqrt{2}$, $-10.5 \leqq t \leqq 10.5$ in $[-10.5, 10.5]$ by $[-6.1, 6.1]$.

4. $x = \cos t$, $y = \sqrt{3}\cos t (= \sqrt{3}x)$, $t = \frac{2\pi}{3}$. $y' = \sqrt{3}$, $y'' = 0$. Tangent line: $y + \frac{\sqrt{3}}{2} = \sqrt{3}(x + \frac{1}{2})$ or $y = \sqrt{3}x$ (a line is its own tangent).

5. $x = \sec^2 t - 1$, $y = \tan t$, $t = -\frac{\pi}{4}(x = y^2)$. $\frac{dy}{dx} = \frac{\sec^2 t}{2\sec^2 t\tan t} = \frac{1}{2\tan t} (= -\frac{1}{2}$ at $t = -\frac{\pi}{4})$. Tangent line: $y + 1 = -\frac{1}{2}(x - 1)$ or $y = -\frac{1}{2}x - \frac{1}{2}$. $\frac{d^2y}{dx^2} = \frac{-\frac{1}{2}\csc^2 t}{2\sec^2 t\tan t} = -\frac{1}{4}\cot^3 t (= \frac{1}{4}$ at $t = -\frac{\pi}{4})$. Graph $x_1 = \sec^2 t - 1$, $y_1 = \tan t$, $x_2 = t$, $y_2 = -\frac{1}{2}t - \frac{1}{2}$, $-7 \leqq t \leqq 15$ in $[-6.2, 14.2]$ by $[-8, 4]$.

6. $\frac{dy}{dx} = \frac{\sec^2 t}{\sec t\tan t} = \csc t$. $m = \frac{dy}{dx}\big|_{t=\pi/6} = 2$. Tangent line: $y - \frac{1}{\sqrt{3}} = 2(x - \frac{2}{\sqrt{3}})$ or $y = 2x - \sqrt{3}$. $\frac{d^2y}{dx^2} = \frac{dy'/dt}{dx/dt} = -\frac{\csc t\cot t}{\sec t\tan t} = -\cot^3 t$. $\frac{d^2y}{dx^2}\big|_{t=\pi/6} = -3\sqrt{3}$. Graph

$x_1 = (\cos t)^{-1}$, $y_1 = \tan t$, $x_2 = t$, $y_2 = 2t - \sqrt{3}$, $-\pi \leqq t \leqq \pi$ in $[-4.1, 4.4]$ by $[-2.3, 2.7]$ in Dot Format.

7. $\frac{dy}{dx} = \frac{1/(2\sqrt{t})}{1} = \frac{1}{2\sqrt{t}} = \frac{1}{2}t^{-1/2}$. $m = \frac{dy}{dx}|_{t=1/4} = 1$. Tangent line: $y - \frac{1}{2} = (1)(x - \frac{1}{4})$ or $y = x + \frac{1}{4}$. $\frac{d^2y}{dx^2} = -\frac{1}{4}t^{-3/2}$, $\frac{d^2y}{dx^2}|_{t=1/4} = -2$. Graph $x_1 = t$, $y_1 = \sqrt{t}$, $x_2 = t$, $y_2 = t + 0.25$, $-10.5 \leqq t \leqq 10.5$ in $[-10.5, 10.5]$ by $[-6.1, 6.1]$.

8. $x = -\sqrt{t+1}$, $y = \sqrt{3t}$, $t = 3$. $\frac{dy}{dx} = \frac{3/2\sqrt{3t}}{(-1/2\sqrt{t+1})} = -\sqrt{3}\sqrt{\frac{t+1}{t}}$. At $t = 3$, $m = -\sqrt{3}\sqrt{\frac{4}{3}} = -2$. Tangent line: $y - 3 = -2(x+2)$ or $y = -2x - 1$. $\frac{d^2y}{dx^2} = -(\sqrt{3}/2)\sqrt{\frac{t}{t+1}}\frac{t-(t+1)}{t^2}/(-1/2\sqrt{t+1}) = -\frac{\sqrt{3}}{t^{3/2}}$. $\frac{d^2y}{dx^2}|_{t=3} = -\frac{\sqrt{3}}{3^{3/2}} = -\frac{1}{3}$. Graph $x_1 = -\sqrt{t+1}$, $y_1 = \sqrt{3t}$, $x_2 = t$, $y_2 = -2t - 1$, $-10 \leqq t \leq 10$ in $[-9.5, 7.5]$ by $[-5, 5]$.

9. $\frac{dy}{dx} = \frac{4t^3}{4t} = t^2$. $m = \frac{dy}{dx}|_{t=-1} = 1$. Tangent line: $y - 1 = (1)(x-5)$ or $y = x - 4$. $\frac{d^2y}{dx^2} = \frac{dy'/dt}{dx/dt} = \frac{2t}{4t} = \frac{1}{2}$ for all $t \neq 0$. Graph $x_1 = 2t^2 + 3$, $y_1 = t^4$, $x_2 = t$, $y_2 = t - 4$, $-2 \leqq t \leqq 20$ in $[-13.2, 24.2]$ by $[-6, 16]$.

10. $x = \frac{1}{t}$, $y = -2 + \ln t$, $t = 1$. $\frac{dy}{dx} = \frac{(1/t)}{(-1/t^2)} = -t$. At $t = 1$, $m = -1$. Tangent line: $y + 2 = -(x-1)$ or $y = -x - 1$. $\frac{d^2y}{dx^2} = \frac{-1}{(-1/t^2)} = t^2$. $\frac{d^2y}{dx^2}|_{t=1} = 1$. Graph $x_1 = \frac{1}{t}$, $y_1 = -2 + \ln t$, $x_2 = t$, $y_2 = -t - 1$, $-10 \leqq t \leqq 10$ in $[-10, 10]$ by $[-10, 10]$.

11. $\frac{dy}{dx} = \frac{\sin t}{1 - \cos t}$. $m = \frac{dy}{dx}|_{t=\pi/3} = \sqrt{3}$. Tangent line: $y - \frac{1}{2} = \sqrt{3}[x - (\frac{\pi}{3} - \frac{\sqrt{3}}{2})]$ or $y = \sqrt{3}x + 2 - \frac{\sqrt{3}}{3}\pi$. $\frac{d^2y}{dx^2} = -\frac{1}{(1-\cos t)^2}$, $\frac{d^2y}{dx^2}|_{t=\pi/3} = -4$. Graph $x_1 = t - \sin t$, $y_1 = 1 - \cos t$, $x_2 = t$, $y_2 = \sqrt{3}\ t + 2 - \frac{\sqrt{3}}{3}\ \pi$, $-17 \leqq t \leqq 17$ in $[-17, 17]$ by $[-10, 10]$.

12. $x = \cos t$, $y = 1 + \sin t$, $t = \frac{\pi}{2}$. $\frac{dy}{dx} = \frac{\cos t}{-\sin t} = -\cot t$. At $t = \frac{\pi}{2}$, $m = 0$. Tangent line: $y = 2$. $\frac{d^2y}{dx^2} = \frac{\csc^2 t}{-\sin t} = -\csc^3 t$. $\frac{d^2y}{dx^2}|_{t=\pi/2} = -1$. Graph $x_1 = \cos t$, $y_1 = 1 + \sin t$, $x_2 = t$, $y_2 = 2$, $-\pi \leqq t \leqq \pi$ in $[-2.55, 2.55]$ by $[0, 3]$.

13. Graph in $[-1, 1]$ by $[0, \pi]$. $(\frac{dx}{dt})^2 + (\frac{dy}{dt})^2 = \sin^2 t + (1 + \cos t)^2 = 2 + 2\cos t = 2(1 + \cos t)\frac{(1-\cos t)}{1 - \cos t} = \frac{2\sin^2 t}{1 - \cos t}$. On $0 \leqq t \leqq \pi$, $\sqrt{(\frac{dx}{dt})^2 + (\frac{dy}{dt})^2} = \frac{\sqrt{2}\sin t}{\sqrt{1-\cos t}}$. Let $u = 1 - \cos t$, $du = \sin t\, dt$. $\sqrt{2}\int_0^\pi \frac{\sin t\, dt}{\sqrt{1-\cos t}} = \sqrt{2}\int_0^2 u^{-1/2} du = 2\sqrt{2}u^{1/2}|_0^2 = 4$.

14. Graph in $[0, 3\sqrt{3}]$ by $[0, 4.5]$. $(\frac{dx}{dt})^2 + (\frac{dy}{dt})^2 = (3t^2)^2 + (3t)^2 = 9t^2(t^2 + 1)$. $\int_0^{\sqrt{3}} 3t\sqrt{t^2+1} = (t^2+1)^{3/2}]_0^{\sqrt{3}} = 8 - 1 = 7$.

15. Graph in $[0,8]$ by $[1/3,9]$. $(\frac{dx}{dt})^2+(\frac{dy}{dt})^2=t^2+[\frac{1}{2}(2t+1)^{1/2}2]^2=t^2+2t+1=(t+1)^2$. $\int_0^4(t+1)dt=\frac{t^2}{2}+t|_0^4=12$.

16. Graph in $[\sqrt{3},9]$ by $[0,7.5]$. $(\frac{dx}{dt})^2+(\frac{dy}{dt})^2=(2t+3)+(1+t)^2=(t+2)^2$. $\int_0^3(t+2)dt=\frac{t^2}{2}+2t]_0^3=\frac{9}{2}+6=\frac{21}{2}$.

17. Graph in $[8,4\pi]$ by $[0,8]$. $\frac{dx}{dt}=8(-\sin t+\sin t+t\cos t)=8t\cos t$. Similarly, $\frac{dy}{dt}=8t\sin t$. $\int_0^{\pi/2}\sqrt{8^2t^2(\cos^2 t+\sin^2 t)}dt=8\int_0^{\pi/2}t\,dt=4t^2|_0^{\pi/2}=\pi^2$.

18. Graph, $0\leqq t\leqq\pi/3$, in $[0,0.5]$ by $[0.5,1]$ or, for a bigger picture, $-\pi/2<t<\pi/2$, in $[-3,3]$ by $[0,1]$. $(\frac{dx}{dt})^2+(\frac{dy}{dt})^2=(\sec t-\cos t)^2+\sin^2 t=\tan^2 t$. $\int_0^{\pi/3}\tan t\,dt=-\ln\cos t]_0^{\pi/3}=-(\ln(\frac{1}{2})-\ln 1)=\ln 2$.

19. $\frac{dx}{dt}=-\sin t$, $\frac{dy}{dt}=\cos t$. $2\pi\int_0^{2\pi}y\sqrt{(\frac{dx}{dt})^2+(\frac{dy}{dt})^2}dt=2\pi\int_0^{2\pi}(2+\sin t)dt=2\pi(2t-\cos t)|_0^{2\pi}=8\pi^2$.

20. $(\frac{dx}{dt})^2+(\frac{dy}{dt})^2=t+\frac{1}{t}$. $2\pi\int x\,ds=2\pi\int_0^{\sqrt{3}}(2/3)t^{3/2}\sqrt{t+(1/t)}dt=\frac{4\pi}{3}\int_0^{\sqrt{3}}t\sqrt{t^2+1}dt=\frac{4\pi}{3}\frac{1}{3}(t^2+1)^{3/2}]_0^{\sqrt{3}}=\frac{4\pi}{9}(8-1)=\frac{28\pi}{9}$.

21. $\frac{dx}{dt}=1$, $\frac{dy}{dt}=t+\sqrt{2}$. $2\pi\int_{-\sqrt{2}}^{\sqrt{2}}x\sqrt{(\frac{dx}{dt})^2+(\frac{dy}{dt})^2}dt=2\pi\int_{-\sqrt{2}}^{\sqrt{2}}(t+\sqrt{2})\sqrt{t^2+2\sqrt{2}t+3}dt=\pi\int_1^9u^{1/2}du$ (where $u=t^2+2\sqrt{2}t+3$, $du=2(t+\sqrt{2})dt$) $=\frac{2}{3}\pi u^{3/2}]_1^9=\frac{52\pi}{3}$.

22. $(\frac{dx}{dt})^2+(\frac{dy}{dt})^2=(\sec t-\cos t)^2+\sin^2 t=\sec^2 t-1=\tan^2 t$. $2\pi\int y\,ds=2\pi\int_0^{\pi/3}\cos t\tan t\,dt=2\pi\int_0^{\pi/3}\sin t\,dt=2\pi[-\cos t]_0^{\pi/3}=\pi$.

23. $\frac{dx}{dt}=2$, $\frac{dy}{dt}=1$. $2\pi\int_0^1 y\sqrt{(\frac{dx}{dt})^2+(\frac{dy}{dt})^2}dt=2\pi\int_0^1(t+1)\sqrt{5}dt=2\pi\sqrt{5}[\frac{t^2}{2}+t]_0^1=3\sqrt{5}\pi$. Check: $r_1=1$, $r_2=2$, slant height $=$ distance between $(0,1)$ and $(2,2)=\sqrt{5}$, area $=\pi(1+2)\sqrt{5}=3\sqrt{5}\pi$.

24. $(\frac{dx}{dt})^2+(\frac{dy}{dt})^2=h^2+r^2$. $2\pi\int y\,dt=2\pi\int_0^1\sqrt{h^2+r^2}r\,t\,dt=\pi r\sqrt{h^2+r^2}t^2]_0^1=\pi r\sqrt{h^2+r^2}$. The formula checks because $\sqrt{h^2+r^2}=$ slant height.

25. a) $\frac{dx}{dt}=-2\sin 2t$, $\frac{dy}{dt}=2\cos 2t$. $\int_0^{\pi/2}\sqrt{(\frac{dx}{dt})^2+(\frac{dy}{dt})^2}dt=\int_0^{\pi/2}\sqrt{4(\sin^2 2t+\cos^2 2t)}dt=\pi$ b) $\frac{dx}{dt}=\pi\cos\pi t$, $\frac{dy}{dt}=-\pi\sin\pi t$. $\int_{-1/2}^{1/2}\sqrt{\pi^2(\cos^2\pi t+\sin^2\pi t)}dt=\pi\int_{-1/2}^{1/2}dt=\pi$.

26. a) With $a=1$, $e=0.5$, we use $4\sqrt{1-0.25\cos^2 t}=g(t)$ for the integrand. With $n=10$ and endpoints 0, $\frac{\pi}{2}$, TRAP yields 5.7566. NINT$(g(t),t,0,\frac{\pi}{2})=5.7556$. b) $|E_T|\leqq\frac{b-a}{12}h^2M\leqq\frac{(\pi/2)}{12}(\frac{\pi/2}{10})^2(4\cdot 1)<0.013$.

27. Let $x_1 = (8\sin 2t)\cos t$, $y_1 = (8\sin 2t)\sin t$, $x_2 = \text{NDER}(x_1, t)$, $y_2 = \text{NDER}(y_1, t)$. $\text{NINT}(\sqrt{x_2^2 + y_2^2}, t, 0, \pi/2) = 19.377$.

28. Graph $x_1 = (6\sin 2.5t)\cos t$, $y_1 = (6\sin 2.5t)\sin t$, $0 \leqq t \leqq a = \pi/2.5$ in $[0, 6]$ by $[0, 5]$. Let $x_2 = \text{NDER}(x_1, t)$, $y_2 = \text{NDER}(y_1, t)$. Length $= \text{NINT}(\sqrt{x_2^2 + y_2^2}, t, 0, \pi/2.5) = 13.808$.

29. Let $x_1 = 2t\cos t$, $y_1 = 2t\sin t$, $x_2 = \text{NDER}(x_1, t)$, $y_2 = \text{NDER}(y_1, t)$. $\text{NINT}(\sqrt{x_2^2 + y_2^2}, t, 0, 50) = 2505.105$.

30. Graph $x_1 = 8t\sin t$, $y_1 = 8\cos t$, $0 \leqq t \leqq 3$, Tstep $= 0.1$ in $[0, 15]$ by $[-8, 8]$. Let $x_2 = \text{NDER}(x_1, t)$, $y = \text{NDER}(y_1, t)$. Length $= \text{NINT}(\sqrt{x_2^2 + y_2^2}, t, 0, 3) = 31.967$.

31. Let $x_1 = 3\sin t$, $y_1 = 5 + 3\sin 2t$, $0 \leqq t \leqq 2\pi$. The graph in $[-4, 4]$ by $[0, 9]$ suggests that the curve is symmetric with respect to the y-axis. Indeed if we replace t by $t + \pi$, (x, y) is replaced by $(-x, y)$. The right half of the curve which is all we need is traced out as t goes from 0 to π. Let $x_2 = \text{NDER}(x_1, t)$, $y_2 = \text{NDER}(y_1, t)$. S. area $= 2\pi\text{NINT}(x_1\sqrt{x_2^2 + y_2^2}, t, 0, \pi) = 159.485$.

32. Graph $x_1 = 5 + 3\sin t$, $y_1 = 3\sin 2t$, $0 \leqq t \leqq 2\pi$. Tstep$=$ 0.05 in $[0, 9]$ by $[-4, 4]$. Let $x_2 = \text{NDER}(x_1, t)$, $y_2 = \text{NDER}(y_1, t)$. Area of solid $= 2\pi\,\text{NINT}(x_1\sqrt{x_2^2 + y_2^2}, t, 0, 2\pi) = 888.703$.

33. Let $x_1 = \sin t$, $y_1 = \sin 2t$. Graph x_1, y_1, $0 \leqq t \leqq 2\pi$ in $[-1.86, 1.86]$ by $[-1.1, 1.1]$. $\frac{dy}{dx} = \frac{dy/dt}{dx/dt} = \frac{2\cos 2t}{\cos t}$. $\frac{dy}{dx} = 0$ when $t = \frac{\pi}{4}$ (first quadrant) and at that point $(x, y) = (\frac{\sqrt{2}}{2}, 1)$. For suitable values at the origin we take $t = 0$ and $t = \pi$. $\frac{dy}{dx}|_{t=0} = 2$, $\frac{dy}{dx}|_{t=\pi} = -2$. The tangent lines at the origin are $y = \pm 2x$. To confirm, we graph x_1, y_1, $x_2 = t$, $y_2 = 2t$, $x_3 = t$, $y_3 = -2t$, $-2\pi \leqq t \leqq 2\pi$ in the viewing rectangle given above.

34. Graph $x_1 = \sin 2t$, $y_1 = \sin 3t$, $0 \leqq t \leqq 2\pi$ in $[-1.86, 1.86]$ by $[-1.1, 1.1]$. $\frac{dy}{dx} = \frac{3\cos 3t}{2\cos 2t} \cdot \frac{dy}{dx} = 0$ first when $3t = \frac{\pi}{2}$ or $t = \frac{\pi}{6}$ and at that point $(x, y) = (\frac{\sqrt{3}}{2}, 1)$. For suitable t-values at the origin we take $t = 0$ and $t = \pi$. $\frac{dy}{dx}|_{t=0} = \frac{3}{2}$, $\frac{dy}{dx}|_{t=\pi} = -\frac{3}{2}$. The tangent lines at the origin are $y = \pm\frac{3}{2}x$. To confirm, we graph x_1, y_1, $x_2 = t$, $y_2 = 1.5t$, $x_3 = t$, $y_3 = -1.5t$, $-2\pi \leqq t \leqq 2\pi$ in the rectangle above.

35 through 41. For each of these we may use $0 \leqq t \leqq 2\pi$ in $[-1.86, 1.86]$ by $[-1.1, 1.1]$. For 38 and 39, $\frac{\pi}{2} \leqq t \leqq \frac{3\pi}{2}$ suffices.

42. Graph $x = 5\cos t + 10\cos(ct)$, $y = 5\sin t + 10\sin(ct)$, $0 \leqq t \leqq 2\pi$ in $[-34, 34]$ by $[-20, 20]$ for each $c = 2, 4, 6, 8$.

43. Graph $x = 12\cos t + 6\cos(ct)$, $y = 12\sin t + 6\sin(ct)$, $0 \leqq t \leqq 2\pi$ in $[-34, 34]$ by $[-20, 20]$ for each $c = 2, 4, 6, 8$.

44. Let $x_1 = 5\cos t + 10\cos(8t)$, $y_1 = 5\sin t + 10\sin(8t)$, $x_2 = \text{NDER}(x_1, t)$, $y_2 = \text{NDER}(y_1, t)$. Length $= \text{NINT}(\sqrt{x_2^2 + y_2^2}, t, 0, 2\pi) = 503.14$.

45. Let $x_1 = 12\cos t + 6\cos(8t)$, $y_1 = 12\sin t + 6\sin(8t)$, $x_2 = \text{NDER}(x_1, t)$, $y_2 = \text{NDER}(y_1, t)$. Length $= \int_0^{2\pi} \sqrt{(\frac{dx_1}{dt})^2 + (\frac{dy_1}{dt})^2} dt = \text{NINT}(\sqrt{x_2^2 + y_2^2}, t, 0, 2\pi) = 306.324$.

10.5 POLAR COORDINATES

1. {a,c}, {b,d}, {e,k}, {f,h}, {g,j}, {i,l}, {m,o}, {n,p}

2. a) $(3, 0)$ b) $(-3, 0)$ c) $(3, 0)$ d) $(-3, 0)$ e) $(-1, \sqrt{3})$ f) $(-1, -\sqrt{3})$ g) $(1, \sqrt{3})$ h) $(-1, -\sqrt{3})$ i) $(1, -\sqrt{3})$ j) $(1, \sqrt{3})$ k) $(-1, \sqrt{3})$ l) $(1, -\sqrt{3})$

3.

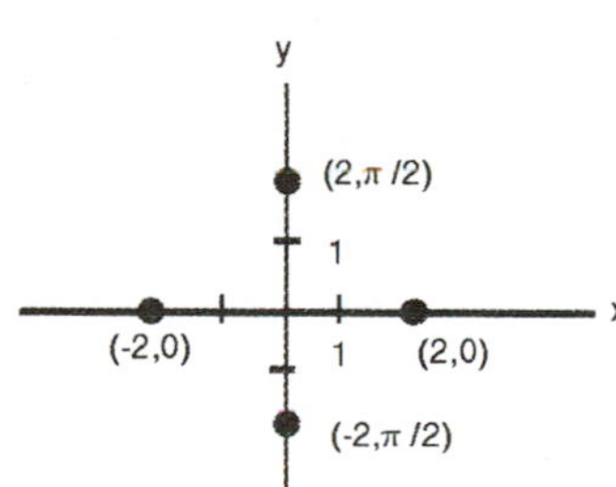

a) $(2, \frac{\pi}{2} + 2n\pi)$, $(-2, -\frac{\pi}{2} + 2n\pi)$ b) $(2, 2n\pi)$, $(-2, (2n+1)\pi)$ c) $(-2, \frac{\pi}{2} + 2n\pi)$, $(2, -\frac{\pi}{2} + 2n\pi)$ d) $(-2, 2n\pi)$, $(2, (2n+1)\pi)$. $n = 0, \pm 1, \pm 2, \ldots$

4.

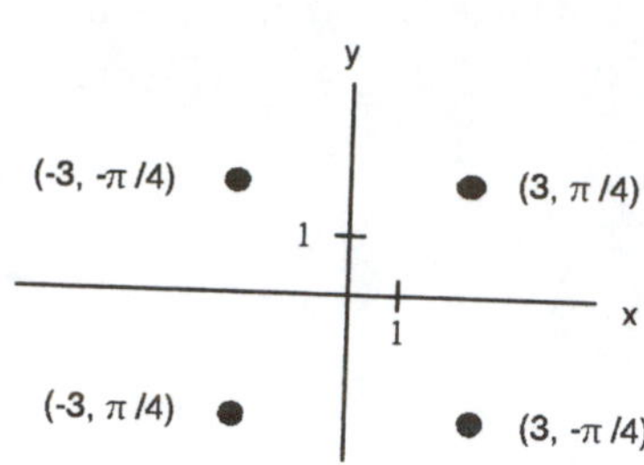

a) $(3, \frac{\pi}{4}+2n\pi)$, $(-3, \frac{5\pi}{4}+2n\pi)$ b) $(-3, \frac{\pi}{4}+2n\pi)$, $(3, \frac{5\pi}{4}+2n\pi)$ c) $(3, -\frac{\pi}{4}+2n\pi)$, $(-3, \frac{3\pi}{4}+2n\pi)$ d) $(-3, -\frac{\pi}{4}+2n\pi)$, $(3, \frac{3\pi}{4}+2n\pi)$

5. In each case we use $x = r\cos\theta$, $y = r\sin\theta$. a) $x = \sqrt{2}\cos\frac{\pi}{4} = 1$, $y = \sqrt{2}\sin\frac{\pi}{4} = 1$. $(1,1)$ b) $x = 1\cdot\cos 0 = 1$, $y = 1\cdot\sin 0 = 0$. $(1,0)$ c) $(0,0)$ d) $x = -\sqrt{2}\cos\frac{\pi}{4} = -1$, $y = -\sqrt{2}\sin\frac{\pi}{4} = -1$. $(-1,-1)$ e) $x = -3\cos\frac{5\pi}{6} = \frac{3\sqrt{3}}{2}$, $y = -3\sin\frac{5\pi}{6} = -\frac{3}{2}$. $(\frac{3\sqrt{3}}{2}, -\frac{3}{2})$ f) $x = 5\cos\theta = 5(\frac{3}{5}) = 3$, $y = 5\sin\theta = 5(\frac{4}{5}) = 4$. $(3,4)$ g) $x = -\cos 7\pi = 1$, $y = 0$. $(1,0)$ h) $x = 2\sqrt{3}\cos\frac{2\pi}{3} = -\sqrt{3}$, $y = 2\sqrt{3}\sin\frac{2\pi}{3} = 2\sqrt{3}(\frac{\sqrt{3}}{2}) = 3$. $(-\sqrt{3}, 3)$.

6. $(0,\theta)$, any real θ

7. Graph $r = 2$, $0 \leqq \theta \leqq 2\pi$ is a square window containing $[-2,2]$ by $[-2,2]$ in polar mode.

8.

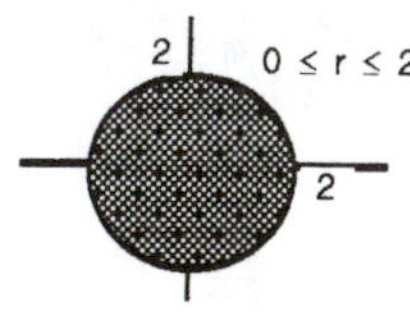

9.

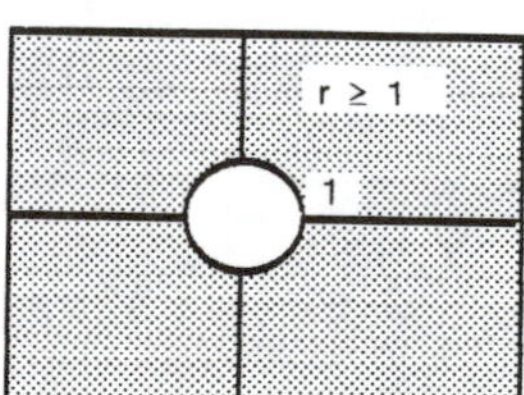

10.

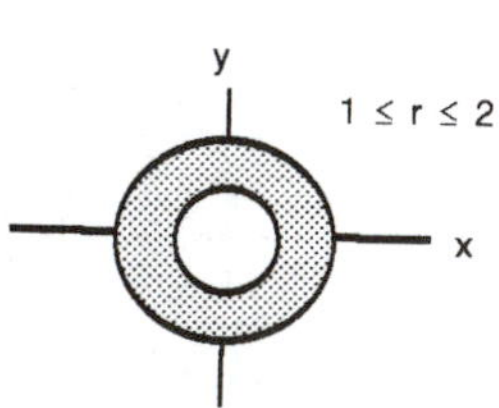

11.

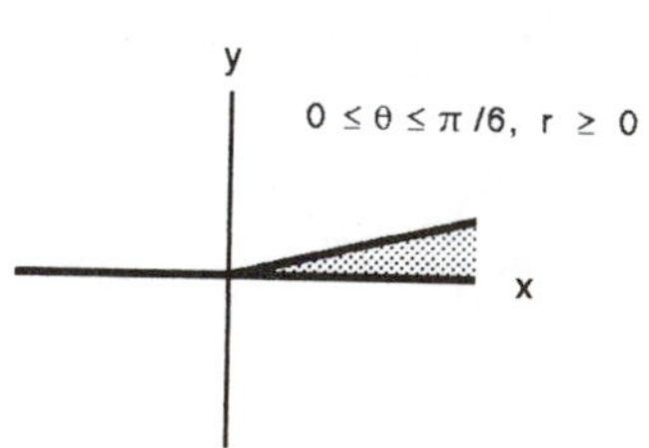

12.

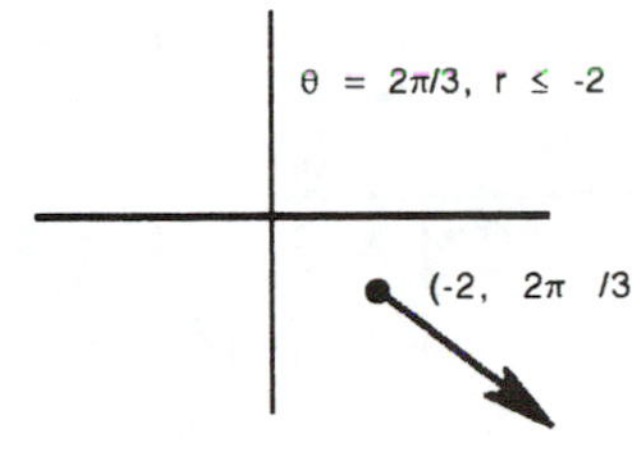

13.

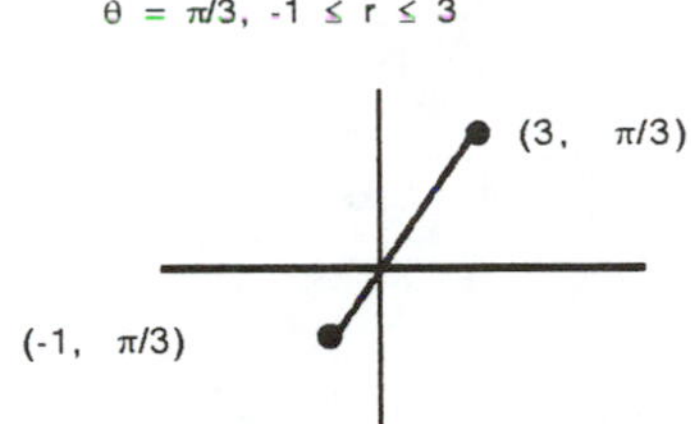

14.

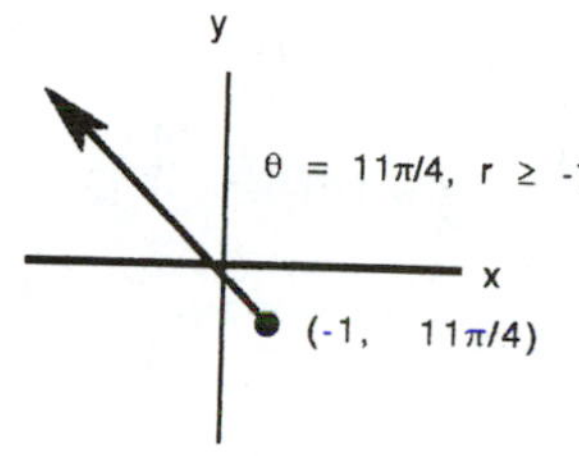

15. The graph consists of the origin and the positive y-axis.

16. The graph consists of the origin and the negative y-axis.

17. The graph consists of the upper half of the unit circle including $(-1,0)$ and $(1,0)$.

18. The graph consists of the lower half of the unit circle including $(-1,0)$ and $(1,0)$.

19.

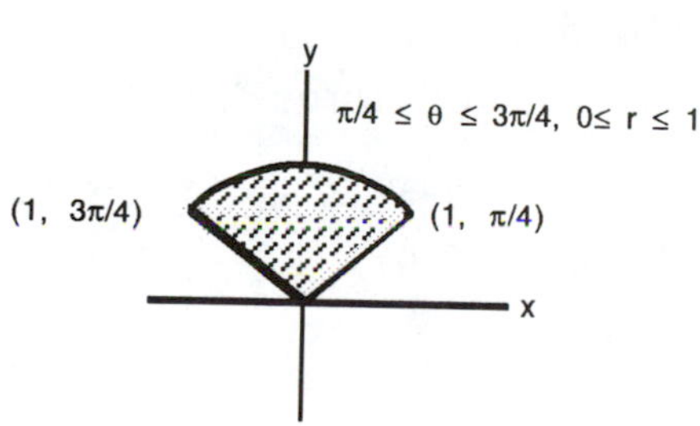

20.

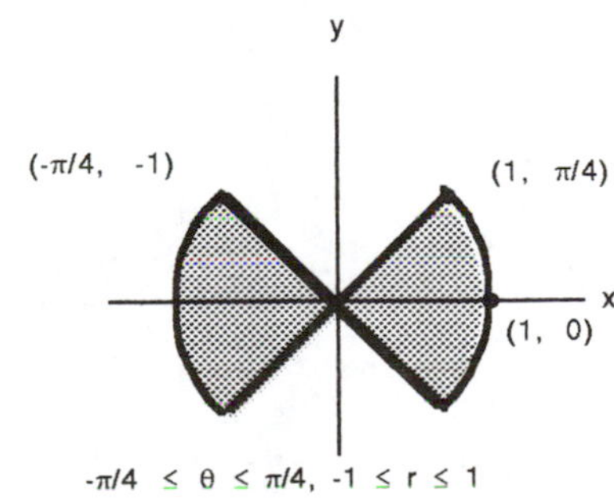

21.

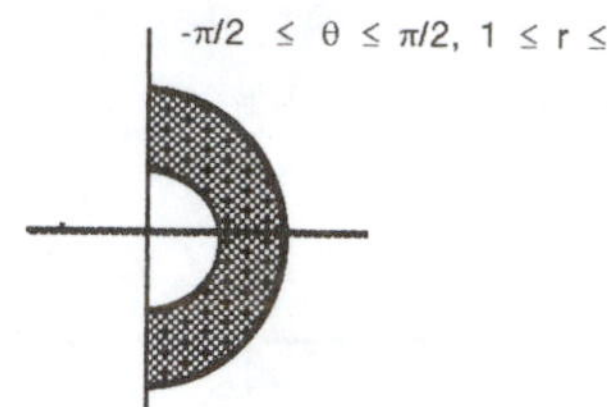

22. 

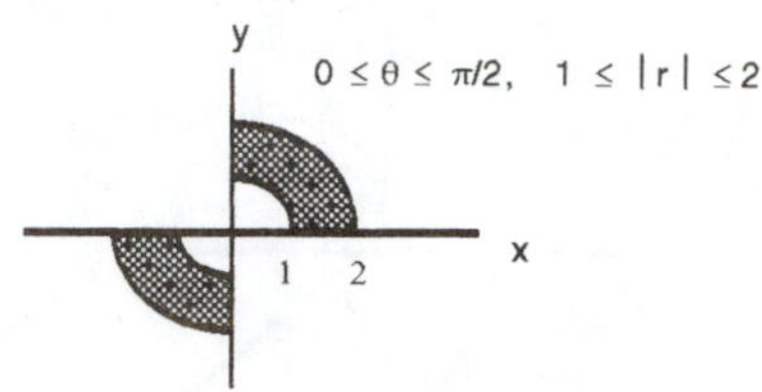

23. $r\cos\theta = 2 \Rightarrow x = 2$. Vertical line consisting of all points with x-coordinate 2.

24. $r\sin\theta = -1 \Rightarrow y = -1$, horizontal line through $(0,-1)$.

25. $r\sin\theta = 4$ is equivalent to $y = 4$, the horizontal line through $(1,4)$.

26. $r\cos\theta = 0$. $x = 0$, the y-axis.

27. $r\sin\theta = 0$. $y = 0$, the x-axis.

28. $r\cos\theta = -3$. $x = -3$, the vertical line through $(-3,0)$.

29. $r\cos\theta + r\sin\theta = 1$. $x + y = 1$, the line through $(1,0)$ and $(0,1)$.

30. $r\sin\theta = r\cos\theta$. $y = x$.

31. $r^2 = 1$. $x^2 + y^2 = 1$, the unit circle.

32. $r^2 = 4r\sin\theta$. $x^2 + y^2 = 4y$, $x^2 + y^2 - 4y + 4 = 0 + 4$, $x^2 + (y-2)^2 = 2^2$. Circle with center $(0,2)$ and radius 2.

33. $r = \frac{5}{\sin\theta - 2\cos\theta}$. $r\sin\theta - 2r\cos\theta = 5$, $y - 2x = 5$, $y = 2x + 5$. Line with slope 2 through $(0,5)$.

34. $r = 4\tan\theta\sec\theta = 4\frac{\sin\theta}{\cos^2\theta}$. $r\cos^2\theta = 4\sin\theta$, $(r\cos\theta)^2 = 4r\sin\theta$, $x^2 = 4y$. This is the parabola with vertex $(0,0)$ which passes through $(\pm 2, 1)$.

35. $x = 7$. $r\cos\theta = 7$, $r = 7\sec\theta$. Graph $r = 7(\cos\theta)^{-1}$, $-\frac{\pi}{2} \leqq \theta \leqq \frac{\pi}{2}$ in $[0,10]$ by $[-50,50]$ obtaining the vertical line through $(7,0)$.

36. $y = 1$. $r\sin\theta = 1$, $r = \csc\theta$.

37. $x = y$. $r\cos\theta = r\sin\theta$, $\cos\theta = \sin\theta$, $\tan\theta = 1$, $\theta = \frac{\pi}{4}$.

38. $x - y = 3$. $r\cos\theta - r\sin\theta = 3$, $r = \frac{3}{\cos\theta - \sin\theta}$.

39. $x^2 + y^2 = 4$, $r = 2$, circle.

40. $x^2 - y^2 = 1$, hyperbola. $r^2\cos^2\theta - r^2\sin^2\theta = 1$, $r^2 = \frac{1}{\cos^2\theta - \sin^2\theta} = \frac{1}{\cos 2\theta} = \sec 2\theta$. $r = \pm\sqrt{\sec 2\theta}$.

41. $\frac{x^2}{9} + \frac{y^2}{4} = 1$. $\frac{r^2\cos^2\theta}{9} + \frac{r^2\sin^2\theta}{4} = 1$, $(4\cos^2\theta + 9\sin^2\theta)r^2 = 36$, $r = \frac{\pm 6}{\sqrt{4\cos^2\theta + 9\sin^2\theta}}$, ellipse.

42. $xy = 2$. $r\cos\theta r\sin\theta = 2$, $r^2 = \frac{2}{\cos\theta\sin\theta}$, $r = \frac{\pm\sqrt{2}}{\sqrt{\cos\theta\sin\theta}}$, hyperbola.

43. $y^2 = 4x$. $r^2\sin^2\theta = 4r\cos\theta$, $r = \frac{4\cos\theta}{\sin^2\theta}$. To confirm, graph r for $0 \leqq \theta \leqq 2\pi$ in $[-8.5, 25.5]$ by $[-10, 10]$. Parabola.

44. $x^2 - y^2 = 25\sqrt{x^2 + y^2}$. $r^2(\cos^2\theta - \sin^2\theta) = 25r$, $r^2\cos 2\theta - 25r = 0$, $r(r\cos 2\theta - 25) = 0$. This yields $r = 0$ (the origin) plus $r = 25\sec 2\theta$.

45. $x = r\cos\theta = f(\theta)\cos\theta$, $y = r\sin\theta = f(\theta)\sin\theta$.

10.6 GRAPHING IN POLAR COORDINATES

1. In Exercises 1 through 12 the student should use the method of Example 1 including a table of values, use of symmetries and "slope at $(0, \theta_0) = \tan\theta_0$" as a guide as to how the curve goes into and out of the origin. The student's results can then be checked on a grapher. One way of carrying out this check is given in the answers. It is assumed that a graphing utility with a polar graphing mode and with a screen "squaring" function is being used.

Graph $r = 1 + \cos\theta$, $0 \leqq \theta \leqq 2\pi$, θStep $= 0.1$ in a "squared" rectangle containing $[-0.25, 2]$ by $[-1.3, 1.3]$, for example $[-1.33, 3.08]$ by $[-1.3, 1.3]$.

2. Graph $r = 2 - 2\cos\theta$, $0 \leqq \theta \leqq 2\pi$ in $[-6.8, 3.7]$ by $[-3.1, 3.1]$.

3. Graph $r = 1 - \sin\theta$, $0 \leqq \theta \leqq 2\pi$ in $[-2.4, 2.4]$ by $[-2.2, 0.7]$.

4. Graph $r = 1 + \sin\theta$, $0 \leqq \theta \leqq 2\pi$ in $[-2.1, 2.1]$ by $[-0.45, 2]$.

5. Graph $r = 2 + \sin\theta$, $0 \leqq \theta \leqq 2\pi$ in $[-3.5, 3.6]$ by $[-1.2, 3]$.

6. Graph $r = 1 + 2\sin\theta$, $0 \leqq \theta \leqq 2\pi$ in $[-3, 3]$ by $[-0.45, 3]$.

7. Graph $r = 2\sqrt{\cos 2\theta}$ and $r = -2\sqrt{\cos 2\theta}$ in $[-2, 2]$ by $[-1.2, 1.2]$. Use θStep $= 0.01$, $-\frac{\pi}{4} \leqq \theta \leqq \frac{\pi}{4}$.

8. $r^2 = 4\sin\theta$. Graph $r_1 = 2\sqrt{\sin\theta}$, $r_2 = -r_1$, $0 \leqq \theta \leqq \pi$ in $[-3.6, 3.6]$ by $[-2.13, 2.13]$.

9. Graph $r = \theta$ in $[-33, 33]$ by $[-18.4, 20]$ first using $0 \leqq \theta \leqq 20$ then $-20 \leqq \theta \leqq 0$ and then $-20 \leqq \theta \leqq 20$.

10. Graph $r = \sin(\theta/2)$, $0 \leqq \theta \leqq 4\pi$ in $[-1.7, 1.7]$ by $[-1, 1]$.

11. Graph $r = 8\cos 2\theta$, $0 \leqq \theta \leqq 2\pi$ in $[-13.6, 13.6]$ by $[-8, 8]$. A complete graph requires a minimum of 2π for the range of θ. The factor 8 stretches the graph of $r = \cos 2\theta$ away from the origin by a factor of 8. The range, $\frac{\pi}{4} \leqq \theta \leqq \frac{3\pi}{4}$, for example, produces a closed curve. Replacing θ by 2θ produces 3 more leaves.

12. Graph $r = 8\cos 3\theta$, $0 \leqq \theta \leqq \pi$ in $[-13.6, 13.6]$ by $[-8, 8]$. A complete graph requires a minimum of π units for the range of θ. The factor 8 stretches the graph of $r = \cos 3\theta$ away from the origin by a factor of 8. The range $\frac{\pi}{6} \leqq \theta \leqq \frac{\pi}{2}$, for example, produces a closed curve. Replacing θ by 3θ has tripled the number of leaves.

13. $r = -1 + \cos\theta$. Slope $= \frac{r'\sin\theta + r\cos\theta}{r'\cos\theta - r\sin\theta} = \frac{-\sin^2\theta - \cos\theta + \cos^2\theta}{-\sin\theta\cos\theta + \sin\theta - \cos\theta\sin\theta}$. At $\theta = \frac{\pi}{2}$, $r = -1$ and slope $= \frac{-1-0+0}{1} = -1$ and the tangent at the rectangular point $(0, -1)$ is $y = -x - 1$ or $r(\sin\theta + \cos\theta) = -1$, $r = \frac{-1}{\sin\theta + \cos\theta}$. At $\theta = -\frac{\pi}{2}$, $r = -1$ and slope $= \frac{-1}{-1} = 1$ and the tangent at the rectangular point $(0, 1)$ is $y = x + 1$ or $r(\sin\theta - \cos\theta) = 1$, $r = \frac{1}{\sin\theta - \cos\theta}$. Graph $r = -1 + \cos\theta$ and the two tangents, $-\pi \leqq \theta \leqq \pi$ in $[-2.6, 4.3]$ by $[-2, 2]$.

14. $r = -1 + \sin\theta$; $\theta = 0, \frac{\pi}{2}, \pi$. Slope $= \frac{r'\sin\theta + r\cos\theta}{r'\cos\theta - r\sin\theta} = \frac{\cos\theta\sin\theta - \cos\theta + \sin\theta\cos\theta}{\cos^2\theta + \sin\theta - \sin^2\theta}$. At $\theta = 0$, $r = -1$, slope $= -1$ and the tangent at $(-1, 0)$ is $y = -x - 1$ or $r\sin\theta = -r\cos\theta - 1$, $r = \frac{-1}{\sin\theta + \cos\theta}$. At $\theta = \frac{\pi}{2}$, $r = 0$, $m = \tan\frac{\pi}{2}$ is undefined so the y-axis is the tangent line. At $\theta = \pi$, $r = -1$, slope $= 1$ and the tangent at $(1, 0)$ is $y = x - 1$ or $r = \frac{-1}{\sin\theta - \cos\theta}$. Graph $r_1 = -1 + \cos\theta$, $r_2 = \frac{-1}{\sin\theta + \cos\theta}$, $r_3 = \frac{-1}{\sin\theta - \cos\theta}$, $0 \leqq \theta \leqq 2\pi$ in $[-3.4, 3.4]$ by $[-2, 2]$. Regard the y-axis as the tangent at the origin.

15. At the origin $r = \sin 2\theta = 0$ which implies $2\theta = n\pi$, $\theta = n\pi/2$. But $\tan(n\pi/2) = \pm\infty$ or 0 so the tangent line at the origin is either horizontal or vertical, i.e., it is either the x-axis or the y-axis. Slope $= \frac{r'\sin\theta + r\cos\theta}{r'\cos\theta - r\sin\theta} = \frac{2\cos 2\theta\sin\theta + \sin 2\theta\cos\theta}{2\cos 2\theta\cos\theta - \sin 2\theta\sin\theta}$. We give the details for the case $\theta = -3\pi/4$. $r = \sin[2(-\frac{3\pi}{4})] = \sin(-\frac{3\pi}{2}) = 1$: $(1, -3\pi/4)$ is the point on the curve ($x = r\cos\theta = -\frac{\sqrt{2}}{2}$, $y = r\sin\theta = \frac{-\sqrt{2}}{2}$). When $\theta = -\frac{3\pi}{4}$, slope $= \frac{2(0)(-\frac{\sqrt{2}}{2}) + (1)(-\frac{\sqrt{2}}{2})}{2(0)(-\frac{\sqrt{2}}{2}) - (1)(-\frac{\sqrt{2}}{2})} =$

-1. Tangent line: $y+\frac{\sqrt{2}}{2} = -1(x+\frac{\sqrt{2}}{2})$, $x+y = -\sqrt{2}$, $r(\cos\theta+\sin\theta) = -\sqrt{2}$, $r = -\sqrt{2}/(\sin\theta + \cos\theta)$. The remaining cases are similar: $\theta = \frac{\pi}{4}$, $m = -1$; $\theta = -\frac{\pi}{4}$, $m = 1$; $\theta = \frac{3\pi}{4}$, $m = 1$. Graph $r = \sin 2\theta$, $r = \pm\sqrt{2}/(\sin\theta + \cos\theta)$, $r = \pm\sqrt{2}/(\sin\theta - \cos\theta)$, $0 \leqq \theta \leqq 2\pi$ in $[-1.9, 1.9]$ by $[-1.2, 1.2]$ and regard the x- and y-axes as tangent lines also.

16. $r = \cos 2\theta$; $\theta = 0, \pm\frac{\pi}{2}, \pi$ and θ for which $r = 0$. $\cos 2\theta = 0$ leads to $2\theta = (2n+1)\frac{\pi}{2}$, $\theta = (2n+1)\frac{\pi}{4}$. $\tan[(2n+1)\frac{\pi}{4}] = \pm 1$. Going into the origin in the first and third quadrant, the slope is 1 while in the second and fourth it is -1. Thus $y = \pm x$ or $\theta = \pm\frac{\pi}{4}$ are tangents at the origin. Slope $= \frac{r'\sin\theta + r\cos\theta}{r'\cos\theta - r\sin\theta} = \frac{-2\sin 2\theta\sin\theta + \cos 2\theta\cos\theta}{-2\sin 2\theta\cos\theta - \cos 2\theta\sin\theta}$. At $\theta = 0$, $r = 1$ and the slope is undefined. The tangent line is $x = 1$ or $r = \sec\theta$. At $\theta = \pm\frac{\pi}{2}$, $r = -1$, slope $= 0$ and the tangent lines are $y = \pm 1$ or $r = \pm\csc\theta$. At $\theta = \pi$, $r = 1$, the slope is undefined and $x = -1$ or $r = -\sec\theta$ is the tangent line. Graph $r = \cos 2\theta$, $r = \pm\sec\theta$, $r = \pm\csc\theta$, $0 \leqq \theta \leqq 2\pi$ in $[-3.4, 3.4]$ by $[-2, 2]$. Use the line-drawing feature of your grapher to draw in $y = \pm x$.

17. Graph $r = \pm 2\sqrt{\cos 2\theta}$, $-\frac{\pi}{4} \leqq \theta \leqq \frac{\pi}{4}$ in $[-2, 2]$ by $[-1.2, 1.2]$. Use θStep $=$ 0.01. Because of the $\pm$ sign, $\frac{\pi}{2}$ is a minimum range of θ, but it must be over an interval in which $\cos 2\theta$ is non-negative.

18. Graph $r = \pm 2\sqrt{\sin 2\theta}$, $0 \leqq \theta \leqq \frac{\pi}{2}$ in $[-3.4, 3.4]$ by $[-2, 2]$. $\frac{\pi}{2}$ is a minimum range of θ.

19. a) Graph $r = \frac{1}{2}+\cos\theta$, $0 \leqq \theta \leqq 2\pi$, θStep $= 0.1$ in $[-0.9, 2.3]$ by $[-0.94, 0.94]$.
b) Graph $r = 0.5 + \sin\theta$, $0 \leqq \theta \leqq 2\pi$ in $[-1.5, 1.5]$ by $[-0.2, 1.6]$.

20. a) Graph $r = 1 - \cos\theta$, $0 \leqq \theta \leqq 2\pi$ in $[-3.3, 1.55]$ by $[-1.43, 1.42]$.
b) Graph $r = -1 + \sin\theta$, $0 \leqq \theta \leqq 2\pi$ in $[-2.1, 2.1]$ by $[-0.5, 2]$.

21. a) Graph $r = 1.5 + \cos\theta$, $0 \leqq \theta \leqq 2\pi$ in $[-2, 4]$ by $[-1.8, 1.9]$. b) Graph $1.5 - \sin\theta$, $0 \leqq \theta \leqq 2\pi$ in $[-3.1, 3.1]$ by $[-2.7, 1]$.

22. a) Graph $r = 2 + \cos\theta$, $0 \leqq \theta \leqq 2\pi$ in $[-4.1, 6.1]$ by $[-3, 3]$. b) Graph $r = -2 + \sin\theta$, $0 \leqq \theta \leqq 2\pi$ in $[-4.2, 4.2]$ by $[-1.5, 3.5]$.

23. Graph $r = 2 - 2\cos\theta$, $0 \leqq \theta \leqq 2\pi$ in $[-6.5, 3.2]$ by $[-2.8, 2.8]$. The region consists of this closed curve (a cardioid) and every point inside it.

24. Graph $r_1 = \sqrt{\cos\theta}$, $r_2 = -\sqrt{\cos\theta}$, $-\frac{\pi}{2} \leqq \theta \leqq \frac{\pi}{2}$ in $[-1.7, 1.7]$ by $[-1, 1]$. The region consists of this closed curve and every point inside it.

25. $(2, 3\pi/4)$ also has coordinates $(-2, -\pi/4)$ and these coordinates satisfy the equation.

26. The point also has coordinates $(-\frac{1}{2}, \frac{\pi}{2})$ and these satisfy the equation.

27. We first solve the equations simultaneously: $r = 1 + \cos\theta = 1 - \cos\theta \Rightarrow \cos\theta = 0 \Rightarrow \theta = (2k+1)\frac{\pi}{2}$. We obtain the points $(1, \frac{\pi}{2})$, $(1, \frac{3\pi}{2})$. We graph both curves, $0 \leqq \theta \leqq 2\pi$ in $[-2.5, 2.5]$ by $[-1.5, 1.5]$ and find that the origin is the only other point of intersection.

28. $r = 1 + \sin\theta = 1 - \sin\theta \Rightarrow \sin\theta = 0 \Rightarrow \theta = n\pi$ and we obtain the points $(1, 0)$ and $(1, \pi)$. A graph of the two curves shows that the origin is the only other point of intersection.

29. $r^2 = (1 - \sin\theta)^2 = 4\sin\theta \Rightarrow 1 - 2\sin\theta + \sin^2\theta = 4\sin\theta$, $\sin^2\theta - 6\sin\theta + 1 = 0$, $\sin\theta = \frac{6\pm\sqrt{36-4}}{2} = 3 \pm 2\sqrt{2}$. Since $\sin\theta \leqq 1$, $\sin\theta = 3 - 2\sqrt{2}$, $\theta = \sin^{-1}(3-2\sqrt{2})$ or $\pi - \sin^{-1}(3-2\sqrt{2})$; $r = 1 - \sin\theta = 1 - (3 - 2\sqrt{2}) = 2(\sqrt{2}-1)$. Thus $(2(\sqrt{2}-1), \sin^{-1}(3-2\sqrt{2}))$ and $(2(\sqrt{2}-1), \pi - \sin^{-1}(3-2\sqrt{2}))$ are points of intersection. A careful study of the graphs shows that the origin and $(2, \frac{3\pi}{2})$ are the only other points of intersection.

30. $r^2 = \sqrt{2}\sin\theta = \sqrt{2}\cos\theta$, $\sin\theta = \cos\theta$, $\tan\theta = 1$, $\theta = \frac{\pi}{4}$ only since $r^2 \geqq 0$. When $\theta = \frac{\pi}{4}$, $r^2 = \sqrt{2}(\frac{1}{\sqrt{2}}) = 1$, $r = \pm 1$. Thus $(\pm 1, \frac{\pi}{4})$ are points of intersection. A study of the graphs $r = \pm\sqrt{\sqrt{2}\sin\theta}$, $r = \pm\sqrt{\sqrt{2}\cos\theta}$ shows that $(\pm 1, \frac{3\pi}{4})$ and the origin are the remaining points of intersection.

31. $r^2 = \sin 2\theta = \cos 2\theta$, $\tan 2\theta = 1$ and $\sin 2\theta > 0$. This yields the possibilities $2\theta = \frac{\pi}{4} + 2n\pi$, $\theta = \frac{\pi}{8} + n\pi$. This yields only the distinct points $(\pm 2^{-1/4}, \frac{\pi}{8})$. From the graph of the two curves we see that $(0, 0)$ is the only other point of intersection.

32. $r = 1 + \cos\frac{\theta}{2} = 1 - \sin\frac{\theta}{2}$ leads to $\tan\frac{\theta}{2} = -1$, $\frac{\theta}{2} = -\frac{\pi}{4} + n\pi$, $\theta = -\frac{\pi}{2} + 2n\pi$. From this with $n = 0, 1$, we get the points $(1 + \frac{1}{\sqrt{2}}, -\frac{\pi}{2})$ and $(1 - \frac{1}{\sqrt{2}}, \frac{3\pi}{2})$. The graph of the curves, $0 \leqq \theta \leqq 4\pi$ in $[-2, 2]$ by $[-2, 2]$ reveals that $(1 - \frac{1}{\sqrt{2}}, \frac{\pi}{2})$, $(1 + \frac{1}{\sqrt{2}}, \frac{\pi}{2})$ and the origin are the remaining points of intersection.

33. Graph $r = 1$, $r = 2\sin 2\theta$, $0 \leqq \theta \leqq 2\pi$ in $[-3.4, 3.4]$ by $[-2, 2]$. We see that there are two points of intersection in each quadrant. Four can be found using the above system and four can be found using $r = -1$, $r = 2\sin 2\theta$. $1 = 2\sin 2\theta$ leads to $\sin 2\theta = \frac{1}{2}$, $2\theta = \frac{\pi}{6} + 2n\pi$ or $\frac{5\pi}{6} + 2n\pi$, $\theta = \frac{\pi}{12} + n\pi$ and $\theta = \frac{5\pi}{12} + n\pi$. Similarly, $-1 = 2\sin 2\theta$ leads to $\theta = \frac{7\pi}{12} + n\pi$ and $\frac{11\pi}{12} + n\pi$. The points are: $(1, \frac{\pi}{12}), (1, \frac{5\pi}{12}), (-1, \frac{19\pi}{12}), (-1, \frac{23\pi}{12}), (1, \frac{13\pi}{12}), (1, \frac{17\pi}{12}), (-1, \frac{7\pi}{12}), (-1, \frac{11\pi}{12})$.

34. $r = 1$, $r^2 = \sin 2\theta$. $r^2 = \sin 2\theta = 1$ leads to $2\theta = \frac{\pi}{2} + 2n\pi$, $\theta = \frac{\pi}{4} + n\pi$. This yields the points $(1, \frac{\pi}{4}), (1, \frac{5\pi}{4})$. Graphing $r = \pm\sqrt{\sin 2\theta}$ and $r = 1$ together shows that these are the only points of intersection.

35. Graph $y = 5\sin\theta$, $0 \leqq \theta \leqq \pi$ in $[-5, 5]$ by $[-0.44, 5.44]$. Period π.

36. Graph $r = 5\cos\theta$, $0 \leqq \theta \leqq \pi$ in $[-3.1, 7.1]$ by $[-3, 3]$. Period of graph $= \pi$.

37. Graph $r = 5\sin 2\theta$, $0 \leqq \theta \leqq 2\pi$ in $[-8.5, 8.5]$ by $[-5, 5]$. Period 2π.

38. Graph $r = 5\cos 2\theta$, $0 \leqq \theta \leqq 2\pi$ in $[-8.5, 8.5]$ by $[-5, 5]$. Period $= 2\pi$.

39. Graph $r = 5\sin 5\theta$, $0 \leqq \theta \leqq \pi$ in $[-8.5, 8.5]$ by $[-5, 5]$. Period π.

40. Graph $r = 5\cos 5\theta$, $0 \leqq \theta \leqq \pi$ in $[-8.5, 8.5]$ by $[-5, 5]$. Period $= \pi$.

41. Graph $r = 5\sin(2.5\theta)$, $0 \leqq \theta \leqq 4\pi$ in $[-8.5, 8.5]$ by $[-5, 5]$. Period 4π.

42. Graph $r = 5\cos 2.5\theta$, $0 \leqq \theta \leqq 4\pi$ in $[-8.5, 8.5]$ by $[-5, 5]$. Period $= 4\pi$.

43. Graph $r = 5\sin 1.5\theta$, $0 \leqq \theta \leqq 4\pi$ in $[-8.5, 8.5]$ by $[-5, 5]$. Period 4π.

44. Graph $r = 5\cos 1.5\theta$, $0 \leqq \theta \leqq 4\pi$ in $[-8.5, 8.5]$ by $[-5, 5]$. Period $= 4\pi$.

45. Graph $r = 1 - 2\sin 3\theta$, $0 \leqq \theta \leqq 2\pi$ in $[-4.7, 4.7]$ by $[-2.1, 3.4]$. Period 2π.

46. Graph $r = 1 + 2\sin\frac{\theta}{2}$, $0 \leqq \theta \leqq 4\pi$ in $[-5.1, 3.7]$ by $[-2.6, 2.6]$. Period $= 4\pi$. Kidney.

47. a) Graph $r = e^{\theta/10}$, $-20 \leqq \theta \leqq 10$ in $[-4.3, 3.6]$ by $[-2, 2.7]$. b) Graph $r = 8/\theta$, $-20 \leqq \theta \leqq 20$ in the same window. c) Graph $r = 10/\sqrt{\theta}$ and $r = -10/\sqrt{\theta}$, $0 \leqq \theta \leqq 200$, θStep $= 0.5$ in the same window.

48. Suppose first the graph is symmetric with respect to the x- and y-axes and let (r, θ) be on the graph. Since the graph is symmetric with respect to the x-axis, $(r, -\theta)$ must be on the graph. Since $(r, -\theta)$ is on the graph and the graph is symmetric with respect to the y-axis, $(r, \pi - (-\theta)) = (r, \theta + \pi)$ is on the graph. So if (r, θ) is on the graph, $(r, \theta + \pi)$ must be on the graph. Therefore the graph is symmetric with respect to the origin. The proofs for the remaining two cases are similar.

49. Infinite period

50. Here is a possible conjecture for a special case: if m and n are odd with no common factor, $r = 5\sin(\frac{m}{n}\theta)$ is a rose with m overlapping petals having period $n\pi$.

51. Graph $r = 1.75 + (0.06/2\pi)\theta$, $0 \leqq \theta \leqq 10\pi$ in $[-3, 3]$ by $[-3, 3]$.

52. $r = r_0 + (\frac{\theta}{2\pi})b$, replacing α by θ. $\frac{dr}{d\theta} = \frac{b}{2\pi}$. $S = \int_0^\alpha \sqrt{r^2 + (\frac{dr}{d\theta})^2 d\theta} = \int_0^\alpha \sqrt{r^2 + (\frac{b}{2\pi})^2 d\theta}$.

53. $r = r_0 + (\frac{\theta}{2\pi})b = 1.75 + (0.06/2\pi)\theta$. $\frac{dr}{d\theta} = 0.06/2\pi$. $S = \int_0^{80\pi} \sqrt{(1.75 + (0.06/2\pi)\theta)^2 + (0.06/2\pi)^2 d\theta} =$ $\text{NINT}(\sqrt{(1.75 + (0.06/2\pi)x)^2 + (0.06/2\pi)^2}, x, 0, 80\pi) = 741.420\text{cm}$.

54. a) Since b is very small in comparison with r, $(b/2\pi)^2$ is even smaller in comparison with r^2. Thus $S = \int_0^\alpha \sqrt{r^2 + (\frac{b}{2\pi})^2 d\theta} \approx \int_0^\alpha \sqrt{r^2 + 0 d\theta} = \int_0^\alpha r\, d\theta$. b) $S_a = \int_0^{80\pi} r\, d\theta = \int_0^{80\pi} [1.75 + (0.06/2\pi)\theta] d\theta = 741.416\text{cm}$, compared with $S = 741.420\text{cm}$.

55. a) We use the approximation S_a of Exercise 54. $S_a = \int_0^\alpha r\, d\theta = \int_0^\alpha [r_0 + (\theta/2\pi)b] d\theta = r_0\alpha + (b/4\pi)\alpha^2$. Letting $\alpha = 2\pi n$ in this equation, we can arrive at the equation $bn^2 + 4\pi n - 2S_a = 0$. Using the appropriate solution from the quadratic formula, we obtain $n = (-2\pi + \sqrt{4\pi^2 + 2bS_a})/b$. b) The speed of the take-up reel steadily decreases.

56. The counter value must be a decreasing function of time.

58. a) $r = f(\theta - \alpha)$ b) Graph $r = 1 - \cos\theta$, $r = 1 - \cos(\theta - \pi/6)$, $r = 1 - \cos(\theta - \pi/2)$ and $r = 1 - \cos(\theta - 2\pi/3)$, $0 \leqq \theta \leqq 2\pi$ in $[-3.4, 3.4]$ by $[-2, 2]$.

10.7 POLAR EQUATIONS OF CONIC SECTIONS

1. $r\cos(\theta - \theta_0) = r_0$ becomes $r\cos(\theta - \frac{\pi}{6}) = 5$. $5 = r[\cos\theta\cos\frac{\pi}{6} + \sin\theta\sin\frac{\pi}{6}] = \frac{\sqrt{3}}{2}(r\cos\theta) + \frac{1}{2}(r\sin\theta)$, $5 = \frac{\sqrt{3}}{2}x + \frac{1}{2}y$, $\sqrt{3}x + y = 10$.

2. $r\cos(\theta - \theta_0) = r_0$ becomes $r\cos(\theta - \frac{3\pi}{4}) = 2$. $2 = r[\cos\theta\cos\frac{3\pi}{4} + \sin\theta\sin\frac{3\pi}{4}] = -\frac{\sqrt{2}}{2}x + \frac{\sqrt{2}}{2}y$, $y = x + 2\sqrt{2}$.

3. Graph $r = \frac{\sqrt{2}}{\cos(\theta - \frac{\pi}{4})}$, $0 \leqq \theta \leqq \pi$ in $[-5.8, 7.8]$ by $[-3, 5]$. $r[\cos\theta\cos\frac{\pi}{4} + \sin\theta\sin\frac{\pi}{4}] = \sqrt{2}$ leads to $x + y = 2$.

4. Graph $r = \frac{3}{\cos(\theta - \frac{2\pi}{3})}$, $0 \leqq \theta \leqq \pi$ in $[-8.5, 8.5]$ by $[-5, 5]$. $r[\cos\theta\cos\frac{2\pi}{3} + \sin\theta\sin\frac{2\pi}{3}] = 3$ leads to $y = \frac{\sqrt{3}}{3}x + 2\sqrt{3}$.

5. $r = 2a\cos\theta = 8\cos\theta$

6. $r = 2a\sin\theta = 2\sqrt{2}\sin\theta$

7. $r = 4\cos\theta = 2(2)\cos\theta$. Center: $(2,0)$, radius $= 2$. Check your sketch by graphing for $0 \leqq \theta \leqq \pi$ in $[-1.4, 5.4]$ by $[-2, 2]$.

8. $r = 6\sin\theta = 2(3)\sin\theta$. Center: $(3, \frac{\pi}{2})$, radius 3. Check your sketch by graphing $r = 6\sin\theta$, $0 \leqq \theta \leqq \pi$ in $[-5.1, 5.1]$ by $[0, 6]$.

9. $r = \frac{ke}{1+e\cos\theta} = \frac{2}{1+\cos\theta}$

10. $r = \frac{ke}{1+e\sin\theta} = \frac{2}{1+\sin\theta}$

11. $r = \frac{ke}{1+e\cos\theta} = \frac{8}{1+2\cos\theta}$

12. $r = \frac{ke}{1-e\sin\theta} = \frac{30}{1-5\sin\theta}$

13. $r = \frac{ke}{1+e\cos\theta} = \frac{\frac{1}{2}}{1+\frac{1}{2}\cos\theta} = \frac{1}{2+\cos\theta}$

14. $r = \frac{2(1/4)}{1-(1/4)\cos\theta} = \frac{2}{4-\cos\theta}$

15. $r = \frac{ke}{1-e\sin\theta} = \frac{10(\frac{1}{5})}{1-(\frac{1}{5})\sin\theta} = \frac{10}{5-\sin\theta}$

16. $r = \frac{ke}{1+e\sin\theta} = \frac{2}{1+(\frac{1}{3})\sin\theta} = \frac{6}{3+\sin\theta}$

17. $r = \frac{1}{1+\cos\theta}$. Directrix: $x = 1$, vertex: $(\frac{1}{2}, 0)$. Graph for $-\pi \leqq \theta \leqq \pi$ in $[-7.4, 2.8]$ by $[-3, 3]$. Include the directrix $r = 1/\cos\theta$.

18. $r = \frac{6}{2+\cos\theta} = \frac{3}{1+(1/2)\cos\theta}$. $e = 1/2$, $ke = 3$, $k = 6$. Directrix: $x = 6$. Let $P(v, 0)$, $v > 0$, be the vertex concerned. $PF = ePD \Rightarrow v = (1/2)(6-v)$, $v = 2$. (See Table 10.4-1). Let $Q(w, \pi)$ be the other vertex. $QF = eQD$ becomes $w = \frac{1}{2}(w+6)$, $w = 6$. On the x-axis the vertices are located at $x = -6$, $x = 2$ and so the center is at $x = (-6+2)/2 = -2$ and has polar coordinates $(2, \pi)$. Graph $r = \frac{6}{2+\cos\theta}$, $0 \leqq \theta \leqq 2\pi$ in $[-8.5, 8.5]$ by $[-5, 5]$. Also include the directrix $r = 6/\cos\theta$.

19. $r = \frac{25}{10-5\cos\theta} = \frac{5}{2-\cos\theta} = \frac{5(\frac{1}{2})}{1-(\frac{1}{2})\cos\theta}$. $e = \frac{1}{2}$, directrix: $x = -5$. Let $P(u, \pi)$, $u > 0$, be the vertex concerned. $PF = ePD \Rightarrow u = \frac{1}{2}(5-u) \Rightarrow u = \frac{5}{3}$, vertex: $(\frac{5}{3}, \pi)$. $1 - e^2 = \frac{3}{4}$. $5(\frac{1}{2}) = ke = a(1-e^2) = \frac{3a}{4}$, $a = \frac{10}{3}$, center at $x = ea = \frac{5}{3}$, vertices at $x = \frac{5}{3} \pm \frac{10}{3}$, i.e., at $x = -\frac{5}{3}$ and 5. Vertices: $(\frac{5}{3}, \pi)$ and $(5, 0)$. Center at $(\frac{5}{3}, 0)$. Graph $r = \frac{5}{2-\cos\theta}$, $-\pi \leqq \theta \leqq \pi$ in $[-8.5, 8.5]$ by $[-5, 5]$. Include the directrix $r = -5\sec\theta$.

20. $r = \frac{4}{2-2\cos\theta} = \frac{2(1)}{1-\cos\theta}$. Directrix: $x = -2$, vertex: $(-1, 0)$. Graph $r_1 = 2/(1-\cos\theta)$, $r_2 = -2/\cos\theta$, $-\pi \leqq \theta \leqq \pi$ in $[-8.5, 8.5]$ by $[-5, 5]$.

21. $r = \frac{400}{16+8\sin\theta} = \frac{25}{1+\frac{1}{2}\sin\theta} = \frac{50(\frac{1}{2})}{1+\frac{1}{2}\sin\theta}$. $e = \frac{1}{2}$, directrix $y = 50$. Let $P(v, \frac{\pi}{2})$ be the corresponding vertex. $PF = ePD$, $v = (\frac{1}{2})(50 - v)$, $2v = 50 - v$, $v = \frac{50}{3}$. (See Table 10.4-3). Let $Q(w, \frac{\pi}{2})$ be the other vertex. $QF = eQD$ becomes $-w = \frac{1}{2}(-w + 50)$, $-2w = -w + 50$, $w = -50$. If $(r, \frac{\pi}{2})$ is the center, it is midway between the vertices: $r = \frac{1}{2}(\frac{50}{3} - 50) = -\frac{50}{3}$. Graph $r = 25/(1 + 0.5\sin\theta)$, $0 \leqq \theta \leqq 2\pi$ in $[-93, 93]$ by $[-50, 60]$. Also include $r = 50/\sin\theta$.

22. $r = \frac{12}{3+3\sin\theta} = \frac{4(1)}{1+\sin\theta}$. Directrix: $y = 4$, vertex: $(2, \frac{\pi}{2})$. Check your sketch by graphing $r_1 = \frac{4}{1+\sin\theta}$ and $r_2 = 4/\sin\theta$, $-\pi \leqq \theta \leqq \pi$ in $[-8.5, 8.5]$ by $[-5, 5]$.

23. $r = \frac{8}{2-2\sin\theta} = \frac{4\cdot 1}{1-(1)\sin\theta}$. $e = 1$, directrix $y = -4$. Graph $r = \frac{4}{1-\sin\theta}$ and $r = \frac{-4}{\sin\theta}$, $0 \leqq \theta \leqq 2\pi$ in $[-13, 13]$ by $[-5, 12]$. Vertex: $(2, \frac{3\pi}{2})$.

24. $r = \frac{4}{2-\sin\theta} = \frac{2}{1-(1/2)\sin\theta} = \frac{4(1/2)}{1-(1/2)\sin\theta}$. $e = \frac{1}{2}$, $k = 4$. Directrix: $y = -4$. Let $P(v, \frac{3\pi}{2})$ be the vertex closer to the focus at the origin. $PF = ePD \Rightarrow v = (\frac{1}{2})(4 - v)$, $v = \frac{4}{3}$. $ke = a(1 - e^2)$ leads to $2 = a(1 - \frac{1}{4})$, $a = \frac{8}{3}$. Since the vertex is at $(\frac{4}{3}, \frac{3\pi}{2})$, the center is at $(-\frac{4}{3}, \frac{3\pi}{2})$ or $(\frac{4}{3}, \frac{\pi}{2})$. The higher vertex is at $(a + \frac{4}{3}, \frac{\pi}{2}) = (4, \frac{\pi}{2})$. Graph $r_1 = \frac{4}{2-\sin\theta}$, $r_2 = -4/\sin\theta$, $-\pi \leqq \theta \leqq \pi$ in $[-8.5, 8.5]$ by $[-5, 5]$.

25. Graph the ellipses sequentially in $[-2, 1.4]$ by $[-1, 1]$, $0 \leqq \theta \leqq 2\pi$, θStep = 0.1. The last two require a larger rectangle. As e increases, the center moves to the left, the ellipse increases in size. The ellipse also flattens out horizontally as can be seen by graphing in $[-11.3, 2.3]$ by $[-4, 4]$.

26. Graph r, $0 \leqq \theta \leqq 2\pi$ in $[-8.7, 17]$ by $[-7.5, 7.5]$ in dot format. In each case the graph is a hyperbola with the x-axis being the focal axis. The point $(1, 0)$ lies between the two vertices, the left vertex starting out at $(0.524, 0)$. As e increases both vertices (and the center) converge to $(1, 0)$.

27. Graph these sequentially in $[-27, 27]$ by $[-16, 16]$. As k becomes more negative, the parabola opens up wider and wider to the right. As k becomes more and more positive, the parabola opens to the left wider and wider.

28. $0 \leqq \theta \leqq 2\pi$, $[-17, 17]$ by $[-10, 10]$ may be used.

29. Graph $r = 3\sec(\theta - \pi/3) = 3/\cos(\theta - \pi/3)$, $0 \leqq \theta \leqq 2\pi$ in $[-11, 19]$ by $[-6, 11.7]$. $x + \sqrt{3}y = 6$ in rectangular coordinates.

30. $r = 4\sin\theta$. (Circle) Graph $r = 4\sin\theta$, $0 \leqq \theta \leqq \pi$ in $[-3.4, 3.4]$ by $[0, 4]$.

31. Graph $r = 8/(4 + \cos\theta)$, $0 \leqq \theta \leqq 2\pi$ in $[-4.9, 3.9]$ by $[-2.6, 2.6]$.

32. Graph $r = 1/(1 - \sin\theta)$, $0 \leqq \theta \leqq 2\pi$ in $[-5.95, 5.95]$ by $[-2, 5]$.

33. Graph $r = 1/(1 + 2\sin\theta)$, $0 \leqq \theta \leqq 2\pi$ in $[-2.7, 2.7]$ by $[-0.93, 2.27]$ in dot format, θStep $= 0.01$.

34. $r = 4\sec(\theta + \frac{\pi}{6})$. $r\cos(\theta + \frac{\pi}{6}) = 4$ leads to $y = \sqrt{3}x - 8$. Graph $r = 4\sec(\theta + \frac{\pi}{6})$, $-\frac{2\pi}{3} \leqq \theta \leqq \frac{\pi}{3}$ in $[-5, 10]$ by $[-12, 5]$.

35. Graph $r = -2\cos\theta$, $0 \leqq \theta \leqq \pi$ in $[-2.7, 0.7]$ by $[-1, 1]$.

36. Graph $8/(4 + \sin\theta)$, $0 \leqq \theta \leqq 2\pi$ in $[-4.1, 4.1]$ by $[-3, 2]$.

37. Graph $r = 1/(1 + \cos\theta)$, $0 \leqq \theta \leqq 2\pi$, θStep $= 0.1$ in $[-7.7, 5.9]$ by $[-4, 4]$.

38. Graph $r = 1/(1 + 2\cos\theta)$, $0 \leqq \theta \leqq 2\pi$ in $[-2.7, 4.04]$ by $[-4, 4]$. Use dot format.

39. Graph $r = 2\cos\theta$, $0 \leqq \theta \leqq \pi$ in $[-0.7, 2.7]$ by $[-1, 1]$. The region consists of the circle and all points within it.

40. Graph $r = -3\cos\theta$, $0 \leqq \theta \leqq \pi$ in $[-4.1, 1.1]$ by $[-1.5, 1.5]$. The region is the circle and all points inside the circle.

41. a) Without loss of generality we may assume that the major axis coincides with part of the x-axis and that a focus (the sun) is located at the origin. By Equation (15) an equation is $r = \frac{a(1-e^2)}{1+e\cos\theta}$. r is minimized when $\cos\theta = 1$, $\theta = 0$, $r = \frac{a(1-e^2)}{1+e} = a(1-e)$. r is maximized when $\cos\theta = -1$, $\theta = \pi$, $r = \frac{a(1-e^2)}{1-e} = a(1+e)$.

b)

Planet	$a(1-e)AU$	$a(1+e)AU$
Mercury	0.3075	0.4667
Venus	0.7184	0.7282
Earth	0.9833	1.017
Mars	1.382	1.666
Jupiter	4.951	5.455
Saturn	9.021	10.057
Uranus	18.30	20.06
Neptune	29.81	30.31
Pluto	29.65	49.23

42. a) We use the formula $r = \frac{a(1-e^2)}{1+e\cos\theta}$. For Mercury we have $r = \frac{0.3871(1-(0.0256)^2)}{1+0.2056\cos\theta} = \frac{0.3707}{1+0.2056\cos\theta}$. Here is a list of the results.

Mercury:	$r = \dfrac{0.3707}{1+0.2056\cos\theta}$	Jupiter:	$r = \dfrac{5.191}{1+0.0484\cos\theta}$
Venus:	$r = \dfrac{0.7233}{1+0.0068\cos\theta}$	Saturn:	$r = \dfrac{9.511}{1+0.0543\cos\theta}$
Earth:	$r = \dfrac{0.9997}{1+0.0617\cos\theta}$	Uranus:	$r = \dfrac{19.14}{1+0.0460\cos\theta}$
Mars:	$r = \dfrac{1.511}{1+0.0934\cos\theta}$	Neptune:	$r = \dfrac{30.06}{1+0.0082\cos\theta}$
		Pluto:	$r = \dfrac{37.012}{1+0.2481\cos\theta}$

Using $0 \leqq \theta \leqq 2\pi$, graph the first four in $[-2.72, 2.72]$ by $[-1.6, 1.6]$ and the second four in $[-52.7, 52.7]$ by $[-40, 40]$. If a viewing window includes the largest orbits, the smallest orbits can't be seen due to the limited size and number of pixels.

43. a) $r = 2\sin\theta$ leads to $r^2 = 2r\sin\theta$, $x^2+y^2 = 2y$, $x^2+y^2-2y+1 = 1$, $x^2+(y-1)^2 = 1$. $r = \csc\theta$ yields $r\sin\theta = 1$, $y = 1$. b) Graph $r = 2\sin\theta$, $r = 1/\sin\theta$, $0 \leqq \theta \leqq \pi$ in $[-1.7, 1.7]$ by $[0, 2]$. Label the points of intersection $(1,1)$, $(\sqrt{2}, \frac{\pi}{4})$ and $(-1,1)$, $(\sqrt{2}, \frac{3\pi}{4})$.

44. a) $r = 2\cos\theta$ leads to $r^2 = 2r\cos\theta$, $x^2+y^2 = 2x$, $x^2-2x+1+y^2 = 1$, $(x-1)^2+y^2 = 1$. $r = \sec\theta$ yields $r\cos\theta = 1$, $x = 1$. b) Graph $r = 2\cos\theta$, $r = 1/\cos\theta$, $0 \leqq \theta \leqq \pi$ in $[-2.9, 2.9]$ by $[-2, 2]$. Label the points of intersection $(1,1)$, $(\sqrt{2}, \frac{\pi}{4})$ and $(1,-1)$, $(-\sqrt{2}, \frac{3\pi}{4})$.

45. Use Fig. 10.69 with $r\cos\theta = 4$, i.e., $x = k = 4$. In the parabola $FP = PD$ or $r = 4 - FB = 4 - r\cos\theta$, $r + r\cos\theta = 4$, $r = \frac{4}{1+\cos\theta}$.

46. The directrix is the line $2 = r\cos(\theta - \pi/2) = r\cos((\pi/2) - \theta) = r\sin\theta = y$, $y = 2 = k$. Using $r = \frac{ke}{1+e\sin\theta}$, we get $r = \frac{2}{1+\sin\theta}$.

47. We restrict ourselves here to the case where a and b are both nonzero. Let L be the line with equation $ax + by = c$ or $y = -\frac{a}{b}x + \frac{c}{b}$. L has slope $-\frac{a}{b}$ so the line through the origin perpendicular to L has equation $y = \frac{b}{a}x$. Solving simultaneously, we find that the point of intersection of L and $y = \frac{b}{a}x$ is $P_0(\frac{ca}{a^2+b^2}, \frac{cb}{a^2+b^2})$. $r^2 = x^2 + y^2$ and $\theta = \tan^{-1}\frac{y}{x}$ but since we must have

$r_0 \geqq 0$, we find $r_0 = \frac{|c|}{\sqrt{a^2+b^2}}$ and $\theta_0 = \tan^{-1}\frac{b}{a}$ or $\theta_0 = \pi + \tan^{-1}\frac{b}{a}$. (Since $-\frac{\pi}{2} \leqq \tan^{-1}\frac{b}{a} \leqq \frac{\pi}{2}$, we must take $\theta_0 = \pi + \tan^{-1}\frac{b}{a}$ if P_0 is in the 2nd or 3rd quadrant.) $r = \frac{r_0}{\cos(\theta-\theta_0)}$ becomes $r = \frac{|c|}{\sqrt{a^2+b^2}\cos(\theta-\theta_0)}$ where θ_0 is as stated above.

48. $2x-y=-5$. From the x- any y-intercepts we can see that P_0 lies in the second quadrant. From Exercise 47 $r = \frac{5}{\sqrt{5}\cos(\theta-(\pi+\tan^{-1}(-1/2)))} = \frac{\sqrt{5}}{\cos(\theta-\pi+\tan^{-1}0.5)}$.

49. $3x+2y=6$. Here P_0 must be in the first quadrant. From Exercise 47, $r = \frac{6}{\sqrt{13}\cos(\theta-\tan^{-1}\frac{2}{3})}$.

50. $2x-y=4$. From the intercepts $(2,0), (0,-4)$ we see that P_0 is in quadrant 4. From Problem 47, $r = \frac{4}{\sqrt{5}\cos(\theta-\tan^{-1}(-1/2))} = \frac{4}{\sqrt{5}\cos(\theta+\tan^{-1}0.5)}$.

51. $4x+3y=-12$. From the intercepts $(-3,0)$ and $(0,-4)$ we see that P_0 lies in the 3rd quadrant. From Exercise 47, $r = \frac{12}{5\cos(\theta-(\pi+\tan^{-1}(3/4)))}$.

52.

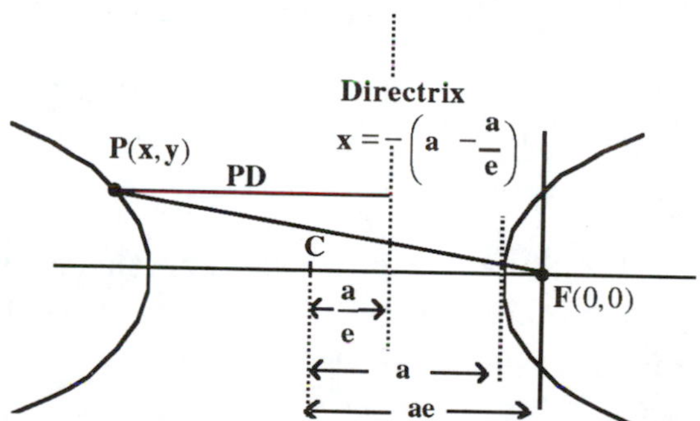

We have $PF = ePD$ or $r = e[-(ae-a/e)-x]$, $r = -ae^2+a-re\cos\theta$, $r(1+e\cos\theta) = a(1-e^2)$, $r = \frac{a(1-e^2)}{1+e\cos\theta}$ which is the required equation. One may show that points on the right branch also are on the graph, but one must use coordinates for $P(r,\theta)$ in which r is negative. Then $PF = ePD$ becomes $-r = e(x+ae-a/e) = re\cos\theta + ae^2 - a$, $-r(1+e\cos\theta) = a(e^2-1)$, $r = \frac{a(1-e^2)}{1+e\cos\theta}$ as required.

53. The first equation "1." has been derived in the text. For "2.", the equation $PF = e\cdot PD$ becomes $r = e[x-(-k)]$ which leads to $r = e(r\cos\theta+k)$, $(1-e\cos\theta)r = ke$, $r = ke/(1-e\cos\theta)$ as required. For "3.", $PF = e\cdot PD$ becomes $r = e(k-y) = e(k-r\sin\theta)$. This leads to $r(1+e\sin\theta) = ke$ and $r = ke/(1+e\sin\theta)$. For "4.", $PF = e\cdot PD$ becomes $r = e[y-(-k)] = e(r\sin\theta+k)$. This leads to $r(1-e\sin\theta) = ke$, $r = ke/(1-e\sin\theta)$.

10.8 INTEGRATION IN POLAR COORDINATES

1. Graph $r = \cos\theta$, $0 \leqq \theta \leqq \frac{\pi}{4}$ in $[0,1]$ by $[0,0.5]$. Then draw a line segment connecting $(0,0)$ and the rectangular point $(\frac{1}{2},\frac{1}{2})$. $A = \frac{1}{2}\int_0^{\pi/4}\cos^2\theta\,d\theta = \frac{1}{2}\int_0^{\pi/4}(\frac{1+\cos 2\theta}{2})d\theta = \frac{1}{4}[\theta + \frac{\sin 2\theta}{2}]_0^{\pi/4} = \frac{1}{4}(\frac{\pi}{4}+\frac{1}{2}) = \frac{\pi+2}{16}$.

2. Graph $r = e^\theta$, $0 \leqq \theta \leqq \ln 25$ in $[-25,2]$ by $[-5.2,10.8]$. Then draw a line from the origin to the terminal point of the graph. $A = \frac{1}{2}\int_0^{\ln 25} e^{2\theta}d\theta = \frac{1}{4}e^{2\theta}]_0^{\ln 25} = \frac{1}{4}[e^{\ln 25^2} - 1] = \frac{624}{4} = 156$.

3. Graph $r = 4 + 2\cos\theta$, $0 \leqq \theta \leqq 2\pi$ in $[-8.2,12.2]$ by $[-6,6]$. $A = \frac{1}{2}\int_0^{2\pi}[2(2+\cos\theta)]^2 d\theta = 2\int_0^{2\pi}(4 + 4\cos\theta + \frac{1}{2} + \frac{\cos 2\theta}{2})d\theta = 2[\frac{9}{2}\theta + 4\sin\theta + \frac{\sin 2\theta}{4}]_0^{2\pi} = 18\pi$.

4. For one example take $a = 2$ and graph $r = 2(1+\cos\theta)$, $0 \leqq \theta \leqq 2\pi$ in $[-3.6,6.6]$ by $[-3,3]$. $A = \frac{a^2}{2}\int_0^{2\pi}(1+\cos\theta)^2 d\theta = \frac{a^2}{2}\int_0^{2\pi}(1 + 2\cos\theta + \cos^2\theta)d\theta = \frac{a^2}{2}[\theta + 2\sin\theta + \frac{1}{2}(\theta - \frac{\sin 2\theta}{2})]_0^{2\pi} = \frac{a^2}{2}[2\pi + 0 + \frac{2\pi}{2}] = \frac{3}{2}a^2\pi$.

5. Graph $r = \cos 2\theta$, $-\frac{\pi}{4} \leqq \theta \leqq \frac{\pi}{4}$ in $[-0.09,1.12]$ by $[-0.35,0.35]$ for one leaf. $A = \frac{1}{2}\int_{-\pi/4}^{\pi/4}\cos^2(2\theta)d\theta = \frac{1}{4}\int_{-\pi/4}^{\pi/4}(1+\cos 4\theta)d\theta = \frac{1}{4}[\theta + \frac{\sin 4\theta}{4}]_{-\pi/4}^{\pi/4} = \frac{\pi}{8}$.

6. For one example take $a = 2$ and graph $r = 4\sin\theta$, $0 \leqq \theta \leqq \pi$ in $[-3.4,3.4]$ by $[0,4]$. $A = \frac{1}{2}\int_0^{\pi}(2a\sin\theta)^2 d\theta = 2a^2\int_0^{\pi}\sin^2\theta\,d\theta = a^2\int_0^{\pi}(1-\cos 2\theta)d\theta = a^2[\theta - \frac{\sin 2\theta}{2}]_0^{\pi} = \pi a^2$.

7. For the purpose of graphing let $a = 2$. For the entire graph use $r = \sqrt{8\cos 2\theta}$, $0 \leqq \theta \leqq 2\pi$ in $[-3,3]$ by $[-1.7,1.7]$. $A = 4$(area in 1st quadrant) $= 4(\frac{1}{2})\int_0^{\pi/4} 2a^2\cos 2\theta\,d\theta = 4a^2\frac{\sin 2\theta}{2}]_0^{\pi/4} = 2a^2$.

8. Graph $r = 2\sqrt{\sin 2\theta}$, $0 \leqq \theta \leqq \frac{\pi}{2}$ in $[-0.7,2.7]$ by $[0,2]$. $A = \frac{1}{2}\int_0^{\pi/2} 4\sin 2\theta\,d\theta = 2[-\frac{\cos 2\theta}{2}]_0^{\pi/2} = 2$.

9. Graph $r_1 = \sqrt{2\sin 3\theta}$ and $r_2 = -r_1$, $0 \leqq \theta \leqq 2\pi$ in $[-2.5,2.5]$ by $[-1.5,1.5]$. Area $= 6$(one leaf) $= 3\int_0^{\pi/3}(2\sin 3\theta)d\theta = -6\frac{\cos 3\theta}{3}]_0^{\pi/3} = 4$.

10. Graph $r_1 = 2\sin\theta$, $r_2 = 2\cos\theta$, $0 \leqq \theta \leqq \pi$ in $[-2.05,3.05]$ by $[-1,2]$ to see the region inside both circles. $A = 2$(area up to $\theta = \frac{\pi}{4}$) $= (2)\frac{1}{2}\int_0^{\pi/4}(2\sin\theta)^2 d\theta = 2\int_0^{\pi/4}(1-\cos 2\theta)d\theta = 2[\theta - \frac{\sin 2\theta}{2}]_0^{\pi/4} = 2[\frac{\pi}{4} - \frac{1}{2}] = \frac{\pi}{2} - 1$.

11. Graph $r = 1$ and $r = 2\sin\theta$, $0 \leqq \theta \leqq 2\pi$ in $[-2.5,2.5]$ by $[-1,2]$. $2\sin\theta = 1$ yields $\sin\theta = \frac{1}{2}$, $\theta = \frac{\pi}{6}, \frac{5\pi}{6}$. $A =$ area of top circle−area of top part of top circle $= \pi(1)^2 - \frac{1}{2}\int_{\pi/6}^{5\pi/6}(4\sin^2\theta - 1)d\theta = \pi - \frac{1}{2}\int_{\pi/6}^{5\pi/6}(2 - 2\cos 2\theta - 1)d\theta = \pi - \frac{1}{2}[\theta - \sin 2\theta]_{\pi/6}^{5\pi/6} = \pi - \frac{1}{2}[\frac{5\pi}{6} - (-\frac{\sqrt{3}}{2}) - (\frac{\pi}{6} - \frac{\sqrt{3}}{2})] = \pi - \frac{\pi}{3} - \frac{\sqrt{3}}{2} = \frac{2\pi}{3} - \frac{\sqrt{3}}{2}$.

12. Graph $r_1 = 2$ and $r_2 = 2(1-\cos\theta)$, $-\frac{\pi}{2} \leqq \theta \leqq \frac{3\pi}{2}$ in $[-7.8, 5.8]$ by $[-4, 4]$. For $-\frac{\pi}{2} \leqq \theta \leqq \frac{\pi}{2}$ the region is bounded by $r = 2(1-\cos\theta)$ and for $\frac{\pi}{2} \leqq \theta \leqq \frac{3\pi}{2}$ the region is bounded by $r = 2$. $A = \frac{1}{2}\int_{-\pi/2}^{\pi/2} 4(1-\cos\theta)^2 d\theta +$ area of semi-circle $= 2\int_{-\pi/2}^{\pi/2}(1-2\cos\theta+\cos^2\theta)d\theta + \pi 2^2/2 = 2[\theta - 2\sin\theta]_{-\pi/2}^{\pi/2} + \int_{-\pi/2}^{\pi/2}(1+\cos 2\theta)d\theta + 2\pi = 5\pi - 8$.

13. Graph $r = 2(1+\cos\theta)$ and $r = 2(1-\cos\theta)$, $0 \leqq \theta \leqq 2\pi$ in $[-6.8, 6.8]$ by $[-4, 4]$. Since both curves are symmetric with respect to the x-axis and one is the reflection of the other through the y-axis, we need only take 4 times the area in the first quadrant. The latter area is determined by the second curve. $A = 4[\frac{1}{2}\int_0^{\pi/2}[2(1-\cos\theta)]^2 d\theta] = 8\int_0^{\pi/2}(1 - 2\cos\theta + \frac{1}{2} + \frac{\cos 2\theta}{2})d\theta = 8[\frac{3}{2}\theta - 2\sin\theta + \frac{\sin 2\theta}{4}]_0^{\pi/2} = 8[\frac{3\pi}{4} - 2] = 6\pi - 16$.

14. Graph $r_1 = (\sqrt{6})\sqrt{\cos 2\theta}$ and $r_2 = \sqrt{3}$, $0 \leqq \theta \leqq 2\pi$ in $[-2.7, 2.7]$ by $[-1.8, 1.8]$. $r^2 = 6\cos 2\theta = 3$, $\cos 2\theta = \frac{1}{2}$. One solution is $2\theta = \pi/3$, $\theta = \pi/6$. $A = 4$ (area in first quadrant) $= 4(\frac{1}{2})\int_0^{\pi/6}[6\cos 2\theta - 3]d\theta = 2[3\sin 2\theta - 3\theta]_0^{\pi/6} = 2(3\sqrt{3}/2 - \pi/2) = 3\sqrt{3} - \pi$.

15. Graph $r = 3a\cos\theta$ and $r = a(1+\cos\theta)$, with $a = 2$, $0 \leqq \theta \leqq 2\pi$ in $[-3, 8.3]$ by $[-3.3, 3.3]$. Points of intersection: $r = 3a\cos\theta = a(1+\cos\theta)$, $3\cos\theta = 1 + \cos\theta$, $\cos\theta = \frac{1}{2}$, $\theta = \pm\frac{\pi}{3}$. $A = \frac{1}{2}\int_{-\pi/3}^{\pi/3}[(3a\cos\theta)^2 - (a(1+\cos\theta))^2]d\theta = a^2\int_0^{\pi/3}[9\cos^2\theta - 1 - 2\cos\theta - \cos^2\theta]d\theta = a^2\int_0^{\pi/3}(4 + 4\cos 2\theta - 1 - 2\cos\theta)d\theta = a^2[3\theta + 2\sin 2\theta - 2\sin\theta]_0^{\pi/3} = a^2[\pi + \sqrt{3} - \sqrt{3}] = \pi a^2$.

16. Graph $r_1 = -2\cos\theta$ and $r_2 = 1$, $0 \leqq \theta \leqq 2\pi$ in $[-2.2, 1.2]$ by $[-1, 1]$. $r = -2\cos\theta = 1$, $\cos\theta = -\frac{1}{2}$ has as one solution $\theta = \frac{2\pi}{3}$. $A = 2$ (area top half) $= (2)[\frac{1}{2}\int_{2\pi/3}^{\pi} 4\cos^2\theta\, d\theta -$ (area of sector of circle)$] = 2[\int_{2\pi/3}^{\pi}(1 + \cos 2\theta)d\theta - \frac{1}{2}(1^2)(\pi - \frac{2\pi}{3})] = 2[(\theta + \frac{\sin 2\theta}{2})_{2\pi/3}^{\pi} - \frac{\pi}{6}] = \frac{\pi}{3} + \frac{\sqrt{3}}{2}$.

17. a) $A_1 = 2$ (area of top half) $= 2[\frac{1}{2}\int_0^{2\pi/3}(2\cos\theta + 1)^2 d\theta] = \int_0^{2\pi/3}(4\cos^2\theta + 4\cos\theta + 1)d\theta = \int_0^{2\pi/3}(2 + 2\cos 2\theta + 4\cos\theta + 1)d\theta = 3\theta + \sin 2\theta + 4\sin\theta]_0^{2\pi/3} = 2\pi - \frac{\sqrt{3}}{2} + 4(\frac{\sqrt{3}}{2}) = 2\pi + \frac{3\sqrt{3}}{2}$. b) $A_2 = A_1 - (\pi - \frac{3\sqrt{3}}{2}) = \pi + 3\sqrt{3}$.

18. Graph $r_1 = 6$ and $r_2 = 3(\sin\theta)^{-1}$, $0 \leqq \theta \leqq 2\pi$ in $[-10.2, 10.2]$ by $[-6, 6]$. $r = 6 = 3/\sin\theta$ leads to $\sin\theta = \frac{1}{2}$ and one solution is $\theta = \frac{\pi}{6}$. $A = 2$ (area right half) $= 2[$area of sector of circle $- \frac{1}{2}\int_{\pi/6}^{\pi/2} 9\csc^2\theta\, d\theta] = 2[\frac{1}{2}(6^2)(\frac{\pi}{2} - \frac{\pi}{6}) + \frac{9}{2}(\cot\theta)_{\pi/6}^{\pi/2}] = 12\pi - 9\sqrt{3}$.

19. Graph $r = \theta^2$, $0 \leqq \theta \leqq \sqrt{5}$ in $[-4.9, 2.5]$ by $[0, 4.4]$. $L = \int_0^{\sqrt{5}} \sqrt{\theta^4 + 4\theta^2}\,d\theta = \int_0^{\sqrt{5}} \theta\sqrt{\theta^2+4}\,d\theta$. Let $u = \theta^2 + 4$, $du = 2\theta\, d\theta$. $L = \frac{1}{2}\int_4^9 u^{1/2} du = \frac{1}{2}(\frac{2}{3})u^{3/2}]_4^9 = \frac{1}{3}(27-8) = \frac{19}{3}$.

20. Graph $r = e^\theta/\sqrt{2}$, $0 \leqq \theta \leqq \pi$ in $[-16.6, 1.57]$ by $[-3.4, 7.4]$. $L = \int_0^\pi \sqrt{e^{2\theta}/2 + e^{2\theta}/2}\,d\theta = \int_0^\pi e^\theta d\theta = e^\pi - 1$.

21. $r = \sec\theta = 1/\cos\theta \Rightarrow r\cos\theta = 1$, $x = 1$, $0 \leqq \theta \leqq \frac{\pi}{4}$. The initial point is $(1,0)$ and the terminal point is $(\sqrt{2}, \pi/4)$. In rectangular coordinates $(1,0)$ to $(1,1)$ on $x = 1$ which has length 1.

22. $r = \csc\theta$ or $r\sin\theta = 1$, $y = 1$. The curve is the line segment from $(2, \pi/6)$ to $(1, \pi/2)$ or, in rectangular coordinates from $(\sqrt{3}, 1)$ to $(0, 1)$ which has length $\sqrt{3}$.

23. Graph $r = 1 + \cos\theta$, $0 \leqq \theta \leqq 2\pi$ in $[-1.7, 3.4]$ by $[-1.5, 1.5]$. $r' = -\sin\theta$. $L = 2\int_0^\pi \sqrt{1 + 2\cos\theta + \cos^2\theta + \sin^2\theta}\,d\theta = 2\int_0^\pi \sqrt{2}\sqrt{1+\cos\theta}\,d\theta = 2\sqrt{2}\int_0^\pi \sqrt{(1+\cos\theta)\frac{(1-\cos\theta)}{1-\cos\theta}}\,d\theta = 2\sqrt{2}\int_0^\pi \frac{\sin\theta\, d\theta}{\sqrt{1-\cos\theta}}$. Let $u = 1 - \cos\theta$, $du = \sin\theta\, d\theta$. $L = 2\sqrt{2}\int_0^2 u^{-1/2} du = 2\sqrt{2}(2u^{1/2})]_0^2 = 8$.

24. a) $L = \int_0^{2\pi} \sqrt{a^2 + 0^2}\,d\theta = 2\pi a$ b) $L = \int_0^\pi \sqrt{a^2\cos^2\theta + a^2\sin^2\theta}\,d\theta = \int_0^\pi a\,d\theta = a\pi$. $r = a\cos\theta$ is the circle $(x - \frac{a}{2})^2 + y^2 = (\frac{a}{2})^2$ of radius $\frac{a}{2}$ so the result is consistent with the formula $C = 2\pi r$. c) The result for $r = a\sin\theta$, $x^2 + (y - \frac{a}{2})^2 = (\frac{a}{2})^2$ is obtained as in b). We assumed $a > 0$.

25. $A = \int_\alpha^\beta 2\pi x\, ds = 2\pi\int_0^{\pi/4} r\cos\theta\sqrt{\cos 2\theta + (\frac{-2\sin 2\theta}{2\sqrt{\cos 2\theta}})^2}\,d\theta = 2\pi\int_0^{\pi/4} \sqrt{\cos 2\theta}\cos\theta\sqrt{\cos 2\theta + \frac{\sin^2 2\theta}{\cos 2\theta}}\,\theta = 2\pi\int_0^{\pi/4} \cos\theta\sqrt{\cos^2 2\theta + \sin^2 2\theta}\,d\theta = 2\pi\int_0^{\pi/4} \cos\theta\, d\theta = 2\pi\sin\theta]_0^{\pi/4} = \sqrt{2}\pi$.

26. $r = (\sqrt{2})e^{\theta/2}$, $0 \leqq \theta \leqq \pi/2$ about the x-axis. $A = 2\pi\int_0^{\pi/2} y\sqrt{2e^\theta + \frac{1}{2}e^\theta}\,d\theta = 2\pi\int_0^{\pi/2} \sqrt{2}e^{\theta/2}(\sin\theta)\sqrt{\frac{5}{2}e^{\theta/2}}\,d\theta = 2\sqrt{5}\pi\int_0^{\pi/2} e^\theta\sin\theta\, d\theta =$ (integration by parts) $2\sqrt{5}\pi[\frac{e^\theta}{2}(\sin\theta - \cos\theta)]_0^{\pi/2} = 2\sqrt{5}\pi[\frac{e^{\pi/2}}{2} - \frac{1}{2}(-1)] = \sqrt{5}\pi(e^{\pi/2} + 1)$.

27. $r^2 = \cos 2\theta$, $2r\frac{dr}{d\theta} = -2\sin 2\theta$, $\frac{dr}{d\theta} = -\frac{\sin 2\theta}{r}$, $(\frac{dr}{d\theta})^2 = \frac{\sin^2 2\theta}{r^2} = \frac{\sin^2 2\theta}{\cos 2\theta}$. The desired area is twice that generated by the arc $r = \sqrt{\cos 2\theta}$, $0 \leqq \theta \leqq \frac{\pi}{4}$. $A = 4\pi\int_0^{\pi/4} r\sin\theta\sqrt{\cos 2\theta + \frac{\sin^2 2\theta}{\cos 2\theta}}\,d\theta = 4\pi\int_0^{\pi/4} \sin\theta\sqrt{\cos^2 2\theta + \sin^2 2\theta}\,d\theta = -4\pi\cos\theta]_0^{\pi/4} = -4\pi[\frac{\sqrt{2}}{2} - 1] = 2(2 - \sqrt{2})\pi$.

28. $r = 2a\cos\theta$ about the y-axis. We assume $a > 0$. $A = 2\pi\int_{-\pi/2}^{\pi/2} r\cos\theta\sqrt{4a^2\cos^2\theta + 4a^2\sin^2\theta}d\theta = 2\pi\int_{-\pi/2}^{\pi/2} 2a\cos^2\theta(2a)d\theta = 8a^2\pi(\frac{1}{2})\int_{-\pi/2}^{\pi/2}(1+\cos 2\theta)d\theta = 4\pi a^2[\theta + \frac{\sin 2\theta}{2}]_{-\pi/2}^{\pi/2} = 4\pi^2a^2$.

29. b) $y_1 \to \infty$ as $x \to 0^+$ c) The integral is improper because y_1 is not defined at the endpoint $x = 0$.

30. a) Graph $f(x)$ in $[0,5]$ by $[0,40]$. (This takes a long time.) $f(x) \to 39.356\ldots$ as $x \to 0^+$. b) $f(x)$ steadily decreases getting closer and closer to 0. c) $39.356\ldots$.

31. $A = a^2\int_0^{2\pi}(1+2\cos\theta+\cos^2\theta)d\theta = a^2\int_0^{2\pi}(1+2\cos\theta+\frac{1}{2}+\frac{\cos 2\theta}{2})d\theta = a^2[\frac{3\theta}{2} + 2\sin\theta + \frac{\sin 2\theta}{4}]_0^{2\pi} = 3\pi a^2$. $\bar{x} = \frac{1}{A}(\frac{2}{3})\int_0^{2\pi} r^3\cos\theta\, d\theta = \frac{2a^3}{3A}\int_0^{2\pi}(1+\cos\theta)^3\cos\theta\, d\theta$. We use NINT and obtain $\bar{x} = (0.8333\ldots)a$ ($5a/6$ converting the decimal answer to a fraction). Since the graph is symmetric with respect to the x-axis, $\bar{y} = 0$.

32. $0 \leqq r \leqq a$, $0 \leqq \theta \leqq \pi$. $\bar{x} = \frac{\frac{2}{3}\int_0^{\pi} a^3\cos\theta\, d\theta}{\int_0^{\pi} a^2 d\theta} = 0$. $\bar{y} = \frac{\frac{2}{3}\int_0^{\pi} a^3\sin\theta\, d\theta}{\int_0^{\pi} a^2 d\theta} = \frac{\frac{4a^3}{3}}{a^2\pi} = \frac{4a}{3\pi}$.

PRACTICE EXERCISES, CHAPTER 10

1. $x = \frac{y^2}{8}$, $y^2 = 4(2)x$. Focus: $(2,0)$, directrix: $x = -2$. Graph $y = \pm 2\sqrt{2x}$ in $[-11.8, 16.8]$ by $[-8.4, 8.4]$. Use the line-drawing feature to include $x = -2$.

2. $y = -x^2/4$, $x^2 = 4(-1)y$. Focus: $(0,-1)$, directrix: $y = 1$. Graph $y_1 = -x^2/4$ and $y_2 = 1$ in $[-3.73, 3.73]$ by $[-2.4, 2]$.

3. $16x^2 + 7y^2 = 112$, $\frac{x^2}{7} + \frac{y^2}{16} = 1$, vertices: $(0, \pm 4)$. $c = \sqrt{16-7} = 3$. Foci: $(0, \pm 3)$. $e = c/a = 3/4$. $a/e = 4(\frac{4}{3}) = \frac{16}{3}$. Directrices: $y = \pm\frac{16}{3}$. Graph $y = \pm 4\sqrt{1 - \frac{x^2}{7}}$, $y = \pm\frac{16}{3}$ in $[-10.2, 10.2]$ by $[-6, 6]$.

4. $x^2 + 2y^2 = 4$, $\frac{x^2}{4} + \frac{y^2}{2} = 1$. $a = 2$, $b = \sqrt{2}$. Vertices: $(\pm 2, 0)$. $c = \sqrt{4-2} = \sqrt{2}$. Foci: $(\pm\sqrt{2}, 0)$. $e = c/a = \sqrt{2}/2$. Directrices: $x = \pm a/e = \pm 2\sqrt{2}$. In polar mode graph $r_1 = 2/\sqrt{\cos^2\theta + 2\sin^2\theta}$, $r_2 = 2(\sqrt{2})(\cos\theta)^{-1}$, $r_3 = -r_2$, $0 \leqq \theta \leqq 2\pi$ in $[-3.4, 3.4]$ by $[-2, 2]$.

5. $3x^2 - y^2 = 3$, $x^2 - \frac{y^2}{3} = 1$. Vertices: $(\pm 1, 0)$. $c = \sqrt{1+3} = 2$. Foci: $(\pm 2, 0)$. $e = \frac{c}{a} = 2$, directrices: $x = \pm\frac{a}{e} = \pm\frac{1}{2}$. Graph $y = \pm\sqrt{3(x^2-1)}$ in $[-15, 15]$ by $[-10, 10]$.

6. $2y^2 - 8x^2 = 16$, $\frac{y^2}{8} - \frac{x^2}{2}$. Vertices: $(0, \pm 2\sqrt{2})$. $c = \sqrt{8+2} = \sqrt{10}$. Foci: $(0, \pm\sqrt{10})$. $e = \frac{c}{a} = \frac{\sqrt{10}}{\sqrt{8}} = \frac{\sqrt{5}}{2}$. Directrices: $y = \pm\frac{a}{e} = \pm\frac{a^2}{c} = \pm\frac{8}{\sqrt{10}} = \pm\frac{4\sqrt{10}}{5}$. Check your sketch by graphing $y = \pm 2\sqrt{x^2+2}$, $y = \pm\frac{4\sqrt{10}}{5}$ in $[-17, 17]$ by $[-10, 10]$.

7. $B^2 - 4AC = -3 < 0$ indicates ellipse but there is no solution and no graph. When we solve the quadratic for y, the discriminant is negative for all x.

8. $x^2+3xy+2y^2+x+y+1 = 0$. $B^2-4AC = 9-4(1)2 = 1 > 0$, hyperbola. $2y^2 + (3x+1)y + (x^2+x+1) = 0$. Graph $y = [-(3x+1) \pm \sqrt{(3x+1)^2 - 8(x^2+x+1)}]/4$ in $[-17, 17]$ by $[-10, 10]$.

9. $B^2 - 4AC = 0$, parabola. $4y^2 + (4x+1)y + (x^2+x+1) = 0$. Graph $y = \frac{-(4x+1)\pm\sqrt{-8x-15}}{8}$ in $[-13.5, 3.5]$ by $[0, 10]$.

10. $x^2 + 2xy - 2y^2 + x + y + 1 = 0$. $B^2 - 4AC = 4 - 4(1)(-2) = 12 > 0$, hyperbola. $-2y^2 + (2x+1)y + (x^2+x+1) = 0$. Graph $y = [-(2x+1) \pm \sqrt{(2x+1)^2 + 8(x^2+x+1)}]/(-4)$ in $[-17, 17]$ by $[-10, 10]$.

11. $2x^2 + xy + 2y^2 - 15 = 0$. $B^2 - 4AC = 1 - 16 = -15 < 0$, ellipse. Since $A = C$, $\theta = \frac{\pi}{4}$, $x' = \frac{\sqrt{2}}{2}(x-y)$, $y' = \frac{\sqrt{2}}{2}(x+y)$. Simplifying $2x'^2 + x'y' + 2y'^2 = 15$, we obtain $\frac{x^2}{6} + \frac{y^2}{10} = 1$. The original equation is $2y^2 + xy + (2x^2 - 15) = 0$. We graph $y = \pm\sqrt{10(1 - \frac{x^2}{6})}$ and $y = \frac{-x\pm\sqrt{x^2-8(2x^2-15)}}{4}$ in $[-7.5, 7.5]$ by $[-4, 5]$.

12. $x^2 + 2\sqrt{3}xy - y^2 = 4$. $B^2 - 4AC = 12 - 4(1)(-1) = 16 > 0$, hyperbola. $\theta = \frac{1}{2}\tan^{-1}(\frac{B}{A-C}) = \frac{1}{2}\tan^{-1}\sqrt{3} = \frac{\pi}{6}$. We substitute $x' = x\cos\frac{\pi}{6} - y\sin\frac{\pi}{6} = \frac{\sqrt{3}x-y}{2}$ and $y' = x\sin\frac{\pi}{6} + y\cos\frac{\pi}{6} = \frac{x+\sqrt{3}y}{2}$ into $x'^2 + 2\sqrt{3}x'y' - y'^2 = 4$, simplify and obtain $x^2 - y^2 = 2$. We apply the quadratic formula to $y^2 - 2\sqrt{3}xy + (4 - x^2) = 0$. Let $y_1 = \sqrt{12x^2 - 4(4 - x^2)} = 4\sqrt{x^2 - 1}$. Graph $y_2 = (2\sqrt{3}x + y_1)/2$, $y_3 = (2\sqrt{3}x - y_1)/2$, $y_4 = \sqrt{x^2 - 2}$, $y_5 = -y_4$ in $[-8.5, 8.5]$ by $[-4, 6]$.

13. Since $A = C (= 0)$, $\theta = \frac{\pi}{4}$. $x'y' = 2$ becomes $\frac{1}{\sqrt{2}}(x-y)\frac{1}{\sqrt{2}}(x+y) = \frac{x^2}{2} - \frac{y^2}{2} = 2$, $\frac{x^2}{4} - \frac{y^2}{4} = 1$. $e = \frac{c}{a} = \frac{\sqrt{4+4}}{2} = \sqrt{2}$.

14. $a = 2$. $e = c/a$, $c = ae = 2(2) = 4$. $b^2 = c^2 - a^2 = 16 - 4 = 12$. $\frac{x^2}{4} - \frac{y^2}{12} = 1$.

15. a) $V = \pi\int_{-2}^{2} y^2 dx = 2\pi\int_0^2 \frac{36-9x^2}{4} dx = \frac{\pi}{2}[36x - 3x^3]_0^2 = \frac{\pi}{2}(72 - 24) = 24\pi$.
b) $V = \pi\int_{-3}^{3} x^2 dy = 2\pi\int_0^3 \frac{36-4y^2}{9} dy = \frac{2\pi}{9}[36y - \frac{4}{3}y^3]_0^3 = 16\pi$.

16. $9x^2 - 4y^2 = 36$, $\frac{x^2}{4} - \frac{y^2}{9} = 1$. Vertices: $(\pm 2, 0)$. $V = \pi \int_2^4 y^2 dx = \frac{9\pi}{4} \int_2^4 (x^2 - 4)dx = 24\pi$.

17. We may minimize the distance squared: $d^2 = (x-0)^2 + (y-3)^2 = 4t^2 + (t^2-3)^2 = t^4 - 2t^2 + 9 = (t^2-1)^2 + 8$. We see that this is minimal when $t = \pm 1$ which corresponds to the points $(\pm 2, 1)$.

18. $A = 2\int_0^{b/2}(h - \frac{4h}{b^2}x^2)dx = 2[hx - \frac{4h}{3b^2}x^3]_0^{b/2} = 2[\frac{bh}{2} - \frac{4bh}{(3)8}] = \frac{2}{3}bh$.

19. $x = t/2$, $y = t+1$, $-\infty < t < \infty$. $t = 2x = y - 1$ or $y = 2x + 1$. The entire line is traced out from left to right.

20. $x = \sqrt{t}$, $y = 1 - \sqrt{t}$; $t \geqq 0$. $y = 1 - x$, $x \geqq 0$. The half-line is traced out from left to right.

21. $x = (1/2)\tan t$, $y = (1/2)\sec t$; $-\pi/2 < t < \pi/2$. $y^2 = \frac{\sec^2 t}{4} = \frac{1+\tan^2 t}{4} = \frac{1}{4} + (\frac{\tan t}{2})^2 = \frac{1}{4} + x^2$, $y^2 - x^2 = \frac{1}{4}$, $\frac{y^2}{(0.5)^2} - \frac{x^2}{(0.5)^2} = 1$, hyperbola. The graph is the upper branch traced out from left to right. Confirm in parametric mode in $[-25.5, 25.5]$ by $[-10, 20]$.

22. $x = -2\cos t$, $y = 2\sin t$; $0 \leqq t \leqq \pi$. $x^2 + y^2 = 4$, the semi-circle of radius 2 with center at the origin is traced out in the clockwise direction from $(-2, 0)$ to $(2, 0)$.

23. $x = -\cos t$, $y = \cos^2 t$, $0 \leqq t \leqq \pi$. $y = x^2$, $-1 \leqq x \leqq 1$. The portion of the parabola $y = x^2$ determined by $-1 \leqq x \leqq 1$ is traced out from left to right.

24. $x = 4\cos t$, $y = 9\sin t$; $0 \leqq t \leqq 2\pi$. $(\frac{x}{4})^2 + (\frac{y}{9})^2 = \cos^2 t + \sin^2 t = 1$. The ellipse $\frac{x^2}{16} + \frac{y^2}{81} = 1$ is traced out in the counterclockwise direction.

25. $16x^2 + 9y^2 = 144$, $\frac{x^2}{9} + \frac{y^2}{16} = 1$. Let $x = 3\cos t$, $y = 4\sin t$, $0 \leqq t \leqq 2\pi$.

26. $x = 2\cos t$, $y = 2\sin t$, $0 \leqq t \leqq 2\pi$ traces out the circle once counterclockwise. We change the direction if we replace t by $-t$: $x = 2\cos t$, $y = -2\sin t$. If $t = \pi$, $(x, y) = (-2, 0)$. Thus a possible answer is $x = 2\cos t$, $y = -2\sin t$, $\pi \leqq t \leqq \pi + 3(2\pi) = 7\pi$.

27. $x = (1/2)\tan t$, $y = (1/2)\sec t$; $t = \pi/3$. $\frac{dy}{dx} = \frac{dy/dt}{dx/dt} = \frac{\sec t \tan t}{\sec^2 t} = \frac{\tan t}{\sec t} = \sin t$. $\frac{dy}{dx}]_{t=\pi/3} = \sqrt{3}/2$. Tangent line: $y - 1 = (\sqrt{3}/2)(x - \sqrt{3}/2)$. $d^2y/dx^2 = (dy'/dt)/(dx/dt) = \cos t/((1/2)\sec^2 t) = 2\cos^3 t$. $[d^2y/dx^2]_{t=\pi/3} = 1/4$.

28. $x = 1 + \frac{1}{t^2}$, $y = 1 - \frac{3}{t}$; $t = 2$. $\frac{dy}{dx} = \frac{3t^{-2}}{-2t^{-3}} = -\frac{3}{2}t$. $\frac{dy}{dx}|_{t=2} = -3$. Tangent line: $y + \frac{1}{2} = -3(x - \frac{5}{4})$. $\frac{d^2y}{dx^2} = \frac{-3/2}{-2t^{-3}} = \frac{3}{4}t^2$. $\frac{d^2y}{dx^2}|_{t=2} = 6$.

29. Graph $x = e^{2t} - t/8$, $y = e^t$; $0 \leqq t \leqq \ln 2$ in $[-0.16, 4.4]$ by $[-0.26, 2.5]$. $L = \int_0^{\ln 2} \sqrt{(dx/dt)^2 + (dy/dt)^2}dt = \int_0^{\ln 2} \sqrt{(2e^{2t} - 1/8)^2 + e^{2t}}dt = \int_0^{\ln 2} \sqrt{(2e^{2t} + 1/8)^2}dt = \int_0^{\ln 2}(2e^{2t} + 1/8)dt = e^{2t} + t/8]_0^{\ln 2} = 4 + \ln 2/8 - 1 = 3 + \ln 2/8$.

30. Graph $x = t^2$, $y = \frac{t^3}{3} - t$, $-2.5 \leqq t \leqq 2.5$ in $[-1, 5]$ by $[-1.8, 1.8]$. $y = \frac{t^3}{3} - t = t(\frac{t^2-3}{3}) = 0$ when $t = 0, \pm\sqrt{3}$. The loop is traced out from $t = -\sqrt{3}$ to $t = \sqrt{3}$. $L = \int_{-\sqrt{3}}^{\sqrt{3}} \sqrt{(2t)^2 + (t^2 - 1)^2}dt = \int_{-\sqrt{3}}^{\sqrt{3}} \sqrt{(t^2 + 1)^2}dt = \int_{-\sqrt{3}}^{\sqrt{3}}(t^2 + 1)dt = 4\sqrt{3}$.

31. $A = 2\pi \int y\, ds = 2\pi \int_0^{\sqrt{5}} 2t\sqrt{t^2 + 2^2}dt$. Let $u = t^2 + 4$, $du = 2tdt$. $A = 2\pi \int_4^9 u^{1/2}du = 2\pi(\frac{2}{3})u^{3/2}]_4^9 = (4\pi/3)(27 - 8) = 76\pi/3$.

32. $x = t^2 + \frac{1}{2t}$, $y = 4\sqrt{t}$, $\frac{1}{\sqrt{2}} \leqq t \leqq 1$; y-axis. $A = 2\pi \int x ds = 2\pi \int_{1/\sqrt{2}}^1 (t^2 + \frac{1}{2t})\sqrt{(2t - \frac{1}{2t^2})^2 + (\frac{2}{\sqrt{t}})^2}dt = 2\pi \int_{1/\sqrt{2}}^1 (t^2 + \frac{1}{2t})\sqrt{(2t + \frac{1}{2t^2})^2}dt = 2\pi \int_{1/\sqrt{2}}^1 (t^2 + \frac{1}{2t})(2t + \frac{1}{2t^2})dt = 2\pi \int_{1/\sqrt{2}}^1 (2t^3 + \frac{3}{2} + t^{-3}/4)dt = 2\pi(2 - \frac{3\sqrt{2}}{4})$.

33. Graph $r = \cos 2\theta$, $0 \leqq \theta \leqq 2\pi$ in $[-1.7, 1.7]$ by $[-1, 1]$. Period $= 2\pi$. Four-leaved rose.

34. $r \cos\theta = 1$. This is the vertical line $x = 1$. $r = \sec\theta$ has period π.

35. Graph $r = 6/(1 - 2\cos\theta)$, $0 \leqq \theta \leqq 2\pi$ in $[-30, 22]$ by $[-13, 18]$ in dot format. Period $= 2\pi$. Hyperbola.

36. Four-leaved rose. Period $= 2\pi$. Graph $r = \sin 2\theta$, $0 \leqq \theta \leqq 2\pi$ in $[-1.7, 1.7]$ by $[-1, 1]$.

37. Graph $r = \theta$, $-4\pi \leqq \theta \leqq 4\pi$ in $[-18.6, 18.6]$ by $[-12.3, 9.7]$. Infinite period. Spiral.

38. $r^2 = \cos 2\theta$. Lemniscate. The period is 2π if we use $r = \sqrt{\cos 2\theta}$; it is π if we use $r = \pm\sqrt{\cos 2\theta}$. Graph $r = \sqrt{\cos 2\pi}$, $0 \leqq \theta \leqq 2\pi$ in $[-1.7, 1.7]$ by $[-1, 1]$.

39. Graph $r = 1 + \cos\theta$, $0 \leqq \theta \leqq 2\pi$ in $[-1.6, 3.2]$ by $[-1.4, 1.4]$. Period 2π. Cardioid.

40. Cardioid. Period $= 2\pi$. Graph $r = 1 - \sin\theta$, $0 \leqq \theta \leqq 2\pi$ in $[-2.5, 2.5]$ in $[-2.35, 0.65]$.

41. Graph $r = 2/(1 - \cos\theta)$, $0 \leqq \theta \leqq 2\pi$ in $[-10, 17]$ by $[-8, 8]$. Period 2π. Parabola.

42. Lemniscate. The period is $\frac{3\pi}{2}$ if $r = \sqrt{\sin 2\theta}$ is used; it is $\frac{\pi}{2}$ if $r = \pm\sqrt{\sin 2\theta}$ is used. Graph $r = \sqrt{\sin 2\theta}$, $0 \leqq \theta \leqq 2\pi$ in $[-1.7, 1.7]$ by $[-1, 1]$.

43. Graph $r = -\sin\theta$, $0 \leqq \theta \leqq 2\pi$ in $[-1, 1]$ by $[-1.1, 0.1]$. Period π. Circle.

44. Limaçon. Period $= 2\pi$. Graph $r = 2\cos\theta + 1$, $0 \leqq \theta \leqq 2\pi$ in $[-2, 4.8]$ by $[-2, 2]$.

45. $2\sqrt{3} = r\cos(\theta - \frac{\pi}{3}) = r(\cos\theta\cos\frac{\pi}{3} + \sin\theta\sin\frac{\pi}{3}) = \frac{1}{2}x + \frac{\sqrt{3}}{2}y$, $x + \sqrt{3}y = 4\sqrt{3}$, $\sqrt{3}y = -x + 4\sqrt{3}$, $y = -\frac{1}{3}x + 4$. Graph $r = 2\sqrt{3}/\cos(\theta - \frac{\pi}{3})$, $0 \leqq \theta \leqq 2\pi$ in $[-17, 17]$ by $[-5, 15]$ and $y = -\frac{1}{\sqrt{3}}x + 4$ in the same window.

46. $\frac{\sqrt{2}}{2} = r\cos(\theta - \frac{3\pi}{4}) = (r\cos\theta)\cos\frac{3\pi}{4} + (r\sin\theta)\sin\frac{3\pi}{4} = x(-\frac{1}{\sqrt{2}}) + y(\frac{1}{\sqrt{2}})$, $y = x + 1$. Graph $r = \sqrt{2}/(2\cos(\theta - \frac{3\pi}{4}))$, $0 \leqq \theta \leqq \pi$ in $[-5.5, 6.5]$ by $[-2, 5]$.

47. $r = 2\sin\theta$, $r^2 = 2r\sin\theta$, $x^2 + y^2 = 2y$, $x^2 + y^2 - 2y + 1 = 1$, $x^2 + (y-1)^2 = 1$. Center $(0, 1)$, radius $= 1$.

48. $r = -4\cos\theta$, $r^2 = -4r\cos\theta$, $x^2 + y^2 = -4x$, $x^2 + 4x + 4 + y^2 = 4$, $(x+2)^2 + y^2 = 4$. Center $(-2, 0)$, radius $= 2$.

49. $r = 6\cos\theta$, $r^2 = 6r\cos\theta$, $x^2 + y^2 = 6x$, $x^2 - 6x + 9 + y^2 = 9$, $(x-3)^2 + y^2 = 9$. The graph of $r = 6\cos\theta$ is the graph of the circle with center $(3, 0)$ and radius 3. The region defined consists of all points on and within this circle. Graph the circle $r = 6\cos\theta$, $0 \leqq \theta \leqq \pi$ in $[-2.1, 8.1]$ by $[-3, 3]$.

50. The graph of $r = -4\sin\theta$ is the circle with center $(0, -2)$ (rectangular coordinates) and radius 2. The region consists of all points on and within this circle. Graph $r = -4\sin\theta$, $0 \leqq \theta \leqq \pi$ in $[-3.4, 3.4]$ by $[-4, 0]$.

51. $r = \sin\theta$, $r = 1 + \sin\theta$. We first solve the equations simultaneously and then check the graphs to see if any points have been missed. $r = \sin\theta = 1 + \sin\theta$, $0 = 1$, no solution. Graph $r = \sin\theta$, $r = 1 + \sin\theta$, $0 \leqq \theta \leqq 2\pi$ in $[-2, 2]$ by $[-0.25, 2]$. We see that $(0, 0)$ is the only point of intersection.

52. $r = \cos\theta$, $r = 1 - \cos\theta$. $\cos\theta = 1 - \cos\theta$, $2\cos\theta = 1$, $\cos\theta = \frac{1}{2}$, $\theta = \frac{\pi}{3}, \frac{5\pi}{3}$. $r = 0$ can be obtained in both equations. Answer: The origin, $(\frac{1}{2}, \frac{\pi}{3})$ and $(\frac{1}{2}, \frac{5\pi}{3})$.

53. $r = 1 + \sin\theta = -1 + \sin\theta$, $1 = -1$, no solution. Graph for $0 \leqq \theta \leqq 2\pi$ in $[-1.9, 1.9]$ by $[-0.25, 2]$. We see that the two graphs are identical. Answer: all points on the graph. Analytically, $r = f(\theta)$ is equivalent to $-r = f(\theta - \pi)$.

So $r = -1 + \sin\theta$ is equivalent to $-r = -1 + \sin(\theta - \pi) = -1 - \sin\theta$ or $r = 1 + \sin\theta$.

54. $r = 1 + \cos\theta = -1 - \cos\theta$, $2\cos\theta = -2$, $\cos\theta = -1$, $\theta = \pi$. This yields the point $(0, \pi)$. The graph shows two more points of intersection. $P_1 = (1, \frac{\pi}{2})$ is on the first curve and $P_1 = (-1, \frac{3\pi}{2})$ is on the second. $P_2 = (1, \frac{3\pi}{2})$ is on the first curve while $(-1, \frac{\pi}{2})$ is on the second. Answer: the origin, P_1 and P_2.

55. Graph the parabola $r = 2/(1 + \cos\theta)$, $0 \leqq \theta \leqq 2\pi$ in $[-26.5, 7.5]$ by $[-10, 10]$. $(1, 0)$ is the vertex.

56. $r = \frac{8}{2+\cos\theta} = \frac{4}{1+(1/2)\cos\theta} = \frac{8(1/2)}{1+(1/2)\cos\theta}$, $e = 1/2$, ellipse, one focus at $(0, 0)$, $k = 8$. $a = ke/(1 - e^2) = 4/(3/4) = 16/3$. Center on x-axis is at $x = -ea = -8/3$. Vertices at $x = -\frac{8}{3} \pm \frac{16}{3}$, i.e., $(-8, 0)$ and $(\frac{8}{3}, 0)$. Graph $r = \frac{8}{2+\cos\theta}$, $0 \leqq \theta \leqq 2\pi$ in $[-10.51, 5.2]$ by $[-4.62, 4.62]$.

57. $r = \frac{6}{1-2\cos\theta} = \frac{3(2)}{1-2\cos\theta}$, $e = 2$, hyperbola, $(0, 0)$ is a focus and $x = -3$ is the corresponding directrix. Vertices: $(-6, 0)$, $(2, \pi)$. Graph r, $0 \leqq \theta \leqq 2\pi$ in $[-13, 4]$ by $[-5, 5]$ in dot format.

58. $r = \frac{12}{3+\sin\theta} = \frac{4}{1+(1/3)\sin\theta} = \frac{12(1/3)}{1+(1/3)\sin\theta}$, $e = 1/3$, $k = 12$. $a = ke/(1 - e^2) = 4/(8/9) = \frac{9}{2}$. Center on y-axis is at $y = -ea = -\frac{3}{2}$. Vertices at $y = -\frac{3}{2} \pm \frac{9}{2}$, i.e., $(-6, \frac{\pi}{2})$ and $(3, \frac{\pi}{2})$. Graph $r = 12/(3 + \sin\theta)$, $0 \leqq \theta \leqq 2\pi$ in $[-8.2, 8.2]$ by $[-6.2, 3.5]$.

59. $e = 2$, $x = r\cos\theta = k = 2$. From Equation (11) or Example 4 of 10.7, $r = \frac{ke}{1+e\cos\theta} = \frac{4}{1+2\cos\theta}$.

60. $e = 1$, $x = r\cos\theta = -k = -4$. From Equation (12) of 10.7, $r = \frac{ke}{1-e\cos\theta} = \frac{4}{1-\cos\theta}$.

61. $e = \frac{1}{2}$, $r\sin\theta = y = k = 2$. From Table 10.4, $r = \frac{ke}{1+e\sin\theta} = \frac{1}{1+\frac{1}{2}\sin\theta} = \frac{2}{2+\sin\theta}$. This may be confirmed by graphing $r = 2/(2+\sin\theta)$, $0 \leqq \theta \leqq 2\pi$ in $[-2.3, 2.3]$ by $[-2, 0.7]$.

62. $e = 1/3$, $r\sin\theta = y = -k = -6$. From Table 10.4, $r = \frac{ke}{1-e\sin\theta} = \frac{2}{1-(1/3)\sin\theta} = \frac{6}{3-\sin\theta}$.

63. We use (4) of Section 10.6. $r^2 = \cos 2\theta$. $r = 0$ if and only if $\cos 2\theta = 0$, $2\theta = \frac{\pi}{2} + n\pi$, $\theta = \frac{\pi}{4} + \frac{n\pi}{2}$. For this curve $\theta = \frac{\pi}{4}$ and $\theta = \frac{3\pi}{4}$ cover all cases. In rectangular coordinates, $y = \pm x$.

64. $r = 2\cos\theta + 1 = 0$, $\cos\theta = -\frac{1}{2}$, $\theta = \frac{2\pi}{3}$ and $\theta = \frac{4\pi}{3}$. Tan $\frac{2\pi}{3} = -\sqrt{3}$, $\tan\frac{4\pi}{3} = \sqrt{3}$ so in rectangular coordinates: $y = \pm\sqrt{3}$.

65. Graph $r = \sin 2\theta$, $0 \leqq \theta \leqq 2\pi$ in $[-1.3, 1.3]$ by $[-0.77, 0.77]$. $\frac{dr}{d\theta} = 2\cos 2\theta$, $\frac{d^2r}{d\theta^2} = -4\sin 2\theta$. From this we see that r is extreme when $\theta = \frac{\pi}{4}, \frac{3\pi}{4}, \frac{5\pi}{4}$ and $\frac{7\pi}{4}$. Slope at $= \frac{r'\sin\theta + r\cos\theta}{r'\cos\theta - r\sin\theta} = \frac{2\cos 2\theta\sin\theta + \sin 2\theta\cos\theta}{2\cos 2\theta\cos\theta - \sin 2\theta\sin\theta}$. Slope at $(1, \frac{\pi}{4}) = \frac{\sqrt{2}/2}{-\sqrt{2}/2} = -1$. The tangent line at $(1, \frac{\pi}{4})$ must be perpendicular to $y = x$ or $\theta = \frac{\pi}{4}$. By Equation (3) of 10.7, $r\cos(\theta - \frac{\pi}{4}) = 1$ is an equation for this tangent line. By symmetry the other tangent lines are: at $(1, \frac{3\pi}{4})$, $r\cos(\theta - \frac{3\pi}{4}) = 1$; at $(1, \frac{5\pi}{4})$, $r\cos(\theta - \frac{5\pi}{4}) = 1$; at $(1, \frac{7\pi}{4})$, $r\cos(\theta - \frac{7\pi}{4}) = 1$.

66. Graph $r = 1 + \sin\theta$, $0 \leqq \theta \leqq 2\pi$ in $[-1.9, 1.9]$ by $[-0.25, 2]$. It crosses the x-axis at $(1, 0)$ and at $(1, \pi)$, $\frac{dy}{dx} = \frac{r'\sin\theta + r\cos\theta}{r'\cos\theta - r\sin\theta} = \frac{\cos\theta\sin\theta + (1+\sin\theta)\cos\theta}{\cos^2\theta - (1+\sin\theta)\sin\theta}$. Slope at $(1, 0) = 1$, slope at $(1, \pi) = -1$, $r\sin\theta - 0 = (1)(r\cos\theta - 1)$, $r(\sin\theta - \cos\theta) = -1$, $r = \frac{-1}{\sin\theta - \cos\theta}$ is an equation of the tangent line at $(1, 0)$. $r\sin\theta - 0 = (-1)(r\cos\theta + 1)$, $r = \frac{-1}{\sin\theta + \cos\theta}$ is an equation of the tangent line at $(1, \pi)$.

67. Graph $r = 2 - \cos\theta$, $0 \leqq \theta \leqq 2\pi$ in $[-4.7, 2.7]$ by $[-2.2, 2.2]$. $A = \frac{1}{2}\int_0^{2\pi}(2 - \cos\theta)^2 d\theta = \frac{1}{2}\int_0^{2\pi}(4 - 4\cos\theta + \frac{1}{2} + \frac{\cos 2\theta}{2})d\theta = \frac{1}{2}[\frac{9\theta}{2} - 4\sin\theta + \frac{\sin 2\theta}{4}]_0^{2\pi} = \frac{9\pi}{2}$.

68. Graph $r = \sin 3\theta$, $0 \leqq \theta \leqq \pi/3$ in $[-0.04, 0.92]$ by $[0, 0.56]$. $A = \frac{1}{2}\int_0^{\pi/3}\sin^2 3\theta\, d\theta = \frac{1}{4}\int_0^{\pi/3}(1 - \cos 6\theta)d\theta = \frac{1}{4}[\theta - \frac{\sin 6\theta}{6}]_0^{\pi/3} = \frac{\pi}{12}$.

69. Graph $r = 1 + \cos 2\theta$ and $r = 1$, $0 \leqq \theta \leqq 2\pi$ in $[-2, 2]$ by $[-1.2, 1.2]$. $r = 1 + \cos 2\theta = 1$, $\cos 2\theta = 0$, $2\theta = \frac{\pi}{2} + n\pi$, $\theta = \frac{\pi}{4} + \frac{n\pi}{2}$. $A = 4$ (area in first quadrant) $= 4(\frac{1}{2})\int_0^{\pi/4}[(1 + \cos 2\theta)^2 - 1]d\theta = 2\int_0^{\pi/4}(2\cos 2\theta + \frac{1}{2} + \frac{1}{2}\cos 4\theta)d\theta = 2[\sin 2\theta + \frac{\theta}{2} + \frac{\sin 4\theta}{8}]_0^{\pi/4} = 2(1 + \pi/8) = 2 + \pi/4$.

70. Graph $r = 2(1 + \sin\theta)$ and $r = 2\sin\theta$, $0 \leqq \theta \leqq 2\pi$ in $[-4, 4]$ by $[-0.5, 4]$. $A =$ area of cardioid$-$area of circle $= \frac{1}{2}\int_0^{2\pi}[2(1 + \sin\theta)]^2 d\theta - \pi(1)^2 = 2\int_0^{2\pi}(1 + 2\sin\theta + \frac{1}{2} - \frac{1}{2}\cos 2\theta)d\theta - \pi = 2[\frac{3\theta}{2} - 2\cos\theta - \frac{\sin 2\theta}{4}]_0^{2\pi} - \pi = 2[3\pi - 2 + 2] - \pi = 5\pi$.

71. $r = \sqrt{\cos 2\theta}$, $r' = \frac{-2\sin 2\theta}{2\sqrt{\cos 2\theta}} = -\frac{\sin 2\theta}{\sqrt{\cos 2\theta}}$. $A = 2\pi\int_0^{\pi/4} r\sin\theta\sqrt{\cos 2\theta + \frac{\sin^2 2\theta}{\cos 2\theta}}d\theta = 2\pi\int_0^{\pi/4}\sin\theta\, d\theta = -2\pi\cos\theta]_0^{\pi/4} = -2\pi(\sqrt{2}/2 - 1) = (2 - \sqrt{2})\pi$.

72. $r^2 = \sin 2\theta$. In the first quadrant $r = \sqrt{\sin 2\theta}$, $r' = \frac{2\cos 2\theta}{2\sqrt{\sin 2\theta}} = \frac{\cos 2\theta}{\sqrt{\sin 2\theta}}$. $A = 2$ (area of top half) $= 4\pi\int x\, ds = 4\pi\int_0^{\pi/2}\sqrt{\sin 2\theta}\cos\theta\sqrt{\sin 2\theta + \cos^2 2\theta/\sin 2\theta}d\theta = 4\pi\int_0^{\pi/2}\cos\theta\, d\theta = 4\pi\sin\theta]_0^{\pi/2} = 4\pi$.

73. Graph $r = -1 + \cos\theta$, $0 \leqq \theta \leqq 2\pi$ in $[-1.3, 3]$ by $[-1.3, 1.3]$. $L = \int_0^{2\pi} \sqrt{r^2 + r'^2}d\theta = 2\int_0^{\pi} \sqrt{(-1+\cos\theta)^2 + \sin^2\theta d\theta} =$ $2\int_0^{\pi} \sqrt{2 - 2\cos\theta}\sqrt{\frac{1+\cos\theta}{1+\cos\theta}}d\theta = 2\sqrt{2}\int_0^{\pi} \frac{\sqrt{1-\cos^2\theta}}{\sqrt{1+\cos\theta}}d\theta = 2\sqrt{2}\int_0^{\pi} \frac{\sin\theta\, d\theta}{\sqrt{1+\cos\theta}} =$ $-4\sqrt{2}\sqrt{1+\cos\theta}]_0^{\pi} = 4\sqrt{2}\sqrt{2} = 8.$

74. Graph $r = \sin\theta$, $0 \leqq \theta \leqq \pi$ in $[-0.85, 0.85]$ by $[0, 1]$. $L = \int_0^{\pi} \sqrt{\sin^2\theta + \cos^2\theta d\theta} = \pi$. This is consistent with circumference $= 2\pi$ (radius) $= 2\pi(\frac{1}{2}) = \pi$.

75. Graph $r = \cos^3(\theta/3)$, $0 \leqq \theta \leqq \pi/4$ in $[0.34, 1.4]$ by $[0, 0.64]$. $L = \int_0^{\pi/4} \sqrt{\cos^6(\theta/3) + (\cos^2(\theta/3)\sin(\theta/3))^2 d\theta} =$ $\int_0^{\pi/4} \sqrt{\cos^4(\theta/3)}\sqrt{\cos^2(\theta/3) + \sin^2(\theta/3)d\theta} = \int_0^{\pi/4} \cos^2(\theta/3)d\theta =$ $\frac{1}{2}\int_0^{\pi/4}(1+\cos(2\theta/3))d\theta = \frac{1}{2}[\theta + \frac{3}{2}\sin(2\theta/3)]_0^{\pi/4} = \frac{1}{2}[\frac{\pi}{4} + \frac{3}{2}\sin(\pi/6)] = \frac{1}{8}(\pi+3).$

76. Graph $r = \sqrt{1+\sin 2\theta}$, $0 \leqq \theta \leqq 2\pi$ in $[-2.2, 2.2]$ by $[-1.3, 1.3]$. $r' = \cos 2\theta/\sqrt{1+\sin 2\theta}$. $L = \int_0^{2\pi} \sqrt{1 + \sin 2\theta + \cos^2 2\theta/(1+\sin 2\theta)d\theta} =$ $\int_0^{2\pi} \sqrt{\frac{1+2\sin 2\theta+\sin^2 2\theta+\cos^2 2\theta}{1+\sin 2\theta}}d\theta = \sqrt{2}\int_0^{2\pi} d\theta = 2\sqrt{2}\pi.$

77. Graph $x_1 = t$, $y_1 = t^2 - 1$, $x_2 = x_1\cos(\pi/6) - y_1\sin(\pi/6)$, $y_2 = x_1\sin(\pi/6) + y_1\cos(\pi/6)$, $-3 \leqq t \leqq 3$ in $[-6.6, 3]$ by $[-1, 8.4]$. Let $P(x, y)$ be a point on the rotated curve. If we rotate P back $30°(\theta = -\pi/6)$, the coordinates of the point obtained, $(\frac{\sqrt{3}}{2}x + \frac{1}{2}y, -\frac{1}{2}x + \frac{\sqrt{3}}{2}y)$ (by (4) of 10.2), will satisfy the original equation, $y = x^2 - 1$. Substituting these in and simplifying, we obtain $3x^2 + 2\sqrt{3}xy + y^2 + 2x - 2\sqrt{3}y - 4 = 0$.

78. Graph $x_1 = 3\cos t$, $y_1 = 2\sin t$, $x_2 = x_1\cos(\pi/3) - y_1\sin(\pi/3)$, $y_2 = x_1\sin(\pi/3) + y_1\cos(\pi/3)$, $0 \leqq t \leqq 2\pi$ in $[-4.7, 4.7]$ by $[-2.8, 2.8]$. Let $P(x, y)$ be a point on the rotated curve. If we rotate P back $(\theta = -\pi/3)$, the coordinates of the point obtained, $(\frac{1}{2}x + \frac{\sqrt{3}}{2}y, -\frac{\sqrt{3}}{2}x + \frac{1}{2}y)$, will satisfy the original equation. Substituting and simplifying, we obtain $31x^2 - 10\sqrt{3}xy + 21y^2 = 144$.

79. a) $\frac{1}{2\pi}\int_0^{2\pi} a(1-\cos\theta)d\theta = \frac{a}{2\pi}[\theta - \sin\theta]_0^{2\pi} = a$. b) $\frac{1}{2\pi}\int_0^{2\pi} a\, d\theta = a$. c) $\frac{1}{(\pi/2)-(-\pi/2)}\int_{-\pi/2}^{\pi/2} a\cos\theta\, d\theta = \frac{1}{\pi}a\sin\theta]_{-\pi/2}^{\pi/2} = \frac{2a}{\pi}$.

80. Let $\theta = \theta_0$ be a fixed ray. $r = a\theta$ will intersect the ray at, say, $\theta = \theta_0 + 2k\pi$ and the next time at $\theta = \theta_0 + 2k\pi + 2\pi$. The distance on $\theta = \theta_0$ between intersection marks is $a(\theta_0 + 2k\pi + 2\pi) - a(\theta_0 + 2k\pi) = 2\pi a$ which is independent of k.

81. We use Fig. 10.13. $\frac{r_{max} - r_{min}}{r_{max} + r_{min}} = \frac{(c+a)-(a-c)}{(c+a)+(a-c)} = \frac{2c}{2a} = \frac{c}{a} = e.$

82. a) a =semimajor axis = $(300 + 8000 + 1000)/2 = 4650$. c = distance from center of ellipse to center of Earth (focus) = $a - (4000 + 300) = 350$. $e = \frac{c}{a} = \frac{350}{4650} = \frac{7}{93}$. b) $r = \frac{a(1-e^2)}{1+e\cos\theta} = \frac{4650(1-(7/93)^2)}{1+(7/93)\cos\theta} = \frac{430000}{93+7\cos\theta}$.

83. $\int_\alpha^\beta \sqrt{(2f(\theta))^2 + (2f'(\theta))^2 d\theta} = \int_\alpha^\beta \sqrt{4(f(\theta))^2 + 4(f'(\theta))^2 d\theta} = \sqrt{4}\int_\alpha^\beta \sqrt{(f(\theta))^2 + (f'(\theta))^2 d\theta} = 2L$.

84. $A_1 = 2\pi \int y\,ds = 2\pi \int_\alpha^\beta r\sin\theta\sqrt{(2f(\theta))^2 + (2f'(\theta))^2 d\theta} =$
$2\pi \int_\alpha^\beta 2f(\theta)(\sin\theta)2\sqrt{f^2(\theta) + (f'(\theta))^2 d\theta} =$
$(4)2\pi \int_\alpha^\beta f(\theta)\sin\theta\sqrt{f^2(\theta + (f'(\theta))^2 d\theta} = 4A_2$.

CHAPTER 11

VECTORS AND ANALYTIC GEOMETRY IN SPACE

11.1 VECTORS IN THE PLANE

1.

a) B, A, A+B

b) A+B+C, C, A, B

c) A-B, A, B

d) A-C, A, C

2.

a) A-B, B, A

b) A-B, A-B+C, C

c) 2A-B, B, A, A

d) C, B, A, A+B+C=0

3. $3\boldsymbol{u} = 9\boldsymbol{i} - 6\boldsymbol{j}$; the scalar components are 9 and 6.

4. $-2\boldsymbol{v} = +4\boldsymbol{i} - 10\boldsymbol{j}$; the scalar components are 4 and -10.

5. $\boldsymbol{u} + \boldsymbol{v} = \boldsymbol{i} + 3\boldsymbol{j}$; the scalar components are 1 and 3.

6. $\boldsymbol{u} - \boldsymbol{v} = 5\boldsymbol{i} - 7\boldsymbol{j}$; the scalar components are 5 and -7.

7. $2\boldsymbol{u} - 3\boldsymbol{v} = (6\boldsymbol{i} - 4\boldsymbol{j}) - (-6\boldsymbol{i} + 15\boldsymbol{j}) = 12\boldsymbol{i} - 19\boldsymbol{j}$; the scalar components are 12 and -19.

8. $-2\boldsymbol{u} + 5\boldsymbol{v} = (-6\boldsymbol{i} + 4\boldsymbol{j}) + (-10\boldsymbol{i} + 25\boldsymbol{j}) = -16\boldsymbol{i} + 29\boldsymbol{j}$; the scalar components are -16 and 29.

9. $\overrightarrow{P_1P_2} = (2-1)\boldsymbol{i} + (-1-3)\boldsymbol{j} = \boldsymbol{i} - 4\boldsymbol{j}$

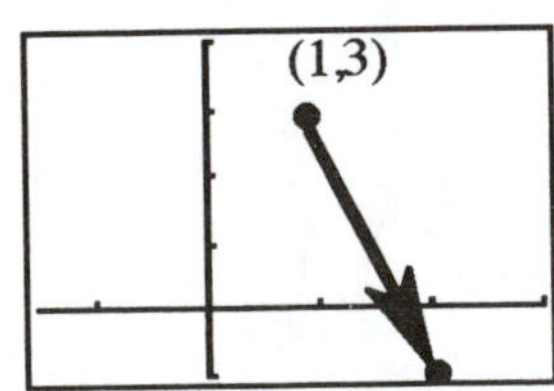

10. $P_3 = (\frac{2-4}{2}, \frac{-1+3}{2}) = (-1, 1)$;
$\vec{0P_3} = -\boldsymbol{i} + \boldsymbol{j}$

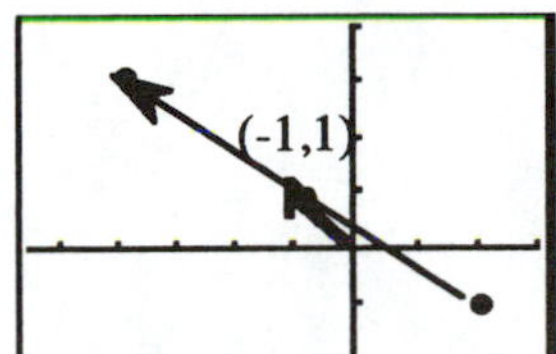

11. $-2\boldsymbol{i} - 3\boldsymbol{j}$

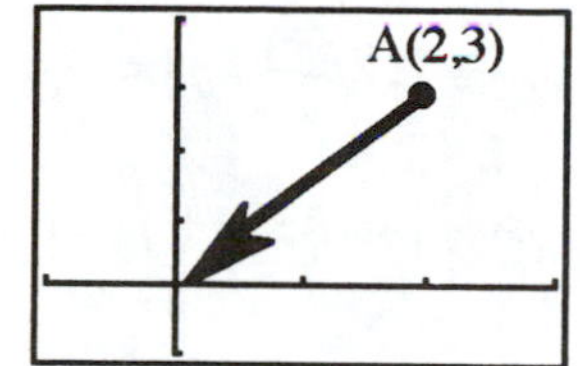

12. $\vec{AB} + \vec{CD} = (2-1)\boldsymbol{i} + (0+1)\boldsymbol{j} + (-2+1)\boldsymbol{i} + (2-3)\boldsymbol{j} = \boldsymbol{i} + \boldsymbol{j} - \boldsymbol{i} - \boldsymbol{j} = \boldsymbol{0}$.

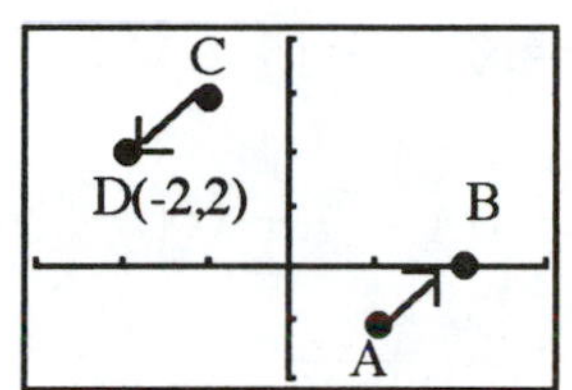

13. $\theta = \frac{\pi}{6}$: $\boldsymbol{u} = \frac{\sqrt{3}}{2}\boldsymbol{i} + \frac{1}{2}\boldsymbol{j}$;
$\theta = \frac{2\pi}{3}$: $\boldsymbol{u} = -\frac{1}{2}\boldsymbol{i} + \frac{\sqrt{3}}{2}\boldsymbol{j}$

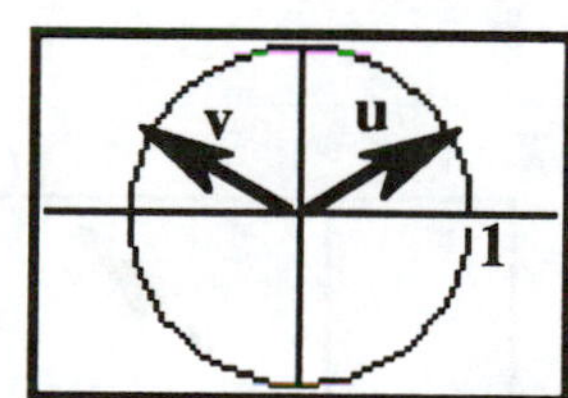

14 $\theta = \frac{\pi}{4}$: $\boldsymbol{u} = \frac{\sqrt{2}}{2}\boldsymbol{i} - \frac{\sqrt{2}}{2}\boldsymbol{j}$;
$\theta = -\frac{3\pi}{4}$: $\boldsymbol{u} = -\frac{\sqrt{2}}{2}\boldsymbol{i} - \frac{\sqrt{2}}{2}\boldsymbol{j}$

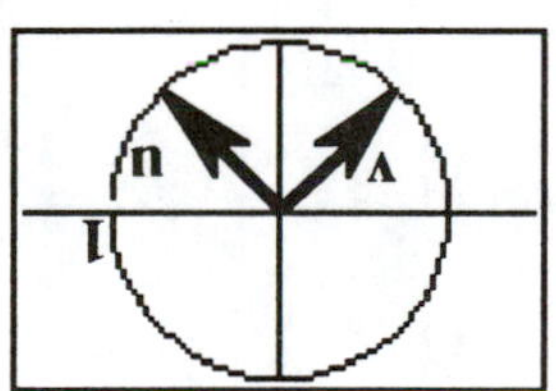

15. $\boldsymbol{j} = \cos 90°\boldsymbol{i} + \sin 90°\boldsymbol{j}$; $\boldsymbol{u} = \cos(90° - 120°)\boldsymbol{i} + \sin(90° - 120°)\boldsymbol{j} = \frac{\sqrt{3}}{2}\boldsymbol{i} - \frac{1}{2}\boldsymbol{j}$.

16. $\boldsymbol{i} = \cos 0°\boldsymbol{i} + \sin 0°\boldsymbol{j}$; $\boldsymbol{u} = \cos 135°\boldsymbol{i} + \sin 135°\boldsymbol{j} = -\frac{\sqrt{2}}{2}\boldsymbol{i} + \frac{\sqrt{2}}{2}\boldsymbol{j}$

17. $|\boldsymbol{v}|^2 = 4 + 9 \Rightarrow |\boldsymbol{v}| = \sqrt{13}$.

18. $|\boldsymbol{v}|^2 = 3^2 + 4^2 \Rightarrow |\boldsymbol{v}| = \sqrt{25} = 5$.

19. $|\boldsymbol{v}|^2 = \frac{9}{25} + \frac{16}{25} \Rightarrow |\boldsymbol{v}| = \sqrt{\frac{25}{25}} = 1$.

20. $|\boldsymbol{v}|^2 = \frac{16}{25} + \frac{9}{25} \Rightarrow |\boldsymbol{v}| = 1$.

21. $|\boldsymbol{v}|^2 = \frac{25}{13^2} + \frac{144}{13^2} \Rightarrow |\boldsymbol{v}| = 1.$

22. $|\boldsymbol{v}|^2 = \frac{64}{17^2} + \frac{225}{17^2} = \frac{289}{289} \Rightarrow |\boldsymbol{v}| = 1.$

23. $\boldsymbol{u} = \frac{1}{\sqrt{3^2+4^2}}[3\boldsymbol{i} + 4\boldsymbol{j}] = \frac{3}{5}\boldsymbol{i} + \frac{4}{5}\boldsymbol{j}.$

24. $\boldsymbol{u} = \frac{1}{\sqrt{16+9}}[4\boldsymbol{i} - 3\boldsymbol{j}] = \frac{4}{5}\boldsymbol{i} - \frac{3}{5}\boldsymbol{j}.$

25. $\boldsymbol{u} = \frac{1}{\sqrt{144+25}}[12\boldsymbol{i} - 5\boldsymbol{j}] = \frac{12}{13}\boldsymbol{i} - \frac{5}{13}\boldsymbol{j}.$

26. $\boldsymbol{u} = \frac{1}{\sqrt{255+64}}[-15\boldsymbol{i} + 8\boldsymbol{j}] = -\frac{15}{17}\boldsymbol{i} + \frac{8}{17}\boldsymbol{j}.$

27. $\boldsymbol{u} = \frac{1}{\sqrt{4+9}}[2\boldsymbol{i} + 3\boldsymbol{j}] = \frac{2}{\sqrt{13}}\boldsymbol{i} + \frac{3}{\sqrt{13}}\boldsymbol{j}.$

28. $\boldsymbol{u} = \frac{1}{\sqrt{25+4}}[5\boldsymbol{i} - 2\boldsymbol{j}] = \frac{5}{\sqrt{29}}\boldsymbol{i} - \frac{2}{\sqrt{29}}\boldsymbol{j}.$

29. length $= \sqrt{1^2 + 1^2} = \sqrt{2}$; direction $= \frac{1}{\sqrt{2}}(\boldsymbol{i} + \boldsymbol{j})$; $\boldsymbol{i} + \boldsymbol{j} = \sqrt{2}\left(\frac{1}{\sqrt{2}}\boldsymbol{i} + \frac{1}{\sqrt{2}}\boldsymbol{j}\right).$

30. length $= \sqrt{4+9} = \sqrt{13}$; direction $= \frac{1}{\sqrt{13}}(2\boldsymbol{i} - 3\boldsymbol{j})$; $2\boldsymbol{i} - 3\boldsymbol{j} = \sqrt{13}\left[\frac{2}{\sqrt{13}}\boldsymbol{i} - \frac{3}{\sqrt{13}}\boldsymbol{j}\right].$

31. length $= \sqrt{3 + 1^2} = 2$; direction $= \frac{1}{2}(\sqrt{3}\boldsymbol{i} + \boldsymbol{j})$; $\sqrt{3}\boldsymbol{i} + \boldsymbol{j} = 2\left(\frac{\sqrt{3}}{2}\boldsymbol{i} + \frac{1}{2}\boldsymbol{j}\right).$

32. length $= \sqrt{4+9} = \sqrt{3}$; direction $= \frac{1}{\sqrt{13}}(-2\boldsymbol{i} + 3\boldsymbol{j})$; $2\boldsymbol{i} + 3\boldsymbol{j} = \sqrt{13}\left[-\frac{2}{\sqrt{13}}\boldsymbol{i} + \frac{3}{\sqrt{13}}\boldsymbol{j}\right].$

33. length $= \sqrt{5^2 + 12^2} = 13$; direction $= \frac{5}{13}(2\boldsymbol{i} + \frac{12}{13}\boldsymbol{j})$; answer $= 13\left[\frac{5}{13}\boldsymbol{i} + \frac{12}{13}\boldsymbol{j}\right].$

34. length $= \sqrt{25 + 144} = 13$; direction $= \frac{1}{13}(-5\boldsymbol{i} - 12\boldsymbol{j})$; $-5\boldsymbol{i} - 12\boldsymbol{j} = 13\left[-\frac{5}{13}\boldsymbol{i} - \frac{12}{13}\boldsymbol{j}\right].$

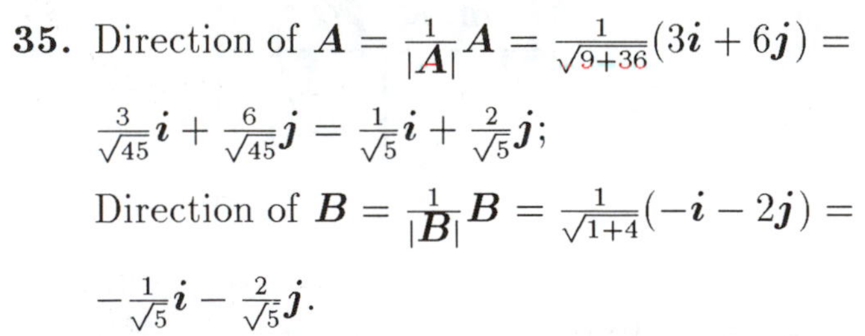

35. Direction of $\boldsymbol{A} = \frac{1}{|\boldsymbol{A}|}\boldsymbol{A} = \frac{1}{\sqrt{9+36}}(3\boldsymbol{i} + 6\boldsymbol{j}) = \frac{3}{\sqrt{45}}\boldsymbol{i} + \frac{6}{\sqrt{45}}\boldsymbol{j} = \frac{1}{\sqrt{5}}\boldsymbol{i} + \frac{2}{\sqrt{5}}\boldsymbol{j}$;
Direction of $\boldsymbol{B} = \frac{1}{|\boldsymbol{B}|}\boldsymbol{B} = \frac{1}{\sqrt{1+4}}(-\boldsymbol{i} - 2\boldsymbol{j}) = -\frac{1}{\sqrt{5}}\boldsymbol{i} - \frac{2}{\sqrt{5}}\boldsymbol{j}.$

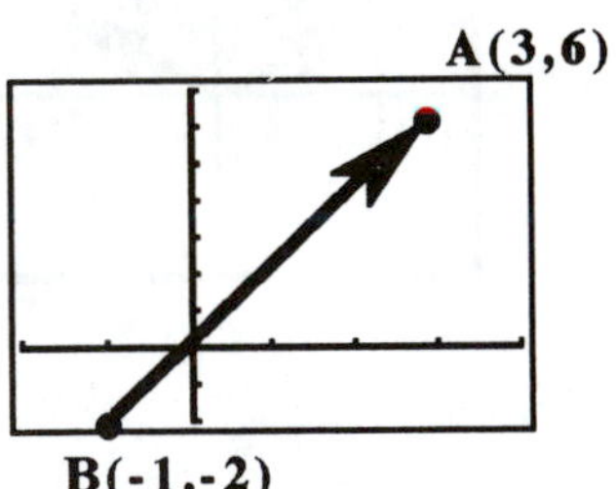

36. $\frac{1}{|\boldsymbol{A}|}\boldsymbol{A} = \frac{1}{\sqrt{9+36}}[3\boldsymbol{i} + 6\boldsymbol{j}] = \frac{1}{3\sqrt{5}}[3\boldsymbol{i} + 6\boldsymbol{j}] = \frac{1}{\sqrt{5}}[\boldsymbol{i} + 2\boldsymbol{j}]$;
$\frac{1}{|\boldsymbol{B}|}\boldsymbol{B} = \frac{1}{\sqrt{\frac{1}{4}+1}}\left[\frac{1}{2}\boldsymbol{i} + \boldsymbol{j}\right] = \frac{2}{\sqrt{5}}\left[\frac{1}{2}\boldsymbol{i} + \boldsymbol{j}\right] = \frac{1}{\sqrt{5}}[\boldsymbol{i} + 2\boldsymbol{j}].$

37. $y' = 2x$; $y'(2) = 4$. $\boldsymbol{i} + 4\boldsymbol{j}$ is tangent $\Rightarrow$ $\pm\frac{1}{\sqrt{5}}(\boldsymbol{i} + 4\boldsymbol{j})$ are unit tangent vectors; $4\boldsymbol{i} - \boldsymbol{j}$ is normal $\Rightarrow \pm\frac{1}{\sqrt{5}}(4\boldsymbol{i} - \boldsymbol{j})$ are unit normal vectors.

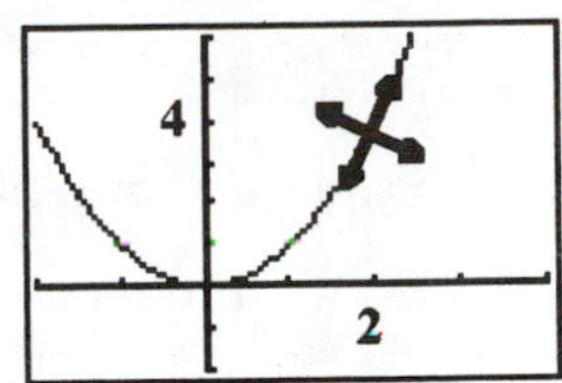

38. $y' = e^x$, $y'(\ln 2) = 2$; $\boldsymbol{i} + 2\boldsymbol{j}$ is a tangent vector; $2\boldsymbol{i} - \boldsymbol{j}$ is normal. Unit tangent vectors are $\pm\frac{1}{\sqrt{5}}(\boldsymbol{i} + 2\boldsymbol{j})$; unit normal vectors are $\pm\frac{1}{\sqrt{5}}(2\boldsymbol{i} - \boldsymbol{j})$.

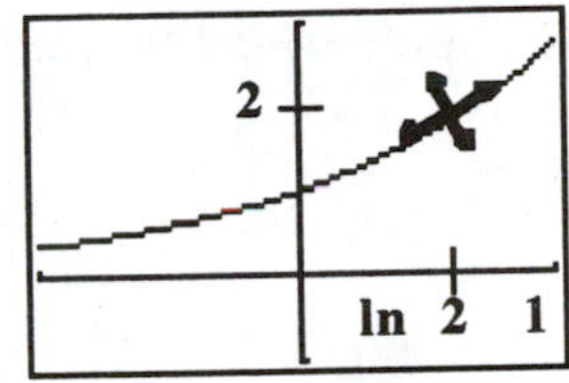

39. $2x + 4yy' = 0 \Rightarrow y'(2) = \frac{-2}{2} = -1$; $\boldsymbol{i} - \boldsymbol{j}$ is tangent and $\boldsymbol{i} + \boldsymbol{j}$ is normal to the curve. Unit tangent vectors are $\pm\frac{1}{\sqrt{2}}(\boldsymbol{i} - \boldsymbol{j})$; unit normal vectors are $\pm\frac{1}{\sqrt{2}}(\boldsymbol{i} + \boldsymbol{j})$.

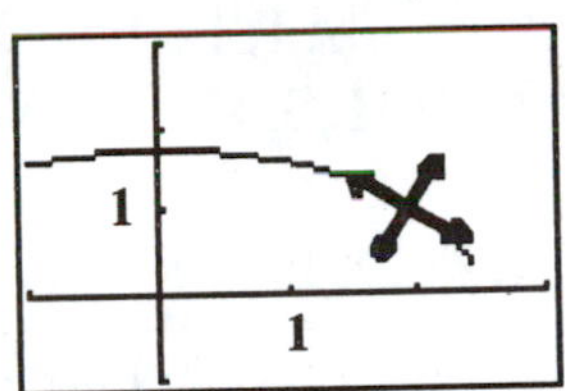

40. $8x + 2yy' = 0 \Rightarrow y'(\sqrt{2}) = 2$; $\boldsymbol{i} + 2\boldsymbol{j}$ is tangent and $2\boldsymbol{i} - \boldsymbol{j}$ are normal to the curve. Unit tangent vectors are $\pm\frac{1}{\sqrt{5}}(\boldsymbol{i} + 2\boldsymbol{j})$; unit normal vectors are $\pm\frac{1}{\sqrt{5}}(2\boldsymbol{i} - \boldsymbol{j})$.

41. $y' = \frac{1}{1+x^2} \Rightarrow y'(1) = \frac{1}{2}$; $2\boldsymbol{i} + \boldsymbol{j}$ is tangent and $\boldsymbol{i} - 2\boldsymbol{j}$ is normal to the curve. Unit tangent vectors are $\pm\frac{1}{\sqrt{5}}(2\boldsymbol{i} + \boldsymbol{j})$; unit normal vectors are $\pm\frac{1}{\sqrt{5}}(\boldsymbol{i} - 2\boldsymbol{j})$.

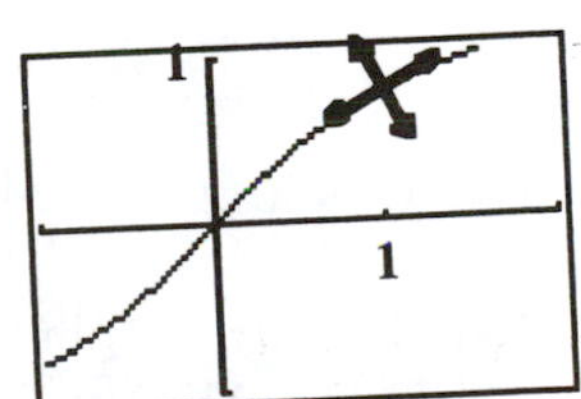

42. $y' = \frac{1}{\sqrt{1-x^2}} \Rightarrow y'(\frac{\sqrt{2}}{2}) = \sqrt{2}$; $\boldsymbol{i} + \sqrt{2}\boldsymbol{j}$ is tangent and $\sqrt{2}\boldsymbol{i} - \boldsymbol{j}$ is normal to the curve. Unit tangent vectors are $\pm\frac{1}{\sqrt{3}}(\boldsymbol{i} + \sqrt{2}\boldsymbol{j})$; unit normal vectors are $\pm\frac{1}{\sqrt{3}}(\sqrt{2}\boldsymbol{i} - \boldsymbol{j})$.

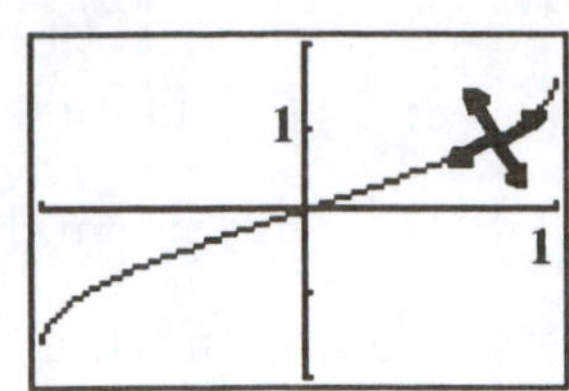

43. The same line segment is used in representing both $\boldsymbol{v}$ and $-\boldsymbol{v}$; the slopes of the vectors are the same.

44. If $\boldsymbol{v} = a\boldsymbol{i} + \boldsymbol{j}$ has length 0, then $a^2 + b^2 = 0$ which means $a = 0$ and $b = 0$; $\boldsymbol{v}$ is the zero vector.

11.2 CARTESIAN (RECTANGULAR) COORDINATES AND VECTORS IN SPACE

1. A line through $(2, 3, 0)$ parallel to the z-axis.

2. A line parallel to the y-axis, passing through $(-1, 2, 0)$.

3. The x-axis.

4. A line through $(1, 0, 6)$ parallel to the z-axis.

5. Circle in xy-plane, center at $(0, 0, 0)$, radius 2.

6. Circle in plane $z = -2$, center at $(0, 0, -2)$, radius 2.

7. Circle in xz-plane, center at $(0, 0, 0)$, radius 2.

8. Circle in yz-plane, center at $(0, 0, 0)$, radius 1.

9. Circle in yz-plane, center at $(0, 0, 0)$, radius 1.

10. Circle in the plane $y = -4$, center at $(0, -4, 0)$, radius $= \sqrt{25 - 16} = 3$.

11. This is a sphere - center at $(0, 0, -3)$, radius 5 - sliced by a plane. The set forms the circle $x^2 + y^2 + 3^2 = 25^2$ or $x^2 + y^2 = 4^2$ in the xy-plane.

12. Circle in xz-plane, center at $(0, 0, 0)$, radius $= \sqrt{4 - 1} = \sqrt{3}$.

13. a) The first quadrant in the xy-plane b) The fourth quadrant in the xy-plane

14. a) Slab parallel to the yz-plane, 1 unit thick, between the planes $x = 0$ and $x = 1$, b) Square column of side 1, parallel to z-axis, c) The unit cube in the corner of the first octant.

15. a) The interior and surface of the unit sphere (center at $(0, 0, 0)$),

b) All of 3-space but the interior and surface of the unit sphere

16. a) The interior and boundary of the circle $x^2 + y^2 = 1$ in the xy-plane,

b) The interior and boundary of the circle $x^2 + y^2 = 1$ in the plane $z = 3$,

c) A cylindrical column with both a) and c) as cross-sections

17. a) The surface of the top half ($z \geq 0$) of the unit sphere, b) The interior and surface of the top half of the unit sphere

18. a) The line $y = x$ in the xy-plane, b) The plane $y = x$

19. a) $x = 3$, b) $y = -1$, c) $z = -2$ **20.** a) $x = 3$, b) $y = -1$, c) $z = 2$

21. a) $z = 1$, b) $x = 3$, c) $y = -1$

22. a) $x^2 + y^2 = 4,\ z = 0$; b) $y^2 + z^2 = 4,\ x = 0$; c) $x^2 + z^2 = 4,\ y = 0$

23. a) $(x - 0)^2 + (y - 2)^2 = 2^2,\ z = 0$; b) $(y - 2)^2 + (z - 0)^2 = 2^2,\ x = 0$; c) $(x - 0)^2 + (z - 0)^2 = 2^2,\ y = 2$

24. a) $(x + 3)^2 + (y - 4)^2 = 1,\ z = 1$; b) $(y - 4)^2 + (z - 1)^2 = 1,\ x = -3$; c) $(x + 3)^2 + (z - 1)^2 = 1,\ y = 4$

25. a) x can be anything; $y = 3,\ z = -1$; b) $x = 1,\ z = -1$, c) $x = 1,\ y = 3$

26. $\sqrt{x^2 + (y - 2)^2 + z^2} = \sqrt{x^2 + y^2 + z^2} \Rightarrow (y - 2)^2 = y^2$; the unique solution is $y = 1$

27. The plane is $z = 3$; the sphere is $x^2 + y^2 + z^2 = 25$. Hence: $x^2 + y^2 + 9 = 25,\ z = 3$.

28. $x^2+y^2+(z-1)^2 = 2^2$ and $x^2+y^2+(z+1)^2 = 2^2 \Rightarrow (z-1) = \pm(z+1) \Rightarrow z = 0$; substituting gives $x^2 + y^2 + 1^2 = 2^2$ or $x^2 + y^2 = 3$.

29. $0 \leq z \leq 1$ **30.** $0 \leq x \leq 2,\ 0 \leq y \leq 2,\ 0 \leq z \leq 2$

31. $z \le 0$ **32.** $x^2 + y^2 + z^2 = 1,\ z \ge 0$

33. a) $(x-1)^2 + (y-1)^2 + (z-1)^2 < 1$; b) $(x-1)^2 + (y-1)^2 + (z-1)^2 > 1$

34. $1 \le x^2 + y^2 + z^2 \le 4$

35. $|\boldsymbol{A}| = \sqrt{2^2 + 1^2 + 2^2} = 3$, direction $\frac{\boldsymbol{A}}{|\boldsymbol{A}|} = \frac{2}{3}\boldsymbol{i} + \frac{1}{3}\boldsymbol{j} - \frac{2}{3}\boldsymbol{k}$.

36. $|\boldsymbol{A}| = \sqrt{3^2 + 6^2 + 2^2} = 7,\ \frac{\boldsymbol{A}}{|\boldsymbol{A}|} = \frac{3}{7}\boldsymbol{i} - \frac{6}{7}\boldsymbol{j} + \frac{2}{7}\boldsymbol{k}$

37. $|\boldsymbol{A}| = \sqrt{1^2 + 4^2 + 8^2} = 9,\ \frac{\boldsymbol{A}}{|\boldsymbol{A}|} = \frac{1}{9}\boldsymbol{i} + \frac{4}{9}\boldsymbol{j} - \frac{8}{9}\boldsymbol{k}$

38. $|\boldsymbol{A}| = \sqrt{9^2 + 2^2 + 6^2} = 11,\ \frac{\boldsymbol{A}}{|\boldsymbol{A}|} = \frac{9}{11}\boldsymbol{i} - \frac{2}{11}\boldsymbol{j} + \frac{6}{11}\boldsymbol{k}$

39. $|\boldsymbol{A}| = 5,\ \frac{\boldsymbol{A}}{|\boldsymbol{A}|} = \boldsymbol{k}$ **40.** $|\boldsymbol{A}| = 6$, direction $= \boldsymbol{i}$ **41.** $|\boldsymbol{A}| = 4,\ \frac{\boldsymbol{A}}{|\boldsymbol{A}|} = -\boldsymbol{j}$

42. $|\boldsymbol{A}| = \sqrt{(\frac{3}{5})^2 + (\frac{4}{5})^2} = 1$, direction $= \boldsymbol{A} = \frac{3}{5}\boldsymbol{i} + \frac{4}{5}\boldsymbol{k}$

43. $|\boldsymbol{A}| = \sqrt{(\frac{1}{3})^2 + (\frac{1}{4})^2} = \frac{5}{12};\ \frac{\boldsymbol{A}}{|\boldsymbol{A}|} = \frac{12}{5}(-\frac{1}{3}\boldsymbol{j} + \frac{1}{4}\boldsymbol{k}) = -\frac{4}{5}\boldsymbol{j} + \frac{3}{5}\boldsymbol{k}$

44. $|\boldsymbol{A}| = \sqrt{\frac{1}{2} + \frac{1}{2}} = 1;\ \frac{\boldsymbol{A}}{|\boldsymbol{A}|} = A = \frac{1}{\sqrt{2}}\boldsymbol{i} - \frac{1}{\sqrt{2}}\boldsymbol{k}$

45. $|\boldsymbol{A}| = \sqrt{3(\frac{1}{6})} = \sqrt{\frac{1}{2}} = \frac{1}{\sqrt{2}};\ \frac{\boldsymbol{A}}{|\boldsymbol{A}|} = \sqrt{2}\left(\frac{1}{\sqrt{6}}\boldsymbol{i} - \frac{1}{\sqrt{6}}\boldsymbol{j} - \frac{1}{\sqrt{6}}\boldsymbol{k}\right) = \frac{1}{\sqrt{3}}\boldsymbol{i} - \frac{1}{\sqrt{3}}\boldsymbol{j} - \frac{1}{\sqrt{3}}\boldsymbol{k}$

46. $|\boldsymbol{A}| = \sqrt{\frac{1}{3} + \frac{1}{3} + \frac{1}{3}} = 1$; direction $= \frac{1}{\sqrt{3}}\boldsymbol{i} + \frac{1}{\sqrt{3}}\boldsymbol{j} + \frac{1}{\sqrt{3}}\boldsymbol{k}$

47. a) Distance $= |\overrightarrow{P_1P_2}| = |2\boldsymbol{i} + 2\boldsymbol{j} - \boldsymbol{k}| = \sqrt{4+4+1} = 3$; b) Direction $= \frac{2}{3}\boldsymbol{i} + \frac{2}{3}\boldsymbol{j} - \frac{1}{3}\boldsymbol{k}$; c) Midpoint is $\left(\frac{1+3}{2}, \frac{1+3}{2}, \frac{1}{2}\right) = \left(2, 2, \frac{1}{2}\right)$

48. a) Distance $= |\overrightarrow{P_1P_2}| = |3\boldsymbol{i} + 4\boldsymbol{j} - 5\boldsymbol{k}| = \sqrt{50} = 5\sqrt{2}$; b) Direction $= \frac{3}{5\sqrt{2}}\boldsymbol{i} + \frac{4}{5\sqrt{2}}\boldsymbol{j} - \frac{5}{5\sqrt{2}}\boldsymbol{k}$; c) Midpoint $= \left(\frac{1}{2}, 3, \frac{5}{2}\right)$

49. a) Distance $= |\overrightarrow{P_1P_2}| = |3\boldsymbol{i} - 6\boldsymbol{j} + 2\boldsymbol{k}| = 7$; b) Direction $= \frac{3}{7}\boldsymbol{i} - \frac{6}{7}\boldsymbol{j} + \frac{2}{7}\boldsymbol{k}$, c) Midpoint $= \left(\frac{5}{2}, 1, 6\right)$

50. a) Distance $= |\overrightarrow{P_1P_2}| = |-\boldsymbol{i} - \boldsymbol{j} - \boldsymbol{k}| = \sqrt{3}$; b) Direction $= -\frac{1}{\sqrt{3}}\boldsymbol{i} - \frac{1}{\sqrt{3}}\boldsymbol{j} - \frac{1}{\sqrt{3}}\boldsymbol{k}$; c) Midpoint $= \left(\frac{5}{2}, \frac{7}{2}, \frac{9}{2}\right)$

51. a) Distance $= |\overrightarrow{P_1P_2}| = |2\boldsymbol{i} - 2\boldsymbol{j} - 2\boldsymbol{k}| = \sqrt{12} = 2\sqrt{3}$; b) Direction $= \frac{1}{\sqrt{3}}\boldsymbol{i} - \frac{1}{\sqrt{3}}\boldsymbol{j} - \frac{1}{\sqrt{3}}\boldsymbol{k}$, c) Midpoint $= (1, -1, -1)$

52. a) Distance $= |\overrightarrow{P_1P_2}| = |-5\boldsymbol{i} - 3\boldsymbol{j} + 2\boldsymbol{k}| = \sqrt{38}$; b) Direction $= -\frac{5}{\sqrt{38}}\boldsymbol{i} - \frac{3}{\sqrt{38}}\boldsymbol{j} - \frac{2}{\sqrt{38}}\boldsymbol{k}$; c) Midpoint $= \left(\frac{5}{2}, \frac{3}{2}, -1\right)$

53. a) $2\boldsymbol{i}$, b) $-\sqrt{3}\boldsymbol{k}$, c) $\frac{3}{10}\boldsymbol{j} + \frac{4}{10}\boldsymbol{k}$, d) $6\boldsymbol{i} - 2\boldsymbol{j} + 3\boldsymbol{k}$

54. a) $-7\boldsymbol{j}$, b) $-\frac{3\sqrt{2}}{5}\boldsymbol{i} - \frac{4\sqrt{2}}{5}\boldsymbol{j}$, c) $\frac{1}{4}\boldsymbol{i} - \frac{1}{3}\boldsymbol{j} - \boldsymbol{k}$, d) $\frac{a}{\sqrt{2}}\boldsymbol{i} + \frac{a}{\sqrt{3}}\boldsymbol{j} - \frac{a}{\sqrt{6}}\boldsymbol{k}$

55. $|\boldsymbol{A}| = \sqrt{12^2 + 5^2} = 13 \Rightarrow \boldsymbol{v} = \frac{7}{13}\boldsymbol{A} = \frac{84}{13}\boldsymbol{i} - \frac{35}{13}\boldsymbol{j}$

56. $|\boldsymbol{A}| = \sqrt{3} \Rightarrow \boldsymbol{v} = \frac{\sqrt{5}}{\sqrt{3}}\boldsymbol{A} = \sqrt{\frac{5}{3}}\boldsymbol{i} + \sqrt{\frac{5}{3}}\boldsymbol{j} + \sqrt{\frac{5}{3}}\boldsymbol{k}$

57. $|\boldsymbol{A}| = \sqrt{4 + 9 + 36} = 7 \Rightarrow \boldsymbol{v} = -\frac{5}{7}\boldsymbol{A} = -\frac{10}{7}\boldsymbol{i} + \frac{15}{7}\boldsymbol{j} - \frac{30}{7}\boldsymbol{k}$

58. $|\boldsymbol{A}| = \frac{\sqrt{3}}{2} \Rightarrow \boldsymbol{v} = \frac{-3}{\frac{\sqrt{3}}{2}}\boldsymbol{A} = -2\sqrt{3}\ \boldsymbol{A} = -\sqrt{3}\boldsymbol{i} + \sqrt{3}\boldsymbol{j} + \sqrt{3}\boldsymbol{k}$

59. a) $C(-2, 0, 2)$, radius $= \sqrt{8} = 2\sqrt{2}$, b) $C(-\frac{1}{2}, -\frac{1}{2}, -\frac{1}{2})$, radius $= \frac{\sqrt{21}}{2}$, c) $C(\sqrt{2}, \sqrt{2}, -\sqrt{2})$, radius $= \sqrt{2}$, d) $C(0, -\frac{1}{3}, \frac{1}{3})$, radius $= \frac{\sqrt{29}}{3}$

60. a) $(x-1)^2 + (y-2)^2 + (z-3)^2 = 14$; b) $x^2 + (y+1)^2 + (z-5)^2 = 4$, c) $(x+2)^2 + y^2 + z^2 = 3$, d) $x^2 + (y+7)^2 + z^2 = 49$

61. $x^2 + 4x + y^2 + z^2 - 4z = 0 \Rightarrow x^2 + 4x + 4 + y^2 + z^2 - 4z + 4 = 8 \Rightarrow (x+2)^2 + y^2 + (z-2)^2 = 8$; Center $(-2, 0, 2)$, radius $\sqrt{8}$

62. $2(x^2 + \frac{1}{2}x) + 2(y^2 + \frac{1}{2}y) + 2(z^2 + \frac{1}{2}z) = 9 \Rightarrow x^2 + \frac{1}{2}x + \frac{1}{16} + y^2 + \frac{1}{2}y + \frac{1}{16} + z^2 + \frac{1}{2}z + \frac{1}{16} = \frac{9}{2} + \frac{3}{16}$; or $(x + \frac{1}{4})^2 + (y + \frac{1}{4})^2 + (z + \frac{1}{4})^2 = \frac{75}{16}$; Center $(-\frac{1}{4}, -\frac{1}{4}, -\frac{1}{4})$, radius $\frac{5}{4}\sqrt{3}$

63. $x^2 + y^2 + z^2 - 2z + 1 = 0 + 1$; Center $(0, 0, 1)$, radius 1

64. $3x^2 + 3(y^2 + \frac{2}{3}y) + 3(z^2 - \frac{2}{3}z) = 9 \Rightarrow x^2 + (y^2 + \frac{2}{3}y + \frac{1}{9}) + (z^2 - \frac{2}{3}z + \frac{1}{9}) = 3 + \frac{2}{9} = \frac{29}{9}$; Center $(0, -\frac{1}{3}, \frac{1}{3})$, radius $\frac{\sqrt{29}}{3}$

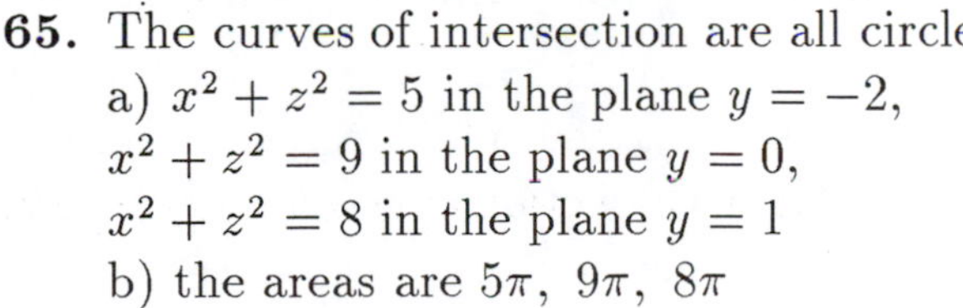

65. The curves of intersection are all circles:
a) $x^2 + z^2 = 5$ in the plane $y = -2$,
$x^2 + z^2 = 9$ in the plane $y = 0$,
$x^2 + z^2 = 8$ in the plane $y = 1$
b) the areas are 5π, 9π, 8π

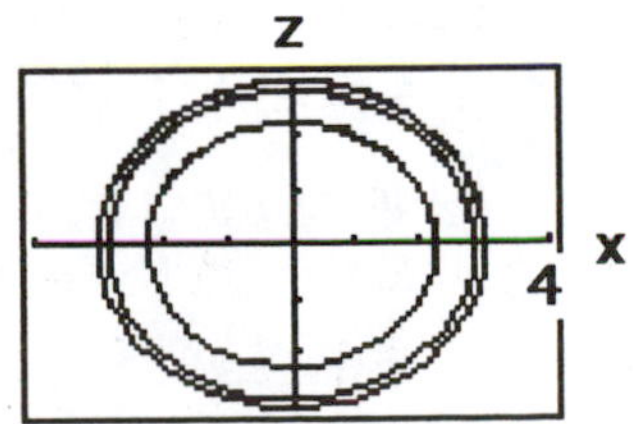

66. a) The intersections are all circles:
$y^2 + z^2 = 8$ in $x = 2$, $y^2 + z^2 = 9$ in
$x = 0$, $y^2 + z^2 = 5$ in $x = 2$
b) areas are 8π, 9π, 5π

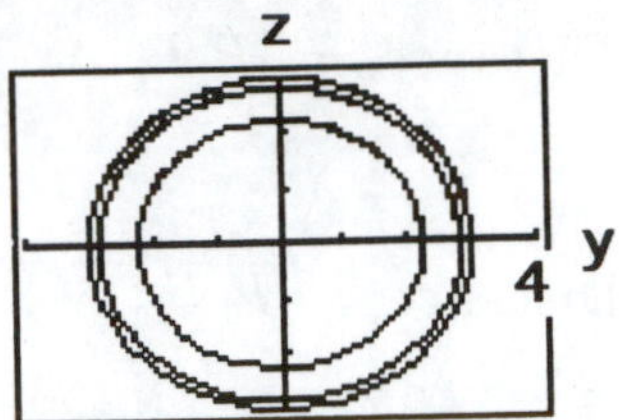

67. a) In $z = 2$, $x^2 + y^2 + 4x = 8 - 4$ or $x^2 + 4x + 4 + y^2 = 8$, i.e., $(x+2)^2 + y^2 = 8$; in $z = 4$, $x^2 + 4x + y^2 = 16 - 16$, or $(x + 2)^2 + y^2 = 4$

b)

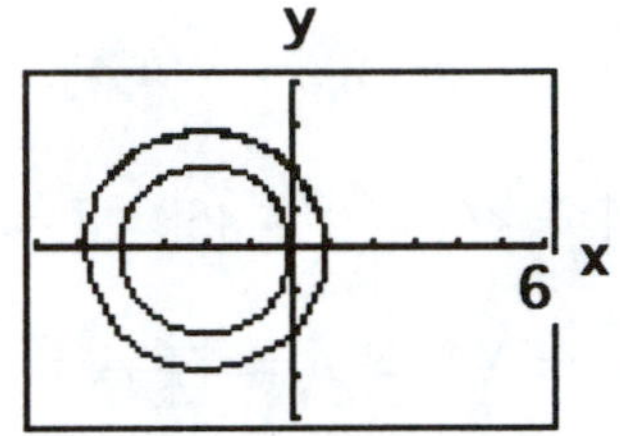

68. a) In $x = -2$, $y^2 + z^2 = 4z = -(-2)^2 - 4(-2) = 4$, or $y^2 + (z - 2)^2 = 8$; in $x = 0$, $y^2 + z^2 - 4z = 0$, or $y^2 + (z - 2)^2 = 4$

b)

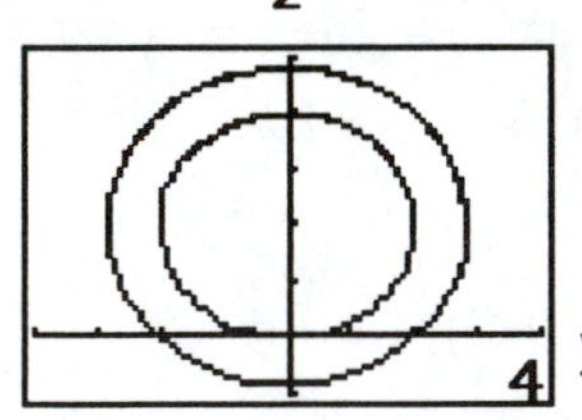

69. a) $\sqrt{y^2 + z^2}$,
b) $\sqrt{x^2 + z^2}$,
c) $\sqrt{x^2 + y^2}$

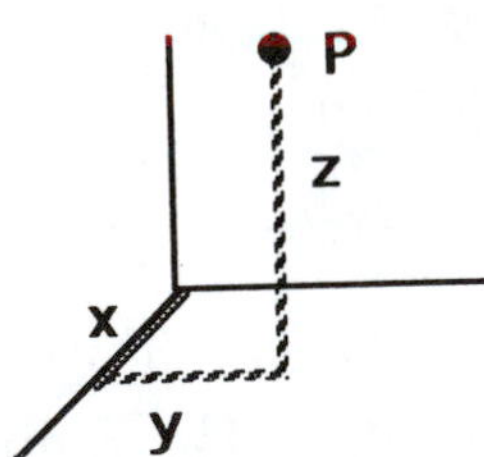

70. a) z, b) x, c) y

11.3 DOT PRODUCTS

1. $\boldsymbol{A}\cdot\boldsymbol{B} = 3\cdot 0 + 2\cdot 5 + 0\cdot 1 = 10$, $|\boldsymbol{A}| = \sqrt{3^2+2^2+0^2} = \sqrt{13}$, $|\boldsymbol{B}| = \sqrt{0^2+5^2+1^2} = \sqrt{26}$, $\cos\theta = \frac{\boldsymbol{A}\cdot\boldsymbol{B}}{|\boldsymbol{A}||\boldsymbol{B}|} = \frac{10}{\sqrt{13}\sqrt{26}} = \frac{10}{13\sqrt{2}}$; $|\boldsymbol{B}|\cos\theta = \frac{\sqrt{26}\cdot 10}{13\sqrt{2}} = \frac{10}{\sqrt{13}}$. $\text{Proj}_{\boldsymbol{A}}\boldsymbol{B} = \frac{\boldsymbol{A}\cdot\boldsymbol{B}}{\boldsymbol{A}\cdot\boldsymbol{A}}\boldsymbol{A} = \frac{10}{13}[3\boldsymbol{i}+2\boldsymbol{j}]$.

Problems 2-12 are done in exactly the same way.

	$\boldsymbol{A}\cdot\boldsymbol{B}$	$\lvert\boldsymbol{A}\rvert$	$\lvert\boldsymbol{B}\rvert$	$\cos\theta$	$\lvert\boldsymbol{B}\rvert\cos\theta$	$\text{Proj}_{\boldsymbol{A}}\boldsymbol{B} = \frac{\boldsymbol{A}\cdot\boldsymbol{B}}{\boldsymbol{A}\cdot\boldsymbol{A}}\boldsymbol{A}$
1.	10	$\sqrt{13}$	$\sqrt{26}$	$\frac{10}{13\sqrt{2}}$	$\frac{10}{\sqrt{13}}$	$\frac{10}{13}[3\boldsymbol{i}+2\boldsymbol{j}]$
2.	0	1	$\sqrt{34}$	0	0	$\boldsymbol{0}$
3.	4	$\sqrt{14}$	2	$\frac{2}{\sqrt{14}}$	$\frac{4}{\sqrt{14}}$	$\frac{2}{7}[3\boldsymbol{i}-2\boldsymbol{j}-\boldsymbol{k}]$
4.	0	$\sqrt{53}$	1	0	0	$\boldsymbol{0}$
5.	2	$\sqrt{34}$	$\sqrt{3}$	$\frac{2}{\sqrt{3}\sqrt{34}}$	$\frac{2}{\sqrt{34}}$	$\frac{1}{17}[5\boldsymbol{j}-3\boldsymbol{k}]$
6.	0	1	$\sqrt{\frac{3}{2}}$	0	0	$\boldsymbol{0}$
7.	$\sqrt{3}-\sqrt{2}$	$\sqrt{2}$	3	$\frac{\sqrt{3}-\sqrt{2}}{3\sqrt{2}}$	$\frac{\sqrt{3}-\sqrt{2}}{2}$	$\frac{\sqrt{3}-\sqrt{2}}{2}[-\boldsymbol{i}+\boldsymbol{j}]$
8.	2	$\sqrt{2}$	$\sqrt{3}$	$\sqrt{\frac{2}{3}}$	$\sqrt{2}$	$\boldsymbol{i}+\boldsymbol{k}$
9.	-25	5	5	-1	-5	$-2\boldsymbol{i}+4\boldsymbol{j}-\sqrt{5}\boldsymbol{k}$
10.	$\sqrt{17}-10$	$\sqrt{26}$	11	$\frac{\sqrt{17}-10}{11\sqrt{26}}$	$\frac{\sqrt{17}-10}{\sqrt{26}}$	$\frac{\sqrt{17}-10}{26}[-5\boldsymbol{i}+\boldsymbol{j}]$
11.	25	15	5	$\frac{1}{3}$	$\frac{5}{3}$	$\frac{1}{9}[10\boldsymbol{i}+11\boldsymbol{j}-2\boldsymbol{k}]$
12.	13	15	3	$\frac{13}{45}$	$\frac{13}{15}$	$\frac{13}{225}[2\boldsymbol{i}+10\boldsymbol{j}-11\boldsymbol{k}]$

13. $\boldsymbol{B} = \frac{\boldsymbol{A}\cdot\boldsymbol{B}}{\boldsymbol{A}\cdot\boldsymbol{A}}\boldsymbol{A}+(\boldsymbol{B}-\frac{\boldsymbol{A}\cdot\boldsymbol{B}}{\boldsymbol{A}\cdot\boldsymbol{A}}\boldsymbol{A}) = \frac{3}{2}\boldsymbol{A}+(\boldsymbol{B}-\frac{3}{2}\boldsymbol{A}) = \frac{3}{2}(\boldsymbol{i}+\boldsymbol{j})+[3\boldsymbol{j}+4\boldsymbol{k}-\frac{3}{2}\boldsymbol{i}+\boldsymbol{j}] = [\frac{3}{2}(\boldsymbol{i}+\boldsymbol{j})]+[-\frac{3}{2}\boldsymbol{i}+\frac{3}{2}\boldsymbol{j}+4\boldsymbol{k}]$

14. $\boldsymbol{B} = \frac{\boldsymbol{A}\cdot\boldsymbol{B}}{\boldsymbol{A}\cdot\boldsymbol{A}}\boldsymbol{A}+[\boldsymbol{B}-\frac{\boldsymbol{A}\cdot\boldsymbol{B}}{\boldsymbol{A}\cdot\boldsymbol{A}}\boldsymbol{A}] = \frac{1}{2}\boldsymbol{A}+[\boldsymbol{B}-\frac{1}{2}\boldsymbol{A}] = [\frac{1}{2}\boldsymbol{i}+\frac{1}{2}\boldsymbol{j}]+[\boldsymbol{j}+\boldsymbol{k}-\frac{1}{2}\boldsymbol{i}-\frac{1}{2}\boldsymbol{j}] = [\frac{1}{2}\boldsymbol{i}+\frac{1}{2}\boldsymbol{j}]+[-\frac{1}{2}\boldsymbol{i}+\frac{1}{2}\boldsymbol{j}+\boldsymbol{k}]$

15. $\boldsymbol{B} = \frac{\boldsymbol{A}\cdot\boldsymbol{B}}{\boldsymbol{A}\cdot\boldsymbol{A}}\boldsymbol{A}+[\boldsymbol{B}-\frac{\boldsymbol{A}\cdot\boldsymbol{B}}{\boldsymbol{A}\cdot\boldsymbol{A}}\boldsymbol{A}] = \frac{28}{6}\boldsymbol{A}+[\boldsymbol{B}-\frac{28}{6}\boldsymbol{A}] = \frac{14}{3}[\boldsymbol{i}+2\boldsymbol{j}-\boldsymbol{k}]+[8\boldsymbol{i}+4\boldsymbol{j}-12\boldsymbol{k}-\frac{14}{3}(\boldsymbol{i}+2\boldsymbol{j}-\boldsymbol{k})] = \frac{14}{3}[\boldsymbol{i}+2\boldsymbol{j}-\boldsymbol{k}]+[\frac{10}{3}\boldsymbol{i}-\frac{16}{3}\boldsymbol{j}-\frac{22}{3}\boldsymbol{k}]$

16. $\frac{\boldsymbol{A}\cdot\boldsymbol{B}}{\boldsymbol{A}\cdot\boldsymbol{A}}A+[\boldsymbol{B}-\frac{\boldsymbol{A}\cdot\boldsymbol{B}}{\boldsymbol{A}\cdot\boldsymbol{A}}\boldsymbol{A}] = \frac{1}{1}\boldsymbol{i}+[\boldsymbol{i}+\boldsymbol{j}+\boldsymbol{k}-\boldsymbol{i}] = \boldsymbol{i}+[\boldsymbol{j}+\boldsymbol{k}]$

17. $\boldsymbol{N} \cdot \overrightarrow{P_0P} = 1(x-2) + 2 \cdot (y-1) = 0$
$\Rightarrow x + 2y = 4$

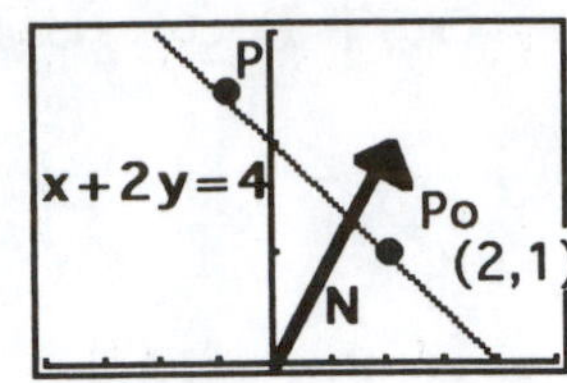

18. $\boldsymbol{N} \cdot \overrightarrow{P_0P} = 1(x-2) - 1(y-1) = 0$
$\Rightarrow x - y = -3$

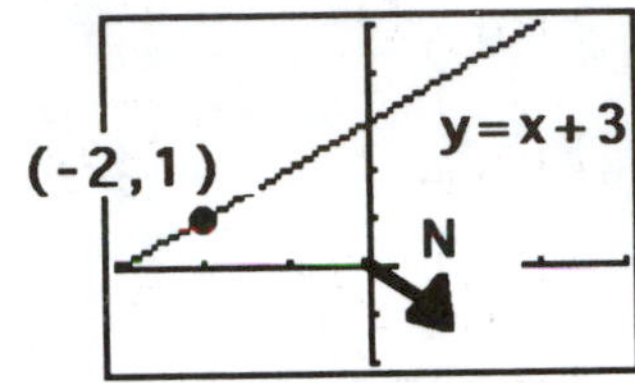

19. $\boldsymbol{N} \cdot \overrightarrow{P_0P} = -2(x+1) - 1(y-2) = 0$
$\Rightarrow -2x - y = 0$

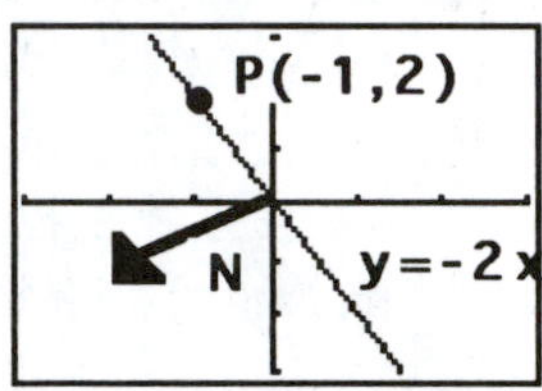

20. $\boldsymbol{N} \cdot \overrightarrow{P_0P} = 2(x+1) - 3(y-2) = 0$
$\Rightarrow 2x - 3y = -8$

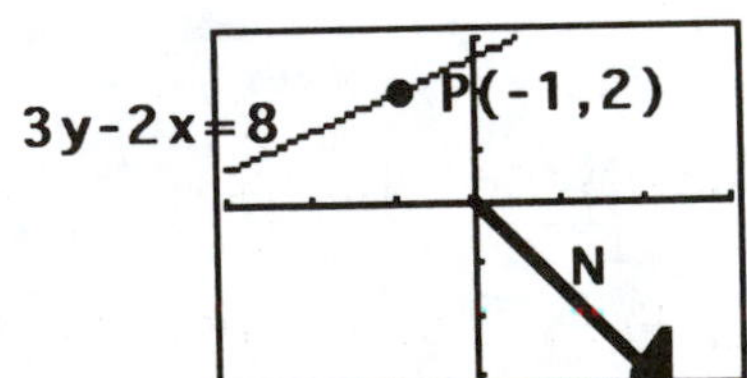

21. The point $P(0,2)$ is on the line. Then $\overrightarrow{PS} = (2-0)\boldsymbol{i} + (8-2)\boldsymbol{j} = 2\boldsymbol{i} + 2\boldsymbol{j}$; $\boldsymbol{N} = \boldsymbol{i} + 3\boldsymbol{j}$. Distance $= |\text{Proj}_{\boldsymbol{N}}\ \overrightarrow{PS}\ | = \left|\overrightarrow{PS} \cdot \frac{\boldsymbol{N}}{|\boldsymbol{N}|}\right| = \frac{2 \cdot 1 + 6 \cdot 3}{\sqrt{1^2+3^2}} = \frac{20}{\sqrt{10}} = 2\sqrt{10}$.

22. Let P be $(0,2)$. $\overrightarrow{PS} = 0\boldsymbol{i} - 2\boldsymbol{j}$; $\boldsymbol{N} = \boldsymbol{i} + 3\boldsymbol{j}$. Distance $= \left|\overrightarrow{PS} \cdot \frac{\boldsymbol{N}}{|\boldsymbol{N}|}\right| = \left|\frac{-6}{\sqrt{10}}\right| = 6/\sqrt{10}$.

23. Let P be $(1,0)$; $\vec{PS}= \boldsymbol{i}+\boldsymbol{j}$; $\boldsymbol{N}=\boldsymbol{i}+\boldsymbol{j}$. Distance $=\left|\frac{1\cdot1+1\cdot1}{\sqrt{2}}\right|=\sqrt{2}$.

24. Let P be $(0,0)$. $\vec{PS}= \boldsymbol{i}+3\boldsymbol{j}$; $\boldsymbol{N}=2\boldsymbol{i}+\boldsymbol{j}$. Distance $=\left|\frac{2+3}{\sqrt{5}}\right|=\sqrt{5}$.

25. $\boldsymbol{A}\cdot\boldsymbol{B}=\frac{1}{\sqrt{3}}\frac{1}{\sqrt{6}}(1(-2)-1(-1)+1(1))=0$; $\boldsymbol{A}\cdot\boldsymbol{B}=\frac{1}{\sqrt{3}}\frac{1}{\sqrt{2}}(0-1(1)+1(1))=0$; $\boldsymbol{B}\cdot\boldsymbol{C}=\frac{1}{\sqrt{2}}\frac{1}{\sqrt{6}}(0-1+1)=0$

26. $\text{Proj}_{\boldsymbol{A}}\boldsymbol{D}=\frac{\boldsymbol{A}.\boldsymbol{D}}{\boldsymbol{A}.\boldsymbol{A}}\boldsymbol{A}=\frac{\frac{1}{\sqrt{3}}-\frac{1}{\sqrt{3}}+\frac{1}{\sqrt{3}}}{\frac{1}{3}(1+1+1)}\boldsymbol{A}=\frac{1}{\sqrt{3}}\boldsymbol{A}=\frac{1}{3}(\boldsymbol{i}-\boldsymbol{j}+\boldsymbol{k})$; $\text{Proj}_{\boldsymbol{B}}\boldsymbol{D}=\frac{\boldsymbol{B}.\boldsymbol{D}}{\boldsymbol{B}.\boldsymbol{B}}\boldsymbol{B}=\frac{\frac{1}{\sqrt{2}}+\frac{1}{\sqrt{2}}}{\frac{1}{2}(1+1)}\boldsymbol{B}=\frac{2}{\sqrt{2}}\boldsymbol{B}=\boldsymbol{j}+\boldsymbol{k}$; $\text{Proj}_{\boldsymbol{C}}\boldsymbol{D}=\frac{\boldsymbol{C}.\boldsymbol{D}}{\boldsymbol{C}.\boldsymbol{C}}\boldsymbol{C}=\frac{\frac{-2}{\sqrt{6}}-\frac{1}{\sqrt{6}}+\frac{1}{\sqrt{6}}}{\frac{1}{6}(4+1+1)}\boldsymbol{C}=\frac{-2}{\sqrt{6}}\boldsymbol{C}=\frac{1}{3}(+2\boldsymbol{i}+\boldsymbol{j}-\boldsymbol{k})$; $\text{Proj}_{\boldsymbol{A}}\boldsymbol{D}+\text{Proj}_{\boldsymbol{B}}\boldsymbol{D}+\text{Proj}_{\boldsymbol{C}}\boldsymbol{D}=\boldsymbol{i}+\boldsymbol{j}+\boldsymbol{k}$.

27. Since $\boldsymbol{i}\cdot\boldsymbol{j}=0$ and $\boldsymbol{i}\cdot\boldsymbol{k}=0$, we have $\boldsymbol{i}\cdot\boldsymbol{j}=\boldsymbol{i}\cdot\boldsymbol{k}$. However, $\boldsymbol{j}\neq\boldsymbol{k}$.

28. $(\boldsymbol{v}_1+\boldsymbol{v}_2)\cdot(\boldsymbol{v}_1-\boldsymbol{v}_2)=\boldsymbol{v}_1\cdot\boldsymbol{v}_1+\boldsymbol{v}_2\cdot\boldsymbol{v}_1-\boldsymbol{v}_1\cdot\boldsymbol{v}_2-\boldsymbol{v}_2\cdot\boldsymbol{v}_2=|\boldsymbol{v}_1|^2-|\boldsymbol{v}_2|^2$. If $\boldsymbol{v}_1$ and $\boldsymbol{v}_2$ have the same lengths their sum is orthogonal to their difference. Example, let $\boldsymbol{v}_1=\boldsymbol{i}$, $\boldsymbol{v}_2=\boldsymbol{j}$; $(\boldsymbol{v}_1+\boldsymbol{v}_2)\cdot(\boldsymbol{v}_1-\boldsymbol{v}_2)=(\boldsymbol{i}+\boldsymbol{j})\cdot(\boldsymbol{i}-\boldsymbol{j})=1-1=0$.

29. $\vec{AB}=3\boldsymbol{i}+\boldsymbol{j}-3\boldsymbol{k}$, $\vec{AC}=2\boldsymbol{i}-2\boldsymbol{j}$, $\vec{BC}=-\boldsymbol{i}-3\boldsymbol{j}+3\boldsymbol{k}$. Recalling that $\vec{BA}=-\vec{AB}$, $\vec{CA}=-\vec{AC}$, we have $< A=\cos^{-1}\left[\frac{\vec{AB}\cdot\vec{AC}}{|\vec{AB}|\cdot|\vec{AC}|}\right]=\cos^{-1}\frac{4}{\sqrt{19}\sqrt{8}}=71.068°$, $< B=\cos^{-1}\left[\frac{\vec{BA}\cdot\vec{BC}}{|\vec{BA}|\cdot|\vec{BC}|}\right]=\cos^{-1}\frac{15}{\sqrt{19}\sqrt{19}}=37.864°$, $< C=180-(< A+< B)=71.068°$

30. $\theta=\cos^{-1}\left[\frac{(\boldsymbol{A}.\boldsymbol{B})}{|\boldsymbol{A}|\cdot|\boldsymbol{B}|}\right]=\cos^{-1}\left[\frac{13}{3\cdot\sqrt{4+100+121}}\right]=\cos^{-1}\left[\frac{13}{45}\right]=73.209°$.

31. For the unit cube, the corner diagonally opposite the origin is at $A(1,1,1)$. The diagonal corner of the face in the xz-plane is at $B(1,0,1)$. The angle between $\vec{OA}$ and $\vec{OB}$ is $\cos^{-1}\left[\frac{1+1}{\sqrt{3}\sqrt{2}}\right]=35.264°$.

32. Use the cube in #31, $\theta=\cos^{-1}\left[\frac{\boldsymbol{i}\cdot(\boldsymbol{i}+\boldsymbol{j}+\boldsymbol{k})}{|\boldsymbol{i}||\boldsymbol{i}+\boldsymbol{j}+\boldsymbol{k}|}\right]=\cos^{-1}\left[\frac{1}{\sqrt{3}}\right]=54.736°$.

33. Work $=\boldsymbol{F}\cdot\vec{PQ}=(-5\boldsymbol{k})\cdot(\boldsymbol{i}+\boldsymbol{j}+\boldsymbol{k})=-5\ N\cdot m$

34. Work $=(60000\boldsymbol{i})\cdot(1000\boldsymbol{i})=6\times10^7\ N\cdot m$.

35. Work $=|\boldsymbol{F}||\vec{PQ}|\cos\theta=(200)(20)\cos41°=3018.838\ N\cdot m$.

36. Work $= (4000)(1000)\cos 60° = 2{,}000{,}000\ N \cdot m$.

37. These curves are straight lines; their tangents (and, thus, their normals) always have the same directions. Find the angle between the normals: $\boldsymbol{N}_1 = 3\boldsymbol{i} + \boldsymbol{j}$, $\boldsymbol{N}_2 = 2\boldsymbol{i} - \boldsymbol{j}$. $\theta = \cos^{-1}\frac{6-1}{\sqrt{10}\sqrt{5}} = \cos^{-1}\frac{1}{\sqrt{2}} = 45°$ or $135°$.

38. $-2x+y=-1$, $5x+y=3 \Rightarrow \boldsymbol{N}_1 = -2\boldsymbol{i}+\boldsymbol{j}$, $\boldsymbol{N}_2 = 5\boldsymbol{i}+\boldsymbol{j}$. $\theta = \cos^{-1}\frac{-10+1}{\sqrt{5}\sqrt{26}} = 142.125°$.

39. The curves intersect at $x = -1.92630321991$ (use technology), at that point, by evaluating the derivative, the slope of the tangent to $y = x^2 - 2$ is $B = -3.85260643982$; the other tangent has slope $C = -0.292287566682$. A set of tangent vectors is $\boldsymbol{i} + B\boldsymbol{j}$ and $\boldsymbol{i} + C\boldsymbol{j}$; $\theta = \cos^{-1}\left[\frac{1+B\cdot C}{\sqrt{B^2+1}\sqrt{C^2+1}}\right] = \cos^{-1} 0.512699775941 = 59.156°$.

40. The curves intersect at $x = 0$ and $x = 1$. At $x = 0$, $(x^3)' = 3x^2 = 0$; $y = \sqrt{x}$ has a vertical tangent $\Rightarrow \theta = 90°$. At $x = 1$, $(x^3)' = 3x^2 = 3$, $(\sqrt{x})' = \frac{1}{2}$; i.e., the tangent lines have slopes 3 and $\frac{1}{2}$. A set of tangent vectors is $3\boldsymbol{i} - \boldsymbol{j}$ and $\boldsymbol{i} - 2\boldsymbol{j}$. $\theta = \cos^{-1}\left[\frac{3+2}{\sqrt{10}\sqrt{5}}\right] = \cos^{-1}\frac{1}{\sqrt{2}} = 45°$ or $135°$.

41. The curves intersect at $x = 0$ and $x = -1.406$; at $x = 0$, the tangents have slopes 0 and 2; $\theta = \cos^{-1}\left[\frac{1}{\sqrt{5}}\right] = 63.4°$. At $x = -1.406$, the tangents have slope -0.811 and 0.986; $\theta = \cos^{-1}\left[\frac{1+(-0.811)(0.986)}{\sqrt{(0.811)^2+1}\sqrt{(0.986)^2+1}}\right] = 1.460$ radians $\approx 83.7°$. See #39 for more details.

42. The curves intersect at $x = 0.386$, where the tangents have slopes -1.228 and 0.926. As in #39, $\theta = \cos^{-1}\left[\frac{1+(-1.228)(0.926)}{\sqrt{1+(1.228)^2}\sqrt{1+(0.926)^2}}\right] = 1.429$ radians $\approx 81.86°$. The curves also intersect at $x = 1.962$ where the tangents have slopes 1.923 and -0.381. $\theta = \cos^{-1}\left[\frac{1+(1.923)(-0.381)}{\sqrt{1+(1.923)^2}\sqrt{(1+(0.381)^2}}\right] = 19.72°$.

43. $\boldsymbol{A}\cdot\boldsymbol{B} = |\boldsymbol{A}||\boldsymbol{B}|\cos\theta = \frac{|\boldsymbol{A}|^2+|\boldsymbol{B}|^2-|\boldsymbol{C}|^2}{2} = \frac{1}{2}\left[a_1^2+a_2^2+a_3^2+b_1^2+b_2^2+b_3^2 - \{(b_1-a_1)^2+(b_2-a_2)^2+(b_3-a_3)^2\}\right] = \frac{1}{2}\left[2a_1b_1+2a_2b_2+2a_3b_3\right] = a_1b_1+a_2b_2+a_3b_3$.

11.4 CROSS PRODUCTS

1. $A \times B = \begin{vmatrix} i & j & k \\ 2 & -2 & -1 \\ 1 & 0 & -1 \end{vmatrix} = i(2-0) + j(-1+2) + k(0+2) = 2i + j + 2k$;
$|A \times B| = \sqrt{4+1+4} = 3$, direction $= [\frac{2}{3}i + \frac{1}{3}j + \frac{2}{3}k]$. $B \times A = -(A \times B)$; length $= 3$, direction $= -[\frac{2}{3}i + \frac{1}{3}j + \frac{2}{3}k]$. In Problems 2-8, $A \times B$ and $B \times A$ have the same length and (when it exists) opposite direction.

2. $A \times B = \begin{vmatrix} i & j & k \\ 2 & 3 & 0 \\ -1 & 1 & 0 \end{vmatrix} = 5k$, length 5, direction is k. $B \times A = -5k$; length 5, direction is $-k$.

3. $A \times B = \begin{vmatrix} i & j & k \\ 2 & -2 & 4 \\ -1 & 1 & -2 \end{vmatrix} = (4-4)i + (-4+4)j + (2-2)k = \mathbf{0}$.
Length $= 0$; $A \times B$ has no direction. $B \times A = 0$, has length 0 and no direction.

4. $A \times B = \begin{vmatrix} i & j & k \\ 1 & 1 & -1 \\ 0 & 0 & 0 \end{vmatrix} = \mathbf{0}$ has length 0, no direction; $B \times A = \mathbf{0}$, length 0, no direction

5. $A \times B = \begin{vmatrix} i & j & k \\ 2 & 0 & 0 \\ 0 & -3 & 0 \end{vmatrix} = -6k$; length $= 6$, direction $= -k$.

6. $A \times B = \begin{vmatrix} i & j & k \\ 1 & 1 & 0 \\ 0 & 1 & 1 \end{vmatrix} = i - j + k$, length $\sqrt{3}$, direction $\frac{1}{\sqrt{3}}(i - j + k)$; $B \times A = -i + j - k$; length $\sqrt{3}$, direction $-\frac{1}{\sqrt{3}}(i - j + k)$.

7. $A \times B = \begin{vmatrix} i & j & k \\ -8 & -2 & -4 \\ 2 & 2 & 1 \end{vmatrix} = (-2+8)i + (-8+8)j + (-16+4)k = 6i - 12k$;
length $= \sqrt{36+144} = \sqrt{180} = 6\sqrt{5}$; direction is $\frac{1}{\sqrt{5}}i - \frac{2}{\sqrt{5}}k$; $B \times A$ has length $6\sqrt{5}$, opposite direction.

8. $\boldsymbol{A}\times\boldsymbol{B} = \begin{vmatrix} \boldsymbol{i} & \boldsymbol{j} & \boldsymbol{k} \\ \frac{3}{2} & -\frac{1}{2} & 1 \\ 1 & 1 & -2 \end{vmatrix} = \boldsymbol{i}(-1-1)+\boldsymbol{j}(1-3)+\boldsymbol{k}(\frac{3}{2}+\frac{1}{2}) = -2\boldsymbol{i}-2\boldsymbol{j}+2\boldsymbol{k};$
length $= 2\sqrt{3}$, direction $\frac{1}{\sqrt{3}}(-\boldsymbol{i}-\boldsymbol{j}+\boldsymbol{k})$; $\boldsymbol{B}\times\boldsymbol{A}$ has same length, opposite direction.

9. $\boldsymbol{A}\times\boldsymbol{B} = \boldsymbol{k}$

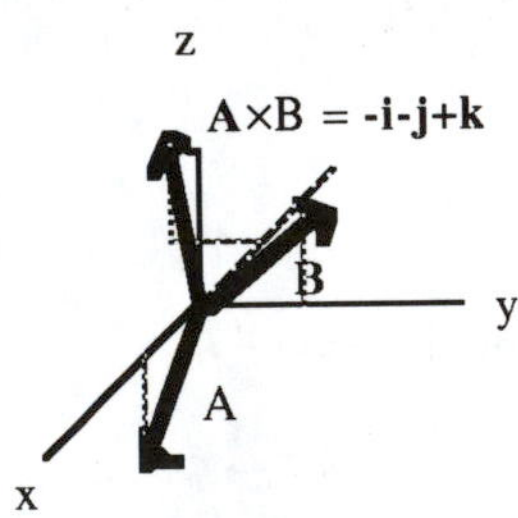

10. $\boldsymbol{A}\times\boldsymbol{B} = \begin{vmatrix} \boldsymbol{i} & \boldsymbol{j} & \boldsymbol{k} \\ 1 & 0 & 1 \\ 0 & 1 & 0 \end{vmatrix} = -\boldsymbol{i}+\boldsymbol{k}$

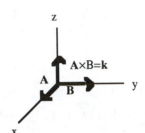

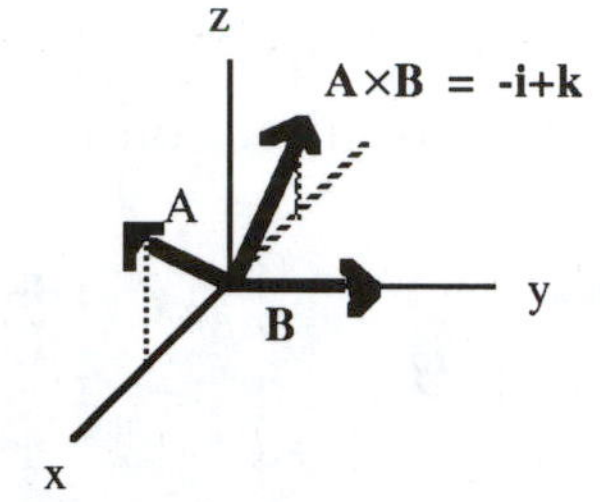

11. $\boldsymbol{A}\times\boldsymbol{B} = \begin{vmatrix} \boldsymbol{i} & \boldsymbol{j} & \boldsymbol{k} \\ 1 & 0 & -1 \\ 1 & 1 & 1 \end{vmatrix} = \boldsymbol{i}-\boldsymbol{j}+\boldsymbol{k}$

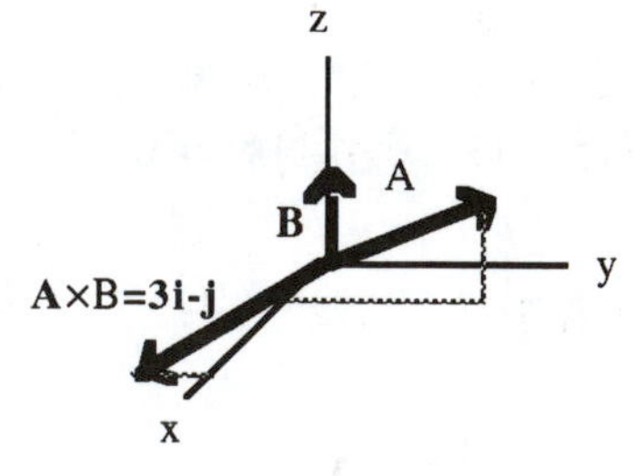

12. $\boldsymbol{A}\times\boldsymbol{B} = \begin{vmatrix} \boldsymbol{i} & \boldsymbol{j} & \boldsymbol{k} \\ 2 & -1 & 0 \\ 1 & 2 & 0 \end{vmatrix} = 5\boldsymbol{k}$

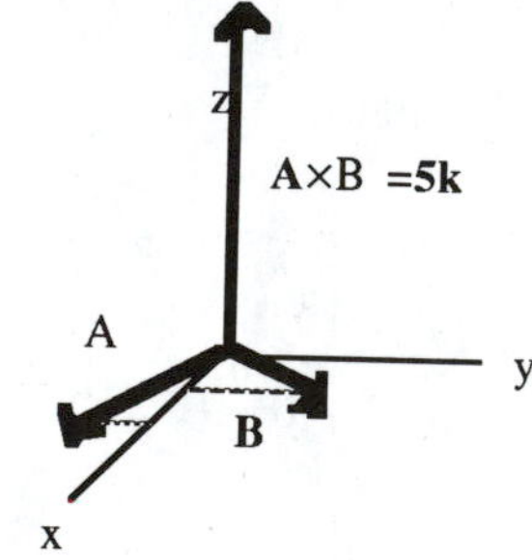

13. $\boldsymbol{A}\times\boldsymbol{B} = \begin{vmatrix} \boldsymbol{i} & \boldsymbol{j} & \boldsymbol{k} \\ 1 & 3 & 2 \\ 0 & 0 & 1 \end{vmatrix} = 3\boldsymbol{i}-\boldsymbol{j}$

14. $\boldsymbol{A}\times\boldsymbol{B} = \begin{vmatrix} \boldsymbol{i} & \boldsymbol{j} & \boldsymbol{k} \\ 1 & 2 & 0 \\ 0 & 2 & 1 \end{vmatrix} = 2\boldsymbol{i}-\boldsymbol{j}+2\boldsymbol{k}$

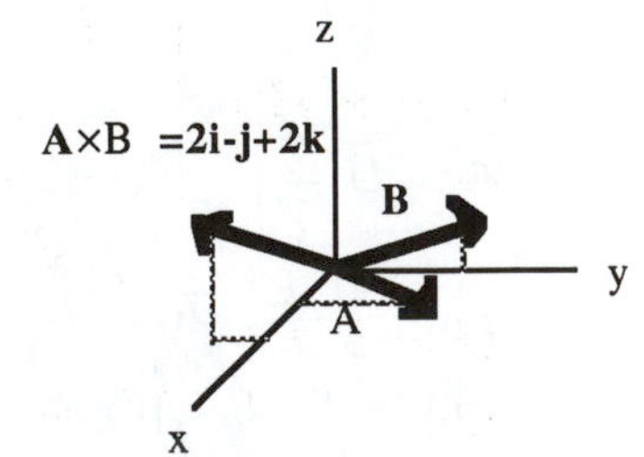

15. a) $\boldsymbol{N} = \pm\, \overrightarrow{PQ} \times \overrightarrow{PR} = \pm \begin{vmatrix} \boldsymbol{i} & \boldsymbol{j} & \boldsymbol{k} \\ 1 & 1 & -3 \\ -1 & 3 & -1 \end{vmatrix} = \pm(8\boldsymbol{i} + 4\boldsymbol{j} + 4\boldsymbol{k})$;

b) $\frac{1}{2}\,|\,\overrightarrow{PQ} \times \overrightarrow{PR}\,| = \frac{1}{2}\sqrt{64+16+16} = \frac{1}{2}(4\sqrt{6}) = 2\sqrt{6}$;

c) $\frac{\boldsymbol{N}}{|\boldsymbol{N}|} = \frac{1}{\sqrt{6}}(2\boldsymbol{i} + \boldsymbol{j} + \boldsymbol{k})$

16. a) $\boldsymbol{N} = \pm\, \overrightarrow{PQ} \times \overrightarrow{PR} = \pm \begin{vmatrix} \boldsymbol{i} & \boldsymbol{j} & \boldsymbol{k} \\ 1 & 0 & 2 \\ 2 & -2 & 0 \end{vmatrix} = \pm(4\boldsymbol{i} + 4\boldsymbol{j} - 2\boldsymbol{k})$,

b) $\frac{1}{2}|\boldsymbol{N}| = \frac{1}{2}\sqrt{16+16+4} = 3$; c) $\frac{\boldsymbol{N}}{|\boldsymbol{N}|} = \frac{1}{6}(4\boldsymbol{i}+4\boldsymbol{j}-2\boldsymbol{k}) = \pm\frac{1}{3}(2\boldsymbol{i}+2\boldsymbol{j}-\boldsymbol{k})$

17. a) $\boldsymbol{N} = \pm\, \overrightarrow{PQ} \times \overrightarrow{PR} = \pm \begin{vmatrix} \boldsymbol{i} & \boldsymbol{j} & \boldsymbol{k} \\ 1 & 1 & 1 \\ 1 & 1 & 0 \end{vmatrix} = \pm(-\boldsymbol{i} + \boldsymbol{j})$; b) $\frac{1}{2}|\boldsymbol{N}| = \frac{1}{2}\sqrt{2} = 1/\sqrt{2}$; c) $\frac{\boldsymbol{N}}{|\boldsymbol{N}|} = \pm\frac{1}{\sqrt{2}}(-\boldsymbol{i} + \boldsymbol{j})$

18. a) $\boldsymbol{N} = \pm\, \overrightarrow{PQ} \times \overrightarrow{PR} = \pm \begin{vmatrix} \boldsymbol{i} & \boldsymbol{j} & \boldsymbol{k} \\ 2 & -1 & -1 \\ 1 & 0 & -2 \end{vmatrix} = \pm(2\boldsymbol{i} + 3\boldsymbol{j} + \boldsymbol{k})$;

b) $\frac{1}{2}|\boldsymbol{N}| = \frac{1}{2}\sqrt{4+9+1} = \frac{1}{2}\sqrt{14}$; c) $\frac{\boldsymbol{N}}{|\boldsymbol{N}|} = \pm\frac{2}{\sqrt{14}}(2\boldsymbol{i} + 3\boldsymbol{j} + \boldsymbol{k})$

19. $\boldsymbol{A} \cdot \boldsymbol{B} = 5 \cdot 0 + (-1)(1) + (1)(-5) = -6$; $\boldsymbol{A} \cdot \boldsymbol{C} = 5 \cdot 15 + (-1)3 + 1(-3) \neq 0$; $\boldsymbol{B} \cdot \boldsymbol{C} = 0(-15) + 1(3) - 5 \cdot 3 \neq 0$. No two are perpendicular. $\boldsymbol{A} \times \boldsymbol{B} = \begin{vmatrix} \boldsymbol{i} & \boldsymbol{j} & \boldsymbol{k} \\ 5 & -1 & 1 \\ 0 & 1 & -5 \end{vmatrix} = 4\boldsymbol{i} + 25\boldsymbol{j} + 5\boldsymbol{k} \neq \boldsymbol{0}$; $\boldsymbol{A} \times \boldsymbol{C} = \begin{vmatrix} \boldsymbol{i} & \boldsymbol{j} & \boldsymbol{k} \\ 5 & -1 & 1 \\ -15 & 3 & -3 \end{vmatrix} = \boldsymbol{0}$;

$\boldsymbol{B} \times \boldsymbol{C} = \begin{vmatrix} \boldsymbol{i} & \boldsymbol{j} & \boldsymbol{k} \\ 0 & 1 & -5 \\ -15 & 3 & -3 \end{vmatrix} = 12\boldsymbol{i} + \cdots$(not $\boldsymbol{0}$). $\boldsymbol{A}$ and $\boldsymbol{C}$ are parallel.

20. a) $\boldsymbol{A} \cdot \boldsymbol{B} = 0$, $\boldsymbol{A} \cdot \boldsymbol{C} = 0$, $\boldsymbol{B} \cdot \boldsymbol{C} = 0 \Rightarrow \boldsymbol{A} \perp \boldsymbol{B}$, $\boldsymbol{A} \perp \boldsymbol{C}$, $\boldsymbol{B} \perp \boldsymbol{C}$,

b) $\boldsymbol{A} \times \boldsymbol{B} = \begin{vmatrix} \boldsymbol{i} & \boldsymbol{j} & \boldsymbol{k} \\ 1 & 2 & -1 \\ -1 & 1 & 1 \end{vmatrix} \neq \boldsymbol{0}$, $\boldsymbol{A} \times \boldsymbol{C} = \begin{vmatrix} \boldsymbol{i} & \boldsymbol{j} & \boldsymbol{k} \\ 5 & -1 & 1 \\ -15 & 3 & -3 \end{vmatrix} \neq \boldsymbol{0}$,

$\boldsymbol{B} \times \boldsymbol{C} = \begin{vmatrix} \boldsymbol{i} & \boldsymbol{j} & \boldsymbol{k} \\ -1 & 1 & 1 \\ -15 & 3 & -3 \end{vmatrix} \neq \boldsymbol{0} \Rightarrow$ no two are parallel.

21. $\boldsymbol{A}\times\boldsymbol{B} = \begin{vmatrix} \boldsymbol{i} & \boldsymbol{j} & \boldsymbol{k} \\ 2 & -1 & 0 \\ 1 & 3 & -2 \end{vmatrix} = 2\boldsymbol{i}+4\boldsymbol{j}+7\boldsymbol{k}$. $(\boldsymbol{A}\times\boldsymbol{B})\cdot\boldsymbol{A} = 2\cdot 2+4(-1)+7(0) = 0$; $(\boldsymbol{A}\times\boldsymbol{B})\cdot\boldsymbol{B} = 2(1)+4(3)+7(-2) = 0$.

22. $\boldsymbol{A}\times\boldsymbol{B}$ is perpendicular to the plane containing $\boldsymbol{A}$ and $\boldsymbol{B} \Rightarrow \boldsymbol{A}\times\boldsymbol{B}\perp\boldsymbol{A}$ and $\boldsymbol{A}\times\boldsymbol{B}\perp\boldsymbol{B} \Rightarrow (\boldsymbol{A}\times\boldsymbol{B})\cdot\boldsymbol{A} = (\boldsymbol{A}\times\boldsymbol{B})\cdot\boldsymbol{B} = 0$.

23. a) $\text{Proj}_{\boldsymbol{B}}\boldsymbol{A} = \frac{\boldsymbol{A}\cdot\boldsymbol{B}}{\boldsymbol{B}\cdot\boldsymbol{B}}\boldsymbol{B}$, b) $\boldsymbol{A}\times\boldsymbol{B}$, c) $\frac{\sqrt{\boldsymbol{A}\cdot\boldsymbol{A}}}{\sqrt{\boldsymbol{B}\cdot\boldsymbol{B}}}\boldsymbol{B}$, d) $(\boldsymbol{A}\times\boldsymbol{B})\times\boldsymbol{C}$,

e) $(\boldsymbol{B}\times\boldsymbol{C})\times\boldsymbol{A}$

24. $\boldsymbol{i}\times(\boldsymbol{i}) = \boldsymbol{0} = \boldsymbol{i}\times(7\boldsymbol{i})$, but $\boldsymbol{i}\neq 7\boldsymbol{i}$.

25. $|\overrightarrow{PQ}| = 2/3$ ft. $|\boldsymbol{F}\times\overrightarrow{PQ}| = |\boldsymbol{F}|\cdot|\overrightarrow{PQ}|\sin 55^\circ = (30)(\frac{2}{3})\sin 55^\circ = 16.383$ foot-pounds.

26. $|\overrightarrow{PQ}| = 2/3$. $|\boldsymbol{F}\times\overrightarrow{PQ}| = |\boldsymbol{F}|\cdot|\overrightarrow{PQ}|\sin 125^\circ = (30)(2/3)\sin 125^\circ = 16.383$ foot-pounds.

27. – 30.

	$\boldsymbol{A}\times\boldsymbol{B}$	$(\boldsymbol{A}\times\boldsymbol{B})\cdot\boldsymbol{C}$	$\boldsymbol{B}\times\boldsymbol{C}$	$(\boldsymbol{B}\times\boldsymbol{C})\cdot\boldsymbol{A}$	$\boldsymbol{C}\times\boldsymbol{A}$	$(\boldsymbol{C}\times\boldsymbol{A})\cdot\boldsymbol{B}$,	Vol.
27.	$4\boldsymbol{k}$	8	$4\boldsymbol{i}$	8	$4\boldsymbol{j}$	8	8
28.	$\boldsymbol{i}+4\boldsymbol{j}+3\boldsymbol{k}$	4	$3\boldsymbol{i}+4\boldsymbol{j}-5\boldsymbol{k}$	4	$\boldsymbol{i}-\boldsymbol{k}$	4	4
29.	$\boldsymbol{i}-2\boldsymbol{j}-4\boldsymbol{k}$	-7	$-2\boldsymbol{i}-3\boldsymbol{j}+\boldsymbol{k}$	-7	$-2\boldsymbol{i}+4\boldsymbol{j}+\boldsymbol{k}$	-7	7
30.	$-\boldsymbol{i}+3\boldsymbol{j}+\boldsymbol{k}$	8	$4(\boldsymbol{i}-\boldsymbol{j}-\boldsymbol{k})$	8	$-6\boldsymbol{i}+2\boldsymbol{j}-2\boldsymbol{k}$	8	8

11.5 LINES AND PLANES IN SPACE

1. $x-3=t,\ y+4=t,\ z+1=t$, or $x=3+1\cdot t,\ y=-4+1\cdot t,\ z=-3+1\cdot t$

2. $x-2=4t,\ y-3=-2t,\ z+1=3t$, or $x=2+4t,\ y=3-2t,\ z=-1+3t$

3. Direction of $\overrightarrow{PQ}$ is $-2\boldsymbol{i}-2\boldsymbol{j}+2\boldsymbol{k}$; using $P: x-1=-2t,\ y-2=-2t,\ z+1=2t$, or $x=1-2t,\ y=2-2t,\ z=-1+2t$.

4. $\vec{PQ} = (3+2)\boldsymbol{i} + (5-0)\boldsymbol{j} + (-2-3)\boldsymbol{k} = 5\boldsymbol{i} + 5\boldsymbol{j} - 5\boldsymbol{k} = 5(\boldsymbol{i} + \boldsymbol{j} - \boldsymbol{k})$ is parallel to $\boldsymbol{i} + \boldsymbol{j} - \boldsymbol{k}$. $x = -2 + 1 \cdot t,\ y = 1 \cdot t,\ z = 3 - 1 \cdot t$ is a set of parametric equations; $x = -2 + 5t,\ y = 5t, z = 3 - 5t$ is another set.

5. $\vec{PQ} = -\boldsymbol{j} - \boldsymbol{k}$; using $P : x - 1 = 0,\ y - 2 = -t,\ z - 0 = -t,$ or $x = 1,\ y = 2 - t,\ z = -t$.

6. $\vec{PQ} = (e-3)\boldsymbol{i} + (\sqrt{2} - \pi)\boldsymbol{j} + 9\boldsymbol{k}$; using $P : x - 3 = (e-3)t,\ y - \pi = (\sqrt{2} - \pi)t,\ z + 2 = 9t,$ or $x = 3 + (e-3)t,\ y = \pi + (\sqrt{2} - \pi)t,\ z = -2 + 9t$.

7. Write the vector as $0\boldsymbol{i} + 2\boldsymbol{j} + \boldsymbol{k}$. The line is: $x = 0 + 0t,\ y = 0 + 2t,\ z = 0 + 1 \cdot t$, or $x = 0,\ y = 2t,\ z = t$.

8. The line passes through $(3, -2, 1)$ instead of through $(1, 2, 0)$: $x = 3 + 2t,\ y = -2 - t,\ z = 1 + 3t$.

9. The line is parallel to $0\boldsymbol{i} + 0\boldsymbol{j} + \boldsymbol{j}$; $x = 1,\ y = 1,\ z = 1 + t$.

10. The line has direction $3\boldsymbol{i} + 7\boldsymbol{j} - 5\boldsymbol{k}$: $x = 2 + 3t,\ y = 4 + 7t,\ z = 5 - 5t$.

11. The line is parallel to the normal of the plane: $\boldsymbol{i} + 2\boldsymbol{j} + 2\boldsymbol{k}$. $x = 0 + 1 \cdot t,\ y = -7 + 2t,\ z = 0 + 2t,$ or $x = t,\ y = -7 + 2t,\ z = 2t$.

12. The line has direction $\boldsymbol{A} \times \boldsymbol{B} = \begin{vmatrix} \boldsymbol{i} & \boldsymbol{j} & \boldsymbol{k} \\ 1 & 2 & 3 \\ 3 & 4 & 5 \end{vmatrix} = -2\boldsymbol{i} + 4\boldsymbol{j} - 2\boldsymbol{k}$. Using $P(2, 3, 0)$: $x = 2 - 2t,\ y = 3 + 4t,\ z = -2t$.

13. The x-axis is parallel to $\boldsymbol{i}$; the origin lies on the axis. $x = 0 + 1 \cdot t,\ y = 0 + 0 \cdot t,\ x = 0 + 0 \cdot t,$ or $x = t,\ y = 0,\ z = 0$.

14. The direction is $\boldsymbol{k}, (0, 0, 0)$ is on the axis: $x = 0,\ y = 0,\ z = t$.

15. Let P be $(0, 0, 0)$, Q be $(1, 1, 1)$. Then $\vec{PQ} = \boldsymbol{i} + \boldsymbol{j} + \boldsymbol{k}$ and the line through P and Q has equations $x = t,\ y = t,\ z = t$. When $t = 0,\ (x, y, z) = (0, 0, 0)$ or P; when $t = 1,\ (x, y, z) = (1, 1, 1)$ or Q

the segment is: $x = t,\ y = t,\ z = t,\ 0 \le t \le 1$.

Problems 16-22 are similar to this one.

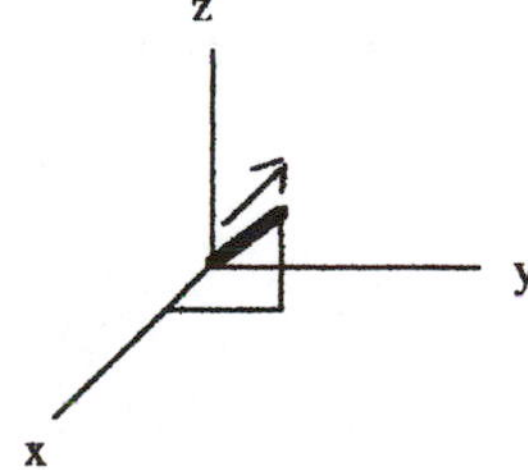

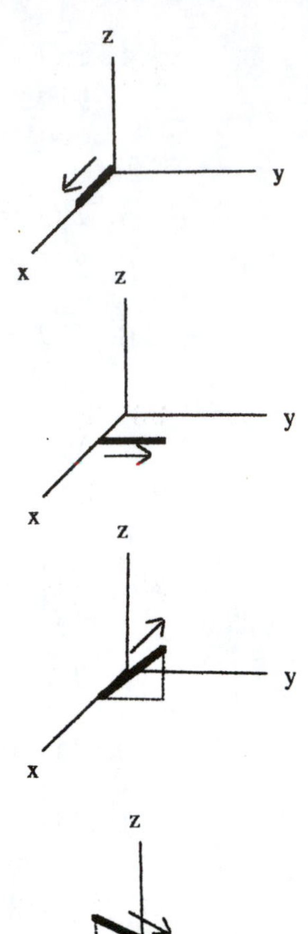

16. $\vec{PQ} = \boldsymbol{i}$: using $P(0,0,0)$,
$x = t,\ y = 0,\ z = 0,\ 0 \le t \le 1$.

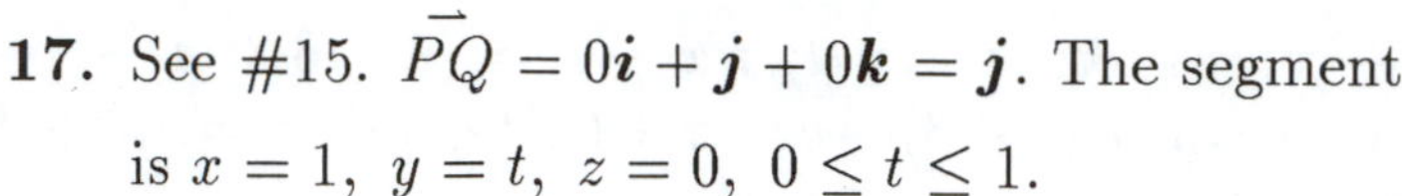

17. See #15. $\vec{PQ} = 0\boldsymbol{i} + \boldsymbol{j} + 0\boldsymbol{k} = \boldsymbol{j}$. The segment is $x = 1,\ y = t,\ z = 0,\ 0 \le t \le 1$.

18. $\vec{PQ} = \boldsymbol{k}$; using $P(1,1,0)$,
$x = 1,\ y = 1,\ z = t,\ 0 \le t \le 1$.

19. See #15. $\vec{PQ} = 2\boldsymbol{j}$. The segment is
$x = 0,\ y = -1 + 2t,\ z = 1,\ 0 \le t \le 1$.

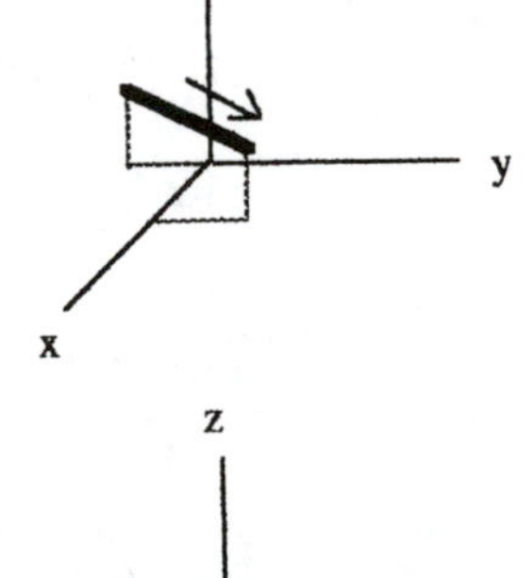

20. $\vec{PQ} = -3\boldsymbol{i} + 2\boldsymbol{j}$; using $P(3,0,0)$,
$x = 3 - 3t,\ y = 2t,\ z = 0,\ 0 \le t \le 1$.

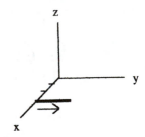

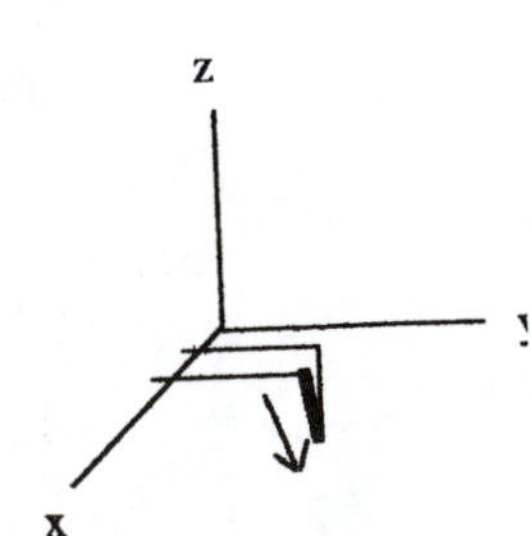

21. See #15. $\vec{PQ} = -\boldsymbol{i} - 2\boldsymbol{k}$. The segment is
$x = 2 - t,\ y = 2,\ z = -2t,\ 0 \le t \le 1$.

22. $\vec{PQ} = -\boldsymbol{i} + 3\boldsymbol{j} + 3\boldsymbol{k}$; using $P(1,-1,-2)$,

$x = 1 - t,\ y = -1 + 3t,$

$z = -2 + 3t,\ 0 \le t \le 1$

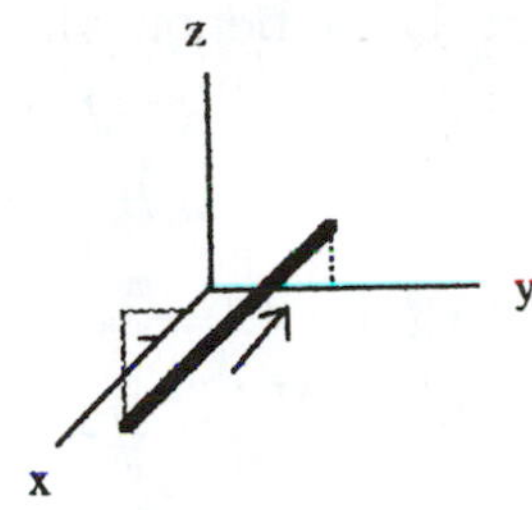

23. $3(x-0) - 2(y-2) - 1(z+1) = 0 \Rightarrow 3x - 2y - z = -3$

24. $3(x-1) + 1\cdot(y+1) + 1\cdot(z-3) = 0 \Rightarrow 3x + y + z = 5$

25. Label the points $A(1,1{-}1)$, $B(2,0,2)$ and $C(0,-2,1)$. $\vec{AB} = \boldsymbol{i}-\boldsymbol{j}+3\boldsymbol{k}$, $\vec{BC} = -2\boldsymbol{i} - 2\boldsymbol{j} - \boldsymbol{k}$. The vector $\vec{AB} \times \vec{BC} = \begin{vmatrix} \boldsymbol{i} & \boldsymbol{j} & \boldsymbol{k} \\ 1 & -1 & 3 \\ -2 & -2 & -1 \end{vmatrix} = 7\boldsymbol{i} - 5\boldsymbol{j} - 4\boldsymbol{k}$ is normal to the plane. Since the plane passes through A, it has equation $7(x-1) - 5(y-1) - 4(z+1) = 0$, or $7x - 5y - 4z = 6$.

26. See#25. $\boldsymbol{N} = (-\boldsymbol{i}+\boldsymbol{j}+2\boldsymbol{k}) \times (-2\boldsymbol{i}+\boldsymbol{j}+\boldsymbol{k}) = \begin{vmatrix} \boldsymbol{i} & \boldsymbol{j} & \boldsymbol{k} \\ -1 & 1 & 2 \\ -2 & 1 & 1 \end{vmatrix} = -\boldsymbol{i} - 3\boldsymbol{j} + \boldsymbol{k}$. Evaulate $-x-3y+z$ at $(2,4,5)$ to obtain $-x-3y+z = -9$ or $x+3y-z = 9$.

27. The vector $1\boldsymbol{i}+3\boldsymbol{j}+4\boldsymbol{k}$ is normal to the plane. $(x-2)+3(y-4)+4(z-5) = 0$, or $x + 3y + 4z = 34$.

28. $\vec{0A} = \boldsymbol{i} - 2\boldsymbol{j} + \boldsymbol{k}$ is normal to the plane. $1\cdot(x-1) - 2(y+2) + 1\cdot(z-1) = 0 \Rightarrow x - 2y + z = 6$.

29. Let $Q(4t,-2t,2t)$ be on the line; let $f(t) = |\ \vec{QP}\ |^2 = (4t)^2 + (-2t)^2 + (2t-12)^2$. Then $f'(t) = 2\cdot 4t\cdot 4 + 2(-2t)(-2) + 2(2t-12)\cdot 2 = 32t + 8t + 8t - 48 = 48(t-1)$. $f'(t) = 0$ at $t = 1$. $\sqrt{f(1)} = \sqrt{16+4+100} = \sqrt{120}$.

30. Let $Q(t)$ be on the line: minimize the square of the distance from P to Q : $f(t) = (5+3t)^2 + (5+4t)^2 + (-3-5t)^2 \Rightarrow f'(t) = 6(5+3t) + 8(5+4t) + 10(3+5t) = 100t + 100$. $f'(-1) = 0 \Rightarrow$ distance $= \sqrt{f(-1)} = 3$.

31. The point is <u>on</u> the line $\Rightarrow$ distance $= 0$.

32. Let $Q(t)$ be on the line; minimize the square of the distance from P to $Q: f(t) = (2t-2)^2+(1+2t-1)^2+(2t+1)^2 \Rightarrow f'(t) = 4(2t-2)+8t+4(2t+1) = 24t - 4 \Rightarrow f'(\frac{1}{6}) = 0$, minimum distance is $\sqrt{f(\frac{1}{6})} = \sqrt{\frac{28}{6}} = \sqrt{\frac{14}{3}}$.

33. Let S be $(2, -3, 4)$; then the point $P(13, 0, 0)$ is on the plane. Distance $= \left|\vec{PS} \cdot \frac{\boldsymbol{N}}{|\boldsymbol{N}|}\right| = \left|(-11\boldsymbol{i} - 3\boldsymbol{j} + 4\boldsymbol{k}) \cdot \frac{(\boldsymbol{i} + 2\boldsymbol{j} + 2\boldsymbol{k})}{\sqrt{9}}\right| = \frac{|-11-6+8|}{3} = 3.$

34. Let $S = (0, 0, 0)$; $P(2, 0, 0)$ is on the plane; $\vec{PS} = -2\boldsymbol{i}$. Distance $= \left|\vec{PS} \cdot \frac{\boldsymbol{N}}{|\boldsymbol{N}|}\right| = \left|(-2\boldsymbol{i}) \cdot \frac{(3\boldsymbol{i} + 2\boldsymbol{j} + 6\boldsymbol{k})}{\sqrt{9 + 4 + 36}}\right| = \frac{|-6|}{\sqrt{49}} = \frac{6}{7}.$

35. See # 33: $S(0, 1, 1)$. Let P be $(0, 0, -4)$. Distance $= \left|\vec{PS} \cdot \frac{\boldsymbol{N}}{|\boldsymbol{N}|}\right| = \left|(\boldsymbol{j} + 5\boldsymbol{k}) \cdot \frac{(4\boldsymbol{j} + 3\boldsymbol{k})}{\sqrt{25}}\right| = \frac{|4+15|}{5} = \frac{19}{5}.$

36. See #33: $S(2, 2, 3)$. Let $P = (2, 0, 0)$. Distance $= \left|\vec{PS} \cdot \frac{\boldsymbol{N}}{|\boldsymbol{N}|}\right| = \left|(2\boldsymbol{j} + 3\boldsymbol{k}) \cdot \frac{(2\boldsymbol{i} + \boldsymbol{j} + 2\boldsymbol{k})}{\sqrt{9}}\right| = \frac{8}{3}.$

37. See #33: $S(0, -1, 0)$. Let $P = (0, 4, 0)$. Distance $= \left|\vec{PS} \cdot \frac{\boldsymbol{N}}{|\boldsymbol{N}|}\right| = \left|(-5\boldsymbol{j}) \cdot \frac{(2\boldsymbol{i} + \boldsymbol{j} + 2\boldsymbol{k})}{\sqrt{9}}\right| = \frac{|-5|}{3} = \frac{5}{3}.$

38. See #33. Let $P = (0, 0, 4)$. Distance $= \left|\vec{PS} \cdot \frac{\boldsymbol{N}}{|\boldsymbol{N}|}\right| = \left|(\boldsymbol{i} - 5\boldsymbol{k}) \cdot \frac{(-4\boldsymbol{i} + \boldsymbol{j} + \boldsymbol{k})}{\sqrt{18}}\right| = \frac{9}{\sqrt{18}} = \frac{3}{\sqrt{2}}.$

39. $2(1-t) - (3t) + 3(1+t) = 6 \Rightarrow -2t = 1 \Rightarrow t = -\frac{1}{2} \Rightarrow x = \frac{3}{2}, \ y = -\frac{3}{2}, \ z = \frac{1}{2}.$

40. $6(2) + 3(3 + 2t) - 4(-2 - 2t) = -12 \Rightarrow t = -\frac{41}{14} \Rightarrow x = 2, \ y = 3 - \frac{41}{7} = -\frac{20}{7}, \ z = -2 + \frac{41}{7} = \frac{27}{7}.$

41. $(1 + 2t) + (1 + 5t) + (3t) = 2 \Rightarrow 10t = 0 \Rightarrow t = 0 \Rightarrow x = 1, \ y = 1, \ z = 0.$

42. $2(-1 + 3t) - 3(5t) = 7 \Rightarrow t = -1 \Rightarrow x = -4, \ y = -2, \ z = -5.$

43. Find the angle between the normals $\boldsymbol{N}_1 = \boldsymbol{i} + \boldsymbol{j}$ and $\boldsymbol{N}_2 = 2\boldsymbol{i} + \boldsymbol{j} - 2\boldsymbol{k}$. $\theta = \cos^{-1}\left[\frac{\boldsymbol{N}_1 \cdot \boldsymbol{N}_2}{|\boldsymbol{N}_1||\boldsymbol{N}_2|}\right] = \cos^{-1}\left[\frac{3}{\sqrt{2}\sqrt{9}}\right] = \cos^{-1}\frac{1}{\sqrt{2}} = 45°.$

44. $\theta = \cos^{-1}\left[\frac{\boldsymbol{N}_1\cdot\boldsymbol{N}_2}{|\boldsymbol{N}_1||\boldsymbol{N}_2|}\right] = \cos^{-1}\left[\frac{(5\boldsymbol{i}+\boldsymbol{j}-\boldsymbol{k})\cdot(\boldsymbol{i}-2\boldsymbol{j}+3\boldsymbol{k})}{\sqrt{27}\sqrt{12}}\right] = \cos^{-1}[0] =$ $90°$.

45. See #43: $\theta = \cos^{-1}\left[\frac{\boldsymbol{N}_1\cdot\boldsymbol{N}_2}{|\boldsymbol{N}_1||\boldsymbol{N}_2|}\right] = \cos^{-1}\left[\frac{4-4-2}{\sqrt{12}\sqrt{9}}\right] = \cos^{-1}\left[\frac{-2}{\sqrt{108}}\right] = 101.096°$.

46. See #43: $\theta = \cos^{-1}\left[\frac{\boldsymbol{N}_1\cdot\boldsymbol{N}_2}{|\boldsymbol{N}_1||\boldsymbol{N}_2|}\right] = \cos^{-1}\left[\frac{(\boldsymbol{i}+\boldsymbol{j}+\boldsymbol{k})\cdot(\boldsymbol{k})}{\sqrt{3}\cdot 1}\right] = \cos^{-1}\frac{1}{\sqrt{3}} =$

$54.736°$.

47. See #43: $\theta = \cos^{-1}\left[\frac{\boldsymbol{N}_1\cdot\boldsymbol{N}_2}{|\boldsymbol{N}_1||\boldsymbol{N}_2|}\right] = \cos^{-1}\left[\frac{2+4-1}{\sqrt{9}\sqrt{6}}\right] = \cos^{-1}\frac{5}{\sqrt{54}} = 47.124°$.

48. See #43: $\theta = \cos^{-1}\left[\frac{(4\boldsymbol{j}+3\boldsymbol{k})\cdot(3\boldsymbol{i}+2\boldsymbol{j}+6\boldsymbol{k})}{\sqrt{25}\sqrt{49}}\right] = \cos^{-1}\left[\frac{26}{35}\right] = 42.025°$.

49. The line is parallel to the cross product $\boldsymbol{v}$ of the normals: $\boldsymbol{v} = \begin{vmatrix} \boldsymbol{i} & \boldsymbol{j} & \boldsymbol{k} \\ 1 & 1 & 1 \\ 1 & 1 & 0 \end{vmatrix} =$ $-\boldsymbol{i}+\boldsymbol{j}$. Find a point on the line by setting $x = 0$ and solving $y+z = 1,\ y = 2$ to get $(0, 2, -1)$ is on the line: $x = -t,\ y = 2+t,\ z = -1$.

50. The line is parallel to $\boldsymbol{N}_1 \times \boldsymbol{N}_2 = \begin{vmatrix} \boldsymbol{i} & \boldsymbol{j} & \boldsymbol{k} \\ 3 & -6 & -2 \\ 2 & 1 & -2 \end{vmatrix} = 14\boldsymbol{i} + 2\boldsymbol{j} + 15\boldsymbol{k}$; set $x = 1$ and solve $-6y - 2z = 0,\ y - 2z = 0$ to find $(1, 0, 0)$ on the line: $x = 1 + 14t,\ y = 2t,\ y = 15t$.

51. $\boldsymbol{v} = \boldsymbol{N}_1 \times \boldsymbol{N}_2 = \begin{vmatrix} \boldsymbol{i} & \boldsymbol{j} & \boldsymbol{k} \\ 1 & -2 & 4 \\ 1 & 1 & -2 \end{vmatrix} = \boldsymbol{i}(0)+\boldsymbol{j}(6)+\boldsymbol{k}3 = 6\boldsymbol{j}+3\boldsymbol{k}$. To find a point on the line, set $z = 0$ and solve $x - 2y = 2,\ x + y = 5$ to get $y = 1,\ x = 4$. Hence $(4, 1, 0)$ is in the line which has equations: $x = 4,\ y = 1 + 6t,\ z = 3t$.

52. The line is parallel to $\boldsymbol{N}_1 \times \boldsymbol{N}_2 = \begin{vmatrix} \boldsymbol{i} & \boldsymbol{j} & \boldsymbol{k} \\ 5 & -2 & 0 \\ 0 & 4 & -5 \end{vmatrix} = 10\boldsymbol{i} + 25\boldsymbol{j} + 20\boldsymbol{k}$ which is parallel to $2\boldsymbol{i} + 5\boldsymbol{j} + 4\boldsymbol{k}$. Set $x = 0$ to obtain $(0, -\frac{11}{2}, -1)$ on the line: $x = 2t,\ y = -\frac{11}{2} + 5t,\ z = -1 + 4t$.

53. Let $A(1,2,3)$; $B(-1,6,2)$ and $C(2,6,3)$ are on the plane; $\boldsymbol{N}$ is parallel to $\vec{AB} \times \vec{AC} = \begin{vmatrix} \boldsymbol{i} & \boldsymbol{j} & \boldsymbol{k} \\ -2 & 4 & -1 \\ 1 & 4 & 0 \end{vmatrix} = 4\boldsymbol{i} - \boldsymbol{j} - 12\boldsymbol{k}$. The plane is

$4(x-1) - 1 \cdot (y-2) - 12(x-3) = 0$, or $4x - y - 12z = -34$.

54. A normal to the plane is given by the cross-product of the direction vectors of L_1 and L_2: $\boldsymbol{N} = \begin{vmatrix} \boldsymbol{i} & \boldsymbol{j} & \boldsymbol{k} \\ 3 & 1 & 1 \\ 5 & 2 & -1 \end{vmatrix} = -3\boldsymbol{i} + 8\boldsymbol{j} + \boldsymbol{k}$. Take an arbitrary point on the plane, say $P(-1,2,-1)$: $-3(x+1) + 8(y-2) + 1(z+1) = 0 \Rightarrow -3x + 8y + z = 18$.

55. The xy-plane has equation $z = 0$; $3t = 0 \Rightarrow t = 0 \Rightarrow x = 1,\ y = -1 : (1,-1,0)$; the yz-plane has equation $x = 0$; $1 + 2t = 0 \Rightarrow t = -\frac{1}{2} \Rightarrow y = -\frac{1}{2},\ z = -\frac{3}{2} : (0, -\frac{1}{2} - \frac{3}{2})$; the xz-plane has equation $y = 0$; $-1 - t = 0 \Rightarrow t = -1 \Rightarrow x = -1,\ z = -3 : (-1, 0 - 3)$.

56. In the xy-plane the line can be represented $x = \sqrt{3}t,\ y = t$. Since $z = 3$ the line is: $x = \sqrt{3}t,\ y = t,\ z = 3$.

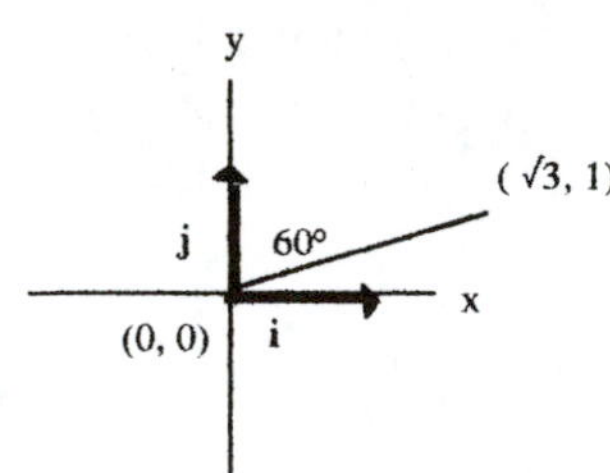

57. $\boldsymbol{v} = -2\boldsymbol{i} + 5\boldsymbol{j} - 3\boldsymbol{k}$ is parallel to the line; $\boldsymbol{N} = 2\boldsymbol{i} + \boldsymbol{j} - \boldsymbol{k}$ is perpendicular to the plane. $\boldsymbol{N} \cdot \boldsymbol{v} \neq 0 \Rightarrow$ the line and plane are not parallel. No.

58. $A_1x + B_1y + C_1z = D_1$ and $A_2x + B_2y + C_2z = D_2$ will be parallel if their normal vectors have the same direction, i.e., $A_1 = \lambda A_2, B_1 = \lambda B_2,\ C_1 = \lambda C_2$; they will be perpendicular if their normal vectors are perpendicular, i.e., $A_1A_2 + B_1B_2 + C_1C_2 = 0$.

11.6 SURFACES IN SPACE

1.

$x^2 + y^2 = 4$

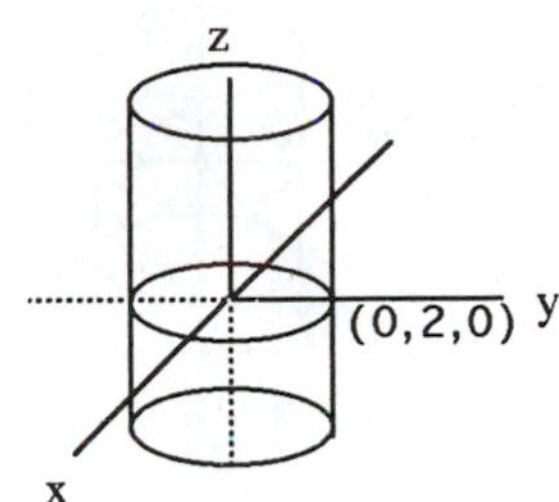

2

$x^2 + z^2 = 4$

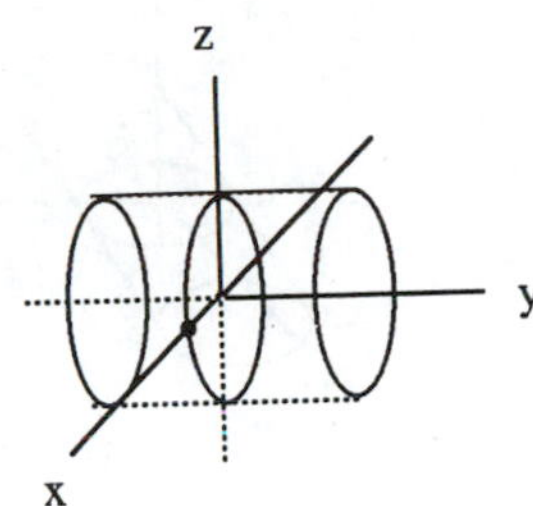

3.

$y^2 + z^2 = 1$

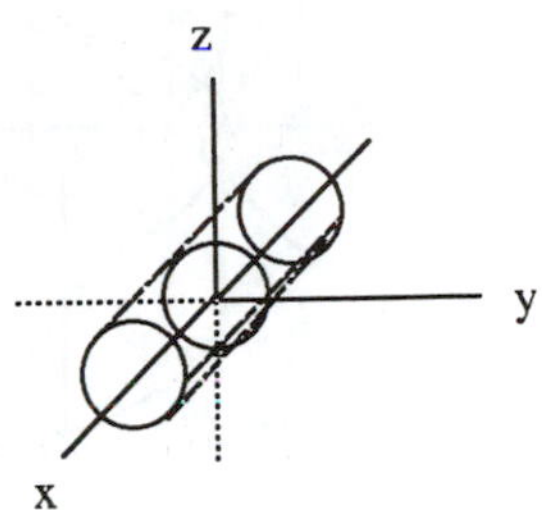

4.

$z = y^2 / 4$

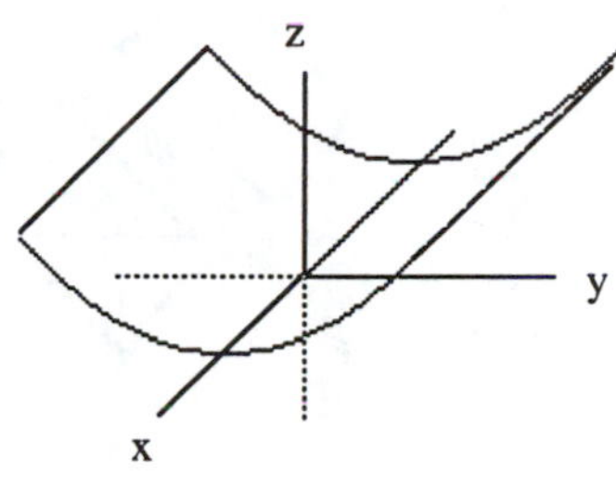

5.

$z = y^2 - 1$

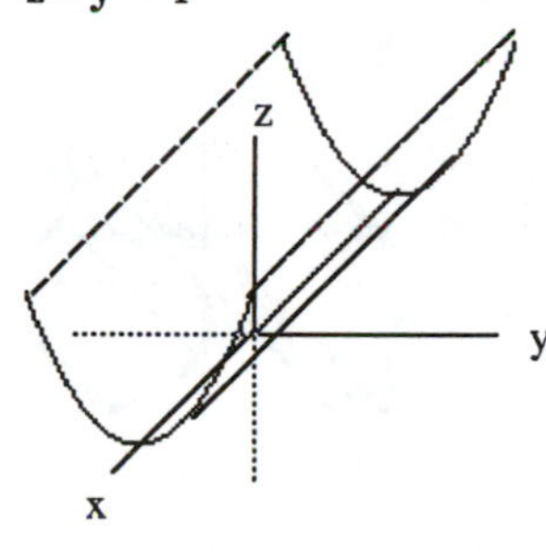

6.

$x = y^2$

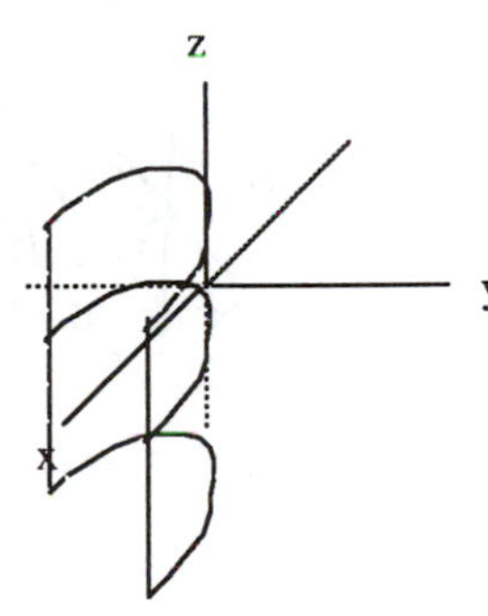

7.

$z = 4 - x^2$

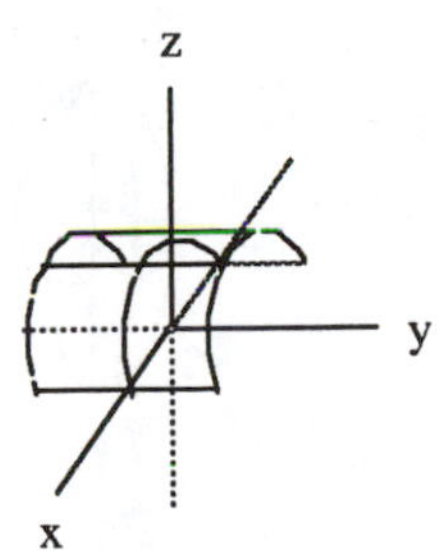

8.

$x = 4 - y^2$

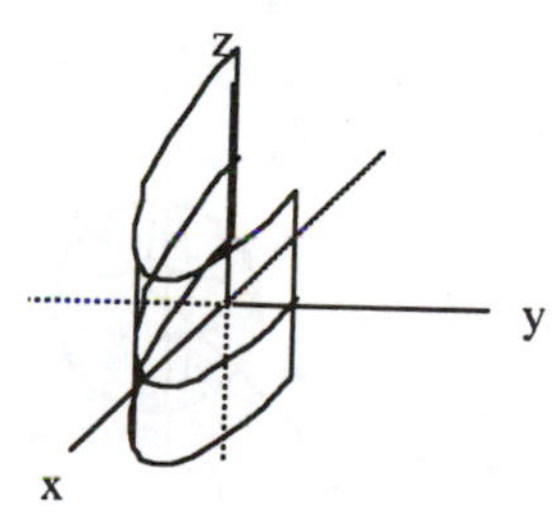

9.

$y = x^2$

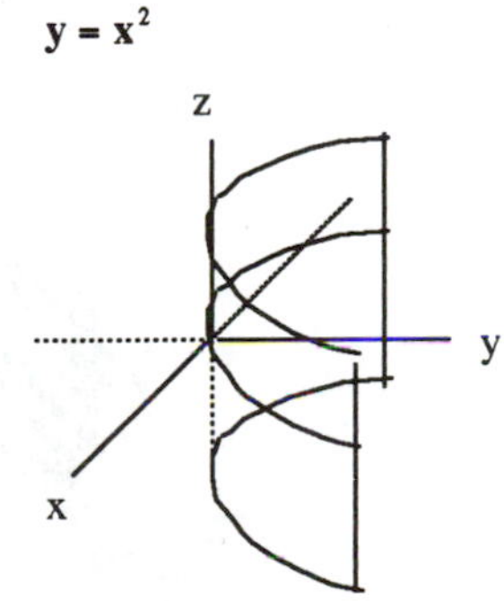

10.

$y = x^2 - 2$

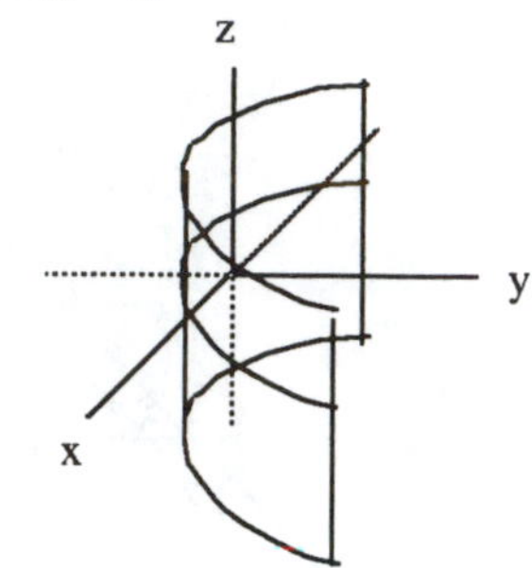

11.

$y^2 + 4z^2 = 16$

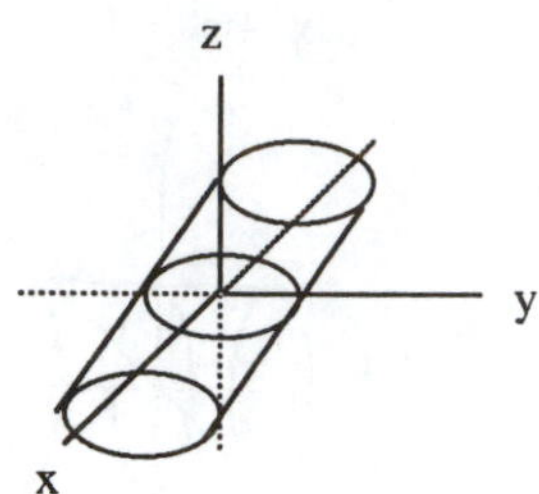

12.

$4x^2 + y^2 = 36$

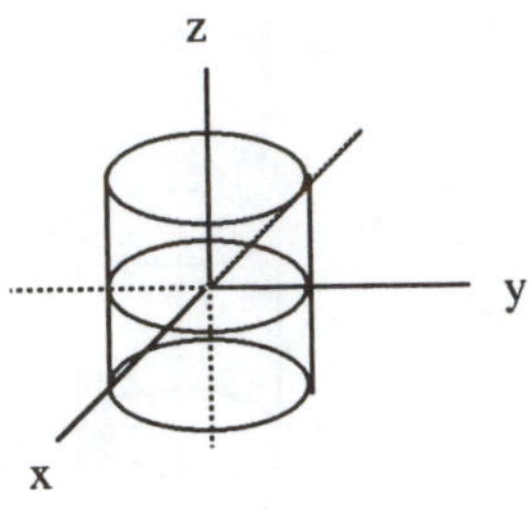

13.

$z^2 + 4y^2 = 9$

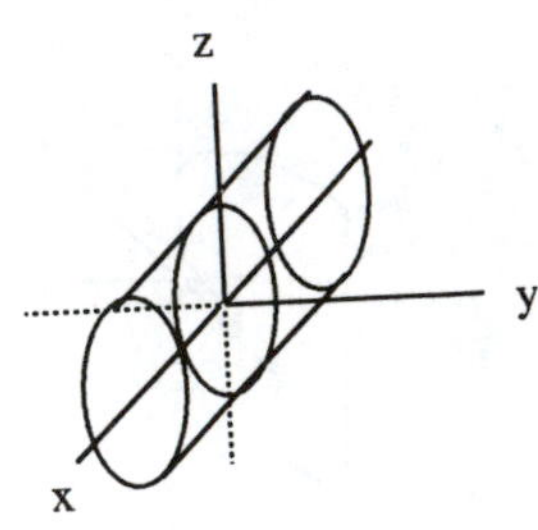

14.

$y^2 - z^2 = 4$

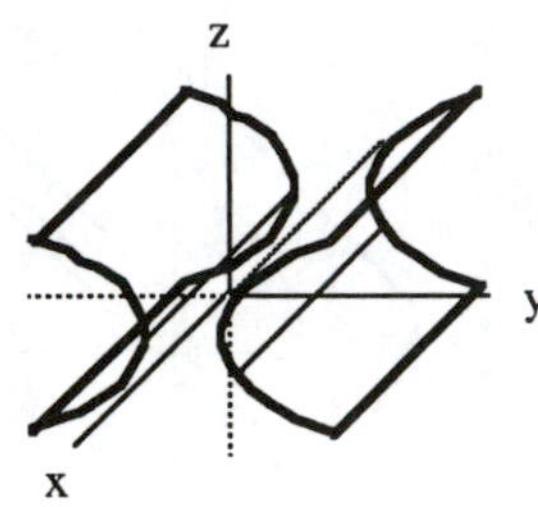

15.

$z^2 - y^2 = 1$

z
y
x

16.

$yz = 1$

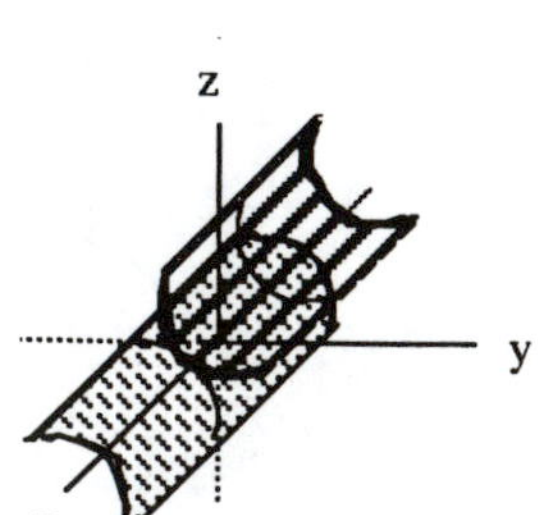

17.

$9x^2 + y^2 + z^2 = 9$

z
y
x

18.

$4x^2 + 4y^2 + z^2 = 16$

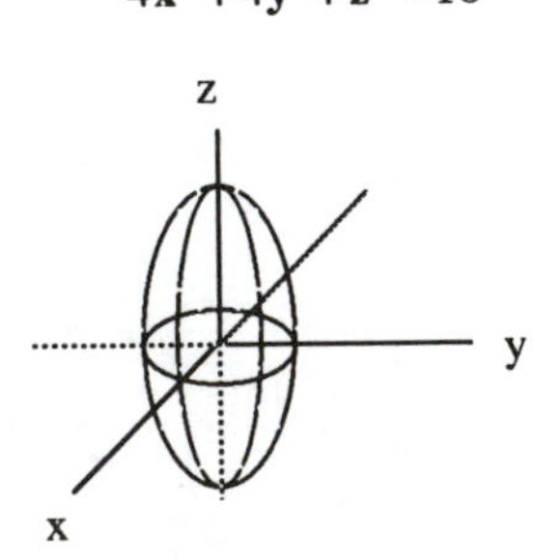

19.

$x^2 + y^2 + z^2 = 4$

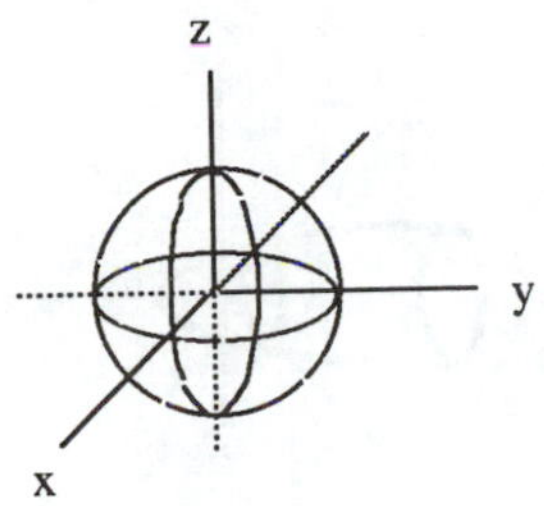

20.

$9x^2 + 4y^2 + z^2 = 36$

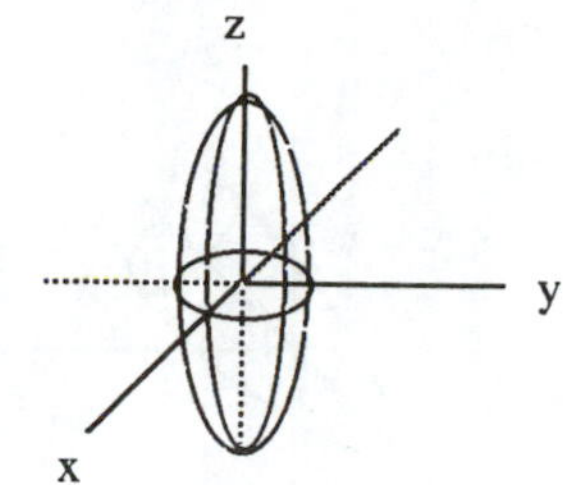

21.

$4x^2 + 9y^2 + 4z^2 = 36$

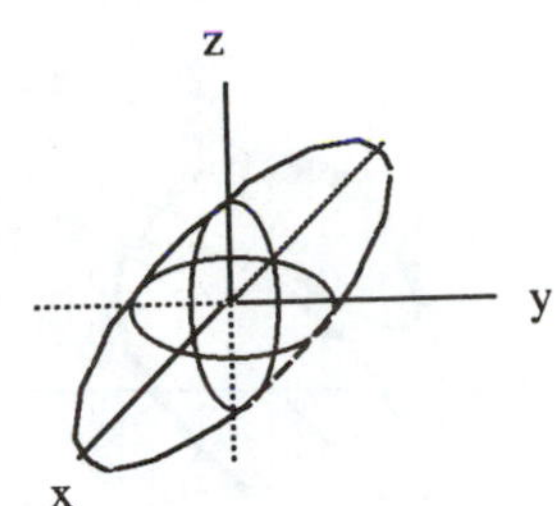

22.

$9x^2 + 4y^2 + 36z^2 = 36$

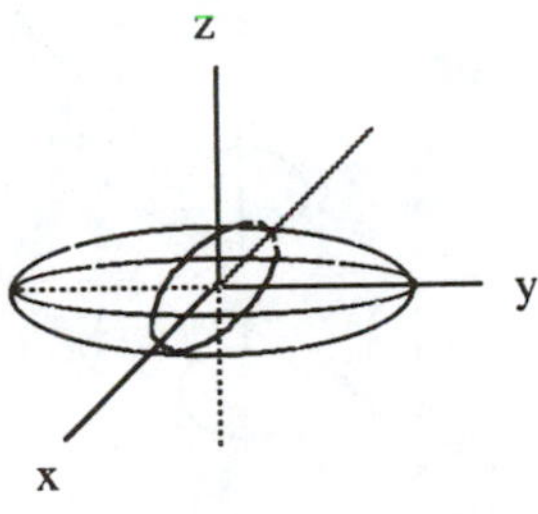

23.

$x^2 + y^2 = z$

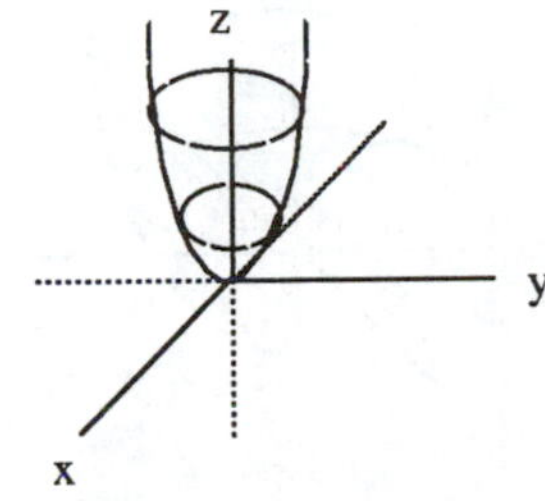

24.

$x^2 + z^2 = y$

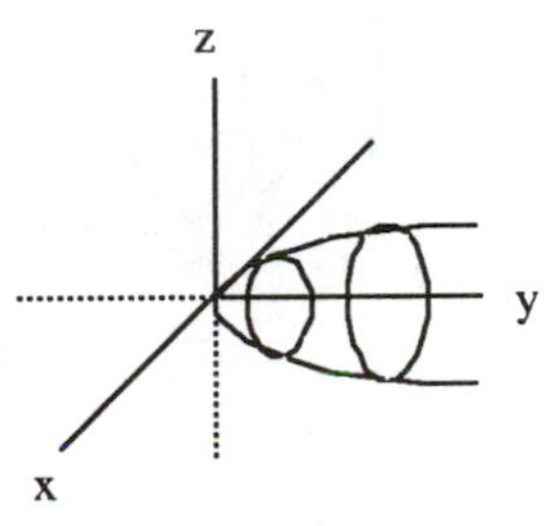

25.

$x^2 + 4y^2 = z$

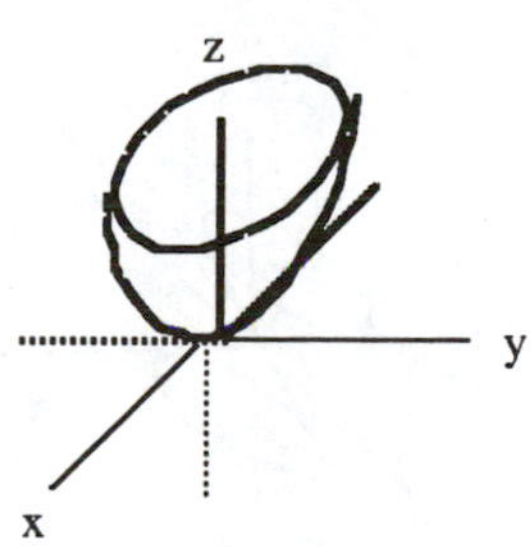

26.

$x^2 + 9y^2 = z$

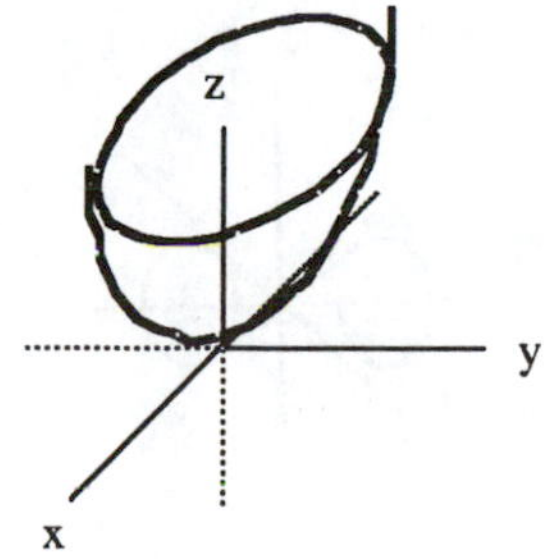

27.

$z = 8 - x^2 - y^2$

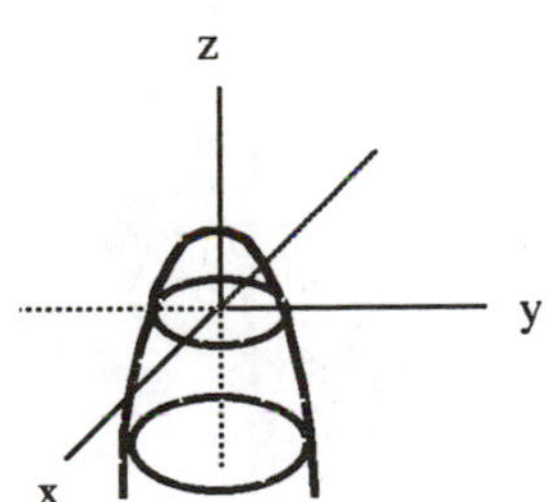

28.

$z = 18 - x^2 - 9y^2$

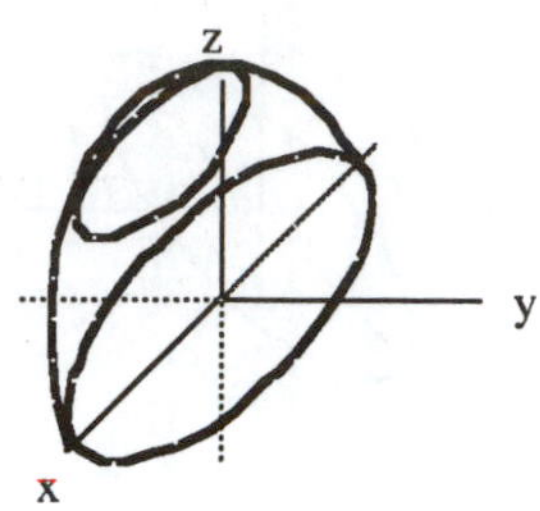

29.

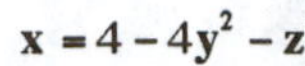

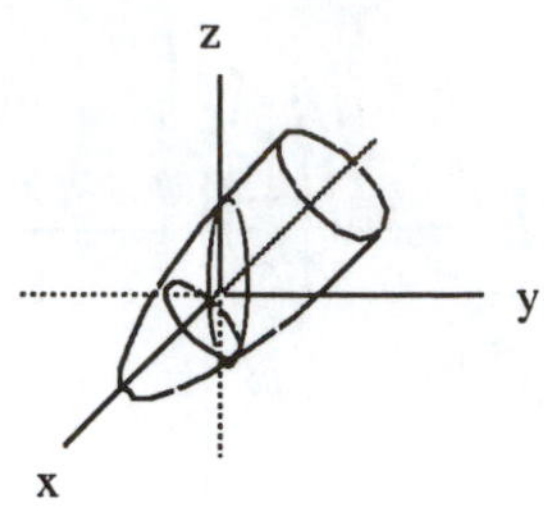

30.

$y = 1 - x^2 - z^2$

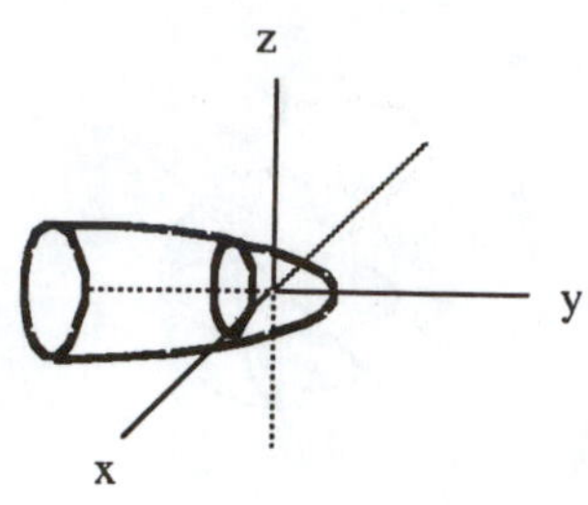

31.

$z = x^2 + y^2 + 1$

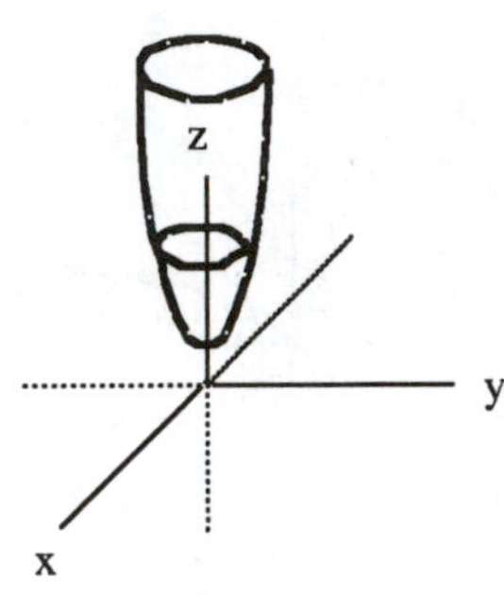

32.

$z = 4x^2 + y^2 - 4$

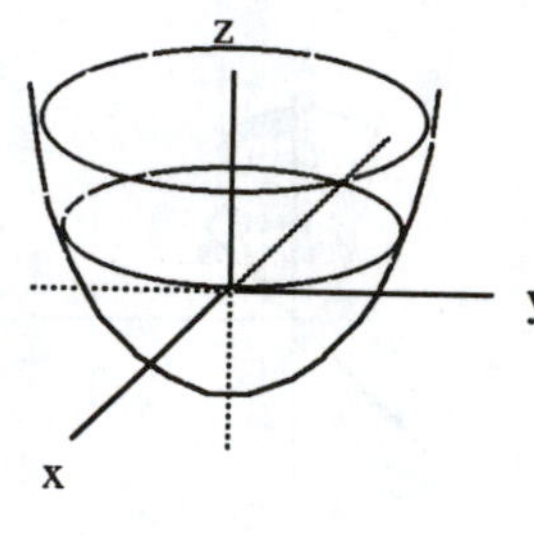

33.

$x^2 + y^2 = z^2$

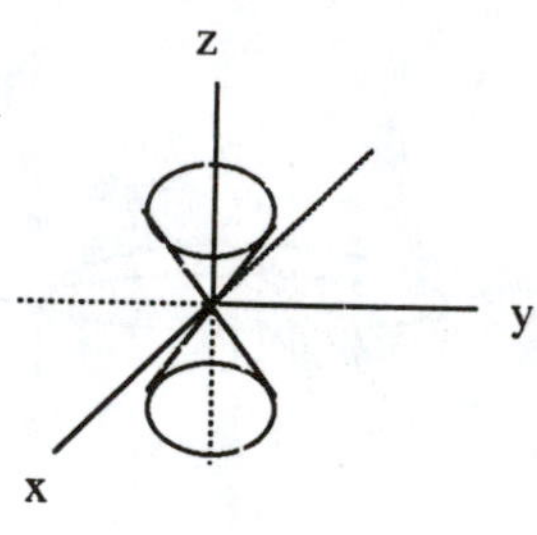

34.

$y^2 + z^2 = x^2$

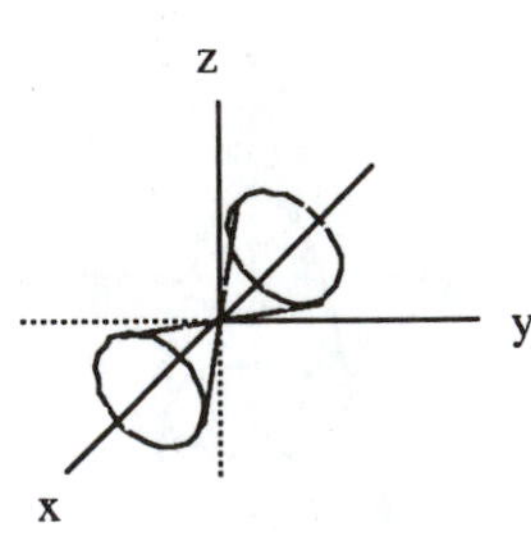

35.

$x^2 + z^2 = y^2$

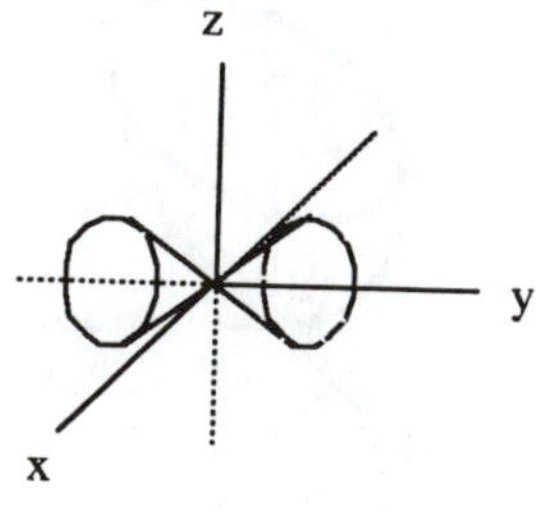

36.

$4x^2 + 9y^2 = z^2$

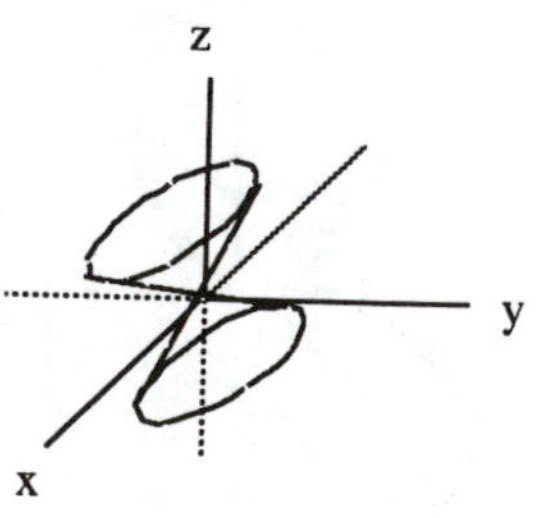

37.

$9x^2 + 4y^2 = 36z^2$

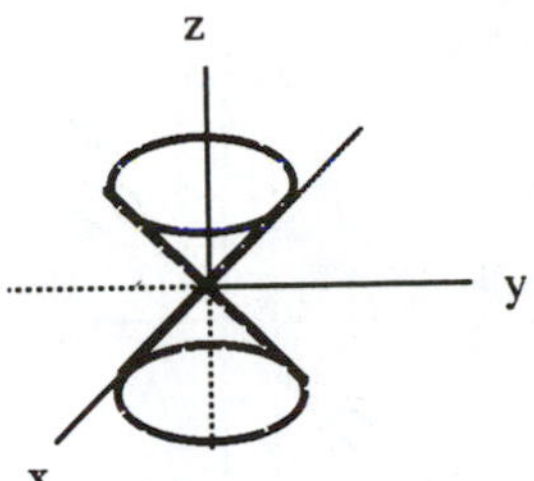

38.

$4x^2 + 9z^2 = 9y^2$

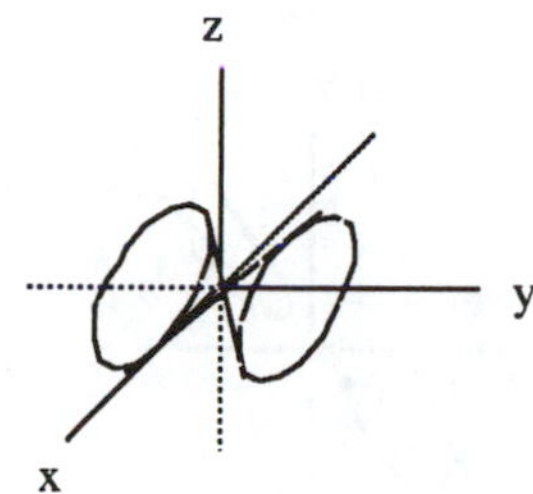

39.

$x^2 + y^2 - z^2 = 1$

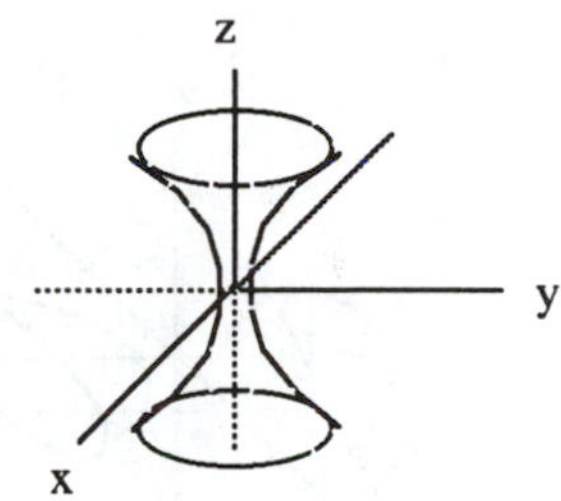

40.

$y^2 + z^2 - x^2 = 1$

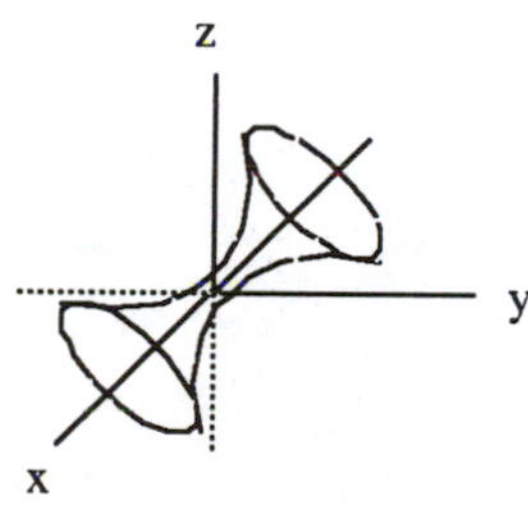

41.

$(y^2 / 4) + (z^2 / 9) - (x^2 / 4) = 1$

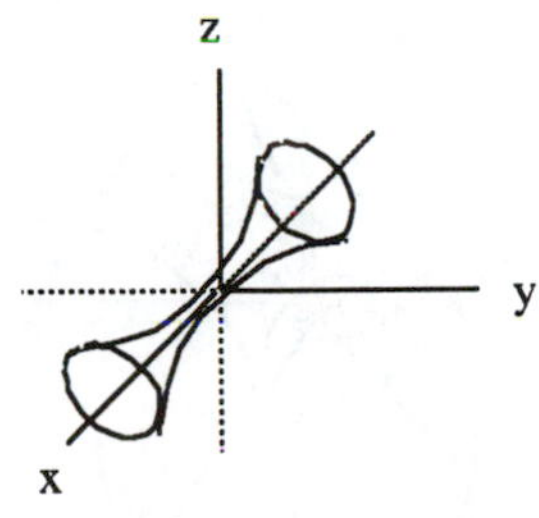

42.

$(x^2 / 4) + (y^2 / 4) - (z^2 / 9) = 1$

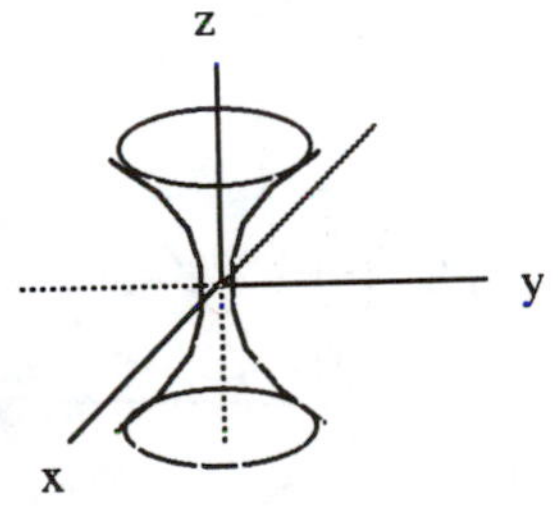

43.

$(x^2 / 4) + y^2 - z^2 = 1$

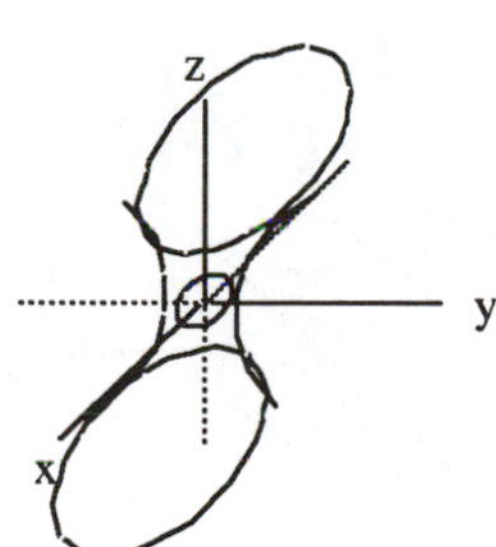

44.

$z^2 - x^2 - y^2 = 1$

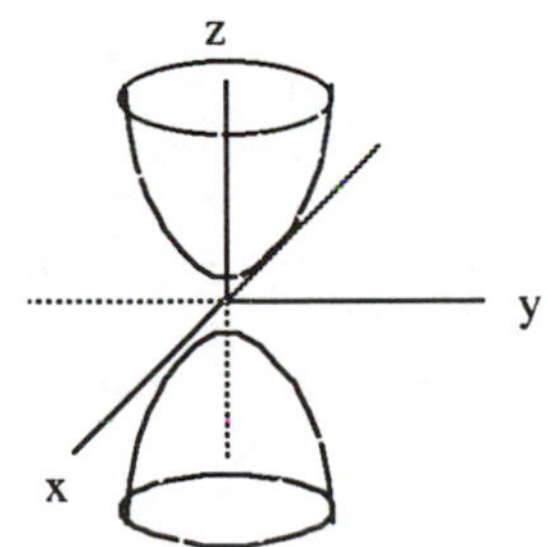

45.

$z^2 - (x^2 / 4) - y^2 = 1$

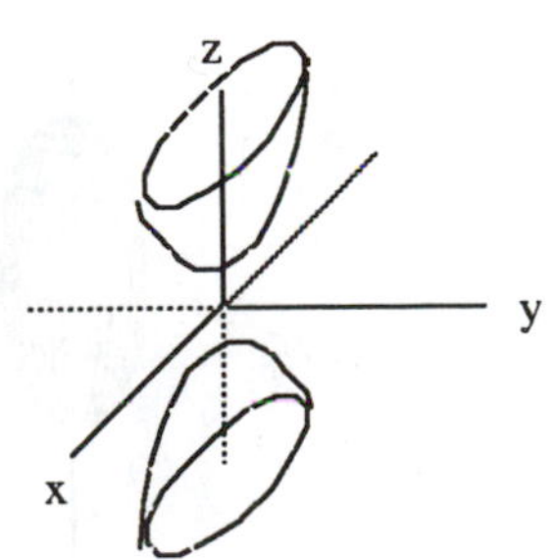

46.

$(y^2/4) - (x^2/4) - z^2 = 1$

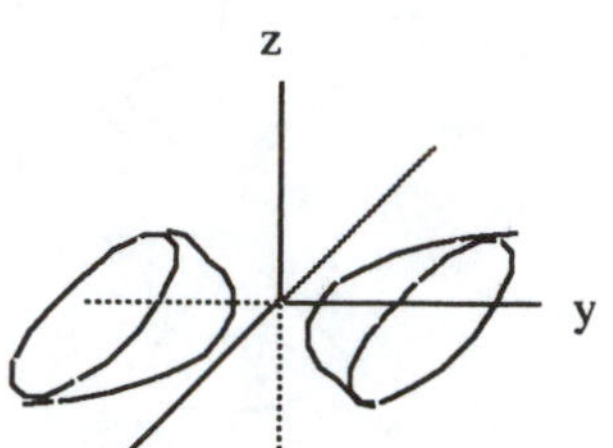

47.

$x^2 - y^2 - (z^2/4) = 1$

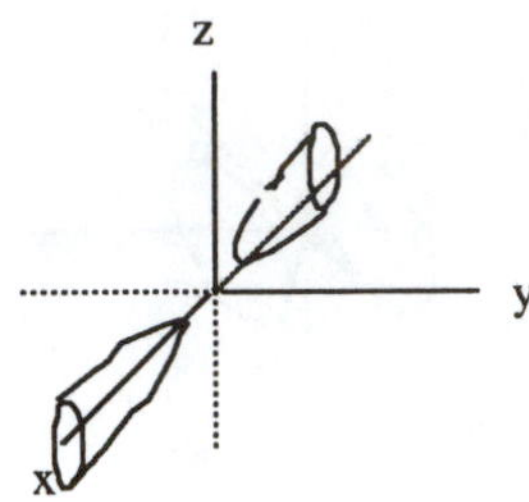

48.

$(x^2/4) - (z^2/4) - y^2 = 1$

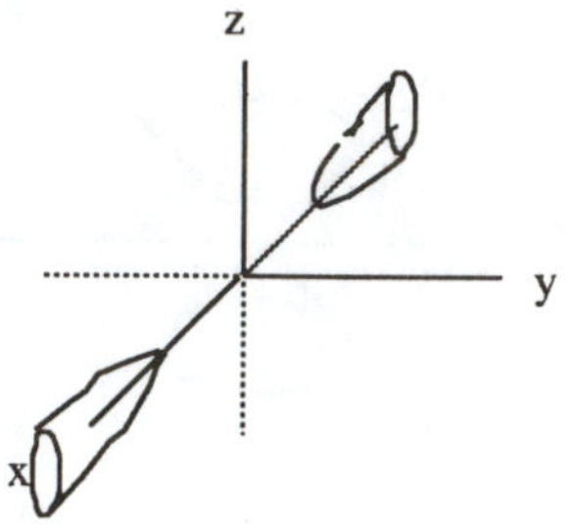

49.

$y^2 - x^2 = z$

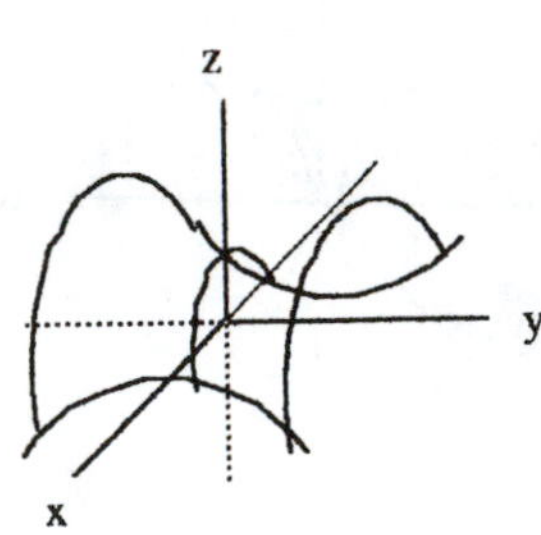

50.

$x^2 - y^2 = z$

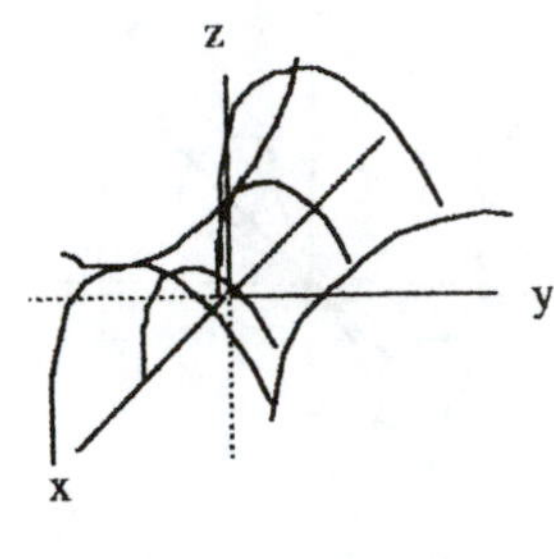

51.

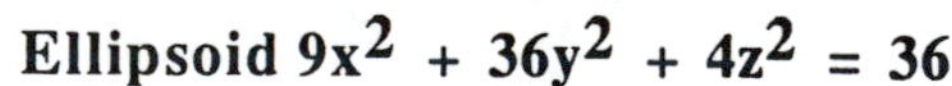

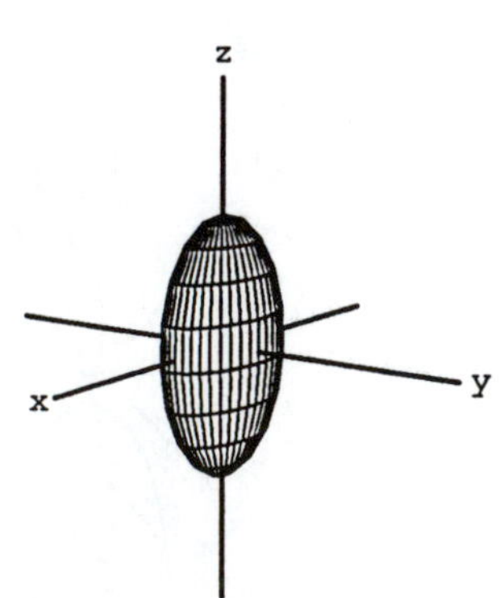

52. **Paraboloid $3y^2 + 3z^2 = 2x$**

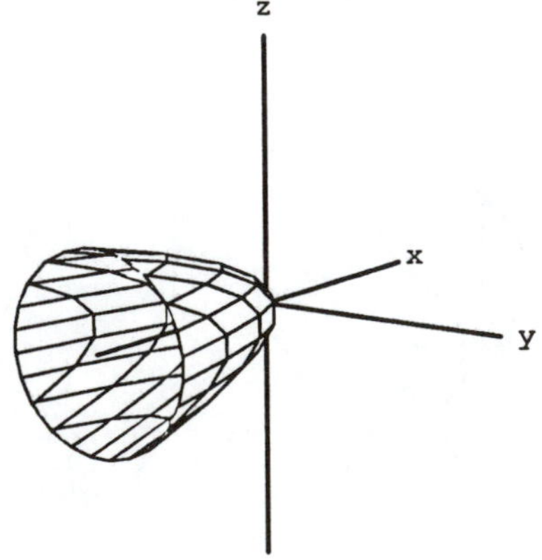

53. Cone $9x^2 + 36z^2 = 4y^2$

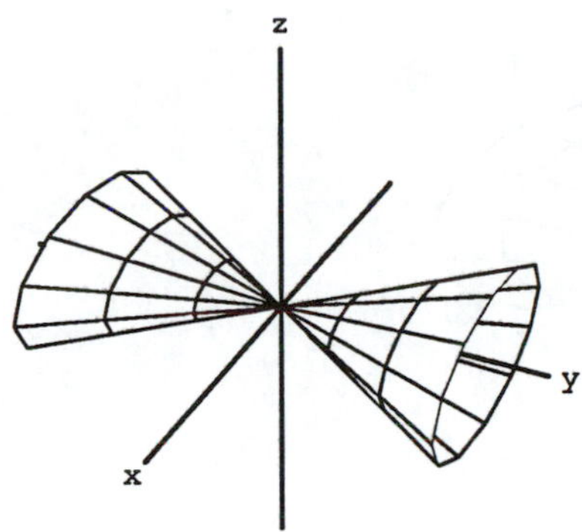

54. Hyperboloid of two sheets $16y^2 - 9x^2 - 144z^2 = 144$

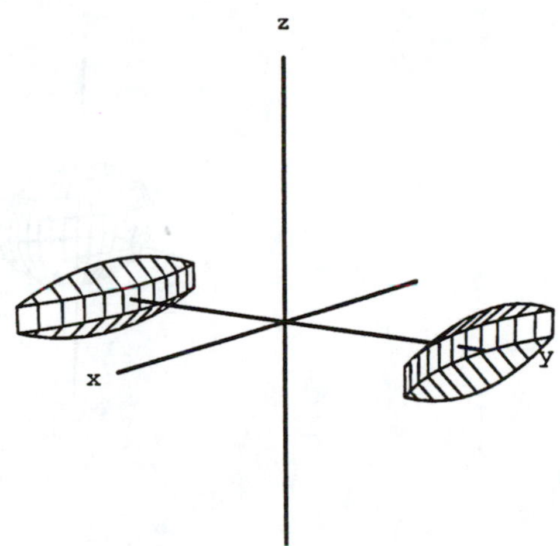

55. Paraboloid $9x^2 + 16z^2 = 72y$

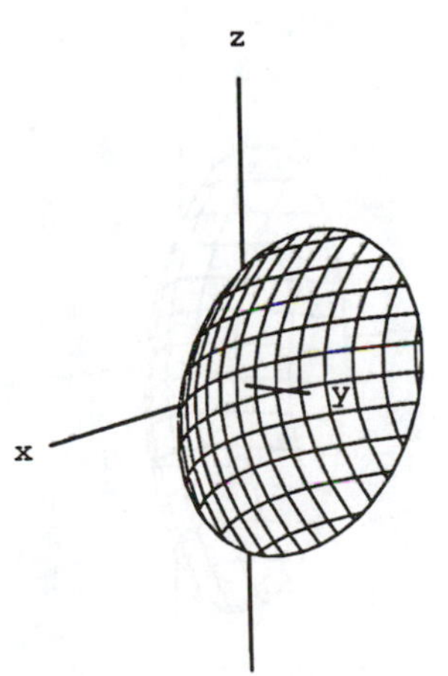

56. **Hyperboloid of one sheet - $4x^2 + 9y^2 + 36z^2 = 36$**

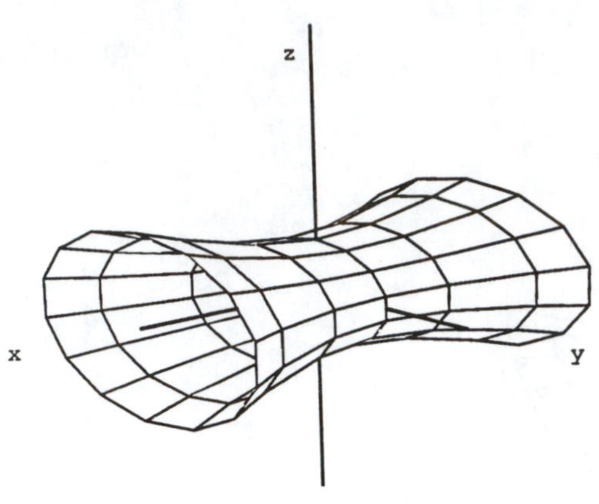

57. **Hyperbolic paraboloid**

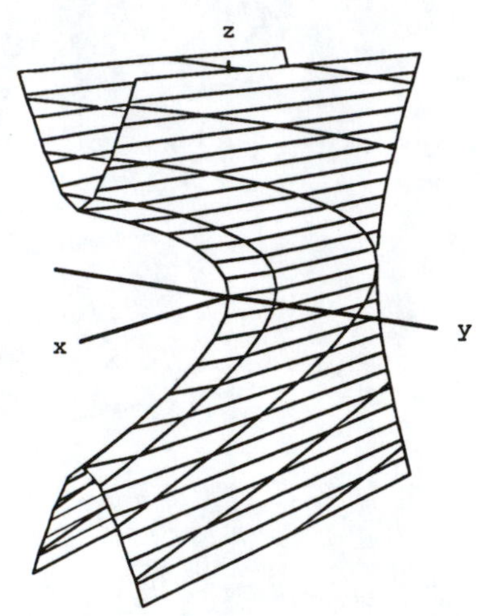

58. **Ellipsoid $36x^2 + 4y^2 + 9z^2 = 36$**

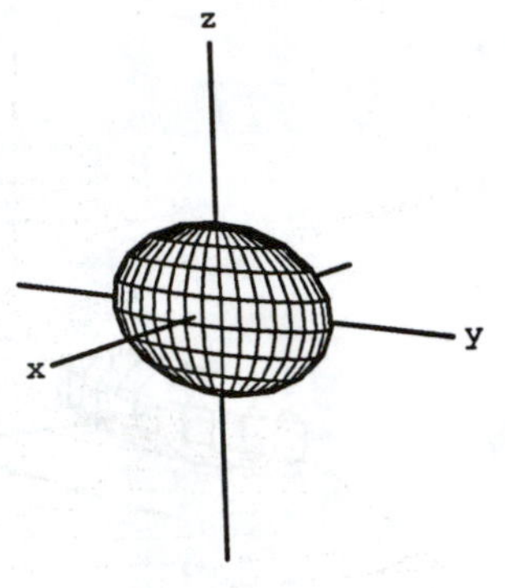

59. **Paraboloid $2x^2 + 2z^2 = 3y$**

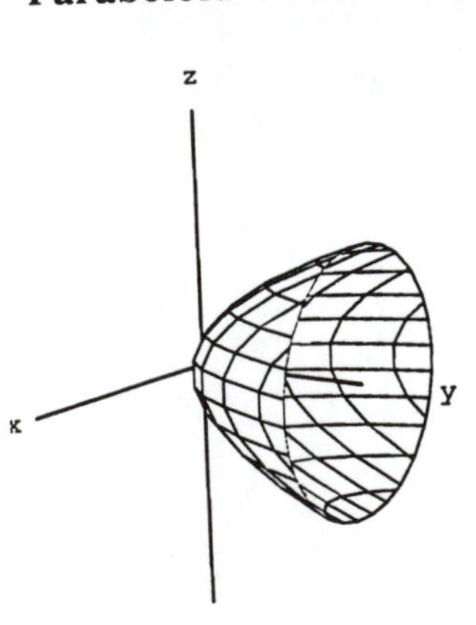

60. **Paraboloid $12y^2 + 3z^2 = 16x$**

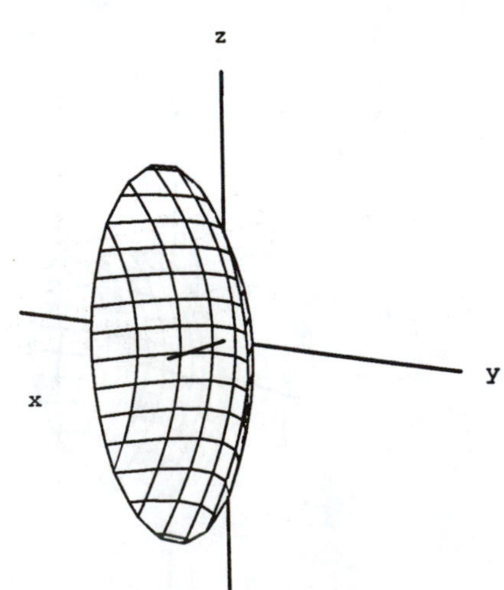

61. **Hyperboloid of two sheets** $9x^2 - 36y^2 - 4z^2 = 36$

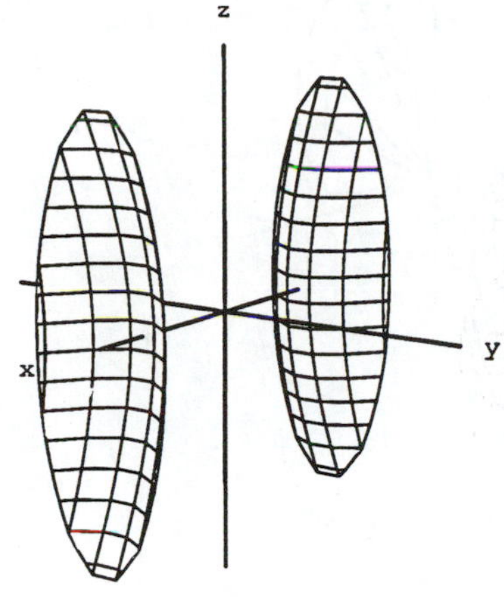

62. **Hyperbolic paraboloid** $x^2 - 4z^2 = 2y$

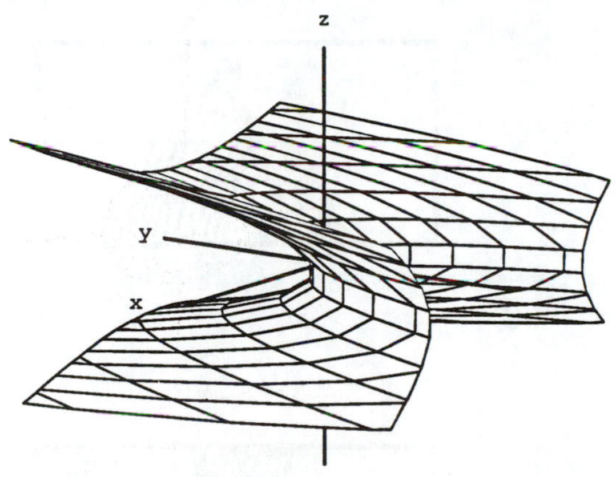

63. **Hyperboloid of one sheet** $25x^2 - 100y^2 + 4z^2 = 100$

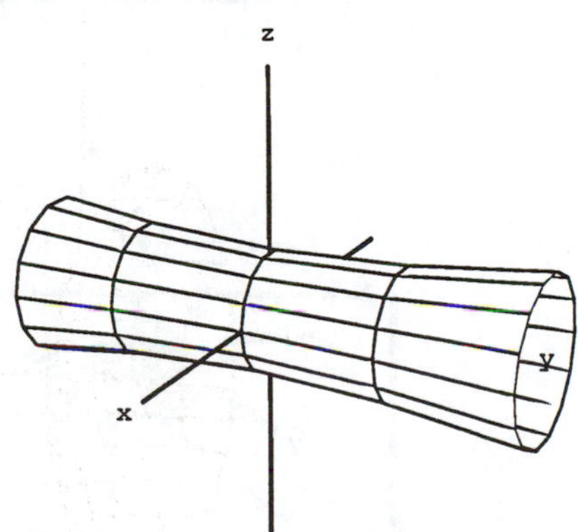

64.

Cone $16y^2 + z^2 = 4x^2$

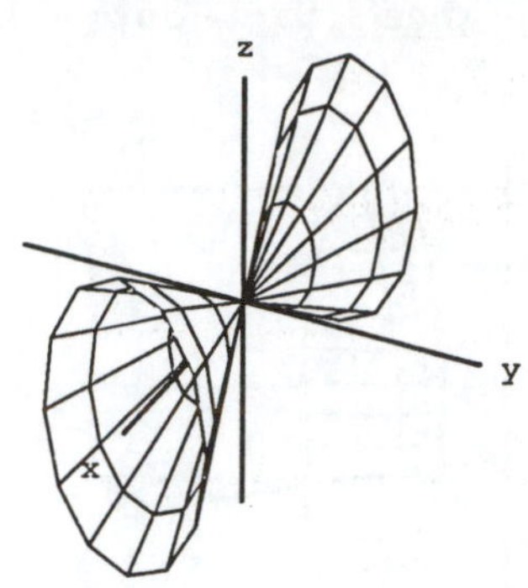

65.

$z = y^2$

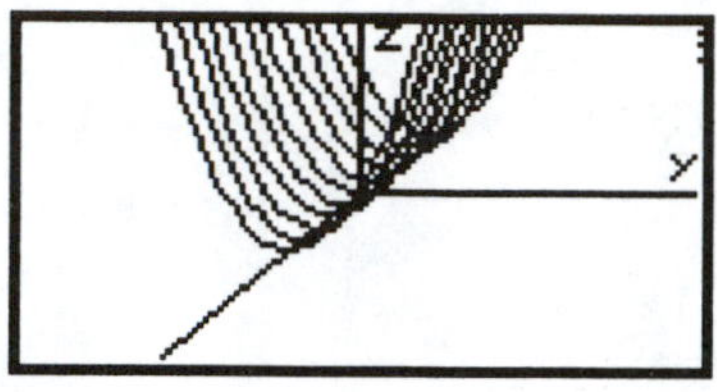

66.

$z = 1 - y^2$

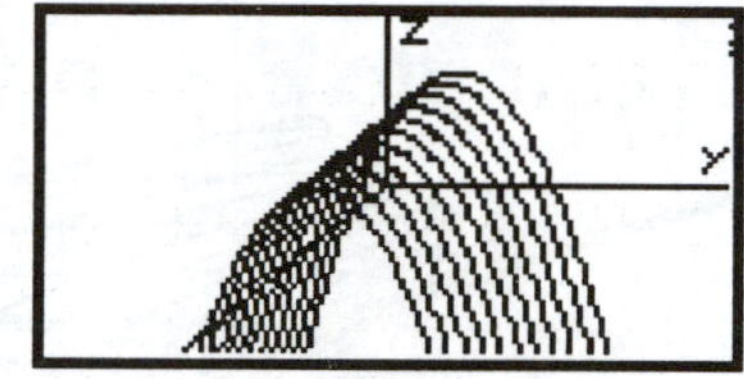

67.

$z = x^2 + y^2$

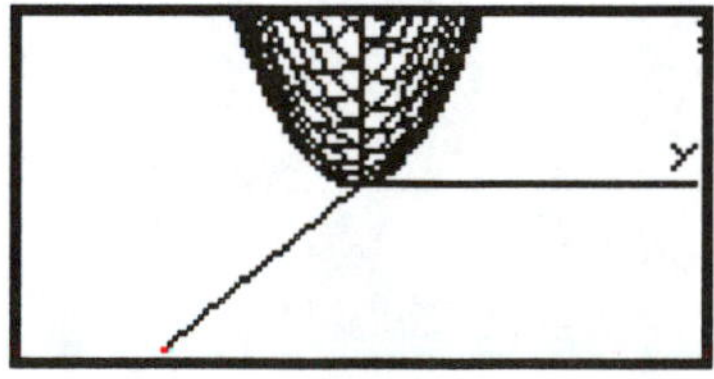

68.

$z = x^2 + 2y^2$

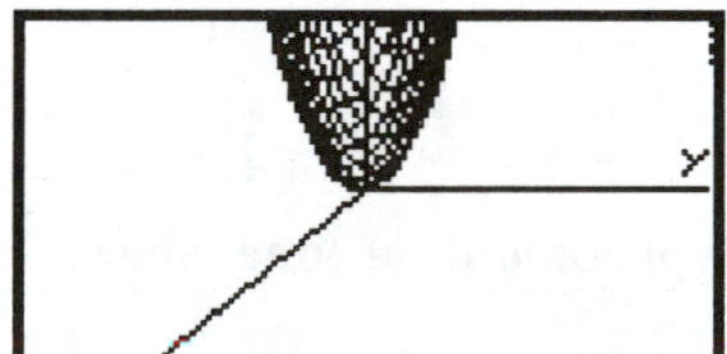

69.

$z = \sqrt{1 - x^2}$

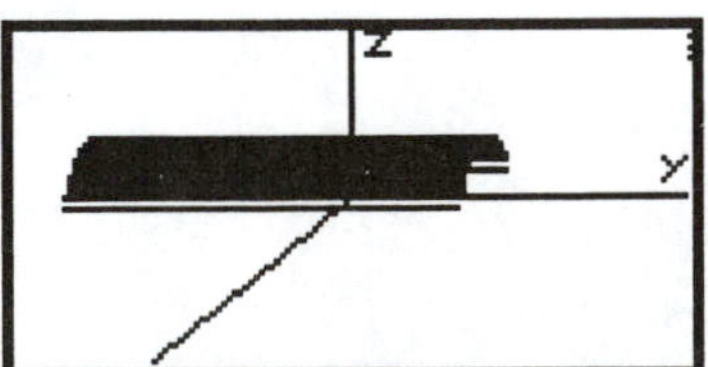

70.

$z = \sqrt{1 - (y^2 / 4)}$

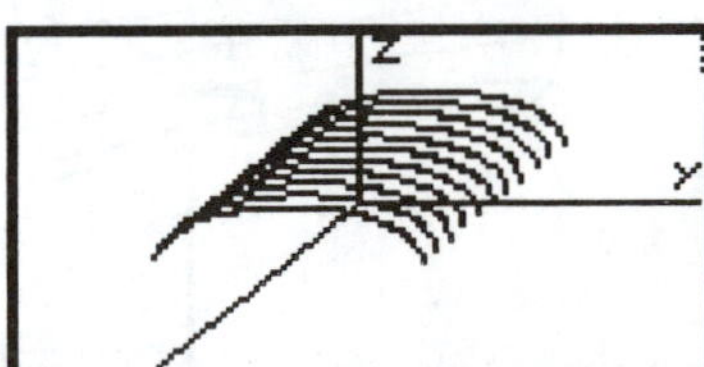

71. $z = \sqrt{x^2 + 2y^2 + 4}$

72. The problem is to find the foci of the ellipsoid. In the xz-plane ($y = 0$) the lithotripter is the ellipse $\frac{x^2}{10^2} + \frac{z^2}{13^2} = 1$. The foci lie on the z axis at $(0, 0, \pm c)$ where $c^2 = 13^2 - 10^2$. The source should be placed $\sqrt{69} \approx 8.3$ inches from the center; the patient 8.3 inches away from the center and on the other side.

11.7 CYLINDRICAL AND SPHERICAL COORDINATES

1. – 10. Use Equations (1) and (2) to find the coordinates:

	Rectangular (x, y, z)	Cylindrical (r, θ, z)	Spherical (ρ, ϕ, θ)
1.	$(0, 0, 0)$	$(0, 0^*, 0)$	$(0, 0^*, 0^*)$
2.	$(1, 0, 0)$	$(1, 0, 0)$	$(1, \pi/2, 0)$
3.	$(0, 1, 0)$	$(1, \pi/2, 0)$	$(1, \pi/2, \pi/2)$
4.	$(0, 0, 1)$	$(0, 0^*, 1)$	$(1, 0, 0^*)$
5.	$(1, 0, 0)$	$(1, 0, 0)$	$(1, \pi/2, 0)$
6.	$(\sqrt{2}, 0, 1)$	$(\sqrt{2}, 0, 1)$	$(\sqrt{3}, \cos^{-1}(1/\sqrt{3}), 0)$
7.	$(0, 1, 1)$	$(1, \pi/2, 1)$	$(\sqrt{2}, \cos^{-1}(1/\sqrt{2}), \pi/2)$
8.	$(0, -3/2, \sqrt{3}/2)$	$(3/2, -\pi/2, \sqrt{3}/2)$	$(\sqrt{3}, \pi/3, -\pi/2)$

0^* can be any angle

9. $\phi = \pi/2 \Rightarrow$ point is in xy-plane ($z = 0$); $\theta = 3\pi/2 \Rightarrow$ point is on negative y-axis ($x = 0$); $\rho = 2\sqrt{2} \Rightarrow r^2 = 8$. Hence $(x, y, z) = (0, -\sqrt{8}, 0) = (0, -2\sqrt{2}, 0)$; $(r, \theta, z) = (2\sqrt{2}, 3\pi/2, 0)$.

10. $\phi = \pi \Rightarrow$ point is on negative z-axis $(x = 0,\ y = 0)$. $\rho = \sqrt{2} \Rightarrow (x, y, z) = (0, 0, -\sqrt{2})$, $(r, \theta, z) = (0, 0^*, -\sqrt{2})$; 0^* can be any angle.

11. $r = 0 \Rightarrow$ rectangular: $x = 0,\ y = 0,\ z = 0$; spherical: $\phi = 0$ or $\phi = \pi, \rho = \rho,\ \theta = \theta$. The z-axis.

12. $x^2 + y^2 = 5 \Rightarrow$ cylindrical: $r = \sqrt{5}$; spherical: $\rho^2 \sin^2 \varphi \cos^2 \theta + \rho^2 \sin^2 \varphi \sin^2 \theta = \rho^2 \sin^2 \varphi = 5$; circular cylinder parallel to the z-axis.

13. $z = 0 \Rightarrow$ rectangular: $z = 0$; cylindrical: $z = 0$; spherical: $\phi = \pi/2$; the xy-plane.

14. $z = -2 \Rightarrow$ cylindrical: $z = -2$; spherical: $\rho \cos \varphi = -2$; the plane $z = -2$.

15. $\rho \cos \varphi = 3 \Rightarrow$ rectangular, cylindrical: $z = 3$; the plane $z = 3$.

16. $\sqrt{x^2 + y^2} = z \Rightarrow$ cylindrical: $r = z$; spherical: $\rho \sin \varphi = \rho \cos \varphi \Rightarrow \sin \varphi = \cos \varphi \Rightarrow \varphi = \pi/4$; a cone.

17. $\rho \sin \varphi \cos \theta = 0 \Rightarrow x = 0$; rectangular: $x = 0$; cylindrical: $\theta = \pi/2$ or $3\pi/2$; the yz-plane.

18. $\tan^2 \varphi = 1 \Rightarrow$ rectangular: $\sin^2 \varphi = \cos^2 \varphi \Rightarrow \rho^2 \sin^2 \varphi = \rho^2 \cos^2 \varphi \Rightarrow x^2 + y^2 = z^2$; cylindrical: $r^2 = z^2$; a cone.

19. Spherical: $\rho^2 = 4$ or $\rho = 2$; cylindrical: $r^2 + z^2 = 4$; sphere of radius 2, center at origin.

20. $x^2 + y^2 + (z - \frac{1}{2})^2 = \frac{1}{4} \Rightarrow$ spherical: $x^2 + y^2 + z^2 - z = 0$, or $\rho^2 - \rho \cos \varphi = 0$, or $\rho = \cos \varphi$; cylindrical: $(z - \frac{1}{2})^2 = \frac{1}{4} - r^2$; a sphere.

21. $\rho = 2 \sin \theta \Rightarrow \rho^2 = 2\rho \sin \theta \Rightarrow x^2 + y^2 + z^2 = 2y$; cylindrical: $r^2 + z^2 = 2r \sin \theta$; a sphere.

22. $\rho = 6 \cos \varphi \Rightarrow \rho^2 = 6\rho \cos \varphi \Rightarrow x^2 + y^2 + z^2 = 6z$; cylindrical: $r^2 + z^2 = 6z$; a sphere.

23. $r = \csc \theta \Rightarrow r \sin \theta = 1 \Rightarrow$ rectangular: $y = 1$; spherical: $\rho \sin \varphi \sin \theta = 1$; the plane $y = 1$.

24. $r = -3 \sec \theta \Rightarrow r \cos \theta = -3 \Rightarrow$ rectangular: $x = 3$; spherical: $\rho \sin \varphi \cos \theta = 3$; the plane $x = 3$.

25. The lower half of the sphere with center at $(0,0,1)$, radius 1; spherical: $\rho^2 - 2\rho\cos\varphi = 0$, or $\rho = 2\cos\varphi$, since $\rho^2 = 2z$, $z \le 1 \Rightarrow \rho < \sqrt{2} \Rightarrow \frac{\pi}{4} < \varphi$; cylindrical: $r^2 = 2z - z^2$, $0 \le z \le 1$.

26. This is a slice of a sphere between the planes $z = 0$ and $z = \frac{\sqrt{3}}{2}$. Spherical: $\rho = \sqrt{3}$, $0 \le \rho\cos\varphi \le \frac{\sqrt{3}}{2} \Rightarrow \pi/3 \le \varphi \le \frac{\pi}{2}$; cylindrical: $z^2 = 3 - r^2$, $0 \le z \le \frac{\sqrt{3}}{2}$.

27. This is the top third of a sphere. Rectangular: $x^2 + y^2 + z^2 = 4$, $2(0.5) \le z \le 2$; cylindrical: $r^2 + z^2 = 4$, $1 \le z \le 2$.

28. See #27. This is the bottom of a sphere; rectangular: $x^2 + y^2 + z^2 = 4$, $z \le -\sqrt{2}$; spherical: $\rho = 2$, $\frac{3\pi}{4} \le \varphi \le \pi$.

29. $\phi = \frac{\pi}{3}$, $0 \le \rho \le 2$ is the truncated top half of a cone; rectangular: $z = \frac{1}{2}\rho \Rightarrow \rho = 2z$, $x = \frac{\sqrt{3}}{2}\rho\cos\theta$, $y = \frac{\sqrt{3}}{2}\rho\sin\theta \Rightarrow x^2 + y^2 = \frac{3}{4}\rho^2 = 3z^2$, $0 \le z \le 1$; cylindrical: $r^2 = 3z^2$, $0 \le z \le 1$.

30. $\varphi = \frac{\pi}{2}$ is the xy-plane; rectangular: $z = 0$, $x^2 + y^2 \le 7$; cylindrical: $r^2 \le 7$, $z = 0$; the interior and boundary of a circle in the xy-plane.

31. A right circular cylinder whose cross-sections perpendicular to the xy-plane are circles of radius 4; The axis of the cylinder is the z-axis.

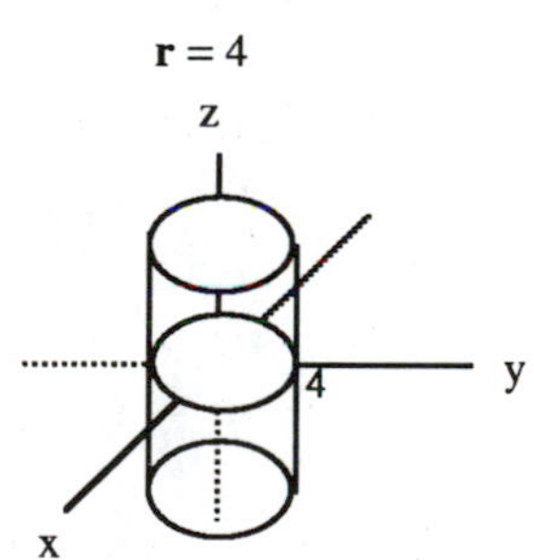

32. A sphere of radius 1, center at $(0,0,0)$

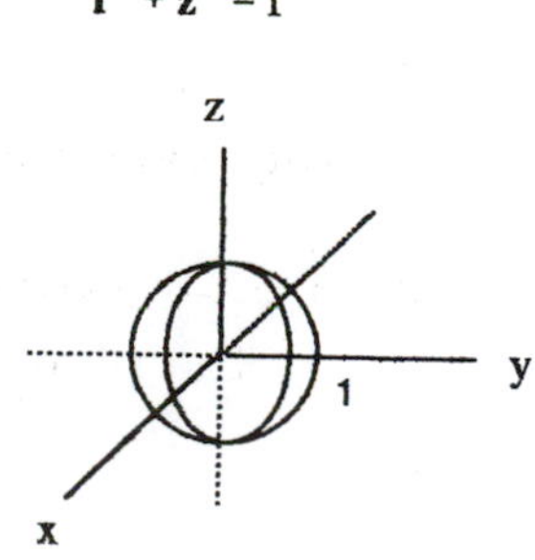

33. Right circular cylinder generated by the cardioid $r = 1 - \cos\theta$

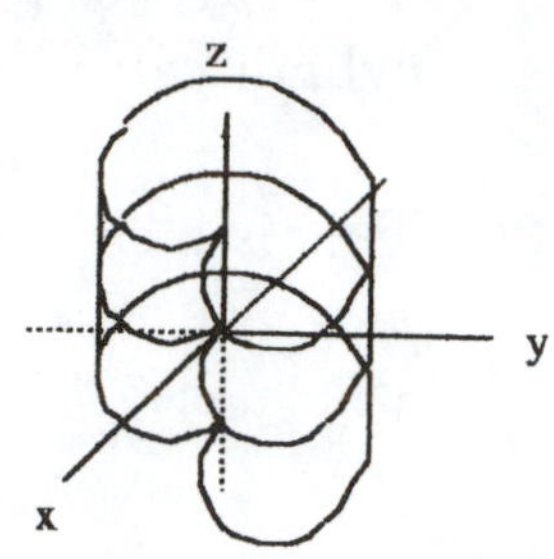

34. $r = 2\cos\theta \Rightarrow r^2 = 2r\cos\theta \Rightarrow$ $x^2 + y^2 = 2x \Rightarrow (x-1)^2 + y^2 = 1$; a cylinder generated by this circle.

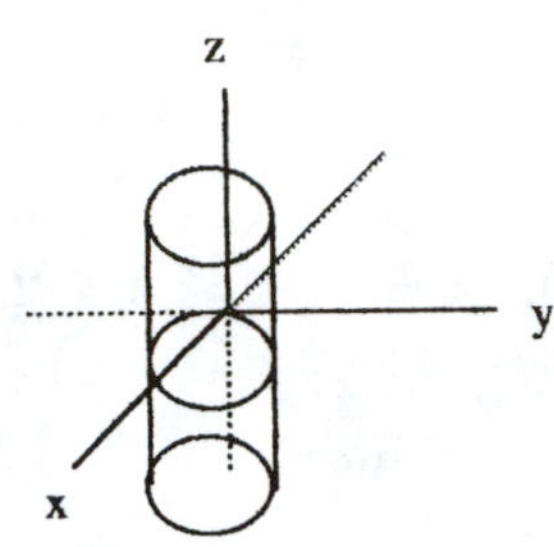

35. Circle of radius 2, center at $(0,0,3)$, parallel to the xy-plane

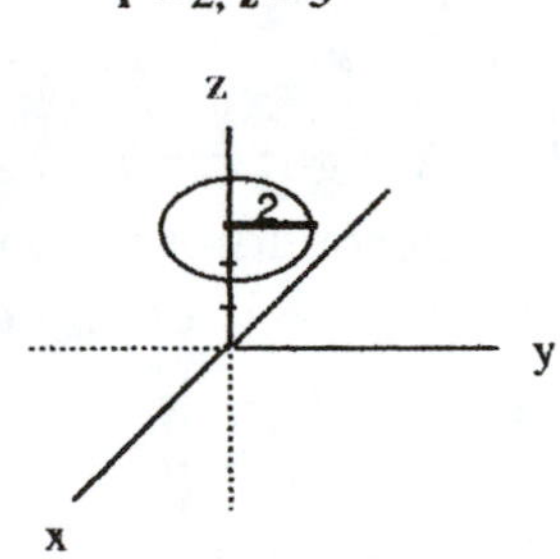

36. The intersection of the plane $\theta = \frac{\pi}{6}$ and the upper nappe of the cone $z = x^2 + y^2$

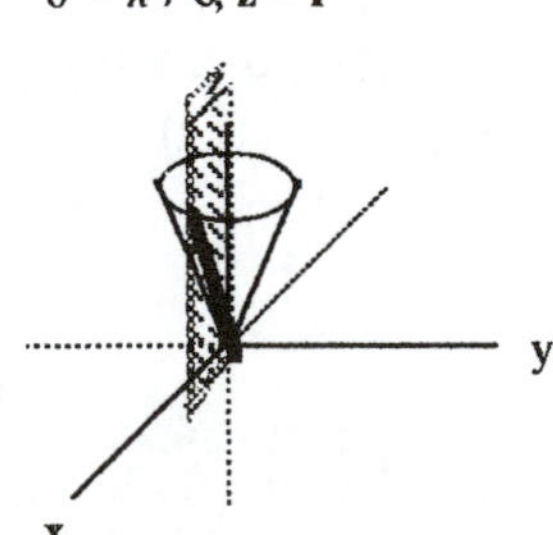

37. A spiral up the side of the cylinder $r = 3$

$r = 3,\quad z = \theta/2$

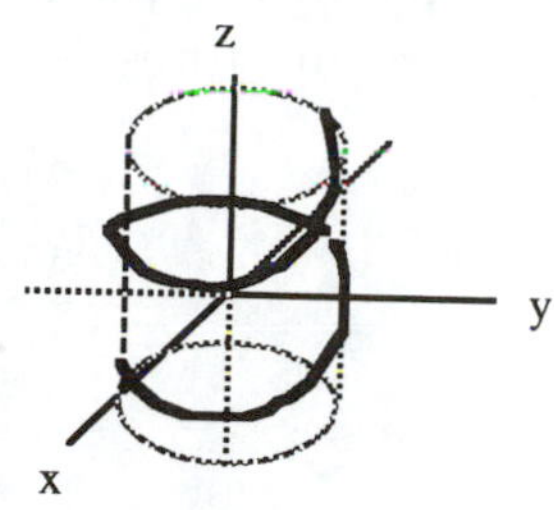

38. A spiral around the cylinder $x^2 + y^2 = 1$

$r = 1,\quad z = 2\cos\theta$

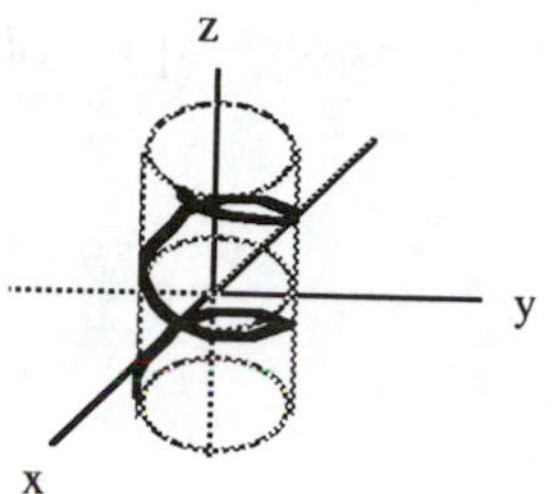

39. A circle in the xy-plane

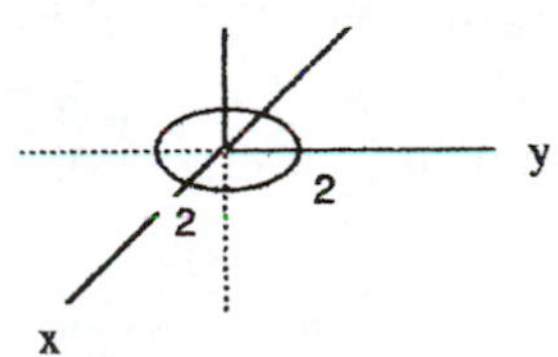

40. A circle of radius $3\sqrt{2}$ in the plane $z = 3\sqrt{2}$

$\rho = 6,\quad \phi = \pi/4$

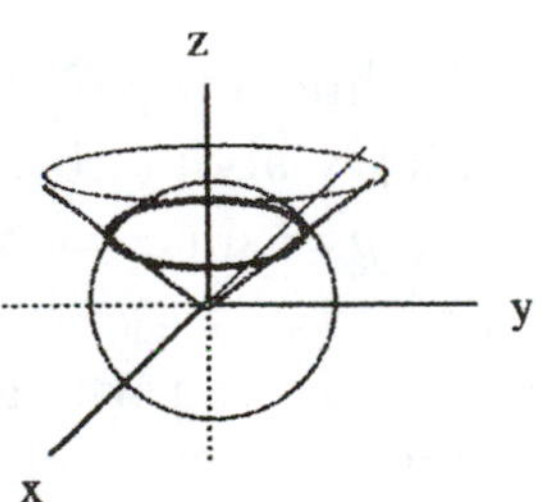

41. The intersection of the cone $\varphi = \frac{\pi}{4}$ and the plane $\theta = \frac{\pi}{4}$; intersecting lines

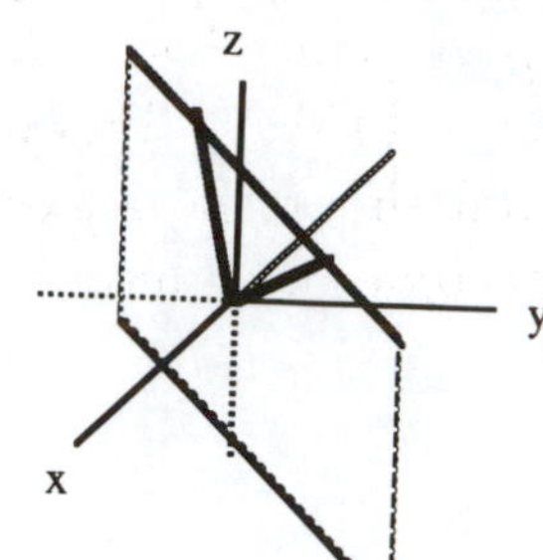

42. The intersection of the yz-plane and the lower nappe of a cone

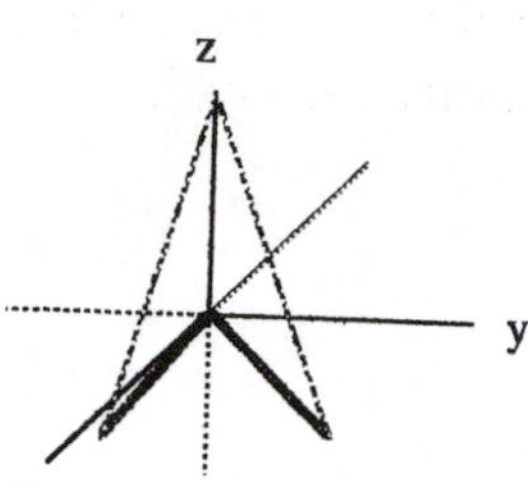

43. The curve lies in the positive half of the yz-plane $(\theta = \frac{\pi}{2})$, where $y = \rho \sin\phi$, $\rho = 4\sin\phi \Rightarrow \rho^2 = 4\rho\sin\phi = 4y$, $y^2 + z^2 = 4y$; $(y-2)^2 + z^2 = 4$. The curve is a circle of radius 2.

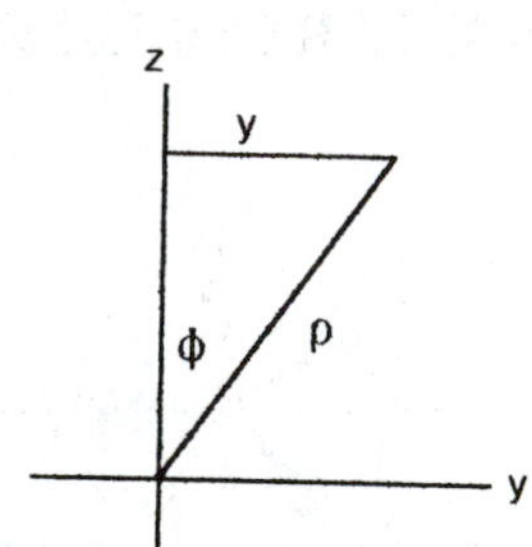

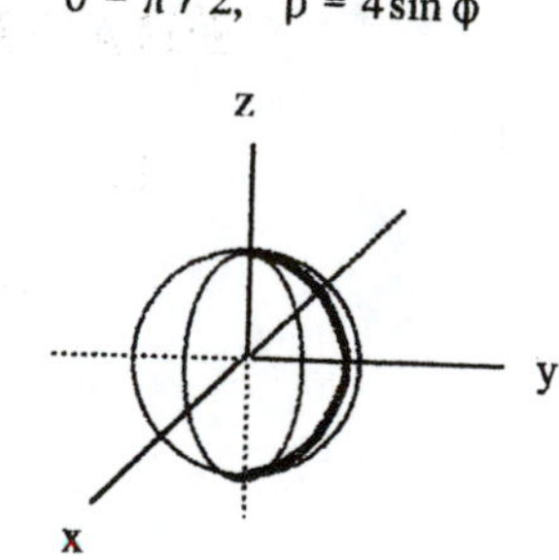

44. By the hint in #46, there will be symmetry about the z-axis. Assume $\theta = \frac{\pi}{2}$, $\rho = \sin\varphi \Rightarrow \rho^2 = \rho\sin\varphi \Rightarrow$ $y^2 + z^2 = y \Rightarrow (y - \frac{1}{2})^2 + z^2 = \frac{1}{4}$. The set is the torus generated by revolving the circle about the z-axis.

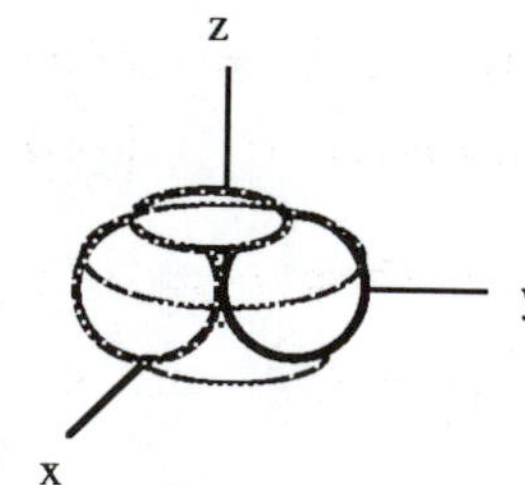

45. There will by symmetry about the z-axis. In the yz-plane, $\rho = \cos\varphi \Rightarrow \rho^2 = \rho\cos\varphi \Rightarrow y^2 + z^2 = z$, or $y^2 + (z - \frac{1}{2})^2 = \frac{1}{4}$. When revolved about the z-axis, this becomes a sphere.

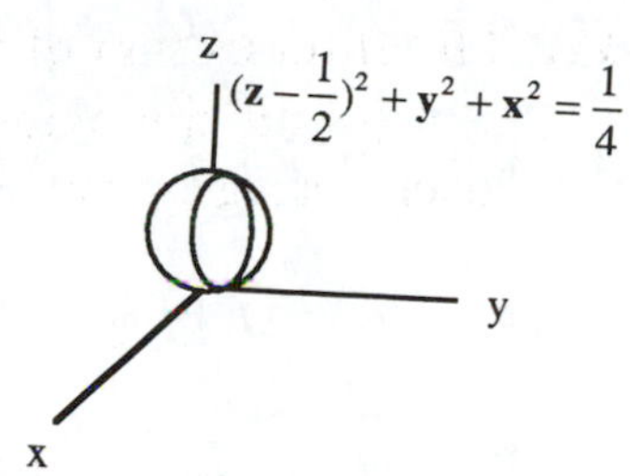

46. In the yz-plane this is half the cardioid $r = 1 - \cos\varphi$; revolved about the z-axis, it resembles an apple.

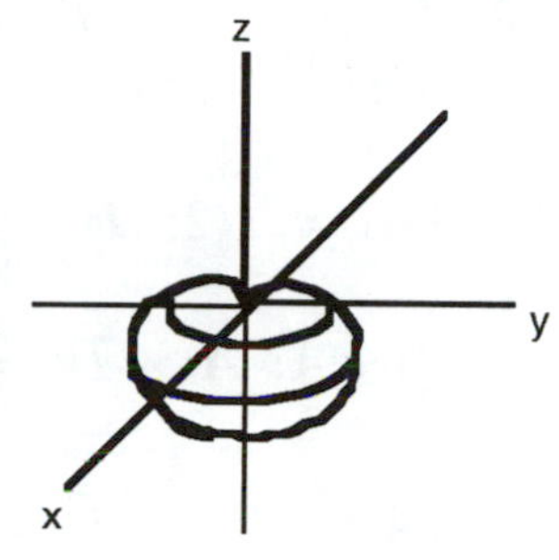

PRACTICE EXERCISES, CHAPTER 11

1.

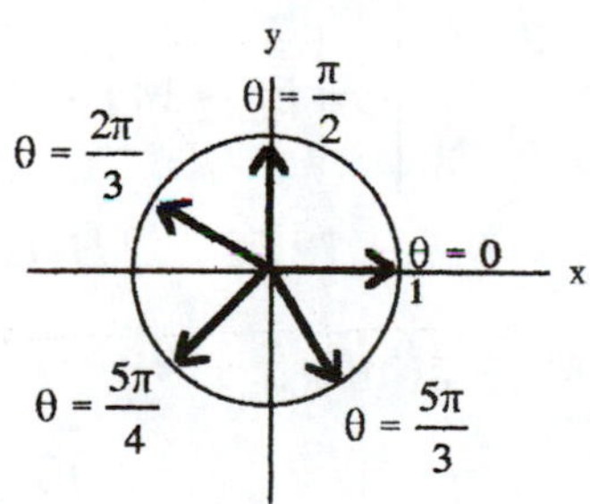

2. a) $\cos 45°\boldsymbol{i} + \sin 45°\boldsymbol{j} = \frac{1}{\sqrt{2}}\boldsymbol{i} + \frac{1}{\sqrt{2}}\boldsymbol{j}$; b) $\cos(-120°)\boldsymbol{i} + \sin(-120°)\boldsymbol{j} = -\frac{\sqrt{3}}{2}\boldsymbol{i} - \frac{1}{2}\boldsymbol{j}$.

3. $y = \tan x \Rightarrow y' = \sec^2 x$; $y'(P) = (\sqrt{2}/1)^2 = 2 \Rightarrow \boldsymbol{T} = \boldsymbol{i} + 2\boldsymbol{j}$; $|\boldsymbol{T}|^2 = 5 \Rightarrow$ unit tangents are $\pm\left(\frac{1}{\sqrt{5}}\boldsymbol{i} + \frac{2}{\sqrt{5}}\boldsymbol{j}\right)$; unit normals are $\pm\left(\frac{2}{\sqrt{5}}\boldsymbol{i} - \frac{1}{\sqrt{5}}\boldsymbol{j}\right)$.

4. $x^2+y^2 = 25 \Rightarrow y' = -\frac{x}{y}$; $y'(P) = \frac{3}{4} \Rightarrow \boldsymbol{T} = 4\boldsymbol{i} - 3\boldsymbol{j}$; $|\boldsymbol{T}| = 5 \Rightarrow$ unit tangents are $\pm\left(\frac{4}{5}\boldsymbol{i} - \frac{3}{5}\boldsymbol{j}\right)$; unit normals are $\pm\left(\frac{3}{5}\boldsymbol{i} + \frac{4}{5}\boldsymbol{j}\right)$.

5. $\boldsymbol{v} = \sqrt{2}(\boldsymbol{i}+\boldsymbol{j}) \Rightarrow |\boldsymbol{v}| = \sqrt{2+2} = 2$; direction is $\frac{1}{|\boldsymbol{v}|}\boldsymbol{v} = \frac{1}{\sqrt{2}}\boldsymbol{i} + \frac{1}{\sqrt{2}}\boldsymbol{j}$.

6. $\boldsymbol{v} = -1(\boldsymbol{i}+\boldsymbol{j}) \Rightarrow |\boldsymbol{v}| = \sqrt{2} \Rightarrow$ direction is $\frac{1}{|\boldsymbol{v}|}\boldsymbol{v} = \frac{-1}{\sqrt{2}}(\boldsymbol{i}+\boldsymbol{j})$.

7. $\boldsymbol{v} = 2\boldsymbol{i} - 3\boldsymbol{j} + 6\boldsymbol{k} \Rightarrow |\boldsymbol{v}| = \sqrt{4+9+36} = 7$; direction $= \frac{1}{|\boldsymbol{v}|}\boldsymbol{v} = \frac{2}{7}\boldsymbol{i} - \frac{3}{7}\boldsymbol{j} + \frac{6}{7}\boldsymbol{k}$.

8. $\boldsymbol{v} = \boldsymbol{i} + 2\boldsymbol{j} - \boldsymbol{k} \Rightarrow |\boldsymbol{v}| = \sqrt{1+2^2+1} = \sqrt{6}$; direction is $\frac{1}{|\boldsymbol{v}|}\boldsymbol{v} = \frac{1}{\sqrt{6}}(\boldsymbol{i}+2\boldsymbol{j}-\boldsymbol{k})$.

9. $\boldsymbol{v} = c(4\boldsymbol{i} - \boldsymbol{j} + 4\boldsymbol{k})$: $|\boldsymbol{v}| = c\sqrt{16+1+16} = c\sqrt{33} = 2 \Rightarrow c = \frac{2}{\sqrt{33}}$; $\boldsymbol{v} = (\frac{2}{\sqrt{33}})(4\boldsymbol{i} - \boldsymbol{j} + 4\boldsymbol{k})$.

10. $\boldsymbol{B} = -\frac{3}{5}\boldsymbol{i} - \frac{4}{5}\boldsymbol{k}$ has the desired direction, $|\boldsymbol{B}| = 1 \Rightarrow 5\boldsymbol{B} = -3\boldsymbol{i} - 4\boldsymbol{k}$ is the answer.

11. $|\boldsymbol{A}| = \sqrt{2}$; $|\boldsymbol{B}| = \sqrt{4+1+4} = 3$; $\boldsymbol{A}\cdot\boldsymbol{B} = 1\cdot 2 + 1\cdot 1 + 0\cdot(-2) = 3 = \boldsymbol{B}\cdot\boldsymbol{A}$; $\boldsymbol{A}\times\boldsymbol{B} = \begin{vmatrix} \boldsymbol{i} & \boldsymbol{j} & \boldsymbol{k} \\ 1 & 1 & 0 \\ 2 & 1 & -2 \end{vmatrix} = -2\boldsymbol{i} + 2\boldsymbol{j} - \boldsymbol{k}$; $\boldsymbol{B}\times\boldsymbol{A} = 2\boldsymbol{i} - 2\boldsymbol{j} + \boldsymbol{k}$, $|\boldsymbol{A}\times\boldsymbol{B}| = \sqrt{9} = 3$; $\cos\theta \frac{\boldsymbol{A}\cdot\boldsymbol{B}}{|\boldsymbol{A}||\boldsymbol{B}|} = \frac{3}{3\sqrt{2}} = \frac{1}{\sqrt{2}} \Rightarrow \theta = \frac{\pi}{4}$; $\text{comp}_{\boldsymbol{A}}\boldsymbol{B} = \frac{\boldsymbol{A}\cdot\boldsymbol{B}}{|\boldsymbol{A}|} = \frac{3}{\sqrt{2}}$; $\text{proj}_{\boldsymbol{A}}\boldsymbol{B} = \frac{\boldsymbol{A}\cdot\boldsymbol{B}}{\boldsymbol{A}\cdot\boldsymbol{A}}\boldsymbol{A} = \frac{3}{\sqrt{2}}(\sqrt{2}\boldsymbol{i} + \sqrt{2}\boldsymbol{j}) = 3(\boldsymbol{i}+\boldsymbol{j})$.

12. $|\boldsymbol{A}| = \sqrt{25+1+1} = \sqrt{27} = 3\sqrt{3}$, $|\boldsymbol{B}| = \sqrt{1+4+9} = \sqrt{14}$, $\boldsymbol{A}\cdot\boldsymbol{B} = \boldsymbol{B}\cdot\boldsymbol{A} = 5 - 2 + 3 = 6$, $\boldsymbol{A}\times\boldsymbol{B} = \begin{vmatrix} \boldsymbol{i} & \boldsymbol{j} & \boldsymbol{k} \\ 5 & 1 & 1 \\ 1 & -2 & 3 \end{vmatrix} = 5\boldsymbol{i} - 14\boldsymbol{j} - 11\boldsymbol{k}$, $\boldsymbol{B}\times\boldsymbol{A} = -5\boldsymbol{i} + 14\boldsymbol{j} + 11\boldsymbol{k}$, $\cos\theta = \frac{\boldsymbol{A}\cdot\boldsymbol{B}}{|\boldsymbol{A}||\boldsymbol{B}|} = \frac{6}{3\sqrt{42}} \Rightarrow \theta = 72.025°$, $|\boldsymbol{B}|\cos\theta = 2\sqrt{\frac{14}{42}} = \frac{2}{\sqrt{3}}$, $\text{proj}_{\boldsymbol{A}}\boldsymbol{B} = \frac{\boldsymbol{A}\cdot\boldsymbol{B}}{\boldsymbol{A}\cdot\boldsymbol{A}}\boldsymbol{A} = \frac{6}{27}(5\boldsymbol{i}+\boldsymbol{j}+\boldsymbol{k}) = \frac{2}{9}(5\boldsymbol{i}+\boldsymbol{j}+\boldsymbol{k})$.

13. $\boldsymbol{B} = \left(\frac{\boldsymbol{A}\cdot\boldsymbol{B}}{\boldsymbol{A}\cdot\boldsymbol{A}}\right)\boldsymbol{A} + \left(\boldsymbol{B} - \frac{\boldsymbol{A}\cdot\boldsymbol{B}}{\boldsymbol{A}\cdot\boldsymbol{A}}\boldsymbol{A}\right) = \frac{8}{6}\boldsymbol{A} + \left(\boldsymbol{B} - \frac{8}{6}\boldsymbol{A}\right) = \frac{4}{3}(2\boldsymbol{i}+\boldsymbol{j}-\boldsymbol{k}) + (\boldsymbol{i}+\boldsymbol{j}-5\boldsymbol{k} - \frac{4}{3}(2\boldsymbol{i}+\boldsymbol{j}-\boldsymbol{k})) = \frac{4}{3}(2\boldsymbol{i}+\boldsymbol{j}-\boldsymbol{k})) + \frac{1}{3}(-5\boldsymbol{i}-\boldsymbol{j}-11\boldsymbol{k})$.

14. $\boldsymbol{B} = \left(\frac{\boldsymbol{A}\cdot\boldsymbol{B}}{\boldsymbol{A}\cdot\boldsymbol{A}}\right)\boldsymbol{A} + \left(\boldsymbol{B} - \frac{\boldsymbol{A}\cdot\boldsymbol{B}}{\boldsymbol{A}\cdot\boldsymbol{A}}\boldsymbol{A}\right) = -\frac{1}{5}(\boldsymbol{i}-2\boldsymbol{j}) + (\boldsymbol{i}+\boldsymbol{j}+\boldsymbol{k} - (-\frac{1}{5})(\boldsymbol{i}-2\boldsymbol{j})) = -\frac{1}{5}(\boldsymbol{i}-2\boldsymbol{j}) + \frac{1}{5}(6\boldsymbol{i}+3\boldsymbol{j}+5\boldsymbol{k})$.

15. $\boldsymbol{A}\times\boldsymbol{B} = \begin{vmatrix} \boldsymbol{i} & \boldsymbol{j} & \boldsymbol{k} \\ 1 & 0 & 0 \\ 1 & 1 & 0 \end{vmatrix} = \boldsymbol{k}$.

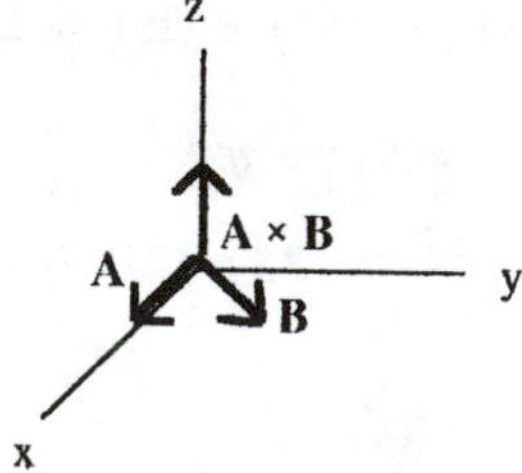

16. $\boldsymbol{A}\times\boldsymbol{B}=\begin{vmatrix} \boldsymbol{i} & \boldsymbol{j} & \boldsymbol{k} \\ 1 & -1 & 0 \\ 1 & 1 & 0 \end{vmatrix}=2\boldsymbol{k}$.

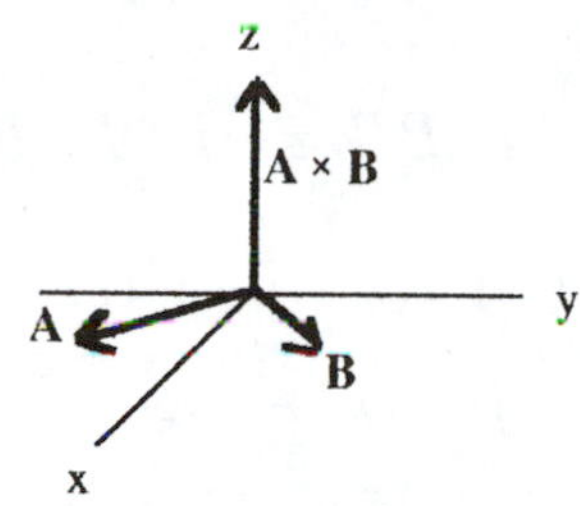

17. $P(3,2)$; $S(0,\frac{1}{2})$ is on the line, $\boldsymbol{N}=3\boldsymbol{i}+4\boldsymbol{j}$ is normal to the line; $\overrightarrow{SP}=3\boldsymbol{i}+\frac{3}{2}\boldsymbol{j}$; distance P to line $=\left|\text{proj}_{\boldsymbol{N}}\,\overrightarrow{SP}\right|=\left|\dfrac{\boldsymbol{N}\cdot\overrightarrow{SP}}{|\boldsymbol{N}|}\right|=\frac{9+6}{\sqrt{25}}=\frac{15}{5}=3$.

18. $P(-1,1)$; $S(\frac{9}{5},0)$ is on the line; $\overrightarrow{SP}=-\frac{14}{5}\boldsymbol{i}+\boldsymbol{j}$; $\boldsymbol{N}=5\boldsymbol{i}-12\boldsymbol{j}\Rightarrow$ distance P to line $=\left|\text{proj}_{\boldsymbol{N}}\,\overrightarrow{SP}\right|=\left|\dfrac{\boldsymbol{N}\cdot\overrightarrow{SP}}{|\boldsymbol{N}|}\right|=\left|\frac{-26}{\sqrt{25+144}}\right|=2$.

19. $P(6,0,-6)$; $S(4,0,-6)$ is on the plane; $\overrightarrow{SP}=2\boldsymbol{i}$, $\boldsymbol{N}=\boldsymbol{i}-\boldsymbol{j}$; distance $=\left|\dfrac{\boldsymbol{N}\cdot\overrightarrow{SP}}{|\boldsymbol{N}|}\right|=\frac{2}{\sqrt{2}}=\sqrt{2}$.

20. $P(3,0,10)$; $S(3,0,-4)$ is on the plane; $\overrightarrow{SP}=14\boldsymbol{k}$, $\boldsymbol{N}=2\boldsymbol{i}+3\boldsymbol{j}+\boldsymbol{k}$; distance $=\left|\dfrac{\boldsymbol{N}\cdot\overrightarrow{SP}}{|\boldsymbol{N}|}\right|=\frac{14}{\sqrt{4+9+1}}=\sqrt{14}$.

21. $2x+y-z=$ constant; $(3,-2,1)$ on plane $\Rightarrow 6-2-1=3=$ constant; $2x+y-z=3$.

22. $\boldsymbol{N}=1\boldsymbol{i}-2\boldsymbol{j}+3\boldsymbol{k}\Rightarrow x-2y+3z=$ constant. $(-1,6,0)$ on the plane $\Rightarrow x-2y+3z=-13$.

23. $\overrightarrow{PQ}=\boldsymbol{i}+2\boldsymbol{j}+\boldsymbol{k}$; $\overrightarrow{PR}=-2\boldsymbol{i}+3\boldsymbol{j}-3\boldsymbol{k}$; $\boldsymbol{N}=\overrightarrow{PQ}\times\overrightarrow{PR}=\begin{vmatrix} \boldsymbol{i} & \boldsymbol{j} & \boldsymbol{k} \\ 1 & 2 & 1 \\ -2 & 3 & -3 \end{vmatrix}=$ $-9\boldsymbol{i}+\boldsymbol{j}+7\boldsymbol{k}$. The plane has equation $-9(x-1)+1(y+1)+7(z-2)=0$, or $-9x+y+7z=4$.

24. $\vec{PQ} = -\boldsymbol{i}+\boldsymbol{j}$, $\vec{PR} = -\boldsymbol{i}+\boldsymbol{k}$; $\vec{PQ} \times \vec{PR} = \begin{vmatrix} \boldsymbol{i} & \boldsymbol{j} & \boldsymbol{k} \\ -1 & 1 & 0 \\ -1 & 0 & 1 \end{vmatrix} = \boldsymbol{i}+\boldsymbol{j}+\boldsymbol{k}$; plane is $1 \cdot (x-1) + 1(y-0) + 1(z-0) = 0$, or $x+y+z=1$.

25. $\boldsymbol{v} = -3\boldsymbol{i} + 0\boldsymbol{j} + 7\boldsymbol{k} \Rightarrow x = 1-3t,\ y = 2,\ z = 3+7t$.

26. xy-plane $\Rightarrow z = 0 \Rightarrow t = 0 \Rightarrow$ point is $(1,-1,0)$; yz-plane $\Rightarrow x = 0 \Rightarrow t = -\frac{1}{2} \Rightarrow$ point is $(0,-\frac{1}{2},-\frac{3}{2})$; xz-plane $\Rightarrow y = 0 \Rightarrow t = -1 \Rightarrow$ point is $(-1,0,-3)$.

27. $x = 2t,\ y = -t,\ z = -t$, is the line; it intersects the plane when $3(2t) - 5(-t) + 2(-t) = 6$, or $t = \frac{2}{3}$. The points is $(\frac{4}{3},-\frac{2}{3},-\frac{2}{3})$.

28. $\vec{PQ} = 0 \cdot \boldsymbol{i} + \boldsymbol{j} - \boldsymbol{k} \Rightarrow x = 1 + 0 \cdot t,\ y = 2 + 1 \cdot t,\ z = 0 - 1 \cdot t:\quad x = 1,\ y = 2+t,\ z = -t$.

29. Set $y - 0$, then $z = 0$ to obtain points $P(10,0,-9)$ and $Q(-5,3,0)$ on the line; $\vec{PQ} = -15\boldsymbol{i} + 3\boldsymbol{j} + 9\boldsymbol{k}$ is the direction of the line. Using P, the line has equations $x = 10 - 15t,\ y = 3t,\ z = -9 + 9t$.

30. Find the angle between the normals $\boldsymbol{N}_1 = \boldsymbol{i} + \boldsymbol{j}$ and $\boldsymbol{N}_2 = \boldsymbol{j} + \boldsymbol{k}$: $\theta = \cos^{-1}\left[\frac{\boldsymbol{N}_1 \cdot \boldsymbol{N}_2}{|\boldsymbol{N}_1||\boldsymbol{N}_2|}\right] = \cos^{-1}\frac{1}{\sqrt{2}\sqrt{2}} = \cos^{-1}\frac{1}{2} = 60°$.

31. Minimizing $d^2 = (4-2)^2 + (4+2t+1)^2 + (4t+10)^2$, we have $0 = 2(5+2t)2 + 2(4t+10)4$ or $t = -\frac{5}{2} \Rightarrow d^2 = 4 + (0)^2 + (0)^2 \Rightarrow d = 2$.

32. $d^2 = (10+4t+1)^2 + (-3-4)^2 + (4t-3)^2$; $2dd' = 2 \cdot 4(11+4t) + 2 \cdot 4(4t-3)$. Set $d' = 0 \Rightarrow 0 = 11 + 4t + 4t - 3 \Rightarrow t = -1$; $d(-1) = \sqrt{7^2+7^2+7^2} = 7\sqrt{3}$.

33. Work $= \boldsymbol{F} \cdot \vec{PQ} = |\boldsymbol{F}||PQ|\cos\theta = (40 \text{ lb})(800 \text{ ft})\cos 28° = 28254.323$ ft · lb.

34. Following Example 1, if $\boldsymbol{F}$ is the force, $|PQ| \cdot |\boldsymbol{F}|\sin 90° = 15$ ft · lb. $\frac{3(\text{ft})}{4} \cdot |\boldsymbol{F}| \cdot 1 = 15$ ft · lb $\Rightarrow \boldsymbol{F} = 20$ lb.

35. Area $= |\boldsymbol{A} \times \boldsymbol{B}| = \begin{vmatrix} \boldsymbol{i} & \boldsymbol{j} & \boldsymbol{k} \\ 1 & 1 & -1 \\ 2 & 1 & 1 \end{vmatrix} = |2\boldsymbol{i} - 3\boldsymbol{j} - \boldsymbol{k}| = \sqrt{14}$; Volume $\boldsymbol{C} \cdot (\boldsymbol{A} \times \boldsymbol{B}) = -2 + 6 - 3 = 1$.

36. Area $= |\boldsymbol{A} \times \boldsymbol{B}| = \begin{vmatrix} \boldsymbol{i} & \boldsymbol{j} & \boldsymbol{k} \\ 1 & 1 & 0 \\ 0 & 1 & 0 \end{vmatrix} = |\boldsymbol{k}| = 1$; Volume $= \boldsymbol{C} \cdot (\boldsymbol{A} \times \boldsymbol{B}) = 1$.

37. Not always true: (b) since if: $\boldsymbol{A} = 2\boldsymbol{i}$, $\boldsymbol{A} \cdot \boldsymbol{A} = 4$, $|\boldsymbol{A}| = 2$; and (e) since: $\boldsymbol{A} \times \boldsymbol{B} = -\boldsymbol{B} \times \boldsymbol{A}$. (a), (c), (d), (f), (g), (h) are always true.

38. All are always true.

39. In plane: y-axis; in 3-space: yz-plane.

40. Line in the plane, plane in 3-space.

41. In plane: circle with center at origin, radius 2; in 3-space, a right circular cylinder with axis parallel to z-axis generated by the circle.

42. Ellipse in the plane, in 3-space an elliptical cylinder, generated by the ellipse.

43. In the plane: parabola opening to the right with vertex at $(0, 0)$; in 3-space: the cylinder generated by the parabola with axis parallel to the z-axis.

44. In the plane: hyperbola; in 3-space, the cylinder generated by the hyperbola.

45. In the plane: cardioid with dimple at right; in 3-space a cylinder generated by the cardioid with axis parallel to the z-axis.

46. In the plane: circle of radius $\frac{1}{2}$, center at $(0, \frac{1}{2})$; in 3-space, the cylinder generated by the circle.

47. In the plane a horizontal lemniscate (∞ sign); in 3-space a cylinder generated by the lemniscate.

48. Four-leafed rose (or propeller) in the plane; cylinder generated by the rose in 3-space.

49. Surface of sphere of radius 2, centered at the origin.

50. Plane parallel to z-axis, generated by the line $y = x$.

51. The upper nappe of a cone whose surface makes an angle of $\frac{\pi}{6}$ with the z-axis.

52. Circle of radius 1, center at $(0, 0, 0)$ in the xy-plane.

53. The upper hemisphere of the unit sphere.

54. The region between the spheres $x^2 + y^2 + z^2 = 1$ and $x^2 + y^2 + z^2 = 2$ as well as the surfaces of these spheres.

	Rectangular	Cylindrical	Spherical
55.	$(1,0,0)$	$(1,0,0)$	$(1,\frac{\pi}{2},0)$
56.	$(0,1,0)$	$(1,\frac{\pi}{2},0)$	$(1,\frac{\pi}{2},\frac{\pi}{2})$
57.	$(0,1,1)$	$(1,\frac{\pi}{2},1)$	$(\sqrt{2},\frac{\pi}{4},\frac{\pi}{2})$
58.	$(1,0,-\sqrt{3})$	$(1,0,-\sqrt{3})$	$(2,\frac{5\pi}{6},0)$
59.	$(1,1,1)$	$(\sqrt{2},\frac{\pi}{4},1)$	$(\sqrt{3},\cos^{-1}(\frac{1}{\sqrt{3}}),\frac{\pi}{4})$
60.	$(0,-1,1)$	$(1,\frac{3\pi}{2},1)$	$(\sqrt{2},\frac{\pi}{4},\frac{3\pi}{2})$

61. Rectangular $z = 2 \Rightarrow$ cylindrical $z = 2 \Rightarrow$ spherical $\rho\cos\phi = 2$; this is a plane parallel to xy-plane.

62. Cylindrical: $z = r$; spherical: $\varphi = \frac{\pi}{4}$, $0 \le \varphi \le \pi/2$; upper nappe of a cone.

63. Cylindrical: $z = r^2 \Rightarrow$ rectangular: $z = x^2 + y^2 \Rightarrow$ spherical: $\rho\cos\phi = \rho^2\sin^2\varphi\cos^2\theta + \rho^2\sin^2\varphi\sin^2\theta = \rho^2\sin^2\varphi$ or $\rho = \frac{\cos\varphi}{\sin^2\varphi}$, a paraboloid symmetric about the z-axis, opening up.

64. $r = \cos\theta \Rightarrow r^2 = r\cos\theta \Rightarrow x^2 + y^2 = x$, rectangular; spherical: $x^2 + y^2 = x \Rightarrow \rho^2\sin^2\varphi = \rho\sin\varphi\cos\theta \Rightarrow \rho\sin\varphi = \cos\theta$; a circular cylinder, parallel to z axis, generated by $(x - \frac{1}{2})^2 + y^2 = (\frac{1}{2})^2$.

65. Spherical $\rho = 4 \Rightarrow \sqrt{r^2 + z^2} = 4$ or $r^2+z^2 = 16$, cylindrical; $x^2+y^2+z^2 = 16$, rectangular; sphere of radius 4 centered at the origin.

66. Rectangular: $z = 1$; cylindrical; $z = 1$; the plane $z = 1$.

67. Sphere

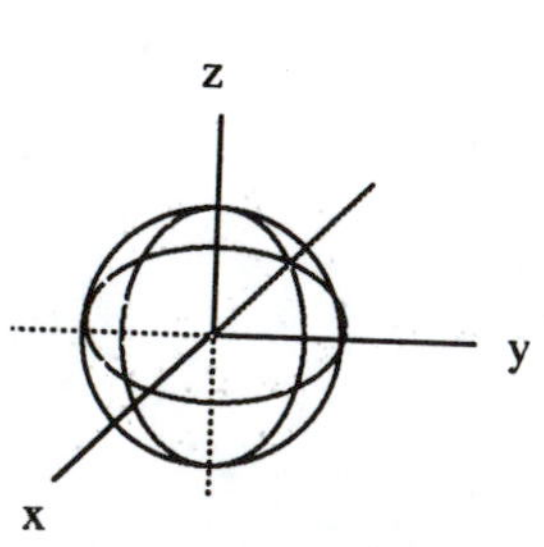

68. Sphere

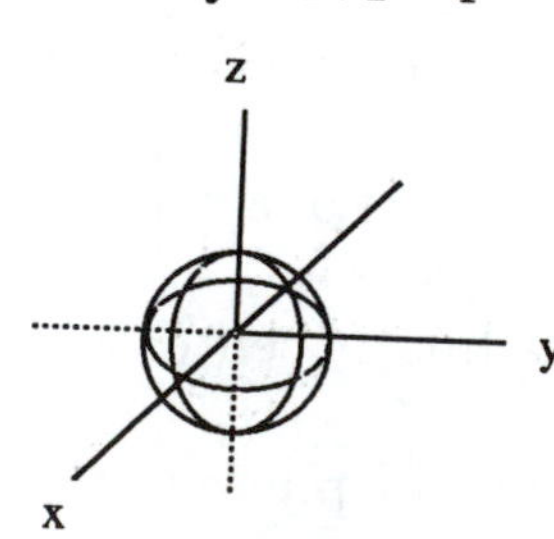

69. Ellipsoid

$4x^2 + 4y^2 + z^2 = 4$

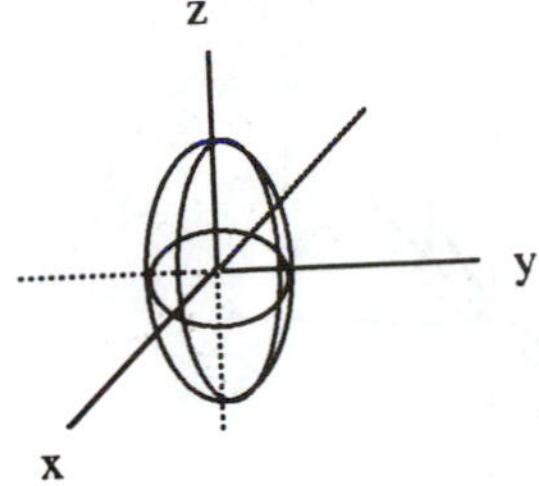

70. Ellipsoid

$36x^2 + 9y^2 + 4z^2 = 36$

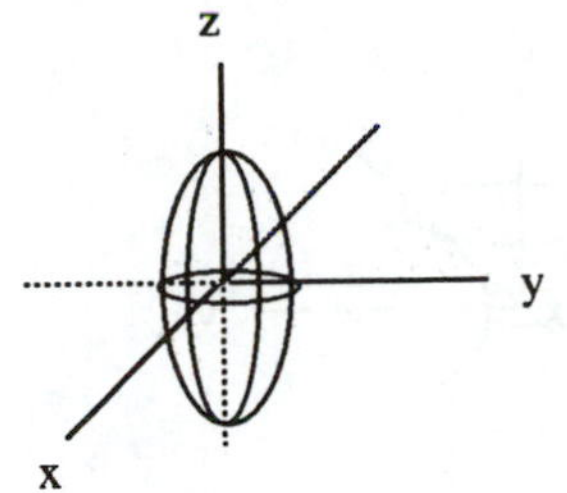

71. Circular Paraboloid

$z = -(x^2 + y^2)$

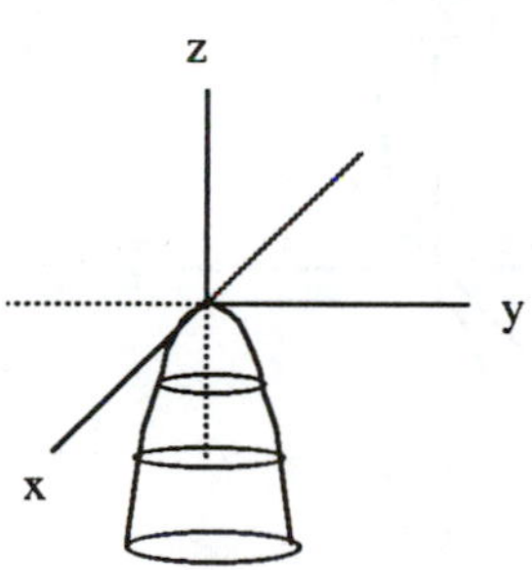

72. Circular Paraboloid

$y = -(x^2 + z^2)$

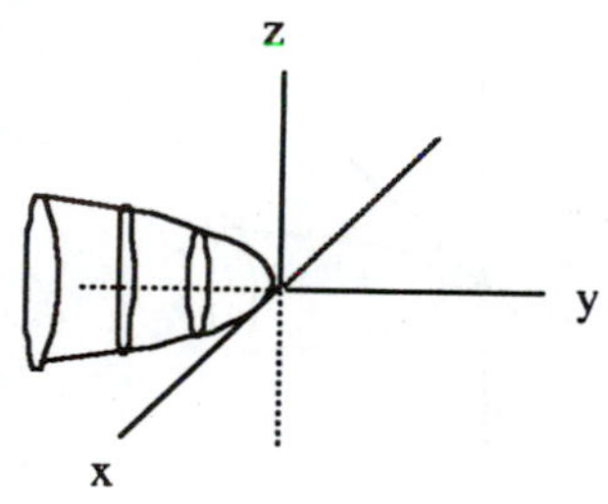

73. Cone about z-axis

$x^2 + y^2 = z^2$

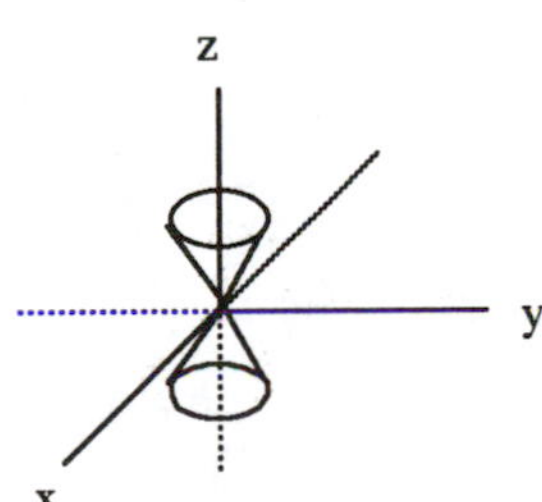

74. Cone about y-axis

$x^2 + z^2 = y^2$

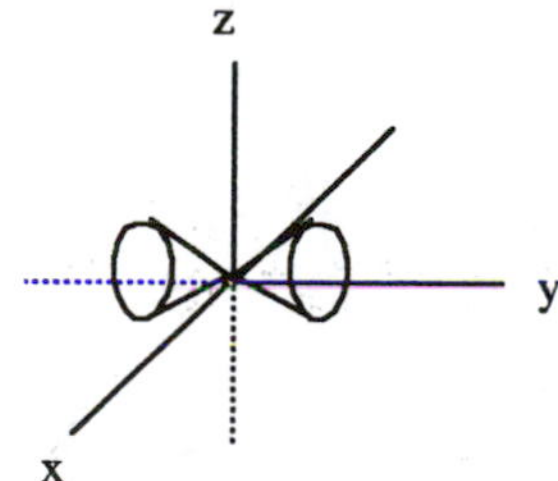

75. Hyperboloid of one sheet

$x^2 + y^2 - z^2 = 4$

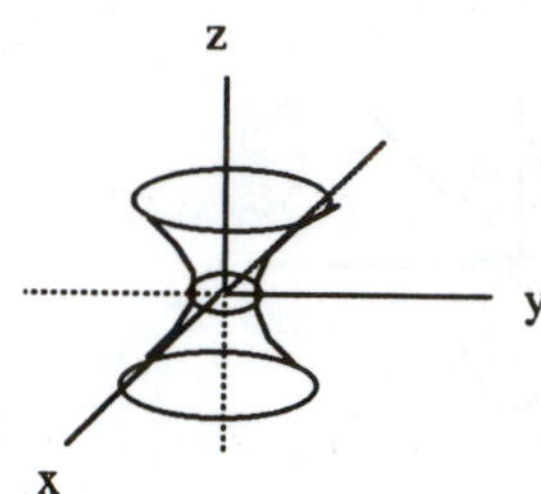

76. Hyperboloid of one sheet

$4y^2 + z^2 - 4x^2 = 4$

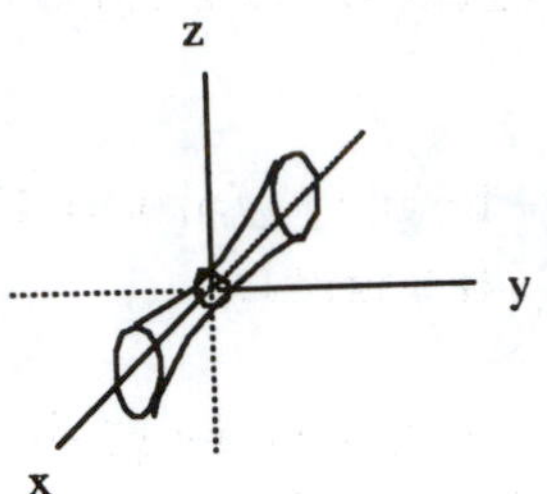

77. Hyperboloid of two sheets

$y^2 - x^2 - z^2 = 1$

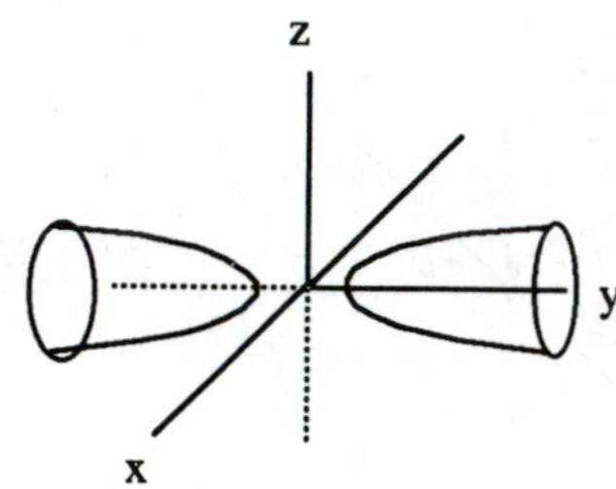

78. Hyperboloid of two sheets

$z^2 - x^2 - y^2 = 1$

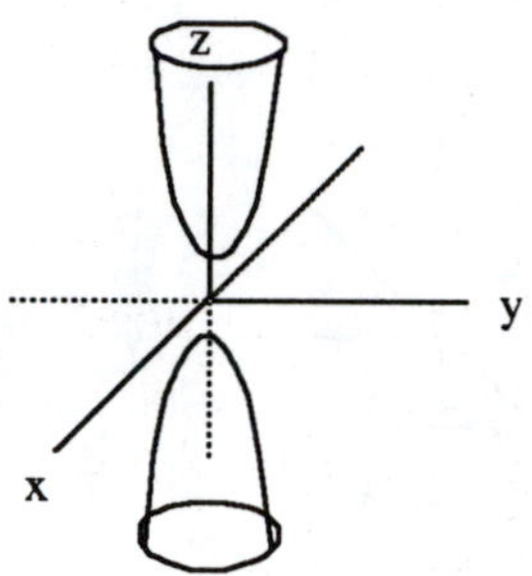

CHAPTER 12

VECTOR-VALUED FUNCTIONS, PARAMETRIZATIONS, AND MOTION IN SPACE

12.1 Vector-Valued Functions and Curves in Space; Derivatives and Integrals

1. $\boldsymbol{r} = (2\cos t)\boldsymbol{i} + (3\sin t)\boldsymbol{j} + 4t\boldsymbol{k}$, $t = \frac{\pi}{2}$. a) $\boldsymbol{v} = \frac{d\boldsymbol{r}}{dt} = (-2\sin t)\boldsymbol{i} + (3\cos t)\boldsymbol{j} + 4\boldsymbol{k}$ b) $\boldsymbol{a} = \frac{d\boldsymbol{v}}{dt} = (-2\cos t)\boldsymbol{i} + (-3\sin t)\boldsymbol{j}$ c) Speed at $t = \pi/2 = |\boldsymbol{v}(\pi/2)| = \sqrt{4\sin^2(\pi/2) + 9\cos^2(\pi/2) + 4^2} = \sqrt{20} = 2\sqrt{5}$
d) (**direction** at $t = \pi/2) = \boldsymbol{v}(\pi/2)/|\boldsymbol{v}(\pi/2)| = (-2\boldsymbol{i} + 4\boldsymbol{k})/2\sqrt{5}$
e) $\boldsymbol{v}(\pi/2) = 2\sqrt{5}\left[(-2\boldsymbol{i} + 4\boldsymbol{k})/2\sqrt{5}\right] = 2\sqrt{5}\left[-(1/\sqrt{5})\boldsymbol{i} + (2/\sqrt{5})\boldsymbol{k}\right]$.

2. a) $\boldsymbol{v} = \frac{d\boldsymbol{r}}{dt} = \boldsymbol{i} + 2t\boldsymbol{j} + 2\boldsymbol{k}$ b) $\boldsymbol{a} = \frac{d\boldsymbol{v}}{dt} = 2\boldsymbol{j}$ c) At $t = 1$: speed $= \sqrt{1 + 2^2 + 2^2} = 3$ d) **direction** $= \frac{1}{3}\boldsymbol{i} + \frac{2}{3}\boldsymbol{j} + \frac{2}{3}\boldsymbol{k}$ e) $\boldsymbol{v} = 3(\frac{1}{3}\boldsymbol{i} + \frac{2}{3}\boldsymbol{j} + \frac{2}{3}\boldsymbol{k})$.

3. a) $\boldsymbol{v} = \frac{d\boldsymbol{r}}{dt} = (-2\sin 2t)\boldsymbol{j} + (2\cos t)\boldsymbol{k}$ b) $\boldsymbol{a} = \frac{d\boldsymbol{v}}{dt} = (-4\cos 2t)\boldsymbol{j} + (-2\sin t)\boldsymbol{k}$ c) At $t = 0$: speed $= \sqrt{0^2 + 2^2} = 2$ d) **direction** $= 2\boldsymbol{k}/2 = \boldsymbol{k}$ e) $\boldsymbol{v} = 2\boldsymbol{k}$.

4. a) $\boldsymbol{v} = e^t\boldsymbol{i} + \frac{4}{9}e^{2t}\boldsymbol{j}$ b) $\boldsymbol{a} = e^t\boldsymbol{i} + \frac{8}{9}e^{2t}\boldsymbol{j}$
c) At $t = \ln 3$: speed $= \sqrt{3^2 + (\frac{4}{9}9)^2} = 5$ d) **direction** $= (3\boldsymbol{i} + 4\boldsymbol{j})/5$
e) $\boldsymbol{v} = 5\,[3\boldsymbol{i} + 4\boldsymbol{j}/5] = 3\boldsymbol{i} + 4\boldsymbol{j}$.

5. a) $\boldsymbol{v} = (\sec t\tan t)\boldsymbol{i} + (\sec^2 t)\boldsymbol{j} + (4/3)\boldsymbol{k}$ b) $\boldsymbol{a} = (\sec t\tan^2 t + \sec^3 t)\boldsymbol{i} + (2\sec^2 t\tan t)\boldsymbol{j}$ c) At $t = \pi/6$, speed $= \sqrt{(\frac{2}{\sqrt{3}}\frac{1}{\sqrt{3}})^2 + (\frac{4}{3})^2 + (\frac{4}{3})^2} = 2$
d) **direction** $= [(2/3)\boldsymbol{i} + (4/3)\boldsymbol{j} + (4/3)\boldsymbol{k}]/2 = (1/3)\boldsymbol{i} + (2/3)\boldsymbol{j} + (2/3)\boldsymbol{k}$
e) $\boldsymbol{v} = 2\,[(1/3)\boldsymbol{i} + (2/3)\boldsymbol{j} + (2/3)\boldsymbol{k}]$.

6. a) $\boldsymbol{v} = \frac{2}{t+1}\boldsymbol{i} + 2t\boldsymbol{j} + t\boldsymbol{k}$ b) $\boldsymbol{a} = -\frac{2}{(t+1)^2}\boldsymbol{i} + 2\boldsymbol{j} + \boldsymbol{k}$ c) At $t = 1$: $\boldsymbol{v} = \boldsymbol{i} + 2\boldsymbol{j} + \boldsymbol{k}$, speed $= \sqrt{1 + 2^2 + 1} = \sqrt{6}$ d) **direction** $= \frac{1}{\sqrt{6}}\boldsymbol{i} + \frac{2}{\sqrt{6}}\boldsymbol{j} + \frac{1}{\sqrt{6}}\boldsymbol{k}$
e) $\boldsymbol{v} = \sqrt{6}\left(\frac{1}{\sqrt{6}}\boldsymbol{i} + \frac{2}{\sqrt{6}}\boldsymbol{j} + \frac{1}{\sqrt{6}}\boldsymbol{k}\right)$.

7. a) $\boldsymbol{v} = (-e^{-t})\boldsymbol{i} + (-6\sin 3t)\boldsymbol{j} + (6\cos 3t)\boldsymbol{k}$ b) $\boldsymbol{a} = (e^{-t})\boldsymbol{i} + (-18\cos 3t)\boldsymbol{j} + (-18\sin 3t)\boldsymbol{k}$ c) At $t = 0$: speed $= \sqrt{(-1)^2 + 0^2 + 6^2} = \sqrt{37}$
d) **direction** $= (-\boldsymbol{i} + 6\boldsymbol{k})/\sqrt{37}$ e) $\boldsymbol{v} = \sqrt{37}\left[(-\boldsymbol{i} + 6\boldsymbol{k})/\sqrt{37}\right]$.

8. a) $\boldsymbol{v} = \boldsymbol{i} + \sqrt{2}t\boldsymbol{j} + t^2\boldsymbol{k}$ b) $\boldsymbol{a} = \sqrt{2}\boldsymbol{j} + 2t\boldsymbol{k}$
c) At $t = 1$: speed $= \sqrt{1+2+1} = 2$ d) **direction** $= \frac{1}{2}\boldsymbol{i} + \frac{\sqrt{2}}{2}\boldsymbol{j} + \frac{1}{2}\boldsymbol{k}$
e) $\boldsymbol{v} = 2(\frac{1}{2}\boldsymbol{i} + \frac{\sqrt{2}}{2}\boldsymbol{j} + \frac{1}{2}\boldsymbol{k})$.

9. $\boldsymbol{r} = (3t+1)\boldsymbol{i} + \sqrt{3}t\boldsymbol{j} + t^2\boldsymbol{k}$, $\boldsymbol{v} = 3\boldsymbol{i} + \sqrt{3}\boldsymbol{j} + 2t\boldsymbol{k}$, $\boldsymbol{a} = 2\boldsymbol{k}$, $\boldsymbol{v}(0) = 3\boldsymbol{i} + \sqrt{3}\boldsymbol{j}$, $\boldsymbol{a}(0) = 2\boldsymbol{k}$. $\theta = \cos^{-1}[\boldsymbol{v}(0)\cdot\boldsymbol{a}(0)/|\boldsymbol{v}(0)||\boldsymbol{a}(0)|] = \cos^{-1} 0 = \pi/2$.

10. $\boldsymbol{v} = \frac{\sqrt{2}}{2}\boldsymbol{i} + \left(\frac{\sqrt{2}}{2} - 32t\right)\boldsymbol{j}$, $\boldsymbol{a} = -32\boldsymbol{j}$, $\boldsymbol{v}(0) = \frac{\sqrt{2}}{2}\boldsymbol{i} + \frac{\sqrt{2}}{2}\boldsymbol{j}$, $\boldsymbol{a}(0) = -32\boldsymbol{j}$. $\theta = \cos^{-1}[\boldsymbol{v}(0)\cdot\boldsymbol{a}(0)/(|\boldsymbol{v}(0)||\boldsymbol{a}(0)|)] = \cos^{-1}\left[-16\sqrt{2}/(1\cdot 32)\right] = \cos^{-1}(-\sqrt{2}/2) = \frac{3\pi}{4}$.

11. $\boldsymbol{v} = \left(\frac{2t}{t^2+1}\right)\boldsymbol{i} + \left(\frac{1}{t^2+1}\right)\boldsymbol{j} + \left(\frac{t}{\sqrt{t^2+1}}\right)\boldsymbol{k}$, $\boldsymbol{a} = \frac{2(1-t^2)}{(t^2+1)^2}\boldsymbol{i} + \frac{-2t}{(t^2+1)^2}\boldsymbol{j} + \left(\frac{1}{(t^2+1)^{3/2}}\right)\boldsymbol{k}$, $\boldsymbol{v}(0) = \boldsymbol{j}$, $\boldsymbol{a}(0) = 2\boldsymbol{i} + \boldsymbol{k}$. $\theta = \cos^{-1}[\boldsymbol{v}(0)\cdot\boldsymbol{a}(0)/(|\boldsymbol{v}(0)||\boldsymbol{a}(0)|)] = \cos^{-1} 0 = \pi/2$.

12. $\boldsymbol{v} = \frac{2}{3}(1+t)^{1/2}\boldsymbol{i} - \frac{2}{3}(1-t)^{1/2}\boldsymbol{j} + \frac{1}{3}\boldsymbol{k}$, $\boldsymbol{a} = \frac{1}{3}(1+t)^{-1/2}\boldsymbol{i} + \frac{1}{3}(1-t)^{-1/2}\boldsymbol{j}$, $\boldsymbol{v}(0) = \frac{2}{3}\boldsymbol{i} - \frac{2}{3}\boldsymbol{j} + \frac{1}{3}\boldsymbol{k}$, $\boldsymbol{a}(0) = \frac{1}{3}\boldsymbol{i} + \frac{1}{3}\boldsymbol{j}$. $\theta = \cos^{-1}[0] = \frac{\pi}{2}$.

13. $\boldsymbol{v} = (1-\cos t)\boldsymbol{i} + (\sin t)\boldsymbol{j}$, $\boldsymbol{a} = (\sin t)\boldsymbol{i} + (\cos t)\boldsymbol{j}$, $\boldsymbol{v}\cdot\boldsymbol{a} = 0$ yields $(1-\cos t)\sin t + \sin t\cos t = \sin t = 0$. Answer: $t = 0, \pi, 2\pi$.

14. $\boldsymbol{v} = (\cos t)\boldsymbol{i} + \boldsymbol{j} - (\sin t)\boldsymbol{k}$, $\boldsymbol{a} = (-\sin t)\boldsymbol{i} - (\cos t)\boldsymbol{k}$. $\boldsymbol{v}\cdot\boldsymbol{a} = -\cos t(\sin t) + \sin t(\cos t) = 0$. $\boldsymbol{v}$ and $\boldsymbol{a}$ are orthogonal for all $t \geq 0$.

15. $\int_0^1 [t^3\boldsymbol{i} + 7\boldsymbol{j} + (t+1)\boldsymbol{k}]\,dt = \frac{t^4}{4}\Big]_0^1 \boldsymbol{i} + 7t]_0^1\boldsymbol{j} + \left[\frac{t^2}{2} + t\right]_0^1 \boldsymbol{k} = \frac{1}{4}\boldsymbol{i} + 7\boldsymbol{j} + \frac{3}{2}\boldsymbol{k}$.

16. $\int_1^2 \left[(6-6t)\boldsymbol{i} + 3\sqrt{t}\boldsymbol{j} + (4t^{-2})\boldsymbol{k}\right]dt = [6t - 3t^2]_1^2\,\boldsymbol{i} + \left[2t^{3/2}\right]_1^2 \boldsymbol{j} - [4t^{-1}]_1^2\,\boldsymbol{k} = -3\boldsymbol{i} + 2(2\sqrt{2}-1)\boldsymbol{j} + 2\boldsymbol{k}$.

17. $\int_{-\pi/4}^{\pi/4} [(\sin t)\boldsymbol{i} + (1+\cos t)\boldsymbol{j} + (\sec^2 t)\boldsymbol{k}]\,dt = [-\cos t]_{-\pi/4}^{\pi/4}\,\boldsymbol{i} + [t + \sin t]_{-\pi/4}^{\pi/4}\,\boldsymbol{j} + \tan t]_{-\pi/4}^{\pi/4}\,\boldsymbol{k} = (\pi/2 + \sqrt{2})\boldsymbol{j} + 2\boldsymbol{k}$.

18. $\int_0^{\pi/3} [(\sec t\tan t)\boldsymbol{i} + (\tan t)\boldsymbol{j} + (2\sin t\cos t)\boldsymbol{k}]\,dt = [\sec t]_0^{\pi/3}\,\boldsymbol{i} - [\ln|\cos t|]_0^{\pi/3}\,\boldsymbol{j} + \left[\sin^2 t\right]_0^{\pi/3} \boldsymbol{k} = (2-1)\boldsymbol{i} - \left[\ln\frac{1}{2} - \ln 1\right]\boldsymbol{j} + \left[\left(\frac{\sqrt{3}}{2}\right)^2 - 0\right]\boldsymbol{k} = \boldsymbol{i} + (\ln 2)\boldsymbol{j} + \frac{3}{4}\boldsymbol{k}$.

19. $\int_1^4 \left[\frac{1}{t}\boldsymbol{i} + \frac{1}{5-t}\boldsymbol{j} + \frac{1}{2t}\boldsymbol{k}\right]dt = \ln|t|]_1^4\,\boldsymbol{i} - \ln|5-t|]_1^4\,\boldsymbol{j} + \frac{1}{2}\ln|t|\Big]_1^4 \boldsymbol{k} = (\ln 4)\boldsymbol{i} + (\ln 4)\boldsymbol{j} + (\ln 2)\boldsymbol{k}$.

20. $\int_0^1 \left[\frac{2}{\sqrt{1-t^2}}\boldsymbol{i} + \frac{\sqrt{3}}{1+t^2}\boldsymbol{k}\right]dt = 2\left[\sin^{-1} t\right]_0^1 \boldsymbol{i} + \sqrt{3}\,[\tan^{-1} t]_0^1\,\boldsymbol{k} = \pi\boldsymbol{i} + \sqrt{3}\frac{\pi}{4}\boldsymbol{k}$.

21. $\boldsymbol{r} = (\sin t)\boldsymbol{i} + (\cos t)\boldsymbol{j}$. $\boldsymbol{v} = (\cos t)\boldsymbol{i} - (\sin t)\boldsymbol{j}$, $\boldsymbol{a} = -(\sin t)\boldsymbol{i} - (\cos t)\boldsymbol{j} = -\boldsymbol{r}$. $\boldsymbol{v}(\pi/4) = (\sqrt{2}/2)(\boldsymbol{i} - \boldsymbol{j})$, $\boldsymbol{a}(\pi/4) = -(\sqrt{2}/2)(\boldsymbol{i} + \boldsymbol{j})$, $\boldsymbol{v}(\pi/2) = -\boldsymbol{j}$, $\boldsymbol{a}(\pi/2) = -\boldsymbol{i}$.

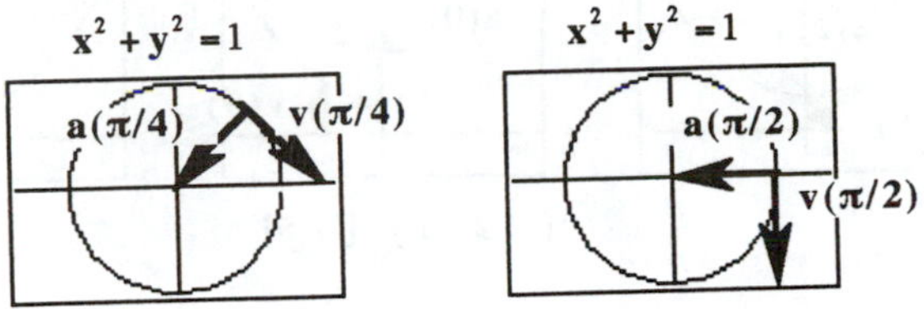

22. $\boldsymbol{r} = (4\cos\frac{t}{2})\boldsymbol{i} + (4\sin\frac{t}{2})\boldsymbol{j}$, $\boldsymbol{v} = -2(\sin\frac{t}{2})\boldsymbol{i} + 2(\cos\frac{t}{2})\boldsymbol{j}$, $\boldsymbol{a} = -(\cos\frac{t}{2})\boldsymbol{i} - (\sin\frac{t}{2})\boldsymbol{j}$. $\boldsymbol{v}(\pi) = -2\boldsymbol{i}$, $\boldsymbol{a}(\pi) = -\boldsymbol{j}$, $\boldsymbol{v}(3\pi/2) = -\sqrt{2}\boldsymbol{i} - \sqrt{2}\boldsymbol{j}$, $\boldsymbol{a}(3\pi/2) = \frac{\sqrt{2}}{2}\boldsymbol{i} - \frac{\sqrt{2}}{2}\boldsymbol{j}$.

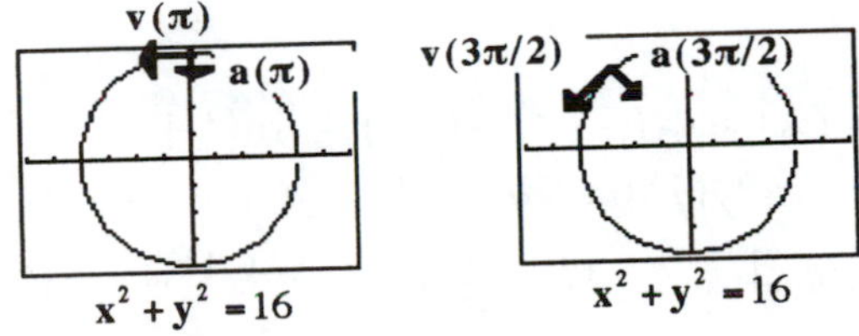

23. $\boldsymbol{v} = (1 - \cos t)\boldsymbol{i} + (\sin t)\boldsymbol{j}$, $\boldsymbol{a} = (\sin t)\boldsymbol{i} + (\cos t)\boldsymbol{j}$. $\boldsymbol{v}(\pi) = 2\boldsymbol{i}$, $\boldsymbol{a}(\pi) = -\boldsymbol{j}$, $\boldsymbol{v}(3\pi/2) = \boldsymbol{i} - \boldsymbol{j}$, $\boldsymbol{a}(3\pi/2) = -\boldsymbol{i}$.

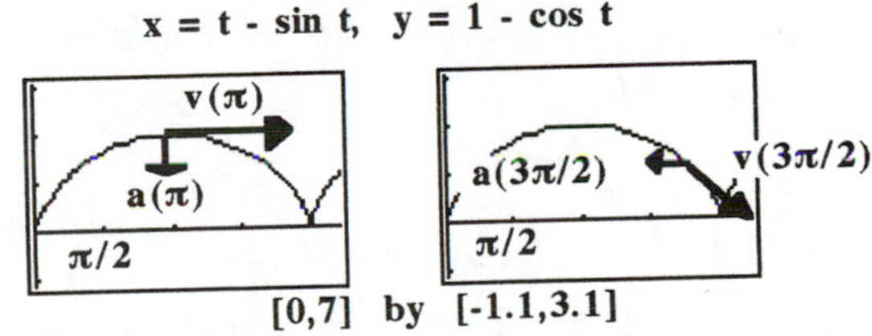

24. $\boldsymbol{r} = t\boldsymbol{i} + (t^2+1)\boldsymbol{j}$, $\boldsymbol{v} = \boldsymbol{i} + 2t\boldsymbol{j}$, $\boldsymbol{a} = 2\boldsymbol{j}$. $\boldsymbol{v}(-1) = \boldsymbol{i} - 2\boldsymbol{j}$, $\boldsymbol{a}(-1) = 2\boldsymbol{j}$, $\boldsymbol{v}(0) = \boldsymbol{i}$, $\boldsymbol{a}(0) = 2\boldsymbol{j}$, $\boldsymbol{v}(1) = \boldsymbol{i} + 2\boldsymbol{j}$, $\boldsymbol{a}(1) = 2\boldsymbol{j}$.

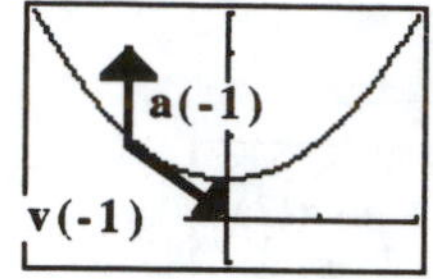

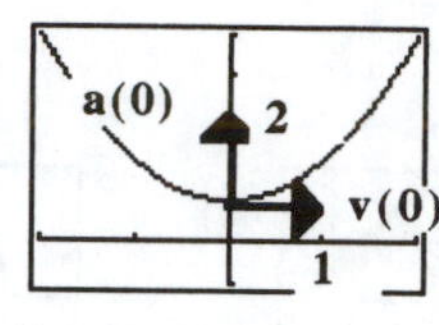

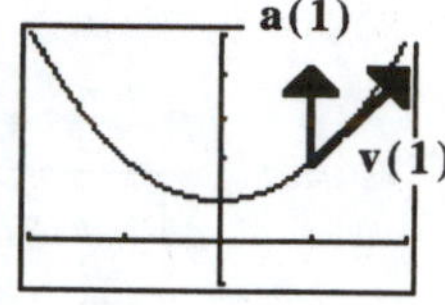

[-2,2] by [-1,5]

25. Integrating, we obtain $\boldsymbol{r} = \left(-\frac{t^2}{2} + C_1\right)\boldsymbol{i} + \left(-\frac{t^2}{2} + C_2\right)\boldsymbol{j} + \left(-\frac{t^2}{2} + C_3\right)\boldsymbol{k} = -\frac{t^2}{2}(\boldsymbol{i} + \boldsymbol{j} + \boldsymbol{k}) + C_1\boldsymbol{i} + C_2\boldsymbol{j} + C_3\boldsymbol{k}$. Setting $t = 0$, we obtain $\boldsymbol{C} = C_1\boldsymbol{i} + C_2\boldsymbol{j} + C_3\boldsymbol{k} = \boldsymbol{i} + 2\boldsymbol{j} + 3\boldsymbol{k}$. Hence $\boldsymbol{r} = -\frac{t^2}{2}(\boldsymbol{i} + \boldsymbol{j} + \boldsymbol{k}) + \boldsymbol{i} + 2\boldsymbol{j} + 3\boldsymbol{k} = \left(1 - \frac{t^2}{2}\right)\boldsymbol{i} + \left(2 - \frac{t^2}{2}\right)\boldsymbol{j} + \left(3 - \frac{t^2}{2}\right)\boldsymbol{k}$.

26. $\boldsymbol{r} = 90t^2\boldsymbol{i} + (90t^2 - \frac{16}{3}t^3)\boldsymbol{j} + \boldsymbol{C}$. $\boldsymbol{r}(0) = \boldsymbol{C} = 100\boldsymbol{j}$. $\boldsymbol{r} = 90t^2\boldsymbol{i} + (90t^2 - \frac{16}{3}t^3 + 100)\boldsymbol{j}$.

27. $\boldsymbol{r} = \left[(t+1)^{3/2} + C_1\right]\boldsymbol{k} + (-e^{-t} + C_2)\boldsymbol{j} + [\ln(t+1) + C_3]\,\boldsymbol{k}$. When $t = 0$, $\boldsymbol{r} = (1 + C_1)\boldsymbol{i} + (-1 + C_2)\boldsymbol{j} + C_3\boldsymbol{k} = \boldsymbol{k}$ so $C_1 = -1$, $C_2 = 1$, $C_3 = 1$. $\boldsymbol{r} = \left[(t+1)^{3/2} - 1\right]\boldsymbol{i} + (1 - e^{-t})\boldsymbol{j} + [\ln(t+1) + 1]\,\boldsymbol{k}$

28. $\boldsymbol{r} = \left(\frac{t^4}{4} + 2t^2\right)\boldsymbol{i} + \frac{t^2}{2}\boldsymbol{j} + \left(\frac{2}{3}t^3\boldsymbol{k}\right) + \boldsymbol{C}$. $\boldsymbol{r}(0) = \boldsymbol{C} = \boldsymbol{i} + \boldsymbol{j}$. $\boldsymbol{r} = \left(\frac{t^4}{4} + 2t^2 + 1\right)\boldsymbol{i} + \left(\frac{t^2}{2} + 1\right)\boldsymbol{j} + \frac{2}{3}t^3\boldsymbol{k}$

29. $\dfrac{d\boldsymbol{r}}{dt} = C_1\boldsymbol{i} + C_2\boldsymbol{j} + (-32t + C_3)\boldsymbol{k}$. When $t = 0$, this must equal $8\boldsymbol{i} + 8\boldsymbol{j}$ so $\dfrac{d\boldsymbol{r}}{dt} = 8\boldsymbol{i} + 8\boldsymbol{j} - 32t\boldsymbol{k}$. $\boldsymbol{r} = (8t + A_1)\boldsymbol{i} + (8t + A_2)\boldsymbol{j} + (-16t^2 + A_3)\boldsymbol{k}$. When $t = 0$: $\boldsymbol{r} = A_1\boldsymbol{i} + A_2\boldsymbol{j} + A_3\boldsymbol{k} = 100\boldsymbol{k}$ so our final answer is $\boldsymbol{r} = 8t\boldsymbol{i} + 8t\boldsymbol{j} + (100 - 16t^2)\boldsymbol{k}$.

30. $\dfrac{d\boldsymbol{r}}{dt} = -(\boldsymbol{i} + \boldsymbol{j} + \boldsymbol{k})t + \boldsymbol{C}_1$. When $t = 0$, $\dfrac{d\boldsymbol{r}}{dt} = C_1 = 0$, so $\dfrac{d\boldsymbol{r}}{dt} = -(\boldsymbol{i} + \boldsymbol{j} + \boldsymbol{k})t$. $\boldsymbol{r} = -(\boldsymbol{i} + \boldsymbol{j} + \boldsymbol{k})\frac{t^2}{2} + \boldsymbol{C}$. When $t = 0$, $\boldsymbol{r} = \boldsymbol{C} = 10(\boldsymbol{i} + \boldsymbol{j} + \boldsymbol{k})$ so $\boldsymbol{r}(t) = (10 - \frac{t^2}{2})(\boldsymbol{i} + \boldsymbol{j} + \boldsymbol{k})$.

31. $\boldsymbol{v} = (1-\cos t)\boldsymbol{i}+\sin t\boldsymbol{j}$, $\boldsymbol{a} = \sin t\boldsymbol{i}+\cos t\boldsymbol{j}$. $|\boldsymbol{v}|^2 = 1-2\cos t+\cos^2 t+\sin^2 t = 2(1-\cos t)$. This has maximum value when $\cos t = -1$ and minimum value when $\cos t = 1$. We conclude that the maximum and minimum values of $|\boldsymbol{v}|$ are, respectively, 2 and 0. Since $|\boldsymbol{a}| = 1$ is constant, 1 is both the maximum and minimum value of $|\boldsymbol{a}|$.

32. $\boldsymbol{r} = (3\cos t)\boldsymbol{j} + (2\sin t)\boldsymbol{k}$, $\boldsymbol{v} = (-3\sin t)\boldsymbol{j} + (2\cos t)\boldsymbol{k}$, $\boldsymbol{a} = (-3\cos t)\boldsymbol{j} + (-2\sin t)\boldsymbol{k}$. $|\boldsymbol{v}|^2 = 9\sin^2 t+4\cos^2 t = 5\sin^2 t+4(\sin^2 t+\cos^2 t) = 5\sin^2 t+4$. Since $\sin^2 t$ has maximum and minimum values 1 and 0, $|\boldsymbol{v}|^2$ has maximum and minimum values 9 and 4 and $|\boldsymbol{v}|$ has maximum and minimum values 3 and 2. $|\boldsymbol{a}|^2 = 9\cos^2 t + 4\sin^2 t$ so we will get the same maximum and minimum values for $|\boldsymbol{a}|$.

33. $\boldsymbol{f} = C_1\boldsymbol{i} + C_2\boldsymbol{j} + C_3\boldsymbol{k}$ where C_1, C_2, C_3 are constants. Hence $\dfrac{d\boldsymbol{f}}{dt} = 0i + 0j + 0k = \mathbf{0}$.

34. $\boldsymbol{f} = f_1\boldsymbol{i} + f_2\boldsymbol{j} + f_3\boldsymbol{k}$ where f_1, f_2, f_3 are differentiable scalar functions of t. $\dfrac{d(c\boldsymbol{f})}{dt} = \dfrac{d}{dt}(cf_1\boldsymbol{i}+cf_2\boldsymbol{j}+cf_3\boldsymbol{k}) = \dfrac{d(cf_1)}{dt}\boldsymbol{i}+\dfrac{d(cf_2)}{dt}\boldsymbol{j}+\dfrac{d(cf_3)}{dt}\boldsymbol{k} = c\dfrac{df_1}{dt}\boldsymbol{i}+c\dfrac{df_2}{dt}\boldsymbol{j}+c\dfrac{df_3}{dt}\boldsymbol{k} = c(\dfrac{df_1}{dt}\boldsymbol{i}+\dfrac{df_2}{dt}\boldsymbol{j}+\dfrac{df_3}{dt}\boldsymbol{k}) = c\dfrac{d\boldsymbol{f}}{dt}$.

35. $\boldsymbol{u} = u_1\boldsymbol{i}+u_2\boldsymbol{j}+u_3\boldsymbol{k}$ and $\boldsymbol{v} = v_1\boldsymbol{i}+v_2\boldsymbol{j}+v_3\boldsymbol{k}$ where u_i and v_i are certain differentiable functions of t for $i = 1, 2, 3$. $\frac{d}{dt}(\boldsymbol{u}+\boldsymbol{v}) = \frac{d}{dt}\left[(u_1+v_1)\boldsymbol{i}+(u_2+v_2)\boldsymbol{j}+(u_3+v_3)\boldsymbol{k}\right] = \left(\frac{du_1}{dt}+\frac{dv_1}{dt}\right)\boldsymbol{i} + \left(\frac{du_2}{dt}+\frac{dv_2}{dt}\right)\boldsymbol{j} + \left(\frac{du_3}{dt}+\frac{dv_3}{dt}\right)\boldsymbol{k} = \left(\frac{du_1}{dt}\boldsymbol{i}+\frac{du_2}{dt}\boldsymbol{j}+\frac{du_3}{dt}\boldsymbol{k}\right)+\left(\frac{dv_1}{dt}\boldsymbol{i}+\frac{dv_2}{dt}\boldsymbol{j}+\frac{dv_3}{dt}\boldsymbol{k}\right) = \dfrac{d\boldsymbol{u}}{dt}+\dfrac{d\boldsymbol{v}}{dt}$. We omit the proof of the second formula which is quite similar: it may be obtained by replacing certain +'s by −'s in the above proof.

36. $\boldsymbol{u} = u_1\boldsymbol{i} + u_2\boldsymbol{j} + u_3\boldsymbol{k}$ and $\boldsymbol{v} = v_1\boldsymbol{i} + v_2\boldsymbol{j} + v_3\boldsymbol{k}$ where u_i and v_i are certain differentiable functions of t for $i = 1, 2, 3$. $\frac{d}{dt}(\boldsymbol{u}\times\boldsymbol{v}) = \frac{d}{dt}\left[(u_2v_3 - v_2u_3)\boldsymbol{i} + (v_1u_3 - u_1v_3)\boldsymbol{j} + (u_1v_2 - v_1u_2)\boldsymbol{k}\right] = (u_2v_3' + u_2'v_3 - v_2u_3' - v_2'u_3)\boldsymbol{i} + (v_1u_3' + v_1'u_3 - u_1v_3' - u_1'v_3)\boldsymbol{j} + (u_1v_2' + u_1'v_2 - v_1u_2' - v_1'u_2)\boldsymbol{k}$. On the other hand, $\dfrac{d\boldsymbol{u}}{dt}\times\boldsymbol{v}+\boldsymbol{u}\times\dfrac{d\boldsymbol{v}}{dt} = \left[(u_2'v_3 - v_2u_3')\boldsymbol{i} + (v_1u_3' - u_1'v_3)\boldsymbol{j} + (u_1'v_2 - v_1u_2')\boldsymbol{k}\right] + \left[(u_2v_3' - v_2'u_3)\boldsymbol{i} + (u_3v_1' - u_1v_3')\boldsymbol{j} + (u_1v_2' - u_2v_1')\boldsymbol{k}\right]$. Completing the component-wise addition, we see that $\frac{d}{dt}(\boldsymbol{u}\times\boldsymbol{v}) = \dfrac{d\boldsymbol{u}}{dt}\times\boldsymbol{v}+\boldsymbol{u}\times\dfrac{d\boldsymbol{v}}{dt}$.

12.2 MODELING PROJECTILE MOTION

1. $x = (v_0 \cos\alpha)t$, $21 = (0.840\,\text{km/sec})(\cos 60°)t$, $t = 50\,\text{sec}$. Graph $x = 840(\cos(\pi/3))t$, $y = 840(\sin(\pi/3))t - \frac{1}{2}(9.8)t^2$, $0 \leqq t \leqq 150$, tStep = 0.5 in $[0, 70000]$ by $[0, 30000]$. Use TRACE to see that when $t = 50\,\text{sec}$, $x = 21000$ m. Caution: on our calculator, $\sin\pi/3$ gave 0; one can use $\sin(\pi/3)$ for the correct value.

2. $R = \frac{v_0^2}{g}\sin 2\alpha$ is maximal when $\sin 2\alpha = 1$, $\alpha = \pi/4$ and $R = v_0^2/g$. This leads to $24500 = v_0^2/9.8$, $v_0 = \sqrt{(9.8)24500} = 490$ m/sec.

3. a) $t = \frac{2v_0 \sin\alpha}{g} = \frac{2(500\,\text{m/sec})\sin 45°}{9.8\,\text{m/sec}^2} = 72.15\,\text{sec}$. $R = \frac{v_0^2}{g}\sin 2\alpha = \frac{500^2 \sin 90°}{9.8} = 25510\,\text{m} = 25.51$ km b) $x = (v_0\cos\alpha)t$, $t = (0.5\,\text{km/sec})(\cos 45°)t$, $t = 10\sqrt{2}\,\text{sec}$. $y = (v_0 \sin\alpha)t - \frac{1}{2}gt^2 = 500\frac{\sqrt{2}}{2}10\sqrt{2} - \frac{1}{2}9.8(10\sqrt{2})^2 = 4020$ m $= 4.02$ km c) $y_{\max} = \frac{(v_0\sin\alpha)^2}{2g} = \frac{(500\sin 45°)^2}{2(9.8)} = 6377.55$ m.

4. $y = y_0 + t(v_0 \sin\alpha) - \frac{1}{2}gt^2 = 32 + t(32\sin 30°) - 16t^2 = 32 + 16t - 16t^2 = 16(2 + t - t^2) = 16(2 - t)(1 + t) = 0$ when $t = 2\,\text{sec}$. $x(t) = (v_0\cos\alpha)t = 32(\sqrt{3}/2)t = 16\sqrt{3}t$. $x(2) = 32\sqrt{3} = 55.426$ ft.

5. $R = \frac{v_0^2}{g}\sin 2\alpha$. If we replace α by $90° - \alpha$, the new range is $\frac{v_0^2}{g}\sin[2(90° - \alpha)] = \frac{v_0^2}{g}\sin(180° - 2\alpha) = \frac{v_0^2}{g}\sin 2\alpha = R$.

6. $R = \frac{v_0^2}{g}\sin 2\alpha = \frac{400^2}{9.8}\sin 2\alpha = 16000$, $\sin 2\alpha = 0.98$, $2\alpha = \sin^{-1} 0.98$ and $2\alpha = 180° - \sin^{-1} 0.98$. Using degree mode, we obtain $\alpha = 39.26°$ and $50.74°$.

7. $R = \frac{v_0^2}{g}\sin 2\alpha$, $10 = \frac{v_0^2}{9.8}\sin 90°$, $v_0 = \sqrt{98} = 9.9$ m/sec. $6 = \frac{98}{9.8}\sin 2\alpha$, $\sin 2\alpha = 0.6$, $2\alpha = \sin^{-1} 0.6$ and $\pi - \sin^{-1} 0.6$. Solving for α and using degree measure, we obtain $\alpha = 18.43°$ and $71.57°$.

8. Here $\alpha = 0°$. $x = (v_0\cos\alpha)t = 5 \times 10^6 t = 0.4$, $t = \frac{0.4}{5\times 10^6} = \frac{40}{5\times 10^8} = 8 \times 10^{-8}$ sec. $y = (v_0 \sin\alpha)t - \frac{1}{2}gt^2 = -\frac{9.8}{2}t^2 = -4.9t^2$. $y(8 \times 10^{-8}) = -4.9(64 \times 10^{-16}) = -313.6 \times 10^{-16}$ m $= -3.136 \times 10^{-12}$ cm. It will drop 3.136×10^{-12} cm.

9. $R = \frac{v_0^2}{g}\sin 2\alpha$, $(248.8)3$ ft $= \frac{v_0^2}{32}\sin 18°$, $v_0 = \sqrt{\frac{32(248.8)3}{\sin 18°}} = 278.02$ ft/sec $= 278.02\frac{\text{ft}}{\text{sec}}\frac{1\,\text{mile}}{5280\,\text{ft}}\frac{3600\,\text{sec}}{1\,\text{hr}} = 189.56$ mph.

10. New $y_{max} = \frac{((2v_0)\sin\alpha)^2}{2g} = 4\frac{(v_0 \sin\alpha)^2}{2g} = 4(\text{old } y_{max})$. New range $= \frac{(2v_0)^2}{g}\sin 2\alpha = 4\frac{v_0^2}{g}\sin 2\alpha = 4$ (old range). To double the height and range the new initial speed must be $\sqrt{2}v_0 \approx 1.41v_0$ and so the initial speed must be increased by about 41%.

11. We solve $y = \frac{3}{4}y_{max}$ for t: $(v_0 \sin\alpha)t - \frac{1}{2}gt^2 = \frac{3(v_0 \sin\alpha)^2}{8g}$, (multiplying by $-8g$) $4g^2t^2 - 8gv_0 \sin\alpha + 3(v_0\sin\alpha)^2 = 0$. Solving for the shorter (first) time, we get $t = \frac{8gv_0 \sin\alpha - \sqrt{64g^2(v_0\sin\alpha)^2 - 48g^2(v_0\sin\alpha)^2}}{8g^2} = \frac{4gv_0\sin\alpha}{8g^2} = \frac{1}{2}\frac{v_0\sin\alpha}{g} = \frac{1}{2}$ (time for y_{max}).

12. $200 = R = \frac{v_0^2}{g}\sin 2\alpha = \frac{(80\sqrt{10/3})^2}{32}\sin 2\alpha$, $\sin 2\alpha = \frac{32(200)}{(80\sqrt{10/3})^2}$, $\alpha = \frac{1}{2}\sin^{-1}(\frac{32(200)}{(80\sqrt{10/3})^2}) = 32.079°$. Using this value of α, we find $y_{max} = \frac{(v_0\sin\alpha)^2}{2g} = \frac{((80\sqrt{10/3})\sin\alpha)^2}{2(32)} = 31.339$ ft. The cannon's angle of elevation should be 32.079°. The performer will have maximum height 31.339 ft so will not strike the ceiling.

13. Let us find the height of the ball when $x = 369$ ft. We first have to find t such that $x(t) = 369$. $x = t(v_0\cos\alpha) = t(116\cos 45°) = 58\sqrt{2}t = 369$, $t = \frac{369}{58\sqrt{2}}$ sec. $y = (v_0 \sin\alpha)t - \frac{1}{2}gt^2 = 58\sqrt{2}t - 16t^2$. $y(\frac{369}{58\sqrt{2}}) = 45.193$ ft. So in flight, the ball passes just above the pin.

14. The time for $x = 135$ we find from $x = (v_0\cos\alpha)t = 135$. $90\cos 30°t = 135$, $t = 135/(90\sqrt{3}/2) = \sqrt{3}$. At that time $y = (90\sin 30°)\sqrt{3} - \frac{1}{2}(32)(\sqrt{3})^2 = (45\sqrt{3} - 48)$ ft $= 29.942$ ft. It won't clear the tree.

15. We use the equations for launching from (x_0, y_0). $x = (v_0\cos 20°)t = 315$, $y = 3 + v_0(\sin 20°)t - 16t^2 = 37$. We solve the first equation for v_0 and substitute into the second equation. $v_0 = 315/(t\cos 20°)$, $315\tan 20° - 16t^2 = 34$. The positive solution is $t_1 = \sqrt{315\tan 20° - 34/4} \approx 2.245$ sec. It takes t_1 seconds to get to the wall. The initial speed is $v_0 = 315/(t_1\cos 20°) \approx 149.31$ ft/sec. But $\frac{dx}{dt} = v_0\cos 20°$, $\frac{dy}{dt} = v_0\sin 20° - 32t$. The speed at t_1 is $|v(t_1)| = \sqrt{(v_0\cos 20°)^2 + (v_0\sin 20° - 32t_1)^2} \approx 141.83$ ft/sec neglecting forces other than gravity.

16. $\frac{10\text{ mi}}{\text{hr}} = \frac{10\text{ mi}}{\text{hr}}\frac{1\text{ hr}}{3600\text{ sec}}\frac{5280\text{ ft}}{1\text{ mi}} = 14.66\ldots$ ft/sec. a) $\boldsymbol{r}(t) = (145t\cos 23° - 14.67t)\boldsymbol{i} + (2.5 + 145t\sin 23° - 16t^2)\boldsymbol{j}$. $x(t) = 145t\cos 23° - 14.67t$, $y(t) = 2.5 + 145t\sin 23° - 16t^2$ b) In parametric mode graph $x(t)$, $y(t)$, $0 \leq t \leq 3.6$ in $[0, 428]$ by $[0, 52.66]$. c) We use TRACE and tStep = 0.1 for the following estimates. i) $y_{max} = 52.64$ ft at $t = 1.8$ sec. ii) It travels up to about $x = 427$ ft in about $t = 3.6$ sec when it hits the ground. iii) The ball is 18.06 ft high at $t = 0.3$ sec when it is 35.64 ft from home plate. It is 19.96 ft high at $t = 3.2$ sec, 380.17 ft from home plate. iv) By drawing in the fence, LINE$(300, 0, 300, 15)$, we see it is easily a home run. In fact when $x = 300$ ft, $y > 40$ ft.

17. a) $x = x_0 + (v_0 \cos\alpha)t = 13 + (35\cos 27°)t$ (The net will stand at $x = 25$ ft.) $y = y_0 + (v_0 \sin\alpha)t - \frac{1}{2}gt^2 = 4 + (35 \sin 27°)t - 16t^2$. $\boldsymbol{r} = [13 + (35\cos 27°)t]\boldsymbol{i} + [4 + (35 \sin 27°)t - 16t^2]\boldsymbol{j}$. b) Graph x, y above, $0 \leqq t \leqq 1.2$, Tstep $= 0.1$ or 0.05 in $[0, 51]$ by $[-9.7, 19.7]$. Also draw in the net: Line$(25, 0, 25, 6)$. c) i) $y_{\max} \approx 7.9$ ft occurs at $t \approx 0.5$ sec. (We are using TRACE). ii) It travels $50.4 - 13 = 37.4$ ft and hits the ground at about 1.2 sec. iii) $y = 7$ ft at $t \approx 0.25$ sec and is $x - 13 = 20.8 - 13 = 7.8$ ft from the point of impact. Also at $t \approx 0.75$ sec and $x \approx 36.4 - 13 = 23.4$ ft from the point of impact. iv) The ball hits the net by i).

18. We enter x, y in parametric form as determined in Example 3. (Caution: $\cos 20(u)$ is not the same as $(\cos 20)(u)$). We use $0 \leqq t \leqq 3.5$, tStep $= 0.01$ in $[0, 500]$ by $[0, 100]$ and we simply use TRACE for the approximations. (c-1) $y_{\max} = 43.07$ ft at $t = 1.56$ sec (c-2) $y = 35$ ft at about $t = 0.86$ sec and $t = 2.28$ sec, 120.23 ft and 307.78 ft from home plate, respectively. c-3) The ball travels 426.02 ft hitting the ground at $t = 3.23$ sec. When $t = 2.91$ sec, $x = 386.82$ ft and $y = 14.64$ ft; so a 15 ft-fence placed 386.82 ft from home plate would prevent a home run. When $t = 2.98$ sec, $x = 395.45$ ft and $y = 11.65$ ft; so a 12 ft-fence placed 395.45 ft from home plate would prevent a home run.

19. We graph $x = \frac{152(\cos 20°)}{0.12}(1 - e^{-0.12t})$, $y = 3 + \frac{152(\sin 20°)}{0.12}(1 - e^{-0.12t}) + (\frac{32}{0.12^2})(1 - 0.12t - e^{-0.12t})$, $0 \leqq t \leqq 3.2$, tStep $= 0.01$ in $[0, 500]$ by $[0, 100]$ and we use TRACE for the approximations. a) $y_{\max} = 40.435$ ft at $t = 1.48$ sec b) It travels about 373 ft hitting the ground at $t = 3.13$ sec c) $y = 30$ ft at about $t = 0.69$ sec and $t = 2.3$ sec, 94.59 ft and 287.08 ft from home plate, respectively. d) When x is 340 ft, y is between 13.7 ft and 14.2 ft so the ball is a home run.

20. We first convert: $\frac{12\,\text{mi}}{\text{hr}} = \frac{12\,\text{mi}}{\text{hr}} \frac{1\,\text{hr}}{3600\,\text{sec}} \frac{5280\,\text{ft}}{1\,\text{mi}} = 17.6$ ft/sec. a) $x = \frac{152(\cos 20°)}{0.08}(1 - e^{-0.08t}) - 17.6t$, $y = 3 + \frac{152(\sin 20°)}{0.08}(1 - e^{-0.08t}) + \frac{32}{0.08^2}(1 - 0.08t - e^{-0.08t})$. Using these values of x and y, the vector form of the flight function is $\boldsymbol{r}(t) = x\boldsymbol{i} + y\boldsymbol{j}$. b) We use x, y above. In parametric mode graph $x_1 = x$, $y_1 = y$, tstep $= 0.1$, $0 \leqq t \leqq 3.2$ in $[0, 410]$ by $[0, 5]$ to simulate the motion with the wind and use $x_2 = x_1 + 17.6t$, $y_2 = y_1$ without wind. c) i) With and without the wind (using TRACE) we obtain $y_{\max} = 41.88$ at $t = 1.5$. ii) With the wind the ball travels in the air about 345 ft and without the wind about 401 ft. In each case it hits the ground at approximately $t = 3.2$ sec. iii) With the wind: $y = 35.489$ ft when $x = 108.191$ ft, $t = 0.9$ sec and $y = 34.782$ ft when $x = 249.414$ ft and $t = 2.2$ sec. Without the wind: $y = 35.489$ ft when $x = 124.031$ ft and $t = 0.9$ sec and $y = 34.782$ ft when $x = 288.134$ ft and $t = 2.2$ sec iv) With the wind by ii) the ball hits the ground before 380 ft so it certainly is not a home run. Without the wind the ball is only at $y \approx 8.5$ ft when $x = 380.985$ ft so it is still not a home run. By trial and error we find that we need a wind speed of about 12 ft/sec (8.182 mi/h) *in the direction of the flight of the ball* for the hit to be a home run.

21. See the solution to Exercise 19 for the set up with $k = 0.12$. Here we use the TRACE function with tStep = 0.1, $0 \leqq t \leqq 3.5$.

k	$y_{\max}$	Time for y_{max}	Flight Distance	Flight Time
0.01	44.77 ft	1.6 sec	463.66 ft	3.3 sec
0.02	44.34	1.6	456.13	3.3
0.05	43.05	1.6	422.38	3.2
0.1	41.15	1.5	391.15	3.2
0.15	39.39	1.5	354.10	3.1
0.20	37.85	1.4	322.22	3.0
0.25	36.41	1.4	294.62	2.9

c) Using l'Hospital's rule, we can show $\frac{1-e^{-kt}}{k} \to t$ and $\frac{1-kt-e^{-kt}}{k^2} \to -\frac{t^2}{2}$ as $k \to 0$. Thus the limit of Eq. (13) as $k \to 0$ is Eq. (6). This makes sense: as the air density diminishes to 0, the air resistance to the motion of the projectile diminishes to 0, as was assumed in Eq. (6).

22. a) We graph $x_1 = 95(\cos 42°)t$, $y_1 = 95(\sin 42°)t - 16t^2$, $x_2 = 100(\cos 25°)t$, $y_2 = 100(\sin 25°)t$, $0 \leqq t \leqq 2$, tStep $= 0.05$ in $[0, 200]$ by $[0, 100]$ and use the TRACE function. $t_f = 1.9\,\text{sec}$. b) $r_f = \frac{y}{\sin 25°} = \frac{63.01807}{\sin 25°} = 149.11$ ft c) If $t = t_f$, the time of impact, then $\frac{y}{x} = \tan\beta$ or $\frac{(v_0 \sin\alpha)t - \frac{1}{2}gt^2}{(v_0\cos\alpha)t} = \frac{\sin\beta}{\cos\beta}$. After algebraic manipulation, this reduces to $t[gt - 2v_0(\sin\alpha\cos\beta - \cos\alpha\sin\beta)/\cos\beta] = 0$. This yields $t = 0$ or $t = \frac{2v_0\sin(\alpha-\beta)}{g\cos\beta}$. Using the specified values of v_0, α, β and $g = 32$, we get $t_f = 1.9154\,\text{sec}$ compared to the approximation $1.9\,\text{sec}$. d) $r_f = \frac{y}{\sin\beta} = [(v_0\sin\alpha)t_f - \frac{1}{2}gt_f^2]/\sin\beta$. Substituting and reducing, we finally obtain $r_f = \frac{2v_0^2(\sin(\alpha-\beta))\cos\alpha}{g\cos^2\beta}$. With the specified values $r_f = 149.21$ ft compared with the approximation 149.11 ft.

23. Marble A will be R units downrange at time t_1 where $x = (v_0 \cos\alpha)t_1 = R$, $t_1 = R/(v_0\cos\alpha)$. At that time $y(= (v_0 \sin\alpha)t - \frac{1}{2}gt^2$ for marble A) will be $(v_0\sin\alpha)R/(v_0\cos\alpha) - \frac{1}{2}g(R/(v_0\cos\alpha))^2$. But the height of marble B is given by $R\tan\alpha - \frac{1}{2}gt^2$ so both marbles will be at the same height at $t = t_1$ and will collide. The result is independent of the initial velocity v_0.

24. For convenience we change notation:

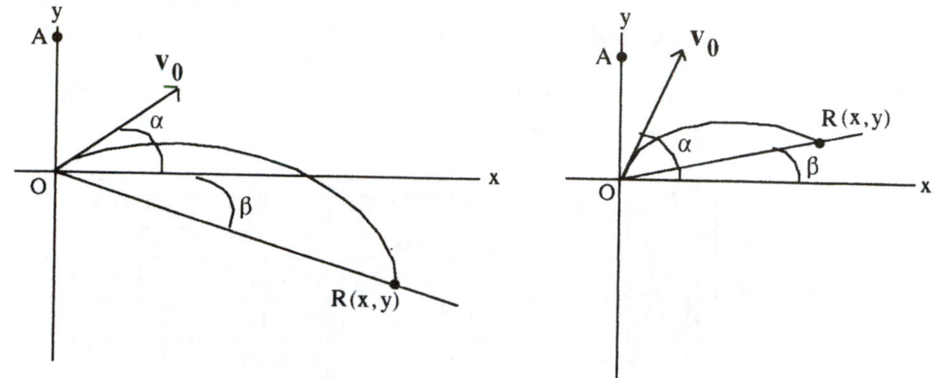

$\frac{y}{x} = -\tan\beta$, $y = -x\tan\beta$, $(v_0\sin\alpha)t - \frac{1}{2}gt^2 = -v_0\cos\alpha(\tan\beta)t$, $t[\frac{1}{2}gt - v_0(\cos\alpha\tan\beta + \sin\alpha)] = 0$ leads to $t = \frac{2v_0(\cos\alpha\tan\beta+\sin\alpha)}{g}$. OR is maximized when x is maximized. $x = (v_0\cos\alpha)t = (\frac{2v_0^2}{g})(\cos^2\alpha\tan\beta + \sin\alpha\cos\alpha)$. $\frac{dx}{d\alpha} = (\frac{2v_0^2}{g})(-2\sin\alpha\cos\alpha\tan\beta + \cos^2\alpha - \sin^2\alpha) = (\frac{2v_0^2}{g})(\cos 2\alpha - \sin 2\alpha\tan\beta)$. x is maximal when $\cos 2\alpha - \sin 2\alpha \tan\beta = 0$, $1 - \tan 2\alpha\tan\beta = 0$, $\tan 2\alpha = \cot\beta$, $\tan 2\alpha = \tan(\frac{\pi}{2} - \beta)$, $2\alpha = \frac{\pi}{2} - \beta$, $\alpha = \frac{\pi}{4} - \frac{\beta}{2}$. Therefore $\alpha + \beta = \frac{\pi}{4} - \frac{\beta}{2} + \beta = \frac{\pi}{4} + \frac{\beta}{2} = \frac{1}{2}(\frac{\pi}{2} + \beta) = \frac{1}{2}(\angle AOR)$ as required.

In the second diagram $\frac{y}{x} = \tan\beta$. Proceeding as in the first part, we get $t = (\frac{2v_0)}{g}(\sin\alpha - \cos\alpha\tan\beta)$. $\frac{dx}{d\alpha} = 0$ leads to $\alpha = \frac{\pi}{4} + \frac{\beta}{2}$. Hence $\alpha - \beta = \frac{\pi}{4} - \frac{\beta}{2} = \frac{1}{2}(\frac{\pi}{2} - \beta) = \frac{1}{2}(\angle AOR)$, i.e., $\boldsymbol{v}_0$ bisects $\angle AOR$.

25. $\frac{d^2\boldsymbol{r}}{dt^2} = -g\boldsymbol{j}$ yields $\frac{d\boldsymbol{r}}{dt} = -gt\boldsymbol{j} + \boldsymbol{C}_1$. When $t = 0$, $\frac{d\boldsymbol{r}}{dt} = \boldsymbol{C}_1 = (v_0\cos\alpha)\boldsymbol{i} + (v_0\sin\alpha)\boldsymbol{j}$. So for arbitrary t, $\frac{d\boldsymbol{r}}{dt} = (v_0\cos\alpha)\boldsymbol{i} + (v_0\sin\alpha - gt)\boldsymbol{j}$. In turn this yields $\boldsymbol{r} = (v_0\cos\alpha)t\boldsymbol{i} + ((v_0\sin\alpha)t - \frac{1}{2}gt^2)\boldsymbol{j} + \boldsymbol{C}_2$. When $t = 0$, $\boldsymbol{r} = \boldsymbol{C}_2 = x_0\boldsymbol{i} + y_0\boldsymbol{j}$. So for arbitrary t, $\boldsymbol{r} = (x_0 + (v_0\cos\alpha)t)\boldsymbol{i} + (y_0 + (v_0\sin\alpha)t - \frac{1}{2}gt^2)\boldsymbol{j}$. This is equivalent to the desired parametric equations.

26. Integrating both sides of the equation, we obtain $-gt - ky = (-g - kv_0 \sin\alpha)e^{-kt}/(-k) + C$. At $t = 0$, this becomes $0 = (-g - kv_0 \sin\alpha)/(-k) + C$. We solve the last equation for C, substitute it into the first equation, then solve for y. After manipulating terms, we obtain the desired expression for y.

12.3 DIRECTED DISTANCE AND THE UNIT TANGENT VECTOR T

1. $\boldsymbol{v} = \frac{d\boldsymbol{r}}{dt} = (-2\sin t)\boldsymbol{i} + (2\cos t)\boldsymbol{j} + \sqrt{5}\boldsymbol{k}$. $|\boldsymbol{v}| = \sqrt{4\sin^2 t + 4\cos^2 t + 5} = \sqrt{4+5} = 3$. $\boldsymbol{T} = \frac{\boldsymbol{v}}{|\boldsymbol{v}|} = -\frac{2}{3}\sin t\boldsymbol{i} + \frac{2}{3}\cos t\boldsymbol{j} + \frac{\sqrt{5}}{3}\boldsymbol{k}$. $L = \int_0^\pi |\boldsymbol{v}|dt = 3\pi$.

2. $\boldsymbol{v} = \frac{d\boldsymbol{r}}{dt} = 12\cos 2t\boldsymbol{i} - 12\sin 2t\boldsymbol{j} + 5\boldsymbol{k}$. $|\boldsymbol{v}| = \sqrt{144(\cos^2 2t + \sin^2 2t) + 25} = 13$. $\boldsymbol{T} = (12/13)\cos 2t\boldsymbol{i} - (12/13)\sin 2t\boldsymbol{j} + (5/13)\boldsymbol{k}$. $L = \int_0^\pi 13dt = 13\pi$.

3. $\boldsymbol{v} = \frac{d\boldsymbol{r}}{dt} = \boldsymbol{i} + t^{1/2}\boldsymbol{k}$. $|\boldsymbol{v}| = \sqrt{1+t}$. $\boldsymbol{T} = \frac{1}{\sqrt{1+t}}\boldsymbol{i} + \sqrt{\frac{t}{1+t}}\boldsymbol{k}$. $L = \int_0^8 \sqrt{1+t}dt = \frac{2}{3}(1+t)^{3/2}]_0^8 = \frac{2}{3}[9^{3/2} - 1] = \frac{52}{3}$.

4. $\boldsymbol{v} = \frac{d\boldsymbol{r}}{dt} = -3\cos^2 t\sin t\boldsymbol{j} + 3\sin^2 t\cos t\boldsymbol{k}$. $|\boldsymbol{v}|^2 = 9(\cos^4 t\sin^2 t + \sin^4 t\cos^2 t) = 9\sin^2 t\cos^2 t(\cos^2 t + \sin^2 t) = 9\sin^2 t\cos^2 t$. $|\boldsymbol{v}| = 3|\sin t||\cos t|$. $\boldsymbol{T} = \frac{\boldsymbol{v}}{|\boldsymbol{v}|} = -|\cos t|\frac{\sin t}{|\sin t|}\boldsymbol{j} + |\sin t|\frac{\cos t}{|\cos t|}\boldsymbol{k}$. For $0 \leqq t \leq \frac{\pi}{2}$, $\boldsymbol{T} = -\cos t\boldsymbol{j} + \sin t\boldsymbol{k}$. $L = \int_0^{\pi/2} 3\sin t\cos t\, dt = \frac{3}{2}\sin^2 t]_0^{\pi/2} = \frac{3}{2}$.

5. $\boldsymbol{v} = \boldsymbol{i} - \boldsymbol{j} + \boldsymbol{k}$, $|\boldsymbol{v}| = \sqrt{3}$. $\boldsymbol{T} = \frac{1}{\sqrt{3}}(\boldsymbol{i} - \boldsymbol{j} + \boldsymbol{k})$. $L = \int_0^3 \sqrt{3}dt = 3\sqrt{3}$.

6. $\boldsymbol{v} = \frac{d\boldsymbol{r}}{dt} = 18t^2\boldsymbol{i} - 6t^2\boldsymbol{j} - 9t^2\boldsymbol{k} = 3t^2(6\boldsymbol{i} - 2\boldsymbol{j} - 3\boldsymbol{k})$. $|\boldsymbol{v}| = 3t^2\sqrt{6^2+4+9} = 21t^2$. $\boldsymbol{T} = \frac{1}{7}(6\boldsymbol{i} - 2\boldsymbol{j} - 3\boldsymbol{k})$. $L = \int_{-1}^1 21t^2dt = 42\int_0^1 t^2dt = \frac{42}{3} = 14$.

7. $\boldsymbol{v} = (\cos t - t\sin t)\boldsymbol{i} + (\sin t + t\cos t)\boldsymbol{j} + \sqrt{2}t^{1/2}\boldsymbol{k}$. $|\boldsymbol{v}| = \sqrt{\cos^2 t - 2t\cos t\sin t + t^2\sin^2 t + \sin^2 t + 2t\sin t\cos t + t^2\cos^2 t + 2t} = \sqrt{t^2 + 2t + 1} = t + 1$. $\boldsymbol{T} = (\frac{\cos t - t\sin t}{t+1})\boldsymbol{i} + (\frac{\sin t + t\cos t}{t+1})\boldsymbol{j} + (\frac{\sqrt{2t}}{t+1})\boldsymbol{k}$. $L = \int_0^\pi (t+1)dt = \frac{t^2}{2} + t]_0^\pi = \frac{\pi^2}{2} + \pi$.

8. $\boldsymbol{v} = (t\cot t + \sin t - \sin t)\boldsymbol{i} + (-t\sin t + \cos t - \cos t)\boldsymbol{j} = t\cos t\boldsymbol{i} - t\sin t\boldsymbol{j}$. $|\boldsymbol{v}| = \sqrt{t^2(\cos^2 t + \sin^2 t)} = |t|$. $\boldsymbol{T} = \frac{\boldsymbol{v}}{|\boldsymbol{v}|} = \frac{t}{|t|}(\cos t\boldsymbol{i} - \sin t\boldsymbol{j})$. For $0 \leqq t$, $\boldsymbol{T} = \cos t\boldsymbol{i} - \sin t\boldsymbol{j}$. $L = \int_0^{\sqrt{2}} t\, dt = \frac{t^2}{2}]_0^{\sqrt{2}} = 1$.

9. $\boldsymbol{v} = -4\sin t\boldsymbol{i} + 4\cos t\boldsymbol{j} + 3\boldsymbol{k}$, $|\boldsymbol{v}| = \sqrt{16\sin^2 t + 16\cos^2 t + 9} = 5$. $s(t) = \int_{t_0}^t 5d\tau = 5(t - t_0)$. $L = \frac{5\pi}{2}$.

10. $\boldsymbol{v} = \frac{d\boldsymbol{r}}{dt} = (-\sin t + t\cos t + \sin t)\boldsymbol{i} + (\cos t + t\sin t - \cos t)\boldsymbol{j} = t\cos t\boldsymbol{i} + t\sin t\boldsymbol{j}$. $|\boldsymbol{v}|^2 = t^2(\cos^2 t + \sin^2 t) = t^2$. $|\boldsymbol{v}| = t$ for $t \geqq 0$. $s(t) = \int_{t_0}^t \tau\, d\tau = \frac{1}{2}(t^2 - t_0^2)$. Length of section $= \frac{1}{2}(\pi^2 - \frac{\pi^2}{4}) = \frac{3\pi^2}{8}$.

11. $\boldsymbol{v} = (e^t\cos t - e^t\sin t)\boldsymbol{i} + (e^t\sin t + e^t\cos t)\boldsymbol{j} + e^t\boldsymbol{k}$. $|\boldsymbol{v}| = \sqrt{e^{2t}(\cos^2 t + \sin^2 t) + e^{2t}(\sin^2 t + \cos^2 t) + e^{2t}} = \sqrt{3}e^t$. $s(t) = \int_{t_0}^t \sqrt{3}e^\tau d\tau = \sqrt{3}(e^t - e^{t_0})$. $L = \sqrt{3}(e^{\ln 4} - e^0) = \sqrt{3}(4-1) = 3\sqrt{3}$.

12. $\boldsymbol{v} = 2\boldsymbol{i} + 3\boldsymbol{j} - 6\boldsymbol{k}$. $|\boldsymbol{v}| = \sqrt{4+9+36} = 7$. $s(t) = \int_{t_0}^t 7\, d\tau = 7(t - t_0)$. $L = 7$.

13. $\boldsymbol{v} = \sqrt{2}\boldsymbol{i} + \sqrt{2}\boldsymbol{j} = 2t\boldsymbol{k}$. $|\boldsymbol{v}| = \sqrt{2+2+4t^2} = 2\sqrt{1+t^2}$. The points correspond to $t = 0$ to $t = 1$. From a table of integrals we use $\int \sqrt{x^2+a^2}dx = \frac{x}{2}\sqrt{x^2+a^2} + \frac{a^2}{2}\ln|x + \sqrt{x^2+a^2}| + C$. $L = 2\int_0^1 \sqrt{t^2+1}dt = 2[\frac{t}{2}\sqrt{t^2+1} + \frac{1}{2}\ln|t + \sqrt{t^2+1}|]_0^1 = \sqrt{2} + \ln(1+\sqrt{2})$.

14. Cut the cylinder by the planes $z = 0$ and $z = 2\pi$ and work with the portion between these two planes. Cut this portion along a line parallel to the z-axis starting at $(1,0,0)$. Since the circumference is 2π, if we now open and flatten, we obtain a 2π by 2π square. The portion of the helix, $0 \leqq t \leqq 2\pi$, has now become the diagonal of the square.

15. a) $\boldsymbol{v} = -4\sin 4t\boldsymbol{i} + 4\cos 4t\boldsymbol{j} + 4\boldsymbol{k}$. $|\boldsymbol{v}| = \sqrt{16+16} = 4\sqrt{2}$. $L = \int_0^{\pi/2} 4\sqrt{2}dt = 2\sqrt{2}\pi$. b) $\boldsymbol{v} = -\frac{1}{2}\sin(\frac{t}{2})\boldsymbol{i} + \frac{1}{2}\cos(\frac{t}{2})\boldsymbol{j} + \frac{1}{2}\boldsymbol{k}$. $|\boldsymbol{v}| = \sqrt{\frac{1}{4}+\frac{1}{4}} = \frac{\sqrt{2}}{2}$. $L = \int_0^{4\pi}\frac{\sqrt{2}}{2}dt = 2\sqrt{2}\pi$. c) $\boldsymbol{v} = -\sin t\boldsymbol{i} - \cos t\boldsymbol{j} - \boldsymbol{k}$. $|\boldsymbol{v}| = \sqrt{2}$. $L = \int_{-2\pi}^0 \sqrt{2}dt = 2\sqrt{2}\pi$.

12.4 CURVATURE, TORSION, AND THE TNB FRAME

1. $\boldsymbol{v} = \boldsymbol{i} - \tan t\boldsymbol{j}$. $|\boldsymbol{v}| = \sqrt{1+\tan^2 t} = \sec t$ (because $\sec t > 0$ for $-\pi/2 < t < \pi/2$). $\boldsymbol{T} = \frac{\boldsymbol{v}}{|\boldsymbol{v}|} = \frac{1}{\sec t}\boldsymbol{i} - \frac{\tan t}{\sec t}\boldsymbol{j} = \cos t\boldsymbol{i} - \sin t\boldsymbol{j}$. $\dfrac{d\boldsymbol{T}}{dt} = -\sin t\boldsymbol{i} - \cos t\boldsymbol{j}$. $\boldsymbol{N} = \dfrac{(d\boldsymbol{T}/dt)}{|d\boldsymbol{T}/dt|} = -\sin t\boldsymbol{i} - \cos t\boldsymbol{j}$. $\kappa = \left|\dfrac{d\boldsymbol{T}}{ds}\right| = \left|\dfrac{d\boldsymbol{T}}{dt}\dfrac{dt}{ds}\right| = \left|\dfrac{d\boldsymbol{T}}{dt}\right|\dfrac{1}{|\boldsymbol{v}|} = \dfrac{1}{\sec t} = \cos t$.

2. $\boldsymbol{v} = \frac{\sec t\tan t}{\sec t}\boldsymbol{i} + \boldsymbol{j} = \tan t\boldsymbol{i} + \boldsymbol{j}$. $|\boldsymbol{v}| = \sqrt{\tan^2 t + 1} = \sec t$ (on $-\pi/2 < t < \pi/2$). $\boldsymbol{T} = \dfrac{\boldsymbol{v}}{|\boldsymbol{v}|} = \sin t\boldsymbol{i} + \cos t\boldsymbol{j}$. $\dfrac{d\boldsymbol{T}}{dt} = \cos t\boldsymbol{i} - \sin t\boldsymbol{j} = \boldsymbol{N}$ (it's already a unit vector). $\kappa = \left|\dfrac{d\boldsymbol{T}}{ds}\right| = \left|\dfrac{d\boldsymbol{T}}{dt}\dfrac{dt}{ds}\right| = \left|\dfrac{d\boldsymbol{T}}{dt}\right|\dfrac{1}{|\boldsymbol{v}|} = 1\cdot\dfrac{1}{\sec t} = \cos t$.

3. $\boldsymbol{v} = 2\boldsymbol{i} - 2t\boldsymbol{j}$, $|\boldsymbol{v}| = \sqrt{4+4t^2} = 2\sqrt{1+t^2}$. $\boldsymbol{T} = \dfrac{2\boldsymbol{i} - 2t\boldsymbol{j}}{2\sqrt{1+t^2}} = \dfrac{\boldsymbol{i} - t\boldsymbol{j}}{\sqrt{1+t^2}}$. $\dfrac{d\boldsymbol{T}}{dt} = -(1+t^2)^{-3/2}t\boldsymbol{i} - \frac{\sqrt{1+t^2} - t^2/\sqrt{1+t^2}}{1+t^2}\boldsymbol{j} = \frac{-t}{(1+t^2)^{3/2}}\boldsymbol{i} - \frac{1}{(1+t^2)^{3/2}}\boldsymbol{j}$. $\left|\dfrac{d\boldsymbol{T}}{dt}\right| = \sqrt{\frac{t^2}{(1+t^2)^3} + \frac{1}{(1+t^2)^3}} = \frac{1}{1+t^2}$. $\boldsymbol{N} = \dfrac{d\boldsymbol{T}/dt}{|d\boldsymbol{T}/dt|} = -\dfrac{t}{\sqrt{1+t^2}}\boldsymbol{i} - \dfrac{1}{\sqrt{1+t^2}}\boldsymbol{j}$. $\kappa = \left|\dfrac{d\boldsymbol{T}}{ds}\right| = \left|\dfrac{d\boldsymbol{T}}{dt}\dfrac{dt}{ds}\right| = \dfrac{1}{1+t^2}\dfrac{1}{|\boldsymbol{v}|} = \dfrac{1}{2(1+t^2)^{3/2}}$.

4. $\boldsymbol{v} = (-\sin t + t\cos t + \sin t)\boldsymbol{i} + (\cos t + t\sin t - \cos t)\boldsymbol{j} = t\cos t\boldsymbol{i} + t\sin t\boldsymbol{j}$, $t > 0$. $|\boldsymbol{v}|^2 = t^2(\cos^2 t + \sin^2 t) = t^2$, $|\boldsymbol{v}| = t$, $t > 0$. $\boldsymbol{T} = \dfrac{\boldsymbol{v}}{|\boldsymbol{v}|} = \cos t\boldsymbol{i} + \sin t\boldsymbol{j}$. $\dfrac{d\boldsymbol{T}}{dt} = -\sin t\boldsymbol{i} + \cos t\boldsymbol{j} = \boldsymbol{N}$ (already a unit vector). $\kappa = \left|\dfrac{d\boldsymbol{T}}{ds}\right| = \left|\dfrac{d\boldsymbol{T}}{dt}\dfrac{dt}{ds}\right| = \left|\dfrac{d\boldsymbol{T}}{dt}\right|\dfrac{1}{|\boldsymbol{v}|} = 1(\dfrac{1}{t}) = \dfrac{1}{t}$.

5. $\boldsymbol{v} = (3\cos t)\boldsymbol{i} - 3\sin t\boldsymbol{j} + 4\boldsymbol{k}$, $\boldsymbol{a} = -3\sin t\boldsymbol{i} - 3\cos t\boldsymbol{j}$, $|\boldsymbol{v}| = \sqrt{9\cos^2 t + 9\sin^2 t + 16} = 5$. $\boldsymbol{T} = \dfrac{\boldsymbol{v}}{|\boldsymbol{v}|} = \dfrac{3}{5}\cos t\boldsymbol{i} - \dfrac{3}{5}\sin t\boldsymbol{j} + \dfrac{4}{5}\boldsymbol{k}$. $\dfrac{d\boldsymbol{T}}{dt} = -\dfrac{3}{5}\sin t\boldsymbol{i} - \dfrac{3}{5}\cos t\boldsymbol{j}$, $\left|\dfrac{d\boldsymbol{T}}{dt}\right| = \dfrac{3}{5}$, $\boldsymbol{N} = -\sin t\boldsymbol{i} - \cos t\boldsymbol{j}$. $\boldsymbol{B} = \boldsymbol{T}\times\boldsymbol{N} = \begin{vmatrix} \boldsymbol{i} & \boldsymbol{j} & \boldsymbol{k} \\ \frac{3}{5}\cos t & -\frac{3}{5}\sin t & \frac{4}{5} \\ -\sin t & -\cos t & 0 \end{vmatrix} = \dfrac{4}{5}\cos t\boldsymbol{i} - \dfrac{4}{5}\sin t\boldsymbol{j} - \dfrac{3}{5}\boldsymbol{k}$. $\kappa = \left|\dfrac{d\boldsymbol{T}}{ds}\right| = \left|\dfrac{d\boldsymbol{T}}{dt}\dfrac{dt}{ds}\right| = \dfrac{3}{5}\dfrac{1}{|\boldsymbol{v}|} = \dfrac{3}{25}$. $\tau = \left\|\begin{matrix} 3\cos t & -3\sin t & 4 \\ -3\sin t & -3\cos t & 0 \\ -3\cos t & 3\sin t & 0 \end{matrix}\right\| \div |\boldsymbol{v}\times\boldsymbol{a}|^2 = 36/|\boldsymbol{v}\times\boldsymbol{a}|^2$. $\boldsymbol{v}\times\boldsymbol{a} = \begin{vmatrix} \boldsymbol{i} & \boldsymbol{j} & \boldsymbol{k} \\ 3\cos t & -3\sin t & 4 \\ -3\sin t & -3\cos t & 0 \end{vmatrix} = 12\cos t\boldsymbol{i} - 12\sin\boldsymbol{j} - 9\boldsymbol{k}$. $|\boldsymbol{v}\times\boldsymbol{a}|^2 = 12^2 + 9^2 = 225$. $\tau = 36/225 = 4/25$.

6. We use the work of Exercise 4. For $t > 0$, $\boldsymbol{v} = t\cos t\boldsymbol{i} + t\sin t\boldsymbol{j}$, $\boldsymbol{T} = \cos t\boldsymbol{i} + \sin t\boldsymbol{j}$, $\boldsymbol{N} = -\sin t\boldsymbol{i} + \cos t\boldsymbol{j}$, $\kappa = \frac{1}{t}$. $\boldsymbol{B} = \boldsymbol{T}\times\boldsymbol{N} = \begin{vmatrix} \boldsymbol{i} & \boldsymbol{j} & \boldsymbol{k} \\ \cos t & \sin t & 0 \\ -\sin t & \cos t & 0 \end{vmatrix} = \boldsymbol{k}$. $\tau = \left|\dfrac{d\boldsymbol{B}}{ds}\right| = 0$, since $\boldsymbol{B}$ is a constant vector.

7. $\boldsymbol{v} = (e^t \cos t - e^t \sin t)\boldsymbol{i} + (e^t \cos t + e^t \sin t)\boldsymbol{j}$, $\boldsymbol{a} = (-e^t \sin t + e^t \cos t - e^t \cos t - e^t \sin t)\boldsymbol{i} + (e^t \cos t - e^t \sin t + e^t \sin t + e^t \cos t)\boldsymbol{j} = -2e^t \sin t\boldsymbol{i} + 2e^t \cos t\boldsymbol{j}$. $|\boldsymbol{v}| = \sqrt{e^{2t}(\cos t - \sin t)^2 + e^{2t}(\cos t + \sin t)^2} = \sqrt{2e^{2t}} = \sqrt{2}e^t$. $\boldsymbol{T} = \frac{\boldsymbol{v}}{|\boldsymbol{v}|} = \left(\frac{\cos t - \sin t}{\sqrt{2}}\right)\boldsymbol{i} + \left(\frac{\cos t + \sin t}{\sqrt{2}}\right)\boldsymbol{j}$. $\dfrac{d\boldsymbol{T}}{dt} = \dfrac{-\sin t - \cos t}{\sqrt{2}}\boldsymbol{i} + \dfrac{-\sin t + \cos t}{\sqrt{2}}\boldsymbol{j}$. $\left|\dfrac{d\boldsymbol{T}}{dt}\right| = \frac{1}{\sqrt{2}}\sqrt{(\sin t + \cos t)^2 + (\cos t - \sin t)^2} = 1$. $\boldsymbol{N} = -\frac{(\sin t + \cos t)}{\sqrt{2}}\boldsymbol{i} + \frac{(\cos t - \sin t)}{\sqrt{2}}\boldsymbol{j}$. $\boldsymbol{B} = \boldsymbol{T} \times \boldsymbol{N} = \begin{vmatrix} \boldsymbol{i} & \boldsymbol{j} & \boldsymbol{k} \\ \frac{\cos t - \sin t}{\sqrt{2}} & \frac{\cos t + \sin t}{\sqrt{2}} & 0 \\ -\frac{(\cos t + \sin t)}{\sqrt{2}} & \frac{\cos t - \sin t}{\sqrt{2}} & 0 \end{vmatrix} = \frac{1}{2}\left[(\cos t - \sin t)^2 + (\cos t + \sin t)^2\right]\boldsymbol{k} = \boldsymbol{k}$. $\kappa = \left|\dfrac{d\boldsymbol{T}}{ds}\right| = \left|\dfrac{d\boldsymbol{T}}{dt}\dfrac{dt}{ds}\right| = 1(\dfrac{1}{|\boldsymbol{v}|}) = \dfrac{1}{\sqrt{2}e^t}$. $\tau = 0$ because the last column of the determinant is all 0's.

8. $\boldsymbol{v} = 12\cos 2t\boldsymbol{i} - 12\sin 2t\boldsymbol{j} + 5\boldsymbol{k}$, $|\boldsymbol{v}|^2 = 144(\cos^2 2t + \sin^2 2t) + 25 = 169$, $|\boldsymbol{v}| = 13$. $\boldsymbol{T} = \dfrac{\boldsymbol{v}}{|\boldsymbol{v}|} = \dfrac{12}{13}\cos 2t\boldsymbol{i} - \dfrac{12}{13}\sin 2t\boldsymbol{j} + \dfrac{5}{13}\boldsymbol{k}$. $\dfrac{d\boldsymbol{T}}{dt} = -\dfrac{24}{13}\sin 2t\boldsymbol{i} - \dfrac{24}{13}\cos 2t\boldsymbol{j}$, $\left|\dfrac{d\boldsymbol{T}}{dt}\right| = \dfrac{24}{13}$. $\boldsymbol{N} = \dfrac{d\boldsymbol{T}/dt}{|d\boldsymbol{T}/dt|} = -\sin 2t\boldsymbol{i} - \cos 2t\boldsymbol{j}$. $\boldsymbol{B} = \boldsymbol{T} \times \boldsymbol{N} = \begin{vmatrix} \boldsymbol{i} & \boldsymbol{j} & \boldsymbol{k} \\ \frac{12}{13}\cos 2t & -\frac{12}{13}\sin 2t & \frac{5}{13} \\ -\sin 2t & -\cos 2t & 0 \end{vmatrix} = \frac{5}{13}\cos 2t\boldsymbol{i} - \frac{5}{13}\sin 2t\boldsymbol{j} - \frac{12}{13}\boldsymbol{k}$. $\kappa = \left|\dfrac{d\boldsymbol{T}}{ds}\right| = \left|\dfrac{d\boldsymbol{T}}{dt}\dfrac{dt}{ds}\right| = \left|\dfrac{d\boldsymbol{T}}{dt}\right|\frac{1}{|\boldsymbol{v}|} = \frac{24}{13}\frac{1}{13} = \frac{24}{169}$. $\tau = \left|\dfrac{d\boldsymbol{B}}{ds}\right| = \left|\dfrac{d\boldsymbol{B}}{dt}\dfrac{dt}{ds}\right| = \left|\dfrac{d\boldsymbol{B}}{dt}\right|\frac{1}{|\boldsymbol{v}|} = \frac{1}{13}\left|-\frac{10}{13}\sin 2t\boldsymbol{i} - \frac{10}{13}\cos 2t\boldsymbol{j}\right| = \frac{1}{13}\frac{10}{13} = \frac{10}{169}$.

9. $\boldsymbol{v} = 2\boldsymbol{i} + 2t\boldsymbol{j}$, $|\boldsymbol{v}| = \sqrt{4 + 4t^2} = 2\sqrt{1 + t^2}$. $a_T = \frac{d}{dt}|\boldsymbol{v}| = \frac{2t}{\sqrt{1+t^2}}$. $\boldsymbol{a} = 2\boldsymbol{j}$, $|\boldsymbol{a}| = 2$. $a_N = \sqrt{|\boldsymbol{a}|^2 - a_T^2} = \sqrt{4 - \frac{4t^2}{1+t^2}} = \frac{2}{\sqrt{1+t^2}}$. $\boldsymbol{a} = a_T\boldsymbol{T} + a_N\boldsymbol{N} = \frac{2t}{\sqrt{1+t^2}}\boldsymbol{T} + \frac{2}{\sqrt{1+t^2}}\boldsymbol{N}$.

10. $\boldsymbol{v} = \frac{2t}{t^2+1}\boldsymbol{i} + (1 - \frac{2}{t^2+1})\boldsymbol{j} = \frac{1}{t^2+1}(2t\boldsymbol{i} + (t^2 - 1)\boldsymbol{j})$. $|\boldsymbol{v}| = \frac{1}{t^2+1}\sqrt{4t^2 + t^4 - 2t^2 + 1} = \frac{t^2+1}{t^2+1} = 1$. $a_T = \dfrac{d|\boldsymbol{v}|}{dt} = 0$. $\boldsymbol{a} = \dfrac{(t^2+1)2 - 2t(2t)}{(t^2+1)^2}\boldsymbol{i} + \dfrac{(t^2+1)2t - (t^2-1)2t}{(t^2+1)^2}\boldsymbol{j} = \dfrac{2 - 2t^2}{(t^2+1)^2}\boldsymbol{i} + \dfrac{4t}{(t^2+1)^2}\boldsymbol{j}$. Since $a_T = 0$, $a_N = \sqrt{|\boldsymbol{a}|^2 - a_T^2} = |\boldsymbol{a}| = \frac{1}{(t^2+1)^2}\sqrt{(2 - 2t^2)^2 + 16t^2} = \frac{1}{(t^2+1)^2}\sqrt{4(1 - 2t^2 + t^4 + 4t^2)} = \frac{2(t^2+1)}{(t^2+1)^2} = \frac{2}{t^2+1}$. $\boldsymbol{a} = a_T\boldsymbol{T} + a_N\boldsymbol{N} = \frac{2}{t^2+1}\boldsymbol{N}$.

11. $\boldsymbol{v} = -a \sin t\boldsymbol{i} + a \cos t\boldsymbol{j} + b\boldsymbol{k}$, $|\boldsymbol{v}| = \sqrt{a^2+b^2}$. $a_T = \frac{d}{dt}|\boldsymbol{v}| = 0$. $\boldsymbol{a} = -a \cos t\boldsymbol{i} - a \sin t\boldsymbol{j}$, $|\boldsymbol{a}| = a$. $a_N = \sqrt{|\boldsymbol{a}|^2 - a_T^2} = \sqrt{a^2} = a$, assuming $a > 0$. $\boldsymbol{a} = a_T\boldsymbol{T} + a_N\boldsymbol{N} = a\boldsymbol{N}$.

12. $\boldsymbol{v} = 3\boldsymbol{i} + \boldsymbol{j} - 3\boldsymbol{k}$, $|\boldsymbol{v}| = \sqrt{19}$. $a_T = \dfrac{d|\boldsymbol{v}|}{dt} = 0$. $\boldsymbol{a} = \boldsymbol{0}$. $a_N = \sqrt{|\boldsymbol{a}|^2 - a_T^2} = 0$. $\boldsymbol{a} = 0\boldsymbol{T} + 0\boldsymbol{N}$.

13. $\boldsymbol{v} = \boldsymbol{i} + 2\boldsymbol{j} + 2t\boldsymbol{k}$, $\boldsymbol{a} = 2\boldsymbol{k}$, $|\boldsymbol{v}| = \sqrt{1+4+4t^2} = \sqrt{5+4t^2}$, $\frac{d}{dt}|\boldsymbol{v}| = \frac{4t}{\sqrt{5+4t^2}}$. When $t = 1$, $a_T = \frac{4}{\sqrt{9}} = \frac{4}{3}$. $a_N = \sqrt{|\boldsymbol{a}|^2 - a_T^2} = \sqrt{4 - \frac{16}{9}} = \frac{\sqrt{20}}{3} = \frac{2\sqrt{5}}{3}$. $\boldsymbol{a} = \frac{4}{3}\boldsymbol{T} + \frac{2\sqrt{5}}{3}\boldsymbol{N}$.

14. $\boldsymbol{v} = (-t \sin t + \cos t)\boldsymbol{i} + (t \cos t + \sin t)\boldsymbol{j} + 2t\boldsymbol{k}$, $\boldsymbol{a} = (-t \cos t - \sin t - \sin t)\boldsymbol{i} + (-t \sin t + \cos t + \cos t)\boldsymbol{j} + 2\boldsymbol{k} = -(t \cos t + 2 \sin t)\boldsymbol{i} + (2 \cos t - t \sin t)\boldsymbol{j} + 2\boldsymbol{k}$. $|\boldsymbol{v}|^2 = t^2 \sin^2 t - 2t \sin t \cos t + \cos^2 t + t^2 \cos^2 t + 2t \sin t \cos t + \sin^2 t + 4t^2 = 5t^2 + 1$, $|\boldsymbol{v}| = \sqrt{5t^2+1}$. $\dfrac{d|\boldsymbol{v}|}{dt} = \dfrac{10t}{2\sqrt{5t^2+1}} = \dfrac{5t}{\sqrt{5t^2+1}}$. Thus $a_T = 0$ when $t = 0$. $\boldsymbol{a}(0) = 2\boldsymbol{j} + 2\boldsymbol{k}$. $a_N = \sqrt{|\boldsymbol{a}(0)|^2 + a_T^2} = \sqrt{8} = 2\sqrt{2}$. $\boldsymbol{a}(0) = 2\sqrt{2}\boldsymbol{N}$.

15. $\boldsymbol{v} = 2t\boldsymbol{i} + (1+t^2)\boldsymbol{j} + (1-t^2)\boldsymbol{k}$, $\boldsymbol{a} = 2\boldsymbol{i} + 2t\boldsymbol{j} - 2t\boldsymbol{k}$. $|\boldsymbol{v}| = \sqrt{4t^2+1+2t^2+t^4+1-2t^2+t^4} = \sqrt{2t^4+4t^2+2} = \sqrt{2}(t^2+1)$. $\dfrac{d|\boldsymbol{v}|}{dt} = 2\sqrt{2}t$. At $t = 0$, $a_T = 0$, $\boldsymbol{a} = 2\boldsymbol{i}$, $a_N = \sqrt{|a|^2 - a_T^2} = \sqrt{4-0} = 2$, $\boldsymbol{a} = 0\boldsymbol{T} + 2\boldsymbol{N} = 2\boldsymbol{N}$.

16. $\boldsymbol{v} = (-e^t \sin t + e^t \cos t)\boldsymbol{i} + (e^t \cos t + e^t \sin t)\boldsymbol{j} + \sqrt{2}e^t\boldsymbol{k}$. $\boldsymbol{a} = (-e^t \cos t - e^t \sin t - e^t \sin t + e^t \cos t)\boldsymbol{i} + (-e^t \sin t + e^t \cos t + e^t \cos t + e^t \sin t)\boldsymbol{j} + \sqrt{2}e^t\boldsymbol{k} = (-2e^t \sin t)\boldsymbol{i} + 2e^t \cos t\boldsymbol{j} + \sqrt{2}e^t\boldsymbol{k}$. $|\boldsymbol{v}|^2 = e^{2t}[(\cos t - \sin t)^2 + (\cos t + \sin t)^2 + 2] = 4e^{2t}$, $|\boldsymbol{v}| = 2e^t$. $\dfrac{d|\boldsymbol{v}|}{dt} = 2e^t = a_T$. At $t = 0$, $a_T = 2$. $\boldsymbol{a}(0) = 2\boldsymbol{j} + \sqrt{2}\boldsymbol{k}$, $|\boldsymbol{a}(0)|^2 = 4 + 2 = 6$. At $t = 0$, $a_N = \sqrt{6-4} = \sqrt{2}$. $\boldsymbol{a}(0) = 2\boldsymbol{T}(0) + \sqrt{2}\boldsymbol{N}(0)$.

17. $\boldsymbol{r} = (\cos t)\boldsymbol{i} + (\sin t)\boldsymbol{j} - \boldsymbol{k}$. $\boldsymbol{r}(\frac{\pi}{4}) = \frac{\sqrt{2}}{2}\boldsymbol{i} + \frac{\sqrt{2}}{2}\boldsymbol{j} - \boldsymbol{k}$. $\boldsymbol{v} = -\sin t\boldsymbol{i} + \cos t\boldsymbol{j}$, $|\boldsymbol{v}| = 1$, so $\boldsymbol{T} = \boldsymbol{v}$. $\boldsymbol{T}(\frac{\pi}{4}) = -\frac{\sqrt{2}}{2}\boldsymbol{i} + \frac{\sqrt{2}}{2}\boldsymbol{j}$. $\dfrac{d\boldsymbol{T}}{dt} = \dfrac{d\boldsymbol{v}}{dt} = -\cos t\boldsymbol{i} - \sin t\boldsymbol{j}$, a unit vector, so $\boldsymbol{N} = -\cos t\boldsymbol{i} - \sin t\boldsymbol{j}$, $\boldsymbol{N}(\frac{\pi}{4}) = -\frac{\sqrt{2}}{2}\boldsymbol{i} - \frac{\sqrt{2}}{2}\boldsymbol{j}$. $\boldsymbol{B}(\frac{\pi}{4}) = \boldsymbol{T}(\frac{\pi}{4}) \times \boldsymbol{N}(\frac{\pi}{4}) = \begin{vmatrix} \boldsymbol{i} & \boldsymbol{j} & \boldsymbol{k} \\ -\frac{\sqrt{2}}{2} & \frac{\sqrt{2}}{2} & 0 \\ -\frac{\sqrt{2}}{2} & -\frac{\sqrt{2}}{2} & 0 \end{vmatrix} = \boldsymbol{k}$. The position vector, $\boldsymbol{r}(\frac{\pi}{4})$, has terminal point $(\frac{\sqrt{2}}{2}, \frac{\sqrt{2}}{2}, -1)$. For the equation of a plane we use $A(x-x_0)+B(y-y_0)+C(z-z_0)=0$. Osculating plane: normal $\boldsymbol{B}$, $1\cdot(z+1)=0$ or $z=-1$. Normal plane: normal vector $\boldsymbol{T}$, $-\frac{\sqrt{2}}{2}(x-\frac{\sqrt{2}}{2})+\frac{\sqrt{2}}{2}(y-\frac{\sqrt{2}}{2})=0$ or $y=x$. Rectifying plane: normal vector $\boldsymbol{N}, -\frac{\sqrt{2}}{2}(x-\frac{\sqrt{2}}{2})-\frac{\sqrt{2}}{2}(y-\frac{\sqrt{2}}{2})=0$ which reduces to $y=-x$.

18. $\boldsymbol{r}=\cos t\boldsymbol{i}+\sin t\boldsymbol{j}+t\boldsymbol{k}$. $\boldsymbol{r}(0)=\boldsymbol{i}$, $\boldsymbol{v}=-\sin t\boldsymbol{i}+\cos t\boldsymbol{j}+\boldsymbol{k}$, $|\boldsymbol{v}|=\sqrt{2}$, so $\boldsymbol{T}=-\frac{1}{\sqrt{2}}\sin t\boldsymbol{i}+\frac{1}{\sqrt{2}}\cos t\boldsymbol{j}+\frac{1}{\sqrt{2}}\boldsymbol{k}$. $\boldsymbol{T}(0)=\frac{1}{\sqrt{2}}\boldsymbol{j}+\frac{1}{\sqrt{2}}\boldsymbol{k}$. $\dfrac{d\boldsymbol{T}}{dt}=-\dfrac{1}{\sqrt{2}}\cos t\boldsymbol{i}-\dfrac{1}{\sqrt{2}}\sin t\boldsymbol{j}$ so $\boldsymbol{N}=-\cos t\boldsymbol{i}-\sin t\boldsymbol{j}$, $\boldsymbol{N}(0)=-\boldsymbol{i}$. $\boldsymbol{B}(0)=\boldsymbol{T}(0)\times\boldsymbol{N}(0)=$ $\begin{vmatrix}\boldsymbol{i} & \boldsymbol{j} & \boldsymbol{k}\\ 0 & \frac{1}{\sqrt{2}} & \frac{1}{\sqrt{2}}\\ -1 & 0 & 0\end{vmatrix}=-\frac{1}{\sqrt{2}}\boldsymbol{j}+\frac{1}{\sqrt{2}}\boldsymbol{k}$. $\boldsymbol{r}(0)=\boldsymbol{i}$ has terminal point $(1,0,0)$. Osculating plane: normal $\boldsymbol{B}$, $-\frac{1}{\sqrt{2}}y+\frac{1}{\sqrt{2}}z$ or $z=y$. Normal plane: normal $\boldsymbol{T}$, $\frac{1}{\sqrt{2}}y+\frac{1}{\sqrt{2}}z=0$ or $z=-y$. Rectifying plane: normal $\boldsymbol{N}$, $-(x-1)=0$ or $x=1$.

19. If $|\boldsymbol{v}|$ is constant, so is $|\boldsymbol{v}|^2=\boldsymbol{v}\cdot\boldsymbol{v}$. Hence $\frac{d}{dt}(\boldsymbol{v}\cdot\boldsymbol{v})=\dfrac{d\boldsymbol{v}}{dt}\cdot\boldsymbol{v}+\boldsymbol{v}\cdot\dfrac{d\boldsymbol{v}}{dt}=2(\boldsymbol{a}\cdot\boldsymbol{v})=0$. Therefore $\boldsymbol{a}$ is normal to $\boldsymbol{v}$ whose direction is the direction of the path. Alternatively, $a_T=\frac{d}{dt}|\boldsymbol{v}|=0$ so $\boldsymbol{a}=a_T\boldsymbol{T}+a_N\boldsymbol{N}=a_N\boldsymbol{N}$ which is normal to $\boldsymbol{T}$.

20. a) $\boldsymbol{r}=x\boldsymbol{i}+f\boldsymbol{j}$, $\boldsymbol{v}=\boldsymbol{i}+f'\boldsymbol{j}$, $\boldsymbol{a}=f''\boldsymbol{j}$. $\kappa=\dfrac{|\boldsymbol{v}\times\boldsymbol{a}|}{|\boldsymbol{v}|^3}=\begin{vmatrix}\boldsymbol{i} & \boldsymbol{j} & \boldsymbol{k}\\ 1 & f' & 0\\ 0 & f'' & 0\end{vmatrix}/|\boldsymbol{v}|^3=$ $|f''\boldsymbol{k}|/|\boldsymbol{v}|^3=\dfrac{|f''|}{(1+(f')^2)^{3/2}}$. b) $y=f(x)=\ln(\cos x)$, $-\pi/2<x<\pi/2$. $f'(x)=-\frac{\sin x}{\cos x}=-\tan x$, $f''(x)=-\sec^2 x$. $\kappa=\frac{\sec^2 x}{(1+\tan^2 x)^{3/2}}=\frac{\sec^2 x}{\sec^3 x}=\cos x$. c) Graph $x_1=t$, $y_1=\kappa=\cos t$, $x_2=t$, $y_2=\ln\cos t$, $-\pi/2<t<\pi/2$ in $[-\frac{\pi}{2},\frac{\pi}{2}]$ by $[-3,1]$. At the endpoints of the interval where $\kappa\to 0$, the angle ϕ between $\boldsymbol{T}$ and $\boldsymbol{i}$ is changing at a slower and slower rate. It is changing fastest at $x=0$ where κ is maximal.

21. $\boldsymbol{r}=t\boldsymbol{i}+\sin t\boldsymbol{j}$, $\boldsymbol{v}=\boldsymbol{i}+\cos t\boldsymbol{j}$, $\boldsymbol{a}=-\sin t\boldsymbol{j}$. At $t=\frac{\pi}{2}$, $\boldsymbol{v}=\boldsymbol{i}$, $|\boldsymbol{v}|=1$, $\boldsymbol{a}=-\boldsymbol{j}$. $\kappa=\dfrac{|\boldsymbol{v}\times\boldsymbol{a}|}{|\boldsymbol{v}|^3}=\left\|\begin{matrix}\boldsymbol{i} & \boldsymbol{j} & \boldsymbol{k}\\ 1 & 0 & 0\\ 0 & -1 & 0\end{matrix}\right\|=|-\boldsymbol{k}|=1$. Radius $=1/\kappa=1$. Thus the center of the circle of curvature is one unit below $(\pi/2,1)$ so is $(\pi/2,0)$. The circle has equation $(x-\pi/2)^2+y^2=1$. Graph $x_1(t)=t$, $y_1(t)=\sin t$

and $x_2(t) = \frac{\pi}{2} + \cos t$, $y_2(t) = \sin t$, $-2\pi \leqq t \leqq 2\pi$, tstep $= 0.1$ in $[-2\pi, 2\pi]$ by $[-4.2, 3.2]$.

22. $\boldsymbol{r} = (2\ln t)\boldsymbol{i} - (t + \frac{1}{t})\boldsymbol{j}$, $\boldsymbol{v} = \frac{2}{t}\boldsymbol{i} - (1 - \frac{1}{t^2})\boldsymbol{j}$, $\boldsymbol{a} = -\frac{2}{t^2}\boldsymbol{i} - \frac{2}{t^3}\boldsymbol{j}$. $\boldsymbol{v}(1) = 2\boldsymbol{i}$, $\boldsymbol{a}(1) = -2\boldsymbol{i} - 2\boldsymbol{j}$. $\boldsymbol{v}(1) \times \boldsymbol{a}(1) = \begin{vmatrix} \boldsymbol{i} & \boldsymbol{j} & \boldsymbol{k} \\ 2 & 0 & 0 \\ -2 & -2 & 0 \end{vmatrix} = 4\boldsymbol{k}$. At $t = 1$, $\kappa = \dfrac{|-4\boldsymbol{k}|}{|2\boldsymbol{i}|^3} = \dfrac{4}{8} = \dfrac{1}{2}$. The radius of curvature is $\frac{1}{\kappa} = 2$. Since $\boldsymbol{T}(1) = \boldsymbol{v}(1)/|\boldsymbol{v}(1)| = \boldsymbol{i}$, $\boldsymbol{N} = -\boldsymbol{j}$ taking the concave down direction. The center of the circle of curvature is $1/\kappa = 2$ units from $(0, -2)$ along the normal, so it is $(0, -4)$. Equation: $(x - 0)^2 + (y + 4)^2 = 4$ or $x^2 + (y + 4)^2 = 4$. To support this graph $x_2 = 2\cos t$, $y_2 = -4 + 2\sin t$ along with the preceding graph. Graph $x_1(t) = 2\ln t$, $y_1(t) = -(t + \frac{1}{t})$ and $x_2(t) = 2\cos t$, $y_2(t) = -4 + 2\sin t$, $0 \leqq t \leqq 3\pi$, tstep $= 0.1$ in $[-9.3, 9.3]$ by $[-10, 1]$. Remark: One can eliminate t and find $y = -2\cosh(x/2)$ for an equation of the original curve.

23. $\boldsymbol{v} \times \boldsymbol{a} = (ds/dt)\boldsymbol{T} \times \left[\frac{d^2s}{dt^2}\boldsymbol{T} + \kappa(ds/dt)^2\boldsymbol{N}\right] = (\frac{ds}{dt})(\frac{d^2s}{dt^2})\boldsymbol{T} \times \boldsymbol{T} + \kappa(\frac{ds}{dt})^3\boldsymbol{T} \times \boldsymbol{N} = \boldsymbol{0} + \kappa(\frac{ds}{dt})^3\boldsymbol{T} \times \boldsymbol{N} = \kappa(\frac{ds}{dt})^3\boldsymbol{T} \times \boldsymbol{N}$. $|\boldsymbol{v} \times \boldsymbol{a}| = \kappa(\frac{ds}{dt})^3|\boldsymbol{T}||\boldsymbol{N}| = \kappa(\frac{ds}{dt})^3 = \kappa|\boldsymbol{v}|^3$. Therefore $\kappa = \dfrac{|\boldsymbol{v} \times \boldsymbol{a}|}{|\boldsymbol{v}|^3}$.

24. $\boldsymbol{r} = \cos t\boldsymbol{i} + \sin t\boldsymbol{j} + t\boldsymbol{k}$, $\boldsymbol{v} = -\sin t\boldsymbol{i} + \cos t\boldsymbol{j} + \boldsymbol{k}$, $\boldsymbol{a} = -\cos t\boldsymbol{i} - \sin t\boldsymbol{j}$. $|\boldsymbol{v}| = \sqrt{2}$, $\boldsymbol{T} = \frac{1}{\sqrt{2}}\boldsymbol{v}$, $d\boldsymbol{T}/dt = \frac{1}{\sqrt{2}}\boldsymbol{a}$, $|d\boldsymbol{T}/dt| = \frac{1}{\sqrt{2}}$, $\boldsymbol{N} = (d\boldsymbol{T}/dt)/|d\boldsymbol{T}/dt| = \boldsymbol{a}$. $\boldsymbol{B} = \boldsymbol{T} \times \boldsymbol{N} = \begin{vmatrix} \boldsymbol{i} & \boldsymbol{j} & \boldsymbol{k} \\ -\frac{\sin t}{\sqrt{2}} & \frac{\cos t}{\sqrt{2}} & \frac{1}{\sqrt{2}} \\ -\cos t & -\sin t & 0 \end{vmatrix} = \frac{\sin t}{\sqrt{2}}\boldsymbol{i} - \frac{\cos t}{\sqrt{2}}\boldsymbol{j} + \frac{1}{\sqrt{2}}\boldsymbol{k}$. $\tau = \dfrac{|d\boldsymbol{B}/dt|}{|\boldsymbol{v}|} = \dfrac{1}{\sqrt{2}}\left|\dfrac{\cos t}{\sqrt{2}}\boldsymbol{i} + \dfrac{\sin t}{\sqrt{2}}\boldsymbol{j}\right| = \dfrac{1}{2}$.

25. $\kappa = \frac{a}{a^2+b^2}$. We wish to find the maximum value of the function $f(x) = \frac{x}{x^2+b^2}$, $x \geqq 0$. $f'(x) = \frac{(x^2+b^2)-x(2x)}{(x^2+b^2)^2} = \frac{b^2-x^2}{x^2+b^2}$. $f'(x) = 0$, $x \geqq 0$ yields $x = b$. Since $f'(x) > 0$ for $0 \leqq x < b$ and $f'(x) < 0$ for $x > b$, $f(b) = \frac{b}{2b^2} = \frac{1}{2b}$ is the absolute maximum of f and of κ. $b > 0$ or else the helix degenerates into a circle.

12.5 PLANETARY MOTION AND SATELLITES

1. $T^2 = \frac{4\pi a^3}{GM}$. $T = \sqrt{\frac{4(6.808\times10^6)^3\pi^2}{(6.6720\times10^{-11})(5.975\times10^{24})}} = 5590.00524462\,\text{sec} = 93.17\,\text{min}$ compared to 93.11 min. in the Table.

2. $e = \frac{r_0 v_0^2}{GM} - 1$, $v_0 = \sqrt{\frac{GM}{r_0}(e+1)} =$ $\sqrt{(6.6720 \times 10^{-11})(1.99 \times 10^{30})(0.0167+1)/149577000000} = 30041$ m/sec.

3. $\frac{T^2}{a^3} = \frac{4\pi^2}{GM}$, $\frac{a^3}{T^2} = \frac{GM}{4\pi^2}$, $a = (\frac{GMT^2}{4\pi^2})^{1/3} =$ $((6.6720 \times 10^{-11})(5.975 \times 10^{24})((92.25)60)^2/(4\pi^2))^{1/3} = 6.763 \times 10^6$ m $=$ 6763 km. $a =$ (diameter of Earth + perigee height + apogee height)/2 $=$ $(2(6378.533) + 183 + 589)/2 = 6765$ km.

4. a) $\frac{T^2}{a^3} = \frac{4\pi^2}{GM}$, $\frac{a^3}{T^2} = \frac{GM}{4\pi^2}$, $a = (GMT^2/(4\pi)^2)^{1/3} = ((6.6720 \times 10^{-11})(6.418 \times 10^{23})((1639)60)^2/(4\pi^2))^{1/3} = 2.189 \times 10^7$ m $= 2.189 \times 10^4$ km $= 21890$ km
b) Diameter of Mars $= 2a -$perigee $h -$apogee $ht = 2(21890) - 1499 - 35800 =$ 6481 km.

5. $T = (4\pi^2 a^3/(GM))^{1/2} = (4\pi^2(2.2030 \times 10^7)^3/((6.6720 \times 10^{-11})(6.418 \times 10^{23}))^{1/2} = 9.9283 \times 10^4\,\text{sec} = 1655\,\text{min}$.

6. If a satellite holds a geostationary orbit, then its period = Earth's rotational period = 1436.1 min = 86166 sec. $\frac{T^2}{a^3} = \frac{4\pi^2}{GM}$, $\frac{a^3}{T^2} = \frac{GM}{4\pi^2}$, $a = (GMT^2/(4\pi^2))^{1/3} =$ $((6.6720\times10^{-11})(5.975\times10^{24})86166^2/(4\pi^2))^{1/3}\text{m} = 42167$ km.

7. $\frac{T^2}{a^3} = \frac{4\pi^2}{GM}$, $\frac{a^3}{T^2} = \frac{GM}{4\pi^2}$, $a = (GMT^2/(4\pi^2))^{1/3} = ((6.6720 \times 10^{-11})(6.418 \times 10^{23})(1477.4(60))^2/(4\pi^2))^{1/3} = 2.042 \times 10^7$ m $= 20420$ km.

8. $a = (GMT^2/(4\pi^2))^{1/3} = ((6.6720\times10^{-11})(5.975\times10^{24})(2.36055\times10^6)^2/(4\pi^2))^{1/3} =$ $3.831953\times10^8\text{m} = 3.831953\times10^5$ km $=$ mean distance to center of Earth. Distance to surface of Earth $= a -$ radius of Earth $= (3.831953\times10^5 - 6378.533)$ km $= 376800$ km.

9. From Eq. (24), $|\boldsymbol{v}| = \sqrt{GM/r} = ((6.6720 \times 10^{-11}(5.975 \times 10^{24})/r)^{1/2} =$ $(1.9967 \times 10^7)/\sqrt{r}$ m/sec.

10. $\frac{T^2}{a^3} = \frac{4\pi^2}{GM}$. For planets in our solar system: $\frac{T^2}{a^3} = 4\pi^2/((6.6720\times10^{-11})(1.99\times 10^{30})) = 2.97 \times 10^{-19}\,\text{sec}^2/\text{m}^3$. For satellites orbiting the earth: $\frac{T^2}{a^3} =$ $4\pi^2/((6.6720 \times 10^{-11})(5.975 \times 10^{24})) = 9.903 \times 10^{-14}\,\text{sec}^2/\text{m}^3$. For satelites orbiting the moon: $T^2/a^3 = 4\pi^2/((6.6720 \times 10^{-11})(7.354 \times 10^{22})) = 8.046 \times 10^{-12}\,\text{sec}^2/\text{m}^3$.

11. $e = \frac{r_0 v_0^2}{GM} - 1$. Circle: $e = 0$, $\frac{r_0 v_0^2}{GM} = 1$, $v_0 = \sqrt{\frac{GM}{r_0}}$. Ellipse: $e < 1$, $\frac{r_0 v_0^2}{GM} - 1 < 1$, $\frac{r_0 v_0^2}{GM} < 2$, $v_0 < \sqrt{\frac{2GM}{r_0}}$. Hyperbola: $e > 1$, $v_0 > \sqrt{\frac{2GM}{r_0}}$. Since $e \geq 0$, $v_0 \geqq \sqrt{\frac{GM}{r_0}}$ in all cases.

12. By Eq. (24), $\frac{GM}{|v|^2}$, $|\boldsymbol{v}| = \sqrt{\frac{GM}{r}}$. Since G, M and r are constants so is $|\boldsymbol{v}|$.

13. $\boldsymbol{u_r} = \cos\theta \boldsymbol{i} + \sin\theta \boldsymbol{j}$. The $\boldsymbol{u_r}$-component of $\boldsymbol{a} = \frac{d^2r}{dt^2} - r(\frac{d\theta}{dt})^2$. $\frac{dr}{dt} = (\sinh\theta)\frac{d\theta}{dt} = 2\sinh\theta$, $\frac{d^2r}{dt^2} = 2(\cosh\theta)\frac{d\theta}{dt} = 4\cosh\theta$. Hence $\frac{d^2r}{dt^2} - r(\frac{d\theta}{dt})^2 = 4\cosh\theta - (\cosh\theta)(2)^2 = 0$.

PRACTICE EXERCISES, CHAPTER 12

1. $\boldsymbol{r} = (4\cos t)\boldsymbol{i} + (\sqrt{2}\sin t)\boldsymbol{j}$, $\boldsymbol{v} = (-4\sin t)\boldsymbol{i} + (\sqrt{2}\cos t)\boldsymbol{j}$, $\boldsymbol{a} = (-4\cos t)\boldsymbol{i} + (-\sqrt{2}\sin t)\boldsymbol{j}$. $\boldsymbol{r}(0) = 4\boldsymbol{i}$, $\boldsymbol{v}(0) = \sqrt{2}\boldsymbol{j}$, $\boldsymbol{a}(0) = -4\boldsymbol{i}$. $\boldsymbol{r}(\pi/4) = 2\sqrt{2}\boldsymbol{i} + \boldsymbol{j}$, $\boldsymbol{v}(\pi/4) = -2\sqrt{2}\boldsymbol{i} + \boldsymbol{j}$, $\boldsymbol{a}(\pi/4) = -2\sqrt{2}\boldsymbol{i} - \boldsymbol{j}$.

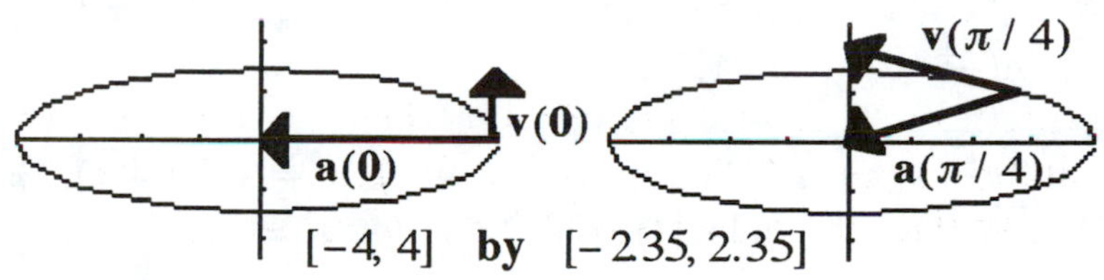

2. $\boldsymbol{r} = (\sqrt{3}\sec t)\boldsymbol{i} + (\sqrt{3}\tan t)\boldsymbol{j}$, $\boldsymbol{v} = \sqrt{3}\sec t \tan t\boldsymbol{i} + \sqrt{3}\sec^2 t\boldsymbol{j}$, $\boldsymbol{a} = (\sqrt{3}\sec^3 t + \sqrt{3}\sec t\tan^2 t)\boldsymbol{i} + 2\sqrt{3}\sec^2 t\tan t\boldsymbol{j}$. $\boldsymbol{r}(0) = \sqrt{3}\boldsymbol{i}$, $\boldsymbol{v}(0) = \sqrt{3}\boldsymbol{j}$, $\boldsymbol{a}(0) = \sqrt{3}\boldsymbol{i}$. $\boldsymbol{r}(\pi/6) = 2\boldsymbol{i} + \boldsymbol{j}$, $\boldsymbol{v}(\pi/6) = \frac{2}{\sqrt{3}}\boldsymbol{i} + \frac{4}{\sqrt{3}}\boldsymbol{j}$, $\boldsymbol{a}(\pi/6) = (\frac{8}{3} + \frac{2}{3})\boldsymbol{i} + 2\sqrt{3}(\frac{4}{3})\frac{1}{\sqrt{3}}\boldsymbol{j} = \frac{10}{3}\boldsymbol{i} + \frac{8}{3}\boldsymbol{j}$.

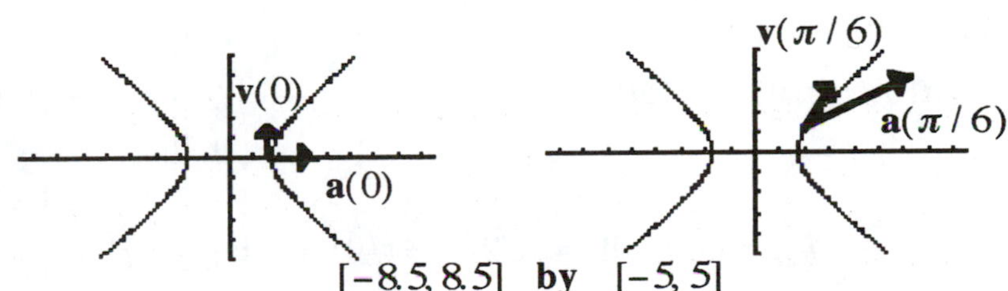

3. $\int_0^1[(3+6t)\boldsymbol{i}+(4+8t)\boldsymbol{j}+(6\pi\cos\pi t)\boldsymbol{k}]dt = [3t+3t^2]_0^1\boldsymbol{i}+[4t+4t^2]_0^1\boldsymbol{j}+[6\sin\pi t]_0^1\boldsymbol{k} = 6\boldsymbol{i}+8\boldsymbol{j}$.

4. $\int_e^{e^2}[\frac{2\ln t}{t}\boldsymbol{i}+\frac{1}{t\ln t}\boldsymbol{j}+\frac{1}{t}\boldsymbol{k}]dt = [(\ln t)^2]_e^{e^2}\boldsymbol{i}+[\ln(\ln t)]_e^{e^2}\boldsymbol{j}+[\ln t]_e^{e^2}\boldsymbol{k} = 3\boldsymbol{i}+(\ln 2)\boldsymbol{j}+\boldsymbol{k}$.

5. $\frac{d\boldsymbol{r}}{dt} = -(\sin t)\boldsymbol{i}+(\cos t)\boldsymbol{j}+\boldsymbol{k}$, $\boldsymbol{r}=\boldsymbol{j}$ when $t=0$. $\boldsymbol{r} = \cos t\boldsymbol{i}+\sin t\boldsymbol{j}+t\boldsymbol{k}+\boldsymbol{C}$. $\boldsymbol{r}(0) = \boldsymbol{i}+\boldsymbol{C} = \boldsymbol{j}$, $\boldsymbol{C} = -\boldsymbol{i}+\boldsymbol{j}$, $\boldsymbol{r} = (\cos t-1)\boldsymbol{i}+(\sin t+1)\boldsymbol{j}+t\boldsymbol{k}$.

6. $\boldsymbol{r} = \tan^{-1}t\boldsymbol{i}-\sin^{-1}t\boldsymbol{j}+\sqrt{t^2+1}\boldsymbol{k}+\boldsymbol{C}$. When $t=0$, $\boldsymbol{r} = \boldsymbol{k}+\boldsymbol{C} = \boldsymbol{j}+\boldsymbol{k}$. Hence $\boldsymbol{C}=\boldsymbol{j}$ and $\boldsymbol{r} = \tan^{-1}t\boldsymbol{i}+(1-\sin^{-1}t)\boldsymbol{j}+\sqrt{t^2+1}\boldsymbol{k}$.

7. $\frac{d^2\boldsymbol{r}}{dt^2} = 2\boldsymbol{j}$, $\frac{d\boldsymbol{r}}{dt} = \boldsymbol{k}$ and $\boldsymbol{r}=\boldsymbol{i}$ when $t=0$. $\frac{d\boldsymbol{r}}{dt} = 2t\boldsymbol{j}+\boldsymbol{C}_1$, $\frac{d\boldsymbol{r}}{dt}(0) = \boldsymbol{C}_1 = \boldsymbol{k}$, $\frac{d\boldsymbol{r}}{dt} = 2t\boldsymbol{j}+\boldsymbol{k}$. $\boldsymbol{r} = t^2\boldsymbol{j}+t\boldsymbol{k}+\boldsymbol{C}_2$, $\boldsymbol{r}(0) = \boldsymbol{C}_2 = \boldsymbol{i}$, $\boldsymbol{r} = \boldsymbol{i}+t^2\boldsymbol{j}+t\boldsymbol{k}$.

8. $\frac{d\boldsymbol{r}}{dt} = -2t\boldsymbol{i}-4t\boldsymbol{j}+\boldsymbol{C}_1$. At $t=1$, $\frac{d\boldsymbol{r}}{dt} = -2\boldsymbol{i}-4\boldsymbol{j}+\boldsymbol{C}_1 = 4\boldsymbol{i}$. $\boldsymbol{C}_1 = 6\boldsymbol{i}+4\boldsymbol{j}$, $\frac{d\boldsymbol{r}}{dt} = (6-2t)\boldsymbol{i}+(4-4t)\boldsymbol{j}$. $\boldsymbol{r} = (6t-t^2)\boldsymbol{i}+(4t-2t^2)\boldsymbol{j}+\boldsymbol{C}_2$. At $t=1$, $\boldsymbol{r} = 5\boldsymbol{i}+2\boldsymbol{j}+\boldsymbol{C}_2 = 3\boldsymbol{i}+3\boldsymbol{j}$, $\boldsymbol{C}_2 = -2\boldsymbol{i}+\boldsymbol{j}$. $\boldsymbol{r} = (6t-t^2-2)\boldsymbol{i}+(4t-2t^2+1)\boldsymbol{j}$.

9. $L = \int_0^{\pi/4}\sqrt{(\frac{dx}{dt})^2+(\frac{dy}{dt})^2+(\frac{dz}{dt})^2} = \int_0^{\pi/4}\sqrt{(-2\sin t)^2+(2\cos t)^2+(2t)^2dt} = \int_0^{\pi/4}2\sqrt{1+t^2} = t\sqrt{1+t^2}+\ln|t+\sqrt{t^2+1}|]_0^{\pi/4} = \frac{\pi}{4}\frac{\sqrt{16+\pi^2}}{4}+\ln(\frac{\pi}{4}+\frac{\sqrt{16+\pi^2}}{4}) = \frac{\pi\sqrt{16+\pi^2}}{16}+\ln(\frac{\pi+\sqrt{16+\pi^2}}{4}) = 1.7199\ldots$.

10. $\boldsymbol{v} = -3\sin t\boldsymbol{i}+3\cos t\boldsymbol{j}+3t^{1/2}\boldsymbol{k}$, $|\boldsymbol{v}|^2 = 9(\sin^2 t+\cos^2 t+t)$, $|\boldsymbol{v}| = 3\sqrt{1+t}$. $L = \int_0^3 3\sqrt{1+t}dt = 3(\frac{2}{3})(1+t)^{3/2}]_0^3 = 2[4^{3/2}-1] = 14$.

11. $\boldsymbol{r} = \frac{4}{9}(1+t)^{3/2}\boldsymbol{i}+\frac{4}{9}(1-t)^{3/2}\boldsymbol{j}+\frac{1}{3}t\boldsymbol{k}$, $\boldsymbol{v} = \frac{2}{3}(1+t)^{1/2}\boldsymbol{i}-\frac{2}{3}(1-t)^{1/2}\boldsymbol{j}+\frac{1}{3}\boldsymbol{k}$, $\boldsymbol{a} = \frac{1}{3}(1+t)^{-1/2}\boldsymbol{i}+\frac{1}{3}(1-t)^{-1/2}\boldsymbol{j}$. $|\boldsymbol{v}| = \sqrt{\frac{4}{9}(1+t)+\frac{4}{9}(1-t)+\frac{1}{9}} = 1$ so $\boldsymbol{T} = \boldsymbol{v}/|\boldsymbol{v}| = \boldsymbol{v}$. $\boldsymbol{T}(0) = \frac{2}{3}\boldsymbol{i}-\frac{2}{3}\boldsymbol{j}+\frac{1}{3}\boldsymbol{k}$. $d\boldsymbol{T}/dt = \boldsymbol{a}$, $|\frac{d\boldsymbol{T}}{dt}| = \sqrt{\frac{1}{9}\frac{1}{1+t}+\frac{1}{9}\frac{1}{1-t}} = \frac{1}{3}\sqrt{\frac{2}{1-t^2}}$. $\boldsymbol{N} = (d\boldsymbol{T}/dt)/|d\boldsymbol{T}/dt|$, $\boldsymbol{N}(0) = (\frac{1}{3}\boldsymbol{i}+\frac{1}{3}\boldsymbol{j})/(\sqrt{2}/3) = (\boldsymbol{i}+\boldsymbol{j})/\sqrt{2}$.

$$\boldsymbol{B} = \boldsymbol{T}\times\boldsymbol{N},\ \boldsymbol{B}(0) = \begin{vmatrix} \boldsymbol{i} & \boldsymbol{j} & \boldsymbol{k} \\ \frac{2}{3} & -\frac{2}{3} & \frac{1}{3} \\ \frac{1}{\sqrt{2}} & \frac{1}{\sqrt{2}} & 0 \end{vmatrix} = -\frac{1}{3\sqrt{2}}\boldsymbol{i}+\frac{1}{3\sqrt{2}}\boldsymbol{j}+\frac{4}{3\sqrt{2}}\boldsymbol{k}.$$ At $t=0$, $\kappa =$

$$|\boldsymbol{v}(0)\times\boldsymbol{a}(0)|/|\boldsymbol{v}(0)|^3 = |\boldsymbol{v}(0)\times\boldsymbol{a}(0)| = \left\|\begin{matrix} \boldsymbol{i} & \boldsymbol{j} & \boldsymbol{k} \\ \frac{2}{3} & -\frac{2}{3} & \frac{1}{3} \\ \frac{1}{3} & \frac{1}{3} & 0 \end{matrix}\right\| = |-\tfrac{1}{9}\boldsymbol{i}+\tfrac{1}{9}\boldsymbol{j}+\tfrac{4}{9}\boldsymbol{k}| =$$

$$\tfrac{1}{9}|-\boldsymbol{i}+\boldsymbol{j}+4\boldsymbol{k}| = \sqrt{18}/9 = \sqrt{2}/3.\quad \tau(0) = \pm\begin{vmatrix} \dot{x} & \dot{y} & \dot{z} \\ \ddot{x} & \ddot{y} & \ddot{z} \\ \dddot{x} & \dddot{y} & \dddot{z} \end{vmatrix}\ |\boldsymbol{v}\times\boldsymbol{a}|_{t=0}^2 =$$

$$\pm\begin{vmatrix} \frac{2}{3} & -\frac{2}{3} & \frac{1}{3} \\ \frac{1}{3} & \frac{1}{3} & 0 \\ -\frac{1}{6} & \frac{1}{6} & 0 \end{vmatrix} /(\sqrt{2}/3)^2 = (1/27)/(2/9) = 1/6.$$

12. $\boldsymbol{r} = \sin t\boldsymbol{i} + \sqrt{2}\cos t\boldsymbol{j} + \sin t\boldsymbol{k}$, $\boldsymbol{v} = \cos t\boldsymbol{i} - \sqrt{2}\sin t\boldsymbol{j} + \cos t\boldsymbol{k}$, $\boldsymbol{a} = -\sin t\boldsymbol{i} - \sqrt{2}\cos t\boldsymbol{j} - \sin t\boldsymbol{k}$, $|\boldsymbol{v}| = \sqrt{\cos^2 t + 2\sin^2 t + \cos^2 t} = \sqrt{2}$. $\boldsymbol{T} = \boldsymbol{v}/|\boldsymbol{v}| = \frac{1}{\sqrt{2}}\cos t\boldsymbol{i} - \sin t\boldsymbol{j} + \frac{1}{\sqrt{2}}\cos t\boldsymbol{k}$. $\frac{d\boldsymbol{T}}{dt} = -\frac{1}{\sqrt{2}}\sin t\boldsymbol{i} - \cos t\boldsymbol{j} - \frac{1}{\sqrt{2}}\sin t\boldsymbol{k}$, $|\frac{d\boldsymbol{T}}{dt}| = \sqrt{\frac{1}{2}\sin^2 t + \cos^2 t + \frac{1}{2}\sin^2 t} = 1$. So $\boldsymbol{N} = \frac{d\boldsymbol{T}}{dt}$. $\boldsymbol{B} = \boldsymbol{T} \times \boldsymbol{N} =$
$$\begin{vmatrix} \boldsymbol{i} & \boldsymbol{j} & \boldsymbol{k} \\ \frac{1}{\sqrt{2}}\cos t & -\sin t & \frac{1}{\sqrt{2}}\cos t \\ -\frac{1}{\sqrt{2}}\sin t & -\cos t & -\frac{1}{\sqrt{2}}\sin t \end{vmatrix} = \frac{\boldsymbol{i}-\boldsymbol{k}}{\sqrt{2}}.\ \kappa = |\tfrac{d\boldsymbol{T}}{ds}| = |\tfrac{d\boldsymbol{T}}{dt}\tfrac{dt}{ds}| = |\tfrac{d\boldsymbol{T}}{dt}|\tfrac{1}{|\boldsymbol{v}|} = \tfrac{1}{\sqrt{2}}.$$
$\tau = |\frac{d\boldsymbol{B}}{ds}| = 0$ since $\boldsymbol{B}$ is constant.

13. $\boldsymbol{v} = (3+6t)\boldsymbol{i} + (4+8t)\boldsymbol{j} + 6\sin t\boldsymbol{k}$, $|\boldsymbol{v}| = \sqrt{[3(1+2t)]^2 + [4(1+2t)]^2 + 36\sin^2 t} = \sqrt{25(1+2t)^2 + 36\sin^2 t}$. $a_T = \frac{d|\boldsymbol{v}|}{dt} = \frac{100(1+2t)+72\sin t\cos t}{2\sqrt{25(1+2t)^2+36\sin^2 t}}$. When $t = 0$, $a_T = \frac{100}{10} = 10$. $\boldsymbol{a} = 6\boldsymbol{i} + 8\boldsymbol{j} + 6\cos t\boldsymbol{k}$, $\boldsymbol{a}(0) = 6\boldsymbol{i} + 8\boldsymbol{j} + 6\boldsymbol{k}$. $a_N = \sqrt{|\boldsymbol{a}|^2 - a_T^2} = \sqrt{6^2 + 8^2 + 6^2 - 10^2} = 6$. $\boldsymbol{a}(0) = 10\boldsymbol{T}(0) + 6\boldsymbol{N}(0)$.

14. $\boldsymbol{v} = \boldsymbol{i} + (1+4t)\boldsymbol{j} + 2t\boldsymbol{k}$. $|\boldsymbol{v}| = \sqrt{1+1+8t+16t^2+4t^2} = \sqrt{2+8t+20t^2}$. $a_T = \frac{1}{2}(2+8t+20t^2)^{-1/2}(8+40t) = (4+10t)/\sqrt{2+8t+20t^2}$. When $t = 0$, $a_T = 4/\sqrt{2} = 2\sqrt{2}$. $\boldsymbol{a} = 4\boldsymbol{j} + 2\boldsymbol{k} = \boldsymbol{a}(0)$. $a_N = \sqrt{|\boldsymbol{a}(0)|^2 - a_T^2} = \sqrt{16+4-8} = 2\sqrt{3}$. $\boldsymbol{a}(0) = 2\sqrt{2}\boldsymbol{T} + 2\sqrt{3}\boldsymbol{N}$.

15. $\boldsymbol{v} = -\frac{1}{2}(1+t^2)^{-3/2}2t\boldsymbol{i} + \frac{\sqrt{1+t^2} - t[\frac{1}{2}(1+t^2)^{-1/2}2t]}{1+t^2}\frac{\sqrt{1+t^2}}{\sqrt{1+t^2}}\boldsymbol{j} = -\frac{t}{(1+t^2)^{3/2}}\boldsymbol{i} + \frac{1+t^2-t^2}{(1+t^2)^{3/2}}\boldsymbol{j}$. $|\boldsymbol{v}|^2 = \frac{t^2+1}{(1+t^2)^3} = \frac{1}{(1+t^2)^2}$, $|\boldsymbol{v}| = \frac{1}{1+t^2}$. Maximal speed $= |v(0)| = 1$, when denominator is smallest.

16. $\boldsymbol{r} = (e^t\cos t)\boldsymbol{i} + (e^t\sin t)\boldsymbol{j}$, $\boldsymbol{v} = (e^t\cos t - e^t\sin t)\boldsymbol{i} + (e^t\cos t + e^t\sin t)\boldsymbol{j}$, $\boldsymbol{a} = (e^t\cos t - e^t\sin t - e^t\sin t - e^t\cos t)\boldsymbol{i} + (e^t\cos t - e^t\sin t + e^t\sin t + e^t\cos t)\boldsymbol{j} = -2e^t\sin t\boldsymbol{i} + 2e^t\cos t\boldsymbol{j}$. We see that $\boldsymbol{r}\cdot\boldsymbol{a} = 0$ for all t. $\boldsymbol{r}$ and $\boldsymbol{a}$ are always orthogonal and $\theta = \frac{\pi}{2}$.

17. $\boldsymbol{v} = -5\sin t\boldsymbol{j} + 3\cos t\boldsymbol{k}$, $\boldsymbol{a} = -5\cos t\boldsymbol{j} - 3\sin t\boldsymbol{k}$. $\boldsymbol{v}\cdot\boldsymbol{a} = 25\sin t\cos t - 9\sin t\cos t = 8\sin 2t = 0$ when $2t = n\pi$, $t = \frac{n\pi}{2}$. In $0 \leqq t \leqq \pi$ when $t = 0, \frac{\pi}{2}, \pi$.

18. $\boldsymbol{r}\cdot(\boldsymbol{i} - \boldsymbol{j}) = 2 - 4\sin\frac{t}{2} = 0$ leads to $\sin\frac{t}{2} = \frac{1}{2}$; the smallest positive solution is $t = \frac{\pi}{3}$.

19. $x = (v_0 \cos\alpha)t = (44\cos 45°)3 = 66\sqrt{2}$ ft. $y = 8 + (v_0 \sin\alpha)t - 16t^2 = 8 + 66\sqrt{2} - 144 = -42.66$ ft. So it must be on the level ground.

20. $y_{\max} = y_0 + \frac{(v_0 \sin\alpha)^2}{2g} = 7 + \frac{(80\sqrt{2}/2)^2}{64} = 57$ ft.

21. $x = (v_0 \cos 45°)t = (4325)3$ ft. $y = (v_0 \sin 45°)t - 16t^2 = 0$. From the first equation $t = 3(4325)\sqrt{2}/v_0$. From the second equation $v_0 = 16\sqrt{2}t = 16(6)(4325)/v_0$, $v_0^2 = 16(2)(4325)$, $v_0 = 4\sqrt{6(4325)} = 644.36$ ft/sec. Replacing 4325 by 4752, we get $v_0 = 675.42$ ft/sec.

22. $109.5 = R = \frac{v_0^2}{g}\sin 2\alpha = \frac{v_0^2}{32}$, $v_0 = \sqrt{32(109.5)} = 59.19$ ft/sec.

23. a) $\frac{8 \text{ mi}}{\text{hr}} = \frac{8 \text{ mi}}{\text{hr}}\frac{1 \text{ hr}}{3600 \text{ sec}}\frac{5280 \text{ ft}}{1 \text{mi}} = \frac{176}{15}$ ft/sec (to avoid round off error). $x = (155\cos 18°)t - (176/15)t$, $y = 4 + (155\sin 18°)t - 16t^2$ and with the same x, y, $\boldsymbol{r} = x\boldsymbol{i} + y\boldsymbol{j}$. b) Graph x, y in a), $0 \leqq t \leqq 4$, tstep $= 0.05$ in $[0, 500]$ by $[0, 100]$. Also include LINE $(380, 0, 380, 10)$. c) With the setup in b) we use the TRACE function, (c-i) $y_{\text{max}} = 39.85$ ft at $t = 1.5$ sec (c-ii) $y = 0$ at about $t = 3.1$ sec and $x = 420.61$ ft. (c-iii) $y = 25$ ft at about $t = 0.55$ sec and $t = 2.45$ sec, when $x = 74.62$ ft and 332.42 ft, respectively. (c-4) Zooming-in and again including LINE $(380, 0, 380, 10)$ shows it is a home run.

24. 8 mi/hr = (176/15) ft/sec. Graph $x = \frac{155}{0.09}(1 - e^{-0.09t})\cos 18° - (176/15)t$, $y = 4 + \frac{155}{0.09}(1 - e^{-0.09t})(\sin 18°) + \frac{32}{(0.09)^2}(1 - 0.09t - e^{-0.09t})$, $0 \leqq t \leqq 3$, tstep $= 0.05$ in $[0, 500]$ by $[0, 100]$. We use the TRACE function for the following approximations. a) $y_{\text{max}} = 36.92$ ft at $t = 1.4$ sec. b) $y = 30$ ft at about $t = 0.75$ sec and $t = 2.05$ sec, 98.11 ft and 251.91 ft from home plate, respectively. c) $y = 0$ at about $t = 2.95$ sec when x is about 347.32 ft. d) No, since by c) the ball hits the ground short of the fence. It is a home run if $k = 0.01$.

25. $\boldsymbol{r} = e^t\boldsymbol{i} + \sin t\boldsymbol{j} + \ln(1-t)\boldsymbol{k}$. When $t = 0$, we get the point $(1, 0, 0)$. $\boldsymbol{v} = e^t\boldsymbol{i} + \cos t\boldsymbol{j} - \frac{1}{1-t}\boldsymbol{k}$, $\boldsymbol{v}(0) = \boldsymbol{i} + \boldsymbol{j} - \boldsymbol{k}$. The tangent line has direction numbers $1, 1, -1$ and parametric equations $x = 1 + t$, $y = t$, $z = -t$.

26. $\boldsymbol{v} = -\sqrt{2}\sin t\boldsymbol{i} + \sqrt{2}\cos t\boldsymbol{j} + \boldsymbol{k}$, $\boldsymbol{v}(\frac{\pi}{4}) = -\boldsymbol{i} + \boldsymbol{j} + \boldsymbol{k}$. $\boldsymbol{r}(\frac{\pi}{4}) = \boldsymbol{i} + \boldsymbol{j} + \frac{\pi}{4}\boldsymbol{k}$. The tangent line has equations $x = 1 - t$, $y = 1 + t$, $z = \frac{\pi}{4} + t$.

27. $\boldsymbol{v} \times \boldsymbol{a} = \begin{vmatrix} \boldsymbol{i} & \boldsymbol{j} & \boldsymbol{k} \\ 3 & 4 & 0 \\ 5 & 15 & 0 \end{vmatrix} = 25\boldsymbol{k}$. $|\boldsymbol{v}| = \sqrt{3^2 + 4^2} = 5$. $\kappa = \frac{|\boldsymbol{v}\times\boldsymbol{a}|}{|\boldsymbol{v}|^3} = \frac{25}{125} = \frac{1}{5}$.

28. The point $(1, 1, \frac{2}{3})$ occurs for $t = 1$. $\boldsymbol{v} = \boldsymbol{i}+\boldsymbol{j}+\sqrt{t}\boldsymbol{k}$. $\boldsymbol{v}(1) = \boldsymbol{i}+\boldsymbol{j}+\boldsymbol{k}$ which is parallel to $\boldsymbol{T}$ and hence normal to the normal plane: $(x-1)+(y-1)+(z-\frac{2}{3}) = 0$ or $x + y + z = \frac{8}{3}$.

29. $\begin{vmatrix} -a\sin t & a\cos t & b \\ -a\cos t & -a\sin t & 0 \\ a\sin t & -a\cos t & 0 \end{vmatrix} = b(a^2\cos^2 t + a^2\sin^2 t) = ba^2$. $\boldsymbol{v} \times \boldsymbol{a} =$

$\begin{vmatrix} \boldsymbol{i} & \boldsymbol{j} & \boldsymbol{k} \\ -a\sin t & a\cos t & b \\ -a\cos t & -a\sin t & 0 \end{vmatrix} = ab\sin t\boldsymbol{i} - ab\cos t\boldsymbol{j} + a^2\boldsymbol{k}$. $|\boldsymbol{v}\times\boldsymbol{a}|^2 = a^2b^2(\sin^2 t + \cos^2 t) + a^4 = a^2(a^2+b^2)$. $\tau = ba^2/[a^2(a^2+b^2)] = b/(a^2+b^2)$. For a given value of a this has maximum value $1/(2a)$. (See the solution of 12.4 #25 and interchange a and b.)

30. Here $x = \ln(\sec t)$, $y = t$, $-\frac{\pi}{2} < t < \frac{\pi}{2}$. $\dot{x} = \tan t$, $\ddot{x} = \sec^2 t$. $\dot{y} = 1$, $\ddot{y} = 0$. $\kappa = \frac{|\dot{x}\ddot{y}-\dot{y}\ddot{x}|}{(\dot{x}^2+\dot{y}^2)^{3/2}} = \frac{|0-\sec^2 t|}{(\tan^2 t+1)^{3/2}} = \frac{\sec^2 t}{(\sec^2 t)^{3/2}} = \frac{\sec^2 t}{\sec^3 t} = \cos t$.

31. a) Let R be the point $(0, y_0)$ and let $\alpha = \angle ROT$. In triangle ROT, $\cos\alpha = y_0/6380$. In triangle SOT, $\cos\alpha = 6380/(6380+437)$. Hence $y_0 = \frac{6380^2}{(6380+437)} = 5971$ km.

b) $ds = \sqrt{1+(\frac{dx}{dy})^2}dy$. $x = \sqrt{(6380)^2 - y^2}$, $\frac{dx}{dy} = \frac{-y}{\sqrt{(6380)^2-y^2}}$, $1 + (\frac{dx}{dy})^2 = \frac{6380^2}{6380^2-y^2}$. $VA = \int_{5971}^{6380} 2\pi x\,ds = 2\pi\int_{5971}^{6380} 6380dy = 2\pi 6380(6380-5971) = (5.2188\times 10^6)\pi$ km^2.

c) $(VA/4\pi r^2)100\% = (5.2188\times 10^8)/(4(6380)^2)\% = 3.21\%$.

32. $r = \frac{(1+e)r_0}{1+e\cos\theta}$ is periodic of period 2π. We may confine our attention to the period where $0 \leqq \theta \leqq 2\pi$. $\frac{dr}{d\theta} = \frac{-(1+e)r_0(-e\sin\theta)}{(1+e\cos\theta)^2}$. $\theta = 0, \pi, 2\pi$ are the only critical points and include the endpoints. $r(0) = \frac{(1+e)r_0}{1+e\cos 0} = r_0 = r(2\pi)$, $r(\pi) = \frac{(1+e)r_0}{(1-e)} > r_0$ since $0 < e < 1$. Thus $r(0) = r_0$ is the minimum value of r; it occurs at the time when $\theta = 0$. We have used the fact that a continuous function on a closed interval attains its absolute extreme values either at endpoints or critical points.

33. a) $f(0) = -1$, $f(2) = 2 - 1 - \frac{1}{2}\sin 2 = 1 - \frac{1}{2}\sin 2 > 1 - \frac{1}{2} = \frac{1}{2} > 0$. By the Intermediate Value Property $f(x) = 0$ has a solution between $x = 0$ and $x = 2$. b) $f'(x) = 1 - 0.5\cos x$. We use $x - \frac{f(x)}{f'(x)} = x - \frac{x-1-0.5\sin x}{1-0.5\cos x}$ for our recursion formula, starting with $x_0 = 0$. The sequence of numbers we obtain converges to 1.49870113352. Graph f in the window $[-2, 3]$ by $[-2, 2]$.

CHAPTER 13

FUNCTIONS OF TWO OR MORE VARIABLES AND THEIR DERIVATIVES

13.1 FUNCTIONS OF TWO OR MORE INDEPENDENT VARIABLES

1. Domain: All points in the xy–plane; Range: All Real Numbers
   Level curves are straight lines parallel to the line $y = x$.

2. Domain: Set of all (x,y) so that $y - x \geq 0 \Rightarrow y \geq x$; Range: $z \geq 0$
   Level curves are straight lines of the form $y - x = c$ where $c \geq 0$.

3. Domain: Set of all (x,y) so that $(x,y) \neq (0,0)$; Range: All Real Numbers
   Level curves are circles with center $(0,0)$ and radii ≥ 0.

4. Domain: All points in the xy–plane; Range: $-1 \leq z \leq 1$
   Level curves are circles with center $(0,0)$ and radii ≥ 0.

5. Domain: All point in the xy–plane; Range: All Real Numbers
   Level curves are hyperbolas with the x and y axes as asymptotes when $f(x,y) \neq 0$, and the x and y axes when $f(x,y) = 0$.

6. Domain: All $(x,y) \neq (0,y)$; Range: All Real Numbers
   Level curves are parabolas with vertex $(0,0)$ and the y–axis as axis of the parabola.

7. Domain: All points in the xy–plane; Range: $z \geq 0$
   Level curves are ellipses with center $(0,0)$ and major and minor axes along the x and y axes.

8. Domain: All points on the xy–plane; Range: All Real Numbers
   Level curves are hyperbolas centered at $(0,0)$ and transverse and conjugate axes along the x and y axes.

9. 
a)

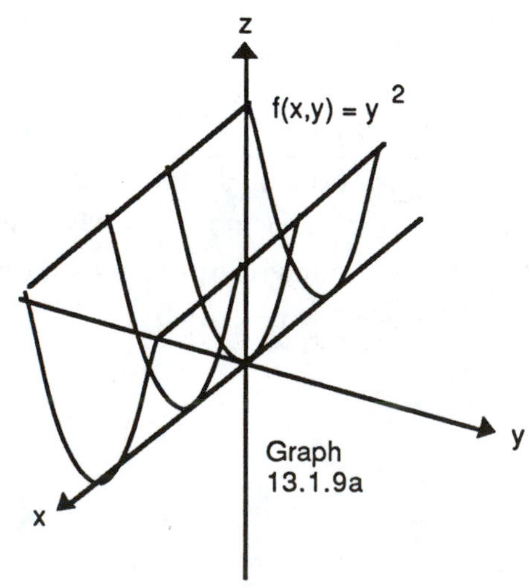

Graph 13.1.9a

b)

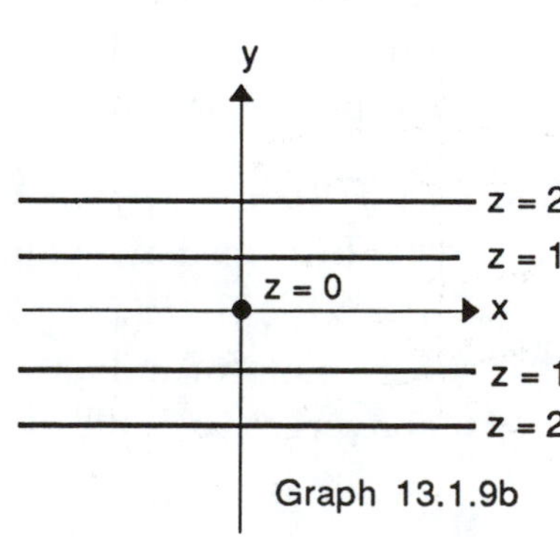

Graph 13.1.9b

10. a)

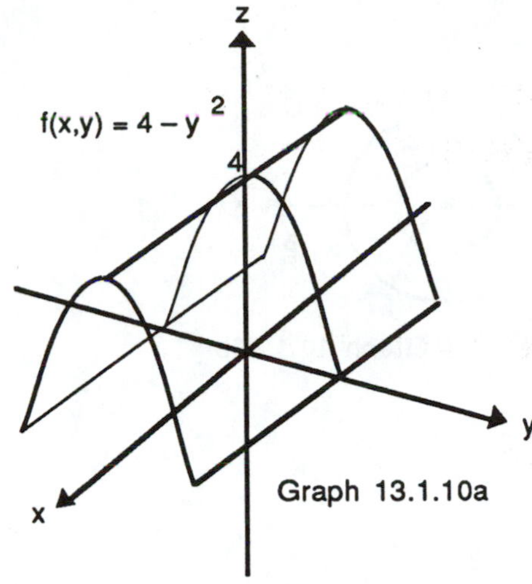

Graph 13.1.10a

b)

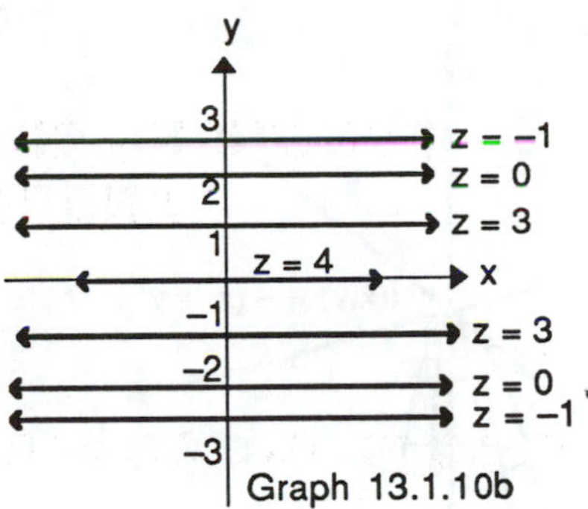

Graph 13.1.10b

11. a)

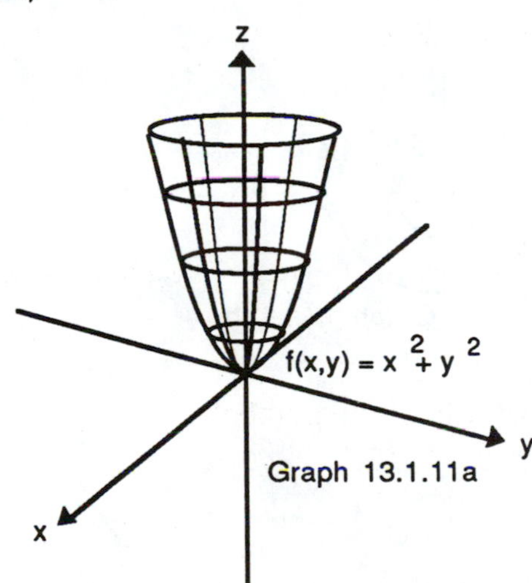

Graph 13.1.11a

b)

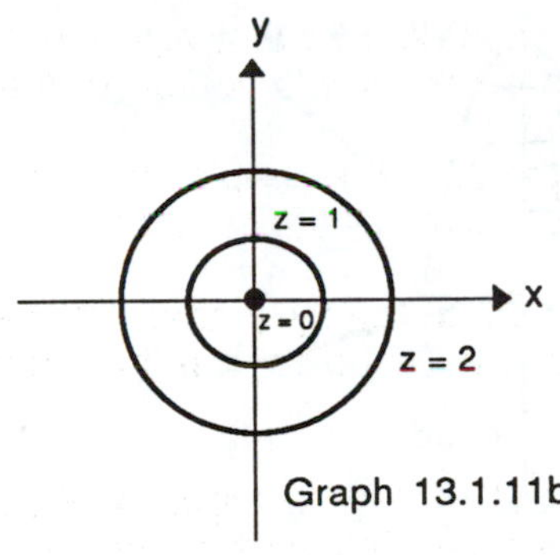

Graph 13.1.11b

12. a)

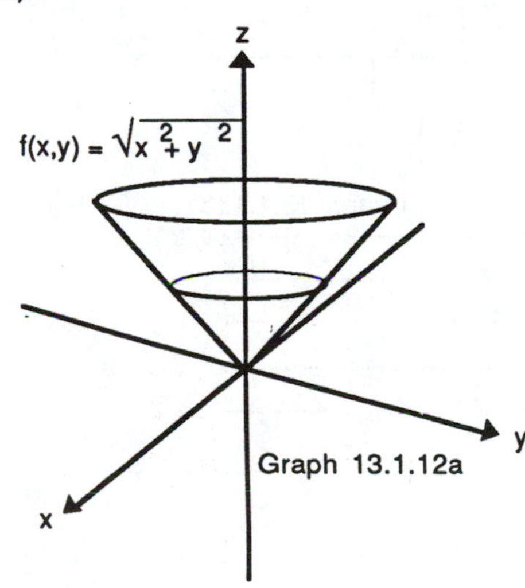

Graph 13.1.12a

b)

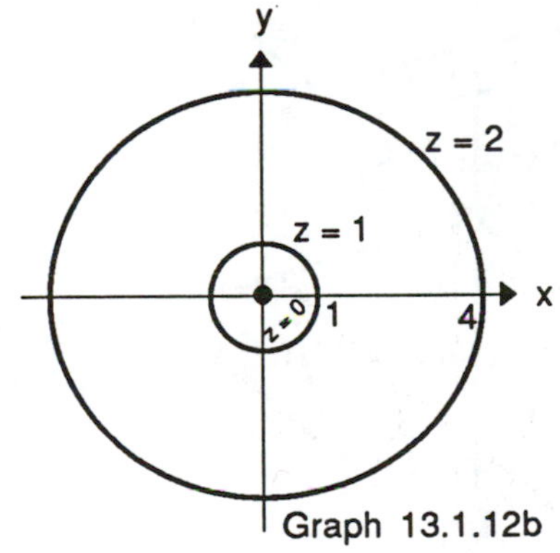

Graph 13.1.12b

13. a)

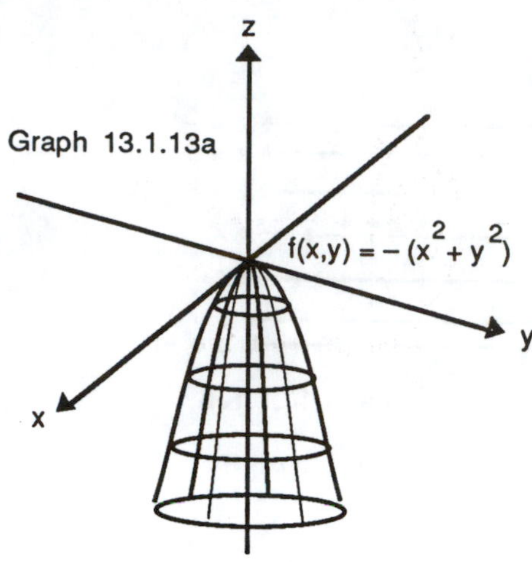

b)

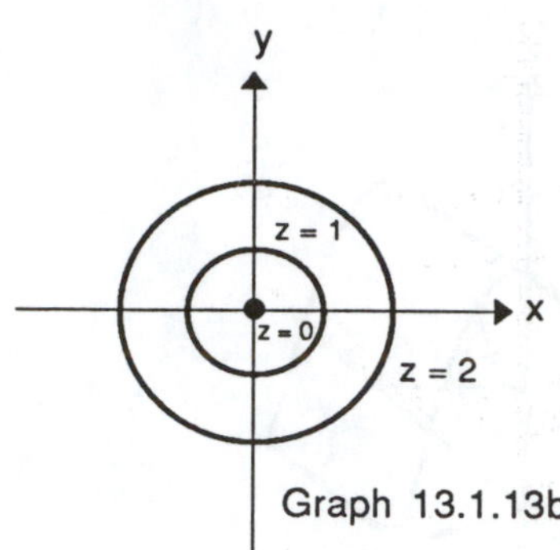

14. a)

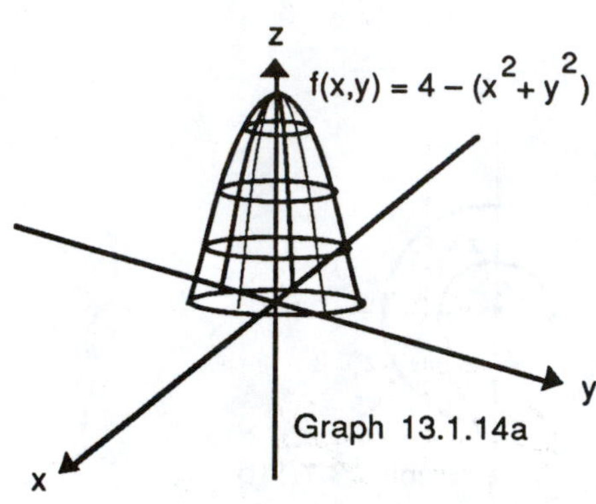

b)

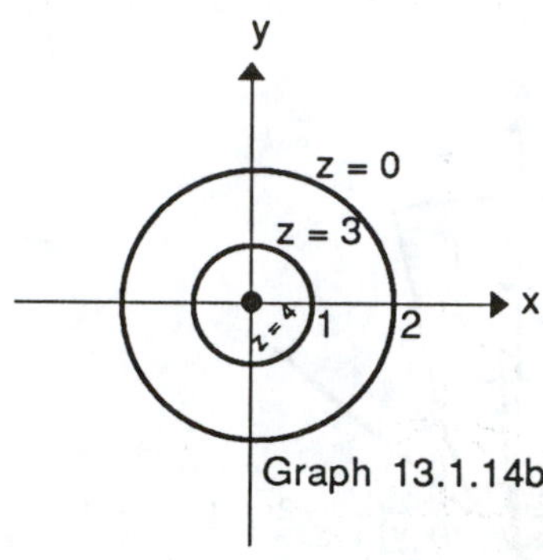

15. a)

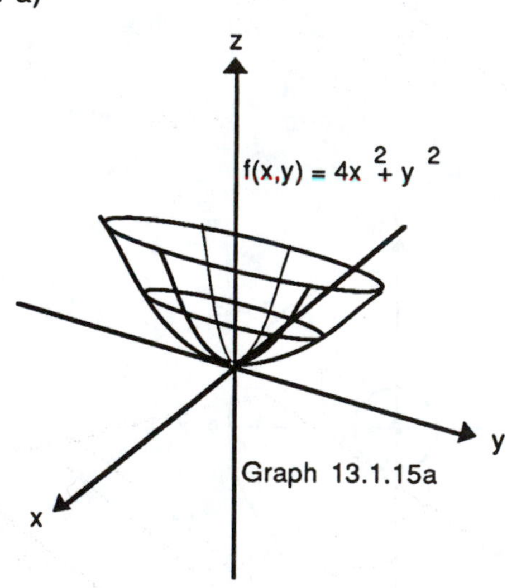

b)

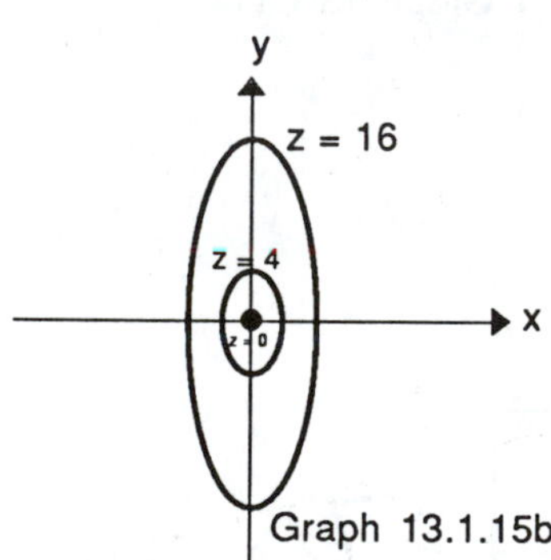

16. a)

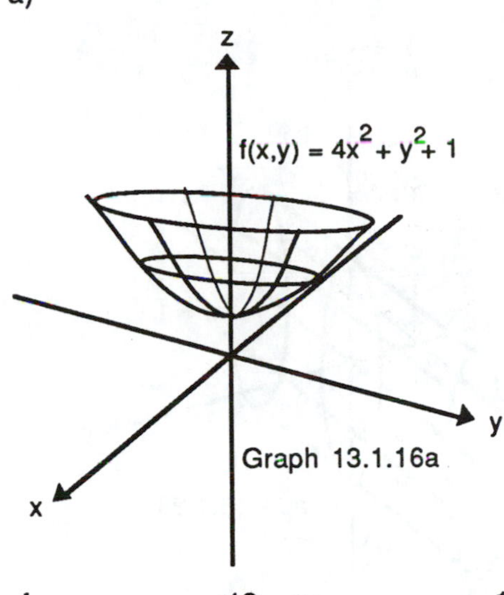

Graph 13.1.16a

b)

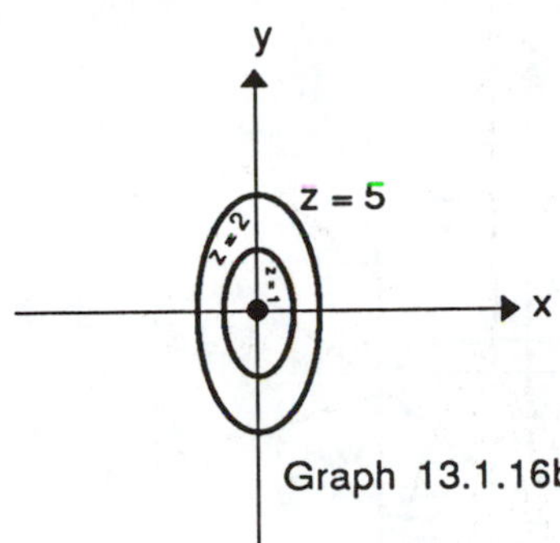

Graph 13.1.16b

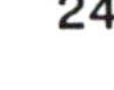

17. f   18. e   19. a   20. c   21. d   22. b

23.

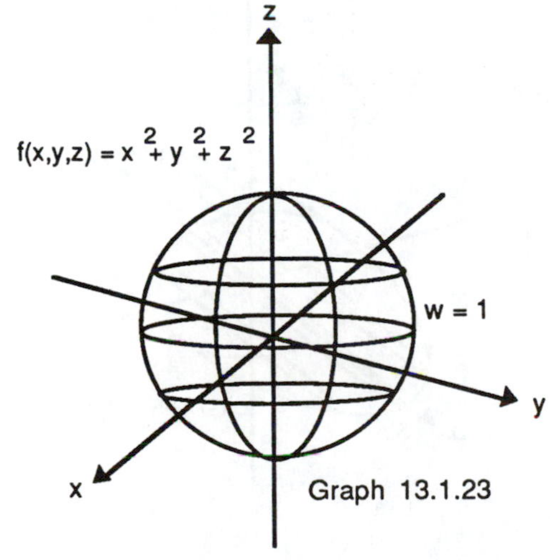

Graph 13.1.23

24.

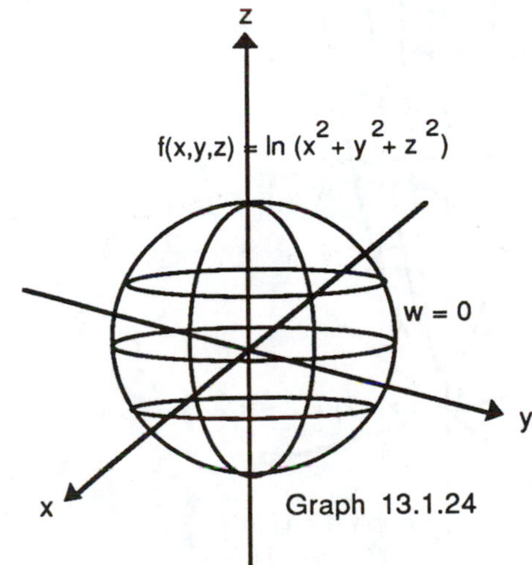

Graph 13.1.24

25.

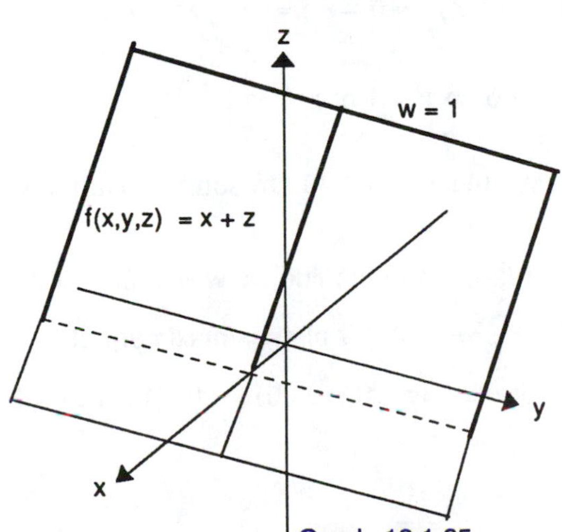

Graph 13.1.25

26.

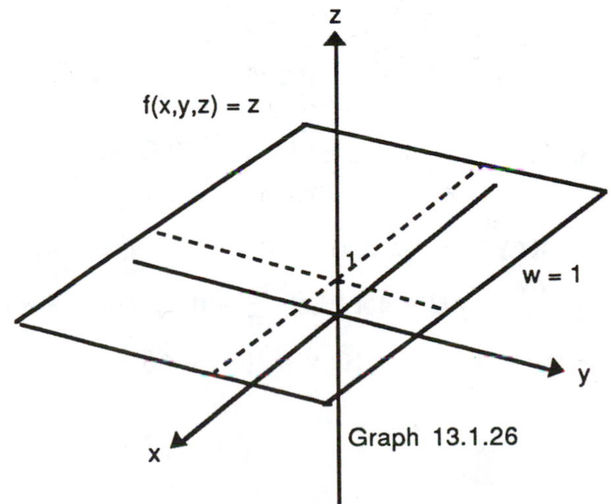

Graph 13.1.26

27.

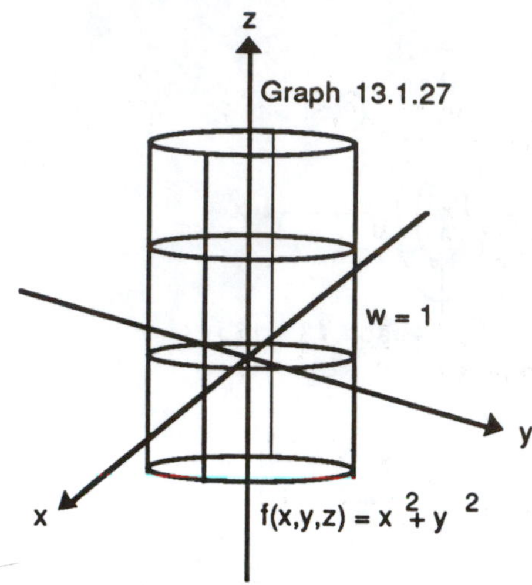

28.

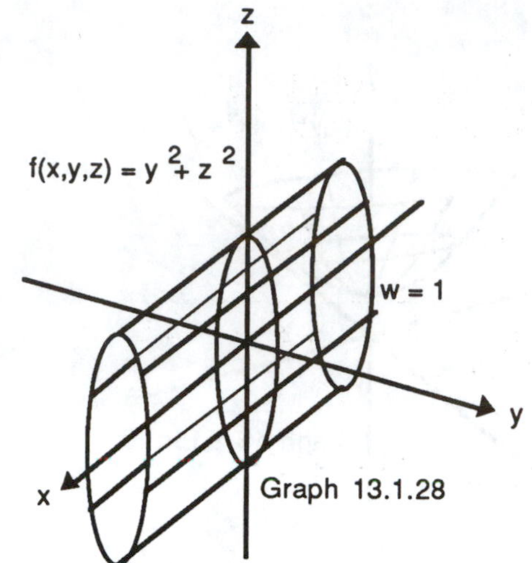

29.

w = 1

1

f(x,y,z) = z − x 2 − y 2

y

x

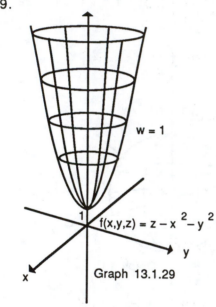

Graph 13.1.29

30.

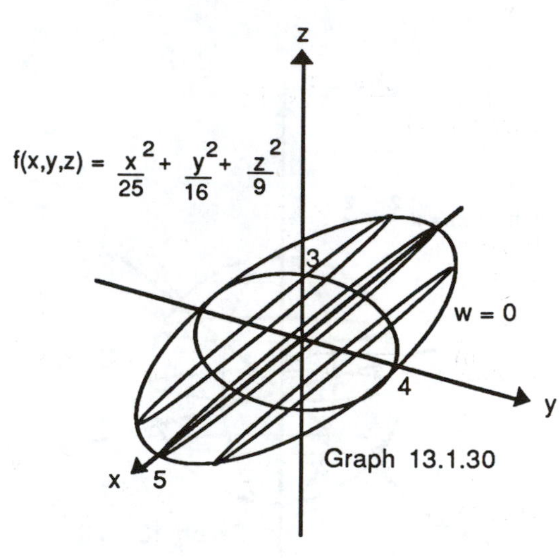

31. $f(x,y) = 16 - x^2 - y^2$ and $(2\sqrt{2},\sqrt{2}) \Rightarrow z = 16 - \left(2\sqrt{2}\right)^2 - \left(\sqrt{2}\right)^2 = 6 \Rightarrow 6 = 16 - x^2 - y^2 \Rightarrow x^2 + y^2 = 10$

32. $f(x,y) = \sqrt{x^2 - 1}$ and $(1,0) \Rightarrow z = \sqrt{1^2 - 1} = 0 \Rightarrow x^2 - 1 = 0 \Rightarrow x = 1$ or $x = -1$

33. $w = 4\left(\frac{Th}{d}\right)^{1/2} = 4\left(\frac{(290\text{ k})(16.8\text{ km})}{5\text{ k/km}}\right)^{1/2} = 124.86$ km. $\therefore$ must be 62.43 km south of Nantucket.

34. $f(x,y,z) = xyz$ and $x = 20 - t$, $y = t$, $z = 20 \Rightarrow w = (20 - t)(t)(20)$ along the line $\Rightarrow w = 400t - 20t^2 \Rightarrow \frac{dw}{dt} = 400 - 40t$. Let $\frac{dw}{dt} = 0 \Rightarrow 400 - 40t = 0 \Rightarrow t = 10$. $\frac{d^2w}{dt^2} = -40$ for all $t \Rightarrow$ maximum at $t = 10 \Rightarrow x = 20 - 10 = 10$, $y = 10$, $z = 20 \Rightarrow$ maximum of f along the line is $f(10,10,20) = (10)(10)(20) = 2000$.

SECTION 13.2 LIMITS AND CONTINUITY

1. $\lim_{(x,y)\to(0,0)} \frac{3x^2 - y^2 + 5}{x^2 + y^2 + 2} = \frac{5}{2}$

2. $\lim_{(x,y)\to(1,1)} \ln|1 + x^2y^2| = \ln 2$

3. $\lim_{(x,y)\to(0,\ln 2)} e^{x-y} = \frac{1}{2}$

4. $\lim_{(x,y)\to(0,4)} \frac{x}{\sqrt{y}} = 0$

5. $\lim_{P\to(1,3,4)} \sqrt{x^2 + y^2 + z^2 - 1} = 5$

6. $\lim_{P\to(1,2,6)} \left(\frac{1}{x} + \frac{1}{y} + \frac{1}{z}\right) = \frac{5}{3}$

7. $\lim_{(x,y)\to(0,\pi/4)} \sec x \tan y = 1$

8. $\lim_{(x,y)\to(0,0)} \cos\left(\frac{x^2 + y^2}{x + y + 1}\right) = 1$

9. $\lim_{(x,y)\to(1,1)} \cos\left(\sqrt[3]{|xy| - 1}\right) = 1$

10. $\lim_{(x,y)\to(1,0)} \frac{x \sin y}{x^2 + 1} = 0$

11. $\lim_{(x,y)\to(0,0)} \frac{e^y \sin x}{x} = 1$

12. $\lim_{(x,y)\to(0,0)} \tan^{-1}\left(\frac{1}{\sqrt{x^2 + y^2}}\right) = \frac{\pi}{2}$

13. $\lim_{\substack{(x,y)\to(1,1)\\ x \neq y}} \frac{x^2 - 2xy + y^2}{x - y} = \lim_{(x,y)\to(1,1)} \frac{(x - y)^2}{x - y} = \lim_{(x,y)\to(1,1)} (x - y) = 0$

14. $\lim_{\substack{(x,y)\to(1,1)\\ x \neq y}} \frac{x^2 - y^2}{x - y} = \lim_{(x,y)\to(1,1)} (x + y) = 2$

15. $\lim_{\substack{(x,y)\to(1,1)\\ x \neq 1}} \frac{xy - y - 2x + 2}{x - 1} = \lim_{(x,y)\to(1,1)} \frac{(x - 1)(y - 2)}{x - 1} = \lim_{(x,y)\to(1,1)} (y - 2) = -1$

16. $\lim_{\substack{(x,y)\to(2,-4)\\ y \neq 4,\ x \neq x^2}} \frac{y + 4}{x^2y - xy + 4x^2 - 4x} = \lim_{(x,y)\to(2,-4)} \frac{y + 4}{x(x - 1)(y + 4)} = \lim_{(x,y)\to(2,-4)} \frac{1}{x(x - 1)} = \frac{1}{2}$

17. $\lim_{P\to(2,3,-6)} \sqrt{x^2 + y^2 + z^2} = 7$

18. $\lim_{P\to(0,-2,0)} \ln\sqrt{x^2 + y^2 + z^2} = \ln 2$

19. $\lim_{P\to(3,3,0)} \left(\sin^2 x + \cos^2 y + \sec^2 z\right) = \sin^2 3 + \cos^2 3 + \sec^2 0 = 2$

20. $\lim_{P\to(\pi,0,3)} ze^{-2y} \cos 2x = (3)e^{-2(0)} \cos 2\pi = 3$

21. $\lim_{P\to(-1/4,\pi/2,2)} \tan^{-1}(xyz) = \tan^{-1}\left(-\frac{\pi}{4}\right)$

22. $\lim_{P\to(1,-1,-1)} \frac{2xy + yz}{x^2 + z^2} = -\frac{1}{2}$

23. a) Continuous at all (x,y)
    b) Continuous at all (x,y) except (0,0)

24. a) Continuous at all (x,y) so that $x \neq y$
    b) Continuous at all (x,y)

25. a) Continuous at all (x,y) except where x = 0 or y = 0
    b) Continuous at all (x,y)

26. a) Continuous at all (x,y) so that $x^2 - 3x + 2 \neq 0 \Rightarrow (x-2)(x-1) \neq 0 \Rightarrow x \neq 2$ and $x \neq 1$

b) Continuous at all (x,y) so that $y \neq x^2$

27. a) Continuous at all (x,y,z)

b) Continuous at all (x,y,z) except the interior of the cylinder $x^2 + y^2 = 1$

28. a) Continuous at all (x,y,z) so that $xyz > 0$

b) Continuous at all (x,y,z)

29. a) Continuous at all (x,y,z) so that $(x,y,z) \neq (x,y,0)$

b) Continuous at all (x,y,z) except those on the sphere $x^2 + y^2 + z^2 = 1$

30. a) Continuous at all (x,y,z) except (0,0,0)

b) Continuous at all (x,y,z) except those of the form (0,y,0) or (x,0,0), that is all (x,y,z) except those on the x and y axes

31. $$\lim_{\substack{(x,y)\to(0,0)\\ \text{along } y=x,\\ x>0}} \frac{x}{\sqrt{x^2+y^2}} = \lim_{(x,y)\to(0,0)} \frac{x}{\sqrt{x^2+x^2}} = \lim_{(x,y)\to(0,0)} \frac{x}{\sqrt{2}\,|x|} = \lim_{(x,y)\to(0,0)} \frac{x}{\sqrt{2}\,x} = \lim_{(x,y)\to(0,0)} \frac{1}{\sqrt{2}} = \frac{1}{\sqrt{2}}$$

$$\lim_{\substack{(x,y)\to(0,0)\\ \text{along } y=x,\\ x<0}} \frac{x}{\sqrt{x^2+y^2}} = \lim_{(x,y)\to(0,0)} \frac{x}{\sqrt{2}\,|x|} = \lim_{(x,y)\to(0,0)} \frac{x}{\sqrt{2}(-x)} = \lim_{(x,y)\to(0,0)} -\frac{1}{\sqrt{2}} = -\frac{1}{\sqrt{2}}$$

$\therefore$ consider paths along $y = x$ where $x > 0$ or $x < 0$.

32. $$\lim_{\substack{(x,y)\to(0,0)\\ \text{along } y=0}} \frac{x^4}{x^4+y^2} = \lim_{(x,y)\to(0,0)} \frac{x^4}{x^4+0^2} = 1. \quad \lim_{\substack{(x,y)\to(0,0)\\ \text{along } y=x^2}} \frac{x^4}{x^4+y^2} = \lim_{(x,y)\to(0,0)} \frac{x^4}{x^4+\left(x^2\right)^2} = \frac{1}{2}$$

$\therefore$ consider paths along $y = 0$ and $y = x^2$

33. $$\lim_{\substack{(x,y)\to(0,0)\\ \text{along } y=kx^2}} \frac{x^4-y^2}{x^4+y^2} = \lim_{(x,y)\to(0,0)} \frac{x^4-\left(kx^2\right)^2}{x^4+\left(kx^2\right)^2} = \lim_{(x,y)\to(0,0)} \frac{x^4-k^2x^4}{x^4+k^2x^4} = \frac{1-k^2}{1+k^2} \Rightarrow$$ different limits for different values of k. $\therefore$ consider paths along $y = kx^2$, k a constant.

34. $$\lim_{\substack{(x,y)\to(0,0)\\ \text{along } y=kx,\ k\neq 0}} \frac{xy}{|xy|} = \lim_{(x,y)\to(0,0)} \frac{x(kx)}{|x(kx)|} = \lim_{(x,y)\to(0,0)} \frac{kx^2}{|kx^2|} = \lim_{(x,y)\to(0,0)} \frac{k}{|k|}.$$ If $k > 0$, the limit is 1; if $k < 0$, the limit is -1. $\therefore$ consider paths along $y = kx$, $k \neq 0$, k a constant.

35. $$\lim_{\substack{(x,y)\to(0,0)\\ \text{along } y=kx,\ k\neq -1}} \frac{x-y}{x+y} = \lim_{(x,y)\to(0,0)} \frac{x-kx}{x+kx} = \frac{1-k}{1+k} \Rightarrow$$ different limits for different values of k. $\therefore$ consider paths along $y = kx$, k a constant, $k \neq -1$.

36. $$\lim_{\substack{(x,y)\to(0,0)\\ \text{along } y=kx,\ k\neq 1}} \frac{x+y}{x-y} = \lim_{(x,y)\to(0,0)} \frac{x+kx}{x-kx} = \frac{1+k}{1-k} \Rightarrow$$ different limits for different values of k. $\therefore$ consider paths along $y = kx$, k a constant, $k \neq 1$

37. $\lim_{(x,y)\to(0,0)} \frac{x^2+y}{y} = \lim_{\substack{(x,y)\to(0,0) \\ \text{along } y = kx^2,\ k \neq 0}} \frac{x^2+kx^2}{kx^2} = \frac{1+k}{k} \Rightarrow$ different limits for different values of k. $\therefore$ consider paths along $y = kx^2$, k a constant, $k \neq 0$.

38. $\lim_{(x,y)\to(0,0)} \frac{x^2}{x^2-y} = \lim_{\substack{(x,y)\to(0,0) \\ \text{along } y = kx^2,\ k \neq 1}} \frac{x^2}{x^2-kx^2} = \frac{1}{1-k} \Rightarrow$ different limits for different values of k. $\therefore$ consider paths along $y = kx^2$, k a constant, $k \neq 0$.

39. a) $f(x,y)\big|_{y = mx} = \frac{2m}{1+m^2} = \frac{2\tan\theta}{1+\tan^2\theta} = \sin 2\theta$. The value of f(x,y) is $\sin 2\theta$ where $\tan\theta = m$ along $y = mx$.

b) Since $f(x,y)\big|_{y = mx} = \sin 2\theta$ and since $-1 \le \sin 2\theta \le 1$ for every θ, $\lim_{(x,y)\to(0,0)} f(x,y)$ varies from -1 to 1 along $y = mx$.

SECTION 13.3 PARTIAL DERIVATIVES

1. $\frac{\partial f}{\partial x} = 2, \frac{\partial f}{\partial y} = 0$

2. $\frac{\partial f}{\partial x} = 0, \frac{\partial f}{\partial y} = -3$

3. $\frac{\partial f}{\partial x} = 0, \frac{\partial f}{\partial y} = 0$

4. $\frac{\partial f}{\partial x} = 2, \frac{\partial f}{\partial y} = -3$

5. $\frac{\partial f}{\partial x} = y-1, \frac{\partial f}{\partial y} = x$

6. $\frac{\partial f}{\partial x} = 2x, \frac{\partial f}{\partial y} = 2y$

7. $\frac{\partial f}{\partial x} = 2x-y, \frac{\partial f}{\partial y} = -x+2y$

8. $\frac{\partial f}{\partial x} = y+3, \frac{\partial f}{\partial y} = x+2$

9. $\frac{\partial f}{\partial x} = 5y-14x+3,$ $\frac{\partial f}{\partial y} = 5x-2y-6$

10. $\frac{\partial f}{\partial x} = \frac{-1}{(x+y)^2}, \frac{\partial f}{\partial y} = \frac{-1}{(x+y)^2}$

11. $\frac{\partial f}{\partial x} = x(x^2+y^2)^{-1/2}, \frac{\partial f}{\partial y} = y(x^2+y^2)^{-1/2}$

12. $\frac{\partial f}{\partial x} = \frac{y^2-x^2}{(x^2+y^2)^2}, \frac{\partial f}{\partial y} = \frac{-2xy}{(x^2+y^2)^2}$

13. $\frac{\partial f}{\partial x} = \frac{-y^2-1}{(xy-1)^2}, \frac{\partial f}{\partial y} = \frac{-x^2-1}{(xy-1)^2}$

14. $\frac{\partial f}{\partial x} = -e^x + e^x(y-x), \frac{\partial f}{\partial y} = e^x$

15. $\frac{\partial f}{\partial x} = e^x \ln y, \frac{\partial f}{\partial y} = \frac{e^x}{y}$

16. $\frac{\partial f}{\partial x} = \cos(x+y), \frac{\partial f}{\partial y} = \cos(x+y)$

17. $\frac{\partial f}{\partial x} = e^x \sin(y+1), \frac{\partial f}{\partial y} = e^x \cos(y+1)$

18. $\frac{\partial f}{\partial x} = \frac{-y}{y^2+x^2}, \frac{\partial f}{\partial y} = \frac{x}{y^2+x^2}$

19. $f_x(x,y,z) = y+z$, $f_y(x,y,z) = x+z$, $f_z(x,y,z) = y+x$

20. $f_x(x,y,z) = -x(x^2+y^2+z^2)^{-3/2}$, $f_y(x,y,z) = -y(x^2+y^2+z^2)^{-3/2}$, $f_z(x,y,z) = -z(x^2+y^2+z^2)^{-3/2}$

21. $f_x(x,y,z) = 0$, $f_y(x,y,z) = 2y$, $f_z(x,y,z) = 4z$

22. $f_x(x,y,z) = 1$, $f_y(x,y,z) = -y(y^2+z^2)^{-1/2}$,$f_z(x,y,z) = -z(y^2+z^2)^{-1/2}$

23. $f_x(x,y,z) = \cos(x + yz)$, $f_y(x,y,z) = z\cos(x + yz)$, $f_z(x,y,z) = y\cos(x + yz)$

24. $f_x(x,y,z) = \sec^2(x + y + z)$, $f_y(x,y,z) = \sec^2(x + y + z)$, $f_z(x,y,z) = \sec^2(x + y + z)$

25. $\frac{\partial f}{\partial t} = -2\pi\sin(2\pi t - \alpha)$, $\frac{\partial f}{\partial \alpha} = \sin(2\pi t - \alpha)$

26. $\frac{\partial g}{\partial u} = 2v^2 e^{2u/v}$, $\frac{\partial g}{\partial v} = 2ve^{2u/v} - 2ue^{2u/v}$

27. $\frac{\partial h}{\partial \rho} = \sin\phi\cos\theta$, $\frac{\partial h}{\partial \phi} = \rho\cos\phi\cos\theta$, $\frac{\partial h}{\partial \theta} = -\rho\sin\phi\sin\theta$

28. $\frac{\partial g}{\partial r} = 1 - \cos\theta$, $\frac{\partial g}{\partial \theta} = r\sin\theta$, $\frac{\partial g}{\partial z} = -1$

29. $\frac{\partial W}{\partial P} = v$, $\frac{\partial W}{\partial V} = P + \frac{\rho v^2}{2g}$, $\frac{\partial W}{\partial \rho} = \frac{Vv^2}{2g}$, $\frac{\partial W}{\partial v} = \frac{V\rho v}{g}$, $\frac{\partial W}{\partial g} = -\frac{V\rho v^2}{2g^2}$

30. $\frac{\partial A}{\partial c} = m$, $\frac{\partial A}{\partial h} = \frac{q}{2}$, $\frac{\partial A}{\partial k} = \frac{m}{q}$, $\frac{\partial A}{\partial m} = \frac{k}{q} + c$, $\frac{\partial A}{\partial q} = -\frac{km}{q^2} + \frac{h}{2}$

31. $\frac{\partial f}{\partial x} = y + z$, $\frac{\partial f}{\partial y} = x + z$, $\frac{\partial^2 f}{\partial x^2} = 0$, $\frac{\partial^2 f}{\partial y^2} = 0$, $\frac{\partial^2 f}{\partial y \partial x} = \frac{\partial^2 f}{\partial x \partial y} = 1$

32. $\frac{\partial f}{\partial x} = y\cos xy$, $\frac{\partial f}{\partial y} = x\cos xy$, $\frac{\partial^2 f}{\partial x^2} = -y^2\sin xy$, $\frac{\partial^2 f}{\partial y^2} = -x^2\sin xy$, $\frac{\partial^2 f}{\partial y \partial x} = \frac{\partial^2 f}{\partial x \partial y} = \cos xy - xy\sin xy$

33. $\frac{\partial g}{\partial x} = 2xy + y\cos x$, $\frac{\partial g}{\partial y} = x^2 - \sin y + \sin x$, $\frac{\partial^2 g}{\partial x^2} = 2y - y\sin x$, $\frac{\partial^2 g}{\partial y^2} = -\cos y$, $\frac{\partial^2 g}{\partial y \partial x} = \frac{\partial^2 g}{\partial x \partial y} = 2x + \cos x$

34. $\frac{\partial h}{\partial x} = e^y$, $\frac{\partial h}{\partial y} = xe^y + 1$, $\frac{\partial^2 h}{\partial x^2} = 0$, $\frac{\partial^2 h}{\partial y^2} = xe^y$, $\frac{\partial^2 h}{\partial y \partial x} = \frac{\partial^2 h}{\partial x \partial y} = e^y$

35. $\frac{\partial r}{\partial x} = \frac{1}{x + y}$, $\frac{\partial r}{\partial y} = \frac{1}{x + y}$, $\frac{\partial^2 r}{\partial x^2} = \frac{-1}{(x + y)^2}$, $\frac{\partial^2 r}{\partial y^2} = \frac{-1}{(x + y)^2}$, $\frac{\partial^2 r}{\partial y \partial x} = \frac{\partial^2 r}{\partial x \partial y} = \frac{-1}{(x + y)^2}$

36. $\frac{\partial s}{\partial x} = \frac{-y}{y^2 + x^2}$, $\frac{\partial s}{\partial y} = \frac{x}{y^2 + x^2}$, $\frac{\partial^2 s}{\partial x^2} = \frac{2xy}{(y^2 + x^2)^2}$, $\frac{\partial^2 s}{\partial y^2} = \frac{-2xy}{(y^2 + x^2)^2}$, $\frac{\partial^2 s}{\partial y \partial x} = \frac{\partial^2 s}{\partial x \partial y} = \frac{y^2 - x^2}{(y^2 + x^2)^2}$

37. $\frac{\partial w}{\partial x} = \frac{2}{2x + 3y}$, $\frac{\partial w}{\partial y} = \frac{3}{2x + 3y}$, $\frac{\partial^2 w}{\partial y \partial x} = \frac{-6}{(2x + 3y)^2}$ and $\frac{\partial^2 w}{\partial x \partial y} = \frac{-6}{(2x + 3y)^2}$

38. $\frac{\partial w}{\partial x} = e^x + \ln y + \frac{y}{x}$, $\frac{\partial w}{\partial y} = \frac{x}{y} + \ln x$, $\frac{\partial^2 w}{\partial y \partial x} = \frac{1}{y} + \frac{1}{x}$ and $\frac{\partial^2 w}{\partial x \partial y} = \frac{1}{y} + \frac{1}{x}$

39. $\frac{\partial w}{\partial x} = y^2 + 2xy^3 + 3x^2y^4$, $\frac{\partial w}{\partial y} = 2xy + 3x^2y^2 + 4x^3y^3$, $\frac{\partial^2 w}{\partial y \partial x} = 2y + 6xy^2 + 12x^2y^3$ and
$\frac{\partial^2 w}{\partial x \partial y} = 2y + 6xy^2 + 12x^2y^3$

40. $\frac{\partial w}{\partial x} = \sin y + y\cos x + y$, $\frac{\partial w}{\partial y} = x\cos y + \sin x + x$, $\frac{\partial^2 w}{\partial y \partial x} = \cos y + \cos x + 1$ and $\frac{\partial^2 w}{\partial x \partial y} = \cos y + \cos x + 1$

41. a) x first b) y first c) x first d) x first e) y first f) y first

42. a) y first three times b) y first three times c) y first twice d) x first twice

43. $xy + z^3x - 2yz = 0 \Rightarrow y + 3z^2x\frac{\partial z}{\partial x} - 2y\frac{\partial z}{\partial x} = 0 \Rightarrow (3z^2x - 2y)\frac{\partial z}{\partial x} = -y - z^3 \Rightarrow \frac{\partial z}{\partial x} = \frac{-y - z^3}{3z^2x - 2y} \Rightarrow$

$\frac{\partial z}{\partial x}(1,1,1) = -2$

44. $xz + y\ln x + x^2 + 4 = 0 \Rightarrow z\frac{\partial x}{\partial z} + x + y\left(\frac{1}{x}\right)\frac{\partial x}{\partial z} + 2x\frac{\partial x}{\partial z} = 0 \Rightarrow \left(z + \frac{y}{x} + 2x\right)\frac{\partial x}{\partial z} = -x \Rightarrow$

$\frac{\partial x}{\partial z} = \frac{-x}{z + \frac{y}{x} + 2x} \Rightarrow \frac{\partial x}{\partial z}(1,-1,-3) = \frac{1}{2}$

45. $\frac{\partial f}{\partial x} = 2x, \frac{\partial f}{\partial y} = 2y, \frac{\partial f}{\partial z} = -4z \Rightarrow \frac{\partial^2 f}{\partial x^2} = 2, \frac{\partial^2 f}{\partial y^2} = 2, \frac{\partial^2 f}{\partial z^2} = -4 \Rightarrow \frac{\partial^2 f}{\partial x^2} + \frac{\partial^2 f}{\partial y^2} + \frac{\partial^2 f}{\partial z^2} = 2 + 2 + (-4) = 0$

46. $\frac{\partial f}{\partial x} = -6xz, \frac{\partial f}{\partial y} = -6yz, \frac{\partial f}{\partial z} = 6z^2 - 3(x^2 + y^2), \frac{\partial^2 f}{\partial x^2} = -6z, \frac{\partial^2 f}{\partial y^2} = -6z, \frac{\partial^2 f}{\partial z^2} = 12z$

$\therefore \frac{\partial^2 f}{\partial x^2} + \frac{\partial^2 f}{\partial y^2} + \frac{\partial^2 f}{\partial z^2} = -6z - 6z + 12z = 0$

47. $\frac{\partial f}{\partial x} = -2e^{-2y}\sin 2x, \frac{\partial f}{\partial y} = -2e^{-2y}\cos 2x, \frac{\partial^2 f}{\partial x^2} = -4e^{-2y}\cos 2x, \frac{\partial^2 f}{\partial y^2} = 4e^{-2y}\cos 2x$

$\therefore \frac{\partial^2 f}{\partial x^2} + \frac{\partial^2 f}{\partial y^2} = -4e^{-2y}\cos 2x + 4e^{-2y}\cos 2x = 0$

48. $\frac{\partial f}{\partial x} = \frac{x}{x^2 + y^2}, \frac{\partial f}{\partial y} = \frac{y}{x^2 + y^2}, \frac{\partial^2 f}{\partial x^2} = \frac{y^2 - x^2}{(x^2 + y^2)^2}, \frac{\partial^2 f}{\partial y^2} = \frac{x^2 - y^2}{(x^2 + y^2)^2}$

$\therefore \frac{\partial^2 f}{\partial x^2} + \frac{\partial^2 f}{\partial y^2} = \frac{y^2 - x^2}{(x^2 + y^2)^2} + \frac{x^2 - y^2}{(x^2 + y^2)^2} = 0$

49. $\frac{\partial f}{\partial x} = -\frac{1}{2}(x^2 + y^2 + z^2)^{-3/2}(2x) = -x(x^2 + y^2 + x^2)^{-3/2}, \frac{\partial f}{\partial y} = -\frac{1}{2}(x^2 + y^2 + z^2)^{-3/2}(2y) = -y(x^2 + y^2 + z^2)^{-3/2}$

$\frac{\partial f}{\partial z} = -\frac{1}{2}(x^2 + y^2 + z^2)^{-3/2}(2z) = -z(x^2 + y^2 + z^2)^{-3/2}. \frac{\partial^2 f}{\partial x^2} = -(x^2 + y^2 + z^2)^{-3/2} + 3x^2(x^2 + y^2 + z^2)^{-5/2}$

$\frac{\partial^2 f}{\partial y^2} = -(x^2 + y^2 + z^2)^{-3/2} + 3y^2(x^2 + y^2 + z^2)^{-5/2}, \frac{\partial^2 f}{\partial z^2} = -(x^2 + y^2 + z^2)^{-3/2} + 3z^2(x^2 + y^2 + z^2)^{-5/2}$

$\therefore \frac{\partial^2 f}{\partial x^2} + \frac{\partial^2 f}{\partial y^2} + \frac{\partial^2 f}{\partial z^2} = \left(-(x^2 + y^2 + z^2)^{-3/2} + 3x^2(x^2 + y^2 + z^2)^{-5/2}\right) +$

$\left(-(x^2 + y^2 + z^2)^{-3/2} + 3y^2(x^2 + y^2 + z^2)^{-5/2}\right) + \left(-(x^2 + y^2 + z^2)^{-3/2} + 3z^2(x^2 + y^2 + z^2)^{-5/2}\right) =$

$-3(x^2 + y^2 + z^2)^{-3/2} + (3x^2 + 3y^2 + 3z^2)(x^2 + y^2 + z^2)^{-5/2} = 0$

50. $\frac{\partial f}{\partial x} = 3e^{3x+4y}\cos 5z, \frac{\partial f}{\partial y} = 4e^{3x+4y}\cos 5z, \frac{\partial f}{\partial z} = -5e^{3x+4y}\sin 5z. \frac{\partial^2 f}{\partial x^2} = 9e^{3x+4y}\cos 5z,$

$\frac{\partial^2 f}{\partial y^2} = 16e^{3x+4y}\cos 5z, \frac{\partial^2 f}{\partial z^2} = -25e^{3x+4y}\cos 5z \Rightarrow \frac{\partial^2 f}{\partial x^2} + \frac{\partial^2 f}{\partial y^2} + \frac{\partial^2 f}{\partial z^2} = 9e^{3x+4y}\cos 5z + 16e^{3x+4y}\cos 5z -$

$25e^{3x+4y}\cos 5z = 0$

51. $\frac{\partial w}{\partial x} = \cos(x + ct)$, $\frac{\partial w}{\partial t} = c\cos(x + ct)$. $\frac{\partial^2 w}{\partial x^2} = -\sin(x + ct)$, $\frac{\partial^2 w}{\partial t^2} = -c^2\sin(x + ct)$
$\therefore \frac{\partial^2 w}{\partial t^2} = c^2(-\sin(x + ct)) = c^2\frac{\partial^2 w}{\partial x^2}$

52. $\frac{\partial w}{\partial x} = -2\sin(2x + 2ct)$, $\frac{\partial w}{\partial t} = -2c\sin(2x + 2ct)$. $\frac{\partial^2 w}{\partial x^2} = -4\cos(2x + 2ct)$, $\frac{\partial^2 w}{\partial t^2} = -4c^2\cos(2x + 2ct)$
$\therefore \frac{\partial^2 w}{\partial t^2} = c^2(-4\cos(2x + 2ct)) = c^2\frac{\partial^2 w}{\partial x^2}$

53. $\frac{\partial w}{\partial x} = \cos(x + ct) - 2\sin(2x + 2ct)$, $\frac{\partial w}{\partial t} = c\cos(x + ct) - 2c\sin(2x + 2ct)$. $\frac{\partial^2 w}{\partial x^2} = -\sin(x + ct) -$
$4\cos(2x + 2ct)$, $\frac{\partial^2 w}{\partial t^2} = -c^2\sin(x + ct) - 4c^2\cos(2x + 2ct)$ $\therefore$ $\frac{\partial^2 w}{\partial t^2} = c^2(-\sin(x + ct) - 4\cos(2x + 2ct)) =$
$c^2\frac{\partial^2 w}{\partial x^2}$

54. $\frac{\partial w}{\partial x} = \frac{1}{x + ct}$, $\frac{\partial w}{\partial t} = \frac{c}{x + ct}$. $\frac{\partial^2 w}{\partial x^2} = \frac{-1}{(x + ct)^2}$, $\frac{\partial^2 w}{\partial t^2} = \frac{-c^2}{(x + ct)^2}$. $\therefore$ $\frac{\partial^2 w}{\partial t^2} = c^2\left(\frac{-1}{(x + ct)^2}\right) = c^2\frac{\partial^2 w}{\partial x^2}$

55. $\frac{\partial w}{\partial x} = 2\sec^2(2x - 2ct)$, $\frac{\partial w}{\partial t} = -2c\sec^2(2x - 2ct)$. $\frac{\partial^2 w}{\partial x^2} = 8\sec^2(2x - 2ct)\tan(2x - 2ct)$,
$\frac{\partial^2 w}{\partial t^2} = 8c^2\sec^2(2x - 2ct)\tan(2x - 2ct)$ $\therefore$ $\frac{\partial^2 w}{\partial t^2} = c^2\left(8\sec^2(2x - 2ct)\tan(2x - 2ct)\right) = c^2\frac{\partial^2 w}{\partial x^2}$

56. $\frac{\partial w}{\partial x} = -15\sin(3x + 3ct) + e^{x+ct}$, $\frac{\partial w}{\partial t} = -15c\sin(3x + 3ct) + ce^{x+ct}$. $\frac{\partial^2 w}{\partial x^2} = -45\cos(3x + 3ct) + e^{x+ct}$
$\frac{\partial^2 w}{\partial t^2} = -45c^2\cos(3x + 3ct) + c^2e^{x+ct}$ $\therefore$ $\frac{\partial^2 w}{\partial t^2} = c^2\left(-45\cos(3x + 3ct) + e^{x+ct}\right) = c^2\frac{\partial^2 w}{\partial x^2}$

SECTION 13.4 THE CHAIN RULE

1. $\frac{\partial w}{\partial x} = 2x$, $\frac{\partial w}{\partial y} = 2y$, $\frac{dx}{dt} = -\sin t$, $\frac{dy}{dt} = \cos t \Rightarrow \frac{dw}{dt} = -2x\sin t + 2y\cos t = -2\cos t\sin t + 2\sin t\cos t = 0$
$\Rightarrow \frac{dw}{dt}(0) = 0$

2. $\frac{\partial w}{\partial x} = 2x$, $\frac{\partial w}{\partial y} = 2y$, $\frac{dx}{dt} = -\sin t + \cos t$, $\frac{dy}{dt} = -\sin t - \cos t \Rightarrow \frac{dw}{dt} = 2x(-\sin t + \cos t) +$
$2y(-\sin t - \cos t) = 2(\cos t + \sin t)(\cos t - \sin t) - 2(\cos t - \sin t)(\sin t + \cos t) = 0 \Rightarrow \frac{dw}{dt}(0) = 0.$

3. $\frac{\partial w}{\partial x} = \frac{1}{z}$, $\frac{\partial w}{\partial y} = \frac{1}{z}$, $\frac{\partial w}{\partial z} = \frac{-(x + y)}{z^2}$, $\frac{dx}{dt} = -2\cos t\sin t$, $\frac{dy}{dt} = 2\sin t\cos t$, $\frac{dz}{dt} = -\frac{1}{t^2} \Rightarrow$
$\frac{dw}{dt} = -\frac{2}{z}\cos t\sin t + \frac{2}{z}\sin t\cos t + \frac{x + y}{z^2t^2} = \frac{\cos^2 t + \sin^2 t}{\frac{1}{t^2}(t^2)} = 1 \Rightarrow \frac{dw}{dt}(3) = 1$

4. $\frac{\partial w}{\partial x} = \frac{2x}{x^2 + y^2 + z^2}, \frac{\partial w}{\partial y} = \frac{2y}{x^2 + y^2 + z^2}, \frac{\partial w}{\partial z} = \frac{2z}{x^2 + y^2 + z^2}, \frac{dx}{dt} = -\sin t, \frac{dy}{dt} = \cos t, \frac{dz}{dt} = 2t^{-1/2} \Rightarrow$

$\frac{dw}{dt} = \frac{-2x \sin t}{x^2 + y^2 + z^2} + \frac{2y \cos t}{x^2 + y^2 + z^2} + \frac{4zt^{-1/2}}{x^2 + y^2 + z^2} = \frac{-2 \sin t \cos t + 2 \sin t \cos t + 4\left(4t^{1/2}\right)t^{-1/2}}{\cos^2 t + \sin^2 t + 16t} = \frac{16}{1 + 16t} \Rightarrow$

$\frac{dw}{dt}(3) = \frac{16}{49}$

5. $\frac{\partial w}{\partial x} = 2ye^x, \frac{\partial w}{\partial y} = 2e^x, \frac{\partial w}{\partial z} = -\frac{1}{z}, \frac{dx}{dt} = \frac{2t}{t^2 + 1}, \frac{dy}{dt} = \frac{1}{t^2 + 1}, \frac{dz}{dt} = e^t \Rightarrow \frac{dw}{dt} = \frac{4yte^x}{t^2 + 1} + \frac{2e^x}{t^2 + 1} - \frac{e^t}{z} =$

$\frac{4t \tan^{-1} t\, e^{\ln(t^2+1)}}{t^2 + 1} + \frac{2(t^2 + 1)}{t^2 + 1} - \frac{e^t}{e^t} = 4t \tan^{-1} t + 1 \Rightarrow \frac{dw}{dt}(1) = \pi + 1$

6. $\frac{\partial w}{\partial x} = -y \cos xy, \frac{\partial w}{\partial y} = -x \cos xy, \frac{\partial w}{\partial z} = 1, \frac{dx}{dt} = 1, \frac{dy}{dt} = \frac{1}{t}, \frac{dz}{dt} = e^{t-1} \Rightarrow \frac{dw}{dt} = -y \cos xy - \frac{x \cos xy}{t} + e^{t-1} =$

$-\ln t(\cos(t \ln t)) - \frac{t \cos(t \ln t)}{t} + e^{t-1} = -\ln t(\cos(t \ln t)) - \cos(t \ln t) + e^{t-1} \Rightarrow \frac{dw}{dt}(1) = 0$

7. $\frac{dz}{dt} = \frac{\partial z}{\partial x}\frac{dx}{dt} + \frac{\partial z}{\partial y}\frac{dy}{dt}$

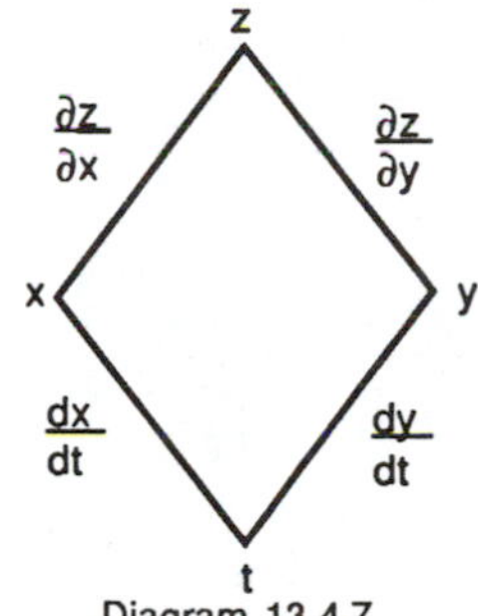

Diagram 13.4.7

8. $\frac{dz}{dt} = \frac{\partial z}{\partial u}\frac{du}{dt} + \frac{\partial z}{\partial v}\frac{dv}{dt} + \frac{\partial z}{\partial w}\frac{dw}{dt}$

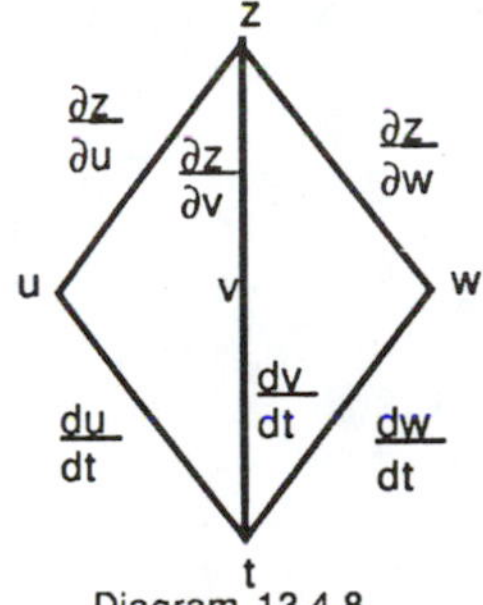

Diagram 13.4.8

9. $\frac{\partial w}{\partial u} = \frac{\partial w}{\partial x}\frac{\partial x}{\partial u} + \frac{\partial w}{\partial y}\frac{\partial y}{\partial u} + \frac{\partial w}{\partial z}\frac{\partial z}{\partial u}$

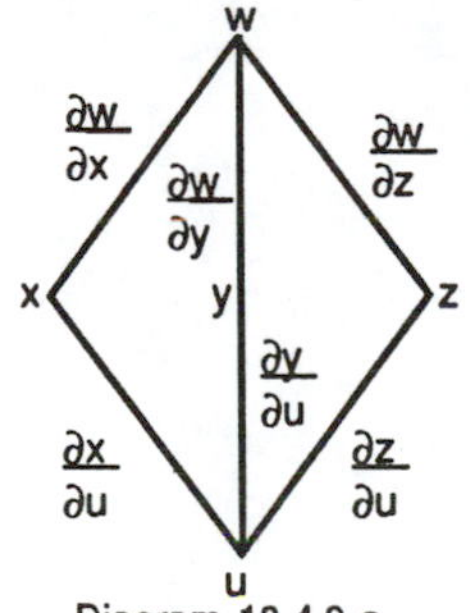

Diagram 13.4.9 a

$\frac{\partial w}{\partial v} = \frac{\partial w}{\partial x}\frac{\partial x}{\partial v} + \frac{\partial w}{\partial y}\frac{\partial y}{\partial v} + \frac{\partial w}{\partial z}\frac{\partial z}{\partial v}$

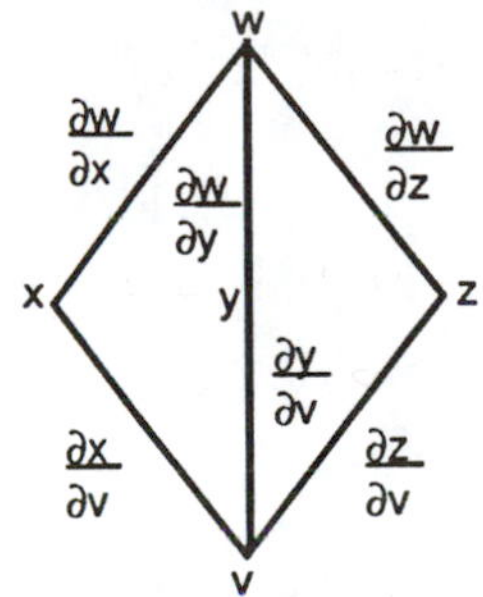

Diagram 13.4.9 b

10. $\frac{\partial w}{\partial x} = \frac{\partial w}{\partial r}\frac{\partial r}{\partial x} + \frac{\partial w}{\partial s}\frac{\partial s}{\partial x} + \frac{\partial w}{\partial t}\frac{\partial t}{\partial x}$

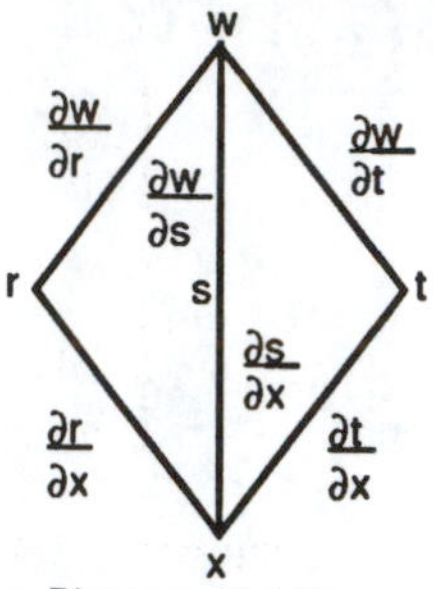
Diagram 13.4.10 a

$\frac{\partial w}{\partial y} = \frac{\partial w}{\partial r}\frac{\partial r}{\partial y} + \frac{\partial w}{\partial s}\frac{\partial s}{\partial y} + \frac{\partial w}{\partial t}\frac{\partial t}{\partial y}$

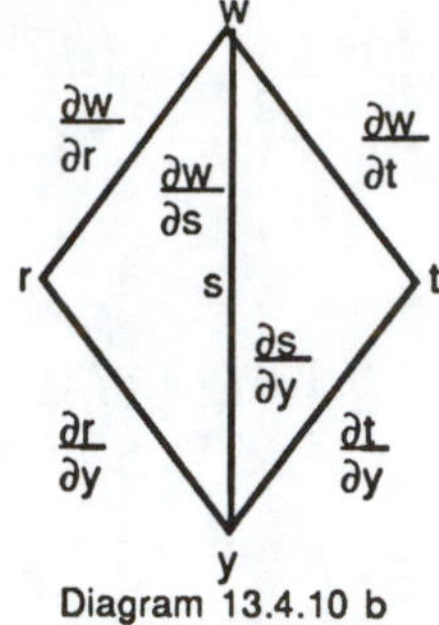
Diagram 13.4.10 b

11. $\frac{\partial w}{\partial u} = \frac{\partial w}{\partial x}\frac{\partial x}{\partial u} + \frac{\partial w}{\partial y}\frac{\partial y}{\partial u}$

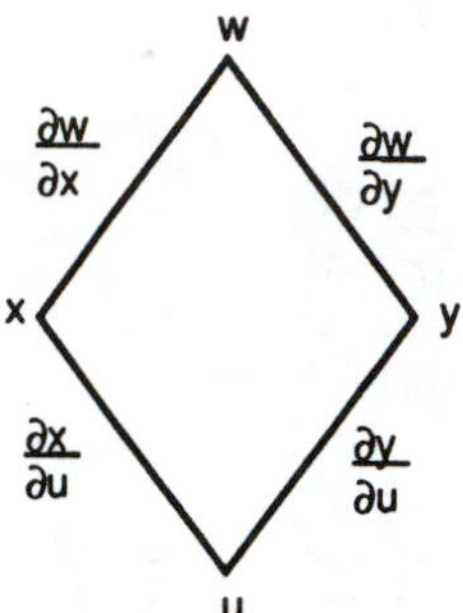
Diagram 13.4.11 a

$\frac{\partial w}{\partial v} = \frac{\partial w}{\partial x}\frac{\partial x}{\partial v} + \frac{\partial w}{\partial y}\frac{\partial y}{\partial v}$

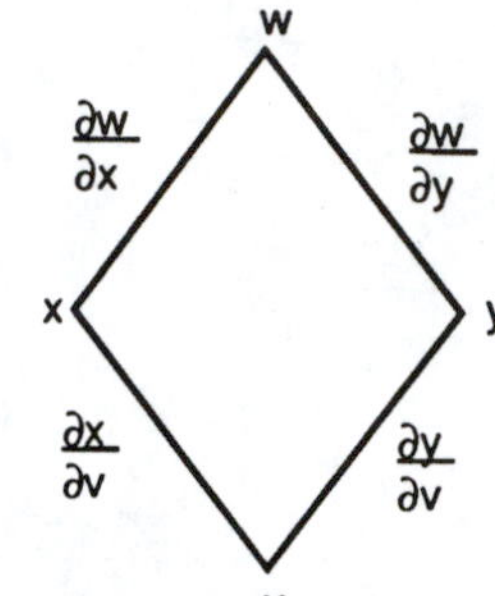
Diagram 13.4.11 b

12. $\frac{\partial w}{\partial x} = \frac{\partial w}{\partial u}\frac{\partial u}{\partial x} + \frac{\partial w}{\partial v}\frac{\partial v}{\partial x}$

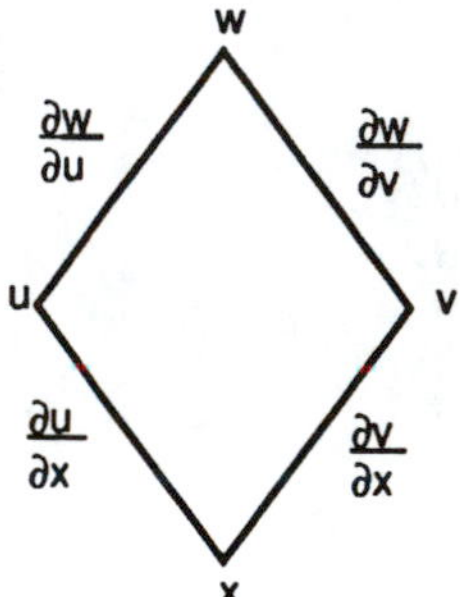
Diagram 13.4.12 a

$\frac{\partial w}{\partial y} = \frac{\partial w}{\partial u}\frac{\partial u}{\partial y} + \frac{\partial w}{\partial v}\frac{\partial v}{\partial y}$

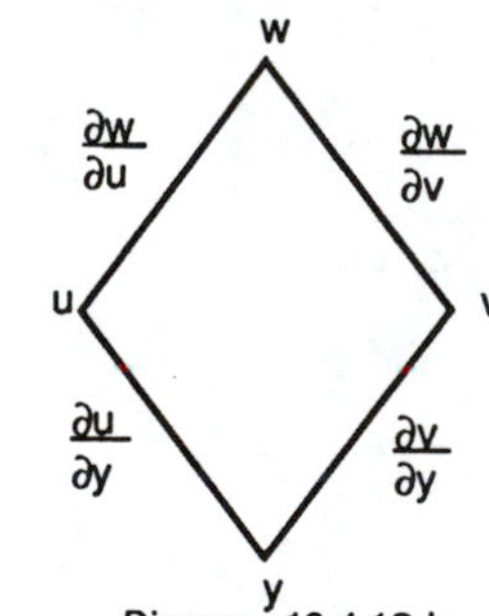
Diagram 13.4.12 b

13. $\frac{\partial z}{\partial t} = \frac{\partial z}{\partial x}\frac{\partial x}{\partial t} + \frac{\partial z}{\partial y}\frac{\partial y}{\partial t}$

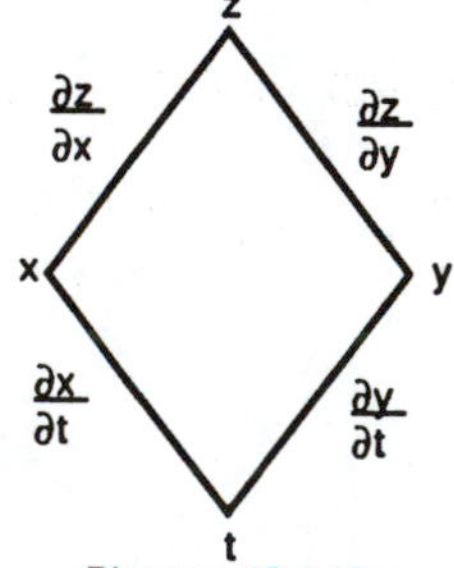
Diagram 13.4.13 a

$\frac{\partial z}{\partial s} = \frac{\partial z}{\partial x}\frac{\partial x}{\partial s} + \frac{\partial z}{\partial y}\frac{\partial y}{\partial s}$

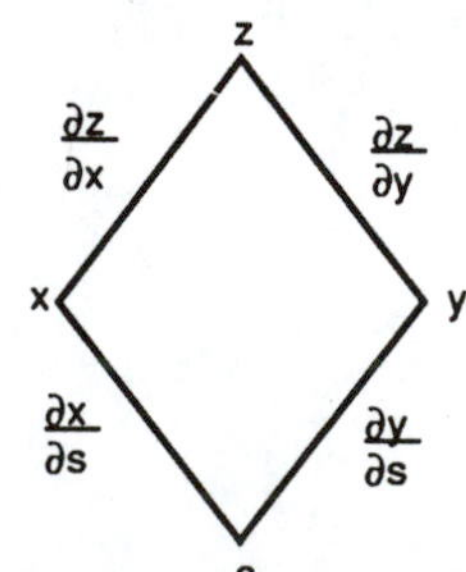
Diagram 13.4.13 b

14. $\frac{\partial y}{\partial r} = \frac{dy}{du}\frac{\partial u}{\partial r}$

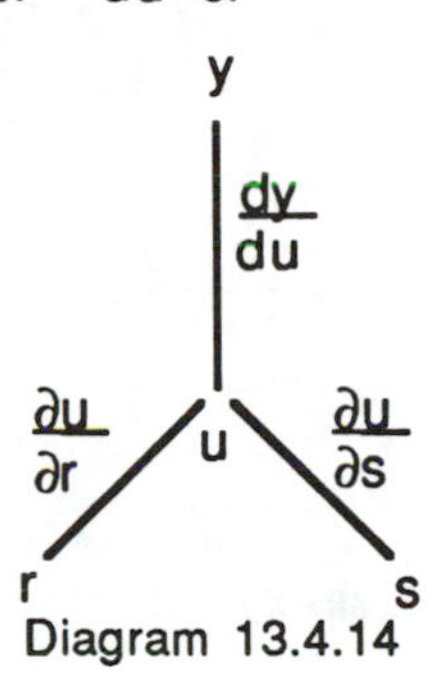

Diagram 13.4.14

15. $\frac{\partial w}{\partial s} = \frac{dw}{du}\frac{\partial u}{\partial s}$

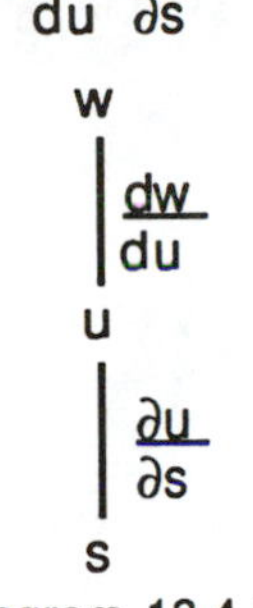

Diagram 13.4.15 a

$\frac{\partial w}{\partial t} = \frac{dw}{du}\frac{\partial u}{\partial t}$

w
dw/du
u
∂u/∂t
t

Diagram 13.4.15 b

16. $\frac{\partial w}{\partial p} = \frac{\partial w}{\partial x}\frac{\partial x}{\partial p} + \frac{\partial w}{\partial y}\frac{\partial y}{\partial p} + \frac{\partial w}{\partial z}\frac{\partial z}{\partial p} + \frac{\partial w}{\partial v}\frac{\partial v}{\partial p}$

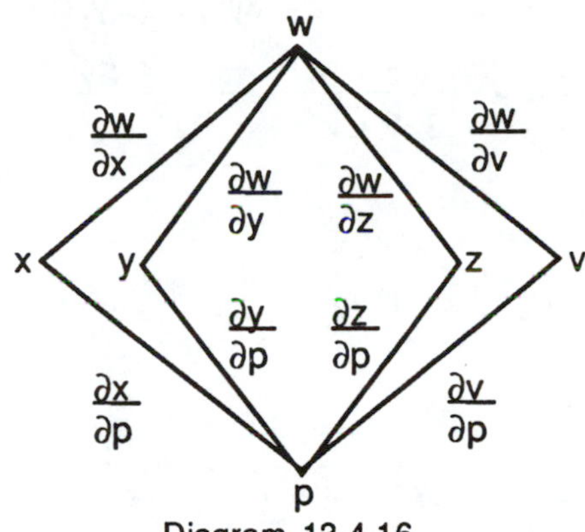

Diagram 13.4.16

17. $\frac{\partial w}{\partial r} = \frac{\partial w}{\partial x}\frac{dx}{dr} + \frac{\partial w}{\partial y}\frac{dy}{dr} = \frac{\partial w}{\partial x}\frac{dx}{dr}$

since $\frac{dy}{dr} = 0$

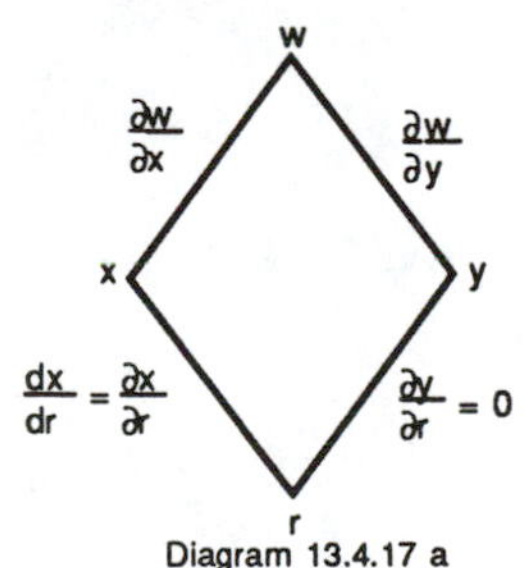

Diagram 13.4.17 a

$\frac{\partial w}{\partial s} = \frac{\partial w}{\partial x}\frac{dx}{ds} + \frac{\partial w}{\partial y}\frac{dy}{ds} = \frac{\partial w}{\partial y}\frac{dy}{ds}$

since $\frac{dx}{ds} = 0$

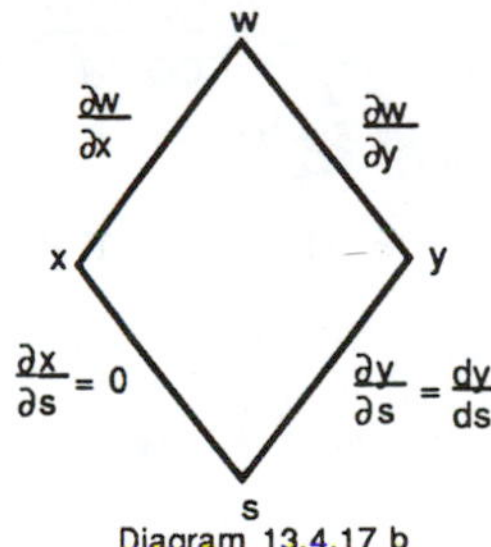

Diagram 13.4.17 b

18. $\frac{\partial w}{\partial s} = \frac{\partial w}{\partial x}\frac{\partial x}{\partial s} + \frac{\partial w}{\partial y}\frac{\partial y}{\partial s}$

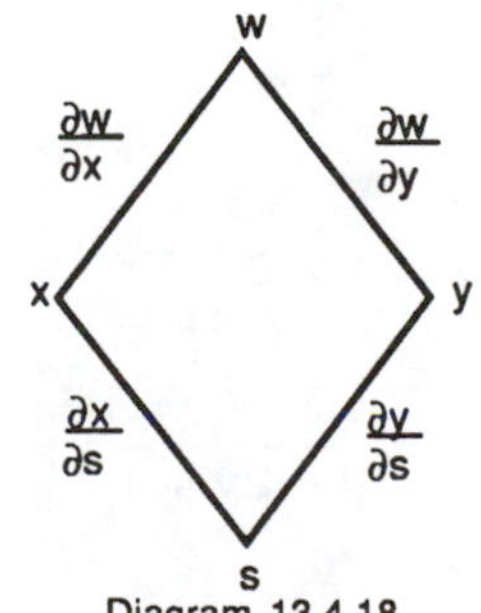

Diagram 13.4.18

19. Let $F(x,y) = x^3 - 2y^2 + xy = 0 \Rightarrow F_x(x,y) = 3x^2 + y$

$F_y(x,y) = -4y + x \Rightarrow \frac{dy}{dx} = -\frac{F_x}{F_y} = -\frac{3x^2 + y}{-4y + x} \Rightarrow$

$\frac{dy}{dx}(1,1) = \frac{4}{3}$

20. Let $F(x,y) = xy + y^2 - 3x - 3 = 0 \Rightarrow F_x(x,y) = y - 3$, $F_y(x,y) = x + 2y \Rightarrow \frac{dy}{dx} = -\frac{F_x}{F_y} = -\frac{y-3}{x+2y} \Rightarrow$
$\frac{dy}{dx}(-1,1) = 2$

21. Let $F(x,y) = x^2 + xy + y^2 - 7 = 0 \Rightarrow F_x(x,y) = 2x + y$, $F_y(x,y) = x + 2y \Rightarrow \frac{dy}{dx} = -\frac{F_x}{F_y} = -\frac{2x+y}{x+2y} \Rightarrow$
$\frac{dy}{dx}(1,2) = -\frac{4}{5}$

22. Let $F(x,y) = xe^y + \sin xy + y - \ln 2 = 0 \Rightarrow F_x(x,y) = e^y + y\cos xy$, $F_y(x,y) = xe^y + x\sin xy + 1 \Rightarrow$
$\frac{dy}{dx} = -\frac{F_x}{F_y} = -\frac{e^y + y\cos xy}{xe^y + x\sin xy + 1} \Rightarrow \frac{dy}{dx}(0,\ln 2) = -(2 + \ln 2)$

23. Let $F(x,y,z)) = z^3 - xy + yz + y^3 - 2 = 0 \Rightarrow F_x(x,y,z) = -y$, $F_y(x,y,z) = -x + z + 3y^2$, $F_z(x,y,z) = 3z^2 + y \Rightarrow$
$\frac{\partial z}{\partial x} = -\frac{F_x}{F_z} = -\frac{-y}{3z^2+y} = \frac{y}{3z^2+y} \Rightarrow \frac{\partial z}{\partial x}(1,1,1) = \frac{1}{4}$. $\frac{\partial z}{\partial y} = -\frac{F_y}{F_z} = -\frac{-x+z+3y^2}{3z^2+y} = \frac{x-z-3y^2}{3z^2+y} \Rightarrow$
$\frac{\partial z}{\partial y}(1,1,1) = -\frac{3}{4}$

24. Let $F(x,y,z) = \frac{1}{x} + \frac{1}{y} + \frac{1}{z} - 1 = 0 \Rightarrow F_x(x,y,z) = -\frac{1}{x^2}$, $F_y(x,y,z) = -\frac{1}{y^2}$, $F_z(x,y,z) = -\frac{1}{z^2} \Rightarrow$
$\frac{\partial z}{\partial x} = -\frac{F_x}{F_z} = -\frac{-\frac{1}{x^2}}{-\frac{1}{z^2}} = -\frac{z^2}{x^2} \Rightarrow \frac{\partial z}{\partial x}(2,3,6) = -9$. $\frac{\partial z}{\partial y} = -\frac{F_y}{F_z} = -\frac{-\frac{1}{y^2}}{-\frac{1}{z^2}} = -\frac{z^2}{y^2} \Rightarrow \frac{\partial z}{\partial y}(2,3,6) = -4$

25. Let $F(x,y,z) = \sin(x+y) + \sin(y+z) + \sin(x+z) = 0 \Rightarrow F_x(x,y,z) = \cos(x+y) + \cos(x+z)$,
$F_y(x,y,z) = \cos(x+y) + \cos(y+z)$, $F_z(x,y,z) = \cos(y+z) + \cos(x+z) \Rightarrow \frac{\partial z}{\partial x} = -\frac{F_x}{F_z} = -\frac{\cos(x+y)+\cos(x+z)}{\cos(y+z)+\cos(x+z)}$
$\Rightarrow \frac{\partial z}{\partial x}(\pi,\pi,\pi) = -1$. $\frac{\partial z}{\partial y} = -\frac{F_y}{F_z} = -\frac{\cos(x+y)+\cos(y+z)}{\cos(y+z)+\cos(x+z)} \Rightarrow \frac{\partial z}{\partial y}(\pi,\pi,\pi) = -1$

26. Let $F(x,y,z) = xe^y + ye^z + 2\ln x - 2 - 3\ln 2 = 0 \Rightarrow F_x(x,y,z) = e^y + \frac{2}{x}$, $F_y(x,y,z) = xe^y + e^z$, $F_z(x,y,z) = y$
$\Rightarrow \frac{\partial z}{\partial x} = -\frac{F_x}{F_z} = -\frac{e^y + \frac{2}{x}}{y} \Rightarrow \frac{\partial z}{\partial x}(1,\ln 2,\ln 3) = -\frac{4}{\ln 2}$. $\frac{\partial z}{\partial y} = -\frac{F_y}{F_z} = -\frac{xe^y + e^z}{y} \Rightarrow \frac{\partial z}{\partial y}(1,\ln 2,\ln 3) = -\frac{5}{\ln 2}$

27. $\frac{\partial w}{\partial r} = \frac{\partial w}{\partial x}\frac{\partial x}{\partial r} + \frac{\partial w}{\partial y}\frac{\partial y}{\partial r} + \frac{\partial w}{\partial z}\frac{\partial z}{\partial r} = 2(x+y+z)(1) + 2(x+y+z)(-\sin(r+s)) + 2(x+y+z)(\cos(r+s)) =$
$2(x+y+z)(1 - \sin(r+s) + \cos(r+s)) = 2(r - s + \cos(r+s) + \sin(r+s))(1 - \sin(r+s) + \cos(r+s)) \Rightarrow$
$\left.\frac{\partial w}{\partial r}\right|_{r=1,s=1} = 12$

28. $\frac{\partial w}{\partial v} = \frac{\partial w}{\partial x}\frac{\partial x}{\partial v} + \frac{\partial w}{\partial y}\frac{\partial y}{\partial v} + \frac{\partial w}{\partial z}\frac{\partial z}{\partial v} = y\left(\frac{2v}{u}\right) + x(1) + \frac{1}{z}(0) = (u+v)\left(\frac{2v}{u}\right) + \frac{v^2}{u} \Rightarrow \left.\frac{\partial w}{\partial v}\right|_{u=-1,v=2} = -5$

29. $\frac{\partial w}{\partial v} = \frac{\partial w}{\partial x}\frac{\partial x}{\partial v} + \frac{\partial w}{\partial y}\frac{\partial y}{\partial v} = \left(2x - \frac{y}{x^2}\right)(-2) + \frac{1}{x}(1) = \left(2(u-2v+1) - \frac{2u+v-2}{(u-2v+1)^2}\right)(-2) + \frac{1}{u-2v+1} \Rightarrow$
$\left.\frac{\partial w}{\partial v}\right|_{u=0,v=0} = -7$

30. $\frac{\partial z}{\partial u} = \frac{\partial z}{\partial x}\frac{\partial x}{\partial u} + \frac{\partial z}{\partial y}\frac{\partial y}{\partial u} = (y \cos xy + \sin y)2u + (x \cos xy + x \cos y)v = \left(uv \cos\left(\left(u^2 + v^2\right)uv\right) + \sin uv\right)2u + \left(\left(u^2 + v^2\right)\cos\left(u^2 + v^2\right)uv + \left(u^2 + v^2\right)\cos uv\right)v \Rightarrow \left.\frac{\partial z}{\partial u}\right|_{u=0,v=1} = 2$

31. $\frac{\partial z}{\partial u} = \frac{dz}{dx}\frac{\partial x}{\partial u} = \frac{5}{1 + x^2}\left(e^u + \ln v\right) = \frac{5}{1 + \left(e^u + \ln v\right)^2}\left(e^u + \ln v\right) \Rightarrow \left.\frac{\partial z}{\partial u}\right|_{u=\ln 2,v=1} = 2$

$\frac{\partial z}{\partial v} = \frac{dz}{dx}\frac{\partial x}{\partial v} = \frac{5}{1 + x^2}\left(\frac{1}{v}\right) = \frac{5}{1 + \left(e^u + \ln v\right)^2}\left(\frac{1}{v}\right) \Rightarrow \left.\frac{\partial z}{\partial v}\right|_{u=\ln 2,v=1} = 5$

32. $\frac{\partial w}{\partial q} = \frac{\partial w}{\partial x}\frac{\partial x}{\partial q} + \frac{\partial w}{\partial y}\frac{\partial y}{\partial q} + \frac{\partial w}{\partial z}\frac{\partial z}{\partial q} + \frac{\partial w}{\partial u}\frac{\partial u}{\partial q} + \frac{\partial w}{\partial v}\frac{\partial v}{\partial q} = yp + (x + z)\tan^{-1} p + (y + u)\frac{1}{q - 3p} + z\left(\frac{1}{2}\right)(q + 5p)^{-1/2} + u(1) = qp \tan^{-1} p + (pq + \ln(q - 3p))\tan^{-1} p + \frac{q \tan^{-1} p + \sqrt{q + 5p}}{q - 3p} + \frac{\ln(q - 3p)}{2(q + 5p)^{1/2}} + \sqrt{q + 5p} \Rightarrow$

$\left.\frac{\partial w}{\partial q}\right|_{p=1,q=4} = 3\pi + 6$

33. $\frac{dV}{dt} = \frac{\partial V}{\partial I}\frac{dI}{dt} + \frac{\partial V}{\partial R}\frac{dR}{dt}$. $V = IR \Rightarrow \frac{\partial V}{\partial I} = R, \frac{\partial V}{\partial R} = I \Rightarrow \frac{dV}{dt} = R\frac{dI}{dt} + I\frac{dR}{dt} \Rightarrow -0.01$ volts/sec = (600 ohms)$\frac{dI}{dt}$ + (0.04 amps)(0.5 ohms/sec) $\Rightarrow \frac{dI}{dt} = -0.00005$ amps/sec.

34. a) $V = abc \Rightarrow \frac{dV}{dt} = \frac{\partial V}{\partial a}\frac{da}{dt} + \frac{\partial V}{\partial b}\frac{db}{dt} + \frac{\partial V}{\partial c}\frac{dc}{dt} = bc\left(\frac{da}{dt}\right) + ac\left(\frac{db}{dt}\right) + ab\left(\frac{dc}{dt}\right) \Rightarrow$

$\left.\frac{dV}{dt}\right|_{a=13,b=9,c=5}$ = (9 cm)(5 cm)(2 cm/sec) + (13 cm)(5 cm)(–5 cm/sec) + (13 cm)(9 cm)(2 cm/sec) = –1 cm^3/sec $\Rightarrow$ Volume is decreasing.

$S = 4ac + 2bc \Rightarrow \frac{dS}{dt} = \frac{\partial S}{\partial a}\frac{da}{dt} + \frac{\partial S}{\partial b}\frac{db}{dt} + \frac{\partial S}{\partial c}\frac{dc}{dt} = 4c\left(\frac{da}{dt}\right) + 2c\left(\frac{db}{dt}\right) + (4a + 2b)\left(\frac{dc}{dt}\right) \Rightarrow$

$\left.\frac{dS}{dt}\right|_{a=13,b=9,c=5}$ = 4(5 cm)(2 cm/sec) + 2(5 cm)(–5 cm/sec) + (4(13 cm) + 2(9 cm))(2 cm/sec) = 130 cm^2/sec. S is increasing.

b) The diagonal is $D = \sqrt{a^2 + b^2 + c^2}$. Then $\frac{dD}{dt} = \frac{\partial D}{\partial a}\frac{da}{dt} + \frac{\partial D}{\partial b}\frac{db}{dt} + \frac{\partial D}{\partial c}\frac{dc}{dt} = \frac{1}{2}(a^2 + b^2 + c^2)^{-1/2}(2a)\frac{da}{dt} + \frac{1}{2}(a^2 + b^2 + c^2)^{-1/2}(2b)\frac{db}{dt} + \frac{1}{2}(a^2 + b^2 + c^2)^{-1/2}(2c)\frac{dc}{dt} = (a^2 + b^2 + c^2)^{-1/2}\left(a\frac{da}{dt} + b\frac{db}{dt} + c\frac{dc}{dt}\right)$

$\Rightarrow \left.\frac{dD}{dt}\right|_{a=13,b=9,c=5}$ =

$\left(13^2\ \text{cm}^2 + 9^2\ \text{cm}^2 + 5^2\ \text{cm}^2\right)^{-1/2}$((13 cm)(2 cm/sec) + (9 cm)(–5 cm/sec) + (5 cm)(2 cm/sec))

$= \frac{-9}{\sqrt{275}}$ cm/sec. The diagonal is decreasing.

35. $f_x(x,y,z) = \cos t$, $f_y(x,y,z) = \sin t$, $f_z(x,y,z) = t^2 + t - 2$. $\frac{df}{dt} = \frac{\partial f}{\partial x}\frac{dx}{dt} + \frac{\partial f}{\partial y}\frac{dy}{dt} + \frac{\partial f}{\partial z}\frac{dz}{dt} = (\cos t)(-\sin t) + (\sin t)(\cos t) + (t^2 + t - 2)(1) = t^2 + t - 2$. $\frac{df}{dt} = 0 \Rightarrow t^2 + t - 2 = 0 \Rightarrow t = -2$ or $t = 1$

$t = -2 \Rightarrow x = \cos(-2), y = \sin(-2), z = -2$; $t = 1 \Rightarrow x = \cos 1, y = \sin 1, z = 1$

36. $\frac{dw}{dt} = \frac{\partial w}{\partial x}\frac{dx}{dt} + \frac{\partial w}{\partial y}\frac{dy}{dt} + \frac{\partial w}{\partial z}\frac{dz}{dt} = \left(2xe^{2y}\cos 3z\right)(-\sin t) + \left(2x^2e^{2y}\cos 3z\right)\left(\frac{1}{t+2}\right) +$
$\left(-3x^2e^{2y}\sin 3z\right)(1) = -2xe^{2y}\cos 3z\sin t + \frac{2x^2e^{2y}\cos 3z}{t+2} - 3x^2e^{2y}\sin 3z$. Now z = 0 and z = t ⇒ t = 0
$\therefore \left.\frac{dw}{dt}\right|_{(1,\ln 2,0);t=0} = 4$

37. a) $\frac{\partial T}{\partial x} = 8x - 4y,\ \frac{\partial T}{\partial y} = 8y - 4x.\ \frac{dT}{dt} = \frac{\partial T}{\partial x}\frac{dx}{dt} + \frac{\partial T}{\partial y}\frac{dy}{dt} = (8x-4y)(-\sin t) + (8y-4x)(\cos t) =$
$(8\cos t - 4\sin t)(-\sin t) + (8\sin t - 4\cos t)(\cos t) = 4\sin^2 t - 4\cos^2 t \Rightarrow \frac{d^2T}{dt^2} = 16\sin t\cos t$
$\frac{dT}{dt} = 0 \Rightarrow 4\sin^2 t - 4\cos^2 t = 0 \Rightarrow \sin^2 t = \cos^2 t \Rightarrow \sin t = \cos t$ or $\sin t = -\cos t \Rightarrow$
$t = \frac{\pi}{4}, \frac{5\pi}{4}$ or $\frac{3\pi}{4}, \frac{7\pi}{4}$ on the interval $0 \le t \le 2\pi$.
$\left.\frac{d^2T}{dt^2}\right|_{t=\pi/4} = 16\sin\frac{\pi}{4}\cos\frac{\pi}{4} > 0 \Rightarrow$ T has a minimum at $t = \frac{\pi}{4}$
$\left.\frac{d^2T}{dt^2}\right|_{t=3\pi/4} = 16\sin\frac{3\pi}{4}\cos\frac{3\pi}{4} < 0 \Rightarrow$ T has a maximum at $t = \frac{3\pi}{4}$
$\left.\frac{d^2T}{dt^2}\right|_{t=5\pi/4} = 16\sin\frac{5\pi}{4}\cos\frac{5\pi}{4} > 0 \Rightarrow$ T has a minimum at $t = \frac{5\pi}{4}$
$\left.\frac{d^2T}{dt^2}\right|_{t=7\pi/4} = 16\sin\frac{7\pi}{4}\cos\frac{7\pi}{4} < 0 \Rightarrow$ T has a maximum at $t = \frac{7\pi}{4}$

b) $T = 4x^2 - 4xy + 4y^2 \Rightarrow \frac{\partial T}{\partial x} = 8x - 4y,\ \frac{\partial T}{\partial y} = 8y - 4x$ (See part a above.)
$t = \frac{\pi}{4} \Rightarrow x = \cos\frac{\pi}{4} = \frac{\sqrt{2}}{2},\ y = \sin\frac{\pi}{4} = \frac{\sqrt{2}}{2} \Rightarrow T\left(\frac{\pi}{4}\right) = 2$
$t = \frac{3\pi}{4} \Rightarrow x = \cos\frac{3\pi}{4} = -\frac{\sqrt{2}}{2},\ y = \sin\frac{3\pi}{4} = \frac{\sqrt{2}}{2} \Rightarrow T\left(\frac{3\pi}{4}\right) = 6$
$t = \frac{5\pi}{4} \Rightarrow x = \cos\frac{5\pi}{4} = -\frac{\sqrt{2}}{2},\ y = \sin\frac{5\pi}{4} = -\frac{\sqrt{2}}{2} \Rightarrow T\left(\frac{5\pi}{4}\right) = 2$
$t = \frac{7\pi}{4} \Rightarrow x = \cos\frac{7\pi}{4} = \frac{\sqrt{2}}{2},\ y = \sin\frac{7\pi}{4} = -\frac{\sqrt{2}}{2} \Rightarrow T\left(\frac{7\pi}{4}\right) = 6$
$\therefore T_{max} = 6$ and $T_{min} = 2$.

SECTION 13.5 DIRECTIONAL DERIVATIVES AND GRADIENT VECTORS

1. $\frac{\partial f}{\partial x} = 2x \Rightarrow \frac{\partial f}{\partial x}(1,1,1) = 2.\ \frac{\partial f}{\partial y} = 2y \Rightarrow \frac{\partial f}{\partial y}(1,1,1) = 2.\ \frac{\partial f}{\partial z} = -4z \Rightarrow \frac{\partial f}{\partial z}(1,1,1) = -4 \Rightarrow \nabla f = 2\mathbf{i} + 2\mathbf{j} - 4\mathbf{k}$

2. $\frac{\partial f}{\partial x} = -6xz \Rightarrow \frac{\partial f}{\partial x}(1,1,1) = -6.\ \frac{\partial f}{\partial y} = -6yz \Rightarrow \frac{\partial f}{\partial y}(1,1,1) = -6.\ \frac{\partial f}{\partial z} = 6z^2 - 3(x^2+y^2) \Rightarrow \frac{\partial f}{\partial z}(1,1,1) = 0 \Rightarrow$
$\nabla f = -6\mathbf{i} - 6\mathbf{j}$

3. $\frac{\partial f}{\partial x} = -x(x^2+y^2+z^2)^{-3/2} \Rightarrow \frac{\partial f}{\partial x}(1,2,-2) = -\frac{1}{27}.\ \frac{\partial f}{\partial y} = -y(x^2+y^2+z^2)^{-3/2} \Rightarrow \frac{\partial f}{\partial y}(1,2,-2) = -\frac{2}{27}.$
$\frac{\partial f}{\partial z} = -z(x^2+y^2+z^2)^{-3/2} \Rightarrow \frac{\partial f}{\partial z}(1,2,-2) = \frac{2}{27} \Rightarrow \nabla f = -\frac{1}{27}\mathbf{i} - \frac{2}{27}\mathbf{j} + \frac{2}{27}\mathbf{k}$

4. $\frac{\partial f}{\partial x} = e^{x+y}\cos z \Rightarrow \frac{\partial f}{\partial x}\left(0,0,\frac{\pi}{6}\right) = \frac{\sqrt{3}}{2}$. $\frac{\partial f}{\partial y} = e^{x+y}\cos z \Rightarrow \frac{\partial f}{\partial y}\left(0,0,\frac{\pi}{6}\right) = \frac{\sqrt{3}}{2}$. $\frac{\partial f}{\partial z} = -e^{x+y}\sin z \Rightarrow$ $\frac{\partial f}{\partial z}\left(0,0,\frac{\pi}{6}\right) = -\frac{1}{2} \Rightarrow \nabla f = \frac{\sqrt{3}}{2}\mathbf{i} + \frac{\sqrt{3}}{2}\mathbf{j} - \frac{1}{2}\mathbf{k}$

5. $\frac{\partial f}{\partial x} = -1$, $\frac{\partial f}{\partial y} = 1 \Rightarrow \nabla f = -\mathbf{i} + \mathbf{j}$

$-1 = y - x$ is the level curve.

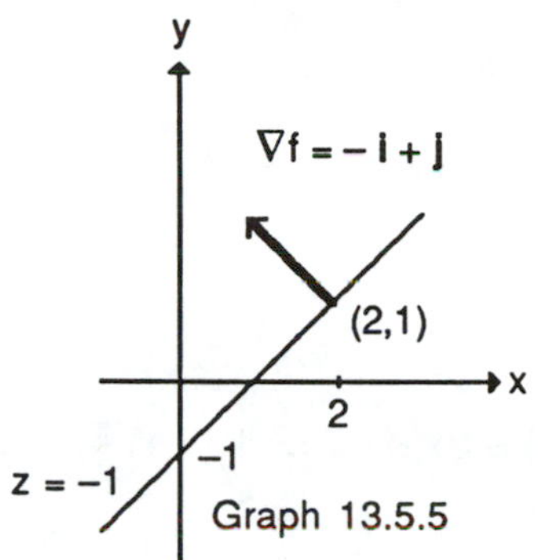

Graph 13.5.5

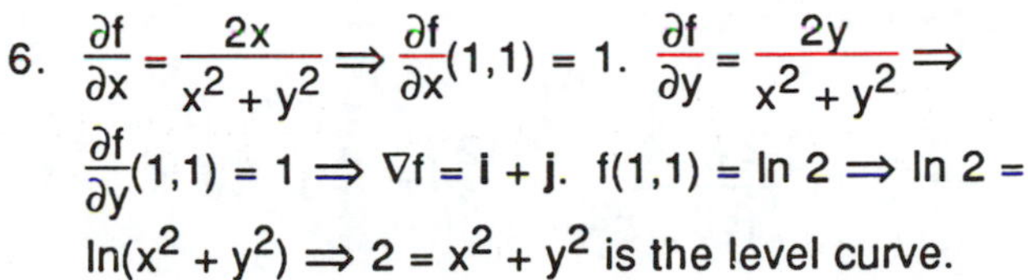

6. $\frac{\partial f}{\partial x} = \frac{2x}{x^2 + y^2} \Rightarrow \frac{\partial f}{\partial x}(1,1) = 1$. $\frac{\partial f}{\partial y} = \frac{2y}{x^2 + y^2} \Rightarrow$ $\frac{\partial f}{\partial y}(1,1) = 1 \Rightarrow \nabla f = \mathbf{i} + \mathbf{j}$. $f(1,1) = \ln 2 \Rightarrow \ln 2 =$ $\ln(x^2 + y^2) \Rightarrow 2 = x^2 + y^2$ is the level curve.

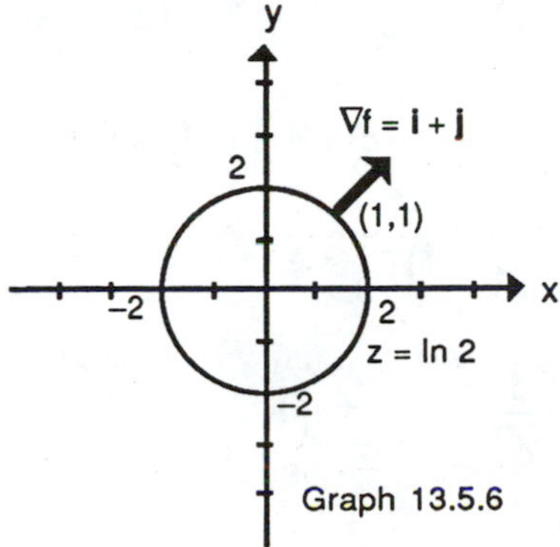

Graph 13.5.6

7. $\frac{\partial f}{\partial x} = -2x \Rightarrow \frac{\partial f}{\partial x}(-1,0) = 2$. $\frac{\partial f}{\partial y} = 1 \Rightarrow$ $\nabla f = 2\mathbf{i} + \mathbf{j}$. $-1 = y - x^2$ is the level curve.

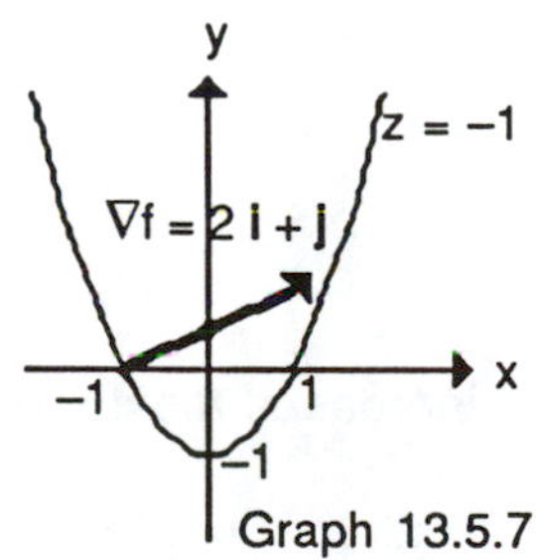

Graph 13.5.7

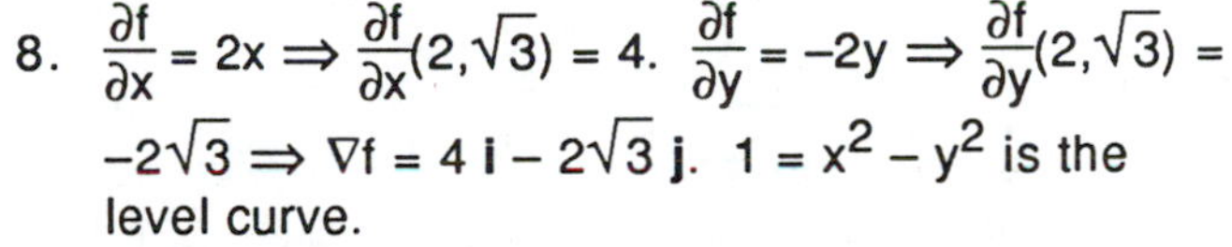

8. $\frac{\partial f}{\partial x} = 2x \Rightarrow \frac{\partial f}{\partial x}(2,\sqrt{3}) = 4$. $\frac{\partial f}{\partial y} = -2y \Rightarrow \frac{\partial f}{\partial y}(2,\sqrt{3}) =$ $-2\sqrt{3} \Rightarrow \nabla f = 4\mathbf{i} - 2\sqrt{3}\,\mathbf{j}$. $1 = x^2 - y^2$ is the level curve.

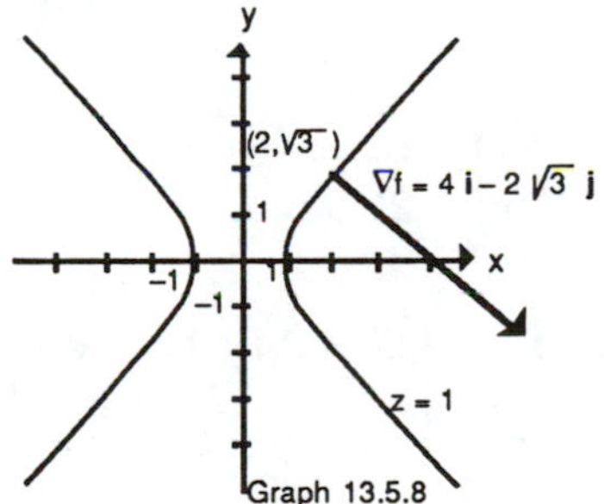

Graph 13.5.8

9. $\mathbf{u} = \frac{\mathbf{A}}{|\mathbf{A}|} = \frac{3\mathbf{i} + 6\mathbf{j} - 2\mathbf{k}}{\sqrt{3^2 + 6^2 + (-2)^2}} = \frac{3}{7}\mathbf{i} + \frac{6}{7}\mathbf{j} - \frac{2}{7}\mathbf{k}$. $f_x(x,y,z) = y + z \Rightarrow f_x(1,-1,2) = 1$, $f_y(x,y,z) = x + z \Rightarrow$ $f_y(1,-1,2) = 3$, $f_z(x,y,z) = y + x \Rightarrow f_z(1,-1,2) = 0 \Rightarrow \nabla f = \mathbf{i} + 3\mathbf{j} \Rightarrow \left(D_{\mathbf{u}}f\right)_{P_0} = \nabla f \cdot \mathbf{u} = \frac{3}{7} + \frac{18}{7} = 3$

10. $\mathbf{u} = \frac{\mathbf{A}}{|\mathbf{A}|} = \frac{\mathbf{i} + \mathbf{j} + \mathbf{k}}{\sqrt{1^2 + 1^2 + 1^2}} = \frac{1}{\sqrt{3}}\mathbf{i} + \frac{1}{\sqrt{3}}\mathbf{j} + \frac{1}{\sqrt{3}}\mathbf{k}$. $f_x(x,y,z) = 2x \Rightarrow f_x(1,1,1) = 2$, $f_y(x,y,z) = 4y \Rightarrow$ $f_y(1,1,1) = 4$, $f_z(x,y,z) = -6z \Rightarrow f_z(1,1,1) = -6 \Rightarrow \nabla f = 2\mathbf{i} + 4\mathbf{j} - 6\mathbf{k} \Rightarrow \left(D_{\mathbf{u}}f\right)_{P_0} = \nabla f \cdot \mathbf{u} = 2\left(\frac{1}{\sqrt{3}}\right) +$ $4\left(\frac{1}{\sqrt{3}}\right) - 6\left(\frac{1}{\sqrt{3}}\right) = 0$

11. $\mathbf{u} = \frac{\mathbf{A}}{|\mathbf{A}|} = \frac{2\mathbf{i} + \mathbf{j} - 2\mathbf{k}}{\sqrt{2^2 + 1^2 + (-2)^2}} = \frac{2}{3}\mathbf{i} + \frac{1}{3}\mathbf{j} - \frac{2}{3}\mathbf{k}$. $f_x(x,y,z) = 3e^x\cos yz \Rightarrow f_x(0,0,0) = 3$, $f_y(x,y,z) =$ $-3ze^x\sin yz \Rightarrow f_y(0,0,0) = 0$, $f_z(x,y,z) = -3ye^x\sin yz \Rightarrow f_z(0,0,0) = 0 \Rightarrow \nabla f = 3\mathbf{i} \Rightarrow \left(D_{\mathbf{u}}f\right)_{P_0} = \nabla f \cdot \mathbf{u} = 2$

12. $\mathbf{u} = \frac{\mathbf{A}}{|\mathbf{A}|} = \frac{\mathbf{i} + 2\mathbf{j} + 2\mathbf{k}}{\sqrt{1^2 + 2^2 + 2^2}} = \frac{1}{3}\mathbf{i} + \frac{2}{3}\mathbf{j} + \frac{2}{3}\mathbf{k}$. $f_x(x,y,z) = y \cos xy + \frac{1}{x} \Rightarrow f_x\left(1,0,\frac{1}{2}\right) = 1$, $f_y(x,y,z) = x \cos xy + ze^{yz} \Rightarrow f_y\left(1,0,\frac{1}{2}\right) = \frac{3}{2}$, $f_z(x,y,z) = y\,e^{yz} + \frac{1}{z} \Rightarrow f_z\left(1,0,\frac{1}{2}\right) = 2 \Rightarrow \nabla f = \mathbf{i} + \frac{3}{2}\mathbf{j} + 2\,\mathbf{k} \Rightarrow \left(D_{\mathbf{u}}f\right)_{P_0} = \nabla f \cdot \mathbf{u} = \frac{1}{3} + 1 + \frac{4}{3} = \frac{8}{3}$

13. $\mathbf{u} = \frac{\mathbf{A}}{|\mathbf{A}|} = \frac{12\,\mathbf{i} + 5\,\mathbf{j}}{\sqrt{12^2 + 5^2}} = \frac{12}{13}\mathbf{i} + \frac{5}{13}\mathbf{j}$. $f_x(x,y,z) = 1 + \frac{y^2}{x^2} \Rightarrow f_x(1,1) = 2$, $f_y(x,y,z) = -\frac{2y}{x} \Rightarrow f_y(1,1) = -2 \Rightarrow \nabla f = 2\,\mathbf{i} - 2\,\mathbf{j} \Rightarrow \left(D_{\mathbf{u}}f\right)_{P_0} = \nabla f \cdot \mathbf{u} = \frac{24}{13} - \frac{10}{13} = \frac{14}{13}$

14. $\mathbf{u} = \frac{\mathbf{A}}{|\mathbf{A}|} = \frac{3\,\mathbf{i} - 4\,\mathbf{j}}{\sqrt{3^2 + (-4)^2}} = \frac{3}{5}\mathbf{i} - \frac{4}{5}\mathbf{j}$. $f_x(x,y,z) = 4x \Rightarrow f_x(-1,1) = -4$, $f_y(x,y,z) = 2y \Rightarrow f_y(-1,1) = 2 \Rightarrow \nabla f = -4\,\mathbf{i} + 2\,\mathbf{j} \Rightarrow \left(D_{\mathbf{u}}f\right)_{P_0} = \nabla f \cdot \mathbf{u} = -\frac{12}{5} - \frac{8}{5} = -4$

15. $\mathbf{u} = \frac{\mathbf{A}}{|\mathbf{A}|} = \frac{3\,\mathbf{i} + 4\,\mathbf{j}}{\sqrt{3^2 + 4^2}} = \frac{3}{5}\mathbf{i} + \frac{4}{5}\mathbf{j}$. $f_x(x,y,z) = 2x + 2y \Rightarrow f_x(1,1) = 4$, $f_y(x,y,z) = 2x = 6y \Rightarrow f_y(1,1) = -4 \Rightarrow \nabla f = 4\,\mathbf{i} - 4\,\mathbf{j} \Rightarrow \left(D_{\mathbf{u}}f\right)_{P_0} = \nabla f \cdot \mathbf{u} = \frac{12}{5} - \frac{16}{5} = -\frac{4}{5}$

16. $\mathbf{u} = \frac{\mathbf{A}}{|\mathbf{A}|} = \frac{2\,\mathbf{i} - \mathbf{j}}{\sqrt{2^2 + (-1)^2}} = \frac{2}{\sqrt{5}}\mathbf{i} - \frac{1}{\sqrt{5}}\mathbf{j}$. $f_x(x,y,z) = \tan^{-1}\left(\frac{y}{x}\right) - \frac{xy}{x^2 + y^2} \Rightarrow f_x(1,1) = \frac{\pi}{4} - \frac{1}{2}$, $f_y(x,y,z) = \frac{1}{\left(\frac{y}{x}\right)^2 + 1} \Rightarrow f_y(1,1) = \frac{1}{2} \Rightarrow \nabla f = \left(\frac{\pi}{4} - \frac{1}{2}\right)\mathbf{i} + \frac{1}{2}\mathbf{j} \Rightarrow \left(D_{\mathbf{u}}f\right)_{P_0} = \nabla f \cdot \mathbf{u} = \left(\frac{\pi}{4} - \frac{1}{2}\right)\frac{2}{\sqrt{5}} + \frac{1}{2}\left(-\frac{1}{\sqrt{5}}\right) = \frac{\pi - 3}{2\sqrt{5}}$

17. $\nabla f = 2x\,\mathbf{i} + \cos y\,\mathbf{j} \Rightarrow \nabla f(1,0) = 2\,\mathbf{i} + \mathbf{j} \Rightarrow \mathbf{u} = \frac{\nabla f}{|\nabla f|} = \frac{2\,\mathbf{i} + \mathbf{j}}{\sqrt{2^2 + 1^2}} = \frac{2}{\sqrt{5}}\mathbf{i} + \frac{1}{\sqrt{5}}\mathbf{j}$. f increases most rapidly in the direction of $\mathbf{u} = \frac{2}{\sqrt{5}}\mathbf{i} + \frac{1}{\sqrt{5}}\mathbf{j}$; decreases most rapidly in the direction $-\mathbf{u} = -\frac{2}{\sqrt{5}}\mathbf{i} - \frac{1}{\sqrt{5}}\mathbf{j}$. $\left(D_{\mathbf{u}}f\right)_{P_0} = \nabla f \cdot \mathbf{u} = \sqrt{5}$, $\left(D_{-\mathbf{u}}f\right)_{P_0} = -\sqrt{5}$.

18. $\nabla f = (2x + y)\,\mathbf{i} + (x + 2y)\,\mathbf{j} \Rightarrow \nabla f(-1,1) = -\mathbf{i} + \mathbf{j} \Rightarrow \mathbf{u} = \frac{\nabla f}{|\nabla f|} = \frac{-\mathbf{i} + \mathbf{j}}{\sqrt{(-1)^2 + 1^2}} = -\frac{1}{\sqrt{2}}\mathbf{i} + \frac{1}{\sqrt{2}}\mathbf{j}$. f increases most rapidly in the direction $\mathbf{u} = -\frac{1}{\sqrt{2}}\mathbf{i} + \frac{1}{\sqrt{2}}\mathbf{j}$ and decreases most rapidly in the direction $-\mathbf{u} = \frac{1}{\sqrt{2}}\mathbf{i} - \frac{1}{\sqrt{2}}\mathbf{j}$. $\left(D_{\mathbf{u}}f\right)_{P_0} = \nabla f \cdot \mathbf{u} = \sqrt{2}$, $\left(D_{-\mathbf{u}}f\right)_{P_0} = -\sqrt{2}$

19. $\nabla f = e^y\,\mathbf{i} + xe^y\,\mathbf{j} + 2z\,\mathbf{k} \Rightarrow \nabla f\left(1, \ln 2, \frac{1}{2}\right) = 2\,\mathbf{i} + 2\,\mathbf{j} + \mathbf{k} \Rightarrow \mathbf{u} = \frac{\nabla f}{|\nabla f|} = \frac{2\,\mathbf{i} + 2\,\mathbf{j} + \mathbf{k}}{\sqrt{2^2 + 2^2 + 1^2}} = \frac{2}{3}\mathbf{i} + \frac{2}{3}\mathbf{j} + \frac{1}{3}\mathbf{k}$. f increases most rapidly in the direction $\mathbf{u} = \frac{2}{3}\mathbf{i} + \frac{2}{3}\mathbf{j} + \frac{1}{3}\mathbf{k}$; decreases most rapidly in the direction $-\mathbf{u} = -\frac{2}{3}\mathbf{i} - \frac{2}{3}\mathbf{j} - \frac{1}{3}\mathbf{k}$. $\left(D_{\mathbf{u}}f\right)_{P_0} = \nabla f \cdot \mathbf{u} = 3$; $\left(D_{-\mathbf{u}}f\right)_{P_0} = -3$

20. $\nabla f = \frac{1}{y}\mathbf{i} - \left(\frac{x}{y^2} + z\right)\mathbf{j} - y\,\mathbf{k} \Rightarrow \nabla f(4,1,1) = \mathbf{i} - 5\,\mathbf{j} - \mathbf{k}$. $\mathbf{u} = \frac{\nabla f}{|\nabla f|} = \frac{\mathbf{i} - 5\,\mathbf{j} - \mathbf{k}}{\sqrt{1^2 + (-5)^2 + (-1)^2}} = \frac{1}{3\sqrt{3}}\mathbf{i} - \frac{5}{3\sqrt{3}}\mathbf{j} - \frac{1}{3\sqrt{3}}\mathbf{k}$. f increases most rapidly in the direction of $\mathbf{u} = \frac{1}{3\sqrt{3}}\mathbf{i} - \frac{5}{3\sqrt{3}}\mathbf{j} - \frac{1}{3\sqrt{3}}\mathbf{k}$; decreases most rapidly in the direction $-\mathbf{u} = -\frac{1}{3\sqrt{3}}\mathbf{i} + \frac{5}{3\sqrt{3}}\mathbf{j} + \frac{1}{3\sqrt{3}}\mathbf{k}$. $\left(D_{\mathbf{u}}f\right)_{P_0} = \nabla f \cdot \mathbf{u} = 3\sqrt{3}$; $\left(D_{-\mathbf{u}}f\right)_{P_0} = -3\sqrt{3}$

21. $\nabla f = (\cos x)\,\mathbf{i} + (\cos y)\,\mathbf{j} + (\cos z)\,\mathbf{k} \Rightarrow \nabla f(0,0,0) = \mathbf{i} + \mathbf{j} + \mathbf{k} \Rightarrow \mathbf{u} = \frac{\nabla f}{|\nabla f|} = \frac{\mathbf{i} + \mathbf{j} + \mathbf{k}}{\sqrt{1^2 + 1^2 + 1^2}} = \frac{1}{\sqrt{3}}\mathbf{i} + \frac{1}{\sqrt{3}}\mathbf{j} + \frac{1}{\sqrt{3}}\mathbf{k}$. f increases most rapidly in the direction $\mathbf{u} = \frac{1}{\sqrt{3}}\mathbf{i} + \frac{1}{\sqrt{3}}\mathbf{j} + \frac{1}{\sqrt{3}}\mathbf{k}$; decreases most rapidly in the direction $-\mathbf{u} = -\frac{1}{\sqrt{3}}\mathbf{i} - \frac{1}{\sqrt{3}}\mathbf{j} - \frac{1}{\sqrt{3}}\mathbf{k}$. $\left(D_{\mathbf{u}}f\right)_{P_0} = \nabla f \cdot \mathbf{u} = \sqrt{3}$, $\left(D_{-\mathbf{u}}f\right)_{P_0} = -\sqrt{3}$

22. $\nabla f = \frac{2x}{x^2 + y^2 - 1}\mathbf{i} + \left(\frac{2y}{x^2 + y^2 - 1} + 1\right)\mathbf{j} + 6\,\mathbf{k} \Rightarrow \nabla f(1,1,0) = 2\,\mathbf{i} + 3\,\mathbf{j} + 6\,\mathbf{k}. \Rightarrow \mathbf{u} = \frac{\nabla f}{|\nabla f|} = \frac{2\,\mathbf{i} + 3\,\mathbf{j} + 6\,\mathbf{k}}{\sqrt{2^2 + 3^2 + 6^2}} = \frac{2}{7}\mathbf{i} + \frac{3}{7}\mathbf{j} + \frac{6}{7}\mathbf{k}$. f increases most rapidly in the direction $\mathbf{u} = \frac{2}{7}\mathbf{i} + \frac{3}{7}\mathbf{j} + \frac{6}{7}\mathbf{k}$; decreases most rapidly in the direction $-\mathbf{u} = -\frac{2}{7}\mathbf{i} - \frac{3}{7}\mathbf{j} - \frac{6}{7}\mathbf{k}$. $\left(D_{\mathbf{u}}f\right)_{P_0} = \nabla f \cdot \mathbf{u} = 7$; $\left(D_{-\mathbf{u}}f\right)_{P_0} = -7$

23. $\nabla f = (-\pi y \sin(\pi xy) + y^2)\,\mathbf{i} + (-\pi x \sin(\pi xy) + 2xy)\,\mathbf{j} \Rightarrow \nabla f(-1,-1) = \mathbf{i} + 2\,\mathbf{j}$. $\mathbf{u} = \frac{\mathbf{A}}{|\mathbf{A}|} = \frac{\mathbf{i} + \mathbf{j}}{\sqrt{1^2 + 1^2}} = \frac{1}{\sqrt{2}}\mathbf{i} + \frac{1}{\sqrt{2}}\mathbf{j}$

$\nabla f \cdot \mathbf{u} = \frac{1}{\sqrt{2}} + \frac{2}{\sqrt{2}} = \frac{3}{\sqrt{2}}$. $\therefore$ $df = (\nabla f \cdot \mathbf{u})ds = \frac{3}{\sqrt{2}}(0.1) = 0{,}15\sqrt{2} \approx 0.212132$

24. $\nabla f = \frac{x}{x^2 + y^2 + z^2}\mathbf{i} + \frac{y}{x^2 + y^2 + z^2}\mathbf{j} + \frac{z}{x^2 + y^2 + z^2}\mathbf{k} \Rightarrow \nabla f(3,4,12) = \frac{3}{169}\mathbf{i} + \frac{4}{169}\mathbf{j} + \frac{12}{169}\mathbf{k}$. $\mathbf{u} = \frac{\mathbf{A}}{|\mathbf{A}|} = \frac{3\,\mathbf{i} + 6\,\mathbf{j} - 2\,\mathbf{k}}{\sqrt{3^2 + 6^2 + (-2)^2}} = \frac{3}{7}\mathbf{i} + \frac{6}{7}\mathbf{j} - \frac{2}{7}\mathbf{k} \Rightarrow \nabla f \cdot \mathbf{u} = \frac{9}{1183}$. $\therefore$ $df = (\nabla f \cdot \mathbf{u})ds = \frac{9}{1183}(0.1) \approx 0.000760$

25. $\nabla f = \left(e^x \cos yz\right)\mathbf{i} - \left(ze^x \sin yz\right)\mathbf{j} - \left(ye^x \sin yz\right)\mathbf{k} \Rightarrow \nabla f(0,0,0) = \mathbf{i}$. $\mathbf{u} = \frac{\mathbf{A}}{|\mathbf{A}|} = \frac{2\,\mathbf{i} + \mathbf{j} - 2\,\mathbf{k}}{\sqrt{2^2 + 1^2 + (-2)^2}} = \frac{2}{3}\mathbf{i} + \frac{1}{3}\mathbf{j} - \frac{2}{3}\mathbf{k} \Rightarrow \nabla f \cdot \mathbf{u} = \frac{2}{3}$. $\therefore$ $df = (\nabla f \cdot \mathbf{u})ds = \frac{2}{3}(0.1) = \frac{1}{15}$ or 0.067

26. $\mathbf{A} = \overrightarrow{P_0P_1} = -2\,\mathbf{i} + 2\,\mathbf{j} + 2\,\mathbf{k}$. $\nabla f = (1 + \cos z)\,\mathbf{i} + (1 - \sin z)\,\mathbf{j} + (-x \sin z - y \cos z)\,\mathbf{k} \Rightarrow \nabla f(2,-1,0) = 2\,\mathbf{i} + \mathbf{j} + \mathbf{k}$. $\mathbf{u} = \frac{\mathbf{A}}{|\mathbf{A}|} = \frac{-2\,\mathbf{i} + 2\,\mathbf{j} + 2\,\mathbf{k}}{\sqrt{(-2)^2 + 2^2 + 2^2}} = -\frac{1}{\sqrt{3}}\mathbf{i} + \frac{1}{\sqrt{3}}\mathbf{j} + \frac{1}{\sqrt{3}}\mathbf{k} \Rightarrow \nabla f \cdot \mathbf{u} = 0$. $\therefore$ $df = (\nabla f \cdot \mathbf{u})ds = 0(0.2) = 0$

27. $\nabla f = y\,\mathbf{i} + (x + 2y)\,\mathbf{j} \Rightarrow \nabla f(3,2) = 2\,\mathbf{i} + 7\,\mathbf{j}$. $\mathbf{A}$, orthogonal to ∇f, is $\mathbf{A} = 7\,\mathbf{i} - 2\,\mathbf{j} \Rightarrow \mathbf{u} = \frac{\mathbf{A}}{|\mathbf{A}|} = \frac{7\,\mathbf{i} - 2\,\mathbf{j}}{\sqrt{7^2 + (-2)^2}} = \frac{7}{\sqrt{53}}\mathbf{i} - \frac{2}{\sqrt{53}}\mathbf{j} \Rightarrow -\mathbf{u} = -\frac{7}{\sqrt{53}}\mathbf{i} + \frac{2}{\sqrt{53}}\mathbf{j}$

28. $\nabla f = \frac{4xy^2}{(x^2 + y^2)^2}\mathbf{i} - \frac{4x^2y}{(x^2 + y^2)^2}\mathbf{j} \Rightarrow \nabla f(1,1) = \mathbf{i} - \mathbf{j}$. $\mathbf{A}$, orthogonal to ∇f, is $\mathbf{A} = \mathbf{i} + \mathbf{j} \Rightarrow$ $\mathbf{u} = \frac{\mathbf{A}}{|\mathbf{A}|} = \frac{\mathbf{i} + \mathbf{j}}{\sqrt{1^2 + 1^2}} = \frac{1}{\sqrt{2}}\mathbf{i} + \frac{1}{\sqrt{2}}\mathbf{j} \Rightarrow -\mathbf{u} = -\frac{1}{\sqrt{2}}\mathbf{i} - \frac{1}{\sqrt{2}}\mathbf{j}$

29. $\nabla f = 2x\,\mathbf{i} + 2y\,\mathbf{j} + 2z\,\mathbf{k} = (2\cos t)\,\mathbf{i} + (2\sin t)\,\mathbf{j} + 2t\,\mathbf{k}$. $\mathbf{v} = (-\sin t)\,\mathbf{i} + (\cos t)\,\mathbf{j} + \mathbf{k} \Rightarrow \mathbf{T} = \frac{\mathbf{v}}{|\mathbf{v}|} =$
$\frac{(-\sin t)\,\mathbf{i} + (\cos t)\,\mathbf{j} + \mathbf{k}}{\sqrt{(\sin t)^2 + (\cos t)^2 + 1^2}} = \left(\frac{-\sin t}{\sqrt{2}}\right)\mathbf{i} + \left(\frac{\cos t}{\sqrt{2}}\right)\mathbf{j} + \frac{1}{\sqrt{2}}\mathbf{k}.$
$\left(D_{\mathbf{T}}f\right)_{P_0} = \nabla f \cdot \mathbf{T} = (2\cos t)\left(\frac{-\sin t}{\sqrt{2}}\right) + (2\sin t)\left(\frac{\cos t}{\sqrt{2}}\right) + 2t\left(\frac{1}{\sqrt{2}}\right) = \frac{2t}{\sqrt{2}} \Rightarrow \left(D_{\mathbf{T}}f\right)\left(\frac{-\pi}{4}\right) = \frac{-\pi}{2\sqrt{2}}$,
$\left(D_{\mathbf{T}}f\right)(0) = 0$, $\left(D_{\mathbf{T}}f\right)\left(\frac{\pi}{4}\right) = \frac{\pi}{2\sqrt{2}}$

30. $\nabla f = 2x\,\mathbf{i} + 2y\,\mathbf{j} = 2(\cos t + t\sin t)\,\mathbf{i} + 2(\sin t - t\cos t)\,\mathbf{j}$. $\mathbf{v} = (t\cos t)\,\mathbf{i} + (t\sin t)\,\mathbf{j} \Rightarrow$
$\mathbf{T} = \frac{\mathbf{v}}{|\mathbf{v}|} = \frac{(t\cos t)\,\mathbf{i} + (t\sin t)\,\mathbf{j}}{\sqrt{(t\cos t)^2 + (t\sin t)^2}} = (\cos t)\,\mathbf{i} + (\sin t)\,\mathbf{j}$ since $t > 0$. $\left(D_{\mathbf{T}}f\right)_{P_0} = \nabla f \cdot \mathbf{T} = 2(\cos t + t\sin t)\cos$
$2(\sin t - t\cos t)\sin t = 2$

31. $\nabla f = f_x(1,2)\,\mathbf{i} + f_y(1,2)\,\mathbf{j}$. $\mathbf{u_1} = \frac{\mathbf{i}+\mathbf{j}}{\sqrt{1^2+1^2}} = \frac{1}{\sqrt{2}}\mathbf{i} + \frac{1}{\sqrt{2}}\mathbf{j}$. $\left(D_{\mathbf{u_1}}f\right)(1,2) =$
$f_x(1,2)\left(\frac{1}{\sqrt{2}}\right) + f_y(1,2)\left(\frac{1}{\sqrt{2}}\right) = 2\sqrt{2} \Rightarrow f_x(1,2) + f_y(1,2) = 4$. $\mathbf{u_2} = -\mathbf{j}$. $\left(D_{\mathbf{u_2}}f\right)(1,2) = f_x(1,2)(0) + f_y(1,2)(-$
$= -3 \Rightarrow -f_y(1,2) = -3 \Rightarrow f_y(1,2) = 3$. $\therefore$ $f_x(1,2) + 3 = 4 \Rightarrow f_x(1,2) = 1$. Then $\nabla f(1,2) = \mathbf{i} + 3\,\mathbf{j}$.
$\mathbf{u} = \frac{\mathbf{A}}{|\mathbf{A}|} = \frac{-\mathbf{i} - 2\,\mathbf{j}}{\sqrt{(-1)^2 + (-2)^2}} = -\frac{1}{\sqrt{5}}\mathbf{i} - \frac{2}{\sqrt{5}}\mathbf{j} \Rightarrow \left(D_{\mathbf{u}}f\right)_{P_0} = \nabla f \cdot \mathbf{u} = -\frac{1}{\sqrt{5}} - \frac{6}{\sqrt{5}} = -\frac{7}{\sqrt{5}}$

32. a) $\left(D_{\mathbf{u}}f\right)_P = 2\sqrt{3} \Rightarrow |\nabla f| = 2\sqrt{3}$. $\mathbf{u} = \frac{\mathbf{A}}{|\mathbf{A}|} = \frac{\mathbf{i}+\mathbf{j}-\mathbf{k}}{\sqrt{1^2+1^2+(-1)^2}} = \frac{1}{\sqrt{3}}\mathbf{i} + \frac{1}{\sqrt{3}}\mathbf{j} - \frac{1}{\sqrt{3}}\mathbf{k}$. $\mathbf{u} = \frac{\nabla f}{|\nabla f|} \Rightarrow$
$\nabla f = |\nabla f|\mathbf{u} \Rightarrow \nabla f = 2\sqrt{3}\left(\frac{1}{\sqrt{3}}\mathbf{i} + \frac{1}{\sqrt{3}}\mathbf{j} - \frac{1}{\sqrt{3}}\mathbf{k}\right) = 2\,\mathbf{i} + 2\,\mathbf{j} - 2\,\mathbf{k}$

b) $\mathbf{A} = \mathbf{i} + \mathbf{j} \Rightarrow \mathbf{u} = \frac{\mathbf{A}}{|\mathbf{A}|} = \frac{\mathbf{i}+\mathbf{j}}{\sqrt{1^2+1^2}} = \frac{1}{\sqrt{2}}\mathbf{i} + \frac{1}{\sqrt{2}}\mathbf{j}$. $\left(D_{\mathbf{u}}f\right)_P = \nabla f \cdot \mathbf{u} = 2\left(\frac{1}{\sqrt{2}}\right)\mathbf{i} + 2\left(\frac{1}{\sqrt{2}}\right)\mathbf{j} = \sqrt{2}\,\mathbf{i} + \sqrt{2}\,\mathbf{j}$

33. $x = g(t)$, $y = h(t) \Rightarrow \mathbf{r} = g(t)\,\mathbf{i} + h(t)\,\mathbf{j} \Rightarrow \mathbf{v} = g'(t)\,\mathbf{i} + h'(t)\,\mathbf{j} \Rightarrow \mathbf{T} = \frac{\mathbf{v}}{|\mathbf{v}|} = \frac{g'(t)\,\mathbf{i} + h'(t)\,\mathbf{j}}{\sqrt{(g'(t))^2 + (h'(t))^2}}$
$z = f(x,y) \Rightarrow \frac{df}{dt} = \frac{\partial f}{\partial x}\frac{dx}{dt} + \frac{\partial f}{\partial y}\frac{dy}{dt} = \frac{\partial f}{\partial x}g'(t) + \frac{\partial f}{\partial y}h'(t)$. If $f(g(t),h(t)) = c$, then $\frac{df}{dt} = 0 \Rightarrow \frac{\partial f}{\partial x}g'(t) + \frac{\partial f}{\partial y}h'(t) =$
Now $\nabla f = \frac{\partial f}{\partial x}\mathbf{i} + \frac{\partial f}{\partial y}\mathbf{j}$. $\therefore$ $\left(D_{\mathbf{T}}f\right) = \nabla f \cdot \mathbf{T} = \frac{\frac{\partial f}{\partial x}g'(t) + \frac{\partial f}{\partial y}h'(t)}{\sqrt{(g'(t))^2 + (h'(t))^2}} = 0 \Rightarrow \nabla f$ is normal to $\mathbf{T}$

34. a) $\nabla(kf) = \frac{\partial(kf)}{\partial x}\mathbf{i} + \frac{\partial(kf)}{\partial y}\mathbf{j} + \frac{\partial(kf)}{\partial z}\mathbf{k} = k\left(\frac{\partial f}{\partial x}\right)\mathbf{i} + k\left(\frac{\partial f}{\partial y}\right)\mathbf{j} + k\left(\frac{\partial f}{\partial z}\right)\mathbf{k} = k\left(\frac{\partial f}{\partial x}\mathbf{i} + \frac{\partial f}{\partial y}\mathbf{j} + \frac{\partial f}{\partial z}\mathbf{k}\right) = k\,\nabla f$

b) $\nabla(f+g) = \frac{\partial(f+g)}{\partial x}\mathbf{i} + \frac{\partial(f+g)}{\partial y}\mathbf{j} + \frac{\partial(f+g)}{\partial z}\mathbf{k} = \left(\frac{\partial f}{\partial x} + \frac{\partial g}{\partial x}\right)\mathbf{i} + \left(\frac{\partial f}{\partial y} + \frac{\partial g}{\partial y}\right)\mathbf{j} + \left(\frac{\partial f}{\partial z} + \frac{\partial g}{\partial z}\right)\mathbf{k} =$
$\frac{\partial f}{\partial x}\mathbf{i} + \frac{\partial g}{\partial x}\mathbf{i} + \frac{\partial f}{\partial y}\mathbf{j} + \frac{\partial g}{\partial y}\mathbf{j} + \frac{\partial f}{\partial z}\mathbf{k} + \frac{\partial g}{\partial z}\mathbf{k} = \left(\frac{\partial f}{\partial x}\mathbf{i} + \frac{\partial f}{\partial y}\mathbf{j} + \frac{\partial f}{\partial z}\mathbf{k}\right) + \left(\frac{\partial g}{\partial x}\mathbf{i} + \frac{\partial g}{\partial y}\mathbf{j} + \frac{\partial g}{\partial z}\mathbf{k}\right) = \nabla f + \nabla g$

c) $\nabla(f - g) = \nabla f - \nabla g$ (Substitute $-g$ in for g in part b above)

d) $\nabla(fg) = \frac{\partial(fg)}{\partial x}\mathbf{i} + \frac{\partial(fg)}{\partial y}\mathbf{j} + \frac{\partial(fg)}{\partial z}\mathbf{k} = \left(\frac{\partial f}{\partial x}g + \frac{\partial g}{\partial x}f\right)\mathbf{i} + \left(\frac{\partial f}{\partial y}g + \frac{\partial g}{\partial y}f\right)\mathbf{j} + \left(\frac{\partial f}{\partial z}g + \frac{\partial g}{\partial z}f\right)\mathbf{k} =$
$\frac{\partial f}{\partial x}g\,\mathbf{i} + \frac{\partial g}{\partial x}f\,\mathbf{i} + \frac{\partial f}{\partial y}g\,\mathbf{j} + \frac{\partial g}{\partial y}f\,\mathbf{j} + \frac{\partial f}{\partial z}g\,\mathbf{k} + \frac{\partial g}{\partial z}f\,\mathbf{k} = f\left(\frac{\partial g}{\partial x}\mathbf{i} + \frac{\partial g}{\partial y}\mathbf{j} + \frac{\partial g}{\partial z}\mathbf{k}\right) + g\left(\frac{\partial f}{\partial x}\mathbf{i} + \frac{\partial f}{\partial y}\mathbf{j} + \frac{\partial f}{\partial z}\mathbf{k}\right) =$
$f\,\nabla g + g\,\nabla f$

SECTION 13.6 TANGENT PLANES AND NORMAL LINES

1. $\nabla f = 2x\,\mathbf{i} + 2y\,\mathbf{j} + 2z\,\mathbf{k} \Rightarrow \nabla f(1,1,1) = 2\,\mathbf{i} + 2\,\mathbf{j} + 2\,\mathbf{k} \Rightarrow$ Tangent plane: $2(x-1) + 2(y-1) + 2(z-1) = 0$ $\Rightarrow x + y + z = 3$; Normal line: $x = 1 + 2t$, $y = 1 + 2t$, $z = 1 + 2t$

2. $\nabla f = 2x\,\mathbf{i} + 2y\,\mathbf{j} - 2z\,\mathbf{k} \Rightarrow \nabla f(3,4,-5) = 6\,\mathbf{i} + 8\,\mathbf{j} + 10\,\mathbf{k} \Rightarrow$ Tangent plane: $6(x-3) + 8(y-4) + 10(z+5)$ $= 0 \Rightarrow 6x + 8y + 10z = 0$; Normal line: $x = 3 + 6t$, $y = 4 + 8t$, $z = -5 + 10t$

3. $\nabla f = -2x\,\mathbf{i} - 2y\,\mathbf{j} + 2z\,\mathbf{k} \Rightarrow \nabla f(3,4,-5) = -6\,\mathbf{i} - 8\,\mathbf{j} - 10\,\mathbf{k} \Rightarrow$ Tangent plane: $-6(x-3) - 8(y-4) -$ $10(z+5) = 0 \Rightarrow 3x + 4y + 5z = 0$; Normal line: $x = 3 - 6t$, $y = 4 - 8t$, $z = -5 - 10t$

4. $\nabla f = -2x\,\mathbf{i} + 2\,\mathbf{k} \Rightarrow \nabla f(2,0,2) = -4\,\mathbf{i} + 2\,\mathbf{k} \Rightarrow$ Tangent plane: $-4(x-2) + 2(z-2) = 0 \Rightarrow -4x + 2z + 4 = 0$; Normal line: $x = 2 - 4t$, $y = 0$, $z = 2 + 2t$

5. $\nabla f = \frac{-2x}{x^2+y^2}\,\mathbf{i} - \frac{2y}{x^2+y^2}\,\mathbf{j} \Rightarrow \nabla f(1,0,0) = -2\,\mathbf{i} + \mathbf{k} \Rightarrow$ Tangent plane: $-2(x-1) + z = 0 \Rightarrow -2x + z = -2$; Normal line: $x = 1 - 2t$, $y = 0$, $z = t$

6. $\nabla f = 2(x+y)\,\mathbf{i} + 2(x+y)\,\mathbf{j} + 2z\,\mathbf{k} \Rightarrow \nabla f(1,2,4) = 6\,\mathbf{i} + 6\,\mathbf{j} + 8\,\mathbf{k} \Rightarrow$ Tangent plane: $6(x-1) + 6(y-2) +$ $8(z-4) = 0 \Rightarrow 6x + 6y + 8z = 50$; Normal line: $x = 1 + 6t$, $y = 2 + 6t$, $z = 4 + 8t$

7. $\nabla f = (2x+2y)\,\mathbf{i} + (2x-2y)\,\mathbf{j} + 2z\,\mathbf{k} \Rightarrow \nabla f(1,-1,3) = 4\,\mathbf{j} + 6\,\mathbf{k} \Rightarrow$ Tangent plane: $4(y+1) + 6(z-3) = 0$ $\Rightarrow 2y + 3z = 7$; Normal line: $x = 1$, $y = -1 + 4t$, $z = 3 + 6t$

8. $\nabla f = \left(-\pi \sin \pi x + 2xy + ze^{xz}\right)\mathbf{i} + \left(-x^2 + z\right)\mathbf{j} + \left(xe^{xz} + y\right)\mathbf{k} \Rightarrow \nabla f(0,1,2) = 2\,\mathbf{i} + 2\,\mathbf{j} + \mathbf{k} \Rightarrow$ Tangent plane: $2(x-0) + 2(y-1) + 1(z-2) = 0 \Rightarrow 2x + 2y + z - 4 = 0$; Normal line: $x = 2t$, $y = 1 + 2t$, $z = 2 + t$

9. $\nabla f = (2x-y)\,\mathbf{i} - (x+2y)\,\mathbf{j} - \mathbf{k} \Rightarrow \nabla f(1,1,-1) = \mathbf{i} - 3\,\mathbf{j} - \mathbf{k} \Rightarrow$ Tangent plane: $1(x-1) - 3(y-1) -$ $(z+1) = 0 \Rightarrow x - 3y - z = -1$; Normal line: $x = 1 + t$, $y = 1 - 3t$, $z = -1 - t$

10. $\nabla f = \mathbf{i} + \mathbf{j} + \mathbf{k}$ for all points $\Rightarrow$ Tangent plane: $1(x-0) + 1(y-1) + 1(z-0) = 0 \Rightarrow x + y + z - 1 = 0$; Normal line: $x = t$, $y = 1 + t$, $z = t$

11. $\nabla f = (2x - 2y - 1)\,\mathbf{i} + (2y - 2x + 3)\,\mathbf{j} - \mathbf{k} \Rightarrow \nabla f(2,-3,18) = 9\,\mathbf{i} - 7\,\mathbf{j} - \mathbf{k} \Rightarrow$ Tangent plane: $9(x-2) -$ $7(y+3) - (z-18) = 0 \Rightarrow 9x - 7y - z = 21$; Normal line: $x = 2 + 9t$, $y = -3 - 7t$, $z = 18 - t$

12. $\nabla f = \mathbf{i} + \left(z - 3y^2\right)\mathbf{j} + (y - 4z)\,\mathbf{k} \Rightarrow \nabla f(-4,-2,1) = \mathbf{i} - 11\,\mathbf{j} - 6\,\mathbf{k} \Rightarrow$ Tangent plane: $1(x+4) -$ $11(y+2) - 6(z-1) = 0 \Rightarrow x - 11y - 6z - 12 = 0$; Normal line: $x = -4 + t$, $y = -2 - 11t$, $z = 1 - 6t$

13. $\nabla f = 2x\,\mathbf{i} + 2y\,\mathbf{j} \Rightarrow \nabla f(1,\sqrt{3},1) = 2\,\mathbf{i} + 2\sqrt{3}\,\mathbf{j}$

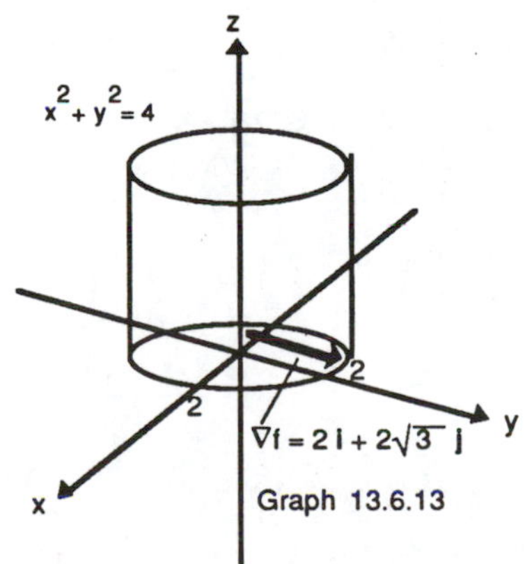

Graph 13.6.13

14. $\nabla f = 2x\,\mathbf{i} + 2z\,\mathbf{k} \Rightarrow \nabla f(1,-1,1) = 2\,\mathbf{i} + 2\,\mathbf{k}$

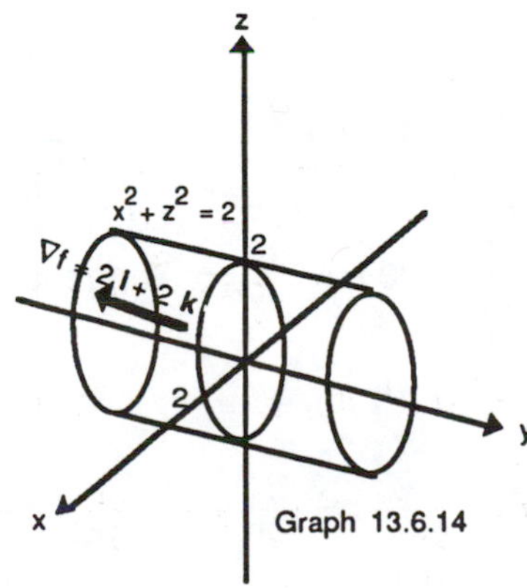

Graph 13.6.14

15. $\nabla f = 2x\,\mathbf{i} + 2y\,\mathbf{j} - \mathbf{k} \Rightarrow \nabla f(1,-1,1) = 2\,\mathbf{i} - 2\,\mathbf{j} - \mathbf{k}$

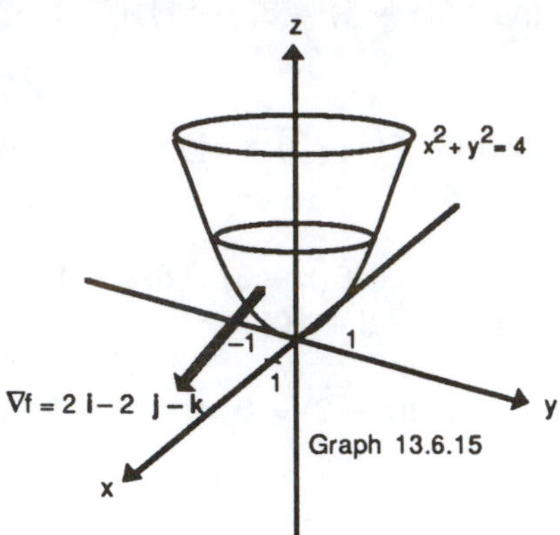

Graph 13.6.15

16. $\nabla f = 2x\,\mathbf{i} + 2y\,\mathbf{j} + \mathbf{k} \Rightarrow \nabla f(-1,-1,2) = -2\,\mathbf{i} - 2\,\mathbf{j} + \mathbf{k}$

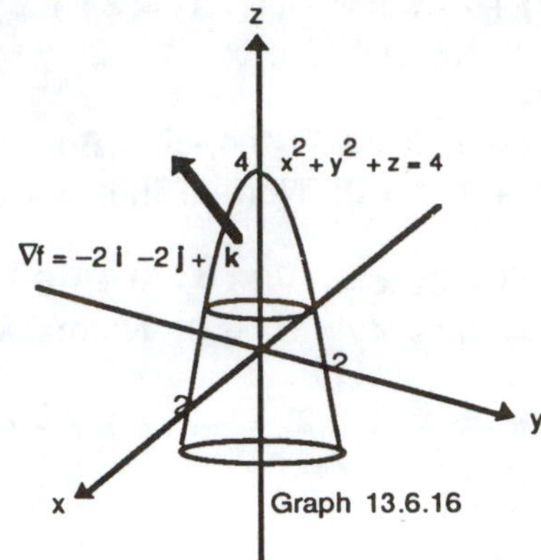

Graph 13.6.16

17. $\nabla f = 2x\,\mathbf{i} + 2y\,\mathbf{j} - 2z\,\mathbf{k} \Rightarrow \nabla f(2,-1,-\sqrt{5}) = 4\,\mathbf{i} - 2\,\mathbf{j} + 2\sqrt{5}\,\mathbf{k}$

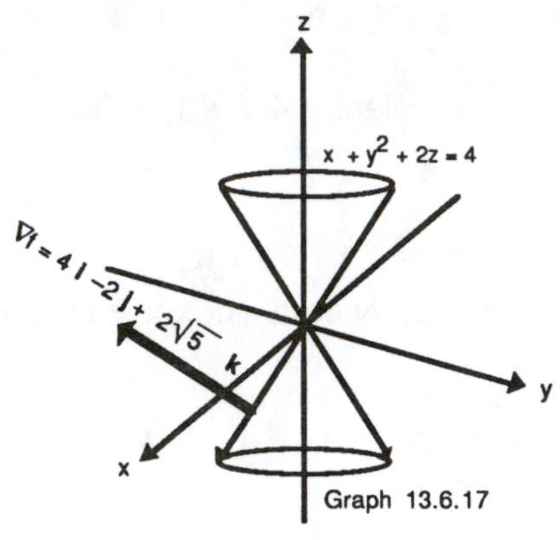

Graph 13.6.17

18. $\nabla f = 8y\,\mathbf{j} + 18z\,\mathbf{k} \Rightarrow \nabla f(0,\frac{3}{2},\sqrt{3}) = 12\,\mathbf{j} + 18\sqrt{3}\,\mathbf{k}$

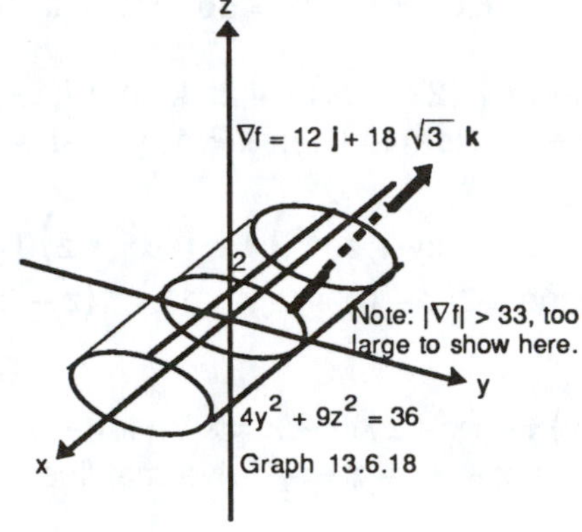

Graph 13.6.18

19. $\nabla f = \mathbf{i} + 2y\,\mathbf{j} + 2\,\mathbf{k} \Rightarrow \nabla f(1,1,1) = \mathbf{i} + 2\,\mathbf{j} + 2\,\mathbf{k}$. $\nabla g = \mathbf{i}$ for all points P. $\mathbf{v} = \nabla f \times \nabla g \Rightarrow$

$$\mathbf{v} = \begin{vmatrix} \mathbf{i} & \mathbf{j} & \mathbf{k} \\ 1 & 2 & 2 \\ 1 & 0 & 0 \end{vmatrix} = 2\,\mathbf{j} - 2\,\mathbf{k} \Rightarrow \text{Tangent line: } x = 1,\ y = 1 + 2t,\ z = 1 - 2t$$

20. $\nabla f = yz\,\mathbf{i} + xz\,\mathbf{j} + xy\,\mathbf{k} \Rightarrow \nabla f(1,1,1) = \mathbf{i} + \mathbf{j} + \mathbf{k}$. $\nabla g = 2x\,\mathbf{i} + 4y\,\mathbf{j} + 6z\,\mathbf{k} \Rightarrow \nabla g(1,1,1) = 2\,\mathbf{i} + 4\,\mathbf{j} + 6\,\mathbf{k}$.

$$\mathbf{v} = \nabla f \times \nabla g \Rightarrow \mathbf{v} = \begin{vmatrix} \mathbf{i} & \mathbf{j} & \mathbf{k} \\ 1 & 1 & 1 \\ 2 & 4 & 6 \end{vmatrix} = 2\,\mathbf{i} - 4\,\mathbf{j} + 2\,\mathbf{k} \Rightarrow \text{Tangent line: } x = 1 + 2t,\ y = 1 - 4t,\ z = 1 + 2t$$

21. $\nabla f = 2x\,\mathbf{i} + 2\,\mathbf{j} + 2\,\mathbf{k} \Rightarrow \nabla f(1,1,\frac{1}{2}) = 2\,\mathbf{i} + 2\,\mathbf{j} + 2\,\mathbf{k}$. $\nabla g = \mathbf{j}$ for all points P. $\mathbf{v} = \nabla f \times \nabla g \Rightarrow$

$$\mathbf{v} = \begin{vmatrix} \mathbf{i} & \mathbf{j} & \mathbf{k} \\ 2 & 2 & 2 \\ 0 & 1 & 0 \end{vmatrix} = -2\,\mathbf{i} + 2\,\mathbf{k} \Rightarrow \text{Tangent line: } x = 1 - 2t,\ y = 1,\ z = \frac{1}{2} + 2t$$

22. $\nabla f = \mathbf{i} + 2y\,\mathbf{j} + \mathbf{k} \Rightarrow \nabla f\left(\frac{1}{2}, 1, \frac{1}{2}\right) = \mathbf{i} + 2\,\mathbf{j} + \mathbf{k}$. $\nabla g = \mathbf{j}$ for all points P. $\mathbf{v} = \nabla f \times \nabla g \Rightarrow$

$$\mathbf{v} = \begin{vmatrix} \mathbf{i} & \mathbf{j} & \mathbf{k} \\ 1 & 2 & 1 \\ 0 & 1 & 0 \end{vmatrix} = -\,\mathbf{i} + \mathbf{k} \Rightarrow \text{Tangent line: } x = \frac{1}{2} - t,\ y = 1,\ z = \frac{1}{2} + t$$

23. $\nabla f = \left(3x^2 + 6xy^2 + 4y\right)\mathbf{i} + \left(6x^2y + 3y^2 + 4x\right)\mathbf{j} - 2z\,\mathbf{k} \Rightarrow \nabla f(1,1,3) = 13\,\mathbf{i} + 13\,\mathbf{j} - 6\,\mathbf{k}$.

$\nabla g = 2x\,\mathbf{i} + 2y\,\mathbf{j} + 2z\,\mathbf{k} \Rightarrow \nabla g(1,1,3) = 2\,\mathbf{i} + 2\,\mathbf{j} + 6\,\mathbf{k}$. $\mathbf{v} = \nabla f \times \nabla g \Rightarrow \mathbf{v} = \begin{vmatrix} \mathbf{i} & \mathbf{j} & \mathbf{k} \\ 13 & 13 & -6 \\ 2 & 2 & 6 \end{vmatrix} =$

$90\,\mathbf{i} - 90\,\mathbf{j} \Rightarrow$ Tangent line: $x = 1 + 90t,\ y = 1 - 90t,\ z = 3$

24. $\nabla f = 2x\,\mathbf{i} + 2y\,\mathbf{j} \Rightarrow \nabla f\left(\sqrt{2}, \sqrt{2}, 4\right) = 2\sqrt{2}\,\mathbf{i} + 2\sqrt{2}\,\mathbf{j}$. $\nabla g = 2x\,\mathbf{i} + 2y\,\mathbf{j} - \mathbf{k} \Rightarrow \nabla g\left(\sqrt{2}, \sqrt{2}, 4\right) = 2\sqrt{2}\,\mathbf{i} +$

$2\sqrt{2}\,\mathbf{j} - \mathbf{k}$. $\mathbf{v} = \nabla f \times \nabla g \Rightarrow \mathbf{v} = \begin{vmatrix} \mathbf{i} & \mathbf{j} & \mathbf{k} \\ 2\sqrt{2} & 2\sqrt{2} & 0 \\ 2\sqrt{2} & 2\sqrt{2} & -1 \end{vmatrix} = -2\sqrt{2}\,\mathbf{i} + 2\sqrt{2}\,\mathbf{j} \Rightarrow$ Tangent line: $x = \sqrt{2} - 2\sqrt{2}\,t$,

$y = \sqrt{2} + 2\sqrt{2}\,t,\ z = 4$

25. $\nabla f = 2x\,\mathbf{i} + 2y\,\mathbf{j} \Rightarrow \nabla f(\sqrt{2}, \sqrt{2}) = 2\sqrt{2}\,\mathbf{i} + 2\sqrt{2}\,\mathbf{j} \Rightarrow$ Tangent line: $2\sqrt{2}\left(x - \sqrt{2}\right) + 2\sqrt{2}\left(y - \sqrt{2}\right) = 0 \Rightarrow$
$\sqrt{2}\,x + \sqrt{2}\,y = 4$

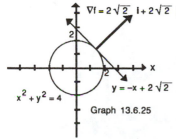

Graph 13.6.25

26. $\nabla f = 2x\,\mathbf{i} + y\,\mathbf{j} \Rightarrow \nabla f(1,2) = 2\,\mathbf{i} + 2\,\mathbf{j} \Rightarrow$ Tangent line: $2(x - 1) + 2(y - 2) = 0 \Rightarrow x + y = 3$

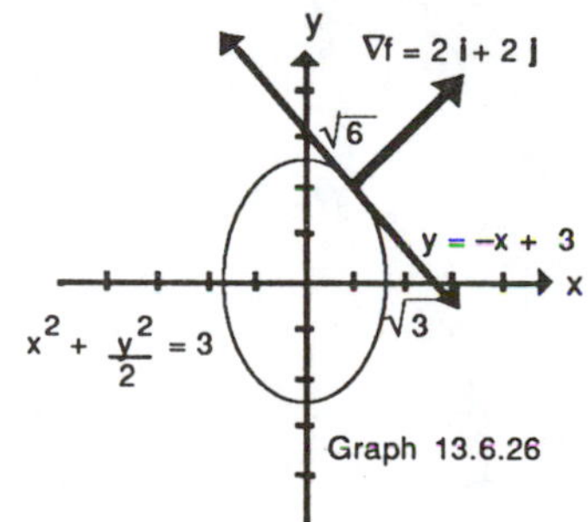

Graph 13.6.26

27. $\nabla f = 8x\,\mathbf{i} + 18y\,\mathbf{j} \Rightarrow \nabla f\left(2, \frac{2\sqrt{5}}{3}\right) = 16\,\mathbf{i} + 12\sqrt{5}\,\mathbf{j} \Rightarrow$ Tangent line: $16(x-2) + 12\sqrt{5}\left(y - \frac{2\sqrt{5}}{3}\right) = 0 \Rightarrow$ $4x + 3\sqrt{5}\,y = 18.$

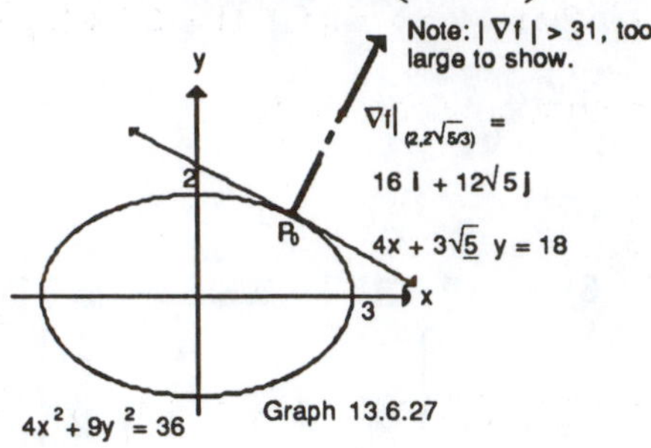

Graph 13.6.27

28. $\nabla f = 2x\,\mathbf{i} - \mathbf{j} \Rightarrow \nabla f(\sqrt{2}, 1) = 2\sqrt{2}\,\mathbf{i} - \mathbf{j} \Rightarrow$ Tangent line: $2\sqrt{2}\left(x - \sqrt{2}\right) - (y-1) = 0 \Rightarrow y = 2\sqrt{2}\,x - 3$

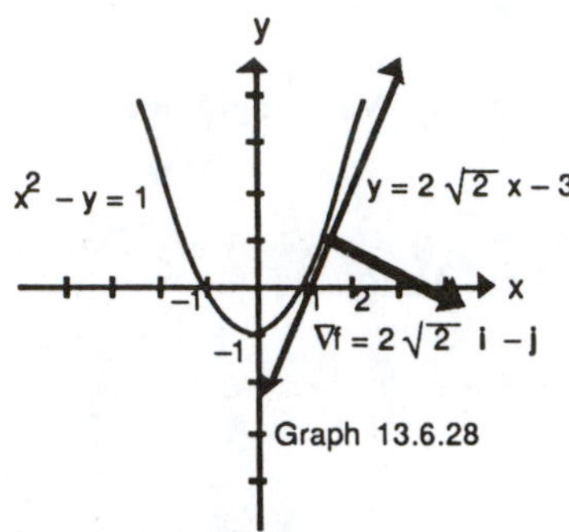

Graph 13.6.28

29. $\nabla f = y\,\mathbf{i} + x\,\mathbf{j} \Rightarrow \nabla f(2,-2) = -2\,\mathbf{i} + 2\,\mathbf{j} \Rightarrow$ Tangent line: $-2(x-2) + 2(y+2) = 0 \Rightarrow y = x - 4$

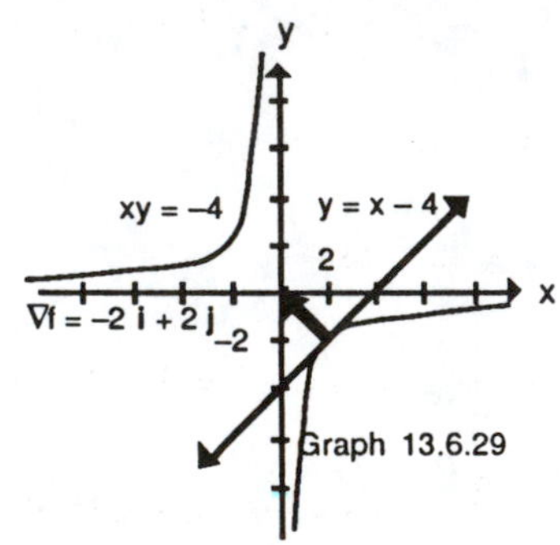

Graph 13.6.29

30. $\nabla f = (2x - y)\,\mathbf{i} + (2y - x)\,\mathbf{j} \Rightarrow \nabla f(-1,2) = -4\,\mathbf{i} + 5\,\mathbf{j} \Rightarrow$ Tangent line: $-4(x+1) + 5(y-2) = 0 \Rightarrow$
$-4x + 5y - 14 = 0$

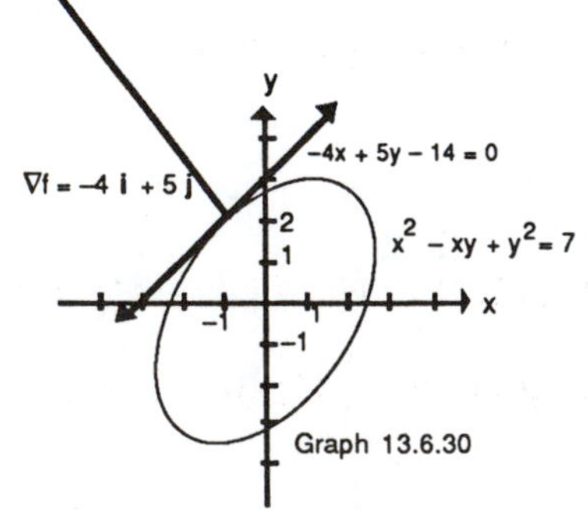

Graph 13.6.30

31. $\mathbf{r} = \sqrt{t}\,\mathbf{i} + \sqrt{t}\,\mathbf{j} - \frac{1}{4}(t+3)\,\mathbf{k} \Rightarrow \mathbf{v} = \frac{1}{2}t^{-1/2}\,\mathbf{i} + \frac{1}{2}t^{-1/2}\,\mathbf{j} - \frac{1}{4}\mathbf{k}$. $t = 1 \Rightarrow x = 1,\ y = 1,\ z = -1 \Rightarrow P_0 = (1,1,-1)$.

Also $t = 1 \Rightarrow \mathbf{v}(1) = \frac{1}{2}\mathbf{i} + \frac{1}{2}\mathbf{j} - \frac{1}{4}\mathbf{k}$. $f(x,y,z) = x^2 + y^2 - z - 3 = 0 \Rightarrow \nabla f = 2x\,\mathbf{i} + 2y\,\mathbf{j} - \mathbf{k} \Rightarrow \nabla f(1,1,-1) =$ $2\,\mathbf{i} + 2\,\mathbf{j} - \mathbf{k}$. $\therefore\ \mathbf{v} = \frac{1}{4}(\nabla f) \Rightarrow$ The curve is normal to the surface.

32. $\mathbf{r} = \sqrt{t}\,\mathbf{i} + \sqrt{t}\,\mathbf{j} + (2t - 1)\,\mathbf{k} \Rightarrow \mathbf{v} = \frac{1}{2}t^{-1/2}\,\mathbf{i} + \frac{1}{2}t^{-1/2}\,\mathbf{j} + 2\,\mathbf{k}$. $t = 1 \Rightarrow x = 1, y = 1, z = 1$. Also $t = 1 \Rightarrow$ $\mathbf{v}(1) = \frac{1}{2}\,\mathbf{i} + \frac{1}{2}\,\mathbf{j} + 2\,\mathbf{k}$. $f(x,y,z) = x^2 + y^2 - z - 1 = 0 \Rightarrow \nabla f = 2x\,\mathbf{i} + 2y\,\mathbf{j} - \mathbf{k} \Rightarrow \nabla f(1,1,1) = 2\,\mathbf{i} + 2\,\mathbf{j} - \mathbf{k} \Rightarrow$ $\mathbf{v} \cdot \nabla f = \frac{1}{2}(2) + \frac{1}{2}(2) + 2(-1) = 0 \Rightarrow$ The curve is tangent to the surface when $t = 1$.

SECTION 13.7 LINEARIZATION AND DIFFERENTIALS

1. a) $f(0,0) = 1$, $f_x(x,y) = 2x \Rightarrow f_x(0,0) = 0$, $f_y(x,y) = 2y \Rightarrow f_y(0,0) = 0 \Rightarrow L(x,y) = 1 + 0(x - 0) + 0(y - 0) = 1$
   b) $f(1,1) = 3$, $f_x(1,1) = 2$, $f_y(1,1) = 2 \Rightarrow L(x,y) = 3 + 2(x - 1) + 2(y - 1) = 2x + 2y - 1$

2. a) $f(1,1) = 1$, $f_x(x,y) = 3x^2y^4 \Rightarrow f_x(1,1) = 3$, $f_y(x,y) = 4x^3y^3 \Rightarrow f_y(1,1) = 4 \Rightarrow L(x,y) = 1 + 3(x - 1) + 4(y - 1)$ $\Rightarrow 3x + 4y - 6$
   b) $f(0,0) = 0$, $f_x(0,0) = 0$, $f_y(0,0) = 0 \Rightarrow L(x,y) = 0$

3. a) $f(0,0) = 1$, $f_x(x,y) = e^x \cos y \Rightarrow f_x(0,0) = 1$, $f_y(x,y) = -e^x \sin y \Rightarrow f_y(0,0) = 0 \Rightarrow L(x,y) = 1 + 1(x - 0) +$ $0(y - 0) = 1 + x$
   b) $f\left(0,\frac{\pi}{2}\right) = 0$, $f_x\left(0,\frac{\pi}{2}\right) = 0$, $f_y\left(0,\frac{\pi}{2}\right) = -1 \Rightarrow L(x,y) = 0 + 0(x - 0) - 1\left(y - \frac{\pi}{2}\right) = -y + \frac{\pi}{2}$

4. a) $f(0,0) = 4$, $f_x(x,y) = 2(x + y + 2) \Rightarrow f_x(0,0) = 4$, $f_y(x,y) = 2(x + y + 2) \Rightarrow f_y(0,0) = 4 \Rightarrow L(x,y) =$ $4 + 4(x - 0) + 4(y - 0) = 4 + 4x + 4y$
   b) $f(1,2) = 25$, $f_x(1,2) = 10$, $f_y(1,2) = 10 \Rightarrow L(x,y) = 25 + 10(x - 1) + 10(y - 2) = 10x + 10y - 5$

5. a) $f(0,0) = 5$, $f_x(x,y) = 3$ for all (x,y), $f_y(x,y) = -4$ for all $(x,y) \Rightarrow L(x,y) = 5 + 3(x - 0) - 4(y - 0) =$ $5 + 3x - 4y$
   b) $f(1,1) = 4$, $f_x(1,1) = 3$, $f_y(1,1) = -4 \Rightarrow L(x,y) = 4 + 3(x - 1) - 4(y - 1) = 3x - 4y + 5$

6. a) $f(0,0) = 1$, $f_x(x,y) = -e^{2y-x} \Rightarrow f_x(0,0) = -1$, $f_y(x,y) = 2e^{2y-x} \Rightarrow f_y(0,0) = 2 \Rightarrow L(x,y) = 1 - 1(x - 0) +$ $2(y - 0) = 1 - x + 2y$
   b) $f(1,2) = e^3$, $f_x(1,2) = -e^3$, $f_y(1,2) = 2e^3 \Rightarrow L(x,y) = e^3 - e^3(x - 1) + 2e^3(y - 2) = -e^3\,x + 2e^3\,y - 2e^3$

7. $f(2,1) = 3$, $f_x(x,y) = 2x - 3y \Rightarrow f_x(2,1) = 1$, $f_y(x,y) = -3x \Rightarrow f_y(2,1) = -6 \Rightarrow L(x,y) = 3 + 1(x - 2) - 6(y - 1)$ $= 7 + x - 6y$. $f_{xx}(x,y) = 2$, $f_{yy}(x,y) = 0$, $f_{xy}(x,y) = -3 \Rightarrow M = 3$. $\therefore$ $|E(x,y)| \le \frac{1}{2}(3)(|x - 2| + |y - 1|)^2 \le$ $\frac{3}{2}(0.1 + 0.1)^2 = 0.06$

8. $f(2,2) = 11$, $f_x(x,y) = x + y + 3 \Rightarrow f_x(2,2) = 7$, $f_y(x,y) = x + \frac{1}{2}y - 3 \Rightarrow f_y(2,2) = 0 \Rightarrow L(x,y) = 11 + 7(x - 2) +$ $0(y - 2) = 7x - 3$. $f_{xx}(x,y) = 1$, $f_{yy}(x,y) = \frac{1}{2}$, $f_{xy}(x,y) = 1 \Rightarrow M = 1$. $\therefore$ $|E(x,y)| \le \frac{1}{2}(1)(|x - 2| + |y - 2|)^2$ $\le \frac{1}{2}(0.1 + 0.1)^2 = 0.02$

9. $f(0,0) = 1$, $f_x(x,y) = \cos y \Rightarrow f_x(0,0) = 1$, $f_y(x,y) = 1 - x\sin y \Rightarrow f_y(0,0) = 1 \Rightarrow L(x,y) = 1 + 1(x - 0) +$ $1(y - 0) = x + y + 1$. $f_{xx}(x,y) = 0$, $f_{yy}(x,y) = 0$, $f_{xy}(x,y) = -\sin y \Rightarrow M = 1$. $\therefore$ $|E(x,y)| \le \frac{1}{2}(1)(|x| + |y|)^2 \le \frac{1}{2}(0.2 + 0.2)^2 = 0.08$

10. $f(1,1) = 0$, $f_x(x,y) = \frac{1}{x} \Rightarrow f_x(1,1) = 1$, $f_y(x,y) = \frac{1}{y} \Rightarrow f_y(1,1) = 1 \Rightarrow L(x,y) = 0 + 1(x-1) + 1(y-1) = x + y - 2$. $f_{xx}(x,y) = -\frac{1}{x^2}$, $f_{yy}(x,y) = -\frac{1}{y^2}$, $f_{xy}(x,y) = 0$. $|x-1| \le 0.1 \Rightarrow 0.99 \le x \le 1.1 \Rightarrow$ max of $|f_{xx}(x,y)|$ on R is $\frac{1}{(0.99)^2} \le 1.03$. $|y-1| \le 0.1 \Rightarrow 0.99 \le y \le 1.1 \Rightarrow$ max of $|f_{yy}(x,y)|$ on R is $\frac{1}{(0.99)^2} \le 1.03. \Rightarrow M = 1.03$. $\therefore$ $|E(x,y)| \le \frac{1}{2}(1.03)(|x-1| + |y-1|)^2 \le 0.515(0.1 + 0.1)^2 = 0.0206$

11. $f(0,0) = 1$, $f_x(x,y) = e^x \cos y \Rightarrow f_x(0,0) = 1$, $f_y(x,y) = -e^x \sin y \Rightarrow f_y(0,0) = 0 \Rightarrow L(x,y) = 1 + 1(x-0) + 0(y-0) = 1 + x$. $f_{xx}(x,y) = e^x \cos y$, $f_{yy}(x,y) = -e^x \cos y$, $f_{xy}(x,y) = -e^x \sin y$. $|x| \le 0.1 \Rightarrow -0.1 \le x \le 0.1$, $|y| \le 0.1 \Rightarrow -0.1 \le y \le 0.1. \Rightarrow$ max of $|f_{xx}(x,y)|$ on R is $e^{0.1}\cos(0.1) \le 1.11$, max of $|f_{yy}(x,y)|$ on R is $e^{0.1}\cos(0.1) \le 1.11$, max of $|f_{xy}(x,y)|$ on R is $e^{0.1}\sin(0.1) \le 0.002 \Rightarrow M = 1.11$.
$\therefore$ $|E(x,y)| \le \frac{1}{2}(1.11)(|x| + |y|)^2 \le 0.555(0.1 + 0.1)^2 = 0.0222$

12. $f_x(x,y) = 2x(y+1) \Rightarrow f_x(1,0) = 2$, $f_y(x,y) = x^2 \Rightarrow f_y(1,0) = 1 \Rightarrow df = 2\,dx + 1\,dy \Rightarrow df$ is more sensitive to changes in x since $f_x > f_y$.

13. Let the width, w, be the long side. Then $A = lw \Rightarrow dA = A_l\,dl + A_w\,dw \Rightarrow dA = w\,dl + l\,dw$. Since $w > l$, dA is more sensitive to a change in w than l. $\therefore$ pay more attention to the width.

14. a) $f_x(x,y) = 2x(y+1) \Rightarrow f_x(1,0) = 2$, $f_y(x,y) = x^2 \Rightarrow f_y(1,0) = 1$. Then $df = 2\,dx + 1\,dy \Rightarrow df$ is more sensitive to changes in x since $f_x > f_y$.
b) $df = 2\,dx + dy \Rightarrow \frac{df}{dy} = 2\frac{dx}{dy} + 1$. $df = 0 \Rightarrow \frac{df}{dy} = 0 \Rightarrow 2\frac{dx}{dy} + 1 = 0 \Rightarrow \frac{dx}{dy} = -\frac{1}{2}$

15. $T_x(x,y) = e^y + e^{-y}$, $T_y(x,y) = x(e^y - e^{-y}) \Rightarrow dT = T_x(x,y)\,dx + T_y(x,y)\,dy = (e^y + e^{-y})dx + x(e^y - e^{-y})dy \Rightarrow dT|_{(2,\ln 2)} = 2.5\,dx + 3.0\,dy$. If $|dx| \le 0.1$, $|dy| \le 0.02$, then the maximum possible error (estimate) $\le 2.5(0.1) + 3.0(0.02) = 0.31$ in magnitude.

16. $V_r = 2\pi rh$, $V_h = \pi r^2 \Rightarrow dV = V_r\,dr + V_h\,dh \Rightarrow \frac{dV}{V} = \frac{2\pi rh\,dr + \pi r^2\,dh}{\pi r^2 h} = \frac{2}{r}dr + \frac{1}{h}dh$. Now, $\left|\frac{dr}{r} \times 100\right| \le 1$, $\left|\frac{dh}{h} \times 100\right| \le 1$. $\therefore$ $\left|\frac{dV}{V} \times 100\right| \le \left|2\frac{dr}{r} \times 100 + \frac{dh}{h} \times 100\right| \le 2\left|\frac{dr}{r} \times 100\right| + \left|\frac{dh}{h} \times 100\right| \le 2(1) + 1 = 3$ or 3%

17. $V_r = 2\pi rh$, $V_h = \pi r^2 \Rightarrow dV = V_r\,dr + V_h\,dh \Rightarrow dV = 2\pi rh\,dr + \pi r^2\,dh \Rightarrow dV|_{(5,12)} = 120\pi\,dr + 25\pi\,dh$. Since $|dr| \le 0.1$ cm, $|dh| \le 0.1$ cm, $dV \le 120\pi(0.1) + 25\pi(0.1) = 14.5\pi\text{ cm}^3$. $V(5,12) = 300\pi\text{ cm}^3 \Rightarrow$ Maximum percentage error $= \pm\frac{14.5\pi}{300\pi} \times 100 = \pm 4.83\%$

18. $V_r = 2\pi rh$, $V_h = \pi r^2 \Rightarrow dV = V_r\,dr + V_h\,dh \Rightarrow dV = 2\pi rh\,dr + \pi r^2\,dh$. Assume $dr = dh \Rightarrow dV = 2\pi rh\,dr + \pi r^2\,dr = (2\pi rh + \pi r^2)\,dr$. For $dV \le 0.1\text{ m}^3$ when $r = 2$ m, $h = 3$ m, $(2\pi(2)(3) + \pi(2)^2) \le 0.1 \Rightarrow dr \le \frac{0.1}{16\pi} \approx 0.001$ m (rounded down). Thus, the error in measuring r and h should be less than or equal to 0.002 m.

19. $df = f_x(x,y)\,dx + f_y(x,y)\,dy = 3x^2y^4dx + 4x^3y^3dy \Rightarrow df|_{(1,1)} = 3\,dx + 4\,dy$. Let $dx = dy \Rightarrow df = 7\,dx$. $|df| \le 0.1 \Rightarrow 7|dx| \le 0.1 \Rightarrow |dx| \le \frac{0.1}{7} \approx 0.014$. $\therefore$ for the square, let $|x-1| \le 0.014$, $|y-1| \le 0.014$

20. a) $\frac{1}{R} = \frac{1}{R_1} + \frac{1}{R_2} \Rightarrow R = \frac{R_1R_2}{R_1+R_2} \Rightarrow dR = R_{R_1}dR_1 + R_{R_2}dR_2 = \frac{(R_2+R_1)R_2 - (R_1R_2)}{(R_1+R_2)^2}dR_1 + \frac{(R_2+R_1)R_1 - (R_1R_2)}{(R_2+R_1)^2}dR_2 = \frac{R_2^2}{(R_2+R_1)^2}dR_1 + \frac{R_1^2}{(R_2+R_1)^2} = \left(\frac{R}{R_1}\right)^2 dR_1 + \left(\frac{R}{R_2}\right)^2 dR_2$

b) $dR = R^2\left[\left(\frac{1}{R_1^2}\right)dR_1 + \left(\frac{1}{R_2^2}\right)dR_2\right] \Rightarrow dR|_{(100,400)} = R^2\left[\frac{1}{(100)^2}dR_1 + \frac{1}{(400)^2}dR_2\right]$

$\therefore$ R will be more sensitive to a change in R_1 since $\frac{1}{(100)^2} > \frac{1}{(400)^2}$.

21. $dR = \left(\frac{R}{R_1}\right)^2 dR_1 + \left(\frac{R}{R_2}\right)^2 dR_2$ (See Exercise 20 above). R_1 changes from 20 to 20.1 ohms $\Rightarrow dR_1 =$ 0.1 ohms, R_2 changes from 25 to 24.9 ohms $\Rightarrow dR_2 = -0.1$ ohms. $\frac{1}{R} = \frac{1}{R_1} + \frac{1}{R_2} \Rightarrow R = \frac{100}{9}$ ohms. $dR|_{(20,25)} = \frac{(100/9)^2}{(20)^2}(0.1) + \frac{(100/9)^2}{(25)^2}(-0.1) = 0.011$ ohms $\Rightarrow$ Percentage change $= \frac{dR}{R}\Big|_{(20,25)} \times 100 = \frac{0.011}{100/9} \times 100 \approx 0.099\%$

22. a) $r^2 = x^2 + y^2 \Rightarrow r = \sqrt{x^2+y^2} \Rightarrow dr = \frac{x}{\sqrt{x^2+y^2}}dx + \frac{y}{\sqrt{x^2+y^2}}dy \Rightarrow dr|_{(3,4)} = \frac{3}{\sqrt{3^2+4^2}}(\pm 0.01) + \frac{4}{\sqrt{3^2+4^2}}(\pm 0.01) = \pm 0.014$. $x = 3, y = 4 \Rightarrow r = 5$. $\therefore \left|\frac{dr}{r} \times 100\right| = \left|\frac{\pm 0.014}{5} \times 100\right| = 0.28\%$

$d\theta = \frac{-y/x^2}{\left(\frac{y}{x}\right)^2 + 1}dx + \frac{1/x}{\left(\frac{y}{x}\right)^2 + 1}dy = \frac{-y}{y^2+x^2}dx + \frac{x}{y^2+x^2}dy \Rightarrow d\theta|_{(3,4)} = \frac{-4}{25}(\pm 0.01) + \frac{3}{25}(\pm 0.01) = \frac{\mp 0.04}{25} + \frac{\pm 0.03}{25} \Rightarrow$ Maximum change in $d\theta$ occurs when dx and dy have opposite signs ($dx = 0.01$ and $dy = -0.01$ or vice versa) $\therefore d\theta = \frac{\pm 0.07}{25} \approx \pm 0.0028$. $\theta = \tan^{-1}\left(\frac{4}{3}\right) \approx 0.927255218 \Rightarrow$ $\left|\frac{d\theta}{\theta} \times 100\right| = \left|\frac{\pm 0.0028}{0.927255218} \times 100\right| \approx 0.30\%$

b) r is more sensitive to changes in y; θ is more sensitive to changes in x.

23. a) $f(1,1,1) = 3$, $f_x(1,1,1) = y + z|_{(1,1,1)} = 2$, $f_y(1,1,1) = x + z|_{(1,1,1)} = 2$, $f_z(1,1,1) = y + x|_{(1,1,1)} = 2 \Rightarrow$ $L(x,y,z) = 2x + 2y + 2z - 3$

b) $f(1,0,0) = 0$, $f_x(1,0,0) = 0$, $f_y(1,0,0) = 1$, $f_z(1,0,0) = 1 \Rightarrow L(x,y,z) = y + z$

c) $f(0,0,0) = 0$, $f_x(0,0,0) = 0$, $f_y(0,0,0) = 0$, $f_z(0,0,0) = 0 \Rightarrow L(x,y,z) = 0$

24. a) $f(1,1,1) = 3$, $f_x(1,1,1) = 2x\big|_{(1,1,1)} = 2$, $f_y(1,1,1) = 2y\big|_{(1,1,1)} = 2$, $f_z(1,1,1) = 2z\big|_{(1,1,1)} = 2 \Rightarrow$ $L(x,y,z) = 2x + 2y + 2z - 3$

b) $f(0,1,0) = 1$, $f_x(0,1,0) = 0$, $f_y(0,1,0) = 2$, $f_z(0,1,0) = 0 \Rightarrow L(x,y,z) = 2y - 1$

c) $f(1,0,0) = 1$, $f_x(1,0,0) = 2$, $f_y(1,0,0) = 0$, $f_z(1,0,0) = 0 \Rightarrow L(x,y,z) = 2x - 1$

25. a) $f(1,0,0) = 1$, $f_x(1,0,0) = \left.\frac{x}{\sqrt{x^2+y^2+z^2}}\right|_{(1,0,0)} = 1$, $f_y(1,0,0) = \left.\frac{y}{\sqrt{x^2+y^2+z^2}}\right|_{(1,0,0)} = 0$, $f_z(1,0,0) = \left.\frac{z}{\sqrt{x^2+y^2+z^2}}\right|_{(1,0,0)} = 0 \Rightarrow L(x,y,z) = x$

b) $f(1,1,0) = \sqrt{2}$, $f_x(1,1,0) = \frac{1}{\sqrt{2}}$, $f_y(1,1,0) = \frac{1}{\sqrt{2}}$, $f_z(1,1,0) = 0 \Rightarrow L(x,y,z) = \frac{1}{\sqrt{2}}x + \frac{1}{\sqrt{2}}y$

c) $f(1,2,2) = 3$, $f_x(1,2,2) = \frac{1}{3}$, $f_y(1,2,2) = \frac{2}{3}$, $f_z(1,2,2) = \frac{2}{3} \Rightarrow L(x,y,z) = \frac{1}{3}x + \frac{2}{3}y + \frac{2}{3}z$

26. a) $f\left(\frac{\pi}{2},1,1\right) = 1$, $f_x\left(\frac{\pi}{2},1,1\right) = \left.\frac{y\cos xy}{z}\right|_{(\pi/2,1,1)} = 0$, $f_y\left(\frac{\pi}{2},1,1\right) = \left.\frac{x\cos xy}{z}\right|_{(\pi/2,1,1)} = 0$, $f_z\left(\frac{\pi}{2},1,1\right) = \left.\frac{-\sin xy}{z^2}\right|_{(\pi/2,1,1)} = -1 \Rightarrow L(x,y,z)) = 2 - z$

b) $f(2,0,1) = 0$, $f_x(2,0,1) = 0$, $f_y(2,0,1) = 2$, $f_z(2,0,1) = 0 \Rightarrow L(x,y,z) = 2y$

27. a) $f(0,0,0) = 2$, $f_x(0,0,0) = \left.e^x\right|_{(0,0,0)} = 1$, $f_y(0,0,0) = \left.-\sin(y+z)\right|_{(0,0,0)} = 0$, $f_z(0,0,0) = \left.-\sin(y+z)\right|_{(0,0,0)} = 0 \Rightarrow L(x,y,z) = 2 + x$

b) $f\left(0,\frac{\pi}{2},0\right) = 1$, $f_x\left(0,\frac{\pi}{2},0\right) = 1$, $f_y\left(0,\frac{\pi}{2},0\right) = -1$, $f_z\left(0,\frac{\pi}{2},0\right) = -1 \Rightarrow L(x,y,z) = x - y - z + \frac{\pi}{2} + 1$

c) $f\left(0,\frac{\pi}{4},\frac{\pi}{4}\right) = 1$, $f_x\left(0,\frac{\pi}{4},\frac{\pi}{4}\right) = 1$, $f_y\left(0,\frac{\pi}{4},\frac{\pi}{4}\right) = -1$, $f_z\left(0,\frac{\pi}{4},\frac{\pi}{4}\right) = -1 \Rightarrow L(x,y,z) = x - y - z + \frac{\pi}{2} + 1$

28. a) $f(1,0,0) = 0$, $f_x(1,0,0) = \left.\frac{yz}{(xyz)^2+1}\right|_{(1,0,0)} = 0$, $f_y(1,0,0) = \left.\frac{xz}{(xyz)^2+1}\right|_{(1,0,0)} = 0$, $f_z(1,0,0) = \left.\frac{xy}{(xyz)^2+1}\right|_{(1,0,0)} = 0 \Rightarrow L(x,y,z) = 0$

b) $f(1,1,0) = 0$, $f_x(1,1,0) = 0$, $f_y(1,1,0) = 0$, $f_z(1,1,0) = 1 \Rightarrow L(x,y,z) = z$

c) $f(1,1,1) = \frac{\pi}{4}$, $f_x(1,1,1) = 1$, $f_y(1,1,1) = 1$, $f_z(1,1,1) = 1 \Rightarrow L(x,y,z) = x + y + z + \frac{\pi}{4} - 3$

29. $f(a,b,c,d) = \begin{vmatrix} a & b \\ c & d \end{vmatrix} = ad - bc \Rightarrow f_a = d$, $f_b = -c$, $f_c = -b$, $f_d = a \Rightarrow df = d\,da - c\,db - b\,dc + a\,dd$. Since $|a|$ is much greater than $|b|$, $|c|$, and $|d|$, f is most sensitive to a change in d.

30. $p(a,b,c) = abc \Rightarrow p_a = bc$, $p_b = ac$, $p_c = ab \Rightarrow dp = bc\,da + ac\,db + ab\,dc \Rightarrow \frac{dp}{p} = \frac{bc\,da + ac\,db + ab\,d}{abc} = \frac{da}{a} + \frac{db}{b} + \frac{dc}{c}$. Since $\left|\frac{da}{a} \times 100\right| = 2$, $\left|\frac{db}{b} \times 100\right| = 2$, $\left|\frac{dc}{c} \times 100\right| = 2$, $\left|\frac{dp}{p} \times 100\right| = \left|\frac{da}{a} \times 100 + \frac{db}{b} \times 100 + \frac{dc}{c} \times 100\right| \le \left|\frac{da}{a} \times 100\right| + \left|\frac{db}{b} \times 100\right| + \left|\frac{dc}{c} \times 100\right| = 2 + 2 + 2 = 6$ or 6%

31. $V = lwh \Rightarrow V_l = wh$, $V_w = lh$, $V_h = lw \Rightarrow dV = wh\,dl + lh\,dw + lw\,dh \Rightarrow \left.dV\right|_{(5,3,2)} = 6\,dl + 10\,dw + 15\,dh$
$dl = 1 \text{ in} = \frac{1}{12} \text{ ft}$, $dw = 1 \text{ in} = \frac{1}{12} \text{ ft}$, $dh = \frac{1}{2} \text{ in} = \frac{1}{24} \text{ ft} \Rightarrow dV = 6\left(\frac{1}{12}\right) + 10\left(\frac{1}{12}\right) + 15\left(\frac{1}{24}\right) = \frac{47}{24} \text{ ft}^3$

32. $A = \frac{1}{2}ab\sin C \Rightarrow A_a = \frac{1}{2}b\sin C$, $A_b = \frac{1}{2}a\sin C$, $A_C = \frac{1}{2}ab\cos C \Rightarrow dA = \frac{1}{2}b\sin C\,da + \frac{1}{2}a\sin C\,db + \frac{1}{2}ab\cos C\,dC$. $dC = \pm 2° = \pm 0.0349$ radians, $da = \pm 0.5$ ft, $db = \pm 0.5$ ft. At $a = 150$ ft, $b = 200$ ft, $C = 60°$, $dA = \frac{1}{2}(200)\sin 60°\,(\pm 0.5) + \frac{1}{2}(150)\sin 60°\,(\pm 0.5) + \frac{1}{2}(200)(150)\cos 60°\,(\pm 0.0349) = \pm 319.23 \text{ ft}^2$

33. $u_x = e^y$, $u_y = xe^y + \sin z$, $u_z = y \cos z \Rightarrow du = e^y\,dx + (xe^y + \sin z)dy + (y \cos z)dz \Rightarrow$
$du\big|_{(2,\ln 3,\pi/2)} = 3\,dx + 7\,dy + 0\,dz = 3\,dx + 7\,dy \Rightarrow$ magnitude of the maximum possible error $\le$
$3(0.2) + 7(0.6) = 4.8$

34. $Q_k = \frac{1}{2}(2kM/h)^{-1/2}(2M/h)$, $Q_M = \frac{1}{2}(2kM/h)^{-1/2}(2k/h)$, $Q_h = \frac{1}{2}(2kM/h)^{-1/2}(-2kM/h^2) \Rightarrow$
$dQ = \frac{1}{2}(2kM/h)^{-1/2}(2M/h)\,dk + \frac{1}{2}(2kM/h)^{-1/2}(2k/h)\,dM + \frac{1}{2}(2kM/h)^{-1/2}(-2kM/h^2)\,dh =$
$\frac{1}{2}(2kM/h)^{-1/2}\left[\frac{2M}{h}\,dk + \frac{2k}{h}\,dM - \frac{2kM}{h^2}\,dh\right] \Rightarrow$
$dQ\big|_{(2,20,0.05)} = \frac{1}{2}\left(2(2)(20)/0.05\right)^{-1/2}\left[\frac{2(20)}{0.05}\,dk + \frac{2(2)}{0.05}\,dM - \frac{2(2)(20)}{(0.05)^2}\,dh\right] =$
$(0.0125)(800\,dk + 80\,dM - 32000\,dh)$ $\therefore$ Q is most sensitive to changes in h.

SECTION 13.8 MAXIMA, MINIMA, AND SADDLE POINTS

1. $f_x(x,y) = 2x + y + 3 = 0$ and $f_y(x,y) = x + 2y - 3 = 0 \Rightarrow x = -3, y = 3 \Rightarrow$ critical point is $(-3,3)$. $f_{xx}(-3,3) = 2$, $f_{yy}(-3,3) = 2$, $f_{xy}(-3,3) = 1 \Rightarrow f_{xx}f_{yy} - f_{xy}{}^2 = 3 > 0$ and $f_{xx} > 0 \Rightarrow$ minimum. $f(-3,3) = -5$, absolute minimum.

2. $f_x(x,y) = 2x + 3y - 6 = 0$ and $f_y(x,y) = 3x + 6y + 3 = 0 \Rightarrow x = 15, y = -8 \Rightarrow$ critical point is $(15,-8)$.
$f_{xx}(15,-8) = 2$, $f_{yy}(15,-8) = 6$, $f_{xy}(15,-8) = 3 \Rightarrow f_{xx}f_{yy} - f_{xy}{}^2 = 3 > 0$ and $f_{xx} > 0 \Rightarrow$ minimum.
$f(15,-8) = -63$, absolute minimum.

3. $f_x(x,y) = 5y - 14x + 3 = 0$ and $f_y(x,y) = 5x - 6 = 0 \Rightarrow x = \frac{6}{5}, y = \frac{69}{25} \Rightarrow$ critical point is $\left(\frac{6}{5},\frac{69}{25}\right)$.
$f_{xx}\left(\frac{6}{5},\frac{69}{25}\right) = -14$, $f_{yy}\left(\frac{6}{5},\frac{69}{25}\right) = 0$, $f_{xy}\left(\frac{6}{5},\frac{69}{25}\right) = 5 \Rightarrow f_{xx}f_{yy} - f_{xy}{}^2 = -25 < 0 \Rightarrow$ saddle point.

4. $f_x(x,y) = 2y - 10x + 4 = 0$ and $f_y(x,y) = 2x - 4y + 4 = 0 \Rightarrow x = \frac{2}{3}, y = \frac{4}{3} \Rightarrow$ critical point is $\left(\frac{2}{3},\frac{4}{3}\right)$.
$f_{xx}\left(\frac{2}{3},\frac{4}{3}\right) = -10$, $f_{yy}\left(\frac{2}{3},\frac{4}{3}\right) = -4$, $f_{xy}\left(\frac{2}{3},\frac{4}{3}\right) = 2 \Rightarrow f_{xx}f_{yy} - f_{xy}{}^2 = 36 > 0$and $f_{xx} < 0 \Rightarrow$
maximum (absolute). $f\left(\frac{2}{3},\frac{4}{3}\right) = 0$

5. $f_x(x,y) = 2x + y + 3 = 0$ and $f_y(x,y) = x + 2 = 0 \Rightarrow x = -2, y = 1 \Rightarrow$ critical point is $(-2,1)$.
$f_{xx}(-2,1) = 2$, $f_{yy}(-2,1) = 0$, $f_{xy}(-2,1) = 1 \Rightarrow f_{xx}f_{yy} - f_{xy}{}^2 = -1 \Rightarrow$ saddle point.

6. $f_x(x,y) = y - 2 = 0$ and $f_y(x,y) = 2y + x - 2 = 0 \Rightarrow x = -2, y = 2 \Rightarrow$ critical point is $(-2,2)$.
$f_{xx}(-2,2) = 0$, $f_y(-2,2) = 2$, $f_{xy}(-2,2) = 1 \Rightarrow f_{xx}f_{yy} - f_{xy}{}^2 = -1 < 0$ saddle point.

7 $f_x(x,y) = 2y - 10x + 4 = 0$ and $f_y(x,y) = 2x - 4y = 0 \Rightarrow x = \frac{4}{9}, y = \frac{2}{9} \Rightarrow$ critical point is $\left(\frac{4}{9},\frac{2}{9}\right)$.
$f_{xx}\left(\frac{4}{9},\frac{2}{9}\right) = -10$, $f_{yy}\left(\frac{4}{9},\frac{2}{9}\right) = -4$, $f_{xy}\left(\frac{4}{9},\frac{2}{9}\right) = 2 \Rightarrow f_{xx}f_{yy} - f_{xy}{}^2 = 36 > 0$ and $f_{xx} < 0 \Rightarrow$
maximum (absolute). $f\left(\frac{4}{9},\frac{2}{9}\right) = -\frac{252}{81}$

8. $f_x(x,y) = 2y - 2x + 3 = 0$ and $f_y(x,y) = 2x - 4y = 0 \Rightarrow x = 3,\ y = \frac{3}{2} \Rightarrow$ critical point is $\left(3,\frac{3}{2}\right)$.
$f_{xx}\left(3,\frac{3}{2}\right) = -2,\ f_{yy}\left(3,\frac{3}{2}\right) = -4,\ f_{xy}\left(3,\frac{3}{2}\right) = 2 \Rightarrow f_{xx}f_{yy} - f_{xy}^2 = 4 > 0$ and $f_{xx} < 0 \Rightarrow$
maximum (absolute). $f\left(3,\frac{3}{2}\right) = \frac{17}{2}$

9. $f_x(x,y) = 2x - 4y = 0$ and $f_y(x,y) = -4x + 2y + 6 = 0 \Rightarrow x = 2,\ y = 1 \Rightarrow$ critical point is $(2,1)$.
$f_{xx}(2,1) = 2,\ f_{yy}(2,1) = 2,\ f_{xy}(2,1) = -4 \Rightarrow f_{xx}f_{yy} - f_{xy}^2 = -12 \Rightarrow$ saddle point.

10. $f_x(x,y) = 6x + 6y - 2 = 0$ and $f_y(x,y) = 6x + 14y + 4 = 0 \Rightarrow x = \frac{13}{12},\ y = -\frac{3}{4} \Rightarrow$ critical point is $\left(\frac{13}{12},-\frac{3}{4}\right)$.
$f_{xx}\left(\frac{13}{12},-\frac{3}{4}\right) = 6,\ f_{yy}\left(\frac{13}{12},-\frac{3}{4}\right) = 14,\ f_{xy}\left(\frac{13}{12},-\frac{3}{4}\right) = 6 \Rightarrow f_{xx}f_{yy} - f_{xy}^2 = 48 > 0$ and $f_{xx} > 0 \Rightarrow$
minimum (absolute). $f\left(\frac{13}{12},-\frac{3}{4}\right) = -\frac{161}{12}$

11. $f_x(x,y) = 4x + 3y - 5 = 0$ and $f_y(x,y) = 3x + 8y + 2 = 0 \Rightarrow x = 2,\ y = -1 \Rightarrow$ critical point is $(2,-1)$.
$f_{xx}(2,-1) = 4,\ f_{yy}(2,-1) = 8,\ f_{xy}(2,-1) = 3 \Rightarrow f_{xx}f_{yy} - f_{xy}^2 = 29 > 0$ and $f_{xx} > 0 \Rightarrow$
minimum (absolute). $f(2,-1) = -6$.

12. $f_x(x,y) = 8x - 6y - 20 = 0$ and $f_y(x,y) = -6x + 10y + 26 = 0 \Rightarrow x = 1,\ y = -2 \Rightarrow$ critical point is $(1,-2)$.
$f_{xx}(1,-2) = 8,\ f_{yy}(1,-2) = 10,\ f_{xy}(1,-2) = -6 \Rightarrow f_{xx}f_{yy} - f_{xy}^2 = 44 > 0$ and $f_{xx} > 0 \Rightarrow$ minimum (absolute)
$f(1,-2) = -36$.

13. $f_x(x,y) = 2x - 4y + 5 = 0$ and $f_y(x,y) = -4x + 2y - 2 = 0 \Rightarrow x = \frac{1}{6},\ y = \frac{4}{3} \Rightarrow$ critical point is $\left(\frac{1}{6},\frac{4}{3}\right)$.
$f_{xx}\left(\frac{1}{6},\frac{4}{3}\right) = 2,\ f_{yy}\left(\frac{1}{6},\frac{4}{3}\right) = 2,\ f_{xy}\left(\frac{1}{6},\frac{4}{3}\right) = -4 \Rightarrow f_{xx}f_{yy} - f_{xy}^2 = -12 \Rightarrow$ saddle point.

14. $f_x(x,y) = 2x - 2 = 0$ and $f_y(x,y) = -2y + 4 = 0 \Rightarrow x = 1,\ y = 2 \Rightarrow$ critical point is $(1,2)$. $f_{xx}(1,2) = 2$,
$f_{yy}(1,2) = -2,\ f_{xy}(1,2) = 0 \Rightarrow f_{xx}f_{yy} - f_{xy}^2 = -4 \Rightarrow$ saddle point.

15. $f_x(x,y) = 2x - 2y - 2 = 0$ and $f_y(x,y) = -2x + 4y + 2 = 0 \Rightarrow x = 1,\ y = 0 \Rightarrow$ critical point is $(1,0)$.
$f_{xx}(1,0) = 2,\ f_{yy}(1,0) = 4,\ f_{xy}(1,0) = -2 \Rightarrow f_{xx}f_{yy} - f_{xy}^2 = 4 > 0$ and $f_{xx} > 0 \Rightarrow$ minimum (absolute).
$f(1,0) = 0$

16. $f_x(x,y) = 2x + 2y = 0$ and $f_y(x,y) = 2x = 0 \Rightarrow x = 0,\ y = 0 \Rightarrow$ critical point is $(0,0)$. $f_{xx}(0,0) = 2,\ f_{yy}(0,0) = 0$,
$f_{xy}(0,0) = 2 \Rightarrow f_{xx}f_{yy} - f_{xy}^2 = -4 \Rightarrow$ saddle point.

17. $f_x(x,y) = 2 - 4x - 2y = 0$ and $f_y(x,y) = 2 - 2x - 2y = 0 \Rightarrow x = 0,\ y = 1 \Rightarrow$ critical point is $(0,1)$. $f_{xx}(0,1) =$
$-4,\ f_{yy}(0,1) = -2,\ f_{xy}(0,1) = -2 \Rightarrow f_{xx}f_{yy} - f_{xy}^2 = 4 > 0$ and $f_{xx} < 0 \Rightarrow$ maximum (absolute).
$f(0,1) = 4$

18. $f_x(x,y) = 2x + y + 1 = 0$ and $f_y(x,y) = x + 2y - 4 = 0 \Rightarrow x = -2,\ y = 3 \Rightarrow$ critical point is $(-2,3)$. $f_{xx}(-2,3) =$
$2,\ f_{yy}(-2,3) = 2,\ f_{xy}(-2,3) = 1 \Rightarrow f_{xx}f_{yy} - f_{xy}^2 = 3 > 0$ and $f_{xx} > 0 \Rightarrow$ minimum (absolute) .
$f(-2,3) = -2$

19. $f_x(x,y) = 3x^2 - 2y = 0$ and $f_y(x,y) = -3y^2 - 2x = 0 \Rightarrow x = 0, y = 0$ or $x = -\frac{2}{3}, y = \frac{2}{3} \Rightarrow$ critical points are $(0,0)$ and $\left(-\frac{2}{3}, \frac{2}{3}\right)$. For $(0,0)$: $f_{xx}(0,0) = 6x\big|_{(0,0)} = 0$, $f_{yy}(0,0) = -6y\big|_{(0,0)} = 0$, $f_{xy}(0,0) = -2 \Rightarrow$ $f_{xx}f_{yy} - f_{xy}^2 = -4 \Rightarrow$ saddle point. For $\left(-\frac{2}{3}, \frac{2}{3}\right)$: $f_{xx}\left(-\frac{2}{3}, \frac{2}{3}\right) = -4$, $f_{yy}\left(-\frac{2}{3}, \frac{2}{3}\right) = -4$, $f_{xy}\left(-\frac{2}{3}, \frac{2}{3}\right) = -2$ $\Rightarrow f_{xx}f_{yy} - f_{xy}^2 = 12 > 0$ and $f_{xx} < 0 \Rightarrow$ local maximum. $f\left(-\frac{2}{3}, \frac{2}{3}\right) = \frac{170}{27}$

20. $f_x(x,y) = 3x^2 + 6x = 0$ and $f_y(x,y) = 3y^2 - 6y = 0 \Rightarrow x = 0, y = 0$ or $x = 0, y = 2$ or $x = -2, y = 0$ or $x = -2, y = 2 \Rightarrow$ critical points are $(0,0)$, $(0,2)$, $(-2,0)$, and $(-2,2)$. For $(0,0)$: $f_{xx}(0,0) = 6x + 6\big|_{(0,0)} = 6$, $f_{yy}(0,0) = 6y - 6\big|_{(0,0)} -6$, $f_{xy}(0,0) = 0 \Rightarrow f_{xx}f_{yy} - f_{xy}^2 = -36 \Rightarrow$ saddle point. For $(0,2)$: $f_{xx}(0,2) = 6$, $f_{yy}(0,2) = 6$, $f_{xy}(0,2) = 0 \Rightarrow f_{xx}f_{yy} - f_{xy}^2 = 36 > 0$ and $f_{xx} > 0 \Rightarrow$ local minimum. $f(0,2) = -12$. For $(-2,0)$: $f_{xx}(-2,0) = -6$, $f_{yy}(-2,0) = -6$, $f_{xy}(-2,0) = 0 \Rightarrow f_{xx}f_{yy} - f_{xy}^2 = 36 > 0$ and $f_{xx} < 0 \Rightarrow$ local maximum. $f(-2,0) = -4$.
For $(-2,2)$: $f_{xx}(-2,2) = -6$, $f_{yy}(-2,2) = 6$, $f_{xy}(-2,2) = 0 \Rightarrow f_{xx}f_{yy} - f_{xy}^2 = -36 \Rightarrow$ saddle point. No absolute extrema.

21. $f_x(x,y) = 12x - 6x^2 + 6y = 0$ and $f_y(x,y) = 6y + 6x = 0 \Rightarrow x = 0, y = 0$ or $x = 1, y = -1 \Rightarrow$ critical points are $(0,0)$ and $(1,-1)$. For $(0,0)$: $f_{xx}(0,0) = 12 - 12x\big|_{(0,0)} = 12$, $f_{yy}(0,0)= 6$, $f_{xy}(0,0) = 6 \Rightarrow$ $f_{xx}f_{yy} - f_{xy}^2 = 36 > 0$ and $f_{xx} > 0 \Rightarrow$ local minimum. $f(0,0) = 0$. For $(1,-1)$: $f_{xx}(1,-1) = 0$, $f_{yy}(1,-1) = 6$, $f_{xy}(1,-1) = 6 \Rightarrow f_{xx}f_{yy} - f_{xy}^2 = -36 \Rightarrow$ saddle point.

22. $f_x(x,y) = 27x^2 - 4y = 0$ and $f_y(x,y) = y^2 - 4x = 0 \Rightarrow x = 0, y = 0$ or $x = \frac{4}{9}, y = \frac{4}{3} \Rightarrow$ critical points are $(0,0)$ and $\left(\frac{4}{9}, \frac{4}{3}\right)$. For $(0,0)$: $f_{xx}(0,0) = 54x\big|_{(0,0)} = 0$, $f_{yy}(0,0) = 2y\big|_{(0,0)} = 0$, $f_{xy}(0,0) = -4 \Rightarrow$ $f_{xx}f_{yy} - f_{xy}^2 = -16 \Rightarrow$ saddle point. For $\left(\frac{4}{9}, \frac{4}{3}\right)$: $f_{xx}\left(\frac{4}{9}, \frac{4}{3}\right) = 24$, $f_{yy}\left(\frac{4}{9}, \frac{4}{3}\right) = \frac{4}{3}$, $f_{xy}\left(\frac{4}{9}, \frac{4}{3}\right) = -4 \Rightarrow$ $f_{xx}f_{yy} - f_{xy}^2 = 16 > 0$ and $f_{xx} > 0 \Rightarrow$ local minimum. $f\left(\frac{4}{9}, \frac{4}{3}\right) = -\frac{8}{27}$

23. $f_x(x,y) = 3x^2 + 3y = 0$ and $f_y(x,y) = 3x + 3y^2 = 0 \Rightarrow x = 0, y = 0$ or $x = -1, y = -1 \Rightarrow$ critical points are $(0,0)$ and $(-1,-1)$. For $(0,0)$: $f_{xx}(0,0) = 6x\big|_{(0,0)} = 0$, $f_{yy}(0,0) = 6y\big|_{(0,0)} = 0$, $f_{xy}(0,0) = 3 \Rightarrow$ $f_{xx}f_{yy} - f_{xy}^2 = -9 \Rightarrow$ saddle point. For $(-1,-1)$: $f_{xx}(-1,-1) = -6$. $f_{yy}(-1,-1) = -6$, $f_{xy}(-1,-1) = 3 \Rightarrow$ $f_{xx}f_{yy} - f_{xy}^2 = 27 > 0$ and $f_{xx} < 0 \Rightarrow$ local maximum. $f(-1,-1) = 1$

24. $f_x(x,y) = 4y - 4x^3 = 0$ and $f_y(x,y) = 4x - 4y^3 = 0 \Rightarrow x = 0, y = 0$ or $x = 1, y = 1 \Rightarrow$ critical points are $(0,0)$ and $(1,1)$. For $(0,0)$: $f_{xx}(0,0) = -12x^2\big|_{(0,0)} = 0$, $f_{yy}(0,0) = -12y^2\big|_{(0,0)} = 0$, $f_{xy}(0,0) = 4 \Rightarrow$ $f_{xx}f_{yy} - f_{xy}^2 = -16 \Rightarrow$ saddle point. For $(1,1)$: $f_{xx}(1,1) = -12$, $f_{yy}(1,1) = -12$, $f_{xy}(1,1) = 4 \Rightarrow$ $f_{xx}f_{yy} - f_{xy}^2 = 128 > 0$ and $f_{xx} < 0 \Rightarrow$ absolute maximum

25.

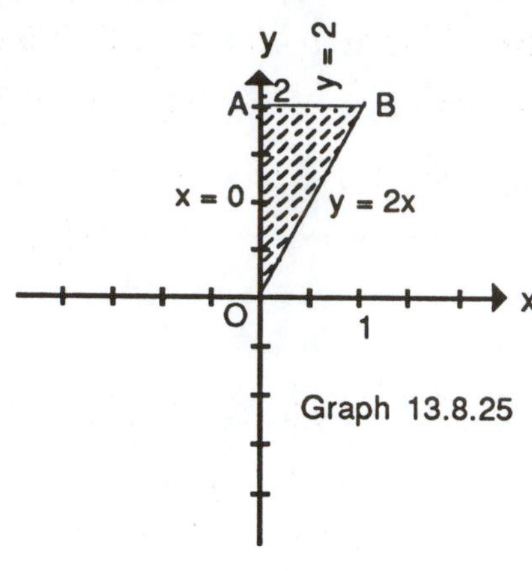

Graph 13.8.25

1. On OA, $f(x,y) = y^2 - 4y + 1 = f(0,y)$ on $0 \le y \le 2$. $y = 0 \Rightarrow f(0,0) = 1$. $y = 2 \Rightarrow f(0,2) = -3$. $f'(0,y) = 2y - 4 = 0 \Rightarrow y = 2 \Rightarrow f(0,2) = -3$.
2. On AB, $f(x,y) = 2x^2 - 4x - 3 = f(x,2)$ on $0 \le x \le 2$. $x = 0 \Rightarrow f(0,2) = -3$. $x = 2 \Rightarrow f(2,2) = -3$. $f'(x,2) = 4x - 4 = 0 \Rightarrow x = 1 \Rightarrow f(1,2) = -5$.
3. On OB, $f(x,y) = 6x^2 - 12x + 1 = f(x,2x)$ on $0 \le x \le 1$. Endpoint values have been found above. $f'(x,2x) = 12x - 12 = 0 \Rightarrow x = 1, y = 2 \Rightarrow (1,2)$, not an interior point of OB.
4. For interior points of the triangular region, $f_x(x,y) = 4x - 4 = 0$ and $f_y(x,y) = 2y - 4 = 0 \Rightarrow x = 1, y = 2 \Rightarrow (1,2)$, not an interior point of the region.

$\therefore$ absolute maximum is 1 at (0,0); absolute minimum is –5 at (1,2)

26.

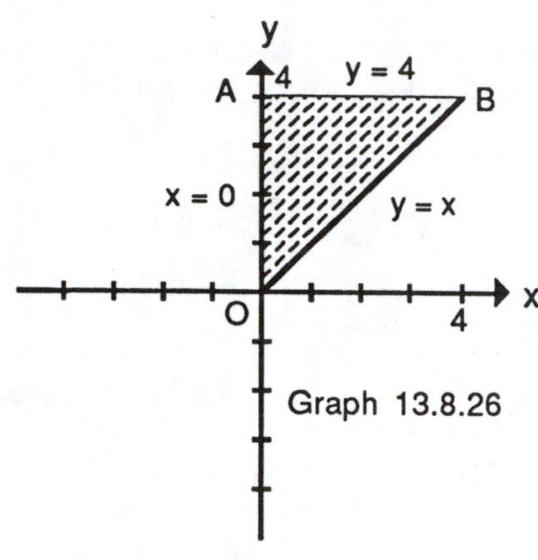

Graph 13.8.26

1. On OA, $D(x,y) = y^2 + 1 = D(0,y)$ on $0 \le y \le 4$. $D'(0,y) = 2y = 0 \Rightarrow y = 0$. $D(0,0) = 1$. $D(0,4) = 17$
2. On AB, $D(x,y) = x^2 - 4x + 17 = D(x,4)$ on $0 \le x \le 4$. $D(4,4) = 17$. $D'(x,4) = 2x - 4 = 0 \Rightarrow x = 2 \Rightarrow (2,4)$ is an interior point of OA. $D(2,4) = 13$
3. On OB, $D(x,y) = x^2 + 1 = D(x,x)$. Endpoint values have been found above. $D'(x,x) = 2x + 1 = 0 \Rightarrow x = -\frac{1}{2}, y = -\frac{1}{2}$, not an interior point of OB.
4. For interior points of the triangular region, $f_x(x,y) = 2x - y = 0$ and $f_y(x,y) = -x + 2y = 0 \Rightarrow x = 0, y = 0$, not an interior point of the region.

$\therefore$ absolute maximum is 17 at (0,4) and (4,4); absolute minimum is 1 at (0,0)

27.

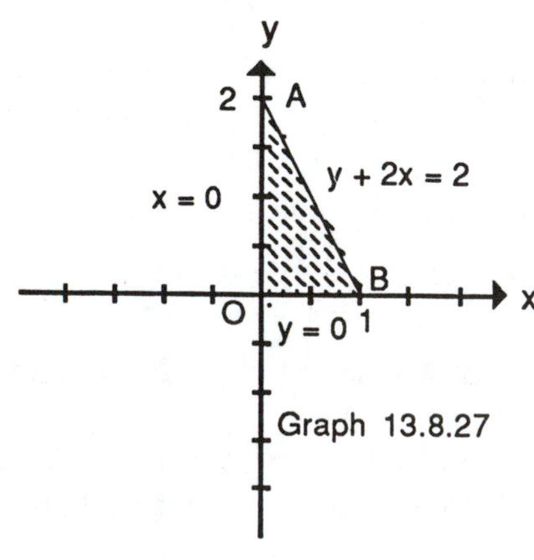

Graph 13.8.27

1. On OA, $f(x,y) = y^2 = f(0,y)$ on $0 \le y \le 2$. $f(0,0) = 0$. $f(0,2) = 4$. $f'(0,y) = 2y = 0 \Rightarrow y = 0, x = 0 \Rightarrow (0,0)$
2. On OB, $f(x,y) = x^2 = f(x,0)$ on $0 \le x \le 1$. $f(1,0) = 1$. $f'(x,0) = 2x = 0 \Rightarrow x = 0, y = 0 \Rightarrow (0,0)$
3. On AB, $f(x,y) = 5x^2 - 8x + 4 = f(x,-2x + 2)$ on $0 \le x \le 1$. $f(0,2) = 4$. $f'(x,-2x + 2) = 10x - 8 = 0 \Rightarrow x = \frac{4}{5}, y = \frac{2}{5}$. $f\left(\frac{4}{5}, \frac{2}{5}\right) = \frac{4}{5}$
4. For interior points of the triangular region, $f_x(x,y) = 2x = 0$ and $f_y(x,y) = 2y = 0 \Rightarrow x = 0, y = 0 \Rightarrow (0,0)$, not an interior point of the region.

$\therefore$ absolute maximum is 4 at (0,2); absolute minimum is 0 at (0,0)

28.

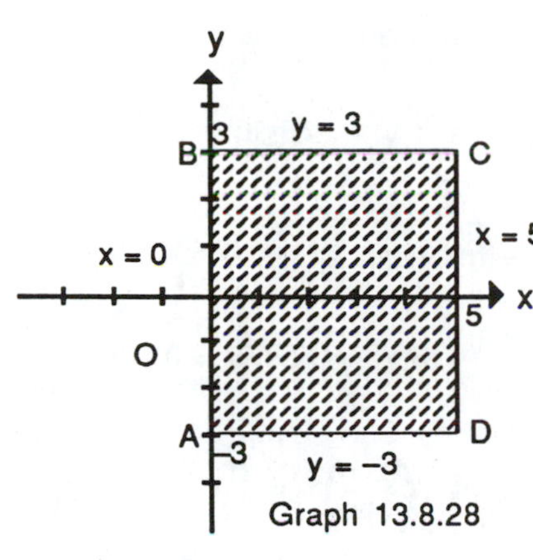

Graph 13.8.28

1. On AB, $T(x,y) = y^2 = T(0,y)$ on $-3 \le y \le 3$. $T(0,-3) = 9$. $T(0,3) = 9$. $T'(0,y) = 2y = 0 \Rightarrow y = 0$, $x = 0$. $T(0,0) = 0$
2. On BC, $T(x,y) = x^2 - 3x + 9 = T(x,3)$ on $0 \le x \le 5$. $T(5,3) = 19$. $T'(x,3) = 2x - 3 = 0 \Rightarrow x = \frac{3}{2}$, $y = 3$. $T\left(\frac{3}{2},3\right) = \frac{27}{4}$
3. On CD, $T(x,y) = y^2 + 5y - 5 = T(5,y)$ on $-3 \le y \le 3$. $T(5,-3) = -11$. $T(5,3) = 19$. $T'(5,y) = 2y + 5 = 0 \Rightarrow y = -\frac{5}{2}$, $x = 5$ $T\left(5,-\frac{5}{2}\right) = -\frac{45}{4}$
4. On AD, $T(x,y) = x^2 - 9x + 9 = T(x,-3)$ on $0 \le x \le 5$. $T'(x,-3) = 2x - 9 = 0 \Rightarrow x = \frac{9}{2}$, $y = -3$. $T\left(\frac{9}{2},-3\right) = -\frac{45}{4}$
5. For interior points of the rectangular region, $T_x(x,y) = 2x + y - 6 = 0$ and $T_y(x,y) = x + 2y = 0 \Rightarrow x = 4$, $y = -2 \Rightarrow (-2,4)$ is an interior critical point. $T(4,-2) = -12$. $\therefore$ absolute maximum is 19 at (5,3); absolute minimum is -12 at $(4,-2)$

29.

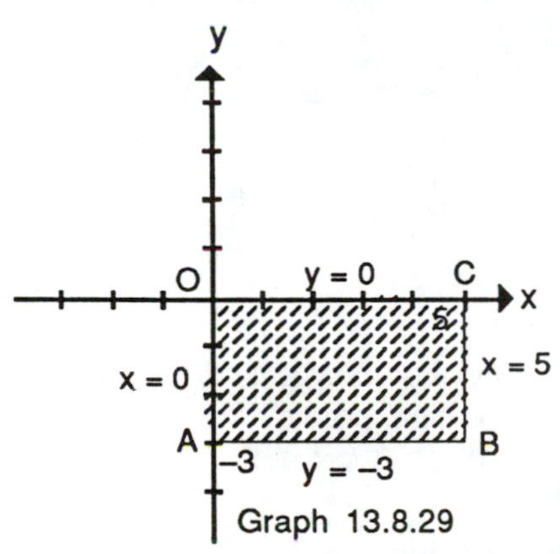

Graph 13.8.29

1. On OC, $T(x,y) = x^2 - 6x + 2 = T(x,0)$ on $0 \le x \le 5$. $T(0,0) = 2$. $T(5,0) = -3$. $T'(x,0) = 2x - 6 = 0 \Rightarrow x = 3$, $y = 0$. $T(3,0) = -7$
2. On CB, $T(x,y) = y^2 + 5y - 3 = T(5,y)$ on $-3 \le y \le 0$. $T(5,-3) = -9$ $T'(5,y) = 2y + 5 = 0 \Rightarrow y = -\frac{5}{2}$, $x = 5$. $T\left(5,-\frac{5}{2}\right) = -\frac{37}{4}$
3. On AB, $T(x,y) = x^2 - 9x + 11 = T(x,-3)$ on $0 \le x \le 5$. $T(0,-3) = 11$. $T'(x,-3) = 2x - 9 = 0 \Rightarrow x = \frac{9}{2}$, $y = -3$. $T\left(\frac{9}{2},-3\right) = -\frac{37}{4}$
4. On AO, $T(x,y) = y^2 + 2 = T(0,y)$ on $-3 \le y \le 0$. $T'(0,y) = 2y = 0 \Rightarrow y = 0$, $x = 0$. (0,0) not an interior point of AO.
5. For interior points of the rectangular region, $T_x(x,y) = 2x + y - 6 = 0$ and $T_y(x,y) = x + 2y = 0 \Rightarrow x = 4$, $y = -2 \Rightarrow T(4,-2) = -10$.

$\therefore$ absolute maximum is 11 at $(0,-3)$; absolute minimum is -10 at $(4,-2)$.

30.

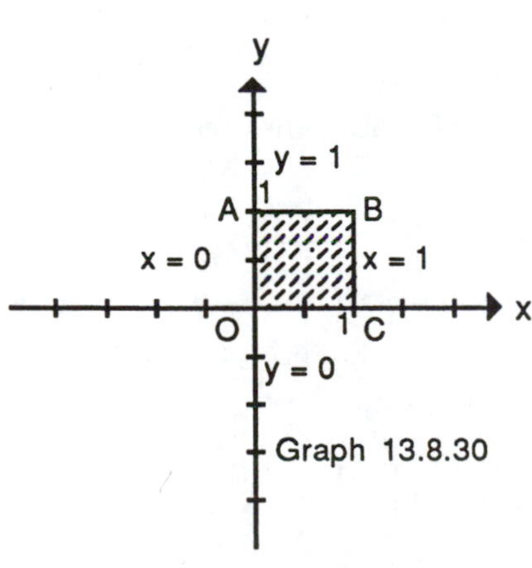

Graph 13.8.30

1. On OA, $f(x,y) = -24y^2 = f(0,y)$ on $0 \le y \le 1$. $f(0,0) = 0$. $f(0,1) = -24$. $f'(0,y) = -48y = 0 \Rightarrow y = 0$, $x = 0$. (0,0) not an interior point of OA.
2. On AB, $f(x,y) = 48x - 32x^3 - 24 = f(x,1)$ on $0 \le x \le 1$. $f(1,1) = -12$. $f'(x,1) = 48 - 96x^2 = 0 \Rightarrow x = \frac{1}{\sqrt{2}}$, $y = 1$ or $x = -\frac{1}{\sqrt{2}}$, $y = 1$ $\left(-\frac{1}{\sqrt{2}},1\right)$ not in the interior of AB. $f\left(\frac{1}{\sqrt{2}},1\right) = 16\sqrt{2} - 24$.
3. On BC, $f(x,y) = 48y - 32 - 24y^2 = f(1,y)$ on $0 \le y \le 1$. $f(1,0) = -32$. $f'(1,y) = 48 - 48y = 0 \Rightarrow y = 1$, $x = 1$. (1,1) not an interior point of BC.
4. On OC, $f(x,y) = -32x^3 = f(x,0)$ on $0 \le x \le 1$. $f'(x,0) = -96x^2 = 0$ $x = 0$, $y = 0$. (0,0) not an interior point of OC
5. For interior points of the rectangular region, $f_x(x,y) = 48y - 96x^2 = 0$ and $f_y(x,y) = 48x - 48y^2 = 0 \Rightarrow$ $x = 0$, $y = 0$ or $x = \frac{1}{\sqrt[3]{4}}$, $y = \frac{1}{\sqrt[3]{2}}$. (0,0) is not an interior point of the region. $f\left(\frac{1}{\sqrt[3]{4}},\frac{1}{\sqrt[3]{2}}\right) = 8 - \frac{12}{\sqrt[3]{2}}$.

$\therefore$ absolute maximum is 0 at (0,0); absolute minimum is -32 at (1,0).

31.

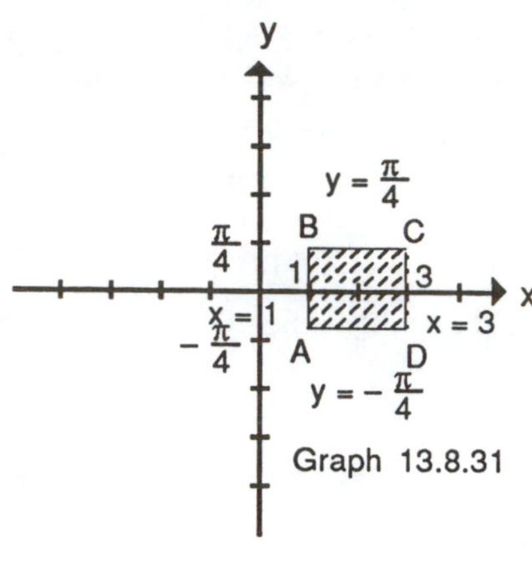

Graph 13.8.31

1. On AB, $f(x,y) = -3\cos y = f(1,y)$ on $-\frac{\pi}{4} \le y \le \frac{\pi}{4}$. $f\left(1,-\frac{\pi}{4}\right) = -\frac{3\sqrt{2}}{2}$. $f\left(1,\frac{\pi}{4}\right) = -\frac{3\sqrt{2}}{2}$. $f'(1,y) = -3\sin y = 0 \Rightarrow y = 0,\ x = 1$ $f(1,0) = -3$.
2. On CD, $f(x,y) = -3\cos y = f(3,y)$ on $-\frac{\pi}{4} \le y \le \frac{\pi}{4}$. $f\left(3,-\frac{\pi}{4}\right) = -\frac{3\sqrt{2}}{2}$. $f\left(3,\frac{\pi}{4}\right) = -\frac{3\sqrt{2}}{2}$. $f'(3,y) = 3\sin y = 0 \Rightarrow y = 0,\ x = 3$. $f(3,0) = -3$.
3. On BC, $f(x,y) = \frac{\sqrt{2}}{2}(x^2 - 4x) = f\left(x,\frac{\pi}{4}\right)$ on $1 \le x \le 3$. $f'\left(x,\frac{\pi}{4}\right) = \frac{\sqrt{2}}{2}(2x - 4) = 0 \Rightarrow x = 2,\ y = \frac{\pi}{4}$. $f\left(2,\frac{\pi}{4}\right) = -2\sqrt{2}$.
4. On AD, $f(x,y) = \frac{\sqrt{2}}{2}(x^2 - 4x) = f\left(x,-\frac{\pi}{4}\right)$ on $1 \le x \le 3$. $f'\left(x,-\frac{\pi}{4}\right) = \frac{\sqrt{2}}{2}(2x - 4) = 0 \Rightarrow x = 2,\ y = -\frac{\pi}{4}$. $f\left(2,-\frac{\pi}{4}\right) = -2\sqrt{2}$.
5. For interior points of the rectangular region, $f_x(x,y) = (2x - 4)\cos y = 0$ and $f_y(x,y) = -(x^2 - 4x)\sin y = 0 \Rightarrow x = 2,\ y = 0$. $f(2,0) = -4$.

$\therefore$ absolute maximum is $-\frac{3\sqrt{2}}{2}$ at $\left(3,-\frac{\pi}{4}\right)$, $\left(3,\frac{\pi}{4}\right)$, $\left(1,-\frac{\pi}{4}\right)$, and $\left(1,\frac{\pi}{4}\right)$; absolute minimum is -4 at $(2,0)$.

32.

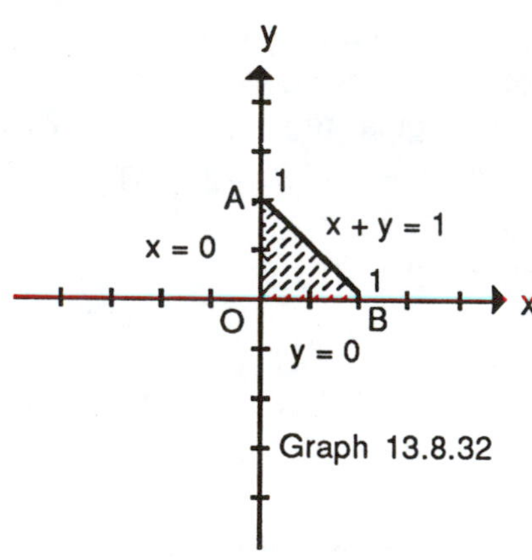

Graph 13.8.32

1. On OA, $f(x,y) = 2y + 1 = f(0,y)$ on $0 \le y \le 1$. $f(0,0) = 1$. $f(0,1) = 3$. $f'(0,y) = 2 \Rightarrow$ no interior critical points.
2. On OB, $f(x,y) = 4x + 1 = f(x,0)$ on $0 \le x \le 1$. $f(1,0) = 5$. $f'(x,0) = 4 \Rightarrow$ no interior critical points.
3. On AB, $f(x,y) = 8x^2 - 6x - 1 = f(x,-x + 1)$ on $0 \le x \le 1$. $f'(x,-x + 1) = 16x - 6 = 0 \Rightarrow x = \frac{3}{8},\ y = \frac{5}{8}$. $f\left(\frac{3}{8},\frac{5}{8}\right) = -\frac{17}{8}$.
4. For interior points of the triangular region, $f_x(x,y) = 4 - 8y = 0$ and $f_y(x,y) = -8x + 2 = 0 \Rightarrow y = \frac{1}{2},\ x = \frac{1}{4}$. $f\left(\frac{1}{4},\frac{1}{2}\right) = 2$.

$\therefore$ absolute maximum is 5 at $(1,0)$; absolute minimum is $-\frac{17}{8}$ at $\left(\frac{3}{8},\frac{5}{8}\right)$.

33. $T_x(x,y) = 2x - 1 = 0$ and $T_y(x,y) = 4y = 0 \Rightarrow x = \frac{1}{2},\ y = 0 \Rightarrow T\left(\frac{1}{2},0\right) = -\frac{1}{4}$. On $x^2 + y^2 = 1$, $T(x,y) = -x^2 - x + 2$ on $-1 \le x \le 1$. $T(-1,0) = 2$, $T(1,0) = 0$. $T'(x,y) = -2x - 1 = 0 \Rightarrow x = -\frac{1}{2},\ y = \pm\frac{\sqrt{3}}{2}$. $T\left(-\frac{1}{2},\frac{\sqrt{3}}{2}\right) = \frac{9}{4}$, $T\left(-\frac{1}{2},-\frac{\sqrt{3}}{2}\right) = \frac{9}{4}$. $\therefore$ hottest is $2\frac{1}{4}{}^\circ$ at $\left(-\frac{1}{2},\frac{\sqrt{3}}{2}\right)$ and $\left(-\frac{1}{2},-\frac{\sqrt{3}}{2}\right)$; coldest is $-\frac{1}{4}{}^\circ$ at $\left(\frac{1}{2},0\right)$.

34. $f_x(x,y) = y + 2 - \frac{2}{x} = 0$ and $f_y(x,y) = x - \frac{1}{y} = 0 \Rightarrow x = 0$ (does not give a point in the open first quadrant) or $x = \frac{1}{2}$, $y = 2$. $f_{xx}\left(\frac{1}{2},2\right) = \frac{2}{x^2}\Big|_{(1/2,2)} = 8$, $f_{yy}\left(\frac{1}{2},2\right) = \frac{1}{y^2}\Big|_{(1/2,2)} = \frac{1}{4}$, $f_{xy}\left(\frac{1}{2},2\right) = 1 \Rightarrow f_{xx}f_{yy} - f_{xy}^2 = 1 > 0$ and $f_{xx} > 0 \Rightarrow$ minimum.

35. a) $f_x(x,y) = 2x - 4y = 0$ and $f_y(x,y) = 2y - 4x = 0 \Rightarrow x = 0, y = 0$. $f_{xx}(0,0) = 2$, $f_{yy}(0,0) = -4$, $f_{xy}(0,0) = -4$ $\Rightarrow f_{xx}f_{yy} - f_{xy}^2 = -24 \Rightarrow$ saddle point.

b) $f_x(x,y) = 2x - 2 = 0$ and $f_y(x,y) = 2y - 4 = 0 \Rightarrow x = 1, y = 2$. $f_{xx}(1,2) = 2$, $f_{yy}(1,2) = 2$, $f_{xy}(1,2) = 0 \Rightarrow$ $f_{xx}f_{yy} - f_{xy}^2 = 4 > 0$ and $f_{xx} > 0 \Rightarrow$ local minimum at (1,2).

c) $f_x(x,y) = 9x^2 - 9 = 0$ and $f_y(x,y) = 2y + 4 = 0 \Rightarrow x = \pm 1, y = -2$. For (1,–2), $f_{xx}(1,-2) = 18x\big|_{(1,-2)} = 18$, $f_{yy}(1,-2) = 2$, $f_{xy}(1,-2) = 0 \Rightarrow f_{xx}f_{yy} - f_{xy}^2 = 36 > 0$ and $f_{xx} > 0 \Rightarrow$ local minimum at (1,–2). For (–1,–2), $f_{xx}(-1,-2) = -18$, $f_{yy}(-1,-2) = 2$, $f_{xy}(-1,-2) = 0 \Rightarrow f_{xx}f_{yy} - f_{xy}^2 = -36 \Rightarrow$ saddle point.

36. a) Minimum at (0,0) since $f(x,y) > 0$ for all other (x,y).

b) Maximum of 1 at (0,0) since $f(x,y) < 1$ for all other (x,y).

c) Neither since $f(x,y) < 0$ for $x < 0$ and $f(x,y) > 0$ for $x > 0$.

d) Neither since $f(x,y) < 0$ for $x < 0, y > 0$ and $f(x,y) > 0$ for $x > 0, y > 0$.

e) Neither since $f(x,y) < 0$ for $x < 0, y > 0$ and $f(x,y) > 0$ for $x > 0, y > 0$.

f) Minimum since $f(x,y) > 0$ for all other (x,y).

37. a) $x = 2\cos t$, $y = 2\sin t \Rightarrow f(t) = 4\cos t \sin t \Rightarrow \frac{df}{dt} = y(-2\sin t) + x(2\cos t) = -4\sin^2 t + 4\cos^2 t$.
$\frac{df}{dt} = 0 \Rightarrow 4\cos^2 t - 4\sin^2 t = 0 \Rightarrow \cos t = \sin t$ or $\cos t = -\sin t$

i) On the quarter circle $x^2 + y^2 = 4$ in the first quadrant, $\frac{df}{dt} = 0$ at $t = \frac{\pi}{4}$. $f(0) = 0$, $f\left(\frac{\pi}{2}\right) = 0$, $f\left(\frac{\pi}{4}\right) = 2$

$\therefore$ absolute minimum is 0 at $t = 0, \frac{\pi}{2}$; absolute maximum is 2 at $t = \frac{\pi}{4}$.

ii) On the half circle $x^2 + y^2 = 4$, $y \ge 0$, $\frac{df}{dt} = 0 \Rightarrow t = \frac{\pi}{4}, \frac{3\pi}{4}$. $f(0) = 0$, $f\left(\frac{\pi}{4}\right) = 2$, $f\left(\frac{3\pi}{4}\right) = -2$, $f(\pi) = 0$

$\therefore$ absolute minimum is –2 at $t = \frac{3\pi}{4}$; absolute maximum is 2 at $t = \frac{\pi}{4}$.

iii) On the full circle $x^2 + y^2 = 4$, $\frac{df}{dt} = 0 \Rightarrow t = \frac{\pi}{4}, \frac{3\pi}{4}, \frac{5\pi}{4}, \frac{7\pi}{4}$. $f(0) = 0$, $f\left(\frac{\pi}{4}\right) = f\left(\frac{5\pi}{4}\right) = 2$, $f\left(\frac{3\pi}{4}\right) = f\left(\frac{7\pi}{4}\right) = -2$, $f(2\pi) = 0$. $\therefore$ absolute minimum is –2 at $t = \frac{3\pi}{4}, \frac{7\pi}{4}$; absolute maximum is 2 at $t = \frac{\pi}{4}, \frac{5\pi}{4}$.

b) $x = 2\cos t$, $y = 2\sin t \Rightarrow f(t) = 2\cos t + 2\sin t \Rightarrow \frac{df}{dt} = -2\sin t + 2\cos t$. $\frac{df}{dt} = 0 \Rightarrow \cos t = \sin t$

i) On $0 \le t \le \frac{\pi}{2}$, $f(0) = 2$, $f\left(\frac{\pi}{2}\right) = 2$. $\frac{df}{dt} = 0 \Rightarrow t = \frac{\pi}{4} \Rightarrow f\left(\frac{\pi}{4}\right) = 2\sqrt{2}$. $\therefore$ absolute minimum is 2 at $t = 0, \frac{\pi}{2}$; absolute maximum is $2\sqrt{2}$ at $t = \frac{\pi}{4}$.

37. (Continued)

ii) On $0 \le t \le \pi$, $f(0) = 2$, $f(\pi) = -2$. $\frac{df}{dt} = 0 \Rightarrow t = \frac{\pi}{4}, \frac{3\pi}{4} \Rightarrow f\left(\frac{\pi}{4}\right) = 2\sqrt{2}$, $f\left(\frac{3\pi}{4}\right) = 0$ $\therefore$ absolute minimum is -2 at $t = \pi$; absolute maximum is $2\sqrt{2}$ at $t = \frac{\pi}{4}$.

iii) On $0 \le t \le 2\pi$, $f(0) = 2$, $f(2\pi) = 2$. $\frac{df}{dt} = 0 \Rightarrow t = \frac{\pi}{4}, \frac{5\pi}{4}$. $f\left(\frac{\pi}{4}\right) = 2\sqrt{2}$, $f\left(\frac{5\pi}{4}\right) = -2\sqrt{2}$. $\therefore$ absolute minimum is $-2\sqrt{2}$ at $t = \frac{5\pi}{4}$; absolute maximum is $2\sqrt{2}$ at $t = \frac{\pi}{4}$.

c) $x = 2\cos t$, $y = 2\sin t \Rightarrow f(t) = 8\cos^2 t + \sin^2 t = 7\cos^2 t + 1 \Rightarrow \frac{df}{dt} = -8\cos t \sin t$. $\frac{df}{dt} = 0 \Rightarrow$ $\cos t = 0$ or $\sin t = 0$

i) On $0 \le t \le \frac{\pi}{2}$, $f(0) = 8$, $f\left(\frac{\pi}{2}\right) = 1$. $\frac{df}{dt} = 0 \Rightarrow t = 0, \frac{\pi}{2}$. $\therefore$ absolute minimum is 1 at $t = \frac{\pi}{2}$; absolute maximum is 8 at $t = 0$.

ii) On $0 \le t \le \pi$, $f(0) = 8$, $f(\pi) = 8$. $\frac{df}{dt} = 0 \Rightarrow t = 0, \pi, \frac{\pi}{2}$. $f\left(\frac{\pi}{2}\right) = 1$. $\therefore$ absolute minimum is 1 at $t = \frac{\pi}{2}$; absolute maximum is 8 at $t = 0, \pi$.

iii) On $0 \le t \le 2\pi$, $f(0) = 8$, $f(2\pi) = 8$. $\frac{df}{dt} = 0 \Rightarrow t = 0, \frac{\pi}{2}, \pi, \frac{3\pi}{2}, 2\pi \Rightarrow f\left(\frac{\pi}{2}\right) = 1$, $f(\pi) = 8$, $f\left(\frac{3\pi}{2}\right) = 1$. $\therefore$ absolute minimum is 1 at $t = \frac{\pi}{2}, \frac{3\pi}{2}$; absolute maximum is 8 at $t = 0, \pi, 2\pi$.

38. a) $x = 2t$, $y = t + 1 \Rightarrow f(t) = 2t^2 + 2t \Rightarrow \frac{df}{dt} = 4t + 2$. $\frac{df}{dt} = 0 \Rightarrow t = -\frac{1}{2}$. $f''(t) = 4 \Rightarrow f\left(-\frac{1}{2}\right)$ is a minimum. $f\left(-\frac{1}{2}\right) = -\frac{1}{2}$ is an absolute minimum since $f(t)$ is an upward parabola. No absolute maximum.

b) $x = 2t$, $y = t + 1$ on $-1 \le t \le 0 \Rightarrow t = -\frac{1}{2}$ is a critical number on the interval (see part a) above). $f\left(-\frac{1}{2}\right) = -\frac{1}{2}$. $f(0) = 0$, $f(-1) = 0 \Rightarrow$ absolute minimum is $-\frac{1}{2}$ at $t = -\frac{1}{2}$; absolute maximum is 0 at $t = 0, -1$.

c) $x = 2t$, $y = t + 1$ on $0 \le t \le 1 \Rightarrow$ no critical numbers on the interval (see part a) above). $f(0) = 0$, $f(1) = 4 \Rightarrow$ absolute minimum is 0 at $t = 0$; absolute maximum is 4 at $t = 1$.

39. a) $x = 3\cos t$, $y = 2\sin t \Rightarrow f(t) = 9 + 3\sin^2 t \Rightarrow \frac{df}{dt} = 6\sin t \cos t$. $\frac{df}{dt} = 0 \Rightarrow \sin t = 0$ or $\cos t = 0$

i) On the quarter ellipse, $\frac{x^2}{9} + \frac{y^2}{4} = 1$, in the first quadrant, $f(0) = 9$, $f\left(\frac{\pi}{2}\right) = 12$. $\frac{df}{dt} = 0 \Rightarrow t = 0, \frac{\pi}{2}$ $\therefore$ absolute maximum is 12 at $t = \frac{\pi}{2}$; absolute minimum is 9 at $t = 0$.

ii) On the half ellipse, $\frac{x^2}{9} + \frac{y^2}{4} = 1$, $y \ge 0$, $f(0) = 9$, $f(\pi) = 9$. $\frac{df}{dt} = 0 \Rightarrow t = 0, \frac{\pi}{2}, \pi$. $f\left(\frac{\pi}{2}\right) = 12$. $\therefore$ absolute minimum is 9 at $t = 0, \pi$; absolute maximum is 12 at $t = \frac{\pi}{2}$.

39. (Continued)

iii) On the full ellipse, $\frac{x^2}{9} + \frac{y^2}{4} = 1$, $f(0) = 9$, $f(2\pi) = 9$. $\frac{df}{dt} = 0 \Rightarrow t = 0, \frac{\pi}{2}, \pi, \frac{3\pi}{2}, 2\pi$. $f\left(\frac{\pi}{2}\right) = 12$, $f(\pi) = 9$, $f\left(\frac{3\pi}{2}\right) = 12$. $\therefore$ absolute maximum is 12 at $t = \frac{\pi}{2}, \frac{3\pi}{2}$; absolute minimum is 9 at $t = 0, \pi, 2\pi$.

b) $x = 3 \cos t$, $y = 2 \sin t \Rightarrow f(t) = 6 \cos t + 6 \sin t \Rightarrow \frac{df}{dt} = -6 \sin t + 6 \cos t$. $\frac{df}{dt} = 0 \Rightarrow \cos t = \sin t$.

i) On the quarter ellipse, $\frac{x^2}{9} + \frac{y^2}{4} = 1$, $f(0) = 6$, $f\left(\frac{\pi}{2}\right) = 6$. $\frac{df}{dt} = 0 \Rightarrow t = \frac{\pi}{4} \Rightarrow f\left(\frac{\pi}{4}\right) = 6\sqrt{2} \Rightarrow$ absolute maximum is $6\sqrt{2}$ at $t = \frac{\pi}{4}$; absolute minimum is 6 at $t = 0, \frac{\pi}{2}$.

ii) On the half ellipse, $\frac{x^2}{9} + \frac{y^2}{4} = 1$, $y \geq 0$, $f(0) = 6$, $f(\pi) = -6$. $\frac{df}{dt} = 0 \Rightarrow t = \frac{\pi}{4} \Rightarrow f\left(\frac{\pi}{4}\right) = 6\sqrt{2} \Rightarrow$ absolute maximum is $6\sqrt{2}$; absolute minimum is -6 at $t = \pi$.

iii) On the full ellipse, $\frac{x^2}{9} + \frac{y^2}{4} = 1$, $f(0) = 6$, $f(2\pi) = 6$. $\frac{df}{dt} = 0 \Rightarrow t = \frac{\pi}{4}, \frac{5\pi}{4} \Rightarrow f\left(\frac{\pi}{4}\right) = 6\sqrt{2}$, $f\left(\frac{5\pi}{4}\right) = -6\sqrt{2} \Rightarrow$ absolute maximum is $6\sqrt{2}$ at $t = \frac{\pi}{4}$; absolute minimum is $-6\sqrt{2}$ at $t = \frac{5\pi}{4}$.

40. $x = \cos t$, $y = \sin t \Rightarrow f(t) = \cos t \sin t \Rightarrow \frac{df}{dt} = -\sin^2 t + \cos^2 t$. On the ellipse $x^2 + 4y^2 = 8$, $0 \leq t \leq 2\pi$, $\frac{df}{dt} = 0 \Rightarrow -\sin^2 t + \cos^2 t = 0 \Rightarrow \cos t = \sin t$ or $\cos t = -\sin t \Rightarrow t = \frac{\pi}{4}, \frac{5\pi}{4}$ or $t = \frac{3\pi}{4}, \frac{7\pi}{4}$. $f\left(\frac{\pi}{4}\right) = f\left(\frac{5\pi}{4}\right) = \frac{1}{2}$, $f\left(\frac{3\pi}{4}\right) = f\left(\frac{7\pi}{4}\right) = -\frac{1}{2}$, $f(0) = f(2\pi) = 0 \Rightarrow$ absolute maximum is $\frac{1}{2}$ t at $= \frac{\pi}{4}, \frac{5\pi}{4}$; absolute minimum is $-\frac{1}{2}$ at $t = \frac{3\pi}{4}, \frac{7\pi}{4}$.

41.

k	x_k	y_k	x_k^2	$x_k y_k$
1	−1	2	1	−2
2	0	1	0	0
3	3	−4	9	−12
Σ	2	−1	10	−14

$m = \frac{2(-1) - 3(-14)}{2^2 - 3(10)} \approx -1.5$, $b = \frac{1}{3}(-1 - (-1.5)2) \approx 0.7$

$\therefore$ $y = -1.5x + 0.7$. $y\big|_{x=4} = -5.3$

42.

k	x_k	y_k	x_k^2	$x_k y_k$
1	−2	0	4	0
2	0	2	0	0
3	2	3	4	6
Σ	0	5	8	6

$m = \frac{0(5) - 3(6)}{0^2 - 3(8)} = \frac{3}{4}$, $b = \frac{1}{3}\left(5 - \frac{3}{4}(0)\right) = \frac{5}{3}$

$\therefore$ $y = \frac{3}{4}x + \frac{5}{3}$. $y\big|_{x=4} = \frac{14}{3}$

43.

k	x_k	y_k	x_k^2	$x_k y_k$
1	0	0	0	0
2	1	2	1	2
3	2	3	4	6
Σ	3	5	5	8

$m = \frac{3(5) - 3(8)}{3^2 - 3(5)} = 1.5$, $b = \frac{1}{3}(5 - 1.5(3)) \approx 0.2$

$\therefore$ $y = 1.5x + 0.2$. $y\big|_{x=4} = 6.2$

44.

k	x_k	y_k	x_k^2	x_ky_k
1	0	1	0	0
2	2	2	4	4
3	3	2	9	6
Σ	5	5	13	10

$m = \frac{5(5) - 3(10)}{5^2 - 3(13)} = \frac{5}{14}$, $b = \frac{1}{3}\left(5 - \frac{5}{14}(5)\right) = \frac{15}{14}$

$\therefore\ y = \frac{5}{14}x + \frac{15}{14}$. $y\big|_{x=4} = \frac{35}{14}$

45.

k	x_k	y_k	x_k^2	x_ky_k
1	12	5.27	144	63.24
2	18	5.68	324	102.24
3	24	6.25	576	150
4	30	7.21	900	216.3
5	36	8.20	1296	295.2
6	42	8.71	1764	365.82
Σ	162	41.32	5004	1192.8

$m = \frac{162(41.32) - 6(1192.8)}{162^2 - 6(5004)} \approx 0.122,$

$b = \frac{1}{6}(41.32 - (0.122)(162)) \approx 3.59$

$\therefore\ y = 0.122\,x + 3.59$

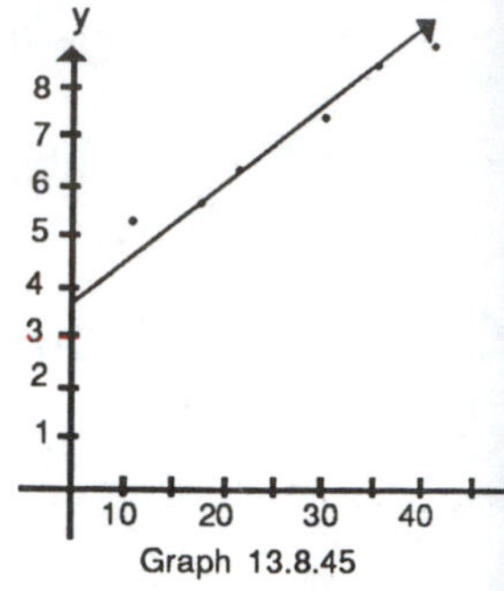

Graph 13.8.45

46.

k	$\left(\frac{1}{D^2}\right)_k$	F_k	$\left(\frac{1}{D^2}\right)_k^2$	$\left(\frac{1}{D^2}\right)_k F_k$
1	0.001	53	0.000001	0.053
2	0.0005	22	0.00000025	0.011
3	0.00025	14	0.000000062	0.0035
4	0.000125	3	0.000000015	0.000375
Σ	0.001875	92	0.000001328	0.067875

$m = \frac{(0.001875)(92) - 4(0.06875)}{(0.001875)^2 - 4(0.000001328)} \approx 57000$, $b = \frac{1}{4}(92 - 57000(0.001875)) \approx -3.72.$

$\therefore\ F = 57000\,\frac{1}{D^2} - 3.72.$

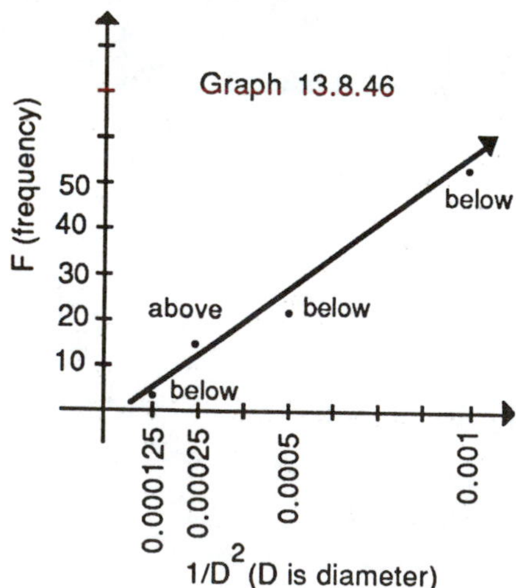

SECTION 13.9 LAGRANGE MULTIPLIERS

1. $\nabla f = y\,\mathbf{i} + x\,\mathbf{j}$, $\nabla g = 2x\,\mathbf{i} + 4y\,\mathbf{j}$. $\nabla f = \lambda \nabla g \Rightarrow y\,\mathbf{i} + x\,\mathbf{j} = \lambda(2x\,\mathbf{i} + 4y\,\mathbf{j}) \Rightarrow y = 2x\lambda$ and $x = 4y\lambda \Rightarrow$ $\lambda = \pm\frac{\sqrt{2}}{4}$ or $x = 0$. CASE 1: If $x = 0$, then $y = 0$ but $(0,0)$ not on the ellipse. $\therefore\ x \neq 0$. CASE 2: $x \neq 0 \Rightarrow \lambda = \pm\frac{\sqrt{2}}{4} \Rightarrow x = \pm\sqrt{2}\,y \Rightarrow (\pm\sqrt{2}\,y)^2 + 2y^2 = 1 \Rightarrow y = \pm\frac{1}{2}$. $\therefore$ f takes on its extreme values at $\left(\pm\sqrt{2}, \frac{1}{2}\right)$ and $\left(\pm\sqrt{2}, \frac{1}{2}\right) \Rightarrow$ the extreme values of f are $\pm\frac{\sqrt{2}}{2}$.

2. $\nabla f = y\,\mathbf{i} + x\,\mathbf{j}$, $\nabla g = 2x\,\mathbf{i} + 2y\,\mathbf{j}$. $\nabla f = \lambda \nabla g \Rightarrow y\,\mathbf{i} + x\,\mathbf{j} = \lambda(2x\,\mathbf{i} + 2y\,\mathbf{j}) \Rightarrow y = 2x\lambda$ and $x = 2y\lambda \Rightarrow x = 0$ or $\lambda = \pm\frac{1}{2}$. CASE 1: $x = 0 \Rightarrow y = 0$. But $(0,0)$ not on $x^2 + y^2 - 10 = 0$. $\therefore\ x \neq 0$. CASE 2: $x \neq 0 \Rightarrow$ $\lambda = \pm\frac{1}{2} \Rightarrow y = 2x\left(\pm\frac{1}{2}\right) = \pm x$. $\therefore\ x^2 + (\pm x)^2 - 10 = 0 \Rightarrow x = \pm\sqrt{5} \Rightarrow y = \pm\sqrt{5}$. $\therefore$ f takes on its extreme values at $(\pm\sqrt{5}, \sqrt{5})$ and $(\pm\sqrt{5}, -\sqrt{5}) \Rightarrow$ the extreme values of f are 5 and -5.

3. $\nabla f = -2x\,\mathbf{i} - 2y\,\mathbf{j}$, $\nabla g = \mathbf{i} + 3\,\mathbf{j}$. $\nabla f = \lambda \nabla g \Rightarrow -2x\,\mathbf{i} - 2y\,\mathbf{j} = \lambda(\mathbf{i} + 3\,\mathbf{j}) \Rightarrow x = -\frac{\lambda}{2}$ and $y = -\frac{3\lambda}{2} \Rightarrow \lambda = -2$ $\Rightarrow x = 1$ and $y = 3 \Rightarrow$ f takes on its extreme value at $(1,3) \Rightarrow$ the extreme value of f is 39.

4. Let (x,y) be on the parabola. Then the distance from the point to the line is $d(x,y) = \frac{|x - y + 1|}{\sqrt{2}}$. We want to minimize d subject to $y^2 = x \Rightarrow d(y) = \frac{|y^2 - y + 1|}{\sqrt{2}} = \frac{\left|\left(y - \frac{1}{2}\right)^2 + \frac{3}{4}\right|}{\sqrt{2}} = \frac{1}{\sqrt{2}}\left(y^2 - y + 1\right)$. $d'(y) = \frac{1}{\sqrt{2}}(2y - 1) \Rightarrow y = \frac{1}{2}$. $d''(y) = 2 \Rightarrow y = \frac{1}{2}$ yields a minimum. $y = \frac{1}{2} \Rightarrow x = \frac{1}{4} \Rightarrow$ the minimum distance is $d\left(\frac{1}{4}, \frac{1}{2}\right) = \frac{\left|\frac{1}{4} - \frac{1}{2} + 1\right|}{\sqrt{2}} = \frac{3\sqrt{2}}{8}$. Note: Lagrange multipliers not used.

5. $\nabla f = 2xy\,\mathbf{i} + x^2\,\mathbf{j}$, $\nabla g = \mathbf{i} + \mathbf{j}$. $\nabla f = \lambda \nabla g \Rightarrow 2xy\,\mathbf{i} + x^2\,\mathbf{j} = \lambda(\mathbf{i} + \mathbf{j}) \Rightarrow 2xy = \lambda$ and $x^2 = \lambda \Rightarrow 2xy = x^2 \Rightarrow$ $x = 0, y = 3$ or $x = 2, y = 1$. $\therefore$ f takes on its extreme values at $(0,3)$ and $(2,1)$. $\therefore$ the extreme values of f are $f(0,3) = 0$ and $f(2,1) = 4$.

6. Let $f(x,y) = x^2 + y^2$ be the square of the distance from the origin. The minimize f subject to $g(x,y) = x^2y - 2$. $\nabla f = 2x\,\mathbf{i} + 2y\,\mathbf{j}$, $\nabla g = 2xy\,\mathbf{i} + x^2\,\mathbf{j}$. $\nabla f = \lambda \nabla g \Rightarrow 2x = 2xy\lambda$ and $2y = x^2\lambda \Rightarrow \lambda = \frac{2y}{x^2} \Rightarrow$ $2x = 2xy\left(\frac{2y}{x^2}\right) \Rightarrow x^2 = 2y^2 \Rightarrow 2y^2y - 2 = 0 \Rightarrow y = 1 \Rightarrow x = \pm\sqrt{2}$. $\therefore$ $(\pm\sqrt{2}, 1)$ are the closest points.

7. a) $\nabla f = \mathbf{i} + \mathbf{j}$, $\nabla g = y\,\mathbf{i} + x\,\mathbf{j}$. $\nabla f = \lambda \nabla g \Rightarrow \mathbf{i} + \mathbf{j} = \lambda(y\,\mathbf{i} + x\,\mathbf{j}) \Rightarrow 1 = \lambda y$ and $1 = \lambda x \Rightarrow y = \frac{1}{\lambda}$ and $x = \frac{1}{\lambda}$ $\Rightarrow \frac{1}{\lambda^2} = 16 \Rightarrow \lambda = \pm 4$. Use $\lambda = 4$ since $x > 0$, $y > 0$. Then $x = 4$, $y = 4 \Rightarrow$ the minimum value is 8 at $x = 4$, $y = 4$.
   b) $\nabla f = y\,\mathbf{i} + x\,\mathbf{j}$, $\nabla g = \mathbf{i} + \mathbf{j}$. $\nabla f = \lambda \nabla g \Rightarrow y\,\mathbf{i} + x\,\mathbf{j} = \lambda(\mathbf{i} + \mathbf{j}) \Rightarrow y = \lambda = x \Rightarrow y = x \Rightarrow y + y = 16 \Rightarrow y = 8$ $\Rightarrow x = 8 \Rightarrow f(8,8) = 64$ is the maximum value.

8. Let $f(x,y) = x^2 + y^2$ be the square of the distance from the origin. Then $\nabla f = 2x\,\mathbf{i} + 2y\,\mathbf{j}$, $\nabla g = (2x + y)\,\mathbf{i} + (2y + x)\,\mathbf{j}$. $\nabla f = \lambda \nabla g \Rightarrow 2x = \lambda(2x + y)$ and $2y = \lambda(2y + x) \Rightarrow \frac{2y}{2y + x} = \lambda \Rightarrow 2x = \frac{2y}{2y + x}(2x + y) \Rightarrow$ $y = \pm x$. $y = x \Rightarrow x^2 + x(x) + x^2 - 1 = 0 \Rightarrow x = \pm\frac{1}{\sqrt{3}} \Rightarrow y = \pm\frac{1}{\sqrt{3}}$. $y = -x \Rightarrow x^2 + x(-x) + (-x)^2 - 1 = 0$ $\Rightarrow x = \pm 1$. $x = 1 \Rightarrow y = -1$, $x = -1 \Rightarrow y = 1$. $f\left(\frac{1}{\sqrt{3}},\frac{1}{\sqrt{3}}\right) = \frac{2}{3} = f\left(-\frac{1}{\sqrt{3}},-\frac{1}{\sqrt{3}}\right)$. $f(1,-1) = 2 = f(-1,1)$.
$\therefore$ $(1,-1)$ and $(-1,1)$ are the farthest; $\left(\frac{1}{\sqrt{3}},\frac{1}{\sqrt{3}}\right)$ and $\left(-\frac{1}{\sqrt{3}},-\frac{1}{\sqrt{3}}\right)$ are the closest.

9. $V = \pi r^2 h \Rightarrow 16\pi = \pi r^2 h \Rightarrow 16 = r^2 h \Rightarrow g(r,h) = r^2 h - 16$. $\nabla S = (2\pi h + 4\pi r)\,\mathbf{i} + 2\pi r\,\mathbf{j}$, $\nabla g = 2rh\,\mathbf{i} + r^2\,\mathbf{j}$. $\nabla S = \lambda \nabla g \Rightarrow (2\pi rh + 4\pi r)\,\mathbf{i} + 2\pi r\,\mathbf{j} = \lambda(2rh\,\mathbf{i} + r^2\,\mathbf{j}) \Rightarrow 2\pi h + 4\pi r = 2rh\lambda$ and $2\pi r = \lambda r^2 \Rightarrow 0 = \lambda r^2 - 2\pi r$ $\Rightarrow r = 0$ or $\lambda = \frac{2\pi}{r}$. Now $r \neq 0 \Rightarrow \lambda = \frac{2\pi}{r} \Rightarrow 2\pi h + 4\pi r = 2rh\left(\frac{2\pi}{r}\right) \Rightarrow 2r = h \Rightarrow 16 = r^2(2r) \Rightarrow r = 2 \Rightarrow$ $h = 4$. $\therefore$ $r = 2$ cm, $h = 4$ cm give the smallest surface area.

10. $g(x,y) = \frac{x^2}{16} + \frac{y^2}{9} - 1 = 0$. $A = (2x)(2y) = 4xy$. $\nabla A = 4y\,\mathbf{i} + 4x\,\mathbf{j}$, $\nabla g = \frac{x}{8}\,\mathbf{i} + \frac{2y}{9}\,\mathbf{j}$. $\nabla A = \lambda \nabla g \Rightarrow$ $4y\,\mathbf{i} + 4x\,\mathbf{j} = \lambda\left(\frac{x}{8}\,\mathbf{i} + \frac{2y}{9}\,\mathbf{j}\right) \Rightarrow 4y = \frac{x}{8}\lambda$ and $4x = \frac{2y}{9}\lambda \Rightarrow \lambda = \frac{32y}{x} \Rightarrow 4x = \frac{2y}{9}\left(\frac{32y}{x}\right) \Rightarrow y = \pm\frac{3}{4}x \Rightarrow$ $\frac{x^2}{16} + \frac{\left(\frac{\pm 3}{4}x\right)^2}{9} = 1 \Rightarrow x^2 = 8 \Rightarrow x = \pm 2\sqrt{2}$. Use $x = 2\sqrt{2}$ since x represents distance. Then $y = \frac{3}{4}(2\sqrt{2})$ $= \frac{3\sqrt{2}}{2}$. $\therefore$ the length is $2x = 4\sqrt{2}$ and the width is $2y = 3\sqrt{2}$.

11. $\nabla T = (8x - 4y)\,\mathbf{i} + (-4x + 2y)\,\mathbf{j}$, $\nabla g = 2x\,\mathbf{i} + 2y\,\mathbf{j}$. $\nabla T = \lambda \nabla g \Rightarrow (8x - 4y)\,\mathbf{i} + (-4x + 2y)\,\mathbf{j} = \lambda(2x\,\mathbf{i} + 2y\,\mathbf{j}) \Rightarrow$ $8x - 4y = 2\lambda x$ and $-4x + 2y = 2\lambda y \Rightarrow y = \frac{-2x}{\lambda - 1}$, $\lambda \neq 1 \Rightarrow 8x - 4\left(\frac{-2x}{\lambda - 1}\right) = 2\lambda x \Rightarrow x = 0$ or $\lambda = 0$ or $\lambda = 5$. $x = 0 \Rightarrow y = 0$. But $(0,0)$ not on $x^2 + y^2 = 25$. $\therefore$ $x \neq 0 \Rightarrow \lambda = 0$ or $\lambda = 5$. $\lambda = 0 \Rightarrow y = 2x \Rightarrow$ $x^2 + (2x)^2 = 25 \Rightarrow x = \pm\sqrt{5} \Rightarrow y = \pm 2\sqrt{5}$. $\lambda = 5 \Rightarrow y = \frac{-2x}{4} = -\frac{1}{2}x \Rightarrow x^2 + \left(-\frac{1}{2}x\right)^2 = 25 \Rightarrow x = \pm 2\sqrt{5}$. $x = 2\sqrt{5} \Rightarrow y = -\sqrt{5}$, $x = -2\sqrt{5} \Rightarrow y = \sqrt{5}$. $T(\sqrt{5},2\sqrt{5}) = 0° = T(-\sqrt{5},-2\sqrt{5})$, the minimum value; $T(2\sqrt{5},-\sqrt{5}) = 125° = T(-2\sqrt{5},\sqrt{5})$, the maximum value.

12. The surface area is given by $S = 4\pi r^2 + 2\pi rh$ subject to the constraint $V(r,h) = \frac{4}{3}\pi r^3 + \pi r^2 h = 8000$. $\nabla S = (8\pi r + 2\pi h)\,\mathbf{i} + 2\pi r\,\mathbf{j}$, $\nabla V = (4\pi r^2 + 2\pi rh)\,\mathbf{i} + \pi r^2\,\mathbf{j}$. $\nabla S = \lambda \nabla V \Rightarrow (8\pi r + 2\pi h)\,\mathbf{i} + 2\pi r\,\mathbf{j} =$ $\lambda((4\pi r^2 + 2\pi rh)\,\mathbf{i} + \pi r^2\,\mathbf{j}) \Rightarrow 8\pi r + 2\pi h = \lambda(4\pi r^2 + 2\pi rh)$ and $2\pi r = \lambda\,\pi r^2 \Rightarrow r = 0$ or $2 = r\lambda$. $r \neq 0 \Rightarrow$ $2 = r\lambda \Rightarrow \lambda = \frac{2}{r} \Rightarrow 4r + h = \frac{2}{r}(2r^2 + rh) \Rightarrow h = 0 \Rightarrow$ The tank is a sphere, there is no cylindrical part.

13. $\nabla f = 2x\,\mathbf{i} + 2y\,\mathbf{j}$, $\nabla g = (2x - 2)\,\mathbf{i} + (2y - 4)\,\mathbf{j}$. $\nabla f = \lambda \nabla g \Rightarrow 2x\,\mathbf{i} + 2y\,\mathbf{j} = \lambda((2x - 2)\,\mathbf{i} + (2y - 4)\,\mathbf{j}) \Rightarrow$ $2x = \lambda(2x - 2)$ and $2y = \lambda(2y - 4) \Rightarrow x = \frac{\lambda}{\lambda - 1}$ and $y = \frac{2\lambda}{\lambda - 1}$, $\lambda \neq 1 \Rightarrow y = 2x \Rightarrow x^2 - 2x + (2x)^2 - 4(2x)$ $= 0 \Rightarrow x = 0, y = 0$ or $x = 2, y = 4$. $\therefore$ $f(0,0) = 0$ is the minimum value, $f(2,4) = 20$ is the maximum value.

14. Let $f(x,y,z) = (x-1)^2 + (y-1)^2 + (z-1)^2$ be the square of the distance from (1,1,1). $\nabla f = 2(x-1)\,\mathbf{i} + 2(y-1)\,\mathbf{j} + 2(z-1)\,\mathbf{k}$, $\nabla g = \mathbf{i} + 2\,\mathbf{j} + 3\,\mathbf{k}$. $\nabla f = \lambda\,\nabla g \Rightarrow 2(x-1)\,\mathbf{i} + 2(y-1)\,\mathbf{j} + 2(z-1)\,\mathbf{k} = \lambda(\mathbf{i} + 2\,\mathbf{j} + 3\,\mathbf{k}) \Rightarrow 2(x-1) = \lambda,\ 2(y-1) = 2\lambda,\ 2(z-1) = 3\lambda \Rightarrow 2(y-1) = 2(2(x-1))$ and $2(z-1) = 3(2(x-1)) \Rightarrow x = \frac{y+1}{2} \Rightarrow z + 2 = 3\left(\frac{y+1}{2}\right) \Rightarrow z = \frac{3y-1}{2} \Rightarrow \frac{y+1}{2} + 2y + 3\left(\frac{3y-1}{2}\right) - 13 = 0 \Rightarrow y = 2 \Rightarrow x = \frac{3}{2},\ z = \frac{5}{2}$. $\therefore$ the point $\left(\frac{3}{2}, 2, \frac{5}{2}\right)$ is the closest.

15. $\nabla f = \mathbf{i} - 2\,\mathbf{j} + 5\,\mathbf{k}$, $\nabla g = 2x\,\mathbf{i} + 2y\,\mathbf{j} + 2z\,\mathbf{k}$. $\nabla f = \lambda\,\nabla g \Rightarrow \mathbf{i} - 2\,\mathbf{j} + 5\,\mathbf{k} = \lambda(2x\,\mathbf{i} + 2y\,\mathbf{j} + 2z\,\mathbf{k}) \Rightarrow 1 = 2x\,\lambda$, $-2 = 2y\,\lambda$, and $5 = 2z\,\lambda \Rightarrow x = \frac{1}{2\lambda},\ y = -\frac{1}{\lambda} = -2x,\ z = \frac{5}{2\lambda} = 5x \Rightarrow x^2 + (-2x)^2 + (5x)^2 = 30 \Rightarrow x = \pm 1$. $x = 1 \Rightarrow y = -2,\ z = 5$. $x = -1 \Rightarrow y = 2,\ z = -5$. $f(1,-2,5) = 30$, the maximum value; $f(-1,2,-5) = -30$, the minimum value.

16. Let $f(x,y,z) = x^2 + y^2 + z^2$ be the square of the distance from the origin. Then $\nabla f = 2x\,\mathbf{i} + 2y\,\mathbf{j} + 2z\,\mathbf{k}$, $\nabla g = 2x\,\mathbf{i} - 2y\,\mathbf{j} - 2z\,\mathbf{k}$. $\nabla f = \lambda\,\nabla g \Rightarrow 2x\,\mathbf{i} + 2y\,\mathbf{j} + 2z\,\mathbf{k} = \lambda(2x\,\mathbf{i} - 2y\,\mathbf{j} - 2z\,\mathbf{k}) \Rightarrow 2x = 2x\lambda,\ 2y = -2y\,\lambda$, and $2z = -2z\,\lambda \Rightarrow x = 0$ or $\lambda = 1$. CASE 1: $\lambda = 1 \Rightarrow 2y = -2y \Rightarrow y = 0$; $2z = -2z \Rightarrow z = 0 \Rightarrow x^2 - 1 = 0 \Rightarrow x = \pm 1$. CASE 2: $x = 0 \Rightarrow -y^2 - z^2 = 1$, which has no solution. $\therefore$ the points closest are $(\pm 1,0,0)$

17. Let $f(x,y,z) = x^2 + y^2 + z^2$ be the square of the distance to the origin. Then $\nabla f = 2x\,\mathbf{i} + 2y\,\mathbf{j} + 2z\,\mathbf{k}$, $\nabla g = y\,\mathbf{i} + x\,\mathbf{j} - \mathbf{k}$. $\nabla f = \lambda\,\nabla g \Rightarrow 2x\,\mathbf{i} + 2y\,\mathbf{j} + 2z\,\mathbf{k} = \lambda(y\,\mathbf{i} + x\,\mathbf{j} - \mathbf{k}) \Rightarrow 2x = \lambda y,\ 2y = \lambda x$, and $2z = -\lambda \Rightarrow x = \frac{\lambda y}{2} \Rightarrow 2y = \lambda\left(\frac{\lambda y}{2}\right) \Rightarrow y = 0$ or $\lambda = \pm 2$. $y = 0 \Rightarrow x = 0 \Rightarrow -z + 1 = 0 \Rightarrow z = 1$. $\lambda = 2 \Rightarrow x = y$, $z = -1 \Rightarrow x^2 - (-1) + 1 = 0 \Rightarrow x^2 + 2 = 0$, no solution. $\lambda = -2 \Rightarrow x = -y,\ z = 1 \Rightarrow (-y)y - 1 + 1 = 0 \Rightarrow y = 0$, again. $\therefore$ (0,0,1) is the point on the surface closest to the origin.

18. Let $f(x,y,z) = x^2 + y^2 + z^2$ be the square of the distance to the origin. Then $\nabla f = 2x\,\mathbf{i} + 2y\,\mathbf{j} + 2z\,\mathbf{k}$, $\nabla g = -y\,\mathbf{i} - x\,\mathbf{j} + 2z\,\mathbf{k}$. $\nabla f = \lambda\,\nabla g \Rightarrow 2x\,\mathbf{i} + 2y\,\mathbf{j} + 2z\,\mathbf{k} = \lambda(-y\,\mathbf{i} - x\,\mathbf{j} + 2z\,\mathbf{k}) \Rightarrow 2x = -y\lambda,\ 2y = -x\lambda$, and $2z = 2z\lambda \Rightarrow \lambda = 1$ or $z = 0$. CASE 1: $\lambda = 1 \Rightarrow 2x = -y$ and $2y = -x \Rightarrow y = 0 \Rightarrow x = 0 \Rightarrow z^2 - 4 = 0 \Rightarrow z = \pm 2$. CASE 2: $z = 0 \Rightarrow -xy - 4 = 0 \Rightarrow y = -\frac{4}{x}$. Then $2x = \frac{4}{x}\lambda \Rightarrow \lambda = \frac{x^2}{2}$ and $-\frac{8}{x} = -x\lambda \Rightarrow -\frac{8}{x} = -x\left(\frac{x^2}{2}\right) \Rightarrow x^4 = 16 \Rightarrow x = \pm 2$. $x = 2 \Rightarrow y = -4$, $x = -2 \Rightarrow y = 4$. We get four points, (2,–4,0), (–2,4,0), (0,0,2), and (0,0,–2). But the points closest are (0,0,2) and (0,0,–2) since they are 2 units fror the origin and the others are $\sqrt{20}$ away.

19. $\nabla f = \mathbf{i} + 2\,\mathbf{j} + 3\,\mathbf{k}$, $\nabla g = 2x\,\mathbf{i} + 2y\,\mathbf{j} + 2z\,\mathbf{k}$. $\nabla f = \lambda\,\nabla g \Rightarrow \mathbf{i} + 2\,\mathbf{j} + 3\,\mathbf{k} = \lambda(2x\,\mathbf{i} + 2y\,\mathbf{j} + 2z\,\mathbf{k}) \Rightarrow 1 = 2x\,\lambda$ $2 = 2y\lambda$, and $3 = 2z\lambda \Rightarrow x = \frac{1}{2\lambda},\ y = \frac{1}{\lambda}$, and $z = \frac{3}{2\lambda} \Rightarrow y = 2x$ and $z = 3x \Rightarrow x^2 + (2x)^2 + (3x)^2 = 25 \Rightarrow x = \pm\frac{5}{\sqrt{14}}$. $x = \frac{5}{\sqrt{14}} \Rightarrow y = \frac{10}{\sqrt{14}},\ z = \frac{15}{\sqrt{14}}$. $x = -\frac{5}{\sqrt{14}} \Rightarrow y = -\frac{10}{\sqrt{14}},\ z = -\frac{15}{\sqrt{14}}$. $f\left(\frac{5}{\sqrt{14}}, \frac{10}{\sqrt{14}}, \frac{15}{\sqrt{14}}\right)$ $5\sqrt{14}$, the maximum value; $f\left(-\frac{5}{\sqrt{14}}, -\frac{10}{\sqrt{14}}, -\frac{15}{\sqrt{14}}\right) = -5\sqrt{14}$, the minimum value.

20. $\nabla f = 2x\,\mathbf{i} + 2y\,\mathbf{j} + 2z\,\mathbf{k}$, $\nabla g = \mathbf{i} + \mathbf{j} + \mathbf{k}$. $\nabla f = \lambda\,\nabla g \Rightarrow 2x\,\mathbf{i} + 2y\,\mathbf{j} + 2z\,\mathbf{k} = \lambda(\mathbf{i} + \mathbf{j} + \mathbf{k}) \Rightarrow 2x = \lambda,\ 2y = \lambda$, $2z = \lambda \Rightarrow x = y = z \Rightarrow x + x + x - 9 = 0 \Rightarrow x = 3 \Rightarrow y = 3,\ z = 3$. $\therefore$ Each of the numbers is 3.

21. $\nabla f = yz\,\mathbf{i} + xz\,\mathbf{j} + xy\,\mathbf{k}$, $\nabla g = \mathbf{i} + \mathbf{j} + 2z\,\mathbf{k}$. $\nabla f = \lambda\,\nabla g \Rightarrow yz\,\mathbf{i} + xz\,\mathbf{j} + xy\,\mathbf{k} = \lambda(\mathbf{i} + \mathbf{j} + 2z\,\mathbf{k}) \Rightarrow yz = \lambda$, $xz = \lambda$, and $xy = \lambda \Rightarrow yz = xz \Rightarrow z = 0$ or $y = x$. But $z > 0 \Rightarrow y = x \Rightarrow x^2 = 2z\lambda$ and $xz = \lambda$. Then $x^2 = 2z(xz)$ $\Rightarrow x = 0$ or $x = 2z^2$. But $x > 0 \Rightarrow x = 2z^2 \Rightarrow y = 2z^2 \Rightarrow 2z^2 + 2z^2 + z^2 = 16 \Rightarrow z = \pm\frac{4}{\sqrt{5}}$. Use $z = \frac{4}{\sqrt{5}}$ since $z > 0 \Rightarrow x = \frac{32}{5}$, $y = \frac{32}{5}$. $f\left(\frac{32}{5}, \frac{32}{5}, \frac{4}{\sqrt{5}}\right) = \frac{4096}{25\sqrt{5}}$

22. $\nabla T = 16x\,\mathbf{i} + 4z\,\mathbf{j} + (4y - 16)\,\mathbf{k}$, $\nabla g = 8x\,\mathbf{i} + 2y\,\mathbf{j} + 8z\,\mathbf{k}$. $\nabla T = \lambda\,\nabla g \Rightarrow 16x\,\mathbf{i} + 4z\,\mathbf{j} + (4y - 16)\,\mathbf{k} = \lambda(8x\,\mathbf{i} + 2y\,\mathbf{j} + 8z\,\mathbf{k}) \Rightarrow 16x = 8x\lambda$, $4z = 2y\lambda$, and $4y - 16 = 8z\lambda \Rightarrow \lambda = 2$ or $x = 0$. CASE 1: $\lambda = 2 \Rightarrow 4z = 2y(2) \Rightarrow z = y$. Then $4z - 16 = 16z \Rightarrow z = -\frac{4}{3} \Rightarrow y = -\frac{4}{3}$. Then $4x^2 + \left(-\frac{4}{3}\right)^2 + 4\left(-\frac{4}{3}\right)^2 = 16 \Rightarrow x = \pm\frac{4}{3}$. CASE 2: $x = 0 \Rightarrow \lambda = \frac{2z}{y} \Rightarrow 4y - 16 = 8z\left(\frac{2z}{y}\right) \Rightarrow y^2 - 4y = 4z^2 \Rightarrow 4(0)^2 + y^2 + y^2 - 4y - 16 = 0 \Rightarrow y = 4$ or $y = -2$. $y = 4 \Rightarrow 4z^2 = 4^2 - 4(4) \Rightarrow z = 0$. $y = -2 \Rightarrow 4z^2 = (-2)^2 - 4(-2) \Rightarrow z = \pm\sqrt{3}$ $T\left(\pm\frac{4}{3}, -\frac{4}{3}, -\frac{4}{3}\right) = 642\frac{2}{3}^\circ$; $T(0,4,0) = 600^\circ$; $T(0,-2,\sqrt{3}) = (600 - 24\sqrt{3})^\circ$; $T(0,-2,-\sqrt{3}) = (600 + 24\sqrt{3})^\circ \approx 641.6^\circ$. $\therefore \left(\pm\frac{4}{3}, -\frac{4}{3}, -\frac{4}{3}\right)$ are the hottest places.

23. $\nabla U = (y + 2)\,\mathbf{i} + x\,\mathbf{j}$, $\nabla g = 2\,\mathbf{i} + \mathbf{j}$. $\nabla U = \lambda\,\nabla g \Rightarrow (y + 2)\,\mathbf{i} + x\,\mathbf{j} = \lambda(2\,\mathbf{i} + \mathbf{j}) \Rightarrow y + 2 = 2\lambda$ and $x = \lambda \Rightarrow y + 2 = 2x \Rightarrow y = 2x - 2 \Rightarrow 2x + 2x - 2 = 30 \Rightarrow x = 8 \Rightarrow y = 14$. $\therefore$ $U(8,14) = \$128$, the maximum value of U under the constraint.

24. $\nabla M = (6 + z)\,\mathbf{i} - 2y\,\mathbf{j} + x\,\mathbf{k}$, $\nabla g = 2x\,\mathbf{i} + 2y\,\mathbf{j} + 2z\,\mathbf{k}$. $\nabla M = \lambda\,\nabla g \Rightarrow (6 + z)\,\mathbf{i} - 2y\,\mathbf{j} + x\,\mathbf{k} = \lambda(2x\,\mathbf{i} + 2y\,\mathbf{j} + 2z\,\mathbf{k}) \Rightarrow 6 + z = 2x\lambda$, $-2y = 2y\lambda$, $x = 2z\lambda \Rightarrow \lambda = -1$ or $y = 0$. CASE 1: $\lambda = -1 \Rightarrow 6 + z = -2x$ and $x = -2z \Rightarrow 6 + z = -2(-2z) \Rightarrow z = 2$, $x = -4$. Then $(-4)^2 + y^2 + 2^2 - 36 = 0 \Rightarrow y = \pm 4$.
CASE 2: $y = 0$. Then $6 + z = 2x\lambda$ and $x = 2z\lambda \Rightarrow \lambda = \frac{x}{2z} \Rightarrow 6 + z = 2x\left(\frac{x}{2z}\right) \Rightarrow 6z + z^2 = x^2 \Rightarrow$ $6z + z^2 + 0^2 + z^2 = 36 \Rightarrow z = -6$ or $z = 3$. $z = -6 \Rightarrow x^2 = 0 \Rightarrow x = 0$. $z = 3 \Rightarrow x^2 = 9 \Rightarrow x = \pm 3$. $\therefore$ we have the points $(\pm 3,0,3)$, $(0,0,-6)$, and $(-4,2,\pm 4)$. $M(3,0,3) = 87$; $M(-3,0,3) = 33$; $M(0,0,-6) = 60$; $M(-4,2,4) = 16$; and $M(-4,2,-4) = 48$. $\therefore$ the weakest field will be at $(-4,2,4)$.

25. $\nabla f = \mathbf{i} + \mathbf{j}$, $\nabla g = y\,\mathbf{i} + x\,\mathbf{j}$. $\nabla f = \lambda\,\nabla g \Rightarrow \mathbf{i} + \mathbf{j} = \lambda(y\,\mathbf{i} + x\,\mathbf{j}) \Rightarrow 1 = y\lambda$ and $1 = x\lambda \Rightarrow y = x \Rightarrow y^2 = 16 \Rightarrow y = \pm 4 \Rightarrow x = \pm 4$. But as $x \to \infty$, $y \to 0$ and $f(x,y) \to \infty$; as $x \to -\infty$, $y \to 0$ and $f(x,y) \to -\infty$.

PRACTICE EXERCISES

1.

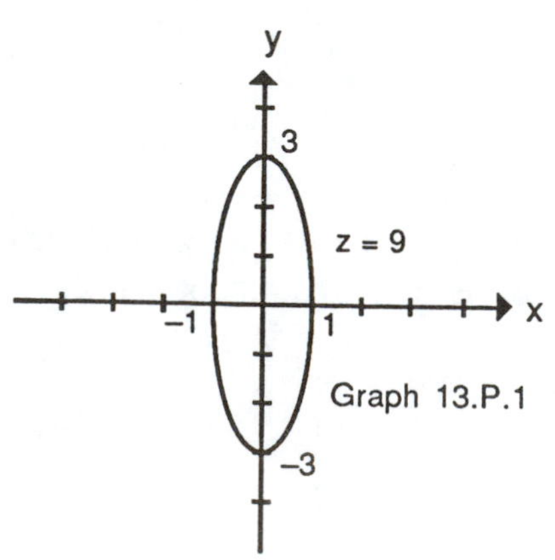

Graph 13.P.1

Domain: All points in the xy–plane
Range: $f(x,y) \geq 0$

Level curves are ellipses with major axis along the y–axis and minor axis along the x–axis.

2.

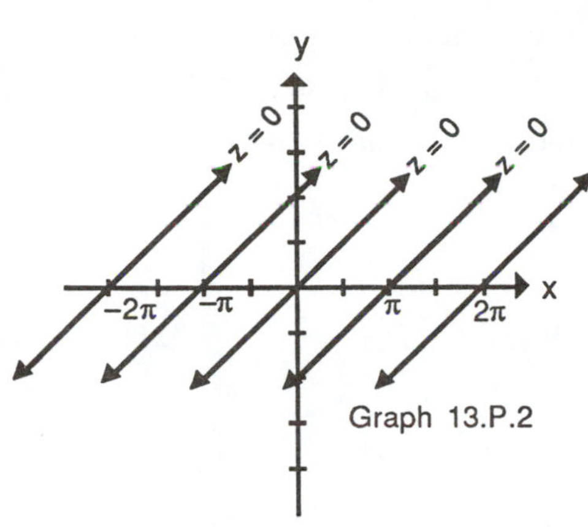

Graph 13.P.2

Domain: All points in the xy–plane
Range: $-1 \le f(x,y) \le 1$

Level curves are lines with m = 1.

3.

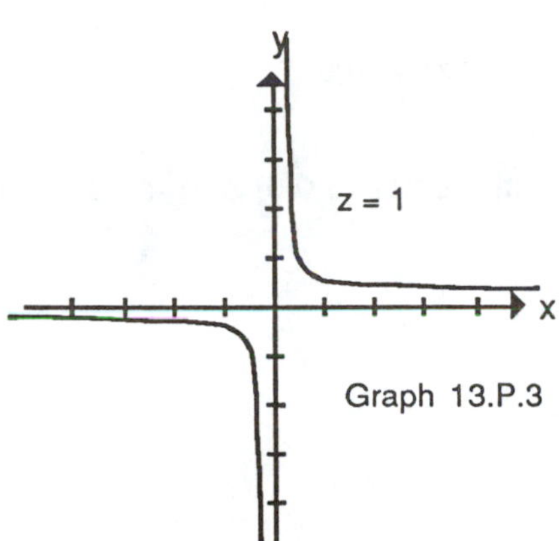

Graph 13.P.3

Domain: All (x,y) such that $x \neq 0$ or $y \neq 0$
Range: $f(x,y) \neq 0$

Level curves are hyperbolas rotated 45° or 135°.

4.

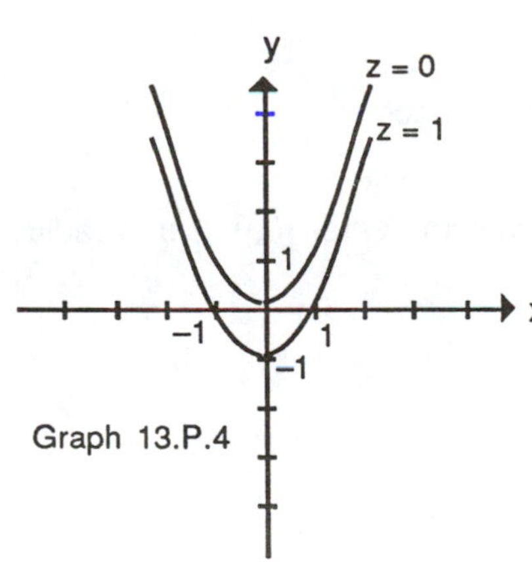

Graph 13.P.4

Domain: All (x,y) so that $x^2 - y \ge 0$
Range: $f(x,y) \ge 0$

Level curves are parabolas with vertex on y–axis.

5.

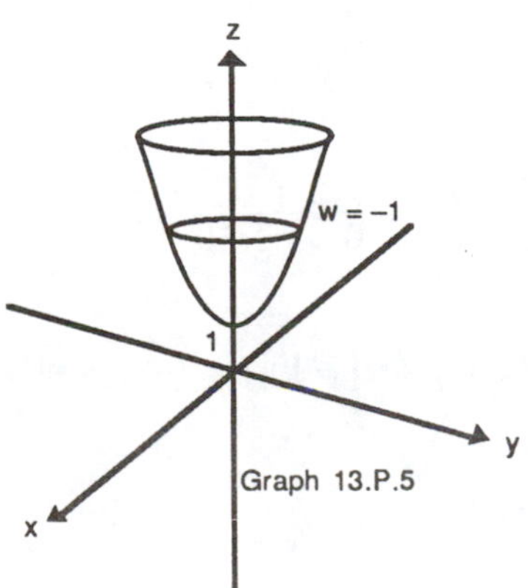

Graph 13.P.5

Domain: All (x,y,z) such (x,y,z $\neq$ (0,0,0)
Range: All Real Numbers

Level surfaces are paraboloids of revolution with the z–axis as axis.

6.

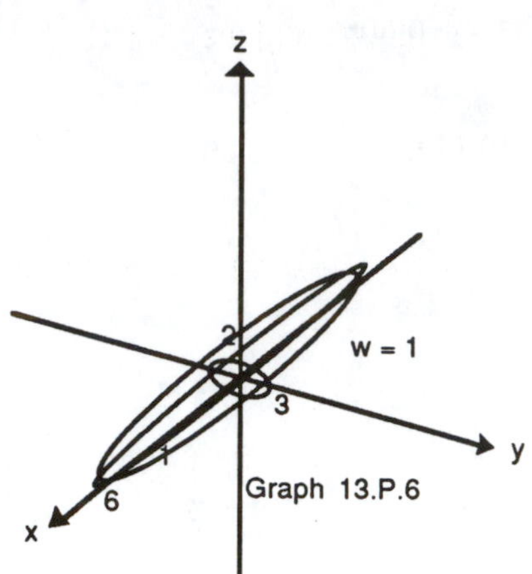

Graph 13.P.6

Domain: All points (x,y,z) in space
Range: $f(x,y,z) \geq 0$

Level surfaces are ellipsoids with center (0,0,0)

7.

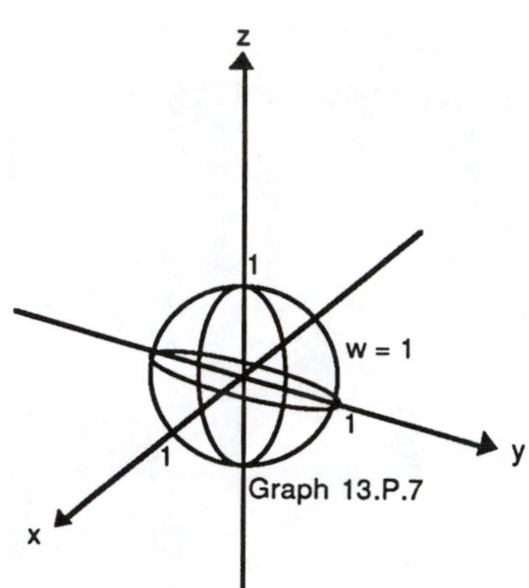

Graph 13.P.7

Domain: All (x,y,z) such that $(x,y,z) \neq (0,0,0)$
Range: $f(x,y,z) > 0$

Level surfaces are spheres with center (0,0,0) and radius $r > 0$.

8.

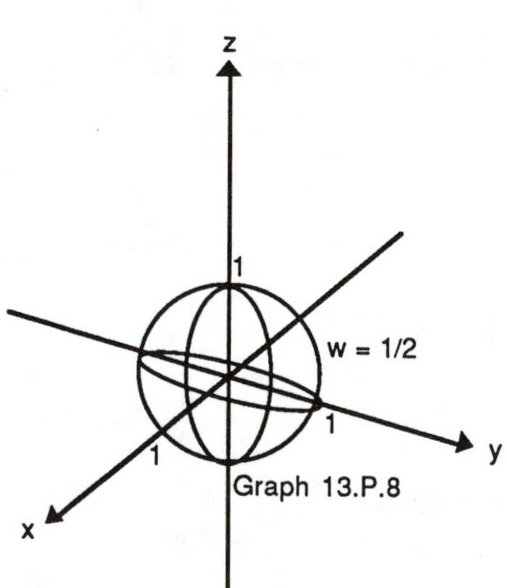

Graph 13.P.8

Domain: All points (x,y,z) in space
Range: $0 < f(x,y,z) \leq 1$

Level surfaces are spheres with center (0,0,0) and radius $r > 0$

9. $\lim_{(x,y)\to(\pi,\ln 2)} e^y \cos x = e^{\ln 2} \cos \pi = -2$

10. $\lim_{(x,y)\to(0,0)} \frac{2+y}{x+\cos y} = \frac{2+0}{0+\cos 0} = 2$

11. $\lim_{(x,y)\to(1,1)} \frac{x^2-y^2}{x-y} = \lim_{(x,y)\to(1,1)} (x+y) = 2$

12. $\lim_{P\to(1,-1,e)} \ln|x+y+z| = \ln|1+(-1)+e| = 1$

13. Let $y = kx^2$, $k \neq 1$. Then $\lim\limits_{\substack{(x,y)\to(0,0)\\ y\neq x^2}} \frac{y}{x^2 - y} = \lim\limits_{(x,kx^2)\to(0,0)} \frac{kx^2}{x^2 - kx^2} = \frac{k}{1 - k^2}$ which gives different limits for different values of k. ∴ the limit doesn't exist.

14. Let $y = kx$. Then $\lim\limits_{\substack{(x,y)\to(0,0)\\ xy\neq 0}} \frac{x^2 + y^2}{xy} = \lim\limits_{(x,kx)\to(0,0)} \frac{x^2 + (kx)^2}{x(kx)} = \frac{1 + k^2}{k}$ which gives different limits for different values of k. ∴ the limit does not exist.

15. $\frac{\partial g}{\partial r} = \cos\theta + \sin\theta$, $\frac{\partial g}{\partial\theta} = -r\sin\theta + r\cos\theta$

16. $f_x(x,y) = \frac{x - y}{x^2 + y^2}$, $f_y(x,y) = \frac{x + y}{x^2 + y^2}$

17. $\frac{\partial f}{\partial R_1} = -\frac{1}{R_1{}^2}$, $\frac{\partial f}{\partial R_2} = -\frac{1}{R_2{}^2}$, $\frac{\partial f}{\partial R_3} = -\frac{1}{R_3{}^2}$

18. $h_x(x,y,z) = 2\pi\cos(2\pi x + y - 3z)$, $h_y(x,y,z) = \cos(2\pi x + y - 3z)$, $h_z(x,y,z) = -3\cos(2\pi x + y - 3z)$

19. $\frac{\partial P}{\partial n} = \frac{RT}{V}$, $\frac{\partial P}{\partial R} = \frac{nT}{V}$, $\frac{\partial P}{\partial T} = \frac{nR}{V}$, $\frac{\partial P}{\partial V} = -\frac{nRT}{V^2}$

20. $f_r(r,l,T,d) = -\frac{1}{2r^2l}\sqrt{\frac{T}{\pi d}}$, $f_l(r,l,T,d) = -\frac{1}{2rl^2}\sqrt{\frac{T}{\pi d}}$, $f_T(r,l,T,d) = \frac{1}{4\pi rld}\sqrt{\frac{T}{\pi d}}$, $f_d(r,l,T,d) = \frac{-T}{2\pi rld^2}\sqrt{\frac{\pi d}{T}}$

21. $\frac{\partial f}{\partial x} = \frac{1}{y}$, $\frac{\partial f}{\partial y} = 1 - \frac{x}{y^2} \Rightarrow \frac{\partial^2 f}{\partial x^2} = 0$, $\frac{\partial^2 f}{\partial y^2} = \frac{2x}{y^3}$, $\frac{\partial^2 f}{\partial y\partial x} = \frac{\partial^2 f}{\partial x\partial y} = -\frac{1}{y^2}$

22. $f_x(x,y) = e^x + y\cos x$, $f_y(x,y) = \sin x \Rightarrow f_{xx}(x,y) = e^x - y\sin x$, $f_{yy}(x,y) = 0$, $f_{xy}(x,y) = f_{yx}(x,y) = \cos x$

23. $\frac{\partial f}{\partial x} = 1 + y - 15x^2 + \frac{2x}{x^2 + 1}$, $\frac{\partial f}{\partial y} = x \Rightarrow \frac{\partial^2 f}{\partial x^2} = -30x + \frac{2 - 2x^2}{(x^2 + 1)^2}$, $\frac{\partial^2 f}{\partial y^2} = 0$, $\frac{\partial^2 f}{\partial y\partial x} = \frac{\partial^2 f}{\partial x\partial y} = 1$

24. $f_x(x,y) = -3y$, $f_y(x,y) = 2y - 3x - \sin y + 7e^y \Rightarrow f_{xx}(x,y) = 0$, $f_{yy}(x,y) = 2 - \cos y + 7e^y$, $f_{xy}(x,y) = f_{yx}(x,y) = -3$.

25. $\frac{\partial w}{\partial x} = y\cos(xy + \pi)$, $\frac{\partial w}{\partial y} = x\cos(xy + \pi)$, $\frac{dx}{dt} = e^t$, $\frac{dy}{dt} = \frac{1}{t + 1} \Rightarrow \frac{dw}{dt} = y\cos(xy + \pi)\,e^t + x\cos(xy + \pi)\left(\frac{1}{t + 1}\right)$ $= e^t\ln(t + 1)\cos(e^t\ln(t + 1) + \pi) + \frac{e^t}{t + 1}\cos(e^t\ln(t + 1) + \pi) \Rightarrow \left.\frac{dw}{dt}\right|_{t=0} = -1$

26. $\frac{\partial w}{\partial x} = e^y$, $\frac{\partial w}{\partial y} = xe^y + \sin z$, $\frac{\partial w}{\partial z} = y\cos z + \sin z$, $\frac{dx}{dt} = t^{-1/2}$, $\frac{dy}{dt} = 1 + \frac{1}{t}$, $\frac{dz}{dt} = \pi \Rightarrow \frac{dw}{dt} = t^{-1/2}e^y + \left(xe^y + \sin z\right)\left(1 + \frac{1}{t}\right) + (y\cos z + \sin z)\pi$. $t = 1 \Rightarrow x = 2$, $y = 0$, $z = \pi \Rightarrow \left.\frac{dw}{dt}\right|_{t=1} = 5$

27. $\frac{\partial w}{\partial x} = 2\cos(2x - y)$, $\frac{\partial w}{\partial y} = -\cos(2x - y)$, $\frac{\partial x}{\partial r} = 1$, $\frac{\partial x}{\partial s} = \cos s$, $\frac{\partial y}{\partial r} = s$, $\frac{\partial y}{\partial s} = r \Rightarrow \frac{\partial w}{\partial r} = 2\cos(2x - y)(1) + (-\cos(2x - y)(s)) = 2\cos(2r + 2\sin s - rs) - s\cos(2r + 2\sin s - rs) \Rightarrow \left.\frac{\partial w}{\partial r}\right|_{r=\pi,s=0} = 2$.

$\frac{\partial w}{\partial s} = 2\cos(2x - y)(\cos s) + (-\cos(2x - y)(r) = 2\cos(2r + 2\sin s - rs)(\cos s) - r\cos(2r + 2\sin s - rs) \Rightarrow \left.\frac{\partial w}{\partial s}\right|_{r=\pi,s=0} = 2 - \pi$

28. $\frac{\partial w}{\partial u} = \frac{dw}{du}\frac{\partial x}{\partial u} = \left(\frac{x}{1+x^2} - \frac{1}{x^2+1}\right)2e^u \cos v$. $u = v = 0 \Rightarrow x = 2 \Rightarrow \frac{\partial w}{\partial u}\Big|_{u=v=0} = \frac{2}{5}$. $\frac{\partial w}{\partial v} = \frac{dw}{dx}\frac{\partial x}{\partial v} =$
$\left(\frac{x}{1+x^2} - \frac{1}{x^2+1}\right)(-2e^u \sin v)$. $\frac{\partial w}{\partial v}\Big|_{u=v=0} = 0$

29. $F_x = -1 - y\cos xy$, $F_y = -2y - x\cos xy$. $\frac{dy}{dx} = -\frac{F_x}{F_y} = -\frac{-1-y\cos xy}{-2y - x\cos xy} = \frac{1+y\cos xy}{-2y - x\cos xy} \Rightarrow$
$\frac{dy}{dx}\Big|_{(x,y)=(0,1)} = -1$

30. $F_x = 2y + e^{x+y}$, $F_y = 2x + e^{x+y} \Rightarrow \frac{dy}{dx} = -\frac{F_x}{F_y} = -\frac{2y + e^{x+y}}{2x + e^{x+y}} \Rightarrow \frac{dy}{dx}\Big|_{(0,\ln 2)} = -(\ln(2) + 1)$

31. $\frac{\partial f}{\partial x} = y + z$, $\frac{\partial f}{\partial y} = x + z$, $\frac{\partial f}{\partial z} = y + x$, $\frac{dx}{dt} = -\sin t$, $\frac{dy}{dt} = \cos t$, $\frac{dz}{dt} = -2\sin 2t \Rightarrow \frac{df}{dt} = -(y+z)\sin t + (x+z)\cos t$
$- 2(y+x)\sin 2t = -(\sin t + \cos 2t)\sin t + (\cos t + \cos 2t)\cos t - 2(\sin t + \cos t)\sin 2t \Rightarrow \frac{df}{dt}\Big|_{t=1} =$
$-(\sin 1 + \cos 2)\sin 1 + (\cos 1 + \cos 2)\cos 1 - 2(\sin 1 + \cos 1)\sin 2$

32. $\frac{\partial w}{\partial x} = \frac{dw}{ds}\frac{\partial s}{\partial x} = 5\frac{dw}{ds}$. $\frac{\partial w}{\partial y} = \frac{dw}{ds}\frac{\partial s}{\partial y} = (1)\frac{dw}{ds} = \frac{dw}{ds}$. $\therefore$ $\frac{\partial w}{\partial x} - 5\frac{\partial w}{\partial y} = 5\frac{dw}{ds} - 5\left(\frac{dw}{ds}\right) = 0$

33. $\nabla f = (-\sin x\cos y)\,\mathbf{i} - (\cos x\sin y)\,\mathbf{j} \Rightarrow \nabla f\Big|_{(\pi/4,\pi/4)} = -\frac{1}{2}\mathbf{i} - \frac{1}{2}\mathbf{j} \Rightarrow |\nabla f| = \sqrt{\left(-\frac{1}{2}\right)^2 + \left(-\frac{1}{2}\right)^2} = \frac{1}{\sqrt{2}}$
$\mathbf{u} = \frac{\nabla f}{|\nabla f|} = \frac{-\frac{1}{2}\mathbf{i} - \frac{1}{2}\mathbf{j}}{\frac{1}{\sqrt{2}}} = -\frac{\sqrt{2}}{2}\mathbf{i} - \frac{\sqrt{2}}{2}\mathbf{j}$. f increases most rapidly in the direction $\mathbf{u} = -\frac{\sqrt{2}}{2}\mathbf{i} - \frac{\sqrt{2}}{2}\mathbf{j}$;
decreases most rapidly in the direction $-\mathbf{u} = \frac{\sqrt{2}}{2}\mathbf{i} + \frac{\sqrt{2}}{2}\mathbf{j}$. $\left(D_{\mathbf{u}}f\right)_{P_0} = \frac{\sqrt{2}}{2}$, $\left(D_{-\mathbf{u}}f\right)_{P_0} = -\frac{\sqrt{2}}{2}$.
$\mathbf{u_1} = \frac{\mathbf{A}}{|\mathbf{A}|} = \frac{3\,\mathbf{i} + 4\,\mathbf{j}}{\sqrt{3^2 + 4^2}} = \frac{3}{5}\mathbf{i} + \frac{4}{5}\mathbf{j}$. $\left(D_{\mathbf{u_1}}f\right)_{P_0} = \nabla f \cdot \mathbf{u_1} = -\frac{7}{10}$.

34. $\nabla f = 2xe^{-2y}\,\mathbf{i} - 2x^2e^{-2y}\,\mathbf{j} \Rightarrow \nabla f\Big|_{(1,0)} = 2\,\mathbf{i} - 2\,\mathbf{j} \Rightarrow |\nabla f| = \sqrt{2^2 + (-2)^2} = 2\sqrt{2}$. $\mathbf{u} = \frac{\nabla f}{|\nabla f|} = \frac{2\,\mathbf{i} - 2\,\mathbf{j}}{2\sqrt{2}} =$
$\frac{1}{\sqrt{2}}\mathbf{i} - \frac{1}{\sqrt{2}}\mathbf{j}$. f increases most rapidly in the direction $\mathbf{u} = \frac{1}{\sqrt{2}}\mathbf{i} - \frac{1}{\sqrt{2}}\mathbf{j}$; decreases most rapidly in the direction $-\mathbf{u} = -\frac{1}{\sqrt{2}}\mathbf{i} + \frac{1}{\sqrt{2}}\mathbf{j}$. $\left(D_{\mathbf{u}}f\right)_{P_0} = \nabla f \cdot \mathbf{u} = 2\sqrt{2}$, $\left(D_{-\mathbf{u}}f\right)_{P_0} = -2\sqrt{2}$. $\mathbf{u_1} = \frac{\mathbf{A}}{|\mathbf{A}|} = \frac{\mathbf{i} + \mathbf{j}}{\sqrt{1^2 + 1^2}}$
$= \frac{1}{\sqrt{2}}\mathbf{i} + \frac{1}{\sqrt{2}}\mathbf{j}$. $\left(D_{\mathbf{u_1}}f\right)_{P_0} = \nabla f \cdot \mathbf{u_1} = 2\left(\frac{1}{\sqrt{2}}\right) + (-2)\left(\frac{1}{\sqrt{2}}\right) = 0$.

35. $\nabla f = \left(\frac{2}{2x + 3y + 6z}\right)\mathbf{i} + \left(\frac{3}{2x + 3y + 6z}\right)\mathbf{j} + \left(\frac{6}{2x + 3y + 6z}\right)\mathbf{k} \Rightarrow \nabla f\Big|_{(-1,-1,1)} = 2\,\mathbf{i} + 3\,\mathbf{j} + 6\,\mathbf{k}$.
$\mathbf{u} = \frac{\nabla f}{|\nabla f|} = \frac{2\,\mathbf{i} + 3\,\mathbf{j} + 6\,\mathbf{k}}{\sqrt{2^2 + 3^2 + 6^2}} = \frac{2}{7}\mathbf{i} + \frac{3}{7}\mathbf{j} + \frac{6}{7}\mathbf{k}$. f increases most rapidly in the direction $\mathbf{u} = \frac{2}{7}\mathbf{i} + \frac{3}{7}\mathbf{j} + \frac{6}{7}\mathbf{k}$;
decreases most rapidly in the direction $-\mathbf{u} = -\frac{2}{7}\mathbf{i} - \frac{3}{7}\mathbf{j} - \frac{6}{7}\mathbf{k}$. $\left(D_{\mathbf{u}}f\right)_{P_0} = \nabla f \cdot \mathbf{u} = 7$,
$\left(D_{-\mathbf{u}}f\right)_{P_0} = -7$. $\mathbf{u_1} = \frac{\mathbf{A}}{|\mathbf{A}|} = \frac{2}{7}\mathbf{i} + \frac{3}{7}\mathbf{j} + \frac{6}{7}\mathbf{k}$ since $\mathbf{A} = \nabla f$. $\Rightarrow$ $\left(D_{\mathbf{u_1}}f\right)_{P_0} = 7$.

36. $\nabla f = (2x + 3y)\,\mathbf{i} + (3x + 2)\,\mathbf{j} + (1 - 2z)\,\mathbf{k} \Rightarrow \nabla f\big|_{(0,0,0)} = 2\,\mathbf{j} + \mathbf{k}$. $\mathbf{u} = \frac{\nabla f}{|\nabla f|} = \frac{2\,\mathbf{j} + \mathbf{k}}{\sqrt{2^2 + 1^2}} = \frac{2}{\sqrt{5}}\mathbf{j} + \frac{1}{\sqrt{5}}\mathbf{k}$.

f increases most rapidly in the direction $\mathbf{u} = \frac{2}{\sqrt{5}}\mathbf{i} + \frac{1}{\sqrt{5}}\mathbf{k}$; decreases most rapidly in the direction $-\mathbf{u} = -\frac{2}{\sqrt{5}}\mathbf{j} - \frac{1}{\sqrt{5}}\mathbf{k}$. $\left(D_{\mathbf{u}}f\right)_{P_0} = \nabla f \cdot \mathbf{u} = \sqrt{5}$; $\left(D_{-\mathbf{u}}f\right)_{P_0} = -\sqrt{5}$. $\mathbf{u}_1 = \frac{\mathbf{A}}{|\mathbf{A}|} = \frac{\mathbf{i} + \mathbf{j} + \mathbf{k}}{\sqrt{1^2 + 1^2 + 1^2}} = \frac{1}{\sqrt{3}}\mathbf{i} + \frac{1}{\sqrt{3}}\mathbf{j} + \frac{1}{\sqrt{3}}\mathbf{k}$

$\left(D_{\mathbf{u}_1}f\right)_{P_0} = \nabla f \cdot \mathbf{u}_1 = 0\left(\frac{1}{\sqrt{3}}\right) + 2\left(\frac{1}{\sqrt{3}}\right) + (1)\left(\frac{1}{\sqrt{3}}\right) = \sqrt{3}$

37.

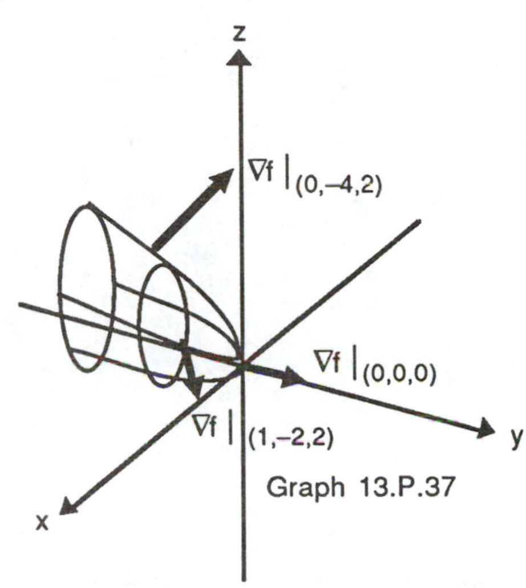

Graph 13.P.37

$\nabla f = 2x\,\mathbf{i} + \mathbf{j} + 2z\,\mathbf{k} \Rightarrow$
$\nabla f\big|_{(1,-2,1)} = 2\,\mathbf{i} + \mathbf{j} + 2\,\mathbf{k}$,
$\nabla f\big|_{(0,0,0)} = \mathbf{j}$,
$\nabla f\big|_{(0,-4,2)} = \mathbf{j} + 4\,\mathbf{k}$

38.

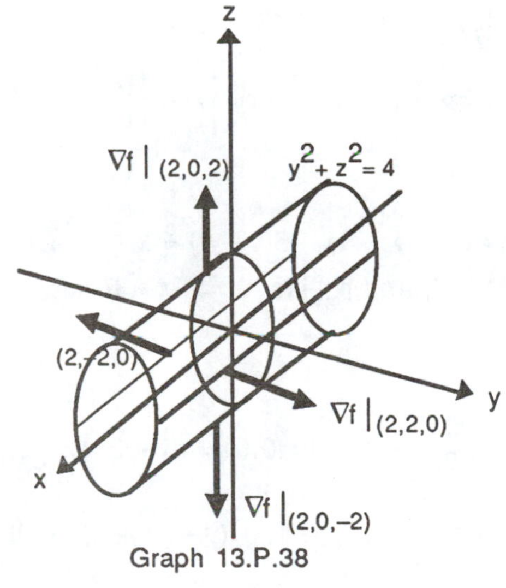

Graph 13.P.38

$\nabla f = 2y\,\mathbf{j} + 2z\,\mathbf{k} \Rightarrow$
$\nabla f\big|_{(2,2,0)} = 4\,\mathbf{j}$,
$\nabla f\big|_{(2,-2,0)} = -4\,\mathbf{j}$,
$\nabla f\big|_{(2,0,2)} = 4\,\mathbf{k}$,
$\nabla f\big|_{(2,0,-2)} = -4\,\mathbf{k}$

39. $\nabla f = 2x\,\mathbf{i} - \mathbf{j} - 5\,\mathbf{k} \Rightarrow \nabla f\big|_{(2,-1,1)} = 4\,\mathbf{i} - \mathbf{j} - 5\,\mathbf{k} \Rightarrow$ Tangent Plane: $4(x - 2) - (y + 1) - 5(z - 1) = 0 \Rightarrow 4x - y - 5z = 4$; Normal Line: $x = 2 + 4t$, $y = -1 - t$, $z = 1 - 5t$

40. $\nabla f = 2x\,\mathbf{i} + 2y\,\mathbf{j} + \mathbf{k} \Rightarrow \nabla f\big|_{(1,1,2)} = 2\,\mathbf{i} + 2\,\mathbf{j} + \mathbf{k} \Rightarrow$ Tangent Plane: $2(x - 1) + 2(y - 1) + (z - 2) = 0 \Rightarrow 2x + 2y + z - 6 = 0$; Normal Line: $x = 1 + 2t$, $y = 1 + 2t$, $z = 2 + t$

41.

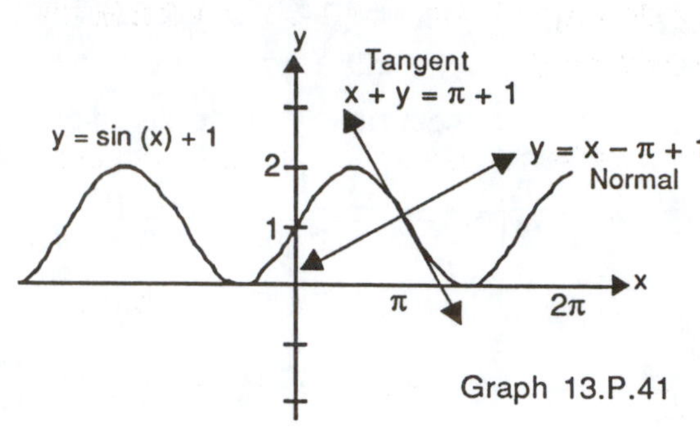

Graph 13.P.41

$\nabla f = (-\cos x)\,\mathbf{i} + \mathbf{j} \Rightarrow \nabla f\big|_{(\pi,1)} = \mathbf{i} + \mathbf{j} \Rightarrow$ Tangent Line: $(x - \pi) + (y - 1) = 0 \Rightarrow x + y = \pi + 1$; Normal Line: $y - 1 = 1(x - \pi) \Rightarrow y = x - \pi + 1$

42.

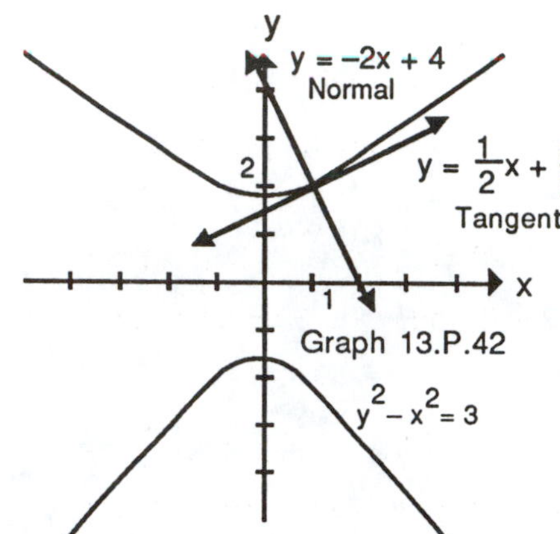

Graph 13.P.42

$\nabla f = -x\,\mathbf{i} + y\,\mathbf{j} \Rightarrow \nabla f\big|_{(1,2)} = -\mathbf{i} + 2\,\mathbf{j} \Rightarrow$ Tangent Line: $-(x - 1) + 2(y - 2) = 0 \Rightarrow y = \frac{1}{2}x + \frac{3}{2}$; Normal Line: $y - 2 = -2(x - 1) \Longrightarrow y = -2x + 4$

43. $f\left(\frac{\pi}{4},\frac{\pi}{4}\right) = \frac{1}{2}$, $f_x\left(\frac{\pi}{4},\frac{\pi}{4}\right) = \cos x \cos y\big|_{(\pi/4,\pi/4)} = \frac{1}{2}$, $f_y\left(\frac{\pi}{4},\frac{\pi}{4}\right) = -\sin x \sin y\big|_{(\pi/4,\pi/4)} = -\frac{1}{2} \Rightarrow L(x,y) = \frac{1}{2} + \frac{1}{2}\left(x - \frac{\pi}{4}\right) - \frac{1}{2}\left(y - \frac{\pi}{4}\right) = \frac{1}{2} + \frac{1}{2}x - \frac{1}{2}y$. $f_{xx}(x,y) = -\sin x \cos y$, $f_{yy}(x,y) = -\sin x \cos y$, $f_{xy}(x,y) = -\cos x \sin y$. $\therefore$ maximum of $|f_{xx}|$, $|f_{yy}|$, and $|f_{xy}|$ is $1 \Rightarrow M = 1 \Rightarrow |E(x,y)| \le \frac{1}{2}(1)\left(\left|x - \frac{\pi}{4}\right| + \left|y - \frac{\pi}{4}\right|\right)^2 \le 0.02$.

44. $f(1,1) = 0$, $f_x(1,1) = y\big|_{(1,1)} = 1$, $f_y(1,1) = x - 6y\big|_{(1,1)} = -5 \Rightarrow L(x,y) = (x - 1) - 5(y - 1) = x - 5y + 4$. $f_{xx}(x,y) = 0$, $f_{yy}(x,y) = -6$, and $f_{xy}(x,y) = 1 \Rightarrow$ maximum of $|f_{xx}|$, $|f_{yy}|$, and $|f_{xy}|$ is $6 \Rightarrow M = 6 \Rightarrow |E(x,y)| \le \frac{1}{2}(6)\left(|x - 1| + |y - 1|\right)^2 \le 0.12$

45. a) $f(1,0,0) = 0$, $f_x(1,0,0) = y - 3z\big|_{(1,0,0)} = 0$, $f_y(1,0,0) = x + 2z\big|_{(1,0,0)} = 1$, $f_z(1,0,0) = 2y - 3x\big|_{(1,0,0)} = -3 \Rightarrow L(x,y,z) = 0(x - 1) + (y - 0) - 3(z - 0) = y - 3z$.

b) $f(1,1,0) = 1$, $f_x(1,1,0) = 1$, $f_y(1,1,0) = 1$, $f_z(1,1,0) = 2 \Rightarrow L(x,y,z) = 1 + (x - 1) + (y - 1) + 2(z - 0) = x + y + 2z - 1$

46. a) $f\left(0,0,\frac{\pi}{4}\right) = 1$, $f_x\left(0,0,\frac{\pi}{4}\right) = -\sqrt{2}\sin x \sin(y + z)\big|_{(0,0,\pi/4)} = 0$, $f_y\left(0,0,\frac{\pi}{4}\right) = \sqrt{2}\cos x \cos(y + z)\big|_{(0,0,\pi/4)}$ 1, $f_z\left(0,0,\frac{\pi}{4}\right) = \sqrt{2}\cos x \cos(y + z)\big|_{(0,0,\pi/4)} = 1 \Rightarrow L(x,y,z) = 1 + 1(y - 0) + 1\left(z - \frac{\pi}{4}\right) = 1 - \frac{\pi}{4} + z$

b) $f\left(\frac{\pi}{4},\frac{\pi}{4},0\right) = \frac{\sqrt{2}}{2}$, $f_x\left(\frac{\pi}{4},\frac{\pi}{4},0\right) = -\frac{\sqrt{2}}{2}$, $f_y\left(\frac{\pi}{4},\frac{\pi}{4},0\right) = \frac{\sqrt{2}}{2}$, $f_z\left(\frac{\pi}{4},\frac{\pi}{4},0\right) = \frac{\sqrt{2}}{2} \Rightarrow L(x,y,z) = \frac{\sqrt{2}}{2} - \frac{\sqrt{2}}{2}\left(x - \frac{\pi}{4}\right) + \frac{\sqrt{2}}{2}\left(y - \frac{\pi}{4}\right) + \frac{\sqrt{2}}{2}(z - 0) = -\frac{\sqrt{2}}{2}x + \frac{\sqrt{2}}{2}y + \frac{\sqrt{2}}{2}z + \frac{\sqrt{2}}{2}$

47. $dV = 2\pi rh\ dr + \pi r^2\ dh \Rightarrow dV\big|_{(1.5,5280)} = 2\pi(1.5)(5280)\ dr + \pi(1.5)^2\ dh = 15840\pi\ dr + 2.25\pi\ dh$. Be more careful with the diameter since it has a greater effect on dV.

48. $df = (2x - y)dx + (-x + 2y)dy \Rightarrow df\big|_{(1,2)} = 3\ dy$. $\therefore$ f is more sensitive to changes in y because a change in x does not change f, but a change in y does change f.

49. $dI = \frac{1}{R}\,dV - \frac{V}{R^2}\,dR \Rightarrow dI\big|_{(24,100)} = \frac{1}{100}\,dV - \frac{24}{100^2}\,dR \Rightarrow dI\big|_{dV=-1,dR=-20} = 0.038$. % change in V = $-\frac{1}{24} = -4.17\%$; % change in R = $-\frac{20}{100} = -20\%$. $I = \frac{24}{100} = 0.24 \Rightarrow$ Estimated % change in I = $\frac{dI}{I} \times 100 = \frac{0.038}{0.24} \times 100 = 15.83\%$

50. $A = \pi ab \Rightarrow dA = \pi b\ da + \pi a\ db \Rightarrow dA\big|_{(10,16);da=\pm0.05,db=\pm0.05} = \pi(16)(\pm0.05) + \pi(10)(\pm0.05) = \pm4.0841\ \text{cm}^2$. $A = \pi(10)(16) = 160\pi\ \text{cm}^2$. $\left|\frac{dA}{A} \times 100\right| = \left|\frac{26\pi(\pm0.05)}{160\pi} \times 100\right| = 0.8125\%$

51. $f_x(x,y) = 2x - y + 2 = 0$ and $f_y(x,y) = -x + 2y + 2 = 0 \Rightarrow x = -2,\ y = -2 \Rightarrow (-2,-2)$ is the critical point. $f_{xx}(-2,-2) = 2,\ f_{yy}(-2,-2) = 2,\ f_{xy}(-2,-2) = -1 \Rightarrow f_{xx}f_{yy} - f_{xy}^2 = 3 > 0$ and $f_{xx} > 0 \Rightarrow$ Minimum (absolute). $f(-2,-2) = -8$

52. $f_x(x,y) = 10x + 4y + 4 = 0$ and $f_y(x,y) = 4x - 4y - 4 = 0 \Rightarrow x = 0,\ y = -1 \Rightarrow$ critical point is $(0,-1)$. $f_{xx}(0,-1) = 10,\ f_{yy}(0,-1) = -4,\ f_{xy}(0,-1) = 4 \Rightarrow f_{xx}f_{yy} - f_{xy}^2 = -56 \Rightarrow$ Saddle Point. $f(0,-1) = 2$

53. $f_x(x,y) = 6y - 3x^2 = 0$ and $f_y(x,y) = 6x - 2y = 0 \Rightarrow x = 0,\ y = 0$ or $x = 6,\ y = 18 \Rightarrow$ critical points are (0,0) and (6,18). For (0,0): $f_{xx}(0,0) = -6x\big|_{(0,0)} = 0,\ f_{yy}(0,0) = -2,\ f_{xy}(0,0) = 6 \Rightarrow f_{xx}f_{yy} - f_{xy}^2 = -36 \Rightarrow$ Saddle Point. $f(0,0) = 0$. For (6,18): $f_{xx}(6,18) = -36,\ f_{yy}(6,18) = -2,\ f_{xy}(6,18) = 6 \Rightarrow f_{xx}f_{yy} - f_{xy}^2 = 36 > 0$ and $f_{xx} < 0 \Rightarrow$ maximum (local since $y = 0$ and $x < 0 \Rightarrow f(x,y)$ increases without bound). $f(6,18) = 108$.

54. $f_x(x,y) = 6x^2 + 3y = 0$ and $f_y(x,y) = 3x + 3y^2 = 0 \Rightarrow x = 0,\ y = 0$ or $x = -\frac{1}{\sqrt[3]{4}},\ y = -\frac{1}{\sqrt[3]{2}} \Rightarrow$ critical points are (0,0) and $\left(-\frac{1}{\sqrt[3]{4}}, -\frac{1}{\sqrt[3]{2}}\right)$. For (0,0): $f_{xx}(0,0) = 12x\big|_{(0,0)} = 0,\ f_{yy}(,0) = 6y\big|_{(0,0)} = 0,\ f_{xy}(0,0) = 3 \Rightarrow f_{xx}f_{yy} - f_{xy}^2 = -9 \Rightarrow$ Saddle Point. $f(0,0) = 0$. For $\left(-\frac{1}{\sqrt[3]{4}}, -\frac{1}{\sqrt[3]{2}}\right)$: $f_{xx}\left(-\frac{1}{\sqrt[3]{4}}, -\frac{1}{\sqrt[3]{2}}\right) = -\frac{12}{\sqrt[3]{4}}$, $f_{yy}\left(-\frac{1}{\sqrt[3]{4}}, -\frac{1}{\sqrt[3]{2}}\right) = -\frac{6}{\sqrt[3]{2}}$, $f_{xy}\left(-\frac{1}{\sqrt[3]{4}}, -\frac{1}{\sqrt[3]{2}}\right) = 3 \Rightarrow f_{xx}f_{yy} - f_{xy}^2 = 27 > 0$ and $f_{xx} < 0 \Rightarrow$ Maximum (local) $f\left(-\frac{1}{\sqrt[3]{4}}, -\frac{1}{\sqrt[3]{2}}\right) = \frac{1}{2}$

55. $f_x(x,y) = 3x^2 - 3y = 0$ and $f_y(x,y) = 3y^2 - 3x = 0 \Rightarrow x = 0,\ y = 0$ or $x = 1,\ y = 1 \Rightarrow$ critical points are (0,0) and (1,1). For (0,0): $f_{xx}(0,0) = 6x\big|_{(0,0)} = 0,\ f_y(0,0) = 6y\big|_{(0,0)} = 0,\ f_{xy}(,0) = -3 \Rightarrow f_{xx}f_{yy} - f_{xy}^2 = -9 \Rightarrow$ Saddle Point. $f(0,0) = 15$. For (1,1): $f_{xx}(1,1) = 6,\ f_{yy}(1,1) = 6,\ f_{xy}(1,1) = -3 \Rightarrow f_{xx}f_{yy} - f_{xy}^2 = 27 > 0$ and $f_{xx} > 0 \Rightarrow$ Minimum (local since $y = 0,\ x < 0 \Rightarrow f(x,y)$ decreases without bound). $f(1,1) = 14$.

56. $f_x(x,y) = 3x^2 + 6x = 0$ and $f_y(x,y) = 3y^2 - 6y = 0 \Rightarrow x = 0, y = 0; x = 0, y = 2; x = -2, y = 0$; or $x = -2, y = 2 \Rightarrow$ critical points are (0,0), (0,2), (–2,0), and (–2,2). For (0,0): $f_{xx}(0,0) = 6x + 6\big|_{(0,0)} = 6$, $f_{yy}(0,0) = 6y - 6\big|_{(0,0)} = -6$, $f_{xy}(0,0) = 0 \Rightarrow f_{xx}f_{yy} - f_{xy}^2 = -36 \Rightarrow$ Saddle Point. $f(0,0) = 0$.
For (0,2): $f_{xx}(0,2) = 6$, $f_{yy}(0,2) = 6$, $f_{xy}(0,2) = 0 \Rightarrow f_{xx}f_{yy} - f_{xy}^2 = 36 > 0$ and $f_{xx} > 0 \Rightarrow$ Minimum (local). $f(0,2) = -4$. For (–2,0): $f_{xx}(-2,0) = -6$, $f_{yy}(-2,0) = -6$, $f_{xy}(-2,0) = 0 \Rightarrow f_{xx}f_{yy} - f_{xy}^2 = 36 > 0$ and $f_{xx} < 0 \Rightarrow$ Maximum (local). $f(-2,0) = 4$. For (–2,2): $f_{xx}(-2,2) = -6$, $f_{yy}(-2,2) = 6$, $f_{xy}(-2,2) = 0 \Rightarrow f_{xx}f_{yy} - f_{xy}^2 = -36 \Rightarrow$ Saddle Point. $f(-2,2) = 0$

57.

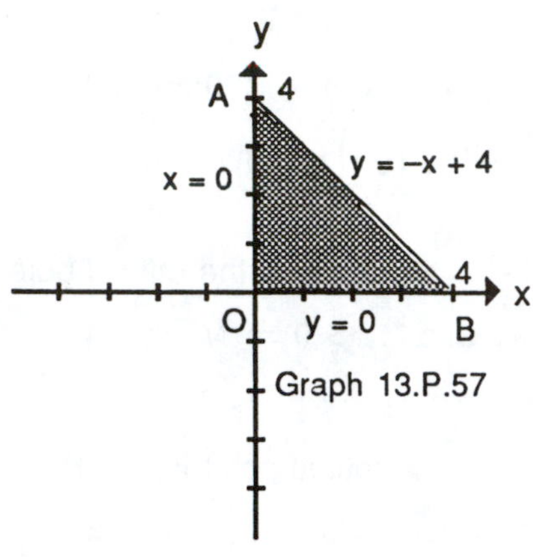

Graph 13.P.57

1. On OA, $f(x,y) = y^2 + 3y = f(0,y)$ for $0 \le y \le 4$. $f(0,0) = 0$, $f(0,4) = 28$. $f'(0,y) = 2y + 3 = 0 \Rightarrow y = -\frac{3}{2}$. But $\left(0, -\frac{3}{2}\right)$ is not in the region.
2. On AB, $f(x,y) = x^2 - 10x + 28 = f(x, -x + 4)$ for $0 \le x \le 4$. $f(4,0) = 4$. $f'(x, -x + 4) = 2x - 10 = 0 \Rightarrow x = 5, y = -1$. But (5,–1) not in the region.
3. On OB, $f(x,y) = x^2 - 3x = f(x,0)$ for $0 \le x \le 4$. $f'(x,0) = 2x - 3 \Rightarrow x = \frac{3}{2}, y = 0 \Rightarrow \left(\frac{3}{2}, 0\right)$ is a critical point. $f\left(\frac{3}{2}, 0\right) = -\frac{9}{4}$
4. For the interior of the triangular region, $f_x(x,y) = 2x + y - 3 = 0$ and $f_y(x,y) = x + 2y + 3 = 0 \Rightarrow x = 3, y = -3$. But (3,–3) is not in the region.

$\therefore$ the absolute maximum is 28 at (0,4); the absolute minimum is $-\frac{9}{4}$ at $\left(\frac{3}{2}, 0\right)$.

58.

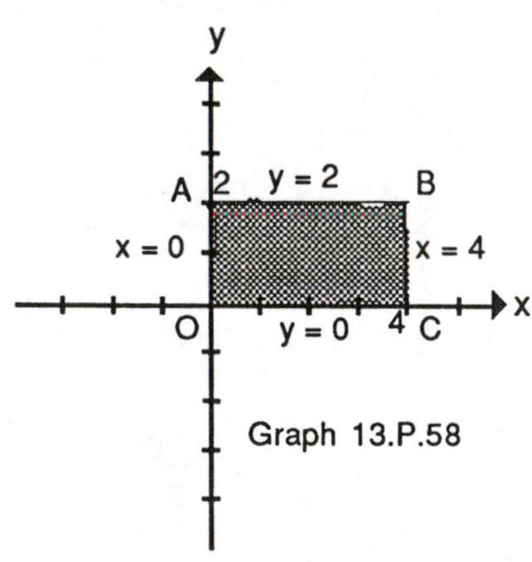

Graph 13.P.58

1. On OA, $f(x,y) = -y^2 + 4y + 1 = f(0,y)$ for $0 \le y \le 2$. $f(0,0) = 1$, $f(0,2) = 5$. $f'(0,y) = -2y + 4 = 0 \Rightarrow y = 2, x = 0 \Rightarrow$ (0,2) which is not in the interior of OA.
2. On AB, $f(x,y) = x^2 - 2x + 5 = f(x,2)$ for $0 \le x \le 4$. $f(4,2) = 13$. $f'(x,2) = 2x - 2 = 0 \Rightarrow x = 1, y = 2 \Rightarrow$ (1,2) is a critical point. $f(1,2) = 4$
3. On BC, $f(x,y) = -y^2 + 4y + 9 = f(4,y)$ for $0 \le y \le 2$. $f(4,0) = 9$. $f'(4,y) = -2y + 4 = 0 \Rightarrow y = 2, x = 4 \Rightarrow$ (4,2) which is not in the interior of BC.
4. On OC, $f(x,y) = x^2 - 2x + 1 = f(x,0)$ for $0 \le x \le 4$. $f'(x,0) = 2x - 2 = 0 \Rightarrow x = 1, y = 0 \Rightarrow$ (1,0) is a critical point. $f(1,0) = 0$.
5. For the interior of the rectangular region, $f_x(x,y) = 2x - 2 = 0$ and $f_y(x,y) = -2y + 4 = 0 \Rightarrow x = 1, y = 2 \Rightarrow$ (1,2) which is not in the interior of the region. $\therefore$ the absolute maximum is 13 at (4,2); the absolute minimum is 0 at (1,0).

59.

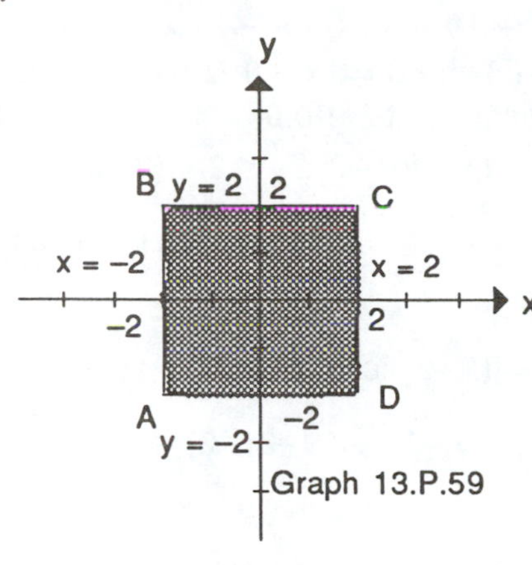

Graph 13.P.59

1. On AB, $f(x,y) = y^2 - y - 4 = f(-2,y)$ for $-2 \le y \le 2$. $f(-2,-2) = 2$, $f(-2,2) = -2$. $f'(-2,y) = 2y - 1 \Rightarrow y = \frac{1}{2}$, $x = -2 \Rightarrow \left(-2,\frac{1}{2}\right)$ is a critical point. $f\left(-2,\frac{1}{2}\right) = -\frac{17}{4}$.
2. On BC, $f(x,y) = -2 = f(x,2)$ for $-2 \le x \le 2$. $f(2,2) = -2$. $f'(x,2) = 0 \Rightarrow$ no critical points in the interior of BC.
3. On CD, $f(x,y) = y^2 - 5y + 4 = f(2,y)$ for $-2 \le y \le 2$. $f(2,-2) = 18$. $f'(2,y) = 2y - 5 = 0 \Rightarrow y = \frac{5}{2}$, $x = 2 \Rightarrow \left(2,\frac{5}{2}\right)$ which is not in the region.
4. On AD, $f(x,y) = 4x + 10 = f(x,-2)$ for $-2 \le x \le 2$. $f'(x,-2) = 4 \Rightarrow$ no critical points in the interior of AD.
5. For the interior of the square, $f_x(x,y) = -y = 2 = 0$ and $f_y(x,y) = 2y - x - 3 = 0 \Rightarrow x = 1$, $y = 2 \Rightarrow (1,2)$ is a critical point. $f(1,2) = -2$

$\therefore$ the absolute maximum is 18 at $(2,-2)$; the absolute minimum is $-\frac{17}{4}$ at $\left(-2,\frac{1}{2}\right)$.

60.

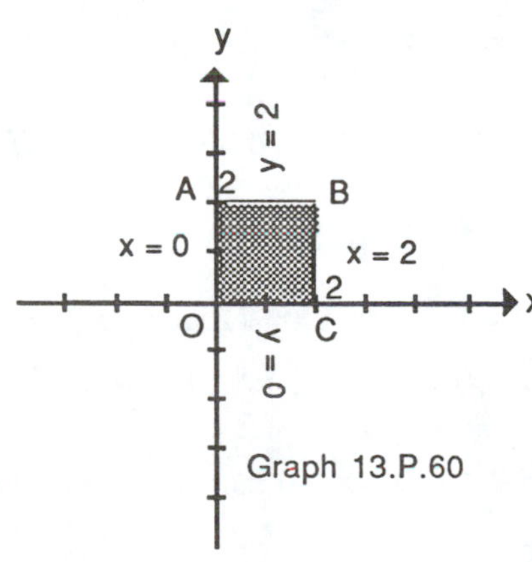

Graph 13.P.60

1. On OA, $f(x,y) = 2y - y^2 = f(0,y)$ for $0 \le y \le 2$. $f(0,0) = 0$, $f(0,2) = 0$. $f'(0,y) = 2 - 2y = 0 \Rightarrow y = 1$, $x = 0 \Rightarrow (0,1)$ is a critical point. $f(0,1) = 1$.
2. On AB, $f(x,y) = 2x - x^2 = f(x,2)$ for $0 \le x \le 2$. $f(2,2) = 0$. $f'(x, 2)$ $2 - 2x = 0 \Rightarrow x = 1$, $y = 2 \Rightarrow (1,2)$ is a critical point. $f(1,2) = 1$
3. On BC, $f(x,y) = 2y - y^2 = f(2,y)$ for $0 \le y \le 2$. $f(2,0) = 0$. $f'(2,y) = 2 - 2y = 0 \Rightarrow y = 1$, $x = 2 \Rightarrow (2,1)$ is a critical point. $f(2,1) = 1$
4. On OC, $f(x,y) = 2x - x^2 = f(x,0)$ for $0 \le x \le 2$. $f'(x,0) = 2 - 2x = 0 \Rightarrow x = 1$, $y = 0 \Rightarrow (1,0)$ is a critical point. $f(1,0) = 1$.
5. For the interior of the rectangular region, $f_x(x,y) = 2 - 2x = 0$ and $f_y(x,y) = 2 - 2y = 0 \Rightarrow x = 1$, $y = 1 \Rightarrow (1,1)$ is a critical point. $f(1,1) = 2$.

$\therefore$ the absolute maximum is 2 at $(1,1)$; the absolute minimum is 0 at $(0,0)$, $(0,2)$, $(2,2)$, and $(2,0)$.

61.

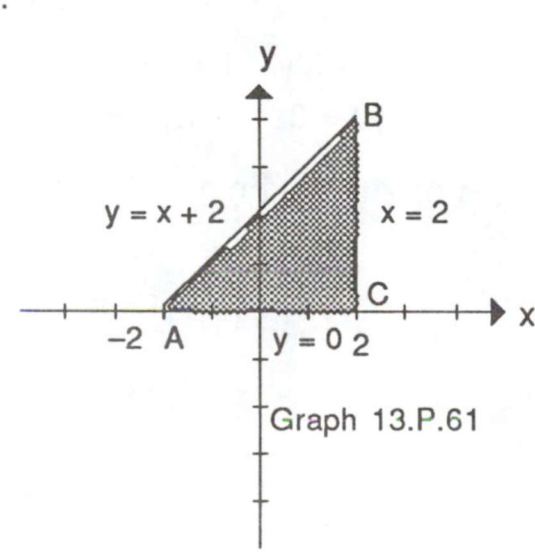

Graph 13.P.61

1. On AB, $f(x,y) = -2x + 4 = f(x,x + 2)$ for $-2 \le x \le 2$. $f(-2,0) = 8$, $f(2,4) = 0$. $f'(x,x + 2) = -2 \Rightarrow$ no critical points in the interior of AB.
2. On BC, $f(x,y) = -y^2 + 4y = f(2,y)$ for $0 \le y \le 4$. $f(2,0) = 0$. $f'(2,y) = -2y + 4 = 0 \Rightarrow y = 2$, $x = 2 \Rightarrow (2,2)$ is a critical point. $f(2,2) = 4$.
3. On AC, $f(x,y) = x^2 - 2x = f(x,0)$ for $-2 \le x \le 2$. $f'(x,0) = 2x - 2 = 0 \Rightarrow x = 1$, $y = 0 \Rightarrow (1,0)$ is a critical point. $f(1,0) = -1$.
4. For the interior of the triangular region, $f_x(x,y) = 2x - 2 = 0$ and $f_y(x,y) = -2y + 4 = 0 \Rightarrow x = 1$, $y = 2 \Rightarrow (1,2)$ is a critical point. $f(1,2) = 3$.

$\therefore$ the absolute maximum is 8 at $(-2,0)$; the absolute minimum is -1 at $(1,0)$.

62.

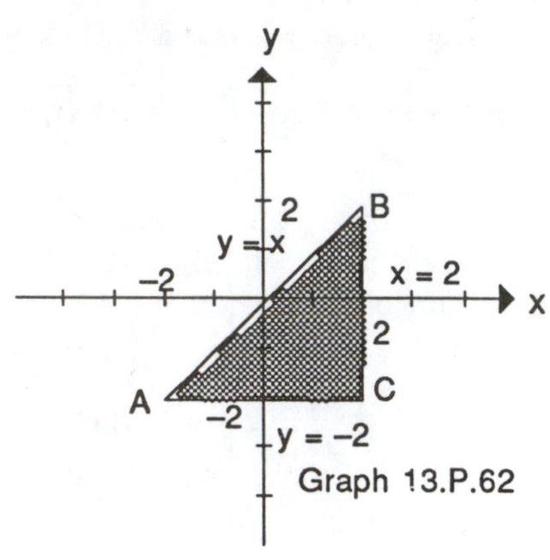

Graph 13.P.62

1. On AB, $f(x,y) = 4x^2 - 2x^4 + 16 = f(x,x)$ for $-2 \le x \le 2$. $f(-2,-2) = 0$, $f(2,2) = 0$. $f'(x,x) = 8x - 8x^3 = 0 \Rightarrow x = 0, y = 0$ or $x = 1, y = 1$ $\Rightarrow (0,0)$ and $(1,1)$ are critical points. $f(0,0) = 16$, $f(1,1) = 18$.
2. On BC, $f(x,y) = 8y - y^4 = f(2,y)$ for $-2 \le y \le 2$. $f(2,-2) = -32$. $f'(2,y) = 8 - 4y^3 = 0 \Rightarrow y = \sqrt[3]{2}$, $x = 2 \Rightarrow (2,\sqrt[3]{2})$ is a critical point. $f(2,\sqrt[3]{2}) = 6\sqrt[3]{2}$.
3. On AC, $f(x,y) = -8x - x^4 = f(x,-2)$ for $-2 \le x \le 2$. $f'(x,-2) = -8 - 4x^3 = 0 \Rightarrow x = \sqrt[3]{-2}$, $y = -2 \Rightarrow (\sqrt[3]{-2},-2)$ is a critical point. $f(\sqrt[3]{-2},-2) = 6\sqrt[3]{2}$.
4. For the interior of the triangular region, $f_x(x,y) = 4y - 4x^3 = 0$ and $f_y(x,y) = 4x - 4y^3 = 0 \Rightarrow$ $x = 0, y = 0$ or $x = 1, y = 1 \Rightarrow (0,0)$ and $(1,1)$, neither of which are interior to the region.

$\therefore$ the absolute maximum is 18 at $(1,1)$; the absolute minimum is -32 at $(2,-2)$.

63. Let $f(x,y) = x^2 + y^2$ be the square of the distance to the origin. $\nabla f = 2x\,\mathbf{i} + 2y\,\mathbf{j}$, $\nabla g = y^2\,\mathbf{i} + 2xy\,\mathbf{j}$. $\nabla f = \lambda\,\nabla g \Rightarrow 2x\,\mathbf{i} + 2y\,\mathbf{j} = \lambda(y^2\,\mathbf{i} + 2xy\,\mathbf{j}) \Rightarrow 2x = \lambda y^2$ and $2y = 2xy\lambda \Rightarrow 2y = \lambda y^2(y\lambda) \Rightarrow y = 0$ (not on $xy^2 = 54$) or $\lambda^2 y^2 - 2 = 0 \Rightarrow y^2 = \frac{2}{\lambda^2}$. $2y = 2xy\lambda \Rightarrow 1 = x\lambda$ since $y \ne 0 \Rightarrow x = \frac{1}{\lambda}$.
$\therefore \frac{1}{\lambda}\left(\frac{2}{\lambda^2}\right) = 54 \Rightarrow \lambda^3 = \frac{1}{27} \Rightarrow \lambda = \frac{1}{3} \Rightarrow x = 3$, $y^2 = 18 \Rightarrow y = \pm 3\sqrt{2} \Rightarrow$ the points nearest to the origin are $(3,\pm 3\sqrt{2})$.

64. $\nabla f = 3x^2\,\mathbf{i} + 2y\,\mathbf{j}$, $\nabla g = 2x\,\mathbf{i} + 2y\,\mathbf{j}$. $\nabla f = \lambda\,\nabla g \Rightarrow 3x^2\,\mathbf{i} + 2y\,\mathbf{j} = \lambda(2x\,\mathbf{i} + 2y\,\mathbf{j}) \Rightarrow 3x^2 = 2x\lambda$ and $2y = 2y\lambda$ $\Rightarrow \lambda = 1$ or $y = 0$. CASE 1: $\lambda = 1 \Rightarrow 3x^2 = 2x \Rightarrow x = 0$ or $x = \frac{2}{3}$. $x = 0 \Rightarrow y = \pm 1 \Rightarrow (0,1)$ and $(0,-1)$. $x = \frac{2}{3} \Rightarrow y = \pm\frac{\sqrt{5}}{3} \Rightarrow \left(\frac{2}{3},\frac{\sqrt{5}}{3}\right)$ and $\left(\frac{2}{3},-\frac{\sqrt{5}}{3}\right)$. CASE 2: $y = 0 \Rightarrow x^2 - 1 = 0 \Rightarrow x = \pm 1 \Rightarrow (1,0)$ and $(-1,0)$ $f(0,\pm 1) = 1$, $f\left(\frac{2}{3},\pm\frac{\sqrt{5}}{3}\right) = \frac{23}{27}$, $f(1,0) = 1$, $f(-1,0) = -1$. $\therefore$ the absolute maximum is 1 at $(0,\pm 1)$ and $(1,0)$; the absolute minimum is -1 at $(-1,0)$.

65. $\nabla T = 400yz^2\,\mathbf{i} + 400xz^2\,\mathbf{j} + 800xyz\,\mathbf{k}$, $\nabla g = 2x\,\mathbf{i} + 2y\,\mathbf{j} + 2z\,\mathbf{k}$. $\nabla T = \lambda\,\nabla g \Rightarrow 400yz^2\,\mathbf{i} + 400xz^2\,\mathbf{j} + 800xyz\,\mathbf{k} = \lambda(2x\,\mathbf{i} + 2y\,\mathbf{j} + 2z\,\mathbf{k}) \Rightarrow 400yz^2 = 2x\lambda$, $400xz^2 = 2y\lambda$, and $800xyz = 2z\lambda$. Solving this system yields the following points: $(0,\pm 1,0)$, $(\pm 1,0,0)$, $\left(\pm\frac{1}{2},\pm\frac{1}{2},\pm\frac{\sqrt{2}}{2}\right)$. $T(0,\pm 1,0) = 0$, $T(\pm 1,0,0) = 0$, $T\left(\pm\frac{1}{2},\pm\frac{1}{2},\pm\frac{\sqrt{2}}{2}\right) = \pm 50$. $\therefore$ 50 is the maximum at $\left(\frac{1}{2},\frac{1}{2},\pm\frac{\sqrt{2}}{2}\right)$ and $\left(-\frac{1}{2},-\frac{1}{2},\pm\frac{\sqrt{2}}{2}\right)$; -50 is the minimum at $\left(\frac{1}{2},-\frac{1}{2},\pm\frac{\sqrt{2}}{2}\right)$ and $\left(-\frac{1}{2},\frac{1}{2},\pm\frac{\sqrt{2}}{2}\right)$.

66. 1. $f(x,y) = x^2 + 3y^2 + 2y$ on $x^2 + y^2 = 1 \Rightarrow \nabla f = 2x\,\mathbf{i} + (6y + 2)\,\mathbf{j}$, $\nabla g = 2x\,\mathbf{i} + 2y\,\mathbf{j}$. $\nabla f = \lambda\,\nabla g \Rightarrow$ $2x\,\mathbf{i} + (6y + 2)\,\mathbf{j} = \lambda(2x\,\mathbf{i} + 2y\,\mathbf{j}) \Rightarrow 2x = 2x\lambda$ and $6y + 2 = 2y\lambda \Rightarrow \lambda = 1$ or $x = 0$. CASE 1: $\lambda = 1 \Rightarrow$ $6y + 2 = 2y \Rightarrow y = -\frac{1}{2}$, $x = \pm\frac{\sqrt{3}}{2} \Rightarrow \left(\pm\frac{\sqrt{3}}{2}, -\frac{1}{2}\right)$. CASE 2: $x = 0 \Rightarrow y^2 = 1 \Rightarrow y = \pm 1 \Rightarrow (0,\pm 1)$. $f\left(\pm\frac{\sqrt{3}}{2}, -\frac{1}{2}\right) = \frac{1}{2}$, $f(0,1) = 5$, $f(0,-1) = 1$. $\therefore$ $\frac{1}{2}$ and 5 are the extreme values on the boundary of the disk.

2. For the interior of the disk, $f_x(x,y) = 2x = 0$ and $f_y(x,y) = 3y + 2 = 0 \Rightarrow x = 0$, $y = -\frac{2}{3}$. $f\left(0,-\frac{2}{3}\right) = 0$.

$\therefore$ the maximum of f on the disk is 5 at (0,1); the absolute minimum of f on the disk is 0 at $\left(0,-\frac{2}{3}\right)$.

CHAPTER 14

MULTIPLE INTEGRALS

14.1 DOUBLE INTEGRALS

1. $\int_0^3 \int_0^2 \left(4 - y^2\right) dy\, dx = \int_0^3 \left[4y - \frac{y^3}{3}\right]_0^2 dx = \frac{16}{3}\int_0^3 dx = 16$

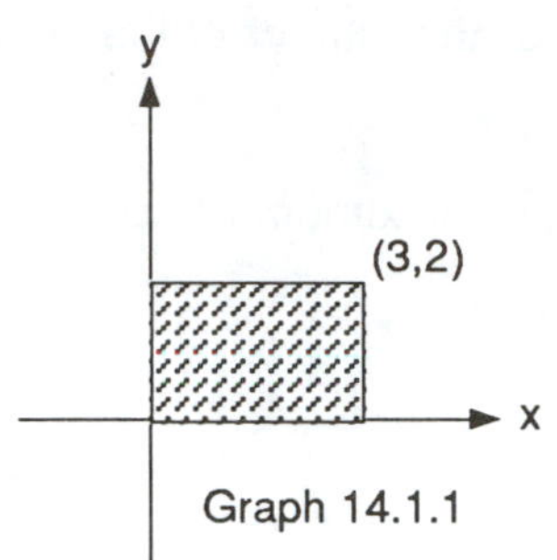

Graph 14.1.1

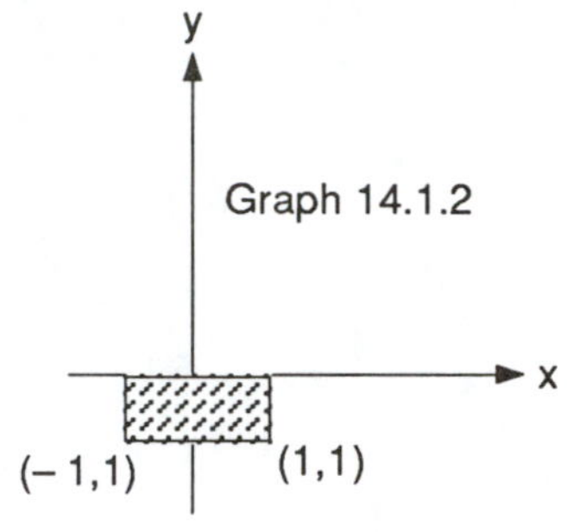

Graph 14.1.2

2. $\int_{-1}^0 \int_{-1}^1 (x + y + 1)\, dx\, dy = \int_{-1}^0 \left[\frac{x^2}{2} + yx + x\right]_{-1}^1 dy =$
$\int_{-1}^0 2y + 2\, dy = 1$

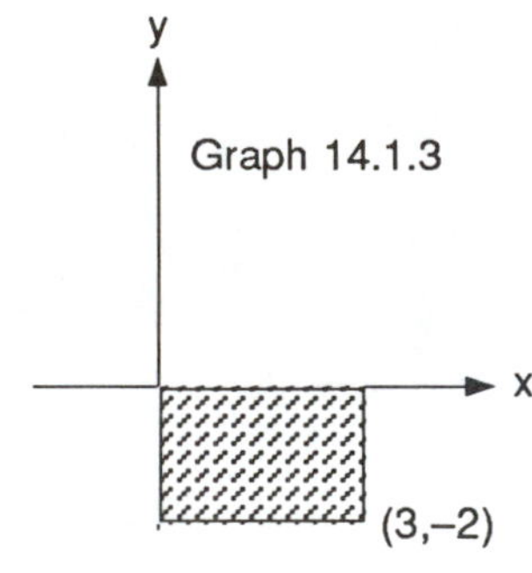

Graph 14.1.3

3. $\int_0^3 \int_{-2}^0 \left(x^2y - 2xy\right) dy\, dx = \int_0^3 \left[\frac{x^2y^2}{2} - xy^2\right]_{-2}^0 dx =$
$\int_0^3 \left(4x - 2x^2\right) dx = \left[2x^2 - \frac{2x^3}{3}\right]_0^3 = 0$

4. $\int_\pi^{2\pi} \int_0^\pi \sin x + \cos y\, dx\, dy = \int_\pi^{2\pi} \left[-\cos x + (\cos y)x\right]_0^\pi dy =$
$\int_\pi^{2\pi} \pi \cos y + 2\, dy = \left[\pi \sin y + 2y\right]_\pi^{2\pi} = 2\pi$

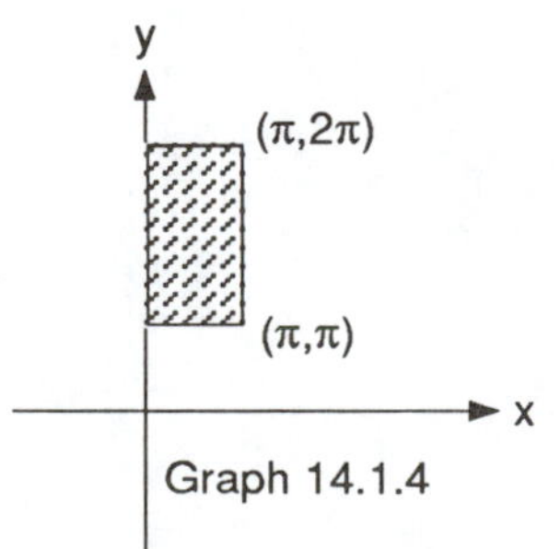

Graph 14.1.4

5. $$\int_0^{\pi}\int_0^{x}(x\sin y)\,dy\,dx = \int_0^{\pi}\big[-x\cos y\big]_0^{x}\,dx = \int_0^{\pi}(x - x\cos x)\,dx = \frac{\pi^2}{2} + 2$$

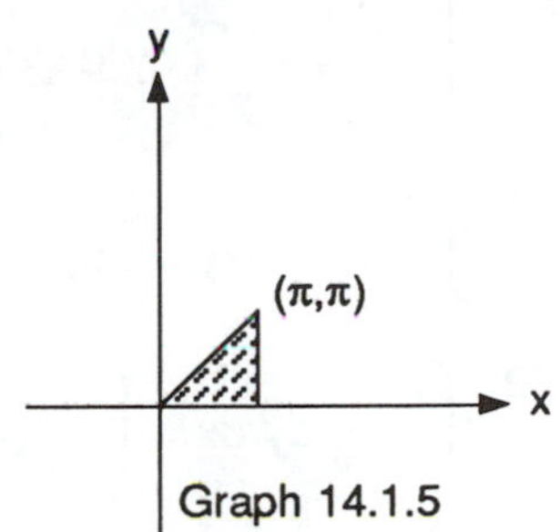

Graph 14.1.5

6. $$\int_1^{\ln 8}\int_0^{\ln y} e^{x+y}\,dx\,dy = \int_1^{\ln 8}\Big[e^{x+y}\Big]_0^{\ln y}\,dy = \int_1^{\ln 8} y\,e^y - e^y\,dy = \Big[(y-1)e^y - e^y\Big]_1^{\ln 8} = 8\ln(8) - 16 + e$$

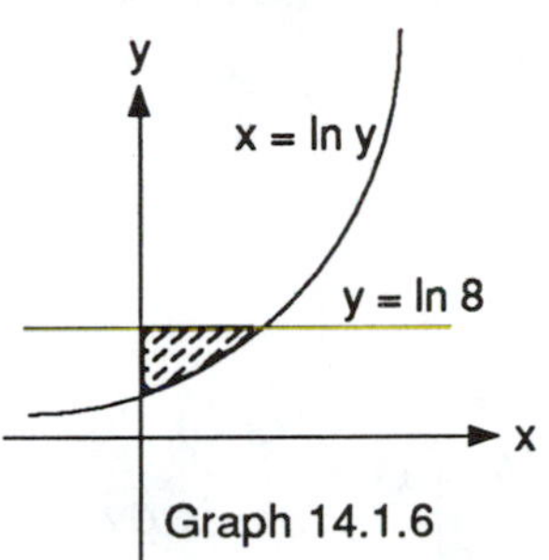

Graph 14.1.6

7. $$\int_0^{\pi}\int_0^{\sin x} y\,dy\,dx = \int_0^{\pi}\left[\frac{y^2}{2}\right]_0^{\sin x}dx = \frac{1}{4}\int_0^{\pi}(1 - \cos 2x)\,dx = \frac{\pi}{4}$$

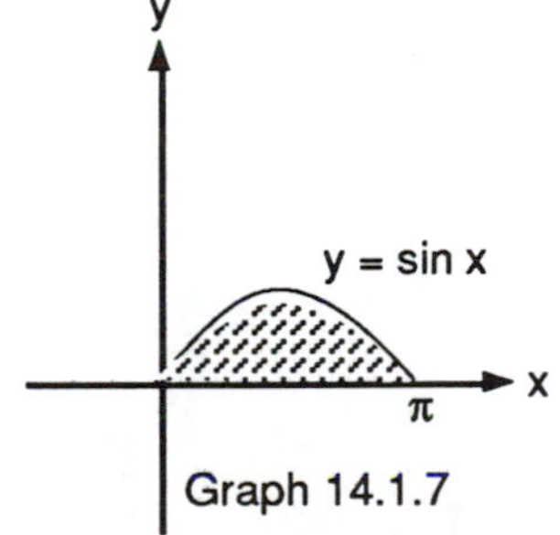

Graph 14.1.7

8. 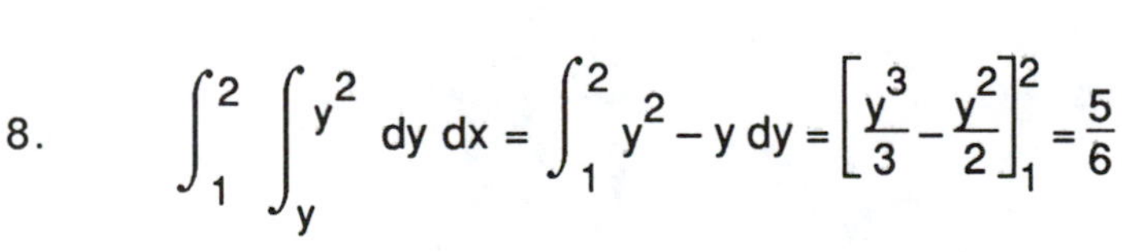

$$\int_1^{2}\int_y^{y^2} dy\,dx = \int_1^{2} y^2 - y\,dy = \left[\frac{y^3}{3} - \frac{y^2}{2}\right]_1^{2} = \frac{5}{6}$$

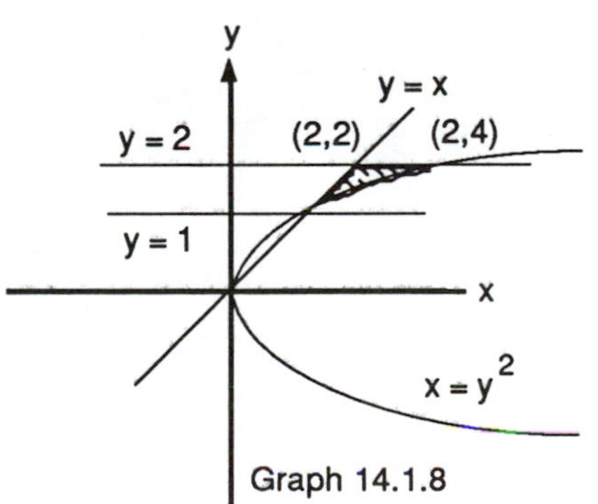

Graph 14.1.8

9. $$\int_1^{2}\int_x^{2x}\frac{x}{y}\,dy\,dx = \int_1^{2}\big[x\ln y\big]_x^{2x}\,dx = \ln(2)\int_1^{2} x\,dx = \frac{\ln 8}{2}$$

10. $$\int_0^{1}\int_0^{1-y} x^2 + y^2\,dx\,dy = \int_0^{1}\left[\frac{x^3}{3} + y^2x\right]_0^{1-y}dy = \int_0^{1}\frac{(1-y)^3}{3} + y^2 - y^3\,dy = \frac{1}{6}$$

11. $\displaystyle\int_0^1 \int_0^{1-x} y - \sqrt{x}\, dy\, dx = \int_0^1 \left[\frac{y^2}{2} - y\sqrt{x}\right]_0^{1-x} dx = \int_0^1 \frac{1 - 2x + x^2}{2} - \sqrt{x}(1 - x)\, dx = -\frac{1}{10}$

12. $\displaystyle\int_0^1 \int_0^1 x^2 + 3xy\, dx\, dy = \int_0^1 \left[\frac{x^3}{3} + \frac{3x^2y}{2}\right]_0^1 dy = \int_0^1 \frac{1}{3} + \frac{3y}{2}\, dy = \frac{13}{12}$

13. $\displaystyle\int_1^2 \int_1^2 \frac{1}{xy}\, dy\, dx = \int_1^2 \frac{1}{x}(\ln 2 - \ln 1)\, dx = \ln 2 \int_1^2 \frac{1}{x}\, dx = (\ln 2)^2$

14. $\displaystyle\int_0^1 \int_0^{\pi} y \cos xy\, dx\, dy = \int_0^1 [\sin xy]_0^{\pi}\, dy = \int_0^1 \sin \pi y\, dy = \frac{2}{\pi}$

15. $\displaystyle\int_0^4 \int_0^{(4-y)/2} dx\, dy$

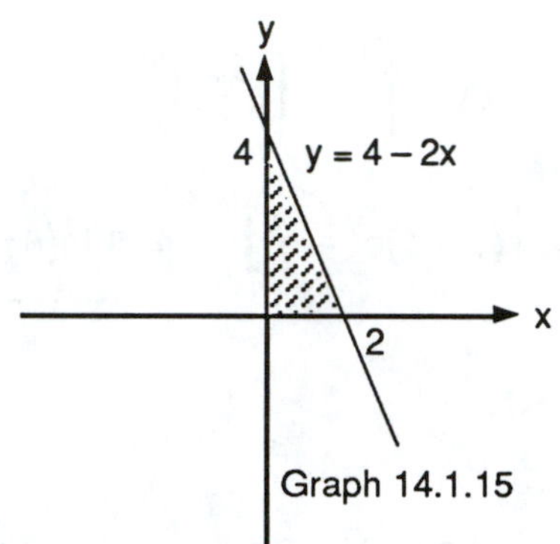

Graph 14.1.15

16. $\displaystyle\int_2^4 \int_0^{(4-y)/2} dx\, dy$

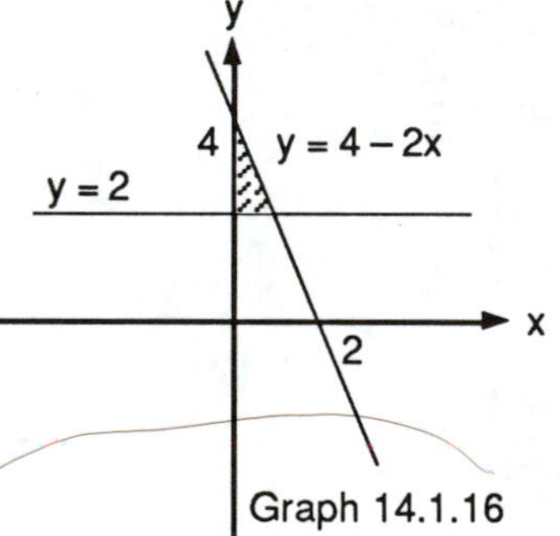

Graph 14.1.16

17. 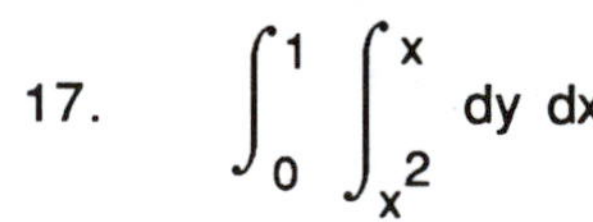

$\displaystyle\int_0^1 \int_{x^2}^{x} dy\, dx$

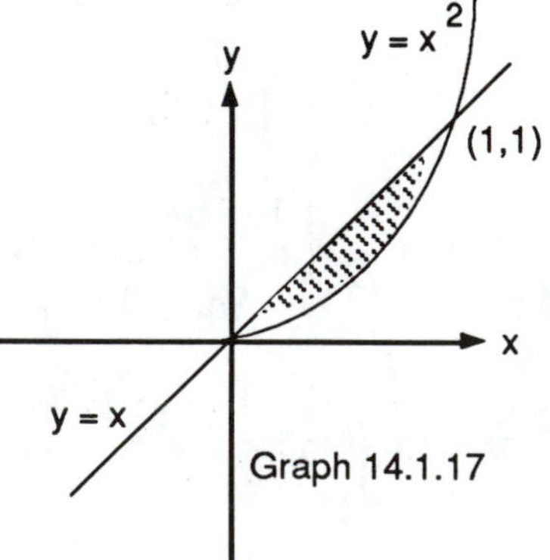

Graph 14.1.17

18. $\int_0^2 \int_0^{x^2} dy\, dx$

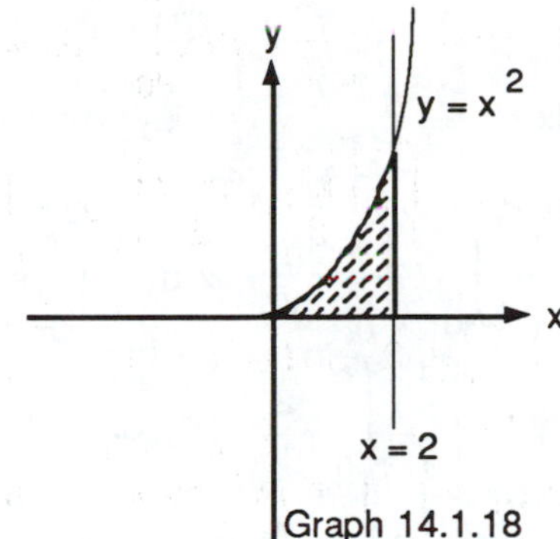

Graph 14.1.18

19. $\int_1^{\exp(2)} \int_{\ln y}^2 dx\, dy$

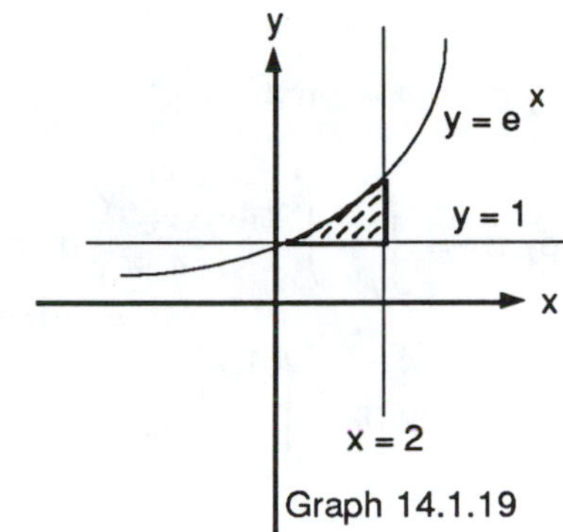

Graph 14.1.19

20. $\int_0^1 \int_0^{x^2} dy\, dx$

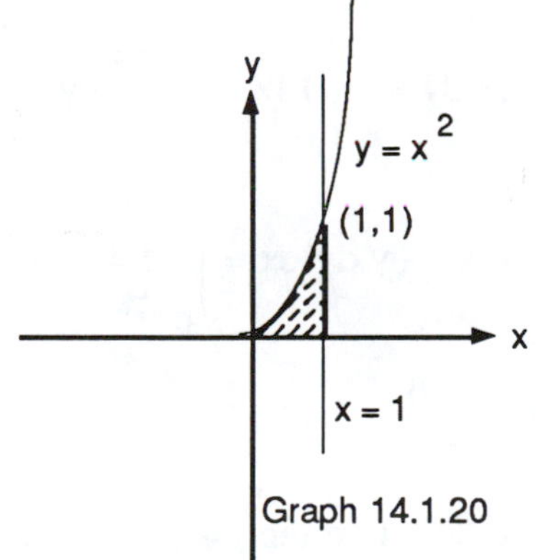

Graph 14.1.20

21. $\int_0^1 \int_1^{\exp(x)} dy\, dx = \int_1^e \int_{\ln y}^1 dx\, dy$

22. $\int_0^1 \int_{\sqrt{x}}^1 \cos(x + y)\, dy\, dx = \int_0^1 \int_0^{y^2} \cos(x + y)\, dx\, dy$

23. $\int_0^2 \int_0^{x^3} f(x,y)\, dy\, dx = \int_0^8 \int_{\sqrt[3]{y}}^2 f(x,y)\, dx\, dy$

24. $\int_0^1 \int_{-\sqrt{y}}^{\sqrt{y}} f(x,y)\,dx\,dy = \int_{-1}^{1} \int_{x^2}^{1} f(x,y)\,dy\,dx$

25. $\int_0^\pi \int_x^\pi \frac{\sin y}{y}\,dy\,dx = \int_0^\pi \int_0^y \frac{\sin y}{y}\,dx\,dy = \int_0^\pi \sin y\,dy = 2$

26. $\int_0^1 \int_y^1 x^2 e^{xy}\,dx\,dy = \int_0^1 \int_0^x x^2 e^{xy}\,dy\,dx = \int_0^1 x\left[e^{xy}\right]_0^x dx = \int_0^1 x\left(\exp(x^2) - 1\right)dx = \frac{e-2}{2}$

27. $\int_0^2 \int_x^2 2y^2 \sin xy\,dy\,dx = \int_0^2 \int_0^y 2y^2 \sin xy\,dx\,dy = \int_0^2 [-2y\cos xy]_0^y\,dy =$

$\int_0^2 -\cos y^2(2y) + 2y\,dy = 4 - \sin 4$

28. $\int_0^2 \int_0^{4-x^2} \frac{xe^{2y}}{4-y}\,dy\,dx = \int_0^4 \int_0^{\sqrt{4-y}} \frac{xe^{2y}}{4-y}\,dx\,dy = \int_0^4 \left[\frac{x^2e^{2y}}{2(4-y)}\right]_0^{\sqrt{4-y}} dx = \frac{1}{2}\int_0^4 e^{2y}\,dx = \frac{e^8-1}{4}$

29. $V = \int_{-4}^{1} \int_{3x}^{4-x^2} (x+4)\,dy\,dx = \int_{-4}^{1} \left[(xy+4y)\right]_{3x}^{4-x^2} dx = \int_{-4}^{1} \left(-x^3 - 7x^2 - 8x + 16\right)dx = \frac{625}{12}$

30. $V = \int_0^2 \int_0^{\sqrt{4-x^2}} (3-y)\,dy\,dx = \int_0^2 \left[3y - \frac{y^2}{2}\right]_0^{\sqrt{4-x^2}} dx = \int_0^2 3\sqrt{4-x^2} - \left(\frac{4-x^2}{2}\right)dx = \frac{9\pi - 8}{3}$

31. $V = \int_0^2 \int_0^3 4 - y^2\,dx\,dy = \int_0^2 \left[4x - y^2x\right]_0^3 dy = \int_0^2 \left(12 - 3y^2\right)dy = 16$

32. $V = \int_0^2 \int_0^{4-x^2} 4 - x^2 - y\,dy\,dx = \int_0^2 \left[4y - x^2y - \frac{y^2}{2}\right]_0^{4-x^2} dx =$

$\int_0^2 8 - 4x^2 + \frac{x^4}{2}\,dx = \frac{128}{15}$

33. $\int_1^3 \int_1^x \frac{1}{xy}\,dy\,dx = 0.603$, use the Calculus Toolkit

34. $\int_0^1 \int_0^1 \exp\left(-\left(x^2+y^2\right)\right)dy\,dx = 0.557$, use the Calculus Toolkit

35. $\int_0^1 \int_0^1 \tan^{-1} xy\,dy\,dx = 0.233$, use the Calculus Toolkit

36. $\int_{-1}^{1} \int_0^{\sqrt{1-x^2}} 3\sqrt{1-x^2-y^2}\,dy\,dx = 3.142$, use the Calculus Toolkit

14.2 AREA, MOMENTS, AND CENTERS OF MASS

1. $\displaystyle\int_0^2 \int_2^{2-x} dy\,dx = \int_0^2 2 - x\,dx = 2$

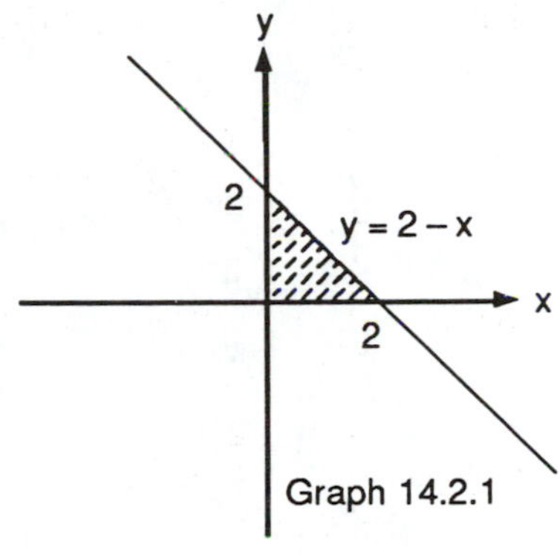

Graph 14.2.1

2. $\displaystyle\int_0^{\ln 2} \int_0^{\exp(x)} dy\,dx = \int_0^{\ln 2} e^x\,dx = 1$

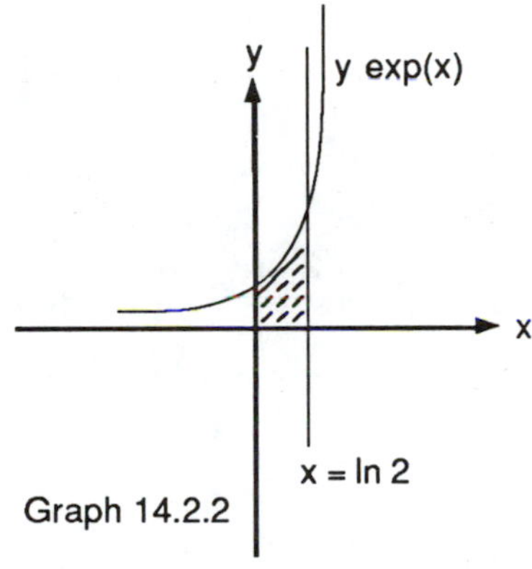

Graph 14.2.2

3. $\displaystyle\int_0^2 \int_{2x}^4 dy\,dx = \int_0^2 4 - 2x\,dx = 4$

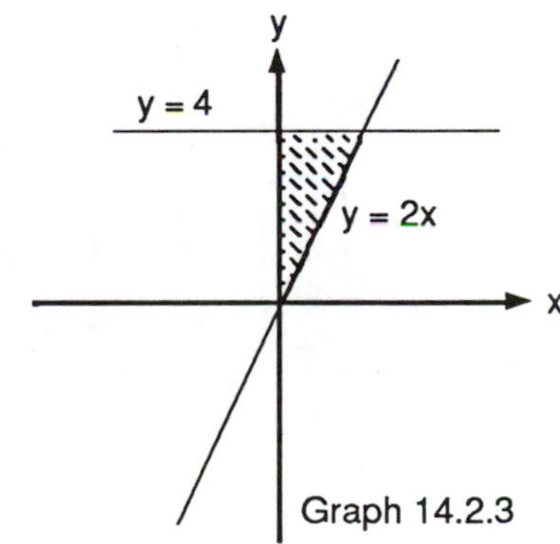

Graph 14.2.3

4. $\displaystyle\int_{-2}^1 \int_{y-2}^{-y^2} dx\,dy = \int_{-2}^1 -y^2 - y + 2\,dy = \frac{9}{2}$

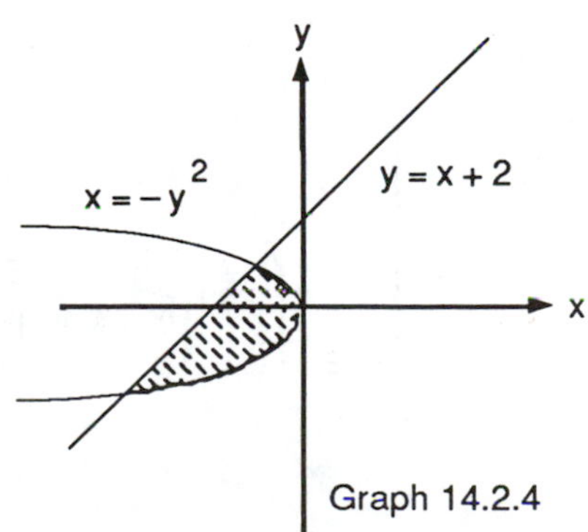

Graph 14.2.4

5. $\int_0^1 \int_{y^2}^{2y-y^2} dx\, dy = \int_0^1 2y - 2y^2\, dy = \frac{1}{3}$

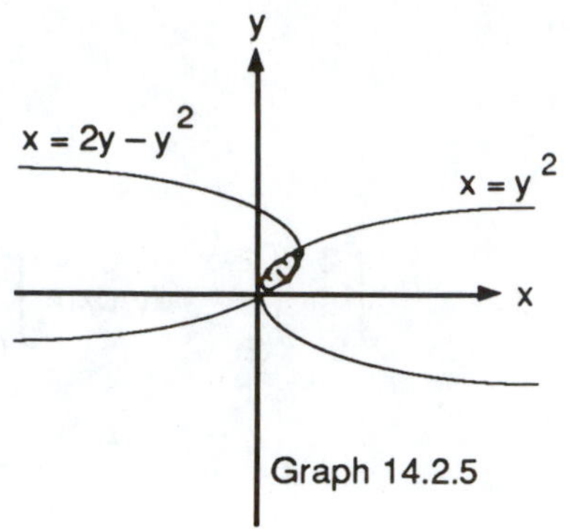

Graph 14.2.5

6. $\int_0^2 \int_{-y}^{y-y^2} dx\, dy = \int_0^2 2y - y^2\, dy = \frac{4}{3}$

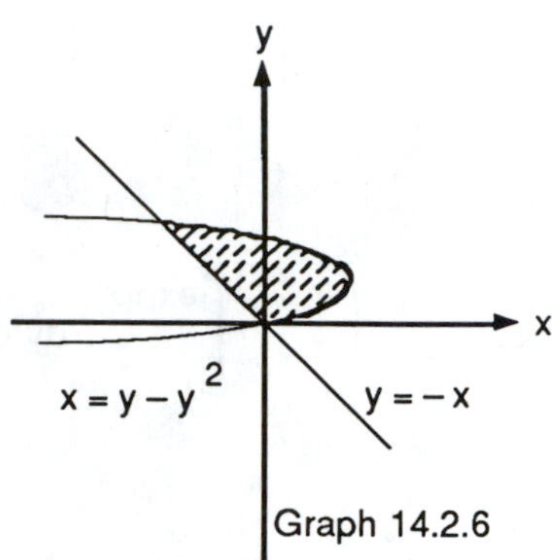

Graph 14.2.6

7. $2\int_0^1 \int_{-1}^{2\sqrt{1-x^2}} dy\, dx = 2\int_0^1 2\sqrt{1-x^2} + 1\, dx = \pi + 2$

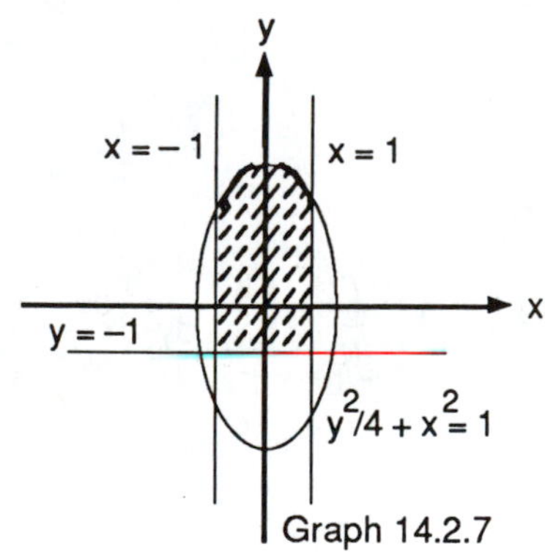

Graph 14.2.7

8. $\int_{-2}^0 \int_{-1}^{x^2} dy\, dx + \int_0^1 \int_{2x-1}^{x^2} dy\, dx =$

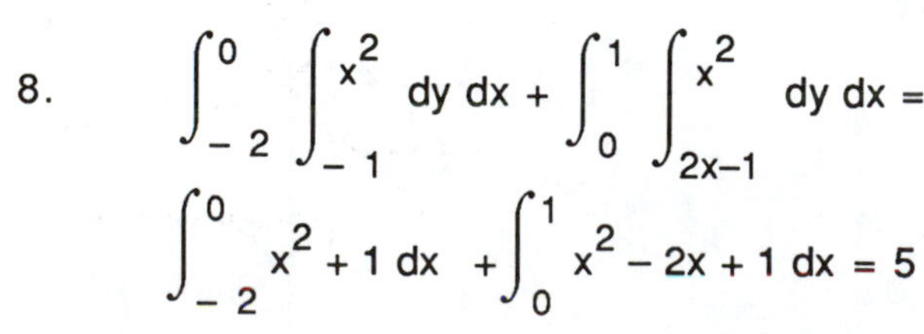

$\int_{-2}^0 x^2 + 1\, dx + \int_0^1 x^2 - 2x + 1\, dx = 5$

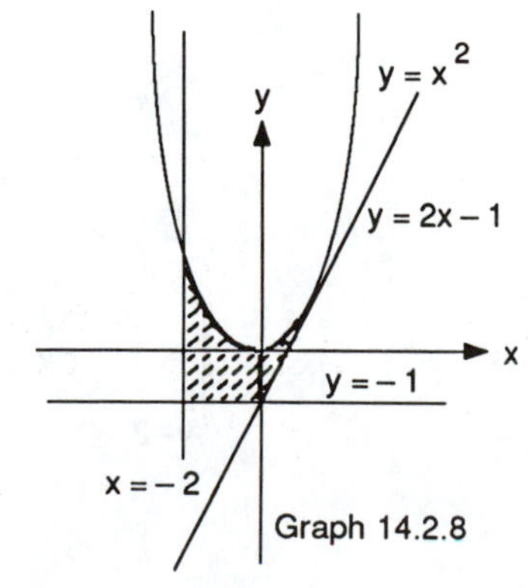

Graph 14.2.8

9. $\displaystyle\int_0^6 \int_{y^2/3}^{2y} dx\, dy = \int_0^6 \left(2y - y^2/3\right) dy = 12$

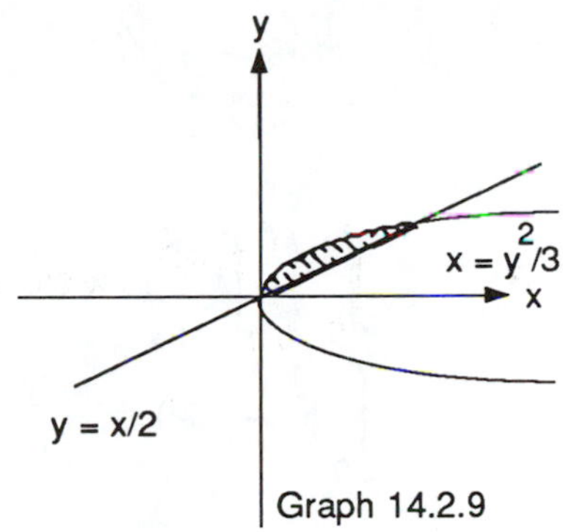

Graph 14.2.9

10. $\displaystyle\int_0^3 \int_{-x}^{2x-x^2} dy\, dx = \int_0^3 3x - x^2\, dx = \frac{9}{2}$

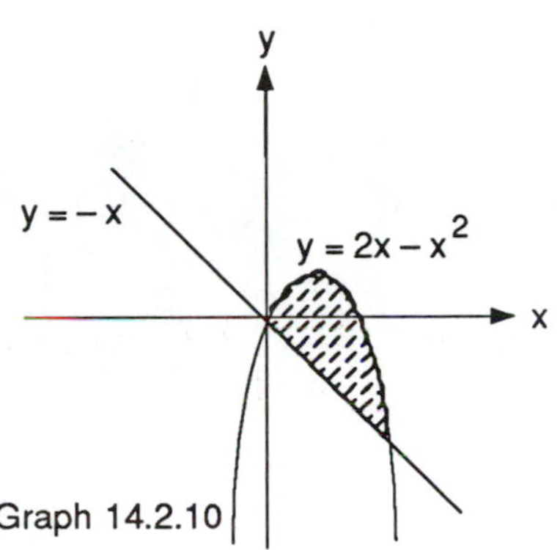

Graph 14.2.10

11. $\displaystyle\int_0^{\pi/4} \int_{\sin x}^{\cos x} dy\, dx = \int_0^{\pi/4} (\cos x - \sin x)\, dx = \sqrt{2} - 1$

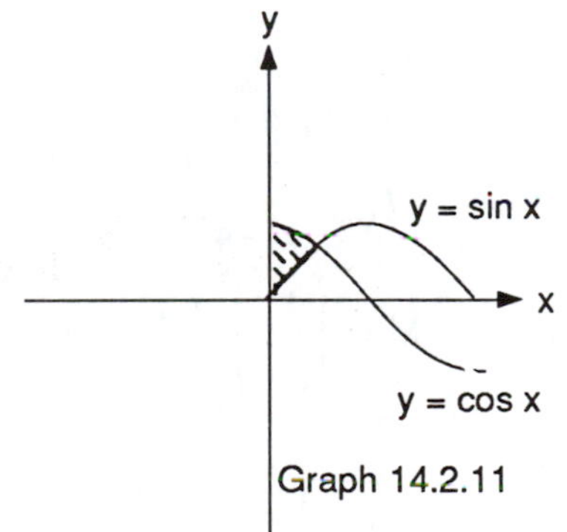

Graph 14.2.11

12. $\displaystyle\int_{-1}^2 \int_{y^2}^{y+2} dx\, dy = \int_{-1}^2 y + 2 - y^2\, dx = 6$

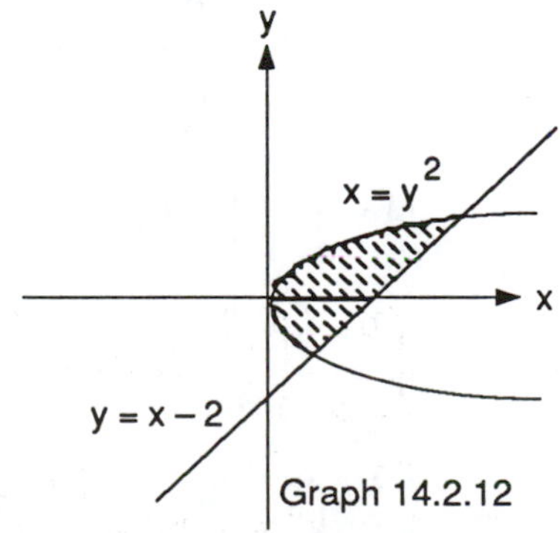

Graph 14.2.12

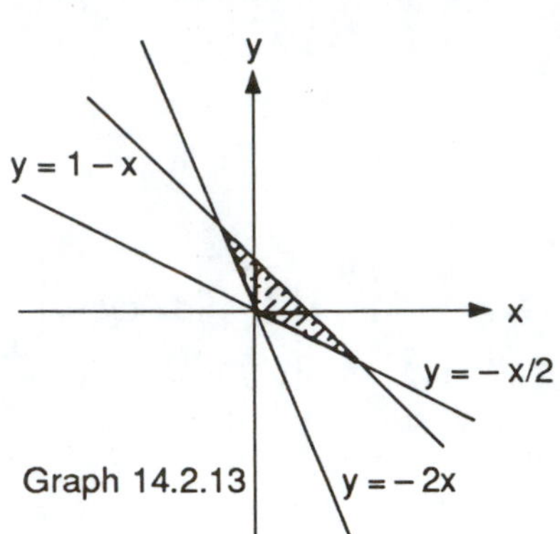

Graph 14.2.13

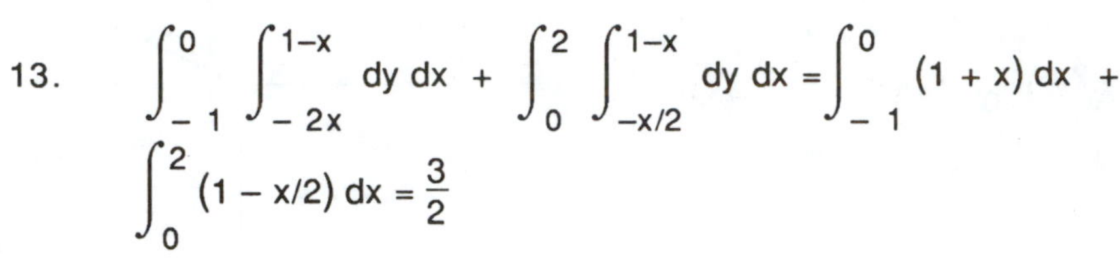

13. $\int_{-1}^{0}\int_{-2x}^{1-x} dy\,dx + \int_{0}^{2}\int_{-x/2}^{1-x} dy\,dx = \int_{-1}^{0}(1 + x)\,dx + \int_{0}^{2}(1 - x/2)\,dx = \frac{3}{2}$

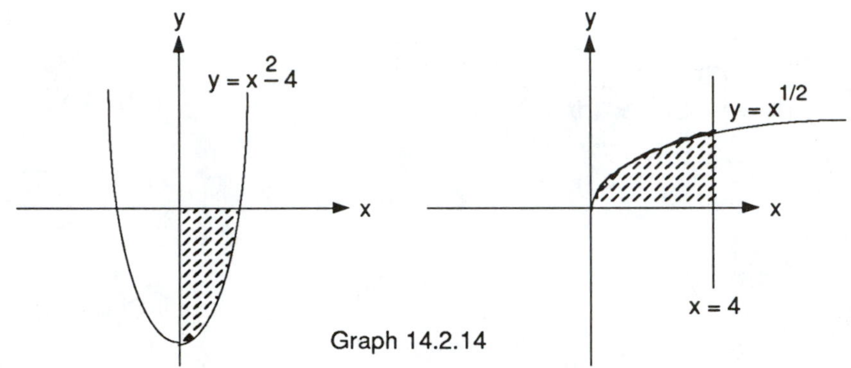

Graph 14.2.14

14. $\int_{0}^{2}\int_{x^2-4}^{0} dy\,dx + \int_{0}^{4}\int_{0}^{\sqrt{x}} dy\,dx = \int_{0}^{2} 4 - x^2\,dx + \int_{0}^{4} x^{1/2}\,dx = \frac{32}{3}$

15. $M = \int_{0}^{1}\int_{x}^{2-x^2} 3\,dy\,dx = 3\int_{0}^{1} 2 - x^2 - x\,dx = \frac{7}{2}$

$M_y = \int_{0}^{1}\int_{x}^{2-x^2} 3x\,dy\,dx = 3\int_{0}^{1} [xy]_{x}^{2-x^2}\,dx = 3\int_{0}^{1}\left(2x - x^3 - x^2\right)dx = \frac{5}{4}$

$M_x = \int_{0}^{1}\int_{x}^{2-x^2} 3y\,dy\,dx = \frac{3}{2}\int_{0}^{1}\left[y^2\right]_{x}^{2-x^2}dx = \frac{3}{2}\int_{0}^{1} 4 - 5x^2 + x^4\,dx = \frac{19}{5} \Rightarrow \overline{x} = \frac{5}{14},\ \overline{y} = \frac{38}{35}$

16. $M = \int_{0}^{3}\int_{0}^{3} dy\,dx = \int_{0}^{3} 3\,dx = 9$

$I_x = \int_{0}^{3}\int_{0}^{3} y^2\,dy\,dx = \int_{0}^{3}\left[\frac{y^3}{3}\right]_{0}^{3}dx = 27\ ,\ R_x = \sqrt{\frac{I_x}{M}} = \sqrt{3}$

$I_y = \int_{0}^{3}\int_{0}^{3} x^2\,dy\,dx = \int_{0}^{3}\left[x^2y\right]_{0}^{3}dx = \int_{0}^{3} x^2\,dx = 27\ ,\ R_y = \sqrt{\frac{I_y}{M}} = \sqrt{3}$

17. $M = \int_0^2 \int_{y^2/2}^{4-y} dx\, dy = \int_0^2 4 - y - \frac{y^2}{2}\, dy = \frac{14}{3}$

$M_y = \int_0^2 \int_{y^2/2}^{4-y} x\, dx\, dy = \frac{1}{2}\int_0^2 \left[x^2\right]_{y^2/2}^{4-y} dx = \frac{1}{2}\int_0^2 16 - 8y + y^2 - \frac{y^4}{4}\, dy = \frac{128}{15}$

$M_x = \int_0^2 \int_{y^2/2}^{4-y} y\, dx\, dy = \int_0^2 y\left(4 - y - \frac{y^2}{2}\right) dy = \frac{10}{3} \Rightarrow \overline{x} = \frac{64}{35},\ \overline{y} = \frac{5}{7}$

18. $M = \int_0^3 \int_3^{3-x} dy\, dx = \int_0^3 3 - x\, dx = \frac{9}{2}$

$M_y = \int_0^3 \int_0^{3-x} x\, dy\, dx = \int_0^3 [xy]_0^{3-x} dx = \int_0^3 3x - x^2\, dx = \frac{9}{2} \Rightarrow \overline{x} = 1,\ \overline{y} = 1$, by symmetry

19. $M = 2\int_0^1 \int_0^{\sqrt{1-x^2}} dy\, dx = 2\int_0^1 \sqrt{1 - x^2}\, dx = \frac{\pi}{2}$

$M_x = 2\int_0^1 \int_0^{\sqrt{1-x^2}} y\, dy\, dx = \int_0^1 \left[y^2\right]_0^{\sqrt{1-x^2}} dx = \int_0^1 1 - x^2\, dx = \frac{2}{3} \Rightarrow \overline{x} = 0$, by symmetry; $\overline{y} = \frac{4}{3\pi}$

20. $M = \frac{125\delta}{6}$; $M_y = \delta \int_0^5 \int_x^{6x-x^2} x\, dy\, dx = \delta\int_0^5 [xy]_x^{6x-x^2} dx = \delta\int_0^5 5x^2 - x^3\, dx = \frac{625\delta}{12}$

$M_x = \delta \int_0^5 \int_x^{6x-x^2} y\, dy\, dx = \delta\int_0^5 \left[y^2\right]_x^{6x-x^2} dx = \delta\int_0^5 35x^2 - 12x^3 + x^4\, dx = \frac{625\delta}{3} \Rightarrow$

$\overline{x} = \frac{5}{2},\ \overline{y} = 10$

21. $M = \int_0^a \int_0^{\sqrt{a^2-x^2}} dy\, dx = \frac{\pi a^2}{4}$; $M_y = \int_0^a \int_0^{\sqrt{a^2-x^2}} x\, dy\, dx =$

$\int_0^a [xy]_0^{\sqrt{a^2-x^2}} dx = -\frac{1}{2}\int_0^a \sqrt{a^2 - y^2}(-2x)\, dx = \frac{a^3}{3} \Rightarrow \overline{x} = \overline{y} = \frac{4a}{3\pi}$, by symmetry

22. $I_x = \int_{-2}^2 \int_{-\sqrt{4-x^2}}^{\sqrt{4-x^2}} y^2\, dy\, dx = \int_{-2}^2 \left[\frac{y^3}{3}\right]_{-\sqrt{4-x^2}}^{\sqrt{4-x^2}} dx = \frac{2}{3}\int_{-2}^2 \left(4 - x^2\right)^{3/2} dx = 4\pi;$

$I_y = 4\pi$, by symmetry; $I_o = I_x + I_y = 8\pi$

23. $M = \int_0^\pi \int_0^{\sin x} dy\, dx = \int_0^\pi \sin x\, dx = 2$; $M_x = \int_0^\pi \int_0^{\sin x} y\, dy\, dx = \frac{1}{2}\int_0^\pi \left[y^2\right]_0^{\sin x} dx =$

$\frac{1}{4}\int_0^\pi 1 - \cos 2x\, dx = \frac{\pi}{4} \Rightarrow \overline{x} = \frac{\pi}{2},\ \overline{y} = \frac{\pi}{8}$

24. $I_y = \int_{\pi}^{2\pi} \int_0^{(\sin^2 x)/x^2} x^2\, dy\, dx = \frac{1}{2}\int_{\pi}^{2\pi} (1 - \cos 2x)\, dx = \frac{\pi}{2}$

25. $M = \int_0^2 \int_{-y}^{y-y^2} (x + y)\, dx\, dy = \int_0^2 \left[\frac{x^2}{2} + xy\right]_{-y}^{y-y^2} dy =$

$\int_0^2 \left(\frac{y^4}{2} - 2y^3 + 2y^2\right) dy = \frac{8}{15}$; $I_x = \int_0^2 \int_{-y}^{y-y^2} y^2(x + y)\, dx\, dy = \int_0^2 \left[\frac{x^2y^2}{2} + xy^3\right]_{-y}^{y-y^2} dy =$

$\int_0^2 \left(\frac{y^6}{2} - 2y^5 + 2y^4\right) dy = \frac{64}{105}$; $R_x = \sqrt{\frac{I_x}{M}} = \sqrt{\frac{8}{7}}$

26. $M = \int_{-\sqrt{3}/2}^{\sqrt{3}/2} \int_{4y^2}^{\sqrt{12-4y^2}} 5x\, dx\, dy = \frac{5}{2}\int_{-\sqrt{3}/2}^{\sqrt{3}/2} \left(12 - 4y^2 - 16y^4\right) dy = 23\sqrt{3}$

27. $M = \int_0^1 \int_x^{2-x} (6x + 3y + 3)\, dy\, dx = \int_0^1 - 12x^2 + 12\, dx = 8$; $M_y = \int_0^1 \int_x^{2-x} x(6x + 3y + 3)\, dy\, dx =$

$\int_0^1 12x - 12x^3\, dx = 3$; $M_x = \int_0^1 \int_x^{2-x} y(6x + 3y + 3)\, dy\, dx =$

$\int_0^1 14 - 6x - 6x^2 - 2x^3\, dx = \frac{17}{2} \Rightarrow \overline{x} = \frac{3}{8}, \overline{y} = \frac{17}{16}$

28. $M = \int_0^1 \int_{y^2}^{2y-y^2} y + 1\, dx\, dy = \int_0^1 2y - 2y^3\, dy = \frac{1}{2}$; $M_x = \int_0^1 \int_{y^2}^{2y-y^2} y(y + 1)\, dx\, dy =$

$\int_0^1 2y^2 - 2y^4\, dy = \frac{4}{15}$; $M_y = \int_0^1 \int_{y^2}^{2y-y^2} x(y + 1)\, dx\, dy = \int_0^1 2y^2 - 2y^4\, dy = \frac{4}{15} \Rightarrow$

$\overline{x} = \frac{8}{15}, \overline{y} = \frac{8}{15}$; $I_x = \int_0^1 \int_{y^2}^{2y-y^2} y^2(y+1)\, dx\, dy = 2\int_0^1 y^3 - y^5\, dy = \frac{1}{6}$

29. $M = \int_0^1 \int_0^6 (x + y + 1)\, dx\, dy = \int_0^1 (6y + 24)\, dy = 27$

$M_x = \int_0^1 \int_0^6 y(x + y + 1)\, dx\, dy = \int_0^1 y(6y + 24)\, dy = 14$

$M_y = \int_0^1 \int_0^6 x(x + y + 1)\, dx\, dy = \int_0^1 (18y + 90)\, dy = 99 \Rightarrow \overline{x} = \frac{11}{3}, \overline{y} = \frac{14}{27}$

$I_y = \int_0^1 \int_0^6 x^2(x + y + 1)\, dx\, dy = 216\int_0^1 \left(\frac{y}{3} + \frac{11}{6}\right) dy = 432 \Rightarrow R_y = \sqrt{\frac{I_y}{M}} = 4$

30. $M = \int_{-1}^{1}\int_{x^2}^{1}(y+1)\,dy\,dx = -\int_{-1}^{1}\left(\frac{x^4}{2}+x^2-\frac{3}{2}\right)dx = \frac{32}{15}$

$M_x = \int_{-1}^{1}\int_{x^2}^{1} y(y+1)\,dy\,dx = \int_{-1}^{1}\left(\frac{5}{6}-\frac{x^6}{3}-\frac{x^4}{2}\right)dx = \frac{48}{35}$

$M_y = \int_{-1}^{1}\int_{x^2}^{1} x(y+1)\,dy\,dx = \int_{-1}^{1}\left(\frac{3x}{2}-\frac{x^5}{2}-x^3\right)dx = 0 \Rightarrow \overline{x} = 0,\ \overline{y} = \frac{9}{14}$

$I_y = \int_{-1}^{1}\int_{x^2}^{1} x^2(y+1)\,dy\,dx = \int_{-1}^{1}\left(\frac{3x^2}{2}-\frac{x^6}{2}-x^4\right)dx = \frac{16}{35} \Rightarrow R_y = \sqrt{\frac{I_y}{M}} = \sqrt{\frac{3}{14}}$

31. $M = \int_{-1}^{1}\int_{0}^{x^2}(7y+1)\,dy\,dx = \int_{-1}^{1}\left(\frac{7x^4}{2}+x^2\right)dx = \frac{31}{15}$

$M_x = \int_{-1}^{1}\int_{0}^{x^2} y(7y+1)\,dy\,dx = \int_{-1}^{1}\left(\frac{7x^6}{3}+\frac{x^4}{2}\right)dx = \frac{13}{15}$

$M_y = \int_{-1}^{1}\int_{0}^{x^2} x(7y+1)\,dy\,dx = \int_{-1}^{1}\left(\frac{7x^5}{2}+x^3\right)dx = 0 \Rightarrow \overline{x} = 0,\ \overline{y} = \frac{13}{31}$

$I_y = \int_{-1}^{1}\int_{0}^{x^2} x^2(7y+1)\,dy\,dx = \int_{-1}^{1} x^2\left(\frac{7x^4}{2}+x^2\right)dx = \frac{7}{5} \Rightarrow R_y = \sqrt{\frac{I_y}{M}} = \sqrt{\frac{21}{31}}$

32. $M = \int_{0}^{20}\int_{-1}^{1}\left(1+\frac{x}{20}\right)dy\,dx = \int_{0}^{20}\left(2+\frac{x}{10}\right)dx = 60$

$M_x = \int_{0}^{20}\int_{-1}^{1} y\left(1+\frac{x}{20}\right)dy\,dx = \int_{0}^{20}\left[\left(1+\frac{x}{20}\right)\left(\frac{y^2}{2}\right)\right]_{-1}^{1}dx = 0$

$M_y = \int_{0}^{20}\int_{-1}^{1} x\left(1+\frac{x}{20}\right)dy\,dx = \int_{0}^{20} x\left(2+\frac{x}{10}\right)dx = \frac{2000}{3} \Rightarrow \overline{x} = \frac{100}{9},\ \overline{y} = 0$

$I_x = \int_{0}^{20}\int_{-1}^{1} y^2\left(1+\frac{x}{20}\right)dy\,dx = \frac{2}{3}\int_{0}^{20}\left(1+\frac{x}{20}\right)dx = 20,\ R_x = \sqrt{\frac{I_x}{M}} = \sqrt{\frac{1}{3}}$

33. $M = \int_0^1 \int_{-y}^{y} (1 + y)\,dx\,dy = \int_0^1 \left(2y^2 + 2y\right) dy = \frac{5}{3}$

$M_x = \int_0^1 \int_{-y}^{y} y(1 + y)\,dx\,dy = 2\int_0^1 \left(y^3 + y^2\right) dy = \frac{7}{6}$

$M_y = \int_0^1 \int_{-y}^{y} x(1 + y)\,dx\,dy = \int_0^1 0\,dy = 0 \Rightarrow \overline{x} = 0, \overline{y} = \frac{7}{10}$

$I_x = \int_0^1 \int_{-y}^{y} y^2(1 + y)\,dx\,dy = \int_0^1 y^2\left(2y^2 + 2y\right) dy = \frac{9}{10} \Rightarrow R_x = \sqrt{\frac{I_x}{M}} = \frac{3\sqrt{6}}{10}$

$I_y = \int_0^1 \int_{-y}^{y} x^2(1 + y)\,dx\,dy = \frac{1}{3}\int_0^1 y^2\left(2y^2 + 2y\right) dy = \frac{3}{10} \Rightarrow R_y = \sqrt{\frac{I_y}{M}} = \frac{3\sqrt{2}}{10}$

$I_o = I_x + I_y = \frac{6}{5}$ and $R_o = \sqrt{\frac{I_o}{M}} = \frac{3\sqrt{2}}{5}$

34. $M = \int_0^1 \int_{-y}^{y} \left(3x^2 + 1\right) dx\,dy = \int_0^1 \left(2y^3 + 2y\right) dy = \frac{3}{2}$

$M_x = \int_0^1 \int_{-y}^{y} y\left(3x^2 + 1\right) dx\,dy = \int_0^1 y\left(2y^3 + 2y\right) dy = \frac{16}{15}$

$M_y = \int_0^1 \int_{-y}^{y} x\left(3x^2 + 1\right) dx\,dy = 0 \Rightarrow \overline{x} = 0, \overline{y} = \frac{32}{45}$

$I_x = \int_0^1 \int_{-y}^{y} y^2\left(3x^2 + 1\right) dx\,dy = \int_0^1 y^2\left(2y^3 + 2y\right) dy = \frac{5}{6} \Rightarrow R_x = \sqrt{\frac{I_x}{M}} = \frac{\sqrt{5}}{3}$

$I_y = \int_0^1 \int_{-y}^{y} x^2\left(3x^2 + 1\right) dx\,dy = 2\int_0^1 \left(\frac{3}{5}y^5 + \frac{1}{3}y^3\right) dy = \frac{11}{30} \Rightarrow R_y = \sqrt{\frac{I_y}{M}} = \sqrt{\frac{11}{45}}$

$I_o = I_x + I_y = \frac{6}{5}$ and $R_o = \sqrt{\frac{I_o}{M}} = \frac{2}{\sqrt{5}}$

35. $f(a) = I_a = \int_0^4 \int_0^2 (y - a)^2\,dy\,dx = \int_0^4 \left(\frac{(2-a)^3}{3} + \frac{a^3}{3}\right) dx = \frac{4}{3}\left[(2-a)^3 + a^3\right] \Rightarrow f'(a) = 0 \Rightarrow$

$a = 1$, $f''(a) = 16 > 0 \Rightarrow f(1)$ is minimum

36. $M = \int_0^1 \int_{-1/\sqrt{1-x^2}}^{1/\sqrt{1-x^2}} dy\,dx = \int_0^1 \frac{2}{\sqrt{1 - x^2}}\,dx = \pi$

$M_y = \int_0^1 \int_{-1/\sqrt{1-x^2}}^{1/\sqrt{1-x^2}} x\,dy\,dx = \int_0^1 \frac{2x}{\sqrt{1 - x^2}}\,dx = 2 \Rightarrow \overline{x} = \frac{2}{\pi}, \overline{y} = 0$ by symmetry

14.3 DOUBLE INTEGRALS IN POLAR FORM

1. $\int_{-1}^{1}\int_{0}^{\sqrt{1-x^2}} dy\,dx = \int_{0}^{\pi}\int_{0}^{1} r\,dr\,d\theta = \frac{1}{2}\int_{0}^{\pi} d\theta = \frac{\pi}{2}$

2. $\int_{-1}^{1}\int_{-\sqrt{1-x^2}}^{\sqrt{1-x^2}} dy\,dx = \int_{0}^{2\pi}\int_{0}^{1} r\,dr\,d\theta = \frac{1}{2}\int_{0}^{2\pi} d\theta = \pi$

3. $\int_{0}^{1}\int_{0}^{\sqrt{1-y^2}}\left(x^2+y^2\right) dx\,dy = \int_{0}^{\pi/2}\int_{0}^{1} r^3\,dr\,d\theta = \frac{1}{4}\int_{0}^{\pi/2} d\theta = \frac{\pi}{8}$

4. $\int_{-1}^{1}\int_{-\sqrt{1-y^2}}^{\sqrt{1-y^2}}\left(x^2+y^2\right) dx\,dy = \int_{0}^{2\pi}\int_{0}^{1} r^3\,dr\,d\theta = \frac{1}{4}\int_{0}^{2\pi} d\theta = \frac{\pi}{2}$

5. $\int_{-a}^{a}\int_{-\sqrt{a^2-x^2}}^{\sqrt{a^2-x^2}} dy\,dx = \int_{0}^{2\pi}\int_{0}^{a} r\,dr\,d\theta = \frac{a^2}{2}\int_{0}^{2\pi} d\theta = \pi a^2$

6. $\int_{0}^{2}\int_{0}^{\sqrt{4-y^2}}\left(x^2+y^2\right) dx\,dy = \int_{0}^{\pi/2}\int_{0}^{2} r^3\,dr\,d\theta = 4\int_{0}^{\pi/2} d\theta = 2\pi$

7. $\int_{0}^{\pi/4}\int_{0}^{\sqrt{2}} \cos\theta\, r^2\,dr\,d\theta = \frac{2\sqrt{2}}{3}\int_{0}^{\pi/4}\cos\theta\,d\theta = \frac{2}{3}$

8. $\int_{0}^{2}\int_{0}^{x} y\,dy\,dx = \int_{0}^{\pi/4}\int_{0}^{2\sec\theta} \sin\theta\, r^2\,dr\,d\theta = \frac{8}{3}\int_{0}^{\pi/4}\tan\theta\sec^2\theta\,d\theta = \frac{4}{3}$

9. $\int_{0}^{3}\int_{0}^{\sqrt{3}x}\frac{1}{\sqrt{x^2+y^2}}\,dy\,dx = \int_{0}^{\pi/3}\int_{0}^{3\sec\theta} dr\,d\theta = \int_{9}^{\pi/3} 3\sec\theta\,d\theta = 3\ln\left(2+\sqrt{3}\right)$

10. $\int_{0}^{2}\int_{0}^{\sqrt{4-x^2}}\frac{xy}{\sqrt{x^2+y^2}}\,dy\,dx = \int_{0}^{\pi/2}\int_{0}^{2}\sin\theta\cos\theta\, r^2\,dr\,d\theta = \frac{8}{3}\int_{0}^{\pi/2}\sin\theta\cos\theta\,d\theta = \frac{4}{3}$

11. $\int_{0}^{1}\int_{0}^{\sqrt{1-x^2}} 5\sqrt{x^2+y^2}\,dy\,dx = \int_{0}^{\pi/2}\int_{0}^{1} 5\,r^2\,dr\,d\theta = \frac{5}{3}\int_{0}^{\pi/2} d\theta = \frac{5\pi}{6}$

12. $\int_{0}^{\infty}\int_{0}^{\infty}\exp\left[-\left(x^2+y^2\right)\right] dx\,dy = \int_{0}^{\pi/2}\int_{0}^{\infty}\exp\left(-r^2\right) r\,dr\,d\theta =$

$\int_{0}^{\pi/2}\lim_{t\to\infty}\int_{0}^{t}\exp\left(-r^2\right) r\,dr\,d\theta = \frac{1}{2}\int_{0}^{\pi/2} d\theta = \frac{\pi}{4}$

13. $\int_{0}^{\pi/2}\int_{0}^{2\sqrt{2-\sin 2\theta}} r\,dr\,d\theta = 2\int_{0}^{\pi/2}\left(2-\sin 2\theta\right) d\theta = 2(\pi-1)$

14. $A = 2\int_{0}^{\pi/2}\int_{1}^{1+\cos\theta} r\,dr\,d\theta = \int_{0}^{\pi/2} 2\cos\theta + \cos^2\theta\,d\theta = \frac{8+\pi}{4}$

15. $$2\int_0^{\pi/6}\int_0^{12\cos 3\theta} r\,dr\,d\theta = 144\int_0^{\pi/6}\cos^2 3\theta\,d\theta = 12\pi$$

16. $$A = \int_0^{2\pi}\int_0^{4\theta/3} r\,dr\,d\theta = \frac{8}{9}\int_0^{2\pi}\theta^2\,d\theta = \frac{64\pi^3}{27}$$

17. $$A = \int_0^{\pi/2}\int_0^{1+\sin\theta} r\,dr\,d\theta = \frac{1}{2}\int_0^{\pi/2}\frac{3}{2} + 2\sin\theta - \frac{\cos 2\theta}{2}\,d\theta = \frac{3\pi+8}{8}$$

18. $$A = 4\int_0^{\pi/2}\int_0^{1-\cos\theta} r\,dr\,d\theta = 2\int_0^{\pi/2}\frac{3}{2} - 2\cos\theta + \frac{\cos 2\theta}{2}\,d\theta = \frac{3\pi-8}{2}$$

19. $$\int_0^{2\pi}\int_0^{\sqrt{3}/2}\frac{1}{1-r^2}\,r\,dr\,d\theta = \ln(2)\int_0^{2\pi}d\theta = \pi\ln 4$$

20. $$4\int_0^{\pi/2}\int_1^{e}\frac{\ln(r^2)}{r}\,dr\,d\theta = 2\int_0^{\pi/2}d\theta = 2\pi$$

21. $$M_x = \int_0^{\pi}\int_0^{1-\cos\theta} 3r^2\sin\theta\,dr\,d\theta = \int_0^{\pi}(1-\cos\theta)^3\sin\theta\,d\theta = 4$$

22. $$I_x = \int_{-a}^{a}\int_{-\sqrt{a^2-x^2}}^{\sqrt{a^2-x^2}} y^2(k)(x^2+y^2)\,dy\,dx = k\int_0^{2\pi}\int_0^{a}\sin^2\theta\,r^5\,dr\,d\theta =$$
$$\frac{ka^6}{6}\int_0^{2\pi}\frac{1-\cos 2\theta}{2}\,d\theta = \frac{ka^6\pi}{6}$$

23. $$M = 2\int_0^{\pi}\int_0^{1+\cos\theta} r\,dr\,d\theta = \int_0^{\pi}(1+\cos\theta)^2\,d\theta = \frac{3\pi}{2}$$
$$M_y = \int_0^{2\pi}\int_0^{1+\cos\theta}\cos\theta\,r^2\,dr\,d\theta = \int_0^{2\pi}\frac{4\cos\theta}{3} + \frac{15}{24} + \cos 2\theta - \sin^2\theta\cos\theta + \frac{\cos 4\theta}{4}\,d\theta =$$
$$\frac{5\pi}{4} \Rightarrow \overline{x} = \frac{5}{6},\ \overline{y} = 0 \text{ by symmetry}$$

24. $$I_o = \int_0^{2\pi}\int_0^{1+\cos\theta} r^3\,dr\,d\theta = \frac{1}{4}\int_0^{2\pi}(1+\cos\theta)^4\,d\theta = \frac{35\pi}{16}$$

25. $$M = \int_{-\infty}^{0}\int_0^{\exp(x)} dy\,dx = \int_{-\infty}^{0} e^x\,dx = \lim_{t\to-\infty}\int_t^0 e^x\,dx = 1$$
$$M_y = \int_{-\infty}^{0}\int_0^{\exp(x)} x\,dy\,dx = \int_{-\infty}^{0} x\,e^x\,dx = \lim_{t\to-\infty}\int_t^0 x\,e^x\,dx = -1$$
$$M_x = \int_{-\infty}^{0}\int_0^{\exp(x)} y\,dy\,dx = \frac{1}{2}\int_{-\infty}^{0} e^{2x}\,dx = \frac{1}{2}\lim_{t\to-\infty}\int_t^0 e^{2x}\,dx = \frac{1}{4} \Rightarrow \overline{x} = -1,\ \overline{y} = \frac{1}{4}$$

26. $$M_y = \int_{-\infty}^{\infty}\int_0^{\exp(-x^2/2)} x\,dy\,dx = 2\int_0^{\infty}\exp(-x^2/2)(x)\,dx = 2\lim_{t\to\infty}\int_0^t \exp(-x^2/2)(x)\,dx = 2$$

14.4 TRIPLE INTEGRALS IN RECTANGULAR COORDINATES

1. $\int_0^1 \int_0^{1-z} \int_0^2 dx\,dy\,dz = 2\int_0^1 \int_0^{1-z} dy\,dz = 2\int_0^1 1 - z\,dz = 1$

2. $\int_0^1 \int_0^2 \int_0^3 dz\,dy\,dx = \int_0^1 \int_0^2 3\,dy\,dx = \int_0^1 6\,dx = 6$, $\int_0^2 \int_0^1 \int_0^3 dz\,dx\,dy$,
$\int_0^3 \int_0^2 \int_0^1 dx\,dy\,dz$, $\int_0^2 \int_0^3 \int_0^1 dx\,dz\,dy$, $\int_0^3 \int_0^1 \int_0^2 dy\,dx\,dz$, $\int_0^1 \int_0^3 \int_0^2 dy\,dz\,dx$

3. $\int_0^1 \int_0^{2-2x} \int_0^{3-3x-3y/2} dz\,dy\,dx = \int_0^1 \int_0^{2-2x} 3 - 3x - \frac{3}{2}y\,dy\,dx = \int_0^1 3 - 6x + 3x^2\,dx = 1$,
$\int_0^2 \int_0^{x-y/2} \int_0^{3-3x-3y/2} dz\,dx\,dy$, $\int_0^1 \int_0^{3-3x} \int_0^{2-2x-2z/3} dy\,dz\,dx$,
$\int_0^3 \int_0^{1-z/3} \int_0^{2-2x-2z/3} dy\,dx\,dz$, $\int_0^2 \int_0^{3-3y/2} \int_0^{1-y/2-z/3} dx\,dz\,dy$,
$\int_0^3 \int_0^{2-2z/3} \int_0^{1-y/2-z/3} dx\,dy\,dz$

4. $\int_0^2 \int_0^3 \int_0^{\sqrt{4-x^2}} dz\,dy\,dx = \int_0^2 \int_0^3 \sqrt{4-x^2}\,dy\,dx = \int_0^2 3\sqrt{4-x^2}\,dx = 3\pi$,
$\int_0^3 \int_0^2 \int_0^{\sqrt{4-x^2}} dz\,dx\,dy$, $\int_0^2 \int_0^{\sqrt{4-x^2}} \int_0^3 dy\,dz\,dx$,
$\int_0^2 \int_0^{\sqrt{4-z^2}} \int_0^3 dy\,dx\,dz$, $\int_0^2 \int_0^3 \int_0^{\sqrt{4-z^2}} dx\,dy\,dz$,
$\int_0^3 \int_0^2 \int_0^{\sqrt{4-z^2}} dx\,dz\,dy$

5. $\int_0^1 \int_0^1 \int_0^1 x^2 + y^2 + z^2\,dz\,dy\,dx = \int_0^1 \int_0^1 \left(x^2 + y^2 + \frac{1}{3}\right) dy\,dx = \int_0^1 x^2 + \frac{2}{3}\,dx = 1$

6. $\int_0^{\sqrt{2}} \int_0^{3y} \int_{x^2+3y^2}^{8-x^2-y^2} dz\,dx\,dy = \int_0^{\sqrt{2}} \int_0^{3y} 8 - 2x^2 - 4y^2\,dx\,dy = \int_0^{\sqrt{2}} 24y - 18y^3 - 12y^3\,dx = -6$

7. $\int_1^e \int_1^e \int_1^e \frac{1}{xyz}\,dx\,dy\,dz = \int_1^e \int_1^e \frac{\ln x}{yz}\,dy\,dz = \int_1^e \frac{1}{z}\,dz = 1$

8. $\int_0^1 \int_0^{3-3x} \int_0^{3-3x-y} dz\,dy\,dx = \int_0^1 \int_0^{3-3x} 3 - 3x - y\,dy\,dx = \int_0^1 \frac{9}{2} - 9x + \frac{9}{2}x^2\,dx = \frac{3}{2}$

9. $\int_0^1 \int_0^{\pi} \int_0^{\pi} y \sin z\,dx\,dy\,dz = \int_0^1 \int_0^{\pi} \pi y \sin z\,dy\,dz = \frac{\pi^3}{2}\int_0^1 \sin z\,dz = \frac{\pi^3}{2}(1 - \cos 1)$

10. $\int_{-1}^{1}\int_{-1}^{1}\int_{-1}^{1}(x + y + z)\,dy\,dx\,dz = \int_{-1}^{1}\int_{-1}^{1} 2x + 2z\,dx\,dz = \int_{-1}^{1} 4z\,dz = 0$

11. $\int_{0}^{3}\int_{0}^{\sqrt{9-x^2}}\int_{0}^{\sqrt{9-x^2}} dz\,dy\,dx = \int_{0}^{3}\int_{0}^{\sqrt{9-x^2}}\sqrt{9-x^2}\,dy\,dx = \int_{0}^{3}\left(9-x^2\right)dx = 18$

12. $\int_{0}^{2}\int_{-\sqrt{4-y^2}}^{\sqrt{4-y^2}}\int_{0}^{2x+y} dz\,dx\,dy = \int_{0}^{2}\int_{-\sqrt{4-y^2}}^{\sqrt{4-y^2}}(2x + y)\,dx\,dy = \int_{0}^{2}\left(4-y^2\right)^{1/2}(2y)\,dy = \frac{16}{3}$

13. $\int_{0}^{1}\int_{0}^{2-x}\int_{0}^{2-x-y} dz\,dy\,dx = \int_{0}^{1}\int_{0}^{2-x} 2 - x - y\,dy\,dx = \int_{0}^{1}\frac{x^2}{2} - 2x + 2\,dx = \frac{7}{6}$

14. $\int_{0}^{1}\int_{0}^{1-x^2}\int_{3}^{4-x^2-y} x\,dz\,dy\,dx = \int_{0}^{1}\int_{0}^{1-x^2} x - x^3 - xy\,dy\,dx = \int_{0}^{1}\frac{x}{2} - x^3 + \frac{x^5}{2}\,dx = \frac{1}{12}$

15. a) $\int_{-1}^{1}\int_{0}^{1-x^2}\int_{x^2}^{1-z} dy\,dz\,dx$ b) $\int_{0}^{1}\int_{-\sqrt{1-z}}^{\sqrt{1-z}}\int_{x^2}^{1-z} dy\,dx\,dz$

c) $\int_{0}^{1}\int_{0}^{1-z}\int_{-\sqrt{y}}^{\sqrt{y}} dx\,dy\,dz$ d) $\int_{0}^{1}\int_{0}^{1-y}\int_{-\sqrt{y}}^{\sqrt{y}} dx\,dz\,dy$

e) $\int_{0}^{1}\int_{-\sqrt{y}}^{\sqrt{y}}\int_{0}^{1-y} dz\,dx\,dy$

16. a) $\int_{0}^{1}\int_{0}^{1}\int_{-1}^{\sqrt{z}} dy\,dz\,dx$ b) $\int_{0}^{1}\int_{0}^{1}\int_{-1}^{\sqrt{y}} dy\,dx\,dz$

c) $\int_{0}^{1}\int_{0}^{\sqrt{z}}\int_{0}^{1} dx\,dy\,dz$ d) $\int_{-1}^{0}\int_{0}^{y^2}\int_{0}^{1} dx\,dz\,dy$

e) $\int_{-1}^{0}\int_{0}^{1}\int_{0}^{y^2} dz\,dx\,dy$

17. $V = \int_{0}^{1}\int_{-1}^{1}\int_{0}^{y^2} dz\,dy\,dx = \int_{0}^{1}\int_{-1}^{1} y^2\,dy\,dx = \frac{2}{3}\int_{0}^{1} dx = \frac{2}{3}$

18. $V = \int_{0}^{1}\int_{0}^{1-x}\int_{0}^{2-2z} dy\,dz\,dx = \int_{0}^{1}\int_{0}^{1-x} 2-2z\,dz\,dx = \int_{0}^{1} 1 - x^2\,dx = \frac{2}{3}$

19. $V = \int_0^4 \int_0^{\sqrt{4-x}} \int_0^{2-y} dz\,dy\,dx = \int_0^4 \int_0^{\sqrt{4-x}} (2-y)\,dy\,dx = \int_0^4 2\sqrt{4-x} - \left(\frac{4-x}{2}\right) dx = \frac{20}{3}$

20. $V = 2\int_0^1 \int_0^{\sqrt{1-x^2}} \int_0^{-y} dz\,dy\,dx = -2\int_0^1 \int_0^{\sqrt{1-x^2}} y\,dy\,dx = -\int_0^1 1 - x^2\,dx = \frac{2}{3}$

21. $V = \int_0^1 \int_0^{2-2x} \int_0^{3-3x-3y/2} dz\,dy\,dx = \int_0^1 \int_0^{2-2x} 3 - 3x - \frac{3}{2}y\,dy\,dx = \int_0^1 3 - 6x + 3x^2\,dx = 1$

22. $V = \int_0^1 \int_0^{1-x} \int_0^{\cos(\pi x/2)} dz\,dy\,dx = \int_0^1 \int_0^{1-x} \left(\cos\left(\frac{\pi x}{2}\right)\right) dy\,dx = \int_0^1 \left(\cos\frac{\pi x}{2}\right)(1-x)\,dx = \frac{4}{\pi^2}$

23. $V = 8\int_0^1 \int_0^{\sqrt{1-x^2}} \int_0^{\sqrt{1-x^2}} dz\,dy\,dx = 8\int_0^1 \int_0^{\sqrt{1-x^2}} \sqrt{1-x^2}\,dy\,dx = 8\int_0^1 1 - x^2\,dx = \frac{16}{3}$

24. $V = \int_0^2 \int_0^{4-x^2} \int_0^{4-x^2-y} dz\,dy\,dx = \int_0^2 \int_0^{4-x^2} 4 - x^2 - y\,dy\,dx = \int_0^2 8 - 4x^2 + \frac{x^4}{2}\,dx = \frac{32}{15}$

25. $V = \int_0^4 \int_0^{\left(\sqrt{16-y^2}\right)/2} \int_0^{4-y} dx\,dz\,dy = \int_0^4 \int_0^{\left(\sqrt{16-y^2}\right)/2} (4-y)\,dz\,dy =$

$\int_0^4 \frac{\sqrt{16-y^2}}{2}(4-y)\,dy = 8\pi + \frac{32}{3}$

26. $V = \int_{-2}^2 \int_{-\sqrt{4-x^2}}^{\sqrt{4-x^2}} \int_0^{3-x} dz\,dy\,dx = \int_{-2}^2 \int_{-\sqrt{4-x^2}}^{\sqrt{4-x^2}} (3-x)\,dy\,dx =$

$\int_{-2}^2 (3-x)\left(2\sqrt{4-x^2}\right) dx = 12\pi$

27. $\text{average} = \frac{1}{8}\int_0^2 \int_0^2 \int_0^2 x^2 + 9\,dz\,dy\,dx = \frac{1}{8}\int_0^2 \int_0^2 \left(2x^2 + 18\right) dy\,dx = \frac{1}{8}\int_0^2 \left(4x^2 + 36\right) dx = \frac{31}{3}$

28. $\text{average} = \frac{1}{2}\int_0^1 \int_0^1 \int_0^2 x + y - z\,dz\,dy\,dx = \frac{1}{2}\int_0^1 \int_0^1 2x + 2y - 2\,dy\,dx = \frac{1}{2}\int_0^1 2x - 1\,dx = 0$

29. $\text{average} = \int_0^1 \int_0^1 \int_0^1 x^2 + y^2 + z^2\,dz\,dy\,dx = \int_0^1 \int_0^1 \left(x^2 + y^2 + \frac{1}{3}\right) dy\,dx = \int_0^1 \left(x^2 + \frac{2}{3}\right) dx = 1$

30. $\text{average} = \frac{1}{8}\int_0^2 \int_0^2 \int_0^2 xyz\,dz\,dy\,dx = \frac{1}{4}\int_0^2 \int_0^2 xy\,dy\,dx = \frac{1}{2}\int_0^2 x\,dx = 1$

14.5 MASSES AND MOMENTS IN THREE DIMENSIONS

1. $I_x = \int_{-c/2}^{c/2} \int_{-b/2}^{b/2} \int_{-a/2}^{a/2} \left(y^2 + z^2\right) dx\, dy\, dz = 4a \int_0^{c/2} \int_0^{b/2} \left(y^2 + z^2\right) dy\, dz =$
$4a \int_0^{c/2} \left(\frac{b^3}{24} + \frac{z^2 b}{2}\right) dx = \frac{abc}{12}\left(b^2 + c^2\right) \Rightarrow I_x = \frac{M}{12}\left(b^2 + c^2\right);\ R_x = \sqrt{\frac{b^2 + c^2}{12}},$
$R_y = \sqrt{\frac{a^2 + c^2}{12}},\ R_z = \sqrt{\frac{a^2 + b^2}{12}}$

2. The plane $z = \frac{4 - 2y}{3}$ is the top of the wedge. $I_x = \int_{-3}^{3} \int_{-2}^{4} \int_{-4/3}^{(4-2y)/3} \left(y^2 + z^2\right) dz\, dy\, dx =$
$\int_{-3}^{3} \int_{-2}^{4} \frac{8y^2}{3} - \frac{2y^3}{3} + \frac{8(2-y)^3}{81} + \frac{64}{81} dy\, dx = \int_{-3}^{3} \frac{104}{3} dx = 208;$
$I_y = \int_{-3}^{3} \int_{-2}^{4} \int_{-4/3}^{2(2-y)/3} \left(x^2 + z^2\right) dz\, dy\, dx = \int_{-3}^{3} \int_{-2}^{4} \frac{8x^2}{3} - \frac{2x^2 y}{3} - \frac{8y}{81} + \frac{80}{81} dy\, dx =$
$\int_{-3}^{3} 12x^2 + \frac{432}{81} dx = 248;\ I_z = \int_{-3}^{3} \int_{-2}^{4} \int_{-4/3}^{2(2-y)/3} \left(x^2 + y^2\right) dz\, dy\, dx =$
$\int_{-3}^{3} \int_{-2}^{4} \left(x^2 + y^2\right)\left(\frac{8}{3} - \frac{2y}{3}\right) dy\, dx = 12 \int_{-3}^{3} x^2 + 2\, dx = 360$

3. $I_x = \int_0^a \int_0^b \int_0^c \left(y^2 + z^2\right) dz\, dy\, dx = \int_0^a \int_0^b cy^2 + \frac{c^3}{3} dy\, dx = \int_0^a \frac{cb^3}{3} + \frac{c^3 b}{3} dx = \frac{abc\left(b^2 + c^2\right)}{3} =$
$\frac{M}{3}\left(b^2 + c^2\right);\ I_y = \frac{M}{3}\left(a^2 + c^2\right)$ and $I_z = \frac{M}{3}\left(a^2 + b^2\right)$ by symmetry

4. a) $M = \int_0^1 \int_0^{1-x} \int_0^{1-x-y} dz\, dy\, dx = \int_0^1 \int_0^{1-x} 1 - x - y\, dy\, dx = \int_0^1 \frac{x^2}{2} - x + \frac{1}{2} dx = \frac{1}{6}$
$M_{yz} = \int_0^1 \int_0^{1-x} \int_0^{1-x-y} x\, dz\, dy\, dx = \int_0^1 \int_0^{1-x} x(1 - x - y)\, dy\, dx = \frac{1}{2}\int_0^1 x^3 - 2x^2 + x\, dx = \frac{1}{24} \Rightarrow$
$\overline{x} = \overline{y} = \overline{z} = \frac{1}{4}$ by symmetry; $I_z = \int_0^1 \int_0^{1-x} \int_0^{1-x-y} \left(y^2 + z^2\right) dz\, dy\, dx =$
$\int_0^1 \int_0^{1-x} y^2 - xy^2 - y^3 + \frac{(1 - x - y)^3}{3} dy\, dx = \frac{1}{6}\int_0^1 (1 - x)^4 dx = \frac{1}{30},\ I_y = I_x = \frac{1}{30}$ by symmetry

b) $R_x = \sqrt{\frac{I_x}{M}} = \sqrt{\frac{1}{5}} \approx 0.4472$, distance from the centroid to the x–axis is $\sqrt{0^2 + \frac{1}{16} + \frac{1}{16}} =$
$\sqrt{\frac{1}{8}} \approx 0.3536$

5. $M = 4\int_0^1\int_0^1\int_{4y^2}^4 dz\,dy\,dx = 4\int_0^1\int_0^1 4-4y^2\,dy\,dx = 16\int_0^1 \frac{2}{3}dx = \frac{32}{3}$

$M_{xy} = 4\int_0^1\int_0^1\int_{4y^2}^4 z\,dz\,dy\,dx = 2\int_0^1\int_0^1 \left(16-16y^4\right)dy\,dx = \frac{128}{5}\int_0^1 dx = \frac{128}{5} \Rightarrow$

$\overline{z} = \frac{12}{5}$, $\overline{x} = \overline{y} = 0$ by symmetry; $I_z = 4\int_0^1\int_0^1\int_{4y^2}^4 \left(x^2+y^2\right)dz\,dy\,dx =$

$16\int_0^1\int_0^1 x^2 - x^2y^2 + y^2 - y^4\,dy\,dx = 16\int_0^1 \frac{2x^2}{3} + \frac{2}{15}dx = \frac{256}{45}$;

$I_x = 4\int_0^1\int_0^1\int_{4y^2}^4 \left(y^2+z^2\right)dz\,dy\,dx = 4\int_0^1\int_0^1 \left(4y^2+\frac{64}{3}\right) - \left(4y^4+\frac{64y^6}{3}\right)dy\,dx =$

$4\int_0^1 \frac{1976}{105}dx = \frac{7904}{105}$; $I_y = 4\int_0^1\int_0^1\int_{4y^2}^4 \left(x^2+z^2\right)dz\,dy\,dx =$

$4\int_0^1\int_0^1 \left(4x^2+\frac{64}{3}\right) - \left(4x^2y^2+\frac{64y^6}{3}\right)dy\,dx = 4\int_0^1 \frac{8}{3}x^2 + \frac{128}{7}dx = \frac{4832}{63}$

6. a) $M = \int_{-2}^2\int_{\left(-\sqrt{4-x^2}\right)/2}^{\left(\sqrt{4-x^2}\right)/2}\int_0^{2-x} dz\,dy\,dx = \int_{-2}^2\int_{\left(-\sqrt{4-x^2}\right)/2}^{\left(\sqrt{4-x^2}\right)/2} (2-x)\,dy\,dx =$

$\int_{-2}^2 (2-x)\left(\sqrt{4-x^2}\right)dx = 4\pi$; $M_{yz} = \int_{-2}^2\int_{\left(-\sqrt{4-x^2}\right)/2}^{\left(\sqrt{4-x^2}\right)/2}\int_0^{2-x} x\,dz\,dy\,dx =$

$\int_{-2}^2\int_{\left(-\sqrt{4-x^2}\right)/2}^{\left(\sqrt{4-x^2}\right)/2} x(2-x)\,dy\,dx = \int_{-2}^2 x(2-x)\left(\sqrt{4-x^2}\right)dx = -2\pi$;

$M_{xz} = \int_{-2}^2\int_{\left(-\sqrt{4-x^2}\right)/2}^{\left(\sqrt{4-x^2}\right)/2}\int_0^{2-x} y\,dz\,dy\,dx = \int_{-2}^2\int_{\left(-\sqrt{4-x^2}\right)/2}^{\left(\sqrt{4-x^2}\right)/2} y(2-x)\,dy\,dx =$

$\frac{1}{2}\int_{-2}^2 (2-x)\left[\frac{4-x^2}{4} - \frac{4-x^2}{4}\right]dx = 0 \Rightarrow \overline{x} = -\frac{1}{2}$, $\overline{y} = 0$

b) $M_{xy} = \int_{-2}^2\int_{\left(-\sqrt{4-x^2}\right)/2}^{\left(\sqrt{4-x^2}\right)/2}\int_0^{2-x} z\,dz\,dy\,dx = \frac{1}{2}\int_{-2}^2\int_{\left(-\sqrt{4-x^2}\right)/2}^{\left(\sqrt{4-x^2}\right)/2} (2-x)^2\,dy\,dx =$

$\frac{1}{2}\int_{-2}^2 (2-x)^2\left(\sqrt{4-x^2}\right)dx = 5\pi \Rightarrow \overline{z} = \frac{5}{4}$

7. a) $M = 4\int_0^2 \int_0^{\sqrt{4-x^2}} \int_{x^2+y^2}^4 dz\,dy\,dx = 4\int_0^{\pi/2} \int_0^2 \int_{r^2}^4 r\,dz\,dr\,d\theta =$

$4\int_0^{\pi/2} \int_0^2 4r - r^3\,dr\,d\theta = 4\int_0^{\pi/2} 4\,d\theta = 8\pi$; $M_{xy} = \int_0^{2\pi} \int_0^2 \int_{r^2}^4 r\,dz\,dr\,d\theta =$

$\int_0^{2\pi} \int_0^2 \frac{r}{2}\left(16 - r^4\right) dr\,d\theta = \frac{32\pi}{3}\int_0^{2\pi} d\theta = \frac{64\pi}{3} \Rightarrow \overline{z} = \frac{8}{3}$, $\overline{x} = \overline{y} = 0$ by symmetry

b) $M = 8\pi$; $4\pi = \int_0^{2\pi} \int_0^{\sqrt{c}} \int_{r^2}^c r\,dz\,dr\,d\theta = \int_0^{2\pi} \int_0^{\sqrt{c}} cr - r^3\,dr\,d\theta = \int_0^{2\pi} \left(\frac{c^2}{4}\right) d\theta = \frac{c^2\pi}{2} \Rightarrow$

$c^2 = 8 \Rightarrow c = 2\sqrt{2}$, since $c > 0$

8. $M = 8$, $M_{xy} = \int_{-1}^1 \int_3^5 \int_{-1}^1 z\,dz\,dy\,dx = \int_{-1}^1 \int_3^5 \left[z^2/2\right]_{-1}^1 dy\,dx = 0$

$M_{yz} = \int_{-1}^1 \int_3^5 \int_{-1}^1 x\,dz\,dy\,dx = 2\int_{-1}^1 \int_3^5 x\,dy\,dx = 4\int_{-1}^1 x^2\,dx = 0$

$M_{xz} = \int_{-1}^1 \int_3^5 \int_{-1}^1 y\,dz\,dy\,dx = 2\int_{-1}^1 \int_3^5 y\,dy\,dx = 16\int_{-1}^1 dx = 32 \Rightarrow \overline{x} = 0,\ \overline{y} = 4,\ \overline{z} = 0$

$I_x = \int_{-1}^1 \int_3^5 \int_{-1}^1 \left(y^2 + z^2\right) dz\,dy\,dx = \int_{-1}^1 \int_3^5 2y^2 + \frac{2}{3}\,dy\,dx = \frac{2}{3}\int_{-1}^1 100\,dx = \frac{400}{3}$

$I_y = \int_{-1}^1 \int_3^5 \int_{-1}^1 \left(x^2 + z^2\right) dz\,dy\,dx = \int_{-1}^1 \int_3^5 2x^2 + \frac{2}{3}\,dy\,dx = \frac{200}{3}\int_{-1}^1 dx = \frac{400}{3}$

$I_z = \int_{-1}^1 \int_3^5 \int_{-1}^1 \left(x^2 + y^2\right) dz\,dy\,dx = 2\int_{-1}^1 \int_3^5 \left(x^2 + y^2\right) dy\,dx = 2\int_{-1}^1 2x^2 + \frac{98}{3}\,dx = \frac{400}{3} \Rightarrow$

$R_x = R_y = R_z = \sqrt{\frac{50}{3}}$; The y–moment will not change, since the y–component of the center of mass is on the y–axis.

9. $I_L = \int_{-2}^2 \int_{-2}^4 \int_{-1}^{1-y/2} \left((y-6)^2 + z^2\right) dz\,dy\,dx = \int_{-2}^2 \int_{-2}^4 \frac{(y-6)^2(4-y)}{2} + \frac{(2-y)^3}{24} + \frac{1}{3}\,dy\,dx =$

$4\int_{-2}^4 \frac{13t^3}{24} + 5t^2 + 16t + \frac{49}{3}\,dt = 1386$, where $t = 2 - y$; $M = 36$, $R_L = \sqrt{\frac{I_L}{M}} = \sqrt{\frac{77}{2}}$

10. The plane $y + 2z = 2$ is the top of the wedge. $I_L = \int_{-2}^2 \int_{-2}^4 \int_{-1}^{(2-y)/2} \left((x-4)^2 + y^2\right) dz\,dy\,dx =$

$\frac{1}{2}\int_{-2}^2 \int_{-2}^4 \left(x^2 - 8x + 16 + y^2\right)(4 - y)\,dy\,dx = \int_{-2}^2 9x^2 - 72x - 162\,dx = 696 \Rightarrow R_L = \sqrt{\frac{I_L}{M}} = \sqrt{\frac{29}{6}}$

11. $M = 8$, $I_L = \int_0^4 \int_0^2 \int_0^1 \left(z^2 + (y-2)^2\right) dz\,dy\,dx = \int_0^4 \int_0^2 y^2 - 4y + \frac{13}{3}\,dy\,dx = \frac{10}{3}\int_0^4 dx = \frac{40}{3} \Rightarrow$

$R_L = \sqrt{\frac{I_L}{M}} = \sqrt{\frac{5}{3}}$

12. $M = 8$, $I_L = \int_0^4 \int_0^2 \int_0^1 \left((x-4)^2 + y^2\right) dz\, dy\, dx = \int_0^4 \int_0^2 \left((x-4)^2 + y^2\right) dy\, dx =$

$\int_0^4 \left(2(x-4)^2 + \frac{8}{3}\right) dx = \frac{160}{3} \Rightarrow R_L = \sqrt{\frac{I_L}{M}} = \sqrt{\frac{20}{3}}$

13. $M = \int_0^2 \int_0^{2-x} \int_0^{2-x-y} 2x\, dz\, dy\, dx = \int_0^2 \int_0^{2-x} 4x - 2x^2 - 2xy\, dy\, dx = \int_0^2 x^3 - 4x^2 + 4x\, dx = \frac{4}{3}$

$M_{xy} = \int_0^2 \int_0^{2-x} \int_0^{2-x-y} 2xz\, dz\, dy\, dx = \int_0^2 \int_0^{2-x} x(2-x-y)^2\, dy\, dx = \int_0^2 \frac{x(2-x)^3}{3}\, dx = \frac{8}{15}$;

$M_{xz} = \frac{8}{15}$ by symmetry; $M_{yz} = \int_0^2 \int_0^{2-x} \int_0^{2-x-y} 2x^2\, dz\, dy\, dx = \int_0^2 \int_0^{2-x} 2x^2(2-x-y)\, dy\, dx =$

$\int_0^2 \left(2x - x^2\right)^2 dx = \frac{16}{15} \Rightarrow \overline{x} = \frac{4}{5}$, $\overline{y} = \overline{z} = \frac{2}{5}$

14. The plane $y + 2z = 2$ is the top of the wedge. $M = \int_{-2}^2 \int_{-2}^4 \int_{-1}^{1-y/2} (x+1)\, dz\, dy\, dx =$

$\int_{-2}^2 \int_{-2}^4 (x+1)\left(2 - \frac{y}{2}\right) dy\, dx = 36$ by the Calculus Toolkit;

$M_{yz} = \int_{-2}^2 \int_{-2}^4 \int_{-1}^{1-y/2} x(x+1)\, dz\, dy\, dx = \int_{-2}^2 \int_{-2}^4 x(x+1)\left(2 - \frac{y}{2}\right) dy\, dx = 48$ by the Calculus Toolkit; $M_{xz} = \int_{-2}^2 \int_{-2}^4 \int_{-1}^{1-y/2} y(x+1)\, dz\, dy\, dx = \int_{-2}^2 \int_{-2}^4 y(x+1)\left(2 - \frac{y}{2}\right) dy\, dx = 0$ by the Calculus Toolkit; $M_{xy} = \int_{-2}^2 \int_{-2}^4 \int_{-1}^{1-y/2} z(x+1)\, dz\, dy\, dx = \frac{1}{2}\int_{-2}^2 \int_{-2}^4 (x+1)\left(2 - \frac{y}{2}\right) dy\, dx = 0$ by the Calculus Toolkit $\Rightarrow \overline{x} = \frac{4}{3}$, $\overline{y} = \overline{z} = 0$; $I_x = \int_{-2}^2 \int_{-2}^4 \int_{-1}^{1-y/2} (x+1)\left(y^2 + z^2\right) dz\, dy\, dx =$

$\int_{-2}^2 \int_{-2}^4 (x+1)\left(2y^2 + \frac{1}{3} - \frac{y^3}{2} + \frac{1}{3}\left(1 - \frac{y}{2}\right)^3\right) dy\, dx = 90$ by the Calculus Toolkit; $I_y =$

$\int_{-2}^2 \int_{-2}^4 \int_{-1}^{1-y/2} (x+1)\left(x^2 + z^2\right) dz\, dy\, dx = \int_{-2}^2 \int_{-2}^4 (x+1)\left(2x^2 + \frac{1}{3} - \frac{x^2 y}{2} + \frac{1}{3}\left(1 - \frac{y}{2}\right)^3\right) dy\, dx =$

66 by the Calculus Toolkit; $I_z = \int_{-2}^2 \int_{-2}^4 \int_{-1}^{1-y/2} (x+1)\left(x^2 + y^2\right) dz\, dy\, dx =$

$\int_{-2}^2 \int_{-2}^4 (x+1)\left(2 - \frac{y}{2}\right)\left(x^2 + y^2\right) dy\, dx = 120$ by the Calculus Toolkit $\Rightarrow R_x = \sqrt{\frac{I_x}{M}} = \sqrt{\frac{5}{2}}$,

$R_y = \sqrt{\frac{I_y}{M}} = \sqrt{\frac{11}{6}}$, $R_z = \sqrt{\frac{I_z}{M}} = \sqrt{\frac{10}{3}}$

15. $M = \int_0^1 \int_0^1 \int_0^1 (x + y + z + 1)\, dz\, dy\, dx = \int_0^1 \int_0^1 \left(x + y + \frac{3}{2}\right) dy\, dx = \int_0^1 (x + 2)\, dx = \frac{5}{2}$

$M_{xy} = \int_0^1 \int_0^1 \int_0^1 (x + y + z + 1)z\, dz\, dy\, dx = \frac{1}{2}\int_0^1 \int_0^1 \left(x + y + \frac{5}{3}\right) dy\, dx =$

$\frac{1}{2}\int_0^1 \left(x + \frac{13}{6}\right) dx = \frac{4}{3} \Rightarrow M_{xy} = M_{yz} = M_{xz} = \frac{4}{3}$ by symmetry $\therefore\ \overline{x} = \overline{y} = \overline{z} = \frac{8}{15}$

$I_z = \int_0^1 \int_0^1 \int_0^1 (x + y + z + 1)\left(x^2 + y^2\right) dz\, dy\, dx = \int_0^1 \int_0^1 \left(x + y + \frac{3}{2}\right)\left(x^2 + y^2\right) dy\, dx =$

$\int_0^1 x^3 + 2x^2 + \frac{1}{3}x + \frac{3}{4} dx = \frac{11}{6} \Rightarrow I_x = I_y = I_z = \frac{11}{6}$ by symmetry $\Rightarrow R_x = R_y = R_z = \sqrt{\frac{I_z}{M}} = \sqrt{\frac{11}{15}}$

16. a) $M = \int_0^2 \int_0^{\sqrt{x}} \int_0^{4-x^2} k\, xy\, dz\, dy\, dx = k\int_0^2 \int_0^{\sqrt{x}} xy\left(4 - x^2\right) dy\, dx =$

$\frac{k}{2}\int_0^2 \left(4x - x^3\right) dx = \frac{32k}{15}$

b) $M_{yz} = \int_0^2 \int_0^{\sqrt{x}} \int_0^{4-x^2} k\, x^2y\, dz\, dy\, dx = k\int_0^2 \int_0^{\sqrt{x}} x^2y\left(4 - x^2\right) dy\, dx = (2.666666)k$, by the Calculus Toolkit $\Rightarrow \overline{x} = 1.25$

c) $M_{xz} = \int_0^2 \int_0^{\sqrt{x}} \int_0^{4-x^2} k\, xy^2\, dz\, dy\, dx = k\int_0^2 \int_0^{\sqrt{x}} xy^2\left(4 - x^2\right) dy\, dx = (1.567267)k$, by the Calculus Toolkit $\Rightarrow \overline{y} = 0.734656$

d) $M_{xy} = \int_0^2 \int_0^{\sqrt{x}} \int_0^{4-x^2} k\, xyz\, dz\, dy\, dx = k\int_0^2 \int_0^{\sqrt{x}} \left(\frac{xy\left(4 - x^2\right)^2}{2}\right) dy\, dx = (2.438095)k$, by the Calculus Toolkit $\Rightarrow \overline{z} = 1.142857$

14.6 TRIPLE INTEGRALS IN CYLINDRICAL AND SPHERICAL COORDINATES

1. $\int_0^{2\pi} \int_0^1 \int_r^{\sqrt{2-r^2}} r\, dz\, dr\, d\theta = \int_0^{2\pi} \int_0^1 \left(\left(2 - r^2\right)^{1/2} r - r^2\right) dr\, d\theta = \int_0^{2\pi} \left(\frac{2^{3/2}}{3} - \frac{2}{3}\right) d\theta = \frac{4\pi\left(\sqrt{2} - 1\right)}{3}$

2. $\int_0^{2\pi} \int_0^3 \int_{r^2/3}^{\sqrt{18-r^2}} r\, dz\, dr\, d\theta = \int_0^{2\pi} \int_0^3 \left(\sqrt{18 - r^2} - \frac{r^2}{3}\right) r\, dr\, d\theta =$

$\int_0^{2\pi} \left[-\frac{\left(18 - r^2\right)^{3/2}}{3} - \frac{r^4}{12}\right]_0^3 d\theta = \frac{9\pi\left(8\sqrt{2} - 1\right)}{2}$

3. $\int_0^{2\pi}\int_0^{\theta/2\pi}\int_0^{3+24r^2} r\,dz\,dr\,d\theta = \int_0^{2\pi}\int_0^{\theta/2\pi}\left(3+24r^2\right) r\,dr\,d\theta = \frac{3}{2}\int_0^{2\pi}\frac{\theta^2}{4\pi^2}+\frac{4\theta^4}{16\pi^4}\,d\theta = \frac{17\pi}{5}$

4. $\int_0^{\pi}\int_0^{\theta/\pi}\int_{-\sqrt{4-r^2}}^{3\sqrt{4-r^2}} r\,dz\,dr\,d\theta = 4\int_0^{\pi}\int_0^{\theta/\pi} 4r-r^3\,dr\,d\theta = 4\int_0^{\pi}\frac{2\theta^2}{\pi^2}-\frac{\theta^4}{4\pi^4}\,d\theta = \frac{37\pi}{15}$

5. $\int_0^{2\pi}\int_0^{1}\int_r^{\left(2-r^2\right)^{-1/2}} 3\,r\,dz\,dr\,d\theta = 3\int_0^{2\pi}\int_0^{1}\left(2-r^2\right)^{-1/2}r - r^2\,dr\,d\theta =$

$3\int_0^{2\pi}\left(\sqrt{2}-\frac{4}{3}\right)d\theta = \pi\left(6\sqrt{2}-8\right)$

6. $\int_0^{2\pi}\int_0^{1}\int_{-1/2}^{1/2}\left(r^2\sin^2\theta+z^2\right) r\,dz\,dr\,d\theta = \int_0^{2\pi}\int_0^{1}\left(r^3\sin^2\theta+\frac{r}{12}\right)dr\,d\theta =$

$\int_0^{2\pi}\frac{\sin^2\theta}{4}+\frac{1}{24}\,d\theta = \frac{\pi}{3}$

7. $\int_0^{\pi}\int_0^{\pi}\int_0^{2\sin\phi} \rho^2\sin\phi\,d\rho\,d\phi\,d\theta = \frac{8}{3}\int_0^{\pi}\int_0^{\pi}\sin^4\phi\,d\phi\,d\theta = \frac{2}{3}\int_0^{\pi}\frac{3\pi}{2}\,d\theta = \pi^2$

8. $\int_0^{2\pi}\int_0^{\pi/4}\int_0^{2} \rho\cos\phi\,\rho^2\sin\phi\,d\rho\,d\phi\,d\theta = 2\int_0^{2\pi}\int_0^{2\pi}\sin 2\phi\,d\phi\,d\theta = \int_0^{2\pi}d\theta = 2\pi$

9. $\int_0^{2\pi}\int_0^{\pi}\int_0^{(1-\cos\phi)/2} \rho^2\sin\phi\,d\rho\,d\phi\,d\theta = \frac{1}{24}\int_0^{2\pi}\int_0^{\pi}\left(1-\cos\phi\right)^3\sin\phi\,d\phi\,d\theta =$

$\frac{1}{6}\int_0^{2\pi}d\theta = \frac{\pi}{3}$

10. $\int_0^{3\pi/2}\int_0^{\pi}\int_0^{1} 5\rho^3\sin^3\phi\,d\rho\,d\phi\,d\theta = \frac{5}{4}\int_0^{3\pi/2}\int_0^{\pi}\sin^3\phi\,d\phi\,d\theta = \frac{5}{3}\int_0^{3\pi/2}d\theta = \frac{5\pi}{2}$

11. $\int_0^{2\pi}\int_0^{\pi/3}\int_{\sec\phi}^{2} 3\rho^2\sin\phi\,d\rho\,d\phi\,d\theta = \int_0^{2\pi}\int_0^{\pi/3}\left(8-\sec^3\phi\right)\sin\phi\,d\phi\,d\theta = \frac{5}{2}\int_0^{2\pi}d\theta = 5\pi$

12. $\int_0^{2\pi}\int_0^{\pi/4}\int_0^{\sec\phi} \rho^3\sin\phi\cos\phi\,d\rho\,d\phi\,d\theta = \frac{1}{4}\int_0^{2\pi}\int_0^{\pi/4}\tan\phi\sec^2\phi\,d\phi\,d\theta =$

$\frac{1}{16}\int_0^{2\pi}d\theta = \frac{\pi}{8}$

13. a) $8\int_0^{\pi/2}\int_0^{\pi/2}\int_0^{2} \rho^2\sin\phi\,d\rho\,d\phi\,d\theta$

b) $8\int_0^{\pi/2}\int_0^{2}\int_0^{\sqrt{4-r^2}} r\,dz\,dr\,d\theta$

c) $8\int_0^{2}\int_0^{\sqrt{4-x^2}}\int_0^{\sqrt{4-x^2-y^2}} dz\,dy\,dx$

14. a) $$\int_0^{\pi/2}\int_0^{3/\sqrt{2}}\int_r^{\sqrt{9-r^2}} zr\,dz\,dr\,d\theta$$

b) $$\int_0^{\pi/2}\int_0^{\pi/4}\int_0^{3} \rho^2\sin\phi\,d\rho\,d\phi\,d\theta$$

c) $$\int_0^{\pi/2}\int_0^{\pi/4}\int_0^{3} \rho^2\sin\phi\,d\rho\,d\phi\,d\theta = 9\int_0^{\pi/2}\int_0^{\pi/4}\sin\phi\,d\phi\,d\theta =$$

$$-9\int_0^{\pi/2}\left(\frac{1}{\sqrt{2}}-1\right)d\theta = \frac{9\pi\left(2-\sqrt{2}\right)}{4}$$

15. $$\int_{-\pi/2}^{\pi/2}\int_0^{\cos\theta}\int_0^{3r^2} f(r,\theta,z)\,r\,dz\,dr\,d\theta$$

16. $$\int_{-\pi/2}^{\pi/2}\int_0^{1}\int_0^{r\cos\theta} r^3\,dz\,dr\,d\theta = \int_{-\pi/2}^{\pi/2}\int_0^{1} r^4\cos\theta\,dr\,d\theta = \frac{1}{5}\int_{-\pi/2}^{\pi/2}\cos\theta\,d\theta = \frac{2}{5}$$

17. $$v = 4\int_0^{\pi/2}\int_0^{1}\int_0^{r^2} r\,dz\,dr\,d\theta = 4\int_0^{\pi/2}\int_0^{1} r^3\,dr\,d\theta = \int_0^{\pi/4} d\theta = \frac{\pi}{2}$$

18. $$v = 4\int_0^{\pi/2}\int_0^{1}\int_{x^2+y^2}^{x^2+y^2+1} r\,dz\,dr\,d\theta = 4\int_0^{\pi/2}\int_0^{1} r\,dr\,d\theta = 2\int_0^{\pi/2} d\theta = \pi$$

19. $$V = 4\int_0^{\pi/2}\int_0^{2}\int_0^{4-r^2} r\,dz\,dr\,d\theta = 4\int_0^{\pi/2}\int_0^{2}\left(4r-r^3\right)dr\,d\theta = 16\int_0^{\pi/2} d\theta = 8\pi$$

20. $$V = \int_0^{2\pi}\int_0^{2}\int_0^{4-r\sin\theta} r\,dz\,dr\,d\theta = \int_0^{2\pi}\int_0^{2}\left(4-r\sin\theta\right)r\,dr\,d\theta = 8\int_0^{2\pi} 1-\frac{\sin\theta}{3}\,d\theta = 16\pi$$

21. $$V = 4\int_0^{\pi/2}\int_0^{1}\int_{4r^2}^{5-r^2} r\,dz\,dr\,d\theta = 4\int_0^{\pi/2}\int_0^{1}\left(5-5r^2\right)r\,dr\,d\theta = 5\int_0^{\pi/2} d\theta = \frac{5\pi}{2}$$

22. $$V = 4\int_0^{\pi/2}\int_1^{3}\int_0^{9-x^2-y^2} r\,dz\,dr\,d\theta = 4\int_0^{\pi/2}\int_1^{3} 9r-r^3\,dr\,d\theta = 64\int_0^{\pi/2} d\theta = 32\pi$$

23. $$V = 8\int_0^{\pi/2}\int_0^{1}\int_0^{\sqrt{4-r^2}} r\,dz\,dr\,d\theta = 8\int_0^{\pi/2}\int_0^{1}\left(4-r^2\right)^{1/2} r\,dr\,d\theta =$$

$$-\frac{8}{3}\int_0^{\pi/2}\left(3^{3/2}-8\right)d\theta = \frac{4\pi\left(8-3\sqrt{3}\right)}{3}$$

24. $$V = 4\int_0^{\pi/2}\int_0^{1}\int_{x^2+y^2}^{\sqrt{2-\left(x^2+y^2\right)}} r\,dz\,dr\,d\theta = 4\int_0^{\pi/2}\int_0^{1}\left(2-r^2\right)^{1/2}(r)-r^3\,dr\,d\theta =$$

$$4\int_0^{\pi/2}\left(\frac{2\sqrt{2}}{3}-\frac{7}{12}\right)d\theta = \frac{\pi\left(8\sqrt{3}-7\right)}{6}$$

25. $$\text{average} = \frac{1}{2\pi}\int_0^{2\pi}\int_0^{1}\int_{-1}^{1} r\,dz\,dr\,d\theta = \frac{1}{2\pi}\int_0^{2\pi}\int_0^{1} 2r^2\,dr\,d\theta = \frac{1}{3\pi}\int_0^{2\pi} d\theta = \frac{2}{3}$$

26. $\text{average} = \frac{1}{4\pi/3}\int_0^{2\pi}\int_0^1\int_{-\sqrt{1-r^2}}^{\sqrt{1-r^2}} r^2\,dz\,dr\,d\theta = \frac{3}{4\pi}\int_0^{2\pi}\int_0^1 \left(1-r^2\right)^{1/2}2r^2\,dr\,d\theta = \frac{3}{32}\int_0^{2\pi} d\theta = \frac{3\pi}{16}$

27. $M = 4\int_0^{\pi/2}\int_0^1\int_0^r r\,dz\,dr\,d\theta = 4\int_0^{\pi/2}\int_0^1 r^2\,dr\,d\theta = \frac{4}{3}\int_0^{\pi/2} d\theta = \frac{2\pi}{3}$

$M_{xy} = \int_0^{2\pi}\int_0^1\int_0^r r\,z\,dz\,dr\,d\theta = \frac{1}{2}\int_0^{2\pi}\int_0^1 r^3\,dr\,d\theta = \frac{1}{8}\int_0^{2\pi} d\theta = \frac{\pi}{4} \Rightarrow$

$\overline{z} = \frac{3}{8}$, $\overline{x} = \overline{y} = 0$ by symmetry

28. $M = 4\int_0^{\pi/2}\int_0^2\int_{\sqrt{r}}^2 r\,dz\,dr\,d\theta = 4\int_0^{\pi/2}\int_0^2 \left(2-\sqrt{r}\right) r\,dr\,d\theta = 4\int_0^{\pi/2} 4 - \frac{2^{7/2}}{5}\,d\theta = \frac{8\pi}{5}\left(5-2\sqrt{2}\right)$

$M_{xy} = 4\int_0^{\pi/2}\int_0^2\int_{\sqrt{r}}^2 r\,z\,dz\,dr\,d\theta = 2\int_0^{\pi/2}\int_0^2 4r - r^2\,dr\,d\theta = 2\int_0^{\pi/2}\frac{16}{3}\,d\theta = \frac{16\pi}{3} \Rightarrow$

$\overline{z} = \frac{10}{3\left(5-2\sqrt{2}\right)}$, $\overline{x} = \overline{y} = 0$ by symmetry

29. $M = 12\pi$, $I_z = \int_0^{2\pi}\int_1^2\int_0^4 r^3\,dz\,dr\,d\theta = 4\int_0^{2\pi}\int_1^2 r^3\,dr\,d\theta = 15\int_0^{2\pi} d\theta = 30\pi \Rightarrow$

$R_z = \sqrt{\frac{I_z}{M}} = \sqrt{\frac{5}{2}}$

30. $M = \int_0^{\pi/2}\int_0^2\int_0^{\sqrt{x^2+y^2}} r\,dz\,dr\,d\theta = \int_0^{\pi/2}\int_0^2 r^{3/2}\,dr\,d\theta = \frac{2}{5}\int_0^{\pi/2} 2^{5/2}\,d\theta = \frac{4\sqrt{2}\pi}{5}$

$M_{yz} = \int_0^{\pi/2}\int_0^2\int_0^{\sqrt{x^2+y^2}} r^2\cos\theta\,dz\,dr\,d\theta = \int_0^{\pi/2}\int_0^2 r^{5/2}\cos\theta\,dr\,d\theta = \frac{2^{9/2}}{7}\int_0^{\pi/2}\cos\theta\,d\theta =$

$\frac{16\sqrt{2}}{7} \Rightarrow \overline{x} = \overline{y} = \frac{20}{7\pi}$ by symmetry

$M_{xy} = = \int_0^{\pi/2}\int_0^2\int_0^{\sqrt{x^2+y^2}} r\,z\,dz\,dr\,d\theta = \frac{1}{2}\int_0^{\pi/2}\int_0^2 r^2\,dr\,d\theta = \frac{4}{3}\int_0^{\pi/2} d\theta = \frac{2\pi}{3} \Rightarrow \overline{z} = \frac{5}{6\sqrt{2}}$

31. a) $I_z = \int_0^{2\pi}\int_0^1\int_{-1}^1 r^2\,dz\,dr\,d\theta = 2\int_0^{2\pi}\int_0^1 r^3\,dr\,d\theta = \frac{1}{2}\int_0^{2\pi} d\theta = \pi$

b) $I_x = \int_0^{2\pi}\int_0^1\int_{-1}^1 \left(r^2\sin^2\theta + z^2\right)\,dz\,dr\,d\theta = \int_0^{2\pi}\int_0^1 \left(2r^2\sin^2\theta + \frac{2}{3}\right) r\,dr\,d\theta =$

$\int_0^{2\pi}\left(\frac{\sin^2\theta}{2} + \frac{1}{3}\right)d\theta = \frac{7\pi}{6}$

32. We orient the cone with its vertex at the origin and note that $\phi = \frac{\pi}{4}$.

$I_x = \int_0^{2\pi}\int_0^1\int_r^1 \left(r^2\sin^2\theta + z^2\right) r\,dz\,dr\,d\theta = \int_0^{2\pi}\int_0^1 r^3\sin^2\theta + \frac{r}{3} - r^4\sin^2\theta - \frac{r^4}{3}\,dr\,d\theta =$
$\int_0^{2\pi} \frac{\sin^2\theta}{20} + \frac{1}{10}\,d\theta = \frac{\pi}{4}$

33. $I_z = \int_0^{2\pi}\int_0^a\int_{-\sqrt{a^2-r^2}}^{\sqrt{a^2-r^2}} r^3\,dz\,dr\,d\theta = -\int_0^{2\pi}\int_0^a r^2\left(a^2-r^2\right)^{1/2}(-2r)\,dr\,d\theta = \int_0^{2\pi}\frac{4a^5}{15}\,d\theta = \frac{8\pi a^5}{15}$

34. $M = \frac{4\pi}{9}$, $M_{yz} = \int_{-\pi/3}^{\pi/3}\int_0^1\int_{-\sqrt{1-r^2}}^{\sqrt{1-r^2}} \left(r^2\cos\theta\right)\,dz\,dr\,d\theta = 2\int_{-\pi/3}^{\pi/3}\int_0^1 \left(r^2\cos\theta\right)\sqrt{1-r^2}\,dr\,d\theta =$
$\frac{\pi}{8}\int_{-\pi/3}^{\pi/3}\cos\theta\,d\theta = \frac{\pi\sqrt{3}}{8} \Rightarrow \overline{x} = \frac{9\sqrt{3}}{32}$, $\overline{x} = \overline{y} = 0$ by symmetry

35. $\int_0^{2\pi}\int_{\pi/3}^{2\pi/3}\frac{a^3}{3}\sin\phi\,d\phi\,d\theta = \frac{a^3}{3}\int_0^{2\pi}d\theta = \frac{2\pi a^3}{3} \Rightarrow V = \frac{4}{3}\pi a^3 - \frac{2\pi a^3}{3} = \frac{2\pi a^3}{3}$

36. $V = \int_0^{\pi/6}\int_0^{\pi/2}\int_0^a \rho^2\sin\phi\,d\rho\,d\phi\,d\theta = \frac{a^3}{3}\int_0^{\pi/6}\int_0^{\pi/2}\sin\phi\,d\phi\,d\theta = \frac{a^3}{3}\int_0^{\pi/6}d\theta = \frac{a^3\pi}{18}$

37. $V = \int_0^{2\pi}\int_0^{\pi/3}\int_{\sec\phi}^2 \rho^2\sin\phi\,d\rho\,d\phi\,d\theta = \frac{1}{3}\int_0^{2\pi}\int_0^{\pi/3} 8\sin\phi - \tan\phi\sec^2\phi\,d\phi\,d\theta =$
$\frac{5}{6}\int_0^{2\pi}d\theta = \frac{5\pi}{3}$

38. $V = 4\int_0^{\pi/2}\int_0^{\pi}\int_0^{1-\cos\phi} \rho^2\sin\phi\,d\rho\,d\phi\,d\theta = \frac{4}{3}\int_0^{\pi/2}\int_0^{\pi}\left(1-\cos\phi\right)^3\sin\phi\,d\phi\,d\theta =$
$\frac{16}{3}\int_0^{\pi/2}d\theta = \frac{8\pi}{3}$

39. $\text{average} = \frac{3}{4\pi}\int_0^{2\pi}\int_0^{\pi}\int_0^1 \rho^3\sin\phi\,d\rho\,d\phi\,d\theta = \frac{3}{16\pi}\int_0^{2\pi}\int_0^{\pi}\sin\phi\,d\phi\,d\theta = \frac{3}{8\pi}\int_0^{2\pi}d\theta = \frac{3}{4}$

40. $M = \int_0^{2\pi}\int_{\pi/3}^{\pi/2}\int_0^2 \rho^2\sin\phi\,d\rho\,d\phi\,d\theta = \frac{8}{3}\int_0^{2\pi}\int_{\pi/3}^{\pi/2}\sin\phi\,d\phi\,d\theta = \frac{4}{3}\int_0^{2\pi}d\theta = \frac{8\pi}{3}$

$M_{xy} = \int_0^{2\pi}\int_{\pi/3}^{\pi/2}\int_0^2 \rho^3\sin\phi\cos\phi\,d\rho\,d\phi\,d\theta = 2\int_0^{2\pi}\int_{\pi/3}^{\pi/2}\sin 2\phi\,d\phi\,d\theta = \frac{1}{2}\int_0^{2\pi}d\theta = \pi \Rightarrow$
$\overline{z} = \frac{3}{8}$, $\overline{x} = \overline{y} = 0$ by symmetry

41. $M = \int_0^{2\pi}\int_0^{\pi/4}\int_0^a \rho^2 \sin\phi \, d\rho \, d\phi \, d\theta = \frac{a^3}{3}\int_0^{2\pi}\int_0^{\pi/4} \sin\phi \, d\phi \, d\theta =$

$\frac{a^3}{3}\int_0^{2\pi} \frac{\sqrt{2}-1}{\sqrt{2}} \, d\theta = \frac{\pi a^3\left(2-\sqrt{2}\right)}{3}$

$M_{xy} = \int_0^{2\pi}\int_0^{\pi/4}\int_0^a \rho^3 \sin\phi \cos\phi \, d\rho \, d\phi \, d\theta = \frac{a^3}{4}\int_0^{2\pi}\int_0^{\pi/4} \sin\phi \cos\phi \, d\phi \, d\theta =$

$\frac{a^4}{16}\int_0^{2\pi} d\theta = \frac{\pi a^4}{8} \Rightarrow \overline{z} = \frac{3\left(2+\sqrt{2}\right)a}{16}$, $\overline{x} = \overline{y} = 0$ by symmetry

42. a) $M = \int_0^{2\pi}\int_0^{\pi}\int_0^a \rho^4 \sin\phi \, d\rho \, d\phi \, d\theta = \frac{a^5}{5}\int_0^{2\pi}\int_0^{\pi} \sin\phi \, d\phi \, d\theta = \frac{2a^5}{5}\int_0^{2\pi} d\theta = \frac{4\pi a^5}{5}$

$I_z = \int_0^{2\pi}\int_0^{\pi}\int_0^a \rho^6 \sin^3\phi \, d\rho \, d\phi \, d\theta = \frac{a^7}{7}\int_0^{2\pi}\int_0^{\pi} \left(1-\cos^2\phi\right)\sin\phi \, d\phi \, d\theta = \frac{4a^7}{21}\int_0^{2\pi} d\theta = \frac{8a^7\pi}{21} \Rightarrow$

$R_z = \sqrt{\frac{I_z}{M}} = \sqrt{\frac{10}{21}}\, a$

b) $M = \int_0^{2\pi}\int_0^{\pi}\int_0^a \rho^3 \sin^2\phi \, d\rho \, d\phi \, d\theta = \frac{a^4}{4}\int_0^{2\pi}\int_0^{\pi} \frac{(1-\cos 2\phi)}{2} \, d\phi \, d\theta =$

$\frac{\pi a^4}{8}\int_0^{2\pi} d\theta = \frac{\pi^2 a^4}{4}$; $I_z = \int_0^{2\pi}\int_0^{\pi}\int_0^a \rho^5 \sin^4\phi \, d\rho \, d\phi \, d\theta = \frac{a^6}{6}\int_0^{2\pi}\int_0^{\pi} \sin^4\phi \, d\phi \, d\theta =$

$\frac{a^6}{16}\int_0^{2\pi} \pi \, d\theta = \frac{a^6\pi^2}{8} \Rightarrow R_z = \sqrt{\frac{I_z}{M}} = \frac{a}{\sqrt{2}}$

14.7 SUBSTITUTIONS IN MULTIPLE INTEGRALS

1. $\int_0^4\int_{y/2}^{1+y/2} \frac{2x-y}{2} \, dx \, dy = \int_0^4 \left[\frac{x^2}{2} - \frac{xy}{2}\right]_{y/2}^{1+y/2} dy = \frac{1}{2}\int_0^4 dy = 2$

2. a) $\begin{vmatrix} \cos v & -u\sin v \\ \sin v & u\cos v \end{vmatrix} = u\cos^2 v + u\sin^2 v = u$

b) $\begin{vmatrix} \sin v & u\cos v \\ \cos v & -u\sin v \end{vmatrix} = -u\sin^2 v - u\cos^2 v = -u$

3. a) $x = \frac{u+v}{3}$, $y = \frac{v-2u}{3}$, $J(u,v) = \begin{vmatrix} 1/3 & 1/3 \\ -2/3 & 1/3 \end{vmatrix} = \frac{1}{9} + \frac{2}{9} = \frac{1}{3}$

b) $\int_4^7 \int_{-1}^2 \frac{vu}{3}\, dv\, du = \frac{1}{2}\int_4^7 u\, du = \frac{33}{4}$

4. a) $J(u,v) = \begin{vmatrix} 1 & 0 \\ v & u \end{vmatrix} = u$

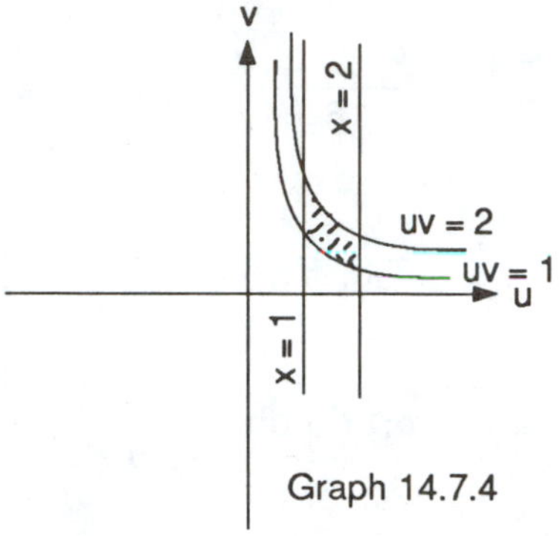

Graph 14.7.4

b) $\int_1^2 \int_{1/u}^{2/u} \left(\frac{uv}{u}\right) u\, dv\, du = \frac{3}{2}\int_a^b \frac{1}{u}\, du = \frac{3 \ln 2}{2}$; $\int_1^2 \int_1^2 \frac{y}{x}\, dy\, dx = \frac{3}{2}\int_1^2 \frac{1}{x}\, dx = \frac{3 \ln 2}{2}$

5. $J(u,v) = \begin{vmatrix} v^{-1} & -uv^{-2} \\ v & u \end{vmatrix} = v^{-1}u + v^{-1}u = \frac{2u}{v}$; $\int_1^1 \int_1^2 (v+u)\left(\frac{2u}{v}\right) dv\, du =$

$\int_1^3 2u + (\ln 4)\, u^2\, du = 8 + \frac{26 \ln 4}{3}$

6. $J(u,v) = \begin{vmatrix} a & 0 \\ 0 & b \end{vmatrix} = ab$; $A = \int_{-a}^{a} \int_{(-b/a)\sqrt{a^2-x^2}}^{(b/a)\sqrt{a^2-x^2}} dy\, dx = \int_{-1}^{1} \int_{-\sqrt{1-u^2}}^{\sqrt{1-u^2}} ab\, du\, dv =$

$2ab\int_{-1}^{1} \sqrt{1-u^2}\, du = ab\pi$

7. $J(r,\theta) = \begin{vmatrix} a\cos\theta & -ar\sin\theta \\ b\sin\theta & br\cos\theta \end{vmatrix} = abr\cos^2\theta + abr\sin^2\theta = abr$

$I_o = \int_{-a}^{a} \int_{(-b/a)\sqrt{a^2-x^2}}^{(b/a)\sqrt{a^2-x^2}} \left(x^2 + y^2\right) dy\, dx = \int_0^{2\pi} \int_0^1 r^3\left(a^2\cos^2\theta + b^2\sin^2\theta\right) abr\, dr\, d\theta =$

$\frac{ab}{4}\int_a^b a^2\cos^2\theta + b^2\sin^2\theta\, d\theta = \frac{\pi ab\left(a^2 + b^2\right)}{4}$

8. $J(u,v) = \begin{vmatrix} 1 & 1/2 \\ 0 & 1 \end{vmatrix} = 1; \int_0^2 \int_{y/2}^{(y+4)/2} y^3(2x-y)\exp\left((2x-y)^2\right) dx\,dy =$

$\int_0^2 \int_0^2 v^3(2u)\exp\left((2u)^2\right) dv\,du = 8\int_0^2 \exp\left(4u^2\right) u\,du = e^{16} - 1$

9. $\begin{vmatrix} \sin\phi\cos\theta & \rho\cos\phi\cos\theta & -\rho\sin\phi\sin\theta \\ \sin\phi\sin\theta & \rho\cos\phi\sin\theta & \rho\sin\phi\cos\theta \\ \cos\phi & -\rho\sin\phi & 0 \end{vmatrix} = \cos\phi \begin{vmatrix} \rho\cos\phi\cos\theta & -\rho\sin\phi\sin\theta \\ \rho\cos\phi\sin\theta & \rho\sin\phi\cos\theta \end{vmatrix} +$

$\rho\sin\phi \begin{vmatrix} \sin\phi\cos\theta & -\rho\sin\phi\sin\theta \\ \sin\phi\sin\theta & \rho\sin\phi\cos\theta \end{vmatrix} = \rho^2\cos\phi\left(\sin\phi\cos\phi\cos^2\theta + \sin\phi\cos\phi\sin^2\theta\right) +$

$\rho^2\sin\phi\left(\sin^2\phi\cos^2\theta + \sin^2\phi\sin^2\theta\right) = \rho^2\sin\phi\cos^2\phi + \rho^2\sin^3\phi =$

$\rho^2\sin\phi\left(\cos^2\phi + \sin^2\phi\right) = \rho^2\sin\phi$

10. $\int_0^3 \int_0^4 \int_{y/2}^{1+y/2} \left(\frac{2x-y}{2} + \frac{z}{3}\right) dx\,dy\,dz = \int_0^3 \int_0^4 \frac{1}{2}(y+1) + \frac{z}{3} - \frac{y}{2}\,dy\,dz = 4\int_0^3 \frac{1}{2} + \frac{z}{3}\,dz = 12$

11. $J(u,v,w) = \begin{vmatrix} a & 0 & 0 \\ 0 & b & 0 \\ 0 & 0 & c \end{vmatrix} = abc$. The transformation takes the $\frac{x^2}{a^2} + \frac{y^2}{b^2} + \frac{z^2}{c^2} = 1$ region in the xyz–space into the $u^2 + v^2 + w^2 = 1$ region in the uvw–space which is a unit sphere with $V = \frac{4}{3}\pi$.

$\therefore\ V = \int\int_R\int dx\,dy\,dz = \int\int_G\int abc\,du\,dv\,dw = \frac{4\pi abc}{3}$

12. $J(u,v,w) = \begin{vmatrix} a & 0 & 0 \\ 0 & b & 0 \\ 0 & 0 & c \end{vmatrix} = abc; \int\int\int |xyz|\,dx\,dy\,dz = 8\int_0^1 \int_0^1 \int_0^1 a^2b^2c^2uvw\,dw\,dv\,du =$

$4a^2b^2c^2 \int_0^1 \int_0^1 uv\,du\,dv = 2a^2b^2c^2\int_0^1 u\,du = a^2b^2c^2$

13. $J(u,v,w) = \begin{vmatrix} 1 & 0 & 0 \\ -v/u^2 & 1/u & 0 \\ 0 & 0 & 1/3 \end{vmatrix} = \frac{1}{3u}; \int\int_R\int x^2y + 3xyz\,dx\,dy\,dz =$

$\int\int_G\int u^2\left(\frac{v}{u}\right) + 3u\frac{v}{u}\frac{w}{3} J(u,v,w)\,du\,dv\,dw = \frac{1}{3}\int_0^3 \int_0^2 \int_1^2 v + \frac{vw}{u}\,du\,dv\,dw =$

$\frac{1}{3}\int_0^3 \int_0^2 v + vw\ln 2\,dv\,dw = \frac{1}{3}\int_0^3 2 + (\ln 4)w\,dw = 2 + \ln 8$

PRACTICE EXERCISES

1. $$\int_1^{10}\int_0^{1/y} ye^{xy}\,dy\,dx = \int_1^{10}\left[e^{xy}\right]_0^{1/y} dx = \int_1^{10}(e-1)\,dx = 9e-9$$

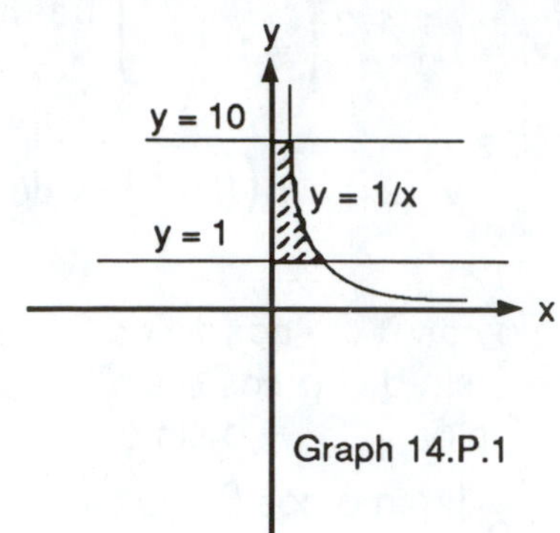

Graph 14.P.1

2. $$\int_0^1\int_0^{x^3} e^{y/x}\,dy\,dx = \int_0^1 x\left[e^{y/x}\right]_0^{x^3} dx = \int_0^1 x\exp\left(x^2\right) - x\,dx = \frac{e-2}{2}$$

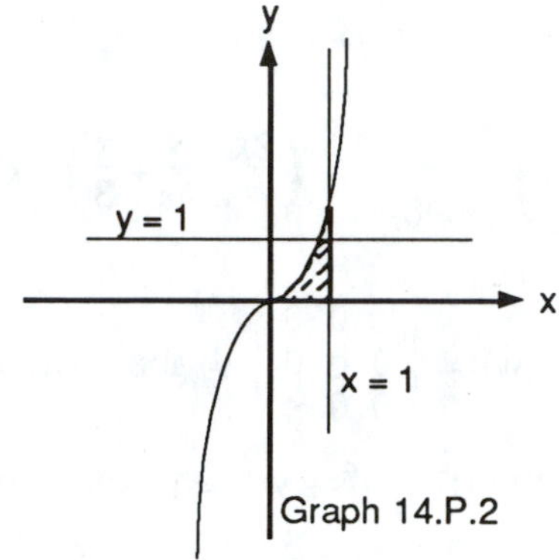

Graph 14.P.2

3. 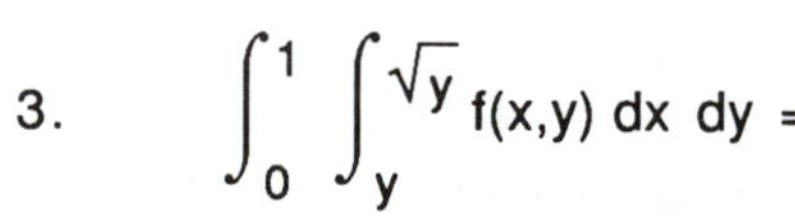
$$\int_0^1\int_y^{\sqrt{y}} f(x,y)\,dx\,dy =$$

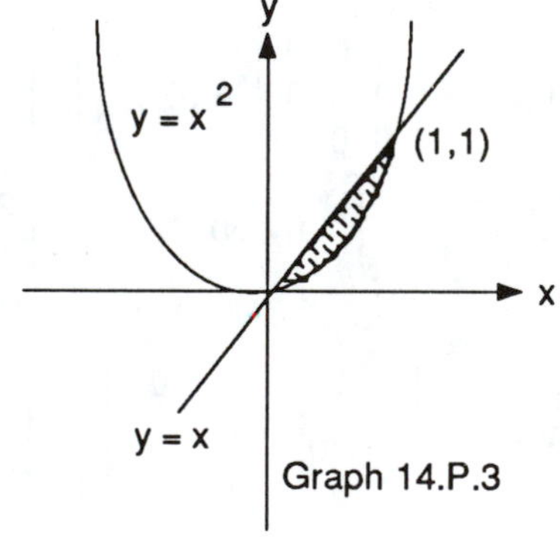

Graph 14.P.3

4. $$\int_0^4\int_{x^2/4}^{x} f(x,y)\,dy\,dx =$$

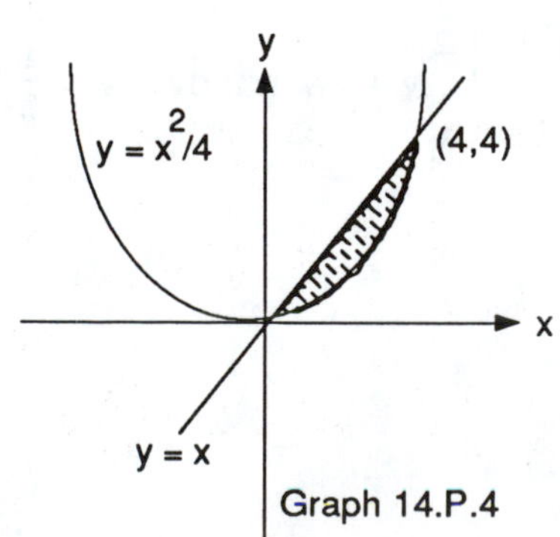

Graph 14.P.4

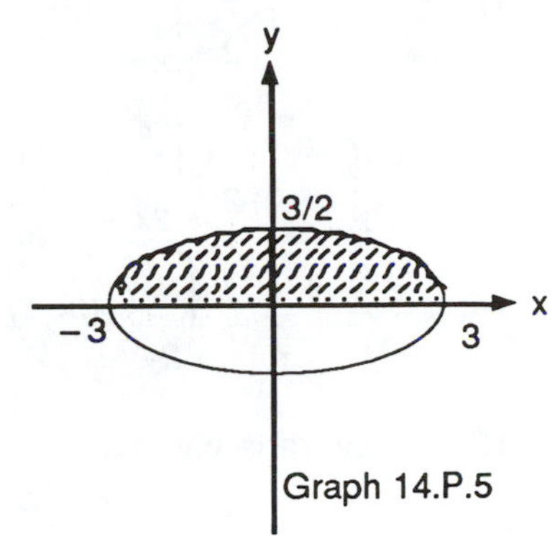

Graph 14.P.5

5. a) $\int_0^{3/2} \int_{-\sqrt{9-4y^2}}^{\sqrt{9-4y^2}} y\,dx\,dy = \int_0^{3/2} \left[yx \right]_{-\sqrt{9-4y^2}}^{\sqrt{9-4y^2}} dy =$

$\int_0^{3/2} 2y\sqrt{9-4y^2}\,dy = \frac{9}{2}$

b) $\int_{-3}^{3} \int_0^{(9-x^2)^{1/2}/2} y\,dy\,dx = \frac{1}{2}\int_{-3}^{3} \left[y^2\right]_0^{(9-x^2)^{1/2}/2} dx =$

$\frac{1}{8}\int_{-3}^{3} \left(9-x^2\right) dx = \frac{9}{2}$

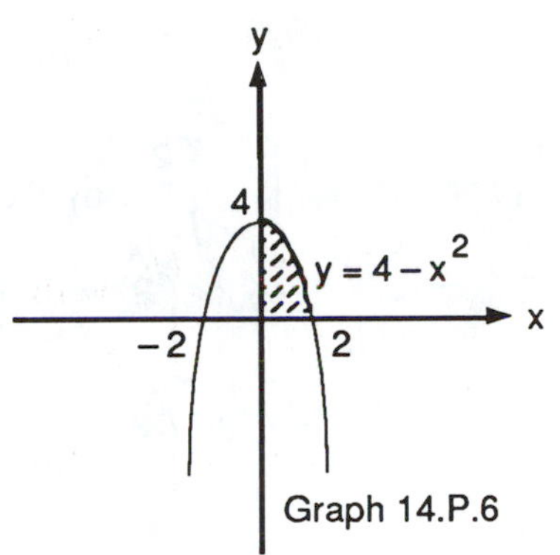

Graph 14.P.6

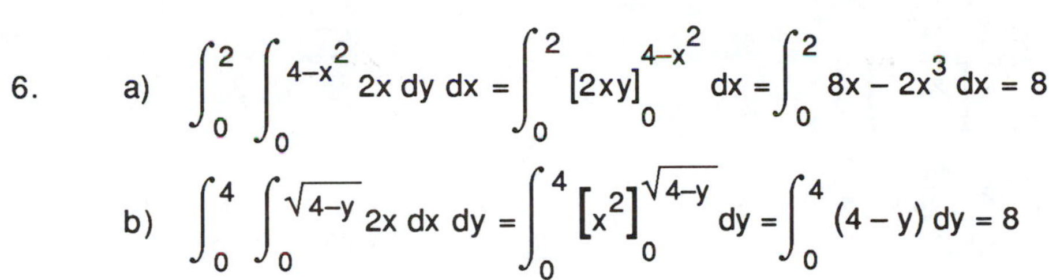

6. a) $\int_0^2 \int_0^{4-x^2} 2x\,dy\,dx = \int_0^2 \left[2xy\right]_0^{4-x^2} dx = \int_0^2 8x - 2x^3\,dx = 8$

b) $\int_0^4 \int_0^{\sqrt{4-y}} 2x\,dx\,dy = \int_0^4 \left[x^2\right]_0^{\sqrt{4-y}} dy = \int_0^4 (4-y)\,dy = 8$

7. $\int_0^1 \int_{2y}^2 4\cos x^2\,dx\,dy = \int_0^2 \int_0^{x/2} 4\cos x^2\,dy\,dx = \int_0^2 \left[\left(4\cos x^2\right)y\right]_0^{x/2} dx =$

$\int_0^2 \left(\cos x^2\right)(2x)\,dx = \sin 4$

8. $\int_0^2 \int_{y/2}^1 \exp\left(x^2\right) dx\,dy = \int_0^1 \int_0^{2x} \exp\left(x^2\right) dy\,dx = \int_0^1 \exp\left(x^2\right)(2x)\,dx = e - 1$

9. $\int_0^8 \int_{\sqrt[3]{x}}^2 \frac{1}{y^4+1}\,dy\,dx = \int_0^2 \int_0^{y^3} \frac{1}{y^4+1}\,dx\,dy = \frac{1}{4}\int_0^2 \frac{4y^3}{y^4+1}\,dy = \frac{\ln 17}{4}$

10. $\int_0^1 \int_{\sqrt[3]{y}}^1 \frac{2\pi \sin \pi x^2}{x^2}\,dx\,dy = \int_0^1 \int_0^{x^3} \left(2\pi \sin \pi x^2\right)/x^2\,dy\,dx = \int_0^1 \left(\frac{2\pi \sin \pi x^2}{x^2}\right) x^3\,dx =$

$\int_0^1 \left(\sin \pi x^2\right)(2\pi x)\,dx = 2$

11. $V = \int_0^1 \int_x^{2-x} x^2 + y^2\,dy\,dx = \int_0^1 \left[x^2 y + \frac{x^3}{3}\right]_x^{2-x} dx =$

$\int_0^1 \left(-\frac{8}{3}x^3 + 4x^2 - 4x + \frac{8}{3}\right) dx = \frac{4}{3}$

12. $V = \int_{-3}^{2} \int_{x}^{6-x^2} x^2\, dy\, dx = \int_{-3}^{2} \left[x^2 y\right]_{x}^{6-x^2} dx = \int_{-3}^{2} 6x^2 - x^4 - x^3\, dx = \frac{125}{4}$

13. $A = \int_{-2}^{0} \int_{2x+4}^{4-x^2} dy\, dx = \int_{-2}^{0} -x^2 - 2x\, dx = \frac{4}{3}$

14. $A = \int_{1}^{4} \int_{2-y}^{\sqrt{y}} dx\, dy = \int_{1}^{4} \sqrt{y} - 2 + y\, dy = \frac{13}{6}$

15. average value $= \int_{0}^{1} \int_{0}^{1} xy\, dy\, dx = \int_{0}^{1} \left[\frac{xy^2}{2}\right]_{0}^{1} dx = \int_{0}^{1} \frac{x}{2}\, dx = \frac{1}{4}$

16. average value $= \frac{1}{\pi/4} \int_{0}^{1} \int_{0}^{\sqrt{1-x^2}} xy\, dy\, dx = \frac{4}{\pi} \int_{0}^{1} \left[\frac{xy^2}{2}\right]_{0}^{\sqrt{1-x^2}} dx = \frac{2}{\pi} \int_{a}^{b} x - x^3\, dx = \frac{1}{2\pi}$

17. $M = \int_{1}^{2} \int_{2/x}^{2} dy\, dx = \int_{1}^{2} 2 - \frac{2}{x}\, dx = 2 - \ln 4$

$M_y = \int_{1}^{2} \int_{2/x}^{2} x\, dy\, dx = \int_{1}^{2} x\left[2 - \frac{2}{x}\right] dx = 1$

$M_x = \int_{1}^{2} \int_{2/x}^{2} y\, dy\, dx = \int_{1}^{2} 2 - \frac{2}{x^2}\, dx = 1 \Rightarrow \overline{x} = \frac{1}{2 - \ln 4}, \overline{y} = \frac{1}{2 - \ln 4}$

18. $I_o = \int_{0}^{2} \int_{2x}^{4} \left(x^2 + y^2\right)(3)\, dy\, dx = 3\int_{0}^{2} 4x^2 - \frac{14x^3}{3} + \frac{64}{3}\, dx = 104$

19. a) $I_o = \int_{-2}^{2} \int_{-1}^{1} \left(x^2 + y^2\right) dy\, dx = \int_{-2}^{2} 2x^2 + \frac{2}{3}\, dx = \frac{40}{3}$

b) $I_x = \int_{-a}^{a} \int_{-b}^{b} y^2\, dy\, dx = \int_{-a}^{a} \frac{2b^3}{3}\, dx = \frac{4ab^3}{3}, I_y = \int_{-b}^{b} \int_{-a}^{a} x^2\, dx\, dy =$

$\int_{-b}^{b} \frac{2a^3}{3}\, dy = \frac{4a^3 b}{3} \Rightarrow I_o = I_x + I_y = \frac{4ab^3}{3} + \frac{4a^3 b}{3} = \frac{4ab\left(b^2 + a^2\right)}{3}$

20. $M = \int_{0}^{3} \int_{0}^{2x/3} dy\, dx = \int_{0}^{3} \frac{2x}{3}\, dx = 3; I_x = \int_{0}^{3} \int_{0}^{2x/3} y^2\, dy\, dx = \frac{8}{81}\int_{0}^{3} x^3\, dx = 2 \Rightarrow R_x = \sqrt{\frac{2}{3}}$

21. $M = \int_{-1}^{1} \int_{-1}^{1} \left(x^2 + y^2 + \frac{1}{3}\right) dy\, dx = \int_{-1}^{1} 2x^2 + \frac{4}{3}\, dx = 4$

$M_x = \int_{-1}^{1} \int_{-1}^{1} y\left(x^2 + y^2 + \frac{1}{3}\right) dy\, dx = \int_{-1}^{1} 0\, dx = 0$

$M_y = \int_{-1}^{1} \int_{-1}^{1} x\left(x^2 + y^2 + \frac{1}{3}\right) dy\, dx = \int_{-1}^{1} x\left(2x^2 + \frac{4}{3}\right) dx = 0$

22. $M = \int_0^1 \int_{x^2}^{x} (x + 1)\, dy\, dx = \int_0^1 \left(x - x^3\right) dx = \frac{1}{4}$

$M_x = \int_0^1 \int_{x^2}^{x} y(x + 1)\, dy\, dx = \frac{1}{2}\int_0^1 \left(x^3 - x^5 + x^2 - x^4\right) dx = \frac{13}{120}$

$M_y = \int_0^1 \int_{x^2}^{x} x(x + 1)\, dy\, dx = \int_0^1 x\left(x - x^3\right) dx = \frac{2}{15} \Rightarrow \overline{x} = \frac{8}{15},\ \overline{y} = \frac{13}{30}$

$I_x = \int_0^1 \int_{x^2}^{x} y^2(x + 1)\, dy\, dx = \frac{1}{3}\int_0^1 \left(x^4 - x^7 + x^3 - x^6\right) dx = \frac{17}{280} \Rightarrow R_x = \sqrt{\frac{I_x}{M}} = \sqrt{\frac{17}{70}}$

$I_y = \int_0^1 \int_{x^2}^{x} x^2(x + 1)\, dy\, dx = \int_0^1 x^2\left(x - x^3\right) dx = \frac{1}{12} \Rightarrow R_y = \sqrt{\frac{I_y}{M}} = \sqrt{\frac{1}{3}}$

23. $\int_{-1}^{1} \int_{-\sqrt{1-x^2}}^{\sqrt{1-x^2}} \frac{2}{\left(1 + x^2 + y^2\right)^2} dy\, dx = \int_0^{2\pi} \int_0^1 \frac{2r}{\left(1 + r^2\right)^2} dr\, d\theta = \frac{1}{2}\int_0^{2\pi} d\theta = \pi$

24. $\int_{-1}^{1} \int_{-\sqrt{1-y^2}}^{\sqrt{1-y^2}} \ln\left(x^2 + y^2 + 1\right) dx\, dy = \int_0^{2\pi} \int_0^1 \ln\left(r^2 + 1\right) r\, dr\, d\theta = \frac{1}{2}\int_0^{2\pi} \ln(4) - 1\, d\theta =$
$\left(\ln(4) - 1\right)\pi$

25. $M = 2\int_0^{\pi/2} \int_1^{1+\cos\theta} r\, dr\, d\theta = \int_0^{\pi/2} 2\cos\theta + \frac{1 + \cos 2\theta}{2} d\theta = \frac{8 + \pi}{4}$

$M_y = \int_{-\pi/2}^{\pi/2} \int_1^{1+\cos\theta} \cos\theta\, r^2\, dr\, d\theta = \int_{-\pi/2}^{\pi/2} \left(\cos^2\theta + \cos^3\theta + \frac{\cos^4\theta}{3}\right) d\theta = \frac{32 + 15\pi}{24} \Rightarrow$
$\overline{x} = \frac{15\pi + 32}{6\pi + 48},\ \overline{y} = 0$ by symmetry

26. $M = \int_{-\pi/3}^{\pi/3} \int_0^3 r\, dr\, d\theta = \frac{9}{2}\int_{-\pi/3}^{\pi/3} d\theta = 3\pi$

$M_y = \int_{-\pi/3}^{\pi/3} \int_0^3 r^2 \cos\theta\, dr\, d\theta = 9\int_{-\pi/3}^{\pi/3} \cos\theta\, d\theta = 9\sqrt{3} \Rightarrow \overline{x} = \frac{3\sqrt{3}}{\pi},\ \overline{y} = 0$ by symmetry

27. $M = \int_0^{\pi/2} \int_1^3 r\, dr\, d\theta = 4\int_0^{\pi/2} d\theta = 2\pi$

$M_y = \int_0^{\pi/2} \int_1^3 r^2 \cos\theta\, dr\, d\theta = \frac{26}{3}\int_0^{\pi/2} \cos\theta\, d\theta = \frac{26}{3} \Rightarrow \overline{x} = \frac{13}{3\pi},\ \overline{y} = \frac{13}{3\pi}$ by symmetry

28. $I^2 = \left(\int_0^{\infty} \exp\left(-x^2\right) dx\right)\left(\int_0^{\infty} \exp\left(-y^2\right) dy\right) = \int_0^{\infty}\int_0^{\infty} \exp\left[-\left(x^2 + y^2\right)\right] dx\, dy =$
$\int_0^{\pi/2} \int_0^{\infty} \exp\left(-r^2\right) r\, dr\, d\theta = \int_0^{\pi/2} \left(\lim_{t \to \infty} \left[-\frac{1}{2}\exp\left(-r^2\right)\right]_0^t\right) d\theta = \frac{1}{2}\int_0^{\pi/2} d\theta = \frac{\pi}{4} \Rightarrow$

$I = \int_0^{\infty} \exp\left(-x^2\right) dx = \frac{\sqrt{\pi}}{2}$

29. $\int_0^{\pi}\int_0^{\pi}\int_0^{\pi}\cos(x+y+z)\,dx\,dy\,dz = \int_0^{\pi}\int_0^{\pi}\sin(z+y+\pi)-\sin(z+y)\,dy\,dz = 0$

$\int_0^{\pi}\left(-\cos(z+2\pi)+\cos(z+\pi)+\cos(z)-\cos(z+\pi)\right)dz =$

30. $\int_0^1\int_0^{1-x^2}\int_{4x^2}^{5-x^2}(x-y+1)\,dz\,dy\,dx = 5\int_0^1\int_0^{1-x^2}(x-y+1)\left(1-x^2\right)dy\,dx =$

$\frac{5}{2}\int_0^1\left(1-x^2\right)^2\left(1+2x+x^2\right)dx = \frac{33}{14}$

31. $V = 2\int_0^{\pi/2}\int_{-\cos y}^{0}\int_0^{-2x}dz\,dx\,dy = -4\int_0^{\pi/2}\int_{-\cos y}^{0}x\,dx\,dy = 2\int_0^{\pi/2}\cos^2 y\,dy = \frac{\pi}{2}$

32. $V = 4\int_0^2\int_0^{\sqrt{4-x^2}}\int_0^{4-x^2}dz\,dy\,dx = 4\int_0^2\int_0^{\sqrt{4-x^2}}4-x^2\,dy\,dx = 4\int_0^2\left(4-x^2\right)^{3/2}dx = 4\pi$

33. a) $4\int_0^{\sqrt{3}}\int_0^{\sqrt{3-x^2}}\int_0^{\sqrt{4-x^2-y^2}}dz\,dy\,dx$

b) $4\int_0^{\pi/2}\int_0^{\sqrt{3}}\int_0^{\sqrt{4-r^2}}r\,dz\,dr\,d\theta$

c) $4\int_0^{\pi/2}\int_0^{\pi/3}\int_{\sec\phi}^{2}\rho^2\sin\phi\,d\rho\,d\phi\,d\theta$

34. a) $I_z = 4\int_0^1\int_0^{\sqrt{1-y^2}}\int_0^{\sqrt{4-x^2-y^2}}\left(x^2+y^2\right)dz\,dy\,dx$

b) $I_z = 4\int_0^{\pi/2}\int_0^1\int_0^{\sqrt{1-r^2}}r^3\,dz\,dr\,d\theta$

c) $I_z = 4\int_0^{\pi/2}\int_0^{\pi/2}\int_0^1\rho^4\sin^3\phi\,d\rho\,d\phi\,d\theta$

35. $V = \int_0^{\pi/2}\int_1^2\int_0^{r^2\sin\theta\cos\theta}r\,dz\,dr\,d\theta = \int_0^{\pi/2}\int_1^2 r^3\sin\theta\cos\theta\,dr\,d\theta =$

$\frac{15}{4}\int_0^{\pi/2}\sin\theta\cos\theta\,d\theta = \frac{15}{8}$

36. $V = 4\int_0^{\pi/2}\int_0^1\int_{r^2}^{\sqrt{2-r^2}}r\,dz\,dr\,d\theta = 4\int_0^{\pi/2}\int_0^1\left(r\sqrt{2-r^2}-r^3\right)dr\,d\theta =$

$\left(\frac{2^{7/2}-7}{3}\right)\int_0^{\pi/2}d\theta = \frac{\pi\left(8\sqrt{2}-7\right)}{6}$

37. $V = 4\int_0^1 \int_0^{\sqrt{1-x^2}} \int_{2x^2+2y^2}^{3-x^2-y^2} dz\,dy\,dx = 4\int_0^{\pi/2} \int_0^1 \int_{2r^2}^{3-r^2} r\,dz\,dr\,d\theta =$

$4\int_0^{\pi/2} \int_0^1 \left(3r - 3r^3\right) dr\,d\theta = 3\int_0^{\pi/2} d\theta = \frac{3\pi}{2}$

38. a) The radius of the sphere is 2; the radius of the hole is 1.

b) $V = 2\int_0^{2\pi} \int_0^{\sqrt{3}} \int_1^{\sqrt{4-z^2}} r\,dr\,dz\,d\theta = \int_0^{2\pi} \int_0^{\sqrt{3}} 3 - z^2\,dz\,d\theta = 2\sqrt{3}\int_0^{2\pi} d\theta = 4\sqrt{3}\pi$

39. a) $M = 4\int_0^{\pi/2} \int_0^1 \int_0^{r^2} z\,r\,dz\,dr\,d\theta = 2\int_0^{\pi/2} \int_0^1 r^5\,dr\,d\theta = \frac{1}{3}\int_0^{\pi/2} d\theta = \frac{\pi}{6}$

$M_{xy} = \int_0^{2\pi} \int_0^1 \int_0^{r^2} z^2\,r\,dz\,dr\,d\theta = \frac{1}{3}\int_0^{2\pi} \int_0^1 r^7\,dr\,d\theta = \frac{1}{24}\int_0^{2\pi} d\theta = \frac{\pi}{12} \Rightarrow \overline{z} = \frac{1}{2},$

$\overline{x} = \overline{y} = 0$ by symmetry; $I_z = \int_0^{2\pi} \int_0^1 \int_0^{r^2} z\,r^3\,dz\,dr\,d\theta = \frac{1}{2}\int_0^{2\pi} \int_0^1 r^7\,dr\,d\theta =$

$\frac{1}{16}\int_0^{2\pi} d\theta = \frac{\pi}{8} \Rightarrow R_z = \sqrt{\frac{I_z}{M}} = \frac{\sqrt{3}}{2}$

b) $M = 4\int_0^{\pi/2} \int_0^1 \int_0^{r^2} r^2\,dz\,dr\,d\theta = 4\int_0^{\pi/2} \int_0^1 r^4\,dr\,d\theta = \frac{4}{5}\int_0^{\pi/2} d\theta = \frac{2\pi}{5}$

$M_{xy} = \int_0^{2\pi} \int_0^1 \int_0^{r^2} z\,r^2\,dz\,dr\,d\theta = \frac{1}{2}\int_0^{2\pi} \int_0^1 r^6\,dr\,d\theta = \frac{1}{14}\int_0^{2\pi} d\theta = \frac{\pi}{7} \Rightarrow \overline{z} = \frac{5}{14}$

$\overline{x} = \overline{y} = 0$ by symmetry; $I_z = \int_0^{2\pi} \int_0^1 \int_0^{r^2} r^4\,dz\,dr\,d\theta = \int_0^{2\pi} \int_0^1 r^6\,dr\,d\theta =$

$\frac{1}{7}\int_0^{2\pi} d\theta = \frac{2\pi}{7} \Rightarrow R_z = \sqrt{\frac{I_z}{M}} = \sqrt{\frac{5}{7}}$

40. a) $M = \int_0^{2\pi} \int_0^1 \int_r^1 z\,r\,dz\,dr\,d\theta = \frac{1}{2}\int_0^{2\pi} \int_0^1 r - r^3\,dr\,d\theta = \frac{1}{8}\int_0^{2\pi} d\theta = \frac{\pi}{4}$

$M_{xy} = \int_0^{2\pi} \int_0^1 \int_r^1 z^2\,r\,dz\,dr\,d\theta = \frac{1}{3}\int_0^{2\pi} \int_0^1 r - r^4\,dr\,d\theta = \frac{1}{10}\int_0^{2\pi} d\theta = \frac{\pi}{5} \Rightarrow \overline{z} = \frac{4}{5},$

$\overline{x} = \overline{y} = 0$ by symmetry; $I_z = \int_0^{2\pi} \int_0^1 \int_r^1 z\,r^3\,dz\,dr\,d\theta = \frac{1}{2}\int_0^{2\pi} \int_0^1 r^3 - r^5\,dr\,d\theta =$

$\frac{1}{24}\int_0^{2\pi} d\theta = \frac{\pi}{12} \Rightarrow R_z = \sqrt{\frac{I_z}{M}} = \sqrt{\frac{1}{3}}$

b) $M = \int_0^{2\pi}\int_0^1\int_r^1 z^2\,dz\,dr\,d\theta = \frac{\pi}{5}$, see part (a) for details

$M_{xy} = \int_0^{2\pi}\int_0^1\int_r^1 z^3 r\,dz\,dr\,d\theta = \frac{1}{4}\int_0^{2\pi}\int_0^1 r - r^5\,dr\,d\theta = \frac{1}{12}\int_0^{2\pi} d\theta = \frac{\pi}{6} \Rightarrow \overline{z} = \frac{5}{6}$,

$\overline{x} = \overline{y} = 0$ by symmetry; $I_z = \int_0^{2\pi}\int_0^1\int_r^1 z^2 r^3\,dz\,dr\,d\theta = \frac{1}{3}\int_0^{2\pi}\int_0^1 r^3 - r^6\,dr\,d\theta =$

$\frac{1}{28}\int_0^{2\pi} d\theta = \frac{\pi}{14} \Rightarrow R_z = \sqrt{\frac{I_z}{M}} = \sqrt{\frac{5}{14}}$

41. $V = 8\int_0^{\pi/2}\int_0^{\pi/2}\int_0^{2\sin\phi} \rho^2\sin\phi\,d\rho\,d\phi\,d\theta = \frac{64}{3}\int_0^{\pi/2}\int_0^{\pi/2} \sin^4\phi\,d\phi\,d\theta = 4\pi\int_0^{\pi/2} d\theta = 2\pi^2$

42. $M = 4\int_0^{\pi/2}\int_0^{\pi/3}\int_0^2 \rho^2\sin\phi\,d\rho\,d\phi\,d\theta = \frac{32}{3}\int_0^{\pi/2}\int_0^{\pi/3} \sin\phi\,d\phi\,d\theta = \frac{16}{3}\int_0^{\pi/2} d\theta = \frac{8\pi}{3}$

$I_z = \int_0^{2\pi}\int_0^{\pi/3}\int_0^2 \rho^4\sin^3\phi\,d\rho\,d\phi\,d\theta = \frac{32}{5}\int_0^{2\pi}\int_0^{\pi/3} \sin^3\phi\,d\phi\,d\theta =$

$\frac{4}{3}\int_0^{2\pi} d\theta = \frac{8\pi}{3}$

43. $M = \frac{2}{3}\pi a^3$; $M_{xy} = \int_0^{2\pi}\int_0^{\pi/2}\int_0^a \rho^3\sin\phi\cos\phi\,d\rho\,d\phi\,d\theta = \frac{a^4}{4}\int_0^{2\pi}\int_0^{\pi/2} \sin\phi\cos\phi\,d\phi\,d\theta =$

$\frac{a^4}{8}\int_0^{2\pi} d\theta = \frac{a^4\pi}{4} \Rightarrow \overline{z} = \frac{3a}{8}$, $\overline{x} = \overline{y} = 0$ by symmetry

44. $M = \frac{4}{3}\pi a^3$; $I_z = \int_0^{2\pi}\int_0^{\pi}\int_0^a \rho^4\sin^3\phi\,d\rho\,d\phi\,d\theta = \frac{a^5}{5}\int_0^{2\pi}\int_0^{\pi} \sin^3\phi\,d\phi\,d\theta = \frac{4a^5}{15}\int_0^{2\pi} d\theta =$

$\frac{8a^5\pi}{15} \Rightarrow R_z = \sqrt{\frac{I_z}{M}} = a\sqrt{\frac{2}{5}}$

CHAPTER 15

VECTOR ANALYSIS

SECTION 15.1 LINE INTEGRALS

1. $\mathbf{r} = t\,\mathbf{i} + (1-t)\,\mathbf{j} \Rightarrow x = t,\ y = 1 - t \Rightarrow y = 1 - x \Rightarrow$ c

2. $\mathbf{r} = \mathbf{i} + \mathbf{j} + t\,\mathbf{k} \Rightarrow x = 1,\ y = 1,\ z = t \Rightarrow$ e

3. $\mathbf{r} = (2\cos t)\,\mathbf{i} + (2\sin t)\,\mathbf{j} \Rightarrow x = 2\cos t,\ y = 2\sin t \Rightarrow x^2 + y^2 = 4 \Rightarrow$ g

4. $\mathbf{r} = t\,\mathbf{i} \Rightarrow x = t,\ y = 0,\ z = 0 \Rightarrow$ a

5. $\mathbf{r} = t\,\mathbf{i} + t\,\mathbf{j} + t\,\mathbf{k} \Rightarrow x = t,\ y = t,\ z = t \Rightarrow$ d

6. $\mathbf{r} = t\,\mathbf{j} + (2 - 2t)\,\mathbf{k} \Rightarrow y = t,\ z = 2 - 2t \Rightarrow z = 2 - 2y \Rightarrow$ b

7. $\mathbf{r} = (t^2 - 1)\,\mathbf{j} + 2t\,\mathbf{k} \Rightarrow y = t^2 - 1,\ z = 2t \Rightarrow y = \frac{z^2}{4} - 1 \Rightarrow$ f

8. $\mathbf{r} = (2\cos t)\,\mathbf{i} + (2\sin t)\,\mathbf{k} \Rightarrow x = 2\cos t,\ z = 2\sin t \Rightarrow x^2 + z^2 = 4 \Rightarrow$ h

9. $\mathbf{r} = t\,\mathbf{i} + (1-t)\,\mathbf{j} \Rightarrow x = t,\ y = 1 - t,\ z = 0 \Rightarrow f(g(t),h(t),k(t)) = 1.\ \frac{dx}{dt} = 1,\ \frac{dy}{dt} = -1,\ \frac{dz}{dt} = 0 \Rightarrow$
$$\sqrt{\left(\frac{dx}{dt}\right)^2 + \left(\frac{dy}{dt}\right)^2 + \left(\frac{dz}{dt}\right)^2}\ dt = \sqrt{2}\ dt \Rightarrow \int_C f(x,y,z)\ ds = \int_0^1 \sqrt{2}\ dt = \sqrt{2}$$

10. $\mathbf{r} = t\,\mathbf{i} + (1-t)\,\mathbf{j} + \mathbf{k} \Rightarrow x = t,\ y = 1 - t,\ z = 1 \Rightarrow f(g(t),h(t),k(t)) = 2t - 2.\ \frac{dx}{dt} = 1,\ \frac{dy}{dt} = -1,\ \frac{dz}{dt} = 0 \Rightarrow$
$$\sqrt{\left(\frac{dx}{dt}\right)^2 + \left(\frac{dy}{dt}\right)^2 + \left(\frac{dz}{dt}\right)^2}\ dt = \sqrt{2}\ dt \Rightarrow \int_C f(x,y,z)\ ds = \int_0^1 (2t - 2)\sqrt{2}\ dt = -\sqrt{2}$$

11. $\mathbf{r} = 2t\,\mathbf{i} + t\,\mathbf{j} + (2 - 2t)\,\mathbf{k} \Rightarrow x = 2t,\ y = t,\ z = 2 - 2t \Rightarrow f(g(t),h(t),k(t)) = 2t^2 - t + 2.\ \frac{dx}{dt} = 2,\ \frac{dy}{dt} = 1,\ \frac{dz}{dt} = -2$
$$\Rightarrow \sqrt{\left(\frac{dx}{dt}\right)^2 + \left(\frac{dy}{dt}\right)^2 + \left(\frac{dz}{dt}\right)^2}\ dt = 3\ dt \Rightarrow \int_C f(x,y,z)\ ds = \int_0^1 (2t^2 - t + 2)3\ dt = \frac{13}{2}$$

12. $\mathbf{r} = (4\cos t)\,\mathbf{i} + (4\sin t)\,\mathbf{j} + 3t\,\mathbf{k} \Rightarrow x = 4\cos t,\ y = 4\sin t,\ z = 3t \Rightarrow f(g(t),h(t),k(t)) = \sqrt{(4\cos t)^2 + (4\sin t)^2}$
$= 4.\ \frac{dx}{dt} = -4\sin t,\ \frac{dy}{dt} = 4\cos t,\ \frac{dz}{dt} = 3 \Rightarrow \sqrt{\left(\frac{dx}{dt}\right)^2 + \left(\frac{dy}{dt}\right)^2 + \left(\frac{dz}{dt}\right)^2}\ dt = 5\ dt \Rightarrow$
$$\int_C f(x,y,z)\ ds = \int_{-2\pi}^{2\pi} 4(5)\ dt = 80\pi$$

13. $\mathbf{r} = \mathbf{i} + \mathbf{j} + t\,\mathbf{k} \Rightarrow x = 1,\ y = 1,\ z = t \Rightarrow f(g(t),h(t),k(t)) = 3t\sqrt{4+t^2}$. $\frac{dx}{dt} = 0,\ \frac{dy}{dt} = 0,\ \frac{dz}{dt} = 1 \Rightarrow$

$$\sqrt{\left(\frac{dx}{dt}\right)^2 + \left(\frac{dy}{dt}\right)^2 + \left(\frac{dz}{dt}\right)^2}\ dt = 1\ dt = dt \Rightarrow \int_C f(x,y,z)\ ds = \int_{-1}^{1} 3t\sqrt{4+t^2}\ dt = 0$$

14. $\mathbf{r} = t\,\mathbf{i} + t\,\mathbf{j} + t\,\mathbf{k} \Rightarrow x = t,\ y = t,\ z = t \Rightarrow f(g(t),h(t),k(t)) = \frac{\sqrt{3}}{t^2+t^2+t^2} = \frac{\sqrt{3}}{3t^2}$. $\frac{dx}{dt} = 1,\ \frac{dy}{dt} = 1,\ \frac{dz}{dt} = 1 \Rightarrow$

$$\sqrt{\left(\frac{dx}{dt}\right)^2 + \left(\frac{dy}{dt}\right)^2 + \left(\frac{dz}{dt}\right)^2}\ dt = \sqrt{3}\ dt \Rightarrow \int_C f(x,y,z)\ ds = \int_1^2 \left(\frac{\sqrt{3}}{3t^2}\right)\sqrt{3}\ dt = \frac{1}{2}$$

15. C_1: $\mathbf{r} = t\,\mathbf{i} + t^2\,\mathbf{j} \Rightarrow x = t,\ y = t^2,\ z = 0 \Rightarrow f(g(t),h(t),k(t)) = t + \sqrt{t^2} = 2t$ since $0 \le t \le 1$. $\frac{dx}{dt} = 1,\ \frac{dy}{dt} = 2t,$

$$\frac{dz}{dt} = 0 \Rightarrow \sqrt{\left(\frac{dx}{dt}\right)^2 + \left(\frac{dy}{dt}\right)^2 + \left(\frac{dz}{dt}\right)^2}\ dt = \sqrt{1+4t^2}\ dt \Rightarrow \int_{C_1} f(x,y,z)\ ds = \int_0^1 2t\sqrt{1+4t^2}\ dt =$$

$\frac{1}{6}\left(5^{3/2}\right) - \frac{1}{6} = \frac{5}{6}\sqrt{5} - \frac{1}{6}$. C_2: $\mathbf{r} = \mathbf{i} + \mathbf{j} + t\,\mathbf{k} \Rightarrow x = 1,\ y = 1,\ z = t \Rightarrow f(g(t),h(t),k(t)) = 2 - t^2$. $\frac{dx}{dt} = 0,$

$$\frac{dy}{dt} = 0,\ \frac{dz}{dt} = 1 \Rightarrow \sqrt{\left(\frac{dx}{dt}\right)^2 + \left(\frac{dy}{dt}\right)^2 + \left(\frac{dz}{dt}\right)^2}\ dt = 1\ dt = dt \Rightarrow \int_{C_2} f(x,y,z)\ ds = \int_0^1 (2 - t^2)\ dt =$$

$$\frac{5}{3}.\ \therefore \int_C f(x,y,z)\ ds = \int_{C_1} f(x,y,z)\ ds + \int_{C_2} f(x,y,z)\ ds = \frac{5}{6}\sqrt{5} + \frac{3}{2}.$$

16. C_1: $\mathbf{r} = t\,\mathbf{k} \Rightarrow x = 0,\ y = 0,\ z = t \Rightarrow f(g(t),h(t),k(t)) = -t^2$. $\frac{dx}{dt} = 0,\ \frac{dy}{dt} = 0,\ \frac{dz}{dt} = 1 \Rightarrow$

$$\sqrt{\left(\frac{dx}{dt}\right)^2 + \left(\frac{dy}{dt}\right)^2 + \left(\frac{dz}{dt}\right)^2}\ dt = 1\ dt \Rightarrow \int_{C_1} f(x,y,z)\ ds = \int_0^1 -t^2\ dt = -\frac{1}{3}.\ C_2: \mathbf{r} = t\,\mathbf{j} + \mathbf{k} \Rightarrow$$

$$x = 0,\ y = t,\ z = 1 \Rightarrow f(g(t),h(t),k(t)) = \sqrt{t} - 1.\ \frac{dx}{dt} = 0,\ \frac{dy}{dt} = 1,\ \frac{dz}{dt} = 1 \Rightarrow \sqrt{\left(\frac{dx}{dt}\right)^2 + \left(\frac{dy}{dt}\right)^2 + \left(\frac{dz}{dt}\right)^2}\ dt =$$

$$1\ dt = dt \Rightarrow \int_{C_2} f(x,y,z)\ ds = \int_0^1 (\sqrt{t} - 1)\ dt = -\frac{1}{3}.\ C_3: \mathbf{r} = t\,\mathbf{i} + \mathbf{j} + \mathbf{k} \Rightarrow x = t,\ y = 1,\ z = 1 \Rightarrow$$

$$f(g(t),h(t),k(t)) = t.\ \frac{dx}{dt} = 1,\ \frac{dy}{dt} = 0,\ \frac{dz}{dt} = 0 \Rightarrow \sqrt{\left(\frac{dx}{dt}\right)^2 + \left(\frac{dy}{dt}\right)^2 + \left(\frac{dz}{dt}\right)^2}\ dt = 1\ dt = dt \Rightarrow$$

$$\int_{C_3} f(x,y,z)\ ds = \int_0^1 t\ dt = \frac{1}{2}.\ \therefore \int_C f(x,y,z)\ ds = -\frac{1}{3} + \left(-\frac{1}{3}\right) + \frac{1}{2} = -\frac{1}{6}$$

17. $\delta(x,y,z) = 2 - z$, $\mathbf{r} = (\cos t)\,\mathbf{j} + (\sin t)\,\mathbf{k}$, $0 \le t \le \pi$, $ds = dt$, $x = 0$, $y = \cos t$, $z = \sin t$, and $M = 2\pi - 2$ are

all given or found in Example 3 in the text, page 946. $I_x = \int_C (y^2 + z^2)\delta\, ds$

$= \int_0^{\pi} (\cos^2 t + \sin^2 t)(2 - \sin t)\, dt = \int_0^{\pi} (2 - \sin t)\, dt = 2\pi - 2 \Rightarrow R_x = \sqrt{\frac{I_x}{M}} = \sqrt{\frac{2\pi - 2}{2\pi - 2}} = 1$

18. $\mathbf{r} = (t^2 - 1)\,\mathbf{j} + 2t\,\mathbf{k} \Rightarrow x = 0,\ y = t^2 - 1,\ z = 2t.\ \frac{dx}{dt} = 0,\ \frac{dy}{dt} = 2t,\ \frac{dz}{dt} = 2 \Rightarrow$

$\sqrt{\left(\frac{dx}{dt}\right)^2 + \left(\frac{dy}{dt}\right)^2 + \left(\frac{dz}{dt}\right)^2}\, dt = 2\sqrt{t^2 + 1}\, dt.\ \therefore\ M = \int_C \delta(x,y,z)\, ds = \int_0^1 \frac{3}{2}t\left(2\sqrt{t^2 + 1}\right) dt =$

$2\sqrt{2} - 1.$

19. Let δ be constant. Let $x = a\cos t$, $y = a\sin t$. Then $\frac{dx}{dt} = -a\sin t,\ \frac{dy}{dt} = a\cos t,\ 0 \le t \le 2\pi,\ \frac{dz}{dt} = 0 \Rightarrow$

$\sqrt{\left(\frac{dx}{dt}\right)^2 + \left(\frac{dy}{dt}\right)^2 + \left(\frac{dz}{dt}\right)^2}\, dt = a\, dt.\ \therefore\ I_z = \int_C (x^2 + y^2)\delta\, ds = \int_0^{2\pi} (a^2\sin^2 t + a^2\cos^2 t)a\delta\, dt =$

$\int_0^{2\pi} a^3\delta\, dt = 2\pi a^3\delta.\ M = \int_C \delta(x,y,z)\, ds = \int_0^{2\pi} \delta a\, dt = 2\pi\delta a.\ R_z = \sqrt{\frac{I_z}{M}} = \sqrt{\frac{2\pi a^3\delta}{2\pi a\delta}} = a.$

20. Let $\delta = a$. $\mathbf{r} = t\,\mathbf{j} + (2 - 2t)\,\mathbf{k} \Rightarrow x = 0,\ y = t,\ z = 2 - 2t \Rightarrow \frac{dx}{dt} = 0,\ \frac{dy}{dt} = 1,\ \frac{dz}{dt} = -2 \Rightarrow$

$\sqrt{\left(\frac{dx}{dt}\right)^2 + \left(\frac{dy}{dt}\right)^2 + \left(\frac{dz}{dt}\right)^2}\, dt = \sqrt{5}\, dt.\ M = \int_C \delta(x,y,z)\, ds = \int_0^1 a\sqrt{5}\, dt = a\sqrt{5}.\ I_x =$

$\int_C (y^2 + z^2)\delta\, ds = \int_0^1 (t^2 + (2 - 2t)^2)a\sqrt{5}\, dt = \int_0^1 (5t^2 - 8t + 4)a\sqrt{5}\, dt = \frac{5}{3}a\sqrt{5}.$

$R_x = \sqrt{\frac{I_x}{M}} = \sqrt{\frac{\frac{5}{3}a\sqrt{5}}{a\sqrt{5}}} = \sqrt{\frac{5}{3}}.\ I_y = \int_C (x^2 + z^2)\delta\, ds = \int_0^1 (2 - 2t)^2 a\sqrt{5}\, dt =$

$\int_0^1 (4 - 8t + 4t^2)a\sqrt{5}\, dt = \frac{4}{3}a\sqrt{5}.\ R_y = \sqrt{\frac{I_y}{M}} = \sqrt{\frac{\frac{4}{3}a\sqrt{5}}{a\sqrt{5}}} = \frac{2}{\sqrt{3}}.$

$I_z = \int_C (x^2 + y^2)\delta\, ds = \int_0^1 t^2 a\sqrt{5}\, dt = \frac{1}{3}a\sqrt{5}.\ R_z = \sqrt{\frac{I_z}{M}} = \sqrt{\frac{\frac{1}{3}a\sqrt{5}}{a\sqrt{5}}} = \frac{1}{\sqrt{3}}.$

21. a) $\mathbf{r} = (\cos t)\,\mathbf{i} + (\sin t)\,\mathbf{j} + t\,\mathbf{k} \Rightarrow x = \cos t,\ y = \sin t,\ z = t \Rightarrow \frac{dx}{dt} = -\sin t,\ \frac{dy}{dt} = \cos t,\ \frac{dz}{dt} = 1 \Rightarrow$

$\sqrt{\left(\frac{dx}{dt}\right)^2 + \left(\frac{dy}{dt}\right)^2 + \left(\frac{dz}{dt}\right)^2}\ dt = \sqrt{2}\ dt.$ $M = \int_C \delta(x,y,z)\ ds = \int_0^{2\pi} \delta\sqrt{2}\ dt = 2\pi\delta\sqrt{2}.$

$I_z = \int_C (x^2 + y^2)\delta\ ds = \int_0^{2\pi} (\cos^2 t + \sin^2 t)\delta\sqrt{2}\ dt = \int_0^{2\pi} \delta\sqrt{2}\ dt = 2\pi\delta\sqrt{2}.$ $R_z = \sqrt{\frac{I_z}{M}} =$

$\sqrt{\frac{2\pi\delta\sqrt{2}}{2\pi\delta\sqrt{2}}} = 1$

b) $M = \int_C \delta(x,y,z)\ ds = \int_0^{4\pi} \delta\sqrt{2}\ dt = 4\pi\delta\sqrt{2}.$ $I_z = \int_C (x^2 + y^2)\delta\ ds = \int_0^{4\pi} \delta\sqrt{2}\ dt = 4\pi\delta\sqrt{2}.$

$R_z = \sqrt{\frac{I_z}{M}} = \sqrt{\frac{4\pi\delta\sqrt{2}}{4\pi\delta\sqrt{2}}} = 1$

22. $\mathbf{r} = (t^2 - 1)\,\mathbf{j} + 2t\,\mathbf{k} \Rightarrow x = 0,\ y = t^2 - 1,\ z = 2t \Rightarrow \frac{dx}{dt} = 0,\ \frac{dy}{dt} = 2t,\ \frac{dz}{dt} = 2$ and $\delta(g(t),h(t),k(t)) = 15\sqrt{t^2 + 1}$

$\Rightarrow \sqrt{\left(\frac{dx}{dt}\right)^2 + \left(\frac{dy}{dt}\right)^2 + \left(\frac{dz}{dt}\right)^2}\ dt = \sqrt{4t^2 + 4}\ dt = 2\sqrt{t^2 + 1}\ dt.$ $\therefore\ \delta\ ds = \left(15\sqrt{t^2 + 1}\right)\left(2\sqrt{t^2 + 1}\right) =$

$30\left(t^2 + 1\right).$ $M_{yz} = \int_C x\delta\ ds = \int_{-1}^{1} 0\ \delta\ dt = 0.$ $M_{xz} = \int_C y\ \delta\ ds = \int_{-1}^{1} (t^2 - 1)\left(30(t^2 + 1)\right) dt$

$\int_{-1}^{1} 30(t^4 - 1)\ dt = -48.$ $M_{xy} = \int_C z\ \delta\ ds = \int_{-1}^{1} 2t\left(30(t^2 + 1)\right) dt = \int_{-1}^{1} 60\left(t^3 + t\right) dt = 0.$

$M = \int_C \delta(x,y,z)\ ds = \int_{-1}^{1} 30\left(t^2 + 1\right) dt = 80.$

$\therefore\ \bar{x} = \frac{0}{80} = 0,\ \bar{y} = -\frac{48}{80} = -\frac{3}{5},\ \bar{z} = \frac{0}{80} = 0.$ $I_y = \int_C (x^2 + z^2)\delta\ ds = \int_{-1}^{1} 4t^2\left(30(t^2 + 1)\right) dt =$

$\int_{-1}^{1} 120\left(t^4 + t^2\right) dt = 128.$ $I_z = \int_C (x^2 + y^2)\delta\ ds = \int_{-1}^{1} (t^2 - 1)^2\left(30(t^2 + 1)\right) dt =$

$\int_{-1}^{1} 30\left(t^6 - t^4 - t^2 + 1\right) dt = \frac{256}{7}.$

SECTION 15.2 VECTOR FIELDS, WORK, CIRCULATION, AND FLUX

1. a) $\mathbf{F} = 3t\,\mathbf{i} + 2t\,\mathbf{j} + 4t\,\mathbf{k}$, $\frac{d\mathbf{r}}{dt} = \mathbf{i} + \mathbf{j} + \mathbf{k} \Rightarrow \mathbf{F}\cdot\frac{d\mathbf{r}}{dt} = 9t \Rightarrow W = \int_0^1 9t\,dt = \frac{9}{2}$

   b) $\mathbf{F} = 3t^2\,\mathbf{i} + 2t\,\mathbf{j} + 4t^4\,\mathbf{k}$, $\frac{d\mathbf{r}}{dt} = \mathbf{i} + 2t\,\mathbf{j} + 4t^3\,\mathbf{k} \Rightarrow \mathbf{F}\cdot\frac{d\mathbf{r}}{dt} = 7t^2 + 16t^7 \Rightarrow W = \int_0^1 (7t^2 + 16t^7)\,dt = \frac{13}{3}$

   c) $\mathbf{F}_1 = 3t\,\mathbf{i} + 2t\,\mathbf{j}$, $\frac{d\mathbf{r}_1}{dt} = \mathbf{i} + \mathbf{j} \Rightarrow \mathbf{F}_1\cdot\frac{d\mathbf{r}_1}{dt} = 5t \Rightarrow W_1 = \int_0^1 5t\,dt = \frac{5}{2}$. $\mathbf{F}_2 = 3\,\mathbf{i} + 2\,\mathbf{j} + 4t\,\mathbf{k}$, $\frac{d\mathbf{r}_2}{dt} = \mathbf{k} \Rightarrow$

   $\mathbf{F}_2\cdot\frac{d\mathbf{r}_2}{dt} = 4t \Rightarrow W_2 = \int_0^1 4t\,dt = 2$. $\therefore\ W = W_1 + W_2 = \frac{9}{2}$

2. a) $\mathbf{F} = \left(\frac{1}{t^2+1}\right)\mathbf{j}$, $\frac{d\mathbf{r}}{dt} = \mathbf{i} + \mathbf{j} + \mathbf{k} \Rightarrow \mathbf{F}\cdot\frac{d\mathbf{r}}{dt} = \frac{1}{t^2+1} \Rightarrow W = \int_0^1 \frac{1}{t^2+1}\,dt = \frac{\pi}{4}$

   b) $\mathbf{F} = \left(\frac{1}{t^2+1}\right)\mathbf{j}$, $\frac{d\mathbf{r}}{dt} = \mathbf{i} + 2t\,\mathbf{j} + 3t^3\,\mathbf{k} \Rightarrow \mathbf{F}\cdot\frac{d\mathbf{r}}{dt} = \frac{2t}{t^2+1} \Rightarrow W = \int_0^1 \frac{2t}{t^2+1}\,dt = \ln 2$

   c) $\mathbf{F}_1 = \left(\frac{1}{t^2+1}\right)\mathbf{j}$, $\frac{d\mathbf{r}_1}{dt} = \mathbf{i} + \mathbf{j} \Rightarrow \mathbf{F}_1\cdot\frac{d\mathbf{r}_1}{dt} = \frac{1}{t^2+1}$. $\mathbf{F}_2 = \frac{1}{2}\mathbf{j}$, $\frac{d\mathbf{r}_2}{dt} = \mathbf{k} \Rightarrow \mathbf{F}_2\cdot\frac{d\mathbf{r}_2}{dt} = 0$.

   $\therefore\ W = \int_0^1 \frac{1}{t^2+1}\,dt = \frac{\pi}{4}$

3. a) $\mathbf{F} = \sqrt{t}\,\mathbf{i} - 2t\,\mathbf{j} + \sqrt{t}\,\mathbf{k}$, $\frac{d\mathbf{r}}{dt} = \mathbf{i} + \mathbf{j} + \mathbf{k} \Rightarrow \mathbf{F}\cdot\frac{d\mathbf{r}}{dt} = 2\sqrt{t} - 2t \Rightarrow W = \int_0^1 (2\sqrt{t} - 2t)\,dt = \frac{1}{3}$

   b) $\mathbf{F} = t^2\,\mathbf{i} - 2t\,\mathbf{j} + t\,\mathbf{k}$, $\frac{d\mathbf{r}}{dt} = \mathbf{i} + 2t\,\mathbf{j} + 4t^3\,\mathbf{k} \Rightarrow \mathbf{F}\cdot\frac{d\mathbf{r}}{dt} = 4t^4 - 3t^2 \Rightarrow W = \int_0^1 (4t^4 - 3t^2)\,dt = -\frac{1}{5}$

   c) $\mathbf{F}_1 = -2t\,\mathbf{j} + \sqrt{t}\,\mathbf{k}$, $\frac{d\mathbf{r}_1}{dt} = \mathbf{i} + \mathbf{j} \Rightarrow \mathbf{F}_1\cdot\frac{d\mathbf{r}_1}{dt} = -2t \Rightarrow W_1 = \int_0^1 -2t\,dt = -1$. $\mathbf{F}_2 = \sqrt{t}\,\mathbf{i} - 2\,\mathbf{j} + \mathbf{k}$,

   $\frac{d\mathbf{r}_2}{dt} = \mathbf{k} \Rightarrow \mathbf{F}_2\cdot\frac{d\mathbf{r}_2}{dt} = 1 \Rightarrow W_2 = \int_0^1 dt = 1$. $\therefore\ W = W_1 + W_2 = 0$

4. a) $\mathbf{F} = t^2\,\mathbf{i} + t^2\,\mathbf{j} + t^2\,\mathbf{k},\ \frac{d\mathbf{r}}{dt} = \mathbf{i} + \mathbf{j} + \mathbf{k} \Rightarrow \mathbf{F}\cdot\frac{d\mathbf{r}}{dt} = 3t^2 \Rightarrow W = \int_0^1 3t^2\,dt = 1$

b) $\mathbf{F} = t^3\,\mathbf{i} + t^6\,\mathbf{j} + t^5\,\mathbf{k},\ \frac{d\mathbf{r}}{dt} = \mathbf{i} + 2t\,\mathbf{j} + 4t^3\,\mathbf{k} \Rightarrow \mathbf{F}\cdot\frac{d\mathbf{r}}{dt} = t^3 + 2t^7 + 4t^8 \Rightarrow W = \int_0^1 (t^3 + 2t^7 + 4t^8)\,dt = \frac{17}{18}$

c) $\mathbf{F_1} = t^2\,\mathbf{j},\ \frac{d\mathbf{r_1}}{dt} = \mathbf{i} + \mathbf{j} \Rightarrow \mathbf{F_1}\cdot\frac{d\mathbf{r_1}}{dt} = t^2 \Rightarrow W_1 = \int_0^1 t^2\,dt = \frac{1}{3}.\ \mathbf{F_2} = \mathbf{i} + t\,\mathbf{j} + t\,\mathbf{k},\ \frac{d\mathbf{r_2}}{dt} = \mathbf{k} \Rightarrow \mathbf{F_2}\cdot\frac{d\mathbf{r_2}}{dt} = t$

$\Rightarrow W_2 = \int_0^1 t\,dt = \frac{1}{2}.\ \therefore\ W = W_1 + W_2 = \frac{5}{6}$

5. a) $\mathbf{F} = (3t^2 - 3t)\,\mathbf{i} + 3t\,\mathbf{j} + \mathbf{k},\ \frac{d\mathbf{r}}{dt} = \mathbf{i} + \mathbf{j} + \mathbf{k} \Rightarrow \mathbf{F}\cdot\frac{d\mathbf{r}}{dt} = 3t^2 + 1 \Rightarrow W = \int_0^1 (3t^2 + 1)\,dt = 2$

b) $\mathbf{F} = (3t^2 - 3t)\,\mathbf{i} + 3t^4\,\mathbf{j} + \mathbf{k},\ \frac{d\mathbf{r}}{dt} = \mathbf{i} + 2t\,\mathbf{j} + 4t^3\,\mathbf{k} \Rightarrow \mathbf{F}\cdot\frac{d\mathbf{r}}{dt} = 6t^5 + 4t^3 + 3t^2 - 3t \Rightarrow$

$W = \int_0^1 \left(6t^5 + 4t^3 + 3t^2 - 3t\right)dt = \frac{3}{2}$

c) $\mathbf{F_1} = (3t^2 - 3t)\,\mathbf{i} + \mathbf{k},\ \frac{d\mathbf{r_1}}{dt} = \mathbf{i} + \mathbf{j} \Rightarrow \mathbf{F_1}\cdot\frac{d\mathbf{r_1}}{dt} = 3t^2 - 3t \Rightarrow W_1 = \int_0^1 (3t^2 - 3t)\,dt = -\frac{1}{2}$

$\mathbf{F_2} = 3t\,\mathbf{j} + \mathbf{k},\ \frac{d\mathbf{r_2}}{dt} = \mathbf{k} \Rightarrow \mathbf{F_2}\cdot\frac{d\mathbf{r_2}}{dt} = 1 \Rightarrow W_2 = \int_0^1 dt = 1.\ \therefore\ W = W_1 + W_2 = \frac{1}{2}$

6. a) $\mathbf{F} = 2t\,\mathbf{i} + 2t\,\mathbf{j} + 2t\,\mathbf{k},\ \frac{d\mathbf{r}}{dt} = \mathbf{i} + \mathbf{j} + \mathbf{k} \Rightarrow \mathbf{F}\cdot\frac{d\mathbf{r}}{dt} = 6t \Rightarrow W = \int_0^1 6t\,dt = 3$

b) $\mathbf{F} = (t^2 + t^4)\,\mathbf{i} + (t^4 + t)\,\mathbf{j} + (t^2 + t)\,\mathbf{k},\ \frac{d\mathbf{r}}{dt} = \mathbf{i} + 2t\,\mathbf{j} + 4t^3\,\mathbf{k} \Rightarrow \mathbf{F}\cdot\frac{d\mathbf{r}}{dt} = 6t^5 + 5t^4 + 3t^2 \Rightarrow$

$W = \int_0^1 (6t^5 + 5t^4 + 3t^2)\,dt = 3$

c) $\mathbf{F_1} = t\,\mathbf{i} + t\,\mathbf{j} + 2t\,\mathbf{k},\ \frac{d\mathbf{r_1}}{dt} = \mathbf{i} + \mathbf{j} \Rightarrow \mathbf{F_1}\cdot\frac{d\mathbf{r_1}}{dt} = 2t \Rightarrow W_1 = \int_0^1 2t\,dt = 1.\ \mathbf{F_2} = (1 + t)\,\mathbf{i} + (t + 1)\,\mathbf{j} + 2\,\mathbf{k},$

$\frac{d\mathbf{r_2}}{dt} = \mathbf{k} \Rightarrow \mathbf{F_2}\cdot\frac{d\mathbf{r_2}}{dt} = 2 \Rightarrow W_2 = \int_0^1 2\,dt = 2.\ \therefore\ W = W_1 + W_2 = 3$

7. $\mathbf{F} = t^3\,\mathbf{i} + t^2\,\mathbf{j} - t^3\,\mathbf{k},\ \frac{d\mathbf{r}}{dt} = \mathbf{i} + 2t\,\mathbf{j} + \mathbf{k} \Rightarrow \mathbf{F}\cdot\frac{d\mathbf{r}}{dt} = 2t^3 \Rightarrow W = \int_0^1 2t^3\,dt = \frac{1}{2}$

8. $\mathbf{F} = (2\sin t)\,\mathbf{i} + (3\cos t)\,\mathbf{j} + (\cos t + \sin t)\,\mathbf{k},\ \frac{d\mathbf{r}}{dt} = (-\sin t)\,\mathbf{i} + (\cos t)\,\mathbf{j} + \frac{1}{6}\mathbf{k} \Rightarrow \mathbf{F}\cdot\frac{d\mathbf{r}}{dt} = 5\cos^2 t - 2 + \frac{1}{6}\cos t$
$+ \frac{1}{6}\sin t \Rightarrow W = \int_0^{2\pi} \left(5\cos^2 t - 2 + \frac{1}{6}\cos t + \frac{1}{6}\sin t\right) dt = \pi$

9. $\mathbf{F} = t\,\mathbf{i} + (\sin t)\,\mathbf{j} + (\cos t)\,\mathbf{k},\ \frac{d\mathbf{r}}{dt} = (\cos t)\,\mathbf{i} - (\sin t)\,\mathbf{j} + \mathbf{k} \Rightarrow \mathbf{F}\cdot\frac{d\mathbf{r}}{dt} = t\cos t - \sin^2 t + \cos t \Rightarrow$
$W = \int_0^{2\pi} (t\cos t - \sin^2 t + \cos t)\,dt = -\pi$

10. $\mathbf{F} = t\,\mathbf{i} + (\cos^2 t)\,\mathbf{j} + (12\sin t)\,\mathbf{k},\ \frac{d\mathbf{r}}{dt} = (\cos t)\,\mathbf{i} - (\sin t)\,\mathbf{j} + \frac{1}{6}\mathbf{k} \Rightarrow \mathbf{F}\cdot\frac{d\mathbf{r}}{dt} = t\cos t - \sin t\cos^2 t + 2\sin t \Rightarrow$
$W = \int_0^{2\pi} \left(t\cos t - \sin t\cos^2 t + 2\sin t\right) dt = 0$

11. $\mathbf{F} = -4t^3\,\mathbf{i} + 8t^2\,\mathbf{j} + 2\,\mathbf{k},\ \frac{d\mathbf{r}}{dt} = \mathbf{i} + 2t\,\mathbf{j} \Rightarrow \mathbf{F}\cdot\frac{d\mathbf{r}}{dt} = 12t^3 \Rightarrow \text{Flow} = \int_0^2 12t^3\,dt = 48$

12. $\mathbf{F} = 12t^2\,\mathbf{j} + 9t^2\,\mathbf{k},\ \frac{d\mathbf{r}}{dt} = 3\,\mathbf{j} + 4\,\mathbf{k} \Rightarrow \mathbf{F}\cdot\frac{d\mathbf{r}}{dt} = 72t^2 \Rightarrow \text{Flow} = \int_0^1 72t^2\,dt = 24$

13. $\mathbf{F} = (\cos t - \sin t)\,\mathbf{i} + (\cos t)\,\mathbf{k},\ \frac{d\mathbf{r}}{dt} = (-\sin t)\,\mathbf{i} + (\cos t)\,\mathbf{k} \Rightarrow \mathbf{F}\cdot\frac{d\mathbf{r}}{dt} = -\sin t\cos t + 1 \Rightarrow$
$\text{Flow} = \int_0^{\pi} (-\sin t\cos t + 1)\,dt = \pi$

14. $\mathbf{F} = (-2\sin t)\,\mathbf{i} - (2\cos t)\,\mathbf{j} + 2\,\mathbf{k},\ \frac{d\mathbf{r}}{dt} = (2\sin t)\,\mathbf{i} + (2\cos t)\,\mathbf{j} + 2\,\mathbf{k} \Rightarrow \mathbf{F}\cdot\frac{d\mathbf{r}}{dt} = -4\sin^2 t - 4\cos^2 t + 4 = 0$
$\Rightarrow \text{Flow} = 0$

15. a) $\mathbf{F} = (\cos t)\,\mathbf{i} + (\sin t)\,\mathbf{j},\ \frac{d\mathbf{r}}{dt} = (-\sin t)\,\mathbf{i} + (\cos t)\,\mathbf{j} \Rightarrow \mathbf{F}\cdot\frac{d\mathbf{r}}{dt} = 0 \Rightarrow$ Circulation = 0.

$M = \cos t,\ N = \sin t,\ dx = -\sin t\,dt,\ dy = \cos t\,dt \Rightarrow \text{Flux} = \int_C M\,dy - N\,dx =$

$\int_0^{2\pi} (\cos^2 t + \sin^2 t)\,dt = \int_0^{2\pi} dt = 2\pi$

15. b) $\mathbf{F} = (\cos t)\,\mathbf{i} + (4 \sin t)\,\mathbf{j}$, $\frac{dr}{dt} = (-\sin t)\,\mathbf{i} + (4 \cos t)\,\mathbf{j} \Rightarrow \mathbf{F}\cdot\frac{dr}{dt} = 15 \sin t \cos t \Rightarrow$ Circ $=$

$\int_0^{2\pi} 15 \sin t \cos t\, dt = 0$. $M = \cos t$, $N = \sin t$, $dx = -\sin t$, $dy = 4 \cos t \Rightarrow$ Flux $=$

$\int_C M dy - N dx = \int_0^{2\pi} (4 \cos^2 t + 4 \sin^2 t)\, dt = 8\pi$

16. $M_1 = (2a \cos t)$, $N_1 = (3a \sin t)$, $dx = -a \sin t\, dt$, $dy = a \cos t\, dt \Rightarrow$ Flux $= \int_C M dy - N dx =$

$\int_0^{2\pi} (2a^2 \cos^2 t - 3a^2 \sin^2 t)\, dt = -\pi a^2$. $M_2 = (2a \cos t)$, $N_2 = (a \cos t - a \sin t)$, $dx = -a \sin t\, dt$,

$dy = a \cos t\, dt \Rightarrow$ Flux $= \int_C M dy - N dx = \int_0^{2\pi} \left(2a^2 \cos^2 t - (-a^2 \sin t \cos t + a^2 \sin^2 t)\right) dt = \pi a^2$

17. $\mathbf{F_1} = (a \cos t)\,\mathbf{i} + (a \sin t)\,\mathbf{j}$, $\frac{dr_1}{dt} = (-a \sin t)\,\mathbf{i} + (a \cos t)\,\mathbf{j} \Rightarrow \mathbf{F_1}\cdot\frac{dr_1}{dt} = 0 \Rightarrow \text{Circ}_1 = 0$. $M_1 = a \cos t$, $N_1 =$

$a \sin t$, $dx = -a \sin t\, dt$, $dy = a \cos t\, dt \Rightarrow \text{Flux}_1 = \int_C M_1 dy - N_1 dx = \int_0^{\pi} (a^2 \cos^2 t + a^2 \sin^2 t)\, dt =$

$\int_0^{\pi} a^2\, dt = a^2\pi$.

$\mathbf{F_2} = t\,\mathbf{i}$, $\frac{dr_2}{dt} = \mathbf{i} \Rightarrow \mathbf{F_2}\cdot\frac{dr_2}{dt} = t \Rightarrow \text{Circ}_2 = \int_{-a}^{a} t\, dt = 0$. $M_2 = t$, $N_2 = 0$, $dx = dt$, $dy = 0 \Rightarrow \text{Flux}_2 =$

$\int_C M_2 dy - N_2 dx = \int_{-a}^{a} 0\, dt = 0$. $\therefore$ Circ $= \text{Circ}_1 + \text{Circ}_2 = 0$, Flux $= \text{Flux}_1 + \text{Flux}_2 = a^2\pi$

18. $\mathbf{F_1} = (a^2 \cos^2 t)\,\mathbf{i} + (a^2 \sin^2 t)\,\mathbf{j}$, $\frac{dr_1}{dt} = (-a \sin t)\,\mathbf{i} + (a \cos t)\,\mathbf{j} \Rightarrow \mathbf{F_1}\cdot\frac{dr_1}{dt} = -a^3 \sin t \cos^2 t + a^3 \cos t \sin^2 t$

$\Rightarrow \text{Circ}_1 = \int_0^{\pi} (-a^3 \sin t \cos^2 t + a^3 \cos t \sin^2 t)\, dt = -\frac{2a^3}{3}$. $M_1 = a^2 \cos^2 t$, $N_1 = a^2 \sin^2 t$,

$dy = a \cos t$, $dx = -a \sin t\, dt \Rightarrow \text{Flux}_1 = \int_C M_1 dy - N_1 dx = \int_0^{\pi} (a^3 \cos^3 t + a^3 \sin^3 t)\, dt = \frac{4}{3} a^3$.

18. (Continued)

$\mathbf{F_2} = t^2\,\mathbf{i}$, $\frac{d\mathbf{r_2}}{dt} = \mathbf{i} \Rightarrow \mathbf{F_2} \cdot \frac{d\mathbf{r_2}}{dt} = t^2 \Rightarrow \text{Circ}_2 = \int_{-a}^{a} t^2\,dt = \frac{2a^3}{3}$. $M_2 = t^2$, $N_2 = 0$, $dy = 0$, $dx = dt \Rightarrow$

$\text{Flux}_2 = \int_C M_2dy - N_2dx = 0$. $\therefore$ Circ $= \text{Circ}_1 + \text{Circ}_2 = 0$, Flux $= \text{Flux}_1 + \text{Flux}_2 = \frac{4}{3}a^3$

19. $\mathbf{F_1} = (-a \sin t)\,\mathbf{i} + (a \cos t)\,\mathbf{j}$, $\frac{d\mathbf{r_1}}{dt} = (-a \sin t)\,\mathbf{i} + (a \cos t)\,\mathbf{j} \Rightarrow \mathbf{F_1} \cdot \frac{d\mathbf{r_1}}{dt} = a^2 \sin^2 t + a^2 \cos^2 t = a^2 \Rightarrow$

$\text{Circ}_1 = \int_0^{\pi} a^2\,dt = a^2\pi.$

$\mathbf{F_2} = t\,\mathbf{j}$, $\frac{d\mathbf{r_2}}{dt} = \mathbf{i} \Rightarrow \mathbf{F_2} \cdot \frac{d\mathbf{r_2}}{dt} = 0 \Rightarrow \text{Circ}_2 = 0$. $\therefore$ Circ $= \text{Circ}_1 + \text{Circ}_2 = a^2\pi$

$M_1 = -a \sin t$, $N_1 = a \cos t$, $dx = -a \sin t$, $dy = a \cos t \Rightarrow \text{Flux}_1 = \int_C M_1dy - N_1dx =$

$\int_0^{\pi} (-a^2 \sin t \cos t + a^2 \sin t \cos t)\,dt = 0$. $M_2 = 0$, $N_2 = t$, $dx = dt$, $dy = 0 \Rightarrow \text{Flux}_2 =$

$\int_C M_2dy - N_2dx = \int_{-a}^{a} -t\,dt = 0$. $\therefore$ Flux $= \text{Flux}_1 + \text{Flux}_2 = 0$

20. $\mathbf{F_1} = (-a^2 \sin^2 t)\,\mathbf{i} + (a^2 \cos^2 t)\,\mathbf{j}$, $\frac{d\mathbf{r_1}}{dt} = (-a \sin t)\,\mathbf{i} + (a \cos t)\,\mathbf{j} \Rightarrow \mathbf{F_1} \cdot \frac{d\mathbf{r_1}}{dt} = a^3 \sin^3 t + a^3 \cos^3 t \Rightarrow$

$\text{Circ}_1 = \int_0^{\pi} (a^2 \sin^3 t + a^3 \cos^3 t)\,dt = \frac{4}{3}a^3$

$\mathbf{F_2} = t^2\,\mathbf{j}$, $\frac{d\mathbf{r_2}}{dt} = \mathbf{i} \Rightarrow \mathbf{F_2} \cdot \frac{d\mathbf{r_2}}{dt} = 0 \Rightarrow \text{Circ}_2 = 0$. $\therefore$ Circ $= \text{Circ}_1 + \text{Circ}_2 = \frac{4}{3}a^3$

$M_1 = -a^2 \sin^2 t$, $N_1 = a^2 \cos^2 t$, $dy = a \cos t\,dt$, $dx = -a \sin t\,dt \Rightarrow \text{Flux}_1 = \int_C M_1dy - N_1dx =$

$\int_0^{\pi} (-a^3 \cos t \sin^2 t + a^3 \sin t \cos^2 t)\,dt = \frac{2}{3}a^3$. $M_2 = 0$, $N_2 = t^2$, $dy = 0$, $dx = dt \Rightarrow \text{Flux}_2 =$

$\int_C M_2dy - N_2dx = \int_{-a}^{a} -t^2\,dt = -\frac{2}{3}a^3$. $\therefore$ Flux $= \text{Flux}_1 + \text{Flux}_2 = 0$

21. $\mathbf{F} = f(t)\,\mathbf{i}$, $\frac{d\mathbf{r}}{dt} = \mathbf{i} + f'(t)\,\mathbf{j} \Rightarrow \mathbf{F} \cdot \frac{d\mathbf{r}}{dt} = f(t) \Rightarrow \int_{t=a}^{t=b} \mathbf{F} \cdot d\mathbf{r} = \int_a^b f(t)\,dt$. Since $t > 0$ for $a \le t \le b$, this integral yields the desired area.

SECTION 15.3 GREEN'S THEOREM IN THE PLANE

1. Equation 15: $M = -y = -a \sin t$, $N = x = a \cos t$, $dx = -a \sin t\, dt$, $dy = a \cos t\, dt \Rightarrow \frac{\partial M}{\partial x} = 0$, $\frac{\partial M}{\partial y} = -1$, $\frac{\partial N}{\partial x} = 1$, $\frac{\partial N}{\partial y} = 0 \Rightarrow \oint_C M\,dy - N\,dx = \int_0^{2\pi} \left((-a \sin t)(a \cos t)\, dt - (a \cos t)(-a \sin t)\, dt\right) =$ 0. $\iint_R \left(\frac{\partial M}{\partial x} + \frac{\partial N}{\partial y}\right) dx\, dy = \iint_R 0\, dx\, dy = 0$

Equation 16: $\oint_C M\,dx + N\,dy = \int_0^{2\pi} \left((-a \sin t)(-a \sin t) + (a \cos t)(a \cos t)\right) dt = 2\pi a^2$

$$\iint_R \left(\frac{\partial N}{\partial x} - \frac{\partial M}{\partial y}\right) dx\, dy = \int_{-a}^{a} \int_{-\sqrt{a^2+x^2}}^{\sqrt{a^2+x^2}} 2\, dx\, dy = \int_{-a}^{a} 4\sqrt{a^2 - x^2}\, dx = 2a^2\pi$$

2. $M = y = a \sin t$, $N = 0$, $dx = -a \sin t\, dt$, $dy = a \cos t\, dt \Rightarrow \frac{\partial M}{\partial x} = 0$, $\frac{\partial M}{\partial y} = 1$, $\frac{\partial N}{\partial x} = 0$, $\frac{\partial N}{\partial y} = 0$

Equation 15: $\oint_C M\,dy - N\,dx = \int_0^{2\pi} a^2 \sin t \cos t\, dt = 0$. $\iint_R \left(\frac{\partial M}{\partial x} + \frac{\partial N}{\partial y}\right) dx\, dy =$ $\iint_R 0\, dx\, dy = 0$

Equation 16: $\oint_C M\,dx + N\,dy = \int_0^{2\pi} (-a^2 \sin^2 t)\, dt = -a^2\pi$. $\iint_R \left(\frac{\partial N}{\partial x} - \frac{\partial M}{\partial y}\right) dx\, dy =$ $\iint_R -1\, dx\, dy = -a^2\pi$

3. $M = 2x = 2a \cos t$, $N = -3y = -3a \sin t$, $dx = -a \sin t$, $dy = a \cos t \Rightarrow \frac{\partial M}{\partial x} = 2$, $\frac{\partial M}{\partial y} = 0$, $\frac{\partial N}{\partial x} = 0$, $\frac{\partial N}{\partial y} = -3$

Equation 15: $\oint_C M\,dy - N\,dx = \int_0^{2\pi} (2a \cos t(a \cos t) + 3a \sin t(-a \sin t))dt =$ $\int_0^{2\pi} \left(2a^2 \cos^2 t - 3a^2 \sin^2 t\right) dt = -\pi a^2$. $\iint_R \left(\frac{\partial M}{\partial x} + \frac{\partial N}{\partial y}\right) = \iint_R -1\, dx\, dy =$

3. (Continued)

$$\int_{-a}^{a} \int_{-\sqrt{a^2-x^2}}^{\sqrt{a^2-x^2}} -1 \, dx \, dy = -\pi a^2$$

Equation 16: $\oint_C M dx + N dy = \int_0^{2\pi} (2a \cos t(-a \sin t) + (-3a \sin t)(a \cos t)) \, dt =$

$$\int_0^{2\pi} \left(-2a^2 \sin t \cos t - 3a^2 \sin t \cos t\right) dt = 0. \quad \iint_R 0 \, dx \, dy = 0$$

4. $M = -x^2y = -a^3 \cos^2 t \sin t$, $N = xy^2 = a^3 \cos t \sin^2 t$, $dx = -a \sin t \, dt$, $dy = a \cos t \, dt \Rightarrow \frac{\partial M}{\partial x} =$ $-2a^2 \cos t \sin t$, $\frac{\partial M}{\partial y} = -a^2 \cos^2 t$, $\frac{\partial N}{\partial x} = a^2 \sin^2 t$, $\frac{\partial N}{\partial y} = 2a^2 \cos t \sin t$.

Equation 15: $\oint_C M dy - N dx = \int_0^{2\pi} \left(-a^4 \cos^3 t \sin t + a^4 \cos t \sin^3 t\right) dt = 0$

$$\iint_R \left(\frac{\partial M}{\partial x} + \frac{\partial N}{\partial y}\right) dx \, dy = \iint_R (-2xy + 2xy) \, dx \, dy$$

Equation 16: $\oint_C M dx + N dy = \int_0^{2\pi} \left(a^4 \cos^2 t \sin^2 t + a^4 \cos^2 t \sin^2 t\right) dt = \int_0^{2\pi} (2a^4 \cos^2 t \sin^2 t) \, dt$

$$= \frac{a^4\pi}{2}. \quad \iint_R \left(\frac{\partial N}{\partial x} - \frac{\partial M}{\partial y}\right) dx \, dy = \iint_R (y^2 + x^2) \, dx \, dy = \int_{-a}^{a} \int_{-\sqrt{a^2-y^2}}^{\sqrt{a^2-y^2}} (y^2 + x^2) \, dx \, dy = \frac{a^4\pi}{2}$$

5. $M = x - y$, $N = y - x \Rightarrow \frac{\partial M}{\partial x} = 1$, $\frac{\partial M}{\partial y} = -1$, $\frac{\partial N}{\partial x} = -1$, $\frac{\partial N}{\partial y} = 1 \Rightarrow$ Flux $= \iint_R 2 \, dx \, dy = \int_0^1 \int_0^1 2 \, dx \, dy$

$= 2$. Circ $= \iint_R (-1 - (-1)) \, dx \, dy = 0$

6. $M = x^2 + 4y$, $N = x + y^2 \Rightarrow \frac{\partial M}{\partial x} = 2x$, $\frac{\partial M}{\partial y} = 4$, $\frac{\partial N}{\partial x} = 1$, $\frac{\partial N}{\partial y} = 2y \Rightarrow$ Flux $= \iint_R (2x + 2y) \, dx \, dy =$

6. (Continued)

$$\int_0^1\int_0^1 (2x + 2y)\,dx\,dy = 2. \text{ Circ} = \iint_R (1 - 4)\,dx\,dy = \int_0^1\int_0^1 -3\,dx\,dy = -3$$

7. $M = y^2 - x^2,\ N = x^2 + y^2 \Rightarrow \frac{\partial M}{\partial x} = -2x,\ \frac{\partial M}{\partial y} = 2y,\ \frac{\partial N}{\partial x} = 2x,\ \frac{\partial N}{\partial y} = 2y \Rightarrow \text{Flux} = \iint_R (-2x + 2y)\,dx\,dy$

$$= \int_0^3\int_0^x (-2x + 2y)dy\,dx = \int_0^3 (-2x^2 + x^2)\,dx = -9. \text{ Circ} = \iint_R (2x - 2y)\,dx\,dy =$$

$$\int_0^3\int_0^x (2x - 2y)\,dy\,dx = \int_0^3 x^2\,dx = 9$$

8. $M = x + y,\ N = -(x^2 + y^2) \Rightarrow \frac{\partial M}{\partial x} = 1,\ \frac{\partial M}{\partial y} = 1,\ \frac{\partial N}{\partial x} = -2x,\ \frac{\partial N}{\partial y} = -2y \Rightarrow \text{Flux} = \iint_R (1 - 2y)\,dx\,dy =$

$$\int_0^1\int_0^x (1 - 2y)\,dy\,dx = \int_0^1 (x - x^2)\,dx = \frac{1}{6}. \text{ Circ} = \iint_R (-2x - 1)\,dx\,dy = \int_0^1\int_0^x (-2x - 1)\,dy\,dx$$

$$= \int_0^1 (-2x^2 - x)\,dx = -\frac{7}{6}$$

9. $M = xy,\ N = y^2 \Rightarrow \frac{\partial M}{\partial x} = y,\ \frac{\partial M}{\partial y} = x,\ \frac{\partial N}{\partial x} = 0,\ \frac{\partial N}{\partial y} = 2y \Rightarrow \text{Flux} = \iint_R (y + 2y)\,dy\,dx = \int_0^1\int_{x^2}^x 3y\,dy\,dx$

$$= \int_0^1 \left(\frac{3x^2}{2} - \frac{3x^4}{2}\right)dx = \frac{1}{5}. \text{ Circ} = \iint_R -x\,dy\,dx = \int_0^1\int_{x^2}^x -x\,dy\,dx = \int_0^1 (-x^2 + x^3)\,dx = -\frac{1}{12}$$

10. $M = -\sin y,\ N = x\cos y \Rightarrow \frac{\partial M}{\partial x} = 0,\ \frac{\partial M}{\partial y} = -\cos y,\ \frac{\partial N}{\partial x} = \cos y,\ \frac{\partial N}{\partial y} = -x\sin y \Rightarrow$ Flux $=$

$$\iint_R (-x\sin y)\,dx\,dy = \int_0^{\pi/2}\int_0^{\pi/2} (-x\sin y)\,dx\,dy = \int_0^{\pi/2}\left(-\frac{\pi^2}{8}\sin y\right)dy = -\frac{\pi^2}{8}$$

$$\text{Circ} = \iint_R (\cos y - (-\cos y))\,dx\,dy = \int_0^{\pi/2}\int_0^{\pi/2} 2\cos y\,dx\,dy = \int_0^{\pi/2} \pi\cos y\,dy = \pi$$

11. $M = y^2,\ N = x^2 \Rightarrow \frac{\partial M}{\partial y} = 2y,\ \frac{\partial N}{\partial x} = 2y \Rightarrow \oint_C y^2\,dx + x^2\,dy = \iint_R (2x - 2y)dy\,dx =$

$$\int_0^1\int_0^{-x+1} (2x - 2y)\,dy\,dx = \int_0^1 (-3x^2 + 4x - 1)\,dx = 0.$$

12. $M = 3y,\ N = 2x \Rightarrow \frac{\partial M}{\partial y} = 3,\ \frac{\partial N}{\partial x} = 2 \Rightarrow \oint_C (3y\,dx + 2x\,dy) = \iint_R (2 - 3)\,dx\,dy = \int_0^{\pi}\int_0^{\sin x} -1\,dy\,dx$

$= -2$

13. $M = 6y + x,\ N = y + 2x \Rightarrow \frac{\partial M}{\partial y} = 6,\ \frac{\partial N}{\partial x} = 2 \Rightarrow \oint_C (6y + x)dx + (y + 2x)dy = \iint_R (2 - 6)\,dy\,dx =$

-4(Area of the circle) $= -16\pi$

14. $M = 2x + y^2,\ N = 2xy + 3y \Rightarrow \frac{\partial M}{\partial y} = 2y,\ \frac{\partial N}{\partial x} = 2y \Rightarrow \oint_C (2x + y^2)\,dx + (2xy + 3y)\,dy =$

$$\iint_R (2y - 2y)\,dx\,dy = 0$$

15. $M = 2xy^3,\ N = 4x^2y^2 \Rightarrow \frac{\partial M}{\partial y} = 6xy^2,\ \frac{\partial N}{\partial x} = 8xy^2 \Rightarrow \oint_C 2xy^3\,dx + 4x^2y^2\,dy = \iint_R (8xy^2 - 6xy^2)\,dx\,dy$

$$\int_0^1\int_0^{x^3} 2xy^2\,dy\,dx = \int_0^1 \frac{2}{3}x^{10}\,dx = \frac{2}{33}$$

16. $M = 4x - 2y,\ N = 2x - 4y \Rightarrow \frac{\partial M}{\partial y} = -2,\ \frac{\partial N}{\partial x} = 2 \Rightarrow \oint_C (4x - 2y)\,dx + (2x - 4y)\,dy =$

$$\iint_R (2 - (-2))\,dx\,dy = 4 \iint_R dx\,dy = 4(\text{Area of the circle}) = 16\pi$$

17. a) $M = f(x),\ N = g(y) \Rightarrow \frac{\partial M}{\partial y} = 0,\ \frac{\partial N}{\partial x} = 0 \Rightarrow \oint_C f(x)\,dx + g(y)\,dy = \iint_R 0\,dy\,dx = 0$

b) $M = ky,\ N = hx \Rightarrow \frac{\partial M}{\partial y} = k,\ \frac{\partial N}{\partial x} = h \Rightarrow \oint_C ky\,dx + hx\,dy = \iint_R (h - k)\,dx\,dy =$

$(h - k)(\text{Area of the region})$

18. $M = 4x^3y,\ N = x^4 \Rightarrow \frac{\partial M}{\partial y} = 4x^3,\ \frac{\partial N}{\partial x} = 4x^3 \Rightarrow \oint_C 4x^3y\,dx + x^4\,dy = \iint_R (4x^3 - 4x^3)\,dx\,dy = 0$

19. Area $= \frac{1}{2}\oint_C x\,dy - y\,dx$. $M = x = a\cos t,\ N = y = a\sin t \Rightarrow dx = -a\sin t\,dt,\ dy = a\cos t\,dt \Rightarrow$ Area $=$

$$\frac{1}{2}\int_0^{2\pi} \left(a^2\cos^2 t + a^2\sin^2 t\right)dt = \frac{1}{2}\int_0^{2\pi} a^2\,dt = \pi a^2$$

20. Area $= \frac{1}{2}\oint_C x\,dy - y\,dx$. $M = x = a\cos t,\ N = y = b\sin t \Rightarrow dx = -a\sin t\,dt,\ dy = b\cos t\,dt \Rightarrow$

$$\text{Area} = \frac{1}{2}\int_0^{2\pi} (ab\cos^2 t + ab\sin^2 t)\,dt = \frac{1}{2}\int_0^{2\pi} ab\,dt = \pi ab$$

21. Area $= \frac{1}{2}\oint_C x\,dy - y\,dx$. $M = x = \cos^3 t,\ N = y = \sin^3 t \Rightarrow dx = -3\cos^2 t\sin t\,dt,\ dy = 3\sin^2 t\cos t\,dt$

$$\Rightarrow \text{Area} = \frac{1}{2}\int_0^{2\pi} (3\sin^2 t\cos^2 t(\cos^2 t + \sin^2 t))\,dt = \frac{1}{2}\int_0^{2\pi} (3\sin^2 t\cos^2 t)\,dt = \frac{3\pi}{8}$$

22. Area $= \frac{1}{2}\oint_C x\,dy - y\,dx$. $M = x = t^2,\ N = y = \frac{t^3}{3} - t \Rightarrow dx = 2t\,dt,\ dy = (t^2 - 1)\,dt \Rightarrow$

$$\text{Area} = \frac{1}{2}\int_{-\sqrt{3}}^{\sqrt{3}} \left(\frac{1}{3}t^4 + t^2\right)dt = \frac{8}{5}\sqrt{3}$$

SECTION 15.4 SURFACE AREA AND SURFACE INTEGRALS

1. $\mathbf{p} = \mathbf{k}$, $\nabla f = 2x\,\mathbf{i} + 2y\,\mathbf{j} - \mathbf{k} \Rightarrow |\nabla f| = \sqrt{(2x)^2 + (2y)^2 + (-1)^2} = \sqrt{4x^2 + 4y^2 + 1}$. $|\nabla f \cdot \mathbf{p}| = 1 \Rightarrow$

$$S = \iint_R \frac{|\nabla f|}{|\nabla f \cdot \mathbf{p}|}\,dA = \iint_R \sqrt{4x^2 + 4y^2 + 1}\,dx\,dy =$$

$$\iint_R \sqrt{4r^2\cos^2\theta + 4r^2\sin^2\theta + 1}\;r\,dr\,d\theta = \int_0^{2\pi}\int_0^{\sqrt{2}} \sqrt{4r^2 + 1}\;r\,dr\,d\theta = \frac{13}{3}\pi$$

2. $\mathbf{p} = \mathbf{k}$. $\nabla f = 2x\,\mathbf{i} + 2y\,\mathbf{j} - \mathbf{k} \Rightarrow |\nabla f| = \sqrt{4x^2 + 4y^2 + 1}$. $|\nabla f \cdot \mathbf{p}| = 1 \Rightarrow S = \iint_R \frac{|\nabla f|}{|\nabla f \cdot \mathbf{p}|}\,dA =$

$$\iint_R \sqrt{4x^2 + 4y^2 + 1}\,dx\,dy = \iint_R \sqrt{4r^2 + 1}\;r\,dr\,d\theta = \int_0^{2\pi}\int_{\sqrt{2}}^{\sqrt{6}} \sqrt{4r^2 + 1}\;r\,dr\,d\theta =$$

$$\int_0^{2\pi} \frac{98}{12}\,d\theta = \frac{49\pi}{3}$$

3. $\mathbf{p} = \mathbf{k}$. $\nabla f = \mathbf{i} + 2\,\mathbf{j} + 2\,\mathbf{k} \Rightarrow |\nabla f| = 3$. $|\nabla f \cdot \mathbf{p}| = 2 \Rightarrow S = \iint_R \frac{|\nabla f|}{|\nabla f \cdot \mathbf{p}|}\,dA = \iint_R \frac{3}{2}\,dx\,dy =$

$$\int_{-1}^{1}\int_{y^2}^{2-y^2} \frac{3}{2}\,dx\,dy = \int_{-1}^{1} (3 - 3y^2)\,dy = 4$$

4. $\mathbf{p} = \mathbf{k}$. $\nabla f = 2x\,\mathbf{i} - 2\,\mathbf{k} \Rightarrow |\nabla f| = \sqrt{4x^2 + 4} = 2\sqrt{x^2 + 1}$. $|\nabla f \cdot \mathbf{p}| = 2 \Rightarrow S = \iint_R \frac{|\nabla f|}{|\nabla f \cdot \mathbf{p}|}\,dA =$

$$\iint_R \frac{2\sqrt{x^2 + 1}}{2}\,dx\,dy = \int_0^{\sqrt{3}}\int_0^{x} \sqrt{x^2 + 1}\,dy\,dx = \int_0^{\sqrt{3}} x\sqrt{x^2 + 1}\,dx = \frac{7}{3}$$

5. $\mathbf{p} = \mathbf{k}$. $\nabla f = 2x\,\mathbf{i} - 2\,\mathbf{j} - 2\,\mathbf{k} \Rightarrow |\nabla f| = \sqrt{(2x)^2 + (-2)^2 + (-2)^2} = \sqrt{4x^2 + 8}$. $|\nabla f \cdot \mathbf{p}| = 2 \Rightarrow S =$

$$\iint_R \frac{|\nabla f|}{|\nabla f\cdot \mathbf{p}|}\,dA = \iint_R \frac{\sqrt{4x^2+8}}{2}\,dx\,dy = \int_0^2\int_0^{3x} \sqrt{x^2+2}\,dy\,dx = \int_0^2 3x\sqrt{x^2+2}\,dx = 6\sqrt{6}-2\sqrt{2}$$

6. $\mathbf{p}=\mathbf{k}$. $\nabla f = \left(2x-\frac{2}{x}\right)\mathbf{i}+\sqrt{15}\,\mathbf{j}-\mathbf{k} \Rightarrow |\nabla f| = \sqrt{\left(2x-\frac{2}{x}\right)^2+\left(\sqrt{15}\right)^2+(-1)^2} = \sqrt{4x^2+8+\frac{4}{x^2}} =$ $\sqrt{\left(2x+\frac{2}{x}\right)^2} = 2x+\frac{2}{x}$ on $1 \le x \le 2$. $|\nabla f\cdot\mathbf{p}| = 1 \Rightarrow S = \iint_R \frac{|\nabla f|}{|\nabla f\cdot \mathbf{p}|}\,dA = \iint_R (2x+2x^{-1})\,dx\,dy =$

$$\int_0^1\int_1^2 (2x+2x^{-1})\,dx\,dy = \int_0^1 (3+2\ln 2)\,dy = 3+2\ln 2$$

7. $\mathbf{p}=\mathbf{k}$. $\nabla f = 2x\,\mathbf{i}+2y\,\mathbf{j}+2z\,\mathbf{k} \Rightarrow |\nabla f| = \sqrt{4x^2+4y^2+4z^2} = \sqrt{8} = 2\sqrt{2}$. $|\nabla f\cdot\mathbf{p}| = 2z \Rightarrow S =$ $\iint_R \frac{|\nabla f|}{|\nabla f\cdot \mathbf{p}|}\,dA = \iint_R \frac{2\sqrt{2}}{2z}\,dA = \sqrt{2}\iint_R \frac{1}{z}\,dA =$

$$\sqrt{2}\iint_R \frac{1}{\sqrt{2-(x^2+y^2)}}\,dA = \sqrt{2}\int_0^{2\pi}\int_0^1 \frac{r\,dr\,d\theta}{\sqrt{2-r^2}} = \sqrt{2}\int_0^{2\pi}(-1+\sqrt{2})\,d\theta = 2\pi(2-\sqrt{2})$$

8. $\mathbf{p}=\mathbf{k}$. $\nabla f = c\,\mathbf{i}-\mathbf{k} \Rightarrow |\nabla f| = \sqrt{c^2+1}$. $|\nabla f\cdot\mathbf{p}| = 1 \Rightarrow S = \iint_R \frac{|\nabla f|}{|\nabla f\cdot \mathbf{p}|}\,dA = \iint_R \sqrt{c^2+1}\,dx\,dy =$

$$\int_0^{2\pi}\int_0^1 \sqrt{c^2+1}\,r\,dr\,d\theta = \int_0^{2\pi} \frac{\sqrt{c^2+1}}{2}\,d\theta = \pi\sqrt{c^2+1}$$

9. $\mathbf{p}=\mathbf{k}$. $\nabla f = 2x\,\mathbf{i}+2z\,\mathbf{j} \Rightarrow |\nabla f| = \sqrt{(2x)^2+(2z)^2} = 2$. $|\nabla f\cdot\mathbf{p}| = 2z$ for the upper surface, $z \ge 0 \Rightarrow$ $S = \iint_R \frac{|\nabla f|}{|\nabla f\cdot \mathbf{p}|}\,dA = 2\iint_R \frac{2}{2z}\,dA = 2\iint_R \frac{1}{z}\,dA = 2\iint_R \frac{1}{\sqrt{1-x^2}}\,dy\,dx =$

$$4\int_{-1/2}^{1/2}\int_0^{1/2} \frac{1}{\sqrt{1-x^2}}\,dy\,dx = 2\int_{-1/2}^{1/2} \frac{1}{\sqrt{1-x^2}}\,dx = \frac{2\pi}{3}$$

10. $\mathbf{p}=\mathbf{j}$. $\nabla f = 2x\,\mathbf{i}+\mathbf{j}+2z\,\mathbf{k} \Rightarrow |\nabla f| = \sqrt{4x^2+4z^2+1}$. $|\nabla f\cdot\mathbf{p}| = 1 \Rightarrow S = \iint_R \frac{|\nabla f|}{|\nabla f\cdot \mathbf{p}|}\,dA =$

10. (Continued)

$$\iint_R \sqrt{4x^2 + 4z^2 + 1}\, dx\, dz = \int_0^{2\pi}\int_0^1 \sqrt{4r^2 + 1}\, r\, dr\, d\theta = \int_0^{2\pi} \left(\frac{5\sqrt{5} - 1}{12}\right) d\theta = \frac{\pi(5\sqrt{5} - 1)}{6}$$

11. The bottom face of the cube is in the xy–plane $\Rightarrow z = 0 \Rightarrow g(x,y,0) = x + y$ and $f(x,y,z) = z = 0 \Rightarrow$

$\nabla f = \mathbf{k} \Rightarrow |\nabla f| = 1$. $\mathbf{p} = \mathbf{k} \Rightarrow |\nabla f\cdot\mathbf{p}| = 1 \Rightarrow d\sigma = dx\, dy \Rightarrow \displaystyle\iint_{z=0} (x + y)\, dx\, dy = \int_0^1\int_0^1 (x + y)\, dx\, dy =$

1. Because of symmetry, you get 1 over the face of the cube in the xz–plane and 1 over the face of the cube in the yz–plane.
In the top of the cube, $g(x,y,z) = g(x,y,1) = x + y + 1$ and $f(x,y,z) = z = 1 \Rightarrow \nabla f = \mathbf{k} \Rightarrow |\nabla f| = 1$. $\mathbf{p} = \mathbf{k} \Rightarrow$

$|\nabla f\cdot\mathbf{p}| = 1 \Rightarrow d\sigma = dx\, dy \Rightarrow \displaystyle\iint_{z=1} (x + y + 1)\, dx\, dy = \int_0^1\int_0^1 (x + y + 1)\, dx\, dy = 2$. Because of

symmetry, the integral is 2 over each of the other two faces. $\therefore \displaystyle\iint_{\text{cube}} (x + y + z)\, d\sigma = 9$.

12. On the face in the xz–plane, $y = 0 \Rightarrow f(x,y,z) = y = 0$ and $g(x,y,z) = g(x,0,z) = z \Rightarrow \nabla f = \mathbf{j} \Rightarrow |\nabla f| = 1$.

$\mathbf{p} = \mathbf{j} \Rightarrow |\nabla f\cdot\mathbf{p}| = 1 \Rightarrow d\sigma = dx\, dz \Rightarrow \displaystyle\iint_{y=0} (y + z)\, d\sigma = \int_0^1\int_0^2 z\, dx\, dz = 1$

On the face in the xy–plane, $z = 0 \Rightarrow f(x,y,z) = z = 0$ and $g(x,y,z) = g(x,y,0) = y \Rightarrow \nabla f = \mathbf{k} \Rightarrow |\nabla f| = 1$.

$\mathbf{p} = \mathbf{k} \Rightarrow |\nabla f\cdot\mathbf{p}| = 1 \Rightarrow d\sigma = dx\, dy \Rightarrow \displaystyle\iint_{z=0} (y + z)\, d\sigma = \int_0^1\int_0^2 y\, dx\, dy = 1$.

On the triangular face in the plane $x = 2$, $f(x,y,z) = x = 2$ and $g(x,y,z) = g(2,y,z) = y + z \Rightarrow \nabla f = \mathbf{i} \Rightarrow$

$|\nabla f| = 1$. $\mathbf{p} = \mathbf{i} \Rightarrow |\nabla f\cdot\mathbf{p}| = 1 \Rightarrow d\sigma = dz\, dy \Rightarrow \displaystyle\iint_{x=2} (y + z)\, d\sigma = \int_0^1\int_0^{1-y} (y + z)\, dz\, dy = \frac{1}{3}$

On the triangular face in the yz–plane, $x = 0 \Rightarrow f(x,y,z) = x = 0$ and $g(x,y,z) = g(0,y,z) = y + z \Rightarrow \nabla f = \mathbf{i}$

$\Rightarrow |\nabla f| = 1$. $\mathbf{p} = \mathbf{i} \Rightarrow |\nabla f\cdot\mathbf{p}| = 1 \Rightarrow d\sigma = dz\, dy \Rightarrow \displaystyle\iint_{x=0} (y + z)\, d\sigma = \int_0^1\int_0^{1-y} (y + z)\, dz\, dy = \frac{1}{3}$

On the sloped face, $y + z = 1$, $f(x,y,z) = y + z = 1$ and $g(x,y,z) = y + z = 1 \Rightarrow \nabla f = \mathbf{j} + \mathbf{k} \Rightarrow |\nabla f| = \sqrt{2}$.

12. (Continued)

$\mathbf{p} = \mathbf{k} \Rightarrow |\nabla f \cdot \mathbf{p}| = 1 \Rightarrow d\sigma = \sqrt{2}\,dx\,dy \Rightarrow \iint_{y+z=1} (y+z)\,d\sigma = \int_0^1\int_0^2 \sqrt{2}\,dx\,dy = 2\sqrt{2}$

$\therefore \iint_{wedge} g(x,y,z)\,d\sigma = \frac{8}{3} + 2\sqrt{2}$

13. On the faces in the coordinate planes, $g(x,y,z) = 0 \Rightarrow$ the integral over these faces is 0.
On the face, $x = a$, $f(x,y,z) = x = a$ and $g(x,y,z) = g(a,y,z) = ayz \Rightarrow \nabla f = \mathbf{i} \Rightarrow |\nabla f| = 1$. $\mathbf{p} = \mathbf{i} \Rightarrow |\nabla f \cdot \mathbf{p}| = 1$

$\Rightarrow d\sigma = dy\,dz \Rightarrow \iint_{x=a} xyz\,d\sigma = \int_0^c\int_0^b ayz\,dy\,dz = \frac{ab^2c^2}{4}$

On the face, $y = b$, $f(x,y,z) = y = b$ and $g(x,y,z) = g(x,b,z) = bxz \Rightarrow \nabla f = \mathbf{j} \Rightarrow |\nabla f| = 1$. $\mathbf{p} = \mathbf{j} \Rightarrow |\nabla f \cdot \mathbf{p}| = 1$

$\Rightarrow d\sigma = dx\,dz \Rightarrow \iint_{y=b} xyz\,dx\,dz = \int_0^c\int_0^a bxz\,dz\,dx = \frac{a^2bc^2}{4}$

On the face, $z = c$, $f(x,y,z) = z = c$ and $g(x,y,z) = g(x,y,c) = cxy \Rightarrow \nabla f = \mathbf{k} \Rightarrow |\nabla f| = 1$. $\mathbf{p} = \mathbf{k} \Rightarrow |\nabla f \cdot \mathbf{p}| = 1$

$\Rightarrow d\sigma = dy\,dx \Rightarrow \iint_{z=c} xyz\,d\sigma = \int_0^b\int_0^a cxy\,dx\,dy = \frac{a^2b^2c}{4}$

$\therefore \iint_S g(x,y,z)\,d\sigma = \frac{abc(ab + ac + bc)}{4}$

14. On the face, $x = a$, $f(x,y,z) = x = a$ and $g(x,y,z) = g(a,y,z) = ayz \Rightarrow \nabla f = \mathbf{i} \Rightarrow |\nabla f| = 1$. $\mathbf{p} = \mathbf{i} \Rightarrow |\nabla f \cdot \mathbf{p}| =$

$1 \Rightarrow d\sigma = dz\,dy \Rightarrow \iint_{x=a} xyz\,d\sigma = \int_{-b}^{b}\int_{-c}^{c} ayz\,dz\,dy = 0$ Because of the symmetry of g on all

the other faces, $\iint_S g(x,y,z)\,d\sigma = 0$

15. $\nabla f = 2\,\mathbf{i} + 2\,\mathbf{j} + \mathbf{k}$ and $g(x,y,z) = x + y + (2 - 2x - 2y) = 2 - x - y \Rightarrow |\nabla f| = 3$. $\mathbf{p} = \mathbf{k} \Rightarrow |\nabla f \cdot \mathbf{p}| = 1 \Rightarrow$

$d\sigma = 3\,dy\,dx \Rightarrow \iint_S (x + y + z)\,d\sigma = 3\int_0^1\int_0^{1-x} (2 - x - y)\,dy\,dx = 2$

16. $f(x,y,z) = y^2 + 4z = 16 \Rightarrow \nabla f = 2y\,\mathbf{j} + 4\,\mathbf{k} \Rightarrow |\nabla f| = \sqrt{4y^2 + 16} = 2\sqrt{y^2 + 4}$. $\mathbf{p} = \mathbf{k} \Rightarrow |\nabla f \cdot \mathbf{p}| = 4 \Rightarrow$

$$d\sigma = \frac{2\sqrt{y^2+4}}{4}\,dx\,dy \Rightarrow \iint_R g(x,y,z)\,d\sigma = \int_{-4}^{4}\int_0^1 x\sqrt{y^2+4}\,\frac{2\sqrt{y^2+4}}{4}\,dx\,dy =$$

$$\int_{-4}^{4}\int_0^1 \frac{x(y^2+4)}{2}\,dx\,dy = \frac{56}{3}$$

17. $\nabla G = 2x\,\mathbf{i} + 2y\,\mathbf{j} + 2z\,\mathbf{k} \Rightarrow |\nabla G| = \sqrt{4x^2 + 4y^2 + 4z^2} = 2a$. $\mathbf{n} = \dfrac{2x\,\mathbf{i} + 2y\,\mathbf{j} + 2z\,\mathbf{k}}{2\sqrt{x^2+y^2+z^2}} = \dfrac{x\,\mathbf{i} + y\,\mathbf{j} + z\,\mathbf{k}}{a} \Rightarrow$

$\mathbf{F} \cdot \mathbf{n} = \dfrac{z^2}{a}$. $|\nabla G \cdot \mathbf{k}| = 2z \Rightarrow d\sigma = \dfrac{2a}{2z}\,dA = \dfrac{a}{z}\,dA$. $\therefore$ Flux $= \displaystyle\iint_R \frac{z^2}{a}\left(\frac{a}{z}\right) dA = \iint_R z\,dA =$

$$\iint_R \sqrt{a^2 - (x^2+y^2)}\,dx\,dy = \int_0^{\pi/2}\int_0^a \sqrt{a^2 - r^2}\;r\,dr\,d\theta = \frac{a^3\pi}{6}$$

18. $\nabla G = 2x\,\mathbf{i} + 2y\,\mathbf{j} + 2z\,\mathbf{k} \Rightarrow |\nabla G| = \sqrt{4x^2 + 4y^2 + 4z^2} = 2a$. $\mathbf{n} = \dfrac{2x\,\mathbf{i} + 2y\,\mathbf{j} + 2z\,\mathbf{k}}{2\sqrt{x^2+y^2+z^2}} = \dfrac{x\,\mathbf{i} + y\,\mathbf{j} + z\,\mathbf{k}}{a} \Rightarrow$

$\mathbf{F} \cdot \mathbf{n} = \dfrac{-xy}{a} + \dfrac{xy}{a} = 0$. $|\nabla G \cdot \mathbf{k}| = 2z \Rightarrow d\sigma = \dfrac{2a}{2z}\,dA = \dfrac{a}{z}\,dA$. $\therefore$ Flux $= \displaystyle\iint_S \mathbf{F} \cdot \mathbf{n}\,dS =$

$$\iint_S 0\,d\sigma = 0$$

19. $\mathbf{n} = \dfrac{x\,\mathbf{i} + y\,\mathbf{j} + z\,\mathbf{k}}{a}$, $d\sigma = \dfrac{a}{z}\,dA$ (See Exercise 17) and $\mathbf{F} \cdot \mathbf{n} = \dfrac{xy}{a} - \dfrac{xy}{a} + \dfrac{z}{a} = \dfrac{z}{a}$.

$\therefore$ Flux $= \displaystyle\iint_R \frac{z}{a}\left(\frac{a}{z}\right) dA = \iint_R 1\,dA = \frac{\pi a^2}{4}$

20. $\mathbf{n} = \dfrac{x\,\mathbf{i} + y\,\mathbf{j} + z\,\mathbf{k}}{a}$, $d\sigma = \dfrac{a}{z}\,dA$ (See Exercise 17) and $\mathbf{F} \cdot \mathbf{n} = \dfrac{zx^2}{a} + \dfrac{zy^2}{a} + \dfrac{z^3}{a} = z\left(\dfrac{x^2+y^2+z^2}{a}\right) = az \Rightarrow$

Flux $= \displaystyle\iint_R za\left(\frac{a}{z}\right) dx\,dy = \iint_R a^2\,dx\,dy = a^2(\text{Area of R}) = \frac{1}{4}\pi a^4$

21. $\mathbf{n} = \frac{x\,\mathbf{i} + y\,\mathbf{j} + z\,\mathbf{k}}{a}$, $d\sigma = \frac{a}{z}\,dA$ (See Exercise 17) and $\mathbf{F} \cdot \mathbf{n} = \frac{x^2}{a} + \frac{y^2}{a} + \frac{z^2}{a} = a \Rightarrow \text{Flux} = \iint_R a\left(\frac{a}{z}\right) dA$

$$= \iint_R \frac{a^2}{z}\,dA = \iint_R \frac{a^2}{\sqrt{a^2 - (x^2 + y^2)}}\,dA = \int_0^{\pi/2}\int_0^a \frac{a^2}{\sqrt{a^2 - r^2}}\,r\,dr\,d\theta = \frac{a^3\pi}{2}$$

22. $\mathbf{n} = \frac{x\,\mathbf{i} + y\,\mathbf{j} + z\,\mathbf{k}}{a}$, $d\sigma = \frac{a}{z}\,dA$ (See Exercise 17) and $\mathbf{F} \cdot \mathbf{n} = \frac{x^2/a + y^2/a + z^2/a}{\sqrt{x^2 + y^2 + z^2}} = \frac{a^2/a}{a} = 1 \Rightarrow$

$$\text{Flux} = \iint_R \frac{a}{z}\,dx\,dy = \iint_R \frac{a}{\sqrt{a^2 - (x^2 + y^2)}}\,dx\,dy = \int_0^{\pi/2}\int_0^a \frac{a}{\sqrt{a^2 - r^2}}\,r\,dr\,d\theta = \frac{a^2\pi}{2}$$

23. $\nabla G = 2y\,\mathbf{i} + \mathbf{k} \Rightarrow |\nabla G| = \sqrt{4y^2 + 1} \Rightarrow \mathbf{n} = \frac{2y\,\mathbf{i} + \mathbf{k}}{\sqrt{4y^2 + 1}} \Rightarrow \mathbf{F} \cdot \mathbf{n} = \frac{2xy - 3z}{\sqrt{4y^2 + 1}}$. $\mathbf{p} = \mathbf{k} \Rightarrow |\nabla G \cdot \mathbf{k}| = 1 \Rightarrow$

$$d\sigma = \sqrt{4y^2 + 1}\,dA \Rightarrow \text{Flux} = \iint_R \left(\frac{2xy - 3z}{\sqrt{4y^2 + 1}}\right)\sqrt{4y^2 + 1}\,dA = \iint_R (2xy - 3z)\,dA =$$

$$\iint_R (2xy - 3(4 - y^2))\,dA = \int_0^1\int_{-2}^2 (2xy - 12 + 3y^2)\,dy\,dx = -32$$

24. $\nabla G = -2x\,\mathbf{i} - 2y\,\mathbf{j} + \mathbf{k} \Rightarrow |\nabla G| = \sqrt{4x^2 + 4y^2 + 1} = \sqrt{4(x^2 + y^2) + 1}$. $\mathbf{p} = \mathbf{k} \Rightarrow |\nabla G \cdot \mathbf{k}| = 1$

$\mathbf{n} = \frac{2x\,\mathbf{i} + 2y\,\mathbf{j} + \mathbf{k}}{\sqrt{4(x^2 + y^2) + 1}} \Rightarrow \mathbf{F} \cdot \mathbf{n} = \frac{8x^2 + 8y^2 - 2}{\sqrt{4(x^2 + y^2) + 1}}$. $d\sigma = \sqrt{4(x^2 + y^2) + 1}\,dx\,dy \Rightarrow$

$$\text{Flux} = \iint_R \frac{8x^2 + 8y^2 - 2}{\sqrt{4(x^2 + y^2) + 1}}\sqrt{4(x^2 + y^2) + 1}\,dx\,dy = \iint_R (8x^2 + 8y^2 - 2)\,dx\,dy =$$

$$\int_0^{2\pi}\int_0^1 (8r^2 - 2)\,r\,dr\,d\theta = 2\pi$$

25. $\nabla G = -e^x\,\mathbf{i} + \mathbf{j} \Rightarrow |\nabla G| = \sqrt{e^{2x} + 1}$. $\mathbf{p} = \mathbf{i} \Rightarrow |\nabla G \cdot \mathbf{i}| = e^x$. $\mathbf{n} = \frac{e^x\,\mathbf{i} - \mathbf{j}}{\sqrt{e^{2x} + 1}} \Rightarrow \mathbf{F} \cdot \mathbf{n} = \frac{-2e^x - 2y}{\sqrt{e^{2x} + 1}}$.

$$d\sigma = \frac{\sqrt{e^{2x} + 1}}{e^x}\,dA \Rightarrow \text{Flux} = \iint_R \frac{-2e^x - 2y}{\sqrt{e^{2x} + 1}}\left(\frac{\sqrt{e^{2x} + 1}}{e^x}\right) dA = \iint_R -4\,dA = \int_0^1\int_1^2 -4\,dy\,dz$$

$$= -4$$

26. $\nabla G = -\frac{1}{x}\mathbf{i} + \mathbf{j} \Rightarrow |\nabla G| = \sqrt{\frac{1}{x^2} + 1} = \frac{\sqrt{1+x^2}}{x}$ since $1 \le x \le e$. $\mathbf{p} = \mathbf{j} \Rightarrow |\nabla G \cdot \mathbf{j}| = 1 \Rightarrow \mathbf{n} = \dfrac{-\frac{1}{x}\mathbf{i} + \mathbf{j}}{\frac{\sqrt{1+x^2}}{x}} =$

$\frac{-\mathbf{i} + x\mathbf{j}}{\sqrt{1+x^2}} \Rightarrow \mathbf{F} \cdot \mathbf{n} = \frac{2xy}{\sqrt{1+x^2}}$. $d\sigma = \frac{\sqrt{1+x^2}}{x}\,dx\,dz \Rightarrow$ Flux $= \iint\limits_R \frac{2xy}{\sqrt{1+x^2}}\left(\frac{\sqrt{1+x^2}}{x}\right) dx\,dz =$

$\int_0^1 \int_1^e 2y\,dx\,dz = \int_1^e \int_0^1 2\ln x\,dz\,dx = 2$

27. $\nabla F = 2x\,\mathbf{i} + 2y\,\mathbf{j} + 2z\,\mathbf{k} \Rightarrow |\nabla F| = \sqrt{4x^2 + 4y^2 + 4z^2} = 2$. $\mathbf{p} = \mathbf{k} \Rightarrow |\nabla F \cdot \mathbf{k}| = 2z$ since $z \ge 0 \Rightarrow d\sigma =$

$\frac{2}{2z}\,dA = \frac{1}{z}\,dA$. $\therefore$ $M = \iint\limits_S \delta\,d\sigma = \frac{\pi}{2}\delta$. $M_{xy} = \iint\limits_S z\delta\,d\sigma = \delta \iint\limits_S z\left(\frac{1}{z}\right) dA =$

$\delta \int_0^1 \int_0^{\sqrt{1-x^2}} dy\,dx = \frac{\pi}{4}\delta$. $\therefore$ $\bar{z} = \dfrac{\frac{\pi}{4}\delta}{\frac{\pi}{2}\delta} = \frac{1}{2}$. Because of symmetry, $\bar{x} = \bar{y} = \frac{1}{2}$. $\therefore$ Centroid $= \left(\frac{1}{2}, \frac{1}{2}, \frac{1}{2}\right)$

28. $\nabla F = 2y\,\mathbf{i} + 2z\,\mathbf{k} \Rightarrow |\nabla F| = \sqrt{4y^2 + 4z^2} = \sqrt{4(y^2 + z^2)} = 6$. $\mathbf{p} = \mathbf{k} \Rightarrow |\nabla F \cdot \mathbf{k}| = 2z$ since $z \ge 0 \Rightarrow$

$d\sigma = \frac{6}{2z}\,dx\,dy = \frac{3}{z}\,dx\,dy$. $\therefore$ $M = \iint\limits_S 1\,d\sigma = \int_{-3}^{3}\int_0^3 \frac{3}{z}\,dx\,dy = \int_{-3}^{3}\int_0^3 \frac{3}{\sqrt{9-y^2}}\,dx\,dy = 9\pi$.

$M_{xy} = \iint\limits_S z\,d\sigma = \int_{-3}^{3}\int_0^3 z\left(\frac{3}{z}\right) dx\,dy = 54$. $M_{xz} = \iint\limits_S y\,d\sigma = \int_{-3}^{3}\int_0^3 y\left(\frac{3}{z}\right) dx\,dy =$

$\int_{-3}^{3}\int_0^3 \frac{3y}{\sqrt{9-y^2}}\,dx\,dy = 0$. $M_{yz} = \iint\limits_S x\,d\sigma = \int_{-3}^{3}\int_0^3 \frac{3x}{\sqrt{9-y^2}}\,dx\,dy = \frac{27}{2}\pi$

$\therefore$ $\bar{x} = \dfrac{\frac{27}{2}\pi}{9\pi} = \frac{3}{2}$, $\bar{y} = 0$, $\bar{z} = \frac{54}{9\pi} = \frac{6}{\pi}$

29. Because of symmetry, $\bar{x} = \bar{y} = 0$. $M = \iint_S \delta\, d\sigma = \delta \iint_S d\sigma = \delta(\text{Area of S}) = 3\pi\sqrt{2}\,\delta$.

$\nabla F = 2x\,\mathbf{i} + 2y\,\mathbf{j} - 2z\,\mathbf{k} \Rightarrow |\nabla F| = \sqrt{4x^2 + 4y^2 + 4z^2} = 2\sqrt{x^2 + y^2 + z^2}$. $\mathbf{p} = \mathbf{k} \Rightarrow |\nabla F \cdot \mathbf{k}| = 2z \Rightarrow$

$d\sigma = \frac{2\sqrt{x^2 + y^2 + z^2}}{2z}\,dA = \frac{\sqrt{x^2 + y^2 + z^2}}{z}\,dA = \frac{\sqrt{x^2 + y^2 + (x^2 + y^2)}}{z}\,dA = \frac{\sqrt{2}\sqrt{x^2 + y^2}}{z}$.

$\therefore M_{xy} = \delta \iint_S z\left(\frac{\sqrt{2}\sqrt{x^2 + y^2}}{z}\right) dA = \delta \iint_S \sqrt{2}\sqrt{x^2 + y^2}\,dA = \delta \int_0^{2\pi}\int_1^2 \sqrt{2}\,r^2\,dr\,d\theta =$

$\frac{14\pi\sqrt{2}}{3}\delta$. $\bar{z} = \frac{\frac{14\pi\sqrt{2}}{3}\delta}{3\pi\sqrt{2}\,\delta} = \frac{14}{9}$. $\therefore (\bar{x}, \bar{y}, \bar{z}) = \left(0, 0, \frac{14}{9}\right)$. $I_z = \iint_S (x^2 + y^2)\,\delta\,d\sigma =$

$\iint_S (x^2 + y^2)\left(\frac{\sqrt{2}\sqrt{x^2 + y^2}}{z}\right)\delta\,dA = \delta\sqrt{2} \iint_S (x^2 + y^2)\,dA = \delta\sqrt{2} \int_0^{2\pi}\int_1^2 r^3\,dr\,d\theta = \frac{15\pi\sqrt{2}}{2}\delta$.

$R_z = \sqrt{I_z/M} = \frac{\sqrt{10}}{2}$

30. $\nabla f = 8x\,\mathbf{i} + 8y\,\mathbf{j} - 2z\,\mathbf{k} \Rightarrow |\nabla f| = \sqrt{64x^2 + 64y^2 + 4z^2} = 2\sqrt{16x^2 + 16y^2 + z^2} = 2\sqrt{4z^2 + z^2} = 2\sqrt{5}\,z$ since $z \geq 0$. $\mathbf{p} = \mathbf{k} \Rightarrow |\nabla f \cdot \mathbf{k}| = 2z \Rightarrow d\sigma = \frac{2\sqrt{5}\,z}{2z}\,dA = \sqrt{5}\,dA \Rightarrow I_z = \iint_S (x^2 + y^2)\,\delta\,d\sigma =$

$\delta\sqrt{5} \iint_R (x^2 + y^2)\,dx\,dy = \delta\sqrt{5} \int_{-\pi/2}^{\pi/2}\int_0^{2\cos\theta} r^2\,r\,dr\,d\theta = \frac{3\sqrt{5}\,\pi\delta}{2}$

31. $f_x(x,y) = 2x$, $f_y(x,y) = 2y \Rightarrow \sqrt{f_x^2 + f_y^2 + 1} = \sqrt{4x^2 + 4y^2 + 1} \Rightarrow \text{Area} = \iint_R \sqrt{4x^2 + 4y^2 + 1}\,dx\,dy =$

$\int_0^{2\pi}\int_0^{\sqrt{3}} \sqrt{4r^2 + 1}\,r\,dr\,d\theta = \frac{\pi}{6}\left(13\sqrt{13} - 1\right)$

32. $f_x(x,y) = \frac{x}{\sqrt{x^2 + y^2}}$, $f_y(x,y) = \frac{y}{\sqrt{x^2 + y^2}} \Rightarrow \sqrt{f_x^2 + f_y^2 + 1} = \sqrt{\frac{x^2}{x^2 + y^2} + \frac{y^2}{x^2 + y^2} + 1} = \sqrt{2} \Rightarrow$

$\text{Area} = \iint_{R_{xy}} \sqrt{2}\,dx\,dy = \sqrt{2}(\text{Area between the ellipse and the circle}) = \sqrt{2}(6\pi - \pi) = 5\pi\sqrt{2}$

33. $f_z(y,z) = -2y,\ f_y(y,z) = -2z \Rightarrow \sqrt{f_y^2 + f_z^2 + 1} = \sqrt{4y^2 + 4z^2 + 1} \Rightarrow \text{Area} = \iint_R \sqrt{4y^2 + 4z^2 + 1}\ dy\,dz$

$= \int_0^{2\pi}\int_0^1 \sqrt{4r^2 + 1}\ r\,dr\,d\theta = \frac{\pi}{6}(5\sqrt{5} - 1)$

34. Over R_{xy}: $z = 2 - \frac{2}{3}x - 2y$ $\quad f_x(x,y) = -\frac{2}{3},\ f_y(x,y) = -2 \Rightarrow \sqrt{f_x^2 + f_y^2 + 1} = \sqrt{\frac{4}{9} + 4 + 1} = \frac{7}{3} \Rightarrow$

Area $= \iint_{R_{xy}} \frac{7}{3}\,dA = \frac{7}{3}$(Area of the triangle in the xy–plane) $= \frac{7}{2}$. Over R_{xz}: $y = -\frac{1}{3}x - \frac{1}{2}z \Rightarrow$

$f_x(x,z) = -\frac{1}{3},\ f_z(x,z) = -\frac{1}{2} \Rightarrow \sqrt{f_x^2 + f_z^2 + 1} = \sqrt{\frac{1}{9} + \frac{1}{4} + 1} = \frac{7}{6} \Rightarrow$ Area $= \iint_{R_{xz}} \frac{7}{6}\,dA =$

$\frac{7}{6}$(Area of the triangle in the xz–plane) $= \frac{7}{2}$. Over R_{yz}: $x = -3y - \frac{3}{2}z \Rightarrow f_y(y,z) = -3,\ f_z(y,z) = -\frac{3}{2} \Rightarrow$
$\sqrt{f_y^2 + f_z^2 + 1} = \sqrt{9 + \frac{9}{4} + 1} = \frac{7}{2}$

$\Rightarrow$ Area $= \iint_{R_{yz}} \frac{7}{2}\,dA = \frac{7}{2}$(Area of the triangle in the yz–plane) $= \frac{7}{2}$

35. $y = \frac{z^2}{2} \Rightarrow f_x(x,z) = 0,\ f_z(x,z) = z \Rightarrow \sqrt{f_x^2 + f_z^2 + 1} = \sqrt{z^2 + 1} \Rightarrow \text{Area} = \int_0^2\int_0^1 \sqrt{z^2 + 1}\ dx\,dz =$

$\sqrt{5} + \frac{1}{2}\ln\left(\sqrt{5} + 2\right)$ (Note: On integrating the second time with respect to z, use the substitution $z = \tan\theta$ which means the integration will go from 0 to $\tan^{-1}2$.)

36. $y = 4 - z \Rightarrow f_x(x,z) = 0,\ f_z(x,z) = -1 \Rightarrow \sqrt{f_x^2 + f_z^2 + 1} = \sqrt{2} \Rightarrow \text{Area} = \iint_{R_{xz}} \sqrt{2}\,dA =$

$\int_0^2\int_0^{4-z^2} \sqrt{2}\,dx\,dz = \frac{16\sqrt{2}}{3}$

SECTION 15.5 THE DIVERGENCE THEOREM

1. $\frac{\partial}{\partial x}(y - x) = -1, \frac{\partial}{\partial y}(z - y) = -1, \frac{\partial}{\partial z}(y - x) = 0 \Rightarrow \nabla \cdot \mathbf{F} = -2 \Rightarrow$ Flux $= \int_{-1}^{1}\int_{-1}^{1}\int_{-1}^{1} -2 \, dx \, dy \, dz = -16$

2. $\frac{\partial}{\partial x}(x^2) = 2x, \frac{\partial}{\partial y}(y^2) = 2y, \frac{\partial}{\partial x}(z^2) = 2z \Rightarrow \nabla \cdot \mathbf{F} = 2x + 2y + 2z \Rightarrow$

$$\text{Flux} = \int_{0}^{1}\int_{0}^{1}\int_{0}^{1} (2x + 2y + 2z) \, dx \, dy \, dz = 3.$$

3. $\frac{\partial}{\partial x}(x^2) = 2x, \frac{\partial}{\partial y}(y^2) = 2y, \frac{\partial}{\partial x}(z^2) = 2z \Rightarrow \nabla \cdot \mathbf{F} = 2x + 2y + 2z \Rightarrow$

$$\text{Flux} = \int_{-1}^{1}\int_{-1}^{1}\int_{-1}^{1} (2x + 2y + 2z) \, dx \, dy \, dz = 0$$

4. $\frac{\partial}{\partial x}(x^2) = 2x, \frac{\partial}{\partial y}(y^2) = 2y, \frac{\partial}{\partial z}(z^2) = 2z \Rightarrow \nabla \cdot \mathbf{F} = 2x + 2y + 2z \Rightarrow$

$$\text{Flux} = \iiint_D (2x + 2y + 2z) \, dx \, dy \, dz = \int_{0}^{1}\int_{0}^{2\pi}\int_{0}^{2} (2r\cos\theta + 2r\sin\theta + 2z) \, r \, dr \, d\theta \, dz = 4\pi$$

5. $\frac{\partial}{\partial x}(y) = 0, \frac{\partial}{\partial y}(xy) = x, \frac{\partial}{\partial z}(-z) = -1 \Rightarrow \nabla \cdot \mathbf{F} = x - 1 \Rightarrow$ Flux $= \iiint_{\text{solid}} (x - 1) \, dz \, dy \, dx =$

$$\int_{0}^{2\pi}\int_{0}^{2}\int_{0}^{r^2} (r\cos\theta - 1) \, dz \, r \, dr \, d\theta = -8\pi$$

6. $\frac{\partial}{\partial x}(x^2) = 2x, \frac{\partial}{\partial y}(xz) = 0, \frac{\partial}{\partial z}(-3z) = -3 \Rightarrow \nabla \cdot \mathbf{F} = 2x - 3 \Rightarrow \text{Flux} = \iiint_D (2x-3)\, dV =$

$$\int_0^{2\pi}\int_0^{2\pi}\int_0^{4} (2\rho \sin\phi \cos\theta - 3)\, \rho^2 \sin\phi \, d\rho \, d\phi \, d\theta = 0$$

7. $\frac{\partial}{\partial x}(x^2) = 2x, \frac{\partial}{\partial y}(-2xy) = -2x, \frac{\partial}{\partial z}(3xz) = 3x \Rightarrow \text{Flux} = \iiint_D 3x \, dx\, dy\, dz =$

$$\int_0^{\pi/2}\int_0^{\pi/2}\int_0^{2} 3\rho \sin\phi \cos\theta(\rho^2 \sin\phi)\, d\rho \, d\phi \, d\theta = 3\pi$$

8. $\frac{\partial}{\partial x}(6x^2 + 2xy) = 12 + 2y, \frac{\partial}{\partial y}(y + x^2 z) = 2, \frac{\partial}{\partial z}(4x^2y^3) = 0 \Rightarrow \nabla \cdot \mathbf{F} = 14 + 2y \Rightarrow \text{Flux} =$

$$\iiint_D (14 + 2y)\, dV = \int_0^{3}\int_0^{\pi/2}\int_0^{2} (14 + 2r\sin\theta)\, r\, dr\, d\theta\, dz = 42\pi + 16$$

9. $\frac{\partial}{\partial x}(2xz) = 2z, \frac{\partial}{\partial y}(-xy) = -x, \frac{\partial}{\partial z}(-z^2) = -2z \Rightarrow \nabla \cdot \mathbf{F} = -x \Rightarrow \text{Flux} = \iiint_D -x\, dV =$

$$\int_0^{2}\int_0^{\sqrt{16-4x^2}}\int_0^{4-y} -x\, dz\, dy\, dx = -\frac{40}{3}$$

10. $\frac{\partial}{\partial x}(x^3) = 3x^2, \frac{\partial}{\partial y}(y^3) = 3y^2, \frac{\partial}{\partial z}(z^3) = 3z^2 \Rightarrow \nabla \cdot \mathbf{F} = 3x^2 + 3y^2 + 3z^2 = 3a^2$ on the sphere $\Rightarrow$

$$\text{Flux} = \iiint_D 3a^2\, dV = 3a^2\left(\frac{4}{3}\pi a^3\right) = 4\pi a^5$$

11. Let $\rho = \sqrt{x^2 + y^2 + z^2}$. Then $\frac{\partial \rho}{\partial x} = \frac{x}{\rho}, \frac{\partial \rho}{\partial y} = \frac{y}{\rho}, \frac{\partial \rho}{\partial z} = \frac{z}{\rho} \Rightarrow \frac{\partial}{\partial x}(\rho x) = \frac{\partial \rho}{\partial x} x + \rho = \frac{x^2}{\rho} = \rho, \frac{\partial}{\partial y}(\rho y) =$ $\frac{\partial \rho}{\partial y} y + \rho = \frac{y^2}{\rho} + \rho, \frac{\partial}{\partial z}(\rho z) = \frac{\partial \rho}{\partial z} z + \rho = \frac{z^2}{\rho} + \rho \Rightarrow \nabla \cdot \mathbf{F} = \frac{x^2 + y^2 + z^2}{\rho} + 3\rho = 4\rho$ since $\rho = \sqrt{x^2 + y^2 + z^2}$

$$\Rightarrow \text{Flux} = \iiint_D 4\rho \, dV = \iiint_D 4\sqrt{x^2 + y^2 + z^2} \, dx\, dy\, dz =$$

$$\int_0^{2\pi} \int_0^{2\pi} \int_1^{\sqrt{2}} (4\rho)\rho^2 \sin\phi \, d\rho\, d\phi\, d\theta = 0$$

12. Let $\rho = \sqrt{x^2 + y^2 + z^2}$. Then $\frac{\partial \rho}{\partial x} = \frac{x}{\rho}, \frac{\partial \rho}{\partial y} = \frac{y}{\rho}, \frac{\partial \rho}{\partial z} = \frac{z}{\rho} \Rightarrow \frac{\partial}{\partial x}\left(\frac{x}{\rho}\right) = \frac{1}{\rho} - \frac{x}{\rho^2}\frac{\partial \rho}{\partial x} = \frac{1}{\rho} - \frac{x^2}{\rho^3}$. Similarly, $\frac{\partial}{\partial y}\left(\frac{y}{\rho}\right) = \frac{1}{\rho} - \frac{y^2}{\rho^3}, \frac{\partial}{\partial z}\left(\frac{z}{\rho}\right) = \frac{1}{\rho} - \frac{z^2}{\rho^3} \Rightarrow \nabla \cdot \mathbf{F} = \frac{3}{\rho} - \frac{x^2 + y^2 + z^2}{\rho^3} = \frac{2}{\rho} \Rightarrow \text{Flux} = \iiint_D \frac{2}{\rho} dV =$

$$\int_0^{2\pi} \int_0^{2\pi} \int_1^{4} \left(\frac{2}{\rho}\right)\rho^2 \sin\phi \, d\rho\, d\phi\, d\theta = 0$$

13. $\frac{\partial}{\partial x}(x) = 1, \frac{\partial}{\partial y}(-2y) = -2, \frac{\partial}{\partial z}(z + 3) = 1 \Rightarrow \nabla \cdot \mathbf{F} = 0 \Rightarrow$ Flux = 0 over the solid. In the xy–plane, z = 0, $\mathbf{n} = -\mathbf{k}$, and $\mathbf{F} = x\,\mathbf{i} - 2y\,\mathbf{j} + 3\,\mathbf{k} \Rightarrow \mathbf{F} \cdot \mathbf{n} = -3 \Rightarrow \text{Flux} = \iint_{z=0} -3 \, d\sigma = -3(\text{Area of the square}) =$ -3. In the yz–plane, $x = 0$, $\mathbf{n} = -\mathbf{i}$, $\mathbf{F} = -2y\,\mathbf{j} + (z + 3)\,\mathbf{k} \Rightarrow \mathbf{F} \cdot \mathbf{n} = 0 \Rightarrow$ Flux = 0.
In the xz–plane, $y = 0$, $\mathbf{n} = -\mathbf{j}$, $\mathbf{F} = x\,\mathbf{i} + (z + 3)\,\mathbf{k} \Rightarrow \mathbf{F} \cdot \mathbf{n} = 0 \Rightarrow$ Flux = 0. ∴ The total flux = $-3 + 0 + 0 + 1 + (-3) + (\text{Flux of the top}) = 0 \Rightarrow$ Flux of the top = 5

14. $|\mathbf{F} \cdot \mathbf{n}| \le \|\mathbf{F}\|\,\|\mathbf{n}\| \le 1$ since $\|\mathbf{F}\| \le 1$, $\|\mathbf{n}\| = 1$. Then $\iiint_D \nabla \cdot \mathbf{F} \, d\sigma = \iint_S \mathbf{f} \cdot \mathbf{n} \, d\sigma \le$

$$\iint_S |\mathbf{F} \cdot \mathbf{n}| \, d\sigma \le \iint_S 1 \, d\sigma = \text{Area of S}$$

15. a) $\frac{\partial}{\partial x}(x) = 1, \frac{\partial}{\partial y}(y) = 1, \frac{\partial}{\partial z}(z) = 1 \Rightarrow \nabla \cdot \mathbf{F} = 3 \Rightarrow \text{Flux} = \iiint_D 3 \, dV = 3 \iiint_D dV =$

3(Volume of the solid)

b) If **F** is orthogonal to **n** at every point of S, then $\mathbf{F} \cdot \mathbf{n} = 0$ everywhere $\Rightarrow \text{Flux} = \iint_S \mathbf{F} \cdot \mathbf{n} \, d\sigma = 0$.

But the Flux is 3(Volume of the solid) ≠ 0. ∴ **F** is not orthogonal to **n** at every point.

16. a) Flux of ∇f over S = $\iint_S \nabla f \cdot \mathbf{n}\, d\sigma = \iiint_D \nabla \cdot \nabla f\, dV = \iiint_D 0\, dV = 0$ since $\nabla \cdot \nabla f = 0$

b) $\nabla f = \frac{\partial f}{\partial x}\mathbf{i} + \frac{\partial f}{\partial y}\mathbf{j} + \frac{\partial f}{\partial z}\mathbf{k} \Rightarrow f\,\nabla f = f\frac{\partial f}{\partial x}\mathbf{i} + f\frac{\partial f}{\partial y}\mathbf{j} + f\frac{\partial f}{\partial z}\mathbf{k} \Rightarrow \nabla \cdot f\nabla f = \frac{\partial f}{\partial x}\frac{\partial f}{\partial x} + f\frac{\partial^2 f}{\partial x^2} + \frac{\partial f}{\partial y}\frac{\partial f}{\partial y} + f\frac{\partial^2 f}{\partial y^2} + \frac{\partial f}{\partial z}\frac{\partial f}{\partial z} +$ $f\frac{\partial^2 f}{\partial z^2} = \left(\frac{\partial f}{\partial x}\right)^2 + \left(\frac{\partial f}{\partial y}\right)^2 + \left(\frac{\partial f}{\partial z}\right)^2 + f\frac{\partial^2 f}{\partial x^2} + f\frac{\partial^2 f}{\partial y^2} + f\frac{\partial^2 f}{\partial z^2} = |\nabla f|^2 + f(\nabla^2 f) = |\nabla f|^2$ since f is harmonic $\Rightarrow$ $\nabla^2 f = 0$. Then $\iint_S f\nabla f \cdot \mathbf{n}\, d\sigma = \iiint_D (\nabla \cdot f\nabla f)\, dV = \iiint_D |\nabla f|^2\, dV$

SECTION 15.6 STOKES' THEOREM

1. curl $\mathbf{F} = \nabla \times \mathbf{F} = 2\,\mathbf{k}$, $\mathbf{n} = \mathbf{k} \Rightarrow$ curl $\mathbf{F} \cdot \mathbf{n} = 2 \Rightarrow d\sigma = dx\,dy \Rightarrow \oint_C \mathbf{F} \cdot d\mathbf{r} = \iint_R 2\, dA =$ 2(Area of the ellipse) = 4π

2. curl $\mathbf{F} = \nabla \times \mathbf{F} = \mathbf{k}$, $\mathbf{n} = \mathbf{k} \Rightarrow$ curl $\mathbf{F} \cdot \mathbf{n} = 1 \Rightarrow d\sigma = dx\,dy \Rightarrow \oint_C \mathbf{F} \cdot d\mathbf{r} = \iint_R dx\,dy = 9\pi$

3. curl $\mathbf{F} = \nabla \times \mathbf{F} = -x\,\mathbf{i} - 2x\,\mathbf{j} + (z-1)\,\mathbf{k}$, $\mathbf{n} = \frac{\mathbf{i}+\mathbf{j}+\mathbf{k}}{\sqrt{3}} \Rightarrow$ curl $\mathbf{F} \cdot \mathbf{n} = \frac{1}{\sqrt{3}}(-3x + z - 1) \Rightarrow d\sigma = \frac{\sqrt{3}}{1}\, dA \Rightarrow$

$$\oint_C \mathbf{F} \cdot d\mathbf{r} = \iint_R \frac{1}{\sqrt{3}}(-3x + z - 1)\sqrt{3}\, dA = \int_0^1 \int_0^{1-x} (-3x + z - 1)\, dy\, dx =$$

$$\int_0^1 \int_0^{1-x} (-3x + (1 - x - y) - 1)\, dy\, dx = \int_0^1 \int_0^{1-x} (-4x - y)\, dy\, dx = -\frac{5}{6}$$

4. curl $\mathbf{F} = \nabla \times \mathbf{F} = (2y - 2z)\,\mathbf{i} + (2z - 2x)\,\mathbf{j} + (2x - 2y)\,\mathbf{k}$, $\mathbf{n} = \frac{\mathbf{i}+\mathbf{j}+\mathbf{k}}{\sqrt{3}} \Rightarrow$ curl $\mathbf{F} \cdot \mathbf{n} =$ $\frac{1}{\sqrt{3}}(2y - 2z + 2z - 2x + 2x - 2y) = 0 \Rightarrow \oint_C \mathbf{F} \cdot d\mathbf{r} = \iint_R 0\, d\sigma = 0$

5. curl $\mathbf{F} = \nabla \times \mathbf{F} = (2y - 0)\,\mathbf{i} + (2z - 2x)\,\mathbf{j} + (2x - 2y)\,\mathbf{k} = 2y\,\mathbf{i} + (2z - 2x)\,\mathbf{j} + (2x - 2y)\,\mathbf{k}$,

$\mathbf{n} = \mathbf{k} \Rightarrow$ curl $\mathbf{F} \cdot \mathbf{n} = 2x - 2y \Rightarrow d\sigma = dx\,dy \Rightarrow \oint_C \mathbf{F} \cdot d\mathbf{r} = \int_{-1}^1 \int_{-1}^1 (2x - 2y)\, dx\, dy = 0$

6. curl $\mathbf{F} = \nabla \times \mathbf{F} = -3x^2y^2\,\mathbf{k}$, $\mathbf{n} = \mathbf{k} \Rightarrow$ curl $\mathbf{F} \cdot \mathbf{n} = -3x^2y^2 \Rightarrow \oint_C \mathbf{F} \cdot d\mathbf{r} = \iint_R -3x^2y^2\,dA =$

$$\int_0^{2\pi}\int_0^a (-3r^2\cos^2\theta(r^2\sin^2\theta))\,r\,dr\,d\theta = \int_0^{2\pi}\int_0^a (-3r^5\cos^2\theta\sin^2\theta)\,dr\,d\theta = -\frac{\pi a^6}{8}$$

7. $x = 3\cos t$, $y = 2\sin t \Rightarrow \mathbf{F} = (2\sin t)\,\mathbf{i} + (9\cos^2 t)\,\mathbf{j} + (9\cos^2 t + 16\sin^4 t)\sin e^{\sqrt{6\sin t\cos t\,(0)}}$ and $\mathbf{r} = (3\cos t)\,\mathbf{i} + (2\sin t)\,\mathbf{j} \Rightarrow d\mathbf{r} = (-3\sin t)\,dt\,\mathbf{i} + (2\cos t)\,dt\,\mathbf{j} \Rightarrow \mathbf{F} \cdot d\mathbf{r} = -6\sin^2 t\,dt + 18\cos^3 t\,dt \Rightarrow$

$$\iint_S \nabla \times \mathbf{F} \cdot \mathbf{n}\,d\sigma = \int_0^{2\pi} (-6\sin^2 t + 18\cos^3 t)\,dt = -6\pi$$

8. curl $\mathbf{F} = \nabla \times \mathbf{F} = -2x\,\mathbf{j} + 2\,\mathbf{k}$. Flux of $\nabla \times \mathbf{F} = \iint_S \nabla \times \mathbf{F} \cdot \mathbf{n}\,d\sigma = \oint_C \mathbf{F} \cdot d\mathbf{r}$. Let C be $x = a\cos t$, $y = a\sin t$. Then $\mathbf{r} = (a\cos t)\,\mathbf{i} + (a\sin t)\,\mathbf{j} \Rightarrow d\mathbf{r} = (-a\sin t)\,dt\,\mathbf{i} + (a\cos t)\,dt\,\mathbf{j}$. Then $\mathbf{F} \cdot d\mathbf{r} = ay\sin t\,dt + ax\cos t\,dt = a^2\sin^2 t\,dt + a^2\cos^2 t\,dt = a^2 dt$. $\therefore$ Flux of $\nabla \times \mathbf{F} = \oint_C \mathbf{F} \cdot d\mathbf{r} = \int_0^{2\pi} a^2\,dt = 2\pi a^2$

9. Let S_1 and S_2 be oriented surfaces that span C and that induce the same positive direction on C. Then $\iint_{S_1} \nabla \times \mathbf{F} \cdot \mathbf{n}_1\,d\sigma_1 = \int_C \mathbf{F} \cdot d\mathbf{r} = \iint_{S_2} \nabla \times \mathbf{F} \cdot \mathbf{n}_2\,d\sigma_2$

10. a) $\nabla \cdot \mathbf{F} = \left(\frac{\partial}{\partial x}\mathbf{i} + \frac{\partial}{\partial y}\mathbf{j} + \frac{\partial}{\partial z}\mathbf{k}\right) \cdot (x\,\mathbf{i} + y\,\mathbf{j} + z\,\mathbf{k}) = 3$

b) $\nabla \times \mathbf{F} = \begin{vmatrix} \mathbf{i} & \mathbf{j} & \mathbf{k} \\ \frac{\partial}{\partial x} & \frac{\partial}{\partial y} & \frac{\partial}{\partial z} \\ x & y & z \end{vmatrix} = 0$

11. a) $\mathbf{F} = M\,\mathbf{i} + N\,\mathbf{j} + P\,\mathbf{k} \Rightarrow$ curl $\mathbf{F} = \left(\frac{\partial P}{\partial y} - \frac{\partial N}{\partial z}\right)\mathbf{i} + \left(\frac{\partial M}{\partial z} - \frac{\partial P}{\partial x}\right)\mathbf{j} + \left(\frac{\partial N}{\partial x} - \frac{\partial M}{\partial y}\right)\mathbf{k}$.

$\nabla \cdot \nabla \times \mathbf{F} = \text{div(curl }\mathbf{F}) = \frac{\partial}{\partial x}\left(\frac{\partial P}{\partial y} - \frac{\partial N}{\partial z}\right) + \frac{\partial}{\partial y}\left(\frac{\partial M}{\partial z} - \frac{\partial P}{\partial x}\right) + \frac{\partial}{\partial z}\left(\frac{\partial N}{\partial x} - \frac{\partial M}{\partial y}\right) = \frac{\partial^2 P}{\partial x\partial y} - \frac{\partial^2 N}{\partial x\partial z} + \frac{\partial^2 M}{\partial y\partial z} - \frac{\partial^2 P}{\partial y\partial x} + \frac{\partial^2 N}{\partial z\partial x} - \frac{\partial^2 M}{\partial z\partial y} = 0$ if the partial derivatives are continuous.

b) $\iint_S \nabla \times \mathbf{F} \cdot \mathbf{n}\,d\sigma = \iiint_D \nabla \cdot \nabla \times \mathbf{F}\,dV = \iiint_D 0\,dV = 0$ if the divergence theorem applies.

12. a) $\mathbf{F} = 2x\,\mathbf{i} + 2y\,\mathbf{j} + 2z\,\mathbf{k} \Rightarrow \text{curl}\,\mathbf{F} = \mathbf{0}\ \therefore\ \oint_C \mathbf{F}\cdot d\mathbf{r} = \iint_S \nabla \times \mathbf{F}\cdot\mathbf{n}\,d\sigma = \iint_S 0\,d\sigma = 0$

b) $\mathbf{F} = \nabla(x^2y^2z^3) = 2xy^2z^3\,\mathbf{i} + 2x^2yz^3\,\mathbf{j} + 3x^2y^2z^2\,\mathbf{k} \Rightarrow \text{curl}\,\mathbf{F} = (6x^2yz^2 - 6x^2yz^2)\,\mathbf{i} + (6xy^2z^2 - 6xy^2z^2)\,\mathbf{j} +$ $(4xyz^3 - 4xyz^3)\,\mathbf{k} = \mathbf{0}.\ \therefore\ \oint_C \mathbf{F}\cdot d\mathbf{r} = \iint_S \nabla \times \mathbf{F}\cdot\mathbf{n}\,d\sigma = \iint_S 0\,d\sigma = 0$

c) $\mathbf{F} = \nabla \times (x\,\mathbf{i} + y\,\mathbf{j} + z\,\mathbf{k}) = \mathbf{0} \Rightarrow \nabla \times \mathbf{F} = \mathbf{0}.\ \therefore\ \oint_C \mathbf{F}\cdot d\mathbf{r} = \iint_S \nabla \times \mathbf{F}\cdot\mathbf{n}\,d\sigma = \iint_S 0\,d\sigma = 0$

d) $\mathbf{F} = \nabla f \Rightarrow \nabla \times \mathbf{F} = \nabla \times \nabla f = \mathbf{0}.\ \therefore\ \oint_C \mathbf{F}\cdot d\mathbf{r} = \iint_S \nabla \times \mathbf{F}\cdot\mathbf{n}\,d\sigma = \iint_S 0\,d\sigma = 0$

13. $\mathbf{F} = \nabla f = -\frac{1}{2}(x^2+y^2+z^2)^{-3/2}(2x)\,\mathbf{i} - \frac{1}{2}(x^2+y^2+z^2)^{-3/2}(2y)\,\mathbf{j} - \frac{1}{2}(x^2+y^2+z^2)^{-3/2}(2z)\,\mathbf{k} =$ $-x(x^2+y^2+z^2)^{-3/2}\,\mathbf{i} - y(x^2+y^2+z^2)^{-3/2}\,\mathbf{j} - z(x^2+y^2+z^2)^{-3/2}\,\mathbf{k}$

a) $\mathbf{r} = (a\cos t)\,\mathbf{i} + (a\sin t)\,\mathbf{j},\ 0 \le t \le 2\pi \Rightarrow d\mathbf{r} = (-a\sin t)\,dt\,\mathbf{i} + (a\cos t)\,dt\,\mathbf{j} \Rightarrow$
$\mathbf{F}\cdot d\mathbf{r} = -x(x^2+y^2+z^2)^{-3/2}(-a\sin t)\,dt - y(x^2+y^2+z^2)^{-3/2}(a\cos t)\,dt =$
$-\frac{a\cos t}{a^3}(-a\sin t)\,dt - \frac{a\sin t}{a^3}(a\cos t)\,dt = 0 \Rightarrow \int_C \mathbf{F}\cdot d\mathbf{r} = 0$

b) $\oint_C \mathbf{F}\cdot d\mathbf{r} = \iint_S \text{curl}\,\mathbf{F}\cdot\mathbf{n}\,d\sigma = \iint_S \text{curl}(\nabla f)\cdot\mathbf{n}\,d\sigma = \iint_S \mathbf{0}\cdot\mathbf{n}\,d\sigma =$ $\iint_S 0\,d\sigma = 0$

14. a) $\text{div}(g\mathbf{F}) = \nabla\cdot g\mathbf{F} = \frac{\partial}{\partial x}(gM) + \frac{\partial}{\partial y}(gN) + \frac{\partial}{\partial z}(gP) = \left(g\frac{\partial M}{\partial x} + M\frac{\partial g}{\partial x}\right) + \left(g\frac{\partial N}{\partial y} + N\frac{\partial g}{\partial y}\right) + \left(g\frac{\partial P}{\partial z} + P\frac{\partial g}{\partial z}\right) =$ $\left(M\frac{\partial g}{\partial x} + N\frac{\partial g}{\partial y} + P\frac{\partial g}{\partial z}\right) + g\left(\frac{\partial M}{\partial x} + \frac{\partial N}{\partial y} + \frac{\partial P}{\partial z}\right) = \mathbf{F}\cdot\nabla g + g(\nabla\cdot\mathbf{F})$

b) $\nabla \times (g\mathbf{F}) = \left(\frac{\partial}{\partial y}(gP) - \frac{\partial}{\partial z}(gN)\right)\mathbf{i} + \left(\frac{\partial}{\partial z}(gM) - \frac{\partial}{\partial x}(gP)\right)\mathbf{j} + \left(\frac{\partial}{\partial x}(gN) - \frac{\partial}{\partial y}(gM)\right)\mathbf{k} =$
$\left(P\frac{\partial g}{\partial y} + g\frac{\partial P}{\partial y} - N\frac{\partial g}{\partial z} - g\frac{\partial N}{\partial z}\right)\mathbf{i} + \left(M\frac{\partial g}{\partial z} + g\frac{\partial M}{\partial z} - P\frac{\partial g}{\partial x} - g\frac{\partial P}{\partial x}\right)\mathbf{j} + \left(N\frac{\partial g}{\partial x} + g\frac{\partial N}{\partial x} - M\frac{\partial g}{\partial y} - g\frac{\partial M}{\partial y}\right)\mathbf{k}$
$= \left(P\frac{\partial g}{\partial y} - N\frac{\partial g}{\partial z}\right)\mathbf{i} + \left(g\frac{\partial P}{\partial y} - g\frac{\partial N}{\partial z}\right)\mathbf{i} + \left(M\frac{\partial g}{\partial z} - P\frac{\partial g}{\partial x}\right)\mathbf{j} + \left(g\frac{\partial M}{\partial z} - g\frac{\partial P}{\partial x}\right)\mathbf{j} + \left(N\frac{\partial g}{\partial x} - M\frac{\partial g}{\partial y}\right)\mathbf{k} +$ $\left(g\frac{\partial N}{\partial x} - g\frac{\partial M}{\partial y}\right)\mathbf{k} = g(\nabla\times\mathbf{F}) + \nabla g \times \mathbf{F}$

c) $(\mathbf{F}_1 \times \mathbf{F}_2) = \begin{vmatrix} \mathbf{i} & \mathbf{j} & \mathbf{k} \\ M_1 & N_1 & P_1 \\ M_2 & N_2 & P_2 \end{vmatrix} = (N_1P_2 - P_1N_2)\,\mathbf{i} - (M_1P_2 - P_1M_2)\,\mathbf{j} + (M_1N_2 - N_1M_2)\,\mathbf{k} \Rightarrow$

$\nabla\cdot(\mathbf{F}_1 \times \mathbf{F}_2) = \nabla\cdot\left((N_1P_2 - P_1N_2)\,\mathbf{i} - (M_1P_2 - P_1M_2)\,\mathbf{j} + (M_1N_2 - N_1M_2)\,\mathbf{k}\right) =$

14. c) (Continued)

$$\frac{\partial}{\partial x}(N_1P_2 - P_1N_2) - \frac{\partial}{\partial y}(M_1P_2 - P_1M_2) + \frac{\partial}{\partial z}(M_1N_2 - N_1M_2) = P_2\frac{\partial N_1}{\partial x} + N_1\frac{\partial P_2}{\partial x} - N_2\frac{\partial P_1}{\partial x} - P_1\frac{\partial N_2}{\partial x}$$
$$- M_1\frac{\partial P_2}{\partial y} - P_2\frac{\partial M_1}{\partial y} + P_1\frac{\partial M_2}{\partial y} + M_2\frac{\partial P_1}{\partial y} + M_1\frac{\partial N_2}{\partial z} + N_2\frac{\partial M_1}{\partial z} - N_1\frac{\partial M_2}{\partial z} - M_2\frac{\partial N_1}{\partial z} =$$
$$M_2\left(\frac{\partial P_1}{\partial y} - \frac{\partial N_1}{\partial z}\right) + N_2\left(\frac{\partial M_1}{\partial z} - \frac{\partial P_1}{\partial x}\right) + P_2\left(\frac{\partial N_1}{\partial x} - \frac{\partial M_1}{\partial y}\right) + M_1\left(\frac{\partial N_2}{\partial z} - \frac{\partial P_2}{\partial y}\right) + N_1\left(\frac{\partial P_2}{\partial x} - \frac{\partial M_2}{\partial z}\right) +$$
$$P_1\left(\frac{\partial M_2}{\partial y} - \frac{\partial N_2}{\partial x}\right) = \mathbf{F_2} \cdot (\nabla \times \mathbf{F_1}) - \mathbf{F_1} \cdot (\nabla \times \mathbf{F_2})$$

SECTION 15.7 PATH INDEPENDENCE, POTENTIAL FUNCTIONS, AND CONSERVATIVE FIELDS

1. $\frac{\partial P}{\partial y} = x = \frac{\partial N}{\partial z}, \frac{\partial M}{\partial z} = y = \frac{\partial P}{\partial x}, \frac{\partial N}{\partial x} = z = \frac{\partial M}{\partial y} \Rightarrow$ Conservative

2. $\frac{\partial P}{\partial y} = x \cos z = \frac{\partial N}{\partial z}, \frac{\partial M}{\partial z} = y \cos z = \frac{\partial P}{\partial x}, \frac{\partial N}{\partial x} = \sin z = \frac{\partial M}{\partial y} \Rightarrow$ Conservative

3. $\frac{\partial P}{\partial y} = -1 \neq \frac{\partial N}{\partial z} \Rightarrow$ Not Conservative

4. $\frac{\partial N}{\partial x} = 1 \neq \frac{\partial M}{\partial y} \Rightarrow$ Not Conservative

5. $\frac{\partial N}{\partial x} = 0 \neq \frac{\partial M}{\partial y} \Rightarrow$ Not Conservative

6. $\frac{\partial P}{\partial y} = 0 = \frac{\partial N}{\partial z}, \frac{\partial M}{\partial z} = 0 = \frac{\partial P}{\partial x}, \frac{\partial N}{\partial x} = -e^x \sin y = \frac{\partial M}{\partial y} \Rightarrow$ Conservative

7. $\frac{\partial f}{\partial x} = 2x \Rightarrow f(x,y,z) = x^2 + g(y,z)$. $\frac{\partial f}{\partial y} = \frac{\partial g}{\partial y} = 3y \Rightarrow g(y,z) = \frac{3y^2}{2} + h(z) \Rightarrow f(x,y,z) = x^2 + \frac{3y^2}{2} + h(z)$. $\frac{\partial f}{\partial z} = h'(z)$ $= 4z. \Rightarrow h(z) = 2z^2 + C \Rightarrow f(x,y,z) = x^2 + \frac{3y^2}{2} + 2z^2 + C$

8. $\frac{\partial f}{\partial x} = y + z \Rightarrow f(x,y,z) = (y + z)x + g(y,z)$. $\frac{\partial f}{\partial y} = x + \frac{\partial g}{\partial y} = x + z \Rightarrow \frac{\partial g}{\partial y} = z \Rightarrow g(y,z) = zy + h(z)$. Then $f(x,y,z) = (y + z)x + zy + h(z)$. $\frac{\partial f}{\partial z} = x + y + h'(z) = x + y \Rightarrow h'(z) = 0 \Rightarrow h(z) = C$. $\therefore$ $f(x,y,z) = (y + z)x + zy + C$

9. $\frac{\partial f}{\partial x} = e^{y+2z} \Rightarrow f(x,y,z) = x\,e^{y+2z} + g(y,z)$. $\frac{\partial f}{\partial y} = x\,e^{y+2z} + \frac{\partial g}{\partial y} = x\,e^{y+2z} \Rightarrow \frac{\partial g}{\partial y} = 0$. Then $f(x,y,z) = x\,e^{y+2z} + h(z)$. $\frac{\partial f}{\partial z} = 2x\,e^{y+2z} + h'(z) = 2x\,e^{y+2z} \Rightarrow h'(z) = 0 \Rightarrow h(z) = C$. $\therefore$ $f(x,y,z) = x\,e^{y+2z} + C$

10. $\frac{\partial f}{\partial x} = y \sin z \Rightarrow f(x,y,z) = xy \sin z + g(y,z)$. $\frac{\partial f}{\partial y} = x \sin z + \frac{\partial g}{\partial y} = x \sin z \Rightarrow \frac{\partial g}{\partial y} = 0 \Rightarrow g(y,z) = h(z)$. Then $f(x,y,z) = xy \sin z + h(z)$. $\frac{\partial f}{\partial z} = xy \cos z + h'(z) = xy \cos z \Rightarrow h'(z) = 0 \Rightarrow h(z) = C$. $\therefore$ $f(x,y,z) = xy \sin z + C$

11. Let $\mathbf{F}(x,y,z) = 2x\,\mathbf{i} + 2y\,\mathbf{j} + 2z\,\mathbf{k} \Rightarrow \frac{\partial P}{\partial y} = 0 = \frac{\partial N}{\partial z}, \frac{\partial M}{\partial z} = 0 = \frac{\partial P}{\partial x}, \frac{\partial N}{\partial x} = 0 = \frac{\partial M}{\partial y} \Rightarrow M\,dx + N\,dy + P\,dz$ is exact. $\frac{\partial f}{\partial x} = 2x \Rightarrow f(x,y,z) = x^2 + g(y,z)$. $\frac{\partial f}{\partial y} = \frac{\partial g}{\partial y} = 2y \Rightarrow g(y,z) = y^2 + h(z) \Rightarrow f(x,y,z) = x^2 + y^2 + h(z)$. $\frac{\partial f}{\partial z} = h'(z) = 2z \Rightarrow h(z) = z^2 + C$. $\therefore$ $f(x,y,z) = x^2 + y^2 + z^2 + C \Rightarrow \int_{(0,0,0)}^{(2,3,-6)} 2x\,dx + 2y\,dy + 2z\,dz =$ $f(2,3,-6) - f(0,0,0) = 49$

12. Let $\mathbf{F}(x,y,z) = yz\,\mathbf{i} + xz\,\mathbf{j} + xy\,\mathbf{k} \Rightarrow \frac{\partial P}{\partial y} = x = \frac{\partial N}{\partial z}, \frac{\partial M}{\partial z} = y = \frac{\partial P}{\partial x}, \frac{\partial N}{\partial x} = z = \frac{\partial M}{\partial y} \Rightarrow M\,dx + N\,dy + P\,dz$ is exact. $\frac{\partial f}{\partial x} = yz \Rightarrow f(x,y,z) = xyz + g(y,z)$. $\frac{\partial f}{\partial y} = xz + \frac{\partial g}{\partial y} = xz \Rightarrow \frac{\partial g}{\partial y} = 0 \Rightarrow g(y,z) = h(z) \Rightarrow f(x,y,z) = xyz +$ $h(z)$. $\frac{\partial f}{\partial z} = xy + h'(z) = xy \Rightarrow h'(z) = 0 \Rightarrow h(z) = C$. $\therefore$ $f(x,y,z) = xyz + C \Rightarrow \int_{(1,1,2)}^{(3,5,0)} yz\,dx + xz\,dy + xy\,dz$ $= f(3,5,0) - f(1,1,2) = -2$

13. Let $\mathbf{F}(x,y,z) = 2xy\,\mathbf{i} + (x^2 - z^2)\,\mathbf{j} - 2yz\,\mathbf{k} \Rightarrow \frac{\partial P}{\partial y} = -2z = \frac{\partial N}{\partial z}, \frac{\partial M}{\partial z} = 0 = \frac{\partial P}{\partial x}, \frac{\partial N}{\partial x} = 2x = \frac{\partial M}{\partial y} \Rightarrow M\,dx + N\,dy +$ $P\,dz$ is exact. $\frac{\partial f}{\partial x} = 2xy \Rightarrow f(x,y,z) = x^2y + g(y,z)$. $\frac{\partial f}{\partial y} = x^2 + \frac{\partial g}{\partial y} = x^2 - z^2 \Rightarrow \frac{\partial g}{\partial y} = -z^2 \Rightarrow g(y,z) = -yz^2 +$ $h(z) \Rightarrow f(x,y,z) = x^2y - yz^2 + h(z)$. $\frac{\partial f}{\partial z} = -2yz + h'(z) = -2yz \Rightarrow h'(z) = 0 \Rightarrow h(z) = C \Rightarrow f(x,y,z) =$ $x^2y - yz^2 + C \Rightarrow \int_{(0,0,0)}^{(1,2,3)} 2xy\,dx + (x^2 - z^2)\,dy - 2yz\,dz = f(1,2,3) - f(0,0,0) = -16$

14. Let $\mathbf{F}(x,y,z) = 3x^2\,\mathbf{i} + \frac{z^2}{y}\,\mathbf{j} + 2z\ln y\,\mathbf{k} \Rightarrow \frac{\partial P}{\partial y} = \frac{2z}{y} = \frac{\partial N}{\partial z}, \frac{\partial M}{\partial z} = 0 = \frac{\partial P}{\partial x}, \frac{\partial N}{\partial x} = 0 = \frac{\partial M}{\partial y} \Rightarrow M\,dx + N\,dy + P\,dz$ is exact. $\frac{\partial f}{\partial x} = 3x^2 \Rightarrow f(x,y,z) = x^3 + g(y,z)$. $\frac{\partial f}{\partial y} = \frac{\partial g}{\partial y} = \frac{z^2}{y} \Rightarrow g(y,z) = z^2\ln y + h(z) \Rightarrow f(x,y,z) = x^3 + z^2\ln y$ $+ h(z)$. $\frac{\partial f}{\partial z} = 2z\ln y + h'(z) = 2z\ln y \Rightarrow h'(z) = 0 \Rightarrow h(z) = C \Rightarrow f(x,y,z) = x^3 + z^2\ln y + C \Rightarrow$ $\int_{(1,1,1)}^{(1,2,3)} 3x^2\,dx + \frac{z^2}{y}\,dy + 2z\ln y\,dz = f(1,2,3) - f(1,1,1) = 9\ln 2$

15. Let $\mathbf{F}(x,y,z) = (\sin y\cos x)\,\mathbf{i} + (\cos y\sin x)\,\mathbf{j} + \mathbf{k} \Rightarrow \frac{\partial P}{\partial y} = 0 = \frac{\partial N}{\partial z}, \frac{\partial M}{\partial z} = 0 = \frac{\partial P}{\partial x}, \frac{\partial N}{\partial x} = \cos y\cos x = \frac{\partial M}{\partial y} \Rightarrow$ $M\,dx + N\,dy + P\,dz$ is exact. $\frac{\partial f}{\partial x} = \sin y\cos x \Rightarrow f(x,y,z) = \sin y\sin x + g(y,z)$. $\frac{\partial f}{\partial y} = \cos y\sin x + \frac{\partial g}{\partial y} =$ $\cos y\sin x \Rightarrow \frac{\partial g}{\partial y} = 0 \Rightarrow g(y,z) = h(z) \Rightarrow f(x,y,z) = \sin y\sin x + h(z)$. $\frac{\partial f}{\partial z} = h'(z) = 1 \Rightarrow h(z) = z + C \Rightarrow$ $f(x,y,z) = \sin y\sin x + z + C \Rightarrow \int_{(1,0,0)}^{(0,1,1)} \sin y\cos x\,dx + \cos y\sin x\,dy + dz = f(0,1,1) - f(1,0,0) = 1$

16. Let $\mathbf{F}(x,y,z) = 2x\,\mathbf{i} - y^2\,\mathbf{j} - \frac{4}{1+z^2}\mathbf{k} \Rightarrow \frac{\partial P}{\partial y} = 0 = \frac{\partial N}{\partial z}, \frac{\partial M}{\partial z} = 0 = \frac{\partial P}{\partial x}, \frac{\partial N}{\partial x} = 0 = \frac{\partial M}{\partial y} \Rightarrow M\,dx + N\,dy + P\,dz$ is exact. $\frac{\partial f}{\partial x} = 2x \Rightarrow f(x,y,z) = x^2 + g(y,z).$ $\frac{\partial f}{\partial y} = \frac{\partial g}{\partial y} = -y^2 \Rightarrow g(y,z) = -\frac{y^3}{3} + h(z) \Rightarrow f(x,y,z) = x^2 - y^2 + h(z).$ $\frac{\partial f}{\partial z} = h'(z) = -\frac{4}{1+z^2} \Rightarrow h(z) = -4\tan^{-1} z + C \Rightarrow f(x,y,z) = x^2 - y^2 - 4\tan^{-1} z + C \Rightarrow$

$$\int_{(0,0,0)}^{(3,3,1)} 2x\,dx - y^2\,dy - \frac{4}{1+z^2}\,dz = f(3,3,1) - f(0,0,0) = -\pi$$

17. Let $\mathbf{F}(x,y,z) = (2\cos y)\,\mathbf{i} + \left(\frac{1}{y} - 2x\sin y\right)\mathbf{j} + \frac{1}{z}\mathbf{k} \Rightarrow \frac{\partial P}{\partial y} = 0 = \frac{\partial N}{\partial z}, \frac{\partial M}{\partial z} = 0 = \frac{\partial P}{\partial x}, \frac{\partial N}{\partial x} = -2\sin y = \frac{\partial M}{\partial y} \Rightarrow$ $M\,dx + N\,dy + P\,dz$ is exact. $\frac{\partial f}{\partial x} = 2\cos y \Rightarrow f(x,y,z) = 2x\cos y + g(y,z).$ $\frac{\partial f}{\partial y} = -2x\sin y + \frac{\partial g}{\partial y} = \frac{1}{y} - 2x\sin y$ $\Rightarrow \frac{\partial g}{\partial y} = \frac{1}{y} \Rightarrow g(y,z) = \ln y + h(z) \Rightarrow f(x,y,z) = 2x\cos y + \ln y + h(z).$ $\frac{\partial f}{\partial z} = h'(z) = \frac{1}{z} \Rightarrow h(z) = \ln z + C \Rightarrow$

$$f(x,y,z) = 2x\cos y + \ln y + \ln z + C \Rightarrow \int_{(0,2,1)}^{(1,\pi/2,2)} 2\cos y\,dx + \left(\frac{1}{y} - 2x\sin y\right)dy + \frac{1}{z}\,dz =$$

$$f(1,\tfrac{\pi}{2},2) - f(0,2,1) = \ln\frac{\pi}{2}$$

18. Let $\mathbf{F}(x,y,z) = (2x\ln y - y^2)\,\mathbf{i} + \left(\frac{x^2}{y} - xz\right)\mathbf{j} - xy\,\mathbf{k}$ (Note: **j** component should be as given here; integrand in the text is not exact.) $\Rightarrow \frac{\partial P}{\partial y} = -x = \frac{\partial N}{\partial z}, \frac{\partial M}{\partial z} = -y = \frac{\partial P}{\partial x}, \frac{\partial N}{\partial x} = \frac{2x}{y} - z = \frac{\partial M}{\partial y} \Rightarrow M\,dx + N\,dy + P\,dz$ is exact. $\frac{\partial f}{\partial x} = 2x\ln y - yz \Rightarrow f(x,y,z) = x^2\ln y - xyz + g(y,z).$ $\frac{\partial f}{\partial y} = \frac{x^2}{y} - xz + \frac{\partial g}{\partial y} = \frac{x^2}{y} - xz \Rightarrow g(y,z) = h(z) \Rightarrow$ $f(x,y,z) = x^2\ln y - xyz + h(z).$ $\frac{\partial f}{\partial z} = -xy + h'(z) = -xy \Rightarrow h'(z) = 0 \Rightarrow h(z) = C \Rightarrow f(x,y,z) = x^2\ln y - xyz + C$

$$\Rightarrow \int_{(1,2,1)}^{(2,1,1)} (2x\ln y - yz)\,dx + \left(\frac{x^2}{y} - xz\right)dy - xy\,dz = f(2,1,1) - f(1,2,1) = -\ln 2$$

19. Let $\mathbf{F}(x,y,z) = \frac{1}{y}\mathbf{i} + \left(\frac{1}{z} - \frac{x}{y^2}\right)\mathbf{j} - \frac{y}{z^2}\mathbf{k} \Rightarrow \frac{\partial P}{\partial y} = -\frac{1}{z^2} = \frac{\partial N}{\partial z}, \frac{\partial M}{\partial z} = 0 = \frac{\partial P}{\partial x}, \frac{\partial N}{\partial x} = -\frac{1}{y^2} = \frac{\partial M}{\partial y} \Rightarrow$ $M\,dx + N\,dy + P\,dz$ is exact. $\frac{\partial f}{\partial x} = \frac{1}{y} \Rightarrow f(x,y,z) = \frac{x}{y} + g(y,z).$ $\frac{\partial f}{\partial y} = -\frac{x}{y^2} + \frac{\partial g}{\partial y} = \frac{1}{z} - \frac{x}{y^2} \Rightarrow \frac{\partial g}{\partial y} = \frac{1}{z} \Rightarrow$ $g(y,z) = \frac{y}{z} + h(z) \Rightarrow f(x,y,z) = \frac{x}{y} + \frac{y}{z} + h(z).$ $\frac{\partial f}{\partial z} = -\frac{y}{z^2} + h'(z) = -\frac{y}{z^2} \Rightarrow h'(z) = 0 \Rightarrow h(z) = C \Rightarrow$

$$f(x,y,z) = \frac{x}{y} + \frac{y}{z} + C \Rightarrow \int_{(1,1,1)}^{(2,2,2)} \frac{1}{y}\,dx + \left(\frac{1}{z} - \frac{x}{y^2}\right)dy - \frac{y}{z^2}\,dz = f(2,2,2) - f(1,1,1) = 0$$

20. Let $\mathbf{F}(x,y,z) = \frac{2x\,\mathbf{i} + 2y\,\mathbf{j} + 2z\,\mathbf{k}}{x^2 + y^2 + z^2}$ (and let $\rho^2 = x^2 + y^2 + z^2 \Rightarrow \frac{\partial \rho}{\partial x} = \frac{x}{\rho}, \frac{\partial \rho}{\partial y} = \frac{y}{\rho}, \frac{\partial \rho}{\partial z} = \frac{z}{\rho}$) $\Rightarrow$ $\frac{\partial P}{\partial y} = -\frac{4yz}{\rho^4} = \frac{\partial N}{\partial z}, \frac{\partial M}{\partial z} = -\frac{4xz}{\rho^4} = \frac{\partial P}{\partial x}, \frac{\partial N}{\partial x} = -\frac{4xy}{\rho^4} = \frac{\partial M}{\partial y} \Rightarrow M\,dx + N\,dy + P\,dz$ is exact. $\frac{\partial f}{\partial x} = \frac{2x}{x^2 + y^2 + z^2} \Rightarrow f(x,y,z) = \ln(x^2 + y^2 + z^2) + g(y,z).$ $\frac{\partial f}{\partial y} = \frac{2y}{x^2 + y^2 + z^2} + \frac{\partial g}{\partial y} = \frac{2y}{x^2 + y^2 + z^2} \Rightarrow \frac{\partial g}{\partial y} = 0$

20. (Continued)

$\Rightarrow g(y,z) = h(z) \Rightarrow f(x,y,z) = \ln(x^2 + y^2 + z^2) + h(z).\ \frac{\partial f}{\partial z} = \frac{2z}{x^2 + y^2 + z^2} + h'(z) = \frac{2z}{x^2 + y^2 + z^2} \Rightarrow h'(z) = 0 \Rightarrow$

$h(z) = C \Rightarrow f(x,y,z) = \ln(x^2 + y^2 + z^2) + C \Rightarrow \int_{(-1,-1,-1)}^{(2,2,2)} \frac{2x\,dx + 2y\,dy + 2z\,dz}{x^2 + y^2 + z^2} = f(2,2,2) - f(-1,-1,-1) = \ln 4$

21. Let $x - 1 = t$, $y - 1 = 2t$, $z - 1 = -2t$, $0 \le t \le 1 \Rightarrow dx = dt$, $dy = 2\,dt$, $dz = -2\,dt \Rightarrow$
$\int_{(1,1,1)}^{(2,3,-1)} y\,dx + x\,dy + 4\,dz = \int_0^1 (2t + 1)dt + (t + 1)2\,dt + 4(-2\,dt) = \int_0^1 (4t - 5)\,dt = -3$

22. Let $x = 0$, $y = 3t$, $z = 4t$, $0 \le t \le 1 \Rightarrow dx = 0$, $dy = 3\,dt$, $dz = 4\,dt \Rightarrow \int_{(0,0,0)}^{(0,3,4)} x^2\,dx + yz\,dy + y^2\,dz =$

$\int_0^1 12t^2(3\,dt) + 9t^2(4\,dt) = \int_0^1 72t^2\,dt = 24$

23. $\frac{\partial P}{\partial y} = 0 = \frac{\partial N}{\partial z},\ \frac{\partial M}{\partial z} = 2z = \frac{\partial P}{\partial x},\ \frac{\partial N}{\partial x} = 0 = \frac{\partial M}{\partial y} \Rightarrow M\,dx + N\,dy + P\,dz$ is exact $\Rightarrow$ **F** is conservative $\Rightarrow$ path independence.

24. Let $\mathbf{F}(x,y,z) = \frac{x\,\mathbf{i} + y\,\mathbf{j} + z\,\mathbf{k}}{\left(\sqrt{x^2 + y^2 + z^2}\right)^3} \Rightarrow \frac{\partial P}{\partial y} = -\frac{yz}{\left(\sqrt{x^2 + y^2 + z^2}\right)^3} = \frac{\partial N}{\partial z},\ \frac{\partial M}{\partial z} = -\frac{xz}{\left(\sqrt{x^2 + y^2 + z^2}\right)^3} = \frac{\partial P}{\partial x}$,
$\frac{\partial N}{\partial x} = -\frac{xy}{\left(\sqrt{x^2 + y^2 + z^2}\right)^3} = \frac{\partial M}{\partial y}$. $\therefore$ $M\,dx + N\,dy + P\,dz$ is exact and $\mathbf{F} = M\,\mathbf{i} + N\,\mathbf{j} + P\,\mathbf{k}$ is conservative
$\Rightarrow$ path independence

25. $\frac{\partial P}{\partial y} = 0,\ \frac{\partial N}{\partial z} = 0,\ \frac{\partial M}{\partial z} = 0,\ \frac{\partial P}{\partial x} = 0,\ \frac{\partial N}{\partial x} = \frac{y^2 - x^2}{(x^2 + y^2)^2},\ \frac{\partial M}{\partial y} = \frac{y^2 - x^2}{(x^2 + y^2)^2} \Rightarrow \text{curl}\,\mathbf{F} = \left(\frac{y^2 - x^2}{(x^2 + y^2)^2} - \frac{y^2 - x^2}{(x^2 + y^2)^2}\right)\mathbf{k}$
$= \mathbf{0}$. $x^2 + y^2 = 1 \Rightarrow \mathbf{r} = (a\cos t)\,\mathbf{i} + (a\sin t)\,\mathbf{j} \Rightarrow d\mathbf{r} = (-a\sin t)\,\mathbf{i} + (a\cos t)\,\mathbf{j} \Rightarrow \mathbf{F} = \frac{-a\sin t}{a^2}\mathbf{i} + \frac{a\cos t}{a^2}\mathbf{j} +$

$z\,\mathbf{k} \Rightarrow \mathbf{F} \cdot d\mathbf{r} = \frac{a^2\sin^2 t}{a^2} + \frac{a^2\cos^2 t}{a^2} = 1 \Rightarrow \int_C \mathbf{F} \cdot d\mathbf{r} = \int_0^{2\pi} 1\,dt = 2\pi$

PRACTICE EXERCISES

1. Path 1: $\mathbf{r} = t\,\mathbf{i} + t\,\mathbf{j} + t\,\mathbf{k} \Rightarrow x = t,\ y = t,\ z = t,\ 0 \le t \le 1 \Rightarrow f(g(t),h(t),h(t)) = 3 - 3t^2$ and $\frac{dx}{dt} = 1,\ \frac{dy}{dt} = 1,$

$\frac{dz}{dt} = 1 \Rightarrow \sqrt{\left(\frac{dx}{dt}\right)^2 + \left(\frac{dy}{dt}\right)^2 + \left(\frac{dz}{dt}\right)^2}\ dt = \sqrt{3}\ dt \Rightarrow \int_C f(x,y,z)\ ds = \int_0^1 \sqrt{3}\left(3 - 3t^2\right) dt = 2\sqrt{3}$

Path 2: $\mathbf{r_1} = t\,\mathbf{i} + t\,\mathbf{j},\ 0 \le t \le 1 \Rightarrow x = t,\ y = t,\ z = 0 \Rightarrow f(g(t),h(t),h(t)) = 2t - 3t^2 + 3$ and $\frac{dx}{dt} = 1,\ \frac{dy}{dt} = 1,$

$\frac{dz}{dt} = 0 \Rightarrow \sqrt{\left(\frac{dx}{dt}\right)^2 + \left(\frac{dy}{dt}\right)^2 + \left(\frac{dz}{dt}\right)^2}\ dt = \sqrt{2}\ dt \Rightarrow \int_{C_1} f(x,y,z)\ ds = \int_0^1 \sqrt{2}\left(2t - 3t^2 + 3\right) dt =$

$3\sqrt{2}$. $\mathbf{r_2} = \mathbf{i} + \mathbf{j} + t\,\mathbf{k} \Rightarrow x = 1,\ y = 1,\ z = t \Rightarrow f(g(t),h(t),h(t)) = 2 - 2t$ and $\frac{dx}{dt} = 0,\ \frac{dy}{dt} = 0,\ \frac{dz}{dt} = 1 \Rightarrow$

$\sqrt{\left(\frac{dx}{dt}\right)^2 + \left(\frac{dy}{dt}\right)^2 + \left(\frac{dz}{dt}\right)^2}\ dt = dt \Rightarrow \int_{C_2} f(x,y,z)\ ds = \int_0^1 (2 - 2t)\ dt = 1.\ \therefore\ \int_C f(x,y,z)\ ds =$

$\int_{C_1} f(x,y,z)\ ds + \int_{C_2} f(x,y,z)\ ds = 3\sqrt{2} + 1$

2. Path 1: $\mathbf{r_1} = t\,\mathbf{i} \Rightarrow x = 1,\ y = 0,\ z = 0 \Rightarrow f(g(t),h(t),h(t)) = t^2$ and $\frac{dx}{dt} = 1,$

$\frac{dy}{dt} = 0,\ \frac{dz}{dt} = 0 \Rightarrow \sqrt{\left(\frac{dx}{dt}\right)^2 + \left(\frac{dy}{dt}\right)^2 + \left(\frac{dz}{dt}\right)^2}\ dt = dt \Rightarrow \int_{C_1} f(x,y,z)\ ds = \int_0^1 t^2 dt = \frac{1}{3}$

$\mathbf{r_2} = t\,\mathbf{j} \Rightarrow x = 0,\ y = t,\ z = 0 \Rightarrow f(g(t),h(t),h(t)) = t$ and $\frac{dx}{dt} = 0,\ \frac{dy}{dt} = 1,\ \frac{dz}{dt} = 0 \Rightarrow$

$\sqrt{\left(\frac{dx}{dt}\right)^2 + \left(\frac{dy}{dt}\right)^2 + \left(\frac{dz}{dt}\right)^2}\ dt = dt \Rightarrow \int_{C_2} f(x,y,z)\ ds = \int_0^1 t\ dt = \frac{1}{2}$

$\mathbf{r_3} = t\,\mathbf{k} \Rightarrow x = 0,\ y = 0,\ z = t \Rightarrow f(g(t),h(t),h(t)) = -t$ and $\frac{dx}{dt} = 0,\ \frac{dy}{dt} = 0,\ \frac{dz}{dt} = 1 \Rightarrow$

$\sqrt{\left(\frac{dx}{dt}\right)^2 + \left(\frac{dy}{dt}\right)^2 + \left(\frac{dz}{dt}\right)^2}\ dt = dt \Rightarrow \int_{C_3} f(x,y,z)\ ds = \int_0^1 -t\ dt = -\frac{1}{2}.\ \therefore\ \int_{\text{Path 1}} f(x,y,z)\ ds =$

$\int_{C_1} f(x,y,z)\ ds + \int_{C_2} f(x,y,z)\ ds + \int_{C_3} f(x,y,z)\ ds = \frac{1}{3}$

Path 2: $\mathbf{r_4} = t\,\mathbf{i} + t\,\mathbf{j} \Rightarrow x = t,\ y = t,\ z = 0 \Rightarrow f(g(t),h(t),h(t)) = t^2 + t$ and $\frac{dx}{dt} = 1,\ \frac{dy}{dt} = 1,\ \frac{dz}{dt} = 0 \Rightarrow$

$\sqrt{\left(\frac{dx}{dt}\right)^2 + \left(\frac{dy}{dt}\right)^2 + \left(\frac{dz}{dt}\right)^2}\ dt = \sqrt{2}\ dt \Rightarrow \int_{C_4} f(x,y,z)\ ds = \int_0^1 \sqrt{2}\left(t^2 + t\right) dt = \frac{5}{6}\sqrt{2}$

2. (Continued)

$\mathbf{r}_3 = t\,\mathbf{k}$ (see above) $\Rightarrow \int_{C_3} f(x,y,z)\,ds = -\frac{1}{2}$. $\therefore \int_{\text{Path 2}} f(x,y,z)\,ds =$

$\int_{C_3} f(x,y,z)\,ds + \int_{C_4} f(x,y,z)\,ds = \frac{5}{6}\sqrt{2} - \frac{1}{2}$

Path 3: $\mathbf{r}_5 = \mathbf{r}_3 \Rightarrow \int_{C_5} f(x,y,z)\,ds = \int_{C_3} f(x,y,z)\,ds = -\frac{1}{2}$. $\mathbf{r}_6 = \mathbf{r}_2 \Rightarrow \int_{C_6} f(x,y,z)\,ds =$

$\int_{C_2} f(x,y,z)\,ds = \frac{1}{2}$. $\mathbf{r}_7 = \mathbf{r}_1 \Rightarrow \int_{C_7} f(x,y,z)\,ds = \int_{C_1} f(x,y,z)\,ds = \frac{1}{3}$. $\therefore \int_{\text{Path 3}} f(x,y,z)\,ds =$

$\int_{C_5} f(x,y,z)\,ds + \int_{C_6} f(x,y,z)\,ds + \int_{C_7} f(x,y,z)\,ds = \int_{C_3} f(x,y,z)\,ds + \int_{C_2} f(x,y,z)\,ds +$

$\int_{C_1} f(x,y,z)\,ds = \frac{1}{3}$

3. $\mathbf{r} = (a\cos t)\,\mathbf{j} + (a\sin t)\,\mathbf{k} \Rightarrow x = 0,\ y = a\cos t,\ z = a\sin t \Rightarrow f(g(t),h(t),h(t)) = \sqrt{a^2\sin^2 t} = a\,|\sin t|$ and

$\frac{dx}{dt} = 0,\ \frac{dy}{dt} = -a\sin t,\ \frac{dz}{dt} = a\cos t \Rightarrow \sqrt{\left(\frac{dx}{dt}\right)^2 + \left(\frac{dy}{dt}\right)^2 + \left(\frac{dz}{dt}\right)^2}\ dt = a\,dt \Rightarrow \int_C f(x,y,z)\,ds =$

$\int_0^{2\pi} a^2\,|\sin t|\,dt = \int_0^{\pi} a^2\sin t\,dt + \int_{\pi}^{2\pi} -a^2\sin t\,dt = 4a^2$

4. (See Exercise 14, Section 15.1.) $\mathbf{r} = t\,\mathbf{i} + t\,\mathbf{j} + t\,\mathbf{k},\ 1 \le t \Rightarrow \int_C f(x,y,z)\,ds = \int_1^{\infty} t^{-2}\,dt =$

$\lim_{k\to\infty} \int_1^{k} t^{-2}\,dt = -1$

5. a) $\mathbf{r} = \sqrt{2}\,t\,\mathbf{i} + \sqrt{2}\,\mathbf{j} + (4 - t^2)\,\mathbf{k},\ 0 \le t \le 1 \Rightarrow x = \sqrt{2}\,t,\ y = \sqrt{2}\,t,\ z = 4 - t^2 \Rightarrow \frac{dx}{dt} = \sqrt{2},\ \frac{dy}{dt} = \sqrt{2},\ \frac{dz}{dt} = -2t$

$\Rightarrow \sqrt{\left(\frac{dx}{dt}\right)^2 + \left(\frac{dy}{dt}\right)^2 + \left(\frac{dz}{dt}\right)^2}\ dt = \sqrt{4 + t^2}\ dt \Rightarrow M = \int_C \delta(x,y,z)\,ds = \int_0^1 3t\sqrt{4 + 4t^2}\,dt =$

$4\sqrt{2} - 2$

b) $M = \int_C \delta(x,y,z)\,ds = \int_0^1 \sqrt{4 + 4t^2}\,dt = \sqrt{2} + \ln(1 + \sqrt{2}\,)$

6. $\mathbf{r} = t\,\mathbf{i} + 2t\,\mathbf{j} + \frac{2}{3}t^{3/2}\,\mathbf{k},\ 0 \le t \le 2 \Rightarrow x = t,\ y = 2t,\ z = \frac{2}{3}t^{3/2} \Rightarrow \frac{dx}{dt} = 1,\ \frac{dy}{dt} = 2,\ \frac{dz}{dt} = t^{1/2} \Rightarrow$ $\sqrt{\left(\frac{dx}{dt}\right)^2 + \left(\frac{dy}{dt}\right)^2 + \left(\frac{dz}{dt}\right)^2}\,dt = \sqrt{t+5}\,dt \Rightarrow M = \int_C \delta(x,y,z)\,ds = \int_0^2 3\sqrt{5+t}\,\sqrt{t+5}\,dt =$ $\int_0^2 3(t+5)\,dt = 36.$ $M_{yz} = \int_C x\delta\,ds = \int_0^2 3t(t+5)\,dt = 38.$ $M_{xz} = \int_C y\delta\,ds =$ $\int_0^2 6t(t+5)\,dt = 46.$ $M_{xy} = \int_C z\delta\,ds = \int_0^2 2t^{3/2}(t+5)\,dt = \frac{144}{7}\sqrt{2}.$ $\therefore\ \bar{x} = M_{yz}/M = \frac{38}{36} = \frac{19}{18},$ $\bar{y} = M_{xz}/M = \frac{23}{18},\ \bar{z} = M_{xy}/M = \frac{\frac{144}{7}\sqrt{2}}{36} = \frac{4}{7}\sqrt{2}$

7. $\mathbf{r} = t\,\mathbf{i} + \frac{2\sqrt{2}}{3}t^{3/2}\,\mathbf{j} + \frac{t^2}{2}\,\mathbf{k},\ 0 \le t \le 2 \Rightarrow x = t,\ y = \frac{2\sqrt{2}}{3}t^{3/2},\ z = \frac{t^2}{2} \Rightarrow \frac{dx}{dt} = t,\ \frac{dy}{dt} = \sqrt{2}\,t^{1/2},\ \frac{dz}{dt} = t \Rightarrow$ $\sqrt{\left(\frac{dx}{dt}\right)^2 + \left(\frac{dy}{dt}\right)^2 + \left(\frac{dz}{dt}\right)^2}\,dt = \sqrt{(t+1)^2} = |t+1| = t+1$ on the domain given. Then $M_{yz} =$ $\int_C x\delta\,ds = \int_0^2 t\left(\frac{1}{t+1}\right)(t+1)\,dt = \int_0^2 t\,dt = 2.$ $M = \int_C \delta\,ds = \int_0^2 \frac{1}{t+1}(t+1)\,dt = \int_0^2 dt = 2$ $M_{xz} = \int_C y\delta\,ds = \int_0^2 \frac{2\sqrt{2}}{3}t^{3/2}\left(\frac{1}{t+1}\right)(t+1)\,dt = \int_0^2 \frac{2\sqrt{2}}{3}t^{3/2}\,dt = \frac{32}{15}.$ $M_{xy} = \int_C z\delta\,ds =$ $\int_0^2 \frac{t^2}{2}\left(\frac{1}{t+1}\right)(t+1)\,dt = \int_0^2 \frac{t^2}{2}\,dt = \frac{4}{3}.$ $\therefore\ \bar{x} = M_{yz}/M = \frac{2}{2} = 1,\ \bar{y} = M_{xz}/M = \frac{32/15}{2} = \frac{16}{15},\ \bar{z} = M_{xy}/M =$ $\frac{4/3}{2} = \frac{2}{3}.$ $I_x = \int_C (y^2+z^2)\delta\,ds = \int_0^2 \left(\frac{8}{9}t^3 + \frac{t^4}{4}\right)dt = \frac{232}{45}.$ $I_y = \int_C (x^2+z^2)\delta\,ds =$ $\int_0^2 \left(t^2 + \frac{t^4}{4}\right)dt = \frac{64}{15}$ $I_z = \int_C (y^2+x^2)\delta\,ds = \int_0^2 \left(t^2 + \frac{8}{9}t^3\right)dt = \frac{56}{9}.$ $R_x = \sqrt{I_x/M}$ $= \sqrt{\frac{232/45}{2}} = \frac{2}{3}\sqrt{\frac{29}{5}}.$ $R_y = \sqrt{I_y/M} = \sqrt{\frac{64/15}{2}} = 4\sqrt{\frac{2}{15}}.$ $R_z = \sqrt{I_z/M} = \sqrt{\frac{56/9}{2}} = \frac{2}{3}\sqrt{7}$

8. $\mathbf{r} = (e^t \cos t)\,\mathbf{i} + (e^t \sin t)\,\mathbf{j} + e^t\,\mathbf{k},\ 0 \le t \le \ln 2 \Rightarrow x = e^t \cos t,\ y = e^t \sin t,\ z = e^t \Rightarrow \frac{dx}{dt} =$ $(e^t \cos t - e^t \sin t),\ \frac{dy}{dt} = (e^t \sin t + e^t \cos t),\ \frac{dz}{dt} = e^t \Rightarrow \sqrt{\left(\frac{dx}{dt}\right)^2 + \left(\frac{dy}{dt}\right)^2 + \left(\frac{dz}{dt}\right)^2}\ dt =$ $\sqrt{(e^t \cos t - e^t \sin t)^2 + (e^t \sin t + e^t \cos t)^2 + \left(e^t\right)^2}\ dt = \sqrt{3e^{2t}}\ dt = \sqrt{3}\,e^t\,dt.$ $\delta\,ds = \sqrt{3}\,e^t\,dt$

Then $M = \int_C \delta\,ds = \int_0^{\ln 2} \sqrt{3}\,e^t\,dt = \sqrt{3}$. $M_{xy} = \int_C z\delta\,ds = \int_0^{\ln 2} \sqrt{3}\,e^t\,(e^t)\,dt = \int_0^{\ln 2} \sqrt{3}\,e^{2t}\,dt$ $= \frac{3\sqrt{3}}{2} \Rightarrow \bar{z} = M_{xy}/M = \frac{3\sqrt{3}/2}{\sqrt{3}} = \frac{3}{2}$. $I_z = \int_C (x^2 + y^2)\delta\,ds = \int_0^{\ln 2} (e^{2t}\cos^2 t + e^{2t}\sin^2 t)\sqrt{3}\,e^t\,dt =$ $\int_0^{\ln 2} \sqrt{3}\,e^{3t}\,dt = \frac{7\sqrt{3}}{3} \Rightarrow R_z = \sqrt{I_z/M} = \sqrt{\frac{7\sqrt{3}/3}{\sqrt{3}}} = \sqrt{\frac{7}{3}}$

9. a) $x^2 + y^2 = 1 \Rightarrow \mathbf{r} = (\cos t)\,\mathbf{i} + (\sin t)\,\mathbf{j},\ 0 \le t \le \pi \Rightarrow x = \cos t,\ y = \sin t \Rightarrow \mathbf{F} = (\cos t + \sin t)\,\mathbf{i} - \mathbf{j}$ and $\frac{d\mathbf{r}}{dt} = (-\sin t)\,\mathbf{i} + (\cos t)\,\mathbf{j} \Rightarrow \mathbf{F} \cdot \frac{d\mathbf{r}}{dt} = -\sin t \cos t - \sin^2 t - \cos t \Rightarrow$ Flow $=$ $\int_0^{\pi} (-\sin t \cos t - \sin^2 t - \cos t)\,dt = -\frac{1}{2}\pi$

b) $\mathbf{r} = -t\,\mathbf{i},\ -1 \le t \le 1 \Rightarrow x = -t,\ y = 0 \Rightarrow \mathbf{F} = -t\,\mathbf{i} - t^2\,\mathbf{j}$ and $\frac{d\mathbf{r}}{dt} = -\mathbf{i} \Rightarrow \mathbf{F} \cdot \frac{d\mathbf{r}}{dt} = t \Rightarrow$ Flow $= \int_{-1}^{1} t\,dt =$ 0

c) $\mathbf{r}_1 = (1 - t)\,\mathbf{i} - t\,\mathbf{j},\ 0 \le t \le 1 \Rightarrow \mathbf{F}_1 = (1 - 2t)\,\mathbf{i} - (1 - 2t - 2t^2)\,\mathbf{j}$ and $\frac{d\mathbf{r}_1}{dt} = -\mathbf{i} - \mathbf{j} \Rightarrow \mathbf{F}_1 \cdot \frac{d\mathbf{r}_1}{dt} = 2t^2 \Rightarrow$ $\text{Flow}_1 = \int_0^1 2t^2\,dt = \frac{2}{3}$. $\mathbf{r}_2 = -t\,\mathbf{i} + (t - 1)\,\mathbf{j},\ 0 \le t \le 1 \Rightarrow \mathbf{F}_2 = -\mathbf{i} - (2t^2 - 2t + 1)\,\mathbf{j}$ and $\frac{d\mathbf{r}_2}{dt} = -\mathbf{i} +$ $\mathbf{j} \Rightarrow \mathbf{F}_2 \cdot \frac{d\mathbf{r}_2}{dt} = -2t^3 + 4t^2 - 2t + 1 \Rightarrow \text{Flow}_2 = \int_0^1 (-2t^3 + 4t^2 - 2t + 1)\,dt = \frac{1}{3}$. $\therefore$ Flow $= \text{Flow}_1 +$ $\text{Flow}_2 = 1$

10. $\mathbf{r}_1 = (\cos t)\,\mathbf{i} + (\sin t)\,\mathbf{j} + t\,\mathbf{k},\ 0 \le t \le \frac{\pi}{2} \Rightarrow \mathbf{F}_1 = (2\cos t)\,\mathbf{i} + 2t\,\mathbf{j} + (2\sin t)\,\mathbf{k}$ and $\frac{d\mathbf{r}_1}{dt} = (-\sin t)\,\mathbf{i} + (\cos t)\,\mathbf{j} +$ $\mathbf{k} \Rightarrow \mathbf{F}_1 \cdot \frac{d\mathbf{r}_1}{dt} = -2\sin t \cos t + 2t\cos t + 2\sin t \Rightarrow \text{Circ}_1 = \int_0^{\pi/2} (-2\sin t \cos t + 2t\cos t + 2\sin t)\,dt =$ $\pi - 2$. $\mathbf{r}_2 = \mathbf{j} + \frac{\pi}{2}(1 - t)\,\mathbf{k},\ 0 \le t \le 1 \Rightarrow \mathbf{F}_2 = \pi(1 - t)\,\mathbf{j} + 2\,\mathbf{k}$ and $\frac{d\mathbf{r}_2}{dt} = -\frac{\pi}{2}\,\mathbf{k} \Rightarrow \mathbf{F}_2 \cdot \frac{d\mathbf{r}_2}{dt} = -\pi \Rightarrow$ $\text{Circ}_2 = \int_0^1 -\pi\,dt = -\pi$. $\mathbf{r}_3 = t\,\mathbf{i} + (1 - t)\,\mathbf{j},\ 0 \le t \le 1 \Rightarrow \mathbf{F}_3 = 2t\,\mathbf{i} + 2(1 - t)\,\mathbf{k}$ and $\frac{d\mathbf{r}_3}{dt} = \mathbf{i} - \mathbf{j} \Rightarrow$

10. (Continued)

$\mathbf{F_3} \cdot \frac{d\mathbf{r_3}}{dt} = 4t - 2 \Rightarrow \text{Circ}_3 = \int_0^1 (4t-2)\,dt = 0$. $\therefore$ Circ = $\text{Circ}_1 + \text{Circ}_2 + \text{Circ}_3 = -2$

11. $M = 2xy + x,\ N = xy - y \Rightarrow \frac{\partial M}{\partial x} = 2y + 1,\ \frac{\partial M}{\partial y} = 2x,\ \frac{\partial N}{\partial x} = y,\ \frac{\partial N}{\partial y} = x - 1 \Rightarrow$ Flux =

$\iint_R (2y + 1 + x - 1)\,dy\,dx = \int_0^1 \int_0^1 (2y + x)\,dy\,dx = \frac{3}{2}$. Circ = $\iint_R (y - 2x)\,dy\,dx =$

$\int_0^1 \int_0^1 (y - 2x)\,dy\,dx = -\frac{1}{2}$

12. $M = y - 6x^2,\ N = x + y^2 \Rightarrow \frac{\partial M}{\partial x} = -12x,\ \frac{\partial M}{\partial y} = 1,\ \frac{\partial N}{\partial x} = 1,\ \frac{\partial N}{\partial y} = 2y \Rightarrow$ Flux = $\iint_R (-12x + 2y)\,dx\,dy =$

$\int_0^1 \int_y^1 (-12x + 2y)\,dx\,dy = \int_0^1 (4y^2 + 2y - 6)\,dy = -\frac{11}{3}$. Circ = $\iint_R (1 - 1)dx\,dy = 0$

13. Let $M = 4x^3y,\ N = x^4 \Rightarrow \frac{\partial M}{\partial y} = 4x^3,\ \frac{\partial N}{\partial x} = 4x^3 \Rightarrow \oint_C 4x^3y\,dx + x^4\,dy = \iint_R (4x^3 - 4x^3)\,dx\,dy = 0$

14. a) Let $M = x,\ N = y \Rightarrow \frac{\partial M}{\partial x} = 1,\ \frac{\partial M}{\partial y} = 0,\ \frac{\partial N}{\partial x} = 0,\ \frac{\partial N}{\partial y} = 1 \Rightarrow$ Flux = $\iint_R (1 + 1)\,dx\,dy = 2\iint_R dx\,dy$

= 2(Area of the region)

b) Let C be a closed curve to which Green's Theorem applies. Let **n** be the unit normal vector to C. Let $\mathbf{F} = x\,\mathbf{i} + y\,\mathbf{j}$. Assume **F** is orthogonal to **n** at every point of C. Then the flux density of **F** at every point of C is 0 since $\mathbf{F} \cdot \mathbf{n} = 0$ at every point of C $\Rightarrow \frac{\partial M}{\partial x} + \frac{\partial N}{\partial y} = 0$ at every point of C. $\therefore$ Flux

$= \iint_R \left(\frac{\partial M}{\partial x} + \frac{\partial N}{\partial y}\right) dx\,dy = \iint_R 0\,dx\,dy = 0$. But part a above states the flux is

2(Area of the region) $\Rightarrow$ the area of the region would be 0 $\Rightarrow$ Contradiction. $\therefore$ **F** cannot be orthogonal to **n** at every point of C.

15. Let M = 8x sin y, N = −8y cos x $\Rightarrow \frac{\partial M}{\partial y} = 8x\cos y,\ \frac{\partial N}{\partial x} = 8y\sin x \Rightarrow \int_C 8x\sin y\,dx - 8y\cos x\,dy =$

$$\iint_R (8y\sin x - 8x\cos y)\,dy\,dx = \int_0^{\pi/2}\int_0^{\pi/2} (8y\sin x - 8x\cos y)\,dy\,dx = 0$$

16. Let $M = y^2$, $N = x^2 \Rightarrow \frac{\partial M}{\partial y} = 2y,\ \frac{\partial N}{\partial x} = 2x \Rightarrow \int_C y^2\,dx + x^2\,dy = \iint_R (2x - 2y)\,dx\,dy =$

$$\int_{-2}^{2}\int_{-\sqrt{4-x^2}}^{\sqrt{4-x^2}} (2x - 2y)\,dy\,dx = \int_{-2}^{2} 4x\sqrt{4 - x^2}\,dx = 0$$

17. Let $z = 1 - x - y \Rightarrow f_x(x,y) = -1,\ f_y(x,y) = -1 \Rightarrow \sqrt{f_x^{\,2} + f_y^{\,2} + 1} = \sqrt{3} \Rightarrow$ Area $= \iint_R \sqrt{3}\,dx\,dy =$

$\sqrt{3}$(Area of the circular region in the xy–plane) $= \pi\sqrt{3}$

18. $\nabla f = -3\,\mathbf{i} + 2y\,\mathbf{j} + 2z\,\mathbf{k},\ \mathbf{p} = \mathbf{i} \Rightarrow |\nabla f| = \sqrt{9 + 4y^2 + 4z^2}$ and $|\nabla f \cdot \mathbf{p}| = 3 \Rightarrow S =$

$$\iint_R \sqrt{9 + 4y^2 + 4z^2}\,dy\,dz = \int_0^{2\pi}\int_0^{\sqrt{3}} \sqrt{9 + 4r^2}\,r\,dr\,d\theta = \int_0^{2\pi} \left(\frac{7}{4}\sqrt{21} - \frac{9}{4}\right) d\theta = \left(\frac{7}{2}\sqrt{21} - \frac{9}{2}\right)\pi$$

19. $\nabla f = 2x\,\mathbf{i} + 2y\,\mathbf{j} + 2z\,\mathbf{k},\ \mathbf{p} = \mathbf{k} \Rightarrow |\nabla f| = \sqrt{4x^2 + 4y^2 + 4z^2} = 2\sqrt{x^2 + y^2 + z^2} = 2$ and $|\nabla f \cdot \mathbf{p}| = |2z| = 2z$

since $z \ge 0 \Rightarrow S = \iint_R \frac{2}{2z}\,dA = \iint_R \frac{1}{z}\,dA = \iint_R \frac{1}{\sqrt{1 - x^2 - y^2}}\,dx\,dy =$

$$\int_0^{2\pi}\int_0^{1/2} \frac{1}{\sqrt{1 - r^2}}\,r\,dr\,d\theta = \int_0^{2\pi} \left(1 - \frac{\sqrt{3}}{2}\right) d\theta = 2\pi - \pi\sqrt{3}$$

20. $\nabla f = 2x\,\mathbf{i} + 2y\,\mathbf{j} + 2z\,\mathbf{k},\ \mathbf{p} = \mathbf{k} \Rightarrow |\nabla f| = \sqrt{4x^2 + 4y^2 + 4z^2} = \sqrt{16} = 4$ and $|\nabla f \cdot \mathbf{p}| = 2z$ since $z \ge 0 \Rightarrow$

$$S = \iint_R \frac{4}{2z}\,dA = \iint_R \frac{2}{z}\,dA = 2\int_0^{\pi/2}\int_0^{2\cos\theta} \frac{2}{\sqrt{4 - r^2}}\,r\,dr\,d\theta = 4\pi - 8$$

21. a) $\nabla f = 2y\,\mathbf{j} - \mathbf{k}$, $\mathbf{p} = \mathbf{k} \Rightarrow |\nabla f| = \sqrt{4y^2+1}$ and $|\nabla f \cdot \mathbf{p}| = 1 \Rightarrow d\sigma = \sqrt{4y^2+1}\,dx\,dy \Rightarrow$

$$\iint_S g(x,y,z)\,d\sigma = \iint_S \frac{z}{\sqrt{4y^2+1}}\sqrt{4y^2+1}\,dx\,dy = \iint_S y(y^2-1)\,d\sigma =$$

$$\int_{-1}^{1}\int_{0}^{3} (y^3 - y)\,dx\,dy = 0$$

b) $$\iint_S g(x,y,z)\,d\sigma = \iint_S \frac{z}{\sqrt{4y^2+1}}\sqrt{4y^2+1}\,dx\,dy = \int_{-1}^{1}\int_{0}^{3} (y^2-1)\,dx\,dy = -4$$

22. $\nabla f = 2y\,\mathbf{j} + 2z\,\mathbf{k}$, $\mathbf{p} = \mathbf{k} \Rightarrow |\nabla f| = \sqrt{4y^2+4z^2} = 2\sqrt{y^2+z^2} = 10$ and $|\nabla f \cdot \mathbf{p}| = 2z$ since $z \ge 0 \Rightarrow d\sigma =$

$\frac{10}{2z}\,dx\,dy = \frac{5}{z}\,dx\,dy \Rightarrow$ $$\iint_S g(x,y,z)\,d\sigma = \iint_R x^4y(y^2+z^2)\left(\frac{5}{z}\right)dx\,dy =$$

$$\iint_R x^4y(25)\left(\frac{5}{\sqrt{25-y^2}}\right)dx\,dy = \int_0^4\int_0^1 \frac{125y}{\sqrt{25-y^2}}x^4\,dx\,dy = \int_0^4 \frac{25y}{\sqrt{25-y^2}}\,dy = 50$$

23. On the face, $z = 1$: $G(x,y,z) = G(x,y,1) = z \Rightarrow \nabla G = \mathbf{k} \Rightarrow |\nabla G| = 1$. $\mathbf{n} = \mathbf{k} \Rightarrow \mathbf{F} \cdot \mathbf{n} = 2xz = 2x$ since $z = 1$

$d\sigma = dA \Rightarrow$ Flux $= \iint_R 2x\,dx\,dy = \int_0^1\int_0^1 2x\,dx\,dy = 1$. On the face, $z = 0$: $G(x,y,z) = G(x,y,0) =$

$z \Rightarrow \nabla G = \mathbf{k} \Rightarrow |\nabla G| = 1$. $\mathbf{n} = -\mathbf{k} \Rightarrow \mathbf{F} \cdot \mathbf{n} = -2xz = 0$ since $z = 0 \Rightarrow$ Flux $= \iint_R 0\,dx\,dy = 0$

On the face, $x = 1$: $G(x,y,z) = G(1,y,z) = x \Rightarrow \nabla G = \mathbf{i} \Rightarrow |\nabla G| = 1$. $\mathbf{n} = \mathbf{i} \Rightarrow \mathbf{F} \cdot \mathbf{n} = 2xy = 2y$ since $x = 1$

Flux $= \int_0^1\int_0^1 2y\,dy\,dz = 1$ On the face, $x = 0$: $G(x,y,z) = G(0,y,z) = x \Rightarrow \nabla G = \mathbf{i} \Rightarrow |\nabla G| = 1$. $\mathbf{n} = -\mathbf{i}$

$\Rightarrow \mathbf{F} \cdot \mathbf{n} = -2xy = 0$ since $x = 0 \Rightarrow$ Flux $= 0$. On the face, $y = 1$: $G(x,y,z) = G(x,1,z) = y \Rightarrow \nabla G = \mathbf{j} \Rightarrow |\nabla G|$

$= 1$. $\mathbf{n} = \mathbf{j} \Rightarrow \mathbf{F} \cdot \mathbf{n} = 2yz = 2z$ since $y = 1 \Rightarrow$ Flux $= \int_0^1\int_0^1 2z\,dz\,dx = 1$. On the face, $y = 0$: $G(x,y,z) =$

$G(z,0,z) = y \Rightarrow \nabla G = \mathbf{j} \Rightarrow |\nabla G| = 1$. $\mathbf{n} = -\mathbf{j} \Rightarrow \mathbf{F} \cdot \mathbf{n} = -2yz = 0$ since $y = 0 \Rightarrow$ Flux $= 0$.

$\therefore$ Total Flux $= 3$

24. Across the cap, $x^2 + y^2 + z^2 = 25$: Let $g(x,y,z) = x^2 + y^2 + z^2 - 25 \Rightarrow \nabla g = 2x\,\mathbf{i} + 2y\,\mathbf{j} + 2z\,\mathbf{k} \Rightarrow$ $|\nabla g| = \sqrt{4x^2 + 4y^2 + 4z^2} = 10$. $\mathbf{p} = \mathbf{k} \Rightarrow |\nabla g \cdot \mathbf{p}| = 2z$ since $z \ge 0$. $\mathbf{n} = \frac{2x\,\mathbf{i} + 2y\,\mathbf{j} + 2z\,\mathbf{k}}{10} = \frac{x\,\mathbf{i} + y\,\mathbf{j} + z\,\mathbf{k}}{5}$. $d\sigma = \frac{10}{2z}\,dA = \frac{5}{z}\,dA$. $\mathbf{F} \cdot \mathbf{n} = \frac{x^2z}{5} + \frac{y^2z}{5} + \frac{z}{5}$. $\therefore$ $\text{Flux}_{\text{cap}} = \iint_R \left(\frac{x^2z}{5} + \frac{y^2z}{5} + \frac{z}{5}\right)\left(\frac{5}{z}\right) dA$

$$= \iint_R (x^2 + y^2 + 1)\,dx\,dy = \int_0^{2\pi}\int_0^4 (r^2 + 1)\,r\,dr\,d\theta = \int_0^{2\pi} 72\,d\theta = 144\pi$$

Across the bottom, $z = 3$: $g(x,y,z) = z - 3$, $\mathbf{p} = \mathbf{k} \Rightarrow \nabla g = \mathbf{k} \Rightarrow |\nabla g| = 1$ and $|\nabla g \cdot \mathbf{p}| = 1$. $\mathbf{n} = -\mathbf{k} \Rightarrow$ $\mathbf{F} \cdot \mathbf{n} = -1 \Rightarrow \text{Flux}_{\text{bottom}} = \iint_R -1\,d\sigma = \iint_R -1\,dA = -1(\text{Area of the circular region}) = -16\pi$

$\therefore$ $\text{Flux} = \text{Flux}_{\text{cap}} + \text{Flux}_{\text{bottom}} = 128\pi$

25. Because of symmetry $\bar{x} = \bar{y} = 0$. Let $F(x,y,z) = x^2 + y^2 + z^2 = 25 \Rightarrow \nabla F = 2x\,\mathbf{i} + 2y\,\mathbf{j} + 2z\,\mathbf{k} \Rightarrow$ $|\nabla F| = \sqrt{4x^2 + 4y^2 + 4z^2} = 10$, $\mathbf{p} = \mathbf{k} \Rightarrow |\nabla F \cdot \mathbf{p}| = 2z$ since $z \ge 0 \Rightarrow M = \iint_R \delta(x,y,z)\,d\sigma =$ $\iint_R z\left(\frac{10}{2z}\right) dA = \iint_R 5\,dA = 5(\text{Area of the circular region}) = 80\pi$. $M_{xy} = \iint_R z\delta\,d\sigma =$

$$\iint_R z\,dA = \iint_R 5\sqrt{25 - x^2 - y^2}\,dx\,dy = \int_0^{2\pi}\int_0^4 5\sqrt{25 - r^2}\,r\,dr\,d\theta = \int_0^{2\pi} \frac{490}{3}\,d\theta = \frac{980}{3}\pi$$

$\therefore$ $\bar{z} = \frac{\frac{980}{3}\pi}{80\pi} = \frac{49}{12}$. Thus $(\bar{x},\bar{y},\bar{z}) = \left(0,0,\frac{49}{12}\right)$. $I_z = \iint_R (x^2 + y^2)\delta\,d\sigma = \iint_R 5(x^2 + y^2)\,dx\,dy =$

$$\int_0^{2\pi}\int_0^4 5r^3\,dr\,d\theta = \int_0^{2\pi} 320\,d\theta = 640\pi. \quad R_z = \sqrt{I_z/M} = \sqrt{\frac{640\pi}{80\pi}} = 2\sqrt{2}$$

26. On the face $z = 1$, $g(x,y,z) = z - 1 = 0$, $\mathbf{p} = \mathbf{k} \Rightarrow \nabla g = \mathbf{k} \Rightarrow |\nabla g| = 1$ and $|\nabla g \cdot \mathbf{p}| = 1 \Rightarrow d\sigma = dA \Rightarrow$

$I = \iint_R (x^2 + y^2)\,dA = 2\int_0^{\pi/4}\int_0^{\sec\theta} r^3\,dr\,d\theta = \frac{2}{3}$. On the face $z = 0$, $g(x,y,z) = z = 0 \Rightarrow$

$\nabla g = \mathbf{k}$, $\mathbf{p} = \mathbf{k} \Rightarrow |\nabla g| = 1 \Rightarrow |\nabla g \cdot \mathbf{p}| = 1 \Rightarrow d\sigma = dA \Rightarrow I = \iint_R (x^2 + y^2)\,dA = \frac{2}{3}$.

On the face $y = 0$, $g(x,y,z) = y = 0 \Rightarrow \nabla g = \mathbf{j}$, $\mathbf{p} = \mathbf{j} \Rightarrow |\nabla g| = 1 \Rightarrow |\nabla g \cdot \mathbf{p}| = 1 \Rightarrow d\sigma = dA \Rightarrow$

26. (Continued)

$$I = \iint_R (x^2 + 0^2)\, dA = \int_0^1 \int_0^1 x^2\, dx\, dz = \frac{1}{3}.$$ On the face $y = 1$, $g(x,y,z) = y - 1 = 0 \Rightarrow$

$\nabla g = \mathbf{j}$, $\mathbf{p} = \mathbf{j} \Rightarrow |\nabla g| = 1 \Rightarrow |\nabla g \cdot \mathbf{p}| = 1 \Rightarrow d\sigma = dA \Rightarrow I = \iint_R (x^2 + 1^2)\, dA = \int_0^1 \int_0^1 (x^2 + 1)\, dx\, dz$

$= \frac{4}{3}$. On the face $x = 1$, $g(x,y,z) = x - 1 = 0 \Rightarrow \nabla g = \mathbf{i}$, $\mathbf{p} = \mathbf{i} \Rightarrow |\nabla g| = 1 \Rightarrow |\nabla g \cdot \mathbf{p}| = 1 \Rightarrow d\sigma = dA \Rightarrow$

$$I = \iint_R (1^2 + y^2)\, dA = \int_0^1 \int_0^1 (1 + y^2)\, dy\, dz = \frac{4}{3}.$$ On the face $x = 0$, $g(x,y,z) = x = 0 \Rightarrow$

$\nabla g = \mathbf{i}$, $\mathbf{p} = \mathbf{i} \Rightarrow |\nabla g| = 1 \Rightarrow |\nabla g \cdot \mathbf{p}| = 1 \Rightarrow d\sigma = dA \Rightarrow I = \iint_R (0^2 + y^2)\, dA = \int_0^1 \int_0^1 y^2\, dy\, dz = \frac{1}{3}$

$\therefore\ I_z = \frac{14}{3}$

27. $\frac{\partial}{\partial x}(2xy) = 2y$, $\frac{\partial}{\partial y}(2yz) = 2z$, $\frac{\partial}{\partial z}(2xz) = 2x \Rightarrow \nabla \cdot \mathbf{F} = 2x + 2y + 2z \Rightarrow \text{Flux} = \iiint_D (2x + 2y + 2z)\, dV$

$$\int_0^1 \int_0^1 \int_0^1 (2x + 2y + 2z)\, dx\, dy\, dz = \int_0^1 \int_0^1 (1 + 2y + 2z)\, dy\, dz = \int_0^1 (2 + 2z)\, dz = 3$$

28. $\frac{\partial}{\partial x}(xz) = z$, $\frac{\partial}{\partial y}(yz) = z$, $\frac{\partial}{\partial z}(1) = 0 \Rightarrow \nabla \cdot \mathbf{F} = 2z \Rightarrow \text{Flux} = \iiint_D 2z\, r\, dr\, d\theta\, dz =$

$$\int_0^{2\pi} \int_0^4 \int_3^5 2z\, dz\, r\, dr\, d\theta = \int_0^{2\pi} \int_0^4 16r\, dr\, d\theta = \int_0^{2\pi} 128\, d\theta = 256\pi$$

29. $\frac{\partial}{\partial x}(-2x) = -2,\ \frac{\partial}{\partial y}(-3y) = -3,\ \frac{\partial}{\partial z}(z) = 1 \Rightarrow \nabla \cdot \mathbf{F} = -4 \Rightarrow \text{Flux} = \iiint_D -4\, dV =$

$$-4\int_0^{2\pi}\int_0^1\int_{r^2}^{\sqrt{2-r^2}} dz\, r\, dr\, d\theta = -4\int_0^{2\pi}\int_0^1 \left(r\sqrt{2-r^2} - r^3\right) dr\, d\theta = -4\int_0^{2\pi}\left(-\frac{7}{12} + \frac{2}{3}\sqrt{2}\right) d\theta =$$

$$\frac{2}{3}\pi\left(7 - 8\sqrt{2}\right)$$

30. $\frac{\partial}{\partial x}(6x + y) = 6,\ \frac{\partial}{\partial y}(-(x + z)) = 0,\ \frac{\partial}{\partial z}(4yz) = 4y \Rightarrow \nabla \cdot \mathbf{F} = 6 + 4y \Rightarrow \text{Flux} = \iiint_D (6 + 4y)\, dV =$

$$\int_0^1\int_0^{\pi/2}\int_0^1 (6 + 4r\sin\theta)\, r\, dr\, d\theta\, dz = \int_0^1\int_0^{\pi/2}\left(3 + \frac{4}{3}\sin\theta\right) d\theta\, dz = \int_0^1\left(\frac{3\pi}{2} + \frac{4}{3}\right) dz = \frac{3\pi}{2} + \frac{4}{3}$$

31. $\nabla f = 2\mathbf{i} + 6\mathbf{j} - 3\mathbf{k} \Rightarrow \nabla \times \mathbf{F} = -2y\,\mathbf{k}.\ \ \mathbf{n} = \frac{2\mathbf{i} + 6\mathbf{j} - 3\mathbf{k}}{\sqrt{4 + 36 + 9}} = \frac{2\mathbf{i} + 6\mathbf{j} - 3\mathbf{k}}{7} \Rightarrow \nabla \times \mathbf{F} \cdot \mathbf{n} = \frac{6}{7}y.\ \ \mathbf{p} = \mathbf{k} \Rightarrow$

$$|\nabla f \cdot \mathbf{p}| = 3 \Rightarrow d\sigma = \frac{7}{3}dA \Rightarrow \oint_C \mathbf{F} \cdot d\mathbf{r} = \iint_R \frac{6}{7}y\, d\sigma = \iint_R \frac{6}{7}y\left(\frac{7}{3}dA\right) = \iint_R 2y\, dx\, dy =$$

$$\int_0^{2\pi}\int_0^1 2r\sin\theta\ r\, dr\, d\theta = \int_0^{2\pi}\frac{2}{3}\sin\theta\, d\theta = 0$$

correct

32. $\text{curl } \mathbf{F} = (8y - 0)\,\mathbf{i} + (0 - 0)\,\mathbf{j} + (1 - 1)\,\mathbf{k} = 8y\,\mathbf{i}.\ \ g(x,y,z) = z + y = 0 \Rightarrow \nabla g = \mathbf{j} + \mathbf{k} \Rightarrow |\nabla g| = \sqrt{2} \Rightarrow$

$$\mathbf{n} = \frac{\mathbf{j} + \mathbf{k}}{\sqrt{2}} \Rightarrow \nabla \times \mathbf{F} \cdot \mathbf{n} = 0.\ \ \mathbf{p} = \mathbf{k} \Rightarrow |\nabla g \cdot \mathbf{p}| = \frac{1}{\sqrt{2}} \Rightarrow d\sigma = \frac{\sqrt{2}}{1/\sqrt{2}}dA = 2\, dA \Rightarrow \oint_C \mathbf{F} \cdot d\mathbf{r} =$$

$$\iint_R \nabla \times \mathbf{F} \cdot \mathbf{n}\, d\sigma = \iint_R 0\, d\sigma = 0$$

33. $\frac{\partial P}{\partial y} = 0 = \frac{\partial N}{\partial z},\ \frac{\partial M}{\partial z} = 0 = \frac{\partial P}{\partial x},\ \frac{\partial N}{\partial x} = 0 = \frac{\partial M}{\partial y} \Rightarrow$ Conservative

34. $\frac{\partial P}{\partial y} = \frac{-zy}{x^2 + y^2 + z^2} = \frac{\partial N}{\partial z},\ \frac{\partial M}{\partial z} = \frac{-xz}{x^2 + y^2 + z^2} = \frac{\partial P}{\partial x},\ \frac{\partial N}{\partial x} = \frac{-xy}{x^2 + y^2 + z^2} = \frac{\partial M}{\partial y} \Rightarrow$ Conservative

35. $\frac{\partial P}{\partial y} = 0 \neq ye^z = \frac{\partial N}{\partial xz} \Rightarrow$ Not Conservative

36. $\frac{\partial P}{\partial y} = \frac{x}{(x+yz)^2} = \frac{\partial N}{\partial z}, \frac{\partial M}{\partial z} = \frac{-y}{(x+yz)^2} = \frac{\partial P}{\partial x}, \frac{\partial N}{\partial x} = \frac{-z}{(x+yz)^2} = \frac{\partial M}{\partial y} \Rightarrow$ Conservative

37. $\frac{\partial f}{\partial x} = 2 \Rightarrow f(x,y,z) = 2x + g(y,z)$. $\frac{\partial f}{\partial y} = \frac{\partial g}{\partial y} = 2y + z \Rightarrow g(y,z) = y^2 + zy + h(z) \Rightarrow f(x,y,z) = 2x + y^2 + zy + h(z)$. $\frac{\partial f}{\partial z} = y + h'(z) = y + 1 \Rightarrow h'(z) = 1 \Rightarrow h(z) = z + C$. $\therefore$ $f(x,y,z) = 2x + y^2 + zy + z + C$

38. $\frac{\partial f}{\partial x} = z \cos xz \Rightarrow f(x,y,z) = \sin xz + g(y,z)$. $\frac{\partial f}{\partial y} = \frac{\partial g}{\partial y} = e^y \Rightarrow g(y,z) = e^y + h(z) \Rightarrow f(x,y,z) = \sin xz + e^y + h(z)$. $\frac{\partial f}{\partial z} = x \cos xz + h'(z) = x \cos xz \Rightarrow h'(z) = 0 \Rightarrow h(z) = C$. $\therefore$ $f(x,y,z) = \sin xz + e^y + C$

39. $\frac{\partial P}{\partial y} = -\frac{1}{2}(x+y+z)^{-3/2} = \frac{\partial N}{\partial z}, \frac{\partial M}{\partial z} = -\frac{1}{2}(x+y+z)^{-3/2} = \frac{\partial P}{\partial x}, \frac{\partial N}{\partial x} = -\frac{1}{2}(x+y+z)^{-3/2} = \frac{\partial M}{\partial y} \Rightarrow M\,dx +$ $N\,dy + P\,dz$ is exact. $\frac{\partial f}{\partial x} = \frac{1}{\sqrt{x+y+z}} \Rightarrow f(x,y,z) = 2\sqrt{x+y+z} + g(y,z)$. $\frac{\partial f}{\partial y} = \frac{1}{\sqrt{x+y+z}} + \frac{\partial g}{\partial y} = \frac{1}{\sqrt{x+y+z}}$ $\Rightarrow \frac{\partial g}{\partial y} = 0 \Rightarrow g(y,z) = h(z) \Rightarrow f(x,y,z) = 2\sqrt{x+y+z} + h(z)$. $\frac{\partial f}{\partial z} = \frac{1}{\sqrt{x+y+z}} + h'(z) = \frac{1}{\sqrt{x+y+z}} \Rightarrow h'(z) = 0$ $\Rightarrow h(z) = C \Rightarrow f(x,y,z) = 2\sqrt{x+y+z} + C \Rightarrow$ $\int_{(-1,1,1)}^{(4,-3,0)} \frac{dx + dy + dz}{\sqrt{x+y+z}} = f(4,-3,0) - f(-1,1,1) = 0$

40. $\frac{\partial P}{\partial y} = -\frac{1}{2\sqrt{yz}} = \frac{\partial N}{\partial z}, \frac{\partial M}{\partial z} = 0 = \frac{\partial P}{\partial x}, \frac{\partial N}{\partial x} = 0 = \frac{\partial M}{\partial y} \Rightarrow M\,dx + N\,dy + P\,dz$ is exact. $\frac{\partial f}{\partial x} = 1 \Rightarrow f(x,y,z) = x +$ $g(y,z)$. $\frac{\partial f}{\partial y} = \frac{\partial g}{\partial y} = -\sqrt{\frac{z}{y}} \Rightarrow g(y,z) = -2\sqrt{yz} + h(z) \Rightarrow f(x,y,z) = x - 2\sqrt{yz} + h(z)$. $\frac{\partial f}{\partial z} = -\sqrt{\frac{y}{z}} + h'(z) =$ $-\sqrt{\frac{y}{z}} \Rightarrow h'(z) = 0 \Rightarrow h(z) = C \Rightarrow f(x,y,z) = x - 2\sqrt{yz} + C$. $\therefore$ $\int_{(1,0,1)}^{(10,3,3)} dx - \sqrt{\frac{z}{y}}\,dy - \sqrt{\frac{y}{z}}\,dz =$ $f(10,3,3) - f(1,0,1) = 3$

41. Over Path 1: $\mathbf{r} = t\,\mathbf{i} + t\,\mathbf{j} + t\,\mathbf{k} \Rightarrow x = t, y = t, z = t$ and $d\mathbf{r} = (\mathbf{i} + \mathbf{j} + \mathbf{k})\,dt \Rightarrow \mathbf{F} = 2t^2\,\mathbf{i} + \mathbf{j} + t^2\,\mathbf{k} \Rightarrow$ $\mathbf{F} \cdot d\mathbf{r} = (3t^2 + 1)\,dt \Rightarrow \text{Work} = \int_0^1 (3t^2 + 1)\,dt = 2$

Over Path 2: $\mathbf{r_1} = t\,\mathbf{i} + t\,\mathbf{j}, 0 \le t \le 1 \Rightarrow x = t, y = t, z = 0$ and $d\mathbf{r_1} = (\mathbf{i} + \mathbf{j})\,dt \Rightarrow \mathbf{F_1} = 2t^2\,\mathbf{i} + \mathbf{j} + t^2\,\mathbf{k} \Rightarrow$ $\mathbf{F_1} \cdot d\mathbf{r_1} = (2t^2 + 1)\,dt \Rightarrow \text{Work}_1 = \int_0^1 (2t^2 + 1)\,dt = \frac{5}{3}$. $\mathbf{r_2} = \mathbf{i} + \mathbf{j} + t\,\mathbf{k}, 0 \le t \le 1 \Rightarrow x = 1, y = 1, z = t$ and $d\mathbf{r_2} = \mathbf{k}\,dt \Rightarrow \mathbf{F_2} = 2\,\mathbf{i} + \mathbf{j} + \mathbf{k} \Rightarrow \mathbf{F_2} \cdot d\mathbf{r_2} = dt \Rightarrow \text{Work}_2 = \int_0^1 dt = 1$. $\therefore$ $\text{Work} = \text{Work}_1 + \text{Work}_2 = \frac{8}{3}$

42. Over Path 1: $\mathbf{r} = t\,\mathbf{i} + t\,\mathbf{j} + t\,\mathbf{k}, 0 \le t \le 1 \Rightarrow x = t, y = t, z = t$ and $d\mathbf{r} = (\mathbf{i} + \mathbf{j} + \mathbf{k})\,dt \Rightarrow \mathbf{F} = 2t^2\,\mathbf{i} + t^2\,\mathbf{j} + \mathbf{k}$ $\Rightarrow \mathbf{F} \cdot d\mathbf{r} = (2t^2 + t^2 + 1)\,dt = (3t^2 + 1)\,dt \Rightarrow \text{Work} = \int_0^1 (3t^2 + 1)\,dt = 2$

Over Path 2: Since f is conservative, $\int_{curve} \mathbf{F} \cdot d\mathbf{r} = 0$ around any simple closed curve.

$\int_{curve} \mathbf{F} \cdot d\mathbf{r} = \int_{C_1} \mathbf{F} \cdot d\mathbf{r} + \int_{C_2} \mathbf{F} \cdot d\mathbf{r}$ where C_1 is the path from (0,0,0) to (1,1,0) to (1,1,1) and C_2 is the path from (1,1,1) to (0,0,0). $\int_{C_2} \mathbf{F} \cdot d\mathbf{r} = -2$ (See Path 1 above.) $\therefore$ $\int_{curve} \mathbf{F} \cdot d\mathbf{r} =$

$$\int_{C_1} \mathbf{F} \cdot d\mathbf{r} + (-2) = 0 \Rightarrow \int_{C_1} \mathbf{F} \cdot d\mathbf{r} = 2$$

43. $dx = (-2 \sin t + 2 \sin 2t)\, dt$, $dy = (2 \cos t - 2 \cos 2t)\, dt$. Area $= \frac{1}{2}\oint_C x\, dy - y\, dx =$

$$\frac{1}{2}\int_0^{2\pi} \big((2 \cos t - \cos 2t)(2 \cos t - 2 \cos 2t) - (2 \sin t - \sin 2t)(-2 \sin t + 2 \sin 2t)\big)\, dt =$$

$$\frac{1}{2}\int_0^{2\pi} (6 - 6 \cos t)\, dt = 6\pi$$

44. $dx = (-2 \sin t - 2 \sin 2t)\, dt$, $dy = (2 \cos t - 2 \cos 2t)\, dt$. Area $= \frac{1}{2}\oint_C x\, dy - y\, dx =$

$$\frac{1}{2}\int_0^{2\pi} \big((2 \cos t + \cos 2t)(2 \cos t - 2 \cos 2t) - (2 \sin t - \sin 2t)(-2 \sin t - 2 \sin 2t)\big)\, dt =$$

$$\frac{1}{2}\int_0^{2\pi} (1 - 2 \cos^3 t + \cos t + 2 \sin^2 t \cos t)\, dt = 2\pi$$

45. $dx = \cos 2t\, dt$, $dy = \cos t\, dt$. Area $= \frac{1}{2}\oint_C x\, dy - y\, dx = \frac{1}{2}\int_0^{\pi} \left(\frac{1}{2} \sin 2t \cos t - \sin t \cos 2t\right) dt =$

$$\frac{1}{2}\int_0^{\pi} (-\sin t \cos^2 t + \sin t)\, dt = \frac{2}{3}$$

46. $dx = (-2a \sin t - 2a \cos 2t)\,dt$, $dy = (b \cos t)\,dt$. Area $= \frac{1}{2}\oint_C x\,dy - y\,dx =$

$$\frac{1}{2}\int_0^{2\pi}\Big((2ab\cos^2 t - ab\cos t\sin 2t) - (-2ab\sin^2 t - 2ab\sin t\cos 2t)\Big)\,dt =$$

$$\frac{1}{2}\int_0^{2\pi}(2ab - 5ab\cos^2 t\sin t + 2ab\sin t)\,dt = ab\pi$$

CHAPTER 16

PREVIEW OF DIFFERENTIAL EQUATIONS

SECTION 16.1 SEPARABLE FIRST ORDER EQUATIONS

1. First Order, Non–Linear

2. Third Order, Non–Linear

3. Fourth Order, Linear

4. Second Order, Linear

5. a) $y = x^2 \Rightarrow y' = 2x,\ y'' = 2 \Rightarrow y'' - y' = 2x - 2x = 0$

b) $y = 1 \Rightarrow y' = 0,\ y'' = 0 \Rightarrow xy'' - y' = 0$

c) $y = C_1x^2 + C_2 \Rightarrow y' = 2C_1x,\ y'' = 2C_1 \Rightarrow xy'' - y' = x(2C_1) - 2C_1x = 0$

6. a) $y = \frac{x}{2} \Rightarrow y' = \frac{1}{2} \Rightarrow y' + \frac{1}{x}y = \frac{1}{2} + \frac{1}{x}\left(\frac{x}{2}\right) = \frac{1}{2} + \frac{1}{2} = 1$

b) $y = \frac{1}{x} + \frac{x}{2} \Rightarrow y' = -\frac{1}{x^2} + \frac{1}{2} \Rightarrow y' + \frac{1}{x}y = -\frac{1}{x^2} + \frac{1}{2} + \frac{1}{x}\left(\frac{1}{x} + \frac{x}{2}\right) = -\frac{1}{x^2} + \frac{1}{2} + \frac{1}{x^2} + \frac{1}{2} = 1$

c) $y = \frac{C}{x} + \frac{x}{2} \Rightarrow y' = -\frac{C}{x^2} + \frac{1}{2} \Rightarrow y' + \frac{1}{x}y = \Rightarrow -\frac{C}{x^2} + \frac{1}{2} + \frac{1}{x}\left(\frac{C}{x} + \frac{x}{2}\right) = -\frac{C}{x^2} + \frac{1}{2} + \frac{C}{x^2} + \frac{1}{2} = 1$

7. a) $y = e^{-x} \Rightarrow y' = -e^{-x} \Rightarrow 2y' + 3y = 2(-e^{-x}) + 3e^{-x} = e^{-x}$

b) $y = e^{-x} + e^{-3x/2} \Rightarrow y' = -e^{-x} - \frac{3}{2}e^{-3x/2} \Rightarrow 2y' + 3y = 2\left(-e^{-x} - \frac{3}{2}e^{-3x/2}\right) + 3\left(e^{-x} + e^{-3x/2}\right) = e^{-x}$

c) $y = e^{-x} + Ce^{-3x/2} \Rightarrow y' = -e^{-x} - \frac{3}{2}Ce^{-3x/2} \Rightarrow 2y' + 3y = 2\left(-e^{-x} - \frac{3}{2}Ce^{-3x/2}\right) + 3\left(e^{-x} + Ce^{-3x/2}\right) = e^{-x}$

8. a) $y = 1 \Rightarrow y' = 0,\ y'' = 0 \Rightarrow yy'' = 0$ and $2(y')^2 - 2y' = 2(0)^2 - 2(0) = 0 \Rightarrow yy'' = 2(y')^2 - 2y$

b) $y = \tan x \Rightarrow y' = \sec^2 x,\ y'' = 2\sec^2 x \tan x \Rightarrow yy'' = \tan x(2\sec^2 x \tan x) = 2\sec^2 x \tan^2 x$ and $2(y')^2 - 2y' = 2(\sec^2 x)^2 - 2\sec^2 x = 2\sec^2 x \tan^2 x$

9. $y'' = -32 \Rightarrow y' = -32x + C_1.\ y'(5) = 0 \Rightarrow -32(5) + C_1 = 0 \Rightarrow C_1 = 160 \Rightarrow y' = -32x + 160 \Rightarrow y = -16x^2 + 160x + C_2.\ y(5) = 400 \Rightarrow -16(5)^2 + 160(5) + C_2 = 400 \Rightarrow C_2 = 0 \Rightarrow y = 160x - 16x^2$

10. $y = 2\left(1 - e^{-3t/2}\right) \Rightarrow \frac{dy}{dt} = 2\left(\frac{3}{2}e^{-3t/2}\right)$ and $y(0) = 2(1 - 1) = 0 \Rightarrow 2\frac{dy}{dt} + 3y = 2\left(2\left(\frac{3}{2}e^{-3t/2}\right)\right) + 3\left(2\left(1 - e^{-3t/2}\right)\right) = 6\,e^{-3t/2} + 6 - 6\,e^{-3t/2} = 6$

11. $y = 3\cos 2t - \sin 2t \Rightarrow y(0) = 3\cos 0 - \sin 0 = 3$ and $y' = -6\sin 2t - 2\cos 2t \Rightarrow y'(0) = -6\sin 0 -$

$2 \cos 0 = -2$ and $y'' = -12 \cos 2t + 4 \sin 2t \Rightarrow y'' + 4y = -12 \cos 2t + 4 \sin 2t + 4(3 \cos 2t - \sin 2t) = 0$

12. $y = 2 - \ln(\cos(x-1)) \Rightarrow y(1) = 2 - \ln(\cos(1-1)) = 2$ and $y' = -\frac{1}{\cos(x-1)}(-\sin(x-1)) = \tan(x-1) \Rightarrow y'(1) = \tan(1-1) = 0$. $y'' = \sec^2(x-1) \Rightarrow y'' - (y')^2 = \sec^2(x-1) - (\tan(x-1))^2 = \sec^2(x-1) - \tan^2(x-1) = 1$

13. a) $\frac{dx}{dt} = 1000 + 0.10\,x \Rightarrow dx = (1000 + 0.10\,x)\,dt \Rightarrow \frac{dx}{1000 + 0.10\,x} = dt \Rightarrow \int \frac{dx}{1000 + 0.10\,x} = \int dt$ $\Rightarrow 10 \ln(1000 + 0.10\,x) = t + C \Rightarrow 1000 + 0.10\,x = e^{0.10(t+C)} = e^{0.10t + 0.10C} = e^{0.10t}\,e^{0.10C}$ $= C_1 e^{0.10t} \Rightarrow x = 10C_1\,e^{0.10t} - 10000 \Rightarrow x = C_2\,e^{0.10t} - 10000$. $x(0) = 1000 \Rightarrow 1000 = C_2 e^{0.10(0)} - 10000 \Rightarrow C_2 = 11000 \Rightarrow x = 11000\,e^{0.10t} - 10000$

b) $100\,000 = 11000\,e^{0.10t} - 10000 \Rightarrow 10 = e^{0.10t} \Rightarrow \ln 10 = 0.10\,t \Rightarrow t = 10 \ln 10 \approx 23.03$

14. $\frac{dT}{dt} = k(T - T_s) \Rightarrow \frac{dT}{T - T_s} = k\,dt \Rightarrow \int \frac{dT}{T - T_s} = \int k\,dt \Rightarrow \ln(T - T_s) = kt + C \Rightarrow T - T_s = C_1\,e^{kt}$

Given $T_s = 20°$ C and the object cools from 100°C to 40°C in 20 minutes, then $T = 100°\text{C} \Rightarrow t_0 = 0 \Rightarrow 100° - 20° = C_1 e^{k(0)} \Rightarrow C_1 = 80° \Rightarrow T - T_s = 80\,e^{kt}$. $t = 20$ min $\Rightarrow T = 40°\text{C} \Rightarrow 40° - 20° = 80\,e^{k(20)}$ $\Rightarrow \frac{1}{4} = e^{20k} \Rightarrow \ln\frac{1}{4} = 20\,k \Rightarrow k = \frac{\ln\frac{1}{4}}{20} \approx -0.069 \Rightarrow T - T_s = 80\,e^{-0.069\,t}$. $T = 60°\text{C} \Rightarrow 60° - 20° = 80\,e^{-0.069\,t} \Rightarrow \frac{1}{2} = e^{-0.069\,t} \Rightarrow t = \frac{\ln\frac{1}{2}}{-0.069} \Rightarrow t \approx 10.05$ minutes

15. $(x^2 + y^2)\,dx + xy\,dy = 0 \Rightarrow \left(1 + \frac{y^2}{x^2}\right)dx + \frac{xy}{x^2}\,dy = 0 \Rightarrow \left(1 + \left(\frac{y}{x}\right)^2\right)dx + \frac{y}{x}\,dy = 0 \Rightarrow \Rightarrow \frac{y}{x}\,dy = -\left(1 + \left(\frac{y}{x}\right)^2\right) \Rightarrow \frac{dy}{dx} = \frac{-\left(1 + \left(\frac{y}{x}\right)^2\right)}{y/x} \Rightarrow$ Homogeneous. $v = \frac{y}{x} \Rightarrow F(v) = \frac{-(1 + v^2)}{v} \Rightarrow \frac{dx}{x} + \frac{dv}{v + \frac{1 + v^2}{v}} = 0 \Rightarrow \frac{dx}{x} + \frac{v\,dv}{1 + 2v^2} = 0 \Rightarrow \ln|x| + \frac{1}{4}\ln(1 + 2v^2) = C \Rightarrow |x|\left(1 + 2v^2\right)^{1/4} = e^C \Rightarrow x^4(1 + 2v^2) = C_1$ where $C_1 = (e^C)^4 \Rightarrow x^4\left(1 + 2\left(\frac{y^2}{x^2}\right)\right) = C_1 \Rightarrow x^4 + 2x^2y^2 = C_1$

16. $(y^2 - xy)\,dx + x^2\,dy = 0 \Rightarrow \left(\frac{y^2}{x^2} - \frac{xy}{x^2}\right)dx + \frac{x^2}{x^2}\,dy = 0 \Rightarrow \left(\left(\frac{y}{x}\right)^2 - \frac{y}{x}\right)dx + dy = 0 \Rightarrow \frac{dy}{dx} = \left(\frac{y}{x} - \left(\frac{y}{x}\right)^2\right) \Rightarrow$ Homogeneous. $v = \frac{y}{x} \Rightarrow F(v) = v - v^2 \Rightarrow \frac{dx}{x} + \frac{dv}{v - (v - v^2)} = 0 \Rightarrow \frac{dx}{x} + \frac{dv}{v^2} = 0 \Rightarrow \ln x - v^{-1} = C \Rightarrow x = C_1 e^{x/y}$

17. $\left(x\,e^{y/x} + y\right)dx - x\,dy = 0 \Rightarrow \left(e^{y/x} + \frac{y}{x}\right)dx - dy = 0 \Rightarrow \frac{dy}{dx} = e^{y/x} + \frac{y}{x} \Rightarrow$ Homogeneous. $v = \frac{y}{x} \Rightarrow F(v) = e^v + v \Rightarrow \frac{dx}{x} + \frac{dv}{v - (e^v + v)} = 0 \Rightarrow \frac{dx}{x} + \frac{dv}{-e^v} = 0 \Rightarrow \ln|x| + e^{-v} = C \Rightarrow \ln|x| + e^{-y/x} = C$

18. $(x - y)\,dx + (x + y)\,dy = 0 \Rightarrow \left(\frac{x}{x} - \frac{y}{x}\right)dx + \left(\frac{x}{x} + \frac{y}{x}\right)dy = 0 \Rightarrow \frac{dy}{dx} = \frac{\left(\frac{y}{x} - 1\right)}{1 + \frac{y}{x}} \Rightarrow$ Homogeneous

$v = \frac{y}{x} \Rightarrow F(v) = \frac{v-1}{v+1} \Rightarrow \frac{dx}{x} + \frac{dy}{v - \left(\frac{v-1}{v+1}\right)} = 0 \Rightarrow \frac{dx}{x} + \frac{(v+1)dv}{v^2+1} = 0 \Rightarrow \frac{dx}{x} + \frac{v\,dv}{v^2+1} + \frac{dv}{v^2+1} = 0 \Rightarrow$

$\ln x + \frac{1}{2}\ln(v^2+1) + \tan^{-1}(v^2+1) = 0$ if $x > 0 \Rightarrow x\sqrt{v^2+1} = C_1\, e^{-\tan^{-1}(v^2+1)} \Rightarrow x\sqrt{\left(\frac{y}{x}\right)^2 + 1} =$

$C_1\, e^{-\tan^{-1}((y/x)^2+1)} \Rightarrow \sqrt{y^2+x^2} = C_1\, e^{-\tan^{-1}((y^2+x^2)/x^2)}$ if $x > 0$.

19. $\frac{dy}{dx} = \frac{y}{x} + \cos\left(\frac{y-x}{x}\right) \Rightarrow \frac{dy}{dx} = \frac{y}{x} + \cos\left(\frac{y}{x} - 1\right) \Rightarrow$ Homogeneous. $v = \frac{y}{x} \Rightarrow F(v) = v + \cos(v-1) \Rightarrow$

$\frac{dx}{x} + \frac{dv}{v - (v + \cos(v-1))} = 0 \Rightarrow \frac{dx}{x} - \sec(v-1)\,dv = 0 \Rightarrow \ln|x| + \ln|\sec(v-1) + \tan(v-1)| = C \Rightarrow$

$\frac{x}{\sec(v-1) + \tan(v-1)} = \pm e^C \Rightarrow \frac{x}{\sec\left(\frac{y-x}{x}\right) + \tan\left(\frac{y-x}{x}\right)} = C_1 \Rightarrow x = C_1\left(\sec\left(\frac{y-x}{x}\right) + \tan\left(\frac{y-x}{x}\right)\right)$

$y(2) = 2 \Rightarrow 2 = C_1\left(\sec\left(\frac{2-2}{2}\right) + \tan\left(\frac{2-2}{2}\right)\right) \Rightarrow C_1 = 2 \Rightarrow x = 2\left(\sec\left(\frac{y-x}{x}\right) + \tan\left(\frac{y-x}{x}\right)\right)$

20. $\left(x \sin\frac{y}{x} - y\cos\frac{y}{x}\right)dx + x\cos\frac{y}{x}\,dy = 0 \Rightarrow \left(\sin\frac{y}{x} - \frac{y}{x}\cos\frac{y}{x}\right)dx + \cos\frac{y}{x}\,dy = 0 \Rightarrow$

$\frac{dy}{dx} = \frac{\frac{y}{x}\cos\frac{y}{x} - \sin\frac{y}{x}}{\cos\frac{y}{x}} \Rightarrow$ Homogeneous. $v = \frac{y}{x} \Rightarrow F(v) = \frac{v\cos v - \sin v}{\cos v} = v - \tan v \Rightarrow \frac{dx}{x} + \frac{dv}{v - (v - \tan v)}$

$= 0 \Rightarrow \frac{dx}{x} + \frac{dv}{\tan v} = 0 \Rightarrow \frac{dx}{x} + \frac{\cos v\,dv}{\sin v} = 0 \Rightarrow \ln|x| + \ln|\sin v| = C \Rightarrow x\sin v = \pm e^C \Rightarrow x\sin\frac{y}{x} = C_1$

$y(2) = \pi \Rightarrow 2\sin\frac{\pi}{2} = C_1 \Rightarrow C_1 = 2 \Rightarrow x\sin\frac{y}{x} = 2$

21. $x\,dx + 2y\,dy = 0 \Rightarrow \frac{x^2}{2} + y^2 = C$. $2y\,dx - x\,dy = 0 \Rightarrow \frac{2\,dx}{x} - \frac{dy}{y} = 0 \Rightarrow 2\ln|x| - \ln|y| = C \Rightarrow \frac{x^2}{|y|} = e^C \Rightarrow$

$\frac{x^2}{y} = \pm e^C = C_1 \Rightarrow x^2 = C_1 y$

22. $3x\,dx + 2y\,dy = 0 \Rightarrow \frac{3x^2}{2} + y^2 = C$. $2y\,dx - 3x\,dy = 0 \Rightarrow \frac{dx}{3x} = \frac{dy}{2y} \Rightarrow \frac{1}{3}\ln|3x| = \frac{1}{2}\ln|2y| + C \Rightarrow$

$\frac{\sqrt[3]{3x}}{\sqrt{2y}} = \pm e^C = C_1,\ y > 0 \Rightarrow \sqrt[3]{3x} = C_1\sqrt{2y}$

23. $y\,dx + x\,dy = 0 \Rightarrow \frac{dx}{x} + \frac{dy}{y} = 0 \Rightarrow \ln|x| + \ln|y| = C \Rightarrow |xy| = e^C \Rightarrow xy = \pm e^C = C_1 \Rightarrow xy = C_1$, a family of hyperbolas with transverse axis the line $y = x$ or $y = -x$. $x\,dx - y\,dy = 0 \Rightarrow \frac{x^2}{2} - \frac{y^2}{2} = C \Rightarrow x^2 - y^2 = C_1$, a family of hyperbolas with transverse axis the x or y axis.

SECTION 16.2 EXACT DIFFERENTIAL EQUATIONS

1. $\frac{\partial M}{\partial y} = 0,\ \frac{\partial N}{\partial x} = 0 \Rightarrow$ Exact

2. $\frac{\partial M}{\partial y} = 1,\ \frac{\partial N}{\partial x} = -1 \Rightarrow$ Not Exact

3. $\frac{\partial M}{\partial y} = -\frac{1}{y^2},\ \frac{\partial N}{\partial x} = -\frac{1}{y^2} \Rightarrow$ Exact

4. $\frac{\partial M}{\partial y} = 2y,\ \frac{\partial N}{\partial x} = 2y \Rightarrow$ Exact

5. $\frac{\partial M}{\partial y} = e^y,\ \frac{\partial N}{\partial x} = e^y \Rightarrow$ Exact

6. $\frac{\partial M}{\partial y} = \cos xy - yx \sin xy,\ \frac{\partial N}{\partial x} = \cos xy - xy \sin xy$ $\Rightarrow$ Exact

7. $\frac{\partial M}{\partial y} = 2,\ \frac{\partial N}{\partial x} = -2 \Rightarrow$ Not Exact

8. $\frac{\partial M}{\partial y} = 2,\ \frac{\partial N}{\partial x} = -2 \Rightarrow$ Not Exact

9. $\frac{\partial M}{\partial y} = 1 = \frac{\partial N}{\partial x} \Rightarrow$ Exact. $\frac{\partial f}{\partial x} = x + y \Rightarrow f(x,y) = \frac{x^2}{2} + xy + k(y)$. $\frac{\partial f}{\partial y} = x + k'(y) = x + y^2 \Rightarrow k(y) = \frac{y^3}{3} + C \Rightarrow$ $f(x,y) = \frac{x^2}{2} + xy + \frac{y^3}{3} + C = C_1 \Rightarrow 3x^2 + 6xy + 2y^3 = C_2$

10. $\frac{\partial M}{\partial y} = 2x\,e^y = \frac{\partial N}{\partial x} \Rightarrow$ Exact. $\frac{\partial f}{\partial x} = 2x\,e^y + e^x \Rightarrow f(x,y) = x^2e^y + e^x + k(y)$. $\frac{\partial f}{\partial y} = x^2e^y + k'(y) = \left(x^2 + 1\right)e^y \Rightarrow$ $k'(y) = e^y \Rightarrow k(y) = e^y + C \Rightarrow f(x,y) = x^2e^y + e^x + e^y + C = C_1 \Rightarrow x^2e^y + e^x + e^y = C_2$

11. $\frac{\partial M}{\partial y} = 2x + 2y = \frac{\partial N}{\partial x} \Rightarrow$ Exact. $\frac{\partial f}{\partial x} = 2xy + y^2 \Rightarrow f(x,y) = x^2y + xy^2 + k(y)$. $\frac{\partial f}{\partial y} = x^2 + 2xy + k'(y) =$ $x^2 + 2xy - y \Rightarrow k'(y) = -y \Rightarrow k(y) = -\frac{y^2}{2} + C \Rightarrow f(x,y) = x^2y + xy^2 - \frac{y^2}{2} + C = C_1 \Rightarrow 2x^2y + 2xy^2 - y^2 = C_2$

12. $\frac{\partial M}{\partial y} = -1 = \frac{\partial N}{\partial x} \Rightarrow$ Exact. $\frac{\partial f}{\partial x} = x^3 - y \Rightarrow f(x,y) = \frac{x^4}{4} - xy + k(y)$. $\frac{\partial f}{\partial y} = -x + k'(y) = -x \Rightarrow k'(y) = 0 \Rightarrow k(y) = C$ $\Rightarrow f(x,y) = \frac{x^4}{4} - xy + C = C_1 \Rightarrow x^4 - 4xy = C_2$

13. $\frac{\partial M}{\partial y} = \frac{1}{y} + \frac{1}{x} = \frac{\partial N}{\partial x} \Rightarrow$ Exact. $\frac{\partial f}{\partial x} = e^x + \ln y + \frac{y}{x} \Rightarrow f(x,y) = e^x + x \ln y + y \ln x + k(y)$. $\frac{\partial f}{\partial y} = \frac{x}{y} + \ln x + k'(y) =$ $\frac{x}{y} + \ln x + \sin y \Rightarrow k'(y) = \sin y \Rightarrow k(y) = -\cos y + C \Rightarrow f(x,y) = e^x + x \ln y + y \ln x - \cos y + C = C_1 \Rightarrow$ $e^x + x \ln y + y \ln x - \cos y = C_2$

14. $\frac{\partial M}{\partial y} = \cos xy - xy \sin xy = \frac{\partial N}{\partial x} \Rightarrow$ Exact. $\frac{\partial f}{\partial x} = y \cos xy + 1 \Rightarrow f(x,y) = \sin xy + k(y)$. $\frac{\partial f}{\partial y} = x \cos xy + k'(y) =$ $x \cos xy + 2y - 3 \Rightarrow k'(y) = 2y - 3 \Rightarrow k(y) = y^2 - 3y + C \Rightarrow f(x,y) = \sin xy + y^2 - 3y + C = C_1 \Rightarrow$ $\sin xy + y^2 - 3y = C_2$

15. $\rho = \frac{1}{y^2} \Rightarrow \frac{1}{y^2}\left(xy^2 + y\right) dx - \frac{1}{y^2}(x\,dy) = 0 \Rightarrow \left(x + \frac{1}{y}\right) dx - \frac{x}{y^2}\,dy = 0 \Rightarrow \frac{\partial M}{\partial y} = -\frac{1}{y^2} = \frac{\partial N}{\partial x} \Rightarrow$ Exact. $\frac{\partial f}{\partial x} = x + \frac{1}{y} \Rightarrow f(x,y) = \frac{x^2}{2} + \frac{x}{y} + k(y)$. $\frac{\partial f}{\partial y} = -\frac{x}{y^2} + k'(y) = -\frac{x}{y^2} \Rightarrow k'(y) = 0 \Rightarrow k(y) = C \Rightarrow f(x,y) = \frac{x^2}{2} + \frac{x}{y} + C =$ $C_1 \Rightarrow x^2 + \frac{2x}{y} = C_2$

16. $\rho = \frac{1}{x^3} \Rightarrow \frac{1}{x^3}(x + 2y)\,dx - \frac{1}{x^3}(x\,dy) = 0 \Rightarrow \left(\frac{1}{x^2} + \frac{2y}{x^3}\right)dx - \frac{1}{x^2}\,dy = 0 \Rightarrow \frac{\partial M}{\partial y} = \frac{2}{x^3} = \frac{\partial N}{\partial x} \Rightarrow$ Exact.
$\frac{\partial f}{\partial x} = \frac{1}{x^2} + \frac{2y}{x^3} \Rightarrow f(x,y) = -\frac{1}{x} - \frac{y}{x^2} + k(y)$. $\frac{\partial f}{\partial y} = -\frac{1}{x^2} + k'(y) = -\frac{1}{x^2} \Rightarrow k'(y) = 0 \Rightarrow k(y) = C \Rightarrow$
$f(x,y) = -\frac{1}{x} - \frac{y}{x^2} + C = C_1 \Rightarrow -\frac{1}{x} - \frac{y}{x^2} = C_2$

17. a) $\rho = \frac{1}{xy} \Rightarrow \frac{1}{xy}(y\,dx) + \frac{1}{xy}(x\,dy) = 0 \Rightarrow \frac{1}{x}\,dx + \frac{1}{y}\,dy = 0 \Rightarrow \frac{\partial M}{\partial y} = 0 = \frac{\partial N}{\partial x} \Rightarrow$ Exact. $\frac{\partial f}{\partial x} = \frac{1}{x} \Rightarrow f(x,y) =$
$\ln|x| + k(y)$. $\frac{\partial f}{\partial y} = k'(y) = \frac{1}{y} \Rightarrow k(y) = \ln|y| + C \Rightarrow f(x,y) = \ln|x| + \ln|y| + C = C_1 \Rightarrow \ln|xy| = C_2 \Rightarrow$
$xy = \pm e^{C_2} \Rightarrow xy = C_3$

b) $\rho = \frac{1}{(xy)^2} \Rightarrow \frac{1}{(xy)^2}(y\,dx) + \frac{1}{(xy)^2}(x\,dy) = 0 \Rightarrow \frac{1}{x^2y}\,dx + \frac{1}{xy^2}\,dy = 0 \Rightarrow \frac{\partial M}{\partial y} = -\frac{1}{x^2y^2} = \frac{\partial N}{\partial x} \Rightarrow$ Exact.
$\frac{\partial f}{\partial x} = \frac{1}{x^2y} \Rightarrow f(x,y) = -\frac{1}{xy} + k(y)$. $\frac{\partial f}{\partial y} = \frac{1}{xy^2} + k'(y) = \frac{1}{xy^2} \Rightarrow k'(y) = 0 \Rightarrow k(y) = C \Rightarrow f(x,y) = -\frac{1}{xy} + C = C_1$
$\Rightarrow xy = C_2$

18. a) $\rho = \frac{1}{y^2} \Rightarrow \frac{1}{y^2}(y\,dx) - \frac{1}{y^2}(x\,dy) = 0 \Rightarrow \frac{1}{y}\,dx - \frac{x}{y^2}\,dy = 0 \Rightarrow \frac{\partial M}{\partial y} = -\frac{1}{y^2} = \frac{\partial N}{\partial x} \Rightarrow$ Exact. $\frac{\partial f}{\partial x} = \frac{1}{y} \Rightarrow$
$f(x,y) = \frac{x}{y} + k(y)$. $\frac{\partial f}{\partial y} = -\frac{x}{y^2} + k'(y) = -\frac{x}{y^2} \Rightarrow k'(y) = 0 \Rightarrow k(y) = C \Rightarrow f(x,y) = \frac{x}{y} + C = C_1 \Rightarrow \frac{x}{y} = C_2$

b) $\rho = \frac{1}{x^2} \Rightarrow \frac{1}{x^2}(y\,dx) - \frac{1}{x^2}(x\,dy) = 0 \Rightarrow \frac{y}{x^2}\,dx - \frac{1}{x}\,dy = 0 \Rightarrow \frac{\partial M}{\partial y} = \frac{1}{x^2} = \frac{\partial N}{\partial x} \Rightarrow$ Exact.
$\frac{\partial f}{\partial x} = \frac{y}{x^2} \Rightarrow f(x,y) = -\frac{y}{x} + k(y)$. $\frac{\partial f}{\partial y} = -\frac{1}{x} + k'(y) = -\frac{1}{x} \Rightarrow k'(y) = 0 \Rightarrow k(y) = C \Rightarrow f(x,y) = -\frac{y}{x} + C = C_1 \Rightarrow$
$-\frac{y}{x} = C_2 \Rightarrow \frac{x}{y} = C_3$

c) $\rho = \frac{1}{xy} \Rightarrow \frac{1}{xy}(y\,dx) - \frac{1}{xy}(x\,dy) = 0 \Rightarrow \frac{1}{x}\,dx - \frac{1}{y}\,dy = 0 \Rightarrow \frac{\partial M}{\partial y} = 0 = \frac{\partial N}{\partial x} \Rightarrow$ Exact.
$\frac{\partial f}{\partial x} = \frac{1}{x} \Rightarrow f(x,y) = \ln|x| + k(y)$. $\frac{\partial f}{\partial y} = k'(y) = -\frac{1}{y} \Rightarrow k(y) = -\ln|y| + C \Rightarrow f(x,y) = \ln|x| - \ln|y| + C = C_1 \Rightarrow$
$\ln\left|\frac{x}{y}\right| = C_2 \Rightarrow \frac{x}{y} = \pm e^{C_2} \Rightarrow \frac{x}{y} = C_3$

d) $\rho = \frac{1}{x^2 + y^2} \Rightarrow \frac{1}{x^2 + y^2}(y\,dx) - \frac{1}{x^2 + y^2}(x\,dy) = 0 \Rightarrow \frac{y}{x^2 + y^2}\,dx - \frac{x}{x^2 + y^2}\,dy = 0 \Rightarrow \frac{\partial M}{\partial y} = \frac{x^2 - y^2}{(x^2 + y^2)^2} =$
$\frac{\partial N}{\partial x} \Rightarrow$ Exact. $\frac{\partial f}{\partial x} = \frac{y}{x^2 + y^2} \Rightarrow f(x,y) = \tan^{-1}\frac{x}{y} + k(y)$. $\frac{\partial f}{\partial y} = \frac{-x}{x^2 + y^2} + k'(y) = \frac{-x}{x^2 + y^2} \Rightarrow k'(y) = 0 \Rightarrow$
$k(y) = C \Rightarrow f(x,y) = \tan^{-1}\frac{x}{y} + C = C_1 \Rightarrow \tan^{-1}\frac{x}{y} = C_2 \Rightarrow \frac{x}{y} = \tan C_2 \Rightarrow \frac{x}{y} = C_3$

19. $\frac{\partial M}{\partial y} = 2y = \frac{\partial N}{\partial x} \Rightarrow$ Exact. $\frac{\partial f}{\partial x} = x + y^2 \Rightarrow f(x,y) = \frac{x^2}{2} + xy^2 + k(y)$. $\frac{\partial f}{\partial y} = 2xy + k'(y) = 2xy + 1 \Rightarrow k'(y) = 1$
$\Rightarrow k(y) = y + C \Rightarrow f(x,y) = \frac{x^2}{2} + xy^2 + y + C = C_1 \Rightarrow x^2 + 2xy^2 + 2y = C_2$. $y(0) = 2 \Rightarrow C_2 = 4 \Rightarrow$
$x^2 + 2xy^2 + 2y = 4$

20. $\frac{\partial M}{\partial y} = e^y = \frac{\partial N}{\partial x} \Rightarrow$ Exact. $\frac{\partial f}{\partial x} = x + e^y \Rightarrow f(x,y) = \frac{x^2}{2} + x\,e^y + k(y)$. $\frac{\partial f}{\partial y} = x\,e^y + k'(y) = y + x\,e^y \Rightarrow k'(y) = y \Rightarrow$
$k(y) = \frac{y^2}{2} + C \Rightarrow f(x,y) = \frac{x^2 + y^2}{2} + x\,e^y + C = C_1 \Rightarrow \frac{x^2 + y^2}{2} + x\,e^y = C_2$. $y(2) = 0 \Rightarrow C_2 = 4 \Rightarrow$
$\frac{x^2 + y^2}{2} + x\,e^y = 4 \Rightarrow x^2 + y^2 + 2x\,e^y = 8$

21. $\frac{\partial M}{\partial y} = -1 = \frac{\partial N}{\partial x} \Rightarrow$ Exact. $\frac{\partial f}{\partial x} = \frac{1}{x} - y \Rightarrow f(x,y) = \ln|x| - xy + k(y)$. $\frac{\partial f}{\partial y} = -x + k'(y) = \frac{1}{y} - x \Rightarrow k'(y) = \frac{1}{y} \Rightarrow$ $k(y) = \ln|y| + C \Rightarrow f(x,y) = \ln|x| - xy + \ln|y| + C = C_1 \Rightarrow \ln|xy| = xy + C_2 \Rightarrow xy = \pm e^{xy+C_2} = \pm e^{C_2}\, e^{xy} \Rightarrow$ $xy = C_3 e^{xy}$. $y(1) = 1 \Rightarrow C_3 = \frac{1}{e} \Rightarrow xy = \frac{1}{e} e^{xy} \Rightarrow xy = e^{xy-1}$

22. $\frac{\partial M}{\partial y} = 1 = \frac{\partial N}{\partial x} \Rightarrow$ Exact. $\frac{\partial f}{\partial x} = 2x + y + 1 \Rightarrow f(x,y) = x^2 + xy + x + k(y)$. $\frac{\partial f}{\partial y} = x + k'(y) = 2y + x + 1 \Rightarrow$ $k'(y) = 2y + 1 \Rightarrow k(y) = y^2 + y + C \Rightarrow f(x,y) = x^2 + xy + y^2 + x + y + C = C_1 \Rightarrow x^2 + y^2 + xy + x + y = C_2$. $y(1) = 5 \Rightarrow C_2 = 37 \Rightarrow x^2 + y^2 + xy + x + y = 37$.

23. $\frac{\partial M}{\partial y} = 2y$, $\frac{\partial N}{\partial x} = ay \Rightarrow a = 2$. $\therefore$ $(3x^2 + y^2)\,dx + 2xy\,dy = 0$ is exact. $\frac{\partial f}{\partial x} = 3x^2 + y^2 \Rightarrow f(x,y) = x^3 + xy^2 +$ $k(y)$. $\frac{\partial f}{\partial y} = 2xy + k'(y) = 2xy \Rightarrow k'(y) = 0 \Rightarrow k(y) = C \Rightarrow f(x,y) = x^3 + xy^2 + C = C_1 \Rightarrow x^3 + xy^2 = C_2$.

24. $\rho = \frac{1}{x^2} \Rightarrow \frac{1}{x^2}\left(x^2 + by^2\right) dx + \frac{1}{x^2}(6xy\,dy) = 0 \Rightarrow \left(1 + \frac{by^2}{x^2}\right) dx + \frac{6y}{x}\,dy = 0 \Rightarrow \frac{\partial M}{\partial y} = \frac{2by}{x^2}, \frac{\partial N}{\partial x} = -\frac{6y}{x^2} \Rightarrow$ $2b = -6$ since ρ makes the equation exact $\Rightarrow b = -3$. $\therefore$ $\left(1 - \frac{3y^2}{x^2}\right) dx + \frac{6y}{x}\,dy = 0$ is exact $\Rightarrow$ $\frac{\partial f}{\partial x} = 1 - \frac{3y^2}{x^2} \Rightarrow f(x,y) = x + \frac{3y^2}{x} + k(y)$. $\frac{\partial f}{\partial y} = \frac{6y}{x} + k'(y) = \frac{6y}{x} \Rightarrow k'(y) = 0 \Rightarrow k(y) = C \Rightarrow f(x,y) = x + \frac{3y^2}{x} + C$ $= C_1 \Rightarrow x + \frac{3y^2}{x} = C_2$

SECTION 16.3 LINEAR FIRST ORDER EQUATIONS

1. $\frac{dy}{dx} + 2y = e^{-x} \Rightarrow P(x) = 2,\ Q(x) = e^{-x} \Rightarrow \int P(x)\,dx = 2x \Rightarrow \rho(x) = e^{2x} \Rightarrow y = \frac{1}{e^{2x}} \int e^{2x}(e^{-x})\,dx \Rightarrow$ $y = \frac{1}{e^{2x}} \int e^x\,dx = \frac{1}{e^{2x}}\left(e^x + C\right) = e^{-x} + C\,e^{-2x}$

2. $\frac{dy}{dx} - \frac{1}{2}y = \frac{1}{2}e^{x/2} \Rightarrow P(x) = -\frac{1}{2},\ Q(x) = \frac{1}{2}e^{x/2} \Rightarrow \int P(x)\,dx = -\frac{1}{2}x \Rightarrow \rho(x) = e^{-x/2} \Rightarrow$ $y = \frac{1}{e^{-x/2}} \int e^{-x/2}\left(\frac{1}{2}e^{x/2}\right) dx = e^{x/2} \int \frac{1}{2}\,dx \Rightarrow y = e^{x/2}\left(\frac{1}{2}x + C\right) = \frac{1}{2}x\,e^{x/2} + C\,e^{x/2}$

3. $\frac{dy}{dx} + \frac{3}{x}y = \frac{\sin x}{x^3} \Rightarrow P(x) = \frac{3}{x},\ Q(x) = \frac{\sin x}{x^3} \Rightarrow \int P(x)\,dx = 3\ln|x| = \ln|x|^3 \Rightarrow \rho(x) = e^{\ln|x|^3} = |x|^3 = x^3$ if $x > 0 \Rightarrow y = \frac{1}{x^3} \int x^3 \left(\frac{\sin x}{x^3}\right) dx = \frac{1}{x^3} \int \sin x\,dx \Rightarrow y = \frac{1}{x^3}(-\cos x + C) = -\frac{\cos x}{x^3} + \frac{C}{x^3}$

4. $\frac{dy}{dx} + \frac{y}{x} = \cos x \Rightarrow P(x) = \frac{1}{x},\ Q(x) = \cos x \Rightarrow \int P(x)\,dx = \ln|x| \Rightarrow \rho(x) = e^{\ln|x|} = x$ if $x > 0 \Rightarrow$ $y = \frac{1}{x} \int x\cos x\,dx = \frac{1}{x}(x\sin x + \cos x + C) \Rightarrow y = \sin x + \frac{\cos x}{x} + \frac{C}{x}$

5. $\frac{dy}{dx} + \frac{4y}{x-1} = \frac{x+1}{(x-1)^3} \Rightarrow P(x) = \frac{4}{x-1},\ Q(x) = \frac{x+1}{(x-1)^3} \Rightarrow \int P(x)\,dx = 4\ln|x-1| = \ln(x-1)^4 \Rightarrow$
$\rho(x) = e^{\ln(x-1)^4} = (x-1)^4 \Rightarrow y = \frac{1}{(x-1)^4}\int (x-1)^4 \frac{x+1}{(x-1)^3}\,dx = \frac{1}{(x-1)^4}\int (x^2-1)\,dx =$
$\frac{1}{(x-1)^4}\left(\frac{x^3}{3} - x + C\right) \Rightarrow y = \frac{x^3}{3(x-1)^4} - \frac{x}{(x-1)^4} + \frac{C}{(x-1)^4}$

6. $\frac{dy}{dx} + 2y = 2x\,e^{-2x} \Rightarrow P(x) = 2,\ Q(x) = 2x\,e^{-2x} \Rightarrow \int P(x)\,dx = \int 2\,dx = 2x \Rightarrow \rho(x) = e^{2x} \Rightarrow$
$y = \frac{1}{e^{2x}}\int e^{2x}\left(2x\,e^{-2x}\right)dx = \frac{1}{e^{2x}}\int 2x\,dx = e^{-2x}\left(x^2 + C\right) \Rightarrow y = x^2 e^{-2x} + C\,e^{-2x}$

7. $\frac{dy}{dx} + (\cot x)\,y = \sec x \Rightarrow P(x) = \cot x,\ Q(x) = \sec x \Rightarrow \int P(x)\,dx = \ln|\sin x| \Rightarrow \rho(x) = e^{\ln|\sin x|} = \sin x$ if
$\sin x > 0 \Rightarrow y = \frac{1}{\sin x}\int \sin x(\sec x)\,dx = \frac{1}{\sin x}\int \tan x\,dx = \frac{1}{\sin x}(\ln|\sec x| + C) = \csc x(\ln|\sec x| + C)$

8. $\frac{dy}{dx} + (\tanh x)\,y = e^{-x}\left(\frac{1}{\cosh x}\right) \Rightarrow P(x) = \tanh x,\ Q(x) = e^{-x}\left(\frac{1}{\cosh x}\right) \Rightarrow \int P(x)\,dx = \ln(\cosh x) \Rightarrow$
$\rho(x) = e^{\ln(\cosh x)} = \cosh x \Rightarrow y = \frac{1}{\cosh x}\int \cosh x(e^{-x})\left(\frac{1}{\cosh x}\right)dx = \frac{1}{\cosh x}\int e^{-x}\,dx =$
$\operatorname{sech} x\left(-e^{-x} + C\right) \Rightarrow y = -e^{-x}\operatorname{sech} x + C\operatorname{sech} x$

9. $\frac{dy}{dx} + 2y = x \Rightarrow P(x) = 2,\ Q(x) = x \Rightarrow \int P(x)\,dx = 2x \Rightarrow \rho(x) = e^{2x} \Rightarrow y = \frac{1}{e^{2x}}\int e^{2x}(x\,dx) =$
$\frac{1}{e^{2x}}\left(\frac{1}{2}x\,e^{2x} - \frac{1}{4}e^{2x} + C\right) \Rightarrow y = \frac{1}{2}x - \frac{1}{4} + C\,e^{-2x}.\ \ y(0) = 1 \Rightarrow -\frac{1}{4} + C = 1 \Rightarrow C = \frac{5}{4} \Rightarrow$
$y = \frac{1}{2}x - \frac{1}{4} + \frac{5}{4}e^{-2x}$

10. $\frac{dy}{dx} + \frac{2y}{x} = x^2 \Rightarrow P(x) = \frac{2}{x},\ Q(x) = x^2 \Rightarrow \int P(x)\,dx = 2\ln|x| \Rightarrow \rho(x) = e^{\ln x^2} = x^2 \Rightarrow$
$y = \frac{1}{x^2}\int x^2\left(x^2\right)dx = \frac{1}{x^2}\int x^4\,dx = \frac{1}{x^2}\left(\frac{x^5}{5} + C\right) = \frac{x^3}{5} + \frac{C}{x^2}.\ \ y(2) = 1 \Rightarrow \frac{8}{5} + \frac{C}{4} = 1 \Rightarrow C = \frac{8}{5} \Rightarrow$
$y = \frac{x^3}{5} + \frac{8}{5x^2}$

11. $\frac{dy}{dx} + \frac{y}{x} = \frac{\sin x}{x} \Rightarrow P(x) = \frac{1}{x},\ Q(x) = \frac{\sin x}{x} \Rightarrow \int P(x)\,dx = \ln|x| \Rightarrow \rho(x) = e^{\ln|x|} = |x| \Rightarrow$
$y = \frac{1}{|x|}\int |x|\frac{\sin x}{x}\,dx = \frac{1}{x}\int x\frac{\sin x}{x}\,dx$ for $x \neq 0 \Rightarrow y = \frac{1}{x}\int \sin x\,dx = \frac{1}{x}(-\cos x + C) \Rightarrow$
$y = -\frac{1}{x}\cos x + \frac{C}{x}.\ \ y\left(\frac{\pi}{2}\right) = 1 \Rightarrow C = \frac{\pi}{2} \Rightarrow y = -\frac{1}{x}\cos x + \frac{\pi}{2x}$

12. $\frac{dy}{dx} - \frac{2y}{x} = x^2 \sec x \tan x \Rightarrow P(x) = -\frac{2}{x},\ Q(x) = x^2 \sec x \tan x \Rightarrow \int P(x)\,dx = -2\ln|x| \Rightarrow \rho(x) = e^{-2\ln|x|}$

$= x^{-2} \Rightarrow y = \frac{1}{x^{-2}} \int x^{-2}(x^2 \sec x \tan x)\,dx = x^2 \int \sec x \tan x\,dx = x^2(\sec x + C) \Rightarrow$

$y = x^2 \sec x + Cx^2.\ y\left(\frac{\pi}{3}\right) = 2 \Rightarrow C = \frac{18}{\pi^2} - 2 \Rightarrow y = x^2 \sec x + \left(\frac{18}{\pi^2} - 2\right)x^2$

13. $\frac{dy}{dx} - ky = 0 \Rightarrow P(x) = -k,\ Q(x) = 0 \Rightarrow \int P(x)\,dx = -kx \Rightarrow \rho(x) = e^{-kx} \Rightarrow$

$y = \frac{1}{e^{-kx}} \int e^{-kx}(0)\,dx = Ce^{kx}.\ y(0) = y_0 \Rightarrow C = y_0 \Rightarrow y = y_0 e^{kx}$

14. a) $\frac{dc}{dt} = \frac{G}{100V} - kc \Rightarrow \frac{dc}{dt} + kc = \frac{G}{100V} \Rightarrow P(t) = k,\ Q(t) = \frac{G}{100V} \Rightarrow \int P(t)\,dt = kt \Rightarrow \rho(t) = e^{kt} \Rightarrow$

$c = \frac{1}{e^{kt}} \int e^{kt} \frac{G}{100V}\,dt = \frac{1}{e^{kt}}\left(\frac{1}{k}\left(\frac{G}{100V}\right)e^{kt} + C\right) \Rightarrow c = \frac{1}{k}\left(\frac{G}{100V}\right) + C\,e^{-kt}.\ c(0) = c_0 \Rightarrow$

$C = c_0 - \frac{G}{100Vk} \Rightarrow c = \frac{G}{100Vk} + \left(c_0 - \frac{G}{100Vk}\right) e^{-kt}$

b) $\lim_{t\to\infty} c = \lim_{t\to\infty}\left(\frac{G}{100Vk} + \left(c_0 - \frac{G}{100Vk}\right) e^{-kt}\right) = \frac{G}{100Vk}$

15. $\frac{dp}{dx} + p = 0 \Rightarrow \frac{dp}{dx} = -p \Rightarrow dp = -p\,dx \Rightarrow \frac{1}{p}\,dp = -dx \Rightarrow \ln|p| = -x + C \Rightarrow p = C_1 e^{-x} \Rightarrow \frac{dy}{dx} = C_1 e^{-x} \Rightarrow$
$y = -C_1 e^{-x} + C_2 = C_3 e^{-x} + C_2$

16. $x\frac{dp}{dx} + 2p = 1 \Rightarrow \frac{dp}{dx} + \frac{2}{x}p = \frac{1}{x} \Rightarrow P(x) = \frac{2}{x},\ Q(x) = \frac{1}{x} \Rightarrow \int P(x)\,dx = 2\ln|x| \Rightarrow \rho(x) = e^{2\ln|x|} = x^2 \Rightarrow$

$p = \frac{1}{x^2}\int x^2\left(\frac{1}{x}\right)dx = \frac{1}{x^2}\int x\,dx = \frac{1}{x^2}\left(\frac{x^2}{2} + C\right) = \frac{1}{2} + \frac{C}{x^2} \Rightarrow \frac{dy}{dx} = \frac{1}{2} + \frac{C}{x^2} \Rightarrow y = \frac{1}{2}x - \frac{C}{x} + C_1$

17. $x\frac{dp}{dx} + p = 0 \Rightarrow \frac{dp}{dx} + \frac{1}{x}p = 0 \Rightarrow P(x) = \frac{1}{x},\ Q(x) = 0 \Rightarrow \int P(x)\,dx = \ln|x| \Rightarrow \rho(x) = e^{\ln|x|} = |x| \Rightarrow$

$p = \frac{1}{|x|}\int |x|(0)\,dx = \frac{C}{|x|} = \frac{C}{x} \text{ if } x > 0 \Rightarrow \frac{dy}{dx} = \frac{C}{x} \Rightarrow y = C\ln x + C_1$

18. $2\frac{dp}{dx} + p^2 = -1 \Rightarrow 2\,dp + p^2\,dx = -\,dx \Rightarrow 2\,dp = -(p^2 + 1)\,dx \Rightarrow \frac{2\,dp}{p^2 + 1} = -\,dx \Rightarrow 2\tan^{-1} p = -x + C \Rightarrow$

$\tan^{-1} p = -\frac{1}{2}(x - C) \Rightarrow p = \tan\left(-\frac{1}{2}(x + C_1)\right) = -\tan\left(\frac{1}{2}(x + C_1)\right) \Rightarrow \frac{dy}{dx} = -\tan\left(\frac{1}{2}(x + C_1)\right) \Rightarrow$

$y = 2\ln\left|\cos\left(\frac{1}{2}(x + C_1)\right)\right| + C_2$

19. $x\frac{dp}{dx} + p = x^2 \Rightarrow \frac{dp}{dx} + \frac{1}{x}p = x \Rightarrow P(x) = \frac{1}{x},\ Q(x) = x \Rightarrow \int P(x)\,dx = \ln x \text{ if } x > 0 \Rightarrow \rho(x) = e^{\ln x} = x \Rightarrow$

$p = \frac{1}{x}\int x(x)\,dx = \frac{1}{x}\int x^2\,dx = \frac{1}{x}\left(\frac{x^3}{3} + C\right) = \frac{x^2}{3} + \frac{C}{x} \Rightarrow \frac{dy}{dx} = \frac{x^2}{3} + \frac{C}{x}.\ y'(1) = 1 \Rightarrow C = \frac{2}{3} \Rightarrow$

$\frac{dy}{dx} = \frac{x^2}{3} + \frac{2}{3x} \Rightarrow y = \frac{x^3}{9} + \frac{2}{3}\ln x + C_1.\ y(1) = 0 \Rightarrow C_1 = -\frac{1}{9} \Rightarrow y = \frac{x^3}{9} + \frac{2}{3}\ln x - \frac{1}{9}$

20. $x\frac{dq}{dx} - 2q = 0 \Rightarrow \frac{dq}{dx} - \frac{2}{x}q = 0 \Rightarrow P(x) = -\frac{2}{x}, Q(x) = 0 \Rightarrow \int P(x)\,dx = -2\ln|x| \Rightarrow \rho(x) = e^{-2\ln|x|} = x^{-2}$
$\Rightarrow q = \frac{1}{x^{-2}}\int x^{-2}(0)\,dx \Rightarrow q = C\,x^2 \Rightarrow \frac{d^2y}{dx^2} = C\,x^2 \Rightarrow \frac{dy}{dx} = \frac{C\,x^3}{3} + C_1 = C_2\,x^3 + C_1 \Rightarrow y = \frac{C_2\,x^4}{4} + C_1\,x +$
$C_3 \Rightarrow y = C_4\,x^4 + C_1\,x + C_3$

21. Steady State $= \frac{V}{R} \Rightarrow$ we want $i = \frac{1}{2}\left(\frac{V}{R}\right) \Rightarrow \frac{1}{2}\left(\frac{V}{R}\right) = \frac{V}{R}\left(1 - e^{-Rt/L}\right) \Rightarrow \frac{1}{2} = 1 - e^{-Rt/L} \Rightarrow -\frac{1}{2} = -e^{-Rt/L}$
$\Rightarrow \ln\frac{1}{2} = -\frac{Rt}{L} \Rightarrow -\frac{L}{R}\ln\frac{1}{2} = t \Rightarrow t = \frac{L}{R}\ln 2$

22. a) $\frac{di}{dt} + \frac{R}{L}i = 0 \Rightarrow P(t) = \frac{R}{L}, Q(t) = 0 \Rightarrow \int P(t)\,dt = \frac{R}{L}t \Rightarrow \rho(t) = e^{Rt/L} \Rightarrow i = \frac{1}{e^{Rt/L}}\int e^{Rt/L}(0)\,dt =$
$C\,e^{-Rt/L}$ where C is the initial current, $i_0 \Rightarrow i = i_0\,e^{-Rt/L}$

b) $\frac{1}{2}i_0 = i_0\,e^{-Rt/L} \Rightarrow e^{-Rt/L} = \frac{1}{2} \Rightarrow -\frac{Rt}{L} = \ln\frac{1}{2} = -\ln 2 \Rightarrow t = \frac{L}{R}\ln 2$

c) $t = \frac{L}{R} \Rightarrow i = i_0\,e^{(-Rt/L)(L/R)} = i_0\,e^{-1}$

23. $t = \frac{3L}{R} \Rightarrow i = \frac{V}{R}\left(1 - e^{(-R/L)(3L/R)}\right) = \frac{V}{R}\left(1 - e^{-3}\right) \approx 0.9502\,\frac{V}{R}$ or about 95% of the steady state value.

SECTION 16.4 SECOND ORDER LINEAR HOMOGENEOUS EQUATIONS

1. $r^2 + 2r = 0 \Rightarrow r_1 = 0, r_2 = -2 \Rightarrow y = C_1e^{0x} + C_2e^{-2x} = C_1 + C_2e^{-2x}$

2. $r^2 + 5r + 6 = 0 \Rightarrow (r + 2)(r + 3) = 0 \Rightarrow r_1 = -2, r_2 = -3 \Rightarrow y = C_1e^{-2x} + C_2e^{-3x}$

3. $r^2 + 6r + 5 = 0 \Rightarrow (r + 5)(r + 1) = 0 \Rightarrow r_1 = -5, r_2 = -1 \Rightarrow y = C_1e^{-5x} + C_2^{-x}$

4. $r^2 - 2r - 3 = 0 \Rightarrow (r - 3)(r + 1) = 0 \Rightarrow r_1 = 3, r_2 = -1 \Rightarrow y = C_1e^{3x} + C_2\,e^{-x}$

5. $r^2 - 4r + 4 = 0 \Rightarrow (r - 2)^2 = 0 \Rightarrow r_1 = r_2 = 2 \Rightarrow y = \left(C_1x + C_2\right)e^{2x}$

6. $r^2 + 6r + 9 = 0 \Rightarrow (r + 3)^2 = 0 \Rightarrow r_1 = r_2 = -3 \Rightarrow y = \left(C_1x + C_2\right)e^{-3x}$

7. $r^2 - 10r + 25 = 0 \Rightarrow (r - 5)^2 = 0 \Rightarrow r_1 = r_2 = 5 \Rightarrow \left(C_1x + C_2\right)e^{5x}$

8. $r^2 - 2\sqrt{2}\,r + 2 = 0 \Rightarrow (r - \sqrt{2}\,)^2 = 0 \Rightarrow r_1 = r_2 = \sqrt{2} \Rightarrow y = \left(C_1x + C_2\right)e^{\sqrt{2}\,x}$

9. $r^2 + r + 1 = 0 \Rightarrow r = \frac{-1 \pm i\sqrt{3}}{2} \Rightarrow y = e^{-x/2}\left(C_1\cos\frac{\sqrt{3}}{2}x + C_2\sin\frac{\sqrt{3}}{2}x\right)$

10. $r^2 - 6r + 10 = 0 \Rightarrow r = 3 \pm i \Rightarrow y = e^{3x}\left(C_1\cos x + C_2\sin x\right)$

11. $r^2 - 2r + 4 = 0 \Rightarrow r = 1 \pm i\sqrt{3} \Rightarrow y = e^{x}\left(C_1\cos\sqrt{3}\,x + C_2\sin\sqrt{3}\,x\right)$

12. $r^2 + 8r + 25 = 0 \Rightarrow r = -4 \pm 3i \Rightarrow y = e^{-4x}\left(C_1 \cos 3x + C_2 \sin 3x\right)$

13. $r^2 - 1 = 0 \Rightarrow r_1 = 1, r_2 = -1 \Rightarrow y = C_1e^x + C_2e^{-x}$. $y(0) = 1 \Rightarrow C_1 + C_2 = 1$. $y' = C_1e^x - C_2e^{-x}$. $y'(0) = -2$ $\Rightarrow -2 = C_1 - C_2$. $\therefore$ $C_1 = -\frac{1}{2}, C_2 = \frac{3}{2} \Rightarrow y = -\frac{1}{2}e^x + \frac{3}{2}e^{-x}$

14. $2r^2 - r - 1 = 0 \Rightarrow (2r + 1)(r - 1) = 0 \Rightarrow r_1 = -\frac{1}{2}, r_2 = 1 \Rightarrow y = C_1e^{-x/2} + C_2e^x$. $y(0) = -1 \Rightarrow -1 = C_1 + C_2$ $y' = -\frac{1}{2}C_1 e^{-x/2} + C_2e^x$. $y'(0) = 0 \Rightarrow 0 = -\frac{1}{2}C_1 + C_2$. $\therefore$ $C_1 = -\frac{2}{3}, C_2 = -\frac{1}{3} \Rightarrow y = -\frac{2}{3}e^{-x/2} - \frac{1}{3}e^x$

15. $r^2 - 4 = 0 \Rightarrow r_1 = 2, r_2 = -2 \Rightarrow y = C_1e^{2x} + C_2e^{-2x}$. $y(0) = 0 \Rightarrow C_1 + C_2$. $y' = 2C_1e^{2x} - 2C_2e^{-2x}$. $y'(0) = 3 \Rightarrow 3 = 2C_1 - 2C_2$. $\therefore$ $C_1 = \frac{3}{4}, C_2 = -\frac{3}{4} \Rightarrow y = \frac{3}{4}e^{2x} - \frac{3}{4}e^{-2x}$

16. $r^2 - 9 = 0 \Rightarrow r_1 = 3, r_2 = -3 \Rightarrow y = C_1e^{3x} + C_2e^{-3x}$. $y(\ln 2) = 1 \Rightarrow 1 = 8C_1 + \frac{1}{8}C_2$. $y' = 3C_1e^{3x} - 3C_2e^{-3x}$. $y'(\ln 2) = -3 \Rightarrow -1 = 8C_1 - \frac{1}{8}C_2$. $\therefore$ $C_1 = 0, C_2 = 8 \Rightarrow y = 8e^{-3x}$

17. $r^2 + 2r + 1 = 0 \Rightarrow (r + 1)^2 = 0 \Rightarrow r_1 = r_2 = -1 \Rightarrow y = \left(C_1x + C_2\right)e^{-x}$. $y(0) = 0 \Rightarrow C_2 = 0$. $y' = C_1\left(e^{-x} - x\,e^{-x}\right)$. $y'(0) = 1 \Rightarrow C_1 = 1 \Rightarrow y = x\,e^{-x}$

18. $4r^2 - 2r + 1 = 0 \Rightarrow r = \frac{1 \pm i\sqrt{3}}{4} \Rightarrow y = e^{x/4}\left(C_1 \cos\frac{\sqrt{3}}{4}x + C_2 \sin\frac{\sqrt{3}}{4}x\right)$. $y(0) = 4 \Rightarrow C_1 = 4$. $y' = \frac{1}{4}e^{x/4}\left(C_1 \cos\frac{\sqrt{3}}{4}x + C_2 \sin\frac{\sqrt{3}}{4}x\right) + e^{x/4}\left(-\frac{\sqrt{3}}{4}C_1 \sin\frac{\sqrt{3}}{4}x + \frac{\sqrt{3}}{4}C_2 \cos\frac{\sqrt{3}}{4}x\right)$. $y'(0) = 2 \Rightarrow$ $\frac{1}{4}C_1 + \frac{\sqrt{3}}{4}C_2 = 2 \Rightarrow C_2 = \frac{4}{\sqrt{3}} \Rightarrow y = e^{x/4}\left(4 \cos\frac{\sqrt{3}}{4}x + \frac{4}{\sqrt{3}}\sin\frac{\sqrt{3}}{4}x\right)$

19. $4r^2 + 12r + 9 = 0 \Rightarrow (2r + 3)^2 = 0 \Rightarrow r_1 = r_2 = -\frac{3}{2} \Rightarrow y = \left(C_1x + C_2\right)e^{-3x/2}$. $y(0) = 0 \Rightarrow C_2 = 0 \Rightarrow$ $y = C_1x\,e^{-3x/2} \Rightarrow y' = C_1\left(e^{-3x/2} - \frac{3}{2}x\,e^{-3x/2}\right)$. $y'(0) = -1 \Rightarrow C_1 = -1 \Rightarrow y = -x\,e^{-3x/2}$

20. $y'' = 0 \Rightarrow y' = C$. $y'(0) = 5 \Rightarrow C = 5 \Rightarrow y' = 5 \Rightarrow y = 5x + C_1$. $y(0) = -3 \Rightarrow -3 = C_1 \Rightarrow y = 5x - 3$

21. $r^2 + 4 = 0 \Rightarrow r = \pm 2i \Rightarrow y = C_1 \cos 2x + C_2 \sin 2x$. $y(0) = 0 \Rightarrow C_1 = 0 \Rightarrow y = C_2 \sin 2x$. $y' = 2C_2 \cos 2x$. $y'(0) = 2 \Rightarrow C_2 = 1 \Rightarrow y = \sin 2x$

22. $r^2 + 9 = 0 \Rightarrow r = \pm 3i \Rightarrow y = C_1 \cos 3x + C_2 \sin 3x$. $y(0) = 0 \Rightarrow C_1 = 0 \Rightarrow y = C_2 \sin 3x$. $y' = 3C_2 \cos 3x$ $y'(0) = \sqrt{3} \Rightarrow C_2 = \frac{\sqrt{3}.}{3} \Rightarrow y = \frac{\sqrt{3}}{3}\sin 3x$

23. $r^2 - 2r + 3 = 0 \Rightarrow r = 1 \pm i\sqrt{2} \Rightarrow y = e^x\left(C_1 \cos\sqrt{2}\,x + C_2 \sin\sqrt{2}\,x\right)$. $y(0) = 2 \Rightarrow C_1 = 2 \Rightarrow y =$ $e^x\left(2 \cos\sqrt{2}\,x + C_2 \sin\sqrt{2}\,x\right)$. $y' = e^x\left(2 \cos\sqrt{2}\,x + C_2 \sin\sqrt{2}\,x\right) +$ $e^x\left(-2\sqrt{2}\sin\sqrt{2}\,x + \sqrt{2}\,C_2 \cos\sqrt{2}\,x\right)$. $y'(0) = 1 \Rightarrow 1 = 2 + \sqrt{2}\,C_2 \Rightarrow C_2 = -\frac{1}{\sqrt{2}} \Rightarrow$

23. (Continued)

$y = e^x\left(2\cos\sqrt{2}\,x - \dfrac{1}{\sqrt{2}}\sin\sqrt{2}\,x\right)$

24. $r^2 - 6r + 10 = 0 \Rightarrow r = 3 \pm i \Rightarrow y = e^{3x}\left(C_1\cos x + C_2\sin x\right)$. $y(0) = 7 \Rightarrow C_1 = 7 \Rightarrow$ $y = e^{3x}\left(7\cos x + C_2\sin x\right)$ $y' = 3e^{3x}\left(7\cos x + C_2\sin x\right) + e^{3x}\left(-7\sin x + C_2\cos x\right)$. $y'(0) = 1 \Rightarrow$ $C_2 = -20 \Rightarrow y = e^{3x}(7\cos x - 20\sin x)$

SECTION 16.5 SECOND ORDER NONHOMOGENEOUS LINEAR EQUATIONS

1. $\dfrac{d^2y}{dx^2} + \dfrac{dy}{dx} = 0 \Rightarrow r^2 + r = 0 \Rightarrow r_1 = 0, r_2 = -1 \Rightarrow y_h = C_1 + C_2e^{-x} \Rightarrow u_1 = 1, u_2 = e^{-x} \Rightarrow D = \begin{vmatrix} 1 & e^{-x} \\ 0 & -e^{-x} \end{vmatrix}$ $= -e^{-x} \Rightarrow v'_1 = -\dfrac{u_2\,F(x)}{D} = x \Rightarrow v_1 = \displaystyle\int x\,dx = \dfrac{x^2}{2} + C_1$. $v'_2 = \dfrac{u_1\,F(x)}{D} = -x\,e^x \Rightarrow v_2 = \displaystyle\int x\,e^x\,dx =$ $-x\,e^x + e^x + C_2 \Rightarrow y = \left(\dfrac{x^2}{2} + C_1\right) + \left(-x\,e^x + e^x + C_2\right)e^{-x} = \dfrac{x^2}{2} - x + C_3 + C_2\,e^{-x}$

2. $\dfrac{d^2y}{dx^2} + y = 0 \Rightarrow r^2 + 1 = 0 \Rightarrow r = \pm i \Rightarrow y_h = C_1\cos x + C_2\sin x \Rightarrow u_1 = \cos x, u_2 = \sin x \Rightarrow D =$ $\begin{vmatrix} \cos x & \sin x \\ -\sin x & \cos x \end{vmatrix} = 1 \Rightarrow v'_1 = -\dfrac{u_2\,F(x)}{D} = -\sin x\tan x \Rightarrow v_1 = \displaystyle\int -\sin x\tan x\,dx = \sin x -$ $\ln(\sec x + \tan x) + C_1$. $v'_2 = \dfrac{u_1\,F(x)}{D} = \cos x\tan x = \sin x \Rightarrow v_2 = \displaystyle\int \sin x\,dx = -\cos x + C_2 \Rightarrow$ $y = \left(\sin x - \ln(\sec x + \tan x) + C_1\right)\cos x + \left(-\cos x + C_2\right)\sin x = -\cos x\ln(\sec x + \tan x) + C_1\cos x +$ $C_2\sin x$

3. $\dfrac{d^2y}{dx^2} + y = 0 \Rightarrow r^2 + 1 = 0 \Rightarrow r = \pm i \Rightarrow y_h = C_1\cos x + C_2\sin x \Rightarrow u_1 = \cos x, u_2 = \sin x \Rightarrow D =$ $\begin{vmatrix} \cos x & \sin x \\ -\sin x & \cos x \end{vmatrix} = 1 \Rightarrow v'_1 = -\dfrac{u_2\,F(x)}{D} = -\sin^2 x \Rightarrow v_1 = \displaystyle\int -\sin^2 x\,dx = -\dfrac{1}{2}x + \dfrac{\sin 2x}{4} + C_1$. $v'_2 = \dfrac{u_1\,F(x)}{D} = \cos x\sin x \Rightarrow v_2 = \displaystyle\int \sin x\cos x\,dx = \dfrac{\sin^2 x}{2} + C_2 \Rightarrow y = \cos x\left(-\dfrac{1}{2}x + \dfrac{\sin 2x}{4} + C_1\right) +$ $\sin x\left(\dfrac{\sin^2 x}{2} + C_2\right) = -\dfrac{1}{2}x\cos x + C_1\cos x + C_3\sin x$

4. $\dfrac{d^2y}{dx^2} + 2\dfrac{dy}{dx} + y = 0 \Rightarrow r^2 + 2r + 1 = 0 \Rightarrow r_1 = r_2 = -1 \Rightarrow y_h = C_1x\,e^{-x} + C_2\,e^{-x} \Rightarrow u_1 = x\,e^{-x}, u_2 = e^{-x}$ $\Rightarrow D = \begin{vmatrix} x\,e^{-x} & e^{-x} \\ e^{-x} - x\,e^{-x} & -e^{-x} \end{vmatrix} = -e^{-2x} \Rightarrow v'_1 = -\dfrac{u_2\,F(x)}{D} = e^{2x} \Rightarrow v_1 = \displaystyle\int e^{2x}\,dx = \dfrac{1}{2}e^{2x} + C_1$. $v'_2 = \dfrac{u_1\,F(x)}{D} = -x\,e^{2x} \Rightarrow v_2 = \displaystyle\int -x\,e^{2x}\,dx = -\dfrac{e^{2x}}{4}(2x - 1) + C_2 \Rightarrow y = x\,e^{-x}\left(\dfrac{1}{2}e^{2x} + C_1\right) -$

4. (Continued)

$e^{-x}\left(\frac{e^{2x}}{4}(2x-1)+C_2\right)=C_1x\,e^{-x}+\frac{e^x}{4}+C_3e^{-x}$

5. $\frac{d^2y}{dx^2}+2\frac{dy}{dx}+y=0 \Rightarrow r^2+2r+1=0 \Rightarrow r_1=r_2=-1 \Rightarrow y_h=C_1x\,e^{-x}+C_2\,e^{-x} \Rightarrow u_1=x\,e^{-x},\ u_2=e^{-x}$

$\Rightarrow D=\begin{vmatrix} x\,e^{-x} & e^{-x} \\ e^{-x}-x\,e^{-x} & -e^{-x}\end{vmatrix}=-e^{-2x} \Rightarrow v'_1=-\frac{u_2\,F(x)}{D}=1 \Rightarrow v_1=\int 1\,dx=x+C_1.\ v'_2=\frac{u_1\,F(x)}{D}=$

$-x \Rightarrow v_2=\int -x\,dx=-\frac{x^2}{2}+C_2 \Rightarrow y=\left(x+C_1\right)xe^{-x}+\left(-\frac{x^2}{2}+C_2\right)e^{-x}=C_1x\,e^{-x}+C_2e^{-x}+\frac{1}{2}x^2\,e^{-x}$

6. $\frac{d^2y}{dx^2}-y=0 \Rightarrow r^2-1=0 \Rightarrow r_1=1,\ r_2=-1 \Rightarrow y_h=C_1\,e^x+C_2\,e^{-x} \Rightarrow u_1=e^x,\ u_2=e^{-x} \Rightarrow$

$D=\begin{vmatrix} e^x & e^{-x} \\ e^x & -e^{-x}\end{vmatrix}=-2 \Rightarrow v'_1=-\frac{u_2\,F(x)}{D}=\frac{x\,e^{-x}}{2} \Rightarrow v_1=\int \frac{x\,e^x}{2}\,dx=\frac{e^{-x}}{2}(-x-1)+C_1.\ v'_2=\frac{u_1\,F(x)}{D}$

$=-\frac{1}{2}x\,e^x \Rightarrow v_2=\int -\frac{1}{2}x\,e^x\,dx=-\frac{1}{2}e^x(x-1)+C_2 \Rightarrow y=e^x\left(\frac{e^{-x}}{2}\right)(-x-1)+C_1e^x+C_2e^{-x}-$

$\frac{1}{2}e^x(x-1)e^{-x}=C_1\,e^x+C_2\,e^{-x}-x$

7. $\frac{d^2y}{dx^2}-y=0 \Rightarrow r^2-1=0 \Rightarrow r_1=1,\ r_2=-1 \Rightarrow y_h=C_1\,e^x+C_2\,e^{-x} \Rightarrow u_1=e^x,\ u_2=e^{-x} \Rightarrow$

$D=\begin{vmatrix} e^x & e^{-x} \\ e^x & -e^{-x}\end{vmatrix}=-2 \Rightarrow v'_1=-\frac{u_2\,F(x)}{D}=\frac{1}{2} \Rightarrow v_1=\frac{1}{2}x+C_1.\ v'_2=\frac{u_1\,F(x)}{D}=-\frac{1}{2}e^{2x} \Rightarrow v_2=-\frac{1}{4}e^{2x}+$

$C_2 \Rightarrow y=\left(\frac{1}{2}x+C_1\right)e^x+\left(-\frac{1}{4}e^{2x}+C_2\right)e^{-x}=\frac{1}{2}x\,e^x+C_3\,e^x+C_2\,e^{-x}$

8. $\frac{d^2y}{dx^2}-y=0 \Rightarrow r^2-1=0 \Rightarrow r_1=1,\ r_2=-1 \Rightarrow y_h=C_1\,e^x+C_2\,e^{-x} \Rightarrow u_1=e^x,\ u_2=e^{-x} \Rightarrow$

$D=\begin{vmatrix} e^x & e^{-x} \\ e^x & -e^{-x}\end{vmatrix}=-2 \Rightarrow v'_1=-\frac{u_2\,F(x)}{D}=\frac{e^{-x}\sin x}{2} \Rightarrow v_1=\int \frac{e^{-x}\sin x}{2}\,dx=\frac{-e^{-x}\,(\sin x+\cos x)}{4}+C_1$

$v'_2=\frac{u_1\,F(x)}{D}=-\frac{e^x\sin x}{2} \Rightarrow v_2=\int -\frac{e^x\sin x}{2}\,dx=\frac{-e^x\,(\sin x-\cos x)}{4}+C_2 \Rightarrow y=$

$e^x\left(\frac{-e^{-x}\,(\sin x+\cos x)}{4}+C_1\right)+e^{-x}\left(\frac{-e^x\,(\sin x-\cos x)}{4}+C_2\right)=C_1\,e^x+C_2\,e^{-x}-\frac{1}{2}\sin x-\frac{1}{2}\cos x$

9. $\frac{d^2y}{dx^2}+4\frac{dy}{dx}+5y=0 \Rightarrow r^2+4r+5=0 \Rightarrow r=-2\pm i \Rightarrow y_h=e^{-2x}\left(C_1\cos x+C_2\sin x\right) \Rightarrow$

$u_1=e^{-2x}\cos x,\ u_2=e^{-2x}\sin x \Rightarrow D=\begin{vmatrix} e^{-2x}\cos x & e^{-2x}\sin x \\ -2e^{-2x}\cos x-e^{-2x}\sin x & -2e^{-2x}\sin x+e^{-2x}\cos x\end{vmatrix}=$

$e^{-4x} \Rightarrow v'_1=-\frac{u_2\,F(x)}{D}=-10\,e^{2x}\sin x \Rightarrow v_1=\int -10\,e^{2x}\sin x\,dx=-2\,e^{2x}(2\sin x-\cos x)+C_1.$

$v'_2=\frac{u_1\,F(x)}{D}=10\,e^{2x}\cos x \Rightarrow v_2=\int 10\,e^{2x}\cos x\,dx=10\left(\frac{e^{2x}}{5}(2\cos x+\sin x)\right)+C_2 \Rightarrow$

9. (Continued)
$y = \left(-2\,e^{2x}(2\sin x - \cos x) + C_1\right)e^{-2x}\cos x + \left(2\,e^{2x}(2\cos x + \sin x) + C_2\right)e^{-2x}\sin x =$
$2 + C_1\,e^{-2x}\cos x + C_2\,e^{-2x}\sin x$

10. $\frac{d^2y}{dx^2} - \frac{dy}{dx} = 0 \Rightarrow r^2 - r = 0 \Rightarrow r_1 = 0,\ r_2 = 1 \Rightarrow y_h = C_1 + C_2\,e^x \Rightarrow u_1 = 1,\ u_2 = e^x \Rightarrow$
$D = \begin{vmatrix} 1 & e^x \\ 0 & e^x \end{vmatrix} = e^x \Rightarrow v'_1 = -\frac{u_2\,F(x)}{D} = -2^x \Rightarrow v_1 = \int -2^x\,dx = -\frac{2^x}{\ln 2} + C_1.\ v'_2 = \frac{u_1\,F(x)}{D} = \left(\frac{2}{e}\right)^x \Rightarrow$
$v_2 = \int \left(\frac{2}{e}\right)^x dx = \left(\frac{2}{e}\right)^x\left(\frac{1}{\ln 2 - 1}\right) + C_2 \Rightarrow y = \left(-\frac{2^x}{\ln 2} + C_1\right) + e^x\left(\left(\frac{2}{e}\right)^x\left(\frac{1}{\ln 2 - 1}\right) + C_2\right) =$
$-\frac{2^x}{\ln 2} + C_1 + \frac{2^x}{\ln 2 - 1} + C_2\,e^x$

11. $\frac{d^2y}{dx^2} + y = 0 \Rightarrow r^2 + 1 = 0 \Rightarrow r = \pm i \Rightarrow y_h = C_1\cos x + C_2\sin x \Rightarrow u_1 = \cos x,\ u_2 = \sin x \Rightarrow$
$D = \begin{vmatrix} \cos x & \sin x \\ -\sin x & \cos x \end{vmatrix} = 1 \Rightarrow v'_1 = -\frac{u_2\,F(x)}{D} = -\tan x \Rightarrow v_1 = \int -\tan x\,dx = -\ln(\sec x) + C_1.$
$v'_2 = \frac{u_1\,F(x)}{D} = 1 \Rightarrow v_2 = \int dx = x + C_2 \Rightarrow y = \left(-\ln(\sec x) + C_1\right)\cos x + \left(x + C_2\right)\sin x = \cos x\,\ln(\cos x)$
$+ x\sin x + C_1\cos x + C_2\sin x$

12. $\frac{d^2y}{dx^2} - \frac{dy}{dx} = 0 \Rightarrow r^2 - r = 0 \Rightarrow r_1 = 0,\ r_2 = 1 \Rightarrow y_h = C_1 + C_2\,e^x \Rightarrow u_1 = 1,\ u_2 = e^x \Rightarrow$
$D = \begin{vmatrix} 1 & e^x \\ 0 & e^x \end{vmatrix} = e^x \Rightarrow v'_1 = -\frac{u_2\,F(x)}{D} = -e^x\cos x \Rightarrow v_1 = \int -e^x\cos x\,dx = -\frac{e^x(\cos x + \sin x)}{2} + C_1.$
$v'_2 = \frac{u_1\,F(x)}{D} = \cos x \Rightarrow v_2 = \int \cos x\,dx = \sin x + C_2 \Rightarrow y = \left(-\frac{e^x(\cos x + \sin x)}{2} + C_1\right) + e^x\left(\sin x + C_2\right)$
$= \frac{e^x\sin x - e^x\cos x}{2} + C_1 + C_2\,e^x$

13. $\frac{d^2y}{dx^2} - 3\frac{dy}{dx} - 10y = 0 \Rightarrow r^2 - 3r - 10 = 0 \Rightarrow r_1 = 5,\ r_2 = -2 \Rightarrow y_h = C_1\,e^{5x} + C_2\,e^{-2x}.\ y_p = C \Rightarrow$
$\frac{d^2y}{dx^2} = 0,\ \frac{dy}{dx} = 0 \Rightarrow -10C = -3 \Rightarrow C = \frac{3}{10} \Rightarrow y_p = \frac{3}{10} \Rightarrow y = \frac{3}{10} + C_1\,e^{5x} + C_2\,e^{-2x}$

14. $\frac{d^2y}{dx^2} - 3\frac{dy}{dx} - 10y = 0 \Rightarrow r^2 - 3r - 10 = 0 \Rightarrow r_1 = 5,\ r_2 = -2 \Rightarrow y_h = C_1\,e^{5x} + C_2\,e^{-2x}.\ y_p = Dx + E \Rightarrow$
$y'_p = D,\ y''_p = 0 \Rightarrow -3D - 10(Dx + E) = 2x - 3 \Rightarrow D = -\frac{1}{5},\ E = \frac{9}{25} \Rightarrow y_p = -\frac{1}{5}x + \frac{9}{25} \Rightarrow$
$y = C_1\,e^{5x} + C_2\,e^{-2x} - \frac{1}{5}x + \frac{9}{25}$

15. $\frac{d^2y}{dx^2} - \frac{dy}{dx} = 0 \Rightarrow r^2 - r = 0 \Rightarrow r_1 = 0,\ r_2 = 1 \Rightarrow y_h = C_1 + C_2\,e^x.\ y_p = B\cos x + C\sin x \Rightarrow \frac{dy}{dx} = -B\sin x +$
$C\cos x,\ \frac{d^2y}{dx^2} = -B\cos x - C\sin x \Rightarrow -B\cos x - C\sin x + B\sin x - C\cos x = \sin x \Rightarrow B = \frac{1}{2},\ C = -\frac{1}{2} \Rightarrow$
$y_p = \frac{1}{2}\cos x - \frac{1}{2}\sin x \Rightarrow y = \frac{1}{2}\cos x - \frac{1}{2}\sin x + C_1 + C_2\,e^x$

16. $\frac{d^2y}{dx^2} + 2\frac{dy}{dx} + y = 0 \Rightarrow r^2 + 2r + 1 = 0 \Rightarrow r = -1 \Rightarrow y_h = C_1 x\, e^{-x} + C_2\, e^{-x}$. $y_p = Dx^2 + Ex + F \Rightarrow$ $y'_p = 2Dx + E$, $y''_p = 2D \Rightarrow 2D + 2(2Dx + E) + Dx^2 + Ex + F = x^2 \Rightarrow D = 1, E = -4, F = 6 \Rightarrow$ $y_p = x^2 - 4x + 6 \Rightarrow y = C_1 x\, e^{-x} + C_2\, e^{-x} + x^2 - 4x + 6$

17. $\frac{d^2y}{dx^2} + y = 0 \Rightarrow r^2 + 1 = 0 \Rightarrow r = \pm i \Rightarrow y_h = C_1 \cos x + C_2 \sin x$. $y_p = B\cos 3x + C \sin 3x \Rightarrow y'_p =$ $-3B \sin 3x + 3C \cos 3x \Rightarrow y''_p = -9B\cos 3x - 9C \sin 3x \Rightarrow -9B\cos 3x - 9C\sin 3x + B\cos 3x +$ $C \sin 3x = \cos 3x \Rightarrow B = -\frac{1}{8}, C = 0 \Rightarrow y_p = -\frac{1}{8}\cos 3x \Rightarrow y = -\frac{1}{8}\cos 3x + C_1 \cos x + C_2 \sin x$

18. $\frac{d^2y}{dx^2} + y = 0 \Rightarrow r^2 + 1 = 0 \Rightarrow r = \pm i \Rightarrow y_h = C_1 \cos x + C_2 \sin x$. $y_p = A\, e^{2x} \Rightarrow y'_p = 2A\, e^{2x} \Rightarrow y''_p =$ $4A\, e^{2x} \Rightarrow 4A\, e^{2x} + A\, e^{2x} = e^{2x} \Rightarrow A = \frac{1}{5} \Rightarrow y_p = \frac{1}{5} e^{2x} \Rightarrow y = C_1 \cos x + C_2 \sin x + \frac{1}{5} e^{2x}$

19. $\frac{d^2y}{dx^2} - \frac{dy}{dx} - 2y = 0 \Rightarrow r^2 - r - 2 = 0 \Rightarrow r_1 = 2, r_2 = -1 \Rightarrow y_h = C_1\, e^{2x} + C_2\, e^{-x}$. $y_p = B\cos x + C \sin x \Rightarrow$ $y'_p = -B \sin x + C\cos x \Rightarrow y''_p = -B\cos x - C \sin x \Rightarrow -B\cos x - C\sin x - (-B \sin x + C\cos x) -$ $2(B\cos x + C \sin x) = 20 \cos x \Rightarrow B = -6, C = -2 \Rightarrow y_p = -6\cos x -2 \sin x \Rightarrow y = -6 \cos x -2 \sin x +$ $C_1\, e^{2x} + C_2\, e^{-x}$

20. $\frac{d^2y}{dx^2} + y = 0 \Rightarrow r^2 + 1 = 0 \Rightarrow r = \pm i \Rightarrow y_h = C_1 \cos x + C_2 \sin x$. $y_p = D\,x + A\, e^x \Rightarrow y'_p = D + A\, e^x \Rightarrow$ $y''_p = A\, e^x \Rightarrow A\, e^x + D\,x + A\, e^x = 2x + 3\, e^x \Rightarrow A = \frac{3}{2}, D = 2 \Rightarrow y_p = 2x + \frac{3}{2} e^x \Rightarrow y = C_1 \cos x + C_2 \sin x$ $+ 2x + \frac{3}{2} e^x$

21. $\frac{d^2y}{dx^2} - y = 0 \Rightarrow r^2 - 1 = 0 \Rightarrow r_1 = 1, r_2 = -1 \Rightarrow y_h = C_1\, e^x + C_2\, e^{-x}$. $y_p = Ax\, e^x + Dx^2 + Ex + F \Rightarrow$ $y'_p = A\, e^x + Ax\, e^x + 2Dx + E \Rightarrow y''_p = 2A\, e^x + Ax\, e^x + 2D \Rightarrow 2A\, e^x + Ax\, e^x + 2D -$ $\left(Ax\, e^x + Dx^2 + Ex + F\right) = e^x + x^2 \Rightarrow A = \frac{1}{2}, D = -1, E = 0, F = -2 \Rightarrow y_p = \frac{1}{2} x\, e^x - x^2 - 2 \Rightarrow$ $y = \frac{1}{2} x\, e^x - x^2 - 2 + C_1\, e^x + C_2\, e^{-x}$

22. $\frac{d^2y}{dx^2} + 2\frac{dy}{dx} + y = 0 \Rightarrow r^2 + 2r + 1 = 0 \Rightarrow r = -1 \Rightarrow y_h = C_1 x\, e^{-x} + C_2\, e^{-x}$. $y_p = B\cos 2x + C \sin 2x \Rightarrow$ $y'_p = -2B \sin 2x + 2C\cos 2x \Rightarrow y''_p = -4B\cos 2x - 4C \sin 2x \Rightarrow -4B\cos 2x - 4C\sin 2x +$ $2(-2B \sin 2x + 2C \cos 2x) + B\cos 2x + C \sin 2x = 6 \sin 2x \Rightarrow B = -\frac{24}{25}, C = -\frac{18}{25} \Rightarrow y_p = -\frac{24}{25}\cos 2x -$ $\frac{18}{25}\sin 2x \Rightarrow y = C_1 x\, e^{-x} + C_2\, e^{-x} - \frac{24}{25}\cos 2x - \frac{18}{25}\sin 2x$

23. $\frac{d^2y}{dx^2} - \frac{dy}{dx} - 6y = 0 \Rightarrow r^2 - r - 6 = 0 \Rightarrow r_1 = 3, r_2 = -2 \Rightarrow y_h = C_1\, e^{3x} + C_2\, e^{-2x}$. $y_p = A\, e^{-x} + B\cos x +$ $D \sin x \Rightarrow y'_p = -A\, e^{-x} - B \sin x + C\cos x \Rightarrow y''_p = A\, e^{-x} - B\cos x - C\sin x \Rightarrow A\, e^{-x} - B\cos x -$ $C \sin x - \left(-A\, e^{-x} - B\sin x + C\cos x\right) -6\left(A\, e^{-x} + B\cos x + C \sin x\right) = e^{-x} - 7\cos x \Rightarrow A = -\frac{1}{4},$

23. (Continued)
$B = \frac{49}{50}$, $C = \frac{7}{50} \Rightarrow y_p = -\frac{1}{4}e^{-x} + \frac{49}{50}\cos x + \frac{7}{50}\sin x \Rightarrow y = -\frac{1}{4}e^{-x} + \frac{49}{50}\cos x + \frac{7}{50}\sin x + C_1 e^{3x} + C_2 e^{-2x}$

24. $\frac{d^2y}{dx^2} + 3\frac{dy}{dx} + 2y = 0 \Rightarrow r^2 + 3r + 2 = 0 \Rightarrow r_1 = -2, r_2 = -1 \Rightarrow y_h = C_1 e^{-2x} + C_2 e^{-x}$. $y_p = Ax\,e^{-2x} + Bx\,e^{-x} + C\,x + D \Rightarrow y'_p = -2Ax\,e^{-2x} + A\,e^{-2x} + B\,e^{-x} - B\,e^{-x} + C \Rightarrow y''_p = 4Ax\,e^{-2x} - 4A\,e^{-2x} - 2B\,e^{-x} + Bx\,e^{-x} \Rightarrow 4Ax\,e^{-2x} - 4A\,e^{-2x} - 2B\,e^{-x} + Bx\,e^{-x} + 3\left(-2Ax\,e^{-2x} + A\,e^{-2x} + B\,e^{-x} - Bx\,e^{-x} + C\right) + 2\left(Ax\,e^{-2x} + Bx\,e^{-x} + Cx + D\right) = e^{-x} + e^{-2x} - x \Rightarrow A = -1, B = 1, C = -\frac{1}{2}, D = \frac{3}{4} \Rightarrow y_p = -x\,e^{-2x} + x\,e^{-x} - \frac{1}{2}x + \frac{3}{4} \Rightarrow y = -x\,e^{-2x} + x\,e^{-x} - \frac{1}{2}x + \frac{3}{4} + C_1 e^{-2x} + C_2 e^{-x}$

25. $\frac{d^2y}{dx^2} + 5\frac{dy}{dx} = 0 \Rightarrow r^2 + 5r = 0 \Rightarrow r_1 = 0, r_2 = -5 \Rightarrow y_h = C_1 + C_2 e^{-5x}$. $y_p = Dx^3 + Ex^2 + Fx \Rightarrow y'_p = 3Dx^2 + 2Ex + F \Rightarrow y''_p = 6Dx + 2E \Rightarrow 6Dx + 2E + 5\left(3Dx^2 + 2Ex + F\right) = 15x^2 \Rightarrow D = 1, E = -\frac{3}{5}, F = \frac{6}{25} \Rightarrow y_p = x^3 - \frac{3}{5}x^2 + \frac{6}{25}x \Rightarrow y = x^3 - \frac{3}{5}x^2 + \frac{6}{25}x + C_1 + C_2 e^{-5x}$

26. $\frac{d^2y}{dx^2} - \frac{dy}{dx} = 0 \Rightarrow r^2 - r = 0 \Rightarrow r_1 = 0, r_2 = 1 \Rightarrow y_h = C_1 + C_2 e^x$. $y_p = Dx^2 + Ex \Rightarrow y'_p = 2Dx + E \Rightarrow y''_p = 2D \Rightarrow 2D - (2Dx + E) = -8x + 3 \Rightarrow D = 4, E = 5 \Rightarrow y_p = 4x^2 + 5x \Rightarrow y = C_1 + C_2{}^{x} + 4x^2 + 5x$

27. $\frac{d^2y}{dx^2} - 3\frac{dy}{dx} = 0 \Rightarrow r^2 - 3r = 0 \Rightarrow r_1 = 0, r_2 = 3 \Rightarrow y_h = C_1 + C_2 e^{3x}$. $y_p = Ax\,e^{3x} + Dx^2 + Ex \Rightarrow y'_p = A\,e^{3x} + 3Ax\,e^{3x} + 2Dx + E \Rightarrow y''_p = 6A\,e^{3x} + 9Ax\,e^{3x} + 2D \Rightarrow 6A\,e^{3x} + 9Ax\,e^{3x} + 2D - 3\left(A\,e^{3x} + 3Ax\,e^{3x} + 2Dx + E\right) = e^{3x} - 12x \Rightarrow A = \frac{1}{3}, D = 2, E = \frac{4}{3} \Rightarrow y_p = \frac{1}{3}x\,e^{3x} + 2x^2 + \frac{4}{3}x \Rightarrow y = \frac{1}{3}x\,e^{3x} + 2x^2 + \frac{4}{3}x + C_1 + C_2 e^{3x}$

28. $\frac{d^2y}{dx^2} + 7\frac{dy}{dx} = 0 \Rightarrow r^2 + 7r = 0 \Rightarrow r_1 = 0, r_2 = -7 \Rightarrow y_h = C_1 + C_2 e^{-7x}$. $y_p = Dx^3 + Ex^2 + Fx \Rightarrow y'_p = 3Dx^2 + 2Ex + F \Rightarrow y''_p = 6Dx + 2E \Rightarrow 6Dx + 2E + 7\left(3Dx^2 + 2Ex + F\right) = 42x^2 + 5x + 1 \Rightarrow D = 2, E = -\frac{1}{2}, F = \frac{2}{7} \Rightarrow y_p = 2x^3 - \frac{1}{2}x^2 + \frac{2}{7}x \Rightarrow y = 2x^3 - \frac{1}{2}x^2 + \frac{2}{7}x + C_1 + C_2 e^{-7x}$

29. $\frac{d^2y}{dx^2} - 5\frac{dy}{dx} = 0 \Rightarrow r^2 - 5r = 0 \Rightarrow r_1 = 0, r_2 = 5 \Rightarrow y_h = C_1 + C_2 e^{5x}$. $y_p = Ax^2 e^{5x} + Bx\,e^{5x} \Rightarrow y'_p = (2A + 5B)x\,e^{5x} + 5Ax^2 e^{5x} + B\,e^{5x} \Rightarrow y''_p = (2A + 10B)e^{5x} + (20A + 25B)x\,e^{5x} + 25Ax^2 e^{5x} \Rightarrow (2A + 10B)e^{5x} + (20A + 25B)x\,e^{5x} + 25Ax^2 e^{5x} - 5\left((2A + 5B)e^{5x} + 5Ax^2 e^{5x} + B\,e^{5x}\right) = x\,e^{5x} \Rightarrow A = \frac{1}{10}, B = -\frac{1}{25} \Rightarrow y_p = \frac{1}{10}x^2 e^{5x} - \frac{1}{25}x\,e^{5x} \Rightarrow y = \frac{1}{10}x^2 e^{5x} - \frac{1}{25}x\,e^{5x} + C_1 + C_2 e^{5x}$

30. $\frac{d^2y}{dx^2} - \frac{dy}{dx} = 0 \Rightarrow r^2 - r = 0 \Rightarrow r_1 = 0, r_2 = 1 \Rightarrow y_h = C_1 + C_2 e^x$. $y_p = A\cos x + B\sin x \Rightarrow y'_p = -A\sin x +$ $B\cos x \Rightarrow y''_p = -A\cos x - B\sin x \Rightarrow -A\cos x - B\sin x - (-A\sin x + B\cos x) = \cos x + \sin x \Rightarrow$ $A = 0, B = -1 \Rightarrow y_p = -\sin x \Rightarrow y = C_1 + C_2 e^x - \sin x$

31. $\frac{d^2y}{dx^2} + y = 0 \Rightarrow r^2 + 1 = 0 \Rightarrow r = \pm i \Rightarrow y_h = C_1\cos x + C_2\sin x$. $y_p = Ax\cos x + Bx\sin x \Rightarrow y'_p =$ $A\cos x - Ax\sin x + B\sin x + Bx\cos x \Rightarrow y''_p = -2A\sin x - Ax\cos x + 2B\cos x - Bx\sin x \Rightarrow$ $-2A\sin x - Ax\cos x + 2B\cos x - Bx\sin x + Ax\cos x + Bx\sin x = 2\cos x + \sin x$ $\Rightarrow A = -\frac{1}{2}, B = 1 \Rightarrow y_p = -\frac{1}{2}x\cos x + x\sin x \Rightarrow y = -\frac{1}{2}x\cos x + x\sin x + C_1\cos x + C_2\sin x$

32. a) $\frac{d^2y}{dx^2} - 4\frac{dy}{dx} + 4y = 0 \Rightarrow r^2 - 4r + 4 = 0 \Rightarrow r = 2 \Rightarrow y_h = C_1 x e^{2x} + C_2 e^{2x} \Rightarrow u_1 = x e^{2x}, u_2 = e^{2x} \Rightarrow$

$$D = \begin{vmatrix} xe^{2x} & e^{2x} \\ e^{2x} + 2xe^{2x} & 2e^{2x} \end{vmatrix} = -e^{4x} \Rightarrow v'_1 = -\frac{u_2 F(x)}{D} = 2 \Rightarrow v_1 = \int 2\,dx = 2x + C_1.\ \ v'_2 = \frac{u_1 F(x)}{D}$$

$$= -2x \Rightarrow v_2 = \int -2x\,dx = -x^2 + C_2 \Rightarrow y = x e^{2x}\left(2x + C_1\right) + e^{2x}\left(-x^2 + C_2\right) = C_1 x e^{2x} + C_2 e^{2x} + x^2 e^{2x}$$

b) $y_h = C_1 x e^{2x} + C_2 e^{2x}$. $y_p = Ax^2 e^{2x} \Rightarrow y'_p = 2Ax e^{2x} + 2Ax^2 e^{2x} \Rightarrow y''_p = 4Ax^2 e^{2x} + 8Ax e^{2x} +$ $2A e^{2x} \Rightarrow 4Ax^2 e^{2x} + 8Ax e^{2x} + 2A e^{2x} - 4\left(2Ax e^{2x} + 2Ax^2 e^{2x}\right) + 4Ax^2 e^{2x} = 2 e^{2x} \Rightarrow A = 1 \Rightarrow$ $y_p = x^2 e^{2x} \Rightarrow y = C_1 x e^{2x} + C_2 e^{2x} + x^2 e^{2x}$

33. a) $\frac{d^2y}{dx^2} - \frac{dy}{dx} = 0 \Rightarrow r^2 - r = 0 \Rightarrow r_1 = 0, r_2 = 1 \Rightarrow y_h = C_1 + C_2 e^x \Rightarrow u_1 = 1, u_2 = e^x \Rightarrow$

$$D = \begin{vmatrix} 1 & e^x \\ 0 & e^x \end{vmatrix} = e^x \Rightarrow v'_1 = -\frac{u_2 F(x)}{D} = -e^x - e^{-x} \Rightarrow v_1 = \int -e^x - e^{-x}\,dx = -e^x + e^{-x} + C_1.$$

$$v'_2 = \frac{u_1 F(x)}{D} = 1 + e^{-2x} \Rightarrow v_2 = \int \left(1 + e^{-2x}\right) dx = x - \frac{1}{2}e^{-2x} + C_2 \Rightarrow$$

$$y = \left(-e^x + e^{-x} + C_1\right) + \left(x - \frac{1}{2}e^{-2x} + C_2\right)e^x = \frac{1}{2}e^{-x} + x e^x + C_1 + C_3 e^x$$

b) $y_h = C_1 + C_2 e^x$. $y_p = Ax e^x + B e^{-x} \Rightarrow y'_p = A e^x + Ax e^x - B e^{-x} \Rightarrow y''_p = 2A e^x + Ax e^x + B e^{-x} \Rightarrow$ $2A e^x + Ax e^x + B e^{-x} - \left(A e^x + Ax e^x - B e^{-x}\right) = e^x + e^{-x} \Rightarrow A = 1, B = \frac{1}{2} \Rightarrow y_p = x e^x + \frac{1}{2}e^{-x} \Rightarrow$ $y = x e^x + \frac{1}{2}e^{-x} + C_1 + C_2 e^x$

34. a) $\frac{d^2y}{dx^2} - 9\frac{dy}{dx} = 0 \Rightarrow r^2 - 9r = 0 \Rightarrow r_1 = 0, r_2 = 9 \Rightarrow y_h = C_1 + C_2 e^{9x} \Rightarrow u_1 = 1, u_2 = e^{9x} \Rightarrow$

$$D = \begin{vmatrix} 1 & e^{9x} \\ 0 & 9e^{9x} \end{vmatrix} = 9e^{9x} \Rightarrow v'_1 = -\frac{u_2 F(x)}{D} = -e^{9x} \Rightarrow v_1 = \int -e^{9x}\,dx = -\frac{1}{9}e^{9x} + C_1.$$

$$v'_2 = \frac{u_1 F(x)}{D} = 1 \Rightarrow v_2 = \int dx = x + C_2 \Rightarrow y = -\frac{1}{9}e^{9x} + C_1 + e^{9x}\left(x + C_2\right) = C_3 e^{9x} + C_1 + x e^{9x}$$

b) $y_h = C_1 + C_2 e^{9x}$. $y_p = Ax e^{9x} \Rightarrow y'_p = A e^{9x} + 9Ax e^{9x} \Rightarrow y''_p = 18A e^{9x} + 81Ax e^{9x} \Rightarrow$ $18A e^{9x} + 81Ax e^{9x} - 9\left(A e^{9x} + 9Ax e^{9x}\right) = 9 e^{9x} \Rightarrow A = 1 \Rightarrow y_p = x e^{9x} \Rightarrow y = C_1 + C_2 e^{9x} + xe^{9x}$

35. a) $\frac{d^2y}{dx^2} - 4\frac{dy}{dx} - 5y = 0 \Rightarrow r^2 - 4r - 5 = 0 \Rightarrow r_1 = 5, r_2 = -1 \Rightarrow y_h = C_1 e^{5x} + C_2 e^{-x} \Rightarrow u_1 = e^{5x},$

$u_2 = e^{-x} \Rightarrow D = \begin{vmatrix} e^{5x} & e^{-x} \\ 5e^{5x} & -e^{-x} \end{vmatrix} = -6e^{4x} \Rightarrow v'_1 = -\frac{u_2 F(x)}{D} = \frac{1}{6}e^{-4x} + \frac{2}{3}e^{-5x} \Rightarrow$

$v_1 = \int \left(\frac{1}{6}e^{-4x} + \frac{2}{3}e^{-5x}\right) dx = -\frac{1}{24}e^{-4x} - \frac{2}{15}e^{-5x} + C_1.\ v'_2 = \frac{u_1 F(x)}{D} = -\frac{1}{6}e^{2x} - \frac{2}{3}e^{x} \Rightarrow$

$v_2 = \int \left(-\frac{1}{6}e^{2x} - \frac{2}{3}e^{x}\right) dx = -\frac{1}{12}e^{2x} - \frac{2}{3}e^{x} + C_2 \Rightarrow y = \left(-\frac{1}{24}e^{-4x} - \frac{2}{15}e^{-5x} + C_1\right)e^{5x} +$

$\left(-\frac{1}{12}e^{2x} - \frac{2}{3}e^{x} + C_2\right)e^{-x} = -\frac{1}{8}e^{x} - \frac{4}{5} + C_1e^{5x} + C_2e^{-x}$

b) $y_h = C_1 e^{5x} + C_2 e^{-x}.\ y_p = Ae^{x} + Bx + C \Rightarrow y'_p = Ae^{x} + B \Rightarrow y''_p = Ae^{x} \Rightarrow Ae^{x} - 4(Ae^{x} + B) -$

$5(Ae^{x} + Bx + C) = e^{x} + 4 \Rightarrow A = -\frac{1}{8}, B = 0, C = -\frac{4}{5} \Rightarrow y_p = -\frac{1}{8}e^{x} - \frac{4}{5} \Rightarrow\ y = -\frac{1}{8}e^{x} - \frac{4}{5} + C_1 e^{5x} +$

$C_2 e^{-x}$

36. $\frac{d^2y}{dx^2} + y = 0 \Rightarrow r^2 + 1 = 0 \Rightarrow r = \pm i \Rightarrow y_h = C_1 \cos x + C_2 \sin x \Rightarrow u_1 = \cos x, u_2 = \sin x \Rightarrow$

$D = \begin{vmatrix} \cos x & \sin x \\ -\sin x & \cos x \end{vmatrix} = 1 \Rightarrow v'_1 = -\frac{u_2 F(x)}{D} = -1 \Rightarrow v_1 = \int -dx = -x + C_1.\ v'_2 = \frac{u_1 F(x)}{D} = \cot x \Rightarrow$

$v_2 = \int \cot x\, dx = -\ln(\cos x) + C_2 \Rightarrow y = (-x + C_1)\cos x + \sin x(-\ln(\cos x) + C_2) = -x\cos x -$

$\sin x \ln(\cos x)$

37. $\frac{d^2y}{dx^2} + y = 0 \Rightarrow r^2 + 1 = 0 \Rightarrow r = \pm i \Rightarrow y_h = C_1 \cos x + C_2 \sin x \Rightarrow u_1 = \cos x, u_2 = \sin x \Rightarrow$

$D = \begin{vmatrix} \cos x & \sin x \\ -\sin x & \cos x \end{vmatrix} = 1 \Rightarrow v'_1 = -\frac{u_2 F(x)}{D} = -\cos x \Rightarrow v_1 = \int -\cos x\, dx = -\sin x + C_1.$

$v'_2 = \frac{u_1 F(x)}{D} = \csc x - \sin x \Rightarrow v_2 = \int (\csc x - \sin x)\, dx = -\ln|\csc x + \cot x| + \cos x + C_2 \Rightarrow$

$y = (-\sin x + C_1)\cos x + (-\ln|\csc x + \cot x| + \cos x + C_2)\sin x = C_1\cos x + C_2 \sin x -$

$\sin x(\ln|\csc x + \cot x|)$

38. $\frac{d^2y}{dx^2} + 4y = 0 \Rightarrow r^2 + 4 = 0 \Rightarrow r = \pm 2i \Rightarrow y_h = C_1 \cos 2x + C_2 \sin 2x.\ \ y_p = B\cos x + C\sin x \Rightarrow$

$y'_p = -B\sin x + C\cos x \Rightarrow y''_p = -B\cos x - C\sin x \Rightarrow -B\cos x - C\sin x + 4(B\cos x + C\sin x) = \sin x$

$\Rightarrow B = 0, C = \frac{1}{3} \Rightarrow y_p = \frac{1}{3}\sin x \Rightarrow y = C_1 \cos 2x + C_2 \sin 2x + \frac{1}{3}\sin x$

39. $\frac{d^2y}{dx^2} - 8\frac{dy}{dx} = 0 \Rightarrow r^2 - 8r = 0 \Rightarrow r_1 = 0, r_2 = 8 \Rightarrow y_h = C_1 + C_2 e^{8x}\ \ y_p = Axe^{8x} \Rightarrow y'_p = Ae^{8x} + 8Axe^{8x}$

$\Rightarrow y''_p = 16Ae^{8x} + 64Axe^{8x} \Rightarrow 16Ae^{8x} + 64Axe^{8x} - 8(Ae^{8x} + 8Axe^{8x}) = e^{8x} \Rightarrow A = \frac{1}{8} \Rightarrow$

$y_p = \frac{1}{8}xe^{8x} \Rightarrow y = \frac{1}{8}xe^{8x} + C_1 + C_2e^{8x}$

40. $\frac{d^2y}{dx^2} + 4\frac{dy}{dx} + 5y = 0 \Rightarrow r^2 + 4r + 5 = 0 \Rightarrow r = -2 \pm i \Rightarrow y_h = e^{-2x}\left(C_1 \cos x + C_2 \sin x\right)$. $y_p = Ax + B \Rightarrow$ $y'_p = A \Rightarrow y''_p = 0 \Rightarrow 4A + 5(Ax + B) = x + 2 \Rightarrow A = \frac{1}{5}, B = \frac{6}{25} \Rightarrow y_p = \frac{1}{5}x + \frac{6}{25} \Rightarrow$ $y = e^{-2x}\left(C_1 \cos x + C_2 \sin x\right) + \frac{1}{5}x + \frac{6}{25}$

41. $\frac{d^2y}{dx^2} - \frac{dy}{dx} = 0 \Rightarrow r^2 - r = 0 \Rightarrow r_1 = 0, r_2 = 1 \Rightarrow y_h = C_1 + C_2 e^x$. $y_p = Dx^4 + Ex^3 + Fx^2 + Gx \Rightarrow$ $y'_p = 4Dx^3 + 3Ex^2 + 2Fx + G \Rightarrow y''_p = 12Dx^2 + 6Ex + 2F \Rightarrow 12Dx^2 + 6Ex + 2F -$ $\left(4Dx^3 + 3Ex^2 + 2Fx + G\right) = x^3 \Rightarrow D = -\frac{1}{4}, E = -1, F = -3, G = -6 \Rightarrow y_p = -\frac{1}{4}x^4 - x^3 - 3x^2 - 6x \Rightarrow$ $y = -\frac{1}{4}x^4 - x^3 - 3x^2 - 6x + C_1 + C_2 e^x$

42. $\frac{d^2y}{dx^2} + 9y = 0 \Rightarrow r^2 + 9 = 0 \Rightarrow r = \pm 3i \Rightarrow y_h = C_1 \cos 3x + C_2 \sin 3x$. $y_p = B \cos x + C \sin x + Dx + E \Rightarrow$ $y'_p = -B \sin x + C \cos x + D \Rightarrow y''_p = -B \cos x - C \sin x \Rightarrow -B \cos x - C \sin x +$ $9(B \cos x + C \sin x + Dx + E) = 9x - \cos x \Rightarrow B = -\frac{1}{8}, C = 0, D = 1, E = 0 \Rightarrow y_p = -\frac{1}{8}\cos x + x \Rightarrow$ $y = C_1 \cos 3x + C_2 \sin 3x - \frac{1}{8}\cos x + x$

43. $\frac{d^2y}{dx^2} + 2\frac{dy}{dx} = 0 \Rightarrow r^2 + 2r = 0 \Rightarrow r_1 = 0, r_2 = -2 \Rightarrow y_h = C_1 + C_2 e^{-2x}$. $y_p = Ax^3 + Bx^2 + Cx + E e^x \Rightarrow$ $y'_p = 3Ax^2 + 2Bx + C + E e^x \Rightarrow y''_p = 6Ax + 2B + E e^x \Rightarrow 6Ax + 2B + E e^x + 2\left(3Ax^2 + 2Bx + C + E e^x\right)$ $= x^2 - e^x \Rightarrow A = \frac{1}{6}, B = -\frac{1}{4}, C = \frac{1}{4}, E = -\frac{1}{3} \Rightarrow y_p = \frac{1}{6}x^3 - \frac{1}{4}x^2 + \frac{1}{4}x - \frac{1}{3}e^x \Rightarrow y = \frac{1}{6}x^3 - \frac{1}{4}x^2 + \frac{1}{4}x -$ $\frac{1}{3}e^x + C_1 + C_2 e^{-2x}$

44. $\frac{d^2y}{dx^2} - 3\frac{dy}{dx} + 2y = 0 \Rightarrow r^2 - 3r + 2 = 0 \Rightarrow r_1 = 1, r_2 = 2 \Rightarrow y_h = C_1 e^x + C_2 e^{2x}$. $y_p = Ax e^x + Bx e^{2x} \Rightarrow$ $y'_p = A e^x + Ax e^x + B e^{2x} + 2Bx e^{2x} \Rightarrow y''_p = 2A e^x + Ax e^x + 4B e^{2x} + 4Bx e^{2x} \Rightarrow 2A e^x + Ax e^x +$ $4B e^{2x} + 4Bx e^{2x} - 3\left(A e^x + Ax e^x + B e^{2x} + 2Bx e^{2x}\right) + 2\left(Ax e^x + Bx e^{2x}\right) = e^x - e^{2x} \Rightarrow A = -1,$ $B = -1 \Rightarrow y_p = -x e^x - x e^{2x} \Rightarrow y = -x e^x - x e^{2x} + C_1 e^x + C_2 e^{2x}$

45. $\frac{d^2y}{dx^2} + y = 0 \Rightarrow r^2 + 1 = 0 \Rightarrow r = \pm i \Rightarrow y_h = C_1 \cos x + C_2 \sin x \Rightarrow u_1 = \cos x, u_2 = \sin x \Rightarrow$

$$D = \begin{vmatrix} \cos x & \sin x \\ -\sin x & \cos x \end{vmatrix} = 1 \Rightarrow v'_1 = -\frac{u_2 F(x)}{D} = -\tan^2 x \Rightarrow v_1 = \int -\tan^2 x \, dx = -\tan x + x + C_1.$$

$$v'_2 = \frac{u_1 F(x)}{D} = \tan x \Rightarrow v_2 = \int \tan x \, dx = \ln|\sec x| + C_2 \Rightarrow y = \left(-\tan x + x + C_1\right)\cos x +$$

$\left(\ln|\sec x| + C_2\right)\sin x \Rightarrow y = x \cos x + \sin x \ln(\sec x) + C_1 \cos x + C_3 \sin x$

46. $\frac{dy}{dx} + 4y = 0 \Rightarrow r + 4 = 0 \Rightarrow r = -4 \Rightarrow y_h = C_1 e^{-4x}$. $y_p = Dx + E \Rightarrow y'_p = D \Rightarrow D + 4(Dx + E) = x \Rightarrow D =$ $\frac{1}{4}, E = -\frac{1}{16} \Rightarrow y_p = \frac{1}{4}x - \frac{1}{16} \Rightarrow y = C_1 e^{-4x} + \frac{1}{4}x - \frac{1}{16}$

47. $\frac{dy}{dx} - 3y = 0 \Rightarrow r - 3 = 0 \Rightarrow r = 3 \Rightarrow y_h = C_1 e^{3x}$. $y_p = A e^x \Rightarrow y'_p = A e^x \Rightarrow A e^x - 3A e^x = e^x \Rightarrow A = -\frac{1}{2} \Rightarrow y_p = -\frac{1}{2} e^x \Rightarrow y = C_1 e^{3x} - \frac{1}{2} e^x$

48. $\frac{dy}{dx} + y = 0 \Rightarrow r + 1 = 0 \Rightarrow r = -1 \Rightarrow y_h = C_1 e^{-x}$. $y_p = A \cos x + B \sin x \Rightarrow y'_p = -A \sin x + B \cos x \Rightarrow -A \sin x + B \cos x + A \cos x + B \sin x = \sin x \Rightarrow A = -\frac{1}{2}, B = \frac{1}{2} \Rightarrow y_p = -\frac{1}{2} \cos x + \frac{1}{2} \sin x \Rightarrow y = C_1 e^{-x} - \frac{1}{2} \cos x + \frac{1}{2} \sin x$

49. $\frac{dy}{dx} - 3y = 0 \Rightarrow r - 3 = 0 \Rightarrow r = 3 \Rightarrow y_h = C_1 e^{3x}$. $y_p = Ax e^{3x} \Rightarrow y'_p = A e^{3x} + 3Ax e^{3x} \Rightarrow A e^{3x} + 3Ax e^{3x} - 3Ax e^{3x} = 5 e^{3x} \Rightarrow A = 5 \Rightarrow y_p = 5x e^{3x} \Rightarrow y = C_1 e^{3x} + 5x e^{3x}$

50. $\frac{d^2y}{dx^2} + y = 0 \Rightarrow r^2 + 1 = 0 \Rightarrow r = \pm i \Rightarrow y_h = C_1 \cos x + C_2 \sin x$. $y_p = A e^{2x} \Rightarrow y'_p = 2A e^{2x} \Rightarrow y''_p = 4A e^{2x} \Rightarrow 4A e^{2x} + A e^{2x} = e^{2x} \Rightarrow A = \frac{1}{5} \Rightarrow y_p = \frac{1}{5} e^{2x} \Rightarrow y = C_1 \cos x + C_2 \sin x + \frac{1}{5} e^{2x}$. $y(0) = 0 \Rightarrow C_1 \cos 0 + C_2 \sin 0 + \frac{1}{5} e^0 = 0 \Rightarrow C_1 = -\frac{1}{5} \Rightarrow y = -\frac{1}{5} \cos x + C_2 \sin x + \frac{1}{5} e^{2x} \Rightarrow y' = \frac{1}{5} \sin x + C_2 \cos x + \frac{2}{5} e^{2x}$. $y'(0) = \frac{2}{5} \Rightarrow \frac{1}{5} \sin 0 + C_2 \cos 0 + \frac{2}{5} e^0 = \frac{2}{5} \Rightarrow C_2 = 0 \Rightarrow y = -\frac{1}{5} \cos x + \frac{1}{5} e^{2x}$

51. $\frac{d^2y}{dx^2} + y = 0 \Rightarrow r^2 + 1 = 0 \Rightarrow r = \pm i \Rightarrow y_h = C_1 \cos x + C_2 \sin x \Rightarrow u_1 = \cos x, u_2 = \sin x \Rightarrow$
$D = \begin{vmatrix} \cos x & \sin x \\ -\sin x & \cos x \end{vmatrix} = 1 \Rightarrow v'_1 = -\frac{u_2 F(x)}{D} = -\sec x \tan x \Rightarrow v_1 = \int -\sec x \tan x \, dx = -\sec x + C_1$. $v'_2 = \frac{u_1 F(x)}{D} = \sec x \Rightarrow v_2 = \int \sec x \, dx = \ln|\sec x + \tan x| + C_2 \Rightarrow y = C_1 \cos x + C_2 \sin x - 1 + \sin x \ln|\sec x + \tan x|$. $y(0) = 1 \Rightarrow C_1 \cos 0 + C_2 \sin 0 - 1 + \sin 0 \ln|\sec 0 + \tan 0| \Rightarrow C_1 = 2 \Rightarrow y = 2 \cos x + C_2 \sin x - 1 + \sin x \ln|\sec x + \tan x| \Rightarrow y' = -2 \sin x + C_2 \cos x + \sec x \sin x + \cos x \ln|\sec x + \tan x|$. $y'(0) = 1 \Rightarrow -2 \sin 0 + C_2 \cos 0 + \sec 0 \sin 0 + \cos 0 \ln|\sec 0 + \tan 0| \Rightarrow C_2 = 1 \Rightarrow y = 2 \cos x + \sin x - 1 + \sin x \ln|\sec x + \tan x|$

52. Step 1: $\frac{1}{y^2}\frac{dy}{dx} + \frac{1}{y} = x^2$ Step 2: Let $u = \frac{1}{y} \Rightarrow \frac{du}{dx} = -\frac{1}{y^2}\frac{dy}{dx} \Rightarrow -\frac{du}{dx} + u = x^2 \Rightarrow \frac{du}{dx} - u = -x^2 \Rightarrow \frac{du}{dx} - u = 0 \Rightarrow r - 1 = 0 \Rightarrow r = 1 \Rightarrow u_h = C_1 e^x$. $u_p = Ax^2 + Bx + C \Rightarrow u'_p = 2Ax + B \Rightarrow 2Ax + B - (Ax^2 + Bx + C) = -x^2 \Rightarrow A = 1, B = 2, C = 2 \Rightarrow u_p = x^2 + 2x + 2 \Rightarrow u = C_1 e^x + x^2 + 2x + 2$
Step 3: $u = \frac{1}{y} \Rightarrow y = u^{-1} = \left(C_1 e^x + x^2 + 2x + 2\right)^{-1}$

53. $y(x) + \int_0^x y(t)\,dt = x \Rightarrow \frac{dy}{dx} + y = 1 \Rightarrow \frac{dy}{dx} = 1 - y \Rightarrow \frac{dy}{1 - y} = dx \Rightarrow -\ln|1 - y| = x + C \Rightarrow 1 - y = C_1 e^{-x} \Rightarrow y = C_1 e^{-x} + 1$. $y(0) = 0 \Rightarrow 0 = C_1 e^0 + 1 \Rightarrow C_1 = -1 \Rightarrow y = -e^{-x} + 1$

SECTION 16.6 OSCILLATION

1. $m\frac{d^2x}{dt^2} + kx = 0 \Rightarrow \frac{d^2x}{dt^2} + \frac{k}{m}x = 0$. Let $\omega = \sqrt{\frac{k}{m}}$. Then $\frac{d^2x}{dt^2} + \omega^2 x = 0 \Rightarrow r^2 + \omega^2 = 0 \Rightarrow r = \pm\,\omega i \Rightarrow$ $x = C_1 \cos \omega t + C_2 \sin \omega t$. $x(0) = x_0 \Rightarrow C_1 \cos 0 + C_2 \sin 0 = x_0 \Rightarrow C_1 = x_0 \Rightarrow x = x_0 \cos \omega t + C_2 \sin \omega t$ $\Rightarrow x' = -x_0\omega \sin \omega t + C_2\omega \cos \omega t$. $x'(0) = v_0 \Rightarrow -x_0\omega \sin 0 + C_2\omega \cos 0 = v_0 \Rightarrow C_2 = \frac{v_0}{\omega} \Rightarrow$ $x = x_0 \cos \omega t + \frac{v_0}{\omega} \sin \omega t$. $x = C \sin(\omega t + \phi)$ where $C = \sqrt{x_0^2 + \left(\frac{v_0}{\omega}\right)^2} = \frac{\sqrt{\omega^2 x_0^2 + v_0^2}}{\omega}$ and $\phi = \tan^{-1}\left(\frac{\omega x_0}{v_0}\right) \Rightarrow x = \frac{\sqrt{\omega^2 x_0^2 + v_0^2}}{\omega} \sin\left(\omega t + \tan^{-1}\left(\frac{\omega x_0}{v_0}\right)\right)$

2. $f(x) = kx \Rightarrow 5$ lbs $= k(0.5$ ft$) \Rightarrow k = 10$ lbs/ft. $x_0 = 0$ ft, $\frac{dx}{dt} = 4$ ft/sec when $t = 0 \Rightarrow \omega = \sqrt{\frac{k}{m}} =$ $\sqrt{\frac{10 \text{ lbs/ft}}{5 \text{ lbs/32 ft/sec}^2}} = 8$ sec. $x = C_1 \cos \omega t + C_2 \sin \omega t$. $x_0 = 0 \Rightarrow C_1 = 0 \Rightarrow x = C_2 \sin \omega t \Rightarrow$ $\frac{dx}{dt} = \omega C_2 \cos \omega t \Rightarrow 4$ ft/sec $= (8$ sec$)C_2 \Rightarrow C_2 = 0.5$ ft/sec$^2 \Rightarrow x = 0.5 \sin 8t$

3. a) $L\frac{d^2i}{dt^2} + R\frac{di}{dt} + \frac{1}{C}i = \frac{dv}{dt}$. $R = 0$, $\frac{1}{LC} = \omega^2$, $v =$ constant $\Rightarrow L\frac{d^2i}{dt^2} + \frac{1}{C}i = 0 \Rightarrow \frac{d^2i}{dt^2} + \frac{1}{LC}i = 0 \Rightarrow$ $\frac{d^2i}{dt^2} + \omega^2 i = 0 \Rightarrow r = \pm\,\omega i \Rightarrow i = C_1 \cos \omega t + C_2 \sin \omega t$

b) $L\frac{d^2i}{dt^2} + R\frac{di}{dt} + \frac{1}{C}i = \frac{dv}{dt}$. $R = 0$, $\frac{1}{LC} = \omega^2$, $v = V \sin \alpha t$, $\alpha \neq \omega \Rightarrow \frac{d^2i}{dt^2} + \omega^2 i = \frac{V\alpha}{L}\cos \alpha t \Rightarrow$ $i_h = C_1 \cos \omega t + C_2 \sin \omega t$. $i_p = A \cos \alpha t + B \sin \alpha t \Rightarrow i'_p = -A\alpha \sin \alpha t + B\alpha \cos \alpha t \Rightarrow$ $i''_p = -A\alpha^2 \cos \alpha t - B\alpha^2 \sin \alpha t \Rightarrow -A\alpha^2 \cos \alpha t - B\alpha^2 \sin \alpha t + \omega^2\left(A \cos \alpha t + B \sin \alpha t\right) =$ $\frac{V\alpha}{L}\cos \alpha t \Rightarrow A = \frac{V\alpha}{L\left(\omega^2 - \alpha^2\right)}$, $B = 0 \Rightarrow i_p = \frac{V\alpha}{L\left(\omega^2 - \alpha^2\right)}\cos \alpha t \Rightarrow i = C_1 \cos \omega t + C_2 \sin \omega t +$ $\frac{V\alpha}{L\left(\omega^2 - \alpha^2\right)}\cos \alpha t$

c) $L\frac{d^2i}{dt^2} + R\frac{di}{dt} + \frac{1}{C}i = \frac{dv}{dt}$ $R = 0$, $\frac{1}{LC} = \omega^2$, $v = V \sin \omega t$, V constant $\Rightarrow L\frac{d^2i}{dt^2} + \frac{1}{C}i = V\omega \cos \omega t \Rightarrow$ $\frac{d^2i}{dt^2} + \omega^2 i = \frac{V\omega}{L}\cos \omega t \Rightarrow i_h = C_1 \cos \omega t + C_2 \sin \omega t$. $i_p = At \cos \omega t + Bt \sin \omega t \Rightarrow i'_p = A \cos \omega t -$ $A\omega t \sin \omega t + B \sin \omega t + B\omega t \cos \omega t \Rightarrow y''_p = -2A\omega \sin \omega t + 2B\omega \cos \omega t - A\omega^2 t \cos \omega t -$ $B\omega^2 t \sin \omega t \Rightarrow -2A\omega \sin \omega t + 2B\omega \cos \omega t - A\omega^2 t \cos \omega t - B\omega^2 t \sin \omega t +$ $\omega^2\left(At \cos \omega t + Bt \sin \omega t\right) = \frac{V\omega}{L}\cos \omega t \Rightarrow A = 0$, $B = \frac{V}{2L} \Rightarrow i_p = \frac{V}{2L}t \sin \omega t \Rightarrow i = C_1 \cos \omega t +$ $C_2 \sin \omega t + \frac{V}{2L}t \sin \omega t$

d) $L\frac{d^2i}{dt^2} + R\frac{di}{dt} + \frac{1}{C}i = \frac{dv}{dt}$ $R = 50$, $L = 5$, $C = 9 \text{ X } 10^{-6}$, v constant $\Rightarrow 5\frac{d^2i}{dt^2} + 50\frac{di}{dt} + \frac{1}{9} \text{ X } 10^6\, i = 0 \Rightarrow$ $\frac{d^2i}{dt^2} + 10\frac{di}{dt} + \frac{1}{45} \text{ X } 10^6\, i = 0 \Rightarrow r^2 + 10r + \frac{1}{45} \text{ X } 10^6 = 0 \Rightarrow r = -5 \pm 5\sqrt{-\frac{7991}{9}} \approx -5 \pm 148.99\, i \Rightarrow$ $i = e^{-5t}\left(C_1 \cos(148.99)t + C_2 \sin(148.99)t\right)$

4. Assume $\theta \approx 0 \Rightarrow \sin\theta \approx \theta \Rightarrow \frac{d^2\theta}{dt^2} + \frac{g}{l}\theta = 0 \Rightarrow r^2 + \frac{g}{l} = 0 \Rightarrow r = \pm i\sqrt{\frac{g}{l}} \Rightarrow \theta = C_1 \cos\sqrt{\frac{g}{l}}\,t + C_2 \sin\sqrt{\frac{g}{l}}\,t$. $t = 0 \Rightarrow \theta = \theta_0 \Rightarrow \theta_0 = C_1 \Rightarrow \theta = \theta_0 \cos\sqrt{\frac{g}{l}}\,t + C_2 \sin\sqrt{\frac{g}{l}}\,t \Rightarrow \frac{d\theta}{dt} = -\theta_0\sqrt{\frac{g}{l}} \sin\sqrt{\frac{g}{l}}\,t + C_2\sqrt{\frac{g}{l}} \cos\sqrt{\frac{g}{l}}\,t$. $t = 0 \Rightarrow \frac{d\theta}{dt} = 0 \Rightarrow C_2 = 0 \Rightarrow \theta = \theta_0 \cos\sqrt{\frac{g}{l}}\,t$

5. $\frac{d^2\theta}{dt^2} = -\frac{2k\theta}{mr^2} \Rightarrow \frac{d^2\theta}{dt^2} + \frac{2k}{mr^2}\theta = 0 \Rightarrow r^2 + \frac{2k}{mr^2} = 0 \Rightarrow r = \pm\sqrt{\frac{2k}{mr^2}}\,i \Rightarrow \theta = C_1 \cos\sqrt{\frac{2k}{mr^2}}\,t + C_2 \sin\sqrt{\frac{2k}{mr^2}}\,t$. $t = 0 \Rightarrow \theta = \theta_0 \Rightarrow C_1 = \theta_0 \Rightarrow \theta = \theta_0 \cos\sqrt{\frac{2k}{mr^2}}\,t + C_2 \sin\sqrt{\frac{2k}{mr^2}}\,t \Rightarrow \frac{d\theta}{dt} = -\theta_0\sqrt{\frac{2k}{mr^2}} \sin\sqrt{\frac{2k}{mr^2}}\,t + C_2\sqrt{\frac{2k}{mr^2}} \cos\sqrt{\frac{2k}{mr^2}}\,t$. $\frac{d\theta}{dt} = v_0$ at $t = 0 \Rightarrow C_2\sqrt{\frac{2k}{mr^2}} = v_0 \Rightarrow C_2 = v_0\sqrt{\frac{mr^2}{2k}} \Rightarrow \theta = \theta_0 \cos\sqrt{\frac{2k}{mr^2}}\,t + v_0\sqrt{\frac{mr^2}{2k}} \sin\sqrt{\frac{2k}{mr^2}}\,t$

6. $\frac{100}{g}\frac{d^2x}{dt^2} = -16\pi x - c\frac{dx}{dt} \Rightarrow \frac{d^2x}{dt^2} + \frac{gc}{100}\frac{dx}{dt} = -\frac{16\pi g}{100}x \Rightarrow r^2 + \frac{gc}{100}r + \frac{16\pi g}{100} = 0 \Rightarrow 100r^2 + gcr + 16\pi g = 0$

$\Rightarrow r = \frac{-4c \pm 4i\sqrt{200\pi - c^2}}{25} \Rightarrow \omega = \frac{4\sqrt{200\pi - c^2}}{25} \Rightarrow \text{Period} = \frac{2\pi}{\omega} = \frac{2\pi}{\frac{4\sqrt{200\pi - c^2}}{25}} = \frac{25\pi}{2\sqrt{200\pi - c^2}}$

$\therefore\ 1.6 = \frac{25\pi}{2\sqrt{200\pi - c^2}} \Rightarrow c = \sqrt{200\pi - \frac{6.25\pi^2}{10.24}} \approx 5.09$

7. a) $f(t) = A\sin\alpha t,\ \alpha \ne \sqrt{\frac{k}{m}} \Rightarrow \frac{d^2x}{dt^2} + \frac{k}{m}x = \frac{k}{m}\left(A\sin\alpha t\right) \Rightarrow r^2 + \frac{k}{m} = 0 \Rightarrow r = \pm i\sqrt{\frac{k}{m}} \Rightarrow$

$x_h = C_1\cos\sqrt{\frac{k}{m}}\,t + C_2\sin\sqrt{\frac{k}{m}}\,t$. $x_p = B\cos\alpha t + C\sin\alpha t \Rightarrow x'_p = -B\alpha\sin\alpha t + C\alpha\cos\alpha t \Rightarrow x''_p = -B\alpha^2\cos\alpha t - C\alpha^2\sin\alpha t \Rightarrow -B\alpha^2\cos\alpha t - C\alpha^2\sin\alpha t + \frac{k}{m}\left(B\cos\alpha t + C\sin\alpha t\right) = \frac{k}{m}\left(A\sin\alpha t\right) \Rightarrow B = 0,\ C = \frac{Ak}{k - m\alpha^2} \Rightarrow x_p = \frac{Ak}{k - m\alpha^2}\sin\alpha t \Rightarrow x = \frac{Ak}{k - m\alpha^2}\sin\alpha t + C_1\cos\sqrt{\frac{k}{m}}\,t + C_2\sin\sqrt{\frac{k}{m}}\,t$. $x(0) = x_0 \Rightarrow C_1 = x_0 \Rightarrow x = x_0\cos\sqrt{\frac{k}{m}}\,t + C_2\sin\sqrt{\frac{k}{m}}\,t + \frac{Ak}{k - m\alpha^2}\sin\alpha t \Rightarrow$

$\frac{dx}{dt} = -x_0\sqrt{\frac{k}{m}}\sin\sqrt{\frac{k}{m}}\,t + C_2\sqrt{\frac{k}{m}}\cos\sqrt{\frac{k}{m}}\,t + \frac{Ak\alpha}{k - m\alpha^2}\cos\alpha t$. $x'(0) = 0 \Rightarrow C_2\sqrt{\frac{k}{m}} + \frac{Ak\alpha}{k - m\alpha^2} = 0 \Rightarrow C_2 = -\frac{Ak\alpha}{k - m\alpha^2}\sqrt{\frac{m}{k}} \Rightarrow x = x_0\cos\sqrt{\frac{k}{m}}\,t - \frac{A\alpha\sqrt{mk}}{k - m\alpha^2}\sin\sqrt{\frac{k}{m}}\,t + \frac{Ak}{k - m\alpha^2}\sin\alpha t$

b) $f(t) = A\sin\alpha t,\ \alpha = \sqrt{\frac{k}{m}} \Rightarrow \frac{d^2x}{dt^2} + \frac{k}{m}x = \frac{k}{m}\left(A\sin\alpha t\right) \Rightarrow r^2 + \frac{k}{m} = 0 \Rightarrow r = \pm\sqrt{\frac{k}{m}}\,i \Rightarrow$

$x_h = C_1\cos\sqrt{\frac{k}{m}}\,t + C_2\sin\sqrt{\frac{k}{m}}\,t$. $x_p = Bt\cos\sqrt{\frac{k}{m}}\,t + Ct\sin\sqrt{\frac{k}{m}}\,t \Rightarrow x'_p = B\cos\sqrt{\frac{k}{m}}\,t - B\sqrt{\frac{k}{m}}\,t\sin\sqrt{\frac{k}{m}}\,t + C\sin\sqrt{\frac{k}{m}}\,t + C\sqrt{\frac{k}{m}}\,t\cos\sqrt{\frac{k}{m}}\,t \Rightarrow x''_p = -2B\sqrt{\frac{k}{m}}\sin\sqrt{\frac{k}{m}}\,t + 2C\sqrt{\frac{k}{m}}\cos\sqrt{\frac{k}{m}}\,t - B\left(\frac{k}{m}\right)t\cos\sqrt{\frac{k}{m}}\,t - C\left(\frac{k}{m}\right)t\sin\sqrt{\frac{k}{m}}\,t \Rightarrow -2B\sqrt{\frac{k}{m}}\sin\sqrt{\frac{k}{m}}\,t +$

7. b) (Continued)

$2C\sqrt{\frac{k}{m}}\cos\sqrt{\frac{k}{m}}\,t - B\left(\frac{k}{m}\right)t\cos\sqrt{\frac{k}{m}}\,t - C\left(\frac{k}{m}\right)t\sin\sqrt{\frac{k}{m}}\,t + \frac{k}{m}\left(Bt\cos\sqrt{\frac{k}{m}}\,t + Ct\sin\sqrt{\frac{k}{m}}\,t\right)$
$= \frac{kA}{m}\sin\alpha t \Rightarrow B = -\frac{A\alpha}{2},\ C = 0 \Rightarrow x_p = -\frac{A\alpha}{2}t\cos\alpha t \Rightarrow x = C_1\cos\sqrt{\frac{k}{m}}\,t + C_2\sin\sqrt{\frac{k}{m}}\,t -$
$\frac{A\alpha}{2}t\cos\alpha t.$ $x(0) = x_0 \Rightarrow C_1 = x_0 \Rightarrow x = x_0\cos\sqrt{\frac{k}{m}}\,t + C_2\sin\sqrt{\frac{k}{m}}\,t - \frac{A\alpha}{2}t\cos\alpha t \Rightarrow$
$\frac{dx}{dt} = -x_0\alpha\sin\alpha t + C_2\alpha\cos\alpha t - \frac{A\alpha}{2}\cos\alpha t - \frac{A\alpha^2}{2}t\sin\alpha t.$ $x'(0) = 0 \Rightarrow C_2\alpha - \frac{A\alpha}{2} = 0 \Rightarrow$
$C_2 = \frac{A}{2} \Rightarrow x = x_0\cos\alpha t + \frac{A}{2}\sin\alpha t - \frac{A\alpha}{2}t\cos\alpha t$

SECTION 16.7 NUMERICAL METHODS

1.

	x_n	y_n
x_0	0	1
x_1	1/5	1.2
x_2	2/5	1.44
x_3	3/5	1.728
x_4	4/5	2.0736
x_5	1	2.48832

Exact Value: $y' = y \Rightarrow \frac{dy}{dx} = y \Rightarrow \frac{dy}{y} = dx \Rightarrow \ln|y| = x + C \Rightarrow$ $e^{\ln|y|} = e^{x+C} \Rightarrow |y| = e^C e^x \Rightarrow y = C_1 e^x.$ $y(0) = 1 \Rightarrow C_1 = 1 \Rightarrow y = e^x \Rightarrow y(1) = e^1 = 2.718281828...$

2. $y' = y,\ y(0) = 1,\ h = \frac{1}{n} \Rightarrow x_{n+1} = x_n + h = x_n + \frac{1}{n},\ y_{n+1} = y_n + hy_n = y_n + \frac{1}{n}y_n = \left(1 + \frac{1}{n}\right)y_n$

$x_0 = 0 \Rightarrow y_0 = 1;\ x_1 = 0 + \frac{1}{n} = \frac{1}{n} \Rightarrow y_1 = \left(1 + \frac{1}{n}\right)1 = \left(1 + \frac{1}{n}\right);\ x_2 = \frac{1}{n} + \frac{1}{n} = \frac{2}{n} \Rightarrow y_2 = \left(1 + \frac{1}{n}\right)\left(1 + \frac{1}{n}\right) =$
$\left(1 + \frac{1}{n}\right)^2;\ x_3 = \frac{3}{n} \Rightarrow y_3 = \left(1 + \frac{1}{n}\right)^3;$ etc.

$\therefore$ let $n = 1 \Rightarrow y_1 = \left(1 + \frac{1}{n}\right)$. Assume $y_k = \left(1 + \frac{1}{n}\right)^k$ for some $k \ge 1$. Then $y_{k+1} = \left(1 + \frac{1}{n}\right)y_k =$
$\left(1 + \frac{1}{n}\right)\left(1 + \frac{1}{n}\right)^k = \left(1 + \frac{1}{n}\right)^{k+1}$. $\therefore$ by mathematical induction, $y_n = \left(1 + \frac{1}{n}\right)^n$ for $n \ge 1$.

$y(1) = \lim_{n\to\infty} y_n = \lim_{n\to\infty}\left(1 + \frac{1}{n}\right)^n = e$

3.

	x_n	y_n
x_0	0	1
x_1	1/5	1.22
x_2	2/5	1.4884
x_3	3/5	1.815848
x_4	4/5	2.2153346
x_5	1	2.702708163

4.

	x_n		k_i		y_n
x_0	0.2	k_1	0.2		
		k_2	0.22		
		k_3	0.222		
		k_4	0.2444	y_1	1.2214
x_2	0.4	k_1	0.24428		
		k_2	0.268708		
		k_3	0.2711508		
		k_4	0.2985101	y_2	1.4918179
x_3	0.6	k_1	0.2983635		
		k_2	0.3281999		
		k_3	0.3311835		
		k_4	0.3646002	y_3	1.8221063
x_4	0.8	k_1	0.3644212		
		k_2	0.4008633		
		k_3	0.4045075		
		k_4	0.4453227	y_4	2.2255205
x_5	1	k_1	0.4451041		
		k_2	0.4896145		
		k_3	0.4940655		
		k_4	0.5439172	y_5	2.7182507

5. $y' = x^2 + y^2 \Rightarrow x_{n+1} = x_n + h \Rightarrow y_{n+1} = y_n + h\left(x_n^{\,2} + y_n^{\,2}\right)$. $Y' = Y^2 \Rightarrow x_{n+1} = x_n + h \Rightarrow Y_{n+1} = Y_n + h(Y_n^{\,2})$. $y_0 = h(x_0^{\,2} + y_0^{\,2}) = Y_0 + h(Y_0)$ since $x_0 = 0 \Rightarrow y_1 = Y_1$. $y_1 + h\left(x_1^{\,2} + y_1^{\,2}\right) > Y_1 + h(Y_1^{\,2})$ since $x_1 > 0 \Rightarrow y_2 > Y_2$. And from here on up to $x = 1$, $y_{n+1} > Y_{n+1}$.
$Y' = Y^2 \Rightarrow \frac{dY}{dx} = Y^2 \Rightarrow \frac{1}{Y^2}\,dY = dx \Rightarrow -\frac{1}{Y} = x + C \Rightarrow \frac{1}{Y} = -x - C \Rightarrow Y = -\frac{1}{x + C}$. $y(0) = 1 \Rightarrow -\frac{1}{C} = 1 \Rightarrow C = -1 \Rightarrow Y = -\frac{1}{x - 1} \Rightarrow y \to \infty$ as $x \to 1^-$. Since $y_{n+1} > Y_{n+1}$, $y_{n+1} \to \infty$ as $x \to 1^-$

6. a) $\frac{dy}{dx} = 1 + y^2 \Rightarrow \frac{1}{1 + y^2}\,dy = dx \Rightarrow \tan^{-1} y = x + C \Rightarrow y = \tan(x + C)$. $y(0) = 0 \Rightarrow C = 0 \Rightarrow y = \tan x$

b) Let $y = -\frac{u'}{u} \Rightarrow \frac{dy}{dx} = \frac{-uu'' + (u')^2}{u^2}$. Then $\frac{-uu'' + (u')^2}{u^2} = 1 + \left(-\frac{u'}{u}\right)^2 \Rightarrow u'' + u = 0 \Rightarrow r^2 + 1 = 0 \Rightarrow r = \pm i \Rightarrow u = C_1 \cos x + C_2 \sin x \Rightarrow u' = -C_1 \sin x + C_2 \cos x$. Now $y(0) = 0 \Rightarrow -\frac{u'}{u} = 0$ when $x = 0 \Rightarrow u'(0) = 0$. Then $0 = -C_1 \sin 0 + C_2 \cos 0 \Rightarrow C_2 = 0$. Then $u = C_1 \cos x$ and $u' = -C_1 \sin x \Rightarrow$
$y = -\frac{u'}{u} = -\frac{-C_1 \sin x}{C_1 \cos x} = \tan x$

Note: For Exercises 7–12, the Calculus Tookit was used.

7. $y = 0.571428572$

8. $y = -0.175611085$

9. $y = 0.810263855$

10. $y = 0.810317873$

11. a) $y = 0.841470985$ b) $y = 0.841470983$

12. a) $y = 0.367879689$ b) $y = 0.367879688$

PRACTICE EXERCISES

1. $e^{y-2}\,dx - e^{x+2y}\,dy = 0 \Rightarrow e^{-x}\,dx - \frac{e^{2y}}{e^{y-2}}\,dy = 0 \Rightarrow e^{-x}\,dx - e^{y+2}\,dy = 0 \Rightarrow -e^{-x} - e^{y+2} = C.$ $y(0) = -2$ $\Rightarrow -e^0 - e^{-2+2} = C \Rightarrow C = -2 \Rightarrow e^{-x} + e^{y+2} = 2$

2. $y \ln y\,dx + (1 + x^2)\,dy = 0 \Rightarrow \frac{dx}{1 + x^2} + \frac{dy}{y \ln y} = 0 \Rightarrow \tan^{-1} x + \ln(\ln y) = C.$ $y(0) = e \Rightarrow \tan^{-1} 0 + \ln(\ln e) =$ $C \Rightarrow C = 0 \Rightarrow \tan^{-1} x + \ln(\ln y) = 0$

3. $\frac{dy}{dx} = \frac{x^2 + y^2}{2xy} \Rightarrow \frac{dy}{dx} = \frac{1 + \left(\frac{y}{x}\right)^2}{2\left(\frac{y}{x}\right)} \Rightarrow$ Homogeneous. $v = \frac{y}{x} \Rightarrow F(v) = \frac{1 + v^2}{2v} \Rightarrow \frac{dx}{x} + \frac{dv}{v - \left(\frac{1 + v^2}{2v}\right)} = 0 \Rightarrow$ $\frac{dx}{x} + \frac{2v\,dv}{v^2 - 1} = 0 \Rightarrow \ln|x| + \ln|v^2 - 1| = C \Rightarrow x(v^2 - 1) = C_1 \Rightarrow x\left(\left(\frac{y}{x}\right)^2 - 1\right) = C_1 \Rightarrow \frac{y^2}{x} - x = C_1.$ $y(5) = 0$ $\Rightarrow 0 - 5 = C_1 \Rightarrow C_1 = -5 \Rightarrow \frac{y^2}{x} - x = -5$

4. $\frac{dy}{dx} = \frac{y(1 + \ln y - \ln x)}{x(\ln y - \ln x)} \Rightarrow \frac{dy}{dx} = \frac{y}{x}\left(\frac{1 + \ln \frac{y}{x}}{\ln \frac{y}{x}}\right) \Rightarrow$ Homogeneous. $v = \frac{y}{x} \Rightarrow F(v) = v\left(\frac{1 + \ln v}{\ln v}\right) \Rightarrow$ $\frac{dx}{x} + \frac{dv}{v - v\left(\frac{1 + \ln v}{\ln v}\right)} = 0 \Rightarrow \frac{dx}{x} + \frac{\ln v\,dv}{-v} = 0 \Rightarrow \ln|x| - (\ln|v|)^2 + C = 0 \Rightarrow \ln|x| - \left(\ln\left|\frac{y}{x}\right|\right)^2 + C = 0 \Rightarrow$ $\ln(x) - \left(\ln\left(\frac{y}{x}\right)\right)^2 + C = 0$ if $x,y > 0$. $y(1) = 1 \Rightarrow \ln 1 - (\ln 1)^2 + C = 0 \Rightarrow C = 0 \Rightarrow \ln x - \left(\ln\left(\frac{y}{x}\right)\right)^2 = 0$

5. $(x^2 + y)\,dx + \left(e^y + x\right)dy = 0 \Rightarrow \frac{\partial M}{\partial y} = 1 = \frac{\partial N}{\partial x} \Rightarrow$ Exact. $\frac{\partial f}{\partial x} = x^2 + y \Rightarrow f(x,y) = \frac{x^3}{3} + xy + k(y) \Rightarrow$ $\frac{\partial f}{\partial y} = x + k'(y) = e^y + x \Rightarrow k'(y) = e^y \Rightarrow k(y) = e^y + C \Rightarrow f(x,y) = \frac{x^3}{3} + xy + e^y + C \Rightarrow \frac{x^3}{3} + xy + e^y = C_1.$ $y(3) = 0 \Rightarrow 9 + 1 = C_1 \Rightarrow C_1 = 10 \Rightarrow \frac{x^3}{3} + xy + e^y = 10$

6. $(e^x + \ln y)\,dx + \left(\frac{x + y}{y}\right)dy = 0 \Rightarrow \frac{\partial M}{\partial y} = \frac{1}{y} = \frac{\partial N}{\partial x} \Rightarrow$ Exact. $\frac{\partial f}{\partial x} = e^x + \ln y \Rightarrow f(x,y) = e^x + x \ln y + k(y) \Rightarrow$ $\frac{\partial f}{\partial y} = \frac{x}{y} + k'(y) = \frac{x + y}{y} \Rightarrow k'(y) = 1 \Rightarrow k(y) = y + C \Rightarrow f(x,y) = e^x + x \ln y + y + C \Rightarrow e^x + x \ln y + y + C = 0.$ $y(\ln 2) = 1 \Rightarrow 2 + 1 + C = 0 \Rightarrow C = -3 \Rightarrow e^x + x \ln y + y - 3 = 0$

7. $(x+1)\frac{dy}{dx} + 2y = x \Rightarrow P(x) = 2,\ Q(x) = x \Rightarrow \int P(x)\,dx = \int 2\,dx = 2x \Rightarrow \rho(x) = e^{2x} \Rightarrow$
$y = \frac{1}{e^{2x}} \int e^{2x}(x\,dx) = \frac{1}{e^{2x}}\left(\frac{1}{2}x\,e^{2x} - \frac{1}{4}e^{2x} + C\right) \Rightarrow y = \frac{1}{2}x - \frac{1}{4} + C\,e^{-2x}.\ y(0) = 1 \Rightarrow 1 = -\frac{1}{4} + C \Rightarrow$
$C = \frac{5}{4} \Rightarrow y = \frac{1}{2}x - \frac{1}{4} + \frac{5}{4}e^{-2x}$

8. $x\frac{dy}{dx} + 2y = x^2 + 1 \Rightarrow \frac{dy}{dx} + \frac{2}{x}y = x + \frac{1}{x} \Rightarrow P(x) = \frac{2}{x},\ Q(x) = x + \frac{1}{x} \Rightarrow \int P(x)\,dx = \int \frac{2}{x}\,dx = 2\ln|x| =$
$\ln x^2 \Rightarrow \rho(x) = e^{\ln x^2} = x^2 \Rightarrow y = \frac{1}{x^2}\int x^2\left(x + \frac{1}{x}\right)dx = \frac{1}{x^2}\left(\frac{x^4}{4} + \frac{x^2}{2} + C\right) = \frac{x^2}{4} + \frac{1}{2} + \frac{C}{x^2}.\ y(1) = 1 \Rightarrow$
$1 = \frac{1}{4} + \frac{1}{2} + C \Rightarrow C = \frac{1}{4} \Rightarrow y = \frac{x^2}{4} + \frac{1}{2} + \frac{1}{4x^2}$

9. $\frac{d^2y}{dx^2} - \left(\frac{dy}{dx}\right)^2 = 1.$ Let $p = \frac{dy}{dx} \Rightarrow \frac{dp}{dx} - p^2 = 1 \Rightarrow \frac{dp}{1+p^2} = dx \Rightarrow \tan^{-1} p = x + C \Rightarrow p = \tan(x + C) \Rightarrow$
$\frac{dy}{dx} = \tan(x + C) \Rightarrow y = \ln|\sec(x + C)| + C_1.\ y'\left(\frac{\pi}{3}\right) = \sqrt{3} \Rightarrow \sqrt{3} = \tan\left(\frac{\pi}{3} + C\right) \Rightarrow \tan^{-1}\sqrt{3} = \frac{\pi}{3} + C \Rightarrow$
$\frac{\pi}{3} = \frac{\pi}{3} + C \Rightarrow C = 0 \Rightarrow y = \ln|\sec x| + C_1.\ y\left(\frac{\pi}{3}\right) = 0 \Rightarrow 0 = \ln\left|\sec\frac{\pi}{3}\right| + C_1 \Rightarrow C_1 = -\ln 2 \Rightarrow$
$y = \ln|\sec x| - \ln 2$

10. $x^2\frac{d^2y}{dx^2} + x\frac{dy}{dx} = 1.$ Let $p = \frac{dy}{dx} \Rightarrow x^2\frac{dp}{dx} + xp = 1 \Rightarrow \frac{dp}{dx} + \frac{1}{x}p = \frac{1}{x^2} \Rightarrow P(x) = \frac{1}{x},\ Q(x) = \frac{1}{x^2} \Rightarrow \int P(x)\,dx$
$= \ln|x| \Rightarrow \rho(x) = e^{\ln|x|} = |x| \Rightarrow p = \frac{1}{|x|}\int |x|\left(\frac{1}{x^2}\right)dx = \frac{1}{x}\int x\left(\frac{1}{x^2}\right)dx = \frac{1}{x}\int \frac{1}{x}\,dx = \frac{1}{x}(\ln x + C)$ if $x > 0$
$\Rightarrow p = \frac{\ln x}{x} + \frac{C}{x} \Rightarrow \frac{dy}{dx} = \frac{\ln x}{x} + \frac{C}{x} \Rightarrow y = (\ln x)^2 + C\ln x + C_1.\ y'(1) = 1 \Rightarrow 1 = \ln 1 + C \Rightarrow C = 1 \Rightarrow$
$y = (\ln x)^2 + \ln x + C_1.\ y(1) = 1 \Rightarrow 1 = (\ln 1)^2 + \ln 1 + C_1 \Rightarrow C_1 = 1 \Rightarrow y = (\ln x)^2 + \ln x + 1$

11. $\frac{d^2y}{dx^2} - 4\frac{dy}{dx} + 3y = 0 \Rightarrow r^2 - 4r + 3 = 0 \Rightarrow r_1 = 3,\ r_2 = 1 \Rightarrow y = C_1 e^{3x} + C_2 e^{x} \Rightarrow y' = 3C_1 e^{3x} + C_2 e^{x}.$
$y'(0) = -2 \Rightarrow -2 = 3C_1 + C_2.\ y(0) = 2 \Rightarrow 2 = C_1 + C_2 \Rightarrow C_1 = -2,\ C_2 = 4 \Rightarrow y = -2e^{3x} + 4e^{x}$

12. $\frac{d^2y}{dx^2} + 5\frac{dy}{dx} + 6y = 0 \Rightarrow r^2 + 5r + 6 = 0 \Rightarrow r_1 = -2,\ r_2 = -3 \Rightarrow y = C_1 e^{-2x} + C_2 e^{-3x} \Rightarrow y' = -2C_1 e^{-2x} -$
$3C_2{}^{-3x}.\ y'(0) = -2 \Rightarrow -2 = -2C_1 - 3C_2.\ y(0) = \frac{5}{6} \Rightarrow \frac{5}{6} = C_1 + C_2 \Rightarrow C_1 = \frac{1}{2},\ C_2 = \frac{1}{3} \Rightarrow y = \frac{1}{2}e^{-2x} +$
$\frac{1}{3}e^{-3x}$

13. $\frac{d^2y}{dx^2} + 4\frac{dy}{dx} + 4y = 0 \Rightarrow r_1 = r_2 = -2 \Rightarrow y = \left(C_1x + C_2\right)e^{-2x}.\ y(0) = 0 \Rightarrow C_2 = 0 \Rightarrow y = C_1x\,e^{-2x} \Rightarrow$
$y' = C_1\left(e^{-2x} - 2x\,e^{-2x}\right).\ y'(0) = 7 \Rightarrow C_1 = 7 \Rightarrow y = 7x\,e^{-2x}$

14. $\frac{d^2y}{dx^2} - 8\frac{dy}{dx} + 16y = 0 \Rightarrow r^2 - 8r + 16 = 0 \Rightarrow r_1 = r_2 = 4 \Rightarrow y = \left(C_1x + C_2\right)e^{4x}.\ y(0) = 4 \Rightarrow C_2 = 4 \Rightarrow$
$y = \left(C_1x + 4\right)e^{4x} \Rightarrow y' = C_1 e^{4x} + 4\left(C_1x + 4\right)e^{4x}.\ y'(0) = -4 \Rightarrow -4 = C_1 + 4(0 + 4) \Rightarrow C_1 = -20 \Rightarrow$
$y = (-20x + 4)e^{4x}$

15. $\frac{d^2y}{dx^2} + 2\frac{dy}{dx} + 2y = 0 \Rightarrow r^2 + 2r + 2 = 0 \Rightarrow r = -1 \pm i \Rightarrow y = e^{-x}\left(C_1 \cos x + C_2 \sin x\right)$. $y(0) = 1 \Rightarrow$ $e^0\left(C_1 \cos 0 + C_2 \sin 0\right) = 1 \Rightarrow C_1 = 1 \Rightarrow y = e^{-x}\left(\cos x + C_2 \sin x\right) \Rightarrow y' = -e^{-x}\left(\cos x + C_2 \sin x\right) +$ $e^{-x}\left(-\sin x + C_2 \cos x\right)$. $y'(0) = 0 \Rightarrow 0 = -\left(\cos 0 + C_2 \sin 0\right) + \left(-\sin 0 + C_2 \cos 0\right) \Rightarrow C_2 = 1 \Rightarrow$ $y = e^{-x}(\cos x + \sin x)$

16. $\frac{d^2y}{dx^2} - 2\frac{dy}{dx} - 4y = 0 \Rightarrow r^2 - 2r - 4 = 0 \Rightarrow r = 1 \pm \sqrt{5} \Rightarrow y = C_1 e^{(1+\sqrt{5})x} + C_2\, e^{(1-\sqrt{5})x}$. $y(0) = 1 \Rightarrow$ $1 = C_1 + C_2$. $y' = \left(1 + \sqrt{5}\right)C_1\, e^{(1+\sqrt{5})x} + \left(1 - \sqrt{5}\right)C_2\, e^{(1-\sqrt{5})x}$. $y'(0) = -1 \Rightarrow -1 = \left(1 + \sqrt{5}\right)C_1 +$ $\left(1 - \sqrt{5}\right)C_2 \Rightarrow C_1 = \frac{-2 + \sqrt{5}}{2\sqrt{5}},\ C_2 = \frac{2 + \sqrt{5}}{2\sqrt{5}} \Rightarrow y = \left(\frac{-2 + \sqrt{5}}{2\sqrt{5}}\right)e^{(1+\sqrt{5})x} + \left(\frac{2 + \sqrt{5}}{2\sqrt{5}}\right)e^{(1-\sqrt{5})x}$

17. $\frac{d^2y}{dx^2} + 2\frac{dy}{dx} = 4x \Rightarrow r^2 + 2r = 0 \Rightarrow r_1 = 0,\ r_2 = -2 \Rightarrow y_h = C_1 + C_2\, e^{-2x}$. $y_p = Ax^2 + Bx \Rightarrow y'_p = 2Ax + B$. $y''_p = 2A \Rightarrow 2A + 2(2Ax + B) = 4x \Rightarrow A = 1,\ B = -1 \Rightarrow y_p = x^2 - x \Rightarrow y = C_1 + C_2\, e^{-2x} + x^2 - x$. $y(0) = 1 \Rightarrow C_1 + C_2 = 1$. $y' = -2C_2\, e^{-2x} + 2x - 1$. $y'(0) = -3 \Rightarrow -2C_2 - 1 = -3 \Rightarrow C_1 = 0,\ C_2 = 1 \Rightarrow$ $y = e^{-2x} + x^2 - x$

18. $\frac{d^2y}{dx^2} + y = \csc x \Rightarrow r^2 + 1 = 0 \Rightarrow r = \pm i \Rightarrow y_h = C_1 \cos x + C_2 \sin x \Rightarrow u_1 = \cos x,\ u_2 = \sin x \Rightarrow$

$$D = \begin{vmatrix} \cos x & \sin x \\ -\sin x & \cos x \end{vmatrix} = 1 \Rightarrow v'_1 = -\frac{u_2\, F(x)}{D} = -\sin x \csc x = -1 \Rightarrow v_1 = \int -\,dx = -x + C_1.$$

$v'_2 = \frac{u_1\, F(x)}{D} = \cos x \csc x = \cot x \Rightarrow v_2 = \int \cot x\, dx = \ln(\sin x) + C_2$ if $\sin x > 0 \Rightarrow y = \cos x\left(-x + C_1\right)$ $+ \sin x\left(\ln(\sin x) + C_2\right) = -x \cos x + C_1 \cos x + \sin x \ln(\sin x) + C_2 \sin x$. $y\left(\frac{\pi}{2}\right) = 1 \Rightarrow -\frac{\pi}{2}\cos\frac{\pi}{2} +$ $C_1 \cos\frac{\pi}{2} + \sin\frac{\pi}{2}\ln\left(\sin\frac{\pi}{2}\right) + C_2 \sin\frac{\pi}{2} = 1 \Rightarrow C_2 = 1 \Rightarrow y = -x \cos x + C_1 \cos x + \sin x \ln(\sin x) + \sin x$ $\Rightarrow y' = \cos x + x \sin x - C_1 \sin x + \cos x \ln(\sin x)$. $y'\left(\frac{\pi}{2}\right) = \frac{\pi}{2} \Rightarrow \cos\frac{\pi}{2} + \frac{\pi}{2}\sin\frac{\pi}{2} - C_1 \sin\frac{\pi}{2} +$ $\cos\frac{\pi}{2}\ln\left(\sin\frac{\pi}{2}\right) = \frac{\pi}{2} \Rightarrow C_1 = 0 \Rightarrow y = -x \cos x + \sin x \ln(\sin x) + \sin x$

19. $\frac{d^2y}{dx^2} - \frac{dy}{dx} - 2y = 3\, e^{2x} \Rightarrow r^2 - r - 2 = 0 \Rightarrow r_1 = 2,\ r_2 = -1 \Rightarrow y_h = C_1\, e^{2x} + C_2\, e^{-x}$. $y_p = Ax\, e^{2x} \Rightarrow$ $y'_p = A\, e^{2x} + 2Ax\, e^{2x} \Rightarrow y''_p = 4A\, e^{2x} + 4Ax\, e^{2x} \Rightarrow 4A\, e^{2x} + 4Ax\, e^{2x} - \left(A\, e^{2x} + 2Ax\, e^{2x}\right) -$ $2\left(Ax\, e^{2x}\right) = 3\, e^{2x} \Rightarrow A = 1 \Rightarrow y_p = x\, e^{2x} \Rightarrow y = x\, e^{2x} + C_1\, e^{2x} + C_2\, e^{-x}$. $y(0) = -2 \Rightarrow C_1 + C_2 = -2$. $y' = 2C_1\, e^{2x} - C_2\, e^{-x} + e^{2x} + 2x\, e^{2x}$. $y'(0) = 0 \Rightarrow 2C_1 - C_2 + 1 = 0 \Rightarrow C_1 = -1,\ C_2 = -1 \Rightarrow$ $y = -\,e^{2x} - e^{-x} + x\, e^{2x}$

20. $\frac{d^2y}{dx^2} - 2\frac{dy}{dx} + 5y = 4\, e^{-x} \Rightarrow r^2 - 2r + 5 = 0 \Rightarrow r = 1 \pm 2i \Rightarrow y_h = e^x\left(C_1 \cos 2x + C_2 \sin 2x\right)$. $y_p = A\, e^{-x}$ $\Rightarrow y'_p = -A\, e^{-x} \Rightarrow y''_p = A\, e^{-x} \Rightarrow A\, e^{-x} - 2\left(-A\, e^{-x}\right) + 5A\, e^{-x} = 4\, e^{-x} \Rightarrow A = \frac{1}{2} \Rightarrow y_p = \frac{1}{2}e^{-x} \Rightarrow$

20. (Continued)

$y = e^x\left(C_1 \cos 2x + C_2 \sin 2x\right) + \frac{1}{2}e^{-x}$. $y(0) = 1 \Rightarrow C_1 + \frac{1}{2} = 1 \Rightarrow C_1 = \frac{1}{2} \Rightarrow$
$y = e^x\left(\frac{1}{2}\cos 2x + C_2 \sin 2x\right) + \frac{1}{2}e^{-x} \Rightarrow y' = e^x\left(\frac{1}{2}\cos 2x + C_2 \sin 2x\right) + e^x\left(-\sin 2x + 2C_2 \cos 2x\right) -$
$\frac{1}{2}e^{-x}$. $y'(0) = -\frac{1}{2} \Rightarrow \frac{1}{2} + 2C_2 - \frac{1}{2} = -\frac{1}{2} \Rightarrow C_2 = -\frac{1}{4} \Rightarrow y = e^x\left(\frac{1}{2}\cos 2x - \frac{1}{4}\sin 2x\right) + \frac{1}{2}e^{-x}$

SOLUTIONS
TO
EXPLORATIONS

SECTION 1.1 COORDINATES AND GRAPHS IN THE PLANE

Exploration 1

1. Upper right corner, (10,10); upper left corner, (-10,10); lower left corner, (-10,-10); lower right corner, (10, -10).

2. The width of most view screens is about 1.5 times the height. For such screens, the dimensions [-10,10] by [-10,10] do not give a square viewing window.

3. Check your settings for your *x*Scl and *y*Scl values. If either scale setting is 1, the respective scale marks in the viewing window show the scale unit.

4. The number and spacing of the scale marks change.

5. If the width of the view screen is about 1.5 times the height, choose view dimensions so that *x*Max − *x*Min = 1.5(*y*Max − *y*Min) or use the ZOOM-BOX option. (For more information on view dimensions and your grapher, see Exercises 35-40 at the end of this section.

SECTION 1.2 SLOPE, AND EQUATIONS FOR LINES

Exploration 1

1. $y - 4 = 1(x - 3)$ or $y + 1 = 1(x + 2)$.

2. $y = x + 1$.

3. The line should pass through (-2,-1) and (3,4).

Exploration 2

1. All lines appear to pass through (0, 3). If we let $x = 0$ in $y = mx + 3$, then $y = 3$.

2. The lines appear to be parallel. The equations show that all four lines have the same slope $m = -3$, and thus the lines <u>are</u> parallel.

3. Try this for both parts 1 and 2.

Exploration 3

1. $$|-x| = \begin{cases} -x & \text{if } -x > 0 \\ 0 & \text{if } \ x = 0 \\ -(-x) = x & \text{if } -x < 0 \end{cases}$$

$$|-x| = \left.\begin{cases} -x & \text{if } x < 0 \\ 0 & \text{if } x = 0 \\ x & \text{if } x > 0 \end{cases}\right\} = |x|$$

2. $-|x|$ is never positive.

3. When x is positive, $|x| = x > 0$. When x is negative, $|x| = -x > 0$. When $x = 0$, $|x| = 0$.

4. $|x|^2 = x^2$, $|-x|^2 = x^2$.

5. $\sqrt{x^2} = |x|$ because $\sqrt{x^2} = x$ when $x \geq 0$ and $\sqrt{x^2} = -x$ when $x < 0$.

SECTION 1.3: RELATIONS, FUNCTIONS, AND THEIR GRAPHS

Exploration 1

1. Grapher activity.

2. In dot format, at most one pixel illuminates in each pixel column. In connected format, more than one pixel can illuminate in a pixel column in order to give an appearance of connected pixels.

3-6. Grapher activities. Note: On all grapher activities, you are encouraged to experiment with different ranges for the t-, x-, and y-values. (For example, in part 6 of this Exploration, let [tMin, tMax] = [0, 100] with t-step = 0.1, and let [xMin, xMax] = [-100, 100] and [yMin, yMax] = [-10, 10].)

Exploration 2

1. Graphing activity. Note: On all graphing activities you should choose (or adjust) range values as you feel it necessary to give you the best possible views.

2. a) $x_1(t) = t^2$, $y_1(t) = t$. The graph should appear symmetric about the x-axis as in Fig 1.41 (b). To demonstrate this symmetry, let $x_2(t) = x_1(t)$, $y_2(t) = -y_1(t)$ and GRAPH $(x_1(t), y_1(t))$ and $(x_2(t), y_2(t))$, then TRACE and jump from graph to graph.

b) $x_1(t) = t$, $y_1(t) = t^3$. The graph should appear symmetric about the origin as in Fig. 1.41(c). To demonstrate this symmetry, let $x_2(t) = -x_1(t)$, $y_2(t) = y_1(t)$, GRAPH $(x_1(t), y_1(t))$ and $(x_2(t), y_2(t))$, then TRACE and jump from graph to graph.

c) $x_1(t) = t$, $y_1(t) = 1/t^2$. The graph should appear symmetric about the y-axis. Part 1 suggests a procedure for demonstrating this symmetry.

Exploration 3

1. GRAPH, in parametric mode, $x_1(t) = t$, $y_1(t) = 4 - t^2$. If the function is even, the graph appears symmetric about the y-axis. To demonstrate this symmetry let $x_2(t) = -x_1(t)$, $y_2(t) = y_1(t)$, GRAPH $(x_1(t), y_1(t))$ and $(x_2(t), y_2(t))$, then TRACE and jump from graph to graph; the cursor will appear to jump across the y-axis. Note: this test can be applied to any function $y = f(x)$ by letting $y_1(t) = f(t)$.

2. GRAPH, in parametric mode, $x_1(t) = t$, $y_1(t) = \sqrt[3]{t}$. If the function is odd, the graph appears symmetric about the origin. To demonstrate this symmetry let $x_2(t) = -x_1(t)$, $y_2(t) = y_1(t)$, GRAPH $(x_1(t), y_1(t))$ and $(x_2(t), y_2(t))$, then TRACE and jump from graph to graph; the cursor will appear to jump across the origin. Note: this test can be applied to any function $y = f(x)$ by letting $y_1(t) = f(t)$.

Exploration 4

1. Library functions available depend on the grapher in use. consult your <u>Owner's Guide</u>. They might include $\sin x$, $\cos x$, $\tan x$, $1/x$, x^2, $\log x$, $\ln x$, $|x|$, $\sin^{-1} x$, $\cos^{-1} x$, $\tan^{-1} x$, $\sqrt{x}$, 10^x, e^x, $\sinh(x)$, $\cosh(x)$, $\tanh(x)$, $\sinh^{-1} x$, $\cosh^{-1} x$, $\tanh^{-1} x$, $\sqrt[3]{x}$, $n!$, round(x), IPart(x), FPart(x), and Int(x) (the greatest integer function).

2. In connected format the grapher connects the different segments of int(x) which is not a correct representation. In this case dot format is better.

3. GRAPH y_1 = -Int($-x$) to get the ceiling function.

Exploration 5

1. Grapher activity.

2. a)

$$f(x) = \begin{cases} x & 0 \le x < 1 \\ x - 1 & 1 \le x \le 2 \end{cases}$$

b)

$$g(x) = \begin{cases} 0 & 0 \le x < 1 \text{ or } 2 < x \le 3 \\ 1 & 1 \le x \le 2 \end{cases}$$

3. a) $y = x(x < 1) + (x - 1)(1 \le x)$, [$x$Min, xMax] = [0, 2].

b) $y = (x \ge 1)(x \le 2)$, [xMin, xMax] = [0, 3].

c) $y = (x)(x < 0) - x(x \ge 0)$.

d) For [xMin, xMax] = [−2, 2], let $y = -1(-2 \le x)(x < -1) + 0 + 1(0 \le x)(x < 1) + 2(1 \le x)(x < 2) + 3(x = 2)$; similarly for other [$x$Min, xMax].

Exploration 6

1-2. Grapher activities.

3. y_1, y_2, y_4 to see original functions and their difference.
y_1, y_2, y_5 to see original functions and their difference.
y_1, y_2, y_6 to see original functions and their product.
y_1, y_2, y_7 to see original functions and their quotient.
y_1, y_2, y_8 to see original functions and their quotient.
y_4, y_5 to see opposite functions.
y_7, y_8 to see reciprocal functions.

Exploration 7

1. $y_3 = f \circ g(x)$, $y_4 = g \circ f(x)$.

2. Conjectures may vary. for example, $f \circ g(x) \ne g \circ f(x)$, $f \circ g(x) = x^2 - 7$, $g \circ f(x) = (x - 7)^2$. Also, the compositions produce shifts of f: $f \circ g(x) = x^2 - 7$ is a vertical shift and $g \circ f(x) = (x - 7)^2$ is a horizontal shift.

3. a) Yes, many. For example, $f(x) = 2x$ and $g(x) = x/2$.

b) Yes. For example, $g(x) = f(x)$, or $g(x) = x$. Others may be possible.

c) Yes, many. For example, $f(x) = x^3$ and $g(x) = x^{1/3}$.

d) $f \circ g \circ h = (f \circ g) \circ h$; $f \circ g \circ h \circ k = (f \circ g \circ h) \circ k$.

SECTION 1.4 GEOMETRIC TRANSFORMATIONS: SHIFTS, REFLECTIONS, STRETCHES, AND SHRINKS

Exploration 1

1. Grapher activity.

2. The seemingly different shapes of the two graphs is an optical illusion caused, in part, by a smaller portion of the graph of $y_2 = x^2 + 2$ being visible. TRACE shows that visible pairs of points in line vertically on the two graphs are a constant 2 units apart.

Exploration 2

1. Grapher activity.

2. Grapher activity.

These two activities graphically support the discussion following Eq. (3).

Exploration 3

1. A horizontal stretch by a factor of 2 doubles the x-coordinate of a point.

2. This would give the parametric equations $x_5(t) = 2t$, $y_5(t) = t^2$.

3. Eliminate the t to get $y = (x/2)^2$.

4. Similarly, to horizontally shrink the graph we have $x_6(t) = t/2$ and $y_6(t) = t^2$. Combining gives $y = (2x)^2$.

5. The parametric equations $x(t) = t/c$, $y(t) = t^2$ shrinks the graph horizontally by a factor of c if $c > 1$ and stretches it horizontally by a factor of $1/c$ when $0 < c < 1$. Eliminating the t gives $y = (cx)^2$, a horizontal shrink by a factor c $(c > 1)$, or a horizontal stretch by a factor $1/c$ $(0 < c < 1)$.

Exploration 4

1. No. We cannot assume a general rule from a particular example; we can argue, however, that the order of the shifts does not matter. Consider, for example, a function $y = f(x)$, a vertical shift of c and a horizontal shift of d:

Vertical shift of f: $f_1(x) = f(x) + c$ followed by Horizontal shift of f_1: $f_3(x) = f_1(x + d) = f(x + d) + c$	Horizontal shift of f: $f_2(x) = f(x + d)$ followed by Vertical shift of f_2: $f_4(x) = f_2(x) + c = f(x + d) + c$

The two results are the same.

2. Vertical shift of $y = x^3$ down 5 units: $y = x^3 - 5$, followed by vertical stretch by a factor of 2: $y = 2(x^3 - 5) = 2x^3 - 10$.

   This is different from the result $y = 2x^3 - 5$ of Example 6, a vertical stretch by factor of 2 followed by a vertical shift down 5 units.

   In general, a stretch followed by a shift does not give the same result as a shift followed by a stretch.

3. If we write v-shift for vertical shift, v-stretch for vertical stretch, etc., then

The six pairs are:	And, in either order they
v-shift, h-shift	give same results.
v-shift, v-stretch	give different results, in general.
v-shift, h-stretch	give same results.
h-shift, v-stretch	give same results.
h-shift, h-stretch	give different results, in general.
v-stretch, h-stretch	give same results.

Exploration 5

1. Grapher activity. This illustrates reflection about the x-axis.

2. When we TRACE in function mode and jump from one graph to the other, the cursor retains the x-coordinate. Because the graphs of y_1 and y_2 have only one domain value in common, TRACE jumps do not help us relate the two graphs.

   If instead we GRAPH:

   $x_1(t) = t$, $y_1(t) = \sqrt[4]{t}$,

   $x_2(t) = -t$, $y_2(t) = \sqrt[4]{t}$,

   TRACE jumps retain the t-value and thus show reflection about the y-axis.

3. Part 1 shows reflection about the x-axis. Part 2 in parametric mode shows reflection about the y-axis. In the study of optics, we learn that a mirror is the perpendicular bisecting plane of the segment formed by a point and its reflection image. An axis is the perpendicular bisector of a point and its reflection image across the axis, as suggested by TRACE jumps.

Exploration 6

1. a) and c) $a(x + b/2a)^2$ is the only term in the entire expression that changes as x changes. When $x = -b/2a$, $(x + b/2a)^2 = 0$, otherwise $(x + b/2a)^2$ is positive. Thus $a(x + b/2a)^2$ increases as we move away from $x = -b/2a$ when a is positive and decreases as we move away from $x = -b/2a$ when a is negative. In other words, the parabola opens upward with vertex at $x = -b/2a$ when a is positive and opens downward with vertex at $x = -b/2a$ when a is negative.

   b) If $x = -b/2a \pm k$, x-values that are equidistant from and on opposite sides of $-b/2a$ on the x-axis, $y = ak^2 + c - b^2/4a$, a single value, so the points on the graph are symmetric across $x = -b/2a$.

2. One way: GRAPH, in parametric mode,

   $$x_1(t) = t, \quad y_1(t) = -2t^2 + 12t - 5;$$
   $$x_2(t) = 3, \quad y_2(t) = t.$$

3. $y = -2(x - 3)^2 + 13$. One way: flip the parabola $y = x^2$ across the x-axis and stretch it vertically by a factor of 2. Then shift the graph 3 units to the right and 13 units up.

Exploration 7

1. Yes. $a = 1 > 0$ so the graph opens in the positive x direction; the axis of symmetry is $y = 0$ and the vertex of the graph is at (0, 0).

2. $y_1 = \sqrt{x}$ and $y_2 = -\sqrt{x}$.

3. Parametric mode: $x(t) = 3t^2 - 12t + 11$, $y(t) = t$. Function mode: $y_1 = 2 + \sqrt{((x + 1)/3)}$, $y_2 = 2 - \sqrt{((x + 1)/3)}$. $a = 3 > 0$ and the graph opens in the positive x-direction; the axis of symmetry is $y = 2$; the vertex of the graph is at (-1, 2).

SECTION 1.5 SOLVING EQUATIONS AND INEQUALITIES GRAPHICALLY

Exploration 1

1. Grapher activity.

2. $x = 2.208$.

3. $x = 2.2079400$.

4. $x = 2.2079400316$.

SECTION 1.6 RELATIONS, FUNCTIONS, AND THEIR INVERSES

Exploration 1

1. $x_1(t) = \cos t$, $y_1(t) = \sin t$. [tMin, tMax] = [0, 2π]. Your graph will best look like a circle if your x-, y-settings give a square viewing window.

2. For a smaller t-step, more values $(x(t), y(t))$ are evaluated and the graph plots slower. For a larger t-step, fewer values $(x(t), y(t))$ are evaluated and the graph plots faster. If t-step is too large, however, the plotted points may be so far apart that the graph (in connected format) will appear to include some segments rather than suggest a circle.

3. If the width of the view screen is about 1.5 times its height, do the following:

   If the graph appears oval, make the x-dimensions about 1.5 times the y-dimensions. Then the viewing window will be approximately square and the graph will appear circular. If the graph appears circular, make the x-dimensions and y-dimensions equal and the graph will appear oval.

   For speed:

   Most graphers have a key that you can press that will automatically give a square viewing window. (It may be called ZOOM-BOX.) A fast way to get equal x- and y-dimensions is to press the key for the standard viewing window [-10, 10] by [-10, 10].

4. $x_1(t) = 4 \cos t$, $y_1(t) = 4 \sin t$.

5. $x_1(t) = 3 + \cos t$, $y_1(t) = -4 + \sin t$.

6. $x_1(t) = 3 + 4 \cos t$, $y_1(t) = -4 + 4 \sin t$.

Exploration 2

1. Grapher activity.

2. $y_1 = -4 + \sqrt{((16 - (x - 3)^2)}$, $y_2 = -4 - \sqrt{((16 - (x - 3)^2)}$.

3. $y_1 = +3\sqrt{(1 - x^2/4)}$, $y_2 = -3\sqrt{(1 - x^2/4)}$.

4. $y_1 = +3\sqrt{(x^2/16 - 1)}$, $y_2 = -3\sqrt{(x^2/16 - 1)}$.

Exploration 3

1. $x_1(t) = t$, $y_1(t) = t^2$;
$x_2(t) = y_1(t)$, $y_2(t) = x_1(t)$.

2. Also GRAPH $x_3(t) = t$, $y_3(t) = t$. In simultaneous mode we see corresponding points on the function $y = x^2$ and its inverse $x = y^2$ appear symmetrically across the line $y = x$.

Exploration 4

1. For (a) and (c), the inverses are also functions. For (b), the inverse is not a function.

2. a) $x_1(t) = t$, $y_1(t) = t^3$; its inverse is:
$x_2(t) = y_1(t)$, $y_2(t) = x_1(t)$.

   b) $x_1(t) = t$, $y_1(t) = 1/t^2$; its inverse is:
$x_2(t) = y_1(t)$, $y_2(t) = x_1(t)$.

   c) $x_1(t) = t$, $y_1(t) = -2t + 4$; its inverse: $x_2(t) = y_1(t)$, $y_2(t) = x_1(t)$.

   Note: the Vertical Line Test fails in (b).

Exploration 5

1. Yes. $y = ax + b$ is not horizontal and so will pass the Horizontal Line Test.

2. The inverse of a linear function is a linear function.

3. $y = mx + b$ has an inverse that is a function if its graph is not horizontal; that is, $m \neq 0$.

4. The inverse is $y = (x - b)/m$. From part 3, $m \neq 0$ allows us to divide by m. This confirms part 2 as $y = x/m - b/m$ is a linear function.

5. For $f(x) = mx + b$ and $f^{-1}(x) = (x - b)/m$, $f^{-1}(f(x)) = f^{-1}(mx + b) = (mx + b - b)/m = mx/m = x$, and $f(f^{-1}(x)) = f((x - b)/m) = m((x - b)/m) + b = x - b + b = x$.

6. Enter y_3 as the same algebraic form as y_1 with x replaced by y_2. For example, $y_1 = 3x - 4$, $y_2 = (x + 4)/3$, $y_3 = 3y_2 - 4$. The graph of y_3 should be the line $y = x$.

Exploration 6

1. Grapher activity.

2. $x < 0$.

3. $x > 0$.

4. $x = 0$.

5 $(1/2)^x < (1/3)^x < (1/5)^x$ when $x < 0$. The inequalities reverse when $x > 0$ and the functions are equal when $x = 0$.

6. $2^{-x} < 3^{-x} < 5^{-x}$ when $x < 0$. The inequalities reverse when $x > 0$ and the functions are equal when $x = 0$.

7. For $a > 0$, $(1/a)^x = a^{-x}$. a^x is the reflection of $(1/a)^x = a^{-x}$ across the y-axis. If $a > 1$ then a^x looks like 2^x; if $0 < a < 1$ then a^x looks like $(1/2)^x$.

Exploration 7

1. $f(x) = -1.5\ (2^{x+1}) - 3$ looks like 2^x shifted 1 unit left, stretched 1.5 units vertically, reflected across the x-axis and finally shifted 3 units down. Notice that because $2^{x+1} = 2(2^x)$ this can also be viewed as $-3(2^x) - 3$ which is stretched vertically by a factor of 3, reflected across the x-axis, and then shifted down three units. Domain = $(-\infty, \infty)$, Range = $(-\infty, -3)$.

2. $f(x) = \log_3(2 - x)$. If we think of $2 - x = -(x - 2)$ then $f(x)$ can be viewed as transforming $\log_3 x$ by first reflecting across the y-axis and then shifting 2 units right. If instead we think of $2 - x$ as $2 + {}^-x$ then $f(x)$ can be obtained by shifting $\log_3 x$ 2 units to the left and then reflecting it across the x-axis. Domain = $(-\infty, 2)$, Range = $(-\infty, \infty)$.

SECTION 1.7 A REVIEW OF TRIGONOMETRIC FUNCTIONS

Exploration 1

1. Grapher activity. Your TRACE values should match the entries in Table 1-3.

2. For $t = 0^\circ, 15^\circ, 30^\circ, \ldots, 360^\circ$.

3. The unit circle is traced counter-clockwise starting at (-1,0).

4. Yes. Any 360° interval which starts at an angle equivalent to −180°. That is, let *t*Min = −180 + 360*n* and let *t*Max = 180 + 360*n* for *n* an integer.

5. In radian mode set *t*Min = −π and *t*Max = π. More generally, let *t*Min = −π + 2π*n* and *t*Max = π + 2π*n* for any integer *n*. (Note: *t*-step = π/12 in radian mode is approximately *t*-step = 15 in degree mode.)

6. One way: set *t*Min = 0, *t*Max = 4π then GRAPH and TRACE. For each value of *t*, $x = \cos t$ and $y = \sin t$.

Exploration 2

1. $x(t)^2 + y(t)^2 = 16(\cos^2(\pi t/3) + \sin^2(\pi t/3) = 4^2$. The Pythagorean Identity for sines and cosines shows that the graph is a circle with radius 4 centered at (0, 0). $\cos(\pi t/3)$ and $\sin(\pi t/3)$ both have period 6, therefore, as *t* varies from 0 to 6 the graph evolves from (4, 0) counterclockwise around to (4, 0).

2. Grapher activity.

3. A $t = 0$ and $t = 6$ the point is at (4, 0). At $t = 3$ and $t = 9$ it is at (-4, 0).

4. *P* is at (4, 0) when *t* is 0 and every six seconds thereafter, i.e., $t = 6k$, $k = 0, 1, 2, 3, \ldots$.

5. $(2\sqrt{2}, 2\sqrt{2})$ is on the line $y = x$ and, therefore, one-eighth of the way through the first revolution. Therefore, *P* will arrive at $(2\sqrt{2}, 2\sqrt{2})$ at $t = 6/8 = 0.75$ and every six seconds thereafter, i.e., at $t = 0.75 + 6k$, $k = 0, 1, 2, \ldots$.

Exploration 3

1. $(x_1(t), y_1(t))$ will trace the unit circle in a counterclockwise direction; $(x_2(t), y_2(t))$ will trace the

graph of $y = \sin t$. Notice that as you move from one graph to the other the y-values (heights) are the same.

2. The y-values are identical, $y_1(t) = y_2(t) = \sin t$. The y-values around the unit circle are values of the sine.

3. With tMax and xMax = 4π we can trace twice around the circle on the left and move through two periods of $y = \sin t$ on the right.

4. With $y_2(t) = \cos t$, tracing shows that $y_2(t) = x_1(t)$. That is, $\cos t$ is the x-value of coordinates on the unit circle. (You may find it interesting to GRAPH $x_2 = \cos t$, $y_2 = t$.) For $y_2 = \tan t$ or $y_2 = \cot t$ (type 1/tan t) four periods are visible; for $y_2 = \sec t$ (type 1/cos t) and $y_2 = \csc t$ (type 1/sin t) two periods are visible. If you use t-step = $\pi/12$ and move from the circle to the graph you can observe the correspondence between points on the circle and on the graphs for special angles ($\pi/6$, $\pi/4$, etc.).

Exploration 4

1. The two graphs are identical.

2. The two graphs appear as mirror images across the x-axis. Also recall Exploration 3, Section 1.3 to see that a function $f(x)$ is even or odd.

Exploration 5

1. Grapher activity.

2. The amplitude appears to be $\sqrt{2}$. Conjecture: $a = \sqrt{2}$.

3. The period appears to be 2π with a horizontal shift of -0.78 which is approximately $\pi/4$. Conjecture: $b = 1$ and $c = \pi/4$.

4. The graph does not appear to be shifted vertically. Conjecture: $d = 0$.

5. $\sin x + \cos x = \sqrt{2} \sin(x + \pi/4)$.

6. We suspect that $y_1 = y_2$, so we would like to see $y_1 - y_2 + 1$ graph as $y = 1$.

Exploration 6

1. $x_1(t) = t$, $y_1(t) = \sin t$;
$x_2(t) = y_1(t)$, $y_2(t) = x_1(t)$; [tMin, tMax] = $[-\pi/2, \pi/2]$.

2. Grapher activity.

3. $x_1(t) = t,\ y_1(t) = \cos t;$
$x_2(t) = y_1(t),\ y_1(t) = x_1(t);\quad [t\text{Min},\ t\text{Max}] = [0,\ \pi].$

$x_1(t) = t,\ y_1(t) = \tan t;$
$x_2(t) = y_1(t),\ y_1(t) = x_1(t);\quad [t\text{Min},\ t\text{Max}] = [-\pi/2,\ \pi/2].$

SECTION 2.1 LIMITS

Exploration 1

1. 1.

2. 4.

3. 2.

4. 8.

5. 16.

6. 36.

7. $\lim_{x\to c} (x + 4) = c + 4.$

8. $\lim_{x\to c} 2x^2 = 2c^2.$

9. $\lim_{x\to c} 4x^2 = 4c^2.$

Exploration 2

The graph suggests the limit is 10/3. Confirming algebraically,

$$\frac{x^2 - 25}{3(x - 5)} = \frac{(x + 5)(x - 5)}{3(x - 5)} = \frac{x + 5}{3} \text{ when } x \neq 5. \quad \text{So,}$$

$$\lim_{x\to 5} \frac{x^2 - 25}{3(x - 5)} = \lim_{x\to 5} \frac{x + 5}{3} = 10/3.$$

Exploration 3

1. $f(x) = (x^3 - 1)/(x - 1)$ appears to approach 3 as $x \to 1$.

2. $f(x) = (x^2 + 1)/(x + 1)$ does not seem to have a limit as $x \to -1$ since the y-values get smaller and smaller as we approach $x = -1$ from the left and larger and larger as we approach $x = -1$ from the right.

3. $f(x) = \sin(1/x)$ does not seem to approach a fixed value as $x \to 0$. Instead the graph oscillates between -1 and 1, not getting closer and closer to any fixed value.

4. $f(x) = [x]$ approaches 2 as we approach $x = 3$ from the left and it approaches 3 as we approach 3 from the right. Because there are two different values that $f(x)$ gets close to there does not seem to be a limit as $x \to 3$.

Exploration 4

Grapher activity. See Section 7.5, Exploration 2.

SECTION 2.2 CONTINUOUS FUNCTIONS

Exploration 1

1. View the graph of f in a window that includes $x = 2$ and $x = -2$. The graph suggests that the discontinuity at $x = 2$ is removable.

   Write $f(x) = (x - 2)(x + 3)/(x - 2)(x + 2) = (x + 3)/(x + 2)$ when $x \neq 2$. As $x \to 2$, this new fraction approaches 5/4 suggesting that there is removable discontinuity at $x = 2$. Define an extension of f by:

$$g(x) = \begin{cases} f(x) & \text{when } x \neq 2 \\ 5/4 & \text{when } x = 2 \end{cases}$$

2. Neither $x = -1$, nor $x = 0$ are in the domain of f. The graph of f seems to approach $y = 0$ as $x \to 0$ suggesting that this discontinuity is removable. Define an extension of f by

$$g(x) = \begin{cases} f(x) & \text{when } x \neq 0 \\ 0 & \text{when } x = 0 \end{cases}$$

SECTION 2.3 THE SANDWICH THEOREM AND (SINθ)/θ

Exploration 1

1. $$\lim_{x\to 0} \frac{\sin 2x}{\sin 3x} = \lim_{x\to 0} \frac{\frac{2x}{2x}\sin 2x}{\frac{3x}{3x}\sin 3x} = \lim_{x\to 0} \frac{2x}{3x} \frac{\frac{\sin 2x}{2x}}{\frac{\sin 3x}{3x}} = 2/3.$$

2. $$\lim_{x\to 0} \frac{\sin 5x}{\sin 8x} = \lim_{x\to 0} \frac{\frac{5x}{5x}\sin 5x}{\frac{8x}{8x}\sin 8x} = \lim_{x\to 0} \frac{5x}{8x} \frac{\frac{\sin 5x}{5x}}{\frac{\sin 8x}{8x}} = 5/8.$$

3. $$\lim_{x\to 0} \frac{\sin ax}{\sin bx} = \lim_{x\to 0} \frac{\frac{ax}{ax}\sin ax}{\frac{bx}{bx}\sin bx} = \lim_{x\to 0} \frac{ax}{bx} \frac{\frac{\sin ax}{ax}}{\frac{\sin bx}{bx}} = a/b.$$

SECTION 2.4 LIMITS INVOLVING INFINITY

Exploration 1

1. The two functions do get farther apart. Their difference, $y_1 - y_2 = -3x$ continues to increase as x increases.

2. The two functions do get closer together. Their ratio, $y_1/y_2 = 1 + 3/2x$, decreases towards 1 as x increases.

3. One way: An end behavior model of $y = f(x)$ is a function whose values are relatively close to those of f as x increases without bound in the sense that, although their values may grow apart ($y_1 - y_2$ increases), comparatively they grow together (that is, their ratio gets closer to 1).

Exploration 2

1. $$g(x) = \begin{cases} f(x) & \text{when } x \neq 2 \\ -4/5 & \text{when } x = 2 \end{cases}$$

2. As $x \to 3^+$, $\frac{x+2}{x+3} \to \frac{5}{6}$ but $\frac{1}{x-3} \to \infty$ so $\lim_{x\to 3^+} \frac{x+2}{x+3} \cdot \frac{1}{x-3} \to \infty$.

3. As $x \to 3^-$, $\frac{x+2}{x+3} \to \frac{5}{6}$ but $\frac{1}{x-3} \to -\infty$ so $\lim_{x\to 3^+} \frac{x+2}{x+3} \cdot \frac{1}{x-3} \to -\infty$.

4. As $x \to 3^+$, $\frac{x+2}{x-3} \to \frac{1}{6}$ but $\frac{1}{x+3} \to \infty$ so $\lim_{x\to 3^-} \frac{x+2}{x-3} \cdot \frac{1}{x+3} \to \infty$.

5. As $x \to 3^-$, $\frac{x+2}{x-3} \to \frac{1}{6}$ but $\frac{1}{x+3} \to -\infty$ so $\lim_{x\to 3^-} \frac{x+2}{x-3} \cdot \frac{1}{x+3} \to -\infty$.

6. $x = -1$ is the vertical asymptote.

7. $x = \sqrt[3]{\frac{1}{2}}$.

Exploration 3

1. Grapher activity.

2. $q(x) = x^2 - 2x - 2$, $r(x) = 4$, $h(x) = x - 2$. If you ZOOM-OUT the graphs of $y_1 = f(x)$ and $y_2 = q(x)$ will appear to merge into a single graph.

3. end-behavior model: x^2;
end behavior asymptote: $x^2 - 8x - 15$.

4. end-behavior model: $2x$; end behavior asymptote: $2x + 3$.

5. end-behavior model: x^3; end behavior asymptote: $x^3 - 9x$.

SECTION 2.5 CONTROLLING FUNCTION OUTPUTS: TARGET VALUES

Exploration 1

1. Keep x in $(7/4, 9/4)$. Proof: $7/4 < x < 9/4 \Rightarrow 7 < 4x < 9 \Rightarrow 4 < 4x - 3 < 6$.

2. Keep x in $(-2, 0)$. Proof: $-2 < x < 0 \Rightarrow 4 < x + 6 < 6$.

3. Keep x in $(1/2, 3/2)$. Proof: $1/2 < x < 3/2 \Rightarrow -3 < -2x < -1 \Rightarrow 4 < -2x + 7 < 6$.

4. Use ZOOM-BOX choosing y-values just "outside" minimum and maximum of the desired range so that the graph enters and leaves the viewing window at the top and bottom and not the left and right. Repeat a few times.

5. Keep x in $(3.76, 4.20)$. Confirm: $3.76 < x < 4.20 \Rightarrow 2(2.76) + 1 < 2x + 1 < 2(4.20) + 1 \Rightarrow$
$2.92 = \sqrt{2(3.76)+1} < \sqrt{2x+1} < \sqrt{2(4.20)+1} = 3.06$ which means that taking x in the given interval keeps the y-values within the target interval $[2.9, 3.1]$.

6. Keep x in $(-2.17, -1.79)$. Confirm: $-2.17 < x < -1.79 \Rightarrow -2.17 + 3 < x + 3 < -1.79 + 3 \Rightarrow 0.91 = \sqrt{-2.17+3} < \sqrt{x + 3} < \sqrt{-1.79+3} = 1.10$ which means that taking x in the given interval keeps the y-values within the target interval $[0.9, 1.1]$.

7. Keep x in (2.91, 3.07). Confirm: $2.91 < x < 3.07 \Rightarrow$ $2(2.91) < 2x < 2(3.07) \Rightarrow 7 - 2(3.07) < 7 - 2x < 7 - 2(2.91)$ $\Rightarrow 0.93 = \sqrt{7-2(3.20)} < \sqrt{7 - 2x} < \sqrt{7-2(2.87)} = 1.08$ which means that taking x in the given interval keeps the y-values within the target interval [0.9, 1.1].

SECTION 2.6 DEFINING LIMITS FORMALLY WITH EPSILONS AND DELTAS

Exploration 1

1. Any value of δ will work because $|k - k| = 0 < \delta$!

2. Choose $\delta = \epsilon$.

3. $f(x) = k$ is continuous because $\lim_{x \to c} k = k = f(c)$ for all c.

   $g(x) = x$ is continuous because $\lim_{x \to c} x = c = g(c)$ for all c.

SECTION 3.1 SLOPES, TANGENT LINES, AND DERIVATIVES

Exploration 1

1. At (1,1), $y - 1 = 2(x - 1)$; at (-1/2, 1/4), $y - 1/4 = -1(x + 1/2)$. At one possible third point (-2, 4) the tangent is $y - 4 = -4(x + 2)$.

2. We should see the parabola $y = x^2$ and a tangent line at each point in (1).

   GRAPH $y_1 = x^2$ and $y_2 = 2x - 1$.

   GRAPH $y_1 = x^2$ and $y_2 = -x - 1/4$.

   GRAPH $y_1 = x^2$ and $y_2 = -4x - 4$.

Exploration 2

1. The first equation is true by substituting the values of $f(x)$ and $f(x + h)$. We simplify algebraically for the second and third equations. The fourth equation is true because the limit of a constant function is the constant.

2. This shows that the calculus definition of the slope of a linear function as a derivative agrees with the algebra definition of the slope of a line as rise/run.

3. The tangent to a line is the line itself.

Exploration 3

1. $$\lim_{h\to 0}\frac{\sqrt{x+h}-\sqrt{x}}{h} = \lim_{h\to 0}\frac{\sqrt{x+h}-\sqrt{x}}{h}\cdot\frac{\sqrt{x+h}+\sqrt{x}}{\sqrt{x+h}+\sqrt{x}}$$

$$= \lim_{h\to 0}\frac{(x+h)-x}{h(\sqrt{x+h}+\sqrt{x})} = \lim_{h\to 0}\frac{1}{\sqrt{x+h}+\sqrt{x}} = \frac{1}{2\sqrt{x}}.$$

2. At $(1.5, \sqrt{1.5})$ the slope is $1/(2\sqrt{1.5})$ so the tangent line is $y - \sqrt{1.5} = (1/(2\sqrt{1.5})(x - 1.5))$. To support graphically, GRAPH this line and $y = \sqrt{x}$ in the same viewing window.

SECTION 3.2 NUMERICAL DERIVATIVES

Exploration 1

1. Grapher activity. You may wish to GRAPH all four functions in [-5, 5] by [-20, 20].

2. NDER($f(x)$) has degree 2.

3. Grapher activity. You may wish to GRAPH all functions in [-3, 3] by [-10, 20].

4. NDER($f(x)$) has degree 3.

5. NDER($f(x)$) is a polynomial function of degree $n - 1$.

SECTION 3.3 DIFFERENTIATION RULES

Exploration 1

1. $3x^2$.

2. $4x^3$.

3. $f'(x) = 3x^2$; $g'(x) = 4x^3$; $h'(x) = nx^{n-1}$.

Exploration 2

1. The graphs appear to be identical. TRACE and move from one graph to the next to see the slight differences in the y-values. We can conclude that for x in [-0.1, 0.1] $\sin x$ is very close to x. We cannot conclude that $\sin x = x$.

2. The graphs are very misleading. $\sin(95x)$ has a period $2\pi/95$ which is very close to the size of each pixel. The

calculator is plotting just one point from each of the 95 periods. These various points look like a complete sinusoid but do not accurately reflect the actual graph.

Exploration 3

1. $y' = 3x^2 - 6x$; $y'' = 6x - 6$; $y''' = 6$; all higher derivatives are zero.

2. For a polynomial of degree n, all derivatives of order $k \geq n + 1$ are zero.

Exploration 4

1. a) $$y' = \frac{x^4[(x-1)(x^2-2x)]' - (x-1)(x^2-2x)[x^4]'}{x^8} =$$
$$\frac{x^4[(x-1)(2x-2) + (x^2-2x)(1)] - (x^3-3x^2+2x)(4x^3)}{x^8} =$$
$$\frac{x^4[(2x^2-4x+2) + (x^2-2x)] - (4x^2-12x+8)(x^4)}{x^8} =$$
$$\frac{x^4[-x^2+6x-6]}{x^8} = \frac{-x^2+6x-6}{x^4}.$$

   b) $$y = \frac{x^3-3x^2+2x}{x^4} = x^{-1}-3x^{-2}+2x^{-3}, \text{ so } y' = -x^{-2}+6x^{-3}-6x^{-4}.$$

2. Clearly the second method is far simpler. The following shows that the results are the same:
$$-x^{-2}+6x^{-3}-6x^{-4} = \frac{-x^2}{x^4}+\frac{6x}{x^4}-\frac{6}{x^4} = \frac{-x^2+6x-6}{x^4}.$$

Exploration 5

1. a) $$f'(x) = \frac{(x^2+4)(5)-(5x)(2x)}{(x^2+4)^2} = \frac{-5x^2+20}{(x^2+4)^2}, \text{ so}$$
$$f''(x) = \frac{(x^2+4)^2(-10x)-(-5x^2+20)(4x^3+16x)}{(x^2+4)^4} =$$
$$\frac{(-10x^5-80x^3-160x)-(-20x^5+320x)}{(x^2+4)^4} = \frac{10x^5-80x^3-480x}{(x^2+4)^4} =$$
$$\frac{10x(x^2-12)(x^2+4)}{(x^2+4)^4}.$$ Since the denominator is always positive and $x^2 + 4 > 0$, $f''(x) > 0$ when $10x(x^2-12) > 0$. That is, when $x > \sqrt{12}$ or $-\sqrt{12} < x < 0$.

b) NDER2 $f(x) > 0$ for $-3.46 < x < 0$ or $x > 3.46$.

2. The two methods support each other. The algebraic method gives an exact rather than an approximate solution but takes more time and effort.

3. Explain your preference.

4. a) $g'(x) = \dfrac{(3x^2+4)(2)-(2x-5)(6x)}{(3x^2+4)^2} = \dfrac{-6x^2+30x+8}{(3x^2+4)^2}$, so

$g''(x) = \dfrac{(3x^2+4)^2(-12x+30)-(-6x^2+30x+8)(12x)(3x^2+4)}{(3x^2+4)^4} =$

$\dfrac{(-36x^3+90x^2-48x+120)-(-72x^3+360x^2+96x)}{(3x^2+4)^3} =$

$\dfrac{36x^3-270x^2-144x+120}{(3x^2+4)^3}$. Because we cannot factor the numerator there is no way to solve $g''(x) > 0$ algebraically.

b) GRAPH NDER2 $g(x)$ and ZOOM and TRACE to identify the three x-intercepts. $g'' > 0$ for $-0.910 < x < 0.460$ or $x > 7.95$.

SECTION 3.4 VELOCITY, SPEED, AND OTHER RATES OF CHANGE

Exploration 1

1. For $y(t) = 16t^2$, the ball bearing "falls" upward. To simulate the falling ball bearing, turn the viewing window upside down, or use $y(t) = 100 - 16t^2$ (to simulate a fall from 100 ft., Example 2). For $y(t) = 100 - 16t^2$, letting t range from 0 to 2.5 will take the ball from 100 ft. to the ground (again, Example 2). To slow the animation choose a smaller value of t-step; to increase the apparent speed choose larger values of t-step.

   To simultaneously simulate the fall on Earth, the Moon, and Jupiter, GRAPH:

   $x_1 = 1,\ y_1 = 100 - 16t^2$
   $x_2 = 2,\ y_2 = 100 - 2.6t^2$
   $x_3 = 3,\ y_3 = 100 - 37.2t^2$ in simultaneous format.

2. GRAPH: $x_1 = 45t,\ y_1 = 1;$ t in [0, 7]
   $x_2 = 55(t - 1),\ y_2 = 2;$ t-step = 0.05,
   x, y in [−1, 300] by [−1, 3].

3. GRAPH: $x_1 = 45t$, $y_1 = 2.1$; t in [0, 7]
$x_2 = 300 - 55(t - 1)$, $y_2 = 1.9$;
x, y in [-1,300] by [-1, 3].

Exploration 2

1. At $t = 2.05$, $x = -2.0295$, $y = 2$.

2. The velocity is positive and decreasing to 0 for t in [0, 0.6] when the particle moves to the right. The particle then reverses direction and the velocity is negative for t in [0.6, 2] first increasing in absolute value, then decreasing in absolute value to 0. After $t = 2$ the velocity is positive and increasing.

3. Graph 3 shows the position of the particle (y) as a function of time t.

4. $y_3' = 12t^2 - 32t + 15 = 0$ when $t = (8 \pm \sqrt{19})/6$ so the velocity is positive, negative, and then positive. The turning points are the zeros of y_3' which support the values $t = 0.6$ and $t = 2.05$ in parts 1 and 2.

   $y_3'' = 24t - 32 = 0$ when $t = 4/3$. The particle accelerates to the left until $t = 4/3$ and accelerates to the right thereafter. This corresponds to the point where the particle starts to slow down while moving to the left.

SECTION 3.5 DERIVATIVES OF TRIGONOMETRIC FUNCTIONS

Exploration 1

1. $y' = 2x - \cos x$.

2. $y' = (x \cos x - \sin x)/x^2$.

3. $y' = 5 - \sin x$.

4. $y' = \cos^2 x - \sin^2 x$.

Exploration 2

When $x_1 = t$ the graphs simultaneously show the position, y_1 and the velocity, y_2, of the weight.

Exploration 3

1. $y' = 3 - \csc^2 x$.

2. $y' = -2 \csc x \cot x$.

3. $y'' = \sec^3 x + \sec x \tan^2 x$.

4. $y'' = -2 \sin x - 3 \cos x$.

SECTION 3.6 THE CHAIN RULE

Exploration 1

1. $y = 15t - 2$. $dy/dt = 15$; $dy/du = 5$, $du/dt = 3$;
   $\frac{dy}{du}\cdot\frac{du}{dt} = 15$. Notice that $\frac{dy}{dt} = \frac{dy}{du}\cdot\frac{du}{dt}$.

2. $dy/dx = 48x^2 + 48x + 12$; $dy/du = 6u^2$, $du/dx = 2$;
   $\frac{dy}{du}\cdot\frac{du}{dx} = 12(4x^2 + 4x + 1)$. Notice that $\frac{dy}{dx} = \frac{dy}{du}\cdot\frac{du}{dx}$.

Exploration 2

1. Grapher activity.

2. There are 7 horizontal tangents.

3. TRACE on y_3 shows NDER(y_2, 2) = 3.9989... which supports $(f \circ g)'(2) = 4$. The slope of the line tangent to the graph of $(f \circ g)$ at $x = 2$ is 4. A <u>square</u> viewing window suggests that $(f \circ g)$ is quite steep at $x = 2$, thereby supporting a slope of 4. (Note that maximum steepness does <u>not</u> occur at $x = 2$ but just a little bit further on.)

SECTION 3.7 IMPLICIT DIFFERENTIATION AND FRACTIONAL POWERS

Exploration 1

1. GRAPH $x_1 = t$, $y_1 = t^3 - 2t$.

2. Graphs as the unit circle. Confirming: $x^2 + y^2 = 1$.

3. Graphs as a circle of radius 3. Confirming: $x^2 + y^2 = 9$.

4. Graphs as an ellipse. Confirming: $x^2 + 4y^2 = 16$.

5. Graphs as a hyperbola. Confirming: $x^2 - y^2 = 1$.

6. Graphs as a hyperbola. Confirming: $x^2 - y^2 = 9$.

7. Graphs as a hyperbola. Confirming: $x^2 - 4y^2 = 16$.

Exploration 2

1. $dy/dt = -\sin t$ and $dx/dt = \cos t$ so $dy/dx = -\sin t/\cos t = -(x/y)$.

2. The graph that rises from the left to the right is for the derivative of the bottom half of the unit circle. As $x \to 1$ or $x \to -1$, both derived graphs approach infinity, indicating vertical tangents at $x = -1$ and $x = +1$.

   To duplicate the view in Fig. 3.48, let $y_1 = \sqrt{(1 - x^2)}$, $y_2 = -\sqrt{(1 - x^2)}$, y_3 = NDER y_1, y_4 = NDER y_2.

Exploration 3

a) $f(x) = \sqrt{x}$ is not differentiable at $x = 0$. NDER has a vertical asymptote and f has a vertical tangent at $x = 0$.

b) $f(x) = x^{1/5}$ is not differentiable at $x = 0$. NDER has a vertical asymptote and f has a vertical tangent at $x = 0$.

c) $f(x) = \sqrt{(1 - x^2)}$ is not differentiable at $x = \pm 1$. NDER has vertical asymptotes and f has vertical tangents at $x = \pm 1$.

SECTION 3.8 LINEAR APPROXIMATIONS AND DIFFERENTIALS

Exploration 1

As we ZOOM-IN the parabola appears more and more like the tangent line. Eventually, the two graphs appear indistinguishable.

Exploration 2

1. $f(3) = 2$ and $f'(3) = 1/4$ so $y - 2 = (1/4)(x - 3)$, that is, $y = 5/4 + x/4$.

2.

x	0.2	0.05	0.005
$\sqrt{1 + x}$	1.095	1.0247	1.002497
$1 + \frac{x}{2}$	1.1	1.0250	1.002500

x	3.2	3.05	3.005
$\sqrt{1 + x}$	2.049	2.01246	2.0012496
$\frac{5}{4} + \frac{x}{4}$	2.050	2.0125	2.00125

The values for the tangent line get closer and closer to the values for the function as the x-values get closer and closer to 0 and 3. That is, $|L(x) - f(x)| \to 0$.

3. $L_0(3) = 1 + 3/2 = 2.5$ while $f(3) = 2.0$; these values differ by 0.5 while the table above suggests that the values of $f(x)$ and $1 + x/2$ are much closer to each other for x near 0. As we extend L_0 in Fig 3.54 we can see the line and the function separating, supporting the numerical evidence in the table.

Exploration 3

The approximations are not linearizations since they are not linear functions. Your should not be able to TRACE too far from 0 before the values of the two functions separate in the tenths decimal place.

Exploration 4

1. Conjecture: The linearization of $y = \sin x$ is $y = x$; the linearization of $y = \cos x$ is $y = 1$.

2. Conjecture: The linearization of $\cos x$ at $\pi/2$ is $y = -x + \pi/2$. A line with slope -1 has equal x- and y-intercepts. Confirm by noting that $f(\pi/2) = 0$ and $f'(\pi/2) = -\sin(\pi/2) = -1$ so $y - 0 = -1(x - \pi)$, that is, $y = -x + \pi/2$.

SECTION 4.1 MAXIMA, MINIMA, AND THE MEAN VALUE THEOREM

Exploration 1

1. There appear to be extreme values at $x = \pm 2$. This is confirmed as $y' = 3x^2 - 12 = 0$ when $x = \pm 2$.

2. Absolute minima occur at 0, $\pm 2\sqrt{3}$. Local maxima occur at both +2 and -2. At the ±2 the derivative is 0 indicating horizontal tangents. At $x = 0$ and $\pm 2\sqrt{3}$ the derivative appears to have jump discontinuities, in which case $|y|$ is

not differentiable at these points and has no tangent lines at these points.

3. On the restricted domain $-1 < x < 4$, the absolute maximum occurs at $x = 2$ and absolute minima occur at $x = 0$ and $x = 2\sqrt{3}$. There are no other extreme values because the endpoints are not included in the domain. The graph, however, suggests that $(-1, 11)$ is a local maximum and $(4, 16)$ is an absolute maximum.

Exploration 2

1. One way to find c graphically: GRAPH $y_1 = x^2$ and $y_2 = \text{NDER}(y_1)$ then TRACE y_2 until $y_2 = 2$. To find c analytically set the derivative equal to 2. ($2x = 2$ so $x = 1$.)

2. For example, let $f(x) = x^3 - x^2$ on $[0, 2]$. $f(0) = 0$, $f(2) = 4$. To find c algebraically set $3x^2 - 2x = 4/2$, the slope of the line joining $(0,0)$ and $(2,4)$. Solve the quadratic to get $c = (2 \pm \sqrt{30})/6$. The negative root is rejected since it does not lie in the interval $[0, 2]$.

   To find c graphically, GRAPH $y_1 = f(x)$, $y_2 = \text{NDER}(y_1)$ and TRACE y_2 until $y_2 = 2$.

   The three graphs show $y_1 = f(x)$, y_2 the line joining the endpoints of the graph of f on $[a, b]$, and y_3 the tangent line at c. y_2 and y_3 are parallel.

Exploration 3

Where y_1 rises its derivative $y_2 > 0$; where y_1 falls, $y_2 < 0$. Conjecture: Where a function increases its derivative is positive and where it decreases its derivative is negative.

SECTION 4.2 PREDICTING HIDDEN BEHAVIOR

Exploration 1

1. Grapher activity.

2. Grapher activity.

3. a) The concavity changes three times.

   b) Conjecture: The concavity changes at most $n - 2$ times in the graph of an n-th degree polynomial.

Each time the concavity changes the second derivative must change signs. If the degree of f is n then the degree of f is $n - 2$ and a polynomial of degree $n - 2$ has at most $n - 2$ roots. Therefore, y_2 can change sign by crossing the x-axis at most $n - 2$ times.

c) The concavity is not changing at integer values. Practical conjecture: Solving graphically for a point of inflection in the graph of a function f cannot be done with much accuracy by ZOOM-IN on the graph of f.

4. If $y = \sin x$ then $y'' = -\sin x$. The value of $\sin x$ changes signs every time $\sin x$ crosses the x-axis; therefore, the concavity of its graph changes at each of these points. The concavity of $y = \sin x$ changes at $x = n\pi$, n an integer.

Exploration 2

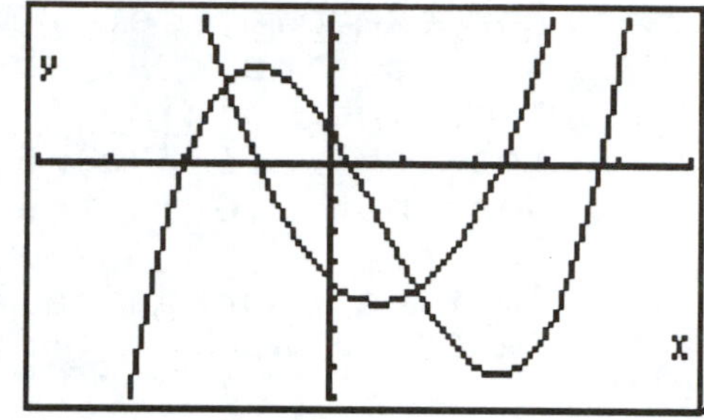

1. $f' = 3x^2 - 4x - 7 = (3x - 7)(x + 1)$ and $f'' = 6x - 4$. f' is negative only between its roots $x = -1, 7/3$, therefore, f increases to a maximum at $x = -1$, decreases to a minimum at $x = 7/3$, and then increases. $f'' < 0$ for $x < 2/3$, $f'' > 0$ for $x > 2/3$ and $f''(2/3) = 0$. Therefore, f is concave down for $x < 2/3$, has a point of inflection at $x = 2/3$ and is concave up for $x > 2/3$.

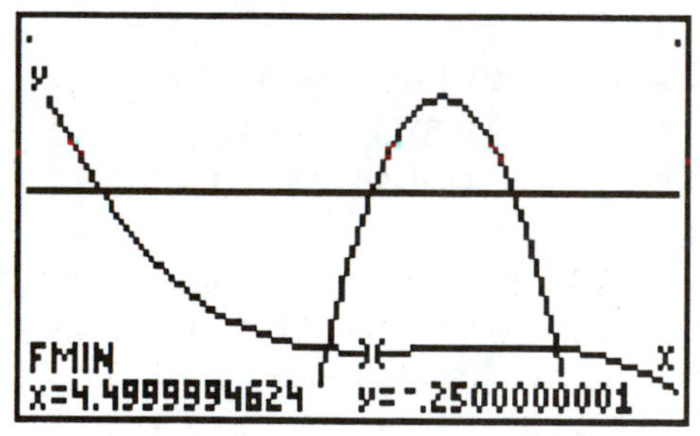

2. a) Examine f first in a standard window and then view it in a [4,5] by [-0.3,-0.1] window to see that there are both a local maximum and minimum. We confirm by seeing that the discriminant of the quadratic $f' = -60x^2 + 546x - 1242$ is positive so that there are two real roots. The x-intercept of $f'' = -120x + 546$ supports a point of inflection at (4.55, -0.245).

b) A [-6,6] by [-25,10] viewing window suggests the important features of the graph. $f' = 4(x + 2)(x - 1)^2$ has zeros at $x = 1$ and $x = -2$ and $f'' = 12(x - 1)(x + 1)$ also has zeros at ±1. f'' changes sign at both zeros so (1,4) and (-1,-12) are both points of inflection of f.

Although f' has two zeros, an extreme value, a relative minimum occurs only at x = -2.

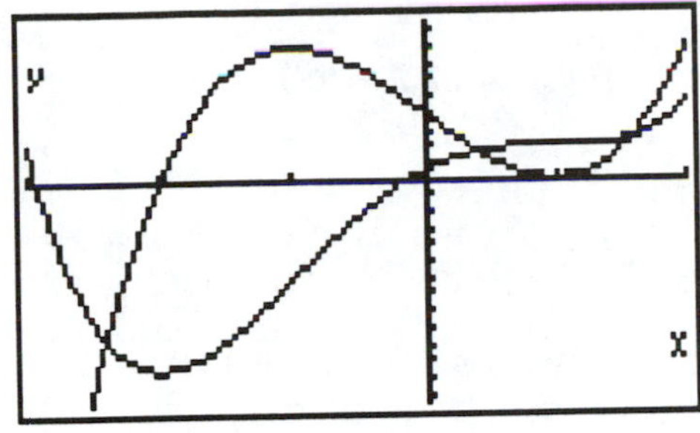

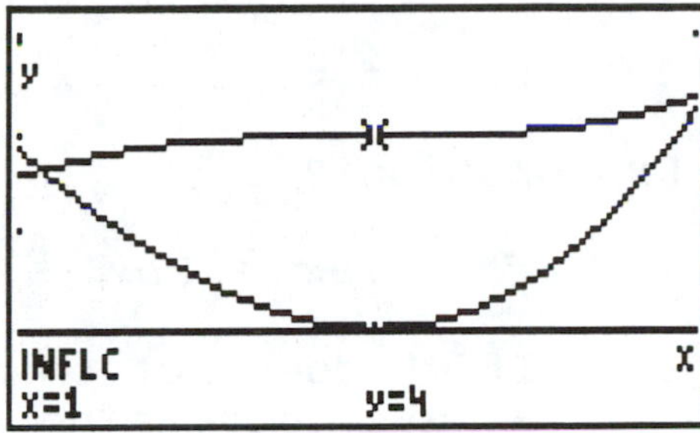

Exploration 3

(a) $f(x) = -x^4$ has a minimum at $x = 0$ since $f'(x) = -4x^3$ changes sign there although $f'(0) = f''(0) = 0$.

(b) $f(x) = x^4$ has a maximum at $x = 0$ since $f'(x) = 4x^3$ changes sign there although $f'(0) = f''(0) = 0$.

(c) $f(x) = x^3$ has neither a maximum nor a minimum at $x = 0$ since $f'(x) = 3x^2$ does not change sign there although $f'(0) = f''(0) = 0$.

SECTION 4.3 POLYNOMIAL FUNCTIONS, NEWTON'S METHOD, AND OPTIMIZATION

Exploration 1

1. $f'' = 6ax + 2b$ is a line with x-intercept $x = -b/3a$. On one side of the intercept f'' is positive (and f is concave up); on the other side f'' is negative (and f is concave down).

2. $f'(x) = 3ax^2 + 3bx + c$ is quadratic. If f' has no real root or if f' has one root (so that f' is tangent to the x-axis) then its graph lies wholly above or below the x-axis. In these cases f' does not change sign and so f will always increase or always decrease. Depending upon the sign of the coefficient $\underline{a}$ the graph will look like 4.27 (a) or (b). If f' has two real roots then it changes sign twice and so f will have two extreme values as in figure 4.27 (c) and (d).

3. (a) $f = x^3 + 3x$ (f' has no real root; Fig 4.27 (a)).

   (b) $f = -x^3 - 3x$ (f' has no real root; Fig 4.27 (b)).

   (c) $f = x^3 + 4$ (f' has one real root; Fig 4.27 (a)).

(d) $f = -x^3 + 4$ (f' has one real root; Fig 4.27 (b)).

(e) $f = x^3 + 4x^2$ (f' has two roots; Fig 4.27 (c)).

(f) $f = -x^3 + 4x^2$ (f' has two roots; Fig 4.27 (d)).

Exploration 2

1. $f' = 4ax^3 + 3bx^2 + 2cx + d$. $f'' = 12ax^2 + 6bx + 2c$.

2. If f'' does not change sign then f' is always increasing or always decreasing as in Fig. 4.27 (a) and (b), and, therefore, has only one root. Consequently, f is always concave up or down and has either an absolute maximum or minimum depending upon the sign of $\underline{a}$. The graph looks like Fig. 4.29 (a) or (b).

3. If f'' changes sign then it changes sign twice. In this case f will have two points of inflection and f' will have two extreme values and so is a cubic of the type shown in Fig. 4.27 (c) or (d). In this case f' either has one, two, or three real roots.

   If f' has one root or two roots (so that f' is tangent to the x-axis), then f' changes sign only once and so f will have only one extreme value. In this case the graph of f will look like Fig. 4.28 (a) or (b) depending upon the leading coefficient $\underline{a}$.

   If f' has three roots then f' will change signs three times and so f will have three extreme values as in 4.28 (c) and (d).

4. (a) $f'' = 12(x - 1)^2$ so f'' does not change sign. Since $a > 0$ f is of the type in Fig. 4.29 (a). Support by graphing in [-3,4] by [-2,25].

   (b) $f'' = 24x^2$ so f'' does not change sign. Since $a > 0$ f is of the type in Fig. 4.29 (a). Support by graphing in [-2,3] by [-2,15].

   (c) $f'' = 12(3x - 2)(x - 2)$ and so f'' changes sign twice. $f' = 12x(x - 2)^2$ has two roots. Since $a > 0$, the graph of f is of the type in Fig. 4.28 (c). Support by graphing in [-1,4] by [-15,10].

   (d) $f'' = 12(3x - 2)(x - 2)$ as in part (c) but now $f' = 6(x - 1)(x^2 - 7x + 1)$ which has three roots. Since $a > 0$ the graph of f is of the type in Fig. 4.28 (c). Support by graphing in [-1,5] by [-5,10].

Exploration 3

1. We get the root -0.3221853546 in just 3 repetitions.

2. Starting with x = 3 gives the same root after 7 repetitions.

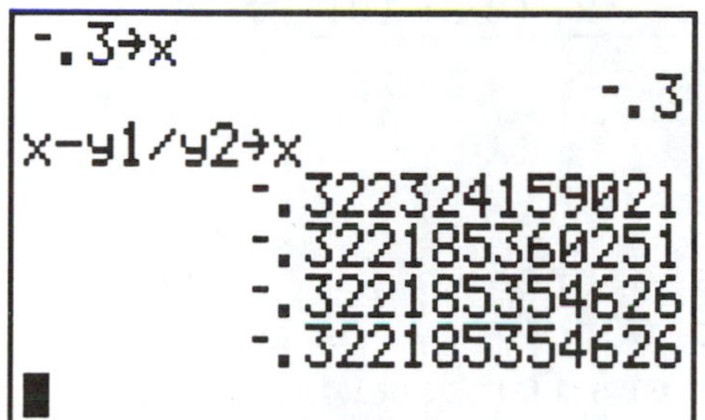

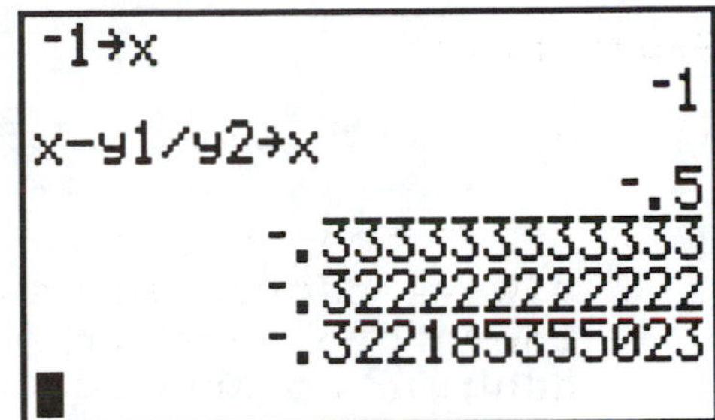

3. Grapher activity.

4. Grapher activity.

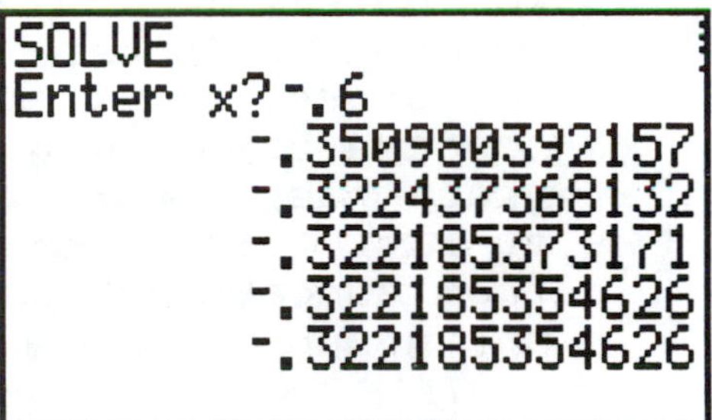

SECTION 4.4 RATIONAL FUNCTIONS AND ECONOMIC APPLICATIONS

Exploration 1

1. Domain: all reals except -3 and 1/2. Intercepts at (0,2) and (-2/3, 0). Vertical asymptotes at x = -3 and x = 1/2. Near x = -3 and x = 1/2 we have the following behaviors:

   $\lim_{x\to -3^-} f(x) = -\infty$, $\lim_{x\to (-1/2)^-} f(x) = -\infty$,

   $\lim_{x\to -3^+} f(x) = \infty$, and $\lim_{x\to (-1/2)^+} f(x) = \infty$.

   Since the degree of the denominator is greater than the degree of the numerator the end behavior asymptote of f is $y = 0$.

2. $$f' = \frac{(2x^2+5x-3)(3) - (3x+2)(4x+5)}{(2x^2+5x-3)^2} = \frac{-(6x^2 + 8x + 19)}{(2x^2 + 5x - 3)^2}.$$

$$f'' = \frac{(2x^2+5x-3)^2(-12x-8) - (-6x^2-8x-19)(2)(2x^2+5x-3)(4x+5)}{(2x^2+5x-3)^4}$$

$$= \frac{(2x^2+5x-3)(-12x-8) - (-6x^2-8x-19)(8x+10)}{(2x^2+5x-3)^3}$$

$$= \frac{24x^3 + 48x^2 + 228x + 214}{(2x^2 + 5x - 3)^3}.$$

3. f' is always negative so f is always decreasing. Therefore f decreases for $x < -3$ from close to 0 to $-\infty$. Between -3 and 1/2, f decreases from $+\infty$ to $-\infty$ and for $x > 1/2$ f decreases from $+\infty$ approaching 0.

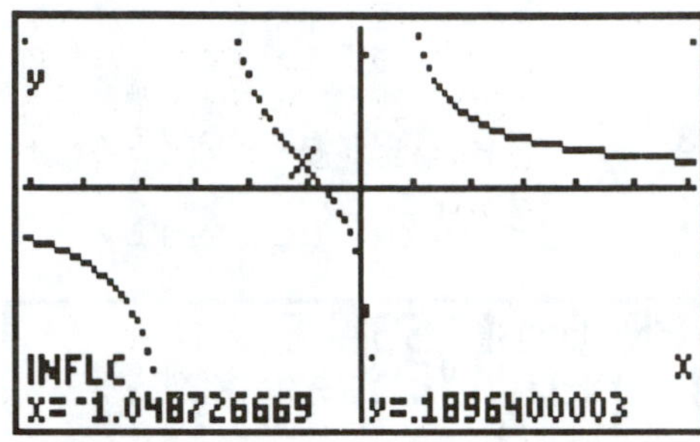

4. f'' is negative to the left of -3. On (-3, 1/2), f'' changes sign from positive to negative at $x = -1.047$. f'' is positive to the right of 1/2.

Therefore, the graph of f is concave down for $x < -3$. The graph has a point of inflection at $x = -1.047$ where it changes from concave up to concave down. To the right of $x = 1/2$ the graph is concave up.

Exploration 2

1. $A'(x) = 0$ when $x\,c'(x) - c(x) = 0$, or when $c'(x) = c(x)/x = A(x)$. Theorem: Average cost is minimized at a production level at which average cost equals marginal cost.

2. i) $A(x) = c'(x)$ when $c(x)/x = c'(x)$, or $x\,c'(x) = c(x)$. Solving $x(3x^2 - 12x + 15) = x^3 - 6x^2 + 15x + 5$, we find $2x^3 - 6x^2 - 5 = 0$, at approximately $x = 3.238$.

   ii) $A'(x) = 0$ when $2x - 6 - 5/x^2 = 0$, or when (approximately), $x = 3.238$.

3. $A''(x) = 2 + \dfrac{10}{x^3} > 0$ for $x > 0$ to confirm that the extreme value at $x = 3.238$ is a minimum.

SECTION 4.5 RADICAL AND TRANSCENDENTAL FUNCTIONS

Exploration 1

1. r is the radius of a circle whose circumference is $8\pi - x$. Find h by the Pythagorean Theorem and then substitute both expressions into the formula for the volume of a cone.

2. The domain of V includes all x for which the square root is nonnegative, that is $0 \le x \le 16\pi$.

3. The radius of the cone, r, must also be positive. This restricts x to the smaller interval $0 \le x \le 8\pi$.

4. Grapher activity. When we ZOOM-IN close enough to read x with an error at most 10^{-3}, that is, between two tick marks 10^{-3} apart horizontally, we are forced to set the vertical dimensions so that we can see consecutive tick marks 10^{-6} apart.

5. $r = (8\pi - x)/2\pi \Rightarrow \dfrac{dr}{dx} = -1/2\pi$ and $h^2 + r^2 = 4^2 \Rightarrow 2h\dfrac{dh}{dx} + 2r\dfrac{dr}{dx} = 0$. Combining gives $\dfrac{dh}{dx} = \dfrac{r}{2\pi h}$. Differentiate V and substitute: $V = \pi r^2 h/3 \Rightarrow \dfrac{dV}{dx} = \dfrac{\pi}{3}(2rh\dfrac{dr}{dx} + r^2\dfrac{dh}{dx}) = \dfrac{\pi}{3}(2rh(-\dfrac{1}{2\pi}) + r^2(\dfrac{r}{2\pi h})) = \dfrac{r}{3}(-h + \dfrac{r^2}{2h})$. The maximum value of V occurs when $\dfrac{dV}{dx} = 0$, that is, when $h = \dfrac{r^2}{2h}$ or $r^2 = 2h^2$ or $r^2 = 2(16 - r^2)$ so $r = 4\sqrt{\dfrac{2}{3}}$. Now solve for x: $\dfrac{8\pi - x}{2\pi} = 4\sqrt{\dfrac{2}{3}}$ and $x = 8\pi(1 - \sqrt{\dfrac{2}{3}}) \approx 4.612$.

Exploration 2

1. 2π. Since $f(x + 2\pi) = f(x)$ 2π is a multiple of the period.

2. GRAPH, for example, over the interval $[-\pi/2, 3\pi/2]$. The graph has asymptotes at $-\pi/2 + n\pi$, n and integer; it increases from $-\infty$, rises to a local maximum, falls slightly to a local minimum, then continues rising to infinity at the vertical asymptote. This increasing-decreasing-increasing pattern repeats in the next interval but the local extreme values are now both positive.

At the same time the graph is concave down and then changes concavity between the first pair of extreme values. This concavity pattern repeats in each succeeding interval of length π. The local extreme values of f and the points of inflection may quite likely be hidden from view in your viewing window -- examples of hidden behavior.

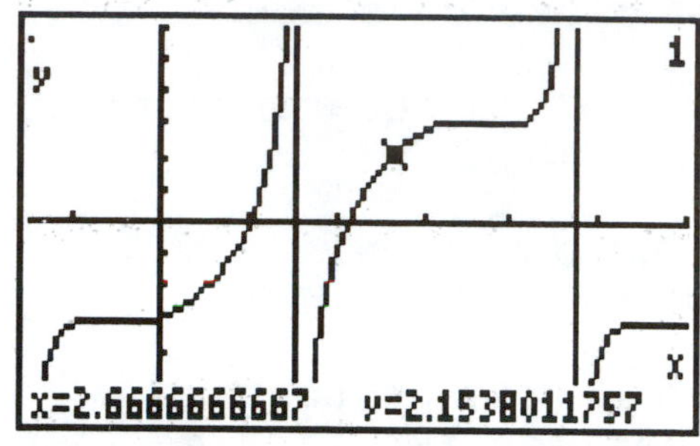

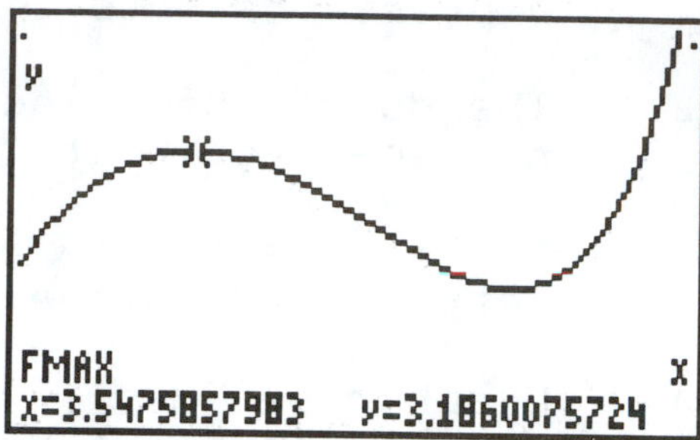

The behavior of f is supported by the derivative graphs. NDER has two roots between each asymptote supporting the pair of extreme values; the second derivative has a single root between the roots of the first derivative. This supports the point of inflection and the single change of concavity within each interval.

3. $f'(x) = \sec^2 x + 3\sin x$. $f''(x) = 2\sec x(\sec x \tan x) - 3\cos x = 2\sec^2 x \tan x - 3\cos x$. A graph of f' shows four sign changes in $[0, 2\pi]$ supporting the increasing-decreasing-increasing pattern observed in f. A graph of f'' shows three sign changes in the same interval supporting the pattern of concavity changes observed in f.

Exploration 3

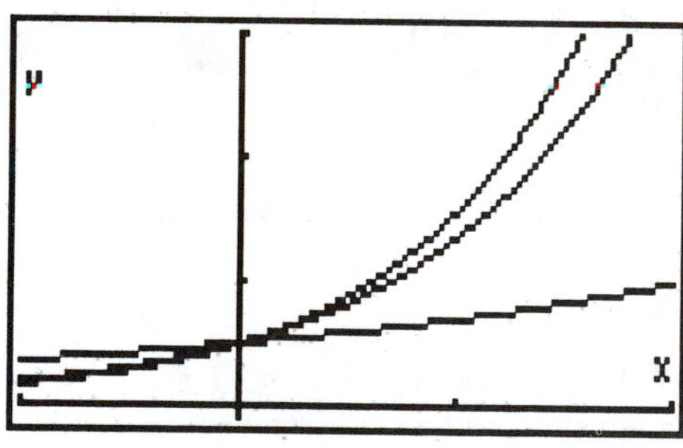

1. It is reasonable to predict that $y = a^x$ for $a = e$, π, and $\sqrt{2}$, will look like $y = 2^x$.

2. When $a = e$, $y = a^x = y'$.

3. (a) A view of the graph in the standard window suggests that $[-2,10]$ by $[-2, 0.5]$ may give a better view. This graph is complete as confirmed by $f' = (1 - x)e^{-x}$ which is zero only at $x = 1$, hence, the only extreme value, and $f'' = (x - 2)e^{x}$ which is zero only at $x = 2$, confirming a single point of inflection on a curve that is first concave down and then concave up.

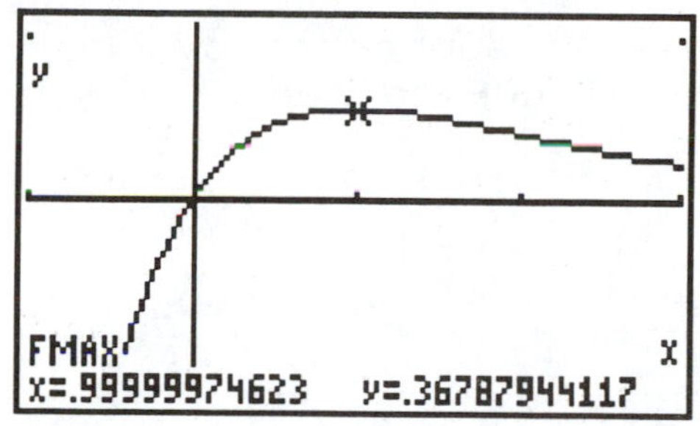

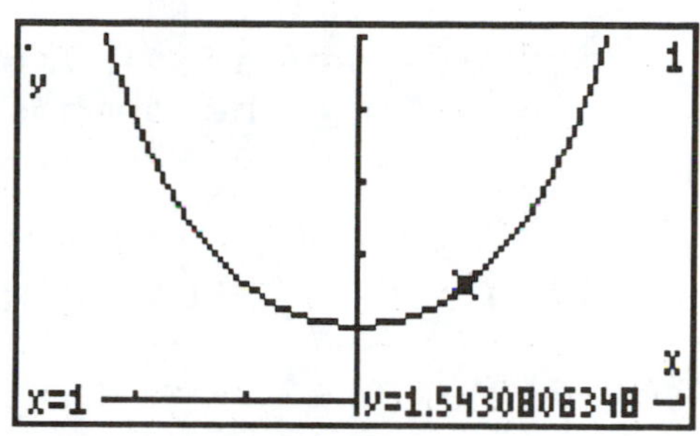

(b) A graph in the standard window shows a minimum at (0, 1), symmetry across the y-axis, and concave up everywhere.

$f' = (e^x - e^{-x})/2$ and $f'' = (e^x + e^{-x})/2$. $f' = 0$ when $e^x = e^{-x}$ or $x = 0$ confirming the extreme value and $f'' > 0$ confirming the concavity. Since $f(-x) = f(x)$ the graph is symmetric around the y-axis, and complete.

Exploration 4

1. Note: to graph $\log_a x$ we use the change of base formula $\log_a x = (\ln x)/(\ln a)$.

   a) GRAPH $y_1 = (\ln x)/(\ln 2)$. The graph appears to be concave down and increasing for all $x > 0$. This is supported by the graphs y_2 = NDER(y_1) which seems to be always positive and y_3 = NDER2(y_1) which appears to be negative.

   b) The graph of $y_1 = (\ln x)/(\ln 0.5)$ appears to be concave up and decreasing for all $x > 0$. This is supported by the graphs of y_2 = NDER(y_1) which appears to be negative y_3 = NDER2(y_1) which appears to positive. Note that $\ln 0.5 = \ln 1/2 = -\ln 2$ so that $\log_{0.5} x = -\log_2 x$.

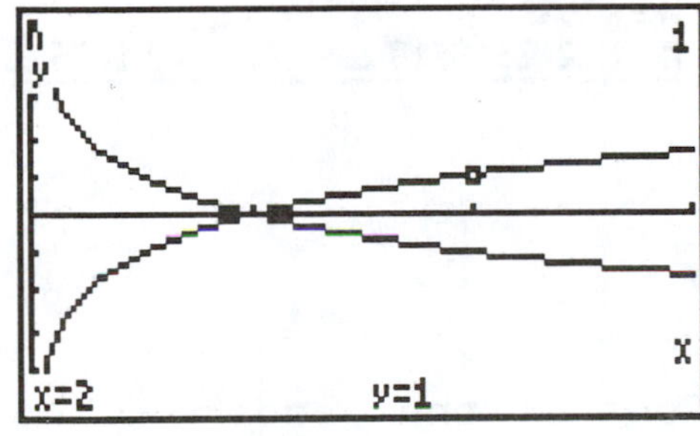

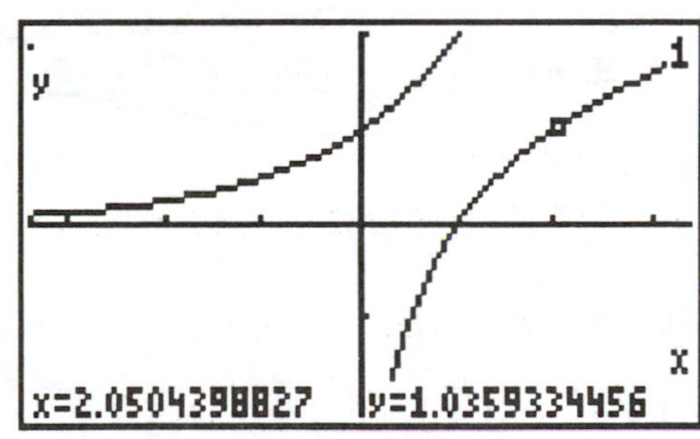

2. GRAPH $y_1 = 2^x$ and $y_2 = \log_2 x$ in a square viewing window. The graphs appear to be symmetrical across $y = x$, suggesting

that they are inverse functions. This is supported by graphing the composition $y_3 = 2y_1(y_2)$. TRACE to see that y_3 is the line $y = x$.

SECTION 4.6 RELATED RATES OF CHANGE

Exploration 1

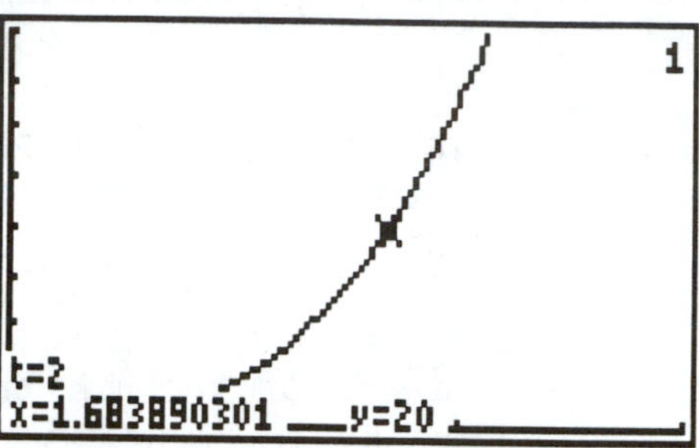

1. $r = \sqrt[3]{\dfrac{15t}{2\pi}}$.

2. As r increases at a constant rate (r is on the horizontal axis), V increases at an increasing rate (V-values are on the vertical axis). (You should understand why this makes sense.)

3. The first is a graph of volume against time. This shows that the volume increases at a constant rate of 10 cm^3/s.

   The second is a graph of the bubble's radius against time. this shows the radius increasing at a decreasing rate.

   The third is a graph of the rate of change of r against time (the same as figure 4.57).

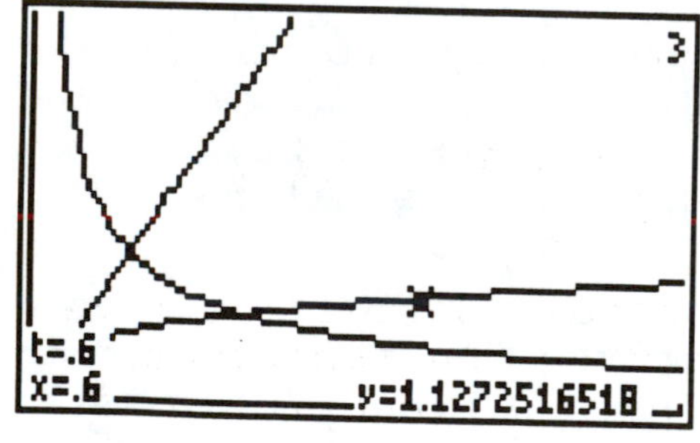

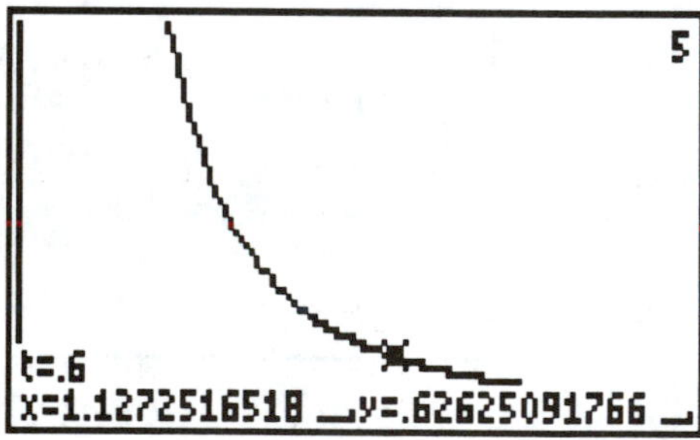

4. One way: $x_5 = x_1$; $y_5 = \text{NDER}(x_1)$.

Exploration 2

1. $x_1(t) = 3t$, $y_1(t) = 0$ simulates the foot of the ladder moving to the right at 3 ft/sec. $x_2(t) = 0$, $y_2(t) = \sqrt{(13^2 - (3t)^2)}$ simulates the motion of the top of the ladder. Note that $y_2(t)$ is computed using the Pythagorean Theorem in a right triangle whose hypotenuse is 13.

2. The ladder slides from vertical to horizontal in t seconds. It makes sense that t begins at 0 and ends when $x = 13$. This greatest x-value, 13, occurs at $t = 13/3$. Therefore, t is in [0, 13/3], x is in [0, 13] and y is in [0, 13].

   However, in order to make the motion visible we do not want the points $(x_1(t), y_1(t))$ and $(x_2(t), y_2(t))$ to lie on the axes. One way is to shift the motion to the right and up 1 unit and GRAPH instead:

   $x_1(t) = 3t + 1$, $y_1(t) = 1$; t in [0, 13/3]
   $x_2(t) = 1$, $y_2(t) = \sqrt{(13^2 - (3t)^2)} + 1$; t-step = 0.111

   and x, y in [0, 14] by [0, 14].

3. Dot format with t-step = 0.5 works well. Observe the top of the ladder falling at an accelerating rate while the foot of the ladder moves with a constant rate.

4. $\dfrac{dy}{dt} = \dfrac{-9t}{\sqrt{169 - 9t^2}}$. So $y_2'(0.5) = -0.3485$ ft/s;
   $y_2'(1) = -0.7115$ ft/s; $y_2'(1.5) = -1.1069$ ft/s;
   $y_2'(2) = -1.5608$ ft/s.

SECTION 4.7 ANTIDERIVATIVES, INITIAL VALUE PROBLEMS, AND MATHEMATICAL MODELING

Exploration 1

1. The graphs of NDER y_1, NDER y_2 and NDER y_3 are identical.

2. The graphs of NDER y_1, NDER y_2 and NDER y_3 are again identical.

3. Conjecture: Functions that have the same derivative have graphs that differ by a vertical shift. The functions differ by a constant. That is, if $f' = g'$ then $f - g$ is constant.

Exploration 2

1. Recall that $x(t) = \cos t$, $y(t) = \sin t$ graphs as the unit circle for [tMin, tMax] = [0, 2π]. The graph of $(x_1(t), y_1(t))$, can be interpreted as a unit circle that has been stretched by a factor of 20 (so that its radius is 20) and shifted vertically 20 units (so that its center is at (0, 20)). The factor of π/6 changes the period from 2π to

12 so these equations can represent Renee's motion on the larger wheel if t is interpreted as time.

Similarly, $(x_2(t), y_2(t))$ graphs as a circle with radius 15 centered at (15, 15). The period of these equations is $t = 8$ so they can represent motion around a wheel which completes a revolution every 8 seconds.

To make the graph more realistic use a square viewing window.

2. TRACE and move between the graphs to see corresponding locations on the two circles. The minimum separation seems to occur at, approximately $t = 21.8$ and the maximum at $t = 16$. Compute the distances from the TRACE coordinates to get the approximate minimum distance 4.2 and an approximate maximum distance of 45.8.

   We thought these equations too complicated to confirm these values through algebraic means. How did you do?

3. The graph of $(x_3(t), y_3(t))$ shows that the distance between Sherrie and Renee varies periodically, repeating every 24 s. The high point of the graph is 45.82 and represents the maximum distance. The low point of the graph is 4.11 and represents the minimum distance. Note that these values support those found in part 2.

4. GRAPH $x_4(t) = t$, $y_4(t) = \text{NDER}(y_3(t))$ to show the rate of change of $y_3(t)$. $y_3(t)$ increases fastest where $y_4(t)$ is greatest, at $t = 3.79$. At this point Renee has just cleared the top of her Ferris wheel while Sherrie is just midway between the top and bottom of her wheel on the way down. $y_3(t)$ decreases fastest at $y_4(t)$'s minimum, $t = 19.58$. At this time Renee is approaching the bottom of her wheel and Sherrie is again approaching the midway point of her wheel on the way down.

SECTION 5.1 CALCULUS AND AREA

Exploration 1

Grapher activity.

Exploration 2

1. $\sum_{k=1}^{3} \sin\left(\frac{k\pi}{3}\right) = \sin\frac{\pi}{3} + \sin\frac{2\pi}{3} + \sin\frac{3\pi}{3} = \frac{\sqrt{3}}{2} + \frac{-\sqrt{3}}{2} + 0 = 0.$

2. a) $\sum_{k=0}^{2} \frac{1}{2^k} = \frac{1}{1} + \frac{1}{2} + \frac{1}{4} = 7/4.$

b) $\sum_{k=-3}^{-1} (k + 1) = (-2) + (-1) + 0 = -3.$

SECTION 5.2 DEFINITE INTEGRALS

Exploration 1

1. a) On [0, 1] $\Delta x = \frac{1}{n}$, $x_k = \frac{k}{n}$ so

$$\text{RRAM}_n = \sum_{k=1}^{n} f(x_k)\Delta x = \sum_{k=1}^{n} \left(\frac{k}{n}\right)^3 \frac{1}{n} = \frac{1}{n^4}\sum_{k=1}^{n} k^3 = \frac{(n+1)^2}{4n^2}.$$

b) On [0, 5] $\Delta x = \frac{5}{n}$, $x_k = \frac{5k}{n}$ so

$$\text{RRAM}_n = \sum_{k=1}^{n} f(x_k)\Delta x = \sum_{k=1}^{n} \left(\frac{5k}{n}\right)^3 \frac{5}{n} = \frac{5^4}{n^4}\sum_{k=1}^{n} k^3 = \frac{5^4(n+1)^2}{4n^2}.$$

c) On [0, x] $\Delta x = \frac{x}{n}$, $x_k = \frac{kx}{n}$ so

$$\text{RRAM}_n = \sum_{k=1}^{n} f(x_k)\Delta x = \sum_{k=1}^{n} \left(\frac{kx}{n}\right)^3 \frac{x}{n} = \frac{x^4}{n^4}\sum_{k=1}^{n} k^3 = \frac{x^4(n+1)^2}{4n^2}.$$

2. $A_0^x t^3 = \lim_{n\to\infty} RRAM_n = \lim_{n\to\infty} \frac{x^4(n+1)^2}{4n^2} = \frac{x^4}{4}.$

Exploration 2

1. 0, because $A_\pi^{2\pi} \sin x = -A_0^\pi \sin x$.

2. 1, because $y = \sin x$ is symmetrical about $x = \pi/2$.

3. $2 + 2\pi$. The vertical shift creates a rectangle underneath the sine curve. The rectangle's height is 2, its width is π, the length of the interval, so its area is $2\cdot\pi$.

4. 4. The vertical stretch doubles the area of each rectangle making up a RAM approximation.

5. 2, because both the graph and the interval of integration are shifted two units to the right.

6. 4. The horizontal stretch doubles the area of each rectangle in a RAM approximation.

7. Some possible conjectures:

(1) $\int_0^b f(x)\,dx = 0$ when the graph encloses areas above and below the x-axis that are equal.

(2) $\int_0^b f(x)\,dx = \frac{1}{2}\int_0^{b/2} f(x)\,dx$ when the graph is symmetrical across the line $x = b/2$.

(3) A vertical shift of h units adds hb to the integral: $\int_0^b f(x) + h\,dx = \int_0^b f(x)\,dx + hb$.

(4) A vertical stretch by k multiplies the integral by the same factor: $\int_0^b k\,f(x)\,dx = k\int_0^b f(x)\,dx$.

(5) If a function is shifted h units horizontally then its integral over $[h, b + h]$ will be the same as the original integral: $\int_0^b f(x + h)\,dx = \int_h^{b+h} f(x)\,dx$.

(6) A horizontal stretch by k of both the function and its interval of integration multiplies the integral by k: $\int_0^{kb} f(x/k)\,dx = k\int_0^b f(x)\,dx$. $k > 1$.

Exploration 3

1. $f(x) = \dfrac{x^2 - 4}{x - 2} = \dfrac{(x - 2)(x + 2)}{x - 2} = x + 2$ when $x \neq 2$, so f has a removable discontinuity at $x = 2$.

2. $\int_0^3 \frac{x^2 - 4}{x - 2}\,dx$ equals the area of a trapezoid with height 3 and bases 2 and 5. Therefore, $\int_0^3 \frac{x^2 - 4}{x - 2}\,dx = \frac{1}{2}(3)(2 + 5) = 10.5$.

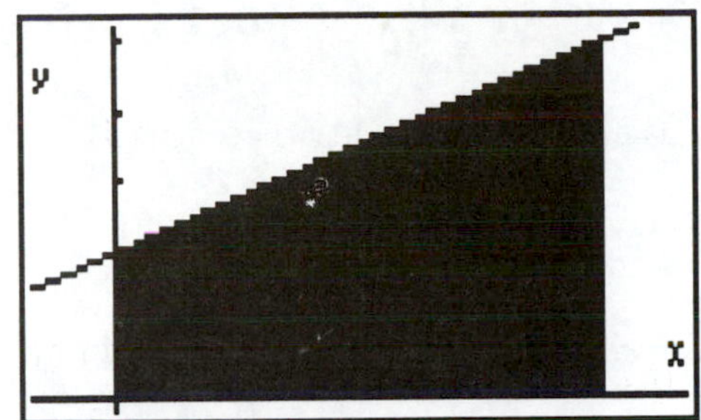

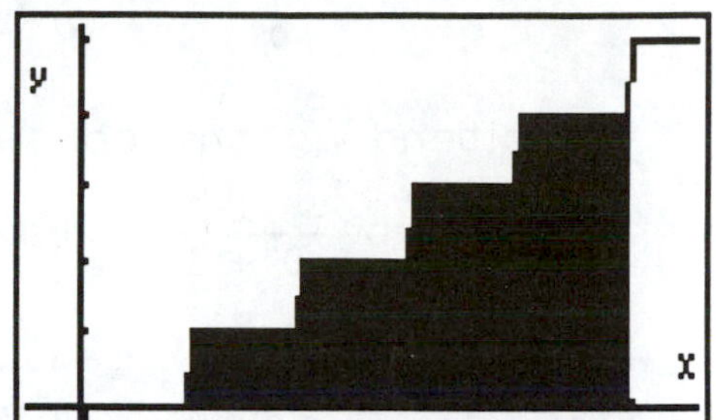

3. $\int_0^5 [x]\,dx$ is the sum of the areas of 5 rectangles each with base 1 and heights of 0, 1, 2, 3, and 4. The integral is, therefore, 0 + 1 + 2 + 3 + 4 = 10.

SECTION 5.3 DEFINITE INTEGRALS AND ANTIDERIVATIVES

Exploration 1

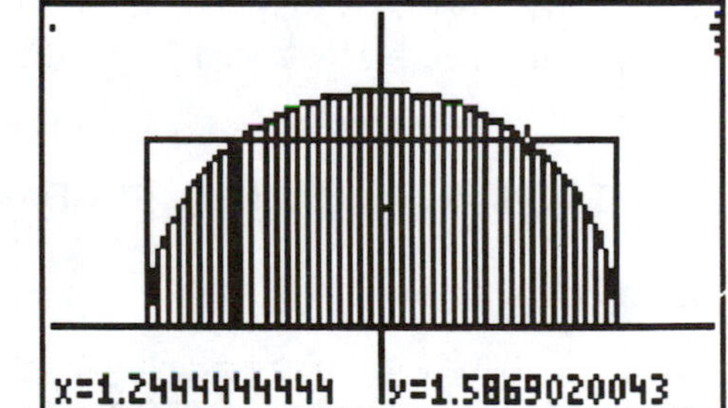

1. $f_{\text{avr}} = \frac{1}{4}\int_{-2}^{2} \sqrt{4 - x^2}\,dx = \pi/2$.

2. Grapher activity. For a parametric view, GRAPH
$x_1(t) = t$, $y_1(t) = \sqrt{(4 - t^2)}$
$x_2(t) = t$, $y_2(t) = \pi/2$
$x_3(t) = -2$, $y_3(t) = t$
$x_4(t) = 2$, $y_4(t) = t$
in a square viewing window with [xMin, xMax] or [yMin, yMax] = [-2, 2]. Regions of equal area are obvious because of symmetry.

3. From the graphs $c = 1.238$. Solving algebraically, $c = \frac{1}{2}\sqrt{16 - \pi^2}$.

Exploration 2

1. The graph of y_1 forms a triangle with base 1, height 1, and therefore area = (1/2)(1)(1) = 1/2. Conjecture: $A(y_2) = 1/3$, $A(y_3) = 1/4$, ..., $A(y_i) = 1/(i + 1)$. In other words, $\mathbf{A}_0^1 x^n = \dfrac{1}{n+1}$. Confirm: $\mathbf{A}_0^1 x^n = \left.\dfrac{x^{n+1}}{n+1}\right|_0^1 = \dfrac{1}{n+1}$.

2. Extending the conjecture from part 1. For $i = 1$, $\mathbf{A}_0^x x^0 = \dfrac{1}{0 + 1} = 1$, which is easily proved true. For $i = -1$, $\mathbf{A}_0^x x^{-1} = \dfrac{1}{-1 + 1}$, which is obviously meaningless with a zero in the denominator. For $i = -2$, $\mathbf{A}_0^x x^{-2} = \dfrac{1}{-2 + 1} =$ -1, which is obviously not true because the area is positive. (Also note that the integrand is undefined at 0.) Similarly, for $i = -3, -4, -5, \ldots$, the conjecture is not true.

   For rational exponents, the conjecture would be $\mathbf{A}_0^x x^{p/q} = \dfrac{1}{p/q + 1}$. This can be shown true for $p/q \geq 0$ and for $-1 < p/q < 0$ as it was done for nonnegative integers above. For $p/q \leq -1$, the conjecture is not true with an argument similar to that used for negative integers above.

SECTION 5.4 THE FUNDAMENTAL THEOREM OF CALCULUS

Exploration 1

The graph of NINT(cos t, 0, x) seems to be a sinusoid with period 2π and amplitude 1. Conjecture: $\int_0^x \cos t\, dt = \sin x + C$.

Exploration 2

1. Predict $y_2 = y_3$

2. TRACE shows that $y_2 - y_3 = 0.0000003$, almost zero.

Exploration 3

1. Grapher activity.

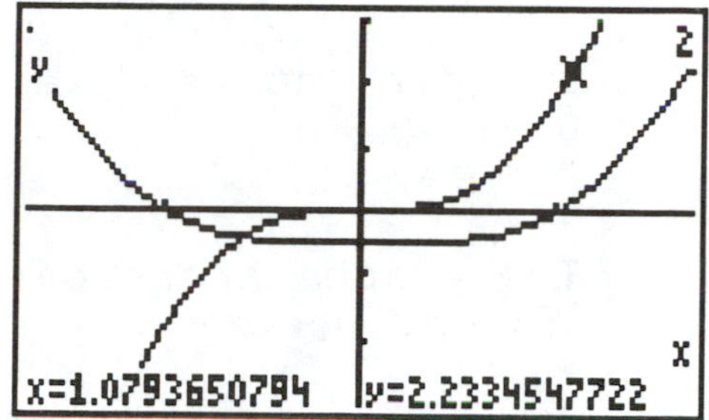

2. Let $u = x^2$. Then $\dfrac{dy}{du} = \sqrt{u \sin u}$ and $\dfrac{du}{dx} = 2x$. Therefore, $\dfrac{dy}{dx} = \dfrac{dy}{du}\cdot\dfrac{du}{dx} = 2x\sqrt{x^2 \sin x}$.

3. The graphs support the extended theorem...they are identical.

Exploration 4

1. $F' = e^{-x^2}$ by the Fundamental Theorem so F always increases since F' is always positive. F has no extreme values because F' is never zero, F' has no singular points, and F' has no endpoints as F is defined for all real numbers.

2. We can determine that F is concave up for negative x and concave down for positive x by observing the signs of F''. $F'' = -2xe^{-x^2}$, so $F'' > 0$ when $x < 0$ and $F'' < 0$ when $x > 0$. Since F changes concavity at $x = 0$, this is a point of inflection. We can confirm this result by interpreting the values of F as areas under f. As x increases towards 0, $f(x)$ increases towards its maximum; F increases as we include additional area and, as x nears zero, the area increases as rapidly as possible. For positive x, F continues to increase but at a diminishing rate as the smaller values of $f(x)$ provide smaller and smaller additional areas.

3. As x increases without bound, F' approaches 0; that is, the graph of F seems to approach a horizontal tangent.

4. We know that $F(0) = 0$. Since F is always increasing this is the only intercept.

Exploration 5

1. The antiderivative $F(x) = \int_0^x f(t)\,dt = 0$ at $x = 0$. Any other antiderivative is a vertical shift of F and hence cannot be 0 at $x = 0$.

2. The graphs are separated by $C = \int_a^b f(t)\,dt$. Proof:

$$\int_a^x f(t)\,dt - \int_b^x f(t)\,dt = \int_a^b f(t)\,dt.$$

3. There is a relationship between a and C but it is not one-to-one it is one-to-many. That is, each value of a determines a unique value of C but each C may be obtained from many different a's.

   (i) $\int_a^x f(t)\,dt = \int_0^x f(t)\,dt + \int_a^0 f(t)\,dt$ so $C = \int_a^0 f(t)\,dt$. This establishes that each a determines a C.

   (ii) Let $f(t) = \sin t$. If $a = 2n\pi$ then $\int_{2n\pi}^x \sin t\,dt = -\cos x + 1$. In this case $C = 1$ is obtained from infinitely many different values of a.

Exploration 6

1. $A = \int_{-2}^0 x^3 - 4x\,dx - \int_0^2 x^3 - 4x\,dx = \frac{x^4}{4} - \frac{4x^2}{2}\Big|_0^{-2} - \frac{x^4}{4} - \frac{4x^2}{2}\Big|_2^C = 8.$

2. (b). Since f is an odd function the area below the x-axis equals the area above and so we can just double the positive area.

   (c) This is the same as Example 9, the area is the area under $|f(x)|$.

   (e) Since the area from 0 to 2 is negative, subtracting it from the positive area is the same as adding its absolute value.

(g) Whether NINT computes a positive or a negative area the sum of its absolute values will give the total area.

The other four choices do not compute the area. (a) fails because it finds the algebraic sum of the two areas over [-2, 2] which is zero. (d) is the same as (a) split into two integrals. (f) fails because it first sums the two signed areas and then takes absolute values. Since the two areas have opposite signs this is the absolute value of a difference not a sum.

3. Note: Other answers are possible.

(a) The area included between y_1 and y_2 is between $x = -2$ and $x = 2$. So $A = \int_{-2}^{2} (4 - x^2) - (x^2 - 4)\,dx$.

(b) Since the two graphs are symmetrical across the x-axis the desired area is twice the area between the x-axis and either y_1 or y_2: $A = 2\int_{-2}^{2} 4 - x^2 dx = 2\int_{-2}^{2} x^2 - 4dx$.

(c) Since y_1 and y_2 are symmetrical across the y-axis, the desired area is twice the area between y_1 and y_2 between $x = 0$ and $x = 2$: $A = 2\int_{0}^{2} (4 - x^2) - (x^2 - 4)\,dx$.

(d) Combining the two types of symmetry mentioned in (b) and (c), the desired area is four times the area between the x-axis and y_1 in the first quadrant:

$$A = 4\int_{0}^{2} 4 - x^2 dx.$$

SECTION 5.5 INDEFINITE INTEGRALS

Exploration 1

1. As you TRACE, move between the graphs. If both F and NINT are antiderivatives they should differ by at most a constant. The TRACE should reveal a constant different in y-values as you move along the curves.

2. This should graph as a horizontal line.

3. GRAPH both and TRACE. The graphs are seen to overlap.

4. The two functions appear to be quite different. Graphing their difference, however, reveals that they differ by the constant 8.6666.

5. With $y_1 = x^2 - 2x + 5$, execute SLOPEFLD. The resulting family of tangent lines closely approximates the antiderivatives $x^3/3 - x^2 + 5x + C$.

Exploration 2

1. GRAPH $x \cos x$ - NDER y_k for $k = 1, 2$, and 3. When $k = 2$ the graph is approximately $y = 0$ showing that $x \sin x - \cos x$ is the antiderivative of $x \cos x$.

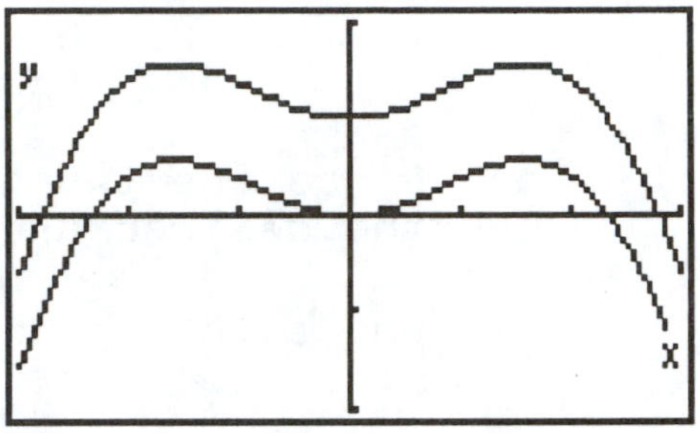

2. GRAPH NINT($x \cos x$) - y_k for $k = 1, 2$, and 3. When $k = 2$ the graph is horizontal showing that $x \sin x + \cos x$ differs from an antiderivative of $x \cos x$ by a constant and is, therefore, an antiderivative too.

3. $[x \sin x + \cos x]' = [x \sin x]' + [\cos x]'$
$= \sin x + x \cos x - \sin x = x \cos x.$

Exploration 3

1. ZOOM-IN gives $x = 1.108$ to three decimal places.

2. The two graphs are nearly identical. Their common y-intercept is -1 as required by the problem's initial value.

SECTION 5.6 INTEGRATION BY SUBSTITUTION -- RUNNING THE CHAIN RULE BACKWARD

Exploration 1

1. $\int \sqrt{1 + x^2}\, 2x\,dx = \int \sqrt{u}\,du = \frac{2}{3}u^{3/2} + C = \frac{2}{3}(1 + x^2)^{3/2} + C.$ The graphs differ by 0.66666 when $x > 0$.

2. $du = 3x^2dx$ so $du/3 = x^2dx$. Therefore, $\int x^2 \sin x^3 dx =$

$\frac{1}{3}\int \sin u\, du = -\frac{1}{3}\cos u = -\frac{1}{3}\cos x^3 + C$. TRACE shows that the two graphs are nearly identical.

3. Grapher activity.

Exploration 2

1. Since the antiderivative is $-\cos x + 1$, a viewing window which includes $[-2\pi, 2\pi]$ by $[0, 2]$ should show the graph $y = -\cos x$ shifted vertically 1 unit.

2. $a = \pi/2$.

Exploration 3

1. If $u = z^2 + 1$ then $du = 2z\, dz$ and
$$\int \frac{2z dz}{\sqrt[3]{z^2 + 1}} = \int \frac{du}{\sqrt[3]{u}} = \frac{3}{2}u^{3/2} + C = \frac{3}{2}(x^2 + 1)^{2/3} + C.$$

2. If $u = \sqrt[3]{z^2 + 1}$ then $du = \dfrac{2z dz}{3(\sqrt[3]{z^2 + 1})^2}$ so
$$\int \frac{2z dz}{\sqrt[3]{z^2 + 1}} = \int 3u\, du = \frac{3}{2}u^2 = \frac{3}{2}(z^2 + 1)^{2/3} + C.$$

Exploration 4

1. $\int_0^2 \sqrt{u}\, du = \frac{2}{3}u^{3/2}\Big|_0^2 = \frac{4}{3}\sqrt{2}$.

2. $\int \sqrt{u}\, du = \frac{2}{3}u^{3/2} = \frac{2}{3}(x^3 + 1)^{3/2}$. Therefore,
$$\frac{2}{3}(x^3 + 1)^{3/2}\Big|_{-1}^{+1} = \frac{4}{3}\sqrt{2}.$$

Exploration 5

1. Numerical activity.

2. Numerical activity.

3. Numerical activity.

SECTION 5.7 NUMERICAL INTEGRATION: THE TRAPEZOIDAL RULE AND SIMPSON'S METHOD

Exploration 1

1. $|E_T| \le \frac{(1-0)}{12}\left(\frac{1}{10}\right)^2 M = \frac{1}{1200}M.$

2. a) $f'' = -x \sin x + 2 \cos x$. Since $x \le 1$ in $[0, 1]$: $|f''| \le |x \sin x| + |\cos x| \le 1 + 2 = 3$. Therefore, we can take $M = 3$.

   b) TRACE shows that $|f''| \le 2.5$ on $[0, 1]$. For support, GRAPH $y_1 = \text{abs}(-x \sin x + 2\cos x)$ and $y_2 = 2.5$ on $[0, 1]$ by $[-1, 4]$.

3. $E_{T_1} \le \frac{3}{1200} = \frac{1}{400}$ while $E_{T_2} \le \frac{2.5}{1200} = \frac{1}{480}$.

4. $E_{T_1} \le \frac{1}{12}\frac{1}{10000}(3) = 0.000025$ and $E_{T_2} \le \frac{1}{12}\frac{1}{10000}(2.5) \approx 0.0000208$. That is $|E_{T_1}| \ge |E_{T_1}|$. Smaller values of h diminish the error estimate as do smaller upper bounds M. However, since h is squared and M is not, reducing h diminishes the error faster than reducing M so both estimates of E_T with $h = 1/100$ have greater accuracy than both estimates with $h = 1/10$.

Exploration 2

1. $f'(x) = 2x \cos(x^2)$ and $f''(x) = 2 \cos(x^2) - 4x^2\sin(x^2)$.

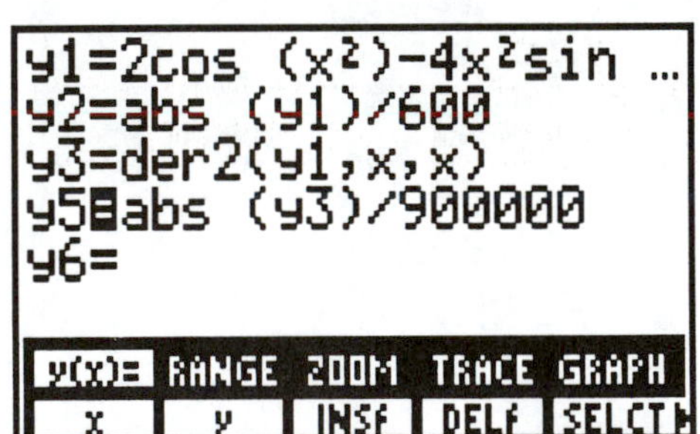

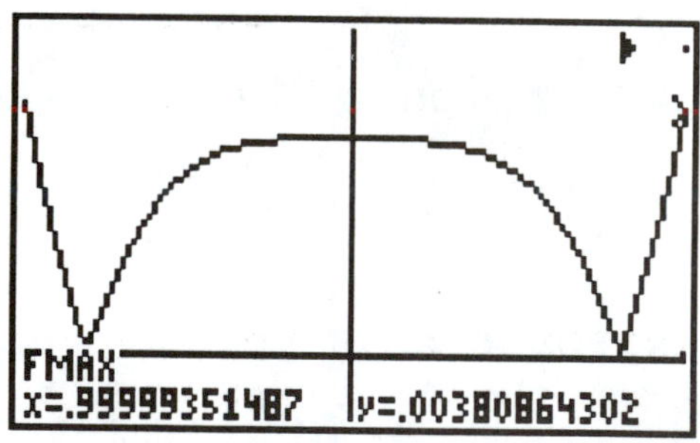

2. The maximum of y_2 on $[-1, 1]$ is 0.0038.

3. The maximum of y_5 on $[-1, 1]$ is 0.0000316.

4. The maximum error with a Simpson approximation is about 1% of the maximum error with a Trapezoidal approximation.

Exploration 3

If S = seq(2/(1 + K*H, K, 0, N, 1) = $\sum_{K=0}^{N} \frac{2}{1 + KH}$. For $y = 1/x$, if $x_k = \frac{1}{1 + KH}$ then $S = 2(y_0 + y_1 + y_2 + .. + y_N)$. Therefore, since $y_0 = 1$ and $y_1 = 0.5$

$$\frac{H}{2}(y_0 + 2y_1 + 2y_2 + \ldots + 2y_{N-1} + y_N) = \frac{H}{2}(S - y_0 - y_1) = \frac{H}{2}(S - 1 - 0.5).$$

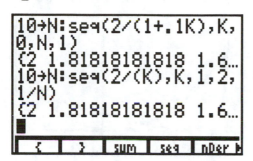

```
10→N:seq(2/(1+.1K),K,
0,N,1)
{2 1.81818181818 1.6…
10→N:seq(2/(K),K,1,2,
1/N)
{2 1.81818181818 1.6…

 {  }  sum  seq  nDer
```

```
For(N,20,100,40):1/N→
H:Disp [N,H/2*(-1-.5+
sum seq(2/(1+H*K),K,0
,N,1))]:End:Pause
     [20 .693303381793]
     [60 .693164541068]
    [100 .693153430482]
```

SECTION 6.1 AREAS BETWEEN CURVES

Exploration 1

1. 4 sin(x/2) ≥ x.

2. The region extends over [0, 3.79].

3. NINT(4 sin(x/2) − x, 0, 3.79) = 3.366.

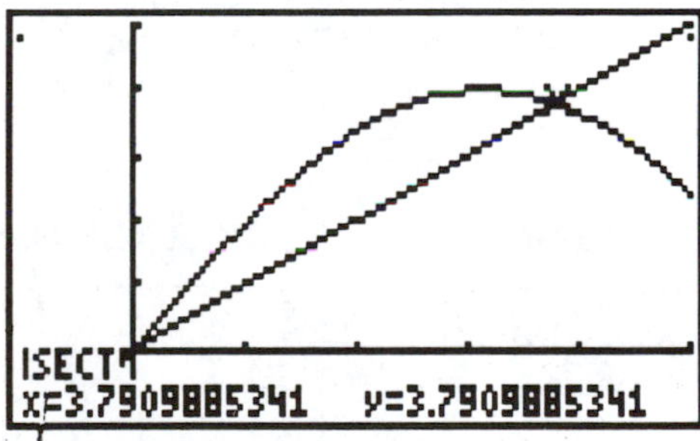

SECTION 6.2 VOLUMES OF SOLIDS OF REVOLUTION -- DISKS AND WASHERS

Exploration 1

1. Grapher activity.

2. $x^2 = 2x$ so $x(x - 2) = 0$. Therefore, $0 \le x \le 2$ and $0 \le y \le 4$.

3. $R(y)$ is the x-coordinate of the parabola $y = x^2$ so $x = \sqrt{y}$. $r(y)$ is the x-coordinate of the line $y = 2x$ so $x = y/2$.

4. $V = \pi\int_0^4 (\sqrt{y})^2 - (\frac{y}{2})^2 dy = 8\pi/3$. This is confirmed by an NINT computation which gives 8.37758041.

Exploration 2

1. $V = \pi\int_{-2}^{3} (x + \frac{6}{\sqrt{x+6}})^2 - (\frac{2x^2+2x+1}{5})^2 dx$ which is complex enough to make technology appealing. NINT gives $30.527\pi = 95.903$.

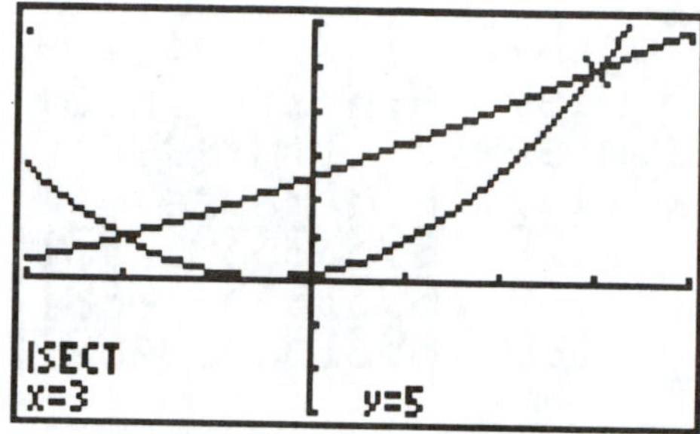

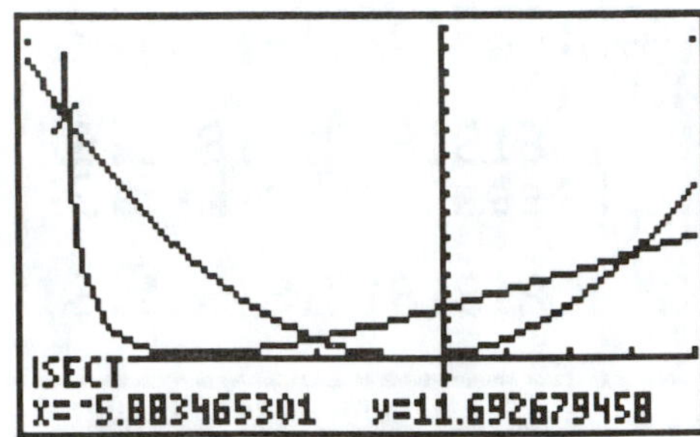

2. ZOOM-OUT reveals a second region bounded by the graphs on [-5.88, -2]. NINT gives the volume of this portion of the solid as $131.214\pi = 412.221$. The total volume is, therefore, $161.741\pi = 508.124$.

SECTION 6.3 CYLINDRICAL SHELLS -- AN ALTERNATIVE TO WASHERS

Exploration 1

1. The limits of integration are $x = 0$ to $x = 1$. The radius of the shell is $2 - x$ and the rectangle's height is $1 - x^2$.

   Therefore, $V = 2\pi\int_0^1 (2 - x)(1 - x^2)\, dx = 13\pi/6$ confirmed to six decimal places by NINT = 6.806784.

2. If the region is shifted left 2 units, the new limits of integration are $x = -2$ to $x = -1$. The radius of the shell is $-x$ and its height is $1 - (x+2)^2$. Therefore,

   $V = 2\pi\int_{-2}^{-1} (-x)(1 - (x+2)^2 dx = 13\pi/6$ which is again confirmed by NINT.

Exploration 2

1. The solid is composed of three parts, swept out by rotating three areas:

   - under the parabola from 0 to 3.5

   - under the line from 3.5 to 4

   - between the line and the parabola from 4 to 4.5

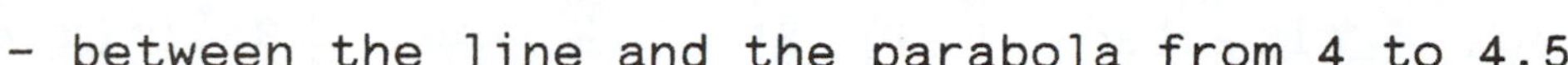

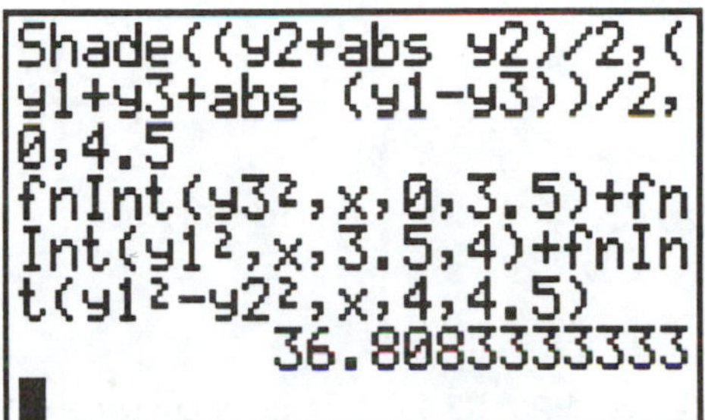
Shade((y2+abs y2)/2,(
y1+y3+abs (y1-y3))/2,
0,4.5
fnInt(y3²,x,0,3.5)+fn
Int(y1²,x,3.5,4)+fnIn
t(y1²-y2²,x,4,4.5)
36.8083333333

```
fnInt(((y1*y2)≥0)*((a
bs y1+abs y2+abs (abs
 y1-abs y2))²-(abs y1
+abs y2-abs (abs y1-a
bs y2))²)/4+((y1*y2)<
0)*(abs y1+abs y2+abs
 (abs y1-abs y2))²/4,
x,A,B)
```

The first two volumes are generated by disks, the last by washers. The total volume is

$$V = \pi\int_0^{3.5} (4x - x^2)^2\,dx + \pi\int_{3.5}^{4} \left(\frac{x}{2}\right)^2 dx + \pi\int_4^{4.5} \left(\frac{x}{2}\right)^2 - (x^2-4x)^2\,dx.$$

An NINT computation gives $33.585\pi + 1.760\pi + 1.463\pi = 36.808\pi \approx 115.637$.

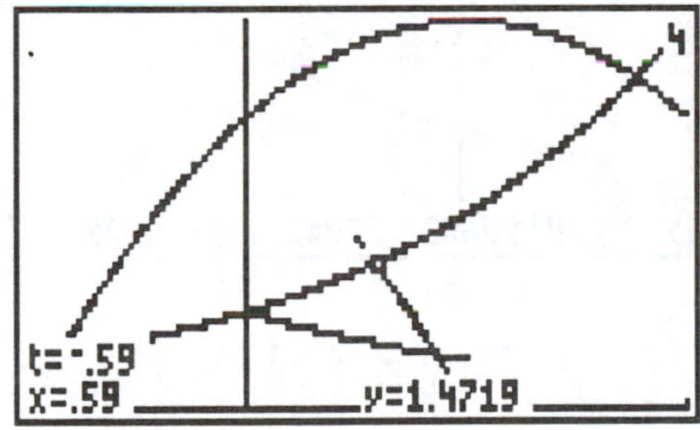
4
t=-.59
x=.59
y=1.4719

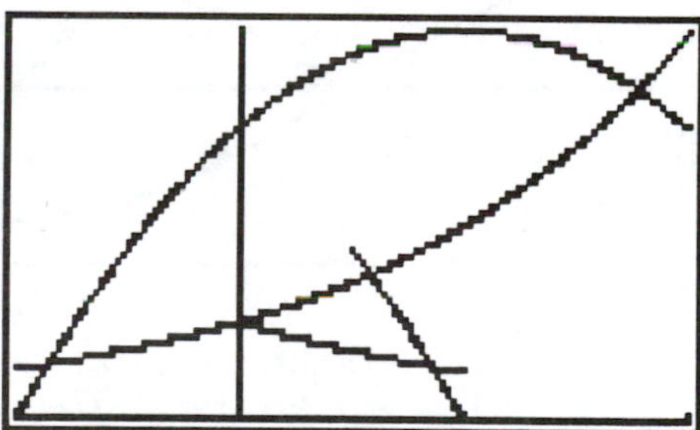

2a) Since $y_2 = 2\cos x$ is negative in this region, the volume can be viewed as formed by the areas under y_1, y_2, and $y_3 = -2\cos x$. The graphs intersect at $x_0 = \tan^{-1}\left(\frac{2}{3}\right) \approx 0.588$. By symmetry, the area under y_1 from 0 to x_0 is the same as the area under y_3 from $\pi - x_0$ to π. Similarly, the area under y_2 from x_0 to $\pi/2$ equals the area under y_2 from $\pi/2$ to $\pi - x_0$. The volume can, therefore, be calculated as twice the volume formed by the first two regions. The volume element for each region is a disk: $dV_1 = 4\cos^2 x\,dx$ and $dV_2 = 9\sin^2 x\,dx$. That is,

$V = 2\left(\int_0^{x_0} 4\cos^2 x\,dx + \int_{x_0}^{\pi/2} 9\sin^2 x\,dx\right)$. An NINT computation gives $V = 2\pi(2.09908 + 6.49949) = 17.197\pi$. We confirm by using the antiderivative formulas #58 and #59 in the Table of Integrals (at the end of the text). These give

$\int_0^{x_0} 4\cos^2 x\,dx = \sin 2(\tan^{-1}(\frac{2}{3})) + 2\tan^{-1}(\frac{2}{3})$ and $\int_{x_0}^{\pi/2} 9\sin^2(x)\,dx =$

$\frac{9\pi}{4} - \frac{9}{2}\tan^{-1}(\frac{2}{3}) + \frac{9}{4}\sin 2(\tan^{-1}(\frac{2}{3}))$.

From a reference triangle for $\theta = \tan^{-1}(\frac{2}{3})$ we find that

$\sin\theta = \frac{2}{\sqrt{13}}$ and $\cos\theta = \frac{3}{\sqrt{13}}$. Since $\sin(2x) = 2\sin x\cos x$ we have $\sin 2\theta = 12/13$. The volume is, therefore,

$V = 2\pi(\frac{12}{13} + 2\tan^{-1}(\frac{2}{3}) + \frac{9\pi}{4} - \frac{9}{2}\tan^{-1}(\frac{2}{3}) + \frac{27}{13})$ or

$2\pi(3 + \frac{9\pi}{4} - \frac{5}{2}\tan^{-1}(\frac{2}{3}) = 17.197\pi \approx 54.026$.

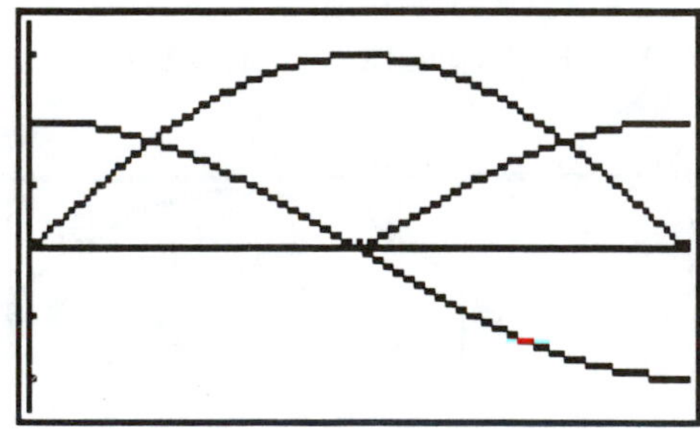

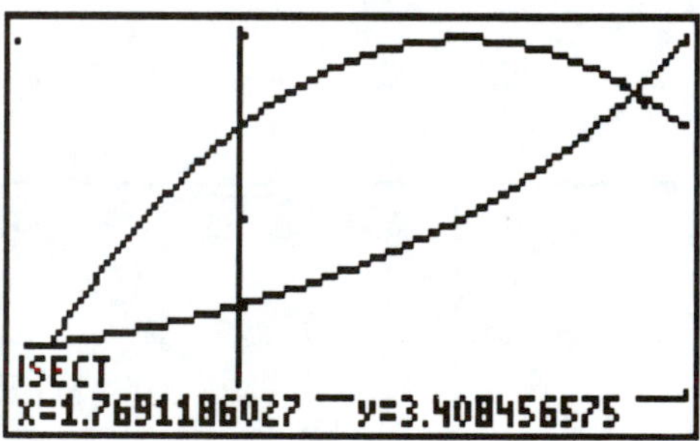

2b) Since both y_1 and y_2 extend to the left of the y-axis, the volume can be viewed as formed by areas in the first quadrant between y_1, y_2, $y_3 = 2^{-x}$ and $y_4 = 4 - (-x-1)^2$ in [-0.2, 3] by [-0.2, 6]. The graphs intersect in five points (which can be found by ZOOM and TRACE): (0.857, 0.552), (0, 1), (0.582, 1.497), (0, 3), and (1.769, 3.408) together with the high point of the parabola (1, 4) separate the region into five horizontal bands each of which rotates around the y-axis to form a portion of the solid.

- the bottom band is between y_4 and y_3, the volume element is a washer from 0.552 to 1

- the next band is from y_4 to the y-axis, the volume element is a disk from 1 to 1.497
- the third band is between y_1 and the y-axis. A volume element is a disk from 1.497 to 3
- the fourth band is between y_1 and y_2. A volume element is a washer taken from 3 to 3.408
- the top band is between y_2 and itself. Volume elements are washers from one side of the parabola to the other from $y = 3.408$ to $y = 4$.

SECTION 6.6 WORK

Exploration 1

1. $k = F/x = 24/.8 = 30$.

2. $x = F/k = 45/30 = 1.5$ m.

3. $W = \int_0^2 30x\,dx = 60$.

4. The work done in stretching or compressing the spring to length L is $W = \int_0^L kx\,dx = \frac{kL^2}{2}$. As L increases the work increases as its square so that it becomes more and more difficult to stretch the spring.

Exploration 2

1. $\Delta V = \pi(\frac{y}{2})^2\Delta y$.

2. $\Delta F = \frac{57}{4}\pi y^2\Delta y$.

3. Distance is $10 - y$.

4. $\Delta W = \frac{57}{4}\pi y^2(10 - y)\Delta y$.

5. $W_n = \sum_{i=0}^{n} \frac{57\pi}{4} y_i^2(10-y_i)\Delta y$.

6. The integral is the limit of the Riemann Sums W_i as n approaches infinity. That is,

$$W_n = \frac{57\pi}{4}\int_0^8 y^2(10 - y)\,dy = 19\pi(2^9) = 30561.413$$ which is confirmed by NINT.

SECTION 6.7 FLUID PRESSURES AND FLUID FORCES

Exploration 1

1. From 0 to 3.

2. $w = 62.5$.

3. $D(y) = 5 - y$.

4. $L(y) = 2y$.

5. $\int_0^3 62.5(5 - y)(2y)\,dy = 1687.5$.

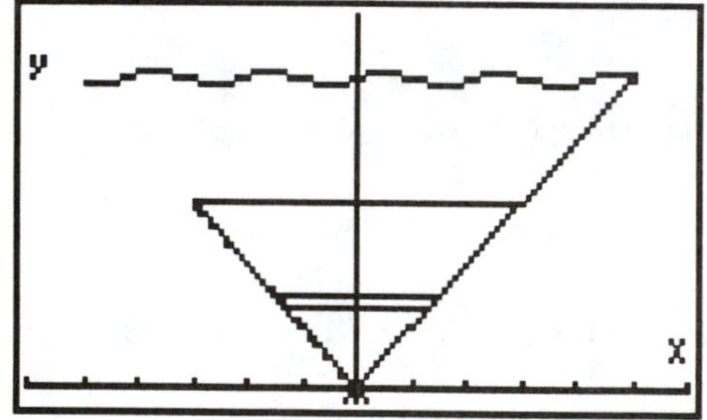

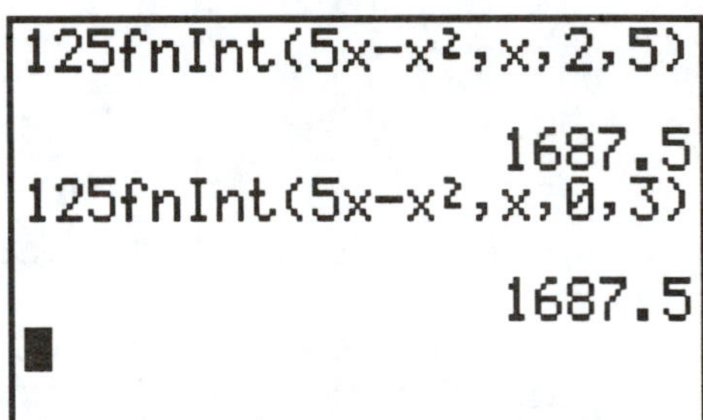

SECTION 6.8 CENTERS OF MASS

Exploration 1

1. The substitution $u^2 = 4 - y$, $2u\,du = -dy$ gives $M_x =$

$$\delta\int_2^0 2(4-u^2)u(-2u\,du) = 256\delta/15 \approx 17.0666,$$ confirmed by NINT.

2. - The center of mass of a vertical strip is $D(x, \frac{4-x^2}{2})$
   - The length of the strip is $4 - x^2$

- The width is dx
- The area is $dA = (4-x^2)dx$
- The mass is $dm = \delta\ dA$

Therefore, $M_x = \delta\int_{-2}^{2}\frac{1}{2}(4-x^2)^2dx \approx 17.0666$, confirmed by NINT.

3. $(\bar{x}, \bar{y}) = (0, 8/5)$ is the center of mass of the plate.

4. By symmetry, $\bar{x}$ is still 0 but $M = \int_{-2}^{2} 8x^2 - 2x^4\,dx = 256/15$ and

$$M_x = \int_{-2}^{2}\frac{4-x^2}{2}2x^2(4-x^4)\,dx = \int_{-2}^{2}16x^2-8x^4+x^6dx = 2048/105.$$

Therefore, the center of mass is $(\bar{x}, \bar{y}) = (0, 8/7)$.

SECTION 6.9 THE BASIC IDEA; OTHER MODELING APPLICATIONS

Exploration 1

1. $(x_1(t), y_1(t))$ graphs the velocity function against time. $(x_2(t), y_2(t))$ gives the position shift against time and so actually simulates the motion. $(x_3(t), y_3(t))$ gives the total distance travelled.

2. The velocity is positive at first and the corresponding shift is also positive. At $t = \pi/2$ the body stops and reverses direction; the position continues to be positive but begins to diminish. At $t = \pi$ the body has moved backwards for as long as it has moved forwards and the net position shift is zero. During the next $\pi/2$ seconds the body moves backwards but is slowing down; its position continues to decrease until, at $t = 3\pi/2$ it is at -5.

3. If $y_2(t)$ is constant the motion is simulated along a line. The advantage is that the x-coordinate now represents the real horizontal displacement of the object. The disadvantage is that when the object changes direction the line segment is invisibly redrawn. However, the motion can be brought out by tracing the graph.

4. Where the dots are close together the velocity has changed little during the time interval, that is, the acceleration is small. Where there are larger gaps between the dots the

object's velocity changes by a larger amount during the same time interval, i.e., the acceleration is greater.

SECTION 7.1 THE NATURAL LOGARITHM FUNCTION

Exploration 1

1. a) GRAPH y_3 and $y_4 = y_1 + y_2$ and TRACE moving from y_3 to y_4.

   b) GRAPH $y_3 = \ln a/x$ and $y_4 = y_1 - y_2$ and TRACE.

   c) GRAPH $y_3 = \ln 1/x$ and $y_4 = -y_2$ and TRACE.

2. Grapher activity.

3. Conjecture: $\ln(f(x)\,x) = \ln f(x) + \ln x$. Test, for example, with $y_1 = |\sin x|$, $y_2 = \ln y_1 x$ and $y_3 = \ln y_1 + \ln x$. GRAPH y_2 and y_3 in $[-1, 4\pi]$ by $[-0.4, 4]$.

   Note, that for $f(x) = x^{2n}$ (n an integer) that $\ln(f(x)x) = \ln(x^{2n}x) = \ln x^{2n+1} = (2n + 1)\ln x$ and so the conjecture is certainly true since $(2n + 1)\ln x = 2n \ln x + \ln x = \ln x^{2n} + \ln x = \ln f(x) + \ln x$.

Exploration 2

1. $\dfrac{d}{dx}(\ln 2x) = \dfrac{1}{2x}\dfrac{d}{dx}(2x) = \dfrac{1}{2x}(2) = \dfrac{1}{x} = \dfrac{d}{dx}(\ln x)$.

2. The proof of the general theorem is identical to the proof above: $\dfrac{d}{dx}(\ln kx) = \dfrac{1}{kx}\dfrac{d}{dx}(kx) = \dfrac{1}{kx}k = \dfrac{1}{x} = \dfrac{d}{dx}(\ln x)$.

3. By the corollary to the Mean Value Theorem these two antiderivatives can differ by a constant, that is, $\ln(kx) - \ln(x) = C$. The natural question is, "What constant?" To answer the question let $x = 1$ to find $\ln k = C$.

SECTION 7.2 THE EXPONENTIAL FUNCTION

Exploration 1

1. Grapher activity.

2. TRACE to see that the two graphs are identical.

3. The two graphs appear to be shifts of each other. For support, GRAPH $y1 - y_3$. The new graph is horizontal. Conjecture that $\int e^x dx = e^x + 1$.

Exploration 2

1. $f'(x) = 2x$ for $x > 0$ and $g'(x) = \frac{d}{dx} f^{-1}(x) = \frac{1}{2\sqrt{x}}$, also $x > 0$.

2.

x	0.5	1.	1.5	2.	3.	4.
$f(x)$	0.25	1.	2.25	4.	9.	16.
$f'(x)$	1.	2.	3.	4.	6.	8.
$g'(f'(x))$	1.	0.5	0.333333	0.25	0.166667	0.125

The numbers in last two rows are reciprocals of each other.

3. $g'(25) = 1/f'(5)$. This is confirmed by $f'(5) = 10$ and $g'(25) = 0.1$.

4. $g'(f(x)) = \frac{1}{f'(x)}$ when g and f are inverses.

5. $f'(x) = \frac{3}{(x+3)^2}$. $f(x) = f^{-1}(x) = \frac{3 - 3x}{x - 2}$ so
$g'(x) = \frac{3}{(x-2)^2}$ and $g'(f(x)) = \frac{(x+3)^2}{3} = \frac{1}{f'(x)}$..

Exploration 3

y_0 is a vertical stretch suggesting that growth is proportional, i.e., if the initial population y_0 is doubled then the population at time t is also doubled.

The positive constant k is a horizontal stretch or shrink and has the opposite effect on population size. Doubling k halves the time it takes for the population to reach a given level.

Exploration 4

Grapher activity.

Exploration 5

1. GRAPH $y_1 = e^{0.04x}$ and $y_2 = 1.5$ in [0, 20] by [0, 2]. TRACE and ZOOM to find $x \approx 10.14$.

   Confirm algebraically, $e^{0.04t} = 1.5$ so $0.04t = \ln(1.5)$ and $t = \dfrac{\ln 1.5}{.04}$ so $t = 10.137$ years.

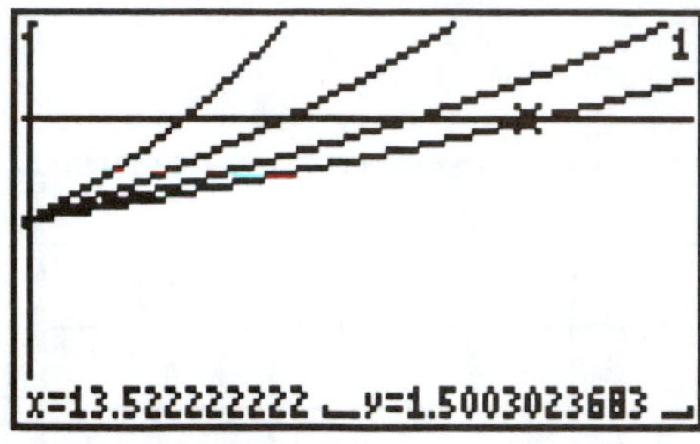

2. The four graphs show that the greater the inflation rate r, the less time it takes for the cost of living to increase 50%. Tracing the graphs gives the approximate values below for the time it takes the cost of living to increase 50%:

r	0.03	0.04	0.06	0.10
t	13.5	10.2	6.8	4.1

SECTION 7.3 OTHER EXPONENTIAL AND LOGARITHMIC FUNCTIONS

Exploration 1

1. GRAPH in [0, 15] by [0, 10].

2. NDER $f > 0$ suggesting that f is always increasing. NDER2 seems to be positive and asymptotic to the x-axis. This suggests that f is always concave up. Therefore, f has no extreme points and no points of inflection.

3. $f'(x) = \sqrt{3}x^{\sqrt{3}-1} > 0$ and $f''(x) = \sqrt{3}(\sqrt{3} - 1)x^{\sqrt{3}-2}$. Since $1 < \sqrt{3} < 2$ $\lim_{x\to\infty} f''(x) = 0$.

Exploration 2

1. Tracing suggests that $y_1 > y_2$.

2. $y_4 = 2$.

3. $y_4 = a$.

4. If $e^{\frac{y_2}{y_1}} = a$ then $\ln a = \dfrac{y_2}{y_1}$ so $y_2 = y_1 \ln a$. Therefore,

$y_2 = a^x \ln a$ so $\dfrac{d}{dx}(a^x) = a^x \ln a$.

Exploration 3

1. The change of base formula gives $\log_a x$ so GRAPH $y_1 = \dfrac{1}{\ln a}\ln x$ or store ln a in C and GRAPH $C \ln x$.

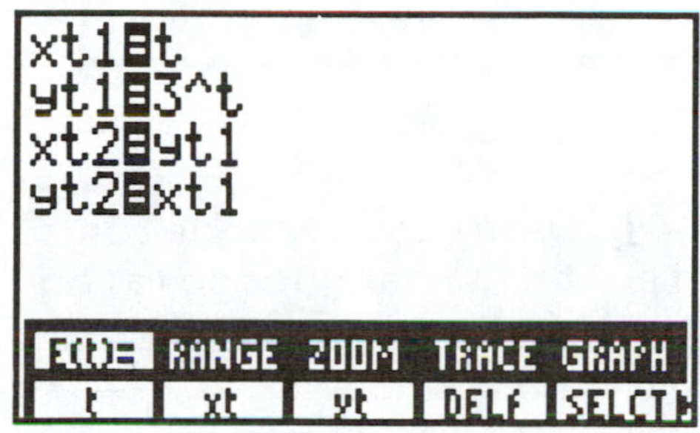

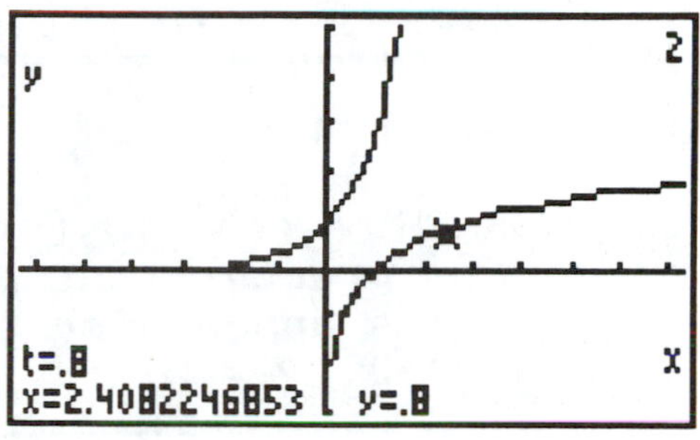

2. GRAPH:

$x_1(t) = a^t$, $y_1(t) = t$;
$x_2(t) = t$, $y_2(t) = x_1(t)$.

If $x_1 = a^x$ then y_2 will graph the inverse function, that is, $y_2 = \log_a x$.

3. By the change of base formula $y = \log_a x = \left(\dfrac{1}{\ln a}\right)\ln x$ so y is a vertical stretch or shrink of $y = \ln x$ depending upon whether $|1/\ln a|$ is greater or less than 1. If $\ln a < 0$ there is also a flip across the x-axis. The table below summarizes the possibilities.

a	$\ln a$	transformation of $\ln x$
$a > e$	$\ln a > 1$	vertical shrink
$1 < a < e$	$0 < \ln a < 1$	vertical stretch
$1/e < a < 1$	$-1 < \ln a < 0$	flip and a stretch
$0 < a < 1/e$	$\ln a < -1$	flip and a shrink

4. Since $a = e^{\ln a}$, $a^x = e^{(\ln a)x}$. Therefore, a^x is a horizontal stretch or shrink of e^x depending upon whether $|\ln a|$ is

greater or less than 1. If $\ln a < 0$ there is also a flip across the y-axis. The table below summarizes the possibilities.

a	$\ln a$	transformation of $\ln x$
$a > e$	$\ln a > 1$	horizontal shrink
$1 < a < e$	$0 < \ln a < 1$	horizontal stretch
$1/e < a < 1$	$-1 < \ln a < 0$	flip and a stretch
$0 < a < 1/e$	$\ln a < -1$	flip and a shrink

Exploration 4

1. GRAPH $y_1 = x^x$ in [-4.7, 4.8] by [-2, 4] (or an x-range on your grapher that makes $\Delta x = .1$. With this range isolated points appear for some negative values of x, e.g., when $x = -2$, $y = 0.25$.

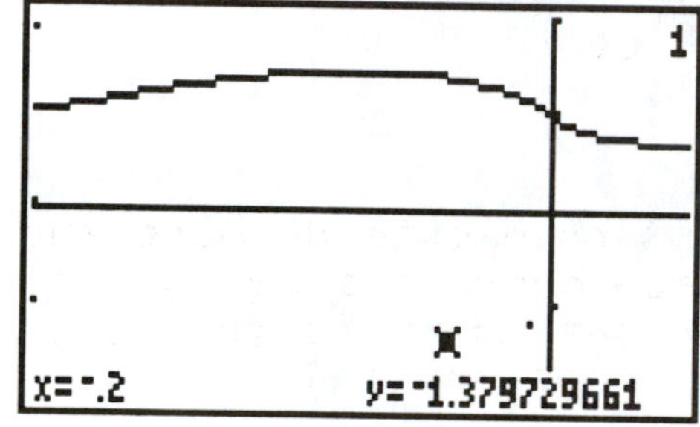

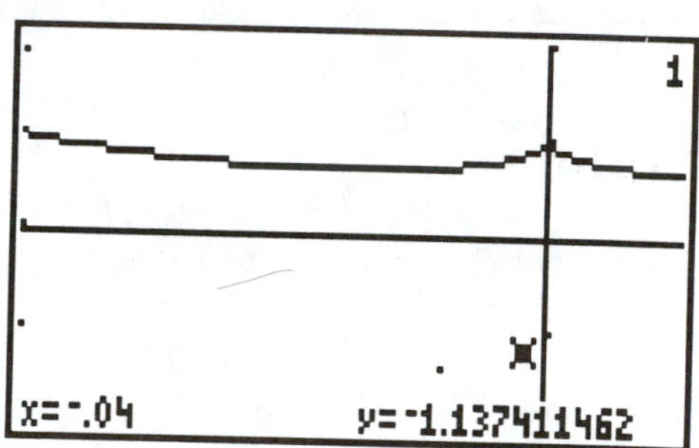

2. GRAPH $g(x) = |x|^x$ in [-4, 5] by [0, 6]. For $x > 0$, $g(x) = x^x$ so g is concave up and has a local minimum at $x = 1/e$. For $x < 0$, $g(-x) = \dfrac{1}{x^x}$, that is, g is a vertical reflection of the reciprocal of f.

   Since f is concave up, decreases to a minimum at $x = 1/e$ and then increases, its reciprocal is concave down, increases to a maximum at $x = 1/e$ and then decreases, approaching the x-axis as a horizontal asymptote. When this is reflected across the y-axis to complete the graph of g the discontinuity at $x = 0$ marks a change of concavity and is, therefore, a point of inflection of the extended function.

3. $h(x) = |x|^{|x|}$ equals y_1 for $x > 0$. Since h is clearly an even function ($h(x) = h(-x)$) its graph reflects across the y-axis so that h has local minima at $\pm 1/e$. (0,1) is still a

removable discontinuity but since the left-hand and right-hand derivatives are unequal, the extended function has a cusp at (1,0) which is also a local maximum.

SECTION 7.4 THE LAW OF EXPONENTIAL CHANGE REVISITED

Exploration 1

Use ZOOM or SOLVE to find that the graphs intersect at, approximately, 140 days. Changing the initial value y_0 changes the y-coordinate of the intersection but not its x-value. That is, the half-life is independent of the initial quantity present.

Exploration 2

1. Grapher activity.

2. View in [0, 20] by [0, 1600] to see the behaviors of the GRAPH for $m = 1$, 2, and 4. To see the graph for $m = 10$ we need a larger window such as [0, 160] by [0, 2400].

Exploration 3

1. The line is drawn rapidly at first and then extends more and more slowly to the right, showing that Jenny is slowing down. Since the line continues to extend to the right (rather than stopping and reversing direction) the acceleration is negative.

2. In this representation of Jenny's motion the spacing of dots along $y = 100$ shows that Jenny is losing speed. In the time vs. position graph this appears as a gradual levelling off of the curve: the slopes are positive but decrease. Since the graph remains concave down the second derivative (Jenny's acceleration) is negative. The second graph also appears to approach a horizontal asymptote suggesting that Jenny never stops, she approaches a limiting distance.

SECTION 7.5 INDETERMINATE FORMS AND L'HOPITAL'S RULE

Exploration 1

1. The two graphs appear to have a common tangent at $x = 0$, the line $y = 1$.

2. NDER y_3 passes through the origin. L'Hopital's Rule, therefore, does not say that $\lim_{x\to 0} \frac{f(x)}{g(x)} = \lim_{x\to 0} \frac{d}{dx}\left(\frac{f(x)}{g(x)}\right)$. Furthermore, since NDER y_3 is not constant, L'Hopital's Rule also does not say that $\lim_{x\to 0} \frac{f(x)}{g(x)} = \frac{d}{dx}\left(\lim_{x\to 0} \frac{f(x)}{g(x)}\right)$.

Exploration 2

1. Both 1 - cos x^6 and x^{12} approach 0 as x does so $\lim_{x\to 0} f(x)$ is the indeterminate form 0/0.

2. L'Hopital's Rule gives $\lim_{x\to 0} \frac{1-\cos x^6}{x^{12}} = \lim_{x\to 0} \frac{6x^5 \sin x^6}{12x^{11}} = \frac{1}{2}\lim_{x\to 0} \frac{\sin x^6}{x^6} = 1/2$.

Exploration 3

1. It appears that $\lim_{x\to \pi/2} \frac{\tan x}{1 + \tan x} = 1$.

2. Whether we approach from the left or the right $\frac{d}{dx}\tan x = \sec^2 x$ so $\lim_{x\to \pm\pi/2} \frac{\tan x}{1 + \tan x} = \lim_{x\to \pm\pi/2} \frac{\sec^2 x}{\sec^2 x} = 1$.

SECTION 7.6 THE RATES AT WHICH FUNCTIONS GROW

Exploration 1

$(x^1)' = 1 > e^x$ for $x < 0$. Therefore, x^1 grows faster than e^x on $x < 0$ and slower on $x > 0$.

$(x^2)' = 2x < e^x$ for all x. Therefore, x^2 grows slower than e^x.

$(x^3)' = 3x^2$ which intersects $y = e^x$ at three points, $x = -0.45$, $x = 0.91$, and $x = 3.73$. For $x < -0.45$ the power function grows faster, on (-0.45, 0.91) the exponential function grows faster, on (0.91, 3.73) the power function grows faster and for $x > 3.73$ the exponential function grows faster.

$(x^4)' = 4x^3$ which intersects $y = e^x$ at $x = 0.83$ and at 7.38. For x in (0.83, 7.38) the power function grows faster while outside this interval e^x grows faster.

In general, it appears that if n is even, the derivative of x^n will intersect e^x at two points. Within the interval created by these intersections the power function grows faster while for all other values the exponential function grows faster.

If n is odd, the derivative of x^n intersects e^x at three points $a < b < c$. For $x < a$ the power function grows faster, for x in (a, b) the exponential function that grows faster, in (b, c) the power function again grows faster while the exponential function grows faster for all $x > c$.

Exploration 2

1a) a^x grows faster than x^2 because
$\lim_{x\to\infty} \frac{a^x}{x^2} = \lim_{x\to\infty} \frac{a^x(\ln a)}{2x} = \lim_{x\to\infty} \frac{a^x(\ln a)^2}{2} = \infty$, using two applications of L'Hopital's Rule.

1b) e^x grows faster than any power function x^n because
$\lim_{x\to\infty} \frac{e^x}{x^n} = \infty$ because each derivative of x^n reduces its power.
After n applications of L'Hopital's Rule the power function has reduced to a constant while the exponential function remains. That is, after n applications of L'Hopital's Rule we have: $\lim_{x\to\infty} \frac{e^x}{x^n} = \lim_{x\to\infty} \frac{e^x}{n!} = \infty$.

2. A similar argument shows that after n applications of L'Hopital's Rule $\lim_{x\to\infty} \frac{a^x}{x^n} = \lim_{x\to\infty} \frac{a^x(\ln a)^n}{n!} = \infty$.

3. $\lim_{x\to\infty} \frac{3^x}{2^x} = \lim_{x\to\infty} 1.5^x = \infty$ so 3^x grows faster than 2^x.

4. In general, if $a > b > 1$ then $\lim_{x\to\infty} \frac{a^x}{b^x} = \lim_{x\to\infty} \left(\frac{a}{b}\right)^x = \infty$ since $a/b > 1$.

5. $\lim_{x\to\infty} \frac{x^m}{x^n} = \lim_{x\to\infty} x^{m-n}$. This limit is infinite if $m > n$ and zero if $m < n$. Therefore, x^m grows faster than x^n if $m > n$.

Exploration 3

$$\lim_{x\to\infty}\frac{\ln x}{x^n} = \lim_{x\to\infty}\frac{1/x}{nx^{n-1}} = \lim_{x\to\infty}\frac{1}{nx^n} = 0 \text{ whenever } n > 0.$$

Therefore, ln x grows more slowly than x^n.

SECTION 7.7 THE INVERSE TRIGONOMETRIC FUNCTIONS

Exploration 1

1. GRAPH $y = \sin^{-1}x$ in [-1, 1] by [-π/2, π/2].
   GRAPH $y = \cos^{-1}x$ in [-1, 1] by [0, π].
   GRAPH $y = \tan^{-1}x$ in [-10, 10] by [-π/2, π/2].

2. In larger [*t*Min, *t*Max] windows $(x_2(t), y_2(t))$ is not a function because each additional period of $y_1(t)$ produce multiple *y*-values for each $x_2(t)$.

Exploration 2

1. Let $x_3(t) = -x_2(t)$, $y_3(t) = -y_2(t)$ and GRAPH $(x_2(t), y_2(t))$ and $(x_3(t), y_3(t))$. TRACE and move between the two graphs to see the cursor reflect through the origin. This is a visual demonstration of the fact that

$$x_2(-t) = \sin(-t) = -\sin t = -x_2(t) \text{ and } y_2(-t) = -t = -y_2(t).$$

2. With $y_1(t) = \tan t$ we see again that $(x_2(t), y_2(t))$ is the graph of $y = \tan^{-1}x$. As in part (1) GRAPH and TRACE $(x_2(t), y_2(t))$ and $(x_3(t), y_3(t))$ to show that the $\tan^{-1}x$ is an odd function.

3. When $y_1(t) = \cos t$ we need to change the viewing window. A convenient one is [*t*Min, *t*Max] = [0, π] and (x, y) in [-1, π] by [-1, π]. The function $\cos^{-1}x$ is neither odd nor even because its graph is not symmetrical across the *y*-axis or the origin.

   The graphs of $\cos^{-1}(-t)$ and $-\cos^{-1}t$ seem to be shifts of each other. For support, GRAPH $-\cos^{-1}t - \cos^{-1}(-t)$ and note that the graph is the horizontal line $y = -\pi$.

Exploration 3

In each case draw reference triangles with the given values:

1. $x = \sqrt{3}$ $r = 1$ to show $\cos^{-1}\frac{\sqrt{3}}{2} = \pi/6$.

   $x = 1$ $r = \sqrt{2}$ to show $\cos^{-1}\frac{1}{\sqrt{2}} = \pi/4$.

   $x = -1$ $r = 2$ to show $\cos^{-1}\frac{\sqrt{3}}{2} = 2\pi/3$.

   $x = -\sqrt{3}$ $r = 2$ to show $\cos^{-1}\frac{\sqrt{3}}{2} = 5\pi/6$.

2. $x = 1$ $y = \sqrt{3}$ to show $\tan^{-1}\sqrt{3} = \pi/3$.

   $x = 1$ $y = 1$ to show $\tan^{-1}(1) = \pi/4$.

   $x = \sqrt{3}$ $y = -1$ to show $\tan^{-1}\frac{-1}{\sqrt{3}} = -\pi/6$.

   $x = 1$ $y = -\sqrt{3}$ to show $\tan^{-1}(-\sqrt{3}) = -\pi/3$.

3. $y = 1$ $r = \sqrt{2}$ to show $csc^{-1}(\sqrt{2}) = \pi/4$.

   $y = \sqrt{3}$ $r = 2$ to show $csc^{-1}\frac{2}{\sqrt{3}} = \pi/3$.

   $y = -1$ $r = 2$ to show $csc^{-1}(-2) = -\pi/6$.

   $y = -\sqrt{3}$ $r = 2$ to show $csc^{-1}\frac{-2}{\sqrt{3}} = -\pi/3$.

4. $x = 1$ $y = 1$ to show $\cot^{-1}(1) = \pi/4$.

   $x = 1$ $y = \sqrt{3}$ to show $\cot^{-1}\frac{1}{\sqrt{3}} = \pi/3$.

   $x = 1$ $y = -\sqrt{3}$ to show $\cot^{-1}\frac{-1}{\sqrt{3}} = 2\pi/3$.

   $x = \sqrt{3}$ $y = -1$ to show $\cot^{-1}(-\sqrt{3}) = 5\pi/6$.

SECTION 7.8 DERIVATIVES OF INVERSE TRIGONOMETRIC FUNCTIONS; RELATED INTEGRALS

Exploration 1

1. $\sin^{-1}x$ is symmetric through the origin because it is an odd function.

2. y_2 should be symmetric across the y-axis since the derivative of an odd function is even. The graph shows this symmetry.

3. y_1 has positive slopes; the graph of y_2 is in agreement.

4. The slopes of y_1 decrease until $x = 0$ and then increase. The graph of y_2 is in agreement since it has a minimum at $x = 0$.

5. y_1 seems to have vertical tangents at its endpoints (which makes sense if we think of $y = \sin x$ which has horizontal tangents at its extreme values.) The slopes of y_1 should increase without bound. y_2 seems to support this with vertical asymptotes at the endpoints.

6. $y = \dfrac{1}{\sqrt{1-x^2}}$ is clearly even and positive (parts 2 and 3). Its minimum occurs at the greatest value of the denominator which is when x^2 is as small as possible, i.e., $x=0$ (part 4). Finally, y has vertical tangents at $x = -1$ and $x = 1$ (part 5).

Exploration 2

1. y_1 should be reflected across the y-axis and then shifted vertically up $\pi/2$. That is, $\sin^{-1}(-x) + \pi/2 = \cos^{-1}(x)$.

2. Since $\dfrac{d}{dx}f(-x) = -f'(x)$, the slopes of the reflected y_1 are opposites. Therefore, its derivative is reflected across the x-axis. The vertical shift will not affect the derivative.

3. $y'_2 = -\dfrac{1}{\sqrt{1 - x^2}}$.

Exploration 3

1. Grapher activity.

2. The graph of y_2 changes sign at $x = 0$ predicting a single minimum. y_3 is always positive predicting that y_1 is always concave up. Further, y_2 seems to rise from negative infinity to positive infinity predicting that y_1 has vertical tangents at its endpoints $x = -1$ and $x = 1$.

   We can confirm these observations analytically by noting that $\dfrac{2x}{\sqrt{1-x^2}}$ changes sign when x does and has vertical asymptotes at $x = -1$ and $x = 1$. At $x = 0$ $y_1 = 0$ so the minimum of y_1 is also its only intercept.

2. $u = x$, $dv = \cos x\,dx$,
$du = 1$, $v = \sin x$. This works.

3. $u = \cos x$, $dv = x\,dx$,
$du = -\sin x$, $v = x^2/2$.

Integration by Parts gives $-x^2\sin x/2 + \int x^2 \sin x\,dx$ which is more complex than the original integral.

4. $u = x\cos x$, $dv = dx$,
$du = -x\sin x + \cos x$ $v = x$.

The new integral is more complex than the original.

Exploration 2

1. $u = 1$ is useless since dv is just the original integral. The second possibility, $u = \ln x$ and $dv = dx$ works fine. We get $du = 1/x\,dx$ and $v = x$. Integration by Parts gives $\int \ln x\,dx = x\ln x - \int 1\,dx = x\ln x - x + C$.

2. $C = 1$. This can be seen in Fig 8.6 by examining the y-intercepts, or TRACE your own graphs.

3. $\int x^2e^x dx = x^2e^x - \int 2xe^x dx$ using $u = x^2$ and $dv = e^x\,dx$. The second integral is simpler and can also be evaluated by Integration by Parts with $u = x$ and $dv = e^x\,dx$. The final evaluation gives $\int x^2e^x dx = x^2e^x - 2x\,e^x + e^x + C =$ $(x^2 - 2x + 1)e^x + C$.

4. Use y_1 = NINT(x^2e^x,0,x), $y_2 = (x^2-2x+2)e^x$ and GRAPH $y_1 - y_2$ and TRACE to find that $C = 2$.

SECTION 8.3 INTEGRALS INVOLVING TRIGONOMETRIC FUNCTIONS

Exploration 1

$NINT(\sqrt{1+\cos(4x)}, 0, \pi/4) = 0.707$ which looks like $\pi/2$. To confirm this analytically use $1 + \cos(4x) = 2\cos^2(2x)$ to get

$$\int_0^{\pi/2} \sqrt{1+\cos(4x)}\,dx = \int_0^{\pi/2} \sqrt{2\cos^2(2x)}\,dx = \sqrt{2}\int_0^{\pi/2} |\cos(2x)|\,dx =$$

$$\sqrt{2}\int_0^{\pi/2} \cos(2x)\,dx = \frac{\sqrt{2}}{2}\sin(2x)\Big|_0^{\pi/2} = \frac{\sqrt{2}}{2}.$$

Exploration 2

Let y_1 = NINT(sin(3x) cos(5x), 0, x) and y_2 = (-cos 8x)/16 + (cos 2x)/4 then GRAPH $y_1 - y_2$ to see that these functions differ by a constant. We can get analytic confirmation by using m=3 and n=5 in identity (2). This gives $\sin 3x \cos 5x = \frac{1}{2}(\sin 8x - \sin 2x)$ so $\int \sin 3x \cos 5x\, dx =$ $\frac{1}{2}\int \sin 8x - \sin 2x dx = \frac{1}{2}(-\frac{1}{8}\cos 8x + \frac{1}{2}\cos 2x) =$ $-\frac{1}{16}\cos 8x + \frac{1}{4}\cos 2x + C.$

Exploration 3

1.

	f	Type	a	$\int_{-a}^{a} f$	$\int_{0}^{a} f$
(a)	$x^2 \cos x$	even	$\pi/2$	0.9348	0.4674
(b)	$\sin^3 x$	odd	$\pi/2$	0	0.6666
(c)	$\frac{\sin^2 x}{\cos x}$	even	$\pi/4$	0.3485	0.1743
(d)	$\sqrt{\sin^2 x + 1}$	even	$\pi/2$	3.8202	1.9101

2. a) f does not have to be an even function if, for some a, $\int_{-a}^{a} f(x)\,dx = 2\int_{0}^{a} f(x)\,dx$. As a counter-example, let

$$f(x) = \begin{cases} x^2 & x \ge 0 \\ 2x & x < 0 \end{cases}$$

which is obviously not an even function. If $a = 3$ then $\int_{-3}^{3} f(x)\,dx = \int_{-3}^{0} 2x\,dx + \int_{0}^{3} x^2\,dx = 18$ and $\int_{0}^{3} f(x)\,dx = 9$.

Note, however, that if $\int_{-a}^{a} f(x)\,dx = 2\int_{0}^{a} f(x)\,dx$ for all a then f must be an even function. The proof is a direct application of the Fundamental Theorem of Calculus.

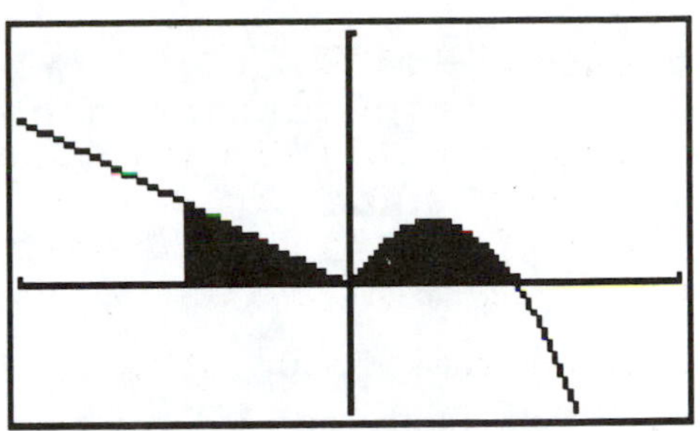

Rewrite the given condition as $\int_{-x}^{x} f(t)\,dt = 2\int_{0}^{x} f(t)\,dt$ then separate the first integral so that $\int_{-x}^{0} f(t)\,dt + \int_{0}^{x} f(t)\,dt = 2\int_{0}^{x} f(t)\,dt$ and differentiate to get $-f(-x) + f(x) = 2f(x)$. Simplify to show that $f(x) = -f(-x)$ for all x, that is, f is even.

b) Similarly, f does not have to be an odd function if it satisfies $\int_{-a}^{a} f(x)\,dx = 0$ for some a. For example, let

$$f(x) = \begin{cases} x^2 & x \geq 0 \\ -2x & x < 0 \end{cases}$$

which is not odd. If $a = 3$ then $\int_{-3}^{3} f(x)\,dx = \int_{-3}^{0} -2x\,dx + \int_{0}^{3} x^2\,dx = 0$.

Once again, however, if $\int_{-a}^{a} f(x)\,dx = 0$ for all a then f must be an odd function. Rewrite the given condition as $\int_{-x}^{x} f(t)\,dt = 0$. Rewrite as $\int_{-x}^{0} f(t)\,dt + \int_{0}^{x} f(t)\,dt = 0$ and differentiate to get $-f(-x) + f(x) = 0$. That is, $f(x) = f(-x)$ for all x so that f is odd.

SECTION 8.4 TRIGONOMETRIC SUBSTITUTIONS

Exploration 1

1. GRAPH $y_1 - y_2$ on [-10, 10] by [-4, 4]. TRACE to see that the functions differ by 3.2188758. (Choose xMin and xMax to include (-5, 5) since the domain of these functions excludes that interval).

2 After graphing, try evaluating $e^{y_1 - y_2}$. You should get 25. Therefore, $y_1 - y_2 = \ln 25$.

3. $y_1 - y_2 = \ln|x + \sqrt{x^2-25}| + \ln|x - \sqrt{x^2-25}|$. Use the multiplication property of logarithms to simplify: $y_1 - y_2 = \ln|(x + \sqrt{x^2-25})(x - \sqrt{x^2-25})| = \ln|x^2 - (x^2 - 25)| = \ln 25$.

Exploration 2

1. NINT(y_1, 0, x) = 1.57079 which looks like $\pi/2$. Confirm analytically by using $u = x$ and $a = \sqrt{2}$ in (2) so that
$$4\int_0^2 \frac{dx}{x^2 + 4} = \frac{1}{2}\tan^{-1}\left(\frac{x}{2}\right)\Big|_0^2 = \pi/2.$$

2. a) $V = \displaystyle\int_0^2 \pi \frac{16}{(x^2 + 4)^2}dx = 16\pi\int_0^2 \frac{dx}{(x^2 + 4)^2}$.

b) Let $x = 2\tan\theta$ so that $x^2+4 = 4(1 + \tan^2\theta) = 4\sec^2\theta$ which simplifies the denominator since $dx = 2\sec^2\theta\, d\theta$. When $x = 0$ $\theta = 0$ and when $x = 2$ $\theta = \pi/2$. The integral becomes $V = 16\pi\displaystyle\int_0^{\pi/2} \frac{2\sec^2\theta\, d\theta}{16\sec^4\theta} = 2\pi\int_0^{\pi/2} \cos^2\theta\, d\theta$.

c) $V = 2\pi\left(\frac{\theta}{2} + \frac{1}{4}\sin 2\theta\right)\Big|_0^{\pi/2} = \pi\left(\frac{\pi + 2}{4}\right) \approx 4.0382$.

d) $NINT(\pi y_1^2, 0, 2) \approx 4.0382$.

Exploration 3

$4x^2 + 4x + 2 = (2x + 1)^2 + 1$ so, let $u = 2x + 1$, $du = 2\,dx$ and $\displaystyle\int \frac{dx}{4x^2 + 4x + 2} = \frac{1}{2}\int \frac{du}{u^2 + 1} = \frac{1}{2}\tan^{-1}(2x+1) + C$. For graphic support let $y_1 = NINT\left(\frac{1}{4x^2+4x+2}, 0, x\right)$ and $y_2 = \frac{1}{2}\tan^{-1}(2x + 1)$ and GRAPH $y_1 - y_2$ to see that the graphs differ by a constant. Or GRAPH $y_3 = \frac{1}{4x^2 + 4x + 2}$ and $y_4 = NDER\left(\frac{1}{2}\tan^{-1}(2x + 1)\right)$.

SECTION 8.5 RATIONAL FUNCTIONS AND PARTIAL FRACTIONS

Exploration 1

1. GRAPH y_4 and y_3 and TRACE to see that the functions are identical.

2. One possibility: "Show that the area between y_4 and y_1 equals the area between y_2 and the x-axis for $x = 0$ to $x = 3$."

Since $y_1 > y_4$, the first area is $\int_0^3 y_1 - y_4 dx =$

$\int_0^3 \frac{6}{x+2} - \frac{6x+7}{(x+2)^2} dx = \int_0^3 \frac{-5}{(x+2)^2} dx = \left.\frac{-5}{x+2}\right|_0^3 = 3/2$ while

the second area is $-\int_0^3 \frac{-5}{(x+2)^2} dx$ since the area is below the x-axis. This also computes to 3/2.

3. Long division gives $\frac{2x^3 - 4x^2 - x - 3}{x^2 - 2x - 3} = 2x + \frac{5x - 3}{x^2 - 2x - 3}$ and a partial fraction decomposition gives $y_5 = 2x + \frac{3}{x-3} + \frac{2}{x+1}$.

Therefore, $I(x) = \int y_5 dx = x^2 + 3 \ln |x-3| + 2 \ln|x + 1| + C$.

For numerical support write $\int_a^b y_5 dx = I(b) - I(a)$. The table below compares $I(b) - I(a)$ and NINT(y_5, a, b) for several choices of a and b. (NOTE: the interval [a, b] must not include $x = -1$ or $x = 3$ where the integral is improper.)

a	b	$I(b) - I(a)$	NINT(y_5, a, b)
4	10	91.4146	91.4146
-6	-2	-36.9822	-36.9822
0	2	2.9014	2.9014

For graphical support, let $y_6 = I(x)$,

y_7 = NINT(y_5, -10, x). GRAPH $y_6 - y_7$ in [-10, -1.1] by [0, 12]
y_7 = NINT(y_5, -.9, x). GRAPH $y_6 - y_7$ in [-.9, 2.9] by [0, 1]
y_7 = NINT(y_5, 3.1, x). GRAPH $y_6 - y_7$ in [3.1, 10] by [0, 6]

In each interval, tracing shows that $y_6 - y_7$ is constant (112.0893, 0.28776, 5.5242 respectively). The constants are different in each interval demonstrating that $\int_a^x y_5 dt$ represents a different antiderivative on each interval. (It is necessary to examine the graphs separately in each interval as an NINT calculation will fail over any interval that includes $x = -1$ or $x = 3$ where y_6 and y_7 are undefined.)

SECTION 8.6 IMPROPER INTEGRALS

Exploration 1

Support Based on Solution 1 We can use the values of NINT(y_1, 0, a) for support by choosing values of a approaching 1 from the left:

a	NINT(y_1, 0, a)
0.99	2.288
0.9999	2.545
0.999999	2.56797
0.99999999	2.57051

For graphical support note that the graph of NINT(y_1, 0, x) rises to, approximately, 2.57 (GRAPH in [0,1] by 0, 3].)

Support Based on Solution 2 We can use the values of NINT(y_2, 1, a) for support by choosing larger and larger values of a.

a	NINT(y_2, 1, a)
1000	1.5688
10^5	1.5708
10^6	1.5708

For graphical support note that the graph of NINT(y_2, 1, x) rises to, approximately, 1.57 (GRAPH in [1,10000] by [0, 2].)

Exploration 2

1. Using properties of logs write $\ln (1/x) = -\ln x$. Therefore, $\lim\limits_{x\to 0^+} \ln x = \lim\limits_{x\to 0^+} -\ln(1/x) = -\infty$.

2. **Investigating $\lim_{x\to 0}$ ln x**

This table of values confirms that $\ln x \to -\infty$.

x	$\ln x$
10^{-2}	-4.605
10^{-20}	-46.052
10^{-40}	-92.103
10^{-60}	-138.155

The values decrease. The limit can be confirmed by observing that $\ln(10^n) = n \ln 10$; as n decreases, becoming more negative, the logarithm also decreases.

The graph of $y_1 = \ln x$ in $[0, 10^{-40}]$ by $[-100, 0]$ supports the table and the limit.

Since $\int_0^1 \frac{dx}{x} = \lim_{b\to 0^+}(-\ln b)$, the divergence of the limit implies the divergence of the integral.

Investigating $\lim_{x\to 0}$ ln 1/x

This table of values confirms that $\ln 1/x \to \infty$

x	$\ln 1/x$
10^{-2}	4.605
10^{-20}	46.052
10^{-40}	92.103
10^{-60}	138.155

The values increase. The limit can be confirmed by observing that $1/10^n = 10^{-n}$ so $\ln(1/10^n) = -n \ln 10$; as n decreases, becoming more negative, the logarithm increases.

The graph of $y_2 = \ln 1/x$ in $[0, 10^{-40}]$ by $[0, 100]$ supports the table and the limit.

Exploration 3

1. Let y_1 = NINT($1/t^2$, t, 0, x) and GRAPH $y_1(1/x)$. If your grapher does not permit composition of functions then GRAPH y_2 = NINT($1/t^2$, t, 0, $1/x$) in [0, .001] by [0, 2]. The graph shows a y-intercept of 1 supporting that $y_2(0) = 1$.

2. To evaluate $V = \lim_{x\to-\infty} \frac{\pi}{8}(4 - e^{2b})$ GRAPH $y_1 = \frac{\pi}{8}(4 - e^{\frac{1}{2x}})$. Since the integral is unbounded below we let $x \to 0^-$ so GRAPH in [−0.001, 0]. The graph appears nearly horizontal with $y_1 = 1.57079$, i.e., $y_1 = \pi/2$.

Exploration 4

1. $\int_1^\infty \frac{dx}{x^2} = \lim_{b\to\infty}\int_1^b \frac{dx}{x^2} = -1/b + 1 = 1$ which agrees with Exploration 3, part 1.

2. For $p \le 0$ let $1/x^p = x^q$ where $q \ge 0$ then

$$\int_0^1 x^q\,dx = \left.\frac{x^{q+1}}{q+1}\right|_0^1 = \frac{1}{q+1} \text{ so } \int_0^1 \frac{dx}{x^p} \text{ converges.}$$

$$\int_1^\infty x^q\,dx = \lim_{b\to\infty} \left.\frac{x^{q+1}}{q+1}\right|_0^b = \lim_{b\to\infty}\frac{b^{q+1}}{q+1} = \infty \text{ so } \int_1^\infty \frac{dx}{x^p} \text{ diverges.}$$

Exploration 5

1. Since $\lim_{x\to\infty} \dfrac{\frac{1}{x^2}}{\frac{1}{(1+x^2)}} = \lim_{x\to\infty} \dfrac{1+x^2}{x^2} = 1$ the two integrals converge. GRAPH $y_1 = \dfrac{\frac{1}{x^2}}{\frac{1}{1+x^2}}$ for support that the ratio approaches 1. The value of NINT($1/(1+x^2)$, 1, 1000) also confirms the convergence of the integral.

2. Exploration 3, part 1 showed that $\int_1^{\infty} \frac{dx}{x^2} = 1$ while

$$\lim_{b\to\infty} \int_1^b \frac{dx}{1 + x^2} = \tan^{-1}x \Big|_1^b = \lim_{b\to\infty} \tan^{-1}b - \pi/4 = \pi/2 - \pi/4 = \pi/4.$$

3. $\lim_{x\to\infty} \dfrac{\frac{1}{e^x}}{\frac{3}{e^x+5}} = \lim_{x\to\infty} \dfrac{1}{3} + \dfrac{5}{3e^x} = 1/3$ so, by the Limit Comparison Test both integrals converge or both diverge.

Since $\int \frac{1}{e^x} dx = \int e^{-x} dx = -e^{-x} + C$,

$$\int_1^{\infty} \frac{1}{e^x} dx = \lim_{b\to\infty} -e^{-x} \Big|_1^b = \lim_{b\to\infty} -e^{-b} + \frac{1}{e} = 1/e.$$

Therefore, both integrals converge.

Exploration 6

1. y_1 appears to approach y_4 as a horizontal asymptote. This end-behavior supports $\int_0^{\infty} \frac{dx}{1 + x^2} = \pi/2$ and, since y_1 is an even function $\int_{-\infty}^0 \frac{dx}{1 + x^2} = \pi/2$ so $\int_{-\infty}^{\infty} \frac{dx}{1 + x^2} = \pi$.

SECTION 8.7 DIFFERENTIAL EQUATIONS

```
NO. OF POINTS=10
           [0 1]
         [.1 1.2]
        [.2 1.42]
       [.3 1.662]
      [.4 1.9282]
     [.5 2.22102]
```

```
                    [0 1]
                [.1 1.21]
             [.2 1.44205]
          [.3 1.69846525]
     [.4 1.98180410125]
     [.5 2.29489353188]
     [.6 2.64085735273]
```

Exploration 1

1. Use $y_1 = 1 + y$, $h = 0.1$, $n = 10$.

2. Grapher activity.

3. Grapher activity.

Exploration 2

1. Differentiate $y = x - 1 + 2e^{-x}$ to get $y' = 1 + e^{-x}$ and note that $y(0) = 1$ to confirm this particular solution to the initial value problem.

2. If your grapher allows the use of lists you can view all graphs simultaneously. GRAPH: $y_1 = x - 1 + \{-6,-5,-4,-3,-2,-1,1,2,3,4,5,6\}e^{-x}$ to extend Figure 8.37.

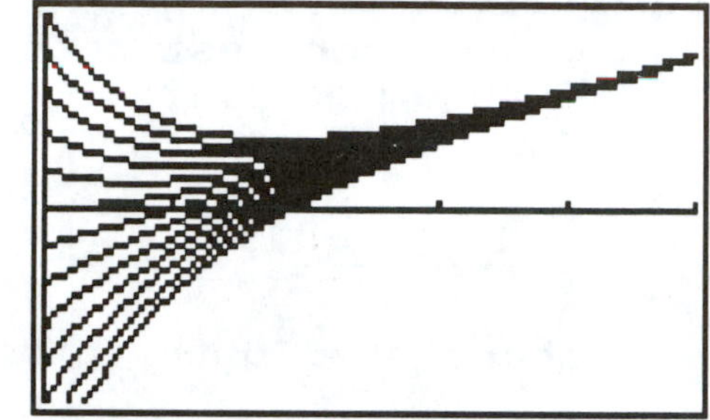

3. $y = x - 1 + Ce^{-x}$ must be the general solution the differential equation $dy/dx = x - y$ because, its derivative, $y' = 1 - Ce^{-x}$ satisfies the equation and it has the one arbitrary constant required for the solution of a first order-differential equation.

SECTION 8.8 COMPUTER ALGEBRA SYSTEMS (CAS)

Exploration 1

1. These two forms of the antiderivative differ by a constant (0.1328125).

2. The Derive solution gives 1.6005 - 1.13799 = 0.46252 while the Mathematica solution gives 1.73333 - 1.27080 = 0.46252. An NINT computation also gives 0.46252.

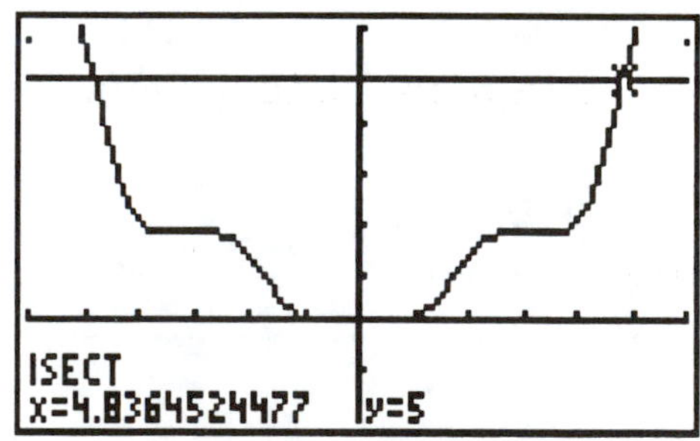

3. $\frac{d}{dx}\int_0^x t\sin^4 t\,dt = x\sin^4 x$ by the Fundamental Theorem of Calculus, so $f'(x) > 0$ for $x > 0$ and $f'(x) < 0$ for $x < 0$. Therefore, f decreases for $x < 0$ and increases for $x > 0$ and so $f(x)$ will intersect the line $y = 5$ once to the left of the origin and once more to the right. GRAPH $y_1 = f(x)$ and $y_2 = 5$ in [-10,10] by [0, 8] to support the two

intersections. TRACE and ZOOM or use SOLVE to identify the solutions $x = -4.836$ and $x = 4.836$.

4. Since the integrand is odd, f must be even. To solve $\int_1^x t\sin^4 t\,dt = 0$, note that $x = 1$ is a solution, by symmetry so is $x = -1$. From (3) $f' > 0$ for positive x so that f increases; therefore, there is no other positive root and, by symmetry, no other negative root.

Exploration 2

1. Grapher activity.

2. $f'(x) = \sin x^2$. On $[0, 3]$ $f'(x) = 0$ when $x^2 = 0, \pi, 2\pi$ (the next solution, 3π is outside the interval). Therefore, $x = 0, \sqrt{\pi}, \sqrt{2\pi}$ are critical points of f. The graph of $y = \sin x^2$ changes sign at $\sqrt{\pi}$ and $\sqrt{2\pi}$ so f has a maximum at $x = \sqrt{\pi}$ and a minimum at $x = \sqrt{2\pi}$. An NINT computation gives the coordinates of these critical points: $(\sqrt{\pi}, 0.895)$ and $(\sqrt{2\pi}, 0.430)$. f' does not change sign at $x = 0$ but 0 and 3 are endpoints of the domain so (0, 0) is a relative minimum of f and (3, 0.774) is a relative maximum.

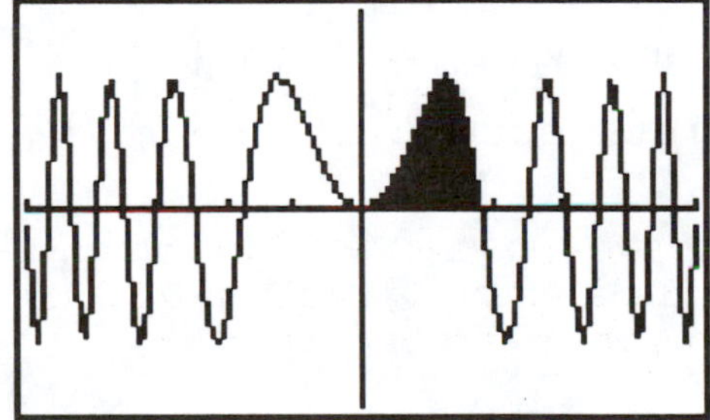

3. The portion of Figure 8.39 with $0 \le x \le 3$ agrees with part (2).

SECTION 9.1 LIMITS OF SEQUENCES OF NUMBERS

Exploration 1

1. a) GRAPH in [0,1] by [0, c+1].

   b) Let $x_2(t) = c$ and $y_2(t) = (t-1)/t$. GRAPH in [0, c+1] by [0, 1] and note how the points seems to accumulate near $(c,1)$.

2. Here the points are spread out, equally spaced horizontally. The accumulation that we observed in (1) now appears as convergence towards the line $y = 1$.

3. The points in the sequence in (2) are now embedded in the graph of $y_1(t)$ at the integer values of x. The convergence of the sequence now appears as the graphs end-behavior asymptote, the line $y = 1$.

Exploration 2

Draw each GRAPH in parametric mode and dot format.

1. $x_1(t) = 3$ and $y_1(t) = 1$ in [0, 20] by [0, 2]. All terms of the sequence plot as the single point (3,1). This visualization does not show a sequence accumulating near a limit, nevertheless, the sequence converges to the constant value 3.

2. $x_2(t) = t - 1$ and $y_2(t) = 1$ in [0, 20] by [0, 2]. The points appear equally spaced. Since they do not accumulate there is no convergence: a_n diverges.

3. $x_3(t) = 1/t$ and $y_3(t) = 1$ in [0, 1] by [0, 2]. The points accumulate at (0, 1) suggesting convergence: $a_n \to 0$.

4. $x_4(t) = \dfrac{(-1)^{t+1}}{t}$ and $y_4(t) = 1$ in [-1, 1] by [0, 2]. TRACE to watch the points accumulate near (0, 1) suggesting $a_n \to 0$.

Exploration 3

1. Grapher activity.

2. $n = 11$

3. With tMax = 1200, GRAPH in [900, 1100] by [0.0008, 0.0012] and TRACE to see $n = 1001$.

Exploration 4

1. Use tMin = 1, tMax = 10, t-step = 1 in [0, 10] by [-1, 2]. The graph of the sequence suggests that the sequence converges to 1.

2. The graph has the new sequence embedded in it. It appears to increase from a y-intercept of (1,0). However, since it is drawn in function mode we can change the viewing window

to [0, 0.001] by [0, 2] to give us a better view of function values near $x = 0$. In this window the graph seems to approach (0, 1); this corresponds to looking at values of the original sequence for $t > 1000$. This provides strong support that the limit of the original sequence is also 1.

SECTION 9.2 INFINITE SERIES

Exploration 1

1. $5.232323\ldots = 5 + \left(\frac{23}{100}\right) + \left(\frac{23}{100^2}\right) + \left(\frac{23}{100^3}\right) + \ldots$ or

   $5.232323\ldots = 5 + 23\sum_{n=1}^{\infty} 100^{-n}$.

2. The geometric series is $\sum_{n=1}^{\infty} (100)^{-n}$ with $a = r = 1/100$, and $a/(1-r) = 1/99$.

3. $5.232323\ldots = 5 + 23/99$ or $518/99$.

4. Answers will vary.

Exploration 2

1. The contrapositive of a statement "If p is then q" is the statement "If not q then not p" which can be rewritten as "If q is not true then p is not true." The statement of the n-th Term Test for Divergence is "If a_n does not converge to 0 then the series Σa_n diverges". The contrapositive is "If Σa_n does not diverge then a_n converges to 0." Removing the double negative gives the boxed statement.

2. In the series $\sum_{n=1}^{\infty}\sum_{i=1}^{n}\frac{1}{n}$ the terms $a_n = \sum_{i=1}^{n}\frac{1}{n} = \frac{1}{n} + \frac{1}{n} + \ldots + \frac{1}{n} = n(1/n) = 1$. The given series then is $\sum_{n=1}^{\infty} 1 = 1 + 1 + \ldots$ which diverges.

SECTION 9.3 SERIES WITHOUT NEGATIVE TERMS: COMPARISON AND INTEGRAL TESTS

Exploration 1

1. $s_{20} = 2.718281828$ which looks like e.

2. The graph rises quickly and then levels off. It seems to approach the horizontal asymptote 2.718281828.

3. Since $1/n! < 1/n$ for $n > 1$, the second series graphs above the first. Therefore, $1/n!$ does not bound the terms of $1/n$ and so offers no information about the convergence of $1/n$.

Exploration 2

1. Partsumg suggests that the graph has a horizontal asymptote. Note that the convergence is very slow; partsumt gives $s_{500} = 1.642936066$ while $s_{1000} = 1.643934567$. $\sqrt{6s_{1000}}$ = 3.140638056 which looks like π.

2. Part 1 suggests that the exact value is $s = \pi^2/6 \approx$ 1.64493... . The table below suggests just how slowly these series converges.

n	s_n
22	1.600496933
203	1.640020072
1071	1.644000796
29354	1.644900001

Exploration 3

1. $\displaystyle\int_1^\infty \frac{1}{x^p}dx = \lim_{b\to\infty}\int_1^b \frac{1}{x^p}dx = \lim_{b\to\infty}\frac{1}{(1-p)b^{1-p}} - \frac{1}{1-p} = \frac{1}{p-1}$ and so converges. Since $1/x^p$ is a continuous, positive, and decreasing function for $x \geq 1$, the series $\sum \frac{1}{n^p}$ also converges.

2. If $p = 1$ the p-series is the harmonic series which diverges.

3. If $p < 1$ then $1/n^p > 1/n$ for $n > 1$. Since the series $\{1/n\}$ diverges so does $\{1/n^p\}$ for $p < 1$.

Exploration 4

a) An end-behavior model for $a_n = \dfrac{2n+1}{(n+1)^2}$ is $b_1 = \dfrac{2}{n}$ which diverges as it is a multiple of the harmonic series.

$\frac{a_n}{b_n} = \frac{(2n + 1)n}{2(n + 1)^2}$ so GRAPH $y_1 = \frac{(2x + 1)x}{2(x + 1)^2}$ in [0, 30] by [0, 1.5]. y_1 approaches the horizontal asymptote $y = 1$. We confirm this analytically by noting that $\lim_{x\to\infty} \frac{(2x + 1)x}{2(x + 1)^2} = \lim_{x\to\infty} 1 - \frac{3x + 2}{2x^2 + 4x + 2} = 1$. Therefore, the limit of a_n/b_n is finite and positive so, by the Limit Comparison Test part (b), a_n diverges.

b) An end-behavior model for $a_n = \frac{100 + n}{n^3 + 2}$ is $b_n = \frac{1}{n^2}$ which converges as it is a p-series with $p = 2$. $\frac{a_n}{b_n} = \frac{100n^2 + n^3}{n^3 + 2}$ so GRAPH $y_1 = \frac{100x^2 + x^3}{x^3 + 2}$ in [0, 30] by [0, 50] to see that the graph rises to a maximum and then seems to approach a horizontal asymptote. Tracing in [0, 1000] by [0, 2] suggests that x approaches 1. We confirm this behavior by observing that $\lim_{x\to\infty} \frac{100x^2 + x^3}{x^3 + 2} = \lim_{x\to\infty} \frac{\frac{100}{x} + 1}{1 + \frac{2}{x^3}} = 1$.
Therefore, the limit of a_n/b_n is finite so, by the Limit Comparison Test part (a), a_n converges.

c) An end-behavior model for $a_n = \frac{1}{2^n - 1}$ is $b_n = \frac{1}{2^n}$ which converges since it is a geometric series with $r = 1/2$. GRAPH $y_1 = \frac{2^x}{2^x - 1}$ in [0, 10] by [0, 2] to see that the graph seems to approach $y = 1$ as a horizontal asymptote. We confirm this behavior by noting that

$$\lim_{x\to\infty} \frac{2^x}{2^x - 1} = \lim_{x\to\infty} \frac{1}{1 - \frac{1}{2^x}} = 1.$$

Therefore, the limit of a_n/b_n is finite so, by the Limit Comparison Test part (a), a_n converges.

SECTION 9.5 ALTERNATING SERIES AND ABSOLUTE CONVERGENCE

Exploration 1

1.

n	s_n	n	s_n
10	0.6456349206	60	0.684883282
20	0.6687714032	70	0.6860553386
30	0.6767581377	80	0.68693624
40	0.6808033818	90	0.6876224873
50	0.6832471606		

Even partial sums increase to the limit and, therefore, are underestimates of L.

2.

n	s_n	n	s_n
9	0.7456349206	59	0.7015499487
19	0.7187714032	69	0.7003410529
29	0.710091471	79	0.69943624
39	0.7058033818	89	0.6987335984
49	0.7032471606		

Odd partial sums decrease to the limit and, therefore, are overestimates of L.

3. S_{200} = 0.6906534305 and S_{500} = 0.6921481806. The series appears to converge slowly; the additional 300 terms in this comparison maintained only the first two digits so the partial sums are quite unstable, that is, their 'early' digits may be affected by quite distant terms in the series.

Exploration 2

1. $|s_{100} - L| < |a_{101}| = 0.0099 < 0.01$

2. $|s_{99} - L| < |a_{100}| = 0.01$, so $s_{99} = 0.6981721793$ is the desired overestimate.

3. Since $|a_{1000}| = 0.001$, $n = 1000$ will give an underestimate and $n = 999$ an overestimate with error at most 0.001.

4. s_{1000} = 0.6926474306 underestimates the sum and s_{999} = 0.6936474306 overestimates the sum with an error no greater than 0.001.

5. Similarly, n = 10000 will give the first underestimate and n = 9999 will give the first overestimate with an error no greater than 0.0001 because $|a_{10000}|$ = 1/10000 = 0.0001. s_{10000} = 0.6930971831 and s_{9999} = 0.6940971831.

SECTION 9.6 POWER SERIES

Exploration 1

1. P_0 = 1 and P_1 = 1 + x. Both of these graphs pass through (0,1) as does y_1. P_1 is tangent to y_1 at (0,1) and so these two functions have the same slope there.

2. The three graphs seems to overlap larger and larger portions of y_1 giving better and better approximations. All the graphs lie above y_1.

3. P_3, P_7, and P_{11} also overlap larger and larger portions of y_1 providing improving approximations. They all lie below y_1.

4.

x	$f(x)$	$P_3(x)$	$P_{10}(x)$
-0.2	0.83333333	0.832	0.8333333504
-0.5	0.666666667	0.625	0.6669921875
0.2	1.25	1.248	1.249999974
0.5	2.	1.875	1.999023438

5. n = 43. If your grapher has sequences built-in you can observe the convergence as follows. Let y_1 = 1/(1-x), y_2 = sum seq(x^K, K, 0, N, 1), and let y_3 = abs(y_1 - y_2). y_2 is the calculator version of $P_N(x)$ and y_3 represents the error in approximating y_1 with y_2.

On the home screen type .9→X:10→N. This sets up the starting values of X and N. (You can choose other starting values if you wish, the process is the same.)

Now type: N+1→N:{N, y_3}. This increases N by 1 and prints out the values of N and y_3. Each time you press enter you will see a new pair of values, N increasing by 1 and y_3, the error, decreasing. When y_3 < 0.1 for the first time,

N = 43. The chart below captures the flavor of the output starting with N = 38:

{38, .1642320327}
{39, .1478088294}
{40, .1330279465}
{41, .1197251518}
{42, .1077526366}
{43, .096977373}

```
.9→x
                .9
For(N,42,47,1):Disp [
N,abs (y1-y2):Pause :
END
   [42 .10775263664 3]
   [43 .096977372979]
```

Exploration 2

1. If $x \neq 0$ then

$$\lim_{n\to\infty} \frac{|a_{n+1}|}{|a_n|} = \lim_{n\to\infty} \frac{\left|\frac{x^{2n+3}}{(2n+3)!}\right|}{\left|\frac{x^{2n+1}}{(2n+1)!}\right|} = \lim_{n\to\infty} \frac{x^2(2n+1)!}{(2n+3)!} = x^2 \lim_{n\to\infty} \frac{1}{(2n+2)(2n+3)}$$

and since this limit is 0 the series converges absolutely. When x = 0, every term of the series is zero so $f(0)$ = 0. Therefore, the series converges absolutely everywhere.

2. If your grapher has sequences enter y_1 = sum seq{$(-1)^N x^{2N+1}/(2N+1)!$, N,0,10,1}. $f(x)$ looks like sin x.

3. The calculator reports an error if you try to GRAPH NDER(y_1, X). Instead, GRAPH y_2 = sum seq{$(-1)^N x^{2N}/(2N)!$, N,0,10,1} which is the term-by-term derivative of y_1. The new graph looks like cos x.

4. Since (cos x) = -sin x, the derivative of y_2 should equal $-y_1$. If you enter y_3 = sum seq{$(-1)^N x^{2N-1}/(2N-1)!$, N,0,10,1} you will not get any graph at all. That is because the first term, with N = 0, now has (-1)! in the denominator. Instead, change the limits so that y_3 = sum seq{$(-1)^N x^{2N-1}/(2N-1)!$, N,1,10,1}. This is not exactly $-y_1$ because it has a different number of terms. Its graph, however, still looks like -sin x.

5. The two graphs are nearly identical from -2π to 2π but differ dramatically outside that interval. This should not shake our conjecture at all since P_{21} is just one approximation of the infinite series $f(x)$.

SECTION 9.7 TAYLOR SERIES AND MACLAURIN SERIES

Exploration 1

1. $P(x) = \sum_{k=0}^{N} C_k(x-a)^k = C_0 + C_1(x-a) + C_2(x-a)^2 + C_3(x-a)^3 + \ldots$ so $P(a) = C_0 + C_1(a-a) + C_2(a-a)^2 + C_3(a-a)^3 + \ldots = C_0$. Therefore, $C_0 = f(a)$.

2. $P'(x) = C_1 + 2C_2(x-a) + 3C_3(x-a)^2 + \ldots$ so $P'(a) = C_1 + 2C_2(a-a) + 3C_3(a-a)^2 + \ldots = C_1$. therefore, $C_1 = f'(a)$.

3. The process can be continued indefinitely by differentiating and evaluating the new derivative at $x = a$. The constant term of P is C_0, the constant term of P' is C_1, the constant term of P'' is $2C_2$ and, in general, the constant term of $P^{(k)}$ is $k!C_k$. Substituting $x=a$ in $P^{(k)}$ eliminates all but the constant term. Therefore, $P^{(k)}(a) = k!\ C_k$. Since we are given that $P^{(k)}(a) = f^{(k)}(a)$ we conclude that $f^{(k)}(a) = k!\ C_k$ and $C_k = f^{(k)}(a)/k!$.

Exploration 2

Answers will vary.

Exploration 3

$$f(x) = \cos x = 1 - \frac{x^2}{2!} + \frac{x^4}{4!} - \frac{x^6}{6!} + \ldots + (-1)^n\frac{x^{2n}}{2n!} + R_{2n}(x)$$

where $R_{2n} = \dfrac{f^{(2n+1)}(c)}{(2n+1)!}x^{2n+1}$. $|f^{2n+1}(x)| \le 1$ since every derivative of $\cos x$ has amplitude 1. Therefore, we can take $M = 1$ and $r = 1$ in the remainder theorem. To show that $\dfrac{|x^{2n+1}|}{(2n+1)!} \to 0$ observe that $\dfrac{x}{2n+1} \to 0$ and, consequently, so does the product. Therefore, the Maclaurin series for $\cos x$ converges to $\cos x$ for every x.

Exploration 4

1. If your grapher has sequences enter y_1 = sum seq{$(-1)^K x^{2K+1}/(2K+1)!$, K, 0, N, 1}. And GRAPH for N = 0, 1, 2, ...

2. GRAPH y_1 as above and $y_2 = \sin x$. The two graphs appear identical for N = 8 although tracing shows differences. For N=15 the graphs trace identically too.

3. GRAPH $y_3 = y_2 - y_1$. With N = 13, and yMin = -10^{-7} and yMax = 10^{-7}, for example, tracing the graph of y_3 shows a maximum error of 1.5226 E-8

Exploration 5

1. Define $e^{x+iy} = e^x e^{iy} = e^x (\cos y + i \sin y)$.

2. $e^{-i\theta} = \cos(-\theta) + i\sin(-\theta) = \cos\theta - i \sin\theta$.

3. $\dfrac{e^{iz}+e^{-iz}}{2} = \dfrac{(\cos z + i\sin z) + (\cos z - i\sin z)}{2} = \cos z$.

4. $\dfrac{e^{iz}-e^{-iz}}{2i} = \dfrac{(\cos z + i\sin z) - (\cos z - i\sin z)}{2i} = \sin z$.

SECTION 10.1 CONIC SECTIONS AND QUADRATIC EQUATIONS

Exploration 1

1. Multiply through by 16 to get $\dfrac{16x^2}{9} + y^2 = 16$ and then solve for y^2: $y^2 = 16 - \dfrac{16x^2}{9}$. Take the square root to confirm the result.

2. No. The ellipse that is graphed looks stretched out in the x-direction more than in the y-direction. Even stranger, graph in [-4,4] by [-4,4] and now it looks like a circle.

3. Now the graph is undistorted and the y-intercepts appear correctly proportioned to the x-intercepts. It is important to square the viewing window with conics because a distorted window can make an ellipse appear like a circle and vice-versa.

Exploration 2

1. No. y_1 gives the top half of each branch and y_2 gives the bottom half. That is because each branch has both positive and negative y-coordinates but y produces only positive y-values and y_2 only negative values.

2. Grapher activity.

3. $y = \pm(3/2)x$ give the two asymptotes. I prefer these to y_3 and y_4 because they give the equations of lines rather than absolute value functions.

4. Plot the vertices of the hyperbola $(\pm a, 0)$ and the asymptotes $y = \pm(b/a)x$. From each vertex sketch a smooth curve approaching each asymptote.

SECTION 10.2 THE GRAPHS OF QUADRATIC EQUATIONS IN *X* AND *Y*

Exploration 1

1. Grapher activity.

2. Tracing gives major axis: 9, minor axis: 6, center: (2.5, -1.75). These values can be confirmed analytically by completing the squares $(8x - 20)^2 + (12y + 21)^2 = 1296$ and then writing in standard form: $\frac{(x-5/2)^2}{(9/2)^2} + \frac{(y+7/4)^2}{3^2} = 1$. This gives the center exactly as (5/2, -7/4) and confirms our values for the major and minor axes.

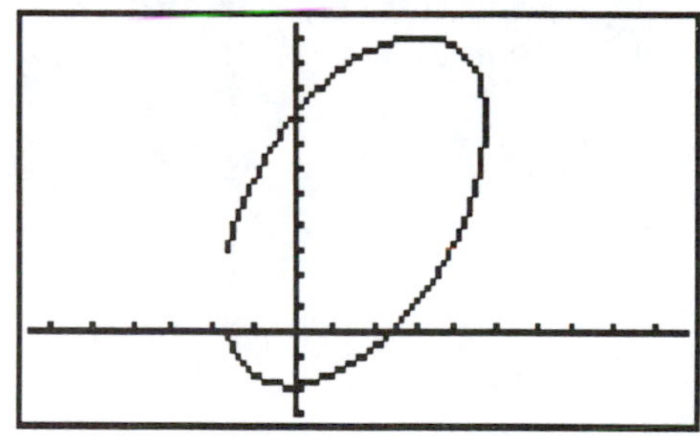

3. A rotated ellipse with vertices at, approximately, (-1, 2) and (-0.25, -8).

Exploration 2

This table suggests some possible answers.

A	B	C	D	E	F	conic
1	0	1	0	0	$-r^2$	circle with center (0,0) and radius r
1	0	c^2 $c>1$	0	0	$-r^2$	ellipse with center (0,0), axes 2 and $2c$
c^2 $c>1$	0	1	0	0	$-r^2$	ellipse with center (0,0), axes $2c$ and 2
a $a>0$	0	0	0	e $e>0$	0	parabola opening up
a $a<0$	0	0	0	e $e>0$	0	parabola opening down
0	0	0	d	e	f	line, degenerate conic
0	b	0	0	0	f	hyperbola with asymptotes $x=0$ and $y=0$.

Exploration 3

1. The graph is a cubic curve equivalent to the explicit function $y = x^3 - x$.

2. The cubic graph is rotated $\pi/6$ radians counter-clockwise. (Note: Enter the equations with parentheses: $x_2(t) = x_1\cos(\pi/6) - y_1\sin(\pi/6)$ or your grapher may interpret $\sin \pi/6$ as $(\sin \pi)/6 = 0$).

3. Grapher activity.

4. Grapher activity.

Exploration 4

1. $x_1(t)$, $y_1(t)$ the graph of the top half of the ellipse in Example 1; $x_2(t)$, $y_2(t)$ is the parametric equation of the top half of the rotated ellipse.

2. If your grapher has list capabilities GRAPH $x_1(t) = t$, $y_1(t) = \{-1, 1\}\sqrt{12 - 3t^2}$ to see both the top and bottom halves of the ellipse and the rotated ellipse drawn. If you do not have list capabilities then GRAPH:

$$x_3(t) = t, \qquad y_3(t) = -y_1(t)$$
$$x_4(t) = x_3(t)\cos\theta - y_3(t)\sin\theta$$
$$y_4(t) = x_3(t)\sin\theta + y_3(t)\cos\theta.$$

3. The ellipse has been rotated counter-clockwise around the origin and shifted down and to the left.

SECTION 10.3 PARAMETRIC EQUATIONS FOR PLANE CURVES

Exploration 1

1. t in $[0, 2\pi]$ for a circle and t in $[0, \pi]$.

2.

t_0	0	$\pi/2$	π	$3\pi/2$
Initial Point	$(a, 0)$	$(0, a)$	$(-a, 0)$	$(0, -a)$

```
For(θ,0,3π/2,π/2):Dis
p [cos (2π/3+θ),sin (
2π/3+θ)]:Pause:End
   [-.5 .866025403784]
 [-.866025403784 -.5]
  [.5 -.866025403784]
   [.866025403784 .5]
```

```
For(θ,0,3π/2,π/2):Dis
p [cos (4π/3+θ),sin (
4π/3+θ)]:Pause :End
  [-.5 -.866025403784]
   [.866025403784 -.5]
    [.5 .866025403784]
  [-.866025403784 .5]
```

3.

Initial Point	Interval Length	Terminal Point
$(a, 0)$	$2\pi/3$	$(-a/2, a\sqrt{3}/2)$
$(a, 0)$	$4\pi/3$	$(-a/2, -a\sqrt{3}/2)$
$(0, a)$	$2\pi/3$	$(-a\sqrt{3}/2, -a/2)$
$(0, a)$	$4\pi/3$	$(a\sqrt{3}/2, -a/2)$
$(-a, 0)$	$2\pi/3$	$(a/2, -a\sqrt{3}/2)$
$(-a, 0)$	$4\pi/3$	$(a/2, a\sqrt{3}/2)$
$(0, -a)$	$2\pi/3$	$(a\sqrt{3}/2, a/2)$
$(0, -a)$	$4\pi/3$	$(-a\sqrt{3}/2, a/2)$

4. For initial point (1, 0) and terminal point (−1, 0) the radius is 1 so use $a = 1$, tMin = 0, tMax = π. For initial point (0, 3) and terminal point (0, −3) use $a = 3$, tMin = π/2 and tMax = 3π/2.

5. To complete a circle with the same initial and terminal point use tMax − tMin = 2π. The smaller the value of t-step the slower the circle is traced, the larger t-step the faster the circle is traced. The price of speed, however, is detail. When t-step is too large the circle appears to have corners. In fact, with t-step = 1 only 6 points are plotted between 0 and 2π and the "circle" doesn't close.

6. Reverse the direction by choosing

   - tMin = 2π, tMax = 0, t-step = −0.1
   - tMin = −π, tMax = π, $a = -1$
   - tMin = −3π/2, tMax = π/2, $a = 1$,
   - $x_1(t) = a \sin t$, $y_1(t) = a \cos t$.

Exploration 2

1. GRAPH $x_1(t) = A \cos t$, $y_1(t) = B \sin t$, t-step = 0.1. Choose a square window including $[-A, A]$ by $[-B, B]$. As the graph is drawn we can "see" the point moving counter-clockwise around the ellipse.

2. When $a > b$ the major axis lies along the x-axis, when $a = b$ the ellipse is a circle, when $a < b$ the major axis lies along the y-axis.

3. If t-step > 0 the graph is traced counter-clockwise; if t-step < 0 it is traced clockwise. The smaller t-step the slower the curve is drawn, the larger t-step the graph is drawn faster but less accurately.

4. GRAPH: $x_1(t) = \cos t$, $y_1(t) = \sin t$;
   $x_2(t) = 2\cos t$, $y_2(t) = \sin t$;
   $x_3(t) = 2\cos t$, $y_3(t) = 2\sin t$ with t in [0, 2π].

   It is interesting to watch the graphs being drawn in simultaneous format because if the three curves represented motion we can watch where the points are on each curve at the same time.

Exploration 3

1. The right branch of the hyperbola only is traced when [tMin, tMax] is [−π/2, π/2]. However, if this closed interval is used a line approximating an asymptote is drawn

because sec t and tan t are not defined at tMin and tMax. To avoid drawing the extra line use a value of tMin slightly larger than $-\pi/2$ and a value of tMax slightly smaller than $\pi/2$, for example, tMin = −1.570796 and tMax = 1.570796. We could graph with tMin = $-\pi$ and tMax = π but this also includes t-values where sec t and tan t are undefined. Any interval whose length is 2π will necessarily include such points.

To avoid this, keep the t-interval tMin = −1.570796 and tMax = 1.570796 but also GRAPH $x_2(t) = x_1(t)$, $y_2(t) = y_1(t)$.

3. GRAPH: $x_1(t) = A \sec t$, $y_1(t) = B \tan t$
$x_2(t) = x_1(t)$, $y_2(t) = y_1(t)$

with tMin = −1.570796 and tMax = 1.570796.

4. The slowing down near a vertex can be observed with t-step = 0.02, x in [1, 2] and y in [−1, 1]. The motion can be explained by examining the t-values that draw this portion of the graph. x is near 1 when cos t is near 1, that is, for t near 0. As t changes from $-\pi/4$ to $\pi/4$ the x-values change from −2 to 2 and the y-values from −1 to 1. But when t changes from $\pi/4$ to $\pi/2$ both x and y approach infinity.

5. Such lines are drawn in connected format when x and y are both large and suddenly change signs. For example, suppose that we are graphing in a t-interval that includes $\pi/2$. Suppose p is the last value of t plotted with $p < \pi/2$ and that q is the first t-value plotted with $q > \pi/2$. In this case, the grapher will connect (sec p, tan p) and (sec q, tan q). But the first point is in the first quadrant and the second point is in the third quadrant so a line will be plotted joining these two points. This can be avoided either by careful choice of the t-interval or by plotting in DOT format.

Exploration 4

1. To see all three cycloids, GRAPH in [−10, 10] by [−3.6, 9.6] (a square window) with t in $[-2\pi, 2\pi]$.

2. The largest y-value of $a(1 - \cos t)$ occurs when $\cos t = -1$, that is, when $y = 2a$. This is the diameter of the wheel so 'a' is the radius of the wheel. For each complete turn of the wheel t changes by 2π so three turns will require a t-interval of length 6π. As usual, a smaller t-step slows down the drawing so that the wheel appears to be rolling at a slower speed; larger t-steps appear to increase the wheels

angular speed. Finally, in a square window the grapher includes a relatively large portion of the y-axis. The more arches of the cycloid that are drawn the smaller the cycloid appears.

3. In order to fit n arches inside 1 large arch, the smaller wheel completes n rotations when the large wheel completes only 1. Therefore, the smaller circumference and, hence, the smaller radius, is $1/n$ times the larger circumference. In order to have the smaller wheel reach complete its n rotations at the same time the larger wheel completes 1, it must be turning n times as fast and, hence, have a period $1/n$ that of the larger wheel. The general parametric equations governing this pair of motions are:

$$x_1(t) = a(t - \sin t), \qquad y_1(t) = a(1 - \cos t)$$
$$x_2(t) = (a/n)(nt - \sin(nt), \qquad y_2(t) = (a/n)(1 - \cos(nt)).$$

Figure 10.27 can be reproduced with $a = 1$, $n = 4$, t in $[0, 2\pi]$ and a square viewing window such as $[-2, 7.2]$ by $[-3, 3]$.

4. a) $x_1(t) = a(t - \sin t)$, $y_1(t) = a(1 - \cos t)$ models the motion of a point P_1 on a moving wheel. To model the motion of the point diametrically opposed to P_1 examine Figure 10.25 and draw a reference triangle by extending the radius through the center in the opposite direction. This gives

$$x = at - a\cos\theta$$
$$y = a - a\sin\theta.$$

Using the transformations of θ into t gives the location of P_2 by:

$$x_2(t) = a(t + \sin t)$$
$$y_2(t) = a(1 + \cos t).$$

```
xt1=t-sin t
yt1=1-cos t
xt2=t+sin t
yt2=1+cos t

F(t)= RANGE ZOOM TRACE GRAPH
t     xt    yt    DELf  SELCT
```

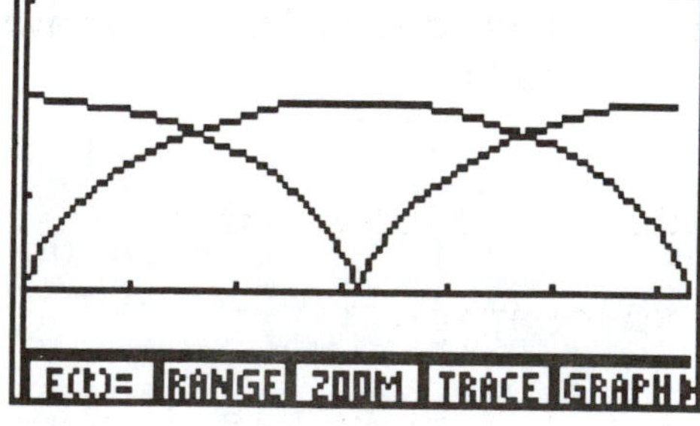

b) P_3 is now at the same relative position but on a wheel a fixed distance behind the first. This can be modelled with a simple horizontal shift:

$$x_3(t) = a(t - \sin t) + b, \quad y_3(t) = a(1 - \cos t).$$

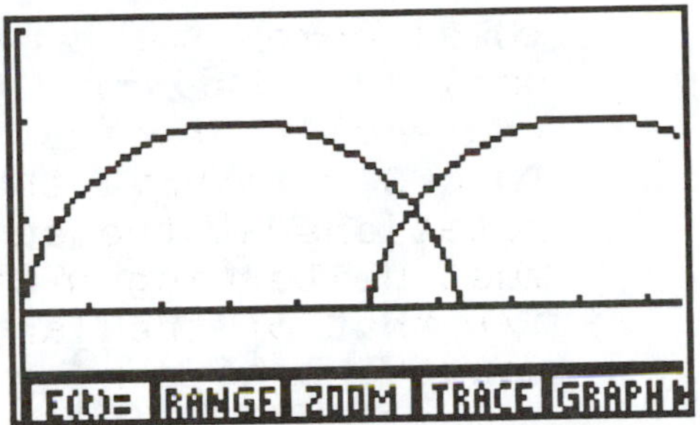

c) Point P_4 is at the same relative position but on a wheel of smaller radius. Refer again to Figure 10.25 and imagine a smaller reference triangle drawn within the triangle at $C(at, a)$. Suppose that the hubcap has radius b, then the location of the new point is given by $x = bt + b\cos\theta$ and $y = b + b\sin\theta$ so that the parametric equations for P_4 are

$$x_4(t) = b(t - \sin t), \qquad y_4(t) = b(1 - \cos t).$$

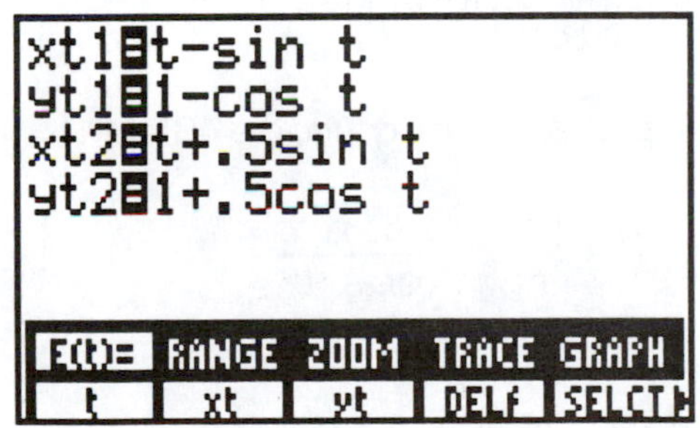

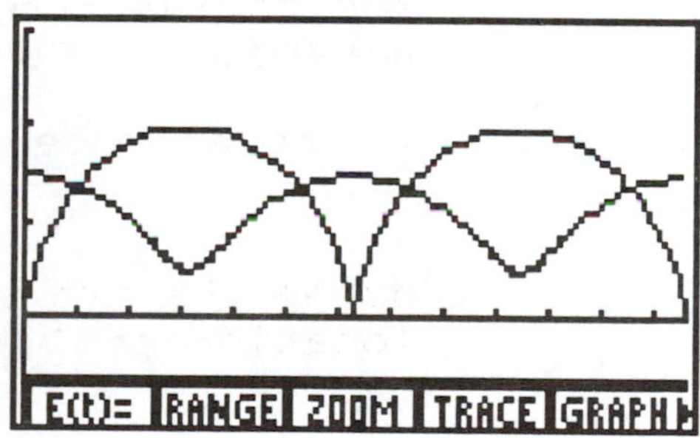

5. The cycloid is drawn in the third quadrant moving to the left.

SECTION 10.5 POLAR COORDINATES

Exploration 1

1. The graph is a circle with radius 6 and center (0,0) traced counter-clockwise.

2. $r = 6$.

Exploration 2

1. a) $x = 2$. Substitute $r \cos\theta$ for x to get $r \cos\theta = 2$.

   b) $xy = 4$ gives $(r \cos\theta)(r \sin\theta) = 4$ so $r^2\cos\theta \sin\theta = 4$.

   c) $x^2 - y^2 = 1$ gives $(r \cos\theta)^2 - (r \sin\theta)^2 = 1$ so $(r^2\cos^2\theta) - (r^2\sin^2\theta)^2 = 1$.

   d) $y^2 - 3x^2 - 4x - 1 = 0$ so $(r \sin\theta)^2 - 3(r \cos\theta)^2 - 4(r \cos\theta) - 1 = 0$ or $r^2\sin^2\theta - 3r^2\cos^2\theta - 4r \cos\theta = 1$. Now replace $\sin^2\theta$ with $1 - \cos^2\theta$ to get $r^2 - 4r^2\cos^2\theta - 4r \cos\theta = 1$ so $r^2 = 4r^2\cos^2\theta + 4r \cos\theta + 1 = (2r \cos\theta + 1)^2$. Take square roots to give the desired polar equation. (The other root, $r = -(1 + 2r \cos\theta)$ traces the same graph and so does not describe an additional part of the curve.)

   e) $x^4 + y^4 + 2x^2y^2 + 2x^3 + 2xy^2 - y^2 = 0$ gives $(x^2 + y^2)^2 + 2x(x^2 + y^2) - y^2 = 0$. Since $x^2 + y^2 = r^2$ we get $(r^2)^2 + 2r^3\cos\theta - r^2\sin^2\theta = 0$. Therefore, $r^4 + 2r^3\cos\theta - r^2 + r^2\cos^2\theta = 0$. So $r^2(r^2 + 2r \cos\theta + \cos^2\theta) = r^2$. That is, $r^2(r + \cos\theta)^2 = r^2$. If r is non-zero then $r + \cos\theta = \pm 1$ or $r = \pm 1 - \cos\theta$. Either sign may be taken for 1 since the two equations give the same cardioid traced in opposite directions.

2. a) In polar mode, $r = 2/\cos\theta$. In parametric mode GRAPH:

$$x_1(t) = \frac{2\cos t}{\cos t}, \quad y_1(t) = \frac{2\sin t}{\cos t}$$
$$x_2(t) = 2, \quad y_2(t) = t$$

   to support that the graphs are the same.

   b) Solve the polar equation for 4 to get $r = \dfrac{2}{\sqrt{\sin\theta \cos\theta}}$. Note that in this form the vertical and horizontal asymptotes ($\theta = 0$ and $\theta = \pi/2$ are obvious). In parametric mode GRAPH:

$$x_1(t) = \frac{2\cos t}{\sqrt{\cos t \sin t}}$$
$$y_1(t) = \frac{2\sin t}{\sqrt{\cos t \sin t}} \quad \text{for } t \text{ in } [0, 3\pi/2].$$

Note that nothing is drawn for t in $[\pi/2, \pi]$ here $(\cos t)(\sin t)$ is negative. Also GRAPH:

$$x_2(t) = t, \quad y_2(t) = 4/t$$

to support that the two graphs are identical.

c) Solve the polar equation for r: $r = \dfrac{1}{\sqrt{\cos^2\theta - \sin^2\theta}}$ and then GRAPH in parametric mode both

$$x_1(t) = \frac{\cos t}{\sqrt{\cos^2 t - \sin^2 t}}, \quad y_1(t) = \frac{\sin t}{\sqrt{\cos^2 t - \sin^2 t}}$$

$$x_2(t) = t, \quad y_2(t) = \sqrt{t^2 - 1}$$

$$x_3(t) = t, \quad y_3(t) = -y_2(t), \qquad t \text{ in } [-\pi, \pi]$$

in a viewing window containing $[-3, 3]$ by $[-2, 2]$.

d) Solve the polar equation for r: $r - 2r\cos\theta = 1$ so $r = \dfrac{1}{1 - 2\cos\theta}$. In parametric mode GRAPH:

$$x_1(t) = \frac{\cos t}{1 - 2\cos t}, \quad y_1(t) = \frac{\sin t}{1 - 2\cos t}.$$

To graph the Cartesian equation solve for y: $y = \pm\sqrt{3x^2 + 4x + 1}$. These can be graphed parametrically with:

$$x_2(t) = t, \quad y_2(t) = \sqrt{3t^2 + 4t + 1};$$

$$x_3(t) = t, \quad y_3(t) = -y_2(t).$$

The graphs appear to be identical. Tracing, however, is not very useful as they are parametized differently and so reach different locations with the same t-values.

e) To graph the polar equation in parametric mode GRAPH:

$$x_1(t) = (1 - \cos t)\cos t$$

$$y_1(t) = (1 - \cos t)\sin t.$$

To graph the Cartesian equation in parametric form we must solve for y. Write $(y^2)^2 + (2x^2 + 2x - 1)y^2 + 2x^3 + x^4 = 0$ which is a quadratic equation in y^2. Use the quadratic formula to get $y^2 = \dfrac{-(2x^2 + 2x - 1) \pm \sqrt{(2x^2+2x-1)^2 - 4(2x^3 + x^4)}}{2}$. We can now solve for y producing four terribly complex equations each of which can be graphed parametrically by letting $x = t$.

3. We can support each equation in parametric mode since both the Polar and the Cartesian equations can be written parametrically. However, the Cartesian equation cannot be written in polar form and so we cannot graph both forms simultaneously in polar mode to obtain graphical support.

4. As the algebra in 2e suggests, it is far more complex to express the Cartesian equation parametrically (so that it can be graphed) than it is to express in Polar form. Sometimes polar form is simply more convenient.

Exploration 3

1. The graph appears to be the vertical line $x = -4$. We can confirm this by noting that $r \cos \theta = x$.

2. The graph appears to be a circle centered at (2, 0) with radius 2. Confirm by writing $r^2 = x^2 + y^2$ and $r \cos\theta = x$ giving $x^2 + y^2 = 4x$ or $(x - 2)^2 + y^2 = 4$.

3. The graph appears to be the line $y = 2x - 4$ but the graph is discontinuous when $2\cos\theta - \sin\theta = 0$, that is, when $\tan\theta = 2$. For θ in $[0, 2\pi]$, there are two such discontinuities, both holes. For other values of θ, we have $2r \cos\theta - r \sin\theta = 4$ or $2x - y = 4$.

SECTION 10.6 GRAPHING IN POLAR COORDINATES

Exploration 1

1. Over $[0, \pi]$ loops of the rose are drawn in the first and fourth quadrants, over $[0, 2\pi]$ all four loops are drawn. The reason that additional parts of the graph appear for θ in $[\pi, 2\pi]$ is that θ determines not just the values of $\sin 2\theta$ but also the direction in which points are plotted.

2. Over $[0, \pi/2]$ and $[0, \pi]$ only the first quadrant loop is drawn. Over both $[0, 3\pi/2]$ and $[0, 2\pi]$ both the first and third quadrant loops are drawn.

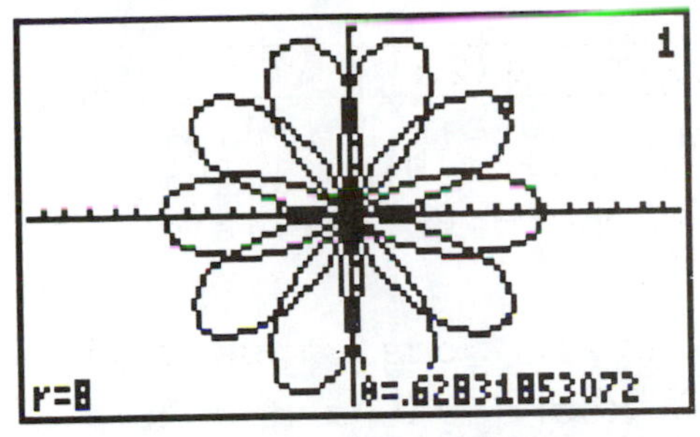

3. The period of $8\sin(2.5\theta)$ is $4\pi/5$ yet only half the graph is drawn in the larger interval $[0, 2\pi]$. For a complete graph $[0, 4\pi]$ is required. The larger interval is necessary because, $\sin(2.5\theta) = -\sin(2.5(\theta + 2\pi))$. In a polar graph this amounts to points in the first and second periods of $r = 8\sin(2.5\theta)$ being plotted on opposite sides of the origin.

4. Many answers are possible. Any function whose period is smaller than 2π and which changes sign every 2π (so that $f(\theta) = -f(\theta + 2\pi)$) will work. For example, $r = 5\sin(3.5\theta)$.

5. The graph of $r = \cos\theta^2$ is quite messy in $0 < \theta < 100$. To better see what is happening reduce the interval and the θ-step and graph in $0 < \theta < 2\pi$ with θ-step = 0.01. This permits a better visualization of the curve. We can see a rose-type curve unfolding with the petals becoming thinner and more frequent. This is supported by the Cartesian graph of $y = \cos x^2$ where we observe the cosine wave's cycles becoming more frequent as x increases.

Exploration 2

If you do not have polar form available on your grapher use parametric mode and GRAPH:

$$x_1(t) = \cos(2t)\cos t,\ y_1(t) = \cos(2t)\sin t;$$
$$x_2(t) = \sin(2t)\cos t,\ y_2(t) = \sin(2t)\sin t,$$

with t in $[0, 2\pi]$ and t-step = 0.1.

1. There are four simultaneous solutions, one in each quadrant. These correspond to simultaneous solutions of the two equations, where the same (r, θ) satisfy each equation. The five other intersections, one in each quadrant and one at the origin correspond to different values of θ, e.g., $r = \cos(8\pi/3)$ and $r = \sin(2\pi/3)$ both plot at $(\sqrt{3}/2, \pi/3)$.

2. Altogether there are 9 intersections, two in each quadrant and one at the origin.

3. The Cartesian graphs intersect at just four points, each corresponding to a simultaneous intersection on the polar graphs.

Exploration 3

1. The first 'collision' occurs at, approximately, (2.41,1.08) and the second at $(-2, \pi)$. They will not collide a third time before the graphs are complete. The next collision will be a repetition of the first.

2. The distance graph appears to repeat every 2π units and shows that as the particles move they approach each other and collide at $t = 1.1$ then move apart reaching a maximum separation at about $t = 5.3$ where the particles are 8.82 units part). They then approach each other and repeat the process.

   The two x-intercepts at the base of the 'W' correspond to a separation of zero, that is, a collision point. These occur at $t = 1.1$ and $t = 3.1$ supporting (approximately) $\theta = 1.08$ and $\theta = \pi$.

   The objects are 4 units apart at $t = 0.3$ and $t = 3.8$. To support these values, GRAPH $x_4(t) = t$, $y_4(t) = 4$ and TRACE.

3. The accuracy could be improved by using a smaller t-step. Of course, this would slow down the graphing considerably.

4. At $\theta = -\pi$, $r_1 = 1 - 3 = -2$ and $r_2 = -2$, confirming the second 'collision' point.

SECTION 10.7 POLAR EQUATIONS OF CONIC SECTIONS

Exploration 1

1. Since L_1 has slope 3 the perpendicular line has slope -1/3. Since it passes through the origin its equation is $y = (-1/3)x$.

2. The two lines intersect at $x = -2.7$, $y = 0.9$ giving $r_0 = \sqrt{(-2.7)^2 + .9^2} = 0.9\sqrt{10}$ and $\theta_0 = \pi + \tan^{-1}(-\frac{1}{3}) = 2.8198$.

3. The graphs are identical.

4. (1) To find the equation of L_2 graphically GRAPH $y_1 = 3x + 9$ and $y_2 = \sqrt{x^2 + (3x+9)^2}$ in the same window. y_2 represents the distance from (x, y_1) to the origin. The perpendicular from the origin intersects L_1 at a point whose distance from the origin is minimal. To locate that point TRACE y_2 to find its minimum and then switch graphs. This

gives an approximate intersection at (2.68, 0.96). Knowing this point we arrive at the equation for the perpendicular $y = (-0.96/2.68)x$.

(2) r_0 is the y-coordinate y_2 found in (1) above. To find θ_0 TRACE and locate the cursor on L_1 as above.

(3) To draw L_2, with the cursor as above select DRAW Line(. Press ENTER to select the current point and then move the cursor to the origin and press ENTER again. A line will join the two selected points.

Exploration 2

1. According to the table this is a type (2) ellipse with one focus at the origin and directrix $x = -5$.

2. You may want to increase the viewing window to [-18, 18] by [-18, 18].

3. The ellipse appears to rotate around the origin counter-clockwise by A units.

SECTION 10.8 INTEGRATION IN POLAR COORDINATES

Exploration 1

1. [2π/3, 4π/3]. Support your guess by graphing with θMin = 2π/3 and θMax = 4π/3. For analytic confirmation notice that $2\cos(2\pi/3) + 1 = 2\cos(4\pi/3) + 1 = 0$.

2. $$A = \int_{2\pi/3}^{4\pi/3} \frac{1}{2}(2\cos\theta + 1)^2 d\theta.$$

   Use $NINT(\frac{1}{2}(r_2)^2, \theta, 2\pi/3, 4\pi/3) = 0.5435$.

3. $$A = \frac{1}{2}\int_{2\pi/3}^{4\pi/3} (2\cos\theta+1)^2 d\theta = \frac{1}{2}\int_{2\pi/3}^{4\pi/3} 4\cos^2\theta + 4\cos\theta + 1\, d\theta =$$

   $$\frac{1}{2}\int_{2\pi/3}^{4\pi/3} 3 + 2\cos 2\theta + 4\cos\theta\, d\theta = \frac{1}{2}(3\theta + \sin 2\theta + 4\sin\theta)\Big|_{2\pi/3}^{4\pi/3} =$$

   $$\pi - \frac{3\sqrt{3}}{2} = 0.5435.$$

4. Since the cosine is even, $2\cos(\theta) + 1 = 2\cos(-\theta) + 1$ and the graph is symmetrical around the x-axis. Therefore, the x-axis should divide the area of the inner loop in half

suggesting that we can calculate the area by doubling the area inside the inner loop above or below the x-axis. We confirm with an NINT computation $NINT((2\cos\theta + 1)^2, \theta, 2\pi/3, \pi) = 0.5435$. Similarly,

$$A = \int_{2\pi/3}^{\pi} (2\cos\theta + 1)^2 d\theta = \int_{2\pi/3}^{\pi} 3 + 2\cos 2\theta + 4\cos\theta \, d\theta = \pi - \frac{3\sqrt{3}}{2}.$$

Exploration 2

1. The limits are 0 to 2π.

2. $\frac{dr}{d\theta} = \sin\theta$ so $\sqrt{r^2 + (\frac{dr}{d\theta})^2} = \sqrt{r^2 + \sin^2\theta}$
$= \sqrt{1 - 2\cos\theta + \cos^2\theta + \sin^2\theta} = \sqrt{2 - 2\cos\theta}$.

3. $NINT(\sqrt{2 - \cos\theta}, \theta, 0, 2\pi) = 8$.

4. $L = \int_0^{2\pi} \sqrt{2(1 - \cos\theta)}\, d\theta = \int_0^{2\pi} \sqrt{4\sin^2(\frac{\theta}{2})}\, d\theta$
$= \int_0^{2\pi} 2\sin(\frac{\theta}{2})\, d\theta = -4\cos\frac{\theta}{2}\Big|_0^{2\pi} = 8.$

Exploration 3

1. TRACE the lemniscate to see that when θ is in $[-\pi/4, \pi/4]$ the right-hand loop is traced once.

2. $r^2 + (\frac{dr}{d\theta})^2 = \cos 2\theta + (\frac{\sin 2\theta}{\sqrt{\cos 2\theta}})^2 = \frac{1}{\cos^2 2\theta}$. Therefore the integrand in (6) is $2\pi \cos\theta$.

3. $NINT(2\pi\cos\theta, \theta, -\pi/4, \pi/4) = 8.8858$

4. $2\pi \int_{-\pi/4}^{\pi/4} \cos\theta \, d\theta = 2\pi\sin\theta\Big|_{-\pi/4}^{\pi/4} = 2\pi\sqrt{2} = 8.8858.$

SECTION 11.1 VECTORS IN THE PLANE

Exploration 1

1. Yes. Since opposite sides of a parallelogram are equal and parallel, the Parallelogram Law implies that vector addition is commutative.

2. Yes.

3. Yes. A 'null' vector, that is, one whose length is zero, acts as an identity.

4. Yes. The vector-addition inverse of v is a vector with the same length but opposite direction.

SECTION 11.2 CARTESIAN (RECTANGULAR) COORDINATES AND VECTORS IN SPACE

Exploration 1

1. - When $z = 0$ the equation of the sphere becomes $x^2 + y^2 = 16$ which is the equation of the sphere's cross-section in the xy-plane (a circle, centered at the (0,0,0) with radius 4).
   - Substitute $z = -2$ to get $x^2 + y^2 = 12$, the equation in the plane $z = -2$ of a circle with radius $\sqrt{12}$ and center (0,0,-2).
   - Let $z = 3$ to get $x^2 + y^2 = 7$, the equation in the plane $z = 7$ of a circle with radius $\sqrt{7}$ and center (0,0,7).

2,3. A circle with radius r_0, can be graphed:

- in function mode by: $y_1 = \sqrt{r_0^2 - x^2}$, $y_2 = -y_1$
- in parametric mode by: $x_1(t) = t$, $y_1(t) = \sqrt{r_0^2 - t^2}$;
  $x_2(t) = t$, $y_2(t) = -y_1(t)$
- in polar mode by: $r = r_0$.

If your calculator has list capabilities replace r_0 with a list variable L_1, STORE values in L_1: {-3.5,-2.5,-1.5,0,2,3,3.3,3.7,3.8,3.9,3.95,3.99} → L_1 and GRAPH.

4. For each z, the cross-section is a circle with radius $r = \sqrt{16 - z^2}$ with z in [-4, 4]. The area of these circles are $A = \pi(16 - z^2)$. In function mode GRAPH $y = \pi(16 - x^2)$ in a window including [-4, 4] by [0, 16π].

SECTION 11.5 LINES AND PLANES IN SPACE

Exploration 1

1. $N_1 \times N_2 = 14\mathbf{i} + 2\mathbf{j} + 15\mathbf{k}$ is perpendicular to each normal vector and, therefore, parallel to each plane.

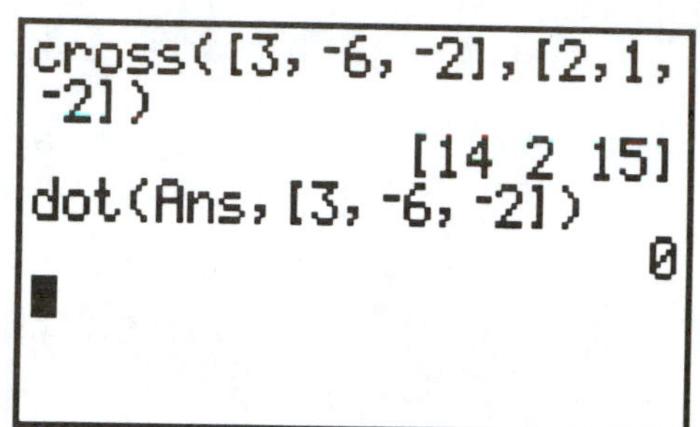

2. Any point on the intersection of the planes is common to both planes. Choose any value for z, say $z = 0$. Substitute in each equation to get $3x - 6y = 15$ and $2x + y - 5$ and solve to get $x = 3$, $y = -1$ giving the point (3,-1,0) lying in both planes.

3. The point (3,-1,0) and the vector $14\mathbf{i} + 2\mathbf{j} + 15\mathbf{k}$ give the parametric equations of the line of intersection:

$$x(t) = 3 + 14t,\quad y(t) = -1 + 2t,\quad z(t) = 15t.$$

SECTION 11.6 SURFACES IN SPACE

Exploration 1

1a) For $y = 0$, the cross-section is $x^2/9 = z^2/4$ or $z = \pm(2/3)x$. GRAPH $y_1 = (2/3)x$, $y_2 = -y_1$. The intersecting plane $y = 0$ passes through the vertex of the cone giving as cross section a pair of lines intersecting at the cone's vertex (0, 0, 0).

For $y = \pm 4$, the cross-section is $\dfrac{z^2}{4} - \dfrac{x^2}{9} = 1$. GRAPH the hyperbola in parametric mode:

$$x_1(t) = t,\quad y_1 = 2\sqrt{1 + \frac{t^2}{9}}$$

$$x_2(t) = t,\quad y_2(t) = -y_1(t).$$

For $y = \pm 10$, the cross-section is $\frac{z^2}{4} - \frac{x^2}{9} = \frac{100}{6}$ or, in standard form, $\frac{z^2}{25} - \frac{x^2}{(15/2)^2} = 1$. This intersection is also a hyperbola. GRAPH in parametric mode,

$$x_1(t) = t,\ y_1(t) = 5\sqrt{1 + \frac{x^2}{7.5^2}}$$

$$x_2(t) = t,\ y_2(t) = -y_1(t).$$

Graphing both hyperbolic intersections simultaneously reveals that as $|y|$ increases, the vertices of the hyperbolas move away from their centers.

Note that all the hyperbolas share the same asymptotes, $z = \pm(2/3)x$.

1b) For $x = 0$, the cross-section is $\frac{y^2}{16} = \frac{z^2}{4}$ or $z = \pm(1/2)y$. To be consistent with the previous graphs use x for y and y for z and GRAPH: $y_1 = (1/2)x$, $y_2 = -y_1$. The intersecting plane $x = 0$ passes through the vertex of the cone perpendicular to the plane $y = 0$, giving as cross section a pair of lines intersecting at the cone's vertex (0, 0, 0).

For $x = \pm 3$, the cross-section is $\frac{z^2}{4} - \frac{y^2}{16} = 1$. GRAPH the hyperbola in parametric mode, again using x for y and y for z:

$$x_1(t) = t,\ y_1(t) = 2\sqrt{1 + \frac{t^2}{16}}$$

$$x_2(t) = t,\ y_2(t) = -y_1(t).$$

For $x = \pm 10$, the cross-section is $\frac{z^2}{4} - \frac{y^2}{16} = \frac{100}{9}$ or, in standard form, $\frac{z^2}{(20/3)^2} - \frac{y^2}{16} = 1$. This intersection is also a hyperbola. GRAPH in parametric mode,

$$x_1(t) = t, \quad y_1(t) = \frac{20}{3}\sqrt{1 + \frac{y^2}{16}}$$

$$x_2(t) = t, \quad y_2(t) = -y_1(t).$$

Graphing both hyperbolic intersections simultaneously reveals that as $|x|$ increases, the vertices of the hyperbolas again move away from their centers.

Note that all the hyperbolas share the same asymptotes, $z = \pm(1/2)x$.

1c) For $z = 0$ the equation is $\frac{x^2}{9} + \frac{y^2}{16} - 0$ which reduces to the single point (0, 0, 0). The plane intersects the cone at its vertex.

When $z = \pm 2$, the cross-section is $\frac{x^2}{9} + \frac{y^2}{16} = 1$, an ellipse with axes of 8 and 6. When $z = \pm 8$ the cross-section is $\frac{x^2}{9} + \frac{y^2}{16} = 16$ or $\frac{x^2}{12^2} + \frac{y^2}{16^2} = 1$, an ellipse with axes 32 and 24. In general, as $|z|$ increases moving the intersecting plane further away from the cone's vertex, the cross-section will be larger and larger ellipses.

2a) $\frac{x^2}{9} + \frac{y^2}{16} + \frac{z^2}{25} = 1$. For $x = 0$ the cross-section is

$\frac{y^2}{16} + \frac{z^2}{25} = 1$, an ellipse with major axis 10, and minor axis 8. For $|x| = \pm 3$ the intersection reduces to a point $(\pm 3, 0, 0)$. There is no intersection if $|x| > 3$. For $|x| < 3$, the intersection is a smaller ellipse. For example, with $x = \pm 1$ the intersections are:

$$\frac{y^2}{16} + \frac{z^2}{25} = \frac{8}{9} \quad \text{or} \quad \frac{y^2}{(128/9)} + \frac{z^2}{(200/9)} = 1.$$

For $y = 0$ the cross-section is $\frac{x^2}{9} + \frac{z^2}{25} = 1$, an ellipse with axes 10 and 6. For $|y| = \pm 4$ the intersections are the points $(0, \pm 4, 0)$. There is no intersection if $|y| > \pm 4$. For $|y| < \pm 4$ as $|y|$ decreases the intersections are smaller and smaller ellipses, that is, as the intersecting plane moves away from $y = 0$ the cross-sections shrink to a point and then vanish.

For $z = 0$ the same pattern emerges. The cross-section at $z = 0$ is the ellipse $\frac{x^2}{9} + \frac{y^2}{16} = 1$ with axes 8 and 6. As $|z|$ increases the elliptical cross-sections diminish in size until, at $z = \pm 5$ they have reduced to the points (0, 0, ±5). For $|z| > 5$ there is no cross-section.

2a) To graph any series of intersections treat one variable as a constant and solve the resulting quadratic equation for one of the other variables. For example, to examine cross-sections with z = constant STORE a value in z, write

$\frac{x^2}{9} + \frac{y^2}{16} = 1 - \frac{z^2}{25}$ so that $y = \pm 4\sqrt{1 - \frac{z^2}{25} - \frac{x^2}{9}}$ and

GRAPH: $x_1(t) = t,\ y_1(t) = \pm 4\sqrt{1 - \frac{z^2}{25} - \frac{t^2}{9}}$

$x_2(t) = t,\ y_2(t) = -y_1(t)$

If you have list capabilities available STORE a list of constants in L_1 and GRAPH:

$x_1(t) = t,\ y_1(t) = \pm 4\sqrt{1 - \frac{L_1^2}{25} - \frac{t^2}{9}}$

$x_2(t) = t,\ y_2(t) = -y_1(t).$

2b) $\frac{x^2}{25} + \frac{y^2}{9} - \frac{z^2}{16} = 1$. For $x = \pm k$ the cross-sections are hyperbolas with asymptotes $y = \pm(3/4)z$. When $x = 0$ the cross-section has vertices at (0, ±3, 0); as $|x|$ increases the vertices move away from the origin.

For $y = \pm k$ the cross-sections are also hyperbolas. These asymptotes are $z = \pm(2/3)x$. The cross-section at $y = 0$ has vertices (±5, 0, 0) and, as $|y|$ increases, the vertices of the cross-sections move further away from the origin.

For $z = \pm k$ the cross-sections are ellipses, centered at the origin, with axes $10\sqrt{1 + \frac{k^2}{16}}$ and $6\sqrt{1 + \frac{k^2}{16}}$. The smallest cross-section occurs at $z = 0$.

2c) $\frac{x^2}{16} + \frac{y^2}{25} = \frac{z}{2}$. For $z = \pm k$ the cross-sections are ellipses, centered at the origin, with axes $4\sqrt{\frac{k}{2}}$ and $5\sqrt{\frac{z}{2}}$. When $z = 0$ the cross section reduces to a point; as $|z|$ increases the axes increases.

For $y = \pm k$ the cross-sections have equations $z = \frac{x^2}{16} + \frac{k^2}{25}$. These parabolas all have the same focus and directrix (in parallel planes $y = k$) but as $|y|$ increases they move away from the origin.

The same pattern holds for $x = \pm k$. The cross-sections are parabolas which are identical in size and shape but move further away from the origin as $|x|$ increases.

2d) $\frac{x^2}{9} + \frac{y^2}{9} = \frac{z}{4}$. This is similar to (c). For constant z the cross-sections are circles; when $z = 0$ the cross-section is a single point, as $|z|$ increases the radii of the circles increase.

For constant x or y the cross-sections are parabolas shifted farther and farther away from the origin in the direction of the intersecting plane.

2e) $\frac{z^2}{16} - \frac{x^2}{25} - \frac{y^2}{9} = 1$. For $z = \pm 4$ the intersection is a single point $(0, 0, \pm 4)$. There are no cross-sections for $|z| < 4$ but for $|z| = k > 4$ the cross-sections are ellipse $\frac{x^2}{25} + \frac{y^2}{9} = \frac{k^2}{16} - 1$ with major and minor axes $10\sqrt{\frac{k^2}{16} - 1}$ and $6\sqrt{\frac{k^2}{16} - 1}$. As k increases the ellipses increase.

For $x = \pm k$ the intersections are hyperbolas all sharing (in their respective planes) asymptotes $z = \pm(4/3)y$. As k increases the vertices of the hyperbolas recede from the origin.

Similarly, for $y = \pm k$, the intersections are hyperbolas which now share the asymptotes $z = \pm(4/5)x$. When $y = 0$ the vertices are at (0, 0, ±4) and, as k increases, they move farther and farther away from the origin.

2f) $\dfrac{y^2}{4} - \dfrac{x^2}{25} = \dfrac{z}{3}$. For $x = \pm k$ the cross-sections are parabolas each shifted down the z-axis relative to parabolas with smaller k. The limiting case $x = 0$ gives a parabola with vertex at the origin. Each parabola opens upwards towards positive z's.

Similarly, when $y = \pm k$ the cross-sections are parabolas now shifted up the z-axis relative to parabolas with smaller k. The limiting case $y = 0$ again gives a parabola with vertex at the origin. These parabolas, however, all open downward toward the negative z-axis.

For $z = 0$, the cross-section is the pair of lines $y = \pm(2/5)x$. For $z = k > 0$ the cross-sections are hyperbolas whose vertices $\left(0, 2\sqrt{\frac{k}{3}}, k\right)$ move away from the origin as k increases. There are no cross-sections for $z < 0$.

SECTION 12.1 VECTOR-VALUED FUNCTIONS AND CURVES IN SPACE; DERIVATIVES AND INTEGRALS

Exploration 1

1,2. $|u(t)| = \sqrt{(\sin t)^2 + (\cos t)^2 + (\sqrt{3})^2} = 2$. Since the length is constant the particle is always two units from the origin, that is, it is moving on the surface of a sphere. Since the k component is also constant the particle is always in the $z = \sqrt{3}$ plane. Therefore, the particle moves in a circle on a sphere centered at (0, 0, 0) with radius 2 so that it is always $\sqrt{3}$ units above the xy-plane.

The motion is that of a yo-yo being twirled in a circle, hence the name of the exploration "The Tethered Particle".

3. $\dfrac{du}{dt} = (\cos t)i - (\sin t)j$ to

$$u \cdot \frac{du}{dt} = \sin t \cos t - \sin t \cos t = 0.$$

4. The position vector is perpendicular to the velocity vector. Since the particle moves on the surface of a sphere, the velocity is tangent to the sphere.

Exploration 2

1. $r(t) = ti + t^2 j.$

2. $v(t) = i + 2tj;\ a(t) = 2j.$

3. The particle moves along the parabola $y = x^2$. At time t it is at (t, t^2). It moves with constant horizontal velocity of 1 unit/sec; the vertical component of velocity is $2t$. Its acceleration is constant ($|a| = 2$) and directed vertically.

SECTION 12.2 MODELING PROJECTILE MOTION

Exploration 1

```
xt1=■00
yt1=15-4t
xt2=(152*cos (20)-8.…
yt2=3+152*sin (20)*t…

F(t)= RANGE ZOOM TRACE GRAPH
t     xt    yt   DELf  SELCT
```

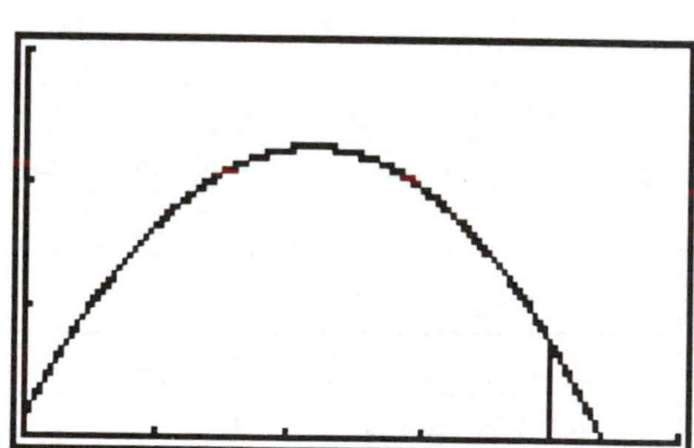

1. To simulate the baseball's flight GRAPH:

$$x_1(t) = (152 \cos 20 - 8.8)\ t,$$
$$y_1(t) = 3 + (152 \sin 20)\ t - 16t^2.$$

To simulate the 15 foot wall 400 feet away GRAPH

$x_2(t) = 400$, $y_2(t) = (15/4)t$ with t in [0, 4] in a [0, 500] by [0, 100] window. Be sure to change the calculator mode to degrees.

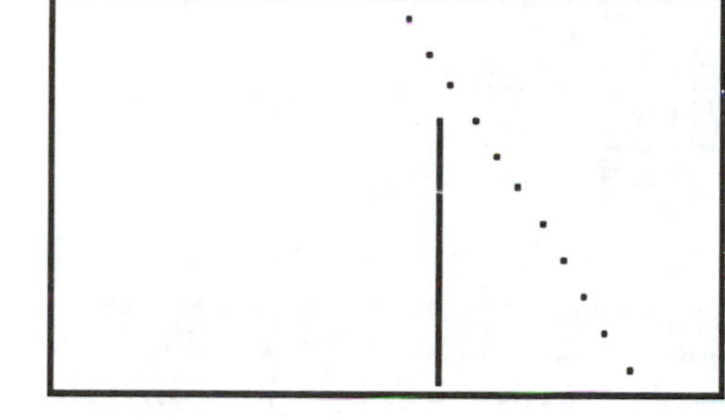

2. The graph seems to show the baseball hitting the top of the 15 foot fence. ZOOM-IN around the top of the fence (or change the window to [390, 410] by [14.5, 15.5] to see that the trajectory of the ball clearly passes over the wall. It is a home run.

For analytic confirmation solve for t with $x_1 = 400$ so

$t = \dfrac{400}{152\cos 20 - 8.8} = 2.984$. Store this value in t and compute $y_1(t) = 15.646$. The ball clears the fence by 0.64 feet . . . a little more than half a foot!

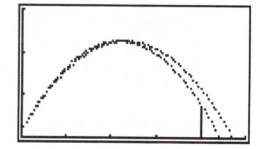

3. Without wind the flight equations are

$x_3(t) = (152 \cos 20)\, t$ and
$y_3(t) = 3 + (152 \sin 20)\, t - 16t^2$.

GRAPH: $(x_1(t), y_1(t))$ and $(x_3(t), y_3(t))$ in [0, 500] by [0, 100] in simultaneous format. The balls seem to separate during the second half of the trajectory. Reducing yMax to 50 shows the balls separating about 1/3 of the way into the flight. The table below gives several comparisons:

time (t)	0.5	1	1.5	2	2.5
distance (x_1) with wind	67.02	134.03	201.05	268.07	335.08
distance (x_3) no wind	71.42	142.83	214.25	285.67	357.08
height (y_1) with wind	24.99	38.99	44.97	42.97	32.97
height (y_3) no wind	24.99	38.99	44.97	42.97	32.97

4. Answers will vary.

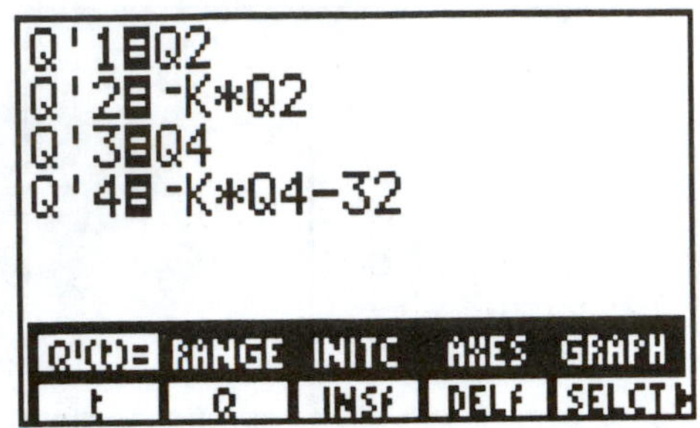

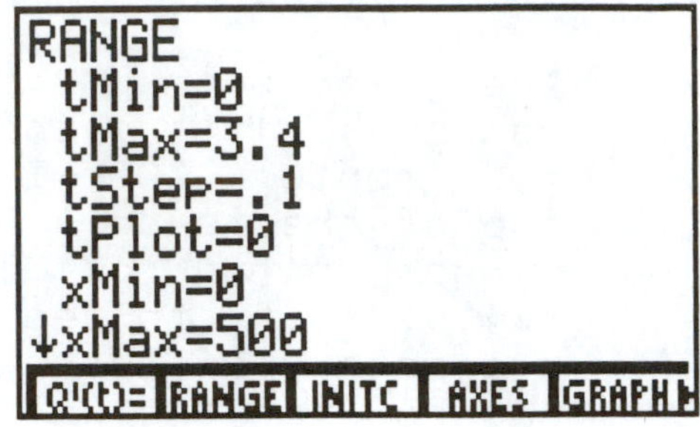

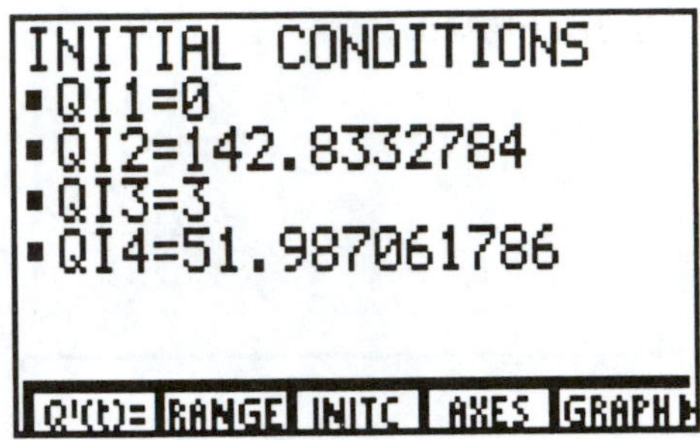

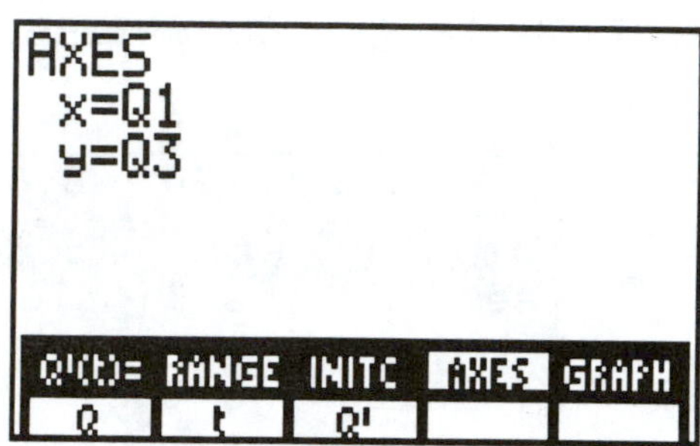

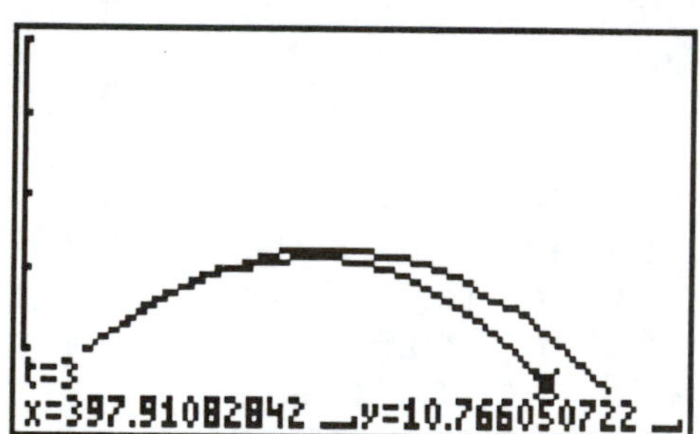

Exploration 2

1. To model the golf ball's motion we use a coordinate system with the x-axis marking distance on the green and the y-axis height above the green. The ball starts at $(x_0, y_0) = (0, 20)$. Since the ball is hit horizontally, $\alpha = 0$. Equation (11) gives the parametric equations:

$$x(t) = v_0 t, \quad y(t) = 20 - 16t^2.$$

It remains to determine the initial velocity.

2,3. Since the ball hits the green 120 feet away, we can find the time t when $y = 0$ and use that time to compute v_0 from $120 = v_0 t$. Since $20 - 16t^2 = 0$, $t = \sqrt{1.25} = 1.118$ so

$v_0 = \dfrac{120}{\sqrt{1.25}} = 107.33$. So the equations of motion are $x(t) = 107.33t$, $y(t) = 20 - 16t^2$ and the ball is in flight for 1.118 seconds.

4. GRAPH $(x(t), y(t))$ with t in [0, 1.2] and t-step = 0.05. Tracing supports that the ball hits the ground between 1.1 and 1.15 seconds between 118 and 123 feet away.

5. Grapher activity.

SECTION 12.4 CURVATURE, TORSION, AND THE TNB FRAME

Exploration 1

1. To find a we need to find the components a_T and a_N. We use equation (7) to find a_T. $\boldsymbol{v} = (t \cos t)\boldsymbol{i} + (t \sin t)\boldsymbol{j}$ so

$|\boldsymbol{v}| = \sqrt{t^2\cos^2 t + t^2\sin^2 t} = \sqrt{t^2}$ and, since $t > 0$, $|\boldsymbol{v}| = t$.

Therefore, $a_T = \dfrac{d}{dt}(|\boldsymbol{v}|) = 1$.

To find a_N we use equation (9). Differentiate $\boldsymbol{v}$ to get $\boldsymbol{a} = (\cos t - t \sin t)\boldsymbol{i} + (\sin t + t \cos t)\boldsymbol{j}$ so that

$|\boldsymbol{a}| = \sqrt{(\cos t - t\sin t)^2 + (\sin t + t\cos t)^2} = \sqrt{1 + t^2}$.

Thus, $a_N = \sqrt{|\boldsymbol{a}|^2 - a_T^2} = \sqrt{(t^2 + 1) - 1} = t$ so $a_N = t$.

Therefore, $\boldsymbol{a} = a_T\mathbf{T} + a_N\mathbf{N} = \mathbf{T} + t\,\mathbf{N}$.

2. GRAPH

- the unit circle $(x_1(t), y_1(t))$,
- the radius vector $(x_2(t), y_2(t))$,
- the acceleration vector:

$x_3(t) = \cos t - t \sin t$,
$y_3(t) = \sin t + t \cos t$.

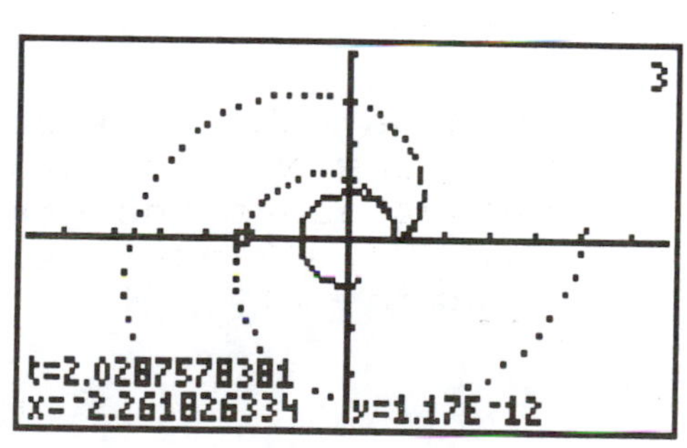

3. To draw the vectors in Figure 12.31 we will need the three sets of equations from (2) and parametric equations for **T** and **N**. Since $t = \frac{\mathbf{v}}{|\mathbf{v}|}$, $\mathbf{T} = (\cos t)\mathbf{i} + (\sin t)\mathbf{j}$.

$$\frac{d\mathbf{T}}{dt} = (-\sin t)\mathbf{i} + (\cos t)\mathbf{j} \text{ and } |d\mathbf{T}/dt| = 1 \text{ so}$$

$\mathbf{N} = (-\sin t)\mathbf{i} + (\cos t)\mathbf{j}$. Enter these equations as

$$x_4(t) = \cos t, \quad y_4(t) = \sin t$$
$$x_5(t) = -\sin t, \quad y_5(t) = \cos t.$$

With these equations in place we can draw the vectors using the LINE command on the DRAW menu. To begin GRAPH $(x_1(t), y_1(t))$ and $(x_2(t), y_2(t))$ and then, from the home screen, enter the following:

$2 \rightarrow t$	or any other value of t
LINE(x_2, y_2, $x_2 + x_3$, $y_2 + y_3$)	draws **a**
LINE(x_2, y_2, $x_2 + x_4$, $y_2 + y_4$)	draws **T**
LINE(x_2, y_2, $x_2 + x_5$, $y_2 + y_5$)	draws **N**

If your grapher permits it enter these commands on a single line so that you can recall them, change t and draw a new set of vectors. In this way you draw the set **a**, **T**, **N** at any $(x_2(t), y_2(t))$.